Competitive Inhibitor

Uncompetitive Inhibitor

Mixed Inhibitor

Lineweaver-Burk Equation

Chapter 8 Nucleotides and Nucleic Acids
UPDATED Molecular Structure Tutorial:

Nucleotides

NEW Simulations:

Nucleotide Structure

DNA/RNA Structure

Sanger Sequencing

Polymerase Chain Reaction

NEW *Nature* Article with Assessment:

LAMP: Adapting PCR for Use in the Field

Animated Biochemical Techniques:

Dideoxy Sequencing of DNA

Polymerase Chain Reaction

Chapter 9 DNA-Based Information Technologies
UPDATED Molecular Structure Tutorial:

Restriction Endonucleases

NEW Simulation:

CRISPR

NEW *Nature* Articles with Assessment:

Assessing Untargeted DNA Cleavage by
   CRISPR/Cas9

Genome Dynamics during Experimental Evolution

Animated Biochemical Techniques:

Plasmid Cloning

Reporter Constructs

Synthesizing an Oligonucleotide Array

Screening an Oligonucleotide Array for Patterns
   of Gene Expression

Yeast Two-Hybrid Systems

Chapter 11 Biological Membranes and Transport
Living Graphs:

Free-Energy Change for Transport (graph)

Free-Energy Change for Transport (equation)

Free-Energy Change for Transport of an Ion

Chapter 12 Biosignaling
UPDATED Molecular Structure Tutorial:

Trimeric G Proteins

Chapter 13 Bioenergetics and Biochemical Reaction Types
Living Graphs:

Free-Energy Change

Free-Energy of Hydrolysis of ATP (graph)

Free-Energy of Hydrolysis of ATP (equation)

Chapter 14 Glycolysis, Gluconeogenesis, and the Pentose Phosphate Pathway
NEW Interactive Metabolic Map:

Glycolysis

NEW Case Study:

Sudden Onset—Introduction to Metabolism

UPDATED Animated Mechanism Figures:

Phosphohexose Isomerase Mechanism

The Class I Aldolase Mechanism

Glyceraldehyde 3-Phosphate Dehydrogenase
   Mechanism

Phosphoglycerate Mutase Mechanism

Alcohol Dehydrogenase Mechanism

Pyruvate Decarboxylase Mechanism

Chapter 16 The Citric Acid Cycle
NEW Interactive Metabolic Map:

The citric acid cycle

NEW Case Study:

An Unexplained Death—Carbohydrate
   Metabolism

UPDATED Animated Mechanism Figures:

Citrate Synthase Mechanism

Isocitrate Dehydrogenase Mechanism

Pyruvate Carboxylase Mechanism

Chapter 17 Fatty Acid Catabolism
NEW Interactive Metabolic Map:

β-Oxidation

NEW Case Study:

A Day at the Beach—Lipid Metabolism

UPDATED Animated Mechanism Figure:

Fatty Acyl-CoA Synthetase Mechanism

Chapter 18 Amino Acid Oxidation and the Production of Urea
UPDATED Animated Mechanism Figures:

Pyridoxal Phosphate Reaction Mechanisms (3)

Carbamoyl Phosphate Synthetase Mechanism

Argininosuccinate Synthetase Mechanism

Serine Dehydratase Mechanism

Serine Hydroxymethyltransferase Mechanism

Glycine Cleavage Enzyme Mechanism

Chapter 19 Oxidative Phosphorylation
Living Graph:

Free-Energy Change for Transport of an Ion

Chapter 20 Photosynthesis and Carbohydrate Synthesis in Plants
UPDATED Molecular Structure Tutorial:

Bacteriorhodopsin

UPDATED Animated Mechanism Figure:

Rubisco Mechanism

Chapter 22 Biosynthesis of Amino Acids, Nucleotides, and Related Molecules
UPDATED Animated Mechanism Figures:

Tryptophan Synthase Mechanism

Thymidylate Synthase Mechanism

Chapter 23 Hormonal Regulation and Integration of Mammalian Metabolism
NEW Case Study:

A Runner's Experiment—Integration of Metabolism (Chs 14–18)

Chapter 24 Genes and Chromosomes
Animation:

Three-Dimensional Packaging of Nuclear Chromosomes

Chapter 25 DNA Metabolism
UPDATED Molecular Structure Tutorial:

Restriction Endonucleases

NEW Simulations:

DNA Replication

DNA Polymerase

Mutation and Repair

NEW *Nature* Article with Assessment:

Looking at DNA Polymerase III Up Close

Animations:

Nucleotide Polymerization by DNA Polymerase

DNA Synthesis

Chapter 26 RNA Metabolism
UPDATED Molecular Structure Tutorial:

Hammerhead Ribozyme

NEW Simulations:

Transcription

mRNA Processing

NEW Animated Mechanism Figure:

RNA Polymerase

NEW *Nature* Article with Assessment:

Alternative RNA Cleavage and Polyadenylation

Animations:

mRNA Splicing

Life Cycle of an mRNA

Chapter 27 Protein Metabolism
NEW Simulation:

Translation

NEW *Nature* Article with Assessment:

Expanding the Genetic Code in the Laboratory

Chapter 28 Regulation of Gene Expression
UPDATED Molecular Structure Tutorial:

Lac Repressor

Lehninger
Principles of Biochemistry

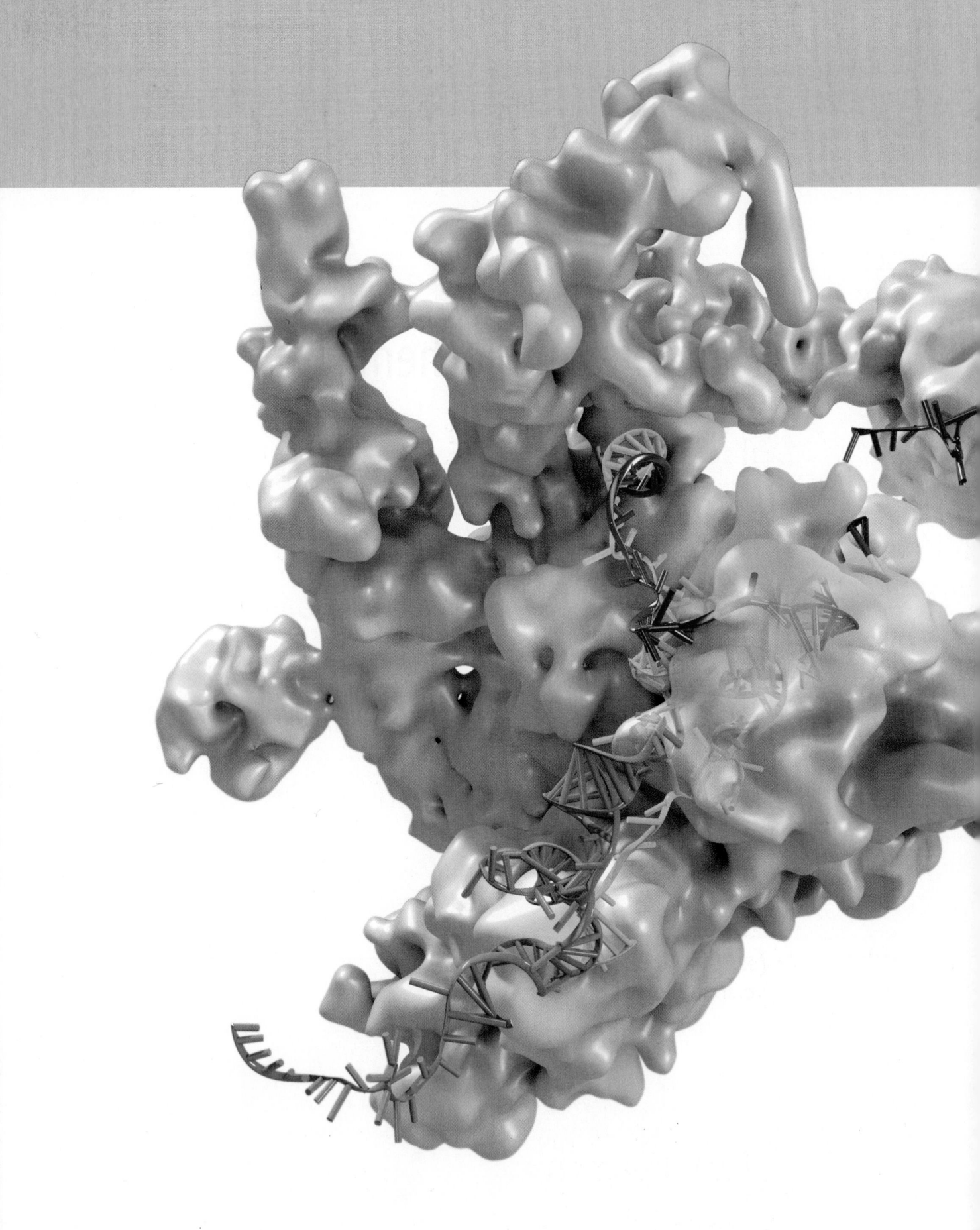

Lehninger
Principles of Biochemistry

SEVENTH EDITION

David L. Nelson

Professor Emeritus of Biochemistry
University of Wisconsin–Madison

Michael M. Cox

Professor of Biochemistry
University of Wisconsin–Madison

w.h.freeman
Macmillan Learning
New York

Vice President, STEM:	Ben Roberts
Senior Acquisitions Editor:	Lauren Schultz
Senior Developmental Editor:	Susan Moran
Assistant Editor:	Shannon Moloney
Marketing Manager:	Maureen Rachford
Marketing Assistant:	Cate McCaffery
Director of Media and Assessment:	Amanda Nietzel
Media Editor:	Lori Stover
Director of Content (Sapling Learning):	Clairissa Simmons
Lead Content Developer, Biochemistry (Sapling Learning):	Richard Widstrom
Content Development Manager for Chemistry (Sapling Learning):	Stacy Benson
Visual Development Editor (Media):	Emiko Paul
Director, Content Management Enhancement:	Tracey Kuehn
Managing Editor:	Lisa Kinne
Senior Project Editor:	Liz Geller
Copyeditor:	Linda Strange
Photo Editor:	Christine Buese
Photo Researcher:	Roger Feldman
Text and Cover Design:	Blake Logan
Illustration Coordinator:	Janice Donnola
Illustrations:	H. Adam Steinberg
Molecular Graphics:	H. Adam Steinberg
Production Manager:	Susan Wein
Composition:	Aptara, Inc.
Printing and Binding:	RR Donnelley
Front Cover Image:	H. Adam Steinberg and Quade Paul
Back Cover Photo:	Yigong Shi

Front cover: An active spliceosome from the yeast *Schizosaccharomyces pombe*. The structure, determined by cryo-electron microscopy, captures a molecular moment when the splicing reaction is nearing completion. It includes the snRNAs U2, U5, and U6, a spliced intron lariat, and many associated proteins. Structure determined by Yigong Shi and colleagues, Tsinghua University, Beijing, China (PDB ID 3JB9, C. Yan et al., *Science* 349:1182, 2015). *Back cover:* Randomly deposited individual spliceosome particles, viewed by electron microscopy. The structure on the front cover was obtained by computationally finding the orientations that are superposable, to reduce the noise and strengthen the signal—the structure of the spliceosome. Photo courtesy of Yigong Shi.

Library of Congress Control Number: 2016943661

North American Edition

ISBN-13: 978-1-4641-2611-6
ISBN-10: 1-4641-2611-9

Printed in the United States of America

First printing

W. H. Freeman and Company
One New York Plaza
Suite 4500
New York, NY 10004-1562
www.macmillanlearning.com

International Edition
Macmillan Higher Education
Houndmills, Basingstoke
RG21 6XS, England
www.macmillanhighered.com/international

To Our Teachers

Paul R. Burton

Albert Finholt

William P. Jencks

Eugene P. Kennedy

Homer Knoss

Arthur Kornberg

I. Robert Lehman

Earl K. Nelson

Wesley A. Pearson

David E. Sheppard

Harold B. White

About the Authors

David L. Nelson, born in Fairmont, Minnesota, received his BS in chemistry and biology from St. Olaf College in 1964 and earned his PhD in biochemistry at Stanford Medical School, under Arthur Kornberg. He was a postdoctoral fellow at the Harvard Medical School with Eugene P. Kennedy, who was one of Albert Lehninger's first graduate students. Nelson joined the faculty of the University of Wisconsin–Madison in 1971 and became a full professor of biochemistry in 1982. He was for eight years Director of the Center for Biology Education at the University of Wisconsin–Madison. He became Professor Emeritus in 2013.

Nelson's research focused on the signal transductions that regulate ciliary motion and exocytosis in the protozoan *Paramecium*. He has a distinguished record as a lecturer and research supervisor. For 43 years he taught (with Mike Cox) an intensive survey of biochemistry for advanced biochemistry undergraduates in the life sciences. He has also taught a survey of biochemistry for nursing students, as well as graduate courses on membrane structure and function and on molecular neurobiology. He has received awards for his outstanding teaching, including the Dreyfus Teacher–Scholar Award, the Atwood Distinguished Professorship, and the Underkofler Excellence in Teaching Award from the University of Wisconsin System. In 1991–1992 he was a visiting professor of chemistry and biology at Spelman College. Nelson's second love is history, and in his dotage he teaches the history of biochemistry and collects antique scientific instruments for use in the Madison Science Museum, of which he is the founding president.

Michael M. Cox was born in Wilmington, Delaware. In his first biochemistry course, the first edition of Lehninger's *Biochemistry* was a major influence in refocusing his fascination with biology and inspiring him to pursue a career in biochemistry. After graduating from the University of Delaware in 1974, Cox went to Brandeis University to do his doctoral work with William P. Jencks, and then to Stanford in 1979 for postdoctoral study with I. Robert Lehman. He moved to the University of Wisconsin–Madison in 1983 and became a full professor of biochemistry in 1992.

Cox's doctoral research was on general acid and base catalysis as a model for enzyme-catalyzed reactions. At Stanford, he began work on the enzymes involved in genetic recombination. The work focused particularly on the RecA protein, designing purification and assay methods that are still in use, and illuminating

David L. Nelson and Michael M. Cox. [Source: Robin Davies, UW-Madison Biochemistry MediaLab.]

the process of DNA branch migration. Exploration of the enzymes of genetic recombination has remained a central theme of his research.

Mike Cox has coordinated a large and active research team at Wisconsin, investigating the enzymology, topology, and energetics of the recombinational DNA repair of double-strand breaks in DNA. The work has focused on the bacterial RecA protein, a wide range of proteins that play auxiliary roles in recombinational DNA repair, the molecular basis of extreme resistance to ionizing radiation, directed evolution of new phenotypes in bacteria, and the applications of all of this work to biotechnology.

For more than three decades he has taught a survey of biochemistry to undergraduates and has lectured in graduate courses on DNA structure and topology, protein-DNA interactions, and the biochemistry of recombination. More recent projects are the organization of a new course on professional responsibility for first-year graduate students and establishment of a systematic program to draw talented biochemistry undergraduates into the laboratory at an early stage of their college career. He has received awards for both his teaching and his research, including the Dreyfus Teacher–Scholar Award, the 1989 Eli Lilly Award in Biological Chemistry, and the 2009 Regents Teaching Excellence Award from the University of Wisconsin. He is also highly active in national efforts to provide new guidelines for undergraduate biochemistry education. Cox's hobbies include turning 18 acres of Wisconsin farmland into an arboretum, wine collecting, and assisting in the design of laboratory buildings.

A Note on the Nature of Science

In this twenty-first century, a typical science education often leaves the philosophical underpinnings of science unstated, or relies on oversimplified definitions. As you contemplate a career in science, it may be useful to consider once again the terms **science**, **scientist**, and **scientific method**.

Science is both a way of thinking about the natural world and the sum of the information and theory that result from such thinking. The power and success of science flow directly from its reliance on ideas that can be tested: information on natural phenomena that can be observed, measured, and reproduced and theories that have predictive value. The progress of science rests on a foundational assumption that is often unstated but crucial to the enterprise: that the laws governing forces and phenomena existing in the universe are not subject to change. The Nobel laureate Jacques Monod referred to this underlying assumption as the "postulate of objectivity." The natural world can therefore be understood by applying a process of inquiry—the scientific method. Science could not succeed in a universe that played tricks on us. Other than the postulate of objectivity, science makes no inviolate assumptions about the natural world. A useful scientific idea is one that (1) has been or can be reproducibly substantiated, (2) can be used to accurately predict new phenomena, and (3) focuses on the natural world or universe.

Scientific ideas take many forms. The terms that scientists use to describe these forms have meanings quite different from those applied by nonscientists. A *hypothesis* is an idea or assumption that provides a reasonable and testable explanation for one or more observations, but it may lack extensive experimental substantiation. A *scientific theory* is much more than a hunch. It is an idea that has been substantiated to some extent and provides an explanation for a body of experimental observations. A theory can be tested and built upon and is thus a basis for further advance and innovation. When a scientific theory has been repeatedly tested and validated on many fronts, it can be accepted as a fact.

In one important sense, what constitutes science or a scientific idea is defined by whether or not it is published in the scientific literature after peer review by other working scientists. As of late 2014, about 34,500 peer-reviewed scientific journals worldwide were publishing some 2.5 million articles each year, a continuing rich harvest of information that is the birthright of every human being.

Scientists are individuals who rigorously apply the scientific method to understand the natural world. Merely having an advanced degree in a scientific discipline does not make one a scientist, nor does the lack of such a degree prevent one from making important scientific contributions. A scientist must be willing to challenge any idea when new findings demand it. The ideas that a scientist accepts must be based on measurable, reproducible observations, and the scientist must report these observations with complete honesty.

The **scientific method** is a collection of paths, all of which may lead to scientific discovery. In the *hypothesis and experiment* path, a scientist poses a hypothesis, then subjects it to experimental test. Many of the processes that biochemists work with every day were discovered in this manner. The DNA structure elucidated by James Watson and Francis Crick led to the hypothesis that base pairing is the basis for information transfer in polynucleotide synthesis. This hypothesis helped inspire the discovery of DNA and RNA polymerases.

Watson and Crick produced their DNA structure through a process of *model building and calculation*. No actual experiments were involved, although the model building and calculations used data collected by other scientists. Many adventurous scientists have applied the process of *exploration and observation* as a path to discovery. Historical voyages of discovery (Charles Darwin's 1831 voyage on H.M.S. *Beagle* among them) helped to map the planet, catalog its living occupants, and change the way we view the world. Modern scientists follow a similar path when they explore the ocean depths or launch probes to other planets. An analog of hypothesis and experiment is *hypothesis and deduction*. Crick reasoned that there must be an adaptor molecule that facilitated translation of the information in messenger RNA into protein. This adaptor hypothesis led to the discovery of transfer RNA by Mahlon Hoagland and Paul Zamecnik.

Not all paths to discovery involve planning. *Serendipity* often plays a role. The discovery of penicillin by Alexander Fleming in 1928 and of RNA catalysts by Thomas Cech in the early 1980s were both chance discoveries, albeit by scientists well prepared to exploit them. *Inspiration* can also lead to important advances. The polymerase chain reaction (PCR), now a central part of biotechnology, was developed by Kary Mullis after a flash of inspiration during a road trip in northern California in 1983.

These many paths to scientific discovery can seem quite different, but they have some important things in common. They are focused on the natural world. They rely on *reproducible observation* and/or *experiment*. All of the ideas, insights, and experimental facts that arise from these endeavors can be tested and reproduced by scientists anywhere in the world. All can be used by other scientists to build new hypotheses and make new discoveries. All lead to information that is properly included in the realm of science. Understanding our universe requires hard work. At the same time, no human endeavor is more exciting and potentially rewarding than trying, with occasional success, to understand some part of the natural world.

With the advent of increasingly robust technologies that provide cellular and organismal views of molecular processes, progress in biochemistry continues apace, providing both new wonders and new challenges. The image on our cover depicts an active spliceosome, one of the largest molecular machines in a eukaryotic cell, and one that is only now yielding to modern structural analysis. It is an example of our current understanding of life at the level of molecular structure. The image is a snapshot from a highly complex set of reactions, in better focus than ever before. But in the cell, this is only one of many steps linked spatially and temporally to many other complex processes that remain to be unraveled and eventually described in future editions. Our goal in this seventh edition of *Lehninger Principles of Biochemistry*, as always, is to strike a balance: to include new and exciting research findings without making the book overwhelming for students. The primary criterion for inclusion of an advance is that the new finding helps to illustrate an important *principle of biochemistry*.

With every revision of this textbook, we have striven to maintain the qualities that made the original Lehninger text a classic: clear writing, careful explanations of difficult concepts, and insightful communication to students of the ways in which biochemistry is understood and practiced today. We have coauthored this text and taught introductory biochemistry together for three decades. Our thousands of students at the University of Wisconsin–Madison over those years have been an endless source of ideas on how to present biochemistry more clearly; they have enlightened and inspired us. We hope that this seventh edition of *Lehninger* will, in turn, enlighten current students of biochemistry everywhere, and inspire all of them to love biochemistry as we do.

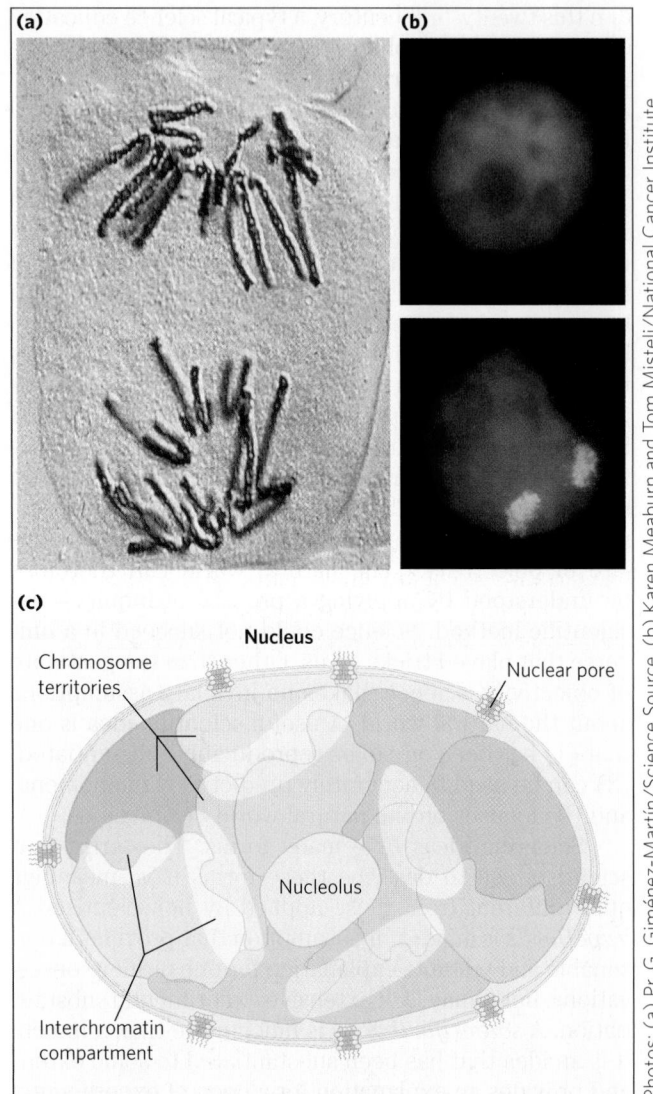

Chromosomal organization in the eukaryotic nucleus

Photos: (a) Pr. G. Giménez-Martín/Science Source. (b) Karen Meaburn and Tom Misteli/National Cancer Institute.

NEW Leading-Edge Science

Among the new or substantially updated topics in this edition are:

■ Synthetic cells and disease genomics (Chapter 1)

■ Intrinsically disordered protein segments (Chapter 4) and their importance in signaling (Chapter 12)

■ Pre–steady state enzyme kinetics (Chapter 6)

■ Gene annotation (Chapter 9)

■ Gene editing with CRISPR (Chapter 9)

■ Membrane trafficking and dynamics (Chapter 11)

■ Additional roles for NADH (Chapter 13)

■ Cellulose synthase complex (Chapter 20)

■ Specialized pro-resolving mediators (Chapter 21)

■ Peptide hormones: incretins and blood glucose; irisin and exercise (Chapter 23)

■ Chromosome territories (Chapter 24)

■ New details of eukaryotic DNA replication (Chapter 25)

■ Cap-snatching; spliceosome structure (Chapter 26)

■ Ribosome rescue; RNA editing update (Chapter 27)

■ New roles for noncoding RNAs (Chapters 26, 28)

■ RNA recognition motif (Chapter 28)

NEW Tools and Technology

The emerging tools of systems biology continue to transform our understanding of biochemistry. These include both new laboratory methods and large, public databases that have become indispensable to researchers. New to this edition of *Lehninger Principles of Biochemistry*:

■ Next-generation DNA sequencing now includes ion semiconductor sequencing (Ion Torrent) and single-molecule real-time (SMRT) sequencing platforms, and the text discussion now follows the description of classical Sanger sequencing (Chapter 8).

■ Gene editing by CRISPR is one of many updates to the discussion of genomics (Chapter 9).

■ LIPID MAPS database and system of classifying lipids is included in the discussion of lipidomics (Chapter 10).

■ Cryo-electron microscopy is described in a new box (Chapter 19).

■ Ribosome profiling to determine which genes are being translated at any given moment, and many related technologies, are included to illustrate the versatility and power of deep DNA sequencing (Chapter 27).

■ Online data resources such as NCBI, PDB, SCOP2, KEGG, and BLAST, mentioned in the text, are listed in the back endpapers for easy reference.

NEW Consolidation of Plant Metabolism

All of plant metabolism is now consolidated into a single chapter, Chapter 20, separate from the discussion of oxidative phosphorylation in Chapter 19. Chapter 20 includes light-driven ATP synthesis, carbon fixation, photorespiration, the glyoxylate cycle, starch and cellulose synthesis, and regulatory mechanisms that ensure integration of all of these activities throughout the plant.

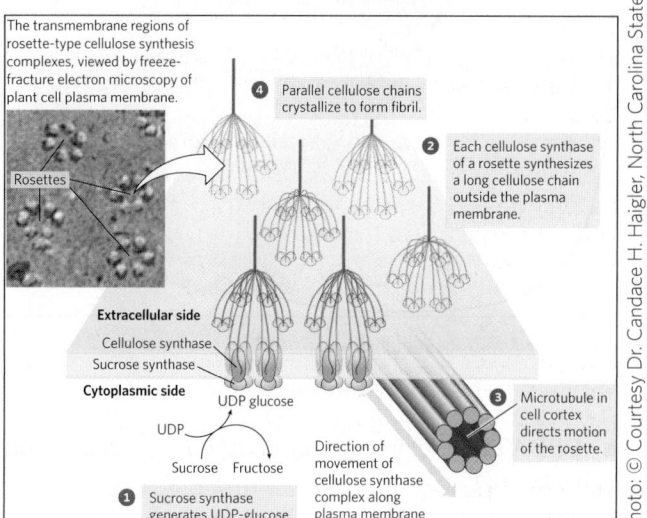

Model for the synthesis of cellulose

Medical Insights and Applications

This icon is used throughout the book to denote material of special medical interest. As teachers, our goal is for students to learn biochemistry and to understand its relevance to a healthier life and a healthier planet. Many sections explore what we know about the molecular mechanisms of disease. The new and updated medical topics in this edition are:

■ **UPDATED** Lactase and lactose intolerance (Chapter 7)

■ **NEW** Guillain-Barré syndrome and gangliosides (Chapter 10)

■ **NEW** Golden Rice Project to prevent diseases of vitamin A deficiency (Chapter 10)

■ **UPDATED** Multidrug resistance transporters and their importance in clinical medicine (Chapter 11)

■ **NEW** Insight into cystic fibrosis and its treatment (Chapter 11)

Structure of the GroEL chaperone protein, as determined by cryo-EM

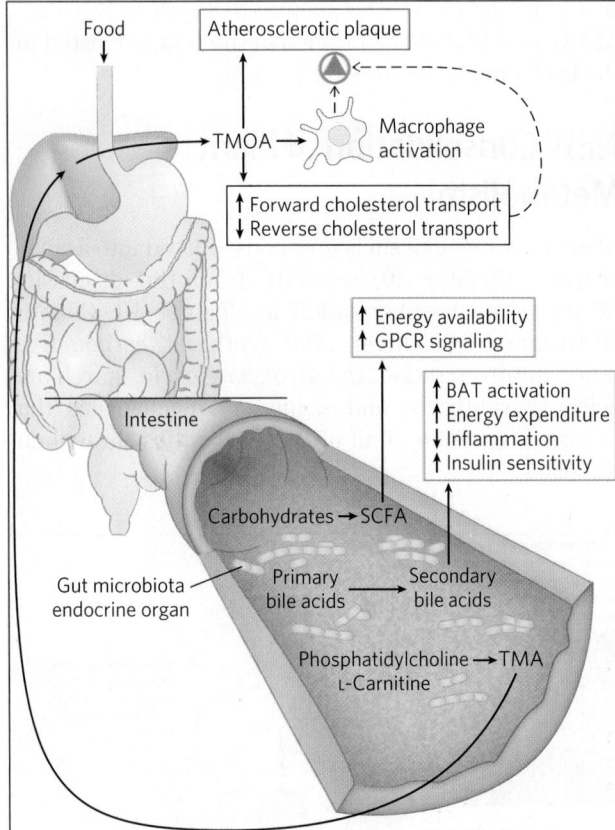

Effects of gut microbe metabolism on health

topics that highlight the interplay of metabolism, obesity, and diabetes are:

- Acidosis in untreated diabetes (Chapter 2)

- Defective protein folding, amyloid deposition in the pancreas, and diabetes (Chapter 4)

- **UPDATED** Blood glucose and glycated hemoglobin in the diagnosis and treatment of diabetes (Box 7-1)

- Advanced glycation end products (AGEs): their role in the pathology of advanced diabetes (Box 7-1)

- Defective glucose and water transport in two forms of diabetes (Box 11-1)

- **NEW** Na^+-glucose transporter and the use of gliflozins in the treatment of type 2 diabetes (Chapter 11)

- Glucose uptake deficiency in type 1 diabetes (Chapter 14)

- MODY: a rare form of diabetes (Box 15-3)

- Ketone body overproduction in diabetes and starvation (Chapter 17)

- **NEW** Breakdown of amino acids: methylglyoxal as a contributor to type 2 diabetes (Chapter 18)

- A rare form of diabetes resulting from defects in mitochondria of pancreatic β cells (Chapter 19)

- Thiazolidinedione-stimulated glyceroneogenesis in type 2 diabetes (Chapter 21)

- Role of insulin in countering high blood glucose (Chapter 23)

- Secretion of insulin by pancreatic β cells in response to changes in blood glucose (Chapter 23)

- How insulin was discovered and purified (Box 23-1)

- **NEW** AMP-activated protein kinase in the hypothalamus in integration of hormonal inputs from gut, muscle, and adipose tissues (Chapter 23)

- **UPDATED** Role of mTORC1 in regulating cell growth (Chapter 23)

- **NEW** Brown and beige adipose as thermogenic tissues (Chapter 23)

- **NEW** Exercise and the stimulation of irisin release and weight loss (Chapter 23)

- **NEW** Short-term eating behavior influenced by ghrelin, PYY_{3-36}, and cannabinoids (Chapter 23)

- **NEW** Role of microbial symbionts in the gut in influencing energy metabolism and adipogenesis (Chapter 23)

- Tissue insensitivity to insulin in type 2 diabetes (Chapter 23)

- **UPDATED** Management of type 2 diabetes with diet, exercise, medication, and surgery (Chapter 23)

- **UPDATED** Colorectal cancer: multistep progression (Chapter 12)

- **NEW** Newborn screening for acyl-carnitine to diagnose mitochondrial disease (Chapter 17)

- **NEW** Mitochondrial diseases, mitochondrial donation, and "three-parent babies" (Chapter 19)

- **UPDATED** Cholesterol metabolism, plaque formation, and atherosclerosis (Chapter 21)

- **UPDATED** Cytochrome P-450 enzymes and drug interactions (Chapter 21)

- **UPDATED** Ammonia toxicity in the brain (Chapter 22)

- **NEW** Xenobiotics as endocrine disruptors (Chapter 23)

Special Theme: Metabolic Integration, Obesity, and Diabetes

Obesity and its medical consequences, including cardiovascular disease and diabetes, are fast becoming epidemic in the industrialized world, and throughout this edition we include new material on the biochemical connections between obesity and health. Our focus on diabetes provides an integrating theme throughout the chapters on metabolism and its control. Some of the

Special Theme: Evolution

Every time a biochemist studies a developmental pathway in nematodes, identifies key parts of an enzyme active site by determining which parts are conserved among species, or searches for the gene underlying a human genetic disease, he or she is relying on evolutionary theory. Funding agencies support work on nematodes with the expectation that the insights gained will be relevant to humans. The conservation of functional residues in an enzyme active site telegraphs the shared history of all organisms on the planet. More often than not, the search for a disease gene is a sophisticated exercise in phylogenetics. Evolution is thus a foundational concept for our discipline. Some of the many areas that discuss biochemistry from an evolutionary viewpoint:

■ Changes in hereditary instructions that allow evolution (Chapter 1)

■ Origins of biomolecules in chemical evolution (Chapter 1)

■ RNA or RNA precursors as the first genes and catalysts (Chapters 1, 26)

■ Timetable of biological evolution (Chapter 1)

■ Use of inorganic fuels by early cells (Chapter 1)

■ Evolution of eukaryotes from simpler cells (endosymbiont theory) (Chapters 1, 19, 20)

■ Protein sequences and evolutionary trees (Chapter 3)

■ Role of evolutionary theory in protein structure comparisons (Chapter 4)

■ Evolution of antibiotic resistance in bacteria (Chapter 6)

■ Evolutionary explanation for adenine nucleotides being components of many coenzymes (Chapter 8)

■ Comparative genomics and human evolution (Chapter 9)

■ Using genomics to understand Neanderthal ancestry (Box 9-3)

■ Evolutionary relationships between V-type and F-type ATPases (Chapter 11)

■ Universal features of GPCR systems (Chapter 12)

■ Evolutionary divergence of β-oxidation enzymes (Chapter 17)

■ Evolution of oxygenic photosynthesis (Chapter 20)

■ **NEW** Presence of organelles, including nuclei, in planctomycete bacteria (Box 22-1)

■ Role of transposons in evolution of the immune system (Chapter 25)

■ Common evolutionary origin of transposons, retroviruses, and introns (Chapter 26)

■ Consolidated discussion of the RNA world hypothesis (Chapter 26)

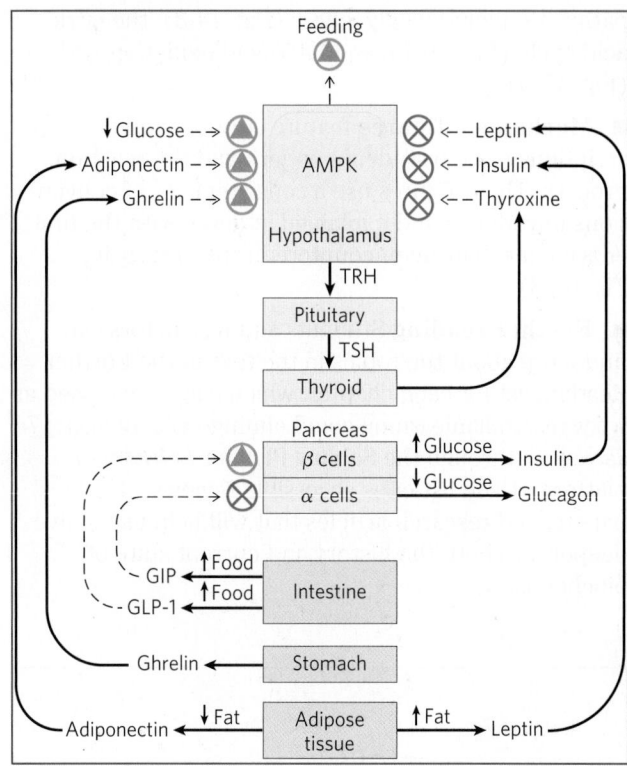

Regulation of feeding behavior

■ Natural variations in the genetic code—exceptions that prove the rule (Box 27-1)

■ Natural and experimental expansion of the genetic code (Box 27-2)

■ Regulatory genes in development and speciation (Box 28-1)

Lehninger Teaching Hallmarks

Students encountering biochemistry for the first time often have difficulty with two key aspects of the course: approaching quantitative problems and drawing on what they have learned in organic chemistry to help them understand biochemistry. These same students must also learn a complex language, with conventions that are often unstated. To help students cope with these challenges, we provide the following study aids:

Focus on Chemical Logic

■ Section 13.2, **Chemical Logic and Common Biochemical Reactions**, discusses the common biochemical reaction types that underlie all metabolic reactions, helping students to connect organic chemistry with biochemistry.

■ **Chemical logic figures** highlight the conservation of mechanism and illustrate patterns that make learning pathways easier. Chemical logic figures are provided for each of the central metabolic pathways, including glycolysis (Fig. 14-3), the citric acid cycle (Fig. 16-7), and fatty acid oxidation (Fig. 17-9).

■ **Mechanism figures** feature step-by-step descriptions to help students understand the reaction process. These figures use a consistent set of conventions introduced and explained in detail with the first enzyme mechanism encountered (chymotrypsin, Fig. 6-23).

■ **Further reading** Students and instructors can find more about the topics in the text in the Further Reading list for each chapter, which can be accessed at www.macmillanlearning.com/LehningerBiochemistry7e as well as through the Sapling Plus for *Lehninger* platform. Each list cites accessible reviews, classic papers, and research articles that will help users dive deeper into both the history and current state of biochemistry.

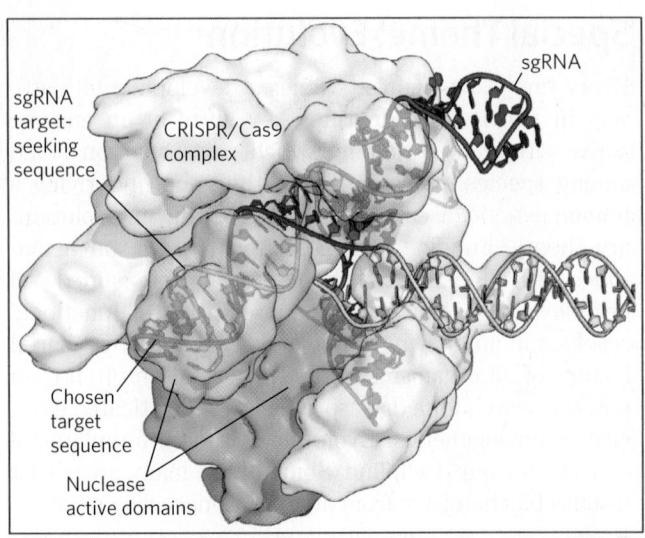

CRISPR/Cas9 structure

■ Figures with **numbered, annotated steps** help explain complex processes.

■ **Summary figures** help students keep the big picture in mind while learning the specifics.

Problem-Solving Tools

■ **In-text Worked Examples** help students improve their quantitative problem-solving skills, taking them through some of the most difficult equations.

■ **More than 600 end-of-chapter problems** give students further opportunity to practice what they have learned.

■ **Data Analysis Problems** (one at the end of each chapter), contributed by Brian White of the University of Massachusetts Boston, encourage students to synthesize what they have learned and apply their knowledge to interpretation of data from the research literature.

Key Conventions

Many of the conventions that are so necessary for understanding each biochemical topic and the biochemical literature are broken out of the text and highlighted. These **Key Conventions** include clear statements of many assumptions and conventions that students are often expected to assimilate without being told (for example, peptide sequences are written from amino- to carboxyl-terminal end, left to right; nucleotide sequences are written from 5′ to 3′ end, left to right).

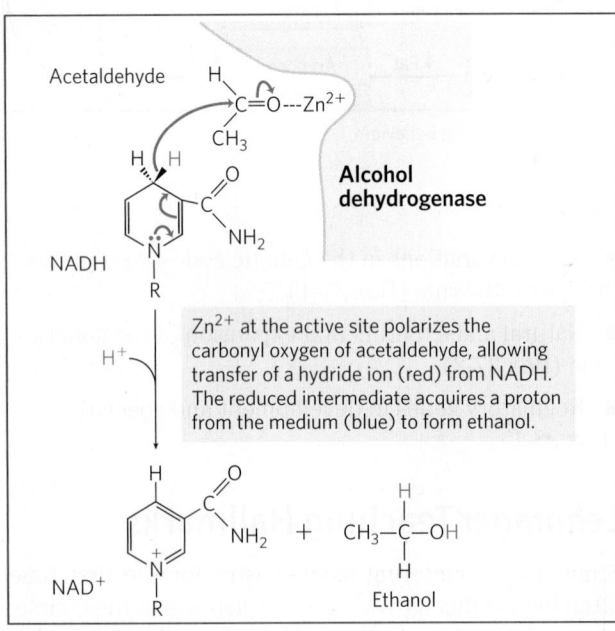

Alcohol dehydrogenase reaction mechanism

Clear Art

■ **Smarter renditions of classic figures** are easier to interpret and learn from.

■ **Molecular structures** are created specifically for this book, using shapes and color schemes that are internally consistent.

Media and Supplements

For this edition of *Lehninger Principles of Biochemistry*, we have thoroughly revised and refreshed the extensive set of online learning tools. In particular, we are moving to a well-established platform that, for the first time, allows us to provide a comprehensive online homework system.

NEW 🅜 Sapling Plus for *Lehninger*

This comprehensive and robust online teaching and learning platform incorporates the e-Book, all instructor and student resources, and instructor assignment and gradebook functionality.

NEW Student Resources in 🅜 Sapling Plus for *Lehninger*

Students are provided with media designed to enhance their understanding of biochemical principles and improve their problem-solving ability.

NEW Online Homework

Sapling Plus for *Lehninger* offers robust, high-level homework questions, with hints and wrong-answer feedback targeted to students' misconceptions, as well as detailed worked-out solutions to reinforce concepts.

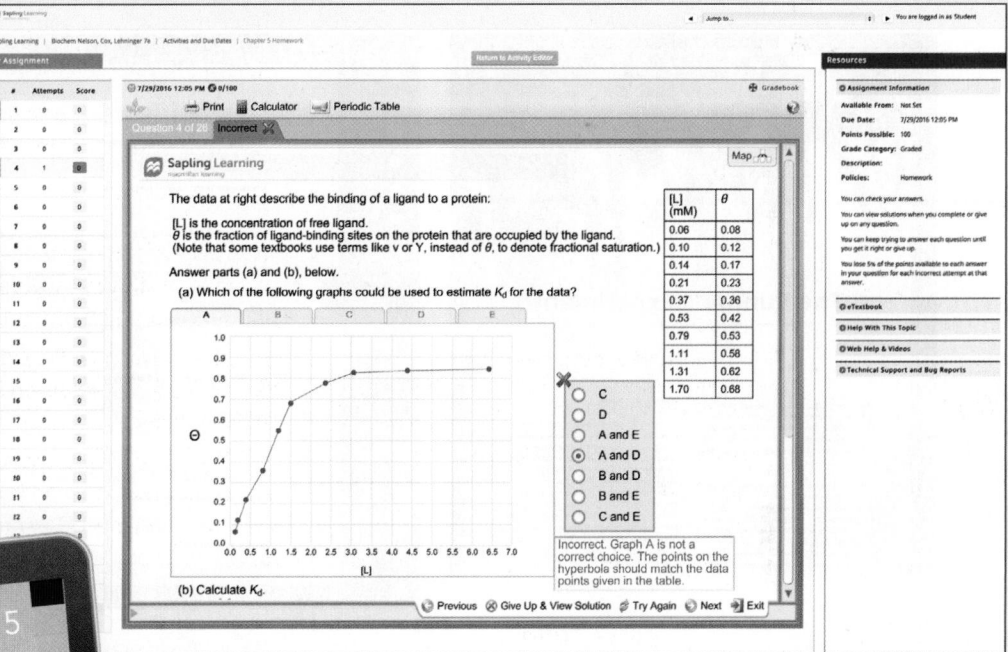

e-Book

The e-Book contains the full contents of the text and embedded links to important media assets (listed on the next two pages).

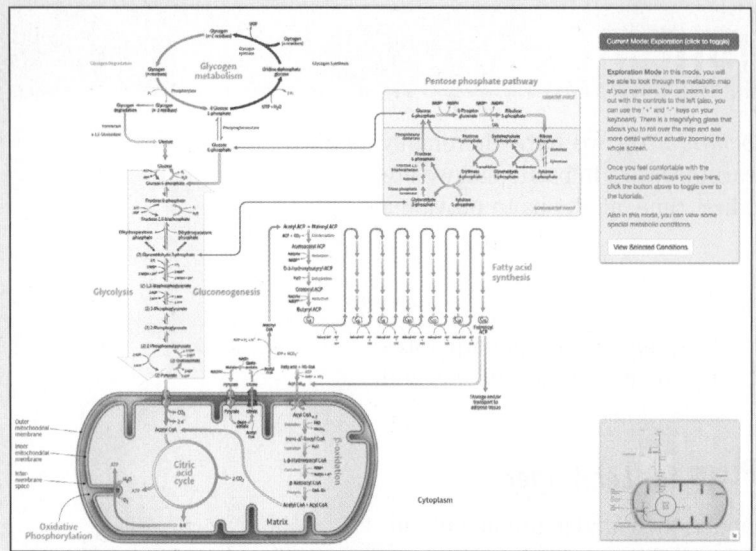

NEW Interactive Metabolic Map

The Interactive Metabolic Map guides students through the most commonly taught metabolic pathways: glycolysis, the citric acid cycle, and β-oxidation. Students can navigate and zoom between overview and detailed views of the map, allowing them to integrate the big-picture connections and fine-grain details of the pathways. Tutorials guide students through the pathways to achieve key learning outcomes. Concept check questions along the way confirm understanding.

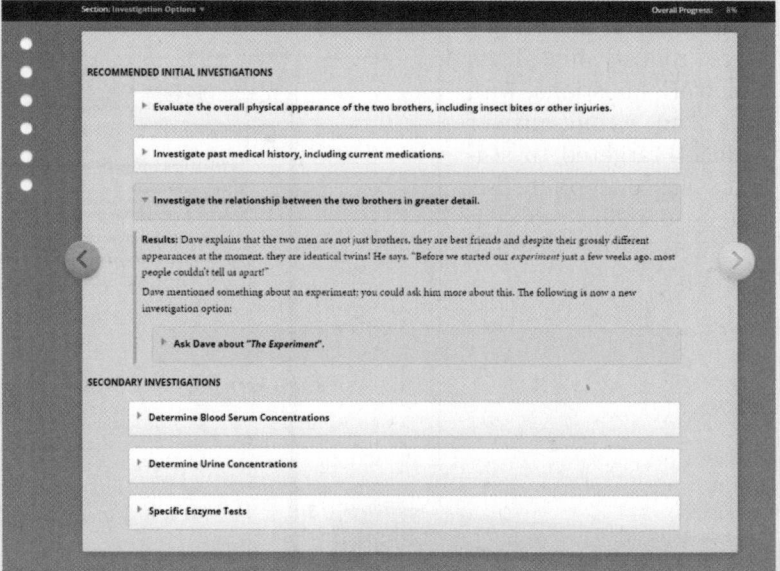

NEW Case Studies

By Justin Hines (Lafayette College) Each of several online case studies introduces students to a biochemical mystery and allows them to determine what investigations to complete as they search for a solution. Final assessments ensure that students have fully completed and understood each case study.

- A Likely Story: Enzyme Inhibition
- An Unexplained Death: Carbohydrate Metabolism
- A Day at the Beach: Lipid Metabolism
- The Runner's Experiment: Integration of Metabolism

- Sudden Onset: Introduction to Metabolism
- Toxic Alcohols: Enzyme Function

More case studies will be added over the course of this edition.

UPDATED Molecular Structure Tutorials

For the seventh edition, these tutorials are updated to JSmol, and now include assessment with targeted feedback to ensure that students grasp key concepts learned from examining various molecular structures in depth.

- Protein Architecture (Chapter 3)
- Oxygen-Binding Proteins (Chapter 5)
- Major Histocompatibility Complex (MHC) Molecules (Chapter 5)
- Nucleotides: Building Blocks of Nucleic Acids (Chapter 8)
- Trimeric G Proteins (Chapter 12)
- Bacteriorhodopsin (Chapter 20)
- Restriction Endonucleases (Chapter 25)
- The Hammerhead Ribozyme: An RNA Enzyme (Chapter 26)
- Lac Repressor: A Gene Regulator (Chapter 28)

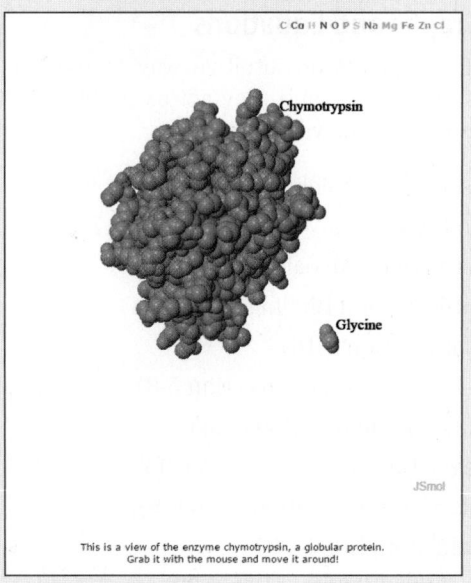

This is a view of the enzyme chymotrypsin, a globular protein. Grab it with the mouse and move it around!

NEW Simulations

Created using art from the text, these biochemical simulations reinforce students' understanding by allowing them to interact with the structures and processes they have encountered. A gamelike format guides students through the simulations. Multiple-choice questions after each simulation ensure that instructors can assess whether students have thoroughly understood each topic.

- Nucleotide Structure
- DNA/RNA Structure
- PCR
- Sanger Sequencing
- CRISPR
- DNA Replication
- DNA Polymerase
- Mutation and Repair
- Transcription
- mRNA Processing
- Translation

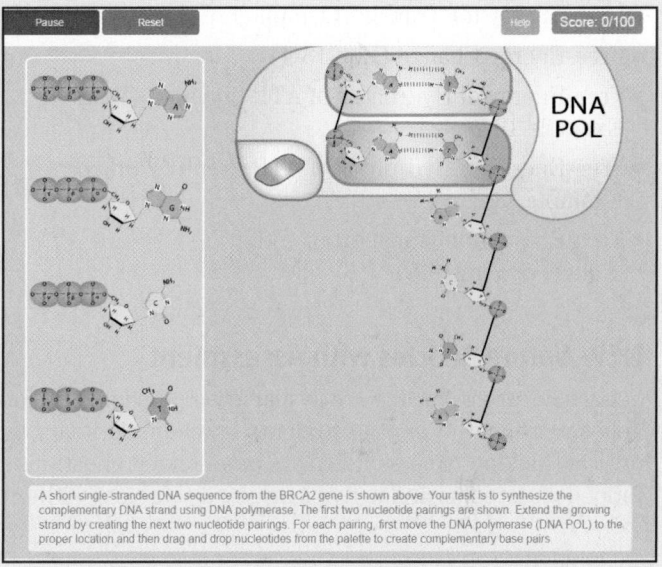

A short single-stranded DNA sequence from the BRCA2 gene is shown above. Your task is to synthesize the complementary DNA strand using DNA polymerase. The first two nucleotide pairings are shown. Extend the growing strand by creating the next two nucleotide pairings. For each pairing, first move the DNA polymerase (DNA POL) to the proper location and then drag and drop nucleotides from the palette to create complementary base pairs.

UPDATED Animated Mechanism Figures

Many mechanism figures from the text are available as animations, accompanied by assessment with targeted feedback. These animations help students learn about key mechanisms at their own pace.

Living Graphs and Equations

These offer students an intuitive way to explore the equations in the text, and they act as problem-solving tools for online homework.

- Henderson-Hasselbalch Equation (Eqn 2-9)
- Titration Curve for a Weak Acid (Fig. 2-17)
- Binding Curve for Myoglobin (Eqn 5-11)
- Cooperative Ligand Binding (Eqn 5-14)
- Hill Equation (Eqn 5-16)
- Protein-Ligand Interactions (Eqn 5-8)
- Competitive Inhibitor (Eqn 6-28)
- Lineweaver-Burk Equation (Box 6-1)
- Michaelis-Menten Equation (Eqn 6-9)
- Mixed Inhibitor (Eqn 6-30)
- Uncompetitive Inhibitor (Eqn 6-29)
- Free-Energy for Transport Equation (Eqn 11-3)
- Free-Energy for Transport Graph (Eqn 11-3)
- Free-Energy Change (Eqn 13-4)
- Free-Energy of Hydrolysis of ATP Equation (Worked Example 13-2)
- Free-Energy of Hydrolysis of ATP Graph (Worked Example 13-2)
- Free-Energy for Transport of an Ion (Eqn 11-4, Eqn 19-8)

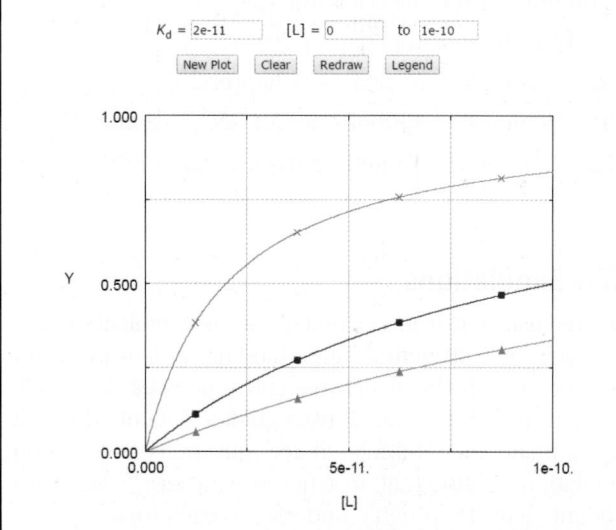

Protein-Ligand Interactions (Equation 5.8)

Proteins function by interacting with other molecules. Molecules that interact reversibly with proteins, without being altered by the interaction, are called ligands. Protein-ligand interactions are at the heart of countless biochemical processes. For this reason, understanding how these interactions are studied and described quantitatively is fundamental to Biochemistry. As described in the text, the key parameter used to describe these interactions is the dissociation constant, K_d. The exercise below will help you understand how the K_d relates the ligand concentration to the fraction of protein binding sites that are occupied by ligand (a parameter called Y). A protein with a high affinity for its ligand will have a low K_d, meaning it will bind to the ligand at quite low ligand concentrations. Similarly, a protein with a low affinity for its ligand will have a high K_d.

$$Y = \frac{[L]}{[L] + K_d}$$

Use the default values and have the computer draw a ligand binding curve by clicking on **New Plot**. Then decrease the K_d by a factor of 10, and draw a new plot. Note how the curve moves to the left as the K_d decreases. Decrease the K_d by another factor of 10, and draw a new plot again. Insert your own values of K_d (see table), and the range over which the X axis will plot the parameter Y, until you have familiarized yourself with how the equation works.

Up to 5 plots can be displayed at one time. The **Clear** button will remove all plots. To see K_d for each plot hit the **Legend** button. The **Redraw** button will refresh the graph. This is useful when the function domain (i.e., **[L]**) has been changed. To see the value of each plot at a given point, move your cursor to the desired location then click.

K_d = 2e-11   [L] = 0   to 1e-10

New Plot   Clear   Redraw   Legend

NEW *Nature* Articles with Assessment

Six articles from *Nature* are available accompanied by tailored, automatically gradable assessment to engage students in reading primary literature and to encourage critical thinking. Also included are open-ended questions that are suitable for use in flipped classrooms and active learning discussions either in class or online.

Animated Biochemical Techniques

Nine animations illustrate the principles behind some of the most commonly used laboratory methods.

Problem-Solving Videos

Created by Scott Ensign of Utah State University, these videos provide students with 24/7 online problem-solving help. Through a two-part approach, each 10-minute video covers a key textbook problem representing a topic that students traditionally struggle to master. Dr. Ensign first describes a proven problem-solving strategy and then applies the strategy to the problem at hand, in clear, concise steps. Students can easily pause, rewind, and review any steps until they firmly grasp, not just the solution, but also the reasoning behind it. Working through the problem in this way is designed to make students better and more confident at applying key strategies as they solve other textbook and exam problems.

Student Print Resources: *The Absolute, Ultimate Guide to Lehninger Principles of Biochemistry*

The Absolute, Ultimate Guide to Lehninger Principles of Biochemistry, Seventh Edition, Study Guide and Solutions Manual, by Marcy Osgood (University of New Mexico School of Medicine) and Karen Ocorr (Sanford-Burnham Medical Research Institute); ISBN 1-4641-8797-5

This guide combines an innovative study guide with a reliable solutions manual (providing extended solutions to end-of-chapter problems) in one convenient volume. Thoroughly class-tested, the study guide includes, for each chapter:

- **Major Concepts**: a road map through the chapter
- **What to Review**: questions that recap key points from previous chapters
- **Discussion Questions**: provided for each section; designed for individual review, study groups, or classroom discussion
- **Self-Test**: "Do you know the terms?"; crossword puzzles; multiple-choice, fact-driven questions; and questions that ask students to apply their new knowledge in new directions—plus answers!

Instructor Resources

Instructors are provided with a comprehensive set of teaching tools, each developed to support the text, lecture presentations, and individual teaching styles. All of these resources are available for download from Sapling Plus for *Lehninger* and from the catalog page at www.MacmillanLearning.com.

Test Bank

A comprehensive test bank, in editable Microsoft Word and Diploma formats, includes 30 to 50 new multiple-choice and short-answer problems per chapter, for a total of 100 questions or more per chapter, each rated by Bloom's level and level of difficulty.

Lecture Slides

Editable lecture slides are tailored to the content of this new edition, with updated, optimized art and text.

Clicker Questions

These dynamic multiple-choice questions can be used with iClicker or other classroom response systems. The clicker questions are written specifically to foster active learning in the classroom and to better inform instructors on students' misunderstandings.

Fully Optimized Art Files

Fully optimized files are available for every figure, photo, and table in the text, featuring enhanced color, high resolution, and enlarged fonts. These files are available as JPEGs or are preloaded into PowerPoint format for each chapter.

Acknowledgments

This book is a team effort, and producing it would be impossible without the outstanding people at W. H. Freeman and Company who have supported us at every step along the way. Susan Moran, Senior Developmental Editor, and Lauren Schultz, Executive Editor, helped develop the revision plan for this edition, made many helpful suggestions, encouraged us, and tried valiantly (if not always successfully) to keep us on schedule. Our outstanding Project Editor, Liz Geller, showed remarkable patience as we regularly failed to meet her deadlines. We thank Design Manager Blake Logan for her artistry in designing the text for the book. We thank Photo Researcher Roger Feldman and Photo Editor Christine Buese for their help in locating images and obtaining permission to use them, and Shannon Moloney for help in orchestrating reviews and providing administrative assistance at many turns. We also thank Lori Stover, Media Editor, Amanda Nietzel, Director of Media and Assessment, and Elaine Palucki, Senior Educational Technology Advisor, for envisioning and overseeing the increasingly important media components to supplement the text. Our gratitude also goes to Maureen Rachford, Marketing Manager, for coordinating the sales and marketing effort. We also wish to thank Kate Parker, whose work on previous editions is still visible in this one. In Madison, Brook Soltvedt is, and has been for all the editions we have worked on, our invaluable first-line editor and critic. She is the first to see manuscript chapters, aids in manuscript and art development, ensures internal consistency in content and nomenclature, and keeps us on task with more-or-less gentle prodding. The deft hand of Linda Strange, who has copyedited all but one edition of this textbook (including the first), is evident in the clarity of the text. She has encouraged and inspired us with her high scientific and literary standards. As she did for the three previous editions, Shelley Lusetti, of New Mexico State University, read every word of the text in proofs, caught numerous mistakes, and made many suggestions that improved the book. The new art and molecular graphics were created by Adam Steinberg of Art for Science, who often made valuable suggestions that led to better and clearer illustrations. We feel very fortunate to have such gifted partners as Brook, Linda, Shelley, and Adam on our team.

We are also deeply indebted to Brian White of the University of Massachusetts Boston, who wrote the data analysis problems at the end of each chapter.

Many others helped us shape this seventh edition with their comments, suggestions, and criticisms. To all of them, we are deeply grateful:

Rebecca Alexander, *Wake Forest University*
Richard Amasino, *University of Wisconsin–Madison*
Mary Anderson, *Texas Woman's University*
Steve Asmus, *Centre College*
Kenneth Balazovich, *University of Michigan*
Rob Barber, *University of Wisconsin–Parkside*

David Bartley, *Adrian College*
Johannes Bauer, *Southern Methodist University*
John Bellizzi, *University of Toledo*
Chris Berndsen, *James Madison University*
James Blankenship, *Cornell University*
Kristopher Blee, *California State University, Chico*
William Boadi, *Tennessee State University*
Sandra Bonetti, *Colorado State University–Pueblo*
Rebecca Bozym, *La Roche College*
Mark Brandt, *Rose-Hulman Institute of Technology*
Ronald Brosemer, *Washington State University*
Donald Burden, *Middle Tennessee State University*
Samuel Butcher, *University of Wisconsin–Madison*
Jeffrey Butikofer, *Upper Iowa University*
Colleen Byron, *Ripon College*
Patricia Canaan, *Oklahoma State University*
Kevin Cannon, *Pennsylvania State Abington College*
Weiguo Cao, *Clemson University*
David Casso, *San Francisco State University*
Brad Chazotte, *Campbell University College of Pharmacy & Health Sciences*
Brooke Christian, *Appalachian State University*
Jeff Cohlberg, *California State University, Long Beach*
Kathryn Cole, *Christopher Newport University*
Jeannie Collins, *University of Southern Indiana*
Megen Culpepper, *Appalachian State University*
Tomas T. Ding, *North Carolina Central University*
Cassidy Dobson, *St. Cloud State University*
Justin Donato, *University of St. Thomas*
Dan Edwards, *California State University, Chico*
Shawn Ellerbroek, *Wartburg College*
Donald Elmore, *Wellesley College*
Ludeman Eng, *Virginia Tech*
Scott Ensign, *Utah State University*
Megan Erb, *George Mason University*
Brent Feske, *Armstrong State University*
Emily Fisher, *Johns Hopkins University*
Marcello Forconi, *College of Charleston*
Wilson Francisco, *Arizona State University*
Amy Gehring, *Williams College*
Jack Goldsmith, *University of South Carolina*
Donna Gosnell, *Valdosta State University*
Lawrence Gracz, *MCPHS University*
Steffen Graether, *University of Guelph*
Michael Griffin, *Chapman University*
Marilena Hall, *Stonehill College*
Prudence Hall, *Hiram College*
Marc Harrold, *Duquesne University*
Mary Hatcher-Skeers, *Scripps College*
Pam Hay, *Davidson College*
Robin Haynes, *Harvard University Extension School*
Deborah Heyl-Clegg, *Eastern Michigan University*
Julie Himmelberger, *DeSales University*
Justin Hines, *Lafayette College*
Charles Hoogstraten, *Michigan State University*
Lori Isom, *University of Central Arkansas*
Roberts Jackie, *DePauw University*
Blythe Janowiak, *Mulligan Saint Louis University*
Constance Jeffery, *University of Illinois at Chicago*
Gerwald Jogl, *Brown University*
Kelly Johanson, *Xavier University of Louisiana*
Jerry Johnson, *University of Houston–Downtown*
Warren Johnson, *University of Wisconsin–Green Bay*
David Josephy, *University of Guelph*
Douglas Julin, *University of Maryland*

Jason Kahn, *University of Maryland*
Marina Kazakevich, *University of Massachusetts Dartmouth*
Mark Kearley, *Florida State University*
Michael Keck, *Keuka College*
Sung-Kun Kim, *Baylor University*
Janet Kirkley, *Knox College*
Robert Kiss, *McGill University*
Michael Koelle, *Yale University*
Dmitry Kolpashchikov, *University of Central Florida*
Andrey Krasilnikov, *Pennsylvania State University*
Amanda Krzysiak, *Bellarmine University*
Terrance Kubiseski, *York University*
Maria Kuhn, *Madonna University*
Min-Hao Kuo, *Michigan State University*
Charles Lauhon, *University of Wisconsin*
Paul Laybourn, *Colorado State University*
Scott Lefler, *Arizona State University*
Brian Lemon, *Brigham Young University–Idaho*
Aime Levesque, *University of Hartford*
Randy Lewis, *Utah State University*
Hong Li, *Florida State University*
Pan Li, *University at Albany, SUNY*
Brendan Looyenga, *Calvin College*
Argelia Lorence, *Arkansas State University*
John Makemson, *Florida International University*
Francis Mann, *Winona State University*
Steven Mansoorabadi, *Auburn University*
Lorraine Marsh, *Long Island University*
Tiffany Mathews, *Pennsylvania State University*
Douglas McAbee, *California State University, Long Beach*
Diana McGill, *Northern Kentucky University*
Karen McPherson, *Delaware Valley College*
Michael Mendenhall, *University of Kentucky*
Larry Miller, *Westminster College*
Rakesh Mogul, *California State Polytechnic University, Pomona*
Judy Moore, *Lenoir-Rhyne University*
Trevor Moraes, *University of Toronto*
Graham Moran, *University of Wisconsin–Milwaukee*
Tami Mysliwiec, *Penn State Berks*
Jeffry Nichols, *Worcester State University*
Brent Nielsen, *Brigham Young University*
James Ntambi, *University of Wisconsin–Madison*
Edith Osborne, *Angelo State University*
Pamela Osenkowski, *Loyola University Chicago*
Gopal Periyannan, *Eastern Illinois University*
Michael Pikaart, *Hope College*
Deborah Polayes, *George Mason University*
Gary Powell, *Clemson University*
Gerry Prody, *Western Washington University*
Elizabeth Prusak, *Bishop's University*
Ramin Radfar, *Wofford College*
Gregory Raner, *University of North Carolina at Greensboro*
Madeline Rasche, *California State University, Fullerton*
Kevin Redding, *Arizona State University*
Cruz-Aguado Reyniel, *Douglas College*
Lisa Rezende, *University of Arizona*
John Richardson, *Austin College*
Jim Roesser, *Virginia Commonwealth University*
Douglas Root, *University of North Texas*
Gillian Rudd, *Georgia Gwinnett College*
Theresa Salerno, *Minnesota State University, Mankato*
Brian Sato, *University of California, Irvine*
Jamie Scaglione, *Carroll University*
Ingeborg Schmidt-Krey, *Georgia Institute of Technology*
Kimberly Schultz, *University of Maryland, Baltimore County*

Jason Schwans, *California State University, Long Beach*
Rhonda Scott, *Southern Adventist University*
Allan Scruggs, *Arizona State University*
Michael Sehorn, *Clemson University*
Edward Senkbei, *Salisbury University*
Amanda Sevcik, *Baylor University*
Robert Shaw, *Texas Tech University*
Nicholas Silvaggi, *University of Wisconsin–Milwaukee*
Jennifer Sniegowski, *Arizona State University Downtown Phoenix Campus*
Narasimha Sreerama, *Colorado State University*
Andrea Stadler, *St. Joseph's College*
Scott Stagg, *Florida State University*
Boris Steipe, *University of Toronto*
Alejandra Stenger, *University of Illinois at Urbana-Champaign*
Steven Theg, *University of California, Davis*
Jeremy Thorner, *University of California, Berkeley*
Kathryn Tifft, *Johns Hopkins University*
Michael Trakselis, *Baylor University*
Bruce Trieselmann, *Durham College*
C.-P. David Tu, *Pennsylvania State University*
Xuemin Wang, *University of Missouri*
Yuqi Wang, *Saint Louis University*
Paul Weber, *Briar Cliff University*
Rodney Weilbaecher, *Southern Illinois University School of Medicine*
Emily Westover, *Brandeis University*
Susan White, *Bryn Mawr College*
Enoka Wijekoon, *University of Guelph*
Kandatege Wimalasena, *Wichita State University*
Adrienne Wright, *University of Alberta*
Chuan Xiao, *University of Texas at El Paso*
Laura Zapanta, *University of Pittsburgh*
Brent Znosko, *Saint Louis University*

We lack the space here to acknowledge all the other individuals whose special efforts went into this book. We offer instead our sincere thanks—and the finished book that they helped guide to completion. We, of course, assume full responsibility for errors of fact or emphasis.

We want especially to thank our students at the University of Wisconsin–Madison for their numerous comments and suggestions. If something in the book does not work, they are never shy about letting us know it. We are grateful to the students and staff of our past and present research groups, who helped us balance the competing demands on our time; to our colleagues in the Department of Biochemistry at the University of Wisconsin–Madison, who helped us with advice and criticism; and to the many students and teachers who have written to suggest ways of improving the book. We hope our readers will continue to provide input for future editions.

Finally, we express our deepest appreciation to our wives, Brook and Beth, and our families, who showed extraordinary patience with, and support for, our book writing.

David L. Nelson
Michael M. Cox
Madison, Wisconsin
June 2016

Contents in Brief

Contents

III   INFORMATION PATHWAYS   955

CHAPTER 1

The Foundations of Biochemistry

Self-study tools that will help you practice what you've learned and reinforce this chapter's concepts are available online. Go to www.macmillanlearning.com/LehningerBiochemistry7e.

About fourteen billion years ago, the universe arose as a cataclysmic explosion of hot, energy-rich subatomic particles. Within seconds, the simplest elements (hydrogen and helium) were formed. As the universe expanded and cooled, material condensed under the influence of gravity to form stars. Some stars became enormous and then exploded as supernovae, releasing the energy needed to fuse simpler atomic nuclei into the more complex elements. Atoms and molecules formed swirling masses of dust particles, and their accumulation led eventually to the formation of rocks, planetoids, and planets. Thus were produced, over billions of years, Earth itself and the chemical elements found on Earth today. About four billion years ago, life arose—simple microorganisms with the ability to extract energy from chemical compounds and, later, from sunlight, which they used to make a vast array of more complex **biomolecules** from the simple elements and compounds on the Earth's surface. We and all other living organisms are made of stardust.

Biochemistry asks how the remarkable properties of living organisms arise from the thousands of different biomolecules. When these molecules are isolated and examined individually, they conform to all the physical and chemical laws that describe the behavior of inanimate matter—as do all the processes occurring in living organisms. The study of biochemistry shows how the collections of inanimate molecules that constitute living organisms interact to maintain and perpetuate life governed solely by the physical and chemical laws that govern the nonliving universe.

Yet organisms possess extraordinary attributes, properties that distinguish them from other collections of matter. What are these distinguishing features of living organisms?

A high degree of chemical complexity and microscopic organization. Thousands of different molecules make up a cell's intricate internal structures **(Fig. 1-1a)**. These include very long polymers, each with its characteristic sequence of subunits, its unique three-dimensional structure, and its highly specific selection of binding partners in the cell.

Systems for extracting, transforming, and using energy from the environment (Fig. 1-1b), enabling organisms to build and maintain their intricate structures and to do mechanical, chemical, osmotic, and electrical work. This counteracts the tendency of all matter to decay toward a more disordered state, to come to equilibrium with its surroundings.

Defined functions for each of an organism's components and regulated interactions among them. This is true not only of macroscopic structures, such as leaves and stems or hearts and lungs, but also of microscopic intracellular structures and individual chemical compounds. The interplay among the chemical components of a living organism is dynamic; changes in one component cause coordinating or compensating changes in another,

(a)   **(b)**

(c)

FIGURE 1-1 Some characteristics of living matter. (a) Microscopic complexity and organization are apparent in this colorized image of a thin section of several secretory cells from the pancreas, viewed with the electron microscope. **(b)** A prairie falcon acquires nutrients and energy by consuming a smaller bird. **(c)** Biological reproduction occurs with near-perfect fidelity. [Sources: (a) SPL/Science Source. (b) W. Perry Conway/Corbis. (c) F1online digitale Bildagentur GmbH/Alamy.]

FIGURE 1-2 Diverse living organisms share common chemical features. Birds, beasts, plants, and soil microorganisms share with humans the same basic structural units (cells) and the same kinds of macromolecules (DNA, RNA, proteins) made up of the same kinds of monomeric subunits (nucleotides, amino acids). They utilize the same pathways for synthesis of cellular components, share the same genetic code, and derive from the same evolutionary ancestors. [Source: The Garden of Eden, 1659 (oil on canvas) by Jan van Kessel the Elder (1626-79)/Johnny van Haeften Gallery, London, UK/ Bridgeman Images.]

with the whole ensemble displaying a character beyond that of its individual parts. The collection of molecules carries out a program, the end result of which is reproduction of the program and self-perpetuation of that collection of molecules—in short, life.

Mechanisms for sensing and responding to alterations in their surroundings. Organisms constantly adjust to these changes by adapting their internal chemistry or their location in the environment.

A capacity for precise self-replication and self-assembly (Fig. 1-1c). A single bacterial cell placed in a sterile nutrient medium can give rise to a billion identical "daughter" cells in 24 hours. Each cell contains thousands of different molecules, some extremely complex; yet each bacterium is a faithful copy of the original, its construction directed entirely by information contained in the genetic material of the original cell. On a larger scale, the progeny of a vertebrate animal share a striking resemblance to their parents, also the result of their inheritance of parental genes.

A capacity to change over time by gradual evolution. Organisms change their inherited life strategies, in very small steps, to survive in new circumstances. The result of eons of evolution is an enormous diversity

of life forms, superficially very different **(Fig. 1-2)** but fundamentally related through their shared ancestry. This fundamental unity of living organisms is reflected at the molecular level in the similarity of gene sequences and protein structures.

Despite these common properties and the fundamental unity of life they reveal, it is difficult to make generalizations about living organisms. Earth has an enormous diversity of organisms. The range of habitats, from hot springs to Arctic tundra, from animal intestines to college dormitories, is matched by a correspondingly wide range of specific biochemical adaptations, achieved within a common chemical framework. For the sake of clarity, in this book we sometimes risk certain generalizations, which, though not perfect, remain useful; we also frequently point out the exceptions to these generalizations, which can prove illuminating.

Biochemistry describes in molecular terms the structures, mechanisms, and chemical processes shared by all organisms and provides organizing principles that underlie life in all its diverse forms. Although biochemistry provides important insights and practical applications in medicine, agriculture, nutrition, and industry, its ultimate concern is with the wonder of life itself.

In this introductory chapter we give an overview of the cellular, chemical, physical, and genetic backgrounds to biochemistry and the overarching principle of evolution—how life emerged and evolved into the diversity of organisms we see today. As you read through the book, you may find it helpful to refer back to this chapter at intervals to refresh your memory of this background material.

1.1 Cellular Foundations

The unity and diversity of organisms become apparent even at the cellular level. The smallest organisms consist of single cells and are microscopic. Larger, multicellular organisms contain many different types of cells, which vary in size, shape, and specialized function. Despite these obvious differences, all cells of the simplest and most complex organisms share certain fundamental properties, which can be seen at the biochemical level.

Cells Are the Structural and Functional Units of All Living Organisms

Cells of all kinds share certain structural features **(Fig. 1-3)**. The **plasma membrane** defines the periphery of the cell, separating its contents from the surroundings. It is composed of lipid and protein molecules that form a thin, tough, pliable, hydrophobic barrier around the cell. The membrane is a barrier to the free passage of inorganic ions and most other charged or polar compounds. Transport proteins in the plasma membrane allow the passage of certain ions and molecules; receptor proteins transmit signals into the cell; and membrane enzymes participate in some reaction pathways. Because the individual lipids and proteins of the plasma membrane are not covalently linked, the entire structure is remarkably flexible, allowing changes in the shape and size of the cell. As a cell grows, newly made lipid and protein molecules are inserted into its plasma membrane; cell division produces two cells, each with its own membrane. This growth and cell division (fission) occurs without loss of membrane integrity.

The internal volume enclosed by the plasma membrane, the **cytoplasm** (Fig. 1-3), is composed of an aqueous solution, the **cytosol**, and a variety of suspended particles with specific functions. These particulate components (membranous organelles such as mitochondria and chloroplasts; supramolecular structures such as **ribosomes** and **proteasomes**, the sites of protein synthesis and degradation) sediment when cytoplasm is centrifuged at 150,000 g (g is the gravitational force of Earth). What remains as the supernatant fluid is the cytosol, a highly concentrated solution containing enzymes and the RNA molecules that encode them; the components (amino acids and nucleotides) from which these macromolecules are assembled; hundreds of small organic molecules called **metabolites**, intermediates in biosynthetic and degradative pathways; **coenzymes**, compounds essential to many enzyme-catalyzed reactions; and inorganic ions (K^+, Na^+, Mg^{2+}, and Ca^{2+}, for example).

All cells have, for at least some part of their life, either a **nucleoid** or a **nucleus**, in which the **genome**—the complete set of genes, composed of DNA—is replicated and stored, with its associated proteins. The nucleoid, in bacteria and archaea, is not separated from the cytoplasm by a membrane; the nucleus, in **eukaryotes**, is enclosed within a double membrane, the nuclear envelope. Cells with nuclear envelopes make up the large domain Eukarya (Greek *eu*, "true," and *karyon*, "nucleus"). Microorganisms without nuclear membranes, formerly grouped together as **prokaryotes** (Greek *pro*, "before"), are now recognized as comprising two very distinct groups: the domains Bacteria and Archaea, described below.

Cellular Dimensions Are Limited by Diffusion

Most cells are microscopic, invisible to the unaided eye. Animal and plant cells are typically 5 to 100 μm in diameter, and many unicellular microorganisms are only 1 to 2 μm long (see the inside of the back cover for information on units and their abbreviations). What limits the dimensions of a cell? The lower limit is probably set by the minimum number of each type of biomolecule required by the cell. The smallest cells, certain bacteria known as mycoplasmas, are 300 nm in diameter and have a volume of about 10^{-14} mL. A single bacterial ribosome is about 20 nm in its longest dimension, so a few ribosomes take up a substantial fraction of the volume in a mycoplasmal cell.

The upper limit of cell size is probably set by the rate of diffusion of solute molecules in aqueous systems. For example, a bacterial cell that depends on oxygen-consuming reactions for energy extraction must obtain molecular oxygen by diffusion from the surrounding medium through its plasma membrane. The cell is so small, and the ratio of its surface area to its volume is so large, that every part of its cytoplasm is easily reached by O_2 diffusing into the cell. With increasing cell size, however, surface-to-volume ratio decreases, until metabolism consumes O_2 faster than diffusion can supply it. Metabolism that requires O_2 thus becomes impossible as cell size increases beyond a certain point, placing a theoretical upper limit on the size of cells. Oxygen is only one of many low molecular weight species that must diffuse from outside the cell to various regions of

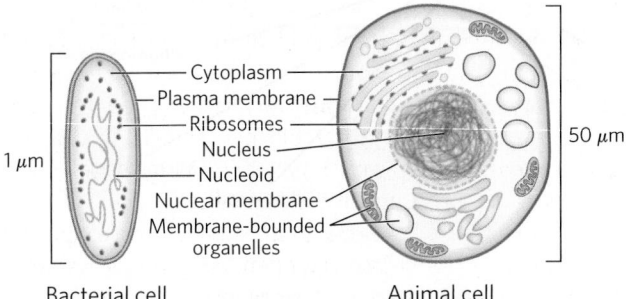

FIGURE 1-3 The universal features of living cells. All cells have a nucleus or nucleoid containing their DNA, a plasma membrane, and cytoplasm. The cytosol is defined as that portion of the cytoplasm that remains in the supernatant after gentle breakage of the plasma membrane and centrifugation of the resulting extract at 150,000 g for 1 hour. Eukaryotic cells contain a variety of membrane-bounded organelles (including mitochondria, chloroplasts) and large particles (ribosomes, for example), which are sedimented by this centrifugation and can be recovered from the pellet.

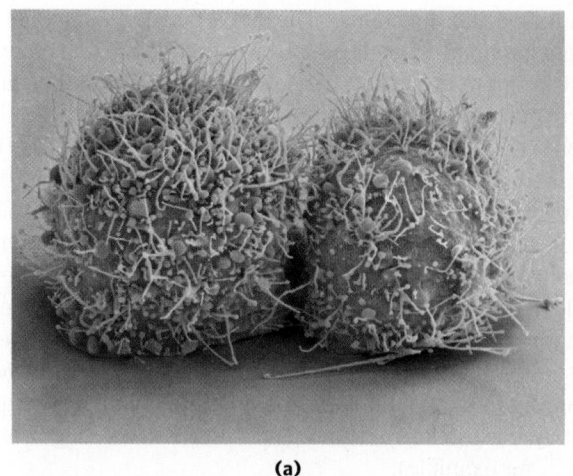

(a)

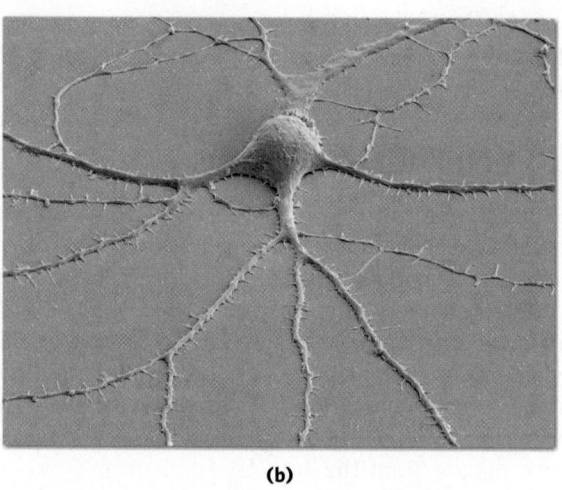

(b)

FIGURE 1-4 Most animal cells have intricately folded surfaces. Colorized scanning electron micrographs show **(a)** the highly convoluted surface of two HeLa cells, a line of human cancer cells cultured in the laboratory, and **(b)** a neuron with its many extensions, each capable of making connections with other neurons. [Sources: (a) NIH National Institute of General Medical Sciences. (b) 2012 National Center for Microscopy & Imaging Research.]

its interior, and the same surface-to-volume argument applies to each of them as well. Many types of animal cells have a highly folded or convoluted surface that increases their surface-to-volume ratio and allows higher rates of uptake of materials from their surroundings **(Fig. 1-4)**.

Organisms Belong to Three Distinct Domains of Life

The development of techniques for determining DNA sequences quickly and inexpensively has greatly improved our ability to deduce evolutionary relationships among organisms. Similarities between gene sequences in various organisms provide deep insight into the course of evolution. In one interpretation of sequence similarities, all living organisms fall into one of three large groups (domains) that define three branches of the evolutionary tree of life originating from a common progenitor **(Fig. 1-5)**. Two large groups of single-celled microorganisms can be distinguished on genetic and biochemical

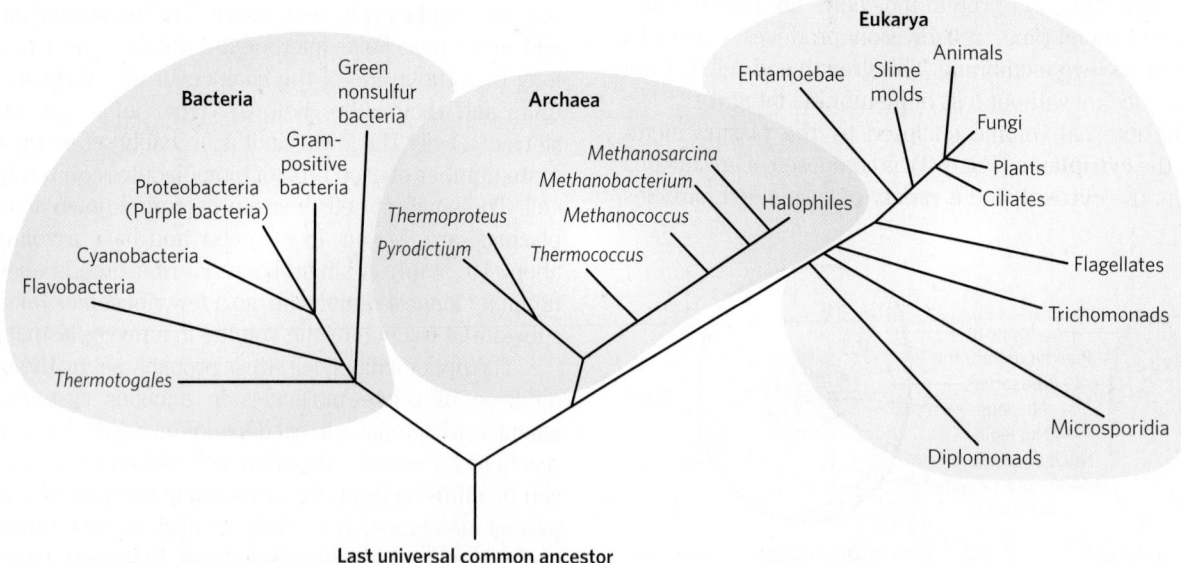

FIGURE 1-5 Phylogeny of the three domains of life. Phylogenetic relationships are often illustrated by a "family tree" of this type. The basis for this tree is the similarity in nucleotide sequences of the ribosomal RNAs of each group; the more similar the sequences, the closer the location of the branches, with the distance between branches representing the degree of difference between two sequences. Phylogenetic trees can also be constructed from similarities across species of the amino acid sequences of a single protein. For example, sequences of the protein GroEL (a bacterial protein that assists in protein folding) were compared to generate the tree in Figure 3-35. The tree in Figure 3-36 is a "consensus" tree, which uses several comparisons such as these to derive the best estimates of evolutionary relatedness among a group of organisms. Genomic sequences from a wide range of bacteria, archaea, and eukaryotes also are consistent with a two-domain model in which eukaryotes are subsumed under the Archaea domain. As more genomes are sequenced, one model may emerge as the clear best fit for the data. [Source: Information from C. R. Woese, *Microbiol. Rev.* 51:221, 1987, Fig. 4.]

grounds: **Bacteria** and **Archaea**. Bacteria inhabit soils, surface waters, and the tissues of other living or decaying organisms. Many of the Archaea, recognized as a distinct domain by Carl Woese in the 1980s, inhabit extreme environments—salt lakes, hot springs, highly acidic bogs, and the ocean depths. The available evidence suggests that the Archaea and Bacteria diverged early in evolution. All eukaryotic organisms, which make up the third domain, **Eukarya**, evolved from the same branch that gave rise to the Archaea; eukaryotes are therefore more closely related to archaea than to bacteria.

Within the domains of Archaea and Bacteria are subgroups distinguished by their habitats. In **aerobic** habitats with a plentiful supply of oxygen, some resident organisms derive energy from the transfer of electrons from fuel molecules to oxygen within the cell. Other environments are **anaerobic**, devoid of oxygen, and microorganisms adapted to these environments obtain energy by transferring electrons to nitrate (forming N_2), sulfate (forming H_2S), or CO_2 (forming CH_4). Many organisms that have evolved in anaerobic environments are *obligate* anaerobes: they die when exposed to oxygen. Others are *facultative* anaerobes, able to live with or without oxygen.

Organisms Differ Widely in Their Sources of Energy and Biosynthetic Precursors

We can classify organisms according to how they obtain the energy and carbon they need for synthesizing cellular material (as summarized in **Fig. 1-6**). There are two broad categories based on energy sources: **phototrophs** (Greek *trophē*, "nourishment") trap and use sunlight, and **chemotrophs** derive their energy from oxidation of a chemical fuel. Some chemotrophs oxidize inorganic fuels—HS^- to S^0 (elemental sulfur), S^0 to SO_4^-, NO_2^- to NO_3^-, or Fe^{2+} to Fe^{3+}, for example. Phototrophs and chemotrophs may be further divided into those that can synthesize all of their biomolecules directly from CO_2 (**autotrophs**) and those that require some preformed organic nutrients made by other organisms (**heterotrophs**). We can describe an organism's mode of nutrition by combining these terms. For example, cyanobacteria are photoautotrophs; humans are chemoheterotrophs. Even finer distinctions can be made, and many organisms can obtain energy and carbon from more than one source under different environmental or developmental conditions.

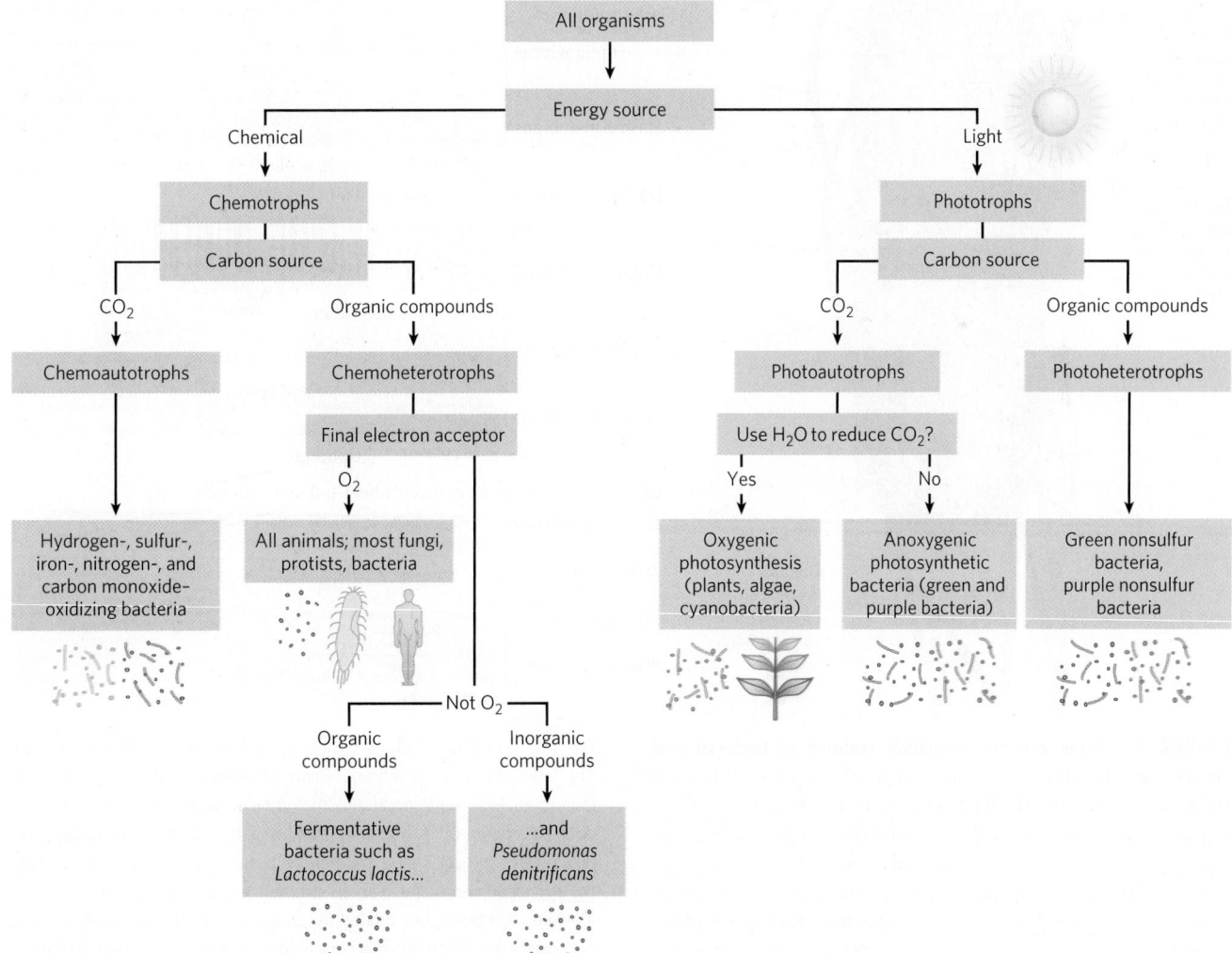

FIGURE 1-6 All organisms can be classified according to their source of energy (sunlight or oxidizable chemical compounds) and their source of carbon for the synthesis of cellular material.

Bacterial and Archaeal Cells Share Common Features but Differ in Important Ways

The best-studied bacterium, *Escherichia coli*, is a usually harmless inhabitant of the human intestinal tract. The *E. coli* cell **(Fig. 1-7a)** is an ovoid about 2 μm long and a little less than 1 μm in diameter, but other bacteria may be spherical or rod-shaped, and some are substantially larger. *E. coli* has a protective outer membrane and an inner plasma membrane that encloses the cytoplasm and the nucleoid. Between the inner and outer membranes is a thin but strong layer of a high molecular weight polymer (peptidoglycan) that gives the cell its shape and rigidity. The plasma membrane and the layers outside it constitute the **cell envelope**. The plasma membranes of bacteria consist of a thin bilayer of lipid molecules penetrated by proteins. Archaeal plasma membranes have a similar architecture, but the lipids can be strikingly different from those of bacteria (see Fig. 10-11). Bacteria and archaea have group-specific specializations of their cell envelopes (Fig. 1-7b–d). Some bacteria, called gram-positive because they are colored by Gram's stain (introduced by Hans Peter Gram in 1882), have a thick layer of peptidoglycan outside their plasma membrane but lack an outer membrane. Gram-negative bacteria have an outer membrane composed of a lipid bilayer into which are inserted complex lipopolysaccharides and proteins called porins that provide transmembrane channels for the diffusion of low molecular weight compounds and ions across this outer membrane. The structures outside the plasma membrane of archaea differ from organism to organism, but they, too, have a layer of peptidoglycan or protein that confers rigidity on their cell envelopes.

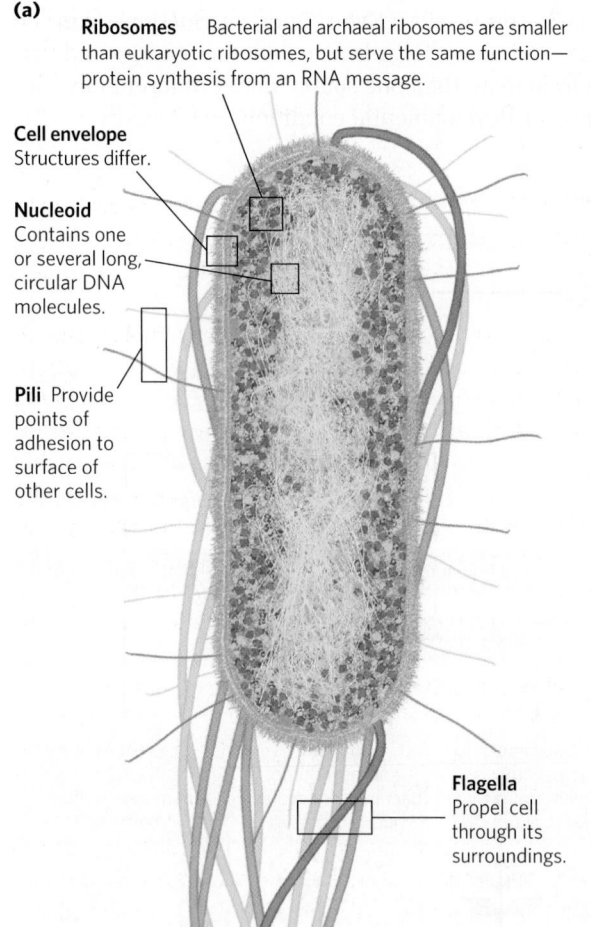

(a)

Ribosomes Bacterial and archaeal ribosomes are smaller than eukaryotic ribosomes, but serve the same function—protein synthesis from an RNA message.

Cell envelope
Structures differ.

Nucleoid
Contains one or several long, circular DNA molecules.

Pili Provide points of adhesion to surface of other cells.

Flagella
Propel cell through its surroundings.

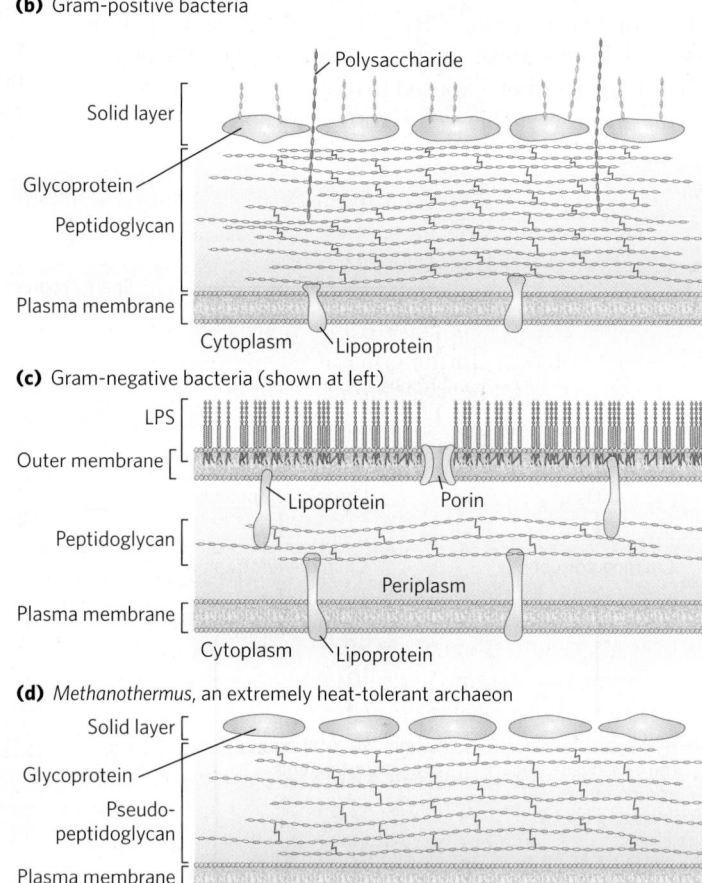

(b) Gram-positive bacteria

Polysaccharide
Solid layer
Glycoprotein
Peptidoglycan
Plasma membrane
Cytoplasm   Lipoprotein

(c) Gram-negative bacteria (shown at left)

LPS
Outer membrane
Lipoprotein   Porin
Peptidoglycan
Periplasm
Plasma membrane
Cytoplasm   Lipoprotein

(d) *Methanothermus*, an extremely heat-tolerant archaeon

Solid layer
Glycoprotein
Pseudo-peptidoglycan
Plasma membrane
Cytoplasm

FIGURE 1-7 Some common structural features of bacterial and archaeal cells. (a) This correct-scale drawing of *E. coli* serves to illustrate some common features. **(b)** The cell envelope of gram-positive bacteria is a single membrane with a thick, rigid layer of peptidoglycan on its outside surface. A variety of polysaccharides and other complex polymers are interwoven with the peptidoglycan, and surrounding the whole is a porous "solid layer" composed of glycoproteins. **(c)** *E. coli* is gram-negative and has a double membrane. Its outer membrane has a lipopolysaccharide (LPS) on the outer surface and phospholipids on the inner surface. This outer membrane is studded with protein channels (porins) that allow small molecules, but not proteins, to diffuse through. The inner (plasma) membrane, made of phospholipids and proteins, is impermeable to both large and small molecules. Between the inner and outer membranes, in the periplasm, is a thin layer of peptidoglycan, which gives the cell shape and rigidity, but does not retain Gram's stain. **(d)** Archaeal membranes vary in structure and composition, but all have a single membrane surrounded by an outer layer that includes either a peptidoglycanlike structure, a porous protein shell (solid layer), or both. [Sources: (a) David S. Goodsell. (b, c, d) Information from S.-V. Albers and B. H. Meyer, *Nature Rev. Microbiol.* 9:414, 2011, Fig. 2.]

The cytoplasm of *E. coli* contains about 15,000 ribosomes, various numbers (10 to thousands) of copies of each of 1,000 or so different enzymes, perhaps 1,000 organic compounds of molecular weight less than 1,000 (metabolites and cofactors), and a variety of inorganic ions. The nucleoid contains a single, circular molecule of DNA, and the cytoplasm (like that of most bacteria) contains one or more smaller, circular segments of DNA called **plasmids**. In nature, some plasmids confer resistance to toxins and antibiotics in the environment. In the laboratory, these DNA segments are especially amenable to experimental manipulation and are powerful tools for genetic engineering (see Chapter 9).

Other species of bacteria, as well as archaea, contain a similar collection of biomolecules, but each species has physical and metabolic specializations related to its environmental niche and nutritional sources. Cyanobacteria, for example, have internal membranes specialized to trap energy from light (see Fig. 20-27). Many archaea live in extreme environments and have biochemical adaptations to survive in extremes of temperature, pressure, or salt concentration. Differences in ribosomal structure gave the first hints that Bacteria and Archaea constituted separate domains. Most bacteria (including *E. coli*) exist as individual cells, but often associate in biofilms or mats, in which large numbers of cells adhere to each other and to some solid substrate beneath or at an aqueous surface. Cells of some bacterial species (the myxobacteria, for example) show simple social behavior, forming many-celled aggregates in response to signals between neighboring cells.

Eukaryotic Cells Have a Variety of Membranous Organelles, Which Can Be Isolated for Study

Typical eukaryotic cells **(Fig. 1-8)** are much larger than bacteria—commonly 5 to 100 μm in diameter, with cell volumes a thousand to a million times larger than those of bacteria. The distinguishing characteristics of eukaryotes are the nucleus and a variety of membrane-enclosed organelles with specific functions. These organelles include **mitochondria**, the site of most of the energy-extracting reactions of the cell; the **endoplasmic reticulum** and **Golgi complexes**, which play central roles in the synthesis and processing of lipids and membrane proteins; **peroxisomes**, in which very long-chain fatty acids are oxidized; and **lysosomes**, filled with digestive enzymes to degrade unneeded cellular debris. In addition to these, plant cells also contain **vacuoles** (which store large quantities of organic acids) and **chloroplasts** (in which sunlight drives the synthesis of ATP in the process of photosynthesis) (Fig. 1-8). Also present in the cytoplasm of many cells are granules or droplets containing stored nutrients such as starch and fat.

In a major advance in biochemistry, Albert Claude, Christian de Duve, and George Palade developed methods for separating organelles from the cytosol and from each other—an essential step in investigating their structures and functions. In a typical cell fractionation **(Fig. 1-9)**, cells or tissues in solution are gently disrupted by physical shear. This treatment ruptures the plasma membrane but leaves most of the organelles intact. The homogenate is then centrifuged; organelles such as nuclei, mitochondria, and lysosomes differ in size and therefore sediment at different rates.

These methods were used to establish, for example, that lysosomes contain degradative enzymes, mitochondria contain oxidative enzymes, and chloroplasts contain photosynthetic pigments. The isolation of an organelle enriched in a certain enzyme is often the first step in the purification of that enzyme.

The Cytoplasm Is Organized by the Cytoskeleton and Is Highly Dynamic

Fluorescence microscopy reveals several types of protein filaments crisscrossing the eukaryotic cell, forming an interlocking three-dimensional meshwork, the **cytoskeleton**. Eukaryotes have three general types of cytoplasmic filaments—actin filaments, microtubules, and intermediate filaments **(Fig. 1-10)**—differing in width (from about 6 to 22 nm), composition, and specific function. All types provide structure and organization to the cytoplasm and shape to the cell. Actin filaments and microtubules also help to produce the motion of organelles or of the whole cell.

Each type of cytoskeletal component consists of simple protein subunits that associate noncovalently to form filaments of uniform thickness. These filaments are not permanent structures; they undergo constant disassembly into their protein subunits and reassembly into filaments. Their locations in cells are not rigidly fixed but may change dramatically with mitosis, cytokinesis, amoeboid motion, or changes in cell shape. The assembly, disassembly, and location of all types of filaments are regulated by other proteins, which serve to link or bundle the filaments or to move cytoplasmic organelles along the filaments. (Bacteria contain actin-like proteins that serve similar roles in those cells.)

The picture that emerges from this brief survey of eukaryotic cell structure is of a cell with a meshwork of structural fibers and a complex system of membrane-enclosed compartments (Fig. 1-8). The filaments disassemble and then reassemble elsewhere. Membranous vesicles bud from one organelle and fuse with another. Organelles move through the cytoplasm along protein filaments, their motion powered by energy-dependent motor proteins. The **endomembrane system** segregates specific metabolic processes and provides surfaces on which certain enzyme-catalyzed reactions occur. **Exocytosis** and **endocytosis**, mechanisms of transport (out of and into cells, respectively) that involve membrane fusion and fission, provide paths between the cytoplasm and surrounding medium, allowing the secretion of substances produced in the cell and uptake of extracellular materials.

(a) Animal cell

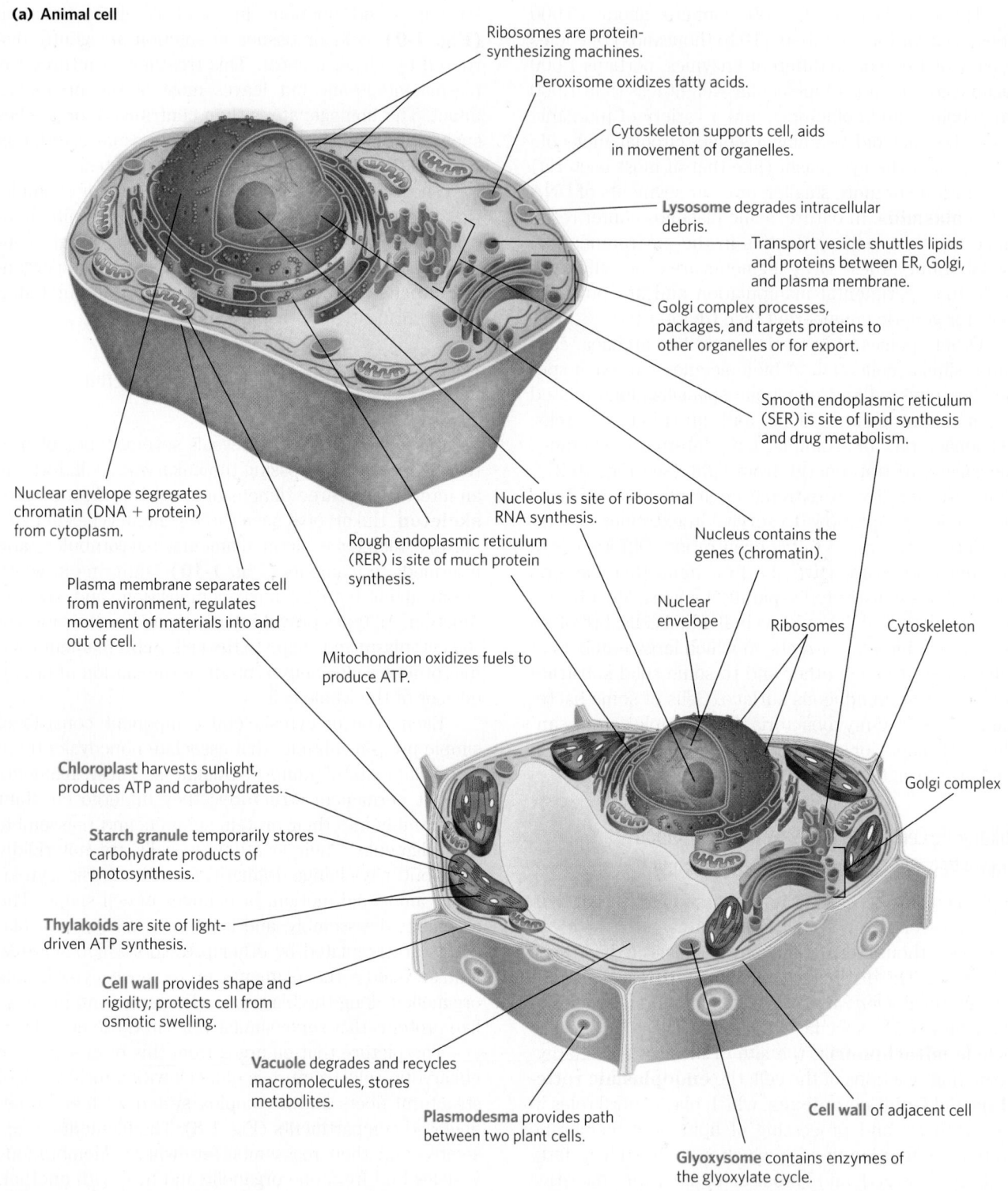

Ribosomes are protein-synthesizing machines.

Peroxisome oxidizes fatty acids.

Cytoskeleton supports cell, aids in movement of organelles.

Lysosome degrades intracellular debris.

Transport vesicle shuttles lipids and proteins between ER, Golgi, and plasma membrane.

Golgi complex processes, packages, and targets proteins to other organelles or for export.

Smooth endoplasmic reticulum (SER) is site of lipid synthesis and drug metabolism.

Nuclear envelope segregates chromatin (DNA + protein) from cytoplasm.

Nucleolus is site of ribosomal RNA synthesis.

Nucleus contains the genes (chromatin).

Plasma membrane separates cell from environment, regulates movement of materials into and out of cell.

Rough endoplasmic reticulum (RER) is site of much protein synthesis.

Nuclear envelope

Ribosomes

Cytoskeleton

Mitochondrion oxidizes fuels to produce ATP.

Chloroplast harvests sunlight, produces ATP and carbohydrates.

Golgi complex

Starch granule temporarily stores carbohydrate products of photosynthesis.

Thylakoids are site of light-driven ATP synthesis.

Cell wall provides shape and rigidity; protects cell from osmotic swelling.

Vacuole degrades and recycles macromolecules, stores metabolites.

Plasmodesma provides path between two plant cells.

Cell wall of adjacent cell

Glyoxysome contains enzymes of the glyoxylate cycle.

(b) Plant cell

FIGURE 1-8 Eukaryotic cell structure. Schematic illustrations of two major types of eukaryotic cell: **(a)** a representative animal cell and **(b)** a representative plant cell. Plant cells are usually 10 to 100 μm in diameter—larger than animal cells, which typically range from 5 to 30 μm. Structures labeled in red are unique to animal cells; those labeled in green are unique to plant cells. Eukaryotic microorganisms (such as protists and fungi) have structures similar to those in plant and animal cells, but many also contain specialized organelles not illustrated here.

This structural organization of the cytoplasm is far from random. The motion and positioning of organelles and cytoskeletal elements are under tight regulation, and at certain stages in its life, a eukaryotic cell undergoes dramatic, finely orchestrated reorganizations, such as the events of mitosis. The interactions between the cytoskeleton and organelles are noncovalent, reversible, and subject to regulation in response to various intracellular and extracellular signals.

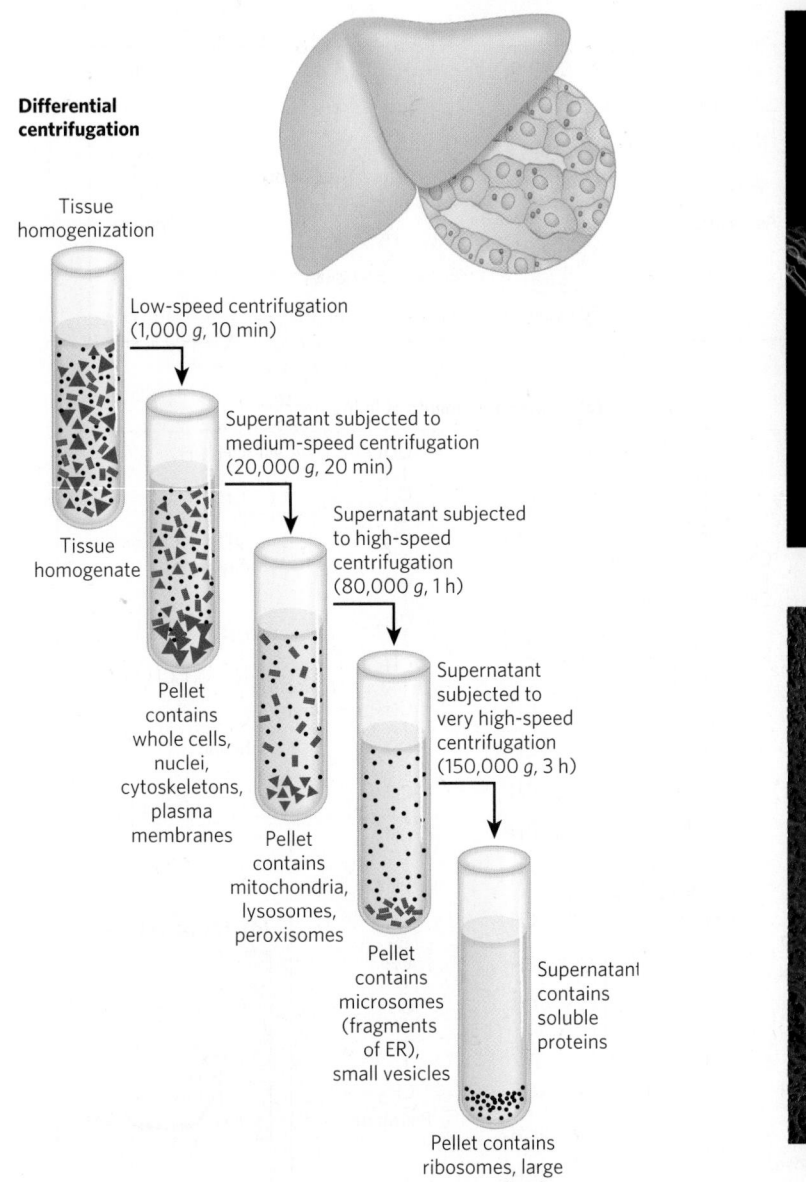

Differential centrifugation

Tissue homogenization

Tissue homogenate

Low-speed centrifugation (1,000 *g*, 10 min)

Pellet contains whole cells, nuclei, cytoskeletons, plasma membranes

Supernatant subjected to medium-speed centrifugation (20,000 *g*, 20 min)

Pellet contains mitochondria, lysosomes, peroxisomes

Supernatant subjected to high-speed centrifugation (80,000 *g*, 1 h)

Pellet contains microsomes (fragments of ER), small vesicles

Supernatant subjected to very high-speed centrifugation (150,000 *g*, 3 h)

Supernatant contains soluble proteins

Pellet contains ribosomes, large macromolecules

FIGURE 1-9 Subcellular fractionation of tissue. A tissue such as liver is first mechanically homogenized to break cells and disperse their contents in an aqueous buffer. The sucrose medium has an osmotic pressure similar to that in organelles, thus balancing diffusion of water into and out of the organelles, which would swell and burst in a solution of lower osmolarity (see Fig. 2-13). The large and small particles in the suspension can be separated by centrifugation at different speeds. Larger particles sediment more rapidly than small particles, and soluble material does not sediment. By careful choice of the conditions of centrifugation, subcellular fractions can be separated for biochemical characterization. [Source: Information from B. Alberts et al., *Molecular Biology of the Cell*, 2nd edn, Garland Publishing, Inc., 1989, p. 165.]

Cells Build Supramolecular Structures

Macromolecules and their monomeric subunits differ greatly in size **(Fig. 1-11)**. An alanine molecule is less than 0.5 nm long. A molecule of hemoglobin, the oxygen-carrying protein of erythrocytes (red blood cells), consists of nearly 600 amino acid subunits in four long

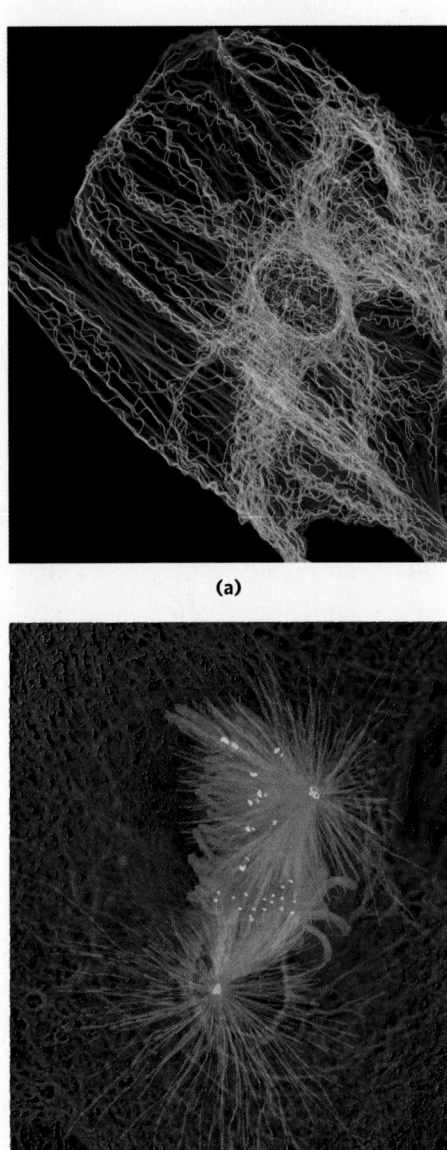

(a)

(b)

FIGURE 1-10 The three types of cytoskeletal filaments: actin filaments, microtubules, and intermediate filaments. Cellular structures can be labeled with an antibody (that recognizes a characteristic protein) covalently attached to a fluorescent compound. The stained structures are visible when the cell is viewed with a fluorescence microscope. **(a)** In this cultured fibroblast cell, bundles of actin filaments are stained red; microtubules, radiating from the cell center, are stained green; and chromosomes (in the nucleus) are stained blue. **(b)** A newt lung cell undergoing mitosis. Microtubules (green), attached to structures called kinetochores (yellow) on the condensed chromosomes (blue), pull the chromosomes to opposite poles, or centrosomes (magenta), of the cell. Intermediate filaments, made of keratin (red), maintain the structure of the cell. [Sources: (a) James J. Faust and David G. Capco, Arizona State University/NIH National Institute of General Medical Sciences. (b) Dr. Alexey Khodjakov, Wadsworth Center, New York State Department of Health.]

chains, folded into globular shapes and associated in a structure 5.5 nm in diameter. In turn, proteins are much smaller than ribosomes (about 20 nm in diameter), which are much smaller than organelles such as

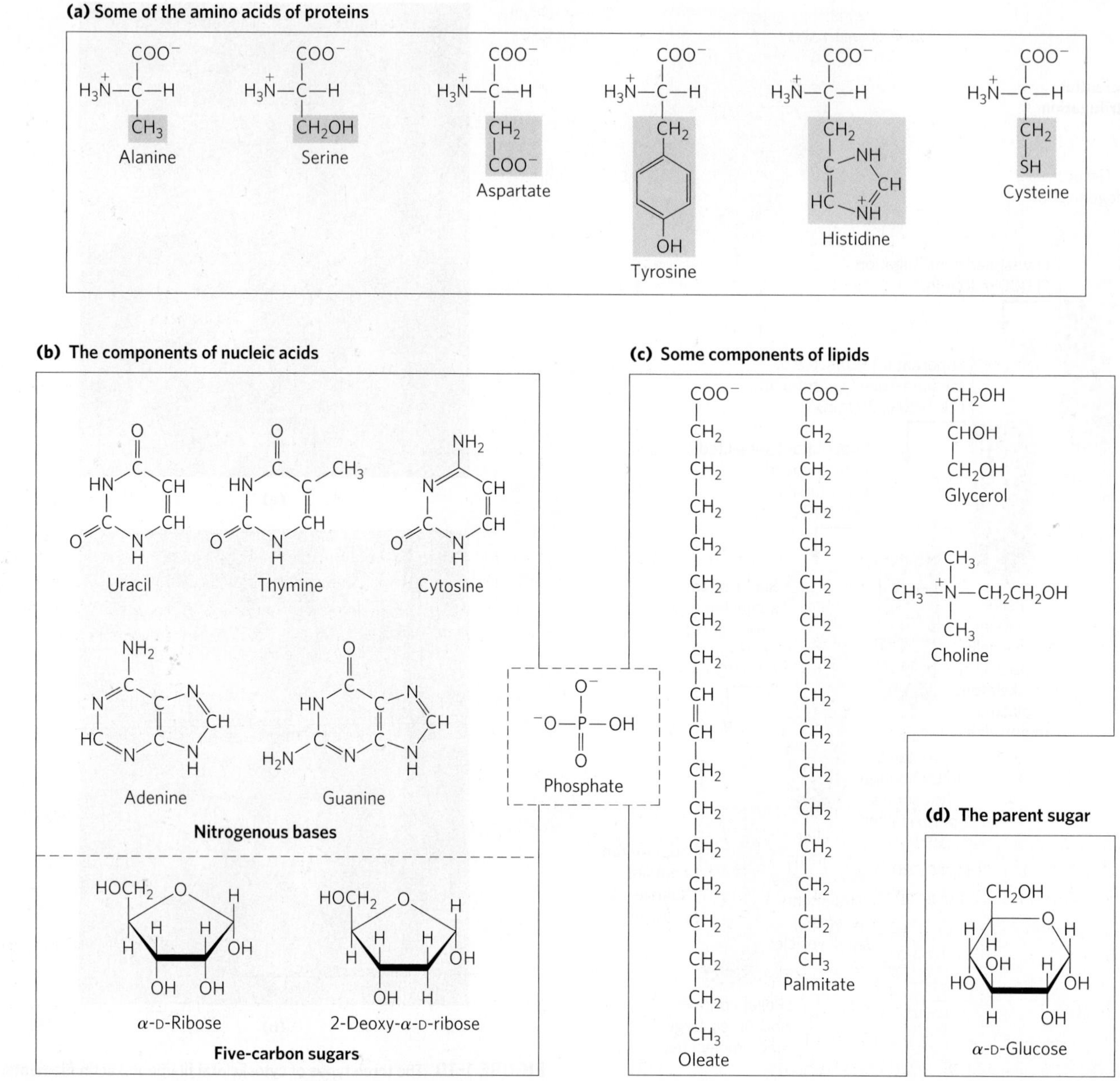

(a) Some of the amino acids of proteins

Alanine • Serine • Aspartate • Tyrosine • Histidine • Cysteine

(b) The components of nucleic acids

Uracil • Thymine • Cytosine • Adenine • Guanine • Phosphate

Nitrogenous bases

α-D-Ribose • 2-Deoxy-α-D-ribose

Five-carbon sugars

(c) Some components of lipids

Glycerol • Choline • Oleate • Palmitate

(d) The parent sugar

α-D-Glucose

FIGURE 1-11 The organic compounds from which most cellular materials are constructed: the ABCs of biochemistry. Shown here are **(a)** six of the 20 amino acids from which all proteins are built (the side chains are shaded light red); **(b)** the five nitrogenous bases, two five-carbon sugars, and phosphate ion from which all nucleic acids are built; **(c)** five components of membrane lipids (including phosphate); and **(d)** D-glucose, the simple sugar from which most carbohydrates are derived.

mitochondria, typically 1,000 nm in diameter. It is a long jump from simple biomolecules to cellular structures that can be seen with the light microscope. **Figure 1-12** illustrates the structural hierarchy in cellular organization.

The monomeric subunits of proteins, nucleic acids, and polysaccharides are joined by covalent bonds. In supramolecular complexes, however, macromolecules are held together by noncovalent interactions—much weaker, individually, than covalent bonds. Among these noncovalent interactions are hydrogen bonds (between polar groups), ionic interactions (between charged groups), aggregations of nonpolar groups in aqueous solution brought about by the hydrophobic effect (sometimes called hydrophobic interactions), and van der Waals interactions (also called London forces)—all of which have energies much smaller than those of covalent bonds. These noncovalent interactions are described in Chapter 2. The large numbers of weak interactions between macromolecules in supramolecular complexes stabilize these assemblies, producing their unique structures.

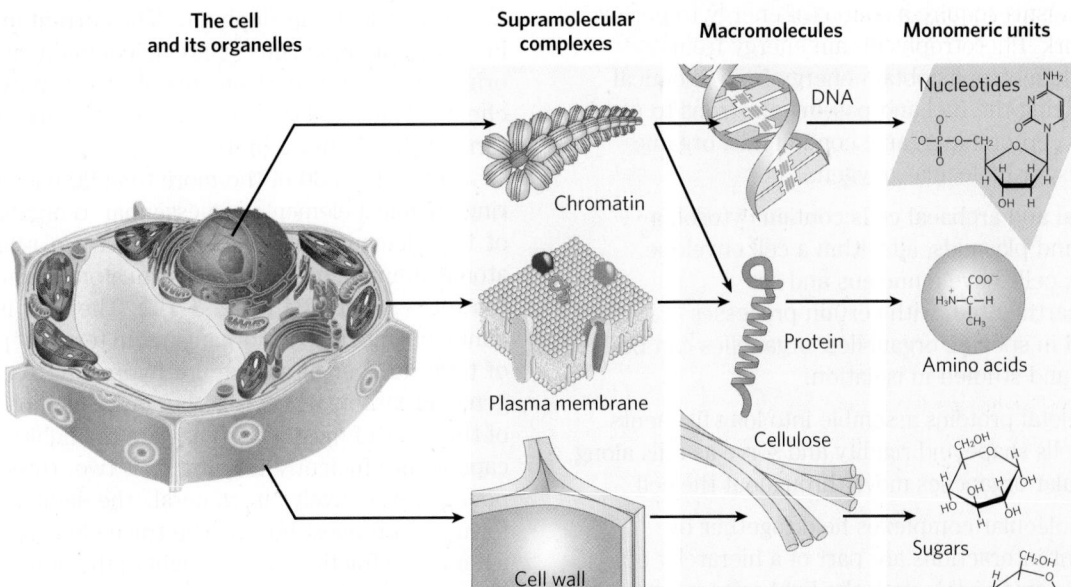

The cell and its organelles

Supramolecular complexes

Macromolecules

Monomeric units

Chromatin

Plasma membrane

Cell wall

DNA

Protein

Cellulose

Nucleotides

Amino acids

Sugars

FIGURE 1-12 Structural hierarchy in the molecular organization of cells. The organelles and other relatively large components of cells are composed of supramolecular complexes, which in turn are composed of smaller macromolecules and even smaller molecular subunits. For example, the nucleus of this plant cell contains chromatin, a supramolec-ular complex that consists of DNA and basic proteins (histones). DNA is made up of simple monomeric subunits (nucleotides), as are proteins (amino acids). [Source: Information from W. M. Becker and D. W. Deamer, *The World of the Cell*, 2nd edn, Benjamin/Cummings Publishing Company, 1991, Fig. 2-15.]

In Vitro Studies May Overlook Important Interactions among Molecules

One approach to understanding a biological process is to study purified molecules in vitro ("in glass"—in the test tube), without interference from other molecules present in the intact cell—that is, in vivo ("in the living"). Although this approach has been remarkably revealing, we must keep in mind that the inside of a cell is quite different from the inside of a test tube. The "interfering" components eliminated by purification may be critical to the biological function or regulation of the molecule purified. For example, in vitro studies of pure enzymes are commonly done at very low enzyme concentrations in thoroughly stirred aqueous solutions. In the cell, an enzyme is dissolved or suspended in the gel-like cytosol with thousands of other proteins, some of which bind to that enzyme and influence its activity. Some enzymes are components of multi-enzyme complexes in which reactants are channeled from one enzyme to another, never entering the bulk solvent. When all of the known macromolecules in a cell are represented in their known dimensions and concentrations (Fig. 1-13), it is clear that the cytosol is very crowded and that diffusion of macromolecules within the cytosol must be slowed by collisions with other large structures. In short, a given molecule may behave quite differently in the cell and in vitro. A central challenge of biochemistry is to understand the influences of cellular organization and macromolecular associations on the function of individual enzymes and other biomolecules—to understand function in vivo as well as in vitro.

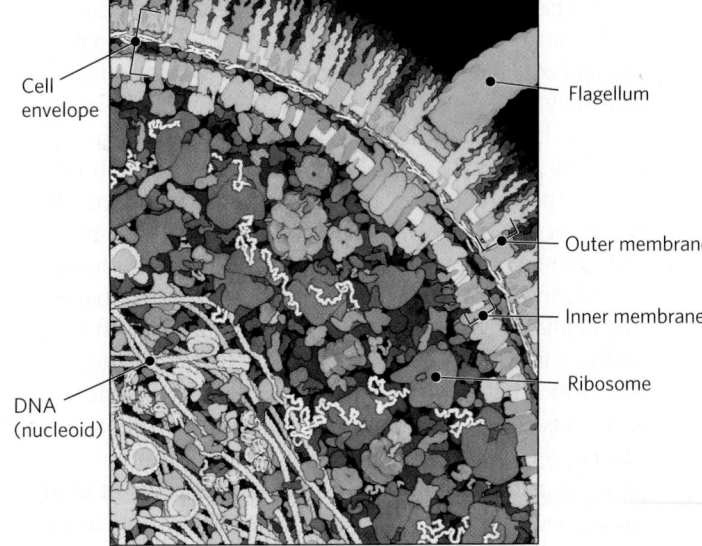

Cell envelope

DNA (nucleoid)

Flagellum

Outer membrane

Inner membrane

Ribosome

FIGURE 1-13 The crowded cell. This drawing by David Goodsell is an accurate representation of the relative sizes and numbers of macromole-cules in one small region of an *E. coli* cell. This concentrated cytosol, crowded with proteins and nucleic acids, is very different from the typical extract of cells used in biochemical studies, in which the cytosol has been diluted manyfold and the interactions between diffusing macromolecules have been strongly altered. [Source: © David S. Goodsell 1999.]

SUMMARY 1.1 Cellular Foundations

■ All cells are bounded by a plasma membrane; have a cytosol containing metabolites, coenzymes, inorganic ions, and enzymes; and have a set of genes contained within a nucleoid (bacteria and archaea) or nucleus (eukaryotes).

■ All organisms require a source of energy to perform cellular work. Phototrophs obtain energy from sunlight; chemotrophs obtain energy from chemical fuels, oxidizing the fuel and passing electrons to good electron acceptors: inorganic compounds, organic compounds, or molecular oxygen.

■ Bacterial and archaeal cells contain cytosol, a nucleoid, and plasmids, all within a cell envelope. Eukaryotic cells have a nucleus and are multicompartmented, with certain processes segregated in specific organelles; organelles can be separated and studied in isolation.

■ Cytoskeletal proteins assemble into long filaments that give cells shape and rigidity and serve as rails along which cellular organelles move throughout the cell.

■ Supramolecular complexes held together by noncovalent interactions are part of a hierarchy of structures, some visible with the light microscope. When individual molecules are removed from these complexes to be studied in vitro, interactions important in the living cell may be lost.

1.2 Chemical Foundations

Biochemistry aims to explain biological form and function in chemical terms. By the late eighteenth century, chemists had concluded that the composition of living matter is strikingly different from that of the inanimate world. Antoine-Laurent Lavoisier (1743–1794) noted the relative chemical simplicity of the "mineral world" and contrasted it with the complexity of the "plant and animal worlds"; the latter, he knew, were composed of compounds rich in the elements carbon, oxygen, nitrogen, and phosphorus.

During the first half of the twentieth century, parallel biochemical investigations of glucose breakdown in yeast and in animal muscle cells revealed remarkable chemical similarities between these two apparently very different cell types; the breakdown of glucose in yeast and in muscle cells involved the same 10 chemical intermediates and the same 10 enzymes. Subsequent studies of many other biochemical processes in many different organisms have confirmed the generality of this observation, neatly summarized in 1954 by Jacques Monod: "What is true of

E. coli is true of the elephant." The current understanding that all organisms share a common evolutionary origin is based in part on this observed universality of chemical intermediates and transformations, often termed "biochemical unity."

Fewer than 30 of the more than 90 naturally occurring chemical elements are essential to organisms. Most of the elements in living matter have a relatively low atomic number; only three have an atomic number above that of selenium, 34 **(Fig. 1-14)**. The four most abundant elements in living organisms, in terms of percentage of total number of atoms, are hydrogen, oxygen, nitrogen, and carbon, which together make up more than 99% of the mass of most cells. They are the lightest elements capable of efficiently forming one, two, three, and four bonds, respectively; in general, the lightest elements form the strongest bonds. The trace elements represent a miniscule fraction of the weight of the human body, but all are essential to life, usually because they are essential to the function of specific proteins, including many enzymes. The oxygen-transporting capacity of the hemoglobin molecule, for example, is absolutely dependent on four iron ions that make up only 0.3% of its mass.

Biomolecules Are Compounds of Carbon with a Variety of Functional Groups

The chemistry of living organisms is organized around carbon, which accounts for more than half of the dry weight of cells. Carbon can form single bonds with hydrogen atoms, and both single and double bonds with oxygen and nitrogen atoms **(Fig. 1-15)**. Of greatest significance in biology is the ability of carbon atoms to form very stable single bonds with up to four other carbon atoms. Two carbon atoms also can share two (or three) electron pairs, thus forming double (or triple) bonds.

The four single bonds that can be formed by a carbon atom project from the nucleus to the four apices of a tetrahedron **(Fig. 1-16)**, with an angle of about 109.5° between any two bonds and an average bond length of 0.154 nm. There is free rotation around each single bond, unless very large or highly charged groups are attached to both carbon atoms, in which case rotation may be restricted. A double bond is shorter (about 0.134 nm) and rigid, and allows only limited rotation about its axis.

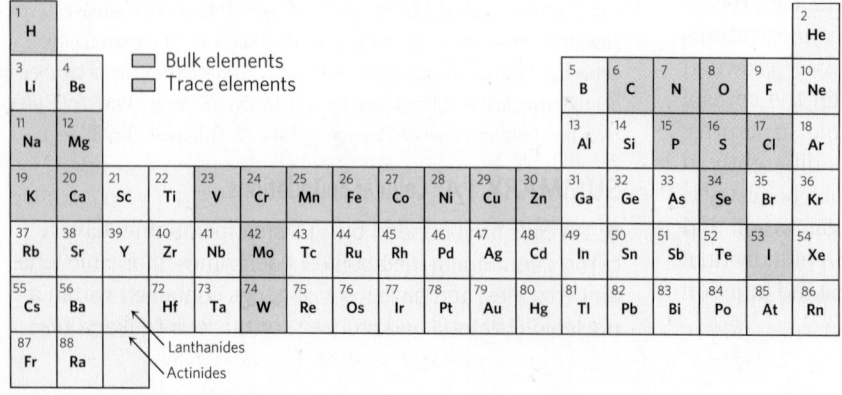

FIGURE 1-14 Elements essential to animal life and health. Bulk elements (shaded light red) are structural components of cells and tissues and are required in the diet in gram quantities daily. For trace elements (shaded yellow), the requirements are much smaller: for humans, a few milligrams per day of Fe, Cu, and Zn, even less of the others. The elemental requirements for plants and microorganisms are similar to those shown here; the ways in which they acquire these elements vary.

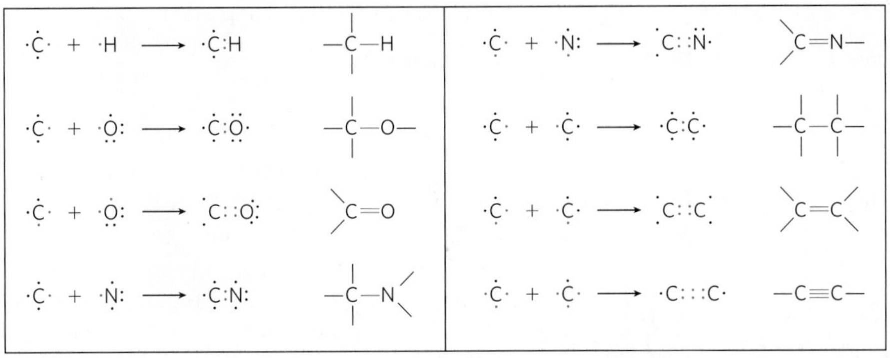

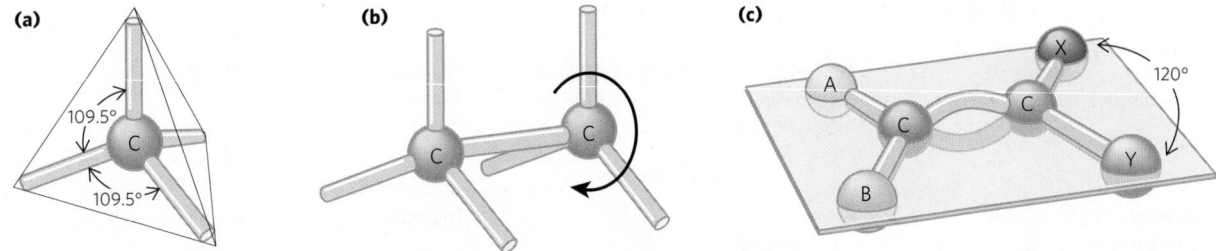

FIGURE 1-15 Versatility of carbon bonding. Carbon can form covalent single, double, and triple bonds (all bonds in red), particularly with other carbon atoms. Triple bonds are rare in biomolecules.

(a)   **(b)**   **(c)**

FIGURE 1-16 Geometry of carbon bonding. (a) Carbon atoms have a characteristic tetrahedral arrangement of their four single bonds. **(b)** Carbon–carbon single bonds have freedom of rotation, as shown for the compound ethane (CH_3—CH_3). **(c)** Double bonds are shorter and do not allow free rotation. The two doubly bonded carbons and the atoms designated A, B, X, and Y all lie in the same rigid plane.

Covalently linked carbon atoms in biomolecules can form linear chains, branched chains, and cyclic structures. It seems likely that the bonding versatility of carbon, with itself and with other elements, was a major factor in the selection of carbon compounds for the molecular machinery of cells during the origin and evolution of living organisms. No other chemical element can form molecules of such widely different sizes, shapes, and composition.

Most biomolecules can be regarded as derivatives of hydrocarbons, with hydrogen atoms replaced by a variety of functional groups that confer specific chemical properties on the molecule, forming various families of organic compounds. Typical of these are alcohols, which have one or more hydroxyl groups; amines, with amino groups; aldehydes and ketones, with carbonyl groups; and carboxylic acids, with carboxyl groups **(Fig. 1-17)**. Many biomolecules are polyfunctional, containing two or more types of functional groups **(Fig. 1-18)**, each with its own chemical characteristics and reactions. The chemical "personality" of a compound is determined by the chemistry of its functional groups and their disposition in three-dimensional space.

Cells Contain a Universal Set of Small Molecules

Dissolved in the aqueous phase (cytosol) of all cells is a collection of perhaps a thousand different small organic molecules (M_r ~100 to ~500), with intracellular concentrations ranging from nanomolar to millimolar (see Fig. 15-4). (See Box 1-1 for an explanation of the various

BOX 1-1   Molecular Weight, Molecular Mass, and Their Correct Units

There are two common (and equivalent) ways to describe molecular mass; both are used in this text. The first is *molecular weight,* or *relative molecular mass,* denoted M_r. The molecular weight of a substance is defined as the ratio of the mass of a molecule of that substance to one-twelfth the mass of an atom of carbon-12 (^{12}C). Since M_r is a ratio, it is dimensionless— it has no associated units. The second is *molecular mass,* denoted m. This is simply the mass of one molecule, or the molar mass divided by Avogadro's number. The molecular mass, m, is expressed in daltons (abbreviated Da). One dalton is equivalent to one-twelfth the mass of an atom of carbon-12; a kilodalton (kDa) is 1,000 daltons; a megadalton (MDa) is 1 million daltons.

Consider, for example, a molecule with a mass 1,000 times that of water. We can say of this molecule either $M_r = 18,000$ or $m = 18,000$ daltons. We can also describe it as an "18 kDa molecule." However, the expression $M_r = 18,000$ daltons is incorrect.

Another convenient unit for describing the mass of a single atom or molecule is the atomic mass unit (formerly amu, now commonly denoted u). One atomic mass unit (1 u) is defined as one-twelfth the mass of an atom of carbon-12. Since the experimentally measured mass of an atom of carbon-12 is 1.9926×10^{-23} g, $1 u = 1.6606 \times 10^{-24}$ g. The atomic mass unit is convenient for describing the mass of a peak observed by mass spectrometry (see Chapter 3, p. 100).

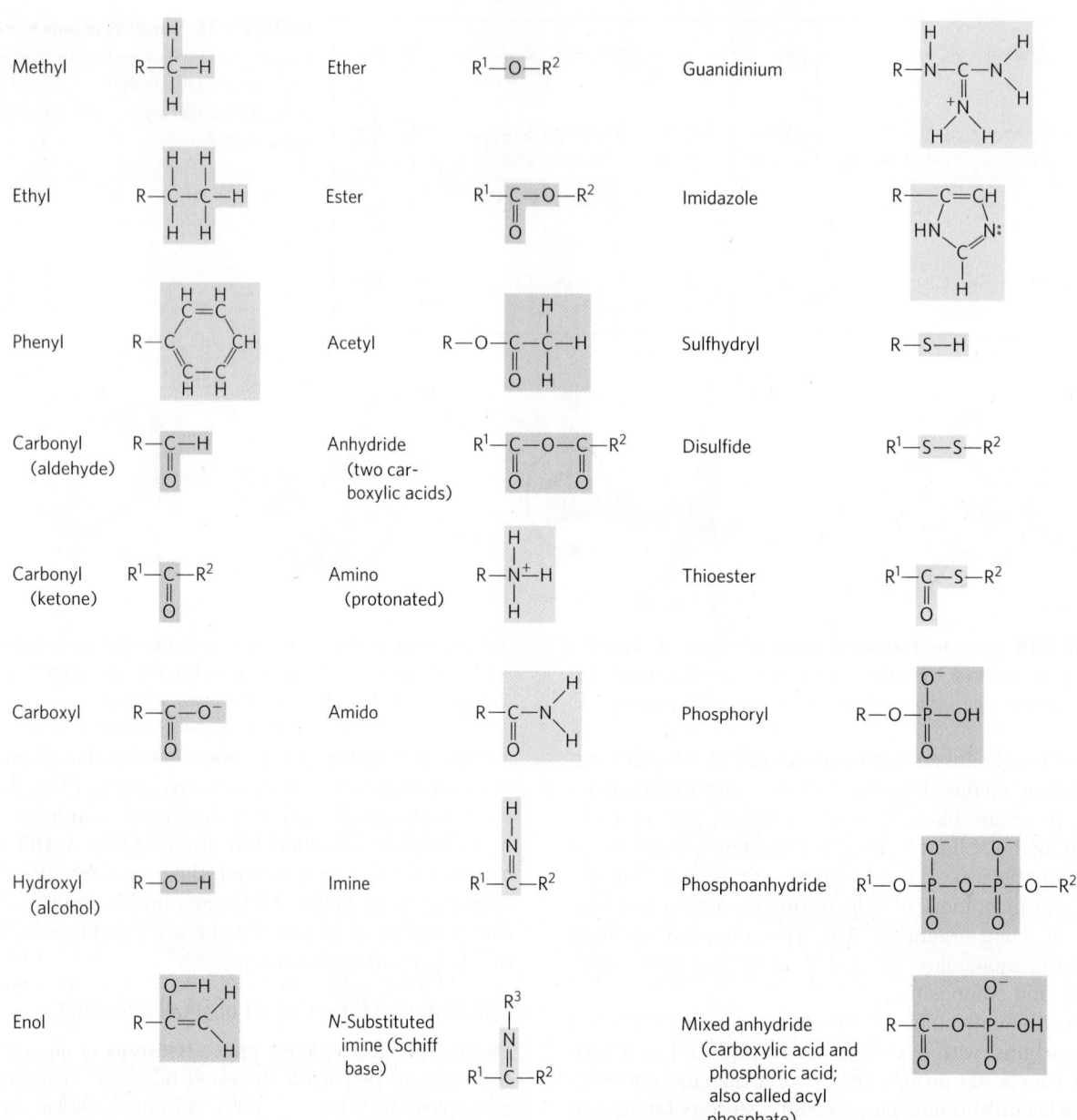

FIGURE 1-17 Some common functional groups of biomolecules. Functional groups are screened with a color typically used to represent the element that characterizes the group: gray for C, red for O, blue for N, yellow for S, and orange for P. In this figure and throughout the book, we use R to represent "any substituent." It may be as simple as a hydrogen atom, but typically it is a carbon-containing group. When two or more substituents are shown in a molecule, we designate them R^1, R^2, and so forth.

ways of referring to molecular weight.) These are the central metabolites in the major pathways occurring in nearly every cell—the metabolites and pathways that have been conserved throughout the course of evolution. This collection of molecules includes the common amino acids, nucleotides, sugars and their phosphorylated derivatives, and mono-, di-, and tricarboxylic acids. The molecules may be polar or charged and are water-soluble. They are trapped in the cell because the plasma membrane is impermeable to them, although specific membrane transporters can catalyze the movement of some molecules into and out of the cell or between compartments in eukaryotic cells. The universal occurrence of the same set of compounds in living cells reflects the evolutionary conservation of metabolic pathways that developed in the earliest cells.

There are other small biomolecules, specific to certain types of cells or organisms. For example, vascular plants contain, in addition to the universal set, small molecules called **secondary metabolites**, which play roles specific to plant life. These metabolites include compounds that give plants their characteristic scents and colors, and compounds such as morphine, quinine, nicotine, and caffeine that are valued for their physiological effects on humans but have other purposes in plants.

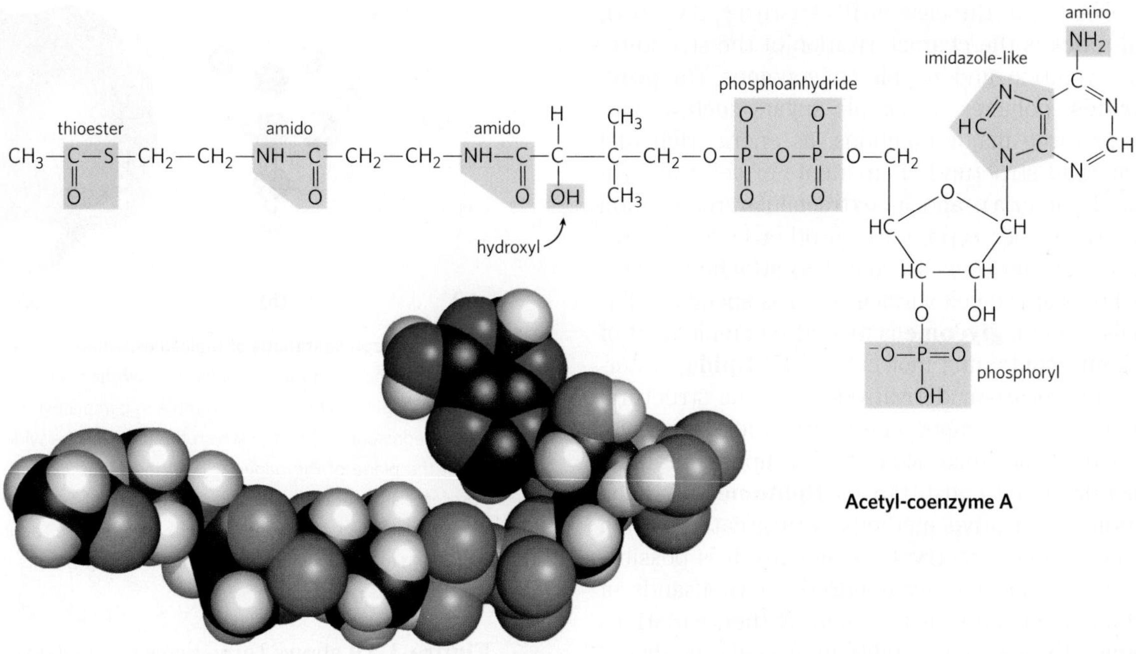

FIGURE 1-18 Several common functional groups in a single biomolecule. Acetyl-coenzyme A (often abbreviated as acetyl-CoA) is a carrier of acetyl groups in some enzymatic reactions. Its functional groups are screened in the structural formula. As we will see in Chapter 2, several of these functional groups can exist in protonated or unprotonated forms, depending on the pH. In the space-filling model, N is blue, C is black, P is orange, O is red, and H is white. The yellow atom at the left is the sulfur of the critical thioester bond between the acetyl moiety and coenzyme A. [Source: Acetyl-CoA extracted from PDB ID 1DM3, Y. Modis and R. K. Wierenga, *J. Mol. Biol.* 297:1171, 2000.]

The entire collection of small molecules in a given cell under a specific set of conditions has been called the **metabolome**, in parallel with the term "genome." **Metabolomics** is the systematic characterization of the metabolome under very specific conditions (such as following administration of a drug or a biological signal such as insulin).

Macromolecules Are the Major Constituents of Cells

Many biological molecules are **macromolecules**, polymers with molecular weights above ~5,000 that are assembled from relatively simple precursors. Shorter polymers are called **oligomers** (Greek *oligos*, "few"). Proteins, nucleic acids, and polysaccharides are macromolecules composed of monomers with molecular weights of 500 or less. Synthesis of macromolecules is a major energy-consuming activity of cells. Macromolecules themselves may be further assembled into supramolecular complexes, forming functional units such as ribosomes. Table 1-1 shows the major classes of biomolecules in an *E. coli* cell.

Proteins, long polymers of amino acids, constitute the largest fraction (besides water) of a cell. Some proteins have catalytic activity and function as enzymes; others serve as structural elements, signal receptors, or transporters that carry specific substances into or out of cells. Proteins are perhaps the most versatile of all biomolecules; a catalog of their many functions would be very long. The sum of all the proteins functioning in a given cell is the cell's **proteome**, and **proteomics** is the systematic characterization of this protein complement under a specific set of conditions. The **nucleic acids**, DNA and RNA, are polymers of nucleotides. They store and transmit genetic information, and some RNA molecules have structural and catalytic roles in supramolecular complexes. The **genome** is the entire sequence of

TABLE 1-1	Molecular Components of an *E. coli* Cell	
	Percentage of total weight of cell	Approximate number of different molecular species
Water	70	1
Proteins	15	3,000
Nucleic acids		
DNA	1	1–4
RNA	6	>3,000
Polysaccharides	3	20
Lipids	2	50[a]
Monomeric subunits and intermediates	2	2,600
Inorganic ions	1	20

Source: A. C. Guo et al., *Nucleic Acids Res.* 41:D625, 2013.
[a]If all permutations and combinations of fatty acid substituents are considered, this number is much larger.

a cell's DNA (or in the case of RNA viruses, its RNA), and **genomics** is the characterization of the structure, function, evolution, and mapping of genomes. The **polysaccharides**, polymers of simple sugars such as glucose, have three major functions: as energy-rich fuel stores, as rigid structural components of cell walls (in plants and bacteria), and as extracellular recognition elements that bind to proteins on other cells. Shorter polymers of sugars (oligosaccharides) attached to proteins or lipids at the cell surface serve as specific cellular signals. A cell's **glycome** is its entire complement of carbohydrate-containing molecules. The **lipids**, water-insoluble hydrocarbon derivatives, serve as structural components of membranes, energy-rich fuel stores, pigments, and intracellular signals. The lipid-containing molecules in a cell constitute its **lipidome**. With the application of sensitive methods with great resolving power (mass spectrometry, for example), it is possible to distinguish and quantify hundreds or thousands of these components and thus to quantify their variations in response to changing conditions, signals, or drugs. Systems biology is an approach that tries to integrate the information from genomics, proteomics, and metabolomics to give a molecular picture of all the activities of a cell under a given set of conditions and the changes that occur when the system is perturbed by external signals or circumstances or by mutations.

Proteins, polynucleotides, and polysaccharides have large numbers of monomeric subunits and thus high molecular weights—in the range of 5,000 to more than 1 million for proteins, up to several *billion* for nucleic acids, and in the millions for polysaccharides such as starch. Individual lipid molecules are much smaller (M_r 750 to 1,500) and are not classified as macromolecules, but they can associate noncovalently into very large structures. Cellular membranes are built of enormous noncovalent aggregates of lipid and protein molecules.

Given their characteristic information-rich subunit sequences, proteins and nucleic acids are often referred to as **informational macromolecules**. Some oligosaccharides, as noted above, also serve as informational molecules.

Three-Dimensional Structure Is Described by Configuration and Conformation

The covalent bonds and functional groups of a biomolecule are, of course, central to its function, but so also is the arrangement of the molecule's constituent atoms in three-dimensional space—its stereochemistry. Carbon-containing compounds commonly exist as **stereoisomers**, molecules with the same chemical bonds and same chemical formula but different **configuration**, the fixed spatial arrangement of atoms. Interactions between biomolecules are invariably **stereospecific**, requiring specific configurations in the interacting molecules.

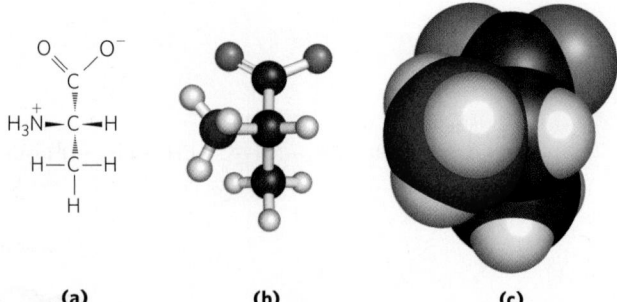

(a)   **(b)**   **(c)**

FIGURE 1-19 Representations of molecules. Three ways to represent the structure of the amino acid alanine (shown here in the ionic form found at neutral pH). **(a)** Structural formula in perspective form: a solid wedge (──) represents a bond in which the atom at the wide end projects out of the plane of the paper, toward the reader; a dashed wedge (┄┄) represents a bond extending behind the plane of the paper. **(b)** Ball-and-stick model, showing bond angles and relative bond lengths. **(c)** Space-filling model, in which each atom is shown with its correct relative van der Waals radius.

Figure 1-19 shows three ways to illustrate the stereochemistry, or configuration, of simple molecules. The perspective diagram specifies stereochemistry unambiguously, but bond angles and center-to-center bond lengths are better represented with ball-and-stick models. In space-filling models, the radius of each "atom" is proportional to its van der Waals radius, and the contours of the model define the space occupied by the molecule (the volume of space from which atoms of other molecules are excluded).

Configuration is conferred by the presence of either (1) double bonds, around which there is little or no freedom of rotation, or (2) chiral centers, around which substituent groups are arranged in a specific orientation. The identifying characteristic of stereoisomers is that they cannot be interconverted without temporarily breaking one or more covalent bonds. **Figure 1-20a** shows the configurations of maleic acid and its isomer, fumaric acid. These compounds are **geometric isomers**, or **cis-trans isomers**; they differ in the arrangement of their substituent groups with respect to the nonrotating double bond (Latin *cis,* "on this side"—groups on the same side of the double bond; *trans,* "across"—groups on opposite sides). Maleic acid (maleate at the neutral pH of cytoplasm) is the cis isomer and fumaric acid (fumarate) the trans isomer; each is a well-defined compound that can be separated from the other, and each has its own unique chemical properties. A binding site (on an enzyme, for example) that is complementary to one of these molecules would not be complementary to the other, which explains why the two compounds have distinct biological roles despite their similar chemical makeup.

In the second type of stereoisomer, four different substituents bonded to a tetrahedral carbon atom may be arranged in two different ways in space—that is, have two configurations—yielding two stereoisomers that have similar or identical chemical properties but

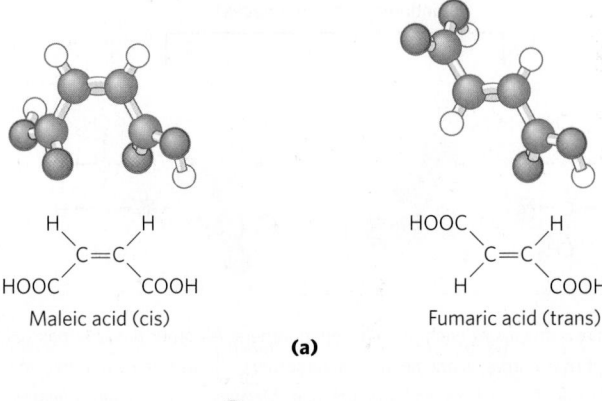

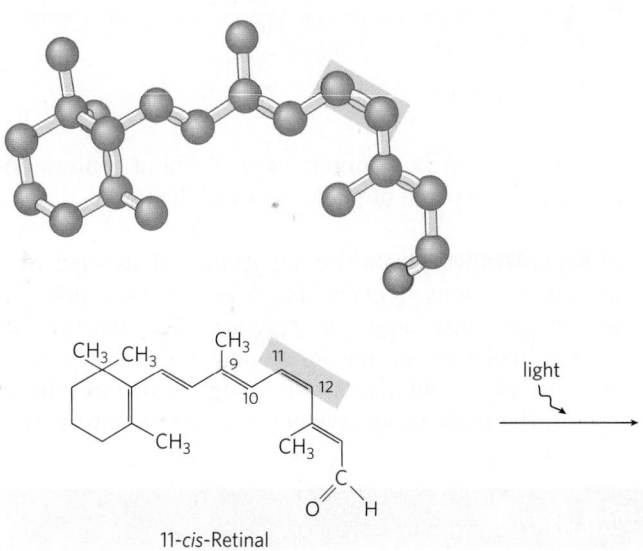

FIGURE 1-20 Configurations of geometric isomers. (a) Isomers such as maleic acid (maleate at pH 7) and fumaric acid (fumarate) cannot be interconverted without breaking covalent bonds, which requires the input of much more energy than the average kinetic energy of molecules at physiological temperatures. **(b)** In the vertebrate retina, the initial event in light detection is the absorption of visible light by 11-*cis*-retinal. The energy of the absorbed light (about 250 kJ/mol) converts 11-*cis*-retinal to all-*trans*-retinal, triggering electrical changes in the retinal cell that lead to a nerve impulse. (Note that the hydrogen atoms are omitted from the ball-and-stick models of the retinals.)

differ in certain physical and biological properties. A carbon atom with four different substituents is said to be asymmetric, and asymmetric carbons are called **chiral centers** (Greek *chiros*, "hand"; some stereoisomers are related structurally as the right hand is to the left).

A molecule with only one chiral carbon can have two stereoisomers; when two or more (n) chiral carbons are present, there can be 2^n stereoisomers. Stereoisomers that are mirror images of each other are called **enantiomers (Fig. 1-21)**. Pairs of stereoisomers that are not

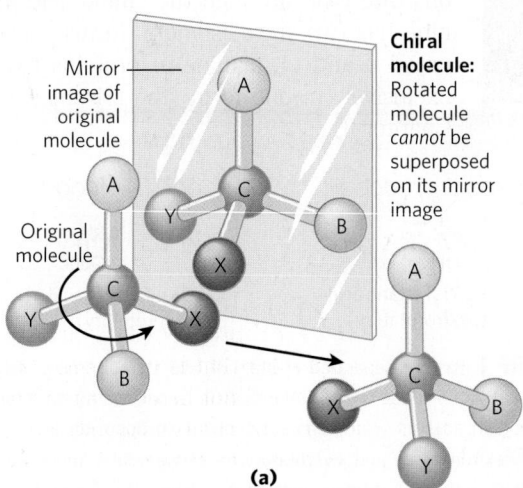

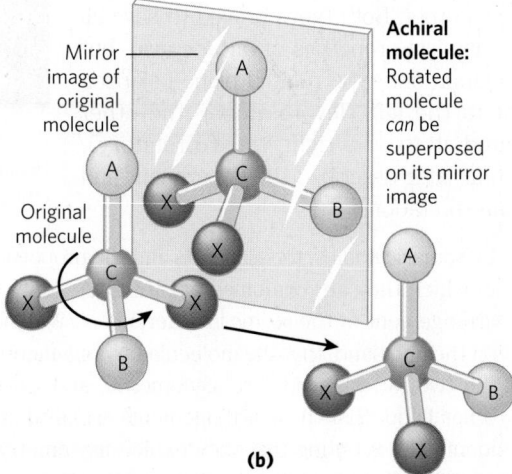

FIGURE 1-21 Molecular asymmetry: chiral and achiral molecules. (a) When a carbon atom has four different substituent groups (A, B, X, Y), they can be arranged in two ways that represent nonsuperposable mirror images of each other (enantiomers). This asymmetric carbon atom is called a chiral atom or chiral center. **(b)** When a tetrahedral carbon has only three dissimilar groups (i.e., the same group occurs twice), only one configuration is possible and the molecule is symmetric, or achiral. In this case the molecule is superposable on its mirror image: the molecule on the left can be rotated counterclockwise (when looking down the vertical bond from A to C) to create the molecule in the mirror.

Enantiomers (mirror images)   Enantiomers (mirror images)

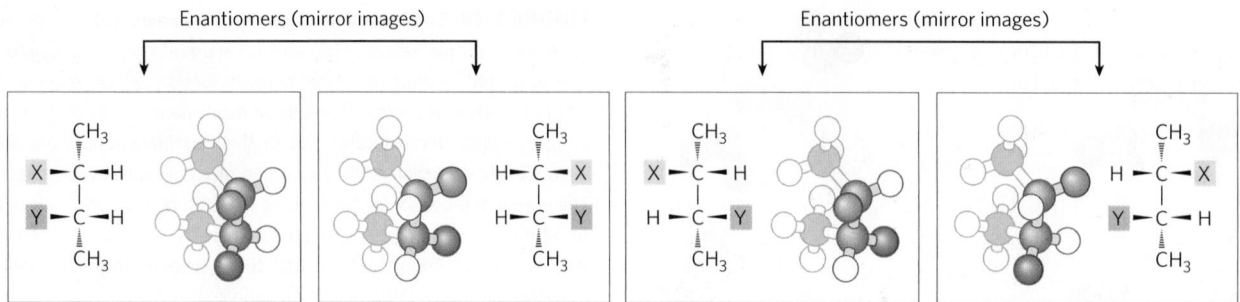

FIGURE 1-22 Enantiomers and diastereomers. There are four different stereoisomers of 2,3-disubstituted butane ($n = 2$ asymmetric carbons, hence $2^n = 4$ stereoisomers). Each is shown in a box as a perspective formula and a ball-and-stick model, which has been rotated to allow the reader to view all the groups. Two pairs of stereoisomers are mirror images of each other, or enantiomers. All other possible pairs are not mirror images and so are diastereomers. [Source: Information from F. Carroll, *Perspectives on Structure and Mechanism in Organic Chemistry*, Brooks/Cole Publishing Co., 1998, p. 63.]

mirror images of each other are called **diastereomers (Fig. 1-22).**

As Louis Pasteur first observed in 1843 (Box 1-2), enantiomers have nearly identical chemical reactivities but differ in a characteristic physical property: their interaction with plane-polarized light. In separate solutions, two enantiomers rotate the plane of plane-polarized light in opposite directions, but an equimolar solution of the two enantiomers (a **racemic mixture**) shows no optical rotation. Compounds without chiral centers do not rotate the plane of plane-polarized light.

>> Key Convention: Given the importance of stereochemistry in reactions between biomolecules (see below), biochemists must name and represent the structure of each biomolecule so that its stereochemistry is unambiguous. For compounds with more than one chiral center, the most useful system of nomenclature is the

---

BOX 1-2   Louis Pasteur and Optical Activity: *In Vino, Veritas*

Louis Pasteur encountered the phenomenon of **optical activity** in 1843, during his investigation of the crystalline sediment that accumulated in wine casks (a form of tartaric acid called paratartaric acid—also called racemic acid, from Latin *racemus*, "bunch of grapes"). He used fine forceps to separate two types of crystals identical in shape but mirror images of each other. Both types proved to have all the chemical properties of tartaric acid, but in solution one type rotated plane-polarized light to the left (levorotatory), the other rotated it to the right (dextrorotatory). Pasteur later described the experiment and its interpretation:

In isomeric bodies, the elements and the proportions in which they are combined are the same, only the arrangement of the atoms is different . . . We know, on the one hand, that the molecular arrangements of the two tartaric acids are asymmetric, and, on the other hand, that these arrangements are absolutely identical, excepting that they exhibit asymmetry in opposite directions. Are the atoms of the dextro acid grouped in the form of a right-handed spiral, or are they placed at the apex of an irregular tetrahedron, or are they disposed according to this or that asymmetric arrangement? We do not know.*

Louis Pasteur 1822–1895
[Source: The Granger Collection.]

Now we do know. X-ray crystallographic studies in 1951 confirmed that the levorotatory and dextrorotatory forms of tartaric acid are mirror images of each other at the molecular level and established the absolute configuration of each (Fig. 1). The same approach has been used to demonstrate that although the amino acid alanine has two stereoisomeric forms (designated D and L), alanine in proteins exists exclusively in one form (the L isomer; see Chapter 3).

(2R,3R)-Tartaric acid
(dextrorotatory)

(2S,3S)-Tartaric acid
(levorotatory)

FIGURE 1 Pasteur separated crystals of two stereoisomers of tartaric acid and showed that solutions of the separated forms rotated plane-polarized light to the same extent but in opposite directions. These dextrorotatory and levorotatory forms were later shown to be the (R,R) and (S,S) isomers represented here. The RS system of nomenclature is explained in the text.

*From Pasteur's lecture to the Société Chimique de Paris in 1883, quoted in R. DuBos, *Louis Pasteur: Free Lance of Science*, p. 95, Charles Scribner's Sons, New York, 1976.

RS system. In this system, each group attached to a chiral carbon is assigned a *priority*. The priorities of some common substituents are

$$—OCH_3 > —OH > —NH_2 > —COOH >$$
$$—CHO > —CH_2OH > —CH_3 > —H$$

For naming in the RS system, the chiral atom is viewed with the group of lowest priority (4 in the following diagram) pointing away from the viewer. If the priority of the other three groups (1 to 3) decreases in clockwise order, the configuration is (*R*) (Latin *rectus*, "right"); if counterclockwise, the configuration is (*S*) (Latin *sinister*, "left"). In this way, each chiral carbon is designated either (*R*) or (*S*), and the inclusion of these designations in the name of the compound provides an unambiguous description of the stereochemistry at each chiral center.

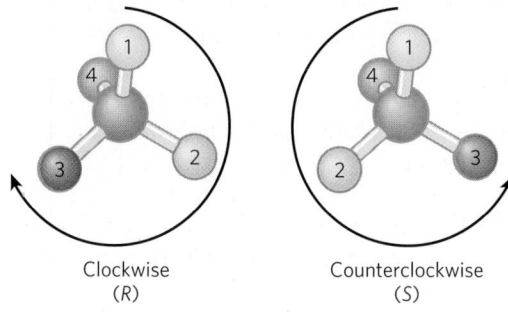

| Clockwise | Counterclockwise |
| (*R*) | (*S*) |

Another naming system for stereoisomers, the D and L system, is described in Chapter 3. A molecule with a single chiral center (the two isomers of glyceraldehyde, for example) can be named unambiguously by either system, as shown here.

L-Glyceraldehyde ≡ (*S*)-Glyceraldehyde **《**

Distinct from configuration is molecular **conformation**, the spatial arrangement of substituent groups that, without breaking any bonds, are free to assume different positions in space because of the freedom of rotation about single bonds. In the simple hydrocarbon ethane, for example, there is nearly complete freedom of rotation around the C—C bond. Many different, interconvertible conformations of ethane are possible, depending on the degree of rotation **(Fig. 1-23)**. Two conformations are of special interest: the staggered, which is more stable than all others and thus predominates, and the eclipsed, which is the least stable. We cannot isolate either of these conformational forms, because they are freely interconvertible. However, when one or more of the hydrogen atoms on each carbon is replaced by a functional group that is either very large or electrically charged, freedom of rotation around the C—C bond is hindered. This limits the number of stable conformations of the ethane derivative.

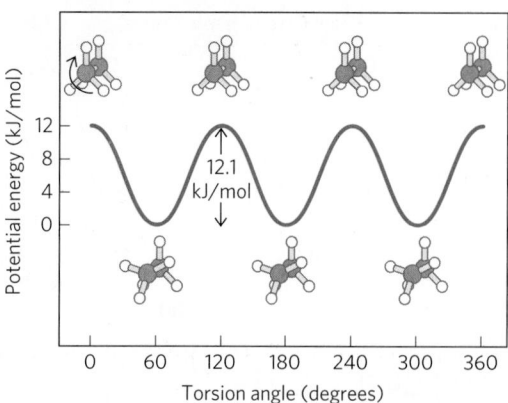

FIGURE 1-23 Conformations. Many conformations of ethane are possible because of freedom of rotation around the C—C bond. In the ball-and-stick model, when the front carbon atom (as viewed by the reader) with its three attached hydrogens is rotated relative to the rear carbon atom, the potential energy of the molecule rises to a maximum in the fully eclipsed conformation (torsion angle 0°, 120°, etc.), then falls to a minimum in the fully staggered conformation (torsion angle 60°, 180°, etc.). Because the energy differences are small enough to allow rapid interconversion of the two forms (millions of times per second), the eclipsed and staggered forms cannot be separately isolated.

Interactions between Biomolecules Are Stereospecific

When biomolecules interact, the "fit" between them must be stereochemically correct. The three-dimensional structure of biomolecules large and small—the combination of configuration and conformation—is of the utmost importance in their biological interactions: reactant with its enzyme, hormone with its receptor on a cell surface, antigen with its specific antibody, for example **(Fig. 1-24)**. The study of biomolecular

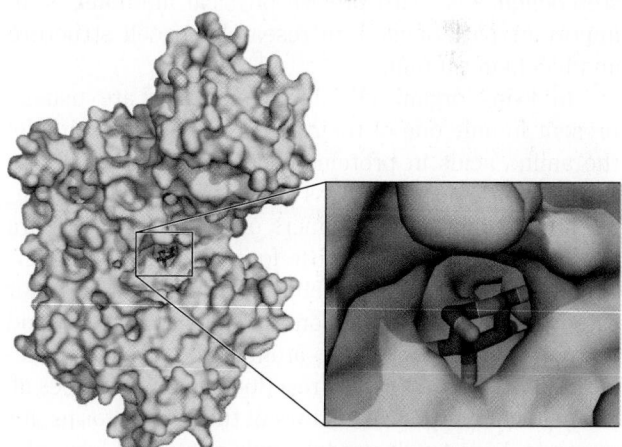

FIGURE 1-24 Complementary fit between a macromolecule and a small molecule. A glucose molecule fits into a pocket on the surface of the enzyme hexokinase and is held in this orientation by several noncovalent interactions between the protein and the sugar. This representation of the hexokinase molecule is produced with software that can calculate the shape of the outer surface of a macromolecule, defined either by the van der Waals radii of all the atoms in the molecule or by the "solvent exclusion volume," the volume that a water molecule cannot penetrate. [Source: PDB ID 3B8A, P. Kuser et al., *Proteins* 72:731, 2008.]

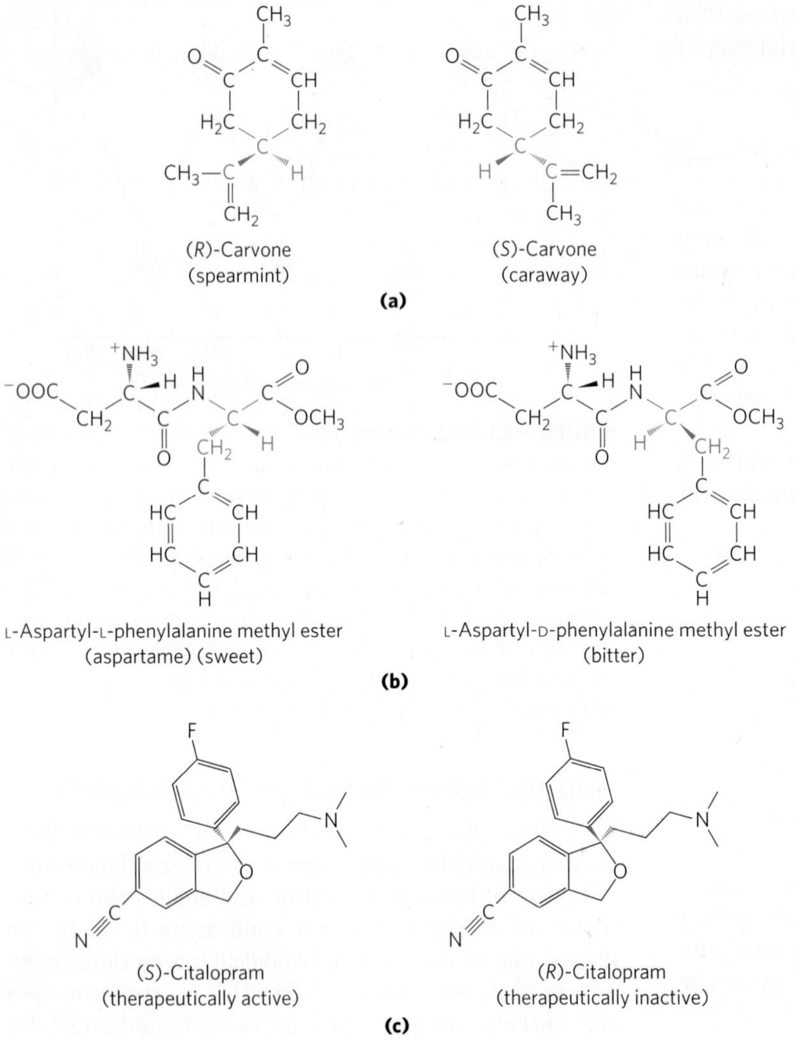

(R)-Carvone
(spearmint)

(S)-Carvone
(caraway)

(a)

L-Aspartyl-L-phenylalanine methyl ester
(aspartame) (sweet)

L-Aspartyl-D-phenylalanine methyl ester
(bitter)

(b)

(S)-Citalopram
(therapeutically active)

(R)-Citalopram
(therapeutically inactive)

(c)

FIGURE 1-25 Stereoisomers have different effects in humans. (a) Two stereoisomers of carvone: (R)-carvone (isolated from spearmint oil) has the characteristic fragrance of spearmint; (S)-carvone (from caraway seed oil) smells like caraway. **(b)** Aspartame, the artificial sweetener sold under the trade name NutraSweet, is easily distinguishable by taste receptors from its bitter-tasting stereoisomer, although the two differ only in the configuration at one of the two chiral carbon atoms. **(c)** The antidepressant medication citalopram (trade name Celexa), a selective serotonin reuptake inhibitor, is a racemic mixture of these two stereoisomers, but only (S)-citalopram has the therapeutic effect. A stereochemically pure preparation of (S)-citalopram (escitalopram oxalate) is sold under the trade name Lexapro. As you might predict, the effective dose of Lexapro is one-half the effective dose of Celexa.

stereochemistry, with precise physical methods, is an important part of modern research on cell structure and biochemical function.

In living organisms, chiral molecules are usually present in only one of their chiral forms. For example, the amino acids in proteins occur only as their L isomers; glucose occurs only as its D isomer. (The conventions for naming stereoisomers of the amino acids are described in Chapter 3; those for sugars, in Chapter 7. The RS system, described above, is the most useful for some biomolecules.) In contrast, when a compound with an asymmetric carbon atom is chemically synthesized in the laboratory, the reaction usually produces all possible chiral forms: a mixture of the D and L forms, for example. Living cells produce only one chiral form of a biomolecule because the enzymes that synthesize that molecule are also chiral.

Stereospecificity, the ability to distinguish between stereoisomers, is a property of enzymes and other proteins and a characteristic feature of biochemical interactions. If the binding site on a protein is complementary to one isomer of a chiral compound, it will not be complementary to the other isomer, for the same reason that a left glove does not fit a right hand. Some striking examples of the ability of biological systems to distinguish stereoisomers are shown in **Figure 1-25**.

The common classes of chemical reactions encountered in biochemistry are described in Chapter 13, as an introduction to the reactions of metabolism.

SUMMARY 1.2 Chemical Foundations

■ Because of its bonding versatility, carbon can produce a broad array of carbon–carbon skeletons with a variety of functional groups; these groups give biomolecules their biological and chemical personalities.

■ A nearly universal set of about a thousand small molecules is found in living cells; the interconversions of these molecules in the central metabolic pathways have been conserved in evolution.

■ Proteins and nucleic acids are linear polymers of simple monomeric subunits; their sequences contain the information that gives each molecule its three-dimensional structure and its biological functions.

■ Molecular configuration can be changed only by breaking covalent bonds. For a carbon atom with four different substituents (a chiral carbon), the substituent groups can be arranged in two different ways, generating stereoisomers with distinct properties. Only one stereoisomer is biologically active. Molecular conformation is the position of atoms in space that can be changed by rotation about single bonds, without breaking covalent bonds.

■ Interactions between biological molecules are almost invariably stereospecific: they require a close fit between complementary structures in the interacting molecules.

1.3 Physical Foundations

Living cells and organisms must perform work to stay alive and to reproduce themselves. The synthetic reactions that occur within cells, like the synthetic processes in any factory, require the input of energy. Energy input is also needed in the motion of a bacterium or an Olympic sprinter, in the flashing of a firefly or the electrical discharge of an eel. And the storage and expression of information require energy, without which structures rich in information inevitably become disordered and meaningless.

In the course of evolution, cells have developed highly efficient mechanisms for coupling the energy obtained from sunlight or chemical fuels to the many energy-requiring processes they must carry out. One goal of biochemistry is to understand, in quantitative and chemical terms, the means by which energy is extracted, stored, and channeled into useful work in living cells. We can consider cellular energy conversions—like all other energy conversions—in the context of the laws of thermodynamics.

Living Organisms Exist in a Dynamic Steady State, Never at Equilibrium with Their Surroundings

The molecules and ions contained within a living organism differ in kind and in concentration from those in the organism's surroundings. A paramecium in a pond, a shark in the ocean, a bacterium in the soil, an apple tree in an orchard—all are different in composition from their surroundings and, once they have reached maturity, maintain a more or less constant composition in the face of a constantly changing environment.

Although the characteristic composition of an organism changes little through time, the population of molecules within the organism is far from static. Small molecules, macromolecules, and supramolecular complexes are continuously synthesized and broken down in chemical reactions that involve a constant flux of mass and energy through the system. The hemoglobin molecules carrying oxygen from your lungs to your brain at this moment were synthesized within the past month; by next month they will have been degraded and entirely replaced by new hemoglobin molecules. The glucose you ingested with your most recent meal is now circulating in your bloodstream; before the day is over these particular glucose molecules will have been converted into something else—carbon dioxide or fat, perhaps—and will have been replaced with a fresh supply of glucose, so that your blood glucose concentration is more or less constant over the whole day. The amounts of hemoglobin and glucose in the blood remain nearly constant because the rate of synthesis or intake of each just balances the rate of its breakdown, consumption, or conversion into some other product. The constancy of concentration is the result of a *dynamic steady state,* a steady state that is far from equilibrium. Maintaining this steady state requires the constant investment of energy; when a cell can no longer obtain energy, it dies and begins to decay toward equilibrium with its surroundings. We consider below exactly what is meant by "steady state" and "equilibrium."

Organisms Transform Energy and Matter from Their Surroundings

For chemical reactions occurring in solution, we can define a **system** as all the constituent reactants and products, the solvent that contains them, and the immediate atmosphere—in short, everything within a defined region of space. The system and its surroundings together constitute the **universe**. If the system exchanges neither matter nor energy with its surroundings, it is said to be **isolated**. If the system exchanges energy but not matter with its surroundings, it is a **closed** system; if it exchanges both energy and matter with its surroundings, it is an **open** system.

A living organism is an open system; it exchanges both matter and energy with its surroundings. Organisms obtain energy from their surroundings in two ways: (1) they take up chemical fuels (such as glucose) from the environment and extract energy by oxidizing them (see Box 1-3, Case 2); or (2) they absorb energy from sunlight.

The first law of thermodynamics describes the principle of the conservation of energy: *in any physical or chemical change, the total amount of energy in the universe remains constant, although the form of the energy may change.* This means that while energy is "used" by a system, it is not "used up"; rather, it is converted from one form into another—from potential energy in chemical bonds, say, into kinetic energy of heat and motion. Cells are consummate transducers of energy, capable of interconverting

BOX 1-3 Entropy: Things Fall Apart

The term "entropy," which literally means "a change within," was first used in 1851 by Rudolf Clausius, one of the formulators of the second law of thermodynamics. It refers to the randomness or disorder of the components of a chemical system. Entropy is a central concept in biochemistry; life requires continual maintenance of order in the face of nature's tendency to increase randomness. A rigorous quantitative definition of entropy involves statistical and probability considerations. However, its nature can be illustrated qualitatively by three simple examples, each demonstrating one aspect of entropy. The key descriptors of entropy are *randomness* and *disorder*, manifested in different ways.

Case 1: The Teakettle and the Randomization of Heat

We know that steam generated from boiling water can do useful work. But suppose we turn off the burner under a teakettle full of water at 100 °C (the "system") in the kitchen (the "surroundings") and allow the teakettle to cool. As it cools, no work is done, but heat passes from the teakettle to the surroundings, raising the temperature of the surroundings (the kitchen) by an infinitesimally small amount until complete equilibrium is attained. At this point all parts of the teakettle and the kitchen are at precisely the same temperature. The free energy that was once concentrated in the teakettle of hot water at 100 °C, *potentially* capable of doing work, has disappeared. Its equivalent in heat energy is still present in the teakettle + kitchen (i.e., the "universe") but has become completely randomized

throughout. This energy is no longer available to do work because there is no temperature differential within the kitchen. Moreover, the increase in entropy of the kitchen (the surroundings) is irreversible. We know from everyday experience that heat never spontaneously passes back from the kitchen into the teakettle to raise the temperature of the water to 100 °C again.

Case 2: The Oxidation of Glucose

Entropy is a state not only of energy but of matter. Aerobic (heterotrophic) organisms extract free energy from glucose obtained from their surroundings by oxidizing the glucose with O_2, also obtained from the surroundings. The end products of this oxidative metabolism, CO_2 and H_2O, are returned to the surroundings. In this process the surroundings undergo an increase in entropy, whereas the organism itself remains in a steady state and undergoes no change in its internal order. Although some entropy arises from the dissipation of heat, entropy also arises from another kind of disorder, illustrated by the equation for the oxidation of glucose:

$$C_6H_{12}O_6 + 6O_2 \longrightarrow 6CO_2 + 6H_2O$$

We can represent this schematically as

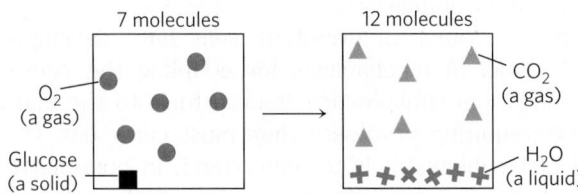

chemical, electromagnetic, mechanical, and osmotic energy with great efficiency **(Fig. 1-26)**.

The Flow of Electrons Provides Energy for Organisms

Nearly all living organisms derive their energy, directly or indirectly, from the radiant energy of sunlight. In the photoautotrophs, light-driven splitting of water during photosynthesis releases its electrons for the reduction of CO_2 and the release of O_2 into the atmosphere:

$$6CO_2 + 6H_2O \xrightarrow{\text{light}} C_6H_{12}O_6 + 6O_2$$
(light-driven reduction of CO_2)

Nonphotosynthetic organisms (chemotrophs) obtain the energy they need by oxidizing the energy-rich products of photosynthesis stored in plants, then passing the electrons thus acquired to atmospheric O_2 to form water, CO_2, and other end products, which are recycled in the environment:

$$C_6H_{12}O_6 + 6O_2 \longrightarrow 6CO_2 + 6H_2O + \text{energy}$$
(energy-yielding oxidation of glucose)

Thus autotrophs and heterotrophs participate in global cycles of O_2 and CO_2, driven ultimately by sunlight, making

these two large groups of organisms interdependent. Virtually all energy transductions in cells can be traced to this flow of electrons from one molecule to another, in a "downhill" flow from higher to lower electrochemical potential; as such, this is formally analogous to the flow of electrons in a battery-driven electric circuit. All these reactions involved in electron flow are **oxidation-reduction reactions**: one reactant is oxidized (loses electrons) as another is reduced (gains electrons).

Creating and Maintaining Order Requires Work and Energy

As we've noted, DNA, RNA, and proteins are informational macromolecules; the precise sequence of their monomeric subunits contains information, just as the letters in this sentence do. In addition to using chemical energy to form the covalent bonds between these subunits, the cell must invest energy to order the subunits in their correct sequence. It is extremely improbable that amino acids in a mixture would spontaneously condense into a single type of protein, with a unique sequence. This would represent increased order in a population of molecules; but according to the second law of thermodynamics, the tendency in nature is toward

The atoms contained in 1 molecule of glucose plus 6 molecules of oxygen, a total of 7 molecules, are more randomly dispersed by the oxidation reaction and are now present in a total of 12 molecules ($6CO_2 + 6H_2O$).

Whenever a chemical reaction results in an increase in the number of molecules—or when a solid substance is converted into liquid or gaseous products, which allow more freedom of molecular movement than solids—molecular disorder, and thus entropy, increases.

Case 3: Information and Entropy

The following short passage from *Julius Caesar,* Act IV, Scene 3, is spoken by Brutus, when he realizes that he must face Mark Antony's army. It is an information-rich nonrandom arrangement of 125 letters of the English alphabet:

> There is a tide in the affairs of men,
> Which, taken at the flood, leads on to fortune;
> Omitted, all the voyage of their life
> Is bound in shallows and in miseries.

In addition to what this passage says overtly, it has many hidden meanings. It not only reflects a complex sequence of events in the play, but also echoes the play's ideas on conflict, ambition, and the demands of leadership. Permeated with Shakespeare's understanding of human nature, it is very rich in information.

However, if the 125 letters making up this quotation were allowed to fall into a completely random, chaotic pattern, as shown in the following box, they would have no meaning whatsoever.

In this form the 125 letters contain little or no information, but they are very rich in entropy. Such considerations have led to the conclusion that information is a form of energy; information has been called "negative entropy." In fact, the branch of mathematics called information theory, which is basic to the programming logic of computers, is closely related to thermodynamic theory. Living organisms are highly ordered, nonrandom structures, immensely rich in information and thus entropy-poor.

---

ever-greater disorder in the universe: *randomness in the universe is constantly increasing*. To bring about the synthesis of macromolecules from their monomeric units, free energy must be supplied to the system (in this case, the cell). We discuss the quantitative energetics of oxidation-reduction reactions in Chapter 13.

>> Key Convention: The randomness or disorder of the components of a chemical system is expressed as **entropy**, **S** (Box 1-3). (We will give a more rigorous definition of entropy in Chapter 13.) Any change in randomness of the system is expressed as entropy change, ΔS, which by convention has a positive value when randomness increases. J. Willard Gibbs, who developed the theory of energy changes during chemical reactions, showed that the **free-energy content**, **G**, of any closed system can be defined in terms of three quantities: **enthalpy**, **H**, reflecting the number and kinds of bonds;

J. Willard Gibbs,1839–1903
[Source: Science Source.]

entropy, S; and the absolute temperature, T (in Kelvin). The definition of free energy is $G = H - TS$. When a chemical reaction occurs at constant temperature, the **free-energy change**, ΔG, is determined by the enthalpy change, ΔH, reflecting the kinds and numbers of chemical bonds and noncovalent interactions broken and formed, and the entropy change, ΔS, describing the change in the system's randomness:

$$\Delta G = \Delta H - T\Delta S$$

where, by definition, ΔH is negative for a reaction that releases heat, and ΔS is positive for a reaction that increases the system's randomness. **<<**

A process tends to occur spontaneously only if ΔG is negative (if free energy is *released* in the process). Yet cell function depends largely on molecules, such as proteins and nucleic acids, for which the free energy of formation is positive: the molecules are less stable and more highly ordered than a mixture of their monomeric components. To carry out these thermodynamically unfavorable, energy-requiring (**endergonic**) reactions, cells couple them to other reactions that liberate free energy (**exergonic** reactions), so that the

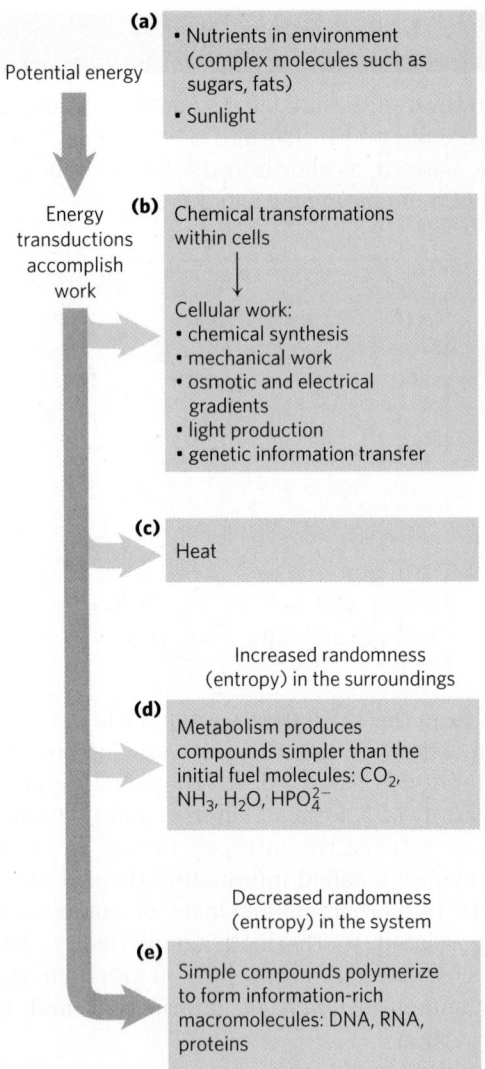

(a)

Potential energy

- Nutrients in environment (complex molecules such as sugars, fats)
- Sunlight

(b)

Energy transductions accomplish work

Chemical transformations within cells

Cellular work:
- chemical synthesis
- mechanical work
- osmotic and electrical gradients
- light production
- genetic information transfer

(c)

Heat

Increased randomness (entropy) in the surroundings

(d)

Metabolism produces compounds simpler than the initial fuel molecules: CO_2, NH_3, H_2O, HPO_4^{2-}

Decreased randomness (entropy) in the system

(e)

Simple compounds polymerize to form information-rich macromolecules: DNA, RNA, proteins

FIGURE 1-26 Some energy transformations in living organisms. As metabolic energy is spent to do cellular work, the randomness of the system plus surroundings (expressed quantitatively as entropy) increases as the potential energy of complex nutrient molecules decreases. **(a)** Living organisms extract energy from their surroundings; **(b)** convert some of it into useful forms of energy to produce work; **(c)** return some energy to the surroundings as heat; and **(d)** release end-product molecules that are less well organized than the starting fuel, increasing the entropy of the universe. One effect of all these transformations is **(e)** increased order (decreased randomness) in the system in the form of complex macromolecules. We return to a quantitative treatment of entropy in Chapter 13.

overall process is exergonic: the *sum* of the free-energy changes is negative.

The usual source of free energy in coupled biological reactions is the energy released by breakage of phosphoanhydride bonds such as those in adenosine triphosphate (ATP; **Fig. 1-27**) and guanosine triphosphate (GTP). Here, each ⓅP represents a phosphoryl group:

Amino acids → protein ΔG_1 is positive (endergonic)
ATP → AMP + Ⓟ—Ⓟ ΔG_2 is negative (exergonic)
   [or ATP → ADP + Ⓟ]

When these reactions are coupled, the sum of ΔG_1 and ΔG_2 is negative—the overall process is exergonic. By this coupling strategy, cells are able to synthesize and maintain the information-rich polymers essential to life.

Energy Coupling Links Reactions in Biology

The central issue in *bioenergetics* (the study of energy transformations in living systems) is the means by which energy from fuel metabolism or light capture is coupled to a cell's energy-requiring reactions. In thinking about energy coupling, it is useful to consider a simple mechanical example, as shown in **Figure 1-28a**. An object at the top of an inclined plane has a certain amount of potential energy as a result of its elevation. It tends to slide down the plane, losing its potential energy of position as it approaches the ground. When an appropriate string-and-pulley device couples the falling object to another, smaller object, the spontaneous downward motion of the larger can lift the smaller, accomplishing a certain amount of work. The amount of energy available to do work is the **free-energy change, ΔG**; this is always somewhat less than the theoretical amount of energy released, because some energy is dissipated as the heat of friction. The greater the elevation of the larger object, the greater the energy released (ΔG) as the object slides downward and the greater the amount of work that can be

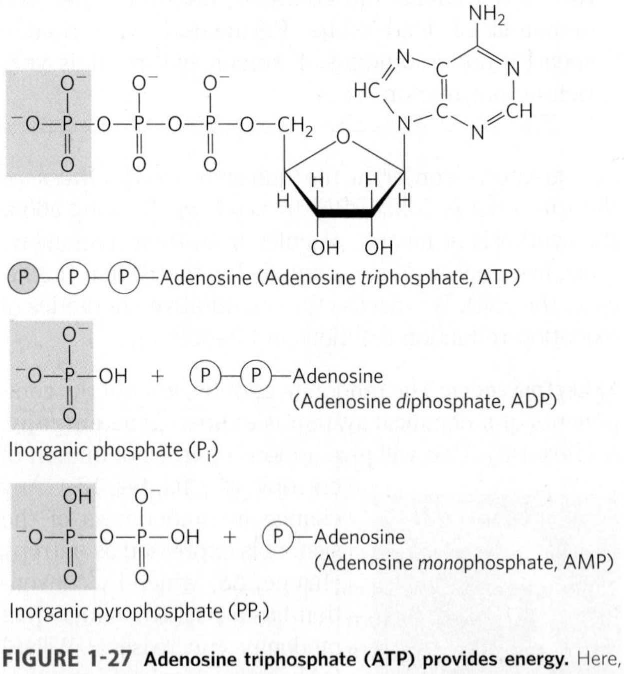

Ⓟ—Ⓟ—Ⓟ—Adenosine (Adenosine *tri*phosphate, ATP)

Inorganic phosphate (P_i)   Ⓟ—Ⓟ—Adenosine (Adenosine *di*phosphate, ADP)

Inorganic pyrophosphate (PP_i)   Ⓟ—Adenosine (Adenosine *mono*phosphate, AMP)

FIGURE 1-27 Adenosine triphosphate (ATP) provides energy. Here, each Ⓟ represents a phosphoryl group. The removal of the terminal phosphoryl group (shaded light red) of ATP, by breakage of a phosphoanhydride bond to generate adenosine diphosphate (ADP) and inorganic phosphate ion (HPO_4^{2-}), is highly exergonic, and this reaction is coupled to many endergonic reactions in the cell (as in the example in Fig. 1-28b). ATP also provides energy for many cellular processes by undergoing cleavage that releases the two terminal phosphates as inorganic pyrophosphate ($H_2P_2O_7^{2-}$), often abbreviated PP_i.

(a) Mechanical example

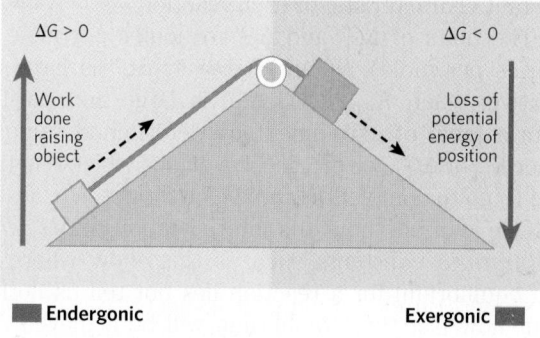

(b) Chemical example

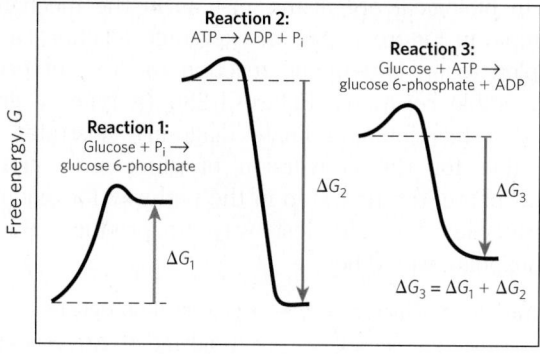

FIGURE 1-28 Energy coupling in mechanical and chemical processes. **(a)** The downward motion of an object releases potential energy that can do mechanical work. The potential energy made available by spontaneous downward motion, an exergonic process (red), can be coupled to the endergonic upward movement of another object (blue). **(b)** In reaction 1, the formation of glucose 6-phosphate from glucose and inorganic phosphate (P_i) yields a product of higher energy than the two reactants. For this endergonic reaction, ΔG is positive. In reaction 2, the exergonic breakdown of adenosine triphosphate (ATP) has a large, negative free-energy change (ΔG_2). The third reaction is the sum of reactions 1 and 2, and the free-energy change, ΔG_3, is the arithmetic sum of ΔG_1 and ΔG_2. Because ΔG_3 is negative, the overall reaction is exergonic and proceeds spontaneously.

accomplished. The larger object can lift the smaller only because, at the outset, the larger object was *far from its equilibrium position*: it had at some earlier point been elevated above the ground, in a process that itself required the input of energy.

How does this apply in chemical reactions? In closed systems, chemical reactions proceed spontaneously until **equilibrium** is reached. When a system is at equilibrium, the rate of product formation exactly equals the rate at which product is converted to reactant. Thus there is no net change in the concentration of reactants and products. The energy change as the system moves from its initial state to equilibrium, with no changes in temperature or pressure, is given by the free-energy change, ΔG. The magnitude of ΔG depends on the particular chemical reaction *and on how far from equilibrium the system is initially*.

Each compound involved in a chemical reaction contains a certain amount of potential energy, related to the kind and number of its bonds. In reactions that occur spontaneously, the products have less free energy than the reactants and thus the reaction releases free energy, which is then available to do work. Such reactions are exergonic; the decline in free energy from reactants to products is expressed as a negative value. Endergonic reactions require an input of energy, and their ΔG values are positive. As in mechanical processes, only part of the energy released in exergonic chemical reactions can be used to accomplish work. In living systems, some energy is dissipated as heat or lost to increasing entropy.

K_{eq} and $\Delta G°$ Are Measures of a Reaction's Tendency to Proceed Spontaneously

The tendency of a chemical reaction to go to completion can be expressed as an equilibrium constant. For the reaction in which a moles of A react with b moles of B to give c moles of C and d moles of D,

$$a\text{A} + b\text{B} \longrightarrow c\text{C} + d\text{D}$$

the equilibrium constant, K_{eq}, is given by

$$K_{eq} = \frac{[\text{C}]^c_{eq}[\text{D}]^d_{eq}}{[\text{A}]^a_{eq}[\text{B}]^b_{eq}}$$

where $[\text{A}]_{eq}$ is the concentration of A, $[\text{B}]_{eq}$ the concentration of B, and so on, *when the system has reached equilibrium*. K_{eq} is dimensionless (that is, has no units of measurement), but, as we explain on page 59, we will include molar units in our calculations to reinforce the point that molar concentrations (represented by the square brackets) must be used in calculating equilibrium constants. A large value of K_{eq} means the reaction tends to proceed until the reactants are almost completely converted into the products.

WORKED EXAMPLE 1-1 **Are ATP and ADP at Equilibrium in Cells?**

The equilibrium constant, K_{eq}, for the following reaction is 2×10^5 M:

$$\text{ATP} \longrightarrow \text{ADP} + \text{HPO}_4^{2-}$$

If the measured cellular concentrations are [ATP] = 5 mM, [ADP] = 0.5 mM, and [P_i] = 5 mM, is this reaction at equilibrium in living cells?

Solution: The definition of the equilibrium constant for this reaction is:

$$K_{eq} = [\text{ADP}][\text{P}_i]/[\text{ATP}]$$

From the measured cellular concentrations given above, we can calculate the mass-action ratio, Q:

$$Q = [\text{ADP}][\text{P}_i]/[\text{ATP}] = (0.5 \text{ mM})(5 \text{ mM})/5 \text{ mM}$$
$$= 0.5 \text{ mM} = 5 \times 10^{-4} \text{ M}$$

This value is *far* from the equilibrium constant for the reaction (2×10^5 M), so the reaction is *very far* from equilibrium in cells. [ATP] is far higher, and [ADP] is far lower, than is expected at equilibrium. How can a cell hold its [ATP]/[ADP] ratio so far from equilibrium? It does so by continuously extracting energy (from nutrients such as glucose) and using it to make ATP from ADP and P_i.

WORKED EXAMPLE 1-2   **Is the Hexokinase Reaction at Equilibrium in Cells?**

For the reaction catalyzed by the enzyme hexokinase:

$$\text{Glucose} + \text{ATP} \longrightarrow \text{glucose 6-phosphate} + \text{ADP}$$

the equilibrium constant, K_{eq}, is 7.8×10^2. In living *E. coli* cells, [ATP] = 5 mM, [ADP] = 0.5 mM, [glucose] = 2 mM, and [glucose 6-phosphate] = 1 mM. Is the reaction at equilibrium in *E. coli*?

Solution: At equilibrium,

$$K_{eq} = 7.8 \times 10^2 = \text{[ADP][glucose 6-phosphate]/[ATP][glucose]}$$

In living cells, [ADP][glucose 6-phosphate]/[ATP][glucose] = (0.5 mM)(1 mM)/(5 mM)(2 mM) = 0.05. The reaction is therefore *far* from equilibrium: the cellular concentrations of the products (glucose 6-phosphate and ADP) are much lower than expected at equilibrium, and those of the reactants are much higher. The reaction therefore tends strongly to go to the right.

Gibbs showed that ΔG (the actual free-energy change) for any chemical reaction is a function of the **standard free-energy change, $\Delta G°$**—a constant that is characteristic of each specific reaction—and a term that expresses the initial concentrations of reactants and products:

$$\Delta G = \Delta G° + RT \ln \frac{[C]_i^c[D]_i^d}{[A]_i^a[B]_i^b} \qquad (1\text{-}1)$$

where $[A]_i$ is the initial concentration of A, and so forth; R is the gas constant; and T is the absolute temperature.

ΔG is a measure of the distance of a system from its equilibrium position. When a reaction has reached equilibrium, no driving force remains and it can do no work: $\Delta G = 0$. For this special case, $[A]_i = [A]_{eq}$, and so on, for all reactants and products, and

$$\frac{[C]_i^c[D]_i^d}{[A]_i^a[B]_i^b} = \frac{[C]_{eq}^c[D]_{eq}^d}{[A]_{eq}^a[B]_{eq}^b}$$

Substituting 0 for ΔG and K_{eq} for $[C]_i^c[D]_i^d/[A]_i^a[B]_i^b$ in Equation 1-1, we obtain the relationship

$$\Delta G° = -RT \ln K_{eq}$$

from which we see that $\Delta G°$ is simply a second way (besides K_{eq}) of expressing the driving force on a reaction. Because K_{eq} is experimentally measurable, we

have a way of determining $\Delta G°$, the thermodynamic constant characteristic of each reaction.

The units of $\Delta G°$ and ΔG are joules per mole (or calories per mole). When $K_{eq} \gg 1$, $\Delta G°$ is large and negative; when $K_{eq} \ll 1$, $\Delta G°$ is large and positive. From a table of experimentally determined values of either K_{eq} or $\Delta G°$, we can see at a glance which reactions tend to go to completion and which do not.

One caution about the interpretation of $\Delta G°$: *thermodynamic* constants such as this show where the final equilibrium for a reaction lies but tell us nothing about how fast that equilibrium will be achieved. The rates of reactions are governed by the parameters of *kinetics*, a topic we consider in detail in Chapter 6.

In biological organisms, just as in the mechanical example in Figure 1-28a, an exergonic reaction can be coupled to an endergonic reaction to drive otherwise unfavorable reactions. Figure 1-28b (a type of graph called a reaction coordinate diagram) illustrates this principle for the conversion of glucose to glucose 6-phosphate, the first step in the pathway for oxidation of glucose. The simplest way to produce glucose 6-phosphate would be:

Reaction 1: $\text{Glucose} + P_i \longrightarrow \text{glucose 6-phosphate}$
(endergonic; ΔG_1 is positive)

(Don't be concerned about the structures of these compounds now; we describe them in detail later in the book.) This reaction does not occur spontaneously; ΔG_1 is positive. A second, highly exergonic reaction can occur in all cells:

Reaction 2: $\text{ATP} \longrightarrow \text{ADP} + P_i$
(exergonic; ΔG_2 is negative)

These two chemical reactions share a common intermediate, P_i, which is consumed in reaction 1 and produced in reaction 2. The two reactions can therefore be coupled in the form of a third reaction, which we can write as the sum of reactions 1 and 2, with the common intermediate, P_i, omitted from both sides of the equation:

Reaction 3: $\text{Glucose} + \text{ATP} \longrightarrow$
$\text{glucose 6-phosphate} + \text{ADP}$

Because more energy is released in reaction 2 than is consumed in reaction 1, the free-energy change for reaction 3, ΔG_3, is negative, and the synthesis of glucose 6-phosphate can therefore occur by reaction 3.

WORKED EXAMPLE 1-3   **Standard Free-Energy Changes Are Additive**

Given that the standard free-energy change for the reaction glucose + $P_i \longrightarrow$ glucose 6-phosphate is 13.8 kJ/mol, and the standard free-energy change for the reaction ATP $\longrightarrow$ ADP + P_i is -30.5 kJ/mol, what is the free-energy change for the reaction glucose + ATP $\longrightarrow$ glucose 6-phosphate + ADP?

Solution: We can write the equation for this reaction as the sum of two other reactions:

(1) Glucose + $P_i \longrightarrow$ glucose 6-phosphate $\Delta G_1^{\circ} = 13.8$ kJ/mol
(2) ATP $\longrightarrow$ ADP + P_i $\Delta G_2^{\circ} = -30.5$ kJ/mol

Sum: Glucose + ATP $\longrightarrow$ glucose 6-phosphate + ADP

$$\Delta G_{Sum}^{\circ} = -16.7 \text{ kJ/mol}$$

The standard free-energy change for two reactions that sum to a third is simply the sum of the two individual reactions. A negative value for ΔG° (−16.7 kJ/mol) indicates that the reaction will tend to occur spontaneously.

The coupling of exergonic and endergonic reactions through a shared intermediate is central to the energy exchanges in living systems. As we shall see, reactions that break down ATP (such as reaction 2 in Fig. 1-28b) release energy that drives many endergonic processes in cells. ATP breakdown in cells is exergonic because *all living cells maintain a concentration of ATP far above its equilibrium concentration*. It is this disequilibrium that allows ATP to serve as the major carrier of chemical energy in all cells. As we describe in detail in Chapter 13, it is not the mere breakdown of ATP that provides energy to drive endergonic reactions; rather, it is the *transfer of a phosphoryl group* from ATP to another small molecule (glucose in the case above) that conserves some of the chemical potential originally in ATP.

WORKED EXAMPLE 1-4 **Energetic Cost of ATP Synthesis**

If the equilibrium constant, K_{eq}, for the reaction

$$\text{ATP} \longrightarrow \text{ADP} + P_i$$

is 2.22×10^5 M, calculate the standard free-energy change, ΔG°, for the *synthesis* of ATP from ADP and P_i at 25 °C.

Solution: First calculate ΔG° for the reaction above.

$$\Delta G^{\circ} = -RT \ln K_{eq}$$
$$= -(8.315 \text{ J/mol} \cdot \text{K})(298 \text{ K})(\ln 2.22 \times 10^5)$$
$$= -30.5 \text{ kJ/mol}$$

This is the standard free-energy change for the *breakdown* of ATP to ADP and P_i. The standard free-energy change for the *reverse* of a reaction has the same absolute value but the opposite sign. The standard free-energy change for the reverse of the above reaction is therefore 30.5 kJ/mol. So, to synthesize 1 mol of ATP under standard conditions (25 °C, 1 M concentrations of ATP, ADP, and P_i), at least 30.5 kJ of energy must be supplied. The actual free-energy change in cells—approximately 50 kJ/mol—is greater than this because the concentrations of ATP, ADP, and P_i in cells are not the standard 1 M (see Worked Example 13-2, p. 509).

WORKED EXAMPLE 1-5 **Standard Free-Energy Change for Synthesis of Glucose 6-Phosphate**

What is the standard free-energy change, ΔG°, under physiological conditions (*E. coli* grows in the human gut, at 37 °C) for the following reaction?

$$\text{Glucose} + \text{ATP} \longrightarrow \text{glucose 6-phosphate} + \text{ADP}$$

Solution: We have the relationship $\Delta G^{\circ} = -RT \ln K_{eq}$, and the value of K_{eq} for this reaction, 7.8×10^2. Substituting the values of R, T, and K_{eq} into this equation gives:

$$\Delta G^{\circ} = -(8.315 \text{ J/mol} \cdot \text{K})(310 \text{ K})(\ln 7.8 \times 10^2) = -17 \text{ kJ/mol}$$

Notice that this value is slightly different from that in Worked Example 1-3. In that calculation we assumed a temperature of 25 °C (298 K), whereas in this calculation we used the physiological temperature of 37 °C (310 K).

Enzymes Promote Sequences of Chemical Reactions

All biological macromolecules are much less thermodynamically stable than their monomeric subunits, yet they are *kinetically stable*: their *uncatalyzed* breakdown occurs so slowly (over years rather than seconds) that, on a time scale that matters for the organism, these molecules are stable. Virtually every chemical reaction in a cell occurs at a significant rate only because of the presence of **enzymes**—biocatalysts that, like all other catalysts, greatly enhance the rate of specific chemical reactions without being consumed in the process.

The path from reactant(s) to product(s) almost invariably involves an energy barrier, called the activation barrier **(Fig. 1-29)**, that must be surmounted

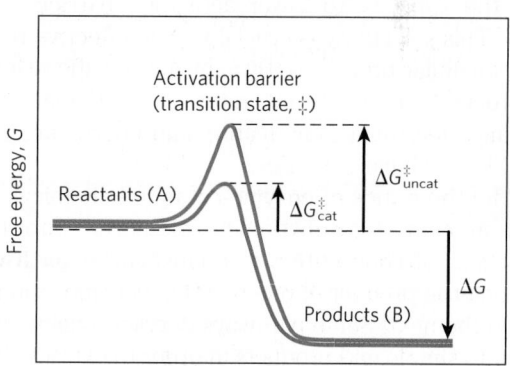

FIGURE 1-29 Energy changes during a chemical reaction. An activation barrier, representing the transition state (see Chapter 6), must be overcome in the conversion of reactants (A) into products (B), even though the products are more stable than the reactants, as indicated by a large, negative free-energy change (ΔG). The energy required to overcome the activation barrier is the activation energy ($\Delta G^{\ddagger}$). Enzymes catalyze reactions by lowering the activation barrier. They bind the transition-state intermediates tightly, and the binding energy of this interaction effectively reduces the activation energy from $\Delta G_{uncat}^{\ddagger}$ (blue curve) to $\Delta G_{cat}^{\ddagger}$ (red curve). (Note that activation energy is *not* related to free-energy change, ΔG.)

for any reaction to proceed. The breaking of existing bonds and formation of new ones generally requires, first, a distortion of the existing bonds to create a **transition state** of higher free energy than either reactant or product. The highest point in the reaction coordinate diagram represents the transition state, and the difference in energy between the reactant in its ground state and in its transition state is the **activation energy**, $\Delta G^{\ddagger}$. An enzyme catalyzes a reaction by providing a more comfortable fit for the transition state: a surface that complements the transition state in stereochemistry, polarity, and charge. The binding of enzyme to the transition state is exergonic, and the energy released by this binding reduces the activation energy for the reaction and greatly increases the reaction rate.

A further contribution to catalysis occurs when two or more reactants bind to the enzyme's surface close to each other and with stereospecific orientations that favor the reaction. This increases by orders of magnitude the probability of productive collisions between reactants. As a result of these factors and several others, discussed in Chapter 6, enzyme-catalyzed reactions commonly proceed at rates greater than 10^{12} times faster than the uncatalyzed reactions. (That is a *million million* times faster!)

Cellular catalysts are, with some notable exceptions, proteins. (Some RNA molecules have enzymatic activity, as discussed in Chapters 26 and 27.) Again with a few exceptions, each enzyme catalyzes a specific reaction, and each reaction in a cell is catalyzed by a different enzyme. Thousands of different enzymes are therefore required by each cell. The multiplicity of enzymes, their specificity (the ability to discriminate between reactants), and their susceptibility to regulation give cells the capacity to lower activation barriers selectively. This selectivity is crucial for the effective regulation of cellular processes. By allowing specific reactions to proceed at significant rates at particular times, enzymes determine how matter and energy are channeled into cellular activities.

The thousands of enzyme-catalyzed chemical reactions in cells are functionally organized into many sequences of consecutive reactions, called **pathways**, in which the product of one reaction becomes the reactant in the next. Some pathways degrade organic nutrients into simple end products in order to extract chemical energy and convert it into a form useful to the cell; together, these degradative, free-energy-yielding reactions are designated **catabolism**. The energy released by catabolic reactions drives the synthesis of ATP. As a result, the cellular concentration of ATP is held far above its equilibrium concentration, so that ΔG for ATP breakdown is large and negative. Similarly, catabolism results in the production of the reduced electron carriers NADH and NADPH, both of which can donate electrons in processes that generate ATP or drive reductive steps in biosynthetic pathways.

Other pathways start with small precursor molecules and convert them to progressively larger and more complex molecules, including proteins and nucleic acids. Such synthetic pathways, which invariably require the input of energy, are collectively designated **anabolism**. The overall network of enzyme-catalyzed pathways, both catabolic and anabolic, constitutes cellular **metabolism**. ATP (and the energetically equivalent nucleoside triphosphates cytidine triphosphate (CTP), uridine triphosphate (UTP), and guanosine triphosphate (GTP)) is the connecting link between the catabolic and anabolic components of this network (shown schematically in **Fig. 1-30**). The pathways of enzyme-catalyzed

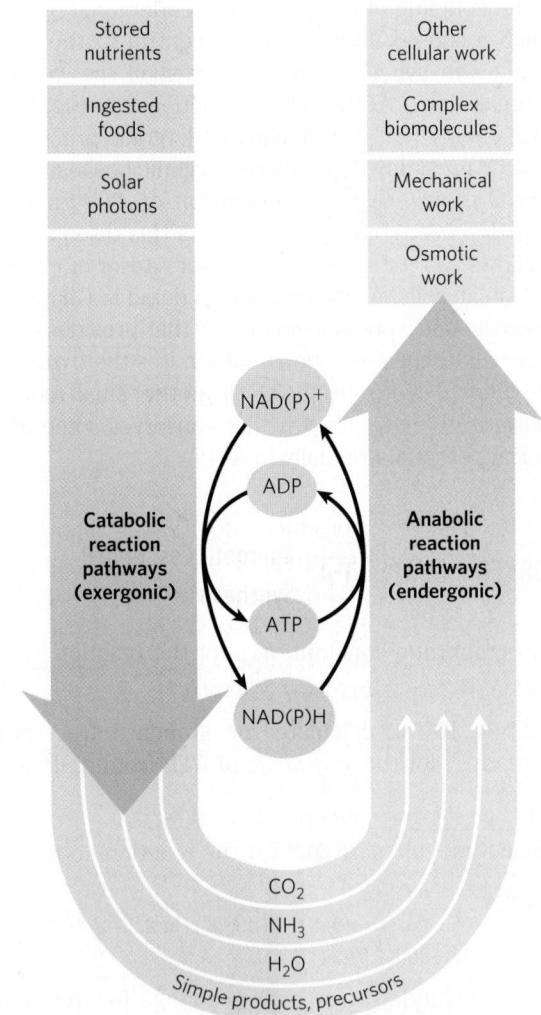

FIGURE 1-30 The central roles of ATP and NAD(P)H in metabolism. ATP is the shared chemical intermediate linking energy-releasing and energy-consuming cellular processes. Its role in the cell is analogous to that of money in an economy: it is "earned/produced" in exergonic reactions and "spent/consumed" in endergonic ones. NAD(P)H (nicotinamide adenine dinucleotide (phosphate)) is an electron-carrying cofactor that collects electrons from oxidative reactions and then donates them in a wide variety of reduction reactions in biosynthesis. Present in relatively low concentrations, these cofactors essential to anabolic reactions must be constantly regenerated by catabolic reactions.

reactions that act on the main constituents of cells—proteins, fats, sugars, and nucleic acids—are virtually identical in all living organisms.

Metabolism Is Regulated to Achieve Balance and Economy

Not only do living cells simultaneously synthesize thousands of different kinds of carbohydrate, fat, protein, and nucleic acid molecules and their simpler subunits, but they do so in the precise proportions required by the cell under any given circumstance. For example, during rapid cell growth the precursors of proteins and nucleic acids must be made in large quantities, whereas in nongrowing cells the requirement for these precursors is much reduced. Key enzymes in each metabolic pathway are regulated so that each type of precursor molecule is produced in a quantity appropriate to the current requirements of the cell.

Consider the pathway in *E. coli* that leads to the synthesis of the amino acid isoleucine, a constituent of proteins. The pathway has five steps catalyzed by five different enzymes (A through F represent the intermediates in the pathway):

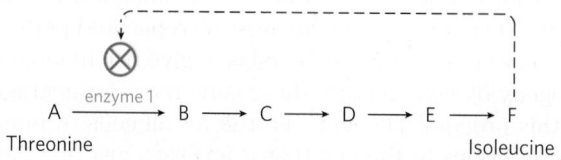

If a cell begins to produce more isoleucine than it needs for protein synthesis, the unused isoleucine accumulates and the increased concentration inhibits the catalytic activity of the first enzyme in the pathway, immediately slowing the production of isoleucine. Such **feedback inhibition** keeps the production and utilization of each metabolic intermediate in balance. (Throughout the book, we use ⊗ to indicate inhibition of an enzymatic reaction.)

Although the concept of discrete pathways is an important tool for organizing our understanding of metabolism, it is an oversimplification. There are thousands of metabolic intermediates in a cell, many of which are part of more than one pathway. Metabolism would be better represented as a web of interconnected and interdependent pathways. A change in the concentration of any one metabolite would start a ripple effect, influencing the flow of materials through other pathways. The task of understanding these complex interactions among intermediates and pathways in quantitative terms is daunting, but **systems biology**, discussed in Chapter 15, has begun to offer important insights into the overall regulation of metabolism.

Cells also regulate the synthesis of their own catalysts, the enzymes, in response to increased or diminished need for a metabolic product; this is the substance of Chapter 28. The regulated expression of genes (the

translation from information in DNA to active protein in the cell) and synthesis of enzymes are other layers of metabolic control in the cell. All layers must be taken into account when describing the overall control of cellular metabolism.

SUMMARY 1.3 Physical Foundations

■ Living cells are open systems, exchanging matter and energy with their surroundings, extracting and channeling energy to maintain themselves in a dynamic steady state distant from equilibrium. Energy is obtained from sunlight or chemical fuels by converting the energy from electron flow into the chemical bonds of ATP.

■ The tendency for a chemical reaction to proceed toward equilibrium can be expressed as the free-energy change, ΔG, which has two components: enthalpy change, ΔH, and entropy change, ΔS. These variables are related by the equation $\Delta G = \Delta H - T\Delta S$.

■ When ΔG of a reaction is negative, the reaction is exergonic and tends to go toward completion; when ΔG is positive, the reaction is endergonic and tends to go in the reverse direction. When two reactions can be summed to yield a third reaction, the ΔG for this overall reaction is the sum of the ΔG values for the two separate reactions.

■ The reactions converting ATP to P_i and ADP or to AMP and PP_i are highly exergonic (large negative ΔG). Many endergonic cellular reactions are driven by coupling them, through a common intermediate, to these highly exergonic reactions.

■ The standard free-energy change for a reaction, $\Delta G°$, is a physical constant that is related to the equilibrium constant by the equation $\Delta G° = -RT \ln K_{eq}$.

■ Most cellular reactions proceed at useful rates only because enzymes are present to catalyze them. Enzymes act in part by stabilizing the transition state, reducing the activation energy, $\Delta G^{\ddagger}$, and increasing the reaction rate by many orders of magnitude. The catalytic activity of enzymes in cells is regulated.

■ Metabolism is the sum of many interconnected reaction sequences that interconvert cellular metabolites. Each sequence is regulated to provide what the cell needs at a given time and to expend energy only when necessary.

1.4 Genetic Foundations

Perhaps the most remarkable property of living cells and organisms is their ability to reproduce themselves for countless generations with nearly perfect fidelity. This continuity of inherited traits implies constancy, over millions of years, in the structure of the molecules

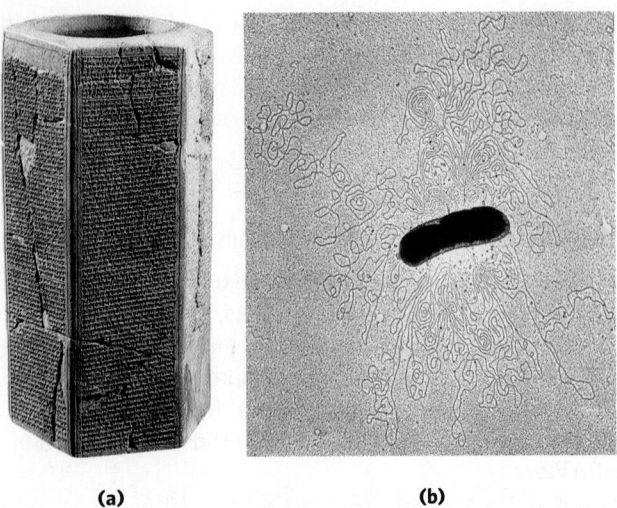

(a) **(b)**

FIGURE 1-31 Two ancient scripts. (a) The Prism of Sennacherib, inscribed in about 700 BCE, describes in characters of the Assyrian language some historical events during the reign of King Sennacherib. The Prism contains about 20,000 characters, weighs about 50 kg, and has survived almost intact for about 2,700 years. **(b)** The single DNA molecule of the bacterium *E. coli*, leaking out of a disrupted cell, is hundreds of times longer than the cell itself and contains all the encoded information necessary to specify the cell's structure and functions. The bacterial DNA contains about 4.6 million characters (nucleotides), weighs less than 10^{-10} g, and has undergone only relatively minor changes during the past several million years. (The yellow spots and dark specks in this colorized electron micrograph are artifacts of the preparation.) [Sources: (a) Erich Lessing/Art Resource, New York. (b) Dr. Gopal Murti-CNRI/Phototake New York.]

that contain the genetic information. Very few historical records of civilization, even those etched in copper or carved in stone **(Fig. 1-31)**, have survived for a thousand years. But there is good evidence that the genetic instructions in living organisms have remained nearly unchanged over very much longer periods; many bacteria have nearly the same size, shape, and internal structure as bacteria that lived almost four billion years ago. This continuity of structure and composition is the result of continuity in the structure of the genetic material.

Among the seminal discoveries in biology in the twentieth century were the chemical nature and the three-dimensional structure of the genetic material, **deoxyribonucleic acid**, **DNA**. The sequence of the monomeric subunits, the nucleotides (strictly, deoxyribonucleotides, as discussed below), in this linear polymer encodes the instructions for forming all other cellular components and provides a template for the production of identical DNA molecules to be distributed to progeny when a cell divides. The perpetuation of a biological species requires that its genetic information be maintained in a stable form, expressed accurately in the form of gene products, and reproduced with a minimum of errors. The effective storage,

expression, and reproduction of the genetic message defines individual species, distinguishes them from one another, and assures their continuity over successive generations.

Genetic Continuity Is Vested in Single DNA Molecules

DNA is a long, thin, organic polymer, the rare molecule that is constructed on the atomic scale in one dimension (width) and the human scale in another (length: a molecule of DNA can be many centimeters long). A human sperm or egg, carrying the accumulated hereditary information of billions of years of evolution, transmits this inheritance in the form of DNA molecules, in which the linear sequence of covalently linked nucleotide subunits encodes the genetic message.

Usually when we describe the properties of a chemical species, we describe the average behavior of a very large number of identical molecules. While it is difficult to predict the behavior of any single molecule in a collection of, say, a picomole (about 6×10^{11} molecules) of a compound, the *average* behavior of the molecules is predictable because so many molecules enter into the average. Cellular DNA is a remarkable exception. The DNA that is the entire genetic material of an *E. coli cell* is a *single molecule* containing 4.64 million nucleotide pairs. That single molecule must be replicated perfectly in every detail if an *E. coli* cell is to give rise to identical progeny by cell division; there is no room for averaging in this process! The same is true for all cells. A human sperm brings to the egg that it fertilizes just one molecule of DNA in each of its 23 different chromosomes, to combine with just one DNA molecule in each corresponding chromosome in the egg. The result of this union is highly predictable: an embryo with all of its ~20,000 genes, constructed of 3 billion nucleotide pairs, intact. An amazing chemical feat.

WORKED EXAMPLE 1-6 Fidelity of DNA Replication

Calculate the number of times the DNA of a modern *E. coli* cell has been copied accurately since its earliest bacterial precursor cell arose about 3.5 billion years ago. Assume for simplicity that over this time period, *E. coli* has undergone, on average, one cell division every 12 hours (this is an overestimate for modern bacteria, but probably an underestimate for ancient bacteria).

Solution:

$$(1 \text{ generation}/12 \text{ hr})(24 \text{ hr/d})(365 \text{ d/yr})(3.5 \times 10^9 \text{ yr})$$
$$= 2.6 \times 10^{12} \text{ generations.}$$

A single page of this book contains about 5,000 characters, so the entire book contains about 5 million characters. The chromosome of *E. coli* also contains about 5 million characters (nucleotide pairs). Imagine

FIGURE 1-32 Complementarity between the two strands of DNA.
DNA is a linear polymer of covalently joined deoxyribonucleotides of four types: deoxyadenylate (A), deoxyguanylate (G), deoxycytidylate (C), and deoxythymidylate (T). Each nucleotide, with its unique three-dimensional structure, can associate very specifically but noncovalently with one other nucleotide in the complementary chain: A always associates with T, and G with C. Thus, in the double-stranded DNA molecule, the entire sequence of nucleotides in one strand is *complementary* to the sequence in the other. The two strands, held together by hydrogen bonds (represented here by vertical light blue lines) between each pair of complementary nucleotides, twist about each other to form the DNA double helix. In DNA replication, the two strands (blue) separate and two new strands (pink) are synthesized, each with a sequence complementary to one of the original strands. The result is two double-helical molecules, each identical to the original DNA.

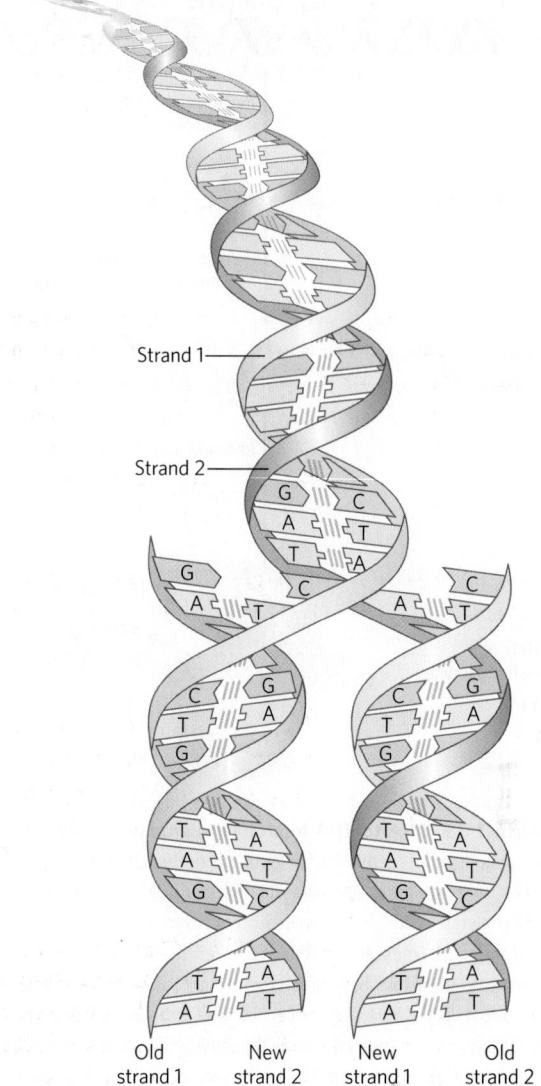

Strand 1

Strand 2

Old strand 1   New strand 2   New strand 1   Old strand 2

making a handwritten copy of this book and passing on the copy to a classmate, who copies it by hand and passes this second copy to a third classmate, who makes a third copy, and so on. How closely would each successive copy of the book resemble the original? Now, imagine the textbook that would result from hand-copying this one a few trillion times!

The Structure of DNA Allows Its Replication and Repair with Near-Perfect Fidelity

The capacity of living cells to preserve their genetic material and to duplicate it for the next generation results from the structural complementarity between the two strands of the DNA molecule **(Fig. 1-32)**. The basic unit of DNA is a linear polymer of four different monomeric subunits, **deoxyribonucleotides**, arranged in a precise linear sequence. It is this linear sequence that encodes the genetic information. Two of these polymeric strands are twisted about each other to form the DNA double helix, in which each deoxyribonucleotide in one strand pairs specifically with a complementary deoxyribonucleotide in the opposite strand. Before a cell divides, the two DNA strands separate and each serves as a template for the synthesis of a new, complementary strand, generating two identical double-helical molecules, one for each daughter cell. If either strand is damaged at any time, continuity of information is assured by the information present in the other strand, which can act as a template for repair of the damage.

The Linear Sequence in DNA Encodes Proteins with Three-Dimensional Structures

The information in DNA is encoded in its linear (one-dimensional) sequence of deoxyribonucleotide subunits, but the expression of this information results in a three-dimensional cell. This change from one to three dimensions occurs in two phases. A linear sequence of deoxyribonucleotides in DNA codes (through an intermediary, RNA) for the production of a protein with a corresponding linear sequence of amino acids **(Fig. 1-33)**. The protein folds into a particular three-dimensional shape, determined by its amino acid sequence and stabilized primarily by noncovalent interactions. Although the final shape of the folded protein is dictated by its amino acid sequence, the folding is aided by "molecular chaperones" (see Fig. 4-30). The precise three-dimensional structure, or **native conformation**, of the protein is crucial to its function.

Once in its native conformation, a protein may associate noncovalently with other macromolecules (other proteins, nucleic acids, or lipids) to form supramolecular complexes such as chromosomes, ribosomes, and membranes. The individual molecules of these complexes have specific, high-affinity binding sites for each other, and within the cell they spontaneously self-assemble into functional complexes.

Although the amino acid sequences of proteins carry all necessary information for achieving the proteins' native conformation, accurate folding and self-assembly also require the right cellular environment—pH, ionic

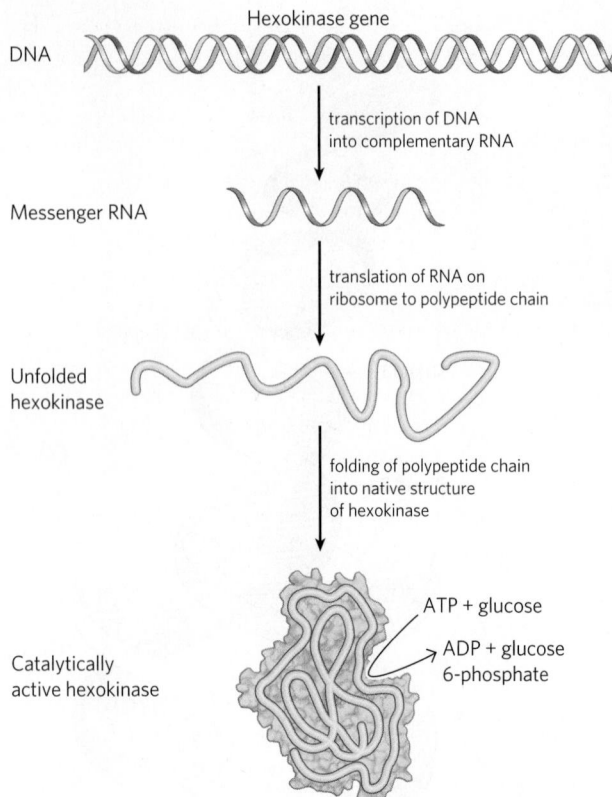

DNA

Hexokinase gene

transcription of DNA
into complementary RNA

Messenger RNA

translation of RNA on
ribosome to polypeptide chain

Unfolded
hexokinase

folding of polypeptide chain
into native structure
of hexokinase

Catalytically
active hexokinase

ATP + glucose

ADP + glucose
6-phosphate

FIGURE 1-33 DNA to RNA to protein to enzyme (hexokinase). The linear sequence of deoxyribonucleotides in the DNA (the gene) that encodes the protein hexokinase is first transcribed into a ribonucleic acid (RNA) molecule with the complementary ribonucleotide sequence. The RNA sequence (messenger RNA) is then translated into the linear protein chain of hexokinase, which folds into its native three-dimensional shape, most likely aided by molecular chaperones. Once in its native form, hexokinase acquires its catalytic activity: it can catalyze the phosphorylation of glucose, using ATP as the phosphoryl group donor.

strength, metal ion concentrations, and so forth. Thus DNA sequence alone is not enough to form and maintain a fully functioning cell.

SUMMARY 1.4 Genetic Foundations

■ Genetic information is encoded in the linear sequence of four types of deoxyribonucleotides in DNA.

■ The double-helical DNA molecule contains an internal template for its own replication and repair.

■ DNA molecules are extraordinarily large, with molecular weights in the millions or billions.

■ Despite the enormous size of DNA, the sequence of its nucleotides is very precise, and the maintenance of this precise sequence over very long times is the basis for genetic continuity in organisms.

■ The linear sequence of amino acids in a protein, which is encoded in the DNA of the gene for that protein, produces a protein's unique three-dimensional structure—a process also dependent on environmental conditions.

■ Individual macromolecules with specific affinity for other macromolecules self-assemble into supramolecular complexes.

1.5 Evolutionary Foundations

Nothing in biology makes sense except in the light of evolution.

—*Theodosius Dobzhansky,* The American Biology Teacher,
March 1973

Great progress in biochemistry and molecular biology in recent decades has amply confirmed the validity of Dobzhansky's striking generalization. The remarkable similarity of metabolic pathways and gene sequences across the three domains of life argues strongly that all modern organisms are derived from a common evolutionary progenitor by a series of small changes (mutations), each of which conferred a selective advantage to some organism in some ecological niche.

Changes in the Hereditary Instructions Allow Evolution

Despite the near-perfect fidelity of genetic replication, infrequent unrepaired mistakes in the DNA replication process lead to changes in the nucleotide sequence of DNA, producing a genetic **mutation** and changing the instructions for a cellular component. Incorrectly repaired damage to one of the DNA strands has the same effect. Mutations in the DNA handed down to offspring—that is, mutations carried in the reproductive cells—may be harmful or even lethal to the new organism or cell; they may, for example, cause the synthesis of a defective enzyme that is not able to catalyze an essential metabolic reaction. Occasionally, however, a mutation *better* equips an organism or cell to survive in its environment **(Fig. 1-34)**. The mutant enzyme might have acquired a slightly different specificity, for example, so that it is now able to use some compound that the cell was previously unable to metabolize. If a population of cells were to find itself in an environment where that compound was the only or the most abundant available source of fuel, the mutant cell would have a selective advantage over the other, unmutated (**wild-type**) cells in the population. The mutant cell and its progeny would survive and prosper in the new environment, whereas wild-type cells would starve and be eliminated. This is what Darwin meant by natural selection—what is sometimes summarized as "survival of the fittest."

Occasionally, a second copy of a whole gene is introduced into the chromosome as a result of defective replication of the chromosome. The second copy is superfluous, and mutations in this gene will not be deleterious; it becomes a means by which the cell may evolve, by producing a new gene with a new function while retaining the original gene and gene function. Seen in this light, the DNA molecules of modern

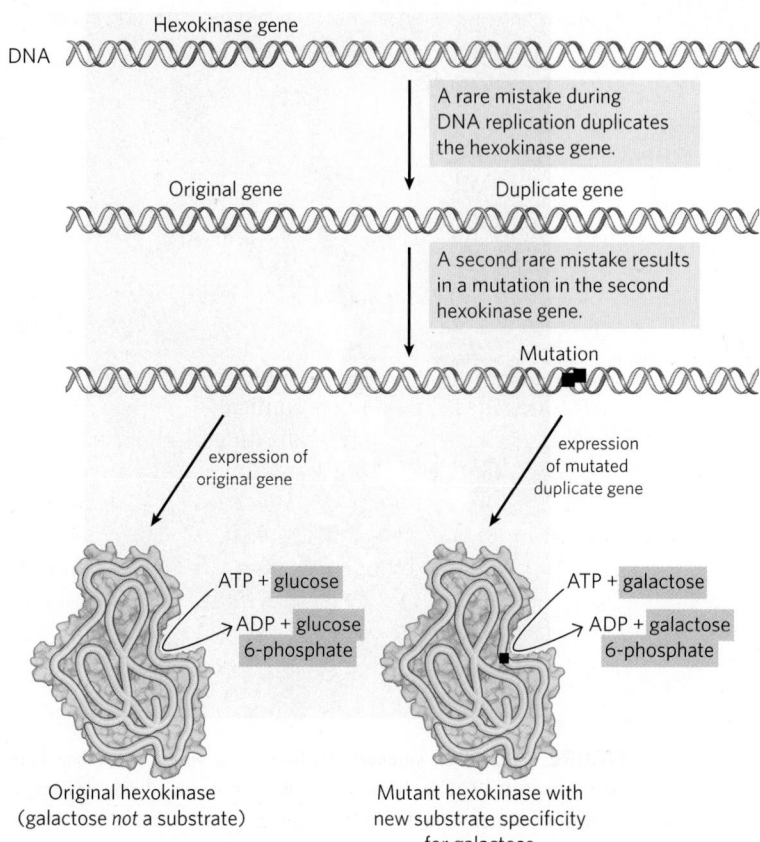

DNA   Hexokinase gene

A rare mistake during DNA replication duplicates the hexokinase gene.

Original gene   Duplicate gene

A second rare mistake results in a mutation in the second hexokinase gene.

Mutation

expression of original gene

expression of mutated duplicate gene

ATP + glucose

ADP + glucose 6-phosphate

ATP + galactose

ADP + galactose 6-phosphate

Original hexokinase (galactose *not* a substrate)

Mutant hexokinase with new substrate specificity for galactose

FIGURE 1-34 Gene duplication and mutation: one path to generate new enzymatic activities. In this example, the single hexokinase gene in a hypothetical organism might occasionally, by accident, be copied twice during DNA replication, such that the organism has two full copies of the gene, one of which is superfluous. Over many generations, as the DNA with two hexokinase genes is repeatedly duplicated, rare mistakes occur, leading to changes in the nucleotide sequence of the superfluous gene and thus of the protein that it encodes. In a few very rare cases, the altered protein produced from this mutant gene can bind a new substrate—galactose in our hypothetical case. The cell containing the mutant gene has acquired a new capability (metabolism of galactose), which may allow the cell to survive in an ecological niche that provides galactose but not glucose. If no gene duplication precedes mutation, the original function of the gene product is lost.

organisms are historical documents, records of the long journey from the earliest cells to modern organisms. The historical accounts in DNA are not complete, however; in the course of evolution, many mutations must have been erased or written over. But DNA molecules are the best source of biological history that we have. The frequency of errors in DNA replication represents a balance between too many errors, which would yield nonviable daughter cells, and too few, which would prevent the genetic variation that allows survival of mutant cells in new ecological niches.

Several billion years of natural selection have refined cellular systems to take maximum advantage of the chemical and physical properties of available raw materials. Chance genetic mutations occurring in individuals in a population, combined with natural selection, have resulted in the evolution of the enormous variety of species we see today, each adapted to its particular ecological niche.

Biomolecules First Arose by Chemical Evolution

In our account thus far, we have passed over the first chapter of the story of evolution: the appearance of the first living cell. Apart from their occurrence in living organisms, organic compounds, including the basic biomolecules such as amino acids and carbohydrates, are found in only trace amounts in the Earth's crust, the sea, and the atmosphere. How did the first living organisms acquire their characteristic organic building

blocks? According to one hypothesis, these compounds were created by the effects of powerful environmental forces—ultraviolet irradiation, lightning, or volcanic eruptions—on the gases in the prebiotic Earth's atmosphere and on inorganic solutes in superheated thermal vents deep in the ocean.

This hypothesis was tested in a classic experiment on the abiotic (nonbiological) origin of organic biomolecules carried out in 1953 by Stanley Miller in the laboratory of Harold Urey. Miller subjected gaseous mixtures such as those presumed to exist on the prebiotic Earth, including NH_3, CH_4, H_2O, and H_2, to electrical sparks produced across a pair of electrodes (to simulate lightning) for periods of a week or more, then analyzed the contents of the closed reaction vessel **(Fig. 1-35)**. The gas phase of the resulting mixture contained CO and CO_2 as well as the starting materials. The water phase contained a variety of organic compounds, including some amino acids, hydroxy acids, aldehydes, and hydrogen cyanide (HCN). This experiment established the possibility of abiotic production of biomolecules in relatively short times under relatively mild conditions. When Miller's carefully stored samples were rediscovered in 2010 and examined with much more sensitive and discriminating techniques (high-performance liquid chromatography and mass spectrometry), his original observations were confirmed and greatly broadened. Previously unpublished experiments by Miller that included H_2S in the gas mixture (mimicking the "smoking" volcanic plumes at the sea

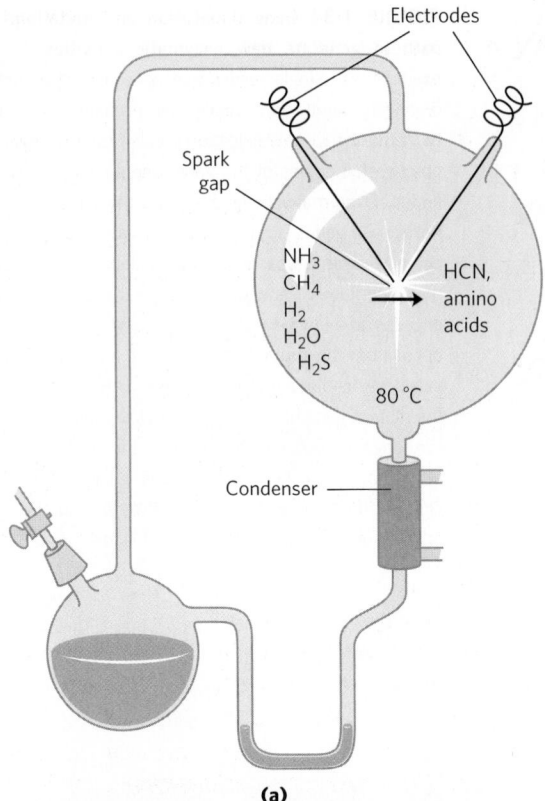

(a)

(b)

FIGURE 1-35 Abiotic production of biomolecules. (a) Spark-discharge apparatus of the type used by Miller and Urey in experiments demonstrating abiotic formation of organic compounds under primitive atmospheric conditions. After subjection of the gaseous contents of the system to electrical sparks, products were collected by condensation. Biomolecules such as amino acids were among the products. **(b)** Stanley L. Miller (1930-2007) using his spark-discharge apparatus. [Source: (b) Bettmann/Corbis.]

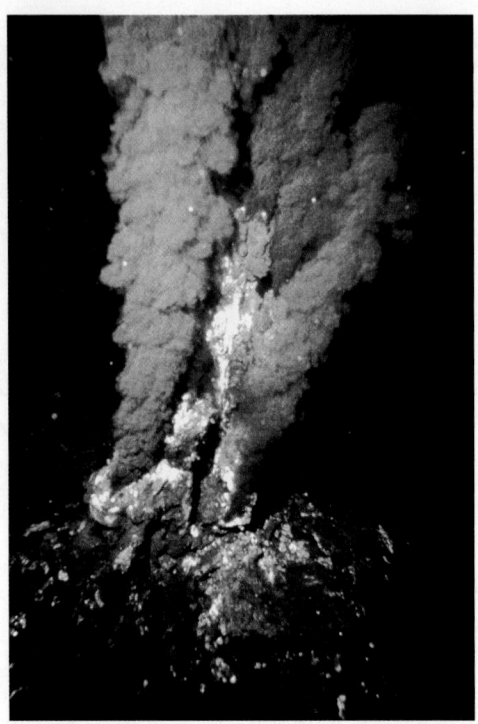

FIGURE 1-36 Black smokers. Hydrothermal vents in the sea floor emit superheated water rich in dissolved minerals. Black "smoke" is formed when the vented solution meets cold sea water and dissolved sulfides precipitate. Diverse life forms, including a variety of archaea and some remarkably complex multicellular organisms, are found in the immediate vicinity of such vents, which may have been the sites of early biogenesis. [Source: P. Rona/OAR/National Undersea Research Program (NURP), NOAA.]

bottom; **Fig. 1-36**) showed the formation of 23 amino acids and 7 organosulfur compounds, as well as a large number of other simple compounds that might have served as building blocks in prebiotic evolution.

More-refined laboratory experiments have provided good evidence that many of the chemical components of living cells can form under these conditions. Polymers of RNA can act as catalysts in biologically significant reactions (see Chapters 26 and 27), and RNA probably played a crucial role in prebiotic evolution, both as catalyst and as information repository. Ribonucleotides, the monomeric units of RNA, have not been formed in the laboratory under prebiotic conditions, so it is possible that prebiotic evolution began with an RNA-like molecule, rather than with RNA itself.

RNA or Related Precursors May Have Been the First Genes and Catalysts

In modern organisms, nucleic acids encode the genetic information that specifies the structure of enzymes, and enzymes catalyze the replication and repair of nucleic acids. The mutual dependence of these two classes of biomolecules brings up the perplexing question: which came first, DNA or protein?

The answer may be that they appeared about the same time, and RNA preceded them both. The discovery that RNA molecules can act as catalysts in their own formation suggests that RNA or a similar molecule may have been the first gene *and* the first catalyst. According to this scenario **(Fig. 1-37)**, one of the earliest stages of biological evolution was the chance formation of an RNA molecule that could catalyze the formation of other RNA molecules of the same sequence—a self-replicating, self-perpetuating RNA. The concentration of

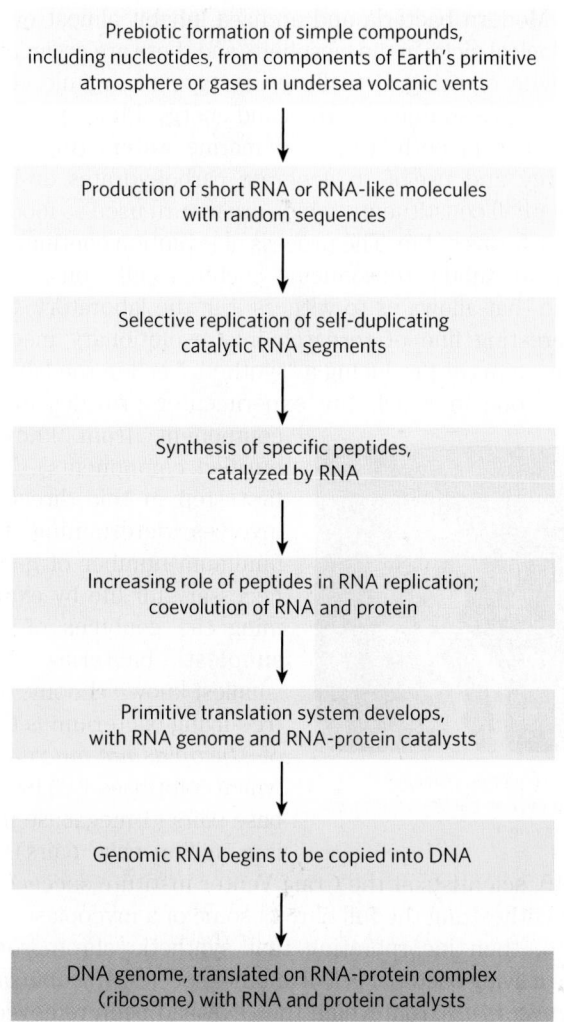

Prebiotic formation of simple compounds, including nucleotides, from components of Earth's primitive atmosphere or gases in undersea volcanic vents

↓

Production of short RNA or RNA-like molecules with random sequences

↓

Selective replication of self-duplicating catalytic RNA segments

↓

Synthesis of specific peptides, catalyzed by RNA

↓

Increasing role of peptides in RNA replication; coevolution of RNA and protein

↓

Primitive translation system develops, with RNA genome and RNA-protein catalysts

↓

Genomic RNA begins to be copied into DNA

↓

DNA genome, translated on RNA-protein complex (ribosome) with RNA and protein catalysts

FIGURE 1-37 A possible "RNA world" scenario.

a self-replicating RNA molecule would increase exponentially, as one molecule formed several, several formed many, and so on. The fidelity of self-replication was presumably less than perfect, so the process would generate variants of the RNA, some of which might be even better able to self-replicate. In the competition for nucleotides, the most efficient of the self-replicating sequences would win, and less efficient replicators would fade from the population.

The division of function between DNA (genetic information storage) and protein (catalysis) was, according to the "RNA world" hypothesis, a later development. New variants of self-replicating RNA molecules developed that had the additional ability to catalyze the condensation of amino acids into peptides. Occasionally, the peptide(s) thus formed would reinforce the self-replicating ability of the RNA, and the pair—RNA molecule and helping peptide—could undergo further modifications in sequence, generating increasingly efficient self-replicating systems. The remarkable discovery that in the protein-synthesizing machinery of modern cells (ribosomes), RNA molecules, not proteins, catalyze the formation of peptide bonds is consistent with the RNA world hypothesis.

Some time after the evolution of this primitive protein-synthesizing system, there was a further development: DNA molecules with sequences complementary to the self-replicating RNA molecules took over the function of conserving the "genetic" information, and RNA molecules evolved to play roles in protein synthesis. (We explain in Chapter 8 why DNA is a more stable molecule than RNA and thus a better repository of inheritable information.) Proteins proved to be versatile catalysts and, over time, took over most of that function. Lipidlike compounds in the primordial mixture formed relatively impermeable layers around self-replicating collections of molecules. The concentration of proteins and nucleic acids within these lipid enclosures favored the molecular interactions required in self-replication.

The RNA world scenario is intellectually satisfying, but it leaves unanswered a vexing question: where did the nucleotides needed to make the initial RNA molecules come from? An alternative to this RNA world scenario supposes that simple metabolic pathways evolved first, perhaps at the hot vents in the ocean floor. A set of linked chemical reactions there might have produced precursors, including nucleotides, before the advent of lipid membranes or RNA. Without more experimental evidence, neither of these hypotheses can be disproved.

Biological Evolution Began More Than Three and a Half Billion Years Ago

Earth was formed about 4.6 billion years ago, and the first evidence of life dates to more than 3.5 billion years ago. In 1996, scientists working in Greenland found chemical evidence of life ("fossil molecules") from as far back as 3.85 billion years ago, forms of carbon embedded in rock that seem to have a distinctly biological origin. Somewhere on Earth during its first billion years the first simple organism arose, capable of replicating its own structure from a template (RNA?) that was the first genetic material. Because the terrestrial atmosphere at the dawn of life was nearly devoid of oxygen, and because there were few microorganisms to scavenge organic compounds formed by natural processes, these compounds were relatively stable. Given this stability and eons of time, the improbable became inevitable: lipid vesicles containing organic compounds and self-replicating RNA gave rise to the first cells, or protocells, and those protocells with the greatest capacity for self-replication became more numerous. The process of biological evolution had begun.

The First Cell Probably Used Inorganic Fuels

The earliest cells arose in a reducing atmosphere (there was no oxygen) and probably obtained energy from inorganic fuels such as ferrous sulfide and ferrous carbonate, both abundant on the early Earth. For example, the reaction

$$FeS + H_2S \longrightarrow FeS_2 + H_2$$

yields enough energy to drive the synthesis of ATP or similar compounds. The organic compounds these early cells required may have arisen by the nonbiological actions of lightning or of heat from volcanoes or thermal vents in the sea on components of the early atmosphere: CO, CO_2, N_2, NH_3, CH_4, and suchlike. An alternative source of organic compounds has been proposed: extraterrestrial space. Space missions in 2006 (Stardust) and 2014 (Philae) found particles of comet dust to contain the simple amino acid glycine and 20 other organic compounds capable of reacting to form biomolecules.

Early unicellular organisms gradually acquired the ability to derive energy from compounds in their environment and to use that energy to synthesize more of their own precursor molecules, thereby becoming less dependent on outside sources. A very significant evolutionary event was the development of pigments capable of capturing the energy of light from the sun, which could be used to reduce, or "fix," CO_2 to form more complex, organic compounds. The original electron donor for these **photosynthetic** processes was probably H_2S, yielding elemental sulfur or sulfate (SO_4^{2-}) as the byproduct. Some hydrothermal vents in the sea bottom (black smokers; Fig. 1-36) emit significant amounts of H_2, which is another possible electron donor in the metabolism of the earliest organisms. Later cells developed the enzymatic capacity to use H_2O as the electron donor in photosynthetic reactions, producing O_2 as waste. Cyanobacteria are the modern descendants of these early photosynthetic oxygen-producers.

Because the atmosphere of Earth in the earliest stages of biological evolution was nearly devoid of oxygen, the earliest cells were anaerobic. Under these conditions, chemotrophs could oxidize organic compounds to CO_2 by passing electrons not to O_2 but to acceptors such as SO_4^{2-}, in this case yielding H_2S as the product. With the rise of O_2-producing photosynthetic bacteria, the atmosphere became progressively richer in oxygen—a powerful oxidant and deadly poison to anaerobes. Responding to the evolutionary pressure of what Lynn Margulis and Dorion Sagan called the "oxygen holocaust," some lineages of microorganisms gave rise to aerobes that obtained energy by passing electrons from fuel molecules to oxygen. Because the transfer of electrons from organic molecules to O_2 releases a great deal of energy, aerobic organisms had an energetic advantage over their anaerobic counterparts when both competed in an environment containing oxygen. This advantage translated into the predominance of aerobic organisms in O_2-rich environments.

Modern bacteria and archaea inhabit almost every ecological niche in the biosphere, and there are organisms capable of using virtually every type of organic compound as a source of carbon and energy. Photosynthetic microbes in both fresh and marine waters trap solar energy and use it to generate carbohydrates and all other cell constituents, which are in turn used as food by other forms of life. The process of evolution continues—and, in rapidly reproducing bacterial cells, on a time scale that allows us to witness it in the laboratory. One interesting line of research into evolutionary mechanisms aims at producing a "synthetic" cell in the laboratory (one in which the experimenter provides every

Lynn Margulis, 1938–2011
[Source: Ben Barnhart/UMass Magazine.]

component from known, purified components). The first step in this direction involves determining the minimum number of genes necessary for life by examining the genomes of the simplest bacteria. The smallest known genome of a free-living bacterium is that of *Mycoplasma mycoides*, which comprises 1.08 megabase pairs (1 megabase pair is a million base pairs). In 2010, scientists at the Craig Venter Institute succeeded in synthesizing the full chromosome of a mycoplasma in vitro, then incorporating that synthetic chromosome into a living bacterial cell of another species, *Mycoplasma capricolum* (from which the DNA had been removed), which thereby acquired the properties of *M. mycoides* (**Fig. 1-38**). This technology opens the way to producing a synthetic cell, with the bare minimum of genes essential to life. With such a cell, one could hope to study, in the laboratory, the evolutionary processes by which protocells gradually diversified and became more complex.

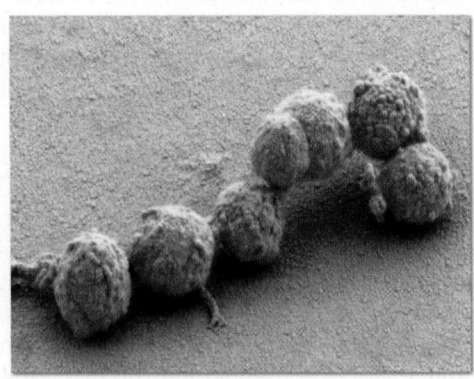

FIGURE 1-38 Synthetic cells. These cells were produced by injecting *Mycoplasma mycoides* DNA synthesized in the laboratory into the enucleated shell of a related organism, *Mycoplasma capricolum*. The synthetic cells reproduce and have properties specific to *M. mycoides*. [Source: ©2012 National Center for Microscopy & Imaging Research.]

Eukaryotic Cells Evolved from Simpler Precursors in Several Stages

Starting about 1.5 billion years ago, the fossil record begins to show evidence of larger and more complex organisms, probably the earliest eukaryotic cells **(Fig. 1-39)**. Details of the evolutionary path from non-nucleated to nucleated cells cannot be deduced from the fossil record alone, but morphological and biochemical comparisons of modern organisms have suggested a sequence of events consistent with the fossil evidence.

Three major changes must have occurred. First, as cells acquired more DNA, the mechanisms required to fold it compactly into discrete complexes with specific proteins and to divide it equally between daughter cells at cell division became more elaborate. Specialized proteins were required to stabilize folded DNA and to pull the resulting DNA-protein complexes (chromosomes) apart during cell division. Second, as cells became larger, a system of intracellular membranes developed, including a double membrane surrounding the DNA. This membrane segregated the nuclear process of RNA synthesis on a DNA template from the cytoplasmic process of protein synthesis on ribosomes. Finally, according to a now widely accepted hypothesis advanced (initially, to

much resistance) by Lynn Margulis, early eukaryotic cells, which were incapable of photosynthesis or aerobic metabolism, enveloped aerobic bacteria or photosynthetic bacteria to form **endosymbiotic** associations that eventually became permanent **(Fig. 1-40)**. Some aerobic bacteria evolved into the mitochondria of modern eukaryotes, and some photosynthetic cyanobacteria became the plastids, such as the chloroplasts of green algae, the likely ancestors of modern plant cells.

At some later stage of evolution, unicellular organisms found it advantageous to cluster together, thereby acquiring greater motility, efficiency, or reproductive success than their free-living single-celled competitors. Further evolution of such clustered organisms led to permanent associations among individual cells and eventually to specialization within the colony—to cellular differentiation.

The advantages of cellular specialization led to the evolution of increasingly complex and highly differentiated organisms, in which some cells carried out the sensory functions, others the digestive, photosynthetic, or reproductive functions, and so forth. Many modern multicellular organisms contain hundreds of different cell types, each specialized for a function that supports the entire organism. Fundamental mechanisms that evolved early have been further refined and embellished through evolution. The same basic structures and mechanisms that underlie the beating motion of cilia in *Paramecium* and of flagella in *Chlamydomonas* are employed by the highly differentiated vertebrate sperm cell, for example.

Molecular Anatomy Reveals Evolutionary Relationships

Biochemists now have an enormously rich, ever increasing treasury of information on the molecular anatomy of cells that they can use to analyze evolutionary relationships and refine evolutionary theory. The sequence of the **genome**, the complete genetic endowment of an organism, has been determined for several thousand bacteria and archaea and for growing numbers of eukaryotic microorganisms, including *Saccharomyces cerevisiae* and *Plasmodium* species; plants, including *Arabidopsis thaliana* and rice; and animals, including *Caenorhabditis elegans* (a roundworm), *Drosophila melanogaster* (the fruit fly), mouse, rat, dog, chimpanzee, and *Homo sapiens* (Table 1-2). It is even possible to recover DNA samples from the tissues of extinct animals such as Neanderthal man and woolly mammoth and sequence it (see Chapter 8). With such sequences in hand, detailed and quantitative comparisons among species can provide deep insight into the evolutionary process. Thus far, the molecular phylogeny derived from gene sequences is consistent with, but in many cases more precise than, the classical phylogeny based on macroscopic structures. Although organisms have continuously diverged at the level of gross anatomy, at the molecular level the basic unity of life is readily apparent; molecular structures and

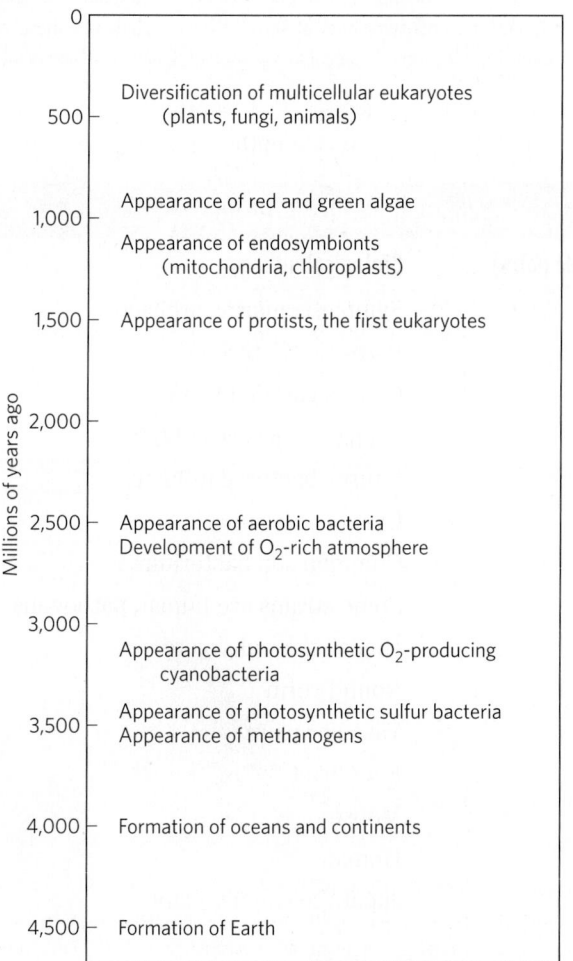

FIGURE 1-39 Landmarks in the evolution of life on Earth.

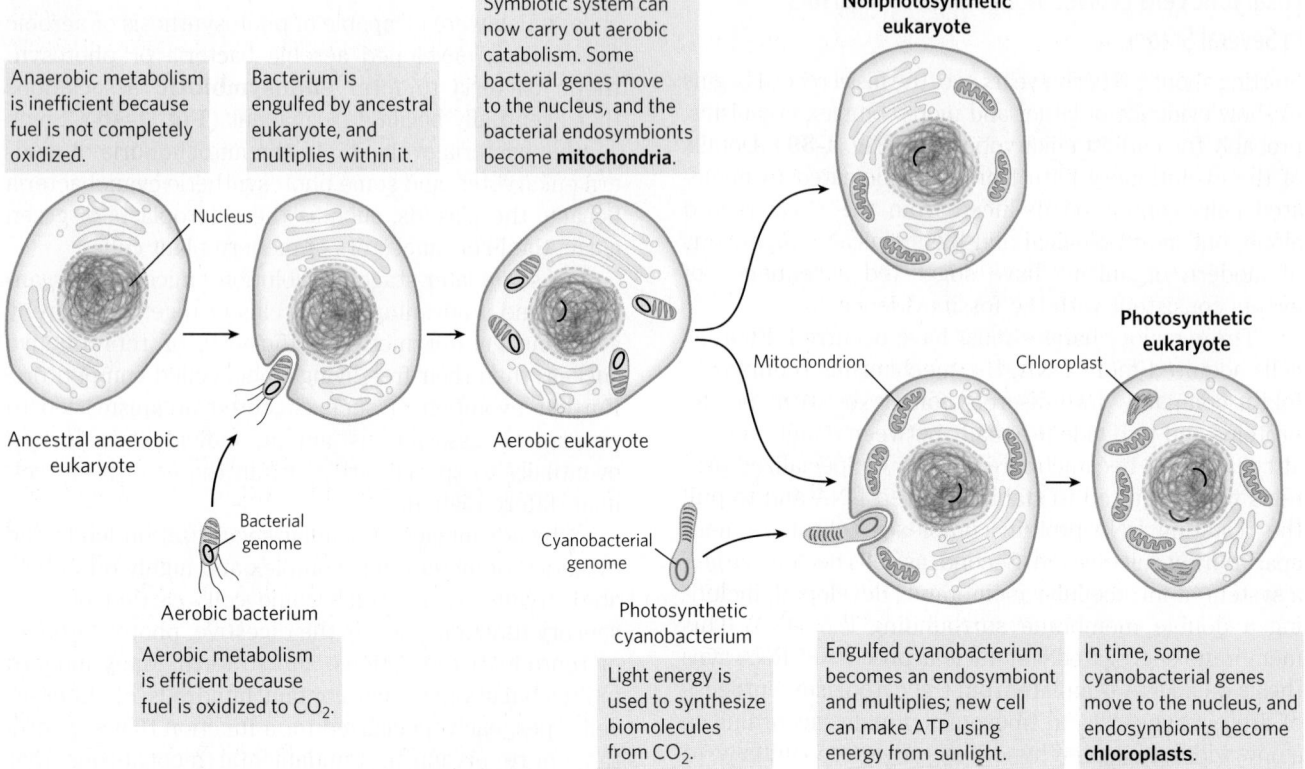

FIGURE 1-40 Evolution of eukaryotes through endosymbiosis. The earliest eukaryote, an anaerobe, acquired endosymbiotic purple bacteria, which carried with them their capacity for aerobic catabolism and became, over time, mitochondria. When photosynthetic cyanobacteria subsequently became endosymbionts of some aerobic eukaryotes, these cells became the photosynthetic precursors of modern green algae and plants.

Labels in figure:
- Anaerobic metabolism is inefficient because fuel is not completely oxidized.
- Bacterium is engulfed by ancestral eukaryote, and multiplies within it.
- Symbiotic system can now carry out aerobic catabolism. Some bacterial genes move to the nucleus, and the bacterial endosymbionts become **mitochondria**.
- Nonphotosynthetic eukaryote
- Nucleus
- Ancestral anaerobic eukaryote
- Aerobic eukaryote
- Mitochondrion
- Chloroplast
- Photosynthetic eukaryote
- Bacterial genome
- Aerobic bacterium
- Aerobic metabolism is efficient because fuel is oxidized to CO_2.
- Cyanobacterial genome
- Photosynthetic cyanobacterium
- Light energy is used to synthesize biomolecules from CO_2.
- Engulfed cyanobacterium becomes an endosymbiont and multiplies; new cell can make ATP using energy from sunlight.
- In time, some cyanobacterial genes move to the nucleus, and endosymbionts become **chloroplasts**.

TABLE 1-2	A Few of the Many Organisms Whose Genomes Have Been Completely Sequenced	
Organism	**Genome size (nucleotide pairs)**	**Biological interest**
Nanoarchaeum equitans	4.9×10^5	Symbiotic marine archaeon
Mycoplasma genitalium	5.8×10^5	Parasitic bacterium
Helicobacter pylori	1.6×10^6	Causes gastric ulcers
Methanocaldococcus jannaschii	1.7×10^6	Archaeon; grows at 85 °C
Haemophilus influenzae	1.9×10^6	Causes bacterial influenza
Synechocystis sp.	3.9×10^6	Cyanobacterium
Bacillus subtilis	4.2×10^6	Common soil bacterium
Escherichia coli	4.6×10^6	Some strains are human pathogens
Saccharomyces cerevisiae	1.2×10^7	Unicellular eukaryote
Caenorhabditis elegans	1.0×10^8	Roundworm
Arabidopsis thaliana	1.2×10^8	Vascular plant
Drosophila melanogaster	1.8×10^8	Fly ("fruit fly")
Mus musculus	2.7×10^9	Mouse
Homo sapiens	3.0×10^9	Human
Paris japonica	1.5×10^{11}	Japanese canopy plant

Sources: www.ncbi.nlm.nih.gov/genome; J. Pellicer et al., *Bot. J. Linn. Soc.* 164:10, 2010.

mechanisms are remarkably similar from the simplest to the most complex organisms. These similarities are most easily seen at the level of sequences, either the DNA sequences that encode proteins or the protein sequences themselves.

When two genes share readily detectable sequence similarities (nucleotide sequence in DNA or amino acid sequence in the proteins they encode), their sequences are said to be homologous and the proteins they encode are **homologs**. If two homologous genes occur in the *same* species, they are said to be paralogous and their protein products are **paralogs**. Paralogous genes are presumed to have been derived by gene duplication followed by gradual changes in the sequences of both copies. Typically, paralogous proteins are similar not only in sequence but also in three-dimensional structure, although they commonly have acquired different functions during their evolution.

Two homologous genes (or proteins) found in *different* species are said to be orthologous, and their protein products are **orthologs**. Orthologs usually have the same function in both organisms, and when a newly sequenced gene in one species is found to be strongly orthologous with a gene in another, this gene is presumed to encode a protein with the same function in both species. By this means, the function of gene products (proteins or RNA molecules) can be deduced from the genomic sequence without any biochemical characterization of the molecules themselves. An **annotated genome** includes, in addition to the DNA sequence itself, a description of the likely function of each gene product, deduced from comparisons with other genomic sequences and established protein functions. Sometimes, by identifying the pathways (sets of enzymes) encoded in a genome, we can deduce from the genomic sequence alone the organism's metabolic capabilities.

The sequence differences between homologous genes may be taken as a rough measure of the degree to which the two species have diverged during evolution—of how long ago their common evolutionary precursor gave rise to two lines with different evolutionary fates. The larger the number of sequence differences, the earlier the divergence in evolutionary history. One can construct a phylogeny (family tree) in which the evolutionary distance between any two species is represented by their proximity on the tree (Fig. 1-5 is an example).

In the course of evolution, new structures, processes, or regulatory mechanisms are acquired, reflections of the changing genomes of the evolving organisms. The genome of a simple eukaryote such as yeast should have genes related to formation of the nuclear membrane, genes not present in bacteria or archaea. The genome of an insect should contain genes that encode proteins involved in specifying a characteristic segmented body plan, genes not present in yeast. The genomes of all vertebrate animals should share genes that specify the development of a spinal column, and those of mammals should have unique genes necessary for the development of the placenta, a characteristic of mammals—and so on. Comparisons of the whole genomes of species in each phylum are leading to the identification of genes critical to fundamental evolutionary changes in body plan and development.

Functional Genomics Shows the Allocations of Genes to Specific Cellular Processes

When the sequence of a genome is fully determined and each gene is assigned a function, molecular geneticists can group genes according to the processes (DNA synthesis, protein synthesis, generation of ATP, and so forth) in which they function and thus find what fraction of the genome is allocated to each of a cell's activities. The largest category of genes in *E. coli, A. thaliana,* and *H. sapiens* consists of those of (as yet) unknown function, which make up more than 40% of the genes in each species. The genes encoding the transporters that move ions and small molecules across plasma membranes make up a significant proportion of the genes in all three species, more in the bacterium and plant than in the mammal (10% of the ~4,400 genes of *E. coli,* ~8% of the ~27,000 genes of *A. thaliana,* and ~4% of the ~20,000 genes of *H. sapiens*). Genes that encode the proteins and RNA required for protein synthesis make up 3% to 4% of the *E. coli* genome, but in the more complex cells of *A. thaliana,* more genes are needed for targeting proteins to their final location in the cell than are needed to synthesize those proteins (about 6% and 2% of the genome, respectively). In general, the more complex the organism, the greater the proportion of its genome that encodes genes involved in the *regulation* of cellular processes and the smaller the proportion dedicated to basic processes, or "housekeeping" functions, such as ATP generation and protein synthesis. The **housekeeping genes** typically are expressed under all conditions and are not subject to much regulation.

Genomic Comparisons Have Increasing Importance in Human Biology and Medicine

The genomes of chimpanzees and humans are 99.9% identical, yet the differences between the two species are vast. The relatively few differences in genetic endowment must explain the possession of language by humans, the extraordinary athleticism of chimpanzees, and myriad other differences. Genomic comparison is allowing researchers to identify candidate genes linked to divergences in the developmental programs of humans and the other primates and to the emergence of complex functions such as language. The picture will become clearer only as more primate genomes become available for comparison with the human genome.

Similarly, the differences in genetic endowment among humans are vanishingly small compared with the differences between humans and chimpanzees, yet these differences account for human variety—including differences in health and in susceptibility to chronic diseases. We have much to learn about the variability in genomic sequence among humans, and the availability of genomic information will almost certainly transform medical diagnosis and treatment. Several monumental studies in which the entire genomic sequence has been determined for hundreds or thousands of people with cancer, type 2 diabetes, schizophrenia, and other diseases or conditions have allowed the identification of many genes in which mutations correlate with the medical condition. Each of those genes codes for a protein that, in principle, might become the target for drugs to treat that condition. We may expect that for some genetic diseases, palliatives will be replaced by cures, and that for disease susceptibilities associated with particular genetic markers, forewarning and perhaps increased preventive measures will prevail. Today's "medical history" may be replaced by a "medical forecast." ■

SUMMARY 1.5 Evolutionary Foundations

■ Occasional inheritable mutations yield organisms that are better suited for survival and reproduction in an ecological niche, and their progeny come to dominate the population in that niche. This process of mutation and selection is the basis for the Darwinian evolution that led from the first cell to all modern organisms. The large number of genes shared by all living organisms explains organisms' fundamental similarities.

■ Life originated about 3.5 billion years ago, most likely with the formation of a membrane-enclosed compartment containing a self-replicating RNA molecule. The components for the first cell may have been produced near thermal vents at the bottom of the sea or by the action of lightning and high temperature on simple atmospheric molecules such as CO_2 and NH_3.

■ The catalytic and genetic roles played by the early RNA genome were, over time, taken over by proteins and DNA, respectively.

■ Eukaryotic cells acquired the capacity for photosynthesis and oxidative phosphorylation from endosymbiotic bacteria. In multicellular organisms, differentiated cell types specialize in one or more of the functions essential to the organism's survival.

■ Knowledge of the complete genomic nucleotide sequences of organisms from different branches of the phylogenetic tree provides insights into evolution and offers great opportunities in human medicine.

Key Terms

All terms are defined in the glossary.

metabolite 3	endergonic reaction 23
nucleus 3	exergonic reaction 23
genome 3	equilibrium 25
eukaryotes 3	standard free-energy
bacteria 5	change, $\Delta G°$ 26
archaea 5	activation energy,
cytoskeleton 7	$\Delta G^{\ddagger}$ 28
stereoisomers 16	catabolism 28
configuration 16	anabolism 28
chiral center 17	metabolism 28
conformation 19	systems biology 29
entropy, S 23	mutation 32
enthalpy, H 23	housekeeping genes 39
free-energy change,	
ΔG 23	

Problems

Some problems related to the contents of the chapter follow. (In solving end-of-chapter problems, you may wish to refer to the tables on the inside of the back cover.) Each problem has a title for easy reference and discussion. For all numerical problems, keep in mind that answers should be expressed with the correct number of significant figures. Brief solutions are provided in Appendix B; expanded solutions are published in the *Absolute Ultimate Study Guide to Accompany Principles of Biochemistry.*

1. The Size of Cells and Their Components

(a) If you were to magnify a cell 10,000-fold (typical of the magnification achieved using an electron microscope), how big would it appear? Assume you are viewing a "typical" eukaryotic cell with a cellular diameter of 50 μm.

(b) If this cell were a muscle cell (myocyte), how many molecules of actin could it hold? Assume the cell is spherical and no other cellular components are present; actin molecules are spherical, with a diameter of 3.6 nm. (The volume of a sphere is $\frac{1}{3}\pi r^3$.)

(c) If this were a liver cell (hepatocyte) of the same dimensions, how many mitochondria could it hold? Assume the cell is spherical; no other cellular components are present; and the mitochondria are spherical, with a diameter of 1.5 μm.

(d) Glucose is the major energy-yielding nutrient for most cells. Assuming a cellular concentration of 1 mM (that is, 1 millimole/L), calculate how many molecules of glucose would be present in our hypothetical (and spherical) eukaryotic cell. (Avogadro's number, the number of molecules in 1 mol of a nonionized substance, is 6.02×10^{23}.)

(e) Hexokinase is an important enzyme in the metabolism of glucose. If the concentration of hexokinase in our eukaryotic cell is 20 μM, how many glucose molecules are present per hexokinase molecule?

2. Components of *E. coli* *E. coli* cells are rod-shaped, about 2 μm long and 0.8 μm in diameter. The volume of a cylinder is $\pi r^2 h$, where h is the height of the cylinder.

(a) If the average density of *E. coli* (mostly water) is 1.1×10^3 g/L, what is the mass of a single cell?

(b) *E. coli* has a protective cell envelope 10 nm thick. What percentage of the total volume of the bacterium does the cell envelope occupy?

(c) *E. coli* is capable of growing and multiplying rapidly because it contains some 15,000 spherical ribosomes (diameter 18 nm), which carry out protein synthesis. What percentage of the cell volume do the ribosomes occupy?

3. Genetic Information in *E. coli* DNA The genetic information contained in DNA consists of a linear sequence of coding units, known as codons. Each codon is a specific sequence of three deoxyribonucleotides (three deoxyribonucleotide pairs in double-stranded DNA), and each codon codes for a single amino acid unit in a protein. The molecular weight of an *E. coli* DNA molecule is about 3.1×10^9 g/mol. The average molecular weight of a nucleotide pair is 660 g/mol, and each nucleotide pair contributes 0.34 nm to the length of DNA.

(a) Calculate the length of an *E. coli* DNA molecule. Compare the length of the DNA molecule with the cell dimensions (see Problem 2). How does the DNA molecule fit into the cell?

(b) Assume that the average protein in *E. coli* consists of a chain of 400 amino acids. What is the maximum number of proteins that can be coded by an *E. coli* DNA molecule?

4. The High Rate of Bacterial Metabolism Bacterial cells have a much higher rate of metabolism than animal cells. Under ideal conditions, some bacteria double in size and divide every 20 min, whereas most animal cells under rapid growth conditions require 24 hours. The high rate of bacterial metabolism requires a high ratio of surface area to cell volume.

(a) Why does surface-to-volume ratio affect the maximum rate of metabolism?

(b) Calculate the surface-to-volume ratio for the spherical bacterium *Neisseria gonorrhoeae* (diameter 0.5 μm), responsible for the disease gonorrhea. Compare it with the surface-to-volume ratio for a globular amoeba, a large eukaryotic cell (diameter 150 μm). The surface area of a sphere is $4\pi r^2$.

5. Fast Axonal Transport Neurons have long thin processes called axons, structures specialized for conducting signals throughout the organism's nervous system. Some axonal processes can be as long as 2 m—for example, the axons that originate in your spinal cord and terminate in the muscles of your toes. Small membrane-enclosed vesicles carrying materials essential to axonal function move along microtubules of the cytoskeleton, from the cell body to the tips of the axons. If the average velocity of a vesicle is 1 μm/s, how long does it take a vesicle to move from a cell body in the spinal cord to the axonal tip in the toes?

6. Is Synthetic Vitamin C as Good as the Natural Vitamin? A claim put forth by some purveyors of health foods is that vitamins obtained from natural sources are more healthful than those obtained by chemical synthesis. For example, pure

L-ascorbic acid (vitamin C) extracted from rose hips is better than pure L-ascorbic acid manufactured in a chemical plant. Are the vitamins from the two sources different? Can the body distinguish a vitamin's source?

7. Identification of Functional Groups Figures 1-17 and 1-18 show some common functional groups of biomolecules. Because the properties and biological activities of biomolecules are largely determined by their functional groups, it is important to be able to identify them. In each of the compounds below, circle and identify by name each functional group.

Ethanolamine
(a)

Glycerol
(b)

Phosphoenolpyruvate, an intermediate in glucose metabolism
(c)

Threonine, an amino acid
(d)

Pantothenate, a vitamin
(e)

D-Glucosamine
(f)

8. Drug Activity and Stereochemistry The quantitative differences in biological activity between the two enantiomers of a compound are sometimes quite large. For example, the D isomer of the drug isoproterenol, used to treat mild asthma, is 50 to 80 times more effective as a bronchodilator than the L isomer. Identify the chiral center in isoproterenol. Why do the two enantiomers have such radically different bioactivity?

Isoproterenol

9. Separating Biomolecules In studying a particular biomolecule (a protein, nucleic acid, carbohydrate, or lipid) in the laboratory, the biochemist first needs to separate it from other biomolecules in the sample—that is, to *purify* it. Specific purification techniques are described later in the book. However, by looking at the monomeric subunits of a biomolecule, you should have some ideas about the characteristics of the molecule that would allow you to separate it from other molecules. For example, how would you separate (a) amino acids from fatty acids and (b) nucleotides from glucose?

10. Silicon-Based Life? Silicon is in the same group of the periodic table as carbon and, like carbon, can form up to four single bonds. Many science fiction stories have been based on the premise of silicon-based life. Is this realistic? What characteristics of silicon make it *less* well adapted than carbon as the central organizing element for life? To answer this question, consider what you have learned about carbon's bonding versatility, and refer to a beginning inorganic chemistry textbook for silicon's bonding properties.

11. Drug Action and Shape of Molecules Some years ago, two drug companies marketed a drug under the trade names Dexedrine and Benzedrine. The structure of the drug is shown below.

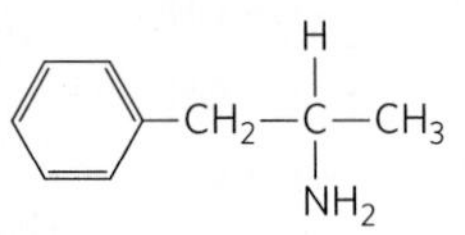

The physical properties (C, H, and N analysis, melting point, solubility, etc.) of Dexedrine and Benzedrine were identical. The recommended oral dosage of Dexedrine (which is still available) was 5 mg/day, but the recommended dosage of Benzedrine (no longer available) was twice that. Apparently it required considerably more Benzedrine than Dexedrine to yield the same physiological response. Explain this apparent contradiction.

12. Components of Complex Biomolecules Figure 1-11 shows the major components of complex biomolecules. For each of the three important biomolecules below (shown in their ionized forms at physiological pH), identify the constituents.

(a) Guanosine triphosphate (GTP), an energy-rich nucleotide that serves as a precursor to RNA:

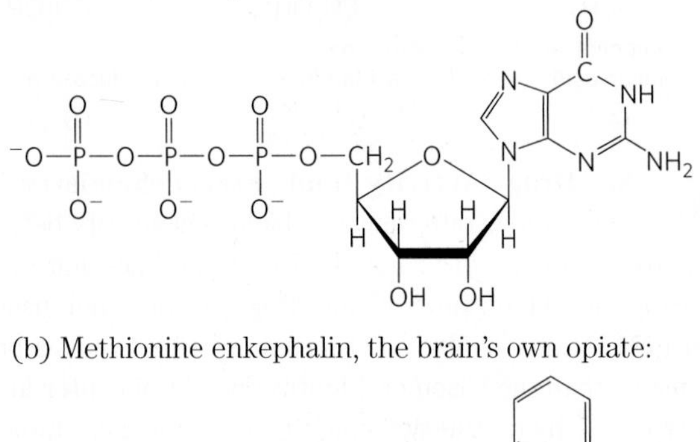

(b) Methionine enkephalin, the brain's own opiate:

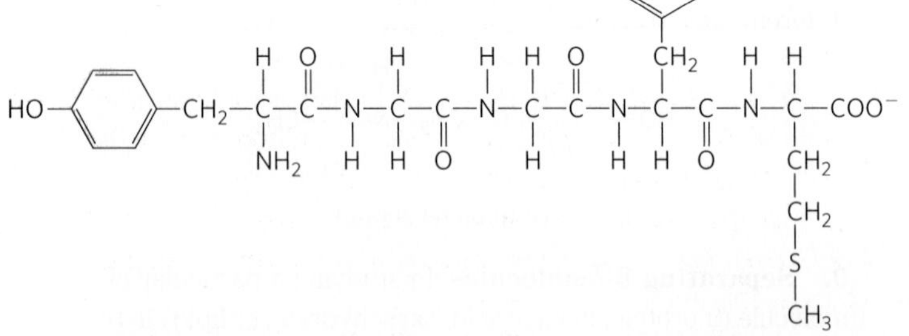

(c) Phosphatidylcholine, a component of many membranes:

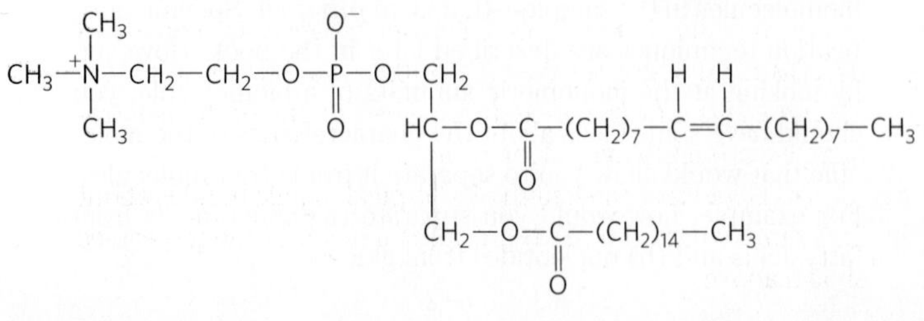

13. Determination of the Structure of a Biomolecule An unknown substance, X, was isolated from rabbit muscle. Its structure was determined from the following observations and experiments. Qualitative analysis showed that X was composed entirely of C, H, and O. A weighed sample of X was completely oxidized, and the H_2O and CO_2 produced were measured; this quantitative analysis revealed that X contained 40.00% C, 6.71% H, and 53.29% O by weight. The molecular mass of X, determined by mass spectrometry, was 90.00 u (atomic mass units; see Box 1-1). Infrared spectroscopy showed that X contained one double bond. X dissolved readily in water to give an acidic solution; the solution demonstrated optical activity when tested in a polarimeter.

(a) Determine the empirical and molecular formula of X.

(b) Draw the possible structures of X that fit the molecular formula and contain one double bond. Consider *only* linear or branched structures and disregard cyclic structures. Note that oxygen makes very poor bonds to itself.

(c) What is the structural significance of the observed optical activity? Which structures in (b) are consistent with the observation?

(d) What is the structural significance of the observation that a solution of X was acidic? Which structures in (b) are consistent with the observation?

(e) What is the structure of X? Is more than one structure consistent with all the data?

14. Naming Stereoisomers with One Chiral Carbon Using the RS System Propranolol is a chiral compound. (*R*)-Propranolol is used as a contraceptive; (*S*)-propranolol is used to treat hypertension. Identify the chiral carbon in the structure below. Is this the (*R*) or the (*S*) isomer? Draw the other isomer.

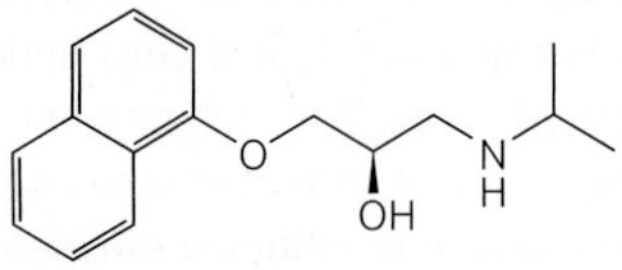

15. Naming Stereoisomers with Two Chiral Carbons Using the RS System The (*R,R*) isomer of methylphenidate (Ritalin) is used to treat attention deficit hyperactivity disorder (ADHD). The (*S,S*) isomer is an antidepressant. Identify the two chiral carbons in the structure below. Is this the (*R,R*) or the (*S,S*) isomer? Draw the other isomer.

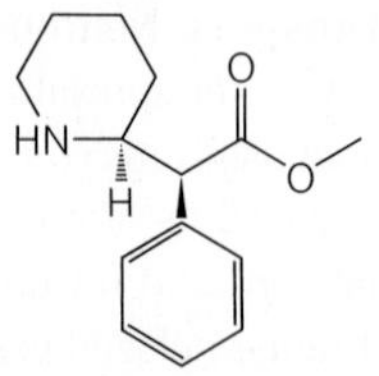

Data Analysis Problem

16. Interaction of Sweet-Tasting Molecules with Taste Receptors Many compounds taste sweet to humans. Sweet taste results when a molecule binds to the sweet receptor, one type of taste receptor, on the surface of certain tongue cells. The stronger the binding, the lower the concentration required to saturate the receptor and the sweeter a given concentration of that substance tastes. The standard free-energy change, $\Delta G°$, of the binding reaction between a sweet molecule and a sweet receptor can be measured in kilojoules or kilocalories per mole.

Sweet taste can be quantified in units of "molar relative sweetness" (MRS), a measure that compares the sweetness of a substance to the sweetness of sucrose. For example, saccharin has an MRS of 161; this means that saccharin is 161 times sweeter than sucrose. In practical terms, this is measured by asking human subjects to compare the sweetness of solutions containing different concentrations of each compound. Sucrose and saccharin taste equally sweet when sucrose is at a concentration 161 times higher than that of saccharin.

(a) What is the relationship between MRS and the $\Delta G°$ of the binding reaction? Specifically, would a more negative $\Delta G°$ correspond to a higher or lower MRS? Explain your reasoning.

Shown below are the structures of 10 compounds, all of which taste sweet to humans. The MRS and $\Delta G°$ for binding to the sweet receptor are given for each substance.

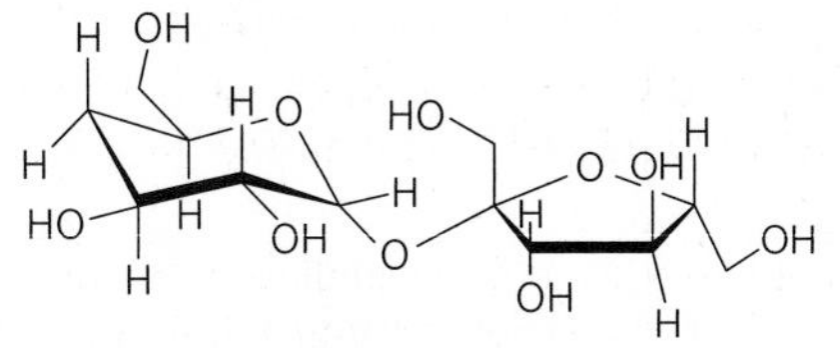

Deoxysucrose
MRS = 0.95
$\Delta G° = -6.67$ kcal/mol

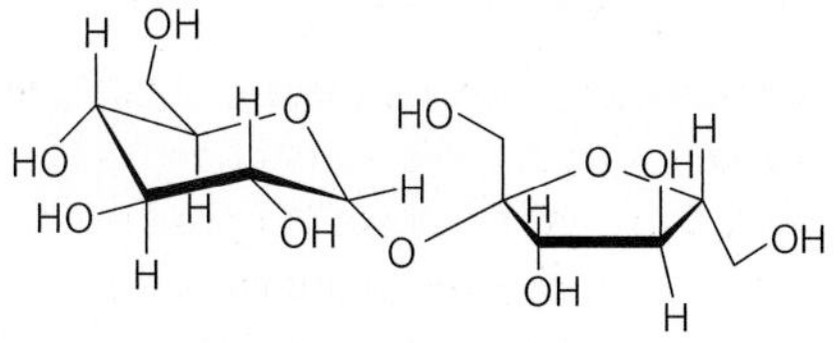

Sucrose
MRS = 1
$\Delta G° = -6.71$ kcal/mol

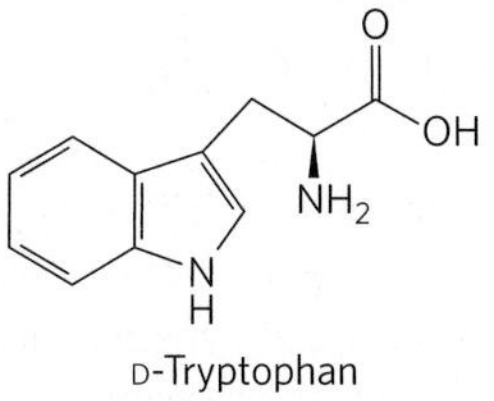

D-Tryptophan
MRS = 21
$\Delta G° = -8.5$ kcal/mol

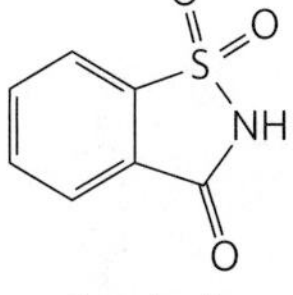

Saccharin
MRS = 161
$\Delta G° = -9.7$ kcal/mol

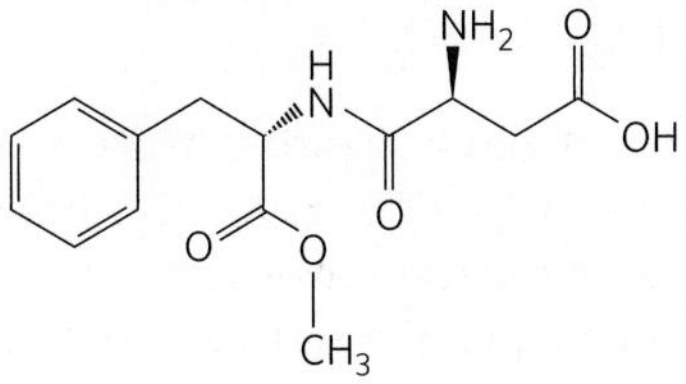

Aspartame
MRS = 172
$\Delta G° = -9.7$ kcal/mol

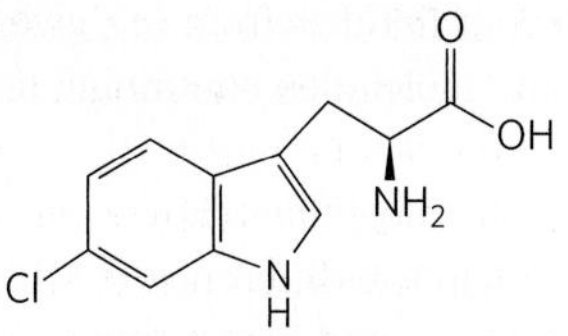

6-Chloro-D-tryptophan
MRS = 906
$\Delta G° = -10.7$ kcal/mol

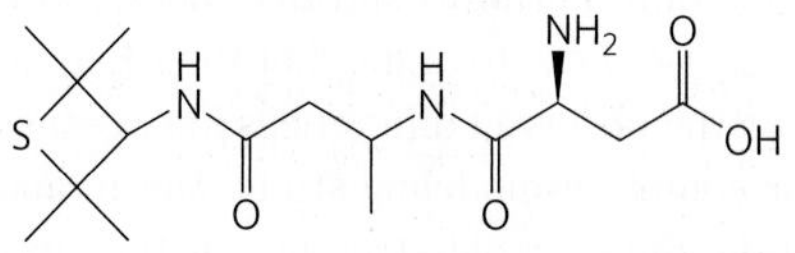

Alitame
MRS = 1,937
$\Delta G° = -11.1$ kcal/mol

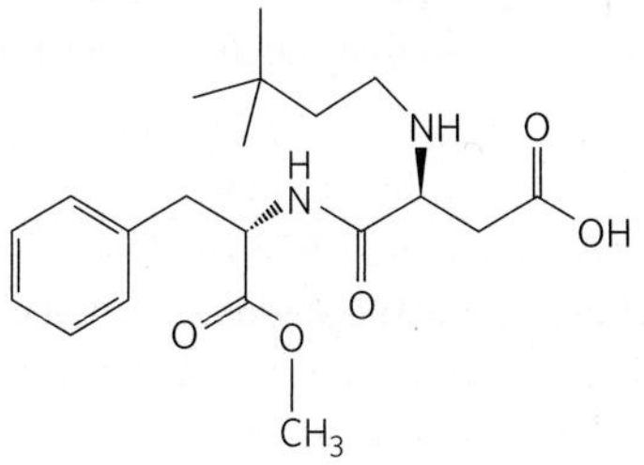

Neotame
MRS = 11,057
$\Delta G° = -12.1$ kcal/mol

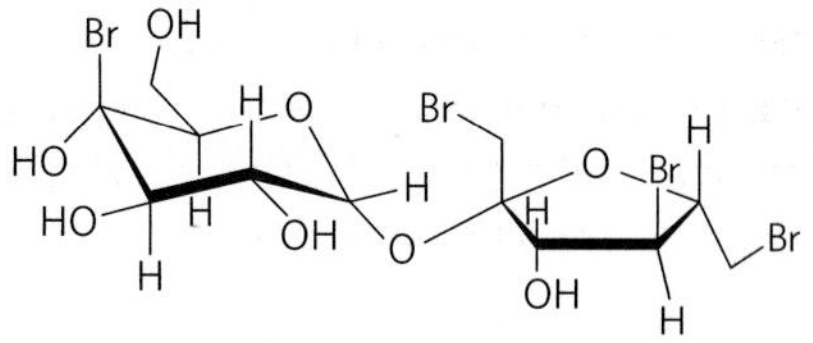

Tetrabromosucrose
MRS = 13,012
$\Delta G° = -12.2$ kcal/mol

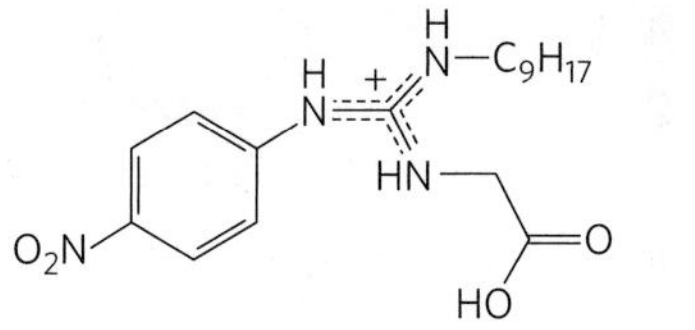

Sucronic acid
MRS = 200,000
$\Delta G° = -13.8$ kcal/mol

Morini, Bassoli, and Temussi (2005) used computer-based methods (often referred to as "in silico" methods) to model the binding of sweet molecules to the sweet receptor.

(b) Why is it useful to have a computer model to predict the sweetness of molecules, instead of a human- or animal-based taste assay?

In earlier work, Schallenberger and Acree (1967) had suggested that all sweet molecules include an "AH-B" structural group, in which "A and B are electronegative atoms separated by a distance of greater than 2.5 Å [0.25 nm] but less than 4 Å [0.4 nm]. H is a hydrogen atom attached to one of the electronegative atoms by a covalent bond."

(c) Given that the length of a "typical" single bond is about 0.15 nm, identify the AH-B group(s) in each of the molecules shown above.

(d) Based on your findings from (c), give two objections to the statement that "molecules containing an AH-B structure will taste sweet."

(e) For two of the molecules shown here, the AH-B model *can* be used to explain the difference in MRS and $\Delta G°$. Which two molecules are these, and how would you use them to support the AH-B model?

(f) Several of the molecules have closely related structures but very different MRS and $\Delta G°$ values. Give two such examples, and use these to argue that the AH-B model is unable to explain the observed differences in sweetness.

In their computer-modeling study, Morini and coauthors used the three-dimensional structure of the sweet receptor and a molecular dynamics modeling program called GRAMM to predict the $\Delta G°$ of binding of sweet molecules to the sweet receptor. First, they "trained" their model—that is, they refined the parameters so that the $\Delta G°$ values predicted by the model matched the known $\Delta G°$ values for one set of sweet molecules (the "training set"). They then "tested" the model by asking it to predict the $\Delta G°$ values for a new set of molecules (the "test set").

(g) Why did Morini and colleagues need to test their model against a different set of molecules from the set it was trained on?

(h) The researchers found that the predicted $\Delta G°$ values for the test set differed from the actual values by, on average, 1.3 kcal/mol. Using the values given with the molecular structures, estimate the resulting error in MRS values.

References

Morini, G., A. Bassoli, and P.A. Temussi. 2005. From small sweeteners to sweet proteins: anatomy of the binding sites of the human T1R2_T1R3 receptor. *J. Med. Chem.* 48:5520–5529.

Schallenberger, R.S., and T.E. Acree. 1967. Molecular theory of sweet taste. *Nature* 216:480–482.

Further Reading is available at www.macmillanlearning.com/LehningerBiochemistry7e.

STRUCTURE AND CATALYSIS

Biochemistry is nothing less than the chemistry of life, and, yes, life can be investigated, analyzed, and understood. To begin, every student of biochemistry needs both a language and some fundamentals; these are provided in Part I.

The chapters of Part I are devoted to the structure and function of the major classes of cellular constituents: water (Chapter 2), amino acids and proteins (Chapters 3 through 6), sugars and polysaccharides (Chapter 7), nucleotides and nucleic acids (Chapter 8), fatty acids and lipids (Chapter 10), and, finally, membranes and membrane signaling proteins (Chapters 11 and 12). We also discuss, in the context of structure and function, the technologies used to study each class of biomolecules. One whole chapter (Chapter 9) is devoted entirely to biotechnologies associated with cloning and genomics.

We begin, in Chapter 2, with water, because its properties affect the structure and function of all other cellular constituents. For each class of organic molecules, we first consider the covalent chemistry of the monomeric units (amino acids, monosaccharides, nucleotides, and fatty acids) and then describe the structure of the macromolecules and supramolecular complexes derived from them. An overarching theme is that the polymeric macromolecules in living systems, though large, are highly ordered chemical entities, with specific sequences of monomeric subunits giving rise to discrete structures and functions. This fundamental theme can be broken down into three interrelated principles: (1) the unique structure of each macromolecule determines its function; (2) noncovalent interactions play a critical role in the structure and thus the function of macromolecules; and (3) the monomeric subunits in polymeric macromolecules occur in specific sequences, representing a form of information on which the ordered living state depends.

The relationship between structure and function is especially evident in proteins, which exhibit an extraordinary diversity of functions. One particular polymeric sequence of amino acids produces a strong, fibrous structure found in hair and wool; another produces a protein that transports oxygen in the blood; a third binds other proteins and catalyzes cleavage of the bonds between their amino acids. Similarly, the special functions of polysaccharides, nucleic acids, and lipids can be understood as resulting directly from their chemical structure, with their characteristic monomeric subunits precisely linked to form functional polymers. Sugars linked together become energy stores, structural fibers, and points of specific molecular recognition; nucleotides strung together in DNA or RNA provide the blueprint for an entire organism; and aggregated lipids form membranes. Chapter 12 unifies the discussion of biomolecule function, describing how specific signaling systems regulate the activities of biomolecules—within a cell, within an organ, and among organs—to keep an organism in homeostasis.

As we move from monomeric units to larger and larger polymers, the chemical focus shifts from covalent bonds to noncovalent interactions. Covalent bonds, at the monomeric and macromolecular level, place constraints on the shapes assumed by large biomolecules. It is the numerous noncovalent interactions, however, that dictate the stable, native conformations of large molecules while permitting the flexibility necessary for their biological function. As we shall see, noncovalent interactions are essential to the catalytic power of enzymes, the critical interaction of complementary base pairs in nucleic acids, and the arrangement and properties of lipids in membranes. The principle that sequences of monomeric subunits are rich in information emerges most fully in the discussion of nucleic acids (Chapter 8). However, proteins and some short polymers of sugars (oligosaccharides) are also information-rich molecules. The amino acid sequence is a form of information that directs the folding of the protein into its unique three-dimensional structure and ultimately determines the function of the protein. Some oligosaccharides also have unique sequences and three-dimensional structures that are recognized by other macromolecules.

Each class of molecules has a similar structural hierarchy: subunits of fixed structure are connected by bonds of limited flexibility to form macromolecules with three-dimensional structures determined by noncovalent interactions. These macromolecules then interact to form the supramolecular structures and organelles that allow a cell to carry out its many metabolic functions. Together, the molecules described in Part I are the stuff of life.

CHAPTER 2

Water

Self-study tools that will help you practice what you've learned and reinforce this chapter's concepts are available online. Go to www.macmillanlearning.com/LehningerBiochemistry7e.

Water is the most abundant substance in living systems, making up 70% or more of the weight of most organisms. The first living organisms on Earth doubtless arose in an aqueous environment, and the course of evolution has been shaped by the properties of the aqueous medium in which life began.

This chapter begins with descriptions of the physical and chemical properties of water, to which all aspects of cell structure and function are adapted. The attractive forces between water molecules and the slight tendency of water to ionize are of crucial importance to the structure and function of biomolecules. We review the topic of ionization in terms of equilibrium constants, pH, and titration curves, and consider how aqueous solutions of weak acids or bases and their salts act as buffers against pH changes in biological systems. The water molecule and its ionization products, H^+ and OH^-, profoundly influence the structure, self-assembly, and properties of all cellular components, including proteins, nucleic acids, and lipids. The noncovalent interactions responsible for the strength and specificity of "recognition" among biomolecules are decisively influenced by water's properties as a solvent, including its ability to form hydrogen bonds with itself and with solutes.

2.1 Weak Interactions in Aqueous Systems

Hydrogen bonds between water molecules provide the cohesive forces that make water a liquid at room temperature and a crystalline solid (ice) with a highly ordered arrangement of molecules at cold temperatures. Polar biomolecules dissolve readily in water because they can replace water-water interactions with more energetically favorable water-solute interactions. In contrast, nonpolar biomolecules are poorly soluble in water because they interfere with water-water interactions but are unable to form water-solute interactions. In aqueous solutions, nonpolar molecules tend to cluster together. Hydrogen bonds and ionic, hydrophobic (Greek, "water-fearing"), and van der Waals interactions are individually weak, but collectively they have a very significant influence on the three-dimensional structures of proteins, nucleic acids, polysaccharides, and membrane lipids.

Hydrogen Bonding Gives Water Its Unusual Properties

Water has a higher melting point, boiling point, and heat of vaporization than most other common solvents (Table 2-1). These unusual properties are a consequence of attractions between adjacent water molecules that give liquid water great internal cohesion. A look at the electron structure of the H_2O molecule reveals the cause of these intermolecular attractions.

Each hydrogen atom of a water molecule shares an electron pair with the central oxygen atom. The geometry of the molecule is dictated by the shapes of the outer electron orbitals of the oxygen atom, which are similar to the sp^3 bonding orbitals of carbon (see Fig. 1-16). These orbitals describe a rough tetrahedron, with a hydrogen atom at each of two corners and unshared

TABLE 2-1 Melting Point, Boiling Point, and Heat of Vaporization of Some Common Solvents

	Melting point (°C)	Boiling point (°C)	Heat of vaporization (J/g)[a]
Water	0	100	2,260
Methanol (CH_3OH)	−98	65	1,100
Ethanol (CH_3CH_2OH)	−117	78	854
Propanol ($CH_3CH_2CH_2OH$)	−127	97	687
Butanol ($CH_3(CH_2)_2CH_2OH$)	−90	117	590
Acetone (CH_3COCH_3)	−95	56	523
Hexane ($CH_3(CH_2)_4CH_3$)	−98	69	423
Benzene (C_6H_6)	6	80	394
Butane ($CH_3(CH_2)_2CH_3$)	−135	−0.5	381
Chloroform ($CHCl_3$)	−63	61	247

[a]The heat energy required to convert 1.0 g of a liquid at its boiling point and at atmospheric pressure into its gaseous state at the same temperature. It is a direct measure of the energy required to overcome attractive forces between molecules in the liquid phase.

electron pairs at the other two corners **(Fig. 2-1a)**. The H—O—H bond angle is 104.5°, slightly less than the 109.5° of a perfect tetrahedron because of crowding by the nonbonding orbitals of the oxygen atom.

The oxygen nucleus attracts electrons more strongly than does the hydrogen nucleus (a proton); that is, oxygen is more electronegative. This means that the shared electrons are more often in the vicinity of the oxygen atom than of the hydrogen. The result of this unequal electron sharing is two electric dipoles in the water molecule, one along each of the H—O bonds; each hydrogen atom bears a partial positive charge ($\delta+$), and the oxygen atom bears a partial negative charge equal in magnitude to the sum of the two partial positives ($2\delta-$). As a result, there is an electrostatic attraction between the oxygen

atom of one water molecule and the hydrogen of another (Fig. 2-1b), called a **hydrogen bond**. Throughout this book, we represent hydrogen bonds with three parallel blue lines, as in Figure 2-1b.

Hydrogen bonds are relatively weak. Those in liquid water have a **bond dissociation energy** (the energy required to break a bond) of about 23 kJ/mol, compared with 470 kJ/mol for the covalent O—H bond in water or 348 kJ/mol for a covalent C—C bond. The hydrogen bond is about 10% covalent, due to overlaps in the bonding orbitals, and about 90% electrostatic. At room temperature, the thermal energy of an aqueous solution (the kinetic energy of motion of the individual atoms and molecules) is of the same order of magnitude as that required to break hydrogen bonds. When water is heated, the increase in temperature reflects the faster motion of individual water molecules. At any given time, most of the molecules in liquid water are hydrogen-bonded, but the lifetime of each hydrogen bond is just 1 to 20 picoseconds (1 ps = 10^{-12} s); when one hydrogen bond breaks, another hydrogen bond forms, with the same partner or a new one, within 0.1 ps. The apt phrase "flickering clusters" has been applied to the short-lived groups of water molecules interlinked by hydrogen bonds in liquid water. The sum of all the hydrogen bonds between H_2O molecules confers great internal cohesion on liquid water. Extended networks of hydrogen-bonded water molecules also form bridges between solutes (proteins and nucleic acids, for example) that allow the larger molecules to interact with each other over distances of several nanometers without physically touching.

The nearly tetrahedral arrangement of the orbitals about the oxygen atom (Fig. 2-1a) allows each water molecule to form hydrogen bonds with as many as four neighboring water molecules. In liquid water at room temperature and atmospheric pressure, however, water molecules are disorganized and in continuous motion, so that each molecule forms hydrogen bonds with an average of only 3.4 other molecules. In ice, on the other hand, each water molecule is fixed in space and forms hydrogen bonds with

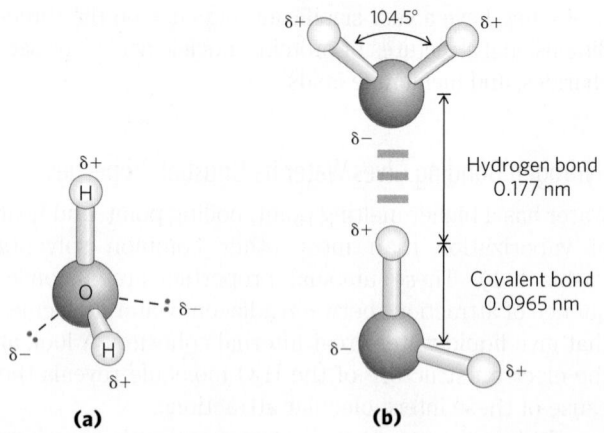

FIGURE 2-1 Structure of the water molecule. (a) The dipolar nature of the H_2O molecule is shown in a ball-and-stick model; the dashed lines represent the nonbonding orbitals. There is a nearly tetrahedral arrangement of the outer-shell electron pairs around the oxygen atom; the two hydrogen atoms have localized partial positive charges ($\delta+$) and the oxygen atom has a partial negative charge ($\delta-$). **(b)** Two H_2O molecules joined by a hydrogen bond (designated here, and throughout this book, by three blue lines) between the oxygen atom of the upper molecule and a hydrogen atom of the lower one. Hydrogen bonds are longer and weaker than covalent O—H bonds.

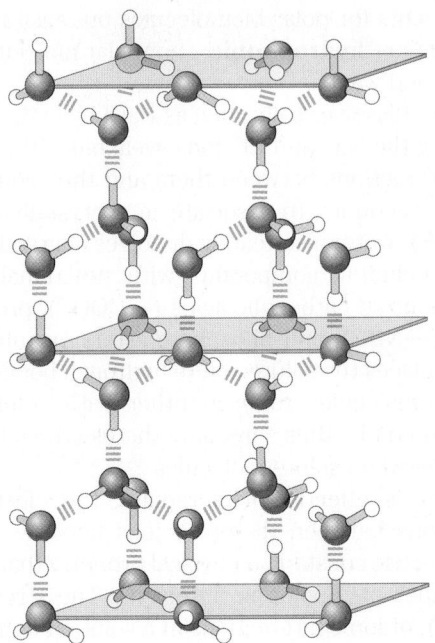

FIGURE 2-2 Hydrogen bonding in ice. In ice, each water molecule forms four hydrogen bonds, the maximum possible for a water molecule, creating a regular crystal lattice. By contrast, in liquid water at room temperature and atmospheric pressure, each water molecule hydrogen-bonds with an average of 3.4 other water molecules. This crystal lattice structure makes ice less dense than liquid water, and thus ice floats on liquid water.

a full complement of four other water molecules to yield a regular lattice structure **(Fig. 2-2)**. Hydrogen bonds account for the relatively high melting point of water, because much thermal energy is required to break a sufficient proportion of hydrogen bonds to destabilize the crystal lattice of ice (Table 2-1). When ice melts or water evaporates, heat is taken up by the system:

$$H_2O\,(\text{solid}) \longrightarrow H_2O\,(\text{liquid}) \qquad \Delta H = +5.9 \text{ kJ/mol}$$
$$H_2O\,(\text{liquid}) \longrightarrow H_2O\,(\text{gas}) \qquad \Delta H = +44.0 \text{ kJ/mol}$$

During melting or evaporation, the entropy of the aqueous system increases as the highly ordered arrays of water molecules in ice relax into the less orderly hydrogen-bonded arrays in liquid water or into the wholly disordered gaseous state. At room temperature, both the melting of ice and the evaporation of water occur spontaneously; the tendency of the water molecules to associate through hydrogen bonds is outweighed by the energetic push toward randomness. Recall that the free-energy change (ΔG) must have a negative value for a process to occur spontaneously: $\Delta G = \Delta H - T\,\Delta S$, where ΔG represents the driving force, ΔH the enthalpy change from making and breaking bonds, and ΔS the change in randomness. Because ΔH is positive for melting and evaporation, it is clearly the increase in entropy (ΔS) that makes ΔG negative and drives these changes.

Water Forms Hydrogen Bonds with Polar Solutes

Hydrogen bonds are not unique to water. They readily form between an electronegative atom (the hydrogen acceptor, usually oxygen or nitrogen) and a hydrogen atom covalently bonded to another electronegative atom (the

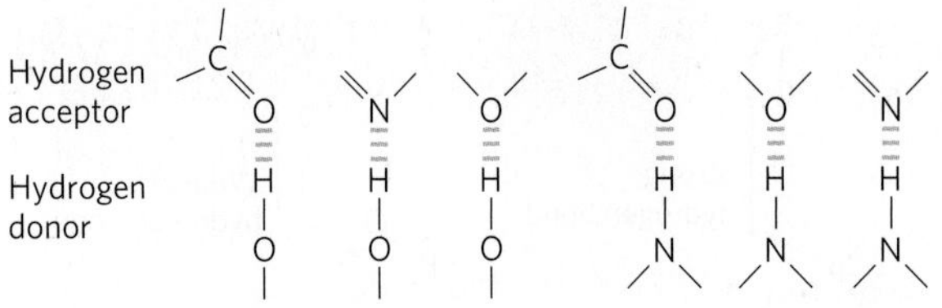

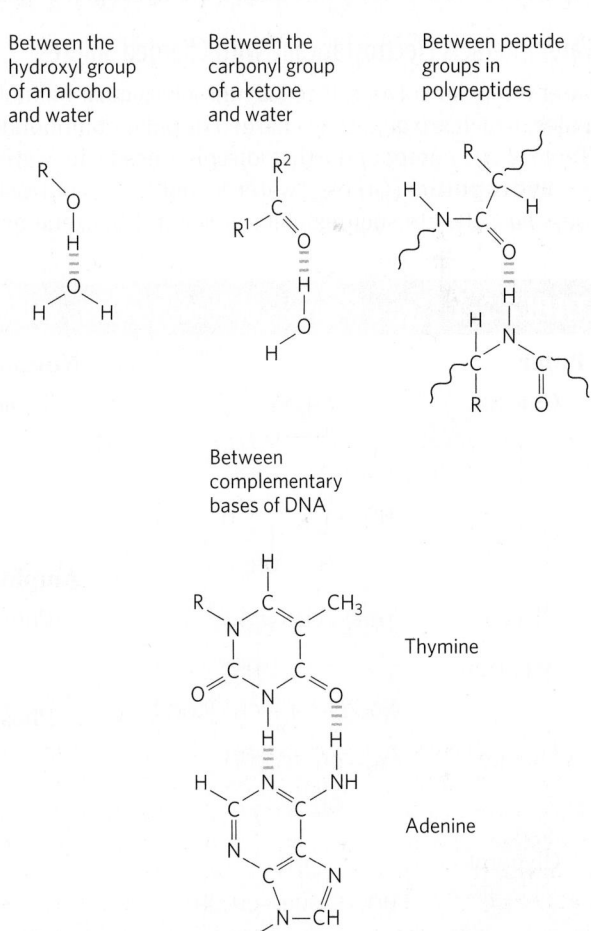

FIGURE 2-3 Common hydrogen bonds in biological systems. The hydrogen acceptor is usually oxygen or nitrogen; the hydrogen donor is another electronegative atom.

hydrogen donor) in the same or another molecule **(Fig. 2-3)**. Hydrogen atoms covalently bonded to carbon atoms do not participate in hydrogen bonding, because carbon is only slightly more electronegative than hydrogen and thus the C—H bond is only very weakly polar. The distinction explains why butane ($CH_3(CH_2)_2CH_3$) has a boiling point of only -0.5 °C, whereas butanol ($CH_3(CH_2)_2CH_2OH$) has a relatively high boiling point of 117 °C. Butanol has a polar hydroxyl group and thus can form intermolecular hydrogen bonds. Uncharged but polar biomolecules such as sugars dissolve readily in water because of the stabilizing effect of hydrogen bonds between the hydroxyl groups or carbonyl oxygen of the sugar and the polar water molecules. Alcohols, aldehydes, ketones, and compounds containing N—H bonds all form hydrogen bonds with water molecules **(Fig. 2-4)** and tend to be soluble in water.

Between the hydroxyl group of an alcohol and water

Between the carbonyl group of a ketone and water

Between peptide groups in polypeptides

Between complementary bases of DNA

Thymine

Adenine

FIGURE 2-4 Some biologically important hydrogen bonds.

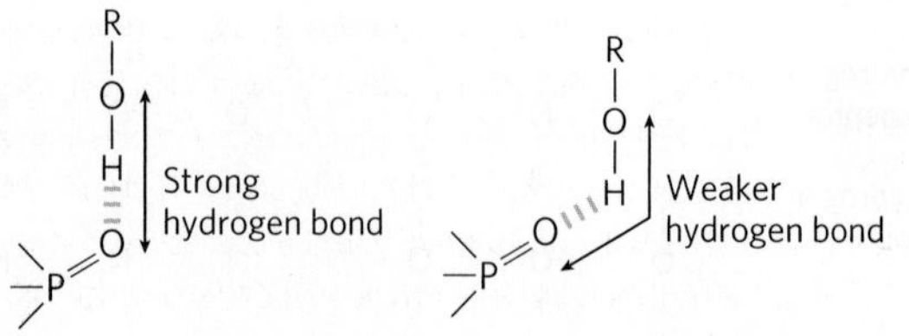

FIGURE 2-5 Directionality of the hydrogen bond. The attraction between the partial electric charges (see Fig. 2-1) is greatest when the three atoms involved in the bond (in this case O, H, and O) lie in a straight line. When the hydrogen-bonded moieties are structurally constrained (when they are parts of a single protein molecule, for example), this ideal geometry may not be possible and the resulting hydrogen bond is weaker.

Hydrogen bonds are strongest when the bonded molecules are oriented to maximize electrostatic interaction, which occurs when the hydrogen atom and the two atoms that share it are in a straight line—that is, when the acceptor atom is in line with the covalent bond between the donor atom and H **(Fig. 2-5)**. This arrangement puts the positive charge of the hydrogen ion directly between the two partial negative charges. Hydrogen bonds are thus highly directional and capable of holding two hydrogen-bonded molecules or groups in a specific geometric arrangement. As we shall see, this property of hydrogen bonds confers very precise three-dimensional structures on protein and nucleic acid molecules, which have many intramolecular hydrogen bonds.

Water Interacts Electrostatically with Charged Solutes

Water is a polar solvent. It readily dissolves most biomolecules, which are generally charged or polar compounds (Table 2-2); compounds that dissolve easily in water are **hydrophilic** (Greek, "water-loving"). In contrast, nonpolar solvents such as chloroform and benzene are poor solvents for polar biomolecules but easily dissolve those that are **hydrophobic**—nonpolar molecules such as lipids and waxes.

Water dissolves salts such as NaCl by hydrating and stabilizing the Na^+ and Cl^- ions, weakening the electrostatic interactions between them and thus counteracting their tendency to associate in a crystalline lattice **(Fig. 2-6)**. Water also readily dissolves charged biomolecules, including compounds with functional groups such as ionized carboxylic acids ($-COO^-$), protonated amines ($-NH_3^+$), and phosphate esters or anhydrides. Water replaces the solute-solute hydrogen bonds linking these biomolecules to each other with solute-water hydrogen bonds, thus screening the electrostatic interactions between solute molecules.

Water is effective in screening the electrostatic interactions between dissolved ions because it has a high dielectric constant, a physical property that reflects the number of dipoles in a solvent. The strength, or force (F), of ionic interactions in a solution depends on the magnitude of the charges (Q), the distance between the charged groups (r), and the dielectric constant (ε, which is dimensionless) of the solvent in which the interactions occur:

$$F = \frac{Q_1 Q_2}{\varepsilon r^2}$$

For water at 25 °C, ε is 78.5, and for the very nonpolar solvent benzene, ε is 4.6. Thus, ionic interactions between dissolved ions are much stronger in less polar environments. The dependence on r^2 is such that ionic attractions or repulsions operate only over short distances—in the range of 10 to 40 nm (depending on the electrolyte concentration) when the solvent is water.

TABLE 2-2	Some Examples of Polar, Nonpolar, and Amphipathic Biomolecules (Shown as Ionic Forms at pH 7)

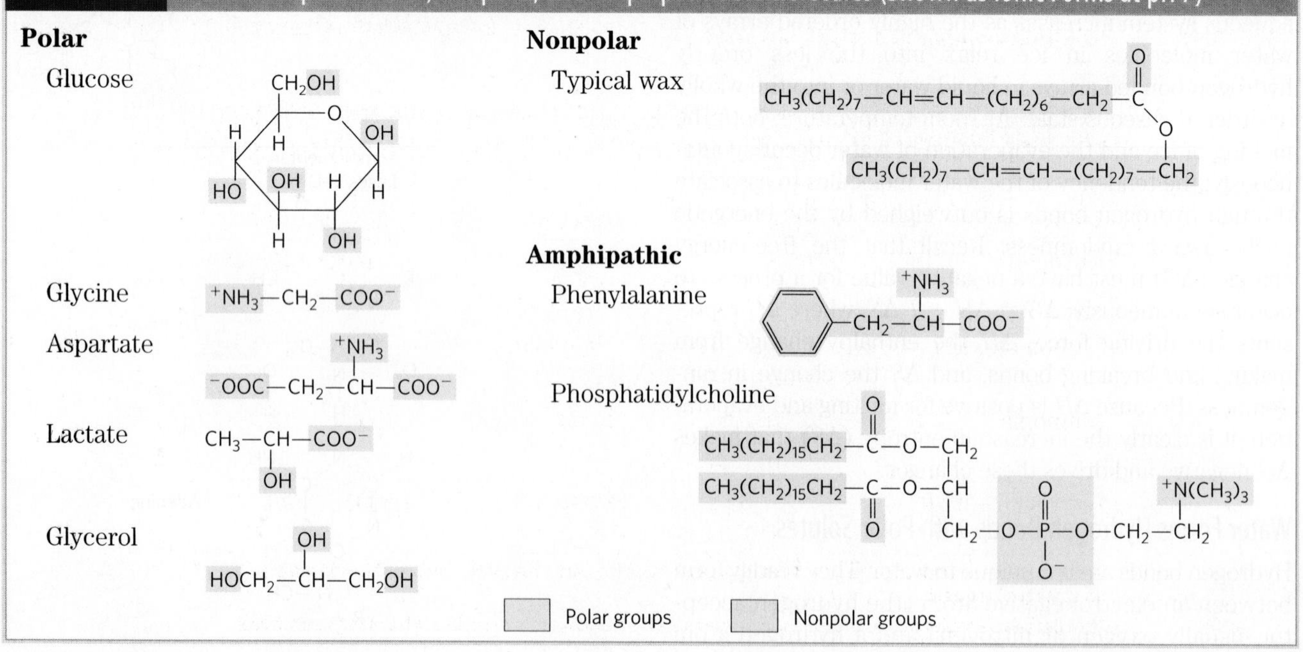

Polar

Glucose

Glycine $^+NH_3-CH_2-COO^-$

Aspartate $^-OOC-CH_2-\overset{\overset{+NH_3}{|}}{CH}-COO^-$

Lactate $CH_3-\overset{\underset{OH}{|}}{CH}-COO^-$

Glycerol $HOCH_2-\overset{\overset{OH}{|}}{CH}-CH_2OH$

Nonpolar

Typical wax

Amphipathic

Phenylalanine

Phosphatidylcholine

☐ Polar groups   ☐ Nonpolar groups

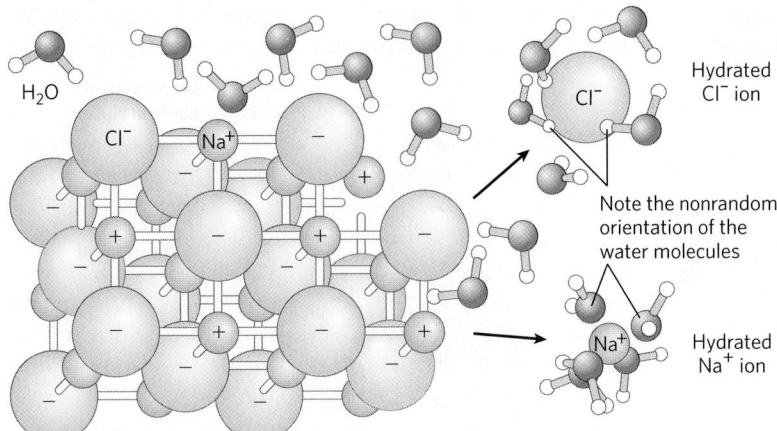

FIGURE 2-6 Water as solvent. Water dissolves many crystalline salts by hydrating their component ions. The NaCl crystal lattice is disrupted as water molecules cluster about the Cl^- and Na^+ ions. The ionic charges are partially neutralized, and the electrostatic attractions necessary for lattice formation are weakened.

Entropy Increases as Crystalline Substances Dissolve

As a salt such as NaCl dissolves, the Na^+ and Cl^- ions leaving the crystal lattice acquire far greater freedom of motion (Fig. 2-6). The resulting increase in entropy (randomness) of the system is largely responsible for the ease of dissolving salts such as NaCl in water. In thermodynamic terms, formation of the solution occurs with a favorable free-energy change: $\Delta G = \Delta H - T\Delta S$, where ΔH has a small positive value and $T\Delta S$ a large positive value; thus ΔG is negative.

Nonpolar Gases Are Poorly Soluble in Water

The molecules of the biologically important gases CO_2, O_2, and N_2 are nonpolar. In O_2 and N_2, electrons are shared equally by both atoms. In CO_2, each C=O bond is polar, but the two dipoles are oppositely directed and cancel each other (Table 2-3). The movement of molecules from the disordered gas phase into aqueous solution constrains their motion and the motion of water molecules and therefore represents a decrease in entropy. The nonpolar nature of these gases and the decrease in entropy when they enter solution combine to make them very poorly soluble in water (Table 2-3). Some organisms have water-soluble "carrier proteins" (hemoglobin and myoglobin, for example) that facilitate the transport of O_2. Carbon dioxide forms carbonic acid (H_2CO_3) in aqueous solution and is transported as the HCO_3^- (bicarbonate) ion, either free—bicarbonate is very soluble in water ($\sim$100 g/L at 25 °C)—or bound to hemoglobin. Three other gases, NH_3, NO, and H_2S, also have biological roles in some organisms; these gases are polar, dissolve readily in water, and ionize in aqueous solution.

Nonpolar Compounds Force Energetically Unfavorable Changes in the Structure of Water

When water is mixed with benzene or hexane, two phases form; neither liquid is soluble in the other. Nonpolar compounds such as benzene and hexane are hydrophobic—they are unable to undergo energetically favorable interactions with water molecules, and they interfere with the hydrogen bonding among water molecules. All molecules or ions in aqueous solution interfere with the hydrogen bonding of some water molecules in their immediate vicinity, but polar or charged

Gas	Structure[a]	Polarity	Solubility in water (g/L)[b]
Nitrogen	N≡N	Nonpolar	0.018 (40 °C)
Oxygen	O=O	Nonpolar	0.035 (50 °C)
Carbon dioxide	$\overset{\delta-}{\longleftarrow}\ \overset{\delta-}{\longrightarrow}$ O=C=O	Nonpolar	0.97 (45 °C)
Ammonia	H H H N ↓$\delta-$	Polar	900 (10 °C)
Hydrogen sulfide	H H S ↓$\delta-$	Polar	1,860 (40 °C)

TABLE 2-3 Solubilities of Some Gases in Water

[a]The arrows represent electric dipoles; there is a partial negative charge ($\delta-$) at the head of the arrow, a partial positive charge ($\delta+$; not shown here) at the tail.

[b]Note that polar molecules dissolve far better even at low temperatures than do nonpolar molecules at relatively high temperatures.

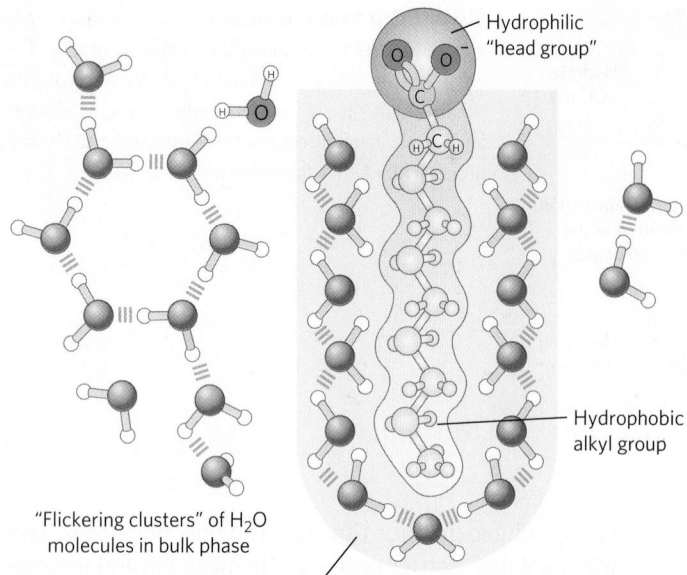

Hydrophilic "head group"

"Flickering clusters" of H₂O molecules in bulk phase

Hydrophobic alkyl group

Highly ordered H₂O molecules form "cages" around the hydrophobic alkyl chains.

(a)

FIGURE 2-7 Amphipathic compounds in aqueous solution. (a) Long-chain fatty acids have very hydrophobic alkyl chains, each of which is surrounded by a layer of highly ordered water molecules. **(b)** By clustering together in micelles, the fatty acid molecules expose the smallest possible hydrophobic surface area to the water, and fewer water molecules are required in the shell of ordered water. The energy gained by freeing immobilized water molecules stabilizes the micelle.

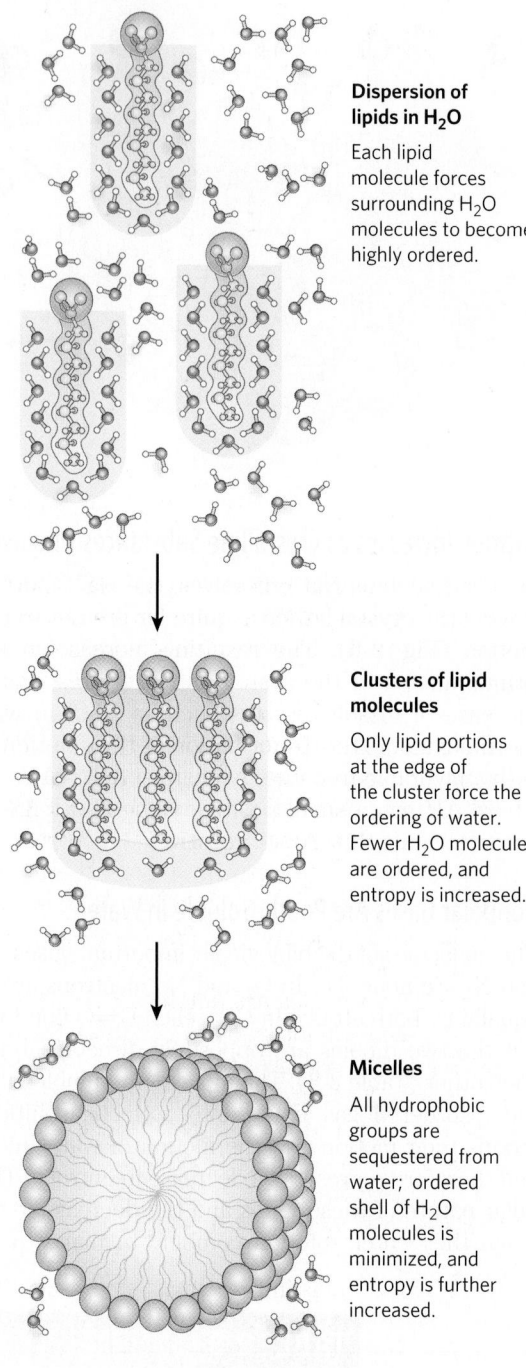

Dispersion of lipids in H₂O

Each lipid molecule forces surrounding H₂O molecules to become highly ordered.

Clusters of lipid molecules

Only lipid portions at the edge of the cluster force the ordering of water. Fewer H₂O molecules are ordered, and entropy is increased.

Micelles

All hydrophobic groups are sequestered from water; ordered shell of H₂O molecules is minimized, and entropy is further increased.

(b)

solutes (such as NaCl) compensate for lost water-water hydrogen bonds by forming new solute-water interactions. The net change in enthalpy (ΔH) for dissolving these solutes is generally small. Hydrophobic solutes, however, offer no such compensation, and their addition to water may therefore result in a small gain of enthalpy; the breaking of hydrogen bonds between water molecules takes up energy from the system, requiring the input of energy from the surroundings. In addition to requiring this input of energy, dissolving hydrophobic compounds in water produces a measurable decrease in entropy. Water molecules in the immediate vicinity of a nonpolar solute are constrained in their possible orientations as they form a highly ordered cagelike shell around each solute molecule. These water molecules are not as highly oriented as those in **clathrates**, crystalline compounds of nonpolar solutes and water, but the effect is the same in both cases: the ordering of water molecules reduces entropy. The number of ordered water molecules, and therefore the magnitude of the entropy decrease, is proportional to the surface area of the hydrophobic solute enclosed within the cage of water molecules. The free-energy change for dissolving a nonpolar solute in water is thus unfavorable: $\Delta G = \Delta H - T\,\Delta S$, where ΔH has a positive value, ΔS has a negative value, and ΔG is positive.

Amphipathic compounds contain regions that are polar (or charged) and regions that are nonpolar

(Table 2-2). When an amphipathic compound is mixed with water, the polar, hydrophilic region interacts favorably with the water and tends to dissolve, but the nonpolar, hydrophobic region tends to avoid contact with the water **(Fig. 2-7a)**. The nonpolar regions of the molecules cluster together to present the smallest hydrophobic area to the aqueous solvent, and the polar regions are arranged to maximize their interaction with the solvent (Fig. 2-7b), a phenomenon called the **hydrophobic effect**. These stable structures of amphipathic compounds in water, called **micelles**, may contain hundreds or thousands of molecules. The forces that hold the nonpolar regions of the molecules together are

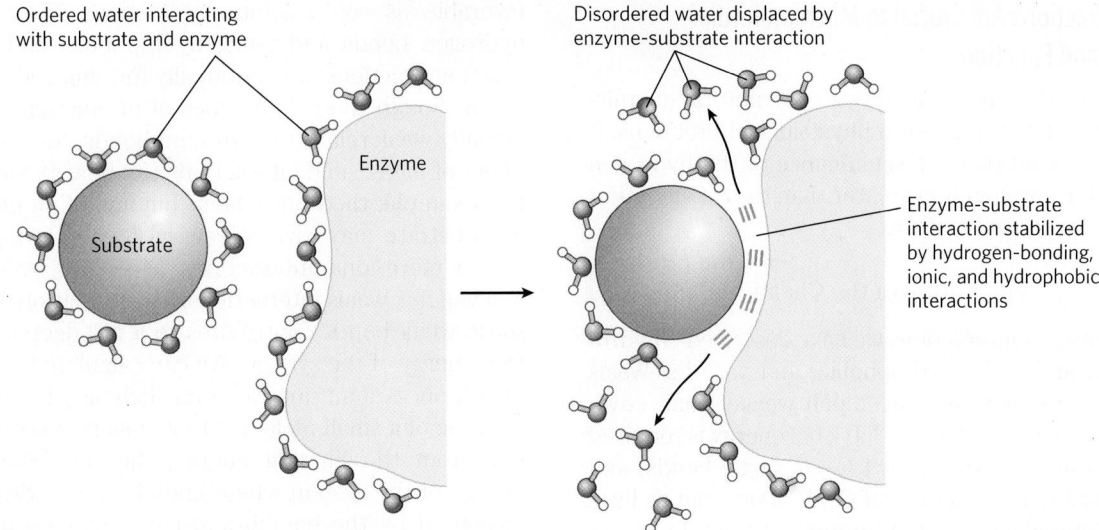

Ordered water interacting with substrate and enzyme

Enzyme

Substrate

Disordered water displaced by enzyme-substrate interaction

Enzyme-substrate interaction stabilized by hydrogen-bonding, ionic, and hydrophobic interactions

FIGURE 2-8 Release of ordered water favors formation of an enzyme-substrate complex. While separate, both enzyme and substrate force neighboring water molecules into an ordered shell. Binding of substrate to enzyme releases some of the ordered water, and the resulting increase in entropy provides a thermodynamic push toward formation of the enzyme-substrate complex (see p. 196).

sometimes referred to as **hydrophobic interactions**, although this terminology can be confusing because the strength of the interactions is not due to any intrinsic attraction between nonpolar moieties. Rather, it results from the system's achieving the greatest thermodynamic stability by minimizing the number of ordered water molecules required to surround hydrophobic portions of the solute molecules.

Many biomolecules are amphipathic; proteins, pigments, certain vitamins, and the sterols and phospholipids of membranes all have both polar and nonpolar surface regions. Structures composed of these molecules are stabilized by the hydrophobic effect, which favors aggregation of the nonpolar regions. The hydrophobic effect on interactions among lipids, and between lipids and proteins, is the most important determinant of structure in biological membranes. The aggregation of nonpolar amino acids in protein interiors, driven by the hydrophobic effect, also stabilizes the three-dimensional structures of proteins.

Hydrogen bonding between water and polar solutes also causes an ordering of water molecules, but the energetic effect is less significant than with nonpolar solutes. Disruption of ordered water molecules is part of the driving force for binding of a polar substrate (reactant) to the complementary polar surface of an enzyme: entropy increases as the enzyme displaces ordered water from the substrate and as the substrate displaces ordered water from the enzyme surface **(Fig. 2-8)**.

van der Waals Interactions Are Weak Interatomic Attractions

When two uncharged atoms are brought very close together, their surrounding electron clouds influence each other. Random variations in the positions of the electrons around one nucleus may create a transient electric dipole, which induces a transient, opposite electric dipole in the nearby atom. The two dipoles weakly attract each other, bringing the two nuclei closer. These weak attractions are called **van der Waals interactions** (also known as London forces). As the two nuclei draw closer together, their electron clouds begin to repel each other. At the point where the net attraction is maximal, the nuclei are said to be in van der Waals contact. Each atom has a characteristic **van der Waals radius**, a measure of how close that atom will allow another to approach (Table 2-4). In the "space-filling" molecular models shown throughout this book, the atoms are depicted in sizes proportional to their van der Waals radii.

TABLE 2-4	van der Waals Radii and Covalent (Single-Bond) Radii of Some Elements	
Element	van der Waals radius (nm)	Covalent radius for single bond (nm)
H	0.11	0.030
O	0.15	0.066
N	0.15	0.070
C	0.17	0.077
S	0.18	0.104
P	0.19	0.110
I	0.21	0.133

Sources: For van der Waals radii, R. Chauvin, *J. Phys. Chem.* 96:9194, 1992. For covalent radii, L. Pauling, *Nature of the Chemical Bond,* 3rd edn, Cornell University Press, 1960.

Note: van der Waals radii describe the space-filling dimensions of atoms. When two atoms are joined covalently, the atomic radii at the point of bonding are shorter than the van der Waals radii because the joined atoms are pulled together by the shared electron pair. The distance between nuclei in a van der Waals interaction or a covalent bond is about equal to the sum of the van der Waals or covalent radii, respectively, for the two atoms. Thus the length of a carbon–carbon single bond is about 0.077 nm + 0.077 nm = 0.154 nm.

Weak Interactions Are Crucial to Macromolecular Structure and Function

I believe that as the methods of structural chemistry are further applied to physiological problems, it will be found that the significance of the hydrogen bond for physiology is greater than that of any other single structural feature.

—*Linus Pauling,*
The Nature of the Chemical Bond, 1939

The noncovalent interactions we have described—hydrogen bonds and ionic, hydrophobic, and van der Waals interactions (Table 2-5)—are much weaker than covalent bonds. An input of about 350 kJ of energy is required to break a mole of (6×10^{23}) C—C single bonds, and about 410 kJ to break a mole of C—H bonds, but as little as 4 kJ is sufficient to disrupt a mole of typical van der Waals interactions. Interactions driven by the hydrophobic effect are also much weaker than covalent bonds, although they are substantially strengthened by a highly polar solvent (a concentrated salt solution, for example). Ionic interactions and hydrogen bonds are variable in strength, depending on the polarity of the solvent and the alignment of the hydrogen-bonded atoms, but they are always significantly weaker than covalent bonds. In aqueous solvent at 25 °C, the available thermal energy can be of the same order of magnitude as the strength of these weak interactions, and the interaction between solute and solvent (water) molecules is nearly as favorable as solute-solute interactions. Consequently, hydrogen bonds and ionic, hydrophobic, and van der Waals interactions are continually forming and breaking.

Although these four types of interactions are individually weak relative to covalent bonds, the cumulative effect of many such interactions can be very significant. For example, the noncovalent binding of an enzyme to its substrate may involve several hydrogen bonds and one or more ionic interactions, as well as hydrophobic and van der Waals interactions. The formation of each of these weak bonds contributes to a net decrease in the free energy of the system. We can calculate the stability of a noncovalent interaction, such as the hydrogen bonding of a small molecule to its macromolecular partner, from the binding energy, the reduction in the energy of the system when binding occurs. Stability, as measured by the equilibrium constant (see below) of the binding reaction, varies *exponentially* with binding energy. To dissociate two biomolecules (such as an enzyme and its bound substrate) that are associated noncovalently through multiple weak interactions, all these interactions must be disrupted at the same time. Because the interactions fluctuate randomly, such simultaneous disruptions are very unlikely. Therefore, 5 or 20 weak interactions bestow much greater molecular stability than would be expected intuitively from a simple summation of small binding energies.

Macromolecules such as proteins, DNA, and RNA contain so many sites of potential hydrogen bonding or ionic, van der Waals, or hydrophobic interactions that the cumulative effect of the many small binding forces can be enormous. For macromolecules, the most stable (that is, the native) structure is usually that in which weak interactions are maximized. The folding of a single polypeptide or polynucleotide chain into its three-dimensional shape is determined by this principle. The binding of an antigen to a specific antibody depends on the cumulative effects of many weak interactions. As noted earlier, the energy released when an enzyme binds noncovalently to its substrate is the main source of the enzyme's catalytic power. The binding of a hormone or a neurotransmitter to its cellular receptor protein is the result of multiple weak interactions. One consequence of the large size of enzymes and receptors (relative to their substrates or ligands) is that their extensive surfaces provide many opportunities for weak interactions. At the molecular level, the complementarity between interacting biomolecules reflects the complementarity and weak interactions between polar and charged groups and the proximity of hydrophobic patches on the surfaces of the molecules.

When the structure of a protein such as hemoglobin **(Fig. 2-9)** is determined by x-ray crystallography (see Box 4-5), water molecules are often found to be bound so tightly that they are part of the crystal structure; the same is true for water in crystals of RNA or DNA. These bound water molecules, which can also be detected in aqueous solutions by nuclear magnetic resonance, have

TABLE 2-5 Four Types of Noncovalent ("Weak") Interactions among Biomolecules in Aqueous Solvent

Hydrogen bonds
  Between neutral groups

  Between peptide bonds

Ionic interactions
  Attraction

  Repulsion

Hydrophobic interactions

van der Waals interactions   Any two atoms in close proximity

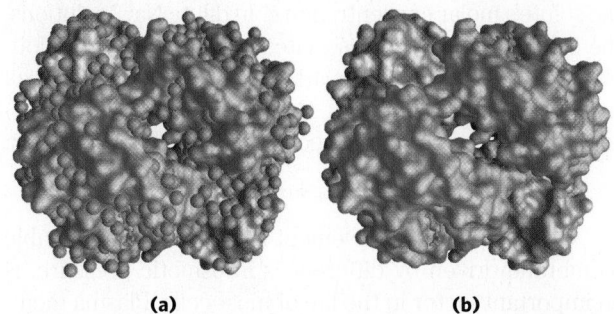

(a) **(b)**

FIGURE 2-9 Water binding in hemoglobin. The crystal structure of hemoglobin, shown **(a)** with bound water molecules (red spheres) and **(b)** without the water molecules. The water molecules are so firmly bound to the protein that they affect the x-ray diffraction pattern as though they were fixed parts of the crystal. The two α subunits of hemoglobin are shown in gray, the two β subunits in blue. Each subunit has a bound heme group (red stick structure), visible only in the β subunits in this view. The structure and function of hemoglobin are discussed in detail in Chapter 5. [Source: PDB ID 1A3N, J. R. H. Tame and B. Vallone, *Acta Crystallogr. D* 56:805, 2000.]

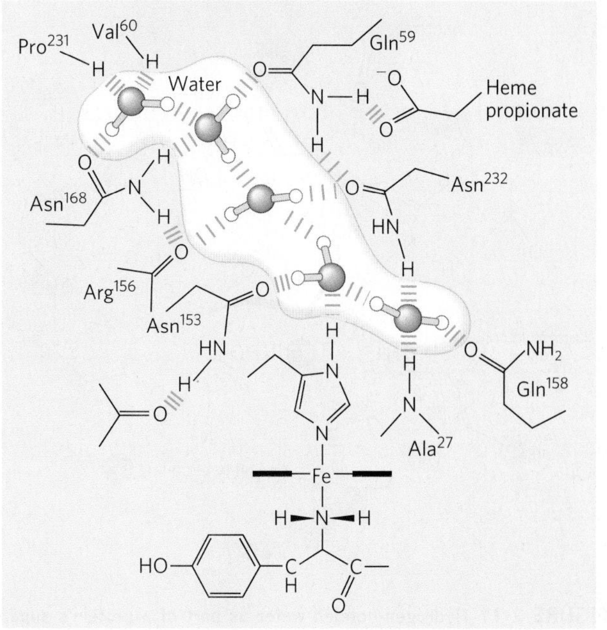

FIGURE 2-10 Water chain in cytochrome *f*. Water is bound in a proton channel of the membrane protein cytochrome *f*, which is part of the energy-trapping machinery of photosynthesis in chloroplasts (see Fig. 20-21). Five water molecules are hydrogen-bonded to each other and to functional groups of the protein: the peptide backbone atoms of valine, proline, arginine, and alanine residues, and the side chains of three asparagine and two glutamine residues. The protein has a bound heme (see Fig. 5-1), its iron ion facilitating electron flow during photosynthesis. Electron flow is coupled to the movement of protons across the membrane, which probably involves "proton hopping" (see Fig. 2-14) through this chain of bound water molecules. [Source: Information from P. Nicholls, *Cell. Mol. Life Sci.* 57:987, 2000, Fig. 6a (redrawn from PDB ID 1HCZ, S. E. Martinez et al., *Prot. Sci.* 5:1081, 1996).]

distinctly different properties from those of the "bulk" water of the solvent. They are, for example, not osmotically active (see below). For many proteins, tightly bound water molecules are essential to their function. In a key reaction in photosynthesis, for example, protons flow across a biological membrane as light drives the flow of electrons through a series of electron-carrying proteins (see Fig. 20-21). One of these proteins, cytochrome *f*, has a chain of five bound water molecules **(Fig. 2-10)** that may provide a path for protons to move through the membrane by a process known as "proton hopping" (described below). Another such light-driven proton pump, bacteriorhodopsin, almost certainly uses a chain of precisely oriented bound water molecules in the transmembrane movement of protons (see Fig. 20-29b). Tightly bound water molecules can also form an essential part of a protein's ligand-binding site. In a bacterial arabinose-binding protein, for example, five water molecules form hydrogen bonds that provide critical cross-links between the sugar (arabinose) and the amino acid residues in the sugar-binding site **(Fig. 2-11)**.

Solutes Affect the Colligative Properties of Aqueous Solutions

Solutes of all kinds alter certain physical properties of the solvent, water: its vapor pressure, boiling point, melting point (freezing point), and osmotic pressure. These are called **colligative properties** (colligative meaning "tied together"), because the effect of solutes on all four properties has the same basis: the concentration of water is lower in solutions than in pure water. The effect of solute concentration on the colligative properties of water is independent of the chemical properties of the solute; it depends only on the *number* of solute particles (molecules or ions) in a given amount

of water. For example, a compound such as NaCl, which dissociates in solution, has an effect on osmotic pressure that is twice that of an equal number of moles of a nondissociating solute such as glucose.

Water molecules tend to move from a region of higher water concentration to one of lower water concentration, in accordance with the tendency in nature for a system to become disordered. When two different aqueous solutions are separated by a semipermeable membrane (one that allows the passage of water but not solute molecules), water molecules diffusing from the region of higher water concentration to the region of lower water concentration produce osmotic pressure **(Fig. 2-12)**. Osmotic pressure, Π, measured as the force necessary to resist water movement (Fig. 2-12c), is approximated by the van't Hoff equation:

$$\Pi = icRT$$

in which R is the gas constant and T is the absolute temperature. The symbol i is the van't Hoff factor, which is a measure of the extent to which the solute dissociates into two or more ionic species. The term ic is the **osmolarity** of the solution, the product of the van't Hoff factor i and

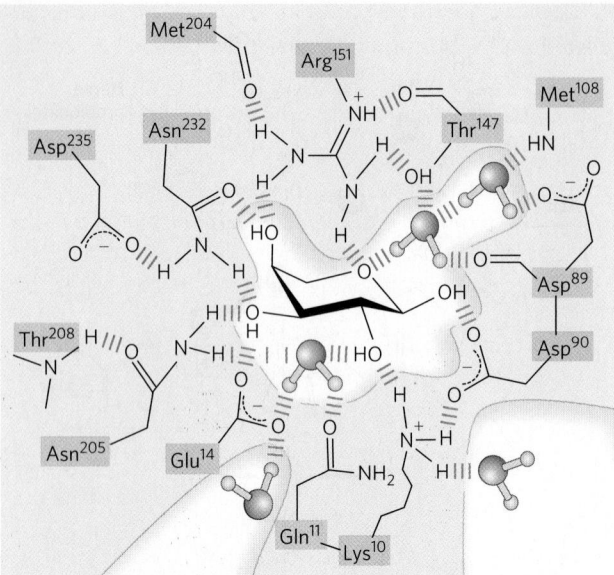

FIGURE 2-11 Hydrogen-bonded water as part of a protein's sugar-binding site. In the L-arabinose-binding protein of the bacterium *E. coli*, five water molecules are essential components of the hydrogen-bonded network of interactions between the sugar arabinose (center) and at least 13 amino acid residues in the sugar-binding site. Viewed in three dimensions, these interacting groups constitute two layers of binding moieties; amino acid residues in the first layer are screened in red, those in the second layer in green. Some of the hydrogen bonds are drawn longer than others for clarity; in reality, all hydrogen bonds are the same length. [Source: Information from P. Ball, *Chem. Rev.* 108:74, 2008, Fig. 16.]

the solute's molar concentration c. In dilute NaCl solutions, the solute completely dissociates into Na^+ and Cl^-, doubling the number of solute particles, and thus $i = 2$. For all nonionizing solutes, $i = 1$. For solutions of several (n) solutes, Π is the sum of the contributions of each species:

$$\Pi = RT(i_1c_1 + i_2c_2 + i_3c_3 + \cdots + i_nc_n)$$

Osmosis, water movement across a semipermeable membrane driven by differences in osmotic pressure, is an important factor in the life of most cells. Plasma membranes are more permeable to water than to most other small molecules, ions, and macromolecules because protein channels (aquaporins; see Fig. 11-43) in the membrane selectively permit the passage of water. Solutions of osmolarity equal to that of a cell's cytosol are said to be **isotonic** relative to that cell. Surrounded by an isotonic solution, a cell neither gains nor loses water (**Fig. 2-13**). In a **hypertonic** solution, one with higher osmolarity than that of the cytosol, the cell shrinks as water moves out. In a **hypotonic** solution, one with a lower osmolarity than the cytosol, the cell swells as water enters. In their natural environments, cells generally contain higher concentrations of biomolecules and ions than their surroundings, so osmotic pressure tends to drive water into cells. If not somehow counterbalanced, this inward movement

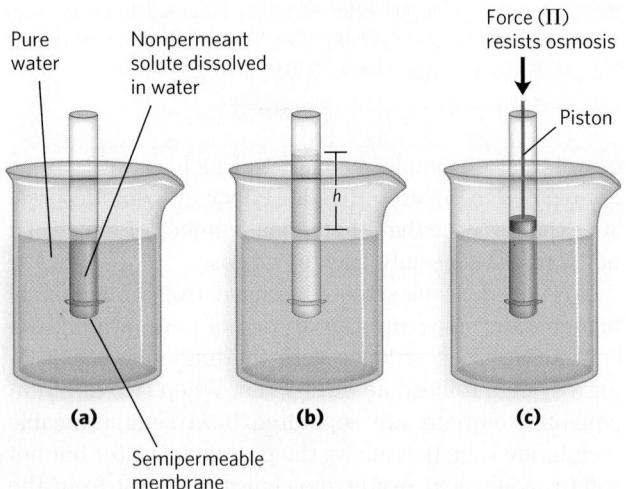

FIGURE 2-12 Osmosis and the measurement of osmotic pressure. **(a)** The initial state. The tube contains an aqueous solution, the beaker contains pure water, and the semipermeable membrane allows the passage of water but not solute. Water flows from the beaker into the tube to equalize its concentration across the membrane. **(b)** The final state. Water has moved into the solution of the nonpermeant compound, diluting it and raising the column of solution within the tube. At equilibrium, the force of gravity operating on the solution in the tube exactly balances the tendency of water to move into the tube, where its concentration is lower. **(c)** Osmotic pressure (Π) is measured as the force that must be applied to return the solution in the tube to the level of the water in the beaker. This force is proportional to the height, h, of the column in (b).

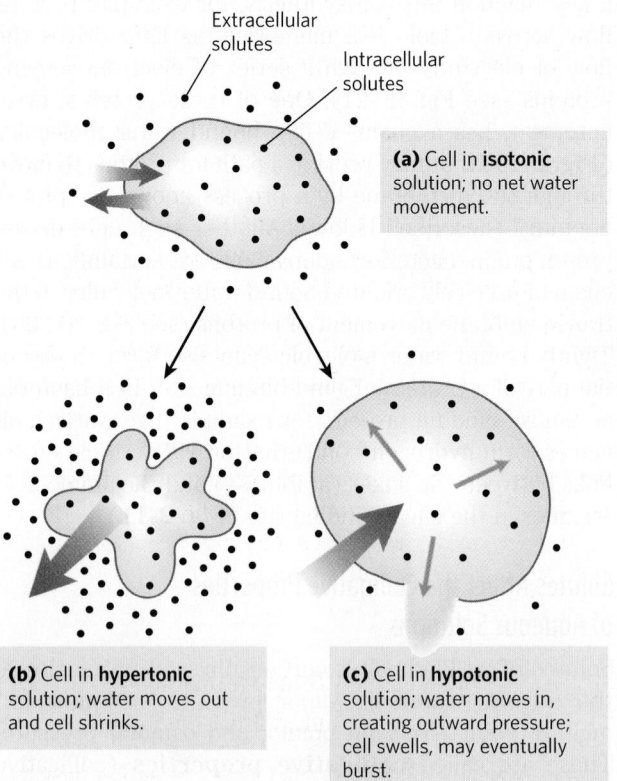

FIGURE 2-13 Effect of extracellular osmolarity on water movement across a plasma membrane. When a cell in osmotic balance with its surrounding medium—that is, a cell in **(a)** an isotonic medium—is transferred into **(b)** a hypertonic solution or **(c)** a hypotonic solution, water moves across the plasma membrane in the direction that tends to equalize osmolarity outside and inside the cell.

of water would distend the plasma membrane and eventually cause bursting of the cell (osmotic lysis).

Several mechanisms have evolved to prevent this catastrophe. In bacteria and plants, the plasma membrane is surrounded by a nonexpandable cell wall of sufficient rigidity and strength to resist osmotic pressure and prevent osmotic lysis. Certain freshwater protists that live in a highly hypotonic medium have an organelle (contractile vacuole) that pumps water out of the cell. In multicellular animals, blood plasma and interstitial fluid (the extracellular fluid of tissues) are maintained at an osmolarity close to that of the cytosol. The high concentration of albumin and other proteins in blood plasma contributes to its osmolarity. Cells also actively pump out Na^+ and other ions into the interstitial fluid to stay in osmotic balance with their surroundings.

Because the effect of solutes on osmolarity depends on the *number* of dissolved particles, not their *mass*, macromolecules (proteins, nucleic acids, polysaccharides) have far less effect on the osmolarity of a solution than would an equal mass of their monomeric components. For example, a *gram* of a polysaccharide composed of 1,000 glucose units has the same effect on osmolarity as a *milligram* of glucose. Storing fuel as polysaccharides (starch or glycogen) rather than as glucose or other simple sugars avoids an enormous increase in osmotic pressure in the storage cell.

Plants use osmotic pressure to achieve mechanical rigidity. The very high solute concentration in the plant cell vacuole draws water into the cell (Fig. 2-13), but the nonexpandable cell wall prevents swelling; instead, the pressure exerted against the cell wall (turgor pressure) increases, stiffening the cell, the tissue, and the plant body. When the lettuce in your salad wilts, it is because loss of water has reduced turgor pressure. Osmosis also has consequences for laboratory protocols. Mitochondria, chloroplasts, and lysosomes, for example, are enclosed by semipermeable membranes. In isolating these organelles from broken cells, biochemists must perform the fractionations in isotonic solutions (see Fig. 1-9) to prevent excessive entry of water into the organelles and the swelling and bursting that would follow. Buffers used in cellular fractionations commonly contain sufficient concentrations of sucrose or some other inert solute to protect the organelles from osmotic lysis.

WORKED EXAMPLE 2-1 Osmotic Strength of an Organelle I

Suppose the major solutes in intact lysosomes are KCl (~ 0.1 M) and NaCl (~ 0.03 M). When isolating lysosomes, what concentration of sucrose is required in the extracting solution at room temperature (25 °C) to prevent swelling and lysis?

Solution: We want to find a concentration of sucrose that gives an osmotic strength equal to that produced by the KCl and NaCl in the lysosomes. The equation for calculating osmotic strength (the van't Hoff equation) is

$$\Pi = RT(i_1c_1 + i_2c_2 + i_3c_3 + \cdots + i_nc_n)$$

where R is the gas constant 8.315 J/mol·K, T is the absolute temperature (Kelvin), c_1, c_2, and c_3 are the molar concentrations of each solute, and i_1, i_2, and i_3 are the numbers of particles each solute yields in solution ($i = 2$ for KCl and NaCl).

The osmotic strength of the lysosomal contents is

$$\begin{aligned}\Pi_{\text{lysosome}} &= RT(i_{\text{KCl}}c_{\text{KCl}} + i_{\text{NaCl}}c_{\text{NaCl}}) \\ &= RT[(2)(0.1 \text{ mol/L}) + (2)(0.03 \text{ mol/L})] \\ &= RT(0.26 \text{ mol/L})\end{aligned}$$

The osmotic strength of a sucrose solution is given by

$$\Pi_{\text{sucrose}} = RT(i_{\text{sucrose}}c_{\text{sucrose}})$$

In this case, $i_{\text{sucrose}} = 1$, because sucrose does not ionize. Thus,

$$\Pi_{\text{sucrose}} = RT(c_{\text{sucrose}})$$

The osmotic strength of the lysosomal contents equals that of the sucrose solution when

$$\begin{aligned}\Pi_{\text{sucrose}} &= \Pi_{\text{lysosome}} \\ RT(c_{\text{sucrose}}) &= RT(0.26 \text{ mol/L}) \\ c_{\text{sucrose}} &= 0.26 \text{ mol/L}\end{aligned}$$

So the required concentration of sucrose (FW 342) is $(0.26 \text{ mol/L})(342 \text{ g/mol}) = 88.92 \text{ g/L}$. Because the solute concentrations are only accurate to one significant figure, $c_{\text{sucrose}} = 0.09 \text{ kg/L}$.

WORKED EXAMPLE 2-2 Osmotic Strength of an Organelle II

Suppose we decided to use a solution of a polysaccharide, say glycogen (p. 254), to balance the osmotic strength of the lysosomes (described in Worked Example 2-1). Assuming a linear polymer of 100 glucose units, calculate the amount of this polymer needed to achieve the same osmotic strength as the sucrose solution in Worked Example 2-1. The M_r of the glucose polymer is $\sim 18,000$, and, like sucrose, it does not ionize in solution.

Solution: As derived in Worked Example 2-1,

$$\Pi_{\text{sucrose}} = RT(0.26 \text{ mol/L})$$

Similarly,

$$\Pi_{\text{glycogen}} = RT(i_{\text{glycogen}}c_{\text{glycogen}}) = RT(c_{\text{glycogen}})$$

For a glycogen solution with the same osmotic strength as the sucrose solution,

$$\begin{aligned}\Pi_{\text{glycogen}} &= \Pi_{\text{sucrose}} \\ RT(c_{\text{glycogen}}) &= RT(0.26 \text{ mol/L}) \\ c_{\text{glycogen}} &= (0.26 \text{ mol/L}) = (0.26 \text{ mol/L})(18,000 \text{ g/mol}) \\ &= 4.68 \text{ kg/L}\end{aligned}$$

Or, when significant figures are taken into account, $c_{glycogen} = 5$ kg/L, an absurdly high concentration.

As we'll see later (p. 254), cells of liver and muscle store carbohydrate not as low molecular weight sugars such as glucose or sucrose but as the high molecular weight polymer glycogen. This allows the cell to contain a large mass of glycogen with a minimal effect on the osmolarity of the cytosol.

SUMMARY 2.1 Weak Interactions in Aqueous Systems

■ The very different electronegativities of H and O make water a highly polar molecule, capable of forming hydrogen bonds with itself and with solutes. Hydrogen bonds are fleeting, primarily electrostatic, and weaker than covalent bonds. Water is a good solvent for polar (hydrophilic) solutes, with which it forms hydrogen bonds, and for charged solutes, with which it interacts electrostatically.

■ Nonpolar (hydrophobic) compounds dissolve poorly in water; they cannot hydrogen-bond with the solvent, and their presence forces an energetically unfavorable ordering of water molecules at their hydrophobic surfaces. To minimize the surface exposed to water, nonpolar and amphipathic compounds such as lipids form aggregates (micelles) in which the hydrophobic moieties are sequestered in the interior, an association driven by the hydrophobic effect, and only the more polar moieties interact with water.

■ Weak, noncovalent interactions, in large numbers, decisively influence the folding of macromolecules such as proteins and nucleic acids. The most stable macromolecular conformations are those in which hydrogen bonding is maximized within the molecule and between the molecule and the solvent, and in which hydrophobic moieties cluster in the interior of the molecule away from the aqueous solvent.

■ The physical properties of aqueous solutions are strongly influenced by the concentrations of solutes. When two aqueous compartments are separated by a semipermeable membrane (such as the plasma membrane separating a cell from its surroundings), water moves across that membrane to equalize the osmolarity in the two compartments. This tendency for water to move across a semipermeable membrane produces the osmotic pressure.

2.2 Ionization of Water, Weak Acids, and Weak Bases

Although many of the solvent properties of water can be explained in terms of the uncharged H_2O molecule, the small degree of ionization of water to hydrogen ions (H^+) and hydroxide ions (OH^-) must also be taken into account. Like all reversible reactions, the ionization of water can be described by an equilibrium constant.

When weak acids are dissolved in water, they contribute H^+ by ionizing; weak bases consume H^+ by becoming protonated. These processes are also governed by equilibrium constants. The total hydrogen ion concentration from all sources is experimentally measurable and is expressed as the pH of the solution. To predict the state of ionization of solutes in water, we must take into account the relevant equilibrium constants for each ionization reaction. We therefore turn now to a brief discussion of the ionization of water and of weak acids and bases dissolved in water.

Pure Water Is Slightly Ionized

Water molecules have a slight tendency to undergo reversible ionization to yield a hydrogen ion (a proton) and a hydroxide ion, giving the equilibrium

$$H_2O \rightleftharpoons H^+ + OH^- \tag{2-1}$$

Although we commonly show the dissociation product of water as H^+, free protons do not exist in solution; hydrogen ions formed in water are immediately hydrated to form **hydronium ions** (H_3O^+). Hydrogen bonding between water molecules makes the hydration of dissociating protons virtually instantaneous:

The ionization of water can be measured by its electrical conductivity; pure water carries electrical current as H_3O^+ migrates toward the cathode and OH^- toward the anode. The movement of hydronium and hydroxide ions in the electric field is extremely fast compared with that of other ions such as Na^+, K^+, and Cl^-. This high ionic mobility results from the kind of "proton hopping" shown in **Figure 2-14**. No individual proton moves very far through the bulk solution, but a series of proton hops between hydrogen-bonded water molecules causes the *net* movement of a proton over a long distance in a remarkably short time. (OH^- also moves rapidly by proton hopping, but in the opposite direction.) As a result of the high ionic mobility of H^+, acid-base reactions in aqueous solutions are exceptionally fast. As noted above, proton hopping very likely also plays a role in biological proton-transfer reactions (Fig. 2-10; see also Fig. 20-29b).

Because reversible ionization is crucial to the role of water in cellular function, we must have a means of expressing the extent of ionization of water in quantitative terms. A brief review of some properties of reversible chemical reactions shows how this can be done.

The position of equilibrium of any chemical reaction is given by its **equilibrium constant, K_{eq}** (sometimes expressed simply as K). For the generalized reaction

$$A + B \rightleftharpoons C + D \tag{2-2}$$

the equilibrium constant K_{eq} can be defined in terms of the concentrations of reactants (A and B) and products (C and D) at equilibrium:

$$K_{eq} = \frac{[C]_{eq}[D]_{eq}}{[A]_{eq}[B]_{eq}}$$

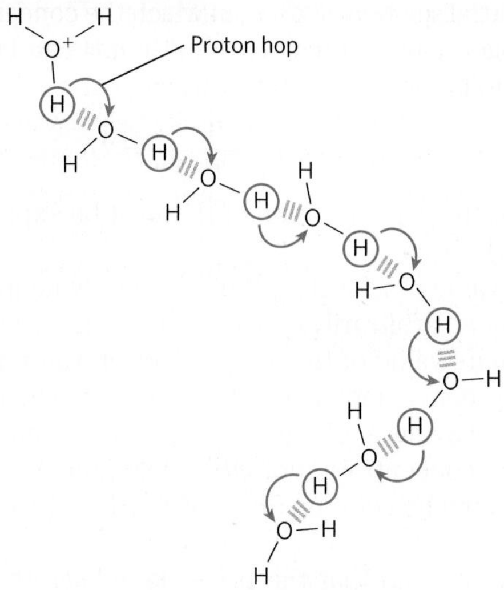

Hydronium ion gives up a proton.

Water accepts proton and
becomes a hydronium ion.

FIGURE 2-14 Proton hopping. Short "hops" of protons between a series of hydrogen-bonded water molecules result in an extremely rapid net movement of a proton over a long distance. As a hydronium ion (upper left) gives up a proton, a water molecule some distance away (bottom) acquires one, becoming a hydronium ion. Proton hopping is much faster than true diffusion and explains the remarkably high ionic mobility of H^+ ions compared with other monovalent cations such as Na^+ and K^+.

Strictly speaking, the concentration terms should be the *activities*, or effective concentrations in nonideal solutions, of each species. Except in very accurate work, however, the equilibrium constant may be approximated by measuring the *concentrations* at equilibrium. For reasons beyond the scope of this discussion, equilibrium constants are dimensionless. Nonetheless, we have generally retained the concentration units (M) in the equilibrium expressions used in this book to remind you that molarity is the unit of concentration used in calculating K_{eq}.

The equilibrium constant is fixed and characteristic for any given chemical reaction at a specified temperature. It defines the composition of the final equilibrium mixture, regardless of the starting amounts of reactants and products. Conversely, we can calculate the equilibrium constant for a given reaction at a given temperature if the equilibrium concentrations of all its reactants and products are known. As we showed in Chapter 1 (p. 26), the standard free-energy change ($\Delta G°$) is directly related to ln K_{eq}.

The Ionization of Water Is Expressed by an Equilibrium Constant

The degree of ionization of water at equilibrium (Eqn 2-1) is small; at 25 °C only about two of every 10^9 molecules in pure water are ionized at any instant. The equilibrium constant for the reversible ionization of water is

$$K_{eq} = \frac{[H^+][OH^-]}{[H_2O]} \tag{2-3}$$

In pure water at 25 °C, the concentration of water is 55.5 M—grams of H_2O in 1 L divided by its gram molecular weight: (1,000 g/L)/(18.015 g/mol)—and is essentially constant in relation to the very low concentrations of H^+ and OH^-, namely 1×10^{-7} M. Accordingly, we can substitute 55.5 M in the equilibrium constant expression (Eqn 2-3) to yield

$$K_{eq} = \frac{[H^+][OH^-]}{[55.5 \text{ M}]}$$

On rearranging, this becomes

$$(55.5 \text{ M})(K_{eq}) = [H^+][OH^-] = K_w \tag{2-4}$$

where K_w designates the product $(55.5 \text{ M})(K_{eq})$, the **ion product of water** at 25 °C.

The value for K_{eq}, determined by electrical-conductivity measurements of pure water, is 1.8×10^{-16} M at 25 °C. Substituting this value for K_{eq} in Equation 2-4 gives the value of the ion product of water:

$$K_w = [H^+][OH^-] = (55.5 \text{ M})(1.8 \times 10^{-16} \text{ M})$$
$$= 1.0 \times 10^{-14} \text{ M}^2$$

Thus the product $[H^+][OH^-]$ in aqueous solutions at 25 °C always equals 1×10^{-14} M². When there are exactly equal concentrations of H^+ and OH^-, as in pure water, the solution is said to be at **neutral pH**. At this pH, the concentrations of H^+ and OH^- can be calculated from the ion product of water as follows:

$$K_w = [H^+][OH^-] = [H^+]^2 = [OH^-]^2$$

Solving for $[H^+]$ gives

$$[H^+] = \sqrt{K_w} = \sqrt{1 \times 10^{-14} \text{ M}^2}$$
$$[H^+] = [OH^-] = 10^{-7} \text{ M}$$

As the ion product of water is constant, whenever $[H^+]$ is greater than 1×10^{-7} M, $[OH^-]$ must be less than 1×10^{-7} M, and vice versa. When $[H^+]$ is very high, as in a solution of hydrochloric acid, $[OH^-]$ must be very low. From the ion product of water we can calculate $[H^+]$ if we know $[OH^-]$, and vice versa.

WORKED EXAMPLE 2-3   **Calculation of [H⁺]**

What is the concentration of H^+ in a solution of 0.1 M NaOH?

Solution: We begin with the equation for the ion product of water:

$$K_w = [H^+][OH^-]$$

With $[OH^-] = 0.1$ M, solving for $[H^+]$ gives

$$[H^+] = \frac{K_w}{[OH^-]} = \frac{1 \times 10^{-14} \text{ M}^2}{0.1 \text{ M}} = \frac{10^{-14} \text{ M}^2}{10^{-1} \text{ M}}$$
$$= 10^{-13} \text{ M}$$

WORKED EXAMPLE 2-4 Calculation of [OH⁻]

What is the concentration of OH⁻ in a solution with an H⁺ concentration of 1.3×10^{-4} M?

Solution: We begin with the equation for the ion product of water:

$$K_w = [H^+][OH^-]$$

With $[H^+] = 1.3 \times 10^{-4}$ M, solving for $[OH^-]$ gives

$$[OH^-] = \frac{K_w}{[H^+]} = \frac{1 \times 10^{-14} \text{ M}^2}{0.00013 \text{ M}} = \frac{10^{-14} \text{ M}^2}{1.3 \times 10^{-4} \text{ M}}$$

$$= 7.7 \times 10^{-11} \text{ M}$$

In all calculations be sure to round your answer to the correct number of significant figures, as here.

The pH Scale Designates the H⁺ and OH⁻ Concentrations

The ion product of water, K_w, is the basis for the **pH scale** (Table 2-6). It is a convenient means of designating the concentration of H⁺ (and thus of OH⁻) in any aqueous solution in the range between 1.0 M H⁺ and 1.0 M OH⁻. The term **pH** is defined by the expression

$$pH = \log \frac{1}{[H^+]} = -\log [H^+]$$

The symbol p denotes "negative logarithm of." For a precisely neutral solution at 25 °C, in which the concentration of hydrogen ions is 1.0×10^{-7} M, the pH can be calculated as follows:

$$pH = \log \frac{1}{1.0 \times 10^{-7}} = 7.0$$

Note that the concentration of H⁺ must be expressed in molar (M) terms.

The value of 7 for the pH of a precisely neutral solution is not an arbitrarily chosen figure; it is derived from the absolute value of the ion product of water at 25 °C, which by convenient coincidence is a round number. Solutions having a pH greater than 7 are alkaline or basic; the concentration of OH⁻ is greater than that of H⁺. Conversely, solutions having a pH less than 7 are acidic.

Keep in mind that the pH scale is logarithmic, not arithmetic. To say that two solutions differ in pH by 1 pH unit means that one solution has ten times the H⁺ concentration of the other, but it does not tell us the absolute magnitude of the difference. **Figure 2-15** gives the pH values of some common aqueous fluids. A cola drink (pH 3.0) or red wine (pH 3.7) has an H⁺ concentration approximately 10,000 times that of blood (pH 7.4).

The pH of an aqueous solution can be approximately measured with various indicator dyes, including

TABLE 2-6	The pH Scale		
[H⁺] (M)	**pH**	**[OH⁻] (M)**	**pOH[a]**
10^0 (1)	0	10^{-14}	14
10^{-1}	1	10^{-13}	13
10^{-2}	2	10^{-12}	12
10^{-3}	3	10^{-11}	11
10^{-4}	4	10^{-10}	10
10^{-5}	5	10^{-9}	9
10^{-6}	6	10^{-8}	8
10^{-7}	7	10^{-7}	7
10^{-8}	8	10^{-6}	6
10^{-9}	9	10^{-5}	5
10^{-10}	10	10^{-4}	4
10^{-11}	11	10^{-3}	3
10^{-12}	12	10^{-2}	2
10^{-13}	13	10^{-1}	1
10^{-14}	14	10^0 (1)	0

[a]The expression pOH is sometimes used to describe the basicity, or OH⁻ concentration, of a solution; pOH is defined by the expression pOH = −log [OH⁻], which is analogous to the expression for pH. Note that in all cases, pH + pOH = 14.

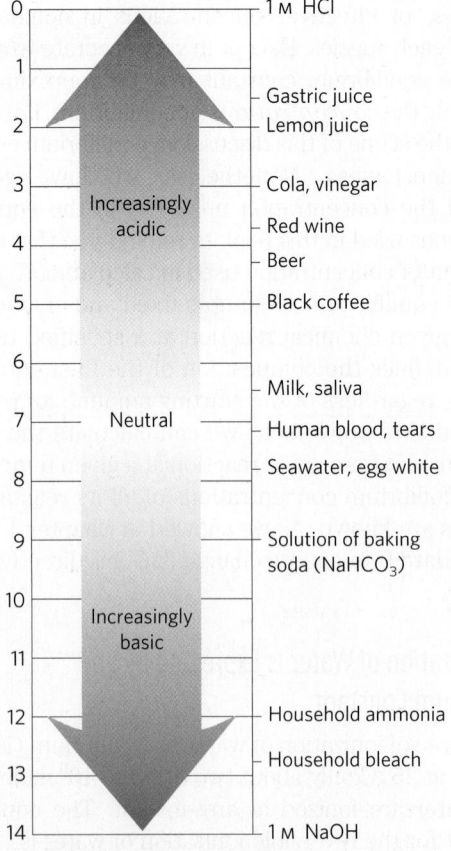

FIGURE 2-15 The pH of some aqueous fluids.

litmus, phenolphthalein, and phenol red. These dyes undergo color changes as a proton dissociates from the dye molecule. Accurate determinations of pH in the chemical or clinical laboratory are made with a glass electrode that is selectively sensitive to H^+ concentration but insensitive to Na^+, K^+, and other cations. In a pH meter, the signal from the glass electrode placed in a test solution is amplified and compared with the signal generated by a solution of accurately known pH.

Measurement of pH is one of the most important and frequently used procedures in biochemistry. The pH affects the structure and activity of biological macromolecules; for example, the catalytic activity of enzymes is strongly dependent on pH (see Fig. 2-22). Measurements of the pH of blood and urine are commonly used in medical diagnoses. The pH of the blood plasma of people with severe, uncontrolled diabetes, for example, is often below the normal value of 7.4; this condition is called **acidosis** (described in more detail below). In certain other diseases the pH of the blood is higher than normal, a condition known as **alkalosis**. Extreme acidosis or alkalosis can be life-threatening. ■

Weak Acids and Bases Have Characteristic Acid Dissociation Constants

Hydrochloric, sulfuric, and nitric acids, commonly called strong acids, are completely ionized in dilute aqueous solutions; the strong bases NaOH and KOH are also completely ionized. Of more interest to biochemists is the behavior of weak acids and bases—those not completely ionized when dissolved in water. These are ubiquitous in biological systems and play important roles in metabolism and its regulation. The behavior of aqueous solutions of weak acids and bases is best understood if we first define some terms.

Acids may be defined as proton donors and bases as proton acceptors. When a proton donor such as acetic acid (CH_3COOH) loses a proton, it becomes the corresponding proton acceptor, in this case the acetate anion (CH_3COO^-). A proton donor and its corresponding proton acceptor make up a **conjugate acid-base pair** (**Fig. 2-16**), related by the reversible reaction

$$CH_3COOH \rightleftharpoons CH_3COO^- + H^+$$

Each acid has a characteristic tendency to lose its proton in an aqueous solution. The stronger the acid,

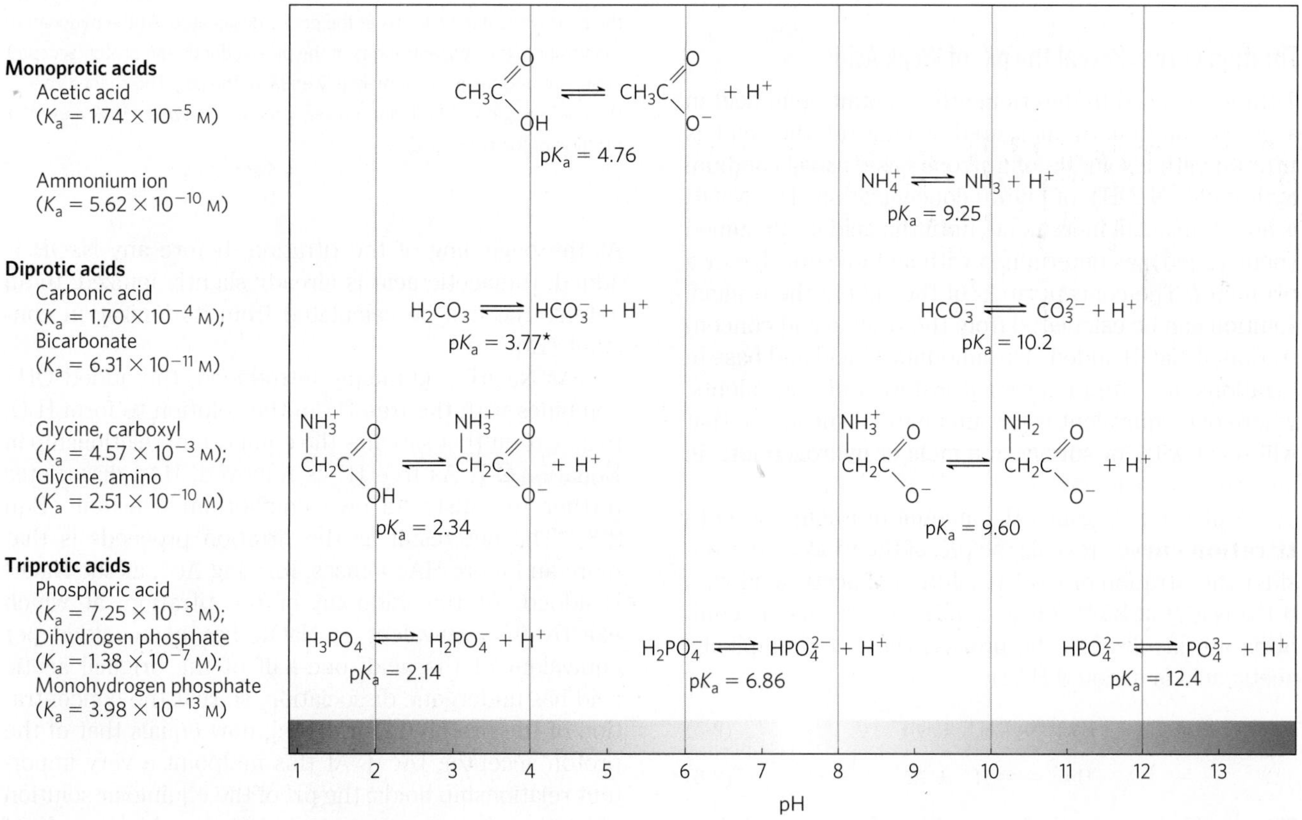

FIGURE 2-16 Conjugate acid-base pairs consist of a proton donor and a proton acceptor. Some compounds, such as acetic acid and ammonium ion, are monoprotic: they can give up only one proton. Others are diprotic (carbonic acid and glycine) or triprotic (phosphoric acid). The dissociation reactions for each pair are shown where they occur along a pH gradient. The equilibrium or dissociation constant (K_a) and its negative logarithm, the pK_a, are shown for each reaction. *For an explanation of apparent discrepancies in pK_a values for carbonic acid (H_2CO_3), see p. 67.

the greater its tendency to lose its proton. The tendency of any acid (HA) to lose a proton and form its conjugate base (A^-) is defined by the equilibrium constant (K_{eq}) for the reversible reaction

$$HA \rightleftharpoons H^+ + A^-$$

for which

$$K_{eq} = \frac{[H^+][A^-]}{[HA]} = K_a$$

Equilibrium constants for ionization reactions are usually called **ionization constants** or **acid dissociation constants**, often designated K_a. The dissociation constants of some acids are given in Figure 2-16. Stronger acids, such as phosphoric and carbonic acids, have larger ionization constants; weaker acids, such as monohydrogen phosphate (HPO_4^{2-}), have smaller ionization constants.

Also included in Figure 2-16 are values of **pK_a**, which is analogous to pH and is defined by the equation

$$pK_a = \log \frac{1}{K_a} = -\log K_a$$

The stronger the tendency to dissociate a proton, the stronger is the acid and the lower its pK_a. As we shall now see, the pK_a of any weak acid can be determined quite easily.

Titration Curves Reveal the pK_a of Weak Acids

Titration is used to determine the amount of an acid in a given solution. A measured volume of the acid is titrated with a solution of a strong base, usually sodium hydroxide (NaOH), of known concentration. The NaOH is added in small increments until the acid is consumed (neutralized), as determined with an indicator dye or a pH meter. The concentration of the acid in the original solution can be calculated from the volume and concentration of NaOH added. The amounts of acid and base in titrations are often expressed in terms of equivalents, where one equivalent is the amount of a substance that will react with, or supply, one mole of hydrogen ions in an acid-base reaction.

A plot of pH against the amount of NaOH added (a **titration curve**) reveals the pK_a of the weak acid. Consider the titration of a 0.1 M solution of acetic acid with 0.1 M NaOH at 25 °C (**Fig. 2-17**). Two reversible equilibria are involved in the process (here, for simplicity, acetic acid is denoted HAc):

$$H_2O \rightleftharpoons H^+ + OH^- \tag{2-5}$$

$$HAc \rightleftharpoons H^+ + Ac^- \tag{2-6}$$

The equilibria must simultaneously conform to their characteristic equilibrium constants, which are, respectively,

$$K_w = [H^+][OH^-] = 1 \times 10^{-14}\ M^2 \tag{2-7}$$

$$K_a = \frac{[H^+][Ac^-]}{[HAc]} = 1.74 \times 10^{-5}\ M \tag{2-8}$$

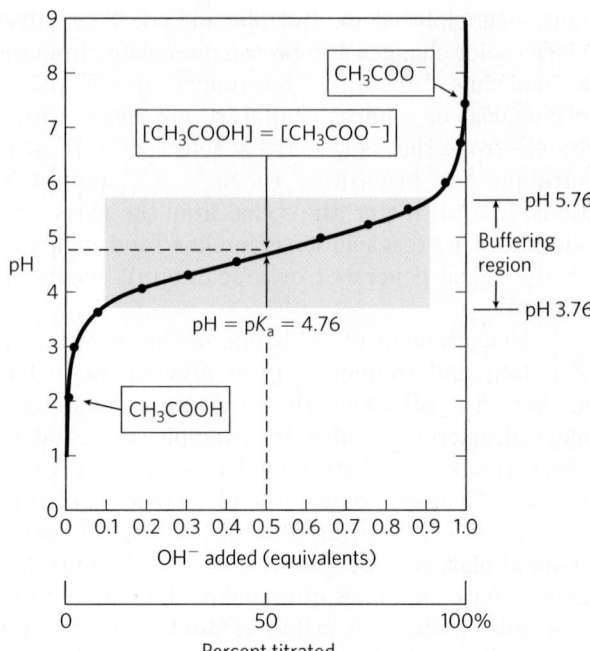

FIGURE 2-17 The titration curve of acetic acid. After addition of each increment of NaOH to the acetic acid solution, the pH of the mixture is measured. This value is plotted against the amount of NaOH added, expressed as a fraction of the total NaOH required to convert all the acetic acid (CH_3COOH) to its deprotonated form, acetate (CH_3COO^-). The points so obtained yield the titration curve. Shown in the boxes are the predominant ionic forms at the points designated. At the midpoint of the titration, the concentrations of the proton donor and proton acceptor are equal, and the pH is numerically equal to the pK_a. The shaded zone is the useful region of buffering power, generally between 10% and 90% titration of the weak acid.

At the beginning of the titration, before any NaOH is added, the acetic acid is already slightly ionized, to an extent that can be calculated from its ionization constant (Eqn 2-8).

As NaOH is gradually introduced, the added OH^- combines with the free H^+ in the solution to form H_2O, to an extent that satisfies the equilibrium relationship in Equation 2-7. As free H^+ is removed, HAc dissociates further to satisfy its own equilibrium constant (Eqn 2-8). The net result as the titration proceeds is that more and more HAc ionizes, forming Ac^-, as the NaOH is added. At the midpoint of the titration, at which exactly 0.5 equivalent of NaOH has been added per equivalent of the acid, one-half of the original acetic acid has undergone dissociation, so that the concentration of the proton donor, [HAc], now equals that of the proton acceptor, [Ac^-]. At this midpoint a very important relationship holds: the pH of the equimolar solution of acetic acid and acetate is exactly equal to the pK_a of acetic acid (pK_a = 4.76; Figs 2-16, 2-17). The basis for this relationship, which holds for all weak acids, will soon become clear.

As the titration is continued by adding further increments of NaOH, the remaining nondissociated

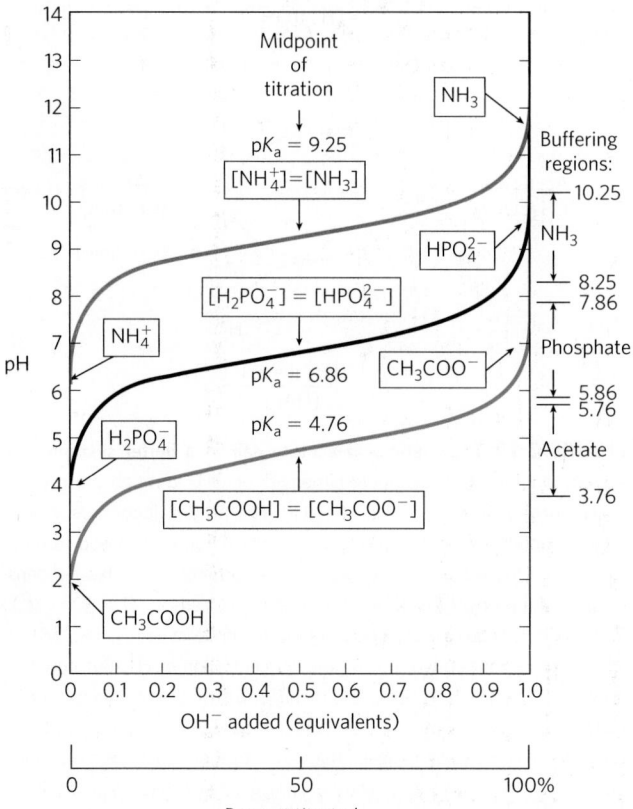

FIGURE 2-18 Comparison of the titration curves of three weak acids. Shown here are the titration curves for CH_3COOH, $H_2PO_4^-$, and NH_4^+. The predominant ionic forms at designated points in the titration are given in boxes. The regions of buffering capacity are indicated at the right. Conjugate acid-base pairs are effective buffers between approximately 10% and 90% neutralization of the proton-donor species.

SUMMARY 2.2 Ionization of Water, Weak Acids, and Weak Bases

■ Pure water ionizes slightly, forming equal numbers of hydrogen ions (hydronium ions, H_3O^+) and hydroxide ions. The extent of ionization is described by an equilibrium constant, $K_{eq} = \dfrac{[H^+][OH^-]}{[H_2O]}$, from which the ion product of water, K_w, is derived. At 25 °C, $K_w = [H^+][OH^-] = (55.5 \text{ M})(K_{eq}) = 10^{-14} \text{ M}^2$.

■ The pH of an aqueous solution reflects, on a logarithmic scale, the concentration of hydrogen ions:
$$pH = \log \frac{1}{[H^+]} = -\log [H^+]$$

■ The greater the acidity of a solution, the lower its pH. Weak acids partially ionize to release a hydrogen ion, thus lowering the pH of the aqueous solution. Weak bases accept a hydrogen ion, increasing the pH. The extent of these processes is characteristic of each particular weak acid or base and is expressed as an acid dissociation constant:
$$K_{eq} = \frac{[H^+][A^-]}{[HA]} = K_a.$$

■ The pK_a expresses, on a logarithmic scale, the relative strength of a weak acid or base:
$$pK_a = \log \frac{1}{K_a} = -\log K_a.$$

■ The stronger the acid, the smaller its pK_a; the stronger the base, the larger its pK_a. The pK_a can be determined experimentally; it is the pH at the midpoint of the titration curve for the acid or base.

2.3 Buffering against pH Changes in Biological Systems

Almost every biological process is pH-dependent; a small change in pH produces a large change in the rate of the process. This is true not only for the many reactions in which the H^+ ion is a direct participant, but also for those reactions in which there is no apparent role for H^+ ions. The enzymes that catalyze cellular reactions, and many of the molecules on which they act, contain ionizable groups with characteristic pK_a values. The protonated amino and carboxyl groups of amino acids and the phosphate groups of nucleotides, for example, function as weak acids; their ionic state is determined by the pH of the surrounding medium. (When an ionizable group is sequestered in the middle of a protein, away from the aqueous solvent, its pK_a, or apparent pK_a, can be significantly different from its pK_a in water.) As we noted above, ionic interactions are among the forces that stabilize a protein molecule and allow an enzyme to recognize and bind its substrate.

Cells and organisms maintain a specific and constant cytosolic pH, usually near pH 7, keeping biomolecules in

acetic acid is gradually converted into acetate. The end point of the titration occurs at about pH 7.0: all the acetic acid has lost its protons to OH^-, to form H_2O and acetate. Throughout the titration the two equilibria (Eqns 2-5, 2-6) coexist, each always conforming to its equilibrium constant.

Figure 2-18 compares the titration curves of three weak acids with very different ionization constants: acetic acid ($pK_a = 4.76$); dihydrogen phosphate, $H_2PO_4^-$ ($pK_a = 6.86$); and ammonium ion, NH_4^+ ($pK_a = 9.25$). Although the titration curves of these acids have the same shape, they are displaced along the pH axis because the three acids have different strengths. Acetic acid, with the highest K_a (lowest pK_a) of the three, is the strongest of the three weak acids (loses its proton most readily); it is already half dissociated at pH 4.76. Dihydrogen phosphate loses a proton less readily, being half dissociated at pH 6.86. Ammonium ion is the weakest acid of the three and does not become half dissociated until pH 9.25.

The titration curve of a weak acid shows graphically that a weak acid and its anion—a conjugate acid-base pair—can act as a buffer, as we describe in the next section.

their optimal ionic state. In multicellular organisms, the pH of extracellular fluids is also tightly regulated. Constancy of pH is achieved primarily by biological buffers: mixtures of weak acids and their conjugate bases.

Buffers Are Mixtures of Weak Acids and Their Conjugate Bases

Buffers are aqueous systems that tend to resist changes in pH when small amounts of acid (H^+) or base (OH^-) are added. A buffer system consists of a weak acid (the proton donor) and its conjugate base (the proton acceptor). As an example, a mixture of equal concentrations of acetic acid and acetate ion, found at the midpoint of the titration curve in Figure 2-17, is a buffer system. Notice that the titration curve of acetic acid has a relatively flat zone extending about 1 pH unit on either side of its midpoint pH of 4.76. In this zone, a given amount of H^+ or OH^- added to the system has much less effect on pH than the same amount added outside the zone. This relatively flat zone is the **buffering region** of the acetic acid–acetate buffer pair. At the midpoint of the buffering region, where the concentration of the proton donor (acetic acid) exactly equals that of the proton acceptor (acetate), the buffering power of the system is maximal; that is, its pH changes least on addition of H^+ or OH^-. The pH at this point in the titration curve of acetic acid is equal to its pK_a. The pH of the acetate buffer system does change slightly when a small amount of H^+ or OH^- is added, but this change is very small compared with the pH change that would result if the same amount of H^+ or OH^- were added to pure water or to a solution of the salt of a strong acid and strong base, such as NaCl, which has no buffering power.

Buffering results from two reversible reaction equilibria occurring in a solution of nearly equal concentrations of a proton donor and its conjugate proton acceptor. **Figure 2-19** explains how a buffer system works. Whenever H^+ or OH^- is added to a buffer, the result is a small change in the ratio of the relative concentrations of the weak acid and its anion and thus a small change in pH. The decrease in concentration of one component of the system is balanced exactly by an increase in the other. The sum of the buffer components does not change, only their ratio changes.

Each conjugate acid-base pair has a characteristic pH zone in which it is an effective buffer (Fig. 2-18). The $H_2PO_4^-/HPO_4^{2-}$ pair has a pK_a of 6.86 and thus can serve as an effective buffer system between approximately pH 5.9 and pH 7.9; the NH_4^+/NH_3 pair, with a pK_a of 9.25, can act as a buffer between approximately pH 8.3 and pH 10.3.

The Henderson-Hasselbalch Equation Relates pH, pK_a, and Buffer Concentration

The titration curves of acetic acid, $H_2PO_4^-$, and NH_4^+ (Fig. 2-18) have nearly identical shapes, suggesting that

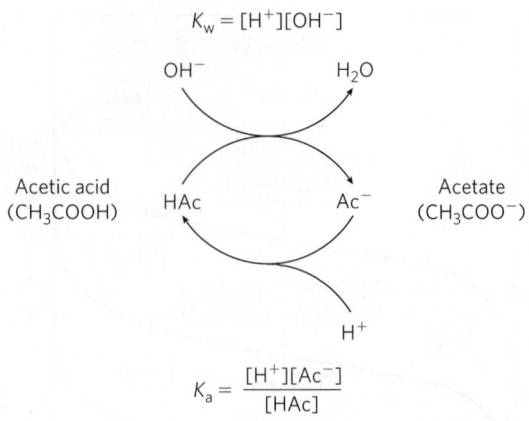

$$K_w = [H^+][OH^-]$$

$$K_a = \frac{[H^+][Ac^-]}{[HAc]}$$

FIGURE 2-19 The acetic acid–acetate pair as a buffer system. The system is capable of absorbing either H^+ or OH^- through the reversibility of the dissociation of acetic acid. The proton donor, acetic acid (HAc), contains a reserve of bound H^+, which can be released to neutralize an addition of OH^- to the system, forming H_2O. This happens because the product $[H^+][OH^-]$ transiently exceeds K_w (1×10^{-14} M^2). The equilibrium quickly adjusts to restore the product to 1×10^{-14} M^2 (at 25 °C), thus transiently reducing the concentration of H^+. But now the quotient $[H^+][Ac^-]/[HAc]$ is less than K_a, so HAc dissociates further to restore equilibrium. Similarly, the conjugate base, Ac^-, can react with H^+ ions added to the system; again, the two ionization reactions simultaneously come to equilibrium. Thus a conjugate acid-base pair, such as acetic acid and acetate ion, tends to resist a change in pH when small amounts of acid or base are added. Buffering action is simply the consequence of two reversible reactions taking place simultaneously and reaching their points of equilibrium as governed by their equilibrium constants, K_w and K_a.

these curves reflect a fundamental law or relationship. This is indeed the case. The shape of the titration curve of any weak acid is described by the Henderson-Hasselbalch equation, which is important for understanding buffer action and acid-base balance in the blood and tissues of vertebrates. This equation is simply a useful way of restating the expression for the ionization constant of an acid. For the ionization of a weak acid HA, the Henderson-Hasselbalch equation can be derived as follows:

$$K_a = \frac{[H^+][A^-]}{[HA]}$$

First solve for $[H^+]$:

$$[H^+] = K_a \frac{[HA]}{[A^-]}$$

Then take the negative logarithm of both sides:

$$-\log[H^+] = -\log K_a - \log \frac{[HA]}{[A^-]}$$

Substitute pH for $-\log[H^+]$ and pK_a for $-\log K_a$:

$$pH = pK_a - \log \frac{[HA]}{[A^-]}$$

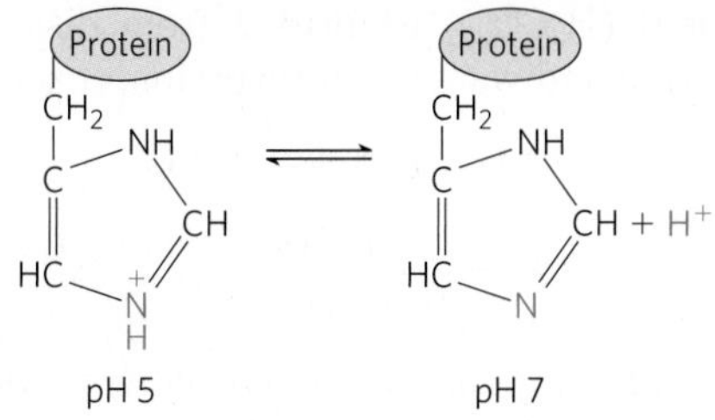

FIGURE 2-20 Ionization of histidine. The amino acid histidine, a component of proteins, is a weak acid. The pK_a of the protonated nitrogen of the side chain is 6.0.

Now invert $-\log$ [HA]/[A$^-$], which involves changing its sign, to obtain the **Henderson-Hasselbalch equation**:

$$pH = pK_a + \log \frac{[A^-]}{[HA]} \qquad (2\text{-}9)$$

This equation fits the titration curve of all weak acids and enables us to deduce some important quantitative relationships. For example, it shows why the pK_a of a weak acid is equal to the pH of the solution at the midpoint of its titration. At that point, [HA] = [A$^-$], and

$$pH = pK_a + \log 1 = pK_a + 0 = pK_a$$

The Henderson-Hasselbalch equation also allows us to (1) calculate pK_a, given pH and the molar ratio of proton donor and acceptor; (2) calculate pH, given pK_a and the molar ratio of proton donor and acceptor; and (3) calculate the molar ratio of proton donor and acceptor, given pH and pK_a.

Weak Acids or Bases Buffer Cells and Tissues against pH Changes

The intracellular and extracellular fluids of multicellular organisms have a characteristic and nearly constant pH. The organism's first line of defense against changes in internal pH is provided by buffer systems. The cytoplasm of most cells contains high concentrations of proteins, and these proteins contain many amino acids with functional groups that are weak acids or weak bases. For example, the side chain of histidine **(Fig. 2-20)** has a pK_a of 6.0 and thus can exist in either the protonated or unprotonated form near neutral pH. Proteins containing histidine residues therefore buffer effectively near neutral pH.

WORKED EXAMPLE 2-5 Ionization of Histidine

Calculate the fraction of histidine that has its imidazole side chain protonated at pH 7.3. The pK_a values for histidine are pK_1 = 1.8, pK_2 (imidazole) = 6.0, and pK_3 = 9.2 (see Fig. 3-12b).

Solution: The three ionizable groups in histidine have sufficiently different pK_a values that the first acid (—COOH) is

completely ionized before the second (protonated imidazole) begins to dissociate a proton, and the second ionizes completely before the third (—NH$_3^+$) begins to dissociate its proton. (With the Henderson-Hasselbalch equation, we can easily show that a weak acid goes from 1% ionized at 2 pH units below its pK_a to 99% ionized at 2 pH units above its pK_a; see also Fig. 3-12b.) At pH 7.3, the carboxyl group of histidine is entirely deprotonated (—COO$^-$) and the α-amino group is fully protonated (—NH$_3^+$). We can therefore assume that at pH 7.3, the only group that is partially dissociated is the imidazole group, which can be protonated (we'll abbreviate as HisH$^+$) or not (His).

We use the Henderson-Hasselbalch equation:

$$pH = pK_a + \log \frac{[A^-]}{[HA]}$$

Substituting pK_2 = 6.0 and pH = 7.3:

$$7.3 = 6.0 + \log \frac{[\text{His}]}{[\text{HisH}^+]}$$

$$1.3 = \log \frac{[\text{His}]}{[\text{HisH}^+]}$$

$$\text{antilog } 1.3 = \frac{[\text{His}]}{[\text{HisH}^+]} = 2.0 \times 10^1$$

This gives us the *ratio* of [His] to [HisH$^+$] (20 to 1 in this case). We want to convert this ratio to the *fraction* of total histidine that is in the unprotonated form (His) at pH 7.3. That fraction is 20/21 (20 parts His per 1 part HisH$^+$, *in a total of 21 parts histidine* in either form), or about 95.2%; the remainder (100% minus 95.2%) is protonated—about 5%.

Nucleotides such as ATP, as well as many metabolites of low molecular weight, contain ionizable groups that can contribute buffering power to the cytoplasm. Some highly specialized organelles and extracellular compartments have high concentrations of compounds that contribute buffering capacity: organic acids buffer the vacuoles of plant cells; ammonia buffers urine.

Two especially important biological buffers are the phosphate and bicarbonate systems. The phosphate buffer system, which acts in the cytoplasm of all cells, consists of $H_2PO_4^-$ as proton donor and HPO_4^{2-} as proton acceptor:

$$H_2PO_4^- \rightleftharpoons H^+ + HPO_4^{2-}$$

The phosphate buffer system is maximally effective at a pH close to its pK_a of 6.86 (Figs 2-16, 2-18) and thus tends to resist pH changes in the range between about 5.9 and 7.9. It is therefore an effective buffer in biological fluids; in mammals, for example, extracellular fluids and most cytoplasmic compartments have a pH in the range of 6.9 to 7.4.

WORKED EXAMPLE 2-6   Phosphate Buffers

(a) What is the pH of a mixture of 0.042 M NaH_2PO_4 and 0.058 M Na_2HPO_4?

Solution: We use the Henderson-Hasselbalch equation, which we'll express here as

$$pH = pK_a + \log \frac{[\text{conjugate base}]}{[\text{acid}]}$$

In this case, the acid (the species that gives up a proton) is $H_2PO_4^-$, and the conjugate base (the species that gains a proton) is HPO_4^{2-}. Substituting the given concentrations of acid and conjugate base and the pK_a (6.86),

$$pH = 6.86 + \log \frac{0.058}{0.042} = 6.86 + 0.14 = 7.0$$

We can roughly check this answer. When more conjugate base than acid is present, the acid is more than 50% titrated and thus the pH is above the pK_a (6.86), where the acid is exactly 50% titrated.

(b) If 1.0 mL of 10.0 M NaOH is added to a liter of the buffer prepared in (a), how much will the pH change?

Solution: A liter of the buffer contains 0.042 mol of NaH_2PO_4. Adding 1.0 mL of 10.0 M NaOH (0.010 mol) would titrate an equivalent amount (0.010 mol) of NaH_2PO_4 to Na_2HPO_4, resulting in 0.032 mol of NaH_2PO_4 and 0.068 mol of Na_2HPO_4. The new pH is

$$pH = pK_a + \log \frac{[HPO_4^{2-}]}{[H_2PO_4^-]}$$

$$= 6.86 + \log \frac{0.068}{0.032} = 6.86 + 0.33 = 7.2$$

(c) If 1.0 mL of 10.0 M NaOH is added to a liter of pure water at pH 7.0, what is the final pH? Compare this with the answer in (b).

Solution: The NaOH dissociates completely into Na^+ and OH^-, giving $[OH^-] = 0.010$ mol/L $= 1.0 \times 10^{-2}$ M. The pOH is the negative logarithm of $[OH^-]$, so pOH = 2.0. Given that in all solutions, pH + pOH = 14, the pH of the solution is 12.

So, an amount of NaOH that increases the pH of water from 7 to 12 increases the pH of a buffered solution, as in (b), from 7.0 to just 7.2. Such is the power of buffering!

Blood plasma is buffered in part by the bicarbonate system, consisting of carbonic acid (H_2CO_3) as proton donor and bicarbonate (HCO_3^-) as proton acceptor (K_1 is the first of several equilibrium constants in the bicarbonate buffering system):

$$H_2CO_3 \rightleftharpoons H^+ + HCO_3^-$$

$$K_1 = \frac{[H^+][HCO_3^-]}{[H_2CO_3]}$$

This buffer system is more complex than other conjugate acid-base pairs because one of its components, carbonic acid (H_2CO_3), is formed from dissolved (d) carbon dioxide and water, in a reversible reaction:

$$CO_2(d) + H_2O \rightleftharpoons H_2CO_3$$

$$K_2 = \frac{[H_2CO_3]}{[CO_2(d)][H_2O]}$$

Carbon dioxide is a gas under normal conditions, and CO_2 dissolved in an aqueous solution is in equilibrium with CO_2 in the gas (g) phase:

$$CO_2(g) \rightleftharpoons CO_2(d)$$

$$K_3 = \frac{[CO_2(d)]}{[CO_2(g)]}$$

The pH of a bicarbonate buffer system depends on the concentration of H_2CO_3 and HCO_3^-, the proton donor and acceptor components. The concentration of H_2CO_3 in turn depends on the concentration of dissolved CO_2, which in turn depends on the concentration of CO_2 in the gas phase, or the **partial pressure** of CO_2, denoted pCO_2. Thus the pH of a bicarbonate buffer exposed to a gas phase is ultimately determined by the concentration of HCO_3^- in the aqueous phase and by pCO_2 in the gas phase.

The bicarbonate buffer system is an effective physiological buffer near pH 7.4, because the H_2CO_3 of blood plasma is in equilibrium with a large reserve capacity of $CO_2(g)$ in the air space of the lungs. As noted above, this buffer system involves three reversible equilibria, in this case between gaseous CO_2 in the lungs and bicarbonate (HCO_3^-) in the blood plasma **(Fig. 2-21)**.

Blood can pick up H^+, such as from the lactic acid produced in muscle tissue during vigorous exercise. Alternatively, it can lose H^+, such as by protonation of the NH_3 produced during protein catabolism. When H^+ is added to blood as it passes through the tissues, reaction 1 in Figure 2-21 proceeds toward a new equilibrium, in

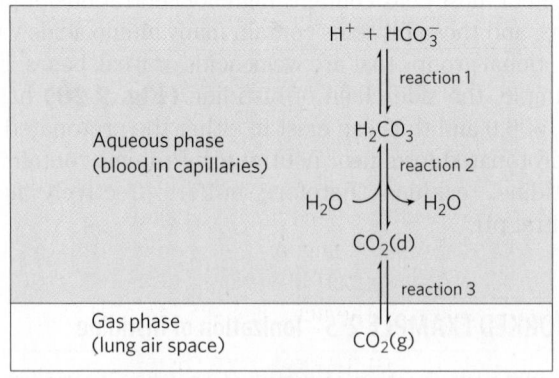

FIGURE 2-21 The bicarbonate buffer system. CO_2 in the air space of the lungs is in equilibrium with the bicarbonate buffer in the blood plasma passing through the lung capillaries. Because the concentration of dissolved CO_2 can be adjusted rapidly through changes in the rate of breathing, the bicarbonate buffer system of the blood is in near-equilibrium with a large potential reservoir of CO_2.

which $[H_2CO_3]$ is increased. This in turn increases $[CO_2(d)]$ in the blood (reaction 2) and thus increases the partial pressure of $CO_2(g)$ in the air space of the lungs (reaction 3); the extra CO_2 is exhaled. Conversely, when H^+ is lost from the blood, the opposite events occur: more H_2CO_3 dissociates into H^+ and HCO_3^- and thus more $CO_2(g)$ from the lungs dissolves in blood plasma. The rate of respiration—that is, the rate of inhaling and exhaling—can quickly adjust these equilibria to keep the blood pH nearly constant. The rate of respiration is controlled by the brain stem, where detection of an increased blood pCO_2 or decreased blood pH triggers deeper and more frequent breathing.

Hyperventilation, the rapid breathing sometimes elicited by stress or anxiety, tips the normal balance of O_2 breathed in and CO_2 breathed out in favor of too much CO_2 breathed out, raising the blood pH to 7.45 or higher. This alkalosis can lead to dizziness, headache, weakness, and fainting. One home remedy for mild alkalosis is to breathe briefly into a paper bag. The air in the bag becomes enriched in CO_2, and inhaling this air increases the CO_2 concentration in the body and blood and decreases blood pH.

At the normal pH of blood plasma (7.4), very little H_2CO_3 is present relative to HCO_3^-, and the addition of just a small amount of base (NH_3 or OH^-) would titrate this H_2CO_3, exhausting the buffering capacity. The important role of H_2CO_3 ($pK_a = 3.57$ at 37 °C) in buffering blood plasma (pH ~7.4) seems inconsistent with our earlier statement that a buffer is most effective in the range of 1 pH unit above and below its pK_a. The explanation for this apparent paradox is the large reservoir of $CO_2(d)$ in blood. Its rapid equilibration with H_2CO_3 results in the formation of additional H_2CO_3:

$$CO_2(d) + H_2O \rightleftharpoons H_2CO_3$$

It is useful in clinical medicine to have a simple expression for blood pH in terms of dissolved CO_2, which is commonly monitored along with other blood gases. We can define a constant, K_h, which is the equilibrium constant for the hydration of CO_2 to form H_2CO_3:

$$K_h = \frac{[H_2CO_3]}{[CO_2(d)]}$$

(The concentration of water is so high (55.5 M) that dissolving CO_2 doesn't change $[H_2O]$ appreciably, so $[H_2O]$ is made part of the constant K_h.) Then, to take the $CO_2(d)$ reservoir into account, we can express $[H_2CO_3]$ as $K_h[CO_2(d)]$ and substitute this expression for $[H_2CO_3]$ in the equation for the acid dissociation of H_2CO_3:

$$K_a = \frac{[H^+][HCO_3^-]}{[H_2CO_3]} = \frac{[H^+][HCO_3^-]}{K_h[CO_2(d)]}$$

Now, the overall equilibrium for dissociation of H_2CO_3 can be expressed in these terms:

$$K_h K_a = K_{combined} = \frac{[H^+][HCO_3^-]}{[CO_2(d)]}$$

We can calculate the value of the new constant, $K_{combined}$, and the corresponding apparent pK, or p$K_{combined}$, from the experimentally determined values of K_h (3.0×10^{-3} M) and K_a (2.7×10^{-4} M) at 37 °C:

$$K_{combined} = (3.0 \times 10^{-3} \text{ M})(2.7 \times 10^{-4} \text{ M})$$
$$= 8.1 \times 10^{-7} \text{ M}^2$$
$$\text{p}K_{combined} = 6.1$$

In clinical medicine, it is common to refer to $CO_2(d)$ as the conjugate acid and to use the apparent, or combined, pK_a of 6.1 to simplify calculation of pH from $[CO_2(d)]$. The concentration of dissolved CO_2 is a function of pCO_2, which in the lung is about 4.8 kilopascals (kPa), corresponding to $[H_2CO_3] \approx 1.2$ mM. Plasma $[HCO_3^-]$ is normally about 24 mM, so $[HCO_3^-]/[H_2CO_3]$ is about 20, and the blood pH is $6.1 + \log 20 \approx 7.4$. ∎

Untreated Diabetes Produces Life-Threatening Acidosis

Human blood plasma normally has a pH between 7.35 and 7.45, and many of the enzymes that function in the blood have evolved to have maximal activity in that pH range. Enzymes typically show maximal catalytic activity at a characteristic pH, called the **pH optimum (Fig. 2-22)**. On either side of this optimum pH, catalytic activity often declines sharply. Thus, a small change in pH can make a large difference in the rate of some crucial enzyme-catalyzed reactions. Biological control of the pH of cells and body fluids is therefore of central importance in all aspects of metabolism and cellular activities, and changes in blood pH have marked physiological consequences, as we know from the alarming experiments described in Box 2-1.

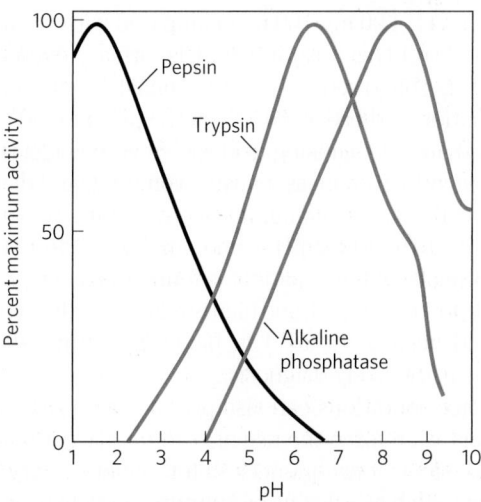

FIGURE 2-22 The pH optima of some enzymes. Pepsin is a digestive enzyme secreted into gastric juice, which has a pH of ~1.5, allowing pepsin to act optimally. Trypsin, a digestive enzyme that acts in the small intestine, has a pH optimum that matches the neutral pH in the lumen of the small intestine. Alkaline phosphatase of bone tissue is a hydrolytic enzyme thought to aid in bone mineralization.

BOX 2-1 🜊 MEDICINE On Being One's Own Rabbit (Don't Try This at Home!)

I wanted to find out what happened to a man when one made him more acid or more alkaline . . . One might, of course, have tried experiments on a rabbit first, and some work had been done along these lines; but it is difficult to be sure how a rabbit feels at any time. Indeed, some rabbits make no serious attempt to cooperate with one.

—J. B. S. Haldane, *Possible Worlds*,
Harper and Brothers, 1928

A century ago, physiologist and geneticist J. B. S. Haldane and his colleague H. W. Davies decided to experiment on themselves, to study how the body controls blood pH. They made themselves alkaline by hyperventilating and ingesting sodium bicarbonate, which left them panting and with violent headaches. They tried to acidify themselves by drinking hydrochloric acid, but calculated that it would take a gallon and a half of dilute HCl to get the desired effect, and a pint was enough to dissolve their teeth and burn their throats. Finally, it occurred to Haldane that if he ate ammonium chloride, it would break down in the body to release hydrochloric acid and ammonia. The ammonia would be converted to harmless urea in the liver. The hydrochloric acid would combine with the sodium bicarbonate present in all tissues, producing sodium chloride and carbon dioxide. In this experiment, the resulting shortness of breath mimicked that in diabetic acidosis or end-stage kidney disease.

Meanwhile, Ernst Freudenberg and Paul György, pediatricians in Heidelberg, were studying tetany—muscle contractions occurring in the hands, arms, feet, and larynx—in infants. They knew that tetany was sometimes seen in patients who had lost large amounts of hydrochloric acid by constant vomiting, and they reasoned that if tissue alkalinity produced tetany, acidity might be expected to cure it. The moment they read Haldane's paper on the effects of ammonium chloride, they tried giving ammonium chloride to babies with tetany, and were delighted to find that the tetany cleared up in a few hours. This treatment didn't remove the primary cause of the tetany, but it did give infant and physician time to deal with that cause.

In individuals with untreated diabetes mellitus, the lack of insulin, or insensitivity to insulin (depending on the type of diabetes), disrupts the uptake of glucose from blood into the tissues and forces the tissues to use stored fatty acids as their primary fuel. For reasons we describe in detail later in the book (see Fig. 23-31), this dependence on fatty acids results in the accumulation of high concentrations of two carboxylic acids, β-hydroxybutyric acid and acetoacetic acid (a combined blood plasma level of 90 mg/100 mL, compared with <3 mg/100 mL in control (healthy) individuals; urinary excretion of 5 g/24 hr, compared with <125 mg/24 hr in controls). Dissociation of these acids lowers the pH of blood plasma to less than 7.35, causing acidosis. Severe acidosis leads to headache, drowsiness, nausea, vomiting, and diarrhea, followed by stupor, coma, and convulsions, presumably because, at the lower pH, some enzyme(s) do not function optimally. When a patient is found to have high blood glucose, low plasma pH, and high levels of β-hydroxybutyric acid and acetoacetic acid in blood and urine, diabetes mellitus is the likely diagnosis.

Other conditions can also produce acidosis. Fasting and starvation force the use of stored fatty acids as fuel, with the same consequences as for diabetes. Very heavy exertion, such as a sprint by runners or cyclists, leads to temporary accumulation of lactic acid in the blood. Kidney failure results in a diminished capacity to regulate bicarbonate levels. Lung diseases (such as emphysema, pneumonia, and asthma) reduce the capacity to dispose of the CO_2 produced by fuel oxidation in the tissues, with the resulting accumulation of H_2CO_3. Acidosis is treated by dealing with the underlying condition—administering insulin to people with diabetes, and steroids or antibiotics to people with lung disease. Severe acidosis can be reversed by administering bicarbonate solution intravenously. ∎

WORKED EXAMPLE 2-7 Treatment of Acidosis with Bicarbonate

Why does intravenous administration of a bicarbonate solution raise the plasma pH?

Solution: The ratio of $[HCO_3^-]$ to $[CO_2(d)]$ determines the pH of the bicarbonate buffer, according to the equation

$$pH = 6.1 + \log \frac{[HCO_3^-]}{[H_2CO_3]}$$

where $[H_2CO_3]$ is directly related to pCO_2, the partial pressure of CO_2. So, if $[HCO_3^-]$ is increased with no change in pCO_2, the pH will rise.

SUMMARY 2.3 Buffering against pH Changes in Biological Systems

■ A mixture of a weak acid (or base) and its salt resists changes in pH caused by the addition of H^+ or OH^-. The mixture thus functions as a buffer.

■ The pH of a solution of a weak acid (or base) and its salt is given by the Henderson-Hasselbalch

equation: $pH = pK_a + \log \dfrac{[A^-]}{[HA]}$.

■ In cells and tissues, phosphate and bicarbonate buffer systems maintain intracellular and extracellular fluids at their optimum (physiological) pH, which is usually close to 7. Enzymes generally work optimally at this pH.

■ Medical conditions that lower the pH of blood, causing acidosis, or raise it, causing alkalosis, can be life threatening.

2.4 Water as a Reactant

Water is not just the solvent in which the chemical reactions of living cells occur; it is very often a direct participant in those reactions. The formation of ATP from ADP and inorganic phosphate is an example of a **condensation reaction** in which the elements of water are eliminated **(Fig. 2-23)**. The reverse of this reaction—cleavage accompanied by the addition of the elements of water—is a **hydrolysis reaction**. Hydrolysis reactions are also responsible for the enzymatic depolymerization of proteins, carbohydrates, and nucleic acids. Hydrolysis reactions, catalyzed by enzymes called **hydrolases**, are almost invariably exergonic; by producing two molecules from one, they lead to an increase in the randomness of the system. The formation of cellular polymers from their subunits by simple reversal of hydrolysis (that is, by condensation reactions) would be endergonic and therefore does not occur. As we shall see, cells circumvent this thermodynamic obstacle by coupling endergonic condensation reactions to exergonic processes, such as breakage of the anhydride bond in ATP.

You are (we hope!) consuming oxygen as you read. Water and carbon dioxide are the end products of the oxidation of fuels such as glucose. The overall reaction can be summarized as

$$\underset{\text{Glucose}}{C_6H_{12}O_6} + 6O_2 \longrightarrow 6CO_2 + 6H_2O$$

The "metabolic water" formed by oxidation of foods and stored fats is actually enough to allow some animals in very dry habitats (gerbils, kangaroo rats, camels) to survive for extended periods without drinking water.

The CO_2 produced by glucose oxidation is converted in erythrocytes to the more soluble HCO_3^-, in a reaction catalyzed by the enzyme carbonic anhydrase:

$$CO_2 + H_2O \rightleftharpoons HCO_3^- + H^+$$

In this reaction, water not only is a substrate but also functions in proton transfer by forming a network of hydrogen-bonded water molecules through which proton hopping occurs (Fig. 2-14).

Green plants and algae use the energy of sunlight to split water in the process of photosynthesis:

$$2H_2O + 2A \xrightarrow{\text{light}} O_2 + 2AH_2$$

In this reaction, A is an electron-accepting species, which varies with the type of photosynthetic organism, and water serves as the electron donor in an oxidation-reduction sequence (see Fig. 20-14) that is fundamental to all life.

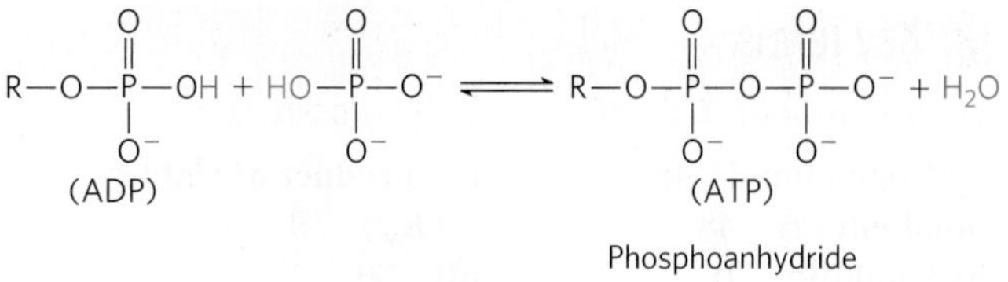

FIGURE 2-23 Participation of water in biological reactions. ATP is a phosphoanhydride formed by a condensation reaction (loss of the elements of water) between ADP and phosphate. R represents adenosine monophosphate (AMP). This condensation reaction requires energy. The hydrolysis of (addition of the elements of water to) ATP to form ADP and phosphate releases an equivalent amount of energy. These condensation and hydrolysis reactions of ATP are just one example of the role of water as a reactant in biological processes.

SUMMARY 2.4 Water as a Reactant

■ Water is both the solvent in which metabolic reactions occur and a reactant in many biochemical processes, including hydrolysis, condensation, and oxidation-reduction reactions.

2.5 The Fitness of the Aqueous Environment for Living Organisms

Organisms have effectively adapted to their aqueous environment and, in the course of evolution, have developed means of exploiting the unusual properties of water. The high specific heat of water (the heat energy required to raise the temperature of 1 g of water by 1 °C) is useful to cells and organisms because it allows water to act as a "heat buffer," keeping the temperature of an organism relatively constant as the temperature of the surroundings fluctuates and as heat is generated as a byproduct of metabolism. Furthermore, some vertebrates exploit the high heat of vaporization of water (Table 2-1) by using (thus losing) excess body heat to evaporate sweat. The high degree of internal cohesion of liquid water, due to hydrogen bonding, is exploited by plants as a means of transporting dissolved nutrients from the roots to the leaves during the process of transpiration. Even the density of ice, lower than that of liquid water, has important biological consequences in the life cycles of aquatic organisms. Ponds freeze from the top down, and the layer of ice at the top insulates the water below from frigid air, preventing the pond (and the organisms in it) from freezing solid. Most fundamental to all living organisms is the fact that many physical and biological properties of cell macromolecules, particularly the proteins and nucleic acids, derive from their interactions with water molecules in the surrounding medium. The influence of water on the course of biological evolution has been profound and determinative. If life forms have evolved elsewhere in the universe, they are unlikely to resemble those of Earth unless liquid water is plentiful in their planet of origin.

Key Terms

Terms in bold are defined in the glossary.

hydrogen bond 48
bond energy 48
hydrophilic 50
hydrophobic 50
amphipathic 52
micelle 52
hydrophobic effect 52
hydrophobic
  interactions 53
van der Waals
  interactions 53
osmolarity 55
osmosis 56
isotonic 56
hypertonic 56
hypotonic 56
equilibrium constant
  (K_{eq}) 58

ion product of water
  (K_w) 59
pH 60
acidosis 61
alkalosis 61
conjugate acid-base
  pair 61
acid dissociation constant
  (K_a) 62
pK_a 62
titration curve 62
buffer 64
buffering region 64
Henderson-Hasselbalch
  equation 65
condensation 69
hydrolysis 69

Problems

1. Effect of Local Environment on Ionic Bond Strength
If the ATP-binding site of an enzyme is buried in the interior of
the enzyme, in a hydrophobic environment, is the ionic interaction between enzyme and substrate stronger or weaker than
that same interaction would be on the surface of the enzyme,
exposed to water? Why?

2. Biological Advantage of Weak Interactions The interactions between biomolecules are often stabilized by weak
interactions such as hydrogen bonds. How might this be an
advantage to the organism?

3. Solubility of Ethanol in Water Explain why ethanol
(CH_3CH_2OH) is more soluble in water than is ethane (CH_3CH_3).

4. Calculation of pH from Hydrogen Ion Concentration
What is the pH of a solution that has an H^+ concentration of
(a) 1.75×10^{-5} mol/L; (b) 6.50×10^{-10} mol/L; (c) 1.0×10^{-4} mol/L;
(d) 1.50×10^{-5} mol/L?

**5. Calculation of Hydrogen Ion Concentration from
pH** What is the H^+ concentration of a solution with pH of
(a) 3.82; (b) 6.52; (c) 11.11?

6. Acidity of Gastric HCl In a hospital laboratory, a
10.0 mL sample of gastric juice, obtained several hours
after a meal, was titrated with 0.1 M NaOH to neutrality; 7.2 mL
of NaOH was required. The patient's stomach contained no
ingested food or drink; thus assume that no buffers were present. What was the pH of the gastric juice?

7. Calculation of the pH of a Strong Acid or Base
(a) Write out the acid dissociation reaction for hydrochloric
acid. (b) Calculate the pH of a solution of 5.0×10^{-4} M HCl. (c)
Write out the acid dissociation reaction for sodium hydroxide.
(d) Calculate the pH of a solution of 7.0×10^{-5} M NaOH.

8. Calculation of pH from Concentration of Strong Acid
Calculate the pH of a solution prepared by diluting 3.0 mL of
2.5 M HCl to a final volume of 100 mL with H_2O.

9. Measurement of Acetylcholine Levels by pH Changes
The concentration of acetylcholine (a neurotransmitter) in a
sample can be determined from the pH changes that accompany its hydrolysis. When the sample is incubated with the
enzyme acetylcholinesterase, acetylcholine is converted to
choline and acetic acid, which dissociates to yield acetate and
a hydrogen ion:

Acetylcholine

Choline   Acetate

In a typical analysis, 15 mL of an aqueous solution containing an unknown amount of acetylcholine had a pH of 7.65.
When incubated with acetylcholinesterase, the pH of the solution decreased to 6.87. Assuming there was no buffer in the
assay mixture, determine the number of moles of acetylcholine in the 15 mL sample.

10. Physical Meaning of pK_a Which of the following aqueous solutions has the lowest pH: 0.1 M HCl; 0.1 M acetic acid
(pK_a = 4.86); 0.1 M formic acid (pK_a = 3.75)?

11. Meanings of K_a and pK_a (a) Does a strong acid have a
greater or lesser tendency to lose its proton than a weak acid?
(b) Does the strong acid have a higher or lower K_a than the
weak acid? (c) Does the strong acid have a higher or lower pK_a
than the weak acid?

12. Simulated Vinegar One way to make vinegar (*not* the
preferred way) is to prepare a solution of acetic acid, the sole
acid component of vinegar, at the proper pH (see Fig. 2-15)
and add appropriate flavoring agents. Acetic acid (M_r 60) is a
liquid at 25 °C, with a density of 1.049 g/mL. Calculate the
volume that must be added to distilled water to make 1 L of
simulated vinegar (see Fig. 2-16).

13. Identifying the Conjugate Base Which is the conjugate base in each of the pairs below?
  (a) RCOOH, $RCOO^-$ (c) $H_2PO_4^-$, H_3PO_4
  (b) RNH_2, RNH_3^+ (d) H_2CO_3, HCO_3^-

**14. Calculation of the pH of a Mixture of a Weak Acid
and Its Conjugate Base** Calculate the pH of a dilute solution that contains a molar ratio of potassium acetate to acetic
acid (pK_a = 4.76) of (a) 2:1; (b) 1:3; (c) 5:1; (d) 1:1; (e) 1:10.

15. Effect of pH on Solubility The strongly polar, hydrogen-
bonding properties of water make it an excellent solvent for
ionic (charged) species. By contrast, nonionized, nonpolar
organic molecules, such as benzene, are relatively insoluble in
water. In principle, the aqueous solubility of any organic acid or

base can be increased by converting the molecules to charged species. For example, the solubility of benzoic acid in water is low. The addition of sodium bicarbonate to a mixture of water and benzoic acid raises the pH and deprotonates the benzoic acid to form benzoate ion, which is quite soluble in water.

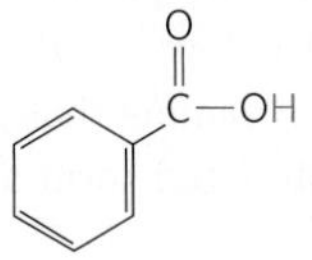

Benzoic acid
$pK_a \approx 5$

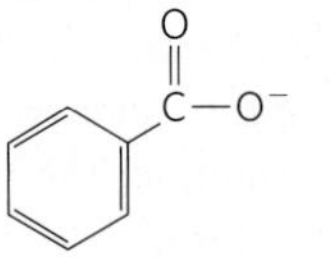

Benzoate ion

Are the following compounds more soluble in an aqueous solution of 0.1 M NaOH or 0.1 M HCl? (The dissociable protons are shown in red.)

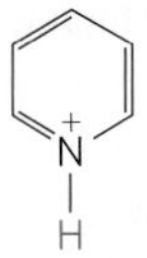

Pyridine ion
$pK_a \approx 5$

(a)

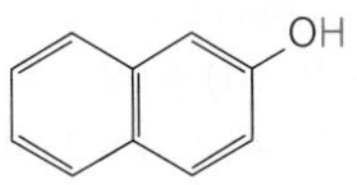

β-Naphthol
$pK_a \approx 10$

(b)

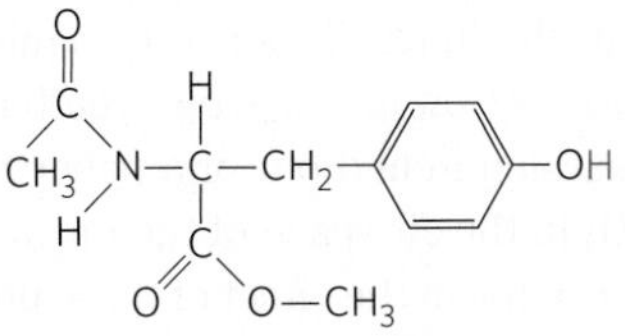

N-Acetyltyrosine methyl ester
$pK_a \approx 10$

(c)

16. Treatment of Poison Ivy Rash The components of poison ivy and poison oak that produce the characteristic itchy rash are catechols substituted with long-chain alkyl groups.

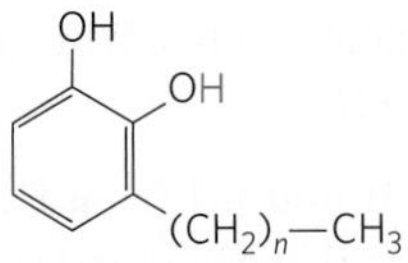

$(CH_2)_n-CH_3$
$pK_a \approx 8$

If you were exposed to poison ivy, which of the treatments below would you apply to the affected area? Justify your choice.

(a) Wash the area with cold water.
(b) Wash the area with dilute vinegar or lemon juice.
(c) Wash the area with soap and water.
(d) Wash the area with soap, water, and baking soda (sodium bicarbonate).

17. pH and Drug Absorption Aspirin is a weak acid with a pK_a of 3.5 (the ionizable H is shown in red):

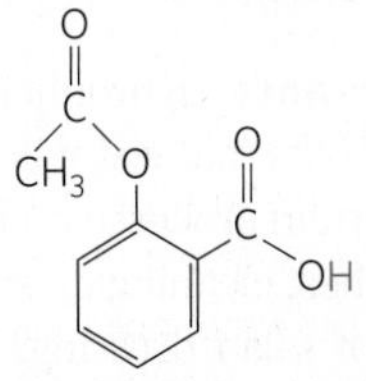

It is absorbed into the blood through the cells lining the stomach and the small intestine. Absorption requires passage through the plasma membrane, the rate of which is determined by the polarity of the molecule: charged and highly polar molecules pass slowly, whereas neutral hydrophobic ones pass rapidly. The pH of the stomach contents is about 1.5, and the pH of the contents of the small intestine is about 6. Is more aspirin absorbed into the bloodstream from the stomach or from the small intestine? Clearly justify your choice.

18. Calculation of pH from Molar Concentrations What is the pH of a solution containing 0.12 mol/L of NH_4Cl and 0.03 mol/L of NaOH (pK_a of NH_4^+/NH_3 is 9.25)?

19. Calculation of pH after Titration of Weak Acid A compound has a pK_a of 7.4. To 100 mL of a 1.0 M solution of this compound at pH 8.0 is added 30 mL of 1.0 M hydrochloric acid. What is the pH of the resulting solution?

20. Properties of a Buffer The amino acid glycine is often used as the main ingredient of a buffer in biochemical experiments. The amino group of glycine, which has a pK_a of 9.6, can exist either in the protonated form ($-NH_3^+$) or as the free base ($-NH_2$), because of the reversible equilibrium

$$R-NH_3^+ \rightleftharpoons R-NH_2 + H^+$$

(a) In what pH range can glycine be used as an effective buffer due to its amino group?
(b) In a 0.1 M solution of glycine at pH 9.0, what fraction of glycine has its amino group in the $-NH_3^+$ form?
(c) How much 5 M KOH must be added to 1.0 L of 0.1 M glycine at pH 9.0 to bring its pH to exactly 10.0?
(d) When 99% of the glycine is in its $-NH_3^+$ form, what is the numerical relation between the pH of the solution and the pK_a of the amino group?

21. Calculation of the pK_a of an Ionizable Group by Titration The pK_a values of a compound with two ionizable groups are $pK_1 = 4.10$ and pK_2 between 7 and 10. A biochemist has 10 mL of a 1.0 M solution of this compound at a pH of 8.00. She adds 10.0 mL of 1.00 M HCl, which changes the pH to 3.20. What is pK_2?

22. Calculation of the pH of a Solution of a Polyprotic Acid Histidine has ionizable groups with pK_a values of 1.8, 6.0, and 9.2, as shown below (His = imidazole group). A biochemist makes up 100 mL of a 0.100 M solution of histidine at a pH of 5.40. She then adds 40 mL of 0.10 M HCl. What is the pH of the resulting solution?

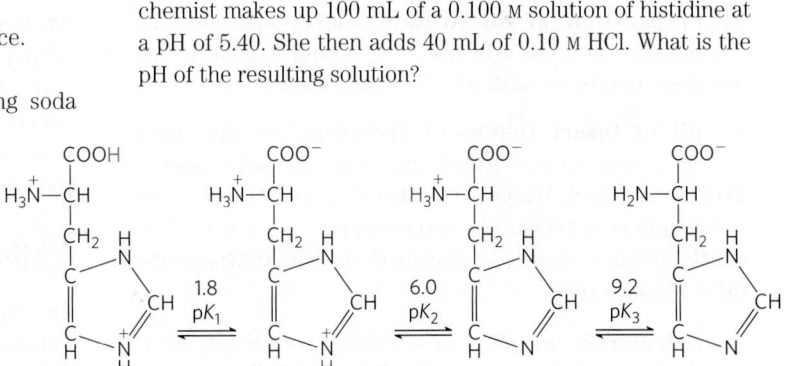

Ionizable group $-COOH \rightleftharpoons -COO^-$
$-HisH^+ \rightleftharpoons -His$
$-NH_3^+ \rightleftharpoons -NH_2$

23. Calculation of Original pH from Final pH after Titration A biochemist has 100 mL of a 0.10 M solution of a weak acid with a pK_a of 6.3. She adds 6.0 mL of 1.0 M HCl, which changes the pH to 5.7. What was the pH of the original solution?

24. Preparation of a Phosphate Buffer What molar ratio of HPO_4^{2-} to $H_2PO_4^-$ in solution would produce a pH of 7.0? Phosphoric acid (H_3PO_4), a triprotic acid, has three pK_a values: 2.14, 6.86, and 12.4. Hint: Only one of the pK_a values is relevant here.

25. Preparation of Standard Buffer for Calibration of a pH Meter The glass electrode used in commercial pH meters gives an electrical response proportional to the concentration of hydrogen ion. To convert these responses to a pH reading, the electrode must be calibrated against standard solutions of known H^+ concentration. Determine the weight in grams of sodium dihydrogen phosphate ($NaH_2PO_4 \cdot H_2O$; FW 138) and disodium hydrogen phosphate (Na_2HPO_4; FW 142) needed to prepare 1 L of a standard buffer at pH 7.00 with a total phosphate concentration of 0.100 M (see Fig. 2-16). See Problem 24 for the pK_a values of phosphoric acid.

26. Calculation of Molar Ratios of Conjugate Base to Weak Acid from pH For a weak acid with a pK_a of 6.0, calculate the ratio of conjugate base to acid at a pH of 5.0.

27. Preparation of Buffer of Known pH and Strength Given 0.10 M solutions of acetic acid (pK_a = 4.76) and sodium acetate, describe how you would go about preparing 1.0 L of 0.10 M acetate buffer of pH 4.00.

28. Choice of Weak Acid for a Buffer Which of these compounds would be the best buffer at pH 5.0: formic acid (pK_a = 3.8), acetic acid (pK_a = 4.76), or ethylamine (pK_a = 9.0)? Briefly justify your answer.

29. Working with Buffers A buffer contains 0.010 mol of lactic acid (pK_a = 3.86) and 0.050 mol of sodium lactate per liter. (a) Calculate the pH of the buffer. (b) Calculate the change in pH when 5 mL of 0.5 M HCl is added to 1 L of the buffer. (c) What pH change would you expect if you added the same quantity of HCl to 1 L of pure water?

30. Use of Molar Concentrations to Calculate pH What is the pH of a solution that contains 0.20 M sodium acetate and 0.60 M acetic acid (pK_a = 4.76)?

31. Preparation of an Acetate Buffer Calculate the cocentrations of acetic acid (pK_a = 4.76) and sodium acetate necessary to prepare a 0.2 M buffer solution at pH 5.0.

32. pH of Insect Defensive Secretion You have been observing an insect that defends itself from enemies by secreting a caustic liquid. Analysis of the liquid shows it to have a total concentration of formate plus formic acid (K_a = 1.8 × 10^{-4}) of 1.45 M; the concentration of formate ion is 0.015 M. What is the pH of the secretion?

33. Calculation of pK_a An unknown compound, X, is thought to have a carboxyl group with a pK_a of 2.0 and another ionizable group with a pK_a between 5 and 8. When 75 mL of 0.1 M NaOH is added to 100 mL of a 0.1 M solution of X at pH 2.0, the

pH increases to 6.72. Calculate the pK_a of the second ionizable group of X.

34. Ionic Forms of Alanine Alanine is a diprotic acid that can undergo two dissociation reactions (see Table 3-1 for pK_a values). (a) Given the structure of the partially protonated form (or zwitterion; see Fig. 3-9) below, draw the chemical structures of the other two forms of alanine that predominate in aqueous solution: the fully protonated form and the fully deprotonated form.

Alanine

Of the three possible forms of alanine, which would be present at the highest concentration in solutions of the following pH: (b) 1.0; (c) 6.2; (d) 8.02; (e) 11.9. Explain your answers in terms of pH relative to the two pK_a values.

35. Control of Blood pH by Respiratory Rate

(a) The partial pressure of CO_2 in the lungs can be varied rapidly by the rate and depth of breathing. For example, a common remedy to alleviate hiccups is to increase the concentration of CO_2 in the lungs. This can be achieved by holding one's breath, by very slow and shallow breathing (hypoventilation), or by breathing in and out of a paper bag. Under such conditions, pCO_2 in the air space of the lungs rises above normal. Qualitatively explain the effect of these procedures on the blood pH.

(b) A common practice of competitive short-distance runners is to breathe rapidly and deeply (hyperventilate) for about half a minute to remove CO_2 from their lungs just before the race begins. Blood pH may rise to 7.60. Explain why the blood pH increases.

(c) During a short-distance run, the muscles produce a large amount of lactic acid ($CH_3CH(OH)COOH$; K_a = 1.38 × 10^{-4} M) from their glucose stores. Why might hyperventilation before a dash be useful?

36. Calculation of Blood pH from CO_2 and Bicarbonate Levels Calculate the pH of a blood plasma sample with a total CO_2 concentration of 26.9 mM and bicarbonate concentration of 25.6 mM. Recall from page 67 that the relevant pK_a of carbonic acid is 6.1.

37. Effect of Holding One's Breath on Blood pH The pH of the extracellular fluid is buffered by the bicarbonate/carbonic acid system. Holding your breath can increase the concentration of $CO_2(g)$ in the blood. What effect might this have on the pH of the extracellular fluid? Explain by showing the relevant equilibrium equation(s) for this buffer system.

Data Analysis Problem

38. "Switchable" Surfactants Hydrophobic molecules do not dissolve well in water. Given that water is a very commonly used solvent, this makes certain processes very difficult: washing oily food residue off dishes, cleaning up spilled oil, keeping the oil and water phases of salad dressings well mixed, and carrying out chemical reactions that involve both hydrophobic and hydrophilic components.

Surfactants are a class of amphipathic compounds that includes soaps, detergents, and emulsifiers. With the use of surfactants, hydrophobic compounds can be suspended in aqueous solution by forming micelles (see Fig. 2-7). A micelle has a hydrophobic core consisting of the hydrophobic compound and the hydrophobic "tails" of the surfactant; the hydrophilic "heads" of the surfactant cover the surface of the micelle. A suspension of micelles is called an emulsion. The more hydrophilic the head group of the surfactant, the more powerful it is—that is, the greater its capacity to emulsify hydrophobic material.

When you use soap to remove grease from dirty dishes, the soap forms an emulsion with the grease that is easily removed by water through interaction with the hydrophilic head of the soap molecules. Likewise, a detergent can be used to emulsify spilled oil for removal by water. And emulsifiers in commercial salad dressings keep the oil suspended evenly throughout the water-based mixture.

There are some situations in which it would be very useful to have a "switchable" surfactant: a molecule that could be reversibly converted between a surfactant and a nonsurfactant.

(a) Imagine such a "switchable" surfactant existed. How would you use it to clean up and then recover the oil from an oil spill?

Liu et al. describe a prototypical switchable surfactant in their 2006 article "Switchable Surfactants." The switching is based on the following reaction:

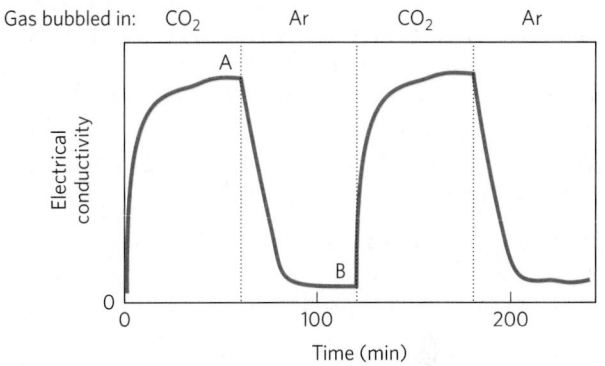

Amidine form   Amidinium form

(b) Given that the pK_a of a typical amidinium ion is 12.4, in which direction (left or right) would you expect the equilibrium of the above reaction to lie? (See Fig. 2-16 for relevant pK_a values.) Justify your answer. Hint: Remember the reaction $H_2O + CO_2 \rightleftharpoons H_2CO_3$.

Liu and colleagues produced a switchable surfactant for which $R = C_{16}H_{33}$. They do not name the molecule in their article; for brevity, we'll call it s-surf.

(c) The amidinium form of s-surf is a powerful surfactant; the amidine form is not. Explain this observation.

Liu and colleagues found that they could switch between the two forms of s-surf by changing the gas that they bubbled through a solution of the surfactant. They demonstrated this switch by measuring the electrical conductivity of the s-surf solution; aqueous solutions of ionic compounds have higher conductivity than solutions of nonionic compounds. They started with a solution of the amidine form of s-surf in water. Their results are shown below; dotted lines indicate the switch from one gas to another.

(d) In which form is the majority of s-surf at point A? At point B?

(e) Why does the electrical conductivity rise from time 0 to point A?

(f) Why does the electrical conductivity fall from point A to point B?

(g) Explain how you would use s-surf to clean up and recover the oil from an oil spill.

Reference

Y. Liu, P.G. Jessop, M. Cunningham, C.A. Eckert, and C.L. Liotta. 2006. Switchable surfactants. *Science* 313:958–960.

Further Reading is available at www.macmillanlearning.com/LehningerBiochemistry7e.

CHAPTER 3

Amino Acids, Peptides, and Proteins

Self-study tools that will help you practice what you've learned and reinforce this chapter's concepts are available online. Go to www.macmillanlearning.com/LehningerBiochemistry7e.

Proteins mediate virtually every process that takes place in a cell, exhibiting an almost endless diversity of functions. To explore the molecular mechanism of a biological process, a biochemist almost inevitably studies one or more proteins. Proteins are the most abundant biological macromolecules, occurring in all cells and all parts of cells. Proteins also occur in great variety; thousands of different kinds may be found in a single cell. As the arbiters of molecular function, proteins are the most important final products of the information pathways discussed in Part III of this book. Proteins are the molecular instruments through which genetic information is expressed.

Relatively simple monomeric subunits provide the key to the structure of the thousands of different proteins. The proteins of every organism, from the simplest of bacteria to human beings, are constructed from the same set of 20 amino acids. Because each of these amino acids has a side chain with distinctive chemical properties, this group of 20 precursor molecules may be regarded as the alphabet in which the language of protein structure is written.

To generate a particular protein, amino acids are covalently linked in a characteristic linear sequence. What is most remarkable is that cells can produce proteins with strikingly different properties and activities by joining the same 20 amino acids in many different combinations and sequences. From these building blocks, different organisms can make such widely diverse products as enzymes, hormones, antibodies, transporters, muscle fibers, the lens protein of the eye, feathers, spider webs, rhinoceros horn, milk proteins, antibiotics, mushroom poisons, and myriad other substances having distinct biological activities (Fig. 3-1). Among these protein products, the enzymes are the most varied and specialized. As the catalysts of almost all cellular reactions, enzymes are one of the keys to understanding the chemistry of life and thus provide a focal point for any course in biochemistry.

Protein structure and function are the topics of this and the next three chapters. Here, we begin with a description of the fundamental chemical properties of amino acids, peptides, and proteins. We also consider how a biochemist works with proteins.

3.1 Amino Acids

Proteins are polymers of amino acids, with each **amino acid residue** joined to its neighbor by a specific type of covalent bond. (The term "residue" reflects the loss of the elements of water when one amino acid is joined to another.) Proteins can be broken down (hydrolyzed) to their constituent amino acids by a variety of methods, and the earliest studies of proteins naturally focused on the free amino acids derived from them. Twenty different amino acids are commonly found in proteins. The first to be discovered was asparagine, in 1806. The last of the 20 to be found, threonine, was not identified until 1938. All the amino acids have trivial or common names, in some cases derived from the source from which they were first isolated. Asparagine was first found in asparagus, and glutamate in wheat gluten; tyrosine was first isolated from cheese (its name is derived from the

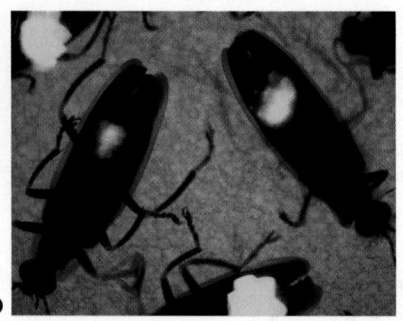

(a) **(b)** **(c)**

FIGURE 3-1 **Some functions of proteins. (a)** The light produced by fireflies is the result of a reaction involving the protein luciferin and ATP, catalyzed by the enzyme luciferase (see Box 13-1). **(b)** Erythrocytes contain large amounts of the oxygen-transporting protein hemoglobin. **(c)** The protein keratin, formed by all vertebrates, is the chief structural component of hair, scales, horn, wool, nails, and feathers. The black rhinoc-eros is extinct in the wild because of the belief prevalent in some parts of the world that a powder derived from its horn has aphrodisiac properties. In reality, the chemical properties of powdered rhinoceros horn are no different from those of powdered bovine hooves or human fingernails. [Sources: (a) Jeff J. Daly/Alamy. (b) Bill Longcore/Science Source. (c) Mary Cooke/Animals Animals.]

Greek *tyros*, "cheese"); and glycine (Greek *glykos*, "sweet") was so named because of its sweet taste.

Amino Acids Share Common Structural Features

All 20 of the common amino acids are α-amino acids. They have a carboxyl group and an amino group bonded to the same carbon atom (the α carbon) **(Fig. 3-2)**. They differ from each other in their side chains, or **R groups**, which vary in structure, size, and electric charge, and which influence the solubility of the amino acids in water. In addition to these 20 amino acids there are many less common ones. Some are residues modified after a protein has been synthesized, others are amino acids present in living organisms but not as constituents of proteins, and two are special cases found in just a few proteins. The common amino acids of proteins have been assigned three-letter abbreviations and one-letter symbols (Table 3-1), which are used as shorthand to indicate the composition and sequence of amino acids polymerized in proteins.

⟩⟩ Key Convention: The three-letter code is transparent, the abbreviations generally consisting of the first three letters of the amino acid name. The one-letter code was devised by Margaret Oakley Dayhoff, considered by many to be the founder of the field of bioinformatics. The one-letter code reflects an attempt to reduce the size of the data files (in an era of punch-card computing) used to describe amino acid sequences. It was designed to be easily memorized, and understanding its origin can help students do just that. For six amino acids (CHIMSV), the first letter of the amino acid name is unique and thus is used as the symbol. For five others (AGLPT), the

Margaret Oakley Dayhoff, 1925–1983

first letter of the name is not unique but is assigned to the amino acid that is most common in proteins (for example, leucine is more common than lysine). For another four, the letter used is phonetically suggestive (RFYW: aRginine, Fenylalanine, tYrosine, tWiptophan). The rest were harder to assign. Four (DNEQ) were assigned letters found within or suggested by their names (asparDic, asparagiNe, glutamEke, Q-tamine). That left lysine. Only a few letters were left, and K was chosen because it was the closest to L. **⟨⟨**

For all the common amino acids except glycine, the α carbon is bonded to four different groups: a carboxyl group, an amino group, an R group, and a hydrogen atom (Fig. 3-2; in glycine, the R group is another hydrogen atom). The α-carbon atom is thus a **chiral center** (p. 17). Because of the tetrahedral arrangement of the bonding orbitals around the α-carbon atom, the four different groups can occupy two unique spatial arrangements, and thus amino acids have two possible stereoisomers. Since they are nonsuperposable mirror images of each other **(Fig. 3-3)**, the two forms represent a class of stereoisomers called **enantiomers** (see Fig. 1-21). All molecules with a chiral center are also **optically active**—that is, they rotate the plane of plane-polarized light (see Box 1-2).

⟩⟩ Key Convention: Two conventions are used to identify the carbons in an amino acid—a practice that can be confusing. The additional carbons in an R group are commonly designated β, γ, δ, ε, and so forth, proceeding out from the α carbon. For most other organic molecules, carbon atoms are simply numbered from one end, giving highest priority (C-1) to the carbon with the substituent containing the atom of highest atomic number. Within this latter convention, the carboxyl carbon

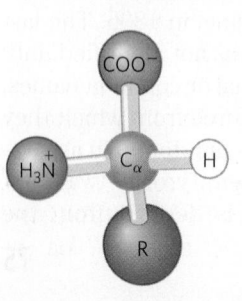

FIGURE 3-2 **General structure of an amino acid.** This structure is common to all but one of the α-amino acids. (Proline, a cyclic amino acid, is the exception.) The R group, or side chain (purple), attached to the α carbon (gray) is different in each amino acid.

TABLE 3-1 Properties and Conventions Associated with the Common Amino Acids Found in Proteins

Amino acid	Abbreviation/ symbol	M_r[a]	pK_a values pK_1 (—COOH)	pK_2 (—NH_3^+)	pK_R (R group)	pI	Hydropathy index[b]	Occurrence in proteins (%)[c]
Nonpolar, aliphatic R groups								
Glycine	Gly G	75	2.34	9.60		5.97	−0.4	7.2
Alanine	Ala A	89	2.34	9.69		6.01	1.8	7.8
Proline	Pro P	115	1.99	10.96		6.48	−1.6[d]	5.2
Valine	Val V	117	2.32	9.62		5.97	4.2	6.6
Leucine	Leu L	131	2.36	9.60		5.98	3.8	9.1
Isoleucine	Ile I	131	2.36	9.68		6.02	4.5	5.3
Methionine	Met M	149	2.28	9.21		5.74	1.9	2.3
Aromatic R groups								
Phenylalanine	Phe F	165	1.83	9.13		5.48	2.8	3.9
Tyrosine	Tyr Y	181	2.20	9.11	10.07	5.66	−1.3	3.2
Tryptophan	Trp W	204	2.38	9.39		5.89	−0.9	1.4
Polar, uncharged R groups								
Serine	Ser S	105	2.21	9.15		5.68	−0.8	6.8
Threonine	Thr T	119	2.11	9.62		5.87	−0.7	5.9
Cysteine[e]	Cys C	121	1.96	10.28	8.18	5.07	2.5	1.9
Asparagine	Asn N	132	2.02	8.80		5.41	−3.5	4.3
Glutamine	Gln Q	146	2.17	9.13		5.65	−3.5	4.2
Positively charged R groups								
Lysine	Lys K	146	2.18	8.95	10.53	9.74	−3.9	5.9
Histidine	His H	155	1.82	9.17	6.00	7.59	−3.2	2.3
Arginine	Arg R	174	2.17	9.04	12.48	10.76	−4.5	5.1
Negatively charged R groups								
Aspartate	Asp D	133	1.88	9.60	3.65	2.77	−3.5	5.3
Glutamate	Glu E	147	2.19	9.67	4.25	3.22	−3.5	6.3

[a]M_r values reflect the structures as shown in Figure 3-5. The elements of water (M_r 18) are deleted when the amino acid is incorporated into a polypeptide.

[b]A scale combining hydrophobicity and hydrophilicity of R groups. The values reflect the free energy (ΔG) of transfer of the amino acid side chain from a hydrophobic solvent to water. This transfer is favorable ($\Delta G < 0$; negative value in the index) for charged or polar amino acid side chains, and unfavorable ($\Delta G > 0$; positive value in the index) for amino acids with nonpolar or more hydrophobic side chains. See Chapter 11. Source: J. Kyte and R. F. Doolittle, *J. Mol. Biol.* 157:105, 1982.

[c]Average occurrence in more than 1,150 proteins. Source: R. F. Doolittle, in *Prediction of Protein Structure and the Principles of Protein Conformation* (G. D. Fasman, ed.), p. 599, Plenum Press, 1989.

[d]As originally composed, the hydropathy index takes into account the frequency with which an amino acid residue appears on the surface of a protein. As proline often appears on the surface in β turns, it has a lower score than its chain of methylene groups would suggest.

[e]Cysteine is generally classified as polar despite having a positive hydropathy index. This reflects the ability of the sulfhydryl group to act as a weak acid and to form a weak hydrogen bond with oxygen or nitrogen.

of an amino acid would be C-1 and the α carbon would be C-2.

$$^-OOC-\underset{\substack{|\\ {}^+NH_3}}{\overset{\overset{\alpha}{2}}{CH}}-\underset{}{\overset{\overset{\beta}{3}}{CH_2}}-\underset{}{\overset{\overset{\gamma}{4}}{CH_2}}-\underset{}{\overset{\overset{\delta}{5}}{CH_2}}-\underset{\substack{|\\ {}^+NH_3}}{\overset{\overset{\epsilon}{6}}{CH_2}}$$

Lysine

In some cases, such as amino acids with heterocyclic R groups (such as histidine), the Greek lettering system is ambiguous and the numbering convention is therefore used. For branched amino acid side chains, equivalent carbons are given numbers after the Greek letters. Leucine thus has δ1 and δ2 carbons (see the structure in Fig. 3-5). **≪**

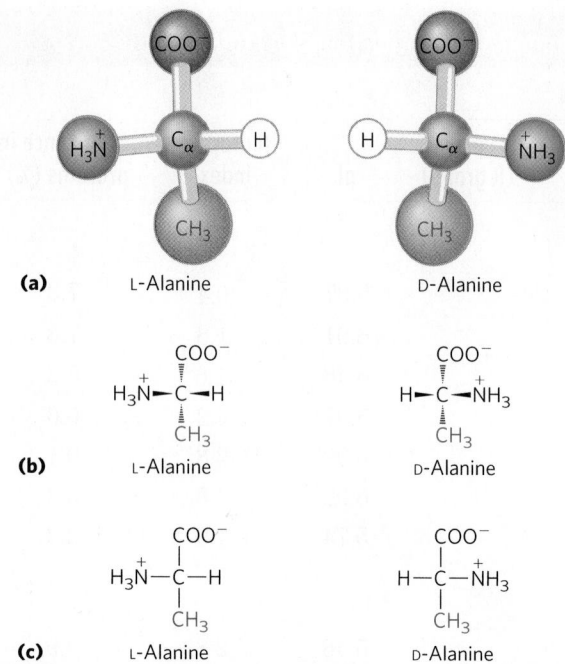

(a)   L-Alanine   D-Alanine

$$H_3\overset{+}{N}-\overset{\overset{\displaystyle COO^-}{|}}{\underset{\underset{\displaystyle CH_3}{|}}{C}}-H \qquad H-\overset{\overset{\displaystyle COO^-}{|}}{\underset{\underset{\displaystyle CH_3}{|}}{C}}-\overset{+}{N}H_3$$

(b)   L-Alanine   D-Alanine

$$H_3\overset{+}{N}-\overset{\overset{\displaystyle COO^-}{|}}{\underset{\underset{\displaystyle CH_3}{|}}{C}}-H \qquad H-\overset{\overset{\displaystyle COO^-}{|}}{\underset{\underset{\displaystyle CH_3}{|}}{C}}-\overset{+}{N}H_3$$

(c)   L-Alanine   D-Alanine

FIGURE 3-3 Stereoisomerism in α-amino acids. (a) The two stereo-isomers of alanine, L- and D-alanine, are nonsuperposable mirror images of each other (enantiomers). **(b, c)** Two different conventions for showing the configurations in space of stereoisomers. In perspective formulas (b), the solid wedge-shaped bonds project out of the plane of the paper, the dashed bonds behind it. In projection formulas (c), the horizontal bonds are assumed to project out of the plane of the paper, the vertical bonds behind. However, projection formulas are often used casually and are not always intended to portray a specific stereochemical configuration.

Special nomenclature has been developed to specify the **absolute configuration** of the four substituents of asymmetric carbon atoms. The absolute configurations of simple sugars and amino acids are specified by the **D, L system (Fig. 3-4)**, based on the absolute configuration of the three-carbon sugar glyceraldehyde, a convention proposed by Emil Fischer in 1891. (Fischer knew what

$$HO-\overset{\overset{\displaystyle {}^1CHO}{|}}{\underset{\underset{\displaystyle {}^3CH_2OH}{|}}{{}^2C}}-H \qquad H-\overset{\overset{\displaystyle CHO}{|}}{\underset{\underset{\displaystyle CH_2OH}{|}}{C}}-OH$$

L-Glyceraldehyde   D-Glyceraldehyde

$$H_3\overset{+}{N}-\overset{\overset{\displaystyle COO^-}{|}}{\underset{\underset{\displaystyle CH_3}{|}}{C}}-H \qquad H-\overset{\overset{\displaystyle COO^-}{|}}{\underset{\underset{\displaystyle CH_3}{|}}{C}}-\overset{+}{N}H_3$$

L-Alanine   D-Alanine

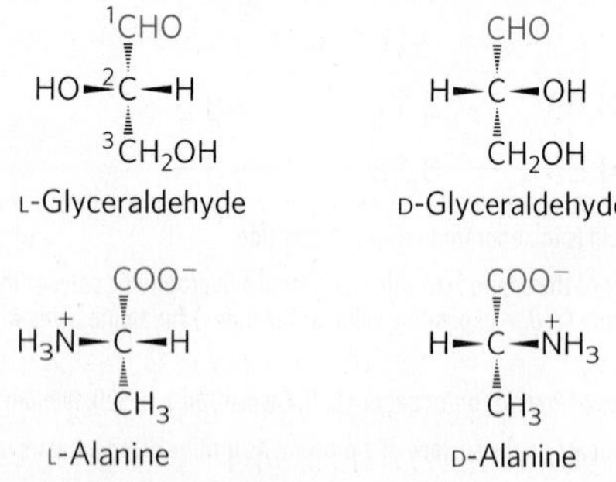

FIGURE 3-4 Steric relationship of the stereoisomers of alanine to the absolute configuration of L- and D-glyceraldehyde. In these perspective formulas, the carbons are lined up vertically, with the chiral atom in the center. The carbons in these molecules are numbered beginning with the terminal aldehyde or carboxyl carbon (red), 1 to 3 from top to bottom as shown. When presented in this way, the R group of the amino acid (in this case the methyl group of alanine) is always below the α carbon. L-Amino acids are those with the α-amino group on the left, and D-amino acids have the α-amino group on the right.

groups surrounded the asymmetric carbon of glyceraldehyde but had to guess at their absolute configuration; he guessed right, as was later confirmed by x-ray diffraction analysis.) For all chiral compounds, stereoisomers having a configuration related to that of L-glyceraldehyde are designated L, and stereoisomers related to D-glyceraldehyde are designated D. The functional groups of L-alanine are matched with those of L-glyceraldehyde by aligning those that can be interconverted by simple, one-step chemical reactions. Thus the carboxyl group of L-alanine occupies the same position about the chiral carbon as does the aldehyde group of L-glyceraldehyde, because an aldehyde is readily converted to a carboxyl group via a one-step oxidation. Historically, the similar L and D designations were used for levorotatory (rotating plane-polarized light to the left) and dextrorotatory (rotating light to the right). However, not all L-amino acids are levorotatory, and the convention shown in Figure 3-4 was needed to avoid potential ambiguities about absolute configuration. By Fischer's convention, L and D refer *only* to the absolute configuration of the four substituents around the chiral carbon, not to optical properties of the molecule.

Another system of specifying configuration around a chiral center is the **RS system**, which is used in the systematic nomenclature of organic chemistry and describes more precisely the configuration of molecules with more than one chiral center (p. 19).

The Amino Acid Residues in Proteins Are L Stereoisomers

Nearly all biological compounds with a chiral center occur naturally in only one stereoisomeric form, either D or L. The amino acid residues in protein molecules are exclusively L stereoisomers. D-Amino acid residues have been found in only a few, generally small peptides, including some peptides of bacterial cell walls and certain peptide antibiotics.

It is remarkable that virtually all amino acid residues in proteins are L stereoisomers. When chiral compounds are formed by ordinary chemical reactions, the result is a racemic mixture of D and L isomers, which are difficult for a chemist to distinguish and separate. But to a living system, D and L isomers are as different as the right hand and the left. The formation of stable, repeating substructures in proteins (Chapter 4) generally requires that their constituent amino acids be of one stereochemical series. Cells are able to specifically synthesize the L isomers of amino acids because the active sites of enzymes are asymmetric, causing the reactions they catalyze to be stereospecific.

Amino Acids Can Be Classified by R Group

Knowledge of the chemical properties of the common amino acids is central to an understanding of biochemistry. The topic can be simplified by grouping the amino acids into five main classes based on the properties of their R groups (Table 3-1), particularly their **polarity**, or tendency to interact with water at biological pH (near pH 7.0). The polarity of the R groups varies widely, from

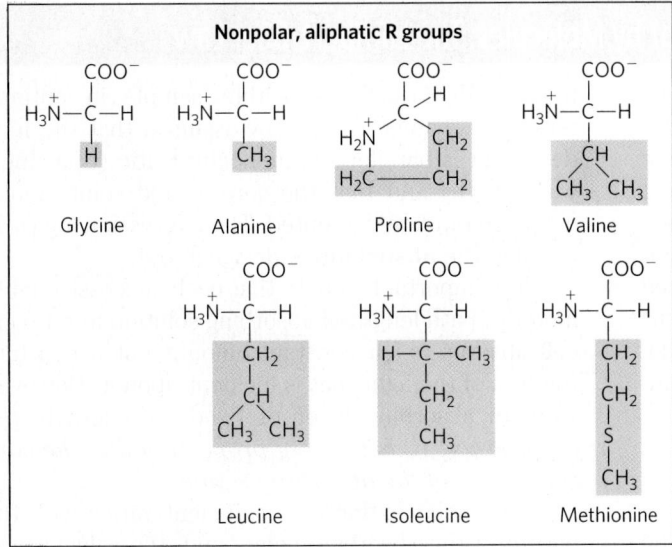

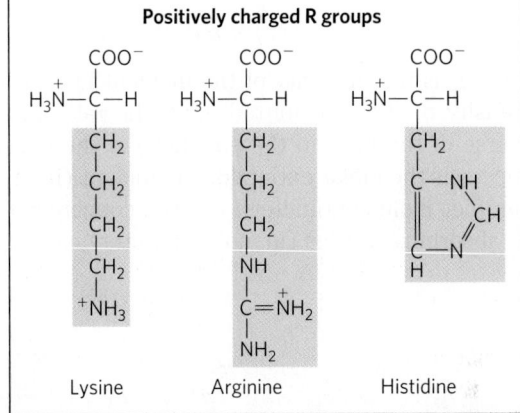

FIGURE 3-5 The 20 common amino acids of proteins. The structural formulas show the state of ionization that would predominate at pH 7.0. The unshaded portions are those common to all the amino acids; the shaded portions are the R groups. Although the R group of histidine is shown uncharged, its pK_a (see Table 3-1) is such that a small but significant fraction of these groups are positively charged at pH 7.0. The protonated form of histidine is shown above the graph in Figure 3-12b.

nonpolar and hydrophobic (water-insoluble) to highly polar and hydrophilic (water-soluble). A few amino acids are somewhat difficult to characterize or do not fit perfectly in any one group, particularly glycine, histidine, and cysteine. Their assignments to particular groupings are the results of considered judgments rather than absolutes.

The structures of the 20 common amino acids are shown in **Figure 3-5**, and some of their properties are listed in Table 3-1. Within each class there are gradations of polarity, size, and shape of the R groups.

Nonpolar, Aliphatic R Groups The R groups in this class of amino acids are nonpolar and hydrophobic. The side chains of **alanine**, **valine**, **leucine**, and **isoleucine** tend to cluster together within proteins, stabilizing protein

structure through the hydrophobic effect. **Glycine** has the simplest structure. Although it is most easily grouped with the nonpolar amino acids, its very small side chain makes no real contribution to interactions driven by the hydrophobic effect. **Methionine**, one of the two sulfur-containing amino acids, has a slightly nonpolar thioether group in its side chain. **Proline** has an aliphatic side chain with a distinctive cyclic structure. The secondary amino (imino) group of proline residues is held in a rigid conformation that reduces the structural flexibility of polypeptide regions containing proline.

Aromatic R Groups Phenylalanine, **tyrosine**, and **tryptophan**, with their aromatic side chains, are relatively nonpolar (hydrophobic). All can contribute to the

BOX 3-1 METHODS Absorption of Light by Molecules: The Lambert-Beer Law

A wide range of biomolecules absorb light at characteristic wavelengths, just as tryptophan absorbs light at 280 nm (see Fig. 3-6). Measurement of light absorption by a spectrophotometer is used to detect and identify molecules and to measure their concentration in solution. The fraction of the incident light absorbed by a solution at a given wavelength is related to the thickness of the absorbing layer (path length) and the concentration of the absorbing species (Fig. 1). These two relationships are combined into the Lambert-Beer law,

$$\log \frac{I_0}{I} = \varepsilon c l$$

where I_0 is the intensity of the incident light, I is the intensity of the transmitted light, the ratio I/I_0 (the inverse of the ratio in the equation) is the transmittance, ε is the molar extinction coefficient (in units of liters per mole-centimeter), c is the concentration of the absorbing species (in moles per liter), and l is the

path length of the light-absorbing sample (in centimeters). The Lambert-Beer law assumes that the incident light is parallel and monochromatic (of a single wavelength) and that the solvent and solute molecules are randomly oriented. The expression $\log (I_0/I)$ is called the **absorbance**, designated A.

It is important to note that each successive millimeter of path length of absorbing solution in a 1.0 cm cell absorbs not a constant amount but a constant fraction of the light that is incident upon it. However, with an absorbing layer of fixed path length, *the absorbance, A, is directly proportional to the concentration of the absorbing solute.*

The molar extinction coefficient varies with the nature of the absorbing compound, the solvent, and the wavelength, and also with pH if the light-absorbing species is in equilibrium with an ionization state that has different absorbance properties.

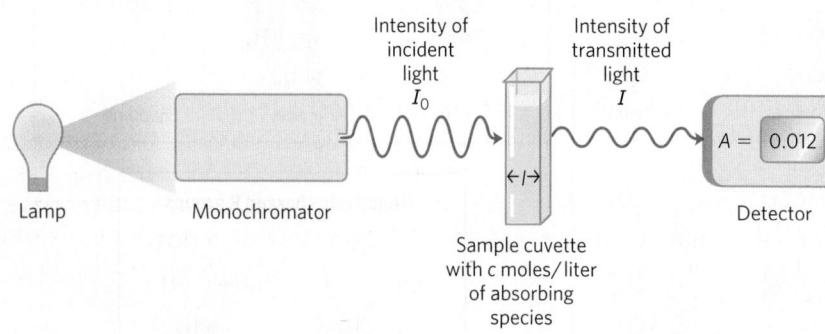

FIGURE 1 The principal components of a spectrophotometer. A light source emits light along a broad spectrum, then the monochromator selects and transmits light of a particular wavelength. The monochromatic light passes through the sample in a cuvette of path length *l*. The absorbance of the sample, log (I_0/I), is proportional to the concentration of the absorbing species. The transmitted light is measured by a detector.

hydrophobic effect. The hydroxyl group of tyrosine can form hydrogen bonds, and it is an important functional group in some enzymes. Tyrosine and tryptophan are significantly more polar than phenylalanine because of the tyrosine hydroxyl group and the nitrogen of the tryptophan indole ring.

Tryptophan and tyrosine, and to a much lesser extent phenylalanine, absorb ultraviolet light (**Fig. 3-6**; see also Box 3-1). This accounts for the characteristic strong absorbance of light by most proteins at a wavelength of 280 nm, a property exploited by researchers in the characterization of proteins.

Polar, Uncharged R Groups The R groups of these amino acids are more soluble in water, or more hydrophilic, than

those of the nonpolar amino acids, because they contain functional groups that form hydrogen bonds with water. This class of amino acids includes **serine**, **threonine**, **cysteine**, **asparagine**, and **glutamine**. The polarity of

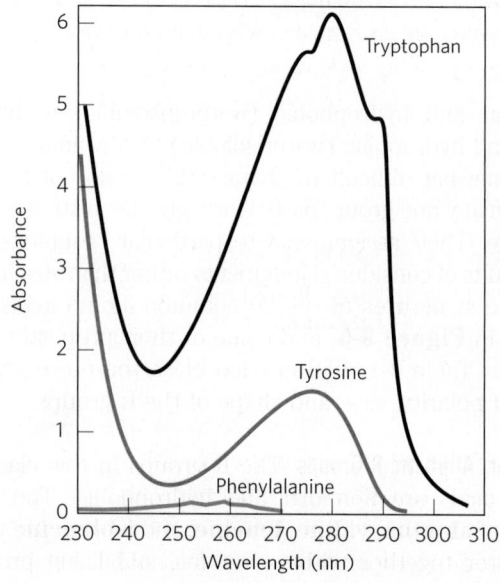

FIGURE 3-6 Absorption of ultraviolet light by aromatic amino acids. Comparison of the light absorption spectra of the aromatic amino acids tryptophan, tyrosine, and phenylalanine at pH 6.0. The amino acids are present in equimolar amounts (10^{-3} M) under identical conditions. The measured absorbance of tryptophan is more than four times that of tyrosine at a wavelength of 280 nm. Note that the maximum light absorption for both tryptophan and tyrosine occurs near 280 nm. Light absorption by phenylalanine generally contributes little to the spectroscopic properties of proteins.

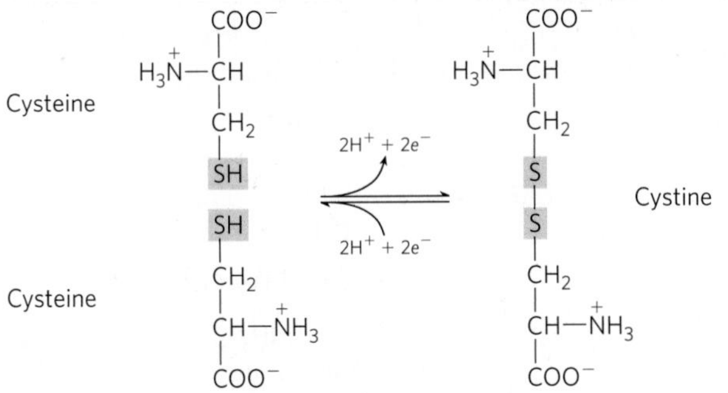

FIGURE 3-7 **Reversible formation of a disulfide bond by the oxidation of two molecules of cysteine.** Disulfide bonds between Cys residues stabilize the structures of many proteins.

serine and threonine is contributed by their hydroxyl groups, and that of asparagine and glutamine by their amide groups. Cysteine is an outlier here because its polarity, contributed by its sulfhydryl group, is quite modest. Cysteine is a weak acid and can make weak hydrogen bonds with oxygen or nitrogen.

Asparagine and glutamine are the amides of two other amino acids also found in proteins—aspartate and glutamate, respectively—to which asparagine and glutamine are easily hydrolyzed by acid or base. Cysteine is readily oxidized to form a covalently linked dimeric amino acid called **cystine**, in which two cysteine molecules or residues are joined by a disulfide bond **(Fig. 3-7)**. The disulfide-linked residues are strongly hydrophobic (nonpolar). Disulfide bonds play a special role in the structures of many proteins by forming covalent links between parts of a polypeptide molecule or between two different polypeptide chains.

Positively Charged (Basic) R Groups The most hydrophilic R groups are those that are either positively or negatively charged. The amino acids in which the R groups have significant positive charge at pH 7.0 are **lysine**, which has a second primary amino group at the ε position on its aliphatic chain; **arginine**, which has a positively charged guanidinium group; and **histidine**, which has an aromatic imidazole group. As the only common amino acid having an ionizable side chain with pK_a near neutrality, histidine may be positively charged (protonated form) or uncharged at pH 7.0. His residues facilitate many enzyme-catalyzed reactions by serving as proton donors/acceptors.

Negatively Charged (Acidic) R Groups The two amino acids having R groups with a net negative charge at pH 7.0 are **aspartate** and **glutamate**, each of which has a second carboxyl group.

Uncommon Amino Acids Also Have Important Functions

In addition to the 20 common amino acids, proteins may contain residues created by modification of common residues already incorporated into a polypeptide—that is, through postsynthetic modification **(Fig. 3-8a)**.

Among these uncommon amino acids are **4-hydroxyproline**, a derivative of proline, and **5-hydroxylysine**, derived from lysine. The former is found in plant cell wall proteins, and both are found in collagen, a fibrous protein of connective tissues. **6-N-Methyllysine** is a constituent of myosin, a contractile protein of muscle. Another important uncommon amino acid is γ-**carboxyglutamate**, found in the blood-clotting protein prothrombin and in certain other proteins that bind Ca^{2+} as part of their biological function. More complex is **desmosine**, a derivative of four Lys residues, which is found in the fibrous protein elastin.

Selenocysteine and **pyrrolysine** are special cases. These rare amino acid residues are not created through a postsynthetic modification. Instead, they are introduced during protein synthesis through an unusual adaptation of the genetic code described in Chapter 27. Selenocysteine contains selenium rather than the sulfur of cysteine. Actually derived from serine, selenocysteine is a constituent of just a few known proteins. Pyrrolysine is found in a few proteins in several methanogenic (methane-producing) archaea and in one known bacterium; it plays a role in methane biosynthesis.

Some amino acid residues in a protein may be modified transiently to alter the protein's function. The addition of phosphoryl, methyl, acetyl, adenylyl, ADP-ribosyl, or other groups to particular amino acid residues can increase or decrease a protein's activity (Fig. 3-8b). Phosphorylation is a particularly common regulatory modification. Covalent modification as a protein regulatory strategy is discussed in more detail in Chapter 6.

Some 300 additional amino acids have been found in cells. They have a variety of functions, but not all are constituents of proteins. **Ornithine** and **citrulline** (Fig. 3-8c) deserve special note because they are key intermediates (metabolites) in the biosynthesis of arginine (Chapter 22) and in the urea cycle (Chapter 18).

Amino Acids Can Act as Acids and Bases

The amino and carboxyl groups of amino acids, along with the ionizable R groups of some amino acids, function as weak acids and bases. When an amino acid lacking an ionizable R group is dissolved in water at neutral pH, it exists in solution as the dipolar ion, or **zwitterion** (German for "hybrid ion"), which can act as either an acid or a base **(Fig. 3-9)**. Substances having this dual (acid-base) nature are **amphoteric** and are often called **ampholytes** (from "amphoteric electrolytes"). A simple monoamino monocarboxylic α-amino acid, such as alanine, is a diprotic acid when fully protonated; it has two groups, the —COOH group and the —NH_3^+ group, that can yield protons:

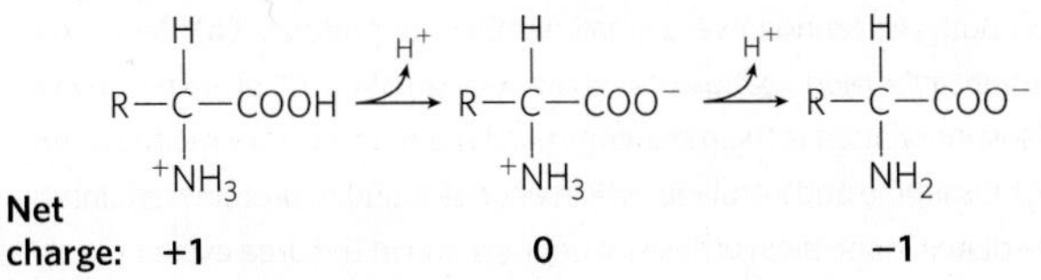

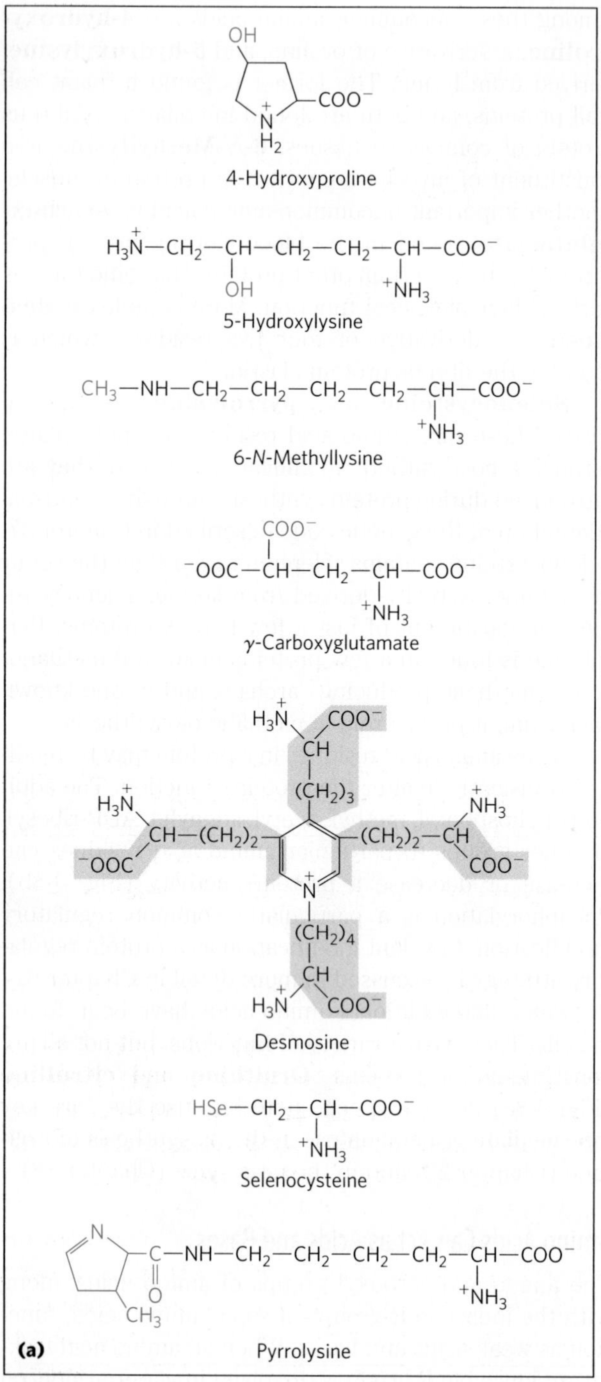

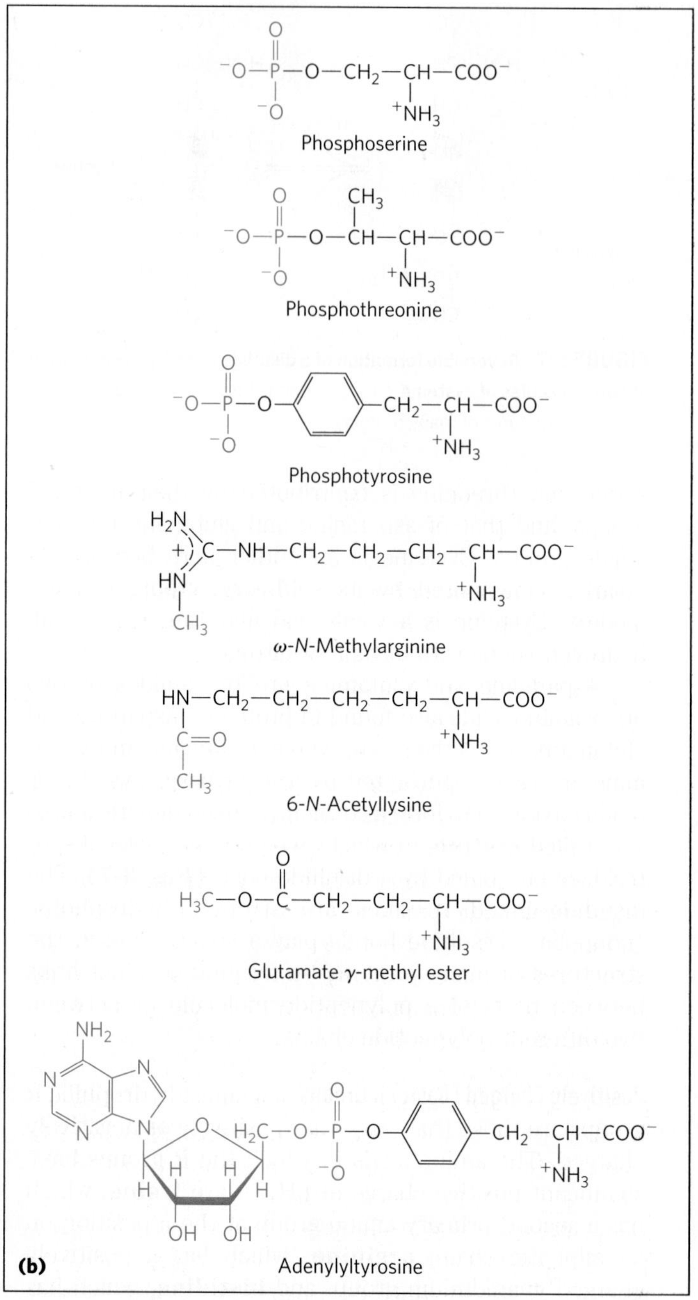

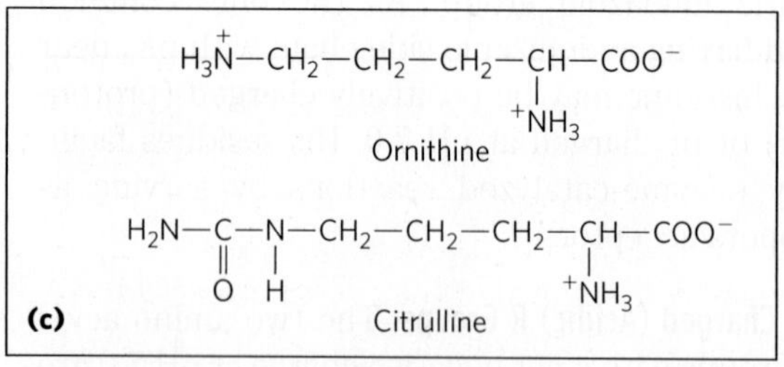

(a)

4-Hydroxyproline

5-Hydroxylysine

6-*N*-Methyllysine

γ-Carboxyglutamate

Desmosine

Selenocysteine

Pyrrolysine

(b)

Phosphoserine

Phosphothreonine

Phosphotyrosine

ω-*N*-Methylarginine

6-*N*-Acetyllysine

Glutamate γ-methyl ester

Adenylyltyrosine

(c)

Ornithine

Citrulline

FIGURE 3-8 Uncommon amino acids. (a) Some uncommon amino acids found in proteins. Most are derived from common amino acids. (Note the use of either numbers or Greek letters in the names of these structures to identify the altered carbon atoms.) Extra functional groups added by modification reactions are shown in red. Desmosine is formed from four Lys residues (the carbon backbones are shaded in light red). Selenocysteine and pyrrolysine are exceptions: these amino acids are added during normal protein synthesis through a highly specialized expansion of the standard genetic code described in Chapter 27. Both are found in very small numbers of proteins. **(b)** Reversible amino acid modifications involved in regulation of protein activity. Phosphorylation is the most common type of regulatory modification. **(c)** Ornithine and citrulline, which are not found in proteins, are intermediates in the biosynthesis of arginine and in the urea cycle.

Amino Acids Have Characteristic Titration Curves

Acid-base titration involves the gradual addition or removal of protons (Chapter 2). **Figure 3-10** shows the titration curve of the diprotic form of glycine. The two ionizable groups of glycine, the carboxyl group and the amino group, are titrated with a strong base such as NaOH. The plot has two distinct stages, corresponding

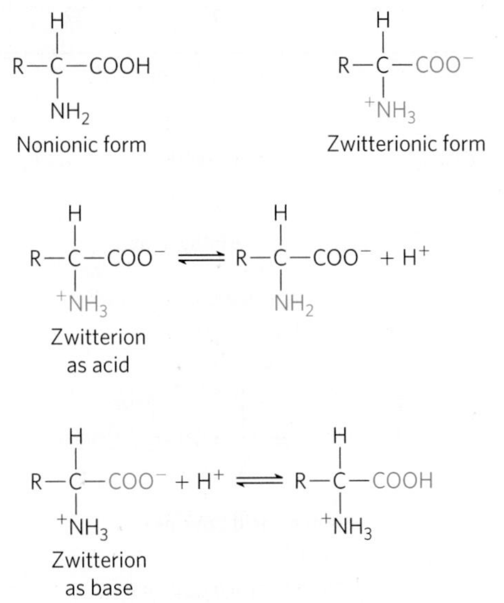

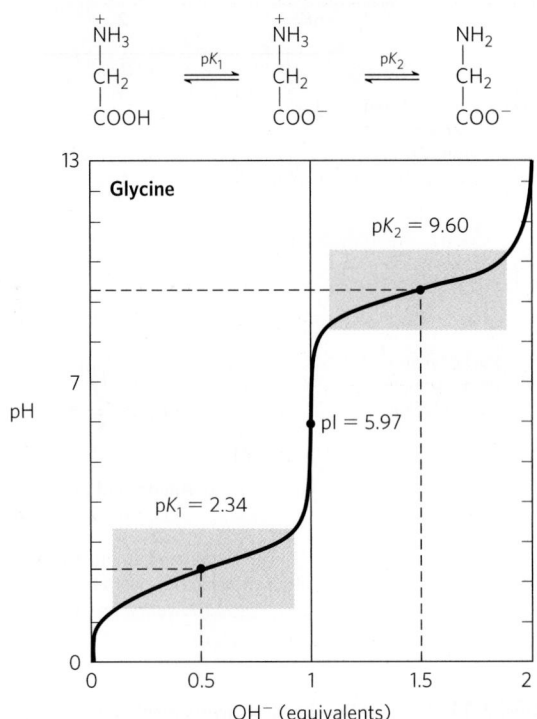

R—C—COOH : Nonionic form
R—C—COO⁻ with ⁺NH₃ : Zwitterionic form

$$R-\overset{H}{\underset{^{+}NH_3}{C}}-COO^- \rightleftharpoons R-\overset{H}{\underset{NH_2}{C}}-COO^- + H^+$$

Zwitterion as acid

$$R-\overset{H}{\underset{^{+}NH_3}{C}}-COO^- + H^+ \rightleftharpoons R-\overset{H}{\underset{^{+}NH_3}{C}}-COOH$$

Zwitterion as base

FIGURE 3-9 Nonionic and zwitterionic forms of amino acids. The nonionic form does not occur in significant amounts in aqueous solutions. The zwitterion predominates at neutral pH. A zwitterion can act as either an acid (proton donor) or a base (proton acceptor).

FIGURE 3-10 Titration of an amino acid. Shown here is the titration curve of 0.1 M glycine at 25 °C. The ionic species predominating at key points in the titration are shown above the graph. The shaded boxes, centered at about $pK_1 = 2.3$ and $pK_2 = 9.60$ indicate the regions of greatest buffering power. Note that 1 equivalent of OH⁻ = 0.1 M NaOH added.

to deprotonation of two different groups on glycine. Each of the two stages resembles in shape the titration curve of a monoprotic acid, such as acetic acid (see Fig. 2-17), and can be analyzed in the same way. At very low pH, the predominant ionic species of glycine is the fully protonated form, ⁺H₃N—CH₂—COOH. In the first stage of the titration, the —COOH group of glycine loses its proton. At the midpoint of this stage, equimolar concentrations of the proton-donor (⁺H₃N—CH₂—COOH) and proton-acceptor (⁺H₃N—CH₂—COO⁻) species are present. As in the titration of any weak acid, a point of inflection is reached at this midpoint where the pH is equal to the pK_a of the protonated group being titrated (see Fig. 2-18). For glycine, the pH at the midpoint is 2.34, thus its —COOH group has a pK_a (labeled pK_1 in Fig. 3-10) of 2.34. (Recall from Chapter 2 that pH and pK_a are simply convenient notations for proton concentration and the equilibrium constant for ionization, respectively. The pK_a is a measure of the tendency of a group to give up a proton, with that tendency decreasing tenfold as the pK_a increases by one unit.) As the titration of glycine proceeds, another important point is reached at pH 5.97. Here there is another point of inflection, at which removal of the first proton is essentially complete and removal of the second has just begun. At this pH glycine is present largely as the dipolar ion (zwitterion) ⁺H₃N—CH₂—COO⁻. We shall return to the significance of this inflection point in the titration curve (labeled pI in Fig. 3-10) shortly.

The second stage of the titration corresponds to the removal of a proton from the —NH₃⁺ group of glycine. The pH at the midpoint of this stage is 9.60, equal to the pK_a (labeled pK_2 in Fig. 3-10) for the —NH₃⁺ group. The titration is essentially complete at a pH of about 12,

at which point the predominant form of glycine is H₂N—CH₂—COO⁻.

From the titration curve of glycine we can derive several important pieces of information. First, it gives a quantitative measure of the pK_a of each of the two ionizing groups: 2.34 for the —COOH group and 9.60 for the —NH₃⁺ group. Note that the carboxyl group of glycine is over 100 times more acidic (more easily ionized) than the carboxyl group of acetic acid, which, as we saw in Chapter 2, has a pK_a of 4.76—about average for a carboxyl group attached to an otherwise unsubstituted aliphatic hydrocarbon. The perturbed pK_a of glycine is caused primarily by the nearby positively charged amino group on the α-carbon atom, an electronegative group that tends to pull electrons toward it (a process called electron withdrawal), as described in **Figure 3-11**. The opposite charges on the resulting zwitterion are also somewhat stabilizing. Similarly, the pK_a of the amino group in glycine is perturbed downward relative to the average pK_a of an amino group. This effect is due largely to electron withdrawal by the electronegative oxygen atoms in the carboxyl groups, increasing the tendency of the amino group to give up a proton. Hence, the α-amino group has a pK_a that is lower than that of an aliphatic amine such as methylamine (Fig. 3-11). In short, the pK_a of any functional group is greatly affected by its chemical environment, a phenomenon sometimes exploited in the active sites of

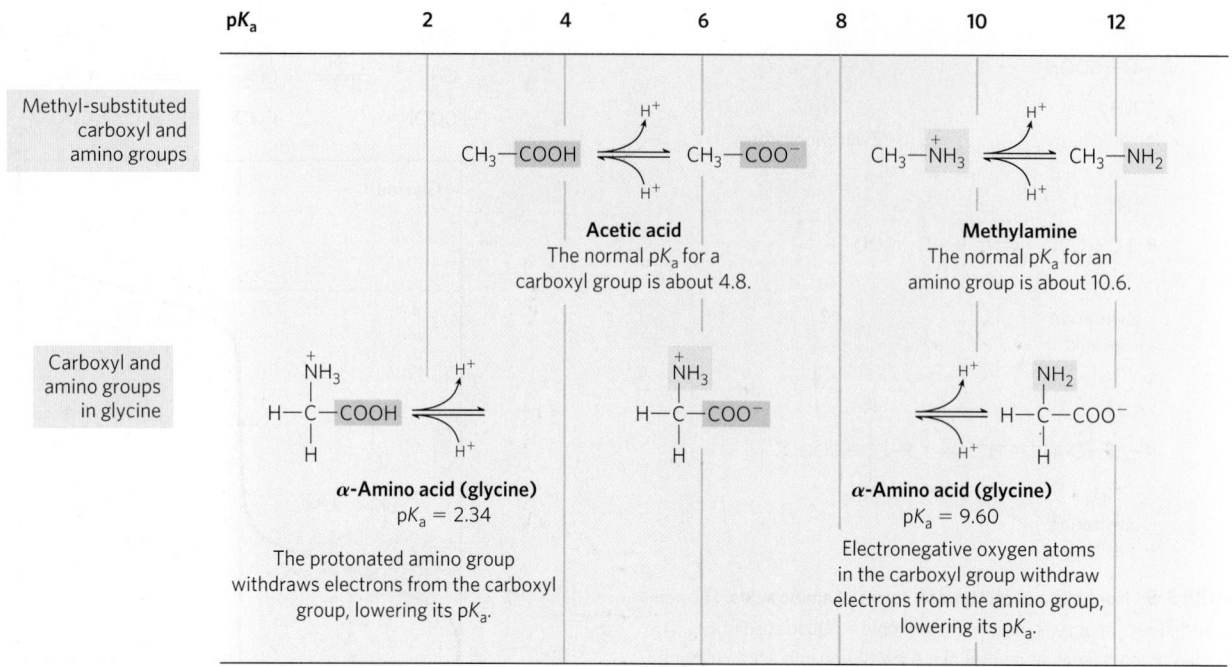

FIGURE 3-11 Effect of the chemical environment on pKₐ. The pKₐ values for the ionizable groups in glycine are lower than those for simple, methyl-substituted amino and carboxyl groups. These downward perturbations of pKₐ are due to intramolecular interactions. Similar effects can be caused by chemical groups that happen to be positioned nearby—for example, in the active site of an enzyme.

enzymes to promote exquisitely adapted reaction mechanisms that depend on the perturbed pK_a values of proton donor/acceptor groups of specific residues.

The second piece of information provided by the titration curve of glycine is that this amino acid has two regions of buffering power. One of these is the relatively flat portion of the curve, extending for approximately 1 pH unit on either side of the first pK_a of 2.34, indicating that glycine is a good buffer near this pH. The other buffering zone is centered around pH 9.60. (Note that glycine is not a good buffer at the pH of intracellular fluid or blood, about 7.4.) Within the buffering ranges of glycine, the Henderson-Hasselbalch equation (p. 65) can be used to calculate the proportions of proton-donor and proton-acceptor species of glycine required to make a buffer at a given pH.

Titration Curves Predict the Electric Charge of Amino Acids

Another important piece of information derived from the titration curve of an amino acid is the relationship between its net charge and the pH of the solution. At pH 5.97, the point of inflection between the two stages in its titration curve, glycine is present predominantly as its dipolar form, fully ionized but with no *net* electric charge (Fig. 3-10). The characteristic pH at which the *net* electric charge is zero is called the **isoelectric point** or **isoelectric pH**, designated **pI**. For glycine, which has no ionizable group in its side chain, the isoelectric point is simply the arithmetic mean of the two pK_a values:

$$pI = \frac{1}{2}(pK_1 + pK_2) = \frac{1}{2}(2.34 + 9.60) = 5.97$$

As is evident in Figure 3-10, glycine has a net negative charge at any pH above its pI and will thus move toward the positive electrode (the anode) when placed in an electric field. At any pH below its pI, glycine has a net positive charge and will move toward the negative electrode (the cathode). The farther the pH of a glycine solution is from its isoelectric point, the greater the net electric charge of the population of glycine molecules. At pH 1.0, for example, glycine exists almost entirely as the form $^+H_3N—CH_2—COOH$ with a net positive charge of 1.0. At pH 2.34, where there is an equal mixture of $^+H_3N—CH_2—COOH$ and $^+H_3N—CH_2—COO^-$, the average or net positive charge is 0.5. The sign and the magnitude of the net charge of any amino acid at any pH can be predicted in the same way.

Amino Acids Differ in Their Acid-Base Properties

The shared properties of many amino acids permit some simplifying generalizations about their acid-base behaviors. First, all amino acids with a single α-amino group, a single α-carboxyl group, and an R group that does not ionize have titration curves resembling that of glycine (Fig. 3-10). These amino acids have very similar, although not identical, pK_a values: pK_a of the —COOH group in the range of 1.8 to 2.4, and pK_a of the —NH₃⁺ group in the range of 8.8 to 11.0 (Table 3-1). The differences in these pK_a values reflect the chemical environments imposed by their R groups. Second, amino acids with an ionizable R group have more complex titration curves, with *three* stages corresponding to the three

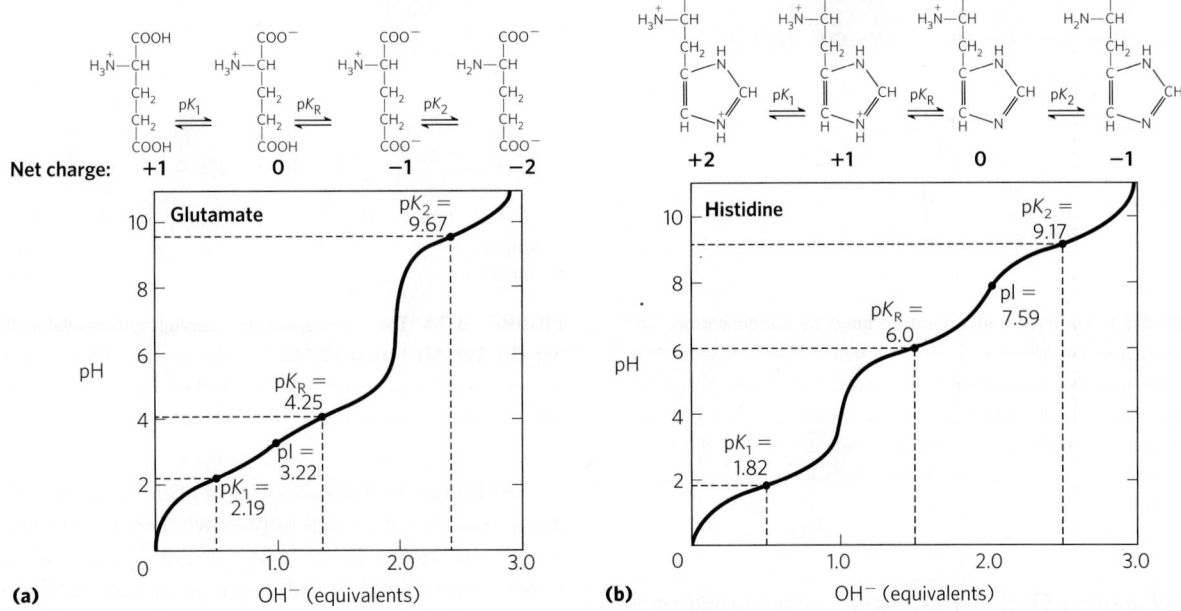

Net charge:

FIGURE 3-12 Titration curves for **(a)** glutamate and **(b)** histidine. The pK_a of the R group is designated here as pK_R.

possible ionization steps; thus they have three pK_a values. The additional stage for the titration of the ionizable R group merges to some extent with that for the titration of the α-carboxyl group, the titration of the α-amino group, or both. The titration curves for two amino acids of this type, glutamate and histidine, are shown in **Figure 3-12**. The isoelectric points reflect the nature of the ionizing R groups present. For example, glutamate has a pI of 3.22, considerably lower than that of glycine. This is due to the presence of two carboxyl groups, which, at the average of their pK_a values (3.22), contribute a net charge of −1 that balances the +1 contributed by the amino group. Similarly, the pI of histidine, with two groups that are positively charged when protonated, is 7.59 (the average of the pK_a values of the amino and imidazole groups), much higher than that of glycine.

Finally, as pointed out earlier, under the general condition of free and open exposure to the aqueous environment, only histidine has an R group ($pK_a = 6.0$) providing significant buffering power near the neutral pH usually found in the intracellular and extracellular fluids of most animals and bacteria (Table 3-1).

SUMMARY 3.1 Amino Acids

■ The 20 amino acids commonly found as residues in proteins contain an α-carboxyl group, an α-amino group, and a distinctive R group substituted on the α-carbon atom. The α-carbon atom of all amino acids except glycine is asymmetric, and thus amino acids can exist in at least two stereoisomeric forms. Only the L stereoisomers, with a configuration related to the absolute configuration of the reference molecule L-glyceraldehyde, are found in proteins.

■ Other, less common amino acids also occur, either as constituents of proteins (usually through modification of common amino acid residues after protein synthesis) or as free metabolites.

■ Amino acids can be classified into five types on the basis of the polarity and charge (at pH 7) of their R groups.

■ Amino acids vary in their acid-base properties and have characteristic titration curves. Monoamino monocarboxylic amino acids (with nonionizable R groups) are diprotic acids ($^+H_3NCH(R)COOH$) at low pH and exist in several different ionic forms as the pH is increased. Amino acids with ionizable R groups have additional ionic species, depending on the pH of the medium and the pK_a of the R group.

3.2 Peptides and Proteins

We now turn to polymers of amino acids, the **peptides** and **proteins**. Biologically occurring polypeptides range in size from small to very large, consisting of two or three to thousands of linked amino acid residues. Our focus is on the fundamental chemical properties of these polymers.

Peptides Are Chains of Amino Acids

Two amino acid molecules can be covalently joined through a substituted amide linkage, termed a **peptide bond**, to yield a dipeptide. Such a linkage is formed by removal of the elements of water (dehydration) from the α-carboxyl group of one amino acid and the α-amino

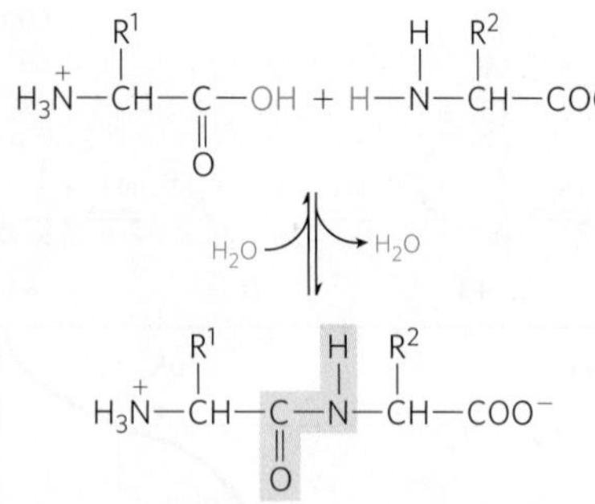

FIGURE 3-13 Formation of a peptide bond by condensation. The α-amino group of one amino acid (with R^2 group) acts as a nucleophile to displace the hydroxyl group of another amino acid (with R^1 group), forming a peptide bond (shaded in light red). Amino groups are good nucleophiles, but the hydroxyl group is a poor leaving group and is not readily displaced. At physiological pH, the reaction shown here does not occur to any appreciable extent.

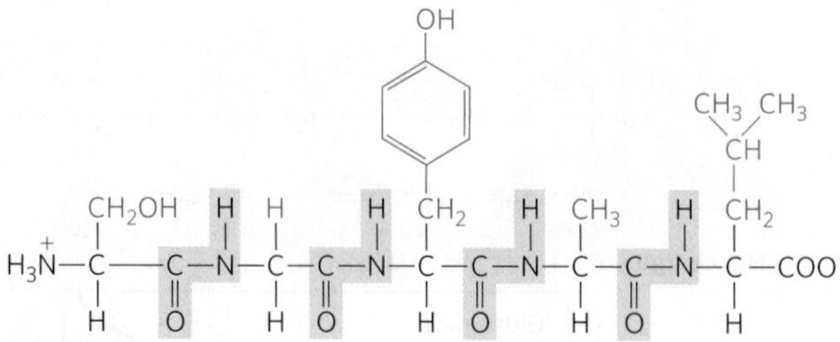

FIGURE 3-14 The pentapeptide serylglycyltyrosylalanylleucine, Ser–Gly–Tyr–Ala–Leu, or SGYAL. Peptides are named beginning with the amino-terminal residue, which by convention is placed at the left. The peptide bonds are shaded in light red; the R groups are in red.

group of another **(Fig. 3-13)**. Peptide bond formation is an example of a condensation reaction, a common class of reactions in living cells. Under standard biochemical conditions, the equilibrium for the reaction shown in Figure 3-13 favors the amino acids over the dipeptide. To make the reaction thermodynamically more favorable, the carboxyl group must be chemically modified or activated so that the hydroxyl group can be more readily eliminated. A chemical approach to this problem is outlined later in this chapter. The biological approach to peptide bond formation is a major topic of Chapter 27.

Three amino acids can be joined by two peptide bonds to form a tripeptide; similarly, four amino acids can be linked to form a tetrapeptide, five to form a pentapeptide, and so forth. When a few amino acids are joined in this fashion, the structure is called an **oligopeptide**. When many amino acids are joined, the product is called a **polypeptide**. Proteins may have thousands of amino acid residues. Although the terms "protein" and "polypeptide" are sometimes used interchangeably, molecules referred to as polypeptides generally have molecular weights below 10,000, and those called proteins have higher molecular weights.

Figure 3-14 shows the structure of a pentapeptide. As already noted, an amino acid unit in a peptide is often called a residue (the part left over after losing the elements of water—a hydrogen atom from its amino group and the hydroxyl moiety from its carboxyl group). In a peptide, the amino acid residue at the end with a free α-amino group is the **amino-terminal** (or *N*-terminal) residue; the residue at the other end, which has a free carboxyl group, is the **carboxyl-terminal** (*C*-terminal) residue.

>> Key Convention: When an amino acid sequence of a peptide, polypeptide, or protein is displayed, the amino-terminal end is placed on the left, the carboxyl-terminal end on the right. The sequence is read left to right, beginning with the amino-terminal end. **<<**

Although hydrolysis of a peptide bond is an exergonic reaction, it occurs only slowly because it has a high activation energy (p. 28). As a result, the peptide bonds in proteins are quite stable, with an average half-life ($t_{1/2}$) of about 7 years under most intracellular conditions.

Peptides Can Be Distinguished by Their Ionization Behavior

Peptides contain only one free α-amino group and one free α-carboxyl group, at opposite ends of the chain **(Fig. 3-15)**. These groups ionize as they do in free amino acids. The α-amino and α-carboxyl groups of all nonterminal amino acids are covalently joined in the peptide bonds, which do not ionize and thus do not contribute to the total acid-base behavior of peptides. However, the R groups of some amino acids can ionize (Table 3-1), and in a peptide these contribute to the overall acid-base properties of the molecule (Fig. 3-15). Thus the acid-base behavior of a peptide can be predicted from its free α-amino and α-carboxyl groups combined with the nature and number of its ionizable R groups.

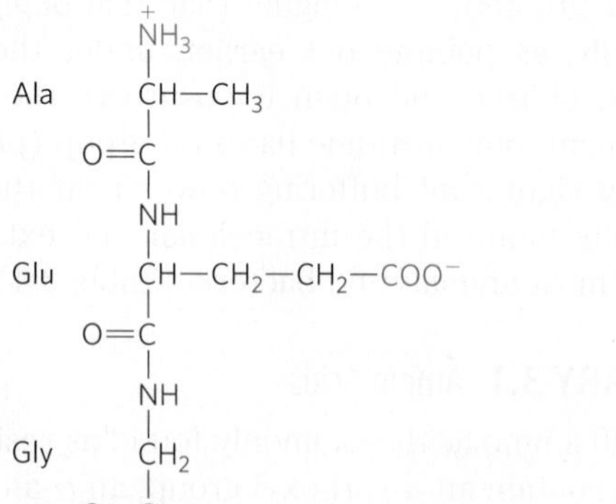

FIGURE 3-15 Alanylglutamylglycyllysine. This tetrapeptide has one free α-amino group, one free α-carboxyl group, and two ionizable R groups. The groups ionized at pH 7.0 are in red.

TABLE 3-2 Molecular Data on Some Proteins

Protein	Molecular weight	Number of residues	Number of polypeptide chains
Cytochrome c (human)	12,400	104	1
Ribonuclease A (bovine pancreas)	13,700	124	1
Lysozyme (chicken egg white)	14,300	129	1
Myoglobin (equine heart)	16,700	153	1
Chymotrypsin (bovine pancreas)	25,200	241	3
Chymotrypsinogen (bovine)	25,700	245	1
Hemoglobin (human)	64,500	574	4
Serum albumin (human)	66,000	609	1
Hexokinase (yeast)	107,900	972	2
RNA polymerase (*E. coli*)	450,000	4,158	5
Apolipoprotein B (human)	513,000	4,536	1
Glutamine synthetase (*E. coli*)	619,000	5,628	12
Titin (human)	2,993,000	26,926	1

Like free amino acids, peptides have characteristic titration curves and a characteristic isoelectric pH (pI) at which they do not move in an electric field. These properties are exploited in some of the techniques used to separate peptides and proteins, as we describe later in the chapter. We should emphasize that the pK_a value for an ionizable R group can change somewhat when an amino acid becomes a residue in a peptide. The loss of charge in the α-carboxyl and α-amino groups, the interactions with other peptide R groups, and other environmental factors can affect the pK_a. The pK_a values for R groups listed in Table 3-1 can be a useful guide to the pH range in which a given group will ionize, but they cannot be strictly applied when an amino acid becomes part of a peptide.

Biologically Active Peptides and Polypeptides Occur in a Vast Range of Sizes and Compositions

No generalizations can be made about the molecular weights of biologically active peptides and proteins in relation to their functions. Naturally occurring peptides range in length from two to many thousands of amino acid residues. Even the smallest peptides can have biologically important effects. Consider the commercially synthesized dipeptide L-aspartyl-L-phenylalanine methyl ester, the artificial sweetener better known as aspartame or NutraSweet.

L-Aspartyl-L-phenylalanine methyl ester
(aspartame)

Many small peptides exert their effects at very low concentrations. For example, a number of vertebrate hormones (Chapter 23) are small peptides. These include oxytocin (nine amino acid residues), which is secreted by the posterior pituitary gland and stimulates uterine contractions, and thyrotropin-releasing factor (three residues), which is formed in the hypothalamus and stimulates the release of another hormone, thyrotropin, from the anterior pituitary gland. Some extremely toxic mushroom poisons, such as amanitin, are also small peptides, as are many antibiotics.

How long are the polypeptide chains in proteins? As Table 3-2 shows, lengths vary considerably. Human cytochrome c has 104 amino acid residues linked in a single chain; bovine chymotrypsinogen has 245 residues. At the extreme is titin, a constituent of vertebrate muscle, which has nearly 27,000 amino acid residues and a molecular weight of about 3,000,000. The vast majority of naturally occurring proteins are much smaller than this, containing fewer than 2,000 amino acid residues.

Some proteins consist of a single polypeptide chain, but others, called **multisubunit** proteins, have two or more polypeptides associated noncovalently (Table 3-2). The individual polypeptide chains in a multisubunit protein may be identical or different. If at least two are identical the protein is said to be **oligomeric**, and the identical units (consisting of one or more polypeptide chains) are referred to as **protomers**. Hemoglobin, for example, has four polypeptide subunits: two identical α chains and two identical β chains, all four held together by noncovalent interactions. Each α subunit is paired in an identical way with a β subunit within the structure of this multisubunit protein, so that hemoglobin can be considered either a tetramer of four polypeptide subunits or a dimer of $\alpha\beta$ protomers.

A few proteins contain two or more polypeptide chains linked covalently. For example, the two polypeptide chains of insulin are linked by disulfide bonds. In such cases, the individual polypeptides are not considered subunits but are commonly referred to simply as chains.

The amino acid composition of proteins is also highly variable. The 20 common amino acids almost never occur in equal amounts in a protein. Some amino acids may occur only once or not at all in a given type of protein; others may occur in large numbers. Table 3-3 shows the amino acid composition of bovine cytochrome c and chymotrypsinogen, the inactive precursor of the digestive enzyme chymotrypsin. These two proteins, with very different functions, also differ significantly in the relative numbers of each kind of amino acid residue.

We can calculate the approximate number of amino acid residues in a simple protein containing no other chemical constituents by dividing its molecular weight by 110. Although the average molecular weight of the 20 common amino acids is about 138, the smaller amino acids predominate in most proteins. If we take into account the proportions in which the various amino acids occur in an average protein (Table 3-1; the averages are determined by surveying the amino acid compositions of more than 1,000 different proteins), the average molecular weight of protein amino acids is nearer to 128. Because a molecule of water (M_r 18) is removed to create each peptide bond, the average molecular weight of an amino acid residue in a protein is about $128 - 18 = 110$.

Some Proteins Contain Chemical Groups Other Than Amino Acids

Many proteins, for example the enzymes ribonuclease A and chymotrypsin, contain only amino acid residues and no other chemical constituents; these are considered simple proteins. However, some proteins contain permanently

TABLE 3-3	Amino Acid Composition of Two Proteins			
	Bovine cytochrome c		**Bovine chymotrypsinogen**	
Amino acid	Number of residues per molecule	Percentage of total[a]	Number of residues per molecule	Percentage of total[a]
Ala	6	6	22	9
Arg	2	2	4	1.6
Asn	5	5	14	5.7
Asp	3	3	9	3.7
Cys	2	2	10	4
Gln	3	3	10	4
Glu	9	9	5	2
Gly	14	13	23	9.4
His	3	3	2	0.8
Ile	6	6	10	4
Leu	6	6	19	7.8
Lys	18	17	14	5.7
Met	2	2	2	0.8
Phe	4	4	6	2.4
Pro	4	4	9	3.7
Ser	1	1	28	11.4
Thr	8	8	23	9.4
Trp	1	1	8	3.3
Tyr	4	4	4	1.6
Val	3	3	23	9.4
Total	104	102	245	99.7

Note: In some common analyses, such as acid hydrolysis, Asp and Asn are not readily distinguished from each other and are together designated Asx (or B). Similarly, when Glu and Gln cannot be distinguished, they are together designated Glx (or Z). In addition, Trp is destroyed by acid hydrolysis. Additional procedures must be employed to obtain an accurate assessment of complete amino acid content.

[a]Percentages do not total to 100%, due to rounding.

TABLE 3-4 Conjugated Proteins

Class	Prosthetic group	Example
Lipoproteins	Lipids	β_1-Lipoprotein of blood
Glycoproteins	Carbohydrates	Immunoglobulin G
Phosphoproteins	Phosphate groups	Casein of milk
Hemoproteins	Heme (iron porphyrin)	Hemoglobin
Flavoproteins	Flavin nucleotides	Succinate dehydrogenase
Metalloproteins	Iron	Ferritin
	Zinc	Alcohol dehydrogenase
	Calcium	Calmodulin
	Molybdenum	Dinitrogenase
	Copper	Plastocyanin

associated chemical components in addition to amino acids; these are called **conjugated proteins**. The non–amino acid part of a conjugated protein is usually called its **prosthetic group**. Conjugated proteins are classified on the basis of the chemical nature of their prosthetic groups (Table 3-4); for example, **lipoproteins** contain lipids, **glycoproteins** contain sugar groups, and **metalloproteins** contain a specific metal. Some proteins contain more than one prosthetic group. Usually the prosthetic group plays an important role in the protein's biological function.

SUMMARY 3.2 Peptides and Proteins

■ Amino acids can be joined covalently through peptide bonds to form peptides and proteins. Cells generally contain thousands of different proteins, each with a different biological activity.

■ Proteins can be very long polypeptide chains of 100 to several thousand amino acid residues. However, some naturally occurring peptides have only a few amino acid residues. Some proteins are composed of several noncovalently associated polypeptide chains, called subunits.

■ Simple proteins yield only amino acids on hydrolysis; conjugated proteins contain in addition some other component, such as a metal or organic prosthetic group.

3.3 Working with Proteins

Biochemists' understanding of protein structure and function has been derived from the study of many individual proteins. To study a protein in detail, the researcher must be able to separate it from other proteins in pure form and must have the techniques to determine its properties. The necessary methods come from protein chemistry, a discipline as old as biochemistry itself and one that retains a central position in biochemical research.

Proteins Can Be Separated and Purified

A pure preparation is essential before a protein's properties and activities can be determined. Given that cells contain thousands of different kinds of proteins, how can one protein be purified? Classical methods for separating proteins take advantage of properties that vary from one protein to the next, including size, charge, and binding properties. These have been supplemented in recent decades by methods involving DNA cloning and genome sequencing that can simplify the process of protein purification. The newer methods, presented in Chapters 8 and 9, often artificially modify the protein being purified, adding a few or many amino acid residues to one or both ends. Convenience thus comes at the price of potentially altering the activity of the purified protein. The purification of proteins in their native state (the form in which they function in the cell) usually relies on methods described here.

The source of a protein is generally tissue or microbial cells. The first step in any protein purification procedure is to break open these cells, releasing their proteins into a solution called a **crude extract**. If necessary, differential centrifugation can be used to prepare subcellular fractions or to isolate specific organelles (see Fig. 1-9).

Once the extract or organelle preparation is ready, various methods are available for purifying one or more of the proteins it contains. Commonly, the extract is subjected to treatments that separate the proteins into different **fractions** based on a property such as size or charge, a process referred to as **fractionation**. Early fractionation steps in a purification utilize differences in protein solubility, which is a complex function of pH, temperature, salt concentration, and other factors. The solubility of proteins is lowered in the presence of some salts, an effect called "salting out." The addition of certain salts in the right amount can selectively precipitate some proteins, while others remain in solution. Ammonium sulfate ($(NH_4)_2SO_4$) is particularly effective and is often used to salt out proteins. The proteins thus precipitated are removed from those remaining in solution by low-speed centrifugation.

A solution containing the protein of interest usually must be further altered before subsequent purification steps are possible. For example, **dialysis** is a procedure that separates proteins from small solutes by taking advantage of the proteins' larger size. The partially purified extract is placed in a bag or tube made of a semipermeable membrane. When this is suspended in a much larger volume of buffered solution of appropriate ionic strength, the membrane allows the exchange of salt and buffer but not proteins. Thus dialysis retains large proteins within the membranous bag or tube while allowing the concentration of other solutes in the protein preparation to change until they come into equilibrium with the solution outside the membrane. Dialysis might be used, for example, to remove ammonium sulfate from the protein preparation.

The most powerful methods for fractionating proteins make use of **column chromatography**, which takes advantage of differences in protein charge, size, binding affinity, and other properties **(Fig. 3-16)**. A porous solid material with appropriate chemical properties (the stationary phase) is held in a column, and a buffered solution (the mobile phase) migrates through it. The protein, dissolved in the same buffered solution that was used to establish the mobile phase, is layered on the top of the column. The protein then percolates through the solid matrix as an ever-expanding band within the larger mobile phase. Individual proteins migrate faster or more slowly through the column, depending on their properties.

Ion-exchange chromatography exploits differences in the sign and magnitude of the net electric charge of proteins at a given pH **(Fig. 3-17a)**. The column matrix is a synthetic polymer (resin) containing bound charged groups; those with bound anionic groups are called **cation exchangers**, and those with bound cationic groups are called **anion exchangers**. The affinity of each protein for the charged groups on the column is affected by the pH (which determines the ionization state of the molecule) and the concentration of competing free salt ions in the surrounding solution. Separation can be optimized by gradually changing the pH and/or salt concentration of the mobile phase so as to create a pH or salt gradient. In **cation-exchange chromatography**, the solid matrix has negatively charged groups. In the mobile phase, proteins with a net positive charge migrate through the matrix more slowly than those with a net negative charge, because the migration of the former is retarded more by interaction with the stationary phase.

In ion-exchange columns, the expansion of the protein band in the mobile phase (the protein solution) is caused both by separation of proteins with different properties and by diffusional spreading. As the length of the column increases, the resolution of two types of protein with different net charges generally improves.

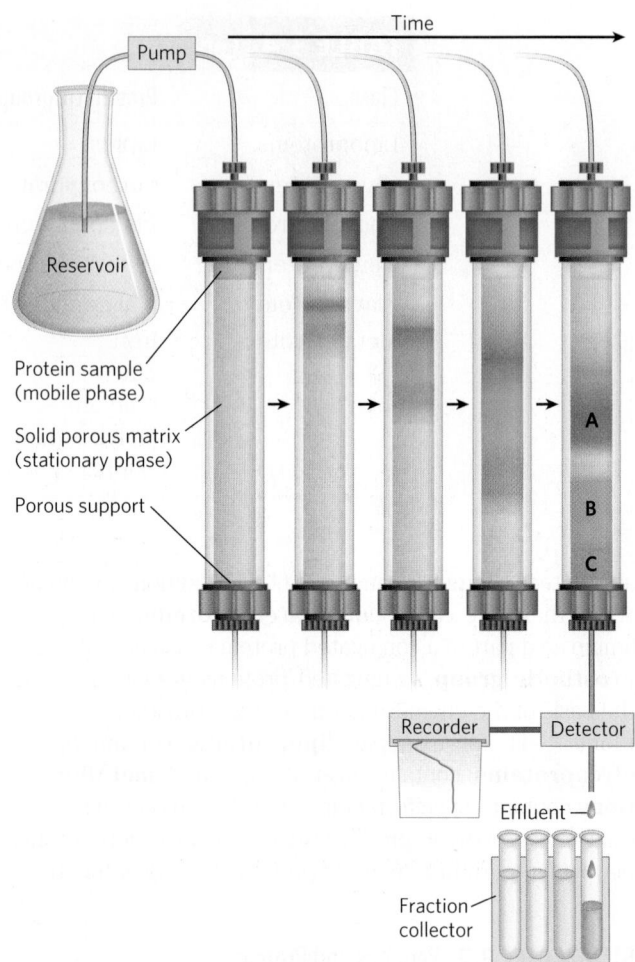

FIGURE 3-16 Column chromatography. The standard elements of a chromatographic column include a solid, porous material (matrix) supported inside a column, generally made of plastic or glass. A solution, the mobile phase, flows through the matrix, the stationary phase. The solution that passes out of the column at the bottom (the effluent) is constantly replaced by solution supplied from a reservoir at the top. The protein solution to be separated is layered on top of the column and allowed to percolate into the solid matrix. Additional solution is added on top. The protein solution forms a band within the mobile phase that is initially the depth of the protein solution applied to the column. As proteins migrate through the column (shown here at five different times), they are retarded to different degrees by their different interactions with the matrix material. The overall protein band thus widens as it moves through the column. Individual types of proteins (such as A, B, and C, shown in blue, red, and green) gradually separate from each other, forming bands within the broader protein band. Separation improves (i.e., resolution increases) as the length of the column increases. However, each individual protein band also broadens with time due to diffusional spreading, a process that decreases resolution. In this example, protein A is well separated from B and C, but diffusional spreading prevents complete separation of B and C under these conditions.

However, the rate at which the protein solution can flow through the column usually decreases with column length. And as the length of time spent on the column increases, the resolution can decline as a result of diffusional spreading within each protein band. As the

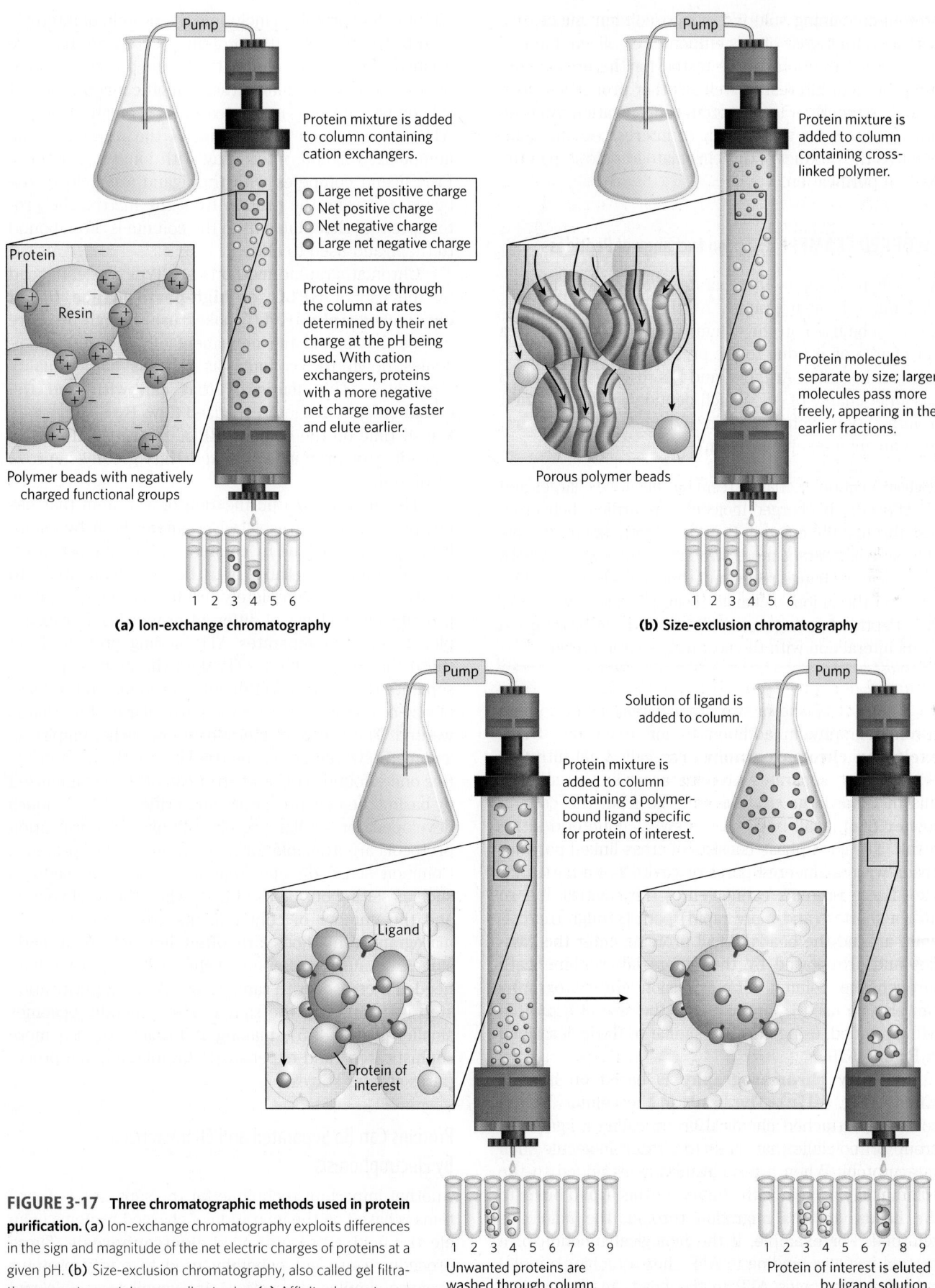

FIGURE 3-17 Three chromatographic methods used in protein purification. (a) Ion-exchange chromatography exploits differences in the sign and magnitude of the net electric charges of proteins at a given pH. **(b)** Size-exclusion chromatography, also called gel filtration, separates proteins according to size. **(c)** Affinity chromatography separates proteins by their binding specificities. Further details of these methods are given in the text.

Protein mixture is added to column containing cation exchangers.

- Large net positive charge
- Net positive charge
- Net negative charge
- Large net negative charge

Proteins move through the column at rates determined by their net charge at the pH being used. With cation exchangers, proteins with a more negative net charge move faster and elute earlier.

Protein

Resin

Polymer beads with negatively charged functional groups

(a) Ion-exchange chromatography

Protein mixture is added to column containing cross-linked polymer.

Protein molecules separate by size; larger molecules pass more freely, appearing in the earlier fractions.

Porous polymer beads

(b) Size-exclusion chromatography

Solution of ligand is added to column.

Protein mixture is added to column containing a polymer-bound ligand specific for protein of interest.

Ligand

Protein of interest

Unwanted proteins are washed through column.

Protein of interest is eluted by ligand solution.

(c) Affinity chromatography

protein-containing solution exits a column, successive portions (fractions) of this effluent are collected in test tubes. Each fraction can be tested for the presence of the protein of interest as well as other properties, such as ionic strength or total protein concentration. All fractions positive for the protein of interest can be combined as the product of this chromatographic step of the protein purification.

WORKED EXAMPLE 3-1   **Ion Exchange of Peptides**

A biochemist wants to separate two peptides by ion-exchange chromatography. At the pH of the mobile phase to be used on the column, one peptide (A) has a net charge of −3 due to the presence of more Glu and Asp residues than Arg, Lys, and His residues. Peptide B has a net charge of +1. Which peptide would elute first from a cation-exchange resin? Which would elute first from an anion-exchange resin?

Solution: A cation-exchange resin has negative charges and binds positively charged molecules, retarding their progress through the column. Peptide B, with its net positive charge, will interact more strongly than peptide A with the cation-exchange resin, and thus peptide A will elute first. On the anion-exchange resin, peptide B will elute first. Peptide A, being negatively charged, will be retarded by its interaction with the positively charged resin.

Figure 3-17 shows two other variations of column chromatography in addition to ion exchange. **Size-exclusion chromatography**, also called gel filtration (Fig. 3-17b), separates proteins according to size. In this method, large proteins emerge from the column sooner than small ones—a somewhat counterintuitive result. The solid phase consists of cross-linked polymer beads with engineered pores or cavities of a particular size. Large proteins cannot enter the cavities and so take a shorter (and more rapid) path through the column, around the beads. Small proteins enter the cavities and are slowed by their more labyrinthine path through the column. Size-exclusion chromatography can also be used to approximate the size of a protein being purified, using methods similar to those described in Figure 3-19.

Affinity chromatography is based on binding affinity (Fig. 3-17c). The beads in the column have a covalently attached chemical group called a ligand—a group or molecule that binds to a macromolecule such as a protein. When a protein mixture is added to the column, any protein with affinity for this ligand binds to the beads, and its migration through the matrix is retarded. For example, if the biological function of a protein involves binding to ATP, then attaching a molecule that resembles ATP to the beads in the column creates an affinity matrix that can help purify the protein. As the protein solution moves through the column,

ATP-binding proteins (including the protein of interest) bind to the matrix. After proteins that do not bind are washed through the column, the bound protein is eluted by a solution containing either a high concentration of salt or free ligand—in this case, ATP or an analog of ATP. Salt weakens the binding of the protein to the immobilized ligand, interfering with ionic interactions. Free ligand competes with the ligand attached to the beads, releasing the protein from the matrix; the protein product that elutes from the column is often bound to the ligand used to elute it.

Chromatographic methods are typically enhanced by the use of **HPLC**, or **high-performance liquid chromatography**. HPLC makes use of high-pressure pumps that speed the movement of the protein molecules down the column, as well as higher-quality chromatographic materials that can withstand the crushing force of the pressurized flow. By reducing the transit time on the column, HPLC can limit diffusional spreading of protein bands and thus greatly improve resolution.

The approach to purification of a protein that has not previously been isolated is guided both by established precedents and by common sense. In most cases, several different methods must be used sequentially to purify a protein completely, each method separating proteins on the basis of different properties. For example, if one step separates ATP-binding proteins from those that do not bind ATP, then the next step must separate the various ATP-binding proteins on the basis of size or charge to isolate the particular protein that is wanted. The choice of methods is somewhat empirical, and many strategies may be tried before the most effective one is found. Trial and error can often be minimized by basing the new procedure on purification techniques developed for similar proteins. Published purification protocols are available for many thousands of proteins. Common sense dictates that inexpensive procedures such as salting out be used first, when the total volume and the number of contaminants are greatest. Chromatographic methods are often impractical at early stages, because the amount of chromatographic medium needed increases with sample size. As each purification step is completed, the sample size generally becomes smaller (Table 3-5), making it feasible to use more sophisticated (and expensive) chromatographic procedures at later stages.

Proteins Can Be Separated and Characterized by Electrophoresis

Another important technique for the separation of proteins is based on the migration of charged proteins in an electric field, a process called **electrophoresis**. These procedures are not generally used to purify proteins because simpler alternatives are usually available and electrophoretic methods often adversely affect the structure and thus the function of proteins. However, as

TABLE 3-5 A Purification Table for a Hypothetical Enzyme

Procedure or step	Fraction volume (mL)	Total protein (mg)	Activity (units)	Specific activity (units/mg)
1. Crude cellular extract	1,400	10,000	100,000	10
2. Precipitation with ammonium sulfate	280	3,000	96,000	32
3. Ion-exchange chromatography	90	400	80,000	200
4. Size-exclusion chromatography	80	100	60,000	600
5. Affinity chromatography	6	3	45,000	15,000

Note: All data represent the status of the sample *after* the designated procedure has been carried out. Activity and specific activity are defined on page 95. After step 5, the enzyme in question has been purified by a factor of 1,500, as reflected in the increase in specific activity relative to that in the crude extract, and the yield of the enzyme is 45%, as reflected in the recovery of total activity.

an analytical method, electrophoresis is extremely important. Its advantage is that proteins can be visualized as well as separated, permitting a researcher to estimate quickly the number of different proteins in a mixture or the degree of purity of a particular protein preparation. Also, electrophoresis can be used to determine crucial properties of a protein such as its isoelectric point and approximate molecular weight.

Electrophoresis of proteins is generally carried out in gels made up of the cross-linked polymer polyacrylamide **(Fig. 3-18)**. The polyacrylamide gel acts as a molecular sieve, slowing the migration of proteins approximately in proportion to their charge-to-mass ratio. Migration may also be affected by protein shape. In electrophoresis, the force moving the macromolecule is the electrical potential, E. The electrophoretic

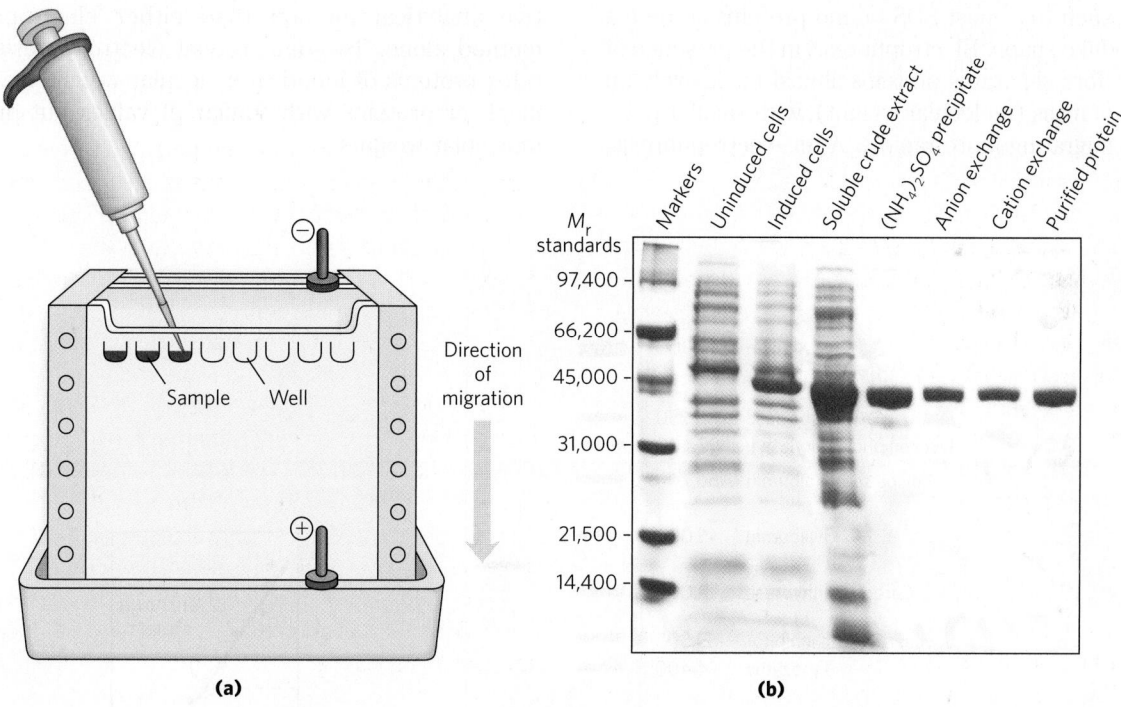

(a)   **(b)**

FIGURE 3-18 Electrophoresis. (a) Different samples are loaded in wells or depressions at the top of the SDS polyacrylamide gel. The proteins move into the gel when an electric field is applied. The gel minimizes convection currents caused by small temperature gradients, as well as protein movements other than those induced by the electric field. **(b)** Proteins can be visualized after electrophoresis by treating the gel with a stain such as Coomassie blue, which binds to the proteins but not to the gel itself. Each band on the gel represents a different protein (or protein subunit); smaller proteins move through the gel more rapidly than larger proteins and therefore are found nearer the bottom of the gel.

This gel illustrates purification of the RecA protein of *Escherichia coli* (described in Chapter 25). The gene for the RecA protein was cloned (Chapter 9) so that its expression (synthesis of the protein) could be controlled. The first lane shows a set of standard proteins (of known M_r), serving as molecular weight markers. The next two lanes show proteins from *E. coli* cells before and after synthesis of RecA protein was induced. The fourth lane shows the proteins in a crude cellular extract. Subsequent lanes (left to right) show the proteins present after successive purification steps. The purified protein is a single polypeptide chain (M_r ~38,000), as seen in the rightmost lane. [Source: (b) Dr. Julia Cox.]

mobility, μ, of a molecule is the ratio of its velocity, V, to the electrical potential. Electrophoretic mobility is also equal to the net charge, Z, of the molecule divided by the frictional coefficient, f, which reflects in part a protein's shape. Thus:

$$\mu = \frac{V}{E} = \frac{Z}{f}$$

The migration of a protein in a gel during electrophoresis is therefore a function of its size and its shape.

An electrophoretic method commonly employed for estimation of purity and molecular weight makes use of the detergent **sodium dodecyl sulfate (SDS)** ("dodecyl" denoting a 12-carbon chain).

$$Na^+ \ ^-O-\overset{\displaystyle O}{\underset{\displaystyle O}{\overset{\|}{\underset{\|}{S}}}}-O-(CH_2)_{11}CH_3$$

Sodium dodecyl sulfate
(SDS)

A protein will bind about 1.4 times its weight of SDS, nearly one molecule of SDS for each amino acid residue. The bound SDS contributes a large net negative charge, rendering the intrinsic charge of the protein insignificant and conferring on each protein a similar charge-to-mass ratio. In addition, SDS binding partially unfolds proteins, such that most SDS-bound proteins assume a similar rodlike shape. Electrophoresis in the presence of SDS therefore separates proteins almost exclusively on the basis of mass (molecular weight), with smaller polypeptides migrating more rapidly. After electrophoresis,

the proteins are visualized by adding a dye such as Coomassie blue, which binds to proteins but not to the gel itself (Fig. 3-18b). Thus, a researcher can monitor the progress of a protein purification procedure as the number of protein bands visible on the gel decreases after each new fractionation step. When compared with the positions to which proteins of known molecular weight migrate in the gel, the position of an unidentified protein can provide a good approximation of its molecular weight **(Fig. 3-19)**. If the protein has two or more different subunits, the subunits are generally separated by the SDS treatment, and a separate band appears for each.

Isoelectric focusing is a procedure used to determine the isoelectric point (pI) of a protein **(Fig. 3-20)**. A pH gradient is established by allowing a mixture of low molecular weight organic acids and bases (ampholytes; p. 81) to distribute themselves in an electric field generated across the gel. When a protein mixture is applied, each protein migrates until it reaches the pH that matches its pI. Proteins with different isoelectric points are thus distributed differently throughout the gel.

Combining isoelectric focusing and SDS electrophoresis sequentially in a process called **two-dimensional electrophoresis** permits the resolution of complex mixtures of proteins **(Fig. 3-21)**. This is a more sensitive analytical method than either electrophoretic method alone. Two-dimensional electrophoresis separates proteins of identical molecular weight that differ in pI, or proteins with similar pI values but different molecular weights.

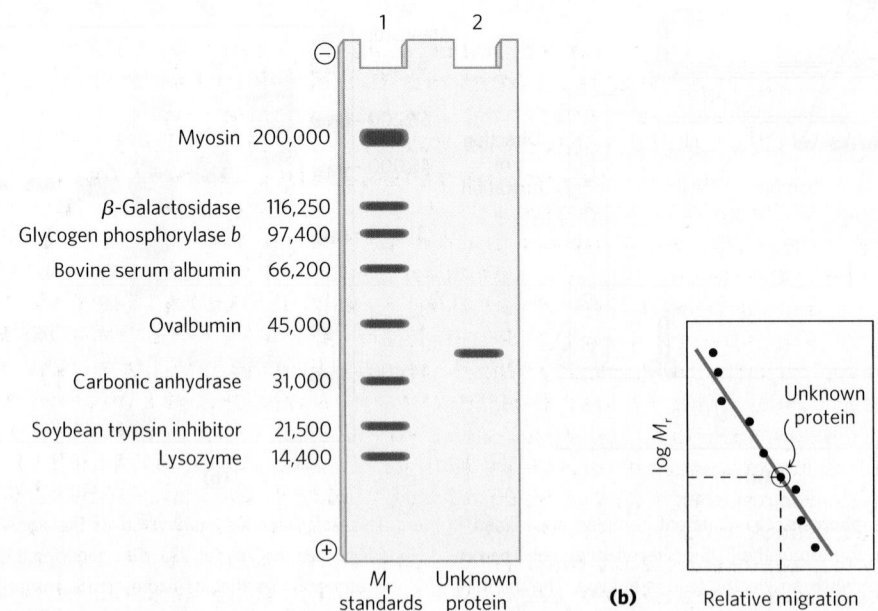

(a) **(b)**

FIGURE 3-19 Estimating the molecular weight of a protein. The electrophoretic mobility of a protein on an SDS polyacrylamide gel is related to its molecular weight, M_r. **(a)** Standard proteins of known molecular weight are subjected to electrophoresis (lane 1). These marker proteins can be used to estimate the molecular weight of an unknown protein (lane 2). **(b)** A plot of log M_r of the marker proteins versus relative migration during electrophoresis is linear, which allows the molecular weight of the unknown protein to be read from the graph. (In similar fashion, a set of standard proteins with reproducible retention times on a size-exclusion column can be used to create a standard curve of retention time versus log M_r. The retention time of an unknown substance on the column can be compared with this standard curve to obtain an approximate M_r.)

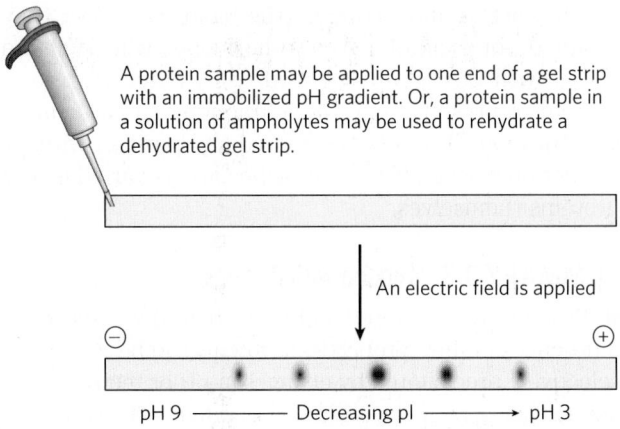

A protein sample may be applied to one end of a gel strip with an immobilized pH gradient. Or, a protein sample in a solution of ampholytes may be used to rehydrate a dehydrated gel strip.

An electric field is applied

⊖   ⊕

pH 9 ———————→ Decreasing pI ———————→ pH 3

After staining, proteins are shown to be distributed along pH gradient according to their pI values.

FIGURE 3-20 Isoelectric focusing. This technique separates proteins according to their isoelectric points. A protein mixture is placed on a gel strip containing an immobilized pH gradient. With an applied electric field, proteins enter the gel and migrate until each reaches a pH equivalent to its pI. Remember that when pH = pI, the net charge of a protein is zero.

Unseparated Proteins Can Be Quantified

To purify a protein, it is essential to have a way of detecting and quantifying that protein in the presence of many other proteins at each stage of the procedure. Often, purification must proceed in the absence of any information about the size and physical properties of the protein or about the fraction of the total protein mass it represents in the extract. For proteins that are enzymes, the amount in a given solution or tissue extract can be measured, or assayed, in terms of the catalytic effect the enzyme produces—that is, the *increase* in the rate at which its substrate is converted to reaction products when the enzyme is present. For this purpose the researcher must know (1) the overall equation of the reaction catalyzed, (2) an analytical procedure for determining the disappearance of the substrate or the appearance of a reaction product, (3) whether the enzyme requires cofactors such as metal ions or coenzymes, (4) the dependence of the enzyme activity on substrate concentration, (5) the optimum pH, and (6) a temperature zone in which the enzyme is stable and has high activity. Enzymes are usually assayed at their optimum pH and at some convenient temperature within the range 25 to 38 °C. Also, very high substrate concentrations are generally used so that the initial reaction rate, measured experimentally, is proportional to enzyme concentration (Chapter 6).

By international agreement, 1.0 unit of enzyme activity for most enzymes is defined as the amount of enzyme causing transformation of 1.0 μmol of substrate to product per minute at 25 °C under optimal conditions of measurement (for some enzymes, this definition is inconvenient, and a unit may be defined differently). The term **activity** refers to the total units of enzyme in a solution. The **specific activity** is the number of enzyme

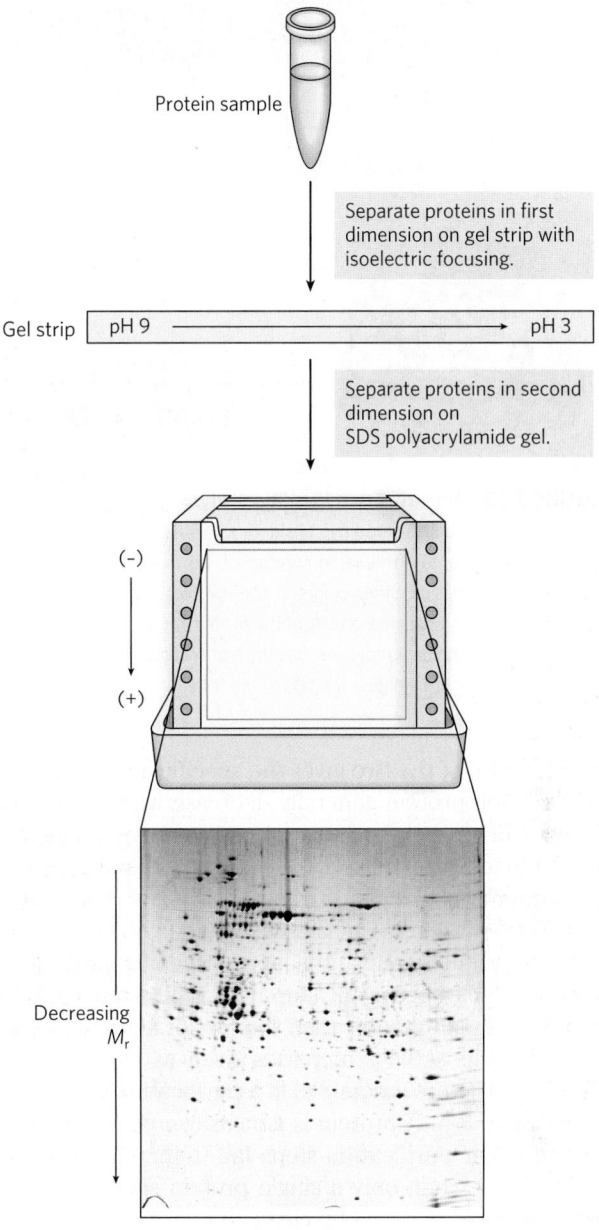

Protein sample

Separate proteins in first dimension on gel strip with isoelectric focusing.

Gel strip   pH 9 ———————————————→ pH 3

Separate proteins in second dimension on SDS polyacrylamide gel.

(−)

Decreasing M_r

(+)

←——————— Decreasing pI ———————→

FIGURE 3-21 Two-dimensional electrophoresis. Proteins are first separated by isoelectric focusing in a thin strip gel. The gel is then laid horizontally on a second, slab-shaped gel, and the proteins are separated by SDS polyacrylamide gel electrophoresis. Horizontal separation reflects differences in pI; vertical separation reflects differences in molecular weight. The original protein complement is thus spread in two dimensions. Thousands of cellular proteins can be resolved using this technique. Individual protein spots can be cut out of the gel and identified by mass spectrometry (see Figs 3-30 and 3-31). [Source: Courtesy of Axel Mogk, from A. Mogk et al., *EMBO J.* 18:6934, 1999, Fig. 7A.]

units per milligram of total protein **(Fig. 3-22)**. The specific activity is a measure of enzyme purity: it increases during purification of an enzyme and becomes maximal and constant when the enzyme is pure (Table 3-5.).

After each purification step, the activity of the preparation (in units of enzyme activity) is assayed, the total amount of protein is determined independently,

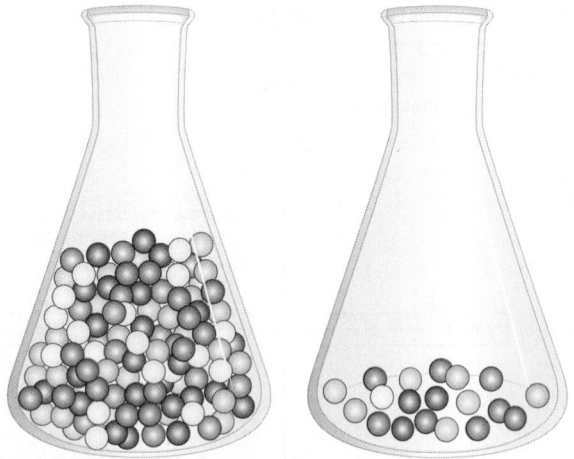

FIGURE 3-22 Activity versus specific activity. The difference between these terms can be illustrated by considering two flasks containing marbles. The flasks contain the same number of red marbles, but different numbers of marbles of other colors. If the marbles represent proteins, both flasks contain the same *activity* of the protein represented by the red marbles. The second flask, however, has the higher *specific activity* because red marbles represent a higher fraction of the total.

and the ratio of the two gives the specific activity. Activity and total protein generally decrease with each step. Activity decreases because there is always some loss due to inactivation or nonideal interactions with chromatographic materials or other molecules in the solution. Total protein decreases because the objective is to remove as much unwanted or nonspecific protein as possible. In a successful step, the loss of nonspecific protein is much greater than the loss of activity; therefore, specific activity increases even as total activity falls. The data are assembled in a purification table similar to Table 3-5. A protein is generally considered pure when further purification steps fail to increase specific activity and when only a single protein species can be detected (for example, by electrophoresis).

For proteins that are not enzymes, other quantification methods are required. Transport proteins can be assayed by their binding to the molecule they transport, and hormones and toxins by the biological effect they produce; for example, growth hormones will stimulate the growth of certain cultured cells. Some structural proteins represent such a large fraction of a tissue mass that they can be readily extracted and purified without a functional assay. The approaches are as varied as the proteins themselves.

SUMMARY 3.3 Working with Proteins

■ Proteins are separated and purified on the basis of differences in their properties. Proteins can be selectively precipitated by changes in pH or temperature, and particularly by the addition of certain salts. A wide range of chromatographic procedures makes use of differences in size, binding affinities, charge, and other properties. These include ion-exchange, size-exclusion, affinity, and high-performance liquid chromatography.

■ Electrophoresis separates proteins on the basis of mass or charge. SDS gel electrophoresis and isoelectric focusing can be used separately or in combination for higher resolution.

■ All purification procedures require a method for quantifying or assaying the protein of interest in the presence of other proteins. Purification can be monitored by assaying specific activity.

3.4 The Structure of Proteins: Primary Structure

Purification of a protein is usually only a prelude to a detailed biochemical dissection of its structure and function. What is it that makes one protein an enzyme, another a hormone, another a structural protein, and still another an antibody? How do they differ chemically? The most obvious distinctions are structural, and to protein structure we now turn.

The structure of large molecules such as proteins can be described at several levels of complexity, arranged in a kind of conceptual hierarchy. Four levels of protein structure are commonly defined **(Fig. 3-23)**.

FIGURE 3-23 Levels of structure in proteins. The *primary structure* consists of a sequence of amino acids linked together by peptide bonds and includes any disulfide bonds. The resulting polypeptide can be arranged into units of *secondary structure*, such as an α helix. The helix is a part of the *tertiary structure* of the folded polypeptide, which is itself one of the subunits that make up the *quaternary structure* of the multisubunit protein, in this case hemoglobin. [Source: PDB ID 1HGA, R. Liddington et al., *J. Mol. Biol.* 228:551, 1992.]

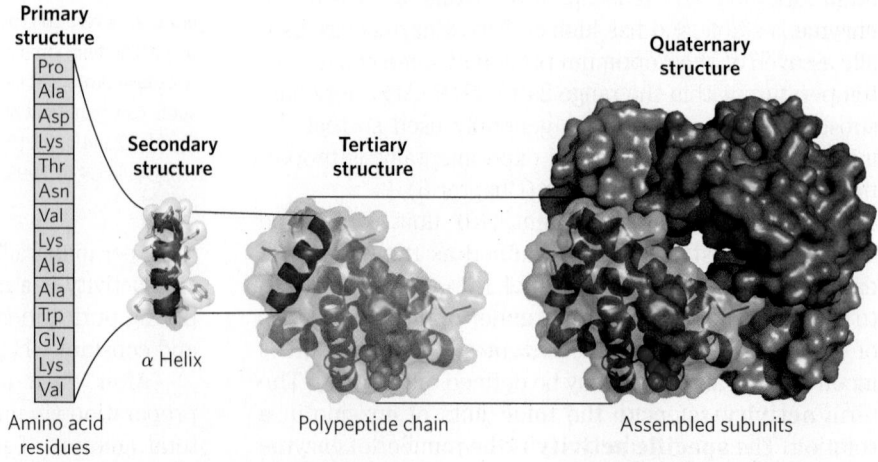

A description of all covalent bonds (mainly peptide bonds and disulfide bonds) linking amino acid residues in a polypeptide chain is its **primary structure**. The most important element of primary structure is the *sequence* of amino acid residues. **Secondary structure** refers to particularly stable arrangements of amino acid residues giving rise to recurring structural patterns. **Tertiary structure** describes all aspects of the three-dimensional folding of a polypeptide. When a protein has two or more polypeptide subunits, their arrangement in space is referred to as **quaternary structure**. Our exploration of proteins will eventually include complex protein machines consisting of dozens to thousands of subunits. Primary structure is the focus of the remainder of this chapter; the higher levels of structure are discussed in Chapter 4.

Differences in primary structure can be especially informative. Each protein has a distinctive number and sequence of amino acid residues. As we shall see in Chapter 4, the primary structure of a protein determines how it folds up into its unique three-dimensional structure, and this in turn determines the function of the protein. We first consider empirical clues that amino acid sequence and protein function are closely linked, then describe how amino acid sequence is determined; finally, we outline the many uses to which this information can be put.

The Function of a Protein Depends on Its Amino Acid Sequence

The bacterium *Escherichia coli* produces more than 3,000 different proteins; a human has ~20,000 genes encoding a much larger number of proteins (through genetic processes discussed in Part III of this text). In both cases, each type of protein has a unique amino acid sequence that confers a particular three-dimensional structure. This structure in turn confers a unique function.

Some simple observations illustrate the importance of primary structure, or the amino acid sequence of a protein. First, as we have already noted, proteins with different functions always have different amino acid sequences. Second, thousands of human genetic diseases have been traced to the production of defective proteins. The defect can range from a single change in the amino acid sequence (as in sickle cell disease, described in Chapter 5) to deletion of a larger portion of the polypeptide chain (as in most cases of Duchenne muscular dystrophy: a large deletion in the gene encoding the protein dystrophin leads to production of a shortened, inactive protein). Finally, on comparing functionally similar proteins from different species, we find that these proteins often have similar amino acid sequences. Thus, a close link between protein primary structure and function is evident.

Is the amino acid sequence absolutely fixed, or invariant, for a particular protein? No; some flexibility is possible. An estimated 20% to 30% of the proteins in humans are **polymorphic**, having amino acid sequence variants in the human population. Many of these variations in sequence have little or no effect on the function of the protein. Furthermore, proteins that carry out a broadly similar function in distantly related species can differ greatly in overall size and amino acid sequence.

Although the amino acid sequence in some regions of the primary structure might vary considerably without affecting biological function, most proteins contain crucial regions that are essential to their function and thus have sequences that are conserved. The fraction of the overall sequence that is critical varies from protein to protein, complicating the task of relating sequence to three-dimensional structure, and structure to function. Before we can consider this problem further, however, we must examine how sequence information is obtained.

The Amino Acid Sequences of Millions of Proteins Have Been Determined

Two major discoveries in 1953 were of crucial importance in the history of biochemistry. In that year, James D. Watson and Francis Crick deduced the double-helical structure of DNA and proposed a structural basis for its precise replication (Chapter 8). Their proposal illuminated the molecular reality behind the idea of a gene. In the same year, Frederick Sanger worked out the sequence of amino acid residues in the polypeptide chains of the hormone insulin **(Fig. 3-24)**, surprising many researchers who had long thought that determining the amino acid sequence of a polypeptide would be a hopelessly difficult task. It quickly became evident that the nucleotide sequence in DNA and the amino acid sequence in proteins were somehow related. Barely a decade after these discoveries, the genetic code relating the nucleotide sequence of DNA to the amino acid sequence of protein molecules was elucidated (Chapter 27). The amino acid sequences of proteins are now most often derived indirectly from the DNA sequences in genome databases. However, an array of techniques derived from traditional methods of polypeptide sequencing still command an important place in protein chemistry. We will summarize the traditional method and mention a few of the techniques derived from it.

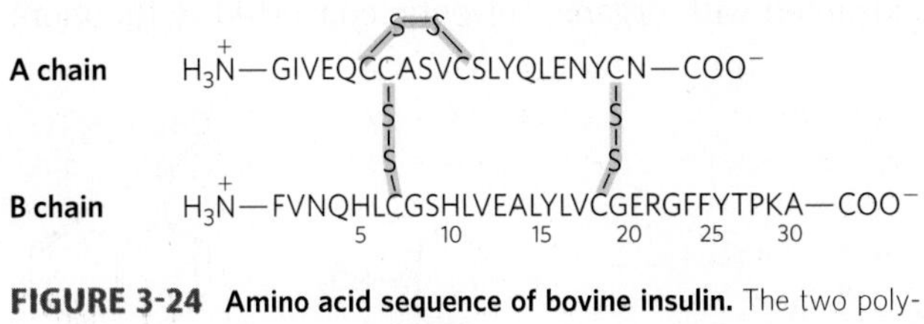

A chain $H_3\overset{+}{N}$—GIVEQCCASVCSLYQLENYCN—COO⁻

B chain $H_3\overset{+}{N}$—FVNQHLCGSHLVEALYLVCGERGFFYTPKA—COO⁻

FIGURE 3-24 Amino acid sequence of bovine insulin. The two polypeptide chains are joined by disulfide cross-linkages (yellow). The A chain of insulin is identical in human, pig, dog, rabbit, and sperm whale insulins. The B chains of the cow, pig, dog, goat, and horse are identical.

Protein Chemistry Is Enriched by Methods Derived from Classical Polypeptide Sequencing

The methods used in the 1950s by Fred Sanger to determine the sequence of the protein insulin are summarized, in their modern form, in **Figure 3-25**. Few proteins are sequenced in this way now, at least in their entirety. As noted above, the sequence of a protein can usually be predicted from the sequence of the gene encoding it, information now readily available in ever-growing genomic databases. However, the traditional sequencing protocols have provided a rich array of tools for biochemists, and almost every step in Figure 3-25 makes use of methods that are widely used in biochemistry labs, sometimes in quite different contexts.

Frederick Sanger, 1918–2013 [Source: UPI/Corbis-Bettmann.]

In the traditional scheme for sequencing large proteins, the amino-terminal amino acid residue was first labeled and its identity determined. The amino-terminal α-amino group can be labeled with 1-fluoro-2,4-dinitrobenzene (FDNB), dansyl chloride, or dabsyl chloride **(Fig. 3-26)**.

The chemical sequencing process itself is based on a two-step process developed by Pehr Edman **(Fig. 3-27)**. The **Edman degradation** procedure labels and removes only the amino-terminal residue from a peptide, leaving all other peptide bonds intact. The peptide is reacted with phenylisothiocyanate under mildly alkaline conditions, which converts the amino-terminal amino acid to a phenylthiocarbamoyl (PTC) adduct. The peptide bond next to the PTC adduct is then cleaved in a step carried out in anhydrous trifluoroacetic acid, with removal of the amino-terminal amino acid as an anilinothiazolinone derivative. The derivatized amino acid is extracted with organic solvents, converted to the more

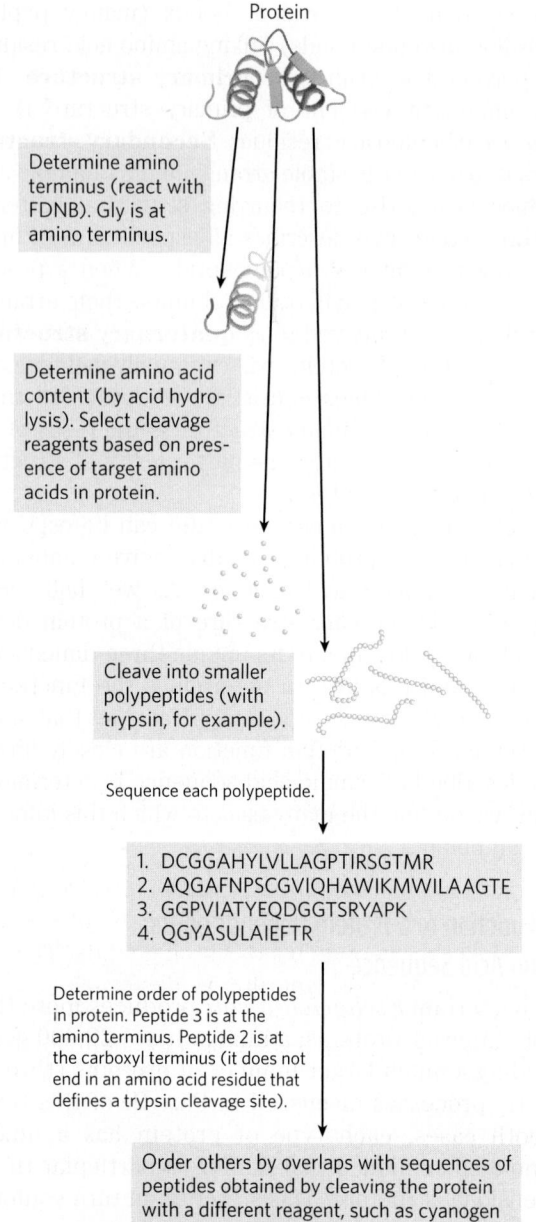

Protein

Determine amino terminus (react with FDNB). Gly is at amino terminus.

Determine amino acid content (by acid hydrolysis). Select cleavage reagents based on presence of target amino acids in protein.

Cleave into smaller polypeptides (with trypsin, for example).

Sequence each polypeptide.

1. DCGGAHYLVLLAGPTIRSGTMR
2. AQGAFNPSCGVIQHAWIKMWILAAGTE
3. GGPVIATYEQDGGTSRYAPK
4. QGYASULAIEFTR

Determine order of polypeptides in protein. Peptide 3 is at the amino terminus. Peptide 2 is at the carboxyl terminus (it does not end in an amino acid residue that defines a trypsin cleavage site).

Order others by overlaps with sequences of peptides obtained by cleaving the protein with a different reagent, such as cyanogen bromide or chymotrypsin.

FIGURE 3-25 Direct protein sequencing. The procedures shown here were developed by Fred Sanger to sequence insulin and have since been used for many additional proteins. FDNB is 1-fluoro-2,4-dinitrobenzene (see text and Fig. 3-26).

stable phenylthiohydantoin derivative by treatment with aqueous acid, and then identified. The use of sequential reactions carried out under first basic and then acidic conditions provides a means of controlling the entire

FDNB   Dansyl chloride   Dabsyl chloride

FIGURE 3-26 Reagents used to modify the α-amino group at the amino terminus.

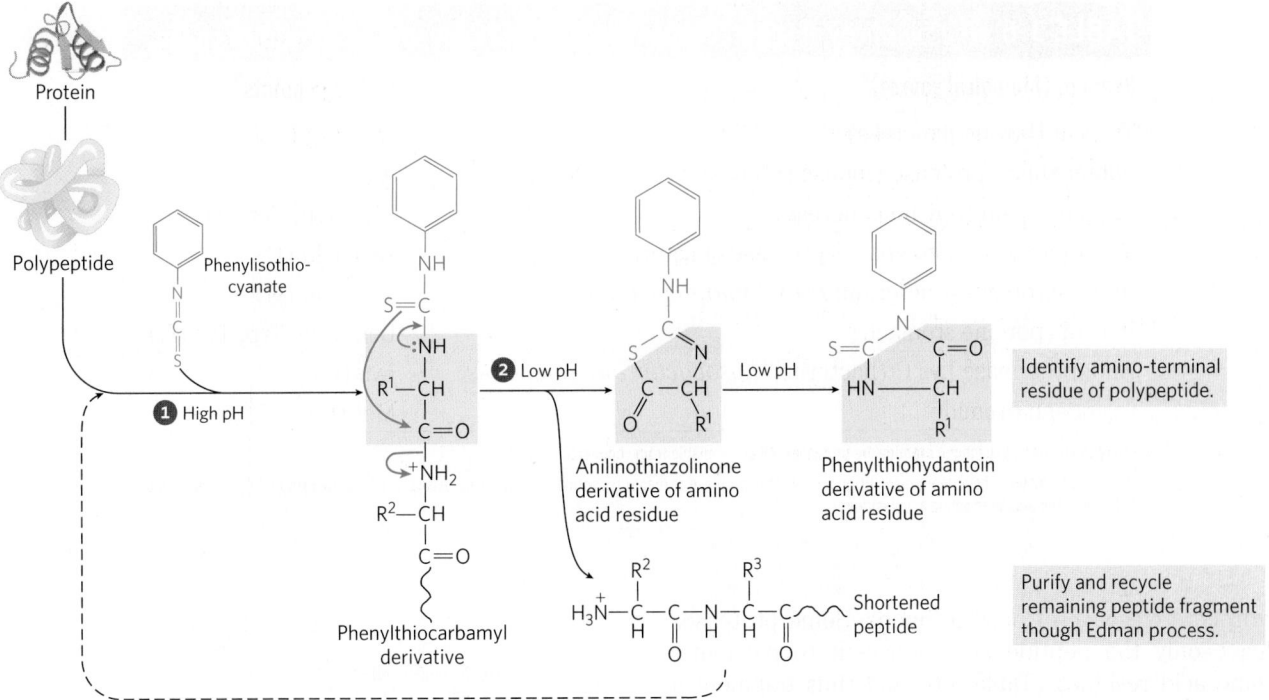

FIGURE 3-27 The protein sequencing chemistry devised by Pehr Edman. The peptide bond nearest to the amino terminus of the protein or polypeptide is cleaved in two steps. The two steps are carried out under very different reaction conditions (basic conditions in step ❶, acidic in step ❷), allowing one step to proceed to completion before the second is initiated.

process. Each reaction with the amino-terminal amino acid can go essentially to completion without affecting any of the other peptide bonds in the peptide. The process is repeated until, typically, as many as 40 sequential amino acid residues are identified. The reactions of the Edman degradation have been automated.

To determine the sequence of large proteins, early developers of sequencing protocols had to devise methods to eliminate disulfide bonds and to cleave proteins precisely into smaller polypeptides. Two approaches to irreversible breakage of disulfide bonds are outlined in **Figure 3-28**. Enzymes called **proteases** catalyze the

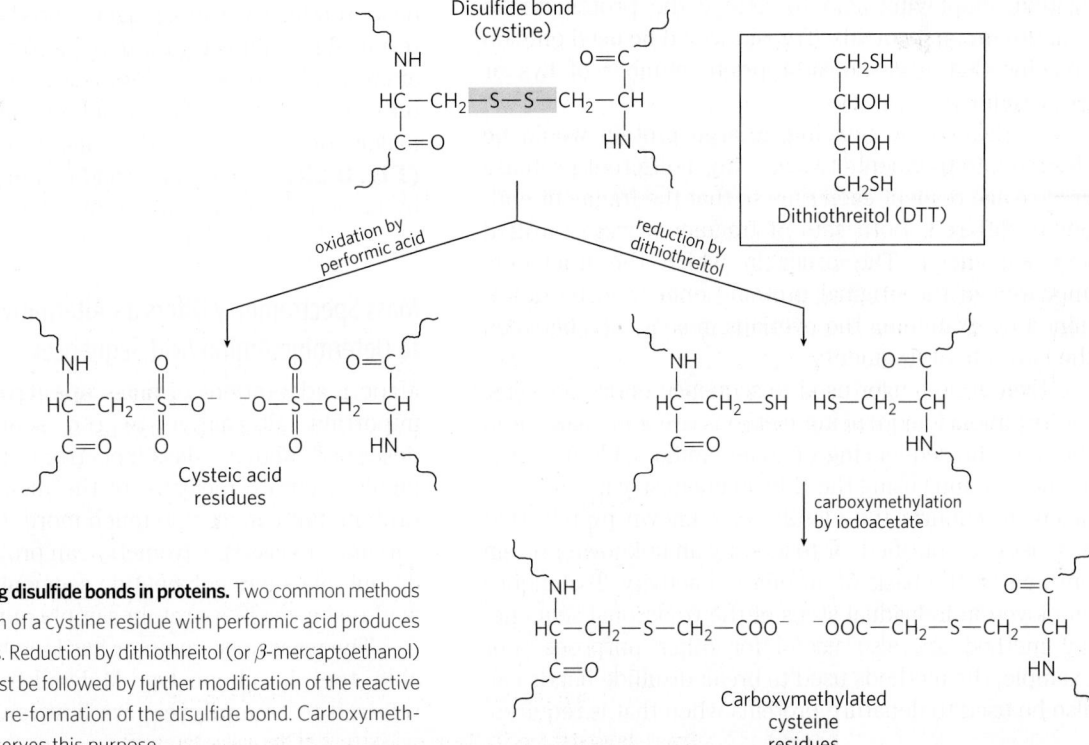

FIGURE 3-28 Breaking disulfide bonds in proteins. Two common methods are illustrated. Oxidation of a cystine residue with performic acid produces two cysteic acid residues. Reduction by dithiothreitol (or β-mercaptoethanol) to form Cys residues must be followed by further modification of the reactive —SH groups to prevent re-formation of the disulfide bond. Carboxymethylation by iodoacetate serves this purpose.

TABLE 3-6 The Specificity of Some Common Methods for Fragmenting Polypeptide Chains

Reagent (biological source)[a]	Cleavage points[b]
Trypsin (bovine pancreas)	Lys, Arg (C)
Submaxillary protease (mouse submaxillary gland)	Arg (C)
Chymotrypsin (bovine pancreas)	Phe, Trp, Tyr (C)
Staphylococcus aureus V8 protease (bacterium *S. aureus*)	Asp, Glu (C)
Asp-*N*-protease (bacterium *Pseudomonas fragi*)	Asp, Glu (N)
Pepsin (porcine stomach)	Leu, Phe, Trp, Tyr (N)
Endoproteinase Lys C (bacterium *Lysobacter enzymogenes*)	Lys (C)
Cyanogen bromide	Met (C)

[a]All reagents except cyanogen bromide are proteases. All are available from commercial sources.

[b]Residues furnishing the primary recognition point for the protease or reagent; peptide bond cleavage occurs on either the carbonyl (C) or the amino (N) side of the indicated amino acid residues.

hydrolytic cleavage of peptide bonds. Some proteases cleave only the peptide bond adjacent to particular amino acid residues (Table 3-6) and thus fragment a polypeptide chain in a predictable and reproducible way. A few chemical reagents also cleave the peptide bond adjacent to specific residues. Among proteases, the digestive enzyme trypsin catalyzes the hydrolysis of only those peptide bonds in which the carbonyl group is contributed by either a Lys or an Arg residue, regardless of the length or amino acid sequence of the chain. A polypeptide with three Lys and/or Arg residues will usually yield four smaller peptides on cleavage with trypsin. Moreover, all except one of these will have a carboxyl-terminal Lys or Arg. The choice of a reagent to cleave the protein into smaller peptides can be aided by first determining the amino acid content of the entire protein, employing acid to reduce the protein to its constituent amino acids. Trypsin would be used only on proteins that have an appropriate number of Lys or Arg residues.

In classical sequencing, a large protein would be cleaved into fragments twice, using a different protease or cleavage reagent each time so that the fragment endpoints differed. Both sets of fragments were purified and sequenced. The order in which the fragments appeared in the original protein could then be determined by examining the overlaps in sequence between the two sets of fragments.

Even if no longer used to sequence entire proteins, the traditional sequencing methods are still valuable in the lab. The sequencing of some amino acids from the amino terminus using the Edman chemistry is often sufficient to confirm the identity of a known protein that has just been purified, or to identify an unknown protein purified on the basis of an unusual activity. Techniques employed in individual steps of the traditional sequencing method are also useful for other purposes. For example, the methods used to break disulfide bonds can also be used to denature proteins when that is required.

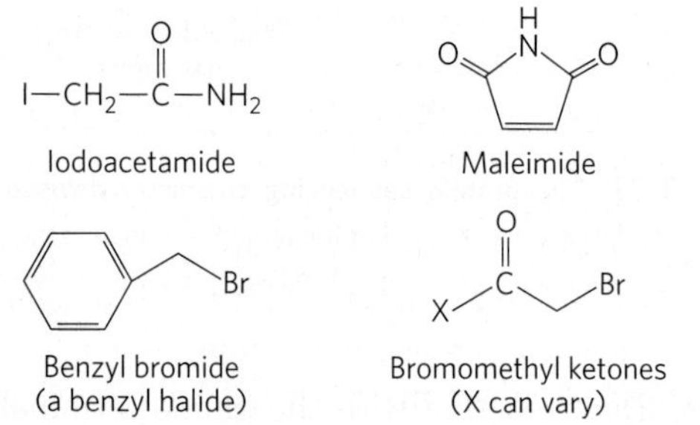

FIGURE 3-29 Reagents used to modify the sulfhydryl groups of Cys residues. (See also Fig. 3-28.)

Furthermore, the effort to label the amino-terminal amino acid residue led eventually to the development of an array of reagents that could react with specific groups on a protein. The same reagents used to label the amino-terminal α-amino group can be used to label the primary amines of Lys residues (Fig. 3-26). The sulfhydryl group on Cys residues can be modified with iodoacetamides, maleimides, benzyl halides, and bromomethyl ketones **(Fig. 3-29)**. Other amino acid residues can be modified by reagents linked to a dye or other molecule to aid in protein detection or functional studies.

Mass Spectrometry Offers an Alternative Method to Determine Amino Acid Sequences

Modern adaptations of **mass spectrometry** provide an important alternative to the sequencing methods described above. Mass spectrometry can provide a highly accurate measure of the molecular weight of a protein, but can also do much more. In particular, some variants of mass spectrometry can provide the sequences of multiple short polypeptide segments (20 to 30 amino acid residues) in a protein sample quite rapidly.

The mass spectrometer has long been an indispensable tool in chemistry. Molecules to be analyzed, referred to as **analytes**, are first ionized in a vacuum.

When the newly charged molecules are introduced into an electric and/or magnetic field, their paths through the field are a function of their mass-to-charge ratio, m/z. This measured property of the ionized species can be used to deduce the mass (m) of the analyte with very high precision.

Although mass spectrometry has been in use for many years, it could not be applied to macromolecules such as proteins and nucleic acids. The m/z measurements are made on molecules in the gas phase, and the heating or other treatment needed to transfer a macromolecule to the gas phase usually caused its rapid decomposition. In 1988, two different techniques were developed to overcome this problem. In one, proteins are placed in a light-absorbing matrix. With a short pulse of laser light, the proteins are ionized and then desorbed from the matrix into the vacuum system. This process, known as **matrix-assisted laser desorption/ionization mass spectrometry**, or **MALDI MS**, has been successfully used to measure the mass of a wide range of macromolecules. In a second and equally successful method, macromolecules in solution are forced directly from the liquid to the gas phase. A solution of analytes is passed through a charged needle that is kept at a high electrical potential, dispersing the solution into a fine mist of charged microdroplets. The solvent surrounding the macromolecules rapidly evaporates, leaving multiply charged macromolecular ions in the gas phase. This technique is called **electrospray ionization mass spectrometry**, or **ESI MS**. Protons added during passage through the needle give additional charge to the macromolecule. The m/z of the molecule can be analyzed in the vacuum chamber.

Mass spectrometry provides a wealth of information for proteomics research, enzymology, and protein chemistry in general. The techniques require only miniscule amounts of sample, so they can be readily applied to the small amounts of protein that can be extracted from a two-dimensional electrophoretic gel. The accurately measured molecular mass of a protein is critical to its identification. Once the mass of a protein is accurately known, mass spectrometry is a convenient and accurate method for detecting changes in mass due to the presence of bound cofactors, bound metal ions, covalent modifications, and so on.

The process for determining the molecular mass of a protein with ESI MS is illustrated in **Figure 3-30**. As it is injected into the gas phase, a protein acquires a variable number of protons, and thus positive charges, from the solvent. The variable addition of these charges creates a spectrum of species with different mass-to-charge ratios. Each successive peak corresponds to a species that differs from that of its neighboring peak by a charge difference of 1 and a mass difference of 1 (one proton). The mass of the protein can be determined from any two neighboring peaks.

Mass spectrometry can also be used to sequence short stretches of polypeptide, an application that has

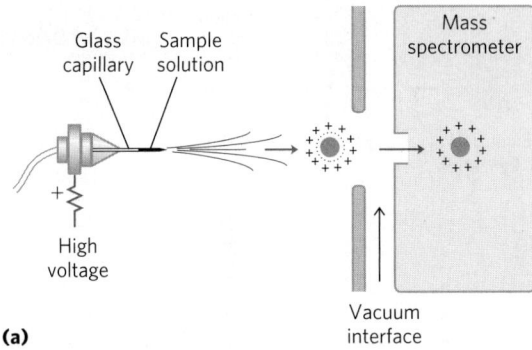

(a)

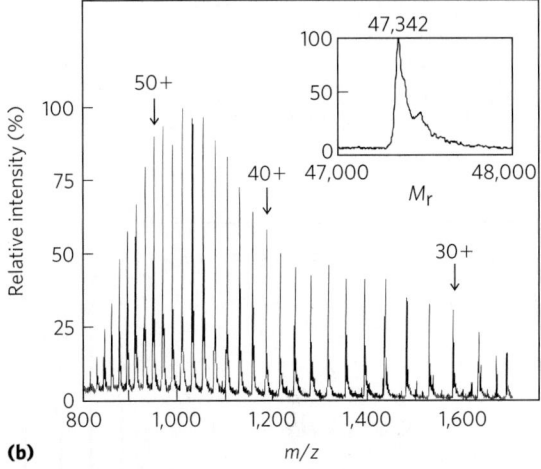

(b)

FIGURE 3-30 Electrospray ionization mass spectrometry of a protein. (a) A protein solution is dispersed into highly charged droplets by passage through a needle under the influence of a high-voltage electric field. The droplets evaporate, and the ions (with added protons in this case) enter the mass spectrometer for m/z measurement. (b) The spectrum generated is a family of peaks, with each successive peak (from right to left) corresponding to a charged species with both mass and charge increased by 1. The inset shows a computer-generated transformation of this spectrum. [Source: Information from M. Mann and M. Wilm, *Trends Biochem. Sci.* 20:219, 1995.]

emerged as an invaluable tool for quickly identifying unknown proteins. Sequence information is extracted using a technique called **tandem MS**, or **MS/MS**. A solution containing the protein under investigation is first treated with a protease or chemical reagent to hydrolyze it to a mixture of shorter peptides. The mixture is then injected into a device that is essentially two mass spectrometers in tandem (**Fig. 3-31a**, top). In the first, the peptide mixture is sorted so that only one of the several types of peptides produced by cleavage emerges at the other end. The sample of the selected peptide, each molecule of which has a charge somewhere along its length, then travels through a vacuum chamber between the two mass spectrometers. In this collision cell, the peptide is further fragmented by high-energy impact with a "collision gas" such as helium or argon that is bled into the vacuum chamber. Each individual peptide is broken in only one place, on average.

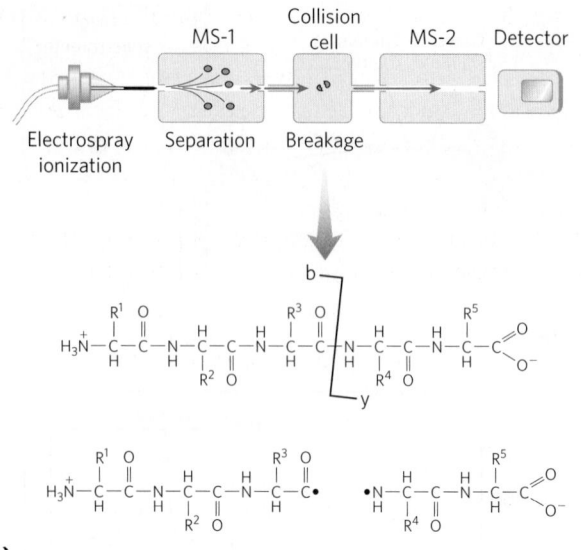

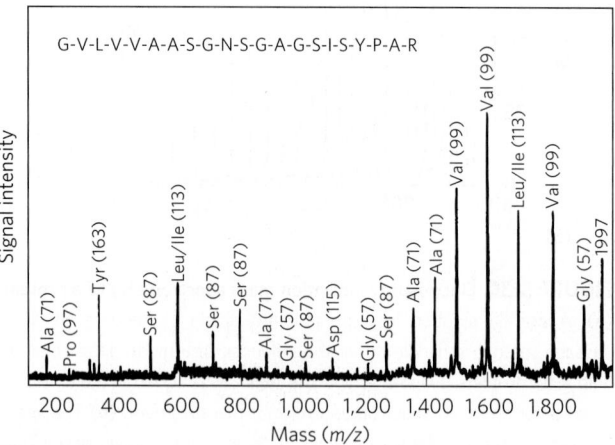

(a)

(b)

FIGURE 3-31 Obtaining protein sequence information with tandem MS.
(a) After proteolytic hydrolysis, a protein solution is injected into a mass spectrometer (MS-1). The different peptides are sorted so that only one type is selected for further analysis. The selected peptide is further fragmented in a chamber between the two mass spectrometers, and m/z for each fragment is measured in the second mass spectrometer (MS-2). Many of the ions generated during this second fragmentation result from breakage of the peptide bond, as shown. These are called b-type or y-type ions, depending on whether the charge is retained on the amino- or carboxyl-terminal side, respectively. **(b)** A typical spectrum with peaks representing the peptide fragments generated from a sample of one small peptide (21 residues). The labeled peaks are y-type ions derived from amino acid residues. The number in parentheses over each peak is the molecular weight of the amino acid ion. The successive peaks differ by the mass of a particular amino acid in the original peptide. The deduced sequence is shown at the top. [Source: Information from T. Keough et al., *Proc. Natl. Acad. Sci. USA* 96:7131, 1999, Fig. 3.]

Although the breaks are not hydrolytic, most occur at the peptide bonds.

The second mass spectrometer then measures the m/z ratios of all the charged fragments. This process generates one or more sets of peaks. A given set of peaks (Fig. 3-31b) consists of all the charged fragments that were generated by breaking the same type of bond (but at different points in the peptide). One set of peaks includes only the fragments in which the charge was retained on the amino-terminal side of the broken bonds; another includes only the fragments in which the charge was retained on the carboxyl-terminal side of the broken bonds. Each successive peak in a given set has one less amino acid than the peak before. The difference in mass from peak to peak identifies the amino acid that was lost in each case, thus revealing the sequence of the peptide. The only ambiguities involve leucine and isoleucine, which have the same mass. Although multiple sets of peaks are usually generated, the two most prominent sets generally consist of charged fragments derived from breakage of the peptide bonds. The amino acid sequence derived from one set can be confirmed by the other, improving the confidence in the sequence information obtained.

The various methods for obtaining protein sequence information complement one another. The Edman degradation procedure is sometimes convenient to get sequence information uniquely from the amino terminus of a protein or peptide. However, it is relatively slow and requires a larger sample than does mass spectrometry. Mass spectrometry can be used for small amounts of sample and for mixed samples. It provides sequence information, but the fragmentation processes can leave unpredictable sequence gaps. Although most protein sequences are now extracted from genomic DNA sequences (Chapter 8) by employing our understanding of the genetic code (Chapter 27), direct protein sequencing is often necessary to identify unknown protein samples. Both protein sequencing methods permit the unambiguous identification of newly purified proteins. Mass spectrometry is the method of choice to identify proteins that are present in small amounts. For example, the technique is sensitive enough to analyze the few hundred nanograms of protein that might be extracted from a single protein band on a polyacrylamide gel. Direct sequencing by mass spectrometry also can reveal the addition of phosphoryl groups or other modifications (Chapter 6). Sequencing by either method can reveal changes in protein sequence that result from the editing of messenger RNA in eukaryotes (Chapter 26). Thus, these methods are all part of a robust toolbox used to investigate proteins and their functions.

Small Peptides and Proteins Can Be Chemically Synthesized

Many peptides are potentially useful as pharmacologic agents, and their production is of considerable commercial importance. There are three ways to obtain a peptide: (1) purification from tissue, a task often made difficult by the vanishingly low concentrations of some peptides; (2) genetic engineering (Chapter 9); or (3) direct chemical synthesis. Powerful techniques now make direct chemical synthesis an attractive option in many cases. In addition to commercial applications, the synthesis of specific peptide portions of larger proteins is an increasingly important tool for the study of protein structure and function.

The complexity of proteins makes the traditional synthetic approaches of organic chemistry impractical for peptides with more than four or five amino acid residues. One problem is the difficulty of purifying the product after each step.

The major breakthrough in this technology was provided by R. Bruce Merrifield in 1962. His innovation was to synthesize a peptide while keeping it attached at one end to a solid support. The support is an insoluble polymer (resin) contained within a column, similar to that used for chromatographic procedures. The peptide is built up on this support one amino acid at a time, through a standard set of reactions in a repeating cycle **(Fig. 3-32)**. At each successive step in the cycle, protective chemical groups block unwanted reactions.

The technology for chemical peptide synthesis is now automated. An important limitation of the process

FIGURE 3-32 Chemical synthesis of a peptide on an insoluble polymer support. Reactions ❶ through ❹ are necessary for the formation of each peptide bond. The 9-fluorenylmethoxycarbonyl (Fmoc) group (shaded blue) prevents unwanted reactions at the α-amino group of the residue (shaded light red). Chemical synthesis proceeds from the carboxyl terminus to the amino terminus, the reverse of the direction of protein synthesis in vivo (Chapter 27).

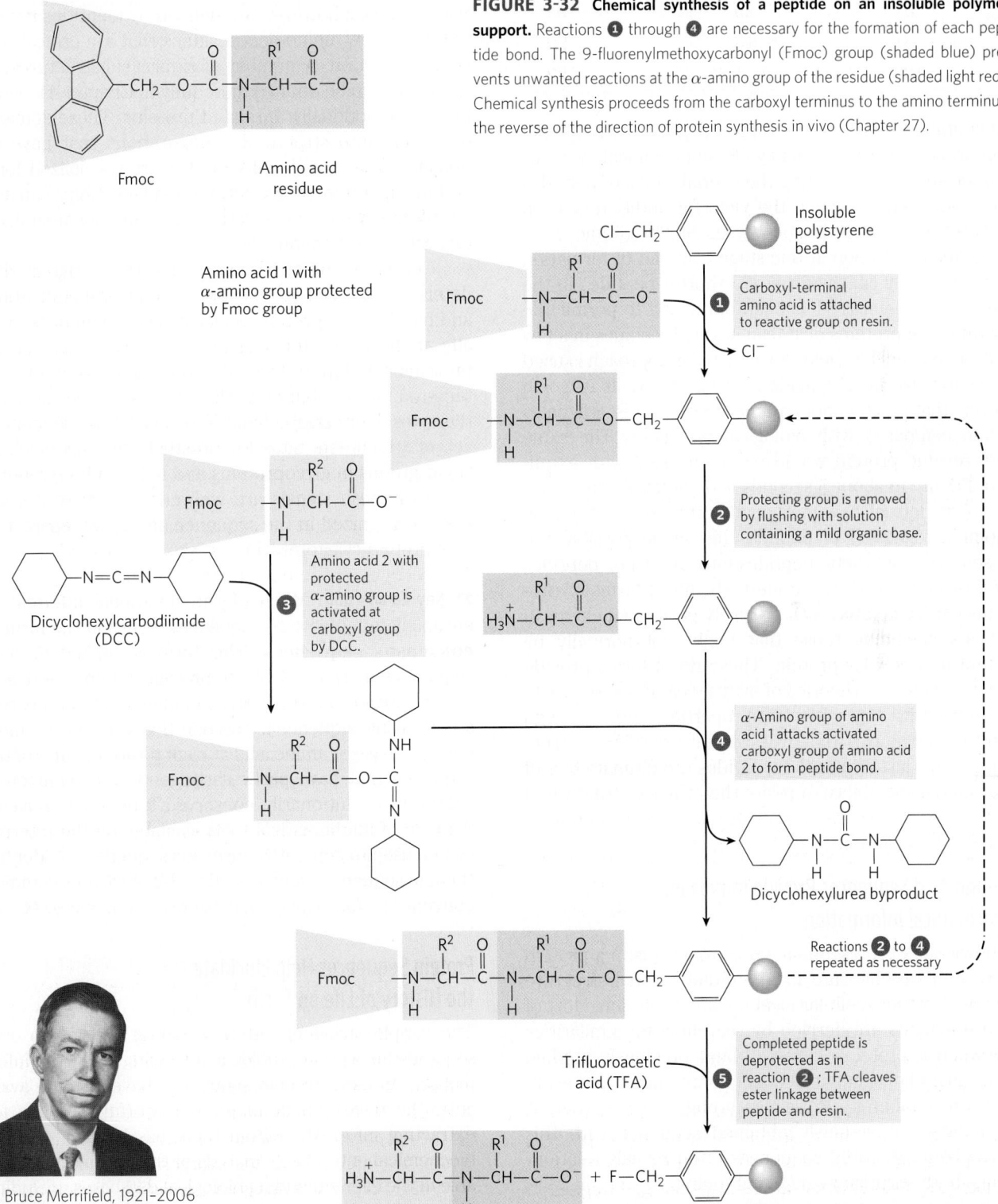

R. Bruce Merrifield, 1921–2006
[Source: Bettmann/Corbis.]

TABLE 3-7 Effect of Stepwise Yield on Overall Yield in Peptide Synthesis

Number of residues in the final polypeptide	Overall yield of final peptide (%) when the yield of each step is:	
	96.0%	99.8%
11	66	98
21	44	96
31	29	94
51	13	90
100	1.8	82

(a limitation shared by the Edman degradation sequencing process) is the efficiency of each chemical cycle, as can be seen by calculating the overall yields of peptides of various lengths when the yield for addition of each new amino acid is 96.0% versus 99.8% (Table 3-7). Incomplete reaction at one stage can lead to formation of an impurity (in the form of a shorter peptide) in the next. The chemistry has been optimized to permit the synthesis of proteins of 100 amino acid residues in a few days in reasonable yield. A very similar approach is used to synthesize nucleic acids (see Fig. 8-33). It is worth noting that this technology, impressive as it is, still pales when compared with biological processes. The same 100-residue protein would be synthesized with exquisite fidelity in about 5 seconds in a bacterial cell.

A variety of new methods for the efficient ligation (joining together) of peptides has made possible the assembly of synthetic peptides into larger polypeptides and proteins. With these methods, novel forms of proteins can be created with precisely positioned chemical groups, including those that might not normally be found in a cellular protein. These novel forms provide new ways to test theories of enzyme catalysis, to create proteins with new chemical properties, and to design protein sequences that will fold into particular structures. This last application provides the ultimate test of our increasing ability to relate the primary structure of a peptide to the three-dimensional structure that it takes up in solution.

Amino Acid Sequences Provide Important Biochemical Information

Knowledge of the sequence of amino acids in a protein can offer insights into its three-dimensional structure and its function, cellular location, and evolution. Most of these insights are derived by searching for similarities between a protein of interest and previously studied proteins. Thousands of sequences are known and available in databases accessible through the Internet. A comparison of a newly obtained sequence with this large bank of stored sequences often reveals relationships both surprising and enlightening.

Exactly how the amino acid sequence determines three-dimensional structure is not understood in detail, nor can we always predict function from sequence. However, protein families that have some shared structural or functional features can be readily identified on the basis of amino acid sequence similarities. Individual proteins are assigned to families based on the degree of similarity in amino acid sequence. Members of a family are usually identical across 25% or more of their sequences, and proteins in these families generally share at least some structural and functional characteristics. Some families, however, are defined by identities involving only a few amino acid residues that are critical to a certain function. A number of similar substructures, or "domains" (to be defined more fully in Chapter 4), occur in many functionally unrelated proteins. These domains often fold into structural configurations that have an unusual degree of stability or that are specialized for a certain environment. Evolutionary relationships can also be inferred from the structural and functional similarities within protein families.

Certain amino acid sequences serve as signals that determine the cellular location, chemical modification, and half-life of a protein. Special signal sequences, usually at the amino terminus, are used to target certain proteins for export from the cell; other proteins are targeted for distribution to the nucleus, the cell surface, the cytosol, or other cellular locations. Other sequences act as attachment sites for prosthetic groups, such as sugar groups in glycoproteins and lipids in lipoproteins. Some of these signals are well characterized and are easily recognized in the sequence of a newly characterized protein (Chapter 27).

≫ Key Convention: Much of the functional information encapsulated in protein sequences comes in the form of **consensus sequences**. This term is applied to such sequences in DNA, RNA, or protein. When a series of related nucleic acid or protein sequences are compared, a consensus sequence is the one that reflects the most common base or amino acid at each position. Parts of the sequence that have particularly good agreement often represent evolutionarily conserved functional domains. A range of mathematical tools available on the Internet can be used to generate consensus sequences or identify them in sequence databases. Box 3-2 illustrates common conventions for displaying consensus sequences. **≪**

Protein Sequences Help Elucidate the History of Life on Earth

The simple string of letters denoting the amino acid sequence of a protein holds a surprising wealth of information. As more protein sequences have become available, the development of more powerful methods for extracting information from them has become a major biochemical enterprise. Analysis of the information available in the ever-expanding biological databases, including

BOX 3-2   Consensus Sequences and Sequence Logos

Consensus sequences can be represented in several ways. To illustrate two types of conventions, we use two examples of consensus sequences (Fig. 1): an ATP-binding structure called a P loop (see Box 12-2) and a Ca^{2+}-binding structure called an EF hand (see Fig. 12-12). The rules described here are adapted from those used by the sequence comparison website PROSITE (http://prosite.expasy.org/sequence_logo.html), using the standard one-letter codes for the amino acids.

[AG]-x(4)-G-K-[ST].

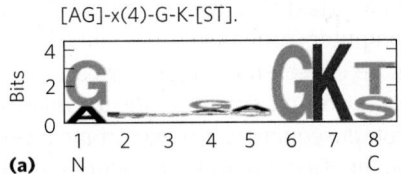

(a)

D-{W}-[DNS]-{ILVFYW}-[DENSTG]-[DNQGHRK]-{GP}-[LIVMC]-[DENQSTAGC]-x(2)-[DE]-[LIVMFYW].

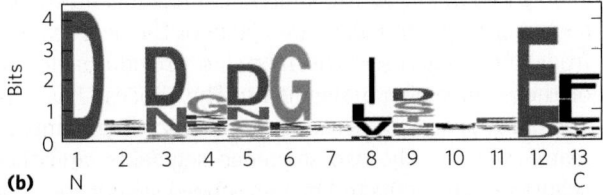

(b)

FIGURE 1 Representations of two consensus sequences. **(a)** P loop, an ATP-binding structure; **(b)** EF hand, a Ca^{2+}-binding structure. [Sources: Sequence data for (a) from document ID PDOC00017 and for (b) from document ID PDOC00018, www.expasy.org/prosite, N. Hulo et al., *Nucleic Acids Res.* 34:D227, 2006. Sequence logos created with WebLogo, http://weblogo.berkeley.edu, G. E. Crooks et al., *Genome Res.* 14:1188, 2004.]

In one type of consensus sequence designation, shown at the top of (a) and (b) in Figure 1, each position is separated from its neighbor by a hyphen. A position where any amino acid is allowed is designated x. Ambiguities are indicated by listing the acceptable amino acids for a given position between square brackets. For example, in (a), [AG] means Ala or Gly.

If all but a few amino acids are allowed at one position, the amino acids that are *not* allowed are listed between curly brackets. For example, in (b), {W} means any amino acid except Trp. Repetition of an element of the pattern is indicated by following that element with a number or range of numbers between parentheses. In (a), for example, x(4) means x-x-x-x; x(2,4) would mean x-x, or x-x-x, or x-x-x-x. When a pattern is restricted to either the amino or carboxyl terminus of a sequence, that pattern starts with < or ends with >, respectively (not so for either example here). A period ends the pattern. Applying these rules to the consensus sequence in (a), either A or G can be found at the first position. Any amino acid can occupy the next four positions, followed by an invariant G and an invariant K. The last position is either S or T.

Sequence logos provide a more informative and graphic representation of an amino acid (or nucleic acid) multiple sequence alignment. Each logo consists of a stack of symbols for each position in the sequence. The overall height of the stack (in bits) indicates the degree of sequence conservation at that position, while the height of each symbol (letter) in the stack indicates the relative frequency of that amino acid (or nucleotide). For amino acid sequences, the colors denote the characteristics of the amino acid: polar (G, S, T, Y, C, Q, N), green; basic (K, R, H), blue; acidic (D, E), red; and hydrophobic (A, V, L, I, P, W, F, M), black. The classification of amino acids in this scheme is somewhat different from that in Table 3-1 and Figure 3-5. The amino acids with aromatic side chains are subsumed into the nonpolar (F, W) and polar (Y) classifications. Glycine, always hard to classify, is assigned to the polar group. Note that when multiple amino acids are acceptable at a particular position, they rarely occur with equal probability. One or a few usually predominate. The logo representation makes the predominance clear, and a conserved sequence in a protein is made obvious. However, the logo obscures some amino acid residues that may be allowed at a position, such as the Cys that occasionally occurs at position 8 of the EF hand in (b).

gene and protein sequences and macromolecular structures, has given rise to the new field of **bioinformatics**. One outcome of this discipline is a growing suite of computer programs, many readily available on the Internet, that can be used by any scientist, student, or knowledgeable layperson. Each protein's function relies on its three-dimensional structure, which in turn is determined largely by its primary structure. Thus, the biochemical information conveyed by a protein sequence is limited only by our understanding of structural and functional principles. The constantly evolving tools of bioinformatics

make it possible to identify functional segments in new proteins and help establish both their sequence and their structural relationships to proteins already in the databases. On a different level of inquiry, protein sequences are beginning to tell us how the proteins evolved and, ultimately, how life evolved on this planet.

The field of molecular evolution is often traced to Emile Zuckerkandl and Linus Pauling, whose work in the mid-1960s advanced the use of nucleotide and protein sequences to explore evolution. The premise is deceptively straightforward. If two organisms are closely related, the sequences of their genes and proteins should be similar. The sequences increasingly diverge as the evolutionary distance between two organisms increases. The promise of this approach began to be realized in the 1970s, when Carl Woese used ribosomal RNA sequences to define the Archaea as a group of living organisms distinct from the Bacteria and Eukarya (see Fig. 1-5). Protein sequences offer an opportunity to greatly refine the available information. With the advent of genome projects investigating organisms from bacteria to humans, the number of available sequences is growing at an enormous rate. This information can be used to trace biological history. The challenge is in learning to read the genetic hieroglyphics.

Evolution has not taken a simple linear path. Complexities abound in any attempt to mine the evolutionary information stored in protein sequences. For a given protein, the amino acid residues essential for the activity of the protein are conserved over evolutionary time. The residues that are less important to function may vary over time—that is, one amino acid may substitute for another—and these variable residues can provide the information to trace evolution. Amino acid substitutions are not always random, however. At some positions in the primary structure, the need to maintain protein function may mean that only particular amino acid substitutions can be tolerated. Some proteins have more variable amino acid residues than others. For these and other reasons, different proteins evolve at different rates.

Another complicating factor in tracing evolutionary history is the rare transfer of a gene or group of genes from one organism to another, a process called **horizontal gene transfer**. The transferred genes may be similar to the genes they were derived from in the original organism, whereas most other genes in the same two organisms may be only distantly related. An example of horizontal gene transfer is the recent rapid spread of antibiotic-resistance genes in bacterial populations. The proteins derived from these transferred genes would not be good candidates for the study of bacterial evolution because they share only a very limited evolutionary history with their "host" organisms.

The study of molecular evolution generally focuses on families of closely related proteins. In most cases, the families chosen for analysis have essential functions in cellular metabolism that must have been present in the earliest viable cells, thus greatly reducing the chance that they were introduced relatively recently by horizontal gene transfer. For example, a protein called EF-1α (elongation factor 1α) is involved

in the synthesis of proteins in all eukaryotes. A similar protein, EF-Tu, with the same function, is found in bacteria. Similarities in sequence and function indicate that EF-1α and EF-Tu are members of a family of proteins that share a common ancestor. The members of protein families are called **homologous proteins**, or **homologs**. The concept of a homolog can be further refined. If two proteins in a family (that is, two homologs) are present in the same species, they are referred to as **paralogs**. Homologs from different species are called **orthologs**. The process of tracing evolution involves first identifying suitable families of homologous proteins and then using them to reconstruct evolutionary paths.

Homologs are identified through the use of increasingly powerful computer programs that can directly compare two or more chosen protein sequences, or can search vast databases to find the evolutionary relatives of one selected protein sequence. The electronic search process can be thought of as sliding one sequence past the other until a section with a good match is found. Within this sequence alignment, a positive score is assigned for each position where the amino acid residues in the two sequences are identical—the value of the score varying from one program to the next—to provide a measure of the quality of the alignment. The process has some complications. Sometimes the proteins being compared match well at, say, two sequence segments, and these segments are connected by less related sequences of different lengths. Thus the two matching segments cannot be aligned at the same time. To handle this, the computer program introduces "gaps" in one of the sequences to bring the matching segments into register **(Fig. 3-33)**. Of course, if a sufficient number of gaps are introduced, almost any two sequences could be brought into some sort of alignment. To avoid uninformative alignments, the programs include penalties for each gap introduced, thus lowering the overall alignment score. With electronic trial and error, the program selects the alignment with the optimal score that maximizes identical amino acid residues while minimizing the introduction of gaps.

Finding identical amino acids is often inadequate in attempts to identify related proteins or, more importantly, to determine how closely related the proteins are on an evolutionary time scale. A more useful analysis also considers the chemical properties of substituted amino acids. Many of the amino acid differences within a protein family may be conservative—that is, an amino acid residue is replaced by a residue having similar chemical properties. For example, a Glu residue may substitute in

Escherichia coli	T G N R T I A V Y D L G G G T F D I S I I E I D E V D G E K T F E V L A T N G D T H L G G E D F D S R L I N Y L
Bacillus subtilis	D E D Q T I L L Y D L G G G T F D V S I L E L G D G V F E V R S T A G D N R L G G D D F D Q V I I D H L

Gap

FIGURE 3-33 Aligning protein sequences with the use of gaps. Shown here is the sequence alignment of a short section of the Hsp70 proteins (a widespread class of protein-folding chaperones) from two well-studied bacterial species, *E. coli* and *Bacillus subtilis*. Introduction of a gap in the *B. subtilis* sequence allows a better alignment of amino acid residues on either side of the gap. Identical amino acid residues are shaded. [Source: Information from R. S. Gupta, *Microbiol. Mol. Biol. Rev.* 62:1435, 1998, Fig. 2.]

one family member for the Asp residue found in another; both amino acids are negatively charged. Such a conservative substitution should logically receive a higher score in a sequence alignment than does a nonconservative substitution, such as replacement of the Asp residue with a hydrophobic Phe residue.

For most efforts to find homologies and explore evolutionary relationships, protein sequences (derived either directly from protein sequencing or from the sequencing of the DNA encoding the protein) are superior to nongenic nucleic acid sequences (those that do not encode a protein or functional RNA). For a nucleic acid, with its four different types of residues, random alignment of nonhomologous sequences will generally yield matches for at least 25% of the positions. Introduction of a few gaps can often increase the fraction of matched residues to 40% or more, and the probability of chance alignment of unrelated sequences becomes quite high. The 20 different amino acid residues in proteins greatly lower the probability of uninformative chance alignments of this type.

The programs used to generate a sequence alignment are complemented by methods that test the reliability of the alignments. A common computerized test is to shuffle the amino acid sequence of one of the proteins being compared to produce a random sequence, then to instruct the program to align the shuffled sequence with the other, unshuffled one. Scores are assigned to the new alignment, and the shuffling and alignment process is repeated many times. The original alignment, before shuffling, should have a score significantly higher than any of those within the distribution of scores generated by the random alignments; this increases the confidence that the sequence alignment has identified a pair of homologs. Note that the absence of a significant alignment score does not necessarily mean that no evolutionary relationship exists between two proteins. As we shall see in Chapter 4, three-dimensional structural similarities sometimes reveal evolutionary relationships where sequence homology has been wiped away by time.

To use a protein family to explore evolution, researchers identify family members with similar molecular functions in the widest possible range of organisms. Information from the family can then be used to trace the evolution of those organisms. By analyzing the sequence divergence in selected protein families, investigators can segregate organisms into classes based on their evolutionary relationships. This information must be reconciled with more classical examinations of the physiology and biochemistry of the organisms.

Certain segments of a protein sequence may be found in the organisms of one taxonomic group but not in other groups; these segments can be used as **signature sequences** for the group in which they are found. An example of a signature sequence is an insertion of 12 amino acids near the amino terminus of the EF-1α/EF-Tu proteins in all archaea and eukaryotes but not in bacteria **(Fig. 3-34)**. This particular signature is one of many biochemical clues that can help establish the evolutionary relatedness of eukaryotes and archaea. Signature sequences have been used to establish evolutionary relationships among groups of organisms at many different taxonomic levels.

By considering the entire sequence of a protein, researchers can now construct more elaborate evolutionary trees with many species in each taxonomic group. **Figure 3-35** presents one such tree for bacteria, based on sequence divergence in the protein GroEL (a protein present in all bacteria that assists in the proper folding of proteins). The tree can be refined by basing it on the sequences of multiple proteins and by supplementing the sequence information with data on the unique biochemical and physiological properties of each species. There are many methods for generating trees, each method with its own advantages and shortcomings, and many ways to represent the resulting evolutionary relationships. In Figure 3-35, the free end points of lines are called "external nodes"; each represents an extant species, and each is so labeled. The points where two lines come together, the "internal nodes," represent extinct ancestor species. In most representations (including Fig. 3-35), the lengths of the lines connecting the nodes are proportional to the number of amino acid substitutions separating one species from another. If we trace two extant species to a common internal node (representing the common ancestor of the two species), the length of the branch connecting each external node to the internal node represents the number of amino acid substitutions separating one extant species from this ancestor. The sum of the lengths of all the line segments that connect an extant species to another extant species through a common ancestor reflects the number of substitutions separating the two extant species. To determine how much time was needed for the various species to diverge, the tree must be calibrated by comparing it with information from the fossil record and other sources.

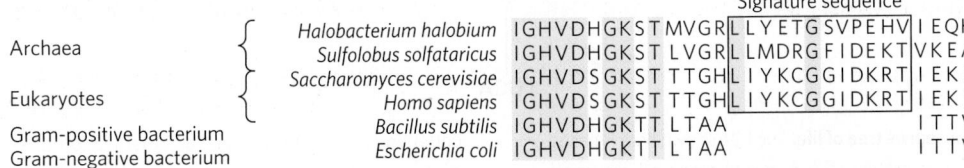

FIGURE 3-34 A signature sequence in the EF-1αEF-Tu protein family. The signature sequence (boxed) is a 12-residue insertion near the amino terminus of the sequence. Residues that align in all species are shaded. Both archaea and eukaryotes have the signature, although the sequences of the insertions are distinct for the two groups. The variation in the signature sequence reflects the significant evolutionary divergence that has occurred at this site since it first appeared in a common ancestor of both groups. [Source: Information from R. S. Gupta, *Microbiol. Mol. Biol. Rev.* 62:1435, 1998, Fig. 7.]

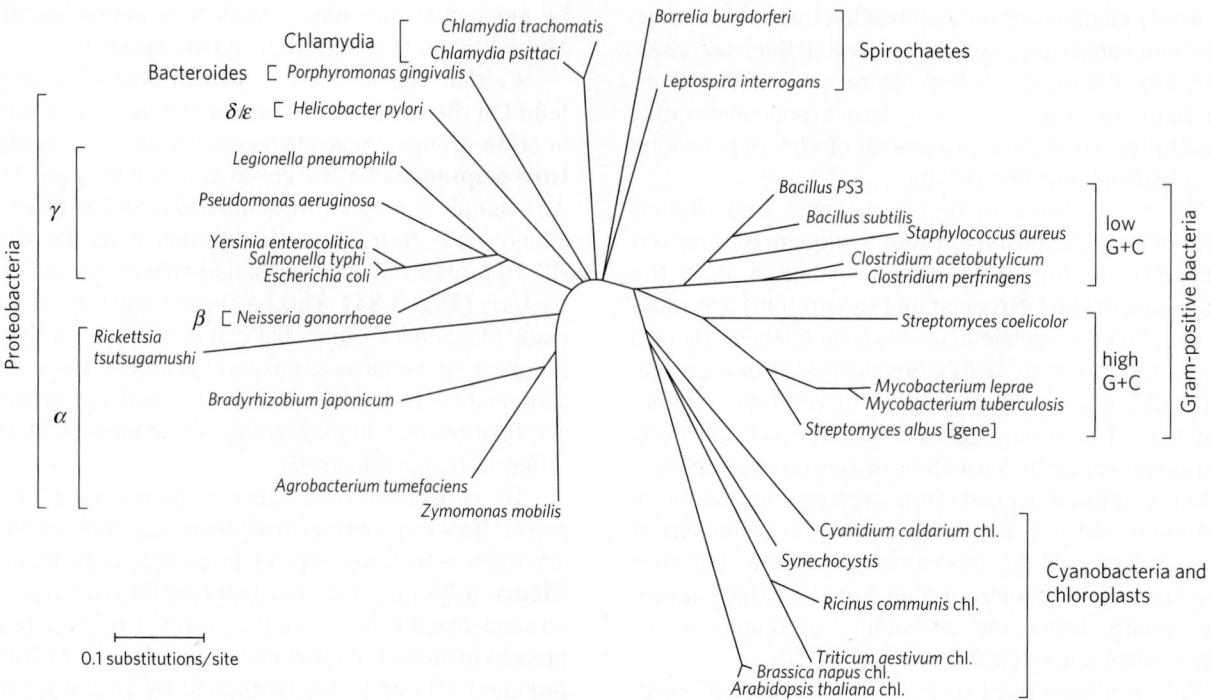

FIGURE 3-35 Evolutionary tree derived from amino acid sequence comparisons. A bacterial evolutionary tree, based on the sequence divergence observed in the GroEL family of proteins. Also included in this tree (lower right) are the chloroplasts (chl.) of some nonbacterial species. [Source: Information from R. S. Gupta, *Microbiol. Mol. Biol. Rev.* 62:1435, 1998, Fig. 11.]

As more sequence information is made available in databases, we can generate evolutionary trees based on multiple proteins. And we can refine these trees as additional genomic information emerges from increasingly sophisticated methods of analysis. All of this work moves us toward the goal of creating a detailed tree of life that describes the evolution and relationship of every organism on Earth. The story is a work in progress, of course **(Fig. 3-36)**. The questions being asked and answered are fundamental to how humans view themselves and

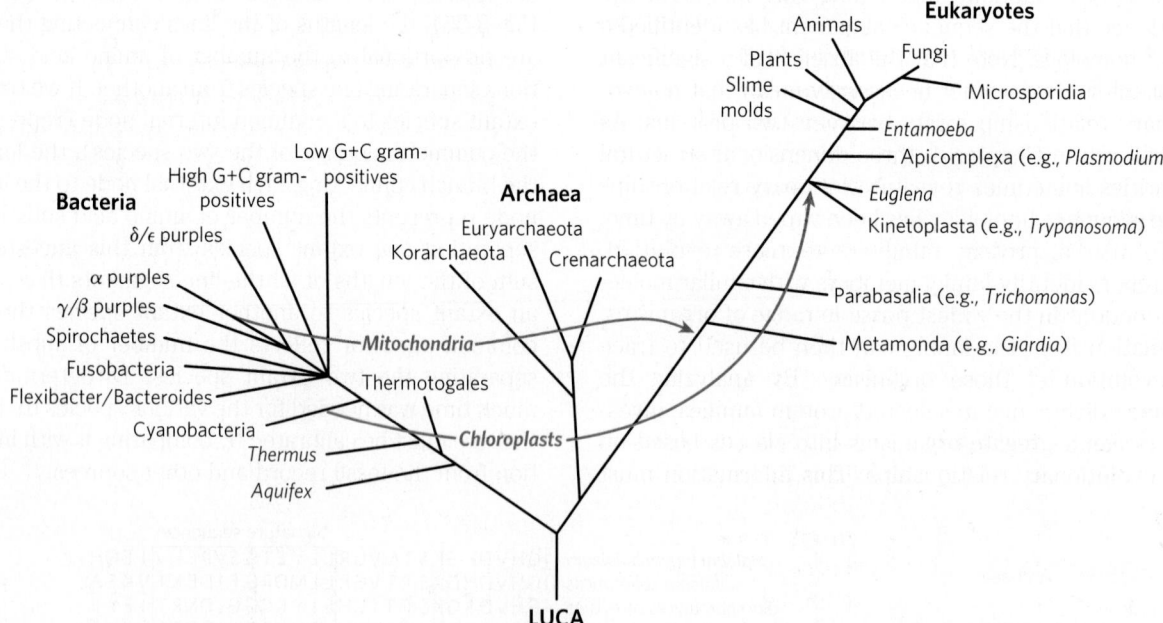

FIGURE 3-36 A consensus tree of life. The tree shown here is based on analyses of many different protein sequences and additional genomic features. The tree presents only a fraction of the available information, as well as only a fraction of the issues remaining to be resolved. Each extant group shown is a complex evolutionary story unto itself. LUCA is the last universal common ancestor from which all other life forms evolved. The blue and green arrows indicate the endosymbiotic assimilation of particular types of bacteria into eukaryotic cells to become mitochondria and chloroplasts, respectively (see Fig. 1-40). [Source: Information from F. Delsuc et al., *Nature Rev. Genet.* 6:363, 2005, Fig. 1.]

the world around them. The field of molecular evolution promises to be among the most vibrant of the scientific frontiers in the twenty-first century.

SUMMARY 3.4 The Structure of Proteins: Primary Structure

■ Differences in protein function result from differences in amino acid composition and sequence. Some variations in sequence may occur in a particular protein, with little or no effect on its function.

■ Amino acid sequences are deduced by fragmenting polypeptides into smaller peptides with reagents known to cleave specific peptide bonds, determining the amino acid sequence of each fragment by the automated Edman degradation procedure, and then ordering the peptide fragments by finding sequence overlaps between fragments generated by different reagents. A protein sequence can also be deduced from the nucleotide sequence of its corresponding gene in DNA or by mass spectrometry.

■ Short proteins and peptides (up to about 100 residues) can be chemically synthesized. The peptide is built up, one amino acid residue at a time, while tethered to a solid support.

■ Protein sequences are a rich source of information about protein structure and function, as well as the evolution of life on Earth. Sophisticated methods are being developed to trace evolution by analyzing the slow changes in amino acid sequences of homologous proteins.

Key Terms

Terms in bold are defined in the glossary.

Problems

1. Absolute Configuration of Citrulline The citrulline isolated from watermelons has the structure shown below. Is it a D- or L-amino acid? Explain.

$$CH_2(CH_2)_2NH-\overset{\displaystyle \underset{\|}{O}}{C}-NH_2$$
$$H-\overset{\displaystyle |}{\underset{\displaystyle |}{C}}-\overset{+}{N}H_3$$
$$COO^-$$

2. Relationship between the Titration Curve and the Acid-Base Properties of Glycine A 100 mL solution of 0.1 M glycine at pH 1.72 was titrated with 2 M NaOH solution. The pH was monitored and the results were plotted as shown in the graph. The key points in the titration are designated I to V. For each of the statements (a) to (o), *identify* the appropriate key point in the titration and *justify* your choice.

(a) Glycine is present predominantly as the species $^+H_3N-CH_2-COOH$.

(b) The *average* net charge of glycine is $+\frac{1}{2}$.

(c) Half of the amino groups are ionized.

(d) The pH is equal to the pK_a of the carboxyl group.

(e) The pH is equal to the pK_a of the protonated amino group.

(f) Glycine has its maximum buffering capacity.

(g) The *average* net charge of glycine is zero.

(h) The carboxyl group has been completely titrated (first equivalence point).

(i) Glycine is completely titrated (second equivalence point).

(j) The predominant species is $^+H_3N-CH_2-COO^-$.

(k) The *average* net charge of glycine is -1.

(l) Glycine is present predominantly as a 50:50 mixture of $^+H_3N-CH_2-COOH$ and $^+H_3N-CH_2-COO^-$.

(m) This is the isoelectric point.

(n) This is the end of the titration.

(o) These are the *worst* pH regions for buffering power.

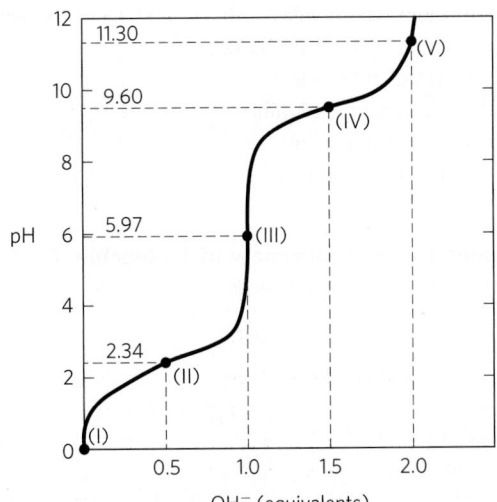

3. How Much Alanine Is Present as the Completely Uncharged Species? At a pH equal to the isoelectric point of alanine, the *net* charge on alanine is zero. Two structures can be drawn that have a net charge of zero, but the predominant form of alanine at its pI is zwitterionic.

Zwitterionic   Uncharged

(a) Why is alanine predominantly zwitterionic rather than completely uncharged at its pI?

(b) What fraction of alanine is in the completely uncharged form at its pI? Justify your assumptions.

4. Ionization State of Histidine Each ionizable group of an amino acid can exist in one of two states, charged or neutral. The electric charge on the functional group is determined by the relationship between its pK_a and the pH of the solution. This relationship is described by the Henderson-Hasselbalch equation.

(a) Histidine has three ionizable functional groups. Write the equilibrium equations for its three ionizations and assign the proper pK_a for each ionization. Draw the structure of histidine in each ionization state. What is the net charge on the histidine molecule in each ionization state?

(b) Draw the structures of the predominant ionization state of histidine at pH 1, 4, 8, and 12. Note that the ionization state can be approximated by treating each ionizable group independently.

(c) What is the net charge of histidine at pH 1, 4, 8, and 12? For each pH, will histidine migrate toward the anode (+) or cathode (−) when placed in an electric field?

5. Separation of Amino Acids by Ion-Exchange Chromatography Mixtures of amino acids can be analyzed by first separating the mixture into its components through ion-exchange chromatography. Amino acids placed on a cation-exchange resin (see Fig. 3-17a) containing sulfonate ($-SO_3^-$) groups flow down the column at different rates because of two factors that influence their movement: (1) ionic attraction between the sulfonate residues on the column and positively charged functional groups on the amino acids, and (2) aggregation of nonpolar amino acid side chains with the hydrophobic backbone of the polystyrene resin. For each pair of amino acids listed, determine which will be eluted first from the cation-exchange column by a pH 7.0 buffer.

(a) Aspartate and lysine
(b) Arginine and methionine
(c) Glutamate and valine
(d) Glycine and leucine
(e) Serine and alanine

6. Naming the Stereoisomers of Isoleucine The structure of the amino acid isoleucine is

(a) How many chiral centers does it have?

(b) How many optical isomers?

(c) Draw perspective formulas for all the optical isomers of isoleucine.

7. Comparing the pK_a Values of Alanine and Polyalanine The titration curve of alanine shows the ionization of two functional groups with pK_a values of 2.34 and 9.69, corresponding to the ionization of the carboxyl and the protonated amino groups, respectively. The titration of di-, tri-, and larger oligopeptides of alanine also shows the ionization of only two functional groups, although the experimental pK_a values are different. The trend in pK_a values is summarized in the table.

Amino acid or peptide	pK_1	pK_2
Ala	2.34	9.69
Ala–Ala	3.12	8.30
Ala–Ala–Ala	3.39	8.03
Ala$-$(Ala)$_n-$Ala, $n \geq 4$	3.42	7.94

(a) Draw the structure of Ala–Ala–Ala. Identify the functional groups associated with pK_1 and pK_2.

(b) Why does the value of pK_1 *increase* with each additional Ala residue in the oligopeptide?

(c) Why does the value of pK_2 *decrease* with each additional Ala residue in the oligopeptide?

8. The Size of Proteins What is the approximate molecular weight of a protein with 682 amino acid residues in a single polypeptide chain?

9. The Number of Tryptophan Residues in Bovine Serum Albumin A quantitative amino acid analysis reveals that bovine serum albumin (BSA) contains 0.58% tryptophan (M_r 204) by weight.

(a) Calculate the *minimum* molecular weight of BSA (i.e., assume there is only one Trp residue per protein molecule).

(b) Size-exclusion chromatography of BSA gives a molecular weight estimate of 70,000. How many Trp residues are present in a molecule of serum albumin?

10. Subunit Composition of a Protein A protein has a molecular mass of 400 kDa when measured by size-exclusion chromatography. When subjected to gel electrophoresis in the presence of sodium dodecyl sulfate (SDS), the protein gives three bands with molecular masses of 180, 160, and 60 kDa. When electrophoresis is carried out in the presence of SDS and dithiothreitol, three bands are again formed, this time with molecular masses of 160, 90, and 60 kDa. Determine the subunit composition of the protein.

11. Net Electric Charge of Peptides A peptide has the sequence

Glu−His−Trp−Ser−Gly−Leu−Arg−Pro−Gly

(a) What is the net charge of the molecule at pH 3, 8, and 11? (Use pK_a values for side chains and terminal amino and carboxyl groups as given in Table 3-1.)

(b) Estimate the pI for this peptide.

12. Isoelectric Point of Pepsin Pepsin is the name given to a mix of several digestive enzymes secreted (as larger precursor proteins) by glands that line the stomach. These glands also secrete hydrochloric acid, which dissolves the particulate matter in food, allowing pepsin to enzymatically cleave individual protein molecules. The resulting mixture of food, HCl, and digestive enzymes is known as chyme and has a pH near 1.5. What pI would you predict for the pepsin proteins? What functional groups must be present to confer this pI on pepsin? Which amino acids in the proteins would contribute such groups?

13. Isoelectric Point of Histones Histones are proteins found in eukaryotic cell nuclei, tightly bound to DNA, which has many phosphate groups. The pI of histones is very high, about 10.8. What amino acid residues must be present in relatively large numbers in histones? In what way do these residues contribute to the strong binding of histones to DNA?

14. Solubility of Polypeptides One method for separating polypeptides makes use of their different solubilities. The solubility of large polypeptides in water depends on the relative polarity of their R groups, particularly on the number of ionized groups: the more ionized groups there are, the more soluble the polypeptide. Which of each pair of polypeptides that follow is more soluble at the indicated pH?

(a) $(Gly)_{20}$ or $(Glu)_{20}$ at pH 7.0
(b) $(Lys–Ala)_3$ or $(Phe–Met)_3$ at pH 7.0
(c) $(Ala–Ser–Gly)_5$ or $(Asn–Ser–His)_5$ at pH 6.0
(d) $(Ala–Asp–Gly)_5$ or $(Asn–Ser–His)_5$ at pH 3.0

15. Purification of an Enzyme A biochemist discovers and purifies a new enzyme, generating the purification table below.

Procedure	Total protein (mg)	Activity (units)
1. Crude extract	20,000	4,000,000
2. Precipitation (salt)	5,000	3,000,000
3. Precipitation (pH)	4,000	1,000,000
4. Ion-exchange chromatography	200	800,000
5. Affinity chromatography	50	750,000
6. Size-exclusion chromatography	45	675,000

(a) From the information given in the table, calculate the specific activity of the enzyme after each purification procedure.

(b) Which of the purification procedures used for this enzyme is most effective (i.e., gives the greatest relative increase in purity)?

(c) Which of the purification procedures is least effective?

(d) Is there any indication based on the results shown in the table that the enzyme after step 6 is now pure? What else could be done to estimate the purity of the enzyme preparation?

16. Dialysis A purified protein is in a Hepes (*N*-(2-hydroxy-ethyl)piperazine-*N'*-(2-ethanesulfonic acid)) buffer at pH 7 with 500 mM NaCl. A sample (1 mL) of the protein solution is placed in a tube made of dialysis membrane and dialyzed against 1 L of the same Hepes buffer with 0 mM NaCl. Small molecules and ions (such as Na^+, Cl^-, and Hepes) can diffuse across the dialysis membrane, but the protein cannot.

(a) Once the dialysis has come to equilibrium, what is the concentration of NaCl in the protein sample? Assume no volume changes occur in the sample during the dialysis.

(b) If the original 1 mL sample were dialyzed twice, successively, against 100 mL of the same Hepes buffer with 0 mM NaCl, what would be the final NaCl concentration in the sample?

17. Peptide Purification At pH 7.0, in what order would the following three peptides (described by their amino acid composition) be eluted from a column filled with a cation-exchange polymer?

Peptide A: Ala 10%, Glu 5%, Ser 5%, Leu 10%, Arg 10%, His 5%, Ile 10%, Phe 5%, Tyr 5%, Lys 10%, Gly 10%, Pro 5%, and Trp 10%.

Peptide B: Ala 5%, Val 5%, Gly 10%, Asp 5%, Leu 5%, Arg 5%, Ile 5%, Phe 5%, Tyr 5%, Lys 5%, Trp 5%, Ser 5%, Thr 5%, Glu 5%, Asn 5%, Pro 10%, Met 5%, and Cys 5%.

Peptide C: Ala 10%, Glu 10%, Gly 5%, Leu 5%, Asp 10%, Arg 5%, Met 5%, Cys 5%, Tyr 5%, Phe 5%, His 5%, Val 5%, Pro 5%, Thr 5%, Ser 5%, Asn 5%, and Gln 5%.

18. Sequence Determination of the Brain Peptide Leucine Enkephalin A group of peptides that influence nerve transmission in certain parts of the brain have been isolated from normal brain tissue. These peptides are known as opioids because they bind to specific receptors that also bind opiate drugs, such as morphine and naloxone. Opioids thus mimic some of the properties of opiates. Some researchers consider these peptides to be the brain's own painkillers. Using the information below, determine the amino acid sequence of the opioid leucine enkephalin. Explain how your structure is consistent with each piece of information.

(a) Complete hydrolysis by 6 M HCl at 110 °C followed by amino acid analysis indicated the presence of Gly, Leu, Phe, and Tyr, in a 2:1:1:1 molar ratio.

(b) Treatment of the peptide with 1-fluoro-2,4-dinitrobenzene followed by complete hydrolysis and chromatography indicated the presence of the 2,4-dinitrophenyl derivative of tyrosine. No free tyrosine could be found.

(c) Complete digestion of the peptide with chymotrypsin followed by chromatography yielded free tyrosine and leucine, plus a tripeptide containing Phe and Gly in a 1:2 ratio.

19. Structure of a Peptide Antibiotic from *Bacillus brevis* Extracts from the bacterium *Bacillus brevis* contain a peptide with antibiotic properties. This peptide forms complexes with metal ions and seems to disrupt ion transport across the cell membranes of other bacterial species, killing them. The structure of the peptide has been determined from the following observations.

(a) Complete acid hydrolysis of the peptide followed by amino acid analysis yielded equimolar amounts of Leu, Orn, Phe, Pro, and Val. Orn is ornithine, an amino acid not present in proteins but present in some peptides. It has the structure

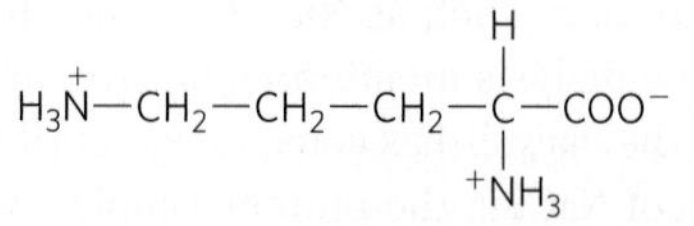

$$H_3\overset{+}{N}-CH_2-CH_2-CH_2-\underset{\underset{+NH_3}{|}}{\overset{\overset{H}{|}}{C}}-COO^-$$

(b) The molecular weight of the peptide was estimated as ~1,200.

(c) The peptide failed to undergo hydrolysis when treated with the enzyme carboxypeptidase. This enzyme catalyzes the hydrolysis of the carboxyl-terminal residue of a polypeptide unless the residue is Pro or, for some reason, does not contain a free carboxyl group.

(d) Treatment of the intact peptide with 1-fluoro-2,4-dinitrobenzene, followed by complete hydrolysis and chromatography, yielded only free amino acids and the following derivative:

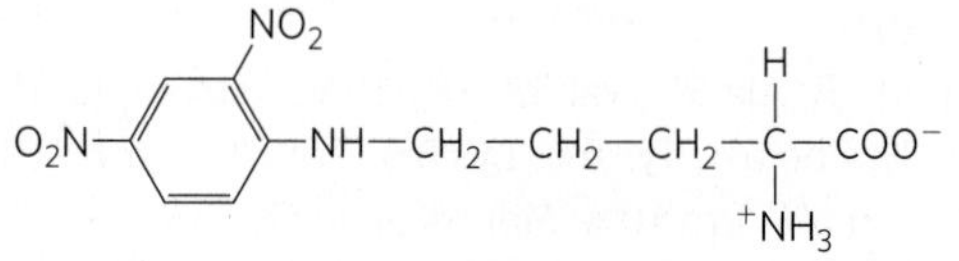

$$O_2N-\!\!\!\!\!\overset{\overset{NO_2}{|}}{\bigcirc}\!\!\!\!\!-NH-CH_2-CH_2-CH_2-\underset{\underset{+NH_3}{|}}{\overset{\overset{H}{|}}{C}}-COO^-$$

(Hint: The 2,4-dinitrophenyl derivative involves the amino group of a side chain rather than the α-amino group.)

(e) Partial hydrolysis of the peptide followed by chromatographic separation and sequence analysis yielded the following di- and tripeptides (the amino-terminal amino acid is always at the left):

Leu–Phe	Phe–Pro	Orn–Leu	Val–Orn
Val–Orn–Leu	Phe–Pro–Val	Pro–Val–Orn	

Given the above information, deduce the amino acid sequence of the peptide antibiotic. Show your reasoning. When you have arrived at a structure, demonstrate that it is consistent with *each* experimental observation.

20. Efficiency in Peptide Sequencing A peptide with the primary structure Lys–Arg–Pro–Leu–Ile–Asp–Gly–Ala is sequenced by the Edman procedure. If each Edman cycle is 96% efficient, what percentage of the amino acids liberated in the fourth cycle will be leucine? Do the calculation a second time, but assume a 99% efficiency for each cycle.

21. Sequence Comparisons Proteins called molecular chaperones (described in Chapter 4) assist in the process of protein folding. One class of chaperones found in organisms from bacteria to mammals is heat shock protein 90 (Hsp90). All Hsp90 chaperones contain a 10 amino acid "signature sequence" that allows ready identification of these proteins in sequence databases. Two representations of this signature sequence are shown below.

Y-x-[NQHD]-[KHR]-[DE]-[IVA]-F-[LM]-R-[ED].

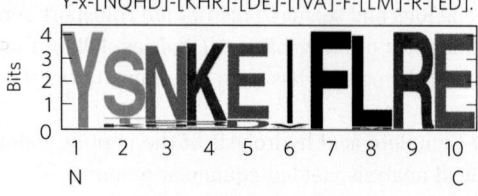

(a) In this sequence, which amino acid residues are invariant (conserved across all species)?

(b) At which position(s) are amino acids limited to those with positively charged side chains? For each position, which amino acid is more commonly found?

(c) At which positions are substitutions restricted to amino acids with negatively charged side chains? For each position, which amino acid predominates?

(d) There is one position that can be any amino acid, although one amino acid appears much more often than any other. What position is this, and which amino acid appears most often?

22. Chromatographic Methods Three polypeptides, the sequences of which are represented below using the one-letter code for their amino acids, are present in a mixture:

1. ATKNRASCLVPKHGALMFWRHKQLVSDPIL QKRQHILVCRNAAG
2. GPYFGDEPLDVHDEPEEG
3. PHLLSAWKGMEGVGKSQSFAALIVILA

Of the three, which one would migrate most slowly during chromatography through:

(a) an ion-exchange resin, beads coated with positively charged groups?

(b) an ion-exchange resin, beads coated with negatively charged groups?

(c) a size-exclusion (gel-filtration) column designed to separate small peptides such as these?

(d) Which peptide contains the ATP-binding motif shown in the following sequence logo?

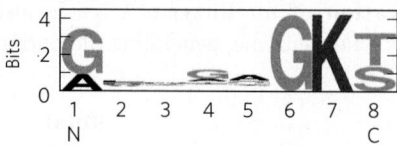

Data Analysis Problem

23. Determining the Amino Acid Sequence of Insulin Figure 3-24 shows the amino acid sequence of bovine insulin. This structure was determined by Frederick Sanger and his coworkers. Most of this work is described in a series of articles published in the *Biochemical Journal* from 1945 to 1955.

When Sanger and colleagues began their work in 1945, it was known that insulin was a small protein consisting of two or four polypeptide chains linked by disulfide bonds. Sanger's team had developed a few simple methods for studying protein sequences.

Treatment with FDNB. FDNB (1-fluoro-2,4-dinitrobenzene) reacted with free amino (but not amide or guanidinium) groups in proteins to produce dinitrophenyl (DNP) derivatives of amino acids:

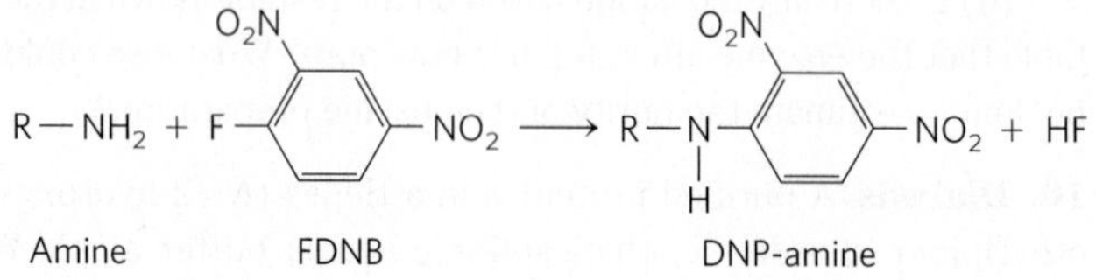

$$R-NH_2 + F-\!\!\!\!\!\overset{\overset{O_2N}{|}}{\bigcirc}\!\!\!\!\!-NO_2 \longrightarrow R-\underset{\underset{H}{|}}{N}-\!\!\!\!\!\overset{\overset{O_2N}{|}}{\bigcirc}\!\!\!\!\!-NO_2 + HF$$

Amine   FDNB   DNP-amine

Acid Hydrolysis. Boiling a protein with 10% HCl for several hours hydrolyzed all of its peptide and amide bonds. Short treatments produced short polypeptides; the longer the treatment, the more complete the breakdown of the protein into its amino acids.

Oxidation of Cysteines. Treatment of a protein with performic acid cleaved all the disulfide bonds and converted all Cys residues to cysteic acid residues (see Fig. 3-28).

Paper Chromatography. This more primitive version of thin-layer chromatography (see Fig. 10-25) separated compounds based on their chemical properties, allowing identification of single amino acids and, in some cases, dipeptides. Thin-layer chromatography also separates larger peptides.

As reported in his first paper (1945), Sanger reacted insulin with FDNB and hydrolyzed the resulting protein. He found many free amino acids, but only three DNP–amino acids: α-DNP-glycine (DNP group attached to the α-amino group), α-DNP-phenylalanine, and ε-DNP-lysine (DNP attached to the ε-amino group). Sanger interpreted these results as showing that insulin had two protein chains: one with Gly at its amino terminus and one with Phe at its amino terminus. One of the two chains also contained a Lys residue, not at the amino terminus. He named the chain beginning with a Gly residue "A" and the chain beginning with Phe "B."

(a) Explain how Sanger's results support his conclusions.

(b) Are the results consistent with the known structure of bovine insulin (see Fig. 3-24)?

In a later paper (1949), Sanger described how he used these techniques to determine the first few amino acids (amino-terminal end) of each insulin chain. To analyze the B chain, for example, he carried out the following steps:

1. Oxidized insulin to separate the A and B chains.
2. Prepared a sample of pure B chain with paper chromatography.
3. Reacted the B chain with FDNB.
4. Gently acid-hydrolyzed the protein so that some small peptides would be produced.
5. Separated the DNP-peptides from the peptides that did not contain DNP groups.
6. Isolated four of the DNP-peptides, which were named B1 through B4.
7. Strongly hydrolyzed each DNP-peptide to give free amino acids.
8. Identified the amino acids in each peptide with paper chromatography.

The results were as follows:

B1: α-DNP-phenylalanine only

B2: α-DNP-phenylalanine; valine

B3: aspartic acid; α-DNP-phenylalanine; valine

B4: aspartic acid; glutamic acid; α-DNP-phenylalanine; valine

(c) Based on these data, what are the first four (aminoterminal) amino acids of the B chain? Explain your reasoning.

(d) Does this result match the known sequence of bovine insulin (Fig. 3-24)? Explain any discrepancies.

Sanger and colleagues used these and related methods to determine the entire sequence of the A and B chains. Their sequence for the A chain was as follows:

$$\overset{1}{\text{Gly}}-\text{Ile}-\text{Val}-\overset{5}{\text{Glx}}-\text{Glx}-\text{Cys}-\text{Cys}-\text{Ala}-\text{Ser}-\overset{10}{\text{Val}}-$$
$$\text{Cys}-\text{Ser}-\text{Leu}-\text{Tyr}-\overset{15}{\text{Glx}}-\text{Leu}-\text{Glx}-\text{Asx}-\text{Tyr}-\overset{20}{\text{Cys}}-\text{Asx}$$

Because acid hydrolysis had converted all Asn to Asp and all Gln to Glu, these residues had to be designated Asx and Glx, respectively (exact identity in the peptide unknown). Sanger solved this problem by using protease enzymes that cleave peptide bonds, but not the amide bonds in Asn and Gln residues, to prepare short peptides. He then determined the number of amide groups present in each peptide by measuring the NH_4^+ released when the peptide was acid-hydrolyzed. Some of the results for the A chain are shown below. The peptides may not have been completely pure, so the numbers were approximate—but good enough for Sanger's purposes.

Peptide name	Peptide sequence	Number of amide groups in peptide
Ac1	Cys–Asx	0.7
Ap15	Tyr–Glx–Leu	0.98
Ap14	Tyr–Glx–Leu–Glx	1.06
Ap3	Asx–Tyr–Cys–Asx	2.10
Ap1	Glx–Asx–Tyr–Cys–Asx	1.94
Ap5pa1	Gly–Ile–Val–Glx	0.15
Ap5	Gly–Ile–Val–Glx–Glx–Cys–Cys– Ala–Ser–Val–Cys–Ser–Leu	1.16

(e) Based on these data, determine the amino acid sequence of the A chain. Explain how you reached your answer. Compare it with Figure 3-24.

References

Sanger, F. 1945. The free amino groups of insulin. *Biochem. J.* 39:507–515.

Sanger, F. 1949. The terminal peptides of insulin. *Biochem. J.* 45:563–574.

Further Reading is available at www.macmillanlearning.com/LehningerBiochemistry7e.

CHAPTER 4

The Three-Dimensional Structure of Proteins

Self-study tools that will help you practice what you've learned and reinforce this chapter's concepts are available online. Go to www.macmillanlearning.com/LehningerBiochemistry7e.

Proteins are big molecules. The covalent backbone of a typical protein contains hundreds of individual bonds. Because free rotation is possible around many of these bonds, the protein can, in principle, assume a virtually uncountable number of conformations. However, each protein has a specific chemical or structural function, suggesting that each has a unique three-dimensional structure **(Fig. 4-1)**. How stable is this structure, what factors guide its formation, and what holds it together? By the late 1920s, several proteins had been crystallized, including hemoglobin (M_r 64,500) and the enzyme urease (M_r 483,000). Given that, generally, the ordered array of molecules in a crystal can form only if the molecular units are identical, the finding that many proteins could be crystallized was evidence that even very large proteins are discrete chemical entities with unique structures. This conclusion revolutionized thinking about proteins and their functions, but the insight it provided was incomplete. Protein structure is always malleable in sometimes surprising ways. Changes in structure can be as important to a protein's function as the structure itself.

In this chapter, we examine the structure of proteins. We emphasize six themes. First, the three-dimensional structure or structures taken up by a protein are determined by its amino acid sequence. Second, the function of a typical protein depends on its structure. Third, most isolated proteins exist in one or a small number of stable structural forms. Fourth, the most important forces stabilizing the specific structures maintained by a given protein are noncovalent; the hydrophobic effect is particularly important. Fifth, amid the huge number of unique protein structures, we can recognize some common structural patterns that help to organize our understanding of protein architecture. Finally, protein structures are not static. All proteins undergo changes in conformation ranging from subtle to dramatic. Parts of many proteins have no discernible structure. For some proteins or parts of proteins, a lack of definable structure is critical to their function.

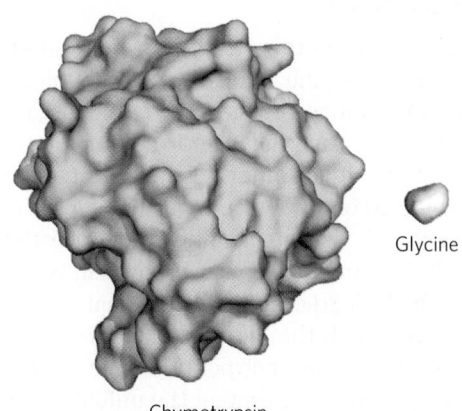

Glycine

Chymotrypsin

FIGURE 4-1 Structure of the enzyme chymotrypsin, a globular protein. A molecule of glycine is shown for size comparison. The known three-dimensional structures of proteins are archived in the Protein Data Bank, or PDB (see Box 4-4). The image shown here was made using data from the entry with PDB ID 6GCH. [Source: PDB ID 6GCH, K. Brady et al., *Biochemistry* 29:7600, 1990.]

4.1 Overview of Protein Structure

The spatial arrangement of atoms in a protein or any part of a protein is called its **conformation**. The possible conformations of a protein or protein segment include any structural state it can achieve without breaking covalent bonds. A change in conformation could occur, for example, by rotation about single bonds. Of the many conformations that are theoretically possible in a protein containing hundreds of single bonds, one or (more commonly) a few generally predominate under biological conditions. The need for multiple stable conformations reflects the changes that must take place in most proteins as they bind to other molecules or catalyze reactions. The conformations existing under a given set of conditions are usually the ones that are thermodynamically the most stable—that is, having the lowest Gibbs free energy (G). Proteins in any of their functional, folded conformations are called **native** proteins.

For the vast majority of proteins, a particular structure or small set of structures is critical to function. However, in many cases, parts of proteins lack discernible structure. These protein segments are intrinsically disordered. In a few cases, entire proteins are intrinsically disordered, yet are fully functional.

What principles determine the most stable conformations of a typical protein? An understanding of protein conformation can be built stepwise from the discussion of primary structure in Chapter 3 through a consideration of secondary, tertiary, and quaternary structures. To this traditional approach we must add the newer emphasis on common and classifiable folding patterns, variously called supersecondary structures, folds, or motifs, which provide an important organizational context to this complex endeavor. We begin by introducing some guiding principles.

A Protein's Conformation Is Stabilized Largely by Weak Interactions

In the context of protein structure, the term **stability** can be defined as the tendency to maintain a native conformation. Native proteins are only marginally stable; the ΔG separating the folded and unfolded states in typical proteins under physiological conditions is in the range of only 20 to 65 kJ/mol. A given polypeptide chain can theoretically assume countless conformations, and as a result, the unfolded state of a protein is characterized by a high degree of conformational entropy. This entropy, along with the hydrogen-bonding interactions of many groups in the polypeptide chain with the solvent (water), tends to maintain the unfolded state. The chemical interactions that counteract these effects and stabilize the native conformation include disulfide (covalent) bonds and the weak (noncovalent) interactions and forces described in Chapter 2: hydrogen bonds, the hydrophobic effect, and ionic interactions.

Many proteins do not have disulfide bonds. The environment within most cells is highly reducing due to

high concentrations of reductants such as glutathione, and most sulfhydryls will thus remain in the reduced state. Outside the cell, the environment is often more oxidizing, and disulfide formation is more likely to occur. In eukaryotes, disulfide bonds are found primarily in secreted, extracellular proteins (for example, the hormone insulin). Disulfide bonds are also uncommon in bacterial proteins. However, thermophilic bacteria, as well as the archaea, typically have many proteins with disulfide bonds, which stabilize proteins; this is presumably an adaptation to life at high temperatures.

For all proteins of all organisms, weak interactions are especially important in the folding of polypeptide chains into their secondary and tertiary structures. The association of multiple polypeptides to form quaternary structures also relies on these weak interactions.

About 200 to 460 kJ/mol are required to break a single covalent bond, whereas weak interactions can be disrupted by a mere 0.4 to 30 kJ/mol. Individual covalent bonds, such as disulfide bonds linking separate parts of a single polypeptide chain, are clearly much stronger than individual weak interactions. Yet, because they are so numerous, the weak interactions predominate as a stabilizing force in protein structure. In general, the protein conformation with the lowest free energy (that is, the most stable conformation) is the one with the maximum number of weak interactions.

The stability of a protein is not simply the sum of the free energies of formation of the many weak interactions within it. For every hydrogen bond formed in a protein during folding, a hydrogen bond (of similar strength) between the same group and water was broken. The net stability contributed by a given hydrogen bond, or the *difference* in free energies of the folded and unfolded states, may be close to zero. Ionic interactions may be either stabilizing or destabilizing. We must therefore look elsewhere to understand why a particular native conformation is favored.

On carefully examining the contribution of weak interactions to protein stability, we find that the **hydrophobic effect** generally predominates. Pure water contains a network of hydrogen-bonded H_2O molecules. No other molecule has the hydrogen-bonding potential of water, and the presence of other molecules in an aqueous solution disrupts the hydrogen bonding of water. When water surrounds a hydrophobic molecule, the optimal arrangement of hydrogen bonds results in a highly structured shell, or **solvation layer**, of water around the molecule (see Fig. 2-7). The increased order of the water molecules in the solvation layer correlates with an unfavorable decrease in the entropy of the water. However, when nonpolar groups cluster together, the extent of the solvation layer decreases, because each group no longer presents its entire surface to the solution. The result is a favorable increase in entropy. As described in Chapter 2, this increase in entropy is the major thermodynamic driving force for the association of hydrophobic groups in aqueous solution. Hydrophobic amino acid side chains therefore tend to cluster in a protein's interior, away from water (think of an oil droplet in water). The amino acid sequences of most proteins thus include a significant content of hydrophobic amino acid side chains (especially Leu, Ile, Val, Phe, and Trp). These are positioned so that they are clustered when the protein is folded, forming a hydrophobic protein core.

Under physiological conditions, the formation of hydrogen bonds in a protein is driven largely by this same entropic effect. Polar groups can generally form hydrogen bonds with water and hence are soluble in water. However, the number of hydrogen bonds per unit mass is generally greater for pure water than for any other liquid or solution, and there are limits to the solubility of even the most polar molecules as their presence causes a net decrease in hydrogen bonding per unit mass. Therefore, a solvation layer forms to some extent even around polar molecules. Although the energy of formation of an intramolecular hydrogen bond between two polar groups in a macromolecule is largely canceled by the elimination of such interactions between these polar groups and water, the release of structured water as intramolecular associations form provides an entropic driving force for folding. Most of the net change in free energy as nonpolar amino acid side chains aggregate within a protein is therefore derived from the increased entropy in the surrounding aqueous solution resulting from the burial of hydrophobic surfaces. This more than counterbalances the large loss of conformational entropy as a polypeptide is constrained into its folded conformation.

The hydrophobic effect is clearly important in stabilizing conformation; the interior of a structured protein is generally a densely packed core of hydrophobic amino acid side chains. It is also important that any polar or charged groups in the protein interior have suitable partners for hydrogen bonding or ionic interactions. One hydrogen bond seems to contribute little to the stability of a native structure, but the presence of hydrogen-bonding groups without partners in the hydrophobic core of a protein can be so *destabilizing* that conformations containing these groups are often thermodynamically untenable. The favorable free-energy change resulting from the combination of several such groups with partners in the surrounding solution can be greater than the free-energy difference between the folded and unfolded states. In addition, hydrogen bonds between groups in a protein form cooperatively (formation of one makes formation of the next one more likely) in repeating secondary structures that optimize hydrogen bonding, as described below. In this way, hydrogen bonds often have an important role in guiding the protein-folding process.

The interaction of oppositely charged groups that form an ion pair, or salt bridge, can have either a stabilizing or destabilizing effect on protein structure. As in the case of hydrogen bonds, charged amino acid side chains interact with water and salts when the protein is unfolded, and the loss of those interactions must be considered when evaluating the effect of a salt bridge on the overall stability of a folded protein. However, the strength of a salt bridge increases as it moves to an environment of lower dielectric constant, ε (p. 50): from the polar aqueous solvent (ε near 80) to the nonpolar protein interior (ε near 4). Salt bridges, especially those that are partly or entirely buried, can thus provide significant stabilization to a protein structure. This trend explains the increased occurrence of buried salt bridges in the proteins of thermophilic organisms. Ionic interactions also limit structural flexibility and confer a uniqueness to a particular protein structure that the clustering of nonpolar groups via the hydrophobic effect cannot provide.

In the tightly packed atomic environment of a protein, one more type of weak interaction can have a significant effect: van der Waals interactions (p. 53). Van der Waals interactions are dipole-dipole interactions involving the permanent electric dipoles in groups such as carbonyls, transient dipoles derived from fluctuations of the electron cloud surrounding any atom, and dipoles induced by interaction of one atom with another that has a permanent or transient dipole. As atoms approach each other, these dipole-dipole interactions provide an attractive intermolecular force that operates only over a limited intermolecular distance (0.3 to 0.6 nm). Van der Waals interactions are weak and, individually, contribute little to overall protein stability. However, in a well-packed protein, or in an interaction between a protein and another protein or other molecule at a complementary surface, the number of such interactions can be substantial.

Most of the structural patterns outlined in this chapter reflect two simple rules: (1) hydrophobic residues are largely buried in the protein interior, away from water, and (2) the number of hydrogen bonds and ionic interactions within the protein is maximized, thus reducing the number of hydrogen-bonding and ionic groups that are not paired with a suitable partner. Proteins within membranes (which we examine in Chapter 11) and proteins that are intrinsically disordered or have intrinsically disordered segments follow different rules. This reflects their particular function or environment, but weak interactions are still critical structural elements. For example, soluble but intrinsically disordered protein segments are enriched in amino acid side chains that are charged (especially Arg, Lys, Glu) or small (Gly, Ala), providing little or no opportunity for the formation of a stable hydrophobic core.

The Peptide Bond Is Rigid and Planar

Covalent bonds, too, place important constraints on the conformation of a polypeptide. In the late 1930s, Linus Pauling and Robert Corey embarked on a series of studies that laid the foundation for our current understanding of protein structure. They began with a careful analysis of the peptide bond.

The α carbons of adjacent amino acid residues are separated by three covalent bonds, arranged as C_α—C—N—C_α. X-ray diffraction studies of crystals of amino acids and of simple dipeptides and tripeptides showed that the peptide C—N bond is somewhat shorter

Linus Pauling, 1901–1994
[Source: Nancy R. Schiff/Getty Images.]

Robert Corey, 1897–1971
[Source: Courtesy California Institute of Technology Archives.]

than the C—N bond in a simple amine and that the atoms associated with the peptide bond are coplanar. This indicated a resonance or partial sharing of two pairs of electrons between the carbonyl oxygen and the amide nitrogen **(Fig. 4-2a)**. The oxygen has a partial negative charge and the hydrogen bonded to the nitrogen has a net partial positive charge, setting up a small electric

dipole. The six atoms of the **peptide group** lie in a single plane, with the oxygen atom of the carbonyl group trans to the hydrogen atom of the amide nitrogen. From these findings Pauling and Corey concluded that the peptide C—N bonds, because of their partial double-bond character, cannot rotate freely. Rotation is permitted about the N—C_α and the C_α—C bonds. The backbone of a polypeptide chain can thus be pictured as a series of rigid planes, with consecutive planes sharing a common point of rotation at C_α (Fig. 4-2b). The rigid peptide bonds limit the range of conformations possible for a polypeptide chain.

Peptide conformation is defined by three dihedral angles (also known as torsion angles) called ϕ (phi), ψ (psi), and ω (omega), reflecting rotation about each of the three repeating bonds in the peptide backbone. A dihedral angle is the angle at the intersection of two planes. In the case of peptides, the planes are defined by bond vectors in the peptide backbone. Two successive bond vectors describe a plane. Three successive bond vectors describe two planes (the central bond vector is common to both; Fig. 4-2c), and the angle between these two planes is what we measure to describe peptide conformation.

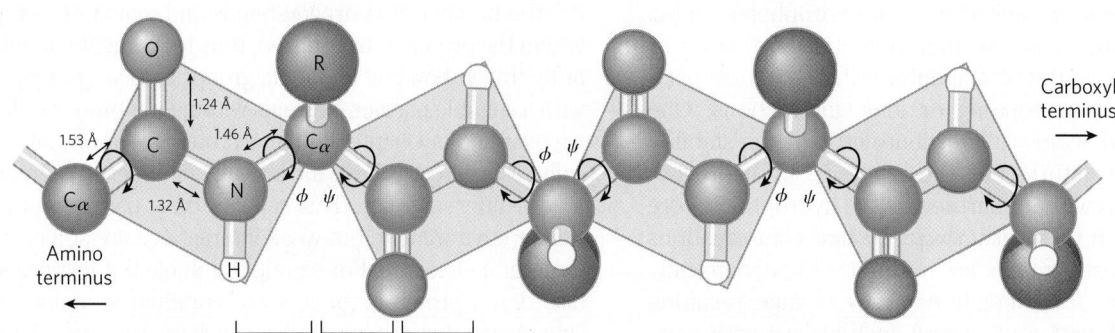

FIGURE 4-2 The planar peptide group. (a) Each peptide bond has some double-bond character due to resonance and cannot rotate. Although the N atom in a peptide bond is often represented with a partial positive charge, careful consideration of bond orbitals and quantum mechanics indicates that the N has a net charge that is neutral or slightly negative. (b) Three bonds separate sequential α carbons in a polypeptide chain. The N—C_α and C_α—C bonds can rotate, described by dihedral angles designated ϕ and ψ, respectively. The peptide C—N bond is not free to rotate. Other single bonds in the backbone may also be rotationally hindered, depending on the size and charge of the R groups. (c) The atoms and planes defining ψ. (d) By convention, ϕ and ψ are 180° (or −180°) when the first and fourth atoms are farthest apart and the peptide is fully extended. As the viewer looks out along the bond undergoing rotation (from either direction), the ϕ and ψ angles increase as the fourth atom rotates clockwise relative to the first. In a protein, some of the conformations shown here (e.g., 0°) are prohibited by steric overlap of atoms. In (b) through (d), the balls representing atoms are smaller than the van der Waals radii for this scale.

>> Key Convention: The important dihedral angles in a peptide are defined by the three bond vectors connecting four consecutive main-chain (peptide backbone) atoms (Fig. 4-2c): ϕ involves the C—N—C$_\alpha$—C bonds (with the rotation occurring about the N—C$_\alpha$ bond), and ψ involves the N—C$_\alpha$—C—N bonds. Both ϕ and ψ are defined as ±180° when the polypeptide is fully extended and all peptide groups are in the same plane (Fig. 4-2d). As one looks down the central bond vector in the direction of the vector arrow (as depicted in Fig. 4-2c for ψ), the dihedral angles increase as the distal (fourth) atom is rotated clockwise (Fig. 4-2d). From the ±180° position, the dihedral angle increases from −180° to 0°, at which point the first and fourth atoms are eclipsed. The rotation can be continued from 0° to +180° (same position as −180°) to bring the structure back to the starting point. The third dihedral angle, ω, is not often considered. It involves the C$_\alpha$—C—N—C$_\alpha$ bonds. The central bond in this case is the peptide bond, where rotation is constrained. The peptide bond is almost always (99.6% of the time) in the trans configuration, constraining ω to a value of ±180°. For a rare cis peptide bond, $\omega = 0°$. **<<**

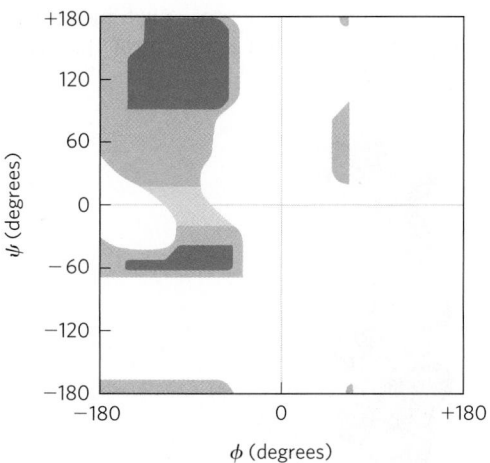

FIGURE 4-3 Ramachandran plot for L-Ala residues. Peptide conformations are defined by the values of ϕ and ψ. Conformations deemed possible are those that involve little or no steric interference, based on calculations using known van der Waals radii and dihedral angles. The areas shaded dark blue represent conformations that involve no steric overlap if the van der Waals radii of each atom are modeled as a hard sphere and that are thus fully allowed. Medium blue indicates conformations permitted if atoms are allowed to approach each other by an additional 0.1 nm, a slight clash. The lightest blue indicates conformations that are permissible if a very modest flexibility (a few degrees) is allowed in the ω dihedral angle that describes the peptide bond itself (generally constrained to 180°). The white regions are conformations that are not allowed. The asymmetry of the plot results from the L stereochemistry of the amino acid residues. The plots for other L residues with unbranched side chains are nearly identical. Allowed ranges for branched residues such as Val, Ile, and Thr are somewhat smaller than for Ala. The Gly residue, which is less sterically hindered, has a much broader range of allowed conformations. The range for Pro residues is greatly restricted because ϕ is limited by the cyclic side chain to the range of −35° to −85°. [Source: Information from T. E. Creighton, *Proteins*, p. 166. © 1984 by W. H. Freeman and Company. Reprinted by permission.]

In principle, ϕ and ψ can have any value between −180° and +180°, but many values are prohibited by steric interference between atoms in the polypeptide backbone and amino acid side chains. The conformation in which both ϕ and ψ are 0° (Fig. 4-2d) is prohibited for this reason; this conformation is merely a reference point for describing the dihedral angles. Backbone angle preferences in a polypeptide represent yet another constraint on the overall folded structure of a protein. Allowed values for ϕ and ψ become evident when ψ is plotted versus ϕ in a **Ramachandran plot (Fig. 4-3)**, introduced by G. N. Ramachandran. As we will see, Ramachandran plots are very useful tools. They are often used to test the quality of three-dimensional protein structures that are deposited in international databases.

SUMMARY 4.1 Overview of Protein Structure

- A typical protein usually has one or more stable three-dimensional structures, or conformations, that reflect its function. Some proteins have segments that are intrinsically disordered.

- Protein structure is stabilized largely by multiple weak interactions. The hydrophobic effect, derived from the increase in entropy of the surrounding water when nonpolar molecules or groups are clustered together, makes the major contribution to stabilizing the globular form of most soluble proteins. Van der Waals interactions also contribute. Hydrogen bonds and ionic interactions are optimized in the thermodynamically most stable structures.

- Nonpeptide covalent bonds, particularly disulfide bonds, play a role in the stabilization of structure in some proteins.

- The nature of the covalent bonds in the polypeptide backbone places constraints on structure. The peptide bond has a partial double-bond character that keeps the entire six-atom peptide group in a rigid planar configuration. The N—C$_\alpha$ and C$_\alpha$—C bonds can rotate to define the dihedral angles ϕ and ψ, respectively, although permitted values of ϕ and ψ are limited by steric and other constraints.

- The Ramachandran plot is a visual description of the combinations of ϕ and ψ dihedral angles that are permitted in a peptide backbone and those that are not permitted due to steric constraints.

4.2 Protein Secondary Structure

The term **secondary structure** refers to any chosen segment of a polypeptide chain and describes the local spatial arrangement of its main-chain atoms, without regard to the positioning of its side chains or its relationship to other segments. A *regular* secondary structure occurs when each dihedral angle, ϕ and ψ, remains the same or nearly the same throughout the segment. There are a few types of secondary structure that are particularly stable and occur widely in proteins. The most prominent are the α-helix and β conformations; another common type is the β turn.

Where a regular pattern is not found, the secondary structure is sometimes referred to as undefined or as a random coil. This last designation, however, does not properly describe the structure of these segments. The path of most of the polypeptide backbone in a typical protein is not random; rather, it is unchanging and highly specific to the structure and function of that particular protein. Our discussion here focuses on the regular structures that are most common.

The α Helix Is a Common Protein Secondary Structure

Pauling and Corey were aware of the importance of hydrogen bonds in orienting polar chemical groups such as the C=O and N—H groups of the peptide bond. They also had the experimental results of William Astbury, who in the 1930s had conducted pioneering x-ray studies of proteins. Astbury demonstrated that the protein that makes up hair and porcupine quills (the fibrous protein α-keratin) has a regular structure that repeats every 5.15 to 5.2 Å. (The angstrom, Å, named after the physicist Anders J. Ångström, is equal to 0.1 nm. Although not an SI unit, it is used universally by structural biologists to describe atomic distances—it is approximately the length of a typical C—H bond.) With this information and their data on the peptide bond, and with the help of precisely constructed models, Pauling and Corey set out to determine the likely conformations of protein molecules.

The first breakthrough came in 1948. Pauling was a visiting lecturer at Oxford University, became ill, and retired to his apartment for several days of rest. Bored with the reading available, Pauling grabbed some paper and pencils to work out a plausible stable structure that could be taken up by a polypeptide chain. The model he developed, and later confirmed in work with Corey and coworker Herman Branson, was the simplest arrangement the polypeptide chain can assume that maximizes the use of internal hydrogen bonding. It is a helical structure, and Pauling and Corey called it the **α helix (Fig. 4-4)**. In this structure, the polypeptide backbone is tightly wound around an imaginary axis drawn longitudinally through the middle of the helix, and the R groups of the amino acid residues protrude outward from the helical backbone. The repeating unit is a single turn of the helix, which extends about 5.4 Å along the long axis, slightly greater than the periodicity Astbury observed on x-ray analysis of hair keratin. The backbone atoms of the amino acid residues in the prototypical α helix have a characteristic set of dihedral angles that define the α helix conformation (Table 4-1), and each helical turn includes 3.6 amino acid residues. The α-helical segments in proteins often deviate slightly from these dihedral angles, and they even vary

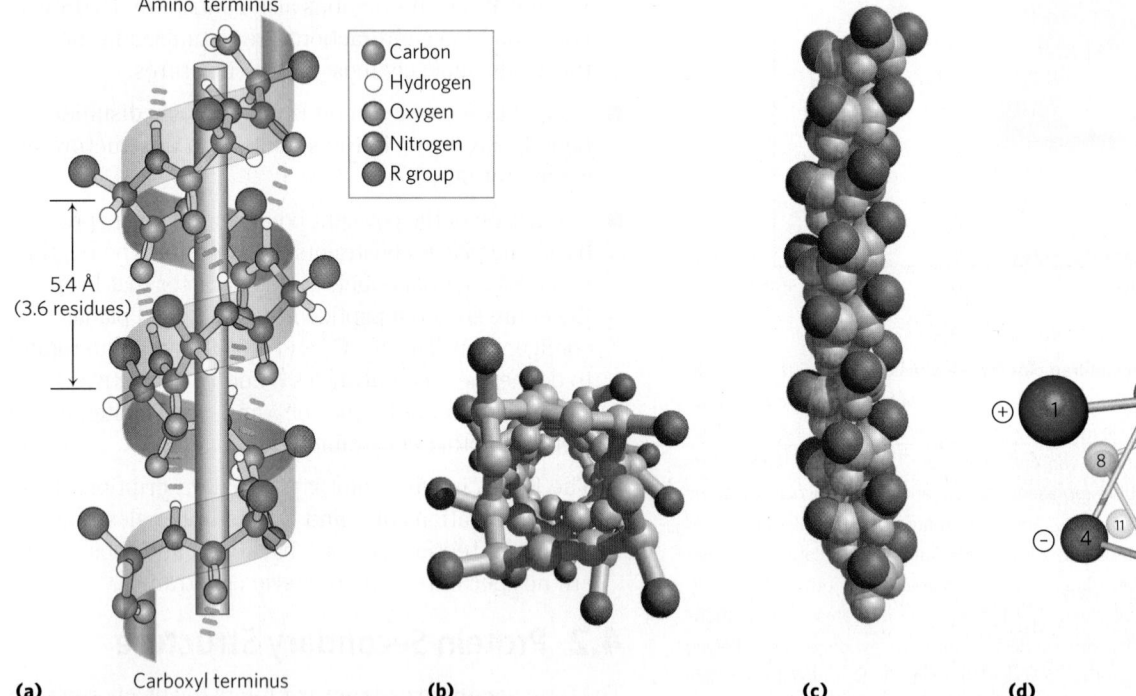

FIGURE 4-4 Models of the α helix, showing different aspects of its structure. (a) Ball-and-stick model showing the intrachain hydrogen bonds. The repeat unit is a single turn of the helix, 3.6 residues. **(b)** The α helix viewed from one end, looking down the longitudinal axis. Note the positions of the R groups, represented by purple spheres. This ball-and-stick model, which emphasizes the helical arrangement, gives the false impression that the helix is hollow, because the balls do not represent the van der Waals radii of the individual atoms. **(c)** As this space-filling model shows, the atoms in the center of the α helix are in very close contact. **(d)** Helical wheel projection of an α helix. This representation can be colored to identify surfaces with particular properties. The yellow residues, for example, could be hydrophobic and conform to an interface between the helix shown here and another part of the same or another polypeptide. The red (negative) and blue (positive) residues illustrate the potential for interaction of oppositely charged side chains separated by two residues in the helix. [Source: (b, c) Derived from PDB ID 4TNC, K. A. Satyshur et al., *J. Biol. Chem.* 263:1628, 1988.]

TABLE 4-1	Idealized ϕ and ψ Angles for Common Secondary Structures in Proteins	
Structure	ϕ	ψ
α Helix	$-57°$	$-47°$
β Conformation		
Antiparallel	$-139°$	$+135°$
Parallel	$-119°$	$+113°$
Collagen triple helix	$-51°$	$+153°$
β Turn type I		
i + 1[a]	$-60°$	$-30°$
i + 2[a]	$-90°$	$0°$
β Turn type II		
i + 1	$-60°$	$+120°$
i + 2	$+80°$	$0°$

Note: In real proteins, dihedral angles often vary somewhat from these idealized values.

[a]The i + 1 and i + 2 angles are those for the second and third amino acid residues in the β turn, respectively.

somewhat within a single, continuous segment so as to produce subtle bends or kinks in the helical axis. Pauling and Corey considered both right- and left-handed variants of the α helix. The subsequent elucidation of the three-dimensional structure of myoglobin and other proteins showed that the right-handed α helix is the common form (Box 4-1). Extended left-handed α helices are theoretically less stable and have not been observed in proteins. The α helix proved to be the predominant structure in α-keratins. More generally, about one-fourth of all amino acid residues in proteins are found in α helices, the exact fraction varying greatly from one protein to another.

Why does the α helix form more readily than many other possible conformations? The answer lies, in part, in its optimal use of internal hydrogen helical bonds. The structure is stabilized by a hydrogen bond between the hydrogen atom attached to the electronegative nitrogen atom of a peptide linkage and the electronegative carbonyl oxygen atom of the fourth amino acid on the amino-terminal side of that peptide bond (Fig. 4-4a). Within the α helix, every peptide bond (except those close to each end of the helix) participates in such hydrogen bonding. Each successive turn of the α helix is held to adjacent turns by three to four hydrogen bonds, conferring significant stability on the overall structure. At the ends of an α-helical segment, there are always three or four amide carbonyl or amino groups that cannot participate in this helical pattern of hydrogen bonding. These may be exposed to the surrounding solvent, where they hydrogen-bond with water, or other parts of the protein may cap the helix to provide the needed hydrogen-bonding partners.

Further experiments have shown that an α helix can form in polypeptides consisting of either L- or D-amino acids. However, all residues must be of one stereoisomeric series; a D-amino acid will disrupt a regular structure consisting of L-amino acids, and vice versa. The most stable form of an α helix consisting of D-amino acids is left-handed.

WORKED EXAMPLE 4-1 **Secondary Structure and Protein Dimensions**

What is the length of a polypeptide with 80 amino acid residues in a single, continuous α helix?

Solution: An idealized α helix has 3.6 residues per turn, and the rise along the helical axis is 5.4 Å. Thus, the rise along the axis for each amino acid residue is 1.5 Å. The length of the polypeptide is therefore 80 residues × 1.5 Å/residue = 120 Å.

Amino Acid Sequence Affects Stability of the α Helix

Not all polypeptides can form a stable α helix. Each amino acid residue in a polypeptide has an intrinsic propensity to

BOX 4-1 METHODS Knowing the Right Hand from the Left

There is a simple method for determining whether a helical structure is right-handed or left-handed. Make fists of your two hands with thumbs outstretched and pointing away from you. Looking at your right hand, think of a helix spiraling up your right thumb in the direction in which the other four fingers are curled as shown (clockwise). The resulting helix is right-handed. Your left hand will demonstrate a left-handed helix, which rotates in the counterclockwise direction as it spirals up your thumb.

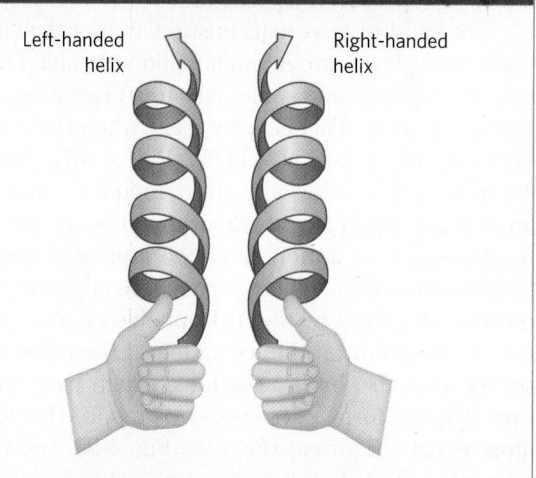

Left-handed helix   Right-handed helix

TABLE 4-2 Propensity of Amino Acid Residues to Take Up an α-Helical Conformation

Amino acid	$\Delta\Delta G°$ (kJ/mol)[a]	Amino acid	$\Delta\Delta G°$ (kJ/mol)[a]
Ala	0	Leu	0.79
Arg	0.3	Lys	0.63
Asn	3	Met	0.88
Asp	2.5	Phe	2.0
Cys	3	Pro	>4
Gln	1.3	Ser	2.2
Glu	1.4	Thr	2.4
Gly	4.6	Tyr	2.0
His	2.6	Trp	2.0
Ile	1.4	Val	2.1

Sources: Data (except proline) from J. W. Bryson et al., *Science* 270:935, 1995. Proline data from J. K. Myers et al., *Biochemistry* 36:10,923, 1997.

[a] $\Delta\Delta G°$ is the difference in free-energy change, relative to that for alanine, required for the amino acid residue to take up the α-helical conformation. Larger numbers reflect greater difficulty taking up the α-helical structure. Data are a composite derived from multiple experiments and experimental systems.

form an α helix (Table 4-2), reflecting the properties of the R group and how they affect the capacity of the adjoining main-chain atoms to take up the characteristic ϕ and ψ angles. Alanine shows the greatest tendency to form α helices in most experimental model systems.

The position of an amino acid residue relative to its neighbors is also important. Interactions between amino acid side chains can stabilize or destabilize the α-helical structure. For example, if a polypeptide chain has a long block of Glu residues, this segment of the chain will not form an α helix at pH 7.0. The negatively charged carboxyl groups of adjacent Glu residues repel each other so strongly that they prevent formation of the α helix. For the same reason, if there are many adjacent Lys and/or Arg residues, with positively charged R groups at pH 7.0, they also repel each other and prevent formation of the α helix. The bulk and shape of Asn, Ser, Thr, and Cys residues can also destabilize an α helix if they are close together in the chain.

The twist of an α helix ensures that critical interactions occur between an amino acid side chain and the side chain three (and sometimes four) residues away on either side of it. This is made clear when the α helix is depicted as a helical wheel (Fig. 4-4d). Positively charged amino acids are often found three residues away from negatively charged amino acids, permitting the formation of an ion pair. Two aromatic amino acid residues are often similarly spaced, resulting in a juxtaposition stabilized by the hydrophobic effect.

A constraint on the formation of the α helix is the presence of Pro or Gly residues, which have the least proclivity to form α helices. In proline, the nitrogen atom is part of a rigid ring (see Fig. 4-8), and rotation

about the N—C$_\alpha$ bond is not possible. Thus, a Pro residue introduces a destabilizing kink in an α helix. In addition, the nitrogen atom of a Pro residue in a peptide linkage has no substituent hydrogen to participate in hydrogen bonds with other residues. For these reasons, proline is only rarely found in an α helix. Glycine occurs infrequently in α helices for a different reason: it has more conformational flexibility than the other amino acid residues. Polymers of glycine tend to take up coiled structures quite different from an α helix.

A final factor affecting the stability of an α helix is the identity of the amino acid residues near the ends of the α-helical segment of the polypeptide. A small electric dipole exists in each peptide bond (Fig. 4-2a). These dipoles are aligned through the hydrogen bonds of the helix, resulting in a net dipole along the helical axis that increases with helix length **(Fig. 4-5)**. The partial positive and negative charges of the helix dipole reside on the peptide amino and carbonyl groups near the amino-terminal and carboxyl-terminal ends, respectively. For this reason, negatively charged amino acids are often found near the amino terminus of the helical segment, where they have a stabilizing interaction with the positive charge of the helix dipole; a positively charged amino acid at the amino-terminal end is destabilizing. The opposite is true at the carboxyl-terminal end of the helical segment.

In summary, five types of constraints affect the stability of an α helix: (1) the intrinsic propensity of an amino acid residue to form an α helix; (2) the interactions between R groups, particularly those spaced three (or four) residues apart; (3) the bulkiness of adjacent R groups; (4) the occurrence of Pro and Gly residues; and (5) interactions between amino acid residues at the ends of the helical segment and the electric dipole

Amino terminus

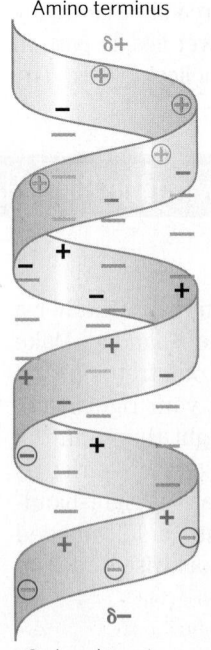

δ+

δ−

Carboxyl terminus

FIGURE 4-5 Helix dipole. The electric dipole of a peptide bond (see Fig. 4-2a) is transmitted along an α-helical segment through the intrachain hydrogen bonds, resulting in an overall helix dipole. In this illustration, the amino and carbonyl constituents of each peptide bond are indicated by + and − symbols, respectively. Non-hydrogen-bonded amino and carbonyl constituents of the peptide bonds near each end of the α-helical region are circled and shown in color.

inherent to the α helix. The tendency of a given segment of a polypeptide chain to form an α helix therefore depends on the identity and sequence of amino acid residues within the segment.

The β Conformation Organizes Polypeptide Chains into Sheets

In 1951, Pauling and Corey predicted a second type of repetitive structure, the β **conformation**. This is a more extended conformation of polypeptide chains, and its structure is again defined by backbone atoms arranged according to a characteristic set of dihedral angles (Table 4-1). In the β conformation, the backbone of the polypeptide chain is extended into a zigzag rather than helical structure **(Fig. 4-6)**. The arrangement of several segments side by side, all of which are in the β conformation, is called a β **sheet**. The zigzag structure of the individual polypeptide segments gives rise to a

pleated appearance of the overall sheet. Hydrogen bonds form between adjacent segments of polypeptide chain within the sheet. The individual segments that form a β sheet are usually nearby on the polypeptide chain but can also be quite distant from each other in the linear sequence of the polypeptide; they may even be in different polypeptide chains. The R groups of adjacent amino acids protrude from the zigzag structure in opposite directions, creating the alternating pattern seen in the side view in Figure 4-6.

The adjacent polypeptide chains in a β sheet can be either parallel or antiparallel (having the same or opposite amino-to-carboxyl orientations, respectively). The structures are somewhat similar, although the repeat period is shorter for the parallel conformation (6.5 Å, vs. 7 Å for antiparallel) and the hydrogen-bonding patterns are different. The interstrand hydrogen bonds are essentially in-line (see Fig. 2-5) in the antiparallel β sheet, whereas they are distorted or not in-line for the parallel variant. The idealized structures exhibit the bond angles given in Table 4-1; these values vary somewhat in real proteins, resulting in structural variation, as seen above for α helices.

β Turns Are Common in Proteins

In globular proteins, which have a compact folded structure, some amino acid residues are in turns or loops where the polypeptide chain reverses direction **(Fig. 4-7)**. These are the connecting elements that link successive runs of α helix or β conformation. Particularly common are β **turns** that connect the ends of two adjacent segments of an antiparallel β sheet. The structure is a 180° turn involving four amino acid residues, with the carbonyl oxygen of the first residue forming a hydrogen bond with the amino-group hydrogen of the fourth. The peptide groups of the central two residues do not participate in any inter-residue hydrogen bonding. Several types of β turns have been described, each defined by the ϕ and ψ angles of the bonds that link the four amino acid residues that make up the particular turn (Table 4-1). Gly and Pro residues often occur in β turns, the former because it is small and flexible, the latter because peptide bonds involving the imino nitrogen of proline readily assume the cis configuration **(Fig. 4-8)**, a form that is particularly amenable to a tight turn. The two types of β turns shown in Figure 4-7 are the most common. Beta turns are often found near the surface of a protein, where the peptide groups of the central two amino acid residues in the turn can hydrogen-bond with water. Considerably less common is the γ turn, a three-residue turn with a hydrogen bond between the first and third residues.

Common Secondary Structures Have Characteristic Dihedral Angles

The α helix and the β conformation are the major repetitive secondary structures in a wide variety of proteins, although other repetitive structures exist in some specialized

(a) β Strand
Side view

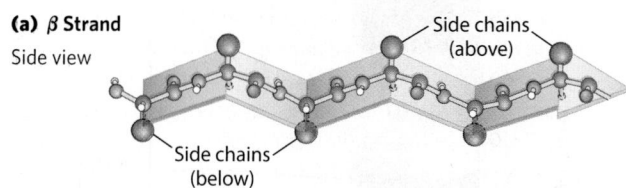

(b) Antiparallel β sheet
Top view

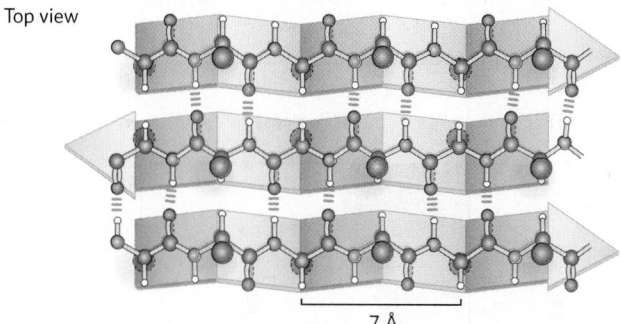

7 Å

(c) Parallel β sheet
Top view

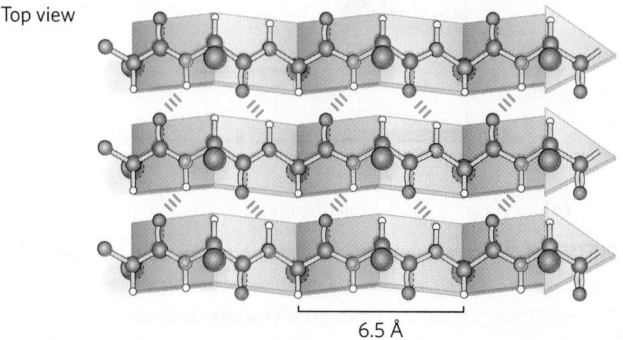

6.5 Å

FIGURE 4-6 The β conformation of polypeptide chains. These **(a)** side and **(b, c)** top views reveal the R groups extending out from the β sheet and emphasize the pleated shape formed by the planes of the peptide bonds. (An alternative name for this structure is β-pleated sheet.) Hydrogen-bond cross-links between adjacent chains are also shown. The amino-terminal to carboxyl-terminal orientations of adjacent chains (arrows) can be the opposite or the same, forming **(b)** an antiparallel β sheet or **(c)** a parallel β sheet.

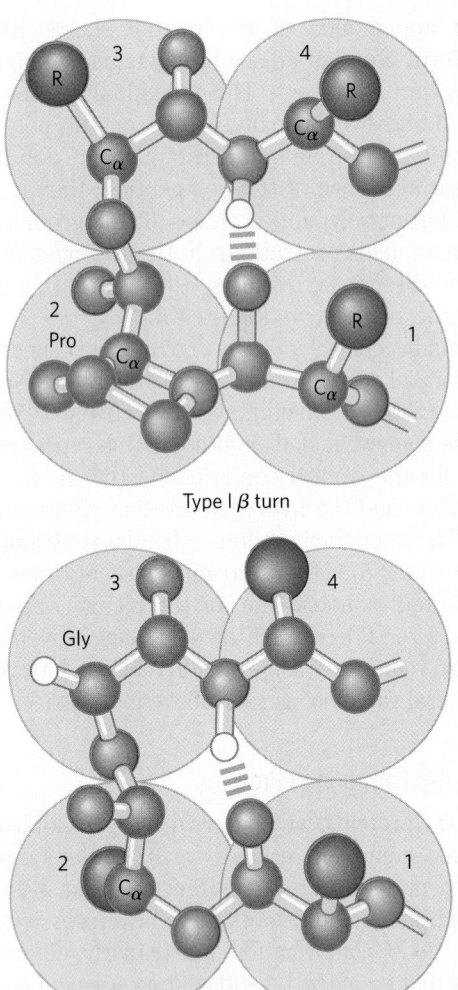

Type I β turn

Type II β turn

FIGURE 4-7 Structures of β turns. Type I and type II β turns are most common, distinguished by the ϕ and ψ angles taken up by the peptide backbone in the turn (see Table 4-1). Type I turns occur more than twice as frequently as type II. Although many amino acid residues are accommodated in these turns, some biases are evident. Pro is the most common residue at position 2 in type I turns, appearing in about 16% of them. Pro is also the most common residue at position 2 in type II turns, appearing about 23% of the time. The most prominent bias is the presence of Gly at position 3 in more than 75% of type II turns. Note the hydrogen bond between the peptide groups of the first and fourth residues of the bends. (Individual amino acid residues are framed by large blue circles. Not all H atoms are shown in these depictions.)

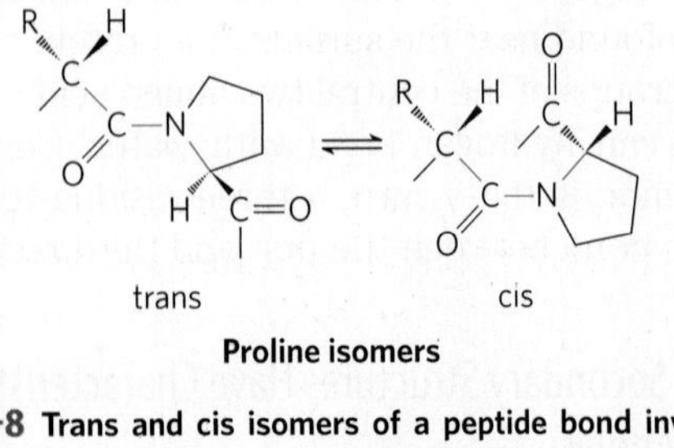

trans   cis

Proline isomers

FIGURE 4-8 Trans and cis isomers of a peptide bond involving the imino nitrogen of proline. Of the peptide bonds between amino acid residues other than Pro, more than 99.95% are in the trans configuration. For peptide bonds involving the imino nitrogen of proline, however, about 6% are in the cis configuration; many of these occur at β turns.

proteins (an example is collagen; see Fig. 4-13). Every type of secondary structure can be completely described by the dihedral angles ϕ and ψ associated with each residue. As shown by a Ramachandran plot, the dihedral angles that define the α helix and the β conformation fall within a relatively restricted range of sterically allowed structures **(Fig. 4-9a)**. Most values of ϕ and ψ taken from known protein structures fall into the expected regions, with high concentrations near the α helix and β conformation values, as predicted (Fig. 4-9b). The only amino acid residue often found in a conformation outside these regions is glycine. Because its side chain is small, a Gly residue can take part in many conformations that are sterically forbidden for other amino acids.

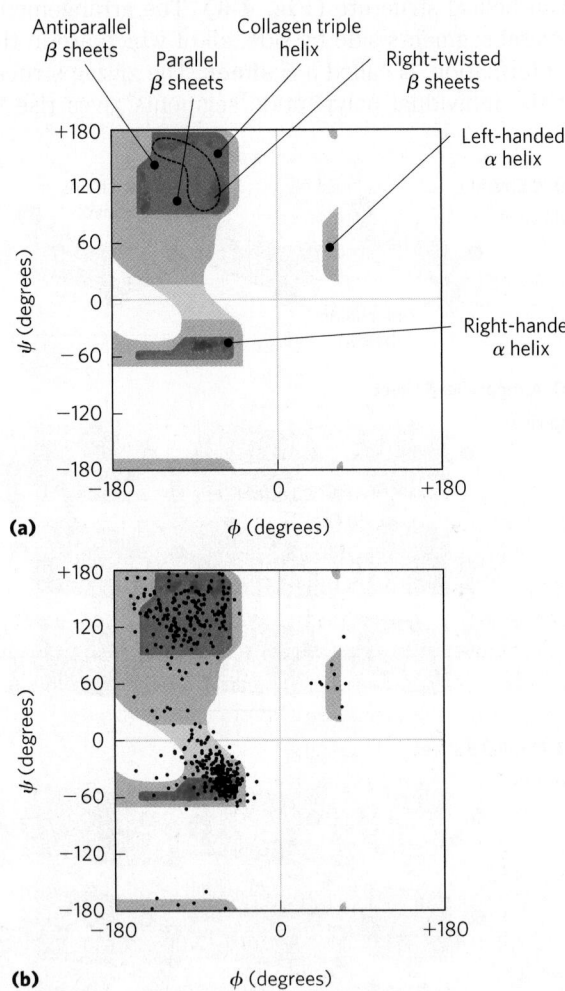

FIGURE 4-9 Ramachandran plots showing a variety of structures. (a) The values of ϕ and ψ for various allowed secondary structures are overlaid on the plot from Figure 4-3. Although left-handed α helices extending over several amino acid residues are theoretically possible, they have not been observed in proteins. **(b)** The values of ϕ and ψ for all the amino acid residues except Gly in the enzyme pyruvate kinase (isolated from rabbit) are overlaid on the plot of theoretically allowed conformations (Fig. 4-3). The small, flexible Gly residues were excluded because they frequently fall outside the expected (blue) ranges. [Sources: (a) Information from T. E. Creighton, *Proteins*, p. 166. © 1984 by W. H. Freeman and Company. (b) Courtesy of Hazel Holden, University of Wisconsin–Madison, Department of Biochemistry.]

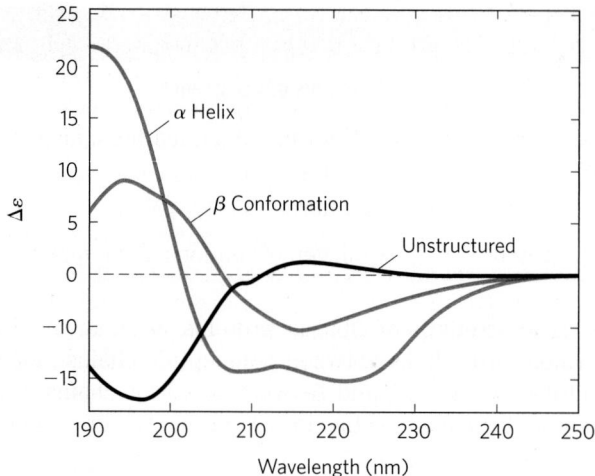

FIGURE 4-10 Circular dichroism spectroscopy. These spectra show polylysine entirely as α helix, as β conformation, or in an unstructured, denatured state. The y axis unit is a simplified version of the units most commonly used in CD experiments. Since the curves are different for α helix, β conformation, and unstructured, the CD spectrum for a given protein can provide a rough estimate for the fraction of the protein made up of the two most common secondary structures. The CD spectrum of the native protein can serve as a benchmark for the folded state, useful for monitoring denaturation or conformational changes brought about by changes in solution conditions.

Common Secondary Structures Can Be Assessed by Circular Dichroism

Any form of structural asymmetry in a molecule gives rise to differences in absorption of left-handed versus right-handed circularly polarized light. Measurement of this difference is called **circular dichroism (CD) spectroscopy.** An ordered structure, such as a folded protein, gives rise to an absorption spectrum that can have peaks or regions with both positive and negative values. For proteins, spectra are obtained in the far UV region (190 to 250 nm). The light-absorbing entity, or chromophore, in this region is the peptide bond; a signal is obtained when the peptide bond is in a folded environment. The difference in molar extinction coefficients (see Box 3-1) for left- and right-handed, circularly polarized light ($\Delta\varepsilon$) is plotted as a function of wavelength. The α-helix and β conformations have characteristic CD spectra **(Fig. 4-10).** Using CD spectra, biochemists can determine whether proteins are properly folded, estimate the fraction of the protein that is folded in either of the common secondary structures, and monitor transitions between the folded and unfolded states.

SUMMARY 4.2 Protein Secondary Structure

- Secondary structure is the local spatial arrangement of the main-chain atoms in a selected segment of a polypeptide chain.

- The most common regular secondary structures are the α helix, the β conformation, and β turns.

- The secondary structure of a polypeptide segment can be completely defined if the ϕ and ψ angles are known for all amino acid residues in that segment.

- Circular dichroism spectroscopy is a method for assessing common secondary structure and monitoring folding in proteins.

4.3 Protein Tertiary and Quaternary Structures

The overall three-dimensional arrangement of all atoms in a protein is referred to as the protein's **tertiary structure.** Whereas the term "secondary structure" refers to the spatial arrangement of amino acid residues that are adjacent in a segment of a polypeptide, tertiary structure includes *longer-range* aspects of amino acid sequence. Amino acids that are far apart in the polypeptide sequence and are in different types of secondary structure may interact within the completely folded structure of a protein. The location of bends (including β turns) in the polypeptide chain and the direction and angle of these bends are determined by the number and location of specific bend-producing residues, such as Pro, Thr, Ser, and Gly. Interacting segments of polypeptide chains are held in their characteristic tertiary positions by several kinds of weak interactions (and sometimes by covalent bonds such as disulfide cross-links) between the segments.

Some proteins contain two or more separate polypeptide chains, or subunits, which may be identical or different. The arrangement of these protein subunits in three-dimensional complexes constitutes **quaternary structure.**

In considering these higher levels of structure, it is useful to designate two major groups into which many proteins can be classified: **fibrous proteins**, with polypeptide chains arranged in long strands or sheets, and **globular proteins**, with polypeptide chains folded into a spherical or globular shape. The two groups are structurally distinct. Fibrous proteins usually consist largely of a single type of secondary structure, and their tertiary structure is relatively simple. Globular proteins often contain several types of secondary structure. The two groups also differ functionally: the structures that provide support, shape, and external protection to vertebrates are made of fibrous proteins, whereas most enzymes and regulatory proteins are globular proteins.

Fibrous Proteins Are Adapted for a Structural Function

α-Keratin, collagen, and silk fibroin nicely illustrate the relationship between protein structure and biological function (Table 4-3). Fibrous proteins share properties that give strength and/or flexibility to the structures in which they occur. In each case, the fundamental structural unit is a simple repeating element of secondary structure. All fibrous proteins are insoluble in water, a property conferred by a high concentration of hydrophobic amino acid residues both in the interior of the protein and on its

TABLE 4-3   Secondary Structures and Properties of Some Fibrous Proteins

Structure	Characteristics	Examples of occurrence
α Helix, cross-linked by disulfide bonds	Tough, insoluble protective structures of varying hardness and flexibility	α-Keratin of hair, feathers, nails
β Conformation	Soft, flexible filaments	Silk fibroin
Collagen triple helix	High tensile strength, without stretch	Collagen of tendons, bone matrix

surface. These hydrophobic surfaces are largely buried, as many similar polypeptide chains are packed together to form elaborate supramolecular complexes. The underlying structural simplicity of fibrous proteins makes them particularly useful for illustrating some of the fundamental principles of protein structure discussed above.

α-Keratin The α-keratins have evolved for strength. Found only in mammals, these proteins constitute almost the entire dry weight of hair, wool, nails, claws, quills, horns, hooves, and much of the outer layer of skin. The α-keratins are part of a broader family of proteins called intermediate filament (IF) proteins. Other IF proteins are found in the cytoskeletons of animal cells. All IF proteins have a structural function and share the structural features exemplified by the α-keratins.

The α-keratin helix is a right-handed α helix, the same helix found in many other proteins. Francis Crick and Linus Pauling, in the early 1950s, independently suggested that the α helices of keratin were arranged as a coiled coil. Two strands of α-keratin, oriented in parallel (with their amino termini at the same end), are wrapped about each other to form a supertwisted coiled coil. The supertwisting amplifies the strength of the overall structure, just as strands are twisted to make a strong rope **(Fig. 4-11)**. The twisting of the axis of an α helix to form a coiled coil explains the discrepancy between the 5.4 Å per turn predicted for an α helix by Pauling and Corey and the 5.15 to 5.2 Å repeating structure observed in the x-ray diffraction of hair (p. 152). The helical path of the supertwists is left-handed, opposite in sense to the α helix. The surfaces where the two α helices touch are made up of hydrophobic amino acid residues, their R groups meshed together in a regular interlocking pattern. This permits a close packing of the polypeptide chains within the left-handed supertwist. Not surprisingly, α-keratin is rich in the hydrophobic residues Ala, Val, Leu, Ile, Met, and Phe.

An individual polypeptide in the α-keratin coiled coil has a relatively simple tertiary structure, dominated by an α-helical secondary structure with its helical axis twisted in a left-handed superhelix. The intertwining of the two α-helical polypeptides is an example of quaternary structure. Coiled coils of this type are common structural elements in filamentous proteins and in the muscle protein myosin (see Fig. 5-27). The quaternary structure of α-keratin can be quite complex. Many coiled coils can be assembled into large supramolecular complexes, such as the arrangement of α-keratin that forms the intermediate filament of hair (Fig. 4-11b).

The strength of fibrous proteins is enhanced by covalent cross-links between polypeptide chains in the multihelical "ropes" and between adjacent chains in a supramolecular assembly. In α-keratins, the cross-links

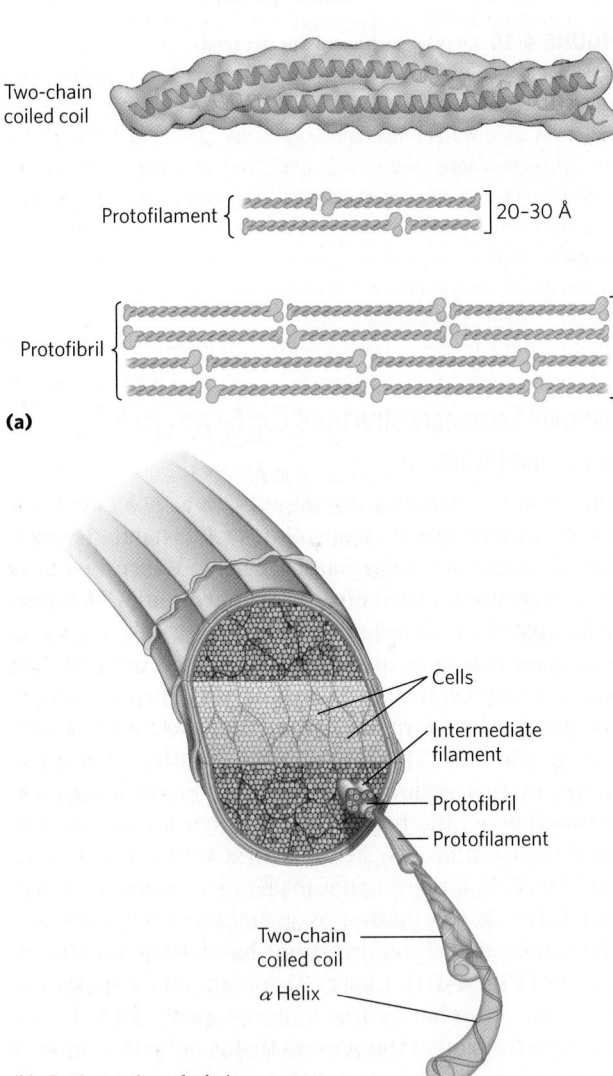

FIGURE 4-11 Structure of hair. (a) Hair α-keratin is an elongated α helix with somewhat thicker elements near the amino and carboxyl termini. Pairs of these helices are interwound in a left-handed sense to form two-chain coiled coils. These then combine in higher-order structures called protofilaments and protofibrils. About four protofibrils—32 strands of α-keratin in all—combine to form an intermediate filament. The individual two-chain coiled coils in the various substructures also seem to be interwound, but the handedness of the interwinding and other structural details are unknown. (b) A hair is an array of many α-keratin filaments, made up of the substructures shown in (a). [Source: (a) PDB ID 3TNU, C. H. Lee et al., *Nature Struct. Mol. Biol.* 19:707, 2012.]

BOX 4-2 Permanent Waving Is Biochemical Engineering

When hair is exposed to moist heat, it can be stretched. At the molecular level, the α helices in the α-keratin of hair are stretched out until they arrive at the fully extended β conformation. On cooling, they spontaneously revert to the α-helix conformation. The characteristic "stretchability" of α-keratins, as well as their numerous disulfide cross-linkages, is the basis of permanent waving. The hair to be waved or curled is first bent around a form of appropriate shape. A solution of a reducing agent, usually a compound containing a thiol or sulfhydryl group (—SH), is then applied with heat. The reducing agent cleaves the cross-linkages by reducing each disulfide bond to form two Cys residues. The moist heat breaks hydrogen bonds and causes the α-helical structure of the polypeptide chains to uncoil. After a time, the reducing solution is removed, and an oxidizing agent is added to establish *new* disulfide bonds between pairs of Cys residues of adjacent polypeptide chains, but not the same pairs as before the treatment. After the hair is washed and cooled, the polypeptide chains revert to their α-helix conformation. The hair fibers now curl in the desired fashion because the new disulfide cross-linkages exert some torsion or twist on the bundles of α-helical coils in the hair fibers. The same process can be used to straighten hair that is naturally curly. A permanent wave (or hair straightening) is not truly permanent, however, because hair grows; in the new hair replacing the old, the α-keratin has the natural pattern of disulfide bonds.

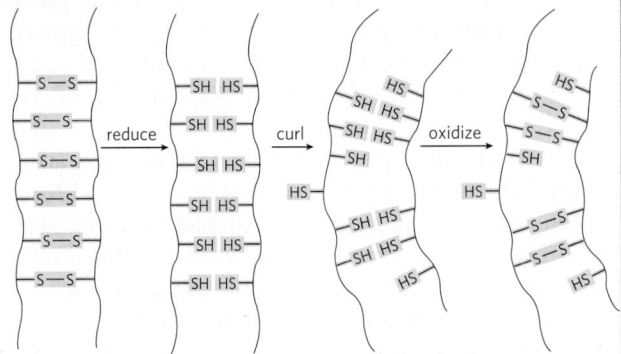

stabilizing quaternary structure are disulfide bonds (Box 4-2). In the hardest and toughest α-keratins, such as those of rhinoceros horn, up to 18% of the residues are cysteines involved in disulfide bonds.

Collagen Like the α-keratins, **collagen** has evolved to provide strength. It is found in connective tissue such as tendons, cartilage, the organic matrix of bone, and the cornea of the eye. The collagen helix is a unique secondary structure, quite distinct from the α helix. It is left-handed and has three amino acid residues per turn (**Fig. 4-12** and Table 4-1). Collagen is also a coiled coil, but one with distinct tertiary and quaternary structures: three separate polypeptides, called α chains (not to be confused with α helices), are supertwisted about each other. The superhelical twisting is right-handed in collagen, opposite in sense to the left-handed helix of the α chains.

There are many types of vertebrate collagen. Typically they contain about 35% Gly, 11% Ala, and 21% Pro and 4-Hyp (4-hydroxyproline, an uncommon amino acid; see Fig. 3-8a). The food product gelatin is derived from collagen. It has little nutritional value as a protein, because collagen is extremely low in many amino acids that are essential in the human diet. The unusual amino acid content of collagen is related to structural constraints unique to the collagen helix. The amino acid sequence in collagen is generally a repeating tripeptide unit, Gly–X–Y, where X is often Pro, and Y is often 4-Hyp. Only Gly residues can be accommodated at the very tight junctions between the individual α chains (Fig. 4-12b). The Pro and 4-Hyp residues permit the sharp twisting of the collagen helix. The amino acid sequence and the supertwisted quaternary structure of collagen allow a very close packing of its three polypeptides. 4-Hydroxyproline has a special role in the structure of collagen—and in human history (Box 4-3).

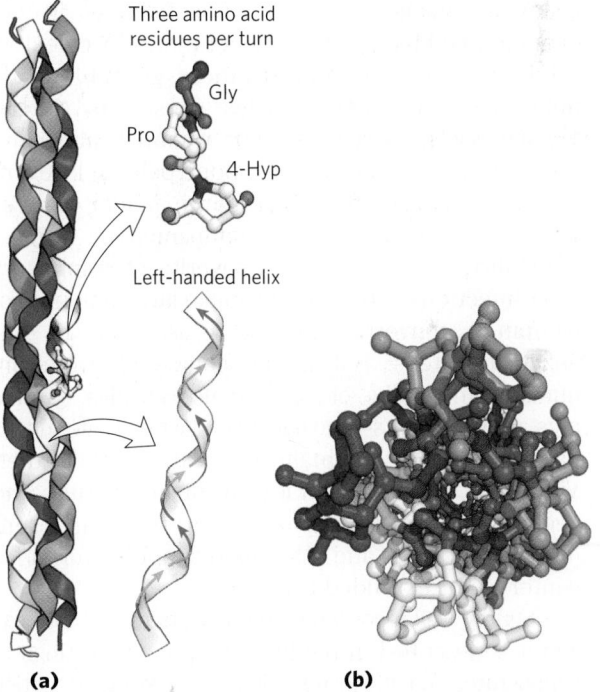

FIGURE 4-12 Structure of collagen. (a) The α chain of collagen has a repeating secondary structure unique to this protein. The repeating tripeptide sequence Gly-X-Pro or Gly-X-4-Hyp adopts a left-handed helical structure with three residues per turn. The repeating sequence used to generate this model is Gly-Pro-4-Hyp. Three of these helices (shown here in white, blue, and purple) wrap around one another with a right-handed twist. (b) The three-stranded collagen superhelix shown from one end, in a ball-and-stick representation. Gly residues are shown in red. Glycine, because of its small size, is required at the tight junction where the three chains are in contact. The balls in this illustration do not represent the van der Waals radii of the individual atoms. The center of the three-stranded superhelix is not hollow, as it appears here, but very tightly packed. [Source: Modified from PDB ID 1CGD, J. Bella et al., *Structure* 3:893, 1995.]

BOX 4-3 MEDICINE Why Sailors, Explorers, and College Students Should Eat Their Fresh Fruits and Vegetables

. . . from this misfortune, together with the unhealthiness of the country, where there never falls a drop of rain, we were stricken with the "camp-sickness," which was such that the flesh of our limbs all shrivelled up, and the skin of our legs became all blotched with black, mouldy patches, like an old jack-boot, and proud flesh came upon the gums of those of us who had the sickness, and none escaped from this sickness save through the jaws of death. The signal was this: when the nose began to bleed, then death was at hand.

—*The Memoirs of the Lord of Joinville*, ca. 1300*

This excerpt describes the plight of Louis IX's army toward the end of the Seventh Crusade (1248–1254), when the scurvy-weakened Crusader army was destroyed by the Egyptians. What was the nature of the malady afflicting these thirteenth-century soldiers?

Scurvy is caused by lack of vitamin C, or ascorbic acid (ascorbate). Vitamin C is required for, among other things, the hydroxylation of proline and lysine in collagen; scurvy is a deficiency disease characterized by general degeneration of connective tissue. Manifestations of advanced scurvy include numerous small hemorrhages caused by fragile blood vessels, tooth loss, poor wound healing and the reopening of old wounds, bone pain and degeneration, and eventually heart failure. Milder cases of vitamin C deficiency are accompanied by fatigue, irritability, and an increased severity of respiratory tract infections. Most animals make large amounts of vitamin C, converting glucose to ascorbate in four enzymatic steps. But in the course of evolution, humans and some other animals—gorillas, guinea pigs, and fruit bats—have lost the last enzyme in this pathway and must obtain ascorbate in their diet. Vitamin C is available in a wide range of fruits and vegetables. Until 1800, however, it was often absent in the dried foods and other food supplies stored for winter or for extended travel.

Scurvy was recorded by the Egyptians in 1500 BCE, and it is described in the fifth century BCE writings of Hippocrates. Yet it did not come to wide public notice until the European voyages of discovery from 1500 to 1800. The first circumnavigation of the globe (1519–1522), led by Ferdinand Magellan, was accomplished only with the loss of more than 80% of his crew to scurvy. During Jacques Cartier's second voyage to explore the St. Lawrence River (1535–1536), his band was threatened with complete disaster until the native Americans taught the men to make a cedar tea that cured

and prevented scurvy (it contained vitamin C). Winter outbreaks of scurvy in Europe were gradually eliminated in the nineteenth century as the cultivation of the potato, introduced from South America, became widespread.

In 1747, James Lind, a Scottish surgeon in the Royal Navy, carried out the first controlled clinical study in recorded history. During an extended voyage on the 50-gun warship *HMS Salisbury*, Lind selected 12 sailors suffering from scurvy and separated them into groups of two. All 12 received the same diet, except that each group was given a different remedy for scurvy from among those recommended at the time. The sailors given lemons and oranges recovered and returned to duty. The sailors given boiled apple juice improved slightly. The remainder continued to deteriorate. Lind's *Treatise on the Scurvy* was published in 1753, but inaction persisted in the Royal Navy for another 40 years. In 1795, the British admiralty finally mandated a ration of concentrated lime or lemon juice for all British sailors (hence the name "limeys"). Scurvy continued to be a problem in some other parts of the world until 1932, when Hungarian scientist Albert Szent-Györgyi, and W. A. Waugh and C. G. King at the University of Pittsburgh, isolated and synthesized ascorbic acid.

James Lind, 1716–1794; naval surgeon, 1739–1748 [Source: Library, Archive and Family History Enquiries, Royal College of Physicians of Edinburgh.]

L-Ascorbic acid (vitamin C) is a white, odorless, crystalline powder. It is freely soluble in water and relatively insoluble in organic solvents. In a dry state, away from light, it is stable for a considerable length of time. The appropriate daily intake of this vitamin is still in dispute. The recommended value in the United States is 90 mg for men, 75 mg for women. The United Kingdom recommends 40 mg, Australia 45 mg, and Russia 50–100 mg. Along with citrus fruits and almost all other fresh fruits, good sources of vitamin C include peppers, tomatoes, potatoes, and broccoli. The vitamin C of fruits and vegetables is destroyed by overcooking or prolonged storage.

So why is ascorbate so necessary to good health? Of particular interest to us here is its role in the formation of collagen. As noted in the text, collagen is constructed of the repeating tripeptide unit Gly–X–Y, where X and Y are generally Pro or 4-Hyp—the proline derivative (4R)-L-hydroxyproline, which plays an essential role in the folding of collagen and in maintaining its structure. The proline ring is normally

*From Ethel Wedgwood, *The Memoirs of the Lord of Joinville: A New English Version*, E. P. Dutton and Company, 1906.

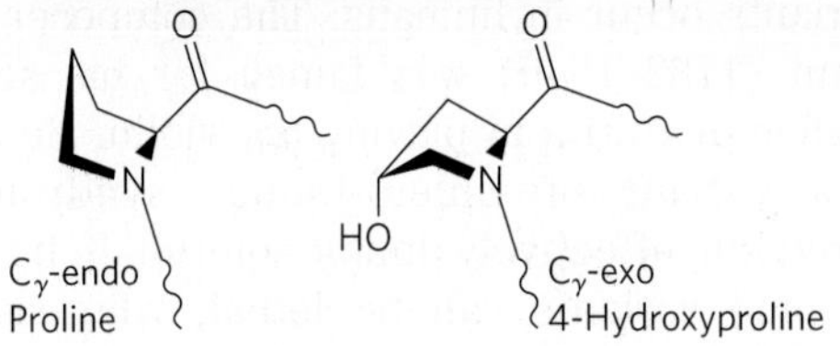

FIGURE 1 The C_γ-endo conformation of proline and the C_γ-exo conformation of 4-hydroxyproline.

found as a mixture of two puckered conformations, called C_γ-endo and C_γ-exo (Fig. 1). The collagen helix structure requires the Pro/4-Hyp residue in the Y positions to be in the C_γ-exo conformation, and it is this conformation that is enforced by the hydroxyl substitution at C-4 in 4-Hyp. The collagen structure also requires that the Pro/4-Hyp residue in the X positions have the C_γ-endo conformation, and introduction of 4-Hyp here can destabilize the helix. In the absence of vitamin C, cells cannot hydroxylate the Pro at the Y positions. This leads to collagen instability and the connective tissue problems seen in scurvy.

The hydroxylation of specific Pro residues in procollagen, the precursor of collagen, requires the action of the enzyme prolyl 4-hydroxylase. This enzyme (M_r 240,000) is an $\alpha_2\beta_2$ tetramer in all vertebrates. The proline-hydroxylating activity is found in the α subunits. Each α subunit contains one atom of nonheme iron (Fe^{2+}), and the enzyme is one of a class of hydroxylases that require α-ketoglutarate in their reactions.

In the normal prolyl 4-hydroxylase reaction (Fig. 2a), one molecule of α-ketoglutarate and one of O_2 bind to the enzyme. The α-ketoglutarate is oxidatively decarboxylated to form CO_2 and succinate. The remaining oxygen atom is then used to hydroxylate an appropriate Pro residue in procollagen. No ascorbate is needed in this reaction. However, prolyl 4-hydroxylase also catalyzes an oxidative decarboxylation of α-ketoglutarate that is not coupled to proline hydroxylation (Fig. 2b). During this reaction, the heme Fe^{2+} becomes oxidized, inactivating the enzyme and preventing the proline hydroxylation. The ascorbate consumed in the reaction is needed to restore enzyme activity—by reducing the heme iron.

Scurvy remains a problem today, not only in remote regions where nutritious food is scarce but, surprisingly, on U.S. college campuses. The only vegetables consumed by some students are those in tossed salads, and days go by without these young adults consuming fruit. A 1998 study of 230 students at Arizona State University revealed that 10% had serious vitamin C deficiencies, and 2 students had vitamin C levels so low that they probably had scurvy. Only half the students in the study consumed the recommended daily allowance of vitamin C.

Eat your fresh fruits and vegetables.

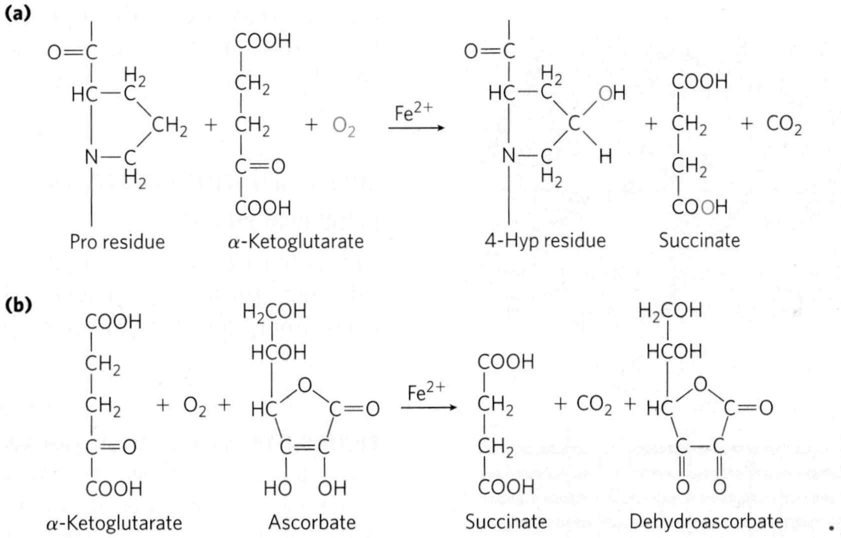

FIGURE 2 Reactions catalyzed by prolyl 4-hydroxylase. **(a)** The normal reaction, coupled to proline hydroxylation, which does not require ascorbate. The fate of the two oxygen atoms from O_2 is shown in red. **(b)** The uncoupled reaction, in which α-ketoglutarate is oxidatively decarboxylated without hydroxylation of proline. Ascorbate is consumed stoichiometrically in this process as it is converted to dehydroascorbate, preventing Fe^{2+} oxidation.

The tight wrapping of the α chains in the collagen triple helix provides tensile strength greater than that of a steel wire of equal cross section. Collagen fibrils **(Fig. 4-13)** are supramolecular assemblies consisting of triple-helical collagen molecules (sometimes referred to as tropocollagen molecules) associated in a variety of ways to provide different degrees of tensile strength. The α chains of collagen molecules and the collagen molecules of fibrils are cross-linked by unusual types of covalent bonds involving Lys, HyLys (5-hydroxylysine; see Fig. 3-8a), or His residues that are present at a few of the X and Y positions. These links create uncommon amino acid residues such as dehydrohydroxylysinonorleucine. The increasingly rigid and brittle character of aging connective tissue results from accumulated covalent cross-links in collagen fibrils.

Polypeptide chain — Lys residue minus ε-amino group (norleucine) — HyLys residue — Polypeptide chain

Dehydrohydroxylysinonorleucine

A typical mammal has more than 30 structural variants of collagen, particular to certain tissues and each somewhat different in sequence and function. Some human genetic defects in collagen structure illustrate the close relationship between amino acid sequence and three-dimensional structure in this protein. Osteogenesis imperfecta is characterized by abnormal bone formation in babies; at least eight variants of this condition, with different degrees of severity, occur in the human population. Ehlers-Danlos syndrome is characterized by loose joints, and at least six variants occur in humans. The composer Niccolò Paganini (1782–1840) was famed for his seemingly impossible dexterity in playing the violin. He suffered from a variant of Ehlers-Danlos syndrome that rendered him effectively double-jointed. In both disorders, some variants can be lethal, whereas others cause lifelong problems.

All of the variants of both conditions result from the substitution of an amino acid residue with a larger R group (such as Cys or Ser) for a single Gly residue in an α chain in one or another of the collagen proteins (a different Gly residue in each disorder). These single-residue substitutions have a catastrophic effect on collagen function because they disrupt the Gly–X–Y repeat that gives collagen its unique helical structure. Given its role in the collagen triple helix (Fig. 4-12), Gly cannot be replaced by another amino acid residue without substantial deleterious effects on collagen structure. ■

Silk Fibroin Fibroin, the protein of silk, is produced by insects and spiders. Its polypeptide chains are predominantly in the β conformation. Fibroin is rich in Ala and Gly residues, permitting a close packing of β sheets and an interlocking arrangement of R groups **(Fig. 4-14)**. The overall structure is stabilized by extensive hydrogen bonding between all peptide linkages in the polypeptides of each β sheet and by the optimization of van der Waals interactions between sheets. Silk does not stretch, because the β conformation is already highly extended (Fig. 4-6). However, the structure is flexible, because the sheets are held together by numerous weak interactions rather than by covalent bonds such as the disulfide bonds in α-keratins.

Structural Diversity Reflects Functional Diversity in Globular Proteins

In a globular protein, different segments of the polypeptide chain (or multiple polypeptide chains) fold back on each other, generating a more compact shape than is

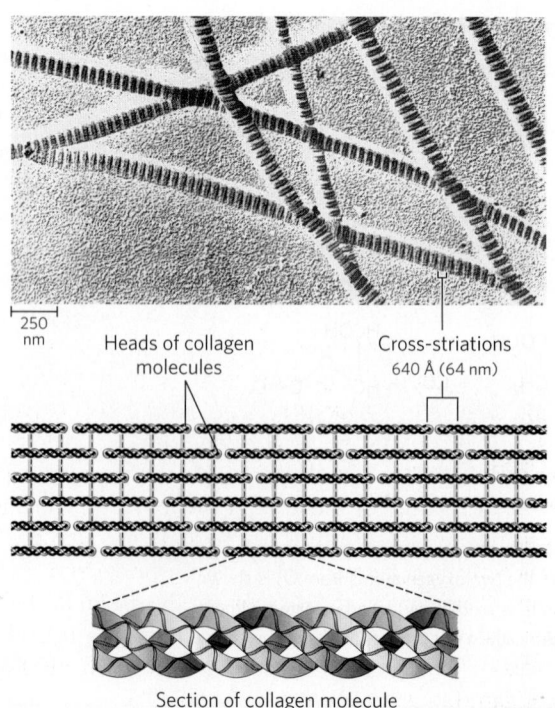

250 nm

Heads of collagen molecules

Cross-striations 640 Å (64 nm)

Section of collagen molecule

FIGURE 4-13 Structure of collagen fibrils. Collagen (M_r 300,000) is a rod-shaped molecule, about 3,000 Å long and only 15 Å thick. Its three helically intertwined α chains may have different sequences; each chain has about 1,000 amino acid residues. Collagen fibrils are made up of collagen molecules aligned in a staggered fashion and cross-linked for strength. The specific alignment and degree of cross-linking vary with the tissue and produce characteristic cross-striations in an electron micrograph. In the example shown here, alignment of the head groups of every fourth molecule produces striations 640 Å (64 nm) apart. [Micrograph source: J. Gross/Biozentrum, University of Basel/Science Source.]

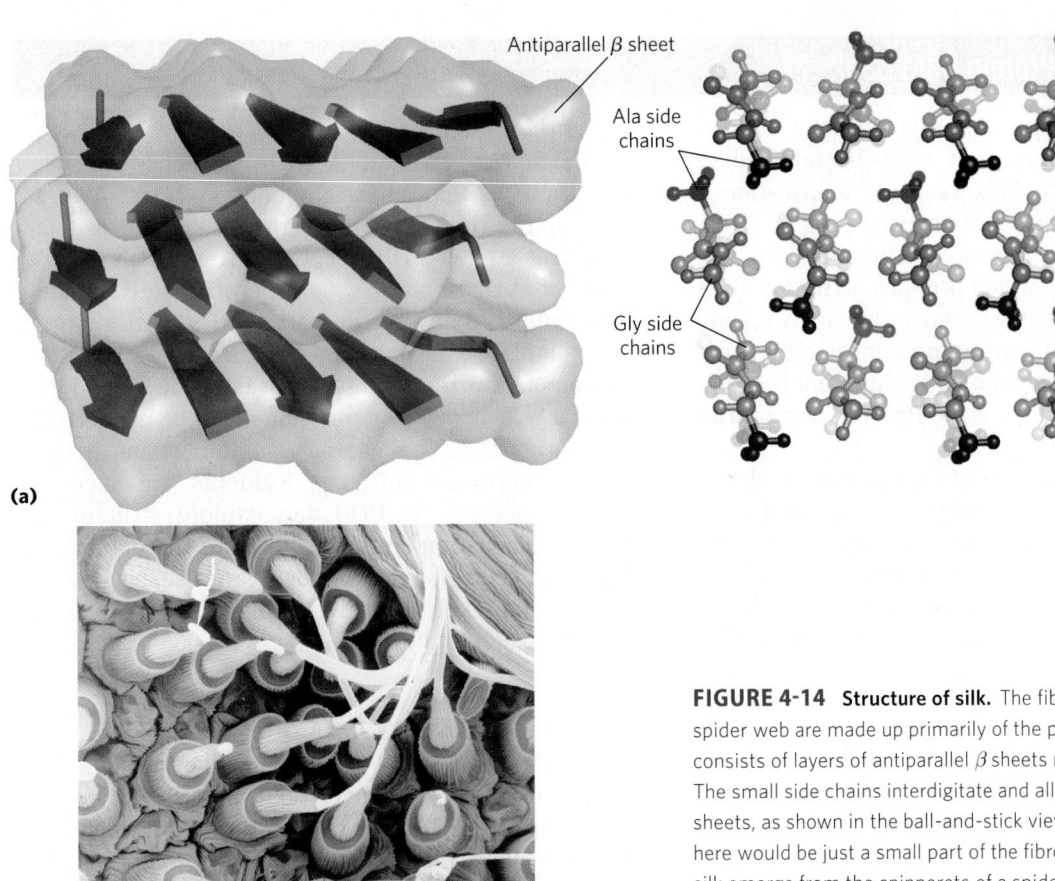

Antiparallel β sheet

Ala side chains

Gly side chains

(a)

(b)   70 μm

FIGURE 4-14   Structure of silk. The fibers in silk cloth and in a spider web are made up primarily of the protein fibroin. **(a)** Fibroin consists of layers of antiparallel β sheets rich in Ala and Gly residues. The small side chains interdigitate and allow close packing of the sheets, as shown in the ball-and-stick view. The segments shown here would be just a small part of the fibroin strand. **(b)** Strands of silk emerge from the spinnerets of a spider in this colorized scanning electron micrograph. [Sources: (a) Model derived from PDB ID 1SLK, S. A. Fossey et al., *Biopolymers* 31:1529, 1991. (b) Tina Weatherby Carvalho/MicroAngela.]

seen in the fibrous proteins **(Fig. 4-15)**. The folding also provides the structural diversity necessary for proteins to carry out a wide array of biological functions. Globular proteins include enzymes, transport proteins, motor proteins, regulatory proteins, immunoglobulins, and proteins with many other functions.

Our discussion of globular proteins begins with the principles gleaned from the first protein structures to be elucidated. This is followed by a detailed description of protein substructure and comparative categorization. Such discussions are possible only because of the vast amount of information available on the Internet from

β Conformation
2,000 × 5 Å

α Helix
900 × 11 Å

Native globular form
100 × 60 Å

FIGURE 4-15   Globular protein structures are compact and varied. Human serum albumin (M_r 64,500) has 585 residues in a single chain. Given here are the approximate dimensions its single polypeptide chain would have if it occurred entirely in extended β conformation or as an α helix. Also shown is the size of the protein in its native globular form, as determined by x-ray crystallography; the polypeptide chain must be very compactly folded to fit into these dimensions.

publicly accessible databases, particularly the Protein Data Bank (Box 4-4).

Myoglobin Provided Early Clues about the Complexity of Globular Protein Structure

The first breakthrough in understanding the three-dimensional structure of a globular protein came from x-ray diffraction studies of myoglobin carried out by John Kendrew and his colleagues in the 1950s. Myoglobin is a relatively small (M_r 16,700), oxygen-binding protein of muscle cells. It functions both to store oxygen and to facilitate oxygen diffusion in rapidly contracting muscle tissue. Myoglobin contains a single polypeptide chain of 153 amino acid residues of known sequence and a single iron protoporphyrin, or heme, group. The same heme group that is found in myoglobin is found in hemoglobin, the oxygen-binding protein of erythrocytes, and is responsible for the deep red-brown color of both myoglobin and hemoglobin. Myoglobin is particularly abundant in the muscles of diving mammals such as whales, seals, and porpoises—so abundant that the muscles of these animals are brown. Storage and distribution of oxygen by muscle myoglobin permits diving mammals to remain submerged for long periods. The activities of myoglobin and other globin molecules are investigated in greater detail in Chapter 5.

BOX 4-4 The Protein Data Bank

The number of known three-dimensional protein structures is now more than 100,000 and doubles every couple of years. This wealth of information is revolutionizing our understanding of protein structure, the relation of structure to function, and the evolutionary paths by which proteins arrived at their present state, which can be seen in the family resemblances that come to light as protein databases are sifted and sorted. One of the most important resources available to biochemists is the **Protein Data Bank** (**PDB**; www.pdb.org).

The PDB is an archive of experimentally determined three-dimensional structures of biological macromolecules, containing virtually all of the macromolecular structures (proteins, RNAs, DNAs, etc.) elucidated to date. Each structure is assigned an identifying label (a four-character identifier

called the PDB ID). Such labels are provided in the figure legends for every PDB-derived structure illustrated in this text so that students and instructors can explore the same structures on their own. The data files in the PDB describe the spatial coordinates of each atom for which the position has been determined (many of the cataloged structures are not complete). Additional data files provide information on how the structure was determined and its accuracy. The atomic coordinates can be converted into an image of the macromolecule by using structure visualization software. Students are encouraged to access the PDB and explore structures, using visualization software linked to the database. Macromolecular structure files can also be downloaded and explored on the desktop, using free software such as JSmol.

Figure 4-16 shows several structural representations of myoglobin, illustrating how the polypeptide chain is folded in three dimensions—its tertiary structure. The red group surrounded by protein is heme. The backbone of the myoglobin molecule consists of eight relatively straight segments of α helix interrupted by bends, some of which are β turns. The longest α helix has 23 amino acid residues and the shortest only 7; all helices are right-handed. More than 70% of the residues in myoglobin are in these α-helical regions. X-ray analysis has revealed the precise position of each of the R groups, which fill up nearly all the space within the folded chain that is not occupied by backbone atoms.

Many important conclusions were drawn from the structure of myoglobin. The positioning of amino acid side chains reflects a structure that is largely stabilized by the hydrophobic effect. Most of the hydrophobic R groups are in the interior of the molecule, hidden from

exposure to water. All but two of the polar R groups are located on the outer surface of the molecule, and all are hydrated. The myoglobin molecule is so compact that its interior has room for only four molecules of water. This dense hydrophobic core is typical of globular proteins. The fraction of space occupied by atoms in an organic liquid is 0.4 to 0.6. In a globular protein the fraction is about 0.75, comparable to that in a crystal (in a typical crystal the fraction is 0.70 to 0.78, near the theoretical maximum). In this packed environment, weak interactions strengthen and reinforce each other. For example, the nonpolar side chains in the core are so close together that short-range van der Waals interactions make a significant contribution to stabilizing hydrophobic interactions.

Deduction of the structure of myoglobin confirmed some expectations and introduced some new elements of secondary structure. As predicted by Pauling and

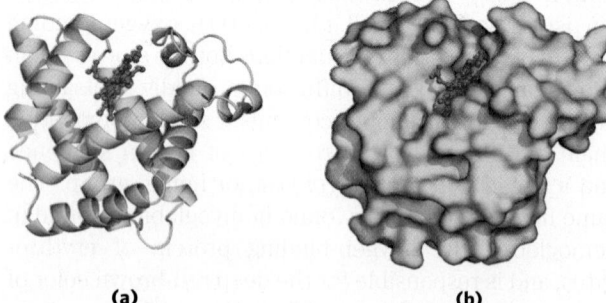 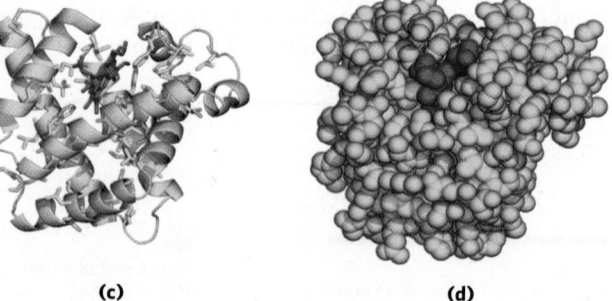

(a)   (b)   (c)   (d)

FIGURE 4-16 Tertiary structure of sperm whale myoglobin. Orientation of the protein is similar in (a) through (d); the heme group is shown in red. In addition to illustrating the myoglobin structure, this figure provides examples of several different ways to display protein structure. **(a)** The polypeptide backbone in a ribbon representation of a type introduced by Jane Richardson, which highlights regions of secondary structure. The α-helical regions are evident. **(b)** Surface contour image; this is useful for visualizing pockets in the protein where other molecules might bind. **(c)** Ribbon representation including side chains (yellow) for the hydrophobic residues Leu, Ile, Val, and Phe. **(d)** Space-filling model with all amino acid side chains. Each atom is represented by a sphere encompassing its van der Waals radius. The hydrophobic residues are again shown in yellow; most are buried in the interior of the protein and thus not visible. [Source: PDB ID 1MBO, S. E. Phillips, *J. Mol. Biol.* 142:531, 1980.]

Corey, all the peptide bonds are in the planar trans configuration. The α helices in myoglobin provided the first direct experimental evidence for the existence of this type of secondary structure. Three of the four Pro residues are found at bends. The fourth Pro residue occurs within an α helix, where it creates a kink necessary for tight helix packing.

The flat heme group rests in a crevice, or pocket, in the myoglobin molecule. The iron atom in the center of the heme group has two bonding (coordination) positions perpendicular to the plane of the heme **(Fig. 4-17)**. One of these is bound to the R group of the His residue at position 93; the other is the site at which an O_2 molecule binds. Within this pocket, the accessibility of the heme group to solvent is highly restricted. This is important for function, because free heme groups in an oxygenated solution are rapidly oxidized from the ferrous (Fe^{2+}) form, which is active in the reversible binding of O_2, to the ferric (Fe^{3+}) form, which does not bind O_2.

As myoglobin structures from many different species were resolved, investigators were able to observe the structural changes that accompany the binding of oxygen or other molecules and thus, for the first time, to understand the correlation between protein structure and function. Hundreds of proteins have now been subjected to similar analysis. Today, nuclear magnetic resonance (NMR) spectroscopy and other techniques supplement x-ray diffraction data, providing more information on a protein's structure (Box 4-5). In addition, the sequencing of the genomic DNA of many organisms (Chapter 9) has identified thousands of genes that encode proteins of known sequence but, as yet, unknown function; this work continues apace.

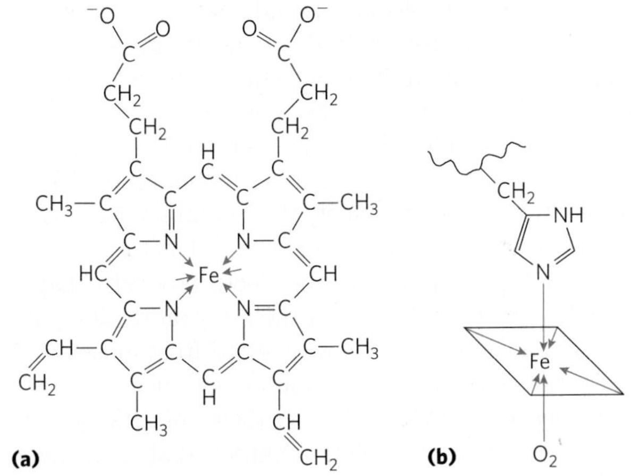

TABLE 4-4	Approximate Proportion of α Helix and β Conformation in Some Single-Chain Proteins	
	Residues (%)[a]	
Protein (total residues)	α Helix	β Conformation
Chymotrypsin (247)	14	45
Ribonuclease (124)	26	35
Carboxypeptidase (307)	38	17
Cytochrome c (104)	39	0
Lysozyme (129)	40	12
Myoglobin (153)	78	0

Source: Data from C. R. Cantor and P. R. Schimmel, *Biophysical Chemistry, Part I: The Conformation of Biological Macromolecules*, p. 100, W. H. Freeman and Company, 1980.

[a] Portions of the polypeptide chains not accounted for by α helix or β conformation consist of bends and irregularly coiled or extended stretches. Segments of α helix and β conformation sometimes deviate slightly from their normal dimensions and geometry.

Globular Proteins Have a Variety of Tertiary Structures

From what we now know about the tertiary structures of thousands of globular proteins, it is clear that myoglobin illustrates just one of many ways in which a polypeptide chain can fold. Table 4-4 shows the proportions of α-helix and β conformations (expressed as percentage of residues in each type) in several small, single-chain, globular proteins. Each of these proteins has a distinct structure, adapted for its particular biological function, but together they share several important properties with myoglobin. Each is folded compactly, and in each case the hydrophobic amino acid side chains are oriented toward the interior (away from water) and the hydrophilic side chains are on the surface. The structures are also stabilized by a multitude of hydrogen bonds and some ionic interactions.

For the beginning student, the very complex tertiary structures of globular proteins—some much larger than myoglobin—are best approached by focusing on common structural patterns, recurring in different and often unrelated proteins. The three-dimensional structure of a typical globular protein can be considered an assemblage of polypeptide segments in the α-helix and β conformations, linked by connecting segments. The structure can then be defined by how these segments stack on one another and how the segments that connect them are arranged.

To understand a complete three-dimensional structure, we need to analyze its folding patterns. We begin by defining two important terms that describe protein structural patterns or elements in a polypeptide chain and then turn to the folding rules.

The first term is **motif**, also called a **fold** or (more rarely) **supersecondary structure**. *A motif or fold is a recognizable folding pattern involving two or more elements of secondary structure and the connection(s) between them.* A motif can be very simple, such as two elements of secondary structure folded against each other,

FIGURE 4-17 The heme group. This group is present in myoglobin, hemoglobin, cytochromes, and many other proteins (the heme proteins). **(a)** Heme consists of a complex organic ring structure, protoporphyrin, which binds an iron atom in its ferrous (Fe^{2+}) state. The iron atom has six coordination bonds, four in the plane of, and bonded to, the flat porphyrin molecule and two perpendicular to it. **(b)** In myoglobin and hemoglobin, one of the perpendicular coordination bonds is bound to a nitrogen atom of a His residue. The other is "open" and serves as the binding site for an O_2 molecule.

BOX 4-5   **METHODS**   Methods for Determining the Three-Dimensional Structure of a Protein

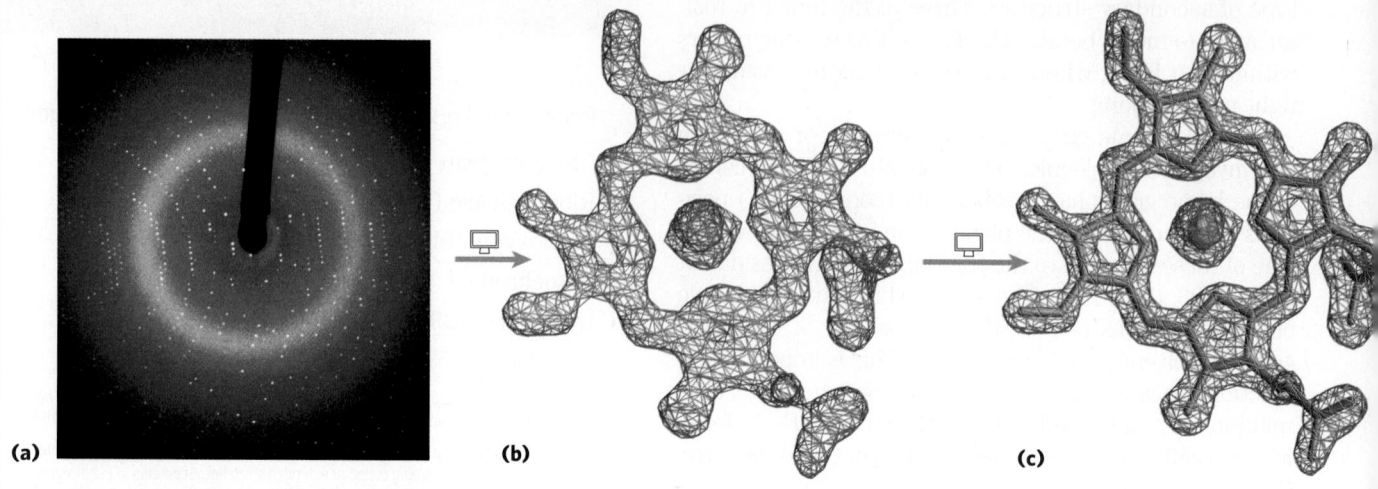

(a)   (b)   (c)

X-Ray Diffraction

The spacing of atoms in a crystal lattice can be determined by measuring the locations and intensities of spots produced on photographic film by a beam of x rays of given wavelength, after the beam has been diffracted by the electrons of the atoms. For example, x-ray analysis of sodium chloride crystals shows that Na^+ and Cl^- ions are arranged in a simple cubic lattice. The spacing of the different kinds of atoms in complex organic molecules, even very large ones such as proteins, can also be analyzed by x-ray diffraction methods. However, the technique for analyzing crystals of complex molecules is far more laborious than for simple salt crystals. When the repeating pattern of the crystal is a molecule as large as, say, a protein, the numerous atoms in the molecule yield thousands of diffraction spots that must be analyzed by computer.

Consider how images are generated in a light microscope. Light from a point source is focused on an object. The object scatters the light waves, and these scattered waves are recombined by a series of lenses to generate an enlarged image of the object. The smallest object whose structure can be determined by such a system—that is, the resolving power of the microscope—is determined by the wavelength of the light, in this case visible light, with wavelengths in the range of 400 to 700 nm. Objects smaller than half the wavelength of the incident light cannot be resolved. To resolve objects as small as proteins we must use x rays, with wavelengths in the range of 0.7 to 1.5 Å (0.07 to 0.15 nm). However, there are no lenses that can recombine x rays to form an image; instead, the pattern of diffracted x rays is collected directly and an image is reconstructed by mathematical techniques.

The amount of information obtained from x-ray crystallography depends on the degree of structural order in the sample. Some important structural parameters were obtained from early studies of the diffraction

patterns of the fibrous proteins arranged in regular arrays in hair and wool. However, the orderly bundles formed by fibrous proteins are not crystals—the molecules are aligned side by side, but not all are oriented in the same direction. More detailed three-dimensional structural information about proteins requires a highly ordered protein crystal. The structures of many proteins are not yet known, simply because they have proved difficult to crystallize. Practitioners have compared making protein crystals to holding together a stack of bowling balls with cellophane tape.

Operationally, there are several steps in x-ray structural analysis (Fig. 1). A crystal is placed in an x-ray beam between the x-ray source and a detector, and a regular array of spots, called reflections, is generated. The spots are created by the diffracted x-ray beam, and each atom in a molecule makes a contribution to each spot. An electron-density map of the protein is reconstructed from the overall diffraction pattern of spots by a mathematical technique called a Fourier transform. In effect, the computer acts as a "computational lens." A model for the structure is then built that is consistent with the electron-density map.

John Kendrew found that the x-ray diffraction pattern of crystalline myoglobin (isolated from muscles of the sperm whale) is highly complex, with nearly 25,000 reflections. Computer analysis of these reflections took place in stages. The resolution improved at each stage until, in 1959, the positions of virtually all the nonhydrogen atoms in the protein had been determined. The amino acid sequence of the protein, obtained by chemical analysis, was consistent with the molecular model. The structures of thousands of proteins, many of them much more complex than myoglobin, have since been determined to a similar level of resolution.

The physical environment in a crystal, of course, is not identical to that in solution or in a living cell. A crystal imposes a space and time average on the structure

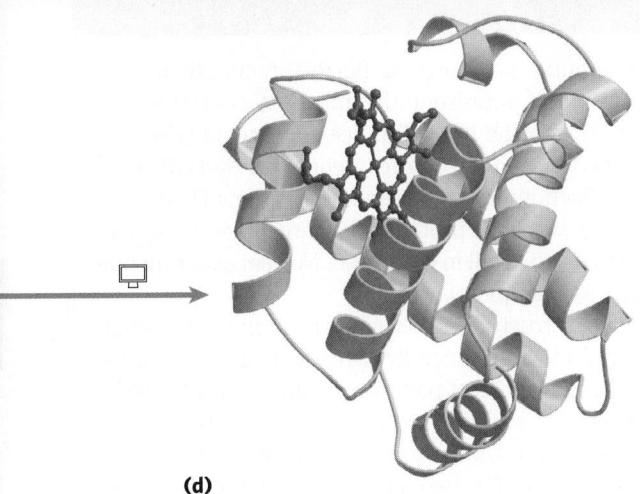

(d)

FIGURE 1 Steps in determining the structure of sperm whale myoglobin by x-ray crystallography. **(a)** X-ray diffraction patterns are generated from a crystal of the protein. **(b)** Data extracted from the diffraction patterns are used to calculate a three-dimensional electron-density map. The electron density of only part of the structure, the heme, is shown here. **(c)** Regions of greatest electron density reveal the location of atomic nuclei, and this information is used to piece together the final structure. Here, the heme structure is modeled into its electron-density map. **(d)** The completed structure of sperm whale myoglobin, including the heme. [Sources: (a, b, c) Courtesy of George N. Phillips, Jr., University of Wisconsin–Madison, Department of Biochemistry. (d) PDB ID 2MBW, E. A. Brucker et al., *J. Biol. Chem.* 271:25,419, 1996.]

deduced from its analysis, and x-ray diffraction studies provide little information about molecular motion within the protein. In principle, the conformation of proteins in a crystal could also be affected by nonphysiological factors such as incidental protein-protein contacts within the crystal. However, when structures derived from the analysis of crystals are compared with structural information obtained by other means (such as NMR, as described below), the crystal-derived structure almost always represents a functional conformation of the protein. X-ray crystallography can be applied successfully to proteins too large to be structurally analyzed by NMR.

Nuclear Magnetic Resonance

An advantage of nuclear magnetic resonance (NMR) studies is that they are carried out on macromolecules in solution, whereas x-ray crystallography is limited to molecules that can be crystallized. NMR can also illuminate the dynamic side of protein structure, including conformational changes, protein folding, and interactions with other molecules.

NMR is a manifestation of nuclear spin angular momentum, a quantum mechanical property of atomic nuclei. Only certain atoms, including ^1H, ^{13}C, ^{15}N, ^{19}F, and ^{31}P, have the kind of nuclear spin that gives rise to an NMR signal. Nuclear spin generates a magnetic dipole. When a strong, static magnetic field is applied to a solution containing a single type of macromolecule, the magnetic dipoles are aligned in the field in one of two orientations, parallel (low energy) or antiparallel (high energy). A short ($\sim$10 μs) pulse of electromagnetic energy of suitable frequency (the resonant frequency, which is in the radio frequency range) is applied at right angles to the nuclei aligned in the magnetic field. Some energy is absorbed as nuclei switch to the high-energy state, and the absorption spectrum that results contains information about the identity of the nuclei and their immediate chemical environment.

The data from many such experiments on a sample are averaged, increasing the signal-to-noise ratio, and an NMR spectrum such as that in Figure 2 is generated.

^1H is particularly important in NMR experiments because of its high sensitivity and natural abundance. For macromolecules, ^1H NMR spectra can become quite complicated. Even a small protein has hundreds of ^1H atoms, typically resulting in a one-dimensional NMR spectrum too complex for analysis. Structural analysis of proteins became possible with the advent of two-dimensional NMR techniques (Fig. 3). These methods allow measurement of distance-dependent coupling of nuclear spins in nearby atoms through space (the nuclear Overhauser effect (NOE), in a method dubbed NOESY) or the coupling of nuclear spins in atoms connected by covalent bonds (total correlation spectroscopy, or TOCSY).

Translating a two-dimensional NMR spectrum into a complete three-dimensional structure can be a laborious process. The NOE signals provide some information about the distances between individual atoms, but for these distance constraints to be useful, the atoms

(Continued on next page)

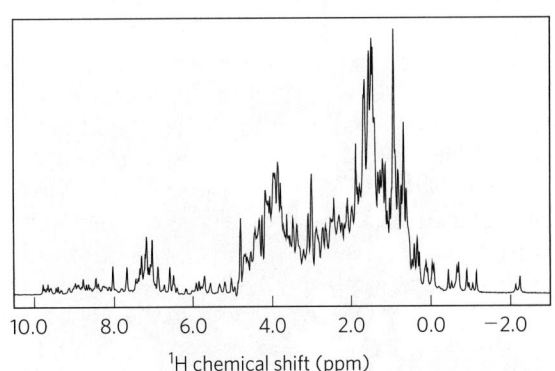

FIGURE 2 One-dimensional NMR spectrum of a globin from a marine blood worm. This protein and sperm whale myoglobin are very close structural analogs, belonging to the same protein structural family and sharing an oxygen-transport function. [Source: Data from B. F. Volkman, National Magnetic Resonance Facility at Madison.]

BOX 4-5 METHODS Methods for Determining the Three-Dimensional Structure of a Protein (*Continued*)

giving rise to each signal must be identified. Complementary TOCSY experiments can help identify which NOE signals reflect atoms that are linked by covalent bonds. Certain patterns of NOE signals have been associated with secondary structures such as α helices. Genetic engineering (Chapter 9) can be used to prepare proteins that contain the rare isotopes ^{13}C or ^{15}N. The new NMR signals produced by these atoms, and the coupling with ^{1}H signals resulting from these substitutions, help in the assignment of individual ^{1}H NOE signals. The process is also aided by a knowledge of the amino acid sequence of the polypeptide.

To generate a three-dimensional structure, researchers feed the distance constraints into a computer along with known geometric constraints such as chirality, van der Waals radii, and bond lengths and angles.

The computer generates a family of closely related structures that represent the range of conformations consistent with the NOE distance constraints (Fig. 3c). The uncertainty in structures generated by NMR is in part a reflection of the molecular vibrations (known as breathing) within a protein structure in solution, discussed in more detail in Chapter 5. Normal experimental uncertainty can also play a role.

Protein structures determined by both x-ray crystallography and NMR generally agree well. In some cases, the precise locations of particular amino acid side chains on the protein exterior are different, often because of effects related to the packing of adjacent protein molecules in a crystal. The two techniques together are at the heart of the rapid increase in the availability of structural information about the macromolecules of living cells.

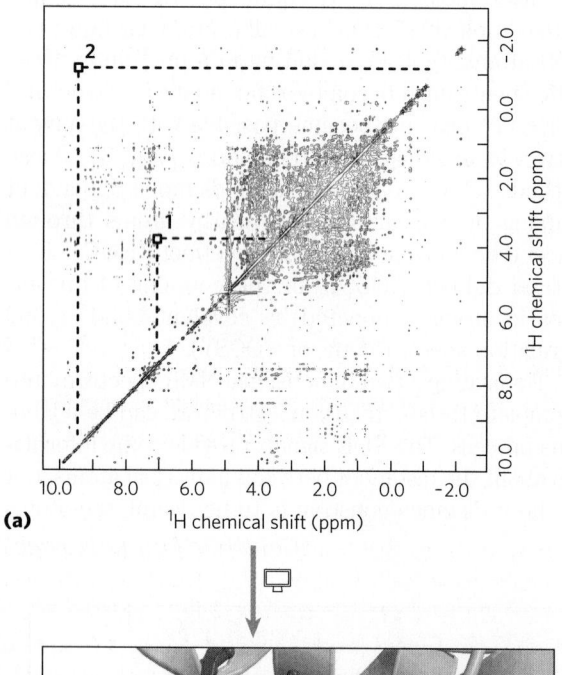

(a)

(b)

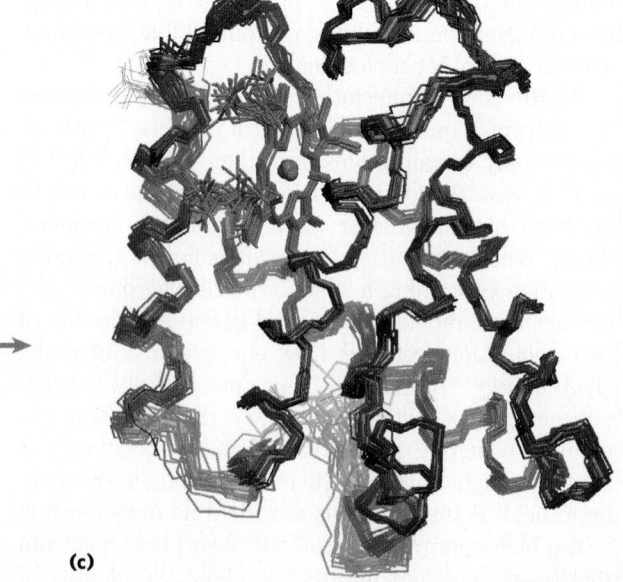

(c)

FIGURE 3 Use of two-dimensional NMR to generate a three-dimensional structure of a globin, the same protein used to generate the data in Figure 2. The diagonal in a two-dimensional NMR spectrum is equivalent to a one-dimensional spectrum. The off-diagonal peaks are NOE signals generated by close-range interactions of ^{1}H atoms that may generate signals quite distant in the one-dimensional spectrum. Two such interactions are identified in **(a)**, and their identities are shown with blue lines in **(b)**. Three lines are drawn for interaction 2 between a methyl group in the protein and a hydrogen on the heme. The methyl group rotates rapidly such that each of its three hydrogens contributes equally to the interaction and the NMR signal. Such information is used to determine the complete three-dimensional structure, as in **(c)**. The multiple lines shown for the protein backbone in (c) represent the family of structures consistent with the distance constraints in the NMR data. The structural similarity with myoglobin (Fig. 1) is evident. The proteins are oriented in the same way in both figures. [Sources: Data and assistance with figure design courtesy of B. F. Volkman, National Magnetic Resonance Facility at Madison. (b) PDB ID 1VRF and (c) PDB ID 1VRE, B. F. Volkman et al., *Biochemistry* 37:10,906, 1998.]

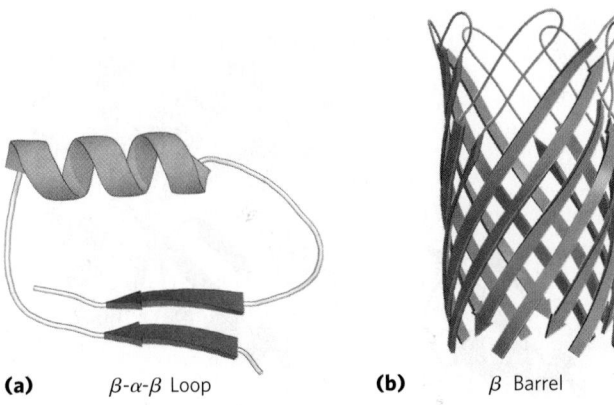

(a) β-α-β Loop   **(b)** β Barrel

FIGURE 4-18 Motifs. (a) A simple motif, the β-α-β loop. **(b)** A more elaborate motif, the β barrel. This β barrel is a single domain of α-hemolysin (a toxin that kills a cell by creating a hole in its membrane) from the bacterium *Staphylococcus aureus*. [Sources: (a) Derived from PDB ID 4TIM, M. E. Noble et al., *J. Med. Chem.*, 34:2709, 1991. (b) Derived from PDB ID 7AHL, L. Song et al., *Science* 274:1859, 1996.]

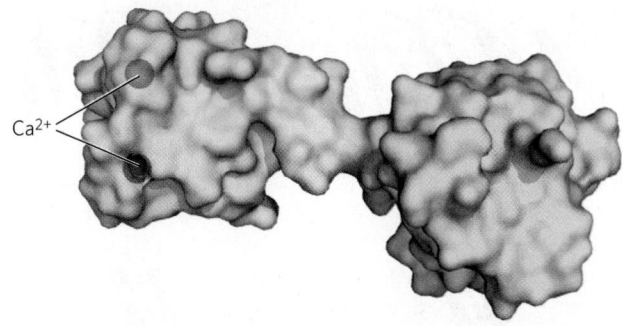

Ca^{2+}

FIGURE 4-19 **Structural domains in the polypeptide troponin C.** This calcium-binding protein, associated with muscle, has two separate calcium-binding domains, shown here in brown and blue. [Source: PDB ID 4TNC, K. A. Satyshur et al., *J. Biol. Chem.* 263:1628, 1988.]

and represent only a small part of a protein. An example is a **β-α-β loop (Fig. 4-18a)**. A motif can also be a very elaborate structure involving scores of protein segments folded together, such as the β barrel (Fig. 4-18b). In some cases, a single large motif may comprise the entire protein. The terms "motif" and "fold" are often used interchangeably, although "fold" is applied more commonly to somewhat more complex folding patterns. The terms encompass any advantageous folding pattern and are useful for describing such patterns. The segment defined as a motif or fold may or may not be independently stable. We have already encountered a well-studied motif, the coiled coil of α-keratin, which is also found in some other proteins. The distinctive arrangement of eight α helices in myoglobin is replicated in all globins and is called the globin fold. Note that a motif is not a hierarchical structural element falling between secondary and tertiary structure. It is simply a folding pattern. The synonymous term "supersecondary structure" is thus somewhat misleading because it suggests hierarchy.

The second term for describing structural patterns is **domain**. A domain, as defined by Jane Richardson in 1981, is a part of a polypeptide chain that is independently stable or could undergo movements as a single entity with respect to the entire protein. Polypeptides with more than a few hundred amino acid residues often fold into two or more domains, sometimes with different functions. In many cases, a domain from a large protein will retain its native three-dimensional structure even when separated (for example, by proteolytic cleavage) from the remainder of the polypeptide chain. In a protein with multiple domains, each domain may appear as a distinct globular lobe **(Fig. 4-19)**; more commonly, extensive contacts between domains make individual domains hard to discern. Different domains often have distinct functions, such as the binding of small molecules or interaction with other proteins. Small proteins usually have only one domain (the domain *is* the protein).

Folding of polypeptides is subject to an array of physical and chemical constraints, and several rules have emerged from studies of common protein folding patterns.

1. The hydrophobic effect makes a large contribution to the stability of protein structures. Burial of hydrophobic amino acid R groups so as to exclude water requires at least two layers of secondary structure. Simple motifs such as the β-α-β loop (Fig. 4-18a) create two such layers.

2. Where they occur together in a protein, α helices and β sheets generally are found in different structural layers. This is because the backbone of a polypeptide segment in the β conformation (Fig. 4-6) cannot readily hydrogen-bond to an α helix that is adjacent to it.

3. Segments adjacent to each other in the amino acid sequence are usually stacked adjacent to each other in the folded structure. Distant segments of a polypeptide may come together in the tertiary structure, but this is not the norm.

4. The β conformation is most stable when the individual segments are twisted slightly in a right-handed sense. This influences both the arrangement of β sheets derived from the twisted segments and the path of the polypeptide connections between them. Two parallel β strands, for example, must be connected by a crossover strand **(Fig. 4-20a)**. In principle, this crossover could have a right- or left-handed conformation, but in proteins it is almost always right-handed. Right-handed connections tend to be shorter than left-handed connections and tend to bend through smaller angles, making them easier to form. The twisting of β sheets also leads to a characteristic twisting of the structure formed by many such segments together, as seen in the β barrel (Fig. 4-18b) and twisted β sheet (Fig. 4-20c), which form the core of many larger structures.

Following these rules, complex motifs can be built up from simple ones. For example, a series of β-α-β loops

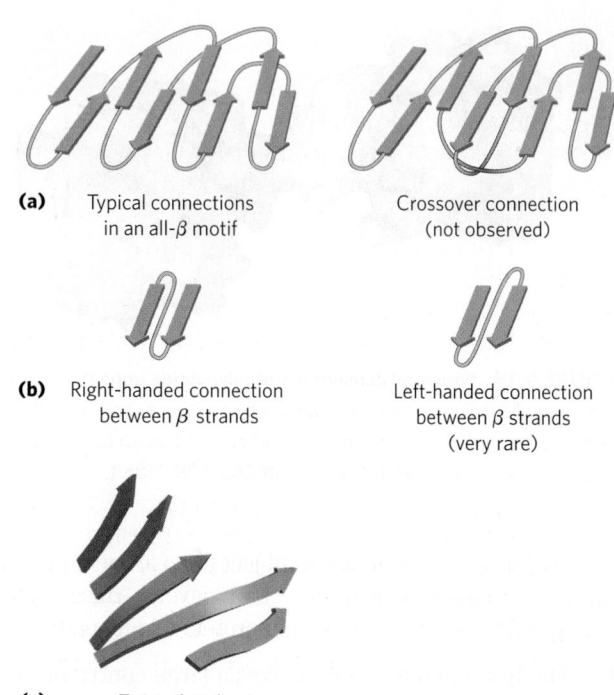

(a)   Typical connections   Crossover connection
           in an all-β motif   (not observed)

(b)   Right-handed connection   Left-handed connection
           between β strands   between β strands
                                       (very rare)

(c)   Twisted β sheet

FIGURE 4-20 Stable folding patterns in proteins. (a) Connections between β strands in layered β sheets. The strands here are viewed from one end, with no twisting. The connections at a given end (e.g., near the viewer) rarely cross one another. An example of such a rare crossover is illustrated by the yellow strand in the structure on the right. **(b)** Because of the right-handed twist in β strands, connections between strands are generally right-handed. Left-handed connections must traverse sharper angles and are harder to form. **(c)** This twisted β sheet is from a domain of photolyase (a protein that repairs certain types of DNA damage) from *E. coli.* Connecting loops have been removed so as to focus on the folding of the β sheet. [Source: (c) Derived from PDB ID 1DNP, H. W. Park et al., *Science* 268:1866, 1995.]

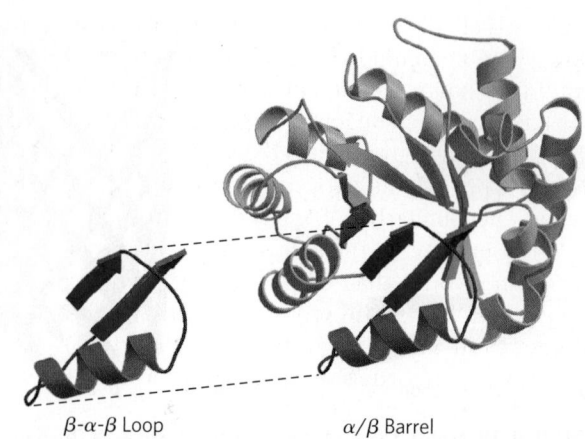

β-α-β Loop   α/β Barrel

FIGURE 4-21 Constructing large motifs from smaller ones. The α/β barrel is a commonly occurring motif constructed from repetitions of the β-α-β loop motif. This α/β barrel is a domain of pyruvate kinase (a glycolytic enzyme) from rabbit. [Source: Derived from PDB ID 1PKN, T. M. Larsen et al., *Biochemistry* 33:6301, 1994.]

arranged so that the β strands form a barrel creates a particularly stable and common motif, the **α/β barrel (Fig. 4-21).** In this structure, each parallel β segment is attached to its neighbor by an α-helical segment. All connections are right-handed. The α/β barrel is found in many enzymes, often with a binding site (for a cofactor or substrate) in the form of a pocket near one end of the barrel. Note that domains with similar folding patterns are said to have the same motif even though their constituent α helices and β sheets may differ in length.

Some Proteins or Protein Segments Are Intrinsically Disordered

In spite of decades of progress in the understanding of protein structure, many proteins cannot be crystallized, making it difficult to determine their three-dimensional structure by methods now considered classical (see Box 4-5). Even where crystallization succeeds, parts of the protein are often so disordered within the crystal that the determined structure does not include those parts. Sometimes, this is due to subtle features of the structure

that render crystallization difficult. However, the reason can be more straightforward: some proteins or protein segments lack an ordered structure in solution.

The concept that some proteins function in the absence of a definable three-dimensional structure comes from reassessment of data from many different proteins. As many as a third of all human proteins may be unstructured or have significant unstructured segments. All organisms have some proteins that fall into this category. **Intrinsically disordered proteins** have properties that are distinct from those of classical, structured proteins. They lack a hydrophobic core and instead are characterized by high densities of charged amino acid residues such as Lys, Arg, and Glu. Pro residues are also prominent, as they tend to disrupt ordered structures.

Structural disorder and high charge density can facilitate the function of some proteins as spacers, insulators, or linkers in larger structures. Other disordered proteins are scavengers, binding up ions and small molecules in solution and serving as reservoirs or garbage dumps. However, many intrinsically disordered proteins are at the heart of important protein interaction networks. The lack of an ordered structure can facilitate a kind of functional promiscuity, allowing one protein to interact with multiple partners. Some intrinsically disordered proteins act to inhibit the action of other proteins by an unusual mechanism: wrapping around their protein targets. One disordered protein may have several or even dozens of protein partners. The structural disorder allows the inhibitor protein to wrap around the multiple targets in different ways. The intrinsically disordered protein p27 plays a key role in controlling mammalian cell division. This protein lacks definable structure in solution. It wraps around and thus inhibits the action of several enzymes called protein kinases (see Chapter 6) that facilitate cell division. The flexible structure of p27 allows it to accommodate itself to its different target proteins. Human tumor cells, which are simply cells that have lost the capacity to control cell division normally, generally have reduced levels of p27; the lower the levels of p27, the poorer the prognosis for the cancer patient. Similarly, intrinsically disordered proteins are often present as hubs or scaffolds at the center of protein networks that constitute signaling pathways (see Fig. 12-26). These proteins, or parts of them, may interact with many different binding partners. They often take on an ordered structure when they interact with other proteins, but the structure they assume may vary with different binding partners. The mammalian protein p53 is also critical in the control of cell division. It contains both structured and unstructured segments, and the different segments interact with dozens of other proteins. An unstructured region of p53 at the carboxyl terminus interacts with at least four different binding partners and assumes a different structure in each of the complexes **(Fig. 4-22)**.

Protein Motifs Are the Basis for Protein Structural Classification

More than 100,000 protein structures are now archived in the Protein Data Bank (PDB). An enormous amount of information about protein structural principles, protein function, and protein evolution is buried in these data. Fortunately, other databases organize this information and make it more readily accessible. In the Structural Classification of Proteins database, or SCOP2 (http://scop2.mrc-lmb.cam.ac.uk), all of the protein information in the PDB can be searched within four different categories: (1) protein relationships, (2) structural classes, (3) protein types, and (4) evolutionary events. The first category provides several options: proteins can be searched with respect to their structural features, evolutionary relationships, or "other" (the latter an attempt to define common motifs and subfolds). The second option organizes all PDB structures according to their secondary structural elements: **all α, all β, α/β** (with

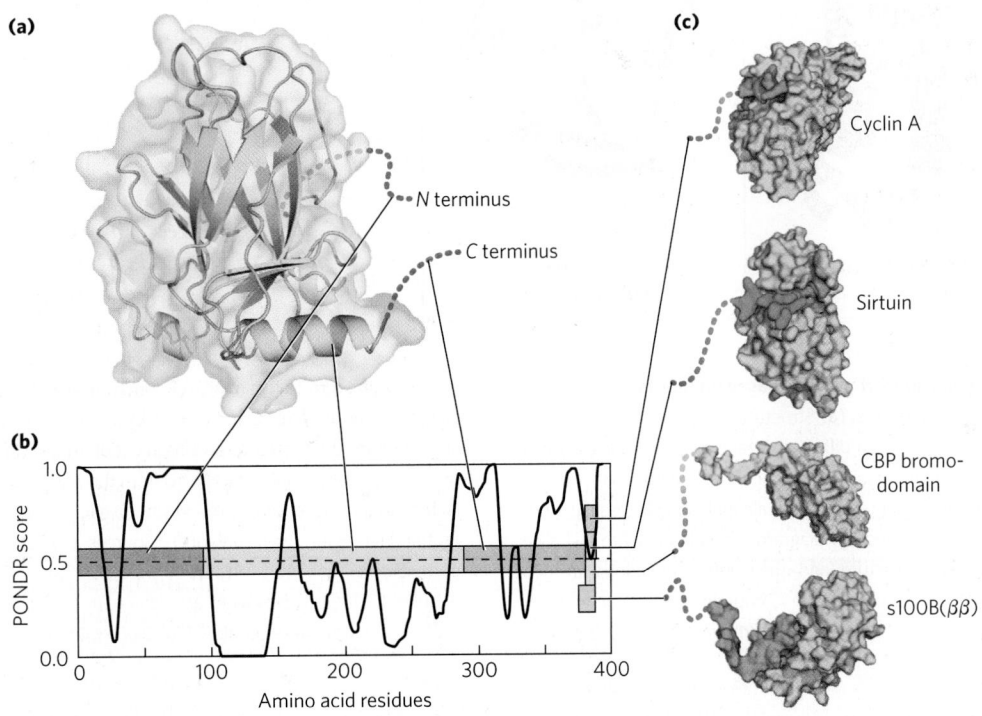

FIGURE 4-22 Binding of the intrinsically disordered carboxyl terminus of p53 protein to its binding partners. (a) The p53 protein is made up of several different segments. Only the central domain is well ordered. **(b)** The linear sequence of the p53 protein is depicted as a colored bar. The overlaid graph presents a plot of the PONDR (Predictor of Natural Disordered Regions) score versus the protein sequence. PONDR is one of the best available algorithms for predicting the likelihood that a given amino acid residue is in a region of intrinsic disorder, based on the surrounding amino acid sequence and amino acid composition. A score of 1.0 indicates a probability of 100% that a protein will be disordered. In the actual protein structure, the tan central domain is ordered. The amino-terminal (blue) and carboxyl-terminal (red) regions are disordered. The very end of the carboxyl-terminal region has multiple binding partners and folds when it binds to each of them; however, the three-dimensional structure that is assumed when binding occurs is different for each of the interactions shown, and thus this carboxyl-terminal segment (11 to 20 residues) is shown in a different color in each complex. [Sources: Information from V. N. Uversky, *Intl. J. Biochem. Cell Biol.* 43:1090, 2011, Fig. 5. (a) Derived from PDB ID 1TUP, Y. Cho et al., *Science* 265:346, 1994. (c) Cyclin A: PDB ID 1H26, E. D. Lowe et al., *Biochemistry* 41:15,625, 2002; sirtuin: PDB ID 1MA3, J. L. Avalos et al., *Mol. Cell* 10:523, 2002; CBP bromodomain: PDB ID 1JSP, S. Mujtaba et al., *Mol. Cell* 13:251, 2004; s100B($\beta\beta$): PDB ID 1DT7, R. R. Rustandi et al., *Nature Struct. Biol.* 7:570, 2000.]

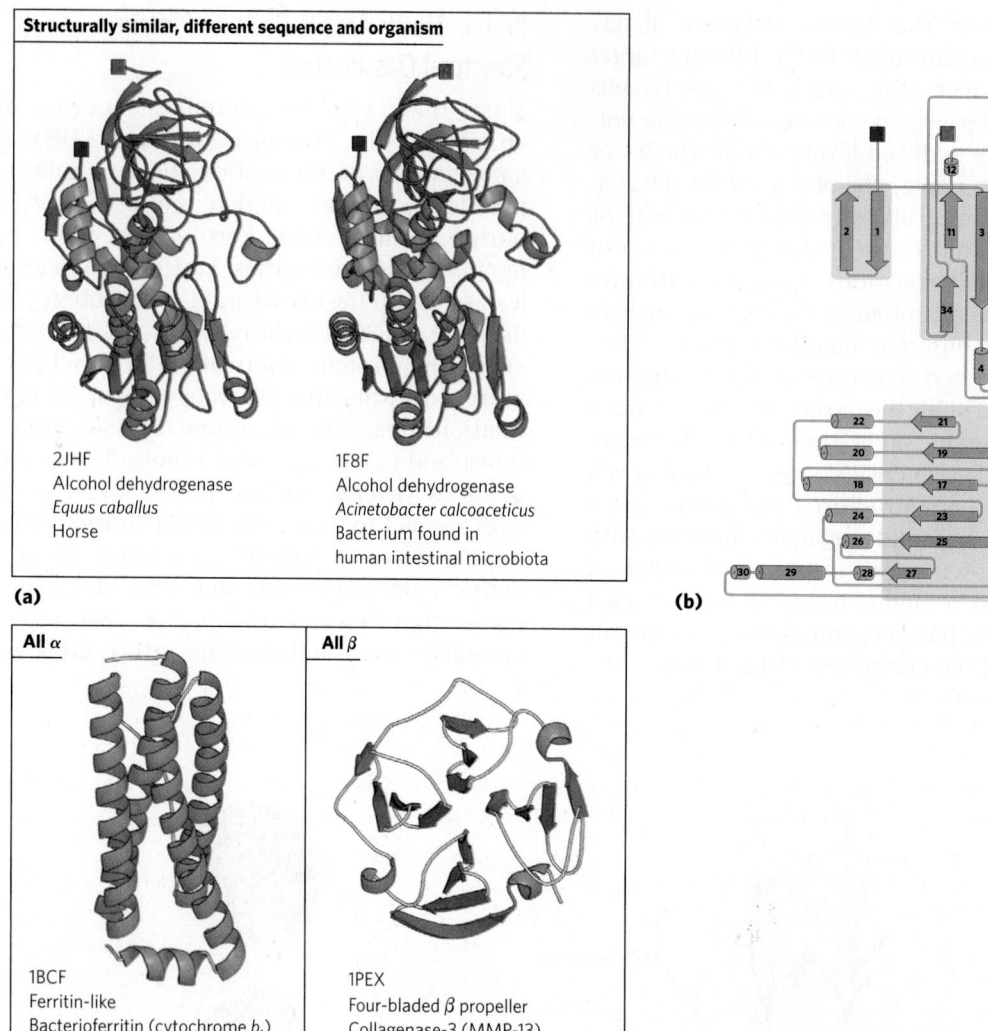

(a)

Structurally similar, different sequence and organism

2JHF
Alcohol dehydrogenase
Equus caballus
Horse

1F8F
Alcohol dehydrogenase
Acinetobacter calcoaceticus
Bacterium found in
human intestinal microbiota

(b)

(c)

All α

1BCF
Ferritin-like
Bacterioferritin (cytochrome b_1)
Escherichia coli

All β

1PEX
Four-bladed β propeller
Collagenase-3 (MMP-13)
Human (*Homo sapiens*)

FIGURE 4-23 Organization of proteins based on motifs. A few of the hundreds of known stable motifs. **(a)** Structural diagrams of the enzyme alcohol dehydrogenase from two different organisms. Such comparisons illustrate evolutionary relationships that conserve structure as well as function. **(b)** A topology diagram for the alcohol dehydrogenase from *Acinetobacter calcoaceticus*. Topology diagrams provide a way to visualize elements of secondary structure and their interconnections in two dimensions, and can be very useful in comparing structural folds or motifs. **(c)** The Structural Classification of Proteins (SCOP2) database (http://scop2.mrc-lmb.cam.ac.uk) organizes protein folds into four classes: all α, all β, α/β, and α + β. Examples of all α and all β folds are shown with their structural classification data (PDB ID, fold name, protein name, and source organism) from the SCOP2 database. The PDB ID is the unique accession code given to each structure archived in the Protein Data Bank (www.pdb.org). [Sources: (a) PDB ID 2JHF, R. Meijers et al., *Biochemistry* 46:5446, 2007; PDB ID 1F8F, J. C. Beauchamp et al. (c) PDB ID 1BCF, F. Frolow et al., *Nature Struct. Biol.* 1:453, 1994; PDB ID 1PEX, F. X. Gomis-Ruth et al., *J. Mol. Biol.* 264:556, 1996.]

α and β segments interspersed or alternating), and **α + β** (with α and β regions somewhat segregated). The third category organizes protein structures by protein type, such as soluble (globular), membrane, fibrous, and intrinsically unstructured proteins. The final category traces structural rearrangements and unusual features of proteins that are evolutionarily related. **Figure 4-23** presents examples of protein motifs taken from SCOP2 to illustrate the potential of searching within each category. The figure also introduces another way to represent elements of secondary structure and the relationships among segments of secondary structure in a protein—the **topology diagram**.

The number of folding patterns is not infinite. Among the more than 80,000 distinct protein structures archived in the PDB, only about 1,200 different folds or motifs are represented. Given the many years of progress in structural biology, new motifs are now only rarely discovered. Many examples of recurring domain or motif structures are available, and these reveal that protein tertiary structure is more reliably conserved than amino acid sequence. The comparison of protein structures can thus provide much information about evolution. Proteins with significant similarity in primary structure and/or with similar tertiary structure and function are said to be in the same **protein family**. The protein structures in the PDB belong to about 4,000 different protein families. A strong evolutionary relationship is usually evident within a protein family. For example, the globin family has many different proteins with both structural and sequence similarities to myoglobin (as seen in the proteins used as examples in Box 4-5 and in Chapter 5). Two or more families that have little similarity in amino acid sequence but make use of the same major structural motif and have functional similarities are grouped into **superfamilies**. An evolutionary relationship among families in a superfamily is considered probable, even though time and functional distinctions—that is, different adaptive pressures—may have erased many of the telltale sequence relationships.

A protein family may be widespread in all three domains of cellular life, the Bacteria, Archaea, and Eukarya, suggesting an ancient origin. Many proteins involved in intermediary metabolism and the metabolism of nucleic acids and proteins fall into this category. Other families may be present in only a small group of organisms, indicating that the structure arose more recently. Tracing the natural history of structural motifs through the use of structural classifications in databases such as SCOP2 provides a powerful complement to sequence analyses in tracing evolutionary relationships. The SCOP2 database is curated manually, with the objective of placing proteins in the correct evolutionary framework based on conserved structural features.

Structural motifs become especially important in defining protein families and superfamilies. Improved protein classification and comparison systems lead inevitably to the elucidation of new functional relationships. Given the central role of proteins in living systems, these structural comparisons can help illuminate every aspect of biochemistry, from the evolution of individual proteins to the evolutionary history of complete metabolic pathways.

Protein Quaternary Structures Range from Simple Dimers to Large Complexes

Many proteins have multiple polypeptide subunits (from two to hundreds). The association of polypeptide chains can serve a variety of functions. Many multisubunit proteins have regulatory roles; the binding of small molecules may affect the interaction between subunits, causing large changes in the protein's activity in response to small changes in the concentration of substrate or regulatory molecules (Chapter 6). In other cases, separate subunits take on separate but related functions, such as catalysis and regulation. Some associations, such as the fibrous

proteins considered earlier in this chapter and the coat proteins of viruses, serve primarily structural roles. Some very large protein assemblies are the site of complex, multistep reactions. For example, each ribosome, the site of protein synthesis, incorporates dozens of protein subunits along with RNA molecules.

A multisubunit protein is also referred to as a **multimer**. A multimer with just a few subunits is often called an **oligomer**. If a multimer has nonidentical subunits, the overall structure of the protein can be asymmetric and quite complicated. However, most multimers have identical subunits or repeating groups of nonidentical subunits, usually in symmetric arrangements. As noted in Chapter 3, the repeating structural unit in such a multimeric protein, whether a single subunit or a group of subunits, is called a **protomer**.

The first oligomeric protein to have its three-dimensional structure determined was hemoglobin (M_r 64,500), which contains four polypeptide chains and four heme prosthetic groups, in which the iron atoms are in the ferrous (Fe^{2+}) state (Fig. 4-17). The protein portion, the globin, consists of two α chains (141 residues each) and two β chains (146 residues each). Note that in this case, α and β do not refer to secondary structures. In a practice that can be confusing to the beginning student, the Greek letters α and β (and γ, δ, and others) are often used to distinguish two different kinds of subunits within a multisubunit protein, regardless of what kinds of secondary structure may predominate in the subunits. Because hemoglobin is four times as large as myoglobin, much more time and effort were required to solve its three-dimensional structure by x-ray analysis, finally achieved by Max Perutz, John Kendrew, and their colleagues in 1959. The subunits of hemoglobin are arranged in symmetric pairs **(Fig. 4-24)**, each pair having one α and one β subunit. Hemoglobin can therefore be described either as a tetramer or as a dimer of $\alpha\beta$ protomers. The role these distinct

Max Perutz, 1914–2002 (left), and John Kendrew, 1917–1997
[Source: Corbis/Hulton Deutsch Collection.]

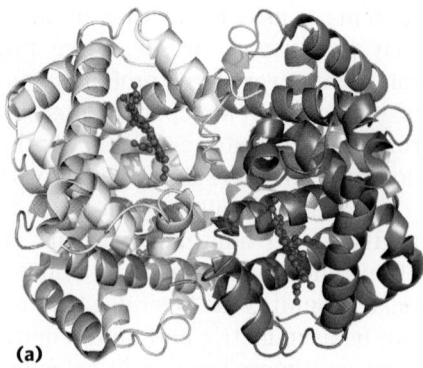

(a)

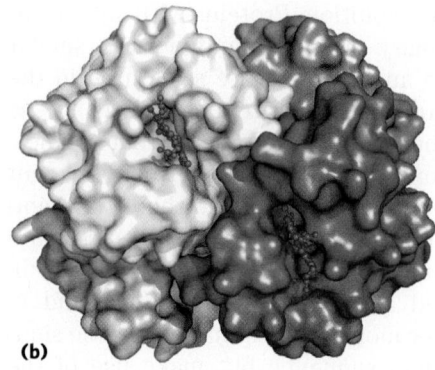

(b)

FIGURE 4-24 Quaternary structure of deoxyhemoglobin. X-ray diffraction analysis of deoxyhemoglobin (hemoglobin without oxygen molecules bound to the heme groups) shows how the four polypeptide subunits are packed together. **(a)** A ribbon representation reveals the secondary structural elements of the structure and the positioning of all the heme prosthetic groups. **(b)** A surface contour model shows the pockets in which the heme prosthetic groups are bound and helps to visualize subunit packing. The α subunits are shown in shades of gray, the β subunits in shades of blue. Note that the heme groups (red) are relatively far apart. [Source: PDB ID 2HHB, G. Fermi et al., *J. Mol. Biol.* 175:159, 1984.]

subunits play in hemoglobin function is discussed extensively in Chapter 5.

SUMMARY 4.3 Protein Tertiary and Quaternary Structures

■ Tertiary structure is the complete three-dimensional structure of a polypeptide chain. Many proteins fall into one of two general classes of proteins based on tertiary structure: fibrous and globular.

■ Fibrous proteins, which serve mainly structural roles, have simple repeating elements of secondary structure.

■ Globular proteins have more complicated tertiary structures, often containing several types of secondary structure in the same polypeptide chain. The first globular protein structure to be determined, by x-ray diffraction methods, was that of myoglobin.

■ The complex structures of globular proteins can be analyzed by examining folding patterns called motifs (also called folds or supersecondary structures). The many thousands of known protein structures are generally assembled from a repertoire of only a few hundred motifs. Domains are regions of a polypeptide chain that can fold stably and independently.

■ Some proteins or protein segments are intrinsically disordered, lacking definable three-dimensional structure. These proteins have distinctive amino acid compositions that allow a more flexible structure. Some of these disordered proteins function as structural components or scavengers; others can interact with many different protein partners, serving as versatile inhibitors or as central components of protein interaction networks. Quaternary structure results from interactions between the subunits of multisubunit (multimeric) proteins or large protein assemblies. Some

multimeric proteins have a repeated unit consisting of a single subunit or a group of subunits, each unit called a protomer.

4.4 Protein Denaturation and Folding

Proteins lead a surprisingly precarious existence. As we have seen, a native protein conformation is only marginally stable. In addition, most proteins must maintain conformational flexibility to function. The continual maintenance of the active set of cellular proteins required under a given set of conditions is called **proteostasis**. Cellular proteostasis requires the coordinated function of pathways for protein synthesis and folding, the refolding of proteins that are partially unfolded, and the sequestration and degradation of proteins that have been irreversibly unfolded or are no longer needed. In all cells, these networks involve hundreds of enzymes and specialized proteins.

As seen in **Figure 4-25**, the life of a protein encompasses much more than its synthesis and later degradation. The marginal stability of most proteins can produce a tenuous balance between folded and unfolded states. As proteins are synthesized on ribosomes (Chapter 27), they must fold into their native conformations. Sometimes this occurs spontaneously, but more often it requires the assistance of specialized enzymes and complexes called chaperones. Many of these same folding helpers function to refold proteins that become transiently unfolded. Proteins that are not properly folded often have exposed hydrophobic surfaces that render them "sticky," leading to the formation of inactive aggregates. These aggregates may lack their normal function but are not inert; their accumulation in cells lies at the heart of diseases ranging from diabetes to Parkinson disease and Alzheimer disease. Not surprisingly, all cells

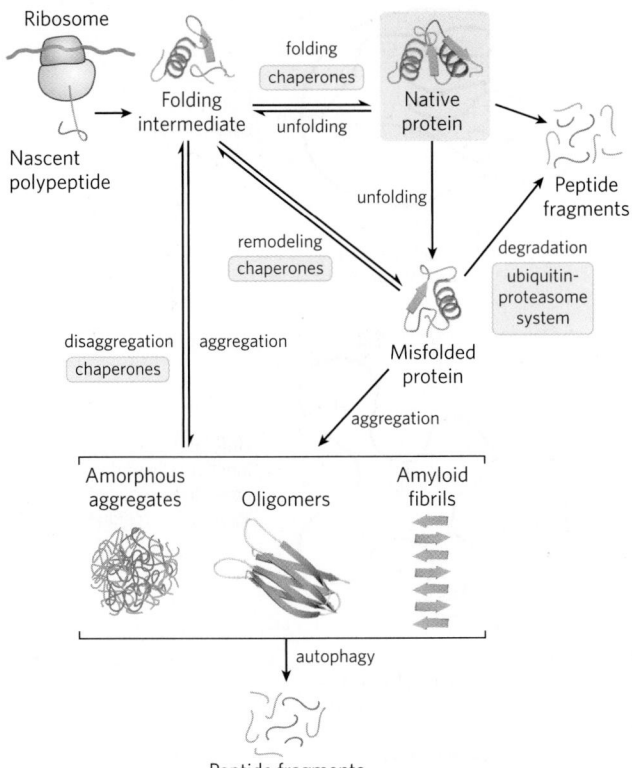

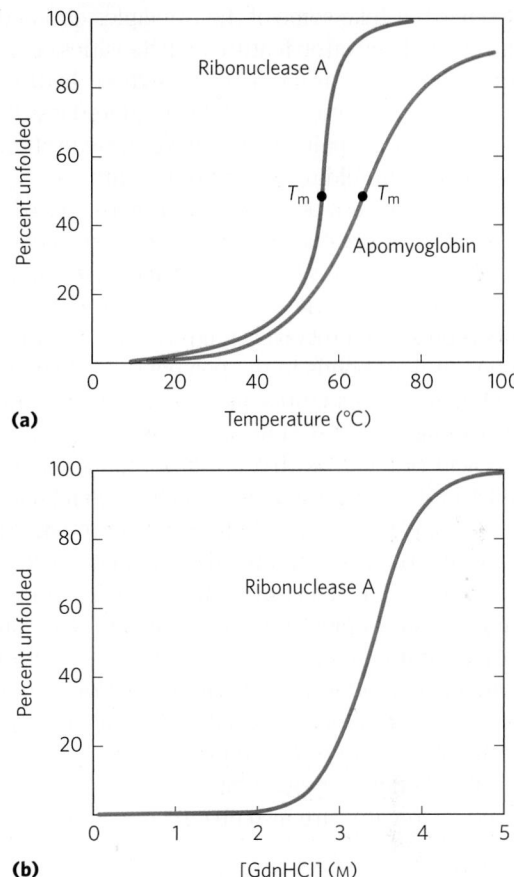

FIGURE 4-25 Pathways that contribute to proteostasis. Three kinds of processes contribute to proteostasis, in some cases with multiple contributing pathways. First, proteins are synthesized on a ribosome. Second, multiple pathways contribute to protein folding, many of which involve the activity of complexes called chaperones. Chaperones (including chaperonins) also contribute to the refolding of proteins that are partially and transiently unfolded. Finally, proteins that are irreversibly unfolded are subject to sequestration and degradation by several additional pathways. Partially unfolded proteins and protein-folding intermediates that escape the quality-control activities of the chaperones and degradative pathways may aggregate, forming both disordered aggregates and ordered amyloid-like aggregates that contribute to disease and aging processes. [Source: Information from F. U. Hartl et al., *Nature* 475:324, 2011, Fig. 6.]

FIGURE 4-26 Protein denaturation. Results are shown for proteins denatured by two different environmental changes. In each case, the transition from the folded to the unfolded state is abrupt, suggesting cooperativity in the unfolding process. **(a)** Thermal denaturation of horse apomyoglobin (myoglobin without the heme prosthetic group) and ribonuclease A (with its disulfide bonds intact; see Fig. 4-27). The midpoint of the temperature range over which denaturation occurs is called the melting temperature, or T_m. Denaturation of apomyoglobin was monitored by circular dichroism (see Fig. 4-10), which measures the amount of helical structure in the protein. Denaturation of ribonuclease A was tracked by monitoring changes in the intrinsic fluorescence of the protein, which is affected by changes in the environment of a Trp residue introduced by mutation. **(b)** Denaturation of disulfide-intact ribonuclease A by guanidine hydrochloride (GdnHCl), monitored by circular dichroism. [Sources: (a) Data from R. A. Sendak et al., *Biochemistry* 35:12,978, 1996; I. Nishii et al., *J. Mol. Biol.* 250:223, 1995. (b) Data from W. A. Houry et al., *Biochemistry* 35:10,125, 1996.]

have elaborate pathways for recycling and/or degrading proteins that are irreversibly misfolded.

The transitions between the folded and unfolded states, and the network of pathways that control these transitions, now become our focus.

Loss of Protein Structure Results in Loss of Function

Protein structures have evolved to function in particular cellular environments. Conditions different from those in the cell can result in protein structural changes, large and small. A loss of three-dimensional structure sufficient to cause loss of function is called **denaturation**. The denatured state does not necessarily equate with complete unfolding of the protein and randomization of conformation. Under most conditions, denatured proteins exist in a set of partially folded states.

Most proteins can be denatured by heat, which has complex effects on many weak interactions in a protein (primarily on the hydrogen bonds). If the temperature is increased slowly, a protein's conformation generally remains intact until an abrupt loss of structure (and function) occurs over a narrow temperature range **(Fig. 4-26)**. The abruptness of the change suggests that unfolding is a cooperative process: loss of structure in one part of the protein destabilizes other parts. The effects of heat on proteins can be mitigated by structure. The very heat-stable proteins of thermophilic bacteria and archaea have evolved to function at the temperature of hot springs (~100 °C). The folded structures of these proteins are often similar to those of proteins in other organisms, but take some of the principles outlined here to extremes. They often feature high densities of charged residues on their surfaces, even tighter hydrophobic packing in their interiors, and folds rendered less flexible by networks of ion pairs, which make these proteins less susceptible to unfolding at high temperatures.

Proteins can also be denatured by extremes of pH, by certain miscible organic solvents such as alcohol or acetone, by certain solutes such as urea and guanidine hydrochloride, or by detergents. Each of these denaturing agents represents a relatively mild treatment in the sense that no covalent bonds in the polypeptide chain are broken. Organic solvents, urea, and detergents act primarily by disrupting the hydrophobic aggregation of nonpolar amino acid side chains that produces the stable core of globular proteins; urea also disrupts hydrogen bonds; and extremes of pH alter the net charge on a protein, causing electrostatic repulsion and the disruption of some hydrogen bonding. The denatured structures resulting from these various treatments are not necessarily the same.

Denaturation often leads to protein precipitation, a consequence of protein aggregate formation as exposed hydrophobic surfaces associate. The aggregates are often highly disordered. The protein precipitate that is seen after boiling an egg white is one example. More-ordered aggregates are also observed in some proteins, as we shall see.

Amino Acid Sequence Determines Tertiary Structure

The tertiary structure of a globular protein is determined by its amino acid sequence. The most important proof of this came from experiments showing that denaturation of some proteins is reversible. Certain globular proteins denatured by heat, extremes of pH, or denaturing reagents will regain their native structure and their biological activity if returned to conditions in which the native conformation is stable. This process is called **renaturation**.

A classic example is the denaturation and renaturation of ribonuclease A, demonstrated by Christian Anfinsen in the 1950s. Purified ribonuclease A denatures completely in a concentrated urea solution in the presence of a reducing agent. The reducing agent cleaves the four disulfide bonds to yield eight Cys residues, and the urea disrupts the stabilizing hydrophobic effect, thus freeing the entire polypeptide from its folded conformation. Denaturation of ribonuclease is accompanied by a complete loss of catalytic activity. When the urea and the reducing agent are removed, the randomly coiled, denatured ribonuclease spontaneously refolds into its correct tertiary structure, with full restoration of its catalytic activity **(Fig. 4-27)**. The refolding of ribonuclease is so accurate that the four intrachain disulfide bonds are re-formed in the same positions in the renatured molecule as in the native ribonuclease. Later, similar results were obtained using chemically synthesized, catalytically active ribonuclease A. This eliminated the

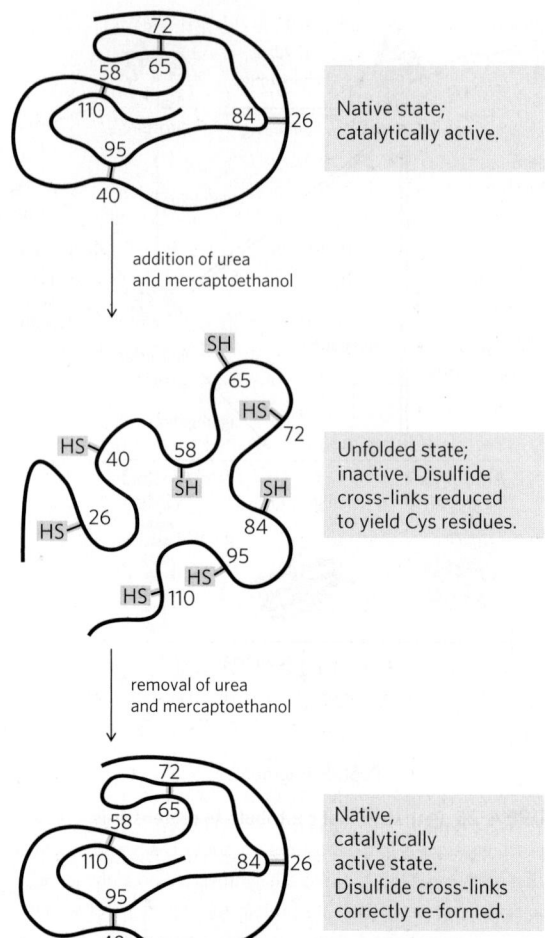

FIGURE 4-27 Renaturation of unfolded, denatured ribonuclease. Urea denatures the ribonuclease, and mercaptoethanol (HOCH$_2$CH$_2$SH) reduces and thus cleaves the disulfide bonds to yield eight Cys residues. Renaturation involves reestablishing the correct disulfide cross-links.

possibility that some minor contaminant in Anfinsen's purified ribonuclease preparation might have contributed to renaturation of the enzyme, thus dispelling any remaining doubt that this enzyme folds spontaneously.

The Anfinsen experiment provided the first evidence that the amino acid sequence of a polypeptide chain contains all the information required to fold the chain into its native, three-dimensional structure. Subsequent work has shown that only a minority of proteins, many of them small and inherently stable, will fold spontaneously into their native form. Even though all proteins have the potential to fold into their native structure, many require some assistance.

Polypeptides Fold Rapidly by a Stepwise Process

In living cells, proteins are assembled from amino acids at a very high rate. For example, *E. coli* cells can make a complete, biologically active protein molecule containing 100 amino acid residues in about 5 seconds at 37 °C. However, the synthesis of peptide bonds on the ribosome is not enough; the protein must fold.

How does the polypeptide chain arrive at its native conformation? Let's assume conservatively that each of the amino acid residues could take up 10 different conformations on average, giving 10^{100} different conformations for the polypeptide. Let's also assume that the protein folds spontaneously by a random process in which it tries out all possible conformations around every single bond in its backbone until it finds its native, biologically active form. If each conformation were sampled in the shortest possible time ($\sim 10^{-13}$ second, or the time required for a single molecular vibration), it would take about 10^{77} years to sample all possible conformations. Clearly, protein folding is not a completely random, trial-and-error process. There must be shortcuts. This problem was first pointed out by Cyrus Levinthal in 1968 and is sometimes called Levinthal's paradox.

The folding pathway of a large polypeptide chain is unquestionably complicated. However, rapid progress has been made in this field, sufficient to produce robust algorithms that can often predict the structure of smaller proteins on the basis of their amino acid sequences. The major folding pathways are hierarchical. Local secondary structures form first. Certain amino acid sequences fold readily into α helices or β sheets, guided by constraints such as those reviewed in our discussion of secondary structure. Ionic interactions, involving charged groups that are often near one another in the linear sequence of the polypeptide chain, can play an important role in guiding these early folding steps. Assembly of local structures is followed by longer-range interactions between, say, two elements of secondary structure that come together to form stable folded structures. The hydrophobic effect plays a significant role throughout the process, as the aggregation of nonpolar amino acid side chains provides an entropic stabilization to intermediates and, eventually, to the final folded structure. The process continues until complete domains form and the entire polypeptide is folded **(Fig. 4-28)**. Notably, proteins dominated by close-range interactions (between pairs of residues generally located near each other in the polypeptide sequence) tend to fold faster than proteins with more complex folding patterns and with many long-range interactions between different segments. As larger proteins with multiple domains are synthesized, domains near the amino terminus (which are synthesized first) may fold before the entire polypeptide has been assembled.

Thermodynamically, the folding process can be viewed as a kind of free-energy funnel **(Fig. 4-29)**. The unfolded states are characterized by a high degree of conformational entropy and relatively high free energy. As folding proceeds, the narrowing of the funnel reflects the decrease in the conformational space that must be searched as the protein approaches its native state. Small depressions along the sides of the free-energy funnel represent semistable intermediates that can briefly slow the folding process. At the bottom of the funnel, an ensemble of folding intermediates has been reduced to a single native conformation (or one of a small set of native conformations). The funnels can have a variety of shapes, depending on the complexity of the folding pathway, the existence of semistable intermediates, and the potential for particular intermediates to assemble into aggregates of misfolded proteins (Fig. 4-29).

Thermodynamic stability is not evenly distributed over the structure of a protein—the molecule has regions of relatively high stability and others of low or negligible stability. For example, a protein may have two stable domains joined by a segment that is entirely disordered. Regions of low stability may allow a protein to alter its conformation between two or more states. As we shall see in the next two chapters, variations in the stability of regions within a protein are often essential to protein function. Intrinsically disordered proteins or protein segments do not fold at all.

As our understanding of protein folding and protein structure improves, increasingly sophisticated computer programs for predicting the structure of proteins from their amino acid sequence are being developed. Prediction of protein structure is a specialty field of bioinformatics, and progress in this area is monitored with a biennial test called the CASP (Critical

Amino acid sequence of a 56-residue peptide

MTYKLILNGKTLKGETTTEAVDAATAEKVFKQYANDNGVDGEWTYDDATKTFTVTE

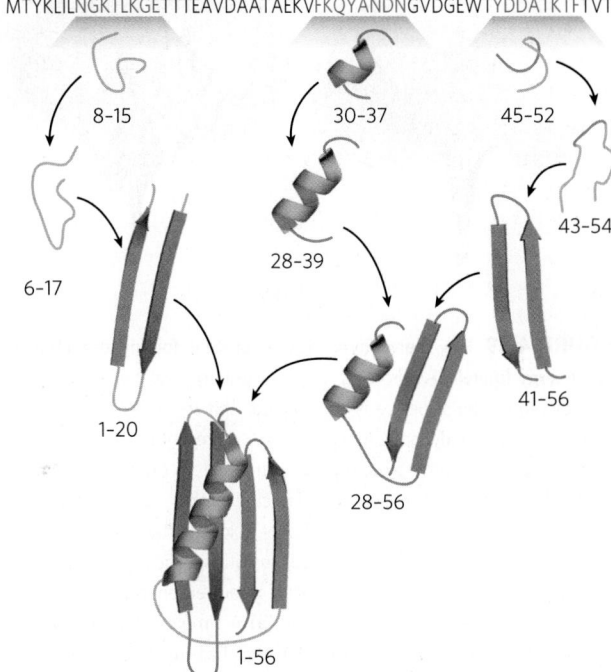

FIGURE 4-28 A protein-folding pathway as defined for a small protein. A hierarchical pathway is shown, based on computer modeling. Small regions of secondary structure are assembled first and then gradually incorporated into larger structures. The program used for this model has been highly successful in predicting the three-dimensional structure of small proteins from their amino acid sequence. The numbers indicate the amino acid residues in this 56 residue peptide that have acquired their final structure in each of the steps shown. [Source: Information from K. A. Dill et al., *Annu. Rev. Biophys.* 37:289, 2008, Fig. 5.]

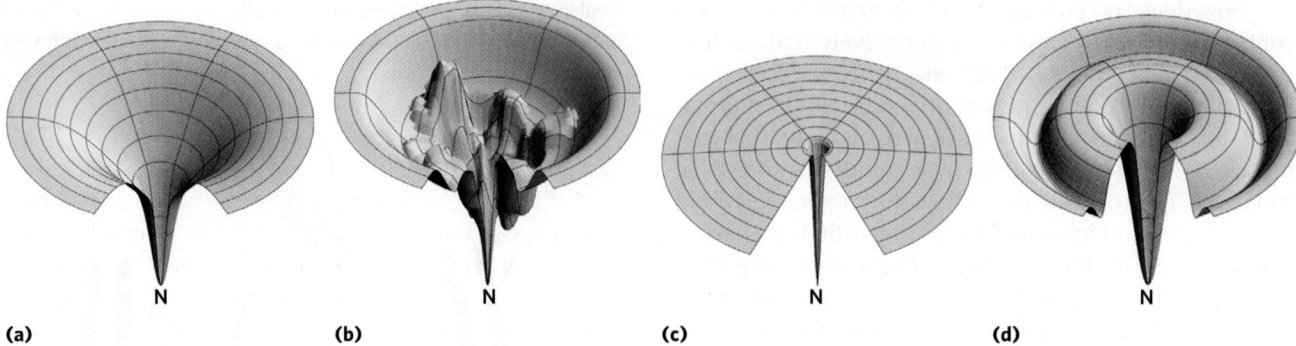

(a)　　　　　　　**(b)**　　　　　　　　**(c)**　　　　　　　**(d)**

FIGURE 4-29 The thermodynamics of protein folding depicted as free-energy funnels. As proteins fold, the conformational space that can be explored by the structure is constrained. This is modeled as a three-dimensional thermodynamic funnel, with ΔG represented by the depth of the funnel and the native structure (N) at the bottom (lowest free-energy point). The funnel for a given protein can have a variety of shapes, depending on the number and types of folding intermediates in the folding pathways. Any folding intermediate with significant stability and a finite lifetime would be represented as a local free-energy minimum—a depression on the surface of the funnel. **(a)** A simple but relatively wide and smooth funnel represents a protein that has multiple folding pathways (that is, the order in which different parts of the protein fold is somewhat random), but it assumes its three-dimensional structure with no folding intermediates that have significant stability. **(b)** This funnel represents a more typical protein that has multiple possible folding intermediates with significant stability on the multiple pathways leading to the native structure. **(c)** A protein with one stable native structure, essentially no other folded intermediates with significant stability, and only one or a very few productive folding pathways is shown as a funnel with one narrow depression leading to the native form. **(d)** A protein with folding intermediates of substantial stability on virtually every pathway leading to the native state (that is, a protein in which a particular motif or domain always folds quickly, but other parts of the protein fold more slowly and in a random order) is depicted by a funnel with a major depression surrounding the depression leading to the native form. [Source: Information from K. A. Dill et al., *Annu. Rev. Biophys.* 37:289, 2008, Fig. 9.]

Assessment of Structural Prediction) competition. Hundreds of research groups from around the world vie to predict the structure of an assigned protein (whose structure has been determined but not yet published). The most successful teams are invited to present their results at a CASP conference. The success of these efforts is improving rapidly, with correct predictions for smaller proteins becoming common.

Some Proteins Undergo Assisted Folding

Not all proteins fold spontaneously as they are synthesized in the cell. Folding for many proteins requires **chaperones**, proteins that interact with partially folded or improperly folded polypeptides, facilitating correct folding pathways or providing microenvironments in which folding can occur. Several types of molecular chaperones are found in organisms ranging from bacteria to humans. Two major families of chaperones, both well studied, are the **Hsp70** family and the **chaperonins**.

The Hsp70 family of proteins generally have a molecular weight near 70,000 and are more abundant in cells stressed by elevated temperatures (hence, *h*eat *s*hock *p*roteins of M_r 70,000, or Hsp70). Hsp70 proteins bind to regions of unfolded polypeptides that are rich in hydrophobic residues. These chaperones thus "protect" both proteins subject to denaturation by heat and new peptide molecules being synthesized (and not yet folded). Hsp70 proteins also block the folding of certain proteins that must remain unfolded until they have been translocated across a membrane (as described in Chapter 27). Some chaperones also facilitate the quaternary assembly of oligomeric proteins. The Hsp70 proteins bind to and release polypeptides in a cycle that uses energy from ATP hydrolysis and involves several other proteins (including a class called Hsp40). **Figure 4-30** illustrates chaperone-assisted folding as elucidated for the eukaryotic Hsp70 and Hsp40 chaperones. The binding of an unfolded polypeptide by an Hsp70 chaperone may break up a protein aggregate or prevent the formation of a new one. When the bound polypeptide is released, it has a chance to resume folding to its native structure. If folding does not occur rapidly enough, the polypeptide may be bound again and the process repeated. Alternatively, the Hsp70-bound polypeptide may be delivered to a chaperonin.

Chaperonins are elaborate protein complexes required for the folding of some cellular proteins that do not fold spontaneously. In *E. coli*, an estimated 10% to 15% of cellular proteins require the resident chaperonin system, called GroEL/GroES, for folding under normal conditions (up to 30% require this assistance when the cells are heat stressed). The analogous chaperonin system in eukaryotes is called Hsp60. The chaperonins first became known when they were found to be necessary for the growth of certain bacterial viruses (hence the designation "Gro"). These chaperone proteins are structured as a series of multisubunit rings, forming two chambers oriented back to back. An unfolded protein is first bound to an exposed hydrophobic surface near the apical end of one GroEL chamber. The protein is then trapped within the chamber when it

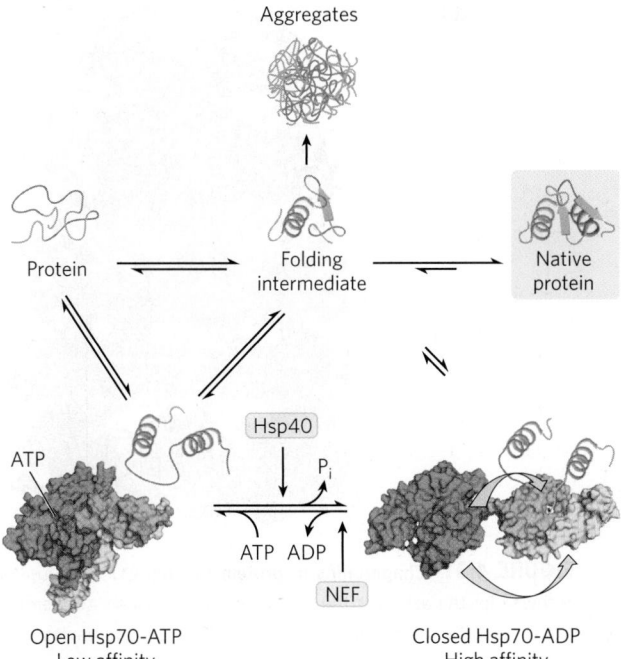

FIGURE 4-30 Chaperones in protein folding. The pathway by which chaperones of the Hsp70 class bind and release polypeptides is illustrated for the eukaryotic chaperones Hsp70 and Hsp40. The chaperones do not actively promote the folding of the substrate protein, but instead prevent aggregation of unfolded peptides. The unfolded or partly folded proteins bind first to the open, ATP-bound form of Hsp70. Hsp40 then interacts with this complex and triggers ATP hydrolysis that produces the closed form of the complex, in which the domains colored orange and yellow come together like the two parts of a jaw, trapping parts of the unfolded protein inside. Dissociation of ADP and recycling of the Hsp70 requires interaction with another protein, nucleotide-exchange factor (NEF). For a population of polypeptide molecules, some fraction of the molecules released after the transient binding of partially folded proteins by Hsp70 will take up the native conformation. The remainder are quickly rebound by Hsp70 or diverted to the chaperonin system (Hsp60; see Fig. 4-31). In bacteria, the Hsp70 and Hsp40 chaperones are called DnaK and DnaJ, respectively. DnaK and DnaJ were first identified as proteins required for in vitro replication of certain viral DNA molecules (hence the "Dna" designation). [Sources: Information from F. U. Hartl et al., *Nature* 475:324, 2011, Fig. 2. Open Hsp70-ATP: PDB ID 2QXL, Q. Liu and W. A. Hendrickson, *Cell* 131:106, 2007. Closed Hsp70-ADP: derived from PDB ID 2KHO, E. B. Bertelson et al., *Proc. Natl. Acad. Sci. USA* 106:8471, 2009, and PDB ID 1DKZ, X. Zhu et al., *Science* 272:1606, 1996.]

is capped transiently by the GroES "lid" **(Fig. 4-31)**. GroEL undergoes substantial conformational changes, coupled to slow ATP hydrolysis, which also regulates the binding and release of GroES. Inside the chamber, a protein has about 10 seconds to fold—the time required for the bound ATP to hydrolyze. Constraining a protein within the chamber prevents inappropriate protein aggregation and also restricts the conformational space that a polypeptide chain can explore as it folds. The protein is released when the GroES cap dissociates, but can rebind rapidly for another round if folding has not been completed. The two chambers in a GroEL complex alternate in binding and releasing unfolded polypeptide

substrates. In eukaryotes, the Hsp60 system utilizes a similar process to fold proteins. However, in place of the GroES lid, protrusions from the apical domains of the subunits flex and close over the chamber. The ATP hydrolytic cycle is also slower in the Hsp60 complexes, giving the constrained proteins more time to fold.

Finally, the folding pathways of some proteins require two enzymes that catalyze isomerization reactions. **Protein disulfide isomerase (PDI)** is a widely distributed enzyme that catalyzes the interchange, or shuffling, of disulfide bonds until the bonds of the native conformation are formed. Among its functions, PDI catalyzes the elimination of folding intermediates with inappropriate disulfide cross-links. **Peptide prolyl cis-trans isomerase (PPI)** catalyzes the interconversion of the cis and trans isomers of peptide bonds formed by Pro residues (Fig. 4-8), which can be a slow step in the folding of proteins that contain some Pro peptide bonds in the cis configuration.

Defects in Protein Folding Provide the Molecular Basis for a Wide Range of Human Genetic Disorders

Despite the many processes that assist in protein folding, misfolding does occur. In fact, protein misfolding is a substantial problem in all cells, and a quarter or more of all polypeptides synthesized may be destroyed because they do not fold correctly. In some cases, the misfolding causes or contributes to the development of serious disease.

Many conditions, including type 2 diabetes, Alzheimer disease, Huntington disease, and Parkinson disease, are associated with a misfolding mechanism: a soluble protein that is normally secreted from the cell is secreted in a misfolded state and converted into an insoluble extracellular **amyloid** fiber. The diseases are collectively referred to as **amyloidoses**. The fibers are highly ordered and unbranched, with a diameter of 7 to 10 nm and a high degree of β-sheet structure. The β segments are oriented perpendicular to the axis of the fiber. In some amyloid fibers the overall structure includes two layers of β sheet, such as that shown for amyloid-β peptide in **Figure 4-32**.

Many proteins can take on the amyloid fibril structure as an alternative to their normal folded conformations, and most of these proteins have a concentration of aromatic amino acid residues in a core region of β sheet or α helix. The proteins are secreted in an incompletely folded conformation. The core (or some part of it) folds into a β sheet before the rest of the protein folds correctly, and the β sheets from two or more incompletely folded protein molecules associate to begin forming an amyloid fibril. The fibril grows in the extracellular space. Other parts of the protein then fold differently, remaining on the outside of the β-sheet core in the growing fibril. The effect of aromatic residues in stabilizing the structure is shown in Figure 4-32c. Because most of the protein molecules fold normally, the onset of symptoms in the amyloidoses is

(a) Folding intermediate delivered by Hsp70-ADP

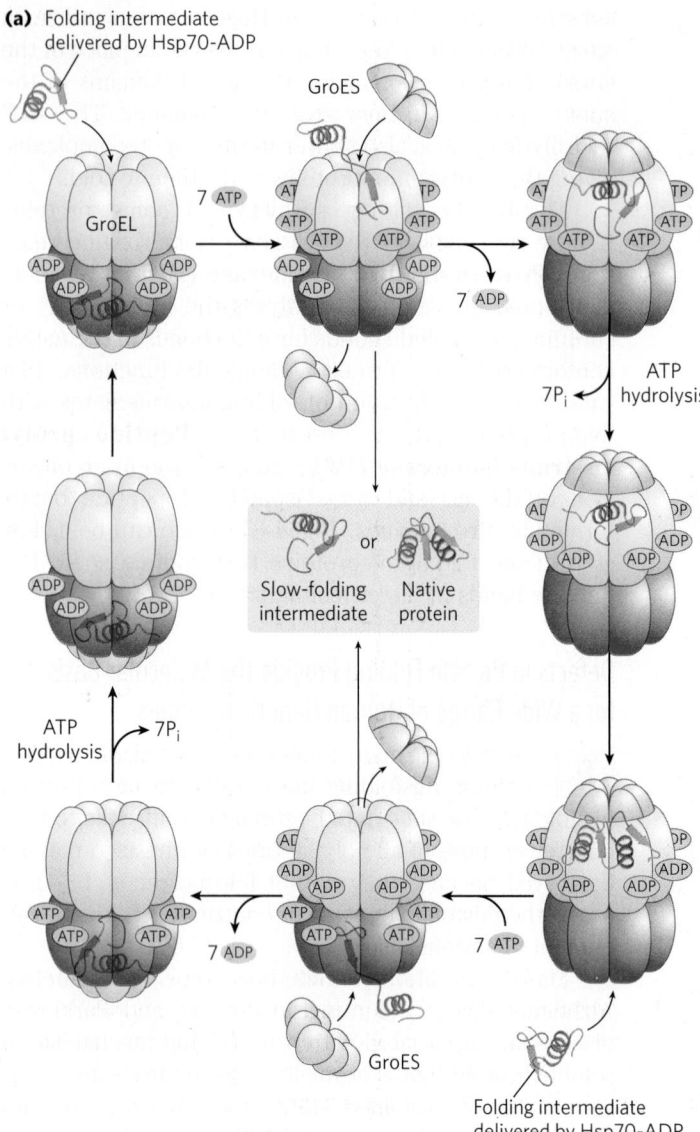

(b)

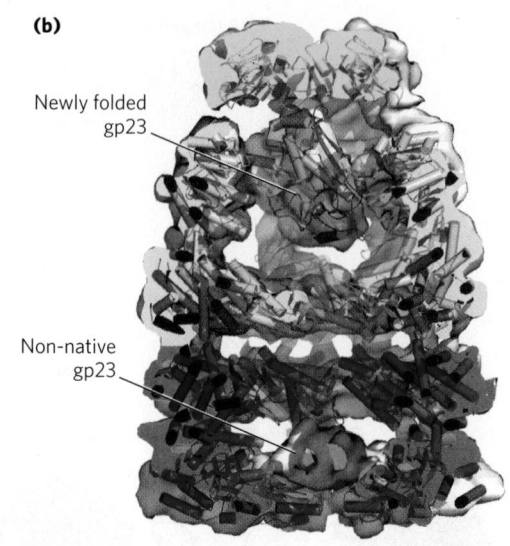

Newly folded gp23

Non-native gp23

FIGURE 4-31 Chaperonins in protein folding. (a) A proposed pathway for the action of the *E. coli* chaperonins GroEL (a member of the Hsp60 protein family) and GroES. Each GroEL complex consists of two large chambers formed by two heptameric rings (each subunit M_r 57,000). GroES, also a heptamer (subunit M_r 10,000), blocks one of the GroEL chambers after an unfolded protein is bound inside. The chamber with the unfolded protein is referred to as cis; the opposite one is trans. Folding occurs within the cis chamber, during the time it takes to hydrolyze the 7 ATP bound to the subunits in the heptameric ring. The GroES and the ADP molecules then dissociate, and the protein is released. The two chambers of the GroEL/Hsp60 systems alternate in the binding and facilitated folding of client proteins. **(b)** A cutaway image of the GroEL/GroES complex. The α-helical secondary structure is represented as cylinders within a transparent surface structure. A folded protein (gp23) is shown within the large interior space of the upper chamber; an unfolded version of gp23 is shown in the lower chamber. [Sources: (a) Information from F. U. Hartl et al., *Nature* 475:324, 2011, Fig. 3. (b) Surface view of GroEL/GroES with unfolded gp23: EMDB-1548, D. K. Clare et al., *Nature* 457:107, 2009; GroEL/GroES: PDB ID 2CGT, D. K. Clare et al., *J. Mol. Biol.* 358:905, 2006; folded gp23: PDB ID 1YUE, A. Fokine et al., *Proc. Natl. Acad. Sci. USA* 102:7163, 2005.]

often very slow. If a person inherits a mutation such as substitution with an aromatic residue at a position that favors formation of amyloid fibrils, disease symptoms may begin at an earlier age.

In eukaryotes, proteins destined for secretion undergo their initial folding in the endoplasmic reticulum (ER; see pathway in Chapter 27). When stress conditions arise, or when protein synthesis threatens to overwhelm the protein-folding capacity of the ER, unfolded proteins can accumulate. These conditions trigger the unfolded protein response (UPR). A set of transcriptional regulators that constitute the UPR bring the various systems into alignment by increasing the concentration of chaperones in the ER or decreasing the rate of overall protein synthesis, or both. Amyloid aggregates that form before the UPR can come into play may be removed. Some are degraded by **autophagy**. In this process, the aggregates are first encapsulated in a membrane, then the contents of the resulting vesicle

are degraded after the vesicle docks with a cytosolic lysosome. Alternatively, misfolded proteins can be degraded by a system of proteases called the ubiquitin-proteasome system (described in Chapter 27). Defects in any of these systems decrease the capacity to deal with misfolded proteins and increase the propensity for development of amyloid-related diseases. The UPR is a complex response involving many protein factors and signals, and inactivation of UPR components may have positive or negative effects on the degree of protein misfolding. This system is an attractive drug target for protein misfolding (amyloid) diseases.

Some amyloidoses are systemic, involving many tissues. Primary systemic amyloidosis is caused by deposition of fibrils consisting of misfolded immunoglobulin light chains (see Chapter 5), or fragments of light chains derived from proteolytic degradation. The mean age of onset is about 65 years. Patients have symptoms including fatigue, hoarseness, swelling, and weight loss, and

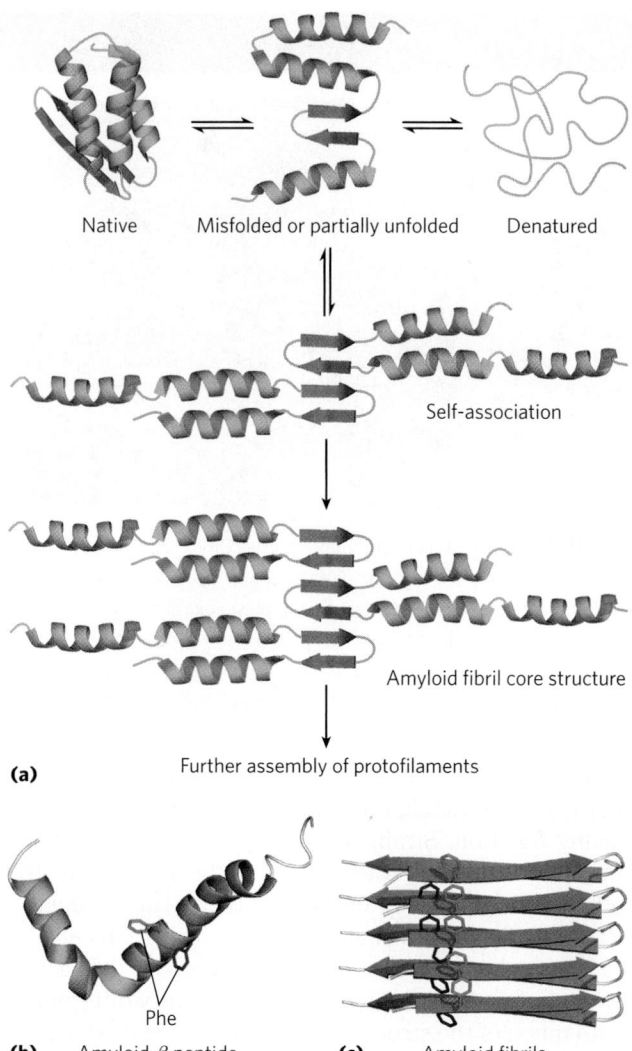

(a)

Native   Misfolded or partially unfolded   Denatured

Self-association

Amyloid fibril core structure

Further assembly of protofilaments

Phe

(b) Amyloid-β peptide

(c) Amyloid fibrils

FIGURE 4-32 Formation of disease-causing amyloid fibrils. (a) Protein molecules whose normal structure includes regions of β sheet undergo partial folding. In a small number of the molecules, before folding is complete, the β-sheet regions of one polypeptide associate with the same region in another polypeptide, forming the nucleus of an amyloid. Additional protein molecules slowly associate with the amyloid and extend it to form a fibril. **(b)** The amyloid-β peptide begins as two α-helical segments of a larger protein. Proteolytic cleavage of this larger protein leaves the relatively unstable amyloid-β peptide, which loses its α-helical structure. It can then assemble slowly into amyloid fibrils **(c)**, which contribute to the characteristic plaques on the exterior of nervous tissue in people with Alzheimer disease. The aromatic side chains shown here play a significant role in stabilizing the amyloid structure. Amyloid is rich in β sheet, with the β strands arranged perpendicular to the axis of the amyloid fibril. Amyloid-β peptide takes the form of two layers of extended parallel β sheet. Some amyloid-forming peptides may fold to form left-handed β helices. [Sources: (a) Information from D. J. Selkoe, *Nature* 426:900, 2003, Fig. 1. (b) PDB ID 1IYT, O. Crescenzi et al., *Eur. J. Biochem.* 269:5642, 2002. (c) PDB ID 2BEG, T. Lührs et al., *Proc. Natl. Acad. Sci. USA* 102:17,342, 2005.]

increase in secretion of an amyloid-prone polypeptide called serum amyloid A (SAA) protein. This protein, or fragments of it, deposits in the connective tissue of the spleen, kidney, and liver, and around the heart. People with this condition, known as secondary systemic amyloidosis, have a wide range of symptoms, depending on the organs initially affected. The disease is generally fatal within a few years. More than 80 amyloidoses are associated with mutations in transthyretin (a protein that binds to and transports thyroid hormones, distributing them throughout the body and brain). A variety of mutations in this protein lead to amyloid deposition concentrated around different tissues, thus producing different symptoms. Amyloidoses are also associated with inherited mutations in the proteins lysozyme, fibrinogen A α chain, and apolipoproteins A-I and A-II; all of these proteins are described in later chapters.

Some amyloid diseases are associated with particular organs. The amyloid-prone protein is generally secreted only by the affected tissue, and its locally high concentration leads to amyloid deposition around that tissue (although some of the protein may be distributed systemically). One common site of amyloid deposition is near the pancreatic islet β cells, responsible for insulin secretion and regulation of glucose metabolism (see Fig. 23-27). Secretion by β cells of a small (37 amino acid) peptide called islet amyloid polypeptide (IAPP), or amylin, can lead to amyloid deposits around the islets, gradually destroying the cells. A healthy human adult has 1 to 1.5 million pancreatic β cells. With progressive loss of these cells, glucose homeostasis is affected and eventually, when 50% or more of the cells are lost, the condition matures into type 2 (non-insulin-dependent) diabetes mellitus.

The amyloid deposition diseases that trigger neurodegeneration, particularly in older adults, are a special class of localized amyloidoses. Alzheimer disease is associated with extracellular amyloid deposition by neurons, involving the amyloid-β peptide (Fig. 4-32b), derived from a larger transmembrane protein (amyloid-β precursor protein) found in most human tissues. When it is part of the larger protein, the peptide is composed of two α-helical segments spanning the membrane. When the external and internal domains are cleaved off by specific proteases, the relatively unstable amyloid-β peptide leaves the membrane and loses its α-helical structure. It can then take the form of two layers of extended parallel β sheet, which can slowly assemble into amyloid fibrils (Fig. 4-32c). Deposits of these amyloid fibers seem to be the primary cause of Alzheimer disease, but a second type of amyloidlike aggregation, involving a protein called tau, also occurs intracellularly (in neurons) in people with Alzheimer disease. Inherited mutations in the tau protein do not result in Alzheimer disease, but they cause a frontotemporal dementia and parkinsonism (a condition with symptoms resembling Parkinson disease) that can be equally devastating.

Several other neurodegenerative conditions involve intracellular aggregation of misfolded proteins. In

many die within the first year after diagnosis. The kidneys or heart are often the most affected organs. Some amyloidoses are associated with other types of disease. People with certain chronic infectious or inflammatory diseases such as rheumatoid arthritis, tuberculosis, cystic fibrosis, and some cancers can experience a sharp

BOX 4-6 ✚ MEDICINE Death by Misfolding: The Prion Diseases

A misfolded brain protein seems to be the causative agent of several rare degenerative brain diseases in mammals. Perhaps the best known of these is bovine spongiform encephalopathy (BSE; also known as mad cow disease). Related diseases include kuru and Creutzfeldt-Jakob disease in humans, scrapie in sheep, and chronic wasting disease in deer and elk. These diseases are also referred to as spongiform encephalopathies, because the diseased brain frequently becomes riddled with holes (Fig. 1). Progressive deterioration of the brain leads to a spectrum of neurological symptoms, including weight loss, erratic behavior, problems with posture, balance, and coordination, and loss of cognitive function. The diseases are fatal.

In the 1960s, investigators found that preparations of the disease-causing agents seemed to lack nucleic acids. At this time, Tikvah Alper suggested that the agent was a protein. Initially, the idea seemed heretical. All disease-causing agents known up to that time—viruses, bacteria, fungi, and so on—contained nucleic acids, and their virulence was related to genetic reproduction and propagation. However, four decades of investigations, pursued most notably by Stanley Prusiner, have provided evidence that spongiform encephalopathies are different.

The infectious agent has been traced to a single protein (M_r 28,000), which Prusiner dubbed **prion** protein (PrP). The name was derived from *prion*-aceous *in*fectious, but Prusiner thought that "prion" sounded better than "proin." Prion protein is a normal

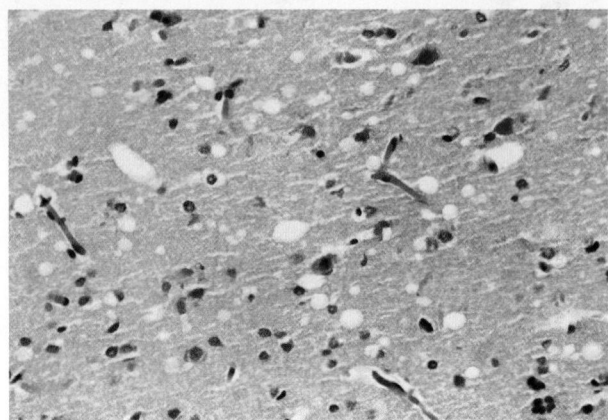

FIGURE 1 Stained section of cerebral cortex from the autopsy of a patient with Creutzfeldt-Jakob disease shows spongiform (vacuolar) degeneration, the most characteristic neurohistological feature. The yellowish vacuoles are intracellular and occur mostly in pre- and postsynaptic processes of neurons. The vacuoles in this section vary in diameter from 20 to 100 μm. [Source: Ralph C. Eagle, Jr./Science Source.]

constituent of brain tissue in all mammals. Its role is not known in detail, but it may have a molecular signaling function. Strains of mice lacking the gene for PrP (and thus the protein itself) suffer no obvious ill effects. Illness occurs only when the normal cellular PrP, or PrP^C, occurs in an altered conformation called PrP^{Sc} (Sc denotes scrapie). The structure of PrP^C has two α helices. The structure of PrP^{Sc} is very different, with much of the structure converted to amyloidlike β

Parkinson disease, the misfolded form of the protein α-synuclein aggregates into spherical filamentous masses called Lewy bodies. Huntington disease involves the protein huntingtin, which has a long polyglutamine repeat. In some individuals, the polyglutamine repeat is longer than normal and a more subtle type of intracellular aggregation occurs. Notably, when the mutant human proteins involved in Parkinson disease and Huntington disease are expressed in *Drosophila melanogaster*, the flies display neurodegeneration expressed as eye deterioration, tremors, and early death. All of these symptoms are highly suppressed if expression of the Hsp70 chaperone is also increased.

Protein misfolding need not lead to amyloid formation to cause serious disease. For example, cystic fibrosis is caused by defects in a membrane-bound protein called *cystic fibrosis transmembrane conductance regulator* (CFTR), which acts as a channel for chloride ions. The most common cystic fibrosis–causing mutation is the deletion of a Phe residue at position 508 in CFTR, which causes improper protein folding. Most of this protein is then degraded and its normal function is lost (see Box 11-2). Many of the disease-related mutations

in collagen (p. 130) also cause defective folding. A particularly remarkable type of protein misfolding is seen in the prion diseases (Box 4-6). ∎

SUMMARY 4.4 Protein Denaturation and Folding

∎ The maintenance of the steady-state collection of active cellular proteins required under a particular set of conditions—called proteostasis—involves an elaborate set of pathways and processes that fold, refold, and degrade polypeptide chains.

∎ The three-dimensional structure and the function of most proteins can be destroyed by denaturation, demonstrating a relationship between structure and function. Some denatured proteins can renature spontaneously to form biologically active protein, showing that tertiary structure is determined by amino acid sequence.

∎ Protein folding in cells is generally hierarchical. Initially, regions of secondary structure may form, followed by folding into motifs and domains. Large ensembles of folding intermediates are rapidly brought to a single native conformation.

sheets (Fig. 2). The interaction of PrP^{Sc} with PrP^C converts the latter to PrP^{Sc}, initiating a domino effect in which more and more of the brain protein converts to the disease-causing form. The mechanism by which the presence of PrP^{Sc} leads to spongiform encephalopathy is not understood.

In inherited forms of prion diseases, a mutation in the gene encoding PrP produces a change in one amino acid residue that is believed to make the conversion of PrP^C to PrP^{Sc} more likely. A complete understanding of prion diseases awaits new information on how prion protein affects brain function. Structural information about PrP is beginning to provide insights into the molecular process that allows the prion proteins to interact so as to alter their conformation (Fig. 2). The significance of prions may extend well beyond spongiform encephalopathies. Evidence is building that prionlike proteins may be responsible for additional neurodegenerative diseases such as multiple system atrophy (MSA), a disease that resembles Parkinson disease.

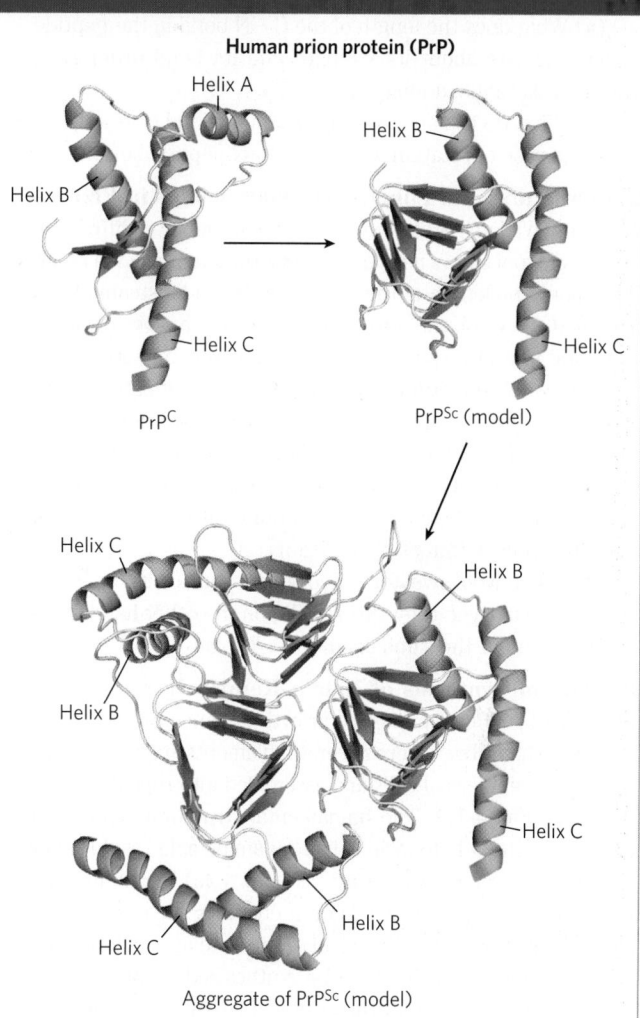

Human prion protein (PrP)

FIGURE 2 Structure of the globular domain of human PrP and models of the misfolded, disease-causing conformation PrP^{Sc}, and an aggregate of PrP^{Sc}. The α helices are labeled to help illustrate the conformational change. Helix A is incorporated into the β-sheet structure of the misfolded conformation. [Sources: Human PrP from PDB ID 1QLX, R. Zahn et al., *Proc. Natl. Acad. Sci. USA* 97:145, 2000. Models from C. Govaerts et al., *Proc. Natl. Acad. Sci. USA* 101:8342, 2004.]

■ For many proteins, folding is facilitated by Hsp70 chaperones and by chaperonins. Disulfide-bond formation and the cis-trans isomerization of Pro peptide bonds are catalyzed by specific enzymes.

■ Protein misfolding is the molecular basis of a wide range of human diseases, including the amyloidoses.

Key Terms

Terms in bold are defined in the glossary.

conformation 116
native conformation 116
hydrophobic effect 116
solvation layer 116
peptide group 118
Ramachandran plot 119
secondary structure 119
α helix 120
β conformation 123
β sheet 123
β turn 123

circular dichroism (CD) spectroscopy 125
tertiary structure 125
quaternary structure 125
fibrous proteins 125
globular proteins 125
α-keratin 126
collagen 127
silk fibroin 130
Protein Data Bank (PDB) 132

motif 133
fold 133
domain 137
intrinsically disordered proteins 138
topology diagram 140
protein family 141
multimer 141
oligomer 141
protomer 141
proteostasis 142
denaturation 143

renaturation 144
chaperone 146
Hsp70 146
chaperonin 146
protein disulfide isomerase (PDI) 147
peptide prolyl cis-trans isomerase (PPI) 147
amyloid 147
amyloidoses 147
autophagy 148
prion 150

Problems

1. Properties of the Peptide Bond In x-ray studies of crystalline peptides, Linus Pauling and Robert Corey found that the C—N bond in the peptide link is intermediate in length (1.32 Å) between a typical C—N single bond (1.49 Å) and a C=N double bond (1.27 Å). They also found that the peptide bond is planar (all four atoms attached to the C—N group are located in the same plane) and that the two α-carbon atoms attached to the C—N are always trans to each other (on opposite sides of the peptide bond).

(a) What does the length of the C—N bond in the peptide linkage indicate about its strength and its bond order (i.e., whether it is single, double, or triple)?

(b) What do the observations of Pauling and Corey tell us about the ease of rotation about the C—N peptide bond?

2. Structural and Functional Relationships in Fibrous Proteins William Astbury discovered that the x-ray diffraction pattern of wool shows a repeating structural unit spaced about 5.2 Å along the length of the wool fiber. When he steamed and stretched the wool, the x-ray pattern showed a new repeating structural unit at a spacing of 7.0 Å. Steaming and stretching the wool and then letting it shrink gave an x-ray pattern consistent with the original spacing of about 5.2 Å. Although these observations provided important clues to the molecular structure of wool, Astbury was unable to interpret them at the time.

(a) Given our current understanding of the structure of wool, interpret Astbury's observations.

(b) When wool sweaters or socks are washed in hot water or heated in a dryer, they shrink. Silk, on the other hand, does not shrink under the same conditions. Explain.

3. Rate of Synthesis of Hair α-Keratin Hair grows at a rate of 15 to 20 cm/yr. All this growth is concentrated at the base of the hair fiber, where α-keratin filaments are synthesized inside living epidermal cells and assembled into ropelike structures (see Fig. 4-11). The fundamental structural element of α-keratin is the α helix, which has 3.6 amino acid residues per turn and a rise of 5.4 Å per turn (see Fig. 4-4a). Assuming that the biosynthesis of α-helical keratin chains is the rate-limiting factor in the growth of hair, calculate the rate at which peptide bonds of α-keratin chains must be synthesized (peptide bonds per second) to account for the observed yearly growth of hair.

4. Effect of pH on the Conformation of α-Helical Secondary Structures The unfolding of the α helix of a polypeptide to a randomly coiled conformation is accompanied by a large decrease in a property called specific rotation, a measure of a solution's capacity to rotate circularly polarized light. Polyglutamate, a polypeptide made up of only L-Glu residues, has the α-helix conformation at pH 3. When the pH is raised to 7, there is a large decrease in the specific rotation of the solution. Similarly, polylysine (L-Lys residues) is an α helix at pH 10, but when the pH is lowered to 7 the specific rotation also decreases, as shown by the following graph.

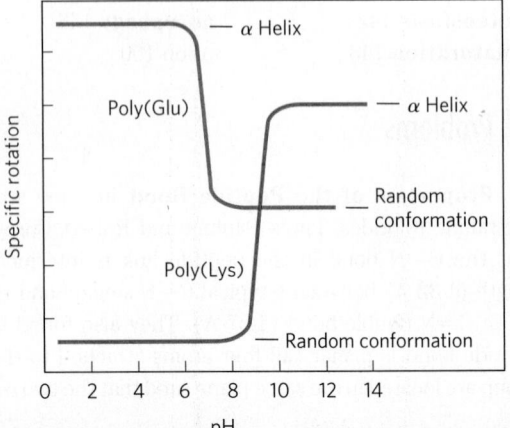

What is the explanation for the effect of the pH changes on the conformations of poly(Glu) and poly(Lys)? Why does the transition occur over such a narrow range of pH?

5. Disulfide Bonds Determine the Properties of Many Proteins Some natural proteins are rich in disulfide bonds, and their mechanical properties (tensile strength, viscosity, hardness, etc.) are correlated with the degree of disulfide bonding.

(a) Glutenin, a wheat protein rich in disulfide bonds, is responsible for the cohesive and elastic character of dough made from wheat flour. Similarly, the hard, tough nature of tortoise shell is due to the extensive disulfide bonding in its α-keratin. What is the molecular basis for the correlation between disulfide-bond content and mechanical properties of the protein?

(b) Most globular proteins are denatured and lose their activity when briefly heated to 65 °C. However, globular proteins that contain multiple disulfide bonds often must be heated longer at higher temperatures to denature them. One such protein is bovine pancreatic trypsin inhibitor (BPTI), which has 58 amino acid residues in a single chain and contains three disulfide bonds. On cooling a solution of denatured BPTI, the activity of the protein is restored. What is the molecular basis for this property?

6. Dihedral Angles A series of torsion angles, ϕ and ψ, that might be taken up by the peptide backbone is shown below. Which of these closely correspond to ϕ and ψ for an idealized collagen triple helix? Refer to Figure 4-9 as a guide.

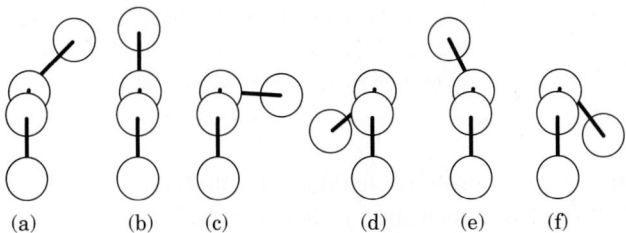

(a)   (b)   (c)   (d)   (e)   (f)

7. Amino Acid Sequence and Protein Structure Our growing understanding of how proteins fold allows researchers to make predictions about protein structure based on primary amino acid sequence data. Consider the following amino acid sequence.

1	2	3	4	5	6	7	8	9	10
Ile–	Ala –	His –	Thr –	Tyr –	Gly –	Pro –	Phe –	Glu –	Ala –

11	12	13	14	15	16	17	18	19	20
Ala–	Met–	Cys –	Lys –	Trp –	Glu –	Ala –	Gln –	Pro –	Asp–

21	22	23	24	25	26	27	28
Gly–	Met–	Glu –	Cys –	Ala –	Phe –	His –	Arg

(a) Where might bends or β turns occur?

(b) Where might intrachain disulfide cross-linkages be formed?

(c) Assuming that this sequence is part of a larger globular protein, indicate the probable location (external surface or interior of the protein) of the following amino acid residues:

Asp, Ile, Thr, Ala, Gln, Lys. Explain your reasoning. (Hint: See the hydropathy index in Table 3-1.)

8. Bacteriorhodopsin in Purple Membrane Proteins Under the proper environmental conditions, the salt-loving archaeon *Halobacterium halobium* synthesizes a membrane protein (M_r 26,000) known as bacteriorhodopsin, which is purple because it contains retinal (see Fig. 10-20). Molecules of this protein aggregate into "purple patches" in the cell membrane. Bacteriorhodopsin acts as a light-activated proton pump that provides energy for cell functions. X-ray analysis of this protein reveals that it consists of seven parallel α-helical segments, each of which traverses the bacterial cell membrane (thickness 45 Å). Calculate the minimum number of amino acid residues necessary for one segment of α helix to traverse the membrane completely. Estimate the fraction of the bacteriorhodopsin protein that is involved in membrane-spanning helices. (Use an average amino acid residue weight of 110.)

9. Protein Structure Terminology Is myoglobin a motif, a domain, or a complete three-dimensional structure?

10. Interpreting Ramachandran Plots Examine the two proteins labeled (a) and (b) below. Which of the two Ramachandran plots, labeled (c) and (d), is more likely to be derived from which protein? Why? [Sources: (a) PDB ID 1GWY, J. M. Mancheno et al., *Structure* 11:1319, 2003. (b) PDB ID 1A6M, J. Vojtechovsky et al., *Biophys. J.* 77:2153, 1999.]

11. Pathogenic Action of Bacteria That Cause Gas Gangrene The highly pathogenic anaerobic bacterium *Clostridium perfringens* is responsible for gas gangrene, a condition in which animal tissue structure is destroyed. This bacterium secretes an enzyme that efficiently catalyzes the hydrolysis of the peptide bond indicated in red:

$$-X-Gly-Pro-Y- \xrightarrow{H_2O}$$
$$-X-COO^- + H_3\overset{+}{N}-Gly-Pro-Y-$$

where X and Y are any of the 20 common amino acids. How does the secretion of this enzyme contribute to the invasiveness of this bacterium in human tissues? Why does this enzyme not affect the bacterium itself?

12. Number of Polypeptide Chains in a Multisubunit Protein A sample (660 mg) of an oligomeric protein of M_r 132,000 was treated with an excess of 1-fluoro-2,4-dinitrobenzene (Sanger's reagent) under slightly alkaline conditions until the chemical reaction was complete. The peptide bonds of the protein were then completely hydrolyzed by heating it with concentrated HCl. The hydrolysate was found to contain 5.5 mg of the following compound:

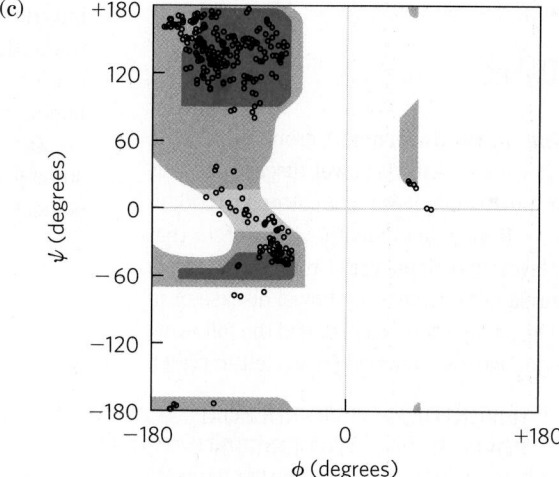

(a)

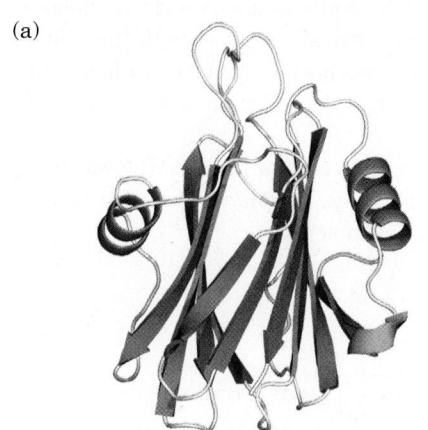

(b)

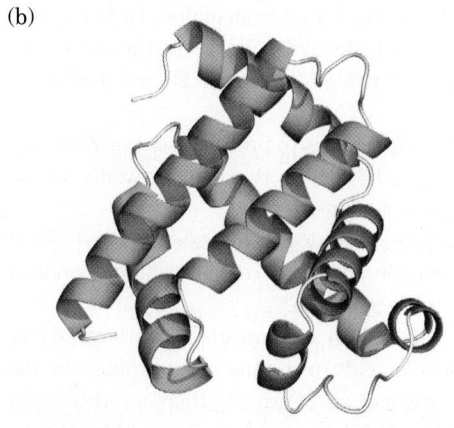

(c)

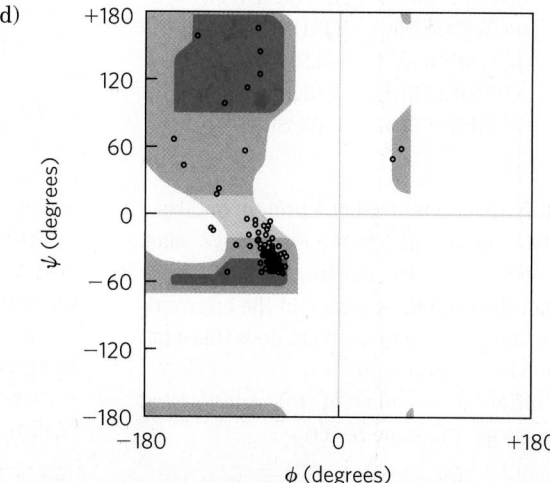

2,4-Dinitrophenyl derivatives of the α-amino groups of other amino acids could not be found.

(a) Explain how this information can be used to determine the number of polypeptide chains in an oligomeric protein.

(b) Calculate the number of polypeptide chains in this protein.

(c) What other analytic technique could you employ to determine whether the polypeptide chains in this protein are similar or different?

13. Predicting Secondary Structure Which of the following peptides is more likely to take up an α-helical structure, and why?

(a) LKAENDEAARAMSEA

(b) CRAGGFPWDQPGTSN

14. Amyloid Fibers in Disease Several small aromatic molecules, such as phenol red (used as a nontoxic drug model), have been shown to inhibit the formation of amyloid in laboratory model systems. A goal of the research on these small aromatic compounds is to find a drug that would efficiently inhibit the formation of amyloid in the brain in people with incipient Alzheimer disease.

(a) Suggest why molecules with aromatic substituents would disrupt the formation of amyloid.

(b) Some researchers have suggested that a drug used to treat Alzheimer disease may also be effective in treating type 2 (non-insulin-dependent) diabetes mellitus. Why might a single drug be effective in treating these two different conditions?

Biochemistry Online

15. Protein Modeling on the Internet A group of patients with Crohn disease (an inflammatory bowel disease) underwent biopsies of their intestinal mucosa in an attempt to identify the causative agent. Researchers identified a protein that was present at higher levels in patients with Crohn disease than in patients with an unrelated inflammatory bowel disease or in unaffected controls. The protein was isolated, and the following *partial* amino acid sequence was obtained (reads left to right):

EAELCPDRCI	HSFQNLGIQC	VKKRDLEQAI
SQRIQTNNNP	FQVPIEEQRG	DYDLNAVRLC
FQVTVRDPSG	RPLRLPPVLP	HPIFDNRAPN
TAELKICRVN	RNSGSCLGGD	EIFLLCDKVQ
KEDIEVYFTG	PGWEARGSFS	QADVHRQVAI
VFRTPPYADP	SLQAPVRVSM	QLRRPSDREL
SEPMEFQYLP	DTDDRHRIEE	KRKRTYETFK
SIMKKSPFSG	PTDPRPPPRR	IAVPSRSSAS
VPKPAPQPYP		

(a) You can identify this protein using a protein database such as UniProt (www.uniprot.org). On the home page, click on the link for a BLAST search. On the BLAST page, enter about 30 residues from the protein sequence in the appropriate search field and submit it for analysis. What does this analysis tell you about the identity of the protein?

(b) Try using different portions of the amino acid sequence. Do you always get the same result?

(c) A variety of websites provide information about the three-dimensional structure of proteins. Find information about the protein's secondary, tertiary, and quaternary structures using database sites such as the Protein Data Bank (PDB; www.pdb.org) or Structural Classification of Proteins (SCOP2; http://scop2.mrc-lmb.cam.ac.uk).

(d) In the course of your Web searches, what did you learn about the cellular function of the protein?

Data Analysis Problem

16. Mirror-Image Proteins As noted in Chapter 3, "The amino acid residues in protein molecules are exclusively L stereoisomers." It is not clear whether this selectivity is necessary for proper protein function or is an accident of evolution. To explore this question, Milton and colleagues (1992) published a study of an enzyme made entirely of D stereoisomers. The enzyme they chose was HIV protease, a proteolytic enzyme made by HIV that converts inactive viral preproteins to their active forms.

Previously, Wlodawer and coworkers (1989) had reported the complete chemical synthesis of HIV protease from L-amino acids (the L-enzyme), using the process shown in Figure 3-32. Normal HIV protease contains two Cys residues, at positions 67 and 95. Because chemical synthesis of proteins containing Cys is technically difficult, Wlodawer and colleagues substituted the synthetic amino acid L-α-amino-n-butyric acid (Aba) for the two Cys residues in the protein. In the authors' words, this was done to "reduce synthetic difficulties associated with Cys deprotection and ease product handling."

(a) The structure of Aba is shown below. Why was this a suitable substitution for a Cys residue? Under what circumstances would it not be suitable?

L-α-Amino-n-butyric acid

Wlodawer and coworkers denatured the newly synthesized protein by dissolving it in 6 M guanidine HCl and then allowed it to fold slowly by dialyzing away the guanidine against a neutral buffer (10% glycerol, 25 mM NaH_2PO_4/Na_2HPO_4, pH 7).

(b) There are many reasons to predict that a protein synthesized, denatured, and folded in this manner would not be active. Give three such reasons.

(c) Interestingly, the resulting L-protease was active. What does this finding tell you about the role of disulfide bonds in the native HIV protease molecule?

In their new study, Milton and coworkers synthesized HIV protease from D-amino acids, using the same protocol as the earlier study (Wlodawer et al.). Formally, there are three possibilities for the folding of the D-protease: it would be (1) the same shape as the L-protease, (2) the mirror image of the L-protease, or (3) something else, possibly inactive.

(d) For each possibility, decide whether or not it is a likely outcome, and defend your position.

In fact, the D-protease was active: it cleaved a particular synthetic substrate and was inhibited by specific inhibitors. To examine the structure of the D- and L-enzymes, Milton and coworkers tested both forms for activity with D and L forms of a chiral peptide substrate and for inhibition by D and L forms of a chiral peptide-analog inhibitor. Both forms were also tested for inhibition by the achiral inhibitor Evans blue. The findings are given in the table.

HIV	Substrate hydrolysis		Inhibition		
			Peptide inhibitor		Evans blue
Protease	D-substrate	L-substrate	D-inhibitor	L-inhibitor	(achiral)
L-protease	−	+	−	+	+
D-protease	+	−	+	−	+

(e) Which of the three models proposed above is supported by these data? Explain your reasoning.

(f) Why does Evans blue inhibit both forms of the protease?

(g) Would you expect chymotrypsin to digest the D-protease? Explain your reasoning.

(h) Would you expect total synthesis from D-amino acids followed by renaturation to yield active enzyme for any enzyme? Explain your reasoning.

References

Milton, R.C., S.C. Milton, and S.B. Kent. 1992. Total chemical synthesis of a D-enzyme: the enantiomers of HIV-1 protease show demonstration of reciprocal chiral substrate specificity. *Science* 256:1445–1448.

Wlodawer, A., M. Miller, M. Jaskólski, B.K. Sathyanarayana, E. Baldwin, I.T. Weber, L.M. Selk, L. Clawson, J. Schneider, and S.B. Kent. 1989. Conserved folding in retroviral proteases: crystal structure of a synthetic HIV-1 protease. *Science* 245:616–621.

Further Reading is available at www.macmillanlearning.com/LehningerBiochemistry7e.

CHAPTER 5

Protein Function

Self-study tools that will help you practice what you've learned and reinforce this chapter's concepts are available online.
Go to www.macmillanlearning.com/LehningerBiochemistry7e.

Proteins function by interacting with other molecules. Knowing the three-dimensional structure of a protein is an important part of understanding protein function, and modern structural biology often includes insights into molecular interactions. However, the protein structures we have examined so far are deceptively static. Proteins are dynamic molecules. Their interactions are affected in physiologically important ways by sometimes subtle, sometimes striking changes in protein conformation. In this chapter and the next, we explore how proteins interact with other molecules and how their interactions are related to dynamic protein structure. We divide these interactions into two types. In some interactions, the result is a reaction that alters the chemical configuration or composition of the interacting molecule, with the protein acting as a reaction catalyst, or **enzyme**; we discuss enzymes and their reactions in Chapter 6. In other interactions, neither the chemical configuration nor the composition of the interacting molecule is changed, and such interactions are the subject of this chapter.

It may seem counterintuitive that a protein's interaction with another molecule could be important if it does not alter the associated molecule. Yet, transient interactions of this type are at the heart of complex physiological processes such as oxygen transport, immune function, and muscle contraction—all topics we examine here. The proteins that carry out these processes illustrate several key principles of protein function, some of which will be familiar from Chapter 4:

The functions of many proteins involve the reversible binding of other molecules. A molecule bound reversibly by a protein is called a **ligand**. A ligand may be any kind of molecule, including another protein. The transient nature of protein-ligand interactions is critical to life, allowing an organism to respond rapidly and reversibly to changing environmental and metabolic circumstances.

A ligand binds at a site on the protein called the **binding site**, which is complementary to the ligand in size, shape, charge, and hydrophobic or hydrophilic character. Furthermore, the interaction is specific: the protein can discriminate among the thousands of different molecules in its environment and selectively bind only one or a few types. A given protein may have separate binding sites for several different ligands. These specific molecular interactions are crucial in maintaining the high degree of order in a living system. (This discussion excludes the binding of water, which may interact weakly and nonspecifically with many parts of a protein. In Chapter 6, we consider water as a specific ligand for many enzymes.)

Proteins are flexible. Changes in conformation may be subtle, reflecting molecular vibrations and small movements of amino acid residues throughout the protein. A protein flexing in this way is sometimes said to "breathe." Changes in conformation may also be more dramatic, with major segments of the protein structure moving as much as several nanometers. Specific conformational changes are frequently essential to a protein's function.

The binding of a protein and ligand is often coupled to a conformational change in the protein that makes the binding site more complementary to the ligand, permitting tighter binding. The structural adaptation that occurs between protein and ligand is called **induced fit**.

In a multisubunit protein, a conformational change in one subunit often affects the conformation of other subunits.

Interactions between ligands and proteins may be regulated, usually through specific interactions with one or more additional ligands. These other ligands may cause conformational changes in the protein that affect the binding of the first ligand.

The enzymes represent a special case of protein function. They bind and chemically transform other molecules. The molecules acted upon by enzymes are called reaction **substrates** rather than ligands, and the ligand-binding site is called the **catalytic site** or **active site**. As you will see, the themes in our discussion of noncatalytic functions of proteins in this chapter—binding, specificity, and conformational change—are continued in Chapter 6, with the added element of proteins participating in chemical transformations.

5.1 Reversible Binding of a Protein to a Ligand: Oxygen-Binding Proteins

Myoglobin and **hemoglobin** may be the most-studied and best-understood proteins. They were the first proteins for which three-dimensional structures were determined, and these two molecules illustrate almost every aspect of that critical biochemical process: the reversible binding of a ligand to a protein. This classic model of protein function tells us a great deal about how proteins work.

Oxygen Can Bind to a Heme Prosthetic Group

Oxygen is poorly soluble in aqueous solutions (see Table 2-3) and cannot be carried to tissues in sufficient quantity if it is simply dissolved in blood serum.

Also, diffusion of oxygen through tissues is ineffective over distances greater than a few millimeters. The evolution of larger, multicellular animals depended on the evolution of proteins that could transport and store oxygen. However, none of the amino acid side chains in proteins are suited for the reversible binding of oxygen molecules. This role is filled by certain transition metals, among them iron and copper, that have a strong tendency to bind oxygen. Multicellular organisms exploit the properties of metals, most commonly iron, for oxygen transport. However, free iron promotes the formation of highly reactive oxygen species such as hydroxyl radicals that can damage DNA and other macromolecules. Iron used in cells is therefore bound in forms that sequester it and/or make it less reactive. In multicellular organisms—especially those in which iron, in its oxygen-carrying capacity, must be transported over large distances—iron is often incorporated into a protein-bound prosthetic group called **heme** (or haem). (Recall from Chapter 3 that a prosthetic group is a compound permanently associated with a protein that contributes to the protein's function.)

Heme consists of a complex organic ring structure, **protoporphyrin**, to which is bound a single iron atom in its ferrous (Fe^{2+}) state **(Fig. 5-1)**. The iron atom has six coordination bonds, four to nitrogen atoms that are part of the flat **porphyrin ring** system and two perpendicular to the porphyrin. The coordinated nitrogen atoms (which have an electron-donating character) help prevent conversion of the heme iron to the ferric (Fe^{3+}) state. Iron in the Fe^{2+} state binds oxygen reversibly; in the Fe^{3+} state it does not bind oxygen. Heme is found in many oxygen-transporting proteins, as well as in some proteins, such as the cytochromes, that participate in oxidation-reduction (electron-transfer) reactions (Chapter 19).

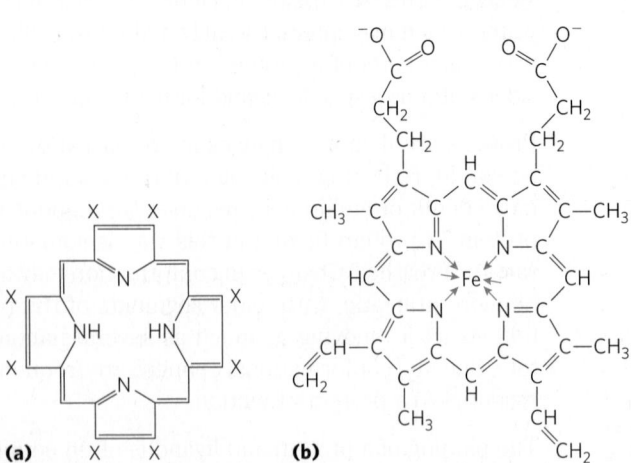

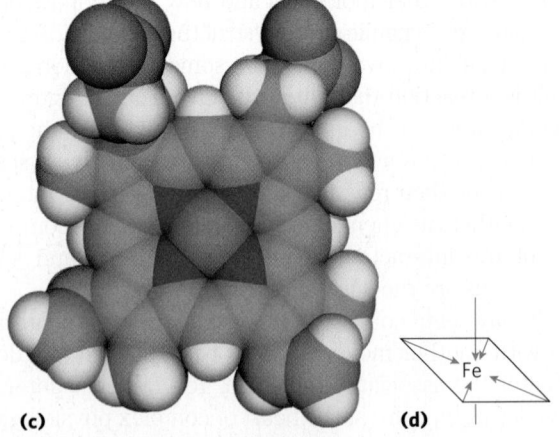

FIGURE 5-1 Heme. The heme group is present in myoglobin, hemoglobin, and many other proteins, designated **heme proteins**. Heme consists of a complex organic ring structure, protoporphyrin IX, with a bound iron atom in its ferrous (Fe^{2+}) state. **(a)** Porphyrins, of which protoporphyrin IX is just one example, consist of four pyrrole rings linked by methene bridges, with substitutions at one or more of the positions denoted X. **(b, c)** Two representations of heme. The iron atom of heme has six coordination bonds: four in the plane of, and bonded to, the flat porphyrin ring system, and **(d)** two perpendicular to it. [Source: (c) Heme extracted from PDB ID 1CCR, H. Ochi et al., *J. Mol. Biol.* 166:407, 1983.]

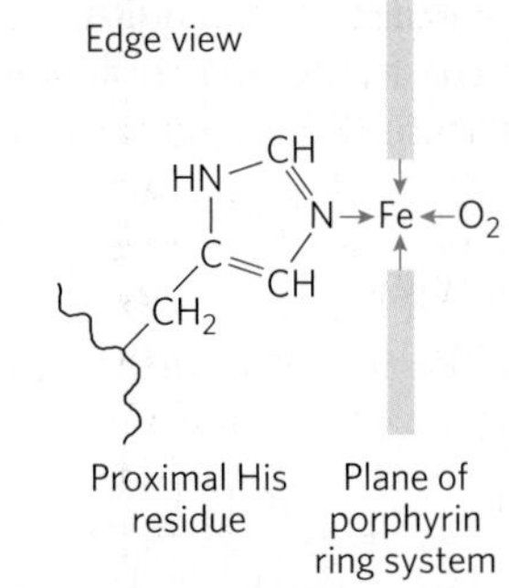

Edge view

Proximal His
residue

Plane of
porphyrin
ring system

FIGURE 5-2 The heme group viewed from the side. This view shows the two coordination bonds to Fe^{2+} that are perpendicular to the porphyrin ring system. One is occupied by a His residue called the proximal His, His^{93} in myoglobin, also designated His F8 (the 8th residue in α helix F; see Fig. 5-3); the other is the binding site for oxygen. The remaining four coordination bonds are in the plane of, and bonded to, the flat porphyrin ring system.

Free heme molecules (heme not bound to protein) leave Fe^{2+} with two "open" coordination bonds. Simultaneous reaction of one O_2 molecule with two free heme molecules (or two free Fe^{2+}) can result in irreversible conversion of Fe^{2+} to Fe^{3+}. In heme-containing proteins, this reaction is prevented by sequestering each heme deep within the protein structure. Thus, access to the two open coordination bonds is restricted. In globins, one of these two coordination bonds is occupied by a side-chain nitrogen of a highly conserved His residue referred to as the **proximal His**. The other is the binding site for molecular oxygen (O_2) **(Fig. 5-2)**. When oxygen binds, the electronic properties of heme iron change; this accounts for the change in color from the dark purple of oxygen-depleted venous blood to the bright red of oxygen-rich arterial blood. Some small molecules, such as carbon monoxide (CO) and nitric oxide (NO), coordinate to heme iron with greater affinity than does O_2. When a molecule of CO is bound to heme, O_2 is excluded, which is why CO is highly toxic to aerobic organisms (a topic explored later, in Box 5-1). By surrounding and sequestering heme, oxygen-binding proteins regulate the access of small molecules to the heme iron.

Globins Are a Family of Oxygen-Binding Proteins

The **globins** are a widespread family of proteins, all having similar primary and tertiary structures. Globins are commonly found in eukaryotes of all classes and even in some bacteria. Most function in oxygen transport or storage, although some play a role in the sensing of oxygen, nitric oxide, or carbon monoxide. The simple nematode worm *Caenorhabditis elegans* has genes encoding 33 different globins. In humans and other mammals, there are at least four kinds of globins. The monomeric myoglobin facilitates oxygen diffusion in muscle tissue. Myoglobin is particularly abundant in the muscles of diving marine mammals such as seals and whales, where it also has an oxygen-storage function for prolonged excursions undersea. The tetrameric hemoglobin is responsible for oxygen transport

in the bloodstream. The monomeric neuroglobin is expressed largely in neurons and helps to protect the brain from hypoxia (low oxygen) or ischemia (restricted blood supply). Cytoglobin, another monomeric globin, is found at high concentrations in the walls of blood vessels, where it functions to regulate levels of nitric oxide (discussed in Chapters 12 and 23).

Myoglobin Has a Single Binding Site for Oxygen

Myoglobin (M_r 16,700; abbreviated Mb) is a single polypeptide of 153 amino acid residues with one molecule of heme. As is typical for a globin polypeptide, myoglobin is made up of eight α-helical segments connected by bends **(Fig. 5-3)**. About 78% of the amino acid residues in the protein are found in these α helices.

Any detailed discussion of protein function inevitably involves protein structure. In the case of myoglobin, we first introduce some structural conventions peculiar to globins. As seen in Figure 5-3, the helical segments are named A through H. An individual amino acid residue is designated either by its position in the amino acid sequence or by its location in the sequence of a particular α-helical segment. For example, the His residue coordinated to the heme in myoglobin—the proximal His—is His^{93} (the 93rd residue from the amino-terminal end of the myoglobin polypeptide sequence) and is also called His F8 (the 8th residue in α helix F). The bends in the structure are designated AB, CD, EF, FG, and so forth, reflecting the α-helical segments they connect.

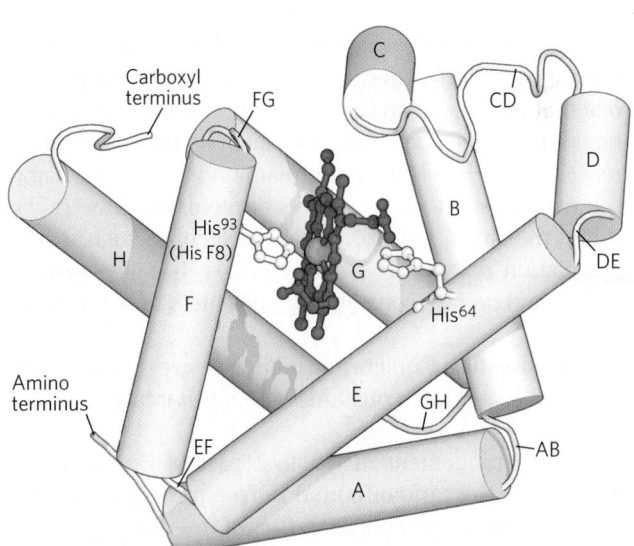

FIGURE 5-3 Myoglobin. The eight α-helical segments (shown here as cylinders) are labeled A through H. Nonhelical residues in the bends that connect them are labeled AB, CD, EF, and so forth, indicating the segments they interconnect. A few bends, including BC and DE, are abrupt and do not contain any residues; these are not normally labeled. The heme is bound in a pocket made up largely of the E and F helices, although amino acid residues from other segments of the protein also participate. [Source: PDB ID 1MBO, S. E. Phillips, *J. Mol. Biol.* 142:531, 1980.]

Protein-Ligand Interactions Can Be Described Quantitatively

The function of myoglobin depends on the protein's ability not only to bind oxygen but also to release it when and where it is needed. Function in biochemistry often revolves around a reversible protein-ligand interaction of this type. A quantitative description of this interaction is a central part of many biochemical investigations.

In general, the reversible binding of a protein (P) to a ligand (L) can be described by a simple **equilibrium expression**:

$$P + L \rightleftharpoons PL \tag{5-1}$$

The reaction is characterized by an equilibrium constant, K_a, such that

$$K_a = \frac{[PL]}{[P][L]} = \frac{k_a}{k_d} \tag{5-2}$$

where k_a and k_d are rate constants (more on these below). The term $\boldsymbol{K_a}$ is an **association constant** (not to be confused with the K_a that denotes an acid dissociation constant; p. 62) that describes the equilibrium between the complex and the unbound components of the complex. The association constant provides a measure of the affinity of the ligand L for the protein. K_a has units of M^{-1}; a higher value of K_a corresponds to a higher affinity of the ligand for the protein.

The equilibrium term K_a is also equivalent to the ratio of the rates of the forward (association) and reverse (dissociation) reactions that form the PL complex. The association rate is described by the rate constant k_a, and dissociation by the rate constant k_d. As discussed further in the next chapter, rate constants are proportionality constants, describing the fraction of a pool of reactant that reacts in a given amount of time. When the reaction involves one molecule, such as the dissociation reaction $PL \longrightarrow P + L$, the reaction is *first order* and the rate constant (k_d) has units of reciprocal time (s^{-1}). When the reaction involves two molecules, such as the association reaction $P + L \longrightarrow PL$, it is called *second order*, and its rate constant (k_a) has units of $M^{-1} s^{-1}$.

≫ Key Convention: Equilibrium constants are denoted with a capital K and rate constants with a lowercase k. **≪**

A rearrangement of the first part of Equation 5-2 shows that the ratio of bound to free protein is directly proportional to the concentration of free ligand:

$$K_a[L] = \frac{[PL]}{[P]} \tag{5-3}$$

When the concentration of the ligand is much greater than the concentration of ligand-binding sites, the binding of the ligand by the protein does not appreciably change the concentration of free (unbound) ligand—that is, [L] remains constant. This condition is broadly applicable to most ligands that bind to proteins in cells and simplifies our description of the binding equilibrium.

We can now consider the binding equilibrium from the standpoint of the fraction, Y, of ligand-binding sites on the protein that are occupied by ligand:

$$Y = \frac{\text{binding sites occupied}}{\text{total binding sites}} = \frac{[PL]}{[PL] + [P]} \tag{5-4}$$

Substituting $K_a[L][P]$ for $[PL]$ (see Eqn 5-3) and rearranging terms gives

$$Y = \frac{K_a[L][P]}{K_a[L][P] + [P]} = \frac{K_a[L]}{K_a[L] + 1} = \frac{[L]}{[L] + \dfrac{1}{K_a}} \tag{5-5}$$

The value of K_a can be determined from a plot of Y versus the concentration of free ligand, [L] (**Fig. 5-4a**). Any equation of the form $x = y/(y + z)$ describes a hyperbola, and Y is thus found to be a hyperbolic function of [L]. The fraction of ligand-binding sites occupied approaches saturation asymptotically as [L] increases. The [L] at which half of the available ligand-binding sites are occupied (that is, $Y = 0.5$) corresponds to $1/K_a$.

It is more common (and intuitively simpler), however, to consider the **dissociation constant, K_d**, which is the reciprocal of K_a ($K_d = 1/K_a$) and has units of molar

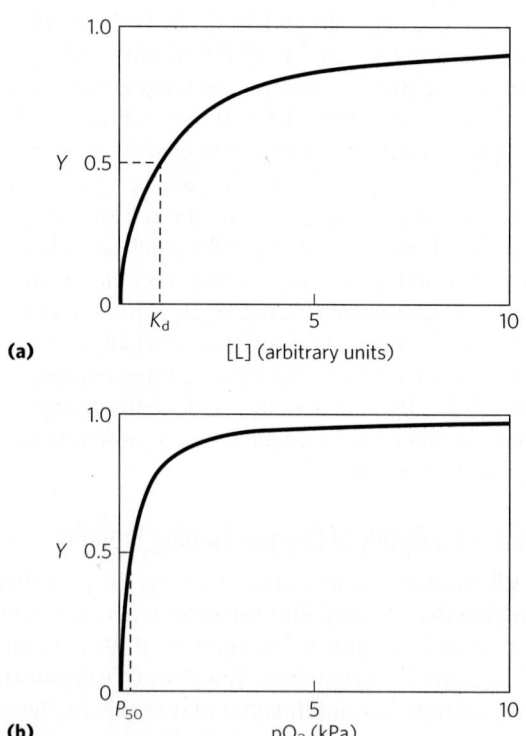

(a)   [L] (arbitrary units)

(b)   pO$_2$ (kPa)

FIGURE 5-4 Graphical representations of ligand binding. The fraction of ligand-binding sites occupied, Y, is plotted against the concentration of free ligand. Both curves are rectangular hyperbolas. **(a)** A hypothetical binding curve for a ligand L. The [L] at which half of the available ligand-binding sites are occupied is equivalent to $1/K_a$, or K_d. The curve has a horizontal asymptote at $Y = 1$ and a vertical asymptote (not shown) at [L] = $-1/K_a$. **(b)** A curve describing the binding of oxygen to myoglobin. The partial pressure of O_2 in the air above the solution is expressed in kilopascals (kPa). Oxygen binds tightly to myoglobin, with a P_{50} of only 0.26 kPa.

TABLE 5-1 Protein Dissociation Constants: Some Examples and Range

Protein	Ligand	K_d (M)[a]
Avidin (egg white)	Biotin	1×10^{-15}
Insulin receptor (human)	Insulin	1×10^{-10}
Anti-HIV immunoglobulin (human)[b]	gp41 (HIV-1 surface protein)	4×10^{-10}
Nickel-binding protein (*E. coli*)	Ni^{2+}	1×10^{-7}
Calmodulin (rat)[c]	Ca^{2+}	3×10^{-6}
		2×10^{-5}

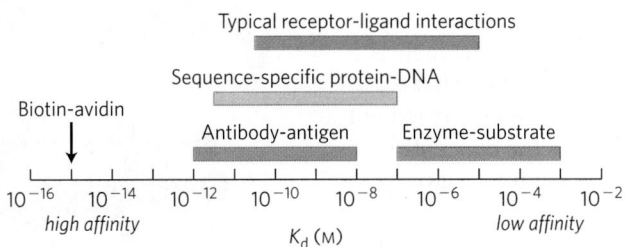

Color bars indicate the range of dissociation constants typical of various classes of interactions in biological systems. A few interactions, such as that between the protein avidin and the enzyme cofactor biotin, fall outside the normal ranges. The avidin-biotin interaction is so tight it may be considered irreversible. Sequence-specific protein-DNA interactions reflect proteins that bind to a particular sequence of nucleotides in DNA, as opposed to general binding to any DNA site.

[a]A reported dissociation constant is valid only for the particular solution conditions under which it was measured. K_d values for a protein–ligand interaction can be altered, sometimes by several orders of magnitude, by changes in the solution's salt concentration, pH, or other variables.

[b]This immunoglobulin was isolated as part of an effort to develop a vaccine against HIV. Immunoglobulins (described later in the chapter) are highly variable, and the K_d reported here should not be considered characteristic of all immunoglobulins.

[c]Calmodulin has four binding sites for calcium. The values shown reflect the highest- and lowest-affinity binding sites observed in one set of measurements.

concentration (M). K_d is the equilibrium constant for the release of ligand. The relevant expressions change to

$$K_d = \frac{[P][L]}{[PL]} = \frac{k_d}{k_a} \tag{5-6}$$

$$[PL] = \frac{[P][L]}{K_d} \tag{5-7}$$

$$Y = \frac{[L]}{[L] + K_d} \tag{5-8}$$

When [L] equals K_d, half of the ligand-binding sites are occupied. As [L] falls below K_d, progressively less of the protein has ligand bound to it. For 90% of the available ligand-binding sites to be occupied, [L] must be nine times greater than K_d.

In practice, K_d is used much more often than K_a to express the affinity of a protein for a ligand. Note that a lower value of K_d corresponds to a higher affinity of ligand for the protein. The mathematics can be reduced to simple statements: K_d is equivalent to the molar concentration of ligand at which half of the available ligand-binding sites are occupied. At this point, the protein is said to have reached half-saturation with respect to ligand binding. The more tightly a protein binds a ligand, the lower the concentration of ligand

required for half the binding sites to be occupied, and thus the lower the value of K_d. Some representative dissociation constants are given in Table 5-1; the scale shows typical ranges for dissociation constants found in biological systems.

WORKED EXAMPLE 5-1 **Receptor-Ligand Dissociation Constants**

Two proteins, A and B, bind to the same ligand, L, with the binding curves shown below.

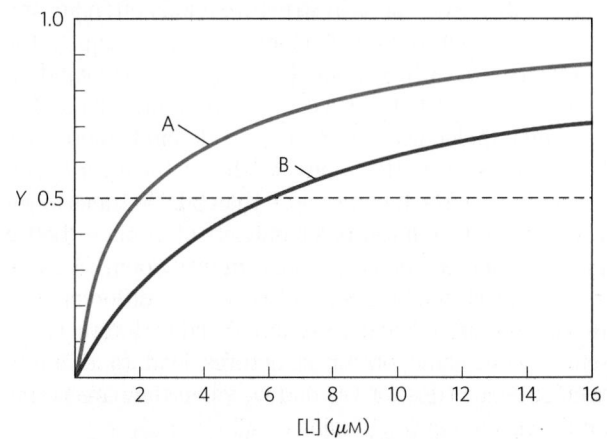

What is the dissociation constant, K_d, for each protein? Which protein (A or B) has a greater affinity for ligand L?

Solution: We can determine the dissociation constants by inspecting the graph. Since Y represents the fraction of binding sites occupied by ligand, the concentration of ligand at which half the binding sites are occupied—that is, the point where the binding curve crosses the line where $Y = 0.5$—is the dissociation constant. For A, $K_d = 2$ μM; for B, $K_d = 6$ μM. Because A is half-saturated at a lower [L], it has a higher affinity for the ligand.

The binding of oxygen to myoglobin follows the patterns discussed above. However, because oxygen is a gas, we must make some minor adjustments to the equations so that laboratory experiments can be carried out more conveniently. We first substitute the concentration of dissolved oxygen for [L] in Equation 5-8 to give

$$Y = \frac{[O_2]}{[O_2] + K_d} \tag{5-9}$$

As for any ligand, K_d equals the $[O_2]$ at which half of the available ligand-binding sites are occupied, or $[O_2]_{0.5}$. Equation 5-9 thus becomes

$$Y = \frac{[O_2]}{[O_2] + [O_2]_{0.5}} \tag{5-10}$$

In experiments using oxygen as a ligand, it is the partial pressure of oxygen (pO_2) in the gas phase above the solution that is varied, because this is easier to measure than the concentration of oxygen dissolved in the solution. The concentration of a volatile substance in solution is always proportional to the local partial pressure of the gas. So, if we define the partial pressure of oxygen at $[O_2]_{0.5}$ as P_{50}, substitution in Equation 5-10 gives

$$Y = \frac{pO_2}{pO_2 + P_{50}} \tag{5-11}$$

A binding curve for myoglobin that relates Y to pO_2 is shown in Figure 5-4b.

Protein Structure Affects How Ligands Bind

The binding of a ligand to a protein is rarely as simple as the above equations would suggest. The interaction is greatly affected by protein structure and is often accompanied by conformational changes. For example, the specificity with which heme binds its various ligands is altered when the heme is a component of myoglobin. For free heme molecules, carbon monoxide binds more than 20,000 times better than does O_2 (that is, the K_d or P_{50} for CO binding to free heme is more than 20,000 times lower than that for O_2), but it binds only about 40 times better than O_2 when the heme is bound in myoglobin. For free heme, the tighter binding by CO reflects differences in the way the orbital structures of CO and O_2 interact with Fe^{2+}. Those same orbital structures lead to different binding geometries for CO and O_2 when they are bound

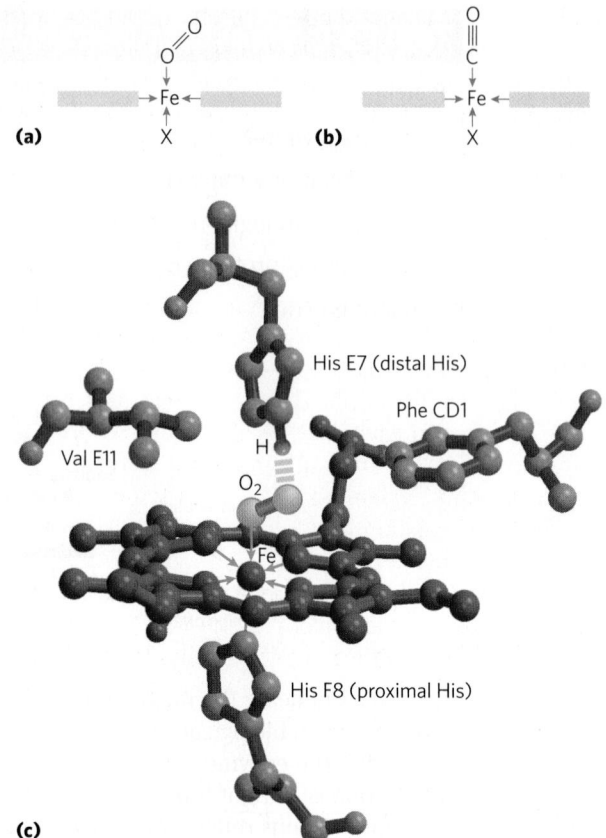

FIGURE 5-5 Steric effects caused by ligand binding to the heme of myoglobin. (a) Oxygen binds to heme with the O_2 axis at an angle, a binding conformation readily accommodated by myoglobin. **(b)** Carbon monoxide binds to free heme with the CO axis perpendicular to the plane of the porphyrin ring. **(c)** Another view of the heme of myoglobin, showing the arrangement of key amino acid residues around the heme. The bound O_2 is hydrogen-bonded to the distal His, His E7 (His[64]), facilitating the binding of O_2 compared with its binding to free heme. [Source: (c) Derived from PDB ID 1MBO, S. E. Phillips, *J. Mol. Biol.* 142:531, 1980.]

to heme **(Fig. 5-5a, b)**. The change in relative affinity of CO and O_2 for heme when the heme is bound to a globin is mediated by the globin structure.

When heme is bound to myoglobin, its affinity for O_2 is selectively increased by the presence of the **distal His** (His[64], or His E7 in myoglobin). The Fe-O_2 complex is much more polar than the Fe-CO complex. There is a partial negative charge distributed across the oxygen atoms in the bound O_2 due to partial oxidation of the interacting iron atom. A hydrogen bond between the imidazole side chain of His E7 and the bound O_2 stabilizes this polar complex electrostatically (Fig. 5-5c). The affinity of myoglobin for O_2 is thus selectively increased by a factor of about 500; there is no such effect for Fe-CO binding in myoglobin. Consequently, the 20,000-fold stronger binding affinity of free heme for CO compared with O_2 declines to approximately 40-fold for heme embedded in myoglobin. This favorable electrostatic effect on O_2 binding is even more dramatic in some invertebrate hemoglobins, where two groups in the binding pocket can form strong hydrogen bonds with O_2, causing the heme group to bind O_2 with greater affinity than CO. This selective enhancement of O_2 affinity in globins is physiologically important and helps prevent poisoning by the CO generated from heme catabolism (see Chapter 22) or other sources.

The binding of O_2 to the heme in myoglobin also depends on molecular motions, or "breathing," in the protein structure. The heme molecule is deeply buried in the folded polypeptide, with limited direct paths for oxygen to move from the surrounding solution to the ligand-binding site. If the protein were rigid, O_2 could not readily enter or leave the heme pocket. However, rapid molecular flexing of the amino acid side chains produces transient cavities in the protein structure, and O_2 makes its way in and out by moving through these cavities. Computer simulations of rapid structural fluctuations in myoglobin suggest there are many such pathways. The distal His acts as a gate to control access to one major pocket near the heme iron. Rotation of that His residue to open and close the pocket occurs on a nanosecond (10^{-9} s) time scale. Even subtle conformational changes can be critical for protein activity.

The distal His functions somewhat differently in some other globins. In neuroglobin, cytoglobin, and some globins found in plants and invertebrates, the distal His is directly coordinated with the heme iron at the location where ligands must bind. In these globins, the O_2 or other ligand must displace the distal His in the process of binding, with a hydrogen bond again forming between the distal His and O_2 after the binding occurs.

Hemoglobin Transports Oxygen in Blood

Nearly all the oxygen carried by whole blood in animals is bound and transported by hemoglobin in erythrocytes (red blood cells). Normal human erythrocytes are small (6 to 9 μm in diameter), biconcave disks. They are formed from precursor stem cells called **hemocytoblasts**. In the maturation process, the stem cell produces daughter cells that form large amounts of hemoglobin and then lose their organelles—nucleus, mitochondria, and endoplasmic reticulum. Erythrocytes are thus incomplete, vestigial cells, unable to reproduce and, in humans, destined to survive for only about 120 days. Their main function is to carry hemoglobin, which is dissolved in the cytosol at a very high concentration (~34% by weight).

In arterial blood passing from the lungs through the heart to the peripheral tissues, hemoglobin is about 96% saturated with oxygen. In the venous blood returning to the heart, hemoglobin is only about 64% saturated. Thus, each 100 mL of blood passing through a tissue releases about one-third of the oxygen it carries, or 6.5 mL of O_2 gas at atmospheric pressure and body temperature.

Myoglobin, with its hyperbolic binding curve for oxygen (Fig. 5-4b), is relatively insensitive to small changes in the concentration of dissolved oxygen and so functions well as an oxygen-storage protein. Hemoglobin, with its multiple subunits and O_2-binding sites, is better suited to

oxygen transport. As we shall see, interactions between the subunits of a multimeric protein can permit a highly sensitive response to small changes in ligand concentration. Interactions among the subunits in hemoglobin cause conformational changes that alter the affinity of the protein for oxygen. The modulation of oxygen binding allows the O_2-transport protein to respond to changes in oxygen demand by tissues.

Hemoglobin Subunits Are Structurally Similar to Myoglobin

Hemoglobin (M_r 64,500; abbreviated Hb) is roughly spherical, with a diameter of nearly 5.5 nm. It is a tetrameric protein containing four heme prosthetic groups, one associated with each polypeptide chain. Adult hemoglobin contains two types of globin, two α chains (141 residues each) and two β chains (146 residues each). Although fewer than half of the amino acid residues are identical in the polypeptide sequences of the α and β subunits, the three-dimensional structures of the two types of subunits are very similar. Furthermore, their structures are very similar to that of myoglobin **(Fig. 5-6)**, even though the amino acid sequences of the three polypeptides are identical at only 27 positions **(Fig. 5-7)**. All three polypeptides are members of the globin family of proteins. The helix-naming convention described for myoglobin is also applied to the hemoglobin polypeptides, except that the α subunit lacks the short D helix. The heme-binding pocket is made up largely of the E and F helices in each of the subunits.

The quaternary structure of hemoglobin features strong interactions between unlike subunits. The $\alpha_1\beta_1$ interface (and its $\alpha_2\beta_2$ counterpart) involves more than 30 residues, and its interaction is sufficiently strong that although mild treatment of hemoglobin with urea tends to disassemble the tetramer into $\alpha\beta$ dimers, these dimers remain intact. The $\alpha_1\beta_2$ (and $\alpha_2\beta_1$) interface involves 19 residues **(Fig. 5-8)**. The hydrophobic effect

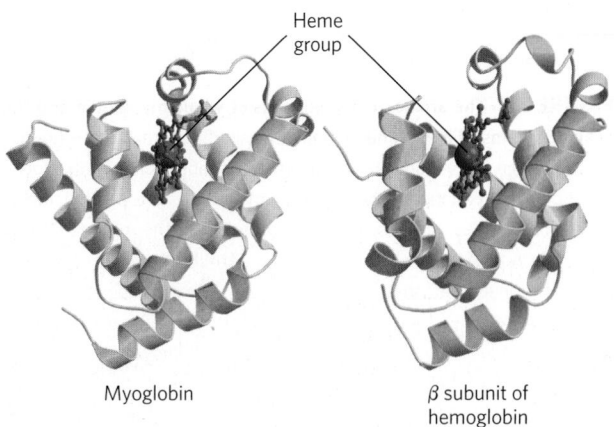

FIGURE 5-6 Comparison of the structures of myoglobin and the β subunit of hemoglobin. [Sources: (left) PDB ID 1MBO, S. E. Phillips, *J. Mol. Biol.* 142:531, 1980. (right) Derived from PDB ID 1HGA, R. Liddington et al., *J. Mol. Biol.* 228:551, 1992.]

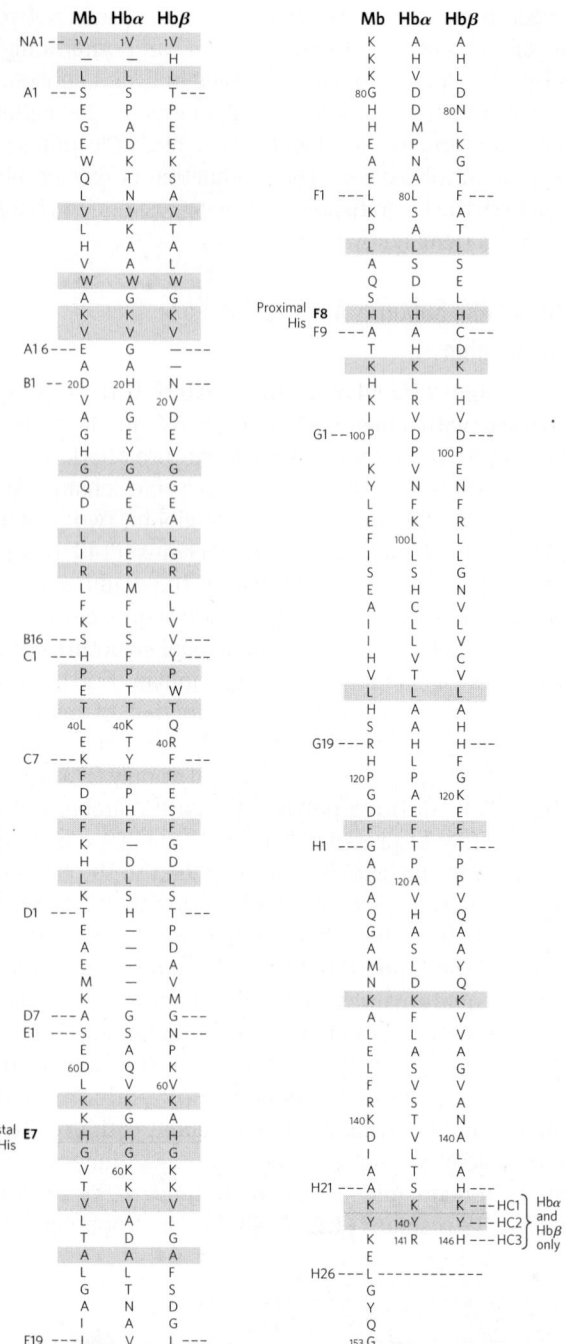

FIGURE 5-7 The amino acid sequences of whale myoglobin and the α and β chains of human hemoglobin. Dashed lines mark helix boundaries. To align the sequences optimally, short gaps must be introduced into both Hb sequences where a few amino acids are present in the other, compared sequences. With the exception of the missing D helix in the Hb α chain (Hbα), this alignment permits the use of the helix lettering convention that emphasizes the common positioning of amino acid residues that are identical in all three structures (shaded). Residues shaded in light red are conserved in all known globins. Note that the common helix-letter-and-number designation for amino acids does not necessarily correspond to a common position in the linear sequence of amino acids in the polypeptides. For example, the distal His residue is His E7 in all three structures, but corresponds to His[64], His[58], and His[63] in the linear sequences of Mb, Hbα, and Hbβ, respectively. Nonhelical residues at the amino and carboxyl termini, beyond the first (A) and last (H) α-helical segments, are labeled NA and HC, respectively.

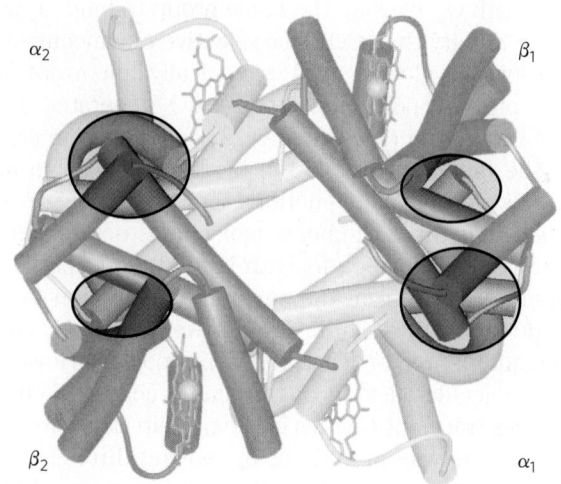

FIGURE 5-8 Dominant interactions between hemoglobin subunits. In this representation, α subunits are light and β subunits are dark. The strongest subunit interactions (highlighted) occur between unlike subunits. When oxygen binds, the $\alpha_1\beta_1$ contact changes little, but there is a large change at the $\alpha_1\beta_2$ contact, with several ion pairs broken. [Source: PDB ID 1HGA, R. Liddington et al., *J. Mol. Biol.* 228:551, 1992.]

plays the major role in stabilizing these interfaces, but there are also many hydrogen bonds and a few ion pairs (or salt bridges), whose importance is discussed below.

Hemoglobin Undergoes a Structural Change on Binding Oxygen

X-ray analysis has revealed two major conformations of hemoglobin: the **R state** and the **T state**. Although oxygen binds to hemoglobin in either state, it has a significantly higher affinity for hemoglobin in the R state. Oxygen binding stabilizes the R state. When oxygen is absent experimentally, the T state is more stable and is thus the predominant conformation of **deoxyhemoglobin**. T and R originally denoted "tense" and "relaxed," respectively, because the T state is stabilized by a greater number of ion pairs, many of which lie at the $\alpha_1\beta_2$ (and $\alpha_2\beta_1$) interface **(Fig. 5-9)**. The binding of O_2 to a hemoglobin subunit in the T state triggers a change in conformation to the R state. When the entire protein undergoes this transition, the structures of the individual subunits change little, but the αβ subunit pairs slide past each other and rotate, narrowing the pocket between the β subunits **(Fig. 5-10)**. In this process, some of the ion pairs that stabilize the T state are broken and some new ones are formed.

Max Perutz proposed that the T → R transition is triggered by changes in the positions of key amino acid side chains surrounding the heme. In the T state, the porphyrin is slightly puckered, causing the heme iron to protrude somewhat on the proximal His (His F8) side. The binding of O_2 causes the heme to assume a more planar conformation, shifting the position of the proximal His and the attached F helix **(Fig. 5-11)**. These changes lead to adjustments in the ion pairs at the $\alpha_1\beta_2$ interface.

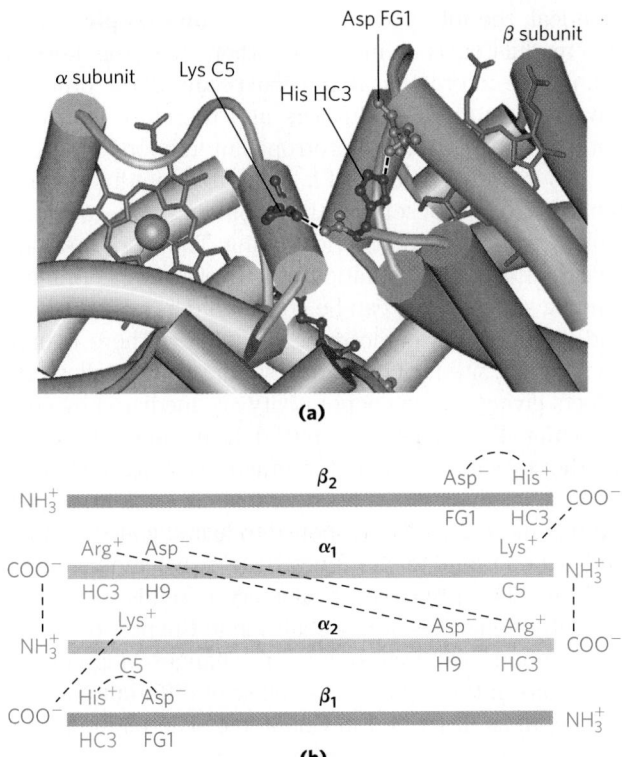

(a)

(b)

FIGURE 5-9 Some ion pairs that stabilize the T state of deoxyhemo-globin. (a) Close-up view of a portion of a deoxyhemoglobin molecule in the T state. Interactions between the ion pairs His HC3 and Asp FG1 of the β subunit (blue) and between Lys C5 of the α subunit (gray) and His HC3 (its α-carboxyl group) of the β subunit are shown with dashed lines. (Recall that HC3 is the carboxyl-terminal residue of the β subunit.) **(b)** Interactions between these ion pairs, and between others not shown in (a), are schematized in this representation of the extended polypeptide chains of hemoglobin. [Source: (a) PDB ID 1HGA, R. Liddington et al., *J. Mol. Biol.* 228:551, 1992.]

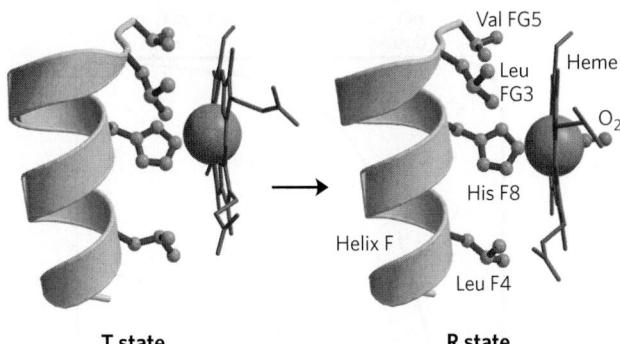

FIGURE 5-11 Changes in conformation near heme on O_2 binding to deoxyhemoglobin. The shift in the position of helix F when heme binds O_2 is thought to be one of the adjustments that triggers the T → R transition. [Sources: T state: derived from PDB ID 1HGA, R. Liddington et al., *J. Mol. Biol.* 228:551, 1992. R state: derived from PDB ID 1BBB, M. M. Silva et al., *J. Biol. Chem.* 267:17,248, 1992; R state modified to represent O_2 instead of CO.]

Hemoglobin Binds Oxygen Cooperatively

Hemoglobin must bind oxygen efficiently in the lungs, where the pO_2 is about 13.3 kPa, and release oxygen in the tissues, where the pO_2 is about 4 kPa. Myoglobin, or any protein that binds oxygen with a hyperbolic binding curve, would be ill-suited to this function, for the reason illustrated in **Figure 5-12**. A protein that bound O_2 with high affinity would bind it efficiently in the lungs but would not release much of it in the tissues. If the protein bound oxygen with a sufficiently low affinity to release it in the tissues, it would not pick up much oxygen in the lungs.

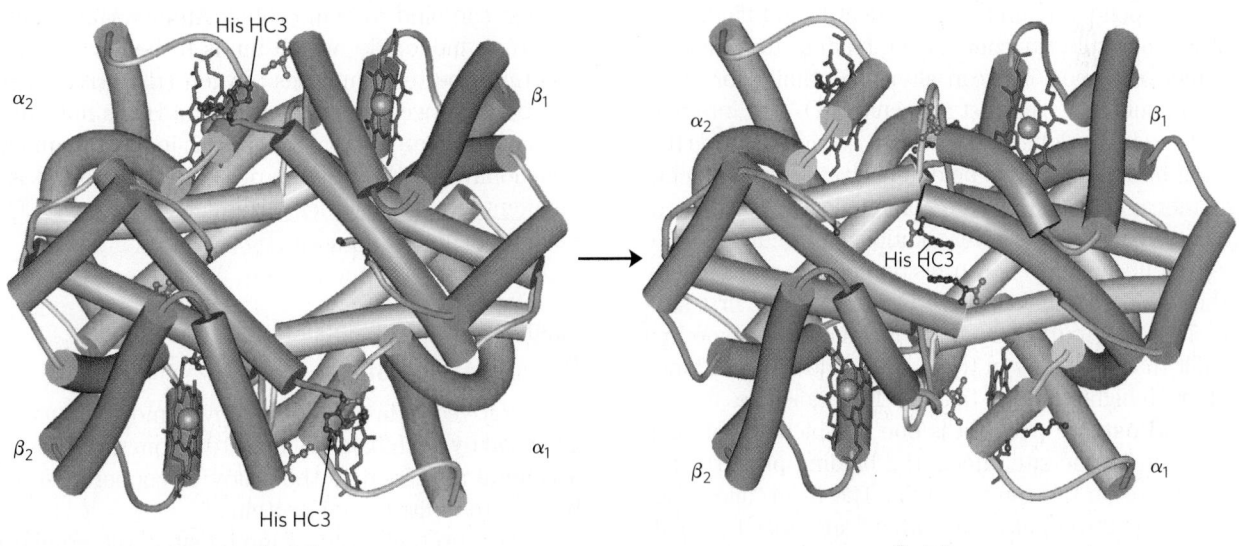

FIGURE 5-10 The T → R transition. In these depictions of deoxyhemo-globin, as in Figure 5-9, the β subunits are blue and the α subunits are gray. Positively charged side chains and chain termini involved in ion pairs are shown in blue, their negatively charged partners in red. The Lys C5 of each α subunit and Asp FG1 of each β subunit are visible but not labeled (compare Fig. 5-9a). Note that the molecule is oriented slightly differently than in Figure 5-9. The transition from the T state to the R state shifts the subunit pairs substantially, affecting certain ion pairs. Most noticeably, the His HC3 residues at the carboxyl termini of the β subunits, which are involved in ion pairs in the T state, rotate in the R state toward the center of the molecule, where they are no longer in ion pairs. Another dramatic result of the T → R transition is a narrowing of the pocket between the β subunits. [Sources: T state: PDB ID 1HGA, R. Liddington et al., *J. Mol. Biol.* 228:551, 1992. R state: PDB ID 1BBB, M. M. Silva et al., *J. Biol. Chem.* 267:17,248, 1992.]

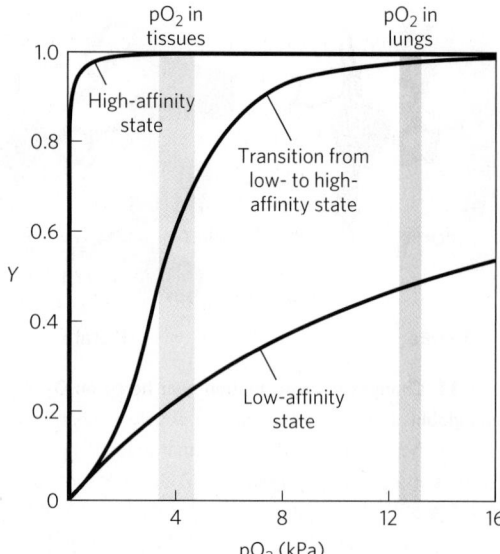

pO$_2$ in tissues

pO$_2$ in lungs

High-affinity state

Transition from low- to high-affinity state

Low-affinity state

Y

pO$_2$ (kPa)

FIGURE 5-12 A sigmoid (cooperative) binding curve. A sigmoid binding curve can be viewed as a hybrid curve reflecting a transition from a low-affinity to a high-affinity state. Because of its cooperative binding, as manifested by a sigmoid binding curve, hemoglobin is more sensitive to the small differences in O$_2$ concentration between the tissues and the lungs, allowing it to bind oxygen in the lungs (where pO$_2$ is high) and release it in the tissues (where pO$_2$ is low).

Hemoglobin solves the problem by undergoing a transition from a low-affinity state (the T state) to a high-affinity state (the R state) as more O$_2$ molecules are bound. As a result, hemoglobin has a hybrid S-shaped, or sigmoid, binding curve for oxygen (Fig. 5-12). A single-subunit protein with a single ligand-binding site cannot produce a sigmoid binding curve—even if binding elicits a conformational change—because each molecule of ligand binds independently and cannot affect ligand binding to another molecule. In contrast, O$_2$ binding to individual subunits of hemoglobin can alter the affinity for O$_2$ in adjacent subunits. The first molecule of O$_2$ that interacts with deoxyhemoglobin binds weakly, because it binds to a subunit in the T state. Its binding, however, leads to conformational changes that are communicated to adjacent subunits, making it easier for additional molecules of O$_2$ to bind. In effect, the T → R transition occurs more readily in the second subunit once O$_2$ is bound to the first subunit. The last (fourth) O$_2$ molecule binds to a heme in a subunit that is already in the R state, and hence it binds with much higher affinity than the first molecule.

An **allosteric protein** is one in which the binding of a ligand to one site affects the binding properties of another site on the same protein. The term "allosteric" derives from the Greek *allos*, "other," and *stereos*, "solid" or "shape." Allosteric proteins are those having "other shapes," or conformations, induced by the binding of ligands referred to as **modulators**. The conformational changes induced by the modulator(s) interconvert more-active and less-active forms of the protein. The modulators for allosteric proteins may be either inhibitors or activators. When the normal ligand and modulator are identical, the interaction is termed **homotropic**. When the modulator is a molecule other than the normal ligand, the interaction is **heterotropic**. Some proteins have two or more modulators and therefore can have both homotropic and heterotropic interactions.

Cooperative binding of a ligand to a multimeric protein, such as we observe with the binding of O$_2$ to hemoglobin, is a form of allosteric binding. The binding of one ligand affects the affinities of any remaining unfilled binding sites, and O$_2$ can be considered as both a ligand and an activating homotropic modulator. There is only one binding site for O$_2$ on each subunit, so the allosteric effects giving rise to cooperativity are mediated by conformational changes transmitted from one subunit to another by subunit-subunit interactions. A sigmoid binding curve is diagnostic of cooperative binding. It permits a much more sensitive response to ligand concentration and is important to the function of many multisubunit proteins. The principle of allostery extends readily to regulatory enzymes, as we shall see in Chapter 6.

Cooperative conformational changes depend on variations in the structural stability of different parts of a protein, as described in Chapter 4. The binding sites of an allosteric protein typically consist of stable segments in proximity to relatively unstable segments, with the latter capable of frequent changes in conformation or intrinsic disorder **(Fig. 5-13)**. When a ligand binds, the moving parts of the protein's binding site may be stabilized in a particular conformation, affecting the conformation of adjacent polypeptide subunits. If the entire binding site were highly stable, then few structural changes could occur in this site or be propagated to other parts of the protein when a ligand bound.

As is the case with myoglobin, ligands other than oxygen can bind to hemoglobin. An important example is carbon monoxide, which binds to hemoglobin about 250 times better than does oxygen (the critical hydrogen bond between O$_2$ and the distal His is not quite as strong in human hemoglobin as it is in most mammalian myoglobins, so the binding of O$_2$ relative to CO is not augmented quite as much). Human exposure to CO can have tragic consequences (Box 5-1).

Cooperative Ligand Binding Can Be Described Quantitatively

Cooperative binding of oxygen by hemoglobin was first analyzed by Archibald Hill in 1910. From this work came a general approach to the study of cooperative ligand binding to multisubunit proteins.

For a protein with n binding sites, the equilibrium of Equation 5-1 becomes

$$P + nL \rightleftharpoons PL_n \qquad (5\text{-}12)$$

and the expression for the association constant becomes

$$K_a = \frac{[PL_n]}{[P][L]^n} \qquad (5\text{-}13)$$

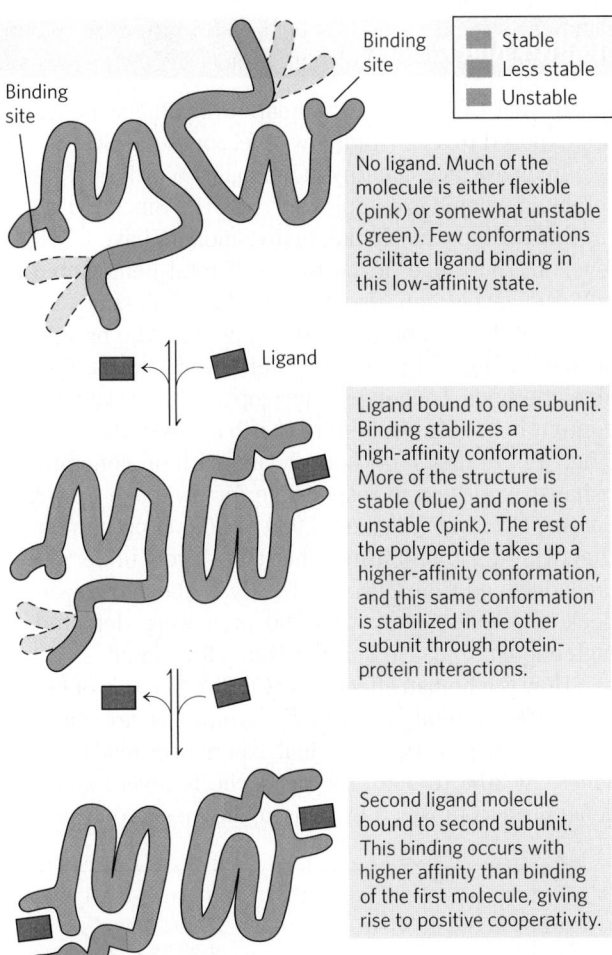

FIGURE 5-13 Structural changes in a multisubunit protein undergoing cooperative binding to ligand. Structural stability is not uniform throughout a protein molecule. Shown here is a hypothetical dimeric protein, with regions of high (blue), medium (green), and low (pink) stability. The ligand-binding sites are composed of both high- and low-stability segments, so affinity for ligand is relatively low. The conformational changes that occur as ligand binds convert the protein from a low- to a high-affinity state, a form of induced fit.

The expression for Y (see Eqn 5-8) is

$$Y = \frac{[L]^n}{[L]^n + K_d} \qquad (5\text{-}14)$$

Rearranging, then taking the log of both sides, yields

$$\frac{Y}{1 - Y} = \frac{[L]^n}{K_d} \qquad (5\text{-}15)$$

$$\log\left(\frac{Y}{1 - Y}\right) = n \log [L] - \log K_d \qquad (5\text{-}16)$$

where $K_d = [L]_{0.5}^n$

Equation 5-16 is the **Hill equation**, and a plot of log $[Y/(1 - Y)]$ versus log [L] is called a **Hill plot**. Based on the equation, the Hill plot should have a slope of n. However, the experimentally determined slope actually reflects not the number of binding sites but the degree of interaction between them. The slope of a Hill plot is therefore

denoted by n_H, the **Hill coefficient**, which is a measure of the degree of cooperativity. If n_H equals 1, ligand binding is not cooperative, a situation that can arise even in a multisubunit protein if the subunits do not communicate. An n_H of greater than 1 indicates positive cooperativity in ligand binding. This is the situation observed in hemoglobin, in which the binding of one molecule of ligand facilitates the binding of others. The theoretical upper limit for n_H is reached when $n_H = n$. In this case the binding would be completely cooperative: all binding sites on the protein would bind ligand simultaneously, and no protein molecules partially saturated with ligand would be present under any conditions. This limit is never reached in practice, and the measured value of n_H is always less than the actual number of ligand-binding sites in the protein.

An n_H of less than 1 indicates negative cooperativity, in which the binding of one molecule of ligand *impedes* the binding of others. Well-documented cases of negative cooperativity are rare.

To adapt the Hill equation to the binding of oxygen to hemoglobin we must again substitute pO_2 for [L] and P_{50}^n for K_d:

$$\log\left(\frac{Y}{1 - Y}\right) = n \log pO_2 - n \log P_{50} \qquad (5\text{-}17)$$

Hill plots for myoglobin and hemoglobin are given in **Figure 5-14**.

Two Models Suggest Mechanisms for Cooperative Binding

Biochemists now know a great deal about the T and R states of hemoglobin, but much remains to be learned about how the T $\rightarrow$ R transition occurs. Two models for

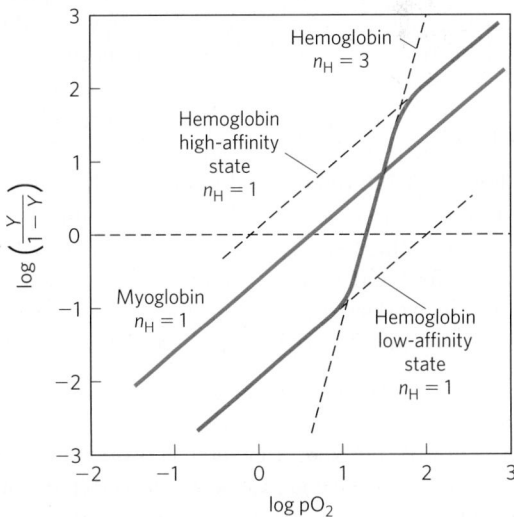

FIGURE 5-14 Hill plots for oxygen binding to myoglobin and hemoglobin. When $n_H = 1$, there is no evident cooperativity. The maximum degree of cooperativity observed for hemoglobin corresponds approximately to $n_H = 3$. Note that while this indicates a high level of cooperativity, n_H is less than n, the number of O_2-binding sites in hemoglobin. This is normal for a protein that exhibits allosteric binding behavior.

BOX 5-1   ♦ MEDICINE   Carbon Monoxide: A Stealthy Killer

Lake Powell, Arizona, August 2000. A family was vacationing on a rented houseboat. They turned on the electrical generator to power an air conditioner and a television. About 15 minutes later, two brothers, aged 8 and 11, jumped off the swim deck at the stern. Situated immediately below the deck was the exhaust port for the generator. Within two minutes, both boys were overcome by the carbon monoxide in the exhaust, which had become concentrated in the space under the deck. Both drowned. These deaths, along with a series of deaths in the 1990s that were linked to houseboats of similar design, eventually led to the recall and redesign of the generator exhaust assembly.

Carbon monoxide (CO), a colorless, odorless gas, is responsible for more than half of yearly deaths due to poisoning worldwide. CO has an approximately 250-fold greater affinity for hemoglobin than does oxygen. Consequently, relatively low levels of CO can have substantial and tragic effects. When CO combines with hemoglobin, the complex is referred to as carboxyhemoglobin, or COHb.

Some CO is produced by natural processes, but locally high levels generally result only from human activities. Engine and furnace exhausts are important sources, as CO is a byproduct of the incomplete combustion of fossil fuels. In the United States alone, nearly 4,000 people succumb to CO poisoning each year, both accidentally and intentionally. Many of the accidental deaths involve undetected CO buildup in enclosed spaces, such as when a household furnace malfunctions or leaks, venting CO into a home. However, CO poisoning can also occur in open spaces, as unsuspecting people at work or play inhale the exhaust from generators, outboard motors, tractor engines, recreational vehicles, or lawn mowers.

Carbon monoxide levels in the atmosphere are rarely dangerous, ranging from less than 0.05 part per million (ppm) in remote and uninhabited areas to 3 to 4 ppm in some cities of the northern hemisphere. In the United States, the government-mandated (Occupational Safety and Health Administration, OSHA) limit for CO at worksites is 50 ppm for people working an eight-hour shift. The tight binding of CO to hemoglobin

means that COHb can accumulate over time as people are exposed to a constant low-level source of CO.

In an average, healthy individual, 1% or less of the total hemoglobin is complexed as COHb. Since CO is a product of tobacco smoke, many smokers have COHb levels in the range of 3% to 8% of total hemoglobin, and the levels can rise to 15% for chain-smokers. COHb levels equilibrate at 50% in people who breathe air containing 570 ppm of CO for several hours. Reliable methods have been developed that relate CO content in the atmosphere to COHb levels in the blood (Fig. 1). In tests of houseboats with a generator exhaust like the one responsible for the Lake Powell deaths, CO levels reached 6,000 to 30,000 ppm under the swim deck, and atmospheric O_2 levels under the deck declined from 21% to 12%. Even above the swim deck, CO levels of up to 7,200 ppm were detected, high enough to cause death within a few minutes.

How is a human affected by COHb? At levels of less than 10% of total hemoglobin, symptoms are rarely observed. At 15%, the individual experiences mild headaches. At 20% to 30%, the headache is severe and is generally accompanied by nausea, dizziness, confusion,

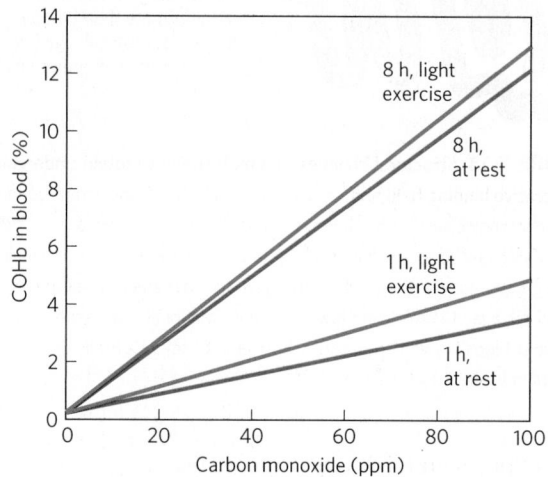

FIGURE 1 Relationship between levels of COHb in blood and concentration of CO in the surrounding air. Four different conditions of exposure are shown, comparing the effects of short versus extended exposure, and exposure at rest versus exposure during light exercise. [Source: Data from R. F. Coburn et al., *J. Clin. Invest.* 44:1899, 1965.]

the cooperative binding of ligands to proteins with multiple binding sites have greatly influenced thinking about this problem.

The first model was proposed by Jacques Monod, Jeffries Wyman, and Jean-Pierre Changeux in 1965, and is called the **MWC model** or the **concerted model (Fig. 5-15a)**. The concerted model assumes that the subunits of a cooperatively binding protein are functionally identical, that each subunit can exist in (at least) two conformations, and that all subunits undergo the transition

from one conformation to the other simultaneously. In this model, no protein has individual subunits in different conformations. The two conformations are in equilibrium. The ligand can bind to either conformation but binds much more tightly to the R state. Successive binding of ligand molecules to the low-affinity conformation (which is more stable in the absence of ligand) makes a transition to the high-affinity conformation more likely.

In the second model, the **sequential model** (Fig. 5-15b), proposed in 1966 by Daniel Koshland and

disorientation, and some visual disturbances; these symptoms are generally reversed if the individual is treated with oxygen. At COHb levels of 30% to 50%, the neurological symptoms become more severe, and at levels near 50%, the individual loses consciousness and can sink into coma. Respiratory failure may follow. With prolonged exposure, some damage becomes permanent. Death normally occurs when COHb levels rise above 60%. Autopsy on the boys who died at Lake Powell revealed COHb levels of 59% and 52%.

Binding of CO to hemoglobin is affected by many factors, including exercise (Fig. 1) and changes in air pressure related to altitude. Because of their higher base levels of COHb, smokers exposed to a source of CO often develop symptoms faster than nonsmokers. Individuals with heart, lung, or blood diseases that reduce the availability of oxygen to tissues may also experience symptoms at lower levels of CO exposure. Fetuses are at particular risk for CO poisoning, because fetal hemoglobin has a somewhat higher affinity for CO than adult hemoglobin. Cases of CO exposure have been recorded in which the fetus died but the woman recovered.

It may seem surprising that the loss of half of one's hemoglobin to COHb can prove fatal—we know that people with any of several anemic conditions manage to function reasonably well with half the usual complement of active hemoglobin. However, the binding of CO to hemoglobin does more than remove protein from the pool available to bind oxygen. It also affects the affinity of the remaining hemoglobin subunits for oxygen. As CO binds to one or two subunits of a hemoglobin tetramer, the affinity for O_2 is increased substantially in the remaining subunits (Fig. 2). Thus, a hemoglobin tetramer with two bound CO molecules can efficiently bind O_2 in the lungs—but it releases very little of it in the tissues. Oxygen deprivation in the tissues rapidly becomes severe. To add to the problem, the effects of CO are not limited to interference with hemoglobin function. CO binds to other heme proteins and a variety of metalloproteins. The effects of these interactions are not yet well understood, but they may be responsible for some of the longer-term effects of acute but nonfatal CO poisoning.

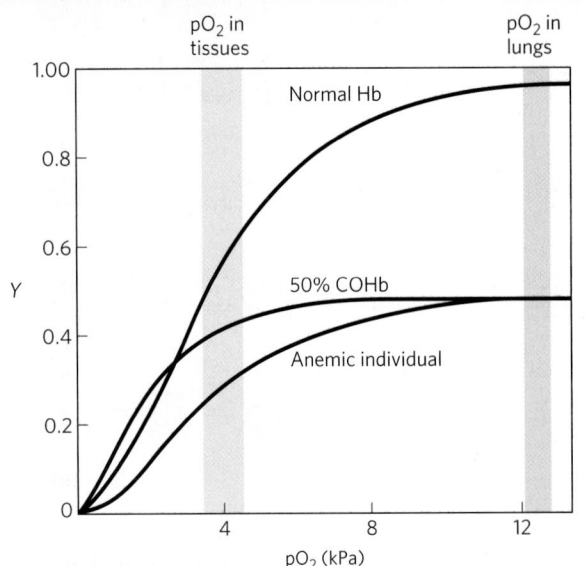

FIGURE 2 Several oxygen-binding curves: for normal hemoglobin, hemoglobin from an anemic individual with only 50% of her hemoglobin functional, and hemoglobin from an individual with 50% of his hemoglobin subunits complexed with CO. The pO_2 in human lungs and tissues is indicated. [Source: Data from F. J. W. Roughton and R. C. Darling, *Am. J. Physiol.* 141:17, 1944.]

When CO poisoning is suspected, rapid removal of the person from the CO source is essential, but this does not always result in rapid recovery. When an individual is moved from the CO-polluted site to a normal, outdoor atmosphere, O_2 begins to replace the CO in hemoglobin—but the COHb level drops only slowly. The half-time is 2 to 6.5 hours, depending on individual and environmental factors. If 100% oxygen is administered with a mask, the rate of exchange can be increased about fourfold; the half-time for O_2-CO exchange can be reduced to tens of minutes if 100% oxygen at a pressure of 3 atm (303 kPa) is supplied. Thus, rapid treatment by a properly equipped medical team is critical.

Carbon monoxide detectors in all homes are highly recommended. This is a simple and inexpensive measure to avoid possible tragedy. After completing the research for this box, we immediately purchased several new CO detectors for our homes.

colleagues, ligand binding can induce a change of conformation in an individual subunit. A conformational change in one subunit makes a similar change in an adjacent subunit, as well as the binding of a second ligand molecule, more likely. There are more potential intermediate states in this model than in the concerted model. The two models are not mutually exclusive; the concerted model may be viewed as the "all-or-none" limiting case of the sequential model. In Chapter 6 we use these models to investigate allosteric enzymes.

Hemoglobin Also Transports H^+ and CO_2

In addition to carrying nearly all the oxygen required by cells from the lungs to the tissues, hemoglobin carries two end products of cellular respiration—H^+ and CO_2—from the tissues to the lungs and the kidneys, where they are excreted. The CO_2, produced by oxidation of organic fuels in mitochondria, is hydrated to form bicarbonate:

$$CO_2 + H_2O \rightleftharpoons H^+ + HCO_3^-$$

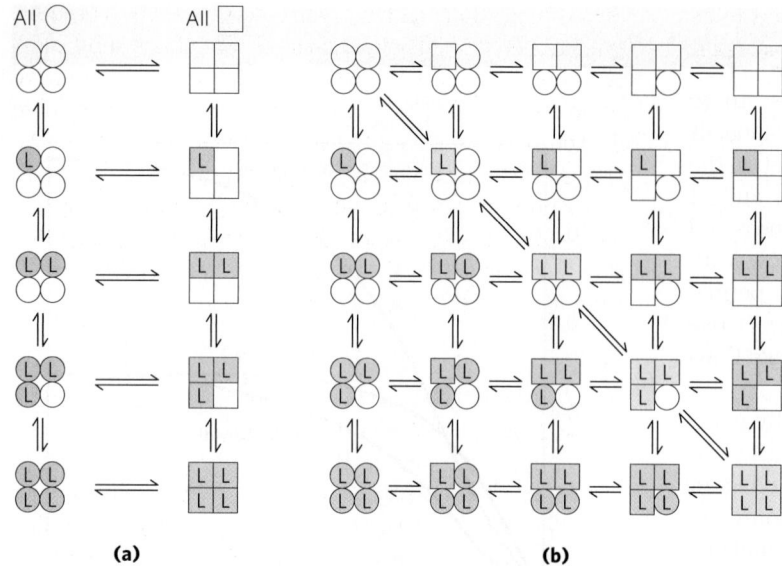

(a)   **(b)**

This reaction is catalyzed by **carbonic anhydrase**, an enzyme particularly abundant in erythrocytes. Carbon dioxide is not very soluble in aqueous solution, and bubbles of CO_2 would form in the tissues and blood if it were not converted to bicarbonate. As you can see from the reaction catalyzed by carbonic anhydrase, the hydration of CO_2 results in an increase in the H^+ concentration (a decrease in pH) in the tissues. The binding of oxygen by hemoglobin is profoundly influenced by pH and CO_2 concentration, so the interconversion of CO_2 and bicarbonate is of great importance to the regulation of oxygen binding and release in the blood.

Hemoglobin transports about 40% of the total H^+ and 15% to 20% of the CO_2 formed in the tissues to the lungs and kidneys. (The remainder of the H^+ is absorbed by the plasma's bicarbonate buffer; the remainder of the CO_2 is transported as dissolved HCO_3^- and CO_2.) The binding of H^+ and CO_2 is inversely related to the binding of oxygen. At the relatively low pH and high CO_2 concentration of peripheral tissues, the affinity of hemoglobin for oxygen decreases as H^+ and CO_2 are bound, and O_2 is released to the tissues. Conversely, in the capillaries of the lung, as CO_2 is excreted and the blood pH consequently rises, the affinity of hemoglobin for oxygen increases and the protein binds more O_2 for transport to the peripheral tissues. This effect of pH and CO_2 concentration on the binding and release of oxygen by hemoglobin is called the **Bohr effect**, after Christian Bohr, the Danish physiologist (and father of physicist Niels Bohr) who discovered it in 1904.

The binding equilibrium for hemoglobin and one molecule of oxygen can be designated by the reaction

$$Hb + O_2 \rightleftharpoons HbO_2$$

but this is not a complete statement. To account for the effect of H^+ concentration on this binding equilibrium, we rewrite the reaction as

$$HHb^+ + O_2 \rightleftharpoons HbO_2 + H^+$$

where HHb^+ denotes a protonated form of hemoglobin. This equation tells us that the O_2-saturation curve of hemoglobin is influenced by the H^+ concentration **(Fig. 5-16)**. Both O_2 and H^+ are bound by hemoglobin, but with inverse affinity. When the oxygen concentration is high, as in the lungs, hemoglobin binds O_2 and releases protons. When the oxygen concentration is low, as in the peripheral tissues, H^+ is bound and O_2 is released.

Oxygen and H^+ are not bound at the same sites in hemoglobin. Oxygen binds to the iron atoms of the hemes, whereas H^+ binds to any of several amino acid residues in the protein. A major contribution to the Bohr effect is made by His^{146} (His HC3) of the β subunits. When protonated, this residue forms one of the ion pairs—to Asp^{94} (Asp FG1)—that helps stabilize deoxyhemoglobin in the T state (Fig. 5-9). The ion pair stabilizes the protonated form of His HC3, giving this residue

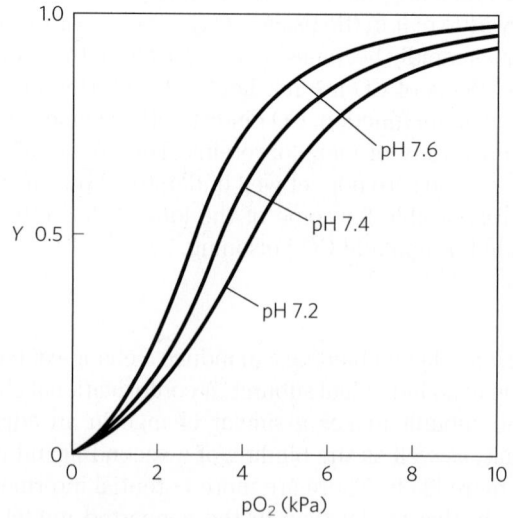

FIGURE 5-16 Effect of pH on oxygen binding to hemoglobin. The pH of blood is 7.6 in the lungs and 7.2 in the tissues. Experimental measurements on hemoglobin binding are often performed at pH 7.4.

an abnormally high pK_a in the T state. The pK_a falls to its normal value of 6.0 in the R state because the ion pair cannot form, and this residue is largely unprotonated in oxyhemoglobin at pH 7.6, the blood pH in the lungs. As the concentration of H^+ rises, protonation of His HC3 promotes release of oxygen by favoring a transition to the T state. Protonation of the amino-terminal residues of the α subunits, certain other His residues, and perhaps other groups has a similar effect.

Thus we see that the four polypeptide chains of hemoglobin communicate with each other not only about O_2 binding to their heme groups but also about H^+ binding to specific amino acid residues. And there is still more to the story. Hemoglobin also binds CO_2, again in a manner inversely related to the binding of oxygen. Carbon dioxide binds as a carbamate group to the α-amino group at the amino-terminal end of each globin chain, forming carbaminohemoglobin:

$$\overset{O}{\underset{O}{\overset{\|}{\underset{\|}{C}}}} + H_2N-\overset{H}{\underset{R}{\overset{|}{\underset{|}{C}}}}-\overset{H}{\underset{O}{\overset{|}{\underset{\|}{C}}}}- \xrightarrow{H^+} \overset{O^-}{\underset{O}{\overset{|}{\underset{\|}{C}}}}-\overset{H}{\underset{}{\overset{|}{\underset{|}{N}}}}-\overset{H}{\underset{R}{\overset{|}{\underset{|}{C}}}}-\overset{H}{\underset{O}{\overset{|}{\underset{\|}{C}}}}-$$

$$\text{Amino-terminal} \qquad \text{Carbamino-terminal}$$
$$\text{residue} \qquad\qquad \text{residue}$$

This reaction produces H^+, contributing to the Bohr effect. The bound carbamates also form additional salt bridges (not shown in Fig. 5-9) that help to stabilize the T state and promote the release of oxygen.

When the concentration of carbon dioxide is high, as in peripheral tissues, some CO_2 binds to hemoglobin and the affinity for O_2 decreases, causing its release. Conversely, when hemoglobin reaches the lungs, the high oxygen concentration promotes binding of O_2 and release of CO_2. It is the capacity to communicate ligand-binding information from one polypeptide subunit to the others that makes the hemoglobin molecule so beautifully adapted to integrating the transport of O_2, CO_2, and H^+ by erythrocytes.

Oxygen Binding to Hemoglobin Is Regulated by 2,3-Bisphosphoglycerate

The interaction of **2,3-bisphosphoglycerate (BPG)** with hemoglobin molecules further refines the function of hemoglobin, and provides an example of heterotropic allosteric modulation.

2,3-Bisphosphoglycerate

BPG is present in relatively high concentrations in erythrocytes. When hemoglobin is isolated, it contains substantial amounts of bound BPG, which can be difficult to remove completely. In fact, the O_2-binding curves for hemoglobin that we have examined to this point were obtained in the presence of bound BPG. 2,3-Bisphosphoglycerate is known to greatly reduce the affinity of hemoglobin for oxygen—there is an inverse relationship between the binding of O_2 and the binding of BPG. We can therefore describe another binding process for hemoglobin:

$$\text{HbBPG} + O_2 \rightleftharpoons \text{HbO}_2 + \text{BPG}$$

BPG binds at a site distant from the oxygen-binding site and regulates the O_2-binding affinity of hemoglobin in relation to the pO_2 in the lungs. BPG is important in the physiological adaptation to the lower pO_2 at high altitudes. For a healthy human at sea level, the binding of O_2 to hemoglobin is regulated such that the amount of O_2 delivered to the tissues is nearly 40% of the maximum that could be carried by the blood (**Fig. 5-17**). Imagine that this person is suddenly transported from sea level to

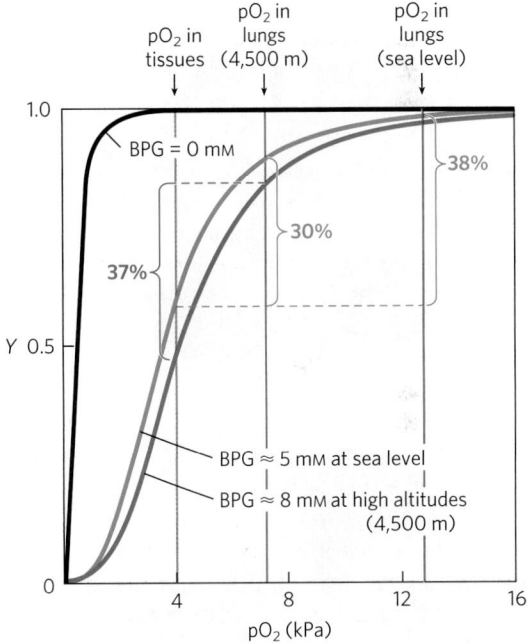

FIGURE 5-17 Effect of 2,3-bisphosphoglycerate on oxygen binding to hemoglobin. The BPG concentration in normal human blood is about 5 mM at sea level and about 8 mM at high altitudes. Note that hemoglobin binds to oxygen quite tightly when BPG is entirely absent, and the binding curve seems to be hyperbolic. In reality, the measured Hill coefficient for O_2-binding cooperativity decreases only slightly (from 3 to about 2.5) when BPG is removed from hemoglobin, but the rising part of the sigmoid curve is confined to a very small region close to the origin. At sea level, hemoglobin is nearly saturated with O_2 in the lungs, but is just over 60% saturated in the tissues, so the amount of O_2 released in the tissues is about 38% of the maximum that can be carried in the blood. At high altitudes, O_2 delivery declines by about one-fourth, to 30% of maximum. An increase in BPG concentration, however, decreases the affinity of hemoglobin for O_2, so approximately 37% of what can be carried is again delivered to the tissues.

an altitude of 4,500 meters, where the pO_2 is considerably lower. The delivery of O_2 to the tissues is now reduced. However, after just a few hours at the higher altitude, the BPG concentration in the blood has begun to rise, leading to a decrease in the affinity of hemoglobin for oxygen. This adjustment in the BPG level has only a small effect on the binding of O_2 in the lungs but a considerable effect on the release of O_2 in the tissues. As a result, the delivery of oxygen to the tissues is restored to nearly 40% of the O_2 that can be transported by the blood. The situation is reversed when the person returns to sea level. The BPG concentration in erythrocytes also increases in people suffering from **hypoxia**, lowered oxygenation of peripheral tissues due to inadequate functioning of the lungs or circulatory system.

The site of BPG binding to hemoglobin is the cavity between the β subunits in the T state **(Fig. 5-18)**. This cavity is lined with positively charged amino acid residues that interact with the negatively charged groups of BPG. Unlike O_2, only one molecule of BPG is bound to each hemoglobin tetramer. BPG lowers hemoglobin's affinity for oxygen by stabilizing the T state. The transition to the

R state narrows the binding pocket for BPG, precluding BPG binding. In the absence of BPG, hemoglobin is converted to the R state more easily.

Regulation of oxygen binding to hemoglobin by BPG has an important role in fetal development. Because a fetus must extract oxygen from its mother's blood, fetal hemoglobin must have greater affinity than the maternal hemoglobin for O_2. The fetus synthesizes γ subunits rather than β subunits, forming $\alpha_2\gamma_2$ hemoglobin. This tetramer has a much lower affinity for BPG than normal adult hemoglobin, and a correspondingly higher affinity for O_2.

Sickle Cell Anemia Is a Molecular Disease of Hemoglobin

The hereditary human disease sickle cell anemia demonstrates strikingly the importance of amino acid sequence in determining the secondary, tertiary, and quaternary structures of globular proteins, and thus their biological functions. Almost 500 genetic variants of hemoglobin are known to occur in the human population; all but a few are quite rare. Most variations consist of differences in a single amino acid residue. The effects on hemoglobin structure and function are often minor but can sometimes be extraordinary. Each hemoglobin variation is the product of an altered gene. Variant genes are called alleles. Because humans generally have two copies of each gene, an individual may have two copies of one allele (thus being homozygous for that gene) or one copy of each of two different alleles (thus heterozygous).

Sickle cell anemia occurs in individuals who inherit the allele for sickle cell hemoglobin from both parents. The erythrocytes of these individuals are fewer and also abnormal. In addition to an unusually large number of immature cells, the blood contains many long, thin, sickle-shaped erythrocytes **(Fig. 5-19)**. When hemoglobin from sickle

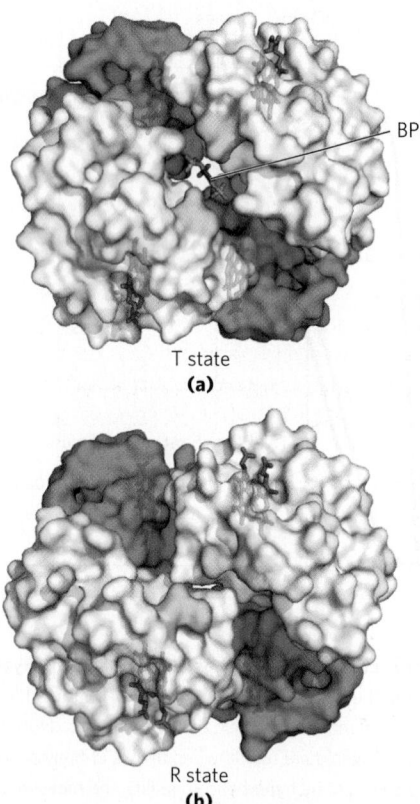

T state
(a)

R state
(b)

FIGURE 5-18 Binding of 2,3-bisphosphoglycerate to deoxyhemoglobin. (a) BPG binding stabilizes the T state of deoxyhemoglobin. The negative charges of BPG interact with several positively charged groups (shown in blue in this surface contour image) that surround the pocket between the β subunits on the surface of deoxyhemoglobin in the T state. (b) The binding pocket for BPG disappears on oxygenation, following transition to the R state. (Compare with Fig. 5-10.) [Sources: (a) PDB ID 1B86, V. Richard et al., *J. Mol. Biol.* 233:270, 1993. (b) PDB ID 1BBB, M. M. Silva et al., *J. Biol. Chem.* 267:17,248, 1992.]

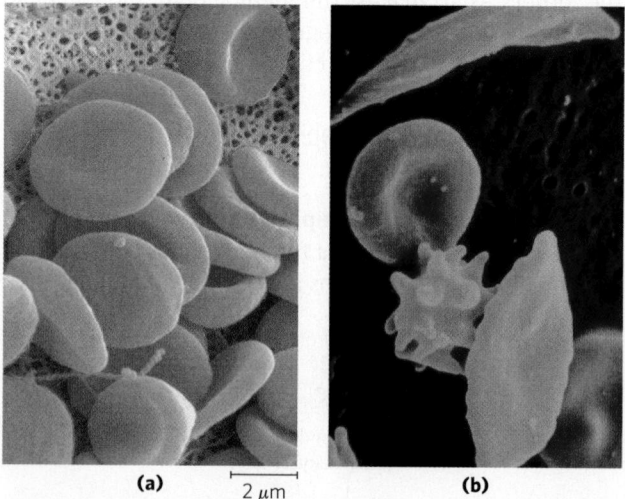

(a)   2 μm   **(b)**

FIGURE 5-19 A comparison of (a) uniform, cup-shaped, normal erythrocytes and (b) the variably shaped erythrocytes seen in sickle-cell anemia, which range from normal to spiny or sickle-shaped. [Sources: (a) A. Syred/Science Source. (b) Jackie Lewin, Royal Free Hospital/Science Source.]

cells (called hemoglobin S, or HbS) is deoxygenated, it becomes insoluble and forms polymers that aggregate into tubular fibers **(Fig. 5-20)**. Normal hemoglobin (hemoglobin A, or HbA) remains soluble on deoxygenation. The insoluble fibers of deoxygenated HbS cause the deformed, sickle shape of the erythrocytes, and the proportion of sickled cells increases greatly as blood is deoxygenated.

The altered properties of HbS result from a single amino acid substitution, a Val instead of a Glu residue at position 6 in the two β chains. The R group of valine has no electric charge, whereas glutamate has a negative charge at pH 7.4. Hemoglobin S therefore has two fewer negative charges than HbA (one fewer on each β chain). Replacement of the Glu residue by Val creates a "sticky" hydrophobic contact point at position 6 of the β chain, which is on the outer surface of the molecule. These sticky spots cause deoxyHbS molecules to associate abnormally with each other, forming the long, fibrous aggregates characteristic of this disorder.

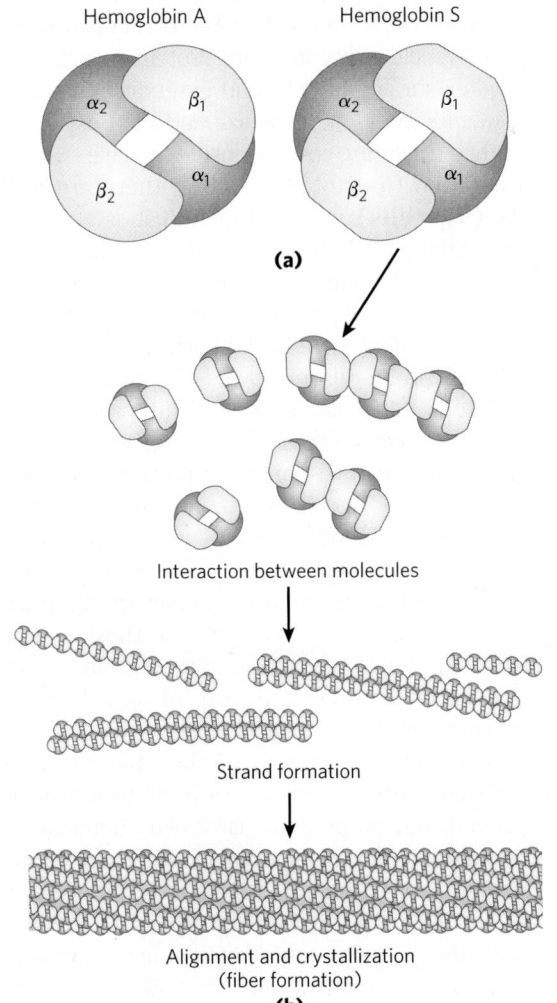

(a)

Interaction between molecules

Strand formation

Alignment and crystallization
(fiber formation)
(b)

FIGURE 5-20 Normal and sickle-cell hemoglobin. (a) Subtle differences between the conformations of HbA and HbS result from a single amino acid change in the β chains. **(b)** As a result of this change, deoxyHbS has a hydrophobic patch on its surface, which causes the molecules to aggregate into strands that align into insoluble fibers.

Sickle cell anemia is life-threatening and painful. People with this disease suffer repeated crises brought on by physical exertion. They become weak, dizzy, and short of breath, and they also experience heart murmurs and an increased pulse rate. The hemoglobin content of their blood is only about half the normal value of 15 to 16 g/100 mL, because sickled cells are very fragile and rupture easily; this results in anemia ("lack of blood"). An even more serious consequence is that capillaries become blocked by the long, abnormally shaped cells, causing severe pain and interfering with normal organ function—a major factor in the early death of many people with the disease.

Without medical treatment, people with sickle cell anemia usually die in childhood. Curiously, the frequency of the sickle cell allele in populations is unusually high in certain parts of Africa. Investigation into this matter led to the finding that when heterozygous, the allele confers a small but significant resistance to lethal forms of malaria. The heterozygous individuals experience a milder condition called sickle cell trait; only about 1% of their erythrocytes become sickled on deoxygenation. These individuals may live completely normal lives if they avoid vigorous exercise and other stresses on the circulatory system. Natural selection has resulted in an allele population that balances the deleterious effects of the homozygous condition against the resistance to malaria afforded by the heterozygous condition. ■

SUMMARY 5.1 Reversible Binding of a Protein to a Ligand: Oxygen-Binding Proteins

■ Protein function often entails interactions with other molecules. A protein binds a molecule, known as a ligand, at its binding site. Proteins may undergo conformational changes when a ligand binds, a process called induced fit. In a multisubunit protein, the binding of a ligand to one subunit may affect ligand binding to other subunits. Ligand binding can be regulated.

■ Myoglobin contains a heme prosthetic group, which binds oxygen. Heme consists of a single atom of Fe^{2+} coordinated within a porphyrin. Oxygen binds to myoglobin reversibly; this simple reversible binding can be described by an association constant K_a or a dissociation constant K_d. For a monomeric protein such as myoglobin, the fraction of binding sites occupied by a ligand is a hyperbolic function of ligand concentration.

■ Normal adult hemoglobin has four heme-containing subunits, two α and two β, similar in structure to each other and to myoglobin. Hemoglobin exists in two interchangeable structural states, T and R. The T state is most stable when oxygen is not bound. Oxygen binding promotes transition to the R state.

■ Oxygen binding to hemoglobin is both allosteric and cooperative. As O_2 binds to one binding site, the hemoglobin undergoes conformational changes that affect the other binding sites—an example of allosteric behavior. Conformational changes between the T and R states, mediated by subunit–subunit interactions, result in cooperative binding; this is described by a sigmoid binding curve and can be analyzed by a Hill plot.

■ Two major models have been proposed to explain the cooperative binding of ligands to multisubunit proteins: the concerted model and the sequential model.

■ Hemoglobin also binds H^+ and CO_2, resulting in the formation of ion pairs that stabilize the T state and lessen the protein's affinity for O_2 (the Bohr effect). Oxygen binding to hemoglobin is also modulated by 2,3-bisphosphoglycerate, which binds to and stabilizes the T state.

■ Sickle cell anemia is a genetic disease caused by a single amino acid substitution (Glu^6 to Val^6) in each β chain of hemoglobin. The change produces a hydrophobic patch on the surface of the hemoglobin that causes the molecules to aggregate into bundles of fibers. This homozygous condition results in serious medical complications.

5.2 Complementary Interactions between Proteins and Ligands: The Immune System and Immunoglobulins

We have seen how the conformations of oxygen-binding proteins affect and are affected by the binding of small ligands (O_2 or CO) to the heme group. However, most protein-ligand interactions do not involve a prosthetic group. Instead, the binding site for a ligand is more often like the hemoglobin binding site for BPG—a cleft in the protein lined with amino acid residues, arranged to make the binding interaction highly specific. Effective discrimination between ligands is the norm at binding sites, even when the ligands have only minor structural differences.

All vertebrates have an immune system capable of distinguishing molecular "self" from "nonself" and then destroying what is identified as nonself. In this way, the immune system eliminates viruses, bacteria, and other pathogens and molecules that may pose a threat to the organism. On a physiological level, the **immune response** is an intricate and coordinated set of interactions among many classes of proteins, molecules, and cell types. At the level of individual proteins, the immune response demonstrates how an acutely sensitive and specific biochemical system is built upon the reversible binding of ligands to proteins.

The Immune Response Includes a Specialized Array of Cells and Proteins

Immunity is brought about by a variety of **leukocytes** (white blood cells), including **macrophages** and **lymphocytes**, all of which develop from undifferentiated stem cells in the bone marrow. Leukocytes can leave the bloodstream and patrol the tissues, each cell producing one or more proteins capable of recognizing and binding to molecules that might signal an infection.

The immune response consists of two complementary systems, the humoral and cellular immune systems. The **humoral immune system** (Latin *humor*, "fluid") is directed at bacterial infections and extracellular viruses (those found in the body fluids), but can also respond to individual foreign proteins. The **cellular immune system** destroys host cells infected by viruses and also destroys some parasites and foreign tissues.

At the heart of the humoral immune response are soluble proteins called **antibodies** or **immunoglobulins**, often abbreviated **Ig**. Immunoglobulins bind bacteria, viruses, or large molecules identified as foreign and target them for destruction. Making up 20% of blood protein, the immunoglobulins are produced by **B lymphocytes**, or **B cells**, so named because they complete their development in the *b*one marrow.

The agents at the heart of the cellular immune response are a class of **T lymphocytes**, or **T cells** (so called because the latter stages of their development occur in the *t*hymus), known as **cytotoxic T cells** (T_C **cells**, also called killer T cells). Recognition of infected cells or parasites involves proteins called **T-cell receptors** on the surface of T_C cells. Receptors are proteins, usually found on the outer surface of cells and extending through the plasma membrane, that recognize and bind extracellular ligands, thus triggering changes inside the cell.

In addition to cytotoxic T cells, there are **helper T cells (T_H cells)**, whose function it is to produce soluble signaling proteins called cytokines, which include the interleukins. T_H cells interact with macrophages. The T_H cells participate only indirectly in the destruction of infected cells and pathogens, stimulating the selective proliferation of those T_C and B cells that can bind to a particular antigen. This process, called **clonal selection**, increases the number of immune system cells that can respond to a particular pathogen. The importance of T_H cells is dramatically illustrated by the epidemic produced by HIV (human immunodeficiency virus), the virus that causes AIDS (acquired immune deficiency syndrome). T_H cells are the primary targets of HIV infection; elimination of these cells progressively incapacitates the entire immune system. Table 5-2 summarizes the functions of some leukocytes of the immune system.

Each recognition protein of the immune system, either a T-cell receptor or an antibody produced by a B cell, specifically binds some particular chemical structure, distinguishing it from virtually all others. Humans are capable of producing more than 10^8 different antibodies with distinct binding specificities. Given this extraordinary diversity, any chemical structure on the surface of a virus or invading cell will most likely be recognized and bound by one or more antibodies. Antibody diversity is derived from random reassembly of a set of immunoglobulin gene segments through genetic recombination mechanisms that are discussed in Chapter 25 (see Fig. 25-43).

A specialized lexicon is used to describe the unique interactions between antibodies or T-cell receptors and

TABLE 5-2	Some Types of Leukocytes Associated with the Immune System
Cell type	**Function**
Macrophages	Ingest large particles and cells by phagocytosis
B lymphocytes (B cells)	Produce and secrete antibodies
T lymphocytes (T cells)	
Cytotoxic (Killer) T cells (T_C)	Interact with infected host cells through receptors on T-cell surface
Helper T cells (T_H)	Interact with macrophages and secrete cytokines (interleukins) that stimulate T_C, T_H, and B cells to proliferate.

the molecules they bind. Any molecule or pathogen capable of eliciting an immune response is called an **antigen**. An antigen may be a virus, a bacterial cell wall, or an individual protein or other macromolecule. A complex antigen may be bound by several different antibodies.

An individual antibody or T-cell receptor binds only a particular molecular structure within the antigen, called its **antigenic determinant** or **epitope**.

It would be unproductive for the immune system to respond to small molecules that are common intermediates and products of cellular metabolism. Molecules of M_r <5,000 are generally not antigenic. However, when small molecules are covalently attached to large proteins in the laboratory, they can be used to elicit an immune response. These small molecules are called **haptens**. The antibodies produced in response to protein-linked haptens will then bind to the same small molecules in their free form. Such antibodies are sometimes used in the development of analytical tests described later in this chapter or as a specific ligand in affinity chromatography (see Fig. 3-17c). We now turn to a more detailed description of antibodies and their binding properties.

Antibodies Have Two Identical Antigen-Binding Sites

Immunoglobulin G (IgG) is the major class of antibody molecule and one of the most abundant proteins in the blood serum. IgG has four polypeptide chains: two large ones, called heavy chains, and two light chains, linked by noncovalent and disulfide bonds into a complex of M_r 150,000. The heavy chains of an IgG molecule interact at one end, then branch to interact separately with the light chains, forming a Y-shaped molecule (**Fig. 5-21**).

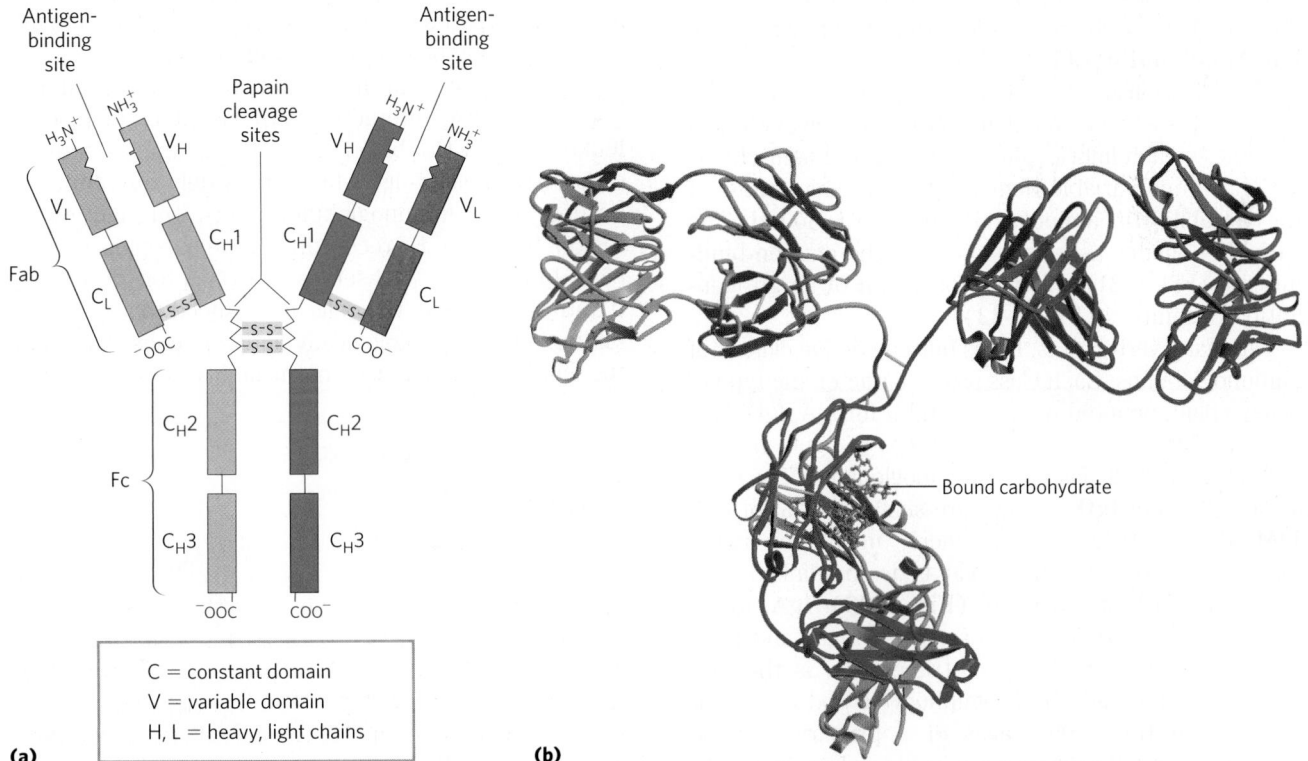

FIGURE 5-21 Immunoglobulin G. (a) Pairs of heavy and light chains combine to form a Y-shaped molecule. Two antigen-binding sites are formed by the combination of variable domains from one light (V_L) and one heavy (V_H) chain. Cleavage with papain separates the Fab and Fc portions of the protein in the hinge region. The Fc portion also contains bound carbohydrate (shown in (b)). (b) A ribbon model of the first complete IgG molecule to be crystallized and structurally analyzed. Although the molecule has two identical heavy chains (two shades of blue) and two identical light chains (two shades of red), it crystallized in the asymmetric conformation shown here. Conformational flexibility may be important to the function of immunoglobulins. [Source: (b) PDB ID 1IGT, L. J. Harris et al., *Biochemistry* 36:1581, 1997.]

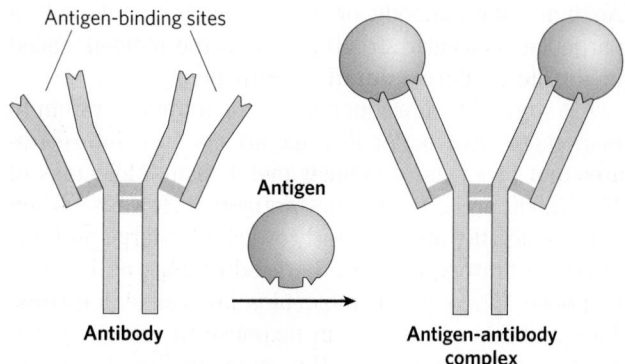

FIGURE 5-22 Binding of IgG to an antigen. To generate an optimal fit for the antigen, the binding sites of IgG often undergo slight conformational changes. Such induced fit is common to many protein-ligand interactions.

At the "hinges" separating the base of an IgG molecule from its branches, the immunoglobulin can be cleaved with proteases. Cleavage with the protease papain liberates the basal fragment, called **Fc** because it usually *c*rystallizes readily, and the two branches, called **Fab**, the *a*ntigen-*b*inding fragments. Each branch has a single antigen-binding site.

The fundamental structure of immunoglobulins was first established by Gerald Edelman and Rodney Porter. Each chain is made up of identifiable domains. Some are constant in sequence and structure from one IgG to the next; others are variable. The constant domains have a characteristic structure known as the **immunoglobulin fold**, a well-conserved structural motif in the all-β class of proteins (Chapter 4). There are three of these constant domains in each heavy chain and one in each light chain. The heavy and light chains also have one variable domain each, in which most of the variability in amino acid sequence is found. The variable domains associate to create the antigen-binding site (Fig. 5-21), allowing formation of an antigen-antibody complex **(Fig. 5-22)**.

In many vertebrates, IgG is but one of five classes of immunoglobulins. Each class has a characteristic type of heavy chain, denoted α, δ, ε, γ, and μ for IgA, IgD, IgE, IgG, and IgM, respectively. Two types of light chain, κ and λ, occur in all classes of immunoglobulins. The overall structures of **IgD** and **IgE** are similar to that of IgG. **IgM** occurs either in a monomeric, membrane-bound form or in a secreted form that is a cross-linked pentamer of this basic structure **(Fig. 5-23)**. **IgA**, found principally in secretions such as saliva, tears, and milk, can be a monomer, dimer, or trimer. IgM is the first antibody to be made by B lymphocytes and the major antibody in the early stages of a primary immune response. Some B cells soon begin to produce IgD (with the same antigen-binding site as the IgM produced by the same cell), but the particular function of IgD is less clear.

The IgG described above is the major antibody in secondary immune responses, which are initiated by a class of B cells called memory B cells. As part of the

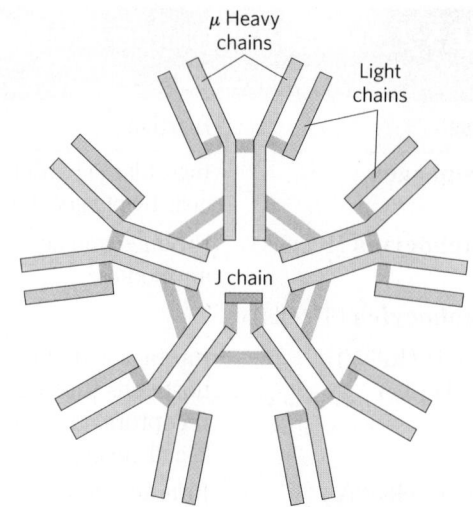

FIGURE 5-23 IgM pentamer of immunoglobulin units. The pentamer is cross-linked with disulfide bonds (yellow). The J chain is a polypeptide of M_r 20,000 found in both IgA and IgM.

organism's ongoing immunity to antigens already encountered and dealt with, IgG is the most abundant immunoglobulin in the blood. When IgG binds to an invading bacterium or virus, it activates certain leukocytes such as macrophages to engulf and destroy the invader, and also activates some other parts of the immune response. Receptors on the macrophage surface recognize and bind the Fc region of IgG. When these Fc receptors bind an antibody-pathogen complex, the macrophage engulfs the complex by phagocytosis **(Fig. 5-24)**.

IgE plays an important role in the allergic response, interacting with basophils (phagocytic leukocytes) in the blood and with histamine-secreting cells called mast cells, which are widely distributed in tissues. This immunoglobulin binds, through its Fc region, to special Fc receptors on the basophils or mast cells. In this form, IgE serves as a receptor for antigen. If antigen is bound, the cells are induced to secrete histamine and other biologically active amines that cause the dilation and increased permeability of blood vessels.

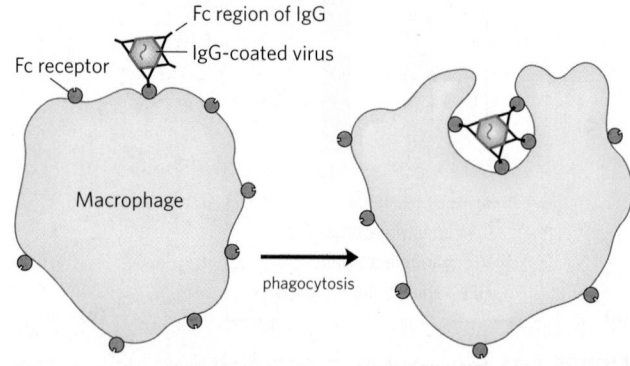

FIGURE 5-24 Phagocytosis of an antibody-bound virus by a macrophage. The Fc regions of antibodies bound to the virus now bind to Fc receptors on the surface of a macrophage, triggering the macrophage to engulf and destroy the virus.

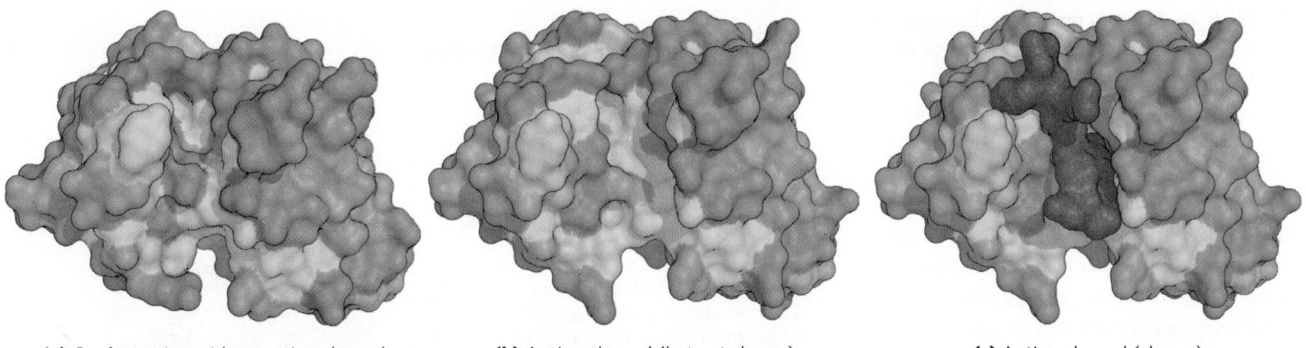

(a) Conformation with no antigen bound   **(b)** Antigen bound (but not shown)   **(c)** Antigen bound (shown)

FIGURE 5-25 Induced fit in the binding of an antigen to IgG. The Fab fragment of an IgG molecule is shown here with the surface contour colored to represent hydrophobicity. Hydrophobic surfaces are yellow, hydrophilic surfaces are blue, with shades of blue to green to yellow in between. **(a)** View of the Fab fragment in the absence of antigen (a small peptide derived from HIV), looking down on the antigen binding site. **(b)** The same view, but with the Fab fragment in the "bound" conformation with the antigen omitted to provide an unobstructed view of the altered binding site. Note how the hydrophobic binding cavity has enlarged and several groups have shifted position. **(c)** The same view as (b) but with the antigen (red) in the binding site. [Sources: (a) PDB ID 1GGC, R. L. Stanfield et al., *Structure* 1:83, 1993. (b, c) PDB ID 1GGI, J. M. Rini et al., *Proc. Natl. Acad. Sci. USA* 90:6325, 1993.]

These effects on the blood vessels are thought to facilitate the movement of immune system cells and proteins to sites of inflammation. They also produce the symptoms normally associated with allergies. Pollen or other allergens are recognized as foreign, triggering an immune response normally reserved for pathogens. ■

Antibodies Bind Tightly and Specifically to Antigen

The binding specificity of an antibody is determined by the amino acid residues in the variable domains of its heavy and light chains. Many residues in these domains are variable, but not equally so. Some, particularly those lining the antigen-binding site, are hypervariable—especially likely to differ. Specificity is conferred by chemical complementarity between the antigen and its specific binding site, in terms of shape and the location of charged, nonpolar, and hydrogen-bonding groups. For example, a binding site with a negatively charged group may bind an antigen with a positive charge in the complementary position. In many instances, complementarity is achieved interactively as the structures of antigen and binding site influence each other as they come closer together. Conformational changes in the antibody and/or the antigen then allow the complementary groups to interact fully. This is an example of induced fit. The complex of a peptide derived from HIV (a model antigen) and an Fab molecule, shown in **Figure 5-25**, illustrates some of these properties. The changes in structure observed on antigen binding are particularly striking in this example.

A typical antibody-antigen interaction is quite strong, characterized by K_d values as low as 10^{-10} M (recall that a lower K_d corresponds to a stronger binding interaction; see Table 5-1). The K_d reflects the energy derived from the hydrophobic effect and the various ionic, hydrogen-bonding, and van der Waals interactions that stabilize the binding. The binding energy required to produce a K_d of 10^{-10} M is about 65 kJ/mol.

The Antibody-Antigen Interaction Is the Basis for a Variety of Important Analytical Procedures

The extraordinary binding affinity and specificity of antibodies make them valuable analytical reagents. Two types of antibody preparations are in use: polyclonal and monoclonal. **Polyclonal antibodies** are those produced by many different B lymphocytes responding to one antigen, such as a protein injected into an animal. Cells in the population of B lymphocytes produce antibodies that bind specific, different epitopes within the antigen. Thus, polyclonal preparations contain a mixture of antibodies that recognize different parts of the protein. **Monoclonal antibodies**, in contrast, are synthesized by a population of identical B cells (a **clone**) grown in cell culture. These antibodies are homogeneous, all recognizing the same epitope. The techniques for producing monoclonal antibodies were developed by Georges Köhler and Cesar Milstein.

The specificity of antibodies has practical uses. A selected antibody can be covalently attached to a resin and used in a chromatography column of the type shown in Figure 3-17c. When a mixture of proteins is added to the column, the antibody specifically binds its

Georges Köhler, 1946–1995
[Source: Bettman/Corbis.]

Cesar Milstein, 1927–2002
[Source: Corbin O'Grady Studio/ Science Source.]

FIGURE 5-26 Antibody techniques. The specific reaction of an antibody with its antigen is the basis of several techniques that identify and quantify a specific protein in a complex sample. **(a)** A schematic representation of the general method. **(b)** An ELISA to test for the presence of herpes simplex virus (HSV) antibodies in blood samples. Wells were coated with an HSV antigen, to which antibodies against HSV will bind. The second antibody is anti-human IgG linked to horseradish peroxidase. Following completion of the steps shown in (a), blood samples with greater amounts of HSV antibody turn brighter yellow. **(c)** An immunoblot. Lanes 1 to 3 are from an SDS gel; samples from successive stages in the purification of a protein kinase were separated and stained with Coomassie blue. Lanes 4 to 6 show the same samples, but these were electrophoretically transferred to a nitrocellulose membrane after separation on an SDS gel. The membrane was then "probed" with antibody against the protein kinase. The numbers between the SDS gel and the immunoblot indicate M_r in thousands. [Sources: (b, c) State of Wisconsin Lab of Hygiene, Madison, WI.]

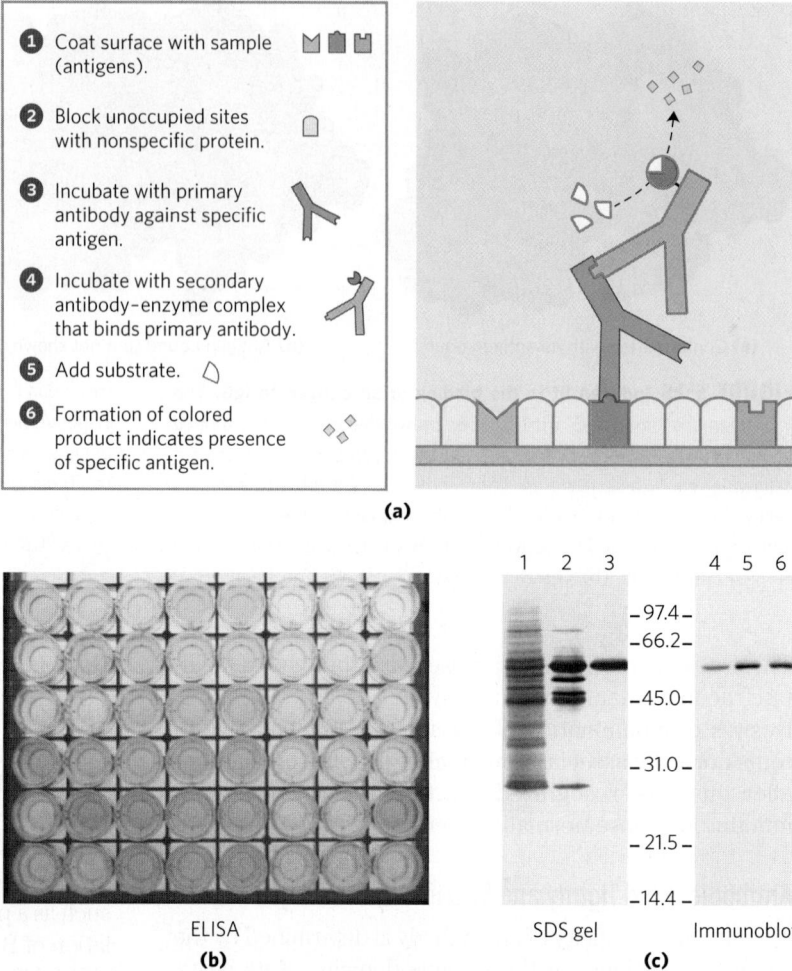

❶ Coat surface with sample (antigens).

❷ Block unoccupied sites with nonspecific protein.

❸ Incubate with primary antibody against specific antigen.

❹ Incubate with secondary antibody-enzyme complex that binds primary antibody.

❺ Add substrate.

❻ Formation of colored product indicates presence of specific antigen.

(a)

ELISA
(b)

1 2 3   4 5 6

−97.4−
−66.2−
−45.0−
−31.0−
−21.5−
−14.4−

SDS gel   Immunoblot
(c)

target protein and retains it on the column while other proteins are washed through. The target protein can then be eluted from the resin by a salt solution or some other agent. This is a powerful protein analytical tool.

In another versatile analytical technique, an antibody is attached to a radioactive label or some other reagent that makes it easy to detect. When the antibody binds the target protein, the label reveals the presence of the protein in a solution or its location in a gel, or even in a living cell. Several variations of this procedure are illustrated in **Figure 5-26**.

An **ELISA** (enzyme-linked immunosorbent assay) can be used to rapidly screen for and quantify an antigen in a sample (Fig. 5-26b). Proteins in the sample are adsorbed to an inert surface, usually a 96-well polystyrene plate. The surface is washed with a solution of an inexpensive nonspecific protein (often casein from nonfat dry milk powder) to block proteins introduced in subsequent steps from adsorbing to unoccupied sites. The surface is then treated with a solution containing the primary antibody— an antibody against the protein of interest. Unbound antibody is washed away, and the surface is treated with a solution containing a secondary antibody—antibody against the primary antibody—linked to an enzyme that catalyzes a reaction that forms a colored product. After unbound secondary antibody is washed away, the substrate of the antibody-linked enzyme is added. Product

formation (monitored as color intensity) is proportional to the concentration of the protein of interest in the sample.

In an **immunoblot assay**, also called a **Western blot** (Fig. 5-26c), proteins that have been separated by gel electrophoresis are transferred electrophoretically to a nitrocellulose membrane. The membrane is blocked (as described above for ELISA), then treated successively with primary antibody, secondary antibody linked to enzyme, and substrate. A colored precipitate forms only along the band containing the protein of interest. Immunoblotting allows the detection of a minor component in a sample and provides an approximation of its molecular weight.

We will encounter other aspects of antibodies in later chapters. They are extremely important in medicine and can tell us much about the structure of proteins and the action of genes.

SUMMARY 5.2 Complementary Interactions between Proteins and Ligands: The Immune System and Immunoglobulins

■ The immune response is mediated by interactions among an array of specialized leukocytes and their associated proteins. T lymphocytes produce T-cell receptors. B lymphocytes produce immunoglobulins. In a process called clonal selection, helper T cells induce the proliferation of B cells and cytotoxic T cells that produce immunoglobulins or proliferation of T-cell receptors that bind to a specific antigen.

■ Humans have five classes of immunoglobulins, each with different biological functions. The most abundant class is IgG, a Y-shaped protein with two heavy and two light chains. The domains near the upper ends of the Y are hypervariable within the broad population of IgGs and form two antigen-binding sites.

■ A given immunoglobulin generally binds to only a part, called the epitope, of a large antigen. Binding often involves a conformational change in the IgG, an induced fit to the antigen.

■ The exquisite binding specificity of immunoglobulins is exploited in analytical techniques such as ELISA and immunoblotting.

5.3 Protein Interactions Modulated by Chemical Energy: Actin, Myosin, and Molecular Motors

Organisms move. Cells move. Organelles and macromolecules within cells move. Most of these movements arise from the activity of a fascinating class of protein-based molecular motors. Fueled by chemical energy, usually derived from ATP, large aggregates of motor proteins undergo cyclic conformational changes that accumulate into a unified, directional force—the tiny force that pulls apart chromosomes in a dividing cell, and the immense force that levers a pouncing, quarter-ton jungle cat into the air.

The interactions among motor proteins, as you might predict, feature complementary arrangements of ionic, hydrogen-bonding, and hydrophobic groups at protein binding sites. In motor proteins, however, the resulting interactions achieve exceptionally high levels of spatial and temporal organization.

Motor proteins underlie the migration of organelles along microtubules, the motion of eukaryotic and bacterial flagella, the movement of some proteins along DNA, and the contraction of muscles. Proteins called kinesins and dyneins move along microtubules in cells, pulling along organelles or reorganizing chromosomes during cell division. An interaction of dynein with microtubules brings about the motion of eukaryotic flagella and cilia. Flagellar motion in bacteria involves a complex rotational motor at the base of the flagellum (see Fig. 19-41). Helicases, polymerases, and other proteins move along DNA as they carry out their functions in DNA metabolism (Chapter 25). Here, we focus on the well-studied example of the contractile proteins of vertebrate skeletal muscle as a paradigm for how proteins translate chemical energy into motion.

The Major Proteins of Muscle Are Myosin and Actin

The contractile force of muscle is generated by the interaction of two proteins, myosin and actin. These proteins are arranged in filaments that undergo transient interactions and slide past each other to bring about contraction. Together, actin and myosin make up more than 80% of the protein mass of muscle.

Myosin (M_r 520,000) has six subunits: two heavy chains (each of M_r 220,000) and four light chains (each of M_r 20,000). The heavy chains account for much of the overall structure. At their carboxyl termini, they are arranged as extended α helices, wrapped around each other in a fibrous, left-handed coiled coil similar to that of α-keratin (**Fig. 5-27a**). At its amino terminus, each heavy chain has a large globular domain containing a site where ATP is hydrolyzed. The light chains are associated with the globular domains. When myosin is treated briefly with the protease trypsin, much of the fibrous tail is cleaved off, dividing the protein into components called light and heavy meromyosin (Fig. 5-27b). The globular domain—called myosin subfragment 1, or S1, or simply the myosin head group—is liberated from heavy meromyosin by cleavage with papain, leaving myosin subfragment 2, or S2. The S1 fragment is the motor domain that makes muscle contraction possible. S1 fragments can be crystallized, and their overall structure, as determined by Ivan Rayment and Hazel Holden, is shown in Figure 5-27c.

In muscle cells, molecules of myosin aggregate to form structures called **thick filaments (Fig. 5-28a)**. These rodlike structures are the core of the contractile unit. Within a thick filament, several hundred myosin molecules are arranged with their fibrous "tails" associated to form a long bipolar structure. The globular domains project from either end of this structure, in regular stacked arrays.

The second major muscle protein, **actin**, is abundant in almost all eukaryotic cells. In muscle, molecules of monomeric actin, called G-actin (*g*lobular actin; M_r 42,000), associate to form a long polymer called F-actin (*f*ilamentous actin). The **thin filament** consists of F-actin (Fig. 5-28b), along with the proteins troponin and tropomyosin (discussed below). The filamentous parts of thin filaments assemble as successive monomeric actin molecules add to one end. On addition, each monomer binds ATP, then hydrolyzes it to ADP, so every actin molecule in the filament is complexed to ADP. This ATP hydrolysis by actin functions only in the assembly of the filaments; it does not contribute directly to the energy expended in muscle contraction. Each actin monomer in the thin filament can bind tightly and specifically to one myosin head group (Fig. 5-28c).

Additional Proteins Organize the Thin and Thick Filaments into Ordered Structures

Skeletal muscle consists of parallel bundles of **muscle fibers**, each fiber a single, very large, multinucleated cell, 20 to 100 μm in diameter, formed from many cells fused together; a single fiber often spans the length of the muscle. Each fiber contains about 1,000 **myofibrils**, 2 μm in diameter, each consisting of a vast number of regularly arrayed thick and thin filaments complexed to

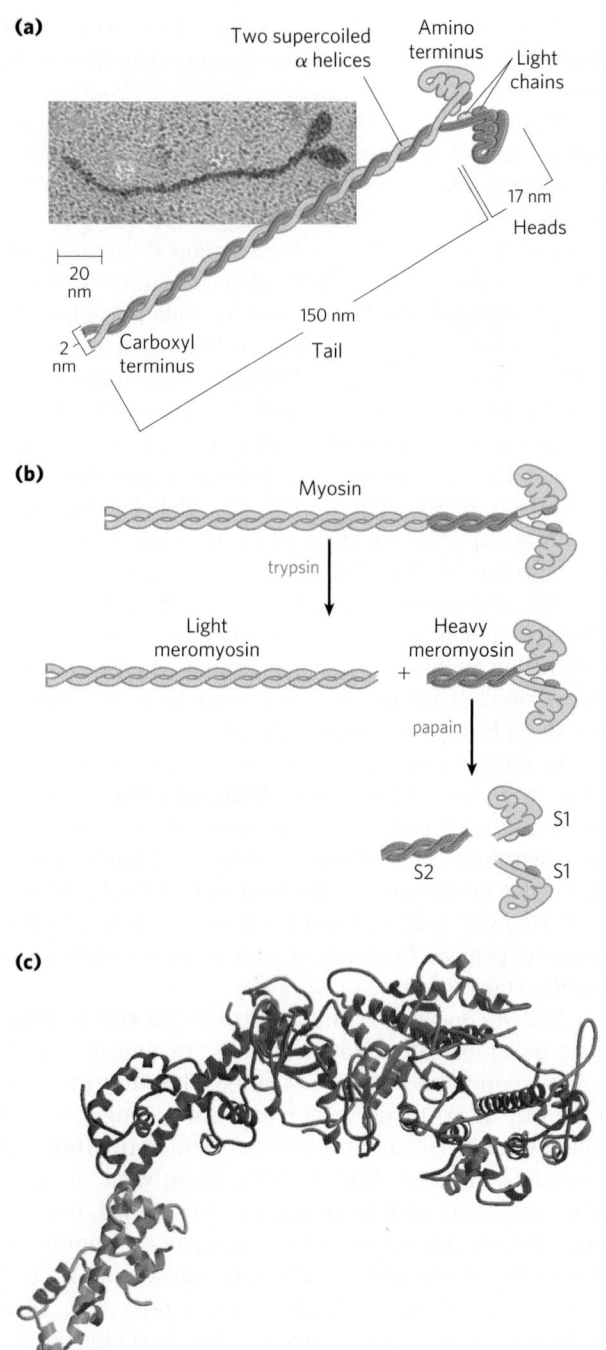

(a)

Two supercoiled α helices

Amino terminus

Light chains

17 nm

Heads

20 nm

2 nm

Carboxyl terminus

150 nm

Tail

(b)

Myosin

trypsin

Light meromyosin

Heavy meromyosin

+

papain

S1

S2

S1

(c)

FIGURE 5-27 Myosin. (a) Myosin has two heavy chains (in two shades of red), the carboxyl termini forming an extended coiled coil (tail) and the amino termini having globular domains (heads). Two light chains (blue) are associated with each myosin head. **(b)** Cleavage with trypsin and papain separates the myosin heads (S1 fragments) from the tails (S2 fragments). **(c)** Ribbon representation of the myosin S1 fragment. The heavy chain is in gray, the two light chains in two shades of blue. [Sources: (a) Takeshi Katayama, et al. "Stimulatory effects of arachidonic acid on myosin ATPase activity and contraction of smooth muscle via myosin motor domain," *Am. J. Physiol. Heart Circ. Physiol.* Vol 298, Issue 2, pp. H505-H514, February 2010, Fig. 6b. **(c)** Courtesy of Ivan Rayment, University of Wisconsin–Madison, Enzyme Institute and Department of Biochemistry; PDB ID 2MYS, I. Rayment et al., *Science* 261:50, 1993.]

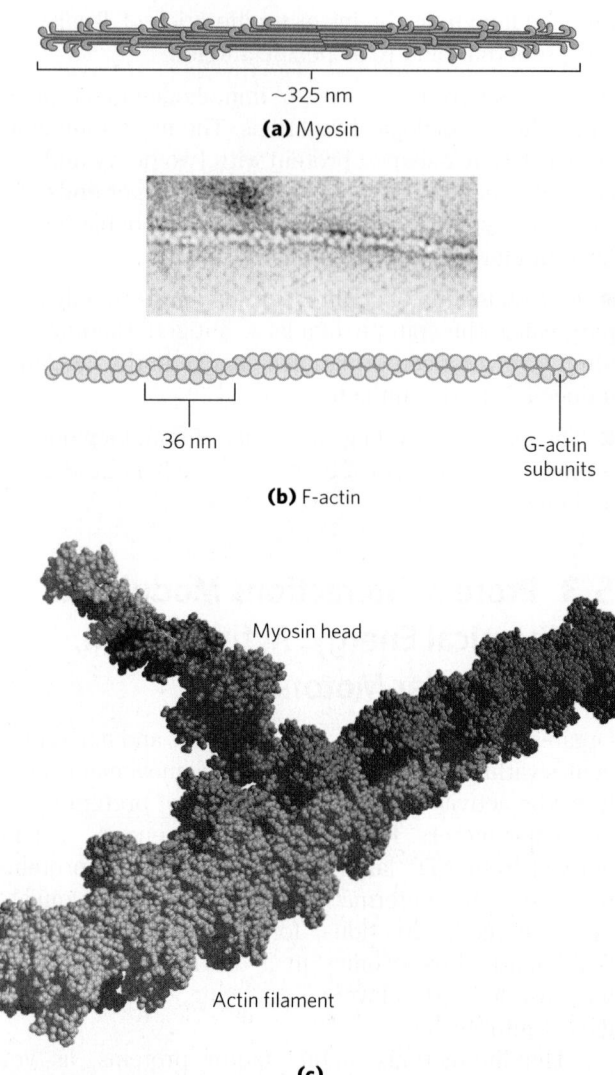

~325 nm

(a) Myosin

36 nm

G-actin subunits

(b) F-actin

Myosin head

Actin filament

(c)

FIGURE 5-28 The major components of muscle. (a) Myosin aggregates to form a bipolar structure called a thick filament. **(b)** F-actin is a filamentous assemblage of G-actin monomers that polymerize two by two, giving the appearance of two filaments spiraling about one another in a right-handed fashion. **(c)** Space-filling model of an actin filament (shades of red) with one myosin head (gray and two shades of blue) bound to an actin monomer within the filament. [Sources: (b) Dr. Roger W. Craig PhD, University of Massachusetts Medical School. (c) Courtesy of Ivan Rayment, University of Wisconsin–Madison, Enzyme Institute and Department of Biochemistry; PDB ID 2MYS, I. Rayment et al., *Science* 261:50, 1993.]

other proteins **(Fig. 5-29)**. A system of flat membranous vesicles called the **sarcoplasmic reticulum** surrounds each myofibril. Examined under the electron microscope, muscle fibers reveal alternating regions of high and low electron density, called the **A bands** and **I bands** (Fig. 5-29b, c). The A and I bands arise from the arrangement of thick and thin filaments, which are aligned and partially overlapping. The I band is the region of the bundle that in cross section would contain only thin filaments. The darker A band stretches the length of the thick filament

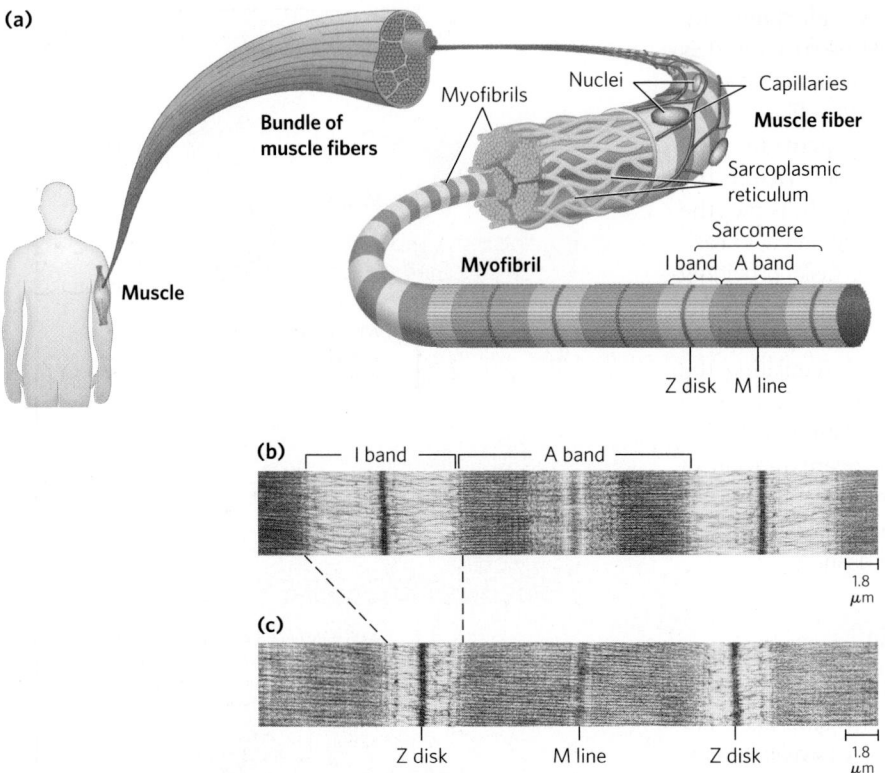

FIGURE 5-29 Skeletal muscle. (a) Muscle fibers consist of single, elongated, multinucleated cells that arise from the fusion of many precursor cells. The fibers are made up of many myofibrils (only six are shown here for simplicity) surrounded by the membranous sarcoplasmic reticulum. The organization of thick and thin filaments in a myofibril gives it a striated appearance. When muscle contracts, the I bands narrow and the Z disks move closer together, as seen in electron micrographs of (b) relaxed and (c) contracted muscle. [Source: (b, c) James E. Dennis/Phototake.]

and includes the region where parallel thick and thin filaments overlap. Bisecting the I band is a thin structure called the **Z disk**, perpendicular to the thin filaments and serving as an anchor to which the thin filaments are attached. The A band, too, is bisected by a thin line, the **M line** or M disk, a region of high electron density in the middle of the thick filaments. The entire contractile unit, consisting of bundles of thick filaments interleaved

at either end with bundles of thin filaments, is called the **sarcomere**. The arrangement of interleaved bundles allows the thick and thin filaments to slide past each other (by a mechanism discussed below), causing a progressive shortening of each sarcomere **(Fig. 5-30)**.

The thin actin filaments are attached at one end to the Z disk in a regular pattern. The assembly includes the minor muscle proteins **α-actinin**, **desmin**, and **vimentin**.

FIGURE 5-30 Muscle contraction. Thick filaments are bipolar structures created by the association of many myosin molecules. (a) Muscle contraction occurs by the sliding of the thick and thin filaments past each other so that the Z disks in neighboring I bands draw closer together. (b) The thick and thin filaments are interleaved such that each thick filament is surrounded by six thin filaments.

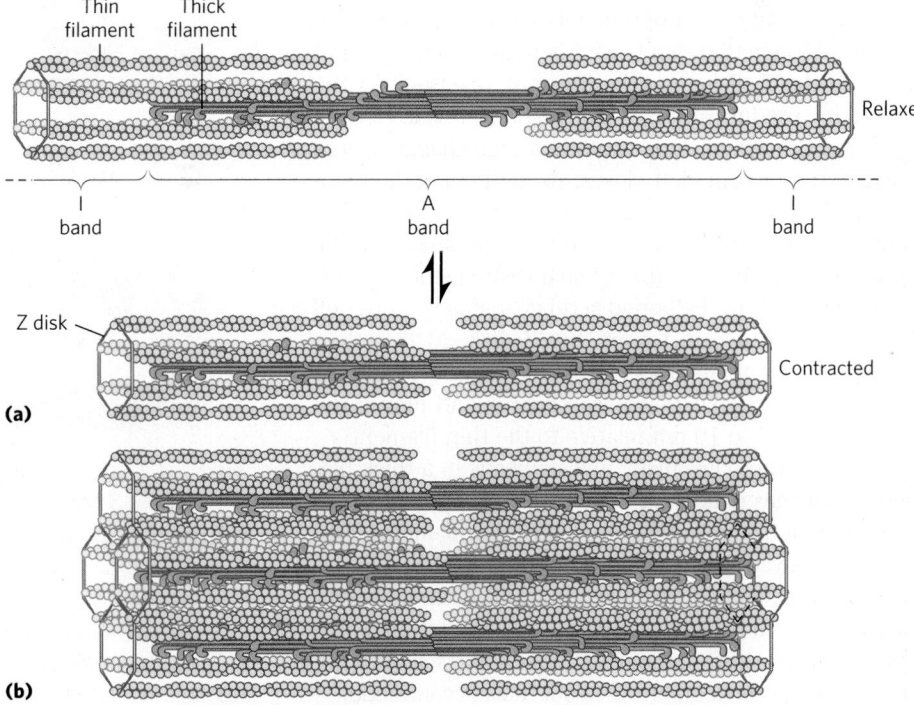

Thin filaments also contain a large protein called **nebulin** (~7,000 amino acid residues), thought to be structured as an α helix that is long enough to span the length of the filament. The M line similarly organizes the thick filaments. It contains the proteins **paramyosin**, **C-protein**, and **M-protein**. Another class of proteins called **titins**, the largest single polypeptide chains discovered thus far (the titin of human cardiac muscle has 26,926 amino acid residues), link the thick filaments to the Z disk, providing additional organization to the overall structure. Among their structural functions, the proteins nebulin and titin are believed to act as "molecular rulers," regulating the length of the thin and thick filaments, respectively. Titin extends from the Z disk to the M line, regulating the length of the sarcomere itself and preventing overextension of the muscle. The characteristic sarcomere length varies from one muscle tissue to the next in a vertebrate, largely due to the different titin variants in the tissues.

Myosin Thick Filaments Slide along Actin Thin Filaments

The interaction between actin and myosin, like that between all proteins and ligands, involves weak bonds. When ATP is not bound to myosin, a face on the myosin head group binds tightly to actin **(Fig. 5-31)**. When ATP binds to myosin and is hydrolyzed to ADP and phosphate, a coordinated and cyclic series of conformational changes occurs in which myosin releases the F-actin subunit and binds another subunit farther along the thin filament.

The cycle has four major steps (Fig. 5-31). In step ❶, ATP binds to myosin and a cleft in the myosin molecule opens, disrupting the actin-myosin interaction so that the bound actin is released. ATP is then hydrolyzed in step ❷, causing a conformational change in the protein to a "high-energy" state that moves the myosin head and changes its orientation in relation to the actin thin filament. Myosin then binds weakly to an F-actin subunit closer to the Z disk than the one just released. As the phosphate product of ATP hydrolysis is released from myosin in step ❸, another conformational change occurs in which the myosin cleft closes, strengthening the myosin-actin binding. This is followed quickly by step ❹, a "power stroke" during which the conformation of the myosin head returns to the original resting state, its orientation relative to the bound actin changing so as to pull the tail of the myosin toward the Z disk. ADP is then released to complete the cycle. Each cycle generates about 3 to 4 pN (piconewtons) of force and moves the thick filament 5 to 10 nm relative to the thin filament.

Because there are many myosin heads in a thick filament, at any given moment some (probably 1% to 3%) are bound to thin filaments. This prevents thick filaments from slipping backward when an individual myosin head releases the actin subunit to which it was bound. The thick filament thus actively slides forward past the adjacent thin filaments. This process, coordinated among the many sarcomeres in a muscle fiber, brings about muscle contraction.

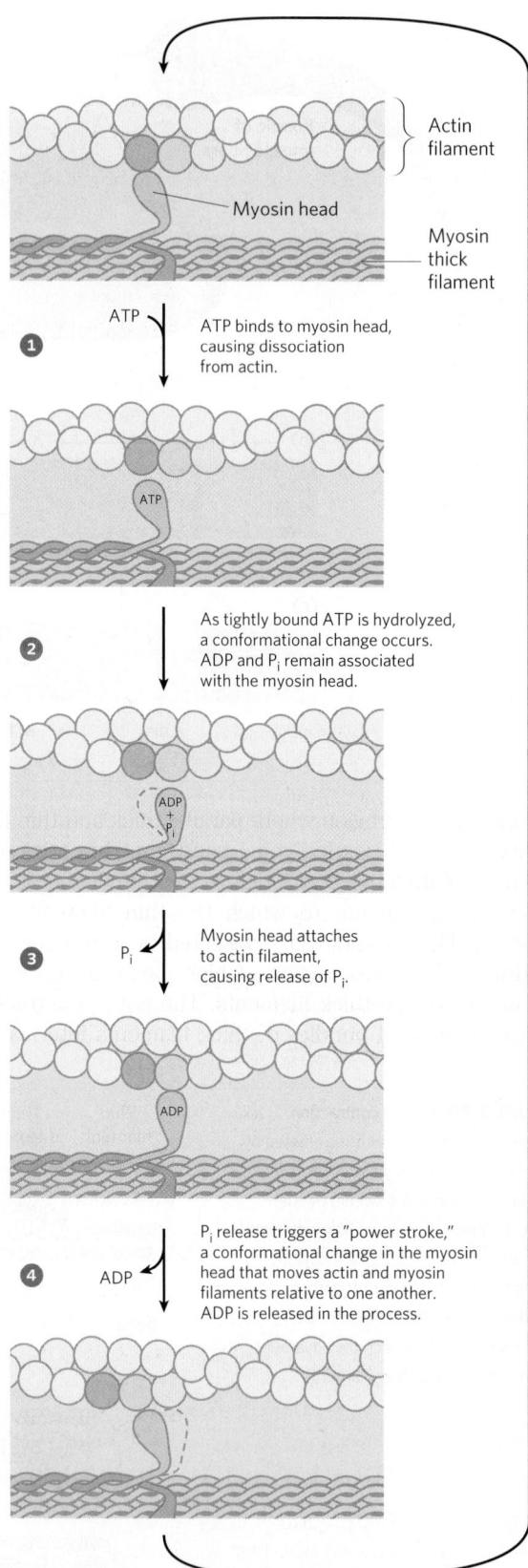

FIGURE 5-31 Molecular mechanism of muscle contraction. Conformational changes in the myosin head that are coupled to stages in the ATP hydrolytic cycle cause myosin to successively dissociate from one actin subunit, then associate with another farther along the actin filament. In this way, the myosin heads slide along the thin filaments, drawing the thick filament array into the thin filament array (see Fig. 5-30).

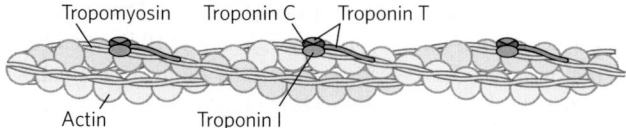

Tropomyosin Troponin C Troponin T

Actin Troponin I

FIGURE 5-32 Regulation of muscle contraction by tropomyosin and troponin. Tropomyosin and troponin are bound to F-actin in the thin filaments. In the relaxed muscle, these two proteins are arranged around the actin filaments so as to block the binding sites for myosin. Tropomyosin is a two-stranded coiled coil of α helices, the same structural motif as in α-keratin (see Fig. 4-11). It forms head-to-tail polymers twisting around the two actin chains. Troponin is attached to the actin-tropomyosin complex at regular intervals of 38.5 nm. Troponin consists of three different subunits: I, C, and T. Troponin I prevents binding of the myosin head to actin; troponin C has a binding site for Ca^{2+}; and troponin T links the entire troponin complex to tropomyosin. When the muscle receives a neural signal to initiate contraction, Ca^{2+} is released from the sarcoplasmic reticulum (see Fig. 5-29a) and binds to troponin C. This causes a conformational change in troponin C, which alters the positions of troponin I and tropomyosin so as to relieve the inhibition by troponin I and allow muscle contraction.

The interaction between actin and myosin must be regulated so that contraction occurs only in response to appropriate signals from the nervous system. The regulation is mediated by a complex of two proteins, **tropomyosin** and **troponin (Fig. 5-32)**. Tropomyosin binds to the thin filament, blocking the attachment sites for the myosin head groups. Troponin is a Ca^{2+}-binding protein. A nerve impulse causes release of Ca^{2+} ions from the sarcoplasmic reticulum. The released Ca^{2+} binds to troponin (another protein-ligand interaction) and causes a conformational change in the tropomyosin-troponin complexes, exposing the myosin-binding sites on the thin filaments. Contraction follows.

Working skeletal muscle requires two types of molecular functions that are common in proteins—binding and catalysis. The actin-myosin interaction, a protein-ligand interaction like that of immunoglobulins with antigens, is reversible and leaves the participants unchanged. When ATP binds myosin, however, it is hydrolyzed to ADP and P_i. Myosin is not only an actin-binding protein, it is also an ATPase—an enzyme. The function of enzymes in catalyzing chemical transformations is the topic of the next chapter.

SUMMARY 5.3 Protein Interactions Modulated by Chemical Energy: Actin, Myosin, and Molecular Motors

■ Protein-ligand interactions achieve a special degree of spatial and temporal organization in motor proteins. Muscle contraction results from choreographed interactions between myosin and actin, coupled to the hydrolysis of ATP by myosin.

■ Myosin consists of two heavy and four light chains, forming a fibrous coiled coil (tail) domain and a globular (head) domain. Myosin molecules are organized into thick filaments, which slide past thin

filaments composed largely of actin. ATP hydrolysis in myosin is coupled to a series of conformational changes in the myosin head, leading to dissociation of myosin from one F-actin subunit and its eventual reassociation with another, farther along the thin filament. The myosin thus slides along the actin filaments.

■ Muscle contraction is stimulated by the release of Ca^{2+} from the sarcoplasmic reticulum. The Ca^{2+} binds to the protein troponin, leading to a conformational change in a troponin-tropomyosin complex that triggers the cycle of actin-myosin interactions.

Key Terms

Terms in bold are defined in the glossary.

ligand 157	**immunoglobulin** 174
binding site 157	**B lymphocyte** or
induced fit 157	**B cell** 174
hemoglobin 158	**T lymphocyte** or
heme 158	**T cell** 174
porphyrin 158	**antigen** 175
heme protein 158	**epitope** 175
globins 159	**hapten** 175
equilibrium expression 160	immunoglobulin fold 176
association constant, K_a 160	**polyclonal**
dissociation constant,	**antibodies** 177
K_d 160	**monoclonal**
allosteric protein 166	**antibodies** 177
modulator 166	**ELISA** 178
Hill equation 167	**immunoblotting** 178
Bohr effect 170	**Western blotting** 178
immune response 174	**myosin** 179
lymphocytes 174	**actin** 179
antibody 174	**sarcomere** 179

Problems

1. Relationship between Affinity and Dissociation Constant Protein A has a binding site for ligand X with a K_d of 10^{-6} M. Protein B has a binding site for ligand X with a K_d of 10^{-9} M. Which protein has a higher affinity for ligand X? Explain your reasoning. Convert the K_d to K_a for both proteins.

2. Negative Cooperativity Which of the following situations would produce a Hill plot with $n_H < 1.0$? Explain your reasoning in each case.

(a) The protein has multiple subunits, each with a single ligand-binding site. Binding of ligand to one site decreases the binding affinity of other sites for the ligand.

(b) The protein is a single polypeptide with two ligand-binding sites, each having a different affinity for the ligand.

(c) The protein is a single polypeptide with a single ligand-binding site. As purified, the protein preparation is heterogeneous, containing some protein molecules that are partially denatured and thus have a lower binding affinity for the ligand.

3. Hemoglobin's Affinity for Oxygen What is the effect of the following changes on the O_2 affinity of hemoglobin? (a) A drop in the pH of blood plasma from 7.4 to 7.2. (b) A decrease in the partial pressure of CO_2 in the lungs from 6 kPa (holding one's breath) to 2 kPa (normal breathing). (c) An increase in the BPG level from 5 mM (normal altitudes) to 8 mM (high altitudes). (d) An increase in CO from 1.0 part per million (ppm) in a normal indoor atmosphere to 30 ppm in a home that has a malfunctioning or leaking furnace.

4. Reversible Ligand Binding I The protein calcineurin binds to the protein calmodulin with an association rate of $8.9 \times 10^3\ M^{-1}s^{-1}$ and an overall dissociation constant, K_d, of 10 nM. Calculate the dissociation rate, k_d, including appropriate units.

5. Reversible Ligand Binding II A binding protein binds to a ligand L with a K_d of 400 nM. What is the concentration of ligand when Y is (a) 0.25, (b) 0.6, (c) 0.95?

6. Reversible Ligand Binding III Three membrane receptor proteins bind tightly to a hormone. Based on the data in the table below, (a) what is the K_d for hormone binding by protein 2? (Include appropriate units.) (b) Which of these proteins binds *most* tightly to this hormone?

Hormone concentration (nM)	Y		
	Protein 1	Protein 2	Protein 3
0.2	0.048	0.29	0.17
0.5	0.11	0.5	0.33
1	0.2	0.67	0.5
4	0.5	0.89	0.8
10	0.71	0.95	0.91
20	0.83	0.97	0.95
50	0.93	0.99	0.98

7. Cooperativity in Hemoglobin Under appropriate conditions, hemoglobin dissociates into its four subunits. The isolated α subunit binds oxygen, but the O_2-saturation curve is hyperbolic rather than sigmoid. In addition, the binding of oxygen to the isolated α subunit is not affected by the presence of H^+, CO_2, or BPG. What do these observations indicate about the source of the cooperativity in hemoglobin?

8. Comparison of Fetal and Maternal Hemoglobins Studies of oxygen transport in pregnant mammals show that the O_2-saturation curves of fetal and maternal blood are markedly different when measured under the same conditions. Fetal erythrocytes contain a structural variant of hemoglobin, HbF, consisting of two α and two γ subunits ($\alpha_2\gamma_2$), whereas maternal erythrocytes contain HbA ($\alpha_2\beta_2$).

(a) Which hemoglobin has a higher affinity for oxygen under physiological conditions, HbA or HbF? Explain.

(b) What is the physiological significance of the different O_2 affinities?

(c) When all the BPG is carefully removed from samples of HbA and HbF, the measured O_2-saturation curves (and conse-

quently the O_2 affinities) are displaced to the left. However, HbA now has a greater affinity for oxygen than does HbF. When BPG is reintroduced, the O_2-saturation curves return to normal, as shown in the graph. What is the effect of BPG on the O_2 affinity of hemoglobin? How can the above information be used to explain the different O_2 affinities of fetal and maternal hemoglobin?

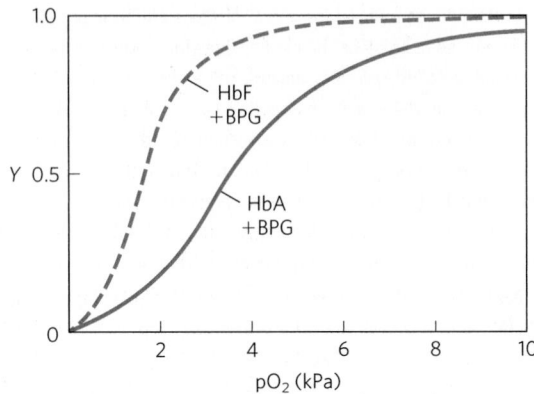

9. Hemoglobin Variants There are almost 500 naturally occurring variants of hemoglobin. Most are the result of a single amino acid substitution in a globin polypeptide chain. Some variants produce clinical illness, though not all variants have deleterious effects. A brief sample follows.

HbS (sickle cell Hb): substitutes a Val for a Glu on the surface

Hb Cowtown: eliminates an ion pair involved in T-state stabilization

Hb Memphis: substitutes one uncharged polar residue for another of similar size on the surface

Hb Bibba: substitutes a Pro for a Leu involved in an α helix

Hb Milwaukee: substitutes a Glu for a Val

Hb Providence: substitutes an Asn for a Lys that normally projects into the central cavity of the tetramer

Hb Philly: substitutes a Phe for a Tyr, disrupting hydrogen bonding at the $\alpha_1\beta_1$ interface

Explain your choices for each of the following:

(a) The Hb variant *least* likely to cause pathological symptoms.

(b) The variant(s) most likely to show pI values different from that of HbA on an isoelectric focusing gel.

(c) The variant(s) most likely to show a decrease in BPG binding and an increase in the overall affinity of the hemoglobin for oxygen.

10. Oxygen Binding and Hemoglobin Structure A team of biochemists uses genetic engineering to modify the interface region between hemoglobin subunits. The resulting hemoglobin variants exist in solution primarily as $\alpha\beta$ dimers (few, if any, $\alpha_2\beta_2$ tetramers form). Are these variants likely to bind oxygen more weakly or more tightly? Explain your answer.

11. Reversible (but Tight) Binding to an Antibody An antibody binds to an antigen with a K_d of 5×10^{-8} M. At what concentration of antigen will Y be (a) 0.2, (b) 0.5, (c) 0.6, (d) 0.8?

12. Using Antibodies to Probe Structure-Function Relationships in Proteins A monoclonal antibody binds to

G-actin but not to F-actin. What does this tell you about the epitope recognized by the antibody?

13. The Immune System and Vaccines A host organism needs time, often days, to mount an immune response against a new antigen, but memory cells permit a rapid response to pathogens previously encountered. A vaccine to protect against a particular viral infection often consists of weakened or killed virus or isolated proteins from a viral protein coat. When injected into a person, the vaccine generally does not cause an infection and illness, but it effectively "teaches" the immune system what the viral particles look like, stimulating the production of memory cells. On subsequent infection, these cells can bind to the virus and trigger a rapid immune response. Some pathogens, including HIV, have developed mechanisms to evade the immune system, making it difficult or impossible to develop effective vaccines against them. What strategy could a pathogen use to evade the immune system? Assume that a host's antibodies and/or T-cell receptors are available to bind to any structure that might appear on the surface of a pathogen and that, once bound, the pathogen is destroyed.

14. How We Become a "Stiff" When a vertebrate dies, its muscles stiffen as they are deprived of ATP, a state called rigor mortis. Using your knowledge of the catalytic cycle of myosin in muscle contraction, explain the molecular basis of the rigor state.

15. Sarcomeres from Another Point of View The symmetry of thick and thin filaments in a sarcomere is such that six thin filaments ordinarily surround each thick filament in a hexagonal array. Draw a cross section (transverse cut) of a myofibril at the following points: (a) at the M line; (b) through the I band; (c) through the dense region of the A band; (d) through the less dense region of the A band, adjacent to the M line (see Fig. 5-29b, c).

Biochemistry Online

16. Lysozyme and Antibodies To fully appreciate how proteins function in a cell, it is helpful to have a three-dimensional view of how proteins interact with other cellular components. Fortunately, this is possible using Web-based protein databases and three-dimensional molecular viewing utilities such as JSmol, a free and user-friendly molecular viewer that is compatible with most browsers and operating systems.

In this exercise, you will examine the interactions between the enzyme lysozyme (Chapter 4) and the Fab portion of the anti-lysozyme antibody. Use the PDB identifier 1FDL to explore the structure of the IgG1 Fab fragment–lysozyme complex (antibody-antigen complex). To answer the following questions, use the information on the Structure Summary page at the Protein Data Bank (www.pdb.org), and view the structure using JSmol or a similar viewer.

(a) Which chains in the three-dimensional model correspond to the antibody fragment and which correspond to the antigen, lysozyme?

(b) What type of secondary structure predominates in this Fab fragment?

(c) How many amino acid residues are in the heavy and light chains of the Fab fragment? In lysozyme? Estimate the percentage of the lysozyme that interacts with the antigen-binding site of the antibody fragment.

(d) Identify the specific amino acid residues in lysozyme and in the variable regions of the Fab heavy and light chains that are situated at the antigen-antibody interface. Are the residues contiguous in the primary sequence of the polypeptide chains?

17. Exploring Antibodies in the Protein Data Bank Use the PDB Molecule of the Month article at www.rcsb.org/pdb/101/motm.do?momID=21 to complete the following exercises.

(a) How many specific antigen-binding sites are there on the first immunoglobulin image on the Web page (image derived from PDB ID 1IGT)?

(b) When a virus enters your lungs, how long does it take for you to produce one or more antibodies that bind to it?

(c) Approximately how many types of different antibodies are present in your blood?

(d) Explore the structure of the immunoglobulin molecule (PDB ID 1IGT) on the Web page by clicking the link in the article or by going directly to www.rcsb.org/pdb/explore/explore.do?structureId=1igt. Use one of the structure viewers provided on the PDB site to create a ribbon structure for this immunoglobulin. Identify the two light chains and two heavy chains, and give them different colors.

Data Analysis Problem

18. Protein Function During the 1980s, the structures of actin and myosin were known only at the resolution shown in Figure 5-28a, b. Although researchers knew that the S1 portion of myosin bound to actin and hydrolyzed ATP, there was a substantial debate about where in the myosin molecule the contractile force was generated. At the time, two competing models were proposed for the mechanism of force generation in myosin.

In the "hinge" model, S1 bound to actin, but the pulling force was generated by contraction of the "hinge region" in the myosin tail. The hinge region is in the heavy meromyosin portion of the myosin molecule, near where trypsin cleaves off light meromyosin (see Fig. 5-27b); this is roughly the point labeled "Two supercoiled α helices" in Figure 5-27a. In the "S1" model, the pulling force was generated in the S1 "head" itself and the tail was just for structural support.

Many experiments were performed but provided no conclusive evidence. Then, in 1987, James Spudich and his colleagues at Stanford University published a study that, although not conclusive, went a long way toward resolving this controversy.

Recombinant DNA techniques were not sufficiently developed to address this issue in vivo, so Spudich and colleagues used an interesting in vitro motility assay. The alga *Nitella* has extremely long cells, often several centimeters long and about 1 mm in diameter. These cells have actin fibers that run along their long axes, and the cells can be cut open along their length to expose the actin fibers. Spudich and his group had observed that plastic beads coated with myosin would "walk" along these fibers in the presence of ATP, just as myosin would do in contracting muscle.

For these experiments, the researchers used a more well-defined method for attaching the myosin to the beads. The "beads" were clumps of killed bacterial (*Staphylococcus aureus*) cells. These cells have a protein on their surface that binds to the Fc region of antibody molecules (Fig. 5-21a). The antibodies, in turn, bind to several (unknown) places along the tail of the myosin molecule. When bead-antibody-myosin complexes were prepared with intact myosin molecules, they would move along *Nitella* actin fibers in the presence of ATP.

(a) Sketch a diagram showing what a bead-antibody-myosin complex might look like at the molecular level.

(b) Why was ATP required for the beads to move along the actin fibers?

(c) Spudich and coworkers used antibodies that bound to the myosin tail. Why would this experiment have failed if they had used an antibody that bound to the part of S1 that normally bound to actin? Why would this experiment have failed if they had used an antibody that bound to actin?

To help focus on the part of myosin responsible for force production, Spudich and colleagues used trypsin to produce two partial myosin molecules (Fig. 5-27b): (1) heavy meromyosin (HMM), made by briefly digesting myosin with trypsin; HMM consists of S1 and the part of the tail that includes the hinge; and (2) short heavy meromyosin (SHMM), made from a more extensive digestion of HMM with trypsin; SHMM consists of S1 and a shorter part of the tail that does not include the hinge. Brief digestion of myosin with trypsin produces HMM and light meromyosin, by cleavage of a single specific peptide bond in the myosin molecule.

(d) Why might trypsin attack this peptide bond first rather than other peptide bonds in myosin?

Spudich and colleagues prepared bead-antibody-myosin complexes with varying amounts of myosin, HMM, and SHMM and measured their speed of movement along *Nitella* actin

fibers in the presence of ATP. The graph below sketches their results.

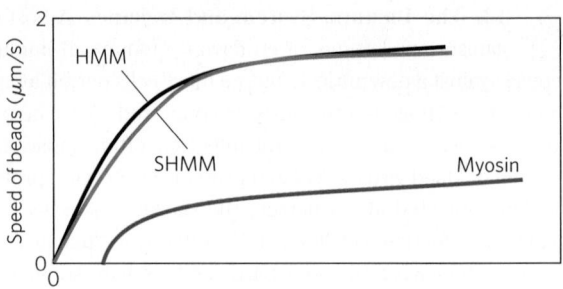

(e) Which model ("S1" or "hinge") is consistent with these results? Explain your reasoning.

(f) Provide a plausible explanation for the increased speed of the beads with increasing myosin density.

(g) Provide a plausible explanation for the plateauing of the speed of the beads at high myosin density.

The more extensive trypsin digestion required to produce SHMM had a side effect: another specific cleavage of the myosin polypeptide backbone in addition to the cleavage in the tail. This second cleavage was in the S1 head.

(h) Why is it surprising that SHMM was still capable of moving beads along the actin fibers?

(i) As it turns out, the tertiary structure of the S1 head remains intact in SHMM. Provide a plausible explanation of how the protein remains intact and functional even though the polypeptide backbone has been cleaved and is no longer continuous.

Reference

Hynes, T.R., S.M. Block, B.T. White, and J.A. Spudich. 1987. Movement of myosin fragments in vitro: domains involved in force production. *Cell* 48:953–963.

Further Reading is available at www.macmillanlearning.com/LehningerBiochemistry7e.

CHAPTER 6

Enzymes

Self-study tools that will help you practice what you've learned and reinforce this chapter's concepts are available online. Go to www.macmillanlearning.com/LehningerBiochemistry7e.

There are two fundamental conditions for life. First, the organism must be able to self-replicate (a topic considered in Part III); second, it must be able to catalyze chemical reactions efficiently and selectively. The central importance of catalysis may seem surprising, but it is easy to demonstrate. As described in Chapter 1, living systems make use of energy from the environment. Many of us, for example, consume substantial amounts of sucrose—common table sugar—as a kind of fuel, usually in the form of sweetened foods and drinks. The conversion of sucrose to CO_2 and H_2O in the presence of oxygen is a highly exergonic process, releasing free energy that we can use to think, move, taste, and see. However, a bag of sugar can remain on the shelf for years without any obvious conversion to CO_2 and H_2O. Although this chemical process is thermodynamically favorable, it is very slow. Yet when sucrose is consumed by a human (or almost any other organism), it releases its chemical energy in seconds. The difference is catalysis. Without catalysis, chemical reactions such as sucrose oxidation could not occur on a useful time scale, and thus could not sustain life.

In this chapter, we turn our attention to the reaction catalysts of biological systems: enzymes, the most remarkable and highly specialized proteins. Enzymes have extraordinary catalytic power, often far greater than that of synthetic or inorganic catalysts. They have a high degree of specificity for their substrates, they accelerate chemical reactions tremendously, and they function in aqueous solutions under very mild conditions of temperature and pH. Few nonbiological catalysts have all these properties.

Enzymes are central to every biochemical process. Acting in organized sequences, they catalyze the hundreds of stepwise reactions that degrade nutrient molecules, conserve and transform chemical energy, and make biological macromolecules from simple precursors.

The study of enzymes has immense practical importance. In some diseases, especially inheritable genetic disorders, there may be a deficiency or even a total absence of one or more enzymes. Other disease conditions may be caused by excessive activity of an enzyme. Measurements of the activities of enzymes in blood plasma, erythrocytes, or tissue samples are important in diagnosing certain illnesses. Many drugs act through interactions with enzymes. Enzymes are also important practical tools in chemical engineering, food technology, and agriculture.

We begin with descriptions of the properties of enzymes and the principles underlying their catalytic power, then introduce enzyme kinetics, a discipline that provides much of the framework for any discussion of enzymes. We then provide specific examples of enzyme mechanisms, illustrating principles introduced earlier in the chapter. We end with a discussion of how enzyme activity is regulated.

6.1 An Introduction to Enzymes

Much of the history of biochemistry is the history of enzyme research. Biological catalysis was first recognized and described in the late 1700s, in studies on the digestion of meat by secretions of the stomach. Research continued in the 1800s with examinations of the conversion of starch to sugar by saliva and various plant extracts. In the 1850s, Louis Pasteur concluded that fermentation of sugar into alcohol by yeast is catalyzed

Eduard Buchner, 1860–1917 [Source: Science Museum/ Science & Society Picture Library.]

James Sumner, 1887–1955 [Source: ©Courtesy Division of Rare and Manuscript Collections, Carl A. Kroch Library, Cornell University, Ithaca, NY. RMC2005_1073.]

J. B. S. Haldane, 1892–1964 [Source: Hans Wild/The LIFE Picture Collection/Getty Images.]

by "ferments." He postulated that these ferments were inseparable from the structure of living yeast cells. This view, called vitalism, prevailed for decades. Then in 1897, Eduard Buchner discovered that cell-free yeast extracts could ferment sugar to alcohol, proving that fermentation was promoted by molecules that continued to function when removed from cells. Buchner's experiment marked the end of vitalistic notions and the dawn of the science of biochemistry. Frederick W. Kühne later gave the name **enzymes** (from the Greek *enzymos*, "leavened") to the molecules detected by Buchner.

The isolation and crystallization of urease by James Sumner in 1926 was a breakthrough in early enzyme studies. Sumner found that urease crystals consisted entirely of protein, and he postulated that all enzymes were proteins. In the absence of other examples, this idea remained controversial for some time. Only in the 1930s was Sumner's conclusion widely accepted, after John Northrop and Moses Kunitz crystallized pepsin, trypsin, and other digestive enzymes and found them also to be proteins. During this period, J. B. S. Haldane wrote a treatise titled *Enzymes*. Although the molecular nature of enzymes was not yet fully appreciated, Haldane made the remarkable suggestion that weak bonding interactions between an enzyme and its substrate might be used to catalyze a reaction. This insight lies at the heart of our current understanding of enzymatic catalysis.

Since the latter part of the twentieth century, thousands of enzymes have been purified, their structures elucidated, and their mechanisms explained.

Most Enzymes Are Proteins

With the exception of a few classes of catalytic RNA molecules (Chapter 26), all enzymes are proteins. Their catalytic activity depends on the integrity of their native protein conformation. If an enzyme is denatured or dissociated into its subunits, catalytic activity is usually lost. If an enzyme is broken down into its component amino acids, its catalytic activity is always destroyed. Thus the primary, secondary, tertiary, and quaternary structures of protein enzymes are essential to their catalytic activity.

Enzymes, like other proteins, have molecular weights ranging from about 12,000 to more than 1 million. Some enzymes require no chemical groups for activity other than their amino acid residues. Others require an additional chemical component called a **cofactor**— either one or more inorganic ions, such as Fe^{2+}, Mg^{2+}, Mn^{2+}, or Zn^{2+} (Table 6-1), or a complex organic or metalloorganic molecule called a **coenzyme**. Coenzymes act as transient carriers of specific functional groups (Table 6-2). Most are derived from vitamins, organic nutrients required in small amounts in the diet. We consider coenzymes in more detail as we encounter them in the metabolic pathways discussed in Part II. Some enzymes require *both* a coenzyme and one or more metal ions for activity. A coenzyme or metal ion that is very tightly or even covalently bound to the enzyme protein is called a **prosthetic group**. A complete, catalytically active enzyme together with its bound coenzyme and/or metal ions is called a **holoenzyme**. The protein part of such an enzyme is called the **apoenzyme** or **apoprotein**. Finally, some enzyme proteins are modified covalently by phosphorylation, glycosylation, and other processes. Many of these alterations are involved in the regulation of enzyme activity.

Enzymes Are Classified by the Reactions They Catalyze

Many enzymes have been named by adding the suffix "-ase" to the name of their substrate or to a word or phrase describing their activity. Thus urease catalyzes hydrolysis of urea, and DNA polymerase catalyzes the polymerization of nucleotides to form DNA. Other enzymes were named by their discoverers for a broad

TABLE 6-1	Some Inorganic Ions That Serve as Cofactors for Enzymes
Ions	**Enzymes**
Cu^{2+}	Cytochrome oxidase
Fe^{2+} or Fe^{3+}	Cytochrome oxidase, catalase, peroxidase
K^+	Pyruvate kinase
Mg^{2+}	Hexokinase, glucose 6-phosphatase, pyruvate kinase
Mn^{2+}	Arginase, ribonucleotide reductase
Mo	Dinitrogenase
Ni^{2+}	Urease
Zn^{2+}	Carbonic anhydrase, alcohol dehydrogenase, carboxypeptidases A and B

TABLE 6-2	Some Coenzymes That Serve as Transient Carriers of Specific Atoms or Functional Groups	
Coenzyme	Examples of chemical groups transferred	Dietary precursor in mammals
Biocytin	CO_2	Biotin
Coenzyme A	Acyl groups	Pantothenic acid and other compounds
5′-Deoxyadenosylcobalamin (coenzyme B_{12})	H atoms and alkyl groups	Vitamin B_{12}
Flavin adenine dinucleotide	Electrons	Riboflavin (vitamin B_2)
Lipoate	Electrons and acyl groups	Not required in diet
Nicotinamide adenine dinucleotide	Hydride ion ($:H^-$)	Nicotinic acid (niacin)
Pyridoxal phosphate	Amino groups	Pyridoxine (vitamin B_6)
Tetrahydrofolate	One-carbon groups	Folate
Thiamine pyrophosphate	Aldehydes	Thiamine (vitamin B_1)

Note: The structures and modes of action of these coenzymes are described in Part II.

function, before the specific reaction catalyzed was known. For example, an enzyme known to act in the digestion of foods was named pepsin, from the Greek *pepsis*, "digestion," and lysozyme was named for its ability to lyse (break down) bacterial cell walls. Still others were named for their source: trypsin, named in part from the Greek *tryein*, "to wear down," was obtained by rubbing pancreatic tissue with glycerin. Sometimes the same enzyme has two or more names, or two different enzymes have the same name. Because of such ambiguities, as well as the ever-increasing number of newly discovered enzymes, biochemists, by international agreement, have adopted a system for naming and classifying enzymes. This system divides enzymes into six classes, each with subclasses, based on the type of reaction catalyzed (Table 6-3). Each enzyme is assigned a four-part classification number and a systematic name, which identifies the reaction it catalyzes. As an example, the formal systematic name of the enzyme catalyzing the reaction

$$ATP + \text{D-glucose} \longrightarrow ADP + \text{D-glucose 6-phosphate}$$

is ATP:D-hexose 6-phosphotransferase, which indicates that it catalyzes the transfer of a phosphoryl group from ATP to glucose. Its Enzyme Commission number (E.C. number)

is 2.7.1.1. The first number (2) denotes the class name (transferase); the second number (7), the subclass (phosphotransferase); the third number (1), a phosphotransferase with a hydroxyl group as acceptor; and the fourth number (1), D-glucose as the phosphoryl group acceptor. For many enzymes, a trivial name is more frequently used—in this case, hexokinase. A complete list and description of the thousands of known enzymes is maintained by the Nomenclature Committee of the International Union of Biochemistry and Molecular Biology (www.chem.qmul.ac.uk/iubmb/enzyme). This chapter is devoted primarily to principles and properties common to all enzymes.

SUMMARY 6.1 An Introduction to Enzymes

■ Life depends on powerful and specific catalysts: enzymes. Almost every biochemical reaction is catalyzed by an enzyme.

■ With the exception of a few catalytic RNAs, all known enzymes are proteins. Many require nonprotein coenzymes or cofactors for their catalytic function.

■ Enzymes are classified according to the type of reaction they catalyze. All enzymes have formal E.C. numbers and names, and most have trivial names.

TABLE 6-3	International Classification of Enzymes	
Class no.	Class name	Type of reaction catalyzed
1	Oxidoreductases	Transfer of electrons (hydride ions or H atoms)
2	Transferases	Group transfer reactions
3	Hydrolases	Hydrolysis reactions (transfer of functional groups to water)
4	Lyases	Cleavage of C—C, C—O, C—N, or other bonds by elimination, leaving double bonds or rings, or addition of groups to double bonds
5	Isomerases	Transfer of groups within molecules to yield isomeric forms
6	Ligases	Formation of C—C, C—S, C—O, and C—N bonds by condensation reactions coupled to cleavage of ATP or similar cofactor

6.2 How Enzymes Work

The enzymatic catalysis of reactions is essential to living systems. Under biologically relevant conditions, uncatalyzed reactions tend to be slow—most biological molecules are quite stable in the neutral-pH, mild-temperature, aqueous environment inside cells. Furthermore, many common chemical processes are unfavorable or unlikely in the cellular environment, such as the transient formation of unstable charged intermediates or the collision of two or more molecules in the precise orientation required for reaction. Reactions required to digest food, send nerve signals, or contract a muscle simply do not occur at a useful rate without catalysis.

An enzyme circumvents these problems by providing a specific environment in which a given reaction can occur more rapidly. The distinguishing feature of an enzyme-catalyzed reaction is that it takes place within the confines of a pocket on the enzyme called the **active site (Fig. 6-1)**. The molecule that is bound in the active site and acted upon by the enzyme is called the **substrate**. The surface of the active site is lined with amino acid residues with substituent groups that bind the substrate and catalyze its chemical transformation. Often, the active site encloses a substrate, sequestering it completely from solution. The enzyme-substrate complex, an entity first proposed by Charles-Adolphe Wurtz in 1880, is central to the action of enzymes. It is also the starting point for mathematical treatments that define the kinetic behavior of enzyme-catalyzed reactions and for theoretical descriptions of enzyme mechanisms.

Enzymes Affect Reaction Rates, Not Equilibria

A simple enzymatic reaction might be written

$$E + S \rightleftharpoons ES \rightleftharpoons EP \rightleftharpoons E + P \quad (6\text{-}1)$$

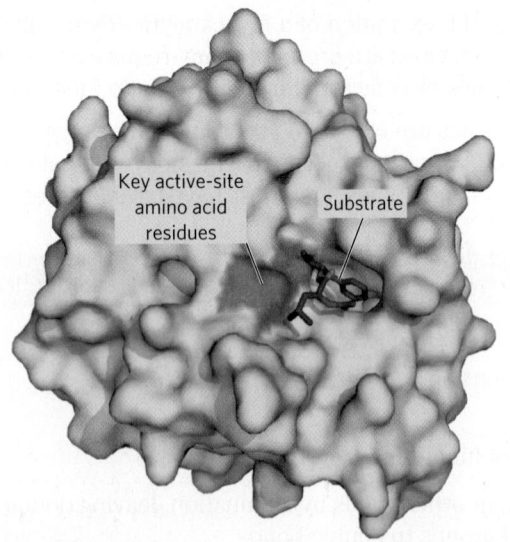

FIGURE 6-1 Binding of a substrate to an enzyme at the active site. The enzyme chymotrypsin with bound substrate. Some key active-site amino acid residues appear as a red splotch on the enzyme surface. [Source: PDB ID 7GCH, K. Brady et al., *Biochemistry* 29:7600, 1990.]

where E, S, and P represent the enzyme, substrate, and product; ES and EP are transient complexes of the enzyme with the substrate and with the product.

To understand catalysis, we must first appreciate the important distinction between reaction equilibria and reaction rates. The function of a catalyst is to increase the *rate* of a reaction. Catalysts do not affect reaction *equilibria*. (Recall that a reaction is at equilibrium when there is no net change in the concentrations of reactants or products.) Any reaction, such as $S \rightleftharpoons P$, can be described by a reaction coordinate diagram **(Fig. 6-2)**, a picture of the energy changes during the reaction. As discussed in Chapter 1, energy in biological systems is described in terms of free energy, G. In the coordinate diagram, the free energy of the system is plotted against the progress of the reaction (the reaction coordinate). The starting point for either the forward or the reverse reaction is called the **ground state**, the contribution to the free energy of the system by an average molecule (S or P) under a given set of conditions.

>> Key Convention: To describe the free-energy changes for reactions, chemists define a standard set of conditions (temperature of 298 K; partial pressure of each gas, 1 atm, or 101.3 kPa; concentration of each solute, 1 M) and express the free-energy change for a reacting system under these conditions as $\Delta G°$, the **standard free-energy change**. Because biochemical systems commonly involve H^+ concentrations far below 1 M, biochemists define a **biochemical standard free-energy change**, $\Delta G'°$, the standard free-energy change *at pH 7.0*; we employ this definition throughout the book. A more complete definition of $\Delta G'°$ is given in Chapter 13. **<<**

The equilibrium between S and P reflects the difference in the free energies of their ground states. In the example shown in Figure 6-2, the free energy of the ground state of P is lower than that of S, so $\Delta G'°$ for the reaction is negative (the reaction is exergonic) and at

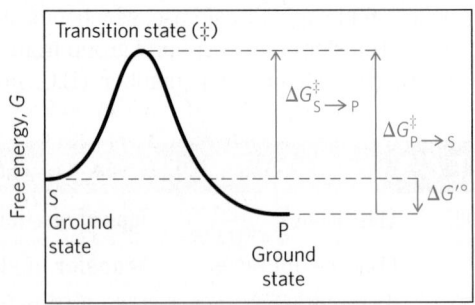

FIGURE 6-2 Reaction coordinate diagram. The free energy of the system is plotted against the progress of the reaction $S \rightarrow P$. A diagram of this kind is a description of the energy changes during the reaction, and the horizontal axis (reaction coordinate) reflects the progressive chemical changes (e.g., bond breakage or formation) as S is converted to P. The activation energies, $\Delta G^{\ddagger}$, for the $S \rightarrow P$ and $P \rightarrow S$ reactions are indicated. $\Delta G'°$ is the overall standard free-energy change in the direction $S \rightarrow P$.

equilibrium there is more P than S (the equilibrium favors P). The position and direction of equilibrium are *not* affected by any catalyst.

A favorable equilibrium does not mean that the S → P conversion will occur at a detectable rate. The *rate* of a reaction is dependent on an entirely different parameter. There is an energy barrier between S and P: the energy required for alignment of reacting groups, formation of transient unstable charges, bond rearrangements, and other transformations required for the reaction to proceed in either direction. This is illustrated by the energy "hill" in Figures 6-2 and 6-3. To undergo reaction, the molecules must overcome this barrier and therefore must be raised to a higher energy level. At the top of the energy hill is a point at which decay to the S or P state is equally probable (it is downhill either way). This is called the **transition state**. The transition state is not a chemical species with any significant stability and should not be confused with a reaction intermediate (such as ES or EP). It is simply a fleeting molecular moment in which events such as bond breakage, bond formation, and charge development have proceeded to the precise point at which decay to substrate or decay to product are equally likely. The difference between the energy levels of the ground state and the transition state is the **activation energy, $\Delta G^{\ddagger}$**. The rate of a reaction reflects this activation energy: a higher activation energy corresponds to a slower reaction. Reaction rates can be increased by raising the temperature and/or pressure, thereby increasing the number of molecules with sufficient energy to overcome the energy barrier. Alternatively, the activation energy can be lowered by adding a catalyst **(Fig. 6-3)**. *Catalysts enhance reaction rates by lowering activation energies.*

Enzymes are no exception to the rule that catalysts do not affect reaction equilibria. The bidirectional arrows in Equation 6-1 make this point: any enzyme that catalyzes the reaction S → P also catalyzes the reaction P → S. The role of enzymes is to *accelerate* the interconversion of S and P. The enzyme is not used up

in the process, and the equilibrium point is unaffected. However, the reaction reaches equilibrium much faster when the appropriate enzyme is present, because the rate of the reaction is increased.

This general principle is illustrated in the conversion of sucrose and oxygen to carbon dioxide and water:

$$C_{12}H_{22}O_{11} + 12O_2 \rightleftharpoons 12CO_2 + 11H_2O$$

This conversion which takes place through a series of separate reactions, has a very large and negative $\Delta G'^{\circ}$, and at equilibrium the amount of sucrose present is negligible. Yet sucrose is a stable compound, because the activation energy barrier that must be overcome before sucrose reacts with oxygen is quite high. Sucrose can be stored in a container with oxygen almost indefinitely without reacting. In cells, however, sucrose is readily broken down to CO_2 and H_2O in a series of reactions catalyzed by enzymes. These enzymes not only accelerate the reactions, they organize and control them so that much of the energy released is recovered in other chemical forms and made available to the cell for other tasks. The reaction pathway by which sucrose (and other sugars) is broken down is the primary energy-yielding pathway for cells, and the enzymes of this pathway allow the reaction sequence to proceed on a biologically useful time scale.

Any reaction may have several steps, involving the formation and decay of transient chemical species called **reaction intermediates.*** A reaction intermediate is any species on the reaction pathway that has a finite chemical lifetime (longer than a molecular vibration, $\sim 10^{-13}$ second). When the S $\rightleftharpoons$ P reaction is catalyzed by an enzyme, the ES and EP complexes can be considered intermediates, even though S and P are stable chemical species (Eqn 6-1); the ES and EP complexes occupy valleys in the reaction coordinate diagram (Fig. 6-3). Additional, less-stable chemical intermediates often exist in the course of an enzyme-catalyzed reaction. The interconversion of two sequential reaction intermediates thus constitutes a reaction step. When several steps occur in a reaction, the overall rate is determined by the step (or steps) with the highest activation energy; this is called the **rate-limiting step**. In a simple case, the rate-limiting step is the highest-energy point in the diagram for interconversion of S and P. In practice, the rate-limiting step can vary with reaction conditions, and for many enzymes several steps may have similar activation energies, which means they are all partially rate-limiting.

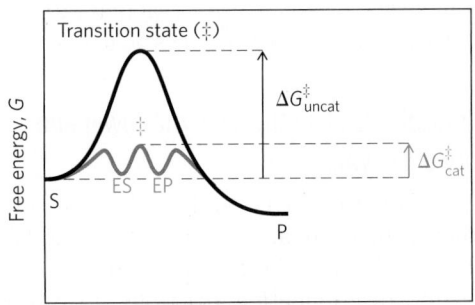

FIGURE 6-3 Reaction coordinate diagram comparing enzyme-catalyzed and uncatalyzed reactions. In the reaction S → P, the ES and EP intermediates occupy minima in the energy progress curve of the enzyme-catalyzed reaction. The terms $\Delta G^{\ddagger}_{uncat}$ and $\Delta G^{\ddagger}_{cat}$ correspond to the activation energy for the uncatalyzed reaction and the overall activation energy for the catalyzed reaction, respectively. The activation energy is lower when the enzyme catalyzes the reaction.

*In this chapter, *step* and *intermediate* refer to chemical reactions and chemical species in the reaction pathway of a single enzyme-catalyzed reaction. In the context of metabolic pathways involving many enzymes (discussed in Part II), these terms are used somewhat differently: an entire enzymatic reaction is often referred to as a "step" in a pathway, and the product of one enzymatic reaction (which is the substrate for the next enzyme in the pathway) is referred to as a pathway "intermediate."

Activation energies are energy barriers to chemical reactions. These barriers are crucial to life itself. The rate at which a molecule undergoes a particular reaction decreases as the activation barrier for that reaction increases. Without such energy barriers, complex macromolecular forms would revert spontaneously to much simpler molecular forms, and the complex and highly ordered structures and metabolic processes of cells could not exist. Over the course of evolution, enzymes have developed to lower activation energies *selectively* for reactions that are needed for cell survival.

Reaction Rates and Equilibria Have Precise Thermodynamic Definitions

Reaction *equilibria* are inextricably linked to the standard free-energy change for the reaction, $\Delta G'^{\circ}$, and reaction *rates* are linked to the activation energy, $\Delta G^{\ddagger}$. A basic introduction to these thermodynamic relationships is the next step in understanding how enzymes work.

An equilibrium such as $S \rightleftharpoons P$ is described by an **equilibrium constant**, K_{eq}, or simply K (p. 25). Under the standard conditions used to compare biochemical processes, an equilibrium constant is denoted K'_{eq} (or K'):

$$K'_{eq} = \frac{[P]}{[S]} \qquad (6\text{-}2)$$

From thermodynamics, the relationship between K'_{eq} and $\Delta G'^{\circ}$ can be described by the expression

$$\Delta G'^{\circ} = -RT \ln K'_{eq} \qquad (6\text{-}3)$$

where R is the gas constant, 8.315 J/mol·K, and T is the absolute temperature, 298 K (25 °C). Equation 6-3 is developed and discussed in more detail in Chapter 13. The important point here is that the equilibrium constant is directly related to the overall standard free-energy change for the reaction (Table 6-4). A large negative value for $\Delta G'^{\circ}$ reflects a favorable reaction equilibrium

(one in which there is much more product than substrate at equilibrium)—but as already noted, this does not mean the reaction will proceed at a rapid rate.

The rate of any reaction is determined by the concentration of the reactant (or reactants) and by a **rate constant**, usually denoted by k. For the unimolecular reaction $S \rightarrow P$, the rate (or velocity) of the reaction, V—representing the amount of S that reacts per unit time—is expressed by a **rate equation**:

$$V = k[S] \qquad (6\text{-}4)$$

In this reaction, the rate depends only on the concentration of S. This is called a first-order reaction. The factor k is a proportionality constant that reflects the probability of reaction under a given set of conditions (pH, temperature, and so forth). Here, k is a first-order rate constant and has units of reciprocal time, such as s^{-1}. If a first-order reaction has a rate constant k of 0.03 s^{-1}, this may be interpreted (qualitatively) to mean that 3% of the available S will be converted to P in 1 second. A reaction with a rate constant of 2,000 s^{-1} will be over in a small fraction of a second. If a reaction rate depends on the concentration of two different compounds, or if the reaction is between two molecules of the same compound, the reaction is second order and k is a second-order rate constant, with units of $M^{-1}s^{-1}$. The rate equation then becomes

$$V = k[S_1][S_2] \qquad (6\text{-}5)$$

From transition-state theory we can derive an expression that relates the magnitude of a rate constant to the activation energy:

$$k = \frac{\mathbf{k}T}{h}e^{-\Delta G^{\ddagger}/RT} \qquad (6\text{-}6)$$

where $\mathbf{k}$ is the Boltzmann constant and h is Planck's constant. The important point here is that the relationship between the rate constant k and the activation energy $\Delta G^{\ddagger}$ is inverse and exponential. In simplified terms, this is the basis for the statement that a lower activation energy means a faster reaction rate.

Now we turn from *what* enzymes do to *how* they do it.

A Few Principles Explain the Catalytic Power and Specificity of Enzymes

Enzymes are extraordinary catalysts. The rate enhancements they bring about are in the range of 5 to 17 orders of magnitude (Table 6-5). Enzymes are also very specific, readily discriminating between substrates with quite similar structures. How can these enormous and highly selective rate enhancements be explained? What is the source of the energy for the dramatic lowering of the activation energies for specific reactions?

The answer to these questions has two distinct but interwoven parts. The first lies in the rearrangement of covalent bonds during an enzyme-catalyzed reaction. Chemical reactions of many types take place between

TABLE 6-4	Relationship between K'_{eq} and $\Delta G'^{\circ}$
K'_{eq}	$\Delta G'^{\circ}$ (kJ/mol)
10^{-6}	34.2
10^{-5}	28.5
10^{-4}	22.8
10^{-3}	17.1
10^{-2}	11.4
10^{-1}	5.7
1	0.0
10^{1}	−5.7
10^{2}	−11.4
10^{3}	−17.1

Note: The relationship is calculated from $\Delta G'^{\circ} = -RT \ln K'_{eq}$ (Eqn 6-3).

TABLE 6-5	Some Rate Enhancements Produced by Enzymes	
Cyclophilin		10^5
Carbonic anhydrase		10^7
Triose phosphate isomerase		10^9
Carboxypeptidase A		10^{11}
Phosphoglucomutase		10^{12}
Succinyl-CoA transferase		10^{13}
Urease		10^{14}
Orotidine monophosphate decarboxylase		10^{17}

substrates and enzymes' functional groups (specific amino acid side chains, metal ions, and coenzymes). Catalytic functional groups on an enzyme may form a transient covalent bond with a substrate and activate it for reaction, or a group may be transiently transferred from the substrate to the enzyme. In many cases, these reactions occur only in the enzyme active site. Covalent interactions between enzymes and substrates lower the activation energy (and thereby accelerate the reaction) by providing an alternative, lower-energy reaction path. The specific types of rearrangements that occur are described in Section 6.4.

The second part of the explanation lies in the *noncovalent* interactions between enzyme and substrate. Recall from Chapter 4 that weak, noncovalent interactions help stabilize protein structure and protein-protein interactions. These same interactions are critical to the formation of complexes between proteins and small molecules, including enzyme substrates. Much of the energy required to lower activation energies is derived from weak, noncovalent interactions between substrate and enzyme. What really sets enzymes apart from most other catalysts is the formation of a specific ES complex. The interaction between substrate and enzyme in this complex is mediated by the same forces that stabilize protein structure, including hydrogen bonds, ionic interactions, and the hydrophobic effect (Chapter 4). Formation of each weak interaction in the ES complex is accompanied by release of a small amount of free energy that stabilizes the interaction. The energy derived from enzyme-substrate interaction is called **binding energy**, ΔG_B. Its significance extends beyond a simple stabilization of the enzyme-substrate interaction. *Binding energy is a major source of free energy used by enzymes to lower the activation energies of reactions.*

Two fundamental and interrelated principles provide a general explanation for how enzymes use noncovalent binding energy:

1. Much of the catalytic power of enzymes is ultimately derived from the free energy released in forming many weak bonds and interactions between an enzyme and its substrate. This binding energy contributes to specificity as well as to catalysis.

2. Weak interactions are optimized in the reaction transition state; enzyme active sites are complementary not to the substrates per se but to the transition states through which substrates pass as they are converted to products during an enzymatic reaction.

These themes are critical to an understanding of enzymes, and they now become our primary focus.

Weak Interactions between Enzyme and Substrate Are Optimized in the Transition State

How does an enzyme use binding energy to lower the activation energy for a reaction? Formation of the ES complex is not the explanation in itself, although some of the earliest considerations of enzyme mechanisms began with this idea. Studies on enzyme specificity carried out by Emil Fischer led him to propose, in 1894, that enzymes were structurally complementary to their substrates, so that they fit together like a lock and key **(Fig. 6-4)**. This elegant idea, that a specific (exclusive)

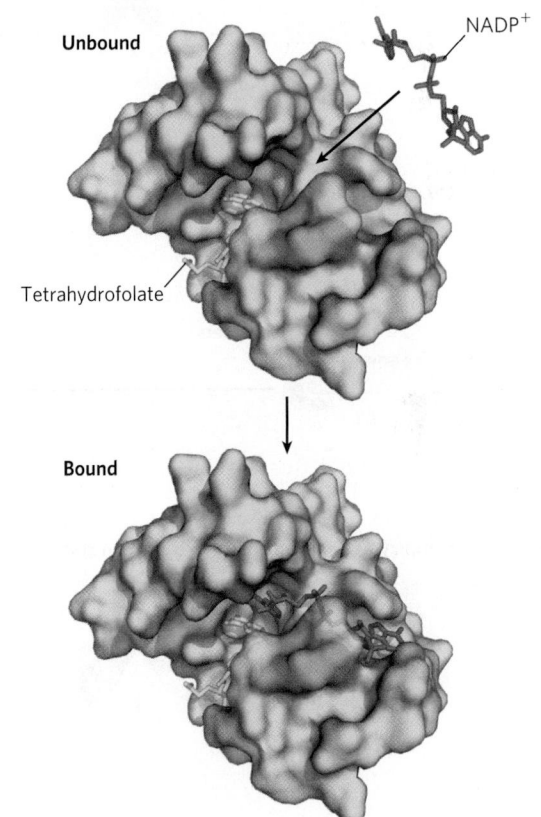

FIGURE 6-4 Complementary shapes of a substrate and its binding site on an enzyme. The enzyme dihydrofolate reductase with its substrate NADP$^+$, unbound and bound; another bound substrate, tetrahydrofolate, is also visible. In this model, the NADP$^+$ binds to a pocket that is complementary to it in shape and ionic properties, an illustration of Emil Fischer's "lock and key" hypothesis of enzyme action. In reality, the complementarity between protein and ligand (in this case, substrate) is rarely perfect, as we saw in Chapter 5. [Source: PDB ID 1RA2, M. R. Sawaya and J. Kraut, *Biochemistry* 36:586, 1997.]

interaction between two biological molecules is mediated by molecular surfaces with complementary shapes, has greatly influenced the development of biochemistry, and such interactions lie at the heart of many biochemical processes. However, the "lock and key" hypothesis can be misleading when applied to enzymatic catalysis. An enzyme completely complementary to its substrate would be a very poor enzyme, as we can demonstrate.

Consider an imaginary reaction, the breaking of a magnetized metal stick. The uncatalyzed reaction is shown in **Figure 6-5a**. Let's examine two imaginary enzymes—two "stickases"—that could catalyze this reaction, both of which employ magnetic forces as a paradigm for the binding energy used by real enzymes. We first design an enzyme perfectly complementary to the substrate (Fig. 6-5b). The active site of this stickase

is a pocket lined with magnets. To react (break), the stick must reach the transition state of the reaction, but the stick fits so tightly in the active site that it cannot bend, because bending would eliminate some of the magnetic interactions between stick and enzyme. Such an enzyme *impedes* the reaction, stabilizing the substrate instead. In a reaction coordinate diagram (Fig. 6-5b), this kind of ES complex would correspond to an energy trough from which the substrate would have difficulty escaping. Such an enzyme would be useless.

The modern notion of enzymatic catalysis, first proposed by Michael Polanyi (1921) and Haldane (1930), was elaborated by Linus Pauling in 1946 and by William P. Jencks in the 1970s: in order to catalyze reactions, an enzyme must be complementary to the *reaction transition state*. This means that optimal interactions

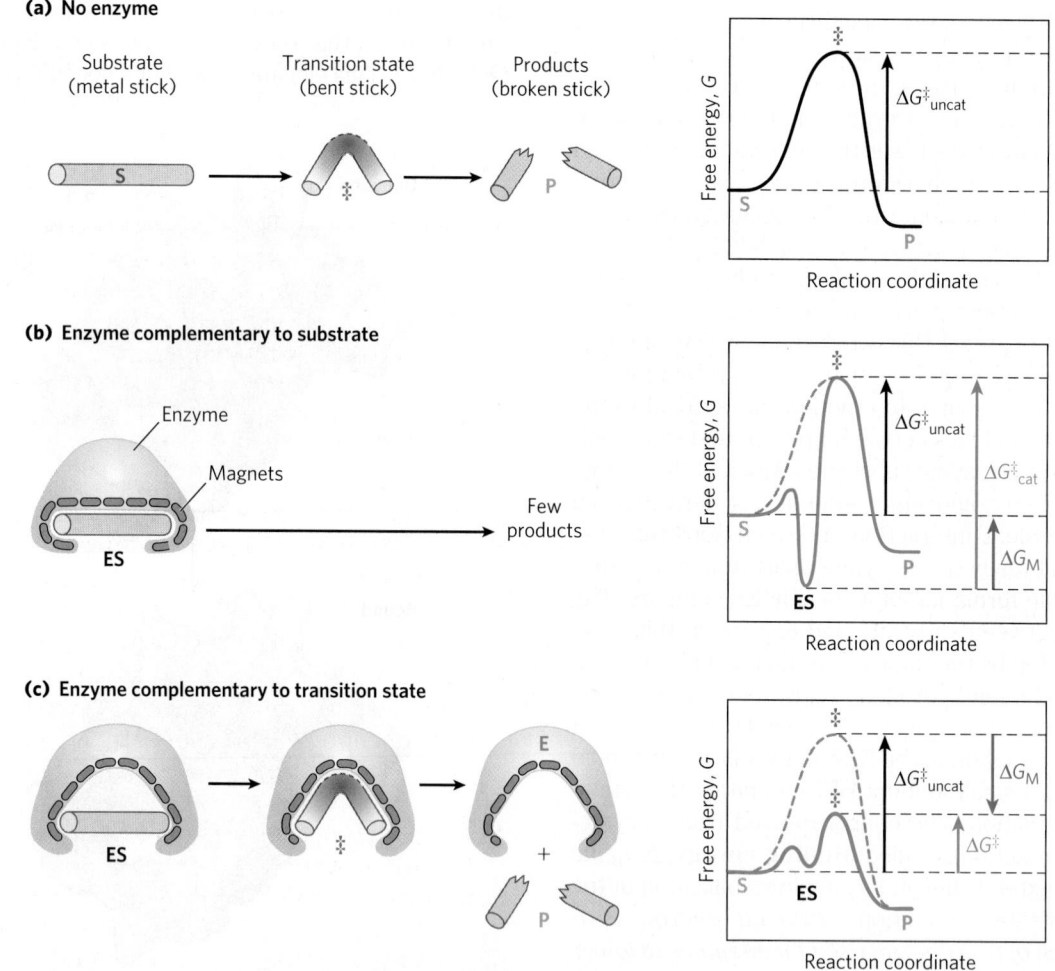

FIGURE 6-5   An imaginary enzyme (stickase) designed to catalyze breakage of a metal stick. (a) Before the stick is broken, it must first be bent (the transition state). In both stickase examples, magnetic interactions take the place of weak bonding interactions between enzyme and substrate. (b) A stickase with a magnet-lined pocket complementary in structure to the stick (the substrate) stabilizes the substrate. Bending is impeded by the magnetic attraction between stick and stickase. (c) An enzyme with a pocket complementary to the reaction transition state helps to destabilize the stick, contributing to catalysis of the reaction. The binding energy of the

magnetic interactions compensates for the increase in free energy required to bend the stick. Reaction coordinate diagrams (right) show the energy consequences of complementarity to substrate versus complementarity to transition state (EP complexes are omitted). ΔG_M, the difference between the transition-state energies of the uncatalyzed and catalyzed reactions, is contributed by the magnetic interactions between the stick and stickase. When the enzyme is complementary to the substrate (b), the ES complex is more stable and has less free energy in the ground state than substrate alone. The result is an *increase* in the activation energy.

between substrate and enzyme occur only in the transition state. Figure 6-5c demonstrates how such an enzyme can work. The metal stick binds to the stickase, but only a subset of the possible magnetic interactions are used in forming the ES complex. The bound substrate must still undergo the increase in free energy needed to reach the transition state. Now, however, the increase in free energy required to draw the stick into a bent and partially broken conformation is offset, or "paid for," by the magnetic interactions that form between our imaginary enzyme and substrate (analogous to the binding energy in a real enzyme) in the transition state. Many of these interactions involve parts of the stick that are distant from the point of breakage; thus interactions between the stickase and nonreacting parts of the stick provide some of the energy needed to catalyze stick breakage. This "energy payment" translates into a lower net activation energy and a faster reaction rate.

Real enzymes work on an analogous principle. Some weak interactions are formed in the ES complex, but the full complement of such interactions between substrate and enzyme is formed only when the substrate reaches the transition state. The free energy (binding energy) released by the formation of these interactions partially offsets the energy required to reach the top of the energy hill. The summation of the unfavorable (positive) activation energy $\Delta G^{\ddagger}$ and the favorable (negative) binding energy ΔG_{B} results in a lower *net* activation energy **(Fig. 6-6)**. Even on the enzyme, the transition state is not a stable species but a brief point in time that the substrate spends atop an energy hill. The enzyme-catalyzed reaction is much faster than the uncatalyzed process, however, because the hill is much smaller. The important principle is that *weak binding interactions between the enzyme and the substrate provide a substantial driving force for enzymatic catalysis.* The groups on the substrate that are involved in these weak interactions can be at some

distance from the bonds that are broken or changed. The weak interactions formed only in the transition state are those that make the primary contribution to catalysis.

The requirement for multiple weak interactions to drive catalysis is one reason why enzymes (and some coenzymes) are so large. An enzyme must provide functional groups for ionic, hydrogen-bond, and other interactions, and also must precisely position these groups so that binding energy is optimized in the transition state. Adequate binding is accomplished most readily by positioning a substrate in a cavity (the active site) where it is effectively removed from water. The size of proteins reflects the need for superstructure to keep interacting groups properly positioned and to keep the cavity from collapsing.

Binding Energy Contributes to Reaction Specificity and Catalysis

Can we demonstrate quantitatively that binding energy accounts for the huge rate accelerations brought about by enzymes? Yes. As a point of reference, Equation 6-6 allows us to calculate that $\Delta G^{\ddagger}$ must be lowered by about 5.7 kJ/mol to accelerate a first-order reaction by a factor of 10, under conditions commonly found in cells. The energy available from formation of a single weak interaction is generally estimated to be 4 to 30 kJ/mol. The overall energy available from many such interactions is therefore sufficient to lower activation energies by the 60 to 100 kJ/mol required to explain the large rate enhancements observed for many enzymes.

The same binding energy that provides energy for catalysis also gives an enzyme its **specificity**, the ability to discriminate between a substrate and a competing molecule. Conceptually, specificity is easy to distinguish from catalysis, but this distinction is much more difficult to make experimentally, because catalysis and specificity arise from the same phenomenon. If an enzyme active site has functional groups arranged optimally to form a variety of weak interactions with a particular substrate in the transition state, the enzyme will not be able to interact to the same degree with any other molecule. For example, if the substrate has a hydroxyl group that forms a hydrogen bond with a specific Glu residue on the enzyme, any molecule lacking a hydroxyl group at that particular position will be a poorer substrate for the enzyme. In addition, any molecule with an extra functional group for which the enzyme has no pocket or binding site is likely to be excluded from the enzyme. In general, *specificity* is derived from the formation of many weak interactions between the enzyme and its specific substrate molecule.

The importance of binding energy to catalysis can be readily demonstrated. For example, the glycolytic enzyme triose phosphate isomerase catalyzes the interconversion

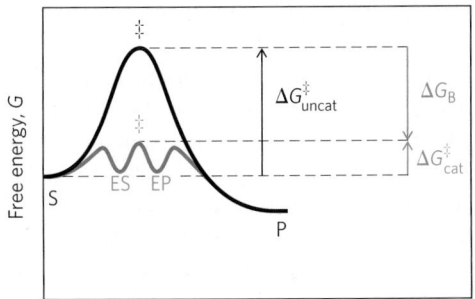

FIGURE 6-6 Role of binding energy in catalysis. To lower the activation energy for a reaction, the system must acquire an amount of energy equivalent to the amount by which $\Delta G^{\ddagger}$ is lowered. Much of this energy comes from binding energy, ΔG_{B}, contributed by formation of weak noncovalent interactions between substrate and enzyme in the transition state. The role of ΔG_{B} is analogous to that of ΔG_{M} in Figure 6-5.

of glyceraldehyde 3-phosphate and dihydroxyacetone phosphate:

Glyceraldehyde 3-phosphate ⇌ (triose phosphate isomerase) Dihydroxyacetone phosphate

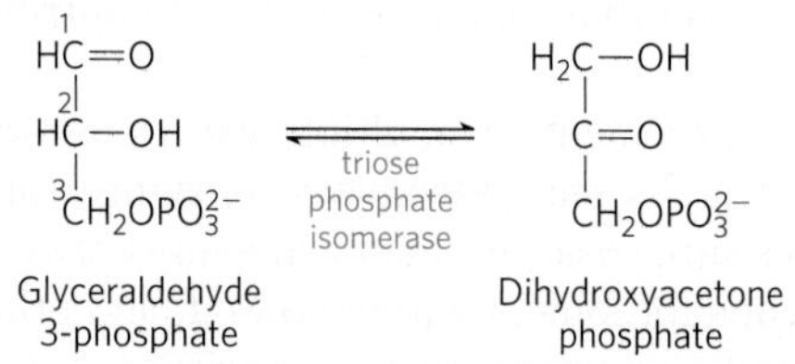

This reaction rearranges the carbonyl and hydroxyl groups on carbons 1 and 2. However, more than 80% of the enzymatic rate acceleration has been traced to enzyme-substrate interactions involving the phosphate group on carbon 3 of the substrate. This was determined by comparing the enzyme-catalyzed reactions with glyceraldehyde 3-phosphate and with glyceraldehyde (no phosphate group at position 3) as substrate.

The general principles outlined above can be illustrated by a variety of recognized catalytic mechanisms. These mechanisms are not mutually exclusive, and a given enzyme might incorporate several types in its overall mechanism of action.

Consider what needs to occur for a reaction to take place. Prominent physical and thermodynamic factors contributing to $\Delta G^{\ddagger}$, the barrier to reaction, might include: (1) the entropy of molecules in solution, which reduces the possibility that they will react together; (2) the solvation shell of hydrogen-bonded water that surrounds and helps to stabilize most biomolecules in aqueous solution; (3) the distortion of substrates that must occur in many reactions; and (4) the need for proper alignment of catalytic functional groups on the enzyme. Binding energy can be used to overcome all these barriers.

First, a large restriction in the relative motions of two substrates that are to react, or **entropy reduction**, is one obvious benefit of binding them to an enzyme. Binding energy constrains the substrates in the proper orientation to react—a substantial contribution to catalysis, because productive collisions between molecules in solution can be exceedingly rare. Substrates can be precisely aligned on the enzyme, with many weak interactions between each substrate and strategically located groups on the enzyme clamping the substrate molecules into the proper positions. Studies have shown that constraining the motion of two reactants can produce rate enhancements of many orders of magnitude (**Fig. 6-7**).

Second, formation of weak bonds between substrate and enzyme results in **desolvation** of the substrate. Enzyme-substrate interactions replace most or all of the hydrogen bonds between the substrate and water that would otherwise impede reaction.

Third, binding energy involving weak interactions formed only in the reaction transition state helps to compensate thermodynamically for the unfavorable free-energy change associated with any distortion, primarily electron redistribution, that the substrate must undergo to react.

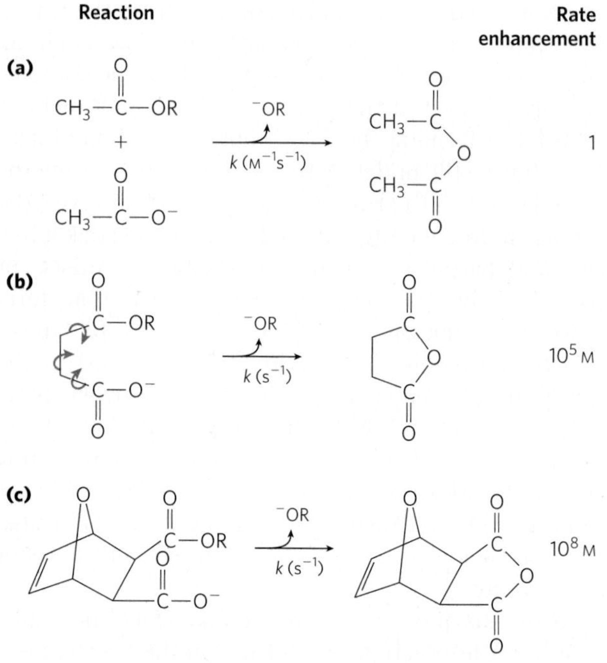

Reaction		Rate enhancement

FIGURE 6-7 Rate enhancement by entropy reduction. Shown here are reactions of an ester with a carboxylate group to form an anhydride. The R group is the same in each case. **(a)** For this bimolecular reaction, the rate constant k is second order, with units of $M^{-1}s^{-1}$. **(b)** When the two reacting groups are in a single molecule, and thus have less freedom of motion, the reaction is much faster. For this unimolecular reaction, k has units of s^{-1}. Dividing the rate constant for (b) by the rate constant for (a) gives a rate enhancement of about 10^5 M. (The enhancement has units of molarity because we are comparing a unimolecular and a bimolecular reaction.) Put another way, if the reactant in (b) were present at a concentration of 1 M, the reacting groups would *behave* as though they were present at a concentration of 10^5 M. Note that the reactant in (b) has freedom of rotation about three bonds (shown with curved arrows), but this still represents a substantial reduction of entropy over (a). If the bonds that rotate in (b) are constrained as in **(c)**, the entropy is reduced further and the reaction exhibits a rate enhancement of 10^8 M relative to (a).

Finally, the enzyme itself usually undergoes a change in conformation when the substrate binds, induced by multiple weak interactions with the substrate. This is referred to as **induced fit**, a mechanism postulated by Daniel Koshland in 1958. The motions can affect a small part of the enzyme near the active site or can involve changes in the positioning of entire domains. Typically, a network of coupled motions occurs throughout the enzyme that ultimately brings about the required changes in the active site. Induced fit serves to bring specific functional groups on the enzyme into the proper position to catalyze the reaction. The conformational change also permits formation of additional weak bonding interactions in the transition state. In either case, the new enzyme conformation has enhanced catalytic properties. As we have seen, induced fit is a common feature of the reversible binding of ligands to proteins (Chapter 5). Induced fit is also important in the interaction of almost every enzyme with its substrate.

Specific Catalytic Groups Contribute to Catalysis

In most enzymes, the binding energy used to form the ES complex is just one of several contributors to the overall catalytic mechanism. Once a substrate is bound to an enzyme, properly positioned catalytic functional groups aid in the cleavage and formation of bonds by a variety of mechanisms, including general acid-base catalysis, covalent catalysis, and metal ion catalysis. These are distinct from mechanisms based on binding energy because they generally involve transient *covalent* interaction with a substrate or group transfer to or from a substrate.

General Acid-Base Catalysis Transfer of a proton is the single most common reaction in biochemistry. One or, often, many proton transfers occur in the course of most reactions that take place in cells. Many biochemical reactions occur through the formation of unstable charged intermediates that tend to break down rapidly to their constituent reactant species, impeding the forward reaction **(Fig. 6-8)**. Charged intermediates can

often be stabilized by the transfer of protons to form a species that breaks down more readily to products. These protons are transferred between an enzyme and a substrate or intermediate.

The effects of catalysis by acids and bases are often studied using nonenzymatic model reactions, in which the proton donors or acceptors are either the constituents of water alone or other weak acids and bases. Catalysis of the type that uses only the H^+ (H_3O^+) or OH^- ions present in water is referred to as **specific acid-base catalysis**. If protons are transferred between the intermediate and water faster than the intermediate breaks down to reactants, the intermediate is effectively stabilized every time it forms. No additional catalysis mediated by other proton acceptors or donors will occur. In many reactions, however, water is not enough to prevent the breakdown to reactants. In these cases, for nonenzymatic reactions in aqueous solutions, weak acids and bases can be added to accelerate the reaction rate. Many weak organic acids can supplement water as proton donors in this situation, or weak organic bases can supplement water as proton acceptors. The term **general acid-base catalysis** refers to proton transfers mediated by weak acids and bases other than water.

General acid-base catalysis becomes crucial in the active site of an enzyme, where water may not be available as a proton donor or acceptor. Several amino acid side chains can and do take on the role of proton donors and acceptors **(Fig. 6-9)**. These groups can be precisely positioned in an enzyme active site to allow proton transfers, providing rate enhancements of the order of 10^2 to 10^5. This type of catalysis occurs on the vast majority of enzymes.

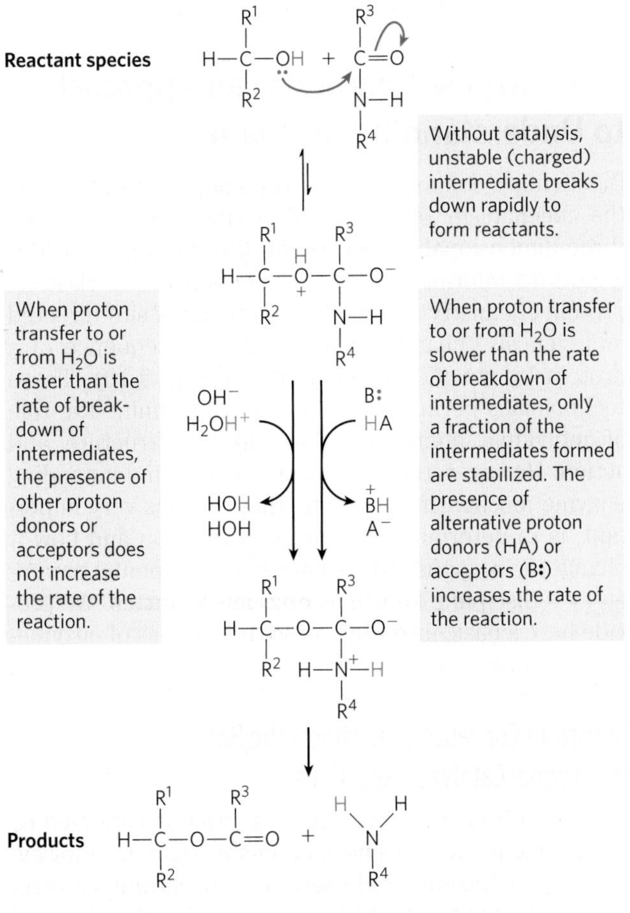

FIGURE 6-8 How a catalyst circumvents unfavorable charge development during cleavage of an amide. The hydrolysis of an amide bond, shown here, is the same reaction as that catalyzed by chymotrypsin and other proteases. Charge development is unfavorable and can be circumvented by donation of a proton by H_3O^+ (specific acid catalysis) or HA (general acid catalysis), where HA represents any acid. Similarly, charge can be neutralized by proton abstraction by OH^- (specific base catalysis) or B: (general base catalysis), where B: represents any base.

FIGURE 6-9 Amino acids in general acid-base catalysis. Many organic reactions that are used to model biochemical processes are promoted by proton donors (general acids) or proton acceptors (general bases). The active sites of some enzymes contain amino acid functional groups, such as those shown here, that can participate in the catalytic process as proton donors or proton acceptors.

Covalent Catalysis In covalent catalysis, a transient covalent bond is formed between the enzyme and the substrate. Consider the hydrolysis of a bond between groups A and B:

$$A\!-\!B \xrightarrow{H_2O} A + B$$

In the presence of a covalent catalyst (an enzyme with a nucleophilic group X:) the reaction becomes

$$A\!-\!B + X: \longrightarrow A\!-\!X + B \xrightarrow{H_2O} A + X: + B$$

Formation and breakdown of a covalent intermediate creates a new pathway for the reaction, but catalysis results *only* when the new pathway has a lower activation energy than the uncatalyzed pathway. Both of the new steps must be faster than the uncatalyzed reaction. Several amino acid side chains, including all those in Figure 6-9, and the functional groups of some enzyme cofactors can serve as nucleophiles in the formation of covalent bonds with substrates. These covalent complexes always undergo further reaction to regenerate the free enzyme. The covalent bond formed between the enzyme and the substrate can activate a substrate for further reaction in a manner that is usually specific to the particular group or coenzyme.

Metal Ion Catalysis Metals, whether tightly bound to the enzyme or taken up from solution along with the substrate, can participate in catalysis in several ways. Ionic interactions between an enzyme-bound metal and a substrate can help orient the substrate for reaction or stabilize charged reaction transition states. This use of weak bonding interactions between metal and substrate is similar to some of the uses of enzyme-substrate binding energy described earlier. Metals can also mediate oxidation-reduction reactions by reversible changes in the metal ion's oxidation state. Nearly a third of all known enzymes require one or more metal ions for catalytic activity.

Most enzymes combine several catalytic strategies to bring about a rate enhancement. A good example is the use of covalent catalysis, general acid-base catalysis, and transition-state stabilization in the reaction catalyzed by chymotrypsin, detailed in Section 6.4.

SUMMARY 6.2 How Enzymes Work

■ Enzymes are highly effective catalysts, commonly enhancing reaction rates by a factor of 10^5 to 10^{17}.

■ Enzyme-catalyzed reactions are characterized by the formation of a complex between substrate and enzyme (an ES complex). Substrate binding occurs in a pocket on the enzyme called the active site.

■ The function of enzymes and other catalysts is to lower the activation energy, $\Delta G^\ddagger$, for a reaction and thereby enhance the reaction rate. The equilibrium of a reaction is unaffected by the enzyme.

■ A significant part of the energy used for enzymatic rate enhancements is derived from weak interactions (hydrogen bonds, aggregation due to the hydrophobic effect, and ionic interactions) between substrate and enzyme. The enzyme active site is structured so that some of these weak interactions occur preferentially in the reaction transition state, thus stabilizing the transition state.

■ The need for multiple interactions is one reason for the large size of enzymes. The binding energy, ΔG_B, is used to offset the energy required for activation, $\Delta G^\ddagger$, in several ways—for example, lowering substrate entropy, causing substrate desolvation, or causing a conformational change in the enzyme (induced fit). Binding energy also accounts for the exquisite specificity of enzymes for their substrates.

■ Additional catalytic mechanisms employed by enzymes include general acid-base catalysis, covalent catalysis, and metal ion catalysis. Catalysis often involves transient covalent interactions between the substrate and the enzyme, or group transfers to and from the enzyme, to provide a new, lower-energy reaction path. In all cases, the enzyme reverts to the unbound state once the reaction is complete.

6.3 Enzyme Kinetics as an Approach to Understanding Mechanism

Biochemists commonly use several approaches to study the mechanism of action of purified enzymes. The three-dimensional structure of the protein provides important information, which is enhanced by classical protein chemistry and modern methods of site-directed mutagenesis (changing the amino acid sequence of a protein by genetic engineering; see Fig. 9-10). These technologies permit enzymologists to examine the role of individual amino acids in enzyme structure and action. However, the oldest approach to understanding enzyme mechanisms, and one that remains very important, is to determine the *rate* of a reaction and how it changes in response to changes in experimental parameters, a discipline known as **enzyme kinetics**. We provide here a basic introduction to the kinetics of enzyme-catalyzed reactions.

Substrate Concentration Affects the Rate of Enzyme-Catalyzed Reactions

A key factor affecting the rate of a reaction catalyzed by an enzyme is the concentration of substrate, [S]. However, studying the effects of substrate concentration is complicated by the fact that [S] changes during the course of an in vitro reaction as substrate is converted to product. One simplifying approach in kinetics experiments is to measure the **initial rate** (or **initial velocity**), designated V_0 (Fig. 6-10). In a typical reaction, the enzyme may be present in nanomolar quantities, whereas [S] may be five or six orders of magnitude higher. If only the beginning of the reaction is monitored, over a period in

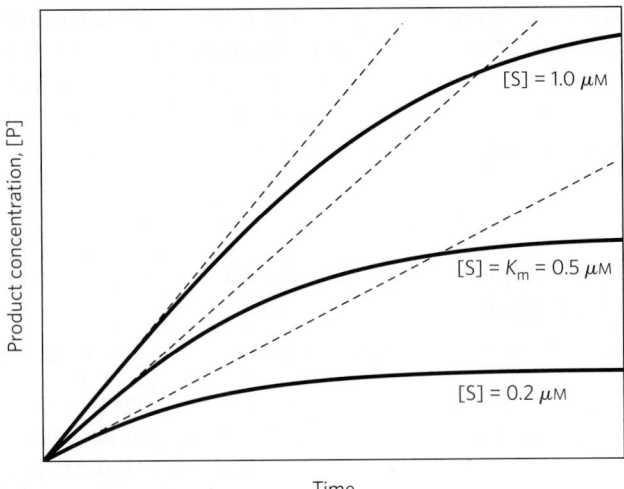

FIGURE 6-10 Initial velocities of enzyme-catalyzed reactions. A theoretical enzyme that catalyzes the reaction S $\rightleftharpoons$ P is present at a concentration sufficient to catalyze the reaction at a maximum velocity, V_{max}, of 1 μM/min. The Michaelis constant, K_m (explained in the text), is 0.5 μM. Progress curves are shown for substrate concentrations below, at, and above the K_m. The rate of an enzyme-catalyzed reaction declines as substrate is converted to product. A tangent to each curve taken at time = 0 defines the initial velocity, V_0, of each reaction.

which only a small percentage of the available substrate is converted to product, [S] can be regarded as constant, to a reasonable approximation. V_0 can then be explored as a function of [S], which is adjusted by the investigator. The effect on V_0 of varying [S] when the enzyme concentration is held constant is shown in **Figure 6-11**. At relatively low concentrations of substrate, V_0 increases

almost linearly with an increase in [S]. At higher substrate concentrations, V_0 increases by smaller and smaller amounts in response to increases in [S]. Finally, a point is reached beyond which increases in V_0 are vanishingly small as [S] increases. This plateau-like V_0 region is close to the **maximum velocity, V_{max}**.

The ES complex is the key to understanding this kinetic behavior, just as it was a starting point for our discussion of catalysis. The kinetic pattern in Figure 6-11 led Victor Henri, following Wurtz's lead, to propose in 1903 that the combination of an enzyme with its substrate molecule to form an ES complex is a necessary step in enzymatic catalysis. This idea was expanded into a general theory of enzyme action, particularly by Leonor Michaelis and Maud Menten in 1913. They postulated that the enzyme first combines reversibly with its substrate to form an enzyme-substrate complex in a relatively fast reversible step:

$$\text{E} + \text{S} \underset{k_{-1}}{\overset{k_1}{\rightleftharpoons}} \text{ES} \qquad (6\text{-}7)$$

The ES complex then breaks down in a slower second step to yield the free enzyme and the reaction product P:

$$\text{ES} \underset{k_{-2}}{\overset{k_2}{\rightleftharpoons}} \text{E} + \text{P} \qquad (6\text{-}8)$$

Because the slower second reaction (Eqn 6-8) must limit the rate of the overall reaction, the overall rate must be proportional to the concentration of the species that reacts in the second step—that is, ES.

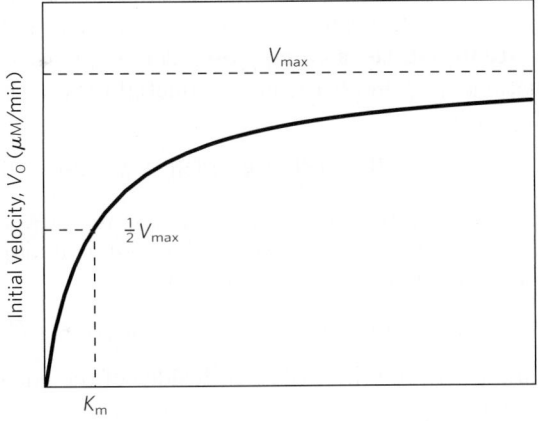

Leonor Michaelis, 1875–1949 [Source: Rockefeller Archive Center.]

Maud Menten, 1879–1960 [Source: Courtesy Archives Service Center, University of Pittsburgh.]

FIGURE 6-11 Effect of substrate concentration on the initial velocity of an enzyme-catalyzed reaction. The maximum velocity, V_{max}, is extrapolated from the plot, because V_0 approaches but never quite reaches V_{max}. The substrate concentration at which V_0 is half maximal is K_m, the Michaelis constant. The concentration of enzyme in an experiment such as this is generally so low that [S] $\gg$ [E] even when [S] is described as low or relatively low. The units shown are typical for enzyme-catalyzed reactions and are given only to help illustrate the meaning of V_0 and [S]. (Note that the curve describes *part* of a rectangular hyperbola, with one asymptote at V_{max}. If the curve were continued below [S] = 0, it would approach a vertical asymptote at [S] = $-K_m$.)

At any given instant in an enzyme-catalyzed reaction, the enzyme exists in two forms, the free or uncombined form E and the combined form ES. At low [S], most of the enzyme is in the uncombined form E. Here, the rate is proportional to [S] because the equilibrium of Equation 6-7 is pushed toward formation of more ES as [S] increases. The maximum initial rate of the catalyzed reaction (V_{max}) is observed when virtually all the enzyme is present as the ES complex and [E] is vanishingly small. Under these conditions, the enzyme is "saturated" with its substrate, so that further increases in [S] have no effect on rate. This condition exists when [S] is sufficiently high that essentially all the free enzyme has been converted to the ES form. After the ES complex breaks down to yield the product P, the enzyme is free to catalyze the reaction of another molecule of substrate (and will do so rapidly under saturating conditions). The saturation effect is a distinguishing characteristic of enzymatic catalysts and is responsible for the plateau observed in Figure 6-11, and the pattern seen in the figure is sometimes referred to as saturation kinetics.

When the enzyme is first mixed with a large excess of substrate, there is an initial transient period, the **pre–steady state**, during which the concentration of ES builds up. For most enzymatic reactions, this period is very brief. It is often too short to be easily observed, lasting just microseconds, and is not evident in Figure 6-10. (We return to the pre–steady state later in this section.) The reaction quickly achieves a **steady state** in which [ES] (and the concentrations of any other intermediates) remains approximately constant over time. The concept of a steady state, introduced by G. E. Briggs and Haldane in 1925, is an approximation based on a simple reality. As noted earlier, enzymes are powerful catalysts that are typically present at concentrations orders of magnitude lower than the concentration of substrate. Once the transient phase or pre–steady state has passed (often after only one enzymatic turnover; that is, conversion of one molecule of substrate to one molecule of product on each molecule of enzyme), P is generated at the same rate that S is consumed only if the concentration of the intermediate ES remains steady. The measured V_0 generally reflects the steady state, even though V_0 is limited to the early part of the reaction, and analysis of these initial rates is referred to as **steady-state kinetics**.

The Relationship between Substrate Concentration and Reaction Rate Can Be Expressed Quantitatively

The curve expressing the relationship between [S] and V_0 (Fig. 6-11) has the same general shape for most enzymes (it approaches a rectangular hyperbola), which can be expressed algebraically by the Michaelis-Menten equation. Michaelis and Menten derived this equation starting from their basic hypothesis that the rate-limiting step in enzymatic reactions is the breakdown of the ES complex to product and free enzyme. The equation is

$$V_0 = \frac{V_{max}[S]}{K_m + [S]} \qquad (6\text{-}9)$$

All these terms—[S], V_0, V_{max}, and a constant, K_m, called the Michaelis constant—are readily measured experimentally.

Here we develop the basic logic and the algebraic steps in a modern derivation of the Michaelis-Menten equation, which includes the steady-state assumption introduced by Briggs and Haldane. The derivation starts with the two basic steps of the formation and breakdown of ES (Eqns 6-7 and 6-8). Early in the reaction,

the concentration of the product, [P], is negligible, and we make the simplifying assumption that the reverse reaction, $P \rightarrow S$ (described by k_{-2}), can be ignored. This assumption is not critical but it simplifies our task. The overall reaction then reduces to

$$E + S \underset{k_{-1}}{\overset{k_1}{\rightleftharpoons}} ES \overset{k_2}{\longrightarrow} E + P \qquad (6\text{-}10)$$

V_0 is determined by the breakdown of ES to form product, which is determined by [ES]:

$$V_0 = k_2[ES] \qquad (6\text{-}11)$$

Because [ES] in Equation 6-11 is not easily measured experimentally, we must begin by finding an alternative expression for this term. First, we introduce the term $[E_t]$, representing the total enzyme concentration (the sum of free and substrate-bound enzyme). Free or unbound enzyme [E] can then be represented by $[E_t] - [ES]$. Also, because [S] is ordinarily far greater than $[E_t]$, the amount of substrate bound by the enzyme at any given time is negligible compared with the total [S]. With these conditions in mind, the following steps lead us to an expression for V_0 in terms of easily measurable parameters.

Step 1 The rates of formation and breakdown of ES are determined by the steps governed by the rate constants k_1 (formation) and $k_{-1} + k_2$ (breakdown to reactants and products, respectively), according to the expressions

$$\text{Rate of ES formation} = k_1([E_t] - [ES])[S] \qquad (6\text{-}12)$$

$$\text{Rate of ES breakdown} = k_{-1}[ES] + k_2[ES] \qquad (6\text{-}13)$$

Step 2 We now make an important assumption: that the initial rate of reaction reflects a steady state in which [ES] is constant—that is, the rate of formation of ES is equal to the rate of its breakdown. This is called the **steady-state assumption**. The expressions in Equations 6-12 and 6-13 can be equated for the steady state, giving

$$k_1([E_t] - [ES])[S] = k_{-1}[ES] + k_2[ES] \qquad (6\text{-}14)$$

Step 3 In a series of algebraic steps, we now solve Equation 6-14 for [ES]. First, the left side is multiplied out and the right side simplified to give

$$k_1[E_t][S] - k_1[ES][S] = (k_{-1} + k_2)[ES] \qquad (6\text{-}15)$$

Adding the term $k_1[ES][S]$ to both sides of the equation and simplifying gives

$$k_1[E_t][S] = (k_1[S] + k_{-1} + k_2)[ES] \qquad (6\text{-}16)$$

We then solve this equation for [ES]:

$$[ES] = \frac{k_1[E_t][S]}{k_1[S] + k_{-1} + k_2} \qquad (6\text{-}17)$$

This can now be simplified further, combining the rate constants into one expression:

$$[ES] = \frac{[E_t][S]}{[S] + (k_{-1} + k_2)/k_1} \qquad (6\text{-}18)$$

The term $(k_{-1} + k_2)/k_1$ is defined as the **Michaelis constant**, K_m. Substituting this into Equation 6-18 simplifies the expression to

$$[ES] = \frac{[E_t][S]}{K_m + [S]} \qquad (6\text{-}19)$$

Step 4 We can now express V_0 in terms of [ES]. Substituting the right side of Equation 6-19 for [ES] in Equation 6-11 gives

$$V_0 = \frac{k_2[E_t][S]}{K_m + [S]} \qquad (6\text{-}20)$$

This equation can be further simplified. Because the maximum velocity occurs when the enzyme is saturated (that is, when $[ES] = [E_t]$), V_{max} can be defined as $k_2[E_t]$. Substituting this in Equation 6-20 gives Equation 6-9:

$$V_0 = \frac{V_{max}[S]}{K_m + [S]}$$

This is the **Michaelis-Menten equation**, the **rate equation** for a one-substrate enzyme-catalyzed reaction. It is a statement of the quantitative relationship between the initial velocity V_0, the maximum velocity V_{max}, and the initial substrate concentration [S], all related through the Michaelis constant K_m. Note that K_m has units of molar concentration. Does the equation fit experimental observations? Yes; we can confirm this by considering the limiting situations where [S] is very high or very low, as shown in **Figure 6-12**.

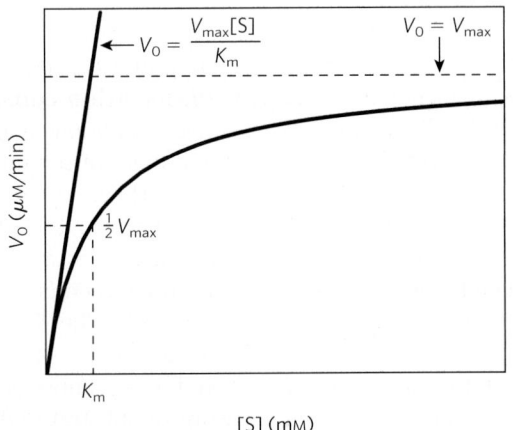

FIGURE 6-12 Dependence of initial velocity on substrate concentration. This graph shows the kinetic parameters that define the limits of the curve at high and low [S]. At low [S], $K_m \gg$ [S], and the [S] term in the denominator of the Michaelis-Menten equation (Eqn 6-9) becomes insignificant. The equation simplifies to $V_0 = V_{max}[S]/K_m$, and V_0 exhibits a linear dependence on [S], as observed here. At high [S], where [S] $\gg K_m$, the K_m term in the denominator of the Michaelis-Menten equation becomes insignificant and the equation simplifies to $V_0 = V_{max}$; this is consistent with the plateau observed at high [S]. The Michaelis-Menten equation is therefore consistent with the observed dependence of V_0 on [S], and the shape of the curve is defined by the terms V_{max}/K_m at low [S] and V_{max} at high [S].

An important numerical relationship emerges from the Michaelis-Menten equation in the special case when V_0 is exactly one-half V_{max} (Fig. 6-12). Then

$$\frac{V_{max}}{2} = \frac{V_{max}[S]}{K_m + [S]} \qquad (6\text{-}21)$$

On dividing by V_{max}, we obtain

$$\frac{1}{2} = \frac{[S]}{K_m + [S]} \qquad (6\text{-}22)$$

Solving for K_m, we get $K_m + [S] = 2[S]$, or

$$K_m = [S], \quad \text{when} \quad V_0 = \tfrac{1}{2}V_{max} \quad (6\text{-}23)$$

This is a very useful, practical definition of K_m: K_m is equivalent to the substrate concentration at which V_0 is one-half V_{max}.

The Michaelis-Menten equation (Eqn 6-9) can be algebraically transformed into versions that are useful in the practical determination of K_m and V_{max} (Box 6-1) and, as we describe later, in the analysis of inhibitor action (see Box 6-2).

Kinetic Parameters Are Used to Compare Enzyme Activities

It is important to distinguish between the Michaelis-Menten equation and the specific kinetic mechanism on which it was originally based. The equation describes the kinetic behavior of a great many enzymes, and all enzymes that exhibit a hyperbolic dependence of V_0 on [S] are said to follow **Michaelis-Menten kinetics**. The practical rule that $K_m = [S]$ when $V_0 = \tfrac{1}{2}V_{max}$ (Eqn 6-23) holds for all enzymes that follow Michaelis-Menten kinetics. (The most important exceptions to Michaelis-Menten kinetics are the regulatory enzymes, discussed in Section 6.5.) However, the Michaelis-Menten equation does not depend on the relatively simple two-step reaction mechanism proposed by Michaelis and Menten (Eqn 6-10). Many enzymes that follow Michaelis-Menten kinetics have quite different reaction mechanisms, and enzymes that catalyze reactions with six or eight identifiable steps often exhibit the same steady-state kinetic behavior. Even though Equation 6-23 holds true for many enzymes, both the magnitude and the real meaning of V_{max} and K_m can differ from one enzyme to the next. This is an important limitation of the steady-state approach to enzyme kinetics. The parameters V_{max} and K_m can be obtained experimentally for any given enzyme, but by themselves they provide little information about the number, rates, or chemical nature of discrete steps in the reaction. Steady-state kinetics nevertheless is the standard language through which biochemists compare and characterize the catalytic efficiencies of enzymes.

Interpreting V_{max} and K_m Figure 6-12 shows a simple graphical method for obtaining an approximate value for K_m. A more convenient procedure, using a **double-reciprocal plot**, is presented in Box 6-1. The K_m can

BOX 6-1   Transformations of the Michaelis-Menten Equation: The Double-Reciprocal Plot

The Michaelis-Menten equation

$$V_0 = \frac{V_{max}[S]}{K_m + [S]}$$

can be algebraically transformed into equations that are more useful in plotting experimental data. One common transformation is derived simply by taking the reciprocal of both sides of the Michaelis-Menten equation:

$$\frac{1}{V_0} = \frac{K_m + [S]}{V_{max}[S]}$$

Separating the components of the numerator on the right side of the equation gives

$$\frac{1}{V_0} = \frac{K_m}{V_{max}[S]} + \frac{[S]}{V_{max}[S]}$$

which simplifies to

$$\frac{1}{V_0} = \frac{K_m}{V_{max}[S]} + \frac{1}{V_{max}}$$

This form of the Michaelis-Menten equation is called the **Lineweaver-Burk equation**. For enzymes obeying the Michaelis-Menten relationship, a plot of $1/V_0$ versus $1/[S]$ (the "double reciprocal" of the V_0 versus [S] plot we have been using to this point) yields a straight line (Fig. 1). This line has a slope of K_m/V_{max}, an intercept of $1/V_{max}$ on the $1/V_0$ axis, and an intercept of $-1/K_m$ on the $1/[S]$ axis. The double-reciprocal presentation, also called a Lineweaver-Burk plot, has the great advantage of allowing a more accurate determination of V_{max}, which can only be *approximated* from a simple plot of V_0 versus [S] (see Fig. 6-12).

Other transformations of the Michaelis-Menten equation have been derived, each with some particular advantage in analyzing enzyme kinetic data. (See Problem 16 at the end of this chapter.)

The double-reciprocal plot of enzyme reaction rates is very useful in distinguishing between certain types of enzymatic reaction mechanisms (see Fig. 6-14) and in analyzing enzyme inhibition (see Box 6-2).

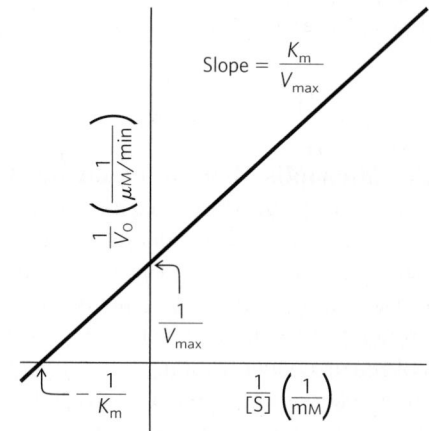

FIGURE 1 A double-reciprocal, or Lineweaver-Burk, plot.

---

vary greatly from enzyme to enzyme, and even for different substrates of the same enzyme (Table 6-6). The term is sometimes used (often inappropriately) as an indicator of the affinity of an enzyme for its substrate. The actual meaning of K_m depends on specific aspects of the reaction mechanism such as the number and relative rates of the individual steps. For reactions with two steps,

$$K_m = \frac{k_2 + k_{-1}}{k_1} \qquad (6\text{-}24)$$

When k_2 is rate-limiting, $k_2 \ll k_{-1}$, and K_m reduces to k_{-1}/k_1, which is defined as the **dissociation constant**, $\mathbf{K_d}$, of the ES complex. Where these conditions hold, K_m does represent a measure of the affinity of the enzyme for its substrate in the ES complex. However, this scenario does not apply for most enzymes. Sometimes $k_2 \gg k_{-1}$, and then $K_m = k_2/k_1$. In other cases, k_2 and k_{-1} are comparable, and K_m remains a more complex function of all three rate constants (Eqn 6-24). The Michaelis-Menten equation and the characteristic saturation behavior of the enzyme still apply, but K_m cannot be considered a simple measure of substrate affinity. Even more common are cases in which the reaction goes through several steps after formation of ES; K_m can then become a very complex function of many rate constants.

The quantity V_{max} also varies greatly from one enzyme to the next. If an enzyme reacts by the two-step Michaelis-Menten mechanism, $V_{max} = k_2[E_t]$, where k_2 is rate-limiting. However, the number of reaction steps and the identity of the rate-limiting step(s) can vary from enzyme to enzyme. For example, consider the common situation where product release, EP → E + P, is rate-limiting. Early in the reaction (when [P] is low),

TABLE 6-6 K_m for Some Enzymes and Substrates

Enzyme	Substrate	K_m (mM)
Hexokinase (brain)	ATP	0.4
	D-Glucose	0.05
	D-Fructose	1.5
Carbonic anhydrase	HCO_3^-	26
Chymotrypsin	Glycyltyrosinylglycine	108
	N-Benzoyltyrosinamide	2.5
β-Galactosidase	D-Lactose	4.0
Threonine dehydratase	L-Threonine	5.0

TABLE 6-7	Turnover Number, k_{cat}, of Some Enzymes	
Enzyme	Substrate	k_{cat} (s^{-1})
Catalase	H_2O_2	40,000,000
Carbonic anhydrase	HCO_3^-	400,000
Acetylcholinesterase	Acetylcholine	14,000
β-Lactamase	Benzylpenicillin	2,000
Fumarase	Fumarate	800
RecA protein (an ATPase)	ATP	0.5

the overall reaction can be adequately described by the scheme

$$E + S \underset{k_{-1}}{\overset{k_1}{\rightleftharpoons}} ES \underset{k_{-2}}{\overset{k_2}{\rightleftharpoons}} EP \overset{k_3}{\longrightarrow} E + P \quad (6\text{-}25)$$

In this case, most of the enzyme is in the EP form at saturation, and $V_{max} = k_3[E_t]$. It is useful to define a more general rate constant, **k_{cat}**, to describe the limiting rate of any enzyme-catalyzed reaction at saturation. If the reaction has several steps and one is clearly rate-limiting, k_{cat} is equivalent to the rate constant for that limiting step. For the simple reaction of Equation 6-10, $k_{cat} = k_2$. For the reaction of Equation 6-25, when product release is clearly rate-limiting, $k_{cat} = k_3$. When several steps are partially rate-limiting, k_{cat} can become a complex function of several of the rate constants that define each individual reaction step. In the Michaelis-Menten equation, $k_{cat} = V_{max}/[E_t]$, and Equation 6-9 becomes

$$V_0 = \frac{k_{cat}[E_t][S]}{K_m + [S]} \quad (6\text{-}26)$$

The constant k_{cat} is a first-order rate constant and hence has units of reciprocal time. It is also called the **turnover number**. It is equivalent to the number of substrate molecules converted to product in a given unit of time on a single enzyme molecule when the enzyme is saturated with substrate. The turnover numbers of several enzymes are given in Table 6-7.

Comparing Catalytic Mechanisms and Efficiencies The kinetic parameters k_{cat} and K_m are useful for the study and comparison of different enzymes, whether their reaction mechanisms are simple or complex. Each enzyme has values of k_{cat} and K_m that reflect the cellular environment, the concentration of substrate normally encountered in vivo by the enzyme, and the chemistry of the reaction being catalyzed.

The parameters k_{cat} and K_m also allow us to evaluate the kinetic efficiency of enzymes, but either parameter alone is insufficient for this task. Two enzymes catalyzing different reactions may have the same k_{cat} (turnover number), yet the rates of the uncatalyzed reactions may be different and thus the rate enhancements brought about by the enzymes may differ greatly. Experimentally, the K_m for an enzyme tends to be similar to the cellular concentration of its substrate. An enzyme that acts on a substrate present at a very low concentration in the cell usually has a lower K_m than an enzyme that acts on a substrate that is more abundant.

The best way to compare the catalytic efficiencies of different enzymes or the turnover of different substrates by the same enzyme is to compare the ratio k_{cat}/K_m for the two reactions. This parameter, sometimes called the **specificity constant**, is the rate constant for the conversion of E + S to E + P. When $[S] \ll K_m$, Equation 6-26 reduces to the form

$$V_0 = \frac{k_{cat}}{K_m}[E_t][S] \quad (6\text{-}27)$$

V_0 in this case depends on the concentration of two reactants, $[E_t]$ and $[S]$, so this is a second-order rate equation, and the constant k_{cat}/K_m is a second-order rate constant with units of $M^{-1}s^{-1}$. There is an upper limit to k_{cat}/K_m, imposed by the rate at which E and S can diffuse together in an aqueous solution. This diffusion-controlled limit is 10^8 to 10^9 $M^{-1}s^{-1}$, and many enzymes have a k_{cat}/K_m near this range (Table 6-8). Such enzymes are said to have achieved catalytic perfection. Note that different values of k_{cat} and K_m can produce the maximum ratio.

TABLE 6-8	Enzymes for Which k_{cat}/K_m Is Close to the Diffusion-Controlled Limit (10^8 to 10^9 $M^{-1}s^{-1}$)			
Enzyme	Substrate	k_{cat} (s^{-1})	K_m (M)	k_{cat}/K_m ($M^{-1}s^{-1}$)
Acetylcholinesterase	Acetylcholine	1.4×10^4	9×10^{-5}	1.6×10^8
Carbonic anhydrase	CO_2	1×10^6	1.2×10^{-2}	8.3×10^7
	HCO_3^-	4×10^5	2.6×10^{-2}	1.5×10^7
Catalase	H_2O_2	4×10^7	1.1×10^0	4×10^7
Crotonase	Crotonyl-CoA	5.7×10^3	2×10^{-5}	2.8×10^8
Fumarase	Fumarate	8×10^2	5×10^{-6}	1.6×10^8
	Malate	9×10^2	2.5×10^{-5}	3.6×10^7
β-Lactamase	Benzylpenicillin	2.0×10^3	2×10^{-5}	1×10^8

Source: A. Fersht, *Structure and Mechanism in Protein Science*, p. 166, W. H. Freeman and Company, 1999.

An enzyme is discovered that catalyzes the chemical reaction

$$\text{SAD} \rightleftharpoons \text{HAPPY}$$

A team of motivated researchers sets out to study the enzyme, which they call happyase. They find that the k_{cat} for happyase is $600\ \text{s}^{-1}$ and carry out several additional experiments.

When $[E_t] = 20$ nM and $[SAD] = 40\ \mu\text{M}$, the reaction velocity, V_0, is $9.6\ \mu\text{M s}^{-1}$. Calculate K_m for the substrate SAD.

Solution: We know k_{cat}, $[E_t]$, $[S]$, and V_0. We want to solve for K_m. Equation 6-26, in which we substitute $k_{cat}[E_t]$ for V_{max} in the Michaelis-Menten equation, is most useful here. Substituting our known values in Equation 6-26 allows us to solve for K_m:

$$V_0 = \frac{k_{cat}[E_t][S]}{K_m + [S]}$$

$$9.6\ \mu\text{M s}^{-1} = \frac{(600\ \text{s}^{-1})(0.020\ \mu\text{M})(40\ \mu\text{M})}{K_m + 40\ \mu\text{M}}$$

$$9.6\ \mu\text{M s}^{-1} = \frac{480\ \mu\text{M}^2\ \text{s}^{-1}}{K_m + 40\ \mu\text{M}}$$

$$9.6\ \mu\text{M s}^{-1}(K_m + 40\ \mu\text{M}) = 480\ \mu\text{M}^2\ \text{s}^{-1}$$

$$K_m + 40\ \mu\text{M} = \frac{480\ \mu\text{M}^2\ \text{s}^{-1}}{9.6\ \mu\text{M s}^{-1}}$$

$$K_m + 40\ \mu\text{M} = 50\ \mu\text{M}$$

$$K_m = 50\ \mu\text{M} - 40\ \mu\text{M}$$

$$K_m = 10\ \mu\text{M}$$

Once you have worked with this equation, you will recognize shortcuts to solve problems like this. For example, one can calculate V_{max} knowing that $k_{cat}[E_t] = V_{max}$ (in this case, $600\ \text{s}^{-1} \times 0.020\ \mu\text{M} = 12\ \mu\text{M s}^{-1}$). A simple rearrangement of Equation 6-26 by dividing both sides by V_{max} gives

$$\frac{V_0}{V_{max}} = \frac{[S]}{K_m + [S]}$$

Thus, the ratio $V_0/V_{max} = 9.6\ \mu\text{M s}^{-1}/12\ \mu\text{M s}^{-1} = [S]/(K_m + [S])$. This simplifies the process of solving for K_m, giving $0.25[S]$, or $10\ \mu\text{M}$.

In a separate happyase experiment using $[E_t] = 10$ mM, the reaction velocity, V_0, is measured as $3\ \mu\text{M s}^{-1}$. What is the [S] used in this experiment?

Solution: Using the same logic as in Worked Example 6-1, we see that the V_{max} for this enzyme concentration is $6\ \mu\text{M s}^{-1}$. Note that the V_0 is exactly half of the V_{max}. Recall that K_m is by definition equal to the [S] at which

$V_0 = \frac{1}{2}V_{max}$. Thus, in this example, the [S] must be the same as the K_m, or $10\ \mu\text{M}$. If V_0 were anything other than $\frac{1}{2}V_{max}$, it would be simplest to use the expression $V_0/V_{max} = [S]/(K_m + [S])$ to solve for [S].

Many Enzymes Catalyze Reactions with Two or More Substrates

We have seen how [S] affects the rate of a simple enzymatic reaction with only one substrate molecule ($S \rightarrow P$). In most enzymatic reactions, however, two (and sometimes more) different substrate molecules bind to the enzyme and participate in the reaction. Nearly two-thirds of all enzymatic reactions have two substrates and two products. These are generally reactions in which a group is transferred from one substrate to the other, or one substrate is oxidized while the other is reduced. For example, in the reaction catalyzed by hexokinase, ATP and glucose are the substrate molecules, and ADP and glucose 6-phosphate are the products:

$$\text{ATP} + \text{glucose} \longrightarrow \text{ADP} + \text{glucose 6-phosphate}$$

A phosphoryl group is transferred from ATP to glucose. The rates of such bisubstrate reactions can also be analyzed by the Michaelis-Menten approach. Hexokinase has a characteristic K_m for each of its substrates (Table 6-6).

Enzymatic reactions with two substrates proceed by one of several different types of pathways. In some cases, both substrates are bound to the enzyme concurrently at some point in the course of the reaction, forming a noncovalent ternary complex (**Fig. 6-13a**); the substrates bind in a random sequence or in a specific order. In other cases, the first substrate is converted to product and dissociates before the second substrate binds, so no ternary complex is formed. An example of this is the Ping-Pong, or double-displacement, mechanism (Fig. 6-13b).

A shorthand notation developed by W. W. Cleland can be helpful in describing reactions with multiple substrates and products. In this system, referred to as **Cleland nomenclature**, substrates are denoted A, B, C, and D, in the order in which they bind to the enzyme, and products are denoted P, Q, S, T, in the order in which they dissociate. Enzymatic reactions with one, two, three, or four substrates are referred to as uni, bi, ter, and quad, respectively. The enzyme is, as usual, denoted E, but if it is modified in the course of the reaction, successive forms are denoted F, G, and so on. The progress of the reaction is indicated with a horizontal line, with successive chemical species indicated below it. If there is an alternative in the reaction path, the horizontal line is bifurcated. Steps involving binding and dissociating substrates and products are indicated with vertical lines.

Common reactions with two substrates and two products (bi bi) are described with the shorthand forms illustrated in Figure 16-13c for an ordered bi bi reaction and a random bi bi reaction. In the latter example, the

(a) Enzyme reaction involving a ternary complex

Ordered

$$E + S_1 \rightleftharpoons ES_1 \overset{S_2}{\rightleftharpoons} ES_1S_2 \longrightarrow E + P_1 + P_2$$

Random order

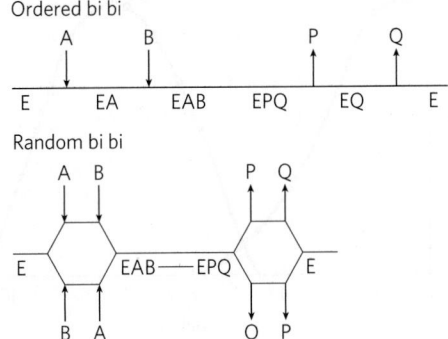

$$ES_1S_2 \longrightarrow E + P_1 + P_2$$

(b) Enzyme reaction in which no ternary complex is formed

$$E + S_1 \rightleftharpoons ES_1 \rightleftharpoons E'P_1 \overset{P_1}{\rightleftharpoons} E' \overset{S_2}{\rightleftharpoons} E'S_2 \longrightarrow E + P_2$$

(c) Cleland nomenclature

Ordered bi bi

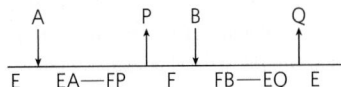

Random bi bi

(d) Ping-Pong in Cleland nomenclature

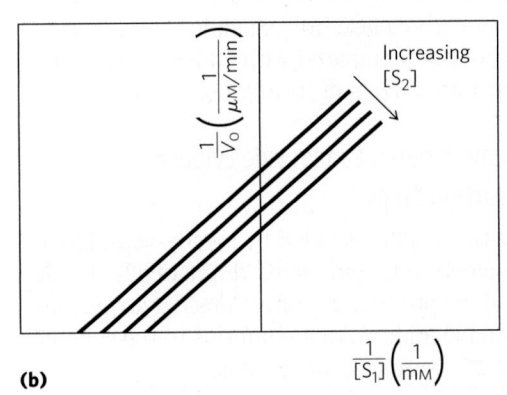

FIGURE 6-13 Common mechanisms for enzyme-catalyzed bisubstrate reactions. (a) The enzyme and both substrates come together to form a ternary complex. In ordered binding, substrate 1 must bind before substrate 2 can bind productively. In random binding, the substrates can bind in either order. **(b)** An enzyme-substrate complex forms, a product leaves the complex, the altered enzyme forms a second complex with another substrate molecule, and the second product leaves, regenerating the enzyme. Substrate 1 may transfer a functional group to the enzyme (to form the covalently modified E′), which is subsequently transferred to substrate 2. This is called a Ping-Pong or double-displacement mechanism. **(c)** Ternary complex formation depicted using Cleland nomenclature. In the ordered bi bi and the random bi bi reactions shown here, the release of product follows the same pattern as the binding of substrate—both ordered or both random. **(d)** The Ping-Pong or double-displacement reaction described with Cleland nomenclature.

release of product is also random, as indicated by the two sets of bifurcations. Rarely, the binding of substrates is ordered and release of products random, or vice versa, eliminating the bifurcation at one end or the other of the progress line. In a Ping-Pong reaction, lacking a ternary complex, the pathway has a transient second form of the enzyme, F (Fig. 6-13d). This is the form in which a group has been transferred from the first substrate, A, to create a transient covalent attachment to the enzyme. As noted above, such reactions are often called double-displacement reactions, as a group is transferred first from substrate A to the enzyme and then from the enzyme to substrate B. Substrates A and B do not encounter each other on the enzyme.

Michaelis-Menten steady-state kinetics can provide only limited information about the number of steps and intermediates in an enzymatic reaction, but it can be used to distinguish between pathways that have a ternary intermediate and pathways—including Ping-Pong pathways—that do not **(Fig. 6-14)**. As we will see when we consider enzyme inhibition, steady-state kinetics can also distinguish between ordered and random binding of substrates and products in reactions with ternary intermediates.

Enzyme Activity Depends on pH

In general, steady-state kinetics provides information required to characterize an enzyme and assess its catalytic efficiency. Additional information can be gained by examining how the key experimental parameters k_{cat} and k_{cat}/K_m change when reaction conditions change, particularly pH. Enzymes have an optimum pH (or pH range) at which their activity is maximal **(Fig. 6-15)**; at higher or lower pH, activity decreases. This is not surprising.

(a)

(b)

FIGURE 6-14 Steady-state kinetic analysis of bisubstrate reactions. In these double-reciprocal plots (see Box 6-1), the concentration of substrate 1 is varied while the concentration of substrate 2 is held constant. This is repeated for several values of [S$_2$], generating several separate lines. **(a)** Intersecting lines indicate that a ternary complex is formed in the reaction; **(b)** parallel lines indicate a Ping-Pong (double-displacement) pathway.

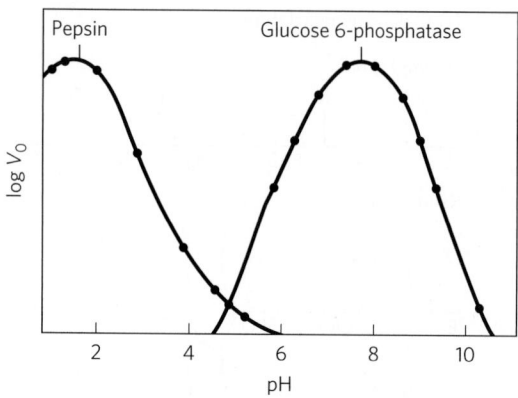

FIGURE 6-15 The pH-activity profiles of two enzymes. These curves are constructed from measurements of initial velocities when the reaction is carried out in buffers of different pH. Because pH is a logarithmic scale reflecting 10-fold changes in [H⁺], the changes in V_0 are also plotted on a logarithmic scale. The pH optimum for the activity of an enzyme is generally close to the pH of the environment in which the enzyme is normally found. Pepsin, a peptidase found in the stomach, has a pH optimum of about 1.6. The pH of gastric juice is between 1 and 2. Glucose 6-phosphatase of hepatocytes (liver cells), with a pH optimum of about 7.8, is responsible for releasing glucose into the blood. The normal pH of the cytosol of hepatocytes is about 7.2.

Amino acid side chains in the active site may act as weak acids and bases only if they maintain a certain state of ionization. Elsewhere in the protein, removing a proton from a His residue, for example, might eliminate an ionic interaction essential for stabilizing the active conformation of the enzyme. A less common cause of pH sensitivity is titration of a group on the substrate.

The pH range over which an enzyme undergoes changes in activity can provide a clue to the type of amino acid residue involved (see Table 3-1). A change in activity near pH 7.0, for example, often reflects titration of a His residue. The effects of pH must be interpreted with some caution, however. In the closely packed environment of a protein, the pK_a of amino acid side chains can be significantly altered. For example, a nearby positive charge can lower the pK_a of a Lys residue, and a nearby negative charge can increase it. Such effects sometimes result in a pK_a that is shifted by several pH units from its value in the free amino acid. In the enzyme acetoacetate decarboxylase, for example, one Lys residue has a pK_a of 6.6 (compared with 10.5 in free lysine) due to electrostatic effects of nearby positive charges.

Pre–Steady State Kinetics Can Provide Evidence for Specific Reaction Steps

The mechanistic insight provided by steady-state kinetics can be augmented, sometimes dramatically, by an examination of the pre–steady state. Consider an enzyme with a reaction mechanism that conforms to the scheme in Equation 6-25, featuring three steps:

$$\text{E} + \text{S} \underset{k_{-1}}{\overset{k_1}{\rightleftharpoons}} \text{ES} \underset{k_{-2}}{\overset{k_2}{\rightleftharpoons}} \text{EP} \overset{k_3}{\longrightarrow} \text{E} + \text{P}$$

Overall catalytic efficiency for this reaction can be assessed with steady-state kinetics, but the rates of the individual steps cannot be determined in this way, and the slow (rate-limiting) step can rarely be identified. To measure the rate constants of individual steps, the reaction must be studied during its pre–steady state. The first turnover of an enzyme-catalyzed reaction often occurs in seconds or milliseconds, so researchers use special equipment that allows mixing and sampling on this timescale (**Fig. 6-16a**). Reactions are stopped and protein-bound products are quantified, after the timed addition and rapid mixing of an acid that denatures the protein and releases all bound molecules. A detailed description of pre–steady state kinetics is beyond the scope of this text, but we can illustrate the power of this approach by a simple example of an enzyme that uses the pathway shown in Equation 6-25. This example also involves an enzyme that catalyzes a relatively slow reaction, so the pre–steady state is more conveniently observed.

For many enzymes, dissociation of product is rate-limiting. In this example (Fig. 6-16b, c), the rate of dissociation of the product (k_3) is slower than the rate of its formation (k_2). Product dissociation therefore dictates the rates observed in the steady state. How do we know that k_3 is rate-limiting? A slow k_3 gives rise to a burst of product formation in the pre–steady state, because the preceding steps are relatively fast. The burst reflects the rapid conversion of one molecule of substrate to one molecule of product at each enzyme active site. The observed rate of product formation slows to the steady-state rate as the bound product is slowly released. Each enzymatic turnover after the first one must proceed through the slow product-release step. However, the rapid generation of product in that first turnover provides much information. The amplitude of the burst—when one molecule of product is generated per molecule of enzyme present (Fig. 6-16c), measured by extrapolating the steady-state progress line back to zero time—is the highest amplitude possible. This provides one piece of evidence that product release is, indeed, rate-limiting. The rate constant for the chemical reaction step, k_2, can be derived from the observed rate of the burst phase. Of course, enzymes do not always conform to the simple reaction scheme of Equation 6-25. Formally, the observation of a burst indicates that a rate-limiting step (typically, product release, or an enzyme conformational change, or another chemical step) occurs after formation of the product being monitored. Additional experiments and analysis can often define the rates of each step in a multistep enzymatic reaction. Some examples of the application of pre–steady state kinetics are included in the descriptions of specific enzymes in Section 6.4.

Enzymes Are Subject to Reversible or Irreversible Inhibition

Enzyme inhibitors are molecules that interfere with catalysis, slowing or halting enzymatic reactions. Enzymes catalyze virtually all cellular processes, so it should not

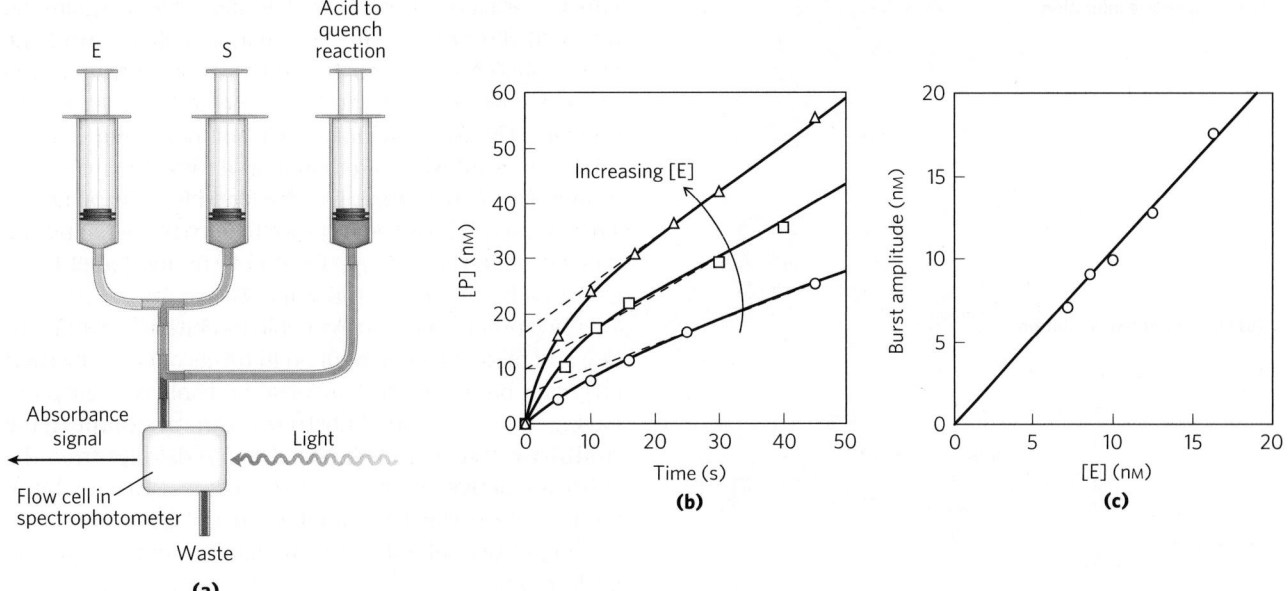

FIGURE 6-16 Pre-steady state kinetics. The transient phase that constitutes the pre-steady state often exists for mere seconds or milliseconds, requiring specialized equipment to monitor it. **(a)** A simple schematic for a rapid-mixing device, called a stopped-flow device. Enzyme (E) and substrate (S) are mixed with the aid of mechanically operated syringes. The reaction is quenched at a programmed time by adding a denaturing acid through another syringe, and the amount of product formed is measured, in this case with a spectrophotometer. **(b)** Experimental data for an enzyme reaction show the pre–steady state occurring in the first 5 to 10 seconds. This is a relatively slow reaction and is used as an example because the steady state can be conveniently monitored. The slope of the lines after 15 seconds reflects the steady state. Extrapolating this slope back to zero time (dashed lines) gives the amplitude of the burst phase. The progress of the reaction during the pre–steady state primarily reflects the chemical steps in the reaction (details of which are not shown). The presence of a burst implies that a step following the chemical step that produces P is rate-limiting—in this case, the product-release step. Notice that the extrapolated intercept at time = 0 increases as [E] increases. **(c)** A plot of burst amplitude (the intercepts from (b)) versus [E] shows that one molecule of P is formed in each active site during the burst (pre–steady state) phase. This provides evidence that step 3, product release, is the rate-limiting step, because it is the only step following product formation in this simple enzymatic reaction. The enzyme used in this experiment was RNase P, one of the catalytic RNAs described in Chapter 26. [Source: (b, c) Data from J. Hsieh et al., *RNA* 15:224, 2009.]

be surprising that enzyme inhibitors are among the most important pharmaceutical agents known. For example, aspirin (acetylsalicylate) inhibits the enzyme that catalyzes the first step in the synthesis of prostaglandins, compounds involved in many processes, including some that produce pain. The study of enzyme inhibitors also has provided valuable information about enzyme mechanisms and has helped define some metabolic pathways. There are two broad classes of enzyme inhibitors: reversible and irreversible.

Reversible Inhibition One common type of **reversible inhibition** is called competitive **(Fig. 6-17a)**. A **competitive inhibitor** competes with the substrate for the active site of an enzyme. While the inhibitor (I) occupies the active site, it prevents the substrate from binding to the enzyme. Many competitive inhibitors are structurally similar to the substrate and combine with the enzyme to form an EI complex, but without leading to catalysis. Even fleeting combinations of this type will reduce the efficiency of an enzyme. By taking into account the molecular geometry of inhibitors, we can reach conclusions about which parts of the normal substrate bind to the enzyme. Competitive inhibition can be analyzed quantitatively by steady-state kinetics.

In the presence of a competitive inhibitor, the Michaelis-Menten equation (Eqn 6-9) becomes

$$V_0 = \frac{V_{max}[S]}{\alpha K_m + [S]} \qquad (6\text{-}28)$$

where

$$\alpha = 1 + \frac{[I]}{K_I} \quad \text{and} \quad K_I = \frac{[E][I]}{[EI]}$$

Equation 6-28 describes the important features of competitive inhibition. The experimentally determined variable αK_m, the K_m observed in the presence of the inhibitor, is often called the "apparent" K_m.

Bound inhibitor does not inactivate the enzyme. When the inhibitor dissociates, substrate can bind and react. Because the inhibitor binds reversibly to the enzyme, the competition can be biased to favor the substrate simply by adding more substrate. When [S] far exceeds [I], the probability that an inhibitor molecule will bind to the enzyme is minimized and the reaction exhibits a normal V_{max}. However, in the presence of inhibitor, higher concentrations of substrate are needed to approach that V_{max}. The [S] at which $V_0 = \frac{1}{2}V_{max}$, the apparent K_m, increases in the presence of inhibitor by the factor α. This effect on apparent K_m, combined with the

(a) Competitive inhibition

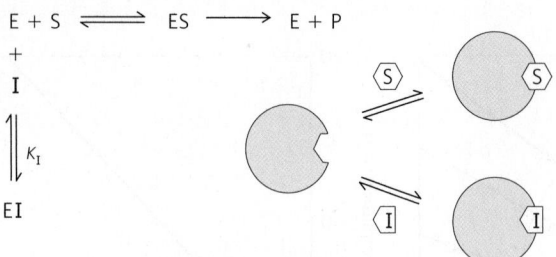

(b) Uncompetitive inhibition

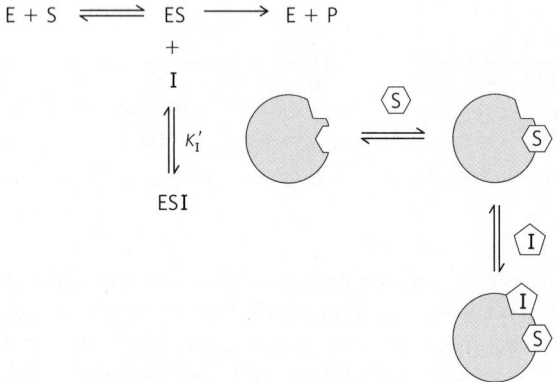

(c) Mixed inhibition

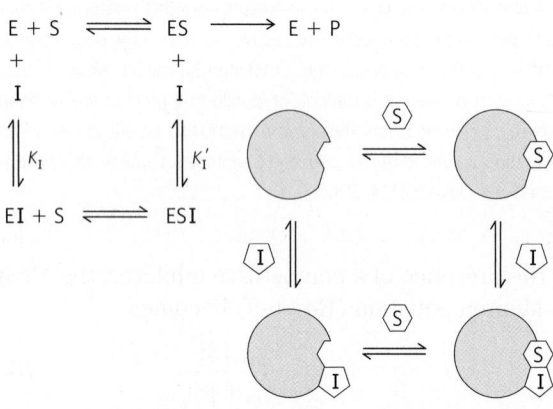

FIGURE 6-17 Three types of reversible inhibition. (a) Competitive inhibitors bind to the enzyme's active site; K_I is the equilibrium dissociation constant for inhibitor binding to E. **(b)** Uncompetitive inhibitors bind at a separate site, but bind only to the ES complex; K'_I is the equilibrium constant for inhibitor binding to ES. **(c)** Mixed inhibitors bind at a separate site, but may bind to either E or ES.

absence of an effect on V_{max}, is diagnostic of competitive inhibition and is readily revealed in a double-reciprocal plot (Box 6-2). The equilibrium constant for inhibitor binding, K_I, can be obtained from the same plot.

A medical therapy based on competition at the active site is used to treat patients who have ingested methanol, a solvent found in gas-line antifreeze. The liver enzyme alcohol dehydrogenase converts methanol to formaldehyde, which is damaging to many tissues. Blindness is a common result of methanol ingestion, because the eyes are particularly sensitive to formaldehyde. Ethanol competes effectively with methanol as an alternative substrate for alcohol dehydrogenase. The

effect of ethanol is much like that of a competitive inhibitor, with the distinction that ethanol is also a substrate for alcohol dehydrogenase and its concentration will decrease over time as the enzyme converts it to acetaldehyde. The therapy for methanol poisoning is slow intravenous infusion of ethanol, at a rate that maintains a controlled concentration in the blood for several hours. This slows the formation of formaldehyde, lessening the danger while the kidneys filter out the methanol to be excreted harmlessly in the urine. ∎

Two other types of reversible inhibition, uncompetitive and mixed, can be defined in terms of one-substrate enzymes, but in practice are observed only with enzymes having two or more substrates. An **uncompetitive inhibitor** (Fig. 6-17b) binds at a site distinct from the substrate active site and, unlike a competitive inhibitor, binds only to the ES complex. In the presence of an uncompetitive inhibitor, the Michaelis-Menten equation is altered to

$$V_0 + \frac{V_{max}[S]}{K_m + \alpha'[S]} \qquad (6\text{-}29)$$

where

$$\alpha' = 1 + \frac{[I]}{K'_I} \qquad \text{and} \qquad K'_I = \frac{[ES][I]}{[ESI]}$$

As described by Equation 6-29, at high concentrations of substrate, V_0 approaches V_{max}/α'. Thus, an uncompetitive inhibitor lowers the measured V_{max}. Apparent K_m also decreases, because the [S] required to reach one-half V_{max} decreases by the factor α'. This behavior can be explained as follows. Because the enzyme is inactive when the uncompetitive inhibitor is bound, but the inhibitor is not competing with substrate for binding, the inhibitor effectively removes some fraction of the enzyme molecules from the reaction. Given that V_{max} depends on [E], the observed V_{max} decreases, and given that inhibitor binds only to the ES complex, only ES (not free enzyme) is deleted from the reaction, so the [S] needed to reach $\frac{1}{2}V_{max}$—that is, K_m—declines by the same amount.

A **mixed inhibitor** (Fig. 6-17c) also binds at a site distinct from the substrate active site, but it binds to either E or ES. The rate equation describing mixed inhibition is

$$V_0 = \frac{V_{max}[S]}{\alpha K_m + \alpha'[S]} \qquad (6\text{-}30)$$

where α and α' are defined as above. A mixed inhibitor usually affects both K_m and V_{max}. V_{max} is affected because the inhibitor renders some fraction of the available enzyme molecules inactive, lowering the effective [E] on which V_{max} depends. The K_m may increase or decrease, depending on which enzyme form, E or ES, the inhibitor binds to most strongly. The special case of $\alpha = \alpha'$, rarely encountered in experiments, classically has been defined as **noncompetitive inhibition**. Examine Equation 6-30 to see why a noncompetitive inhibitor would affect the V_{max} but not the K_m.

BOX 6-2   Kinetic Tests for Determining Inhibition Mechanisms

The double-reciprocal plot (see Box 6-1) offers an easy way of determining whether an enzyme inhibitor is competitive, uncompetitive, or mixed. Two sets of rate experiments are carried out, with the enzyme concentration held constant in each set. In the first set, [S] is also held constant, permitting measurement of the effect of increasing inhibitor concentration [I] on the initial rate V_0 (not shown). In the second set, [I] is held constant but [S] is varied. The results are plotted as $1/V_0$ versus $1/[S]$.

Figure 1 shows a set of double-reciprocal plots, one obtained in the absence of inhibitor and two at different concentrations of a competitive inhibitor. Increasing [I] results in a family of lines with a common intercept on the $1/V_0$ axis but with different slopes. Because the intercept on the $1/V_0$ axis equals $1/V_{max}$,

we know that V_{max} is unchanged by the presence of a competitive inhibitor. That is, regardless of the concentration of a competitive inhibitor, a sufficiently high substrate concentration will always displace the inhibitor from the enzyme's active site. Above the graph is the rearrangement of Equation 6-28 on which the plot is based. The value of α can be calculated from the change in slope at any given [I]. Knowing [I] and α, we can calculate K_I from the expression

$$\alpha = 1 + \frac{[I]}{K_I}$$

For uncompetitive and mixed inhibition, similar plots of rate data give the families of lines shown in Figures 2 and 3. Changes in axis intercepts signal changes in V_{max} and K_m.

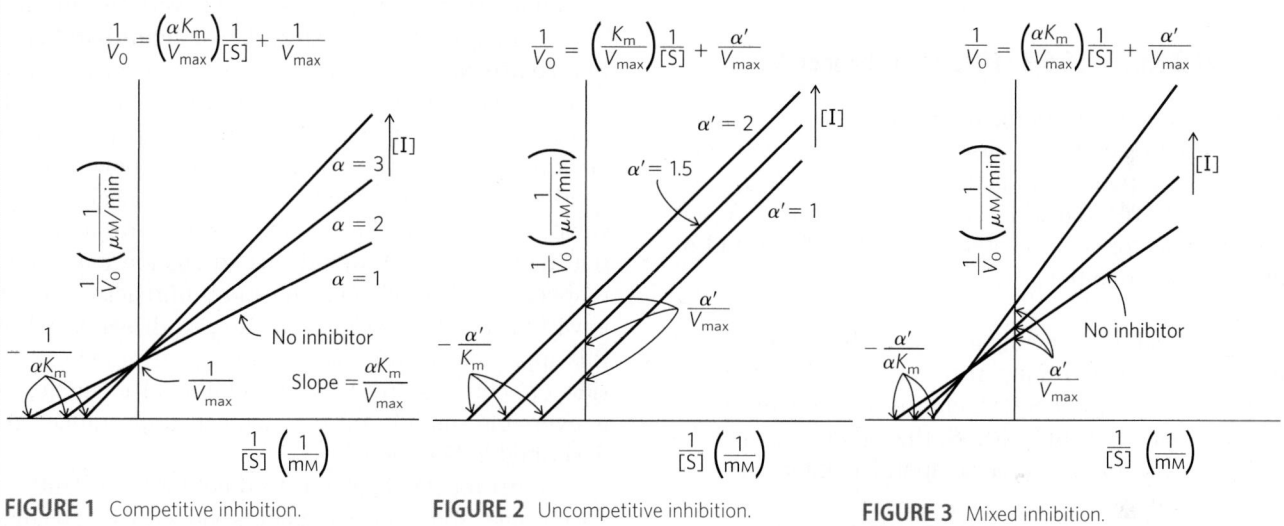

FIGURE 1 Competitive inhibition.   **FIGURE 2** Uncompetitive inhibition.   **FIGURE 3** Mixed inhibition.

Equation 6-30 is a general expression for the effects of reversible inhibitors, simplifying to the expressions for competitive and uncompetitive inhibition when $\alpha' = 1.0$ or $\alpha = 1.0$, respectively. From this expression we can summarize the effects of inhibitors on individual kinetic parameters. For all reversible inhibitors, the apparent $V_{max} = V_{max}/\alpha'$, because the right side of Equation 6-30 always simplifies to V_{max}/α' at sufficiently high substrate concentrations. For competitive inhibitors, $\alpha' = 1.0$ and can thus be ignored. Taking this expression for apparent V_{max}, we can also derive a general expression for apparent K_m to show how this parameter changes in the presence of reversible inhibitors. Apparent K_m, as always, equals the [S] at which V_0 is one-half apparent V_{max} or, more generally, when $V_0 = V_{max}/2\alpha'$. This condition is met when [S] = $\alpha K_m/\alpha'$. Thus, apparent $K_m = \alpha K_m/\alpha'$. The terms α and α' reflect the binding of inhibitor to E and ES, respectively. Thus, the term $\alpha K_m/\alpha'$ is a mathematical expression of the relative affinity of inhibitor for the two

enzyme forms. This expression is simpler when either α or α' is 1.0 (for uncompetitive or competitive inhibitors), as summarized in Table 6-9.

In practice, uncompetitive and mixed inhibition are observed only for enzymes with two or more substrates—say, S_1 and S_2—and are very important in the experimental analysis of such enzymes. If an inhibitor binds to the site normally occupied by S_1, it may act as

TABLE 6-9	Effects of Reversible Inhibitors on Apparent V_{max} and Apparent K_m	
Inhibitor type	Apparent V_{max}	Apparent K_m
None	V_{max}	K_m
Competitive	V_{max}	αK_m
Uncompetitive	V_{max}/α'	K_m/α'
Mixed	V_{max}/α'	$\alpha K_m/\alpha'$

a competitive inhibitor in experiments in which $[S_1]$ is varied. If an inhibitor binds to the site normally occupied by S_2, it may act as a mixed or uncompetitive inhibitor of S_1. The actual inhibition patterns observed depend on whether the S_1- and S_2-binding events are ordered or random, and thus the order in which substrates bind and products leave the active site can be determined. Product inhibition experiments in which one of the reaction products is provided as an inhibitor are often particularly informative. If only one of two reaction products is present, no reverse reaction can take place. However, a product generally binds to some part of the active site and can thus serve as an effective inhibitor when the second product is not present. Enzymologists can combine steady-state kinetic studies involving different combinations and amounts of products and inhibitors with pre–steady state analysis to develop a detailed picture of the mechanism of a bisubstrate reaction.

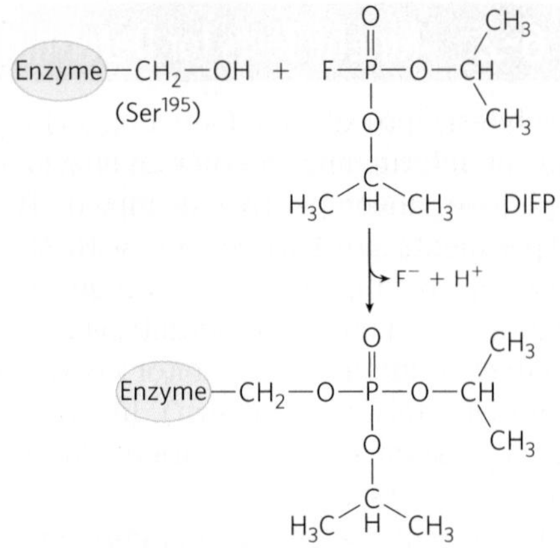

FIGURE 6-18 Irreversible inhibition. Reaction of chymotrypsin with diisopropylfluorophosphate (DIFP), which modifies Ser[195], irreversibly inhibits the enzyme. This has led to the conclusion that Ser[195] is the key active-site Ser residue in chymotrypsin.

WORKED EXAMPLE 6-3 Effect of Inhibitor on K_m

The researchers working on happyase (see Worked Examples 6-1 and 6-2) discover that the compound STRESS is a potent competitive inhibitor of happyase. Addition of 1 nM STRESS increases the measured K_m for SAD by a factor of 2. What are the values for α and α' under these conditions?

Solution: Recall that the apparent K_m, the K_m measured in the presence of a competitive inhibitor, is defined as αK_m. Because K_m for SAD increases by a factor of 2 in the presence of 1 nM STRESS, the value of α must be 2. The value of α' for a competitive inhibitor is 1, by definition.

Irreversible Inhibition The **irreversible inhibitors** bind covalently with or destroy a functional group on an enzyme that is essential for the enzyme's activity, or they form a highly stable noncovalent association. Formation of a covalent link between an irreversible inhibitor and an enzyme is a particularly effective way to inactivate an enzyme. Irreversible inhibitors are another useful tool for studying reaction mechanisms. Amino acids with key catalytic functions in the active site can sometimes be identified by determining which residue is covalently linked to an inhibitor after the enzyme is inactivated. An example is shown in **Figure 6-18**.

A special class of irreversible inhibitors is the **suicide inactivators**. These compounds are relatively unreactive until they bind to the active site of a specific enzyme. A suicide inactivator undergoes the first few chemical steps of the normal enzymatic reaction, but instead of being transformed into the normal product, the inactivator is converted to a very reactive

compound that combines irreversibly with the enzyme. These compounds are also called **mechanism-based inactivators**, because they hijack the normal enzyme reaction mechanism to inactivate the enzyme. Suicide inactivators play a significant role in *rational drug design*, a modern approach to obtaining new pharmaceutical agents in which chemists synthesize novel substrates based on knowledge of substrates and reaction mechanisms. A well-designed suicide inactivator is specific for a single enzyme and is unreactive until it is within that enzyme's active site, so drugs based on this approach can offer the important advantage of few side effects (Box 6-3). Some additional examples of irreversible inhibitors of medical importance are described in Section 6.4.

An irreversible inhibitor need not bind covalently to the enzyme. Noncovalent binding is enough, if that binding is so tight that the inhibitor dissociates only rarely. How does a chemist develop a tight-binding inhibitor? Recall that enzymes evolve to bind most tightly to the transition states of the reactions that they catalyze. In principle, if one can design a molecule that looks like that reaction transition state, it should bind tightly to the enzyme. Even though transition states cannot be observed directly, chemists can often predict the approximate structure of a transition state based on accumulated knowledge about reaction mechanisms. Although the transition state is by definition transient and thus unstable, in some cases, stable molecules can be designed that resemble transition states. These are called **transition-state analogs**. They bind to an enzyme more tightly than does the substrate in the ES complex, because they fit into the active site better (that is, form a greater number of weak interactions) than the substrate itself. The idea of transition-state analogs was suggested by Pauling in the 1940s, and it has been explored using a variety of enzymes. For example, transition-state analogs designed to inhibit the glycolytic enzyme aldolase bind to that enzyme more than four

BOX 6-3   🜲 MEDICINE   Curing African Sleeping Sickness with a Biochemical Trojan Horse

African sleeping sickness, or African trypanosomiasis, is caused by protists (single-celled eukaryotes) called trypanosomes (Fig. 1). This disease (and related trypanosome-caused diseases) is medically and economically significant in many developing nations. Until the late twentieth century, the disease was virtually incurable. Vaccines are ineffective because the parasite has a novel mechanism to evade the host immune system.

The cell coat of trypanosomes is covered with a single protein, which is the antigen to which the human immune system responds. Every so often, however, by a process of genetic recombination (see Table 28-1), a few cells in the population of infecting trypanosomes switch to a new protein coat, not recognized by the immune system. This process of "changing coats" can occur hundreds of times. The result is a chronic cyclic infection: the human host develops a fever, which subsides as the immune system beats back the first infection; trypanosomes with changed coats then become the seed for a second infection, and the fever recurs. This cycle can repeat for weeks, and the weakened person eventually dies.

Some modern approaches to treating African sleeping sickness have been based on an understanding of enzymology and metabolism. In at least one such approach, this involves pharmaceutical agents designed as mechanism-based enzyme inactivators (suicide inactivators). A vulnerable point in trypanosome metabolism is the pathway of polyamine biosynthesis. The polyamines spermine and spermidine, involved in DNA packaging, are required in large amounts in rapidly dividing cells. The first step in their synthesis is catalyzed by ornithine decarboxylase, an

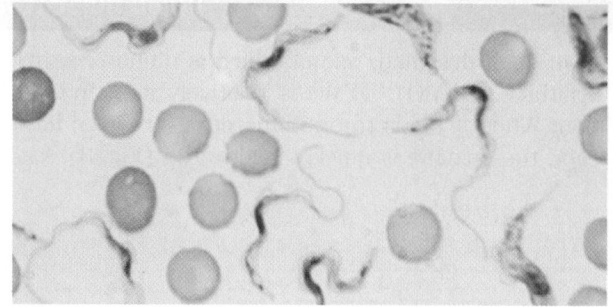

FIGURE 1 *Trypanosoma brucei rhodesiense,* one of several trypanosomes known to cause African sleeping sickness. [Source: John Mansfield, University of Wisconsin–Madison, Department of Bacteriology.]

enzyme that requires for its function a coenzyme called pyridoxal phosphate. Pyridoxal phosphate (PLP), derived from vitamin B_6, forms a covalent bond with the amino acid substrates of the reactions it is involved in and acts as an electron sink to facilitate a variety of reactions (see Fig. 22-32). In mammalian cells, ornithine decarboxylase undergoes rapid turnover—that is, a rapid, constant round of enzyme degradation and synthesis. In some trypanosomes, however, the enzyme (for reasons not well understood) is stable, not readily replaced by newly synthesized enzyme. An inhibitor of ornithine decarboxylase that binds permanently to the enzyme would thus have little effect on human cells, which could rapidly replace inactivated enzyme, but would adversely affect the parasite.

The first few steps of the normal reaction catalyzed by ornithine decarboxylase are shown in Figure 2. Once CO_2 is released, the electron movement is reversed and putrescine is produced (see Fig. 22-32). Based on this mechanism, several suicide inactivators

(Continued on next page)

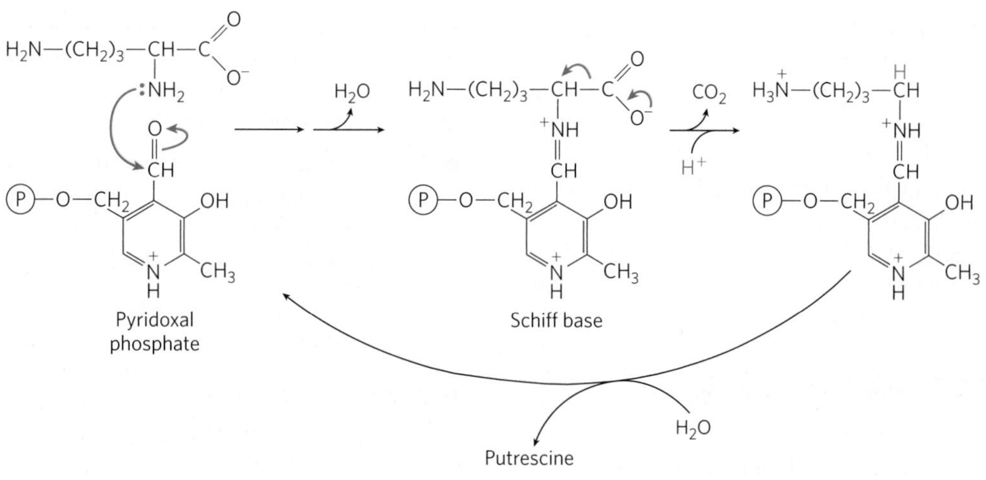

FIGURE 2 Mechanism of ornithine decarboxylase reaction.

BOX 6-3 ⚕ MEDICINE Curing African Sleeping Sickness with a Biochemical Trojan Horse (*Continued*)

have been designed, one of which is difluoromethyl-ornithine (DFMO). DFMO is relatively inert in solution. When it binds to ornithine decarboxylase, however, the enzyme is quickly inactivated (Fig. 3). The

inhibitor acts by providing an alternative electron sink in the form of two strategically placed fluorine atoms, which are excellent leaving groups. Instead of electrons moving into the ring structure of PLP, the reaction results in displacement of a fluorine atom. The —S of a Cys residue at the enzyme's active site then forms a covalent complex with the highly reactive PLP-inhibitor adduct, in an essentially irreversible reaction. In this way, the inhibitor makes use of the enzyme's own reaction mechanisms to kill it.

DFMO has proved highly effective against African sleeping sickness in clinical trials and is now used to treat African sleeping sickness caused by *Trypanosoma brucei gambiense*. Approaches such as this show great promise for treating a wide range of diseases. The design of drugs based on enzyme mechanism and structure can complement the more traditional trial-and-error methods of developing pharmaceuticals.

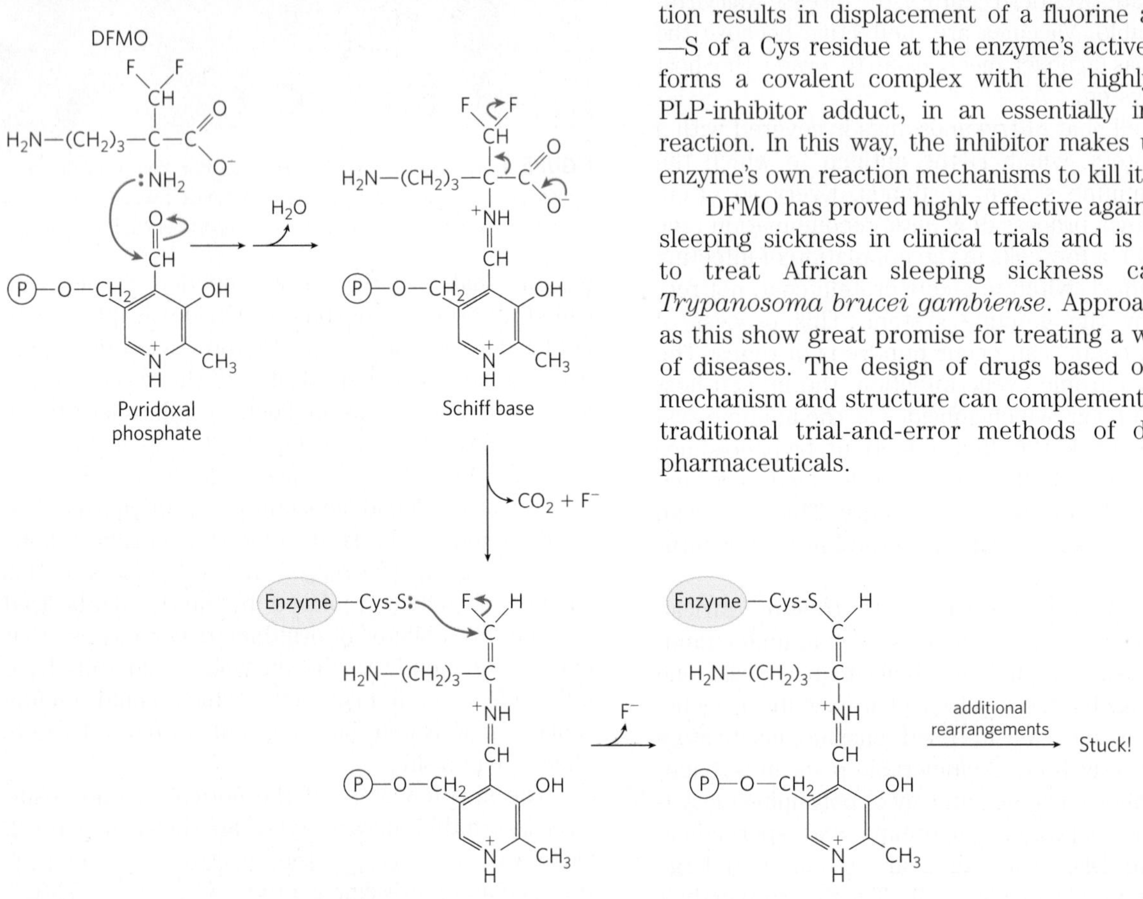

FIGURE 3 Inhibition of ornithine decarboxylase by DFMO.

orders of magnitude more tightly than do its substrates **(Fig. 6-19)**. A transition-state analog cannot perfectly mimic a transition state. Some analogs, however, bind to a target enzyme 10^2 to 10^8 times more tightly than does the normal substrate, providing good evidence that enzyme active sites are indeed complementary to transition states. The concept of transition-state analogs is important to the design of new pharmaceutical agents. As we shall see in Section 6.4, the powerful anti-HIV drugs called protease inhibitors were designed in part as tight-binding transition-state analogs.

SUMMARY 6.3 Enzyme Kinetics as an Approach to Understanding Mechanism

■ Most enzymes have certain kinetic properties in common. When substrate is added to an enzyme, the reaction rapidly achieves a steady state in which the rate at which the ES complex forms balances the rate

at which it breaks down. As [S] increases, the steady-state activity of a fixed concentration of enzyme increases in a hyperbolic fashion to approach a characteristic maximum rate, V_{max}, at which essentially all the enzyme has formed a complex with substrate.

■ The substrate concentration that results in a reaction rate equal to one-half V_{max} is the Michaelis constant K_m, which is characteristic for each enzyme acting on a given substrate. The Michaelis-Menten equation

$$V_0 = \frac{V_{max}[S]}{K_m + [S]}$$

relates initial velocity to [S] and V_{max} through the constant K_m. Michaelis-Menten kinetics is also called steady-state kinetics.

■ K_m and V_{max} have different meanings for different enzymes. The limiting rate of an enzyme-catalyzed reaction at saturation is described by the constant k_{cat},

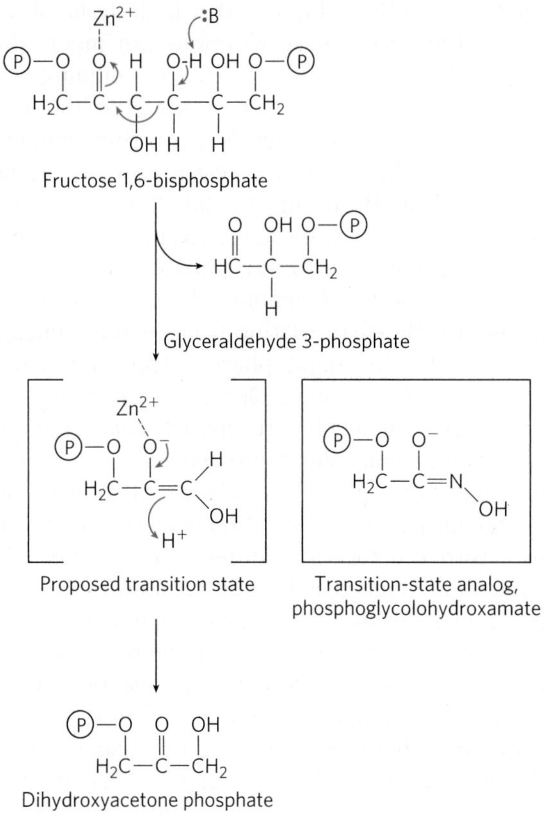

Fructose 1,6-bisphosphate

Glyceraldehyde 3-phosphate

Proposed transition state

Transition-state analog, phosphoglycolohydroxamate

Dihydroxyacetone phosphate

FIGURE 6-19 A transition-state analog. In glycolysis, a class II aldolase (found in bacteria and fungi) catalyzes the cleavage of fructose 1,6-bisphosphate to form glyceraldehyde 3-phosphate and dihydroxyacetone phosphate (see Fig. 14-6 for an example of a class I aldolase reaction, occurring in animals and higher plants). The reaction proceeds via a reverse aldol-condensation-like mechanism. The compound phosphoglycolohydroxamate, which resembles the proposed enediolate transition state, binds to the enzyme nearly 10,000 times better than does the dihydroxyacetone phosphate product.

the turnover number. The ratio k_{cat}/K_m provides a good measure of catalytic efficiency. The Michaelis-Menten equation is also applicable to bisubstrate reactions, which occur by ternary complex or Ping-Pong (double-displacement) pathways.

■ Every enzyme has an optimum pH (or pH range) at which it has maximal activity.

■ Pre–steady state kinetics can provide added insight into enzymatic reaction mechanisms.

■ Reversible inhibition of an enzyme may be competitive, uncompetitive, or mixed. Competitive inhibitors compete with substrate by binding reversibly to the active site, but they are not transformed by the enzyme. Uncompetitive inhibitors bind only to the ES complex, at a site distinct from the active site. Mixed inhibitors bind to either E or ES, again at a site distinct from the active site. In irreversible inhibition, an inhibitor binds permanently to an active site by forming a covalent bond or a very stable noncovalent interaction.

6.4 Examples of Enzymatic Reactions

Thus far we have focused on the general principles of catalysis and on introducing some of the kinetic parameters used to describe enzyme action. We now turn to several examples of specific enzyme reaction mechanisms.

To understand the complete mechanism of action of a purified enzyme, we need to identify all substrates, cofactors, products, and regulators. We also need to know (1) the temporal sequence in which enzyme-bound reaction intermediates form, (2) the structure of each intermediate and each transition state, (3) the rates of interconversion between intermediates, (4) the structural relationship of the enzyme to each intermediate, and (5) the energy contributed by all reacting and interacting groups to the intermediate complexes and transition states. There are still few enzymes for which we have an understanding that meets all these requirements.

We present here the mechanisms for four enzymes: chymotrypsin, hexokinase, enolase, and lysozyme. These examples are not intended to cover all possible classes of enzyme chemistry. They are chosen in part because they are among the best-understood enzymes and in part because they clearly illustrate some general principles outlined in this chapter. The discussion concentrates on selected principles, along with some key experiments that have helped to bring these principles into focus. We use the chymotrypsin example to review some of the conventions used to depict enzyme mechanisms. Much mechanistic detail and experimental evidence is necessarily omitted; no one book could completely document the rich experimental history of these enzymes. In addition, we consider only briefly the special contribution of coenzymes to the catalytic activity of many enzymes. The function of coenzymes is chemically varied, and we describe each coenzyme in detail as it is encountered in Part II.

The Chymotrypsin Mechanism Involves Acylation and Deacylation of a Ser Residue

Bovine pancreatic chymotrypsin (M_r 25,191) is a protease, an enzyme that catalyzes the hydrolytic cleavage of peptide bonds. This protease is specific for peptide bonds adjacent to aromatic amino acid residues (Trp, Phe, Tyr). The three-dimensional structure of chymotrypsin is shown in **Figure 6-20**, with functional groups in the active site emphasized. The reaction catalyzed by this enzyme illustrates the principle of transition-state stabilization and also provides a classic example of general acid-base catalysis and covalent catalysis.

Chymotrypsin enhances the rate of peptide bond hydrolysis by a factor of at least 10^9. It does not catalyze a direct attack of water on the peptide bond; instead, a transient covalent acyl-enzyme intermediate is formed. The reaction thus has two distinct phases. In the acylation phase, the peptide bond is cleaved and an ester linkage is formed between the peptide carbonyl carbon and

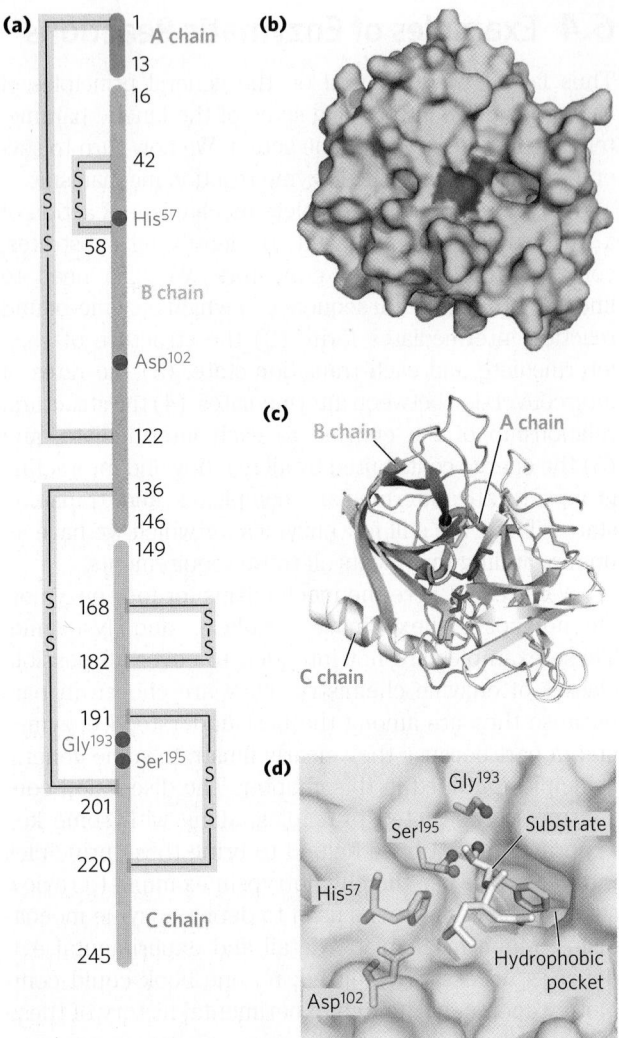

FIGURE 6-20 Structure of chymotrypsin. (a) A representation of primary structure, showing disulfide bonds and the amino acid residues crucial to catalysis. The protein consists of three polypeptide chains linked by disulfide bonds. (The numbering of residues in chymotrypsin, with "missing" residues 14, 15, 147, and 148, is explained in Fig. 6-39.) The active-site amino acid residues are grouped together in the three-dimensional structure. **(b)** A depiction of the enzyme emphasizing its surface. The hydrophobic pocket in which the aromatic amino acid side chain of the substrate is bound is shown in yellow. Key active-site residues, including Ser^{195}, His^{57}, and Asp^{102}, are red. The roles of these residues in catalysis are illustrated in Figure 6-23. **(c)** The polypeptide backbone as a ribbon structure. Disulfide bonds are yellow; the three chains are colored as in part (a). **(d)** A close-up of the active site with a substrate (white and yellow) bound. The hydroxyl of Ser^{195} attacks the carbonyl group of the substrate (the oxygens are red); the developing negative charge on the oxygen is stabilized by the oxyanion hole (amide nitrogens from Ser^{195} and Gly^{193}, in blue), as explained in Figure 6-23. The aromatic amino acid side chain of the substrate (yellow) sits in the hydrophobic pocket. The amide nitrogen of the peptide bond to be cleaved (protruding toward the viewer and projecting the path of the rest of the substrate polypeptide chain) is shown in white. [Source: (b, c, d) PDB ID 7GCH, K. Brady et al., *Biochemistry* 29:7600, 1990.]

the enzyme. In the deacylation phase, the ester linkage is hydrolyzed and the nonacylated enzyme regenerated.

The first evidence for a covalent acyl-enzyme intermediate came from a classic application of pre–steady state kinetics. In addition to its action on polypeptides,

chymotrypsin also catalyzes the hydrolysis of small esters and amides. These reactions are much slower than hydrolysis of peptides because less binding energy is available with smaller substrates (the pre–steady state is also correspondingly longer), thus simplifying the analysis of the resulting reactions. Investigations by B. S. Hartley and B. A. Kilby in 1954 found that chymotrypsin hydrolysis of the ester *p*-nitrophenylacetate, as measured by release of *p*-nitrophenol, proceeds with a rapid burst before leveling off to a slower rate **(Fig. 6-21)**. By extrapolating back to zero time, they concluded that the burst phase corresponded to the release of just under one molecule of *p*-nitrophenol for every enzyme molecule present (a small fraction of their enzyme molecules were inactive). Hartley and Kilby suggested that this release of *p*-nitrophenol occurred during a rapid acylation of all the enzyme molecules, with the rate for subsequent turnover of the enzyme limited by a subsequent, slower deacylation step. Similar results have since been obtained with many other enzymes. The observation of a burst phase provides yet another example of the use of kinetics to break down a reaction into its constituent steps.

Additional features of the chymotrypsin mechanism have been discovered by analyzing the dependence of the reaction on pH. The rate of chymotrypsin-catalyzed

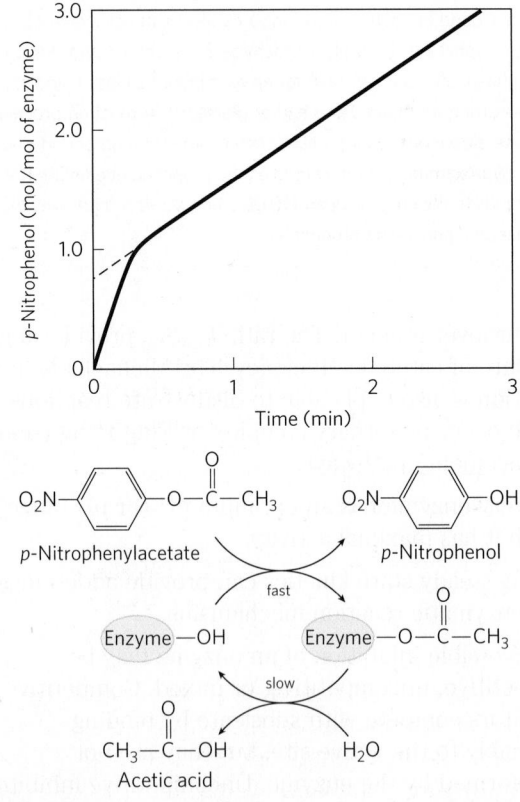

FIGURE 6-21 Pre-steady state kinetic evidence for an acyl-enzyme intermediate. The hydrolysis of *p*-nitrophenylacetate by chymotrypsin is measured by release of *p*-nitrophenol (a colored product). Initially, the reaction releases a rapid burst of *p*-nitrophenol nearly stoichiometric with the amount of enzyme present. This reflects the fast acylation phase of the reaction. The subsequent rate is slower, because enzyme turnover is limited by the rate of the slower deacylation phase.

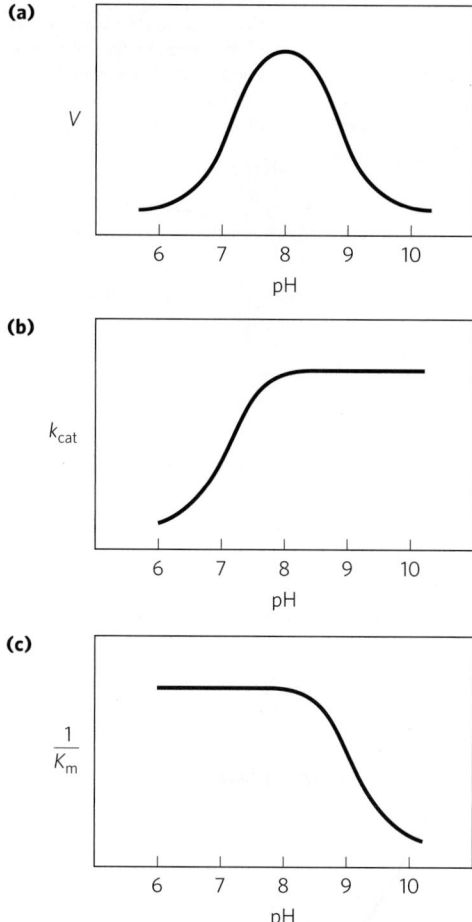

(a)

(b)

(c)

FIGURE 6-22 The pH dependence of chymotrypsin-catalyzed reactions. **(a)** The rates of chymotrypsin-mediated cleavage produce a bell-shaped pH-rate profile with an optimum at pH 8.0. The rate (V) plotted here is that at low substrate concentrations and thus reflects the term k_{cat}/K_m. The plot can be broken down to its components by using kinetic methods to determine the terms k_{cat} and K_m separately at each pH. When this is done **(b, c)**, it becomes clear that the transition just above pH 7 is due to changes in k_{cat}, whereas the transition above pH 8.5 is due to changes in $1/K_m$. Kinetic and structural studies have shown that the transitions illustrated in (b) and (c) reflect the ionization states of the His[57] side chain (when substrate is not bound) and the α-amino group of Ile[16] (at the amino terminus of the B chain), respectively. For optimal activity, His[57] must be unprotonated and Ile[16] must be protonated.

cleavage generally exhibits a bell-shaped pH-rate profile **(Fig. 6-22)**. The rates plotted in Figure 6-22a are obtained at low (subsaturating) substrate concentrations and therefore represent k_{cat}/K_m (see Eqn 6-27, p. 203). A more complete analysis of the rates at different substrate concentrations at each pH allows researchers to determine the individual contributions of the k_{cat} and K_m terms. After obtaining the maximum rates at each pH, one can plot the k_{cat} alone versus pH (Fig. 6-22b); after obtaining the K_m at each pH, researchers can then plot $1/K_m$ versus pH (Fig. 6-22c). Kinetic and structural analyses have revealed that the change in k_{cat} reflects the ionization state of His[57]. The decline in k_{cat} at low pH results from protonation of His[57] (so that it can no longer extract a proton from Ser[195] in step ❷ of the reaction;

see **Fig. 6-23**). This rate reduction illustrates the importance of general acid and general base catalysis in the mechanism for chymotrypsin. The changes in the $1/K_m$ term reflect the ionization of the α-amino group of Ile[16] (at the amino-terminal end of one of the enzyme's three polypeptide chains). This group forms a salt bridge to Asp[194], stabilizing the active conformation of the enzyme. When this group loses its proton at high pH, the salt bridge is eliminated, and a conformational change closes the hydrophobic pocket where the aromatic amino acid side chain of the substrate inserts (Fig. 6-20). Substrates can no longer bind properly, which is measured kinetically as an increase in K_m.

As shown in Figure 6-23, the nucleophile in the acylation phase is the oxygen of Ser[195]. (Proteases with a Ser residue that plays this role in reaction mechanisms are called **serine proteases**.) The pK_a of a Ser hydroxyl group is generally too high for the unprotonated form to be present in significant concentrations at physiological pH. However, in chymotrypsin, Ser[195] is linked to His[57] and Asp[102] in a hydrogen-bonding network referred to as the **catalytic triad**. When a peptide substrate binds to chymotrypsin, a subtle change in conformation compresses the hydrogen bond between His[57] and Asp[102], resulting in a stronger interaction, called a low-barrier hydrogen bond. This enhanced interaction increases the pK_a of His[57] from ~7 (for free histidine) to >12, allowing the His residue to act as an enhanced general base that can remove the proton from the Ser[195] hydroxyl group. Deprotonation prevents development of a highly unstable positive charge on the Ser[195] hydroxyl and makes the Ser side chain a stronger nucleophile. At later reaction stages, His[57] also acts as a proton donor, protonating the amino group in the displaced portion of the substrate (the leaving group).

As the Ser[195] oxygen attacks the carbonyl group of the substrate (Fig. 6-23, step ❷), a very short-lived tetrahedral intermediate is formed in which the carbonyl oxygen acquires a negative charge. This charge, forming within a pocket on the enzyme called the oxyanion hole, is stabilized by hydrogen bonds contributed by the amide groups of two peptide bonds in the chymotrypsin backbone. One of these hydrogen bonds (contributed by Gly[193]) is present only in this intermediate and in the transition states for its formation and breakdown; it reduces the energy required to reach these states. This is an example of the use of binding energy in catalysis through enzyme–transition state complementarity.

An Understanding of Protease Mechanisms Leads to New Treatments for HIV Infections

New pharmaceutical agents are almost always designed to inhibit an enzyme. The extremely successful therapies developed to treat HIV infections provide a case in point. The human immunodeficiency virus (HIV) is the agent that causes acquired immune deficiency syndrome (AIDS). In 2015, an estimated 34 to 41 million people worldwide were living with HIV

How to Read Reaction Mechanisms—A Refresher

Chemical reaction mechanisms, which trace the formation and breakage of covalent bonds, are communicated with dots and curved arrows, a convention known informally as "electron pushing." A covalent bond consists of a shared pair of electrons. Nonbonded electrons important to the reaction mechanism are designated by dots ($^-\ddot{\text{O}}$H). Curved arrows ($\frown$) represent the movement of electron pairs. For movement of a single electron (as in a free radical reaction), a single-headed (fishhook-type) arrow is used ($\frown$). Most reaction steps involve an unshared electron pair (as in the chymotrypsin mechanism).

Some atoms are more electronegative than others; that is, they more strongly attract electrons. The relative electronegativities of atoms encountered in this text are F > O > N > C ≈ S > P ≈ H. For example, the two electron pairs making up a C=O (carbonyl) bond are not shared equally; the carbon is relatively electron-deficient as the oxygen draws away the electrons. Many reactions involve an electron-rich atom (a nucleophile) reacting with an electron-deficient atom (an electrophile). Some common nucleophiles and electrophiles in biochemistry are shown at right.

In general, a reaction mechanism is initiated at an unshared electron pair of a nucleophile. In mechanism diagrams, the base of the electron-pushing arrow originates near the electron-pair dots, and the head of the arrow points directly at the electrophilic center being attacked. Where the unshared electron pair confers a formal negative charge on the nucleophile, the negative charge symbol itself can represent the unshared electron pair and serves as the base of the arrow. In the chymotrypsin mechanism, the nucleophilic electron pair in the ES complex between steps ❶ and ❷ is provided by the oxygen of the Ser¹⁹⁵ hydroxyl group. This electron pair (2 of the 8 valence electrons of the hydroxyl oxygen) provides the base of the curved arrow. The electrophilic center under attack is the carbonyl carbon of the peptide bond to be cleaved. The C, O, and N atoms have a maximum of 8 valence electrons, and H has a maximum of 2. These atoms are occasionally found in unstable states with less than their maximum allotment of electrons, but C, O, and N cannot have more than 8. Thus, when the electron pair from chymotrypsin's Ser¹⁹⁵ attacks the substrate's carbonyl carbon, an electron pair is displaced from the carbon valence shell (you cannot have 5 bonds to carbon!). These electrons move toward the more electronegative carbonyl oxygen. The oxygen has 8 valence electrons both before and after this chemical process, but the number shared with the carbon is reduced from 4 to 2, and the carbonyl oxygen acquires a negative charge. In the next step, the electron pair conferring the negative charge on the oxygen moves back to re-form a bond with carbon and reestablish the carbonyl linkage. Again, an electron pair must be displaced from the carbon, and this time it is the electron pair shared with the amino group of the peptide linkage. This breaks the peptide bond. The remaining steps follow a similar pattern.

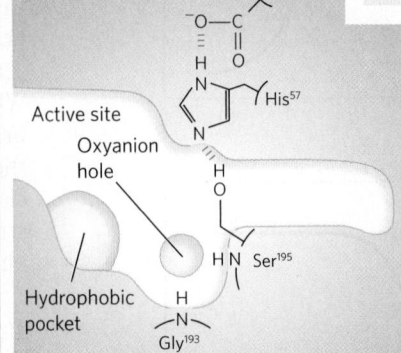

Chymotrypsin (free enzyme)

Active site
Oxyanion hole
Hydrophobic pocket

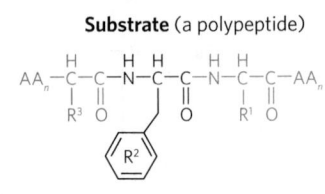

When substrate binds, the side chain of the residue adjacent to the peptide bond to be cleaved nestles in a hydrophobic pocket on the enzyme, positioning the peptide bond for attack.

❶

Substrate (a polypeptide)

Product 2

❼

Enzyme–product 2 complex

Dissociation of the second product from the active site regenerates free enzyme.

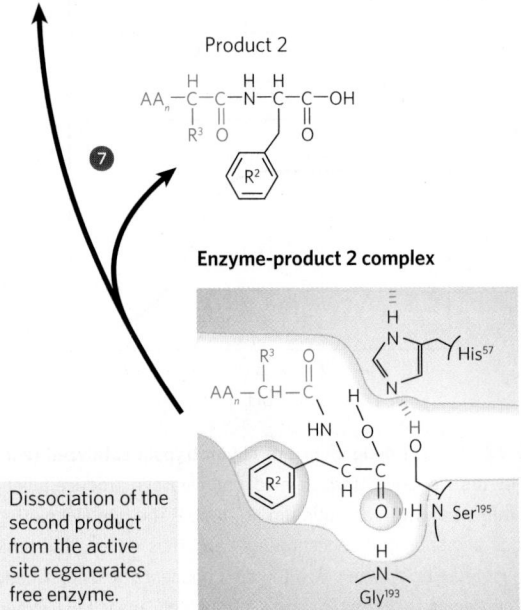

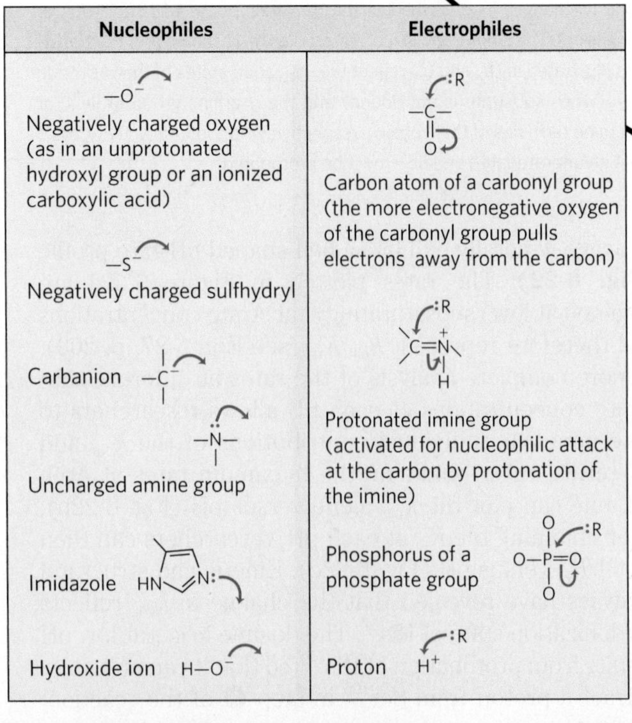

Nucleophiles	Electrophiles
—O⁻ Negatively charged oxygen (as in an unprotonated hydroxyl group or an ionized carboxylic acid)	Carbon atom of a carbonyl group (the more electronegative oxygen of the carbonyl group pulls electrons away from the carbon)
—S⁻ Negatively charged sulfhydryl	Protonated imine group (activated for nucleophilic attack at the carbon by protonation of the imine)
Carbanion	Phosphorus of a phosphate group
Uncharged amine group	Proton
Imidazole	
Hydroxide ion H—O⁻	

Enzyme-substrate complex

Interaction of Ser[195] and His[57] generates a strongly nucleophilic alkoxide ion on Ser[195]; the ion attacks the peptide carbonyl group, forming a tetrahedral acyl-enzyme. This is accompanied by formation of a short-lived negative charge on the carbonyl oxygen of the substrate, which is stabilized by hydrogen bonding in the oxyanion hole.

Short-lived intermediate*
(acylation)

Instability of the negative charge on the substrate carbonyl oxygen leads to collapse of the tetrahedral intermediate; re-formation of a double bond with carbon displaces the bond between carbon and the amino group of the peptide linkage, breaking the peptide bond. The amino leaving group is protonated by His[57], facilitating its displacement.

Product 1

MECHANISM FIGURE 6-23 **Hydrolytic cleavage of a peptide bond by chymotrypsin.** The reaction has two phases. In the acylation phase (steps ❶ to ❹), formation of a covalent acyl-enzyme intermediate is coupled to cleavage of the peptide bond. In the deacylation phase (steps ❺ to ❼), deacylation regenerates the free enzyme; this is essentially the reverse of the acylation phase, with water mirroring, in reverse, the role of the amine component of the substrate.

*The short-lived tetrahedral intermediate following step ❷, and the second tetrahedral intermediate that forms later, are sometimes referred to as transition states, but this terminology can cause confusion. An *intermediate* is any chemical species with a finite lifetime, "finite" being defined as longer than the time required for a molecular vibration ($\sim 10^{-13}$ second). A *transition state* is simply the maximum-energy species formed on the reaction coordinate and does not have a finite lifetime. The tetrahedral intermediates formed in the chymotrypsin reaction closely resemble, both energetically and structurally, the transition states leading to their formation and breakdown. However, the intermediate represents a committed stage of completed bond formation, whereas the transition state is part of the process of reaction. In the case of chymotrypsin, given the close relationship between the intermediate and the actual transition state, the distinction between them is routinely glossed over. Furthermore, the interaction of the negatively charged oxygen with the amide nitrogens in the oxyanion hole, often referred to as transition-state stabilization, also serves to stabilize the intermediate in this case. Not all intermediates are so short-lived that they resemble transition states. The chymotrypsin acyl-enzyme intermediate is much more stable and more readily detected and studied, and it is never confused with a transition state.

Acyl-enzyme intermediate

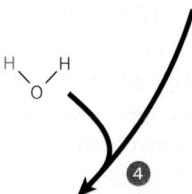

An incoming water molecule is deprotonated by general base catalysis, generating a strongly nucleophilic hydroxide ion. Attack of hydroxide on the ester linkage of the acyl-enzyme generates a second tetrahedral intermediate, with oxygen in the oxyanion hole again taking on a negative charge.

Acyl-enzyme intermediate

Short-lived intermediate* (deacylation)

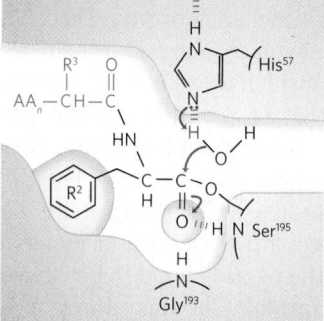

Collapse of the tetrahedral intermediate forms the second product, a carboxylate anion, and displaces Ser[195].

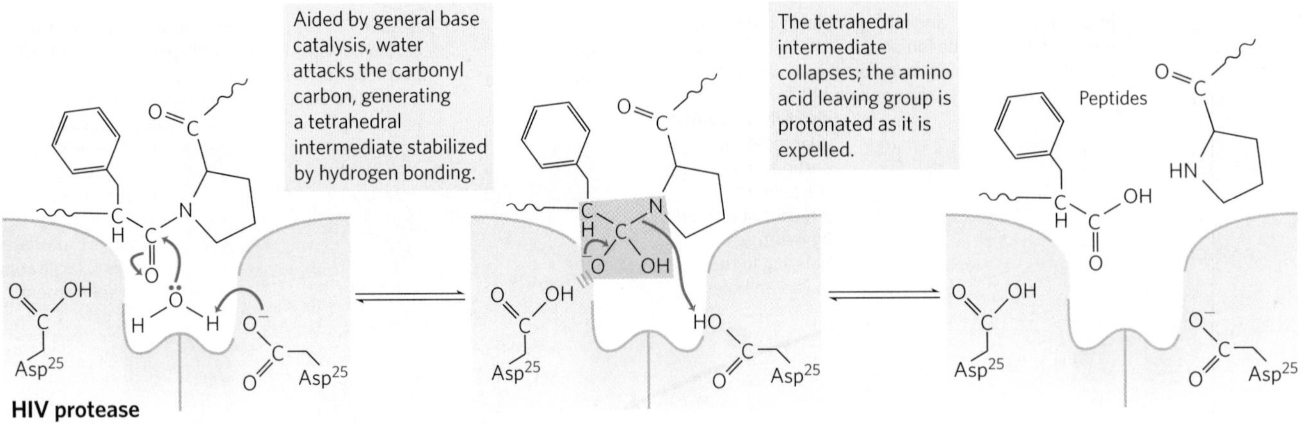

Aided by general base catalysis, water attacks the carbonyl carbon, generating a tetrahedral intermediate stabilized by hydrogen bonding.

The tetrahedral intermediate collapses; the amino acid leaving group is protonated as it is expelled.

HIV protease

FIGURE 6-24 Mechanism of action of HIV protease. Two active-site Asp residues (from different subunits) act as general acid-base catalysts, facilitating the attack of water on the peptide bond. The unstable tetrahedral intermediate in the reaction pathway is shaded light red.

infections, with about 2 million new infections that year and approximately 1.2 million fatalities. AIDS first surfaced as a worldwide epidemic in the 1980s; HIV was discovered soon after and was identified as a **retrovirus**. Retroviruses possess an RNA genome and an enzyme, reverse transcriptase, capable of using RNA to direct the synthesis of a complementary DNA. Efforts to understand HIV and develop therapies for HIV infection benefited from decades of basic research, both on enzyme mechanisms and on the properties of other retroviruses.

A retrovirus such as HIV has a relatively simple life cycle (see Fig. 26-32). Its RNA genome is converted to duplex DNA in several steps catalyzed by the reverse transcriptase (described in Chapter 26). The duplex DNA is then inserted into a chromosome in the nucleus of the host cell by the enzyme integrase (described in Chapter 25). The integrated copy of the viral genome can remain dormant indefinitely. Alternatively, it can be transcribed back into RNA, which can then be translated into proteins to construct new virus particles. Most of the viral genes are translated into large polyproteins, which are cut by an HIV protease into the individual proteins needed to make the virus (see Fig. 26-33). Only three key enzymes operate in this cycle—the reverse transcriptase, the integrase, and the protease. These enzymes thus represent the most promising drug targets.

There are four major subclasses of proteases. The serine proteases, such as chymotrypsin and trypsin, and the cysteine proteases (in which a Cys residue serves a catalytic role similar to that of Ser in the active site) form covalent enzyme-substrate complexes; the aspartyl proteases and metalloproteases do not. The HIV protease is an aspartyl protease. Two active-site Asp residues facilitate the direct attack of a water molecule on the carbonyl group of the peptide bond to be cleaved **(Fig. 6-24)**. The initial product of this attack is an unstable tetrahedral intermediate, much like that in the chymotrypsin reaction. This intermediate is close in structure and energy to the reaction transition state. The drugs that have been developed as HIV protease

inhibitors form noncovalent complexes with the enzyme, but they bind to it so tightly that they can be considered irreversible inhibitors. The tight binding is derived in part from their design as transition-state analogs. The success of these drugs makes a point worth emphasizing: the catalytic principles we have studied in this chapter are not simply abstruse ideas to be memorized—their application saves lives.

The HIV protease is most efficient at cleaving peptide bonds between Phe and Pro residues. The active site has a pocket that binds an aromatic group next to the bond to be cleaved. Several HIV protease inhibitors are shown in **Figure 6-25**. Although the structures appear varied, they all share a core structure: a main chain with a hydroxyl group positioned next to a branch containing a benzyl group. This arrangement targets the benzyl group to an aromatic (hydrophobic) binding pocket. The adjacent hydroxyl group mimics the negatively charged oxygen in the tetrahedral intermediate in the normal reaction, providing a transition-state analog. The remainder of each inhibitor structure was designed to fit into and bind to various crevices along the surface of the enzyme, enhancing overall binding. The availability of these effective drugs has vastly increased the lifespan and quality of life of millions of people with HIV and AIDS. In early 2015, 15 million of the approximately 37 million people living with HIV infection were receiving antiretroviral therapy. ■

Hexokinase Undergoes Induced Fit on Substrate Binding

Yeast hexokinase (M_r 107,862) is a bisubstrate enzyme that catalyzes the reversible reaction

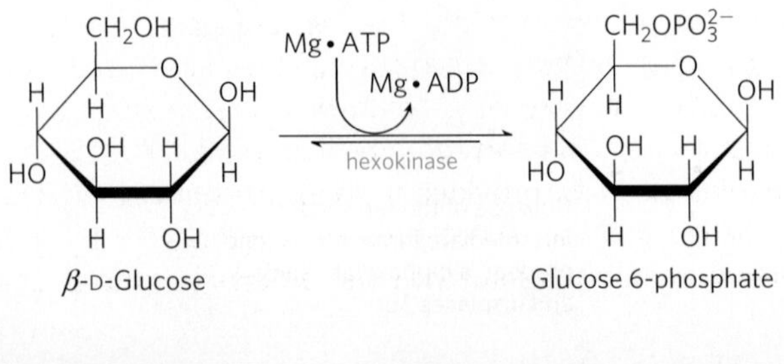

β-D-Glucose   Glucose 6-phosphate

Indinavir

Nelfinavir

Lopinavir

Saquinavir

FIGURE 6-25 HIV protease inhibitors. The hydroxyl group (red) acts as a transition-state analog, mimicking the oxygen of the tetrahedral intermediate. The adjacent benzyl group (blue) helps to properly position the drug in the active site.

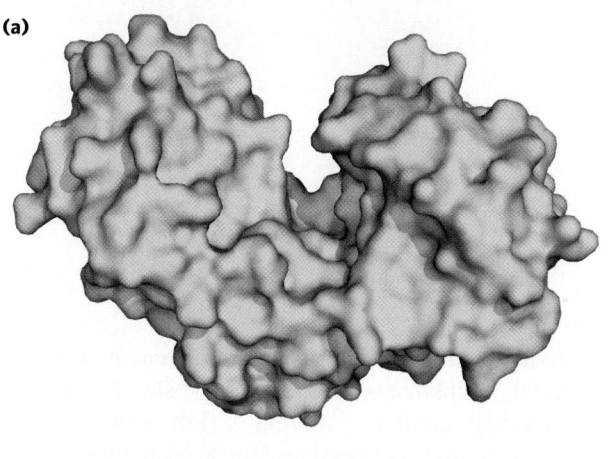

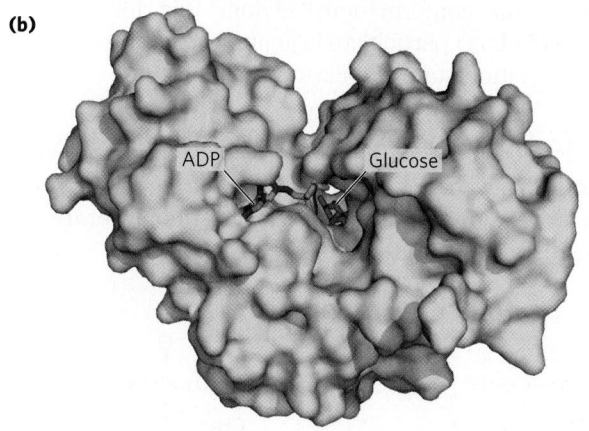

FIGURE 6-26 Induced fit in hexokinase. (a) Hexokinase has a U-shaped structure. **(b)** The ends pinch toward each other in a conformational change induced by binding of D-glucose. [Sources: (a) PDB ID 2YHX, C. M. Anderson et al., *J. Mol. Biol.* 123:15, 1978. (b) PDB ID 2E2O, modeled with ADP derived from PDB ID 2E2Q, H. Nishimasu, et al., *J. Biol. Chem.* 282:9923, 2007.]

ATP and ADP always bind to enzymes as a complex with the metal ion Mg^{2+}.

In the hexokinase reaction, the γ-phosphoryl of ATP is transferred to the hydroxyl at C-6 of glucose. This hydroxyl is similar in chemical reactivity to water, and water freely enters the enzyme active site. Yet hexokinase favors the reaction with glucose by a factor of 10^6. The enzyme can discriminate between glucose and water because of a conformational change in the enzyme when the correct substrate binds **(Fig. 6-26)**. Hexokinase thus provides a good example of induced fit. When glucose is not present, the enzyme is in an inactive conformation, with the active-site amino acid side chains out of position for reaction. When glucose (but not water) and Mg·ATP bind, the binding energy derived from this interaction induces a conformational change in hexokinase to the catalytically active form.

This model has been reinforced by kinetic studies. The five-carbon sugar xylose, stereochemically similar to glucose but one carbon shorter, binds to hexokinase but in a position where it cannot be phosphorylated. Nevertheless, addition of xylose to the reaction mixture increases the rate of ATP hydrolysis. Evidently, the binding of xylose is sufficient to induce a change in hexokinase to its active conformation, and the enzyme is thereby "tricked" into phosphorylating water. The hexokinase reaction also illustrates that enzyme specificity is not always a simple matter of binding one compound but not another. In the case of hexokinase, specificity is observed not in the formation of the ES complex but in the relative rates of subsequent catalytic steps. Reaction rates increase greatly in the presence of a substrate, glucose, that is able to accept a phosphoryl group.

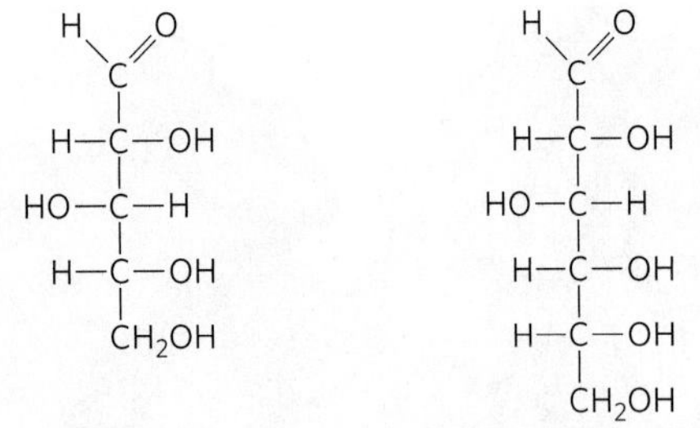

Xylose

Glucose

Induced fit is only one aspect of the catalytic mechanism of hexokinase—like chymotrypsin, hexokinase uses several catalytic strategies. For example, the active-site amino acid residues (those brought into position by the conformational change that follows substrate binding) participate in general acid-base catalysis and transition-state stabilization.

The Enolase Reaction Mechanism Requires Metal Ions

Another glycolytic enzyme, enolase, catalyzes the reversible dehydration of 2-phosphoglycerate to phosphoenolpyruvate:

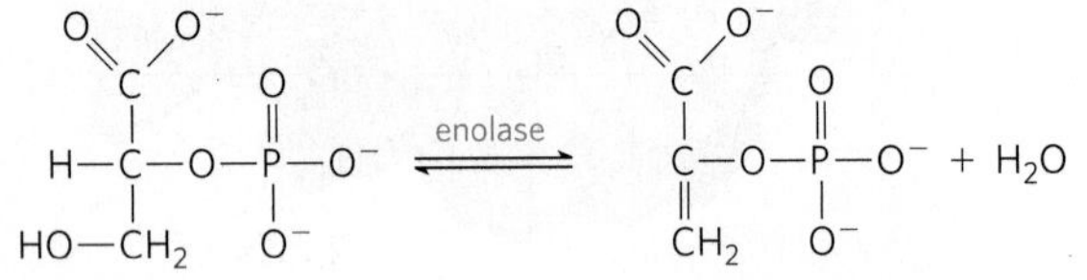

2-Phosphoglycerate ⇌ (enolase) Phosphoenolpyruvate + H_2O

The reaction provides an example of the use of an enzymatic cofactor, in this case a metal ion (an example of coenzyme function is provided in Box 6-3). Yeast enolase

(M_r 93,316) is a dimer with 436 amino acid residues per subunit. The enolase reaction illustrates one type of metal ion catalysis and provides an additional example of general acid-base catalysis and transition-state stabilization. The reaction occurs in two steps **(Fig. 6-27a)**. First, Lys^{345} acts as a general base catalyst, abstracting a proton from C-2 of 2-phosphoglycerate; then Glu^{211} acts as a general acid catalyst, donating a proton to the —OH leaving group. The proton at C-2 of 2-phosphoglycerate is not acidic and thus is quite resistant to its removal by Lys^{345}. However, the electronegative oxygen atoms of the adjacent carboxyl group pull electrons away from C-2, making the attached protons somewhat more labile. In the active site, the carboxyl group of 2-phosphoglycerate undergoes strong ionic interactions with two bound Mg^{2+} ions (Fig. 6-27b), greatly enhancing the electron withdrawal by the carboxyl. Together, these effects render the C-2 protons sufficiently acidic (lowering the pK_a) that one proton can be abstracted to initiate the reaction. As the unstable enolate intermediate is formed, the metal ions further act to shield the two negative charges (on the carboxyl oxygen atoms) that transiently exist in close proximity to each other. Hydrogen bonding to other active-site amino acid residues also contributes to the overall mechanism. The various interactions effectively stabilize both the enolate intermediate and the transition state preceding its formation.

Lysozyme Uses Two Successive Nucleophilic Displacement Reactions

Lysozyme is a natural antibacterial agent found in tears and egg whites. The hen egg white lysozyme (M_r 14,296)

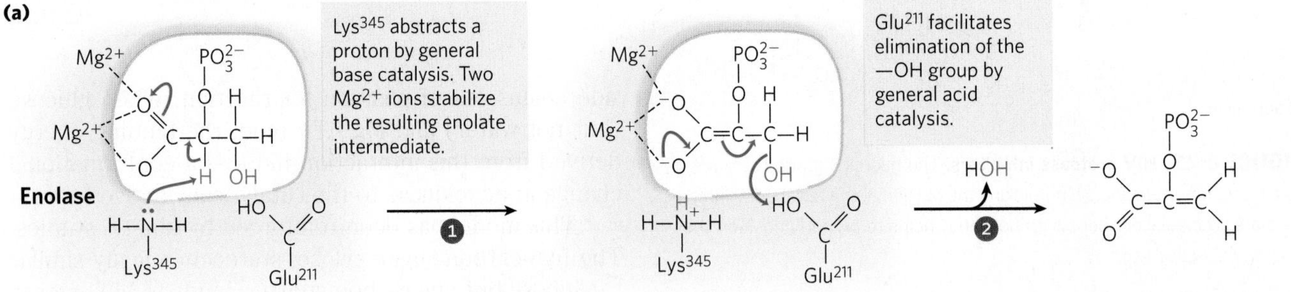

(a)

Enolase

Lys^{345} abstracts a proton by general base catalysis. Two Mg^{2+} ions stabilize the resulting enolate intermediate.

Glu^{211} facilitates elimination of the —OH group by general acid catalysis.

2-Phosphoglycerate bound to enzyme → ① → Enolate intermediate → ② → Phosphoenolpyruvate

(b)

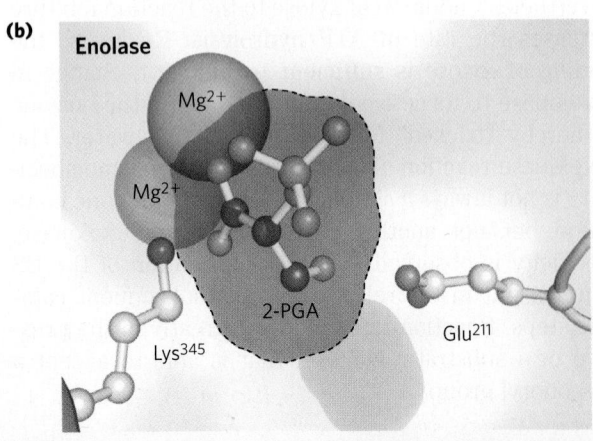

MECHANISM FIGURE 6-27 **Two-step reaction catalyzed by enolase.** **(a)** The mechanism by which enolase converts 2-phosphoglycerate (2-PGA) to phosphoenolpyruvate. The carboxyl group of 2-PGA is coordinated by two magnesium ions at the active site. **(b)** The substrate, 2-PGA, in relation to the Mg^{2+}, Lys^{345}, and Glu^{211} in the enolase active site (gray outline). Nitrogen is shown in blue, phosphorus in orange; hydrogen atoms are not shown. [Source: (b) PDB ID 1ONE, T. M. Larsen et al., *Biochemistry* 35:4349, 1996.]

is a monomer with 129 amino acid residues. This was the first enzyme to have its three-dimensional structure determined, by David Phillips and colleagues in 1965. The structure revealed four stabilizing disulfide bonds and a cleft containing the active site **(Fig. 6-28a)**. More than five decades of investigations have provided a detailed picture of the structure and activity of the enzyme, and an interesting story of how biochemical science progresses.

The substrate of lysozyme is peptidoglycan, a carbohydrate found in many bacterial cell walls. Lysozyme cleaves the $(\beta 1 \rightarrow 4)$ glycosidic C—O bond (p. 258) between the two types of sugar residues in the molecule, N-acetylmuramic acid (Mur2Ac) and

N-acetylglucosamine (GlcNAc) (Fig. 6-28b), often referred to as NAM and NAG, respectively, in the research literature on enzymology. Six residues of the alternating Mur2Ac and GlcNAc in peptidoglycan bind in the active site, in binding sites designated A through F. Model building has shown that the lactyl side chain of Mur2Ac cannot be accommodated in sites C and E, restricting Mur2Ac binding to sites B, D, and F. Only one of the bound glycosidic bonds is cleaved, that between a Mur2Ac residue in site D and a GlcNAc residue in site E. The key catalytic amino acid residues in the active site are Glu[35] and Asp[52] **(Fig. 6-29a)**. The reaction is a nucleophilic substitution, with —OH from water replacing the GlcNAc at C-1 of Mur2Ac.

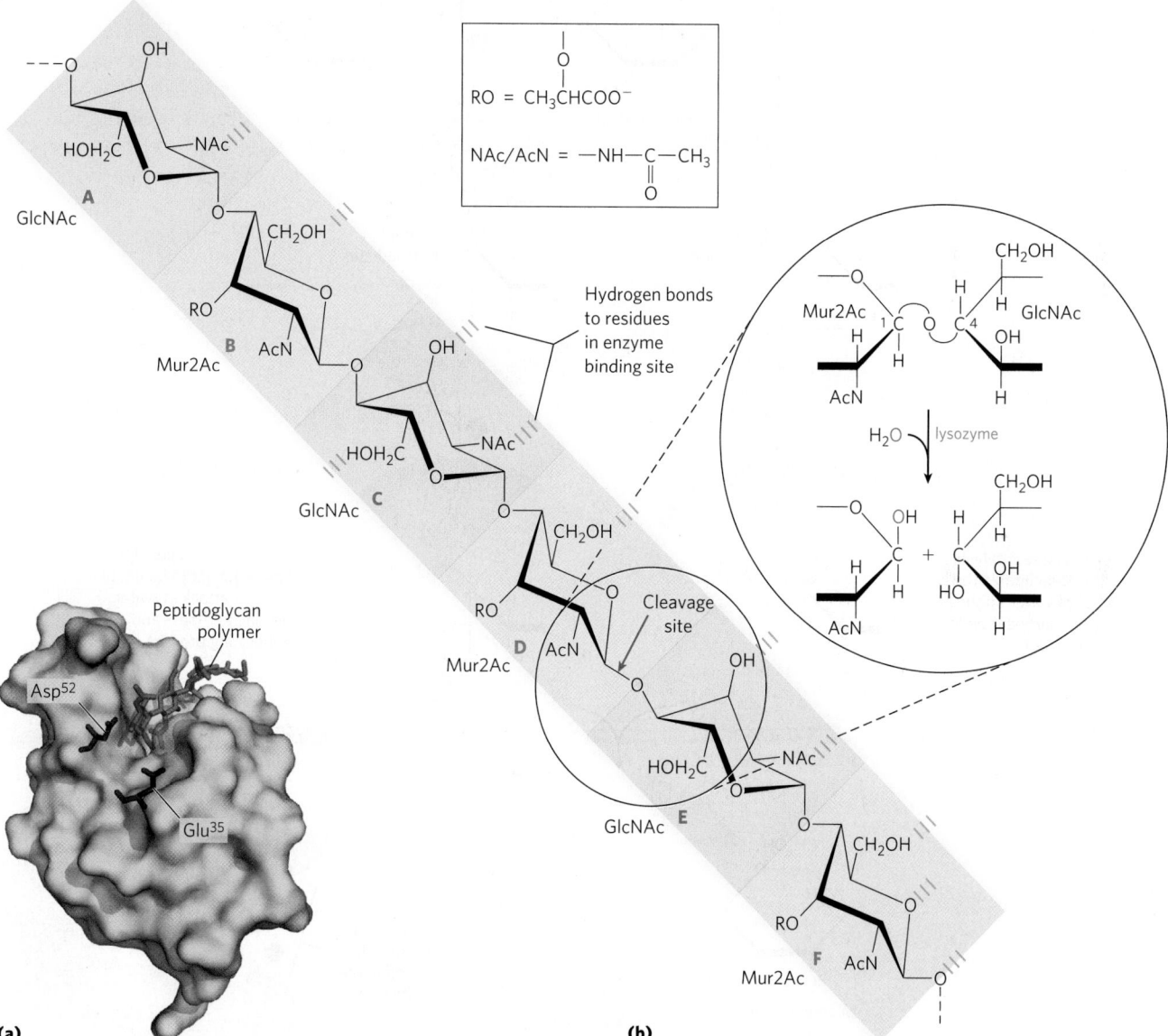

FIGURE 6-28 Hen egg white lysozyme and the reaction it catalyzes.
(a) Surface representation of the enzyme, with the active-site residues Glu[35] and Asp[52] shown as black stick structures and bound substrate shown as a red stick structure. Note that the crystallized enzyme was a mutant, with Gln replacing Glu[35] (see p. 223); the label here refers to the wild-type residue. **(b)** Reaction catalyzed by hen egg white lysozyme. A segment of a peptidoglycan polymer is shown, with the lysozyme binding sites A through F shaded. The glycosidic C—O bond between sugar residues bound to sites D and E is cleaved, as indicated by the red arrow. The hydrolytic reaction is shown in the inset, with the fate of the oxygen in the H_2O traced in red. Mur2Ac is N-acetylmuramic acid; GlcNAc, N-acetylglucosamine. RO— represents a lactyl (lactic acid) group; —NAc and AcN—, an N-acetyl group (see key). [Source: (a) PDB ID 1LZE, K. Maenaka et al., *J. Mol. Biol.* 247:281, 1995.]

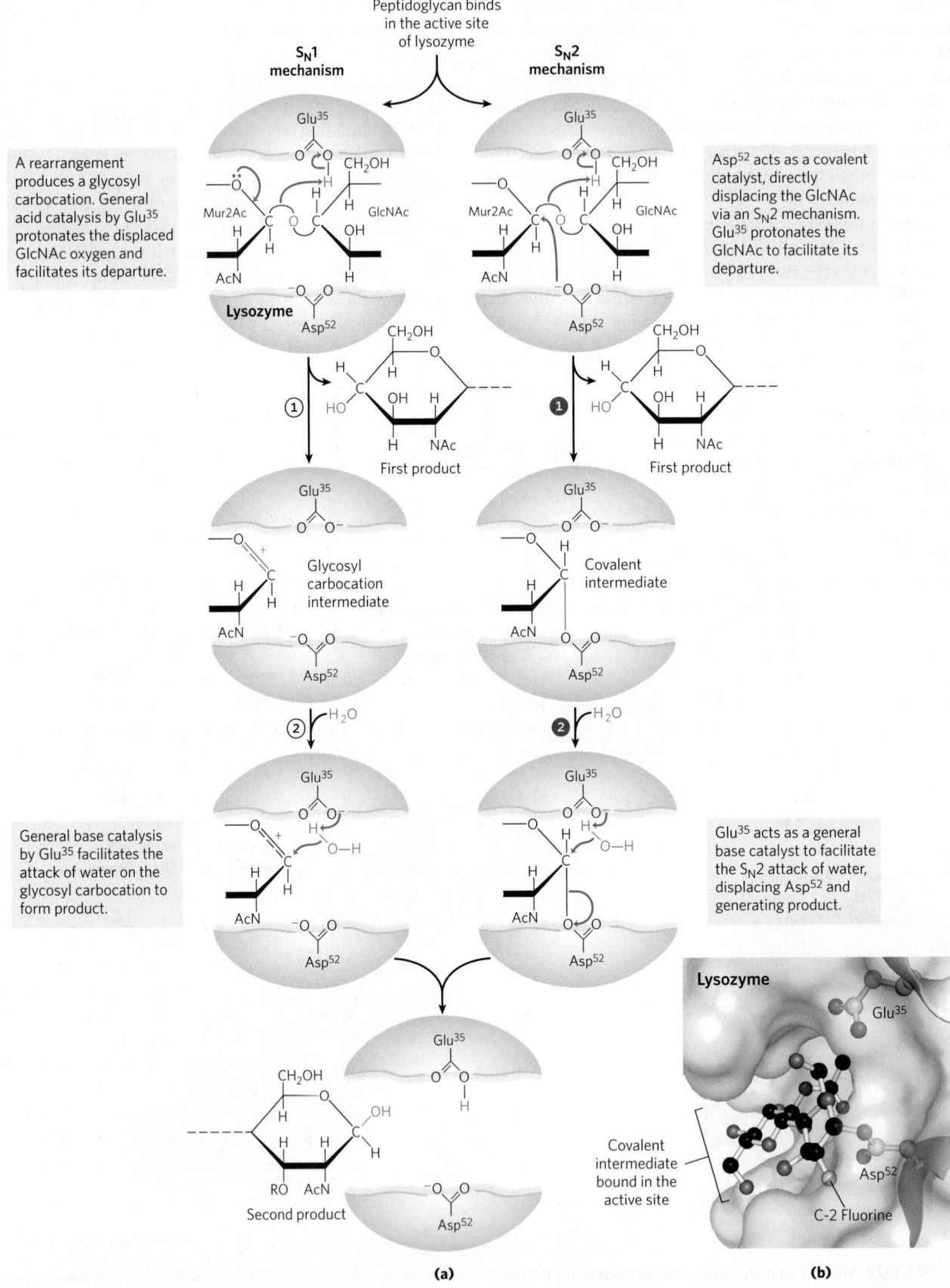

A rearrangement produces a glycosyl carbocation. General acid catalysis by Glu^{35} protonates the displaced GlcNAc oxygen and facilitates its departure.

Asp^{52} acts as a covalent catalyst, directly displacing the GlcNAc via an S$_N$2 mechanism. Glu^{35} protonates the GlcNAc to facilitate its departure.

General base catalysis by Glu^{35} facilitates the attack of water on the glycosyl carbocation to form product.

Glu^{35} acts as a general base catalyst to facilitate the S$_N$2 attack of water, displacing Asp^{52} and generating product.

(a)

(b)

MECHANISM FIGURE 6-29 Lysozyme reaction. In this reaction (described in the text), the water introduced into the product at C-1 of Mur2Ac is in the same configuration as the original glycosidic bond. The reaction is thus a molecular substitution with retention of configuration. **(a)** Two proposed pathways potentially explain the overall reaction and its properties. The S$_N$1 pathway (left) is the original Phillips mechanism. The S$_N$2 pathway (right) is the mechanism most consistent with current data. **(b)** A surface rendering of the lysozyme active site, with the covalent enzyme-substrate intermediate shown as a ball-and-stick structure. (A fluorine-substituted experimental substrate was used; see p. 223.) Side chains of active-site residues are shown as ball-and-stick structures. [Source: (b) PDB ID 1H6M, D. J. Vocadlo et al., *Nature* 412:835, 2001.]

With the active-site residues identified and a detailed structure of the enzyme available, the path to understanding the reaction mechanism seemed open in the 1960s. However, definitive evidence for a particular mechanism eluded investigators for nearly four decades. There are two chemically reasonable mechanisms that could generate the product observed when lysozyme cleaves the glycosidic bond. Phillips and colleagues proposed a dissociative (S_N1-type) mechanism (Fig. 6-29a, left), in which the GlcNAc initially dissociates in step ① to leave behind a glycosyl cation (a carbocation) intermediate. In this mechanism, the departing GlcNAc is protonated by general acid catalysis by Glu[35], located in a hydrophobic pocket that gives its carboxyl group an unusually high pK_a. The carbocation is stabilized by resonance involving the adjacent ring oxygen, as well as by electrostatic interaction with the negative charge on the nearby Asp[52]. In step ②, water attacks at C-1 of Mur2Ac to yield the product. The alternative mechanism (Fig. 6-29a, right) involves two consecutive direct displacement (S_N2-type) steps. In step ❶, Asp[52] attacks C-1 of Mur2Ac to displace the GlcNAc. As in the first mechanism, Glu[35] acts as a general acid to protonate the departing GlcNAc. In step ❷, water attacks at C-1 of Mur2Ac to displace the Asp[52] and generate product.

The Phillips mechanism (S_N1) was widely accepted for more than three decades. However, some controversy persisted and tests continued. The scientific method sometimes advances an issue slowly, and a truly insightful experiment can be difficult to design. Some early arguments against the Phillips mechanism were suggestive but not completely persuasive. For example, the half-life of the proposed glycosyl cation was estimated to be 10^{-12} second, just longer than a molecular vibration and not long enough for the needed diffusion of other molecules. More important, lysozyme is a member of a family of enzymes called "retaining glycosidases," all of which catalyze reactions in which the product has the same anomeric configuration as the substrate (anomeric configurations of carbohydrates are examined in Chapter 7), and all of which are known to have reactive covalent intermediates like that envisioned in the alternative (S_N2) pathway. Hence, the Phillips mechanism ran counter to experimental findings for closely related enzymes.

A compelling experiment tipped the scales decidedly in favor of the S_N2 pathway, as reported by Stephen Withers and colleagues in 2001. Making use of a mutant enzyme (with residue 35 changed from Glu to Gln) and artificial substrates, which combined to slow the rate of key steps in the reaction, these workers were able to stabilize the elusive covalent intermediate. This allowed them to observe the intermediate directly, using both mass spectrometry and x-ray crystallography (Fig. 6-29b).

Is the lysozyme mechanism now proven? No. A key feature of the scientific method, as Albert Einstein once summarized it, is "No amount of experimentation can ever prove me right; a single experiment can prove me wrong." In the case of the lysozyme mechanism, one might argue (and some have) that the artificial substrates, with fluorine substitutions at C-1 and C-2, as were used to stabilize the covalent intermediate, might have altered the reaction pathway. The highly electronegative fluorine could destabilize an already electron-deficient oxocarbenium ion in the glycosyl cation intermediate that might occur in an S_N1 pathway. However, the S_N2 pathway is now the mechanism most in concert with available data.

An Understanding of Enzyme Mechanism Produces Useful Antibiotics

Penicillin was discovered in 1928 by Alexander Fleming, but it was another 15 years before this relatively unstable compound was understood well enough to use it as a pharmaceutical agent to treat bacterial infections. Penicillin interferes with the synthesis of peptidoglycan, the major component of the rigid cell wall that protects bacteria from osmotic lysis. Peptidoglycan consists of polysaccharides and peptides cross-linked in several steps that include a transpeptidase reaction **(Fig. 6-30)**. It is this reaction that is inhibited by penicillin and related compounds **(Fig. 6-31a)**, all of which are irreversible inhibitors of transpeptidase, able to bind its active site through a segment that mimics one conformation of the D-Ala–D-Ala segment of the peptidoglycan precursor. The peptide bond in the precursor is replaced by a highly reactive β-lactam ring in the antibiotic. When penicillin binds to the transpeptidase, an active-site Ser attacks the carbonyl of the β-lactam ring and generates a covalent adduct between penicillin and the enzyme. The leaving group remains attached, however, because it is linked by the remnant of the β-lactam ring (Fig. 6-31b). The covalent complex irreversibly inactivates the enzyme. This, in turn, blocks synthesis of the bacterial cell wall, and most bacteria die as the fragile inner membrane bursts under osmotic pressure.

Human use of penicillin and its derivatives has led to the evolution of strains of pathogenic bacteria that express **β-lactamases (Fig. 6-32a)**, enzymes that cleave β-lactam antibiotics, rendering them inactive. The bacteria thereby become resistant to the antibiotics. The genes for these enzymes have spread rapidly through bacterial populations under the selective pressure imposed by the use (and often overuse) of β-lactam antibiotics. Human medicine responded with the development of compounds such as clavulanic acid, a suicide inactivator, which irreversibly inactivates the β-lactamases (Fig. 6-32b). Clavulanic acid mimics the structure of a β-lactam antibiotic and forms a covalent adduct with a Ser in the β-lactamase active site. This leads to a rearrangement that creates a much more reactive derivative, which is subsequently attacked by another nucleophile in the active site to irreversibly acylate the enzyme and inactivate it. Amoxicillin and clavulanic acid are

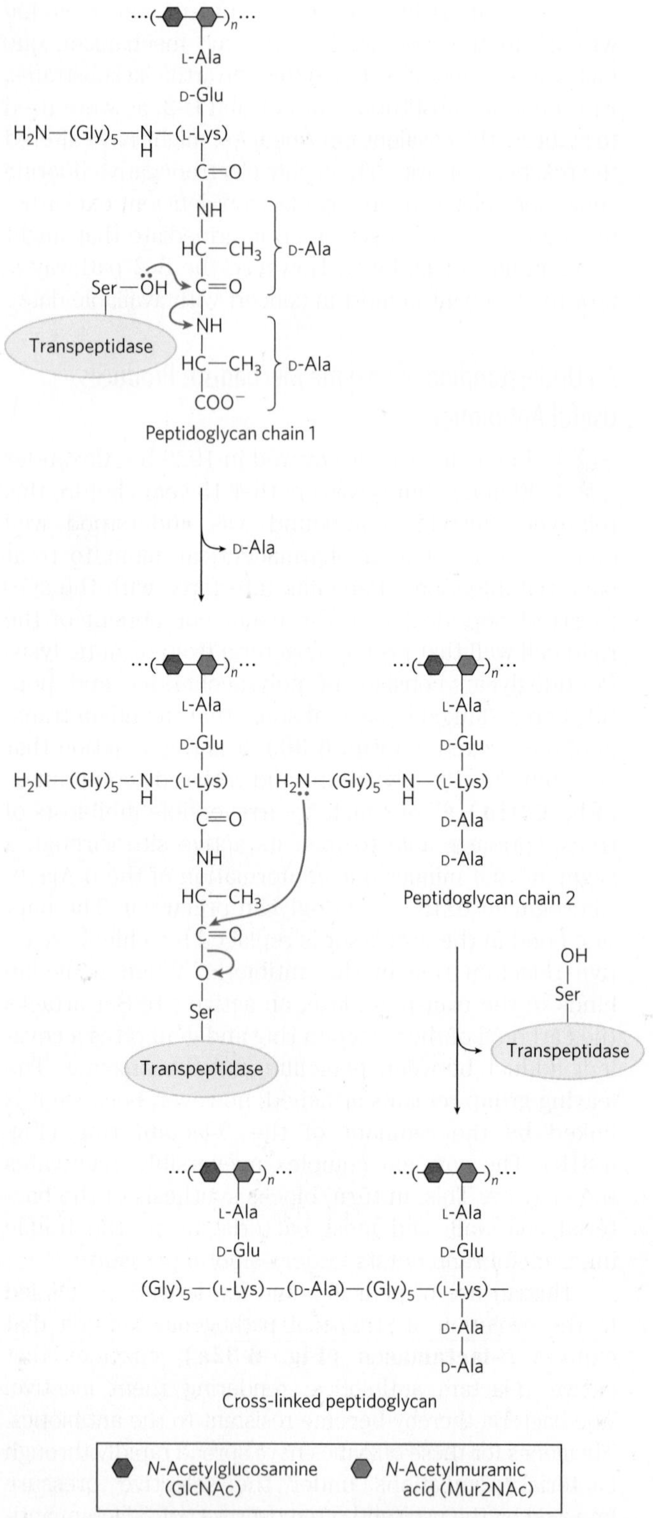

FIGURE 6-30 The transpeptidase reaction. This reaction, which links two peptidoglycan precursors into a larger polymer, is facilitated by an active-site Ser and a covalent catalytic mechanism similar to that of chymotrypsin. Note that peptidoglycan is one of the few places in nature where D-amino acid residues are found. The active-site Ser attacks the carbonyl of the peptide bond between the two D-Ala residues, creating a covalent ester linkage between the substrate and the enzyme, with release of the terminal D-Ala residue. An amino group from the second peptidoglycan precursor then attacks the ester linkage, displacing the enzyme and cross-linking the two precursors.

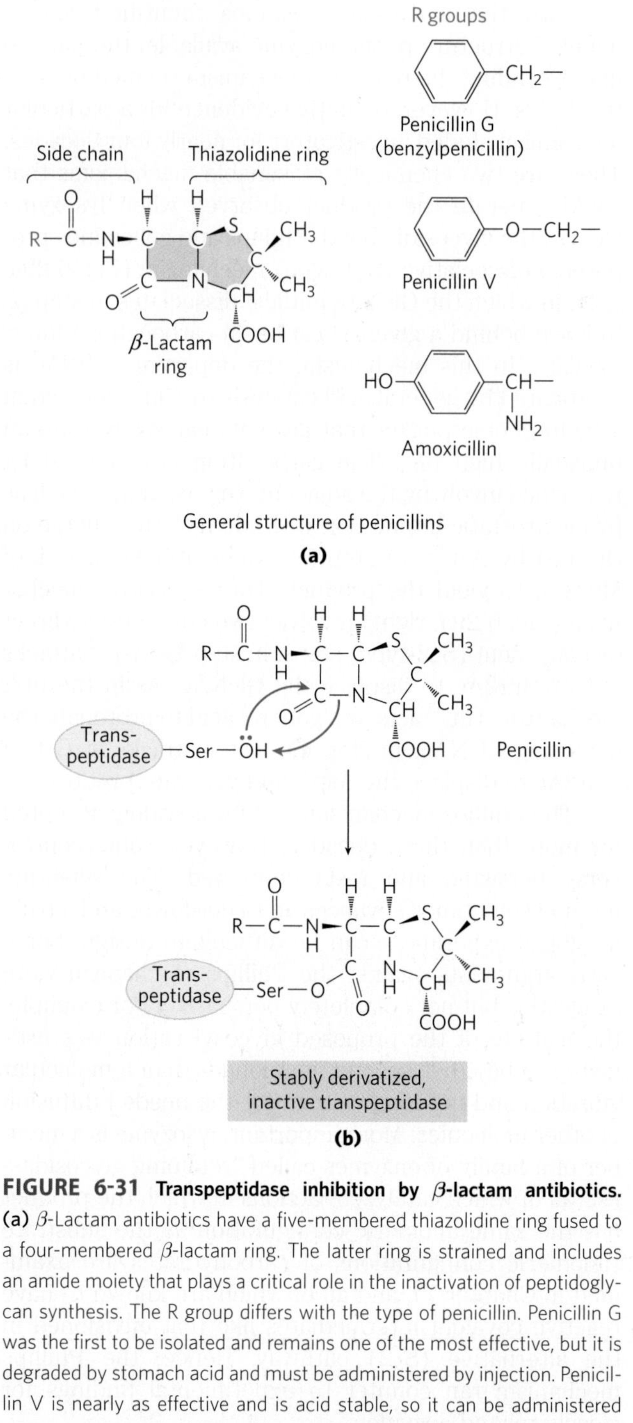

FIGURE 6-31 Transpeptidase inhibition by β-lactam antibiotics. (a) β-Lactam antibiotics have a five-membered thiazolidine ring fused to a four-membered β-lactam ring. The latter ring is strained and includes an amide moiety that plays a critical role in the inactivation of peptidoglycan synthesis. The R group differs with the type of penicillin. Penicillin G was the first to be isolated and remains one of the most effective, but it is degraded by stomach acid and must be administered by injection. Penicillin V is nearly as effective and is acid stable, so it can be administered orally. Amoxicillin has a broad range of effectiveness, is readily administered orally, and is thus the most widely prescribed β-lactam antibiotic. (b) Attack on the amide moiety of the β-lactam ring by a transpeptidase active-site Ser results in a covalent acyl-enzyme product. This is hydrolyzed so slowly that adduct formation is practically irreversible, and the transpeptidase is inactivated.

inactivate it. Amoxicillin and clavulanic acid are combined in a widely used pharmaceutical formulation with the trade name Augmentin. The cycle of chemical warfare between humans and bacteria continues unabated. Strains of disease-causing bacteria that are resistant to both amoxicillin and clavulanic acid have been discovered. Mutations in β-lactamase within these strains

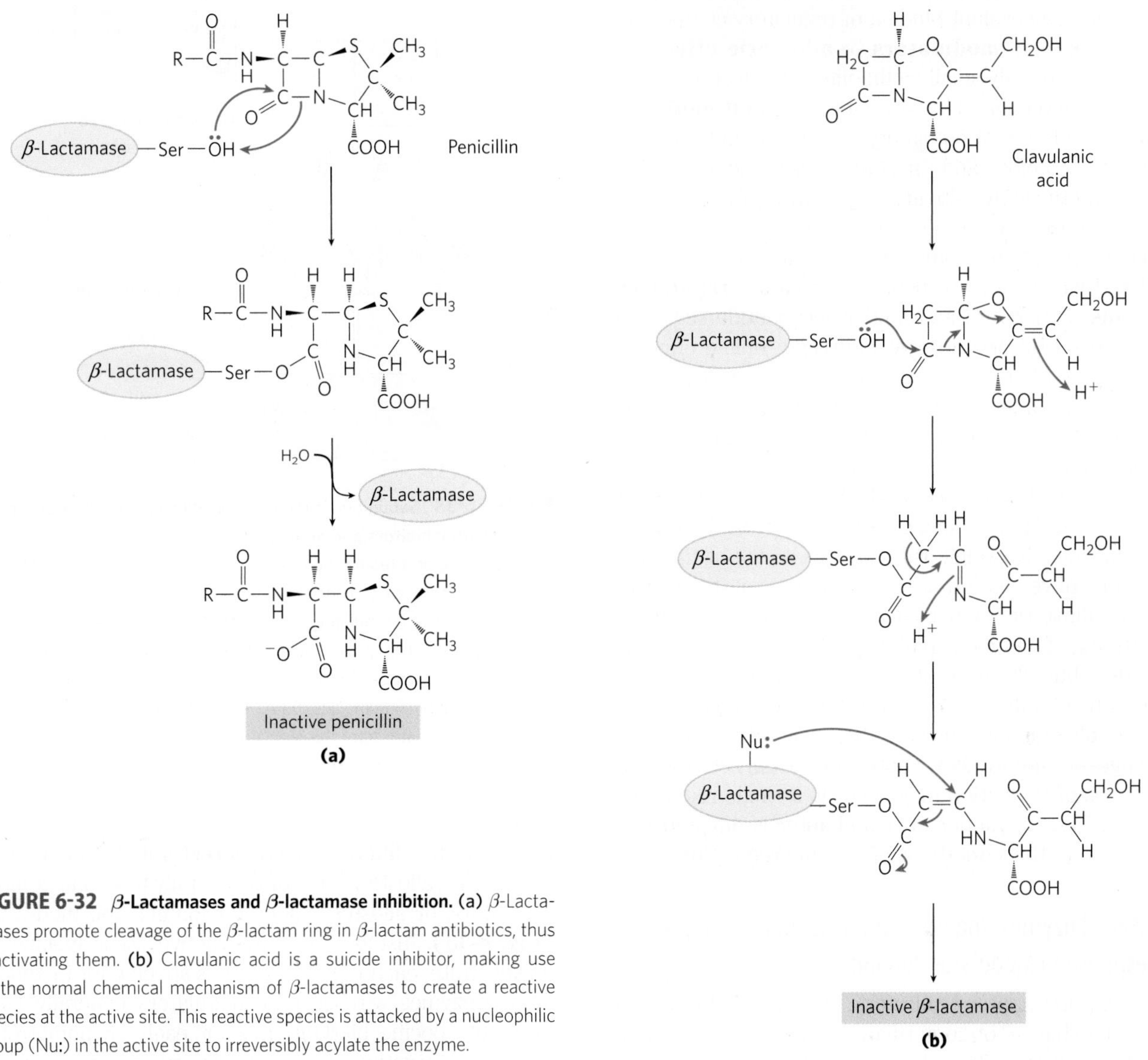

FIGURE 6-32 β-Lactamases and β-lactamase inhibition. (a) β-Lactamases promote cleavage of the β-lactam ring in β-lactam antibiotics, thus inactivating them. **(b)** Clavulanic acid is a suicide inhibitor, making use of the normal chemical mechanism of β-lactamases to create a reactive species at the active site. This reactive species is attacked by a nucleophilic group (Nu:) in the active site to irreversibly acylate the enzyme.

render it unreactive to clavulanic acid. The development of new antibiotics promises to be a growth industry for the foreseeable future. ■

SUMMARY 6.4 Examples of Enzymatic Reactions

■ Chymotrypsin is a serine protease with a well-understood mechanism, featuring general acid-base catalysis, covalent catalysis, and transition-state stabilization.

■ Hexokinase provides an excellent example of induced fit as a means of using substrate binding energy.

■ The enolase reaction proceeds via metal ion catalysis.

■ Lysozyme makes use of covalent catalysis and general acid catalysis as it promotes two successive nucleophilic displacement reactions.

■ Understanding enzyme mechanism allows the development of drugs to inhibit enzyme action.

6.5 Regulatory Enzymes

In cellular metabolism, groups of enzymes work together in sequential pathways to carry out a given metabolic process, such as the multireaction breakdown of glucose to lactate or the multireaction synthesis of an amino acid from simpler precursors. In such enzyme systems, the reaction product of one enzyme becomes the substrate of the next.

Most of the enzymes in each metabolic pathway follow the kinetic patterns we have already described. Each pathway, however, includes one or more enzymes that have a greater effect on the rate of the overall sequence. The catalytic activity of these **regulatory enzymes** increases or decreases in response to certain signals. Adjustments in the rate of reactions catalyzed by regulatory enzymes, and therefore in the rate of entire metabolic sequences, allow the cell to meet changing needs for energy and for biomolecules required in growth and repair.

The activities of regulatory enzymes are modulated in a variety of ways. **Allosteric enzymes** function through reversible, noncovalent binding of regulatory compounds called **allosteric modulators** or **allosteric effectors**, which are generally small metabolites or cofactors. Other enzymes are regulated by reversible **covalent modification**. Both classes of regulatory enzymes tend to be multisubunit proteins, and in some cases the regulatory site(s) and the active site are on separate subunits. Metabolic systems have at least two other mechanisms of enzyme regulation. Some enzymes are stimulated or inhibited when they are bound by separate **regulatory proteins**. Others are activated when peptide segments are removed by **proteolytic cleavage**; unlike effector-mediated regulation, regulation by proteolytic cleavage is irreversible. Important examples of both mechanisms are found in physiological processes such as digestion, blood clotting, hormone action, and vision.

Cell growth and survival depend on efficient use of resources, and this efficiency is made possible by regulatory enzymes. No single rule governs which of the various types of regulation occur in different systems. To a degree, allosteric (noncovalent) regulation may permit fine-tuning of metabolic pathways that are required continuously but at different levels of activity as cellular conditions change. Regulation by covalent modification may be all or none—usually the case with proteolytic cleavage—or it may allow subtle changes in activity. Several types of regulation may occur in a single regulatory enzyme. The remainder of this chapter is devoted to a discussion of these methods of enzyme regulation.

Allosteric Enzymes Undergo Conformational Changes in Response to Modulator Binding

As we saw in Chapter 5, allosteric proteins are those having "other shapes" or conformations induced by the binding of modulators. The same concept applies to certain regulatory enzymes, as conformational changes induced by one or more modulators interconvert more-active and less-active forms of the enzyme. The modulators for allosteric enzymes may be inhibitory or stimulatory. Often the modulator is the substrate itself; regulation in which substrate and modulator are identical is referred to as homotropic. The effect is similar to that of O_2 binding to hemoglobin (Chapter 5): binding of the ligand—or substrate, in the case of enzymes—causes conformational changes that affect the subsequent activity of other sites on the protein. In most cases, the conformational change converts a relatively inactive conformation (often referred to as a T state) to a more active conformation (an R state). When the modulator is a molecule other than the substrate, the enzyme is said to be heterotropic. Note that allosteric modulators should not be confused with uncompetitive and mixed inhibitors. Although the latter bind at a second site on the enzyme, they do not necessarily mediate conformational changes between active and inactive forms, and the kinetic effects are distinct.

The properties of allosteric enzymes are significantly different from those of simple nonregulatory enzymes.

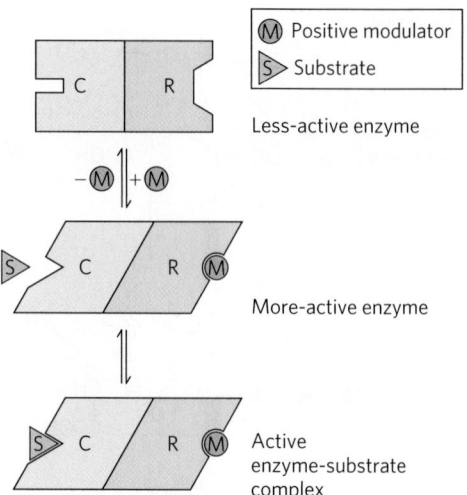

FIGURE 6-33 Subunit interactions in an allosteric enzyme, and interactions with inhibitors and activators. In many allosteric enzymes, the substrate-binding site and the modulator-binding site(s) are on different subunits, the catalytic (C) and regulatory (R) subunits, respectively. Binding of the positive (stimulatory) modulator (M) to its specific site on the regulatory subunit is communicated to the catalytic subunit through a conformational change. This change renders the catalytic subunit active and capable of binding the substrate (S) with higher affinity. On dissociation of the modulator from the regulatory subunit, the enzyme reverts to its inactive or less active form.

Some of the differences are structural. In addition to active sites, allosteric enzymes generally have one or more regulatory, or allosteric, sites for binding the modulator **(Fig. 6-33)**. Just as an enzyme's active site is specific for its substrate, each regulatory site is specific for its modulator. Enzymes with several modulators generally have different specific binding sites for each. In homotropic enzymes, the active site and regulatory site are the same.

Allosteric enzymes are typically larger and more complex than nonallosteric enzymes, with two or more subunits. A classic example is aspartate transcarbamoylase (often abbreviated ATCase), which catalyzes an early step in the biosynthesis of pyrimidine nucleotides, the reaction of carbamoyl phosphate and aspartate to form carbamoyl aspartate:

Carbamoyl phosphate + Aspartate → (aspartate transcarbamoylase, P_i) → *N*-Carbamoylaspar

ATCase has 12 polypeptide chains organized into 6 catalytic subunits (organized as 2 trimeric complexes) and 6 regulatory subunits (organized as 3 dimeric complexes). **Figure 6-34** shows the quaternary structure of this enzyme, deduced from x-ray analysis. The enzyme exhibits allosteric behavior as detailed below, as the

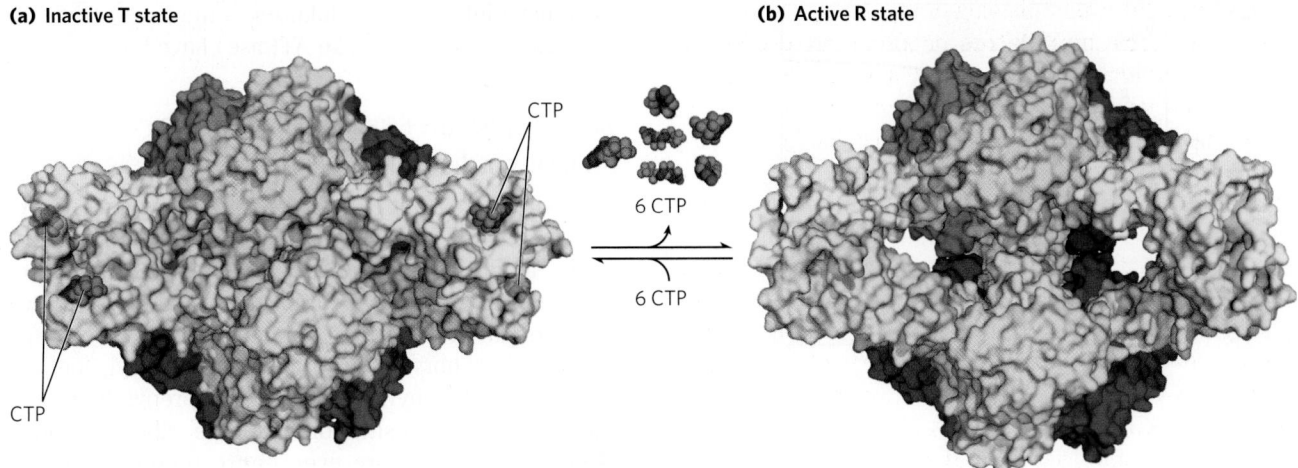

(a) Inactive T state

(b) Active R state

CTP

6 CTP

6 CTP

CTP

CTP

FIGURE 6-34 The regulatory enzyme aspartate transcarbamoylase. (a) The inactive T state and **(b)** the active R state of the enzyme are shown. This allosteric regulatory enzyme has two stacked catalytic clusters, each with three catalytic polypeptide chains (in shades of blue and purple), and three regulatory clusters, each with two regulatory polypeptide chains (in beige and yellow). The regulatory clusters form the points of a triangle (not evident in this side view) surrounding the catalytic subunits. Binding sites for allosteric modulators (including CTP) are on the regulatory subunits. Modulator binding produces large changes in enzyme conformation and activity. The role of this enzyme in nucleotide synthesis, and details of its regulation, are discussed in Chapter 22. [Sources: (a) PDB ID 1RAB, R. P. Kosman et al., *Proteins* 15:147, 1993. (b) PDB ID 1F1B, L. Jin et al., *Biochemistry* 39:8058, 2000.]

catalytic subunits function cooperatively. The regulatory subunits have binding sites for ATP and CTP, which function as positive and negative regulators, respectively. CTP is one of the end products of the pathway, and negative regulation by CTP serves to limit ATCase action under conditions when CTP is abundant. On the other hand, high concentrations of ATP indicate that cellular metabolism is robust, the cell is growing, and additional pyrimidine nucleotides may be needed to support RNA transcription and DNA replication.

The Kinetic Properties of Allosteric Enzymes Diverge from Michaelis-Menten Behavior

Allosteric enzymes show relationships between V_0 and [S] that differ from Michaelis-Menten kinetics. They do exhibit saturation with the substrate when [S] is sufficiently high, but for allosteric enzymes, plots of V_0 versus [S] **(Fig. 6-35)** usually produce a sigmoid saturation curve, rather than the hyperbolic curve typical of nonregulatory enzymes. On the sigmoid saturation curve we can find a value of [S] at which V_0 is half-maximal, but we cannot refer to it with the designation K_m, because the enzyme does not follow the hyperbolic Michaelis-Menten relationship. Instead, the symbol $[S]_{0.5}$ or $K_{0.5}$ is often used to represent the substrate concentration giving half-maximal velocity of the reaction catalyzed by an allosteric enzyme (Fig. 6-35).

Sigmoid kinetic behavior generally reflects cooperative interactions between multiple protein subunits. In other words, changes in the structure of one subunit are translated into structural changes in adjacent subunits, an effect mediated by noncovalent interactions at the interface between subunits. The principles are particularly well illustrated by a nonenzymatic process: O_2 binding to hemoglobin. Sigmoid kinetic behavior is explained by the concerted and sequential models for subunit interactions (see Fig. 5-15).

ATCase effectively illustrates both homotropic and heterotropic allosteric kinetic behavior. Binding of the substrates, aspartate and carbamoyl phosphate, to the enzyme gradually brings about a transition from the relatively inactive T state to the more active R state. This accounts for the sigmoid rather than hyperbolic change in V_0 with increasing [S]. One characteristic of sigmoid kinetics is that small changes in the concentration of a modulator can be associated with large changes in activity. As exemplified in Figure 6-35a, a relatively small increase in [S] in the steep part of the curve causes a comparatively large increase in V_0.

The heterotropic allosteric regulation of ATCase is brought about by its interactions with ATP and CTP. For heterotropic allosteric enzymes, an activator may cause the curve to become more nearly hyperbolic, with a decrease in $K_{0.5}$ but no change in V_{max}, resulting in an increased reaction velocity at a fixed substrate concentration. For ATCase, the interaction with ATP brings this about, and the enzyme exhibits a V_0 versus [S] curve that is characteristic of the active R state at sufficiently high ATP concentrations (V_0 is higher for any value of [S]; Fig. 6-35b). A negative modulator (an inhibitor) may produce a *more* sigmoid substrate-saturation curve, with an increase in $K_{0.5}$, as illustrated by the effects of CTP on ATCase kinetics (see curves for a negative modulater in Fig. 6-35b). Other heterotropic allosteric enzymes respond to an activator by an increase in V_{max} with little change in $K_{0.5}$ (Fig. 6-35c). Heterotropic allosteric enzymes therefore show different kinds of responses in their substrate-activity curves, because

(a)

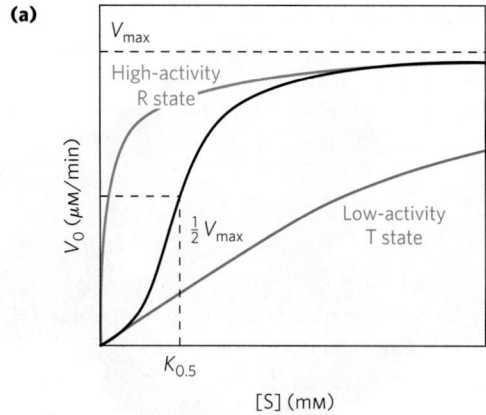

(b)

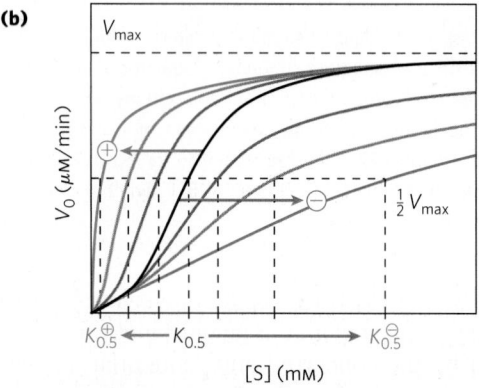

(c)

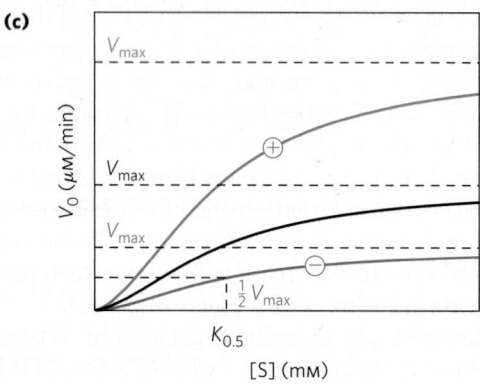

FIGURE 6-35 Substrate-activity curves for representative allosteric enzymes. Three examples of complex responses of allosteric enzymes to their modulators. **(a)** The sigmoid curve of a homotropic enzyme, in which the substrate also serves as a positive (stimulatory) modulator, or activator. Notice the resemblance to the oxygen-saturation curve of hemoglobin (see Fig. 5-12). The sigmoidal curve is a hybrid curve in which the enzyme is present primarily in the relatively inactive T state at low substrate concentration, and primarily in the more active R state at high substrate concentration. The curves for the pure T and R states are plotted separately in color. ATCase exhibits a kinetic pattern similar to this. **(b)** The effects of several different concentrations of a positive modulator (+) or negative modulator (−) on an allosteric enzyme in which $K_{0.5}$ is altered without a change in V_{max}. The central curve shows the substrate-activity relationship without a modulator. For ATCase, CTP is a negative modulator and ATP is a positive modulator. **(c)** A less common type of modulation, in which V_{max} is altered and $K_{0.5}$ is nearly constant.

some have inhibitory modulators, some have activating modulators, and some (like ATCase) have both.

Some Enzymes Are Regulated by Reversible Covalent Modification

In another important class of regulatory enzymes, activity is modulated by covalent modification of one or more of the amino acid residues in the enzyme molecule. Over 500 different types of covalent modification have been found in proteins. Common modifying groups include phosphoryl, acetyl, adenylyl, uridylyl, methyl, amide, carboxyl, myristoyl, palmitoyl, prenyl, hydroxyl, sulfate, and adenosine diphosphate ribosyl groups **(Fig. 6-36)**. There are even entire proteins that are used as specialized modifying groups, including ubiquitin and sumo. All of these groups are generally linked to and removed from a regulated enzyme by separate enzymes. When an amino acid residue in an enzyme is modified, a novel amino acid with altered properties has effectively been introduced into the enzyme. Introduction of a charge can alter the local properties of the enzyme and induce a change in conformation. Introduction of a hydrophobic group can trigger association with a membrane. The changes are often substantial and can be critical to the function of the altered enzyme.

The variety of enzyme modifications is too great to cover in detail, but some examples are instructive. One enzyme that is regulated by methylation is the methyl-accepting chemotaxis protein of bacteria. This protein is part of a system that permits a bacterium to swim toward an attractant (such as a sugar) in solution and away from repellent chemicals. The methylating agent is *S*-adenosyl-methionine (adoMet) (see Fig. 18-18). Acetylation is another common modification, with approximately 80% of the soluble proteins in eukaryotes, including many enzymes, acetylated at their amino terminus. Ubiquitin is added to proteins as a tag that destines them for proteolytic degradation (see Fig. 27-49). Ubiquitination can also have a regulatory function. Sumo is found attached to many eukaryotic nuclear proteins and has roles in the regulation of transcription, chromatin structure, and DNA repair.

ADP-ribosylation is an especially interesting reaction, observed in a number of proteins; the ADP-ribose is derived from nicotinamide adenine dinucleotide (NAD) (see Fig. 8-41). This type of modification occurs for the bacterial enzyme dinitrogenase reductase, resulting in regulation of the important process of biological nitrogen fixation. Diphtheria toxin and cholera toxin are enzymes that catalyze the ADP-ribosylation (and inactivation) of key cellular enzymes or other proteins.

Phosphorylation is the most common type of regulatory modification. It is estimated that one-third of all proteins in a eukaryotic cell are phosphorylated, and one or (often) many phosphorylation events are part of virtually every regulatory process. Some proteins have only one phosphorylated residue, others have several, and a few have dozens of sites for phosphorylation. This

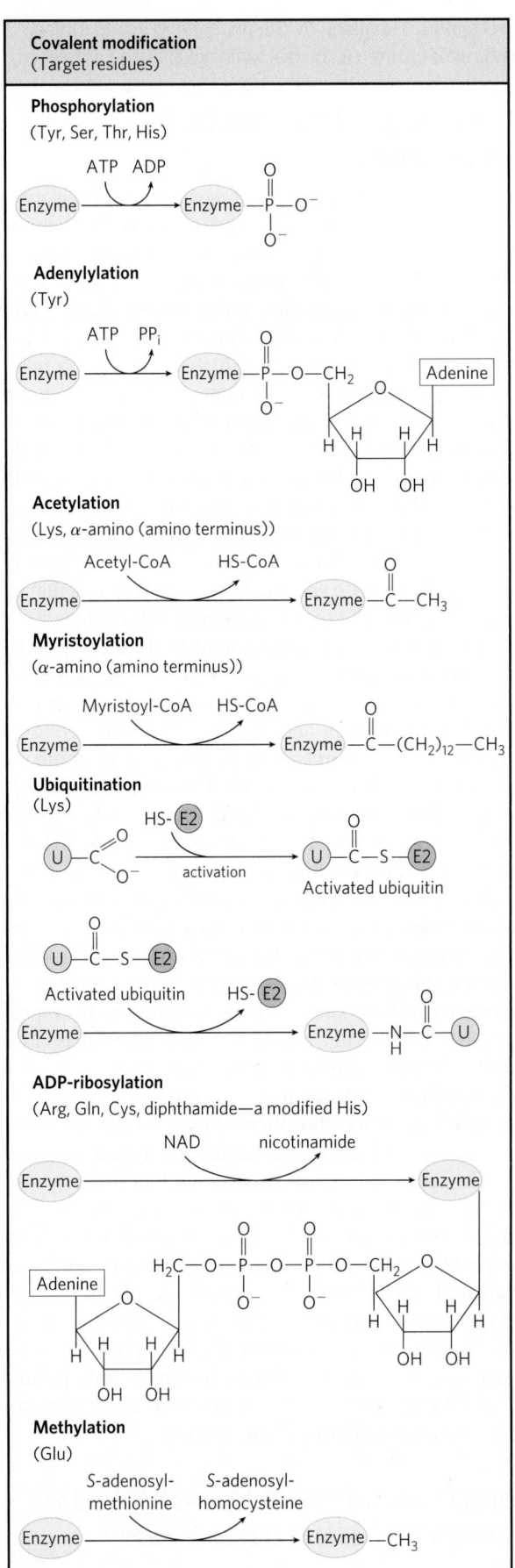

Covalent modification
(Target residues)

Phosphorylation
(Tyr, Ser, Thr, His)

Adenylylation
(Tyr)

Acetylation
(Lys, α-amino (amino terminus))

Myristoylation
(α-amino (amino terminus))

Ubiquitination
(Lys)

Activated ubiquitin

ADP-ribosylation
(Arg, Gln, Cys, diphthamide—a modified His)

Methylation
(Glu)

FIGURE 6-36 Some enzyme modification reactions.

mode of covalent modification is central to a large number of regulatory pathways. We discuss it in some detail here, and again in Chapter 12.

We will encounter all of these types of enzyme modification again in later chapters.

Phosphoryl Groups Affect the Structure and Catalytic Activity of Enzymes

The attachment of phosphoryl groups to specific amino acid residues of a protein is catalyzed by **protein kinases**. More than 500 genes encoding these critical enzymes are found in the human genome. In the reactions they catalyze, the γ-phosphoryl group derived from a nucleoside triphosphate (usually ATP) is transferred to a particular Ser, Thr, or Tyr residue (occasionally His as well) on the target protein. This introduces a bulky, charged group into a region of the target protein that was only moderately polar. The oxygen atoms of a phosphoryl group can hydrogen-bond with one or several groups in a protein, commonly the amide groups of the peptide backbone at the start of an α helix or the charged guanidinium group of an Arg residue. The two negative charges on a phosphorylated side chain can also repel neighboring negatively charged (Asp or Glu) residues. When the modified side chain is located in a region of an enzyme critical to its three-dimensional structure, phosphorylation can have dramatic effects on enzyme conformation and thus on substrate binding and catalysis. Removal of phosphoryl groups from these same target proteins is catalyzed by **phosphoprotein phosphatases**, also called simply **protein phosphatases**.

An important example of enzyme regulation by phosphorylation is the case of glycogen phosphorylase (M_r 94,500) of muscle and liver (Chapter 15), which catalyzes the reaction

$$(\text{Glucose})_n + \text{P}_i \longrightarrow (\text{glucose})_{n-1} + \text{glucose 1-phosphate}$$

Glycogen → Shortened glycogen chain

The glucose 1-phosphate so formed can be used for ATP synthesis in muscle or converted to free glucose in the liver. Note that glycogen phosphorylase, though it adds a phosphate to a substrate, is not itself a kinase, because it does not utilize ATP or any other nucleotide triphosphate as a phosphoryl donor in its catalyzed reaction. It is, however, the substrate for a protein kinase that phosphorylates it. In the discussion below, the phosphoryl groups we are concerned with are those involved in regulation of the enzyme, as distinguished from its catalytic function.

Glycogen phosphorylase occurs in two forms: the more active phosphorylase a and the less active phosphorylase b **(Fig. 6-37)**. Phosphorylase a has two subunits, each with a specific Ser residue that is phosphorylated at its hydroxyl group. These serine phosphate residues are required for maximal activity of the enzyme. The phosphoryl groups can be hydrolytically removed by a separate enzyme called phosphoprotein phosphatase 1 (PP1):

$$\text{Phosphorylase } a + 2\text{H}_2\text{O} \longrightarrow \text{Phosphorylase } b + 2\text{P}_i$$

(more active) → (less active)

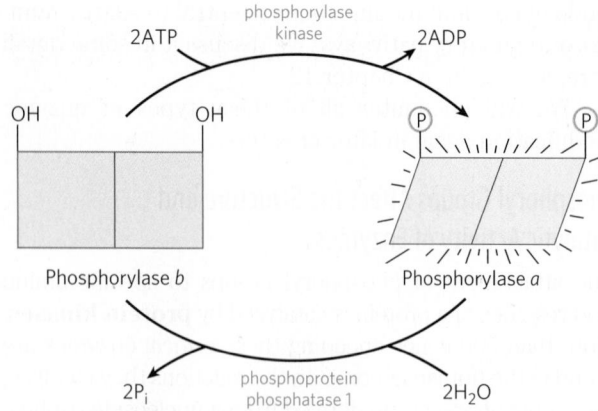

FIGURE 6-37 Regulation of muscle glycogen phosphorylase activity by phosphorylation. In the more active form of the enzyme, phosphorylase *a*, specific Ser residues, one on each subunit, are phosphorylated. Phosphorylase *a* is converted to the less active phosphorylase *b* by enzymatic loss of these phosphoryl groups, promoted by phosphoprotein phosphatase 1 (PP1). Phosphorylase *b* can be reconverted (reactivated) to phosphorylase *a* by the action of phosphorylase kinase.

In this reaction, phosphorylase *a* is converted to phosphorylase *b* by the cleavage of two serine phosphate covalent bonds, one on each subunit of glycogen phosphorylase.

Phosphorylase *b* can, in turn, be reactivated—covalently transformed back into active phosphorylase *a*—by another enzyme, phosphorylase kinase, which catalyzes the transfer of phosphoryl groups from ATP to the hydroxyl groups of the two specific Ser residues in phosphorylase *b*:

$$2\text{ATP} + \text{phosphorylase } b \longrightarrow 2\text{ADP} + \text{phosphorylase } a$$
$$\text{(less active)} \qquad\qquad\qquad \text{(more active)}$$

The breakdown of glycogen in skeletal muscles and the liver is regulated by varying the ratio of the two forms of glycogen phosphorylase. The *a* and *b* forms differ in their secondary, tertiary, and quaternary structures; the active site undergoes changes in structure and, consequently, changes in catalytic activity as the two forms are interconverted.

The regulation of glycogen phosphorylase by phosphorylation illustrates the effects on both structure and catalytic activity of adding a phosphoryl group. In the unphosphorylated state, each subunit of this enzyme is folded so as to bring the 20 residues at its amino terminus, including some basic residues, into a region containing several acidic amino acids; this produces an electrostatic interaction that stabilizes the conformation. Phosphorylation of Ser[14] interferes with this interaction, forcing the amino-terminal domain out of the acidic environment and into a conformation that allows interaction between the Ⓟ-Ser and several Arg side chains. In this conformation, the enzyme is much more active.

Phosphorylation of an enzyme can affect catalysis in another way: by altering substrate-binding affinity. For example, when isocitrate dehydrogenase (an enzyme of the citric acid cycle; see Chapter 16) is phosphorylated,

electrostatic repulsion by the phosphoryl group inhibits the binding of citrate (a tricarboxylic acid) at the active site.

Multiple Phosphorylations Allow Exquisite Regulatory Control

The Ser, Thr, or Tyr residues that are typically phosphorylated in regulated proteins occur within common structural motifs, called consensus sequences, that are recognized by specific protein kinases (Table 6-10). Some kinases are basophilic, preferentially phosphorylating a residue that has basic neighbors; others have different substrate preferences, such as for a residue near a Pro residue. Amino acid sequence is not the only important factor in determining whether a given residue will be phosphorylated, however. Protein folding brings together residues that are distant in the primary sequence; the resulting three-dimensional structure can determine whether a protein kinase has access to a given residue and can recognize it as a substrate. Another factor influencing the substrate specificity of certain protein kinases is the proximity of other phosphorylated residues.

Regulation by phosphorylation is often complicated. Some proteins have consensus sequences recognized by several different protein kinases, each of which can phosphorylate the protein and alter its enzymatic activity. In some cases, phosphorylation is hierarchical: a certain residue can be phosphorylated only if a neighboring residue has already been phosphorylated. For example, glycogen synthase, the enzyme that catalyzes the condensation of glucose monomers to form glycogen (Chapter 15), is inactivated by phosphorylation of specific Ser residues and is also modulated by at least four other protein kinases that phosphorylate four other sites in the enzyme **(Fig. 6-38)**. For example, the enzyme does not become a substrate for glycogen synthase kinase 3 until one site has been phosphorylated by casein kinase II. Some phosphorylations inhibit glycogen synthase more than others, and some combinations of phosphorylations are cumulative. These multiple regulatory phosphorylations provide the potential for extremely subtle modulation of enzyme activity.

To serve as an effective regulatory mechanism, phosphorylation must be reversible. In general, phosphoryl groups are added and removed by different enzymes, and the processes can therefore be separately regulated. Cells contain a family of phosphoprotein phosphatases that hydrolyze specific Ⓟ-Ser, Ⓟ-Thr, and Ⓟ-Tyr esters, releasing P_i. The phosphoprotein phosphatases we know of thus far act on only a subset of phosphorylated proteins, but they show less substrate specificity than protein kinases.

Some Enzymes and Other Proteins Are Regulated by Proteolytic Cleavage of an Enzyme Precursor

For some enzymes, an inactive precursor called a **zymogen** is cleaved to form the active enzyme. Many proteolytic enzymes (proteases) of the stomach and pancreas are

TABLE 6-10 Consensus Sequences for Protein Kinases

Protein kinase	Consensus sequence and phosphorylated residue
Protein kinase A	-x-R-[RK]-x-[ST]-B-
Protein kinase G	-x-R-[RK]-x-[ST]-X-
Protein kinase C	-[RK](2)-x-[ST]-B-[RK](2)-
Protein kinase B	-x-R-x-[ST]-x-K-
Ca^{2+}/calmodulin kinase I	-B-x-R-x(2)-[ST]-x(3)-B-
Ca^{2+}/calmodulin kinase II	-B-x-[RK]-x(2)-[ST]-x(2)-
Mysoin light chain kinase (smooth muscle)	-K(2)-R-x(2)-S-x-B(2)-
Phosphorylase *b* kinase	*-K-R-K-Q-I-S-V-R-*
Extracellular signal-regulated kinase (ERK)	-P-x-[ST]-P(2)-
Cyclin-dependent protein kinase (cdc2)	-x-[ST]-P-x-[KR]-
Casein kinase I	-[SpTp]-x(2)-[ST]-B[a]
Casein kinase II	-x-[ST]-x(2)-[ED]-x-
β-Adrenergic receptor kinase	-[DE](n)-[ST]-x(3)
Rhodopsin kinase	-x(2)-[ST]-E(n)-
Insulin receptor kinase	-x-E(3)-Y-M(4)-*K(2)-S-R-G-D-Y-M-T-M-Q-I-G-K(3)-L-P-A-T-G-D-Y-M-N-M-S-P-V-G-D-*
Epidermal growth factor (EGF) receptor kinase	*-E(4)-Y-F-E-L-V-*

Sources: L. A. Pinna and M. H. Ruzzene, *Biochim. Biophys. Acta* 1314:191, 1996; B. E. Kemp and R. B. Pearson, *Trends Biochem. Sci.* 15:342, 1990; P. J. Kennelly and E. G. Krebs, *J. Biol. Chem.* 266:15,555, 1991.

Note: Shown here are deduced consensus sequences (in roman type) and actual sequences from known substrates (italic). The Ser (S), Thr (T), or Tyr (Y) residue that undergoes phosphorylation is in red; all amino acid residues are shown as their one-letter abbreviations (see Table 3-1). x represents any amino acid; B, any hydrophobic acid; Sp and Tp are Ser and Thr residues that must already be phosphorylated for the kinase to recognize the site.

[a]The best target site has two amino acid residues separating the phosphorylated and target Ser/Thr residues; target sites with one or three intervening residues function at a reduced level.

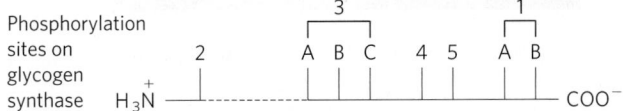

Phosphorylation sites on glycogen synthase

Kinase	Phosphorylation sites	Degree of synthase inactivation
Protein kinase A	1A, 1B, 2, 4	+
Protein kinase G	1A, 1B, 2	+
Protein kinase C	1A	+
Ca^{2+}/calmodulin kinase	1B, 2	+
Phosphorylase *b* kinase	2	+
Casein kinase I	At least nine	+ + + +
Casein kinase II	5	0
Glycogen synthase kinase 3	3A, 3B, 3C	+ + +
Glycogen synthase kinase 4	2	+

regulated in this way. Chymotrypsin and trypsin are initially synthesized as chymotrypsinogen and trypsinogen **(Fig. 6-39)**. Specific cleavage causes conformational changes that expose the enzyme active site. Because this type of activation is irreversible, other mechanisms are needed to inactivate these enzymes. Proteases are inactivated by inhibitor proteins that bind very tightly to the enzyme active site. For example, pancreatic trypsin inhibitor (M_r 6,000) binds to and inhibits trypsin. α_1-Antiproteinase (M_r 53,000) primarily inhibits neutrophil elastase (neutrophils are a type of leukocyte, or white blood cell; elastase is a protease that acts on elastin, a component of some connective tissues). An insufficiency of α_1-antiproteinase, which can be caused by exposure to cigarette smoke, has been associated with lung damage, including emphysema.

FIGURE 6-38 Multiple regulatory phosphorylations. The enzyme glycogen synthase has at least nine separate sites in five designated regions that are susceptible to phosphorylation by one of the cellular protein kinases. Thus, regulation of this enzyme is a matter not of binary (on/off) switching but of finely tuned modulation of activity over a wide range in response to a variety of signals.

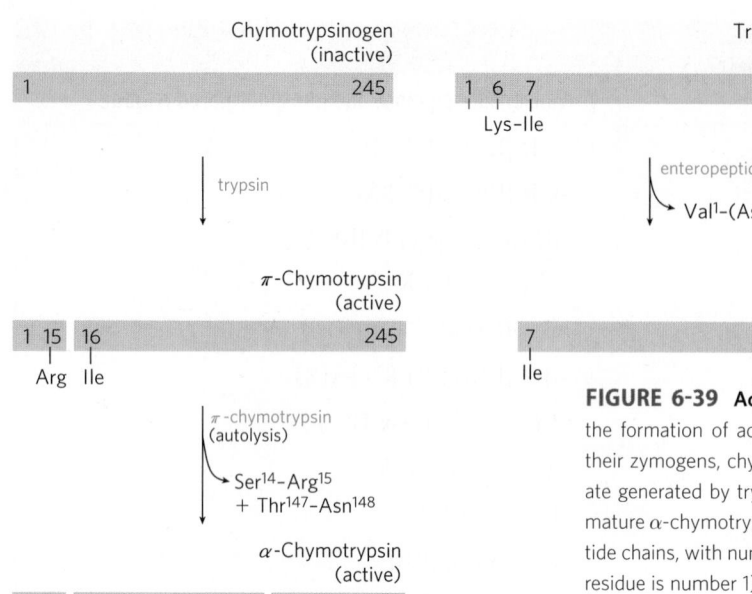

FIGURE 6-39 Activation of zymogens by proteolytic cleavage. Shown here is the formation of active chymotrypsin (formally, α-chymotrypsin) and trypsin from their zymogens, chymotrypsinogen and trypsinogen. The π-chymotrypsin intermediate generated by trypsin cleavage has a somewhat altered specificity relative to the mature α-chymotrypsin. The bars represent the amino acid sequences of the polypeptide chains, with numbers indicating the positions of the residues (the amino-terminal residue is number 1). Residues at the termini of the polypeptide fragments generated by cleavage are indicated below the bars. Note that in the final active forms, some numbered residues are missing. Recall that the three polypeptide chains (A, B, and C) of chymotrypsin are linked by disulfide bonds (see Fig. 6-20).

Proteases are not the only proteins activated by proteolysis. In other cases, however, the precursors are called not zymogens but, more generally, **proproteins** or **proenzymes**, as appropriate. For example, the connective tissue protein collagen is initially synthesized as the soluble precursor procollagen.

A Cascade of Proteolytically Activated Zymogens Leads to Blood Coagulation

A blood clot is an aggregate of cell fragments called platelets, cross-linked and stabilized by proteinaceous fibers consisting mainly of fibrin **(Fig. 6-40a)**. Fibrin is derived from a soluble zymogen called fibrinogen. After albumins and globulins, fibrinogen is usually the third most abundant type of protein in blood plasma. The formation of a blood clot provides a well-studied example of a **regulatory cascade**, a mechanism that allows a very sensitive response to—and amplification of—a molecular signal. The pathways also bring together several other types of regulation.

In a regulatory cascade, a signal leads to the activation of protein X. Protein X catalyzes the activation of protein Y. Protein Y catalyzes the activation of protein Z, and so on. Since proteins X, Y, and Z are catalysts and activate multiple copies of the next protein in the chain, the signal is amplified in each step. In some cases, the activation steps involve proteolytic cleavage and are thus effectively irreversible. In others, activation entails readily reversible protein modification steps such as phosphorylation. Regulatory cascades govern a wide range of biological processes, including, besides blood coagulation, some aspects of cell fate determination during development, the detection of light by retinal rods, and programmed cell death (apoptosis).

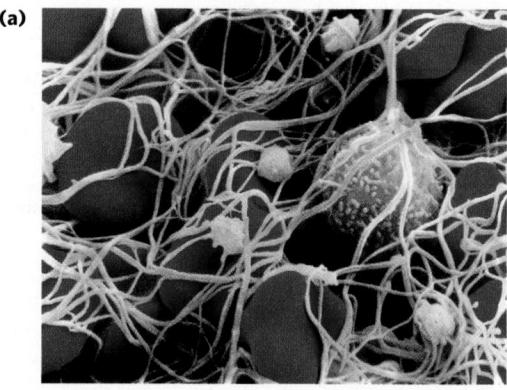

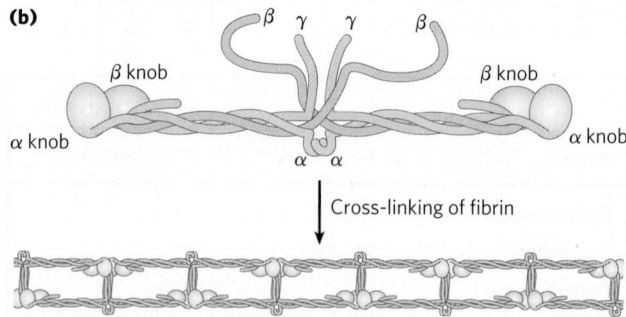

FIGURE 6-40 The function of fibrin in blood clots. (a) A blood clot consists of aggregated platelets (small, light-colored cells) tied together with strands of cross-linked fibrin. Erythrocytes (red in this colorized scanning electron micrograph) are also trapped in the matrix. **(b)** The soluble plasma protein fibrinogen consists of two complexes of α, β, and γ subunits $(\alpha_2\beta_2\gamma_2)$. The removal of amino-terminal peptides from the α and β subunits (not shown) leads to the formation of higher-order complexes and eventual covalent cross-linking that results in the formation of fibrin fibers. The "knobs" are globular domains at the ends of the proteolyzed subunits. [Source: (a) CNRI/Science Source.]

Fibrinogen is a dimer of heterotrimers $(A\alpha_2B\beta_2\gamma_2)$ with three different but evolutionarily related types of subunits (Fig. 6-40b). Fibrinogen is converted to **fibrin** $(\alpha_2\beta_2\gamma_2)$, and thereby activated for blood clotting, by the proteolytic removal of 16 amino acid residues from the amino-terminal end (the A peptide) of each α subunit and 14 amino acid residues from the amino-terminal end (the B peptide) of each β subunit. Peptide removal is catalyzed by the serine protease **thrombin**. The newly exposed amino termini of the α and β subunits fit neatly into binding sites in the carboxyl-terminal globular portions of the γ and β subunits, respectively, of another fibrin protein. Fibrin thus polymerizes into a gel-like matrix to generate a soft clot. Covalent cross-links between the associated fibrins are generated by the condensation of particular Lys residues in one fibrin heterotrimer with Gln residues in another, catalyzed by a transglutaminase, **factor XIIIa**. The covalent cross-links convert the soft clot into a hard clot.

Fibrinogen activation to produce fibrin is the end point of not one but two parallel but intertwined regulatory cascades **(Fig. 6-41)**. One of these is referred to as the contact activation pathway ("contact" refers to interaction of key components of this system with anionic phospholipids presented on the surface of platelets at the site of a wound). As all components of this pathway are found in the blood plasma, it is also called the **intrinsic pathway**. The second path is the tissue factor or **extrinsic pathway**. A major component of this pathway, the protein **tissue factor (TF)**, is not present in the blood. Most of the protein factors in both pathways are designated by roman numerals. Many of these factors are chymotrypsin-like serine proteases, with zymogen precursors that are synthesized in the liver and exported to the blood. Other factors are regulatory proteins that bind to the serine proteases and help to activate them.

Blood clotting begins with the activation of circulating **platelets**—specialized cell fragments that lack nuclei—at the site of a wound. Tissue damage causes collagen molecules present beneath the epithelial cell layer that lines each blood vessel to become exposed to the blood. Platelet activation is primarily triggered by interaction with this collagen. Activation leads to the presentation of anionic phospholipids on the surface of each platelet and the release of signaling molecules such as **thromboxanes** (p. 375) that help stimulate the activation of additional platelets. The activated platelets aggregate at the site of a wound, forming a loose clot. Stabilization of the clot requires the fibrin generated by the coagulation cascades.

The extrinsic pathway comes into play first. Tissue damage exposes the blood plasma to TF embedded largely in the membranes of fibroblasts and smooth muscle cells beneath the endothelial layer. An initiating complex is formed between TF and factor VII, present in the blood plasma. **Factor VII** is a zymogen of a serine protease, and TF is a regulatory protein required for its function. Factor VII is converted to its active form, **factor VIIa**, by proteolytic cleavage carried out by **factor Xa** (another serine

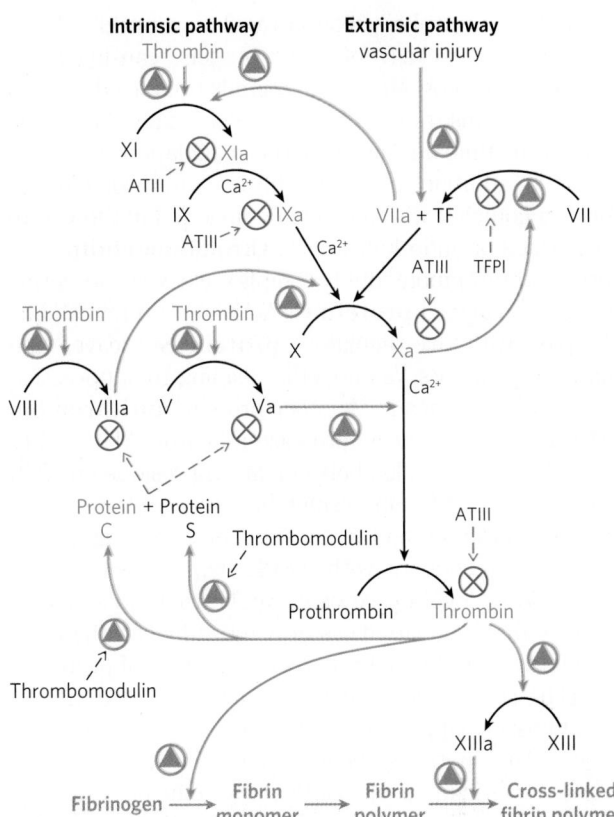

FIGURE 6-41   The coagulation cascades. The interlinked intrinsic and extrinsic pathways leading to the cleavage of fibrinogen to form active fibrin are shown. Active serine proteases in the pathways are shown in blue. Green arrows denote activating steps, and red arrows indicate inhibitory processes.

protease). The TF-VIIa complex then cleaves **factor X**, creating the active form, factor Xa.

If TF-VIIa is needed to cleave X, and Xa is needed to cleave TF-VII, how does the process ever get started? A very small amount of factor VIIa is present in the blood at all times, enough to form a small amount of the active TF-VIIa complex immediately after tissue is damaged. This allows formation of factor Xa and establishes the initiating feedback loop. Once levels of factor Xa begin to build up, Xa (in a complex with regulatory protein factor Va) cleaves prothrombin to form active thrombin, and thrombin cleaves fibrinogen.

The extrinsic pathway thus provides a burst of thrombin. However, the TF-VIIa complex is quickly shut down by the protein **tissue factor protein inhibitor (TFPI)**. Clot formation is sustained by the activation of components of the intrinsic pathway. **Factor IX** is converted to the active serine protease **factor IXa** by the TF-VIIa protease during initiation of the clotting sequence. Factor IXa, in a complex with the regulatory protein **VIIIa**, is relatively stable and provides an alternative enzyme for the proteolytic conversion of factor X to Xa. Activated IXa can also be produced by the serine protease factor XIa. Most of the XIa is generated by cleavage of **factor XI** zymogen by thrombin in a feedback loop.

Left uncontrolled, blood coagulation could eventually lead to blockage of blood vessels, causing heart attacks or strokes. More regulation is thus needed. As a hard clot forms, regulatory pathways are already acting to limit the time during which the coagulation cascade is active. In addition to cleaving fibrinogen, thrombin also forms a complex with a protein embedded in the vascular surface of endothelial cells, **thrombomodulin**. The thrombin-thrombomodulin complex cleaves the serine protease zymogen **protein C**. Activated protein C, in a complex with the regulatory **protein S**, cleaves and inactivates factors Va and VIIIa, leading to suppression of the overall cascade. Another protein, **antithrombin III (ATIII)**, is a serine protease inhibitor. ATIII makes a covalent 1:1 complex between an Arg residue on ATIII and the active-site Ser residue of serine proteases, particularly thrombin and factor Xa. These two regulatory systems, in concert with TFPI, help to establish a threshold or level of exposure to TF that is needed to activate the coagulation cascade. Individuals with genetic defects that eliminate or decrease levels of protein C or ATIII in the blood have a greatly elevated risk of thrombosis (inappropriate formation of blood clots).

The control of blood coagulation has important roles in medicine, particularly in the prevention of blood clotting during surgery and in patients at risk for heart attacks or strokes. Several different medical approaches to anticoagulation are available. The first takes advantage of another feature of several proteins in the coagulation cascade that we have not yet considered. The factors VII, IX, X, and prothrombin, along with proteins C and S, have calcium-binding sites that are critical to their function. In each case, the calcium-binding sites are formed by modification of multiple Glu residues near the amino terminus of each protein to γ-**carboxyglutamate** residues (abbreviated **Gla**; p. 81). The Glu-to-Gla modifications are carried out by enzymes that depend on the function of the fat-soluble vitamin K (p. 380). Bound calcium functions to adhere these proteins to the anionic phospholipids that appear on the surface of activated platelets, effectively localizing the coagulation factors to the area where the clot is to form. Vitamin K antagonists such as **warfarin** (Coumadin) have proven highly effective as anticoagulants. A second approach to anticoagulation is the administration of heparins. **Heparins** are highly sulfated polysaccharides (see Fig. 7-22). They act as anticoagulants by increasing the affinity of ATIII for factor Xa and thrombin, thus facilitating the inactivation of key cascade elements (see Fig. 7-26). Finally, **aspirin** (acetylsalicylate; Fig. 21-15b) is effective as an anticoagulant. Aspirin inhibits the enzyme cyclooxygenase, required for the production of thromboxanes. As aspirin reduces thromboxane release from platelets, the capacity of the platelets to aggregate declines.

Humans born with a deficiency in any one of most components of the clotting cascade have an increased tendency to bleed that varies from mild to essentially uncontrollable, a fatal condition. Genetic defects in the genes encoding the proteins required for blood clotting result in diseases referred to as hemophilias. Hemophilia A is a sex-linked trait resulting from a deficiency in factor VIII. This is the most common human hemophilia, affecting about one in 5,000 males worldwide. The most famous example of the inheritance of hemophilia A occurred among European royalty. Queen Victoria (1819–1901) was evidently a carrier. Prince Leopold, her eighth child, suffered from hemophilia A and died at the age of 31 after a minor fall. At least two of her daughters were carriers and passed the defective gene to other royal families of Europe **(Fig. 6-42)**. ■

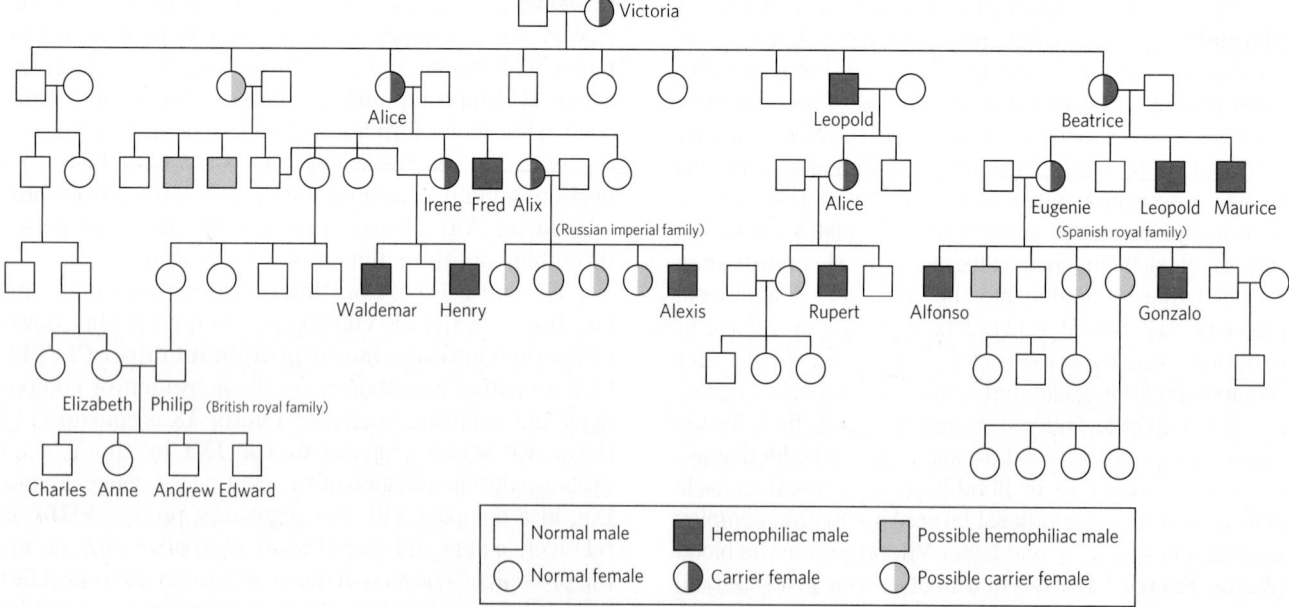

FIGURE 6-42 The royal families of Europe and inheritance of hemophilia A. Males are indicated by squares and females by circles. Males who suffered from hemophilia are represented by red squares, and presumed female carriers by half-red circles.

Some Regulatory Enzymes Use Several Regulatory Mechanisms

Glycogen phosphorylase catalyzes the first reaction in a pathway that feeds stored glucose into energy-yielding carbohydrate metabolism (Chapters 14 and 15). This is an important metabolic pathway, and its regulation is correspondingly complex. Although the primary regulation of glycogen phosphorylase is through covalent modification, as outlined in Figure 6-37, glycogen phosphorylase is also modulated by allosteric binding of AMP,

which is an activator of phosphorylase *b*, and by glucose 6-phosphate and ATP, both inhibitors. In addition, the enzymes that add and remove the phosphoryl groups are themselves regulated by—and so the entire system is sensitive to—the levels of hormones that regulate blood sugar (**Fig. 6-43**; see also Chapters 15 and 23).

Other complex regulatory enzymes are found at key metabolic crossroads. Bacterial glutamine synthetase, which catalyzes a reaction that introduces reduced nitrogen into cellular metabolism (Chapter 22), is among the most complex regulatory enzymes known. It is regulated

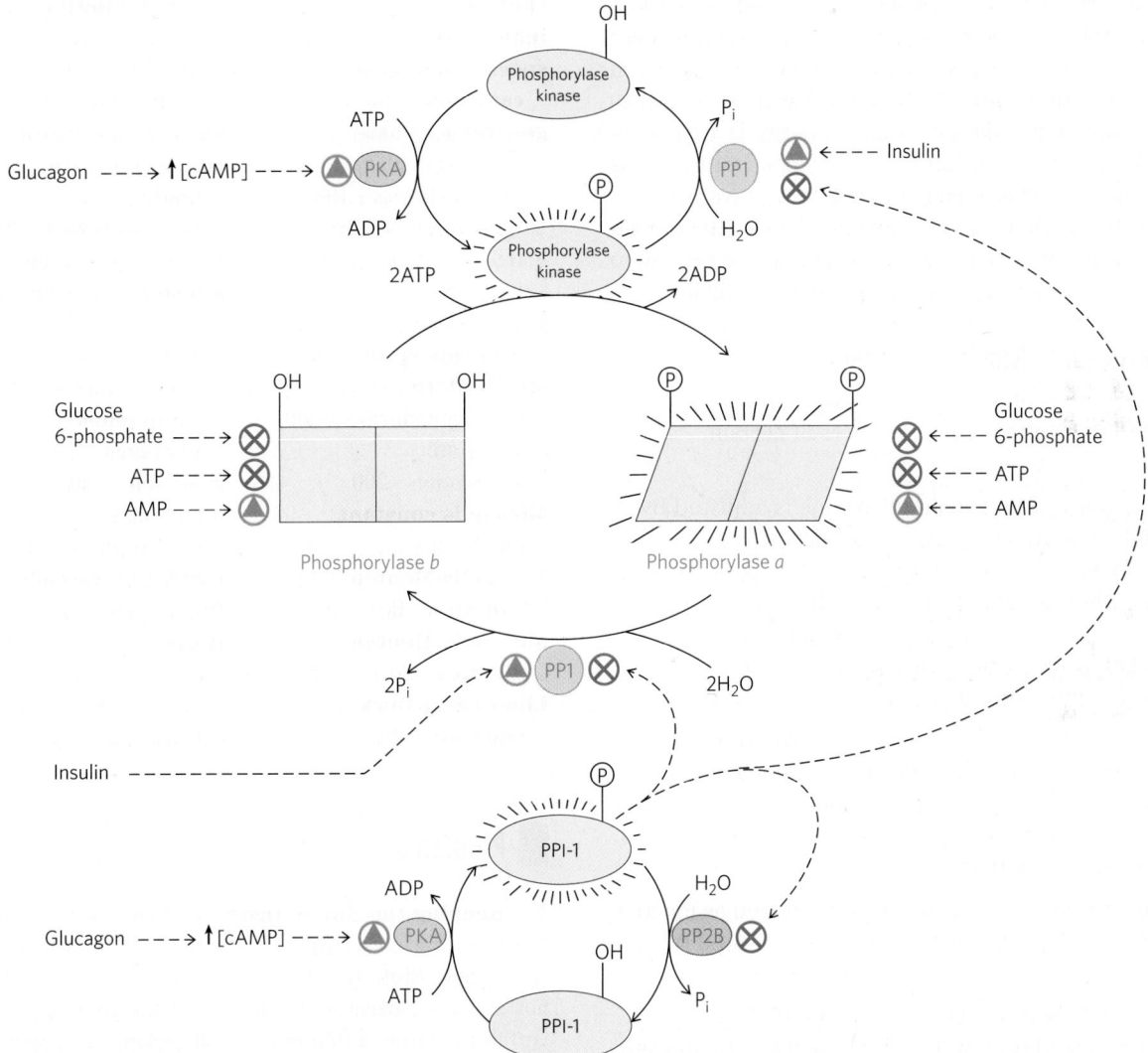

FIGURE 6-43 Regulation of muscle glycogen phosphorylase activity by phosphorylation. The activity of glycogen phosphorylase in muscle is subjected to a multilevel system of regulation involving much more than the covalent modification (phosphorylation) shown in Figure 6-37. Also playing important roles are allosteric regulation and a regulatory cascade sensitive to hormonal status that acts on the enzymes involved in phosphorylation and dephosphorylation. The activity of both forms of the enzyme is allosterically regulated by an activator (AMP) and by inhibitors (glucose 6-phosphate and ATP) that bind to separate sites on the enzyme. The activities of phosphorylase kinase and phosphoprotein phosphatase 1 (PP1) are also regulated by covalent modification, via a short pathway that responds to the hormones glucagon and epinephrine. One path leads to the phosphorylation of phosphorylase kinase and phosphoprotein phosphatase inhibitor 1 (PPI-1). The phosphorylated phosphorylase kinase is activated and, in turn, phosphorylates and activates glycogen phosphorylase. At the same time, the phosphorylated PPI-1 interacts with and inhibits PP1. PPI-1 also keeps itself active (phosphorylated) by inhibiting phosphoprotein phosphatase 2B (PP2B), the enzyme that dephosphorylates (inactivates) it. In this way, the equilibrium between the *a* and *b* forms of glycogen phosphorylase is shifted decisively toward the more active glycogen phosphorylase *a*. Note that both forms of phosphorylase kinase are activated to a degree by Ca^{2+} ion (not shown). This pathway is discussed in more detail in Chapters 14, 15, and 23.

allosterically, with at least eight different modulators; by reversible covalent modification; and by the association of other regulatory proteins, a mechanism examined in detail when we consider the regulation of specific metabolic pathways.

What is the advantage of such complexity in the regulation of enzymatic activity? We began this chapter by stressing the central importance of catalysis to the existence of life. The *control* of catalysis is also critical to life. If all possible reactions in a cell were catalyzed simultaneously, macromolecules and metabolites would quickly be broken down to much simpler chemical forms. Instead, cells catalyze only the reactions they need at a given moment. When chemical resources are plentiful, cells synthesize and store glucose and other metabolites. When chemical resources are scarce, cells use these stores to fuel cellular metabolism. Chemical energy is used economically, parceled out to various metabolic pathways as cellular needs dictate. The availability of powerful catalysts, each specific for a given reaction, makes the regulation of these reactions possible. This, in turn, gives rise to the complex, highly regulated symphony we call life.

SUMMARY 6.5 Regulatory Enzymes

- The activities of metabolic pathways in cells are regulated by control of the activities of certain enzymes.
- The activity of an allosteric enzyme is adjusted by reversible binding of a specific modulator to a regulatory site. A modulator may be the substrate itself or some other metabolite, and the effect of the modulator may be inhibitory or stimulatory. The kinetic behavior of allosteric enzymes reflects cooperative interactions among enzyme subunits.
- Other regulatory enzymes are modulated by covalent modification of a specific functional group necessary for activity. The phosphorylation of specific amino acid residues is a particularly common way to regulate enzyme activity.
- Many proteolytic enzymes are synthesized as inactive precursors called zymogens, which are activated by cleavage to release small peptide fragments.
- Blood clotting is mediated by two interlinked regulatory cascades of proteolytically activated zymogens.
- Enzymes at important metabolic intersections may be regulated by complex combinations of effectors, allowing coordination of the activities of interconnected pathways.

Key Terms

Terms in bold are defined in the glossary.

enzyme 188
cofactor 188
coenzyme 188
prosthetic group 188
holoenzyme 188
apoenzyme 188
apoprotein 188
active site 190
substrate 190
ground state 190
transition state 191
activation energy ($\Delta G^{\ddagger}$) 191
reaction intermediate 191
rate-limiting step 191
equilibrium constant (K_{eq}) 192
rate constant 192
binding energy (ΔG_B) 193
specificity 195
induced fit 196
specific acid-base catalysis 197
general acid-base catalysis 197
covalent catalysis 198
enzyme kinetics 198
initial rate (initial velocity), V_0 198
V_{max} 199
pre–steady state 200
steady state 200
steady-state kinetics 200
steady-state assumption 200
Michaelis constant (K_m) 201
Michaelis-Menten equation 201
Michaelis-Menten kinetics 201
Lineweaver-Burk equation 202
dissociation constant (K_d) 202
k_{cat} 203
turnover number 203
Cleland nomenclature 204
reversible inhibition 207
competitive inhibition 207
uncompetitive inhibition 208
mixed inhibition 208
noncompetitive inhibition 208
irreversible inhibitors 210
suicide inactivator 210
transition-state analog 210
serine proteases 215
regulatory enzyme 225
allosteric enzyme 225
allosteric modulator (allosteric effector) 226
protein kinases 229
protein phosphatases 229
zymogen 230
proproteins (proenzymes) 232
regulatory cascade 232
fibrinogen 233
fibrin 233
thrombin 233
intrinsic pathway 233
extrinsic pathway 233

Problems

1. Keeping the Sweet Taste of Corn The sweet taste of freshly picked corn (maize) is due to the high level of sugar in the kernels. Store-bought corn (several days after picking) is not as sweet, because about 50% of the free sugar is converted to starch within one day of picking. To preserve the sweetness of fresh corn, the husked ears can be immersed in boiling water for a few minutes ("blanched"), then cooled in cold water. Corn processed in this way and stored in a freezer maintains its sweetness. What is the biochemical basis for this procedure?

2. Intracellular Concentration of Enzymes To approximate the concentration of enzymes in a bacterial cell, assume that the cell contains equal concentrations of 1,000 different enzymes in solution in the cytosol and that each protein has a molecular weight of 100,000. Assume also that the bacterial cell is a cylinder (diameter 1.0 μm, height 2.0 μm), that the cytosol (specific gravity 1.20) is 20% soluble protein by weight, and that the soluble protein consists entirely of enzymes. Calculate the *average* molar concentration of each enzyme in this hypothetical cell.

3. Rate Enhancement by Urease The enzyme urease enhances the rate of urea hydrolysis at pH 8.0 and 20 °C by a factor of 10^{14}. If a given quantity of urease can completely hydrolyze a given quantity of urea in 5.0 min at 20 °C and pH 8.0, how long would it take for this amount of urea to be hydrolyzed under the same conditions in the absence of urease? Assume that both reactions take place in sterile systems so that bacteria cannot attack the urea.

4. Protection of an Enzyme against Denaturation by Heat When enzyme solutions are heated, there is a progressive loss of catalytic activity over time due to denaturation of the enzyme. A solution of the enzyme hexokinase incubated at 45 °C lost 50% of its activity in 12 min, but when incubated at 45 °C in the presence of a very large concentration of one of its substrates, it lost only 3% of its activity in 12 min. Suggest why thermal denaturation of hexokinase was retarded in the presence of one of its substrates.

5. Requirements of Active Sites in Enzymes Carboxypeptidase, which sequentially removes carboxyl-terminal amino acid residues from its peptide substrates, is a single polypeptide of 307 amino acids. The two essential catalytic groups in the active site are furnished by Arg^{145} and Glu^{270}.

(a) If the carboxypeptidase chain were a perfect α helix, how far apart (in Å) would Arg^{145} and Glu^{270} be? (Hint: See Fig. 4-4a.)

(b) Explain how the two amino acid residues can catalyze a reaction occurring in the space of a few angstroms.

6. Quantitative Assay for Lactate Dehydrogenase The muscle enzyme lactate dehydrogenase catalyzes the reaction

$$\underset{\text{Pyruvate}}{CH_3-\overset{\overset{\text{O}}{\|}}{C}-COO^-} + NADH + H^+ \longrightarrow \underset{\text{Lactate}}{CH_3-\overset{\overset{\text{OH}}{|}}{\underset{\overset{|}{H}}{C}}-COO^-} + NAD^+$$

NADH and NAD^+ are the reduced and oxidized forms, respectively, of the coenzyme NAD. Solutions of NADH, but *not* NAD^+, absorb light at 340 nm. This property is used to determine the concentration of NADH in solution by measuring spectrophotometrically the amount of light absorbed at 340 nm by the solution. Explain how these properties of NADH can be used to design a quantitative assay for lactate dehydrogenase.

7. Effect of Enzymes on Reactions Which of the listed effects would be brought about by any enzyme catalyzing the following simple reaction?

$$S \underset{k_2}{\overset{k_1}{\rightleftharpoons}} P \quad \text{where} \quad K'_{eq} = \frac{[P]}{[S]}$$

(a) Decreased K'_{eq}; (b) increased k_1; (c) increased K'_{eq}; (d) increased $\Delta G^{\ddagger}$; (e) decreased $\Delta G^{\ddagger}$; (f) more negative $\Delta G'^{\circ}$; (g) increased k_2.

8. Relation between Reaction Velocity and Substrate Concentration: Michaelis-Menten Equation (a) At what substrate concentration would an enzyme with a k_{cat} of 30.0 s^{-1} and a K_m of 0.0050 M operate at one-quarter of its maximum rate?

(b) Determine the fraction of V_{max} that would be obtained at the following substrate concentrations [S]: $\frac{1}{2}K_m$, $2K_m$, and $10K_m$.

(c) An enzyme that catalyzes the reaction X $\rightleftharpoons$ Y is isolated from two bacterial species. The enzymes have the same V_{max} but different K_m values for the substrate X. Enzyme A has a K_m of 2.0 μM, and enzyme B has a K_m of 0.5 μM. The plot below shows the kinetics of reactions carried out with the same concentration of each enzyme and with [X] = 1 μM. Which curve corresponds to which enzyme?

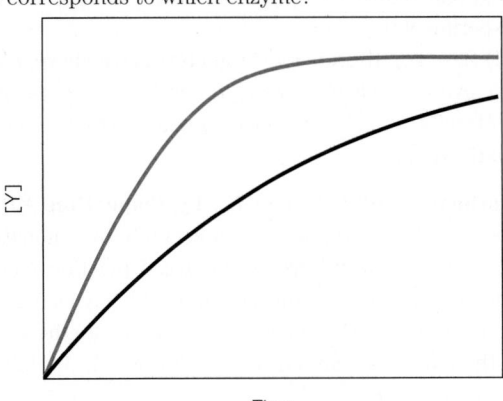

9. Applying the Michaelis-Menten Equation I An enzyme catalyzes the reaction A $\rightleftharpoons$ B. The enzyme is present at a concentration of 2 nM, and the V_{max} is 1.2 $\mu M\ s^{-1}$. The K_m for substrate A is 10 μM. Calculate the initial velocity of the reaction, V_0, when the substrate concentration is (a) 2 μM, (b) 10 μM, (c) 30 μM.

10. Applying the Michaelis-Menten Equation II An enzyme catalyzes the reaction M $\rightleftharpoons$ N. The enzyme is present at a concentration of 1 nM, and the V_{max} is 2 $\mu M\ s^{-1}$. The K_m for substrate M is 4 μM. (a) Calculate k_{cat}. (b) What values of V_{max} and K_m would be observed in the presence of sufficient amounts of an uncompetitive inhibitor to generate an α' of 2.0?

11. Applying the Michaelis-Menten Equation III A research group discovers a new version of happyase, which they call happyase*, that catalyzes the chemical reaction HAPPY $\rightleftharpoons$ SAD. The researchers begin to characterize the enzyme.

(a) In the first experiment, with $[E_t]$ at 4 nM, they find that the V_{max} is 1.6 $\mu M\ s^{-1}$. Based on this experiment, what is the k_{cat} for happyase*? (Include appropriate units.)

(b) In another experiment, with $[E_t]$ at 1 nM and [HAPPY] at 30 μM, the researchers find that $V_0 = 300$ nM s^{-1}. What is the measured K_m of happyase* for its substrate HAPPY? (Include appropriate units.)

(c) Further research shows that the purified happyase* used in the first two experiments was actually contaminated with a reversible inhibitor called ANGER. When ANGER is carefully removed from the happyase* preparation and the two experiments repeated, the measured V_{max} in (a) is increased to 4.8 $\mu M\ s^{-1}$, and the measured K_m in (b) is now 15 μM. For the inhibitor ANGER, calculate the values of α and α'.

(d) Based on the information given above, what type of inhibitor is ANGER?

12. Applying the Michaelis-Menten Equation IV An enzyme is found that catalyzes the reaction X $\rightleftharpoons$ Y. Researchers find that the K_m for the substrate X is 4 μM, and the k_{cat} is 20 min^{-1}.

(a) In an experiment, [X] = 6 mM, and V_0 = 480 nM min^{-1}. What was the [E_t] used in the experiment?

(b) In another experiment, [E_t] = 0.5 μM, and the measured V_0 = 5 μM min^{-1}. What was the [X] used in the experiment?

(c) The compound Z is found to be a very strong competitive inhibitor of the enzyme, with an α of 10. In an experiment with the same [E_t] as in (a), but a different [X], an amount of Z is added that reduces V_0 to 240 nM min^{-1}. What is the [X] in this experiment?

(d) Based on the kinetic parameters given above, has this enzyme evolved to achieve catalytic perfection? Explain your answer briefly, using the kinetic parameter(s) that define catalytic perfection.

13. Estimation of V_{max} and K_m by Inspection Although graphical methods are available for accurate determination of the V_{max} and K_m of an enzyme-catalyzed reaction (see Box 6-1), sometimes these quantities can be quickly estimated by inspecting values of V_0 at increasing [S]. Estimate the V_{max} and K_m of the enzyme-catalyzed reaction for which the following data were obtained:

[S] (M)	V_0 (μM/min)
2.5×10^{-6}	28
4.0×10^{-6}	40
1×10^{-5}	70
2×10^{-5}	95
4×10^{-5}	112
1×10^{-4}	128
2×10^{-3}	139
1×10^{-2}	140

14. Properties of an Enzyme of Prostaglandin Synthesis Prostaglandins are a class of eicosanoids, fatty acid derivatives with a variety of extremely potent actions on vertebrate tissues. They are responsible for producing fever and inflammation and its associated pain. Prostaglandins are derived from the 20-carbon fatty acid arachidonic acid in a reaction catalyzed by the enzyme prostaglandin endoperoxide synthase. This enzyme, a cyclooxygenase, uses oxygen to convert arachidonic acid to PGG$_2$, the immediate precursor of many different prostaglandins (prostaglandin synthesis is described in Chapter 21).

(a) The kinetic data given below are for the reaction catalyzed by prostaglandin endoperoxide synthase. Focusing here on the first two columns, determine the V_{max} and K_m of the enzyme.

[Arachidonic acid] (mM)	Rate of formation of PGG$_2$ (mM min^{-1})	Rate of formation of PGG$_2$ with 10 mg/mL ibuprofen (mM min^{-1})
0.5	23.5	16.67
1.0	32.2	25.25
1.5	36.9	30.49
2.5	41.8	37.04
3.5	44.0	38.91

(b) Ibuprofen is an inhibitor of prostaglandin endoperoxide synthase. By inhibiting the synthesis of prostaglandins, ibuprofen reduces inflammation and pain. Using the data in the first and third columns of the table, determine the type of inhibition that ibuprofen exerts on prostaglandin endoperoxide synthase.

15. Graphical Analysis of V_{max} and K_m The following experimental data were collected during a study of the catalytic activity of an intestinal peptidase with the substrate glycylglycine:

$$\text{Glycylglycine} + H_2O \longrightarrow 2\,\text{glycine}$$

[S] (mM)	Product formed (μmol/min^{-1})
1.5	0.21
2.0	0.24
3.0	0.28
4.0	0.33
8.0	0.40
16.0	0.45

Use graphical analysis (see Box 6-1) to determine the V_{max} and K_m for this enzyme preparation and substrate.

16. The Eadie-Hofstee Equation There are several ways to transform the Michaelis-Menten equation so as to plot data and derive kinetic parameters, each with different advantages depending on the data set being analyzed. One transformation of the Michaelis-Menten equation is the Lineweaver-Burk, or double-reciprocal, equation. Multiplying both sides of the Lineweaver-Burk equation by V_{max} and rearranging gives the Eadie-Hofstee equation:

$$V_0 = (-K_m)\frac{V_0}{[S]} + V_{max}$$

A plot of V_0 versus V_0/[S] for an enzyme-catalyzed reaction is shown below. The blue curve was obtained in the absence of inhibitor. Which of the other curves (A, B, or C) shows the enzyme activity when a competitive inhibitor was added to the reaction mixture? Hint: See Equation 6-30.

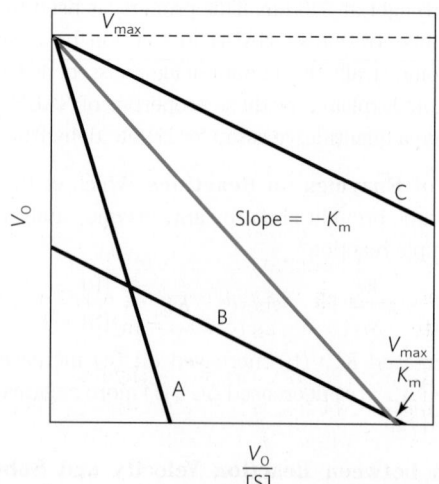

17. The Turnover Number of Carbonic Anhydrase Carbonic anhydrase of erythrocytes (M_r 30,000) has one of the highest turnover numbers known. It catalyzes the reversible hydration of CO_2:

$$H_2O + CO_2 \rightleftharpoons H_2CO_3$$

This is an important process in the transport of CO_2 from the tissues to the lungs. If 10.0 μg of pure carbonic anhydrase catalyzes the hydration of 0.30 g of CO_2 in 1 min at 37 °C at V_{max}, what is the turnover number (k_{cat}) of carbonic anhydrase (in units of min^{-1})?

18. Deriving a Rate Equation for Competitive Inhibition The rate equation for an enzyme subject to competitive inhibition is

$$V_0 = \frac{V_{max}[S]}{\alpha K_m + [S]}$$

Beginning with a new definition of total enzyme as

$$[E_t] = [E] + [ES] + [EI]$$

and the definitions of α and K_I provided in the text, derive the rate equation above. Use the derivation of the Michaelis-Menten equation as a guide.

19. Irreversible Inhibition of an Enzyme Many enzymes are inhibited irreversibly by heavy metal ions such as Hg^{2+}, Cu^{2+}, or Ag^+, which can react with essential sulfhydryl groups to form mercaptides:

$$Enz\text{-}SH + Ag^+ \longrightarrow Enz\text{-}S\text{-}Ag + H^+$$

The affinity of Ag^+ for sulfhydryl groups is so great that Ag^+ can be used to titrate —SH groups quantitatively. To 10.0 mL of a solution containing 1.0 mg/mL of a pure enzyme, an investigator added just enough $AgNO_3$ to completely inactivate the enzyme. A total of 0.342 μmol of $AgNO_3$ was required. Calculate the minimum molecular weight of the enzyme. Why does the value obtained in this way give only the *minimum* molecular weight?

20. Clinical Application of Differential Enzyme Inhibition Human blood serum contains a class of enzymes known as acid phosphatases, which hydrolyze biological phosphate esters under slightly acidic conditions (pH 5.0):

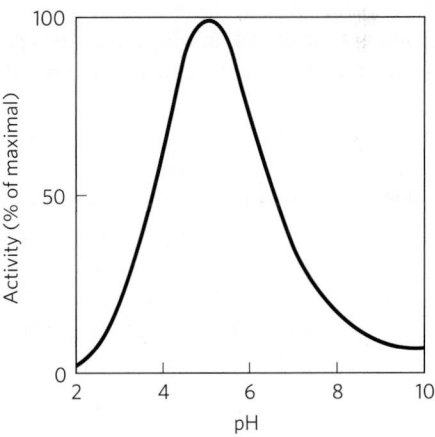

Acid phosphatases are produced by erythrocytes and by the liver, kidney, spleen, and prostate gland. The enzyme of the prostate gland is clinically important, because its increased activity in the blood can be an indication of prostate cancer. The phosphatase from the prostate gland is strongly inhibited by tartrate ion, but acid phosphatases from other tissues are not. How can this information be used to develop a specific procedure for measuring the activity of the acid phosphatase of the prostate gland in human blood serum?

21. Inhibition of Carbonic Anhydrase by Acetazolamide Carbonic anhydrase is strongly inhibited by the drug acetazolamide, which is used as a diuretic (i.e., to increase the production of urine) and to lower excessively high pressure in the eye (due to accumulation of intraocular fluid) in glaucoma. Carbonic anhydrase plays an important role in these and other secretory processes because it participates in

regulating the pH and bicarbonate content of several body fluids. The experimental curve of initial reaction velocity (as percentage of V_{max}) versus [S] for the carbonic anhydrase reaction is illustrated below (upper curve). When the experiment is repeated in the presence of acetazolamide, the lower curve is obtained. From an inspection of the curves and your knowledge of the kinetic properties of competitive and mixed enzyme inhibitors, determine the nature of the inhibition by acetazolamide. Explain your reasoning.

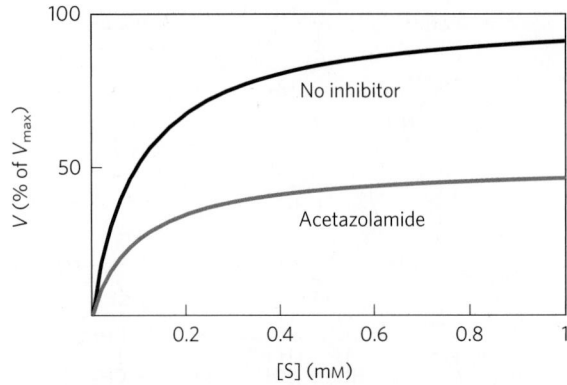

22. The Effects of Reversible Inhibitors Derive the expression for the effect of a reversible inhibitor on observed K_m (apparent $K_m = \alpha K_m/\alpha'$). Start with Equation 6-30 and the statement that apparent K_m is equivalent to the [S] at which $V_0 = V_{max}/2\alpha'$.

23. pH Optimum of Lysozyme The active site of lysozyme contains two amino acid residues essential for catalysis: Glu^{35} and Asp^{52}. The pK_a values of the carboxyl side chains of these residues are 5.9 and 4.5, respectively. What is the ionization state (protonated or deprotonated) of each residue at pH 5.2, the pH optimum of lysozyme? How can the ionization states of these residues explain the pH-activity profile of lysozyme shown below?

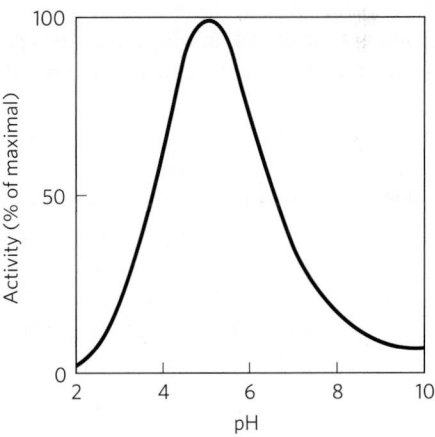

Data Analysis Problem

24. Exploring and Engineering Lactate Dehydrogenase Examining the structure of an enzyme can lead to hypotheses about the relationship between different amino acids in the protein's structure and the protein's function. One way to test these hypotheses is to use recombinant DNA technology to generate mutant versions of the enzyme and then examine the structure and function of these altered forms. The technology used to do this is described in Chapters 8 and 9.

One example of this kind of analysis is the work of A. R. Clarke and colleagues on the enzyme lactate dehydrogenase, published in 1989. Lactate dehydrogenase (LDH) catalyzes the reduction of pyruvate with NADH to form lactate (see Section 14.3). A schematic of the enzyme's active site is shown below; the pyruvate is in the center:

Lactate dehydrogenase

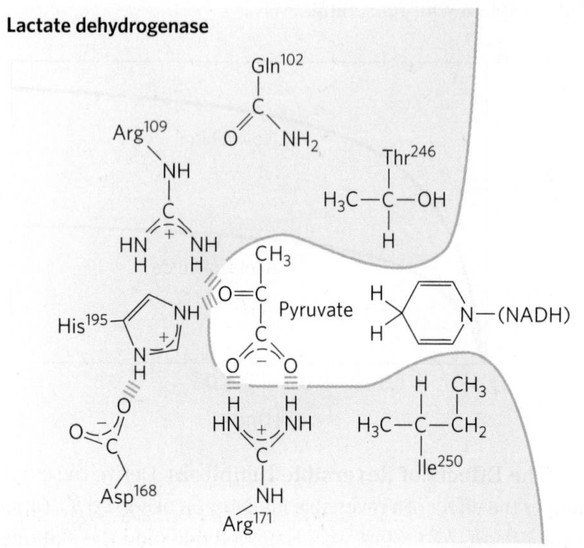

The reaction mechanism is similar to that of many NADH reductions (see Fig. 13-24); it is approximately the reverse of steps ❷ and ❸ of Figure 14-8. The transition state involves a strongly polarized carbonyl group of the pyruvate molecule:

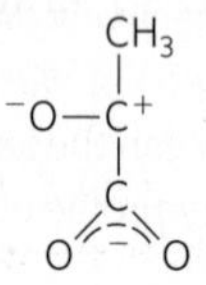

(a) A mutant form of LDH in which Arg[109] is replaced with Gln shows only 5% of the pyruvate binding and 0.07% of the activity of wild-type enzyme. Provide a plausible explanation for the effects of this mutation.

(b) A mutant form of LDH in which Arg[171] is replaced with Lys shows only 0.05% of the wild-type level of substrate binding. Why is this dramatic effect surprising?

(c) In the crystal structure of LDH, the guanidinium group of Arg[171] and the carboxyl group of pyruvate are aligned, as shown above, in a co-planar "forked" configuration. Based on this structure, explain the dramatic effect of substituting Arg[171] with Lys.

(d) A mutant form of LDH in which Ile[250] is replaced with Gln shows reduced binding of NADH. Provide a plausible explanation for this result.

Clarke and colleagues also set out to engineer a mutant version of LDH that would bind and reduce oxaloacetate rather than pyruvate. They made a single substitution, replacing Gln[102] with Arg; the resulting enzyme would reduce oxaloacetate to malate and would no longer reduce pyruvate to lactate. They had therefore converted LDH to malate dehydrogenase.

(e) Sketch the active site of this mutant LDH with oxaloacetate bound.

(f) Why does this mutant enzyme now use oxaloacetate as a substrate instead of pyruvate?

(g) The authors were surprised that substituting a larger amino acid in the active site allowed a larger substrate to bind. Explain this result.

References

Clarke, A.R., T. Atkinson, and J.J. Holbrook. 1989. From analysis to synthesis: new ligand binding sites on the lactate dehydrogenase framework, Part I. *Trends Biochem. Sci.* 14:101–105.

Clarke, A.R., T. Atkinson, and J.J. Holbrook. 1989. From analysis to synthesis: new ligand binding sites on the lactate dehydrogenase framework, Part II. *Trends Biochem. Sci.* 14:145–148.

Further Reading is available at www.macmillanlearning.com/LehningerBiochemistry7e.

CHAPTER 7

Carbohydrates and Glycobiology

Self-study tools that will help you practice what you've learned and reinforce this chapter's concepts are available online. Go to www.macmillanlearning.com/LehningerBiochemistry7e.

Carbohydrates are the most abundant biomolecules on Earth. Each year, photosynthesis converts more than 100 billion metric tons of CO_2 and H_2O into cellulose and other plant products. Certain carbohydrates (sugar and starch) are a dietary staple in most parts of the world, and the oxidation of carbohydrates is the central energy-yielding pathway in most nonphotosynthetic cells. Carbohydrate polymers (also called glycans) serve as structural and protective elements in the cell walls of bacteria and plants and in the connective tissues of animals. Other carbohydrate polymers lubricate skeletal joints and participate in cell-cell recognition and adhesion. Complex carbohydrate polymers covalently attached to proteins or lipids act as signals that determine the intracellular destination or metabolic fate of these hybrid molecules, called **glycoconjugates**. This chapter introduces the major classes of carbohydrates and glycoconjugates and provides a few examples of their many structural and functional roles.

Carbohydrates are polyhydroxy aldehydes or ketones, or substances that yield such compounds on hydrolysis. Many, but not all, carbohydrates have the empirical formula $(CH_2O)_n$; some also contain nitrogen, phosphorus, or sulfur. There are three major size classes of carbohydrates: monosaccharides, oligosaccharides, and polysaccharides (the word "saccharide" is derived from the Greek *sakcharon*, meaning "sugar"). **Monosaccharides**, or simple sugars, consist of a single polyhydroxy aldehyde or ketone unit. The most abundant monosaccharide in nature is the six-carbon sugar D-glucose, sometimes referred to as dextrose. Monosaccharides of four or more carbons tend to have cyclic structures.

Oligosaccharides consist of short chains of monosaccharide units, or residues, joined by characteristic linkages called glycosidic bonds. The most abundant are the **disaccharides**, with two monosaccharide units. Sucrose (cane sugar), for example, consists of the six-carbon sugars D-glucose and D-fructose. All common monosaccharides and disaccharides have names ending with the suffix "-ose." In cells, most oligosaccharides consisting of three or more units do not occur as free entities but are joined to nonsugar molecules (lipids or proteins) in glycoconjugates.

The **polysaccharides** are sugar polymers containing more than 20 or so monosaccharide units; some have hundreds or thousands of units. Some polysaccharides, such as cellulose, are linear chains; others, such as glycogen, are branched. Both cellulose and glycogen consist of recurring units of D-glucose, but they differ in the type of glycosidic linkage and consequently have strikingly different properties and biological roles.

7.1 Monosaccharides and Disaccharides

The simplest of the carbohydrates, the monosaccharides, are either aldehydes or ketones with two or more hydroxyl groups; the six-carbon monosaccharides

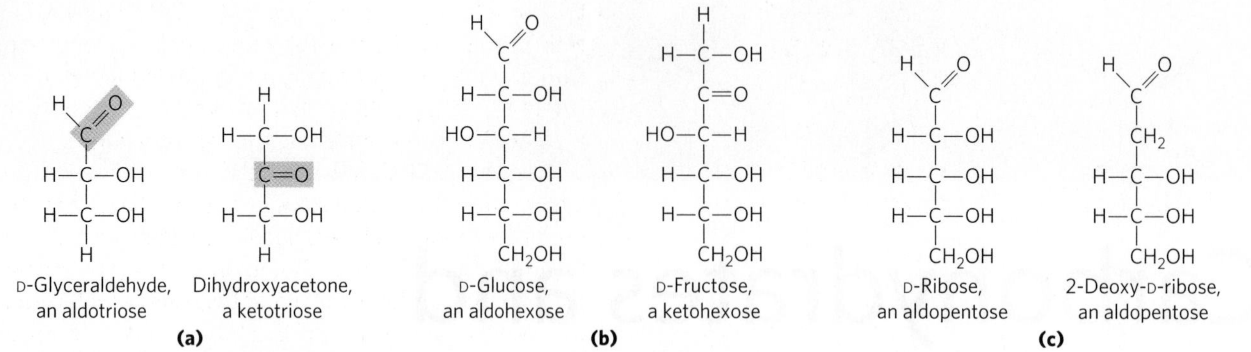

FIGURE 7-1 Representative monosaccharides. (a) Two trioses, an aldose and a ketose. The carbonyl group in each is shaded. **(b)** Two common hexoses. **(c)** The pentose components of nucleic acids. D-Ribose is a component of ribonucleic acid (RNA), and 2-deoxy-D-ribose is a component of deoxyribonucleic acid (DNA).

glucose and fructose have five hydroxyl groups. Many of the carbon atoms to which the hydroxyl groups are attached are chiral centers, which give rise to the many sugar stereoisomers found in nature. Stereoisomerism in sugars is biologically significant because the enzymes that act on sugars are strictly stereospecific, typically preferring one stereoisomer to another by three or more orders of magnitude, as reflected in K_m values or binding constants. It is as difficult to fit the wrong sugar stereoisomer into an enzyme's binding site as it is to put your left glove on your right hand.

We begin by describing the families of monosaccharides with backbones of three to seven carbons—their structure, their stereoisomeric forms, and the means of representing their three-dimensional structures on paper. We then discuss several chemical reactions of the carbonyl groups of monosaccharides. One such reaction, the addition of a hydroxyl group from within the same molecule, generates cyclic forms with four or more backbone carbons (the forms that predominate in aqueous solution). This ring closure creates a new chiral center, adding further stereochemical complexity to this class of compounds. The nomenclature for unambiguously specifying the configuration about each carbon atom in a cyclic form and the means of representing these structures on paper are described in some detail; this information will be useful as we discuss the metabolism of monosaccharides in Part II. We also introduce here some important monosaccharide derivatives encountered in later chapters.

The Two Families of Monosaccharides Are Aldoses and Ketoses

Monosaccharides are colorless, crystalline solids that are freely soluble in water but insoluble in nonpolar solvents. Most have a sweet taste (see Box 7-2, p. 252). The backbones of common monosaccharides are unbranched carbon chains in which all the carbon atoms are linked by single bonds. In this open-chain

form, one of the carbon atoms is double-bonded to an oxygen atom to form a carbonyl group; each of the other carbon atoms has a hydroxyl group. If the carbonyl group is at an end of the carbon chain (that is, in an aldehyde group), the monosaccharide is an **aldose**; if the carbonyl group is at any other position (in a ketone group), the monosaccharide is a **ketose**. The simplest monosaccharides are the two three-carbon trioses: glyceraldehyde, an aldotriose, and dihydroxyacetone, a ketotriose **(Fig. 7-1a)**.

Monosaccharides with four, five, six, and seven carbon atoms in their backbones are called, respectively, tetroses, pentoses, hexoses, and heptoses. There are aldoses and ketoses of each of these chain lengths: aldotetroses and ketotetroses, aldopentoses and ketopentoses, and so on. The hexoses, which include the aldohexose D-glucose and the ketohexose D-fructose (Fig. 7-1b), are the most common monosaccharides in nature—the products of photosynthesis and key intermediates in the central energy-yielding reaction sequence in most organisms. The aldopentoses D-ribose and 2-deoxy-D-ribose (Fig. 7-1c) are components of nucleotides and nucleic acids (Chapter 8).

Monosaccharides Have Asymmetric Centers

All the monosaccharides except dihydroxyacetone contain one or more asymmetric (chiral) carbon atoms and thus occur in optically active isomeric forms (pp. 17–18). The simplest aldose, glyceraldehyde, contains one chiral center (the middle carbon atom) and therefore has two different optical isomers, or **enantiomers (Fig. 7-2)**.

>> **Key Convention:** One of the two enantiomers of glyceraldehyde is, by convention, designated the D isomer; the other is the L isomer. As for other biomolecules with chiral centers, the absolute configurations of sugars are known from x-ray crystallography. To represent three-dimensional sugar structures on paper, we often use **Fischer projection formulas** (Fig. 7-2). In these

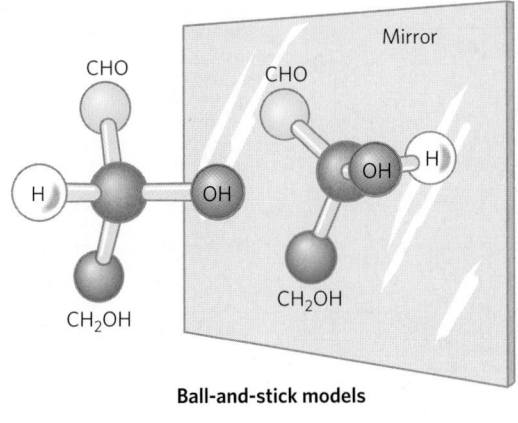

Ball-and-stick models

CHO	CHO
H—C—OH	HO—C—H
CH$_2$OH	CH$_2$OH
D-Glyceraldehyde	L-Glyceraldehyde

Fischer projection formulas

CHO	CHO
H—C—OH	HO—C—H
CH$_2$OH	CH$_2$OH
D-Glyceraldehyde	L-Glyceraldehyde

Perspective formulas

FIGURE 7-2 Three ways to represent the two enantiomers of glyceraldehyde. The enantiomers are mirror images of each other. Ball-and-stick models show the actual configuration of molecules. Recall (see Fig. 1-19) that in perspective formulas, the wide end of a solid wedge projects out of the plane of the paper, toward the reader; a dashed wedge extends behind.

projections, horizontal bonds project out of the plane of the paper, toward the reader; vertical bonds project behind the plane of the paper, away from the reader. ◀◀

In general, a molecule with n chiral centers can have 2^n stereoisomers. Glyceraldehyde has $2^1 = 2$; the aldohexoses, with four chiral centers, have $2^4 = 16$. The stereoisomers of monosaccharides of each carbon-chain length can be divided into two groups that differ in the configuration about the chiral center *most distant* from the carbonyl carbon. Those in which the configuration at this reference carbon is the same as that of D-glyceraldehyde are designated D isomers, and those with the same configuration as L-glyceraldehyde are L isomers. In other words, when the hydroxyl group on the reference carbon is on the right (*dextro*) in a projection formula that has the carbonyl carbon at the top, the sugar is the D isomer; when on the left (*levo*), it is the L isomer. Of the 16 possible aldohexoses, eight are D forms and eight are L. Most of the hexoses of living organisms are D isomers. Why D isomers? An interesting and unanswered question. Recall that all of the amino acids found in proteins are exclusively one of

two possible stereoisomers, L (p. 78). The basis for this initial preference for one isomer during evolution is unknown; however, once one isomer had been selected, it was likely that evolving enzymes would retain their preference for that stereoisomer.

Figure 7-3 shows the structures of the D stereoisomers of all the aldoses and ketoses having three to six carbon atoms. The carbons of a sugar are numbered beginning at the end of the chain nearest the carbonyl group. Each of the eight D-aldohexoses, which differ in the stereochemistry at C-2, C-3, or C-4, has its own name: D-glucose, D-galactose, D-mannose, and so forth (Fig. 7-3a). The four- and five-carbon ketoses are designated by inserting "ul" into the name of a corresponding aldose; for example, D-ribulose is the ketopentose corresponding to the aldopentose D-ribose. (The importance of ribulose will become clear when we discuss the fixation of atmospheric CO_2 by green plants, in Chapter 20.) The ketohexoses are named otherwise: for example, fructose (from the Latin *fructus*, "fruit"; fruits are one source of this sugar) and sorbose (from *Sorbus*, the genus of mountain ash, which has berries rich in the related sugar alcohol sorbitol). Two sugars that differ only in the configuration around one carbon atom are called **epimers;** D-glucose and D-mannose, which differ only in the stereochemistry at C-2, are epimers, as are D-glucose and D-galactose (which differ at C-4) **(Fig. 7-4).**

Some sugars occur naturally in their L form; examples are L-arabinose and the L isomers of some sugar derivatives that are common components of glycoconjugates (Section 7.3).

H   O
C
H—C—OH
HO—C—H
HO—C—H
CH$_2$OH
L-Arabinose

The Common Monosaccharides Have Cyclic Structures

For simplicity, we have thus far represented the structures of aldoses and ketoses as straight-chain molecules (Figs 7-3, 7-4). In fact, in aqueous solution, aldotetroses and all monosaccharides with five or more carbon atoms in the backbone occur predominantly as cyclic (ring) structures in which the carbonyl group has formed a covalent bond with the oxygen of a hydroxyl group along the chain. The formation of these ring structures is the result of a general reaction between alcohols and aldehydes or ketones to form derivatives called **hemiacetals** or **hemiketals.** Two molecules of an alcohol can add to a carbonyl carbon; the product of the first addition is a hemiacetal (for addition to an aldose) or a hemiketal (for addition to a ketose). If the —OH and carbonyl groups are on the same molecule, a five- or six-membered ring results. Addition of the second molecule of alcohol

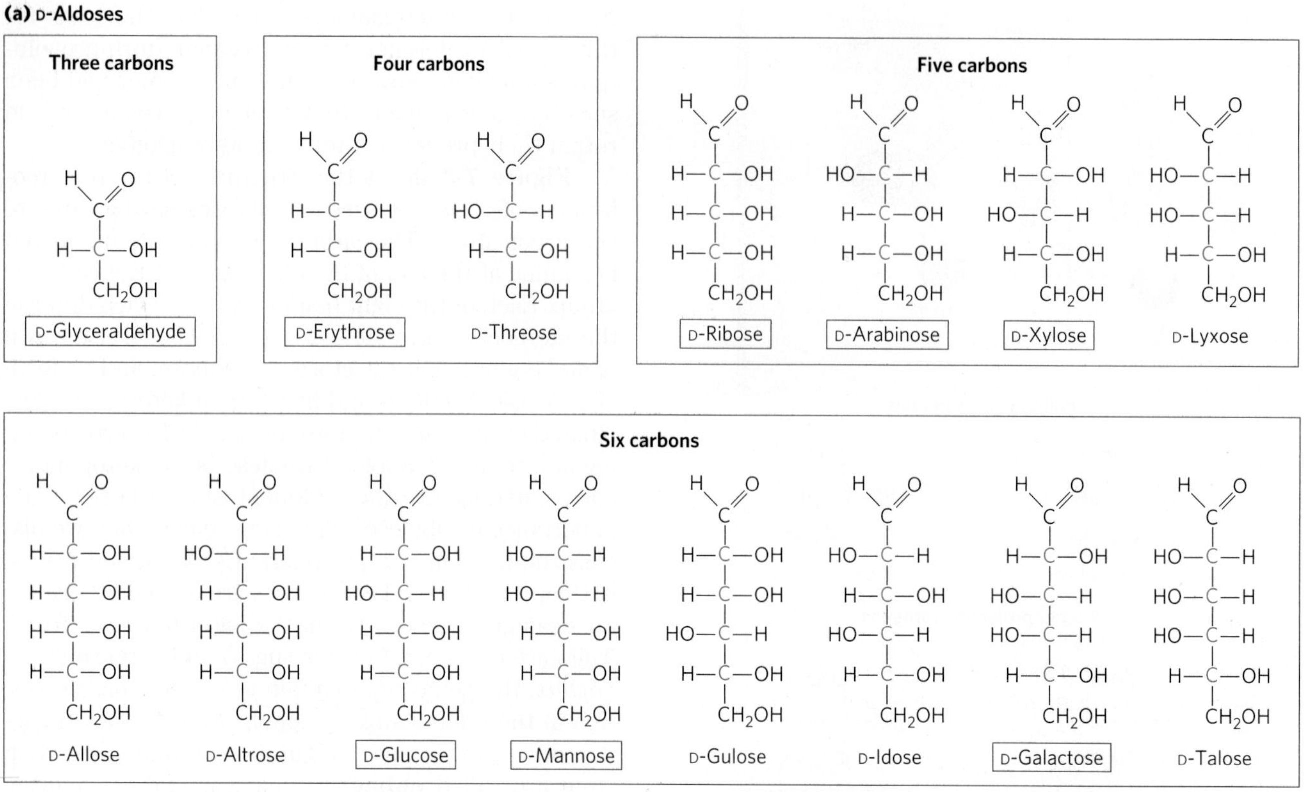

(a) D-Aldoses

Three carbons

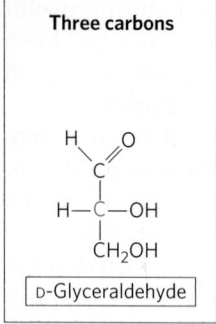

D-Glyceraldehyde

Four carbons

D-Erythrose D-Threose

Five carbons

D-Ribose D-Arabinose D-Xylose D-Lyxose

Six carbons

D-Allose D-Altrose D-Glucose D-Mannose D-Gulose D-Idose D-Galactose D-Talose

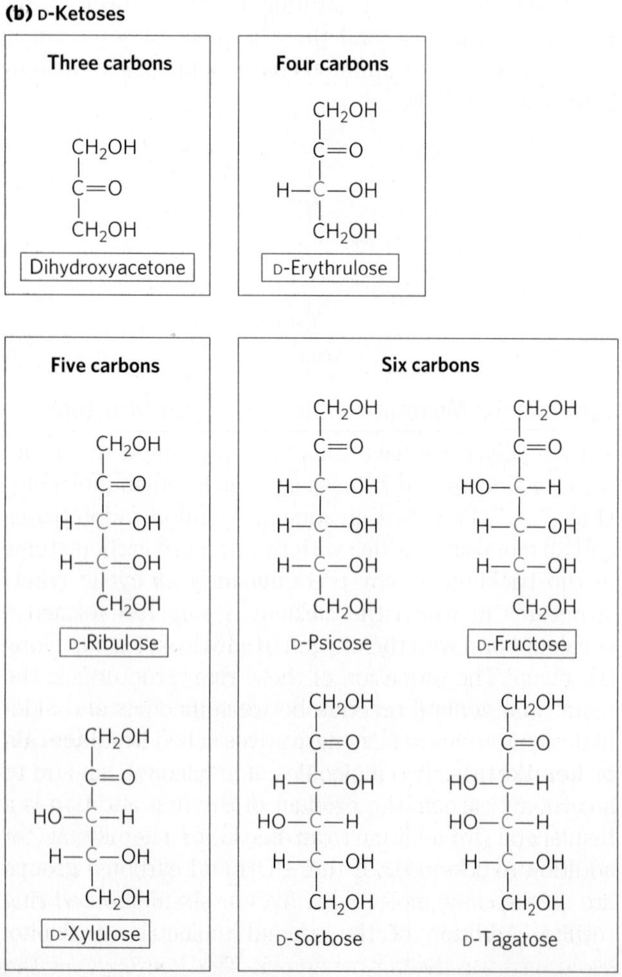

(b) D-Ketoses

Three carbons

Dihydroxyacetone

Four carbons

D-Erythrulose

Five carbons

D-Ribulose

D-Xylulose

Six carbons

D-Psicose D-Fructose

D-Sorbose D-Tagatose

FIGURE 7-3 Aldoses and ketoses. The series of **(a)** D-aldoses and **(b)** D-ketoses having from three to six carbon atoms, shown as projection formulas. The carbon atoms in red are chiral centers. In all these D isomers, the chiral carbon *most distant from the carbonyl carbon* has the same configuration as the chiral carbon in D-glyceraldehyde. The sugars named in boxes are the most common in nature; you will encounter these again in this and later chapters.

produces the full acetal or ketal **(Fig. 7-5)**, and the bond formed is a glycosidic linkage. When the two molecules that react are monosaccharides, the acetal or ketal formed is a disaccharide.

The reaction with the first molecule of alcohol creates an additional chiral center (the carbonyl carbon). Because the alcohol can add in either of two ways, attacking either the "front" or the "back" of the carbonyl carbon, the reaction can produce either of two stereoisomeric configurations, denoted α and β. For example, D-glucose exists in solution as an intramolecular hemiacetal in which the free hydroxyl group at C-5 has reacted with the aldehydic C-1, rendering the latter carbon asymmetric

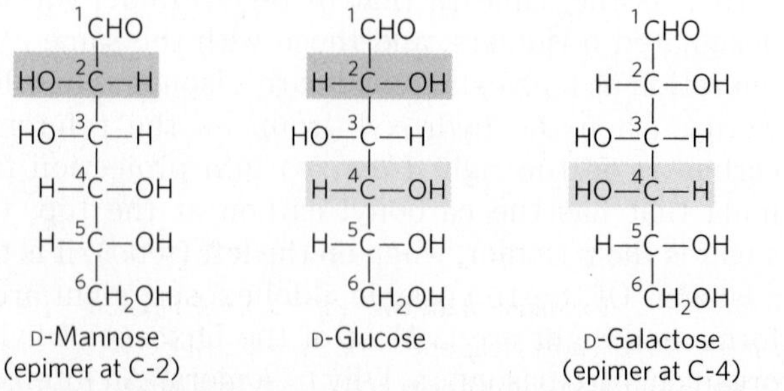

D-Mannose (epimer at C-2) D-Glucose D-Galactose (epimer at C-4)

FIGURE 7-4 Epimers. D-Glucose and two of its epimers are shown as projection formulas. Each epimer differs from D-glucose in the configuration at one chiral center (shaded light red or blue).

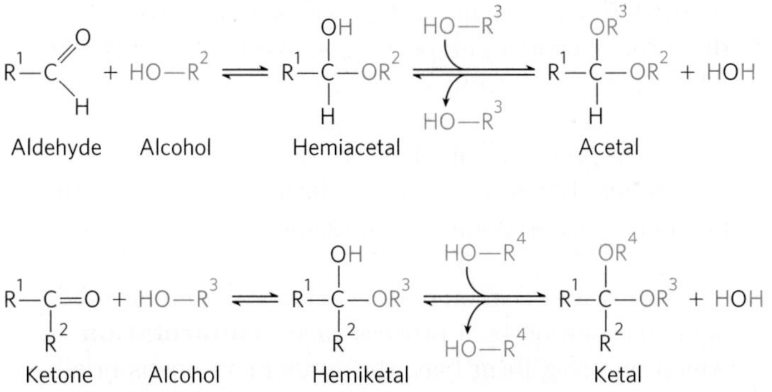

FIGURE 7-5 **Formation of hemiacetals and hemiketals.** An aldehyde or ketone can react with an alcohol in a 1:1 ratio to yield a hemiacetal or hemiketal, respectively, creating a new chiral center at the carbonyl carbon. Substitution of a second alcohol molecule produces an acetal or ketal. When the second alcohol is part of another sugar molecule, the bond produced is a glycosidic bond.

and producing two possible stereoisomers, designated α and β **(Fig. 7-6)**. Isomeric forms of monosaccharides that differ only in their configuration about the hemiacetal or hemiketal carbon atom are called **anomers,** and the carbonyl carbon atom is called the **anomeric carbon**.

Six-membered ring compounds are called **pyranoses** because they resemble the six-membered ring compound

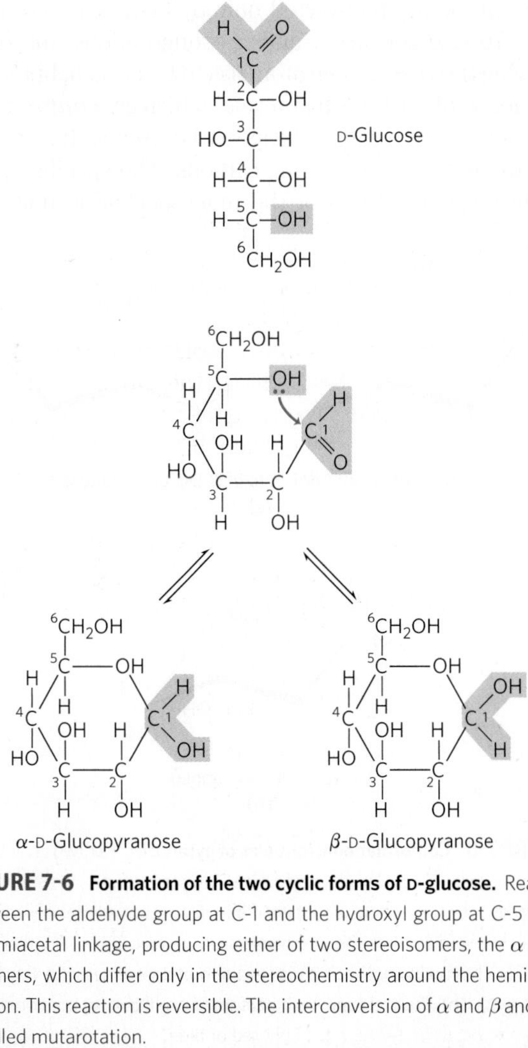

FIGURE 7-6 **Formation of the two cyclic forms of D-glucose.** Reaction between the aldehyde group at C-1 and the hydroxyl group at C-5 forms a hemiacetal linkage, producing either of two stereoisomers, the α and β anomers, which differ only in the stereochemistry around the hemiacetal carbon. This reaction is reversible. The interconversion of α and β anomers is called mutarotation.

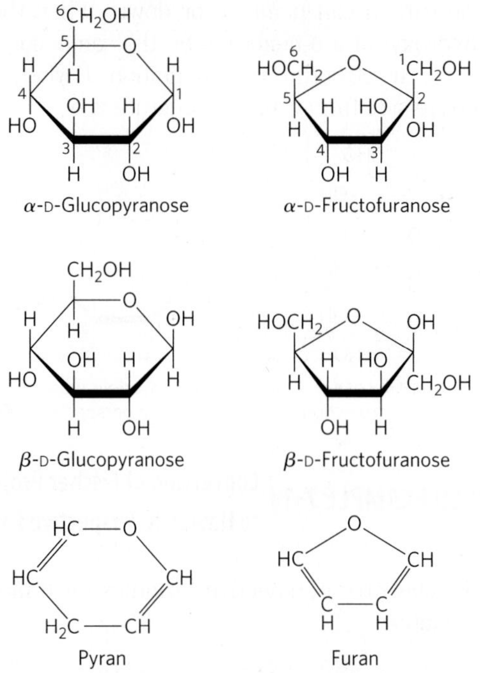

FIGURE 7-7 **Pyranoses and furanoses.** The pyranose forms of D-glucose and the furanose forms of D-fructose are shown here as Haworth perspective formulas. The edges of the ring nearest the reader are represented by bold lines. Hydroxyl groups below the plane of the ring in these Haworth perspectives would appear at the right side of a Fischer projection (compare with Fig. 7-6). Pyran and furan are shown for comparison.

pyran **(Fig. 7-7)**. The systematic names for the two ring forms of D-glucose are therefore α-D-glucopyranose and β-D-glucopyranose. Ketohexoses (such as fructose) also occur as cyclic compounds with α and β anomeric forms. In these compounds, the hydroxyl group at C-5 (or C-6) reacts with the keto group at C-2 to form a **furanose** (or pyranose) ring containing a hemiketal linkage (Fig. 7-5). D-Fructose readily forms the furanose ring (Fig. 7-7); the more common anomer of this sugar in combined forms or in derivatives is β-D-fructofuranose.

Cyclic sugar structures are more accurately represented in **Haworth perspective formulas** than in the Fischer projections commonly used for linear sugar structures. In Haworth perspectives, the six-membered ring is tilted to make its plane almost perpendicular to that of the paper, with the bonds closest to the reader drawn thicker than those farther away, as in Figure 7-7.

>> **Key Convention:** To convert the Fischer projection formula of any linear D-hexose to a Haworth perspective formula showing the molecule's cyclic structure, draw the six-membered ring (five carbons, and one oxygen at the upper right), number the carbons in a clockwise direction beginning with the anomeric carbon, then place the hydroxyl groups. If a hydroxyl group is to the right in the Fischer projection, it is placed pointing down (i.e., below the plane of the ring) in the Haworth perspective; if it is to the left in the Fischer projection, it is placed pointing up (i.e., above the plane) in the Haworth perspective. The terminal —CH₂OH group projects upward for the D enantiomer, downward for the L enantiomer. The hydroxyl on the

anomeric carbon can point up or down. When the anomeric hydroxyl of a D-hexose is on the same side of the ring as C-6, the structure is by definition β; when it is on the opposite side from C-6, the structure is α.

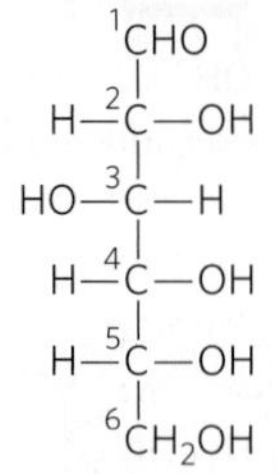

D-Glucose
Fischer projection

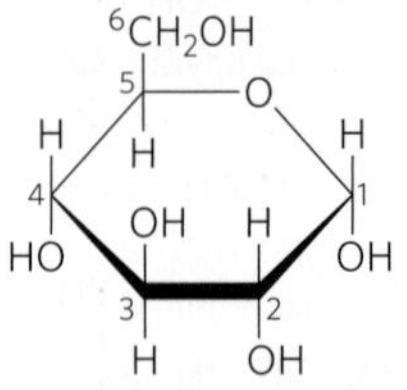

α-D-Glucopyranose
Haworth perspective ≪

WORKED EXAMPLE 7-1 **Conversion of Fischer Projection to Haworth Perspective Formulas**

Draw the Haworth perspective formulas for D-mannose and D-galactose.

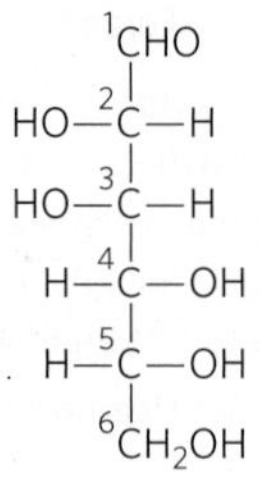

D-Mannose

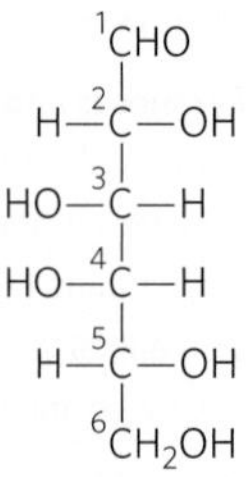

D-Galactose

Solution: Pyranoses are six-membered rings, so start with six-membered Haworth structures with the oxygen atom at the top right. Number the carbon atoms clockwise, starting with the aldose carbon. For mannose, place the hydroxyls on C-2, C-3, and C-4 above, above, and below the ring, respectively (because in the Fischer projection they are on the left, left, and right sides of the mannose backbone). For D-galactose, the hydroxyls are oriented below, above, and above the ring for C-2, C-3, and C-4, respectively. The hydroxyl at C-1 can point either up or down; there are two possible configurations, α and β, at this carbon.

WORKED EXAMPLE 7-2 **Drawing Haworth Perspective Formulas of Sugar Isomers**

Draw the Haworth perspective formulas for α-D-mannose and β-L-galactose.

Solution: The Haworth perspective formula of D-mannose from Worked Example 7-1 can have the hydroxyl group at C-1 pointing either up or down. According to the Key Convention, for the α form, the C-1 hydroxyl is pointing down when C-6 is up, as it is in D-mannose.

For β-L-galactose, use the Fischer representation of D-galactose (see Worked Example 7-1) to draw the correct Fischer representation of L-galactose, which is its mirror image: the hydroxyls at C-2, C-3, C-4, and C-5 are

on the left, right, right, and left sides, respectively. Now draw the Haworth perspective, a six-membered ring in which the —OH groups on C-2, C-3, and C-4 are oriented up, down, and down, respectively, because in the Fischer representation they are on the left, right, and right sides. Because it is the β form, the —OH on the anomeric carbon points down (same side as C-5).

The α and β anomers of D-glucose interconvert in aqueous solution by a process called **mutarotation**, in which one ring form (say, the α anomer) opens briefly into the linear form, then closes again to produce the β anomer (Fig. 7-6). Thus, a solution of β-D-glucose and a solution of α-D-glucose eventually form identical equilibrium mixtures having identical optical properties. This mixture consists of about one-third α-D-glucose, two-thirds β-D-glucose, and very small amounts of the linear form and the five-membered ring (glucofuranose) form.

Haworth perspective formulas like those in Figure 7-7 are commonly used to show the stereochemistry of ring forms of monosaccharides. However, the six-membered pyranose ring is not planar, as Haworth perspectives suggest, but tends to assume either of two "chair" conformations **(Fig. 7-8)**. Recall from Chapter 1 (pp. 16–19) that two *conformations* of a molecule are interconvertible without the breakage of covalent bonds, whereas two *configurations* can be interconverted only by breaking a covalent bond. To interconvert α and β configurations, the bond involving the ring oxygen atom has to be broken, but interconversion of the two chair forms (which are *conformers*) does not require bond breakage and does not change configurations at any of the ring carbons. The specific three-dimensional structures of the monosaccharide units are

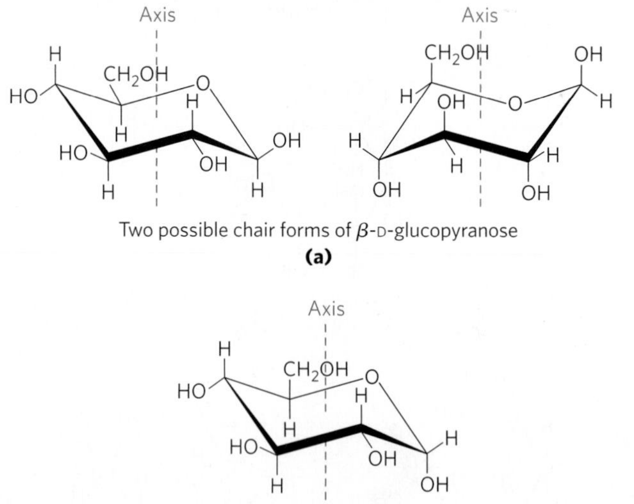

Two possible chair forms of β-D-glucopyranose
(a)

α-D-Glucopyranose
(b)

FIGURE 7-8 Conformational formulas of pyranoses. (a) Two chair forms of the pyranose ring of β-D-glucopyranose. Two *conformers* such as these are not readily interconvertible; an input of about 46 kJ of energy per mole of sugar is required to force the interconversion of chair forms. Another conformation, the "boat" (not shown), is seen only in derivatives with very bulky substituents. **(b)** The preferred chair conformation of α-D-glucopyranose.

important in determining the biological properties and functions of some polysaccharides, as we shall see.

Organisms Contain a Variety of Hexose Derivatives

In addition to simple hexoses such as glucose, galactose, and mannose, there are many sugar derivatives in which a hydroxyl group in the parent compound is replaced with another substituent, or a carbon atom is oxidized to a carboxyl group **(Fig. 7-9)**. In glucosamine, galactosamine, and mannosamine, the hydroxyl at C-2 of the parent compound is replaced with an amino group. The amino group is commonly condensed with acetic acid, as in *N*-acetylglucosamine. This glucosamine derivative is part of many structural polymers, including those of the bacterial cell wall. Substitution of a hydrogen for the hydroxyl group at C-6 of L-galactose or L-mannose produces L-fucose or L-rhamnose, respectively. L-Fucose is found in the complex oligosaccharide components of glycoproteins and glycolipids; L-rhamnose is found in plant polysaccharides.

Oxidation of the carbonyl (aldehyde) carbon of glucose to the carboxyl level produces gluconic acid, used in medicine as an innocuous counterion when administering positively charged drugs (such as quinine) or ions (such as Ca^{2+}). Other aldoses yield other **aldonic acids**. Oxidation of the carbon at the other end of the carbon chain—C-6 of glucose, galactose, or mannose—forms the corresponding **uronic acid**: glucuronic, galacturonic, or mannuronic acid. Both aldonic and uronic acids form stable intramolecular esters called lactones (Fig. 7-9, lower left). The sialic acids are a family of sugars with the same nine-carbon backbone. One of them, *N*-acetylneuraminic acid (often referred to simply as "sialic acid"), is a derivative of *N*-acetylmannosamine that occurs in many glycoproteins and

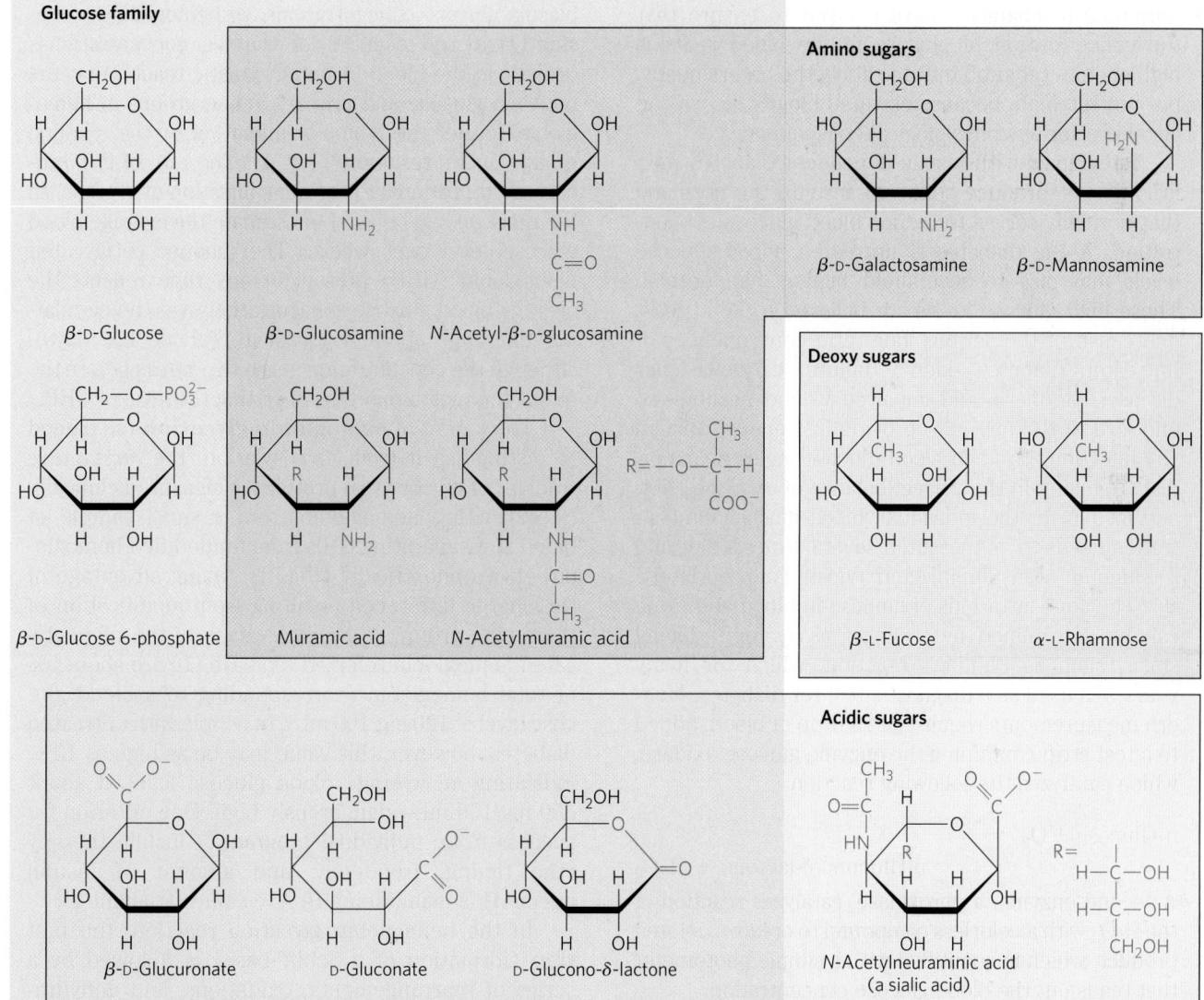

FIGURE 7-9 **Some hexose derivatives important in biology.** In amino sugars, an —NH_2 group replaces one of the —OH groups in the parent hexose. Substitution of —H for —OH produces a deoxy sugar; note that the deoxy sugars shown here occur in nature as the L isomers. The acidic sugars contain a carboxylate group, which confers a negative charge at neutral pH. D-Glucono-δ-lactone results from formation of an ester linkage between the C-1 carboxylate group and the C-5 (also known as the δ carbon) hydroxyl group of D-gluconate.

glycolipids on animal cell surfaces, providing sites of recognition by other cells or extracellular carbohydrate-binding proteins. The carboxylic acid groups of the acidic sugar derivatives are ionized at pH 7, and the compounds are therefore correctly named as the carboxylates—glucuronate, galacturonate, and so forth.

In the synthesis and metabolism of carbohydrates, the intermediates are very often not the sugars themselves but their phosphorylated derivatives. Condensation of phosphoric acid with one of the hydroxyl groups of a sugar forms a phosphate ester, as in glucose 6-phosphate (Fig. 7-9), the first metabolite in the pathway by which most organisms oxidize glucose for energy. Sugar phosphates are relatively stable at neutral pH and bear a negative charge. One effect of sugar phosphorylation within cells is to trap the sugar inside the cell; most cells do not have plasma membrane transporters for phosphorylated sugars. Phosphorylation also activates sugars for

---

BOX 7-1   🔬 MEDICINE   Blood Glucose Measurements in the Diagnosis and Treatment of Diabetes

Glucose is the principal fuel for the brain. When the amount of glucose reaching the brain is too low, the consequences can be dire: lethargy, coma, permanent brain damage, and death (see Fig. 23-25). Complex hormonal mechanisms have evolved to ensure that the concentration of glucose in the blood remains high enough (about 5 mM) to satisfy the brain's needs, but not too high, because elevated blood glucose can also have serious physiological consequences.

Individuals with insulin-dependent diabetes mellitus do not produce sufficient insulin, the hormone that normally serves to reduce blood glucose concentration. If the diabetes is untreated, blood glucose levels may rise to severalfold higher than normal. These high glucose levels are believed to be at least one cause of the serious long-term consequences of untreated diabetes—kidney failure, cardiovascular disease, blindness, and impaired wound healing—so one goal of therapy is to provide just enough insulin (by injection) to keep blood glucose levels near normal. To maintain the correct balance of exercise, diet, and insulin for the individual, blood glucose concentration needs to be measured several times a day, and the amount of insulin injected adjusted appropriately.

The concentrations of glucose in blood and urine can be determined by a simple assay for reducing sugar, such as Fehling's reaction, which for many years was used as a diagnostic test for diabetes. Modern measurements require just a drop of blood, added to a test strip containing the enzyme glucose oxidase, which catalyzes the following reaction:

$$\text{D-Glucose} + O_2 \xrightarrow{\text{glucose oxidase}} \text{D-Glucono-}\delta\text{-lactone} + H_2O_2$$

A second enzyme, a peroxidase, catalyzes reaction of the H_2O_2 with a colorless compound to create a colored product, which is quantified with a simple photometer that reads out the blood glucose concentration.

Because blood glucose levels change with the timing of meals and exercise, single-time measurements do not reflect the *average* blood glucose over hours and days, so dangerous increases may go undetected.

The average glucose concentration can be assessed by looking at its effect on hemoglobin, the oxygen-carrying protein in erythrocytes (p. 163). Transporters in the erythrocyte membrane equilibrate intracellular and plasma glucose concentrations, so hemoglobin is constantly exposed to glucose at whatever concentration is present in the blood. A nonenzymatic reaction occurs between glucose and primary amino groups in hemoglobin (either the amino-terminal Val or the ε-amino groups of Lys residues) (Fig. 1). The rate of this process is proportional to the concentration of glucose, so the reaction can be used to estimate the average blood glucose level over weeks. The amount of glycated hemoglobin (GHB) present at any time reflects the average blood glucose concentration over the circulating "lifetime" of the erythrocyte (about 120 days), although the concentration in the two weeks before the test is the most important in setting the level of GHB.

The extent of **hemoglobin glycation** (so named to distinguish it from glycosylation, the *enzymatic* transfer of glucose to a protein) is measured clinically by extracting hemoglobin from a small sample of blood and separating GHB from unmodified hemoglobin electrophoretically (Fig. 2), taking advantage of the charge difference resulting from modification of the amino group(s). Normal values of the monoglycated hemoglobin referred to as HbA1c are about 5% of total hemoglobin (corresponding to a blood glucose level of 120 mg/100 mL). In people with untreated diabetes, however, this value may be as high as 13%, indicating an average blood glucose level of about 300 mg/100 mL—dangerously high. One criterion for success in an individual program of insulin therapy (the timing, frequency, and amount of insulin injected) is maintaining HbA1c values at about 7%.

In the hemoglobin glycation reaction, the first step (formation of a Schiff base) is followed by a series of rearrangements, oxidations, and dehydrations of the carbohydrate moiety to produce a heterogeneous mixture of AGEs, *advanced glycation end products*. These products can leave the erythrocyte and form covalent cross-links between proteins,

subsequent chemical transformation. Several important phosphorylated derivatives of sugars are components of nucleotides (discussed in the next chapter).

Monosaccharides Are Reducing Agents

Monosaccharides can be oxidized by relatively mild oxidizing agents such as cupric (Cu^{2+}) ion. The carbonyl carbon is oxidized to a carboxyl group. Glucose and other sugars capable of reducing cupric

ion are called **reducing sugars**; the sugars are oxidized to a complex mixture of carboxylic acids. This is the basis of Fehling's reaction, a semiquantitative test for the presence of reducing sugar that for many years was used to detect and measure elevated glucose levels in people with diabetes mellitus. Today, more sensitive methods that involve an immobilized enzyme on a test strip are used; they require only a single drop of blood. ■

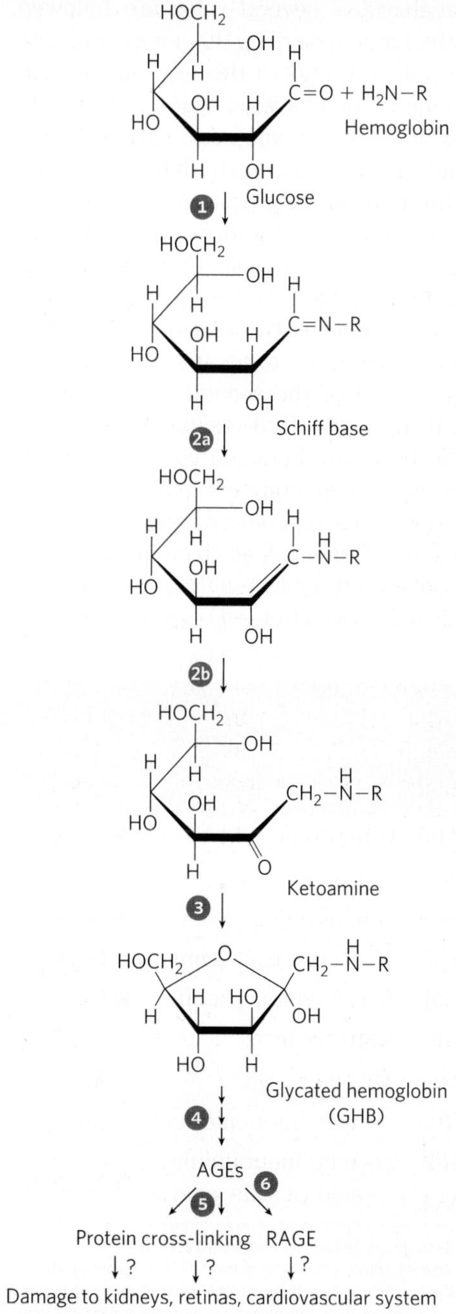

FIGURE 1 The nonenzymatic reaction of glucose with a primary amino group in hemoglobin begins with ❶ formation of a Schiff base, which ❷ undergoes a rearrangement to generate a stable product; ❸ this ketoamine can further cyclize to yield GHB. ❹ Subsequent reactions generate advanced glycation end products (AGEs), such as ε-N-carboxymethyllysine and methylglyoxal, compounds that ❺ can damage other proteins by cross-linking them, causing pathological changes. ❻ The AGE receptor (RAGE), activated by AGE, stimulates downstream events, including inflammation.

interfering with normal protein function (Fig. 1). The accumulation of relatively high concentrations of AGEs in people with diabetes may, by cross-linking critical proteins, cause the damage to the kidneys, retinas, and cardiovascular system that characterizes the disease. This pathogenic process is a potential target for drug action. AGEs also act through transmembrane receptors for AGE, or RAGEs, which trigger the inflammatory response associated with diabetes.

Analyte ID	Percent	Time (min)	Area
Injection	0.0	0.11	17,682
A1a	0.4	0.30	8,051
A1b	1.0	0.45	19,267
A1c	5.9	1.10	110,946
A0	92.6	1.56	1,727,669

Total area 1,883,615

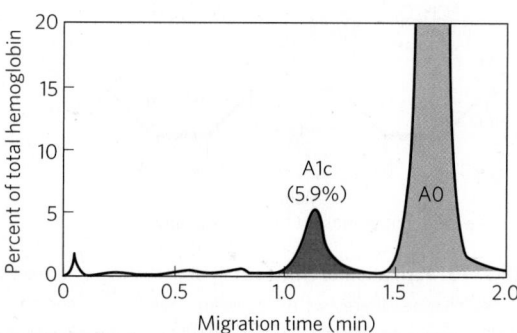

FIGURE 2 Pattern of hemoglobin (detected by its absorption at 415 nm) after electrophoretic separation of nonglycated (A0) and monoglycated (A1c) forms in a thin glass capillary. Integration of the area under the peaks allows calculation of the amount of GHB (HbA1c) as a percentage of total hemoglobin. Shown here is the profile of an individual with a normal level of HbA1c (5.9%).

Disaccharides Contain a Glycosidic Bond

Disaccharides (such as maltose, lactose, and sucrose) consist of two monosaccharides joined covalently by an **O-glycosidic bond**, which is formed when a hydroxyl group of one sugar molecule, typically in its cyclic form, reacts with the anomeric carbon of the other **(Fig. 7-10)**. This reaction represents the formation of an acetal from a hemiacetal (such as glucopyranose) and an alcohol (a hydroxyl group of the second sugar molecule) (Fig. 7-5), and the resulting compound is called a glycoside. Glycosidic bonds are readily hydrolyzed by acid but resist cleavage by base. Thus disaccharides can be hydrolyzed to yield their free monosaccharide components by boiling with dilute acid. **N-glycosyl bonds** join the anomeric carbon of a sugar to a nitrogen atom in glycoproteins (see Fig. 7-30) and nucleotides (see Fig. 8-1).

The oxidation of a sugar by cupric ion (the reaction that defines a reducing sugar) occurs only with the linear form, which exists in equilibrium with the cyclic form(s). When the anomeric carbon is involved in a glycosidic bond (that is, when the compound is a full acetal or ketal; see Fig. 7-5), the easy interconversion of linear and cyclic forms shown in Figure 7-6 is prevented. Because the carbonyl carbon can be oxidized only when the sugar is in its linear form, formation of a glycosidic bond renders a sugar nonreducing. In describing disaccharides or polysaccharides, the end of a chain with a free anomeric carbon (one

not involved in a glycosidic bond) is commonly called the **reducing end**.

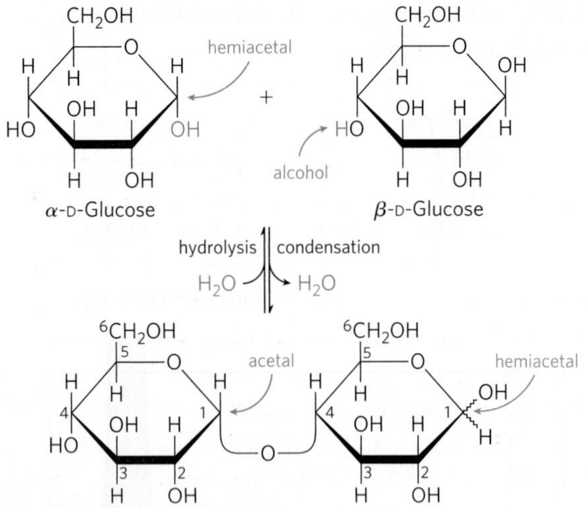

The disaccharide maltose (Fig. 7-10) contains two D-glucose residues joined by a glycosidic linkage between C-1 (the anomeric carbon) of one glucose residue and C-4 of the other. Because the disaccharide retains a free anomeric carbon (C-1 of the glucose residue on the right in Fig. 7-10), maltose is a reducing sugar. The configuration of the anomeric carbon atom in the glycosidic linkage is α. The glucose residue with the free anomeric carbon is capable of existing in α- and β-pyranose forms.

>> **Key Convention:** To name reducing disaccharides such as maltose unambiguously, and especially to name more complex oligosaccharides, several rules are followed. By convention, the name describes the compound written with its nonreducing end to the left, and we can "build up" the name in the following order. (1) Give the configuration (α or β) at the anomeric carbon joining the first monosaccharide unit (on the left) to the second. (2) Name the nonreducing residue; to distinguish five- and six-membered ring structures, insert "furano" or "pyrano" into the name. (3) Indicate in parentheses the two carbon atoms joined by the glycosidic bond, with an arrow connecting the two numbers; for example, (1→4) shows that C-1 of the first-named sugar residue is joined to C-4 of the second. (4) Name the second residue. If there is a third residue, describe the second glycosidic bond by the same conventions. (To shorten the description of complex polysaccharides, three-letter abbreviations or colored symbols for the monosaccharides are often used, as given in Table 7-1.) Following this convention for naming oligosaccharides, maltose is α-D-glucopyranosyl-(1→4)-D-glucopyranose.

α-D-Glucose β-D-Glucose

hemiacetal

+

alcohol

hydrolysis | condensation

H_2O ⇄ H_2O

acetal   hemiacetal

Maltose
α-D-glucopyranosyl-(1→4)-D-glucopyranose

FIGURE 7-10 Formation of maltose. A disaccharide is formed from two monosaccharides (here, two molecules of D-glucose) when an —OH (alcohol) of one monosaccharide molecule (right) condenses with the intramolecular hemiacetal of the other (left), with elimination of H_2O and formation of a glycosidic bond. The reversal of this reaction is hydrolysis— attack by H_2O on the glycosidic bond. The maltose molecule, shown here, retains a reducing hemiacetal at the C-1 not involved in the glycosidic bond. Because mutarotation interconverts the α and β forms of the hemiacetal, the bonds at this position are sometimes depicted with wavy lines to indicate that the structure may be either α or β.

TABLE 7-1	Symbols and Abbreviations for Common Monosaccharides and Some of Their Derivatives		
Abequose	Abe	Glucuronic acid	◆ GlcA
Arabinose	Ara	Galactosamine	☐ GalN
Fructose	Fru	Glucosamine	◩ GlcN
Fucose	▲ Fuc	N-Acetylgalactosamine	☐ GalNAc
Galactose	○ Gal	N-Acetylglucosamine	■ GlcNAc
Glucose	● Glc	Iduronic acid	◇ IdoA
Mannose	◉ Man	Muramic acid	Mur
Rhamnose	Rha	N-Acetylmuramic acid	Mur2Ac
Ribose	Rib	N-Acetylneuraminic acid (a sialic acid)	◆ Neu5Ac
Xylose	★ Xyl		

Note: In a commonly used convention, hexoses are represented as circles, N-acetylhexosamines as squares, and hexosamines as squares divided diagonally. All sugars with the "gluco" configuration are blue, those with the "galacto" configuration are yellow, and "manno" sugars are green. Other substituents can be added as needed: sulfate (S), phosphate (P), O-acetyl (OAc), or O-methyl (OMe).

Because most sugars encountered in this book are the D enantiomers and the pyranose form of hexoses predominates, we generally use a shortened version of the formal name of such compounds, giving the configuration of the anomeric carbon and naming the carbons joined by the glycosidic bond. In this abbreviated nomenclature, maltose is Glc(α1→4)Glc. **«**

The disaccharide lactose **(Fig. 7-11)**, which yields D-galactose and D-glucose on hydrolysis, occurs naturally in milk. The anomeric carbon of the glucose residue is available for oxidation, and thus lactose is a reducing disaccharide. Its abbreviated name is Gal (β1→4)Glc. The enzyme lactase—absent in lactose-intolerant individuals—begins the digestive process in the small intestine by splitting the (β1→4) bond of lactose into monosaccharides, which can be absorbed from the small intestine. Lactose, like other disaccharides, is not absorbed from the small intestine, and in lactose-intolerant individuals, the undigested lactose passes into the large intestine. Here, the increased osmolarity due to dissolved lactose opposes the absorption of water from the intestine into the bloodstream,

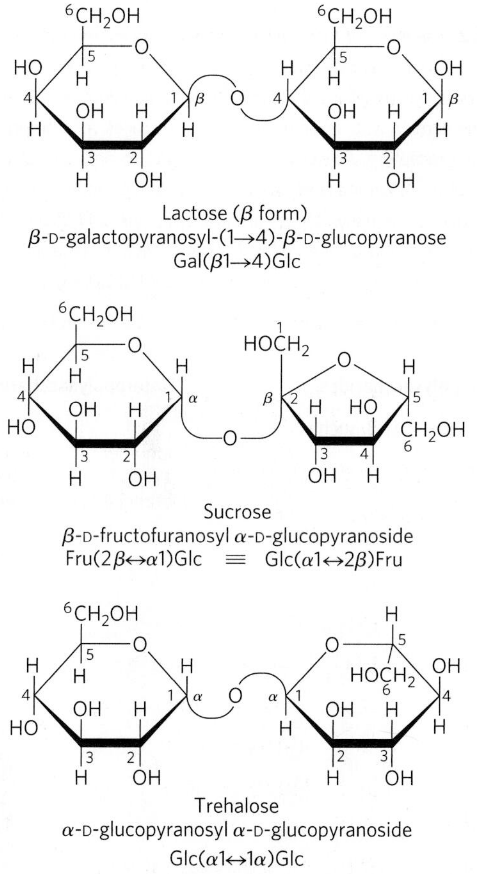

causing watery, loose stools. In addition, fermentation of the lactose by intestinal bacteria produces large volumes of CO_2, which leads to the bloating, cramps, and gas associated with lactose intolerance.

Sucrose is a disaccharide of glucose and fructose. It is formed by plants but not by animals. In contrast to maltose and lactose, sucrose contains no free anomeric carbon atom; the anomeric carbons of both monosaccharide units are involved in the glycosidic bond (Fig. 7-11). Sucrose is therefore a nonreducing sugar, and its stability—its resistance to oxidation—makes it a suitable molecule for the storage and transport of energy in plants. In the abbreviated nomenclature, a double-headed arrow connects the symbols specifying the anomeric carbons and their configurations. Thus the abbreviated name of sucrose is either Glc(α1↔2β)Fru or Fru(β2↔1α)Glc. Sucrose is a major intermediate product of photosynthesis; in many plants it is the principal form in which sugar is transported from the leaves to other parts of the plant body.

Trehalose, Glc(α1↔1α)Glc (Fig. 7-11)—a disaccharide of D-glucose that, like sucrose, is a nonreducing sugar—is a major constituent of the circulating fluid (hemolymph) of insects. It serves as an energy-storage compound.

Lactose gives milk its sweetness, and sucrose, of course, is table sugar. Trehalose is also used commercially as a sweetener. Box 7-2 explains how humans detect sweetness, and how artificial sweeteners such as aspartame act.

SUMMARY 7.1 Monosaccharides and Disaccharides

■ Sugars (also called saccharides) are compounds containing an aldehyde or ketone group and two or more hydroxyl groups.

■ Monosaccharides generally contain several chiral carbons and therefore exist in a variety of stereochemical forms, which may be represented on paper as Fischer projections. Epimers are sugars that differ in configuration at only one carbon atom.

■ Monosaccharides commonly form internal hemiacetals or hemiketals, in which the aldehyde or ketone group joins with a hydroxyl group of the same molecule, creating a cyclic structure; this can be represented as a Haworth perspective formula. The carbon atom originally found in the aldehyde or ketone group (the anomeric carbon) can assume either of two configurations, α and β, which are interconvertible by mutarotation. In the linear form of the monosaccharide, which is in equilibrium with the cyclic forms, the anomeric carbon is easily oxidized, making the compound a reducing sugar.

■ A hydroxyl group of one monosaccharide can add to the anomeric carbon of a second monosaccharide to form an acetal called a glycoside. In this disaccharide, the glycosidic bond protects the anomeric carbon from oxidation, making it a nonreducing sugar.

FIGURE 7-11 Two common disaccharides. Like maltose in Figure 7-10, these are shown as Haworth perspectives. The common name, full systematic name, and abbreviation are given for each disaccharide. Formal nomenclature for sucrose names glucose as the parent glycoside, although it is typically depicted as shown, with glucose on the left. The two abbreviated symbols shown for sucrose are equivalent ($\equiv$).

Lactose (β form)
β-D-galactopyranosyl-(1→4)-β-D-glucopyranose
Gal(β1→4)Glc

Sucrose
β-D-fructofuranosyl α-D-glucopyranoside
Fru(2β↔α1)Glc ≡ Glc(α1↔2β)Fru

Trehalose
α-D-glucopyranosyl α-D-glucopyranoside
Glc(α1↔1α)Glc

BOX 7-2 Sugar Is Sweet, and So Are ... a Few Other Things

Sweetness is one of the five basic flavors that humans can taste (Fig. 1); the others are sour, bitter, salty, and umami. Sweet taste is detected by protein receptors in the plasma membranes of gustatory cells in the taste buds on the surface of the tongue. In humans, two closely related genes (*T1R2* and *T1R3*) encode sweetness receptors (Fig. 2). When a molecule with a compatible structure binds a gustatory receptor's extracellular domain, it triggers a series of events in the cell (including activation of a GTP-binding protein; see Fig. 12-16) that generate an electrical signal to the brain that is interpreted as "sweet." During evolution, there has probably been selection for the ability to taste compounds found in

foods containing important nutrients, such as the carbohydrates that are major fuels for most organisms.

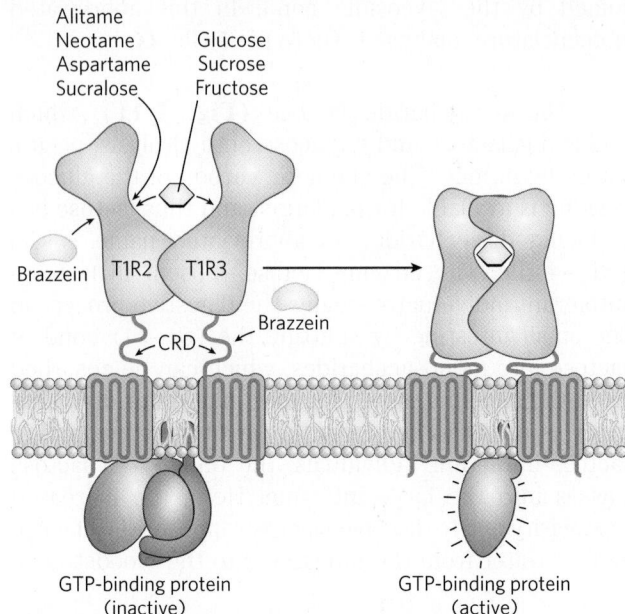

FIGURE 2 The receptor for sweet-tasting substances, showing its regions of interaction (short arrows) with various sweet-tasting compounds. Each receptor has an extracellular domain, a cysteine-rich domain (CRD), and a membrane domain with seven transmembrane helices, a common feature of signaling receptors. Artificial sweeteners bind to only one of the two receptor subunits; natural sugars bind to both. See Chapter 1, Problem 16, for the structures of many of these artificial sweeteners. T1R2 and T1R3 are the proteins encoded by the genes *T1R2* and *T1R3*. [Source: Information from F. M. Assadi-Porter et al., *J. Mol. Biol.* 398:584, 2010, Fig. 1.]

FIGURE 1 A strong stimulus for the sweetness receptors. [Source: David Cook/blueshiftstudios/Alamy.]

■ Oligosaccharides are short polymers of several monosaccharides joined by glycosidic bonds. At one end of the chain, the reducing end, is a monosaccharide unit with its anomeric carbon not involved in a glycosidic bond.

■ The common nomenclature for disaccharides or oligosaccharides specifies the order of monosaccharide units, the configuration at each anomeric carbon, and the carbon atoms involved in the glycosidic linkage(s).

7.2 Polysaccharides

Most carbohydrates found in nature occur as polysaccharides, polymers of medium to high molecular weight (M_r >20,000). Polysaccharides, also called **glycans**, differ from each other in the identity of their recurring monosaccharide units, in the length of their chains, in the types of bonds linking the units, and in the degree of branching. **Homopolysaccharides** contain only a single monomeric species; **heteropolysaccharides** contain two or more different kinds **(Fig. 7-12)**. Some homopolysaccharides serve as storage forms of monosaccharides that are used

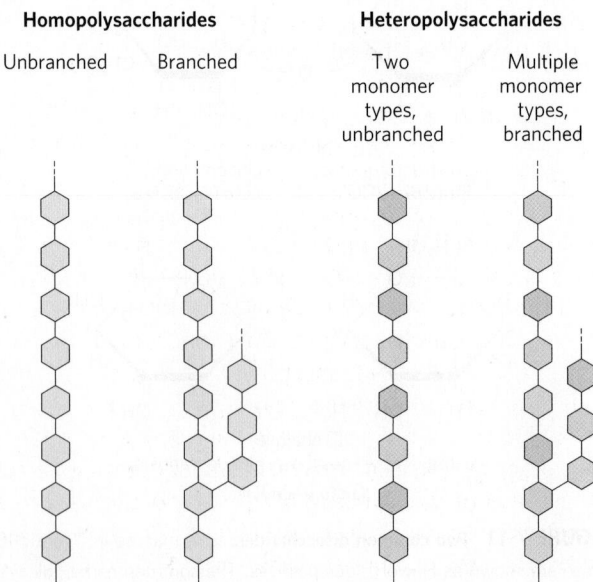

FIGURE 7-12 Homopolysaccharides and heteropolysaccharides. Polysaccharides may be composed of one, two, or several different monosaccharides, in straight or branched chains of varying length.

Most simple sugars, including sucrose, glucose, and fructose, taste sweet, but there are other classes of compounds that also bind the sweet receptors. The amino acids glycine, alanine, and serine are mildly sweet and harmless; nitrobenzene and ethylene glycol have a strong sweet taste, but are toxic. (See Box 18-2 for a remarkable medical mystery involving ethylene glycol poisoning.) Several natural products are extraordinarily sweet. Stevioside, a sugar derivative isolated from the leaves of the stevia plant (*Stevia rebaudiana* Bertoni), is several hundred times sweeter than an equivalent amount of sucrose (table sugar). The small (54 amino acids) protein brazzein, isolated from berries of the Oubli vine (*Pentadiplandra brazzeana* Baillon) in Gabon and Cameroon, is 17,000 times sweeter than sucrose on a molar basis. Presumably, the sweet taste of the berries encourages their consumption by animals that then disperse the seeds so that new plants are established.

There is great interest in the development of artificial sweeteners as weight-reduction aids—compounds that give foods a sweet taste without adding the calories found in sugars. The artificial sweetener aspartame demonstrates the importance of stereochemistry in biology (Fig. 3). According to one simple model of sweetness receptor binding, binding involves three sites on the receptor: AH^+, B^-, and X. Site AH^+ contains a group (an alcohol or amine) that can hydrogen-bond with a group with partial negative charge, such as a carbonyl oxygen, on the sweetener

molecule; the carboxylic acid of aspartame contains such an oxygen. Site B^- contains a group with a partially negative oxygen available to hydrogen-bond with a partially positive atom on the sweetener molecule, such as the amine group of aspartame. Site X is oriented perpendicular to the other two groups and is capable of interacting with a hydrophobic patch on the sweetener molecule, such as the benzene ring of aspartame.

When the steric match is correct, as on the left in Figure 3, the sweet receptor is stimulated and the signal "sweet" is conducted to the brain. When the match is not correct, as on the right, the sweet receptor is not stimulated; in fact, in this case, another receptor (for bitterness) is stimulated by the "wrong" stereoisomer of aspartame. Stereoisomerism really matters!

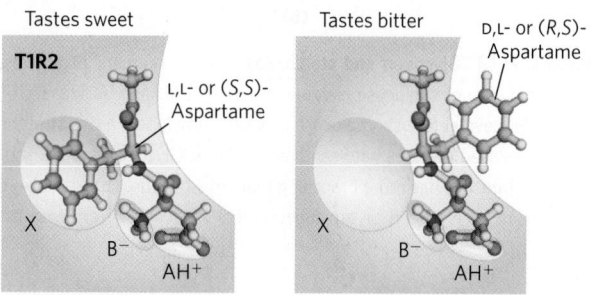

FIGURE 3 Stereochemical basis for the taste of two isomers of aspartame. [Source: Information from http://chemistry.elmhurst.edu/vchembook/549receptor.html, © Charles E. Ophardt, Elmhurst College.]

as fuels; starch and glycogen are homopolysaccharides of this type. Other homopolysaccharides (cellulose and chitin, for example) serve as structural elements in plant cell walls and animal exoskeletons. Heteropolysaccharides provide extracellular support for organisms of all kingdoms. For example, the rigid layer of the bacterial cell envelope (the peptidoglycan) is composed in part of a heteropolysaccharide built from two alternating monosaccharide units (see Fig. 6-28). In animal tissues, the extracellular space is occupied by several types of heteropolysaccharides, which form a matrix that holds individual cells together and provides protection, shape, and support to cells, tissues, and organs.

Unlike proteins, polysaccharides generally do not have defining molecular weights. This difference is a consequence of the mechanisms of assembly of the two types of polymer. As we shall see in Chapter 27, proteins are synthesized on a template (messenger RNA) of defined sequence and length, by enzymes that follow the template exactly. For polysaccharide synthesis there is no template; rather, the program for polysaccharide synthesis is intrinsic to the enzymes that catalyze the polymerization of the

monomeric units, and there is no specific stopping point in the synthetic process; the products thus vary in length.

Some Homopolysaccharides Are Storage Forms of Fuel

The most important storage polysaccharides are starch in plant cells and glycogen in animal cells. Both polysaccharides occur intracellularly as large clusters or granules. Starch and glycogen molecules are heavily hydrated, because they have many exposed hydroxyl groups available to hydrogen-bond with water. Most plant cells have the ability to form starch (see Fig. 20-5), and starch storage is especially abundant in tubers (underground stems), such as potatoes, and in seeds.

Starch contains two types of glucose polymer, amylose and amylopectin **(Fig. 7-13)**. Amylose consists of long, unbranched chains of D-glucose residues connected by $(\alpha 1 \rightarrow 4)$ linkages (as in maltose). Such chains vary in molecular weight from a few thousand to more than a million. Amylopectin also has a high molecular weight (up to 200 million) but unlike amylose is highly branched. The glycosidic linkages joining successive glucose residues

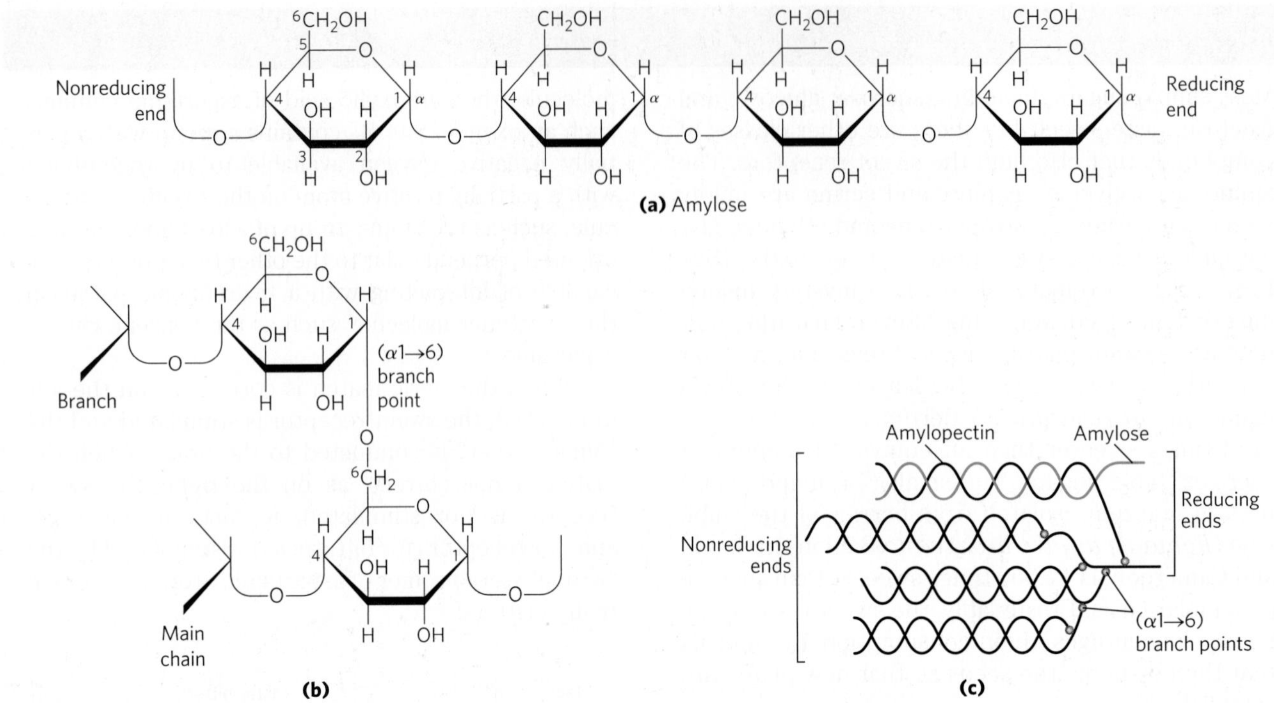

(a) Amylose

(b)

(c)

FIGURE 7-13 Glycogen and starch. (a) A short segment of amylose, a linear polymer of D-glucose residues in ($\alpha1{\rightarrow}4$) linkage. A single chain can contain several thousand glucose residues. Amylopectin has stretches of similarly linked residues between branch points. Glycogen has the same basic structure, but has more branching than amylopectin. **(b)** An ($\alpha1{\rightarrow}6$) branch point of glycogen or amylopectin. **(c)** A cluster of amylose and amylopectin like that believed to occur in starch granules. Strands of amylopectin (black) form double-helical structures with each other or with amylose strands (blue). Amylopectin has frequent ($\alpha1{\rightarrow}6$) branch points (red). Glucose residues at the nonreducing ends of the outer branches are removed enzymatically during the mobilization of starch for energy production. Glycogen has a similar structure but is more highly branched and more compact.

in amylopectin chains are ($\alpha1{\rightarrow}4$); the branch points (occurring every 24 to 30 residues) are ($\alpha1{\rightarrow}6$) linkages.

Glycogen is the main storage polysaccharide of animal cells. Like amylopectin, glycogen is a polymer of ($\alpha1{\rightarrow}4$)-linked glucose subunits, with ($\alpha1{\rightarrow}6$)-linked branches, but glycogen is more extensively branched (on average, a branch every 8 to 12 residues) and more compact than starch. Glycogen is especially abundant in the liver, where it may constitute as much as 7% of the wet weight; it is also present in skeletal muscle. In hepatocytes glycogen is found in large granules (see Fig. 15-26), which are clusters of smaller granules composed of single, highly branched glycogen molecules with an average molecular weight of several million. The large glycogen granules also contain, in tightly bound form, the enzymes responsible for the synthesis and degradation of glycogen (see Fig. 15-42).

Because each branch in glycogen ends with a nonreducing sugar unit, a glycogen molecule with n branches has $n + 1$ nonreducing ends, but only one reducing end. When glycogen is used as an energy source, glucose units are removed one at a time from the nonreducing ends. Degradative enzymes that act only at nonreducing ends can work simultaneously on the many branches, speeding the conversion of the polymer to monosaccharides.

Why not store glucose in its monomeric form? It has been calculated that hepatocytes store glycogen equivalent to a glucose concentration of 0.4 M. The actual concentration of glycogen, which is insoluble and contributes little to the osmolarity of the cytosol, is about 0.01 μM. If the cytosol contained 0.4 M glucose, the osmolarity would be threateningly elevated, leading to osmotic entry of water that might rupture the cell (see Fig. 2-13). Furthermore, with an intracellular glucose concentration of 0.4 M and an external concentration of about 5 mM (the concentration in the blood of a mammal), the free-energy change for glucose uptake into cells against this very high concentration gradient would be prohibitively large.

Dextrans are bacterial and yeast polysaccharides made up of ($\alpha1{\rightarrow}6$)-linked poly-D-glucose; all have ($\alpha1{\rightarrow}3$) branches, and some also have ($\alpha1{\rightarrow}2$) or ($\alpha1{\rightarrow}4$) branches. Dental plaque, formed by bacteria growing on the surface of teeth, is rich in dextrans, which are adhesive and allow the bacteria to stick to teeth and to each other. Dextrans also provide a source of glucose for bacterial metabolism. Synthetic dextrans are components of several commercial products (for example, Sephadex) used in the fractionation of proteins by size-exclusion chromatography (see Fig. 3-17b). The dextrans in these products are chemically cross-linked to form insoluble materials of various sizes.

Some Homopolysaccharides Serve Structural Roles

Cellulose, a tough, fibrous, water-insoluble substance, is found in the cell walls of plants, particularly in stalks, stems, trunks, and all the woody portions of the plant body. Cellulose constitutes much of the mass of wood, and cotton is almost pure cellulose. Like amylose, the cellulose molecule is a linear, unbranched homopolysaccharide,

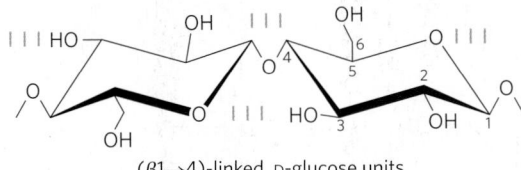

(β1→4)-linked D-glucose units

FIGURE 7-14 Cellulose. Two units of a cellulose chain; the D-glucose residues are in (β1→4) linkage. The rigid chair structures can rotate relative to one another.

(a)

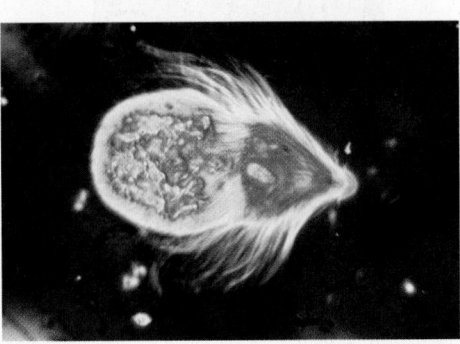

(b)

FIGURE 7-15 Cellulose breakdown by *Trichonympha*. (a) The termite *Cryptotermes domesticus* gnaws off and ingests particles of wood, rich in cellulose. **(b)** *Trichonympha*, a protistan symbiont in the termite gut, produces the enzyme cellulase, which breaks the (β1→4) glycosidic bonds in cellulose, making wood a source of metabolizable sugar (glucose) for the protist and its host termite. Many invertebrates can digest cellulose, but only a few vertebrates (the ruminants, such as cattle, sheep, and goats). The ruminants can use cellulose as food because the first of their four stomach compartments (rumen) teems with bacteria and protists that secrete cellulase. [Sources: (a) David McClenaghan/CSIRO Entomology. (b) Eric V. Grave/Science Source.]

consisting of 10,000 to 15,000 D-glucose units. But there is a very important difference: in cellulose the glucose residues have the β configuration **(Fig. 7-14)**, whereas in amylose the glucose is in the α configuration. The glucose residues in cellulose are linked by (β1→4) glycosidic bonds, in contrast to the (α1→4) bonds of amylose. This difference causes individual molecules of cellulose and amylose to fold differently in space, giving them very different macroscopic structures and physical properties (see below). The tough, fibrous nature of cellulose makes it useful in such commercial products as cardboard and insulation material, and it is a major constituent of cotton and linen fabrics. Cellulose is also the starting material for the commercial production of cellophane, rayon, and lyocell.

Glycogen and starch ingested in the diet are hydrolyzed by α-amylases and glycosidases, enzymes in saliva and the small intestine that break (α1→4) glycosidic bonds between glucose units. Most vertebrate animals cannot use cellulose as a fuel source, because they lack an enzyme to hydrolyze the (β1→4) linkages. Termites readily digest cellulose (and therefore wood), but only because their intestinal tract harbors a symbiotic microorganism, *Trichonympha*, that secretes cellulase, which hydrolyzes the (β1→4) linkages **(Fig. 7-15)**. Molecular genetic studies have revealed that genes encoding cellulose-degrading enzymes are present in the genomes of a wide range of invertebrate animals, including arthropods and nematodes. There is one important exception to the absence of cellulase in vertebrates: ruminant animals such as cattle, sheep, and goats harbor symbiotic microorganisms in the rumen (the first of their four stomach compartments) that can hydrolyze cellulose, allowing the animal to degrade dietary cellulose from soft grasses, but not from woody plants. Fermentation in the rumen yields acetate, propionate, and β-hydroxybutyrate, which the animal uses to synthesize the sugars in milk.

Biomass that is rich in cellulose can be used as starting material for the fermentation of carbohydrates to ethanol, to be used as a gasoline additive (switchgrass is a common biofuel crop). The annual production of biomass on Earth (accomplished primarily by photosynthetic organisms) is the energetic equivalent of nearly a trillion barrels of crude oil, when converted to ethanol by fermentation. Because of their potential use in biomass conversion to bioenergy, cellulose-degrading enzymes such as cellulase are under vigorous investigation. Supramolecular complexes called cellulosomes, found on the outside surface of the bacterium *Clostridium*

cellulolyticum, include the catalytic subunit of cellulase, along with proteins that hold one or more cellulase molecules to the bacterial surface, and a subunit that binds cellulose and positions it in the catalytic site.

A major fraction of photosynthetic biomass is the woody portion of plants and trees, which consists of cellulose plus several other polymers derived from carbohydrates that are not easily digestible, either chemically or biologically. Lignins, for example, make up some 30% of the mass of wood. Synthesized from precursors that include phenylalanine and glucose, lignins are complex polymers with covalent cross-links to cellulose that complicate the digestion of cellulose by cellulase. If woody plants are to be used in the production of ethanol from biomass, better means of digesting wood components will need to be found.

Chitin is a linear homopolysaccharide composed of *N*-acetylglucosamine residues in (β1→4) linkage **(Fig. 7-16)**. The only chemical difference from cellulose is the replacement of the hydroxyl group at C-2 with an acetylated amino group. Chitin forms extended fibers similar to those of cellulose, and like cellulose cannot be

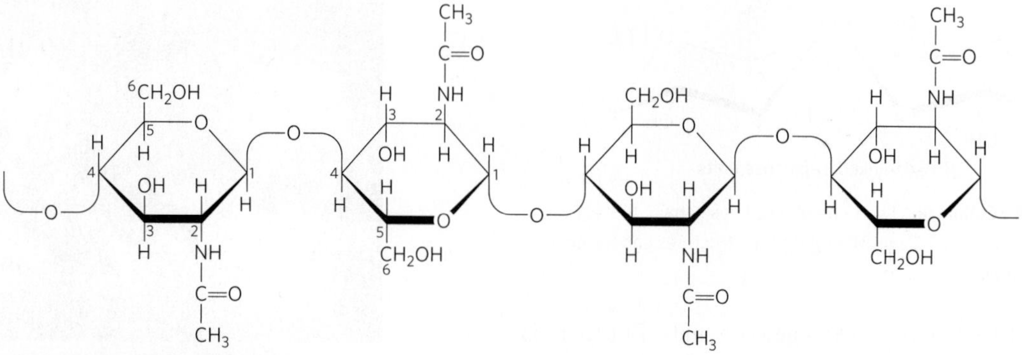

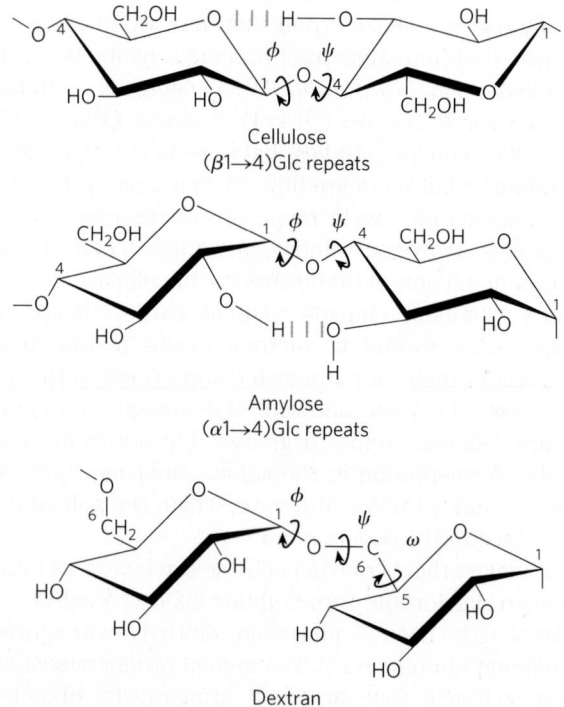

FIGURE 7-16 Chitin. (a) A short segment of chitin, a homopolymer of *N*-acetyl-D-glucosamine units in (β1→4) linkage. **(b)** A spotted June beetle (*Pelidnota punctata*), showing its surface armor (exoskeleton) of chitin. [Source: (b) Paul Whitten/Science Source.]

digested by vertebrates. Chitin is the principal component of the hard exoskeletons of nearly a million species of arthropods—insects, lobsters, and crabs, for example—and is probably the second most abundant polysaccharide, next to cellulose, in nature; an estimated 1 billion tons of chitin are produced in the biosphere each year.

Steric Factors and Hydrogen Bonding Influence Homopolysaccharide Folding

The folding of polysaccharides in three dimensions follows the same principles as those governing polypeptide structure: subunits with a more-or-less rigid structure dictated by covalent bonds form three-dimensional macromolecular structures that are stabilized by weak interactions within or between molecules, such as hydrogen bonds, interactions due to the hydrophobic effect, van der Waals interactions, and, for polymers with charged subunits, electrostatic interactions. Because polysaccharides have so many hydroxyl groups, hydrogen bonding has an especially important influence on their structure. Glycogen, starch, and cellulose are composed of pyranoside (six-membered ring) subunits, as are the oligosaccharides of glycoproteins and glycolipids, to be discussed later. Such molecules can be represented as a series of rigid pyranose rings connected by an oxygen atom bridging two carbon atoms (the glycosidic bond). There is, in principle, free rotation about both C—O bonds linking the residues (Fig. 7-14), but as in polypeptides (see Figs 4-2, 4-9), rotation about each bond is limited by steric hindrance by substituents. The three-dimensional structures of these molecules can be described in terms of the dihedral angles, ϕ and ψ, about the glycosidic bond **(Fig. 7-17)**, analogous to angles ϕ and ψ made by the peptide bond.

The bulkiness of the pyranose ring and its substituents, along with electronic effects at the anomeric carbon, place constraints on the angles ϕ and ψ; thus certain conformations are much more stable than others,

Cellulose
(β1→4)Glc repeats

Amylose
(α1→4)Glc repeats

Dextran
(α1→6)Glc repeats (with (α1→3) branches, not shown)

FIGURE 7-17 Conformation at the glycosidic bonds of cellulose, amylose, and dextran. The polymers are depicted as rigid pyranose rings joined by glycosidic bonds, with free rotation about these bonds. Note that in dextran there is also free rotation about the bond between C-5 and C-6 (torsion angle ω (omega)).

● $\phi,\psi = 30°,-40°$

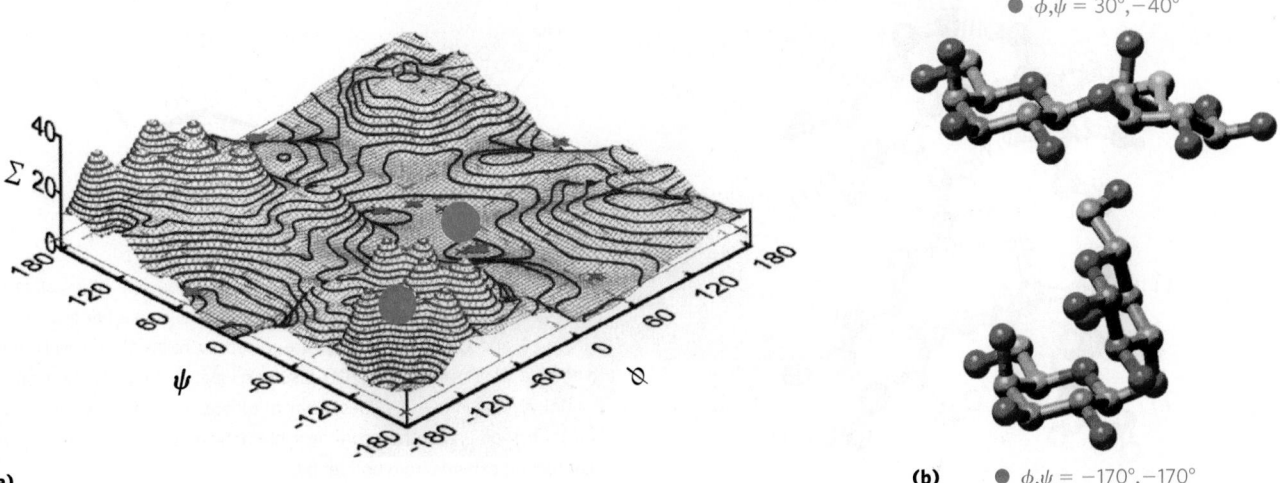

(a)   **(b)**   ● $\phi,\psi = -170°,-170°$

FIGURE 7-18 A map of favored conformations for oligosaccharides and polysaccharides. The torsion angles ψ (psi) and ϕ (phi) (see Fig. 7-17), which define the spatial relationship between adjacent rings, can in principle have any value from 0° to 360°. In fact, some of the torsion angles would give conformations that are sterically hindered, whereas others give conformations that maximize hydrogen bonding. **(a)** When the relative energy (Σ) is plotted for each value of ϕ and ψ, with isoenergy ("same energy") contours drawn at intervals of 1 kcal/mol above the minimum energy state, the result is a map of preferred conformations. This is analogous to the Ramachandran plot for peptides (see Figs 4-3, 4-9). **(b)** Two energetic extremes for the disaccharide Gal($\beta1\rightarrow3$)Gal, which fall on the energy diagram (a) as shown by the red and blue dots. The red dot indicates the least favored conformation; the blue dot, the most favored conformation. The known conformations of the three polysaccharides shown in Figure 7-17 have been determined by x-ray crystallography, and all fall within the lowest-energy regions of the map. [Source: (a) Courtesy of H.-J. Gabius and Herbert Kaltner, University of Munich, from a figure provided by C.-W. von der Lieth, Heidelberg.]

as can be shown on a map of energy as a function of ϕ and ψ **(Fig. 7-18)**.

The most stable three-dimensional structure for the ($\alpha1\rightarrow4$)-linked chains of starch and glycogen is a tightly coiled helix **(Fig. 7-19)**, stabilized by interchain hydrogen bonds. In amylose (with no branches) this structure is regular enough to allow crystallization and thus determination of the structure by x-ray diffraction. The average plane of each residue along the amylose chain forms a 60° angle with the average plane of the preceding residue, so the helical structure has six residues per turn. For amylose, the core of the helix is of precisely the right dimensions to accommodate iodine as complex ions (I_3^- and I_5^-), giving an intensely blue complex. This interaction is a common qualitative test for amylose.

For cellulose, the most stable conformation is that in which each chair is turned 180° relative to its neighbors, yielding a straight, extended chain. All —OH groups are available for hydrogen bonding with neighboring chains. With several chains lying side by side, a stabilizing network of interchain and intrachain hydrogen bonds produces straight, stable supramolecular fibers of great tensile strength **(Fig. 7-20)**. This property of cellulose has made it useful to civilizations for millennia. Many manufactured products, including papyrus, paper, cardboard, rayon, insulating tiles, and a variety of other useful materials, are derived from cellulose. The water content of these materials is low because extensive interchain hydrogen bonding between cellulose molecules satisfies their capacity for hydrogen-bond formation.

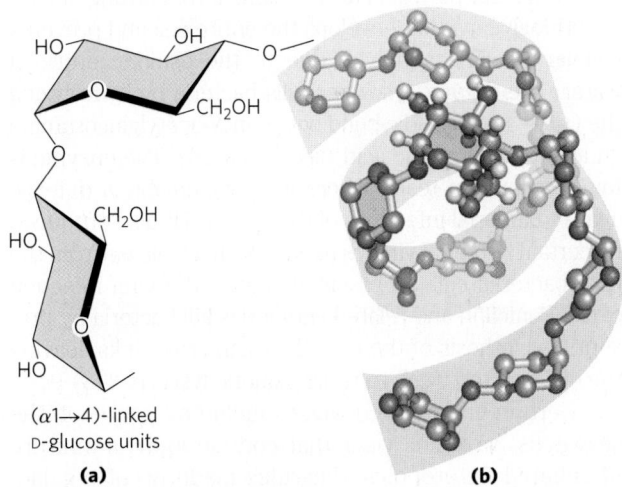

($\alpha1\rightarrow4$)-linked
D-glucose units
(a)   **(b)**

FIGURE 7-19 Helical structure of starch (amylose). (a) In the most stable conformation, with adjacent rigid chairs, the polysaccharide chain is curved, rather than linear as in cellulose (see Fig. 7-14). **(b)** A model of a segment of amylose; for clarity, the hydroxyl groups are omitted from all but one of the glucose residues. Compare the two residues shaded in pink with the chemical structures in (a). The conformation of ($\alpha1\rightarrow4$) linkages in amylose, amylopectin, and glycogen causes these polymers to assume tightly coiled helical structures. These compact structures produce the dense granules of stored starch or glycogen seen in many cells (see Fig. 20-2). [Source: (b) PDB ID 1C58, K. Gessler et al., *Proc. Natl. Acad. Sci. USA* 96:4246, 1999.]

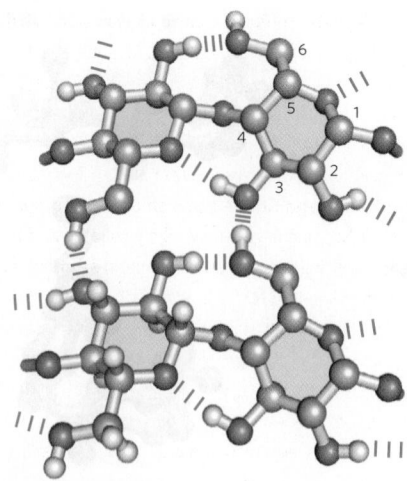

FIGURE 7-20 Cellulose chains. Scale drawing of segments of two parallel cellulose chains, showing the conformation of the D-glucose residues and the hydrogen-bond cross-links. In the hexose unit at the lower left, all hydrogen atoms are shown; in the other three hexose units, the hydrogens attached to carbon are omitted for clarity, as they do not participate in hydrogen bonding.

Bacterial and Algal Cell Walls Contain Structural Heteropolysaccharides

The rigid component of bacterial cell walls (peptidoglycan) is a heteropolymer of alternating ($\beta1\rightarrow4$)-linked N-acetylglucosamine and N-acetylmuramic acid residues (see Fig. 20-30). The linear polymers lie side by side in the cell wall, cross-linked by short peptides, the exact structure of which depends on the bacterial species. The peptide cross-links weld the polysaccharide chains into a strong sheath (peptidoglycan) that envelops the entire cell and prevents cellular swelling and lysis due to the osmotic entry of water. The enzyme lysozyme kills bacteria by hydrolyzing the ($\beta1\rightarrow4$) glycosidic bond between N-acetylglucosamine and N-acetylmuramic acid (see Fig. 6-28). The enzyme is found in human tears, where it is presumably a defense against bacterial infections of the eye, and is also produced by certain bacterial viruses to ensure their release from the host bacterial cell, an essential step of the viral infection cycle. Penicillin and related antibiotics kill bacteria by preventing synthesis of the peptidoglycan cross-links, leaving the cell wall too weak to resist osmotic lysis (p. 223).

Certain marine red algae, including some of the seaweeds, have cell walls that contain **agar**, a mixture of sulfated heteropolysaccharides made up of D-galactose and an L-galactose derivative ether-linked between C-3 and C-6. Agar is a complex mixture of polysaccharides, all with the same backbone structure but substituted to varying degrees with sulfate and pyruvate. **Agarose** (M_r ~150,000) is the agar component with the fewest charged groups (sulfates, pyruvates) **(Fig. 7-21)**. The remarkable gel-forming property of agarose makes it useful in the biochemistry laboratory. When a suspension of agarose in water is heated and cooled, the agarose forms a double helix: two molecules in parallel orientation twist together with a helix repeat of three residues; water molecules are trapped in the central

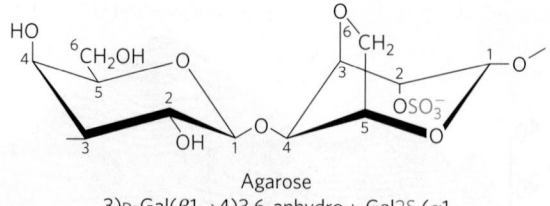

Agarose
3)D-Gal($\beta1\rightarrow4$)3,6-anhydro-L-Gal2S ($\alpha1$
repeating units

FIGURE 7-21 Agarose. The repeating unit consists of D-galactose ($\beta1\rightarrow4$)-linked to 3,6-anhydro-L-galactose (in which an ether bridge connects C-3 and C-6). These units are joined by ($\alpha1\rightarrow3$) glycosidic links to form a polymer 600 to 700 residues long. A small fraction of the 3,6-anhydrogalactose residues have a sulfate ester at C-2 (as shown here). The open parentheses in the systematic name indicate that the repeating unit extends from both ends.

cavity. These structures in turn associate with each other to form a gel—a three-dimensional matrix that traps large amounts of water. Agarose gels are used as inert supports for the electrophoretic separation of nucleic acids. Agar is also used to form a surface for the growth of bacterial colonies. Another commercial use of agar is for the capsules in which some vitamins and drugs are packaged; the dried agar material dissolves readily in the stomach and is metabolically inert.

Glycosaminoglycans Are Heteropolysaccharides of the Extracellular Matrix

The extracellular space in the tissues of multicellular animals is filled with a gel-like material, the **extracellular matrix (ECM)**, also called ground substance, which holds the cells together and provides a porous pathway for the diffusion of nutrients and oxygen to individual cells. The ECM that surrounds fibroblasts and other connective tissue cells is composed of an interlocking meshwork of heteropolysaccharides and fibrous proteins such as fibrillar collagens, elastins, and fibronectins. Basement membrane is a specialized ECM that underlies epithelial cells; it consists of specialized collagens, laminins, and heteropolysaccharides. These heteropolysaccharides, the **glycosaminoglycans**, are a family of linear polymers composed of repeating disaccharide units **(Fig. 7-22)**. They are unique to animals and bacteria and are not found in plants. One of the two monosaccharides is always either N-acetylglucosamine or N-acetylgalactosamine; the other is in most cases a uronic acid, usually D-glucuronic or L-iduronic acid. Some glycosaminoglycans contain esterified sulfate groups. The combination of sulfate groups and the carboxylate groups of the uronic acid residues gives glycosaminoglycans a very high density of negative charge. To minimize the repulsive forces among neighboring charged groups, these molecules assume an extended conformation in solution, forming a rodlike helix in which the negatively charged carboxylate groups occur on alternate sides of the helix (as shown for heparin in Fig. 7-22). The extended rod form also provides maximum separation between the negatively charged sulfate groups. The specific patterns

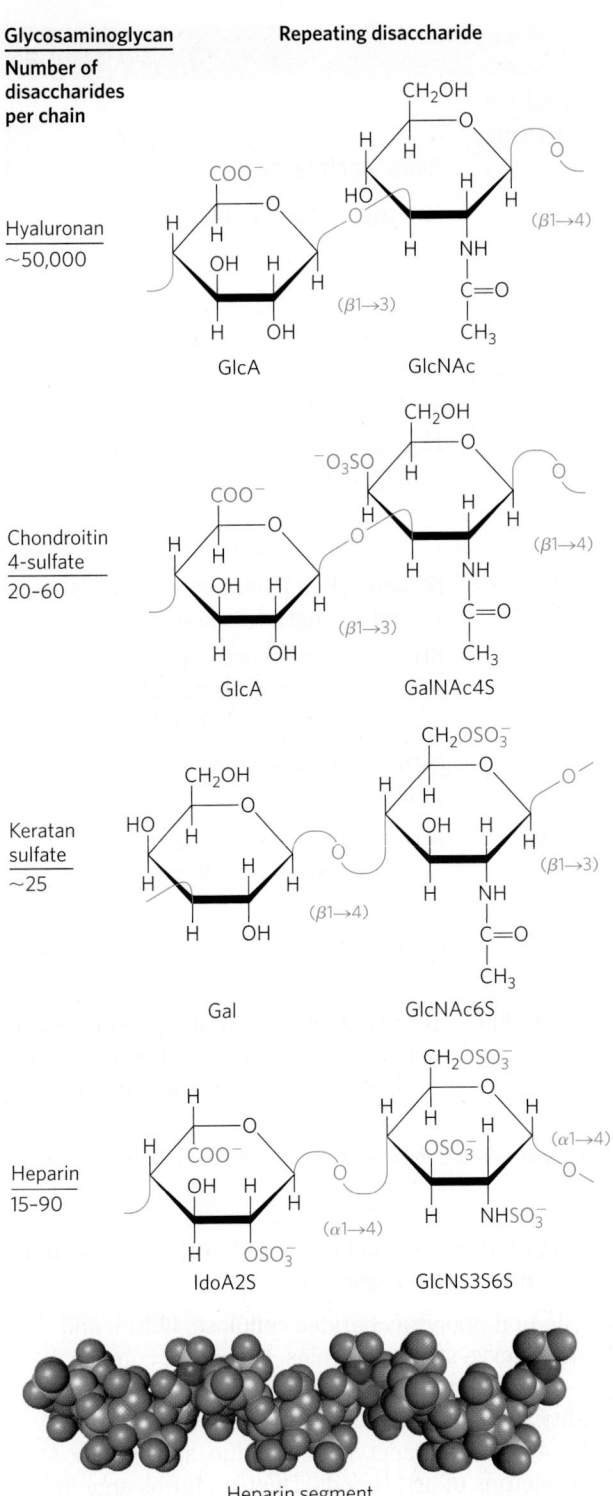

Glycosaminoglycan   **Repeating disaccharide**

Number of disaccharides per chain

Hyaluronan ~50,000

Chondroitin 4-sulfate 20–60

Keratan sulfate ~25

Heparin 15–90

Heparin segment

FIGURE 7-22 Repeating units of some common glycosaminoglycans of extracellular matrix. The molecules are copolymers of alternating uronic acid and amino sugar residues (keratan sulfate is the exception), with sulfate esters in any of several positions, except in hyaluronan. The ionized carboxylate and sulfate groups (red in the perspective formulas) give these polymers their characteristic high negative charge. Therapeutic heparin contains primarily iduronic acid (IdoA) and a smaller proportion of glucuronic acid (GlcA, not shown) and is generally highly sulfated and heterogeneous in length. The space-filling model shows a heparin segment as its structure in solution, as determined by NMR spectroscopy. The carbons in the iduronic acid sulfate are colored blue; those in glucosamine sulfate are green. Oxygen and sulfur atoms are shown in their standard colors of red and yellow, respectively. The hydrogen atoms are not shown (for clarity). Heparan sulfate (not shown) is similar to heparin but has a higher proportion of GlcA and fewer sulfate groups, arranged in a less regular pattern. [Source: Molecular model: PDB ID 1HPN, B. Mulloy et al., *Biochem. J.* 293:849, 1993.]

in the synovial fluid of joints and give the vitreous humor of the vertebrate eye its jellylike consistency (the Greek *hyalos* means "glass"; hyaluronan can have a glassy or translucent appearance). Hyaluronan is also a component of the ECM of cartilage and tendons, to which it contributes tensile strength and elasticity as a result of its strong noncovalent interactions with other components of the matrix. Hyaluronidase, an enzyme secreted by some pathogenic bacteria, can hydrolyze the glycosidic linkages of hyaluronan, rendering tissues more susceptible to bacterial invasion. In many animal species, a similar enzyme in sperm hydrolyzes the outer glycosaminoglycan coat around an ovum, allowing sperm penetration.

Other glycosaminoglycans differ from hyaluronan in three respects: they are generally much shorter polymers, they are covalently linked to specific proteins (proteoglycans), and one or both monomeric units differ from those of hyaluronan. **Chondroitin sulfate** (Greek *chondros,* "cartilage") contributes to the tensile strength of cartilage, tendons, ligaments, heart valves, and the walls of the aorta. **Dermatan sulfate** (Greek *derma,* "skin") contributes to the pliability of skin and is also present in blood vessels and heart valves. In this polymer, many of the glucuronate residues present in chondroitin sulfate are replaced by their C-5 epimer, L-iduronate (IdoA).

α-L-Iduronate (IdoA)   β-D-Glucuronate (GlcA)

Keratan sulfates (Greek *keras,* "horn") have no uronic acid, and their sulfate content is variable. They are present in cornea, cartilage, bone, and a variety of horny structures formed from dead cells: horn, hair, hoofs, nails, and claws. **Heparan sulfate** (Greek *hēpar,* "liver"; it was originally isolated from dog liver) is produced by all animal cells and contains variable arrangements of sulfated and nonsulfated sugars. The sulfated segments of

of sulfated and nonsulfated sugar residues in glycosaminoglycans allow specific recognition by a variety of protein ligands that bind electrostatically to these molecules. The sulfated glycosaminoglycans are attached to extracellular proteins to form proteoglycans (Section 7.3).

The glycosaminoglycan **hyaluronan** (hyaluronic acid) contains alternating residues of D-glucuronic acid and N-acetylglucosamine (Fig. 7-22). With up to 50,000 repeats of the basic disaccharide unit, hyaluronan has a molecular weight of several million; it forms clear, highly viscous, noncompressible solutions that serve as lubricants

TABLE 7-2 Structures and Roles of Some Polysaccharides

Polymer	Type[a]	Repeating unit[b]	Size (number of monosaccharide units)	Roles/significance
Starch				Energy storage: in plants
Amylose	Homo-	($\alpha1\rightarrow4$) Glc, linear	50–5,000	
Amylopectin	Homo-	($\alpha1\rightarrow4$) Glc, with ($\alpha1\rightarrow6$) Glc branches every 24–30 residues	Up to 10^6	
Glycogen	Homo-	($\alpha1\rightarrow4$) Glc, with ($\alpha1\rightarrow6$) Glc branches every 8–12 residues	Up to 50,000	Energy storage: in bacteria and animal cells
Cellulose	Homo-	($\beta1\rightarrow4$) Glc	Up to 15,000	Structural: in plants, gives rigidity and strength to cell walls
Chitin	Homo-	($\beta1\rightarrow4$) GlcNAc	Very large	Structural: in insects, spiders, crustaceans, gives rigidity and strength to exoskeletons
Dextran	Homo-	($\alpha1\rightarrow6$) Glc, with ($\alpha1\rightarrow3$) branches	Wide range	Structural: in bacteria, extracellular adhesive
Peptidoglycan	Hetero-; peptides attached	4)Mur2Ac($\beta1\rightarrow4$) GlcNAc($\beta1$	Very large	Structural: in bacteria, gives rigidity and strength to cell envelope
Agarose	Hetero-	3)D-Gal ($\beta1\rightarrow4$)3,6-anhydro-L-Gal($\alpha1$	1,000	Structural: in algae, cell wall material
Hyaluronan (a glycosamino-glycan)	Hetero-; acidic	4)GlcA ($\beta1\rightarrow3$) GlcNAc($\beta1$	Up to 100,000	Structural: in vertebrates, extracellular matrix of skin and connective tissue; viscosity and lubrication in joints

[a] Each polymer is classified as a homopolysaccharide (homo-) or heteropolysaccharide (hetero-).

[b] The abbreviated names for the peptidoglycan, agarose, and hyaluronan repeating units indicate that the polymer contains repeats of this disaccharide unit. For example, in peptidoglycan, the GlcNAc of one disaccharide unit is ($\beta1\rightarrow4$)-linked to the first residue of the next disaccharide unit.

the chain allow it to interact with a large number of proteins, including growth factors and ECM components, as well as various enzymes and factors present in plasma. Heparin is a highly sulfated, intracellular form of heparan sulfate produced primarily by mast cells (a type of leukocyte, or immune cell). Its physiological role is not yet clear, but purified heparin is used as a therapeutic agent to inhibit coagulation of blood through its capacity to bind the protease inhibitor antithrombin (see Fig. 7-27).

Table 7-2 summarizes the composition, properties, roles, and occurrence of the polysaccharides described in Section 7.2.

SUMMARY 7.2 Polysaccharides

■ Polysaccharides (glycans) serve as stored fuel and as structural components of cell walls and extracellular matrix.

■ The homopolysaccharides starch and glycogen are storage fuels in plant, animal, and bacterial cells. They consist of D-glucose units with ($\alpha1\rightarrow4$) linkages, and both contain some branches.

■ The homopolysaccharides cellulose, chitin, and dextran serve structural roles. Cellulose, composed of ($\beta1\rightarrow4$)-linked D-glucose residues, lends strength and rigidity to plant cell walls. Chitin, a polymer of ($\beta1\rightarrow4$)-linked N-acetylglucosamine, strengthens the exoskeletons of arthropods. Dextran forms an adhesive coat around certain bacteria.

■ Homopolysaccharides fold in three dimensions. The chair form of the pyranose ring is essentially rigid, so the conformation of the polymers is determined by rotation about the bonds from the rings to the oxygen atom of the glycosidic linkage. Starch and glycogen form helical structures with intrachain hydrogen bonding; cellulose and chitin form long, straight strands that interact with neighboring strands.

■ Bacterial and algal cell walls are strengthened by heteropolysaccharides—peptidoglycan in bacteria, agar in red algae. The repeating disaccharide in peptidoglycan is GlcNAc($\beta1\rightarrow4$)Mur2Ac; in agar, it is D-Gal($\beta1\rightarrow4$)3,6-anhydro-L-Gal.

■ Glycosaminoglycans are extracellular heteropolysaccharides in which one of the two monosaccharide units is a uronic acid (keratan sulfate is an exception) and the other is an N-acetylated amino sugar. Sulfate esters on some of the hydroxyl groups and on the amino group of some glucosamine residues in heparin and heparan sulfate give these polymers a high density of negative charge, forcing them to assume extended conformations. These polymers (hyaluronan, chondroitin sulfate, dermatan sulfate, and keratan sulfate) provide viscosity, adhesiveness, and tensile strength to the extracellular matrix.

7.3 Glycoconjugates: Proteoglycans, Glycoproteins, and Glycosphingolipids

In addition to their important roles as fuel stores (starch, glycogen, dextran) and as structural materials (cellulose, chitin, peptidoglycans), polysaccharides and oligosaccharides are information carriers. Some provide communication between cells and their extracellular surroundings; others label proteins for transport to and localization in specific organelles, or for destruction when the protein is malformed or superfluous; and others serve as recognition sites for extracellular signal molecules (growth factors, for example) or extracellular parasites (bacteria or viruses). On almost every eukaryotic cell, specific oligosaccharide chains attached to components of the plasma membrane form a carbohydrate layer (the glycocalyx), several nanometers thick, that serves as an information-rich surface that the cell shows to its surroundings. These oligosaccharides are central players in cell-cell recognition and adhesion, cell migration during development, blood clotting, the immune response, wound healing, and other cellular processes. In most of these cases, the informational carbohydrate is covalently joined to a protein or a lipid to form a **glycoconjugate**, which is the biologically active molecule **(Fig. 7-23).**

Proteoglycans are macromolecules of the cell surface or ECM in which one or more sulfated glycosaminoglycan chains are joined covalently to a membrane protein or a secreted protein. The glycosaminoglycan chain can bind to extracellular proteins through electrostatic interactions between the protein and the negatively charged sugar moieties on the proteoglycan. Proteoglycans are major components of all extracellular matrices.

Glycoproteins have one or several oligosaccharides of varying complexity joined covalently to a protein. They are usually found on the outer face of the plasma membrane (as part of the glycocalyx), in the ECM, and in the blood. Inside cells, they are found in specific organelles such as Golgi complexes, secretory granules, and lysosomes. The oligosaccharide portions of glycoproteins are very heterogeneous and, like glycosaminoglycans, are

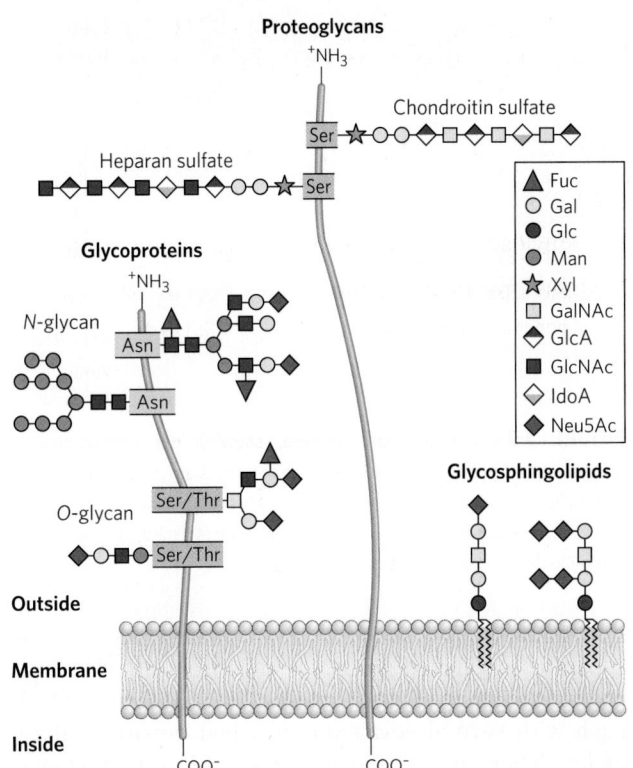

FIGURE 7-23 Glycoconjugates. The structures of some typical proteoglycans, glycoproteins, and glycosphingolipids described in the text.

rich in information, forming highly specific sites for recognition and high-affinity binding by carbohydrate-binding proteins called lectins. Some cytosolic and nuclear proteins can be glycosylated as well.

Glycosphingolipids are plasma membrane components in which the hydrophilic head groups are oligosaccharides. As in glycoproteins, the oligosaccharides act as specific sites for recognition by lectins. Neurons are rich in glycosphingolipids, which help in nerve conduction and myelin formation. Glycosphingolipids also play a role in signal transduction in cells. Sphingolipids are considered in more detail in Chapters 10 and 11.

Proteoglycans Are Glycosaminoglycan-Containing Macromolecules of the Cell Surface and Extracellular Matrix

Mammalian cells can produce at least 40 types of proteoglycans. These molecules act as tissue organizers, and they influence various cellular activities, such as growth factor activation and adhesion. The basic proteoglycan unit consists of a "core protein" with covalently attached glycosaminoglycan(s). The point of attachment is a Ser residue, to which the glycosaminoglycan is joined through a tetrasaccharide bridge **(Fig. 7-24).** The Ser residue is generally in the sequence –Ser–Gly–X–Gly– (where X is any amino acid residue), although not every protein with this sequence has an attached glycosaminoglycan.

Many proteoglycans are secreted into the ECM, but some are integral membrane proteins (see Fig. 11-6).

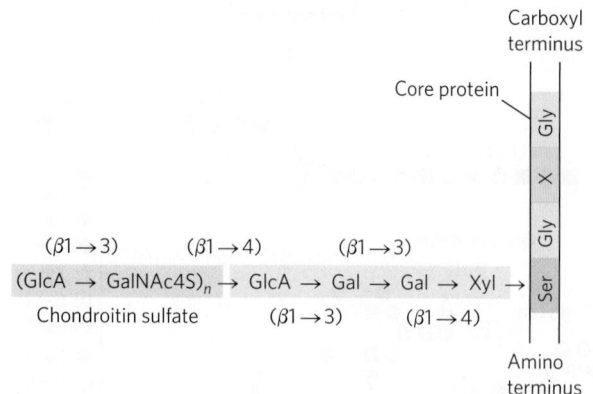

FIGURE 7-24 Proteoglycan structure, showing the tetrasaccharide bridge. A typical tetrasaccharide linker (blue) connects a glycosaminoglycan—in this case, chondroitin sulfate (orange)—to a Ser residue in the core protein. The xylose residue at the reducing end of the linker is joined by its anomeric carbon to the hydroxyl of the Ser residue.

For example, the sheetlike ECM (basal lamina) that separates organized groups of cells from other groups contains a family of core proteins (M_r 20,000 to 40,000), each with several covalently attached heparan sulfate chains. There are two major families of membrane heparan sulfate proteoglycans. **Syndecans** have a single transmembrane domain and an extracellular domain bearing three to five chains of heparan sulfate and, in some cases, chondroitin sulfate **(Fig. 7-25a)**. **Glypicans** are attached to the membrane by a lipid anchor, a derivative of the membrane lipid phosphatidylinositol (see Fig. 11-13). Both syndecans and glypicans can be shed into the extracellular space. A protease in the ECM that cuts proteins close to the membrane surface releases syndecan ectodomains (domains outside the plasma membrane), and a phospholipase that breaks the connection to the membrane lipid releases glypicans. These mechanisms provide a way for a cell to change its surface features quickly. Shedding is highly regulated and is activated in proliferating cells, such as cancer cells. Proteoglycan shedding is involved in cell-cell recognition and adhesion, and in the proliferation and differentiation of cells. Numerous chondroitin sulfate and dermatan sulfate proteoglycans also exist, some as membrane-bound entities, others as secreted products in the ECM.

The glycosaminoglycan chains can bind to a variety of extracellular ligands and thereby modulate the ligands' interaction with specific receptors of the cell surface. Detailed studies of heparan sulfate demonstrate a domain structure that is not random; some domains (typically three to eight disaccharide units long) differ from neighboring domains in sequence and in ability to bind to specific proteins. Highly sulfated domains (called NS domains) alternate with domains having unmodified GlcNAc and GlcA residues (*N*-acetylated, or NA, domains) (Fig. 7-25b). The exact pattern of sulfation in the NS domain depends on the particular proteoglycan; given the number of possible modifications of the GlcNAc–IdoA (iduronic acid) dimer, at least 32 different

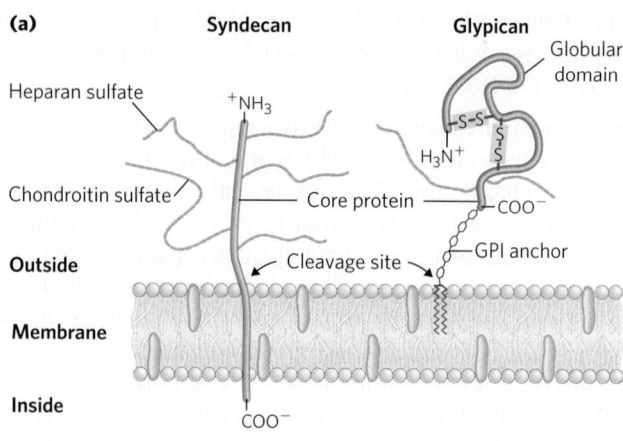

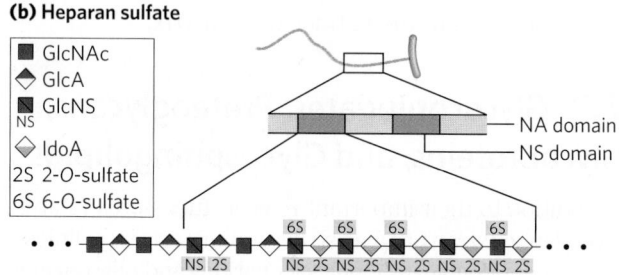

FIGURE 7-25 Two families of membrane proteoglycans. (a) Schematic diagrams of a syndecan and a glypican in the plasma membrane. Syndecans are held in the membrane through the hydrophobic effect by interactions between a sequence of nonpolar amino acid residues and plasma membrane lipids; they can be released by a single proteolytic cut near the membrane surface. In a typical syndecan, the extracellular amino-terminal domain is covalently attached (by tetrasaccharide linkers such as those in Fig. 7-24) to three heparan sulfate chains and two chondroitin sulfate chains. Glypicans are held in the membrane by a covalently attached membrane lipid (GPI anchor; see Fig. 11-13), but are shed if the bond between the lipid portion of the GPI anchor (phosphatidylinositol) and the oligosaccharide linked to the protein is cleaved by a phospholipase. All glypicans have 14 conserved Cys residues, which form disulfide bonds to stabilize the protein moiety, and either two or three glycosaminoglycan chains attached near the carboxyl terminus, close to the membrane surface. **(b)** Along a heparan sulfate chain, regions rich in sulfated sugars, the NS domains (green), alternate with regions with chiefly unmodified residues of GlcNAc and GlcA, the NA domains (gray). One of the NS domains is shown in more detail, revealing a high density of modified residues: GlcNS (*N*-sulfoglucosamine), with a sulfate ester at C-6; and both GlcA and IdoA, with a sulfate ester at C-2. The exact pattern of sulfation in the NS domain differs among proteoglycans. [Sources: (a) Information from U. Häcker et al., *Nature Rev. Mol. Cell Biol.* 6:530, 2005. (b) Information from J. Turnbull et al., *Trends Cell Biol.* 11:75, 2001.]

disaccharide units are possible. Furthermore, the same core protein can display different heparan sulfate structures when synthesized in different cell types.

Heparan sulfate molecules with precisely organized NS domains bind specifically to extracellular proteins and signaling molecules to alter their activities. The change in activity may result from a conformational change in the protein that is induced by the binding **(Fig. 7-26a)**, or it may be due to the ability of adjacent domains of heparan sulfate to bind to two different proteins, bringing them into close proximity and enhancing protein-protein interactions (Fig. 7-26b). A third general mechanism of action is the binding of

(a) Conformational activation

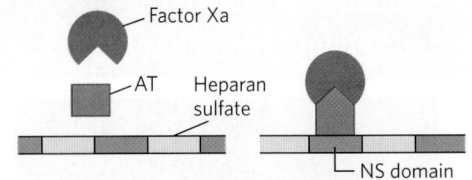

A conformational change induced in the protein antithrombin (AT) on binding a specific pentasaccharide NS domain allows its interaction with blood clotting factor Xa, preventing clotting.

(b) Enhanced protein-protein interaction

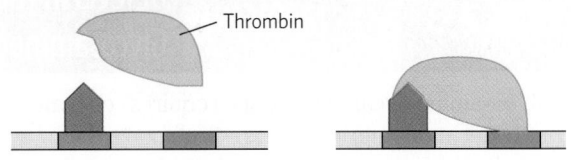

Binding of AT and thrombin to two adjacent NS domains brings the two proteins into close proximity, favoring their interaction, which inhibits blood clotting.

(c) Coreceptor for extracellular ligands

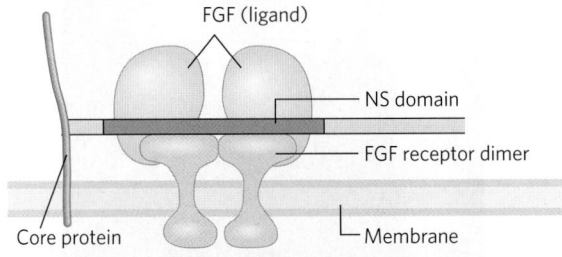

NS domains interact with both the fibroblast growth factor (FGF) and its receptor, bringing the oligomeric complex together and increasing the effectiveness of a low concentration of FGF.

(d) Cell surface localization/concentration

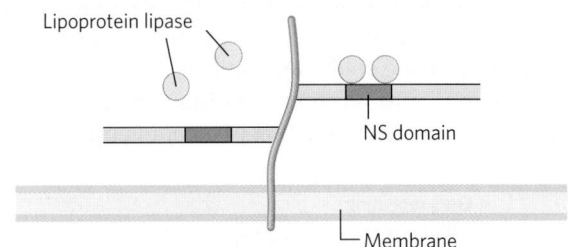

The high density of negative charges in heparan sulfate attracts positively charged lipoprotein lipase molecules and holds them by electrostatic and sequence-specific interactions with NS domains.

FIGURE 7-26 Four types of protein interactions with NS domains of heparan sulfate. [Source: Information from J. Turnbull et al., *Trends Cell Biol.* 11:75, 2001.]

extracellular signal molecules (growth factors, for example) to heparan sulfate, which increases their local concentrations and enhances their interaction with growth factor receptors on the cell surface; in this case, the heparan sulfate acts as a coreceptor (Fig. 7-26c). For example, fibroblast growth factor (FGF), an extracellular protein signal that stimulates cell division, first binds to heparan sulfate moieties of syndecan molecules in the target cell's plasma membrane. Syndecan presents FGF to the FGF plasma membrane receptor, and only then can FGF interact productively with its receptor to trigger cell division. Finally, in another type of mechanism, the NS domains interact—electrostatically and otherwise—with a variety of soluble molecules outside the cell, maintaining high local concentrations at the cell surface (Fig. 7-26d).

The protease thrombin, essential to blood coagulation (see Fig. 6-41), is inhibited by another blood protein, antithrombin, which prevents premature blood clotting. Antithrombin does not bind to or inhibit thrombin in the absence of heparan sulfate. In the presence of heparan sulfate or heparin, the binding affinity of thrombin for antithrombin increases 2,000-fold, and thrombin is strongly inhibited. When thrombin and antithrombin are crystallized in the presence of a short (16 residue) segment of heparan sulfate, the negatively charged heparan sulfate mimic is seen to bridge positively charged regions of the two proteins, causing an allosteric change that inhibits thrombin's protease activity **(Fig. 7-27)**. The

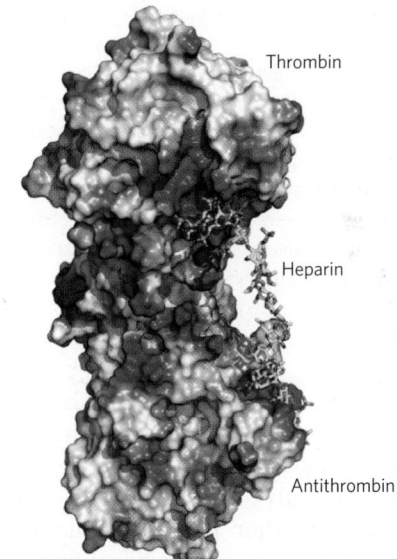

FIGURE 7-27 Molecular basis for heparan sulfate enhancement of the binding of thrombin to antithrombin. In this crystal structure of thrombin, antithrombin, and a 16 residue heparan-sulfate-like polymer, all crystallized together, the binding sites for heparan sulfate in both proteins are rich in Arg and Lys residues. These positively charged regions, shown in blue, allow strong electrostatic interaction with multiple negatively charged sulfates and carboxylates of the heparan sulfate. Consequently, the affinity of antithrombin for thrombin is three orders of magnitude greater in the presence of heparan sulfate than in its absence. Regions of thrombin and antithrombin rich in negatively charged residues are shown in red in this electrostatic representation of the two proteins. [Source: PDB ID 1TB6, W. Li, et al., *Nature Struct. Mol. Biol.* 11:857, 2004.]

BOX 7-3 ⚕ MEDICINE Defects in the Synthesis or Degradation of Sulfated Glycosaminoglycans Can Lead to Serious Human Disease

Glycosaminoglycan synthesis requires enzymes that activate monomeric sugars, transport them across membranes, condense the activated sugars into polysaccharides, and add sulfates. Mutations in any of these enzymes in humans can lead to structural defects in the glycosaminoglycan (or in the proteoglycans formed from them). The result can be any of a wide variety of defects in cell signaling, cell proliferation, tissue morphogenesis, or interactions with growth factors (Fig. 1). For example, failure to extend the disaccharide unit GlcNAc-GlcA leads to a bone abnormality in which multiple, large bone spurs develop (Fig. 2).

When the defect occurs in degradative enzymes, the accumulation of incompletely degraded

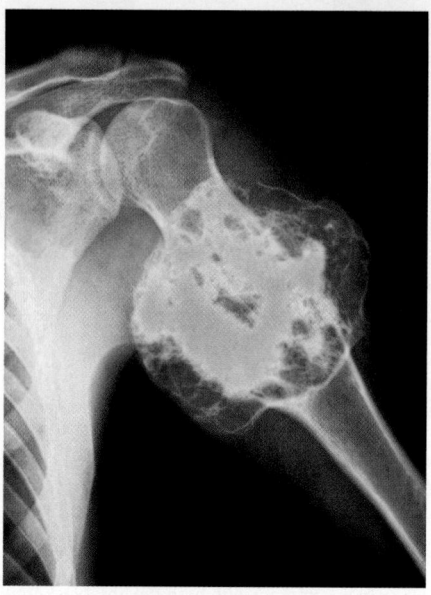

FIGURE 2 Bone deformation characteristic of multiple hereditary exostoses, a disease resulting from a genetic inability to add the GlcNAc-GlcA disaccharide to the growing heparan sulfate or heparin chain (see ❻ in Fig. 1). The extra bone growth is artificially colored red in this x-ray of the humerus (upper arm bone). [Source: CNRI/Science Photo Library/Science Source.]

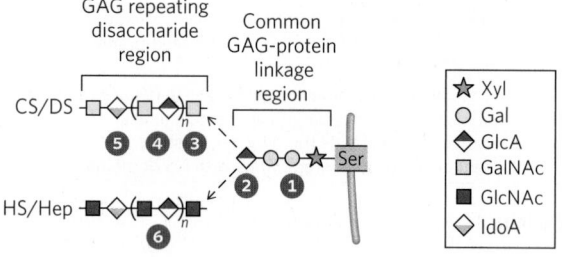

FIGURE 1 A segment of proteoglycan showing the normal structure of the glycosaminoglycans (GAGs) chondroitin sulfate or dermatan sulfate (CS/DS) (top) and heparan sulfate or heparin (HS/Hep) (bottom), attached through the linkage region to a Ser residue in the core protein. When a specific biosynthetic enzyme is absent because of a mutation, the numbered elements cannot be added to the growing oligosaccharide, and the product is truncated. The dysfunctional GAGs result in several types of human disease, depending on the site of truncation: ❶ progeroid-type Ehlers-Danlos syndrome—with hyperextensible joints, fragile skin, and premature aging; ❷ short stature or frequent joint dislocations; ❸ neuropathy (nerve damage); ❹ skeletal defects; ❺ bipolar disorder or diaphragmatic hernia; and ❻ bone deformations in the form of large bone spurs.

glycosaminoglycans can produce diseases ranging from moderate, as in Scheie syndrome, with joint stiffening but normal intelligence and life span, to severe, as in Hurler syndrome, with enlarged internal organs, heart disease, dwarfism, mental retardation, and early death. Glycosaminoglycans were formerly called mucopolysaccharides, and diseases caused by genetic defects in their breakdown are often still called mucopolysaccharidoses.

binding sites for heparan sulfate and heparin in both proteins are rich in Arg and Lys residues; the amino acids' positive charges interact electrostatically with the sulfates of the glycosaminoglycans. Antithrombin also inhibits two other blood coagulation proteins (factors IXa and Xa) in a heparan sulfate–dependent process.

The importance of correctly synthesizing sulfated domains in heparan sulfate is demonstrated in mutant ("knockout") mice lacking the enzyme that sulfates the C-2 hydroxyl of iduronate (IdoA). These animals are born without kidneys and with severe developmental abnormalities of the skeleton and eyes. Other studies demonstrate that membrane proteoglycans are important in the liver for clearing lipoproteins from the blood. Finally, there is growing evidence that proteoglycans containing heparan sulfate and chondroitin

sulfate provide directional cues for axon outgrowth, influencing the path taken by developing axons in the nervous system.

The functional importance of proteoglycans and the glycosaminoglycans associated with them can also be seen in the effects of mutations that block the synthesis or degradation of these polymers in humans (Box 7-3).

Some proteoglycans can form **proteoglycan aggregates**, enormous supramolecular assemblies of many core proteins all bound to a single molecule of hyaluronan. Aggrecan core protein (M_r ~250,000) has multiple chains of chondroitin sulfate and keratan sulfate, joined to Ser residues in the core protein through trisaccharide linkers, to give an aggrecan monomer of M_r ~2 × 10^6. When a hundred or more of these "decorated" core proteins bind a single, extended molecule of hyaluronate

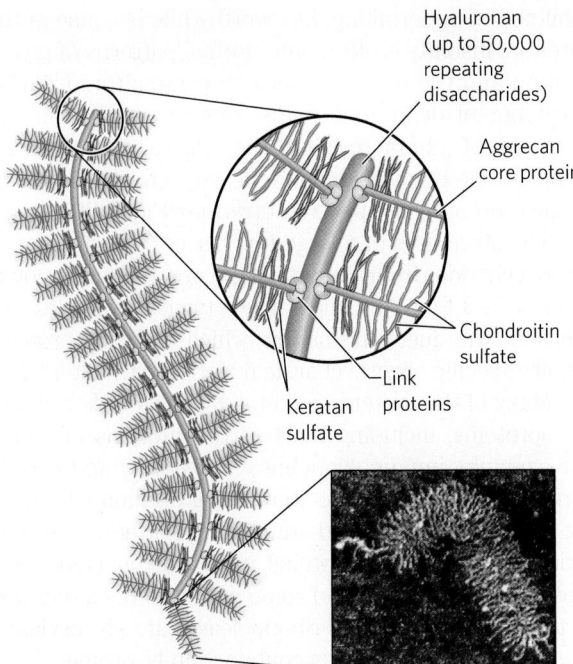

FIGURE 7-28 Proteoglycan aggregate of the extracellular matrix. Schematic drawing of a proteoglycan with many aggrecan molecules. One very long molecule of hyaluronan is associated noncovalently with about 100 molecules of the core protein aggrecan. Each aggrecan molecule contains many covalently bound chondroitin sulfate and keratan sulfate chains. Link proteins at the junction between each core protein and the hyaluronan backbone mediate the core protein–hyaluronan interaction. The micrograph shows a single molecule of aggrecan, viewed with the atomic force microscope (see Box 19-2). [Source: Micrograph courtesy of Laurel Ng. Reprinted with permission from Ng, L., Grodinsky, A., Patwari, P., Sandy, J., Plaas, A. H. K., & Ortiz, C. (2003) Individual cartilage aggrecan macromolecules and their constituent glycosaminoglycans visualized via atomic force microscopy. *J. Struct. Biol.* 143:242–257, Fig. 7a left © Elsevier.]

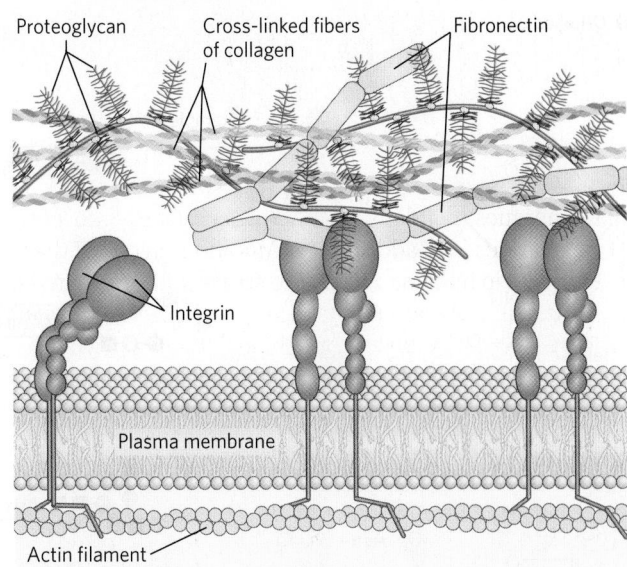

FIGURE 7-29 Interactions between cells and the extracellular matrix. The association between cells and the proteoglycan of the extracellular matrix is mediated by a membrane protein (integrin) and by an extracellular protein (fibronectin in this example) with binding sites for both integrin and the proteoglycan. Note the close association of collagen fibers with the fibronectin and proteoglycan.

(Fig. 7-28), the resulting proteoglycan aggregate ($M_r > 2 \times 10^8$) and its associated water of hydration occupy a volume about equal to that of a bacterial cell! Aggrecan interacts strongly with collagen in the ECM of cartilage, contributing to the development, tensile strength, and resilience of this connective tissue.

Interwoven with these enormous extracellular proteoglycans are fibrous matrix proteins such as collagen, elastin, and fibronectin, forming a cross-linked meshwork that gives the whole ECM strength and resilience. Some of these proteins are multiadhesive, a single protein having binding sites for several different matrix molecules. Fibronectin, for example, has separate domains that bind fibrin, heparan sulfate, collagen, and a family of plasma membrane proteins called integrins that mediate signaling between the cell interior and the ECM. The overall picture of cell-matrix interactions that emerges (Fig. 7-29) shows an array of interactions between cellular and extracellular molecules. These interactions serve not merely to anchor cells to the ECM, providing the strength and elasticity of skin and joints. They also provide paths that direct the migration of cells in developing tissue and serve to convey information in both directions across the plasma membrane.

Glycoproteins Have Covalently Attached Oligosaccharides

Glycoproteins are carbohydrate-protein conjugates in which the glycans are branched and are smaller and more structurally diverse than the huge glycosaminoglycans of proteoglycans. The carbohydrate is attached at its anomeric carbon through a glycosidic link to the —OH of a Ser or Thr residue (*O*-linked), or through an *N*-glycosyl link to the amide nitrogen of an Asn residue (*N*-linked) (Fig. 7-30). Some glycoproteins have a single oligosaccharide chain, but many have more than one; the carbohydrate may constitute from 1% to 70% or more of the glycoprotein by mass. About half of all proteins of mammals are glycosylated, and about 1% of all mammalian genes encode enzymes involved in the synthesis and attachment of these oligosaccharide chains. *N*-linked oligosaccharides are generally found in the consensus sequence N-{P}-[ST]; not all potential sites are used. (See Box 3-2 for the conventions on representing consensus sequences.) There appears to be no specific consensus sequence for *O*-linked oligosaccharides, although regions bearing *O*-linked chains tend to be rich in Gly, Val, and Pro residues.

One class of glycoproteins found in the cytoplasm and the nucleus is unique in that the glycosylated positions in the protein carry only single residues of *N*-acetylglucosamine, in *O*-glycosidic linkage to the hydroxyl group of Ser side chains. This modification is reversible and often occurs on the same Ser residues that

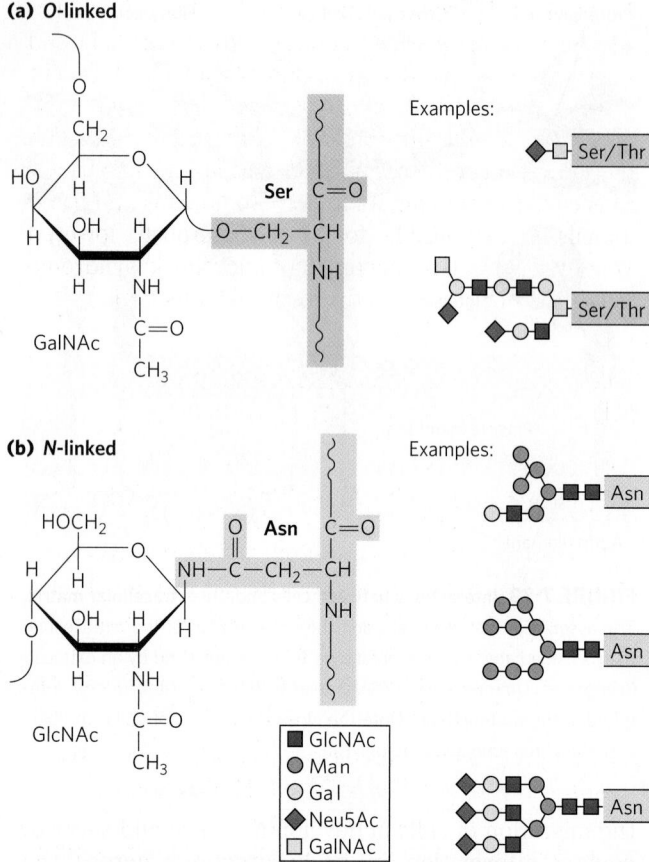

FIGURE 7-30 Oligosaccharide linkages in glycoproteins. (a) O-linked oligosaccharides have a glycosidic bond to the hydroxyl group of Ser or Thr residues (light red), illustrated here with GalNAc as the sugar at the reducing end of the oligosaccharide. One simple chain and one complex chain are shown. **(b)** N-linked oligosaccharides have an N-glycosyl bond to the amide nitrogen of an Asn residue (green), illustrated here with GlcNAc as the terminal sugar. Three common types of oligosaccharide chains that are N-linked in glycoproteins are shown. A complete description of oligosaccharide structure requires specification of the position and stereochemistry (α or β) of each glycosidic linkage.

are phosphorylated at some stage in the protein's activity. The two modifications are mutually exclusive, and this type of glycosylation is important in the regulation of protein activity. We discuss protein phosphorylation at length in Chapter 12.

As we shall see in Chapter 11, the external surface of the plasma membrane has many membrane glycoproteins with arrays of covalently attached oligosaccharides of varying complexity. **Mucins** are secreted or membrane glycoproteins that can contain large numbers of O-linked oligosaccharide chains. Mucins are present in most secretions; they are what gives mucus its characteristic slipperiness.

Glycomics is the systematic characterization of all carbohydrate components of a given cell or tissue, including those attached to proteins and to lipids. For glycoproteins, this also means determining which proteins are glycosylated and where in the amino acid sequence each oligosaccharide is attached. This is a

challenging undertaking, but worthwhile because of the potential insights it offers into normal patterns of glycosylation and the ways in which they are altered during development or in genetic diseases or cancer. Current methods of characterizing the entire carbohydrate complement of cells depend heavily on sophisticated application of mass spectrometry (see Fig. 7-39).

The structures of a large number of O- and N-linked oligosaccharides from a variety of glycoproteins are known; Figures 7-23 and 7-30 show a few typical examples. We consider the mechanisms by which specific proteins acquire specific oligosaccharide moieties in Chapter 27.

Many of the proteins secreted by eukaryotic cells are glycoproteins, including most of the proteins of blood. For example, immunoglobulins (antibodies) and certain hormones, such as follicle-stimulating hormone, luteinizing hormone, and thyroid-stimulating hormone, are glycoproteins. Many milk proteins, including the major whey protein α-lactalbumin, and some of the proteins secreted by the pancreas (such as ribonuclease) are glycosylated, as are most of the proteins contained in lysosomes.

The biological advantages of adding oligosaccharides to proteins are slowly being uncovered. The very hydrophilic carbohydrate clusters alter the polarity and solubility of the proteins with which they are conjugated. Oligosaccharide chains that are attached to newly synthesized proteins in the endoplasmic reticulum (ER) and elaborated in the Golgi complex serve as destination labels and also act in protein quality control, targeting misfolded proteins for degradation (see Figs 27-41, 27-42). When numerous negatively charged oligosaccharide chains are clustered in a single region of a protein, the charge repulsion among them favors the formation of an extended, rodlike structure in that region. The bulkiness and negative charge of oligosaccharide chains also protect some proteins from attack by proteolytic enzymes. Beyond these global physical effects on protein structure, there are also more specific biological effects of oligosaccharide chains in glycoproteins (Section 7.4). The importance of normal protein glycosylation is clear from the finding of at least 40 different genetic disorders of glycosylation in humans, all causing severely defective physical or mental development; some of these disorders are fatal.

Glycolipids and Lipopolysaccharides Are Membrane Components

Glycoproteins are not the only cellular components that bear complex oligosaccharide chains; some lipids, too, have covalently bound oligosaccharides. **Gangliosides** are membrane lipids of eukaryotic cells in which the polar head group, the part of the lipid that forms the outer surface of the membrane, is a complex oligosaccharide containing a sialic acid (Fig. 7-9) and other monosaccharide residues. Some of the oligosaccharide moieties of gangliosides, such as those that determine human blood groups (see Fig. 10-14), are identical with those found in certain glycoproteins, which therefore also contribute to blood group type. Like the oligosaccharide moieties of glycoproteins, those of membrane lipids are generally, perhaps always, found on the outer face of the plasma membrane.

Lipopolysaccharides are the dominant surface feature of the outer membrane of gram-negative bacteria such as *Escherichia coli* and *Salmonella typhimurium*. These molecules are prime targets of the antibodies produced by the vertebrate immune system in response to bacterial infection and are therefore important determinants of the serotype of bacterial strains. (Serotypes are strains that are distinguished on the basis of antigenic properties.) The lipopolysaccharides of *S. typhimurium* contain six fatty acids bound to two glucosamine residues, one of which is the point of attachment for a complex oligosaccharide **(Fig. 7-31)**. *E. coli* has similar but unique lipopolysaccharides. The lipid A portion of the lipopolysaccharides of some bacteria is called endotoxin; its toxicity to humans and other animals is responsible for the dangerously lowered blood pressure that occurs in toxic shock syndrome resulting from gram-negative bacterial infections. ▪

SUMMARY 7.3 Glycoconjugates: Proteoglycans, Glycoproteins, and Glycosphingolipids

■ Proteoglycans are glycoconjugates in which one or more large glycans, called sulfated glycosaminoglycans (heparan sulfate, chondroitin sulfate, dermatan sulfate, or keratan sulfate), are covalently attached to a core protein. Bound to the outside of the plasma membrane by a transmembrane peptide or a covalently attached lipid, proteoglycans provide points of adhesion, recognition, and information transfer between cells, or between a cell and the extracellular matrix.

■ Glycoproteins contain oligosaccharides covalently linked to Asn or Ser/Thr residues. The glycans are typically branched and smaller than glycosaminoglycans. Many cell surface or extracellular proteins are glycoproteins, as are most secreted proteins. The covalently attached oligosaccharides influence the folding and stability of the proteins, provide critical information about the targeting of newly synthesized proteins, and allow specific recognition by other proteins.

■ Glycomics is the determination of the full complement of sugar-containing molecules in a cell or tissue and determination of the function of each such molecule.

■ Glycolipids and glycosphingolipids in plants and animals and lipopolysaccharides in bacteria are components of the cell envelope, with covalently attached oligosaccharide chains exposed on the cell's outer surface.

7.4 Carbohydrates as Informational Molecules: The Sugar Code

Glycobiology, the study of the structure and function of glycoconjugates, is one of the most active and exciting areas of biochemistry and cell biology. It is becoming increasingly clear that cells use specific oligosaccharides to encode important information about intracellular targeting of proteins, cell-cell interactions, cell differentiation and tissue development, and extracellular signals. Our discussion uses just a few examples to illustrate the diversity of structure and the range of biological activity of the glycoconjugates. In Chapter 20 we discuss the biosynthesis of polysaccharides, and in

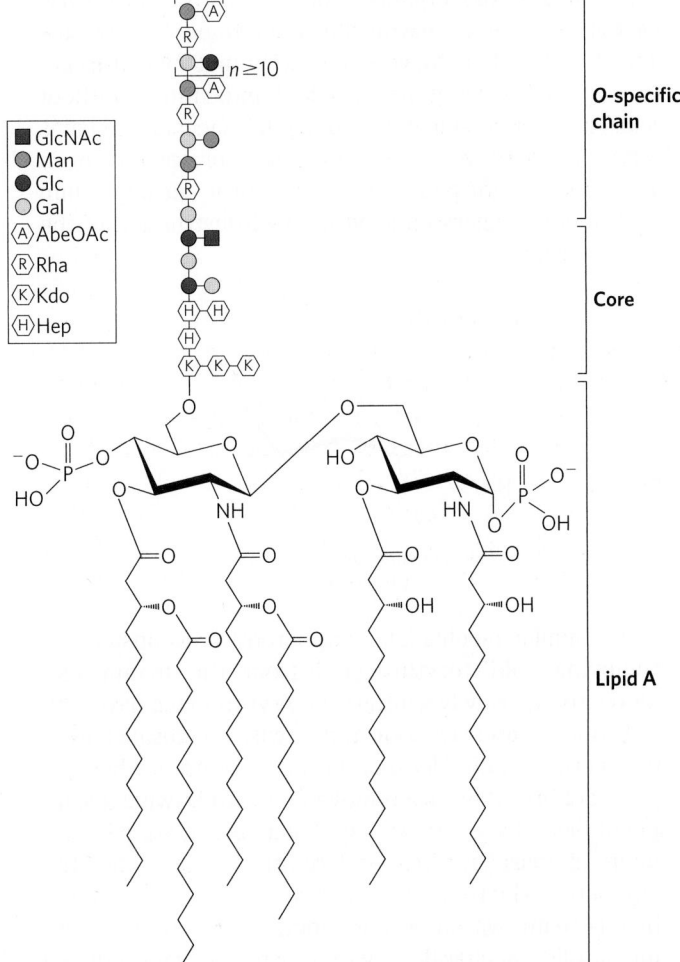

O-specific chain

Core

Lipid A

■ GlcNAc
● Man
● Glc
○ Gal
Ⓐ AbeOAc
Ⓡ Rha
Ⓚ Kdo
Ⓗ Hep

n ≥ 10

FIGURE 7-31 Bacterial lipopolysaccharides. Schematic diagram of the lipopolysaccharide of the outer membrane of *Salmonella typhimurium*. Kdo is 3-deoxy-D-*manno*-octulosonic acid (previously called *ketodeoxyoctonic acid*); Hep is L-*glycero*-D-*manno*-heptose; AbeOAc is abequose (a 3,6-dideoxyhexose) acetylated on one of its hydroxyls. Different bacterial species have subtly different lipopolysaccharide structures, but they have in common a lipid region (lipid A, also known as endotoxin), composed of six fatty acid residues, and two phosphorylated glucosamines, a core oligosaccharide, and an "O-specific" chain, which is the principal determinant of the serotype (immunological reactivity) of the bacterium. The outer membranes of the gram-negative bacteria *S. typhimurium* and *E. coli* contain so many lipopolysaccharide molecules that the cell surface is almost completely covered with O-specific chains.

Chapter 27, the assembly of oligosaccharide chains on glycoproteins.

Improved methods for the analysis of oligosaccharide and polysaccharide structure have revealed remarkable complexity and diversity in the oligosaccharides of glycoproteins and glycolipids. Consider the oligosaccharide chains in Figure 7-30, typical of those found in many glycoproteins. The most complex of those shown contains 14 monosaccharide residues of four different kinds, variously linked as $(1\rightarrow2)$, $(1\rightarrow3)$, $(1\rightarrow4)$, $(1\rightarrow6)$, $(2\rightarrow3)$, and $(2\rightarrow6)$, some with the α and some with the β configuration. Branched structures, not found in nucleic acids or proteins, are common in oligosaccharides. With the reasonable assumption that 20 different monosaccharide subunits are available for construction of oligosaccharides, we can calculate that many billions of different hexameric oligosaccharides are possible; this compares with 6.4×10^7 (20^6) different hexapeptides possible for the 20 common amino acids, and 4,096 (4^6) different hexanucleotides for the four nucleotide subunits. If we also allow for variations in oligosaccharides resulting from sulfation of one or more residues, the number of possible oligosaccharides increases by two orders of magnitude. In reality, only a subset of possible combinations is found, given the restrictions imposed by the biosynthetic enzymes and the availability of precursors. Nevertheless, the enormously rich structural information in glycans does not merely rival but far surpasses that of nucleic acids in the density of information contained in a molecule of modest size. Each of the oligosaccharides represented in Figures 7-23 and 7-30 presents a unique, three-dimensional face—a word in the sugar code—readable by the proteins that interact with it.

Lectins Are Proteins That Read the Sugar Code and Mediate Many Biological Processes

Lectins, found in all organisms, are proteins that bind carbohydrates with high specificity and with moderate to high affinity. Lectins serve in a wide variety of cell-cell recognition, signaling, and adhesion processes and in intracellular targeting of newly synthesized proteins. Plant lectins, abundant in seeds, probably serve as deterrents to insects and other predators. In the laboratory, purified plant lectins are useful reagents for detecting and separating glycans and glycoproteins with different oligosaccharide moieties. Here we discuss just a few examples of the roles of lectins in animal cells.

Some peptide hormones that circulate in the blood have oligosaccharide moieties that strongly influence their circulatory half-life. Luteinizing hormone and thyrotropin (polypeptide hormones produced in the pituitary) have N-linked oligosaccharides that end with the disaccharide GalNAc4S($\beta1\rightarrow4$)GlcNAc, which is recognized by a lectin (receptor) of hepatocytes. (GalNAc4S is N-acetylgalactosamine sulfated on the —OH group at C-4.) Receptor-hormone interaction mediates the

uptake and destruction of luteinizing hormone and thyrotropin, reducing their concentration in the blood. Thus the blood levels of these hormones undergo a periodic rise (due to pulsatile secretion by the pituitary) and fall (due to constant destruction by hepatocytes).

The residues of Neu5Ac (a sialic acid) situated at the ends of the oligosaccharide chains of many plasma glycoproteins (Fig. 7-23) protect these proteins from uptake and degradation in the liver. For example, ceruloplasmin, a copper-containing serum glycoprotein, has several oligosaccharide chains ending in Neu5Ac. The mechanism that removes sialic acid residues from serum glycoproteins is unclear. It may be due to the activity of the enzyme neuraminidase (also called sialidase) produced by invading organisms or to a steady, slow release of the residues by extracellular enzymes. The plasma membrane of hepatocytes has lectin molecules (asialoglycoprotein receptors; "asialo-" indicating "without sialic acid") that specifically bind oligosaccharide chains with galactose residues no longer "protected" by a terminal Neu5Ac residue. Receptor-ceruloplasmin interaction triggers endocytosis and destruction of the ceruloplasmin.

N-Acetylneuraminic acid (Neu5Ac)
(a sialic acid)

A similar mechanism is apparently responsible for removing "old" erythrocytes from the mammalian bloodstream. Newly synthesized erythrocytes have several membrane glycoproteins with oligosaccharide chains that end in Neu5Ac. In the laboratory, when the sialic acid residues are removed by withdrawing a sample of blood from experimental animals, treating it with neuraminidase in vitro, and reintroducing it into the circulation, the treated erythrocytes disappear from the bloodstream within a few hours; erythrocytes with intact oligosaccharides (withdrawn and reintroduced without neuraminidase treatment) continue to circulate for days.

Cell surface lectins—both human lectins and the lectins of infectious agents—are important in the development of some diseases. **Selectins** are a family of plasma membrane lectins that mediate cell-cell recognition and adhesion in a wide range of cellular processes. One such process is the movement of immune cells (leukocytes) through the capillary wall, from blood to tissues, at sites of infection or inflammation **(Fig. 7-32)**. At an infection site, P-selectin on the surface of capillary endothelial cells interacts with a specific oligosaccharide of the surface glycoproteins of circulating leukocytes.

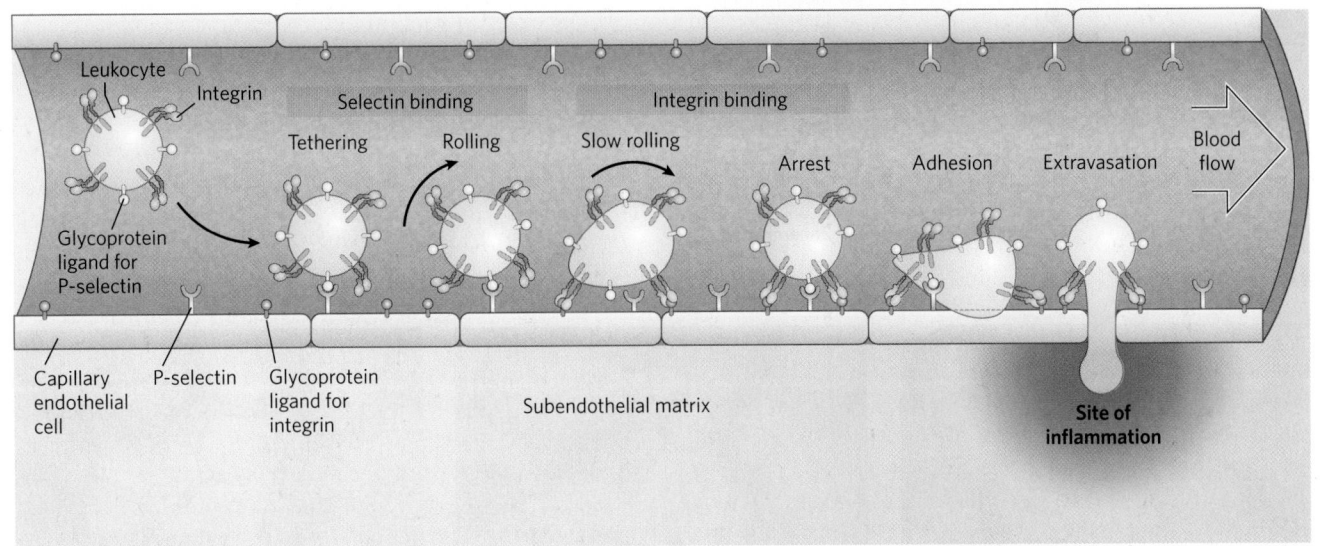

FIGURE 7-32 Role of lectin-ligand interactions in leukocyte movement to the site of an infection or injury. A leukocyte circulating through a capillary is slowed by transient interactions between P-selectin molecules in the plasma membrane of the capillary endothelial cells and glycoprotein ligands for P-selectin on the leukocyte surface. As the leukocyte interacts with successive P-selectin molecules, it rolls along the capillary surface. Near a site of inflammation, stronger interactions between integrin in the leukocyte surface and its ligand in the capillary surface lead to tight adhesion. The leukocyte stops rolling and, under the influence of signals sent from the site of inflammation, begins extravasation—escape through the capillary wall—as it moves toward the region of inflammation.

This interaction slows the leukocytes as they roll along the endothelial lining of the capillaries. A second interaction, between integrin molecules in the leukocyte plasma membrane and an adhesion protein on the endothelial cell surface, now stops the leukocyte and allows it to move through the capillary wall into the infected tissues to initiate the immune attack. Two other selectins participate in this "lymphocyte homing": E-selectin on the endothelial cell and L-selectin on the leukocyte bind their cognate oligosaccharides on the leukocyte and endothelial cell, respectively.

Human selectins mediate the inflammatory responses in rheumatoid arthritis, asthma, psoriasis, multiple sclerosis, and the rejection of transplanted organs, and thus there is great interest in developing drugs that inhibit selectin-mediated cell adhesion. Many carcinomas express an antigen normally present only in fetal cells (sialyl Lewis x, or sialyl Le^x) that, when shed into the circulation, facilitates tumor cell survival and metastasis. Carbohydrate derivatives that mimic the sialyl Le^x portion of sialoglycoproteins or that alter the biosynthesis of the oligosaccharide might prove effective as selectin-specific drugs for treating chronic inflammation or metastatic disease.

Several animal viruses, including the influenza virus, attach to their host cells through interactions with oligosaccharides displayed on the host cell surface. The lectin of the influenza virus, known as the HA (hemagglutinin) protein, is essential for viral entry and infection. After the virus has entered a host cell and has been replicated, the newly synthesized viral particles bud out of the cell, wrapped in a portion of its plasma membrane. A viral sialidase (neuraminidase) trims the terminal sialic acid residue from the host cell's oligosaccharides, releasing the viral particles from their interaction with the cell and preventing their aggregation with one another. Another round of infection can now begin. The antiviral drugs oseltamivir (Tamiflu) and zanamivir (Relenza) are used clinically in the treatment of influenza. These drugs are sugar analogs; they inhibit the viral sialidase by competing with the host cell's oligosaccharides for binding **(Fig. 7-33)**. This prevents the release of viruses from the infected cell and also causes viral particles to aggregate, both of which block another cycle of infection.

Some microbial pathogens have lectins that mediate bacterial adhesion to host cells or the entry of toxin into cells. For example, *Helicobacter pylori* has a surface lectin that adheres to oligosaccharides on the surface of epithelial cells that line the inner surface of the stomach **(Fig. 7-34)**. Among the binding sites recognized by the *H. pylori* lectin is the oligosaccharide Lewis b (Le^b), which is present in the glycoproteins and glycolipids that define the type O blood group determinant (see Fig. 10-14). This observation helps to explain the severalfold greater incidence of gastric ulcers in people of blood type O than in those of type A or B; *H. pylori* attacks their epithelial cells more effectively. Chemically synthesized analogs of the Le^b oligosaccharide may prove useful in treating this type of ulcer. Administered orally, they could prevent bacterial adhesion (and thus infection) by competing with the gastric glycoproteins for binding to the bacterial lectin.

Some of the most devastating of the human parasitic diseases, widespread in much of the developing

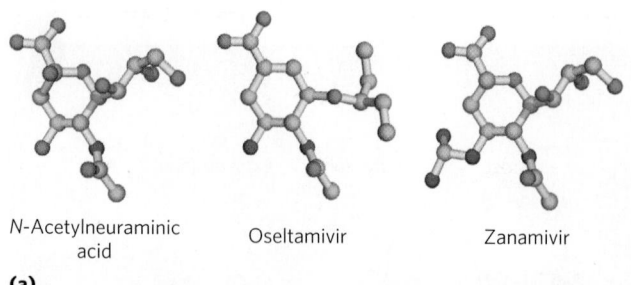

N-Acetylneuraminic
acid   Oseltamivir   Zanamivir

(a)

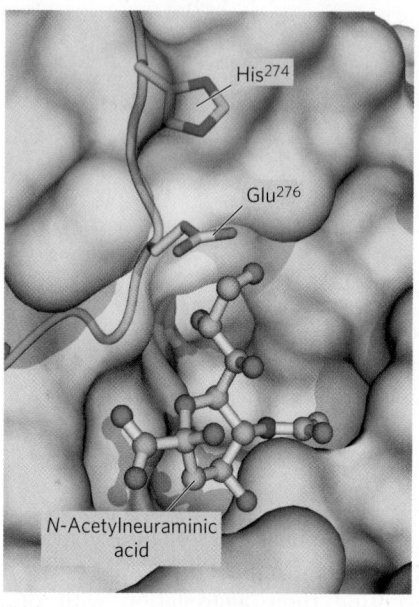

(b)

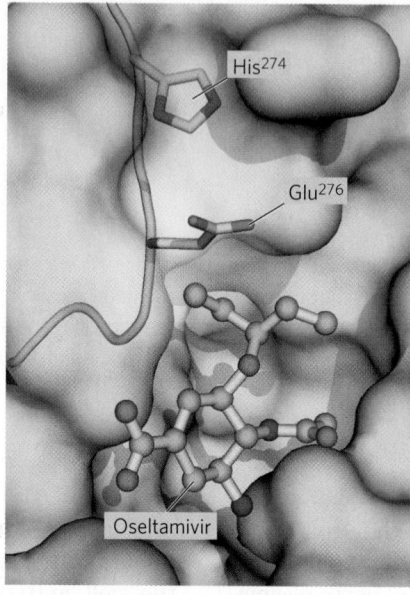

(c)

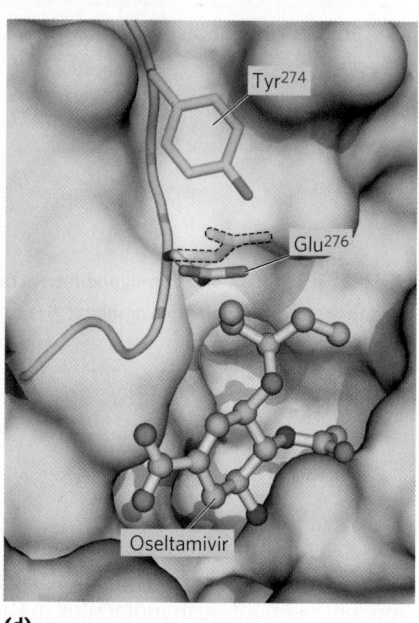

(d)

FIGURE 7-33 Binding site on influenza neuraminidase for N-acetylneuraminic acid and an antiviral drug, oseltamivir. (a) The normal binding ligand for this enzyme is a sialic acid, N-acetylneuraminic acid. The drugs oseltamivir and zanamivir occupy the same site on the enzyme, competitively inhibiting it and blocking viral release from the host cell. **(b)** The normal interaction with N-acetylneuraminic acid in the binding site. **(c)** Oseltamivir can fit into this site by pushing a nearby Glu residue out of the way. **(d)** A mutation in the influenza virus's gene for neuraminidase replaces a His near this Glu residue with the larger side chain of a Tyr. Now, oseltamivir is not as effective at pushing the Glu out of its way, and the drug binds much less well to the binding site, making the mutant virus effectively resistant to oseltamivir. [Sources: (b) PDB ID 2BAT, J. N. Varghese et al., *Proteins* 14:327, 1992. (c) PDB ID 2HU4, R. J. Russell et al., *Nature* 443:45, 2006. (d) PDB ID 3CL0, P. J. Collins et al., *Nature* 453:1258, 2008.]

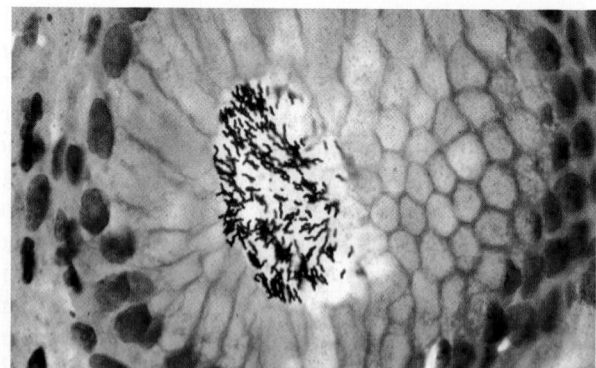

FIGURE 7-34 An ulcer in the making. *Helicobacter pylori* cells adhering to the gastric surface. This bacterium causes ulcers through interactions between a bacterial surface lectin and the Le[b] oligosaccharide (a blood group antigen) of the epithelial cells lining the inside surface of the stomach. [Source: R. M. Genta/Miraca Life Sciences Research Institute, Irving, Texas, and D. Y. Graham/Veterans Affairs Medical Center, Houston, Texas.]

world, are caused by eukaryotic microorganisms that display unusual surface oligosaccharides, which in some cases are known to be protective for the parasites. These organisms include the trypanosomes, responsible for African sleeping sickness and Chagas disease (see Box 6-3); *Plasmodium falciparum*, the malaria parasite; and *Entamoeba histolytica*, the causative agent of amoebic dysentery. The prospect of finding drugs that interfere with the synthesis of these unusual oligosaccharide chains, and therefore with the replication of the parasites, has inspired much recent work on the biosynthetic pathways of these oligosaccharides. ■

Lectins also act intracellularly, in sorting proteins for transportation to specific cellular compartments (see Chapter 27). For example, an oligosaccharide containing mannose 6-phosphate, recognized by a lectin, marks newly synthesized proteins in the Golgi complex for transfer to the lysosome (see Fig. 27-41).

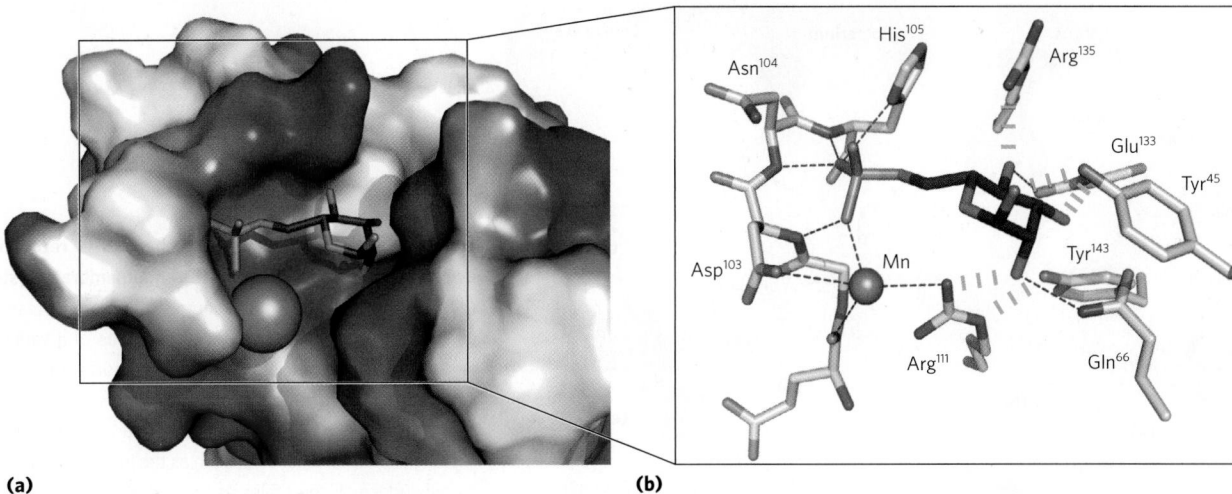

(a)   **(b)**

FIGURE 7-35 Details of a lectin-carbohydrate interaction. (a) Structure of the bovine mannose 6-phosphate receptor complexed with mannose 6-phosphate. The protein is represented as a surface contour image, showing the surface as predominantly negatively charged (red) or positively charged (blue). Mannose 6-phosphate is shown as a stick structure; a manganese ion is shown as a violet sphere. **(b)** An enlarged view of the binding site. Mannose 6-phosphate is hydrogen-bonded to

Arg[111] and coordinated with the manganese ion (shown smaller than its van der Waals radius, for clarity). Each hydroxyl group of mannose is hydrogen-bonded to the protein. The His[105] hydrogen-bonded to a phosphate oxygen of mannose 6-phosphate may be the residue that, when protonated at low pH, causes the receptor to release mannose 6-phosphate into the lysosome. [Source: (a, b) PDB ID 1M6P, D. L. Roberts et al., *Cell* 93:639, 1998.]

Lectin-Carbohydrate Interactions Are Highly Specific and Often Multivalent

The high density of information in the structure of oligosaccharides provides a sugar code with an essentially unlimited number of unique "words" small enough to be read by a single protein. In their carbohydrate-binding sites, lectins have a subtle molecular complementarity that allows interaction only with their correct carbohydrate binding partners. The result is an extraordinarily high specificity in these interactions. The affinity between an oligosaccharide and an individual carbohydrate-binding domain (CBD) of a lectin is sometimes modest (micromolar to millimolar K_d values), but the effective affinity is often greatly increased by lectin multivalency, in which a single lectin molecule has multiple CBDs. In a cluster of oligosaccharides—as is commonly found on a membrane surface, for example—each oligosaccharide can engage one of the lectin's CBDs, strengthening the interaction. When cells express multiple lectin receptors, the avidity of the interaction can be very high, enabling highly cooperative events such as cell attachment and rolling (Fig. 7-32).

X-ray crystallographic studies of the structure of the mannose 6-phosphate receptor/lectin reveal details of its interaction with mannose 6-phosphate that explain the specificity of the binding and the role of a divalent cation in the lectin-sugar interaction **(Fig. 7-35a)**. His[105] is hydrogen-bonded to one of the oxygen atoms of the phosphate (Fig. 7-35b). When the protein tagged with mannose 6-phosphate reaches the lysosome (which

has a lower internal pH than the Golgi complex), the receptor loses its affinity for mannose 6-phosphate. Protonation of His[105] may be responsible for this change in binding.

In addition to such highly specific interactions, there are more general interactions that contribute to the binding of many carbohydrates to their lectins. For example, many sugars have a more polar and a less polar

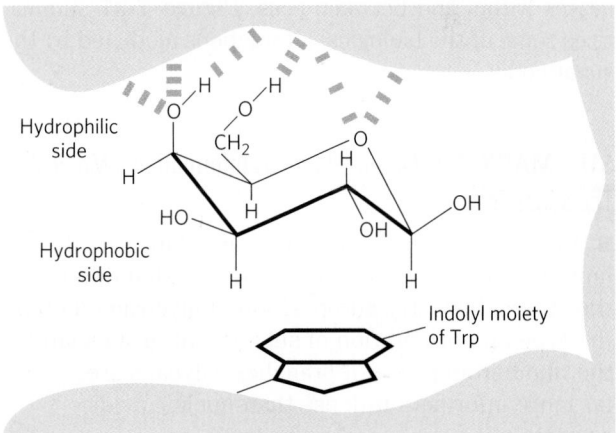

FIGURE 7-36 Interactions of sugar residues due to the hydrophobic effect. Sugar units such as galactose have a more polar side (the top of the chair as shown here, with the ring oxygen and several hydroxyls), available to hydrogen-bond with the lectin, and a less polar side that can interact with nonpolar side chains in the protein, such as the indole ring of Trp residues, through the hydrophobic effect. [Source: Information from a figure provided by Dr. C.-W. von der Lieth, Heidelberg; H.-J. Gabius, *Naturwissenschaften* 87:108, 2000, Fig. 6.]

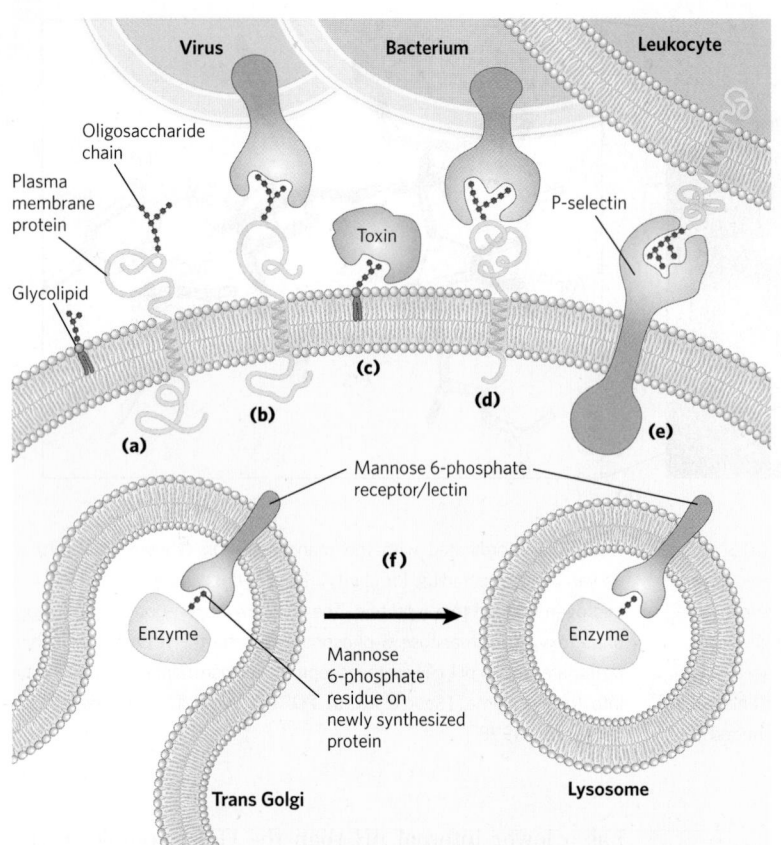

FIGURE 7-37 Role of oligosaccharides in recognition events at the cell surface and in the endomembrane system. (a) Oligosaccharides with unique structures (represented as strings of red hexagons) are components of a variety of glycoproteins or glycolipids on the outer surface of plasma membranes. Their oligosaccharide moieties are bound by extracellular lectins with high specificity and affinity. (b) Viruses that infect animal cells, such as the influenza virus, bind to cell surface glycoproteins as the first step in infection. (c) Bacterial toxins, such as the cholera and pertussis toxins, bind to a surface glycolipid before entering a cell. (d) Some bacteria, such as *H. pylori*, adhere to and then colonize or infect animal cells. (e) Selectins (lectins) in the plasma membrane of certain cells mediate cell-cell interactions, such as those of leukocytes with the endothelial cells of the capillary wall at an infection site. (f) The mannose 6-phosphate receptor/lectin of the trans Golgi complex binds to the oligosaccharide of lysosomal enzymes, targeting them for transfer into the lysosome. [Source: Information from N. Sharon and H. Lis, *Sci. Am.* 268 (January):82, 1993.]

side **(Fig. 7-36)**; the more polar side hydrogen-bonds with the lectin, while the less polar undergoes interactions with nonpolar amino acid residues through the hydrophobic effect. The sum of all these interactions produces high-affinity binding and high specificity of lectins for their carbohydrates. This represents a kind of information transfer that is clearly central in many processes within and between cells. **Figure 7-37** summarizes some of the biological interactions mediated by the sugar code.

SUMMARY 7.4 Carbohydrates as Informational Molecules: The Sugar Code

■ Monosaccharides can be assembled into an almost limitless variety of oligosaccharides, which differ in the stereochemistry and position of glycosidic bonds, the type and orientation of substituent groups, and the number and type of branches. Glycans are far more information-dense than nucleic acids or proteins.

■ Lectins, proteins with highly specific carbohydrate-binding domains, are commonly found on the outer surface of cells, where they initiate interaction with other cells. In vertebrates, oligosaccharide tags "read" by lectins govern the rate of degradation of certain peptide hormones, circulating proteins, and blood cells.

■ Bacterial and viral pathogens and some eukaryotic parasites adhere to their animal cell targets through binding of lectins in the pathogens to oligosaccharides on the target cell surface.

■ X-ray crystallography of lectin-sugar complexes shows the detailed complementarity between the two molecules, which accounts for the strength and specificity of lectin interactions with carbohydrates.

7.5 Working with Carbohydrates

A growing appreciation of the importance of oligosaccharide structure in biological signaling and recognition has been the driving force behind the development of methods for analyzing the structure and stereochemistry of complex oligosaccharides. Oligosaccharide analysis is complicated by the fact that, unlike nucleic acids and proteins, oligosaccharides can be branched and are joined by a variety of linkages. The high charge density of many oligosaccharides and polysaccharides, and the relative lability of the sulfate esters in glycosaminoglycans, present further difficulties.

For simple, linear polymers such as amylose, the positions of the glycosidic bonds are determined by the classical method of exhaustive methylation: treating the intact polysaccharide with methyl iodide in a strongly basic medium to convert all free hydroxyls to acid-stable methyl ethers, then hydrolyzing the methylated

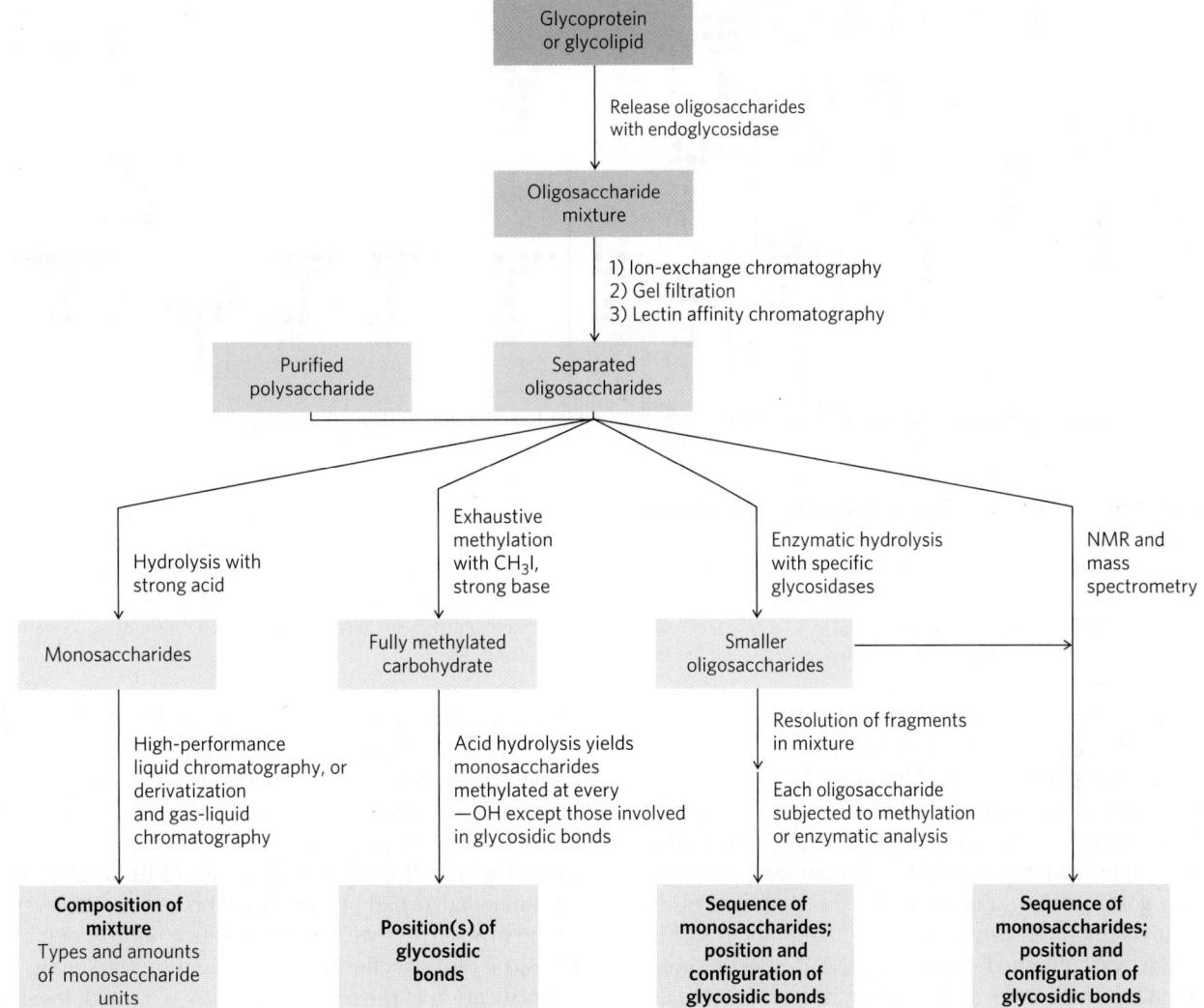

FIGURE 7-38 Methods of carbohydrate analysis. A carbohydrate purified in the first stage of the analysis often requires all four analytical routes for its complete characterization.

polysaccharide in acid. The only free hydroxyls in the monosaccharide derivatives so produced are those that were involved in glycosidic bonds. To determine the sequence of monosaccharide residues, including any branches that are present, exoglycosidases of known specificity are used to remove residues one at a time from the nonreducing end(s). The known specificity of these exoglycosidases often allows deduction of the position and stereochemistry of the linkages.

For analysis of the oligosaccharide moieties of glycoproteins and glycolipids, the oligosaccharides are released by purified enzymes—glycosidases that specifically cleave O- or N-linked oligosaccharides, or lipases that remove lipid head groups. Alternatively, O-linked glycans can be released from glycoproteins by treatment with hydrazine.

The resulting mixtures of carbohydrates are resolved into their individual components by a variety of methods **(Fig. 7-38)**, including the same techniques used in protein and amino acid separation: fractional precipitation by solvents, and ion-exchange and size-exclusion chromatography (see Fig. 3-17). Highly purified lectins, attached covalently to an insoluble support, are commonly used in affinity chromatography of carbohydrates.

Hydrolysis of oligosaccharides and polysaccharides in strong acid yields a mixture of monosaccharides, which can be identified and quantified by chromatographic techniques to yield the overall composition of the polymer.

Oligosaccharide analysis relies increasingly on mass spectrometry and high-resolution NMR spectroscopy. Matrix-assisted laser desorption/ionization mass spectrometry (MALDI MS) and tandem mass spectrometry (MS/MS), both described in Chapter 3, are readily applicable to polar compounds such as oligosaccharides. MALDI MS is a very sensitive method for determining the mass of a molecular ion (in this case, the entire

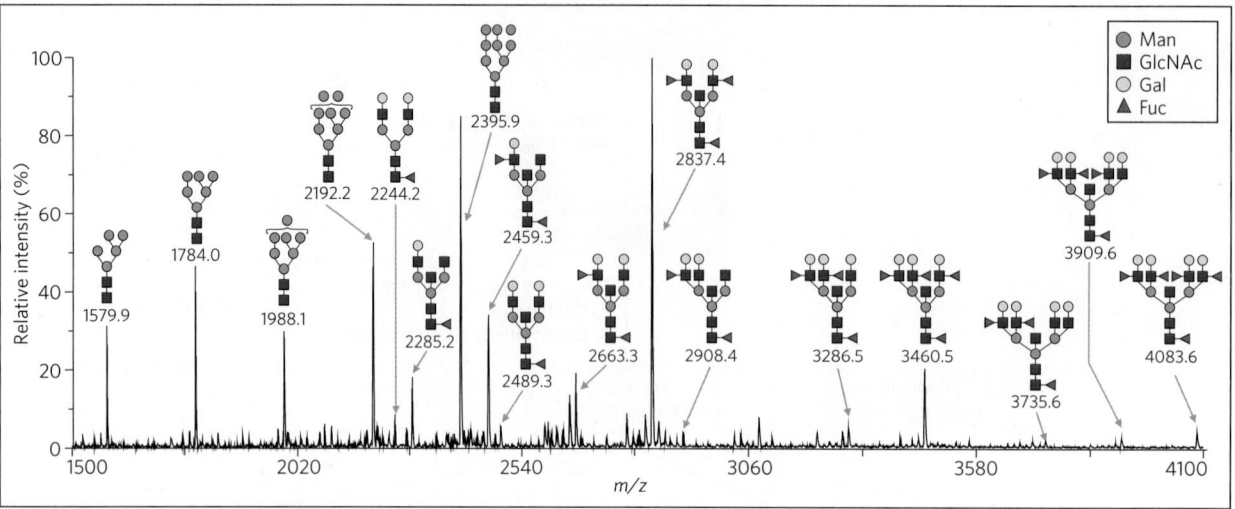

FIGURE 7-39 Separation and quantification of the oligosaccharides in a group of glycoproteins. In this experiment, a mixture of proteins extracted from kidney tissue was treated to release oligosaccharides from glycoproteins, and the oligosaccharides were analyzed by matrix-assisted laser desorption/ionization mass spectrometry (MALDI MS). Each distinct oligosaccharide produces a peak at its molecular mass, and the area under the curve reflects the quantity of that oligosaccharide. The most prominent oligosaccharide here (mass 2837.4 u) is composed of 13 sugar residues; other oligosaccharides, containing as few as 7 and as many as 19 residues, were also resolved by this method. [Source: Courtesy of Anne Dell. Reprinted with permission from E. M. Comelli et al., *Glycobiology* 16:117, 2006, Fig. 3.]

oligosaccharide chain; **Fig. 7-39**). MS/MS reveals the mass of the molecular ion and many of its fragments, which are usually the result of breakage of the glycosidic bonds. NMR analysis alone (see Box 4-5), especially for oligosaccharides of moderate size, can yield much information about sequence, linkage position, and anomeric carbon configuration. For example, the structure of the heparin segment shown as a space-filling model in Figure 7-22 was obtained entirely by NMR spectroscopy. Automated procedures and commercial instruments are used for the routine determination of oligosaccharide structure, but the sequencing of branched oligosaccharides joined by more than one type of bond remains a far more formidable task than determining the linear sequences of proteins and nucleic acids.

Another important tool in working with carbohydrates is chemical synthesis, which has proved to be a powerful approach to understanding the biological functions of glycosaminoglycans and oligosaccharides. The chemistry involved in such syntheses is difficult, but carbohydrate chemists can now synthesize short segments of almost any glycosaminoglycan, with correct stereochemistry, chain length, and sulfation pattern, and oligosaccharides significantly more complex than those shown in Figure 7-30. Solid-phase oligosaccharide synthesis is based on the same principles (and has the same advantages) as peptide synthesis (see Fig. 3-32), but requires a set of tools unique to carbohydrate chemistry: blocking groups and activating groups that allow the synthesis of glycosidic linkages with the correct hydroxyl group. Synthetic approaches of this type currently represent an area of great interest, because it is difficult to purify defined oligosaccharides in adequate quantities from natural sources.

To identify proteins with specific affinity for particular oligosaccharides, **oligosaccharide microarrays** are used. The principle is the same as for DNA microarrays (Figs 9-22, 9-23), but the technical problems are more challenging. Pure oligosaccharides are attached to a glass slide in microdroplets, and the slide is exposed to a potential lectin (glycan-binding protein) that has been tagged with a fluorescent molecule **(Fig. 7-40)**. After all the nonadsorbed protein is washed away, observation of the microarrays with a fluorescence microscope identifies the oligosaccharides recognized by the lectin, and quantification of the fluorescence gives a rough measure of lectin-oligosaccharide affinity.

SUMMARY 7.5 Working with Carbohydrates

■ Establishing the complete structure of oligosaccharides and polysaccharides requires determination of the linear sequence, branching positions, the configuration of each monosaccharide unit, and the positions of the glycosidic linkages—a more complex problem than protein and nucleic acid analysis.

■ The structures of oligosaccharides and polysaccharides are usually determined by a combination of methods: specific enzymatic hydrolysis to determine stereochemistry at the glycosidic bond and to produce smaller fragments for further analysis; methylation to locate glycosidic bonds; and stepwise degradation to determine sequence and configuration of anomeric carbons.

■ Mass spectrometry and high-resolution NMR spectroscopy, applicable to small samples of carbohydrate, yield essential information about sequence, configuration at anomeric and other carbons, and positions of glycosidic bonds.

■ Solid-phase synthetic methods yield defined oligosaccharides that are of great value in exploring

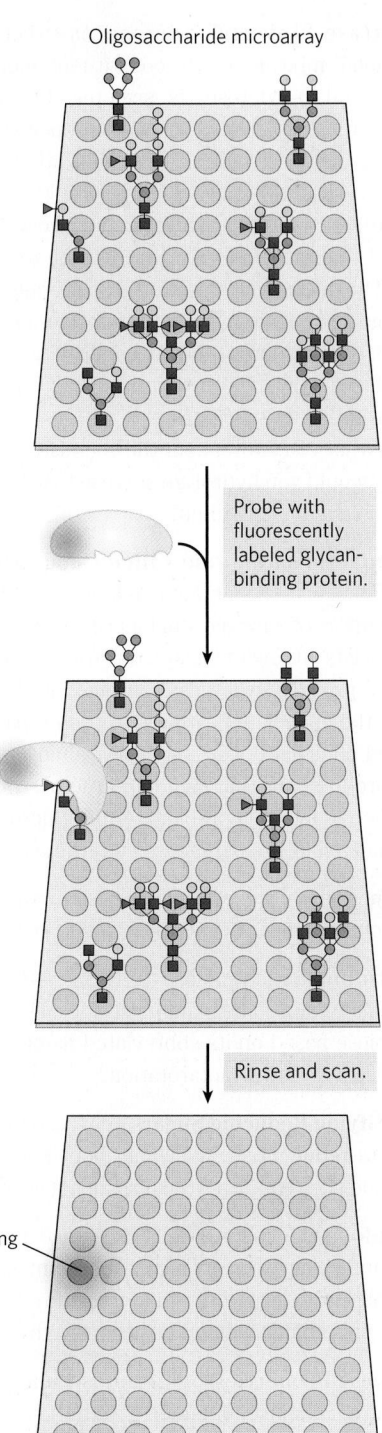

Oligosaccharide microarray

Probe with fluorescently labeled glycan-binding protein.

Rinse and scan.

Binding

FIGURE 7-40 Oligosaccharide microarrays to determine the specificity and affinity of carbohydrate binding by lectins. Solutions of pure samples of oligosaccharides, synthesized or isolated from nature, are placed in microscopic droplets on a glass slide and attached to the glass through an inert spacer. Each spot represents a different oligosaccharide. The protein sample to be tested for its affinity for oligosaccharides is first conjugated with a fluorescent marker, then the sample is poured over the slide and allowed to equilibrate; any nonadsorbed protein is washed away. Observation of the microarray with a fluorescence microscope shows which spots have adsorbed protein (they glow green), and assessment of the fluorescence intensity gives a rough measure of protein-oligosaccharide binding affinity. [Source: Information from P. H. Seeberger, *Nature Chem. Biol.* 5:368, 2009, Fig. 2a.]

lectin-oligosaccharide interactions and may prove clinically useful.

■ Microarrays of pure oligosaccharides are useful in determining the specificity and affinity of lectin binding to specific oligosaccharides.

Key Terms

Terms in bold are defined in the glossary.

Problems

1. Sugar Alcohols In the monosaccharide derivatives known as sugar alcohols, the carbonyl oxygen is reduced to a hydroxyl group. For example, D-glyceraldehyde can be reduced to glycerol. However, this sugar alcohol is no longer designated D or L. Why?

2. Recognizing Epimers Using Figure 7-3, identify the epimers of (a) D-allose, (b) D-gulose, and (c) D-ribose at C-2, C-3, and C-4.

3. Melting Points of Monosaccharide Osazone Derivatives Many carbohydrates react with phenylhydrazine ($C_6H_5NHNH_2$) to form bright yellow crystalline derivatives known as osazones:

$$\text{Glucose} \xrightarrow{\;C_6H_5NHNH_2\;} \text{Osazone derivative of glucose}$$

The melting temperatures of these derivatives are easily determined and are characteristic for each osazone. This information was used to help identify monosaccharides before the development of HPLC or gas chromatography. Listed below are the melting points (MPs) of some aldose-osazone derivatives.

Monosaccharide	MP of anhydrous monosaccharide (°C)	MP of osazone derivative (°C)
Glucose	146	205
Mannose	132	205
Galactose	165–168	201
Talose	128–130	201

As the table shows, certain pairs of derivatives have the same melting points, although the nonderivatized monosaccharides do not. Why do glucose and mannose, and similarly galactose and talose, form osazone derivatives with the same melting points?

4. Configuration and Conformation Which bond(s) in α-D-glucose must be broken to change its configuration to β-D-glucose? Which bond(s) to convert D-glucose to D-mannose? Which bond(s) to convert one "chair" form of D-glucose to the other?

5. Deoxysugars Is D-2-deoxygalactose the same chemical as D-2-deoxyglucose? Explain.

6. Sugar Structures Describe the common structural features and the differences for each of the following pairs: (a) cellulose and glycogen; (b) D-glucose and D-fructose; (c) maltose and sucrose.

7. Reducing Sugars Draw the structural formula for α-D-glucosyl-(1→6)-D-mannosamine, and circle the part of this structure that makes the compound a reducing sugar.

8. Hemiacetal and Glycosidic Linkages Explain the difference between a hemiacetal and a glycoside.

9. A Taste of Honey The fructose in honey is mainly in the β-D-pyranose form. This is one of the sweetest carbohydrates known, about twice as sweet as glucose; the β-D-furanose form of fructose is much less sweet. The sweetness of honey gradually decreases at a high temperature. Also, high-fructose corn syrup (a commercial product in which much of the glucose in corn syrup is converted to fructose) is used for sweetening *cold* but not *hot* drinks. What chemical property of fructose could account for both these observations?

10. Glucose Oxidase in Determination of Blood Glucose The enzyme glucose oxidase isolated from the mold *Penicillium notatum* catalyzes the oxidation of β-D-glucose to D-glucono-δ-lactone. This enzyme is highly specific for the β anomer of glucose and does not affect the α anomer. In spite of this specificity, the reaction catalyzed by glucose oxidase is commonly used in a clinical assay for total blood glucose—that is, for solutions consisting of a mixture of β- and α-D-glucose. What are the circumstances required to make this possible? Aside from allowing the detection of smaller quantities of glucose, what advantage does glucose oxidase offer over Fehling's reagent for measuring blood glucose?

11. Invertase "Inverts" Sucrose As sweet as sucrose is, an equimolar mixture of its constituent monosaccharides, D-glucose and D-fructose, is sweeter. Besides enhancing sweetness, fructose has hygroscopic properties that improve the texture of foods, reducing crystallization and increasing moisture.

In the food industry, hydrolyzed sucrose is called invert sugar, and the yeast enzyme that hydrolyzes it is called invertase. The hydrolysis reaction is generally monitored by measuring the specific rotation of the solution, which is positive (+66.4°) for sucrose, but becomes negative (inverts) as more D-glucose (specific rotation = +52.7°) and D-fructose (specific rotation = −92°) form.

From what you know about the chemistry of the glycosidic bond, how would you hydrolyze sucrose to invert sugar nonenzymatically in a home kitchen?

12. Manufacture of Liquid-Filled Chocolates The manufacture of chocolates containing a liquid center is an interesting application of enzyme engineering. The flavored liquid center consists largely of an aqueous solution of sugars rich in fructose to provide sweetness. The technical dilemma is the following: the chocolate coating must be prepared by pouring hot melted chocolate over a solid (or almost solid) core, yet the final product must have a liquid, fructose-rich center. Suggest a way to solve this problem. (Hint: Sucrose is much less soluble than a mixture of glucose and fructose.)

13. Anomers of Sucrose? Lactose exists in two anomeric forms, but no anomeric forms of sucrose have been reported. Why?

14. Gentiobiose Gentiobiose (D-Glc(β1→6)D-Glc) is a disaccharide found in some plant glycosides. Draw the structure of gentiobiose based on its abbreviated name. Is it a reducing sugar? Does it undergo mutarotation?

15. Identifying Reducing Sugars Is *N*-acetyl-β-D-glucosamine (Fig. 7-9) a reducing sugar? What about D-gluconate? Is the disaccharide GlcN(α1→1α)Glc a reducing sugar?

16. Cellulose Digestion Cellulose could provide a widely available and cheap form of glucose, but humans cannot digest it. Why not? If you were offered a procedure that allowed you to acquire this ability, would you accept? Why or why not?

17. Physical Properties of Cellulose and Glycogen The almost pure cellulose obtained from the seed threads of *Gossypium* (cotton) is tough, fibrous, and completely insoluble in water. In contrast, glycogen obtained from muscle or liver disperses readily in hot water to make a turbid solution. Despite their markedly different physical properties, both substances are (1→4)-linked D-glucose polymers of comparable molecular weight. What structural features of these two polysaccharides underlie their different physical properties? Explain the biological advantages of their respective properties.

18. Dimensions of a Polysaccharide Compare the dimensions of a molecule of cellulose and a molecule of amylose, each of M_r 200,000.

19. Growth Rate of Bamboo The stems of bamboo, a tropical grass, can grow at the phenomenal rate of 0.3 m/day under optimal conditions. Given that the stems are composed almost entirely of cellulose fibers oriented in the direction of growth, calculate the number of sugar residues per second that must be added enzymatically to growing cellulose chains to account for the growth rate. Each D-glucose unit contributes ~0.5 nm to the length of a cellulose molecule.

20. Glycogen as Energy Storage: How Long Can a Game Bird Fly? Since ancient times it has been observed that certain game birds, such as grouse, quail, and pheasants, are easily fatigued. The Greek historian Xenophon wrote: "The bustards . . . can be caught if one is quick in starting them up, for they will fly only a short distance, like partridges, and soon tire; and their flesh is delicious." The flight muscles of game birds rely almost entirely on the use of glucose 1-phosphate for energy, in the form of ATP (Chapter 14). The glucose 1-phosphate is formed by the breakdown of stored muscle glycogen, catalyzed by the enzyme glycogen phosphorylase. The rate of ATP production is limited by the rate at which glycogen can be broken down. During a "panic flight," the game bird's rate of glycogen breakdown is quite high, approximately 120 μmol/min of glucose 1-phosphate produced per gram of fresh tissue. Given that the flight muscles usually contain about 0.35% glycogen by weight, calculate how long a game bird can fly. (Assume the average molecular weight of a glucose residue in glycogen is 162 g/mol.)

21. Relative Stability of Two Conformers Explain why the two structures shown in Figure 7-18b are so different in energy (stability). Hint: See Figure 1-23.

22. Volume of Chondroitin Sulfate in Solution One critical function of chondroitin sulfate is to act as a lubricant in skeletal joints by creating a gel-like medium that is resilient to friction and shock. This function seems to be related to a distinctive property of chondroitin sulfate: the volume occupied by the molecule is much greater in solution than in the dehydrated solid. Why is the volume so much larger in solution?

23. Heparin Interactions Heparin, a highly negatively charged glycosaminoglycan, is used clinically as an anticoagulant. It acts by binding several plasma proteins, including antithrombin III, an inhibitor of blood clotting. The 1:1 binding of heparin to antithrombin III seems to cause a conformational change in the protein that greatly increases its ability to inhibit clotting. What amino acid residues of antithrombin III are likely to interact with heparin?

24. Permutations of a Trisaccharide Think about how one might estimate the number of possible trisaccharides composed of N-acetylglucosamine 4-sulfate (GlcNAc4S) and glucuronic acid (GlcA), and draw 10 of them.

25. Effect of Sialic Acid on SDS Polyacrylamide Gel Electrophoresis Suppose you have four forms of a protein, all with identical amino acid sequence but containing zero, one, two, or three oligosaccharide chains, each ending in a single sialic acid residue. Draw the gel pattern you would expect when a mixture of these four glycoproteins is subjected to SDS polyacrylamide gel electrophoresis (see Fig. 3-18) and stained for protein. Identify any bands in your drawing.

26. Information Content of Oligosaccharides The carbohydrate portion of some glycoproteins may serve as a cellular recognition site. To perform this function, the oligosaccharide moiety must have the potential to exist in a large variety of forms. Which can produce a greater variety of structures: oligopeptides composed of five different amino acid residues, or oligosaccharides composed of five different monosaccharide residues? Explain.

27. Determination of the Extent of Branching in Amylopectin The amount of branching (number of (α1→6) glycosidic bonds) in amylopectin can be determined by the following procedure. A sample of amylopectin is exhaustively methylated—treated with a methylating agent (methyl iodide) that replaces the hydrogen of every sugar hydroxyl with a methyl group, converting —OH to —OCH$_3$. All the glycosidic bonds in the treated sample are then hydrolyzed in aqueous acid, and the amount of 2,3-di-O-methylglucose so formed is determined.

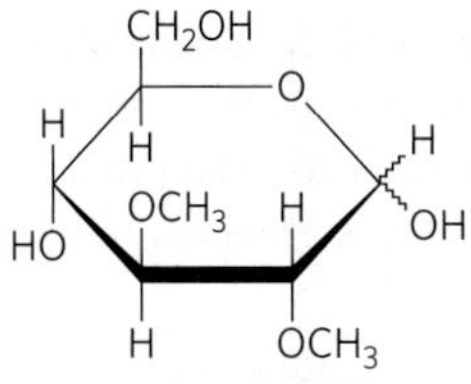

2,3-Di-O-methylglucose

(a) Explain the basis of this procedure for determining the number of (α1→6) branch points in amylopectin. What happens to the unbranched glucose residues in amylopectin during the methylation and hydrolysis procedure?

(b) A 258 mg sample of amylopectin treated as described above yielded 12.4 mg of 2,3-di-O-methylglucose. Determine what percentage of the glucose residues in the amylopectin contained an (α1→6) branch. (Assume that the average molecular weight of a glucose residue in amylopectin is 162 g/mol.)

28. Structural Analysis of a Polysaccharide A polysaccharide of unknown structure was isolated, subjected to exhaustive methylation, and hydrolyzed. Analysis of the products revealed three methylated sugars: 2,3,4-tri-O-methyl-D-glucose, 2,4-di-O-methyl-D-glucose, and 2,3,4,6-tetra-O-methyl-D-glucose, in the ratio 20:1:1. What is the structure of the polysaccharide?

Data Analysis Problem

29. Determining the Structure of ABO Blood Group Antigens The human ABO blood group system was first discovered in 1901, and in 1924 this trait was shown to be inherited at a single gene locus with three alleles. In 1960, W. T. J. Morgan published a paper summarizing what was known at that time about the structure of the ABO antigen molecules. When the paper was published, the complete structures of the A, B, and O antigens were not yet known; this paper is an example of what scientific knowledge looks like "in the making."

In any attempt to determine the structure of an unknown biological compound, researchers must deal with two fundamental problems: (1) If you don't know what *it* is, how do you know if *it* is pure? (2) If you don't know what *it* is, how do you know that your extraction and purification conditions have not changed *its* structure? Morgan addressed problem 1 through several methods. One method is described in his paper as observing "constant analytical values after fractional solubility tests" (p. 312). In this case, "analytical values" are measurements of chemical composition, melting point, and so forth.

(a) Based on your understanding of chemical techniques, what could Morgan mean by "fractional solubility tests"?

(b) Why would the analytical values obtained from fractional solubility tests of a *pure* substance be constant, and those of an *impure* substance not be constant?

Morgan addressed problem 2 by using an assay to measure the immunological activity of the substance present in different samples.

(c) Why was it important for Morgan's studies, and especially for addressing problem 2, that this activity assay be quantitative (measuring a level of activity) rather than simply qualitative (measuring only the presence or absence of a substance)?

The structure of the blood group antigens is shown in Figure 10-14. In his paper, Morgan listed several properties of the three antigens, A, B, and O, that were known at that time (p. 314):

1. Type B antigen has a higher content of galactose than A or O.
2. Type A antigen contains more total amino sugars than B or O.
3. The glucosamine:galactosamine ratio for the A antigen is roughly 1.2; for B, it is roughly 2.5.

(d) Which of these findings is (are) consistent with the known structures of the blood group antigens?

(e) How do you explain the discrepancies between Morgan's data and the known structures?

In later work, Morgan and his colleagues used a clever technique to obtain structural information about the blood group antigens. Enzymes had been found that would specifically degrade the antigens. However, these were available only as crude enzyme preparations, perhaps containing more than one enzyme of unknown specificity. Degradation of the blood type antigens by these crude enzymes could be inhibited by the addition of particular sugar molecules to the reaction. Only sugars found in the blood type antigens would cause this inhibition. One enzyme preparation, isolated from the protozoan *Trichomonas foetus*, would degrade all three antigens and was inhibited by the addition of particular sugars. The results of these studies are summarized in the table below, showing the percentage of substrate remaining unchanged when the *T. foetus* enzyme acted on the blood group antigens in the presence of sugars.

Sugar added	Unchanged substrate (%)		
	A antigen	B antigen	O antigen
Control—no sugar	3	1	1
L-Fucose	3	1	100
D-Fucose	3	1	1
L-Galactose	3	1	3
D-Galactose	6	100	1
N-Acetylglucosamine	3	1	1
N-Acetylgalactosamine	100	6	1

For the O antigen, a comparison of the control and L-fucose results shows that L-fucose inhibits the degradation of the antigen. This is an example of product inhibition, in which an excess of reaction product shifts the equilibrium of the reaction, preventing further breakdown of substrate.

(f) Although the O antigen contains galactose, *N*-acetylglucosamine, and *N*-acetylgalactosamine, none of these sugars inhibited the degradation of this antigen. Based on these data, is the enzyme preparation from *T. foetus* an endoglycosidase or exoglycosidase? (Endoglycosidases cut bonds between interior residues; exoglycosidases remove one residue at a time from the end of a polymer.) Explain your reasoning.

(g) Fucose is also present in the A and B antigens. Based on the structure of these antigens, why does fucose fail to prevent their degradation by the *T. foetus* enzyme? What structure would be produced?

(h) Which of the results in (f) and (g) are consistent with the structures shown in Figure 10-14? Explain your reasoning.

Reference

Morgan, W.T.J. 1960. The Croonian Lecture: a contribution to human biochemical genetics; the chemical basis of blood-group specificity. *Proc. R. Soc. Lond. B Biol. Sci.* 151:308–347.

Further Reading is available at www.macmillanlearning.com/LehningerBiochemistry7e.

CHAPTER 8

Nucleotides and Nucleic Acids

Self-study tools that will help you practice what you've learned and reinforce this chapter's concepts are available online. Go to www.macmillanlearning.com/LehningerBiochemistry7e.

Nucleotides have a variety of roles in cellular metabolism. They are the energy currency in metabolic transactions, the essential chemical links in the response of cells to hormones and other extracellular stimuli, and the structural components of an array of enzyme cofactors and metabolic intermediates. And, last but certainly not least, they are the constituents of nucleic acids: **deoxyribonucleic acid (DNA)** and **ribonucleic acid (RNA)**, the molecular repositories of genetic information. The structure of every protein, and ultimately of every biomolecule and cellular component, is a product of information programmed into the nucleotide sequence of cellular (or viral) nucleic acids. The ability to store and transmit genetic information from one generation to the next is a fundamental condition for life.

This chapter provides an overview of the chemical nature of the nucleotides and nucleic acids found in most cells; a more detailed examination of the function of nucleic acids is the focus of Part III of this text.

8.1 Some Basics

The amino acid sequence of every protein in a cell, and the nucleotide sequence of every RNA, is specified by a nucleotide sequence in the cell's DNA. A segment of a DNA molecule that contains the information required for the synthesis of a functional biological product, whether protein or RNA, is referred to as a **gene**. A cell typically has many thousands of genes, and DNA molecules, not surprisingly, tend to be very large. The storage and transmission of biological information are the only known functions of DNA.

RNAs have a broader range of functions, and several classes are found in cells. **Ribosomal RNAs (rRNAs)** are components of ribosomes, the complexes that carry out the synthesis of proteins. **Messenger RNAs (mRNAs)** are intermediaries, carrying information for the synthesis of a protein from one or a few genes to a ribosome. **Transfer RNAs (tRNAs)** are adapter molecules that faithfully translate the information in mRNA into a specific sequence of amino acids. In addition to these major classes, there are many RNAs with special functions, described in depth in Part III.

Nucleotides and Nucleic Acids Have Characteristic Bases and Pentoses

A **nucleotide** has three characteristic components: (1) a nitrogenous (nitrogen-containing) base, (2) a pentose, and (3) one or more phosphates (**Fig. 8-1**). The molecule without a phosphate group is called a **nucleoside**. The nitrogenous bases are derivatives of two parent compounds, **pyrimidine** and **purine**. The bases and pentoses of the common nucleotides are heterocyclic compounds.

>> **Key Convention:** The carbon and nitrogen atoms in the parent structures are conventionally numbered to facilitate the naming and identification of the many derivative compounds. The convention for the pentose ring follows

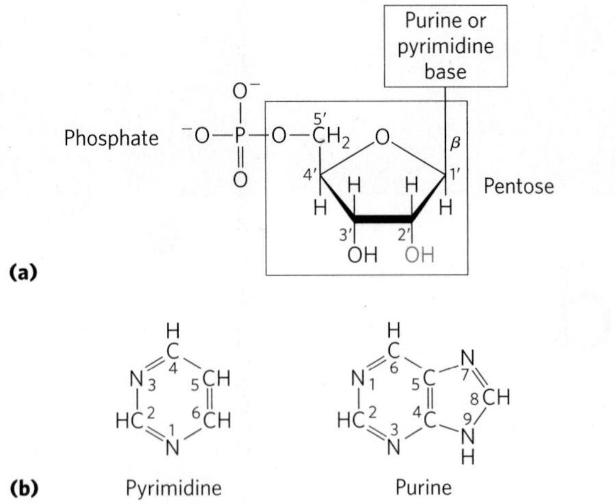

(a)

(b) Pyrimidine   Purine

FIGURE 8-1 Structure of nucleotides. (a) General structure showing the numbering convention for the pentose ring. This is a ribonucleotide. In deoxyribonucleotides the —OH group on the 2′ carbon (in red) is replaced with —H. **(b)** The parent compounds of the pyrimidine and purine bases of nucleotides and nucleic acids, showing the numbering conventions.

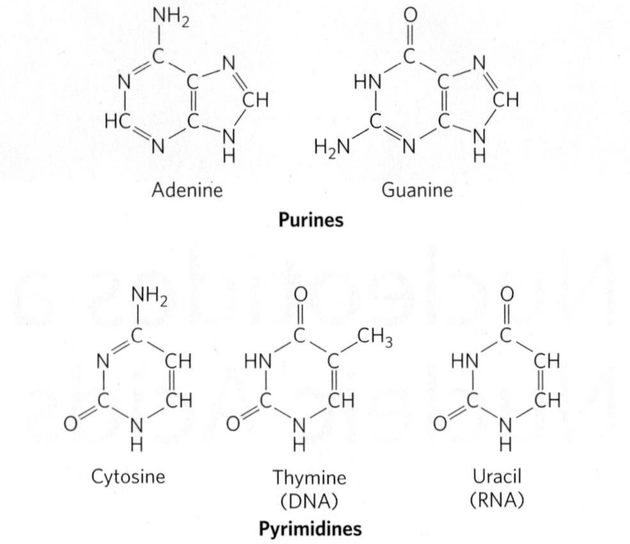

Adenine   Guanine

Purines

Cytosine   Thymine
(DNA)   Uracil
(RNA)

Pyrimidines

FIGURE 8-2 Major purine and pyrimidine bases of nucleic acids. Some of the common names of these bases reflect the circumstances of their discovery. Guanine, for example, was first isolated from guano (bird manure), and thymine was first isolated from thymus tissue.

rules outlined in Chapter 7, but in the pentoses of nucleotides and nucleosides the carbon numbers are given a prime (′) designation to distinguish them from the numbered atoms of the nitrogenous bases. **≪**

The base of a nucleotide is joined covalently (at N-1 of pyrimidines and N-9 of purines) in an N-β-glycosyl bond to the 1′ carbon of the pentose, and the phosphate is esterified to the 5′ carbon. The N-β-glycosyl bond is formed by removal of the elements of water (a hydroxyl group from the pentose and hydrogen from the base), as in O-glycosidic bond formation (see Fig. 7-30).

Both DNA and RNA contain two major purine bases, **adenine** (A) and **guanine** (G), and two major pyrimidines. In both DNA and RNA one of the pyrimidines is **cytosine** (C), but the second common pyrimidine is not the same in both: it is **thymine** (T) in DNA and **uracil** (U)

in RNA. Only occasionally does thymine occur in RNA or uracil in DNA. The structures of the five major bases are shown in **Figure 8-2**, and the nomenclature of their corresponding nucleotides and nucleosides is summarized in Table 8-1.

Nucleic acids have two kinds of pentoses. The recurring deoxyribonucleotide units of DNA contain 2′-deoxy-D-ribose, and the ribonucleotide units of RNA contain D-ribose. In nucleotides, both types of pentoses are in their β-furanose (closed five-membered ring) form. As **Figure 8-3** shows, the pentose ring is not planar but occurs in one of a variety of conformations generally described as "puckered."

≫ Key Convention: Although DNA and RNA seem to have two distinguishing features—different pentoses

TABLE 8-1	Nucleotide and Nucleic Acid Nomenclature		
Base	Nucleoside	Nucleotide	Nucleic acid
Purines			
Adenine	Adenosine	Adenylate	RNA
	Deoxyadenosine	Deoxyadenylate	DNA
Guanine	Guanosine	Guanylate	RNA
	Deoxyguanosine	Deoxyguanylate	DNA
Pyrimidines			
Cytosine	Cytidine	Cytidylate	RNA
	Deoxycytidine	Deoxycytidylate	DNA
Thymine	Thymidine or deoxythymidine	Thymidylate or deoxythymidylate	DNA
Uracil	Uridine	Uridylate	RNA

Note: "Nucleoside" and "nucleotide" are generic terms that include both ribo- and deoxyribo- forms. Also, ribonucleosides and ribonucleotides are here designated simply as nucleosides and nucleotides (e.g., riboadenosine as adenosine), and deoxyribonucleosides and deoxyribonucleotides as deoxynucleosides and deoxynucleotides (e.g., deoxyriboadenosine as deoxyadenosine). Both forms of naming are acceptable, but the shortened names are more commonly used. Thymine is an exception; "ribothymidine" is used to describe its unusual occurrence in RNA.

FIGURE 8-3 Conformations of ribose. (a) In solution, the straight-chain (aldehyde) and ring (β-furanose) forms of free ribose are in equilibrium. RNA contains only the ring form, β-D-ribofuranose. Deoxyribose undergoes a similar interconversion in solution, but in DNA exists solely as β-2'-deoxy-D-ribofuranose. **(b)** Ribofuranose rings in nucleotides can exist in four different puckered conformations. In all cases, four of the five atoms are nearly in a single plane. The fifth atom (C-2' or C-3') is on either the same (endo) or the opposite (exo) side of the plane relative to the C-5' atom.

and the presence of uracil in RNA and thymine in DNA—it is the pentoses that uniquely define the identity of a nucleic acid. If the nucleic acid contains 2'-deoxy-D-ribose, it is DNA by definition, even if it contains uracil. Similarly, if the nucleic acid contains D-ribose, it is RNA, regardless of its base composition. ❮❮

Figure 8-4 gives the structures and names of the four major **deoxyribonucleotides** (deoxyribonucleoside

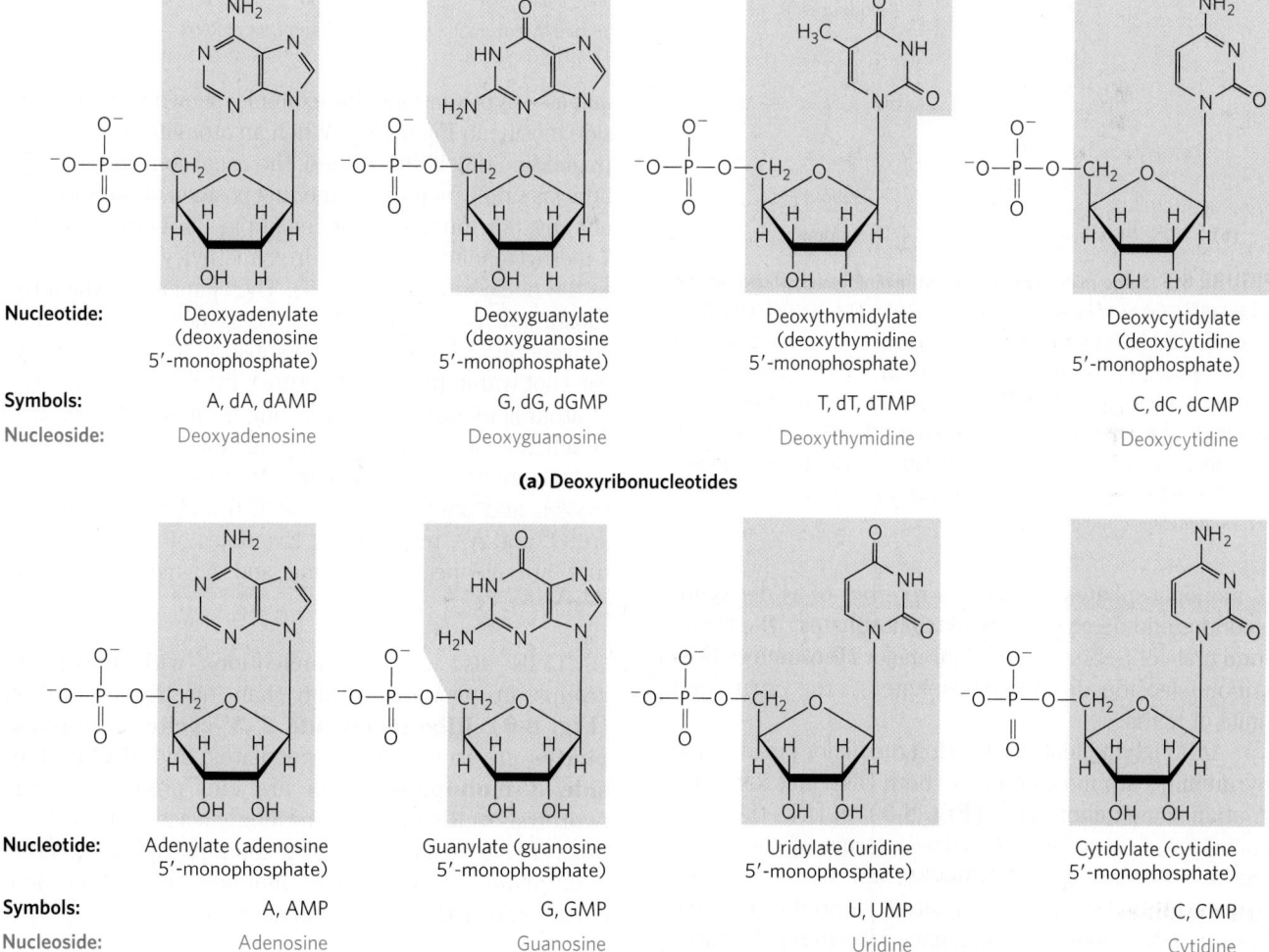

(a) Deoxyribonucleotides

Nucleotide:	Deoxyadenylate (deoxyadenosine 5'-monophosphate)	Deoxyguanylate (deoxyguanosine 5'-monophosphate)	Deoxythymidylate (deoxythymidine 5'-monophosphate)	Deoxycytidylate (deoxycytidine 5'-monophosphate)
Symbols:	A, dA, dAMP	G, dG, dGMP	T, dT, dTMP	C, dC, dCMP
Nucleoside:	Deoxyadenosine	Deoxyguanosine	Deoxythymidine	Deoxycytidine

(b) Ribonucleotides

Nucleotide:	Adenylate (adenosine 5'-monophosphate)	Guanylate (guanosine 5'-monophosphate)	Uridylate (uridine 5'-monophosphate)	Cytidylate (cytidine 5'-monophosphate)
Symbols:	A, AMP	G, GMP	U, UMP	C, CMP
Nucleoside:	Adenosine	Guanosine	Uridine	Cytidine

FIGURE 8-4 Deoxyribonucleotides and ribonucleotides of nucleic acids. All nucleotides are shown in their free form at pH 7.0. The nucleotide units of DNA **(a)** are usually symbolized as A, G, T, and C, sometimes as dA, dG, dT, and dC; those of RNA **(b)** as A, G, U, and C. In their free form the deoxyribonucleotides are commonly abbreviated dAMP, dGMP, dTMP, and dCMP; the ribonucleotides, AMP, GMP, UMP, and CMP. For each nucleotide in the figure, the more common name is followed by the complete name in parentheses. All abbreviations assume that the phosphate group is at the 5' position. The nucleoside portion of each molecule is shaded in light red. In this and the following illustrations, the ring carbons are not shown.

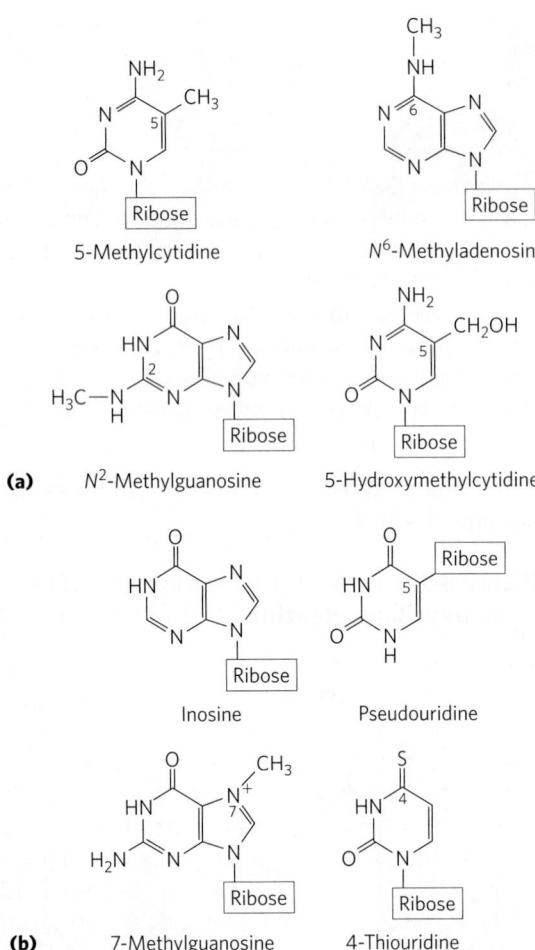

(a) N²-Methylguanosine   5-Hydroxymethylcytidine

Inosine   Pseudouridine

(b)   7-Methylguanosine   4-Thiouridine

FIGURE 8-5 Some minor purine and pyrimidine bases, shown as the nucleosides. (a) Minor bases of DNA. 5-Methylcytidine occurs in the DNA of animals and higher plants, N⁶-methyladenosine in bacterial DNA, and 5-hydroxymethylcytidine in the DNA of animals and of bacteria infected with certain bacteriophages. **(b)** Some minor bases of tRNAs. Inosine contains the base hypoxanthine. Note that pseudouridine, like uridine, contains uracil; they are distinct in the point of attachment to the ribose—in uridine, uracil is attached through N-1, the usual attachment point for pyrimidines; in pseudouridine, through C-5.

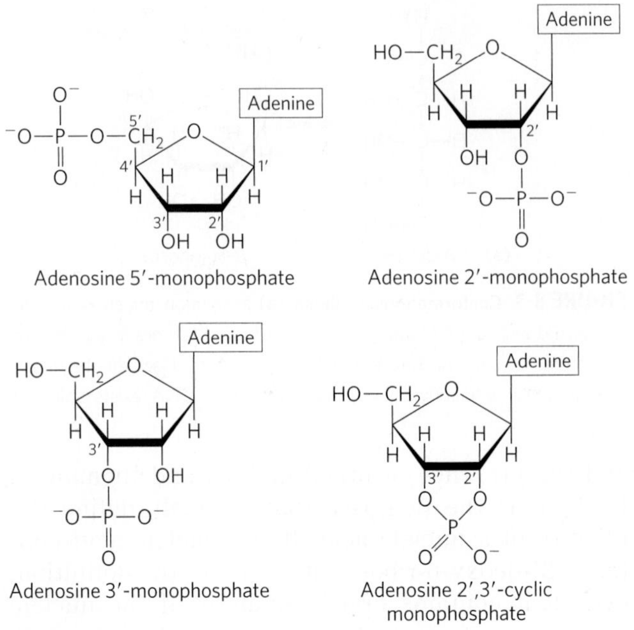

5'-monophosphates; sometimes referred to as deoxynucleotides and deoxynucleoside triphosphates), the structural units of DNAs, and the four major **ribonucleotides** (ribonucleoside 5'-monophosphates), the structural units of RNAs.

Although nucleotides bearing the major purines and pyrimidines are most common, both DNA and RNA also contain some minor bases **(Fig. 8-5)**. In DNA the most common of these are methylated forms of the major bases; in some viral DNAs, certain bases may be hydroxymethylated or glucosylated. Altered or unusual bases in DNA molecules often have roles in regulating or protecting the genetic information. Minor bases of many types are also found in RNAs, especially in tRNAs (see Fig. 8-25 and Fig. 26-22).

>> Key Convention: The nomenclature for the minor bases can be confusing. Like the major bases, many have common

Adenosine 5'-monophosphate   Adenosine 2'-monophosphate

Adenosine 3'-monophosphate   Adenosine 2',3'-cyclic monophosphate

FIGURE 8-6 Some adenosine monophosphates. Adenosine 2'-monophosphate, 3'-monophosphate, and 2',3'-cyclic monophosphate are formed by enzymatic and alkaline hydrolysis of RNA.

names—hypoxanthine, for example, shown as its nucleoside inosine in Figure 8-5. When an atom in the purine or pyrimidine ring is substituted, the usual convention (used here) is simply to indicate the ring position of the substituent by its number—for example, 5-methylcytosine, 7-methylguanine, and 5-hydroxymethylcytosine (shown as the nucleosides in Fig. 8-5). The element to which the substituent is attached (N, C, O) is not identified. The convention changes when the substituted atom is exocyclic (not within the ring structure), in which case the type of atom is identified, and the ring position to which it is attached is denoted with a superscript. The amino nitrogen attached to C-6 of adenine is N^6; similarly, the carbonyl oxygen and amino nitrogen at C-6 and C-2 of guanine are O^6 and N^2, respectively. Examples of this nomenclature are N^6-methyladenosine and N^2-methylguanosine (Fig. 8-5). **◄◄**

Cells also contain nucleotides with phosphate groups in positions other than on the 5' carbon **(Fig. 8-6)**. **Ribonucleoside 2',3'-cyclic monophosphates** are isolatable intermediates, and **ribonucleoside 3'-monophosphates** are end products of the hydrolysis of RNA by certain ribonucleases. Other variations are adenosine 3',5'-cyclic monophosphate (cAMP) and guanosine 3',5'-cyclic monophosphate (cGMP), considered at the end of this chapter.

Phosphodiester Bonds Link Successive Nucleotides in Nucleic Acids

The successive nucleotides of both DNA and RNA are covalently linked through phosphate-group "bridges," in which the 5'-phosphate group of one nucleotide unit is

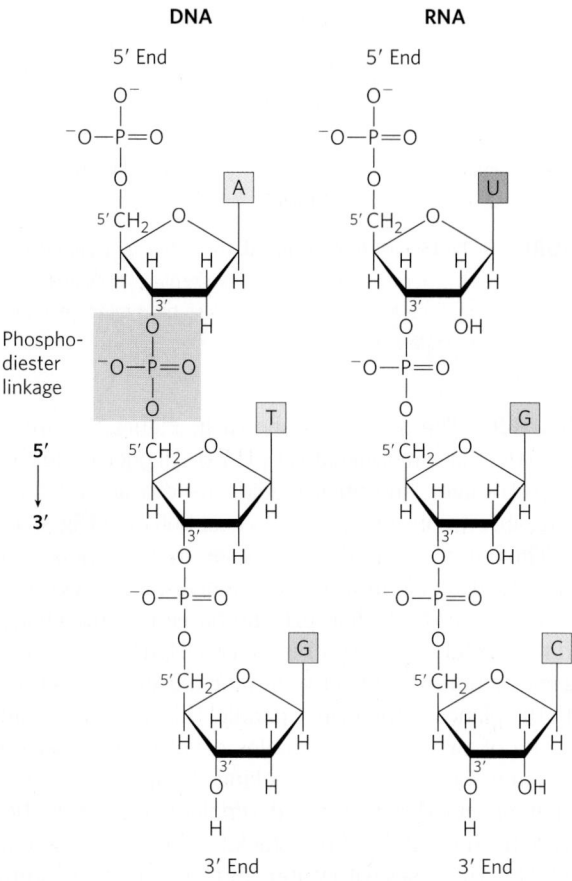

FIGURE 8-7 Phosphodiester linkages in the covalent backbone of DNA and RNA. The phosphodiester bonds (one of which is shaded in the DNA) link successive nucleotide units. The backbone of alternating pentose and phosphate groups in both types of nucleic acid is highly polar. The 5′ and 3′ ends of the macromolecule may be free or may have an attached phosphoryl group.

joined to the 3′-hydroxyl group of the next nucleotide, creating a **phosphodiester linkage (Fig. 8-7)**. Thus the covalent backbones of nucleic acids consist of alternating phosphate and pentose residues, and the

nitrogenous bases may be regarded as side groups joined to the backbone at regular intervals. The backbones of both DNA and RNA are hydrophilic. The hydroxyl groups of the sugar residues form hydrogen bonds with water. The phosphate groups, with a pK_a near 0, are completely ionized and negatively charged at pH 7, and the negative charges are generally neutralized by ionic interactions with positive charges on proteins, metal ions, and polyamines.

>> Key Convention: All the phosphodiester linkages in DNA and RNA have the same orientation along the chain (Fig. 8-7), giving each linear nucleic acid strand a specific polarity and distinct 5′ and 3′ ends. By definition, the **5′ end** lacks a nucleotide attached at the 5′ position, and the **3′ end** lacks a nucleotide attached at the 3′ position. Other groups (most often one or more phosphates) may be present on one or both ends. The 5′→3′ orientation of a strand of nucleic acid refers to the *ends* of the strand and the orientation of individual nucleotides, not the orientation of the individual phosphodiester bonds linking its constituent nucleotides. **<<**

The covalent backbone of DNA and RNA is subject to slow, nonenzymatic hydrolysis of the phosphodiester bonds. In the test tube, RNA is hydrolyzed rapidly under alkaline conditions, but DNA is not; the 2′-hydroxyl groups in RNA (absent in DNA) are directly involved in the process. Cyclic 2′,3′-monophosphate nucleotides are the first products of the action of alkali on RNA and are rapidly hydrolyzed further to yield a mixture of 2′- and 3′-nucleoside monophosphates **(Fig. 8-8)**.

The nucleotide sequences of nucleic acids can be represented schematically, as illustrated below by a segment of DNA with five nucleotide units. The phosphate groups are symbolized by Ⓟ, and each deoxyribose is symbolized by a vertical line, from C-1′ at the top to C-5′ at the bottom (but keep in mind that the sugar is always

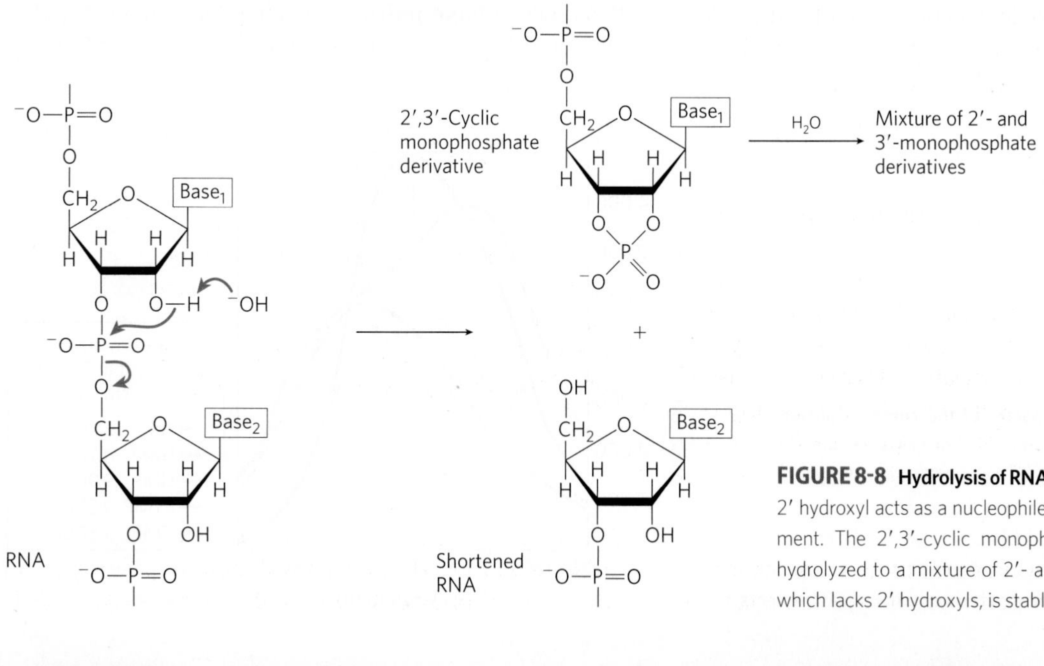

FIGURE 8-8 Hydrolysis of RNA under alkaline conditions. The 2′ hydroxyl acts as a nucleophile in an intramolecular displacement. The 2′,3′-cyclic monophosphate derivative is further hydrolyzed to a mixture of 2′- and 3′-monophosphates. DNA, which lacks 2′ hydroxyls, is stable under similar conditions.

in its closed-ring β-furanose form in nucleic acids). The connecting lines between nucleotides (which pass through ⓟ) are drawn diagonally from the middle (C-3′) of the deoxyribose of one nucleotide to the bottom (C-5′) of the next.

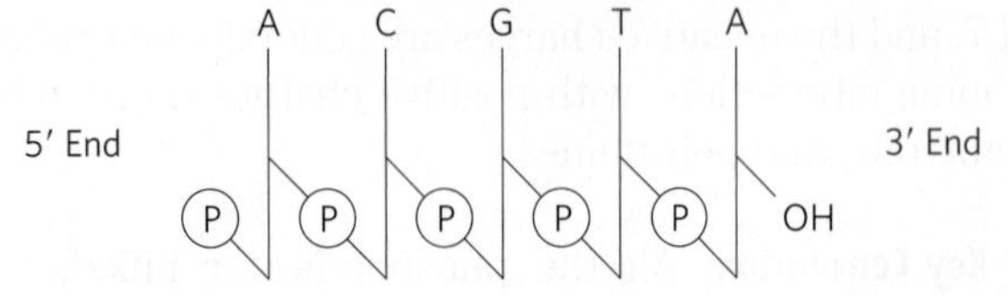

A short nucleic acid is referred to as an **oligonucleotide**. The definition of "short" is somewhat arbitrary, but polymers containing 50 or fewer nucleotides are generally called oligonucleotides. A longer nucleic acid is called a **polynucleotide**.

Some simpler representations of this pentadeoxyribonucleotide are pA-C-G-T-A$_{OH}$, pApCpGpTpA, and pACGTA.

>> **Key Convention:** The sequence of a single strand of nucleic acid is always written with the 5′ end at the left and the 3′ end at the right—that is, in the 5′→3′ direction. ◀◀

The Properties of Nucleotide Bases Affect the Three-Dimensional Structure of Nucleic Acids

Free pyrimidines and purines are weakly basic compounds and thus are called bases. The purines and pyrimidines common in DNA and RNA are aromatic molecules (Fig. 8-2), a property with important consequences for the structure, electron distribution, and light absorption of nucleic acids. Electron delocalization among atoms in the ring gives most of the bonds in the ring partial double-bond character. One result is that pyrimidines are planar molecules and purines are very nearly planar, with a slight pucker. Free pyrimidine and purine bases may exist in two or more tautomeric forms depending on the pH. Uracil, for example, occurs in lactam, lactim, and double lactim forms

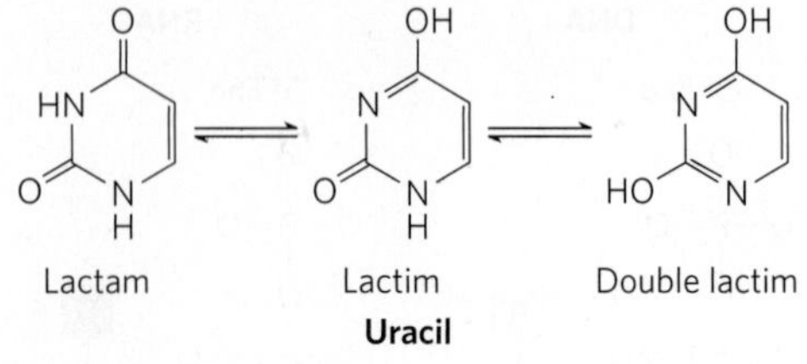

FIGURE 8-9 Tautomeric forms of uracil. The lactam form predominates at pH 7.0; the other forms become more prominent as pH decreases. The other free pyrimidines and the free purines also have tautomeric forms, but they are more rarely encountered.

(Fig. 8-9). The structures shown in Figure 8-2 are the tautomers that predominate at pH 7.0. All nucleotide bases absorb UV light, and nucleic acids are characterized by a strong absorption at wavelengths near 260 nm **(Fig. 8-10)**.

The purine and pyrimidine bases are hydrophobic and relatively insoluble in water at the near-neutral pH of the cell. At acidic or alkaline pH, the bases become charged and their solubility in water increases. Hydrophobic stacking interactions in which two or more bases are positioned with the planes of their rings parallel (like a stack of coins) are one of two important modes of interaction between bases in nucleic acids. The stacking also involves a combination of van der Waals and dipole-dipole interactions between the bases. Base stacking helps to minimize contact of the bases with water, and base-stacking interactions are very important in stabilizing the three-dimensional structure of nucleic acids, as described later.

The functional groups of pyrimidines and purines are ring nitrogens, carbonyl groups, and exocyclic amino groups. Hydrogen bonds involving the amino and carbonyl groups are the most important mode of interaction between two (and occasionally three or four) complementary strands of nucleic acid. The most common hydrogen-bonding patterns are those defined by James D. Watson and Francis Crick in 1953, in which A bonds specifically to T (or U) and G bonds to C **(Fig. 8-11)**. These two types of **base pairs** predominate in double-stranded DNA and RNA, and the tautomers shown in Figure 8-2

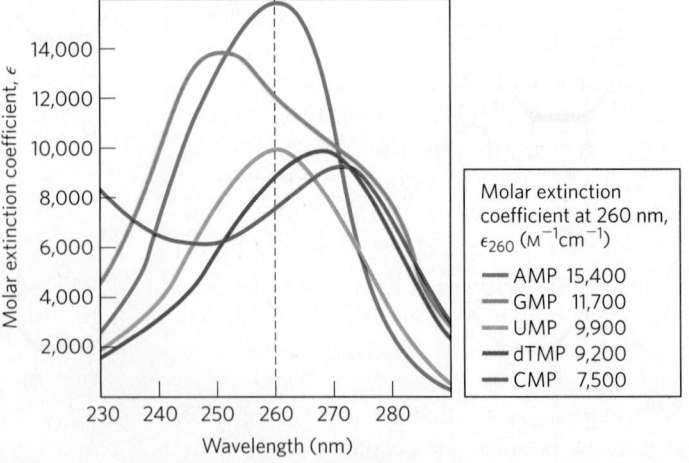

FIGURE 8-10 Absorption spectra of the common nucleotides. The spectra are shown as the variation in molar extinction coefficient with wavelength. The molar extinction coefficients at 260 nm and pH 7.0 (ε_{260}) are listed in the table. The spectra of corresponding ribonucleotides and deoxyribonucleotides, as well as the nucleosides, are essentially identical. For mixtures of nucleotides, a wavelength of 260 nm (dashed vertical line) is used for absorption measurements.

Molar extinction coefficient at 260 nm, ε_{260} (M⁻¹cm⁻¹)

AMP	15,400
GMP	11,700
UMP	9,900
dTMP	9,200
CMP	7,500

FIGURE 8-11 Hydrogen-bonding patterns in the base pairs defined by Watson and Crick. Here as elsewhere, hydrogen bonds are represented by three blue lines.

are responsible for these patterns. It is this specific pairing of bases that permits the duplication of genetic information, as we discuss later in this chapter.

James D. Watson
[Source: UPI/Bettmann/Corbis.]

Francis Crick, 1916–2004
[Source: UPI/Bettmann/Corbis.]

SUMMARY 8.1 Some Basics

■ A nucleotide consists of a nitrogenous base (purine or pyrimidine), a pentose sugar, and one or more phosphate groups. Nucleic acids are polymers of nucleotides, joined together by phosphodiester linkages between the 5′-hydroxyl group of one pentose and the 3′-hydroxyl group of the next.

■ There are two types of nucleic acid: RNA and DNA. The nucleotides in RNA contain ribose, and the common pyrimidine bases are uracil and cytosine. In DNA, the nucleotides contain 2′-deoxyribose, and the common pyrimidine bases are thymine and cytosine. The primary purines are adenine and guanine in both RNA and DNA.

8.2 Nucleic Acid Structure

The discovery of the structure of DNA by Watson and Crick in 1953 gave rise to entirely new disciplines and influenced the course of many established ones. In this section we focus on DNA structure, some of the events that led to its discovery, and more recent refinements in our understanding of DNA. We also introduce RNA structure.

As in the case of protein structure (Chapter 4), it is sometimes useful to describe nucleic acid structure in terms of hierarchical levels of complexity (primary, secondary, tertiary). The primary structure of a nucleic acid is its covalent structure and nucleotide sequence. Any regular, stable structure taken up by some or all of the nucleotides in a nucleic acid can be referred to as secondary structure. Most structures considered in the remainder of this chapter fall under the heading of secondary structure. The complex folding of large chromosomes within eukaryotic chromatin and bacterial nucleoids, or the elaborate folding of large tRNA or rRNA molecules, is generally considered tertiary structure. DNA tertiary structure is discussed in Chapter 24. RNA tertiary structure is considered briefly in this chapter and more thoroughly in Chapter 26.

DNA Is a Double Helix That Stores Genetic Information

DNA was first isolated and characterized by Friedrich Miescher in 1868. He called the phosphorus-containing substance "nuclein." Not until the 1940s, with the work of Oswald T. Avery, Colin MacLeod, and Maclyn McCarty, was there any compelling evidence that DNA was the genetic material. Avery and his colleagues found that an extract of a virulent strain of the bacterium *Streptococcus pneumoniae* (causing disease in mice) could be used to transform a nonvirulent strain of the same bacterium into a virulent strain. They were able to demonstrate through various chemical tests that it was DNA from the virulent strain (not protein, polysaccharide, or RNA, for example) that carried the genetic information for virulence. Then in 1952, experiments by Alfred D. Hershey and Martha Chase, in which they studied the infection of bacterial cells by a virus (bacteriophage) with radioactively labeled DNA or protein, removed any remaining doubt that DNA, not protein, carried the genetic information.

Another important clue to the structure of DNA came from the work of Erwin Chargaff and his colleagues in the late 1940s. They found that the four nucleotide bases of DNA occur in different ratios in the DNAs of different organisms and that the amounts of certain bases are closely related. These data, collected from DNAs of a great many different species, led Chargaff to the following conclusions:

1. The base composition of DNA generally varies from one species to another.

2. DNA specimens isolated from different tissues of the same species have the same base composition.

3. The base composition of DNA in a given species does not change with an organism's age, nutritional state, or changing environment.

4. In all cellular DNAs, regardless of the species, the number of adenosine residues is equal to the number of thymidine residues (that is, A = T), and the number of guanosine residues is equal to the number of cytidine residues (G = C). From these relationships it follows that the sum of the purine residues equals the sum of the pyrimidine residues; that is, A + G = T + C.

These quantitative relationships, sometimes called "Chargaff's rules," were confirmed by many subsequent researchers. They were a key to establishing the three-dimensional structure of DNA and yielded clues to how genetic information is encoded in DNA and passed from one generation to the next.

To shed more light on the structure of DNA, Rosalind Franklin and Maurice Wilkins used the powerful method of x-ray diffraction (see Box 4-5) to analyze DNA fibers in the early 1950s. Although lacking the molecular definition of diffraction from crystals, the x-ray diffraction pattern generated from the fibers was informative **(Fig. 8-12)**. The pattern revealed that DNA molecules are helical, with two periodicities along their long axis, a primary one of 3.4 Å and a secondary one of 34 Å. The problem then was to formulate a three-dimensional model of the DNA molecule that could account not only for the x-ray diffraction data but also

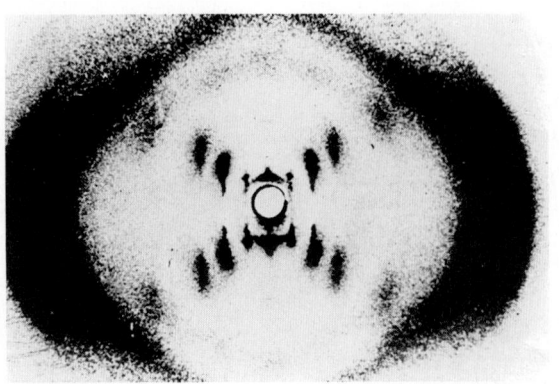

FIGURE 8-12 X-ray diffraction pattern of DNA fibers. The spots forming a cross in the center denote a helical structure. The heavy bands at the left and right arise from the recurring bases. [Source: Science Source.]

for the specific A = T and G = C base equivalences discovered by Chargaff and for the other chemical properties of DNA.

James Watson and Francis Crick relied on this accumulated information about DNA to set about deducing its structure. In 1953 they postulated a three-dimensional model of DNA structure that accounted for all the available data. It consists of two helical DNA chains wound around the same axis to form a right-handed double helix (see Box 4-1 for an explanation of the right- or left-handed sense of a helical structure). The hydrophilic backbones of alternating deoxyribose and phosphate groups are on the outside of the double helix, facing the surrounding water. The furanose ring of each deoxyribose is in the C-2′ endo conformation. The purine and pyrimidine bases of both strands are stacked inside the double helix, with their hydrophobic and nearly planar ring structures very close together and perpendicular to the long axis. The offset pairing of the two strands creates a **major groove** and **minor groove** on the surface of the duplex **(Fig. 8-13)**. Each nucleotide base of one strand is paired in the same plane with a base of the other strand. Watson and Crick found that the hydrogen-bonded base pairs illustrated in Figure 8-11, G with C and A with T, are those that fit best within the structure, providing a rationale for Chargaff's rule that in any DNA, G = C and A = T. It is important to note that

Rosalind Franklin, 1920–1958
[Source: Science Source.]

Maurice Wilkins, 1916–2004
[Source: UPI/Bettmann/Corbis.]

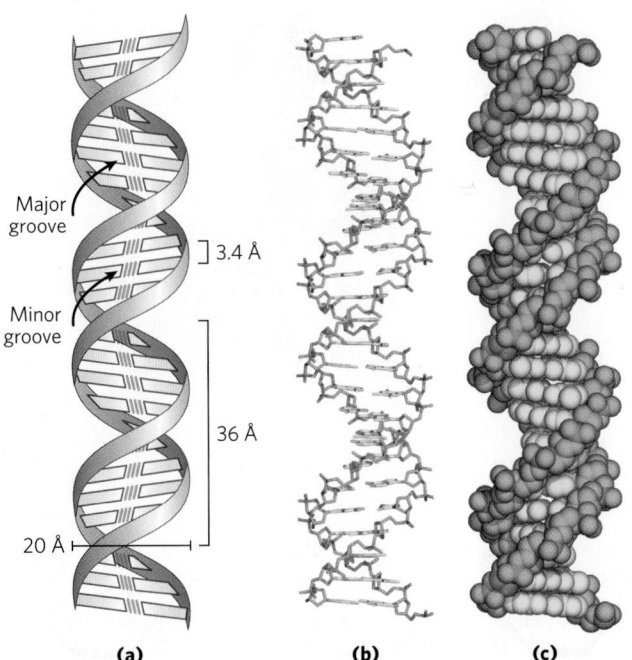

Major groove

Minor groove

3.4 Å

36 Å

20 Å

(a) **(b)** **(c)**

FIGURE 8-13 Watson-Crick model for the structure of DNA. The original model proposed by Watson and Crick had 10 base pairs, or 34 Å (3.4 nm), per turn of the helix; subsequent measurements revealed 10.5 base pairs, or 36 Å (3.6 nm), per turn. **(a)** Schematic representation, showing dimensions of the helix. **(b)** Stick representation showing the backbone and stacking of the bases. **(c)** Space-filling model.

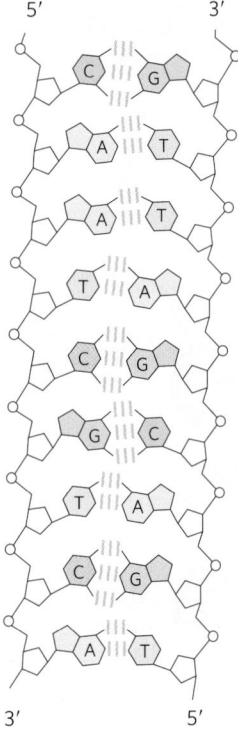

5′   3′

3′   5′

FIGURE 8-14 Complementarity of strands in the DNA double helix. The complementary antiparallel strands of DNA follow the pairing rules proposed by Watson and Crick. The base-paired antiparallel strands differ in base composition: the left strand has the composition $A_3T_2G_1C_3$; the right, $A_2T_3G_3C_1$. They also differ in sequence when each chain is read in the 5′→3′ direction. Note the base equivalences: A = T and G = C in the duplex.

three hydrogen bonds can form between G and C, symbolized G≡C, but only two can form between A and T, symbolized A=T. Pairings of bases other than G with C and A with T tend (to varying degrees) to destabilize the double-helical structure.

When Watson and Crick constructed their model, they had to decide at the outset whether the strands of DNA should be **parallel** or **antiparallel**—whether their 3′,5′-phosphodiester bonds should run in the same or opposite directions. An antiparallel orientation produced the most convincing model, and later work with DNA polymerases (Chapter 25) provided experimental evidence that the strands are indeed antiparallel, a finding ultimately confirmed by x-ray analysis.

To account for the periodicities observed in the x-ray diffraction patterns of DNA fibers, Watson and Crick manipulated molecular models to arrive at a structure in which the vertically stacked bases inside the double helix would be 3.4 Å apart; the secondary repeat distance of about 34 Å was accounted for by the presence of 10 base pairs in each complete turn of the double helix. The structure in aqueous solution differs slightly from that in fibers, having 10.5 base pairs per helical turn (Fig. 8-13).

As **Figure 8-14** shows, the two antiparallel polynucleotide chains of double-helical DNA are not identical in either base sequence or composition. Instead they are **complementary** to each other. Wherever adenine occurs in one chain, thymine is found in the other;

similarly, wherever guanine occurs in one chain, cytosine is found in the other.

The DNA double helix, or duplex, is held together by hydrogen bonding between complementary base pairs (Fig. 8-11) and by base-stacking interactions. The complementarity between the DNA strands is attributable to the hydrogen bonding between base pairs; however, the hydrogen bonds do not contribute significantly to the stability of the structure. The double helix is primarily stabilized by metal cations, which shield the negative charges of backbone phosphates, and by base-stacking interactions between complementary base pairs. Base-stacking interactions between adjacent G≡C pairs are stronger than those between adjacent A=T pairs or adjacent pairs including all four bases. Because of this, DNA duplexes with higher G≡C content are more stable.

The important features of the double-helical model of DNA structure are supported by much chemical and biological evidence. Moreover, the model immediately suggested a mechanism for the transmission of genetic information. The essential feature of the model is the complementarity of the two DNA strands. As Watson and Crick were able to see, well before confirmatory data became available, this structure could logically be replicated by (1) separating the two strands and (2) synthesizing a complementary strand for each. Because nucleotides in each new strand are joined in a sequence specified by the base-pairing rules stated above, each preexisting strand functions as a template to guide the synthesis of one complementary strand **(Fig. 8-15)**. These expectations were experimentally confirmed, inaugurating a revolution in our understanding of biological inheritance.

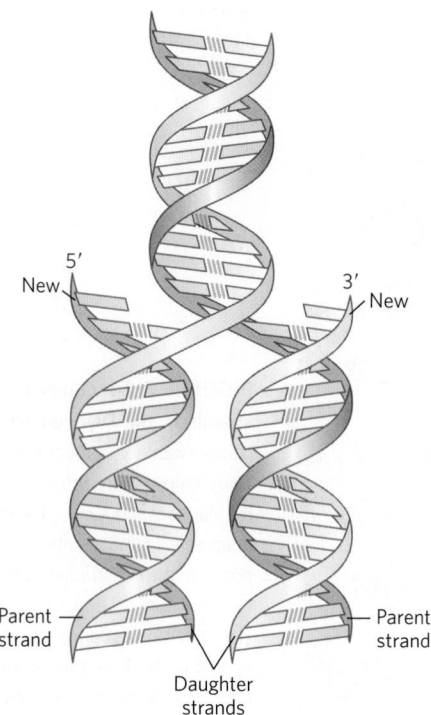

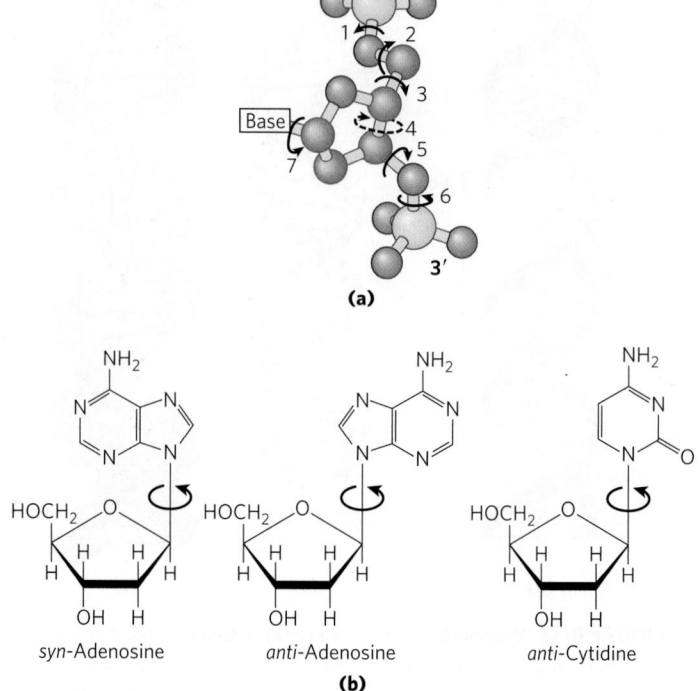

FIGURE 8-15 Replication of DNA as suggested by Watson and Crick. The preexisting or "parent" strands become separated, and each is the template for biosynthesis of a complementary "daughter" strand (in pink).

FIGURE 8-16 Structural variation in DNA. (a) The conformation of a nucleotide in DNA is affected by rotation about seven different bonds. Six of the bonds rotate freely. The limited rotation about bond 4 gives rise to ring pucker. This conformation is endo or exo, depending on whether the atom is displaced to the same side of the plane as C-5' or to the opposite side (see Fig. 8-3b). **(b)** For purine bases in nucleotides, only two conformations with respect to the attached ribose units are sterically permitted, anti or syn. Pyrimidines occur in the anti conformation.

DNA Can Occur in Different Three-Dimensional Forms

DNA is a remarkably flexible molecule. Considerable rotation is possible around several types of bonds in the sugar–phosphate (phosphodeoxyribose) backbone, and thermal fluctuation can produce bending, stretching, and unpairing (melting) of the strands. Many significant deviations from the Watson-Crick DNA structure are found in cellular DNA, some or all of which may be important in DNA metabolism. These structural variations generally do not affect the key properties of DNA defined by Watson and Crick: strand complementarity, antiparallel strands, and the requirement for A=T and G≡C base pairs.

Structural variation in DNA reflects three things: the different possible conformations of the deoxyribose, rotation about the contiguous bonds that make up the phosphodeoxyribose backbone **(Fig. 8-16a)**, and free rotation about the C-1'–N-glycosyl bond (Fig. 8-16b). Because of steric constraints, purines in purine nucleotides are restricted to two stable conformations with respect to deoxyribose, called syn and anti (Fig. 8-16b). Pyrimidines are generally restricted to the anti conformation because of steric interference between the sugar and the carbonyl oxygen at C-2 of the pyrimidine.

The Watson-Crick structure is also referred to as **B-form DNA**, or B-DNA. The B form is the most stable structure for a random-sequence DNA molecule under physiological conditions and is therefore the standard point of reference in any study of the properties of DNA. Two structural variants that have been well characterized in crystal structures are the **A** and **Z forms**. These three DNA conformations are shown in **Figure 8-17**, with a summary of their properties. The A form is favored in many solutions that are relatively devoid of water. The DNA is still arranged in a right-handed double helix, but the helix is wider and the number of base pairs per helical turn is 11, rather than 10.5 as in B-DNA. The plane of the base pairs in A-DNA is tilted about 20° relative to B-DNA base pairs, thus the base pairs in A-DNA are not perfectly perpendicular to the helix axis. These structural changes deepen the major groove while making the minor groove shallower. The reagents used to promote crystallization of DNA tend to dehydrate it, and thus most short DNA molecules tend to crystallize in the A form.

Z-form DNA is a more radical departure from the B structure; the most obvious distinction is the left-handed helical rotation. There are 12 base pairs per helical turn, and the structure appears more slender and elongated. The DNA backbone takes on a zigzag appearance. Certain nucleotide sequences fold into left-handed Z helices much more readily than others.

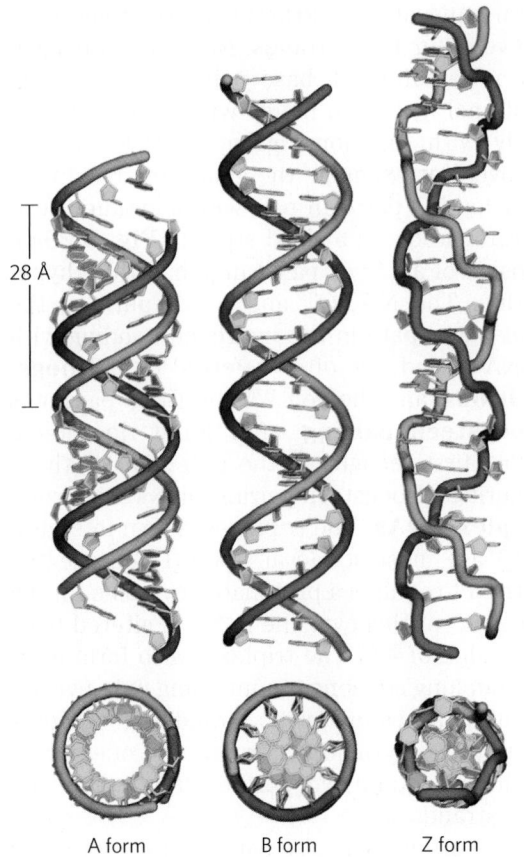

	A form	B form	Z form
Helical sense	Right handed	Right handed	Left handed
Diameter	~26 Å	~20 Å	~18 Å
Base pairs per helical turn	11	10.5	12
Helix rise per base pair	2.6 Å	3.4 Å	3.7 Å
Base tilt normal to the helix axis	20°	6°	7°
Sugar pucker conformation	C-3' endo	C-2' endo	C-2' endo for pyrimidines; C-3' endo for purines
Glycosyl bond conformation	Anti	Anti	Anti for pyrimidines; syn for purines

A form   B form   Z form

FIGURE 8-17 Comparison of A, B, and Z forms of DNA. Each structure shown here has 36 base pairs. The riboses and bases are shown in yellow. The phosphodiester backbone is represented as a blue rope. Blue is the color used to represent DNA strands in later chapters. The table summarizes some properties of the three forms of DNA.

Prominent examples are sequences in which pyrimidines alternate with purines, especially alternating C and G (that is, in the helix, alternating C≡G and G≡C pairs) or 5-methyl-C and G residues. To form the left-handed helix in Z-DNA, the purine residues flip to the syn conformation, alternating with pyrimidines in the anti conformation. The major groove is barely apparent in Z-DNA, and the minor groove is narrow and deep.

Whether A-DNA occurs in cells is uncertain, but there is evidence for some short stretches (tracts) of Z-DNA in both bacteria and eukaryotes. These Z-DNA tracts may play a role (as yet undefined) in regulating the expression of some genes or in genetic recombination.

Certain DNA Sequences Adopt Unusual Structures

Other sequence-dependent structural variations found in larger chromosomes may affect the function and metabolism of the DNA segments in their immediate vicinity. For example, bends occur in the DNA helix wherever four or more adenosine residues appear sequentially in one strand. Six adenosines in a row produce a bend of about 18°. The bending observed with this and other sequences may be important in the binding of some proteins to DNA.

A common type of DNA sequence is a **palindrome**. A palindrome is a word, phrase, or sentence that is spelled identically when read either forward or backward; two examples are ROTATOR and NURSES RUN. In DNA, the term is applied to regions of DNA with **inverted repeats,** such that an inverted, self-complementary sequence in one strand is repeated in the opposite orientation in the paired strand, as in **Figure 8-18.** The self-complementarity within each strand confers the potential to form **hairpin** or

Palindrome

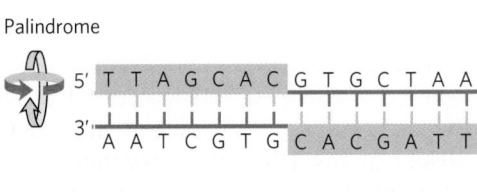

Mirror repeat

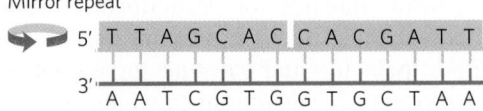

FIGURE 8-18 Palindromes and mirror repeats. Palindromes are sequences of double-stranded nucleic acids with twofold symmetry. To superimpose one repeat (shaded sequence) on the other, it must be rotated 180° about the horizontal axis, then 180° about the vertical axis, as shown by the colored arrows. A mirror repeat, on the other hand, has a symmetric sequence within each strand. Superimposing one repeat on the other requires only a single 180° rotation about the vertical axis.

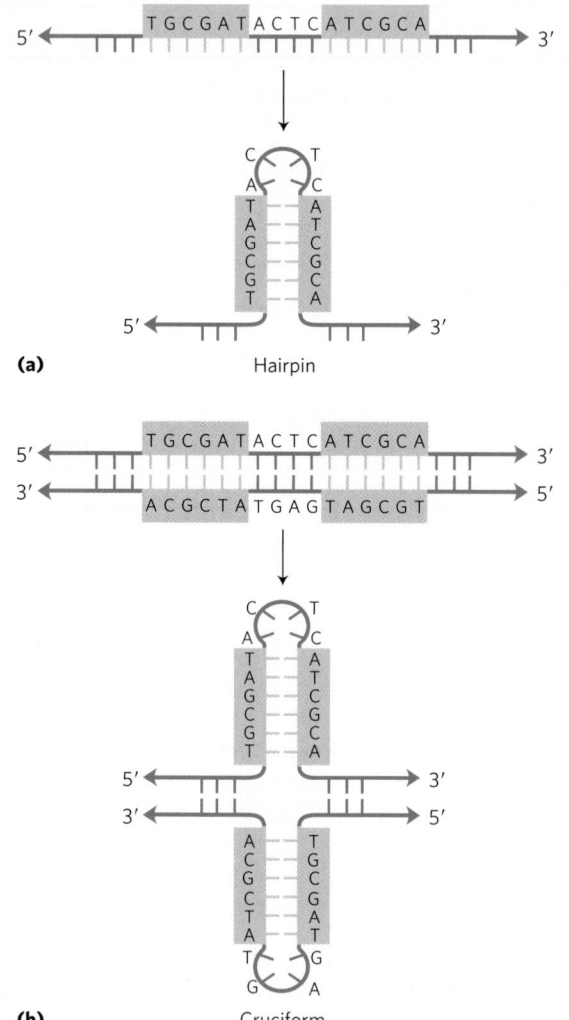

FIGURE 8-19 Hairpins and cruciforms. Palindromic DNA (or RNA) sequences can form alternative structures with intrastrand base pairing. **(a)** When only a single DNA (or RNA) strand is involved, the structure is called a hairpin. **(b)** When both strands of a duplex DNA are involved, it is called a cruciform. Blue shading highlights asymmetric sequences that can pair with the complementary sequence either in the same strand or in the complementary strand.

cruciform (cross-shaped) structures **(Fig. 8-19)**. When the inverted repeat occurs within each individual strand of the DNA, the sequence is called a **mirror repeat**. Mirror repeats do not have complementary sequences within the same strand and thus cannot form hairpin or cruciform structures. Sequences of these types are found in almost every large DNA molecule and can encompass a few base pairs or thousands. The extent to which palindromes occur as cruciforms in cells is not known, although some cruciform structures have been demonstrated in vivo in *Escherichia coli*. Self-complementary sequences cause isolated single strands of DNA (or RNA) in solution to fold into complex structures containing multiple hairpins.

Several unusual DNA structures are formed from three or even four DNA strands. Nucleotides participating in a Watson-Crick base pair (Fig. 8-11) can form additional hydrogen bonds with a third strand, particularly with functional groups arrayed in the major groove. For example, the guanosine residue of a G≡C nucleotide pair can pair with a cytidine residue (if protonated) on a third strand **(Fig. 8-20a)**; the adenosine of an A=T pair can pair with a thymidine residue. The N-7, O^6, and N^6 of purines, the atoms that participate in the hydrogen bonding with a third DNA strand, are often referred to as **Hoogsteen positions**, and the non-Watson-Crick pairing is called **Hoogsteen pairing**, after Karst Hoogsteen, who in 1963 first recognized the potential for these unusual pairings. Hoogsteen pairing allows the formation of **triplex DNAs**. The triplexes shown in Figure 8-20 (a, b) are most stable at low pH because the C≡G·C$^+$ triplet requires a protonated cytosine. In the triplex, the pK_a of this cytosine is >7.5, altered from its normal value of 4.2. The triplexes also form most readily within long sequences containing only pyrimidines or only purines in a given strand. Some triplex DNAs contain two pyrimidine strands and one purine strand; others contain two purine strands and one pyrimidine strand.

Four DNA strands can also pair to form a tetraplex (quadruplex), but this occurs readily only for DNA sequences with a very high proportion of guanosine residues (Fig. 8-20c, d). The guanosine tetraplex, or **G tetraplex**, is quite stable over a broad range of conditions. The orientation of strands in the tetraplex can vary as shown in Figure 8-20e.

In the DNA of living cells, sites recognized by many sequence-specific DNA-binding proteins (Chapter 28) are arranged as palindromes, and polypyrimidine or polypurine sequences that can form triple helices are found within regions involved in the regulation of expression of some eukaryotic genes. In principle, synthetic DNA strands designed to pair with these sequences to form triplex DNA could disrupt gene expression. This approach to controlling cellular metabolism is of commercial interest for its potential application in medicine and agriculture.

Messenger RNAs Code for Polypeptide Chains

We now turn our attention to the expression of the genetic information that DNA contains. RNA, the second major form of nucleic acid in cells, has many functions. In gene expression, RNA acts as an intermediary by carrying the information encoded in DNA to specify the amino acid sequence of a functional protein.

Given that the DNA of eukaryotes is largely confined to the nucleus whereas protein synthesis occurs on ribosomes in the cytoplasm, some molecule other than DNA must carry the genetic message from the

T=A•T   (a)   C≡G•C⁺

(b)

Guanosine tetraplex
(c)

(d)

FIGURE 8-20 DNA structures containing three or four DNA strands.
(a) Base-pairing patterns in one well-characterized form of triplex DNA.
The Hoogsteen pair in each case is shown in red. (b) Triple-helical DNA
containing two pyrimidine strands (red and white; sequence TTCCT)
and one purine strand (blue; sequence AAGGAA). The blue and white
strands are antiparallel and paired by normal Watson-Crick base-pairing
patterns. The third (all-pyrimidine) strand (red) is parallel to the purine
strand and paired through non-Watson-Crick hydrogen bonds. The triplex
is viewed from the side, with six triplets shown. (c) Base-pairing pattern
in the guanosine tetraplex structure. (d) Four successive tetraplets from
a G tetraplex structure. (e) Possible variants in the orientation of strands
in a G tetraplex. [Sources: (b) Modified from PDB ID 1BCE, J. L. Asensio
et al., *Nucleic Acids Res.* 26:3677, 1998. (d) PDB ID 244D, G. Laughlan et al.,
Science 265:520, 1994.]

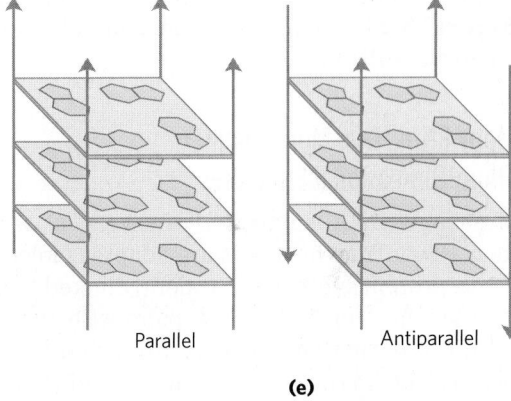

Parallel   Antiparallel
(e)

nucleus to the cytoplasm. As early as the 1950s, RNA was considered the logical candidate: RNA is found in both the nucleus and the cytoplasm, and an increase in protein synthesis is accompanied by an increase in the amount of cytoplasmic RNA and an increase in its rate of turnover. These and other observations led several researchers to suggest that RNA carries genetic information from DNA to the protein-synthesizing machinery of the ribosome. In 1961, François Jacob and Jacques Monod presented a unified (and essentially correct) picture of many aspects of this process. They proposed the name "messenger RNA" (mRNA) for that portion of the total cellular RNA carrying the genetic information from DNA to the ribosomes. The

mRNAs are formed on a DNA template by the process of **transcription**. Once they reach the ribosomes, the messengers provide the templates that specify amino acid sequences in polypeptide chains. Although mRNAs from different genes can vary greatly in length, the mRNAs from a particular gene generally have a defined size.

In bacteria and archaea, a single mRNA molecule may code for one or several polypeptide chains. If it carries the code for only one polypeptide, the mRNA is **monocistronic**; if it codes for two or more different polypeptides, the mRNA is **polycistronic**. In eukaryotes, most mRNAs are monocistronic. (For the purposes of this discussion, "cistron" refers to a

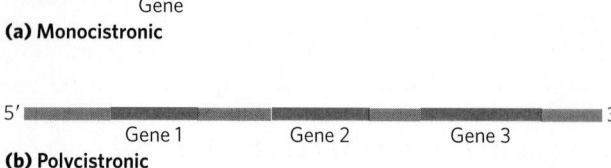

(a) Monocistronic

(b) Polycistronic

FIGURE 8-21 Bacterial mRNA. Schematic diagrams show **(a)** monocistronic and **(b)** polycistronic mRNAs of bacteria. Red segments represent RNA coding for a gene product; gray segments represent noncoding RNA. In the polycistronic transcript, noncoding RNA separates the three genes.

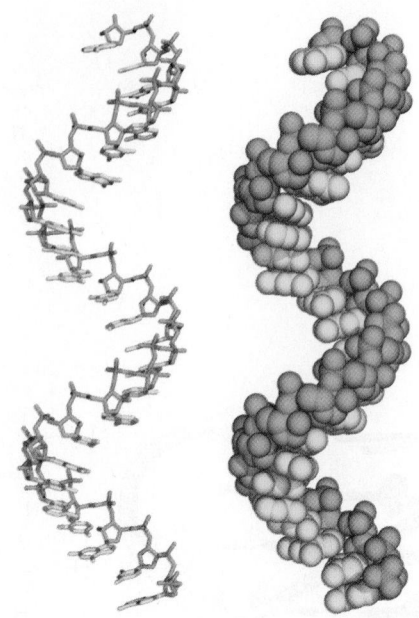

FIGURE 8-22 Typical right-handed stacking pattern of single-stranded RNA. The bases are shown in yellow, the phosphorus atoms in orange, and the riboses and phosphate oxygens in green. Green is used to represent RNA strands in succeeding chapters, just as blue is used for DNA.

gene. The term itself has historical roots in the science of genetics, and its formal genetic definition is beyond the scope of this text.) The minimum length of an mRNA is set by the length of the polypeptide chain for which it codes. For example, a polypeptide chain of 100 amino acid residues requires an RNA coding sequence of at least 300 nucleotides, because each amino acid is coded by a nucleotide triplet (this and other details of protein synthesis are discussed in Chapter 27). However, mRNAs transcribed from DNA are always somewhat longer than the length needed simply to code for a polypeptide sequence (or sequences). The additional, noncoding RNA includes sequences that regulate protein synthesis. **Figure 8-21** summarizes the general structure of bacterial mRNAs.

Many RNAs Have More Complex Three-Dimensional Structures

Messenger RNA is only one of several classes of cellular RNA. Transfer RNAs are adapter molecules that act in protein synthesis; covalently linked to an amino acid at one end, each tRNA pairs with the mRNA in such a way that amino acids are joined to a growing polypeptide in the correct sequence. Ribosomal RNAs are components of ribosomes. There is also a wide variety of special-function RNAs, including some (called ribozymes) that have enzymatic activity. All the RNAs are considered in detail in Chapter 26. The diverse and often complex functions of these RNAs reflect a diversity of structure much richer than that observed in DNA molecules.

The product of transcription of DNA is always single-stranded RNA. The single strand tends to assume a right-handed helical conformation dominated by base-stacking interactions **(Fig. 8-22)**, which are stronger between two purines than between a purine and pyrimidine or between two pyrimidines. The purine-purine interaction is so strong that a pyrimidine separating two purines is often displaced from the stacking pattern so that the purines can interact. Any self-complementary sequences in the molecule produce more complex structures. RNA can base-pair with complementary regions of either RNA or DNA. Base pairing matches the pattern for DNA: G pairs with C and A pairs with U (or with the occasional T residue in some RNAs). One difference is that base pairing between G and U residues is allowed in RNA (see Fig. 8-24) when complementary sequences in two single strands of RNA (or within a single strand of RNA that folds back on itself to align the residues) pair with each other. The paired strands in RNA or RNA-DNA duplexes are antiparallel, as in DNA.

When two strands of RNA with perfectly complementary sequences are paired, the predominant double-stranded structure is an A-form right-handed double helix. However, strands of RNA that are perfectly paired over long regions of sequence are uncommon. The three-dimensional structures of many RNAs, like those of proteins, are complex and unique. Weak interactions, especially base-stacking interactions, help stabilize RNA structures, just as they do in DNA. Z-form helices have been made in the laboratory (under very high-salt or high-temperature conditions). The B form of RNA has not been observed. Breaks in the regular A-form helix caused by mismatched or unmatched bases in one or both strands are common and result in bulges or internal loops **(Fig. 8-23)**. Hairpin loops form between nearby self-complementary (palindromic) sequences. Extensive base-paired helical segments are formed in many RNAs **(Fig. 8-24)**, and the resulting hairpins are the most common type of secondary structure in RNA.

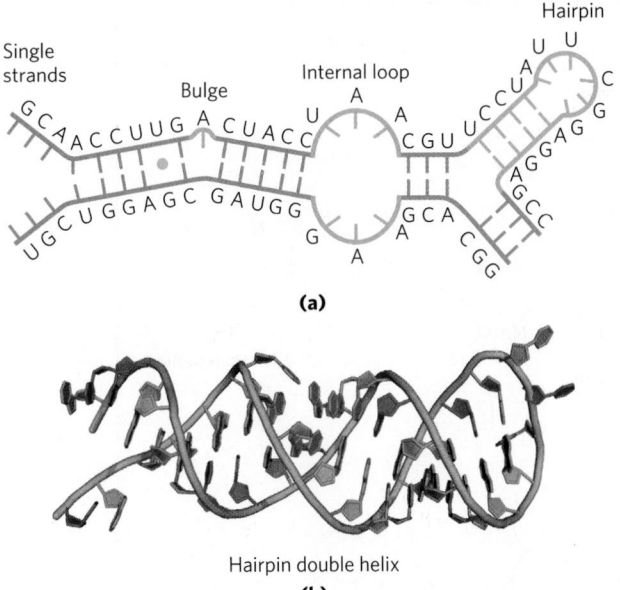

(a)

Hairpin double helix

(b)

FIGURE 8-23 Secondary structure of RNAs. (a) Bulge, internal loop, and hairpin loop. **(b)** The paired regions generally have an A-form right-handed helix, as shown for a hairpin. [Source: (b) Modified from PDB ID 1GID, J. H. Cate et al., *Science* 273:1678, 1996.]

FIGURE 8-24 Base-paired helical structures in an RNA. Shown here is the possible secondary structure of the M1 RNA component of the enzyme RNase P of *E. coli*, with many hairpins. RNase P, which also contains a protein component (not shown), functions in the processing of transfer RNAs (see Fig. 26-26). The two square brackets indicate additional complementary sequences that may be paired in the three-dimensional structure. The blue dots indicate non-Watson-Crick G≡U base pairs (boxed inset). Note that G≡U base pairs are allowed only when presynthesized strands of RNA fold up or anneal with each other. There are no RNA polymerases (the enzymes that synthesize RNAs on a DNA template) that insert a U opposite a template G, or vice versa, during RNA synthesis. [Source: B. D. James et al., *Cell* 52:19, 1988.]

Specific short base sequences (such as UUCG) are often found at the ends of RNA hairpins and are known to form particularly tight and stable loops. Such sequences may act as starting points for the folding of an RNA molecule into its precise three-dimensional structure. Other contributions are made by hydrogen bonds that are not part of standard Watson-Crick base pairs. For example, the 2'-hydroxyl group of ribose can hydrogen-bond with other groups. Some of these properties are evident in the tertiary structure of the phenylalanine transfer RNA of yeast—the tRNA responsible for inserting Phe residues into polypeptides—and in two RNA enzymes, or ribozymes, whose functions, like those of protein enzymes, depend on their three-dimensional structures **(Fig. 8-25)**.

The analysis of RNA structure and the relationship between its structure and its function is an emerging field of inquiry that has many of the same complexities as the analysis of protein structure. The importance of understanding RNA structure grows as we become increasingly aware of the large number of functional roles for RNA molecules.

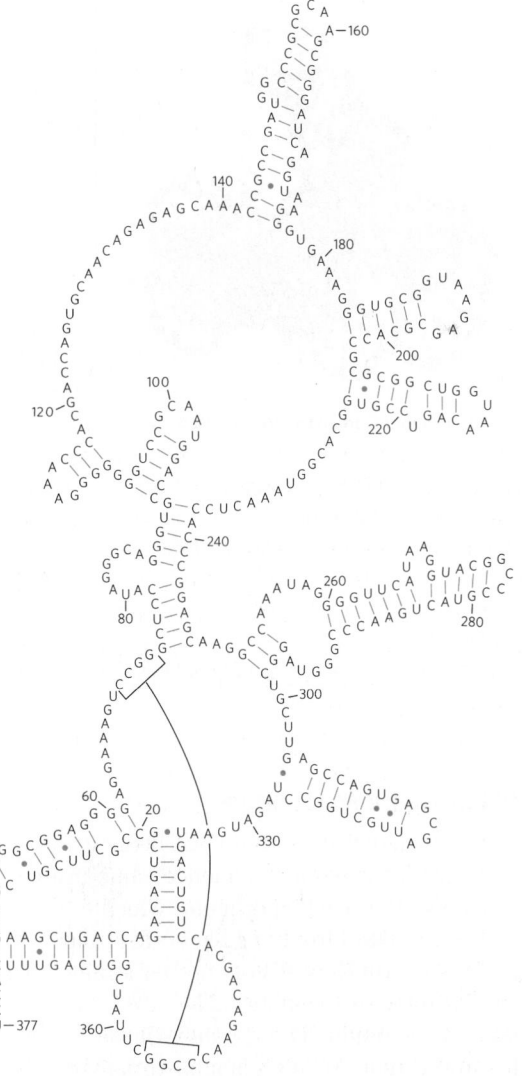

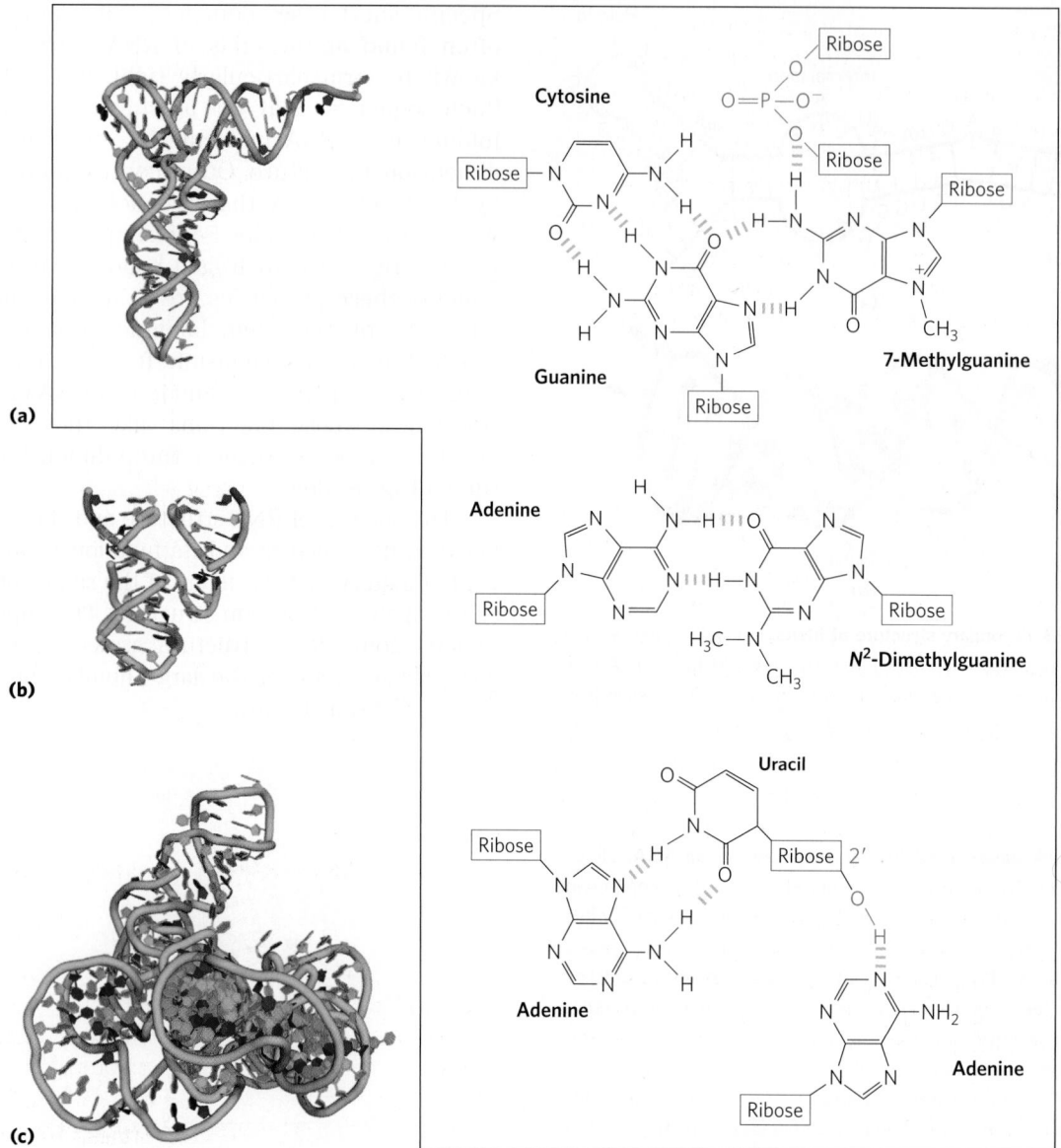

(a)

(b)

(c)

FIGURE 8-25 Three-dimensional structure in RNA. (a) Three-dimensional structure of phenylalanine tRNA of yeast. Some unusual base-pairing patterns found in this tRNA are shown. Note also the involvement of the oxygen of a ribose phosphodiester bond in one hydrogen-bonding arrangement, and a ribose 2'-hydroxyl group in another (both in red). **(b)** A hammerhead ribozyme (so named because the secondary structure at the active site looks like the head of a hammer), derived from certain plant viruses. Ribozymes, or RNA enzymes, catalyze a variety of reactions, primarily in RNA metabolism and protein synthesis. The complex three-dimensional structures of these RNAs reflect the complexity inherent in catalysis, as described for protein enzymes in Chapter 6. **(c)** A segment of mRNA known as an intron, from the ciliated protozoan *Tetrahymena thermophila*. This intron (a ribozyme) catalyzes its own excision from between exons in an mRNA strand (discussed in Chapter 26). [Sources: (a) PDB ID 1TRA, E. Westhof and M. Sundaralingam, *Biochemistry* 25:4868, 1986. (b) Modified from PDB ID 1MME, W. G. Scott et al., *Cell* 81:991, 1995. (c) Modified from PDB ID 1GRZ, B. L. Golden et al., *Science* 282:259, 1998.]

SUMMARY 8.2 Nucleic Acid Structure

■ Many lines of evidence show that DNA bears genetic information. Some of the earliest evidence came from the Avery-MacLeod-McCarty experiment, which showed that DNA isolated from one bacterial strain can enter and transform the cells of another strain, endowing it with some of the inheritable characteristics of the donor. The Hershey-Chase experiment showed that the DNA of a bacterial virus, but not its protein coat, carries the genetic message for replication of the virus in a host cell.

■ Putting together the available data, Watson and Crick postulated that native DNA consists of two antiparallel chains in a right-handed double-helical arrangement. Complementary base pairs, A═T and G≡C, are formed by hydrogen bonding within the helix. The base pairs are stacked perpendicular to the long axis of the double helix, 3.4 Å apart, with 10.5 base pairs per turn.

■ DNA can exist in several structural forms. Two variations of the Watson-Crick form, or B-DNA, are A- and Z-DNA. Some sequence-dependent structural variations cause bends in the DNA molecule. DNA strands with appropriate sequences can form hairpin or cruciform structures or triplex or tetraplex DNA.

■ Messenger RNA transfers genetic information from DNA to ribosomes for protein synthesis. Transfer RNA and ribosomal RNA are also involved in protein synthesis. RNA can be structurally complex; single RNA strands can fold into hairpins, double-stranded regions, or complex loops.

8.3 Nucleic Acid Chemistry

The role of DNA as a repository of genetic information depends in part on its inherent stability. The chemical transformations that do occur are generally very slow in the absence of an enzyme catalyst. The long-term storage of information without alteration is so important to a cell, however, that even very slow reactions that alter DNA structure can be physiologically significant. Processes such as carcinogenesis and aging may be intimately linked to slowly accumulating, irreversible alterations of DNA. Other, nondestructive alterations also occur and are essential to function, such as the strand separation that must precede DNA replication or transcription. In addition to providing insights into physiological processes, our understanding of nucleic acid chemistry has given us a powerful array of technologies that have applications in molecular biology, medicine, and forensic science. We now examine the chemical properties of DNA and a few of these technologies.

Double-Helical DNA and RNA Can Be Denatured

Solutions of carefully isolated, native DNA are highly viscous at pH 7.0 and room temperature (25 °C). When such a solution is subjected to extremes of pH or to temperatures above 80 °C, its viscosity decreases sharply, indicating that the DNA has undergone a physical change. Just as heat and extremes of pH denature globular proteins, they also cause denaturation, or melting, of double-helical DNA. Disruption of the hydrogen bonds between paired bases and of base-stacking interactions causes unwinding of the double helix to form two single strands, completely separate from each other along the entire length or part of the length (partial denaturation) of the molecule. No covalent bonds in the DNA are broken (**Fig. 8-26**).

Renaturation of a partially denatured DNA molecule is a rapid one-step process, as long as a double-helical segment of a dozen or more residues still unites the two strands. When the temperature or pH is returned to the range in which most organisms live, the unwound segments of the two strands spontaneously rewind, or **anneal**, to yield the intact duplex (Fig. 8-26). However, if the two strands are completely separated, renaturation

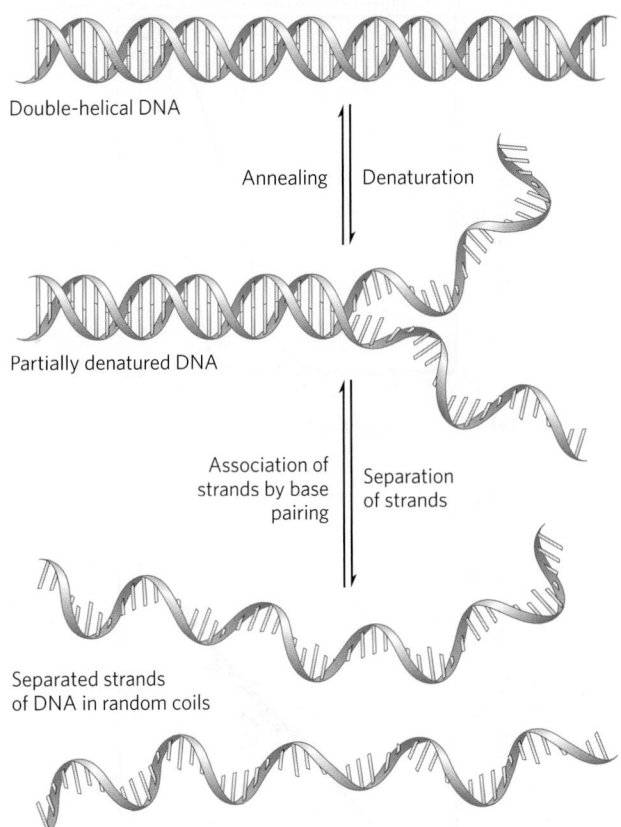

Double-helical DNA

Annealing | Denaturation

Partially denatured DNA

Association of strands by base pairing | Separation of strands

Separated strands of DNA in random coils

FIGURE 8-26 **Reversible denaturation and annealing (renaturation) of DNA.**

occurs in two steps. In the first, relatively slow step, the two strands "find" each other by random collisions and form a short segment of complementary double helix. The second step is much faster: the remaining unpaired bases successively come into register as base pairs, and the two strands "zipper" themselves together to form the double helix.

The close interaction between stacked bases in a nucleic acid has the effect of decreasing its absorption of UV light relative to that of a solution with the same concentration of free nucleotides, and the absorption is decreased further when two complementary nucleic acid strands are paired. This is called the hypochromic effect. Denaturation of a double-stranded nucleic acid produces the opposite result: an increase in absorption called the hyperchromic effect. The transition from double-stranded DNA to the denatured, single-stranded form can thus be detected by monitoring UV absorption at 260 nm.

Viral or bacterial DNA molecules in solution denature when they are heated slowly (**Fig. 8-27**). Each species of DNA has a characteristic denaturation temperature, or melting point (t_m; formally, the temperature at which half the DNA is present as separated single strands): the higher its content of G≡C base pairs, the higher the melting point of the DNA. This is primarily because, as we saw earlier, G≡C base pairs make greater contributions to base stacking than do A=T

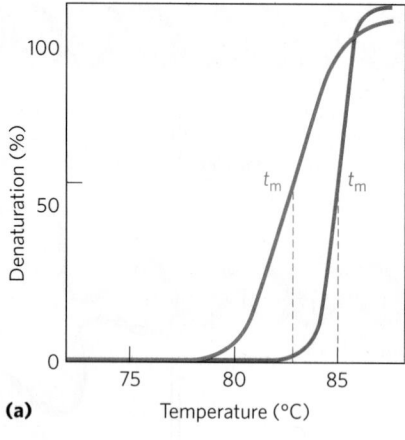

(a)

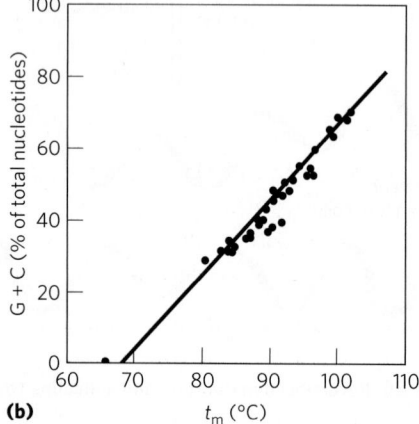

(b)

FIGURE 8-27 Heat denaturation of DNA. (a) The denaturation, or melting, curves of two DNA specimens. The temperature at the midpoint of the transition (t_m) is the melting point; it depends on pH and ionic strength and on the size and base composition of the DNA. **(b)** Relationship between t_m and the G + C content of a DNA. [Source: (b) Adapted from J. Marmur and P. Doty, *J. Mol. Biol.* 5:109, 1962.]

base pairs. Thus the melting point of a DNA molecule, determined under fixed conditions of pH and ionic strength, can yield an estimate of its base composition. If denaturation conditions are carefully controlled, regions that are rich in A═T base pairs will denature while most of the DNA remains double-stranded. Such denatured regions (called bubbles) can be visualized with electron microscopy **(Fig. 8-28)**. In the strand separation of DNA that occurs in vivo during processes such as DNA replication and transcription, the site where strand separation is initiated is often rich in A═T base pairs, as we shall see.

Duplexes of two RNA strands or one RNA strand and one DNA strand (RNA-DNA hybrids) can also be denatured. Notably, RNA duplexes are more stable to heat denaturation than DNA duplexes. At neutral pH, denaturation of a double-helical RNA often requires temperatures 20 °C or more higher than those required for denaturation of a DNA molecule with a comparable sequence, assuming that the strands in each molecule are perfectly complementary. The stability of an RNA-DNA hybrid is

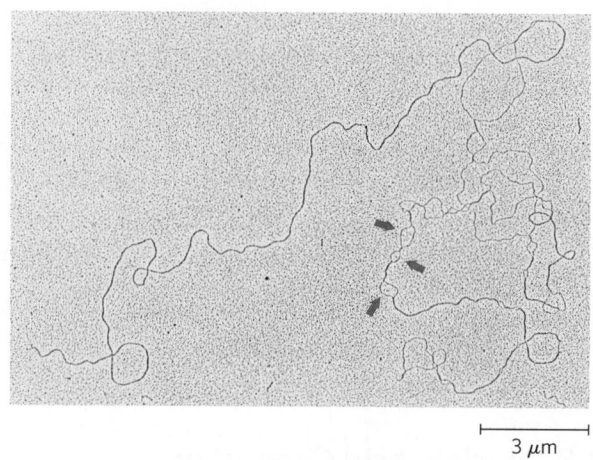

3 μm

FIGURE 8-28 Partially denatured DNA. This DNA was partially denatured, then fixed to prevent renaturation during sample preparation. The shadowing method used to visualize the DNA in this electron micrograph increases its diameter approximately fivefold and obliterates most details of the helix. However, length measurements can be obtained, and single-stranded regions are readily distinguishable from double-stranded regions. The arrows point to some single-stranded bubbles where denaturation has occurred. The regions that denature are highly reproducible and are rich in A═T base pairs. [Source: Ross B. Inman.]

generally intermediate between that of RNA and DNA duplexes. The physical basis for these differences in thermal stability is not known.

WORKED EXAMPLE 8-1 **DNA Base Pairs and DNA Stability**

In samples of DNA isolated from two unidentified species of bacteria, X and Y, adenine makes up 32% and 17%, respectively, of the total bases. What relative proportions of adenine, guanine, thymine, and cytosine would you expect to find in the two DNA samples? What assumptions have you made? One of these species was isolated from a hot spring (64 °C). Which species is most likely the thermophilic bacterium, and why?

Solution: For any double-helical DNA, A = T and G = C. The DNA from species X has 32% A and therefore must contain 32% T. This accounts for 64% of the bases and leaves 36% as G≡C pairs: 18% G and 18% C. The sample from species Y, with 17% A, must contain 17% T, accounting for 34% of the base pairs. The remaining 66% of the bases are thus equally distributed as 33% G and 33% C. This calculation is based on the assumption that both DNA molecules are double-stranded.

The higher the G + C content of a DNA molecule, the higher the melting temperature. Species Y, having the DNA with the higher G + C content (66%), most likely is the thermophilic bacterium; its DNA has a higher melting temperature and thus is more stable at the temperature of the hot spring.

Nucleotides and Nucleic Acids Undergo Nonenzymatic Transformations

Purines and pyrimidines, along with the nucleotides of which they are a part, undergo spontaneous alterations in their covalent structure. The rate of these reactions is generally *very slow*, but they are physiologically significant because of the cell's very low tolerance for alterations in its genetic information. Alterations in DNA structure that produce permanent changes in the genetic information encoded therein are called **mutations**, and much evidence suggests an intimate link between the accumulation of mutations in an individual organism and the process of aging and carcinogenesis.

Several nucleotide bases undergo spontaneous loss of their exocyclic amino groups (deamination) **(Fig. 8-29a)**. For example, under typical cellular conditions, deamination of cytosine (in DNA) to uracil occurs in about one of every 10^7 cytidine residues in 24 hours. This rate of deamination corresponds to about 100 spontaneous events per day, on average, in a mammalian cell. Deamination of adenine and guanine occurs at about 1/100th this rate.

The slow cytosine deamination reaction seems innocuous enough, but it is almost certainly the reason why DNA contains thymine rather than uracil. The product of cytosine deamination (uracil) is readily recognized as foreign in DNA and is removed by a repair system (Chapter 25). If DNA normally contained uracil, recognition of uracils resulting from cytosine deamination would be more difficult, and unrepaired uracils would lead to permanent sequence changes as they were paired with adenines during replication. Cytosine deamination would gradually lead to a decrease in G≡C base pairs and an increase in A=U base pairs in the DNA of all cells. Over the millennia, cytosine deamination could eliminate G≡C base pairs and the genetic code that depends on them. Establishing thymine as one of the four bases in DNA may well have been one of the crucial turning points in evolution, making the long-term storage of genetic information possible.

Another important reaction in deoxyribonucleotides is the hydrolysis of the *N*-β-glycosyl bond between the base and the pentose. The base is lost, creating a DNA lesion called an AP (apurinic, apyrimidinic) site or abasic site (Fig. 8-29b). Purines are lost at a higher rate than pyrimidines. As many as one in 10^5 purines

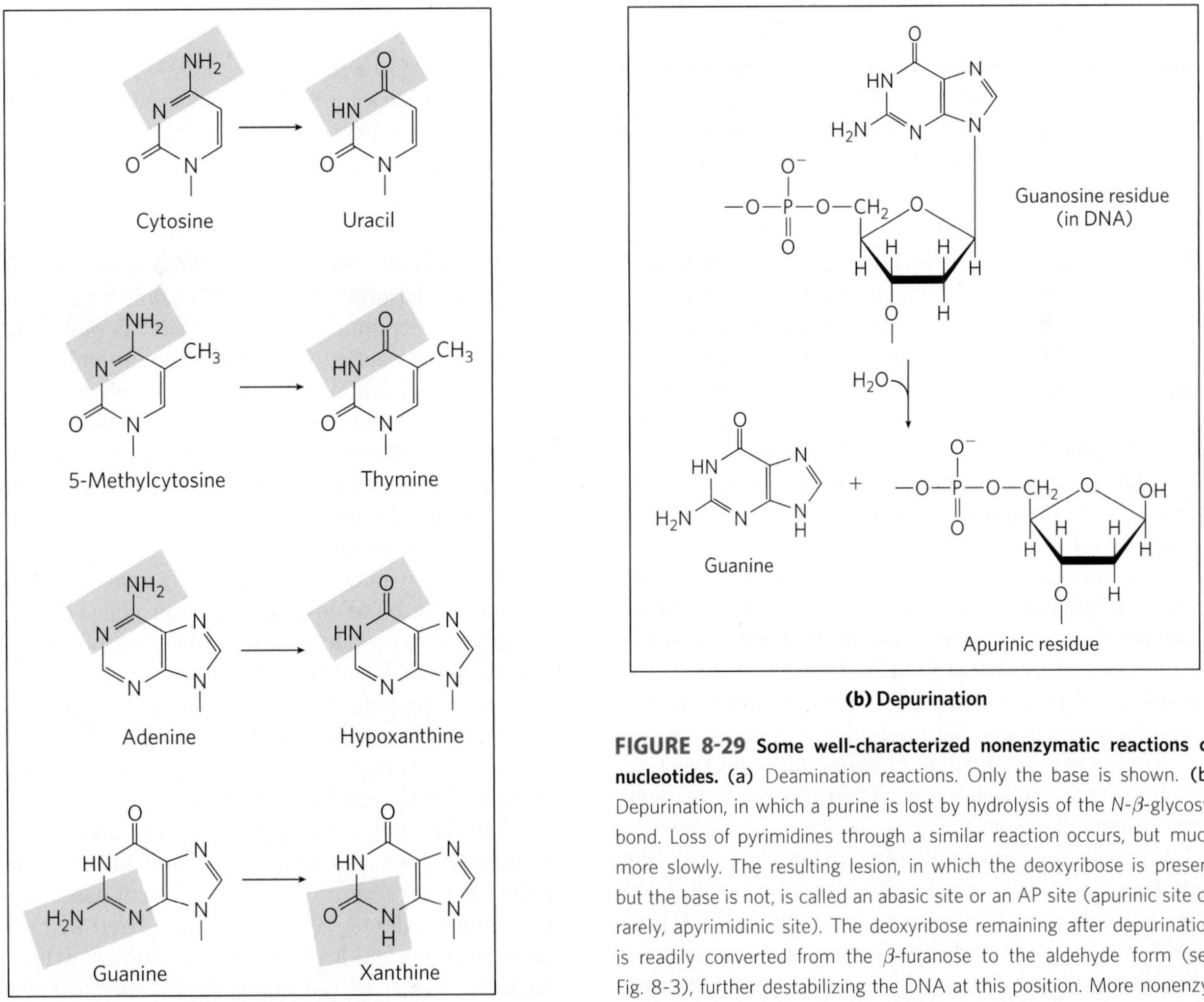

(a) Deamination

(b) Depurination

FIGURE 8-29 Some well-characterized nonenzymatic reactions of nucleotides. (a) Deamination reactions. Only the base is shown. **(b)** Depurination, in which a purine is lost by hydrolysis of the *N*-β-glycosyl bond. Loss of pyrimidines through a similar reaction occurs, but much more slowly. The resulting lesion, in which the deoxyribose is present but the base is not, is called an abasic site or an AP site (apurinic site or, rarely, apyrimidinic site). The deoxyribose remaining after depurination is readily converted from the β-furanose to the aldehyde form (see Fig. 8-3), further destabilizing the DNA at this position. More nonenzymatic reactions are illustrated in Figures 8-30 and 8-31.

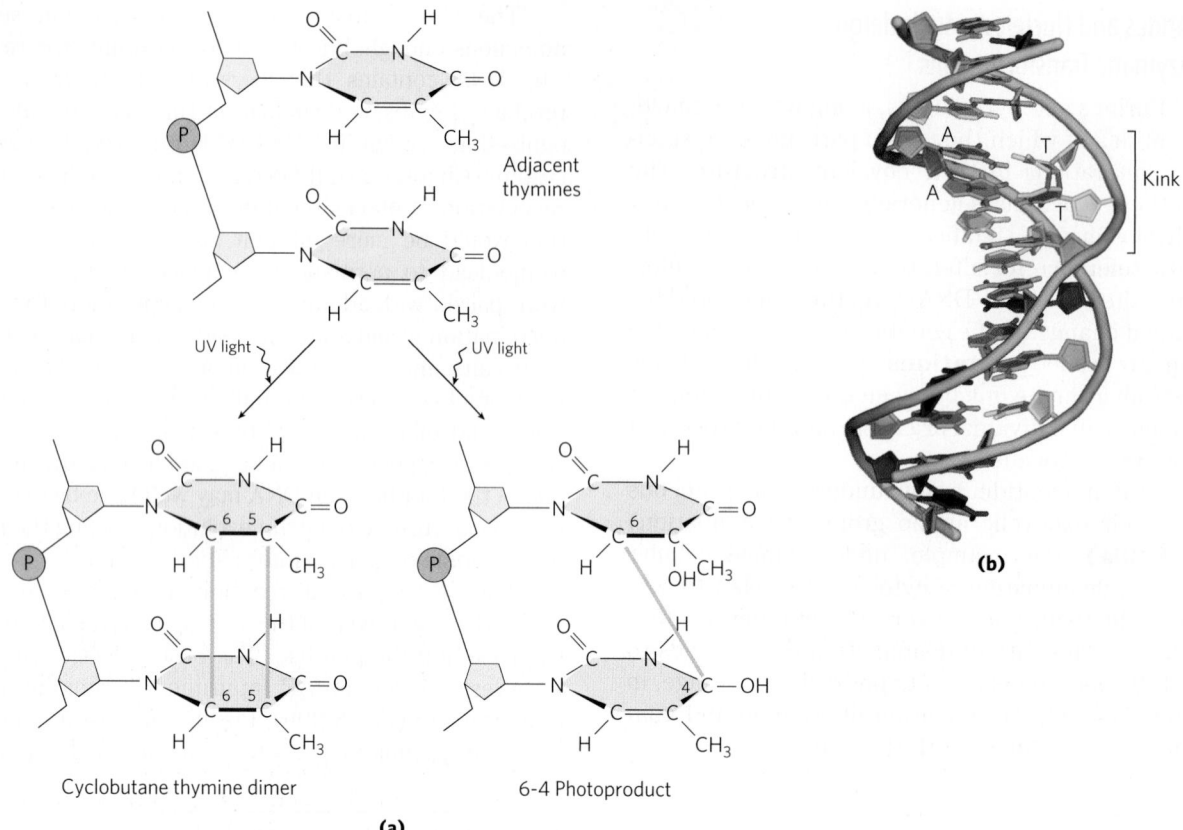

FIGURE 8-30 Formation of pyrimidine dimers induced by UV light. (a) One type of reaction (on the left) results in the formation of a cyclobutyl ring involving C-5 and C-6 of adjacent pyrimidine residues. An alternative reaction (on the right) results in a 6-4 photoproduct, with a linkage between C-6 of one pyrimidine and C-4 of its neighbor. (b) Formation of a cyclobutane pyrimidine dimer introduces a bend or kink into the DNA. [Source: (b) PDB ID 1TTD, K. McAteer et al., *J. Mol. Biol.* 282:1013, 1998.]

(10,000 per mammalian cell) are lost from DNA every 24 hours under typical cellular conditions. Depurination of ribonucleotides and RNA is much slower and less physiologically significant. In the test tube, loss of purines can be accelerated by dilute acid. Incubation of DNA at pH 3 causes selective removal of the purine bases, resulting in a derivative called apurinic acid.

Other reactions are promoted by radiation. UV light induces the condensation of two ethylene groups to form a cyclobutane ring. In the cell, the same reaction between adjacent pyrimidine bases in nucleic acids forms cyclobutane pyrimidine dimers. This happens most frequently between adjacent thymidine residues on the same DNA strand **(Fig. 8-30)**. A second type of pyrimidine dimer, called a 6-4 photoproduct, is also formed during UV irradiation. Ionizing radiation (x rays and gamma rays) can cause ring opening and fragmentation of bases as well as breaks in the covalent backbone of nucleic acids.

Virtually all forms of life are exposed to energy-rich radiation capable of causing chemical changes in DNA. Near-UV radiation (with wavelengths of 200 to 400 nm), which makes up a significant portion of the solar spectrum, is known to cause pyrimidine dimer formation and other chemical changes in the DNA of bacteria and of human skin cells. We are subjected to a constant field of ionizing radiation in the form of cosmic rays, which can penetrate deep into the earth, as well as radiation emitted from radioactive elements, such as radium, plutonium, uranium, radon, ^{14}C, and ^{3}H. X rays used in medical and dental examinations and in radiation therapy of cancer and other diseases are another form of ionizing radiation. It is estimated that UV and ionizing radiations are responsible for about 10% of all DNA damage caused by environmental agents.

DNA also may be damaged by reactive chemicals introduced into the environment as products of industrial activity. Such products may not be injurious per se but may be metabolized by cells into forms that are. There are two prominent classes of such agents **(Fig. 8-31)**: (1) deaminating agents, particularly nitrous acid (HNO_2) or compounds that can be metabolized to nitrous acid or nitrites, and (2) alkylating agents.

Nitrous acid, formed from organic precursors such as nitrosamines and from nitrite and nitrate salts, is a potent accelerator of the deamination of bases. Bisulfite has similar effects. Both agents are used as preservatives in processed foods to prevent the growth of toxic bacteria. They do not seem to increase cancer risks significantly when used in this way, perhaps because

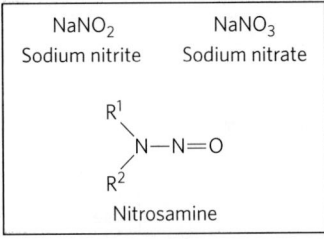

(a) Nitrous acid precursors

FIGURE 8-31 Chemical agents that cause DNA damage. (a) Precursors of nitrous acid, which promotes deamination reactions. **(b)** Alkylating agents. Most generate modified nucleotides nonenzymatically.

(b) Alkylating agents

they are used in only small amounts and make only a minor contribution to the overall levels of DNA damage. (The potential health risk from food spoilage if these preservatives were not used is much greater.)

Alkylating agents can alter certain bases of DNA. For example, the highly reactive chemical dimethylsulfate (Fig. 8-31b) can methylate a guanine to yield O^6-methylguanine, which cannot base-pair with cytosine.

Many similar reactions are brought about by alkylating agents normally present in cells, such as S-adenosyl methionine.

The most important source of mutagenic alterations in DNA is oxidative damage. Reactive oxygen species such as hydrogen peroxide, hydroxyl radicals, and superoxide radicals arise during irradiation or (more commonly) as a byproduct of aerobic metabolism. These species damage DNA through any of a large, complex group of reactions, ranging from oxidation of deoxyribose and base moieties to strand breaks. Of these species, the hydroxyl radicals are responsible for most oxidative DNA damage. Cells have an elaborate defense system to destroy reactive oxygen species, including enzymes such as catalase and superoxide dismutase that convert reactive oxygen species to harmless products. A fraction of these oxidants inevitably escape cellular defenses, however, and are able to damage DNA. Accurate estimates for the extent of this

damage are not yet available, but every day the DNA of each human cell is subjected to thousands of damaging oxidative reactions.

This is merely a sampling of the best-understood reactions that damage DNA. Many carcinogenic compounds in food, water, or air exert their cancer-causing effects by modifying bases in DNA. Nevertheless, the integrity of DNA as a polymer is better maintained than that of either RNA or protein, because DNA is the only macromolecule that has the benefit of extensive biochemical repair systems. These repair processes (described in Chapter 25) greatly lessen the impact of damage to DNA. ■

Some Bases of DNA Are Methylated

Certain nucleotide bases in DNA molecules are enzymatically methylated. Adenine and cytosine are methylated more often than guanine and thymine. Methylation is generally confined to certain sequences or regions of a DNA molecule. In some cases, the function of methylation is well understood; in others, the function remains unclear. All known DNA methylases use S-adenosylmethionine as a methyl group donor (Fig. 8-31b). $E.\ coli$ has two prominent methylation systems. One serves as part of a defense mechanism that helps the cell to distinguish its DNA from foreign DNA by marking its own DNA with methyl groups and destroying DNA (that is, foreign DNA) without the methyl groups (this is known as a restriction-modification system; see p. 322). The other system methylates adenosine residues within the sequence (5′)GATC(3′) to N^6-methyladenosine (Fig. 8-5a). Methyl groups are added by the Dam (*DNA adenine methylation*) methylase, a component of a system that repairs mismatched base pairs formed occasionally during DNA replication (see Fig. 25-21).

In eukaryotic cells, about 5% of cytidine residues in DNA are methylated to 5-methylcytidine (Fig. 8-5a). Methylation is most common at CpG sequences, producing methyl-CpG symmetrically on both strands of the DNA. The extent of methylation of CpG sequences varies by region in large eukaryotic DNA molecules.

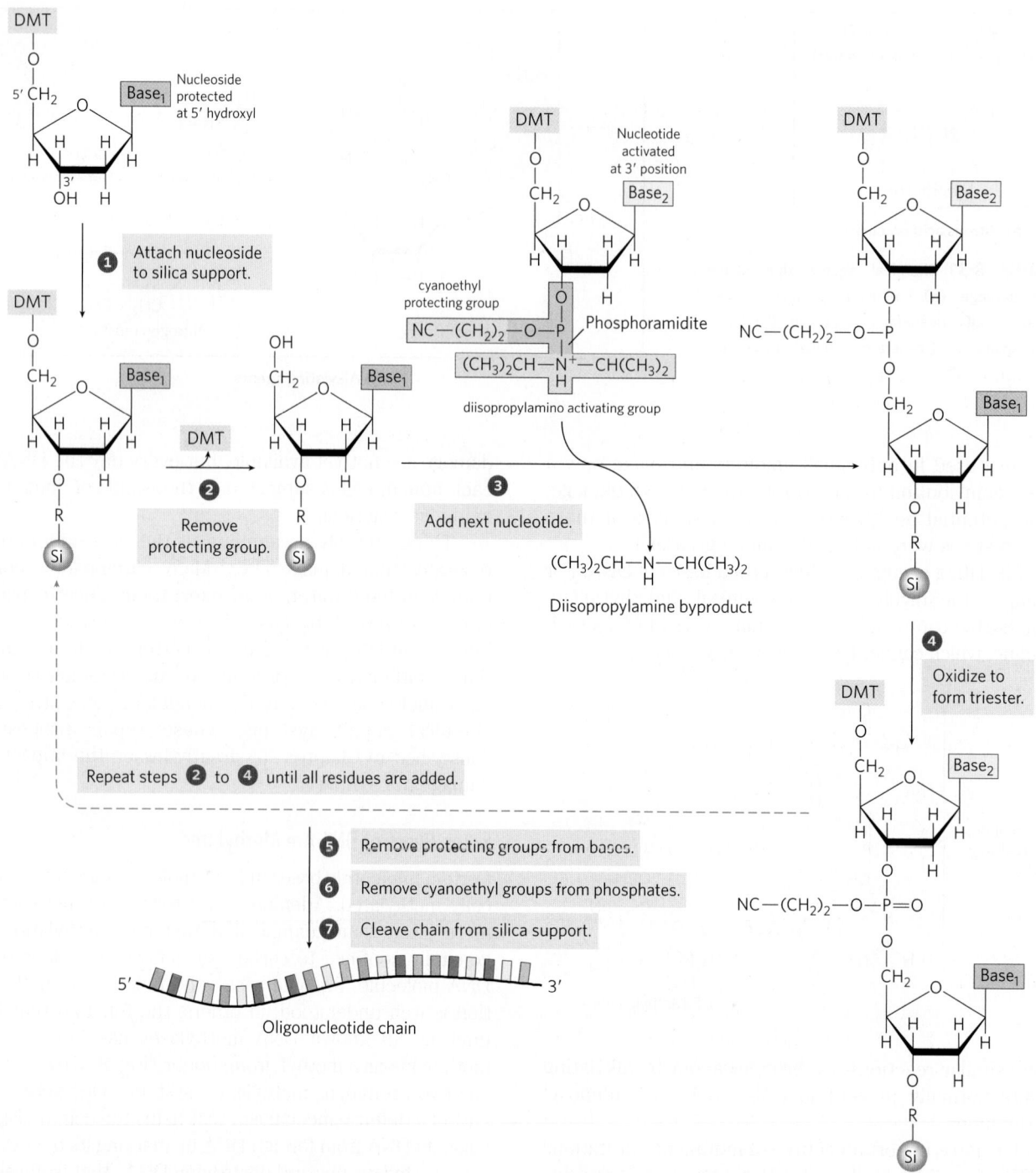

FIGURE 8-32 Chemical synthesis of DNA by the phosphoramidite method. Automated DNA synthesis is conceptually similar to the synthesis of polypeptides on a solid support. The oligonucleotide is built up on the solid support (silica), one nucleotide at a time, in a repeated series of chemical reactions with suitably protected nucleotide precursors. ❶ The first nucleoside (which will be the 3′ end) is attached to the silica support at the 3′ hydroxyl (through a linking group, R) and is protected at the 5′ hydroxyl with an acid-labile dimethoxytrityl group (DMT). The reactive groups on all bases are also chemically protected. ❷ The protecting DMT group is removed by washing the column with acid (the DMT group is colored, so this reaction can be followed spectrophotometrically). ❸ The next nucleotide has a reactive phosphoramidite at its 3′ position: a trivalent phosphite (as opposed to the more oxidized pentavalent phosphate normally present in nucleic acids) with one linked oxygen

replaced by an amino group or substituted amine. In the common variant shown, one of the phosphoramidite oxygens is bonded to the deoxyribose, the other is protected by a cyanoethyl group, and the third position is occupied by a readily displaced diisopropylamino group. Reaction with the immobilized nucleotide forms a 5′,3′ linkage, and the diisopropylamino group is eliminated. In step ❹, the phosphite linkage is oxidized with iodine to produce a phosphotriester linkage. Reactions ❷ through ❹ are repeated until all nucleotides are added. At each step, excess nucleotide is removed before addition of the next nucleotide. In steps ❺ and ❻ the remaining protecting groups on the bases and the phosphates are removed, and in ❼ the oligonucleotide is separated from the solid support and purified. The chemical synthesis of RNA is somewhat more complicated because of the need to protect the 2′ hydroxyl of ribose without adversely affecting the reactivity of the 3′ hydroxyl.

The Chemical Synthesis of DNA Has Been Automated

An important practical advance in nucleic acid chemistry was the rapid and accurate synthesis of short oligonucleotides of known sequence. The methods were pioneered by H. Gobind Khorana and his colleagues in the 1970s. Refinements by Robert Letsinger and Marvin Caruthers led to the chemistry now in widest use, called the phosphoramidite method **(Fig. 8-32)**. The synthesis is carried out with the growing strand attached to a solid support, using principles similar to those used by Merrifield for peptide synthesis (see Fig. 3-32), and is readily automated. The efficiency of each addition step is very high, allowing the routine synthesis of polymers containing 70 or 80 nucleotides and, in some laboratories, much longer strands. The availability of relatively inexpensive DNA polymers with predesigned sequences revolutionized all areas of biochemistry.

Gene Sequences Can Be Amplified with the Polymerase Chain Reaction

Genome projects, as described in Chapter 9, have given rise to online databases containing the complete genome sequences of thousands of organisms. If we know the sequence of at least the end portions of a DNA segment we are interested in, we can hugely amplify the number of copies of that DNA segment with the **polymerase chain reaction (PCR)**, a process conceived by Kary Mullis in 1983. The amplified DNA can then be used for a multitude of purposes, as we shall see.

The PCR procedure, shown in **Figure 8-33**, relies on enzymes called **DNA polymerases**. These enzymes synthesize DNA strands from deoxyribonucleotides (dNTPs), using a DNA template. DNA polymerases do not synthesize DNA de novo, but instead must add nucleotides to preexisting strands, referred to as **primers** (see Chapter 25). In PCR, two synthetic oligonucleotides are prepared for use as replication primers that can be extended by a DNA polymerase. These oligonucleotide primers are complementary to sequences on opposite strands of the target DNA, positioned so that their 5′ ends define the ends of the segment to be amplified, and they become part of the amplified sequence. The 3′ ends of the annealed primers are oriented toward each other and positioned to prime DNA synthesis across the targeted DNA segment.

The PCR procedure has an elegant simplicity. Basic PCR requires four components: a DNA sample containing the segment to be amplified, the pair of

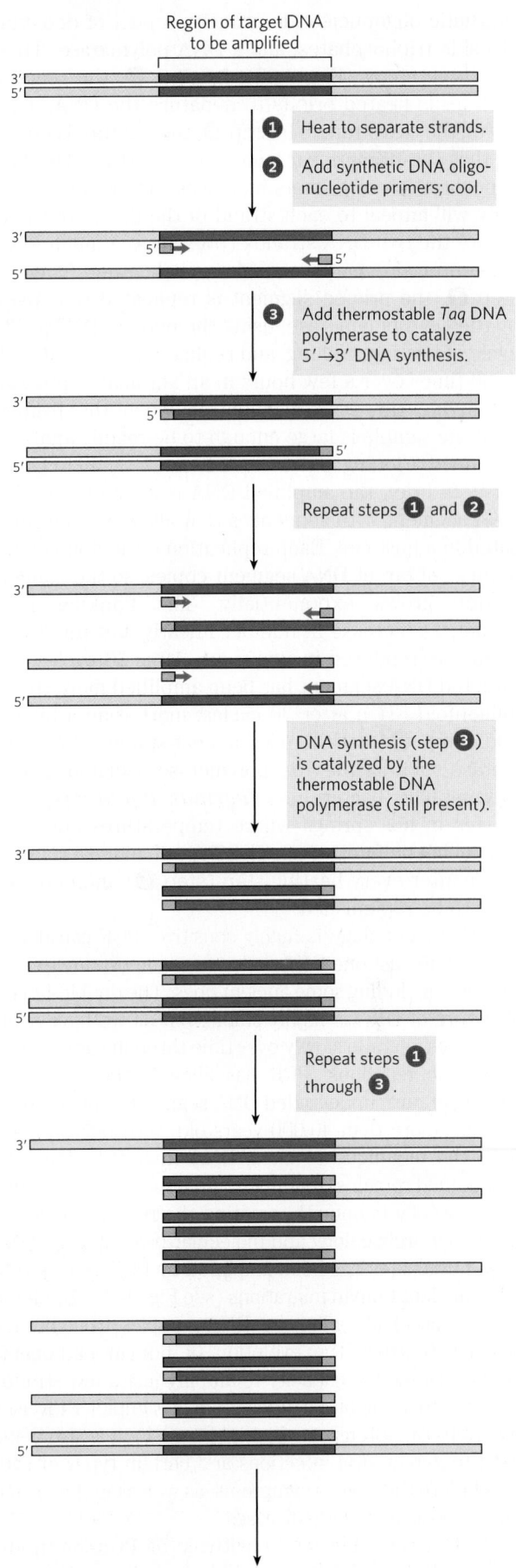

FIGURE 8-33 Amplification of a DNA segment by the polymerase chain reaction (PCR). The PCR procedure has three steps. DNA strands are ❶ separated by heating, then ❷ annealed to an excess of short synthetic DNA primers (orange) that flank the region to be amplified (dark blue); ❸ new DNA is synthesized by polymerization catalyzed by DNA polymerase. The thermostable *Taq* DNA polymerase is not denatured by the heating steps. The three steps are repeated for 25 or 30 cycles in an automated process carried out in a small benchtop instrument called a thermocycler.

Region of target DNA to be amplified

❶ Heat to separate strands.

❷ Add synthetic DNA oligonucleotide primers; cool.

❸ Add thermostable *Taq* DNA polymerase to catalyze 5′→3′ DNA synthesis.

Repeat steps ❶ and ❷.

DNA synthesis (step ❸) is catalyzed by the thermostable DNA polymerase (still present).

Repeat steps ❶ through ❸.

After 20 cycles, the target sequence has been amplified about 10^6-fold.

synthetic oligonucleotide primers, a pool of deoxynucleoside triphosphates, and a DNA polymerase. There are three steps (Fig. 8-33). In step ❶, the reaction mixture is heated briefly to denature the DNA, separating the two strands. In step ❷, the mixture is cooled so that the primers can anneal to the DNA. The high concentration of primers increases the likelihood that they will anneal to each strand of the denatured DNA before the two DNA strands (present at a much lower concentration) can reanneal to each other. Then, in step ❸, the primed segment is replicated selectively by the DNA polymerase, using the pool of dNTPs. The cycle of heating, cooling, and replication is repeated 25 to 30 times over a few hours in an automated process, amplifying the DNA segment between the primers until the sample is large enough to be readily analyzed or cloned. Cloning is described in more detail in Chapter 9. In brief, the amplified DNA is joined to another DNA segment with sequences that allow it to be replicated in a host cell. Each replication cycle doubles the number of target DNA segment copies, so the concentration grows exponentially. The flanking DNA sequences increase in number linearly, but this effect is quickly rendered insignificant. After 20 cycles, the targeted DNA segment has been amplified more than a millionfold (2^{20}); after 30 cycles, more than a billionfold. Step ❸ of PCR uses a heat-stable DNA polymerase such as the *Taq* polymerase, isolated from a thermophilic bacterium (*Thermus aquaticus*) that thrives in hot springs where temperatures approach the boiling point of water. The *Taq* polymerase remains active after every heating step (step ❶) and does not have to be replenished.

This technology is highly sensitive: PCR can detect and amplify just one DNA molecule in almost any type of sample—including some ancient ones. The double-helical structure of DNA is highly stable, but as we have seen, DNA does degrade slowly over time through various nonenzymatic reactions. PCR has allowed the successful cloning of rare, undegraded DNA segments isolated from samples more than 40,000 years old. Investigators have used the technique to clone DNA fragments from the mummified remains of humans and extinct animals, such as the woolly mammoth, creating the research fields of molecular archaeology and molecular paleontology. DNA from burial sites has been amplified by PCR and used to trace ancient human migrations (see Fig. 9-33). Epidemiologists use PCR-enhanced DNA samples from human remains to trace the evolution of human pathogenic viruses. Due to its capacity to amplify just a few strands of DNA that might be present in a sample, PCR is a potent tool in forensic medicine (Box 8-1). It is also being used to detect viral infections and certain types of cancers before they cause symptoms, as well as in the prenatal diagnosis of genetic diseases.

Given the extreme sensitivity of PCR methods, contamination of samples is a serious issue. In many applications, including forensic and ancient DNA tests, controls must be run to make sure the amplified DNA is not derived from the researcher or from contaminating bacteria.

The Sequences of Long DNA Strands Can Be Determined

In its capacity as a repository of information, a DNA molecule's most important property is its nucleotide sequence. Until the late 1970s, determining the sequence of a nucleic acid containing as few as 5 or 10 nucleotides was very laborious. The development of two techniques in 1977 (one by Allan Maxam and Walter Gilbert, the other by Frederick Sanger) made possible the sequencing of larger DNA molecules. The techniques depended on the improved understanding of nucleotide chemistry and DNA metabolism and on improved electrophoretic methods for separating DNA strands that differ in size by only one nucleotide. (See Fig. 3-18 for a description of gel electrophoresis.)

Although the two methods are similar in strategy, **Sanger sequencing**, also known as dideoxy chain-termination sequencing, proved to be technically easier and became the basis of more modern sequencing protocols **(Fig. 8-34)**. It depends upon the construction of a new DNA strand. Like PCR, this method makes use of DNA polymerases and a primer to synthesize a DNA strand complementary to the strand under analysis. Each added deoxynucleotide is complementary, through base pairing, to a base in the template strand. In Sanger sequencing, the sequence obtained is that of the newly synthesized strand complementary to the template strand being analyzed.

In the reaction catalyzed by DNA polymerase, the 3′-hydroxyl group of the primer reacts with an incoming dNTP to form a new phosphodiester bond (Fig. 8-34a). In the Sanger sequencing reaction, nucleotide analogs called dideoxynucleoside triphosphates (ddNTPs) interrupt DNA synthesis because they bind to the template strand but lack the 3′-hydroxyl group needed to add the next nucleotide (Fig. 8-34b). For instance, the addition of ddCTP in small amounts to a reaction system containing a much larger amount of dCTP (along with the other three dNTPs) leads to competition every time the DNA polymerase encounters a G in the template strand. Usually, dC is added, and synthesis of the strand continues. Sometimes, ddC will be added instead, and the strand will be terminated at that position. Thus, a small fraction of the synthesized strands are prematurely terminated at every position where dC would normally be added, opposite each template dG. Given the excess of dCTP over ddCTP, the chance that the analog will be incorporated instead of dC is small. But enough ddCTP is present to ensure that each new strand has a high probability of acquiring at least one ddC at some point (at one or another of the G residues in the template) during synthesis. The result is a solution containing a mixture of fragments, each ending with a ddC residue. Each G residue in the template generates C-terminated fragments of a particular length. The different-sized fragments, separated by electrophoresis, reveal the location of C residues in the synthesized DNA strand.

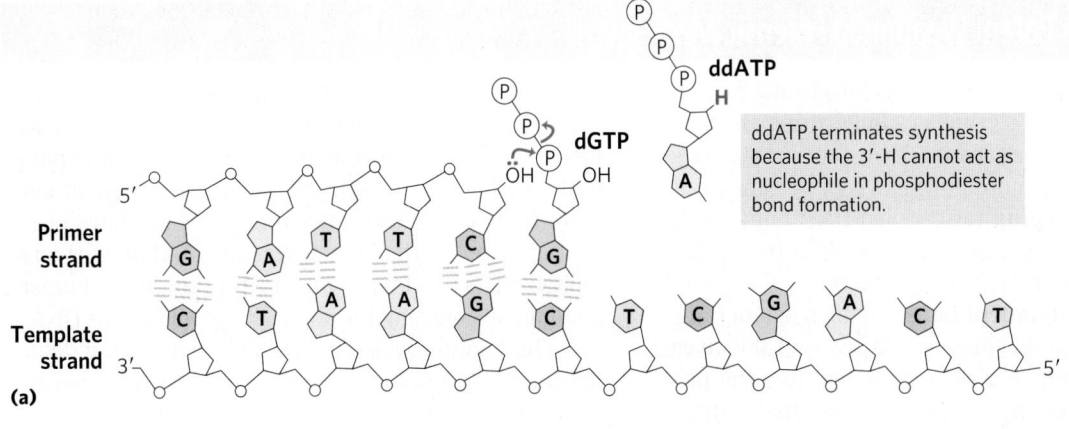

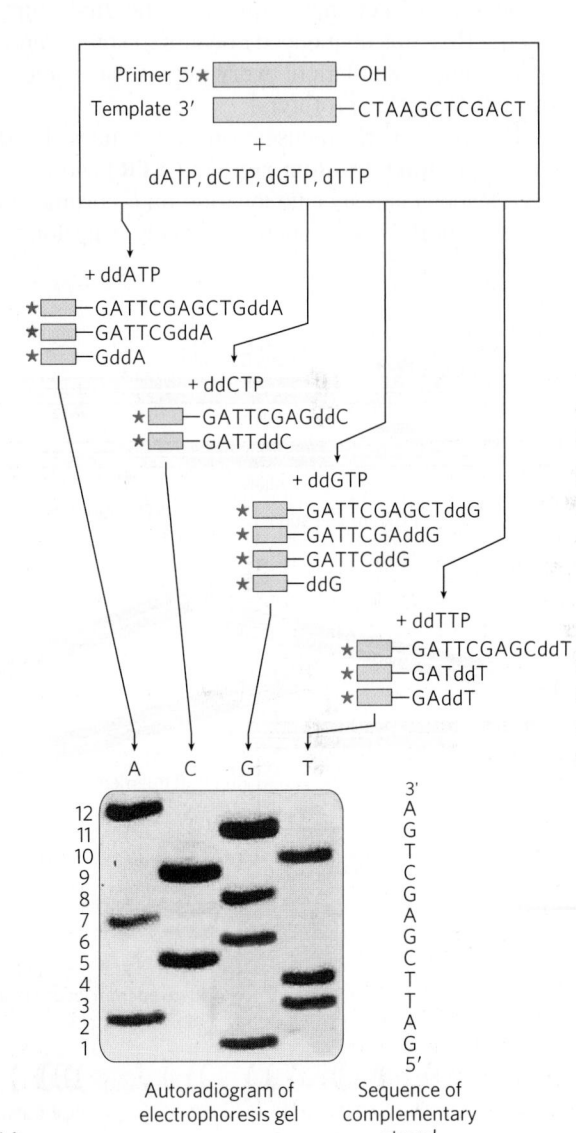

FIGURE 8-34 DNA sequencing by the Sanger method.

This method makes use of the mechanism of DNA synthesis by DNA polymerases (Chapter 25). **(a)** DNA polymerases require both a primer (a short oligonucleotide strand), to which nucleotides are added, and a template strand to guide the selection of each new nucleotide. In cells, the 3′-hydroxyl group of the primer reacts with an incoming deoxynucleoside triphosphate—dGTP in this example—to form a new phosphodiester bond. The Sanger sequencing procedure uses dideoxynucleoside triphosphate (ddNTP) analogs to interrupt DNA synthesis. (The Sanger method is also known as dideoxy chain-termination sequencing.) When a ddNTP—ddATP in this example—is inserted in place of a dNTP, strand elongation is halted after the analog is added, because the analog lacks the 3′-hydroxyl group needed for the next step. **(b)** Dideoxynucleoside triphosphate analogs have —H (red) rather than —OH at the 3′ position of the ribose ring. **(c)** The DNA to be sequenced is used as the template strand, and a short primer, radioactively (in the example here) or fluorescently labeled, is annealed to it. By addition of small amounts of a single ddNTP, for example ddCTP, to an otherwise normal reaction system, the synthesized strands will be prematurely terminated at some locations where dC normally occurs. Given the excess of dCTP over ddCTP, the chance that the analog will be incorporated whenever a dC is to be added is small. However, enough ddCTP is present to ensure that each new strand has a high probability of acquiring at least one ddC at some point during synthesis. The result is a solution containing a mixture of labeled fragments, each ending with a C residue. Each C residue in the sequence generates a set of fragments of a particular length, such that the different-sized fragments, separated by electrophoresis, reveal the location of C residues. This procedure is repeated separately for each of the four ddNTPs, and the sequence can be read directly from an autoradiogram of the gel. Because shorter DNA fragments migrate faster, the fragments near the bottom of the gel represent the nucleotide positions closest to the primer (the 5′ end), and the sequence is read (in the 5′→3′ direction) from bottom to top. Note that the sequence obtained is that of the strand *complementary* to the strand being analyzed. [Source: (c) Dr. Lloyd Smith, University of Wisconsin–Madison, Department of Chemistry.]

When this procedure was first developed, the process was repeated separately for each of the four ddNTPs. Radioactively labeled primers allowed researchers to detect the DNA fragments generated during the DNA synthesis reactions. The sequence of the synthesized DNA strand was read directly from an autoradiogram of the resulting gel (Fig. 8-34c). Because shorter DNA fragments migrate faster, the fragments near the bottom

BOX 8-1 A Potent Weapon in Forensic Medicine

One of the most accurate methods for placing an individual at the scene of a crime is a fingerprint. But with the advent of recombinant DNA technology (see Chapter 9), a much more powerful tool became available: **DNA genotyping** (also called DNA fingerprinting or DNA profiling). As first described by English geneticist Alec Jeffreys in 1985, the method is based on **sequence polymorphisms**, slight sequence differences among individuals—1 in every 1,000 base pairs (bp), on average. Each difference from the prototype human genome sequence (the first human genome that was sequenced) occurs in some fraction of the human population; every person has some differences from this prototype.

Forensic work focuses on differences in the lengths of **short tandem repeat (STR)** sequences. An STR locus is a specific location on a chromosome where a short DNA sequence (usually 4 bp long) is repeated many times in tandem. The loci most often used in STR genotyping are short—4 to 50 repeats (16 to 200 bp for tetranucleotide repeats)—and have multiple length variants in the human population. More than 20,000 tetranucleotide STR loci have been characterized in the human genome. And more than a million STRs of all types may be present in the human genome, accounting for about 3% of all human DNA.

The length of a particular STR in a given individual can be determined with the aid of the polymerase chain reaction (see Fig. 8-33). The use of PCR also makes the procedure sensitive enough to be applied to the very small samples often collected at crime scenes. The DNA sequences flanking STRs are unique to each STR locus and are identical (except for very rare mutations) in all humans. PCR primers are targeted to this flanking DNA and are designed to amplify the DNA across the STR (Fig. 1a). The length

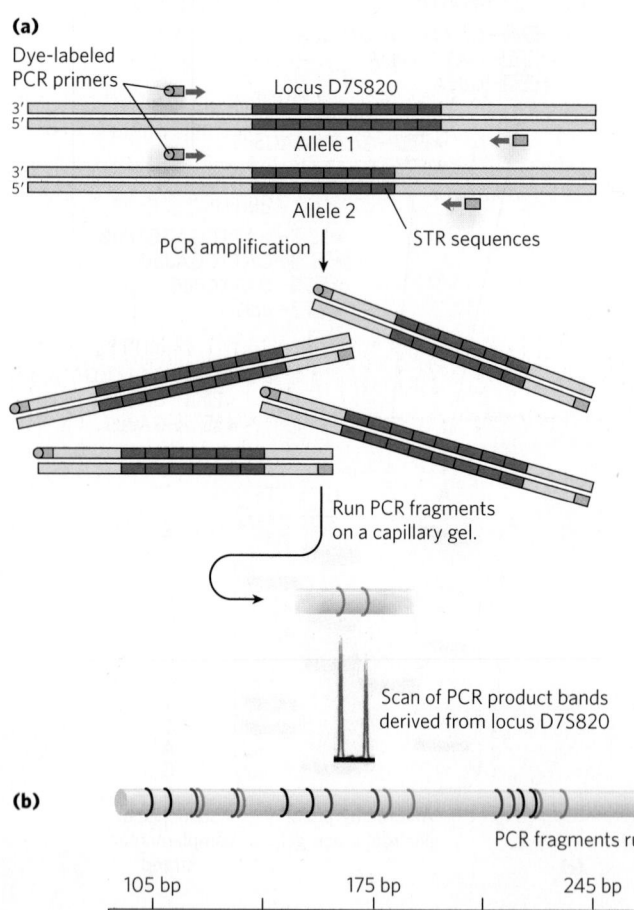

FIGURE 1 (a) STR loci can be analyzed by PCR. Suitable PCR primers (with an attached dye to aid in subsequent detection) are targeted to sequences on each side of the STR, and the region between them is amplified. If the STR sequences have different lengths on the two chromosomes of an individual's chromosome pair, two PCR products of different lengths result. **(b)** The PCR products from amplification of up to 16 STR loci can be run on a single capillary acrylamide gel (a "16-plex" analysis). Determination of which locus corresponds to which signal depends on the color of the fluorescent dye attached to the primers used in the process and on the size range in which the signal appears (the size range can be controlled by which sequences—those closer to or more distant from the STR—are targeted by the designed PCR primers). Fluorescence is given in relative fluorescence units (RFU), as measured against a standard supplied with the kit. [Source: (b) Courtesy of Carol Bingham, Promega Corporation.]

of the PCR product then reflects the length of the STR in that sample. Because each human inherits one chromosome of each chromosome pair from each parent, the STR lengths on the two chromosomes are often different, generating two different STR lengths from one individual. The PCR products are subjected to electrophoresis on a very thin polyacrylamide gel in a capillary tube. The resulting bands are converted into a set of peaks that accurately reveal the size of each PCR fragment and thus the length of the STR in the corresponding allele. Analysis of multiple STR loci can yield a profile that is unique to an individual (Fig. 1b). This is typically done with a commercially available kit that includes PCR primers unique to each locus, linked to colored dyes to help distinguish the different PCR products. PCR amplification enables investigators to obtain STR genotypes from less than 1 ng of partially degraded DNA, an amount that can be obtained from a single hair follicle, a small fraction of a drop of blood, a small semen sample, or samples that might be months or even many years old. When good STR genotypes are obtained, the chance of misidentification is less than 1 in 10^{18} (a quintillion).

The successful forensic use of STR analysis required standardization, first attempted in the United Kingdom in 1995. The U.S. standard, called the Combined DNA Index System (CODIS), established in 1998, is based on 13 well-studied STR loci, which must be present in any DNA-typing experiment carried out in the United States (Table 1). The amelogenin gene is also used as a marker in the analyses. Present on the human sex chromosomes, this gene has a slightly different length on the X and Y chromosomes. PCR amplification across this gene thus generates different-sized products that can reveal the sex of the DNA donor. By mid-2015, the CODIS database contained more than 14 million STR genotypes and had assisted in more than 274,000 forensic investigations.

DNA genotyping has been used to both convict and acquit suspects, and to establish paternity with an extraordinary degree of certainty. In the United States, there have been at least 330 postconviction exonerations based on DNA evidence. The impact of these procedures on court cases will continue to grow as standards are refined and as international STR genotyping databases grow. Even very old mysteries can be solved. In 1996, STR genotyping helped confirm identification of the bones of the last Russian czar and his family, who were assassinated in 1918.

TABLE 1	Properties of the Loci Used for the CODIS Database			
Locus	Chromosome	Repeat motif	Repeat length (range)[a]	Number of alleles seen[b]
CSF1PO	5	TAGA	5–16	20
FGA	4	CTTT	12.2–51.2	80
TH01	11	TCAT	3–14	20
TPOX	2	GAAT	4–16	15
VWA	12	[TCTG][TCTA]	10–25	28
D3S1358	3	[TCTG][TCTA]	8–21	24
D5S818	5	AGAT	7–18	15
D7S820	7	GATA	5–16	30
D8S1179	8	[TCTA][TCTG]	7–20	17
D13S317	13	TATC	5–16	17
D16S539	16	GATA	5–16	19
D18S51	18	AGAA	7–39.2	51
D21S11	21	[TCTA][TCTG]	12–41.2	82
Amelogenin[c]	X, Y	Not applicable		

Source: Data from J. M. Butler, *Forensic DNA Typing*, 2nd edn, Elsevier, 2005, p. 96.

[a]Repeat lengths observed in the human population. Partial or imperfect repeats can be included in some alleles.

[b]Number of different alleles observed as of 2005 in the human population. Careful analysis of a locus in many individuals is a prerequisite to its use in forensic DNA typing.

[c]Amelogenin is a gene, of slightly different size on the X and Y chromosomes, that is used to establish gender.

of the gel represented the nucleotide positions closest to the primer (the 5′ end), and the sequence was read (in the 5′→3′ direction) from bottom to top.

DNA sequencing was first automated by a variation of the Sanger method, in which each of the four ddNTPs used for a reaction was labeled with a different-colored fluorescent tag **(Fig. 8-35)**. With this technology, all four ddNTPs could be introduced into a single reaction. Researchers could sequence DNA molecules containing thousands of nucleotides in a few hours, and the entire genomes of hundreds of organisms were sequenced in this way. For example, in the Human Genome Project, researchers sequenced all 3.2×10^9 base pairs (bp) of the DNA in a human cell (see Chapter 9) in an effort that spanned nearly a decade and included contributions from dozens of laboratories worldwide. This form of Sanger sequencing is still used for routine analysis of short segments of DNA.

DNA Sequencing Technologies Are Advancing Rapidly

DNA sequencing technologies continue to evolve. A complete human genome can now be sequenced in a day or two, a bacterial genome in a few hours. With modest expense, a personal genomic sequence can be routinely included in each individual's medical record. These advances have been made possible by methods sometimes referred to as next-generation, or "next-gen," sequencing. The sequencing strategies are sometimes similar to and sometimes quite different from that used in the Sanger method. Innovations have allowed a miniaturization of the procedure, a massive increase in scale, and a corresponding decrease in cost.

A genomic sequence is determined in several steps. First, the genomic DNA to be sequenced is sheared at random locations to generate fragments a few hundred base pairs long. Synthetic oligonucleotides of known sequence are ligated to each end of all the fragments, providing a point of reference on every DNA molecule. The individual fragments are then immobilized on a solid surface, and each is amplified in place by PCR to form a tight cluster of identical fragments. The solid surface is part of a channel that allows liquid solutions to flow over the samples. The result is a solid surface just a few centimeters wide with millions of attached DNA clusters, each cluster

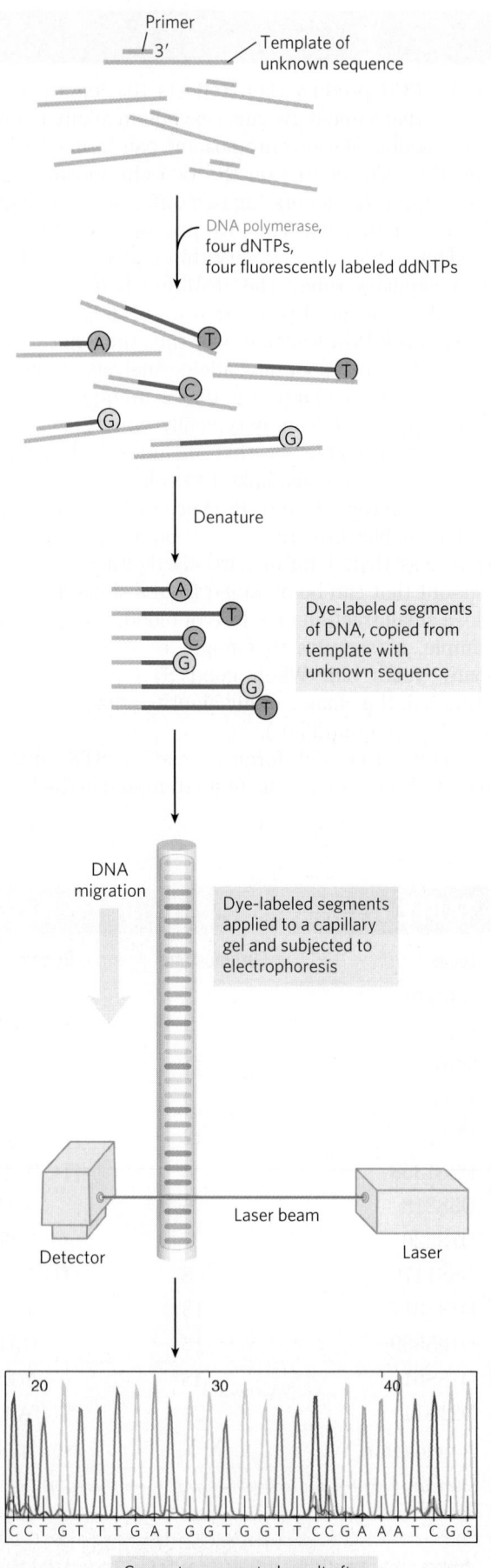

FIGURE 8-35 Automation of DNA sequencing reactions. In the Sanger method, each ddNTP can be linked to a fluorescent (dye) molecule that gives the same color to all the fragments terminating in that nucleotide, with a different color for each nucleotide. All four labeled ddNTPs are added to the reaction mix together. The resulting colored DNA fragments are separated by size in an electrophoretic gel in a capillary tube (a refinement of gel electrophoresis that allows faster separations). All fragments of a given length migrate through the capillary gel together in a single band, and the color associated with each band is detected with a laser beam. The DNA sequence is read by identifying the color sequences in the bands as they pass the detector and feeding this information directly to a computer. The amount of fluorescence in each band is represented as a peak in the computer output. [Source: Data provided by Lloyd Smith, University of Wisconsin–Madison, Department of Chemistry.]

(a)

Pulse in dATP

5′

3′

No flash of light; dATP degraded by apyrase.

Pulse in dGTP

5′

3′

Base is incorporated; pyrophosphate is released.

(P)(P)

Sulfurylase uses pyrophosphate to convert adenosine 5′-phosphosulfate into ATP.

(S)

(P)(P)(P)

In the presence of ATP, luciferase reacts with luciferin.

Flash of light

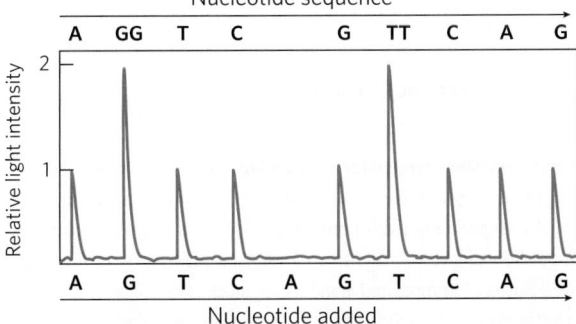

containing multiple copies of a single DNA sequence derived from a random genomic DNA fragment. The efficiency of next-generation sequencing comes from sequencing all of these millions of clusters at the same time, with the data from each cluster captured and stored in a computer.

Two widely used next-generation methods employ different strategies to accomplish the sequencing reactions. In both cases, the sequence one obtains is that of a newly synthesized DNA strand complementary to the DNA template strands being analyzed. One of the methods, known as 454 sequencing (the numbers refer to a code used during development of the technology and have no scientific meaning), uses a strategy called **pyrosequencing** in which the addition of nucleotides is detected by flashes of light **(Fig. 8-36)**. The four dNTPs (unaltered) are pulsed onto the reacting surface one at a time in a repeating sequence. The nucleotide solution is retained on the surface just long enough for DNA polymerase (one of several enzymes present in the medium bathing the surface) to add that nucleotide to any cluster where it is complementary to the next nucleotide in the template sequence. Excess nucleotide is destroyed quickly by the enzyme apyrase before the next nucleotide pulse. When a specific nucleotide is successfully added to the strands of a cluster, pyrophosphate is released as a byproduct. The enzyme sulfurylase uses the pyrophosphate to transform adenosine 5′-phosphosulfate in the medium to ATP. The appearance of ATP provides the signal that a nucleotide has been added to the DNA. Also in the medium are the enzyme luciferase and its substrate luciferin. When ATP is generated, luciferase catalyzes a reaction with luciferin that emits a tiny flash of light. (This reaction gives fireflies their flash; see Box 13-1.) When many tiny flashes occur in a cluster, the emitted light can be recorded in a captured image. For example, when dCTP is added to the solution, flashes occur only at clusters where G is the next base in the template and C is the next nucleotide to be added

FIGURE 8-36 Next-generation pyrosequencing. (a) Pyrosequencing uses flashes of light to detect the addition of complementary nucleotides on the DNA (template) to be sequenced. Each individual segment of the DNA to be sequenced is attached to a tiny DNA capture bead, then amplified on the bead by PCR. Each bead is immersed in an emulsion and placed in a tiny (diameter ~29 μm) well on a picotiter plate. The reaction of luciferin and ATP with luciferase produces light flashes when a nucleotide is added to a particular DNA cluster in a particular well. **(b)** Artist's rendition of a very small part of one cycle of a 454 sequencing run. Each white spot represents a single DNA fragment cluster, with the same clusters shown over multiple cycles. In this example, reading the top (or bottom) red-circled spot from left to right across each row gives the sequence for that cluster.

(b)

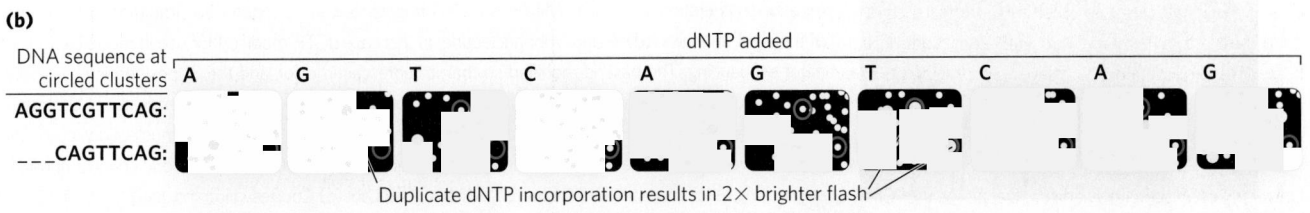

Duplicate dNTP incorporation results in 2× brighter flash

to the growing DNA chain. If there is a string of two, three, or four G residues in the template, a similar number of C residues are added to the growing strand in one cycle. This is recorded as a "flash" amplitude at that cluster that is two, three, or four times greater than when only one C residue is added. Similarly, when dGTP is added, flashes occur at a different set of clusters, marking those as clusters where G is the next nucleotide added to the sequence (where C is present in the template). The length of DNA that can be reliably sequenced in a single cluster by this method—often referred to as the read length, or "read"—is typically 400 to 500 nucleotides, and is rapidly increasing.

The second widely used next-generation method employs a technique known as **reversible terminator sequencing (Fig. 8-37)**, which lies at the heart of the

(a)

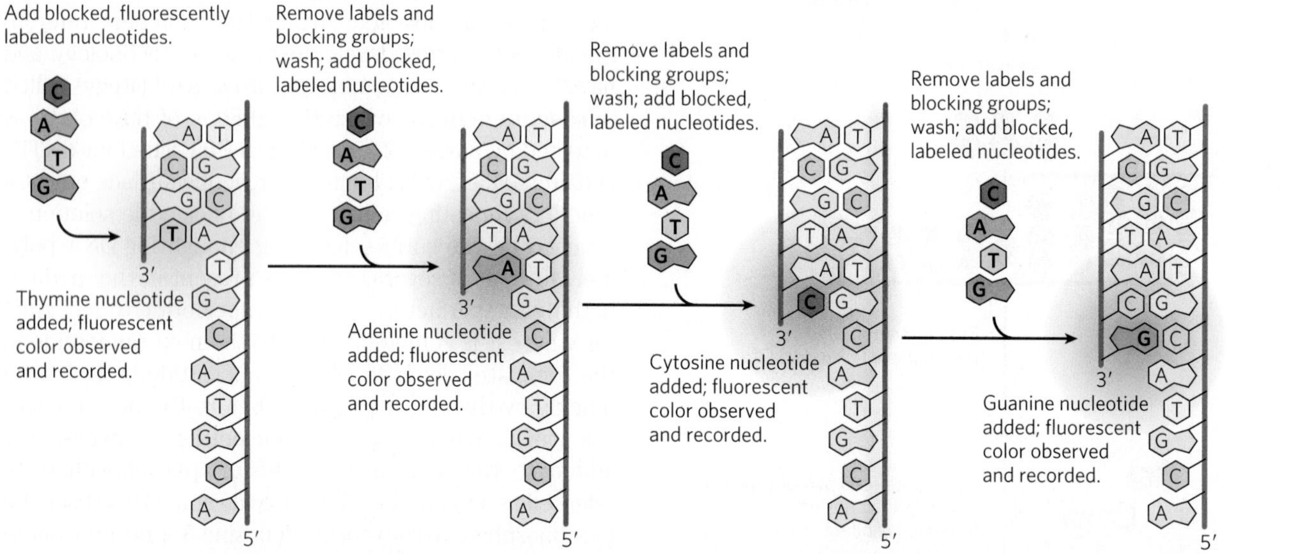

Add blocked, fluorescently labeled nucleotides.

Thymine nucleotide added; fluorescent color observed and recorded.

Remove labels and blocking groups; wash; add blocked, labeled nucleotides.

Adenine nucleotide added; fluorescent color observed and recorded.

Remove labels and blocking groups; wash; add blocked, labeled nucleotides.

Cytosine nucleotide added; fluorescent color observed and recorded.

Remove labels and blocking groups; wash; add blocked, labeled nucleotides.

Guanine nucleotide added; fluorescent color observed and recorded.

(b) dNTP incorporated
TACGGTCTC:
CCCCCCAGT:

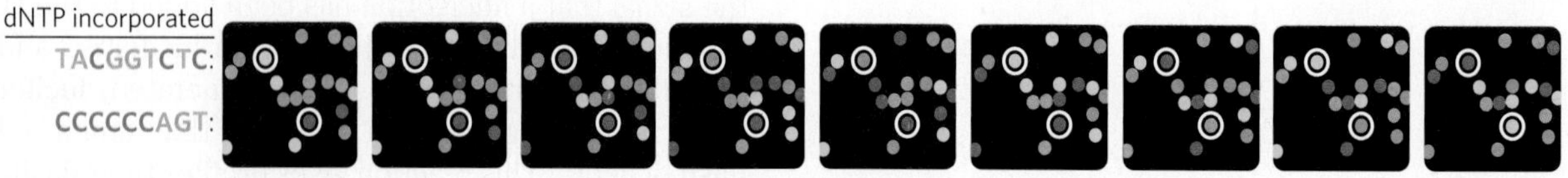

(c)

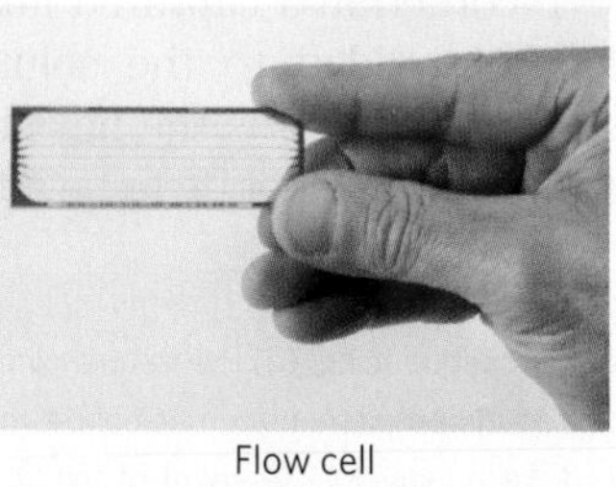

Flow cell

(d)

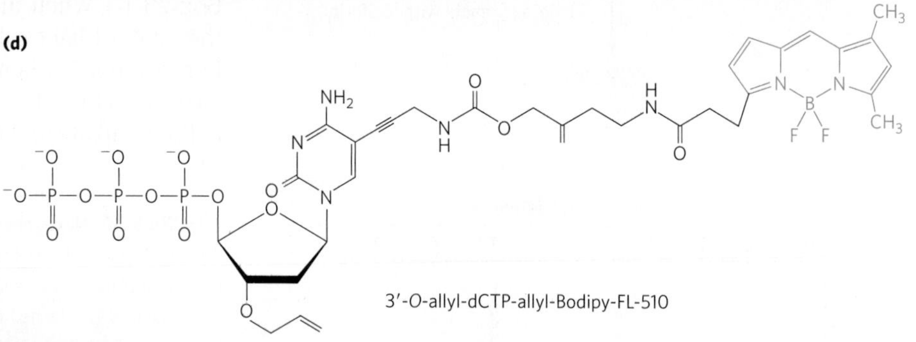

3'-O-allyl-dCTP-allyl-Bodipy-FL-510

(e)

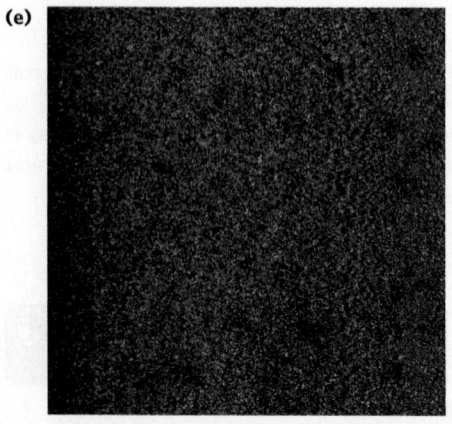

FIGURE 8-37 Next-generation reversible terminator sequencing. (a) The reversible terminator method of sequencing uses fluorescent tags to identify nucleotides. Blocking groups on each fluorescently labeled nucleotide prevent multiple nucleotides from being added in a single cycle. **(b)** Artist's rendition of nine successive cycles from one very small part of an Illumina sequencing run. Each colored spot represents the location of a cluster of immobilized identical oligonucleotides affixed to the surface of the flow cell. The white-circled spots represent the same two clusters on the surface over successive cycles, with the sequences indicated. Data are recorded and analyzed digitally. **(c)** Typical flow cell used for a next-generation sequencer. Millions of DNA fragments can be sequenced simultaneously in each of the eight channels. **(d)** A deoxyribonucleotide, in this case dCTP, modified for use in reversible terminator sequencing. The base is modified so that it fluoresces in color, and the 3' position is chemically blocked. Both the dye and the 3'-end-blocking group can be removed, either chemically or photolytically, leaving a free 3'-OH group for addition of the next nucleotide. The modified nucleotides currently used in reversible terminator sequencing are proprietary. **(e)** Part of the surface of one channel during a sequencing reaction. [Source: (c) Courtesy Michael Cox Lab. (e) Courtesy Illumina, Inc.]

Illumina sequencer. Once the genomic sequences are fragmented and oligonucleotides of known sequence are attached at the ends, the DNA segments are immobilized on a solid surface and amplified in place by PCR. A special sequencing primer is then added that is complementary to the oligonucleotides of known sequence at the segment ends. Four different modified deoxynucleotides (A, T, G, and C), each with a particular fluorescent label that identifies the nucleotide by color, are added, along with DNA polymerase. The labeled nucleotides are special terminator nucleotides with blocking groups attached to their 3′ ends that permit only one nucleotide to be added to each strand. The polymerase adds the appropriate nucleotide to the strands in each cluster. Next, lasers excite all the fluorescent labels, and an image of the entire surface reveals the color (and thus the identity)

of the base added to each cluster. The fluorescent label and the blocking groups are then chemically or photolytically removed, in preparation for adding a new nucleotide to each cluster. The sequencing proceeds stepwise. Read lengths are shorter for this method, typically 100 to 200 nucleotides per cluster, although refinements are continuing.

Using these increasingly powerful methods, determining the complete genomic sequence of an organism is much faster and cheaper. A few hundred base pairs of sequence may have little value unless one knows where on a chromosome the sequence is located. Translating the sequences of millions of short DNA fragments into a complex and contiguous genomic sequence requires the computerized alignment of overlapping fragments **(Fig. 8-38)**. The number of times

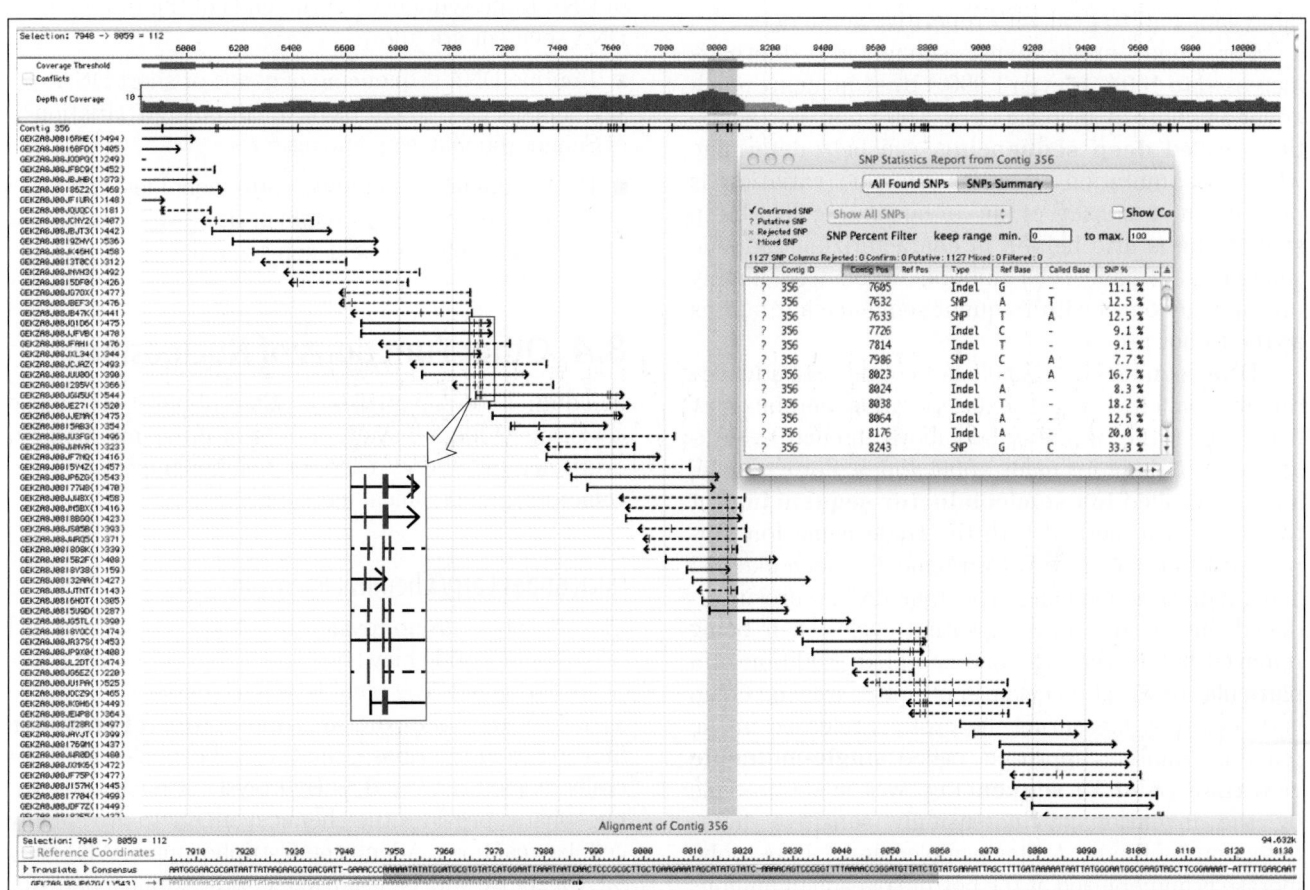

FIGURE 8-38 Sequence assembly. In a genomic sequence, each base pair of the genome is usually represented in multiple sequenced fragments, referred to as reads. Shown here is a small part of the genomic sequence of a new variant species of *E. coli*, with the reads generated by a 454 sequencer. The numbers at the top represent base-pair positions in the genome, relative to an arbitrarily defined reference point. All sequence fragments come from a particular long contig designated 356. The reads are represented by horizontal arrows, with computer-assigned identifiers for each read listed at the left. DNA strand segments are sequenced at random, with sequences obtained from one strand (5′ to 3′, left to right) represented by solid arrows and sequences obtained from the other strand (5′ to 3′, right to left) represented by dashed-line arrows (the latter automatically reported as their

complement when they are merged with the overall dataset). The "coverage threshold" at the top is a measure of sequence quality. The wider green bar indicates sequences that have been obtained enough times to generate high confidence in the results; the depth of the coverage line indicates how many times a given base pair appears in a sequenced read. The vertical blue-shaded line indicates a part of the sequence that is highlighted by shading in the sequence line at the bottom of the figure. The "SNP statistics report" (inset) is a listing of positions where sequence changes called single nucleotide polymorphisms (SNPs; see Chapter 9) seem to be present in some of the reads. These putative SNPs are often checked by additional sequencing. They are indicated in the reads by thin blue vertical slash marks within the horizontal lines for each read.

that a particular nucleotide in a genome is sequenced, on average, is referred to as the **sequencing depth** or sequencing coverage. In most cases, a sufficiently large number of random fragments are sequenced so that each nucleotide in the genome is sequenced an average of 30 to 40 times (30–40× coverage). Although the coverage of particular nucleotides may vary (some will be sequenced 100 times, perhaps a few not at all), this level of coverage ensures that most genomic nucleotides will be sequenced at least 10 times and that most sequencing errors will be detected and eliminated. The overlaps allow the computer to trace the sequence through a chromosome, from one overlapping fragment to another, permitting the assembly of long, contiguous sequences called **contigs**. In a successful genomic sequencing exercise, many contigs can extend over millions of base pairs. Special strategies are needed to fill in the inevitable gaps and to deal with repetitive sequences.

For some applications, sequencing depth is increased to 100× or even 1,000× by sequencing much larger amounts of genomic DNA. This approach, sometimes called **deep sequencing**, can help determine whether a mutation or other genomic variation is present in a subset of an organism's cells. Deep sequencing is also helpful in characterizing genomic sequences in cancerous tumors, which have highly unstable genomes with frequent sequence alterations as the tumor grows.

DNA sequencing technologies continue to advance rapidly, and a few newer next-generation methods now complement the two described above and may eventually replace them for many applications. For example, a method called **ion semiconductor sequencing** (at the heart of a method with the trade name Ion Torrent) uses immobilized DNA fragments, much like 454 and Illumina sequencing. The four dNTPs are introduced one by one in a repeating cycle, each being removed before the next one is added. Addition of a particular dNTP at a certain spot in the growing chain is detected by measuring the protons released in the reaction. Another approach, called **single-molecule real-time (SMRT) sequencing**, was made possible by the invention of increasingly sensitive light-detection methods. A single molecule of DNA polymerase is immobilized at the bottom of each of millions of precisely engineered pores on the flow cell. The polymerase captures fragmented genomic segments as they diffuse into the pore. The labeled dNTPs then diffuse in, each newly added nucleotide releasing its colored fluorescent group as it attaches to the DNA chain. An innovative light-detection system records the color of the resulting light flash at the bottom of the pore, revealing the identity of each added nucleotide. The method is accurate and can generate particularly long read lengths, up to nearly 10,000 base pairs.

SUMMARY 8.3 Nucleic Acid Chemistry

- Native DNA undergoes reversible unwinding and separation of strands (melting) on heating or at extremes of pH. DNAs rich in G≡C pairs have higher melting points than DNAs rich in A=T pairs.

- DNA is a relatively stable polymer. Spontaneous reactions such as deamination of certain bases, hydrolysis of base-sugar N-glycosyl bonds, radiation-induced formation of pyrimidine dimers, and oxidative damage occur at very low rates, yet are important because of a cell's very low tolerance for changes in its genetic material.

- Oligonucleotides of known sequence can be synthesized rapidly and accurately.

- The polymerase chain reaction (PCR) provides a convenient and rapid method for amplifying segments of DNA if the sequences of the ends of the targeted DNA segment are known.

- Routine DNA sequencing of genes or short DNA segments is carried out using an automated variation of Sanger dideoxy sequencing.

- DNA sequences, including entire genomes, can be efficiently determined in hours or days with a range of methods, including next-gen sequencing.

8.4 Other Functions of Nucleotides

In addition to their roles as the subunits of nucleic acids, nucleotides have a variety of other functions in every cell: as energy carriers, components of enzyme cofactors, and chemical messengers.

Nucleotides Carry Chemical Energy in Cells

The phosphate group covalently linked at the 5′ hydroxyl of a ribonucleotide may have one or two additional phosphates attached. The resulting molecules are referred to as nucleoside mono-, di-, and triphosphates (**Fig. 8-39**). Starting from the ribose, the three phosphates are generally labeled α, β, and γ. Hydrolysis of nucleoside triphosphates provides the chemical energy to drive many cellular reactions. Adenosine 5′-triphosphate, ATP, is by far the most widely used nucleoside triphosphate for this purpose, but UTP, GTP, and CTP are also used in some reactions. Nucleoside triphosphates also serve as the activated precursors of DNA and RNA synthesis, as described in Chapters 25 and 26.

The energy released by hydrolysis of ATP and the other nucleoside triphosphates is accounted for by the structure of the triphosphate group. The bond between the ribose and the α phosphate is an ester linkage. The α, β and β, γ linkages are phosphoanhydrides (**Fig. 8-40**). Hydrolysis of the ester linkage yields about 14 kJ/mol under standard conditions,

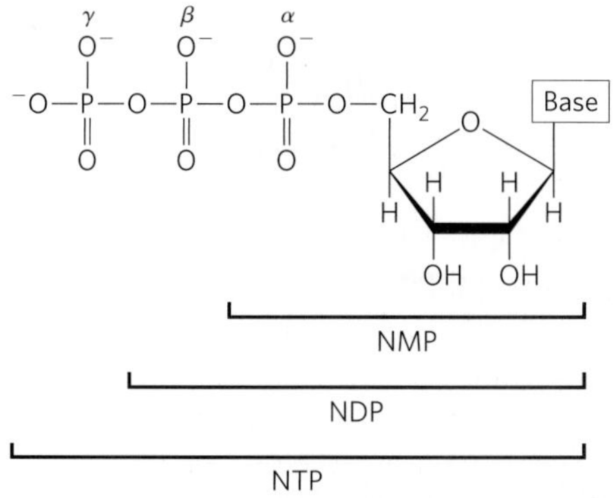

Abbreviations of ribonucleoside 5′-phosphates			
Base	Mono-	Di-	Tri-
Adenine	AMP	ADP	ATP
Guanine	GMP	GDP	GTP
Cytosine	CMP	CDP	CTP
Uracil	UMP	UDP	UTP

Abbreviations of deoxyribonucleoside 5′-phosphates			
Base	Mono-	Di-	Tri-
Adenine	dAMP	dADP	dATP
Guanine	dGMP	dGDP	dGTP
Cytosine	dCMP	dCDP	dCTP
Thymine	dTMP	dTDP	dTTP

FIGURE 8-39 Nucleoside phosphates. General structure of the nucleoside 5′-mono-, di-, and triphosphates (NMPs, NDPs, and NTPs) and their standard abbreviations. In the deoxyribonucleoside phosphates (dNMPs, dNDPs, and dNTPs), the pentose is 2′-deoxy-D-ribose.

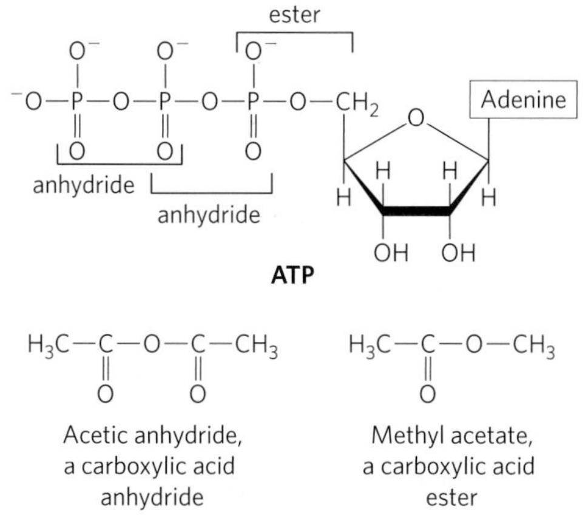

FIGURE 8-40 The phosphate ester and phosphoanhydride bonds of ATP. Hydrolysis of an anhydride bond yields more energy than hydrolysis of the ester. A carboxylic acid anhydride and carboxylic acid ester are shown for comparison.

whereas hydrolysis of each anhydride bond yields about 30 kJ/mol. ATP hydrolysis often plays an important thermodynamic role in biosynthesis. When coupled to a reaction with a positive free-energy change, ATP hydrolysis shifts the equilibrium of the overall process to favor product formation (recall the relationship between the equilibrium constant and free-energy change described by Eqn 6-3 on p. 192).

Adenine Nucleotides Are Components of Many Enzyme Cofactors

A variety of enzyme cofactors serving a wide range of chemical functions include adenosine as part of their structure **(Fig. 8-41)**. They are unrelated structurally except for the presence of adenosine. In none of these cofactors does the adenosine portion participate directly in the primary function, but removal of adenosine generally results in a drastic reduction of cofactor activities. For example, removal of the adenine nucleotide (3′-phosphoadenosine diphosphate) from acetoacetyl-CoA, the coenzyme A derivative of acetoacetate, reduces its reactivity as a substrate for β-ketoacyl-CoA transferase (an enzyme of lipid metabolism) by a factor of 10^6. Although this requirement for adenosine has not been investigated in detail, it must involve the binding energy between enzyme and substrate (or cofactor) that is used both in catalysis and in stabilizing the initial enzyme-substrate complex (Chapter 6). In the case of β-ketoacyl-CoA transferase, the nucleotide moiety of coenzyme A seems to be a binding "handle" that helps to pull the substrate (acetoacetyl-CoA) into the active site. Similar roles may be found for the nucleoside portion of other nucleotide cofactors.

Why is adenosine, rather than some other large molecule, used in these structures? The answer here may involve a form of evolutionary economy. Adenosine is certainly not unique in the amount of potential binding energy it can contribute. The importance of adenosine probably lies not so much in some special chemical characteristic as in the evolutionary advantage of using one compound for multiple roles. Once ATP became the universal source of chemical energy, systems developed to synthesize ATP in greater abundance than the other nucleotides; because it is abundant, it becomes the logical choice for incorporation into a wide variety of structures. The economy extends to protein structure. A single protein domain that binds adenosine can be used in different enzymes. Such a domain, called a **nucleotide-binding fold**, is found in many enzymes that bind ATP and nucleotide cofactors.

Some Nucleotides Are Regulatory Molecules

Cells respond to their environment by taking cues from hormones or other external chemical signals. The interaction of these extracellular chemical signals ("first messengers") with receptors on the cell surface often leads to the production of **second messengers** inside the cell, which in turn leads to adaptive changes in the cell interior (Chapter 12). Often, the second messenger

Coenzyme A

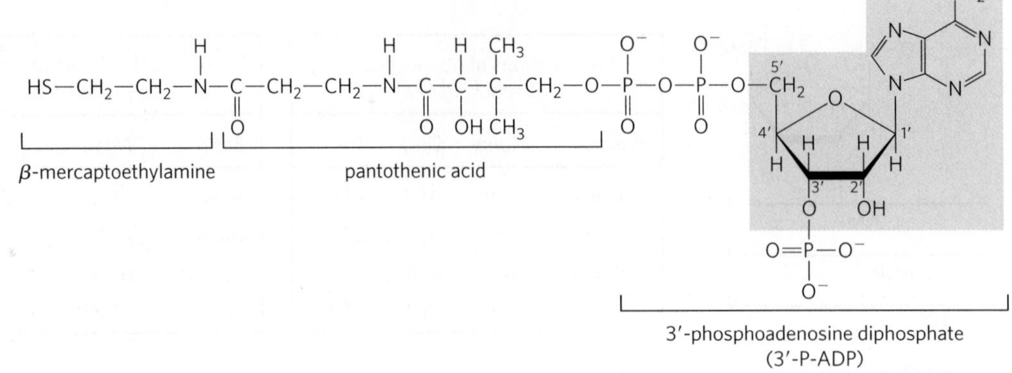

β-mercaptoethylamine   pantothenic acid

3'-phosphoadenosine diphosphate
(3'-P-ADP)

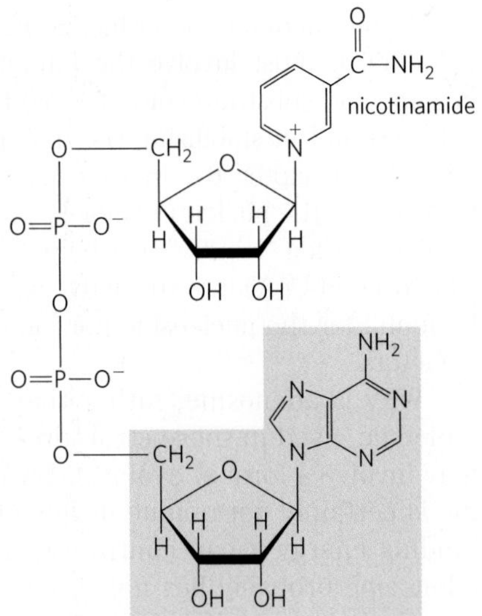

riboflavin

nicotinamide

Nicotinamide adenine dinucleotide (NAD⁺)

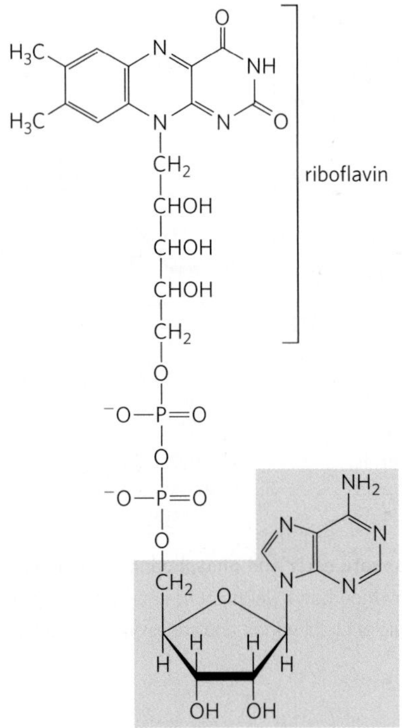

Flavin adenine dinucleotide (FAD)

FIGURE 8-41 Some coenzymes containing adenosine. The adenosine portion is shaded in light red. Coenzyme A (CoA) functions in acyl group transfer reactions; the acyl group (such as the acetyl or acetoacetyl group) is attached to the CoA through a thioester linkage to the β-mercaptoethylamine moiety.

NAD^+ functions in hydride transfers, and FAD, the active form of vitamin B_2 (riboflavin), in electron transfers. Another coenzyme incorporating adenosine is 5'-deoxyadenosylcobalamin, the active form of vitamin B_{12} (see Box 17-2), which participates in intramolecular group transfers between adjacent carbons.

is a nucleotide **(Fig. 8-42)**. One of the most common is **adenosine 3′,5′-cyclic monophosphate (cyclic AMP, or cAMP)**, formed from ATP in a reaction catalyzed by adenylyl cyclase, an enzyme associated with the inner face of the plasma membrane. Cyclic AMP serves regulatory functions in virtually every cell outside the plant kingdom. Guanosine 3′,5′-cyclic monophosphate (cGMP) also has regulatory functions in many cells.

Another regulatory nucleotide, ppGpp (Fig. 8-42), is produced in bacteria in response to a slowdown in protein synthesis during amino acid starvation. This nucleotide inhibits the synthesis of the rRNA and tRNA molecules (see Fig. 28-22) needed for protein

synthesis, preventing the unnecessary production of nucleic acids.

Adenine Nucleotides Also Serve as Signals

ATP and ADP also serve as signaling molecules in many unicellular and multicellular organisms, including humans. In mammals, certain neurons release ATP at synapses, which binds P_{2X} receptors on the postsynaptic cell, triggering changes in membrane potential or the release of an intracellular second messenger that initiates diverse physiological processes, including taste, inflammation, and smooth muscle contraction. One important

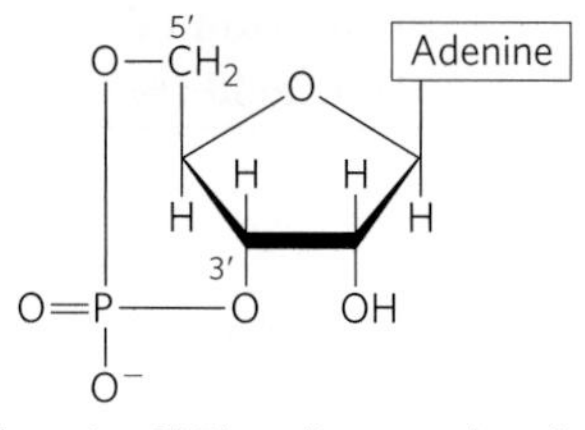

Adenosine 3′,5′-cyclic monophosphate
(cyclic AMP; cAMP)

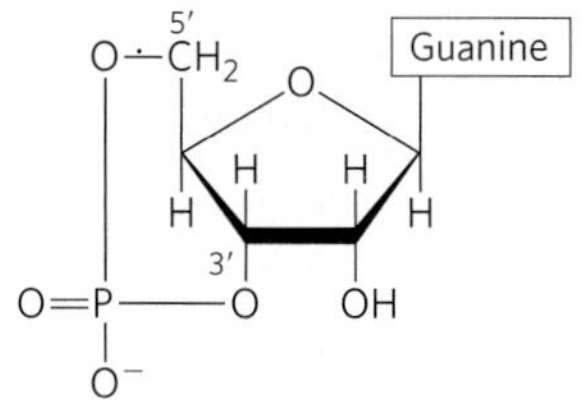

Guanosine 3′,5′-cyclic monophosphate
(cyclic GMP; cGMP)

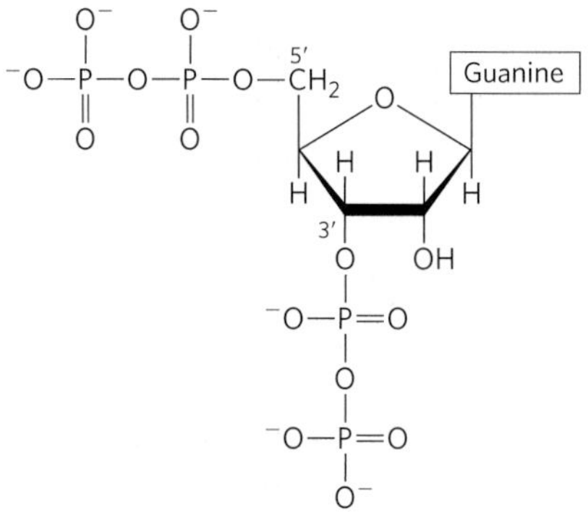

Guanosine 5′-diphosphate,3′-diphosphate
(guanosine tetraphosphate; ppGpp)

FIGURE 8-42 Three regulatory nucleotides.

class of ATP receptors that mediate the sensation of pain is an obvious target for drug development. Extracellular ADP is a signaling molecule that acts through P_{2Y} receptors in sensitive cell types. By preventing ADP from binding the P_{2Y} receptors of platelets, the drug clopidogrel (Plavix) inhibits undesirable blood clotting in patients with cardiac disease. Signaling pathways are discussed in more detail in Chapter 12. ■

SUMMARY 8.4 Other Functions of Nucleotides

■ ATP is the central carrier of chemical energy in cells. The presence of an adenosine moiety in a variety of enzyme cofactors may be related to binding-energy requirements.

■ Cyclic AMP, formed from ATP in a reaction catalyzed by adenylyl cyclase, is a common second messenger produced in response to hormones and other chemical signals.

■ ATP and ADP serve as neurotransmitters in a variety of signaling pathways.

Key Terms

Terms in bold are defined in the glossary.

**deoxyribonucleic acid
(DNA)** 279
**ribonucleic acid
(RNA)** 279
gene 279
**ribosomal RNA
(rRNA)** 279
**messenger RNA
(mRNA)** 279
transfer RNA (tRNA) 279
nucleotide 279
nucleoside 279
pyrimidine 279
purine 279
deoxyribonucleotides 281
ribonucleotide 282
**phosphodiester
linkage** 283
5′ end 283
3′ end 283
oligonucleotide 284
polynucleotide 284
base pair 284
major groove 286
minor groove 286
B-form DNA 288
A-form DNA 288
Z-form DNA 288
palindrome 289
hairpin 289
cruciform 290

triplex DNA 290
G tetraplex 290
transcription 291
**monocistronic
mRNA** 291
polycistronic mRNA 291
mutation 297
**polymerase chain
reaction (PCR)** 301
DNA polymerases 301
Sanger sequencing 302
**sequence polymor-
phisms** 304
**short tandem repeat
(STR)** 304
**DNA sequencing
technologies** 306
pyrosequencing 307
**reversible terminator
sequencing** 308
sequencing depth 310
contig 310
**ion semiconductor
sequencing** 310
**single-molecule
real-time (SMRT)
sequencing** 310
second messenger 311
**adenosine 3′,5′-cyclic
monophosphate (cyclic
AMP, cAMP)** 312

Problems

1. Nucleotide Structure Which positions in the purine ring of a purine nucleotide in DNA have the potential to form hydrogen bonds but are not involved in Watson-Crick base pairing?

2. Base Sequence of Complementary DNA Strands One strand of a double-helical DNA has the sequence (5′)GCGCAATATTTCTCAAAATATTGCGC(3′). Write the base sequence of the complementary strand. What special type of sequence is contained in this DNA segment? Does the double-stranded DNA have the potential to form any alternative structures?

3. DNA of the Human Body Calculate the weight in grams of a double-helical DNA molecule stretching from Earth to the moon (~320,000 km). The DNA double helix weighs about 1×10^{-18} g per 1,000 nucleotide pairs; each base pair extends 3.4 Å. For an interesting comparison, your body contains about 0.5 g of DNA.

4. DNA Bending Assume that a poly(A) tract five base pairs long produces a 20° bend in a DNA strand. Calculate the total (net) bend produced in a DNA if the center base pairs (the third of five) of two successive $(dA)_5$ tracts are located (a) 10 base pairs apart; (b) 15 base pairs apart. Assume 10 base pairs per turn in the DNA double helix.

5. Distinction between DNA Structure and RNA Structure Hairpins may form at palindromic sequences in single strands of either RNA or DNA. How is the helical structure of a long and fully base-paired (except at the end) hairpin in RNA different from that of a similar hairpin in DNA?

6. Nucleotide Chemistry The cells of many eukaryotic organisms have highly specialized systems that specifically repair G–T mismatches in DNA. The mismatch is repaired to form a $G \equiv C$ (not $A = T$) base pair. This G–T mismatch repair mechanism occurs in addition to a more general system that repairs virtually all mismatches. Suggest why cells might require a specialized system to repair G–T mismatches.

7. Denaturation of Nucleic Acids A duplex DNA oligonucleotide in which one of the strands has the sequence TAATAC GACTCACTATAGGG has a melting temperature (t_m) of 59 °C. If an RNA duplex oligonucleotide of identical sequence (substituting U for T) is constructed, will its melting temperature be higher or lower?

8. Spontaneous DNA Damage Hydrolysis of the N-glycosyl bond between deoxyribose and a purine in DNA creates an AP site. An AP site generates a thermodynamic destabilization greater than that created by any DNA mismatched base pair. This effect is not completely understood. Examine the structure of an AP site (see Fig. 8-29b) and describe some chemical consequences of base loss.

9. Prediction of Nucleic Acid Structure from Its Sequence A part of a sequenced chromosome has the sequence (on one strand) ATTGCATCCGCGCGTGCGCGCGC-GATCCCGTTACTTTCCG. Which part of this sequence is most likely to take up the Z conformation?

10. Nucleic Acid Structure Explain why the absorption of UV light by double-stranded DNA increases (the hyperchromic effect) when the DNA is denatured.

11. Determination of Protein Concentration in a Solution Containing Proteins and Nucleic Acids The concentration of protein or nucleic acid in a solution containing both can be estimated by using their different light absorption properties: proteins absorb most strongly at 280 nm and nucleic acids at 260 nm. Estimates of their respective concentrations in a mixture can be made by measuring the absorbance (A) of the solution at 280 and 260 nm and using the table below, which gives $R_{280/260}$, the ratio of absorbances at 280 and 260 nm; the percentage of total mass that is nucleic acid; and a factor, F, that corrects the A_{280} reading and gives a more accurate protein estimate. The protein concentration (in mg/mL) = $F \times A_{280}$ (assuming the cuvette is 1 cm wide). Calculate the protein concentration in a solution of $A_{280} = 0.69$ and $A_{260} = 0.94$.

$R_{280/260}$	Proportion of nucleic acid (%)	F
1.75	0.00	1.116
1.63	0.25	1.081
1.52	0.50	1.054
1.40	0.75	1.023
1.36	1.00	0.994
1.30	1.25	0.970
1.25	1.50	0.944
1.16	2.00	0.899
1.09	2.50	0.852
1.03	3.00	0.814
0.979	3.50	0.776
0.939	4.00	0.743
0.874	5.00	0.682
0.846	5.50	0.656
0.822	6.00	0.632
0.804	6.50	0.607
0.784	7.00	0.585
0.767	7.50	0.565
0.753	8.00	0.545
0.730	9.00	0.508
0.705	10.00	0.478
0.671	12.00	0.422
0.644	14.00	0.377
0.615	17.00	0.322
0.595	20.00	0.278

12. Solubility of the Components of DNA Draw the following structures and rate their relative solubilities in water (most soluble to least soluble): deoxyribose, guanine, phosphate. How are these solubilities consistent with the three-dimensional structure of double-stranded DNA?

13. Polymerase Chain Reaction One strand of a chromosomal DNA sequence is shown below. An investigator wants to amplify and isolate a DNA fragment defined by the segment shown in red, using the polymerase chain reaction (PCR). Design two PCR primers, each 20 nucleotides long, that can be used to amplify this DNA segment. The final PCR product generated with your primers should include no sequences outside the segment in red.

5′ – – – AATGCCGTCAGCCGATCTGCCTCGAGTCAATCGA TGCTGGTAACTTGGGGTATAAAGCTTACCCATGGTATCGTAG TTAGATTGATTGTTAGGTTCTTAGGTTTAGGTTTCTGGTATT GGTTTAGGGTCTTTGATGCTATTAATTGTTTGGTTTTTGATTT GGTCTTTATATGGTTTATGTTTTAAGCCGGGTTTTGTCTGG-GATGGTTCGTCTGATGTGCGCGTAGCGTGCGGCG – – –3′

14. Genomic Sequencing In large-genome sequencing projects, the initial data usually reveal gaps where no sequence information has been obtained. To close the gaps, DNA primers complementary to the 5′-ending strand (that is, identical to the sequence of the 3′-ending strand) at the end of each contig are especially useful. Explain how these primers might be used.

15. Next-Generation Sequencing In reversible terminator sequencing, how would the sequencing process be affected if the 3′-end-blocking group of each nucleotide were replaced with the 3′-H present in the dideoxynucleotides used in Sanger sequencing?

16. Sanger Sequencing Logic In the Sanger (dideoxy) method for DNA sequencing, a small amount of a dideoxynucleoside triphosphate—say, ddCTP—is added to the sequencing reaction along with a larger amount of the corresponding dCTP. What result would be observed if the dCTP were omitted?

17. DNA Sequencing The following DNA fragment was sequenced by the Sanger method. The red asterisk indicates a fluorescent label.

*5′ ━━━━ 3′-OH
3′ ━━━━ ATTACGCAAGGACATTAGAC---5′

A sample of the DNA was reacted with DNA polymerase and each of the nucleotide mixtures (in an appropriate buffer) listed below. Dideoxynucleotides (ddNTPs) were added in relatively small amounts.

1. dATP, dTTP, dCTP, dGTP, ddTTP
2. dATP, dTTP, dCTP, dGTP, ddGTP
3. dATP, dCTP, dGTP, ddTTP
4. dATP, dTTP, dCTP, dGTP

The resulting DNA was separated by electrophoresis on an agarose gel, and the fluorescent bands on the gel were located. The band pattern resulting from nucleotide mixture 1 is shown below. Assuming that all mixtures were run on the same gel, what did the remaining lanes of the gel look like?

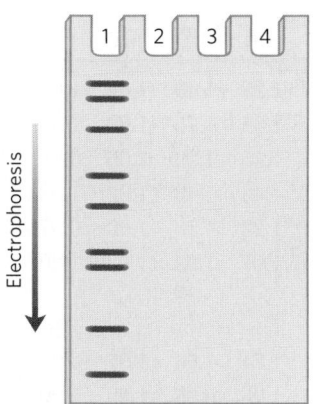

18. Snake Venom Phosphodiesterase An exonuclease is an enzyme that sequentially cleaves nucleotides from the end of a polynucleotide strand. Snake venom phosphodiesterase,

which hydrolyzes nucleotides from the 3′ end of any oligonucleotide with a free 3′-hydroxyl group, cleaves between the 3′ hydroxyl of the ribose or deoxyribose and the phosphoryl group of the next nucleotide. It acts on single-stranded DNA or RNA and has no base specificity. This enzyme was used in sequence determination experiments before the development of modern nucleic acid sequencing techniques. What are the products of partial digestion by snake venom phosphodiesterase of an oligonucleotide with the sequence (5′)GCGCCAUUGC(3′)—OH?

19. Preserving DNA in Bacterial Endospores Bacterial endospores form when the environment is no longer conducive to active cell metabolism. The soil bacterium *Bacillus subtilis*, for example, begins the process of sporulation when one or more nutrients are depleted. The end product is a small, metabolically dormant structure that can survive almost indefinitely with no detectable metabolism. Spores have mechanisms to prevent accumulation of potentially lethal mutations in their DNA over periods of dormancy that can exceed 1,000 years. *B. subtilis* spores are much more resistant than are the organism's growing cells to heat, UV radiation, and oxidizing agents, all of which promote mutations.

(a) One factor that prevents potential DNA damage in spores is their greatly decreased water content. How would this affect some types of mutations?

(b) Endospores have a category of proteins called small acid-soluble proteins (SASPs) that bind to their DNA, preventing formation of cyclobutane-type dimers. What causes cyclobutane dimers, and why do bacterial endospores need mechanisms to prevent their formation?

20. Oligonucleotide Synthesis In the scheme of Figure 8-34, each new base to be added to the growing oligonucleotide is modified so that its 3′ hydroxyl is activated and the 5′ hydroxyl has a dimethoxytrityl (DMT) group attached. What is the function of the DMT group on the incoming base?

Biochemistry Online

21. The Structure of DNA Elucidation of the three-dimensional structure of DNA helped researchers understand how this molecule conveys information that can be faithfully replicated from one generation to the next. To see the secondary structure of double-stranded DNA, go to the Protein Data Bank website (www.pdb.org). Use the PDB identifiers listed below to retrieve the structure summaries for the two forms of DNA. View the 3D structure using JSmol (click the 3D View tab or the JSmol link in the Structure Image window on the summary page). You will need to use both the display menus on the screen and the scripting controls in the JSmol menu (accessed by clicking on the JSmol logo in the lower right corner of the image screen) to complete the following exercises. Refer to the JSmol help links as needed.

(a) Access PDB ID 141D, a highly conserved, repeated DNA sequence from the end of the genome of HIV-1 (the virus that causes AIDS). Set the Style to Ball and Stick. Then use the scripting controls to color by element (Color > Atoms > By Scheme > Element (CPK)). Identify the sugar–phosphate backbone for each strand of the DNA duplex. Locate and identify individual bases. Identify the 5′ end of each strand. Locate the major and minor grooves. Is this a right- or left-handed helix?

(b) Access PDB ID 145D, a DNA with the Z conformation. Set the Style to Ball and Stick. Then use the scripting controls to color by element (Main Menu > Color > Atoms > By Scheme > Element (CPK)). Identify the sugar–phosphate backbone for each strand of the DNA duplex. Is this a right- or left-handed helix?

(c) To fully appreciate the secondary structure of DNA, view the molecules in stereo. From the scripting control Main Menu select Style > Stereographic > Cross-eyed viewing or Wall-eyed viewing. (If you have stereographic glasses available, select the appropriate option.) You will see two images of the DNA molecule. Sit with your nose approximately 10 inches from the monitor and focus on the tip of your nose (cross-eyed) or on the opposite edges of the screen (wall-eyed). In the background you should see three images of the DNA helix. Shift your focus to the middle image, which should appear three-dimensional. (Note that only one of the two authors can make this work.)

▊ Data Analysis Problem

22. Chargaff's Studies of DNA Structure The chapter section "DNA Is a Double Helix That Stores Genetic Information" includes a summary of the main findings of Erwin Chargaff and his coworkers, listed as four conclusions ("Chargaff's rules"; p. 286). In this problem, you will examine the data Chargaff collected in support of these conclusions.

In one paper, Chargaff (1950) described his analytical methods and some early results. Briefly, he treated DNA samples with acid to remove the bases, separated the bases by paper chromatography, and measured the amount of each base with UV spectroscopy. His results are shown in the three tables below. The *molar ratio* is the ratio of the number of moles of each base in the sample to the number of moles of phosphate in the sample—this gives the fraction of the total number of bases represented by each particular base. The *recovery* is the sum of all four bases (the sum of the molar ratios); full recovery of all bases in the DNA would give a recovery of 1.0.

Molar ratios in ox DNA

Base	Thymus Prep. 1	Prep. 2	Prep. 3	Spleen Prep. 1	Prep. 2	Liver Prep. 1
Adenine	0.26	0.28	0.30	0.25	0.26	0.26
Guanine	0.21	0.24	0.22	0.20	0.21	0.20
Cytosine	0.16	0.18	0.17	0.15	0.17	
Thymine	0.25	0.24	0.25	0.24	0.24	
Recovery	*0.88*	*0.94*	*0.94*	*0.84*	*0.88*	

Molar ratios in human DNA

Base	Sperm Prep. 1	Prep. 2	Thymus Prep. 1	Liver Normal	Carcinoma
Adenine	0.29	0.27	0.28	0.27	0.27
Guanine	0.18	0.17	0.19	0.19	0.18
Cytosine	0.18	0.18	0.16		0.15
Thymine	0.31	0.30	0.28		0.27
Recovery	*0.96*	*0.92*	*0.91*		*0.87*

Molar ratios in DNA of microorganisms

Base	Yeast Prep. 1	Prep. 2	Avian tubercle bacilli Prep. 1
Adenine	0.24	0.30	0.12
Guanine	0.14	0.18	0.28
Cytosine	0.13	0.15	0.26
Thymine	0.25	0.29	0.11
Recovery	*0.76*	*0.92*	*0.77*

(a) Based on these data, Chargaff concluded that "no differences in composition have so far been found in DNA from different tissues of the same species." This corresponds to conclusion 2 in this chapter. However, a skeptic looking at the data above might say, "They certainly look different to me!" If you were Chargaff, how would you use the data to convince the skeptic to change her mind?

(b) The base composition of DNA from normal and cancerous liver cells (hepatocarcinoma) was not distinguishably different. Would you expect Chargaff's technique to be capable of detecting a difference between the DNA of normal and cancerous cells? Explain your reasoning.

As you might expect, Chargaff's data were not completely convincing. He went on to improve his techniques, as described in his 1951 paper, in which he reported molar ratios of bases in DNA from a variety of organisms.

Source	A:G	T:C	A:T	G:C	Purine:pyrimidine
Ox	1.29	1.43	1.04	1.00	1.1
Human	1.56	1.75	1.00	1.00	1.0
Hen	1.45	1.29	1.06	0.91	0.99
Salmon	1.43	1.43	1.02	1.02	1.02
Wheat	1.22	1.18	1.00	0.97	0.99
Yeast	1.67	1.92	1.03	1.20	1.0
Haemophilus influenzae type c	1.74	1.54	1.07	0.91	1.0
E. coli K-12	1.05	0.95	1.09	0.99	1.0
Avian tubercle bacillus	0.4	0.4	1.09	1.08	1.1
Serratia marcescens	0.7	0.7	0.95	0.86	0.9
Bacillus schatz	0.7	0.6	1.12	0.89	1.0

(c) According to Chargaff, as stated in conclusion 1 in this chapter, "The base composition of DNA generally varies from one species to another." Provide an argument, based on the data presented so far, that supports this conclusion.

(d) According to conclusion 4, "In *all* cellular DNAs, regardless of the species, . . . A + G = T + C." Provide an argument, based on the data presented so far, that supports this conclusion.

Part of Chargaff's intent was to disprove the "tetranucleotide hypothesis"; this was the idea that DNA was a monotonous tetranucleotide polymer $(AGCT)_n$ and therefore not capable of containing sequence information. Although the data presented above show that DNA cannot be simply a repeating tetranucleotide—if so, all samples would have molar ratios of 0.25 for each base—it was still possible that the DNA from different organisms was a slightly more complex, but still monotonous, repeating sequence.

To address this issue, Chargaff took DNA from wheat germ and treated it with the enzyme deoxyribonuclease for different time intervals. At each time interval, some of the DNA was converted to small fragments; the remaining, larger fragments

he called the "core." In the table below, the "19% core" corresponds to the larger fragments left behind when 81% of the DNA was degraded; the "8% core" corresponds to the larger fragments left after 92% degradation.

Base	Intact DNA	19% Core	8% Core
Adenine	0.27	0.33	0.35
Guanine	0.22	0.20	0.20
Cytosine	0.22	0.16	0.14
Thymine	0.27	0.26	0.23
Recovery	*0.98*	*0.95*	*0.92*

(e) How would you use these data to argue that wheat germ DNA is not a monotonous repeating sequence?

References

Chargaff, E. 1950. Chemical specificity of nucleic acids and mechanism of their enzymatic degradation. *Experientia* 6:201–209.

Chargaff, E. 1951. Structure and function of nucleic acids as cell constituents. *Fed. Proc.* 10:654–659.

Further Reading is available at www.macmillanlearning.com/LehningerBiochemistry7e.

CHAPTER 9

DNA-Based Information Technologies

Self-study tools that will help you practice what you've learned and reinforce this chapter's concepts are available online. Go to www.macmillanlearning.com/LehningerBiochemistry7e.

The complexity of the molecules and systems revealed in this book can sometimes conceal a biochemical reality: what we have learned is just a beginning. Novel proteins and lipids and carbohydrates and nucleic acids are discovered every day, and we often have no clue as to their functions. How many have yet to be encountered, and what might they do? Even well-characterized biomolecules continue to challenge researchers with countless unresolved mechanistic and functional questions. A new era, defined by technologies that provide broad access to the entirety of a cell's DNA, the genome, has accelerated progress.

The word "genome," coined by German botanist Hans Winkler in 1920, was derived simply by combining *gene* and the final syllable of *chromosome*. A **genome** today is defined as the complete haploid genetic complement of an organism. In essence, a genome is one copy of the hereditary information required to specify the organism. For sexually reproducing organisms, the genome includes one set of autosomes and one of each type of sex chromosome. When cells have organelles that also contain DNA, the genetic content of the organelles is not considered part of the nuclear genome. Mitochondria, found in most eukaryotic cells, and chloroplasts, in the light-harvesting cells of photosynthetic organisms, each have their own distinct genome. For viruses, which can have genetic material composed of DNA or RNA, the genome is a complete copy of the nucleic acid required to specify the virus.

The thousands of completed genome sequences in hand have provided one look at the immensity of the task ahead. Simply put, we do not know the function of most of the DNA—often including half or more of the genes—in a typical genome. Those same genomic sequences, however, also provide an unprecedented opportunity. There is no greater source of information about a cell or organism than that buried in its own DNA. The technologies we turn to in this chapter (along with several discussed in Chapter 8) allow us to take advantage of this information resource, and they touch every topic we explore in subsequent chapters.

As objects of study, DNA molecules present a special problem: their size. Chromosomes are far and away the largest biomolecules in any cell. How does a researcher find the information he or she seeks when it is just a small part of a chromosome that can include millions or even billions of contiguous base pairs? Solutions to these problems began to emerge in the 1970s.

Decades of advances by thousands of scientists working in genetics, biochemistry, cell biology, and physical chemistry came together in the laboratories of Paul Berg, Herbert Boyer, and Stanley Cohen to yield the first techniques for locating, isolating, preparing, and studying small segments of DNA derived from much larger chromosomes. Advanced technologies described in Chapter 8, still evolving and improving, followed closely behind. In 1986, Thomas H. Roderick of the Jackson Laboratories in Bar Harbor, Maine, came up with *Genomics* as the name for a new journal, and the

Paul Berg
[Source: NIH National
Library of Medicine.]

Herbert Boyer
[Source: Courtesy Dr. Jane
Gitschier.]

Stanley N. Cohen
[Source: NIH National
Library of Medicine.]

word ended up defining a new field. The modern science of **genomics** is dedicated to the study of DNA on a cellular scale. In turn, genomics contributes to **systems biology**, the study of biochemistry on the scale of whole cells and organisms.

Every student and instructor, when considering the topics we present in this chapter, encounters a conflict. First, the methods we describe were made possible by advances in our understanding of DNA and RNA metabolism. Hence, one must understand some fundamental concepts of DNA replication, RNA transcription, protein synthesis, and gene regulation to appreciate how these methods work. At the same time, however, modern biochemistry relies on these same methods to such an extent that a current treatment of any aspect of the discipline becomes very difficult without a proper introduction to them. By presenting these technologies early in the book, we acknowledge that they are inextricably interwoven with both the advances that gave rise to them and the newer discoveries they now make possible. The background we necessarily provide makes the discussion here not just an introduction to technology but also a preview of many of the fundamentals of DNA and RNA biochemistry encountered in later chapters.

We begin by outlining the principles of DNA cloning, then illustrate the range of applications and the potential of many newer technologies that support and accelerate the advance of biochemistry.

9.1 Studying Genes and Their Products

A researcher has isolated a new enzyme that she knows is the key to a human disease. She hopes to isolate large amounts of the protein to crystallize it for structural analysis and to study it. She wants to alter amino acid residues at its active site so that she can understand the reaction it catalyzes. She plans an elaborate research program to elucidate how this enzyme interacts with, and is regulated by, other proteins in the cell. All of this, and much more, becomes possible if she can obtain the gene encoding her enzyme. Unfortunately, that gene consists of just a few thousand base pairs within a human chromosome with a size measured in hundreds of

millions of base pairs. How does she isolate the small segment that she needs and then study it? The answer lies in DNA cloning and methods developed to manipulate cloned genes.

Genes Can Be Isolated by DNA Cloning

A *clone* is an identical copy. This term originally applied to cells of a single type, isolated and allowed to reproduce to create a population of identical cells. When applied to DNA, a clone represents many identical copies of a particular gene segment. In brief, our researcher must cut the gene out of the larger chromosome, attach it to a much smaller piece of carrier DNA, and allow microorganisms to make many copies of it. This is the process of **DNA cloning**. The result is selective amplification of a particular gene or DNA segment so that it may be isolated and studied. Classically, the cloning of DNA from any organism entails five general procedures:

1. *Obtaining the DNA segment to be cloned.* Enzymes called restriction endonucleases act as precise molecular scissors, recognizing specific sequences in DNA and cleaving genomic DNA into smaller fragments suitable for cloning. Alternatively, genomic DNA can be sheared randomly into fragments of a desired size. Since the sequence of targeted genomic regions is often known (available in databases), some DNA segments to be cloned are amplified by the polymerase chain reaction (PCR) or are simply synthesized (both methods described in Chapter 8).

2. *Selecting a small molecule of DNA capable of autonomous replication.* These small DNAs are called **cloning vectors** (a vector is a carrier or delivery agent). Most cloning vectors used in the laboratory are modified versions of naturally occurring small DNA molecules found in bacteria or lower eukaryotes such as yeast. Small viral DNAs may also play this role.

3. *Joining two DNA fragments covalently.* The enzyme DNA ligase links the cloning vector to the DNA fragment to be cloned. Composite DNA molecules of this type, comprising covalently linked segments from two or more sources, are called **recombinant DNAs**.

4. *Moving recombinant DNA from the test tube to a host organism.* The host organism provides the enzymatic machinery for DNA replication.

5. *Selecting or identifying host cells that contain recombinant DNA.* The cloning vector generally has features that allow the host cells to survive in an environment in which cells lacking the vector would die. Cells containing the vector are thus "selectable" in that environment.

TABLE 9-1	Some Enzymes Used in Recombinant DNA Technology
Enzyme(s)	**Function**
Type II restriction endonucleases	Cleave DNA molecules at specific base sequences
DNA ligase	Joins two DNA molecules or fragments
DNA polymerase I (*E. coli*)	Fills gaps in duplexes by stepwise addition of nucleotides to 3′ ends
Reverse transcriptase	Makes a DNA copy of an RNA molecule
Polynucleotide kinase	Adds a phosphate to the 5′-OH end of a polynucleotide to label it or permit ligation
Terminal transferase	Adds homopolymer tails to the 3′-OH ends of a linear duplex
Exonuclease III	Removes nucleotide residues from the 3′ ends of a DNA strand
Bacteriophage λ exonuclease	Removes nucleotides from the 5′ ends of a duplex to expose single-stranded 3′ ends
Alkaline phosphatase	Removes terminal phosphates from the 5′ or 3′ end (or both)

The methods used to accomplish these and related tasks are collectively referred to as **recombinant DNA technology** or, more informally, **genetic engineering**.

Much of our initial discussion focuses on DNA cloning in the bacterium *Escherichia coli*, the first organism used for recombinant DNA work and still the most common host cell. *E. coli* has many advantages: its DNA metabolism (like many other of its biochemical processes) is well understood; many naturally occurring cloning vectors associated with *E. coli*, such as plasmids and bacteriophages (bacterial viruses; also called phages), are well characterized; and techniques are available for moving DNA expeditiously from one bacterial cell to another. The principles discussed here are broadly applicable to DNA cloning in other organisms, a topic discussed more fully later in the section.

Restriction Endonucleases and DNA Ligases Yield Recombinant DNA

Particularly important to recombinant DNA technology is a set of enzymes (Table 9-1) made available through decades of research on nucleic acid metabolism. Two classes of enzymes lie at the heart of the classic approach to generating and propagating a recombinant DNA molecule **(Fig. 9-1)**. First, **restriction endonucleases** (also called restriction enzymes) recognize and

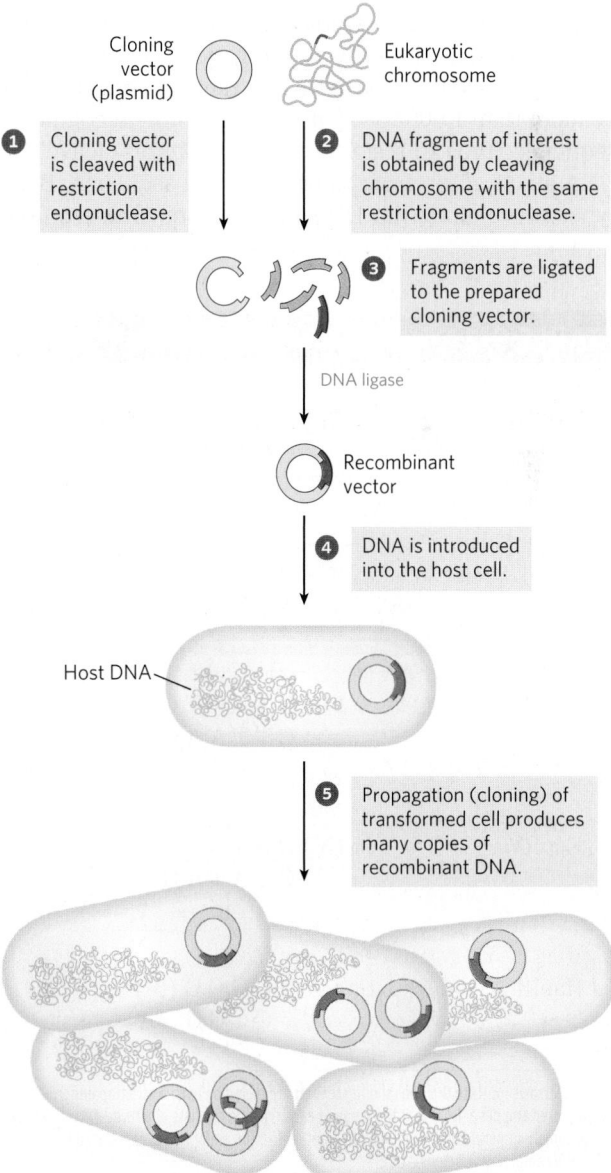

FIGURE 9-1 Schematic illustration of DNA cloning. A cloning vector and eukaryotic chromosomes are separately cleaved with the same restriction endonuclease. (A single chromosome is shown here for simplicity.) The fragments to be cloned are then ligated to the cloning vector. The resulting recombinant DNA (only one recombinant vector is shown here) is introduced into a host cell, where it can be propagated (cloned). Note that this drawing is not to scale: the size of the *E. coli* chromosome relative to that of a typical cloning vector (such as a plasmid) is much greater than depicted here.

Cloning vector (plasmid)

Eukaryotic chromosome

1 Cloning vector is cleaved with restriction endonuclease.

2 DNA fragment of interest is obtained by cleaving chromosome with the same restriction endonuclease.

3 Fragments are ligated to the prepared cloning vector.

DNA ligase

Recombinant vector

4 DNA is introduced into the host cell.

Host DNA

5 Propagation (cloning) of transformed cell produces many copies of recombinant DNA.

cleave DNA at specific sequences (recognition sequences or restriction sites) to generate a set of smaller fragments. Second, the DNA fragment to be cloned is joined to a suitable cloning vector by using **DNA ligases** to link the DNA molecules together. The recombinant vector is then introduced into a host cell, which amplifies the fragment in the course of many generations of cell division.

Restriction endonucleases are found in a wide range of bacterial species. As Werner Arber discovered in the early 1960s, their biological function is to recognize and cleave foreign DNA (the DNA of an infecting virus, for example); such DNA is said to be *restricted*. In the host cell's DNA, the sequence that would be recognized by one of its own restriction endonucleases is protected from digestion by methylation of the DNA, catalyzed by a specific DNA methylase. The restriction endonuclease and the corresponding methylase are sometimes referred to as a **restriction-modification system**.

There are three types of restriction endonucleases, designated I, II, and III. Types I and III are generally large, multisubunit complexes containing both the endonuclease and methylase activities. Type I restriction endonucleases cleave DNA at random sites that can be more than 1,000 base pairs (bp) from the recognition sequence. Type III restriction endonucleases cleave the DNA about 25 bp from the recognition sequence. Both types move along the DNA in a reaction that requires the energy of ATP. **Type II restriction endonucleases**, first isolated by Hamilton Smith in 1970, are simpler, require no ATP, and catalyze the hydrolytic cleavage of particular phosphodiester bonds in the DNA within the recognition sequence itself. The extraordinary utility of this group of restriction endonucleases was demonstrated by Daniel Nathans, who first used them to develop novel methods for mapping and analyzing genes and genomes.

Thousands of type II restriction endonucleases have been discovered in different bacterial species, and more than 100 different DNA sequences are recognized by one or more of these enzymes. The recognition sequences are usually 4 to 6 bp long and are palindromic (see Fig. 8-18). Table 9-2 lists sequences recognized by a few type II restriction endonucleases.

Some restriction endonucleases make staggered cuts on the two DNA strands, leaving two to four nucleotides of one strand unpaired at each resulting end. These unpaired strands are referred to as **sticky ends (Fig. 9-2a)** because they can base-pair with each other or with complementary sticky ends of other DNA fragments. Other restriction endonucleases cleave both strands of DNA straight across, at opposing phosphodiester bonds, leaving no unpaired bases on the ends, often called **blunt ends** (Fig. 9-2b).

The average size of the DNA fragments produced by cleaving genomic DNA with a restriction

TABLE 9-2	Recognition Sequences for Some Type II Restriction Endonucleases		
BamHI	(5′) G G A T C C (3′) C C T A G G	HindIII	(5′) A A G C T T (3′) T T C G A A
ClaI	(5′) A T C G A T (3′) T A G C T A	NotI	(5′) G C G G C C G C (3′) C G C C G G C G
EcoRI	(5′) G A A T T C (3′) C T T A A G	PstI	(5′) C T G C A G (3′) G A C G T C
EcoRV	(5′) G A T A T C (3′) C T A T A G	PvuII	(5′) C A G C T G (3′) G T C G A C
HaeIII	(5′) G G C C (3′) C C G G	Tth111I	(5′) G A C N N N G T C (3′) C T G N N N C A G

Note: Arrows indicate the phosphodiester bonds cleaved by each restriction endonuclease. Asterisks indicate bases that are methylated by the corresponding methylase (where known). N denotes any base. Note that the name of each enzyme consists of a three-letter abbreviation of the bacterial species from which it is derived, sometimes followed by a strain designation and roman numerals to distinguish different restriction endonucleases isolated from the same bacterial species. Thus BamHI is the first (I) restriction endonuclease characterized from *Bacillus amyloliquefaciens*, strain H.

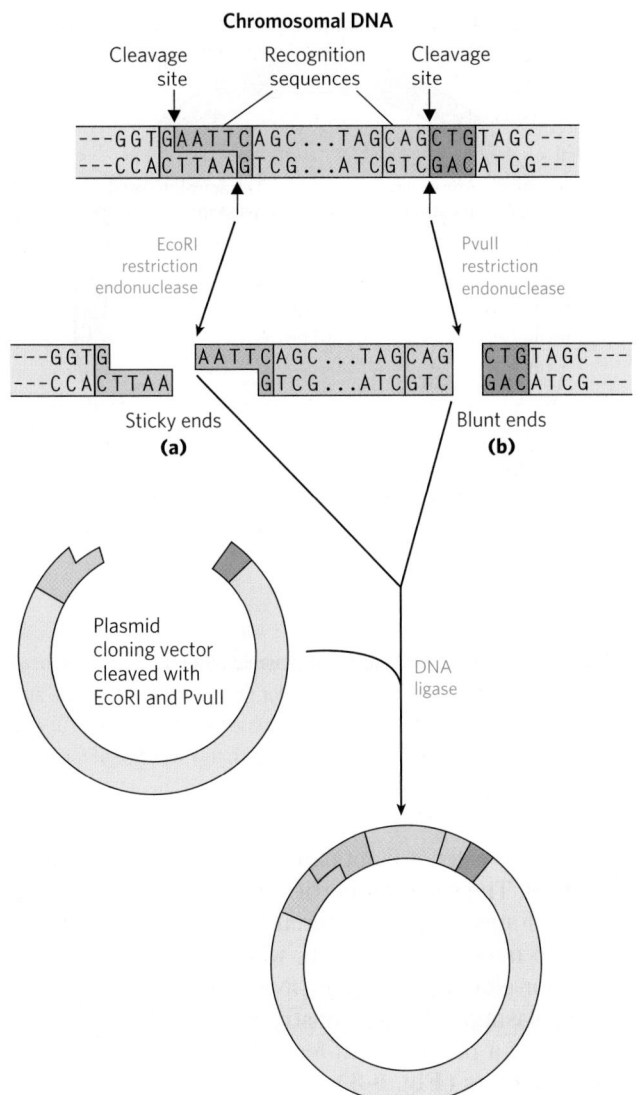

Chromosomal DNA

Cleavage site | Recognition sequences | Cleavage site

```
---GGTGAATTCAGC...TAGCAGCTGTAGC---
---CCACTTAAGTCG...ATCGTCGACATCG---
```

EcoRI restriction endonuclease

PvuII restriction endonuclease

```
---GGTG        AATTCAGC...TAGCAG        CTGTAGC---
---CCACTTAA        GTCG...ATCGTC        GACATCG---
```

Sticky ends **(a)**   Blunt ends **(b)**

Plasmid cloning vector cleaved with EcoRI and PvuII

DNA ligase

Synthetic polylinker

EcoRI sticky end | PstI | HindIII | BamHI | SmaI | EcoRI sticky end

```
5'AATTCCTGCAGAAGCTTCCGGATCCCCGGG
   GGACGTCTTCGAAGGCCTAGGGGCCCTTAA
```

Plasmid cloning vector
(cleaved with EcoRI)

DNA ligase

EcoRI · PstI · HindIII · BamHI · SmaI · EcoRI

Polylinker

(c)

FIGURE 9-2 Cleavage of DNA molecules by restriction endonucleases. Restriction endonucleases recognize and cleave only specific sequences, leaving either **(a)** sticky ends (with protruding single strands) or **(b)** blunt ends. Fragments can be ligated to other DNAs, such as the cleaved cloning vector (a plasmid) shown here. This reaction is facilitated by the annealing of complementary sticky ends. Ligation is less efficient for DNA fragments with blunt ends than for those with complementary sticky ends, and DNA fragments with different (noncomplementary) sticky ends generally are not ligated. **(c)** A synthetic DNA fragment with recognition sequences for several restriction endonucleases can be inserted into a plasmid that has been cleaved by a restriction endonuclease. The insert is called a linker; an insert with multiple restriction sites is called a polylinker.

endonuclease depends on the frequency with which a particular restriction site occurs in the DNA molecule; this in turn depends largely on the size of the recognition sequence.

In a DNA molecule with a random sequence in which all four nucleotides were equally abundant, a 6 bp sequence recognized by a restriction endonuclease such as BamHI would occur, on average, once every 4^6 (4,096) bp. Enzymes that recognize a 4 bp sequence would produce smaller DNA fragments from a random-sequence DNA molecule; a recognition sequence of this size would be expected to occur about once every 4^4 (256) bp. In natural DNA molecules, particular recognition sequences tend to occur less frequently than this because nucleotide sequences in DNA are not random and the four nucleotides are not equally abundant. In laboratory experiments, the average size of the fragments produced by restriction endonuclease cleavage of a large DNA can be increased by simply terminating the reaction before

completion; the result is called a partial digest. Average fragment size can also be increased by using a special class of endonucleases called homing endonucleases (see Fig. 26-37). These recognize and cleave much longer DNA sequences (14 to 20 bp).

Once a DNA molecule has been cleaved into fragments, a particular fragment of known size can be partially purified by agarose or acrylamide gel electrophoresis (p. 302) or by HPLC (p. 92). For a typical mammalian genome, however, cleavage by a restriction endonuclease usually yields too many different DNA fragments to permit convenient isolation of a particular fragment. A common intermediate step in the cloning of a specific gene or DNA segment is the construction of a DNA library (described in Section 9.2).

After the target DNA fragment is isolated, DNA ligase can be used to join it to a similarly digested cloning vector—that is, a vector digested by the *same* restriction endonuclease; a fragment generated by EcoRI, for example, generally will not link to a fragment generated by BamHI. As described in more detail in Chapter 25 (see Fig. 25-16), DNA ligase catalyzes the formation of new phosphodiester bonds in a reaction that uses ATP or a similar cofactor. The base pairing of complementary sticky ends greatly facilitates the ligation reaction (Fig. 9-2a). Blunt ends can also be ligated, albeit less efficiently. Researchers can create new DNA sequences for a wide range of purposes by inserting synthetic DNA fragments, called **linkers**, to bridge the ends that are being ligated. Inserted DNA fragments with multiple recognition sequences for restriction endonucleases (often useful later as points for inserting additional DNA by cleavage and ligation) are called **polylinkers** (Fig. 9-2c).

The effectiveness of sticky ends in selectively joining two DNA fragments was apparent in the earliest recombinant DNA experiments. Before restriction endonucleases were widely available, some workers found they could generate sticky ends by the combined action of the bacteriophage λ exonuclease and terminal transferase (Table 9-1). The fragments to be joined were given complementary homopolymeric tails. Peter Lobban and Dale Kaiser used this method in 1971 in the first experiments to join naturally occurring DNA fragments. Similar methods were used soon after in Paul Berg's laboratory to join DNA segments from simian virus 40 (SV40) to DNA derived from bacteriophage λ, thereby creating the first recombinant DNA molecule with DNA segments from different species.

Cloning Vectors Allow Amplification of Inserted DNA Segments

The principles that govern the delivery of recombinant DNA in clonable form to a host cell, and its subsequent amplification in the host, are well illustrated by considering three popular cloning vectors: plasmids and bacterial artificial chromosomes, used in experiments with *E. coli*, and a vector used to clone large DNA segments in yeast.

Plasmids A plasmid is a circular DNA molecule that replicates separately from the host chromosome. The wide variety of naturally occurring bacterial plasmids range in size from 5,000 to 400,000 bp. Many of the plasmids found in bacterial populations are little more than molecular parasites, similar to viruses but with a more limited capacity to transfer from one cell to another. To survive in the host cell, plasmids incorporate several specialized sequences that enable them to make use of the cell's resources for their own replication and gene expression.

Naturally occurring plasmids usually have a symbiotic role in the cell. They may provide genes that confer resistance to antibiotics or that perform new functions for the cell. For example, the Ti plasmid of *Agrobacterium tumefaciens* allows the host bacterium to

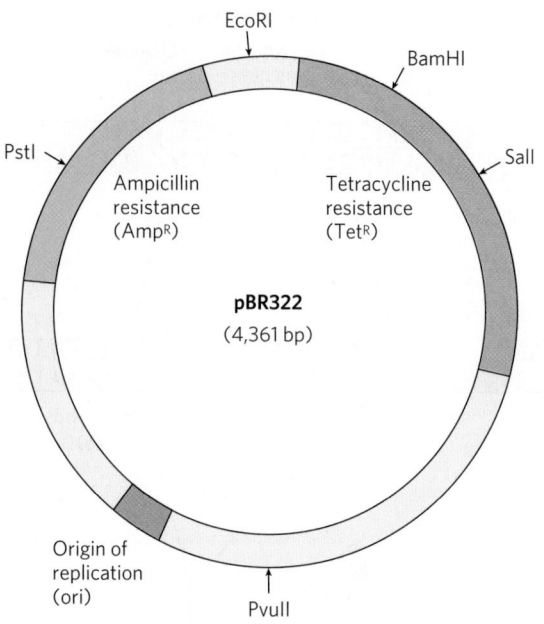

FIGURE 9-3 The constructed *E. coli* plasmid pBR322. Notice the location of some important restriction sites, for PstI, EcoRI, BamHI, SalI, and PvuII; ampicillin- and tetracycline-resistance genes (*amp*^{R} and *tet*^{R}); and the replication origin (ori). Constructed in 1977, this was one of the early plasmids designed expressly for cloning in *E. coli*.

colonize the cells of plants and make use of the plant's resources. The same properties that enable plasmids to grow and survive in a bacterial or eukaryotic host are useful to molecular biologists who want to engineer a vector for cloning a specific DNA segment. The classic *E. coli* plasmid pBR322, constructed in 1977, is a good example of a plasmid with features useful in almost all cloning vectors **(Fig. 9-3)**:

1. The plasmid pBR322 has an **origin of replication**, or **ori**, a sequence where replication is initiated by cellular enzymes (see Chapter 25). This sequence is required to propagate the plasmid. An associated regulatory system is present that limits replication to maintain pBR322 at a level of 10 to 20 copies per cell.

2. The plasmid contains genes that confer resistance to the antibiotics tetracycline (Tet^R) and ampicillin (Amp^R), allowing the selection of cells that contain the intact plasmid or a recombinant version of the plasmid (discussed below).

3. Several unique recognition sequences in pBR322 are targets for restriction endonucleases (PstI, EcoRI, BamHI, SalI, and PvuII), providing sites where the plasmid can be cut to insert foreign DNA.

4. The small size of the plasmid (4,361 bp) facilitates its entry into cells and the biochemical manipulation of the DNA. This small size was the result of trimming away many DNA segments from a larger, parent plasmid—sequences that the molecular biologist does not need.

The replication origins inserted in common plasmid vectors were originally derived from naturally occurring plasmids. As in pBR322, each of these origins is regulated to maintain a particular plasmid copy number. Depending on the origin used, the plasmid copy number can vary from one to hundreds or thousands per cell, providing many options for investigators. Two different plasmids cannot function in the same cell if they use the same origin of replication, because the regulation of one will interfere with the replication of the other. Such plasmids are said to be incompatible. When a researcher wants to introduce two or more different plasmids into a bacterial cell, each plasmid must have a different replication origin.

In the laboratory, small plasmids can be introduced into bacterial cells by a process called **transformation**. The cells (often *E. coli*, but other bacterial species are also used) and plasmid DNA are incubated together at 0 °C in a calcium chloride solution, then are subjected to heat shock by rapidly shifting the temperature to between 37 °C and 43 °C. For reasons not well understood, some of the cells treated in this way take up the plasmid DNA. Some species of bacteria, such as *Acinetobacter baylyi*, are naturally competent for DNA uptake and do not require the calcium chloride–heat shock treatment. In an alternative method, called **electroporation**, cells incubated with the plasmid DNA are subjected to a high-voltage pulse, which transiently renders the bacterial membrane permeable to large molecules.

Regardless of the approach, relatively few cells take up the plasmid DNA, so a method is needed to identify those that do. The usual strategy is to utilize one of two types of genes in the plasmid, referred to as selectable and screenable markers. A **selectable marker** either permits the growth of a cell (positive selection) or kills the cell (negative selection) under a defined set of conditions. The plasmid pBR322 provides markers for both positive and negative selection **(Fig. 9-4)**. A **screenable marker** is a gene encoding a protein that causes the cell to produce a colored or fluorescent molecule. Cells are not harmed when the gene is present, and the cells that carry the plasmid are easily identified by the colored or fluorescent colonies they produce.

Transformation of typical bacterial cells with purified DNA (never a very efficient process) becomes less successful as plasmid size increases, and it is difficult to clone DNA segments longer than about 15,000 bp when plasmids are used as the vector.

To illustrate the use of a plasmid as a cloning vector, consider the bacterial gene encoding a recombinase called the RecA protein (see Chapter 25). In most bacteria, the gene encoding RecA is one of the thousands of genes on a chromosome millions of base pairs long. The *recA* gene is just over 1,000 bp long. A plasmid would be a good choice for cloning a gene of this size. As described later, the cloned gene can be altered in a variety of ways, and the gene variants can be expressed at high levels to enable purification of the encoded protein.

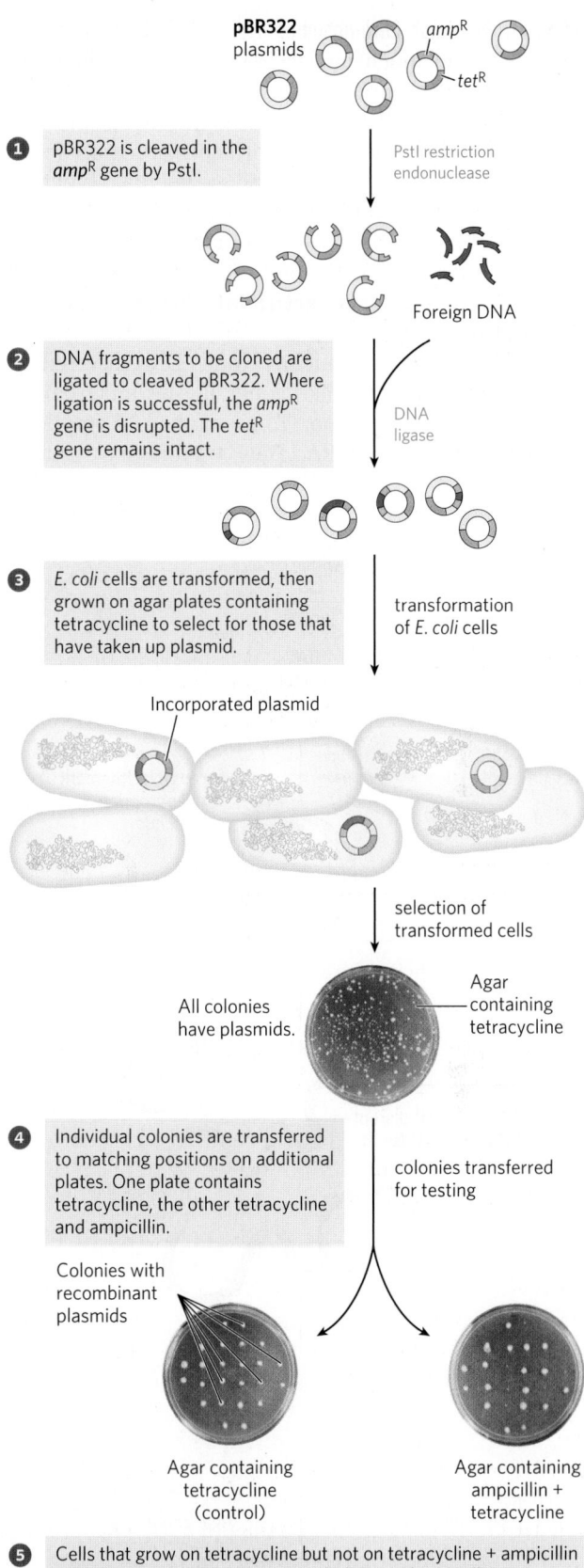

① pBR322 is cleaved in the *amp*^R gene by PstI.

PstI restriction endonuclease

Foreign DNA

② DNA fragments to be cloned are ligated to cleaved pBR322. Where ligation is successful, the *amp*^R gene is disrupted. The *tet*^R gene remains intact.

DNA ligase

③ *E. coli* cells are transformed, then grown on agar plates containing tetracycline to select for those that have taken up plasmid.

transformation of *E. coli* cells

Incorporated plasmid

selection of transformed cells

All colonies have plasmids.

Agar containing tetracycline

④ Individual colonies are transferred to matching positions on additional plates. One plate contains tetracycline, the other tetracycline and ampicillin.

colonies transferred for testing

Colonies with recombinant plasmids

Agar containing tetracycline (control)

Agar containing ampicillin + tetracycline

⑤ Cells that grow on tetracycline but not on tetracycline + ampicillin contain recombinant plasmids with disrupted ampicillin resistance, hence the foreign DNA. Cells with pBR322 without foreign DNA retain ampicillin resistance and grow on both plates.

FIGURE 9-4 Use of pBR322 to clone foreign DNA in *E. coli* and identify cells containing the DNA. [Source: Elizabeth A. Wood, University of Wisconsin-Madison, Department of Biochemistry.]

Bacterial Artificial Chromosomes Researchers sometimes want to clone much longer DNA segments than can typically be incorporated into standard plasmid cloning vectors such as pBR322. To meet this need, plasmid vectors have been developed with special features that allow the cloning of very long segments (typically 100,000 to 300,000 bp) of DNA. Once such large segments of cloned DNA have been added, these vectors are large enough to be thought of as chromosomes and are known as **bacterial artificial chromosomes**, or **BACs (Fig. 9-5)**.

A BAC vector (without any cloned DNA inserted) is a relatively simple plasmid, generally not much larger than other plasmid vectors. To accommodate very long segments of cloned DNA, BAC vectors have stable origins of replication that maintain the plasmid at one or two copies per cell. The low copy number is useful in cloning large segments of DNA, because it limits the opportunities for unwanted recombination reactions that can unpredictably alter large cloned DNAs over time. BACs also include *par* genes, which encode proteins that direct the reliable distribution of the recombinant chromosomes to daughter cells at cell division, thereby increasing the likelihood of each daughter cell carrying one copy, even when few copies are present. The BAC vector includes both selectable and screenable markers. The BAC vector shown in Figure 9-5 contains a gene that confers resistance to the antibiotic chloramphenicol (Cam^R). Vector-containing cells can be selected by growing them on agar plates containing this antibiotic—a positive selection, as the cells with the vector survive. A *lacZ* gene, required for production of the enzyme β-galactosidase, is a screenable marker that can reveal which cells contain plasmids—now chromosomes—that incorporate the cloned DNA segments. The β-galactosidase catalyzes conversion of the colorless molecule 5-bromo-4-chloro-3-indolyl-β-D-galactopyranoside (more simply, X-gal) to a blue product. If the gene is intact and expressed, the colony containing it is blue. If gene expression is disrupted by the introduction of a cloned DNA segment, the colony is white.

Yeast Artificial Chromosomes As with *E. coli*, yeast genetics is a well-developed discipline. The genome of *Saccharomyces cerevisiae* contains only 14×10^6 bp (less than four times the size of the *E. coli* chromosome), and its entire sequence is known. Yeast is also very easy to maintain and grow on a large scale in the laboratory. Plasmid vectors have been constructed for insertions into yeast cells, employing the same principles that govern the use of *E. coli* vectors. Convenient methods for moving DNA into and out of yeast cells permit the study of many aspects of eukaryotic cell biochemistry. Some recombinant plasmids incorporate multiple replication

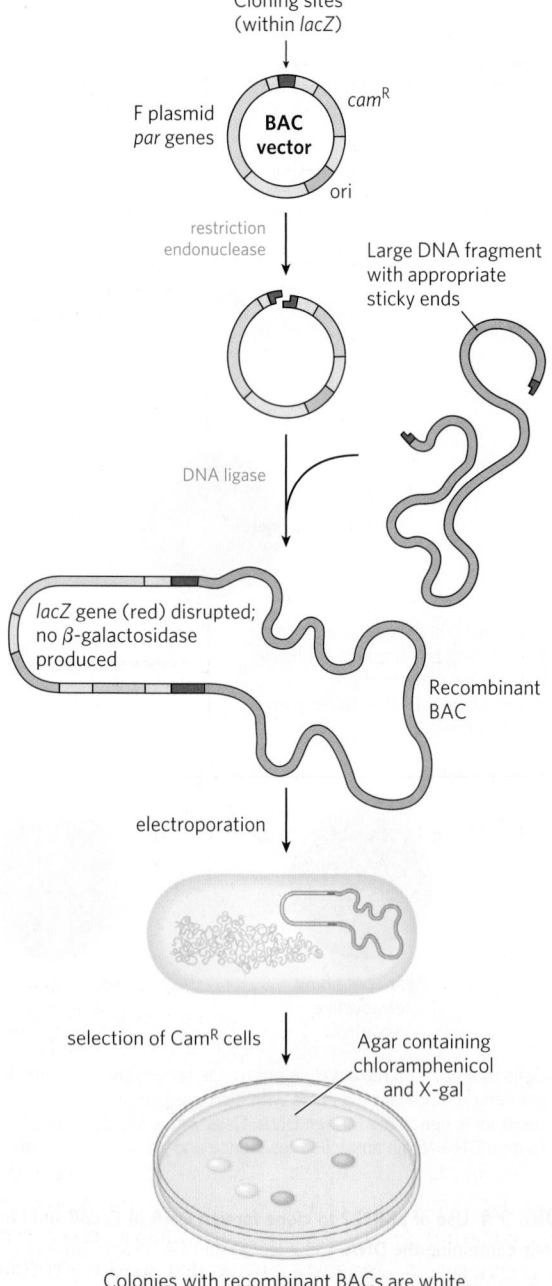

Colonies with recombinant BACs are white.

FIGURE 9-5 Bacterial artificial chromosomes (BACs) as cloning vectors. The vector is a relatively simple plasmid, with a replication origin (ori) that directs replication. The *par* genes, derived from a type of plasmid called an F plasmid, assist in the even distribution of plasmids to daughter cells at cell division. This increases the likelihood of each daughter cell carrying one copy of the plasmid, even when few copies are present. The low number of copies is useful in cloning large segments of DNA, because this limits the opportunities for unwanted recombination reactions that can unpredictably alter large cloned DNAs over time. The BAC includes selectable markers. A *lacZ* gene (required for the production of the enzyme β-galactosidase) is situated in the cloning region such that it is inactivated by cloned DNA inserts. Introduction of recombinant BACs into cells by electroporation is promoted by the use of cells with an altered (more porous) cell wall. Recombinant DNAs are screened for resistance to the antibiotic chloramphenicol (Cam^R). Plates also contain X-gal, a substrate for β-galactosidase that yields a blue product. Colonies with active β-galactosidase, and hence no DNA insert in the BAC vector, turn blue; colonies without β-galactosidase activity, and thus with the desired DNA inserts, are white.

origins and other elements that allow them to be used in more than one species (e.g., in yeast and in *E. coli*). Plasmids that can be propagated in cells of two or more species are called **shuttle vectors**.

Research on large genomes and the associated need for high-capacity cloning vectors led to the development of **yeast artificial chromosomes**, or **YACs (Fig. 9-6)**. YAC vectors contain all the elements needed to maintain a eukaryotic chromosome in the yeast nucleus: a yeast origin of replication, two selectable markers, and specialized sequences (derived from the telomeres and centromere) needed for stability and proper segregation of the chromosomes at cell division (see Chapter 24). In preparation for its use in cloning, the vector is propagated as a circular bacterial plasmid and then isolated and purified. Cleavage with a restriction endonuclease (BamHI in Fig. 9-6) removes a length of DNA between two telomere sequences (TEL), leaving the telomeres at the ends of the linearized DNA. Cleavage at another internal site (by EcoRI in Fig. 9-6) divides the vector into two DNA segments, referred to as vector arms, each with a different selectable marker.

The genomic DNA to be cloned is prepared by partial digestion with restriction endonucleases to obtain a suitable fragment size. Genomic fragments are then separated by **pulsed field gel electrophoresis**, a variation of gel electrophoresis (see Fig. 3-18) that segregates very large DNA segments. DNA fragments of appropriate size (up to about 2×10^6 bp) are mixed with the prepared vector arms and ligated. The ligation mixture is then used to transform yeast cells (pretreated to partially degrade their cell walls) with these very large DNA molecules—which now have the structure and size to be considered yeast chromosomes. Culture on a medium that requires the presence of both selectable marker genes ensures the growth of only those yeast cells that contain an artificial chromosome with a large insert sandwiched between the two vector arms (Fig. 9-6). The stability of YAC clones increases with the length of the cloned DNA segment (up to a point). Those with inserts of more than 150,000 bp are nearly as stable as normal cellular chromosomes, whereas those with inserts of less than 100,000 bp are gradually lost during mitosis (so, generally, there are no yeast cell clones carrying only the two vector ends ligated together or vectors with only short inserts). YACs that lack a telomere at either end are rapidly degraded.

As with BACs, YAC vectors can be used to clone very long segments of DNA. In addition, the DNA cloned in a YAC can be altered to study the function of specialized sequences in chromosome metabolism, mechanisms of gene regulation and expression, and many other aspects of eukaryotic molecular biology.

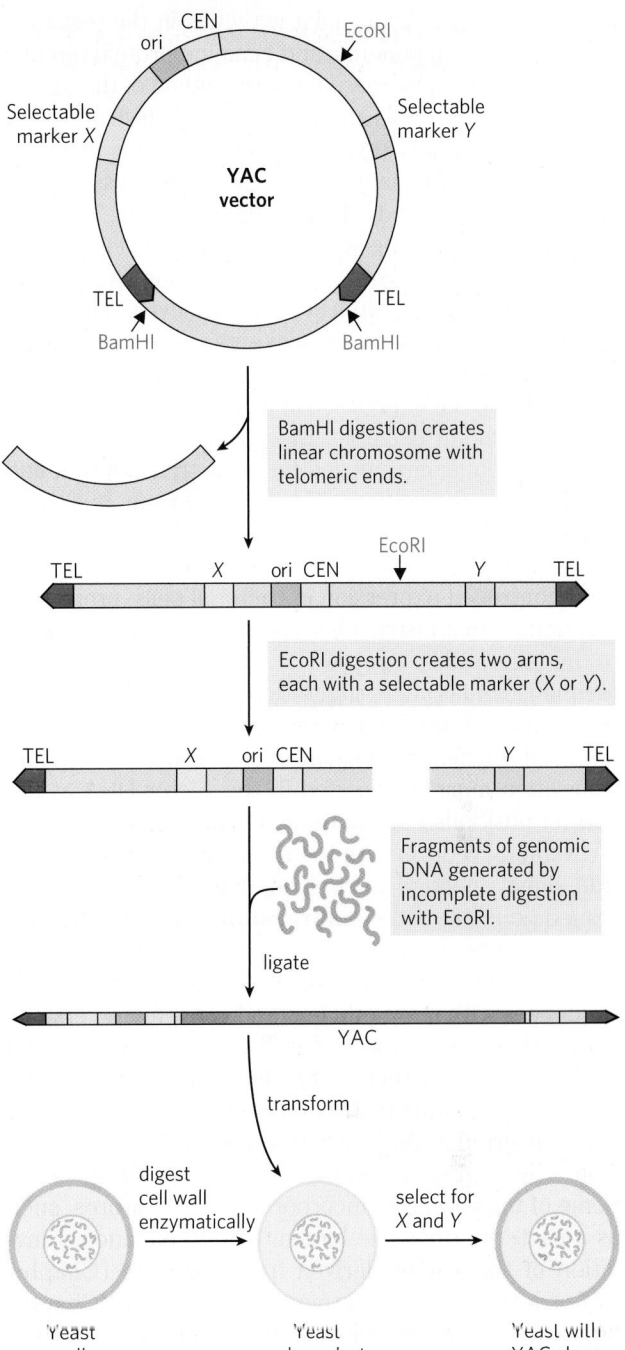

FIGURE 9-6 Construction of a yeast artificial chromosome (YAC). A YAC vector includes an origin of replication (ori), a centromere (CEN), two telomeres (TEL), and selectable markers (*X* and *Y*). Digestion with BamHI and EcoRI generates two separate DNA arms, each with a telomeric end and one selectable marker. A large segment of DNA (e.g., up to 2×10^6 bp from the human genome) is ligated to the two arms to create a yeast artificial chromosome. The YAC transforms yeast cells (prepared by removal of the cell wall to form spheroplasts), and the cells are selected for *X* and *Y*; the surviving cells propagate the DNA insert.

Cloned Genes Can Be Expressed to Amplify Protein Production

Frequently, the product of a cloned gene, rather than the gene itself, is of primary interest—particularly when the protein has commercial, therapeutic, or research value. Biochemists use purified proteins for many purposes, including to elucidate protein function, study reaction mechanisms, generate antibodies to the proteins, reconstitute complex cellular activities in the test tube with purified components, and examine protein binding partners. With an increased understanding of the fundamentals of DNA, RNA, and protein metabolism and their regulation in a host organism such as *E. coli* or yeast, investigators can manipulate cells to express cloned genes in order to study their protein products. The general goal is to alter the sequences around a cloned gene to trick the host organism into producing the protein product of the gene, often at very high levels. This overexpression of a protein can make its subsequent purification much easier.

We'll use the expression of a eukaryotic protein in a bacterium as an example. Eukaryotic genes have surrounding sequences needed for their transcription and regulation in the cells they are derived from, but these sequences do not function in bacteria. Thus, eukaryotic genes lack the DNA sequence elements required for their controlled expression in bacterial cells: promoters (sequences that instruct RNA polymerase where to bind to initiate mRNA synthesis), ribosome-binding sites (sequences that allow translation of the mRNA to protein), and additional regulatory sequences. Appropriate bacterial regulatory sequences for transcription and translation must be inserted in the vector DNA at the correct positions relative to the eukaryotic gene.

Cloning vectors with the transcription and translation signals needed for the regulated expression of a cloned gene are called **expression vectors**. The rate of expression of the cloned gene is controlled by replacing the gene's normal promoter and regulatory sequences with more efficient and convenient versions supplied by the vector. Generally, a well-characterized promoter and its regulatory elements are positioned near several unique restriction sites for cloning, so that genes inserted at the restriction sites will be expressed from the regulated promoter elements **(Fig. 9-7)**. Some of these vectors incorporate other features, such as a bacterial ribosome-binding site to enhance translation of the mRNA derived from the gene (Chapter 27) or a transcription termination sequence (Chapter 26). In some cases, cloned genes are so efficiently expressed that their protein product represents 10% or more of the cellular protein. At these concentrations, some foreign proteins can kill the host cell (usually *E. coli*), so expression of the cloned gene must be limited to the few hours before the planned harvesting of the cells.

Many Different Systems Are Used to Express Recombinant Proteins

Every living organism has the capacity to express genes in its genomic DNA; thus, in principle, any organism can serve as a host to express proteins from a different (heterologous) species. Almost every sort of organism has, indeed, been used for this purpose,

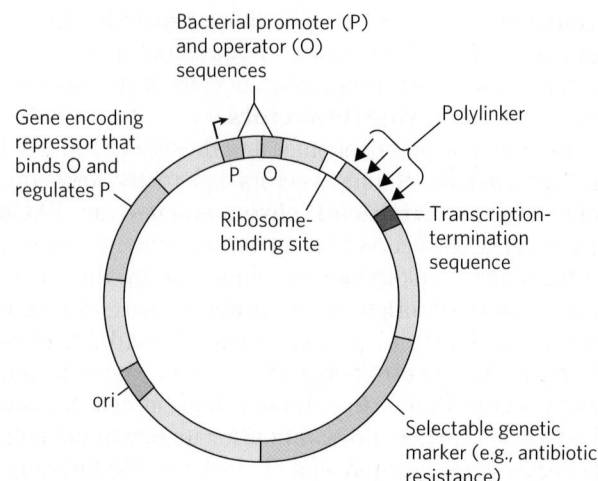

FIGURE 9-7 DNA sequences in a typical *E. coli* expression vector. The gene to be expressed is inserted into one of the restriction sites in the polylinker, near the promoter (P), with the end of the gene encoding the amino terminus of the protein positioned closest to the promoter. The promoter allows efficient transcription of the inserted gene, and the transcription-termination sequence sometimes improves the amount and stability of the mRNA produced. The operator (O) permits regulation by a repressor that binds to it. The ribosome-binding site provides sequence signals for the efficient translation of the mRNA derived from the gene. The selectable marker allows the selection of cells containing the recombinant DNA.

and each host type has a particular set of advantages and disadvantages.

Bacteria Bacteria, especially *E. coli*, remain the most common hosts for protein expression. The regulatory sequences that govern gene expression in *E. coli* and many other bacteria are well understood and can be harnessed to express cloned proteins at high levels. Bacteria are easy to store and grow in the laboratory, on inexpensive growth media. Efficient methods also exist to get DNA into bacteria and extract DNA from them. Bacteria can be grown in huge amounts in commercial fermenters, providing a rich source of the cloned protein. Problems do exist, however. When expressed in bacteria, some heterologous proteins do not fold correctly, and many do not undergo the posttranslational modifications or proteolytic cleavage that may be necessary for their activity. Certain features of a gene sequence also can make a particular gene difficult to express in bacteria. For example, intrinsically disordered regions are more common in eukaryotic proteins. When expressed in bacteria, many eukaryotic proteins aggregate into insoluble cellular precipitates called inclusion bodies. For these and many other reasons, some eukaryotic proteins are inactive when purified from bacteria or cannot be expressed at all. To help address some of these problems, new bacterial host strains are regularly being developed that include enhancements such as the engineered presence of eukaryotic protein chaperones or enzymes that modify eukaryotic proteins.

There are many specialized systems for expressing proteins in bacteria. The promoter and regulatory sequences associated with the lactose operon (see Chapter 28) are often fused to the gene of interest to direct transcription. The cloned gene will be transcribed when lactose is added to the growth medium. However, regulation in the lactose system is "leaky": it is not turned off completely when lactose is absent—a potential problem if the product of the cloned gene is toxic to the host cells. Transcription from the Lac promoter is also not efficient enough for some applications.

An alternative system uses the promoter and RNA polymerase of a bacterial virus called bacteriophage T7. If the cloned gene is fused to a T7 promoter, it is transcribed, not by the *E. coli* RNA polymerase, but by the T7 RNA polymerase. The gene encoding this polymerase is separately cloned into the same cell in a construct that affords tight regulation (allowing controlled production of the T7 RNA polymerase). The polymerase is also very efficient and directs high levels of expression of most genes fused to the T7 promoter. This system has been used to express the RecA protein in bacterial cells **(Fig. 9-8)**.

Yeast *Saccharomyces cerevisiae* is probably the best understood eukaryotic organism and one of the easiest to grow and manipulate in the laboratory. Like bacteria, this yeast can be grown on inexpensive media. Yeast

have tough cell walls that are difficult to breach in order to introduce DNA vectors, so bacteria are more convenient for doing much of the genetic engineering and vector maintenance. This is why the yeast vector was first propagated in bacteria. Several excellent shuttle vectors exist for this purpose.

The principles underlying the expression of a protein in yeast are the same as those for bacteria. Cloned genes must be linked to promoters that can direct high-level expression in yeast cells. For example, the yeast *GAL1* and *GAL10* genes are under cellular regulation such that they are expressed when yeast cells are grown in media with galactose but shut down when the cells are grown in glucose. Thus, if a heterologous gene is expressed using these same regulatory sequences, the expression of that gene can be controlled simply by choosing an appropriate medium for cell growth.

Some of the same problems that accompany protein expression in bacteria also occur with yeast. Heterologous proteins may not fold properly, yeast may lack the enzymes needed to modify the proteins to their active forms, or certain features of the gene sequence may hinder expression of a protein. However, because *S. cerevisiae* is a eukaryote, the expression of eukaryotic genes (especially yeast genes) is sometimes more efficient in this host than in bacteria. The products may also be folded and modified more accurately than are proteins expressed in bacteria.

Insects and Insect Viruses **Baculoviruses** are insect viruses with double-stranded DNA genomes. When they infect their insect larval hosts, they act as parasites, killing the larvae and turning them into factories for virus production. Late in the infection process, the viruses produce large amounts of two proteins (p10 and polyhedrin), neither of which is needed for production of viruses in cultured insect cells. The genes for both of these proteins can be replaced with the gene for a heterologous protein. When the resulting recombinant virus is used to infect insect cells or larvae, the heterologous protein is often produced at very high levels—up to 25% of the total protein present at the end of the infection cycle.

Autographa californica multicapsid nucleopolyhedrovirus (AcMNPV; *A. californica* is a moth species it infects) is the baculovirus most often used for protein expression. It has a large genome (134,000 bp), too large for direct cloning. Virus purification is also cumbersome. These problems have been solved by the creation of **bacmids**, large circular DNAs that include the entire baculovirus genome along with sequences that allow replication of the bacmid in *E. coli* **(Fig. 9-9)**. The gene of interest is cloned into a smaller plasmid and combined with the larger plasmid by site-specific recombination in vivo (see Fig. 25-38). The recombinant bacmid is then isolated and transfected into insect cells (the term **transfection** is used when the DNA

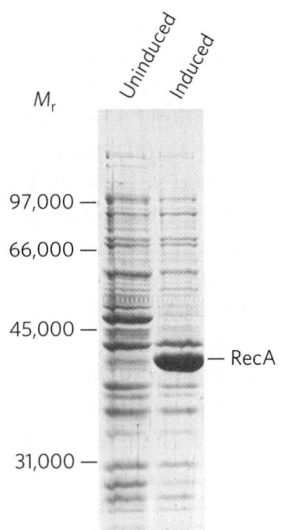

FIGURE 9-8 Regulated expression of RecA protein in a bacterial cell. The gene encoding the RecA protein, fused to a bacteriophage T7 promoter, is cloned into an expression vector. Under normal growth conditions (uninduced), no RecA protein appears. When the T7 RNA polymerase is induced in the cell, the *recA* gene is expressed, and large amounts of RecA protein are produced. The positions of standard molecular weight markers that were run on the same gel are indicated. [Source: Courtesy Rachel Britt, Department of Biochemistry, University of Wisconsin-Madison.]

(a)

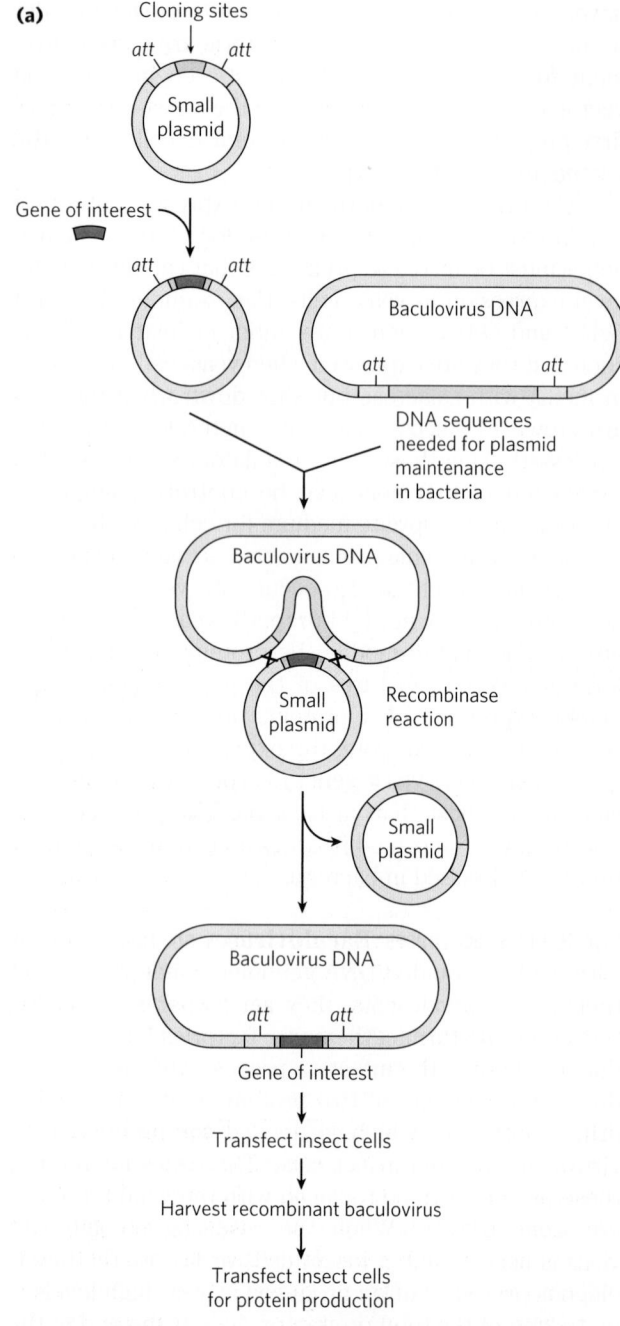

FIGURE 9-9 Cloning with baculoviruses. (a) Shown here is the construction of a typical vector used for protein expression in baculoviruses. The gene of interest is cloned into a small plasmid (top left) between two sites (*att*) recognized by a site-specific recombinase, then is introduced into the baculovirus vector by site-specific recombination (see Fig. 25-38). This generates a circular DNA product that is used to infect the cells of an insect larva. The gene of interest is expressed during the infection cycle, downstream of a promoter that normally expresses a baculovirus coat protein at very high levels. **(b)** The photographs show larvae of the cabbage looper moth. The larva on the left was infected with a recombinant baculovirus vector expressing a protein that produces a red color; on the right, an uninfected larva. [Source: (b) USDA-ARS.]

used for transformation includes viral sequences and leads to viral replication), followed by recovery of the protein once the infection cycle is finished. A wide range of bacmid systems are available commercially. Baculovirus systems are not successful with all proteins. However, with these systems, insect cells sometimes successfully replicate the protein-modification patterns of higher eukaryotes and produce active, correctly modified eukaryotic proteins.

Mammalian Cells in Culture The most convenient way to introduce cloned genes into a mammalian cell is with viruses. This method takes advantage of the natural capacity of a virus to insert its DNA or RNA into a cell, and sometimes into the cellular chromosome. A variety of engineered mammalian viruses are available as vectors, including human adenoviruses and retroviruses. The gene of interest is cloned so that its expression is controlled by a virus promoter. The virus uses its natural infection mechanisms to introduce the recombinant genome into cells, where the cloned protein is expressed. These systems have the advantage that proteins can be expressed either transiently (if the viral DNA is maintained separately from the host cell genome and eventually degraded) or permanently (if the viral DNA is integrated into the host cell genome). With the correct choice of host cell, the proper post-translational modification of the protein to its active form can be ensured. However, the growth of mammalian cells in tissue culture is very expensive, and this technology is generally used to test the function of a protein in vivo rather than to produce a protein in large amounts.

Alteration of Cloned Genes Produces Altered Proteins

Cloning techniques can be used not only to overproduce proteins but to produce proteins that are altered, subtly or dramatically, from their native forms. Specific amino acids may be replaced individually by **site-directed mutagenesis**. This technique has greatly enhanced research on proteins by allowing investigators to make specific changes in the primary structure and examine

(b)

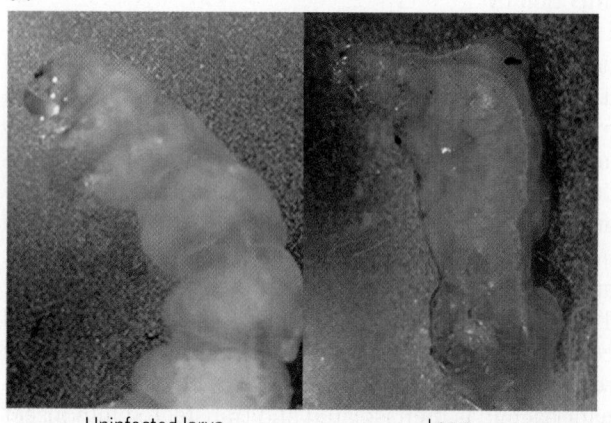

Uninfected larva   Larva
                  infected with a baculovirus

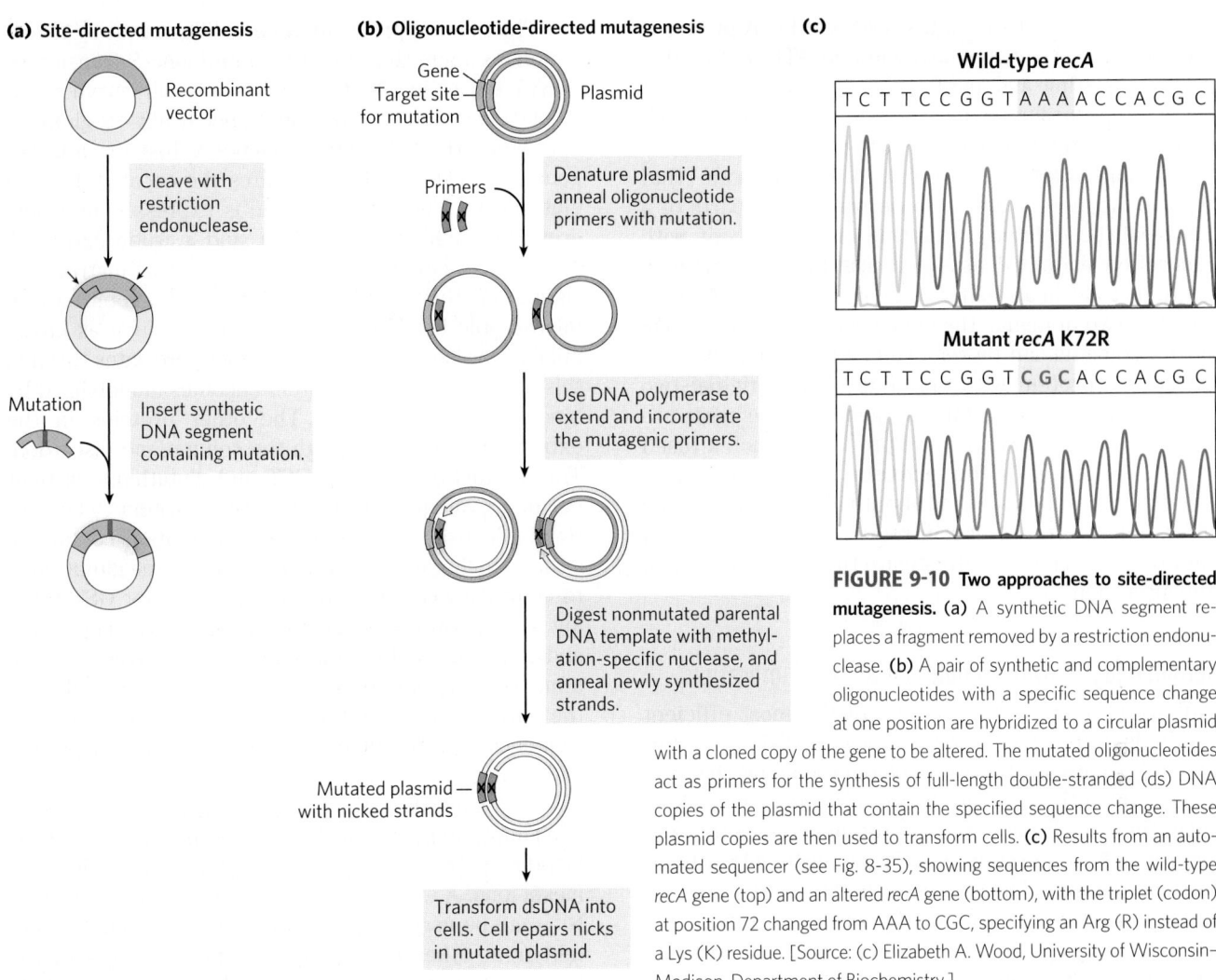

(a) Site-directed mutagenesis

Recombinant vector

Cleave with restriction endonuclease.

Mutation

Insert synthetic DNA segment containing mutation.

(b) Oligonucleotide-directed mutagenesis

Gene
Target site for mutation

Plasmid

Primers

Denature plasmid and anneal oligonucleotide primers with mutation.

Use DNA polymerase to extend and incorporate the mutagenic primers.

Digest nonmutated parental DNA template with methyl-ation-specific nuclease, and anneal newly synthesized strands.

Mutated plasmid — with nicked strands

Transform dsDNA into cells. Cell repairs nicks in mutated plasmid.

(c)

Wild-type *recA*

T C T T C C G G T A A A A C C A C G C

Mutant *recA* K72R

T C T T C C G G T **C G C** A C C A C G C

FIGURE 9-10 Two approaches to site-directed mutagenesis. (a) A synthetic DNA segment replaces a fragment removed by a restriction endonuclease. **(b)** A pair of synthetic and complementary oligonucleotides with a specific sequence change at one position are hybridized to a circular plasmid with a cloned copy of the gene to be altered. The mutated oligonucleotides act as primers for the synthesis of full-length double-stranded (ds) DNA copies of the plasmid that contain the specified sequence change. These plasmid copies are then used to transform cells. **(c)** Results from an automated sequencer (see Fig. 8-35), showing sequences from the wild-type *recA* gene (top) and an altered *recA* gene (bottom), with the triplet (codon) at position 72 changed from AAA to CGC, specifying an Arg (R) instead of a Lys (K) residue. [Source: (c) Elizabeth A. Wood, University of Wisconsin–Madison, Department of Biochemistry.]

the effects of these changes on the protein's folding, three-dimensional structure, and activity. This powerful approach to studying protein structure and function changes the amino acid sequence by altering the DNA sequence of the cloned gene. If appropriate restriction sites flank the sequence to be altered, researchers can simply remove a DNA segment and replace it with a synthetic one, identical to the original except for the desired change **(Fig. 9-10a)**.

When suitably located restriction sites are not present, **oligonucleotide-directed mutagenesis** can create a specific DNA sequence change (Fig. 9-10b). The cloned gene is denatured, separating the strands. Two short, complementary synthetic DNA strands, each with the desired base change, are annealed to opposite strands of the cloned gene within a suitable circular DNA vector. The mismatch of a single base pair in 30 to 40 bp does not prevent annealing. The two annealed oligonucleotides serve to prime DNA synthesis in both directions around the plasmid vector, creating two complementary strands that contain the mutation. After several cycles of selective amplification by the

polymerase chain reaction (PCR; see Fig. 8-33), the mutation-containing DNA predominates in the population and can be used to transform bacteria. Most of the transformed bacteria will have plasmids carrying the mutation. If necessary, the nonmutant template plasmid DNA can be selectively eliminated by cleavage with the restriction enzyme DpnI. The template plasmid, usually isolated from wild-type *E. coli*, has a methylated A residue in every copy of the four-nucleotide palindrome GATC (called a dam site; see Fig. 25-21). The new DNA containing the mutation does not have methylated A residues, because the replication is done in vitro. Given that DpnI selectively cleaves DNA at the sequence GATC only if the A residue in one or both strands is methylated, it breaks down only the template.

For an example, we go back to the bacterial *recA* gene. The product of this gene, the RecA protein, has several activities (see Section 25.3): it binds to and forms a filamentous structure on DNA, aligns two DNAs of similar sequence, and hydrolyzes ATP. A particular amino acid residue in RecA (a 352 residue polypeptide), the Lys residue at position 72, is involved in ATP hydrolysis.

By changing Lys[72] to an Arg, a variant of RecA protein is created that will bind, but not hydrolyze, ATP (Fig. 9-10c). The engineering and purification of this variant RecA protein has facilitated research into the roles of ATP hydrolysis in the functioning of this protein.

Changes can be introduced into a gene that involve far more than one base pair. Large parts of a gene can be deleted by cutting out a segment with restriction endonucleases and ligating the remaining portions to form a smaller gene. For example, if a protein has two domains, the gene segment encoding one of the domains can be removed so that the gene now encodes a protein with only one of the original two domains. Parts of two different genes can be ligated to create new combinations; the product of such a fused gene is called a **fusion protein**. Researchers have ingenious methods to bring about virtually any genetic alteration in vitro. After reintroducing the altered DNA into the cell, they can investigate the consequences of the alteration.

Terminal Tags Provide Handles for Affinity Purification

Affinity chromatography is one of the most efficient methods for purifying proteins (see Fig. 3-17c). Unfortunately, many proteins do not bind a ligand that can be conveniently immobilized on a column matrix. However, the gene for almost any protein can be altered to express a fusion protein that can be purified by affinity chromatography. The gene encoding the target protein is fused to a gene encoding a peptide or protein that binds a simple, stable ligand with high affinity and specificity. The peptide or protein used for this purpose is referred to as a **tag**. Tag sequences can be added to genes such that the resulting proteins have tags at their amino or carboxyl terminus. Table 9-3 lists some of the peptides or proteins commonly used as tags.

The general procedure can be illustrated by focusing on a system that uses the glutathione-*S*-transferase (GST) tag **(Fig. 9-11)**. GST is a small enzyme (M_r 26,000) that binds tightly and specifically to glutathione. When the GST gene sequence is fused to a target gene, the fusion protein acquires the capacity to bind glutathione. The fusion protein is expressed in a host organism such as a bacterium, and a crude extract is prepared. A column is filled with a porous matrix consisting of the ligand (glutathione) immobilized on microscopic beads of a stable polymer such as cross-linked agarose. As the crude extract percolates through this matrix, the fusion protein becomes immobilized by binding the glutathione. The other proteins in the extract are washed through the column and discarded. The interaction between GST and glutathione is tight but noncovalent, allowing the fusion protein to be gently eluted from the column with a solution containing either a higher concentration of salts or free glutathione to compete with the immobilized ligand for GST binding. The fusion protein is often obtained with good yield and high purity. In some commercially available systems, the tag can be entirely or largely removed from the purified fusion protein by a protease that cleaves a sequence near the junction between the target protein and its tag.

A shorter tag with widespread application consists of a simple sequence of six or more His residues. These histidine tags, or His tags, bind tightly and specifically to nickel ions. A chromatography matrix with immobilized Ni^{2+} can be used to quickly separate a His-tagged protein from other proteins in an extract. Some of the larger tags, such as maltose-binding protein, provide added stability and solubility, allowing the purification of cloned proteins that are otherwise inactive due to improper folding or insolubility.

Affinity chromatography using terminal tags is powerful and convenient. The tags have been successfully used in thousands of published studies; in many cases, the protein would be impossible to purify and study without the tag. However, even very small tags can affect the properties of the proteins they are attached to, thereby influencing the study results. For example, the tag may adversely affect protein folding. Even if the tag is removed by a protease, one or a few extra amino acid residues can remain behind on the target protein, which may or may not affect the protein's activity. The types of experiments to be carried out, and the results obtained from them, should always be evaluated with the aid of well-designed controls to assess any effect of a tag on protein function.

The Polymerase Chain Reaction Can Be Adapted for Convenient Cloning

Careful design of the primers used for PCR (see Fig. 8-33) can alter the amplified segment by the inclusion, at each end, of additional DNA not present in the

TABLE 9-3	Commonly Used Protein Tags	
Tag protein/ peptide	Molecular mass (kDa)	Immobilized ligand
Protein A	59	Fc portion of IgG
(His)$_6$	0.8	Ni^{2+}
Glutathione-*S*-transferase (GST)	26	Glutathione
Maltose-binding protein	41	Maltose
β-Galactosidase	116	*p*-Aminophenyl-β-D-thiogalactoside (TPEG)
Chitin-binding domain	5.7	Chitin

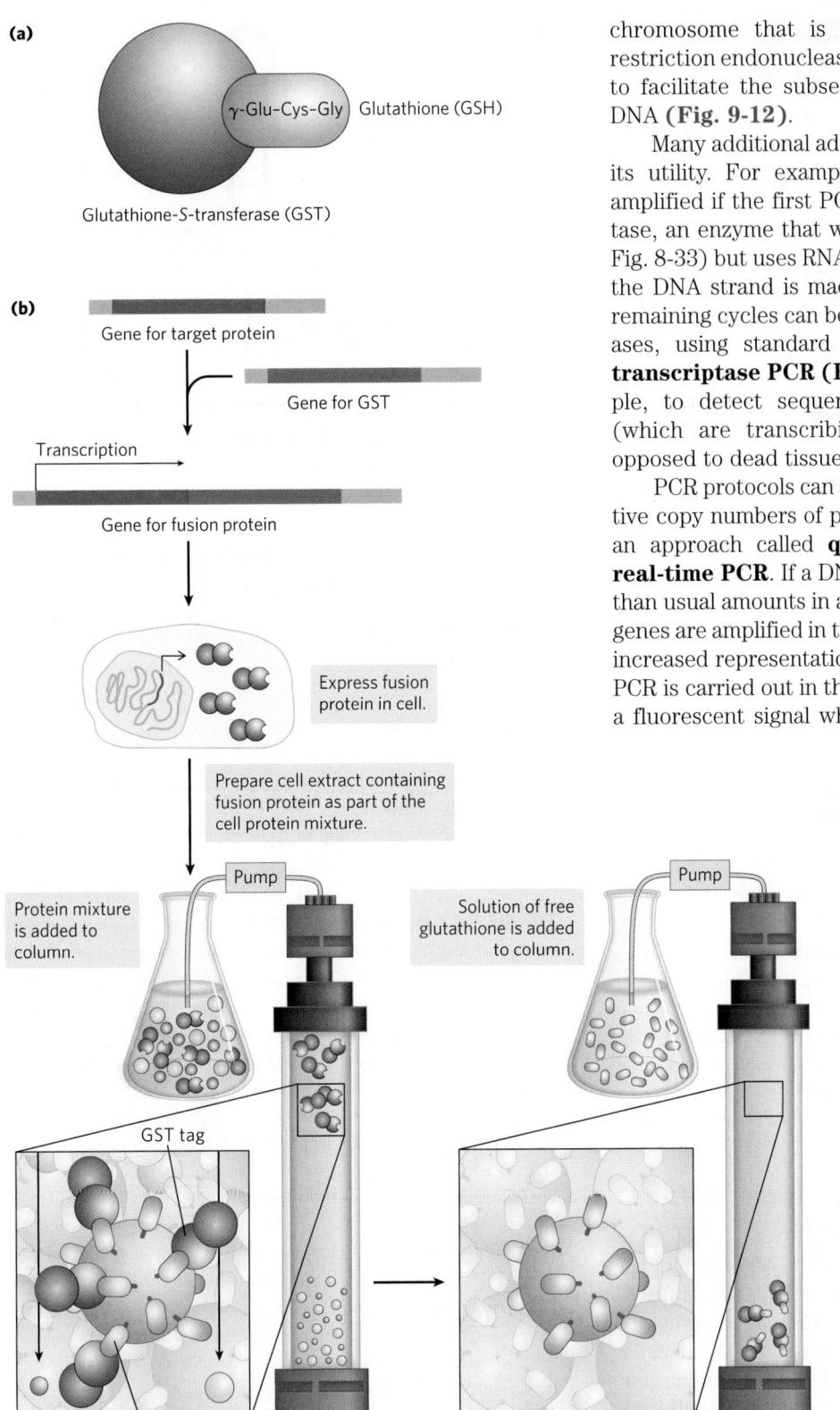

chromosome that is being targeted. For example, restriction endonuclease cleavage sites can be included to facilitate the subsequent cloning of the amplified DNA **(Fig. 9-12)**.

Many additional adaptations of PCR have increased its utility. For example, sequences in RNA can be amplified if the first PCR cycle uses reverse transcriptase, an enzyme that works like DNA polymerase (see Fig. 8-33) but uses RNA as a template (Fig. 9-12). After the DNA strand is made from the RNA template, the remaining cycles can be carried out with DNA polymerases, using standard PCR protocols. This **reverse transcriptase PCR (RT-PCR)** can be used, for example, to detect sequences derived from living cells (which are transcribing their DNA into RNA) as opposed to dead tissues.

PCR protocols can also be used to estimate the relative copy numbers of particular sequences in a sample, an approach called **quantitative PCR (qPCR)** or **real-time PCR**. If a DNA sequence is present in higher than usual amounts in a sample—for example, if certain genes are amplified in tumor cells—qPCR can reveal the increased representation of that sequence. In brief, the PCR is carried out in the presence of a probe that emits a fluorescent signal when the PCR product is present

FIGURE 9-11 Use of tagged proteins in protein purification. (a) Glutathione-*S*-transferase (GST) is a small enzyme that binds glutathione (a glutamate residue to which a Cys–Gly dipeptide is attached at the carboxyl carbon of the Glu side chain, hence the abbreviation GSH). **(b)** The GST tag is fused to the carboxyl terminus of the protein by genetic engineering. The tagged protein is expressed in the cell and is present in the crude extract when the cells are lysed. The extract is subjected to affinity chromatography (see Fig. 3-17c) through a matrix with immobilized glutathione. The GST-tagged protein binds to the glutathione, retarding the protein's migration through the column, while other proteins are washed through rapidly. The tagged protein is subsequently eluted with a solution containing elevated salt concentration or free glutathione.

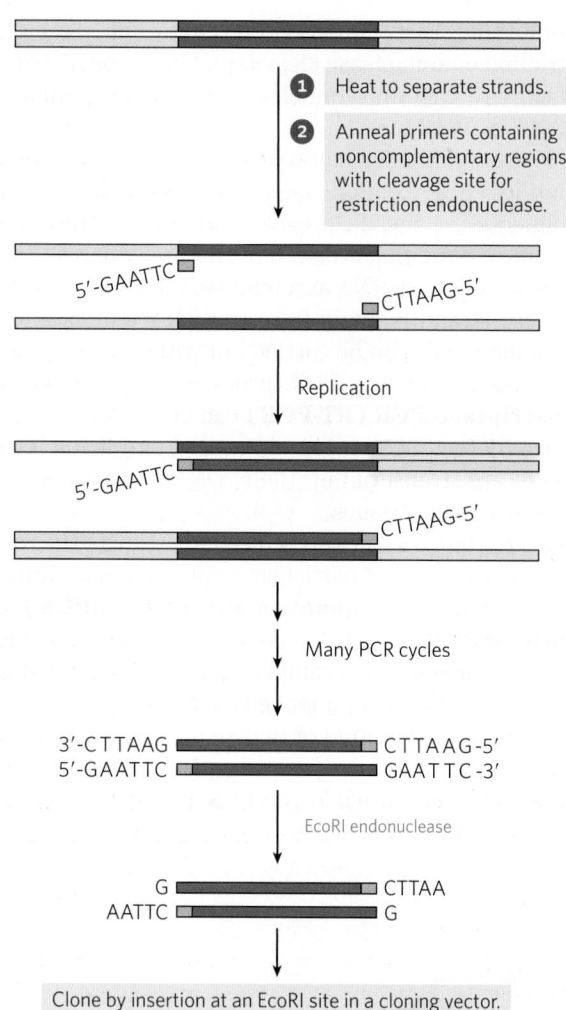

FIGURE 9-12 Cloning of a PCR-amplified DNA segment. DNA that has been amplified by the polymerase chain reaction (see Fig. 8-33) can be cloned. The primers can include noncomplementary ends that have a site for cleavage by a restriction endonuclease. Although these parts of the primers do not anneal to the target DNA, the PCR process incorporates them into the DNA that is amplified. Cleavage of the amplified fragments at these sites creates sticky ends, used in ligation of the amplified DNA to a cloning vector.

(Fig. 9-13). If the sequence of interest is present at higher levels than other sequences in the sample, the PCR signal will reach a predetermined threshold faster. Reverse transcriptase PCR and qPCR can be combined to determine the relative concentrations of a particular mRNA molecule in a cell, and thereby monitor gene expression, under different environmental conditions.

SUMMARY 9.1 Studying Genes and Their Products

■ DNA cloning and genetic engineering involve the cleavage of DNA and assembly of DNA segments in new combinations—recombinant DNA.

■ Cloning entails cutting DNA into fragments with enzymes; selecting and possibly modifying a fragment of interest; inserting the DNA fragment into a suitable

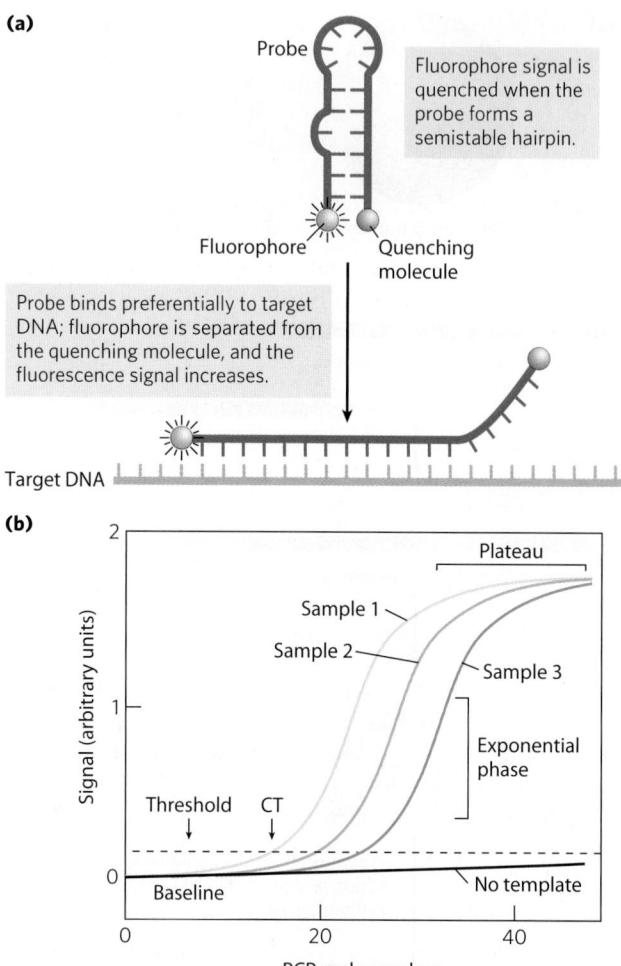

FIGURE 9-13 Quantitative PCR. PCR can be used quantitatively, by carefully monitoring the progress of a PCR amplification and determining when a DNA segment has been amplified to a specified threshold level. **(a)** The amount of PCR product present is determined by measuring the level of a fluorescent probe attached to a reporter oligonucleotide complementary to the DNA segment that is being amplified. Probe fluorescence is not detectable initially, due to a fluorescence quencher attached to the same oligonucleotide. When the reporter oligonucleotide pairs with its complement in a copy of the amplified DNA segment, the fluorophore is separated from the quenching molecule and fluorescence results. **(b)** As the PCR reaction proceeds, the amount of the targeted DNA segment increases exponentially, and the fluorescent signal also increases exponentially as the oligonucleotide probes anneal to the amplified segments. After many PCR cycles, the signal reaches a plateau as one or more reaction components become exhausted. When a segment is present in greater amounts in one sample than another, its amplification reaches a defined threshold level earlier. The "No template" line follows the slow increase in background signal observed in a control that does not include added sample DNA. CT is the cycle number at which the threshold is first surpassed.

cloning vector; transferring the vector with the DNA insert into a host cell for replication; and identifying and selecting cells that contain the DNA fragment.

■ Key enzymes in gene cloning include restriction endonucleases (especially the type II enzymes) and DNA ligase.

- Cloning vectors include plasmids and, for the longest DNA inserts, bacterial artificial chromosomes (BACs) and yeast artificial chromosomes (YACs).

- Genetic engineering techniques manipulate cells to express and/or alter cloned genes.

- Proteins or peptides can be attached to a protein of interest by altering its cloned gene, creating a fusion protein. The additional peptide segments can be used to detect the protein or to purify it, using convenient affinity chromatography methods.

- The polymerase chain reaction (PCR) permits the amplification of chosen segments of DNA or RNA for cloning and can be adapted to determine gene copy number or to monitor gene expression quantitatively.

9.2 Using DNA-Based Methods to Understand Protein Function

Protein function can be described on three levels. **Phenotypic function** describes the effects of a protein on the entire organism. For example, loss of the protein may lead to slower growth of the organism, an altered development pattern, or even death. **Cellular function** is a description of the network of interactions a protein engages in at the cellular level. Identifying interactions with other proteins in the cell can help define the kinds of metabolic processes in which the protein participates. Finally, **molecular function** refers to the precise biochemical activity of a protein, including details such as the reactions an enzyme catalyzes or the ligands a receptor binds. The challenge of understanding the functions of the thousands of uncharacterized or poorly characterized proteins found in a typical cell has given rise to a wide variety of techniques. DNA-based methods make a critical contribution to this effort and can provide information on all three levels. With these technologies, we can determine when a particular protein is expressed, the other proteins it might be related to, its location in the cell, other cellular components it interacts with, and what happens to the cell when the protein is missing.

DNA Libraries Are Specialized Catalogs of Genetic Information

A **DNA library** is a collection of DNA clones, usually gathered for purposes of gene discovery or the determination of gene or protein function. The library can take a variety of forms, depending on the source of the DNA and the ultimate purpose of the library.

The largest is a **genomic library**, produced when the complete genome of an organism is cleaved into thousands of fragments. All the fragments are cloned by insertion of each fragment into a cloning vector. This creates a complex mixture of recombinant vectors, each with a different cloned fragment. Library construction begins with *partial* digestion of the DNA by restriction endonucleases, such that any given sequence appears in fragments of a limited range of sizes—a range compatible with the cloning vector. Fragments that are too large or too small for cloning are removed by centrifugation or electrophoresis. The cloning vector, such as a BAC or YAC, is cleaved with the same restriction endonuclease used to digest the DNA and is ligated to the genomic DNA fragments. The ligated DNA mixture is then used to transform bacteria or yeast cells to produce a library of cells, each harboring a different recombinant DNA molecule. Ideally, all the DNA of the genome under study is represented in the library. Each transformed bacterium or yeast cell grows into a colony, or clone, of identical cells, each cell bearing the same recombinant plasmid, one of many represented in the overall library.

Efforts to define gene or protein function often make use of more specialized libraries. An example is a library that includes only those sequences of DNA that are *expressed*—that is, transcribed into RNA—in a given organism, or even just in certain cells or tissues. Such a library lacks any genomic DNA that is not transcribed. The researcher first extracts mRNA from an organism, or from specific cells of an organism, and then prepares the **complementary DNAs (cDNAs)**. This multistep reaction, shown in **Figure 9-14**, relies on reverse transcriptase, which synthesizes DNA from a template RNA. The resulting double-stranded DNA fragments are inserted into a suitable vector and cloned, creating a population of clones called a **cDNA library**. The presence of a gene for a particular protein in such a library implies that this gene is expressed in the cells and under the conditions used to generate the library.

Sequence or Structural Relationships Provide Information on Protein Function

One important reason to sequence many genomes is to provide a database that can be used to assign gene functions by genome comparisons, an enterprise referred to as **comparative genomics**. This field is deeply rooted in and, indeed, made possible by evolutionary biology. Sometimes a newly discovered gene is related by sequence homologies to a previously studied gene in another or the same species, and its function can be entirely or partly defined by that relationship. Genes that occur in different species but have a clear sequence and functional relationship to each other are called **orthologs**. Genes similarly related to each other within a single species are called **paralogs**. We introduced these terms in Chapter 3 in the context of proteins. As with proteins, information about the function of a gene in one species can be used to at least tentatively assign function to the orthologous gene found in a second species. The correlation is easiest to make when comparing genomes from relatively closely related

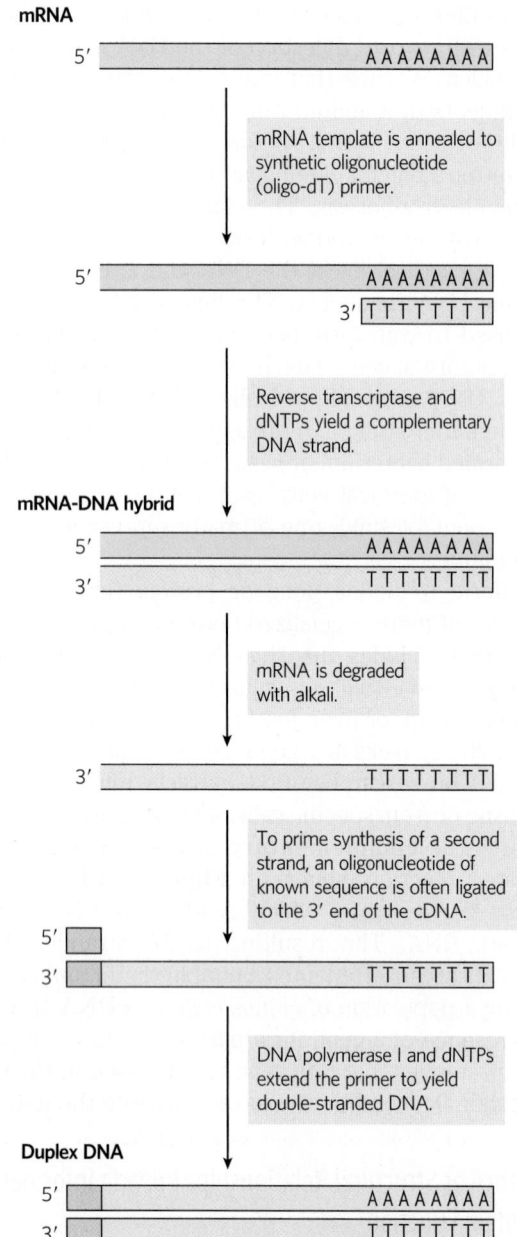

FIGURE 9-14 Building a cDNA library from mRNA. A cell's total mRNA content includes transcripts from thousands of genes, and the cDNAs generated from this mRNA are correspondingly heterogeneous. Reverse transcriptase can synthesize DNA on an RNA or a DNA template (see Fig. 26-32). To prime the synthesis of a second DNA strand, oligonucleotides of known sequence are ligated to the 3′ end of the first strand, and the double-stranded cDNA so produced is cloned into a plasmid.

species, such as mouse and human, although many clearly orthologous genes have been identified in species as distant as bacteria and humans. Sometimes even the order of genes on a chromosome is conserved over large segments of the genomes of closely related species **(Fig. 9-15)**. Conserved gene order, called **synteny**, provides additional evidence for an orthologous relationship between genes at identical locations within the related segments.

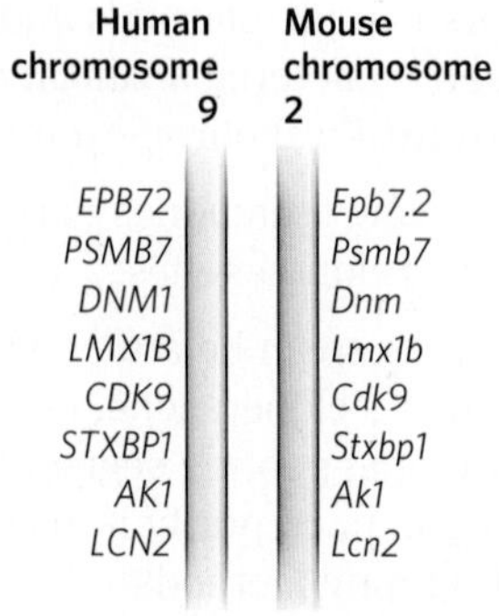

Human chromosome 9	Mouse chromosome 2
EPB72	Epb7.2
PSMB7	Psmb7
DNM1	Dnm
LMX1B	Lmx1b
CDK9	Cdk9
STXBP1	Stxbp1
AK1	Ak1
LCN2	Lcn2

FIGURE 9-15 Synteny in the human and mouse genomes. Large segments of the two genomes have closely related genes aligned in the same order on the chromosomes. In these short segments of human chromosome 9 and mouse chromosome 2, the genes show a very high degree of homology, as well as the same gene order. The different lettering schemes for the gene names simply reflect the different naming conventions for the two species. [Source: Information from T. G. Wolfsberg et al., *Nature* 409:824, 2001, Fig. 1.]

Alternatively, certain amino acid sequences associated with particular structural motifs (Chapter 4) may be identified within a protein. The presence of a structural motif may help to define molecular function by suggesting that a protein, say, catalyzes ATP hydrolysis, binds to DNA, or forms a complex with zinc ions. These relationships are determined with the aid of sophisticated computer programs, limited only by the current information on gene and protein structure and by our capacity to associate sequences with particular structural motifs. Sequences at an enzyme active site that have been highly conserved during evolution are typically associated with catalytic function, and their identification is often a key step in defining an enzyme's reaction mechanism. The reaction mechanism, in turn, provides information needed to develop new enzyme inhibitors that can be used as pharmaceutical agents.

Fusion Proteins and Immunofluorescence Can Reveal the Location of Proteins in Cells

Often, an important clue to the function of a gene product comes from determining its location within the cell. For example, a protein found exclusively in the nucleus could be involved in processes that are unique to that organelle, such as transcription, replication, or chromatin condensation. Researchers often engineer fusion proteins for the purpose of locating a protein in the cell or organism. Some of the most useful fusions are the attachment of marker proteins that signal the location by direct visualization or by immunofluorescence.

A particularly useful marker is the **green fluorescent protein (GFP) (Fig. 9-16)**, discovered by Osamu Shimomura. As subsequently shown by Martin Chalfie, a target gene (encoding the protein of interest) fused to the GFP gene generates a fusion protein that is highly fluorescent—it literally lights up when exposed to blue light—and can be visualized directly in a living

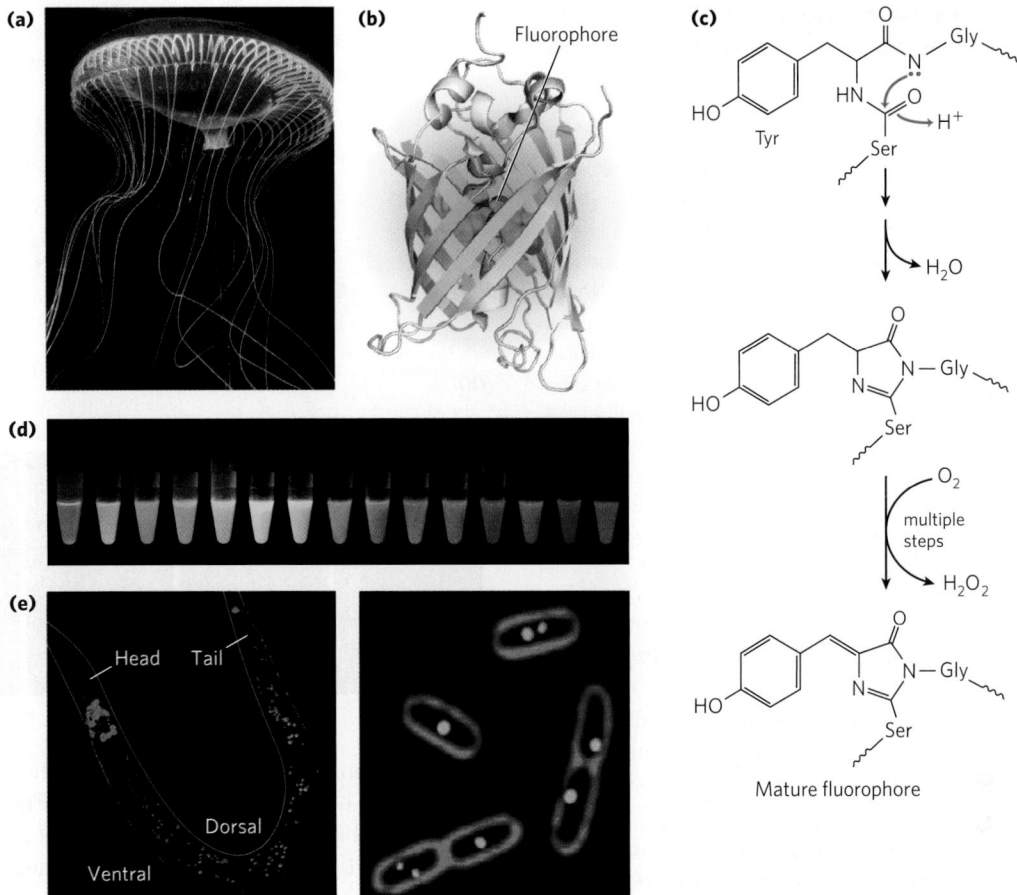

FIGURE 9-16 Green fluorescent protein (GFP). **(a)** GFP is derived from the jellyfish *Aequorea victoria*, which is abundant in Puget Sound, Washington. **(b)** The protein has a β-barrel structure; the fluorophore (shown as a space-filling model) is in the center of the barrel. **(c)** The fluorophore in GFP is derived from a sequence of three amino acids: –Ser^{65}–Tyr^{66}–Gly^{67}–. The fluorophore achieves its mature form through an internal rearrangement, coupled to a multistep oxidation reaction. An abbreviated mechanism is shown here. **(d)** Variants of GFP are now available in almost any color of the visible spectrum. **(e)** A GLR1-GFP fusion protein fluoresces bright green in *Caenorhabditis elegans*, a nematode worm (left). GLR1 is a glutamate receptor of nervous tissue. (In this photograph, autofluorescing

fat droplets are false colored in magenta.) The membranes of *E. coli* cells (right) are stained with a red fluorescent dye. The cells are expressing a protein that binds to a resident plasmid, fused to GFP. The green spots indicate the locations of plasmids. [Sources: (a) Chris Parks/ImageQuest Marine. (b) PDB ID 1GFL, F. Yang et al., *Nature Biotechnol.* 14:1246, 1996. (c, d) Courtesy of Roger Tsien, University of California, San Diego, Department of Pharmacology, and Paul Steinbach. (e) (left) Courtesy Penelope J. Brockie and Andres V. Maricq, Department of Biology, University of Utah; (right) courtesy Joseph A. Pogliano, from J. Pogliano et al. (2001), Multicopy plasmids are clustered and localized in *Escherichia coli*, *Proc. Natl. Acad. Sci. USA* 98:4486–4491.]

cell. GFP is a protein derived from the jellyfish *Aequorea victoria*. The protein has a β-barrel structure with a fluorophore (the fluorescent component of the protein) in the center. The fluorophore is derived from a rearrangement and oxidation of three amino acid residues. Because this reaction is autocatalytic and requires no proteins or cofactors other than molecular oxygen, GFP is readily cloned in an active form in almost any cell. Just a few molecules of this protein can be observed microscopically, allowing the study of its location and movements in a cell. Careful protein engineering by Roger Tsien, coupled with the isolation of related fluorescent proteins from other marine coelenterates, has made variants of these proteins available in an array of colors (Fig. 9-16d) and other characteristics (brightness, stability). If fusion to GFP does not impair the function or properties of a protein one wishes to study,

the fusion protein can be used to reveal the protein's location in the cell under a range of conditions and to detect interactions with other labeled proteins. With this technology, for example, the protein GLR1 (a glutamate

Osamu Shinomura
[Source: Josh Reynolds/AP Images.]

Martin Chalfie
[Source: Diane Bondareff/AP Images.]

Roger Y. Tsien
[Source: HO/ Reuters/Corbis.]

FIGURE 9-17 Indirect immunofluorescence. (a) The protein of interest is bound to a primary antibody, and a secondary antibody is added; this second antibody, with one or more attached fluorescent groups, binds to the first. Multiple secondary antibodies can bind the primary antibody, amplifying the signal. If the protein of interest is in the interior of the cell, the cell is fixed and permeabilized, and the two antibodies are added in succession. **(b)** The end result is an image in which bright spots indicate the location of the protein or proteins of interest in the cell. The images here show a nucleus from a human fibroblast, successively stained with antibodies and fluorescent labels for DNA polymerase ε, for PCNA, an important polymerase accessory protein, and for bromo-deoxyuridine (BrdU), a nucleotide analog. The BrdU, added as a brief pulse, identifies regions undergoing active DNA replication. The patterns of staining show that DNA polymerase ε and PCNA co-localize to regions of active DNA synthesis (rightmost image); one such region is visible in the white box. [Source: (b) Fuss, J. and Linn, S., 2002, "Human DNA polymerase ε co-localizes with proliferating cell nuclear antigen and DNA replication late, but not early, in S phase," *J. Biol. Chem.* 277:8658–8666. Courtesy Jill Fuss, University of California, Berkeley.]

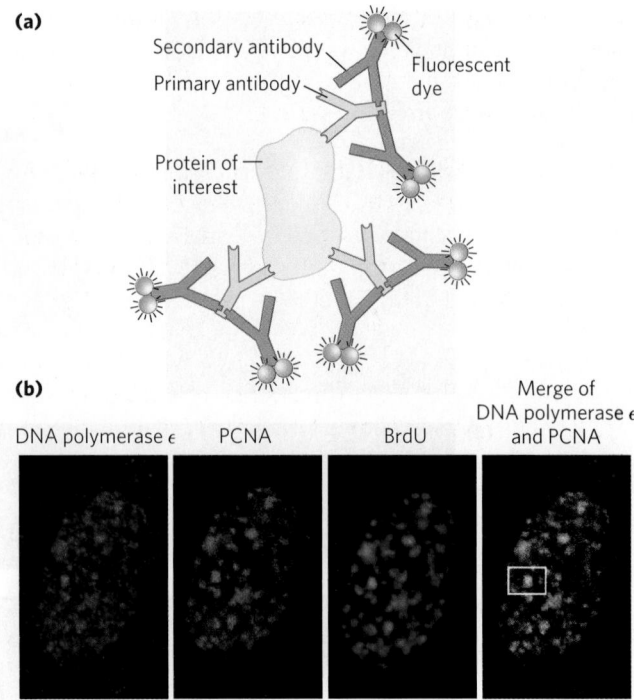

receptor of nervous tissue) has been visualized as a GLR1-GFP fusion protein in the nematode *Caenorhabditis elegans* (Fig. 9-16e). Shimomura, Chalfie, and Tsien received the 2008 Nobel Prize in Chemistry for their work in developing GFP as a tool for biochemical investigation.

In many cases, visualization of a GFP fusion protein in a living cell is not possible or practical or even desirable. The GFP fusion protein may be inactive or may not be expressed at sufficient levels to allow visualization. In this case, **immunofluorescence** is an alternative approach for visualizing the endogenous (unaltered) protein. This approach requires fixation (and thus death) of the cell. The protein of interest is sometimes expressed as a fusion protein with an **epitope tag**, a short protein sequence that is bound tightly by a well-characterized, commercially available antibody. Fluorescent molecules (fluorochromes) are attached to this antibody. More commonly, the target protein is unaltered and is bound by an antibody that is specific for the protein. Next, a second antibody is added that binds specifically to the first one, and it is the second antibody that has the attached fluorochrome(s) **(Fig. 9-17a)**. A variation of this indirect approach to visualization is to attach biotin molecules to the first antibody, then add streptavidin (a bacterial protein closely related to avidin, a protein that binds biotin; see Table 5-1) complexed with fluorochromes. The interaction between biotin and streptavidin is one of the strongest and most specific known, and the potential to add multiple fluorochromes to each target protein gives this method great sensitivity. In all of these cases, the end product is a microscopic view of a cell in which a spot of light (a focus) reveals the location of the protein.

Highly specialized cDNA libraries can be made by cloning cDNAs or cDNA fragments into a vector that fuses each cDNA sequence with the sequence for a marker, called a reporter gene. The fused gene is often called a

reporter construct. For example, all the genes in the library may be fused to the GFP gene **(Fig. 9-18)**. Each cell in the library expresses one of these fused genes. The cellular location of the product of any gene represented in the library is revealed as foci of light in cells that express the gene at sufficient levels—assuming that the fusion protein retains its normal function and location.

Protein-Protein Interactions Can Help Elucidate Protein Function

Another key to defining the function of a particular protein is to determine what other cellular components it binds to. In the case of protein-protein interactions, the

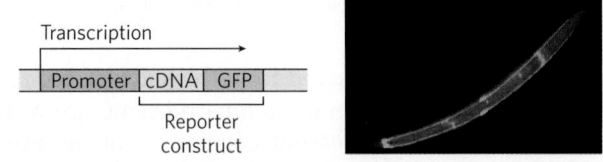

FIGURE 9-18 Specialized DNA libraries. Cloning of a cDNA next to the GFP gene creates a reporter construct. Transcription proceeds through the gene of interest (the inserted cDNA) and the reporter gene (here, GFP), and the mRNA transcript is expressed as a fusion protein. The GFP part of the protein is visible with the fluorescence microscope. Just one example is shown here; thousands of genes can be fused to GFP in similar constructs and stored in libraries in which each cell or organism in the library expresses a different protein fused to GFP. If the fusion protein is properly expressed, researchers can assess its location in the cell or organism. The photograph shows a nematode worm containing a GFP fusion protein expressed only in the four "touch" neurons that run the length of its body. [Source: Courtesy of Kevin Strange, PhD, and Michael Christensen, PhD, Department of Pharmacology, Vanderbilt University Medical Center.]

association of a protein of unknown function with one whose function is known can provide a compelling implication of a functional relationship. The techniques used in this effort are quite varied.

Purification of Protein Complexes By fusing the gene encoding a protein under study with the gene for an epitope tag, investigators can precipitate the protein product of the fusion gene by complexing it with the antibody that binds the epitope. This process is called **immunoprecipitation (Fig. 9-19)**. If the tagged protein is expressed in cells, other proteins that bind to it precipitate with it. Identifying the associated proteins reveals some of the intracellular protein-protein interactions of the tagged protein. There are many variations of this process. For example, a crude extract of cells that express a tagged protein is added to a column containing immobilized antibody (see Fig. 3-17c for a description of affinity chromatography). The tagged protein binds to the antibody, and proteins that interact with the tagged protein are sometimes also retained on the column. The connection between the protein and the tag is cleaved with a specific

protease, and the protein complexes are eluted from the column and analyzed. Researchers can use these methods to define complex networks of interactions within a cell. In principle, the chromatographic approach to analyzing protein-protein interactions can be used with any type of protein tag (His tag, GST, etc.) that can be immobilized on a suitable chromatographic medium.

The selectivity of this approach has been enhanced with **tandem affinity purification (TAP) tags**. Two consecutive tags are fused to a target protein, and the fusion protein is expressed in a cell **(Fig. 9-20)**. The first tag is protein A, a protein found at the surface of the bacterium *Staphylococcus aureus* that binds tightly to mammalian immunoglobulin G (IgG). The second tag is often a calmodulin-binding peptide. A crude extract

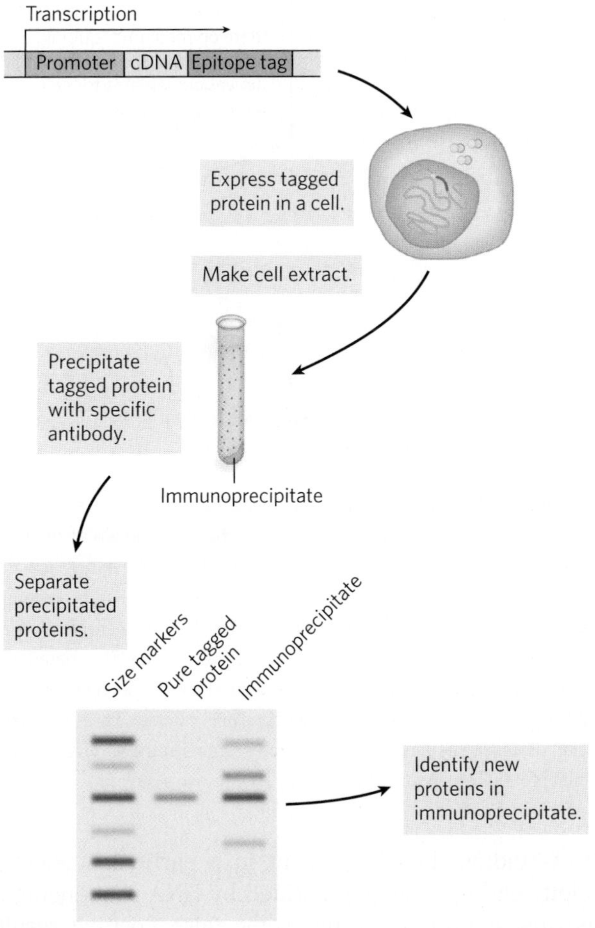

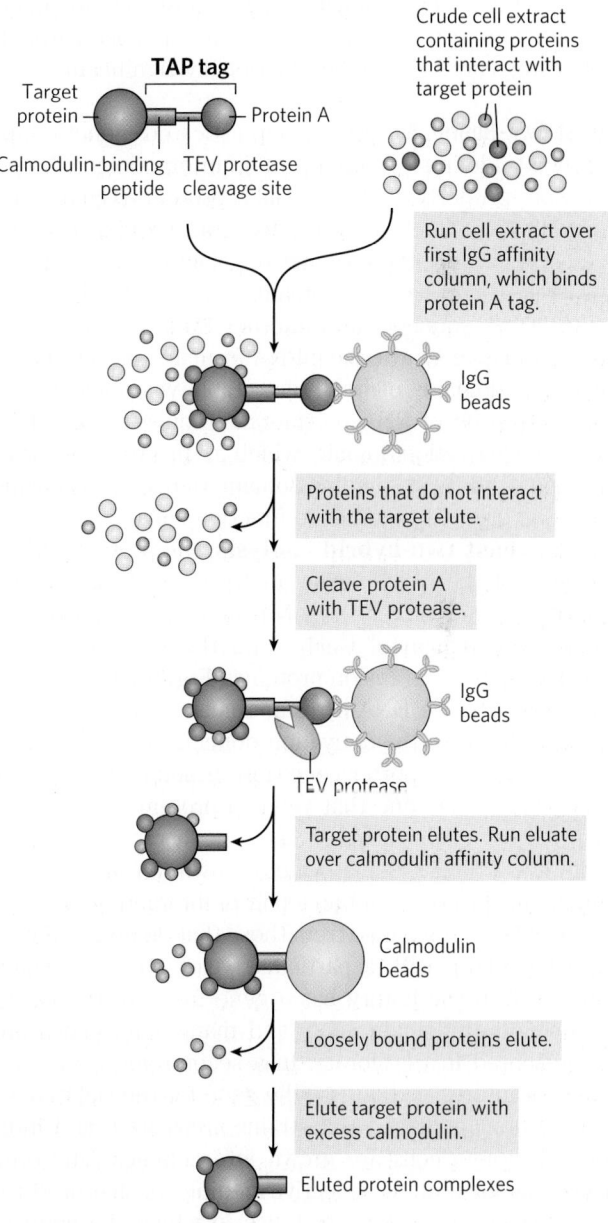

FIGURE 9-20 Tandem affinity purification (TAP) tags. A TAP-tagged protein and associated proteins are isolated by two consecutive affinity purifications, as described in the text.

FIGURE 9-19 The use of epitope tags to study protein-protein interactions. The gene of interest is cloned next to a gene for an epitope tag, and the resulting fusion protein is precipitated by antibodies to the epitope. Any other proteins that interact with the tagged protein also precipitate, thereby helping to elucidate protein-protein interactions.

containing the TAP-tagged fusion protein is passed through a column matrix with attached IgG antibodies that bind protein A. Most of the unbound cellular proteins are washed through the column, but proteins that normally interact with the target protein in the cell are retained. The first tag is then cleaved from the fusion protein with a highly specific protease, TEV protease, and the shortened fusion target protein and any proteins associated noncovalently with the target protein are eluted from the column. The eluent is then passed through a second column containing a matrix with attached calmodulin that binds the second tag. Loosely bound proteins are again washed from the column. After the second tag is cleaved, the target protein is eluted from the column with its associated proteins. The two consecutive purification steps eliminate any weakly bound contaminants. False positives are minimized, and protein interactions that persist through both steps are likely to be functionally significant.

Yeast Two-Hybrid Analysis A sophisticated genetic approach to defining protein-protein interactions is based on the properties of the Gal4 protein (Gal4p; see Fig. 28-31), which activates the transcription of *GAL* genes (encoding the enzymes of galactose metabolism) in yeast. Gal4p has two domains: one that binds a specific DNA sequence, and another that activates RNA polymerase to synthesize mRNA from an adjacent gene. The two domains of Gal4p are stable when separated, but activation of RNA polymerase requires interaction with the activation domain, which in turn requires positioning by the DNA-binding domain. Hence, the domains must be brought together to function correctly.

In **yeast two-hybrid analysis**, the protein-coding regions of the genes to be analyzed are fused to the yeast gene for either the DNA-binding domain or the activation domain of Gal4p, and the resulting genes express a series of fusion proteins **(Fig. 9-21)**. If a protein fused to the DNA-binding domain interacts with a protein fused to the activation domain, transcription is activated. The reporter gene transcribed by this activation is generally one that yields a protein required for growth or an enzyme that catalyzes a reaction with a colored product. Thus, when grown on the proper medium, cells that contain a pair of interacting proteins are easily distinguished from those that do not. A library can be set up with a particular yeast strain in which each cell in the library has a gene fused to the Gal4p DNA-binding domain gene, and many such genes are represented in the library. In a second yeast strain, a gene of interest is fused to the gene for the Gal4p activation domain. The yeast strains are mated, and individual diploid cells are grown into colonies. The only cells that grow on the selective medium, or that produce the appropriate color, are those in which the gene of interest is binding to a partner, allowing transcription of the reporter gene. This allows large-scale screening for cellular proteins that interact with the target protein. The interacting protein that is fused to the Gal4p

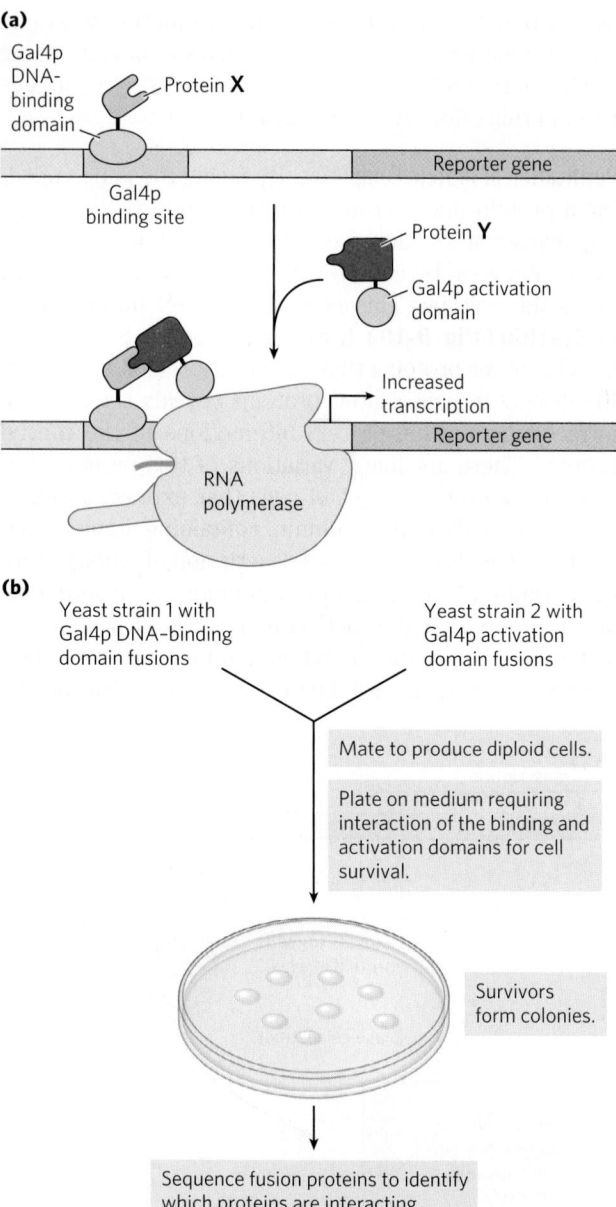

FIGURE 9-21 **Yeast two-hybrid analysis.** (a) The goal is to bring together the DNA-binding domain and the activation domain of the yeast Gal4 protein (Gal4p) through the interaction of two proteins, X and Y, to which one or other of the domains is fused. This interaction is accompanied by the expression of a reporter gene. (b) The two gene fusions are created in separate yeast strains, which are then mated. The mated mixture is plated on a medium on which the yeast cannot survive unless the reporter gene is expressed. Thus, all surviving colonies have interacting fusion proteins. Sequencing of the fusion proteins in the survivors reveals which proteins are interacting.

DNA-binding domain present in a particular selected colony can be quickly identified by DNA sequencing of the fusion protein's gene. Some false positive results occur, due to the formation of multiprotein complexes.

These techniques for determining cellular localization and protein interactions provide important clues to protein function. However, they do not replace classical biochemistry. They simply give researchers an expedited entrée into important new biological problems. When paired with the simultaneously evolving tools of biochemistry and molecular biology, the techniques described here are speeding the discovery not only of new proteins but of new biological processes and mechanisms.

DNA Microarrays Reveal RNA Expression Patterns and Other Information

Major refinements of the technology underlying DNA libraries, PCR, and hybridization have come together in the development of **DNA microarrays**, which allow the rapid and simultaneous screening of many thousands of genes. In the most commonly used technique, DNA segments from genes of known sequence, a few dozen to hundreds of base pairs long, are synthesized directly on a solid surface by a process called photolithography **(Fig. 9-22)**. Thousands of independent sequences are generated, each occupying a tiny part, or spot, of a surface measuring just a few square centimeters. The pattern of sequences is predesigned, with each of many thousands of spots containing sequences derived from a particular gene. The resulting array, or chip, may include sequences

derived from every gene of a bacterial or yeast genome, or selected families of genes from a larger genome. Once constructed, the microarray can be probed with mRNAs or cDNAs from a particular cell type or cell culture to identify the genes being expressed in those cells.

A microarray can provide a snapshot of all the genes in an organism, informing the researcher about the genes that are expressed at a given stage in the organism's development or under a particular set of environmental conditions. For example, the total complement of mRNA can be isolated from cells at two different stages of development and converted to cDNA with reverse transcriptase. With the use of fluorescently labeled deoxyribonucleotides, the two cDNA samples can be made so that one fluoresces red, the other green **(Fig. 9-23)**. The cDNA from the two samples is mixed and used to probe the microarray. Each cDNA anneals to only one spot, corresponding to the gene encoding the mRNA that gave rise to that cDNA. Spots that fluoresce green represent genes that produce mRNAs at higher levels at one developmental stage; those that fluoresce red represent genes expressed at higher levels

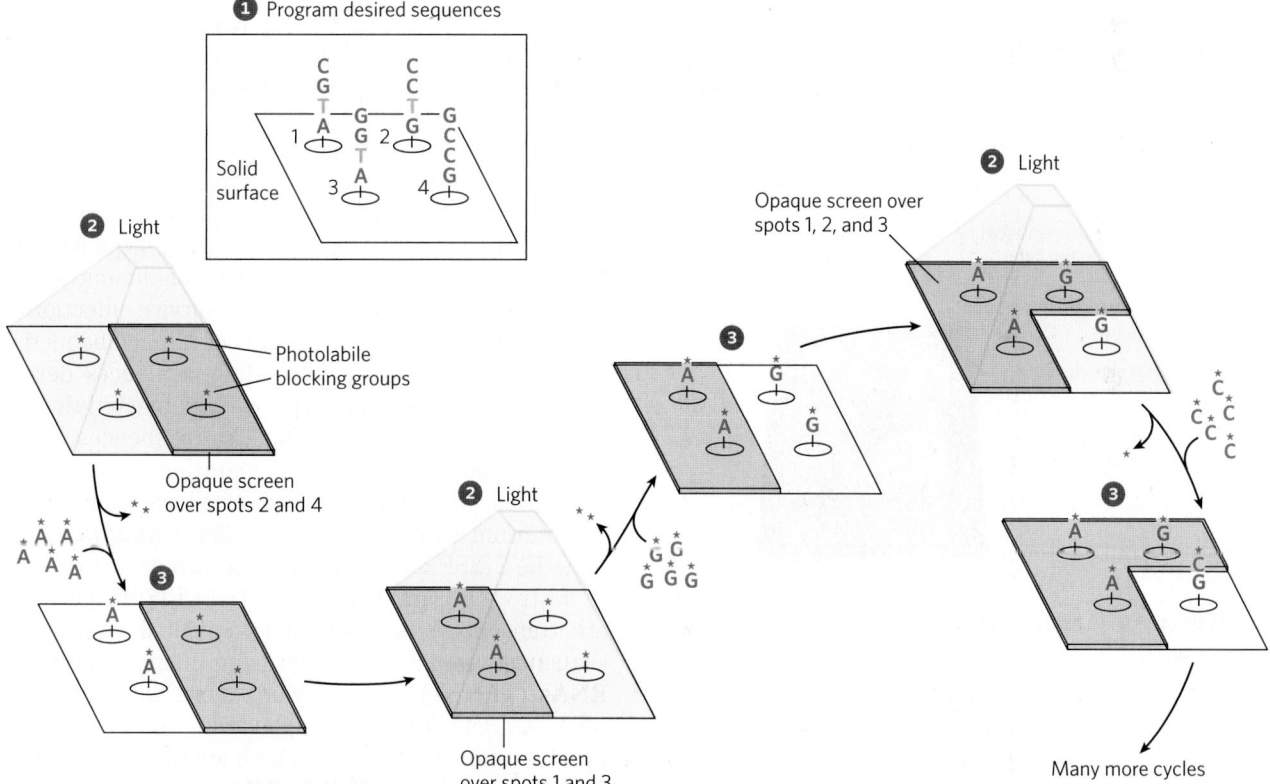

FIGURE 9-22 Photolithography to create a DNA microarray. ❶ A computer is programmed with the desired oligonucleotide sequences. Nucleophilic groups, attached to a solid surface, are initially rendered inactive by photolabile blocking groups (shown here as *). ❷ Before a flash of light, an opaque screen blocks the light from some areas of the surface, preventing their activation. Other areas, or "spots," are exposed. ❸ A solution containing one 5′-photoprotected phosphoramidate nucleotide (e.g., A*) is washed over the spots. The 5′ hydroxyl of the nucleotide is blocked with the photolabile group (*) to prevent unwanted reactions, and the nucleotide links to the exposed surface nucleophilic groups at

the appropriate spots by displacement of its activated 3′ phosphoramidate. The surface is washed successively with solutions containing each remaining nucleotide (G*, C*, T*), with each wash preceded by a flash of light to remove the photolabile blocking groups of nucleotides or surface groups at the appropriate locations (steps ❷ and ❸, repeated). Additional nucleotides are added, one at a time, to extend the nascent oligonucleotide, using screens and light to ensure that the correct nucleotides are added at each spot in the correct sequence. The process is repeated until the required sequences are built up on each of the thousands of spots in a DNA microarray.

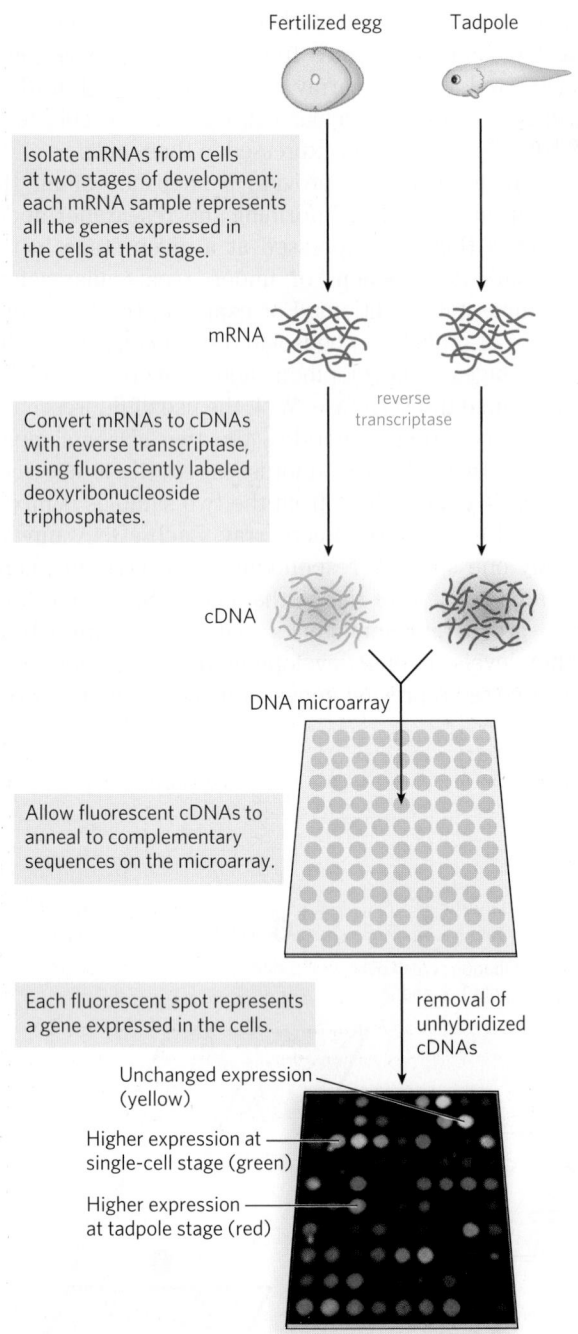

Fertilized egg   Tadpole

Isolate mRNAs from cells at two stages of development; each mRNA sample represents all the genes expressed in the cells at that stage.

mRNA

reverse transcriptase

Convert mRNAs to cDNAs with reverse transcriptase, using fluorescently labeled deoxyribonucleoside triphosphates.

cDNA

DNA microarray

Allow fluorescent cDNAs to anneal to complementary sequences on the microarray.

removal of unhybridized cDNAs

Each fluorescent spot represents a gene expressed in the cells.

Unchanged expression (yellow)

Higher expression at single-cell stage (green)

Higher expression at tadpole stage (red)

FIGURE 9-23 A DNA microarray experiment. A microarray can be prepared from any known DNA sequence, from any source. Once the DNA is attached to a solid support, the microarray can be probed with other fluorescently labeled nucleic acids. Here, mRNA samples are collected from cells of a frog at two different stages of development.

circumstances of its expression can provide important clues about its role in the cell.

Inactivating or Altering a Gene with CRISPR Can Reveal Gene Function

One of the most informative paths to understanding the function of a gene is to change (mutate) the gene or delete it. An investigator can then examine how the genomic alteration affects cell growth or function. The methods available to modify genomes grow more sophisticated every year. One increasingly common strategy is to introduce a highly specific nuclease into a cell to cut the gene of interest at a site that is functionally critical, generating a double-strand break. In eukaryotes, such breaks are most commonly repaired by cellular systems that promote nonhomologous end joining (NHEJ), a process described in Chapter 25. NHEJ seals the double-strand break, but the process is imprecise. Nucleotides are often deleted or added during the repair, inactivating the gene. In bacteria, introduced double-strand breaks are usually repaired more accurately, by homologous recombination systems (Chapter 25), but inactivating mutations can appear. Several nucleases have been engineered that can be precisely targeted to almost any sequence, but the process was expensive until the advent of **CRISPR/Cas systems** in 2011.

"CRISPR" stands for clustered, regularly interspaced short palindromic repeats; as the name suggests, these consist of a series of regularly spaced short repeats in the bacterial genome. A Cas (CRISPR-associated) protein is a nuclease. The CRISPR sequences and Cas protein are components of a kind of immune system that evolved to allow bacteria to survive infection by bacteriophages. CRISPR sequences are embedded in the bacterial genome, surrounding sequences derived from phage pathogens that previously infected the bacterium without killing it. The viral sequences are, in effect, spacer sequences separating the CRISPR sequences. When the same bacteriophage again attacks a bacterium with the corresponding CRISPR/Cas system, the CRISPR sequence and Cas protein act together to destroy the viral DNA. First, the CRISPR sequences are transcribed to RNA, and individual viral spacer sequences are cleaved to form products called **guide RNAs (gRNAs)**, which include some adjacent repeat RNA. A gRNA forms a complex with one or more Cas proteins and, in some cases, with another RNA called a **trans-activating CRISPR RNA**, or **tracrRNA**. The resulting complex binds specifically to the invading bacteriophage DNA, cleaving and destroying it through the nuclease activities associated with the Cas proteins.

The current technology was made possible by discovery of a relatively simple CRISPR/Cas system in *Streptococcus pyogenes*. This system requires only a single Cas protein, Cas9, to cleave DNA. Work in many laboratories, particularly those of Jennifer Doudna and Emmanuelle Charpentier, has produced a streamlined CRISPR/Cas9 system composed of just one protein

at another stage. If a gene produces mRNAs that are equally abundant at both stages of development, the corresponding spot fluoresces yellow. By using a mixture of two samples to measure relative rather than absolute sequence abundance, the method corrects for inconsistencies among spots in the microarray. The spots that fluoresce provide a snapshot of all the genes being expressed in the cells at the moment they were harvested—gene expression examined on a genome-wide scale. For a gene of unknown function, the time and

(a)

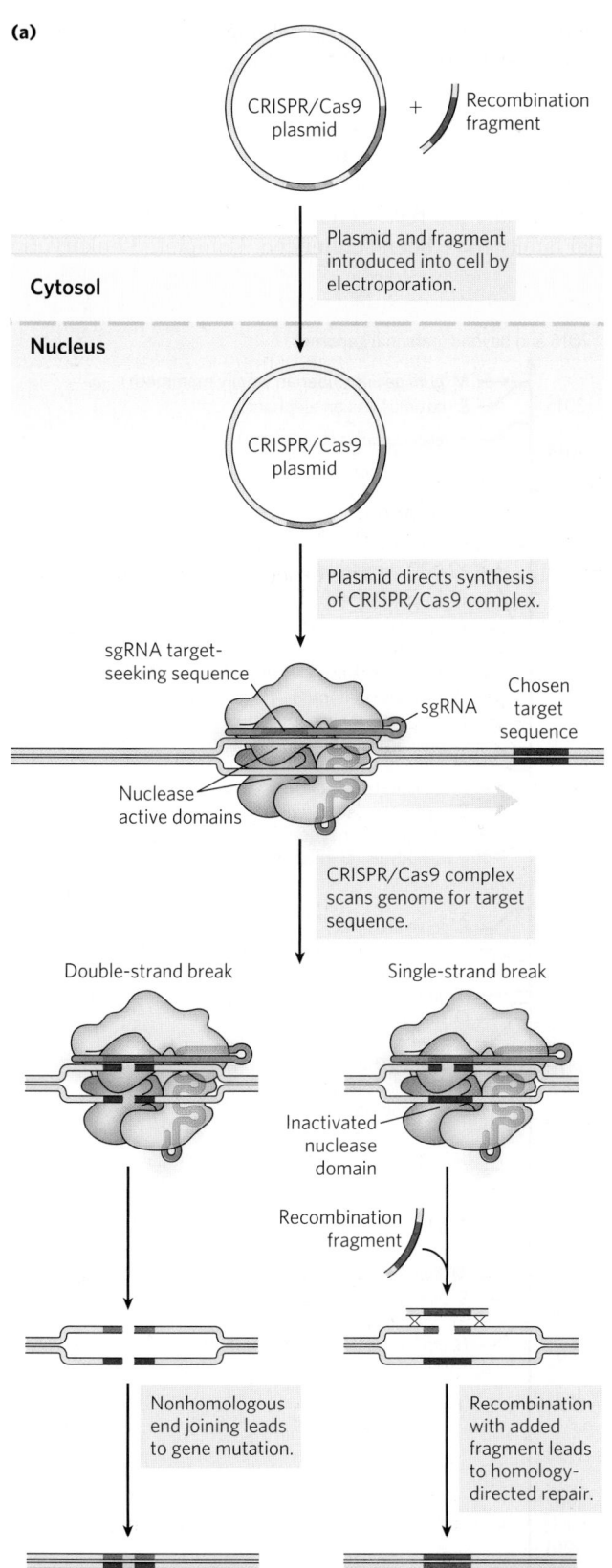

(b)

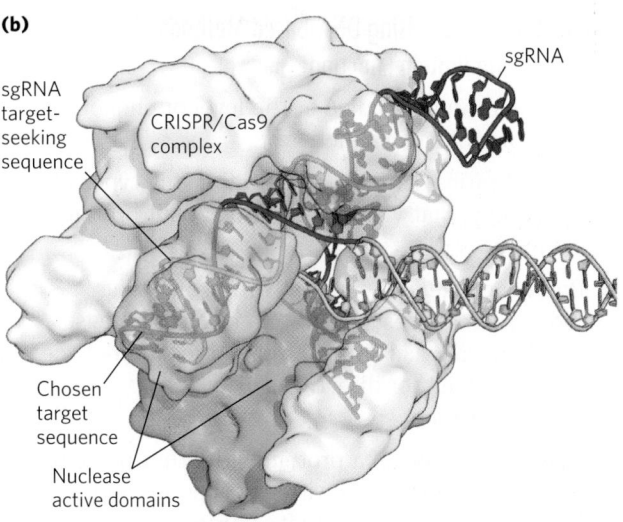

FIGURE 9-24 The CRISPR/Cas9 system for genomic engineering.
(a) The genes encoding the Cas9 protein and sgRNA are introduced into a cell where a targeted genomic change is planned. The sgRNA has a region complementary to the chosen genomic target sequence (purple); this region can be engineered to include any desired sequence. A complex consisting of the CRISPR sgRNA and the Cas9 protein forms within the cell and binds to the chosen target site in the DNA. The structure of the bound complex is shown in **(b)**. In the pathway shown on the left in (a), two nuclease active sites in the Cas9 protein separately cleave each DNA strand in the target, producing a double-strand break. The double-strand break is usually repaired by nonhomologous end joining, which generally deletes or alters the nucleotides at the site where joining occurs. Alternatively, as shown in the pathway on the right, if one nuclease site is inactivated, Cas9 nuclease activity creates a single-strand break in the target sequence. In the presence of a recombination donor DNA fragment, identical to the target sequence but incorporating the desired sequence change (fragment shown in red), homologous DNA recombination will sometimes change the sequence at the site of the break to match that of the donor DNA. [Source: PDB ID 4UN3, C. Anders et al., *Nature* 513:569, 2014.]

enzyme that cleaves just one strand, forming a single-strand break, or nick. The sgRNA is needed both to pair with the target sequence in the DNA and to activate the nuclease domains for cleavage.

Plasmids expressing the required protein and RNA components of CRISPR/Cas9 can be introduced into cells by electroporation (p. 325). In cells from many organisms, the targeted gene is inactivated in 10% to 50% of the treated cells. If a genomic change (mutation) rather than a simple gene inactivation is required, it can be introduced by recombination when a DNA fragment encompassing the cleavage site and including the desired change enters the cell with the CRISPR/Cas9 plasmids. This recombination is often inefficient, but success can be improved somewhat by introducing a nick rather than a double-strand break at the target site (Fig. 9-24).

New applications for CRISPR/Cas9 are being developed rapidly. Potential therapeutic uses are still many years away, but developments are pointing the way to future treatments for genetic diseases, HIV disease, and many other human ailments.

(Cas9) and one associated RNA, consisting of gRNA and tracrRNA fused into a **single guide RNA (sgRNA)**. The guide sequence can be altered to target almost any genomic sequence **(Fig. 9-24)**. Cas9 has two separate nuclease domains: one domain cleaves the DNA strand paired with the sgRNA, and the other cleaves the opposite DNA strand. Inactivating one domain creates an

SUMMARY 9.2 Using DNA-Based Methods to Understand Protein Function

■ Proteins can be studied at the level of phenotypic, cellular, or molecular function.

■ DNA libraries can be a prelude to many types of investigations that yield information about protein function.

■ By fusing a gene of interest with genes that encode green fluorescent protein or epitope tags, researchers can visualize the cellular location of the gene product, either directly or by immunofluorescence.

■ The interactions of a protein with other proteins or RNA can be investigated with epitope tags and immunoprecipitation or affinity chromatography. Yeast two-hybrid analysis probes molecular interactions in vivo.

■ Microarrays can reveal changes in the expression patterns of genes in response to cellular stimuli, developmental stages, or shifting conditions.

■ The CRISPR/Cas9 system provides a powerful and inexpensive way to inactivate genes or to alter their sequence in order to investigate their function.

9.3 Genomics and the Human Story

Automation of the original Sanger DNA sequencing method led to the first complete sequencing of bacterial genomes in the 1990s. Two human genome sequences were completed in 2001. One resulted from a publicly funded effort led first by James Watson and later by Francis Collins. A parallel, private effort was led by Craig Venter. These accomplishments reflected more than a decade of intense effort coordinated in dozens of laboratories around the world, but they were just a beginning. With the advent of new sequencing technologies (Chapter 8), the time required to sequence a human genome has been reduced from years to days.

The human genome is an increasingly small part of the genome sequencing story. The genomes of thousands of other species have now been sequenced and made publicly available, providing a look at genomic complexity throughout the three domains of living organisms: Bacteria, Archaea, and Eukarya **(Fig. 9-25)**. Whereas many early sequencing projects focused on species commonly used in research laboratories, they now include species of practical, medical, agricultural, and evolutionary interest. Genomes from every known bacterial family have been sequenced. Completed eukaryotic

Francis S. Collins
[Source: Alex Wong/
Getty Images.]

J. Craig Venter
[Source: Shawn Thew/
Stringer/AFP/Getty Images.]

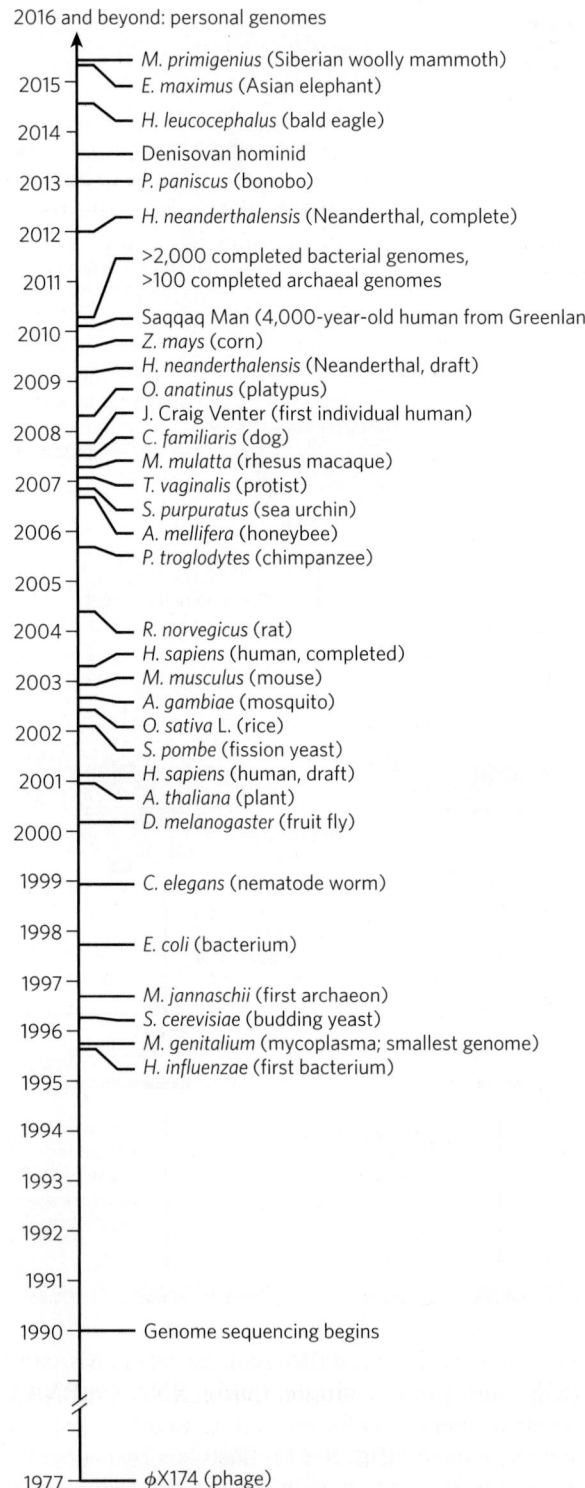

FIGURE 9-25 A genomic sequencing timeline.

genome sequences number in the thousands. Thousands of individual human genomes have been sequenced, and as the number grows, genome-based, personalized medicine is becoming a reality (Box 9-1). Genomes of extinct species such as *Homo neanderthalensis* and of humans who died in past millennia have been sequenced. Each genome sequence becomes an international resource for researchers. Collectively, they provide a source for broad comparisons that help pinpoint both variable and highly conserved gene segments, and allow the identification of genes that are unique to a species or group of species. Efforts to map genes, identify new proteins and disease-related genes, elucidate genetic patterns of medical interest, and trace our evolutionary history are among the many initiatives under way.

BOX 9-1   🜚 MEDICINE   Personalized Genomic Medicine

When twins Noah and Alexis Beery were born in California, they exhibited symptoms that elicited a diagnosis of cerebral palsy. Treatments seemed to have no effect. Not satisfied with the diagnosis or the treatment, the twins' parents, Joe and Retta Beery, took the twins, then age 5, to see a specialist in Michigan, who diagnosed them with a rare genetic condition called DOPA-responsive dystonia. A treatment regimen was devised that successfully suppressed the symptoms and allowed the twins to assume normal lives. However, at age 12, Alexis developed a severe cough and breathing difficulties that again seemed to threaten the child's survival. In one episode, paramedics had to revive her twice. The symptoms did not seem to be related to the dystonia. Might Noah be next? Frustrated and deeply worried, the twins' parents sought a complete genome sequence of both Noah and Alexis. This seemingly unusual step was a natural one for the Beery family. Joe was the chief information officer at Life Technologies, developers of sequencing technologies in use by many large DNA-sequencing centers. The cases of Noah and Alexis were taken up by Matthew Bainbridge and his team at the Baylor College of Medicine Human Genome Sequencing Center in Houston, Texas. The results proved decisive. The twins had mutations in their genomes that produced not only a deficiency in DOPA but also a potential deficiency in production of the hormone serotonin. A small adjustment in Alexis's therapy brought her life-threatening symptoms to an end, and the same therapy was given to her brother. Both siblings now lead normal lives.

The first draft human genome sequence was completed in 2001, after 12 years, at a cost of $3 billion. That cost has plummeted (Fig. 1), and newly completed human genomes are commonplace. The goal of a $1,000 human genome sequencing procedure is on the horizon and promises to make this technology widely available. Since most genomic changes that affect human health are thought to be in protein-coding genes (an assumption that may be challenged in years to come), a cheaper alternative is simply to sequence the 1% of the genome that represents the coding regions (exons) of genes, or the **exome**.

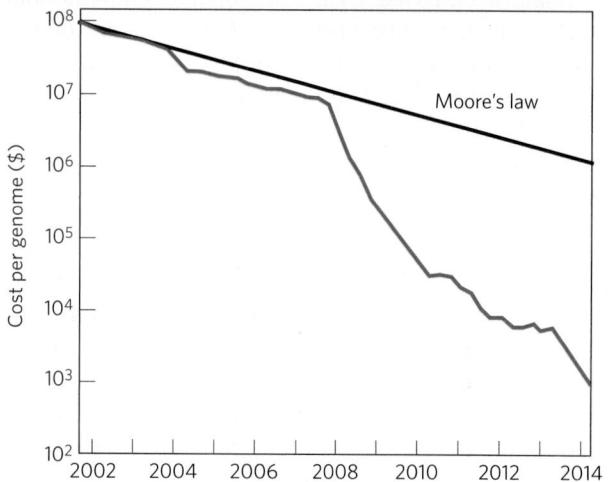

FIGURE 1 Since January 2008, the cost of human genome sequencing has been declining faster than the projected decline in the cost of processing data on computers (Moore's law). [Source: Data from the National Human Genome Research Institute.]

The first human genome sequence came from a haploid genome, derived from a DNA amalgam from several different humans. A high-quality reference genome was completed in 2004. Subsequent completed human genome sequences, many from individual diploid genomes, have demonstrated how much individual genetic variation exists. Relative to the reference sequence, a typical human has about 3.5 million single nucleotide polymorphisms (SNPs; see p. 347) and another few hundred thousand differences in the form of small insertions and deletions and changes in repeat copy numbers. About 60% of the SNPs are heterozygous, present on only one of two paired chromosomes. Only a small portion (5,000 to 10,000) of the SNPs affect the amino acid sequences of proteins encoded by genes.

This complexity ensures that, at least in the short term, successful diagnosis of a condition by whole genome sequencing will be the exception rather than the rule. However, human genomics is advancing rapidly. The number of success stories is increasing as the technology becomes more widely available and the capacity of genomic analysis to recognize causative genetic changes improves.

Annotation Provides a Description of the Genome

A genome sequence is simply a very long string of A, G, T, and C residues, all meaningless until interpreted. The process of **genome annotation** yields information about the location and function of genes and other critical sequences. Genome annotation converts the sequence into information that any researcher can use, and it is typically focused on genomic DNA encompassing genes that encode RNA and protein, the most common targets of scientific investigation. Every newly sequenced genome includes many genes—often 40% or more of the total—about which little or nothing is known.

Using Web-based tools that apply computational power to comparative genomics, scientists can define gene locations and assign tentative gene functions (where possible) based on similarity to genes previously studied in other genomes. The classic BLAST (Basic Local Alignment Search Tool) algorithm allows a rapid search of all genome databases for sequences related to one that a researcher is exploring, and is especially valuable for investigating the function of a particular gene. BLAST is one of many resources available at the NCBI (National Center for Biotechnology Information) site (www.ncbi.nlm.nih.gov), sponsored by the National Institutes of Health, and the Ensembl site (www.ensembl. org), cosponsored by the EMBL-EBI (European Molecular Biology Laboratory–European Bioinformatics Institute) and the Wellcome Trust Sanger Institute.

In every newly described genome sequence, the many genomic segments and genes that have not yet been characterized—that unknown 40% or so of the total—represent a special challenge. Elucidating the function of these genomic elements will probably take many decades. Many of the current experimental approaches again focus on protein-coding genes. A change in growth pattern or in other properties of an organism when a gene is inactivated provides information on the phenotypic function of the protein product of the gene. For several genomes, including those of *S. cerevisiae* and the plant *Arabidopsis thaliana*, gene knockout (inactivation) collections have been developed by genetic engineering. Each clone in an organism's collection has a different inactivated gene, and a large fraction of that organism's genes (except for a core of genes essential for life at all times) are represented in the knockout set. For single-celled organisms such as yeast, these collections are comprehensive. For complex multicellular animals such as mice, knockout collections are built painstakingly over time by many different research groups, one mutation at a time.

The Human Genome Contains Many Types of Sequences

All of these rapidly growing databases have the potential not only to fuel advances in all realms of biochemistry but to change the way we think about ourselves. What does our own genome, and comparisons with those of other organisms, tell us?

In some ways, we are not as complicated as we once imagined. Decades-old estimates that humans had about 100,000 genes within the approximately 3.2×10^9 bp of the human genome have been supplanted by the discovery that we have only about 20,000 protein-coding genes—less than twice the number in a fruit fly (13,600 genes), not many more than in a nematode worm (19,700 genes), and fewer than in a rice plant (38,000 genes).

In other ways, we are more complex than we previously realized. Many, if not most, eukaryotic genes contain one or more segments of DNA that do not code for the amino acid sequence of a polypeptide product. These nontranslated segments interrupt the otherwise colinear relationship between the gene's nucleotide sequence and the amino acid sequence of the encoded polypeptide. Such nontranslated DNA segments are called intervening sequences, or **introns**, and the coding segments are called **exons** (Fig. 9-26); few bacterial genes contain introns. The introns are spliced from a primary RNA transcript to generate a transcript that can be translated contiguously into a protein product (see Chapter 26). An exon often (but not always) encodes a single domain of a larger, multidomain protein. Humans share many protein domain types with plants, worms, and flies, but the domains are mixed and matched in more complex ways, increasing the variety of proteins found in our proteome. Alternative modes of gene expression and RNA splicing permit alternative combinations of exons, leading to the production of more than one protein from a single gene. Alternative splicing (Chapter 26) is far more common in humans and other vertebrates than in worms or bacteria, allowing greater complexity in the number and kinds of proteins generated.

In mammals and some other eukaryotes, the typical gene has a much higher proportion of intron DNA than exon DNA; in most cases, the function of introns is not clear. Less than 1.5% of human DNA is "protein-coding" or exon DNA, carrying information for protein products (Fig. 9-27a). However, when introns are included in the accounting, as much as 30% of the human genome consists of genes that encode proteins. Several efforts

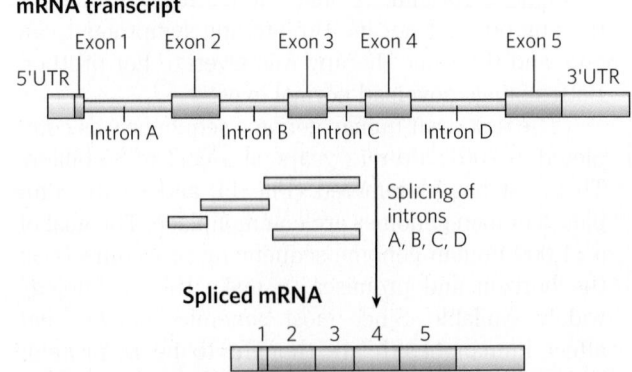

FIGURE 9-26 Introns and exons. This gene transcript contains five exons and four introns, along with 5′ and 3′ untranslated regions (5′UTR and 3′UTR). Splicing removes the introns to create an mRNA product for translation into protein.

(a) Human genome: DNA sequence types

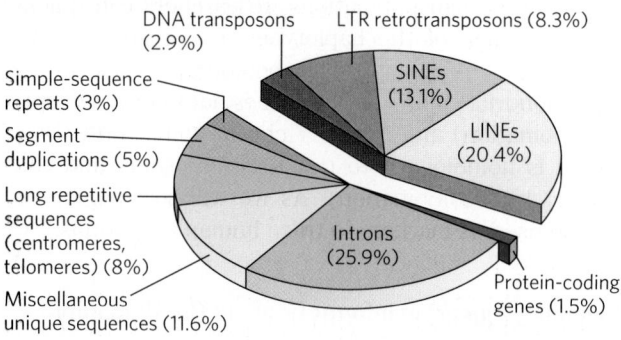

(b) Human genome: Protein-coding genes

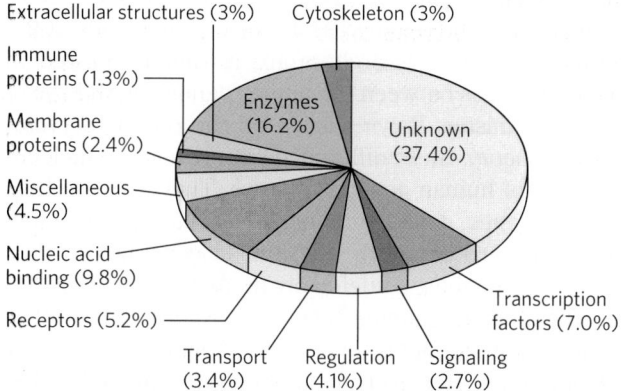

FIGURE 9-27 A snapshot of the human genome. (a) This pie chart shows the proportions of various types of sequences in our genome. The classes of transposons that represent nearly half of the total genomic DNA are indicated in shades of gray. LTR retrotransposons are retrotransposons with long terminal repeats (see Fig. 26-36). Long interspersed nuclear elements (LINEs) and short interspersed nuclear elements (SINEs) are special classes of particularly common DNA transposons. **(b)** The approximately 20,000 protein-coding genes in the human genome can be classified by the type of protein encoded. [Sources: (a) Data from T. R. Gregory, *Nature Rev. Genet.* 6:699, 2005. (b) Data from www.pantherdb.org.]

are under way to categorize protein-coding genes by type of function (Fig. 9-27b).

The relative paucity of protein-coding genes in the human genome leaves a lot of DNA unaccounted for. Much of the DNA that does not encode proteins is in the form of repeated sequences of several kinds. Perhaps most surprising is that about half the human genome is made up of moderately repeated sequences that are derived from **transposons**, segments of DNA, ranging from a few hundred to several thousand base pairs long, that can move from one location to another in the genome. Originally discovered in corn by Barbara McClintock, who called them transposable elements, transposons are a kind of molecular parasite. They make their home in the genomes of essentially every organism. Many transposons contain genes encoding the proteins that catalyze the transposition process itself, as described in more detail in Chapters 25 and 26. There are several classes of transposons in the human genome.

Many are strictly DNA segments, which have slowly increased in number over the millennia as a result of replication events coupled to the transposition process. Some, called retrotransposons, are closely related to retroviruses, transposing from one genomic location to another through RNA intermediates that are reconverted to DNA by reverse transcription. Some transposons in the human genome are active elements, moving at a low frequency, but most are inactive, evolutionary relics altered by mutations. Transposon movement can lead to the redistribution of other genomic sequences, and this has played a major role in human evolution.

Once the protein-coding genes (including exons and introns) and transposons are accounted for, perhaps 25% of the total DNA remains. As a follow-up to the Human Genome Project, the ENCODE initiative was launched by the U.S. National Human Genome Research Institute in 2003 to identify functional elements in the human genome. The work of the worldwide consortium of research groups engaged in the ENCODE initiative has revealed that the vast majority (>80%, including most transposons) of the DNA in the human genome is either transcribed into RNA in at least one type of cell or tissue or is involved in some functional aspect of chromatin structure. Much of the noncoding (nontranscribed) DNA in the remaining 20% contains regulatory elements that affect the expression of the 20,000 protein-coding genes and the many additional genes encoding functional RNAs. Many mutations (SNPs; described below) associated with human genetic diseases lie in this noncoding DNA, probably affecting regulation of one or more genes. As described in Chapters 26 and 27, new classes of functional RNAs are being discovered at a rapid pace. Many of these functional RNAs, now being identified by a variety of screening methods, are produced by RNA-coding genes whose existence was previously unsuspected.

About 3% or so of the human genome consists of highly repetitive sequences referred to as **simple-sequence repeats (SSRs)**. Generally less than 10 bp long, an SSR is sometimes repeated millions of times per cell, distributed in short segments of tandem repeats. The most prominent examples of SSR DNA are found in centromeres and telomeres (see Chapter 24). Human telomeres, for example, consist of up to 2,000 contiguous repeats of the sequence GGTTAG. Additional, shorter repeats of simple sequences also occur throughout the genome. These isolated segments of repeated sequences, often containing up to a few dozen tandem repeats of a simple sequence, are called **short tandem repeats (STR)**. Such sequences are the targets of the technologies used in forensic DNA analysis (see Box 8-1).

What does all this information tell us about the similarities and differences among individual humans? Within the human population there are millions of single-base variations, called **single nucleotide polymorphisms**, or **SNPs** (pronounced "snips"). Each person differs from the next by, on average, 1 in every 1,000 bp.

Many of these variations are in the form of SNPs, but the human population also has a wide range of larger deletions, insertions, and small rearrangements. From these often subtle genetic differences comes the human variety we are all aware of—such as differences in hair color, stature, foot size, eyesight, allergies to medication, and (to some unknown degree) behavior.

The process of genetic recombination during meiosis tends to mix and match these small genetic variations so that different combinations of genes are inherited (see Chapter 25). However, groups of SNPs and other genetic differences that are close together on a chromosome are rarely affected by recombination and are usually inherited together; such a grouping of multiple SNPs is known as a **haplotype**. Haplotypes provide convenient markers for certain human populations and individuals within populations.

Defining a haplotype requires several steps. First, positions that contain SNPs in the human population are identified in genomic DNA samples from multiple individuals **(Fig. 9-28a)**. Each SNP in a prospective haplotype may be separated from the next SNP by several thousand base pairs and still be regarded as "nearby" in the context of chromosomes that extend for millions of base pairs. Second, a set of SNPs typically inherited together is defined as a haplotype (Fig. 9-28b); each haplotype consists of the particular bases found at the various SNP positions within the defined set. Finally, tag SNPs—a subset of SNPs that define an entire haplotype—are chosen to uniquely identify each haplotype (Fig. 9-28c).

By sequencing just these tag positions in genomic samples from human populations, researchers can quickly identify which of the haplotypes are present in each individual. Especially stable haplotypes exist in the mitochondrial genome (which does not undergo meiotic recombination) and on the Y chromosome (only 3% of which is homologous to the X chromosome and thus subject to recombination). As we will see, haplotypes can be used as markers to trace human migrations.

Genome Sequencing Informs Us about Our Humanity

Genome sequencing projects allow researchers to identify conserved genetic elements that are of functional significance, including conserved exon sequences, regulatory regions, and other genomic features (such as centromeres and telomeres). In the ongoing study of the human genome, researchers are further interested in the differences between the human genome and those of other organisms. Relying again on the power of evolutionary theory, these differences can reveal the molecular basis of human genetic diseases. They can also help identify genes, gene alterations, and other genomic features that are unique to the human genome and thus likely to contribute to definably human characteristics.

The human genome is very closely related to other mammalian genomes over large segments of every chromosome. However, for a genome measured in billions of base pairs, differences of just a few percent can add up to millions of genetic distinctions. Searching

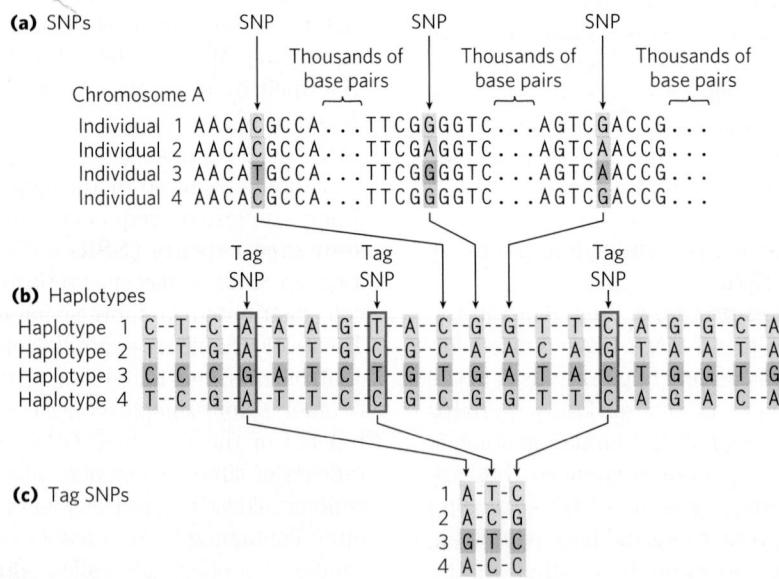

FIGURE 9-28 Haplotype identification. (a) The positions of SNPs in the human genome can be identified in genomic samples. The SNPs can be in any part of the genome, whether or not it is part of a known gene. **(b)** Groups of SNPs are compiled into a haplotype. The SNPs vary in the overall human population, as in the four fictitious individuals shown here, but the SNPs chosen to define a haplotype are often the same in most individuals of a particular population. **(c)** A few SNPs are chosen as haplotype-defining (tag SNPs, outlined in red), and these are used

to simplify the process of identifying an individual's haplotype (by sequencing 3 instead of 20 loci). **(c)** For example, if the positions shown here were sequenced, an A-T-C haplotype might be characteristic of a population native to one location in northern Europe, whereas G-T-C might be the prevailing sequence in a population in Asia. Multiple haplotypes of this kind are used to trace prehistoric human migrations. [Source: Information from International HapMap Consortium, *Nature* 426:789, 2003, Fig. 1.]

among these, and making use of comparative genomics techniques, researchers can begin to explore the molecular basis of our large brain, language skills, tool-making ability, or bipedalism.

The genome sequences of our closest biological relatives, the chimpanzee (*Pan troglodytes*) and bonobo (*Pan paniscus*), offer some important clues, and we can use them to illustrate the comparative process. Human and chimpanzee shared a common ancestor about 7 million years ago. Genomic differences between the species, including SNPs and larger genomic rearrangements such as inversions, deletions, and fusions, can be used to construct a phylogenetic tree **(Fig. 9-29a)**. Over the course of evolution, segments of chromosomes may become inverted as a result of a segmental duplication, transposition of one copy to another arm of the same chromosome, and recombination between them (Fig. 9-29b); such inversions have occurred in the human lineage on chromosomes 1, 12, 15, 16, and 18. Two chromosomes found in other primate lineages have been fused to form human chromosome 2 (Fig. 9-29c). The human lineage thus has 23 chromosome pairs rather than the 24 pairs typical of simians. Once this fusion appeared in the line leading to humans, it would have represented a major barrier to interbreeding with other primates that lacked it.

If we look only at base-pair changes, the published human and chimpanzee genomes differ by only 1.23% (compared with the 0.1% variance from one human to another). Some variations are at positions where there is a known polymorphism in either the human or the chimpanzee population, and these are unlikely to reflect a species-defining evolutionary change. When we ignore these positions, the differences amount to about 1.06%, or about 1 in 100 bp. This small fraction translates into more than 30 million base-pair differences, some of which affect protein function and gene regulation. Humans are approximately as closely related to bonobos as to chimpanzees.

The genomic rearrangements that help distinguish chimpanzee and human include 5 million short insertions or deletions involving a few base pairs each, as well as a substantial number of larger insertions, deletions, inversions, and duplications that can involve many thousands of base pairs. When transposon insertions—a major source of genomic variation—are added to the list, the differences between the human and chimpanzee genomes increase. The chimpanzee genome has two classes of retrotransposons that are not present in the human genome (see Chapter 26). Other types of rearrangements, especially segmental duplications, are also common in primate lineages. Duplications of chromosomal segments can lead to changes in the expression of genes contained in these segments. There are about 90 million bp of such differences between human and chimpanzee, representing another 3% of these genomes. Each species has segments of DNA, constituting 40 to 45 million bp, that are entirely unique to that particular genome,

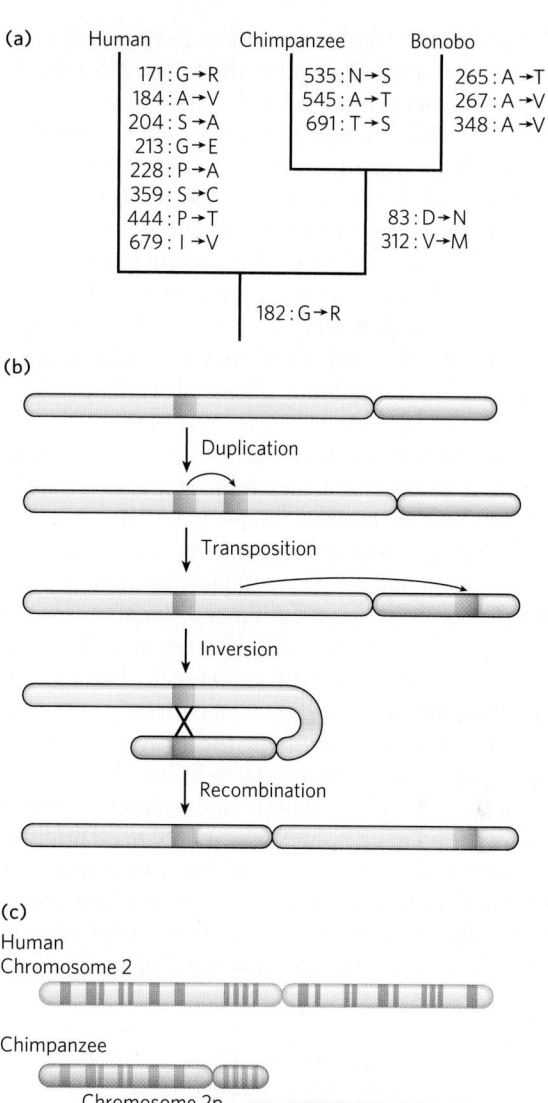

FIGURE 9-29 Genomic alterations in the human lineage. (a) This evolutionary tree is for the progesterone receptor, which helps regulate many events in reproduction. The gene encoding this protein has undergone more evolutionary alterations than most. Amino acid changes associated uniquely with human, chimpanzee, and bonobo are listed beside each branch (with the residue number). **(b)** One of the multistep processes that can lead to the inversion of a chromosome segment. A gene or a chromosome segment is duplicated, then moved to another chromosomal location by transposition. Recombination of the two segments may result in inversion of the DNA between them. **(c)** The genes on chimpanzee chromosomes 2p and 2q are homologous to those on human chromosome 2, implying that two chromosomes fused at some point in the line leading to humans. Homologous regions can be visualized as bands created in metaphase by certain dyes, as shown here. [Source: (a) Information from C. Chen, *Mol. Phylogenet. Evol.* 47:637, 2008.]

with larger chromosomal insertions, duplications, and other rearrangements affecting more base pairs than do single-nucleotide changes. Thus, in all, chimpanzee and human differ over about 4% of their genomes.

Sorting out which genomic distinctions are relevant to features that are uniquely human is a daunting task.

If one assumes a similar rate of evolution in the chimpanzee and human lines after they diverged from their common ancestor, half the changes represent chimpanzee lineage changes and half represent human lineage changes. By comparing both genome sequences with those of more distantly related species referred to as **outgroups**, we can determine which variant was present in the common ancestor. Consider a locus, X, where there is a difference between the human and chimpanzee genomes **(Fig. 9-30)**. The lineage of the orangutan, an outgroup, diverged from that of chimpanzee and human prior to the common ancestor of chimpanzee and human. If the sequence at locus X is identical in orangutan and chimpanzee, this sequence was probably present in the chimpanzee and human ancestor, and the sequence seen in humans is specific to the human lineage. Sequences that are identical in human and orangutan can be eliminated as candidates for human-specific genomic features. The importance of comparisons with closely related outgroups has given rise to new efforts to sequence the genomes of orangutan, macaque, and many other primate species. Comparison of the human and bonobo genomes is refining the analysis of genes and alleles of special significance to humans.

The search for the genetic underpinnings of special human characteristics, such as our enhanced brain function, can benefit from two complementary approaches. The first searches for genomic regions where extreme changes have occurred, such as genes that have been duplicated many times or large genomic segments not present in other primates. The second approach looks at genes known to be involved in relevant human disease conditions. For brain function, for example, one would examine genes that, when mutated, contribute to cognitive or mental disorders.

Observed genetic changes are sometimes concentrated in a particular gene or region, suggesting that these genes or regions played a role in the evolution of special human characteristics. In principle, human-specific traits could reflect changes in protein-coding genes, in regulatory processes, or both. A few classes of protein-coding genes show evidence of accelerated divergence (more amino acid substitutions than in most other genes). These include genes involved in chemosensory perception, immune function, and reproduction. In these cases, rapid evolution is evident in virtually all primate lines, reflecting physiological functions that are critical to all primate species. Another class of genes showing evidence of accelerated evolution is those encoding transcription factors—proteins involved in the expression of other genes (see Chapter 26).

Notably, analyses of the human lineage have not detected an increased rate of genetic change in protein-coding genes involved in brain development or size. In primates, most genes that function uniquely in the brain are even more highly conserved than genes functioning in other tissues, perhaps due to some special constraints related to brain biochemistry. However, some differences in gene expression between humans and other primates are observed. For example, the gene encoding the enzyme glutamate dehydrogenase, which plays an important role in neurotransmitter synthesis, has an increased copy number in humans due to gene duplication. Genomic regions related to gene regulation have disproportionately high numbers of changes in genes involved in neural development and nutrition. A variety of RNA-coding genes, some with expression concentrated in the brain, also show evidence of accelerated evolution **(Fig. 9-31)**. Many of these are probably involved in regulating the expression of other genes. As we continue to discover many new classes of RNA (see Chapter 26), we are likely to radically change our perspective on how evolution alters the workings of living systems.

Genome Comparisons Help Locate Genes Involved in Disease

One of the motivations for the Human Genome Project was its potential for accelerating the discovery of genes underlying genetic diseases. That promise has been fulfilled: more than 4,500 human mutation phenotypes, mostly associated with genetic diseases, have been mapped to particular genes. Some disease-gene hunters caution that, so far, the work may have uncovered mostly the relatively easy cases and that many challenges remain.

The main approach during the past two decades uses **linkage analysis**, yet another approach derived from evolutionary biology. In brief, the gene involved in a

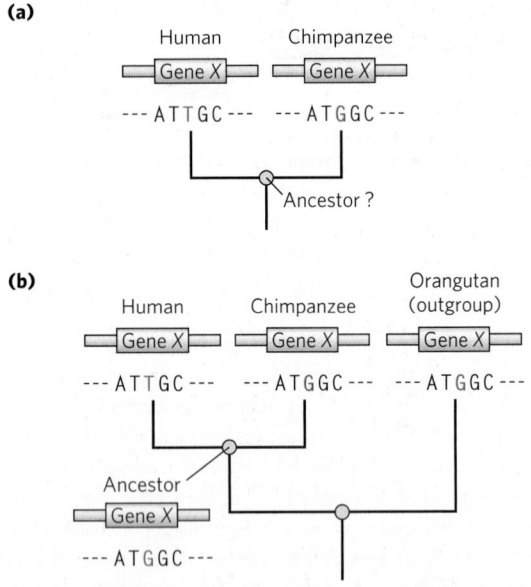

(a)

(b)

FIGURE 9-30 Determination of sequence alterations unique to one ancestral line. (a) Sequences from the same hypothetical gene in human and chimpanzee are compared. The sequence of this gene in the two species' last common ancestor is unknown. **(b)** The orangutan genome is used as an outgroup. The sequence of the orangutan gene is found to be identical to that of the chimpanzee gene. This means that the mutation causing the difference between human and chimpanzee almost certainly occurred in the line leading to modern humans, and the common ancestor of human and chimpanzee (and orangutan) had the variant now found in chimpanzee.

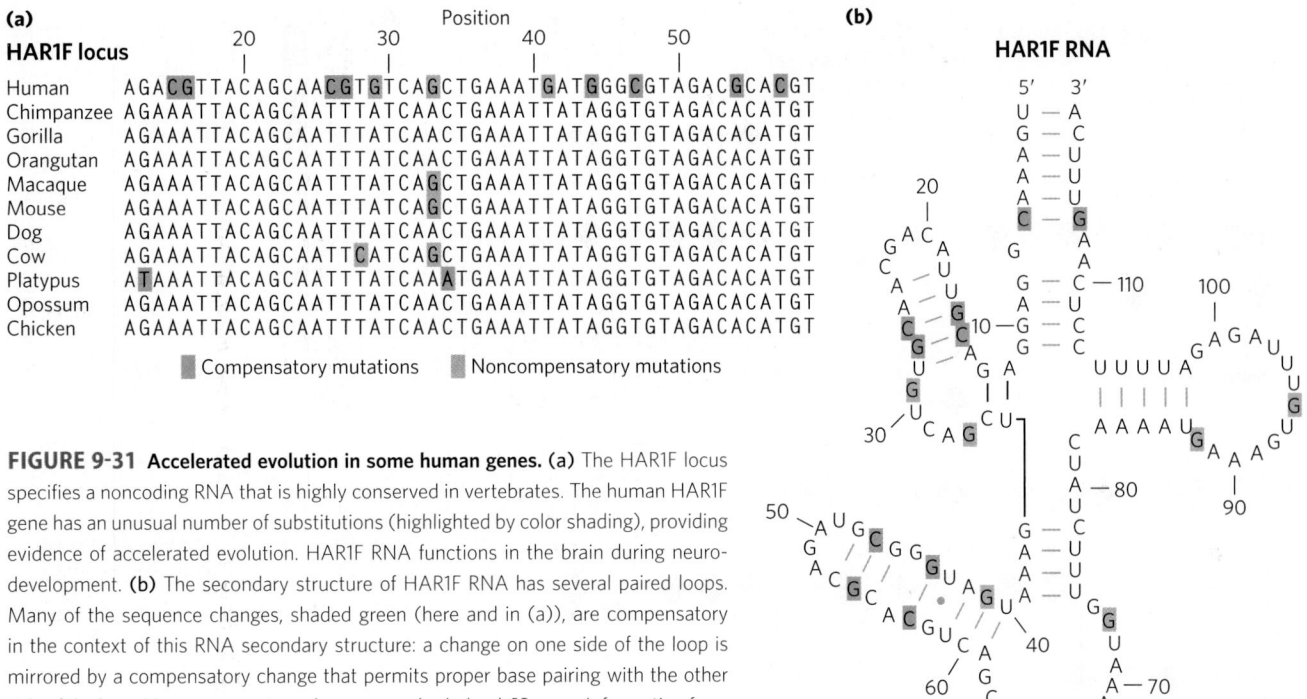

FIGURE 9-31 Accelerated evolution in some human genes. (a) The HAR1F locus specifies a noncoding RNA that is highly conserved in vertebrates. The human HAR1F gene has an unusual number of substitutions (highlighted by color shading), providing evidence of accelerated evolution. HAR1F RNA functions in the brain during neurodevelopment. **(b)** The secondary structure of HAR1F RNA has several paired loops. Many of the sequence changes, shaded green (here and in (a)), are compensatory in the context of this RNA secondary structure: a change on one side of the loop is mirrored by a compensatory change that permits proper base pairing with the other side of the loop. Noncompensatory changes are shaded red. [Source: Information from T. Marques-Bonet, *Annu. Rev. Genomics Hum. Genet.* 10:355, 2009.]

disease condition is mapped relative to well-characterized genetic polymorphisms that occur throughout the human genome. We can illustrate this by describing the search for one gene involved in early-onset Alzheimer disease. About 10% of all cases of Alzheimer disease in the United States result from an inherited predisposition. Several different genes have been discovered that, when mutated, can lead to early onset of the disease. One such gene, *PS1*, encodes the protein presenilin-1, and its discovery made heavy use of linkage analysis. The search begins with large families having multiple individuals affected by a particular disease—in this case, Alzheimer disease. Two of the many family pedigrees used to search for this gene in the early 1990s are shown in **Figure 9-32a**. In studies of this type, DNA samples are collected from both affected and unaffected family members. Researchers first localize the region associated with a disease to a specific chromosome by comparing the genotypes of individuals with and without the disease, focusing especially on close family members. The specific points of comparison are sets of well-characterized SNP loci mapped to each chromosome, as identified by the Human Genome Project. By identifying the SNPs that are most often inherited with the disease-causing gene, investigators can gradually localize the responsible gene to a single chromosome. In the case of the *PS1* gene, coinheritance was strongest with markers on chromosome 14 (Fig. 9-32b).

Chromosomes are very large DNA molecules, and localizing the gene to one chromosome is only a small part of the battle. It is established that this chromosome contains a mutation that gives rise to the disease, but in every individual human genome, every chromosome houses thousands of SNPs and other changes. Simply sequencing the entire chromosome would be unlikely to

reveal the SNP or other change associated with the disease. Instead, investigators rely on statistical methods that correlate the inheritance of additional, more closely spaced polymorphisms with the occurrence of the disease, focusing on a denser panel of polymorphisms known to occur on the chromosome of interest. The more closely a marker is located to a disease gene, the more likely it is to be inherited along with that gene. This process can pinpoint a region of the chromosome that contains the gene. However, the region may still encompass many genes. In our example, linkage analysis indicated that the disease-causing gene, *PS1*, was somewhere near the SNP locus D14S43 (Fig. 9-32c).

The final steps in identifying the gene use the human genome databases. The local region containing the gene is examined, and the genes within it are identified. DNA from many individuals, some who have the disease and some who do not, is sequenced over this region. As the DNA in this region is sequenced from increasing numbers of individuals, gene variants that are consistently present in individuals with the disease and absent in unaffected individuals can be identified. The search can be aided by an understanding of the function of the genes in the target region, because particular metabolic pathways may be more likely than others to produce the disease state. In 1995, the chromosome 14 gene associated with Alzheimer disease was identified as *S182*. The product of this gene was given the name presenilin-1, and the gene was subsequently renamed *PS1*.

Many human genetic diseases are caused by mutations in a single gene or in sequences involved in its regulation. Several different mutations in a particular gene, all leading to the same or related genetic condition, may be present in the human population. For

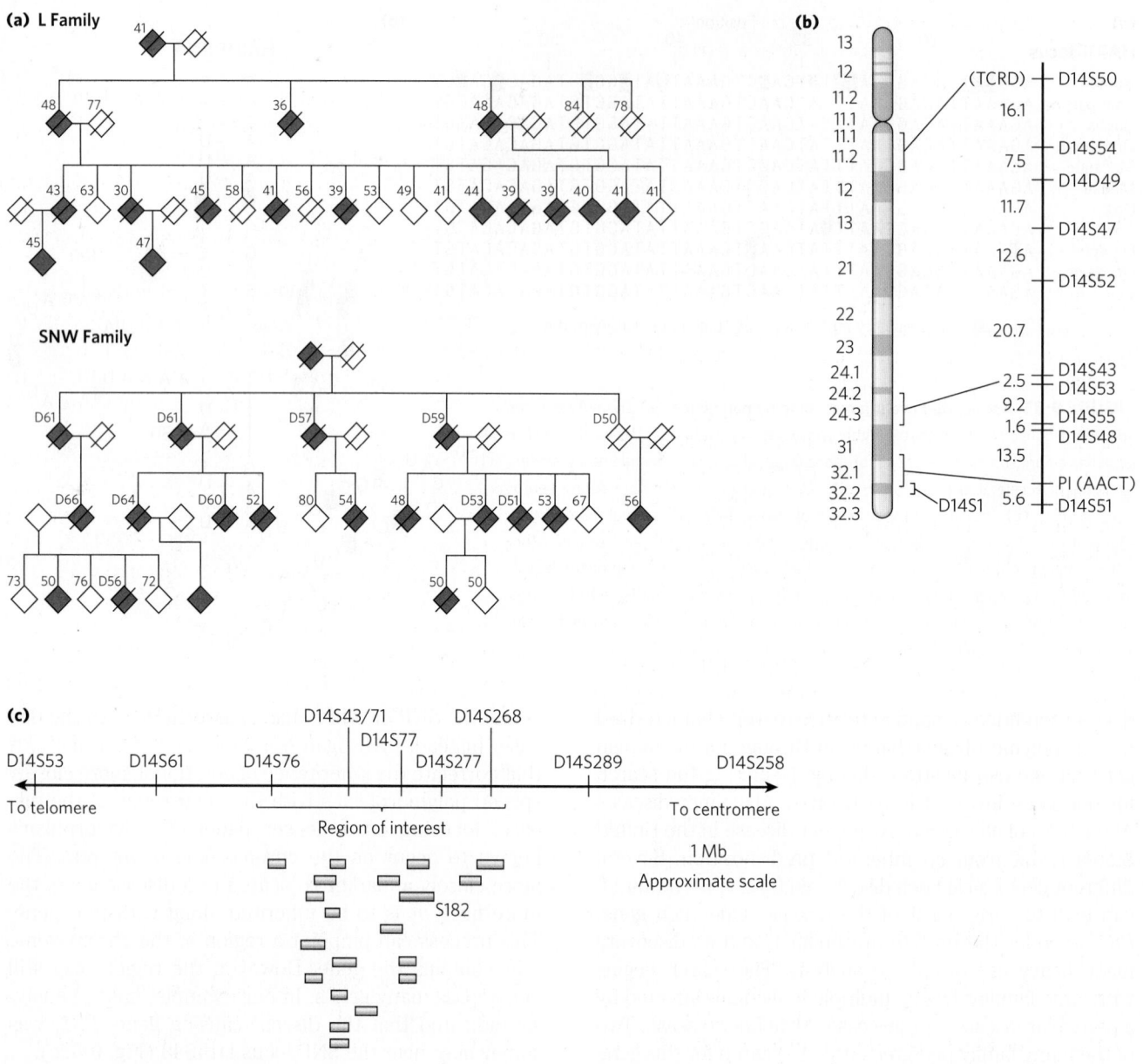

FIGURE 9-32 Linkage analysis in the discovery of disease genes. (a) These pedigrees for two families affected by early-onset Alzheimer disease are based on the data available at the time of the study. Red symbols represent affected individuals; slashes indicate deaths either before or soon after the study. The number above each symbol is the person's age at the time of the study or at time of death (indicated with a D). To protect family privacy, gender is not indicated. **(b)** Chromosome 14, with bands created by certain dyes. Chromosome marker positions are shown at the right, with the genetic distance between them in centimorgans, a genetic distance measure-

ment that reflects the frequency of recombination between markers. *TCRD* (T-cell receptor delta) and *PI* (AACT (α1-antichymotrypsin)), two genes with alterations in the human population, were used along with SNPs as markers in chromosome mapping. **(c)** By comparing DNA from affected and unaffected family members, researchers eventually defined a region of interest near marker D14S43 that contains 19 expressed genes. The gene labeled *S182* (red) encodes presenilin-1. (1 Mb = 10⁶ base pairs.) [Sources: (a, b) Information from G. D. Schellenberg et al., *Science* 258:668, 1992. (c) Information from R. Sherrington et al., *Nature* 375:754, 1995.]

example, there are several variants of *PS1*, all giving rise to a much increased risk of early-onset Alzheimer disease. Another, more extreme example is the several genes encoding different hemoglobins: more than 1,000 known mutational variants are present in the human population. Some of these variants are innocuous; some cause diseases ranging from sickle-cell disease to thalassemias. The inheritance of particular mutant genes may be concentrated in families or in isolated populations.

More complex are cases in which a disease condition is caused by mutations in two different genes

(neither of which, alone, causes the disease), or in which a particular condition is enhanced by an otherwise innocuous mutation in another gene. Identifying the genes and mutations responsible for these digenic diseases is exceedingly difficult, and sometimes such diseases can be documented only within small, isolated, and highly inbred populations.

Modern genome databases are opening up alternative paths to the identification of disease genes. In many cases, we already have biochemical information about the disease. In the case of early-onset Alzheimer disease, an accumulation of the amyloid β-protein in limbic and association cortices of the brain is at least partly responsible for the symptoms. Defects in presenilin-1 (and in a related protein, presenilin-2, encoded by a gene on chromosome 1) lead to the elevated cortical levels of amyloid β-protein. Focused databases are being developed that catalog such functional information on the protein products of genes and on protein-interaction networks and SNP locations, along with other data. The result is a streamlined path to the identification of candidate genes for a particular disease. If a researcher knows a little about the kinds of enzymes or other proteins likely to contribute to disease symptoms, these databases can quickly generate a list of genes known to encode proteins with relevant functions, a list of additional uncharacterized genes with orthologous or paralogous relationships to these genes, a list of proteins known to interact with the target proteins or orthologs in other organisms, and a map of gene positions. Often, with the aid of data from some selected family pedigrees, a short list of potentially relevant genes can be rapidly determined.

These approaches are not limited to human diseases. The same methods can be used to identify the genes involved in diseases—or genes that produce desirable characteristics—in other animals and in plants. Of course, they can also be used to track down genes involved in any observable trait that a researcher might be interested in. ∎

Genome Sequences Inform Us about Our Past and Provide Opportunities for the Future

About 70,000 years ago, a small group of humans in Africa looked out across the Red Sea to Asia. Perhaps encouraged by some innovation in small boat construction, or driven by conflict or famine, or simply curious, they crossed the water barrier. That initial colonization, involving maybe 1,000 individuals, began a journey that did not stop until humans reached Tierra del Fuego (at the southern tip of South America), many thousands of years later. In the process, established populations from previous hominid expansions into Eurasia, including *Homo neanderthalensis*, were displaced. The Neanderthals disappeared, just as other hominid lines had disappeared before them.

The story of how modern humans first appeared in Africa a few hundred thousand years ago, and their migrations as they eventually radiated out of Africa, is written in our DNA. Genomic sequences from multiple species have brought both primate and hominid evolution into sharper focus. Using haplotypes present in extant human populations, we can trace the migrations of our intrepid ancestors across the planet **(Fig. 9-33a)**. The Neanderthals were not simply displaced. Some mingling occurred (Fig. 9-33b). Using sensitive PCR-based methods, we now have a nearly complete sequence of the Neanderthal genome (Box 9-2). We know that about 5% of the genome

---

BOX 9-2 Getting to Know Humanity's Next of Kin

Modern humans and Neanderthals coexisted in Europe and Asia as recently as 30,000 years ago. The human and Neanderthal ancestral populations diverged about 370,000 years ago, before the appearance of anatomically modern humans. Neanderthals used tools, lived in small groups, and buried their dead. Of the known hominid relatives of modern humans, Neanderthals are the closest. For hundreds of millennia, they inhabited large parts of Europe and western Asia (Fig. 1). If the chimpanzee genome can tell us something about what it is to be human, the Neanderthal genome can tell us more. Buried in the bones and other remains taken from burial sites are fragments of Neanderthal genomic DNA.

FIGURE 1 Neanderthals occupied much of Europe and western Asia until about 30,000 years ago. Major Neanderthal archaeological sites are shown here. (Note that the group was named for the site at Neanderthal in Germany.)

Technologies developed for use in forensic science (see Box 8-1) and studies of ancient DNA have been combined to initiate a Neanderthal genome project.

This endeavor is unlike the genome projects aimed at extant species. The Neanderthal DNA is

(Continued on next page)

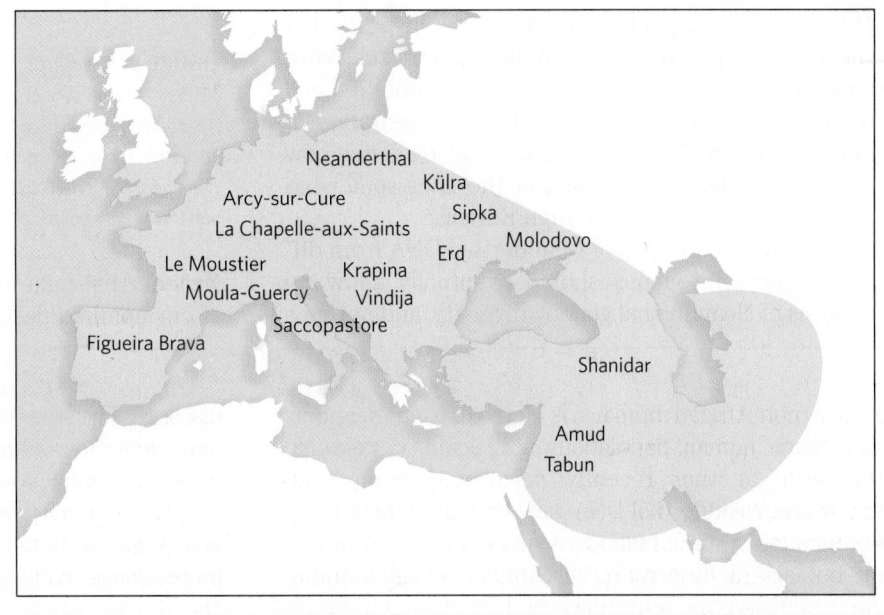

BOX 9-2   Getting to Know Humanity's Next of Kin (*Continued*)

present in small amounts, and it is contaminated with DNA from other animals and bacteria. How does one get at it, and how can one be certain that the sequences really came from Neanderthals? The answers have been revealed by innovative applications of biotechnology. In essence, the small quantities of DNA fragments found in a Neanderthal bone or other remains are cloned into a library, and the cloned DNA segments are sequenced at random, contaminants and all. The sequencing results are compared with the existing human genome and chimpanzee genome databases. Segments derived from Neanderthal DNA are readily distinguished from segments derived from bacteria or insects by computerized analysis, because they have sequences closely related to human and chimpanzee DNA. Once a collection of Neanderthal DNA segments is sequenced, the sequences can be used as probes to identify sequence fragments in ancient samples that overlap with these known fragments. The potential problem of contamination with the closely related modern human DNA can be controlled for by examining mitochondrial DNA. Human populations have readily identifiable haplotypes (distinctive sets of genomic differences; see Fig. 9-28) in their mitochondrial DNA, and analysis of Neanderthal samples has shown that Neanderthals' mitochondrial DNA has its own distinct haplotypes. The presence in the Neanderthal samples of some base-pair differences that are found in the chimpanzee database but not in the human database is more evidence that nonhuman hominid sequences are being found.

A high-quality Neanderthal genomic sequence has been completed, and more are on the way. The data provide evidence that modern humans and the Neanderthals who were the source of this DNA shared a common ancestor about 700,000 years ago (Fig. 2). Analysis of mitochondrial DNA suggests that the two groups continued on the same track, with some gene flow between them, for about 300,000 more years. The lines split with the appearance of anatomically modern humans, although evidence now exists for some intermingling of the lines somewhat later as humans spread through Eurasia.

Expanded libraries of Neanderthal DNA from different sets of remains should eventually allow an analysis of Neanderthal genetic diversity, and perhaps

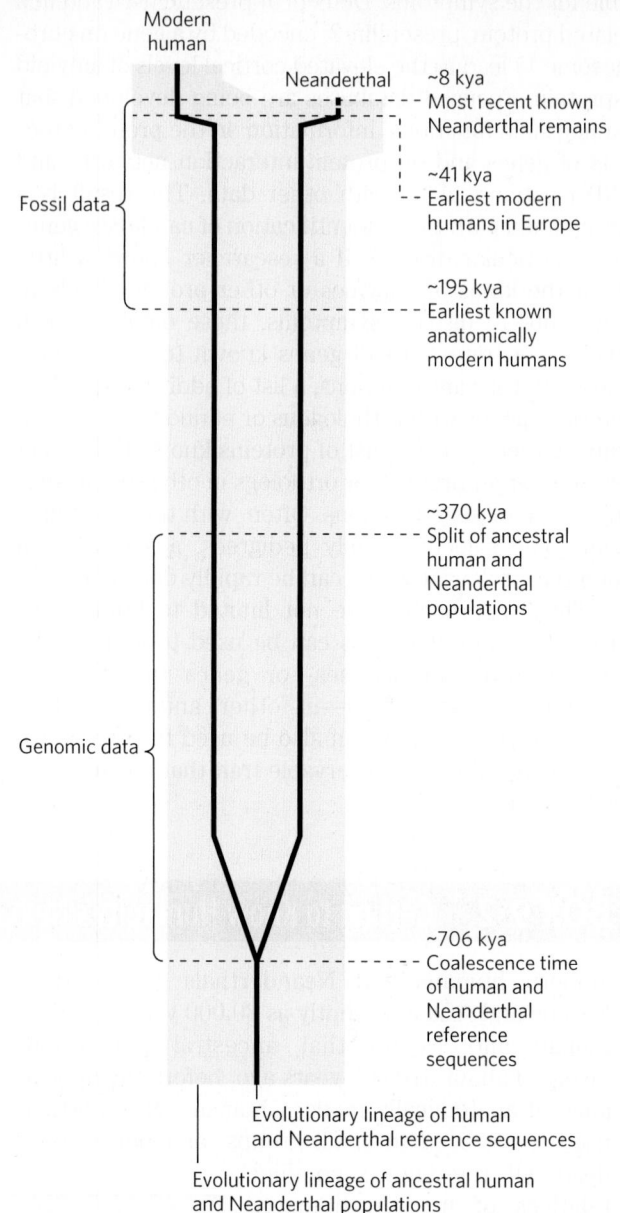

FIGURE 2 This timeline shows the divergence of human and Neanderthal genome sequences (black lines) and of ancestral human and Neanderthal populations (yellow screen). Genomic data provide evidence for some intermingling of the populations up to about 45,000 years ago. Key events in human evolution are noted. [Source: Information from J. P. Noonan et al., *Science* 314:1113, 2006.]

Neanderthal migrations, providing a fascinating look at our hominid past.

of most non-African humans is derived from Neanderthals. Some human populations also acquired genomic DNA from another recently discovered group, the Denisovans. Neanderthal DNA gave humans a more complex immune system, making us more resistant to infection but also a little more susceptible to autoimmune diseases. The story of our past is gradually taking shape as more genomes, of humans alive today and those who lived in past millennia, are being assembled.

The medical promise of personal genomic sequences grows as sequencing costs continue to decline and more genes underlying inherited diseases are defined.

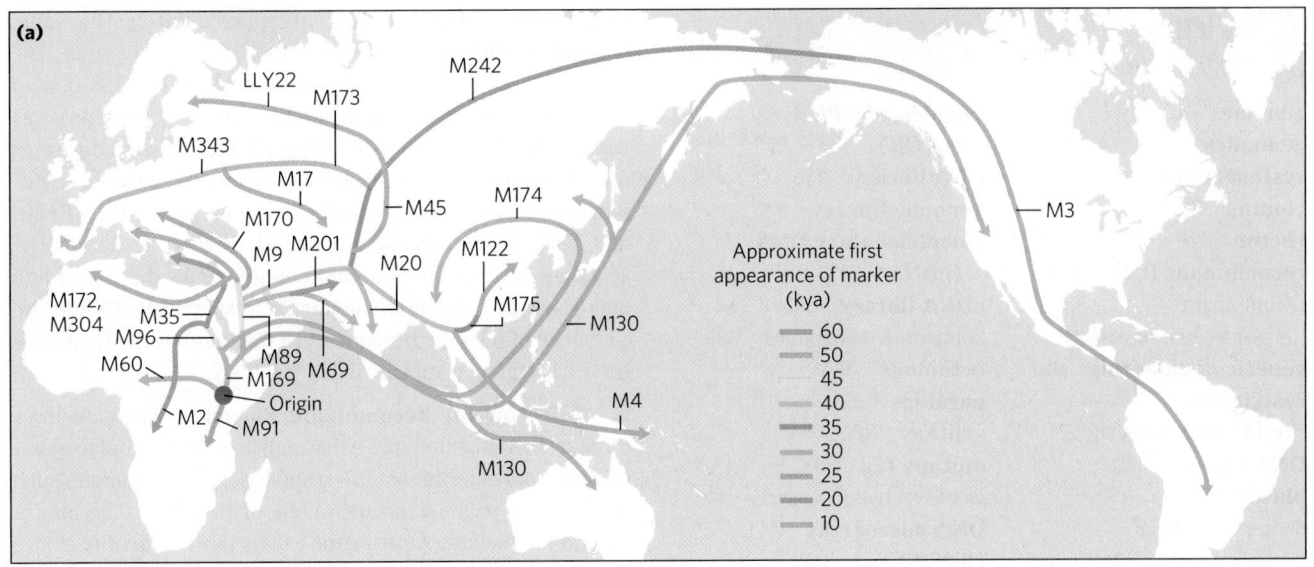

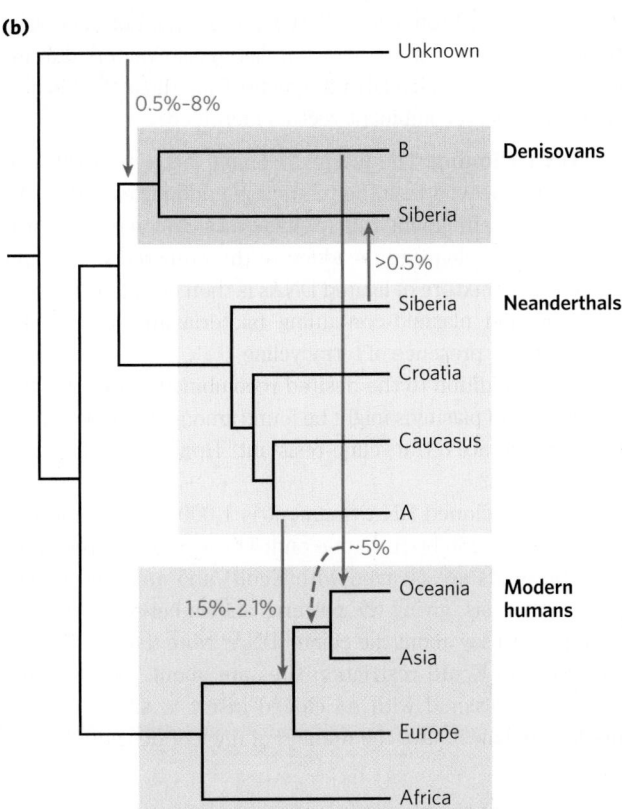

FIGURE 9-33 The paths of human migrations. (a) When a small part of a human population migrates away from a larger group, it takes only part of the population's overall genetic diversity with it. Thus, some haplotypes are present in the migrating group but many are not. At the same time, mutations can create novel haplotypes over time. This map was generated from an analysis of genetic markers (defined haplotypes with M or LLY numbers) on the Y chromosome. The genetic samples were taken from indigenous populations long established at geographic points along the routes shown. Haplotypes that appear suddenly along a migration path, reflecting new changes (mutations) in particular SNP genomic locations in certain isolated populations, are called "founder events." These enable researchers to trace migrations from that point, as other populations with the new haplotype were probably descended from the founder population. The abbreviation *kya* means "thousand years ago." **(b)** Human migrations eventually displaced several closely related hominid groups, but not before some intermingling occurred. This tree illustrates gene-flow events documented from detailed genomic sequences of modern and ancient humans, as well as of Neanderthals and Denisovans. DNA from an unknown group of Neanderthals (A) is recorded in the genomes of all humans with some Eurasian heritage. A transfer of DNA from an unknown ancestor to the Denisovan line (B) contributed to ancestors of present-day individuals native to Australia and Pacific islands (Oceania). [Sources: (a) Information from G. Stix, *Sci. Am.* 299 (July):56, 2008. (b) Information from S. Pääbo, *Cell* 157:216, 2014.]

Knowledge of genomic sequences also provides the prospect of altering them. It is now commonplace to engineer the DNA sequences of organisms ranging from bacteria and yeast to plants and mammals, for research and commercial purposes. Efforts to cure inherited human diseases by human gene therapy have not yet lived up to their potential, but technologies for gene delivery are constantly being improved. Few scientific disciplines will affect the future of our species more than modern genomics.

SUMMARY 9.3 Genomics and the Human Story

■ About 30% of the DNA in the human genome is in the exons and introns of genes that encode proteins.

Nearly half of the DNA is derived from parasitic transposons. Much of the rest encodes RNAs of many types. Simple-sequence repeats make up the centromere and telomeres.

■ The gene alterations that define humanity can be discerned in part through comparative genomics, using other primates.

■ Comparative genomics is also used to locate the gene alterations that define inherited diseases, and the technique can be used to study the evolution and migration of our human ancestors over millennia.

Key Terms

Problems

1. Engineering Cloned DNA When joining two or more DNA fragments, a researcher can adjust the sequence at the junction in a variety of subtle ways, as seen in the following exercises.

(a) Draw the structure of each end of a linear DNA fragment produced by an EcoRI restriction digest (include those sequences remaining from the EcoRI recognition sequence).

(b) Draw the structure resulting from the reaction of this end sequence with DNA polymerase I and the four deoxynucleoside triphosphates (see Fig. 8-34).

(c) Draw the sequence produced at the junction that arises if two ends with the structure derived in (b) are ligated (see Fig. 25-16).

(d) Draw the structure produced if the structure derived in (a) is treated with a nuclease that degrades only single-stranded DNA.

(e) Draw the sequence of the junction produced if an end with structure (b) is ligated to an end with structure (d).

(f) Draw the structure of the end of a linear DNA fragment that was produced by a PvuII restriction digest (include those sequences remaining from the PvuII recognition sequence).

(g) Draw the sequence of the junction produced if an end with structure (b) is ligated to an end with structure (f).

(h) Suppose you can synthesize a short duplex DNA fragment with any sequence you desire. With this synthetic fragment and the procedures described in (a) through (g), design a protocol that would remove an EcoRI restriction site from a DNA molecule and incorporate a new BamHI restriction site at approximately the same location. (See Fig. 9-2.)

(i) Design four different short synthetic double-stranded DNA fragments that would permit ligation of structure (a) with a DNA fragment produced by a PstI restriction digest. In one of these fragments, design the sequence so that the final junction contains the recognition sequences for both EcoRI and PstI. In the second and third fragments, design the sequence so that the junction contains only the EcoRI and only the PstI recognition sequence, respectively. Design the sequence of the fourth fragment so that neither the EcoRI nor the PstI sequence appears in the junction.

2. Selecting for Recombinant Plasmids When cloning a foreign DNA fragment into a plasmid, it is often useful to insert the fragment at a site that interrupts a selectable marker (such as the tetracycline-resistance gene of pBR322). The loss of function of the interrupted gene can be used to identify clones containing recombinant plasmids with foreign DNA. With a yeast artificial chromosome (YAC) vector, it is not necessary to do this; the researcher can still distinguish vectors that incorporate large foreign DNA fragments from those that do not. How are these recombinant vectors identified?

3. DNA Cloning The plasmid cloning vector pBR322 (see Fig. 9-3) is cleaved with the restriction endonuclease PstI. An isolated DNA fragment from a eukaryotic genome (also produced by PstI cleavage) is added to the prepared vector and ligated. The mixture of ligated DNAs is then used to transform bacteria, and plasmid-containing bacteria are selected by growth in the presence of tetracycline.

(a) In addition to the desired recombinant plasmid, what other types of plasmids might be found among the transformed bacteria that are tetracycline-resistant? How can the types be distinguished?

(b) The cloned DNA fragment is 1,000 bp long and has an EcoRI site 250 bp from one end. Three different recombinant plasmids are cleaved with EcoRI and analyzed by gel electrophoresis, giving the patterns shown below. What does each pattern say about the cloned DNA? Note that in pBR322, the PstI and EcoRI restriction sites are about 750 bp apart. The entire plasmid with no cloned insert is 4,361 bp. Size markers in lane 4 have the number of nucleotides noted.

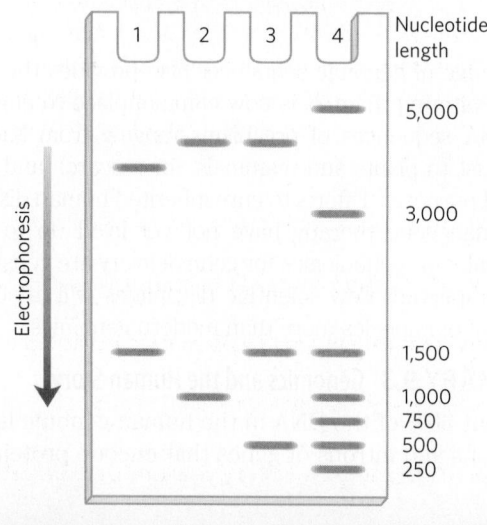

4. Restriction Enzymes The partial sequence of one strand of a double-stranded DNA molecule is

5′ – – – GACGAAGTGCTGCAGAAAGTCCGCGTTATAGGCAT
GAATTCCTGAGG – – – 3′

The cleavage sites for the restriction enzymes EcoRI and PstI are shown below.

EcoRI
          ↓
            *
(5′) G A A T T C (3′)
    C T T A A G
       * ↑

PstI
              ↓
          *
(5′) C T G C A G (3′)
    G A C G T C
    ↑ *

Write the sequence of *both strands* of the DNA fragment created when this DNA is cleaved with both EcoRI and PstI. The top strand of your duplex DNA fragment should be derived from the strand sequence given above.

5. Designing a Diagnostic Test for a Genetic Disease Huntington disease (HD) is an inherited neurodegenerative disorder, characterized by the gradual, irreversible impairment of psychological, motor, and cognitive functions. Symptoms typically appear in middle age, but onset can occur at almost any age. The course of the disease can last 15 to 20 years. The molecular basis of the disease is becoming better understood. The genetic mutation underlying HD has been traced to a gene encoding a protein (M_r 350,000) of unknown function. In individuals who will not develop HD, a region of the gene that encodes the amino terminus of the protein has a sequence of CAG codons (for glutamine) that is repeated 6 to 39 times in succession. In individuals with adult-onset HD, this codon is typically repeated 40 to 55 times. In individuals with childhood-onset HD, this codon is repeated more than 70 times. The length of this simple trinucleotide repeat indicates whether an individual will develop HD, and at approximately what age the first symptoms will occur.

A small portion of the amino-terminal coding sequence of the 3,143-codon HD gene is given below. The nucleotide sequence of the DNA is shown in black, the amino acid sequence corresponding to the gene is shown in blue, and the CAG repeat is shaded. Using Figure 27-7 to translate the genetic code, outline a PCR-based test for HD that could be carried out using a blood sample. Assume the PCR primer must be 25 nucleotides long. By convention, unless otherwise specified, a DNA sequence encoding a protein is displayed with the coding strand—the sequence identical to the mRNA transcribed from the gene (except for U replacing T)—on top, such that it is read 5′ to 3′, left to right.

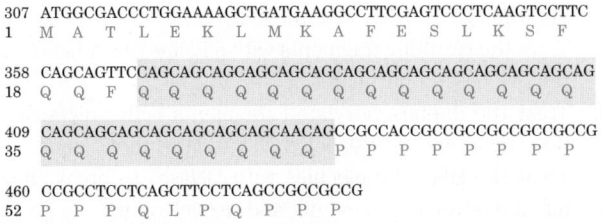

307 ATGGCGACCCTGGAAAAGCTGATGAAGGCCTTCGAGTCCCTCAAGTCCTTC
1   M A T L E K L M K A F E S L K S F

358 CAGCAGTTCCAGCAGCAGCAGCAGCAGCAGCAGCAGCAGCAGCAGCAGCAG
18 Q Q F Q Q Q Q Q Q Q Q Q Q Q Q Q

409 CAGCAGCAGCAGCAGCAGCAGCAACAGCCGCCACCGCCGCCGCCGCCGCCG
35 Q Q Q Q Q Q Q Q Q P P P P P P P

460 CCGCCTCCTCAGCTTCCTCAGCCGCCGCCG
52 P P P Q L P Q P P P

Source: The Huntington's Disease Collaborative Research Group, *Cell* 72:971, 1993.

6. Using PCR to Detect Circular DNA Molecules In a species of ciliated protist, a segment of genomic DNA is sometimes deleted. The deletion is a genetically programmed reaction associated with cellular mating. A researcher proposes that the DNA is deleted in a type of recombination called site-specific recombination, with the DNA at either end of the segment joined together and the deleted DNA ending up as a circular DNA reaction product.

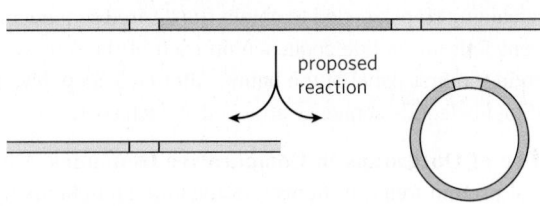

Suggest how the researcher might use the polymerase chain reaction (PCR) to detect the presence of the circular form of the deleted DNA in an extract of the protist.

7. Glowing Plants When grown in ordinary garden soil and watered normally, a plant engineered to express green fluorescent protein (see Fig. 9-16) will glow in the dark, whereas a plant engineered to express firefly luciferase (see Fig. 8-36) will not. Explain these observations.

8. Mapping a Chromosome Segment A group of overlapping clones, designated A through F, is isolated from one region of a chromosome. Each of the clones is separately cleaved by a restriction enzyme, and the pieces are resolved by agarose gel electrophoresis, with the results shown below. There are nine different restriction fragments in this chromosomal region, with a subset appearing in each clone. Using this information, deduce the order of the restriction fragments in the chromosome.

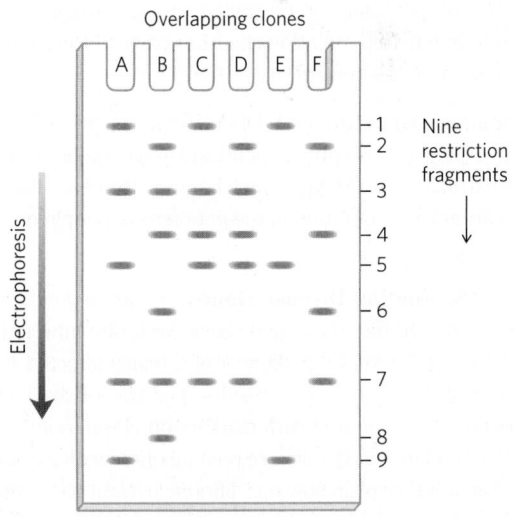

9. Immunofluorescence In the more common protocol for immunofluorescence detection of cellular proteins, an investigator uses two antibodies. The first binds specifically to the protein of interest. The second is labeled with fluorochromes for easy visualization, and it binds to the first antibody. In principle, one could simply label the first antibody and skip one step. Why use two successive antibodies?

10. Yeast Two-Hybrid Analysis You are a researcher who has just discovered a new protein in a fungus. Design a yeast two-hybrid experiment to identify the other proteins in the fungal cell with which your protein interacts and explain how this could help you determine the function of your protein.

11. Use of Photolithography to Make a DNA Microarray Figure 9-22 shows the first steps in the process of making a DNA microarray, or DNA chip, using photolithography. Describe the remaining steps needed to obtain the desired sequences (a different four-nucleotide sequence on each of the four spots) shown in the first panel of the figure. After each step, give the resulting nucleotide sequence attached at each spot.

12. Use of Outgroups in Comparative Genomics A hypothetical protein found in human, orangutan, and chimpanzee has the following sequences (red indicates amino acid residue differences; dashes indicate a deletion—the residues are missing in that sequence):

Human: ATSAAG**Y**DEWEGGK**V**LIHL – – KLQNRGALL
             ELDIGAV

Orangutan: ATSAAG**W**DEWEGGK**V**LIHL**DG**KLQNRGALL
             ELDIGAV

Chimpanzee: ATSAAG**W**DEWEGGK**I**LIHL**DG**KLQNRGALL
             ELDIGAV

What is the most likely sequence of the protein present in the last common ancestor of human and chimpanzee?

13. Human Migrations I Native American populations in North and South America have mitochondrial DNA haplotypes that can be traced to populations in northeast Asia. The Aleut and Eskimo populations in the far northern parts of North America possess a subset of the same haplotypes that link other Native Americans to Asia, and also have several additional haplotypes that can be traced to Asian origins but are not found in native populations in other parts of the Americas. Provide a possible explanation.

14. Human Migrations II DNA (haplotypes) originating from the Denisovans can be found in the genomes of Indigenous Australians and Melanesian Islanders. However, the same DNA markers are not found in the genomes of people native to Africa. Explain.

15. Finding Disease Genes You are a gene hunter, trying to find the genetic basis for a rare inherited disease. Examination of six pedigrees of families affected by the disease provides inconsistent results. For two of the families, the disease is co-inherited with markers on chromosome 7. For the other four families, the disease is co-inherited with markers on chromosome 12. Explain how this difference might have arisen.

Data Analysis Problem

16. HincII: The First Restriction Endonuclease Discovery of the first restriction endonuclease to be of practical use was reported in two papers published in 1970. In the first paper, Smith and Wilcox described the isolation of an enzyme that cleaved double-stranded DNA. They initially demonstrated the enzyme's nuclease activity by measuring the decrease in viscosity of DNA samples treated with the enzyme.

(a) Why does treatment with a nuclease decrease the viscosity of a solution of DNA?

The authors determined whether the enzyme was an endonuclease or exonuclease by treating ^{32}P-labeled DNA with the enzyme, then adding trichloroacetic acid (TCA). Under the conditions used in their experiment, single nucleotides would be TCA-soluble and oligonucleotides would precipitate.

(b) No TCA-soluble ^{32}P-labeled material formed on treatment of the ^{32}P-labeled DNA with the nuclease. Based on this finding, is the enzyme an endonuclease or exonuclease? Explain your reasoning.

When a polynucleotide is cleaved, the phosphate usually is not removed but remains attached to the 5′ or 3′ end of the resulting DNA fragment. Smith and Wilcox determined the location of the phosphate on the fragment formed by the nuclease in the following steps:

1. Treat unlabeled DNA with the nuclease.
2. Treat a sample (A) of the product with γ-^{32}P-labeled ATP and polynucleotide kinase (which can attach the γ-phosphate of ATP to a 5′ OH but not to a 5′ phosphate or to a 3′ OH or 3′ phosphate). Measure the amount of ^{32}P incorporated into the DNA.
3. Treat another sample (B) of the product of step 1 with alkaline phosphatase (which removes phosphate groups from free 5′ and 3′ ends), followed by polynucleotide kinase and γ-^{32}P-labeled ATP. Measure the amount of ^{32}P incorporated into the DNA.

(c) Smith and Wilcox found that sample A had 136 counts/min of ^{32}P; sample B had 3,740 counts/min. Did the nuclease cleavage leave the phosphate on the 5′ or the 3′ end of the DNA fragments? Explain your reasoning.

(d) Treatment of bacteriophage T7 DNA with the nuclease gave approximately 40 specific fragments of various lengths. How is this result consistent with the enzyme's recognizing a specific sequence in the DNA as opposed to making random double-strand breaks?

At this point, there were two possibilities for the site-specific cleavage: the cleavage occurred either (1) at the site of recognition or (2) near the site of recognition but not within the sequence recognized. To address this issue, Kelly and Smith determined the sequence of the 5′ ends of the DNA fragments generated by the nuclease, in the following steps:

1. Treat phage T7 DNA with the enzyme.
2. Treat the resulting fragments with alkaline phosphatase to remove the 5′ phosphates.
3. Treat the dephosphorylated fragments with polynucleotide kinase and γ-^{32}P-labeled ATP to label the 5′ ends.
4. Treat the labeled molecules with DNases to break them into a mixture of mono-, di-, and trinucleotides.

5. Determine the sequence of the labeled mono-, di-, and trinucleotides by comparing them with oligonucleotides of known sequence on thin-layer chromatography.

The labeled products were identified as follows: mononucleotides: A and G; dinucleotides: (5′)ApA(3′) and (5′)GpA(3′); trinucleotides: (5′)ApApC(3′) and (5′)GpApC(3′).

(e) Which model of cleavage is consistent with these results? Explain your reasoning.

Kelly and Smith went on to determine the sequence of the 3′ ends of the fragments. They found a mixture of (5′)TpC(3′) and (5′)TpT(3′). They did not determine the sequence of any trinucleotides at the 3′ end.

(f) Based on these data, what is the recognition sequence for the nuclease, and where in the sequence is the DNA backbone cleaved? Use Table 9-2 as a model for your answer.

References

Kelly, T.J., and H.O. Smith. 1970. A restriction enzyme from *Haemophilus influenzae*: II. Base sequence of the recognition site. *J. Mol. Biol.* 51:393–409.

Smith, H.O., and K.W. Wilcox. 1970. A restriction enzyme from *Haemophilus influenzae*: I. Purification and general properties. *J. Mol. Biol.* 51:379–391.

Further Reading is available at www.macmillanlearning.com/LehningerBiochemistry7e.

CHAPTER 10

Lipids

Self-study tools that will help you practice what you've learned and reinforce this chapter's concepts are available online.
Go to www.macmillanlearning.com/LehningerBiochemistry7e.

iological lipids are a chemically diverse group of compounds, the common and defining feature of which is their insolubility in water. The biological functions of the lipids are as diverse as their chemistry. Fats and oils are the principal stored forms of energy in many organisms. Phospholipids and sterols are major structural elements of biological membranes. Other lipids, although present in relatively small quantities, play crucial roles as enzyme cofactors, electron carriers, light-absorbing pigments, hydrophobic anchors for proteins, "chaperones" to help membrane proteins fold, emulsifying agents in the digestive tract, hormones, and intracellular messengers. This chapter introduces representative lipids of each type, organized according to their functional roles, with emphasis on their chemical structure and physical properties. Although we follow a functional organization for our discussion, the thousands of different lipids can also be organized into eight general categories of chemical structure (see Table 10-2). We discuss the energy-yielding oxidation of lipids in Chapter 17 and their synthesis in Chapter 21.

10.1 Storage Lipids

The fats and oils used almost universally as stored forms of energy in living organisms are derivatives of **fatty acids**. The fatty acids are hydrocarbon derivatives, at about the same low oxidation state (that is, as highly reduced) as the hydrocarbons in fossil fuels. The cellular oxidation of fatty acids (to CO_2 and H_2O), like the controlled, rapid burning of fossil fuels in internal combustion engines, is highly exergonic.

We introduce here the structures and nomenclature of the fatty acids most commonly found in living organisms. Two types of fatty acid–containing compounds, triacylglycerols and waxes, are described to illustrate the diversity of structures and physical properties in this family of compounds.

Fatty Acids Are Hydrocarbon Derivatives

Fatty acids are carboxylic acids with hydrocarbon chains ranging from 4 to 36 carbons long (C_4 to C_{36}). In some fatty acids, this chain is unbranched and fully saturated (contains no double bonds); in others, the chain contains one or more double bonds (Table 10-1). A few contain three-carbon rings, hydroxyl groups, or methyl-group branches.

>> **Key Convention:** A simplified nomenclature for unbranched fatty acids specifies the chain length and number of double bonds, separated by a colon. For example, the 16-carbon saturated palmitic acid is abbreviated 16:0, and the 18-carbon oleic (octadecenoic) acid, with one double bond (shown below), is 18:1. Each line segment of the zigzag in the structure represents a single bond between adjacent carbons. The carboxyl carbon is assigned the number 1 (C-1), and the carbon next to it is C-2. The positions of any double bonds, designated Δ (delta), are specified relative to C-1 by a superscript number indicating the lower-numbered carbon in the double bond. By this convention, oleic acid, with a double bond between C-9 and C-10, is designated $18:1(\Delta^9)$; a 20-carbon fatty acid with one double bond between C-9 and C-10 and another between C-12 and C-13 is designated $20:2(\Delta^{9,12})$.

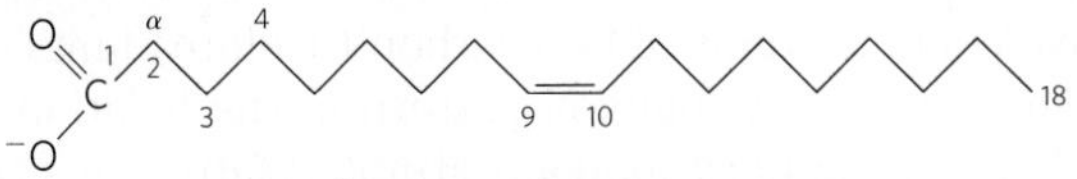

$18:1(\Delta^9)$ *cis*-9-Octadecenoic acid   ≪

TABLE 10-1 Some Naturally Occurring Fatty Acids: Structure, Properties, and Nomenclature

Carbon skeleton	Structure[a]	Systematic name[b]	Common name (derivation)	Melting point (°C)	Solubility at 30 °C (mg/g solvent) Water	Benzene
12:0	$CH_3(CH_2)_{10}COOH$	n-Dodecanoic acid	Lauric acid (Latin *laurus*, "laurel plant")	44.2	0.063	2,600
14:0	$CH_3(CH_2)_{12}COOH$	n-Tetradecanoic acid	Myristic acid (Latin *Myristica*, nutmeg genus)	53.9	0.024	874
16:0	$CH_3(CH_2)_{14}COOH$	n-Hexadecanoic acid	Palmitic acid (Latin *palma* "palm tree")	63.1	0.0083	348
18:0	$CH_3(CH_2)_{16}COOH$	n-Octadecanoic acid	Stearic acid (Greek *stear*, "hard fat")	69.6	0.0034	124
20:0	$CH_3(CH_2)_{18}COOH$	n-Eicosanoic acid	Arachidic acid (Latin *Arachis*, legume genus)	76.5		
24:0	$CH_3(CH_2)_{22}COOH$	n-Tetracosanoic acid	Lignoceric acid (Latin *lignum*, "wood" + *cera*, "wax")	86.0		
16:1(Δ^9)	$CH_3(CH_2)_5CH=$ $CH(CH_2)_7COOH$	*cis*-9-Hexadecenoic acid	Palmitoleic acid	1 to −0.5		
18:1(Δ^9)	$CH_3(CH_2)_7CH=$ $CH(CH_2)_7COOH$	*cis*-9-Octadecenoic acid	Oleic acid (Latin *oleum*, "oil")	13.4		
18:2($\Delta^{9,12}$)	$CH_3(CH_2)_4CH=$ $CHCH_2CH=$ $CH(CH_2)_7COOH$	*cis-,cis*-9,12-Octadecadienoic acid	Linoleic acid (Greek *linon*, "flax")	1–5		
18:3($\Delta^{9,12,15}$)	$CH_3CH_2CH=$ $CHCH_2CH=$ $CHCH_2CH=$ $CH(CH_2)_7COOH$	*cis-,cis-,cis*-9,12,15-Octadecatrienoic acid	α-Linolenic acid	−11		
20:4($\Delta^{5,8,11,14}$)	$CH_3(CH_2)_4CH=$ $CHCH_2CH=$ $CHCH_2CH=$ $CHCH_2CH=$ $CH(CH_2)_3COOH$	*cis-,cis-,cis-,cis*-5,8,11,14-Icosatetraenoic acid	Arachidonic acid	−49.5		

[a]All acids are shown in their nonionized form. At pH 7, all free fatty acids have an ionized carboxylate. Note that numbering of carbon atoms begins at the carboxyl carbon.

[b]The prefix *n-* indicates the "normal" unbranched structure. For instance, "dodecanoic" simply indicates 12 carbon atoms, which could be arranged in a variety of branched forms; "*n*-dodecanic" specifies the linear, unbranched form. For unsaturated fatty acids, the configuration of each double bond is indicated; in biological fatty acids the configuration is almost always cis.

The most commonly occurring fatty acids have even numbers of carbon atoms in an unbranched chain of 12 to 24 carbons (Table 10-1). As we shall see in Chapter 21, the even number of carbons results from the mode of synthesis of these compounds, which involves successive condensations of two-carbon (acetate) units.

There is also a common pattern in the location of double bonds; in most monounsaturated fatty acids the double bond is between C-9 and C-10 (Δ^9), and the other double bonds of polyunsaturated fatty acids are generally Δ^{12} and Δ^{15}. (Arachidonic acid is an exception to this generalization; see Table 10-1.) The double bonds of polyunsaturated fatty acids are almost never conjugated (alternating single and double bonds, as in —CH=CH—CH=CH—), but are separated by a methylene group: —CH=CH—CH$_2$—CH=CH—. In nearly all naturally occurring unsaturated fatty acids, the double bonds are in the cis configuration. Trans fatty acids are produced by fermentation in the rumen of dairy animals and are obtained from dairy products and meat.

» Key Convention: The family of **polyunsaturated fatty acids (PUFAs)** with a double bond between the third and fourth carbon from the methyl end of the chain are of special importance in human nutrition. Because the physiological role of PUFAs is related more to the position of the first double bond near the *methyl* end of the chain than to that near the carboxyl end, an alternative nomenclature is sometimes used for these fatty acids. The carbon of the methyl group—that is, the carbon most distant from the carboxyl group—is called the ω (omega; the last letter in the Greek alphabet) carbon and is given the number 1 (C-1); the carboxyl carbon in this convention has the highest number. The positions of the double bonds are indicated relative to the ω carbon. In this convention, PUFAs with a double bond between C-3 and C-4 are called **omega-3 (ω-3) fatty acids**, and those with a double bond between C-6 and C-7 are **omega-6 (ω-6) fatty acids**. Shown below is eicosapentaenoic acid, which can be designated as $20:5(\Delta^{5,8,11,14,17})$ by the standard nomenclature but is also referred to as an omega-3 fatty acid, emphasizing the biologically important double bond in the omega-3 position.

$20:5(\Delta^{5,8,11,14,17})$ Eicosapentaenoic acid (EPA)   **«**

Humans require the omega-3 PUFA α-linolenic acid (ALA; $18:3(\Delta^{9,12,15})$, in the standard convention), but do not have the enzymatic capacity to synthesize it and must therefore obtain it in the diet. From ALA, humans can synthesize two other omega-3 PUFAs important in cellular function: eicosapentaenoic acid (EPA; $20:5(\Delta^{5,8,11,14,17})$, shown in the Key Convention above) and docosahexaenoic acid (DHA; $22:6(\Delta^{4,7,10,13,16,19})$). An imbalance of omega-6 and omega-3 PUFAs in the diet is associated with an increased risk of cardiovascular disease. The optimal dietary ratio of omega-6 to omega-3 PUFAs is between 1:1 and 4:1, but the ratio in the diets of most North Americans is closer to 10:1 to 30:1. The "Mediterranean diet," which has been associated with lowered cardiovascular risk, is richer in omega-3 PUFAs, obtained in leafy vegetables (salads) and fish oils. The latter oils are especially rich in EPA and DHA, and fish oil supplements are often prescribed for individuals with a history of cardiovascular disease. ■

The physical properties of the fatty acids, and of compounds that contain them, are largely determined by the length and degree of unsaturation of the hydrocarbon chain. The nonpolar hydrocarbon chain accounts for the poor solubility of fatty acids in water. Lauric acid ($12:0$, M_r 200), for example, has a solubility in water of 0.063 mg/g—much less than that of glucose (M_r 180), which is 1,100 mg/g. The longer the fatty acyl chain and the fewer the double bonds, the lower is the solubility in water. The carboxylic acid group is polar (and ionized at neutral pH) and accounts for the slight solubility of short-chain fatty acids in water.

Melting points are also strongly influenced by the length and degree of unsaturation of the hydrocarbon chain. At room temperature (25 °C), the saturated fatty acids from 12:0 to 24:0 have a waxy consistency, whereas unsaturated fatty acids of these lengths are oily liquids. This difference in melting points is due to different degrees of packing of the fatty acid molecules **(Fig. 10-1)**. In the fully saturated compounds, free rotation around each carbon–carbon bond gives the hydrocarbon chain great flexibility; the most stable conformation is the fully extended form, in which the steric hindrance of neighboring atoms is minimized. These molecules can pack together tightly in nearly crystalline arrays, with atoms all along their lengths in van der Waals contact with the atoms of neighboring molecules. In unsaturated fatty acids, a cis double bond forces a kink in the hydrocarbon chain. Fatty acids with one or several such kinks cannot pack together as tightly as fully saturated fatty acids, and their interactions with each other are therefore weaker. Because less thermal energy is needed to disorder these poorly ordered arrays of unsaturated fatty acids, they have markedly lower melting points than saturated fatty acids of the same chain length (Table 10-1).

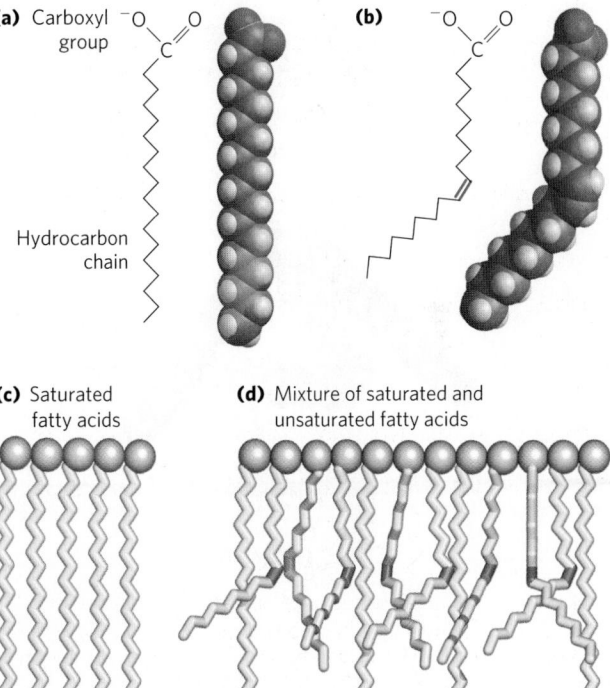

FIGURE 10-1 The packing of fatty acids into stable aggregates. The extent of packing depends on the degree of saturation. **(a)** Two representations of the fully saturated acid stearic acid, 18:0 (stearate at pH 7), in its usual extended conformation. **(b)** The cis double bond (red) in oleic acid, $18:1(\Delta^9)$ (oleate), restricts rotation and introduces a rigid bend in the hydrocarbon tail. All other bonds in the chain are free to rotate. **(c)** Fully saturated fatty acids in the extended form pack into nearly crystalline arrays, stabilized by extensive hydrophobic interaction. **(d)** The presence of one or more fatty acids with cis double bonds (red) interferes with this tight packing and results in less stable aggregates.

In vertebrates, free fatty acids (unesterified fatty acids, with a free carboxylate group) circulate in the blood bound noncovalently to a protein carrier, serum albumin. However, fatty acids are present in blood plasma mostly as carboxylic acid derivatives such as esters or amides. Lacking the charged carboxylate group, these fatty acid derivatives are generally even less soluble in water than are the free fatty acids.

Triacylglycerols Are Fatty Acid Esters of Glycerol

The simplest lipids constructed from fatty acids are the **triacylglycerols**, also referred to as triglycerides, fats, or neutral fats. Triacylglycerols are composed of three fatty acids, each in ester linkage with a single glycerol **(Fig. 10-2)**. Those containing the same kind of fatty acid in all three positions are called simple triacylglycerols and are named after the fatty acid they contain. Simple triacylglycerols of 16:0, 18:0, and 18:1, for example, are tripalmitin, tristearin, and triolein, respectively. Most naturally occurring triacylglycerols are mixed; they contain two or three different fatty acids. To name these compounds unambiguously, the name and position of each fatty acid must be specified.

Because the polar hydroxyls of glycerol and the polar carboxylates of the fatty acids are bound in ester linkages, triacylglycerols are nonpolar, hydrophobic molecules, essentially insoluble in water. Lipids have lower specific gravities than water, which explains why mixtures of oil and water (oil-and-vinegar salad dressing, for example) have two phases: oil, with the lower specific gravity, floats on the aqueous phase.

Triacylglycerols Provide Stored Energy and Insulation

In most eukaryotic cells, triacylglycerols form a separate phase of microscopic, oily droplets in the aqueous cytosol, serving as depots of metabolic fuel. In vertebrates, specialized cells called adipocytes, or fat cells, store large amounts of triacylglycerols as fat droplets that nearly fill the cell **(Fig. 10-3a)**. Triacylglycerols are also stored as oils in the seeds of many types of plants, providing energy and biosynthetic precursors during seed germination (Fig. 10-3b). Adipocytes and germinating seeds contain **lipases**, enzymes that catalyze the hydrolysis of stored triacylglycerols, releasing fatty acids for export to sites where they are required as fuel.

There are two significant advantages to using triacylglycerols as stored fuels, rather than polysaccharides

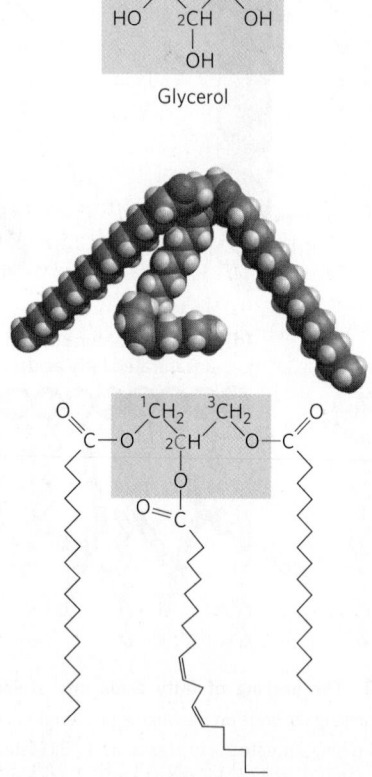

FIGURE 10-2 Glycerol and a triacylglycerol. The mixed triacylglycerol shown here has three different fatty acids attached to the glycerol backbone. When glycerol has different fatty acids at C-1 and C-3, C-2 is a chiral center (p. 17).

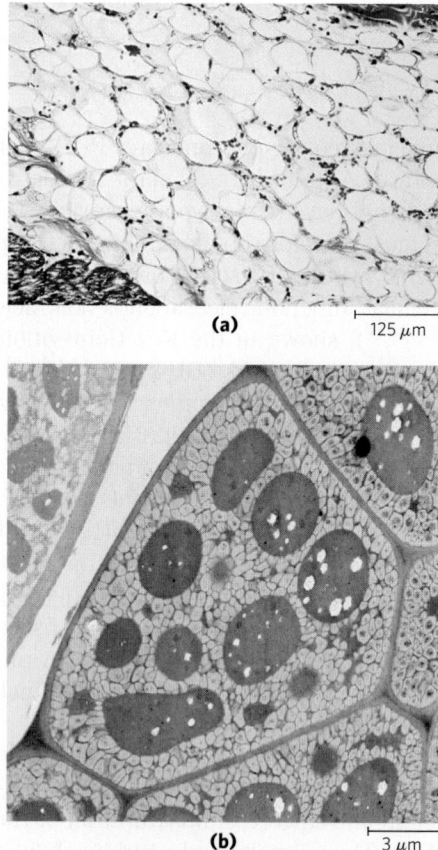

FIGURE 10-3 Fat stores in cells. (a) Cross section of human white adipose tissue. Each cell contains a fat droplet (white) so large that it squeezes the nucleus (stained red) against the plasma membrane. **(b)** Cross section of a cotyledon cell from a seed of the plant *Arabidopsis*. The large dark structures are protein bodies, which are surrounded by stored oils in the light-colored oil bodies. [Sources: (a) Biophoto Associates/Science Source. (b) Courtesy Howard Goodman, Department of Genetics, Harvard Medical School.]

such as glycogen and starch. First, the carbon atoms of fatty acids are more reduced than those of sugars, so oxidation of triacylglycerols yields more than twice as much energy, gram for gram, as the oxidation of carbohydrates. Second, because triacylglycerols are hydrophobic and therefore unhydrated, the organism that carries stored fuel in the form of fat does not have to carry the extra weight of water of hydration that is associated with stored polysaccharides (2 g per gram of polysaccharide). Humans have fat tissue (composed primarily of adipocytes) under the skin, in the abdominal cavity, and in the mammary glands. Moderately obese people with 15 to 20 kg of triacylglycerols deposited in their adipocytes could meet their energy needs for months by drawing on their fat stores. In contrast, the human body can store less than a day's energy supply in the form of glycogen. Carbohydrates such as glucose do offer certain advantages as quick sources of metabolic energy, one of which is their ready solubility in water.

In some animals, triacylglycerols stored under the skin serve not only as energy stores but as insulation against low temperatures. Seals, walruses, penguins, and other warm-blooded polar animals are amply padded with triacylglycerols. In hibernating animals (bears, for example), the huge fat reserves accumulated before hibernation serve the dual purposes of insulation and energy storage (see Box 17-1).

Partial Hydrogenation of Cooking Oils Improves Their Stability but Creates Fatty Acids with Harmful Health Effects

Most natural fats, such as those in vegetable oils, dairy products, and animal fat, are complex mixtures of simple and mixed triacylglycerols. These contain a variety of fatty acids differing in chain length and degree of saturation **(Fig. 10-4)**. Vegetable oils such as corn (maize) oil and olive oil are composed largely of triacylglycerols with unsaturated fatty acids and thus are liquids at room temperature. Triacylglycerols containing only saturated fatty acids, such as tristearin, the major component of beef fat, are white, greasy solids at room temperature.

When lipid-rich foods are exposed too long to the oxygen in air, they may spoil and become rancid. The unpleasant taste and smell associated with rancidity result from the oxidative cleavage of double bonds in unsaturated fatty acids, which produces aldehydes and carboxylic acids of shorter chain length and therefore higher volatility; these compounds pass readily through the air to your nose. Throughout the twentieth century, to improve the shelf life of vegetable oils used in cooking, and to increase their stability at the high temperatures used in deep-frying, commercial vegetable oils were prepared by partial hydrogenation. This process converts many of the cis double bonds in the fatty acids to single bonds and increases the melting temperature of the oils so that they are more nearly solid at room

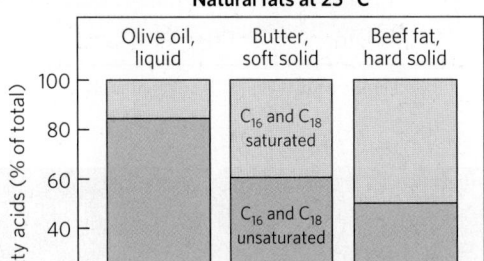

FIGURE 10-4 Fatty acid composition of three food fats. Olive oil, butter, and beef fat consist of mixtures of triacylglycerols differing in their fatty acid composition. The melting points of these fats—and hence their physical state at room temperature (25 °C)—are a direct function of their fatty acid composition. Olive oil has a high proportion of long-chain (C_{16} and C_{18}) unsaturated fatty acids, which accounts for its liquid state at 25 °C. The higher proportion of long-chain (C_{16} and C_{18}) saturated fatty acids in butter increases its melting point, so butter is a soft solid at room temperature. Beef fat, with an even higher proportion of long-chain saturated fatty acids, is a hard solid.

temperature (margarine is produced from vegetable oil in this way). Partial hydrogenation, however, has another, undesirable, effect: some cis double bonds are converted to trans double bonds. There is now strong evidence that dietary intake of trans fatty acids (often referred to simply as "trans fats") leads to a higher incidence of cardiovascular disease, and that avoiding these fats in the diet substantially reduces the risk of coronary heart disease. Dietary trans fatty acids raise the level of triacylglycerols and of LDL ("bad") cholesterol in the blood, and lower the level of HDL ("good") cholesterol, and these changes alone are enough to increase the risk of coronary heart disease. But trans fatty acids may have further adverse effects. They seem, for example, to increase the body's inflammatory response, which is another risk factor for heart disease. (See Chapter 21 for a description of LDL and HDL—low-density and high-density lipoprotein—cholesterol and their health effects.) Regulatory agencies around the world now limit or ban the use of trans fatty acids in prepared and packaged foods. ∎

Waxes Serve as Energy Stores and Water Repellents

Biological waxes are esters of long-chain (C_{14} to C_{36}) saturated and unsaturated fatty acids with long-chain (C_{16} to C_{30}) alcohols **(Fig. 10-5)**. Their melting points (60 to 100 °C) are generally higher than those of triacylglycerols. In plankton, the free-floating microorganisms at the bottom of the food chain for marine animals, waxes are the chief storage form of metabolic fuel.

Waxes also serve a diversity of other functions related to their water-repellent properties and their firm consistency. Certain skin glands of vertebrates secrete waxes to protect hair and skin and keep it pliable, lubricated, and waterproof. Birds, particularly waterfowl, secrete waxes

(a)

$$CH_3(CH_2)_{14} - \overset{\overset{\displaystyle O}{\|}}{C} - O - CH_2 - (CH_2)_{28} - CH_3$$

Palmitic acid   1-Triacontanol

FIGURE 10-5 Biological wax. (a) Triacontanoylpalmitate, the major component of beeswax, is an ester of palmitic acid with the alcohol triacontanol. **(b)** The beeswax of a honeycomb is firm at 25 °C and completely impervious to water. The term "wax" originates in the Old English *weax*, meaning "the material of the honeycomb." [Source: (b) iStockphoto/Thinkstock.]

from their preen glands to keep their feathers water-repellent. The shiny leaves of holly, rhododendrons, poison ivy, and many tropical plants are coated with a thick layer of waxes, which prevents excessive evaporation of water and protects against parasites.

Biological waxes find a variety of applications in the pharmaceutical, cosmetic, and other industries. Lanolin

(from lamb's wool), beeswax (Fig. 10-5), carnauba wax (from a Brazilian palm tree), and wax extracted from the seeds of the jojoba bush are widely used in the manufacture of lotions, ointments, and polishes.

SUMMARY 10.1 Storage Lipids

■ Lipids are water-insoluble cellular components, of diverse structure, that can be extracted from tissues by nonpolar solvents.

■ Almost all fatty acids, the hydrocarbon components of many lipids, have an even number of carbon atoms (usually 12 to 24); they are either saturated or unsaturated, with double bonds almost always in the cis configuration.

■ Triacylglycerols contain three fatty acid molecules esterified to the three hydroxyl groups of glycerol. Simple triacylglycerols contain only one type of fatty acid; mixed triacylglycerols, two or three types. Triacylglycerols are primarily storage fats; they are present in many foods.

■ Because trans fatty acids in the diet are an important risk factor for coronary heart disease, their use in prepared and processed foods has become highly regulated.

■ Waxes are esters of long-chain fatty acids and long-chain alcohols.

10.2 Structural Lipids in Membranes

The central architectural feature of biological membranes is a double layer of lipids, which acts as a barrier to the passage of polar molecules and ions. Membrane lipids are amphipathic: one end of the molecule is hydrophobic, the other hydrophilic. Their hydrophobic interactions with

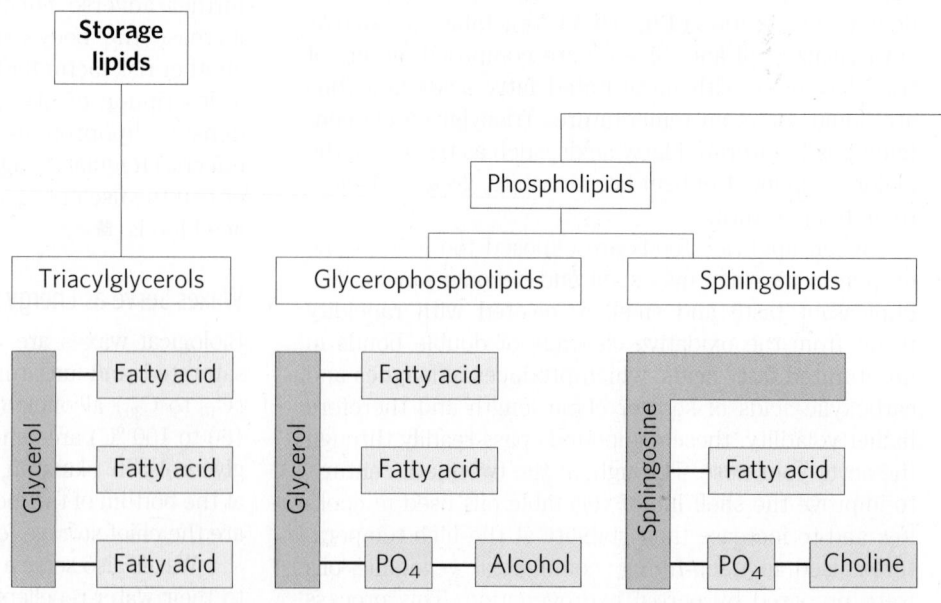

FIGURE 10-6 Some common types of storage and membrane lipids. All the lipid types shown here have either glycerol or sphingosine as the backbone (light red screen), to which are attached one or more long-chain alkyl groups (yellow) and a polar head group (blue). In triacylglycerols, glycerophospholipids, galactolipids, and sulfolipids, the alkyl groups are fatty acids in ester linkage. Sphingolipids contain a single fatty acid, in amide linkage to the sphingosine

each other and their hydrophilic interactions with water direct their packing into sheets called membrane bilayers. In this section, we describe five general types of membrane lipids: glycerophospholipids, in which the hydrophobic regions are composed of two fatty acids joined to glycerol; galactolipids and sulfolipids, which also contain two fatty acids esterified to glycerol, but lack the characteristic phosphate of phospholipids; archaeal tetraether lipids, in which two very long alkyl chains are ether-linked to glycerol at both ends; sphingolipids, in which a single fatty acid is joined to a fatty amine, sphingosine; and sterols, compounds characterized by a rigid system of four fused hydrocarbon rings.

The hydrophilic moieties in these amphipathic compounds may be as simple as a single —OH group at one end of the sterol ring system, or they may be much more complex. In glycerophospholipids and some sphingolipids, a polar head group is joined to the hydrophobic moiety by a phosphodiester linkage; these are the **phospholipids**. Other sphingolipids lack phosphate but have a simple sugar or complex oligosaccharide at their polar ends; these are the **glycolipids (Fig. 10-6)**. Within these groups of membrane lipids, enormous diversity results from various combinations of fatty acid "tails" and polar "heads." The arrangement of these lipids in membranes, and their structural and functional roles therein, are considered in the next chapter.

Glycerophospholipids Are Derivatives of Phosphatidic Acid

Glycerophospholipids, also called phosphoglycerides, are membrane lipids in which two fatty acids are attached in ester linkage to the first and second carbons of glycerol, and a highly polar or charged group is attached through a phosphodiester linkage to the third

L-Glycerol 3-phosphate (*sn*-glycerol 3-phosphate)

FIGURE 10-7 L-Glycerol 3-phosphate, the backbone of phospholipids. Glycerol itself is not chiral, as it has a plane of symmetry through C-2. However, glycerol is prochiral—it can be converted to a chiral compound by adding a substituent such as phosphate to either of the —CH_2OH groups. One unambiguous nomenclature for glycerol phosphate is the D, L system (described on p. 78), in which the isomers are named according to their stereochemical relationships to glyceraldehyde isomers. By this system, the stereoisomer of glycerol phosphate found in most lipids is correctly named either L-glycerol 3-phosphate or D-glycerol 1-phosphate. Another way to specify stereoisomers is the *sn* (stereospecific numbering) system, in which C-1 is, by definition, the group of the prochiral compound that occupies the pro-S position. The common form of glycerol phosphate in phospholipids is, by this system, *sn*-glycerol 3-phosphate (in which C-2 has the R configuration). In archaea, the glycerol in lipids has the other configuration; it is D-glycerol 3-phosphate.

carbon. Glycerol is prochiral; it has no asymmetric carbons, but attachment of phosphate at one end converts it into a chiral compound, which can be correctly named either L-glycerol 3-phosphate, D-glycerol 1-phosphate, or *sn*-glycerol 3-phosphate **(Fig. 10-7)**. Glycerophospholipids are named as derivatives of the parent compound, phosphatidic acid **(Fig. 10-8)**, according to the polar alcohol in the head group. Phosphatidylcholine and phosphatidylethanolamine have choline and ethanolamine as their polar head groups, for example. Cardiolipin is a two-tailed glycerophospholipid in which two phosphatidic acid moieties share the same glycerol as their head group (Fig. 10-8). Cardiolipin is found in

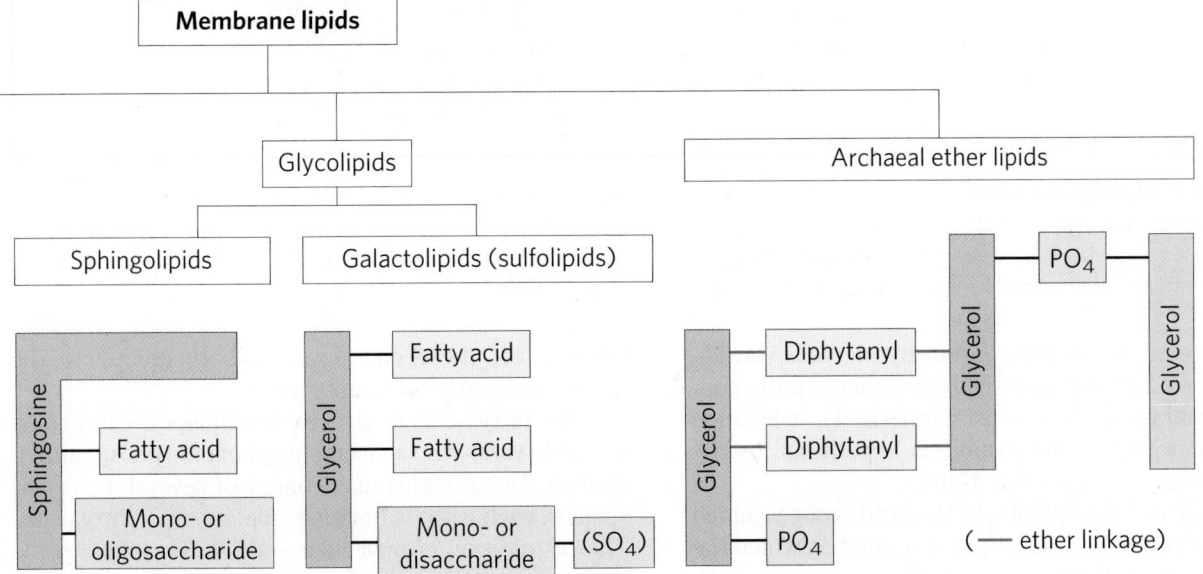

backbone. The membrane lipids of archaea are variable; that shown here has two very long, branched alkyl chains, each end in ether linkage with a glycerol moiety. In phospholipids, the polar head group is joined through a phosphodiester, whereas glycolipids have a direct glycosidic linkage between the head-group sugar and the backbone glycerol.

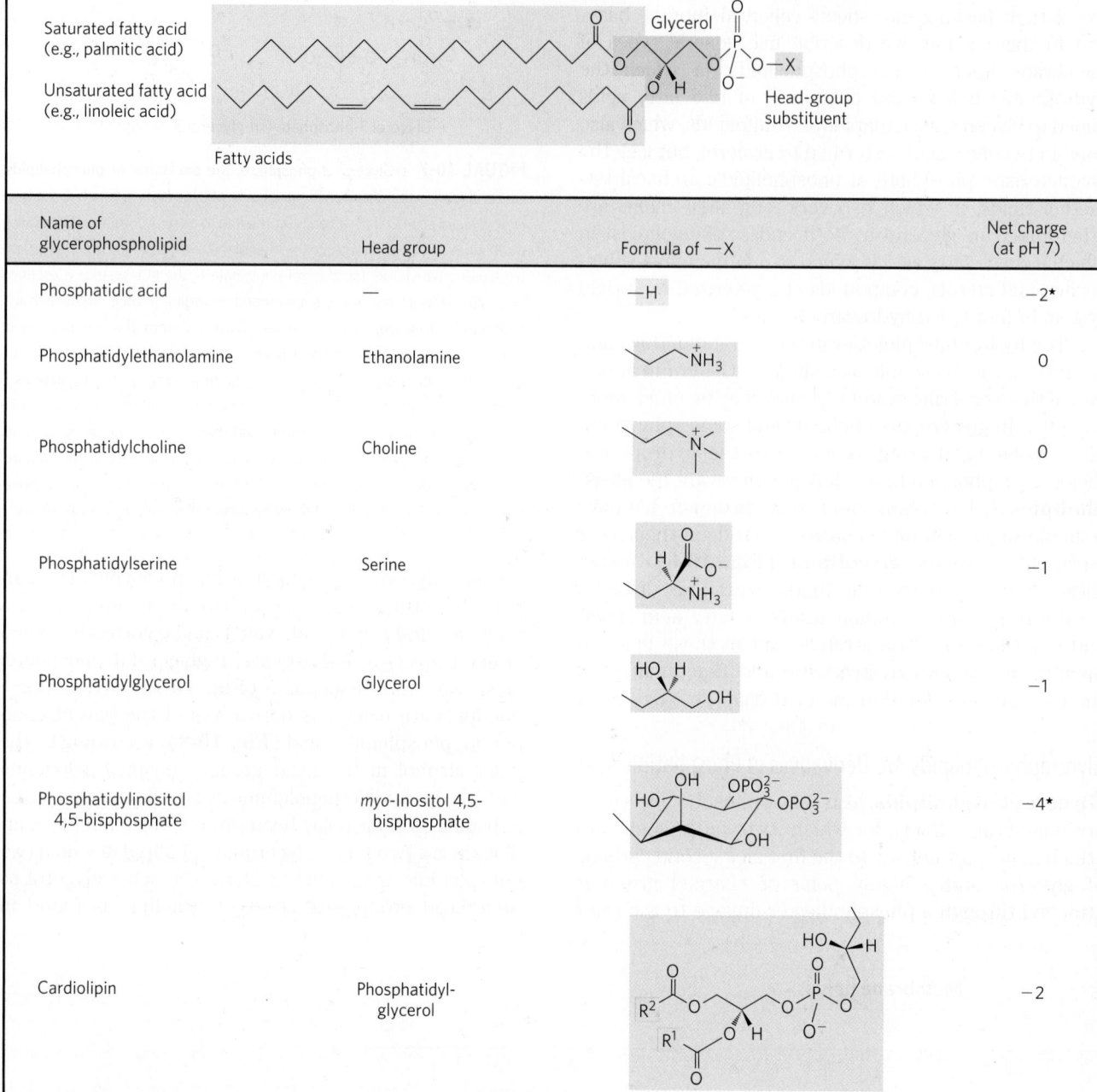

Name of glycerophospholipid	Head group	Formula of —X	Net charge (at pH 7)
Phosphatidic acid	—	—H	−2*
Phosphatidylethanolamine	Ethanolamine		0
Phosphatidylcholine	Choline		0
Phosphatidylserine	Serine		−1
Phosphatidylglycerol	Glycerol		−1
Phosphatidylinositol 4,5-bisphosphate	myo-Inositol 4,5-bisphosphate		−4*
Cardiolipin	Phosphatidylglycerol		−2

FIGURE 10-8 Glycerophospholipids. The common glycerophospholipids are diacylglycerols linked to head-group alcohols through a phosphodiester bond. Phosphatidic acid, a phosphomonoester, is the parent compound. Each derivative is named for the head-group alcohol, with the prefix "phosphatidyl-." In cardiolipin, two phosphatidic acids share a single glycerol (R^1 and R^2 are fatty acyl groups). *Note that phosphate esters each have a charge of about −1.5; one of their —OH groups is only partially ionized at pH 7.

most bacterial membranes; in eukaryotic cells, cardiolipin is located almost exclusively in the inner mitochondrial membrane (where it is synthesized), a location consistent with the endosymbiosis hypothesis for the origin of organelles (see Fig. 1-40).

In all glycerophospholipids, the head group is joined to glycerol through a phosphodiester bond, in which the phosphate group bears a negative charge at neutral pH. The polar alcohol may be negatively charged (as in phosphatidylinositol 4,5-bisphosphate), neutral (phosphatidylserine), or positively charged (phosphatidylcholine, phosphatidylethanolamine). As we shall see in

Chapter 11, these charges contribute greatly to the surface properties of membranes.

The fatty acids in glycerophospholipids can be any of a wide variety, so a given phospholipid (phosphatidylcholine, for example) may consist of several molecular species, each with its unique complement of fatty acids. The distribution of molecular species is specific to the organism, to the particular tissue within the organism, and to the particular glycerophospholipids in the same cell or tissue. In general, glycerophospholipids contain a C_{16} or C_{18} saturated fatty acid at C-1 and a C_{18} or C_{20} unsaturated fatty acid at C-2. With few exceptions, the

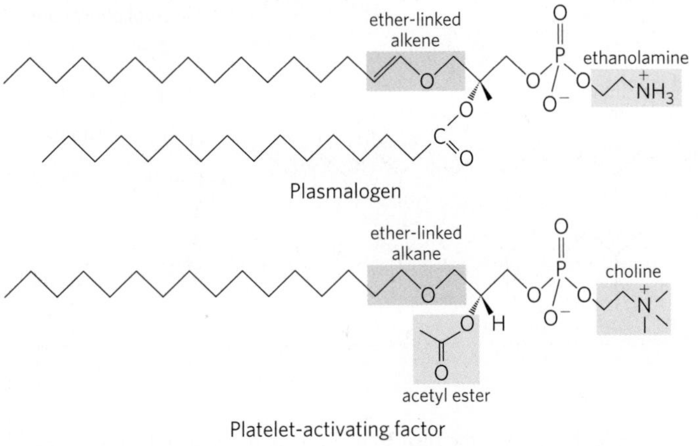

ether-linked alkene

ethanolamine

Plasmalogen

ether-linked alkane

choline

acetyl ester

Platelet-activating factor

FIGURE 10-9 Ether lipids. Plasmalogens have an ether-linked alkenyl chain where most glycerophospholipids have an ester-linked fatty acid (compare Fig. 10-8). Platelet-activating factor has a long ether-linked alkyl chain at C-1 of glycerol, but C-2 is ester-linked to acetic acid, which makes the compound much more water-soluble than most glycerophospholipids and plasmalogens. The head-group alcohol is ethanolamine in plasmalogens and choline in platelet-activating factor.

biological significance of the variation in fatty acids and head groups is not yet understood.

Some Glycerophospholipids Have Ether-Linked Fatty Acids

Some animal tissues and some unicellular organisms are rich in **ether lipids**, in which one of the two acyl chains is attached to glycerol in ether, rather than ester, linkage. The ether-linked chain may be saturated, as in the alkyl ether lipids, or may contain a double bond between C-1 and C-2, as in **plasmalogens (Fig. 10-9)**. Vertebrate heart tissue is uniquely enriched in ether lipids; about half of the heart phospholipids are plasmalogens. The membranes of halophilic bacteria, ciliated protists, and certain invertebrates also contain high proportions of ether lipids. The functional significance of ether lipids in these membranes is unknown; perhaps their resistance to the phospholipases that cleave ester-linked fatty acids from membrane lipids is important in some roles.

At least one ether lipid, **platelet-activating factor**, is a potent molecular signal. It is released from leukocytes called basophils and stimulates platelet aggregation and the release of serotonin (a vasoconstrictor) from platelets. It also exerts a variety of effects on liver, smooth muscle, heart, uterine, and lung tissues

and plays an important role in inflammation and the allergic response. ■

Chloroplasts Contain Galactolipids and Sulfolipids

The second group of membrane lipids includes those that predominate in plant cells: the **galactolipids**, in which one or two galactose residues are connected by a glycosidic linkage to C-3 of a 1,2-diacylglycerol (**Fig. 10-10**; see also Fig. 10-6). Galactolipids are localized in the thylakoid membranes (internal membranes) of chloroplasts; they make up 70% to 80% of the total membrane lipids of a vascular plant, and are therefore probably the most abundant membrane lipids in the biosphere. Phosphate is often the limiting plant nutrient in soil, and perhaps the evolutionary pressure to conserve phosphate for more critical roles favored plants that made phosphate-free lipids. Plant membranes also contain sulfolipids, in which a sulfonated glucose residue is joined to a diacylglycerol in glycosidic linkage. The sulfonate group bears a negative charge like that of the phosphate group in phospholipids.

Archaea Contain Unique Membrane Lipids

Some archaea that live in ecological niches with extreme conditions—high temperatures (boiling water), low pH, high ionic strength, for example—have membrane lipids

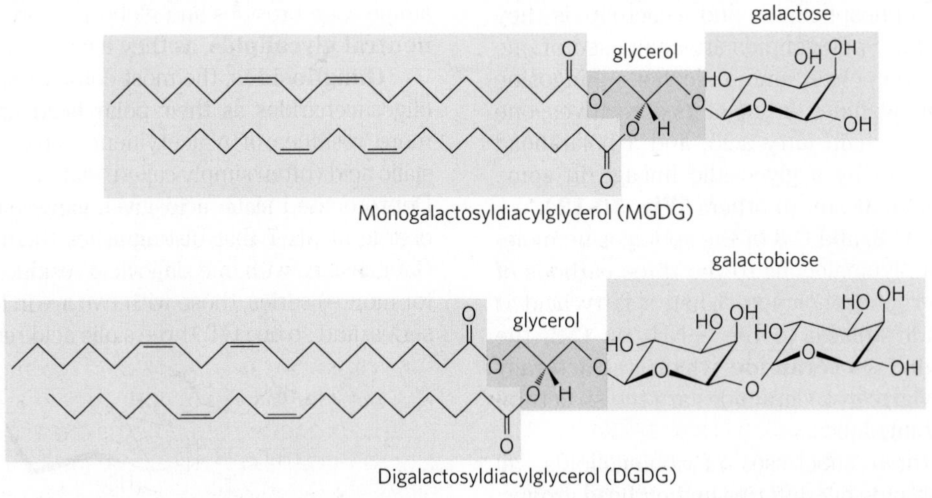

galactose

glycerol

Monogalactosyldiacylglycerol (MGDG)

galactobiose

glycerol

Digalactosyldiacylglycerol (DGDG)

FIGURE 10-10 Two galactolipids of chloroplast thylakoid membranes. In monogalactosyldiacylglycerols (MGDGs) and digalactosyldiacylglycerols (DGDGs), both acyl groups are polyunsaturated and the head groups are uncharged.

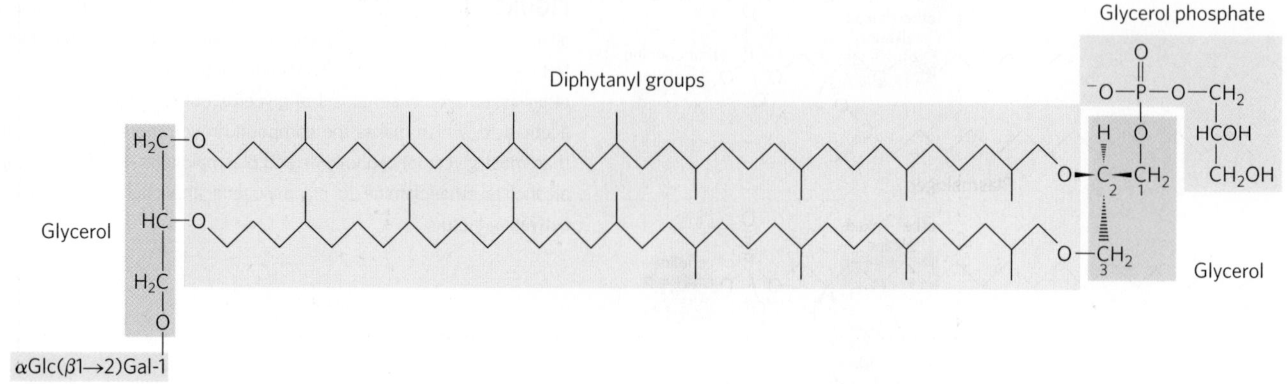

Glycerol phosphate

Diphytanyl groups

Glycerol

αGlc(β1→2)Gal-1

FIGURE 10-11 An unusual membrane lipid found only in some archaea. In this diphytanyl tetraether lipid, the diphytanyl moieties (yellow) are long hydrocarbons composed of eight five-carbon isoprene groups condensed head-to-head. (On the condensation of isoprene units, see Fig. 21-36; also, compare the diphytanyl groups with the 20-carbon phytol side chain of chlorophylls in Fig. 20-8a.) In this extended form, the diphytanyl groups are about twice the length of a 16-carbon fatty acid typically found in the membrane lipids of bacteria and eukaryotes. The glycerol moieties in the archaeal lipids are in the R configuration, in contrast to those of bacteria and eukaryotes, which have the S configuration. Archaeal lipids differ in the substituents on the glycerols. In the molecule shown here, one glycerol is linked to the disaccharide α-glucopyranosyl-(1→2)-β-galactofuranose; the other glycerol is linked to a glycerol phosphate head group.

containing long-chain (32 carbons) branched hydrocarbons linked at each end to glycerol **(Fig. 10-11)**. These linkages are through ether bonds, which are much more stable to hydrolysis at low pH and high temperature than are the ester bonds found in the lipids of bacteria and eukaryotes. In their fully extended form, these archaeal lipids are twice the length of phospholipids and sphingolipids, and can span the full width of the plasma membrane. At each end of the extended molecule is a polar head consisting of glycerol linked to either phosphate or sugar residues. The general name for these compounds, glycerol dialkyl glycerol tetraethers (GDGTs), reflects their unique structure. The glycerol moiety of the archaeal lipids is not the same stereoisomer as that in the lipids of bacteria and eukaryotes; the central carbon is in the R configuration in archaea, but in the S configuration in bacteria and eukaryotes (Fig. 10-7).

Sphingolipids Are Derivatives of Sphingosine

Sphingolipids, the fourth large class of membrane lipids, also have a polar head group and two nonpolar tails, but unlike glycerophospholipids and galactolipids they contain no glycerol. Sphingolipids are composed of one molecule of the long-chain amino alcohol sphingosine (also called 4-sphingenine) or one of its derivatives, one molecule of a long-chain fatty acid, and a polar head group that is joined by a glycosidic linkage in some cases and a phosphodiester in others **(Fig. 10-12)**.

Carbons C-1, C-2, and C-3 of the sphingosine molecule are structurally analogous to the three carbons of glycerol in glycerophospholipids. When a fatty acid is attached in amide linkage to the —NH$_2$ on C-2, the resulting compound is a **ceramide**, which is structurally similar to a diacylglycerol. Ceramides are the structural parents of all sphingolipids.

There are three subclasses of sphingolipids, all derivatives of ceramide but differing in their head groups: sphingomyelins, neutral (uncharged) glycolipids, and gangliosides. **Sphingomyelins** contain phosphocholine

or phosphoethanolamine as their polar head group and are therefore classified along with glycerophospholipids as phospholipids (Fig. 10-6). Indeed, sphingomyelins resemble phosphatidylcholines in their general properties and three-dimensional structure, and in having no net charge on their head groups **(Fig. 10-13)**. Sphingomyelins are present in the plasma membranes of animal cells and are especially prominent in myelin, a membranous sheath that surrounds and insulates the axons of some neurons—thus the name "sphingomyelins."

Glycosphingolipids, which occur largely in the outer face of plasma membranes, have head groups with one or more sugars connected directly to the —OH at C-1 of the ceramide moiety; they do not contain phosphate. **Cerebrosides** have a single sugar linked to ceramide; those with galactose are characteristically found in the plasma membranes of cells in neural tissue, and those with glucose in the plasma membranes of cells in nonneural tissues. **Globosides** are glycosphingolipids with two or more sugars, usually D-glucose, D-galactose, or N-acetyl-D-galactosamine. Cerebrosides and globosides are sometimes called **neutral glycolipids**, as they have no charge at pH 7.

Gangliosides, the most complex sphingolipids, have oligosaccharides as their polar head groups and one or more residues of N-acetylneuraminic acid (Neu5Ac), a sialic acid (often simply called "sialic acid"), at the termini. Deprotonated sialic acid gives gangliosides the negative charge at pH 7 that distinguishes them from globosides. Gangliosides with one sialic acid residue are in the GM (M for mono-) series, those with two are in the GD (D for di-) series, and so on (GT, three sialic acid residues; GQ, four).

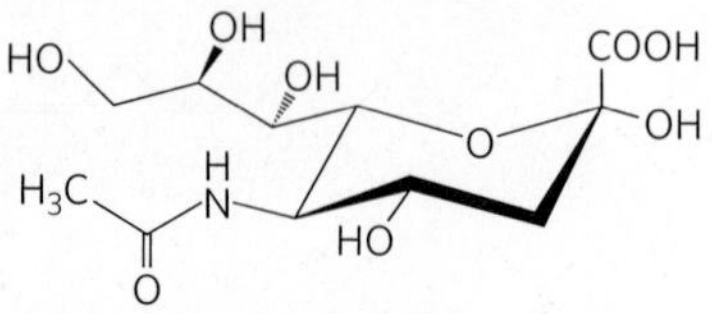

α-N-Acetylneuraminic acid (a sialic acid)
(Neu5Ac)

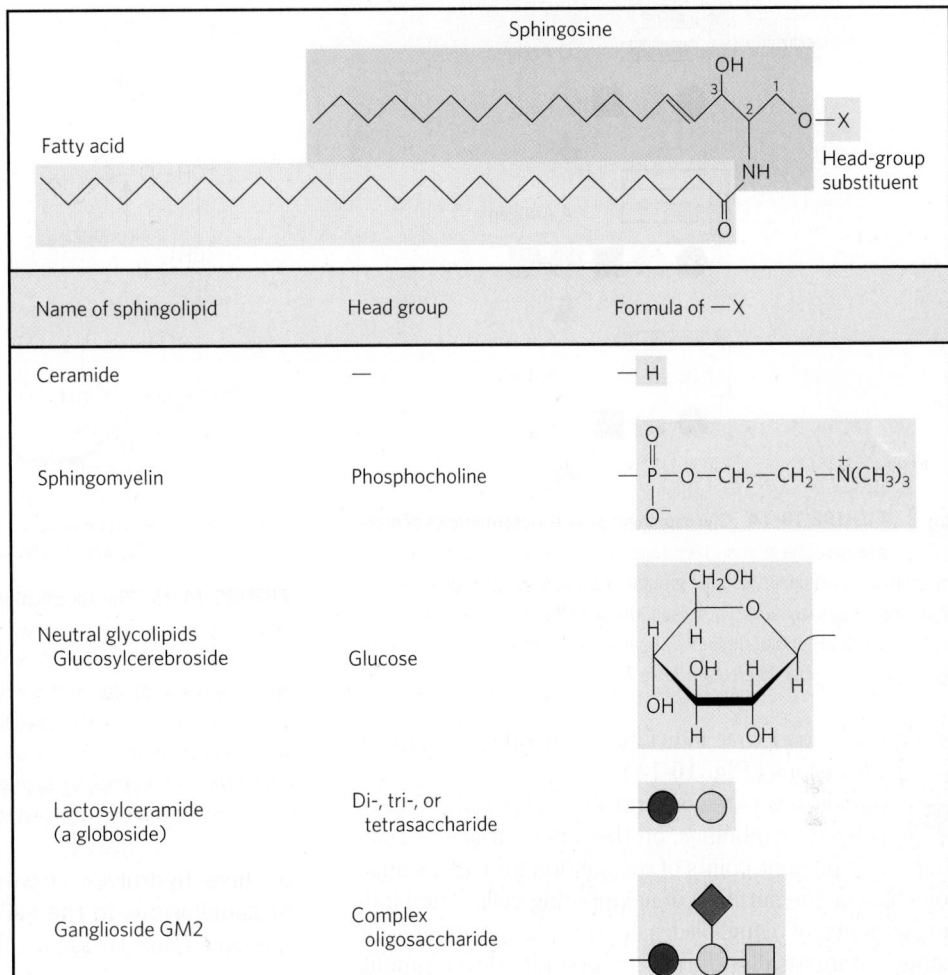

Johann Thudichum, 1829–1901
[Source: J. L. W. Thudichum,
Tubingen, F. Pietzcker (1898).]

FIGURE 10-12 Sphingolipids. The first three carbons at the polar end of sphingosine are analogous to the three carbons of glycerophospholipids. The amino group at C-2 bears a fatty acid in amide linkage. The fatty acid is usually saturated or monounsaturated, with 16, 18, 22, or 24 carbon atoms. Ceramide is the parent compound for this group. Other sphingolipids differ in the polar head group attached at C-1. Gangliosides have very complex oligosaccharide head groups. Standard symbols for sugars are used in this figure, as shown in Table 7-1.

Sphingolipids at Cell Surfaces Are Sites of Biological Recognition

When sphingolipids were discovered more than a century ago by the physician-chemist Johann Thudichum, their biological role seemed as enigmatic as the Sphinx, for which he therefore named them. In humans, at least 60 different sphingolipids have been identified in cellular membranes. Many of these are especially prominent in the plasma membranes of neurons, and some are clearly recognition sites on the cell surface, but a specific function for only a few sphingolipids has been discovered thus far. The carbohydrate moieties of certain sphingolipids define the human blood groups and therefore determine

Phosphatidylcholine

phosphocholine

Sphingomyelin

phosphocholine

FIGURE 10-13 The similar molecular structures of two classes of membrane lipid. Phosphatidylcholine (a glycerophospholipid) and sphingomyelin (a sphingolipid) have similar dimensions and physical properties, but presumably play different roles in membranes.

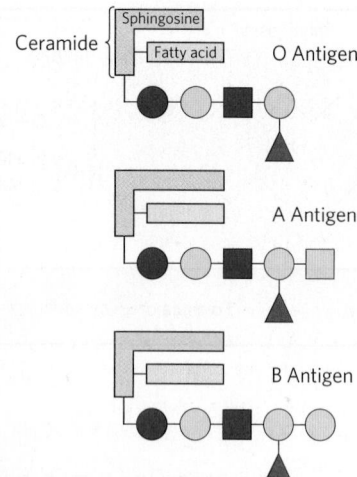

FIGURE 10-14 Glycosphingolipids as determinants of blood groups. The human blood groups (O, A, B) are determined in part by the oligosaccharide head groups of these glycosphingolipids. The same three oligosaccharides are also found attached to certain blood proteins of individuals of blood types O, A, and B, respectively. Standard symbols for sugars are used here (see Table 7-1).

the type of blood that individuals can safely receive in blood transfusions **(Fig. 10-14)**.

[medical symbol icon] Gangliosides are concentrated in the outer face of plasma membranes, on the outer surface of cells, where they present points of recognition for extracellular molecules or the surfaces of neighboring cells. The kinds and amounts of gangliosides in the plasma membrane change dramatically during embryonic development. Tumor formation induces the synthesis of a new complement of gangliosides, and very low concentrations of a specific ganglioside have been found to induce differentiation of cultured neuronal tumor cells. Guillain-Barré syndrome is a serious autoimmune disorder in which the body makes antibodies against its own gangliosides, including those in neurons. The resulting inflammation damages the peripheral nervous system, leading to temporary (or sometimes permanent) paralysis. In cholera, cholera toxin produced by the intestinal bacterium *Vibrio cholerae* enters sensitive cells after attaching to specific gangliosides on the intestinal epithelial cell surface (see Box 12-1). Investigation of the biological roles of diverse gangliosides remains fertile ground for future research. ∎

Phospholipids and Sphingolipids Are Degraded in Lysosomes

Most cells continually degrade and replace their membrane lipids. For each hydrolyzable bond in a glycerophospholipid, there is a specific hydrolytic enzyme in the lysosome **(Fig. 10-15)**. Phospholipases of the A type remove one of the two fatty acids, producing a lysophospholipid. (These esterases do not attack the ether link of plasmalogens.) Lysophospholipases remove the remaining fatty acid.

Gangliosides are degraded by a set of lysosomal enzymes that catalyze the stepwise removal of sugar units, finally yielding a ceramide. A genetic defect in any

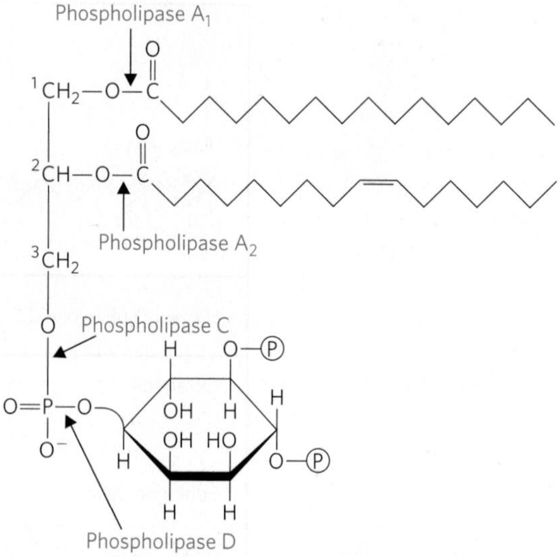

FIGURE 10-15 The specificities of phospholipases. Phospholipases A_1 and A_2 hydrolyze the ester bonds of intact glycerophospholipids at C-1 and C-2 of glycerol, respectively. When one of the fatty acids has been removed by a type A phospholipase, the second fatty acid is removed by a lysophospholipase (not shown). Phospholipases C and D each split one of the phosphodiester bonds in the head group. Some phospholipases act on only one type of glycerophospholipid, such as phosphatidylinositol 4,5-bisphosphate (PIP$_2$, shown here) or phosphatidylcholine; others are less specific.

of these hydrolytic enzymes leads to the accumulation of gangliosides in the cell, with severe medical consequences (Box 10-1).

Sterols Have Four Fused Carbon Rings

Sterols are structural lipids present in the membranes of most eukaryotic cells. The characteristic structure of this fifth group of membrane lipids is the steroid nucleus, consisting of four fused rings, three with six carbons and one with five **(Fig. 10-16)**. The steroid nucleus is almost planar and is relatively rigid; the fused rings do not allow rotation about C—C bonds. **Cholesterol**, the major sterol in animal tissues, is amphipathic, with a polar head group (the hydroxyl group at C-3) and a nonpolar hydrocarbon body (the steroid nucleus and the hydrocarbon side chain at C-17), about as long as a 16-carbon fatty acid

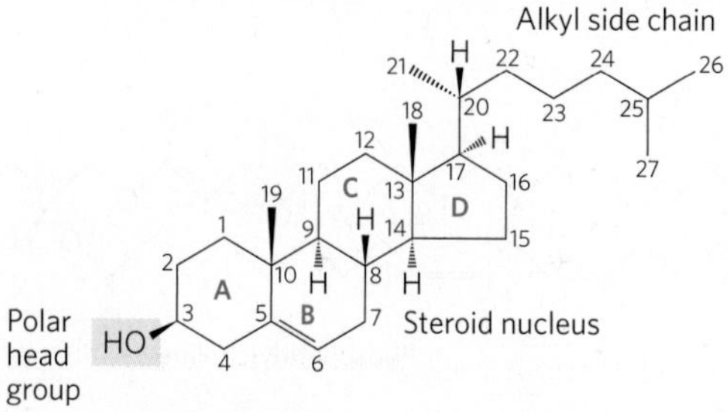

FIGURE 10-16 Cholesterol. In this chemical structure of cholesterol, the rings are labeled A through D to simplify reference to derivatives of the steroid nucleus. The C-3 hydroxyl group (shaded blue) is the polar head group. For storage and transport of the sterol, this hydroxyl group condenses with a fatty acid to form a sterol ester.

BOX 10-1    MEDICINE   Abnormal Accumulations of Membrane Lipids: Some Inherited Human Diseases

The polar lipids of membranes undergo constant metabolic turnover, the rate of their synthesis normally counterbalanced by the rate of breakdown. The breakdown of lipids is promoted by hydrolytic enzymes in lysosomes, each enzyme capable of hydrolyzing a specific bond. When sphingolipid degradation is impaired by a defect in one of these enzymes (Fig. 1), partial breakdown products accumulate in the tissues, causing serious disease. More than 50 distinct lysosomal storage diseases have been discovered, each the result of a single mutation in one of the genes for a lysosomal protein.

For example, Niemann-Pick disease is caused by a rare genetic defect in the enzyme sphingomyelinase, the enzyme that cleaves phosphocholine from sphingomyelin. Sphingomyelin accumulates in the brain, spleen, and liver. The disease becomes evident in infants and causes mental retardation and early death. More common is Tay-Sachs disease, in which ganglioside GM2 accumulates in the brain and spleen (Fig. 2) owing to lack of the enzyme hexosaminidase A. The symptoms of Tay-Sachs disease are progressive developmental retardation, paralysis, blindness, and death by the age of 3 or 4 years.

Genetic counseling can predict and avert many inheritable diseases. Tests on prospective parents can detect abnormal enzymes, then DNA testing can determine the exact nature of the defect and the risk it poses for offspring. Once a pregnancy occurs, fetal cells obtained by sampling a part of the placenta (chorionic villus sampling) or the fluid surrounding the fetus (amniocentesis) can be tested in the same way.

FIGURE 1 Pathways for the breakdown of GM1, globoside, and sphingomyelin to ceramide. A defect in the enzyme hydrolyzing a particular step is indicated by ⊗; the disease that results from accumulation of the partial breakdown product is noted.

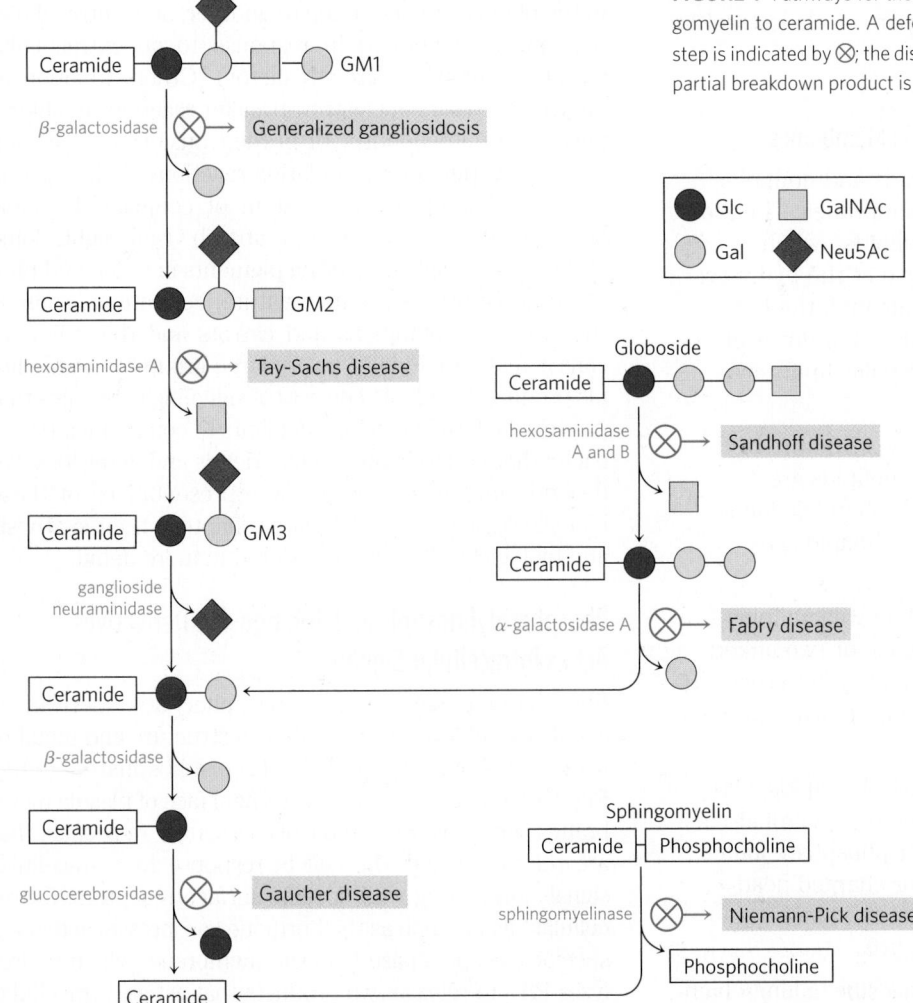

1 μm

FIGURE 2 Electron micrograph of a portion of a brain cell from an infant with Tay-Sachs disease, obtained post mortem, showing abnormal ganglioside deposits in the lysosomes. [Source: Otis Imboden/National Geographic/Getty Images.]

in its extended form. Similar sterols are found in other eukaryotes: stigmasterol in plants and ergosterol in fungi, for example. Bacteria cannot synthesize sterols; a few bacterial species, however, can incorporate exogenous sterols into their membranes. The sterols of all eukaryotes are synthesized from simple five-carbon isoprene subunits, as are the fat-soluble vitamins, quinones, and dolichols described in Section 10.3.

In addition to their roles as membrane constituents, the sterols serve as precursors for a variety of products with specific biological activities. Steroid hormones, for example, are potent biological signals that regulate gene expression. **Bile acids** are polar derivatives of cholesterol that act as detergents in the intestine, emulsifying dietary fats to make them more readily accessible to digestive lipases.

Taurocholic acid
(a bile acid)

We return to cholesterol and other sterols in later chapters, to consider the structural role of cholesterol in biological membranes (Chapter 11), signaling by steroid hormones (Chapter 12), and the remarkable biosynthetic pathway to cholesterol and transport of cholesterol by lipoprotein carriers (Chapter 21).

SUMMARY 10.2 Structural Lipids in Membranes

■ The polar lipids, with polar heads and nonpolar tails, are major components of membranes. The most abundant are the glycerophospholipids, which contain fatty acids esterified to two of the hydroxyl groups of glycerol, and a second alcohol, the head group, esterified to the third hydroxyl of glycerol via a phosphodiester bond. Other polar lipids are the sterols.

■ Glycerophospholipids differ in the structure of their head group; common glycerophospholipids are phosphatidylethanolamine and phosphatidylcholine. The polar heads of the glycerophospholipids are charged at pH near 7.

■ Chloroplast membranes are rich in galactolipids, composed of a diacylglycerol with one or two linked galactose residues, and sulfolipids, diacylglycerols with a linked sulfonated sugar residue and thus a negatively charged head group.

■ Some archaea have unique membrane lipids, with long-chain alkyl groups ether-linked to glycerol at each end and with sugar residues and/or phosphate joined to the glycerol to provide a polar or charged head group. These lipids are stable under the harsh conditions in which these archaea live.

■ The sphingolipids contain sphingosine, a long-chain aliphatic amino alcohol, but no glycerol. Sphingomyelin has, in addition to phosphoric acid and choline, two long hydrocarbon chains, one contributed by a fatty acid and the other by sphingosine. Three other classes of sphingolipids are cerebrosides, globosides, and gangliosides, which contain sugar components.

■ Sterols have four fused rings and a hydroxyl group. Cholesterol, the major sterol in animals, is both a structural component of membranes and precursor to a wide variety of steroids.

10.3 Lipids as Signals, Cofactors, and Pigments

The two functional classes of lipids considered thus far (storage lipids and structural lipids) are major cellular components; membrane lipids make up 5% to 10% of the dry mass of most cells, and storage lipids make up more than 80% of the mass of an adipocyte. With some important exceptions, these lipids play a *passive* role in the cell; lipid fuels are stored until oxidized by enzymes, and membrane lipids form impermeable barriers around cells and cellular compartments. Another group of lipids, present in much smaller amounts, includes those with *active* roles in the metabolic traffic as metabolites and messengers. Some serve as potent signals—as hormones, carried in the blood from one tissue to another, or as intracellular messengers generated in response to an extracellular signal (hormone or growth factor). Others function as enzyme cofactors in electron-transfer reactions in chloroplasts and mitochondria, or in the transfer of sugar moieties in a variety of glycosylation reactions. A third group consists of lipids with a system of conjugated double bonds: pigment molecules that absorb visible light. Some of these act as light-capturing pigments in vision and photosynthesis; others produce natural colorations, such as the orange of pumpkins and carrots and the yellow of canary feathers. Finally, a very large group of volatile lipids produced in plants consists of signaling molecules that pass through the air, allowing plants to communicate with each other and to invite animal friends and deter foes. We describe in this section a few representatives of these biologically active lipids. In later chapters, their synthesis and biological roles are considered in more detail.

Phosphatidylinositols and Sphingosine Derivatives Act as Intracellular Signals

Phosphatidylinositol and its phosphorylated derivatives act at several levels to regulate cell structure and metabolism. Phosphatidylinositol 4,5-bisphosphate (PIP$_2$; Fig. 10-15) in the cytoplasmic (inner) face of plasma membranes serves as a reservoir of messenger molecules that are released inside the cell in response to extracellular signals interacting with specific surface receptors. Extracellular signals such as the hormone vasopressin activate a specific phospholipase C in the membrane, which hydrolyzes PIP$_2$ to release two products that act as intracellular messengers: inositol 1,4,5-trisphosphate (IP$_3$), which is water-soluble, and diacylglycerol, which remains associated with the plasma membrane. IP$_3$ triggers release of Ca^{2+} from the endoplasmic reticulum, and the combination of diacylglycerol and elevated cytosolic Ca^{2+} activates the enzyme protein kinase C. By phosphorylating specific proteins, this enzyme brings about the cell's response to the extracellular signal. This signaling mechanism is described more fully in Chapter 12 (see Fig. 12-11).

Inositol phospholipids also serve as points of nucleation for supramolecular complexes involved in signaling or in exocytosis. Certain signaling proteins bind specifically to phosphatidylinositol 3,4,5-trisphosphate (PIP_3) in the plasma membrane, initiating the formation of multienzyme complexes at the membrane's cytosolic surface. Thus, formation of PIP_3 in response to extracellular signals brings the proteins together in signaling complexes at the surface of the plasma membrane (see Fig. 12-20).

Membrane sphingolipids also can serve as sources of intracellular messengers. Both ceramide and sphingomyelin (Fig. 10-12) are potent regulators of protein kinases, and ceramide or its derivatives are involved in the regulation of cell division, differentiation, migration, and programmed cell death (also called apoptosis; see Chapter 12).

Eicosanoids Carry Messages to Nearby Cells

Eicosanoids are paracrine hormones, substances that act only on cells near the point of hormone synthesis instead of being transported in the blood to act on cells in other tissues or organs. These fatty acid derivatives have a variety of dramatic effects on vertebrate tissues. They are involved in reproductive function; in the inflammation, fever, and pain associated with injury or disease; in the formation of blood clots and the regulation of blood pressure; in gastric acid secretion; and in various other processes important in human health or disease.

Eicosanoids are derived from arachidonate (arachidonic acid; 20:4($\Delta^{5,8,11,14}$)) and eicosapentaenoic acid (EPA; 20:5($\Delta^{5,8,11,14,17}$)), from which they take their general name (Greek *eikosi*, "twenty"). There are four major classes of eicosanoids: prostaglandins, thromboxanes, leukotrienes, and lipoxins **(Fig. 10-17)**. Eicosanoid names include letter designations for the functional groups on the ring and numbers indicating the number of double bonds in the hydrocarbon chain.

Prostaglandins (PG) contain a five-carbon ring. Their name derives from the prostate gland, the tissue from which they were first isolated by Bengt Samuelsson and Sune Bergström. PGE_2 and other series 2 prostaglandins are synthesized from arachidonate; series 3 prostaglandins are derived from EPA (see Fig. 21-12). Prostaglandins have an array of functions. Some stimulate contraction of the smooth muscle of the uterus during menstruation and labor. Others affect blood flow to specific organs, the wake-sleep cycle, and the responsiveness of certain tissues to hormones such as epinephrine and glucagon. Prostaglandins in a third group elevate body temperature (producing fever) and cause inflammation and pain.

The **thromboxanes (TX)** have a six-membered ring containing an ether. They are produced by platelets (also called thrombocytes) and act in the formation of blood clots and reduction of blood flow to the site of a clot. As shown by John Vane, the nonsteroidal antiinflammatory drugs (NSAIDs)—aspirin, ibuprofen, and meclofenamate,

Prostaglandin E_2 (PGE_2)

Arachidonate

NSAIDs

Lipoxin A_4 (LXA_4)

Thromboxane A_2 (TXA_2)

Leukotriene A_4 (LTA_4)

FIGURE 10-17 Arachidonic acid and some eicosanoid derivatives. Arachidonic acid (arachidonate at pH 7) is the precursor of eicosanoids, including the prostaglandins, thromboxanes, leukotrienes, and lipoxins. In prostaglandin E_2, C-8 and C-12 of arachidonate are joined to form the characteristic five-membered ring. In thromboxane A_2, the C-8 and C-12 are joined and an oxygen atom is added to form the six-membered ring. Nonsteroidal antiinflammatory drugs (NSAIDs) such as aspirin and ibuprofen block the formation of prostaglandins and thromboxanes from arachidonate by inhibiting the enzyme cyclooxygenase (prostaglandin H_2 synthase). Leukotriene A_4 has a series of three conjugated double bonds, and no cyclic moiety. Lipoxins are also noncyclic derivatives of arachidonate, with several hydroxyl groups.

John Vane (1927–2004), Sune Bergström (1916–2004), and Bengt Samuelsson [Source: Ira Wyman/Sygma/Corbis.]

for example—inhibit the enzyme prostaglandin H_2 synthase (also called cyclooxygenase2, or COX-2), which catalyzes an early step in the pathway from arachidonate to series 2 prostaglandins and thromboxanes (Fig. 10-17) and from EPA to series 3 prostaglandins and thromboxanes (see Fig. 21-12).

Leukotrienes (LT), first found in leukocytes, contain three conjugated double bonds. They are powerful biological signals. For example, leukotriene D_4, derived from leukotriene A_4, induces contraction of the smooth muscle lining the airways to the lung. Overproduction of leukotrienes causes asthmatic attacks, and leukotriene synthesis is one target of antiasthmatic drugs such as prednisone. The strong contraction of the smooth muscle of the lungs that occurs during anaphylactic shock is part of the potentially fatal allergic reaction in individuals hypersensitive to bee stings, penicillin, or other agents.

Lipoxins (LX), like leukotrienes, are linear eicosanoids. Their distinguishing feature is the presence of several hydroxyl groups along the chain (Fig. 10-17). These compounds are potent antiinflammatory agents. Because their synthesis is stimulated by low doses (81 mg) of aspirin taken daily, this low dose is commonly prescribed for individuals with cardiovascular disease. ■

Steroid Hormones Carry Messages between Tissues

Steroids are oxidized derivatives of sterols; they have the sterol nucleus but lack the alkyl chain attached to ring D of cholesterol, and they are more polar than cholesterol. Steroid hormones move through the bloodstream (on protein carriers) from their site of production to target tissues, where they enter cells, bind to highly specific receptor proteins in the nucleus, and trigger changes in gene expression and thus metabolism. Because hormones have very high affinity for their receptors, very low concentrations of hormones (nanomolar or less) are sufficient to produce responses in target tissues. The major groups of steroid hormones are the male and female sex hormones and the hormones produced by the adrenal cortex, cortisol and aldosterone **(Fig. 10-18)**. Prednisone and prednisolone are steroid drugs with strong antiinflammatory activities, mediated in part by the inhibition of arachidonate release by phospholipase A_2 and consequent inhibition of the synthesis of prostaglandins, thromboxanes, leukotrienes, and lipoxins. These drugs have a variety of medical applications, including the treatment of asthma and rheumatoid arthritis. ■

Vascular plants contain the steroidlike brassinolide (Fig. 10-18), a potent growth regulator that increases the rate of stem elongation and affects the orientation of cellulose microfibrils in the cell wall during growth.

Vascular Plants Produce Thousands of Volatile Signals

Plants produce thousands of different lipophilic compounds, volatile substances that are used to attract pollinators, to repel herbivores, to attract organisms that defend the plant against herbivores, and to communicate with other plants. Jasmonate, for example, derived from

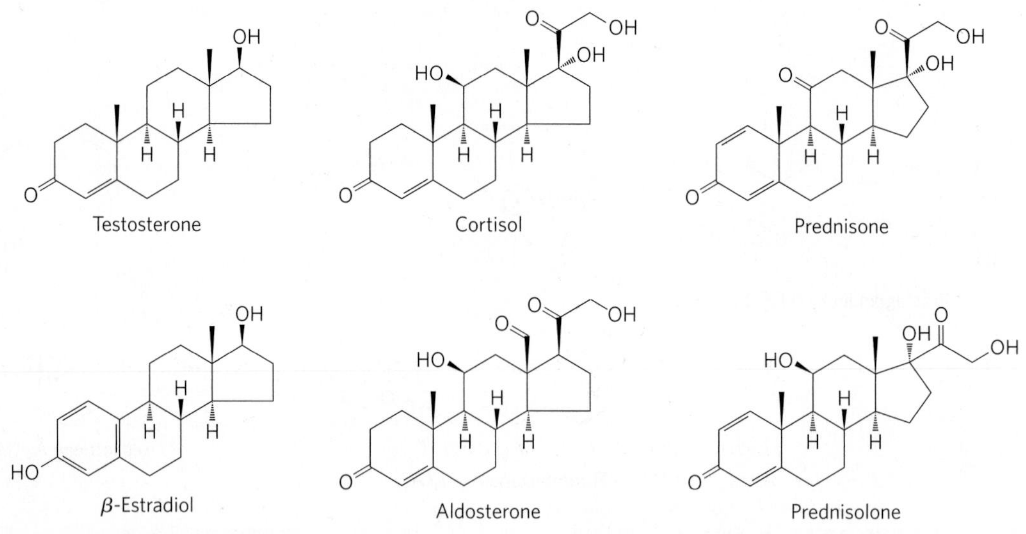

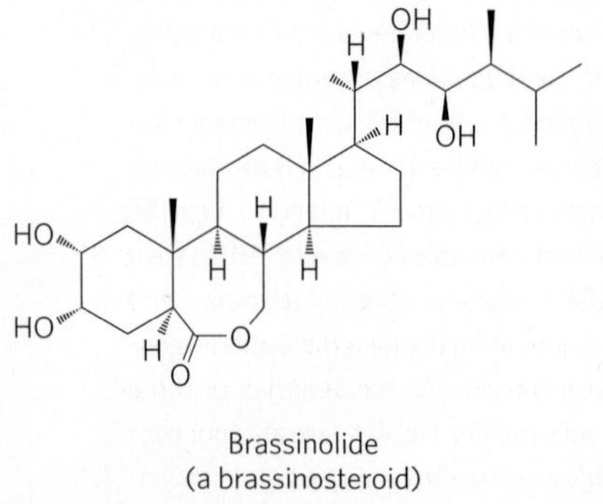

FIGURE 10-18 Steroids derived from cholesterol. Testosterone, the male sex hormone, is produced in the testes. Estradiol, one of the female sex hormones, is produced in the ovaries and placenta. Cortisol and aldosterone are hormones synthesized in the cortex of the adrenal gland; they regulate glucose metabolism and salt excretion, respectively. Prednisone and prednisolone are synthetic steroids used as antiinflammatory agents. Brassinolide is a growth regulator found in vascular plants.

Testosterone

Cortisol

Prednisone

β-Estradiol

Aldosterone

Prednisolone

Brassinolide
(a brassinosteroid)

the fatty acid $18:3(\Delta^{9,12,15})$ in membrane lipids, triggers the plant's defenses in response to insect-inflicted damage. The methyl ester of jasmonate gives the characteristic fragrance of jasmine oil, which is widely used in the perfume industry. Many plant volatiles, including geraniol (the characteristic scent of geraniums), β-pinene (pine trees), limonene (limes), menthol, and carvone (see Fig. 1-25a), are derived from fatty acids or from compounds made by the condensation of five-carbon isoprene units.

$$CH_2=\overset{\overset{\displaystyle CH_3}{|}}{C}-CH=CH_2$$
Isoprene

Vitamins A and D Are Hormone Precursors

During early decades of the twentieth century, a major focus of research in physiological chemistry was the identification of **vitamins**, compounds that are essential to the health of humans and other vertebrates but cannot be synthesized by these animals and must therefore be obtained in the diet. Early nutritional studies identified two general classes of such compounds: those soluble in nonpolar organic solvents (fat-soluble vitamins) and those that could be extracted from foods with aqueous solvents (water-soluble vitamins). Eventually, the fat-soluble group was resolved into the four vitamin groups A, D, E, and K, all of which are isoprenoid compounds synthesized by the condensation of multiple isoprene units. Two of these (D and A) serve as hormone precursors.

Vitamin D₃, also called **cholecalciferol**, is normally formed in the skin from 7-dehydrocholesterol in a photochemical reaction driven by the UV component of sunlight **(Fig. 10-19a)**. Vitamin D₃ is not itself biologically active, but it is converted by enzymes in the liver and kidney to 1α,25-dihydroxyvitamin D₃ (calcitriol), a hormone that regulates calcium uptake in the intestine and calcium levels in kidney and bone. Deficiency of vitamin D leads to defective bone formation and the disease rickets, for which administration of vitamin D produces a dramatic cure (Fig. 10-19b). Vitamin D₂ (ergocalciferol) is a commercial product formed by UV irradiation of the ergosterol of yeast. Vitamin D₂ is structurally similar to D₃, with slight modification to the side chain attached to the sterol D ring. Both have the same biological effects, and D₂ is commonly added to milk and butter as a dietary supplement. The product of vitamin D metabolism, 1α,25-dihydroxyvitamin D₃, regulates gene expression by interacting with specific nuclear receptor proteins (pp. 1156–1157).

Vitamin A₁ (all-*trans*-retinol) and its oxidized metabolites retinoic acid and retinal act in the processes of development, cell growth and differentiation, and vision **(Fig. 10-20)**. Vitamin A₁ or β-carotene in the diet can be converted enzymatically to all-*trans*-retinoic acid, a retinoid hormone that acts through a family of nuclear receptor proteins (RAR, RXR, PPAR) to regulate gene expression central to embryonic development, stem cell differentiation, and cell proliferation. All-*trans*-retinoic

(a)

7-Dehydrocholesterol   UV light, 2 steps (in skin)   Cholecalciferol (vitamin D₃)   1 step in the liver, 1 step in the kidney   1α,25-Dihydroxyvitamin D₃ (calcitriol)

FIGURE 10-19 Vitamin D₃ production and metabolism. (a) Cholecalciferol (vitamin D₃) is produced in the skin by UV irradiation of 7-dehydrocholesterol, which breaks the bond shaded light red. In the liver, a hydroxyl group is added at C-25; in the kidney, a second hydroxylation at C-1 produces the active hormone, 1α,25-dihydroxyvitamin D₃. This hormone regulates the metabolism of Ca^{2+} in kidney, intestine, and bone. **(b)** Dietary vitamin D prevents rickets, a disease once common in cold climates where heavy clothing blocks the UV component of sunlight necessary for the production of vitamin D₃ in skin. In this detail from a large mural by John Steuart Curry, *The Social Benefits of Biochemical Research* (1943), the people and animals on the left show the effects of poor nutrition, including the bowed legs of a boy with classical rickets. On the right are the people and animals made healthier with the "social benefits of research," including the use of vitamin D to prevent and treat rickets. [Source: (b) Courtesy of Media Center, University of Wisconsin–Madison, Department of Biochemistry.]

(b)

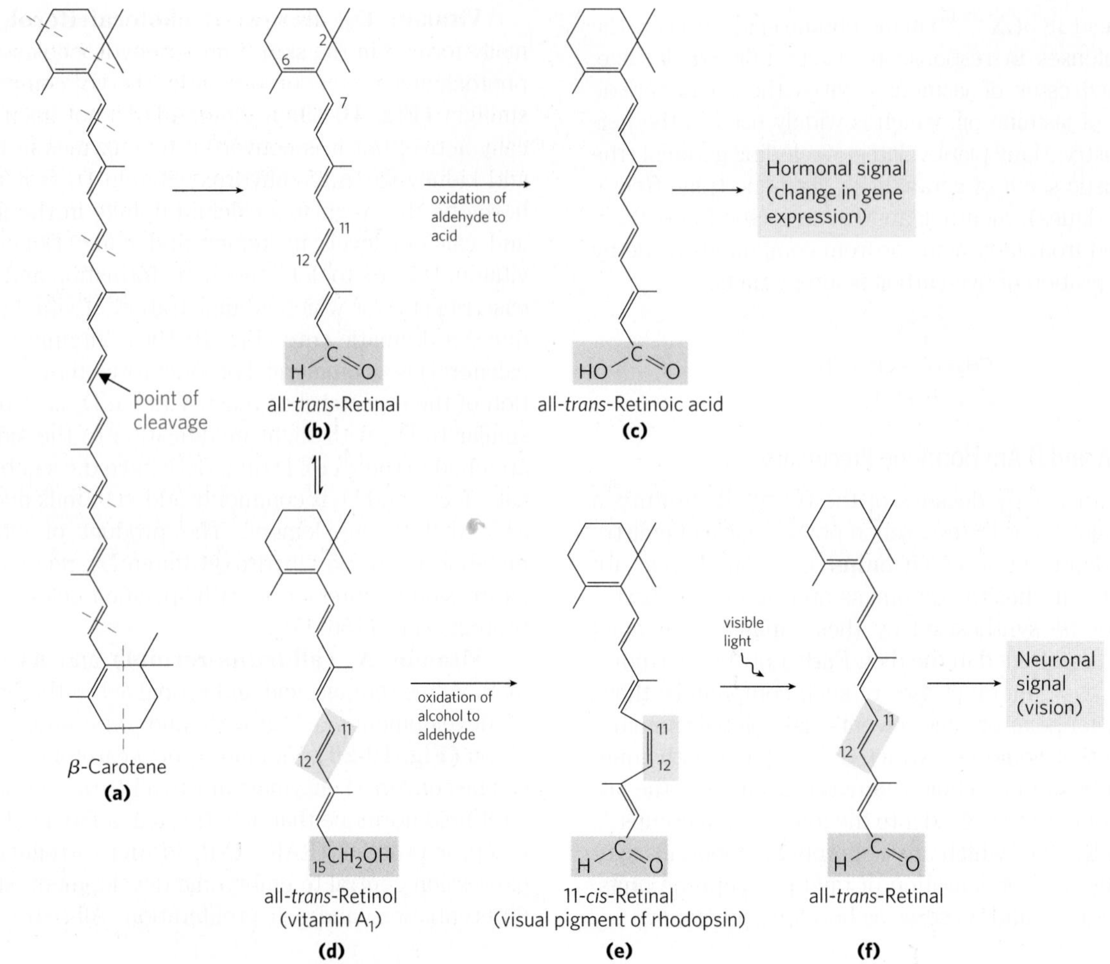

FIGURE 10-20 Structures

Hormonal signal (change in gene expression)

point of cleavage

β-Carotene
(a)

oxidation of aldehyde to acid

all-*trans*-Retinal
(b)

all-*trans*-Retinoic acid
(c)

oxidation of alcohol to aldehyde

all-*trans*-Retinol
(vitamin A₁)
(d)

visible light

11-*cis*-Retinal
(visual pigment of rhodopsin)
(e)

all-*trans*-Retinal
(f)

Neuronal signal (vision)

FIGURE 10-20 Dietary β-carotene and vitamin A₁ as precursors of the retinoids. (a) β-Carotene is shown with its isoprene structural units set off by dashed red lines. Symmetric cleavage of β-carotene yields two molecules of all-*trans*-retinal **(b)**, which can be either further oxidized to all-*trans*-retinoic acid, a retinoid hormone **(c)**, or reduced to all-*trans*-retinol, vitamin A₁ **(d)**. In the visual pathway, all-*trans*-retinol from this reaction, or obtained directly through the diet, can be converted to the aldehyde 11-*cis*-retinal **(e)**. This product combines with the protein opsin to form rhodopsin (not shown), a visual pigment widespread in nature. In the dark, the retinal of rhodopsin is in the 11-*cis* form. When a rhodopsin molecule is excited by visible light, the 11-*cis*-retinal undergoes a series of photochemical reactions that convert it to all-*trans*-retinal **(f)**, forcing a change in the shape of the entire rhodopsin molecule. This transformation in the rod cell of the vertebrate retina sends an electrical signal to the brain that is the basis of visual transduction (see Fig. 12-14).

acid is used to treat certain types of leukemia, and it is the active ingredient in the drug tretinoin (Retin-A), used to treat severe acne and wrinkled skin. In the vertebrate eye, retinal bound to the protein opsin forms the photoreceptor pigment rhodopsin. The photochemical conversion of 11-*cis*-retinal to all-*trans*-retinal is the fundamental event in vision (see Fig. 12-14).

Unlike most vitamins, vitamin A can be stored for some time in the body (primarily as its ester with palmitic acid, in the liver). Vitamin A was first isolated from fish liver oils; eggs, whole milk, and butter are also good dietary sources. Another source is β-carotene (Fig. 10-20), the pigment that gives carrots, sweet potatoes, and other yellow vegetables their characteristic color. Carotene is one of a very large number (>700) of **carotenoids**, natural products with a characteristic extensive system of conjugated double bonds, which makes possible their strong absorption of visible light (450–470 nm).

Vitamin A deficiency in a pregnant woman can lead to congenital malformations and growth retardation in the infant. In adults, vitamin A is also essential to vision, immunity, and reproduction. Deficiency of vitamin A leads to a variety of symptoms, including dryness of the skin, eyes, and mucous membranes, and night blindness, an early symptom commonly used in diagnosing vitamin A deficiency. In the developing world, vitamin A deficiency causes an estimated million or more cases of blindness or death each year. One effective strategy for providing vitamin A is the metabolic engineering of rice strains to overproduce β-carotene. Rice has all the enzymatic machinery to produce β-carotene in its leaves, but these enzymes are less active in the grain. Introduction of two genes into the rice has resulted in "golden rice" having grains much enriched in β-carotene **(Fig. 10-21).** ∎

Vitamins E and K and the Lipid Quinones Are Oxidation-Reduction Cofactors

 Vitamin E is the collective name for a group of closely related lipids called **tocopherols**, all of

FIGURE 10-21 Carotene-enriched rice. Worldwide, about 200 million women and children suffer from vitamin A deficiency, which causes 500,000 cases of irreversible blindness and up to 2 million deaths annually, particularly where rice is a staple food. An international humanitarian effort—the Golden Rice Project—has made great strides in addressing this health crisis. Wild-type rice grains (left) do not produce β-carotene, the metabolic precursor of vitamin A. Rice plants have been genetically engineered to produce β-carotene in the grain, which takes on the yellow color of the carotene (right). A diet supplemented with Golden Rice provides enough β-carotene to prevent vitamin A deficiency and its tragic health consequences. [Source: © Golden Rice Humanitarian Board (www.goldenrice.org).]

which contain a substituted aromatic ring and a long isoprenoid side chain **(Fig. 10-22a)**. Because they are hydrophobic, tocopherols associate with cell membranes, lipid deposits, and lipoproteins in the blood. Tocopherols are biological antioxidants. The aromatic ring reacts with and destroys the most reactive forms of oxygen radicals and other free radicals, protecting unsaturated fatty acids from oxidation and preventing oxidative damage to membrane lipids, which can cause cell fragility. Tocopherols are found in eggs and vegetable oils and are especially abundant in

(a)
Vitamin E: an antioxidant

(b)
Vitamin K_1: a blood-clotting cofactor (phylloquinone)

(c)
Warfarin: a blood anticoagulant

(d)
Ubiquinone: a mitochondrial electron carrier (coenzyme Q)
(n = 4 to 8)

(e)
Plastoquinone: a chloroplast electron carrier (n = 4 to 8)

(f)
Dolichol: a sugar carrier
(n = 9 to 22)

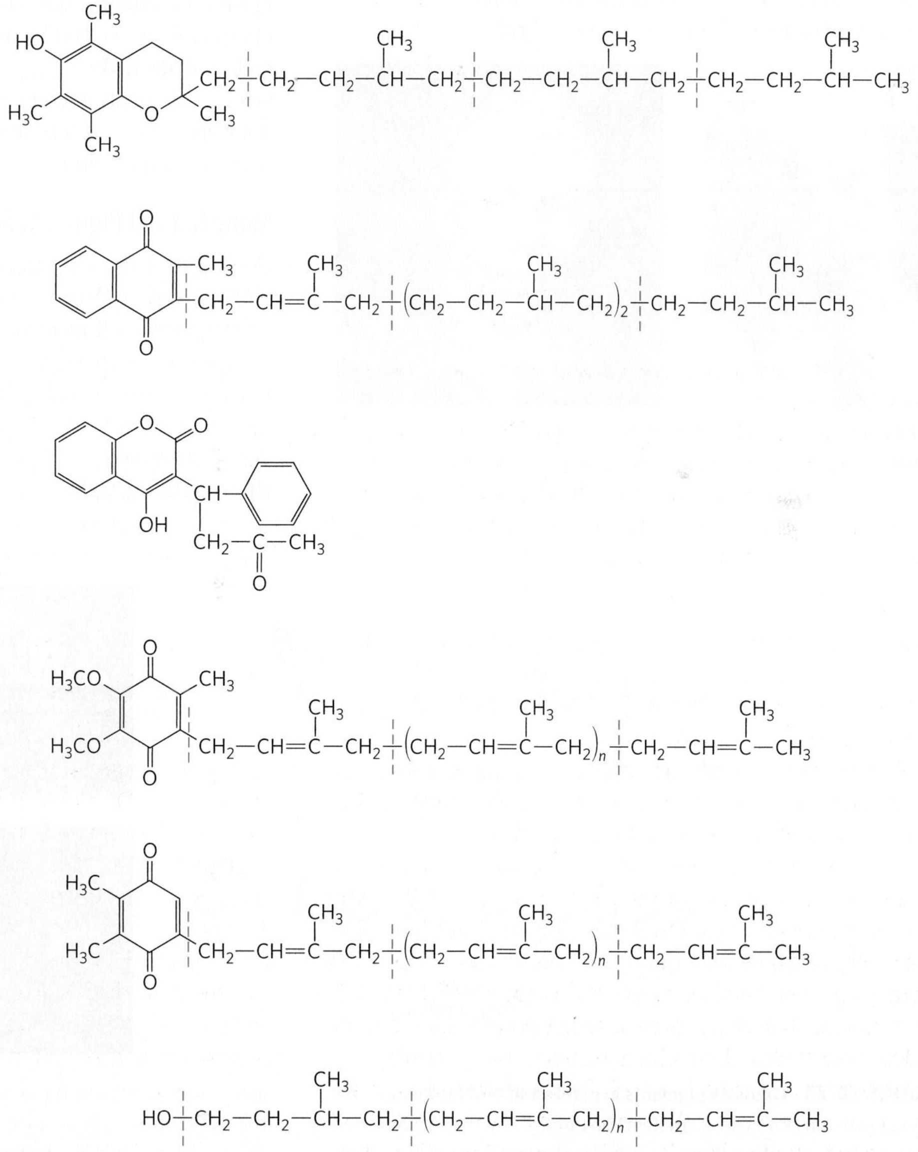

FIGURE 10-22 Some other biologically active isoprenoid compounds or derivatives. Units derived from isoprene are set off by dashed red lines. In most mammalian tissues, ubiquinone (also called coenzyme Q) has 10 isoprene units. Dolichols of animals have 17 to 21 isoprene units (85 to 105 carbon atoms), bacterial dolichols have 11, and those of plants and fungi have 14 to 24.

wheat germ. Laboratory animals fed diets depleted of vitamin E develop scaly skin, muscular weakness and wasting, and sterility. Vitamin E deficiency in humans is very rare; the principal symptom is fragile erythrocytes.

The aromatic ring of **vitamin K** (Fig. 10-22b) undergoes a cycle of oxidation and reduction during the formation of active prothrombin, a blood plasma protein essential in blood clotting. Prothrombin is a proteolytic enzyme that splits peptide bonds in the blood protein fibrinogen to convert it to fibrin, the insoluble fibrous protein that holds blood clots together (see Fig. 6-40). Henrik Dam and Edward A. Doisy independently discovered that vitamin K deficiency slows blood clotting, which can be fatal. Vitamin K deficiency is extremely uncommon in humans, aside from a small percentage of infants who suffer from hemorrhagic disease of the newborn, a potentially fatal disorder. In the United States, newborns are routinely given a 1 mg injection of vitamin K. Vitamin K_1 (phylloquinone) is found in green plant leaves; a related form, vitamin K_2 (menaquinone), is formed by bacteria living in the vertebrate intestine.

Henrik Dam, 1895–1976
[Source: Science Source.]

Edward A. Doisy, 1893–1986
[Source: National Library of Medicine/
Science Photo Library/Science Source.]

Warfarin (Fig. 10-22c) is a synthetic compound that inhibits the formation of active prothrombin. It is particularly poisonous to rats, causing death by internal bleeding. Ironically, this potent rodenticide is also an invaluable anticoagulant drug for treating humans at risk for excessive blood clotting, such as surgical patients and those with coronary thrombosis. ■

Ubiquinone (also called coenzyme Q) and plastoquinone (Fig. 10-22d, e) are isoprenoids that function as lipophilic electron carriers in the oxidation-reduction reactions that drive ATP synthesis in mitochondria and chloroplasts, respectively. Both ubiquinone and plastoquinone can accept either one or two electrons and either one or two protons (see Fig. 19-3).

Dolichols Activate Sugar Precursors for Biosynthesis

During assembly of the complex carbohydrates of bacterial cell walls, and during the addition of polysaccharide units to certain proteins (glycoproteins) and lipids (glycolipids) in eukaryotes, the sugar units to be added are chemically activated by attachment to isoprenoid alcohols called **dolichols** (Fig. 10-22f). These compounds have strong hydrophobic interactions with membrane lipids, anchoring the attached sugars to the membrane, where they participate in sugar-transfer reactions.

Many Natural Pigments Are Lipidic Conjugated Dienes

Conjugated dienes have carbon chains with alternating single and double bonds. Because this structural arrangement allows the delocalization of electrons, the compounds can be excited by low-energy electromagnetic radiation (visible light), giving them colors visible to humans and other animals. Carotene (Fig. 10-20) is yellow-orange; similar compounds give bird feathers their striking reds, oranges, and yellows (**Fig. 10-23**).

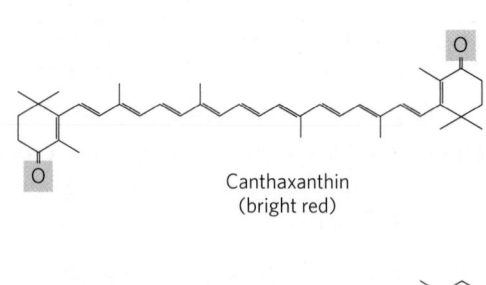

Canthaxanthin
(bright red)

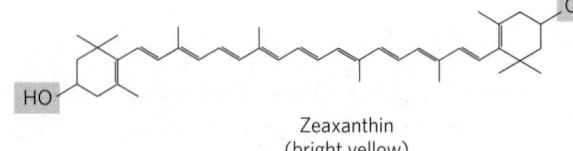

Zeaxanthin
(bright yellow)

FIGURE 10-23 Lipids as pigments in plants and bird feathers. Compounds with long conjugated systems absorb light in the visible region of the spectrum. Subtle differences in the chemistry of these compounds produce pigments of strikingly different colors. Birds acquire the pigments that color their feathers red or yellow by eating plant materials that contain carotenoid pigments, such as canthaxanthin and zeaxanthin. The differences in pigmentation between male and female birds are the result of differences in intestinal uptake and processing of carotenoids. [Sources: Cardinal: Dr. Dan Sudia/Science Source. Goldfinch: Richard Day/VIREO.]

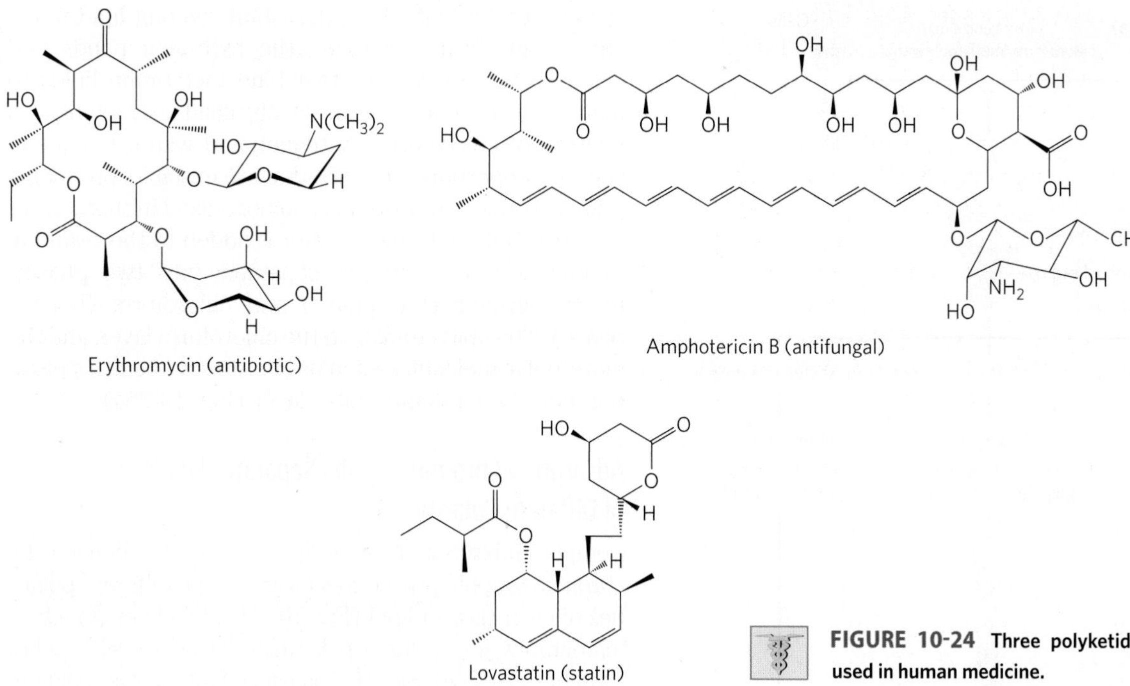

Erythromycin (antibiotic)

Amphotericin B (antifungal)

Lovastatin (statin)

FIGURE 10-24 Three polyketide natural products used in human medicine.

Like sterols, steroids, dolichols, vitamins A, E, D, and K, ubiquinone, and plastoquinone, these pigments are synthesized from five-carbon isoprene derivatives; the biosynthetic pathway is described in detail in Chapter 21.

Polyketides Are Natural Products with Potent Biological Activities

Polyketides are a diverse group of lipids with biosynthetic pathways (Claisen condensations) similar to those for fatty acids. They are **secondary metabolites**, compounds that are not central to an organism's metabolism but serve some subsidiary function that gives the organism an advantage in some ecological niche. Many polyketides find use in medicine as antibiotics (erythromycin), antifungals (amphotericin B), or inhibitors of cholesterol synthesis (lovastatin) **(Fig. 10-24)**. ■

SUMMARY 10.3 Lipids as Signals, Cofactors, and Pigments

■ Some types of lipids, although present in relatively small quantities, play critical roles as cofactors or signals.

■ Phosphatidylinositol bisphosphate is hydrolyzed to yield two intracellular messengers, diacylglycerol and inositol 1,4,5-trisphosphate. Phosphatidylinositol 3,4,5-trisphosphate is a nucleation point for supramolecular protein complexes involved in biological signaling.

■ Prostaglandins, thromboxanes, leukotrienes, and lipoxins, all of which are eicosanoids derived from arachidonate, are extremely potent hormones.

■ Steroid hormones, such as the sex hormones, are derived from sterols. They serve as powerful biological signals, altering gene expression in target cells.

■ Vitamins D, A, E, and K are fat-soluble compounds made up of isoprene units. All play essential roles in

the metabolism or physiology of animals. Vitamin D is precursor to a hormone that regulates calcium metabolism. Vitamin A furnishes the visual pigment of the vertebrate eye and is a regulator of gene expression during epithelial cell growth. Vitamin E functions in the protection of membrane lipids from oxidative damage, and vitamin K is essential in the blood-clotting process.

■ Ubiquinones and plastoquinones, also isoprenoid derivatives, are electron carriers in mitochondria and chloroplasts, respectively.

■ Dolichols activate and anchor sugars to cellular membranes; the sugar groups are then used in the synthesis of complex carbohydrates, glycolipids, and glycoproteins.

■ Lipidic conjugated dienes serve as pigments in flowers and fruits and give bird feathers their striking colors.

■ Polyketides are natural products widely used in medicine.

10.4 Working with Lipids

Because lipids are insoluble in water, their extraction and subsequent fractionation require the use of organic solvents and some techniques not commonly used in the purification of water-soluble molecules such as proteins and carbohydrates. In general, complex mixtures of lipids are separated by differences in polarity or solubility in nonpolar solvents. Lipids that contain ester- or amide-linked fatty acids can be hydrolyzed by treatment with acid or alkali or with specific hydrolytic enzymes (phospholipases, glycosidases) to yield their components for analysis. Some methods commonly used in lipid analysis are shown in **Figure 10-25** and discussed below.

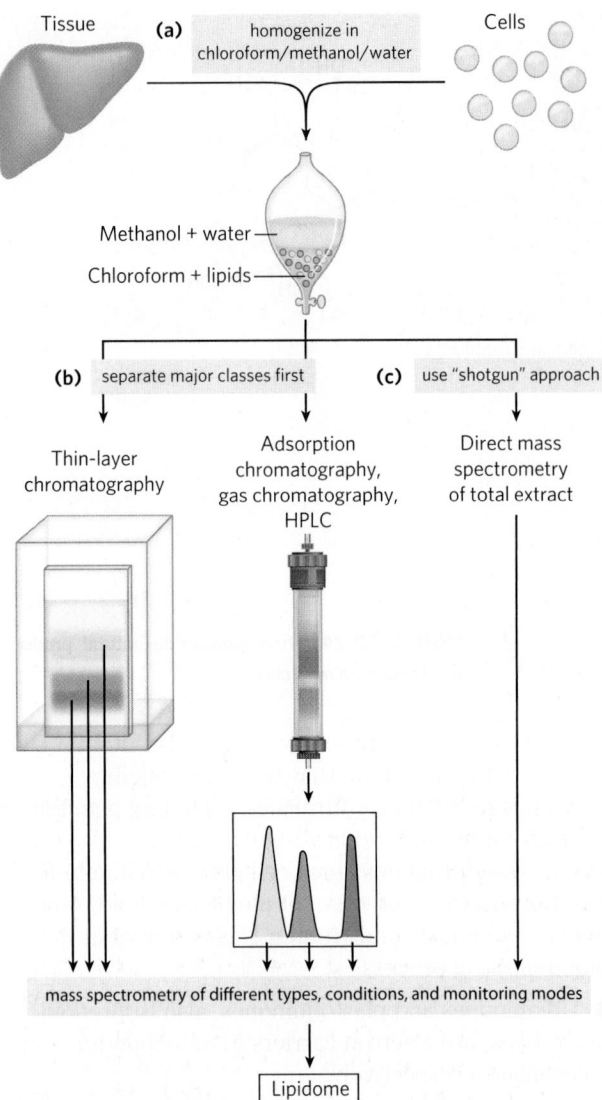

Methanol + water

Chloroform + lipids

(b) separate major classes first

(c) use "shotgun" approach

Thin-layer chromatography

Adsorption chromatography, gas chromatography, HPLC

Direct mass spectrometry of total extract

mass spectrometry of different types, conditions, and monitoring modes

Lipidome

FIGURE 10-25 Common procedures in the extraction, separation, and identification of cellular lipids. (a) Tissue is homogenized in a chloroform/methanol/water mixture, which on addition of water and removal of unextractable sediment by centrifugation yields two phases. **(b)** Major classes of extracted lipids in the chloroform phase may first be separated by thin-layer chromatography (TLC), in which lipids are carried up a silica gel–coated plate by a rising solvent front, less-polar lipids traveling farther than more-polar or charged lipids, or by adsorption chromatography on a column of silica gel, through which solvents of increasing polarity are passed. For example, column chromatography with appropriate solvents can be used to separate closely related lipid species such as phosphatidylserine, phosphatidylglycerol, and phosphatidylinositol. Once separated, each lipid's complement of fatty acids can be determined by mass spectrometry. **(c)** Alternatively, in the "shotgun" approach, an unfractionated extract of lipids can be directly subjected to high-resolution mass spectrometry of different types and under different conditions to determine the total composition of all the lipids—that is, the lipidome.

Lipid Extraction Requires Organic Solvents

Neutral lipids (triacylglycerols, waxes, pigments, and so forth) are readily extracted from tissues with ethyl ether, chloroform, or benzene, solvents that do not permit lipid clustering driven by the hydrophobic effect. Membrane lipids are more effectively extracted by more polar organic solvents, such as ethanol or methanol, which

reduce the hydrophobic interactions among lipid molecules while also weakening the hydrogen bonds and electrostatic interactions that bind membrane lipids to membrane proteins. A commonly used extractant is a mixture of chloroform, methanol, and water, initially in volume proportions (1:2:0.8) that are miscible, producing a single phase. After tissue is homogenized in this solvent to extract all lipids, more water is added to the resulting extract, and the mixture separates into two phases: methanol/water (top phase) and chloroform (bottom phase). The lipids remain in the chloroform layer, and the more polar molecules such as proteins and sugars partition into the methanol/water layer (Fig. 10-25a).

Adsorption Chromatography Separates Lipids of Different Polarity

Complex mixtures of tissue lipids can be fractionated by chromatographic procedures based on the different polarities of each class of lipid (Fig. 10-25b). In adsorption chromatography, an insoluble, polar material such as silica gel (a form of silicic acid, $Si(OH)_4$) is packed into a glass column, and the lipid mixture (in chloroform solution) is applied to the top of the column. (In high-performance liquid chromatography, the column is of smaller diameter and solvents are forced through the column under high pressure.) The polar lipids bind tightly to the polar silicic acid, but the neutral lipids pass directly through the column and emerge in the first chloroform wash. The polar lipids are then eluted, in order of increasing polarity, by washing the column with solvents of progressively higher polarity. Uncharged but polar lipids (cerebrosides, for example) are eluted with acetone, and very polar or charged lipids (such as glycerophospholipids) are eluted with methanol.

Thin-layer chromatography on silicic acid employs the same principle (Fig. 10-25b). A thin layer of silica gel is spread onto a glass plate, to which it adheres. A small sample of lipids dissolved in chloroform is applied near one edge of the plate, which is dipped in a shallow container of an organic solvent or solvent mixture; the entire setup is enclosed in a chamber saturated with the solvent vapor. As the solvent rises on the plate by capillary action, it carries lipids with it. The less polar lipids move farthest, as they have less tendency to bind to the silicic acid. The separated lipids can be detected by spraying the plate with a dye (rhodamine) that fluoresces when associated with lipids, or by exposing the plate to iodine fumes. Iodine reacts reversibly with the double bonds in fatty acids, such that lipids containing unsaturated fatty acids develop a yellow or brown color. Several other spray reagents are also useful in detecting specific lipids. For subsequent analysis, regions containing separated lipids can be scraped from the plate and the lipids recovered by extraction with an organic solvent.

Gas Chromatography Resolves Mixtures of Volatile Lipid Derivatives

Gas chromatography (GC) separates volatile components of a mixture according to their relative tendencies to dissolve in the inert material packed in the chromatography column or to volatilize and move through the column, carried by a current of an inert gas such as helium. Some lipids are naturally volatile, but most must first be derivatized to increase their volatility (that is, lower their boiling point). For an analysis of the fatty acids in a sample of phospholipids, the lipids are first transesterified: heated in a methanol/HCl or methanol/NaOH mixture to convert fatty acids esterified to glycerol into their methyl esters. These fatty acyl methyl esters are then loaded onto the gas chromatography column, and the column is heated to volatilize the compounds. Those fatty acyl esters most soluble in the column material partition into (dissolve in) that material; the less soluble lipids are carried by the stream of inert gas and emerge first from the column. The order of elution depends on the nature of the solid adsorbent in the column and on the boiling point of the components of the lipid mixture. Using these techniques, mixtures of fatty acids of various chain lengths and various degrees of unsaturation can be completely resolved.

Specific Hydrolysis Aids in Determination of Lipid Structure

Certain classes of lipids are susceptible to degradation under specific conditions. For example, all ester-linked fatty acids in triacylglycerols, phospholipids, and sterol esters are released by mild acid or alkaline treatment, and somewhat harsher hydrolysis conditions release amide-bound fatty acids from sphingolipids. Enzymes that specifically hydrolyze certain lipids are also useful in the determination of lipid structure. Phospholipases A, C, and D (Fig. 10-15) each split particular bonds in phospholipids and yield products with characteristic solubilities and chromatographic behaviors. Phospholipase C, for example, releases a water-soluble phosphoryl alcohol (such as phosphocholine from phosphatidylcholine) and a chloroform-soluble diacylglycerol, each of which can be characterized separately to determine the structure of the intact phospholipid. The combination of specific hydrolysis with characterization of the products by thin-layer, gas, or high-performance liquid chromatography often allows determination of a lipid structure.

Mass Spectrometry Reveals Complete Lipid Structure

To establish unambiguously the length of a hydrocarbon chain or the position of double bonds, mass spectrometric analysis of lipids or their volatile derivatives is invaluable. The chemical properties of similar lipids (for example, two fatty acids of similar length unsaturated at different positions, or two isoprenoids with different numbers of isoprene units) are very much alike, and their order of elution from the various chromatographic procedures often does not distinguish between them. When the eluate from a chromatography column is sampled by mass spectrometry, however, the components of a lipid mixture can be simultaneously separated and identified by their unique pattern of fragmentation (**Fig. 10-26**). With the

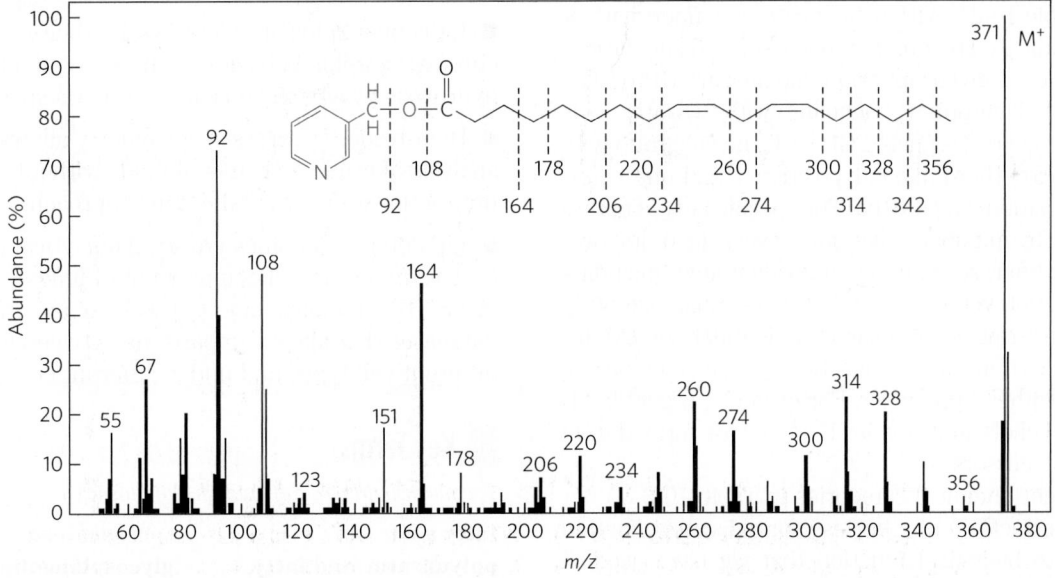

FIGURE 10-26 Determination of fatty acid structure by mass spectrometry. The fatty acid is first converted to a derivative that minimizes migration of the double bonds when the molecule is fragmented by electron bombardment. The derivative shown here is a picolinyl ester of linoleic acid—18:2($\Delta^{9,12}$) (M_r 371)—in which the alcohol is picolinol (red). When bombarded with a stream of electrons, this molecule is volatilized and converted to a parent ion (M$^+$; M_r 371), in which the N atom bears the positive charge, and a series of smaller fragments produced by breakage of C—C bonds in the fatty acid. The mass spectrometer separates these charged fragments according to their mass/charge ratio (m/z). (To review the principles of mass spectrometry, see pp. 100–102.)

The prominent ions at m/z = 92, 108, 151, and 164 contain the pyridine ring of the picolinol and various fragments of the carboxyl group, showing that the compound is indeed a picolinyl ester. The molecular ion, M$^+$ (m/z = 371), confirms the presence of a C$_{18}$ fatty acid with two double bonds. The uniform series of ions 14 atomic mass units (u) apart represents loss of each successive methyl and methylene group from the methyl end of the acyl chain (beginning at C-18; the right end of the molecule as shown here), until the ion at m/z = 300 is reached. This is followed by a gap of 26 u for the carbons of the terminal double bond, at m/z = 274; a further gap of 14 u for the C-11 methylene group, at m/z = 260; and so forth. By this means, the entire structure is determined, although these data alone do not reveal the configuration (cis or trans) of the double bonds. [Source: W. W. Christie, *Lipid Technol.* 8:64, 1996.]

TABLE 10-2	Eight Major Categories of Biological Lipids	
Category	**Category code**	**Examples**
Fatty acids	FA	Oleate, stearoyl-CoA, palmitoylcarnitine
Glycerolipids	GL	Di- and triacylglycerols
Glycerophospholipids	GP	Phosphatidylcholine, phosphatidylserine, phosphatidyethanoloamine
Sphingolipids	SP	Sphingomyelin, ganglioside GM2
Sterol lipids	ST	Cholesterol, progesterone, bile acids
Prenol lipids	PR	Farnesol, geraniol, retinol, ubiquinone
Saccharolipids	SL	Lipopolysaccharide
Polyketides	PK	Tetracycline, erythromycin, aflatoxin B_1

increased resolution of mass spectrometry, it is possible to identify individual lipids in very complex mixtures without first fractionating the lipids in a crude extract. This "shotgun" method (Fig. 10-25c) avoids losses during the preliminary separation of lipid subclasses, and it is faster.

Lipidomics Seeks to Catalog All Lipids and Their Functions

As lipid biochemists have become aware of the thousands of different naturally occurring lipids, they have created a database analogous to the Protein Data Bank. The LIPID MAPS Lipidomics Gateway (www.lipidmaps.org) has its own classification system that places each lipid species in one of eight chemical categories, each designated by two letters (Table 10-2). Within each category, finer distinctions are indicated by numbered classes and subclasses. For example, all glycerophosphocholines are GP01. The subgroup of glycerophosphocholines with two fatty acids in ester linkage is designated GP0101; the subgroup with one fatty acid ether-linked at position 1 and one ester-linked at position 2 is GP0102. The specific fatty acids are designated by numbers that give every lipid its own unique identifier, so that each individual lipid, including lipid types not yet discovered, can be unambiguously described in terms of a 12-character identifier, the LM_ID. One factor used in this classification system is the nature of the biosynthetic precursor. For example, prenol lipids (such as dolichols and vitamins E and K) are formed from isoprenyl precursors.

The eight chemical categories in Table 10-2 do not coincide perfectly with the less formal categorization according to biological function that we have used in this chapter. For example, the structural lipids of membranes include both glycerophospholipids and sphingolipids, which are separate categories in Table 10-2. Each method of classification has its advantages.

The application of mass spectrometric techniques with high throughput and high resolution can provide quantitative catalogs of all the lipids present in a specific cell type under particular conditions—the **lipidome**—and of the ways in which the lipidome changes with differentiation, disease such as cancer, or drug treatment. An animal cell contains more than a thousand different lipid species, each presumably having a specific function.

These functions are known for a growing number of lipids, but the still largely unexplored lipidome offers a rich source of new problems for the next generation of biochemists and cell biologists to solve.

SUMMARY 10.4 Working with Lipids

■ In the determination of lipid composition, the lipids are first extracted from tissues with organic solvents and separated by thin-layer, gas, or high-performance liquid chromatography.

■ Phospholipases specific for one of the bonds in a phospholipid can be used to generate simpler compounds for subsequent analysis.

■ Individual lipids are identified by their chromatographic behavior, their susceptibility to hydrolysis by specific enzymes, or mass spectrometry.

■ High-resolution mass spectrometry allows the analysis of crude mixtures of lipids without prefractionation—the "shotgun" approach.

■ Lipidomics combines powerful analytical techniques to determine the full complement of lipids in a cell or tissue (the lipidome) and to assemble annotated databases that allow comparisons between lipids of different cell types and under different conditions.

Key Terms

Terms in bold are defined in the glossary.

fatty acid 361	sphingomyelin 370
polyunsaturated fatty acid (PUFA) 363	**glycosphingolipid** 370
triacylglycerol 364	**cerebroside** 370
lipases 364	globoside 370
phospholipid 367	**ganglioside** 370
glycolipid 367	**sterol** 372
glycerophospholipid 367	cholesterol 372
ether lipid 369	**prostaglandin (PG)** 375
plasmalogen 369	**thromboxane (TX)** 375
galactolipid 369	**leukotriene (LT)** 376
sphingolipid 370	**lipoxin (LX)** 376
ceramide 370	**vitamin** 377
	vitamin D_3 377
	cholecalciferol 377

Problems

1. Operational Definition of Lipids How is the definition of "lipid" different from the types of definitions used for other biomolecules, such as amino acids, nucleic acids, and proteins?

2. Structure of an Omega-6 Fatty Acid Draw the structure of the omega-6 fatty acid 16:1.

3. Melting Points of Lipids The melting points of a series of 18-carbon fatty acids are: stearic acid, 69.6 °C; oleic acid, 13.4 °C; linoleic acid, −5 °C; and linolenic acid, −11 °C.

(a) What structural aspect of these 18-carbon fatty acids can be correlated with the melting point?

(b) Draw all the possible triacylglycerols that can be constructed from glycerol, palmitic acid, and oleic acid. Rank them in order of increasing melting point.

(c) Branched-chain fatty acids are found in some bacterial membrane lipids. Would their presence increase or decrease the fluidity of the membrane (that is, give the lipids a lower or higher melting point)? Why?

4. Catalytic Hydrogenation of Vegetable Oils Catalytic hydrogenation, used in the food industry, converts double bonds in the fatty acids of the oil triacylglycerols to —CH_2—CH_2—. How does this affect the physical properties of the oils?

5. Impermeability of Waxes What property of the waxy cuticles that cover plant leaves makes the cuticles impermeable to water?

6. Naming Lipid Stereoisomers The two compounds below are stereoisomers of carvone with quite different properties; the one on the left smells like spearmint, and that on the right, like caraway. Name the compounds using the RS system.

Spearmint   Caraway

7. RS Designations for Alanine and Lactate Draw (using wedge-bond notation) and label the (*R*) and (*S*) isomers of 2-aminopropanoic acid (alanine) and 2-hydroxypropanoic acid (lactic acid).

2-Aminopropanoic acid   2-Hydroxypropanoic acid
(alanine)   (lactic acid)

8. Hydrophobic and Hydrophilic Components of Membrane Lipids A common structural feature of membrane lipids is their amphipathic nature. For example, in phosphatidylcholine, the two fatty acid chains are hydrophobic and the phosphocholine head group is hydrophilic. For each of the following membrane lipids, name the components that serve as the hydrophobic and hydrophilic units: (a) phosphatidylethanolamine; (b) sphingomyelin; (c) galactosylcerebroside; (d) ganglioside; (e) cholesterol.

9. Deducing Lipid Structure from Composition Compositional analysis of a certain lipid shows that it has exactly one mole of fatty acid per mole of inorganic phosphate. Could this be a glycerophospholipid? A ganglioside? A sphingomyelin?

10. Deducing Lipid Structure from Molar Ratio of Components Complete hydrolysis of a glycerophospholipid yields glycerol, two fatty acids (16:1(Δ^9) and 16:0), phosphoric acid, and serine in the molar ratio 1:1:1:1:1. Name this lipid and draw its structure.

11. Lipids in Blood Group Determination We note in Figure 10-14 that the structure of glycosphingolipids determines the blood groups A, B, and O in humans. It is also true that glycoproteins determine blood groups. How can both statements be true?

12. The Action of Phospholipases The venom of the Eastern diamondback rattler and the Indian cobra contains phospholipase A₂, which catalyzes the hydrolysis of fatty acids at the C-2 position of glycerophospholipids. The phospholipid breakdown product of this reaction is lysolecithin (lecithin is phosphatidylcholine). At high concentrations, this and other lysophospholipids act as detergents, dissolving the membranes of erythrocytes and lysing the cells. Extensive hemolysis may be life-threatening.

(a) All detergents are amphipathic. What are the hydrophilic and hydrophobic portions of lysolecithin?

(b) The pain and inflammation caused by a snake bite can be treated with certain steroids. What is the basis of this treatment?

(c) Though the high levels of phospholipase A₂ in venom can be deadly, this enzyme is necessary for a variety of normal metabolic processes. What are these processes?

13. Intracellular Messengers from Phosphatidylinositols When the hormone vasopressin stimulates cleavage of PIP₂ by phospholipase C, two products are formed. What are they? Compare their properties and their solubilities in water, and predict whether either would diffuse readily through the cytosol.

14. Isoprene Units in Isoprenoids Geraniol, farnesol, and squalene are called isoprenoids because they are synthesized from five-carbon isoprene units. In each compound, circle the five-carbon units representing isoprene units (see Fig. 10-22).

Geraniol   Farnesol

Squalene

15. Hydrolysis of Lipids Name the products of mild hydrolysis with dilute NaOH of (a) 1-stearoyl-2,3-dipalmitoylglycerol; (b) 1-palmitoyl-2-oleoylphosphatidylcholine.

16. Effect of Polarity on Solubility Rank the following in order of increasing solubility in water: a triacylglycerol, a diacylglycerol, and a monoacylglycerol, all containing only palmitic acid.

17. Chromatographic Separation of Lipids A mixture of lipids is applied to a silica gel column, and the column is then washed with increasingly polar solvents. The mixture consists of phosphatidylserine, phosphatidylethanolamine, phosphatidylcholine, cholesteryl palmitate (a sterol ester), sphingomyelin, palmitate, *n*-tetradecanol, triacylglycerol, and cholesterol. In what order will the lipids elute from the column? Explain your reasoning.

18. Identification of Unknown Lipids Johann Thudichum, who practiced medicine in London about 100 years ago, also dabbled in lipid chemistry in his spare time. He isolated a variety of lipids from neural tissue and characterized and named many of them. His carefully sealed and labeled vials of isolated lipids were rediscovered many years later.

(a) How would you confirm, using techniques not available to Thudichum, that the vials labeled "sphingomyelin" and "cerebroside" actually contain these compounds?

(b) How would you distinguish sphingomyelin from phosphatidylcholine by chemical, physical, or enzymatic tests?

19. Ninhydrin to Detect Lipids on TLC Plates Ninhydrin reacts specifically with primary amines to form a purplish-blue product. A thin-layer chromatogram of rat liver phospholipids is sprayed with ninhydrin, and the color is allowed to develop. Which phospholipids can be detected in this way?

Data Analysis Problem

20. Determining the Structure of the Abnormal Lipid in Tay-Sachs Disease Box 10-1, Figure 1, shows the pathway of breakdown of gangliosides in healthy (normal) individuals and in individuals with certain genetic diseases. Some of the data on which the figure is based were presented in a paper by Lars Svennerholm (1962). Note that the sugar Neu5Ac, *N*-acetylneuraminic acid, represented in the Box 10-1 figure as ◆, is a sialic acid.

Svennerholm reported that "about 90% of the monosialogangliosides isolated from normal human brain" consisted of a compound with ceramide, hexose, *N*-acetylgalactosamine, and *N*-acetylneuraminic acid in the molar ratio 1:3:1:1.

(a) Which of the gangliosides (GM1 through GM3 and globoside) in Box 10-1, Figure 1, fits this description? Explain your reasoning.

(b) Svennerholm reported that 90% of the gangliosides from a patient with Tay-Sachs had a molar ratio (of the same four components given above) of 1:2:1:1. Is this consistent with the Box 10-1 figure? Explain your reasoning.

To determine the structure in more detail, Svennerholm treated the gangliosides with neuraminidase to remove the *N*-acetylneuraminic acid. This resulted in an asialoganglioside that was much easier to analyze. He hydrolyzed it with acid, collected the ceramide-containing products, and determined the molar ratio of the sugars in each product. He did this for both the normal and the Tay-Sachs gangliosides. His results are shown below.

Ganglioside	Ceramide	Glucose	Galactose	Galactosamine
Normal				
Fragment 1	1	1	0	0
Fragment 2	1	1	1	0
Fragment 3	1	1	1	1
Fragment 4	1	1	2	1
Tay-Sachs				
Fragment 1	1	1	0	0
Fragment 2	1	1	1	0
Fragment 3	1	1	1	1

(c) Based on these data, what can you conclude about the structure of the normal ganglioside? Is this consistent with the structure in Box 10-1? Explain your reasoning.

(d) What can you conclude about the structure of the Tay-Sachs ganglioside? Is this consistent with the structure in Box 10-1? Explain your reasoning.

Svennerholm also reported the work of other researchers who "permethylated" the normal asialoganglioside. Permethylation is the same as exhaustive methylation: a methyl group is added to every free hydroxyl group on a sugar. They found the following permethylated sugars: 2,3,6-trimethylglycopyranose; 2,3,4,6-tetramethylgalactopyranose; 2,4,6-trimethylgalactopyranose; and 4,6-dimethyl-2-deoxy-2-aminogalactopyranose.

(e) To which sugar of GM1 does each of the permethylated sugars correspond? Explain your reasoning.

(f) Based on all the data presented so far, what pieces of information about normal ganglioside structure are missing?

Reference

Svennerholm, L. 1962. The chemical structure of normal human brain and Tay-Sachs gangliosides. *Biochem. Biophys. Res. Comm.* 9:436–441.

Biological Membranes and Transport

Self-study tools that will help you practice what you've learned and reinforce this chapter's concepts are available online. Go to www.macmillanlearning.com/LehningerBiochemistry7e.

The first cell probably came into being when a membrane formed, enclosing a small volume of aqueous solution and separating it from the rest of the universe. Membranes define the external boundaries of cells and control the molecular traffic across that boundary (**Fig. 11-1**); in eukaryotic cells, they also divide the internal space into discrete compartments to segregate processes and components. Proteins embedded in and associated with membranes organize complex reaction sequences and are central to both biological energy conservation and cell-to-cell communication. The biological activities of membranes flow from their remarkable physical properties. Membranes are flexible, self-repairing, and selectively permeable to polar solutes. Their flexibility permits the shape changes that accompany cell growth and movement (such as amoeboid movement). With their ability to break and reseal, two membranes can fuse, as in exocytosis, or a single membrane-enclosed compartment can undergo fission to yield two sealed compartments, as in endocytosis or cell division, without creating gross leaks through cellular surfaces. Because membranes are selectively permeable, they retain certain compounds and ions within cells and within specific cellular compartments while excluding others.

Membranes are not merely passive barriers. They include an array of proteins specialized for promoting or catalyzing various cellular processes. At the cell surface, transporters move specific organic solutes and inorganic ions across the membrane; receptors sense extracellular signals and trigger molecular changes in the cell; and adhesion molecules hold neighboring cells together. Within the cell, membranes organize cellular processes such as the synthesis of lipids and certain proteins and the energy transductions in mitochondria and chloroplasts. Because membranes consist of just two layers of molecules, they are very thin—essentially

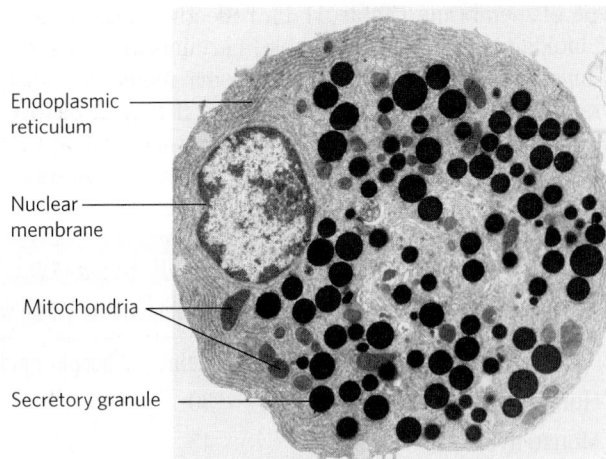

FIGURE 11-1 Biological membranes. This electron micrograph of a thin-sectioned exocrine pancreas cell shows several compartments made up of or bounded by membranes: endoplasmic reticulum, nucleus, mitochondria, and secretory granules. [Source: Don W. Fawcett/Science Source.]

two-dimensional. Intermolecular collisions of membrane proteins and lipids are far more probable in this two-dimensional space than in three-dimensional space, so the efficiency of enzyme-catalyzed processes organized within membranes is vastly increased.

In this chapter we first describe the composition of cellular membranes and their chemical architecture—the molecular structures that underlie their biological functions. Next, we consider the remarkable dynamic features of membranes, in which lipids and proteins move relative to each other. Cell adhesion, endocytosis, and the membrane fusion accompanying neurotransmitter secretion illustrate the dynamic roles of membrane proteins. We then turn to the protein-mediated passage of solutes across membranes via transporters and ion channels. In later chapters we discuss the roles of membranes in signal transduction (Chapters 12 and 23), energy transduction (Chapters 19 and 20), lipid synthesis (Chapter 21), and protein synthesis (Chapter 27).

11.1 The Composition and Architecture of Membranes

One approach to understanding membrane function is to study membrane composition—to determine, for example, which components are common to all membranes and which are unique to membranes with specific functions. So, before describing membrane structure and function, we consider the molecular components of membranes: proteins and polar lipids, which account for almost all the mass of biological membranes, and carbohydrates, present as part of glycoproteins and glycolipids.

Each Type of Membrane Has Characteristic Lipids and Proteins

The relative proportions of protein and lipid vary with the type of membrane (Table 11-1), reflecting the diversity of biological roles. For example, certain neurons have a myelin sheath—an extended plasma membrane that wraps around the cell many times and acts as a passive electrical insulator. The myelin sheath consists primarily of lipids (good insulators), whereas the plasma membranes

of bacteria and the membranes of mitochondria and chloroplasts, the sites of many enzyme-catalyzed processes, contain more protein than lipid (in mass per total mass).

For studies of membrane composition, the first task is to isolate a selected membrane. When eukaryotic cells are subjected to mechanical shear, their plasma membranes are torn and fragmented, releasing cytoplasmic components and membrane-bounded organelles such as mitochondria, chloroplasts, lysosomes, and nuclei. Plasma membrane fragments and intact organelles can be isolated by techniques described in Chapter 1 (see Fig. 1-9) and in Worked Example 2-1 (p. 57).

Cells have mechanisms to control the kinds and amounts of membrane lipid they synthesize and to target specific lipids to particular organelles. Each domain, each species, each tissue or cell type, and the organelles of each cell type have a characteristic set of membrane lipids. Plasma membranes, for example, are enriched in cholesterol and sphingolipids but contain no detectable cardiolipin **(Fig. 11-2)**; mitochondrial membranes are very low in cholesterol and sphingolipids, but they contain most of the cell's phosphatidylglycerol and cardiolipin, which are synthesized within the mitochondria. In all but a few cases, the functional significance of these different combinations is not yet known.

The protein composition of membranes from different sources varies even more widely than their lipid composition, reflecting functional specialization. In addition, some membrane proteins are covalently linked to oligosaccharides. For example, in glycophorin, a glycoprotein of the erythrocyte plasma membrane, 60% of the mass consists of complex oligosaccharides covalently attached to specific amino acid residues. Ser, Thr, and Asn residues are the most common points of carbohydrate attachment (see Fig. 7-30). The sugar moieties of surface glycoproteins influence the folding of the proteins as well as their stability, their intracellular destination, and their orientation in the membrane, and they play a significant role in the specific binding of ligands to glycoprotein surface receptors (see Fig. 7-37).

Some membrane proteins are covalently attached to one or more lipids, which serve as hydrophobic anchors that hold the proteins to the membrane, as we shall see.

TABLE 11-1	Major Components of Plasma Membranes in Various Organisms				
	Components (% by weight)				
	Protein	Phospholipid	Sterol	Sterol type	Other lipids
Human myelin sheath	30	30	19	Cholesterol	Galactolipids, plasmalogens
Mouse liver	45	27	25	Cholesterol	—
Maize leaf	47	26	7	Sitosterol	Galactolipids
Yeast	52	7	4	Ergosterol	Triacylglycerols, steryl esters
Paramecium (ciliated protist)	56	40	4	Stigmasterol	—
E. coli	75	25	0	—	—

Note: Values do not add up to 100% in every case because there are components other than protein, phospholipids, and sterol; plants, for example, have high glycolipid content.

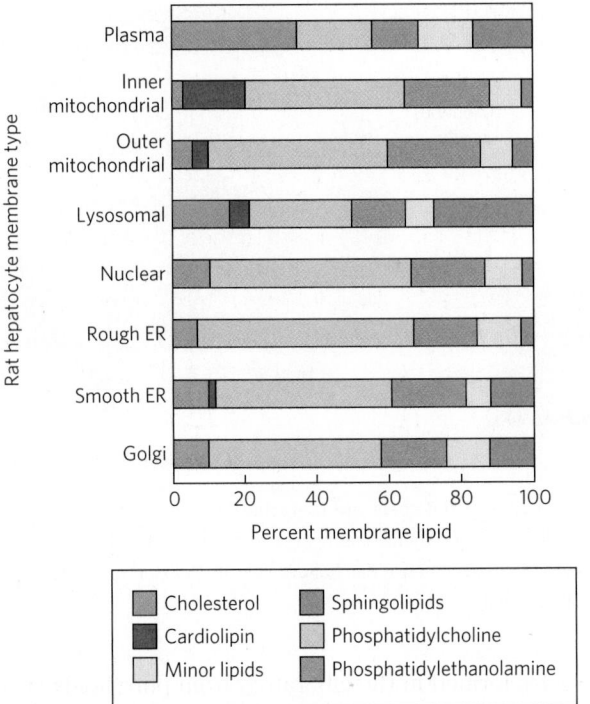

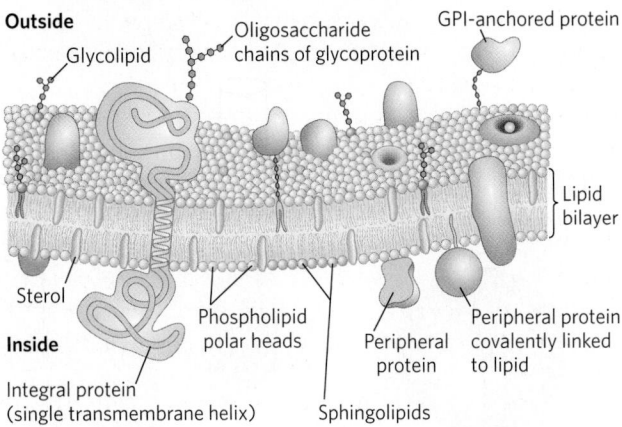

FIGURE 11-3 Fluid mosaic model for plasma membrane structure. The fatty acyl chains in the interior of the membrane form a fluid, hydrophobic region. Integral proteins float in this sea of lipid, held by hydrophobic interactions with their nonpolar amino acid side chains. Both proteins and lipids are free to move laterally in the plane of the bilayer, but movement of either from one leaflet of the bilayer to the other is restricted. The carbohydrate moieties attached to some proteins and lipids of the plasma membrane are exposed on the extracellular surface.

FIGURE 11-2 Lipid composition of the plasma membrane and organelle membranes of a rat hepatocyte. The functional specialization of each membrane type is reflected in its unique lipid composition. Cholesterol is prominent in plasma membranes but barely detectable in mitochondrial membranes. Cardiolipin is a major component of the inner mitochondrial membrane but not of the plasma membrane. Phosphatidylserine, phosphatidylinositol, and phosphatidylglycerol are relatively minor components of most membranes but serve critical functions; phosphatidylinositol and its derivatives, for example, are important in signal transductions triggered by hormones. Sphingolipids, phosphatidylcholine, and phosphatidylethanolamine are present in most membranes but in varying proportions. Glycolipids, which are major components of the chloroplast membranes of plants, are virtually absent from animal cells.

All Biological Membranes Share Some Fundamental Properties

Membranes are impermeable to most polar or charged solutes, but permeable to nonpolar compounds. They are 5 to 8 nm (50 to 80 Å) thick when proteins protruding on both sides are included. The combined evidence from electron microscopy and studies of chemical composition, as well as physical studies of permeability and the motion of individual protein and lipid molecules within membranes, led to the development of the **fluid mosaic model** for the structure of biological membranes **(Fig. 11-3)**. Phospholipids form a bilayer in which the nonpolar regions of the lipid molecules in each layer face the core of the bilayer and their polar head groups face outward, interacting with the aqueous phase on either side. Proteins are embedded in this bilayer sheet, their hydrophobic domains in contact with the fatty acyl chains of membrane lipids. Some proteins protrude from only one side of the membrane; others have domains exposed on both sides. The orientation of proteins in the bilayer

is asymmetric, giving the membrane "sidedness": the protein domains exposed on one side of the bilayer are different from those exposed on the other side, reflecting functional asymmetry. The individual lipid and protein units in a membrane form a fluid mosaic with a pattern that, unlike a mosaic of ceramic tile and mortar, is free to change constantly. The membrane mosaic is fluid because most of the interactions among its components are noncovalent, leaving individual lipid and protein molecules free to move laterally in the plane of the membrane.

We now look at some of these features of membrane structure in more detail and consider the experimental evidence that supports the basic model.

A Lipid Bilayer Is the Basic Structural Element of Membranes

Glycerophospholipids, sphingolipids, and sterols are virtually insoluble in water. When mixed with water, they spontaneously form microscopic lipid aggregates, clustering together, with their hydrophobic moieties in contact with each other and their hydrophilic groups interacting with the surrounding water. The clustering reduces the amount of hydrophobic surface exposed to water and thus minimizes the number of molecules in the shell of ordered water at the lipid-water interface (see Fig. 2-7), resulting in an increase in entropy. This hydrophobic effect provides the thermodynamic driving force for the formation and maintenance of these clusters of lipid molecules. The term **hydrophobic interactions** is sometimes used to describe the clustering of hydrophobic molecular surfaces in an aqueous environment, but it should be clearly understood that the molecules are not interacting chemically; they are simply finding the lowest-energy environment by reducing the hydrophobic, or nonpolar, surface area exposed to water.

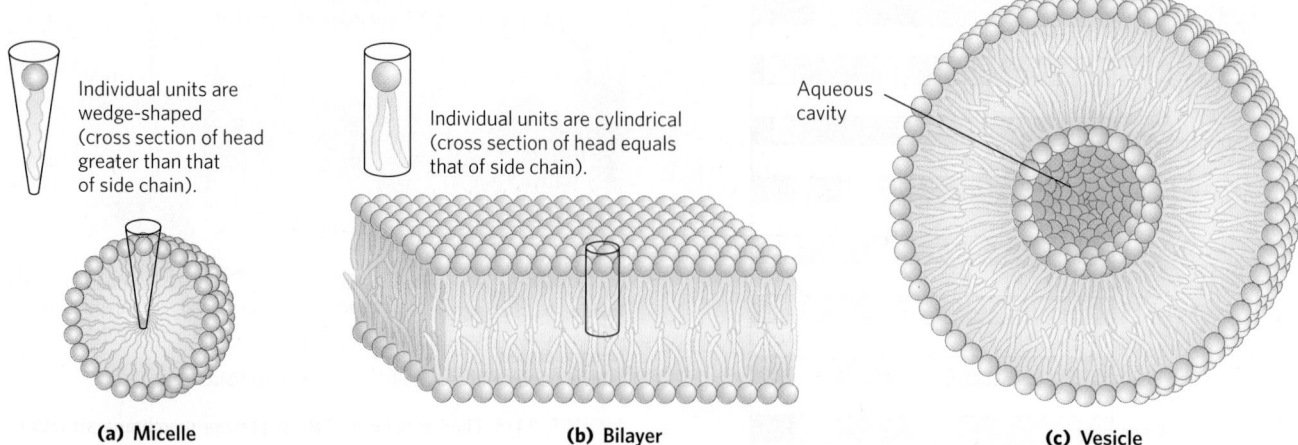

(a) Micelle

Individual units are wedge-shaped (cross section of head greater than that of side chain).

(b) Bilayer

Individual units are cylindrical (cross section of head equals that of side chain).

(c) Vesicle

Aqueous cavity

FIGURE 11-4 Amphipathic lipid aggregates that form in water. (a) In micelles, the hydrophobic chains of the fatty acids are sequestered at the core of the sphere. There is virtually no water in the hydrophobic interior. **(b)** In an open bilayer, all acyl side chains except those at the edges of the sheet are protected from interaction with water. **(c)** When a two-dimensional bilayer folds on itself, it forms a closed bilayer, a three-dimensional hollow vesicle (liposome) enclosing an aqueous cavity.

Depending on the precise conditions and the nature of the lipids, several types of lipid aggregate can form when amphipathic lipids are mixed with water **(Fig. 11-4)**. **Micelles** are spherical structures that contain anywhere from a few dozen to a few thousand amphipathic molecules. These molecules are arranged with their hydrophobic regions aggregated in the interior, where water is excluded, and their hydrophilic head groups at the outer surface, in contact with water. Micelle formation is favored when the cross-sectional area of the head group is greater than that of the acyl side chain(s), as in free fatty acids, lysophospholipids (phospholipids lacking one fatty acid), and many detergents, such as sodium dodecyl sulfate (SDS; p. 94).

A second type of lipid aggregate in water is the **bilayer**, in which two lipid monolayers (leaflets) form a two-dimensional sheet. Bilayer formation is favored if the cross-sectional areas of the head group and acyl side chain(s) are similar, as in glycerophospholipids and sphingolipids. The hydrophobic portions in each monolayer, excluded from water, interact with each other. The hydrophilic head groups interact with water at one or the other surface of the bilayer. Because the hydrophobic regions at its edges (Fig. 11-4b) are in contact with water, a bilayer sheet is relatively unstable and spontaneously folds back on itself to form a hollow sphere, called a **vesicle** or liposome (Fig. 11-4c). The continuous surface of vesicles eliminates exposed hydrophobic regions, allowing bilayers to achieve maximal stability in their aqueous environment. Vesicle formation also creates a separate internal aqueous compartment (the vesicle lumen). It is likely that the antecedents to the first living cells resembled lipid vesicles, their aqueous contents segregated from their surroundings by a hydrophobic shell.

The lipid bilayer is 3 nm (30 Å) thick. The hydrocarbon core, made up of the —CH$_2$— and —CH$_3$ of the fatty acyl groups, is about as nonpolar as decane, and

vesicles formed in the laboratory from pure lipids (liposomes) are essentially impermeable to polar solutes, as is the lipid bilayer of biological membranes (although biological membranes, as we shall see, are permeable to solutes for which they have specific transporters).

Most membrane lipids and proteins are synthesized in the endoplasmic reticulum (ER), and from there they move to their destination organelles or to the plasma membrane **(Fig. 11-5a)**. During this "membrane trafficking," small membrane vesicles bud from the ER, then move to and fuse with the cis Golgi. As lipids and proteins move across the Golgi to its trans side, they undergo a variety of covalent alterations that determine their final location and function in the cell. For example, oligosaccharide chains or fatty acids such as palmitate are covalently linked to specific membrane proteins, and phospholipids undergo reshuffling of their fatty acid components to reach their mature forms. In many cases, these modifications dictate the eventual location of the modified protein. Membrane trafficking is accompanied by striking changes in lipid composition and disposition across the bilayer (Fig. 11-5b). Phosphatidylcholine is the principal phospholipid in the lumenal monolayer of the Golgi membrane, but in transport vesicles leaving the trans Golgi, phosphatidylcholine has largely been replaced by sphingolipids and cholesterol, which, following fusion of the transport vesicles with the plasma membrane, make up the majority of the lipids in the outer monolayer of the cell's membrane. Plasma membrane lipids are asymmetrically distributed between the two monolayers of the bilayer. In the plasma membrane of eukaryotic cells, for example, choline-containing lipids (phosphatidylcholine and sphingomyelin) are typically found in the outer (extracellular, or exoplasmic) leaflet, whereas phosphatidylserine, phosphatidylethanolamine, and the phosphatidylinositols are almost exclusively in the

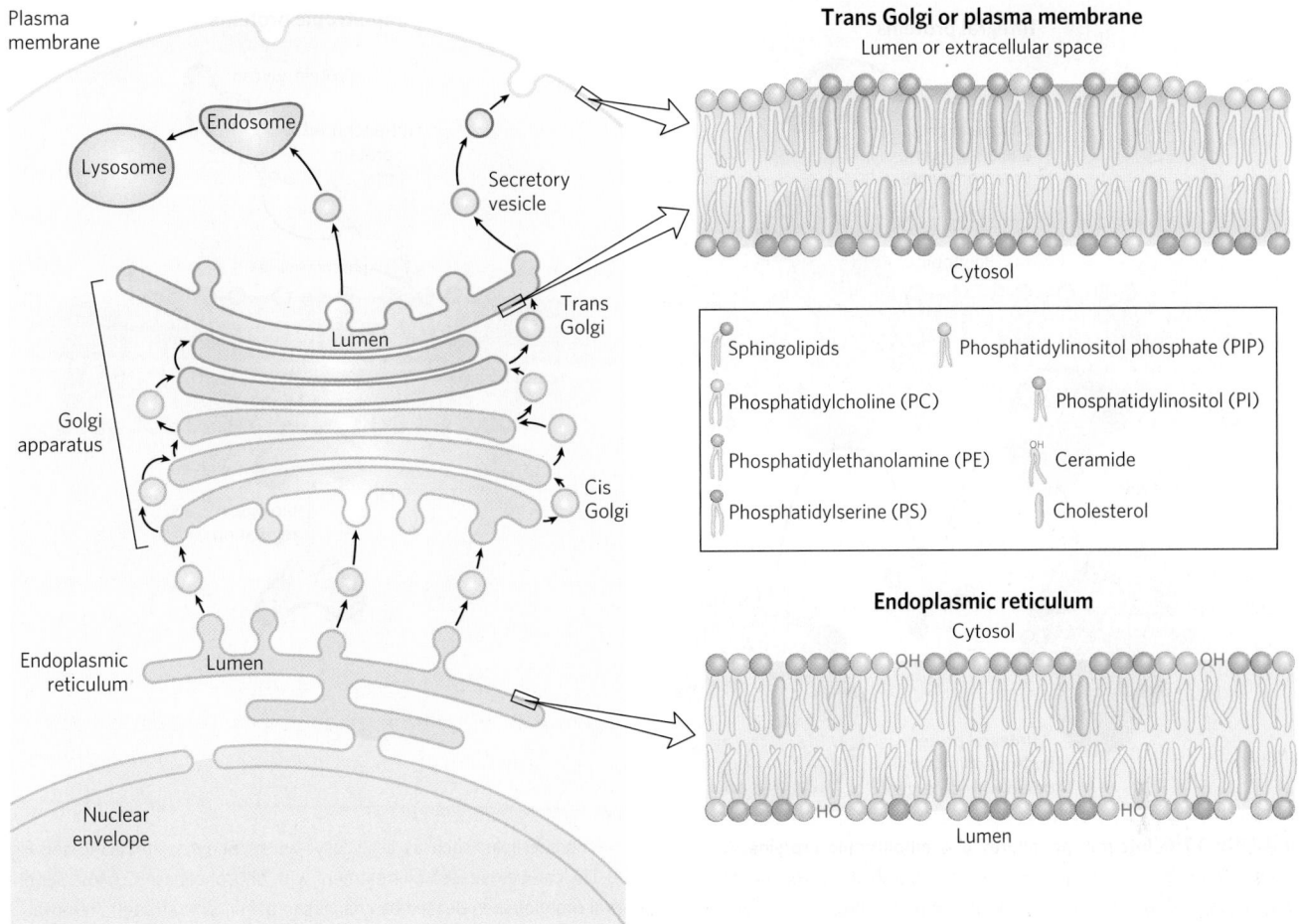

FIGURE 11-5 Compositional changes accompanying membrane trafficking. (a) The path of lipids and proteins during membrane trafficking from the site of their synthesis (ER) through the Golgi apparatus to the cell surface (or to organelles such as lysosomes). Small vesicles bud off the ER, move to and fuse with the cis Golgi, exit the trans Golgi as secretory or transport vesicles, and fuse with the plasma membrane or with endosomes, which give rise to lysosomes. **(b)** During trafficking, both the lipid composition of the bilayer and the disposition of specific lipids between inner and outer leaflets change remarkably. [Source: (b) Information from G. Drin, *Annu. Rev. Biochem.* 83:51, 2014, Fig. 1.]

inner (cytoplasmic) leaflet, where the negatively charged serine and inositol phosphate head groups can interact electrostatically with positively charged regions of peripheral or amphitropic membrane proteins (described below). A second route for redistributing lipids from their site of synthesis to their destination membrane is via specialized protein-mediated conduits referred to as junctions, including ER–plasma membrane junctions and ER-mitochondrial junctions.

An early method for determining the bilayer distribution of a specific phospholipid in the plasma membrane was to treat the intact cell with phospholipase C, which cannot reach lipids in the inner monolayer (leaflet) but removes the head groups of lipids in the outer monolayer. The proportion of each head group released provided an estimate of the fraction of each lipid in the outer monolayer of the plasma membrane. Today, the locations of individual lipids in the plasma membrane or other cellular membranes can be determined, with greater resolution, using methods that employ fluorescent lipid analogs or fluorescent derivatives of antibodies, toxins, or lipid-binding domains that have high

binding affinity and specificity for one lipid type. Location of the bound labeled probes is determined by high-resolution fluorescence microscopy.

Changes over time in the distribution of lipids between plasma membrane monolayers, or leaflets, have biological consequences. For example, in blood platelets, only when the phosphatidylserine in the plasma membrane moves into the outer leaflet is the platelet able to play its role in formation of a blood clot. For many other cell types, exposure of phosphatidylserine on the outer surface marks a cell for destruction by programmed cell death. The movement of phospholipid molecules from one leaflet to another is catalyzed and regulated by specific proteins (see Fig. 11-15).

Three Types of Membrane Proteins Differ in the Nature of Their Association with the Membrane

Integral membrane proteins are embedded within the lipid bilayer and are removable only by agents that overcome the hydrophobic effect, such as detergents, organic solvents, or denaturants **(Fig. 11-6)**. Integral

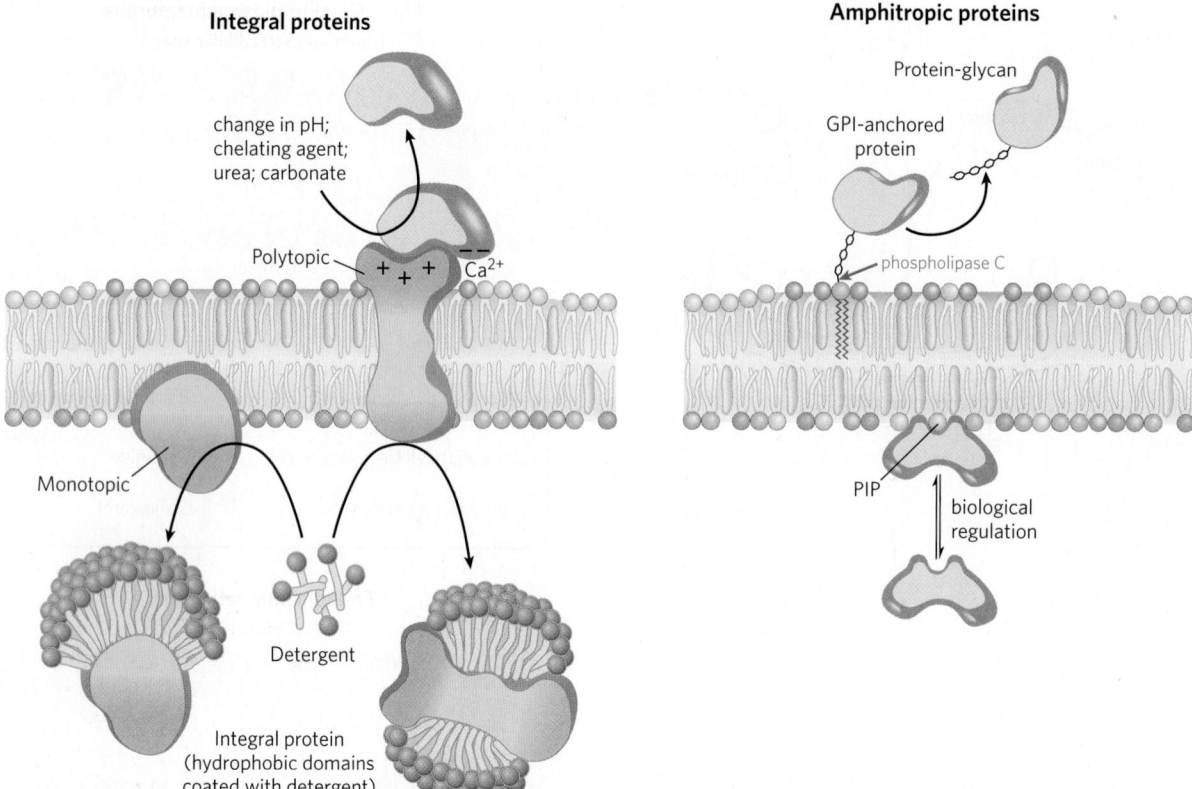

FIGURE 11-6 Integral, peripheral, and amphitropic proteins. Membrane proteins can be operationally distinguished by the conditions required to release them from the membrane. Integral proteins, both monotopic (associated with one leaflet) and polytopic (transmembrane), can be extracted with detergents, which disrupt hydrophobic interactions with the lipid bilayer and form micelle-like clusters around individual protein molecules. Integral proteins covalently attached to a membrane lipid, such as a glycosyl phosphatidylinositol (GPI; see Fig. 11-13), can be released by treatment with phospholipase C. Most peripheral proteins are released by changes in pH or ionic strength, removal of Ca^{2+} by a chelating agent, or addition of urea or carbonate. Amphitropic proteins are sometimes associated with membranes and sometimes not, depending on some type of regulatory process such as reversible palmitoylation.

proteins may be **monotopic**, interacting with just one leaflet of the bilayer, or **polytopic**, having a polypeptide chain that traverses the membrane once or several times. **Peripheral membrane proteins** associate with the membrane through electrostatic interactions and hydrogen bonding with hydrophilic domains of integral proteins and with membrane lipids. In the laboratory, they can be released from their membrane association by relatively mild treatments that interfere with electrostatic interactions or break hydrogen bonds; a commonly used agent is carbonate at high pH. **Amphitropic proteins** associate reversibly with membranes and are therefore found both in the membrane and in the cytosol. Their affinity for membranes results in some cases from the protein's noncovalent interaction with a membrane protein or lipid, and in other cases from the presence of one or more lipids covalently attached to the amphitropic protein (see Fig. 11-13). Generally, the reversible association of amphitropic proteins with the membrane is regulated; for example, phosphorylation or ligand binding can force a conformational change in the protein, exposing a membrane-binding site that was previously inaccessible. Reversible covalent attachment of one or more lipid moieties can also effect a change in the affinity of an amphitropic protein for the membrane.

Many Integral Membrane Proteins Span the Lipid Bilayer

Membrane protein topology (the localization of protein domains relative to the lipid bilayer) can be determined through the use of reagents that react with protein side chains but cannot cross membranes—polar chemical reagents that react with the primary amines of Lys residues, for example, or enzymes such as trypsin that cleave proteins but cannot cross the membrane. If a membrane protein in an intact erythrocyte reacts with a membrane-impermeant reagent, that protein must have at least one domain exposed on the outer (extracellular) face of the membrane. For example, trypsin cleaves extracellular domains but does not affect domains buried within the bilayer or exposed only on the inner surface, unless the plasma membrane is broken to make these domains accessible to the enzyme.

Classic experiments with such topology-specific reagents showed that the erythrocyte glycoprotein **glycophorin** spans the plasma membrane. Its amino-terminal domain (bearing several carbohydrate chains) is on the outer surface and is cleaved by trypsin. The carboxyl terminus protrudes on the inside of the cell, where it cannot react with impermeant reagents. Both the

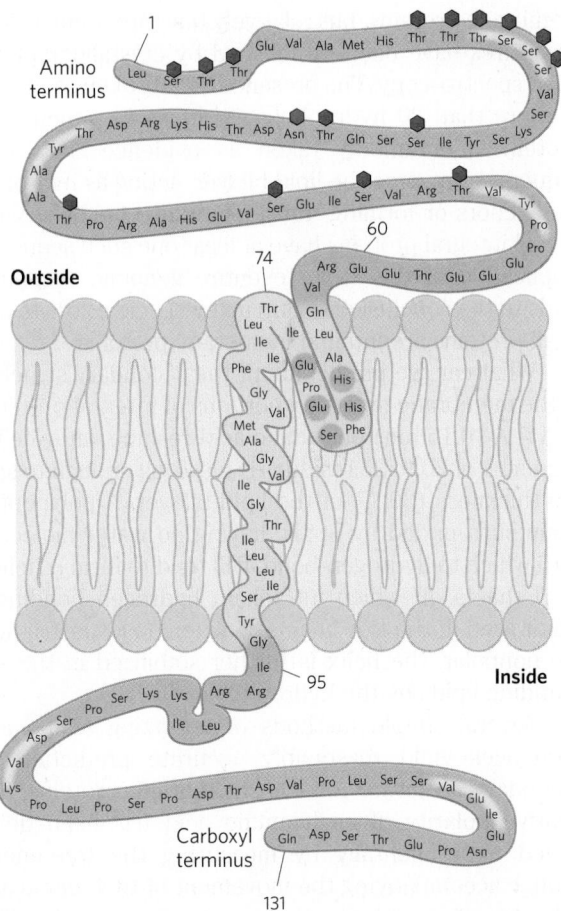

FIGURE 11-7 Transbilayer disposition of glycophorin in an erythrocyte. One hydrophilic domain, containing all the sugar residues, is on the outer surface, and another hydrophilic domain protrudes from the inner face of the membrane. Each red hexagon represents a tetrasaccharide (containing two Neu5Ac (sialic acid), Gal, and GalNAc) O-linked to a Ser or Thr residue; the blue hexagon represents an oligosaccharide N-linked to an Asn residue. The relative size of the oligosaccharide units is larger than shown here. A segment of 19 hydrophobic residues (residues 75 to 93) forms an α helix that traverses the membrane bilayer (see Fig. 11-10a). The segment from residues 64 to 74 has some hydrophobic residues and probably penetrates the outer face of the lipid bilayer, as shown. [Source: Information from V. T. Marchesi et al., *Annu. Rev. Biochem.* 45:667, 1976.]

amino-terminal and carboxyl-terminal domains contain many polar or charged amino acid residues and are therefore hydrophilic. However, a segment in the center of the protein (residues 75 to 93) contains mainly hydrophobic amino acid residues, suggesting that glycophorin has a transmembrane segment arranged as shown in **Figure 11-7**. Rigorous physical and chemical studies have confirmed this topology for glycophorin and have established the topology of many other membrane proteins.

These topological studies have also revealed that the orientation of glycophorin in the membrane is asymmetric: its amino-terminal segment is *always* on the outside. Similar studies of many other integral membrane proteins show that each has a specific orientation in the bilayer, giving the membrane a distinct sidedness. For glycophorin, and for all other glycoproteins of the plasma membrane, the glycosylated domains are invariably found on

the extracellular face of the bilayer. As we shall see, the asymmetric arrangement of membrane proteins results in functional asymmetry. All the molecules of a given ion pump, for example, have the same orientation in the membrane and pump ions in the same direction.

Hydrophobic Regions of Integral Proteins Associate with Membrane Lipids

The firm attachment of integral proteins to membranes is the result of the hydrophobic effect; moving the hydrophobic domains of proteins from contact with membrane lipids to contact with the aqueous environment would have a high thermodynamic cost. Some polytopic proteins have a single hydrophobic sequence in the middle of the molecule (as in glycophorin) or at the amino or carboxyl terminus. Others have multiple hydrophobic sequences, each of which, when in the α-helical conformation, is long enough (about 20 residues) to span the lipid bilayer. (Recall from Worked Example 4-1 that each residue in an α helix adds 1.5 Å to its length.)

One of the best-studied polytopic proteins, bacteriorhodopsin, has seven very hydrophobic internal sequences and spans the lipid bilayer seven times. Bacteriorhodopsin is a light-driven proton pump densely packed in regular arrays in the purple membrane of the bacterium *Halobacterium salinarum*. X-ray crystallography reveals a structure with seven α-helical segments, each traversing the lipid bilayer, connected by nonhelical loops at the inner and outer faces of the membrane **(Fig. 11-8)**. In the amino acid sequence of bacteriorhodopsin, seven segments of about 20 hydrophobic residues can be identified, each forming an α helix that spans the bilayer. The seven helices are clustered together and oriented not quite perpendicular to the bilayer

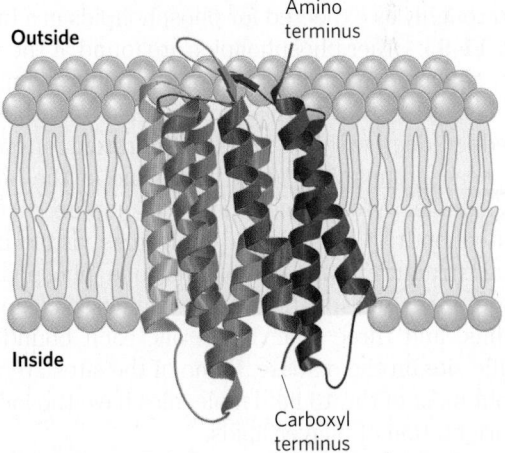

FIGURE 11-8 Bacteriorhodopsin, a membrane-spanning protein. The single polypeptide chain folds into seven hydrophobic α helices, each of which traverses the lipid bilayer roughly perpendicular to the plane of the membrane. The seven transmembrane helices are clustered, and the space around and between them is filled with the acyl chains of membrane lipids. The light-absorbing pigment retinal (see Fig. 10-20) is buried deep within the membrane in contact with several of the helical segments (not shown). The helices are colored to correspond with the hydropathy plot in Figure 11-10b. [Source: PDB ID 2AT9, K. Mitsuoka et al., *J. Mol. Biol.* 286:861, 1999.]

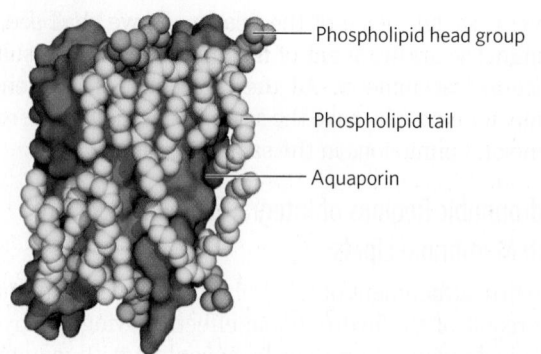

Phospholipid head group

Phospholipid tail

Aquaporin

FIGURE 11-9 Lipid annuli associated with an integral membrane protein. The crystal structure of sheep aquaporin, a transmembrane water channel, includes a shell of phospholipids positioned with their head groups (blue) at the expected positions on the inner and outer membrane surfaces and their hydrophobic acyl chains (gold) intimately associated with the surface of the protein exposed to the bilayer. The lipid forms a "grease seal" around the protein, which is depicted by a dark blue surface representation. [Source: PDB ID 2B6O, T. Gonen et al., *Nature* 438:633, 2005.]

plane, a pattern that (as we shall see in Chapter 12) is a very common motif in membrane proteins involved in signal reception. The hydrophobic effect keeps the nonpolar amino acid residues firmly anchored among the fatty acyl groups of the membrane lipids in the membrane.

Once their structures have been solved by crystallography, many membrane proteins are found to have attached phospholipid molecules, which are presumed to be positioned in the native membranes as they are in the protein crystals. Many of these phospholipid molecules lie on the protein surface, their head groups interacting with polar amino acid residues at the inner and outer membrane–water interfaces and their side chains associated with nonpolar residues. These **annular lipids** form a bilayer shell (annulus) around the protein, oriented roughly as expected for phospholipids in a bilayer **(Fig. 11-9)**. Other phospholipids are found at the interfaces between monomers of multisubunit membrane proteins, where they form a "grease seal." Yet others are embedded deep within a membrane protein, often with their head groups well below the plane of the bilayer. For example, cytochrome oxidase (Complex IV, found in mitochondria) has 13 lipid molecules visible in the crystal structure: two cardiolipins, one phosphatidylcholine, three phosphatidylethanolamines, four prostaglandins, and three triacylglycerols, each bound to a specific site on the oxidase. Some of the sites are internal, but most of the 13 lipid molecules have the location and orientation of bilayer lipids.

The Topology of an Integral Membrane Protein Can Often Be Predicted from Its Sequence

Determination of the three-dimensional structure of a membrane protein—that is, its topology—is generally much more difficult than determining its amino acid sequence, either directly or by gene sequencing. The amino acid sequences are known for thousands of membrane proteins, but relatively few three-dimensional structures have been established by crystallography or NMR spectroscopy. The presence of unbroken sequences of more than 20 hydrophobic residues in a membrane protein is commonly taken as evidence that these sequences traverse the lipid bilayer, acting as hydrophobic anchors or forming transmembrane channels. Virtually all integral proteins have at least one such sequence. Application of this logic to entire genomic sequences leads to the conclusion that in many species, 20% to 30% of all proteins are integral membrane proteins.

What can we predict about the secondary structure of the membrane-spanning portions of integral proteins? At 1.5 Å (0.15 nm) per amino acid residue, an α-helical sequence of 20 to 25 residues is just long enough to span the thickness (30 Å) of the lipid bilayer. A polypeptide chain surrounded by lipids, having no water molecules with which to hydrogen-bond, will tend to form α helices or β sheets, in which intrachain hydrogen bonding is maximized. If the side chains of all amino acids in a helix are nonpolar, the helix is further stabilized in the surrounding lipids by the hydrophobic effect.

Several simple methods of analyzing amino acid sequences yield reasonably accurate predictions of secondary structure for transmembrane proteins. The relative polarity of each amino acid has been determined experimentally by measuring the free-energy change accompanying the movement of that amino acid side chain from a hydrophobic solvent into water. This free energy of transfer, which can be expressed as a **hydropathy index** (see Table 3-1), ranges from highly exergonic for charged or polar residues to highly endergonic for amino acids with aromatic or aliphatic hydrocarbon side chains. The overall hydropathy index (hydrophobicity) of a sequence of amino acids is estimated by summing the free energies of transfer for the residues in the sequence. To scan a polypeptide sequence for potential membrane-spanning segments, an investigator calculates the hydropathy index for successive segments (called windows) of a given size, from 7 to 20 residues. For a window of seven residues, for example, the average indices for residues 1 to 7, 2 to 8, 3 to 9, and so on, are plotted as in **Figure 11-10** (plotted for the middle residue in each window—residue 4 for residues 1 to 7, for example). A region with more than 20 residues of high hydropathy index is presumed to be a transmembrane segment. When the sequences of membrane proteins of known three-dimensional structure are scanned using simple online bioinformatics tools, we find a reasonably good correspondence between predicted and known membrane-spanning segments. Hydropathy analysis predicts a single hydrophobic helix for glycophorin (Fig. 11-10a) and seven transmembrane segments for bacteriorhodopsin (Fig. 11-10b)—in agreement with structures known from x-ray crystallography.

Not all integral membrane proteins are composed of transmembrane α helices. Another structural motif

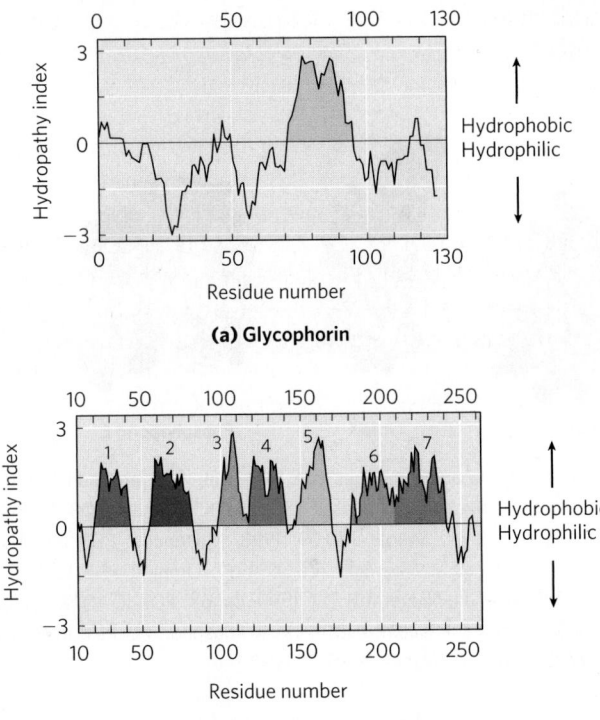

(a) Glycophorin

(b) Bacteriorhodopsin

FIGURE 11-10 Hydropathy plots. Average hydropathy index (see Table 3-1) is plotted against residue number for two integral membrane proteins. The hydropathy index for each amino acid residue in a sequence of defined length, or "window," is used to calculate the average hydropathy for that window. The horizontal axis shows the residue number in the middle of the window. **(a)** Glycophorin from human erythrocytes has a single hydrophobic sequence between residues 75 and 93 (yellow); compare this with Figure 11-7. **(b)** Bacteriorhodopsin, known from independent physical studies to have seven transmembrane helices (see Fig. 11-8), has seven hydrophobic regions. Note, however, that the hydropathy plot is ambiguous in the region of segments 6 and 7. X-ray crystallography has confirmed that this region has two transmembrane segments.

common in bacterial membrane proteins is the **β barrel** (see Fig. 4-18b), in which 20 or more transmembrane segments form β sheets that line a cylinder **(Fig. 11-11)**. The same factors that favor α-helix formation in the hydrophobic interior of a lipid bilayer also stabilize β

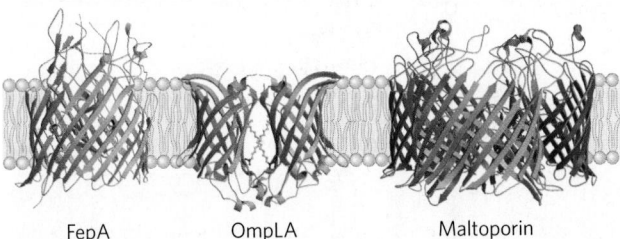

FepA   OmpLA   Maltoporin

FIGURE 11-11 Membrane proteins with β-barrel structure. Three proteins of the *E. coli* outer membrane are shown, viewed in the plane of the membrane. FepA, involved in iron uptake, has 22 membrane-spanning β strands. Outer membrane phospholipase A, or OmpLA, is a 12-stranded β barrel that exists as a dimer in the membrane. Maltoporin, a maltose transporter, is a trimer; each monomer consists of 16 β strands. [Sources: FepA: PDB ID 1FEP, S. K. Buchanan et al., *Nature Struct. Biol.* 6:56, 1999. OmpLA: modified from PDB ID 1QD5, H. J. Snijder et al., *Nature* 401:717, 1999. Maltoporin: modified from PDB ID 1MAL, T. Schirmer et al., *Science* 267:512, 1995.]

barrels: when no water molecules are available to hydrogen-bond with the carbonyl oxygen and nitrogen of the peptide bond, maximal intrachain hydrogen bonding gives the most stable conformation. Planar β sheets do not maximize these interactions and are generally not found in the membrane interior; β barrels allow all possible hydrogen bonds and are common among membrane proteins. **Porins**, proteins that allow certain polar solutes to cross the outer membrane of gram-negative bacteria such as *E. coli*, have many-stranded β barrels lining the polar transmembrane passage. The outer membranes of mitochondria and chloroplasts also contain a variety of β barrels, perhaps a result of the origins of mitochondria and chloroplasts as bacterial endosymbionts (see Fig. 1-40).

A polypeptide is more extended in the β conformation than in an α helix; just seven to nine residues of β conformation are needed to span a membrane. Recall that in the β conformation, alternating side chains project above and below the sheet (see Fig. 4-6). In β strands of membrane proteins, every second residue in the membrane-spanning segment is hydrophobic and interacts with the lipid bilayer; aromatic side chains are commonly found at the lipid-protein interface. The other residues may or may not be hydrophilic.

A further remarkable feature of many transmembrane proteins of known structure is the presence of Tyr and Trp residues at the interface between lipid and water **(Fig. 11-12)**. The side chains of these residues seem to serve as membrane interface anchors, able to interact simultaneously with the central lipid phase and the aqueous phases on either side of the membrane. Another generalization about amino acid location relative to the bilayer is described by the **positive-inside rule**: the positively charged Lys and Arg residues of membrane proteins occur more commonly on the cytoplasmic face of membranes.

Covalently Attached Lipids Anchor Some Membrane Proteins

Some membrane proteins are covalently linked to one or more lipids, which may be of several types: long-chain fatty acids, isoprenoids, sterols, or glycosylated derivatives of *p*hosphatidyl*i*nositol (GPIs; **Fig. 11-13**). The attached lipid provides a hydrophobic anchor that inserts into the lipid bilayer and holds the protein at the membrane surface. The strength of the hydrophobic interaction between a bilayer and a single hydrocarbon chain linked to a protein is barely enough to anchor the protein securely, but many proteins have more than one attached lipid moiety. Furthermore, other interactions, such as ionic attractions between positively charged Lys residues in the protein and negatively charged lipid head groups, can add to the anchoring effect of a covalently bound lipid. For example, the plasma membrane protein MARCKS, which interacts with actin filaments in the process of cell motility, has

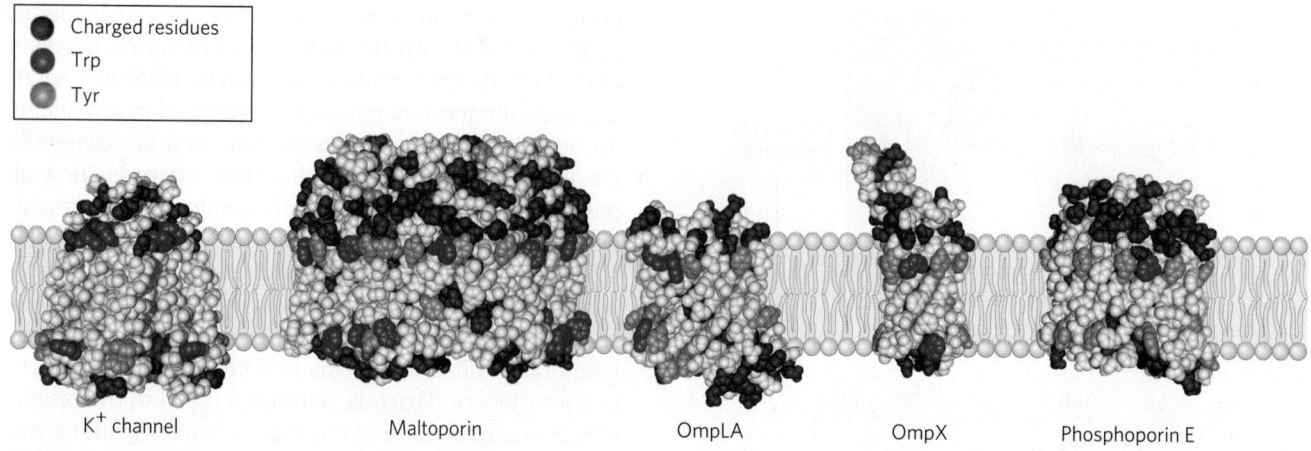

Charged residues
Trp
Tyr

K⁺ channel   Maltoporin   OmpLA   OmpX   Phosphoporin E

FIGURE 11-12 Tyr and Trp residues of membrane proteins clustering at the water-lipid interface. The detailed structures of these five integral membrane proteins are known from crystallographic studies. The K⁺ channel is from the bacterium *Streptomyces lividans* (see Fig. 11-45); maltoporin, OmpLA, OmpX, and phosphoporin E are proteins of the outer membrane of *E. coli.* Residues of Tyr and Trp are found predominantly where the nonpolar region of acyl chains meets the polar head group region. Charged residues (Lys, Arg, Glu, Asp) are found almost exclusively in the aqueous phases. [Sources: K⁺ channel: PDB ID 1BL8, D. A. Doyle et al., *Science* 280:69, 1998. Maltoporin: PDB ID 1AF6, Y. F. Wang et al., *J. Mol. Biol.* 272:56, 1997. OmpLA: PDB ID 1QD5, H. J. Snijder et al., *Nature* 401:717, 1999. OmpX: PDB ID 1QJ9, J. Vogt and G. E. Schulz, *Structure* 7:1301, 1999. Phosphorin E: PDB ID 1PHO, S. W. Cowan et al., *Nature* 358:727, 1992.]

a covalently attached myristoyl moiety, but it also contains the sequence

KKKKKRFSFKKSFKLSGFSFKKNKK
151   175

which adds to the protein's affinity for the membrane. Three clusters of positively charged Lys and Arg residues (screened blue) interact with the negatively charged head group of phosphatidylinositol 4,5-bisphosphate (PIP₂) on the cytoplasmic face of the plasma membrane; five aromatic residues (screened yellow) insert into the lipid bilayer. When the head-group phosphates of PIP₂ are enzymatically removed, MARCKS loses its hold on the plasma membrane and dissociates.

FIGURE 11-13 Lipid-linked membrane proteins. Covalently attached lipids anchor membrane proteins to the lipid bilayer. A palmitoyl group is shown attached by thioester linkage to a Cys residue; an *N*-myristoyl group is generally attached to an amino-terminal Gly residue, typically of a protein that also has a hydrophobic transmembrane segment; the farnesyl and geranylgeranyl groups attached to carboxyl-terminal Cys residues are isoprenoids of 15 and 20 carbons, respectively. The carboxyl-terminal Cys residue is invariably methylated. Glycosyl phosphatidylinositol (GPI) anchors are derivatives of phosphatidylinositol in which the inositol bears a short oligosaccharide covalently joined to the carboxyl-terminal residue of a protein through phosphoethanolamine. GPI-anchored proteins are always on the extracellular face of the plasma membrane. Farnesylated and palmitoylated membrane proteins are found on the inner surface of the plasma membrane, and myristoylated proteins have domains both inside and outside the plasma membrane.

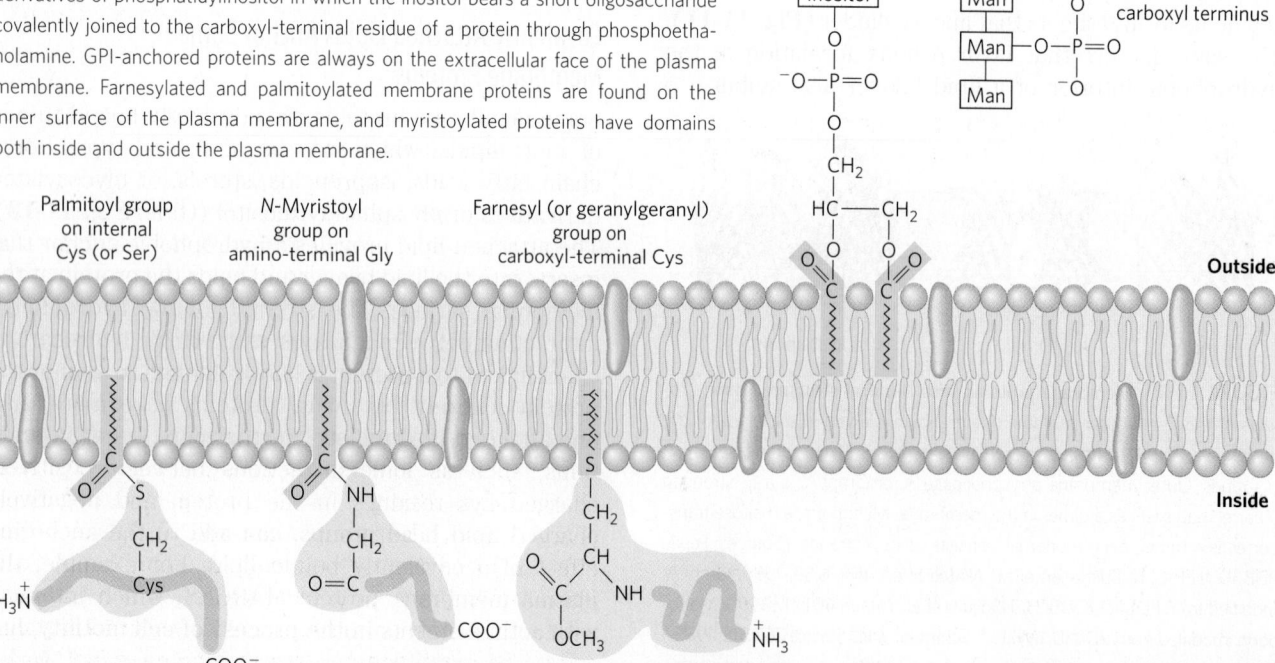

Palmitoyl group on internal Cys (or Ser)

N-Myristoyl group on amino-terminal Gly

Farnesyl (or geranylgeranyl) group on carboxyl-terminal Cys

GPI anchor on carboxyl terminus

Outside

Inside

Beyond merely anchoring a protein to the membrane, the attached lipid may have a more specific role. In the plasma membrane, **GPI-anchored proteins** are exclusively on the outer face and are clustered in certain regions, as discussed later in the chapter (p. 401), whereas other types of lipid-linked proteins (with farnesyl or geranylgeranyl groups attached; Fig. 11-13) are exclusively on the inner face. In polarized epithelial cells (such as intestinal epithelial cells; see Fig. 11-41), in which apical and basal surfaces have different roles, GPI-anchored proteins are directed specifically to the apical surface. Attachment of a specific lipid to a newly synthesized membrane protein therefore has a targeting function, directing the protein to its correct cellular location.

Amphitropic Proteins Associate Reversibly with the Membrane

Some amphitropic proteins contain a PH (pleckstrin homology) domain, a binding pocket that specifically binds phosphatidylinositol 3,4,5-trisphosphate (PIP_3) located on the cytoplasmic face of the plasma membrane. PIP_3 is formed and degraded in response to hormonal and other signals. Another conserved protein domain, SH2 (Src homology), binds membrane proteins with phosphorylated tyrosine (phosphotyrosine) residues, but not their unphosphorylated form. (PH and SH2 domains are discussed in more detail in Chapter 12.) Thus the association of many amphitropic proteins with the plasma membrane can be reversibly controlled by the enzymatic addition or removal of a single phosphoryl group on phosphatidylinositol or a protein Tyr residue. The transient association of specific proteins with the membrane is central to many signaling pathways. When two or more proteins need to interact in a signaling event, confining them to the two-dimensional space of the membrane surface makes their interaction far more likely.

SUMMARY 11.1 The Composition and Architecture of Membranes

■ Biological membranes define cellular boundaries, divide cells into discrete compartments, organize complex reaction sequences, and act in signal reception and energy transformations.

■ Membranes are composed of lipids and proteins in varying combinations particular to each species, cell type, and organelle. The fluid mosaic model, with a lipid bilayer as the basic structural unit, gives a simplified and general picture of membranes.

■ Membrane trafficking is the movement of membrane components from the endoplasmic reticulum into and through the Golgi apparatus, where they are targeted to their final destinations by covalent alterations.

■ Integral membrane proteins are embedded within membranes, their nonpolar amino acid side chains stabilized by contact with the lipid bilayer rather than the surrounding aqueous phase. Peripheral membrane

proteins associate with membranes through electrostatic interactions and hydrogen bonding with membrane phospholipids and integral proteins. Amphitropic proteins associate reversibly with membranes in response to biological signals such as phosphorylation of membrane lipids or proteins or the removal of covalently attached lipids.

■ Many membrane proteins span the lipid bilayer several times, with hydrophobic sequences of about 20 amino acid residues forming transmembrane α helices. Multistranded β barrels are also common in integral proteins in bacterial membranes. Tyr and Trp residues of transmembrane proteins are commonly found at the lipid-water interface.

■ Some membrane proteins have covalently attached lipids that mediate their interaction with the bilayer.

11.2 Membrane Dynamics

One remarkable feature of all biological membranes is their plasticity—their ability to change shape without losing their integrity and becoming leaky. The basis for this property is the noncovalent interactions among lipids in the bilayer and the mobility allowed to individual lipids because they are not covalently anchored to one another. We turn now to the dynamics of membranes: the motions that occur and the transient structures allowed by these motions.

Acyl Groups in the Bilayer Interior Are Ordered to Varying Degrees

Although the lipid bilayer structure is stable, its individual phospholipid molecules have much freedom of motion **(Fig. 11-14)**, depending on the temperature and the lipid composition. Below normal physiological temperatures, the lipids in a bilayer form a gel-like **liquid-ordered (L_o) state**, in which all types of motion of individual lipid molecules are strongly constrained; the bilayer is paracrystalline (Fig. 11-14a). Above physiological temperatures, individual hydrocarbon chains of fatty acids are in constant motion produced by rotation about the carbon–carbon bonds of the long acyl side chains and by lateral diffusion of individual lipid molecules in the plane of the bilayer. This is the **liquid-disordered (L_d) state** (Fig. 11-14b). In the transition from the L_o state to the L_d state, the general shape and dimensions of the bilayer are maintained; what changes is the degree of motion (lateral and rotational) allowed to individual lipid molecules.

At temperatures in the physiological range for a mammal (about 20 to 40 °C), long-chain saturated fatty acids (such as 16:0 and 18:0) tend to pack into an L_o gel phase, but the kinks in unsaturated fatty acids (see Fig. 10-1) interfere with packing, favoring the L_d state. Shorter-chain fatty acyl groups are more mobile than longer-chain fatty acyl groups and thus also favor the L_d state. The sterol content of a membrane, which varies

(a) Liquid-ordered state L₀

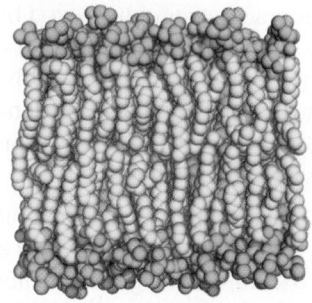

↕ Heat produces thermal motion of side chains (L₀ → L_d transition).

(b) Liquid-disordered state L_d

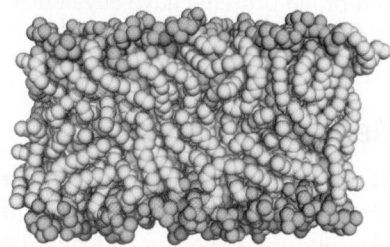

FIGURE 11-14 Two extreme states of bilayer lipids. (a) In the liquid-ordered (L_o) state, polar head groups are uniformly arrayed at the surface, and the acyl chains are nearly motionless and packed with regular geometry. **(b)** In the liquid-disordered (L_d) state, or fluid state, acyl chains undergo much thermal motion and have no regular organization. The state of membrane lipids in biological membranes is maintained somewhere between these extremes. [Source: H. Heller et al., *J. Phys. Chem.* 97:8343, 1993.]

greatly with organism and organelle (Table 11-1), is another important determinant of lipid state. Sterols (such as cholesterol) have paradoxical effects on bilayer fluidity: they interact with phospholipids containing unsaturated fatty acyl chains, compacting them and constraining their motion in bilayers. In contrast, their association with sphingolipids and phospholipids having long, saturated fatty acyl chains tends to make a bilayer fluid that would otherwise, without cholesterol, adopt the L_o state. In biological membranes composed of a variety of phospholipids and sphingolipids, cholesterol tends to associate with sphingolipids and to form regions in the L_o state surrounded by cholesterol-poor regions in the L_d state (see the discussion of membrane rafts below).

Cells regulate their lipid composition to achieve a constant membrane fluidity under various growth conditions. For example, bacteria synthesize more unsaturated fatty acids and fewer saturated ones when cultured at low temperatures than when cultured at higher temperatures (Table 11-2). As a result of this adjustment in lipid composition, membranes of bacteria cultured at high or low temperatures have about the same degree of fluidity. This is presumably essential for the function of many membrane-embedded proteins—enzymes, transporters, and receptors—that act within the lipid bilayer.

Transbilayer Movement of Lipids Requires Catalysis

At physiological temperatures, a lipid molecule diffuses from one leaflet (monolayer) of the bilayer to the other **(Fig. 11-15a)** very slowly, if at all, in most membranes, although lateral diffusion *in the plane* of the bilayer is very rapid (Fig. 11-15b). Transbilayer—or "flip-flop"—movement requires that a polar or charged head group leave its aqueous environment and move into the hydrophobic interior of the bilayer, a process with a large, positive free-energy change. There are, however, situations in which such movement is essential. For example, in the ER, membrane glycerophospholipids are synthesized on the cytosolic face, whereas sphingolipids are synthesized or modified on the lumenal surface. To get from their site of synthesis to their eventual point of deposition, these lipids must undergo flip-flop diffusion.

Proteins called flippases, floppases, and scramblases (Fig. 11-15c) facilitate the transbilayer movement

TABLE 11-2	Fatty Acid Composition of *E. coli* Cells Cultured at Different Temperatures			
	Percentage of total fatty acids[a]			
	10 °C	20 °C	30 °C	40 °C
Myristic acid (14:0)	4	4	4	8
Palmitic acid (16:0)	18	25	29	48
Palmitoleic acid (16:1)	26	24	23	9
Oleic acid (18:1)	38	34	30	12
Hydroxymyristic acid	13	10	10	8
Ratio of unsaturated to saturated[b]	2.9	2.0	1.6	0.38

Source: Data from A. G. Marr and J. L. Ingraham, *J. Bacteriol.* 84:1260, 1962.

[a]The exact fatty acid composition depends not only on growth temperature but on growth stage and growth medium composition.

[b]Ratios calculated as the total percentage of 16:1 plus 18:1 divided by the total percentage of 14:0 plus 16:0. Hydroxymyristic acid was omitted from this calculation.

(a) Uncatalyzed transbilayer ("flip-flop") diffusion

(b) Uncatalyzed lateral diffusion

(c) Catalyzed transbilayer translocations

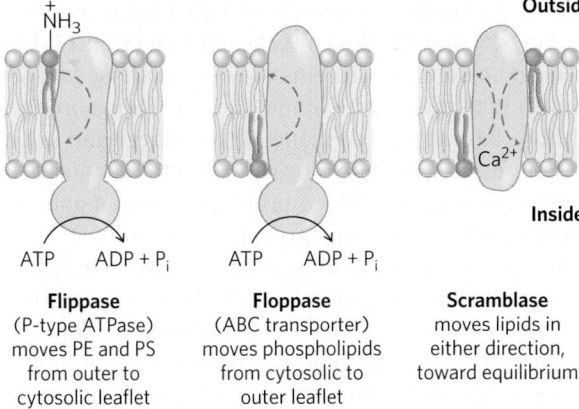

FIGURE 11-15 Motion of single phospholipids in a bilayer. (a) Uncatalyzed movement from one leaflet to the other is very slow, but **(b)** lateral diffusion within the leaflet is very rapid, requiring no catalysis. **(c)** Three types of phospholipid translocaters in the plasma membrane. PE is phosphatidylethanolamine; PS is phosphatidylserine.

(translocation) of individual lipid molecules, providing a path that is energetically more favorable and much faster than the uncatalyzed movement. The combination of asymmetric biosynthesis of membrane lipids, very slow uncatalyzed flip-flop diffusion, and the presence of selective, energy-dependent lipid translocators could account for the transbilayer asymmetry in lipid composition discussed in Section 11.1. Besides contributing to this asymmetry of composition, the energy-dependent transport of lipids to one bilayer leaflet may, by creating a larger surface on one side of the bilayer, be important in generating the membrane curvature essential in the budding of vesicles.

Flippases catalyze translocation of the *amino*-phospholipids phosphatidylethanolamine and phosphatidylserine from the extracellular to the cytosolic leaflet of the plasma membrane, contributing to the asymmetric distribution of phospholipids: phosphatidylethanolamine and phosphatidylserine primarily in the cytosolic leaflet, and the sphingolipids and phosphatidylcholine in the outer leaflet. Keeping phosphatidylserine out of the extracellular leaflet is important: its exposure on the outer surface triggers apoptosis (programmed cell death; see Chapter 12) and engulfment by macrophages that

carry phosphatidylserine receptors. Flippases also act in the ER, where they move newly synthesized phospholipids from their site of synthesis in the cytosolic leaflet to the lumenal leaflet. Flippases consume about one ATP per molecule of phospholipid translocated, and they are structurally and functionally related to the P-type ATPases (active transporters) described on page 413.

Two other types of lipid-translocating activities are known but less well characterized. **Floppases** move plasma membrane phospholipids and sterols from the cytosolic to the extracellular leaflet and, like flippases, are ATP-dependent. Floppases are members of the ABC transporter family, described on page 413; all ABC transporters actively transport hydrophobic substrates outward across the plasma membrane. **Scramblases** are proteins that move any membrane phospholipid across the bilayer down its concentration gradient (from the leaflet where it has a higher concentration to the leaflet where it has a lower concentration); their activity is not dependent on ATP. Scramblase activity leads to controlled randomization of the head-group composition on the two faces of the bilayer. The activity rises sharply with an increase in cytosolic Ca^{2+} concentration, which may result from cell activation, cell injury, or apoptosis; as noted above, exposure of phosphatidylserine on the outer surface marks a cell for apoptosis and engulfment by macrophages. Rhodopsin, the protein that detects light in the vertebrate eye, has a second activity: it is a scramblase that facilitates rapid randomization, exceeding 10,000 phospholipids per protein per second. Finally, a group of proteins that act primarily to move phosphatidylinositol lipids across lipid bilayers, the phosphatidylinositol transfer proteins, are believed to have important roles in lipid signaling and membrane trafficking.

Lipids and Proteins Diffuse Laterally in the Bilayer

Individual lipid molecules can move laterally in the plane of the membrane by changing places with neighboring lipid molecules; that is, they undergo Brownian movement within the bilayer (Fig. 11-15b), which can be quite rapid. A molecule in the outer leaflet of the erythrocyte plasma membrane, for example, can diffuse laterally so fast that it circumnavigates the erythrocyte in seconds. This rapid lateral diffusion in the plane of the bilayer tends to randomize the positions of individual molecules in a few seconds.

Lateral diffusion can be shown experimentally by attaching fluorescent probes to the head groups of lipids and using fluorescence microscopy to follow the probes over time **(Fig. 11-16)**. In one technique, a small region ($5 \ \mu m^2$) of a cell surface with fluorescence-tagged lipids is bleached by intense laser radiation so that the irradiated patch no longer fluoresces when viewed with less-intense (nonbleaching) light in the fluorescence microscope. However, within milliseconds, the region recovers its fluorescence as unbleached lipid

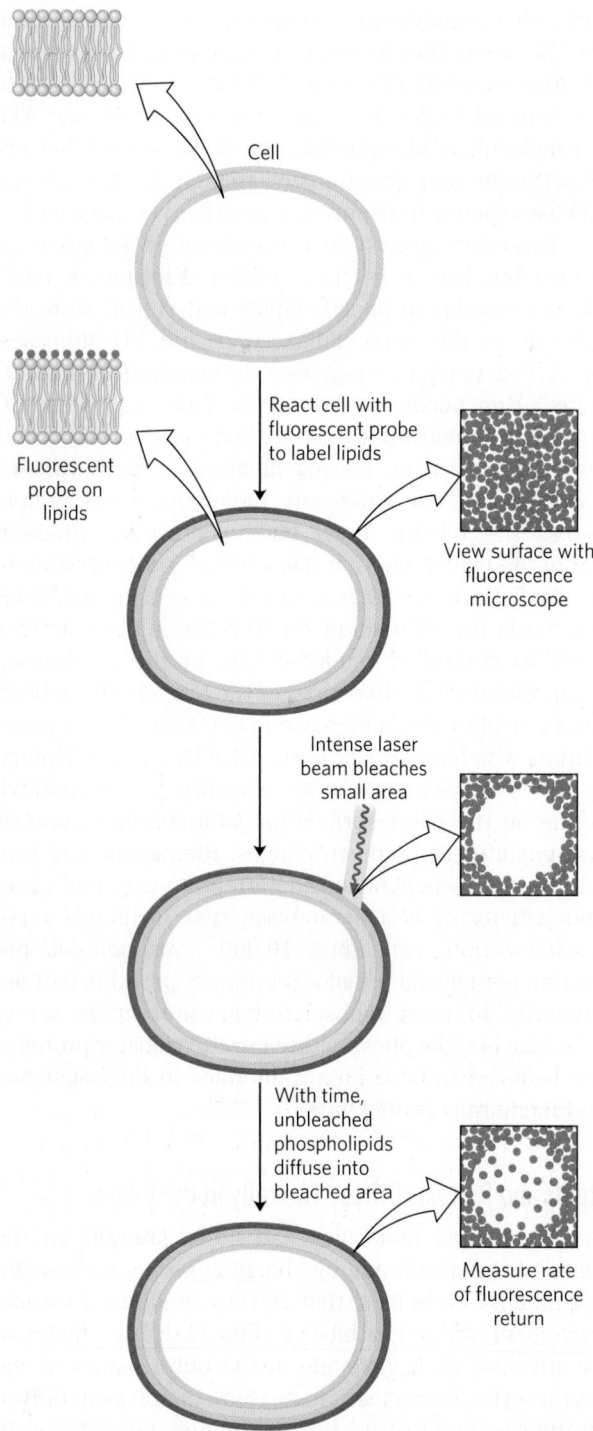

Cell

React cell with fluorescent probe to label lipids

Fluorescent probe on lipids

View surface with fluorescence microscope

Intense laser beam bleaches small area

With time, unbleached phospholipids diffuse into bleached area

Measure rate of fluorescence return

FIGURE 11-16 Measurement of lateral diffusion rates of lipids by fluorescence recovery after photobleaching (FRAP). Lipids in the outer leaflet of the plasma membrane are labeled by reaction with a membrane-impermeant fluorescent probe (red) so that the surface is uniformly labeled when viewed with a fluorescence microscope. A small area is bleached by irradiation with an intense laser beam and becomes nonfluorescent. With the passage of time, labeled lipid molecules diffuse into the bleached region, and it again becomes fluorescent. Researchers can track the time course of fluorescence return and determine a diffusion coefficient for the labeled lipid. The diffusion rates are typically high; a lipid moving at this speed could circumnavigate an *E. coli* cell in one second. (The FRAP method can also be used to measure lateral diffusion of membrane proteins.)

region to a nearby region ("hop diffusion") is rarer; membrane lipids behave as though corralled by fences that they can occasionally cross by hop diffusion **(Fig. 11-17)**.

Many membrane proteins move as if afloat in a sea of lipids. Like membrane lipids, these proteins are free to diffuse laterally in the plane of the bilayer and are in constant motion, as shown by the FRAP technique with fluorescence-tagged surface proteins. Some membrane proteins associate to form large aggregates ("patches") on the surface of a cell or organelle in which individual protein molecules do not move relative to one another; for example, acetylcholine receptors form dense, near-crystalline patches on neuronal plasma membranes at synapses. Other membrane proteins are anchored to internal structures that prevent their free diffusion. In the erythrocyte membrane, both glycophorin and the chloride-bicarbonate exchanger (p. 410) are tethered to

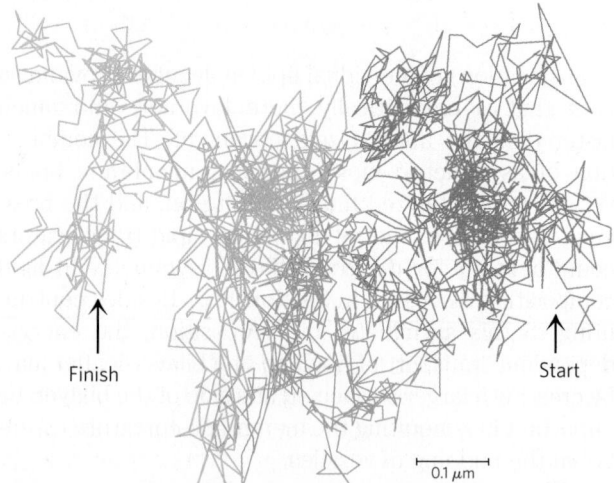

Finish

Start

0.1 μm

FIGURE 11-17 Hop diffusion of individual lipid molecules. The motion of a single fluorescently labeled lipid molecule in a cell surface is recorded on video by fluorescence microscopy, with a time resolution of 25 μs (equivalent to 40,000 frames/s). The track shown here represents a molecule followed for 56 ms (2,250 frames); the trace begins in the purple area and continues through blue, green, and orange. The pattern of movement indicates rapid diffusion within a confined region (about 250 nm in diameter, shown by a single color), with occasional hops into an adjoining region. This finding suggests that the lipids are corralled by molecular fences that they occasionally jump. [Source: Courtesy of Takahiro Fujiwara, Ken Ritchie, Hideji Murakoshi, Ken Jacobson, and Akihiro Kusumi.]

molecules diffuse into the bleached patch and bleached lipid molecules diffuse away from it. The rate of *f*luorescence *r*ecovery *a*fter *p*hotobleaching, or **FRAP**, is a measure of the rate of lateral diffusion of the lipids. Using the FRAP technique, researchers have shown that some membrane lipids diffuse laterally at rates of up to 1 μm/s.

Another technique, single particle tracking, allows one to follow the movement of a *single* lipid molecule in the plasma membrane on a much shorter time scale. Results from these studies confirm that lipid molecules diffuse laterally with rapidity within small, discrete regions of the cell surface but that movement from one such

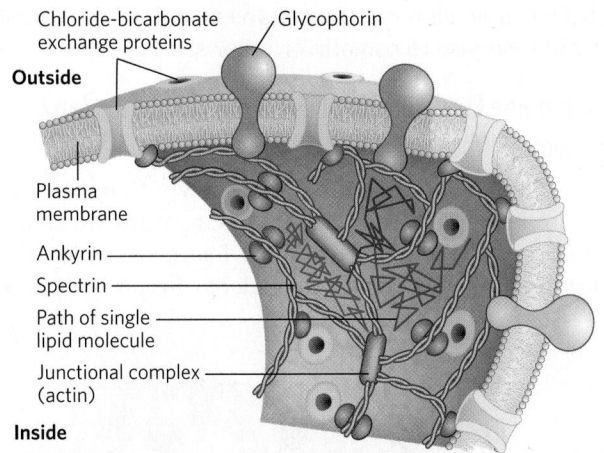

FIGURE 11-18 Restricted motion of the erythrocyte chloride-bicarbonate exchanger and glycophorin. The proteins span the membrane and are tethered to spectrin, a cytoskeletal protein, by another protein, ankyrin, limiting their lateral mobility. Ankyrin is anchored in the membrane by a covalently bound palmitoyl side chain (see Fig. 11-13). Spectrin, a long, filamentous protein, is cross-linked at junctional complexes containing actin. A network of cross-linked spectrin molecules attached to the cytoplasmic face of the plasma membrane stabilizes the membrane, making it resistant to deformation. This network of anchored membrane proteins may form the "corral" suggested by the experiment shown in Figure 11-17; the lipid tracks shown here are confined to different regions defined by the tethered membrane proteins. Occasionally a lipid molecule (green track) jumps from one corral to another (blue track), then another (red track).

spectrin, a filamentous cytoskeletal protein **(Fig. 11-18)**. One possible explanation for the pattern of lateral diffusion of lipid molecules shown in Figure 11-17 is that membrane proteins immobilized by their association with spectrin form the "fences" that define the regions within which relatively unrestricted lipid motion can occur.

Sphingolipids and Cholesterol Cluster Together in Membrane Rafts

We have seen that diffusion of membrane lipids from one bilayer leaflet to the other is very slow unless catalyzed and that the different lipid species of the plasma membrane are asymmetrically distributed in the two leaflets of the bilayer (Fig. 11-5). Even within a single leaflet, the lipid distribution is not uniform. Glycosphingolipids (cerebrosides and gangliosides), which typically contain long-chain saturated fatty acids, form transient clusters in the outer leaflet that largely exclude glycerophospholipids, which typically contain one unsaturated fatty acyl group and a shorter saturated acyl group. The long, saturated acyl groups of sphingolipids can form more compact, more stable associations with the long ring system of cholesterol than can the shorter, often unsaturated, chains of phospholipids. The cholesterol-sphingolipid **microdomains** of the plasma membrane make the bilayer slightly thicker and more ordered (less fluid) than neighboring regions rich in phospholipids, and are more difficult to dissolve with nonionic detergents; they behave like liquid-ordered sphingolipid **rafts** adrift on an ocean of

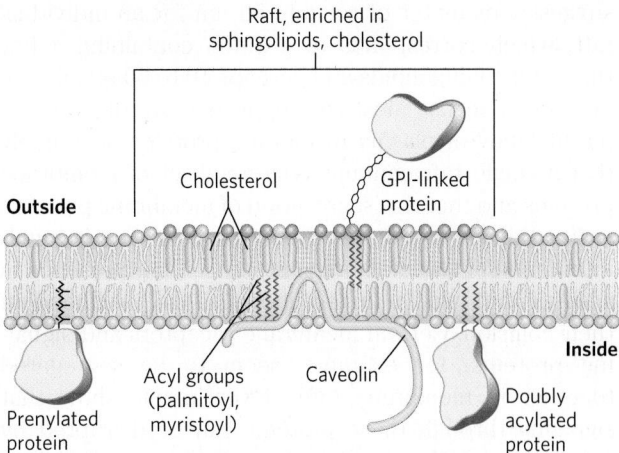

FIGURE 11-19 Membrane microdomains (rafts). Stable associations of sphingolipids and cholesterol in the outer leaflet produce a microdomain, slightly thicker than other membrane regions, that is enriched with specific types of membrane proteins. GPI-anchored proteins are prominent in the outer leaflet of these rafts, and proteins with one or several covalently attached long-chain acyl groups are common in the inner leaflet. Inwardly curved rafts called caveolae are especially enriched in proteins called caveolins (see Fig. 11-20). Proteins with attached prenyl groups (such as Ras; see Box 12-1) tend to be excluded from rafts.

liquid-disordered phospholipids **(Fig. 11-19)**. Proteins with relatively short hydrophobic helical sections (19 to 20 residues) cannot span the thicker bilayer in rafts and thus tend to be excluded. Proteins with longer hydrophobic helices (24 to 25 residues) segregate into the thicker bilayer regions of rafts, where the entire length of the helix is stabilized by the hydrophobic effect.

Lipid rafts are remarkably enriched in two classes of integral membrane proteins, with two specific types of covalently attached lipids. The integral proteins of one class have two long-chain saturated fatty acids (two palmitoyl groups or a palmitoyl and a myristoyl group) covalently attached through Cys residues. Those of the second class, the GPI-anchored proteins, have a glycosyl phosphatidylinositol on their carboxyl-terminal residue (Fig. 11-13). Presumably, these lipid anchors, like the long, saturated acyl chains of sphingolipids, form more stable associations with the cholesterol and long acyl groups in rafts than with the surrounding phospholipids. (It is notable that other lipid-linked proteins, those with covalently attached isoprenyl groups such as farnesyl, are *not* preferentially associated with the outer leaflet of sphingolipid/cholesterol rafts; see Fig. 11-19.) The "raft" and "sea" domains of the plasma membrane are not rigidly separated; membrane proteins can move into and out of lipid rafts in a fraction of a second. But in the shorter time scale (microseconds) more relevant to many membrane-mediated biochemical processes, many of these proteins reside primarily in a raft.

We can estimate the fraction of the cell surface occupied by rafts from the fraction of the plasma membrane that resists dissolution by detergent, which can be as high as 50% in some cases: the rafts cover half of the ocean. Indirect measurements in cultured fibroblasts suggest a diameter of roughly 50 nm for an individual raft, which corresponds to a patch containing a few thousand sphingolipids and perhaps 10 to 50 membrane proteins. Because most cells express more than 50 different kinds of plasma membrane proteins, it is likely that a single raft contains only a subset of membrane proteins and that this segregation of membrane proteins is functionally significant. For a process that involves interaction of two membrane proteins, their presence in a single raft would hugely increase the likelihood of their collision. Certain membrane receptors and signaling proteins, for example, seem to be segregated together in membrane rafts. Experiments show that signaling through these proteins can be disrupted by manipulations that deplete the plasma membrane of cholesterol and destroy lipid rafts.

A **caveolin** is an integral membrane protein with two globular domains connected by a hairpin-shaped hydrophobic domain, which binds the protein to the cytoplasmic leaflet of the plasma membrane. Three palmitoyl groups attached to the carboxyl-terminal globular domain further anchor the protein to the membrane. Caveolins form dimers and associate with cholesterol-rich regions in the membrane. The presence of caveolin dimers forces the associated lipid bilayer to curve inward, forming **caveolae** ("little caves") in the cell surface **(Fig. 11-20)**. Caveolae are unusual rafts: they involve *both* leaflets of the bilayer—the cytoplasmic leaflet, from which the caveolin globular domains project, and the extracellular leaflet, a typical sphingolipid/cholesterol raft with associated GPI-anchored proteins. Caveolae are implicated in a variety of cellular functions, including membrane trafficking within cells and the transduction of external signals into cellular responses. The receptors for insulin and other growth factors, as well as certain GTP-binding proteins and protein kinases associated with transmembrane signaling, seem to be localized in rafts and perhaps in caveolae. We discuss some possible roles of rafts in signaling in Chapter 12.

Caveolae may also provide a means of expanding the cell surface. The lipid bilayer itself is not elastic, but if existing caveolae lose their associated caveolin as the result of a regulatory signal, the caveolae flatten into the plasma membrane (Fig. 11-20c). The effect is to add surface area, allowing the cell to expand without bursting in response to osmotic or other stress.

Membrane Curvature and Fusion Are Central to Many Biological Processes

Caveolins are not unique in their ability to induce curvature in membranes. Changes of curvature are central to one of the most remarkable features of biological membranes: their ability to undergo fusion with other

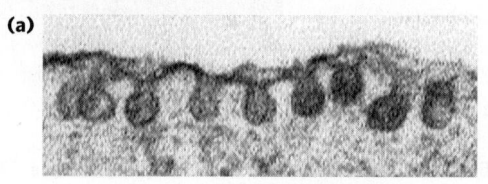
(a)

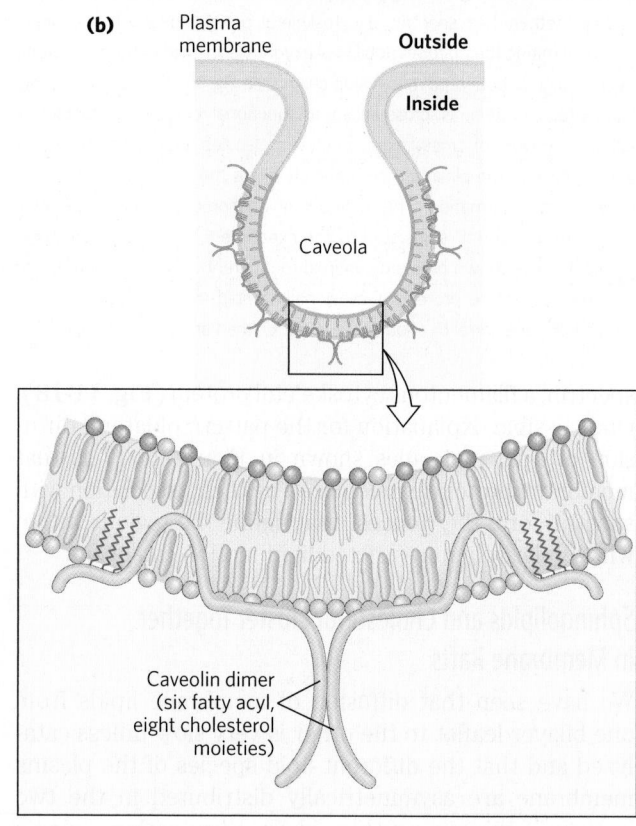
(b)

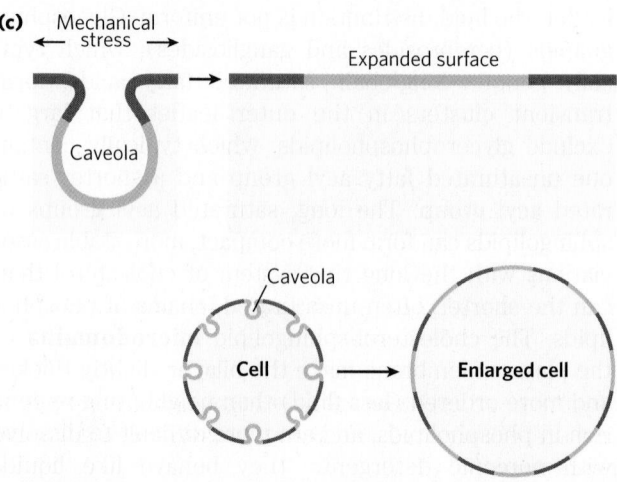
(c)

FIGURE 11-20 A caveolin forces inward curvature of a membrane. Caveolae are small invaginations in the plasma membrane, as seen in **(a)** an electron micrograph of an adipocyte that is surface-labeled with an electron-dense marker. **(b)** Cartoon showing the location and role of a caveolin dimer in causing inward membrane curvature. Each caveolin monomer has a central hydrophobic domain and three long-chain acyl groups (red), which hold the molecule to the inside of the plasma membrane. When several caveolin dimers are concentrated in a small region (a raft), they force a curvature in the lipid bilayer, forming a caveola. Cholesterol molecules in the bilayer are shown in orange. **(c)** Flattening of caveolae allows the plasma membrane to expand in response to various stresses. [Source: (a) Courtesy of R. G. Parton. Reprinted with permission from Macmillan Publishers, Ltd.: *Nature Rev. Mol. Cell Biol.* 8:185-194, Fig. 1a. ©2007]

membranes without losing their continuity. Although membranes are stable, they are by no means static. Within the eukaryotic endomembrane system (which includes the nuclear membrane, endoplasmic reticulum, Golgi complex, and various small vesicles), the membranous compartments constantly reorganize. Vesicles bud from the ER to carry newly synthesized lipids and proteins to other organelles and to the plasma membrane. Exocytosis, endocytosis, cell division, fusion of egg and sperm cells, and entry of a membrane-enveloped virus into a host cell all involve a membrane reorganization that requires the fusion of two membrane segments without loss of continuity **(Fig. 11-21)**. Most of these processes begin with a local increase in membrane curvature. A protein that is intrinsically curved may force a bilayer to curve by binding to it **(Fig. 11-22)**; the binding energy provides the driving force for the increase in bilayer curvature. Alternatively, multiple subunits of a scaffold protein may assemble into curved supramolecular complexes and stabilize curves that spontaneously form in the bilayer. For example, a superfamily of proteins containing **BAR domains** (named for the first three members of the family that were identified: *BIN1*, *a*mphiphysin, and *R*VS167) can assemble into a crescent-shaped scaffold that binds to the membrane surface, forcing or favoring membrane curvature. BAR domains consist of coiled coils that form long, thin, curved dimers with a positively charged concave surface that tends to form ionic interactions with the negatively

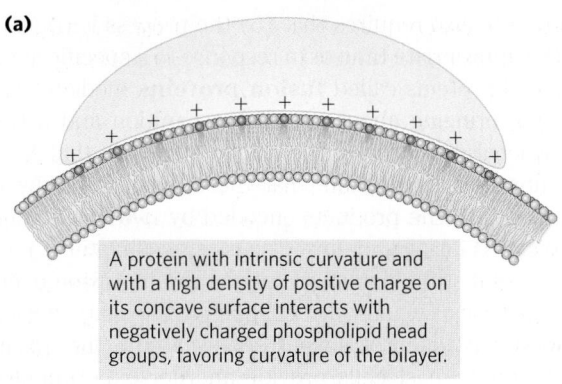

(a) A protein with intrinsic curvature and with a high density of positive charge on its concave surface interacts with negatively charged phospholipid head groups, favoring curvature of the bilayer.

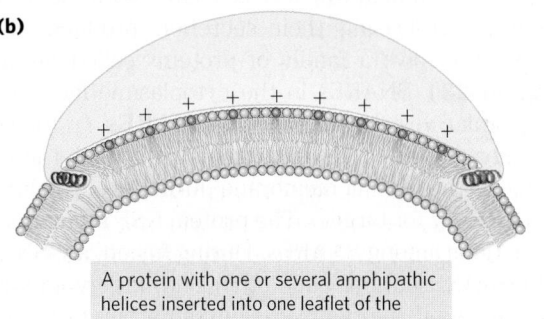

(b) A protein with one or several amphipathic helices inserted into one leaflet of the bilayer crowds the lipids in that leaflet, forcing the membrane to bend.

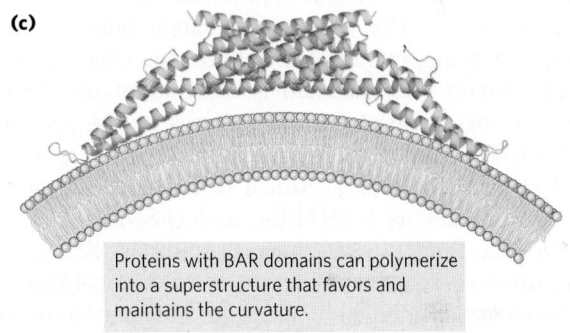

(c) Proteins with BAR domains can polymerize into a superstructure that favors and maintains the curvature.

FIGURE 11-22 Three models for protein-induced curvature of membranes. [Sources: (a, b) Information from B. Qualmann et al., *EMBO J.* 30:3501, 2011, Fig. 1. (c) Information from B. J. Peter et al., *Science* 303:495, 2004, Fig. 1A.]

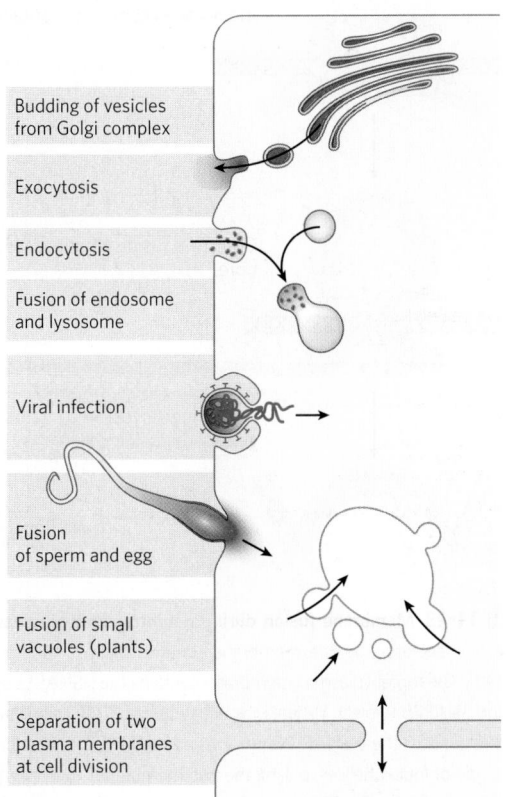

Budding of vesicles from Golgi complex

Exocytosis

Endocytosis

Fusion of endosome and lysosome

Viral infection

Fusion of sperm and egg

Fusion of small vacuoles (plants)

Separation of two plasma membranes at cell division

FIGURE 11-21 Membrane fusion. The fusion of two membranes is central to a variety of cellular processes involving organelles and the plasma membrane.

charged head groups of membrane lipids PIP_2 and PIP_3. The enzymatic formation of these inositol lipids can tag a plasma membrane area for creation of inward curvature by a BAR protein (Fig. 11-22). Some of these BAR proteins also have a helical region that inserts like a wedge into one leaflet of the bilayer, expanding its area relative to the other leaflet and thereby forcing curvature.

Specific fusion of two membranes requires that (1) they recognize each other; (2) their surfaces become closely apposed, which requires removal of the water molecules normally associated with the polar head groups of lipids; (3) their bilayer structures become locally disrupted, resulting in fusion of the outer leaflets of the two membranes (hemifusion); and (4) their bilayers fuse to form a single continuous bilayer. The fusion occurring in receptor-mediated endocytosis, or regulated secretion, also requires that (5) the process is triggered at the appropriate time or in response to a specific signal. Integral proteins called **fusion proteins** mediate these events, bringing about specific recognition and a transient local distortion of the bilayer structure that favors membrane fusion. (Note that these fusion proteins are unrelated to the products encoded by two fused genes, also called fusion proteins, discussed in Chapter 9.)

A well-studied example of membrane fusion occurs at synapses, when intracellular (neuronal) vesicles loaded with neurotransmitter fuse with the plasma membrane. Yeast cells provide another experimentally accessible system in which vesicles fuse with the plasma membrane, releasing their secretion products. Both processes involve a family of proteins called SNAREs **(Fig. 11-23)**. SNAREs in the cytoplasmic face of the intracellular vesicle are called **v-SNAREs** (*v* for vesicle); those in the target membrane with which the vesicle fuses (the plasma membrane during exocytosis) are **t-SNAREs** (*t* for target). The protein NSF regulates the interactions among SNAREs. During fusion, a v-SNARE and t-SNARE bind to each other and undergo a structural change that produces a bundle of long, thin rods made up of helices from both SNAREs and two helices from the protein SNAP25 (Fig. 11-23). The two SNAREs initially interact at their ends, then zip up into the bundle of helices. This structural change pulls the two membranes into contact and initiates the fusion of their lipid bilayers. An alternative designation of SNARE types is based on structural features of the proteins: R-SNAREs have an Arg residue critical to their function, and Q-SNAREs have a critical Gln residue. Typically, R-SNAREs act as v-SNAREs, and Q-SNAREs act as t-SNAREs. James E. Rothman, Randy W. Schekman, and Thomas C. Südhof shared the 2013 Nobel Prize in Physiology or Medicine for their elucidation of the molecular basis of membrane trafficking and fusion.

Thomas C. Südhof, Randy W. Schekman, and James E. Rothman [Source: Alban Wyters/Sipa USA/AP Images.]

The complex of SNAREs and SNAP25 is the target of several powerful neurotoxins. *Clostridium botulinum* toxin is a bacterial protease that cleaves specific bonds in SNARE proteins, preventing

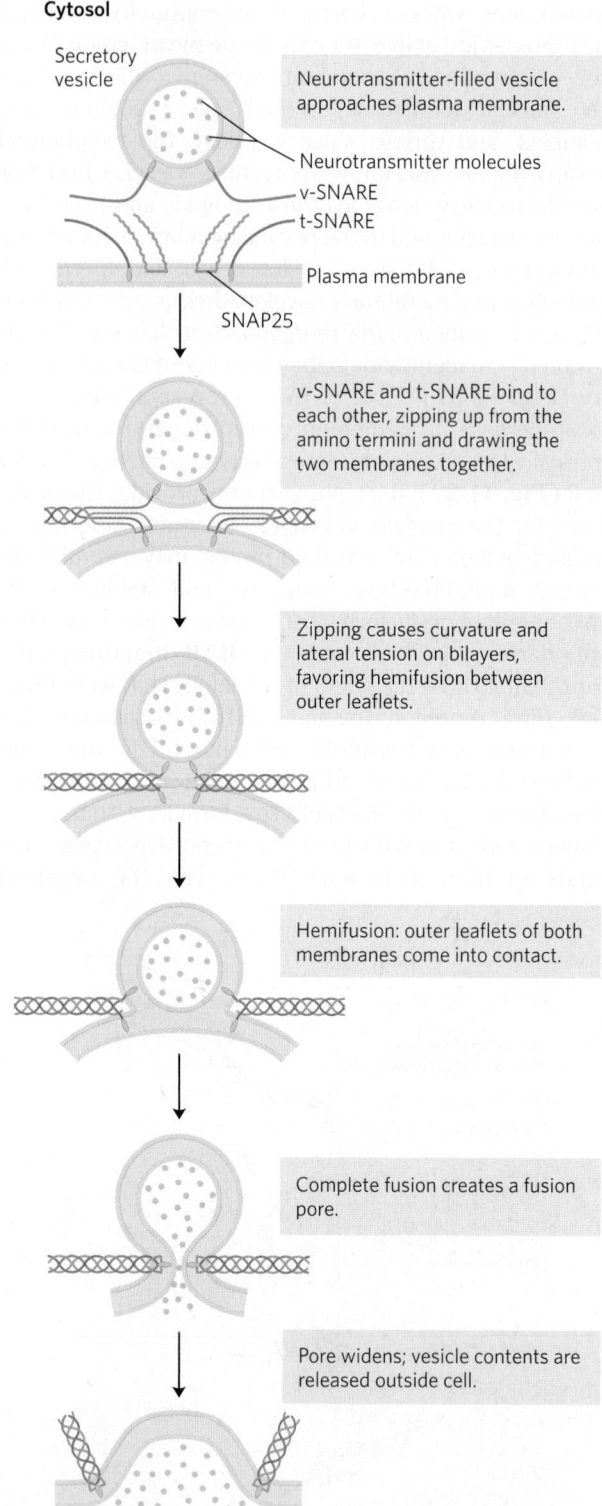

Cytosol

Secretory vesicle

Neurotransmitter-filled vesicle approaches plasma membrane.

Neurotransmitter molecules
v-SNARE
t-SNARE

Plasma membrane

SNAP25

v-SNARE and t-SNARE bind to each other, zipping up from the amino termini and drawing the two membranes together.

Zipping causes curvature and lateral tension on bilayers, favoring hemifusion between outer leaflets.

Hemifusion: outer leaflets of both membranes come into contact.

Complete fusion creates a fusion pore.

Pore widens; vesicle contents are released outside cell.

FIGURE 11-23 Membrane fusion during neurotransmitter release at a synapse. The secretory vesicle membrane contains the v-SNARE synaptobrevin (red). The target (plasma) membrane contains the t-SNAREs syntaxin (blue) and SNAP25 (violet). When a local increase in [Ca^{2+}] signals release of neurotransmitter, the v-SNARE, SNAP25, and t-SNARE interact, forming a coiled bundle of four α helices, pulling the two membranes together and disrupting the bilayer locally. This leads first to hemifusion, joining the outer leaflets of the two membranes, then to complete membrane fusion and neurotransmitter release. NSF (*N*-ethylmaleimide-sensitive *f*usion factor) acts in disassembly of the SNARE complex when fusion is complete. [Source: Information from Y. A. Chen and R. H. Scheller, *Nature Rev. Mol. Cell Biol.* 2:98, 2001.]

neurotransmission and causing paralysis and death. Because of its very high specificity for these proteins, purified botulinum toxin has served as a powerful tool for dissecting the mechanism of neurotransmitter release in vivo and in vitro. Used in small amounts, botulinum toxin (Botox) is used in medicine to treat disorders of eye and neck muscles, as well as cosmetically for the removal of skin wrinkles. Tetanus toxin, produced by the bacterium *Clostridium tetani*, is also a protease with high specificity for SNARE proteins. It causes painful muscle spasms and rigidity of voluntary muscles—hence the characteristic symptom "lockjaw." ■

Integral Proteins of the Plasma Membrane Are Involved in Surface Adhesion, Signaling, and Other Cellular Processes

Several families of integral proteins in the plasma membrane provide specific points of attachment between cells or between a cell and proteins of the extracellular matrix. **Integrins** are surface adhesion proteins that mediate a cell's interaction with the extracellular matrix and with other cells, including some pathogens. Integrins also carry signals in both directions across the plasma membrane, integrating information about the extracellular and intracellular environments. All integrins are heterodimeric proteins composed of two unlike subunits, α and β, each anchored to the plasma membrane by a single transmembrane helix. The large extracellular domains of the α and β subunits combine to form a specific binding site for extracellular proteins such as collagen and fibronectin, which contain a common determinant of integrin binding, the sequence Arg–Gly–Asp (RGD).

Other plasma membrane proteins involved in surface adhesion are the **cadherins**, which undergo homophilic ("with the same kind") interactions with identical cadherins in an adjacent cell. **Selectins** have extracellular domains that, in the presence of Ca^{2+}, bind specific polysaccharides on the surface of an adjacent cell. Selectins are present primarily in the various types of blood cells and in the endothelial cells that line blood vessels (see Fig. 7-32). They are an essential part of the blood-clotting process.

Integral membrane proteins play roles in many other cellular processes. They serve as transporters and ion channels (discussed in Section 11.3) and as receptors for hormones, neurotransmitters, and growth factors (Chapter 12). They are central to oxidative phosphorylation and photophosphorylation (Chapters 19 and 20) and to cell-cell and cell-antigen recognition in the immune system (Chapter 5). Integral proteins are also important players in the membrane fusion that accompanies exocytosis, endocytosis, and the entry of many types of viruses into host cells.

SUMMARY 11.2 Membrane Dynamics

■ Lipids in a biological membrane can exist in liquid-ordered or liquid-disordered states; in the latter state, thermal motion of acyl chains makes the interior of the bilayer fluid. Fluidity is affected by temperature, fatty acid composition, and sterol content.

■ Flip-flop diffusion of lipids between the inner and outer leaflets of a membrane is very slow except when specifically catalyzed by flippases, floppases, or scramblases.

■ Proteins and lipids can diffuse laterally within the plane of the membrane, but this mobility is limited by interactions of membrane proteins with internal cytoskeletal structures and interactions of lipids with lipid rafts. One class of lipid rafts is enriched for sphingolipids and cholesterol with a subset of membrane proteins that are GPI-linked or attached to several long-chain fatty acyl moieties.

■ Caveolins are integral membrane proteins that associate with the inner leaflet of the plasma membrane, forcing it to curve inward to form caveolae, which are involved in membrane transport, signaling, and the expansion of plasma membranes.

■ Specific proteins containing BAR domains cause local membrane curvature and mediate the fusion of two membranes, which accompanies processes such as endocytosis, exocytosis, and viral invasion. Because the inositol phospholipids PIP_2 and PIP_3 are specifically recognized by BAR proteins, their formation may be the signal for the intracellular processes that require membrane curvature.

■ SNAREs are membrane proteins that act in the fusion of vesicles with the plasma membrane, in response to a signal.

■ Integrins, cadherins, and selectins are transmembrane proteins of the plasma membrane that act both to attach cells to each other and to carry messages between the extracellular matrix and the cytoplasm.

11.3 Solute Transport across Membranes

Every living cell must acquire from its surroundings the raw materials for biosynthesis and for energy production, and must release the byproducts of metabolism to its environment; both processes require that small compounds or inorganic ions cross the plasma membrane. Within the eukaryotic cell, different compartments have different concentrations of ions and of metabolic intermediates and products, and these, too, must move across intracellular membranes in tightly regulated processes. A few nonpolar compounds can dissolve in the lipid bilayer and cross a membrane unassisted, but for any polar compound or ion, a specific membrane protein carrier is essential. Approximately 2,000 genes in the human genome encode proteins that function in transporting solutes across membranes. In some cases, a membrane protein simply facilitates the diffusion of a solute down its concentration gradient, but transport can also occur against a gradient of concentration,

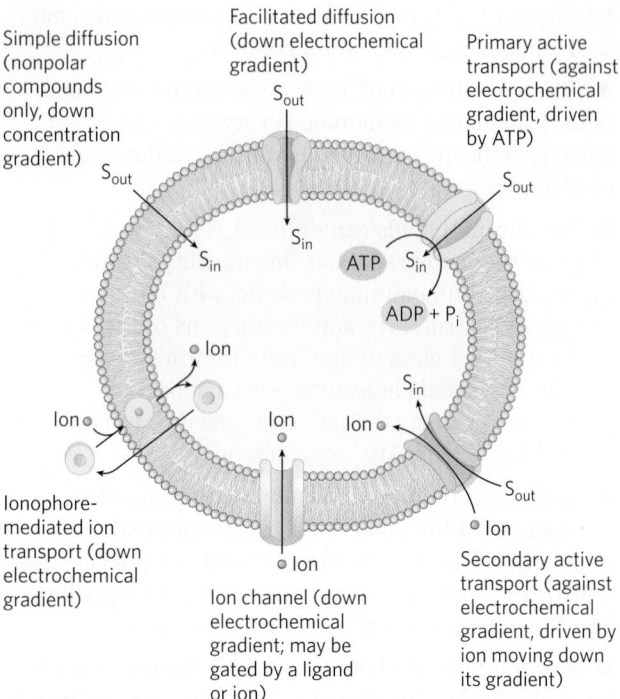

FIGURE 11-24 Summary of transporter types. Some types (ionophores, ion channels, and passive transporters) simply speed transmembrane movement of solutes down their electrochemical gradients, whereas others (active transporters) can pump solutes against a gradient, using ATP or a gradient of a second solute to provide the energy.

electrical potential, or both, and in these cases, as we shall see, the transport process requires energy. Ions may also diffuse across membranes via ion channels formed by proteins, or they may be carried across by ionophores, small molecules that mask the charge of ions and allow them to diffuse through the lipid bilayer. **Figure 11-24** summarizes the various types of transport mechanisms discussed in this section.

Transport May Be Passive or Active

When two aqueous compartments containing unequal concentrations of a soluble compound or ion are separated by a permeable divider (membrane), the solute moves by **simple diffusion** from the region of higher concentration, through the membrane, to the region of lower concentration, until the two compartments have

equal solute concentrations **(Fig. 11-25a)**. When ions of opposite charge are separated by a permeable membrane, there is a transmembrane electrical gradient, a **membrane potential, V_m** (expressed in millivolts). This membrane potential produces a force opposing ion movements that increase V_m and driving ion movements that reduce V_m (Fig. 11-25b). Thus, the direction in which a charged solute tends to move spontaneously across a membrane depends on both the chemical gradient (the difference in solute concentration) and the electrical gradient (V_m) across the membrane. Together these two factors are referred to as the **electrochemical gradient** or **electrochemical potential**. This behavior of solutes is in accord with the second law of thermodynamics: molecules tend to spontaneously assume the distribution of greatest randomness and lowest energy.

Membrane proteins that act by increasing the rate of solute movement across membranes are called transporters or carriers. Transporters are of two general types. **Passive transporters** simply facilitate movement down a concentration gradient, increasing the transport rate. This process is called **passive transport** or **facilitated diffusion**. **Active transporters** (sometimes called pumps) can move substrates across a membrane against a concentration gradient or an electrical potential, a process called **active transport**. **Primary active transporters** use energy provided directly by a chemical reaction; **secondary active transporters** couple uphill transport of one substrate with downhill transport of another.

Transporters and Ion Channels Share Some Structural Properties but Have Different Mechanisms

To pass through a lipid bilayer, a polar or charged solute must first give up its interactions with the water molecules in its hydration shell, then diffuse about 3 nm (30 Å) through a substance (lipid) in which it is poorly soluble **(Fig. 11-26)**. The energy used to strip away the hydration shell and to move the polar compound from water into lipid, then through the lipid bilayer, is regained as the compound leaves the membrane on the other side and is rehydrated. However, the intermediate stage of transmembrane passage is a high-energy state comparable to the transition state in an enzyme-catalyzed chemical reaction. In both cases, an activation barrier

FIGURE 11-25 Movement of solutes across a permeable membrane. (a) Net movement of an electrically neutral solute is toward the side of lower solute concentration until equilibrium is achieved. The solute concentrations on the left and right sides of the membrane, as shown here, are designated C_1 and C_2. The rate of transmembrane solute movement (indicated by the arrows) is proportional to the concentration ratio. **(b)** Net movement of an electrically charged solute is dictated by a combination of the electrical potential (V_m) and the ratio of chemical concentrations (C_2/C_1) across the membrane; net ion movement continues until this electrochemical potential reaches zero.

(a)

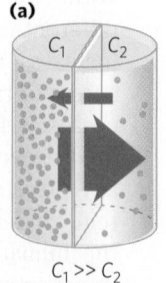

$C_1 \gg C_2$

Before equilibrium

Net flux

$\longrightarrow$

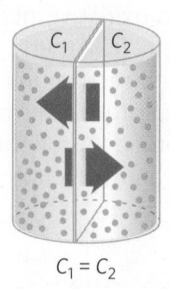

$C_1 = C_2$

At equilibrium

No net flux

(b)

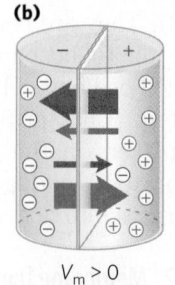

$V_m > 0$

Before equilibrium

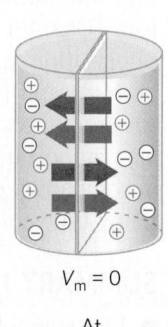

$V_m = 0$

At equilibrium

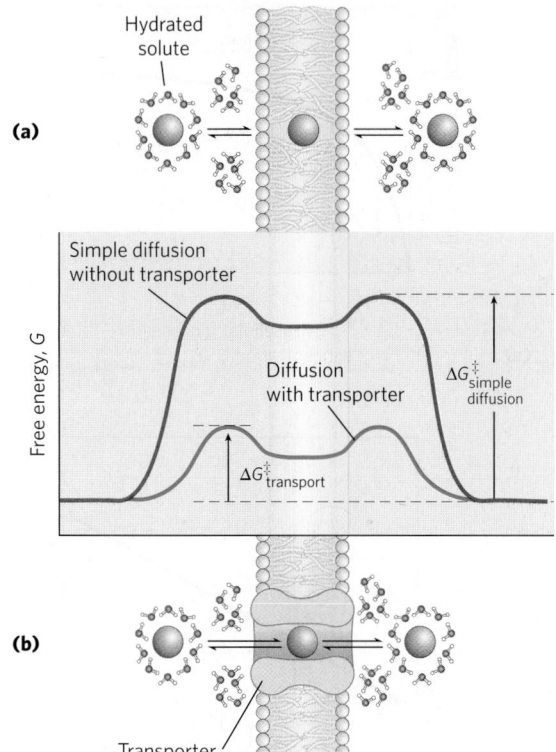

(a)

Hydrated solute

Simple diffusion without transporter

Diffusion with transporter

$\Delta G^{\ddagger}_{\text{simple diffusion}}$

$\Delta G^{\ddagger}_{\text{transport}}$

Free energy, G

(b)

Transporter

FIGURE 11-26 Energy changes accompanying passage of a hydrophilic solute through the lipid bilayer of a biological membrane. (a) In simple diffusion, removal of the hydration shell is highly endergonic, and the energy of activation ($\Delta G^{\ddagger}$) for diffusion through the bilayer is very high. **(b)** A transporter protein reduces the $\Delta G^{\ddagger}$ for transmembrane diffusion of the solute. It does this by forming noncovalent interactions with the dehydrated solute to replace the hydrogen bonding with water and by providing a hydrophilic transmembrane pathway.

must be overcome to reach the intermediate stage (Fig. 11-26; compare with Fig. 6-3). The energy of activation ($\Delta G^{\ddagger}$) for translocation of a polar solute across the bilayer is so large that pure lipid bilayers are virtually impermeable to polar and charged species on time scales relevant to cell growth and division.

Membrane proteins lower the activation energy for transport of polar compounds and ions by providing an alternative path across the membrane for specific solutes. Lowering the activation energy greatly increases the rate of transmembrane movement (recall Eqn 6-6, p. 192). Transporters are not enzymes in the usual sense; their "substrates" are moved from one compartment to another but are not chemically altered. Like enzymes, however, transporters bind their substrates with stereochemical specificity through multiple weak, noncovalent interactions. The negative free-energy change associated with these weak interactions, $\Delta G_{\text{binding}}$, counterbalances the positive free-energy change that accompanies loss of the water of hydration from the substrate, $\Delta G_{\text{dehydration}}$, thereby lowering $\Delta G^{\ddagger}$ for transmembrane passage (Fig. 11-26). Transporter proteins span the lipid bilayer several times, forming a transmembrane pathway lined with hydrophilic amino acid side chains. The pathway provides an alternative route

for a specific substrate to move across the lipid bilayer without its having to dissolve in the bilayer, further lowering $\Delta G^{\ddagger}$ for transmembrane diffusion. The result is an orders-of-magnitude increase in the substrate's rate of passage across the membrane.

Ion channels speed the passage of inorganic ions across membranes by a mechanism different from that of transporters. They provide an aqueous path across the membrane through which inorganic ions can diffuse at very high rates. Most ion channels have a "gate" **(Fig. 11-27a)** regulated by a biological signal. When the gate is open, ions move across the membrane, through the channel, in the direction dictated by the ion's charge and the electrochemical gradient. Movement occurs at rates approaching the limit of unhindered diffusion (tens of millions of ions per second per channel—much higher than typical transporter rates). Ion channels typically show some specificity for an ion, but they are not saturable with their ion substrate. Flow through a channel stops either when the gating mechanism is closed (again, by a biological signal) or when there is no longer an electrochemical gradient providing the driving force for the movement. In contrast, transporters, which bind their "substrates" with high stereospecificity, catalyze transport at rates well below the limits of free diffusion, and they are saturable in the same sense as are enzymes:

(a) Ion channel: single gate

Gate closed

Gate open

(b) Transporter (pump): alternating gates

One gate open

Other gate open

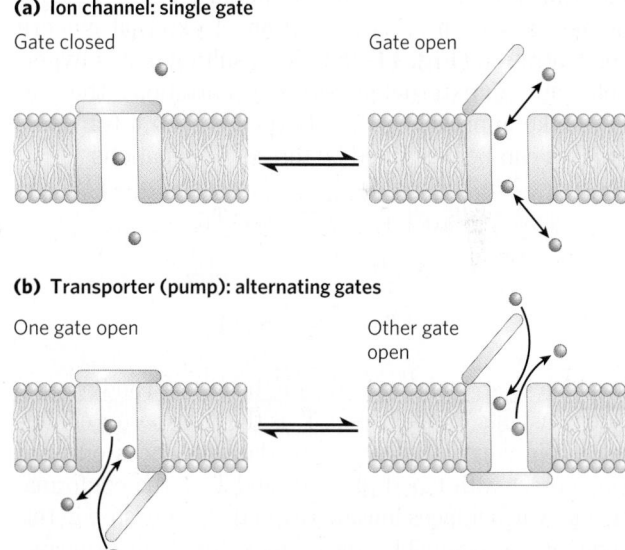

FIGURE 11-27 Differences between channels and transporters. (a) In an ion channel, a transmembrane pore is either open or closed, depending on the position of the single gate. When it is open, ions move through at a rate limited only by the maximum rate of diffusion. **(b)** Transporters have two gates, and both are never open at the same time. Movement of a substrate (an ion or a small molecule) through the membrane is therefore limited by the time needed for one gate to open and close (on one side of the membrane) and the second gate to open. Rates of movement through ion channels can be orders of magnitude greater than rates through transporters, but channels simply allow the ion to flow down the electrochemical gradient, whereas active transporters (pumps) can move a substrate against its concentration gradient. [Source: Information from D. C. Gadsby, *Nature Rev. Mol. Cell Biol.* 10:344, 2009, Fig. 1.]

there is some substrate concentration above which further increases will not produce a greater rate of transport. Transporters have a gate on either side of the membrane, and the two gates are never open at the same time (Fig. 11-27b).

Both transporters and ion channels constitute large families of proteins, defined not only by their primary sequences but by their secondary structures. We next consider some well-studied representatives of the main transporter and channel families. You will also encounter some of these in Chapter 12 when we discuss transmembrane signaling, and some in later chapters in the context of the metabolic pathways in which they participate.

The Glucose Transporter of Erythrocytes Mediates Passive Transport

Energy-yielding metabolism in erythrocytes depends on a constant supply of glucose from the blood plasma, where the glucose concentration is maintained at about 4.5 to 5 mM. Glucose enters the erythrocyte by passive transport via a specific glucose transporter called GLUT1, at a rate about 50,000 times greater than it could cross the membrane unassisted.

The process of glucose transport can be described by analogy with an enzymatic reaction in which the "substrate" is glucose outside the cell (S_{out}), the "product" is glucose inside the cell (S_{in}), and the "enzyme" is the transporter, T. When the initial rate of glucose uptake is measured as a function of external glucose concentration **(Fig. 11-28)**, the resulting plot is hyperbolic: at high external glucose concentrations, the rate of uptake approaches V_{max}. Formally, such a transport process can be described by the set of equations

$$S_{out} + T_1 \underset{k_{-1}}{\overset{k_1}{\rightleftharpoons}} S_{out} \cdot T_1$$

$$k_{-4} \Big\Vert k_4 \qquad\qquad k_{-2} \Big\Vert k_2$$

$$S_{in} + T_2 \underset{k_{-3}}{\overset{k_3}{\rightleftharpoons}} S_{in} \cdot T_2$$

in which k_1, k_{-1}, and so forth, are the forward and reverse rate constants for each step; T_1 is the transporter conformation in which the glucose-binding site faces outward (in contact with blood plasma), and T_2 is the conformation in which it faces inward. Given that every step in this sequence is reversible, the transporter is, in principle, equally able to move glucose into or out of the cell. However, with GLUT1, glucose always moves down its concentration gradient, which normally means *into* the cell. Glucose that enters a cell is generally metabolized immediately, and the intracellular glucose concentration is thereby kept low relative to its concentration in the blood.

The rate equations for glucose transport can be derived exactly as for enzyme-catalyzed reactions (Chapter 6), yielding an expression analogous to the Michaelis-Menten equation

$$V_0 = \frac{V_{max}[S]_{out}}{K_t + [S]_{out}} \tag{11-1}$$

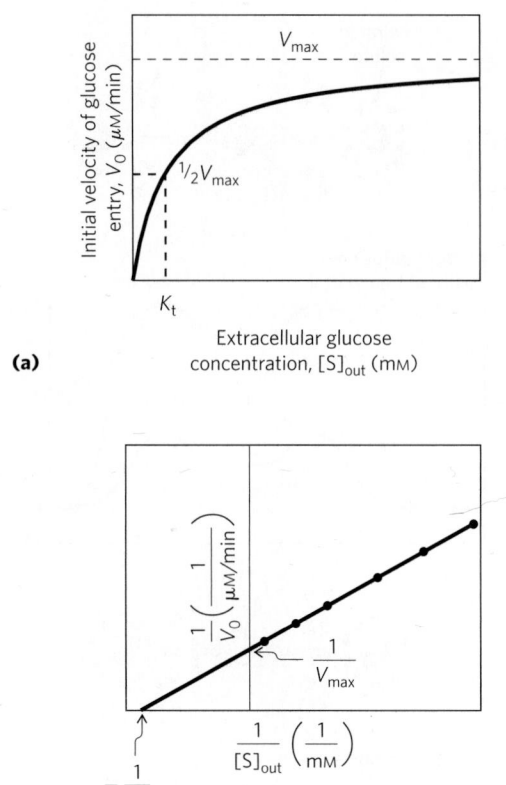

FIGURE 11-28 Kinetics of glucose transport into erythrocytes. (a) The initial rate of glucose entry into an erythrocyte, V_0, depends on the initial concentration of glucose on the outside, $[S]_{out}$. **(b)** Double-reciprocal plot of the data in (a). The kinetics of passive transport is analogous to the kinetics of an enzyme-catalyzed reaction. (Compare these plots with Fig. 6-11 and Box 6-1, Fig. 1.) K_t is analogous to K_m, the Michaelis constant.

in which V_0 is the initial velocity of accumulation of glucose inside the cell when its concentration in the surrounding medium is $[S]_{out}$, and K_t ($K_{transport}$) is a constant analogous to the Michaelis constant, a combination of rate constants that is characteristic of each transport system. This equation describes the *initial* velocity, the rate observed when $[S]_{in} = 0$. As is the case for enzyme-catalyzed reactions, the slope-intercept form of the equation describes a linear plot of $1/V_0$ against $1/[S]_{out}$, from which we can obtain values of K_t and V_{max} (Fig. 11-28b). When $[S]_{out} = K_t$, the rate of uptake is $\frac{1}{2}V_{max}$; the transport process is half-saturated. The concentration of glucose in blood, as noted above, is 4.5 to 5 mM, which is close to the K_t, ensuring that GLUT1 is nearly saturated with substrate and operates near V_{max}.

Because no chemical bonds are made or broken in the conversion of S_{out} to S_{in}, neither "substrate" nor "product" is intrinsically more stable, and the process of entry is therefore fully reversible. As $[S]_{in}$ approaches $[S]_{out}$, the rates of entry and exit become equal. Such a system is therefore incapable of accumulating glucose within a cell at concentrations above that in the surrounding medium; it simply equilibrates glucose on the two sides of the membrane much faster than would occur in the absence of a specific transporter. GLUT1 is specific for D-glucose,

(a)

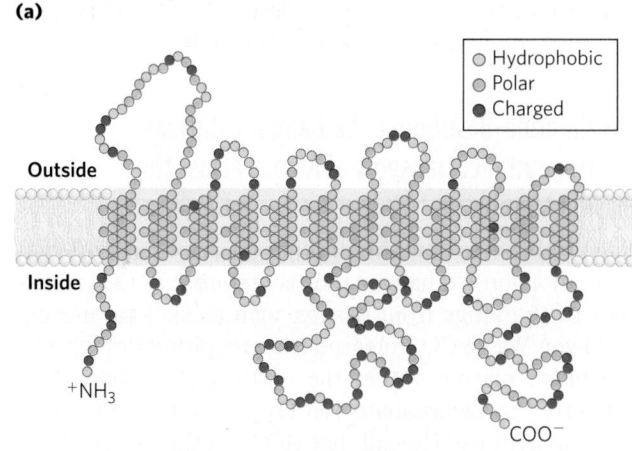

○ Hydrophobic
● Polar
● Charged

Outside

Inside

$+NH_3$

COO^-

(b)

−Ser −Leu −Val −Thr −Asn −Phe −Ile −

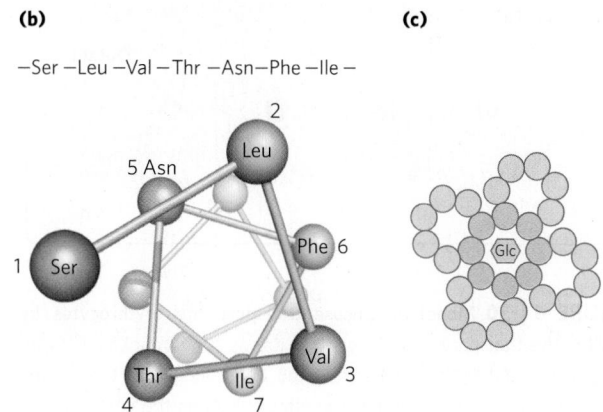

(c)

(d)

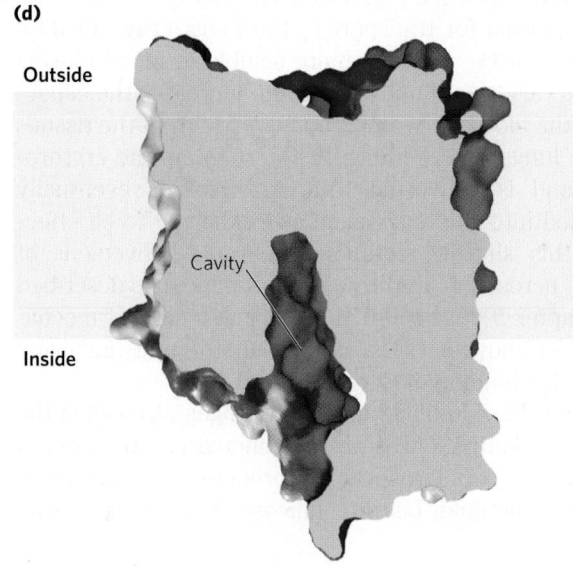

Outside

Cavity

Inside

FIGURE 11-29 Membrane topology of the glucose transporter GLUT1. **(a)** Transmembrane helices are represented here as oblique (angled) rows of three or four amino acid residues, each row depicting one turn of the α helix. Nine of the 12 helices contain three or more polar or charged residues (blue or red), often separated by several hydrophobic residues (yellow). **(b)** A helical wheel diagram shows the distribution of polar and nonpolar residues on the surface of a helical segment. The helix is diagrammed as though observed along its axis from the amino terminus. Adjacent residues in the linear sequence are connected, and each residue is placed around the wheel in the position it occupies in the helix; recall that 3.6 residues are required to make one complete turn of the α helix. In this example, the polar residues (blue) are on one side of the helix, the hydrophobic residues (yellow) on the other. This is, by definition, an amphipathic helix. **(c)** Side-by-side association of amphipathic helices, each with its polar face oriented toward the central cavity, produces a transmembrane channel lined with polar (and charged) residues, available for interaction with glucose. **(d)** The structure of human GLUT1 in the inside-open conformation, as determined by x-ray crystallography. This sliced-open view of the protein shows the long central cavity, open to the inside and lined with many polar side chains (blue). [Sources: (a, c) Information from M. Mueckler, *Eur. J. Biochem.* 219:713, 1994. (d) PDB ID 4PYP, D. Deng et al., *Nature* 510;121, 2014.]

with a measured K_t of about 6 mM. For the close analogs D-mannose and D-galactose, which differ only in the position of one hydroxyl group, the values of K_t are 20 and 30 mM, respectively, and for L-glucose, K_t exceeds 3,000 mM. Thus, GLUT1 shows the three hallmarks of passive transport: high rates of diffusion down a concentration gradient, saturability, and stereospecificity.

GLUT1 is an integral membrane protein ($M_r \sim$56,000) with 12 hydrophobic segments, each forming a membrane-spanning helix **(Fig. 11-29a)**. The helices that line the transmembrane path for glucose are **amphipathic**; for each helix, the residues along one side are predominantly nonpolar, and those on the other side are mainly polar. This amphipathic structure is evident in a helical wheel diagram (Fig. 11-29b). A cluster of amphipathic helices are arranged so that their polar sides face each other and line a hydrophilic pore through which glucose can pass (Fig. 11-29c), while their hydrophobic sides interact with the surrounding membrane lipids such that the hydrophobic effect stabilizes the entire transporter structure.

Structural studies of mammalian GLUT1 and its close analogs from other organisms suggest that the

protein cycles through a series of conformational changes, interconverting a form (T_1) with its glucose-binding site accessible only from the extracellular side, through a form in which the bound glucose is sequestered and inaccessible from either side, to a form (T_2) with the glucose-binding site open only to the intracellular side **(Fig. 11-30)**. The only form of the human GLUT1 protein that has been solved by crystallography (Fig. 11-29d) is the inward-opening form, T_2.

Twelve passive glucose transporters are encoded in the human genome, each with its unique kinetic properties, patterns of tissue distribution, and function (Table 11-3). GLUT1, in addition to supplying glucose to erythrocytes, also transports glucose across the blood-brain barrier, supplying the glucose that is essential for normal brain metabolism. The very rare individuals with defects in GLUT1 have a variety of brain-related symptoms, including seizures, movement and language disorders, and retarded development. Standard care for such individuals includes a ketogenic diet, which provides the ketones that can serve as an alternative energy source for the brain (p. 668). In the liver, GLUT2 transports glucose out of hepatocytes when liver

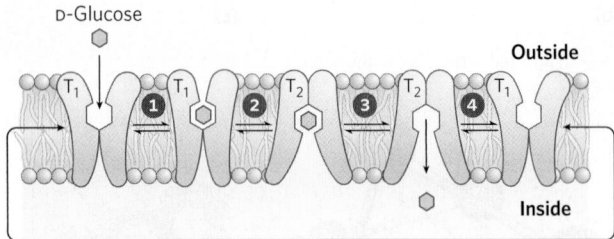

FIGURE 11-30 Model of glucose transport into erythrocytes by GLUT1. The transporter exists in two extreme conformations: T_1, with the glucose-binding site exposed on the outer surface of the plasma membrane, and T_2, with the binding site exposed on the inner surface. Glucose transport occurs in four steps. ❶ Glucose in blood plasma binds to a stereospecific site on T_1; this lowers the activation energy for ❷ a conformational change from $glucose_{out} \cdot T_1$ to $glucose_{in} \cdot T_2$, effecting transmembrane passage of the glucose. ❸ Glucose is released from T_2 into the cytoplasm, and ❹ the transporter returns to the T_1 conformation, ready to transport another glucose molecule. Between the forms T_1 and T_2, there is an intermediate form (not shown here) in which glucose is sequestered within the transporter, with access to neither side.

glycogen is broken down to replenish blood glucose. GLUT2 has a large K_t (≥ 17 mM) and can therefore respond to increased levels of intracellular glucose (produced by glycogen breakdown) by increasing outward transport. Skeletal and heart muscle and adipose tissue have yet another glucose transporter, GLUT4 ($K_t = 5$ mM), which is distinguished by its response to insulin: its activity increases when insulin signals a high blood glucose concentration, thus increasing the rate of glucose

uptake into muscle and adipose tissue. Box 11-1 describes the effect of insulin on this transporter. ∎

The Chloride-Bicarbonate Exchanger Catalyzes Electroneutral Cotransport of Anions across the Plasma Membrane

The erythrocyte contains another passive transport system, an anion exchanger that is essential in CO_2 transport to the lungs from tissues such as skeletal muscle and liver. Waste CO_2 released from respiring tissues into the blood plasma enters the erythrocyte, where it is converted to bicarbonate (HCO_3^-) by the enzyme carbonic anhydrase. (Recall that HCO_3^- is the primary buffer of blood pH; see Fig. 2-21.) The HCO_3^- reenters the blood plasma for transport to the lungs **(Fig. 11-31)**. Because HCO_3^- is much more soluble in blood plasma than is CO_2, this roundabout route increases the capacity of the blood to carry carbon dioxide from the tissues to the lungs. In the lungs, HCO_3^- reenters the erythrocyte and is converted to CO_2, which is eventually released into the lung space and exhaled. To be effective, this shuttle requires very rapid movement of HCO_3^- across the erythrocyte membrane. (As described in Chapter 5 (pp. 169–171), there is a second mechanism for moving CO_2 from tissue to lung, involving reversible binding of CO_2 to hemoglobin.)

The chloride-bicarbonate exchanger, also called the anion exchange (AE) protein, increases the rate of HCO_3^- transport across the erythrocyte membrane more than a millionfold. Like the glucose transporter, it is an

Transporter	Tissue(s) where expressed	K_t (mM)	Role/characteristics[a]
GLUT1	Erythrocytes, blood-brain barrier, placenta, most tissues at a low level	3	Basal glucose uptake; defective in De Vivo disease
GLUT2	Liver, pancreatic islets, intestine, kidney	17	In liver and kidney, removal of excess glucose from blood; in pancreas, regulation of insulin release
GLUT3	Brain (neuron), testis (sperm)	1.4	Basal glucose uptake; high turnover number
GLUT4	Muscle, fat, heart	5	Activity increased by insulin
GLUT5	Intestine (primarily), testis, kidney	6[b]	Primarily fructose transport
GLUT6	Spleen, leukocytes, brain	>5	Possibly no transporter function
GLUT7	Small intestine, colon, testis, prostate	0.3	—
GLUT8	Testis, sperm acrosome	~2	—
GLUT9	Liver, kidney, intestine, lung, placenta	0.6	Urate and glucose transporter in liver, kidney
GLUT10	Heart, lung, brain, liver, muscle, pancreas, placenta, kidney	0.3[c]	Glucose and galactose transporter
GLUT11	Heart, skeletal muscle	0.16	Glucose and fructose transporter
GLUT12	Skeletal muscle, heart, prostate, placenta	—	—

TABLE 11-3   Glucose Transporters in Humans

Sources: Information on localization from M. Mueckler and B. Thorens, *Mol. Aspects Med.* 34:121, 2013. K_t values for glucose from R. Augustin, *IUBMB Life* 62:315, 2010.

[a]Dash indicates role uncertain.

[b]K_m for fructose.

[c]K_m for 2-deoxyglucose.

BOX 11–1 ✚ MEDICINE   Defective Glucose and Water Transport in Two Forms of Diabetes

When ingestion of a carbohydrate-rich meal causes blood glucose to exceed the usual concentration between meals (about 5 mM), excess glucose is taken up by the myocytes of cardiac and skeletal muscle (which store it as glycogen) and by adipocytes (which convert it to triacylglycerols). Glucose uptake into myocytes and adipocytes is mediated by the glucose transporter GLUT4. Between meals, some GLUT4 is present in the plasma membrane, but most (90%) is sequestered in the membranes of small intracellular vesicles (Fig. 1). Insulin released from the pancreas in response to high blood glucose triggers, within minutes, the movement of these vesicles to the plasma membrane, with which they fuse, bringing most of the GLUT4 molecules to the membrane (see Fig. 12-20). With more GLUT4 molecules in action, the rate of glucose uptake increases 15-fold or more. When blood glucose levels return to normal, insulin release slows and most GLUT4 molecules are removed from the plasma membrane and stored in vesicles.

In type 1 (insulin-dependent) diabetes mellitus, the inability to release insulin (and thus to mobilize glucose transporters) results in low rates of glucose uptake into muscle and adipose tissue. One consequence is a prolonged period of high blood glucose after a carbohydrate-rich meal. This condition is the basis for the glucose tolerance test used to diagnose diabetes (Chapter 23).

The water permeability of epithelial cells lining the renal collecting duct in the kidney is due to the presence of an aquaporin (AQP2) in their apical plasma membranes (facing the lumen of the duct). Vasopressin (antidiuretic hormone, ADH) regulates the retention of water by mobilizing AQP2 molecules stored in vesicle membranes within the epithelial cells, much as insulin mobilizes GLUT4 in muscle and adipose tissue. When the vesicles fuse with the epithelial cell plasma membrane, water permeability greatly increases and more water is reabsorbed from the collecting duct and returned to the blood. When the vasopressin level drops, AQP2 is resequestered within vesicles, reducing water retention. In the relatively rare human disease diabetes insipidus, a genetic defect in AQP2 leads to impaired water reabsorption by the kidney. The result is excretion of copious volumes of very dilute urine. If the individual drinks enough water to replace that lost in the urine, there are no serious medical consequences, but insufficient water intake leads to dehydration and imbalances in blood electrolytes, which can lead to fatigue, headache, muscle pain, or even death.

FIGURE 1 **Transport of glucose into a myocyte by GLUT4 is regulated by insulin.** [Source: Information from F. E. Lienhard et al., *Sci. Am.* 266 (January):86, 1992.]

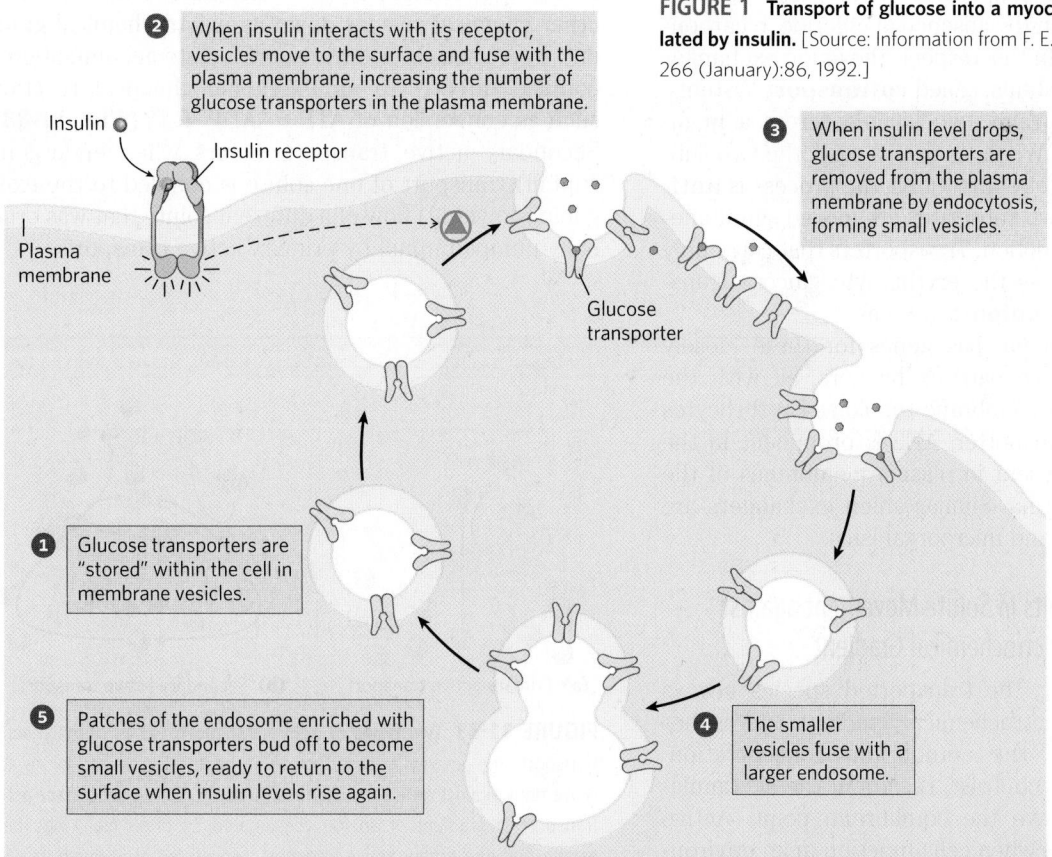

❷ When insulin interacts with its receptor, vesicles move to the surface and fuse with the plasma membrane, increasing the number of glucose transporters in the plasma membrane.

❸ When insulin level drops, glucose transporters are removed from the plasma membrane by endocytosis, forming small vesicles.

Insulin

Insulin receptor

Plasma membrane

Glucose transporter

❶ Glucose transporters are "stored" within the cell in membrane vesicles.

❺ Patches of the endosome enriched with glucose transporters bud off to become small vesicles, ready to return to the surface when insulin levels rise again.

❹ The smaller vesicles fuse with a larger endosome.

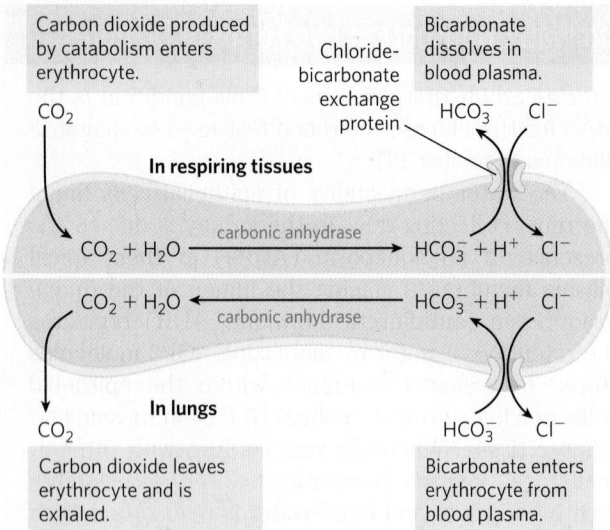

FIGURE 11-31 Chloride-bicarbonate exchanger of the erythrocyte membrane. This cotransport system allows the entry and exit of HCO_3^- without changing the membrane potential. Its role is to increase the CO_2-carrying capacity of the blood. The top half of the figure illustrates the events that take place in respiring tissues; the bottom half, the events in the lungs.

integral protein that probably spans the membrane at least 12 times. This protein mediates the simultaneous movement of two anions: for each HCO_3^- ion that moves in one direction, one Cl^- ion moves in the opposite direction, with no net transfer of charge: the exchange is **electroneutral**. The coupling of Cl^- and HCO_3^- movements is obligatory; in the absence of chloride, bicarbonate transport stops. In this respect, the anion exchanger is typical of those systems, called **cotransport** systems, that simultaneously carry two solutes across a membrane **(Fig. 11-32)**. When, as in this case, the two substrates move in opposite directions, the process is **antiport**. In **symport**, two substrates are moved simultaneously in the same direction. Transporters that carry only one substrate, such as the erythrocyte glucose transporter, are known as **uniport** systems.

The human genome has genes for three closely related chloride-bicarbonate exchangers, all with the same predicted transmembrane topology. Erythrocytes contain the AE1 transporter, AE2 is prominent in the liver, and AE3 is present in plasma membranes of the brain, heart, and retina. Similar anion exchangers are also found in plants and microorganisms.

Active Transport Results in Solute Movement against a Concentration or Electrochemical Gradient

In passive transport, the transported species always moves down its electrochemical gradient and is not accumulated above the equilibrium concentration. Active transport, by contrast, results in the accumulation of a solute above the equilibrium point. Active transport is essential when cells function in an environment in which key substrates are present outside the cell only at very low concentrations. For example, the

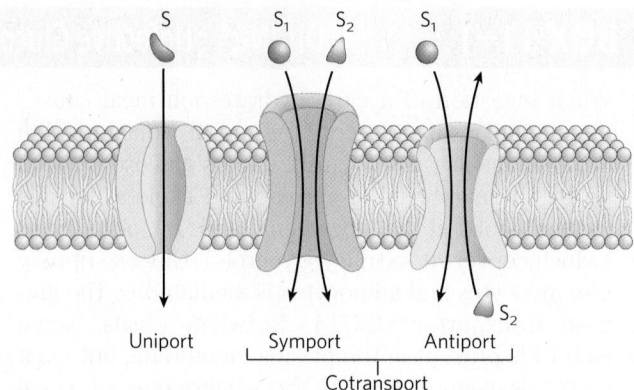

FIGURE 11-32 Three general classes of transport systems. Transporters differ in the number of solutes (substrates) transported and the direction in which each solute moves. Examples of all three types of transporter are discussed in the text. Note that this classification tells us nothing about whether these are energy-requiring (active transport) or energy-independent (passive transport) processes.

bacterium *E. coli* can grow in a medium containing only 1 μM P_i, but the cell must maintain internal P_i levels in the millimolar range. (Worked Example 11-2, below, describes another such situation, which requires cells to pump Ca^{2+} outward across the plasma membrane.) Active transport is thermodynamically unfavorable (endergonic) and takes place only when coupled (directly or indirectly) to an exergonic process such as the absorption of sunlight, an oxidation reaction, the breakdown of ATP, or the concomitant flow of some other chemical species down its electrochemical gradient. In primary active transport, solute accumulation is coupled directly to an exergonic chemical reaction, such as conversion of ATP to ADP + P_i **(Fig. 11-33)**. Secondary active transport occurs when endergonic (uphill) transport of one solute is coupled to the exergonic (downhill) flow of a different solute that was originally pumped uphill by primary active transport.

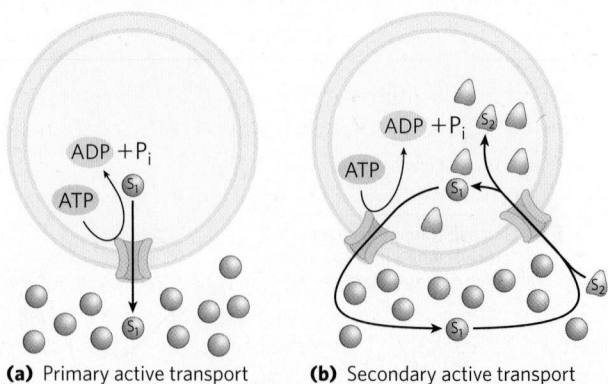

FIGURE 11-33 Two types of active transport. (a) In primary active transport, the energy released by ATP hydrolysis drives solute (S_1) movement against an electrochemical gradient. **(b)** In secondary active transport, a gradient of an ion (designated S_1; often Na^+) has been established by primary active transport. Movement of S_1 down its electrochemical gradient now provides the energy to drive cotransport of a second solute, S_2, against its electrochemical gradient.

The amount of energy needed for the transport of a solute against a gradient can be calculated from the initial concentration gradient. The general equation for the free-energy change in the chemical process that converts substrate (S) to product (P) is

$$\Delta G = \Delta G'^{\circ} + RT \ln \left([P]/[S]\right) \quad (11\text{-}2)$$

where $\Delta G'^{\circ}$ is the standard free-energy change, R is the gas constant (8.315 J/mol·K), and T is the absolute temperature. When the "reaction" is simply transport of a solute from a region where its concentration is C_1 to a region where its concentration is C_2, no bonds are made or broken and $\Delta G'^{\circ}$ is zero. The free-energy change for transport, ΔG_t, is then

$$\Delta G_t = RT \ln \left(C_2/C_1\right) \quad (11\text{-}3)$$

If there is, say, a 10-fold difference in concentration between two compartments, the cost of moving 1 mol of an uncharged solute at 25 °C uphill across a membrane separating the compartments is

$$\Delta G_t = (8.315\ \text{J/mol·K})(298\ \text{K}) \ln (10/1) = 5{,}700\ \text{J/mol}$$
$$= 5.7\ \text{kJ/mol}$$

Equation 11-3 holds for all uncharged solutes.

WORKED EXAMPLE 11-1 | **Energy Cost of Pumping an Uncharged Solute**

Calculate the energy cost (free-energy change) of pumping an uncharged solute against a 10^4-fold concentration gradient at 25 °C.

Solution: Begin with Equation 11-3. Substitute 1.0×10^4 for (C_2/C_1), 8.315 J/mol·K for R, and 298 K for T:

$$\Delta G_t = RT \ln \left(C_2/C_1\right)$$
$$= (8.315\ \text{J/mol·K})(298\ \text{K})(1.0 \times 10^4)$$
$$= 23\ \text{kJ/mol}$$

When the solute is an *ion*, its movement without an accompanying counterion results in the endergonic separation of positive and negative charges, producing an electrical potential; such a transport process is said to be **electrogenic**. The energetic cost of moving an ion depends on the electrochemical potential (Fig. 11-25), the sum of the chemical and electrical gradients:

$$\Delta G_t = RT \ln \left(C_2/C_1\right) + ZF\Delta\psi \quad (11\text{-}4)$$

where Z is the charge on the ion, F is the Faraday constant (96,480 J/V·mol), and $\Delta\psi$ is the transmembrane electrical potential (in volts). Eukaryotic cells typically have plasma membrane potentials of about 0.05 V (with the inside negative relative to the outside), so the second term on the right side of Equation 11-4 can make a significant contribution to the total free-energy change for transporting an ion. Most cells maintain more than a 10-fold difference in ion concentrations across their plasma or intracellular membranes, and for many cells and tissues active transport is therefore a major energy-consuming process.

WORKED EXAMPLE 11-2 | **Energy Cost of Pumping a Charged Solute**

Calculate the energy cost (free-energy change) of pumping Ca^{2+} from the cytosol, where its concentration is about 1.0×10^{-7} M, to the extracellular fluid, where its concentration is about 1.0 mM. Assume a temperature of 37 °C (body temperature in a mammal) and a standard transmembrane potential of 50 mV (inside negative) for the plasma membrane.

Solution: This is a case in which energy must be expended to counter two forces acting on the ion being transported: the membrane potential and the concentration difference across the membrane. These forces are expressed in the two terms on the right side of Equation 11-4:

$$\Delta G_t = RT \ln \left(C_2/C_1\right) + ZF\ \Delta\psi$$

in which the first term describes the chemical gradient and the second describes the electrical potential.

In Equation 11-4, substitute 8.315 J/mol·K for R, 310 K for T, 1.0×10^{-3} for C_2, 1.0×10^{-7} for C_1, +2 (the charge on a Ca^{2+} ion) for Z, 96,500 J/V·mol for F, and 0.050 V for $\Delta\psi$. Note that the transmembrane potential is 50 mV (inside negative), so the change in potential when an ion moves from inside to outside is 50 mV.

$$\Delta G_t = RT \ln (C_2/C_1) + ZF\ \Delta\psi$$
$$= (8.315\ \text{J/mol·K})(310\ \text{K}) \ln \frac{(1.0 \times 10^{-3})}{(1.0 \times 10^{-7})}$$
$$+ 2(96{,}500\ \text{J/V·mol})(0.050\ \text{V})$$
$$= 33\ \text{kJ/mol}$$

The mechanism of active transport is of fundamental importance in biology. As we shall see in Chapters 19 and 20, ATP is formed in mitochondria and chloroplasts by a mechanism that is essentially ATP-driven ion transport operating in reverse. The energy made available by the spontaneous flow of protons across a membrane is calculable from Equation 11-4; remember that ΔG for flow *down* an electrochemical gradient has a negative value, and ΔG for transport of ions *against* an electrochemical gradient has a positive value.

P-Type ATPases Undergo Phosphorylation during Their Catalytic Cycles

The family of active transporters called **P-type ATPases** are cation transporters that are reversibly phosphorylated by ATP (thus the name P-type) as part of the transport cycle. Phosphorylation forces a conformational change that is central to movement of the cation across the membrane. The human genome encodes at least 70 P-type ATPases that share similarities in amino acid sequence and topology, especially near the Asp residue that undergoes phosphorylation. All are integral proteins with 8 or 10 predicted membrane-spanning regions in a single polypeptide, and all are sensitive to inhibition by the transition-state analog **vanadate**, which mimics phosphate when under nucleophilic attack by a water molecule.

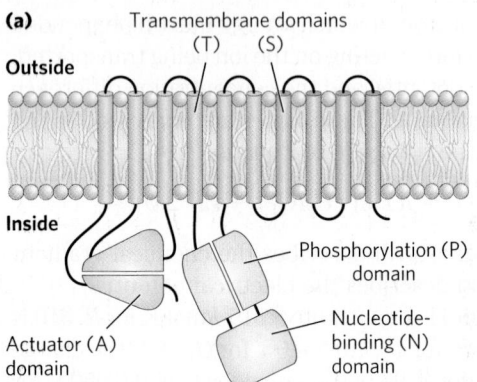

Phosphate   Vanadate

The P-type ATPases are widespread in eukaryotes and bacteria. The Na$^+$K$^+$ ATPase of animal cells (an antiporter for Na$^+$ and K$^+$ ions) and the plasma membrane H$^+$ ATPase of plants and fungi set the transmembrane electrochemical potential in cells by establishing ion gradients across the plasma membrane. These gradients provide the driving force for secondary active transport and are also the basis for electrical signaling in neurons. In animal tissues, the **sarcoplasmic/endoplasmic reticulum Ca^{2+} ATPase (SERCA) pump** and the plasma membrane Ca^{2+} ATPase pump are uniporters for Ca^{2+} ions, which together maintain the cytosolic level of Ca^{2+} below 1 μM. The SERCA pump moves Ca^{2+} from the cytosol into the lumen of the sarcoplasmic reticulum. Parietal cells in the lining of the mammalian stomach have a P-type ATPase that pumps H$^+$ and K$^+$ out of the cells and into the stomach, thereby acidifying the stomach contents. Lipid flippases, as we noted earlier, are structurally and functionally related to P-type transporters. Bacteria and eukaryotes use P-type ATPases to pump toxic heavy metal ions such as Cd^{2+} and Cu^{2+} out of cells.

All P-type pumps have similar structures **(Fig. 11-34)** and similar mechanisms. The mechanism postulated for P-type ATPases takes into account the

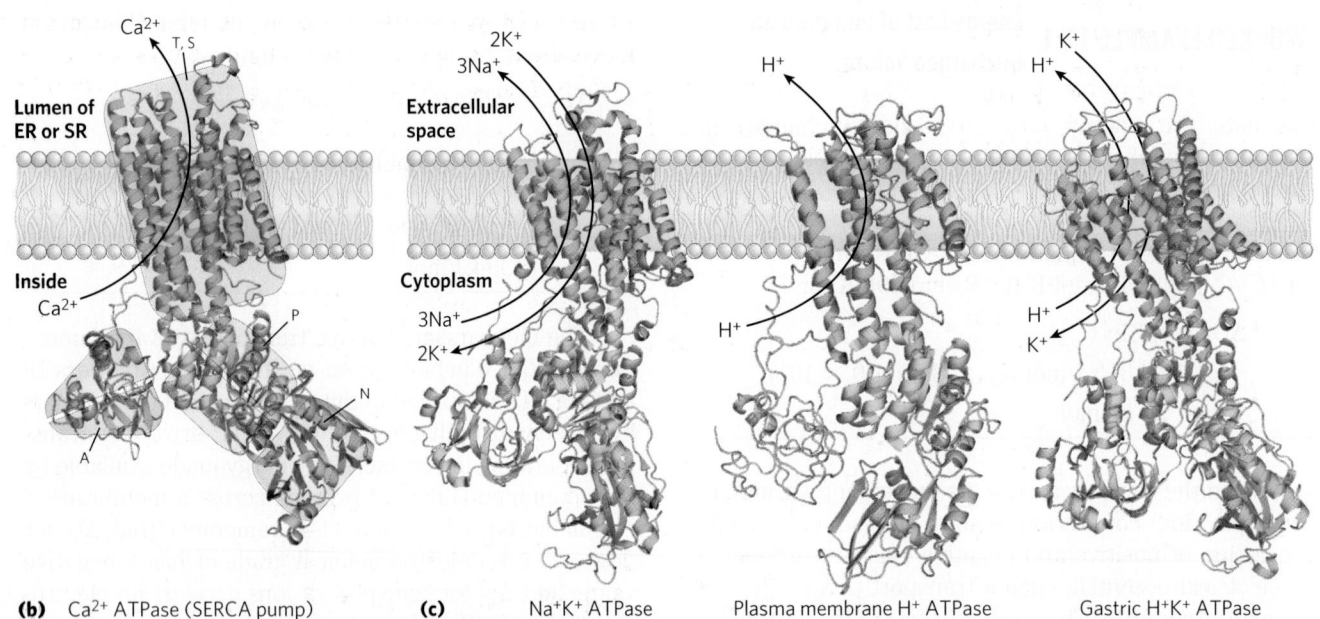

FIGURE 11-34 The general structure of the P-type ATPases. (a) P-type ATPases have three cytoplasmic domains (A, N, and P) and two transmembrane domains (T and S) consisting of multiple helices. The N (nucleotide-binding) domain binds ATP and Mg^{2+}, and it has protein kinase activity that phosphorylates a specific Asp residue found in the P (phosphorylation) domain of all P-type ATPases. The A (actuator) domain has protein phosphatase activity and removes the phosphoryl group from the Asp residue with each catalytic cycle of the pump. A transport (T) domain with six transmembrane helices includes the ion-transporting structure, and four more transmembrane helices make up the support (S) domain, which provides physical support to the transport domain and may have other specialized functions in certain P-type ATPases. The binding sites for the ions to be transported are near the middle of the membrane, 40 to 50 Å from the phosphorylated Asp residue—thus Asp

phosphorylation-dephosphorylation does not *directly* affect ion binding. The A domain communicates movements of the N and P domains to the ion-binding sites. **(b)** A ribbon representation of the Ca^{2+} ATPase (SERCA pump). ATP binds to the N domain, and the Ca^{2+} ions to be transported bind to the T domain. **(c)** Other P-type ATPases have domain structures, and presumably mechanisms, like those of the SERCA pump; shown here are Na$^+$K$^+$ ATPase, the plasma membrane H$^+$ ATPase, and the gastric H$^+$K$^+$ ATPase. [Sources: (a) Information from M. Bublitz et al., *Curr. Opin. Struct. Biol.* 20:431, 2010, Fig. 1. (b) PDB ID 1SU4, C. Toyoshima et al., *Nature* 405:647, 2000. (c) Na$^+$K$^+$ ATPase: PDB ID 3KDP, J. Preben Morth et al., *Nature* 450:1043, 2007; H$^+$ ATPase: PDB ID 3B8C, B. P. Pedersen, et al., *Nature* 450:1111, 2007; H$^+$K$^+$ ATPase: modified from PDB ID 3IXZ, K. Abe et al., *EMBO J.* 28:1637, 2009, modeled following PDB ID 3B8E, J. Preben Morth et al., *Nature* 450:1043, 2007.]

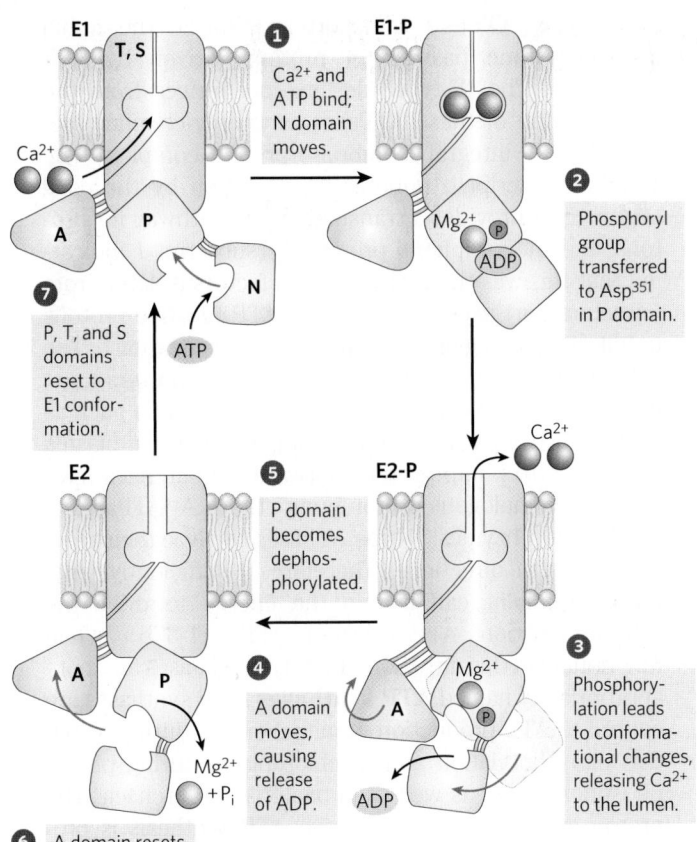

FIGURE 11-35 Postulated mechanism of the SERCA pump. The transport cycle begins with the protein in the E1 conformation, with the Ca^{2+}-binding sites facing the cytosol. Two Ca^{2+} ions bind, then ATP binds to the transporter and phosphorylates Asp^{351}, forming E1-P. Phosphorylation favors the second conformation, E2-P, in which the Ca^{2+}-binding sites, now with a reduced affinity for Ca^{2+}, are accessible on the other side of the membrane (the lumen or extracellular space), and the released Ca^{2+} diffuses away. Finally, E2-P is dephosphorylated, returning the protein to its E1 conformation for another round of transport. [Source: Information from W. Kühlbrandt, *Nature Rev. Mol. Cell Biol.* 5:282, 2004.]

A variation on this basic mechanism is seen in the **Na⁺K⁺ ATPase** of the plasma membrane, discovered by Jens Skou in 1957. This cotransporter couples phosphorylation-dephosphorylation of the critical Asp residue to the simultaneous movement of both Na^+ and K^+ against their electrochemical gradients. The Na^+K^+ ATPase is responsible for maintaining low Na^+ and high K^+ concentrations in the cell relative to the extracellular fluid **(Fig. 11-36)**. For each molecule of ATP converted to ADP and P_i, the transporter moves two K^+ ions inward and three Na^+ ions outward across

Jens Skou
[Source: Lars Moeller/AP Images.]

large conformational changes and the phosphorylation-dephosphorylation of the critical Asp residue in the P (phosphorylation) domain that is known to occur during a catalytic cycle. For the SERCA pump **(Fig. 11-35)**, each catalytic cycle moves two Ca^{2+} ions across the membrane and converts an ATP to ADP and P_i. ATP has two roles in this mechanism, one catalytic and one modulatory. The role of ATP binding and phosphoryl transfer to the enzyme is to bring about the interconversion of two conformations, E1 and E2, of the transporter. In the E1 conformation, the two Ca^{2+}-binding sites are exposed on the cytosolic side of the ER or sarcoplasmic reticulum and bind Ca^{2+} with high affinity. ATP binding and Asp phosphorylation drive a conformational change from E1 to E2 that results in exposure of the Ca^{2+}-binding sites on the lumenal side of the membrane and their greatly reduced affinity for Ca^{2+}, causing release of Ca^{2+} ions into the lumen. By this mechanism, the energy released by hydrolysis of ATP during one phosphorylation-dephosphorylation cycle drives Ca^{2+} across the membrane against a large electrochemical gradient.

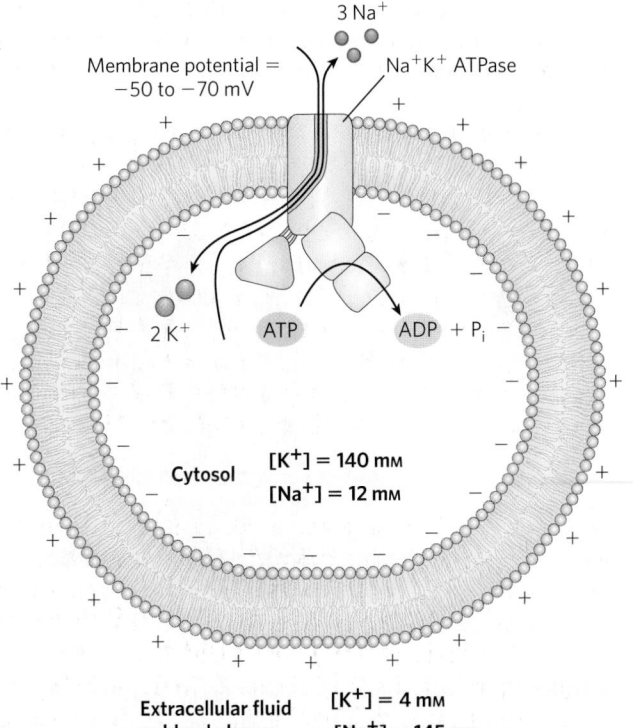

FIGURE 11-36 Role of the Na⁺K⁺ ATPase in animal cells. This active transport system is primarily responsible for setting and maintaining the intracellular concentrations of Na^+ and K^+ in animal cells and for generating the membrane potential. It does this by moving three Na^+ ions out of the cell for every two K^+ ions it moves in. The electrical potential across the plasma membrane is central to electrical signaling in neurons, and the gradient of Na^+ is used to drive the uphill cotransport of solutes in many cell types.

the plasma membrane. Cotransport is therefore electrogenic, creating a net separation of charge across the membrane; in animals, this produces the membrane potential of -50 to -70 mV (inside negative relative to outside) that is characteristic of most cells and is essential to the conduction of action potentials in neurons. The central role of the Na^+K^+ ATPase is reflected in the energy invested in this single reaction: about 25% of the total energy consumption of a human at rest.

V-Type and F-Type ATPases Are ATP-Driven Proton Pumps

V-type ATPases, a class of proton-transporting ATPases, are responsible for acidifying intracellular compartments in many organisms (thus V for $vacuolar$). Proton pumps of this type maintain the vacuoles of fungi and higher plants at a pH between 3 and 6, well below that of the surrounding cytosol (pH 7.5). V-type ATPases are also responsible for the acidification of lysosomes, endosomes, the Golgi complex, and secretory vesicles in animal cells. All V-type ATPases have a similar complex structure, with an integral (transmembrane) domain (V_o) that serves as a proton channel and a peripheral domain (V_1) that contains the ATP-binding site and the ATPase activity **(Fig. 11-37a)**. The structure is similar to that of the well-characterized F-type ATPases.

F-type ATPase transporters catalyze the uphill transmembrane passage of protons, driven by ATP hydrolysis. The "F-type" designation derives from the identification of these ATPases as energy-coupling *fac*tors. The F_o integral membrane protein complex (Fig. 11-37b; subscript o denotes its inhibition by the drug *oligomycin*) provides a transmembrane pathway for protons, and the peripheral protein F_1 (subscript 1 indicating that this was the first of several factors isolated from mitochondria) uses the energy of ATP to drive protons uphill (into a region of higher H^+ concentration). The F_oF_1 organization of proton-pumping transporters must have developed very early in evolution. Bacteria such as *E. coli* use an F_oF_1 ATPase complex in their plasma membrane to pump protons outward, and archaea have a closely homologous proton pump, the A_oA_1 ATPase.

Like all enzymes, F-type ATPases catalyze their reactions in both directions. Therefore, a sufficiently large proton gradient can supply the energy to drive the reverse reaction, ATP synthesis (Fig. 11-37b). When functioning in this direction, the F-type ATPases are more appropriately named **ATP synthases**. ATP synthases are central to ATP production in mitochondria during oxidative phosphorylation and in chloroplasts during photophosphorylation, as well as in bacteria and archaea. The proton gradient needed to drive ATP synthesis is produced by other types of proton pumps powered by substrate oxidation or sunlight. We provide a detailed description of these processes in Chapters 19 and 20.

(a)

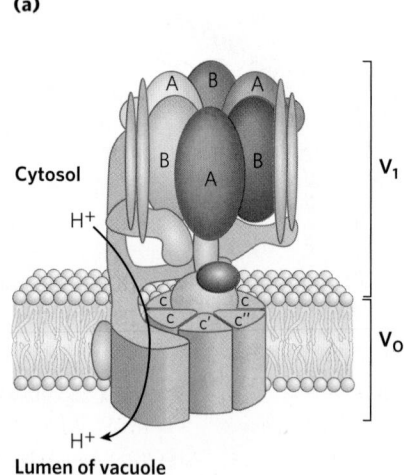

Lumen of vacuole

(b)

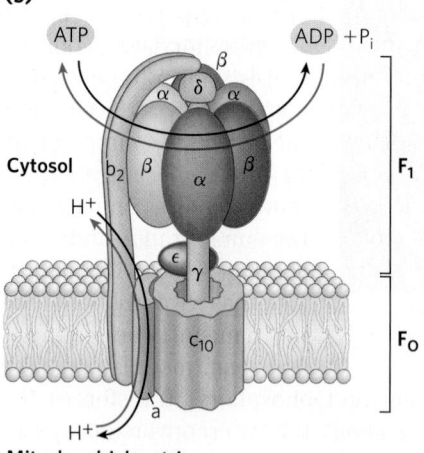

Mitochondrial matrix

FIGURE 11-37 Two proton pumps with similar structures. (a) The V_oV_1 H^+ ATPase uses ATP to pump protons into vacuoles and lysosomes, creating their low internal pH. It has an integral (membrane-embedded) domain, V_o, that includes multiple identical c subunits, and a peripheral domain that projects into the cytosol and contains the ATP-hydrolyzing sites, located on three identical B subunits (purple). **(b)** The F_oF_1 ATPase/ATP synthase of mitochondria has an integral domain, F_o, with multiple copies of the c subunit, and a peripheral domain, F_1, consisting of three α subunits, three β subunits, and a central shaft joined to the integral domain. F_o, and presumably V_o, provides a transmembrane channel through which protons are pumped as

ATP is hydrolyzed on the β subunits of F_1 (B subunits of V_1). The remarkable mechanism by which ATP hydrolysis is coupled to proton movement is described in detail in Chapter 19. It involves rotation of F_o in the plane of the membrane. The structures of the V_oV_1 ATPase and its analogs A_oA_1 ATPase (of archaea) and CF_oCF_1 ATPase (of chloroplasts) are similar to that of F_oF_1, and the mechanisms are also conserved. An ATP-driven proton transporter also can catalyze ATP synthesis (red arrows) as protons flow *down* their electrochemical gradient. This is the central reaction in the processes of oxidative phosphorylation and photophosphorylation, described in detail in Chapters 19 and 20.

ABC Transporters Use ATP to Drive the Active Transport of a Wide Variety of Substrates

ABC transporters constitute a large family of ATP-driven transporters that pump amino acids, peptides, proteins, metal ions, various lipids, bile salts, and many hydrophobic compounds, including drugs, across a membrane against a concentration gradient. Many ABC transporters are located in the plasma membrane, but some are also found in the ER and in the membranes of mitochondria and lysosomes. All members of this family have two ATP-binding domains ("cassettes") that give the family its name—*ATP-binding cassette* transporters—and two transmembrane domains, each containing six transmembrane helices. In some cases, all these domains are in a single, long polypeptide; other ABC transporters have two subunits, each contributing a nucleotide-binding domain (NBD) and a domain with six transmembrane helices. The structures of homologous forms of an ABC transporter from the nematode *Caenorhabditis elegans* and the bacterium *Staphylococcus aureus* have been solved **(Fig. 11-38)** and are believed to represent the two extreme forms that the protein assumes in the course of one transport cycle. One has its substrate-binding site exposed on one side of the membrane, and the other has its substrate-binding site accessible on the other side. Substrates move across the membrane when the two forms interconvert, driven by ATP hydrolysis (Fig. 11-38c). The NBDs of all ABC proteins are similar in sequence and presumably in three-dimensional structure. They constitute the conserved molecular motor that can be coupled to a wide variety of transmembrane domains, each capable of pumping one specific substrate across a membrane. When coupled this way, the ATP-driven motor moves solutes against a concentration gradient, with a stoichiometry of about one ATP hydrolyzed per molecule of substrate transported.

The human genome contains at least 48 genes that encode ABC transporters (Table 11-4). Some of these transporters have very high specificity for a single substrate; others are more promiscuous, able to transport drugs that cells presumably did not encounter during their evolution. Many ABC transporters are involved in maintaining the composition of the lipid bilayer, such as the floppases that move membrane lipids from one leaflet of the bilayer to the other. Many others are needed to transport sterols, sterol derivatives, and fatty acids into the bloodstream for transport throughout the body. For example, the cellular machinery for exporting excess cholesterol includes an ABC transporter (see Fig. 21-47). Mutations in the genes that encode some of these proteins contribute to genetic diseases, including liver failure, retinal degeneration, and Tangier disease. The *cystic fibrosis transmembrane conductance regulator* protein (CFTR) of the plasma membrane is an interesting case of an ABC protein that is an ion channel (for Cl⁻), regulated by ATP hydrolysis, but without the

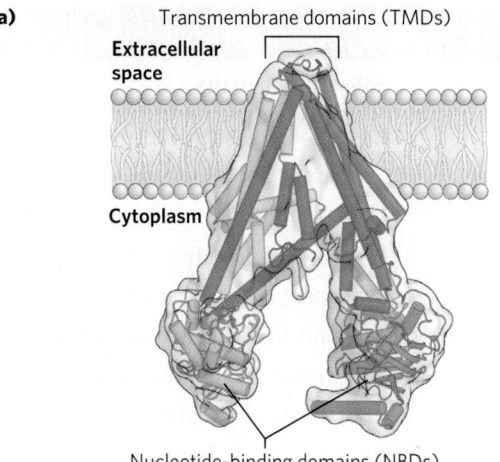

(a) Transmembrane domains (TMDs)

(b)

(c)

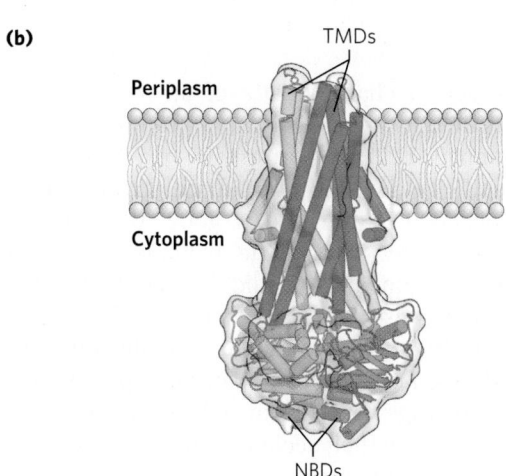
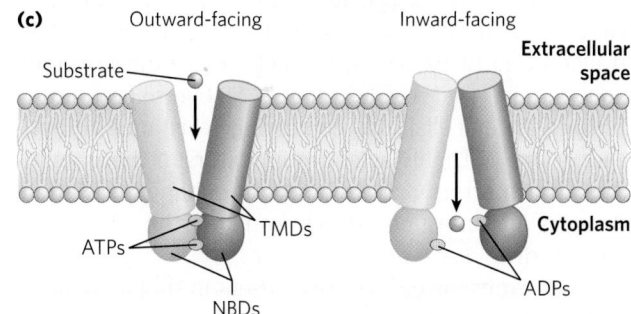

FIGURE 11-38 ABC transporters. (a) The multidrug transporter ABCB1 of *C. elegans*, analogous to MDR1 of humans, in its inward-facing form. The protein has two homologous halves, each with six transmembrane helices in two transmembrane domains (TMDs; blue), and a cytoplasmic nucleotide-binding domain (NBD; red). **(b)** An homologous protein, Sav1866 of *S. aureus*, in its presumed outward-facing form, with its substrate-binding site accessible only from the extracellular space. **(c)** Mechanism proposed for the coupling of ATP hydrolysis to transport. Substrate binds to the transporter on the cytoplasmic side, with ATP bound to the NBD sites. On substrate binding and ATP hydrolysis to ADP, a conformational change exposes the substrate to the outside surface and lowers the affinity of the transporter for its substrate; substrate diffuses away from the transporter and into the extracellular space. Compare this process with the model of glucose transport in Figure 11-30. [Sources: (a) PDB ID 4F4C, M. S. Jin et al., *Nature* 490:566, 2012. (b) PDB ID 2HYD, R. J. Dawson and K. P. Locher, *Nature* 443:180, 2006.]

TABLE 11-4 Some ABC Transporters in Humans

Gene(s)	Role/characteristics	Text discussion
ABCA1	Reverse cholesterol transport; defect causes Tangier disease	pp. 851–852
ABCA4	Only in visual receptors, export of all-*trans* retinal	p. 456, Fig. 12-14
ABCB1	Multidrug resistance P-glycoprotein 1; transport across blood-brain barrier	—
ABCB4	Multidrug resistance; transport of phosphatidylcholine in bile	—
ABCB6	Transports porphyrins into mitochondria for heme synthesis	pp. 880–882
ABCB11	Transports bile salts out of hepatocytes	p. 650, Fig. 17-1
ABCC6	Sulfonylurea receptor; targeted by the drug glipizide in type 2 diabetes	p. 933, Fig. 23-29
ABCG2	Breast cancer resistance protein (BCRP); major exporter of anticancer drugs	p. 418
ABCG5, ABCG8	Act together to limit uptake of sterols from gut	—
ABCC7	CFTR (Cl^- channel); defect causes cystic fibrosis	p. 417; Box 11-2

pumping function characteristic of an active transporter (Box 11-2).

One human ABC transporter with very broad substrate specificity is the **multidrug transporter (MDR1)**, encoded by the *ABCB1* gene. MDR1 in the placental membrane and in the blood-brain barrier ejects toxic compounds that would damage the fetus or the brain. But it is also responsible for the striking resistance of certain tumors to some generally effective antitumor drugs. For example, MDR1 pumps the chemotherapeutic drugs doxorubicin and vinblastine out of cells, thus preventing their accumulation within a tumor and blocking their therapeutic effects. Overexpression of MDR1 is often associated with treatment failure in cancers of the liver, kidney, and colon. A related ABC transporter, BCRP (*b*reast *c*ancer *r*esistance *p*rotein, encoded by the *ABCG2* gene), is overexpressed in breast cancer cells, also conferring resistance to anticancer drugs. Highly selective inhibitors of these multidrug transporters are expected to enhance the effectiveness of antitumor drugs and are the objects of current drug discovery and design.

ABC transporters are also present in simpler animals and in plants and microorganisms. Yeast has 31 genes that encode ABC transporters, *Drosophila* has 56, and *E. coli* has 80, representing 2% of its entire genome. ABC transporters that are used by *E. coli* and other bacteria to import essentials such as vitamin B_{12} are the presumed evolutionary precursors of the MDRs of animal cells. The presence of ABC transporters that confer antibiotic resistance in pathogenic microbes (*Pseudomonas aeruginosa, Staphylococcus aureus, Candida albicans, Neisseria gonorrhoeae,* and *Plasmodium falciparum*) is a serious public health concern and makes these transporters attractive targets for drug design. ∎

Ion Gradients Provide the Energy for Secondary Active Transport

The ion gradients formed by primary transport of Na^+ or H^+ can, in turn, provide the driving force for cotransport of other solutes. Many cell types have transport systems that couple the spontaneous, downhill flow of these ions to the simultaneous uphill pumping of another ion, sugar, or amino acid (Table 11-5).

The **lactose transporter** (**lactose permease**, or **galactoside permease**) of *E. coli* is the well-studied prototype for proton-driven cotransporters. This single polypeptide chain (417 residues) transports one proton and one lactose molecule into the cell, with the net accumulation of lactose (**Fig. 11-39**). *E. coli* normally produces a gradient of protons and charge across its plasma membrane by oxidizing fuels and using the energy of oxidation to pump protons outward. (This mechanism is discussed in detail in Chapter 19.) The plasma membrane is impermeable to protons, but the lactose transporter provides a route for proton reentry into the cell, and as this happens, lactose is simultaneously carried into the cell by symport. The endergonic

TABLE 11-5 Cotransport Systems Driven by Gradients of Na^+ or H^+

Organism/tissue/cell type	Transported solute (moving against its gradient)	Cotransported solute (moving down its gradient)	Type of transport
E. coli	Lactose	H^+	Symport
	Proline	H^+	Symport
	Dicarboxylic acids	H^+	Symport
Intestine, kidney (vertebrates)	Glucose	Na^+	Symport
	Amino acids	Na^+	Symport
Vertebrate cells (many types)	Ca^{2+}	Na^+	Antiport
Higher plants	K^+	H^+	Antiport
Fungi (*Neurospora*)	K^+	H^+	Antiport

BOX 11–2 ⚕ MEDICINE A Defective Ion Channel in Cystic Fibrosis

Cystic fibrosis (CF) is a serious hereditary disease. In the United States, the frequency of CF ranges from 1 in 3,200 live births among whites to 1 in 31,000 live births among Asian Americans. About 5% of whites are carriers, having one defective and one normal copy of the gene. Only individuals with two defective copies show the severe symptoms of the disease: obstruction of the gastrointestinal and respiratory tracts, commonly leading to bacterial infection of the airways.

The defective gene underlying CF was discovered in 1989. It encodes a membrane protein called *cystic fibrosis transmembrane conductance regulator*, or CFTR. This protein has two segments, each containing six transmembrane helices, two nucleotide-binding domains (NBDs), and a regulatory region that connects them (Fig. 1). CFTR is therefore very similar to other ABC transporter proteins, except that it functions as an *ion channel* (for Cl⁻), not as a pump. The channel conducts Cl⁻ across the plasma membrane when both NBDs have bound ATP, and it closes when the ATP on one of the NBDs is broken down to ADP and Pᵢ. The Cl⁻ channel is further regulated by phosphorylation of several Ser residues in the regulatory domain, catalyzed by cAMP-dependent protein kinase (Chapter 12). When the regulatory domain is not phosphorylated, the Cl⁻ channel is closed.

The mutation responsible for CF in 70% of cases results in deletion of a Phe residue at position 508 (a mutation denoted F508del). The mutant protein folds incorrectly, causing it to be degraded in proteasomes. As a result, Cl⁻ movement is reduced across

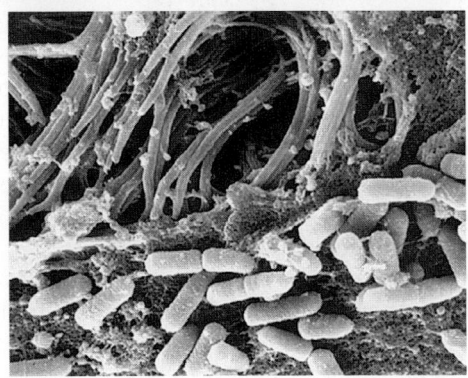

FIGURE 2 Mucus lining the surface of the lungs traps bacteria. In healthy lungs (shown here), these bacteria are killed and swept away by the action of cilia. In CF, this mechanism is impaired, resulting in recurring infections and progressive damage to the lungs. [Source: Tom Moninger, University of Iowa, Iowa City.]

the plasma membranes of epithelial cells that line the airways, digestive tract, exocrine glands (pancreas, sweat glands), bile ducts, and vas deferens. Less-common mutations, such as G551D (Gly⁵⁵¹ changed to Asp), lead to production of CFTR that is correctly folded and inserted into the membrane but is defective in Cl⁻ transfer.

Diminished export of Cl⁻ in individuals with CF is accompanied by diminished export of water from cells, causing the mucus on cell surfaces to become dehydrated, thick, and excessively sticky. In normal circumstances, cilia on the epithelial cells lining the inner surface of the lungs constantly sweep away

(Continued on next page)

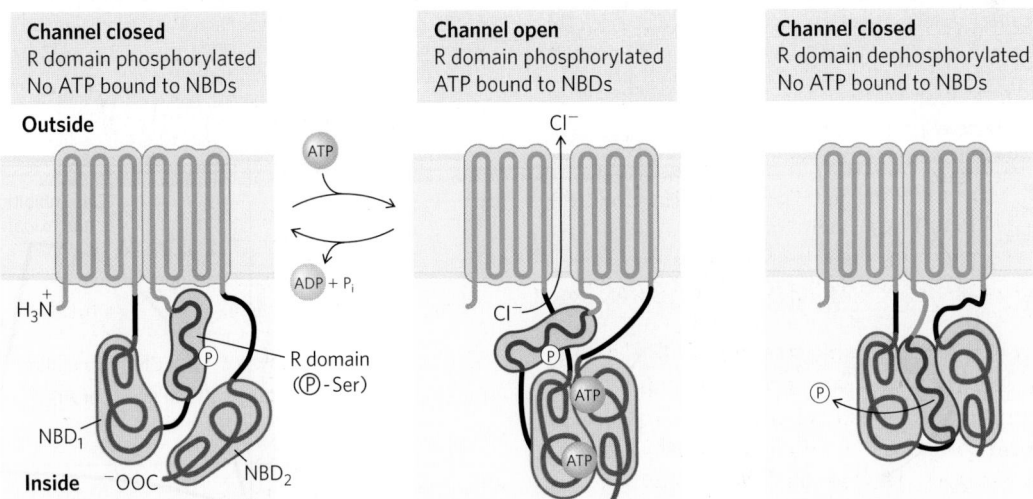

FIGURE 1 Three states of the CFTR protein. The protein has two segments, each with six transmembrane helices, and three functionally significant domains extend from the cytoplasmic surface: NBD₁ and NBD₂ (green) are nucleotide-binding domains that bind ATP, and the regulatory R domain (blue) is the site of phosphorylation by cAMP-dependent protein kinase. When this R domain is phosphorylated but no ATP is bound to the NBDs (left), the channel is closed. The binding of ATP opens the channel (middle) until the bound ATP is hydrolyzed. When the R domain is unphosphorylated (right), it binds the NBD domains and prevents ATP binding and channel opening. CFTR is a typical ABC transporter in all but two respects: most ABC transporters lack the regulatory domain, and CFTR acts as an ion channel (for Cl⁻), not as a typical transporter.

BOX 11–2 ✚ MEDICINE   A Defective Ion Channel in Cystic Fibrosis (*Continued*)

bacteria that settle in this mucus (Fig. 2), but the thick mucus in individuals with CF hinders this process, providing a haven in the lungs for pathogenic bacteria. Frequent infections by bacteria such as *Staphylococcus aureus* and *Pseudomonas aeruginosa* cause progressive damage to the lungs and reduce respiratory efficiency, eventually resulting in death due to inadequate lung function.

Advances in therapy have raised the average life expectancy for people who have CF from just 10 years in 1960 to almost 40 years today. CFTR potentiators such as ivacaftor (VX-770) increase the function of the mutant G551D protein that is properly folded and in place in the plasma membrane. For individuals with the folding defect, F508del, CFTR correctors improve the processing and delivery of the mutant protein to the cell surface; a combination of potentiator and corrector drugs is more effective than the corrector drug alone for these patients (Fig. 3).

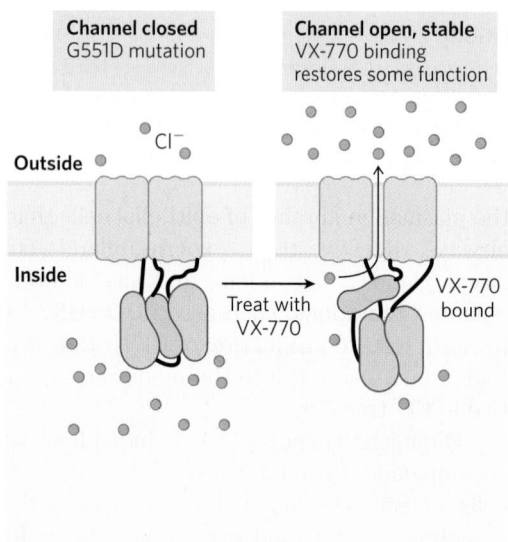

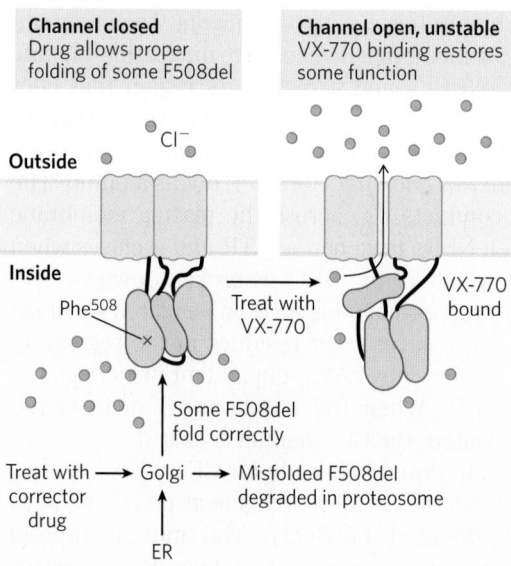

FIGURE 3 **(a)** The CFTR mutation G551D (replacement of Gly[551] with Asp) results in a protein that is inserted into the membrane correctly but is defective as a Cl$^-$ channel. Addition of the potentiator drug VX-770 (ivacaftor) restores partial function to the Cl$^-$ channel. **(b)** The more common mutation F508del (deletion of Phe[508]) prevents proper folding of CFTR, causing it to be degraded in proteasomes. In the presence of a corrector drug, folding and membrane insertion can take place; addition of the potentiator drug results in partial restoration of Cl$^-$ channel activity. The channel is unstable and is degraded over time. [Source: Information from J. P. Clancy, *Sci. Transl. Med.* 6:1, 2014.]

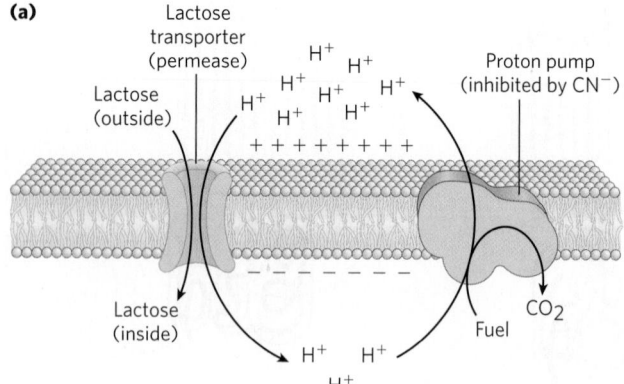

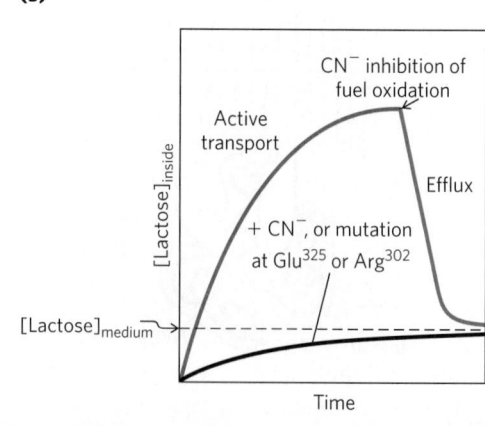

FIGURE 11-39 **Lactose uptake in *E. coli*.** **(a)** The primary transport of H$^+$ out of the cell, driven by the oxidation of a variety of fuels, establishes both a proton gradient and an electrical potential (inside negative) across the membrane. Secondary active transport of lactose into the cell involves symport of H$^+$ and lactose by the lactose transporter. The uptake of lactose against its concentration gradient is entirely dependent on this inflow of protons driven by the electrochemical gradient. **(b)** When the energy-yielding oxidation reactions of metabolism are blocked by cyanide (CN$^-$), the lactose transporter allows equilibration of lactose across the membrane by passive transport. Mutations that affect Glu[325] or Arg[302] have the same effect as cyanide. The dashed line represents the concentration of lactose in the surrounding medium.

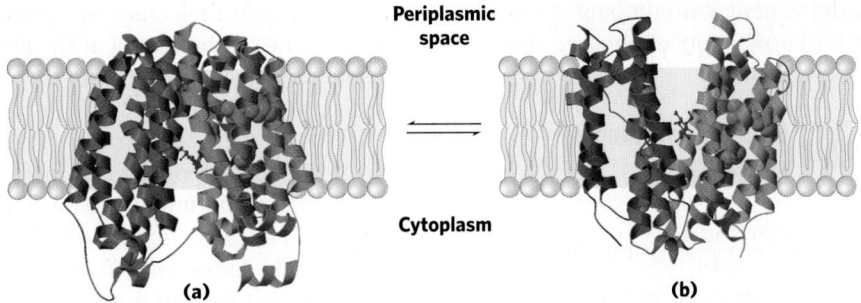

Periplasmic space

Cytoplasm

(a)   (b)

FIGURE 11-40 The lactose transporter (lactose permease) of *E. coli*.
(a) A ribbon representation viewed parallel to the plane of the membrane reveals the 12 transmembrane helices arranged in two nearly symmetric domains, shown in different shades of purple. In the form of the protein for which the crystal structure was determined, the substrate sugar (red) is bound near the middle of the membrane, where the sugar is exposed to the cytoplasm. (b) The postulated second conformation of the transporter, related to the first by a large, reversible conformational change in which the substrate-binding site is exposed first to the periplasm, where lactose is picked up, then to the cytoplasm, where the lactose is released. The interconversion of the two forms is driven by changes in the pairing of charged (protonatable) side chains such as those of Glu^{325} and Arg^{302} (green), which is affected by the transmembrane proton gradient. [Sources: (a) Modified from PDB ID 1PV7, J. Abramson et al., *Science* 301:610, 2003. (b) PDB ID 2CFQ, O. Mirza et al., *EMBO J.* 25:1177, 2006.]

accumulation of lactose is thereby coupled to the exergonic flow of protons into the cell, with a negative overall free-energy change.

The lactose transporter is one member of the **major facilitator superfamily (MFS)** of transporters, which comprises 28 families. Almost all proteins in this superfamily have 12 transmembrane domains (the few exceptions have 14). The proteins share relatively little sequence homology, but the similarity of their secondary structures and topology suggests a common tertiary structure. The crystallographic solution of the *E. coli* lactose transporter provides a glimpse of this general structure **(Fig. 11-40a)**. The protein's 12 transmembrane helices are connected by loops that protrude into the cytoplasm or the periplasmic space (between the plasma membrane and outer membrane or cell wall). The six amino-terminal and six carboxyl-terminal helices form very similar domains to produce a structure with a rough twofold symmetry. In the crystallized form of the protein, a large aqueous cavity is exposed on the cytoplasmic side of the membrane. The substrate-binding site is in this cavity, more or less in the middle of the membrane. The side of the transporter facing outward (the periplasmic face) is closed tightly, with no channel big enough for lactose to enter. The proposed mechanism for transmembrane passage of the substrate (Fig. 11-40b) is that a rocking motion between the two domains, driven by substrate binding and proton movement, alternately exposes the substrate-binding domain to the cytoplasm and to the periplasm. This model is similar to that shown in Figure 11-30 for GLUT1.

In intestinal epithelial cells, glucose and certain amino acids are accumulated by symport with Na^+, down the Na^+ gradient established by the Na^+K^+ ATPase of the plasma membrane **(Fig. 11-41)**. The apical surface of the intestinal epithelial cell (the surface that faces the intestinal contents) is covered with microvilli, long, thin projections of the plasma membrane that greatly increase the surface area exposed to the intestinal contents. The

Na^+-glucose symporter in the apical plasma membrane takes up glucose from the intestine in a process driven by the downhill flow of Na^+:

$$2Na^+_{out} + glucose_{out} \rightarrow 2Na^+_{in} + glucose_{in}$$

The energy required for this process comes from two sources: the greater concentration of Na^+ outside than inside the cell (the chemical potential) and the membrane (electrical) potential, which is inside negative and therefore draws Na^+ inward. The strong thermodynamic tendency for Na^+ to move into the cell provides the energy needed for the transport of glucose into the cell, against its concentration gradient. As in the case of the lactose permease, an ion gradient created and

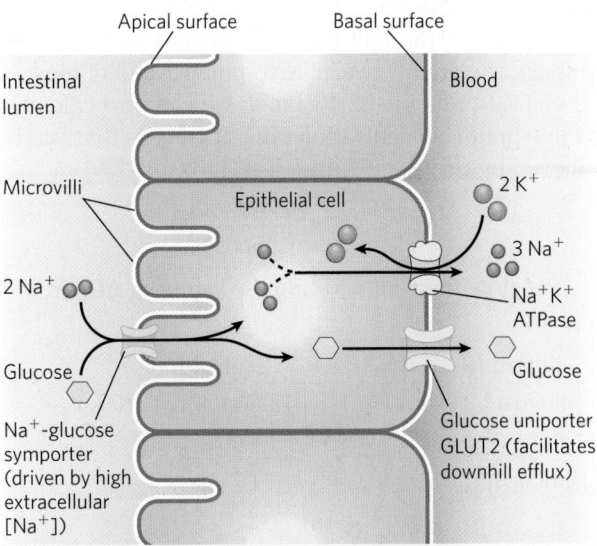

FIGURE 11-41 Glucose transport in intestinal epithelial cells.
Glucose is cotransported with Na^+ across the apical plasma membrane into the epithelial cell. It moves through the cell to the basal surface, where it passes into the blood via GLUT2, a passive glucose uniporter. The Na^+K^+ ATPase continues to pump Na^+ outward to maintain the Na^+ gradient that drives glucose uptake.

sustained by energy-dependent ion pumping serves as the potential energy for cotransport of another species against its concentration gradient.

WORKED EXAMPLE 11-3   **Energetics of Pumping by Symport**

Calculate the maximum $\dfrac{[\text{glucose}]_{\text{in}}}{[\text{glucose}]_{\text{out}}}$ ratio that can be achieved by the plasma membrane Na^+-glucose symporter of an epithelial cell when $[Na^+]_{\text{in}}$ is 12 mM, $[Na^+]_{\text{out}}$ is 145 mM, the membrane potential is -50 mV (inside negative), and the temperature is 37 °C.

Solution: Using Equation 11-4 (p. 413), we can calculate the energy inherent in an electrochemical Na^+ gradient—that is, the cost of moving one Na^+ ion up this gradient:

$$\Delta G_t = RT \ln \frac{[Na^+]_{\text{out}}}{[Na^+]_{\text{in}}} + ZF \, \Delta\psi$$

We then substitute standard values for R, T, and F; the given values for $[Na^+]$ (expressed as molar concentrations); $+1$ for Z (because Na^+ has a positive charge); and 0.050 V for $\Delta\psi$. Note that the membrane potential is -50 mV (inside negative), so the change in potential when an ion moves from inside to outside is 50 mV.

$$\Delta G_t = (8.315 \text{ J/mol·K})(310 \text{ K}) \ln\frac{(1.45 \times 10^{-1})}{(1.2 \times 10^{-2})} +$$
$$1 \, (96{,}500 \text{ J/V·mol})(0.050 \text{ V})$$
$$= 11.2 \text{ kJ/mol}$$

When Na^+ reenters the cell, it releases the electrochemical potential created by pumping it out; ΔG for reentry is -11.2 kJ/mol of Na^+. This is the potential energy per mole of Na^+ that is available to pump glucose. Given that two Na^+ ions pass down their electrochemical gradient and into the cell for each glucose carried in by symport, the energy available to pump 1 mol of glucose is 2×11.2 kJ/mol $= 22.4$ kJ/mol. We can now calculate the maximum concentration ratio of glucose that can be achieved by this pump (from Eqn 11-3, p. 413):

$$\Delta G_t = RT \ln \frac{[\text{glucose}]_{\text{in}}}{[\text{glucose}]_{\text{out}}}$$

Rearranging, then substituting the values of ΔG_t, R, and T, gives

$$\ln \frac{[\text{glucose}]_{\text{in}}}{[\text{glucose}]_{\text{out}}} = \frac{\Delta G_t}{RT} = \frac{22.4 \text{ kJ/mol}}{(8.315 \text{ J/mol·K})(310 \text{ K})} = 8.69$$

$$\frac{[\text{glucose}]_{\text{in}}}{[\text{glucose}]_{\text{out}}} = e^{8.69}$$

$$= 5.94 \times 10^3$$

Thus the cotransporter can pump glucose inward until its concentration inside the epithelial cell is about 6,000 times that outside (in the intestine). (This is the maximum theoretical ratio, assuming a perfectly efficient coupling of Na^+ reentry and glucose uptake.)

As glucose molecules are pumped from the intestine into the epithelial cell at the apical surface, glucose is simultaneously moved from the cell into the blood by passive transport through a glucose transporter (GLUT2) in the basal surface (Fig. 11-41). The crucial role of Na^+ in symport and antiport systems such as this requires the continued outward pumping of Na^+ to maintain the transmembrane Na^+ gradient.

In the kidney, a different Na^+-glucose symporter is the target of drugs used to treat type 2 diabetes. Gliflozins are specific inhibitors of this Na^+-glucose symporter. They lower blood glucose by inhibiting glucose reabsorption in the kidney, thus preventing the damaging effects of elevated blood glucose. Glucose not reabsorbed in the kidney is cleared in the urine.

Because of the essential role of ion gradients in active transport and energy conservation, compounds that collapse ion gradients across cellular membranes are effective poisons, and those that are specific for infectious microorganisms can serve as antibiotics. One such substance is valinomycin, a small cyclic peptide that neutralizes the K^+ charge by surrounding the ion with six carbonyl oxygens **(Fig. 11-42)**. The hydrophobic peptide then acts as a shuttle, carrying K^+ across the membrane down its concentration gradient and deflating that gradient. Compounds that shuttle ions across membranes in this way are called **ionophores** ("ion bearers"). Both valinomycin and monensin (a Na^+-carrying ionophore) are antibiotics; they kill microbial cells by disrupting secondary transport processes and energy-conserving reactions. Monensin is widely used as an antifungal and antiparasitic agent. ∎

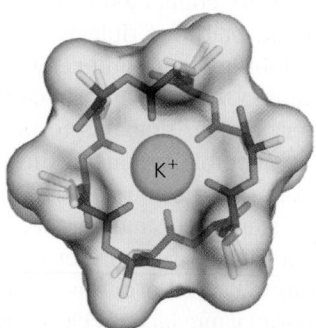

FIGURE 11-42 Valinomycin, a peptide ionophore that binds K^+. In this image, the surface contours are shown as a yellow envelope, through which a stick structure of the peptide and a K^+ ion (green) are visible. The oxygen atoms (red) that bind K^+ are part of a central hydrophilic cavity. Hydrophobic amino acid side chains (yellow) coat the outside of the molecule. Because the exterior of the K^+-valinomycin complex is hydrophobic, the complex readily diffuses through membranes, carrying K^+ down its concentration gradient. The resulting dissipation of the transmembrane ion gradient kills microbial cells, making valinomycin a potent antibiotic. [Source: Coordinates prepared for The Virtual Museum of Minerals and Molecules, http://virtual-museum.soils.wisc.edu/valinomycin/index.html, by Phillip Barak, University of Wisconsin–Madison, Department of Soil Science, using data from K. Neupert-Laves and M. Dobler, *Helv. Chim. Acta* 58:432, 1975.]

Aquaporins Form Hydrophilic Transmembrane Channels for the Passage of Water

A family of integral membrane proteins discovered by Peter Agre, the **aquaporins (AQPs)**, provide channels for rapid movement of water molecules across all plasma membranes. Aquaporins are found in all organisms, and multiple aquaporin genes are generally present, encoding similar but not identical proteins. Eleven aquaporins are known in mammals, each with a specific location and role (Table 11-6). Erythrocytes, which swell or shrink rapidly in response to abrupt changes in extracellular osmolarity as blood travels through the renal medulla, have a high density of aquaporin in their plasma membrane (2×10^5 copies of AQP1 per cell). The exocrine glands that produce sweat, saliva, and tears secrete water through aquaporins. Seven different aquaporins play roles in urine production and water retention in the nephron (the functional

Peter Agre
[Source: Courtesy Dr. Peter Agre, Johns Hopkins University.]

unit of the kidney). Each renal AQP has a specific location in the nephron, and each has specific properties and regulatory features. For example, AQP2 in the epithelial cells of the renal collecting duct is regulated by vasopressin (also called antidiuretic hormone): more water is reabsorbed from the duct into the kidney tissues when the vasopressin level is high. Mutant mice with no AQP2 gene have greater urine output (polyuria) and more dilute urine, the result of the proximal tubule becoming less permeable to water. In humans, genetically defective AQPs are known to be responsible for a variety of diseases, including a relatively rare form of diabetes that is accompanied by polyuria (Box 11-1).

Water molecules flow through an AQP1 channel at a rate of about $10^9 \ s^{-1}$. For comparison, the highest known turnover number for an enzyme is that for catalase, $4 \times 10^7 \ s^{-1}$, and many enzymes have turnover numbers between $1 \ s^{-1}$ and $10^4 \ s^{-1}$ (see Table 6-7). The low activation energy for passage of water through aquaporin channels ($\Delta G^{\ddagger} < 15$ kJ/mol) suggests that water moves through the channels in a continuous stream, in the direction dictated by the osmotic gradient. (For a discussion of osmosis, see p. 56.) Aquaporins do not allow passage of protons (hydronium ions, H_3O^+), which

TABLE 11-6	Permeability Characteristics and Predominant Distribution of Known Mammalian Aquaporins		
Aquaporin	**Permeant (permeability)**	**Tissue distribution**	**Primary subcellular distribution[a]**
AQP0	Water (low)	Lens	Plasma membrane
AQP1	Water (high)	Erythrocyte, kidney, lung, vascular endothelium, brain, eye	Plasma membrane
AQP2	Water (high)	Kidney, vas deferens	Apical plasma membrane, intracellular vesicles
AQP3	Water (high), glycerol (high), urea (moderate)	Kidney, skin, lung, eye, colon	Basolateral plasma membrane
AQP4	Water (high)	Brain, muscle, kidney, lung, stomach, small intestine	Basolateral plasma membrane
AQP5	Water (high)	Salivary gland, lacrimal gland, sweat gland, lung, cornea	Apical plasma membrane
AQP6	Water (low), anions ($NO_3^- > Cl^-$)	Kidney	Intracellular vesicles
AQP7	Water (high), glycerol (high), urea (high)	Adipose tissue, kidney, testis	Plasma membrane
AQP8[b]	Water (high)	Testis, kidney, liver, pancreas, small intestine, colon	Plasma membrane, intracellular vesicles
AQP9	Water (low), glycerol (high), urea (high)	Liver, leukocyte, brain, testis	Plasma membrane
AQP10	Water (low), glycerol (high), urea (high)	Small intestine	Intracellular vesicles

Source: Data from L. S. King et al., *Nature Rev. Mol. Cell Biol.* 5:688, 2004.

[a]The apical plasma membrane faces the lumen of the gland or tissue; the basolateral plasma membrane is along the sides and base of the cell, not facing the lumen of the gland or tissue.

[b]AQP8 might also be permeated by urea.

would collapse membrane electrochemical gradients. What is the basis for this extraordinary selectivity?

We find an answer in the structure of AQP1, as determined by x-ray crystallography. AQP1 **(Fig. 11-43a)** consists of four identical monomers (each M_r 28,000), each of which forms a transmembrane pore with a diameter sufficient to allow passage of water molecules in single file. Each monomer has six transmembrane helical segments and two shorter helices, both of which contain the sequence Asn–Pro–Ala (NPA). The six transmembrane helices form the pore through the monomer, and the two short loops containing the NPA sequences extend toward the middle of the bilayer from opposite sides. Their NPA regions overlap in the middle of the membrane to form part of the specificity filter—the structure that allows only water to pass (Fig. 11-43b).

The water channel narrows to a diameter of 2.8 Å near the center of the membrane, severely restricting the size of molecules that can travel through. The positive charge of a highly conserved Arg residue at this bottleneck discourages the passage of cations such as H_3O^+. The residues that line the channel of each AQP1 monomer are generally nonpolar, but carbonyl oxygens in the peptide backbone, projecting into the narrow part of the channel at intervals, can hydrogen-bond with individual water molecules as they pass through; the two

Asn residues (Asn^{76} and Asn^{192}) in the NPA loops also form hydrogen bonds with the water. The structure of the channel does not permit formation of a chain of water molecules close enough to allow proton hopping (see Fig. 2-14), which would effectively move protons across the membrane. Critical Arg and His residues and electric dipoles formed by the short helices of the NPA loops provide positive charges in positions that repel any protons that might leak through the pore and prevent hydrogen bonding between adjacent water molecules.

An aquaporin isolated from spinach is known to be "gated"—open when two critical Ser residues near the intracellular end of the channel are phosphorylated, and closed when they are dephosphorylated. Both the open and closed structures have been determined by crystallography. Phosphorylation favors a conformation that presses two nearby Leu residues and a His residue into the channel, blocking the movement of water past that point and effectively closing the channel. Other aquaporins are regulated in other ways, allowing rapid changes in membrane permeability to water.

Although generally highly specific for water, some AQPs also allow glycerol or urea to pass at high rates (Table 11-6); these AQPs are believed to be important in the metabolism of glycerol. AQP7, for example, found in the plasma membranes of adipocytes (fat cells),

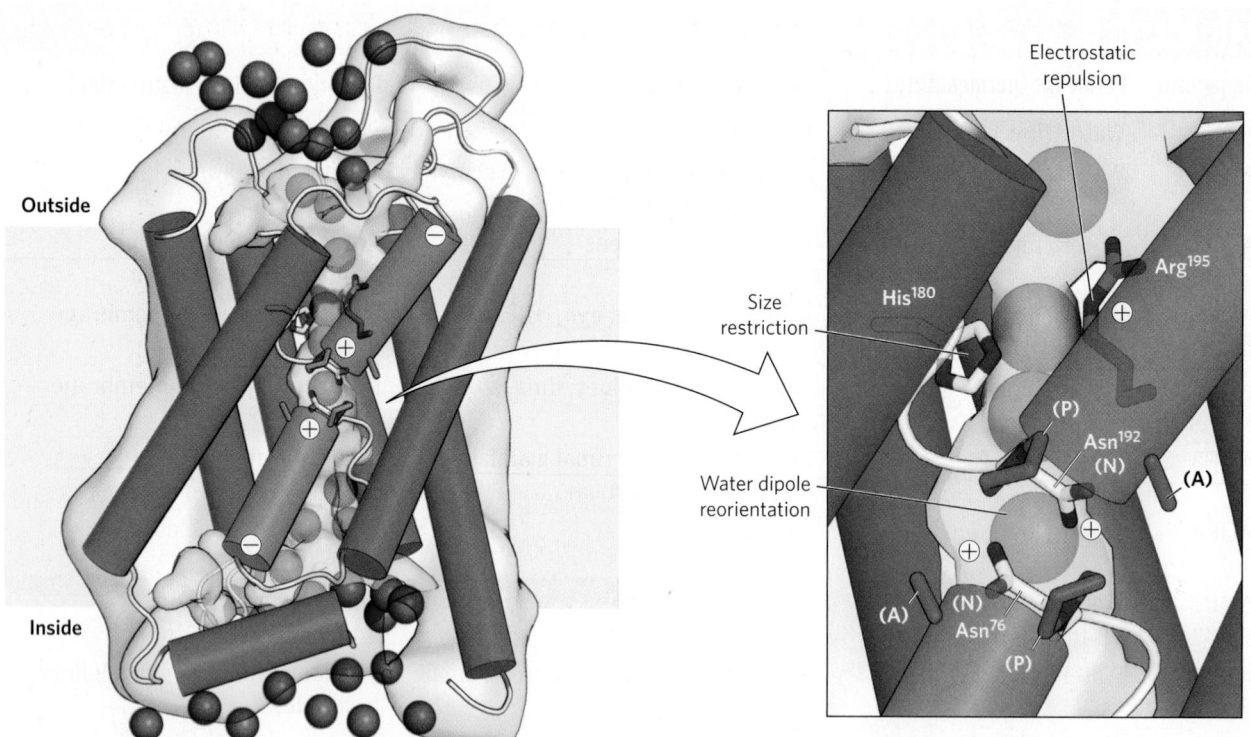

FIGURE 11-43 Aquaporin. The protein is a tetramer of identical subunits, each with a transmembrane pore. **(a)** A monomer of bovine aquaporin, viewed in the plane of the membrane. The helices form a central pore (yellow), through which water (red) passes. **(b)** This closeup view shows that the pore narrows at His^{180} to a diameter of 2.8 Å, limiting passage of molecules larger than H_2O. The positive charge of Arg^{195} repels cations, including H_3O^+, preventing their passage through the pore. The two short helices shown in green contain the Asn–Pro–Ala

(NPA) sequences, found in all aquaporins, that form part of the water channel. These helices are oriented with their positively charged dipoles pointed at the pore in such a way as to force a water molecule to reorient as it passes through. This breaks up hydrogen-bonded chains of water molecules, preventing proton passage by "proton hopping." [Sources: (a) Modified from PDB ID 2B5F, S. Tornroth-Horsefield et al., *Nature* 439:688, 2006. (b) Modified from PDB ID 1J4N, H. Sui et al., *Nature* 414:872, 2001.]

transports glycerol efficiently. This is presumably essential to the import of glycerol for triacylglycerol synthesis, and for its export during triacylglycerol breakdown. Mice with defective AQP7 develop obesity and non-insulin-dependent diabetes.

Ion-Selective Channels Allow Rapid Movement of Ions across Membranes

Ion-selective channels—first recognized in neurons and now known to be present in the plasma membranes of all cells, as well as in the intracellular membranes of eukaryotes—provide another mechanism for moving inorganic ions across membranes. Ion channels, together with ion pumps such as the Na^+K^+ ATPase, determine a plasma membrane's permeability to specific ions and regulate the cytosolic concentration of ions and the membrane potential. In neurons, very rapid changes in the activity of ion channels cause the changes in membrane potential (action potentials) that carry signals from one end of a neuron to the other. In myocytes, rapid opening of Ca^{2+} channels in the sarcoplasmic reticulum releases the Ca^{2+} that triggers muscle contraction. We discuss the signaling functions of ion channels in Chapter 12. Here we describe the structural basis for ion-channel function, using as examples a voltage-gated K^+ channel, the neuronal Na^+ channel, and the acetylcholine receptor ion channel.

Ion channels are distinct from ion transporters in at least three ways. First, the rate of flux through channels can be several orders of magnitude greater than the turnover number for a transporter—10^7 to 10^8 ions/s for an ion channel, approaching the theoretical maximum for unrestricted diffusion. By contrast, the turnover rate of the Na^+K^+ ATPase is about 100 s^{-1}. Second, ion channels are not saturable: rates do not approach a maximum at high substrate concentration. Third, they are gated in response to some type of cellular event. In **ligand-gated channels** (which are generally oligomeric), binding of an extracellular or intracellular small molecule forces an allosteric transition in the protein, which opens or closes the channel. In **voltage-gated ion channels**, a change in transmembrane electrical potential (V_m) causes a charged protein domain to move relative to the membrane, opening or closing the channel. Both types of gating can be very fast. A channel typically opens in a fraction of a millisecond and may remain open for only milliseconds, making these molecular devices effective for very fast signal transmission in the nervous system.

Ion-Channel Function Is Measured Electrically

Because a single ion channel typically remains open for only a few milliseconds, monitoring this process is beyond the limit of most biochemical measurements. Ion fluxes must therefore be measured electrically, either as changes in V_m (in the millivolt range) or as electric current I (in the microampere or picoampere range), using microelectrodes and appropriate amplifiers. In **patch-clamping**, a technique developed by Erwin Neher and Bert Sakmann in 1976, very small currents are measured through a tiny region of the membrane surface containing only one or a few ion-channel molecules **(Fig. 11-44)**. The researcher can measure the size and duration of the current that flows during one opening of an ion channel and can determine how often a channel opens and how that frequency is affected by membrane potential, regulatory ligands,

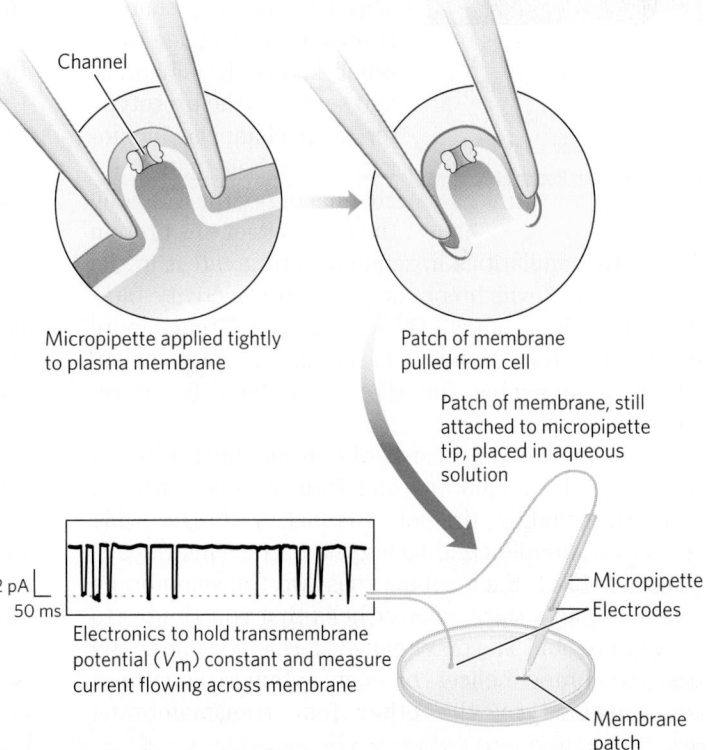

FIGURE 11-44 Electrical measurements of ion-channel function. The "activity" of an ion channel is estimated by measuring the flow of ions through it, using the patch-clamp technique. A finely drawn-out pipette (micropipette) is pressed against the cell surface, and negative pressure in the pipette forms a pressure seal between pipette and membrane. As the pipette is pulled away from the cell, it pulls off a tiny patch of membrane (which may contain one or a few ion channels). After placing the pipette and attached patch in an aqueous solution, the researcher can measure channel activity as the electric current that flows between the contents of the pipette and the aqueous solution. In practice, a circuit is set up that "clamps" the transmembrane potential at a given value and measures the current that must flow to maintain this voltage. With highly sensitive current detectors, researchers can measure the current flowing through a single ion channel, typically a few picoamperes. The trace shows the current through a single acetylcholine receptor channel as a function of time (in milliseconds), revealing how fast the channel opens and closes, how frequently it opens, and how long it stays open. Downward deflection represents channel opening. Clamping the V_m at different values permits determination of the effect of membrane potential on these parameters of channel function. [Source: V. Witzemann et al., *Proc. Natl. Acad. Sci. USA* 93:13,286, 1996.]

Channel

Micropipette applied tightly to plasma membrane

Patch of membrane pulled from cell

Patch of membrane, still attached to micropipette tip, placed in aqueous solution

2 pA
50 ms

Electronics to hold transmembrane potential (V_m) constant and measure current flowing across membrane

Micropipette
Electrodes

Membrane patch

toxins, and other agents. Patch-clamp studies have revealed that as many as 10^4 ions can move through a single ion channel in 1 ms. Such an ion flux represents a huge amplification of the initial signal; for example, only two acetylcholine molecules are needed to open an acetylcholine receptor channel (as described below).

Erwin Neher
[Source: Courtesy Boettcher-Gajewski/Max Planck Institut für Biophysikalische Chemie.]

Bert Sakmann
[Source: Courtesy of Max Planck Institut für Neurobiologie.]

The Structure of a K⁺ Channel Reveals the Basis for Its Specificity

Roderick MacKinnon
[Source: Courtesy Dr. Roderick MacKinnon, Laboratory of Molecular Neurobiology and Biophysics, The Rockefeller University.]

The structure of a potassium channel from the bacterium *Streptomyces lividans*, determined crystallographically by Roderick MacKinnon in 1998, provides important insight into the way ion channels work. This bacterial ion channel is related in sequence to all other known K⁺ channels and serves as the prototype for such channels, including the voltage-gated K⁺ channel of neurons. Among the members of this protein family, the similarities in sequence are greatest in the "pore region," which contains the ion selectivity filter that allows K⁺ (radius 1.33 Å) to pass 10^4 times more readily than Na⁺ (radius 0.95 Å)—at a rate (about 10^8 ions/s) approaching the theoretical limit for unrestricted diffusion.

The K⁺ channel consists of four identical subunits that span the membrane and form a cone within a cone surrounding the ion channel, with the wide end of the double cone facing the extracellular space **(Fig. 11-45a)**. Each subunit has two transmembrane α helices and a third, shorter helix that contributes to the pore region. The outer cone is formed by one of the transmembrane helices of each subunit. The inner cone, formed by the other four transmembrane

helices, surrounds the ion channel and cradles the ion selectivity filter. Viewed perpendicular to the plane of the membrane, the central channel is seen to be just wide enough to accommodate an unhydrated metal ion such as potassium (Fig. 11-45b).

Both the ion specificity and the high flux through the channel are understandable from what we know of the channel's structure (Fig. 11-45c). At the inner and outer plasma membrane surfaces, the entryways to the channel have several negatively charged amino acid residues, which presumably increase the local concentration of cations such as K⁺ and Na⁺. The ion path through the membrane begins (on the inner surface) as a wide, water-filled channel in which the ion can retain its hydration sphere. Further stabilization is provided by the short helices in the pore region of each subunit, with the partial negative charges of their electric dipoles pointed at K⁺ in the channel. About two-thirds of the way through the membrane, this channel narrows in the region of the selectivity filter, forcing the ion to give up its hydrating water molecules. Carbonyl oxygen atoms in the backbone of the selectivity filter replace the water molecules in the hydration sphere, forming a series of perfect coordination shells through which the K⁺ moves. This favorable interaction with the filter is not possible for Na⁺, which is too small to make contact with all the potential oxygen ligands. The preferential stabilization of K⁺ is the basis for the ion selectivity of the filter, and mutations that change residues in this part of the protein eliminate the channel's ion selectivity. The K⁺-binding sites of the filter are flexible enough to collapse to fit any Na⁺ that enters the channel, and this conformational change closes the channel.

There are four potential K⁺-binding sites along the selectivity filter, each composed of an oxygen "cage" that provides ligands for the K⁺ ions (Fig. 11-45c). In the crystal structure, two K⁺ ions are visible within the selectivity filter, about 7.5 Å apart, and two water molecules occupy the unfilled positions. K⁺ ions pass through the filter in single file; their mutual electrostatic repulsion probably just balances the interaction of each ion with the selectivity filter and keeps them moving. Movement of the two K⁺ ions is concerted: first they occupy positions 1 and 3, then they hop to positions 2 and 4. The energetic difference between these two configurations (1, 3 and 2, 4) is very small; energetically, the selectivity pore is not a series of hills and valleys but a flat surface, which is ideal for rapid ion movement through the channel. The structure of the channel seems to have been optimized during evolution to give maximal flow rates and high specificity.

Voltage-gated K⁺ channels are more complex structures than that illustrated in Figure 11-45, but they are variations on the same theme. For example, the mammalian voltage-gated K⁺ channels in the *Shaker* family have an ion channel like that of the bacterial channel shown in

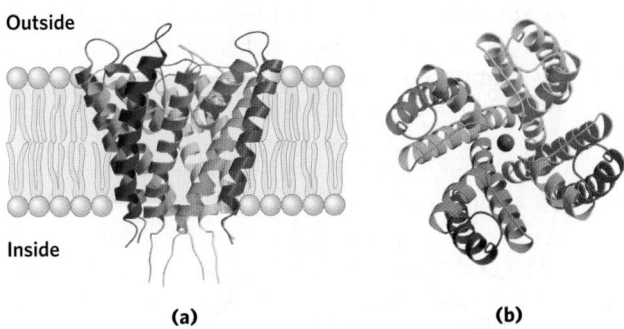

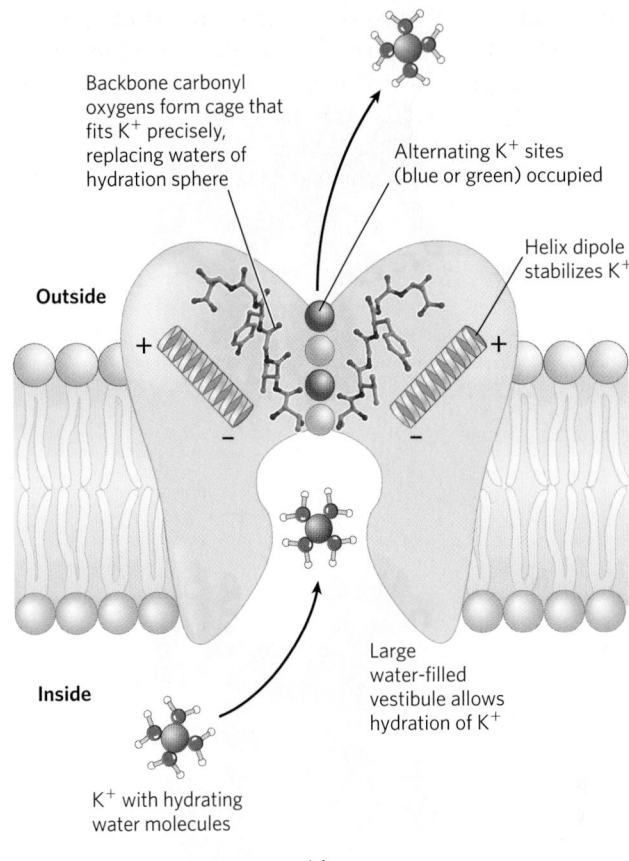

FIGURE 11-45 The K⁺ channel of *Streptomyces lividans.* (a) Viewed in the plane of the membrane, the channel consists of eight transmembrane helices (two from each of four identical subunits), forming a cone with its wide end toward the extracellular space. The inner helices of the cone (lighter colored) line the transmembrane channel, and the outer helices interact with the lipid bilayer. Short segments of each subunit converge in the open end of the cone to make a selectivity filter. (b) This view, perpendicular to the plane of the membrane, shows the four subunits arranged around a central channel just wide enough for a single K⁺ ion to pass. (c) Diagram of a K⁺ channel in cross section, showing the structural features critical to function. Carbonyl oxygens (red) of the peptide backbone in the selectivity filter protrude into the channel, interacting with and stabilizing a K⁺ ion passing through. These ligands are perfectly positioned to interact with each of four K⁺ ions but not with the smaller Na⁺ ions. This preferential interaction with K⁺ is the basis for the ion selectivity. [Sources: (a, b) PDB ID 1BL8, D. A. Doyle et al., *Science* 280:69, 1998. (c) Information from G. Yellen, *Nature* 419:35, 2002, and PDB ID 1J95, M. Zhou et al., *Nature* 411:657, 2001.]

Figure 11-45, but with additional protein domains that sense the membrane potential, move in response to a change in potential, and in moving trigger the opening or closing of the K⁺ channel **(Fig. 11-46)**. The critical transmembrane helix in the voltage-sensing domain of *Shaker* K⁺ channels contains four Arg residues; the positive charges on these residues cause the helix to move relative to the membrane in response to changes in the transmembrane electric field (the membrane potential).

Cells also have channels that specifically conduct Na⁺ or Ca²⁺ and exclude K⁺. In each case, the ability to discriminate among cations requires both a cavity in the binding site of just the right size (neither too large nor too small) to accommodate the ion and a precise positioning within the cavity of carbonyl oxygens that can replace the ion's hydration shell. This fit can be achieved with molecules smaller than proteins; for example, valinomycin (Fig. 11-42) can provide the precise fit that allows high specificity for the binding of one ion rather than another. Chemists have designed small molecules with very high specificity for binding of Li⁺ (radius 0.60 Å), Na⁺ (radius 0.95 Å), K⁺ (radius 1.33 Å), or Rb⁺ (radius 1.48 Å). The biological versions, however—the channel proteins—not only *bind* specifically but *conduct* ions across membranes in a *gated* fashion.

Gated Ion Channels Are Central in Neuronal Function

Virtually all rapid signaling between neurons and their target tissues (such as muscle) is mediated by the rapid opening and closing of ion channels in plasma membranes. For example, Na⁺ channels in neuronal plasma membranes sense the transmembrane electrical gradient and respond to changes by opening or closing. These voltage-gated ion channels are typically very selective for Na⁺ over other monovalent or divalent cations (by factors of 100 or more) and have extremely high flux rates (>10^7 ions/s). Closed in the resting state, Na⁺ channels are opened—activated—by a reduction in the membrane potential. Within milliseconds of opening, a channel closes and remains inactive for many milliseconds. Activation followed by inactivation of Na⁺ channels is the basis for signaling by neurons (see Fig. 12-29).

Another well-studied ion channel is the **nicotinic acetylcholine receptor**, which functions in the passage of an electric signal from a motor neuron to a muscle fiber at the neuromuscular junction (signaling the muscle to contract). Acetylcholine released by the motor neuron diffuses a few micrometers to the plasma membrane of a myocyte, where it binds to an acetylcholine receptor. This forces a conformational change in the receptor, causing its ion channel to open. The resulting inward movement of positively charged ions into the myocyte depolarizes its plasma membrane and triggers contraction. The acetylcholine receptor allows Na⁺, Ca²⁺, and K⁺ to pass through its channel with equal ease, but other cations and all anions are unable to pass. Movement of Na⁺ through an acetylcholine receptor ion channel is unsaturable (its rate is linear

(a)

Outside

S2 S3 S4 K⁺ S6

S1

Inside

S5

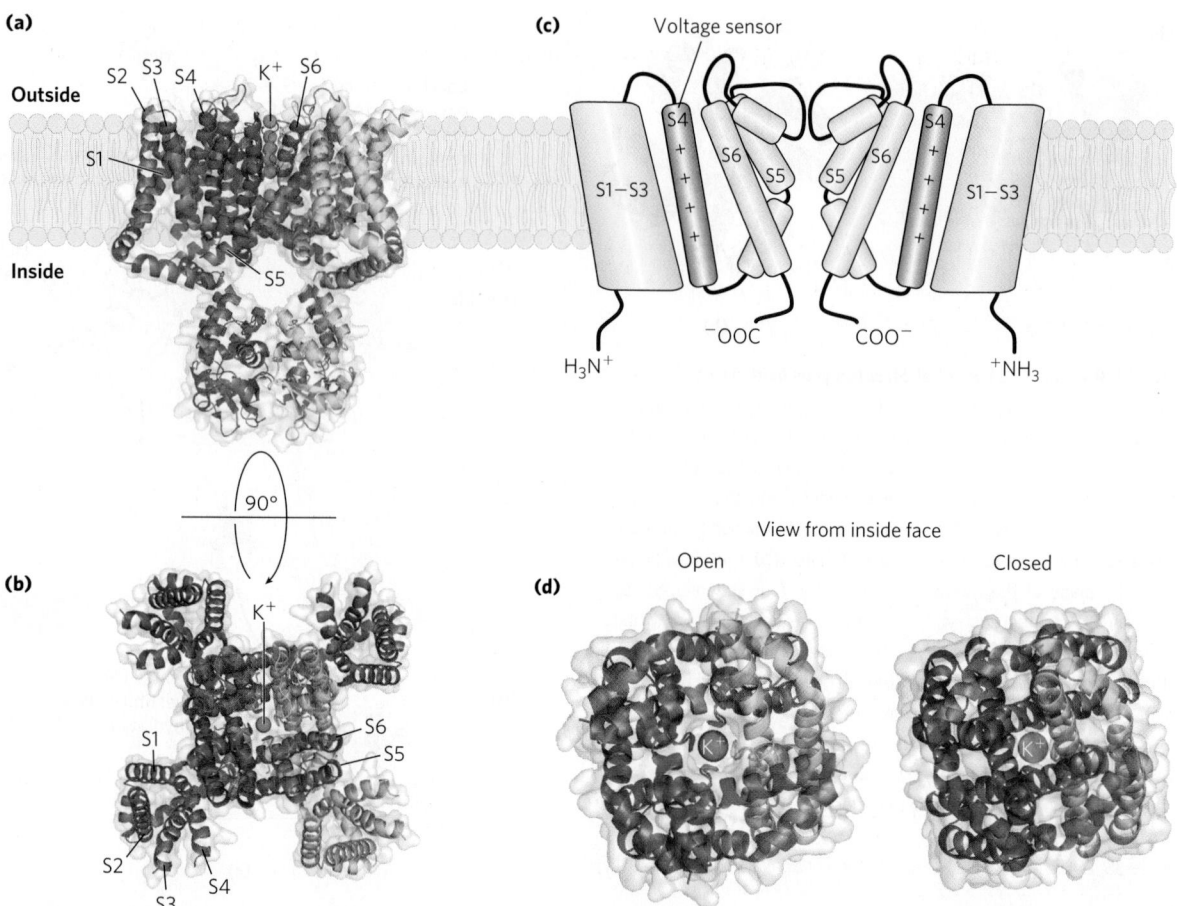

90°

(b)

K⁺

S1

S6

S5

S2

S4

S3

(c)

Voltage sensor

S4

S6

S5 S5

S6

S4

S1–S3

S1–S3

H₃N⁺

$^-$OOC

COO$^-$

$^+$NH₃

View from inside face

Open   Closed

(d)

K⁺   K⁺

FIGURE 11-46 Structural basis for voltage gating in a K⁺ channel of the *Shaker* family. This crystal structure of the Kv1.2-β2 subunit complex from rat brain shows the basic K⁺ channel (corresponding to that shown in Fig. 11-45) with the extra machinery necessary to make the channel sensitive to gating by membrane potential: four transmembrane helical extensions of each subunit and four β subunits. The entire complex, viewed **(a)** in the plane of the membrane and **(b)** perpendicular to the plane (as viewed from outside the membrane), is represented as in Figure 11-45, with each subunit in a different color; each of the four β subunits is the same color as the subunit with which it associates. In (b), each transmembrane helix of one subunit (red) is numbered, S1 to S6. S5 and S6 from each of four subunits form the channel itself and are comparable to the two transmembrane helices of each subunit in Figure 11-45. S1 to S4 are four transmembrane helices. The S4 helix contains the highly conserved Arg residues and is believed to be the chief moving part of the voltage-sensing mechanism. **(c)** A schematic diagram of the voltage-gated channel, showing the basic pore structure (center) and the extra structures that make the channel voltage-sensitive; S4, the Arg-containing helix, is orange. For clarity, the β subunits are not shown in this view. In the resting membrane, the transmembrane electrical potential (inside negative) exerts a pull on positively charged Arg side chains in S4, toward the cytosolic side. When the membrane is depolarized, the pull is lessened, and with complete reversal of the membrane potential, S4 is drawn toward the extracellular side. **(d)** This movement of S4 is physically coupled to opening and closing of the K⁺ channel, which is shown here in its open and closed conformations. Although K⁺ is present in the closed channel, the pore closes on the bottom, near the cytosol, preventing K⁺ passage. [Sources: (a, b, d) PDB ID 2A79, S. B. Long et al., *Science* 309:897, 2005. (c) Information from C. S. Gandhi and E. Y. Isacoff, *Trends Neurosci.* 28:472, 2005.]

with respect to extracellular [Na⁺]) and very fast—about 2×10^7 ions/s under physiological conditions.

CH₃—C
O
‖
O—CH₂—CH₂—$^+$N—CH₃
CH₃
CH₃

Acetylcholine

The acetylcholine receptor channel is typical of many other ion channels that produce or respond to electric signals: it has a "gate" that opens in response to stimulation by a signal molecule (in this case acetylcholine) and an intrinsic timing mechanism that closes the gate after a split second. Thus the acetylcholine signal is transient—an essential feature of all electric signal conduction.

Based on similarities between the amino acid sequences of other ligand-gated ion channels and the acetylcholine receptor, neuronal receptor channels that respond to the extracellular signals γ-aminobutyric acid (GABA), glycine, and serotonin are grouped in the acetylcholine receptor superfamily and probably share three-dimensional structure and gating mechanisms. The GABA$_A$ and glycine receptors are anion channels specific for Cl⁻ or HCO₃⁻, whereas the serotonin receptor, like the acetylcholine receptor, is cation-specific.

Another class of ligand-gated ion channels respond to *intracellular* ligands: 3′,5′-cyclic guanosine mononucleotide (cGMP) in the vertebrate eye, cGMP and cAMP in olfactory neurons, and ATP and inositol 1,4,5-trisphosphate (IP$_3$) in many cell types. These channels are composed of multiple subunits, each with six transmembrane helical domains. We discuss the signaling functions of these ion channels in Chapter 12.

Table 11-7 shows some transporters discussed in other chapters in the context of the pathways in which they act.

TABLE 11-7	Transport Systems Described Elsewhere in This Text	
Transport system and location	**Figure**	**Role**
IP$_3$-gated Ca^{2+} channel of ER	12-11	Allows signaling via changes in cytosolic [Ca^{2+}]
Glucose transporter of animal cell plasma membrane; regulated by insulin	12-20	Increases capacity of muscle and adipose tissue to take up excess glucose from blood
Voltage-gated Na$^+$ channel of neuron	12-29	Creates action potentials in neuronal signal transmission
Fatty acid transporter of myocyte plasma membrane	17-3	Imports fatty acids for fuel
Acyl-carnitine/carnitine transporter of mitochondrial inner membrane	17-6	Imports fatty acids into matrix for β oxidation
Complex I, III, and IV proton transporters of mitochondrial inner membrane	19-16	Act as energy-conserving mechanism in oxidative phosphorylation, converting electron flow into proton gradient
F$_o$F$_1$ ATPase/ATP synthase of mitochondrial inner membrane, chloroplast thylakoid, and bacterial plasma membrane	19-25, 20-20a, 20-24	Interconverts energy of proton gradient and ATP during oxidative phosphorylation and photophosphorylation
Adenine nucleotide antiporter of mitochondrial inner membrane	19-30	Imports substrate ADP for oxidative phosphorylation and exports product ATP
P$_i$-H$^+$ symporter of mitochondrial inner membrane	19-30	Supplies P$_i$ for oxidative phosphorylation
Malate-α-ketoglutarate transporter of mitochondrial inner membrane	19-31	Shuttles reducing equivalents (as malate) from matrix to cytosol
Glutamate-aspartate transporter of mitochondrial inner membrane	19-31	Completes shuttling begun by malate-α-ketoglutarate shuttle
Uncoupling protein UCP1, a proton pore of mitochondrial inner membrane	19-36, 23-35	Allows dissipation of proton gradient in mitochondria as means of thermogenesis and/or disposal of excess fuel
Cytochrome *bf* complex, a proton transporter of chloroplast thylakoid	20-19	Acts as proton pump, driven by electron flow through the Z scheme; source of proton gradient for photosynthetic ATP synthesis
Bacterorhodopsin, a light-driven proton pump	20-27	Is light-driven source of proton gradient for ATP synthesis in halophilic bacterium
P$_i$-triose phosphate antiporter of chloroplast inner membrane	20-42, 20-43	Exports photosynthetic product from stroma; imports P$_i$ for ATP synthesis
Citrate transporter of mitochondrial inner membrane	21-10	Provides cytosolic citrate as source of acetyl-CoA for lipid synthesis
Pyruvate transporter of mitochondrial inner membrane	21-10	Is part of mechanism for shuttling citrate from matrix to cytosol
LDL receptor in animal cell plasma membrane	21-41	Imports, by receptor-mediated endocytosis, lipid-carrying particles
Protein translocase of ER	27-40	Transports into ER proteins destined for plasma membrane, secretion, or organelles
Nuclear pore protein translocase	27-44a	Shuttles proteins between nucleus and cytoplasm
Bacterial protein transporter	27-46	Exports secreted proteins through plasma membrane

Defective Ion Channels Can Have Severe Physiological Consequences

The importance of ion channels to physiological processes is clear from the effects of mutations in specific ion-channel proteins (Table 11-8, Box 11-2). Genetic defects in the voltage-gated Na^+ channel of the myocyte plasma membrane result in diseases in which muscles are periodically either paralyzed (as in hyperkalemic periodic paralysis) or stiff (as in paramyotonia congenita). Cystic fibrosis is the result of a mutation that changes one amino acid in the protein CFTR, a Cl^- ion channel; the defective process in this case is not neurotransmission but secretion by various exocrine gland cells with activities tied to Cl^- ion fluxes.

Many naturally occurring toxins act on ion channels, and the potency of these toxins further illustrates the importance of normal ion-channel function. Tetrodotoxin (produced by the puffer fish, *Sphaeroides rubripes*) and saxitoxin (produced by the marine dinoflagellate *Gonyaulax*, which causes "red tides") act by binding to the voltage-gated Na^+ channels of neurons and preventing normal action potentials. Puffer fish is an ingredient of the Japanese delicacy fugu, which may be prepared only by chefs specially trained to separate succulent morsel from deadly poison. Eating shellfish that have fed on *Gonyaulax* can also be fatal; shellfish are not sensitive to saxitoxin, but they concentrate it in their muscles, which become highly poisonous to organisms higher up the food chain. The venom of the black mamba snake contains dendrotoxin, which interferes with voltage-gated K^+ channels. Tubocurarine, the active component of curare (used as an arrow poison in the Amazon region), and two other toxins from snake venoms, cobrotoxin and bungarotoxin, block the acetylcholine receptor or prevent the opening of its ion channel. By blocking signals from nerves to muscles, all these toxins cause paralysis and possibly death. On the positive side, the extremely high affinity of bungarotoxin for the acetylcholine receptor ($K_d = 10^{-15}$ M) has proved useful experimentally: the radiolabeled toxin was used to quantify the receptor during its purification. ■

SUMMARY 11.3 Solute Transport across Membranes

■ Movement of polar compounds and ions across biological membranes requires transporter proteins. Some transporters simply facilitate passive diffusion of a solute across the membrane, from a higher to a lower concentration. Others transport solutes against an electrochemical gradient; this requires a source of metabolic energy.

■ Carriers, like enzymes, show saturation and stereospecificity for their substrates. Transport via these systems may be passive or active. Primary active transporters are driven by ATP or electron-transfer reactions; secondary active transporters are driven by coupled flow of two solutes, one of which (often H^+ or Na^+) flows down its electrochemical gradient as the other is pulled up its gradient.

■ The GLUT transporters, such as GLUT1 of erythrocytes, carry glucose into cells by passive transport. These transporters are uniporters, carrying only one substrate. Symporters permit simultaneous passage of two substances in the same direction; examples are the lactose transporter of *E. coli*, driven by the energy of a proton gradient (lactose-H^+ symport), and the glucose transporter of intestinal epithelial cells, driven by a Na^+ gradient (glucose-Na^+ symport). Antiporters mediate simultaneous passage of two substances in opposite

TABLE 11-8	Some Diseases Resulting from Ion Channel Defects	
Ion channel	Affected gene	Disease
Na^+ (voltage-gated, skeletal muscle)	*SCN4A*	Hyperkalemic periodic paralysis (or paramyotonia congenita)
Na^+ (voltage-gated, neuronal)	*SCN1A*	Generalized epilepsy with febrile seizures
Na^+ (voltage-gated, cardiac muscle)	*SCN5A*	Long QT syndrome 3
Ca^{2+} (neuronal)	*CACNA1A*	Familial hemiplegic migraine
Ca^{2+} (voltage-gated, retina)	*CACNA1F*	Congenital stationary night blindness
Ca^{2+} (polycystin-1)	*PKD1*	Polycystic kidney disease
K^+ (neuronal)	*KCNQ4*	Dominant deafness
K^+ (voltage-gated, neuronal)	*KCNQ2*	Benign familial neonatal convulsions
Nonspecific cation (cGMP-gated, retinal)	*CNCG1*	Retinitis pigmentosa
Acetylcholine receptor (skeletal muscle)	*CHRNA1*	Congenital myasthenic syndrome
Cl^-	*ABCC7*	Cystic fibrosis

directions; examples are the chloride-bicarbonate exchanger of erythrocytes and the ubiquitous Na^+K^+ ATPase.

■ In animal cells, Na^+K^+ ATPase maintains the differences in cytosolic and extracellular concentrations of Na^+ and K^+, and the resulting Na^+ gradient is used as the energy source for a variety of secondary active transport processes.

■ The Na^+K^+ ATPase of the plasma membrane and the Ca^{2+} transporters of the sarcoplasmic/endoplasmic reticulum (the SERCA pumps) are examples of P-type ATPases; they undergo reversible phosphorylation during their catalytic cycle. F-type ATPase proton pumps (ATP synthases) are central to energy-conserving mechanisms in mitochondria and chloroplasts. V-type ATPases produce gradients of protons across some intracellular membranes, including plant vacuolar membranes.

■ ABC transporters carry a variety of substrates (including many drugs) out of cells, using ATP as the energy source.

■ Ionophores are lipid-soluble molecules that bind specific ions and carry them passively across membranes, dissipating the energy of electrochemical ion gradients.

■ Water moves across membranes through aquaporins. Some aquaporins are regulated; some also transport glycerol or urea.

■ Ion channels provide hydrophilic pores through which select ions can diffuse, moving down their electrical or chemical concentration gradients; these channels characteristically are unsaturable, have very high flux rates, and are highly specific for one ion. Most are voltage- or ligand-gated. The neuronal Na^+ channel is voltage-gated, and the acetylcholine receptor ion channel is gated by acetylcholine, which triggers conformational changes that open and close the transmembrane path.

Key Terms

Terms in bold are defined in the glossary.

Problems

1. Determining the Cross-Sectional Area of a Lipid Molecule When phospholipids are layered gently onto the surface of water, they orient at the air-water interface with their head groups in the water and their hydrophobic tails in the air. An experimental apparatus **(a)** has been devised that reduces the surface area available to a layer of lipids. By measuring the force necessary to push the lipids together, it is possible to determine when the molecules are packed tightly in a continuous monolayer; as that area is approached, the force needed to further reduce the surface area increases sharply **(b)**. How would you use this apparatus to determine the average area occupied by a single lipid molecule in the monolayer?

(a) Force applied here to compress monolayer

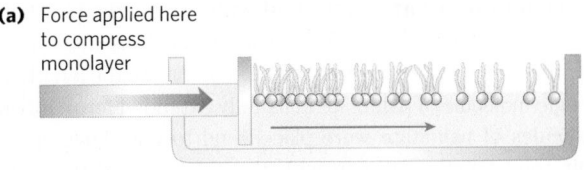

(b)

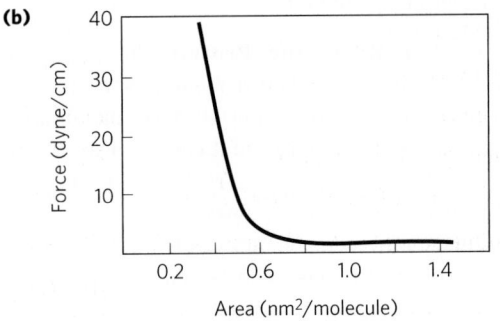

2. Evidence for a Lipid Bilayer In 1925, E. Gorter and F. Grendel used an apparatus like that described in Problem 1 to determine the surface area of a lipid monolayer formed by lipids extracted from erythrocytes of several animal species. They used a microscope to measure the dimensions of individual cells, from which they calculated the average surface area of one erythrocyte. They obtained the data shown in the table below. Were these investigators justified in concluding that "chromocytes [erythrocytes] are covered by a layer of fatty substances that is two molecules thick" (i.e., a lipid bilayer)?

Animal	Volume of packed cells (mL)	Number of cells (per mm^3)	Total surface area of lipid monolayer from cells (m^2)	Total surface area of one cell (μm^2)
Dog	40	8,000,000	62	98
Sheep	10	9,900,000	6.0	29.8
Human	1	4,740,000	0.92	99.4

Source: Data from E. Gorter and F. Grendel, *J. Exp. Med.* 41:439, 1925.

3. Number of Detergent Molecules per Micelle When a small amount of the detergent sodium dodecyl sulfate (SDS; Na$^+$CH$_3$(CH$_2$)$_{11}$OSO$_3^-$) is dissolved in water, the detergent ions enter the solution as monomeric species. As more detergent is added, a concentration is reached (the critical micelle concentration) at which the monomers associate to form micelles. The critical micelle concentration of SDS is 8.2 mM. The micelles have an average particle weight (the sum of the molecular weights of the constituent monomers) of 18,000. Calculate the number of detergent molecules in the average micelle.

4. Properties of Lipids and Lipid Bilayers Lipid bilayers formed between two aqueous phases have this important property: they form two-dimensional sheets, the edges of which close on each other and undergo self-sealing to form vesicles (liposomes).

(a) What properties of lipids are responsible for this property of bilayers? Explain.

(b) What are the consequences of this property for the structure of biological membranes?

5. Length of a Fatty Acid Molecule The carbon–carbon bond distance for single-bonded carbons such as those in a saturated fatty acyl chain is about 1.5 Å. Estimate the length of a single molecule of palmitate in its fully extended form. If two molecules of palmitate were placed end to end, how would their total length compare with the thickness of the lipid bilayer in a biological membrane?

6. Location of a Membrane Protein The following observations are made on an unknown membrane protein, X. It can be extracted from disrupted erythrocyte membranes into a concentrated salt solution, and it can be cleaved into fragments by proteolytic enzymes. Treatment of erythrocytes with proteolytic enzymes followed by disruption and extraction of membrane components yields intact X. However, treatment of erythrocyte "ghosts" (which consist of just plasma membranes, produced by disrupting the cells and washing out the hemoglobin) with proteolytic enzymes, followed by disruption and extraction, yields extensively fragmented X. What do these observations indicate about the location of X in the plasma membrane? Do the properties of X resemble those of an integral or peripheral membrane protein?

7. Predicting Membrane Protein Topology from Sequence You have cloned the gene for a human erythrocyte protein, which you suspect is a membrane protein. From the nucleotide sequence of the gene, you know the amino acid sequence. From this sequence alone, how would you evaluate the possibility that the protein is an integral protein? Suppose the protein proves to be an integral protein with one transmembrane segment. Suggest biochemical or chemical experiments that might allow you to determine whether the protein is oriented with the amino terminus on the outside or the inside of the cell.

8. Surface Density of a Membrane Protein *E. coli* can be induced to make about 10,000 copies of the lactose transporter (M_r 31,000) per cell. Assume that *E. coli* is a cylinder 1 μm in diameter and 2 μm long. What fraction of the plasma membrane surface is occupied by the lactose transporter molecules? Explain how you arrived at this conclusion.

9. Molecular Species in the *E. coli* Membrane The plasma membrane of *E. coli* is about 75% protein and 25% phospholipid by weight. How many molecules of membrane lipid are present for each molecule of membrane protein? Assume an average protein M_r of 50,000 and an average phospholipid M_r of 750. What more would you need to know to estimate the fraction of the membrane surface that is covered by lipids?

10. Temperature Dependence of Lateral Diffusion The experiment described in Figure 11-16 was performed at 37 °C. If the experiment were carried out at 10 °C, what effect would you expect on the rate of diffusion? Why?

11. Membrane Self-Sealing Cellular membranes are self-sealing—if they are punctured or disrupted mechanically, they quickly and automatically reseal. What properties of membranes are responsible for this important feature?

12. Lipid Melting Temperatures Membrane lipids in tissue samples obtained from different parts of a reindeer's leg have different fatty acid compositions. Membrane lipids from tissue near the hooves contain a larger proportion of unsaturated fatty acids than those from tissue in the upper leg. What is the significance of this observation?

13. Flip-Flop Diffusion The inner leaflet (monolayer) of the human erythrocyte membrane consists predominantly of phosphatidylethanolamine and phosphatidylserine. The outer leaflet consists predominantly of phosphatidylcholine and sphingomyelin. Although the phospholipid components of the membrane can diffuse in the fluid bilayer, this sidedness is preserved at all times. How?

14. Membrane Permeability At pH 7, tryptophan crosses a lipid bilayer at about one-thousandth the rate of indole, a closely related compound:

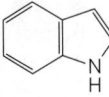

Suggest an explanation for this observation.

15. Use of the Helical Wheel Diagram A helical wheel is a two-dimensional representation of a helix, a view along its central axis (see Fig. 11-29b; see also Fig. 4-4d). Use the helical wheel diagram shown here to determine the distribution of amino acid residues in a helical segment with the sequence –Val–Asp–Arg–Val–Phe–Ser–Asn–Val–Cys–Thr–His–Leu–Lys–Thr–Leu–Gln–Asp–Lys–

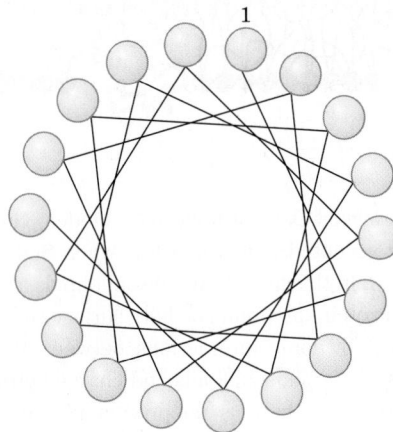

What can you say about the surface properties of this helix? How would you expect the helix to be oriented in the tertiary structure of an integral membrane protein?

16. Synthesis of Gastric Juice: Energetics Gastric juice (pH 1.5) is produced by pumping HCl from blood plasma (pH 7.4) into the stomach. Calculate the amount of free energy required to concentrate the H^+ in 1 L of gastric juice at 37 °C. Under cellular conditions, how many moles of ATP must be hydrolyzed to provide this amount of free energy? The free-energy change for ATP hydrolysis under cellular conditions is about −58 kJ/mol (as explained in Chapter 13). Ignore the effects of the transmembrane electrical potential.

17. Energetics of the Na^+K^+ ATPase For a typical vertebrate cell with a membrane potential of −0.070 V (inside negative), what is the free-energy change for transporting 1 mol of Na^+ from the cell into the blood at 37 °C? Assume the concentration of Na^+ inside the cell is 12 mM and in blood plasma it is 145 mM.

18. Action of Ouabain on Kidney Tissue Ouabain specifically inhibits the Na^+K^+ ATPase activity of animal tissues but is not known to inhibit any other enzyme. When ouabain is added to thin slices of living kidney tissue, it inhibits oxygen consumption by 66%. Why? What does this observation tell us about the use of respiratory energy by kidney tissue?

19. Energetics of Symport Suppose you determined experimentally that a cellular transport system for glucose, driven by symport of Na^+, could accumulate glucose to concentrations 25 times greater than in the external medium, while the external $[Na^+]$ was only 10 times greater than the intracellular $[Na^+]$. Would this violate the laws of thermodynamics? If not, how could you explain this observation?

20. Labeling the Lactose Transporter A bacterial lactose transporter, which is highly specific for lactose, contains a Cys residue that is essential to its transport activity. Covalent reaction of *N*-ethylmaleimide (NEM) with this Cys residue irreversibly inactivates the transporter. A high concentration of lactose in the medium prevents inactivation by NEM, presumably by sterically protecting the Cys residue, which is in or near the lactose-binding site. You know nothing else about the transporter protein. Suggest an experiment that might allow you to determine the M_r of this Cys-containing transporter polypeptide.

21. Intestinal Uptake of Leucine You are studying the uptake of L-leucine by epithelial cells of the mouse intestine. Measurements of the rates of uptake of L-leucine and several of its analogs, with and without Na^+ in the assay buffer, yield the results given in the table below. What can you conclude about the properties and mechanism of the leucine transporter? Would you expect L-leucine uptake to be inhibited by ouabain?

Substrate	Uptake in presence of Na^+		Uptake in absence of Na^+	
	V_{max}	K_t (mM)	V_{max}	K_t (mM)
L-Leucine	420	0.24	23	0.2
D-Leucine	310	4.7	5	4.7
L-Valine	225	0.31	19	0.31

22. Effect of an Ionophore on Active Transport Consider the leucine transporter described in Problem 21. Would V_{max} and/or K_t change if you added a Na^+ ionophore to the assay solution containing Na^+? Explain.

23. Water Flow through an Aquaporin A human erythrocyte has about 2×10^5 AQP1 monomers. If water molecules flow through the plasma membrane at a rate of 5×10^8 per AQP1 tetramer per second, and the volume of an erythrocyte is 5×10^{-11} mL, how rapidly could an erythrocyte halve its volume as it encountered the high osmolarity (1 M) in the interstitial fluid of the renal medulla? Assume that the erythrocyte consists entirely of water.

Biochemistry Online

24. Predicting Membrane Protein Topology I Online bioinformatics tools make hydropathy analysis easy if you know the amino acid sequence of a protein. At the Protein Data Bank (www.pdb.org), the Protein Feature View displays additional information about a protein gleaned from other databases, such as UniProt and SCOP2. A simple graphical view of a hydropathy plot created using a window of 15 residues shows hydrophobic regions in red and hydrophilic regions in blue.

(a) Looking only at the displayed hydropathy plots in the Protein Feature View, what predictions would you make about the membrane topology of these proteins: glycophorin A (PDB ID 1AFO), myoglobin (PDB ID 1MBO), and aquaporin (PDB ID 2B6O)?

(b) Now, refine your information using the ProtScale tools at the ExPASy bioinformatics resource portal. Each of the PDB Protein Feature Views was created with a UniProt Knowledgebase ID. For glycophorin A, the UniProtKB ID is P02724; for myoglobin, P02185; and for aquaporin, Q6J8I9. Go to the ExPASy portal (http://web.expasy.org/protscale) and select the Kyte & Doolittle hydropathy analysis option, with a window of 7 amino acids. Enter the UniProtKB ID for aquaporin (Q6J8I9, which you can also get from the PDB's Protein Feature View page), then select the option to analyze the complete chain (residues 1 to 263). Use the default values for the other options and click Submit to get a hydropathy plot. Save a GIF image of this plot. Now repeat the analysis using a window of 15 amino acids. Compare the results for the 7-residue and 15-residue window analyses. Which one gives you a better signal-to-noise ratio?

(c) Under what circumstances would it be important to use a narrower window?

25. Predicting Membrane Protein Topology II The receptor for the hormone epinephrine in animal cells is an integral membrane protein (M_r 64,000) that is believed to have seven membrane-spanning regions.

(a) Show that a protein of this size is capable of spanning the membrane seven times.

(b) Given the amino acid sequence of this protein, how would you predict which regions of the protein form the membrane-spanning helices?

(c) Go to the Protein Data Bank (www.pdb.org). Use the PDB identifier 1DEP to retrieve the data page for a portion of the β-adrenergic receptor (one type of epinephrine receptor) isolated from turkey. Using JSmol to explore the structure, predict whether this portion of the receptor is located within the membrane or at the membrane surface. Explain your answer. Now use the Protein Feature View to see the hydrophobicity analysis of the sequence. Does this support your answer?

(d) Retrieve the data for a portion of another receptor, the acetylcholine receptor of neurons and myocytes, using the PDB identifier 1A11. As in (c), predict where this portion of the receptor is located and explain your answer.

If you have not used the PDB, see Box 4-4 (p. 132) for more information.

Data Analysis Problem

26. The Fluid Mosaic Model of Biological Membrane Structure Figure 11-3 shows the currently accepted fluid mosaic model of biological membrane structure. This model was presented in detail in a review article by S. J. Singer in

1971. In the article, Singer presented the three models of membrane structure that had been proposed up to that time:

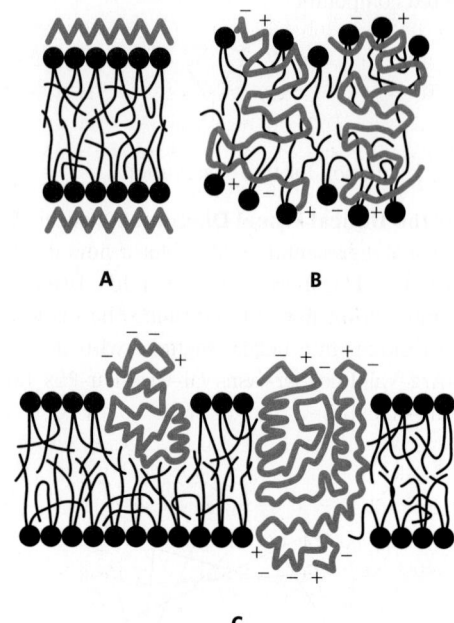

A. The Davson-Danielli-Robertson Model. This was the most widely accepted model in 1971, when Singer's review was published. In this model, the phospholipids are arranged as a bilayer. Proteins are found on both surfaces of the bilayer, attached to it by ionic interactions between the charged head groups of the phospholipids and charged groups of the proteins. Crucially, there is no protein in the interior of the bilayer.

B. The Benson Lipoprotein Subunit Model. Here the proteins are globular and the membrane is a protein-lipid mixture. The hydrophobic tails of the lipids are embedded in the hydrophobic parts of the proteins. The lipid head groups are exposed to the solvent. There is no lipid bilayer.

C. The Lipid-Globular Protein Mosaic Model. This is the model shown in Figure 11-3. The lipids form a bilayer and proteins are embedded in it, some extending through the bilayer and others not. Proteins are anchored in the bilayer by interactions between the hydrophobic tails of the lipids and hydrophobic portions of the protein.

For the data given below, consider how each piece of information aligns with each of the three models of membrane structure. Which model(s) are supported, which are not supported, and what reservations do you have about the data or their interpretation? Explain your reasoning.

(a) When cells were fixed, stained with osmium tetroxide, and examined in the electron microscope, the membranes showed a "railroad track" appearance, with two dark-staining lines separated by a light space.

(b) The thickness of membranes in cells fixed and stained in the same way was found to be 5 to 9 nm. The thickness of a "naked" phospholipid bilayer, without proteins, was 4 to 4.5 nm. The thickness of a single monolayer of proteins was about 1 nm.

(c) Singer wrote in his article: "The average amino acid composition of membrane proteins is not distinguishable from that of soluble proteins. In particular, a substantial fraction of the residues is hydrophobic" (p. 165).

(d) As described in Problems 1 and 2 of this chapter, researchers had extracted membranes from cells, extracted the lipids, and compared the area of the lipid monolayer with the area of the original cell membrane. The interpretation of the results was complicated by the issue illustrated in the graph of Problem 1: the area of the monolayer depended on how hard it was pushed. With very light pressures, the ratio of monolayer area to cell membrane area was about 2.0. At higher pressures—thought to be more like those found in cells—the ratio was substantially lower.

(e) Circular dichroism spectroscopy uses changes in polarization of UV light to make inferences about protein secondary structure (see Fig. 4-10). On average, this technique showed that membrane proteins have a large amount of α helix and little or no β sheet. This finding was consistent with most membrane proteins having a globular structure.

(f) Phospholipase C is an enzyme that removes the polar head group (including the phosphate) from phospholipids. In several studies, treatment of intact membranes with phospholipase C removed about 70% of the head groups without disrupting the "railroad track" structure of the membrane.

(g) Singer described in his article a study in which "a glycoprotein of molecular weight about 31,000 in human red blood cell membranes is cleaved by tryptic treatment of the membranes into soluble glycopeptides of about 10,000 molecular weight, while the remaining portions are quite hydrophobic" (p. 199). Trypsin treatment did not cause gross changes in the membranes, which remained intact.

Singer's review also included many more studies in this area. In the end, though, the data available in 1971 did not conclusively prove Model C was correct. As more data have accumulated, this model of membrane structure has been accepted by the scientific community.

Reference

Singer, S.J. 1971. The molecular organization of biological membranes. In *Structure and Function of Biological Membranes* (L. I. Rothfield, ed.), pp. 145–222. New York: Academic Press, Inc.

Further Reading is available at www.macmillanlearning.com/LehningerBiochemistry7e.

CHAPTER 12

Biosignaling

Self-study tools that will help you practice what you've learned and reinforce this chapter's concepts are available online.
Go to www.macmillanlearning.com/LehningerBiochemistry7e.

The ability of cells to receive and act on signals from beyond the plasma membrane is fundamental to life. Bacterial cells receive constant input from membrane proteins that act as information receptors, sampling the surrounding medium for pH, osmotic strength, the availability of food, oxygen, and light, and the presence of noxious chemicals, predators, or competitors for food. These signals elicit appropriate responses, such as motion toward food or away from toxic substances or the formation of dormant spores in a nutrient-depleted medium. In multicellular organisms, cells with different functions exchange a wide variety of signals. Plant cells respond to growth hormones and to variations in sunlight. Animal cells exchange information about the concentrations of ions and glucose in extracellular fluids, the interdependent metabolic activities taking place in different tissues, and, in an embryo, the correct placement of cells during development. In all these cases, the signal represents *information* that is detected by specific receptors and converted to a cellular response, which always involves a *chemical* process. This conversion of information into a chemical change, **signal transduction**, is a universal property of living cells.

12.1 General Features of Signal Transduction

Signal transductions are remarkably specific and exquisitely sensitive. **Specificity** is achieved by precise molecular complementarity between the signal and receptor molecules **(Fig. 12-1a)**, mediated by the same kinds of weak (noncovalent) forces that mediate enzyme-substrate and antigen-antibody interactions. Multicellular organisms have an additional level of specificity, because the receptors for a given signal, or the intracellular targets of a given signal pathway, are present only in certain cell types. Thyrotropin-releasing hormone, for example, triggers responses in the cells of the anterior pituitary but not in hepatocytes, which lack receptors for this hormone. Epinephrine alters glycogen metabolism in hepatocytes but not in adipocytes; in this case, both cell types have receptors for the hormone, but whereas hepatocytes contain glycogen and the glycogen-metabolizing enzyme that is stimulated by epinephrine, adipocytes contain neither. Adipocytes respond to epinephrine by metabolizing triacylglycerols to release fatty acids, which are then transported to other tissues.

437

(a) Specificity
Signal molecule fits binding site on its complementary receptor; other signals do not fit.

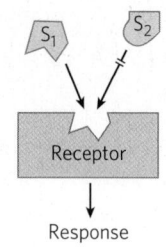

(d) Desensitization/Adaptation
Receptor activation triggers a feedback circuit that shuts off the receptor or removes it from the cell surface.

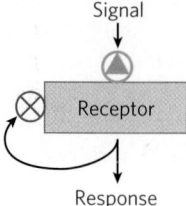

(b) Amplification
When enzymes activate enzymes, the number of affected molecules increases geometrically in an enzyme cascade.

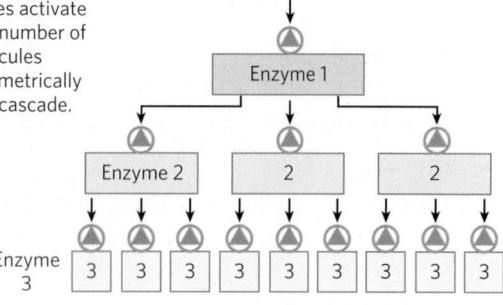

(e) Integration
When two signals have opposite effects on a metabolic characteristic such as the concentration of a second messenger X, or the membrane potential V_m, the regulatory outcome results from the integrated input from both receptors.

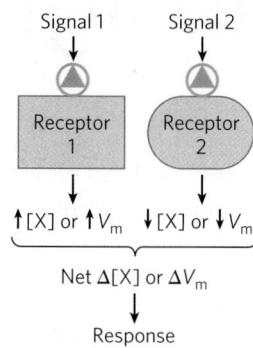

(c) Modularity
Proteins with multivalent affinities form diverse signaling complexes from interchangeable parts. Phosphorylation provides reversible points of interaction.

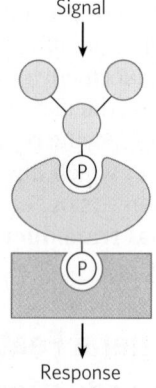

(f) Localized response
When the enzyme that destroys an intracellular message is clustered with the message producer, the message is degraded before it can diffuse to distant points, so the response is only local and brief.

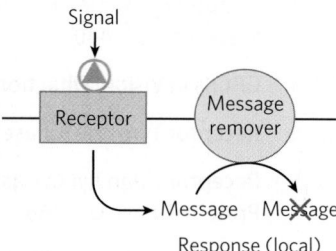

FIGURE 12-1 Six features of signal-transducing systems.

Three factors account for the extraordinary sensitivity of signal transduction: the high affinity of receptors for signal molecules, cooperativity (often but not always) in the ligand-receptor interaction, and amplification of the signal by enzyme cascades. The **affinity** between signal (ligand) and receptor can be expressed as the dissociation constant K_d, commonly 10^{-7} M or less—meaning that the receptor detects micromolar to nanomolar concentrations of a signal molecule.

Cooperativity in receptor-ligand interactions results in large changes in receptor activation with small changes in ligand concentration (recall the effect of cooperativity on oxygen binding to hemoglobin; see Fig. 5-12). **Amplification** results when an enzyme is activated by a signal receptor and, in turn, catalyzes the activation of many molecules of a second enzyme, each of which activates many molecules of a third enzyme, and so on, in a so-called **enzyme cascade** (Fig. 12-1b). Such cascades can produce amplifications of several orders of magnitude within milliseconds. The response

to a signal must also be terminated, such that the downstream effects are in proportion to the strength of the original stimulus.

Interacting signaling proteins are **modular**. Many signaling proteins have multiple domains that recognize specific features in other proteins, or in the cytoskeleton or plasma membrane. This modularity allows a cell to mix and match a set of signaling molecules to create a wide variety of multienzyme complexes with different functions or cellular locations. One common theme in these interactions is the binding of one modular signaling protein to phosphorylated residues in another protein; the resulting interaction can be regulated by phosphorylation or dephosphorylation of the protein partner (Fig. 12-1c). Nonenzymatic **scaffold proteins** with affinity for several enzymes that interact in cascades bring these enzymes together, ensuring that they interact at specific cellular locations and at specific times. Many of the domains involved in protein-protein interactions are intrinsically disordered (see Fig. 4-22), capable of folding differently depending on which protein they interact with. As a result, a single protein can have multiple functions in signaling pathways.

The sensitivity of receptor systems is subject to modification. When a signal is present continuously, the receptor system becomes **desensitized** (Fig. 12-1d), so that it no longer responds to the signal. When the stimulus falls below a certain threshold, the system again becomes sensitive. Think of what happens to your visual transduction system when you walk from bright sunlight into a darkened room or from darkness into the light.

Signal **integration** (Fig. 12-1e) is the ability of the system to receive multiple signals and produce a unified response appropriate to the combined needs of the cell or organism. Different signaling pathways converse with each other at several levels, generating complex cross talk that maintains homeostasis in the cell and the organism.

A final noteworthy feature of signal-transducing systems is **response localization** within a cell (Fig. 12-1f). When the components of a signaling system are confined to a specific subcellular structure (a raft in the plasma membrane, for example), a cell can regulate a process locally, without affecting distant regions of the cell.

One of the revelations of research on signaling is the remarkable degree to which signaling mechanisms have been conserved during evolution. Although the number of different biological signals is probably in the thousands (Table 12-1 lists a few important types), and the kinds of response elicited by these signals are comparably numerous, the machinery for transducing all of these signals is built from about 10 basic types of protein components.

In this chapter we examine some examples of the major classes of signaling mechanisms, looking at how they are integrated in specific biological functions such as responses to hormones and growth factors; the senses of sight, smell, and taste; the transmission of nerve signals; and control of the cell cycle. Often, the end result of a signaling pathway is the phosphorylation of a few specific target-cell proteins, which changes their activities and thus the activities of the cell. Throughout our discussion we emphasize the conservation of fundamental mechanisms for the transduction of biological signals and the adaptation of these basic mechanisms to a wide range of signaling pathways.

TABLE 12-1	Some Signals to Which Cells Respond
Antigens	Light
Cell surface glycoproteins/ oligosaccharides	Mechanical touch
Developmental signals	Microbial, insect pathogens
Extracellular matrix components	Neurotransmitters
	Nutrients
Growth factors	Odorants
Hormones	Pheromones
Hypoxia	Tastants

We consider the molecular details of several representative signal-transduction systems, classified according to the type of receptor. The trigger for each system is different, but the general features of signal transduction are common to all: a signal interacts with a receptor; the activated receptor interacts with cellular machinery, producing a second signal or a change in the activity of a cellular protein; the metabolic activity of the target cell undergoes a change; and finally, the transduction event ends. To illustrate these general features of signaling systems, we look at examples of four basic receptor types **(Fig. 12-2)**.

1. *G protein–coupled receptors* that *indirectly* activate (through GTP-binding proteins, or G proteins) enzymes that generate intracellular second messengers. This type of receptor is illustrated by the β-adrenergic receptor system that detects epinephrine (adrenaline) (Section 12.2). Vision, olfaction, and gustation are sensory systems that also operate through G protein–coupled receptors (Section 12.3).

2. *Receptor enzymes* in the plasma membrane that have an enzymatic activity on the cytoplasmic side, triggered by ligand binding on the extracellular side. Receptors with tyrosine kinase activity, for example, catalyze the phosphorylation of Tyr residues in specific intracellular target proteins. The insulin receptor is one example (Section 12.4); the receptor for epidermal growth factor (EGFR) is another. Receptor guanylyl cyclases also fall in this general class (Section 12.5).

3. *Gated ion channels* of the plasma membrane that open and close (hence the term "gated") in response to the binding of chemical ligands or changes in transmembrane potential. These are the simplest signal transducers.

4. *Nuclear receptors* that bind specific ligands (such as the hormone estrogen) and alter the rate at which specific genes are transcribed and translated into cellular proteins. Because steroid hormones function through mechanisms intimately related to the regulation of gene expression, we consider them only briefly here (Section 12.8) and defer a detailed discussion of their action until Chapter 28.

As we begin this discussion of biological signaling, a word about the nomenclature of signaling proteins is in order. These proteins are typically discovered in one context and named accordingly, then prove to be involved in a broader range of biological functions for which the original name is not helpful. For example, the retinoblastoma protein, pRb, was initially identified as the site of a mutation that contributes to cancer of the retina (retinoblastoma), but it is now known to function in many pathways essential to cell division in all cells, not just those of

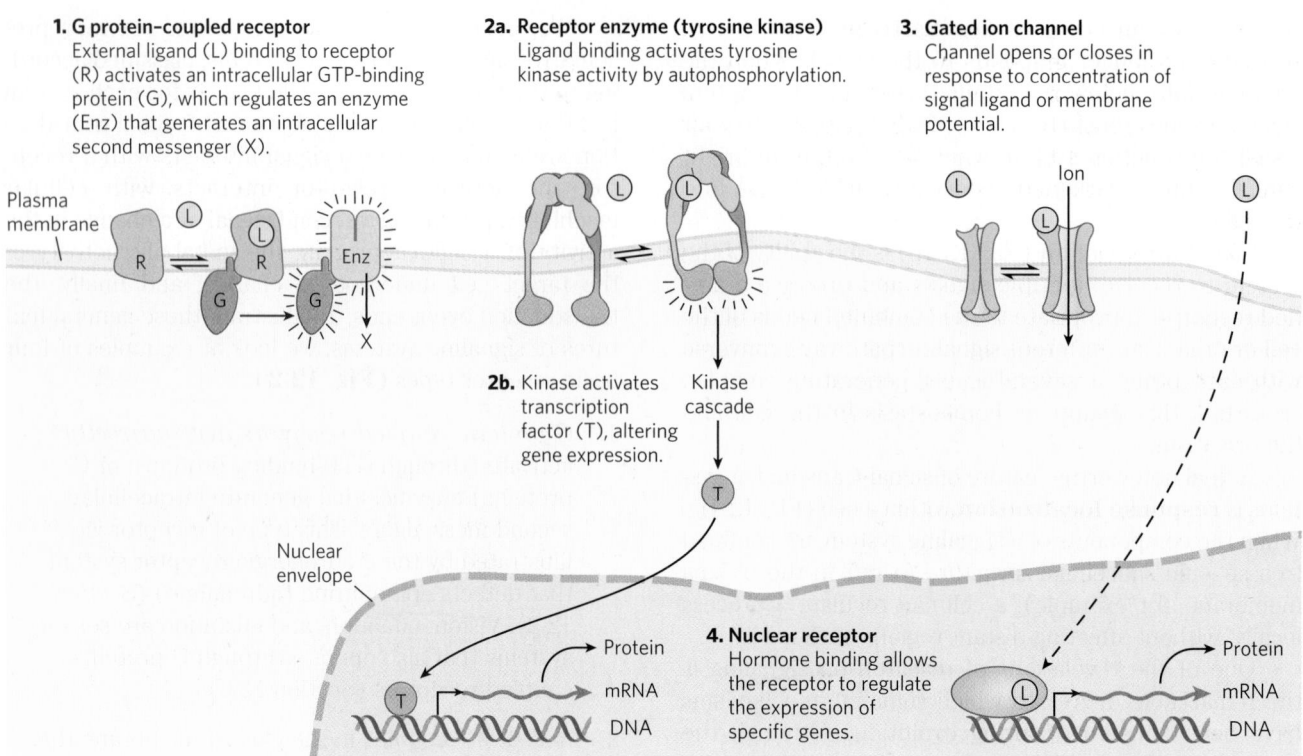

1. G protein–coupled receptor
External ligand (L) binding to receptor (R) activates an intracellular GTP-binding protein (G), which regulates an enzyme (Enz) that generates an intracellular second messenger (X).

2a. Receptor enzyme (tyrosine kinase)
Ligand binding activates tyrosine kinase activity by autophosphorylation.

3. Gated ion channel
Channel opens or closes in response to concentration of signal ligand or membrane potential.

2b. Kinase activates transcription factor (T), altering gene expression.

Kinase cascade

4. Nuclear receptor
Hormone binding allows the receptor to regulate the expression of specific genes.

FIGURE 12-2 Four general types of signal transducers.

the retina. Some genes and proteins are given noncommittal names: the tumor suppressor protein p53, for example, is a *p*rotein of *53* kDa, but its name gives no clue to its great importance in the regulation of cell division and the development of cancer. In this chapter we generally define these protein names as we encounter them, introducing the names commonly used by researchers in the field. Don't be discouraged if you can't get them all straight the first time you encounter them.

SUMMARY 12.1 General Features of Signal Transduction

■ All cells have specific and highly sensitive signal-transducing mechanisms, which have been conserved during evolution.

■ A wide variety of stimuli act through specific protein receptors in the plasma membrane.

■ The receptors bind the signal molecule and initiate a process that amplifies the signal, integrates it with input from other receptors, and transmits the information throughout the cell, or in some cases to a local region of the cell. If the signal persists, receptor desensitization reduces or ends the response.

■ Multicellular organisms have four general types of signaling mechanisms: plasma membrane proteins that act through G proteins, receptors with internal enzyme activity (such as tyrosine kinase), gated ion channels, and nuclear receptors that bind steroids and alter gene expression.

12.2 G Protein–Coupled Receptors and Second Messengers

As their name implies, **G protein–coupled receptors (GPCRs)** are receptors that act through a member of the **guanosine nucleotide–binding protein**, or **G protein**, family. Three essential components define signal transduction through GPCRs: a plasma membrane receptor with seven transmembrane helical segments, a G protein that cycles between active (GTP-bound) and inactive (GDP-bound) forms, and an effector enzyme (or ion channel) in the plasma membrane that is regulated by the activated G protein. An extracellular signal such as a hormone, growth factor, or neurotransmitter is the "first messenger" that activates a receptor from outside the cell. When the receptor is activated, its associated G protein exchanges its bound GDP for a GTP from the cytosol. The G protein then dissociates from the activated receptor and binds to the nearby effector enzyme, altering its activity. The effector enzyme then causes a change in the cytosolic concentration of a low molecular weight metabolite or inorganic ion, which acts as a **second messenger** to activate or inhibit one or more downstream targets, often protein kinases.

The human genome encodes just over 800 GPCRs, about 350 for detecting hormones, growth factors, and other endogenous ligands, and perhaps 500 that serve as olfactory (smell) and gustatory (taste) receptors. GPCRs have been implicated in many common human conditions, including allergies, depression, blindness, diabetes, and various cardiovascular defects, with serious health consequences. GPCR mutations are also found in 20% of all cancers. More than a third of *all* drugs on the market target one GPCR or another. For example, the β-adrenergic receptor, which mediates the effects of epinephrine, is the target of the "beta blockers," prescribed for such diverse conditions as hypertension, cardiac arrhythmia, glaucoma, anxiety, and migraine headache. More than 100 of the GPCRs found in the human genome are still "orphan receptors," meaning that their natural ligands are not yet identified, and so we know nothing about their biology. The β-adrenergic receptor, with well-understood biology and pharmacology, is the prototype for all GPCRs, and our discussion of signal-transducing systems begins there. ■

The β-Adrenergic Receptor System Acts through the Second Messenger cAMP

Epinephrine sounds the alarm when a threat requires the organism to mobilize its energy-generating machinery; it signals the need to fight or flee. Epinephrine action begins when the hormone binds to a protein receptor in the plasma membrane of an epinephrine-sensitive cell. **Adrenergic receptors** ("adrenergic" reflects the alternative name for epinephrine, adrenaline) are of four general types, α_1, α_2, β_1, and β_2, defined by differences in their affinities and responses to a group of agonists and antagonists. **Agonists** are molecules (natural ligands or their structural analogs) that bind to a receptor and produce the effects of the natural ligand; **antagonists** are analogs that bind the receptor without triggering the normal effect and thereby block the effects of agonists, including the natural ligand. In some cases, the affinity of a synthetic agonist or antagonist for the receptor is greater than that of the natural agonist **(Fig. 12-3)**. The four types of adrenergic receptors are found in different target tissues and mediate different responses to epinephrine. Here we focus on the β-**adrenergic receptors** of muscle, liver, and adipose tissue. These receptors mediate changes in fuel metabolism, as described in Chapter 23, including the increased breakdown of glycogen and fat. Adrenergic receptors of the β_1 and β_2 subtypes act through the same mechanism, so in our discussion, "β-adrenergic" applies to both types.

Like all GPCRs, the β-adrenergic receptor is an integral protein with seven hydrophobic, helical regions of 20 to 28 amino acid residues that span the plasma membrane seven times, thus the alternative name for GPCRs: **heptahelical receptors**. The binding of epinephrine to a site on the receptor deep within the plasma membrane **(Fig. 12-4a**, step ❶) promotes a conformational change in the receptor's intracellular domain that affects its interaction with an associated G protein, promoting the dissociation of GDP and

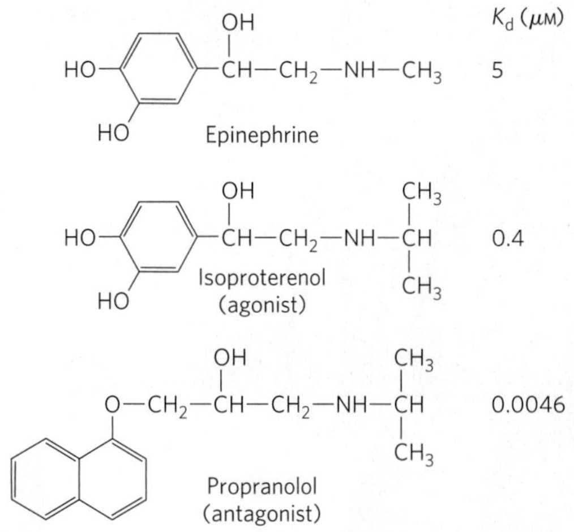

FIGURE 12-3 Epinephrine and its synthetic analogs. Epinephrine, also called adrenaline, is released from the adrenal gland and regulates energy-yielding metabolism in muscle, liver, and adipose tissue. It also serves as a neurotransmitter in adrenergic neurons. Its affinity for its receptor is expressed as a dissociation constant for the receptor-ligand complex. Isoproterenol and propranolol are synthetic analogs, one an agonist with an affinity for the receptor that is higher than that of epinephrine, and the other an antagonist with extremely high affinity.

binding of GTP from the cytosol (step ❷). For all GPCRs, the G protein is heterotrimeric, composed of three different subunits: α, β, and γ. These G proteins are therefore known as **trimeric G proteins**. In this case, it is the α subunit that binds GDP or GTP and transmits the signal from the activated receptor to the effector protein. Because this G protein activates its effector, it is referred to as a **stimulatory G protein**, or **G_s**. Like other G proteins (Box 12-1), G_s functions as a biological "switch": when the nucleotide-binding site of G_s (on the α subunit) is occupied by GTP, G_s is turned on and can activate its effector protein (adenylyl cyclase in the present case); with GDP bound to the site, G_s is switched off. In the active form, the β and γ subunits of G_s dissociate from the α subunit as a $\beta\gamma$ dimer, and $G_{s\alpha}$, with its bound GTP, moves in the plane of the membrane from the receptor to a nearby molecule of adenylyl cyclase (step ❸). $G_{s\alpha}$ is held to the membrane by a covalently attached palmitoyl group (see Fig. 11-13).

Adenylyl cyclase is an integral protein of the plasma membrane, with its active site on the cytoplasmic face. The association of active $G_{s\alpha}$ with adenylyl cyclase stimulates the cyclase to catalyze the synthesis of second messenger cAMP from ATP (Fig. 12-4a, step ❹; Fig. 12-4b), raising the cytosolic [cAMP]. The interaction between $G_{s\alpha}$ and adenylyl cyclase is possible only when $G_{s\alpha}$ is bound to GTP. The mammalian genome encodes nine isozymes of membrane-localized adenylyl cyclase, all with highly conserved sequences but, presumably, with discrete functions.

The stimulation by $G_{s\alpha}$ is self-limiting; $G_{s\alpha}$ *has intrinsic GTPase activity that inactivates $G_{s\alpha}$ by*

(a)

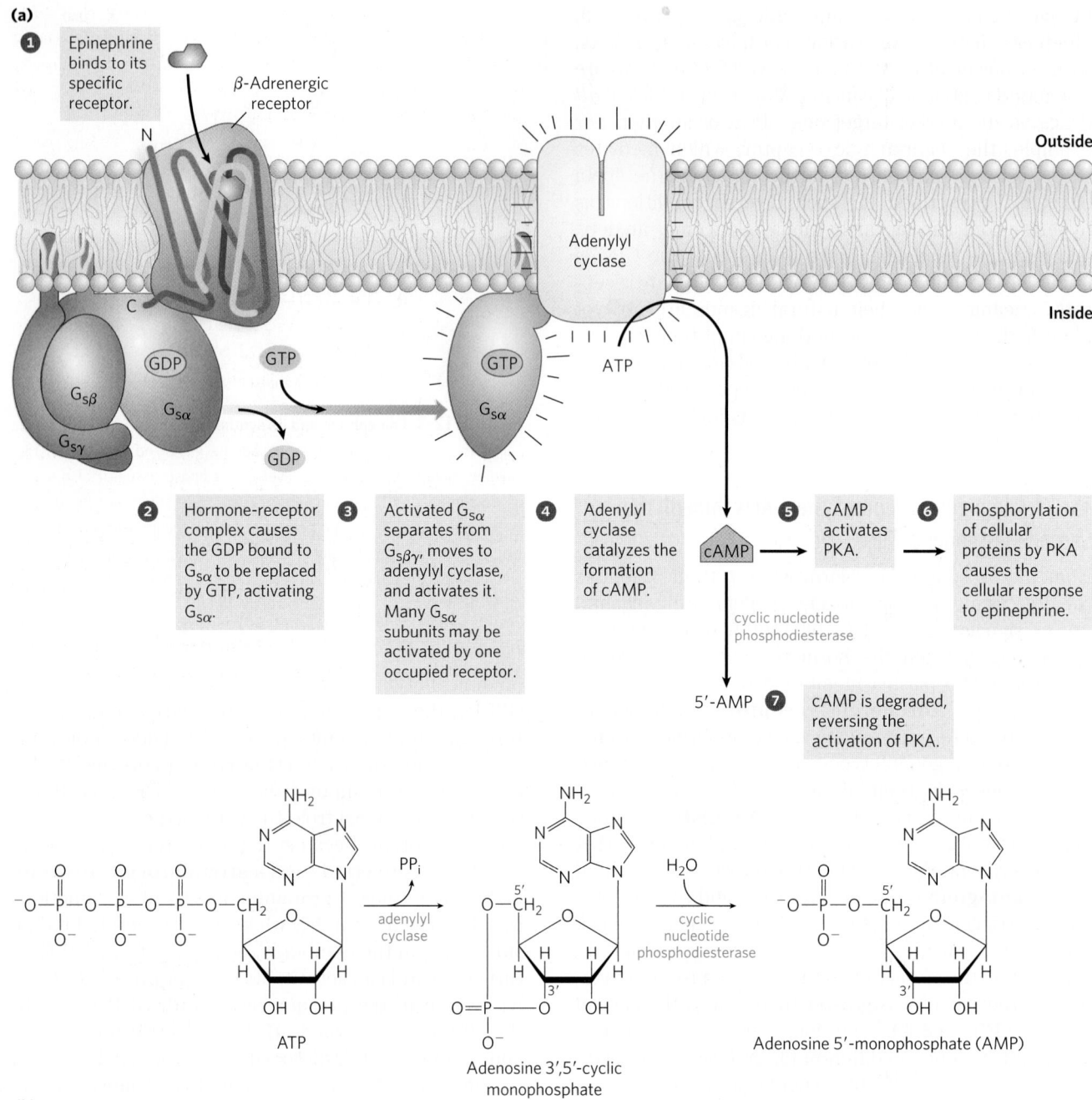

❶ Epinephrine binds to its specific receptor.

β-Adrenergic receptor

N

Outside

C

Inside

Adenylyl cyclase

GDP

GTP

$G_{s\beta}$

$G_{s\alpha}$

$G_{s\gamma}$

GDP

GTP

$G_{s\alpha}$

ATP

❷ Hormone-receptor complex causes the GDP bound to $G_{s\alpha}$ to be replaced by GTP, activating $G_{s\alpha}$.

❸ Activated $G_{s\alpha}$ separates from $G_{s\beta\gamma}$, moves to adenylyl cyclase, and activates it. Many $G_{s\alpha}$ subunits may be activated by one occupied receptor.

❹ Adenylyl cyclase catalyzes the formation of cAMP.

cAMP

❺ cAMP activates PKA.

❻ Phosphorylation of cellular proteins by PKA causes the cellular response to epinephrine.

cyclic nucleotide phosphodiesterase

5'-AMP

❼ cAMP is degraded, reversing the activation of PKA.

(b)

ATP → (PP$_i$, adenylyl cyclase) → Adenosine 3',5'-cyclic monophosphate (cAMP) → (H$_2$O, cyclic nucleotide phosphodiesterase) → Adenosine 5'-monophosphate (AMP)

FIGURE 12-4 Transduction of the epinephrine signal: the β-adrenergic pathway. **(a)** The mechanism that couples binding of epinephrine to its receptor with activation of adenylyl cyclase; the seven steps are discussed in the text. The same adenylyl cyclase molecule in the plasma membrane may be regulated by a stimulatory G protein (G_s), as shown, or by an inhibitory G protein (G_i, not shown). G_s and G_i are under the influence of different hormones. Hormones that induce GTP binding to G_i cause *inhibition* of adenylyl cyclase, resulting in lower cellular [cAMP]. **(b)** The combined action of the enzymes that catalyze steps ❹ and ❼, synthesis and hydrolysis of cAMP by adenylyl cyclase and cAMP phosphodiesterase, respectively.

converting its bound GTP to GDP **(Fig. 12-5)**. The now inactive $G_{s\alpha}$ dissociates from adenylyl cyclase, rendering the cyclase inactive. $G_{s\alpha}$ reassociates with the βγ dimer ($G_{s\beta\gamma}$), and inactive G_s is again available to interact with a hormone-bound receptor.

The role of $G_{s\alpha}$ in serving as a biological "switch" protein is not unique. A variety of G proteins act as binary switches in signaling systems with GPCRs and in many processes that involve membrane fusion or fission (Box 12-1).

Epinephrine exerts its downstream effects through the increase in [cAMP] that results from activation of adenylyl cyclase. Cyclic AMP, the second messenger, allosterically activates **cAMP-dependent protein kinase**, also called **protein kinase A** or **PKA** (Fig. 12-4a, step ❺), which catalyzes the phosphorylation of specific

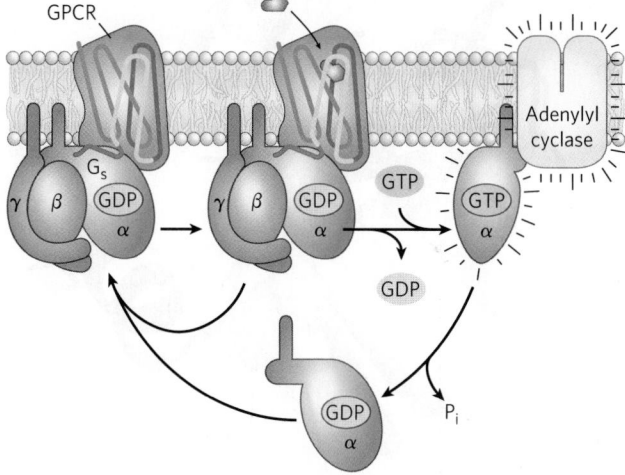

1 G_s with GDP bound is turned off; it cannot activate adenylyl cyclase.

2 Contact of G_s with hormone-receptor complex causes displacement of bound GDP by GTP.

3 G_s with GTP bound dissociates into α and $\beta\gamma$ subunits. $G_{s\alpha}$-GTP is turned on; it can activate adenylyl cyclase.

FIGURE 12-5 The GTPase switch. G proteins cycle between GDP-bound (off) and GTP-bound (on). The protein's intrinsic GTPase activity, in many cases stimulated by RGS proteins (regulators of G-protein signaling; see Box 12-1), determines how quickly bound GTP is hydrolyzed to GDP and thus how long the G protein remains active.

4 GTP bound to $G_{s\alpha}$ is hydrolyzed by the protein's intrinsic GTPase; $G_{s\alpha}$ thereby turns itself off. The inactive α subunit reassociates with the $\beta\gamma$ subunit.

Ser or Thr residues of targeted proteins, including glycogen phosphorylase b kinase. The latter enzyme is active when phosphorylated and can begin the process of mobilizing glycogen stores in muscle and liver in anticipation of the need for energy, as signaled by epinephrine.

The inactive form of PKA has two identical catalytic subunits (C) and two identical regulatory subunits (R) **(Fig. 12-6a)**. The tetrameric R_2C_2 complex is catalytically inactive, because an autoinhibitory domain of each R subunit occupies the substrate-binding cleft of each C subunit. Cyclic AMP is an allosteric activator of PKA. When cAMP binds to the R subunits, they undergo a conformational change that moves the autoinhibitory domain of R out of the catalytic domain of C, and the R_2C_2 complex dissociates to yield two free, catalytically active C subunits. This same basic mechanism—displacement of an autoinhibitory domain—mediates the allosteric activation of many types of protein kinases by their second messengers (as in Figs 12-18 and 12-25, for example).

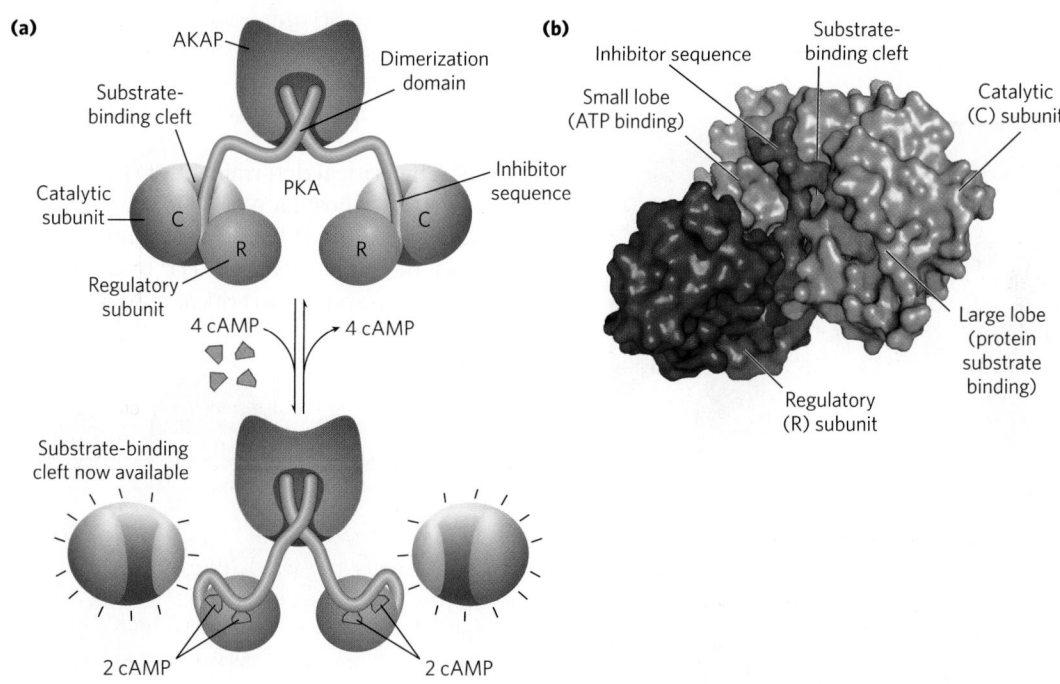

FIGURE 12-6 Activation of cAMP-dependent protein kinase (PKA).
(a) When [cAMP] is low, the two identical regulatory subunits (R; red) associate with the two identical catalytic subunits (C). In this R_2C_2 complex, the inhibitor sequences of the R subunits lie in the substrate-binding cleft of the C subunits and prevent binding of protein substrates; the complex is therefore catalytically inactive. The amino-terminal sequences of the R subunits interact to form an R_2 dimer, the site of binding to an A kinase anchoring protein (AKAP), described later in the text. When [cAMP] rises in response to a hormonal signal, each R subunit binds two cAMP molecules and undergoes a dramatic reorganization that pulls its inhibitory

sequence away from the C subunit, opening up the substrate-binding cleft and releasing each C subunit in its catalytically active form. **(b)** A crystal structure showing part of the R_2C_2 complex—one C subunit and part of one R subunit. The amino-terminal dimerization region of the R subunit is omitted for simplicity. The small lobe of C contains the ATP-binding site, and the large lobe surrounds and defines the cleft where the protein substrate binds and undergoes phosphorylation at a Ser or Thr residue, with a phosphoryl group transferred from ATP. In this inactive form, the inhibitor sequence of R blocks the substrate-binding cleft of C, inactivating it. [Source: (b) PDB ID 3FHI, C. Kim et al., *Science* 307:690, 2005.]

BOX 12-1   G Proteins: Binary Switches in Health and Disease

Alfred G. Gilman and Martin Rodbell discovered the critical roles of guanosine nucleotide–binding proteins (G proteins) in a wide variety of cellular processes, including sensory perception, signaling for cell division, growth and differentiation, intracellular movements of proteins and membrane vesicles, and protein synthesis. The human genome encodes nearly 200 of these proteins, which differ in size and subunit structure, intracellular location, and function. But all G proteins share a common feature: they can become activated and then, after a brief period, can inactivate themselves, thereby serving as molecular binary switches with built-in timers. This superfamily of proteins includes the trimeric G proteins involved in adrenergic signaling (G_s and G_i) and vision (transducin); small G proteins such as that involved in insulin signaling (Ras) and others that function in vesicle trafficking (ARF and Rab), transport into and out of the nucleus (Ran; see Fig. 27-44), and timing of the cell cycle (Rho); and several proteins involved in protein synthesis (initiation factor IF2 and elongation factors EF-Tu and EF-G; see Chapter 27). Many G proteins have covalently bound lipids, which give them an affinity for membranes and dictate their locations in the cell.

Alfred G. Gilman, 1941–2015 [Source: Shelly Katz/Liaison Agency/Getty Images.]

Martin Rodbell, 1925–1998 [Source: Courtesy Andrew M. Rodbell.]

All G proteins have the same core structure and use the same mechanism for switching between an inactive conformation, favored when GDP is bound, and an active conformation, favored when GTP is bound. We can use the Ras protein (~20 kDa), a minimal signaling unit, as a prototype for all members of this superfamily (Fig. 1).

In the GTP-bound conformation, the G protein exposes previously buried regions (called **switch I** and **switch II**) that interact with proteins downstream in the signaling pathway, until the G protein inactivates itself by hydrolyzing its bound GTP to

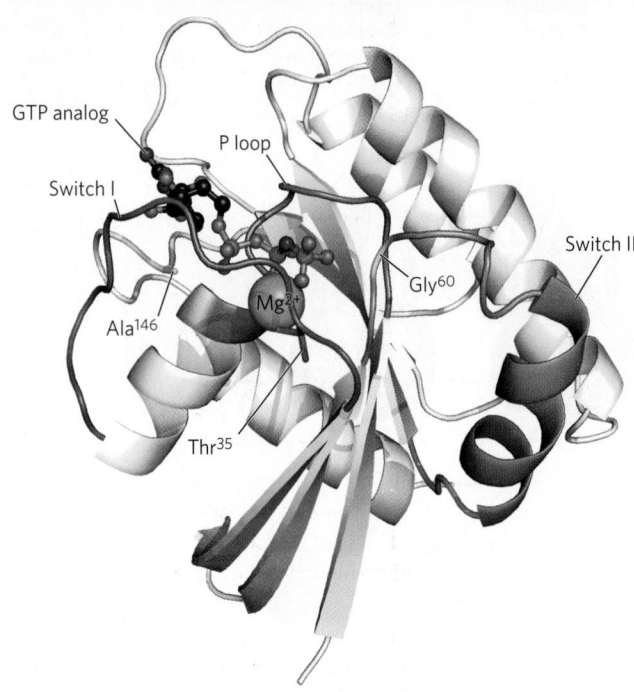

FIGURE 1 The Ras protein, the prototype for all G proteins. Mg^{2+}-GTP is held by critical residues in the phosphate-binding P loop (blue) and by Thr^{35} in the switch I (red) and Gly^{60} in the switch II (green) regions. Ala^{146} gives specificity for GTP over ATP. In the structure shown here, the nonhydrolyzable GTP analog Gpp(NH)p is in the GTP-binding site. [Source: PDB ID 5P21, E. F. Pai et al., *EMBO J.* 9:2351, 1990.]

GDP. The critical determinant of G-protein conformation is the γ phosphate of GTP, which interacts with a region called the **P loop** (*p*hosphate-binding; Fig. 2). In Ras, the γ phosphate of GTP binds to a Lys residue in the P loop and to two critical residues, Thr^{35} in switch

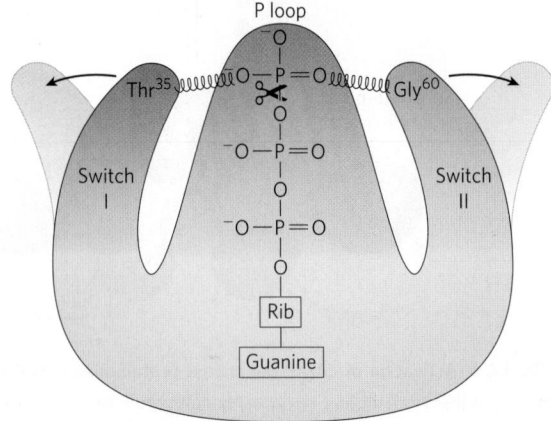

FIGURE 2 When bound GTP is hydrolyzed by the GTPase activities of Ras and its GTPase activator protein (GAP), loss of hydrogen bonds to Thr^{35} and Gly^{60} allows the switch I and switch II regions to relax into a conformation in which they are no longer available to interact with downstream targets. [Source: Information from I. R. Vetter and A. Wittinghofer, *Science* 294:1299, 2001, Fig. 3.]

I and Gly[60] in switch II, that hydrogen-bond with the oxygens of the γ phosphate of GTP. These hydrogen bonds act like a pair of springs holding the protein in its active conformation. When GTP is cleaved to GDP and P_i is released, these hydrogen bonds are lost; the protein then relaxes into its inactive conformation, burying the sites that, in its active state, interact with other partners. Ala[146] hydrogen-bonds to the guanine oxygen, allowing GTP, but not ATP, to bind.

The intrinsic GTPase activity of most G proteins is very weak, but is increased up to 10^5-fold by **GTPase activator proteins (GAPs)**, also called, in the case of heterotrimeric G proteins, **regulators of G protein signaling (RGSs**; Fig. 3). GAPs (and RGSs) thus determine how long the switch remains on. They contribute a critical Arg residue that reaches into the G-protein GTPase active site and assists in catalysis. The intrinsically slow process of replacing bound GDP with GTP, switching the protein on, is catalyzed by **guanosine nucleotide–exchange factors (GEFs)** associated with the G protein (Fig. 3). The ligand-bound β-adrenergic receptor is one of many GEFs, and a broad range of proteins act as GAPs. Their combined effects set the level of GTP-bound G proteins, and thus the strength of the response to signals that arrive at the receptors.

Because G proteins play crucial roles in so many signaling processes, it is not surprising that defects in G proteins lead to a variety of diseases. In about 25% of all human cancers (and in a much higher proportion of certain types of cancer), there is a mutation in a Ras protein—typically in one of the critical residues around the GTP-binding site or in the P loop—that virtually eliminates its GTPase activity. Once activated by GTP binding, this Ras protein remains constitutively active, promoting cell division in cells that should not divide. The tumor suppressor gene *NF1* encodes a GAP that enhances the GTPase activity of normal Ras. Mutations in *NF1* that result in a nonfunctioning GAP leave Ras with only its intrinsic GTPase activity, which is very weak (that is, has a very low turnover number); once activated by GTP binding, Ras stays active for an extended period, continuing to send the signal: divide.

Defective heterotrimeric G proteins can also lead to disease. Mutations in the gene that encodes the α subunit of G_s (which mediates changes in [cAMP] in response to hormonal stimuli) may result in a $G_α$ that is permanently active or permanently inactive. "Activating" mutations generally occur in residues crucial to GTPase activity; they lead to a continuously elevated [cAMP], with significant downstream consequences, including undesirable cell proliferation. For example, such mutations are found in about 40% of pituitary tumors (adenomas). Individuals with "inactivating" mutations in $G_α$ are unresponsive to hormones (such as thyroid hormone) that act through cAMP. Mutation in the gene for the transducin α subunit ($T_α$), which is involved in visual signaling, leads to a type of night blindness, apparently due to defective interaction between the activated $T_α$ subunit and the phosphodiesterase of the rod outer segment (see Fig. 12-14). A sequence variation in the gene encoding the β subunit of a heterotrimeric G protein is commonly found in individuals with hypertension (high blood pressure), and this variant gene is suspected of involvement in obesity and atherosclerosis.

The pathogenic bacterium that causes cholera produces a toxin that targets a G protein, interfering with normal signaling in host cells. **Cholera toxin**, secreted by *Vibrio cholerae* in the intestine of an infected person, is a heterodimeric protein. Subunit B recognizes and binds to specific gangliosides on the surface of intestinal epithelial cells and provides a route for subunit A to enter these cells. After entry, subunit A is broken into two fragments, A1 and A2. A1 associates with the host cell's

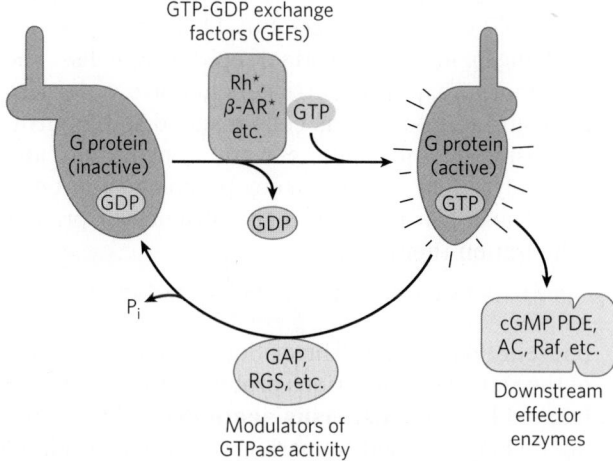

FIGURE 3 Many factors regulate the activity of G proteins (green). Inactive G proteins, both small G proteins such as Ras and heterotrimeric G proteins such as G_s, interact with upstream GTP-GDP exchange factors (red). Often these exchange factors are activated (*) receptors such as rhodopsin (Rh) and β-adrenergic receptors (AR). The G proteins are activated by GTP binding, and in the GTP-bound form, activate downstream effector enzymes (blue), such as cGMP phosphodiesterase (PDE), adenylyl cyclase (AC), and Raf. GTPase activator proteins (GAPs, in the case of small G proteins) and regulators of G protein signaling (RGSs) (yellow), by modulating the GTPase activity of G proteins, determine how long the G protein will remain active.

(Continued on next page)

BOX 12-1 G Proteins: Binary Switches in Health and Disease (*Continued*)

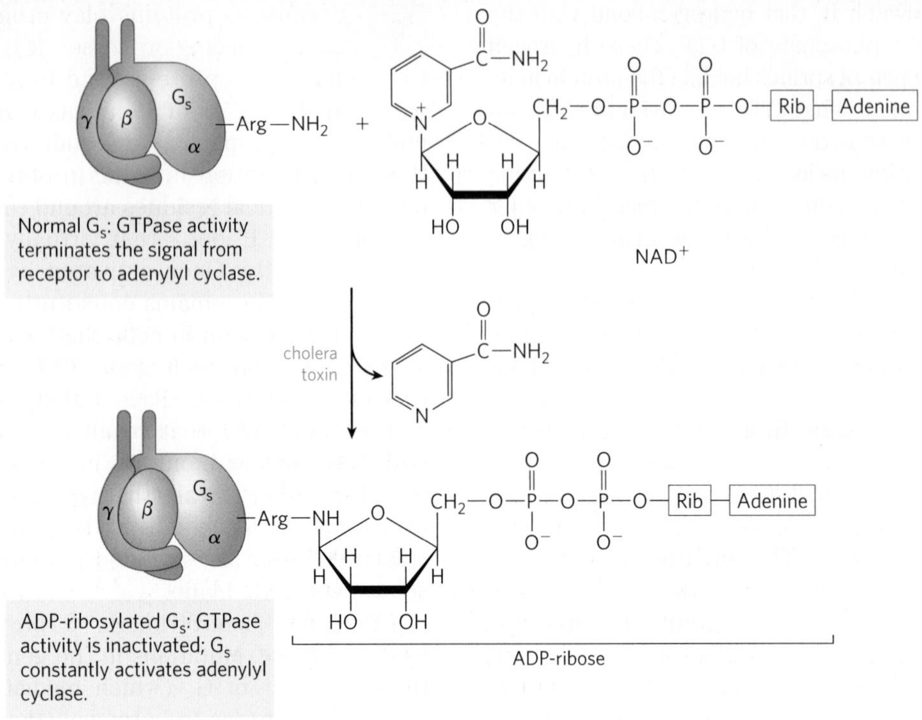

Normal G_s: GTPase activity terminates the signal from receptor to adenylyl cyclase.

NAD^+

cholera toxin

ADP-ribosylated G_s: GTPase activity is inactivated; G_s constantly activates adenylyl cyclase.

ADP-ribose

FIGURE 4 The bacterial toxin that causes cholera is an enzyme that catalyzes transfer of the ADP-ribose moiety of NAD^+ to an Arg residue of G_s. The G proteins thus modified fail to respond to normal hormonal stimuli. The pathology of cholera results from defective regulation of adenylyl cyclase and overproduction of cAMP.

ADP-ribosylation factor ARF6, a small G protein, through residues in its switch I and switch II regions—which are accessible only when ARF6 is in its active (GTP-bound) form. This association with ARF6 activates A1, which catalyzes the transfer of ADP-ribose from NAD^+ to the critical Arg residue in the P loop of the α subunit of G_s (Fig. 4). ADP-ribosylation blocks the GTPase activity of G_s and thereby renders G_s permanently active. This results in continuous activation of the adenylyl cyclase of intestinal epithelial cells, chronically high [cAMP], and chronically active PKA. PKA phosphorylates the CFTR Cl^- channel (see Box 11-2) and a Na^+-H^+ exchanger in the intestinal epithelial cells. The resultant efflux of NaCl triggers massive water loss through the intestine as cells respond to the ensuing osmotic imbalance. Severe dehydration and electrolyte loss are the major pathologies in cholera. These can be fatal in the absence of prompt rehydration therapy. ■

The structure of the substrate-binding cleft in PKA is the prototype for all known protein kinases (Fig. 12-6b); certain residues in this cleft region have identical counterparts in all of the 544 protein kinases encoded in the human genome. The ATP-binding site of each catalytic subunit positions ATP perfectly for the transfer of its terminal (γ) phosphoryl group to the —OH in the side chain of a Ser or Thr residue in the target protein.

As indicated in Figure 12-4a (step ❻), PKA regulates many enzymes downstream in the signaling pathway. Although these downstream targets have diverse functions, they share a region of sequence similarity around the Ser or Thr residue that undergoes phosphorylation, a sequence that marks them for regulation by PKA (Table 12-2). The substrate-binding cleft of PKA recognizes these sequences and phosphorylates their Thr or Ser residue. Comparison of the sequences of various protein substrates for PKA has yielded the **consensus sequence**—the neighboring residues needed to mark a Ser or Thr residue for phosphorylation.

As in many signaling pathways, signal transduction by adenylyl cyclase entails several steps that *amplify* the original hormone signal **(Fig. 12-7)**. First, the binding of one hormone molecule to one receptor molecule catalytically activates many G_s molecules that associate with the activated receptor, one after the other. Next, by activating one molecule of adenylyl cyclase, each active $G_{s\alpha}$ molecule stimulates the catalytic synthesis of *many* molecules of cAMP. The second messenger cAMP now activates PKA, each molecule of which catalyzes the phosphorylation of *many* molecules of the target protein—phosphorylase *b* kinase in Figure 12-7. This

TABLE 12-2 Some Enzymes and Other Proteins Regulated by cAMP-Dependent Phosphorylation (by PKA)

Enzyme/protein	Sequence phosphorylated[a]	Pathway/process regulated
Glycogen synthase	RASCTSSS	Glycogen synthesis

Phosphorylase *b* kinase
  α subunit
  β subunit
	VEFRRLSI
	RTKRSGSV
		Glycogen breakdown

Pyruvate kinase (rat liver)	GVLRRASVAZL	Glycolysis
Pyruvate dehydrogenase complex (type L)	GYLRRASV	Pyruvate to acetyl-CoA
Hormone-sensitive lipase	PMRRSV	Triacylglycerol mobilization and fatty acid oxidation
Phosphofructokinase-2/fructose 2,6-bisphosphatase	LQRRRGSSIPQ	Glycolysis/gluconeogenesis
Tyrosine hydroxylase	FIGRRQSL	Synthesis of L-dopa, dopamine, norepinephrine, and epinephrine
Histone H1	AKRKASGPPVS	DNA condensation
Histone H2B	KKAKASRKESYSVYVYK	DNA condensation
Cardiac phospholamban (cardiac pump regulator)	AIRRAST	Intracellular $[Ca^{2+}]$
Protein phosphatase-1 inhibitor-1	IRRRRPTP	Protein dephosphorylation
PKA consensus sequence[b]	xR[RK]x[ST]B	Many

[a]The phosphorylated S or T residue is shown in red. All residues are given as their one-letter abbreviations (See Table 3-1).

[b]x is any amino acid; B is any hydrophobic amino acid. See Box 3-2 for conventions used in displaying consensus sequences.

kinase activates glycogen phosphorylase *b*, which leads to the rapid mobilization of glucose from glycogen. The net effect of the cascade is amplification of the hormonal signal by several orders of magnitude, which accounts for the very low concentration of epinephrine (or any other hormone) required for hormone activity. This signaling pathway is also rapid: the signal leads to intracellular changes within milliseconds or even microseconds.

Several Mechanisms Cause Termination of the β-Adrenergic Response

To be useful, a signal-transducing system has to *turn off* after the hormonal or other stimulus has ended, and mechanisms for shutting off the signal are intrinsic to all signaling systems. Most systems also adapt to the continued presence of the signal by becoming less sensitive to it, by *desensitizing*. The β-adrenergic system illustrates both. Here, our focus is on termination.

The response to β-adrenergic stimulation will end when the concentration of epinephrine in the blood

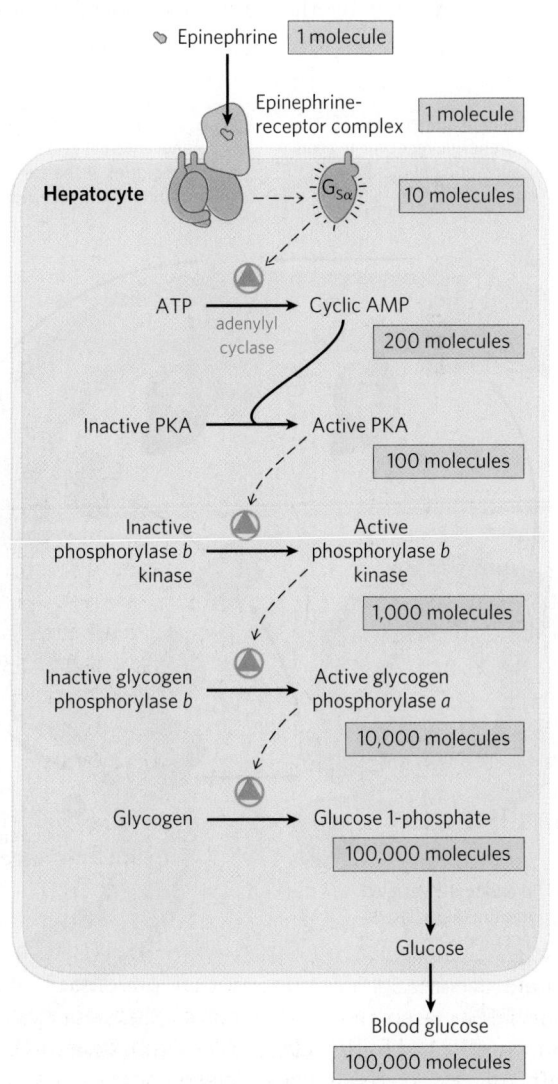

FIGURE 12-7 Epinephrine cascade. Epinephrine triggers a series of reactions in hepatocytes in which catalysts activate catalysts, resulting in great amplification of the original hormone signal. The numbers of molecules shown are simply to illustrate amplification and are almost certainly gross underestimates. Binding of one molecule of epinephrine to one β-adrenergic receptor on the cell surface activates many (possibly hundreds of) G proteins, one after another, each of which goes on to activate a molecule of the enzyme adenylyl cyclase. Adenylyl cyclase acts catalytically, producing many molecules of cAMP for each activated adenylyl cyclase. (Because two molecules of cAMP are required to activate one PKA catalytic subunit, this step does not amplify the signal.)

drops below the K_d for its receptor. The hormone then dissociates from the receptor, and the latter reassumes its inactive conformation, in which it can no longer activate G_s.

A second means of ending the response is the hydrolysis of GTP bound to the G_α subunit, catalyzed by the intrinsic GTPase activity of the G protein. Conversion of bound GTP to GDP favors the return of G_α to the conformation in which it binds the $G_{\beta\gamma}$ subunits—the conformation in which the G protein is unable to interact with or stimulate adenylyl cyclase. This ends the production of cAMP. The rate of inactivation of G_s depends on the GTPase activity, which for G_α alone is very feeble. However, GTPase activator proteins (GAPs) strongly stimulate this GTPase activity, causing more rapid inactivation of the G protein (see Box 12-1). GAPs can themselves be regulated by other factors, providing a fine-tuning of the response to β-adrenergic stimulation. A third mechanism for terminating the response is to remove the second messenger: cAMP is hydrolyzed to 5′-AMP (not active as a second messenger) by **cyclic nucleotide phosphodiesterase** (Fig. 12-4a, step ❼; 12-4b).

Finally, at the end of the signaling pathway, the metabolic effects that result from enzyme phosphorylation are reversed by the action of phosphoprotein phosphatases, which hydrolyze phosphorylated Ser, Thr, or Tyr residues, releasing inorganic phosphate (P_i). About 150 genes in the human genome encode phosphoprotein phosphatases, fewer than the number (544) encoding protein kinases, reflecting the relative promiscuity of the phosphoprotein phosphatase. A single phosphoprotein phosphatase (PP1) dephosphorylates some 200 different phosphoprotein targets. Some phosphatases are known to be regulated; others may act constitutively. When [cAMP] drops and PKA returns to its inactive form (step ❼ in Fig. 12-4a), the balance between phosphorylation and dephosphorylation is tipped toward dephosphorylation by these phosphatases.

The β-Adrenergic Receptor Is Desensitized by Phosphorylation and by Association with Arrestin

The mechanisms for signal termination described above take effect when the stimulus ends. A different mechanism, desensitization, damps the response *even while the signal persists*. Desensitization of the β-adrenergic receptor is mediated by a protein kinase that phosphorylates the receptor on the intracellular domain that normally interacts with G_s **(Fig. 12-8)**. When the receptor remains occupied

❶ Binding of epinephrine (E) to β-adrenergic receptor triggers dissociation of $G_{s\beta\gamma}$ from $G_{s\alpha}$ (not shown).

❷ $G_{s\beta\gamma}$ recruits βARK to the membrane, where it phosphorylates Ser residues at the carboxyl terminus of the receptor.

FIGURE 12-8 Desensitization of the β-adrenergic receptor in the continued presence of epinephrine. This process is mediated by two proteins: β-adrenergic protein kinase (βARK) and β-arrestin (βarr; also known as arrestin 2). Not shown here is the phosphorylation and activation of βARK by PKA. PKA is activated by the rise in [cAMP] in response to the initial signal, epinephrine.

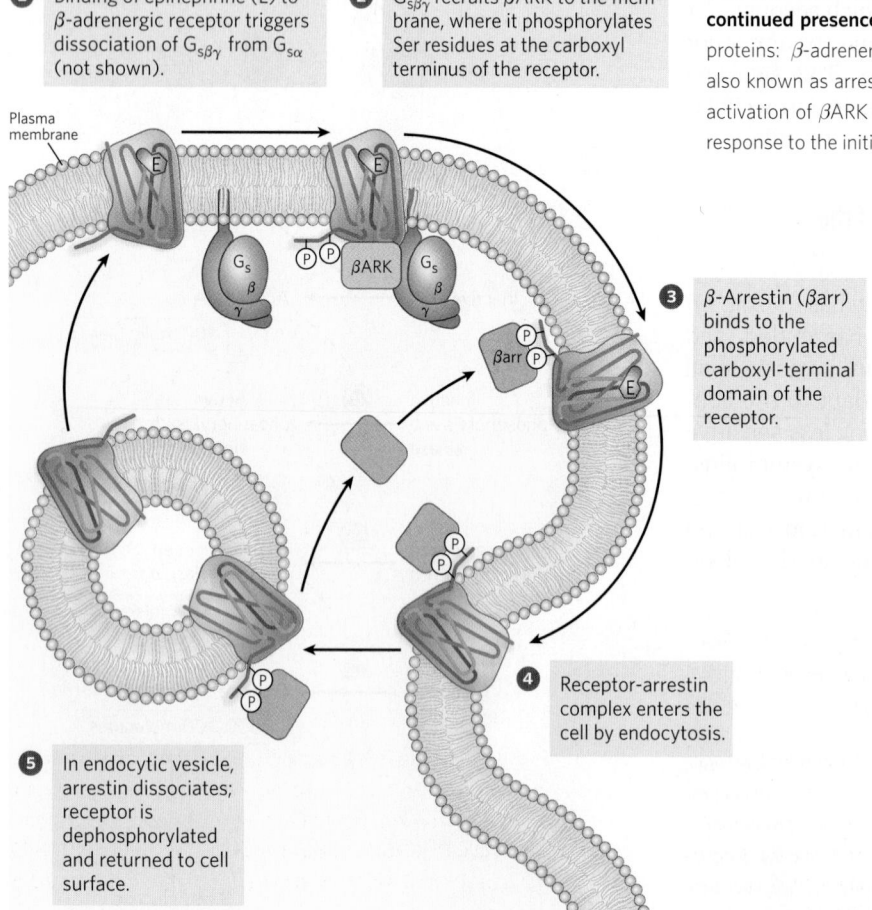

❸ β-Arrestin (βarr) binds to the phosphorylated carboxyl-terminal domain of the receptor.

❹ Receptor-arrestin complex enters the cell by endocytosis.

❺ In endocytic vesicle, arrestin dissociates; receptor is dephosphorylated and returned to cell surface.

with epinephrine, **β-adrenergic receptor kinase**, or **βARK** (also commonly called **GRK2**; see below), phosphorylates several Ser residues near the receptor's carboxyl terminus, which is on the cytoplasmic side of the plasma membrane. PKA, activated by the rise in [cAMP], phosphorylates, and thereby activates, βARK. βARK is then drawn to the plasma membrane by its association with the $G_{s\beta\gamma}$ subunits and is thus positioned to phosphorylate the receptor. Receptor phosphorylation creates a binding site for the protein **β-arrestin**, or **βarr** (also called arrestin 2), and binding of β-arrestin blocks the sites in the receptor that interact with the G protein **(Fig. 12-9)**. The binding of β-arrestin also facilitates the sequestration of receptor molecules, their removal from the plasma membrane by endocytosis into small intracellular vesicles (endosomes). The arrestin-receptor complex recruits clathrin and other proteins involved in vesicle formation (see Fig. 27-27), which initiate membrane invagination, leading to the formation of endosomes containing the adrenergic receptor. In this state, the receptors are inaccessible to epinephrine and therefore inactive. These receptor molecules are eventually dephosphorylated and returned to the plasma membrane, completing the circuit and resensitizing the system to epinephrine. β-Adrenergic receptor kinase is a member of a family of **G protein–coupled receptor kinases (GRKs)**, all of which phosphorylate GPCRs on their carboxyl-terminal cytoplasmic domains and play roles similar to that of βARK in desensitization and resensitization of their receptors. At least five different GRKs and four different arrestins are encoded in the human genome; each GRK is capable of desensitizing a particular subset of GPCRs, and each arrestin can interact with many different types of phosphorylated receptors.

The receptor-arrestin complex has another important role: it initiates signaling by a different pathway, the MAPK cascade described below. Thus, acting through a single GPCR, epinephrine triggers two distinct signaling pathways. The two pathways, one triggered by the receptor's interaction with a G protein and the other by its interaction with arrestin, can be differentially affected by the agonist; in some cases, one agonist favors the G-protein pathway and another favors the arrestin pathway. This bias is an important consideration in the development of a medication that acts through a GPCR. For example, the most addictive of the opioid drugs of abuse act more strongly through G-protein signaling than through arrestin. An ideal opioid pain medication would act through the branch of the pathway that has therapeutic effects and not through the pathway that leads to addiction. ■

Cyclic AMP Acts as a Second Messenger for Many Regulatory Molecules

Epinephrine is just one of many hormones, growth factors, and other regulatory molecules that act by changing the intracellular [cAMP] and thus the activity of PKA (Table 12-3). For example, glucagon binds

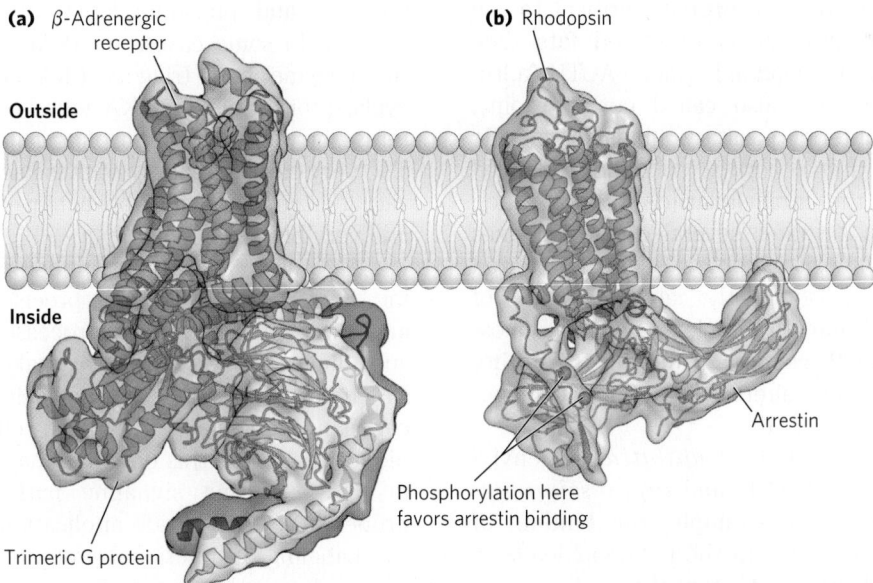

(a) β-Adrenergic receptor   **(b)** Rhodopsin

Outside

Inside

Arrestin

Phosphorylation here favors arrestin binding

Trimeric G protein

FIGURE 12-9 Mutual exclusion of trimeric G protein and arrestin in their interaction with a GPCR. (a) The complex of the β-adrenergic receptor with its trimeric G protein, G_s. **(b)** The complex of β-arrestin with the β-adrenergic receptor has not yet been solved, but the complex with another closely similar GCPR, visual rhodopsin, has, as shown here. (Rhodopsin is discussed later in the chapter.) Comparison of the two structures makes it clear that the binding of arrestin blocks the binding of the G protein and so prevents further activation of G proteins, effectively ending the response to the initial signal (epinephrine). [Sources: (a) PDB ID 3SN6, S. G. F. Rasmussen et al., *Nature* 477:549, 2011, Fig. 2c. (b) PDB ID 4ZWJ, Y. Kang et al., *Nature* 523:561, 2015, Fig. 2b.]

TABLE 12-3	Some Signals That Use cAMP as Second Messenger

Corticotropin (ACTH)
Corticotropin-releasing hormone (CRH)
Dopamine [D_1, D_2]
Epinephrine (β-adrenergic)
Follicle-stimulating hormone (FSH)
Glucagon
Histamine [H_2]
Luteinizing hormone (LH)
Melanocyte-stimulating hormone (MSH)
Odorants (many)
Parathyroid hormone
Prostaglandins E_1, E_2 (PGE_1, PGE_2)
Serotonin [5-HT_1, 5-HT_4]
Somatostatin
Tastants (sweet, bitter)
Thyroid-stimulating hormone (TSH)

Note: Receptor subtypes in square brackets. Subtypes may have different transduction mechanisms. For example, serotonin is detected in some tissues by receptor subtypes 5-HT_1 and 5-HT_4, which act through adenylyl cyclase and cAMP, and in other tissues by receptor subtypes 5-HT_2, acting through the phospholipase C-IP_3 mechanism (see Table 12-4).

to its receptors in the plasma membrane of adipocytes, activating (via a G_s protein) adenylyl cyclase. PKA, stimulated by the resulting rise in [cAMP], phosphorylates and activates two proteins critical to the mobilization of the fatty acids of stored fats (see Fig. 17-3). Similarly, the peptide hormone ACTH (adrenocorticotropic hormone, also called corticotropin), produced by the anterior pituitary, binds to specific receptors in the adrenal cortex, activating adenylyl cyclase and raising the intracellular [cAMP]. PKA then phosphorylates and activates several of the enzymes required for the synthesis of cortisol and other steroid hormones. In many cell types, the catalytic subunit of PKA can also move into the nucleus, where it phosphorylates the **cAMP response element binding protein (CREB)**, which alters the expression of specific genes regulated by cAMP.

Some hormones act by *inhibiting* adenylyl cyclase, thus *lowering* [cAMP] and *suppressing* protein phosphorylation. For example, the binding of somatostatin to its receptor in the pancreas leads to activation of an **inhibitory G protein**, or G_i, structurally homologous to G_s, that inhibits adenylyl cyclase and lowers [cAMP]. In this way, somatostatin inhibits the secretion of several hormones, including glucagon. In adipose tissue, prostaglandin E_2 (PGE_2; see Fig. 10-17) inhibits adenylyl cyclase, thus lowering [cAMP]

and slowing the mobilization of lipid reserves triggered by epinephrine and glucagon. In certain other tissues, PGE_2 stimulates cAMP synthesis: its receptors are coupled to adenylyl cyclase through a stimulatory G protein, G_s. In tissues with α_2-adrenergic receptors, epinephrine lowers [cAMP]; in this case, the receptors are coupled to adenylyl cyclase through an inhibitory G protein, G_i. In short, an extracellular signal such as epinephrine or PGE_2 can have different effects on different tissues or cell types, depending on three factors: the type of receptor in the tissue, the type of G protein (G_s or G_i) with which the receptor is coupled, and the set of PKA target enzymes in the cell. By summing the influences that tend to increase and decrease [cAMP], a cell achieves the integration of signals that is a general feature of signal-transducing mechanisms (Fig. 12-1e).

Another factor that explains how so many types of signals can be mediated by a single second messenger (cAMP) is the confinement of the signaling process to a specific region of the cell by **adaptor proteins**—noncatalytic proteins that hold together other protein molecules that function in concert (further described below). **AKAPs (A kinase anchoring proteins)** have multiple distinct protein-binding domains; they are multivalent adaptor proteins. One domain binds to the R subunits of PKA (see Fig. 12-6a) and another binds to a specific structure in the cell, confining the PKA to the vicinity of that structure. For example, specific AKAPs bind PKA to microtubules, actin filaments, ion channels, mitochondria, or the nucleus. Different types of cells have different complements of AKAPs, so cAMP might stimulate phosphorylation of mitochondrial proteins in one cell and phosphorylation of actin filaments in another. In some cases, an AKAP connects PKA with the enzyme that triggers PKA activation (adenylyl cyclase) or terminates PKA action (cAMP phosphodiesterase or phosphoprotein phosphatase) **(Fig. 12-10)**. The very close proximity of these activating and inactivating enzymes presumably achieves a highly localized, and very brief, response.

As is now clear, to fully understand cellular signaling, researchers need tools precise enough to detect and study where signaling processes take place at the subcellular level and when they take place in real time. In studies of the intracellular localization of biochemical changes, biochemistry meets cell biology, and techniques that cross this boundary have become essential in understanding signaling pathways. Fluorescent probes have found wide application in signaling studies. Labeling of functional proteins with a fluorescent tag such as the green fluorescent protein (GFP) reveals their location within the cell (see Fig. 9-16). Changes in the state of association of two proteins (such as the R and C subunits of PKA) can be seen by measuring the nonradiative transfer of energy between fluorescent probes attached to each protein, a

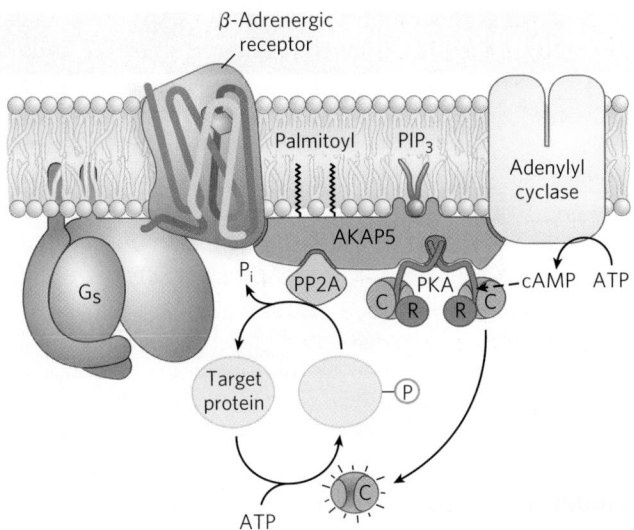

FIGURE 12-10 Nucleation of supramolecular complexes by A kinase anchoring proteins (AKAPs). AKAP5 is one of a family of proteins that act as multivalent scaffolds, holding PKA catalytic subunits—through interaction of the AKAP with the PKA regulatory subunits—in proximity to a particular region or structure in the cell. AKAP5 is targeted to rafts in the cytoplasmic face of the plasma membrane by two covalently attached palmitoyl groups and a site that binds phosphatidylinositol 3,4,5-trisphosphate (PIP_3) in the membrane. AKAP5 also has binding sites for the β-adrenergic receptor, adenylyl cyclase, PKA, and a phosphoprotein phosphatase (PP2A), bringing them all together in the plane of the membrane. When epinephrine binds to the β-adrenergic receptor, adenylyl cyclase produces cAMP, which reaches the nearby PKA quickly and with very little dilution. PKA phosphorylates its target protein, altering its activity, until the phosphoprotein phosphatase removes the phosphoryl group and returns the target protein to its prestimulus state. The AKAPs in this and other cases bring about a high local concentration of enzymes and second messengers, so that the signaling circuit remains highly localized and the duration of the signal is limited.

technique called fluorescence resonance energy transfer (FRET; Box 12-2).

Diacylglycerol, Inositol Trisphosphate, and Ca^{2+} Have Related Roles as Second Messengers

A second broad class of GPCRs are coupled through a G protein to a plasma membrane **phospholipase C (PLC)** that catalyzes cleavage of the membrane phospholipid phosphatidylinositol 4,5-bisphosphate, or PIP_2 (see Fig. 10-15). When one of the hormones that acts by this mechanism (Table 12-4) binds its specific receptor in the plasma membrane (**Fig. 12-11**, step ❶), the receptor-hormone complex catalyzes GTP-GDP exchange on an associated G protein, $\mathbf{G_q}$ (step ❷), activating it in much the same way that the β-adrenergic receptor activates G_s (Fig. 12-4). The activated G_q activates the PIP_2-specific PLC (Fig. 12-11, step ❸), which catalyzes the production of two potent second messengers (step ❹), **diacylglycerol** and **inositol 1,4,5-trisphosphate**, or IP_3 (not to be confused with PIP_3, p. 463).

Inositol 1,4,5-trisphosphate (IP_3)

Inositol trisphosphate, a water-soluble compound, diffuses from the plasma membrane to the endoplasmic reticulum (ER), where it binds to specific IP_3-gated Ca^{2+} channels, causing them to open. The action of the SERCA pump (p. 414) ensures that $[Ca^{2+}]$ in the ER is orders of magnitude higher than that in the cytosol, so when these gated Ca^{2+} channels open, Ca^{2+} rushes into the cytosol (Fig. 12-11, step ❺), and the cytosolic $[Ca^{2+}]$ rises sharply to about 10^{-6} M. One effect of elevated $[Ca^{2+}]$ is the activation of **protein kinase C** (**PKC**; C for Ca^{2+}). Diacylglycerol cooperates with Ca^{2+} in activating PKC, thus also acting as a second messenger (step ❻). Activation involves the movement of a PKC domain (the pseudosubstrate domain) away from its location in the substrate-binding region of the enzyme, allowing the enzyme to bind and phosphorylate proteins that contain a PKC consensus sequence—Ser or Thr residues embedded in an amino acid sequence recognized by PKC (step ❼). There are several isozymes of PKC, each

TABLE 12-4	Some Signals That Act through Phospholipase C, IP_3, and Ca^{2+}	
Acetylcholine [muscarinic M_1]	Gastrin-releasing peptide	Platelet-derived growth factor (PDGF)
α_1-Adrenergic agonists	Glutamate	Serotonin [5-HT_2]
Angiogenin	Gonadotropin-releasing hormone (GRH)	Thyrotropin-releasing hormone (TRH)
Angiotensin II	Histamine [H_1]	Vasopressin
ATP [P_{2x}, P_{2y}]	Light (*Drosophila*)	
Auxin	Oxytocin	

Note: Receptor subtypes are in square brackets; see footnote to Table 12-3.

BOX 12-2 METHODS   FRET: Biochemistry Visualized in a Living Cell

Fluorescent probes are commonly used to detect rapid biochemical changes in single living cells. They can be designed to give an essentially instantaneous report (within nanoseconds) on the changes in intracellular concentration of a second messenger or in the activity of a protein kinase. Furthermore, fluorescence microscopy has sufficient resolution to reveal where in the cell such changes are occurring. In one widely used procedure, the fluorescent probes are derived from a naturally occurring fluorescent protein, the **green fluorescent protein (GFP)**, described in Chapter 9 (see Fig. 9-16), and variants with different fluorescence spectra, produced by genetic engineering or obtained from various marine coelenterates. For example, in the yellow fluorescent protein (YFP), Ala^{206} in GFP is replaced by a Lys residue, changing the wavelength of light absorption and fluorescence. Other variants of GFP fluoresce blue (BFP) or cyan (CFP) light, and a related protein (mRFP1) fluoresces red light (Fig. 1). GFP and its variants are compact structures that retain their ability to fold into their native β-barrel conformation even when fused with another protein. These fluorescent hybrid proteins act as spectroscopic rulers for measuring distances between interacting proteins within a cell and, indirectly, as measures of local concentrations of compounds that change the distance between two proteins.

An excited fluorescent molecule such as GFP or YFP can dispose of the energy from the absorbed photon in either of two ways: (1) by fluorescence, emitting a photon of slightly longer wavelength (lower energy) than the exciting light, or (2) by nonradiative

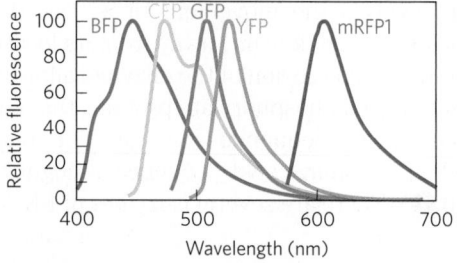

FIGURE 1 Emission spectra of some GFP variants.

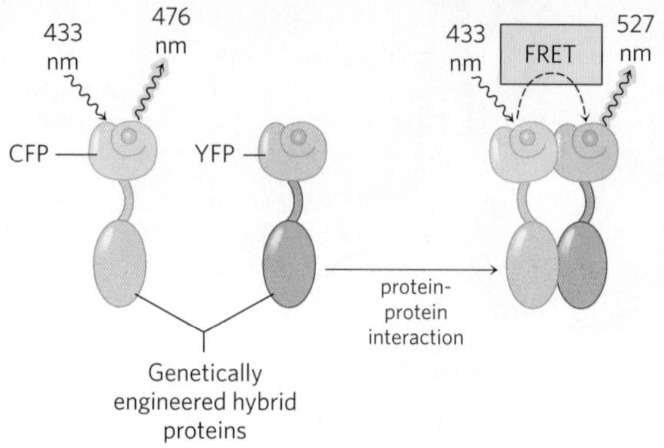

When the donor protein (CFP) is excited with monochromatic light of wavelength 433 nm, it emits fluorescent light at 476 nm (left). When the (red) protein fused with CFP interacts with the (purple) protein fused with YFP, that interaction brings CFP and YFP close enough to allow fluorescence resonance energy transfer (FRET) between them. Now, when CFP absorbs light of 433 nm, instead of fluorescing at 476 nm, it transfers energy directly to YFP, which then fluoresces at its characteristic emission wavelength, 527 nm. The ratio of light emission at 527 and 476 nm is therefore a measure of the extent of interaction between the red and purple proteins.

FIGURE 2 When the donor protein (CFP) is excited with monochromatic light of wavelength 433 nm, it emits fluorescent light at 476 nm (left). When the (red) protein fused with CFP interacts with the (purple) protein fused with YFP, that interaction brings CFP and YFP close enough to allow fluorescence resonance energy transfer (FRET) between them. Now, when CFP absorbs light of 433 nm, instead of fluorescing at 476 nm, it transfers energy directly to YFP, which then fluoresces at its characteristic emission wavelength, 527 nm. The ratio of light emission at 527 and 476 nm is therefore a measure of the extent of interaction between the red and purple proteins.

fluorescence resonance energy transfer (FRET), in which the energy of the excited molecule (the donor) passes directly to a nearby molecule (the acceptor) *without emission of a photon*, exciting the acceptor (Fig. 2). The acceptor can now decay to its ground state by fluorescence; the emitted photon has a longer wavelength (lower energy) than both the original exciting light and the fluorescence emission of the donor. This second mode of decay (FRET) is possible only when donor and acceptor are close to each other (within 1 to 50 Å); the efficiency of FRET is inversely proportional to the *sixth power* of the distance between donor and acceptor. Thus very small changes in the distance between donor and acceptor register as very large changes in FRET, measured as the fluorescence of the acceptor molecule when the donor is excited. With sufficiently sensitive light detectors, this fluorescence signal can be located to specific regions of a single, living cell.

with a characteristic tissue distribution, target protein specificity, and role. Their targets include cytoskeletal proteins, enzymes, and nuclear proteins that regulate gene expression. Taken together, this family of enzymes has a wide range of cellular actions, affecting neuronal and immune function and the regulation of cell division. Compounds that lead to overexpression of PKC or increase its activity to abnormal levels act as tumor

promoters; animals exposed to these substances have increased rates of cancer.

Calcium Is a Second Messenger That Is Localized in Space and Time

There are many variations on this basic scheme for Ca^{2+} signaling. In many cell types that respond to extracellular

FRET has been used to measure [cAMP] in living cells. The gene for BFP is fused with that for the regulatory subunit (R) of cAMP-dependent protein kinase (PKA), and the gene for GFP is fused with that for the catalytic subunit (C) (Fig. 3). When these two hybrid proteins are expressed in a cell, BFP (donor; excitation at 380 nm, emission at 460 nm) and GFP (acceptor; excitation at 475 nm, emission at 545 nm) in the inactive PKA (R_2C_2 tetramer) are close enough to undergo FRET. Wherever in the cell [cAMP] increases, the R_2C_2 complex dissociates into R_2 and 2 C and the FRET signal is lost, because donor and acceptor are now too far apart for efficient FRET. Viewed in the fluorescence microscope, the region of higher [cAMP] has a minimal GFP signal and higher

BFP signal. Measuring the ratio of emission at 460 nm and 545 nm gives a sensitive measure of the change in [cAMP]. By determining this ratio for all regions of the cell, the investigator can generate a false color image of the cell in which the ratio, or relative [cAMP], is represented by the intensity of the color. Images recorded at timed intervals reveal changes in [cAMP] over time.

A variation of this technology has been used to measure the activity of PKA in a living cell (Fig. 4). Researchers create a phosphorylation target for PKA by producing a hybrid protein containing four elements: YFP (acceptor); a short peptide with a Ser residue surrounded by the consensus sequence for PKA; a Ⓟ–Ser-binding domain (called 14-3-3); and CFP (donor). When the Ser residue is not phosphorylated, 14-3-3 has no affinity for the Ser residue and the hybrid protein exists in an extended form, with the donor and acceptor too far apart to generate a FRET signal. Wherever PKA is active in the cell, it phosphorylates the Ser residue of the hybrid protein, and 14-3-3 binds to the Ⓟ–Ser. In doing so, it draws YFP and CFP together and a FRET signal is detected with the fluorescence microscope, revealing the presence of active PKA.

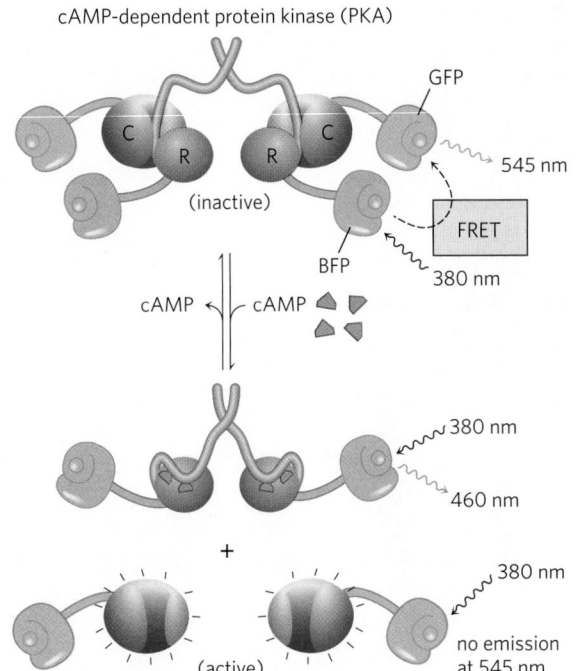

FIGURE 3 Measuring [cAMP] with FRET. Gene fusion creates hybrid proteins that exhibit FRET when the PKA regulatory (R) and catalytic (C) subunits are associated (low [cAMP]). When [cAMP] rises, the subunits dissociate and FRET ceases. The ratio of emission at 460 nm (dissociated) and 545 nm (complexed) thus offers a sensitive measure of [cAMP].

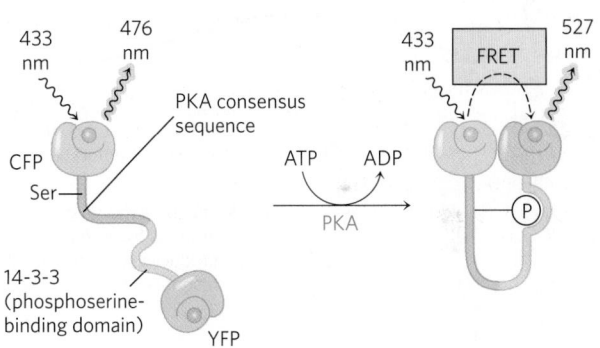

FIGURE 4 Measuring the activity of PKA with FRET. An engineered protein links YFP and CFP via a peptide that contains (1) a Ser residue surrounded by the consensus sequence for phosphorylation by PKA and (2) the 14-3-3 Ⓟ-Ser-binding domain. Active PKA phosphorylates the Ser residue, which docks with the 14-3-3 binding domain, bringing the fluorescence proteins close enough to allow FRET, revealing the presence of active PKA.

signals, Ca^{2+} serves as a second messenger that triggers intracellular responses, such as exocytosis in neurons and endocrine cells, contraction in muscle, and cytoskeletal rearrangements during amoeboid movement. In unstimulated cells, cytosolic $[Ca^{2+}]$ is kept very low ($<10^{-7}$ M) by the action of Ca^{2+} pumps in the ER, mitochondria, and plasma membrane (as further discussed below). Hormonal, neural, or other stimuli

cause either an influx of Ca^{2+} into the cell through specific Ca^{2+} channels in the plasma membrane or the release of sequestered Ca^{2+} from the ER or mitochondria, in either case raising the cytosolic $[Ca^{2+}]$ and triggering a cellular response.

Changes in intracellular $[Ca^{2+}]$ are detected by Ca^{2+}-binding proteins that regulate a variety of Ca^{2+}-dependent enzymes. **Calmodulin** (**CaM**; M_r 17,000) is

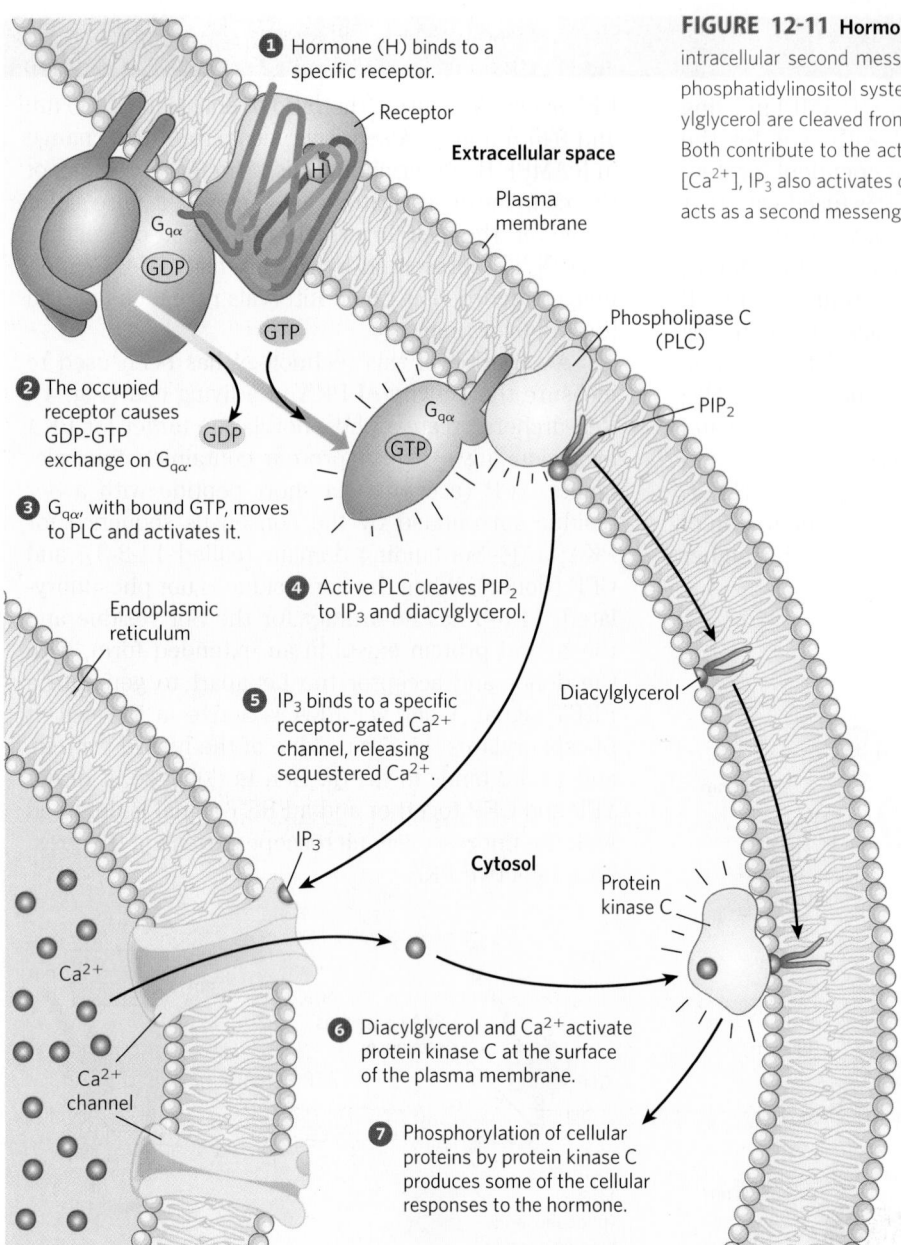

1 Hormone (H) binds to a specific receptor.

Receptor

Extracellular space

Plasma membrane

$G_{q\alpha}$

GDP

GTP

2 The occupied receptor causes GDP-GTP exchange on $G_{q\alpha}$.

GDP

$G_{q\alpha}$

GTP

Phospholipase C (PLC)

PIP$_2$

3 $G_{q\alpha}$, with bound GTP, moves to PLC and activates it.

Endoplasmic reticulum

4 Active PLC cleaves PIP$_2$ to IP$_3$ and diacylglycerol.

Diacylglycerol

5 IP$_3$ binds to a specific receptor-gated Ca^{2+} channel, releasing sequestered Ca^{2+}.

IP$_3$

Cytosol

Protein kinase C

Ca^{2+}

6 Diacylglycerol and Ca^{2+} activate protein kinase C at the surface of the plasma membrane.

Ca^{2+} channel

7 Phosphorylation of cellular proteins by protein kinase C produces some of the cellular responses to the hormone.

FIGURE 12-11 Hormone-activated phospholipase C and IP$_3$. Two intracellular second messengers are produced in the hormone-sensitive phosphatidylinositol system: inositol 1,4,5-trisphosphate (IP$_3$) and diacylglycerol are cleaved from phosphatidylinositol 4,5-bisphosphate (PIP$_2$). Both contribute to the activation of protein kinase C. By raising cytosolic [Ca^{2+}], IP$_3$ also activates other Ca^{2+}-dependent enzymes; thus Ca^{2+} also acts as a second messenger.

an acidic protein with four high-affinity Ca^{2+}-binding sites. When intracellular [Ca^{2+}] rises to about 10^{-6} M (1 μM), the binding of Ca^{2+} to calmodulin drives a conformational change in the protein **(Fig. 12-12a)**. Calmodulin associates with a variety of proteins and, in its Ca^{2+}-bound state, modulates their activities (Fig. 12-12b). It is a member of a family of Ca^{2+}-binding proteins that also includes troponin (see Fig. 5-32), which triggers skeletal muscle contraction in response to increased [Ca^{2+}]. Members of this family share a characteristic Ca^{2+}-binding structure, the EF hand (Fig. 12-12c).

Calmodulin is an integral subunit of the **Ca^{2+}/ calmodulin-dependent protein kinases (CaM kinases**, types I through IV). When intracellular [Ca^{2+}] increases in response to a stimulus, calmodulin binds Ca^{2+}, undergoes a change in conformation, and activates the CaM kinase. The kinase then phosphorylates

target enzymes, regulating their activities. Calmodulin is also a regulatory subunit of phosphorylase b kinase of muscle, which is activated by Ca^{2+}. Thus Ca^{2+} triggers ATP-requiring muscle contractions while also activating glycogen breakdown, providing fuel for ATP synthesis. Many other enzymes are also known to be modulated by Ca^{2+} through calmodulin (Table 12-5). The activity of the second messenger Ca^{2+}, like that of cAMP, can be spatially restricted; after its release triggers a local response, Ca^{2+} is generally removed before it can diffuse to distant parts of the cell.

Commonly, Ca^{2+} level does not simply rise and then fall, but rather oscillates with a period of a few seconds **(Fig. 12-13)**—even when the extracellular concentration of the triggering hormone remains constant. The mechanism underlying [Ca^{2+}] oscillations presumably entails feedback regulation by Ca^{2+} on some part of the

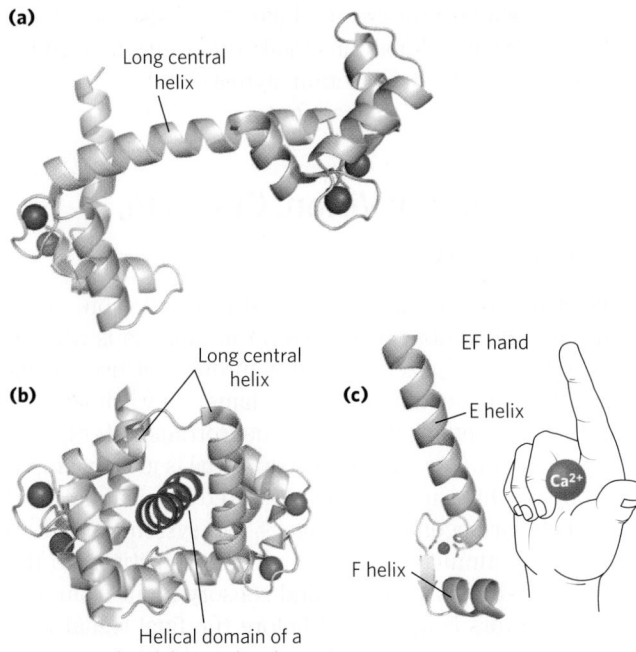

(a)

Long central helix

(b)

Long central helix

Helical domain of a calmodulin-regulated protein

(c)

EF hand

E helix

Ca²⁺

F helix

FIGURE 12-12 Calmodulin. This is the protein mediator of many Ca^{2+}-stimulated enzymatic reactions. Calmodulin has four high-affinity Ca^{2+}-binding sites ($K_d \approx$ 0.1 to 1 μM). **(a)** A ribbon model of the crystal structure of calmodulin. The four Ca^{2+}-binding sites are occupied by Ca^{2+} (purple). The amino-terminal domain is on the left; the carboxyl-terminal domain on the right. **(b)** Calmodulin associated with a helical domain of one of the many enzymes it regulates, calmodulin-dependent protein kinase II. Notice that the long central α helix of calmodulin visible in (a) has bent back on itself in binding to the helical substrate domain. The central helix of calmodulin is clearly more flexible in solution than in crystal. **(c)** Each of the four Ca^{2+}-binding sites occurs in a helix-loop-helix motif called the EF hand, also found in many other Ca^{2+}-binding proteins. [Sources: (a) PDB ID 1CLL, R. Chattopadhyaya et al., *J. Mol. Biol.* 228:1177, 1992. (b, c) PDB ID 1CDL, W. E. Meador et al., *Science* 257:1251, 1992.]

TABLE 12-5	Some Proteins Regulated by Ca^{2+} and Calmodulin
Adenylyl cyclase (brain)	
Ca^{2+}/calmodulin-dependent protein kinases (CaM kinases I to IV)	
Ca^{2+}-dependent Na^+ channel (*Paramecium*)	
Ca^{2+}-release channel of sarcoplasmic reticulum	
Calcineurin (phosphoprotein phosphatase 2B)	
cAMP phosphodiesterase	
cAMP-gated olfactory channel	
cGMP-gated Na^+, Ca^{2+} channels (rod and cone cells)	
Glutamate decarboxylase	
Myosin light-chain kinases	
NAD^+ kinase	
Nitric oxide synthase	
Phosphatidylinositol 3-kinase	
Plasma membrane Ca^{2+} ATPase (Ca^{2+} pump)	
RNA helicase (p68)	

Ca^{2+}-release process. Whatever the mechanism, the effect is that one kind of signal (hormone concentration, for example) is converted into another (frequency and amplitude of intracellular $[Ca^{2+}]$ "spikes"). The Ca^{2+} signal diminishes as Ca^{2+} diffuses away from the initial source (the Ca^{2+} channel), is sequestered in the ER, or is pumped out of the cell.

There is significant cross talk between the Ca^{2+} and cAMP signaling systems. In some tissues, both the enzyme that produces cAMP (adenylyl cyclase) and the enzyme that degrades cAMP (phosphodiesterase) are stimulated by Ca^{2+}. Temporal and spatial changes in $[Ca^{2+}]$ can therefore produce transient, localized changes in [cAMP]. We have noted already that PKA, the enzyme

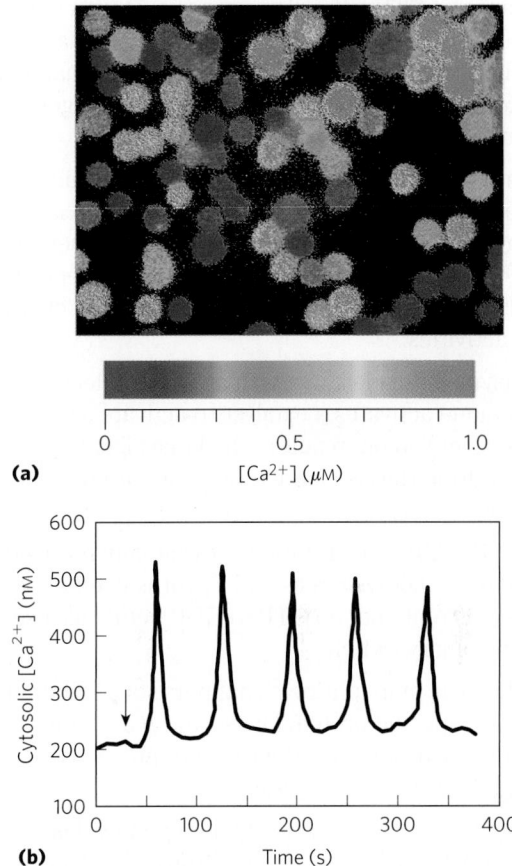

(a)

0   0.5   1.0

$[Ca^{2+}]$ (μM)

(b)

Time (s)

FIGURE 12-13 Triggering of oscillations in intracellular $[Ca^{2+}]$ by extracellular signals. (a) A dye (fura) that undergoes fluorescence changes when it binds Ca^{2+} is allowed to diffuse into cells, and its instantaneous light output is measured by fluorescence microscopy. Fluorescence intensity is represented by color; the color scale relates intensity of color to $[Ca^{2+}]$, allowing determination of the absolute $[Ca^{2+}]$. In this case, thymocytes (cells of the thymus) have been stimulated with extracellular ATP, which raises their internal $[Ca^{2+}]$. The cells are heterogeneous in their responses: some have high intracellular $[Ca^{2+}]$ (red), others have much lower $[Ca^{2+}]$ (blue). **(b)** When such a probe is used in a single hepatocyte, the agonist norepinephrine (added at the arrow) causes oscillations of $[Ca^{2+}]$ from 200 to 500 nM. Similar oscillations are induced in other cell types by other extracellular signals. [Sources: (a) Courtesy Michael D. Cahalan, Department of Physiology and Biophysics, University of California, Irvine. (b) T. A. Rooney et al., *J. Biol. Chem.* 264:17,131, 1989.]

that responds to cAMP, is often part of a highly localized supramolecular complex assembled on scaffold proteins such as AKAPs. This subcellular localization of target enzymes, combined with temporal and spatial gradients in $[Ca^{2+}]$ and [cAMP], allows a cell to respond to one or several signals with subtly nuanced metabolic changes, localized in space and time.

SUMMARY 12.2 G Protein–Coupled Receptors and Second Messengers

■ G protein–coupled receptors (GPCRs) share a common structural arrangement of seven transmembrane helices and act through heterotrimeric G proteins. On ligand binding, GPCRs catalyze the exchange of GTP for GDP on the G protein, causing dissociation of the G_α subunit; G_α then stimulates or inhibits the activity of an effector enzyme, changing the local concentration of its second-messenger product.

■ The β-adrenergic receptor activates a stimulatory G protein, G_s, thereby activating adenylyl cyclase and raising the concentration of the second messenger cAMP. Cyclic AMP stimulates cAMP-dependent protein kinase to phosphorylate key target enzymes, changing their activities.

■ Enzyme cascades, in which a single molecule of hormone activates a catalyst to activate another catalyst, and so on, result in the large signal amplification that is characteristic of hormone receptor systems.

■ Cyclic AMP concentration is eventually reduced by cAMP phosphodiesterase, and G_s turns itself off by hydrolysis of its bound GTP to GDP, acting as a self-limiting binary switch.

■ When the epinephrine signal persists, β-adrenergic receptor–specific protein kinase and β-arrestin temporarily desensitize the receptor and cause it to move into intracellular vesicles.

■ Some receptors *stimulate* adenylyl cyclase through G_s; others *inhibit* it through G_i. Thus cellular [cAMP] reflects the integrated input of two (or more) signals.

■ Noncatalytic adaptor proteins such as AKAPs hold together proteins involved in a signaling process, increasing the efficiency of their interactions and, in some cases, confining the process to a specific subcellular location.

■ Some GPCRs act via a plasma membrane phospholipase C that cleaves PIP_2 to diacylglycerol and IP_3. By opening Ca^{2+} channels in the endoplasmic reticulum, IP_3 raises cytosolic $[Ca^{2+}]$. Diacylglycerol and Ca^{2+} act together to activate protein kinase C, which phosphorylates and changes the activity of

specific cellular proteins. Cellular $[Ca^{2+}]$ also regulates (often through calmodulin) many other enzymes and proteins involved in secretion, cytoskeletal rearrangements, or contraction.

12.3 GPCRs in Vision, Olfaction, and Gustation

The detection of light, odors, and tastes (vision, olfaction, and gustation, respectively) in animals is accomplished by specialized sensory neurons that use signal-transduction mechanisms fundamentally similar to those that detect hormones, neurotransmitters, and growth factors. An initial sensory signal is greatly amplified by mechanisms that include gated ion channels and intracellular second messengers; the system adapts to continued stimulation by changing its sensitivity to the stimulus (desensitization); and sensory input from several receptors is integrated before the final signal goes to the brain.

The Vertebrate Eye Uses Classic GPCR Mechanisms

Visual transduction **(Fig. 12-14)** begins when light falls on **rhodopsin**, a GPCR in the disk membranes of rod cells of the vertebrate eye. (Rod cells do not detect colors; cone cells do (see below).) The light-absorbing pigment (chromophore) 11-*cis*-retinal is covalently attached to **opsin**, the protein component of rhodopsin, which lies near the middle of the disk membrane bilayer. When a photon is absorbed by the retinal component of rhodopsin (step ❶), the energy causes a photochemical change; 11-*cis*-retinal is converted to all-*trans*-retinal (see Figs 1-20b and 10-20). This change in the structure of the chromophore forces conformational changes in the rhodopsin molecule, allowing it to interact with and thus activate its trimeric G protein, transducin. Rhodopsin now stimulates the exchange of bound GDP on transducin for GTP from the cytosol (Fig. 12-14, step ❷), and activated transducin stimulates the membrane protein cyclic GMP (cGMP) phosphodiesterase (PDE) by removing an inhibitory subunit (step ❸). The activated cGMP PDE degrades the second messenger 3′,5′-cGMP to 5′-GMP, lowering the concentration of cGMP (step ❹). A cGMP-dependent Na^+ or Ca^{2+} channel in the plasma membrane closes (step ❺), while a Na^+-Ca^{2+} active antiporter continues to pump Ca^{2+} outward across the plasma membrane (step ❻), making the transmembrane electrical potential more negative inside (that is, hyperpolarizing the rod cell). This electrical change passes through a series of specialized nerve cells to the visual cortex of the brain.

Several steps in the visual-transduction process result in a huge amplification of the signal. Each excited rhodopsin molecule activates at least 500 molecules of

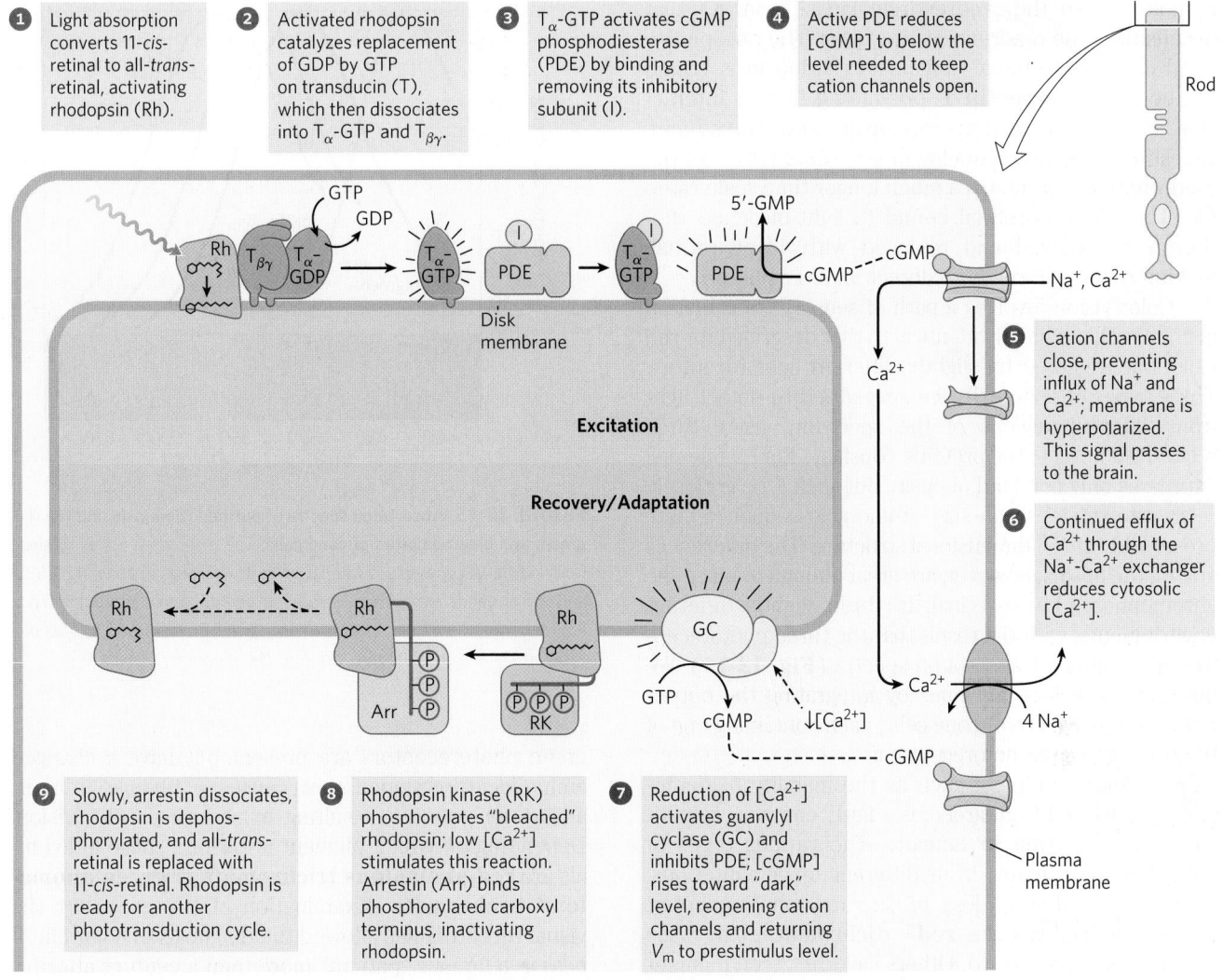

1 Light absorption converts 11-*cis*-retinal to all-*trans*-retinal, activating rhodopsin (Rh).

2 Activated rhodopsin catalyzes replacement of GDP by GTP on transducin (T), which then dissociates into T_α-GTP and $T_{\beta\gamma}$.

3 T_α-GTP activates cGMP phosphodiesterase (PDE) by binding and removing its inhibitory subunit (I).

4 Active PDE reduces [cGMP] to below the level needed to keep cation channels open.

5 Cation channels close, preventing influx of Na^+ and Ca^{2+}; membrane is hyperpolarized. This signal passes to the brain.

6 Continued efflux of Ca^{2+} through the Na^+-Ca^{2+} exchanger reduces cytosolic $[Ca^{2+}]$.

7 Reduction of $[Ca^{2+}]$ activates guanylyl cyclase (GC) and inhibits PDE; [cGMP] rises toward "dark" level, reopening cation channels and returning V_m to prestimulus level.

8 Rhodopsin kinase (RK) phosphorylates "bleached" rhodopsin; low $[Ca^{2+}]$ stimulates this reaction. Arrestin (Arr) binds phosphorylated carboxyl terminus, inactivating rhodopsin.

9 Slowly, arrestin dissociates, rhodopsin is dephosphorylated, and all-*trans*-retinal is replaced with 11-*cis*-retinal. Rhodopsin is ready for another phototransduction cycle.

FIGURE 12-14 Molecular consequences of photon absorption by rhodopsin in the rod outer segment. The top half of the figure (steps **1** to **5**) describes excitation; the bottom shows post-illumination steps: recovery (steps **6** and **7**) and adaptation (steps **8** and **9**).

transducin, and each transducin molecule can activate a molecule of cGMP PDE. This phosphodiesterase has a remarkably high turnover number: each activated molecule hydrolyzes 4,200 molecules of cGMP per second. The binding of cGMP to cGMP-gated ion channels is cooperative, and a relatively small change in [cGMP] therefore registers as a large change in ion conductance. The result of these amplifications is exquisite sensitivity to light. Absorption of a single photon closes 1,000 or more ion channels for Na^+ and Ca^{2+}, hyperpolarizing the cell's membrane potential (V_m) by about 1 mV.

As your eyes move across this line, the retinal images of the first words disappear rapidly—before you see the next series of words. In that short interval, a great deal of biochemistry has taken place. Very soon after illumination of the rod or cone cells stops, the photosensory system shuts off. The α subunit of transducin (T_α, with bound GTP) has intrinsic GTPase activity. Within milliseconds after the decrease in

light intensity, GTP is hydrolyzed and T_α reassociates with $T_{\beta\gamma}$. The inhibitory subunit of PDE, which had been bound to T_α-GTP, is released and reassociates with PDE, strongly inhibiting its activity and thus slowing cGMP breakdown.

At the same time, a second factor that helps to end the response to light is the reduction of intracellular $[Ca^{2+}]$ that results from continued Ca^{2+} efflux through the Na^+-Ca^{2+} exchanger (Fig. 12-14, step **6**). High $[Ca^{2+}]$ inhibits the enzyme that makes cGMP (guanylyl cyclase; step **7**), so cGMP production rises when $[Ca^{2+}]$ falls, quickly reaching its prestimulus level.

In response to prolonged illumination, rhodopsin itself undergoes changes that limit the duration of its signaling activity. The conformational change induced in rhodopsin by light absorption exposes several Thr and Ser residues in its carboxyl-terminal domain, and these residues are phosphorylated by **rhodopsin kinase** (step **8**), which is functionally and structurally homologous to the β-adrenergic kinase (βARK) that desensitizes the β-adrenergic receptor. The phosphorylated carboxyl-terminal domain of rhodopsin is bound by the protein **arrestin 1**, preventing further interaction between activated rhodopsin and transducin. Arrestin 1 is a close homolog of arrestin 2 (βarr) of the β-adrenergic system. On a much longer time scale (step ❾), the all-*trans*-retinal bound to light-bleached rhodopsin is removed and replaced with 11-*cis*-retinal, making rhodopsin ready to detect another photon.

Color vision involves a path of sensory transduction in cone cells essentially identical to that described for rod cells, but triggered by slightly different light receptors. Three types of cone cells are specialized to detect light from different regions of the spectrum, using three related photoreceptor proteins (opsins). Each cone cell expresses only one kind of opsin, but each type is closely related to rhodopsin in size, amino acid sequence, and, presumably, three-dimensional structure. The differences among the opsins, however, are great enough to place the chromophore, 11-*cis*-retinal, in three slightly different environments, with the result that the three photoreceptors have different absorption spectra **(Fig. 12-15)**. We discriminate colors and hues by integrating the output from the three types of cone cells, each containing one of the three types of photoreceptors.

Color blindness, such as the inability to distinguish red from green, is a fairly common, genetically inherited trait in humans. The various types of color blindness result from different opsin mutations. One form is due to loss of the red photoreceptor; affected individuals are **red⁻ dichromats** (they see only two primary colors). Others lack the green pigment and are **green⁻ dichromats**. In some cases, the red and

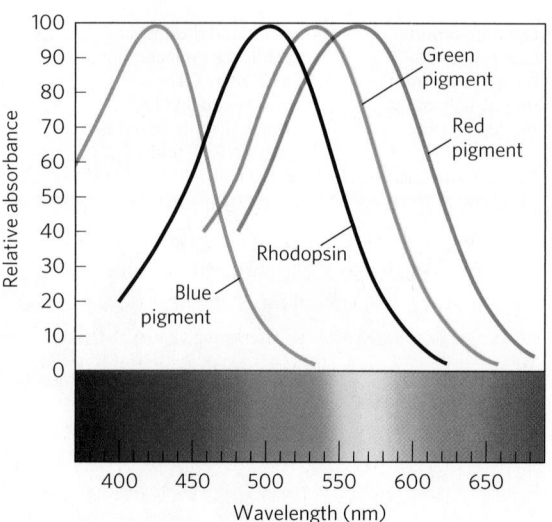

FIGURE 12-15 Absorption spectra of purified rhodopsin and the red, green, and blue receptors of cone cells. The receptor spectra, obtained from individual cone cells isolated from cadavers, peak at about 420, 530, and 560 nm, and the maximum absorption for rhodopsin is at about 500 nm. For reference, the visible spectrum for humans is about 380 to 750 nm. [Source: Data from J. Nathans, *Sci. Am.* 260 (February):42, 1989.]

green photoreceptors are present but have a changed amino acid sequence that causes a change in their absorption spectra, resulting in abnormal color vision. Depending on which pigment is altered, these individuals are **red-anomalous trichromats** or **green-anomalous trichromats**. Examination of the genes for the visual receptors has allowed the diagnosis of color blindness in a famous "patient" more than a century after his death (Box 12-3). ∎

BOX 12-3 🩺 MEDICINE Color Blindness: John Dalton's Experiment from the Grave

The chemist John Dalton (of atomic theory fame) was color-blind. He thought it probable that the vitreous humor of his eyes (the fluid that fills the eyeball behind the lens) was tinted blue, unlike the colorless fluid of normal eyes. He proposed that after his death, his eyes should be dissected and the color of the vitreous humor determined. His wish was honored. The day after Dalton's death in July 1844, Joseph Ransome dissected his eyes and found the vitreous humor to be perfectly colorless. Ransome, like many scientists, was reluctant to throw samples away. He placed Dalton's eyes in a jar of preservative, where they stayed for a century and a half (Fig. 1).

Then, in the mid-1990s, molecular biologists in England took small samples of Dalton's retinas and extracted DNA. Using the known gene sequences for the opsins of the red and green light receptors, they amplified the relevant sequences (using techniques described in Chapter 8) and determined that Dalton

FIGURE 1 Dalton's eyes. [Source: Professor J. D. Mollon, Department of Experimental Psychology, Cambridge University.]

had the opsin gene for the red photopigment but lacked the opsin gene for the green photopigment. Dalton was a green⁻ dichromat. So, 150 years after his death, the experiment Dalton started—by hypothesizing about the cause of his color blindness—was finally finished.

Vertebrate Olfaction and Gustation Use Mechanisms Similar to the Visual System

The sensory cells that detect odors and tastes have much in common with the visual receptor system. Binding of an odorant molecule to its specific GPCR triggers a change in receptor conformation, activating a G protein, G_{olf}, analogous to transducin and to G_s of the β-adrenergic system. The activated G_{olf} activates adenylyl cyclase, raising the local [cAMP]. The cAMP-gated Na^+ and Ca^{2+} channels of the plasma membrane open, and the influx of Na^+ and Ca^{2+} produces a small depolarization called the **receptor potential**. If a sufficient number of odorant molecules encounter receptors, the receptor potential is strong enough to cause the neuron to fire an action potential. This signal is relayed to the brain in several stages and registers as a specific smell. All these events occur within 100 to 200 ms. When the olfactory stimulus is no longer present, the transducing machinery shuts itself off in several ways. A cAMP phosphodiesterase returns [cAMP] to the prestimulus level. G_{olf} hydrolyzes its bound GTP to GDP, thereby inactivating itself. Phosphorylation of the receptor by a specific kinase prevents its interaction with G_{olf}, by a mechanism analogous to that used to desensitize the β-adrenergic receptor and rhodopsin. Some odorants are detected by another mechanism we have seen in other signal transductions: activation of a phospholipase and production of IP_3, leading to a rise in intracellular $[Ca^{2+}]$.

The sense of taste in vertebrates reflects the activity of gustatory neurons clustered in taste buds on the surface of the tongue. For example, sweet-tasting molecules are those that bind receptors in "sweet" taste buds. In taste sensory neurons, GPCRs are coupled to the heterotrimeric G protein **gustducin**. When the tastant molecule binds its receptor, gustducin is activated and stimulates cAMP production by adenylyl cyclase. The resulting elevation of [cAMP] activates PKA, which phosphorylates K^+ channels in the plasma membrane, causing them to close and sending an electrical signal to the brain. Other taste buds specialize in detecting bitter, sour, salty, or umami (savory) tastants, using various combinations of second messengers and ion channels in the transduction mechanisms.

All GPCR Systems Share Universal Features

We have now looked at several types of signaling systems (hormone signaling, vision, olfaction, and gustation) in which membrane receptors are coupled to second messenger–generating enzymes through G proteins. As we have intimated, signaling mechanisms must have arisen early in evolution; genomic studies have revealed hundreds of genes encoding GPCRs in vertebrates, arthropods (*Drosophila* and mosquito), and the roundworm *Caenorhabditis elegans*. Even the common budding yeast *Saccharomyces* uses GPCRs and G proteins to detect the opposite mating type. Overall patterns have been conserved, and the introduction of variety has given modern organisms the ability to respond to a wide range of stimuli

TABLE 12-6	Some Signals That Act through GPCRs
Amines	Tachykinin
Acetylcholine (muscarinic)	Thyrotropin-releasing hormone
Dopamine	Urotensin II
Epinephrine	**Protein hormones**
Histamine	Follicle-stimulating hormone
Serotonin	
Peptides	Gonadotropin
Angiotensin	Lutropin-choriogonadotropic hormone
Bombesin	
Bradykinin	Thyrotropin
Chemokine	**Prostanoids**
Colecystokinin (CCK)	Prostacyclin
Endothelin	Prostaglandin
Gonadotropin-releasing hormone	Thromboxane
Interleukin-8	**Others**
Melanocortin	Cannabinoids
Neuropeptide Y	Lysosphingolipids
Neurotensin	Melatonin
Orexin	Olfactory stimuli
Somatostatin	Opioids
	Rhodopsin

(Table 12-6). Of the approximately 20,000 genes in the human genome, as many as 1,000 encode GPCRs, including hundreds for olfactory stimuli and many orphan receptors, for which the natural ligand is not yet known.

All well-studied signal-transducing systems that act through heterotrimeric G proteins share some common features that reflect their evolutionary relatedness **(Fig. 12-16)**. The receptors have seven transmembrane segments, a domain (generally the loop between transmembrane helices 6 and 7) that interacts with a G protein, and a carboxyl-terminal cytoplasmic domain that undergoes reversible phosphorylation on several Ser or Thr residues. The ligand-binding site (or, in the case of light reception, the light receptor) is buried deep in the membrane and includes residues from several of the transmembrane segments. Ligand binding (or light) induces a conformational change in the receptor, exposing a domain that can interact with a G protein. Heterotrimeric G proteins activate or inhibit effector enzymes (adenylyl cyclase, PDE, or PLC), which change the concentration of a second messenger (cAMP, cGMP, IP_3, or Ca^{2+}). In the hormone-detecting systems, the final output is an activated protein kinase that regulates some cellular process by phosphorylating a protein critical to that process. In sensory neurons, the output is a change in membrane potential and a consequent

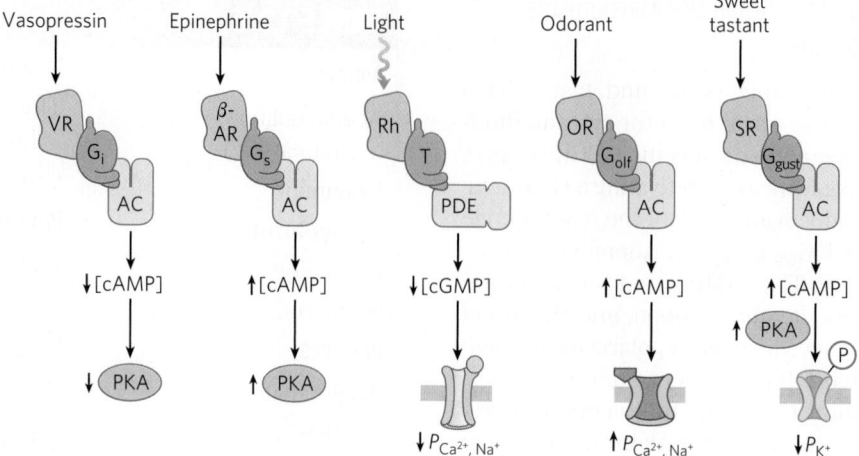

FIGURE 12-16 Common features of signaling systems that detect hormones, light, smells, and tastes. GPCRs provide signal specificity, and their interaction with G proteins provides signal amplification. Heterotrimeric G proteins activate effector enzymes: adenylyl cyclase (AC) and phosphodiesterases (PDEs) that degrade cAMP or cGMP. Changes in concentration of the second messengers (cAMP, cGMP) result in alterations in enzymatic activities via phosphorylation or alterations in the permeability (P) of surface membranes to Ca^{2+}, Na^+, and K^+. The resulting depolarization or hyperpolarization of the sensory cell (the signal) passes through relay neurons to sensory centers in the brain. In the best-studied cases, desensitization includes phosphorylation of the receptor and binding of a protein (arrestin) that interrupts receptor–G protein interactions. (The path of odorant detection by production of IP_3 and increase in intracellular $[Ca^{2+}]$, mentioned in the text, is not shown here.) VR is the vasopressin receptor; β-AR, the β-adrenergic receptor; Rh, rhodopsin; OR, olfactory receptor; SR, sweet-taste receptor.

electrical signal that passes to another neuron in the pathway connecting the sensory cell to the brain.

All these systems self-inactivate. Bound GTP is converted to GDP by the intrinsic GTPase activity of G proteins, often augmented by GTPase-activating proteins (GAPs) or RGS proteins (regulators of G-protein signaling; see Fig. 12-5 and Box 12-1, Fig. 4). In some cases, the effector enzymes that are the targets of modulation by G proteins also serve as GAPs. The desensitization mechanism involving phosphorylation of the carboxyl-terminal region followed by arrestin binding is widespread and may be universal.

Each of the 1,000 GPCRs of vertebrates is expressed selectively, in certain cell types or under certain conditions. Together, they allow cells and tissues to respond to a wide array of stimuli, including various low molecular weight amines, peptides, proteins, eicosanoids and other lipids, as well as light and the many compounds detected by olfaction and gustation. The determination of several GPCR structures by crystallography **(Fig. 12-17),**

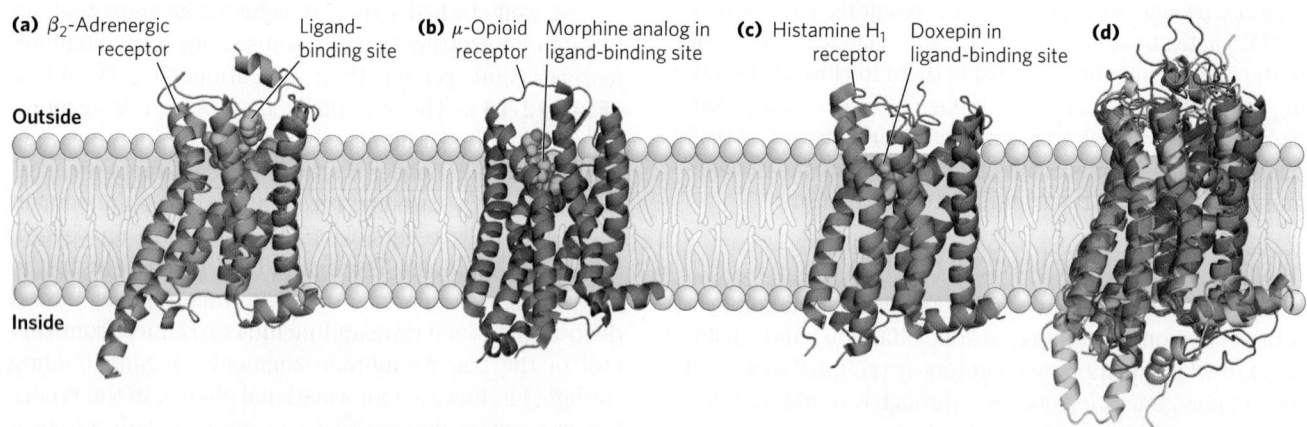

FIGURE 12-17 The β-adrenergic receptor and several other GPCRs. **(a)** The β_2-adrenergic receptor with the agonist epinephrine, shown in yellow, in the ligand-binding site. **(b)** The μ opioid receptor, the target of morphine and codeine, with a morphine analog in the ligand-binding site. **(c)** The histamine H_1 receptor with the bound drug doxepin. **(d)** Five GPCR structures superimposed to show the remarkable conservation of structure. Shown are the human A2A adenosine receptor (orange); turkey β_1-adrenergic receptor (blue), human β_2-adrenergic receptor (green), squid rhodopsin (yellow); and bovine rhodopsin (red). [Sources: (a) PDB ID 3SN6, S. G. F. Rasmussen et al., *Nature* 477:549, 2011. (b) PDB ID 4DKL, A. Manglik et al., *Nature* 485:321, 2012. (c) PDB ID 3RZE, T. Shimamura et al., *Nature* 475:65, 2011. (d) Human A2A adenosine receptor: PDB ID 3EML, V. P. Jaakola et al., *Science* 322:1211, 2008; turkey β_1-adrenergic receptor: PDB ID 2VT4, A. Warne et al., *Nature* 454:486, 2008; human β_2-adrenergic receptor: PDB ID 2RH1, V. Cherezov et al., *Science* 318:1258, 2007; squid rhodopsin: PDB ID 2Z73, M. Murakami and T. Kouyama, *Nature* 453:363, 2008; bovine rhodopsin: PDB ID 1U19, T. Okada et al., *J. Mol. Biol.* 342:571, 2004.]

including the β-adrenergic receptor and the histamine receptor, has stimulated great interest in both the transduction mechanism(s) and the possibilities of altering receptor activity with drugs. These two receptors are the targets of a variety of widely used beta-blocker and antihistamine medications, respectively. The structural similarities among GPCRs go beyond the common seven–transmembrane helix pattern; as Figure 12-17d shows, the structures of five different GPCRs are almost superimposable. Clearly, something about this three-dimensional structure makes it effective as a transducer of many disparate signals.

SUMMARY 12.3 GPCRs in Vision, Olfaction, and Gustation

■ Vision, olfaction, and gustation in vertebrates employ GPCRs, which act through heterotrimeric G proteins to change the membrane potential (V_m) of a sensory neuron.

■ In rod and cone cells of the retina, light activates rhodopsin, which activates the G protein transducin. The freed α subunit of transducin activates a cGMP phosphodiesterase, which lowers [cGMP] and thus closes cGMP-dependent ion channels in the outer segment of the neuron. The resulting hyperpolarization of the rod or cone cell carries the signal to the next neuron in the pathway, and eventually to the brain.

■ In olfactory neurons, olfactory stimuli, acting through GPCRs and G proteins, trigger an increase in [cAMP] (by activating adenylyl cyclase) or [Ca^{2+}] (by activating PLC). These second messengers affect ion channels and thus the V_m.

■ Gustatory neurons have GPCRs that respond to tastants by altering levels of cAMP, which changes V_m by gating ion channels.

■ There is a high degree of conservation of signaling proteins and transduction mechanisms across signaling systems and across species.

12.4 Receptor Tyrosine Kinases

The **receptor tyrosine kinases (RTKs)**, a family of plasma membrane receptors with intrinsic protein kinase activity, transduce extracellular signals by a mechanism fundamentally different from that of GPCRs. RTKs have a ligand-binding domain on the extracellular face of the plasma membrane and an enzyme active site on the cytoplasmic face, connected by a single transmembrane segment. The cytoplasmic domain is a protein kinase that phosphorylates Tyr residues (a Tyr kinase) in specific target proteins. The receptors for insulin and epidermal growth factor are prototypes for the approximately 60 RTKs in humans.

Stimulation of the Insulin Receptor Initiates a Cascade of Protein Phosphorylation Reactions

Insulin regulates both metabolic enzymes and gene expression. Insulin does not enter cells, but initiates a signal that travels a branched pathway from the plasma membrane receptor to insulin-sensitive enzymes in the cytosol, and to the nucleus, where it stimulates the transcription of specific genes. The active insulin receptor protein (INSR) consists of two identical α subunits protruding from the outer face of the plasma membrane and two transmembrane β subunits with their carboxyl termini protruding into the cytosol—a dimer of $\alpha\beta$ monomers **(Fig. 12-18)**. The α subunits contain the insulin-binding domain, and the intracellular domains of the β subunits contain the protein kinase activity that transfers a phosphoryl group from ATP to the hydroxyl group of Tyr residues in specific target proteins. Signaling through INSR begins when the binding of one insulin molecule between the two subunits of the dimer activates the Tyr kinase activity, and each β subunit phosphorylates three critical Tyr residues near the carboxyl terminus of the other β subunit. This **autophosphorylation** opens the active site so that the enzyme can phosphorylate Tyr residues of other target proteins. The mechanism of activation of the INSR protein kinase is similar to that described for PKA and PKC: a region of the cytoplasmic domain (an autoinhibitory sequence) that usually occludes the active site moves out of the active site after being phosphorylated, opening the site for the binding of target proteins (Fig. 12-18).

When INSR is autophosphorylated (**Fig. 12-19**, step ❶) and becomes an active Tyr kinase, one of its targets is insulin receptor substrate-1 (IRS-1; step ❷). Once phosphorylated on several of its Tyr residues, IRS-1 becomes the point of nucleation for a complex of proteins (step ❸) that carry the message from the insulin receptor to end targets in the cytosol and nucleus, through a long series of intermediate proteins. First, a Ⓟ–Tyr residue of IRS-1 binds to the **SH2 domain** of the protein Grb2. (SH2 is an abbreviation of *Src homology 2*, so named because the sequence of an SH2 domain is similar to that of a domain in Src (pronounced *sark*), another protein Tyr kinase.) Many signaling proteins contain SH2 domains, all of which bind Ⓟ–Tyr residues in a protein partner. Grb2 (*growth factor receptor-bound protein 2*) is an adaptor protein, with no intrinsic enzymatic activity. Its function is to bring together two proteins (in this case, IRS-1 and the protein Sos) that must interact to enable signal transduction. In addition to its SH2 (Ⓟ–Tyr-binding) domain, Grb2 contains a second protein-binding domain, SH3, that binds to a proline-rich region of Sos, recruiting Sos to the growing receptor complex. When bound to Grb2, Sos acts as a guanosine nucleotide–exchange factor (GEF), catalyzing the replacement of bound GDP with GTP on Ras, a G protein.

Ras is the prototype of a family of **small G proteins** that mediate a wide variety of signal transductions (see Box 12-1). Like the trimeric G protein that functions with the β-adrenergic system (Fig. 12-5), Ras can exist in either the GTP-bound (active) or GDP-bound (inactive) conformation, but Ras ($\sim$20 kDa)

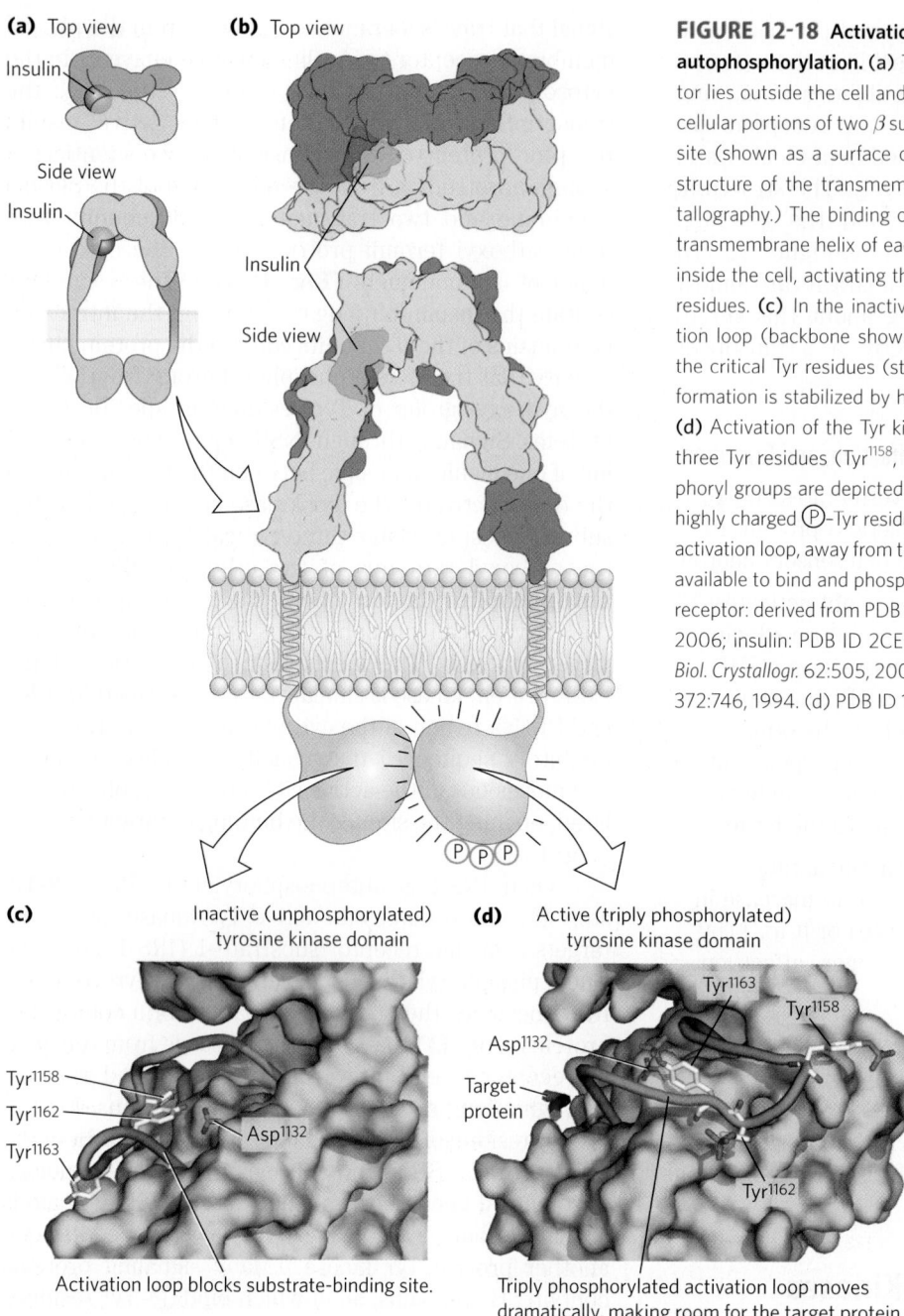

(a) Top view

Insulin

Side view

Insulin

(b) Top view

Insulin

Side view

(c) Inactive (unphosphorylated) tyrosine kinase domain

Tyr¹¹⁵⁸
Tyr¹¹⁶²
Tyr¹¹⁶³
Asp¹¹³²

Activation loop blocks substrate-binding site.

(d) Active (triply phosphorylated) tyrosine kinase domain

Tyr¹¹⁶³
Tyr¹¹⁵⁸
Asp¹¹³²
Target protein
Tyr¹¹⁶²

Triply phosphorylated activation loop moves dramatically, making room for the target protein in the substrate-binding site.

FIGURE 12-18 Activation of the insulin-receptor tyrosine kinase by autophosphorylation. (a) The insulin-binding region of the insulin receptor lies outside the cell and comprises **(b)** two α subunits and the extracellular portions of two β subunits, intertwined to form the insulin-binding site (shown as a surface contour model of the crystal structure). (The structure of the transmembrane domain has not been solved by crystallography.) The binding of insulin is communicated through the single transmembrane helix of each β subunit to the paired Tyr kinase domains inside the cell, activating them to phosphorylate each other on three Tyr residues. **(c)** In the inactive form of the Tyr kinase domain, the activation loop (backbone shown in teal) sits in the active site, and none of the critical Tyr residues (stick structures) are phosphorylated. This conformation is stabilized by hydrogen bonding between Tyr^{1162} and Asp^{1132}. **(d)** Activation of the Tyr kinase allows each β subunit to phosphorylate three Tyr residues (Tyr^{1158}, Tyr^{1162}, Tyr^{1163}) on the other β subunit. (Phosphoryl groups are depicted in red and orange.) The introduction of three highly charged ℗–Tyr residues forces a 30 Å change in the position of the activation loop, away from the substrate-binding site, which thus becomes available to bind and phosphorylate a target protein. [Sources: (b) Insulin receptor: derived from PDB ID 2DTG, N. M. McKern et al., *Nature* 443:218, 2006; insulin: PDB ID 2CEU, J. L. Whittingham et al., *Acta Crystallogr. D Biol. Crystallogr.* 62:505, 2006. (c) PDB ID 1IRK, S. R. Hubbard et al., *Nature* 372:746, 1994. (d) PDB ID 1IR3, S. R. Hubbard, *EMBO J.* 16:5572, 1997.]

acts as a monomer. When GTP binds, Ras can activate a protein kinase, Raf-1 (Fig. 12-19, step ❹), the first of three protein kinases—Raf-1, MEK, and ERK—that form a cascade in which each kinase activates the next by phosphorylation (step ❺). The protein kinases MEK and ERK are activated by phosphorylation of both a Thr and a Tyr residue. When activated, ERK mediates some of the biological effects of insulin by entering the nucleus and phosphorylating transcription factors, including Elk1 (step ❻), that modulate the transcription of about 100 insulin-regulated genes (step ❼), some of which encode proteins essential for cell division. Thus, insulin acts as a growth factor.

The proteins Raf-1, MEK, and ERK are members of three larger families, for which several nomenclatures are used. ERK is in the **MAPK** family (*m*itogen-*a*ctivated *p*rotein *k*inases; mitogens are extracellular signals that induce mitosis and cell division). Soon after discovery of the first MAPK enzyme, that enzyme was found to be activated by another protein kinase, which was named MAP kinase kinase (MEK belongs to this family), and when a third kinase that activated MAP kinase kinase was discovered, it was given the slightly ludicrous family name MAP kinase kinase kinase (Raf-1 is in this family). Somewhat less cumbersome are the abbreviations for these three families: MAPK, MAPKK, and MAPKKK. Kinases in

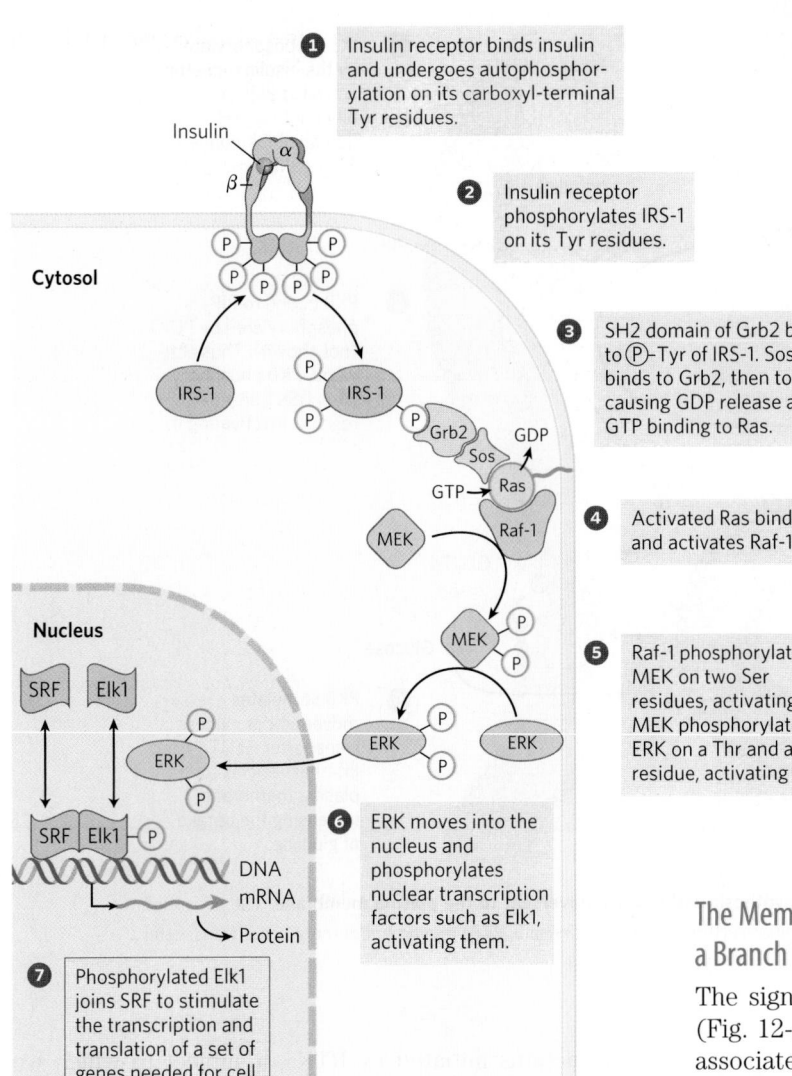

① Insulin receptor binds insulin and undergoes autophosphorylation on its carboxyl-terminal Tyr residues.

② Insulin receptor phosphorylates IRS-1 on its Tyr residues.

③ SH2 domain of Grb2 binds to ℗-Tyr of IRS-1. Sos binds to Grb2, then to Ras, causing GDP release and GTP binding to Ras.

④ Activated Ras binds and activates Raf-1.

⑤ Raf-1 phosphorylates MEK on two Ser residues, activating it. MEK phosphorylates ERK on a Thr and a Tyr residue, activating it.

⑥ ERK moves into the nucleus and phosphorylates nuclear transcription factors such as Elk1, activating them.

⑦ Phosphorylated Elk1 joins SRF to stimulate the transcription and translation of a set of genes needed for cell division.

FIGURE 12-19 Regulation of gene expression by insulin through a MAP kinase cascade. The insulin receptor (INSR) consists of two α subunits on the outer face of the plasma membrane and two β subunits that traverse the membrane and protrude from the cytoplasmic face. Binding of insulin to the α subunits triggers a conformational change that allows the autophosphorylation of Tyr residues in the carboxyl-terminal domain of the β subunits. Autophosphorylation further activates the Tyr kinase domain, which then catalyzes phosphorylation of other target proteins. The signaling pathway by which insulin regulates the expression of specific genes consists of a cascade of protein kinases, each of which activates the next. INSR is a Tyr-specific kinase; the other kinases (all shown in blue) phosphorylate Ser or Thr residues. MEK is a dual-specificity kinase that phosphorylates both a Thr and a Tyr residue in ERK (extracellular regulated kinase). MEK is *mitogen-activated, ERK-activating kinase*; SRF is *serum response factor.*

the MAPK and MAPKKK families are specific for Ser or Thr residues, and MAPKKs (here, MEK) phosphorylate both a Ser and a Tyr residue in their substrate, a MAPK (here, ERK).

Biochemists now recognize this insulin pathway as but one instance of a more general scheme in which hormone signals, via pathways similar to that shown in Figure 12-19, result in a change in the phosphorylation of target enzymes by protein kinases or phosphoprotein phosphatases. The target of phosphorylation is often another protein kinase, which then phosphorylates a third protein kinase, and so on. The result is a cascade of reactions that amplifies the initial signal by many orders of magnitude (see Fig. 12-1b). **MAPK cascades** (Fig. 12-19) mediate signaling initiated by a variety of growth factors, such as platelet-derived growth factor (PDGF) and epidermal growth factor (EGF). Another general scheme exemplified by the insulin receptor pathway is the use of nonenzymatic adaptor proteins to bring together the components of a branched signaling pathway, to which we now turn.

The Membrane Phospholipid PIP₃ Functions at a Branch in Insulin Signaling

The signaling pathway from insulin branches at IRS-1 (Fig. 12-19, step **②**). Grb2 is not the only protein that associates with phosphorylated IRS-1. The enzyme phosphoinositide 3-kinase (PI3K) binds IRS-1 through PI3K's SH2 domain **(Fig. 12-20)**. Thus activated, PI3K converts the membrane lipid phosphatidylinositol 4,5-bisphosphate (PIP₂) to phosphatidylinositol 3,4,5-trisphosphate (PIP₃) by the transfer of a phosphoryl group from ATP. The multiply (negatively) charged head group of PIP₃, protruding on the cytoplasmic side of the plasma membrane, is the starting point for a second signaling branch involving another cascade of protein kinases. When bound to PIP₃, protein kinase B (PKB; also called Akt) is phosphorylated and activated by yet another protein kinase, PDK1. The activated PKB then phosphorylates Ser or Thr residues in its target proteins, one of which is glycogen synthase kinase 3 (GSK3). In its active, nonphosphorylated form, GSK3 phosphorylates glycogen synthase, inactivating it and thereby contributing to the slowing of glycogen synthesis. (This mechanism is only part of the explanation for the effects of insulin on glycogen metabolism; see Fig. 15-42.) When phosphorylated by PKB, GSK3 is inactivated. By thus preventing inactivation of glycogen synthase in liver and muscle, the cascade of protein phosphorylations initiated by insulin stimulates glycogen synthesis (Fig. 12-20). In a third signaling branch in muscle and fat tissue, PKB triggers the clathrin-aided

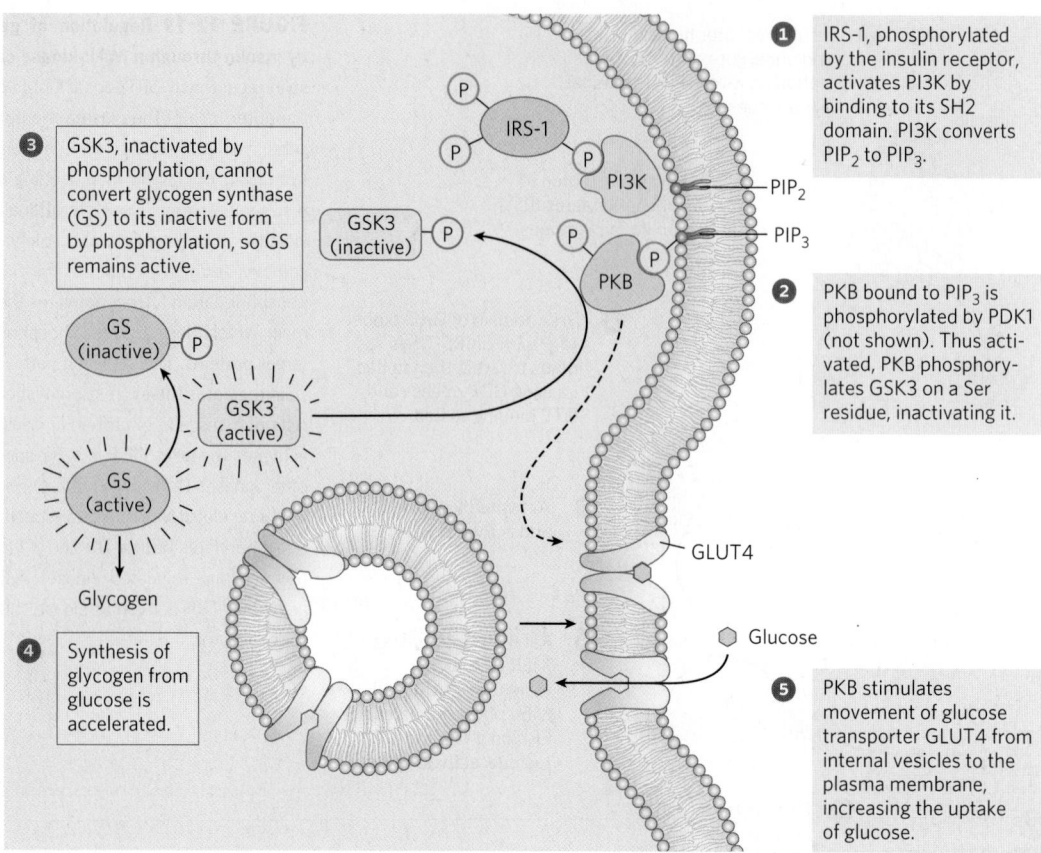

③ GSK3, inactivated by phosphorylation, cannot convert glycogen synthase (GS) to its inactive form by phosphorylation, so GS remains active.

① IRS-1, phosphorylated by the insulin receptor, activates PI3K by binding to its SH2 domain. PI3K converts PIP₂ to PIP₃.

② PKB bound to PIP₃ is phosphorylated by PDK1 (not shown). Thus activated, PKB phosphorylates GSK3 on a Ser residue, inactivating it.

④ Synthesis of glycogen from glucose is accelerated.

⑤ PKB stimulates movement of glucose transporter GLUT4 from internal vesicles to the plasma membrane, increasing the uptake of glucose.

FIGURE 12-20 Insulin action on glycogen synthesis and GLUT4 movement to the plasma membrane. The activation of PI3 kinase (PI3K) by phosphorylated IRS-1 initiates (through protein kinase B, PKB) movement of the glucose transporter GLUT4 to the plasma membrane, and the activation of glycogen synthase.

movement of glucose transporters (GLUT4) from intracellular vesicles to the plasma membrane, stimulating glucose uptake from the blood (Fig. 12-20, step **⑤**; see also Box 11-1).

As in all signaling pathways, there is a mechanism for terminating the activity of the PI3K-PKB pathway. A PIP₃-specific phosphatase (PTEN in humans) removes the phosphoryl group at the 3 position of PIP₃ to produce PIP₂, which no longer serves as a binding site for PKB, and the signaling chain is broken. In various types of cancer, it is often found that the PTEN gene has undergone mutation, resulting in a defective regulatory circuit and abnormally high levels of PIP₃ and of PKB activity. The result is a continuing signal for cell division and thus tumor growth. ∎

The insulin receptor is the prototype for several receptor enzymes with a similar structure and RTK activity **(Fig. 12-21)**. The receptors for EGF and PDGF, for example, have structural and sequence similarities to INSR, and both have a protein Tyr kinase activity that phosphorylates IRS-1. Many of these receptors dimerize after binding ligand; INSR is the exception, as it is already an $(\alpha\beta)_2$ dimer before insulin binds. (The protomer of the insulin receptor is one $\alpha\beta$ unit.) The binding of adaptor proteins such as Grb2 to ⑰–Tyr residues is a common mechanism for promoting protein-protein

interactions initiated by RTKs, a subject to which we return in Section 12.6.

In addition to the many receptors that act as protein Tyr kinases (the RTKs), several receptorlike plasma membrane proteins have protein Tyr phosphatase activity. Based on the structures of these proteins, we can surmise that their ligands are components of the extracellular matrix or are surface molecules on other cells. Although their signaling roles are not yet as well understood as those of the RTKs, they clearly have the potential to reverse the actions of signals that stimulate RTKs.

What spurred the evolution of such complicated regulatory machinery? This system allows one activated receptor to activate several IRS-1 molecules, amplifying the insulin signal, and it provides for the integration of signals from different receptors such as EGFR and PDGFR, each of which can phosphorylate IRS-1. Furthermore, because IRS-1 can activate any of several proteins that contain SH2 domains, a single receptor acting through IRS-1 can trigger two or more signaling pathways; insulin affects gene expression through the Grb2-Sos-Ras-MAPK pathway and affects glycogen metabolism and glucose transport through the PI3K-PKB pathway. Finally, there are several closely related IRS proteins (IRS2, IRS3), each with its own characteristic tissue distribution and

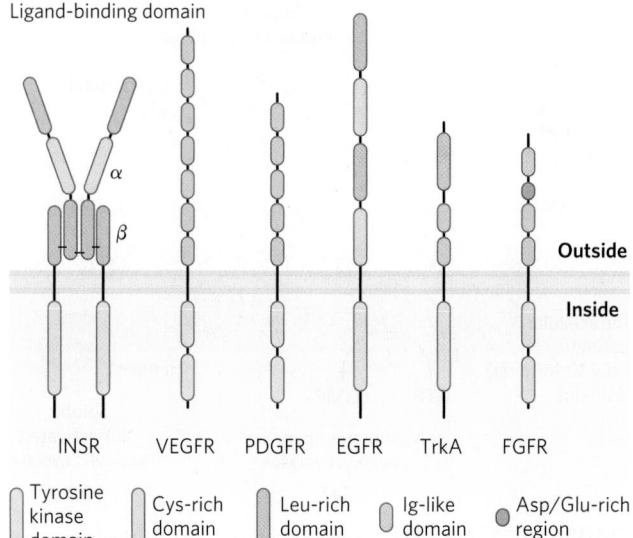

FIGURE 12-21 Receptor tyrosine kinases. Growth factor receptors that initiate signals through Tyr kinase activity include those for insulin (INSR), vascular epidermal growth factor (VEGFR), platelet-derived growth factor (PDGFR), epidermal growth factor (EGFR), high-affinity nerve growth factor (TrkA), and fibroblast growth factor (FGFR). All these receptors have a Tyr kinase domain on the cytoplasmic side of the plasma membrane (blue). The extracellular domain is unique to each type of receptor, reflecting the different growth-factor specificities. These extracellular domains are typically combinations of structural motifs such as cysteine- or leucine-rich segments and segments containing one of several motifs common to immunoglobulins (Ig). Many other receptors of this type are encoded in the human genome, each with a different extracellular domain and ligand specificity.

function, further enriching the signaling possibilities in pathways initiated by RTKs.

Cross Talk among Signaling Systems Is Common and Complex

For simplicity, we have treated individual signaling pathways as separate sequences of events leading to

separate metabolic consequences, but there is, in fact, extensive cross talk among signaling systems. The regulatory circuitry that governs metabolism is richly interwoven and multilayered. We have discussed the signaling pathways for insulin and epinephrine separately, but they do not operate independently. Insulin opposes the metabolic effects of epinephrine in most tissues, and activation of the insulin signaling pathway directly attenuates signaling through the β-adrenergic signaling system. For example, the INSR kinase directly phosphorylates two Tyr residues in the cytoplasmic tail of a β_2-adrenergic receptor, and PKB, activated by insulin **(Fig. 12-22)**, phosphorylates two Ser residues in the same region. Phosphorylation of these four residues triggers clathrin-aided internalization of the β_2-adrenergic receptor, taking it out of service and lowering the cell's sensitivity to epinephrine. A second type of cross talk between these receptors occurs when $\circledP$–Tyr residues on the β_2-adrenergic receptor, phosphorylated by INSR, serve as nucleation points for SH2 domain–containing proteins such as Grb2 (Fig. 12-22, left side). Activation of the MAPK ERK by insulin (Fig. 12-19) is 5- to 10-fold greater in the presence of the β_2-adrenergic receptor, presumably because of this cross talk. Signaling systems that use cAMP and Ca^{2+} also show extensive interaction; each second messenger affects the generation and concentration of the other. One of the major challenges of systems biology is to sort out the effects of such interactions on the overall metabolic patterns in each tissue—a daunting task.

SUMMARY 12.4 Receptor Tyrosine Kinases

■ The insulin receptor, INSR, is the prototype of receptor enzymes with Tyr kinase activity. When insulin binds, each $\alpha\beta$ unit of INSR phosphorylates the β subunit of its partner, activating the receptor's Tyr kinase activity. The kinase catalyzes the phosphorylation of Tyr residues on other proteins, such as IRS-1.

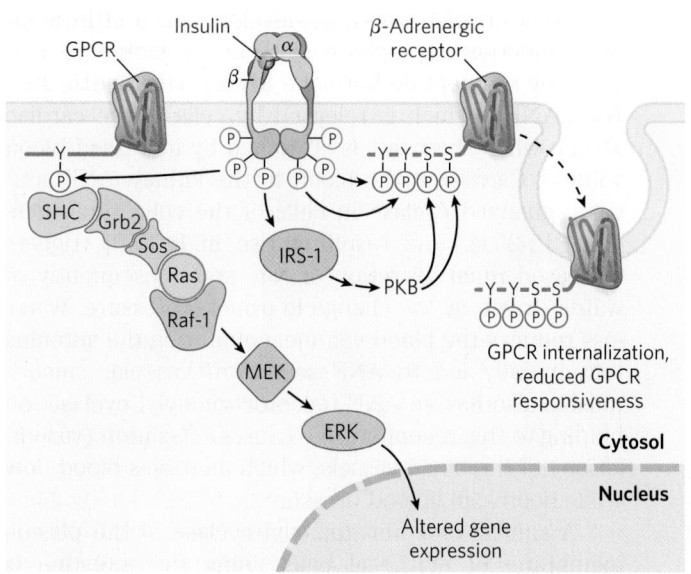

FIGURE 12-22 Cross talk between the insulin receptor and the β_2-adrenergic receptor (or other GPCR). When INSR is activated by insulin binding, its Tyr kinase directly phosphorylates the β_2-adrenergic receptor (right side) on two Tyr residues (Tyr[350] and Tyr[364]) near its carboxyl terminus, and indirectly (through activation of protein kinase B (PKB); see Fig. 12-20) causes phosphorylation of two Ser residues in the same region. The effect of these phosphorylations is internalization of the adrenergic receptor, reducing the response to the adrenergic stimulus. Alternatively (left side), INSR-catalyzed phosphorylation of a GPCR (an adrenergic or other receptor) on a carboxyl-terminal Tyr creates the point of nucleation for activating the MAPK cascade (see Fig. 12-19), with Grb2 serving as the adaptor protein. In this case, INSR has used the GPCR to enhance its own signaling.

- Phosphotyrosine residues in IRS-1 serve as binding sites for proteins with SH2 domains. Some of these proteins, such as Grb2, have two or more protein-binding domains and can serve as adaptors that bring two proteins into proximity.

- Sos bound to Grb2 catalyzes GDP-GTP exchange on Ras (a small G protein), which in turn activates a MAPK cascade that ends with the phosphorylation of target proteins in the cytosol and nucleus. The result is specific metabolic changes and altered gene expression.

- The enzyme PI3K, activated by interaction with IRS-1, converts the membrane lipid PIP_2 to PIP_3, which becomes the point of nucleation for proteins in a second and third branch of insulin signaling.

- There are extensive interconnections among signaling pathways, allowing integration and fine-tuning of multiple hormonal effects.

12.5 Receptor Guanylyl Cyclases, cGMP, and Protein Kinase G

Guanylyl cyclases **(Fig. 12-23)** are receptor enzymes that, when activated, convert GTP to the second messenger 3′,5′-cyclic GMP (cGMP):

GTP

PP$_i$

Guanosine 3′,5′-cyclic monophosphate
(cGMP)

Many of the actions of cGMP in animals are mediated by **cGMP-dependent protein kinase**, also called **protein kinase G (PKG)**. On activation by cGMP, PKG phosphorylates Ser and Thr residues in target proteins.

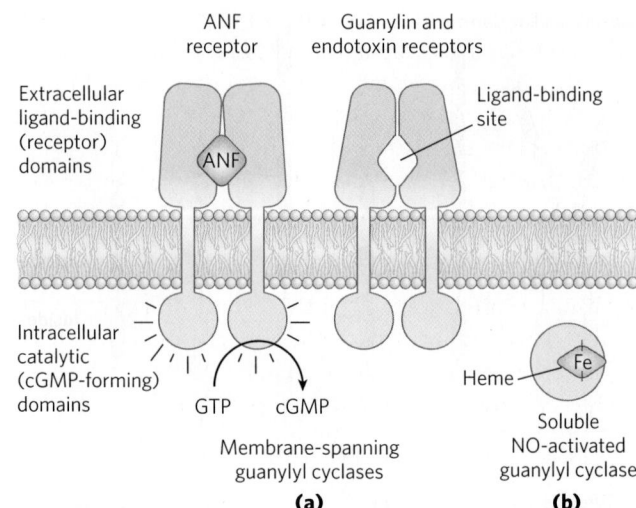

FIGURE 12-23 Two types of guanylyl cyclase that participate in signal transduction. (a) One type is a homodimer with a single membrane-spanning segment in each monomer, connecting the extracellular ligand-binding domain and the intracellular guanylyl cyclase domain. Receptors of this type are used to detect two extracellular ligands: atrial natriuretic factor (ANF; receptors in cells of the renal collecting ducts and vascular smooth muscle) and guanylin (peptide hormone produced in the intestine, with receptors in intestinal epithelial cells). The guanylin receptor is also the target of a bacterial endotoxin that triggers severe diarrhea. **(b)** The other type of guanylyl cyclase is a soluble heme-containing enzyme that is activated by intracellular nitric oxide (NO); this form is present in many tissues, including smooth muscle of the heart and blood vessels.

The catalytic and regulatory domains of this enzyme are in a single polypeptide (M_r ~80,000). Part of the regulatory domain fits snugly in the substrate-binding cleft. Binding of cGMP forces this pseudosubstrate out of the binding site, opening the site to target proteins containing the PKG consensus sequence.

Cyclic GMP carries different messages in different tissues. In the kidney and intestine it triggers changes in ion transport and water retention; in cardiac muscle (a type of smooth muscle) it signals relaxation; in the brain it may be involved both in development and in adult brain function. Guanylyl cyclase in the kidney is activated by the peptide hormone **atrial natriuretic factor (ANF)**, which is released by cells in the cardiac atrium when the heart is stretched by increased blood volume. Carried in the blood to the kidney, ANF activates guanylyl cyclase in cells of the collecting ducts (Fig. 12-23a). The resulting rise in [cGMP] triggers increased renal excretion of Na$^+$ and consequently of water, driven by the change in osmotic pressure. Water loss reduces the blood volume, countering the stimulus that initially led to ANF secretion. Vascular smooth muscle also has an ANF receptor–guanylyl cyclase; on binding to this receptor, ANF causes relaxation (vasodilation) of the blood vessels, which increases blood flow while decreasing blood pressure.

A similar receptor guanylyl cyclase in the plasma membrane of epithelial cells lining the intestine is activated by the peptide **guanylin** (Fig. 12-23a), which regulates Cl^- secretion in the intestine. This receptor is also the target of a heat-stable peptide endotoxin produced by *Escherichia coli* and other gram-negative bacteria. The elevation in [cGMP] caused by the endotoxin increases Cl^- secretion and consequently decreases reabsorption of water by the intestinal epithelium, producing diarrhea.

A distinctly different type of guanylyl cyclase is a cytosolic protein with a tightly associated heme group (Fig. 12-23b), an enzyme activated by nitric oxide (NO). Nitric oxide is produced from arginine by Ca^{2+}-dependent **NO synthase**, present in many mammalian tissues, and diffuses from its cell of origin into nearby cells.

NO is sufficiently nonpolar to cross plasma membranes without a carrier. In the target cell, it binds to the heme group of guanylyl cyclase and activates cGMP production. In the heart, cGMP-dependent protein kinase reduces the forcefulness of contractions by stimulating the ion pump(s) that remove Ca^{2+} from the cytosol.

NO-induced relaxation of cardiac muscle is the same response brought about by nitroglycerin and other nitrovasodilators taken to relieve **angina pectoris**, the pain caused by contraction of a heart deprived of O_2 because of blocked coronary arteries. Nitric oxide is unstable and its action is brief; within seconds of its formation, it undergoes oxidation to nitrite or nitrate. Nitrovasodilators produce long-lasting relaxation of cardiac muscle because they break down over several hours, yielding a steady stream of NO. The value of nitroglycerin as a treatment for angina was discovered serendipitously in factories producing nitroglycerin as an explosive in the 1860s. Workers with angina reported that their condition was much improved during the workweek but worsened on weekends. The physicians treating these workers heard this story so often that they made the connection, and a drug was born.

The effects of increased cGMP synthesis diminish after the stimulus ceases, because a specific phosphodiesterase (cGMP PDE) converts cGMP to the inactive 5'-GMP. Humans have several isoforms of cGMP PDE, with different tissue distributions. The isoform in the blood vessels of the penis is inhibited by the drugs sildenafil (Viagra) and tadalafil (Cialis), which therefore cause [cGMP] to remain elevated once raised by an appropriate stimulus, accounting for the usefulness of this drug in the treatment of erectile dysfunction.

Sildenafil (Viagra)

Tadalafil (Cialis)

Cyclic GMP has another mode of action in the vertebrate eye: it causes ion-specific channels to open in the retinal rod and cone cells, as we discussed in Section 12.3.

SUMMARY 12.5 Receptor Guanylyl Cyclases, cGMP, and Protein Kinase G

■ Several signals, including atrial natriuretic factor and guanylin, act through receptor enzymes with guanylyl cyclase activity. The cGMP so produced is a second messenger that activates cGMP-dependent protein kinase (PKG). This enzyme alters metabolism by phosphorylating specific enzyme targets.

■ Nitric oxide is a short-lived messenger that stimulates a soluble guanylyl cyclase, raising [cGMP] and stimulating PKG.

12.6 Multivalent Adaptor Proteins and Membrane Rafts

Two generalizations have emerged from studies of signaling systems such as those we have discussed so far. First, protein kinases that phosphorylate, and phosphatases that dephosphorylate, Tyr, Ser, and Thr residues are central to signaling, *directly* affecting the activities of a large number of protein substrates by phosphorylation/dephosphorylation. Second, protein-protein interactions brought about by the reversible phosphorylation of Tyr, Ser, and Thr residues in signaling proteins create *docking sites* for other proteins that bring about *indirect* effects on proteins downstream in the signaling pathway. In fact, many signaling proteins are *multivalent*: they can interact with several different proteins simultaneously to form multiprotein signaling complexes. In this section we present a few examples to illustrate the general principles of phosphorylation-dependent protein interactions in signaling pathways.

Protein Modules Bind Phosphorylated Tyr, Ser, or Thr Residues in Partner Proteins

The protein Grb2 in the insulin signaling pathway (Figs 12-19 and 12-22) binds through its SH2 domain to other proteins that have exposed Ⓟ–Tyr residues. The human genome encodes at least 87 SH2-containing proteins, many already known to participate in signaling. The Ⓟ–Tyr residue is bound in a deep pocket in an SH2 domain, with each of its phosphate oxygens participating in hydrogen bonding or electrostatic interactions; the positive charges on two Arg residues figure prominently in the binding. Subtle differences in the structure of SH2 domains account for the specificities of the interactions of SH2-containing proteins with various Ⓟ–Tyr-containing proteins. The SH2 domain typically interacts with a Ⓟ–Tyr (which is assigned the index position 0) and the next three residues toward the carboxyl terminus (designated +1, +2, +3). Some proteins with SH2 domains (Src, Fyn, Hck, Nck) favor negatively charged residues in the +1 and +2 positions; others (PLCγ1, SHP2) have a long hydrophobic groove that binds to aliphatic residues in positions +1 to +5. These differences define subclasses of SH2 domains that have different partner specificities.

Phosphotyrosine-binding domains, or **PTB domains**, are another binding partner for Ⓟ–Tyr proteins **(Fig. 12-24)**, but their critical sequences and three-dimensional structure distinguish them from SH2 domains. The human genome encodes 24 proteins that contain PTB domains, including IRS-1, which we have already encountered in its role as an adaptor protein in insulin-signal transduction (Fig. 12-19). The Ⓟ–Tyr binding sites for SH2 and PTB domains on partner proteins are created by Tyr kinases and eliminated by protein tyrosine phosphatases (PTPs).

As we have seen, other signaling protein kinases, including PKA, PKC, PKG, and members of the MAPK cascade, phosphorylate Ser or Thr residues in their target proteins, which in some cases acquire the ability to interact with partner proteins through the phosphorylated residue, triggering a downstream process. An alphabet soup of domains that bind Ⓟ–Ser or Ⓟ–Thr residues has been identified, and more are sure to be found. Each domain favors a certain sequence around the phosphorylated residue, so proteins with that domain bind to and interact with a specific subset of phosphorylated proteins.

In some cases, the region on a protein that binds Ⓟ–Tyr of a substrate protein is masked by the region's interaction with a Ⓟ–Tyr in the same protein. For example, the soluble protein Tyr kinase Src, when phosphorylated on a critical Tyr residue, is rendered inactive; an SH2 domain needed to bind the substrate protein instead binds the internal Ⓟ–Tyr. When this Ⓟ–Tyr residue is hydrolyzed by a phosphoprotein phosphatase, the Tyr kinase activity of Src is activated **(Fig. 12-25a)**.

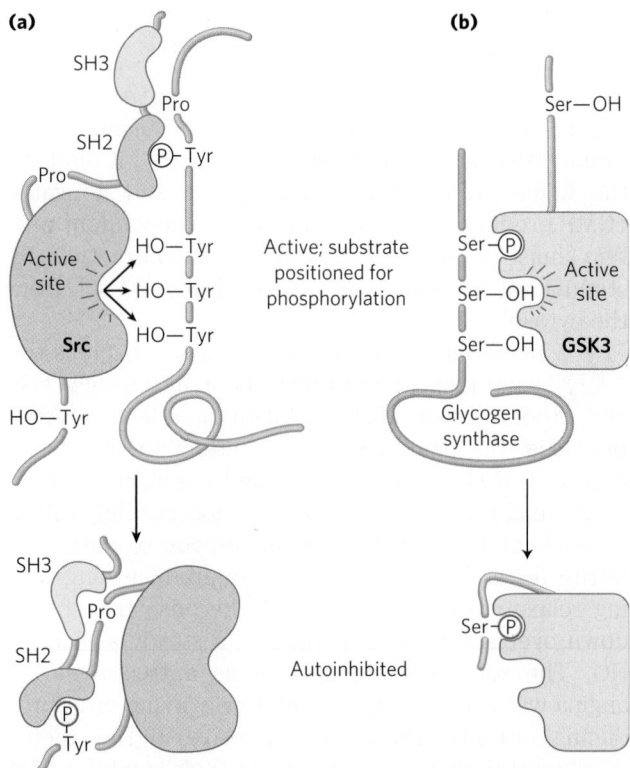

FIGURE 12-25 Mechanism of autoinhibition of Src and GSK3. (a) In the active form of the Tyr kinase Src, an SH2 domain binds a Ⓟ–Tyr in the protein substrate, and an SH3 domain binds a proline-rich region of the substrate, thus lining up the active site of the kinase with several target Tyr residues in the substrate (top). When Src is phosphorylated on a specific Tyr residue (bottom), the SH2 domain binds the internal Ⓟ–Tyr instead of the Ⓟ–Tyr of the substrate, and the SH3 domain binds an internal proline-rich region, preventing productive enzyme-substrate binding; the enzyme is thus autoinhibited. (b) In the active form of glycogen synthase kinase 3 (GSK3), an internal Ⓟ–Ser-binding domain is available to bind Ⓟ–Ser in its substrate (glycogen synthase) and to position the kinase to phosphorylate neighboring Ser residues (top). Phosphorylation of an internal Ser residue allows this internal kinase segment to occupy the Ⓟ–Ser-binding site, blocking substrate binding (bottom).

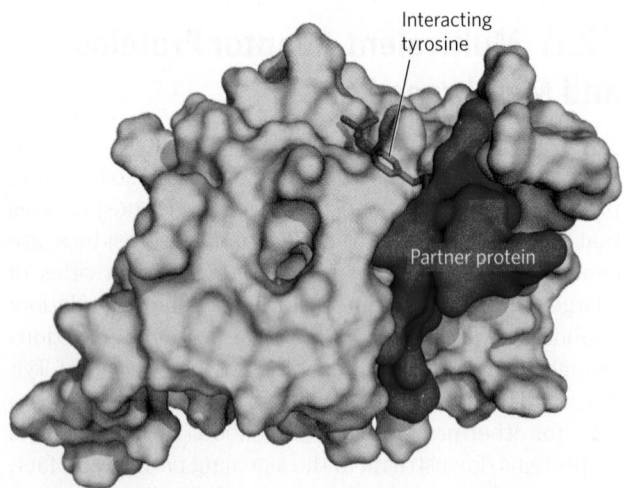

FIGURE 12-24 Interaction of a PTB domain with a Ⓟ-Tyr residue in a partner protein. The PTB domain is represented as a blue surface contour. The partner protein is held to the kinase by multiple noncovalent interactions, which confer specificity on the interaction and position the Ⓟ-Tyr residue in a binding pocket at the enzyme's active site. [Source: PDB ID 1SHC, M. M. Zhou et al., *Nature* 378:584, 1995.]

Similarly, glycogen synthase kinase 3 (GSK3) is inactive when phosphorylated on a Ser residue in its autoinhibitory domain (Fig. 12-25b). Dephosphorylation of that domain frees the enzyme to bind (and then phosphorylate) its target proteins.

In addition to the three commonly phosphorylated residues in proteins, there is a fourth phosphorylated structure that nucleates the formation of supramolecular complexes of signaling proteins: the phosphorylated head group of the membrane phosphatidylinositols. Many signaling proteins contain domains such as SH3 and PH (pleckstrin homology domain) that bind tightly to PIP_3 protruding on the cytoplasmic side of the plasma membrane. Wherever the enzyme PI3K creates this head group (as it does in response to the insulin signal), proteins that bind PIP_3 will cluster at the membrane surface.

Most of the proteins involved in signaling at the plasma membrane have one or more protein- or phospholipid-binding domains; many have three or more, and thus are multivalent in their interactions with other signaling proteins. **Figure 12-26** shows just a few of the multivalent proteins known to participate in signaling.

Many of the complexes include components with membrane-binding domains. Given the location of so many signaling processes at the inner surface of the plasma membrane, the molecules that must collide to produce the signaling response are effectively confined to two-dimensional space—the membrane surface; collisions are far more likely here than in the three-dimensional space of the cytosol.

In summary, a remarkable picture of signaling pathways has emerged from studies of many signaling proteins and their multiple binding domains. An initial signal results in phosphorylation of the receptor or a target protein, triggering the assembly of large multiprotein complexes, held together on scaffolds with multivalent binding capacities. Some of these complexes contain several protein kinases that activate each other in turn, producing a cascade of phosphorylation and a great amplification of the initial signal. The interactions between cascade kinases are not left to the vagaries of random collisions in three-dimensional space. In the MAPK cascade, for example, a scaffold protein, KSR, binds all three kinases (MAPK, MAPKK, and MAPKKK),

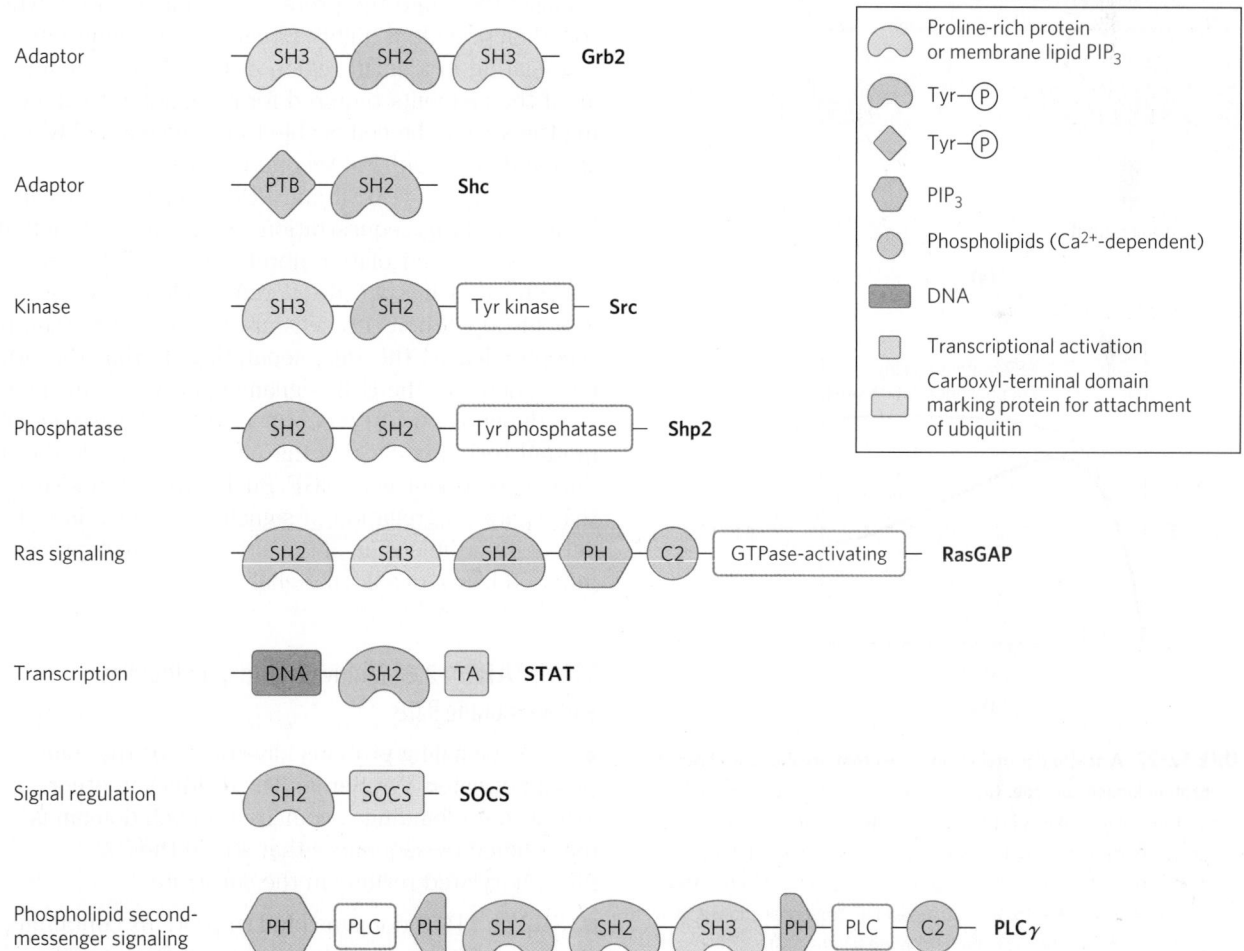

FIGURE 12-26 Some binding modules of signaling proteins. Each protein is represented by a line (with the amino terminus to the left); symbols indicate the location of conserved binding domains (with specificities as listed in the key; abbreviations are explained in the text); green boxes indicate catalytic activities. The name of each protein is given at its carboxyl-terminal end. These signaling proteins interact with phosphorylated proteins or phospholipids in many permutations and combinations to form integrated signaling complexes. [Source: Information from T. Pawson et al., *Trends Cell Biol.* 11:504, 2001, Fig. 5.]

ensuring their proximity and correct orientation and even conferring allosteric properties on the interactions among the kinases, which makes their serial phosphorylation sensitive to very small stimuli **(Fig. 12-27)**.

Phosphotyrosine phosphatases remove the phosphate from ⓟ–Tyr residues, reversing the effect of phosphorylation. There are at least 37 genes encoding protein tyrosine phosphatases (PTPs) in the human genome. About half of these are receptorlike integral proteins with a single transmembrane domain; they are presumably controlled by extracellular factors not yet identified. Other PTPs are soluble and contain SH2 domains that determine their molecular partners and intracellular location. In addition, animal cells have protein ⓟ–Ser

and ⓟ–Thr phosphatases such as PP1 that reverse the effects of Ser- and Thr-specific protein kinases. We can see, then, that signaling occurs in **protein circuits**, which are effectively hardwired from signal receptor to response effector and can be switched off instantly by the hydrolysis of a single upstream phosphate ester bond. In these circuits, protein kinases are the writers, domains such as SH2 are the readers, and PTPs and other phosphatases, the erasers.

The multivalency of signaling proteins allows the assembly of many different combinations of Lego-like signaling modules, each combination suited to particular signals, cell types, and metabolic circumstances, yielding diverse signaling circuits of extraordinary complexity.

Membrane Rafts and Caveolae Segregate Signaling Proteins

Membrane rafts (Chapter 11) are regions of the membrane bilayer enriched in sphingolipids, sterols, and certain proteins, including many proteins attached to the bilayer by GPI (*g*lycosylated derivatives of *p*hosphatidyl*i*nositol) anchors. The β-adrenergic receptor is segregated in rafts that also contain G proteins, adenylyl cyclase, PKA, and the protein phosphatase PP2, which together provide a highly integrated signaling unit. By segregating in a small region of the plasma membrane all of the elements required for responding to and ending the signal, the cell is able to produce a highly localized and brief "puff" of second messenger.

Some RTKs (EGFR and PDGFR) are also localized in rafts, and this sequestration very likely has functional significance. In isolated fibroblasts, EGFR is usually concentrated in specialized rafts called caveolae (see Fig. 11-20). When the cells are treated with EGF, the receptor leaves the raft, separating it from the other components of the EGF signaling pathway. This migration depends on the receptor's protein kinase activity; mutant receptors lacking this activity remain in the raft during treatment with EGF. Such experiments suggest that spatial segregation of signaling proteins in rafts is yet another dimension of the already complex processes initiated by extracellular signals.

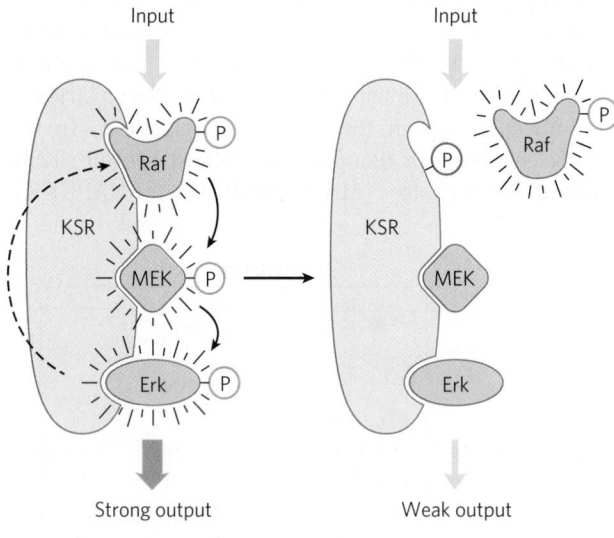

(a)

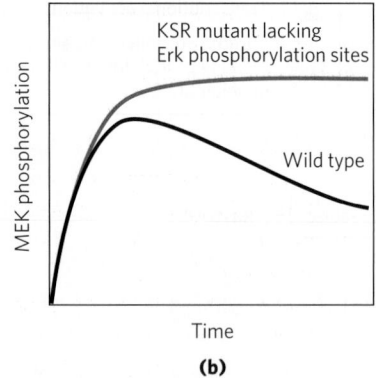

(b)

FIGURE 12-27 A scaffold protein from yeast that organizes and regulates a protein kinase cascade. (a) The scaffold protein KSR has binding sites for all three of the kinases in the Raf-MEK-Erk cascade. With the binding of all three in appropriate orientations, interactions among the proteins are rapid and efficient. When Erk has been activated (left), it phosphorylates the binding site for Raf (right), forcing a conformational change that displaces Raf and thereby prevents the phosphorylation of MEK. The result of this feedback regulation is that MEK phosphorylation is temporary. **(b)** In yeast cells with mutant KSR lacking the phosphorylation sites (red curve), no feedback occurs, producing a different time course of signaling. [Source: Information from M. C. Good et al., *Science* 332:680, 2011, Fig. 2E.]

SUMMARY 12.6 Multivalent Adaptor Proteins and Membrane Rafts

■ Many signaling proteins have domains that bind phosphorylated Tyr, Ser, or Thr residues in other proteins; the binding specificity for each domain is determined by sequences that adjoin the phosphorylated residue in the substrate.

■ SH2 and PTB domains bind to proteins containing ⓟ–Tyr residues; other domains bind ⓟ–Ser and ⓟ–Thr residues in various contexts.

■ SH3 and PH domains bind the membrane phospholipid PIP$_3$.

■ Many signaling proteins are multivalent, with several different binding modules. By combining the substrate specificities of various protein kinases with the specificities of domains that bind phosphorylated Ser, Thr, or Tyr residues, and with phosphatases that can rapidly inactivate a signaling pathway, cells create a large number of multiprotein signaling complexes.

■ Membrane rafts and caveolae sequester groups of signaling proteins in small regions of the plasma membrane, effectively raising their local concentrations and making signaling more efficient.

12.7 Gated Ion Channels

Ion Channels Underlie Electrical Signaling in Excitable Cells

Certain cells in multicellular organisms are "excitable": they can detect an external signal, convert it into an electrical signal (specifically, a change in membrane potential), and pass it on. Excitable cells play central roles in nerve conduction, muscle contraction, hormone secretion, sensory processes, and learning and memory. The excitability of sensory cells, neurons, and myocytes depends on ion channels, signal transducers that provide a regulated path for the movement of inorganic ions such as Na^+, K^+, Ca^{2+}, and Cl^- across the plasma membrane in response to various stimuli. Recall from Chapter 11 that these ion channels are "gated": they may be open or closed, depending on whether the associated receptor has been activated by the binding of its specific ligand (a neurotransmitter, for example) or by a change in the transmembrane electrical potential. The Na^+K^+ ATPase is electrogenic; it creates a charge imbalance across the plasma membrane by carrying 3 Na^+ out of the cell for every 2 K^+ carried in **(Fig. 12-28a)**. The action of the ATPase makes the inside of the cell negative relative to the outside. Inside the cell, $[K^+]$ is much higher and $[Na^+]$ is much lower than outside the cell (Fig. 12-28b). The direction of spontaneous ion flow across a polarized membrane is dictated by the electrochemical potential of that ion across the membrane, which has two components: the difference in concentration of the ion on the two sides of the membrane, and the difference in electrical potential (V_m), typically expressed in millivolts (see Eqn 11-4, p. 413). Given the ion concentration differences and a V_m of about -60 mV (inside negative), opening of a Na^+ or Ca^{2+} channel will result in a spontaneous inward flow of Na^+ or Ca^{2+} (and depolarization), whereas opening of a K^+ channel will result in a spontaneous outward flow of K^+ (and hyperpolarization) (Fig. 12-28b). In this case, K^+ moves out of the cell against the electrical gradient, because the large concentration difference exerts a stronger effect than the V_m. For Cl^-, the membrane potential predominates, so when a Cl^- channel opens, Cl^- flows outward.

The number of ions that must flow to produce a physiologically significant change in the membrane potential is negligible relative to the concentrations of

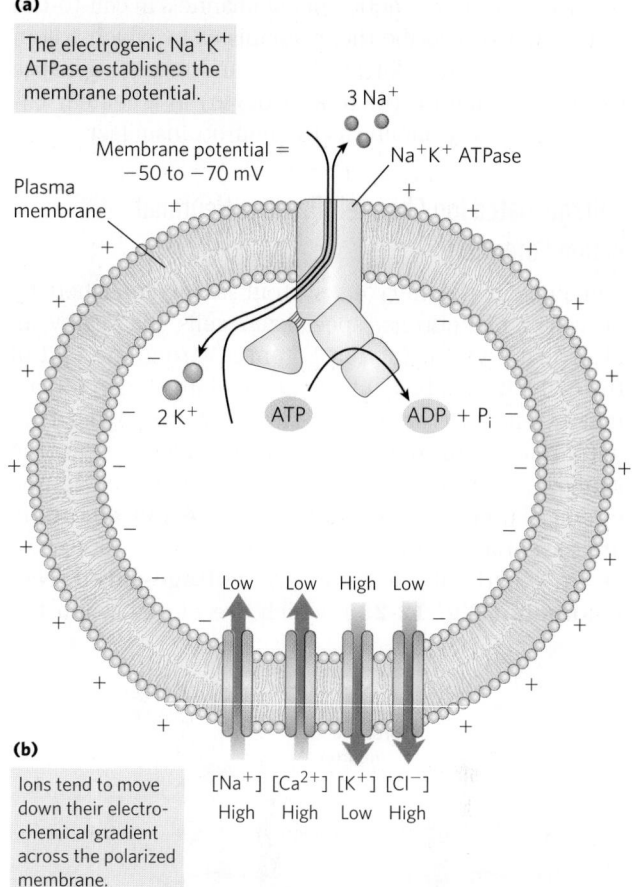

(a)

The electrogenic Na^+K^+ ATPase establishes the membrane potential.

Membrane potential = -50 to -70 mV

3 Na^+

Na^+K^+ ATPase

Plasma membrane

2 K^+ ATP   ADP + P_i

Low Low High Low

(b)

Ions tend to move down their electrochemical gradient across the polarized membrane.

| $[Na^+]$ | $[Ca^{2+}]$ | $[K^+]$ | $[Cl^-]$ |
| High | High | Low | High |

FIGURE 12-28 Transmembrane electrical potential. (a) The electrogenic Na^+K^+ ATPase produces a transmembrane electrical potential of about 60 mV (inside negative). **(b)** Blue arrows show the direction in which ions tend to move spontaneously across the plasma membrane in an animal cell, driven by the combination of chemical and electrical gradients. The chemical gradient drives Na^+ and Ca^{2+} inward (producing depolarization) and K^+ outward, against its electrical gradient (producing hyperpolarization). The electrical gradient drives Cl^- outward, against its concentration gradient (producing depolarization).

Na^+, K^+, and Cl^- in cells and extracellular fluid, so the ion fluxes that occur during signaling in excitable cells have essentially no effect on the concentrations of these ions. With Ca^{2+}, the situation is different; because the intracellular $[Ca^{2+}]$ is generally very low ($\sim10^{-7}$ M), inward flow of Ca^{2+} can significantly alter the cytosolic $[Ca^{2+}]$, allowing it to serve as a second messenger.

The membrane potential of a cell at a given time is the result of the types and numbers of ion channels open at that instant. The precisely timed opening and closing of ion channels and the resulting transient changes in membrane potential underlie the electrical signaling by which the nervous system stimulates the skeletal muscles to contract, the heart to beat, or secretory cells to release their contents. Moreover, many hormones exert their effects by altering the membrane potential of their target cells. These mechanisms are not limited to animals; ion channels play important roles in the responses of bacteria, protists, and plants to environmental signals.

To illustrate the action of ion channels in cell-to-cell signaling, we describe the mechanisms by which a neuron passes a signal along its length and across a synapse to the next neuron (or to a myocyte) in a cellular circuit, using acetylcholine as the neurotransmitter.

Voltage-Gated Ion Channels Produce Neuronal Action Potentials

Signaling in the nervous system is accomplished by networks of neurons, specialized cells that carry an electrical impulse (action potential) from one end of the cell (the cell body) through an elongated cytoplasmic extension (the axon). The electrical signal triggers release of neurotransmitter molecules at the synapse, carrying the signal to the next cell in the circuit. Three types of **voltage-gated ion channels** are essential to this signaling mechanism. Along the entire length of the axon are **voltage-gated Na⁺ channels (Fig. 12-29)**, which are closed when the

membrane is at rest ($V_m = -60$ mV) but open briefly when the membrane is depolarized locally in response to acetylcholine (or some other neurotransmitter). Also distributed along the axon are **voltage-gated K⁺ channels**, which open, a split second later, in response to the depolarization that results when nearby Na⁺ channels open. The depolarizing flow of Na⁺ into the axon (influx) is thus rapidly countered by a repolarizing outward flow of K⁺ (efflux). At the distal end of the axon are **voltage-gated Ca²⁺ channels**, which open when the wave of depolarization (Fig. 12-29, step ❶) and repolarization (step ❷) caused by the activity of Na⁺ and K⁺ channels arrives, triggering release of the neurotransmitter acetylcholine—which carries the signal to another neuron (fire an action potential!) or to a muscle fiber (contract!).

The voltage-gated Na⁺ channels are selective for Na⁺ over other cations by a factor of 100 or more. They also have a very high flux rate of $>10^7$ ions/s. A Na⁺ channel that opens in response to a reduction in transmembrane electrical potential closes within milliseconds and remains unable to reopen for many milliseconds. The influx of Na⁺ through the open Na⁺ channels depolarizes the membrane locally, causing voltage-gated K⁺ channels to open (Fig. 12-29, step ❶). The resulting K⁺ efflux repolarizes the membrane locally, reestablishing the inside-negative membrane potential (step ❷). (We discuss the structure and mechanism of voltage-gated K⁺ channels in some detail in Section 11.3; see Figs 11-45 and 11-46.) A brief pulse of

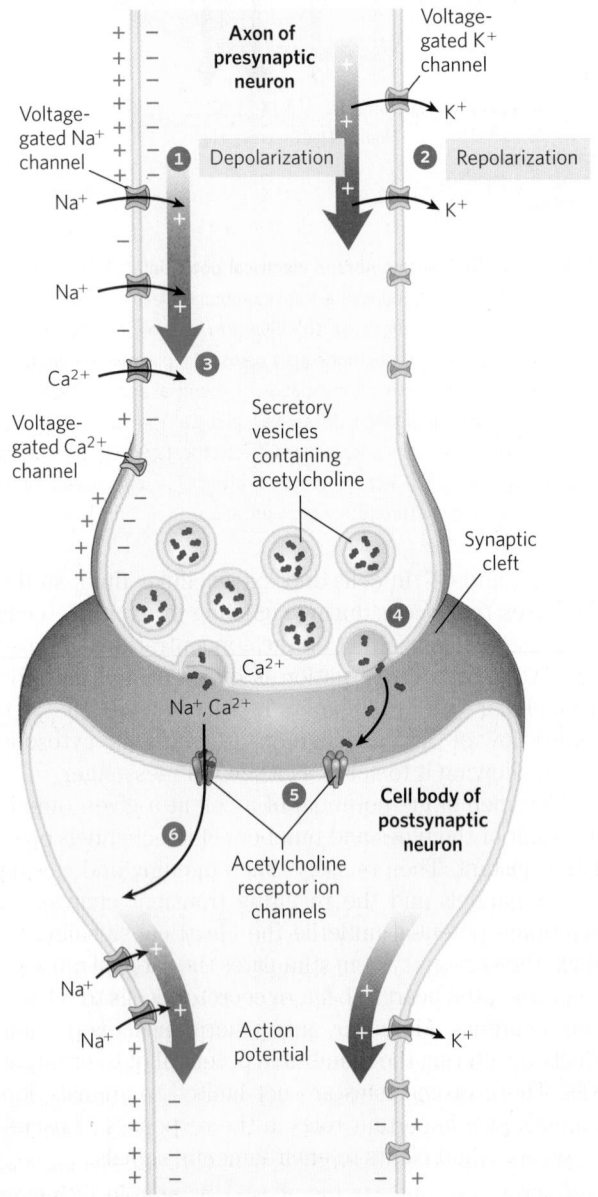

FIGURE 12-29 Role of voltage-gated and ligand-gated ion channels in neural transmission. Initially, the plasma membrane of the presynaptic neuron is polarized (inside negative) through the action of the electrogenic Na⁺K⁺ ATPase, which pumps out 3 Na⁺ for every 2 K⁺ pumped in (see Fig. 12-28). ❶ A stimulus to this neuron (not shown) causes an action potential to move along the axon (blue arrow), away from the cell body. The opening of a voltage-gated Na⁺ channel allows Na⁺ entry, and the resulting local depolarization causes the adjacent Na⁺ channel to open, and so on. The directionality of movement of the action potential is ensured by the brief refractory period that follows the opening of each voltage-gated Na⁺ channel. ❷ A split second after the action potential passes a point in the axon, voltage-gated K⁺ channels open, allowing K⁺ exit, which brings about repolarization of the membrane (red arrow) to make it ready for the next action potential. (Note that, for clarity, Na⁺ channels and K⁺ channels are drawn on opposite sides of the axon, but both types are uniformly distributed in the axonal membrane; also, positive and negative charges are shown only on the left, but as the wave of potential sweeps the axon, the membrane potential is the same at any given point along the axon.) ❸ When the wave of depolarization reaches the axon tip, voltage-gated Ca²⁺ channels open, allowing Ca²⁺ entry. ❹ The resulting increase in internal [Ca²⁺] triggers exocytotic release of the neurotransmitter acetylcholine into the synaptic cleft. ❺ Acetylcholine binds to a receptor on the postsynaptic neuron (or myocyte), causing its ligand-gated ion channel to open. ❻ Extracellular Na⁺ and Ca²⁺ enter through this channel, depolarizing the postsynaptic cell. The electrical signal has thus passed to the cell body of the postsynaptic neuron and will move along its axon to a third neuron (or a myocyte) by this same sequence of events.

depolarization thus traverses the axon as local depolarization triggers the brief opening of neighboring Na^+ channels, then K^+ channels. The short period that follows the opening of each Na^+ channel, during which it cannot open again, ensures that a unidirectional wave of depolarization—the action potential—sweeps from the nerve cell body toward the end of the axon.

When the wave of depolarization reaches the voltage-gated Ca^{2+} channels, they open (step ❸), and Ca^{2+} enters from the extracellular space. The rise in cytoplasmic $[Ca^{2+}]$ then triggers release of acetylcholine by exocytosis into the synaptic cleft (step ❹). Acetylcholine diffuses to the postsynaptic cell (another neuron or a myocyte), where it binds to acetylcholine receptors and triggers depolarization (described below). Thus the message is passed to the next cell in the circuit. We see, then, that gated ion channels convey signals in either of two ways: by changing the cytoplasmic concentration of an ion (such as Ca^{2+}), which then serves as an intracellular second messenger, or by changing V_m and affecting other membrane proteins that are sensitive to V_m. The passage of an electrical signal through one neuron and on to the next illustrates both types of mechanism.

Neurons Have Receptor Channels That Respond to Different Neurotransmitters

Animal cells, especially those of the nervous system, contain a variety of ion channels gated by ligands, voltage, or both. Receptors that are themselves ion channels are classified as **ionotropic**, to distinguish them from receptors that generate a second messenger (**metabotropic** receptors). Acetylcholine acts on an ionotropic receptor in the postsynaptic cell. The acetylcholine receptor is a cation channel. When occupied by acetylcholine, the receptor opens to the passage of cations (Na^+, K^+, and Ca^{2+}), triggering depolarization of the cell. The neurotransmitters serotonin, glutamate, and glycine all can act through ionotropic receptors that are structurally related to the acetylcholine receptor. Serotonin and glutamate trigger the opening of cation (Na^+, K^+, Ca^{2+}) channels, whereas glycine opens Cl^--specific channels.

Depending on which ion passes through a channel, binding of the ligand (neurotransmitter) for that channel results in either depolarization or hyperpolarization of the target cell. A single neuron normally receives input from many other neurons, each releasing its own characteristic neurotransmitter with its characteristic depolarizing or hyperpolarizing effect. The target cell's V_m therefore reflects the *integrated* input (Fig. 12-1e) from multiple neurons. The cell responds with an action potential only if the integrated input adds up to a net depolarization of sufficient size.

The receptor channels for acetylcholine, glycine, glutamate, and γ-aminobutyric acid (GABA) are gated by *extracellular* ligands. *Intracellular* second messengers—such as cAMP, cGMP, IP_3, Ca^{2+}, and ATP—regulate ion channels of the type we saw in the sensory transductions of vision, olfaction, and gustation.

Toxins Target Ion Channels

Many of the most potent toxins found in nature act on ion channels. For example, dendrotoxin (from the black mamba snake) blocks the action of voltage-gated K^+ channels, tetrodotoxin (produced by puffer fish) acts on voltage-gated Na^+ channels, and cobrotoxin disables acetylcholine receptor ion channels. Why, in the course of evolution, have ion channels become the preferred target of toxins, rather than some critical metabolic target such as an enzyme essential in energy metabolism?

Ion channels are extraordinary amplifiers; opening of a single channel can allow the flow of 10 million ions per second. Consequently, relatively few molecules of an ion channel protein are needed per neuron for signaling functions. This means that a relatively small number of toxin molecules with high affinity for ion channels, acting from outside the cell, can have a pronounced effect on neurosignaling throughout the body. A comparable effect by way of a metabolic enzyme, typically present in cells at much higher concentrations than ion channels, would require far greater numbers of the toxin molecule.

SUMMARY 12.7 Gated Ion Channels

■ Ion channels gated by membrane potential or ligands are central to signaling in neurons and other cells.

■ The voltage-gated Na^+ and K^+ channels of neuronal membranes carry the action potential along the axon as a wave of depolarization (Na^+ influx) followed by repolarization (K^+ efflux).

■ Arrival of an action potential at the distal end of a presynaptic neuron triggers neurotransmitter release. The neurotransmitter (acetylcholine, for example) diffuses to the postsynaptic neuron (or the myocyte, at a neuromuscular junction), binds to specific receptors in the plasma membrane, and triggers a change in V_m.

■ Neurotoxins, produced by many organisms, attack neuronal ion channels and are therefore fast-acting and deadly.

12.8 Regulation of Transcription by Nuclear Hormone Receptors

The steroid, retinoic acid (retinoid), and thyroid hormones form a large group of receptor ligands that exert at least part of their effects by a mechanism fundamentally different from that of other hormones: they act directly in the nucleus to alter gene expression. We discuss their mode of action in detail in Chapter 28, along with other mechanisms for regulating gene expression. Here we give a brief overview.

Steroid hormones (estrogen, progesterone, and cortisol, for example), too hydrophobic to dissolve readily in the blood, are transported on specific carrier proteins from their point of release to their target tissues. In target cells, these hormones pass through the plasma membrane and nuclear membrane by simple diffusion

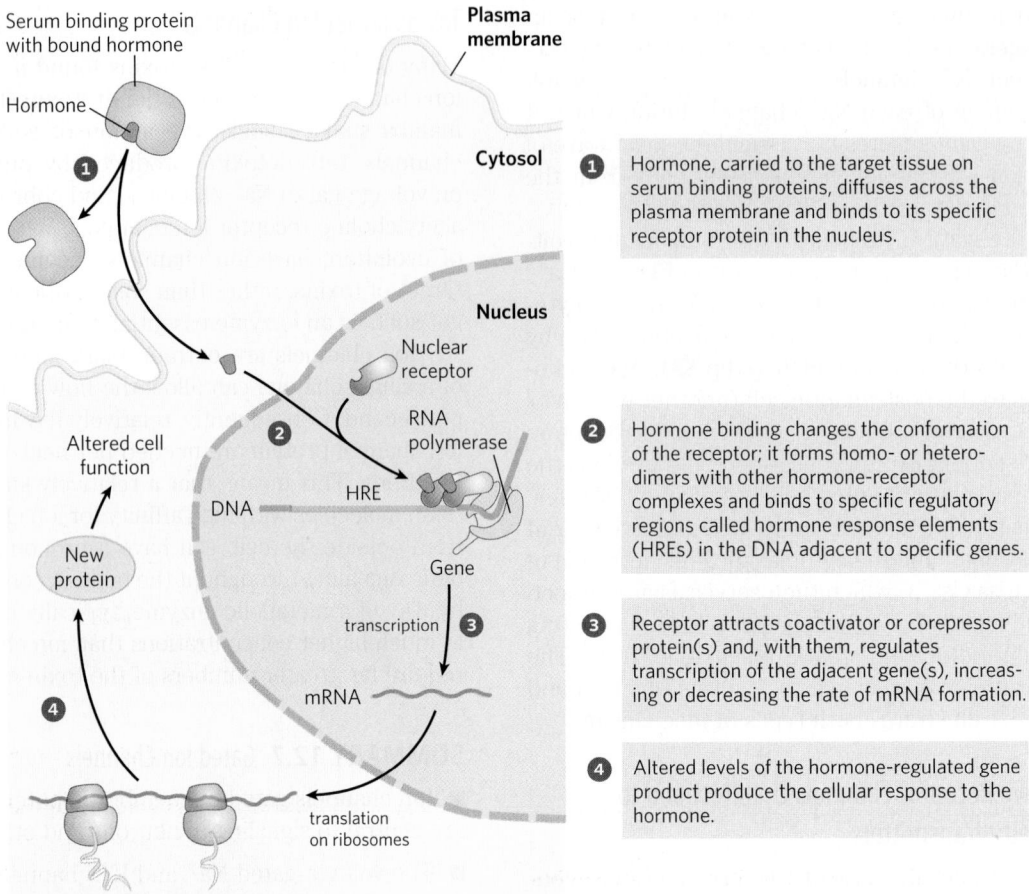

FIGURE 12-30 General mechanism by which steroid and thyroid hormones, retinoids, and vitamin D regulate gene expression. The details of transcription and protein synthesis are discussed in Chapters 26 and 27. Some steroids also act through plasma membrane receptors by a completely different mechanism.

1 Hormone, carried to the target tissue on serum binding proteins, diffuses across the plasma membrane and binds to its specific receptor protein in the nucleus.

2 Hormone binding changes the conformation of the receptor; it forms homo- or hetero-dimers with other hormone-receptor complexes and binds to specific regulatory regions called hormone response elements (HREs) in the DNA adjacent to specific genes.

3 Receptor attracts coactivator or corepressor protein(s) and, with them, regulates transcription of the adjacent gene(s), increasing or decreasing the rate of mRNA formation.

4 Altered levels of the hormone-regulated gene product produce the cellular response to the hormone.

and bind to specific receptor proteins in the nucleus **(Fig. 12-30)**. Hormone binding triggers changes in the conformation of a receptor protein so that it becomes capable of interacting with specific regulatory sequences in DNA called **hormone response elements (HREs)**, thus altering gene expression (see Fig. 28-33). The bound receptor-hormone complex enhances the expression of specific genes adjacent to HREs, with the help of several other proteins essential for transcription. Hours or days are required for these regulators to have their full effect—the time required for the changes in RNA synthesis and subsequent protein synthesis to become evident in altered metabolism.

The specificity of the steroid-receptor interaction is exploited in the use of the drug **tamoxifen** to treat breast cancer. In some types of breast cancer, division of the cancerous cells depends on the continued presence of estrogen. Tamoxifen is an estrogen antagonist; it competes with estrogen for binding to the estrogen receptor, but the tamoxifen-receptor complex has little or no effect on gene expression. Consequently, tamoxifen administered after surgery or during chemotherapy for hormone-dependent breast cancer slows or stops the growth of remaining cancerous cells. Another steroid analog, the drug **mifepristone (RU486)**, binds to

the progesterone receptor and blocks hormone actions essential to implantation of the fertilized ovum in the uterus, and thus functions as a contraceptive.

Tamoxifen

Mifepristone (RU486)

SUMMARY 12.8 Regulation of Transcription by Nuclear Hormone Receptors

■ Steroid hormones enter cells by simple diffusion and bind to specific receptor proteins.

■ The hormone-receptor complex binds specific regions of DNA, the hormone response elements, and interacts with other proteins to regulate the expression of nearby genes.

12.9 Signaling in Microorganisms and Plants

Much of what we have said about signaling relates to mammalian tissues or cultured cells from such tissues. Bacteria, archaea, eukaryotic microorganisms, and vascular plants must also respond to a variety of external signals—O_2, nutrients, light, noxious chemicals, and so on. We turn here to a brief consideration of the kinds of signaling machinery used by microorganisms and plants.

Bacterial Signaling Entails Phosphorylation in a Two-Component System

In pioneering studies of chemotaxis in bacteria, Julius Adler showed that *E. coli* responds to nutrients in its

Julius Adler
[Source: Courtesy Hildegard Wohl Adler.]

environment, including sugars and amino acids, by swimming toward them, propelled by surface flagella. A family of membrane proteins have binding domains on the outside of the plasma membrane to which specific **attractants** (sugars or amino acids) bind **(Fig. 12-31)**. The signal is transmitted by the so-called **two-component system.** The first component is a **receptor histidine kinase** that, in response to ligand binding, phosphorylates a His residue in its cytoplasmic domain, then catalyzes transfer of the phosphoryl group from the His residue to an Asp residue on the second component, a soluble protein called the **response regulator**. This phosphoprotein moves to the base of a flagellum, carrying the signal from the membrane receptor. Each flagellum is driven by a rotary motor that can propel the cell through its medium or cause it to stall, depending on the direction of motor rotation. The change in attractant concentration over time, signaled through the receptor, allows the cell to determine whether it is moving toward or away from the attractant. If its motion is toward the attractant, the response regulator signals the cell to continue in a straight line (a run); if away from it, the cell tumbles momentarily, acquiring a new direction. Repetition of this behavior results in a random path, biased toward movement in the direction of increasing attractant concentration.

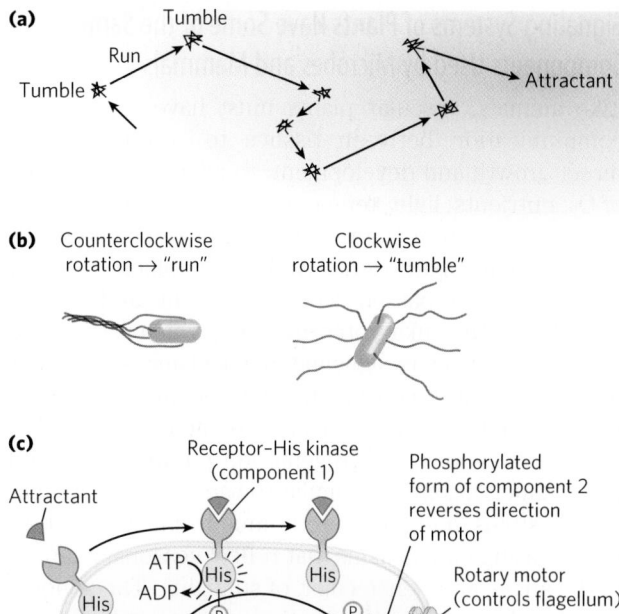

FIGURE 12-31 The two-component signaling mechanism in bacterial chemotaxis. (a) When placed near a source of an attractant solute, *E. coli* performs a random walk, biased toward the attractant. **(b)** Flagella have intrinsic helical structure, and when all flagella rotate counterclockwise, the flagellar helices twist together and move in concert to propel the cell forward in a "run." When the flagella rotate clockwise, the flagellar bundles fly apart, and the cell tumbles briefly until counterclockwise rotation resumes and the cell begins to swim forward again in a new, random direction. When moving toward the attractant, the cell has fewer tumbles and therefore longer runs; when moving away, the frequent tumbles eventually result in movement toward the attractant. **(c)** Flagellar rotation is controlled by a two-component system consisting of a receptor–histidine kinase and an effector protein. When an attractant ligand binds to the receptor domain of the membrane-bound receptor, a protein kinase in the cytosolic domain (component 1) is activated and autophosphorylates a His residue. This phosphoryl group is then transferred to an Asp residue on a response regulator (component 2). After phosphorylation, the response regulator moves to the base of the flagellum, where it causes counterclockwise rotation of the flagella, producing a run.

E. coli detects not only sugars and amino acids but also O_2, extremes of temperature, and other environmental factors, using this basic two-component system. Two-component systems have been detected in many other bacteria, both gram-positive and gram-negative, and in archaea, as well as in protists and fungi. Clearly, this signaling mechanism developed early in the course of cellular evolution and has been conserved.

Various signaling systems used by animal cells also have analogs in bacteria. As the full genomic sequences of more, and increasingly diverse, bacteria become known, researchers have discovered genes that encode proteins similar to protein Ser or Thr kinases, Ras-like proteins regulated by GTP binding, and proteins with SH3 domains. Receptor Tyr kinases have not been detected in bacteria, but $\circledP$–Tyr residues do occur in some bacteria.

Signaling Systems of Plants Have Some of the Same Components Used by Microbes and Mammals

Like animals, vascular plants must have a means of communication between tissues to coordinate and direct growth and development, to adapt to conditions of O_2, nutrients, light, temperature, and water availability, and to warn of the presence of noxious chemicals and damaging pathogens. At least a billion years of evolution have passed since the plant and animal branches of the eukaryotes diverged, which is reflected in the differences in signaling mechanisms: some plant mechanisms are conserved—that is, are similar to those in animals (protein kinases, adaptor proteins, cyclic nucleotides, electrogenic ion pumps, and gated ion channels); some are similar to bacterial two-component systems; and some are unique to plants (such as light-sensing mechanisms that reflect seasonal changes in the angle, and hence color, of sunlight). The genome of the plant *Arabidopsis thaliana* encodes about 1,000 protein Ser/Thr kinases, including about 60 MAPKs and nearly 400 membrane-associated receptor kinases that phosphorylate Ser or Thr residues; a variety of protein phosphatases; enzymes for the synthesis and degradation of cyclic nucleotides; and 100 or more ion channels, including about 20 gated by cyclic nucleotides. Inositol phospholipids are present, as are kinases that interconvert them by phosphorylation of inositol head groups. Even given that *Arabidopsis* has multiple copies of many genes, the presence of this many signaling-related genes certainly reflects a wide array of signaling potential.

Some types of signaling proteins common in animal tissues are not present in plants, or are represented by only a few genes. Protein kinases that are activated by cyclic nucleotides (PKA and PKG) seem to be absent, for example. Heterotrimeric G protein and protein Tyr kinase genes are much less prominent in the plant genome, and the mode of action of these proteins is different from that in animal cells. GPCRs, the largest family of signaling proteins in the human genome, are absent from the plant genome. DNA-binding nuclear steroid receptors are certainly not prominent, and may be absent from plants. Although vascular plants lack the most widely conserved light-sensing mechanism present in animals (rhodopsin, with retinal as pigment), they have a rich collection of other light-detecting mechanisms not found in animal tissues—phytochromes and cryptochromes, for example (Chapter 20).

SUMMARY 12.9 Signaling in Microorganisms and Plants

■ Bacteria and eukaryotic microorganisms have a variety of sensory systems that allow them to sample and respond to their environment. In the two-component system, a receptor His kinase senses the signal and autophosphorylates a His residue, then phosphorylates an Asp residue of the response regulator.

■ Plants respond to many environmental stimuli and employ hormones and growth factors to coordinate the development and metabolic activities of their tissues. Plant genomes encode hundreds of signaling proteins, including some very similar to those of mammals.

■ Plants do not have GPCRs or protein kinases activated by cAMP or cGMP.

12.10 Regulation of the Cell Cycle by Protein Kinases

One of the most dramatic manifestations of signaling pathways is the regulation of the eukaryotic cell cycle. During embryonic growth and later development, cell division occurs in virtually every tissue. In the adult organism, most tissues become quiescent. A cell's "decision" to divide or not is of crucial importance to the organism. When the regulatory mechanisms that limit cell division are defective and cells undergo unregulated division, the result is catastrophic—cancer. Proper cell division requires a precisely ordered sequence of biochemical events that assures every daughter cell a full complement of the molecules required for life. Investigations into the control of cell division in diverse eukaryotic cells have revealed universal regulatory mechanisms. Signaling mechanisms much like those discussed above are central in determining whether and when a cell undergoes cell division, and they also ensure orderly passage through the stages of the cell cycle.

The Cell Cycle Has Four Stages

Cell division accompanying mitosis in eukaryotes occurs in four well-defined stages **(Fig. 12-32)**. In the S (synthesis) phase, the DNA is replicated to produce copies for both daughter cells. In the G2 phase (*G* indicates the gap between divisions), new proteins are synthesized and the cell approximately doubles in size. In the M phase (mitosis), the maternal nuclear envelope breaks down, paired chromosomes are pulled to opposite poles of the cell, each set of daughter chromosomes is surrounded by a newly formed nuclear envelope, and cytokinesis pinches the cell in half, producing two daughter cells (see Fig. 24-23). In embryonic or rapidly proliferating tissue, each daughter cell divides again, but only after a waiting period (G1). In cultured animal cells the entire process takes about 24 hours.

After passing through mitosis and into G1, a cell either continues through another division or ceases to divide, entering a quiescent phase (G0) that may last hours, days, or the lifetime of the cell. When a cell in G0 begins to divide again, it reenters the division cycle through the G1 phase. Differentiated cells such as hepatocytes or adipocytes have acquired their specialized function and form; they remain in the G0 phase. Stem cells retain their potential to divide and to differentiate into any of a number of cell types.

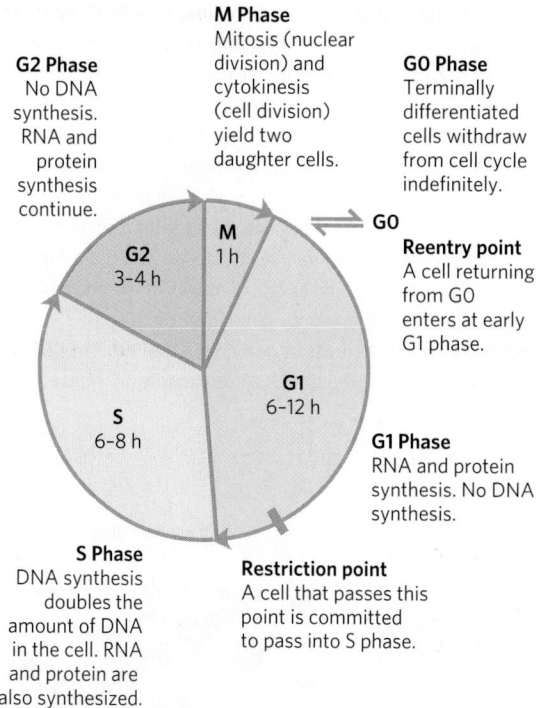

FIGURE 12-32 The eukaryotic cell cycle. The durations (in hours) of the four stages vary, but those shown are typical.

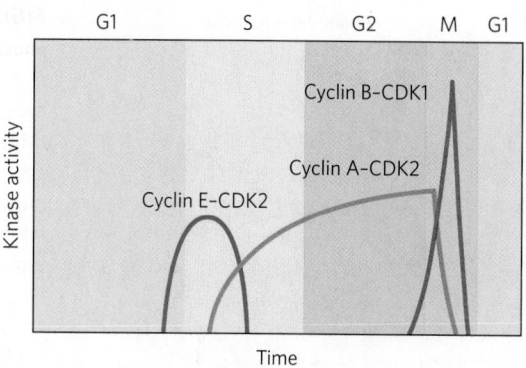

FIGURE 12-33 Variations in the activities of specific CDKs during the cell cycle in animals. Cyclin E–CDK2 activity peaks near the G1 phase–S phase boundary, when the active enzyme triggers synthesis of enzymes required for DNA synthesis (see Fig. 12-37). Cyclin A–CDK2 activity rises during the S and G2 phases, then drops sharply in the M phase, as cyclin B–CDK1 peaks. Cyclin D is active as long as a growth factor is present (not shown). [Source: Data from J. Pines, *Nature Cell Biol.* 1:E73, 1999.]

Levels of Cyclin-Dependent Protein Kinases Oscillate

The timing of the cell cycle is controlled by a family of protein kinases with activities that change in response to cellular signals. By phosphorylating specific proteins at precisely timed intervals, these protein kinases orchestrate the metabolic activities of the cell to produce orderly cell division. The kinases are heterodimers with a regulatory subunit, a **cyclin**, and a catalytic subunit, a **cyclin-dependent protein kinase (CDK)**. In the absence of the cyclin, the catalytic subunit is virtually inactive. When the cyclin binds, the catalytic site opens up, a residue essential to catalysis becomes accessible, and the protein kinase activity of the catalytic subunit increases 10,000-fold. Animal cells have at least 10 different cyclins (designated A, B, and so forth) and at least 8 CDKs (CDK1 through CDK8), which act in various combinations at specific points in the cell cycle. Plants also use a family of CDKs to regulate their cell division in root and shoot meristems, the principal tissues in which division occurs.

In a population of animal cells undergoing synchronous division, some CDK activities show striking oscillations **(Fig. 12-33)**. These oscillations are the result of four mechanisms for regulating CDK activity: phosphorylation or dephosphorylation of the CDK, controlled degradation of the cyclin subunit, periodic synthesis of CDKs and cyclins, and the action of specific CDK-inhibiting proteins. The precisely timed activation and inactivation of a series of CDKs produces signals serving as a master clock that orchestrates the events in normal cell division and ensures that one stage is completed before the next begins.

Regulation of CDKs by Phosphorylation The activity of a CDK is strikingly affected by phosphorylation and dephosphorylation of two critical residues in the protein **(Fig. 12-34)**. Phosphorylation of Thr^{160} of CDK2 stabilizes a conformation in which an autoinhibitory "T loop" is moved away from the substrate-binding cleft in the kinase, opening it to bind protein substrates. Dephosphorylation of $℗-Tyr^{15}$ of CDK2 removes a negative charge that blocks ATP from approaching its binding site. This mechanism for activating a CDK is self-reinforcing; the enzyme (PTP) that dephosphorylates $℗-Tyr^{15}$ is itself a substrate for the CDK and is activated by phosphorylation. The combination of these factors activates the CDK manyfold, allowing it to phosphorylate downstream protein targets critical to progression of the cell cycle **(Fig. 12-35a)**.

The presence of a single-strand break in DNA signals arrest of the cell cycle in G2 by activating two proteins (ATM and ATR; see Fig. 12-37). These proteins trigger a cascade of responses that include inactivation of the PTP that dephosphorylates Tyr^{15} of the CDK. With the CDK inactivated, the cell is arrested in G2, unable to divide until the DNA is repaired and the effects of the cascade are reversed.

Controlled Degradation of Cyclin Highly specific and precisely timed proteolytic breakdown of mitotic cyclins regulates CDK activity throughout the cell cycle (Fig. 12-35b). Progress through mitosis requires first the activation then the destruction of cyclins A and B, which activate the catalytic subunit of the M-phase CDK. These cyclins contain near their amino terminus the sequence –Arg–Thr–Ala–Leu–Gly–Asp–Ile–Gly–Asn–, the "destruction box," which targets the proteins for

(a)

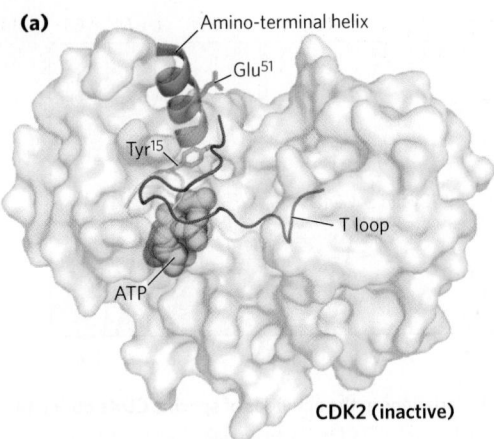

CDK2 (inactive)

FIGURE 12-34 Activation of cyclin-dependent protein kinases (CDKs) by cyclin and phosphorylation. CDKs are active only when associated with a cyclin. The crystal structure of CDK2 with and without a cyclin reveals the basis for this activation. **(a)** Without the cyclin, CDK2 folds so that one segment, the T loop, obstructs the binding site for protein substrates. The binding site for ATP is also near the T loop and is blocked when Tyr^{15} is phosphorylated (not shown). **(b)** When the cyclin binds, it forces conformational changes that move the T loop away from the active site and reorient an amino-terminal helix, bringing a residue critical to catalysis (Glu^{51}) into the active site. **(c)** When a Thr residue in the T loop is phosphorylated, its negative charges are stabilized by interaction with three Arg residues, holding the T loop away from the substrate-binding site. Removal of the phosphoryl group on Tyr^{15} gives ATP access to its binding site, fully activating CDK2 (see Fig. 12-35). [Sources: (a) PDB ID 1HCK, U. Schulze-Gahmen et al., *J. Med. Chem.* 39:4540, 1996. (b) PDB ID 1FIN, P. D. Jeffrey et al., *Nature* 376:313, 1995. (c) PDB ID 1JST, A. A. Russo et al., *Nature Struct. Biol.* 3:696, 1996.]

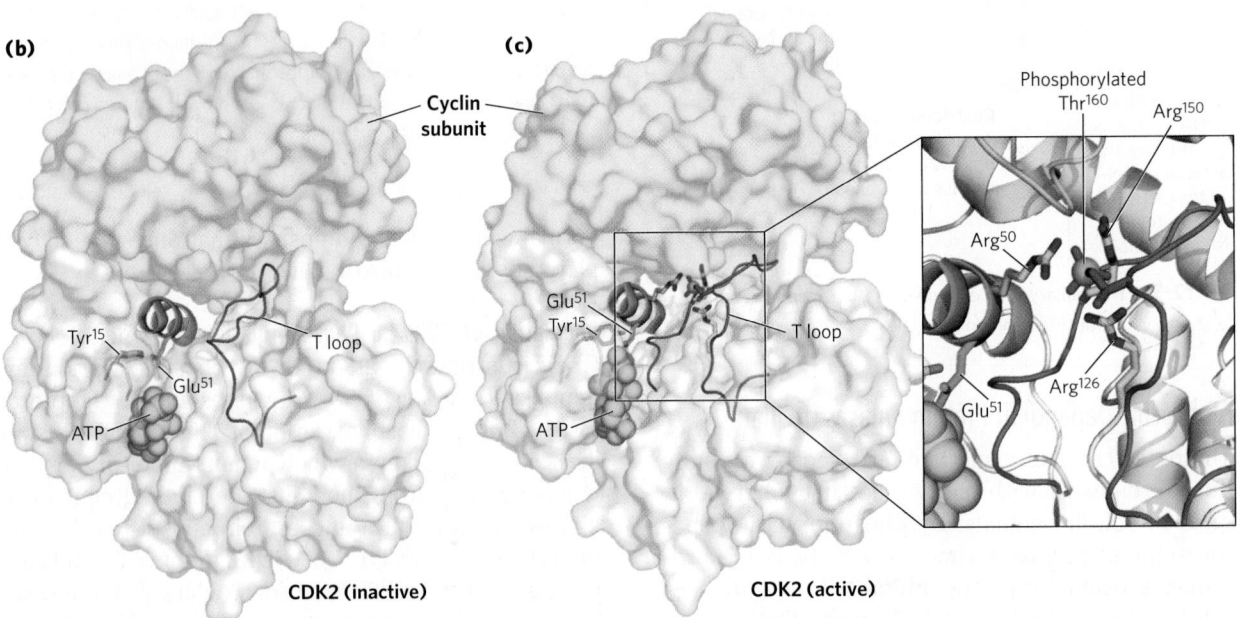

degradation. (This usage of "box" derives from the common practice, in diagramming the sequence of a nucleic acid or protein, of enclosing within a box a short sequence of nucleotide or amino acid residues with some specific function. It does not imply any three-dimensional structure.) The protein DBRP (destruction box recognizing protein) recognizes this sequence and initiates the process of cyclin degradation by bringing together the cyclin and another protein, **ubiquitin**. The cyclin and activated ubiquitin are covalently joined by the enzyme ubiquitin ligase. Several more ubiquitin molecules are then appended, providing the signal for a proteolytic enzyme complex, or **proteasome**, to degrade cyclin.

How is the timing of cyclin breakdown controlled? A feedback loop occurs in the overall process shown in Figure 12-35. Increased CDK activity (step ❹) leads, eventually, to cyclin proteolysis (step ❽). Newly synthesized cyclin associates with and activates the CDK, which phosphorylates and activates DBRP. Active DBRP then causes proteolysis of the cyclin. The lowered cyclin level causes a decline in CDK activity, and the activity of DBRP also drops through slow, constant dephosphorylation and inactivation by a DBRP phosphatase. The cyclin level is ultimately restored by synthesis of new cyclin molecules.

The role of ubiquitin and proteasomes is not limited to the regulation of cyclins; as we shall see in Chapter 27, both also take part in the turnover of cellular proteins, a process fundamental to cellular housekeeping.

Growth Factors Stimulate CDK and Cyclin Synthesis The third mechanism for changing CDK activity is regulation of the rate of synthesis of the cyclin or CDK or both. Extracellular signals such as **growth factors** and cytokines (developmental signals that trigger cell division) activate, by phosphorylation, the nuclear transcription factors Jun and Fos, which promote the synthesis of many gene products, including cyclins, CDKs, and the transcription factor E2F. In turn, E2F stimulates production of several enzymes essential for the synthesis of deoxynucleotides and DNA, and the CDK and cyclin allow the cell to enter the S phase **(Fig. 12-36)**.

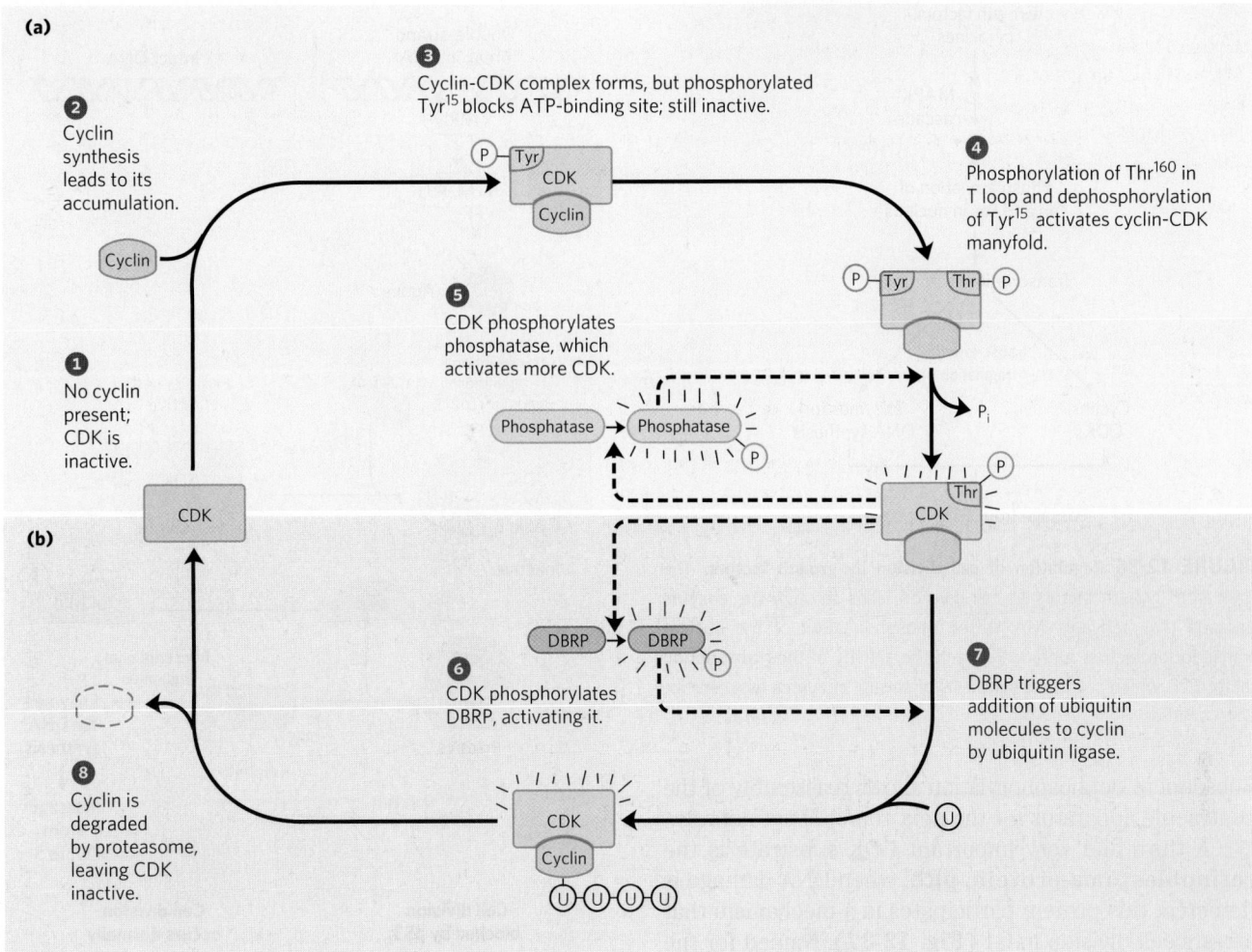

FIGURE 12-35 Regulation of CDK by phosphorylation and proteolysis.
(a) As a cell enters mitosis, the M-phase CDK is inactive (step **❶**). As cyclin is synthesized (step **❷**), the cyclin-CDK complex forms (step **❸**). The T loop lies in the substrate-binding site of CDK, and Ⓟ-Tyr^{15} blocks its ATP-binding site, keeping the complex inactive. When Thr^{160} in the T loop is phosphorylated, the loop moves out of the substrate-binding site, and when Tyr^{15} is dephosphorylated, ATP can bind. These two changes activate

the cyclin-CDK manyfold (step **❹**). Further activation is achieved as CDK also phosphorylates and activates the enzyme that dephosphorylates Ⓟ-Tyr^{15} (step **❺**). **(b)** The active cyclin-CDK complex triggers its own inactivation by phosphorylation of DBRP (destruction box recognizing protein; step **❻**). DBRP and ubiquitin ligase then attach several molecules of ubiquitin (U) to the cyclin (step **❼**), targeting it for destruction by proteolytic enzyme complexes called proteasomes (step **❽**).

Inhibition of CDKs Finally, specific protein inhibitors bind to and inactivate specific CDKs. One such protein is p21, which we discuss below.

These four control mechanisms modulate the activity of specific CDKs that, in turn, control whether a cell will divide, differentiate, become permanently quiescent, or begin a new cycle of division after a period of quiescence. The details of cell cycle regulation, such as the number of different cyclins and kinases and the combinations in which they act, differ from species to species, but the basic mechanism has been conserved in the evolution of all eukaryotic cells.

CDKs Regulate Cell Division by Phosphorylating Critical Proteins

We have examined how cells maintain close control of CDK activity, but how does the activity of CDKs control

the cell cycle? The list of target proteins that CDKs are known to act upon continues to grow, and much remains to be learned. But we can see a general pattern behind CDK regulation by inspecting the effect of CDKs on the structures of lamin and myosin and on the activity of retinoblastoma protein.

The structure of the nuclear envelope is maintained in part by highly organized meshworks of intermediate filaments composed of the protein **lamin**. Breakdown of the nuclear envelope before segregation of the sister chromatids in mitosis is partly due to the phosphorylation of lamin by a CDK, which causes lamin filaments to depolymerize.

A second kinase target is the ATP-driven contractile machinery (actin and myosin) that pinches a dividing cell into two equal parts during cytokinesis. After the division, a CDK phosphorylates a small regulatory subunit of myosin, causing dissociation of myosin from actin filaments and inactivating the contractile machinery.

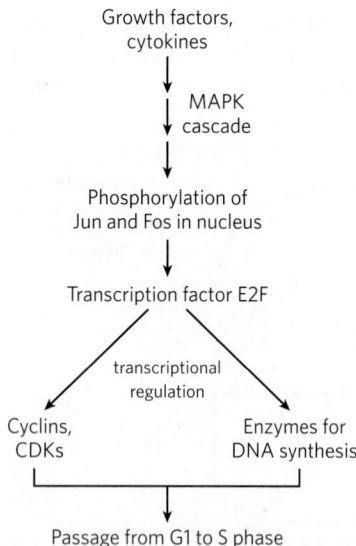

FIGURE 12-36 Regulation of cell division by growth factors. The path from growth factors to cell division leads through the enzyme cascade that activates MAPK, the phosphorylation of the nuclear transcription factors Jun and Fos, and the activity of the transcription factor E2F, which promotes synthesis of several enzymes essential for DNA synthesis.

Subsequent dephosphorylation allows reassembly of the contractile apparatus for the next round of cytokinesis.

A third and very important CDK substrate is the **retinoblastoma protein, pRb**; when DNA damage is detected, this protein participates in a mechanism that arrests cell division in G1 **(Fig. 12-37)**. Named for the retinal tumor cell line in which it was discovered, pRb functions in most, perhaps all, cell types to regulate cell division in response to a variety of stimuli. Unphosphorylated pRb binds the transcription factor E2F; while bound to pRb, E2F cannot promote transcription of a group of genes necessary for DNA synthesis (the genes for DNA polymerase α, ribonucleotide reductase, and other proteins; see Chapter 25). In this state, the cell cycle cannot proceed from the G1 to the S phase, the step that commits a cell to mitosis and cell division. The pRb-E2F blocking mechanism is relieved when pRb is phosphorylated by cyclin E–CDK2, which occurs in response to a signal for cell division to proceed.

When the protein kinases ATM and ATR detect damage to DNA (signaled by the presence of the protein MRN at a double-strand break site), they phosphorylate p53, activating it to serve as a transcription factor that stimulates the synthesis of the protein p21 (Fig. 12-37). This protein inhibits the protein kinase activity of cyclin E–CDK2. In the presence of p21, pRb remains unphosphorylated and bound to E2F, blocking the activity of this transcription factor, and the cell cycle is arrested in G1. This gives the cell time to repair its DNA before entering the S phase, thereby avoiding the potentially disastrous transfer of a defective genome to one or both daughter cells. When the damage is too severe to allow effective repair, this same machinery triggers apoptosis (described

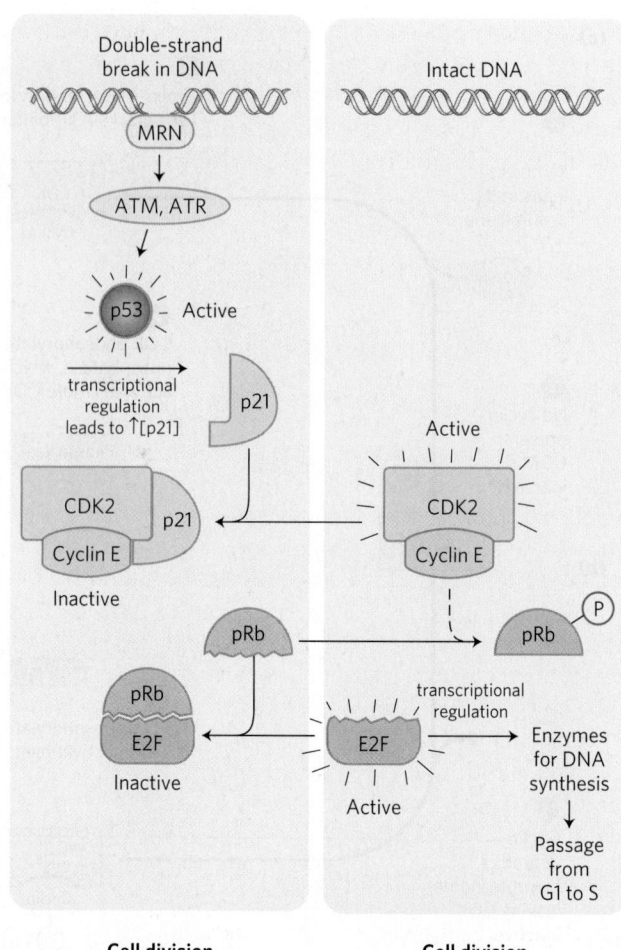

Cell division blocked by p53

Cell division occurs normally

FIGURE 12-37 Regulation of passage from G1 to S by phosphorylation of pRb. Transcription factor E2F promotes transcription of genes for certain enzymes essential to DNA synthesis. The retinoblastoma protein, pRb, can bind E2F (lower left), inactivating it and preventing transcription of these genes. Phosphorylation of pRb by CDK2 prevents it from binding and inactivating E2F, and the genes are transcribed, allowing cell division. Damage to the cell's DNA (upper left) triggers a series of events that inactivate CDK2, blocking cell division. When the protein MRN detects damage to the DNA, it activates two protein kinases, ATM and ATR, and they phosphorylate and activate the transcription factor p53. Active p53 promotes the synthesis of another protein, p21, an inhibitor of CDK2. Inhibition of CDK2 stops the phosphorylation of pRb, which therefore continues to bind and inhibit E2F. With E2F inactivated, genes essential to cell division are not transcribed and cell division is blocked. When DNA has been repaired, this inhibition is released, and the cell divides.

below), a process that leads to the death of the cell, preventing the possible development of a cancer.

SUMMARY 12.10 Regulation of the Cell Cycle by Protein Kinases

■ Progression through the cell cycle is regulated by the cyclin-dependent protein kinases (CDKs), which act at specific points in the cycle, phosphorylating key proteins and modulating their activities. The catalytic subunit of CDKs is inactive unless associated with the regulatory cyclin subunit.

■ The activity of a cyclin-CDK complex changes during the cell cycle through differential synthesis of CDKs, specific degradation of the cyclin, phosphorylation and dephosphorylation of critical residues in CDKs, and binding of inhibitory proteins to specific cyclin-CDKs.

■ Among the targets phosphorylated by cyclin-CDKs are proteins of the nuclear envelope and proteins required for cytokinesis and DNA repair.

12.11 Oncogenes, Tumor Suppressor Genes, and Programmed Cell Death

Tumors and cancer are the result of uncontrolled cell division. Normally, cell division is regulated by a family of extracellular growth factors, proteins that cause resting cells to divide and, in some cases, differentiate. The result is a precise balance between the formation of new cells and cell destruction. Regulation of cell division ensures that skin cells are replaced every few weeks and white blood cells are replaced every few days. When this balance is disturbed by defects in regulatory proteins, the result is sometimes the formation of a clone of cells that divide repeatedly and without regulation (a tumor) until their presence interferes with the function of normal tissues—cancer. The direct cause is almost always a genetic defect in one or more of the proteins that regulate cell division. In some cases, a defective gene is inherited from one parent; in other cases, the mutation occurs when a toxic compound from the environment (a mutagen or carcinogen) or high-energy radiation interacts with the DNA of a single cell to damage it and introduce a mutation. In most cases there is both an inherited and an environmental contribution, and in most cases, more than one mutation is required to cause completely unregulated division and full-blown cancer.

Oncogenes Are Mutant Forms of the Genes for Proteins That Regulate the Cell Cycle

Oncogenes are mutated versions of genes encoding signaling proteins involved in cell cycle regulation. Oncogenes were originally discovered in tumor-causing viruses, then later found to be derived from genes in animal host cells, **proto-oncogenes**, which encode growth-regulating proteins. During a viral infection, the host DNA sequence of a proto-oncogene is sometimes copied into the viral genome, where it proliferates with the virus. In subsequent viral infection cycles, the proto-oncogenes can become defective by truncation or mutation. Viruses, unlike animal cells, do not have effective mechanisms for correcting mistakes during DNA replication, so they accumulate mutations rapidly. When a virus carrying an oncogene infects a new host cell, the viral DNA (and oncogene) can be incorporated into the host cell's DNA, where it can now interfere with the regulation of cell division in the host cell. In an alternative, nonviral mechanism, a single cell

in a tissue exposed to carcinogens may suffer DNA damage that renders one of its regulatory proteins defective, with the same effect as the viral oncogenic mechanism: failed regulation of cell division.

The mutations that produce oncogenes are genetically dominant; if either of a pair of chromosomes contains a defective gene, that gene product sends the signal "divide," and a tumor may result. The oncogenic defect can be in any of the proteins involved in communicating the "divide" signal. Oncogenes discovered thus far include those that encode secreted proteins that act as signaling molecules, growth factors, transmembrane proteins (receptors), cytoplasmic proteins (G proteins and protein kinases), and the nuclear transcription factors that control the expression of genes essential for cell division (Jun, Fos).

Some oncogenes encode surface receptors with defective or missing signal-binding sites, such that their intrinsic Tyr kinase activity is unregulated. For example, the oncoprotein ErbB is essentially identical to the normal receptor for epidermal growth factor, except that ErbB lacks the amino-terminal domain that normally binds EGF **(Fig. 12-38)** and is therefore locked in its activated conformation. As a result, the mutant ErbB protein sends the "divide" signal whether EGF is present or not. Mutations in *erbB2*, the gene for a receptor Tyr kinase related to ErbB, are commonly associated with cancers of the glandular epithelium in breast, stomach, and ovary. (For an explanation of the use of abbreviations in naming genes and their products, see Chapter 25.)

The prominent role played by protein kinases in signaling processes related to normal and abnormal cell division has made these enzymes a prime target in the development of drugs for the treatment of cancer (Box 12-4). Mutant forms of the G protein Ras are

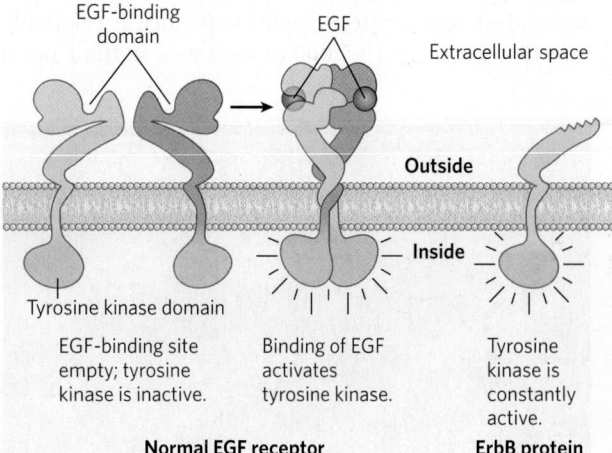

FIGURE 12-38 Oncogene-encoded defective EGF receptor. The product of the *erbB2* oncogene (the ErbB protein) is a truncated version of the normal receptor for epidermal growth factor (EGF). Its intracellular domain has the structure normally induced by EGF binding, but the protein lacks the extracellular binding site for EGF. Unregulated by EGF, ErbB continuously signals cell division.

BOX 12-4 ⚕ MEDICINE Development of Protein Kinase Inhibitors for Cancer Treatment

When a single cell divides without any regulatory limitation, it eventually gives rise to a clone of cells so large that it interferes with normal physiological functions (Fig. 1). This is cancer, a leading cause of death in the developed world, and increasingly so in the developing world. In all types of cancer, the normal regulation of cell division has become dysfunctional due to defects in one or more genes. For example, genes encoding proteins that normally send intermittent signals for cell division become oncogenes, producing constitutively active signaling proteins, or genes encoding proteins that normally restrain cell division (tumor suppressor genes) mutate to produce proteins that lack this braking function. In many tumors, both kinds of mutation have occurred.

Many oncogenes and tumor suppressor genes encode protein kinases or proteins that act in pathways upstream from protein kinases. It is therefore reasonable to hope that specific inhibitors of protein kinases could prove valuable in the treatment of cancer. For example, a mutant form of the EGF receptor is a constantly active receptor Tyr kinase (RTK), signaling cell division whether EGF is present or not (see Fig. 12-38). In about 30% of all women with invasive breast cancer, a mutation in the gene for the receptor HER2/neu yields an RTK with activity increased up to 100-fold. Another RTK, **vascular endothelial growth factor receptor (VEGFR)**, must be activated for the formation of new blood vessels (angiogenesis) to provide a solid tumor with its own blood supply, and inhibition of VEGFR might starve a tumor of essential nutrients. Nonreceptor Tyr kinases can also mutate, resulting in constant signaling and unregulated cell division. For example, the oncogene *Abl* (from the *Ab*elson *l*eukemia virus) is associated with acute myeloid leukemia, a relatively rare blood disease (~5,000 cases a year in the United

States). Another group of oncogenes encode unregulated cyclin-dependent protein kinases. In each of these cases, specific protein kinase inhibitors might be valuable chemotherapeutic agents in the treatment of disease. Not surprisingly, huge efforts are under way to develop such inhibitors. How should one approach this challenge?

Protein kinases of all types show striking conservation of structure at the active site. All share with the prototypical PKA structure the features shown in Figure 2: two lobes that enclose the active site, with a P loop that helps to align and bind the phosphoryl groups of ATP, an activation loop that moves to open the active site to the protein substrate, and a C helix that changes position as the enzyme is activated, bringing the residues in the substrate-binding cleft into their binding positions.

The simplest protein kinase inhibitors are ATP analogs that occupy the ATP-binding site but cannot serve as phosphoryl group donors. Many such compounds are known, but their clinical usefulness is limited by their lack of selectivity—they inhibit virtually all protein kinases and would produce unacceptable side effects. More selectivity is seen with compounds that

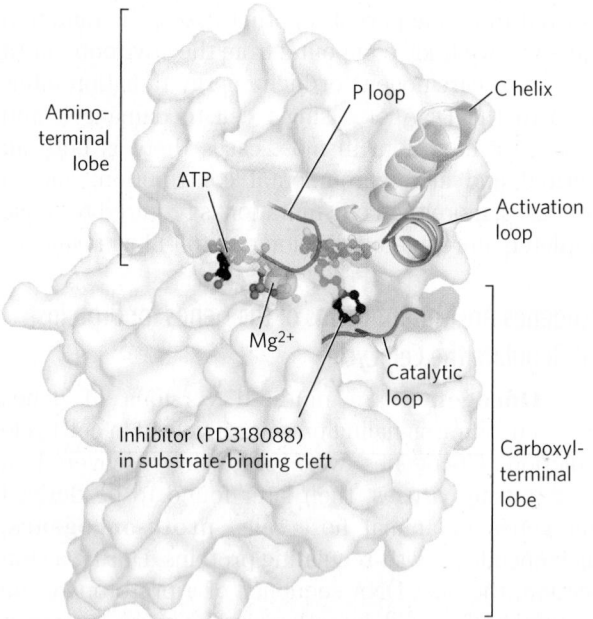

FIGURE 2 Conserved features of the active site of protein kinases. The amino-terminal and carboxyl-terminal lobes surround the active site of the enzyme, near the catalytic loop and the site where ATP binds. The activation loop of this and many other kinases undergoes phosphorylation, then moves away from the active site to expose the substrate-binding cleft, which in this image is occupied by a specific inhibitor of this enzyme, PD318088. The P loop is essential in the binding of ATP, and the C helix must also be correctly aligned for ATP binding and kinase activity. [Source: PDB ID 1S9I, J. F. Ohren et al., *Nature Struct. Mol. Biol.* 11:1192, 2004.]

FIGURE 1 Unregulated division of a single cell in the colon led to a primary cancer that metastasized to the liver. Secondary cancers are seen as white patches in this liver obtained at autopsy. [Source: CNRI/ Science Source.]

fill part of the ATP-binding site but also interact outside this site with parts of the protein unique to the target protein kinase. A third possible strategy is based on the fact that although the active conformations of all protein kinases are similar, their inactive conformations are not. Drugs that target the inactive conformation of a specific protein kinase and prevent its conversion to the active form may have a higher specificity of action. A fourth approach employs the great specificity of antibodies. For example, monoclonal antibodies (p. 177) that bind the extracellular portions of specific RTKs could eliminate the receptors' kinase activity by preventing dimerization or by causing their removal from the cell surface. In some cases, an antibody selectively binding to the surface of cancer cells could cause the immune system to attack those cells.

The search for drugs active against specific protein kinases has yielded encouraging results. For example, imatinib mesylate (Gleevec; Fig. 3a), one of the small-molecule inhibitors, has proved nearly 100% effective in bringing about remission in patients with early-stage chronic myeloid leukemia. Erlotinib (Tarceva; Fig. 3b), which targets EGFR, is effective against advanced non-small-cell lung cancer (NSCLC). Because many cell-division signaling systems involve more than one protein kinase, inhibitors that act on several protein kinases may be useful in the treatment of cancer. Sunitinib (Sutent) and sorafenib (Nexavar) target several protein kinases, including VEGFR and PDGFR. These two drugs are in clinical use for patients with gastrointestinal stromal tumors and advanced renal cell carcinoma, respectively. Trastuzumab (Herceptin), cetuximab (Erbitux), and bevacizumab (Avastin) are monoclonal antibodies that target HER2/neu, EGFR, and VEGFR, respectively; all three drugs are in clinical use for certain types of cancer. Detailed knowledge of the structure around the ATP-binding site makes it possible to design drugs that inhibit a *specific* protein kinase by (1) blocking the critical ATP-binding site, while (2) interacting with residues around that site that are *unique* to that particular protein kinase.

At least a hundred more compounds are in preclinical trials. Among the drugs being evaluated are some obtained from natural sources and some produced by synthetic chemistry. Indirubin is a component of a Chinese herbal preparation traditionally used to treat certain leukemias; it inhibits CDK2 and CDK5. Roscovatine (Fig. 3d), a substituted adenine, has a benzyl ring that makes it highly specific as an inhibitor of CDK2. With several hundred potential anticancer drugs heading toward clinical testing, it is realistic to hope that some will prove more effective or more target-specific than those now in use.

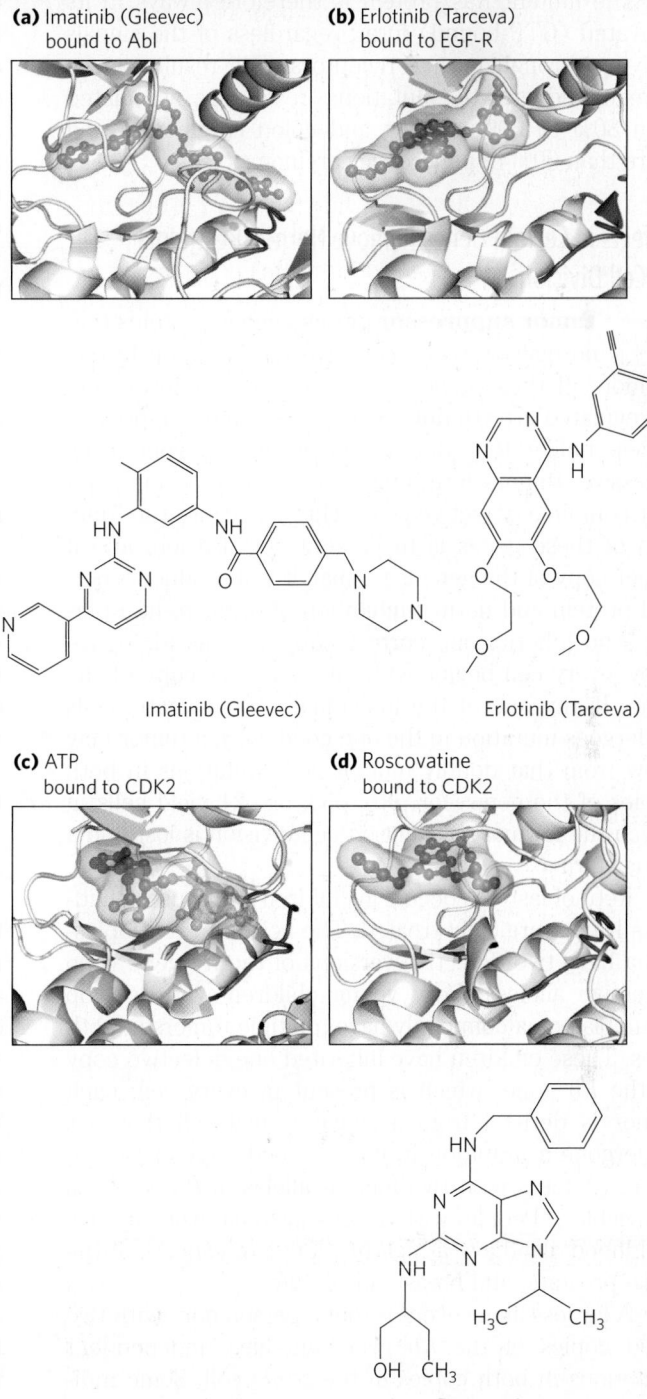

(a) Imatinib (Gleevec) bound to Abl

(b) Erlotinib (Tarceva) bound to EGF-R

Imatinib (Gleevec)   Erlotinib (Tarceva)

(c) ATP bound to CDK2

(d) Roscovatine bound to CDK2

Roscovatine

FIGURE 3 Some protein kinase inhibitors now in clinical trials or clinical use, showing their binding to the target protein. **(a)** Imatinib binds to the Abl kinase (an oncogene product) active site; it occupies both the ATP-binding site and a region adjacent to that site. **(b)** Erlotinib binds to the active site of EGFR. **(c)**, **(d)** Roscovatine is an inhibitor of the cyclin-dependent kinases CDK2, CDK7, and CDK9; shown here are normal Mg^{2+}-ATP binding at the active site (c) and roscovatine binding (d), which prevents the binding of ATP. [Sources: (a) PDB ID 1IEP, B. Nagar et al., *Cancer Res.* 62:4236, 2002. (b) PDB ID 1M17, J. Stamos et al., *J. Biol. Chem.* 277:46,265, 2002. (c) PDB ID 1S9I, J. F. Ohren et al., *Nature Struct. Mol. Biol.* 11:1192, 2004. (d) PDB ID 2A4L, W. F. De Azevedo et al., *Eur. J. Biochem.* 243:518, 1997.]

common in tumor cells. The *ras* oncogene encodes a protein with normal GTP binding but no GTPase activity. The mutant Ras protein is therefore always in its activated (GTP-bound) form, regardless of the signals arriving through normal receptors. The result can be unregulated growth. Mutations in *ras* are associated with 30% to 50% of lung and colon carcinomas and more than 90% of pancreatic carcinomas. ∎

Defects in Certain Genes Remove Normal Restraints on Cell Division

Tumor suppressor genes encode proteins that normally restrain cell division. Mutation in one or more of these genes can lead to tumor formation. Unregulated growth due to defective tumor suppressor genes, unlike that due to oncogenes, is genetically recessive; tumors form only if *both* chromosomes of a pair contain a defective gene. This is because the function of these genes is to prevent cell division, and if either copy of the gene is normal, it will produce a normal protein and normal inhibition of division. In a person who inherits one correct copy and one defective copy, every cell begins with one defective copy of the gene. If any one of the individual's 10^{12} somatic cells undergoes mutation in the one good copy, a tumor may grow from that doubly mutant cell. Mutations in both copies of the genes for pRb, p53, or p21 yield cells in which the normal restraint on cell division is lost and a tumor forms.

Retinoblastoma occurs in children and causes blindness if not surgically treated. The cells of a retinoblastoma have two defective versions of the *Rb* gene (two defective alleles). Very young children who develop retinoblastoma commonly have multiple tumors in both eyes. These children have inherited one defective copy of the *Rb* gene, which is present in every cell; each tumor is derived from a single retinal cell that has undergone a mutation in its one good copy of the *Rb* gene. (A fetus with two mutant alleles in every cell is nonviable.) People with retinoblastoma who survive childhood also have a high incidence of cancers of the lung, prostate, and breast later in life.

A far less likely event is that a person born with two good copies of the *Rb* gene will have independent mutations in both copies in the *same* cell. Some individuals do develop retinoblastomas later in childhood, usually with only one tumor in one eye. These individuals, presumably, were born with two good copies (alleles) of *Rb* in every cell, but both *Rb* alleles in a single retinal cell have undergone mutation, leading to a tumor. After about age three, retinal cells stop dividing, and retinoblastomas at later ages are quite rare.

Stability genes (also called caretaker genes) encode proteins that function in the repair of major genetic defects that result from aberrant DNA replication, ionizing radiation, or environmental carcinogens. Mutations in these genes lead to a high frequency of unrepaired damage (mutations) in other genes, including proto-oncogenes and tumor suppressor genes, and thus to cancer. Among the stability genes are *ATM* (see Fig. 12-37); the *XP* gene family, in which mutations lead to xeroderma pigmentosum; and the *BRCA1* genes associated with some types of breast cancer (see Box 25-1). Mutations in the gene for p53 also cause tumors; in more than 90% of human cutaneous squamous cell carcinomas (skin cancers) and in about 50% of all other human cancers, *p53* is defective. Those very rare individuals who *inherit* one defective copy of *p53* commonly have the Li-Fraumeni cancer syndrome, with multiple cancers (of the breast, brain, bone, blood, lung, and skin) occurring at high frequency and at an early age. The explanation for multiple tumors in this case is the same as that for *Rb* mutations: an individual born with one defective copy of *p53* in every somatic cell is likely to suffer a second *p53* mutation in more than one cell during his or her lifetime.

In summary, then, three classes of defects can contribute to the development of cancer: (1) oncogenes, in which the defect is the equivalent of a car's accelerator pedal being stuck down, with the engine racing; (2) mutated tumor suppressor genes, in which the defect leads to the equivalent of brake failure; and (3) mutated stability genes, with the defect leading to unrepaired damage to the cell's replication machinery—the equivalent of an unskilled car mechanic.

Mutations in oncogenes and tumor suppressor genes do not have an all-or-none effect. In some cancers, perhaps in all, the progression from a normal cell to a malignant tumor requires an accumulation of mutations (sometimes over several decades), none of which, alone, is responsible for the end effect. For example, the development of colorectal cancer has several recognizable stages, each associated with a mutation **(Fig. 12-39)**. If an epithelial cell in the colon undergoes mutation of both copies of the tumor suppressor gene *APC* (adenomatous polyposis coli), it begins to divide faster than normal and produces a clone of itself, a benign polyp (early adenoma). For reasons not yet known, the *APC* mutation results in chromosomal instability, and whole regions of a chromosome are lost or rearranged during cell division. This instability can lead to another mutation, commonly in *ras*, that converts the clone into an intermediate adenoma. A third mutation (often in the tumor suppressor gene *DCC*) leads to a late adenoma. Only when both copies of *p53* become defective does this cell mass become a carcinoma—a malignant, life-threatening tumor. The full sequence therefore requires at least seven genetic "hits": two on each of three tumor suppressor genes (*APC*, *DCC*, and *p53*) and one on the proto-oncogene *ras*. There are probably several other routes to colorectal cancer as well, but the principle that full malignancy results only from multiple mutations is likely to hold true for all of them. Because mutations accumulate over time, the chances of developing full-blown metastatic cancer rise with age (Fig. 12-39).

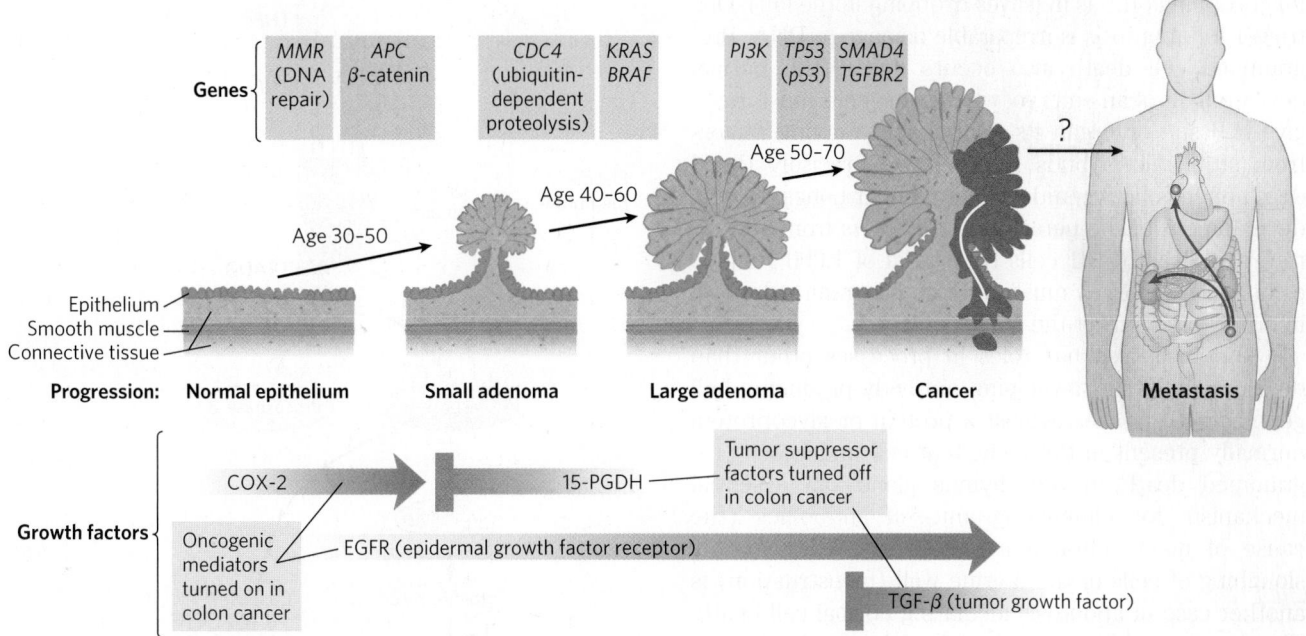

FIGURE 12-39 Multistep transition from normal epithelial cell to colorectal cancer. Serial mutations in oncogenes (green) or tumor suppressor genes (red) lead to progressively less control of cell division, until finally an active tumor forms, which can sometimes metastasize (spread from the initial site to other regions of the body). Mutation of the *MMR* gene leads to defective DNA repair and consequently to a higher rate of mutation. Mutations in both copies of the tumor suppressor gene *APC* lead to benign clusters of epithelial cells that multiply too rapidly (early adenoma). The *CDC4* oncogene results in defective ubiquitination, which is essential to the regulation of cyclin-dependent kinases (see Fig. 12-35). The oncogenes *KRAS* and *BRAF* encode Ras and Raf proteins (see Fig. 12-19), and this further disruption of signaling leads to the formation of a large adenoma, which may be detected by colonoscopy as a benign polyp. Oncogenic mutations in the *PI3K* gene, which encodes the enzyme phosphoinositide-3 kinase, or in *PTEN*, which regulates the synthesis of this enzyme, lead to a further strengthening of the signal: divide now. When a cell in one of the polyps undergoes further mutations, such as in the tumor suppressor genes *DCC* and *p53* (see Fig. 12-37), increasingly aggressive tumors form. Finally, mutations in other tumor suppressor genes such as *SMAD4* lead to a malignant tumor and sometimes to a metastatic tumor that can spread to other tissues. A second type of mutation that can add to the deleterious effects is one that affects the production or action of growth factors or their receptors (bottom). Mutations in EGFR (epidermal growth factor receptor) or TGF-β (transforming growth factor-β) favor uncontrolled growth, as do mutations in the enzymes that produce certain prostaglandins (COX-2; cyclooxygenase; see Fig. 10-17) or the enzyme 15-PGDH (15-hydroxyprostaglandin dehydrogenase). Most malignant tumors of other tissues probably result from a series of mutations such as this, although not necessarily these particular genes, or in this order. [Source: Information from S. D. Markowitz and M. M. Bertagnolli, *N. Engl. J. Med.* 361:2449, 2009, Fig. 2.]

When a polyp is detected in the early adenoma stage and the cells containing the first mutations are removed surgically, late adenomas and carcinomas will not develop; hence the importance of early detection. Cells and organisms, too, have their early detection systems. For example, the ATM and ATR proteins described in Section 12.10 can detect DNA damage too extensive to be repaired effectively. They then trigger, through a pathway that includes p53, the process of apoptosis, in which a cell that has become dangerous to the organism kills itself.

The development of fast and inexpensive sequencing methods has opened a new window on the process by which cancer develops. In a typical study of cancers in humans, the sequences of all 20,000 genes were determined in about 3,300 different tumors, and then compared with the gene sequences in noncancerous tissue from the same patient. Almost 300,000 mutations were detected in all. Only a small fraction of these mutations, the **driver mutations**, were the *cause* of unregulated cell division; the vast majority (>99.9%) were "passenger mutations," which occurred randomly and did not confer a selective growth advantage on the tissue in which they occurred. Among the driver mutations were those in about 75 tumor suppressor genes and about 65 oncogenes. These 140 driver mutations fell in three general categories: those that affect cell survival (in genes encoding Ras, PI3K, MAPK, for example), those that affect cells' ability to maintain an intact genome (ATM, ATR), and those that affect cell fate, causing cells to divide, differentiate, or become quiescent (APC is one example). A relatively small number of mutations were very common in multiple types of cancer, in the genes for Ras, p53, and pRb, for example. ∎

Apoptosis Is Programmed Cell Suicide

Many cells can precisely control the time of their own death by the process of **programmed cell death**, or **apoptosis** (pronounced app′-a-toe′-sis; from the Greek for "dropping off," as in leaves dropping in the fall). One trigger for apoptosis is irreparable damage to DNA. Programmed cell death also occurs during the normal development of an embryo, when some cells must die to give a tissue or organ its final shape. Carving fingers from stubby limb buds requires the precisely timed death of cells between developing finger bones. During development of the nematode *C. elegans* from a fertilized egg, exactly 131 cells (of a total of 1,090 somatic cells in the embryo) must undergo programmed death in order to construct the adult body.

Apoptosis also has roles in processes other than development. If a developing antibody-producing cell generates antibodies against a protein or glycoprotein normally present in the body, that cell undergoes programmed death in the thymus gland—an essential mechanism for eliminating anti-self antibodies (the cause of many autoimmune diseases). The monthly sloughing of cells of the uterine wall (menstruation) is another case of apoptosis mediating normal cell death. The dropping of leaves in the fall is the result of apoptosis in specific cells of the stem. Sometimes cell suicide is not programmed but occurs in response to biological circumstances that threaten the rest of the organism. For example, a virus-infected cell that dies before completion of the infection cycle prevents spread of the virus to nearby cells. Severe stresses such as heat, hyperosmolarity, UV light, and gamma irradiation also trigger cell suicide; presumably the organism is better off with any aberrant, potentially mutated cells dead.

The regulatory mechanisms that trigger apoptosis involve some of the same proteins that regulate the cell cycle. The signal for suicide often comes from outside, through a surface receptor. Tumor necrosis factor (TNF), produced by cells of the immune system, interacts with cells through specific TNF receptors. These receptors have TNF-binding sites on the outer face of the plasma membrane and a "death domain" (~80 amino acid residues) that carries the self-destruct signal through the membrane to cytosolic proteins such as TRADD (TNF receptor–associated death domain) **(Fig. 12-40)**.

When caspase 8, an "initiator" caspase, is activated by an apoptotic signal carried through TRADD, it further self-activates by cleaving its own proenzyme form. Mitochondria are one target of active caspase 8. The protease causes the release of certain proteins contained between the inner and outer mitochondrial membranes: cytochrome *c* and several "effector" caspases (see Fig. 19-39). Cytochrome *c* binds to the proenzyme form of the effector enzyme caspase 9 and stimulates its proteolytic activation. The activated caspase 9, in turn, catalyzes wholesale destruction of cellular proteins—a major cause of apoptotic cell death. One specific target of caspase action is a caspase-activated deoxyribonuclease.

In apoptosis, the monomeric products of protein and DNA degradation (amino acids and nucleotides) are released in a controlled process that allows them

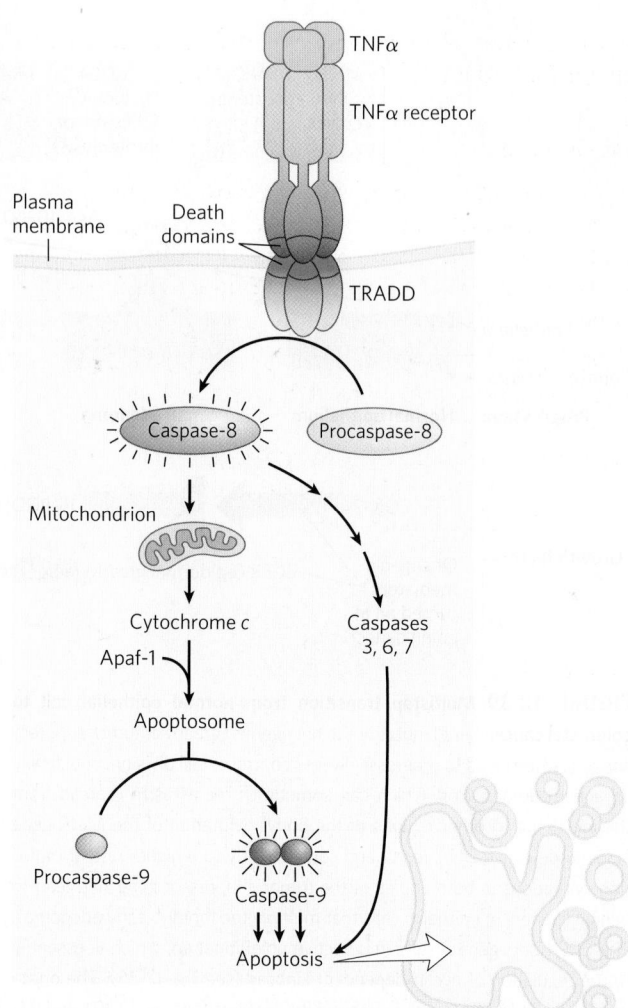

FIGURE 12-40 Initial events of apoptosis. An apoptosis-triggering signal from outside the cell (TNFα) binds to its specific receptor in the plasma membrane. The occupied receptor interacts with the cytosolic protein TRADD through "death domains" (80-residue domains on both TNFα receptor and TRADD), activating TRADD. Activated TRADD initiates a proteolytic cascade that leads to apoptosis: TRADD activates caspase-8, which acts to release cytochrome *c* from mitochondria, which, in concert with protein Apaf-1, activates caspase-9, triggering apoptosis (see Fig. 19-39).

to be taken up and reused by neighboring cells. Apoptosis thus allows the organism to eliminate a cell that is unneeded or potentially dangerous without wasting its components.

SUMMARY 12.11 Oncogenes, Tumor Suppressor Genes, and Programmed Cell Death

■ Oncogenes encode defective signaling proteins. By continually giving the signal for cell division, they lead to tumor formation. Oncogenes are genetically dominant and may encode defective growth factors, receptors, G proteins, protein kinases, or nuclear regulators of transcription.

■ Tumor suppressor genes encode regulatory proteins that normally inhibit cell division; mutations in these genes are genetically recessive but can lead to tumor formation.

■ Cancer is generally the result of an accumulation of mutations in oncogenes and tumor suppressor genes.

■ When stability genes, which encode proteins necessary for the repair of genetic damage, are mutated, other mutations go unrepaired, including mutations in proto-oncogenes and tumor suppressor genes that can lead to cancer.

■ Apoptosis is programmed and controlled cell death that functions during normal development and adulthood to get rid of unnecessary, damaged, or infected cells. Apoptosis can be triggered by extracellular signals such as TNF, acting through plasma membrane receptors.

Key Terms

Terms in bold are defined in the glossary.

signal transduction   437
specificity   437
cooperativity   438
amplification   438
enzyme cascade   438
modularity   438
scaffold proteins   438
desensitization   439
integration   439
response localization 439
G protein–coupled receptors (GPCRs)   440
guanosine nucleotide–binding proteins   440
G proteins   440
second messenger   440
agonist   441
antagonist   441
β-adrenergic receptors   441
heptahelical receptors   441
stimulatory G protein (G$_s$)   441
adenylyl cyclase   442
cAMP-dependent protein kinase (protein kinase A; PKA)   442
P loop   444
GTPase activator protein (GAP)   445
regulator of G protein signaling (RGS)   445
guanosine nucleotide–exchange factor (GEF)   445
consensus sequence   446

β-arrestin (βarr; arrestin 2) 449
G protein–coupled receptor kinases (GRKs)   449
cAMP response element binding protein (CREB)   450
inhibitory G protein (G$_i$)   450
adaptor proteins   450
AKAPs (A kinase anchoring proteins)   450
phospholipase C (PLC)   451
inositol 1,4,5-trisphosphate (IP$_3$)   451
protein kinase C (PKC)   451
green fluorescent protein (GFP)   452
fluorescence resonance energy transfer (FRET)   452
calmodulin (CaM)   453
Ca^{2+}/calmodulin-dependent protein kinases (CaM kinases)   454
rhodopsin   456
opsin   456
rhodopsin kinase   457
receptor potential   459
gustducin   459
receptor Tyr kinase (RTK)   461
autophosphorylation 461
SH2 domain   461
Ras   461

small G proteins   461
MAPKs   462
guanosine 3′,5′-cyclic monophosphate (cyclic GMP; cGMP)   466
cGMP-dependent protein kinase (protein kinase G; PKG)   466
atrial natriuretic factor (ANF)   466
NO synthase   467
PTB domains   468
voltage-gated ion channels   472
ionotropic   473
metabotropic   473
hormone response element (HRE)   474
two-component signaling systems   475

receptor histidine kinase   475
response regulator   475
cyclin   477
cyclin-dependent protein kinase (CDK)   477
ubiquitin   478
proteasome   478
growth factors   478
retinoblastoma protein (pRb)   480
oncogene   481
proto-oncogene   481
tumor suppressor gene   484
programmed cell death   485
apoptosis   485

Problems

1. Hormone Experiments in Cell-Free Systems In the 1950s, Earl W. Sutherland, Jr., and his colleagues carried out pioneering experiments to elucidate the mechanism of action of epinephrine and glucagon. Given what you have learned in this chapter about hormone action, interpret each of the experiments described below. Identify substance X and indicate the significance of the results.

(a) Addition of epinephrine to a homogenate of normal liver resulted in an increase in the activity of glycogen phosphorylase. However, when the homogenate was first centrifuged at a high speed and epinephrine or glucagon was added to the clear supernatant fraction that contains phosphorylase, no increase in the phosphorylase activity occurred.

(b) When the particulate fraction from the centrifugation in (a) was treated with epinephrine, substance X was produced. The substance was isolated and purified. Unlike epinephrine, substance X activated glycogen phosphorylase when added to the clear supernatant fraction of the centrifuged homogenate.

(c) Substance X was heat stable; that is, heat treatment did not affect its capacity to activate phosphorylase. (Hint: Would this be the case if substance X were a protein?) Substance X was nearly identical to a compound obtained when pure ATP was treated with barium hydroxide. (Fig. 8-6 will be helpful.)

2. Effect of Dibutyryl cAMP versus cAMP on Intact Cells The physiological effects of epinephrine should in principle be mimicked by addition of cAMP to the target cells. In practice, addition of cAMP to intact target cells elicits only a minimal physiological response. Why? When the structurally related derivative dibutyryl cAMP (shown below) is added to intact cells, the expected physiological response is readily apparent. Explain the basis for the difference in cellular response to these two substances. Dibutyryl cAMP is widely used in studies of cAMP function.

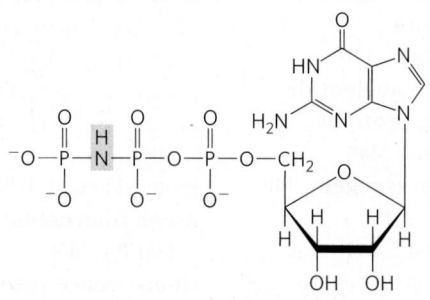

Dibutyryl cAMP
($N^6,O^{2'}$-Dibutyryl adenosine 3',5'-cyclic monophosphate)

3. Effect of Cholera Toxin on Adenylyl Cyclase
The gram-negative bacterium *Vibrio cholerae* produces a protein, cholera toxin (M_r 90,000), that is responsible for the characteristic symptoms of cholera: extensive loss of body water and Na^+ through continuous, debilitating diarrhea. If body fluids and Na^+ are not replaced, severe dehydration results; untreated, the disease is often fatal. When the cholera toxin gains access to the human intestinal tract, it binds tightly to specific sites in the plasma membrane of the epithelial cells lining the small intestine, causing adenylyl cyclase to undergo prolonged activation (hours or days).

(a) What is the effect of cholera toxin on [cAMP] in the intestinal cells?

(b) Based on the information above, suggest how cAMP normally functions in intestinal epithelial cells.

(c) Suggest a possible treatment for cholera.

4. Mutations in PKA Explain how mutations in the R or C subunit of cAMP-dependent protein kinase (PKA) might lead to (a) a constantly active PKA or (b) a constantly inactive PKA.

5. Therapeutic Effects of Albuterol The respiratory symptoms of asthma result from constriction of the bronchi and bronchioles of the lungs, caused by contraction of the smooth muscle of their walls. This constriction can be reversed by raising [cAMP] in the smooth muscle. Explain the therapeutic effects of albuterol, a β-adrenergic agonist taken (by inhalation) for asthma. Would you expect this drug to have any side effects? If so, how might one design a better drug that does not have these effects?

6. Termination of Hormonal Signals Signals carried by hormones must eventually be terminated. Describe several different mechanisms for signal termination.

7. Using FRET to Explore Protein-Protein Interactions In Vivo Figure 12-8 shows the interaction between β-arrestin and the β-adrenergic receptor. How would you use FRET (see Box 12-2) to demonstrate this interaction in living cells? Which proteins would you fuse? Which wavelengths would you use to illuminate the cells, and which wavelengths would you monitor? What would you expect to

observe if the interaction occurred? If it did not occur? How might you explain the failure of this approach to demonstrate this interaction?

8. EGTA Injection EGTA (ethylene glycol-bis(β-amino ethyl ether)-N,N,N',N'-tetraacetic acid) is a chelating agent with high affinity and specificity for Ca^{2+}. By microinjecting a cell with an appropriate Ca^{2+}-EGTA solution, an experimenter can prevent cytosolic [Ca^{2+}] from rising above 10^{-7} M. How would EGTA microinjection affect a cell's response to vasopressin (see Table 12-4)? To glucagon?

9. Amplification of Hormonal Signals Describe all the sources of amplification in the insulin receptor system.

10. Mutations in *ras* How would a mutation in *ras* that leads to formation of a Ras protein with no GTPase activity affect a cell's response to insulin?

11. Differences among G Proteins Compare the G protein G_s, which acts in transducing the signal from β-adrenergic receptors, and the G protein Ras. What properties do they share? How do they differ? What is the functional difference between G_s and G_i?

12. Mechanisms for Regulating Protein Kinases Identify eight general types of protein kinases found in eukaryotic cells, and explain what factor is *directly* responsible for activating each type.

13. Nonhydrolyzable GTP Analogs Many enzymes can hydrolyze GTP between the β and γ phosphates. The GTP analog β,γ-imidoguanosine 5'-triphosphate (Gpp(NH)p), shown below, cannot be hydrolyzed between the β and γ phosphates.

Gpp(NH)p
(β,γ-imidoguanosine 5'-triphosphate)

Predict the effect of microinjection of Gpp(NH)p into a myocyte on the cell's response to β-adrenergic stimulation.

14. Use of Toxin Binding to Purify a Channel Protein
α-Bungarotoxin is a powerful neurotoxin found in the venom of a poisonous snake (*Bungarus multicinctus*). It binds with high specificity to the acetylcholine receptor (AChR; an integral membrane protein) and prevents its ion channel from opening. This interaction was used to purify AChR from the electric organ of torpedo fish.

(a) Outline a strategy for using α-bungarotoxin covalently bound to chromatography beads to purify the AChR protein. (Hint: See Fig. 3-17c.)

(b) Outline a strategy for the use of [^{125}I]α-bungarotoxin to purify the AChR protein.

15. Excitation Triggered by Hyperpolarization In most neurons, membrane *depolarization* leads to the opening of voltage-dependent ion channels, generation of an action potential, and, ultimately, an influx of Ca^{2+}, which causes release of neurotransmitter at the axon terminus. Devise a cellular strategy by which *hyperpolarization* in rod cells could produce excitation of the visual pathway and passage of visual signals to the brain. (Hint: The neuronal signaling pathway in higher organisms consists of a *series* of neurons that relay information to the brain. The signal released by one neuron can be either excitatory or inhibitory to the following, postsynaptic neuron.)

16. Visual Desensitization Oguchi disease is an inherited form of night blindness. Affected individuals are slow to recover vision after a flash of bright light against a dark background, such as the headlights of a car on the freeway. Suggest what the molecular defect(s) might be in Oguchi disease. Explain in molecular terms how this defect would account for night blindness.

17. Effect of a Permeant cGMP Analog on Rod Cells An analog of cGMP, 8-Br-cGMP, will permeate cellular membranes, is only slowly degraded by a rod cell's PDE activity, and is as effective as cGMP in opening the gated channel in the cell's outer segment. If you suspended rod cells in a buffer containing a relatively high [8-Br-cGMP], then illuminated the cells while measuring their membrane potential, what would you observe?

18. Hot and Cool Taste Sensations The sensations of heat and cold are transduced by a group of temperature-gated cation channels. For example, TRPV1, TRPV3, and TRPM8 are usually closed, but open under the following conditions: TRPV1 at $\geq 43\ °C$; TRPV3 at $\geq 33\ °C$; and TRPM8 at $<25\ °C$. These channel proteins are expressed in sensory neurons known to be responsible for temperature sensation.

(a) Propose a reasonable model to explain how exposing a sensory neuron containing TRPV1 to high temperature leads to a sensation of heat.

(b) Capsaicin, one of the active ingredients in "hot" peppers, is an agonist of TRPV1. Capsaicin shows 50% activation of the TRPV1 response at a concentration of 32 nM—a property known as EC_{50}. Explain why even a very few drops of hot pepper sauce can taste very "hot" without actually burning you.

(c) Menthol, one of the active ingredients in mint, is an agonist of TRPM8 ($EC_{50} = 30\ \mu M$) and TRPV3 ($EC_{50} = 20\ mM$). What sensation would you expect from contact with low levels of menthol? With high levels?

19. Oncogenes, Tumor Suppressor Genes, and Tumors For each of the following situations, provide a plausible explanation for how it could lead to unrestricted cell division.

(a) Colon cancer cells often contain mutations in the gene encoding the prostaglandin E_2 receptor. PGE_2 is a growth factor required for the division of cells in the gastrointestinal tract.

(b) Kaposi sarcoma, a common tumor in people with untreated AIDS, is caused by a virus carrying a gene for a protein similar to the chemokine receptors CXCR1 and CXCR2. Chemokines are cell-specific growth factors.

(c) Adenovirus, a tumor virus, carries a gene for the protein E1A, which binds to the retinoblastoma protein, pRb. (Hint: See Fig. 12–37.)

(d) An important feature of many oncogenes and tumor suppressor genes is their cell-type specificity. For example, mutations in the PGE_2 receptor are not typically found in lung tumors. Explain this observation. (Note that PGE_2 acts through a GPCR in the plasma membrane.)

20. Mutations in Tumor Suppressor Genes and Oncogenes Explain why mutations in tumor suppressor genes are recessive (both copies of the gene must be defective for the regulation of cell division to be defective), whereas mutations in oncogenes are dominant.

21. Retinoblastoma in Children Explain why some children with retinoblastoma develop multiple tumors of the retina in both eyes, whereas others have a single tumor in only one eye.

22. Specificity of a Signal for a Single Cell Type Discuss the validity of the following proposition. A signaling molecule (hormone, growth factor, or neurotransmitter) elicits identical responses in different types of target cells if they contain identical receptors.

Data Analysis Problem

23. Exploring Taste Sensation in Mice Pleasing tastes are an evolutionary adaptation to encourage animals to consume nutritious foods. Zhao and coauthors (2003) examined the two major pleasurable taste sensations: sweet and umami. Umami is a "distinct savory taste" triggered by amino acids, especially aspartate and glutamate, and probably encourages animals to consume protein-rich foods. Monosodium glutamate (MSG) is a flavor enhancer that exploits this sensitivity.

At the time the article was published, specific taste receptor proteins for sweet and umami had been tentatively characterized. Three such proteins were known—T1R1, T1R2, and T1R3—which function as heterodimeric receptor complexes: T1R1-T1R3 was tentatively identified as the umami receptor, and T1R2-T1R3 as the sweet receptor. It was not clear how taste sensation was encoded and sent to the brain, and two possible models had been suggested. In the cell-based model, individual taste-sensing cells express only one kind of receptor; that is, there are "sweet cells," "bitter cells," "umami cells," and so on, and each type of cell sends its information to the brain via a different nerve. The brain "knows" which taste is detected by the identity of the nerve fiber that transmits the message. In the receptor-based model, individual taste-sensing cells have several kinds of receptors and send different messages along the same nerve fiber to the brain, the message depending on which receptor is activated. Also unclear at the time was whether there was any interaction between the different taste sensations, or whether parts of one taste-sensing system were required for other taste sensations.

(a) Previous work had shown that different taste receptor proteins are expressed in non-overlapping sets of taste receptor cells. Which model does this support? Explain your reasoning.

Zhao and colleagues constructed a set of "knockout mice"—mice homozygous for loss-of-function alleles for one of the three receptor proteins, T1R1, T1R2, or T1R3—and double-knockout mice with nonfunctioning T1R2 and T1R3. The researchers measured the taste perception of these mice by measuring their "lick rate" of solutions containing different taste molecules. Mice will lick the spout of a feeding bottle with a pleasant-tasting solution more often than one with an unpleasant-tasting solution. The researchers measured relative lick rates: how often the mice licked a sample solution compared with water. A relative lick rate of 1 indicated no preference; <1, an aversion; and >1, a preference.

(b) All four types of knockout strains had the same responses to salt and bitter tastes as did wild-type mice. Which of the above issues did this experiment address? What do you conclude from these results?

The researchers then studied umami taste reception by measuring the relative lick rates of the different mouse strains with different quantities of MSG in the feeding solution. Note that the solutions also contained inosine monophosphate (IMP), a strong potentiator of umami taste reception (and a common ingredient in ramen soups, along with MSG), and ameloride, which suppresses the pleasant salty taste imparted by the sodium of MSG. The results are shown in the graph.

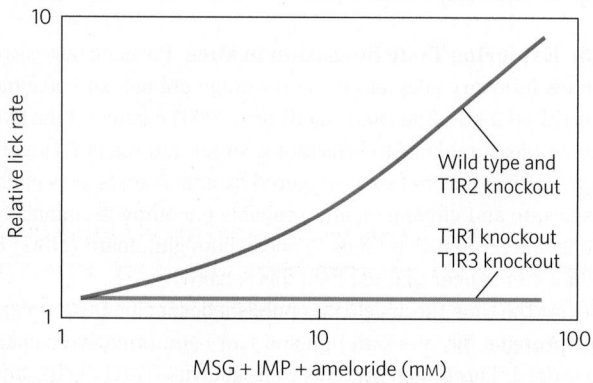

(c) Are these data consistent with the umami taste receptor consisting of a heterodimer of T1R1 and T1R3? Why or why not?

(d) Which model(s) of taste encoding does this result support? Explain your reasoning.

Zhao and coworkers then performed a series of similar experiments using sucrose as a sweet taste. These results are shown below.

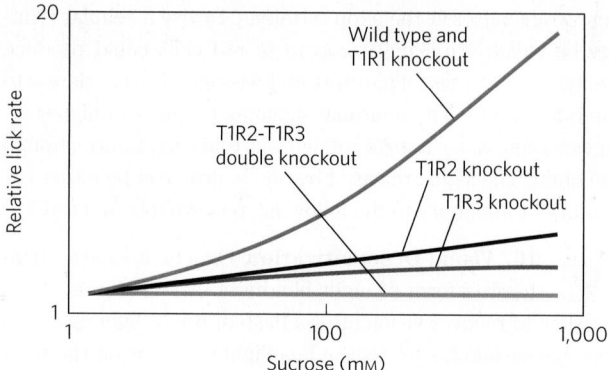

(e) Are these data consistent with the sweet taste receptor consisting of a heterodimer of T1R2 and T1R3? Why or why not?

(f) There were some unexpected responses at very high sucrose concentrations. How do these complicate the idea of a heterodimeric system as presented above?

In addition to sugars, humans also taste other compounds (e.g., saccharin and the peptides monellin and aspartame) as sweet; mice do not taste these as sweet. Zhao and coworkers inserted into TIR2-knockout mice a copy of the human T1R2 gene under the control of the mouse T1R2 promoter. These modified mice now tasted monellin and saccharin as sweet. The researchers then went further, adding to T1R1-knockout mice the RASSL protein—a G protein–linked receptor for the synthetic opiate spiradoline; the RASSL gene was under the control of a promoter that could be induced by feeding the mice tetracycline. These mice did not prefer spiradoline in the absence of tetracycline; in the presence of tetracycline, they showed a strong preference for nanomolar concentrations of spiradoline.

(g) Do these results strengthen your conclusions about the mechanism of taste sensation?

Reference

Zhao, G.Q., Y. Zhang, M.A. Hoon, J. Chandrashekar, I. Erlenbach, N.J.P. Ryba, and C. Zuker. 2003. The receptors for mammalian sweet and umami taste. *Cell* 115:255–266.

Further Reading is available at www.macmillanlearning.com/LehningerBiochemistry7e.

BIOENERGETICS AND METABOLISM

Metabolism is a highly coordinated cellular activity in which many multienzyme systems (metabolic pathways) cooperate to (1) obtain chemical energy by capturing solar energy or degrading energy-rich nutrients from the environment; (2) convert nutrient molecules into the cell's own characteristic molecules, including precursors of macromolecules; (3) polymerize monomeric precursors into macromolecules: proteins, nucleic acids, and polysaccharides; and (4) synthesize and degrade biomolecules required for specialized cellular functions, such as membrane lipids, intracellular messengers, and pigments.

Although metabolism embraces many hundreds of different enzyme-catalyzed reactions, our major concern in Part II is the central metabolic pathways, which are few in number and remarkably similar in all forms of life. Living organisms can be divided into two large groups according to the chemical form in which they obtain carbon from the environment. **Autotrophs** (such as photosynthetic bacteria, green algae, and vascular plants) can use carbon dioxide from the atmosphere as their sole source of carbon, from which they construct all their carbon-containing biomolecules (see Fig. 1-6). **Heterotrophs** cannot use atmospheric carbon dioxide and must obtain carbon from their environment in the form of relatively complex organic molecules such as glucose. Multicellular animals and most microorganisms are heterotrophic. Autotrophic cells and organisms are relatively self-sufficient, whereas heterotrophic cells and organisms, with their requirements for carbon in more complex forms, must subsist on the products of other organisms.

Many autotrophic organisms are photosynthetic and obtain their energy from sunlight, whereas heterotrophic organisms obtain their energy from the degradation of organic nutrients produced by autotrophs. In our biosphere, autotrophs and heterotrophs live together in a vast, interdependent cycle in which autotrophic organisms use atmospheric carbon dioxide to build their organic biomolecules, some of them generating oxygen from water in the process. Heterotrophs, in turn, use the organic products of autotrophs as nutrients and return carbon dioxide to the atmosphere. Some of the oxidation reactions that produce carbon dioxide also consume oxygen, converting it to water. Thus carbon, oxygen, and water are constantly cycled between the heterotrophic and autotrophic worlds, with solar energy as the driving force for this global process **(Fig. 1)**.

All living organisms also require a source of nitrogen, which is necessary for the synthesis of amino acids, nucleotides, and other compounds. Bacteria and plants can generally use either ammonia or nitrate as their sole source of nitrogen, but vertebrates must obtain nitrogen in the form of amino acids or other organic compounds. Only a few organisms—the cyanobacteria and many species of soil bacteria that live symbiotically on the roots of some plants—are capable of converting ("fixing") atmospheric nitrogen (N_2) into ammonia. Other bacteria (the nitrifying bacteria) oxidize ammonia to nitrites and nitrates; yet others convert nitrate to N_2. The anammox bacteria convert ammonia and nitrite to N_2. Thus, in addition to the global carbon and oxygen cycles, a nitrogen cycle operates in the biosphere, turning over huge amounts of nitrogen **(Fig. 2)**. The cycling of carbon, oxygen, and nitrogen, which ultimately involves all species, depends on a proper balance between the activities of the producers (autotrophs) and consumers (heterotrophs) in our biosphere.

These cycles of matter are driven by an enormous flow of energy into and through the biosphere, beginning with the capture of solar energy by photosynthetic

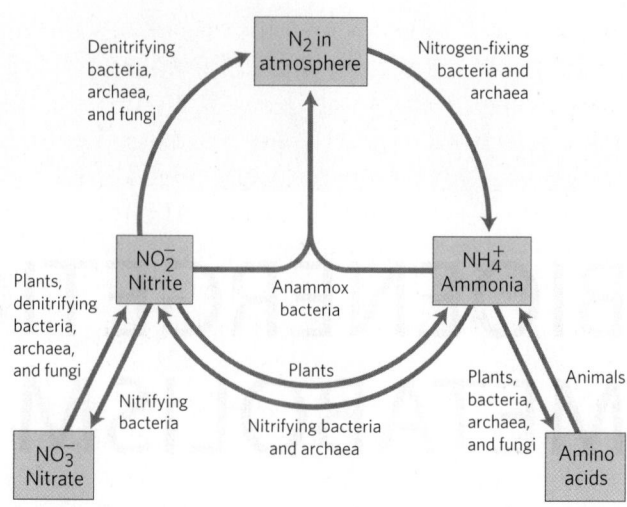

FIGURE 2 Cycling of nitrogen in the biosphere. Gaseous nitrogen (N_2) makes up 80% of the earth's atmosphere.

organisms and use of this energy to generate energy-rich carbohydrates and other organic nutrients; these nutrients are then used as energy sources by heterotrophic organisms. In metabolic processes, and in all energy transformations, there is a loss of useful energy (free energy) and an inevitable increase in the amount of unusable energy (heat and entropy). In contrast to the cycling of matter, therefore, energy flows one way through the biosphere; organisms cannot regenerate useful energy from energy dissipated as heat and entropy. Carbon, oxygen, and nitrogen recycle continuously, but energy is constantly transformed into unusable forms such as heat.

Metabolism, the sum of all the chemical transformations taking place in a cell or organism, occurs through a series of enzyme-catalyzed reactions that constitute **metabolic pathways**. Each of the consecutive steps in a metabolic pathway brings about a specific, small chemical change, usually the removal, transfer, or addition of a particular atom or functional group. The precursor is converted into a product through a series of metabolic intermediates called **metabolites**. The term **intermediary metabolism** is often applied to the combined activities of all the metabolic pathways that interconvert precursors, metabolites, and products of low molecular weight (generally, M_r <1,000).

Catabolism is the degradative phase of metabolism in which organic nutrient molecules (carbohydrates, fats, and proteins) are converted into smaller, simpler end products (such as lactic acid, CO_2, and NH_3). Catabolic pathways release energy, some of which is conserved in the formation of ATP and reduced electron carriers (NADH, NADPH, and $FADH_2$); the rest is

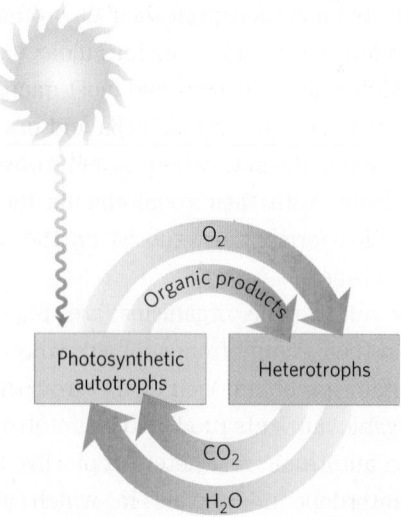

FIGURE 1 Cycling of carbon dioxide and oxygen between the autotrophic (photosynthetic) and heterotrophic domains in the biosphere. The flow of mass through this cycle is enormous; about 4×10^{11} metric tons of carbon are turned over in the biosphere annually.

lost as heat. In **anabolism**, also called biosynthesis, small, simple precursors are built up into larger and more complex molecules, including lipids, polysaccharides, proteins, and nucleic acids. Anabolic reactions require an input of energy, generally in the form of the phosphoryl group transfer potential of ATP and the reducing power of NADH, NADPH, and $FADH_2$ **(Fig. 3)**.

Some metabolic pathways are linear, and some are branched, yielding multiple useful end products from a single precursor or converting several starting materials into a single product. In general, catabolic pathways are *convergent* and anabolic pathways *divergent* **(Fig. 4)**. Some pathways are cyclic: one starting component of the pathway is regenerated in a series of reactions that converts another starting component into a product. We shall see examples of each type of pathway in the following chapters.

Most cells have the enzymes to carry out both the degradation and the synthesis of the important categories of biomolecules—fatty acids, for example. The simultaneous synthesis and degradation of fatty acids would be wasteful, however, and this is prevented by reciprocally regulating the anabolic and catabolic reaction sequences: when one sequence is active, the other is suppressed. Such regulation could not occur if anabolic and catabolic pathways were catalyzed by exactly the same set of enzymes, operating in one direction for anabolism, the opposite direction for catabolism: inhibition of an enzyme involved in catabolism would also inhibit the reaction sequence in the anabolic direction. Catabolic and anabolic pathways that connect the same two end points (glucose $\longrightarrow$ pyruvate, and pyruvate $\longrightarrow$ glucose, for example) may employ many of the same enzymes, but, invariably, at least one of the steps is catalyzed by different enzymes in the catabolic and anabolic directions, and these enzymes are the sites of separate regulation. Moreover, for both anabolic and catabolic pathways to be essentially irreversible, the reactions unique to each direction must include at least one that is thermodynamically very favorable—in other words, a reaction for which the reverse reaction is very unfavorable. As a further contribution to the separate regulation of catabolic and anabolic reaction sequences, paired catabolic and anabolic pathways commonly take place in different cellular compartments: for example, fatty acid catabolism in animal mitochondria, fatty acid synthesis in the cytosol. The concentrations of intermediates, enzymes, and regulators can be maintained at different levels in these different compartments. Because metabolic pathways are subject to kinetic control by substrate concentration, separate pools of anabolic and catabolic intermediates also contribute to the control of metabolic rates. Devices that separate anabolic and catabolic processes will be of particular interest in our discussions of metabolism.

Metabolic pathways are regulated at several levels, from within the cell and from outside. The most immediate regulation is by the availability of substrate; when the intracellular concentration of an enzyme's substrate is near or below K_m (as is commonly the case), the rate of the reaction depends strongly on substrate concentration (see Fig. 6-11). A second type of rapid control from within is allosteric regulation (p. 225) by a metabolic intermediate or coenzyme—an amino acid or ATP, for example—that signals the cell's internal metabolic state. When the cell contains an amount of, say, aspartate sufficient for its immediate needs, or when the cellular level of ATP indicates that further fuel consumption is unnecessary at the moment, these signals allosterically inhibit

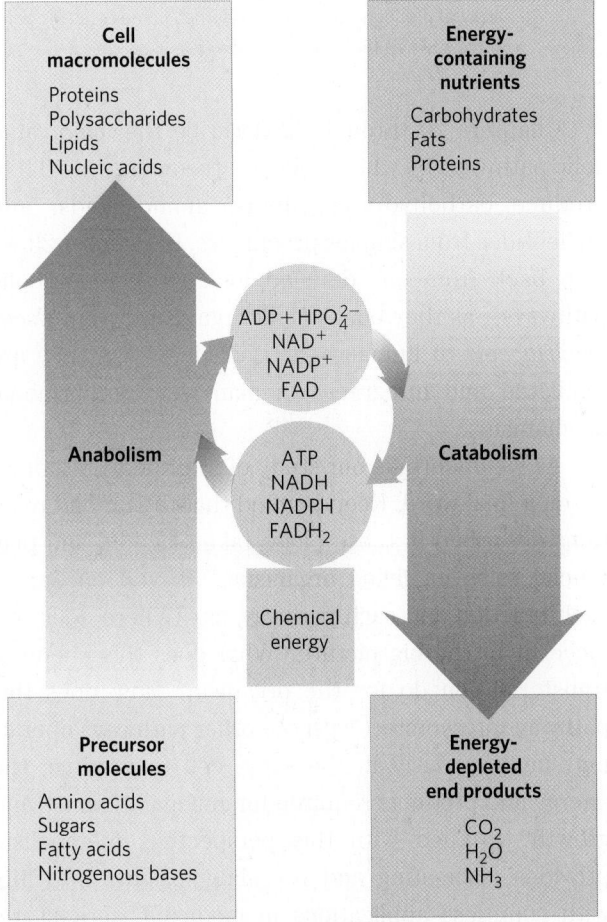

FIGURE 3 **The big picture: energy relationships between catabolic and anabolic pathways.** Catabolic pathways deliver chemical energy in the form of ATP, NADH, NADPH, and $FADH_2$. These energy carriers are used in anabolic pathways to convert small precursor molecules into cellular macromolecules.

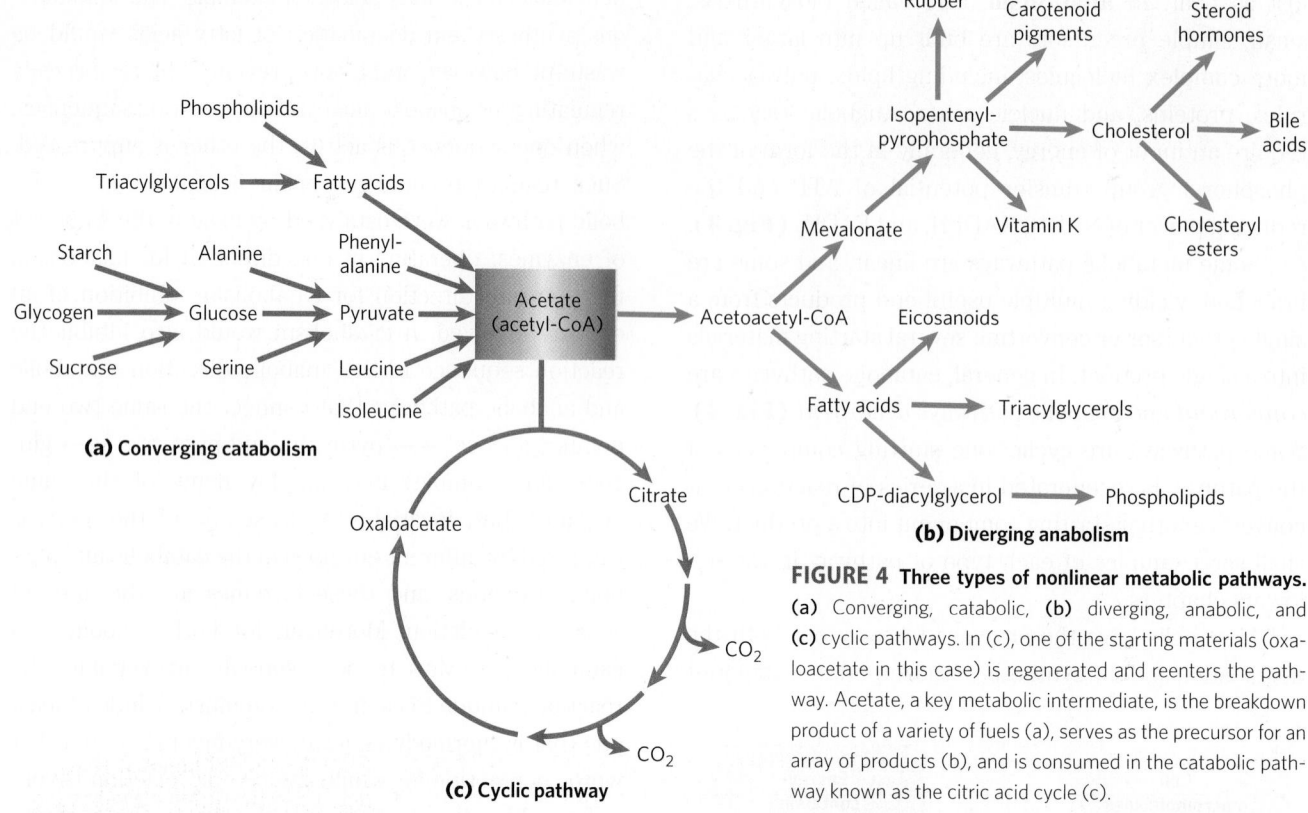

FIGURE 4 Three types of nonlinear metabolic pathways. **(a)** Converging, catabolic, **(b)** diverging, anabolic, and **(c)** cyclic pathways. In (c), one of the starting materials (oxaloacetate in this case) is regenerated and reenters the pathway. Acetate, a key metabolic intermediate, is the breakdown product of a variety of fuels (a), serves as the precursor for an array of products (b), and is consumed in the catabolic pathway known as the citric acid cycle (c).

the activity of one or more enzymes in the relevant pathway. In multicellular organisms, the metabolic activities of different tissues are regulated and integrated by growth factors and hormones that act from outside the cell. In some cases, this regulation occurs virtually instantaneously (sometimes in less than a millisecond) through changes in the levels of intracellular messengers that modify the activity of existing enzyme molecules by allosteric mechanisms or by covalent modification such as phosphorylation. In other cases, the extracellular signal changes the cellular concentration of an enzyme by altering the rate of its synthesis or degradation, so the effect is seen only after minutes or hours.

We begin Part II with a discussion of the basic energetic principles that govern all metabolism (Chapter 13). We then consider the major catabolic pathways by which cells obtain energy from the oxidation of various fuels (Chapters 14 through 20). Chapters 19 and 20 represent the pivotal point of our discussion of metabolism; they concern chemiosmotic energy coupling, a universal mechanism in which a transmembrane electrochemical potential, produced either by substrate oxidation or by light absorption, drives the synthesis of ATP.

Chapters 20 through 22 describe the major anabolic pathways by which cells use the energy in ATP to produce carbohydrates, lipids, amino acids, and nucleotides from simpler precursors. In Chapter 23 we step back from our detailed look at the metabolic pathways—as they occur in all organisms, from *Escherichia coli* to humans—and consider how they are regulated and integrated in mammals by hormonal mechanisms.

As we undertake our study of intermediary metabolism, a final word. Keep in mind that the myriad reactions described in these pages take place in, and play crucial roles in, living organisms. As you encounter each reaction and each pathway, ask: Where does this piece fit in the big picture? What does this chemical transformation do for the organism? How does this pathway interconnect with the other pathways operating simultaneously in the same cell to produce the energy and products required for cell maintenance and growth? Studied with this perspective, metabolism provides fascinating and revealing insights into life, with countless applications in medicine, agriculture, and biotechnology.

CHAPTER 13

Bioenergetics and Biochemical Reaction Types

Self-study tools that will help you practice what you've learned and reinforce this chapter's concepts are available online. Go to www.macmillanlearning.com/LehningerBiochemistry7e.

iving cells and organisms must perform work to stay alive, to grow, and to reproduce. The ability to harness energy and to channel it into biological work is a fundamental property of all living organisms; it must have been acquired very early in cellular evolution. Modern organisms carry out a remarkable variety of energy transductions, conversions of one form of energy to another. They use the chemical energy in fuels to bring about the synthesis of complex, highly ordered macromolecules from simple precursors. They also convert the chemical energy of fuels into concentration gradients and electrical gradients, into motion and heat, and, in a few organisms such as fireflies and some deep-sea fish, into light. Photosynthetic organisms transduce light energy into all these other forms of energy.

The chemical mechanisms that underlie biological energy transductions have fascinated and challenged biologists for centuries. The French chemist Antoine Lavoisier recognized that animals somehow transform chemical fuels (foods) into heat and that this process of respiration is essential to life. He observed that

> . . . in general, respiration is nothing but a slow combustion of carbon and hydrogen, which is entirely similar to that which occurs in a lighted lamp or

candle, and that, from this point of view, animals that respire are true combustible bodies that burn and consume themselves. . . . One may say that this analogy between combustion and respiration has not escaped the notice of the poets, or rather the philosophers of antiquity, and which they had expounded and interpreted. This fire stolen from heaven, this torch of Prometheus, does not only represent an ingenious and poetic idea, it is a faithful picture of the operations of nature, at least for animals that breathe; one may therefore say, with the ancients, that the torch of life lights itself at the moment the infant breathes for the first time, and it does not extinguish itself except at death.*

Antoine Lavoisier, 1743–1794 [Source: INTERFOTO/Alamy.]

In the twentieth century, we began to understand much of the chemistry underlying that "torch of life." Biological energy transductions obey the same chemical and physical laws that govern all other natural processes. It is therefore essential for a student of biochemistry to understand these laws and how they apply to the flow of energy in the biosphere.

In this chapter we first review the laws of thermodynamics and the quantitative relationships among free

*From a memoir by Armand Seguin and Antoine Lavoisier, dated 1789, quoted in A. Lavoisier, *Oeuvres de Lavoisier,* Imprimerie Impériale, Paris, 1862.

energy, enthalpy, and entropy. We then review the common types of biochemical reactions that occur in living cells, reactions that harness, store, transfer, and release the energy taken up by organisms from their surroundings. Our focus then shifts to reactions that have special roles in biological energy exchanges, particularly those involving ATP. We finish by considering the importance of oxidation-reduction reactions in living cells, the energetics of biological electron transfers, and the electron carriers commonly employed as cofactors in these processes.

13.1 Bioenergetics and Thermodynamics

Bioenergetics is the quantitative study of **energy transductions**—changes of one form of energy into another—that occur in living cells, and of the nature and function of the chemical processes underlying these transductions. Although many of the principles of thermodynamics have been introduced in earlier chapters and may be familiar to you, a review of the quantitative aspects of these principles is useful here.

Biological Energy Transformations Obey the Laws of Thermodynamics

Many quantitative observations made by physicists and chemists on the interconversion of different forms of energy led, in the nineteenth century, to the formulation of two fundamental laws of thermodynamics. The first law is the principle of the conservation of energy: *for any physical or chemical change, the total amount of energy in the universe remains constant; energy may change form or it may be transported from one region to another, but it cannot be created or destroyed.* The second law of thermodynamics, which can be stated in several forms, says that the universe always tends toward increasing disorder: *in all natural processes, the entropy of the universe increases.*

Living organisms consist of collections of molecules much more highly organized than the surrounding materials from which they are constructed, and organisms maintain and produce order, seemingly immune to the second law of thermodynamics. But living organisms do not violate the second law; they operate strictly within it. To discuss the application of the second law to biological systems, we must first define those systems and their surroundings.

The *reacting system* is the collection of matter that is undergoing a particular chemical or physical process; it may be an organism, a cell, or two reacting compounds. The reacting system and its *surroundings* together constitute the *universe*. In the laboratory, some chemical or physical processes can be carried out in isolated or closed systems, in which no material or energy is exchanged with the surroundings. Living cells and organisms, however, are open systems, exchanging both material and energy with their surroundings; living systems are never at equilibrium with their surroundings, and the constant transactions between system and surroundings explain how organisms can create order within themselves while operating within the second law of thermodynamics.

In Chapter 1 (p. 23) we defined three thermodynamic quantities that describe the energy changes occurring in a chemical reaction:

Gibbs free energy, *G*, expresses the amount of energy capable of doing work during a reaction at constant temperature and pressure. When a reaction proceeds with the release of free energy (that is, when the system changes so as to possess less free energy), the free-energy change, ΔG, has a negative value and the reaction is said to be **exergonic**. In **endergonic** reactions, the system gains free energy and ΔG is positive.

Enthalpy, *H*, is the heat content of the reacting system. It reflects the number and kinds of chemical bonds in the reactants and products. When a chemical reaction releases heat, it is said to be **exothermic**; the heat content of the products is less than that of the reactants, and the change in enthalpy, ΔH, has, by convention, a negative value. Reacting systems that take up heat from their surroundings are **endothermic** and have positive values of ΔH.

Entropy, *S*, is a quantitative expression for the randomness or disorder in a system (see Box 1-3). When the products of a reaction are less complex and more disordered than the reactants, the reaction is said to proceed with a gain in entropy.

The units of ΔG and ΔH are joules/mole or calories/mole (recall that 1 cal = 4.184 J); units of entropy are joules/mole·Kelvin (J/mol·K) (Table 13-1).

"Now, in the *second* law of thermodynamics . . ."

[Source: Sidney Harris.]

TABLE 13-1	Some Physical Constants and Units Used in Thermodynamics

Boltzmann constant, $\mathbf{k} = 1.381 \times 10^{-23}$ J/K
Avogadro's number, $N = 6.022 \times 10^{23}$ mol^{-1}
Faraday constant, $F = 96{,}480$ J/V·mol
Gas constant, $R = 8.315$ J/mol·K
  ($=1.987$ cal/mol·K)

Units of ΔG and ΔH are J/mol (or cal/mol)
Units of ΔS are J/mol·K (or cal/mol·K)
1 cal $= 4.184$ J

Units of absolute temperature, T, are Kelvin, K
25 °C $= 298$ K
At 25 °C, $RT = 2.478$ kJ/mol
  ($=0.592$ kcal/mol)

Under the conditions existing in biological systems (including constant temperature and pressure), changes in free energy, enthalpy, and entropy are related to each other quantitatively by the equation

$$\Delta G = \Delta H - T\Delta S \qquad (13\text{-}1)$$

in which ΔG is the change in Gibbs free energy of the reacting system, ΔH is the change in enthalpy of the system, T is the absolute temperature, and ΔS is the change in entropy of the system. By convention, ΔS has a positive sign when entropy increases and ΔH, as noted above, has a negative sign when heat is released by the system to its surroundings. Either of these conditions, both of which are typical of energetically favorable processes, tends to make ΔG negative. In fact, ΔG of a spontaneously reacting system is always negative.

The second law of thermodynamics states that the entropy *of the universe* increases during all chemical and physical processes, but it does not require that the entropy increase take place *in the reacting system* itself. The order produced within cells as they grow and divide is more than compensated for by the disorder they create in their surroundings in the course of growth and division (see Box 1-3, case 2). In short, living organisms preserve their internal order by taking from the surroundings free energy in the form of nutrients or sunlight, and returning to their surroundings an equal amount of energy as heat and entropy.

Cells Require Sources of Free Energy

Cells are isothermal systems—they function at essentially constant temperature (and also function at constant pressure). Heat flow is not a source of energy for cells, because heat can do work only as it passes to a zone or object at a lower temperature. The energy that cells can and must use is free energy, described by the Gibbs free-energy function G, which allows prediction of the direction of chemical reactions, their exact equilibrium position, and the amount of work they can (in theory) perform at constant temperature and pressure.

Heterotrophic cells acquire free energy from nutrient molecules, and photosynthetic cells acquire it from absorbed solar radiation. Both kinds of cells transform this free energy into ATP and other energy-rich compounds capable of providing energy for biological work at constant temperature.

Standard Free-Energy Change Is Directly Related to the Equilibrium Constant

The composition of a reacting system (a mixture of chemical reactants and products) tends to continue changing until equilibrium is reached. At the equilibrium concentration of reactants and products, the rates of the forward and reverse reactions are exactly equal and no further net change occurs in the system. The concentrations of reactants and products *at equilibrium* define the equilibrium constant, K_{eq} (p. 25). In the general reaction

$$a\text{A} + b\text{B} \rightleftharpoons c\text{C} + d\text{D}$$

where a, b, c, and d are the number of molecules of A, B, C, and D participating, the equilibrium constant is given by

$$K_{eq} = \frac{[\text{C}]^c[\text{D}]^d}{[\text{A}]^a[\text{B}]^b} \qquad (13\text{-}2)$$

where [A], [B], [C], and [D] are the molar concentrations of the reaction components *at the point of equilibrium*.

When a reacting system is not at equilibrium, the tendency to move toward equilibrium represents a driving force, the magnitude of which can be expressed as the free-energy change for the reaction, ΔG. Under standard conditions (298 K $= 25$ °C), when reactants and products are initially present at 1 M concentrations or, for gases, at partial pressures of 101.3 kilopascals (kPa), or 1 atm, the force driving the system toward equilibrium is defined as the standard free-energy change, $\Delta G°$. By this definition, the standard state for reactions that involve hydrogen ions is [H$^+$] $= 1$ M, or pH 0. Most biochemical reactions, however, occur in well-buffered aqueous solutions near pH 7; both the pH and the concentration of water (55.5 M) are essentially constant.

>> Key Convention: For convenience of calculations, biochemists define a standard state different from that used in chemistry and physics: in the biochemical standard state, [H$^+$] is 10^{-7} M (pH 7) and [H$_2$O] is 55.5 M. For reactions that involve Mg^{2+} (which include most of those with ATP as a reactant), [Mg^{2+}] in solution is commonly taken to be constant at 1 mM. **<<**

Physical constants based on this biochemical standard state are called **standard transformed constants** and are written with a prime (such as $\Delta G'°$ and K'_{eq}) to distinguish them from the untransformed constants used by chemists and physicists. (Note that most other textbooks use the symbol $\Delta G°'$ rather than $\Delta G'°$.

Our use of $\Delta G'^\circ$, recommended by an international committee of chemists and biochemists, is intended to emphasize that the transformed free-energy change, $\Delta G'^\circ$, is the criterion for equilibrium.) For simplicity, we will hereafter refer to these transformed constants as **standard free-energy changes** and **standard equilibrium constants**.

>> Key Convention: In another simplifying convention used by biochemists, when H_2O, H^+, and/or Mg^{2+} are reactants or products, their concentrations are not included in equations such as Equation 13-2 but are instead incorporated into the constants K'_{eq} and $\Delta G'^\circ$. **<<**

Just as K'_{eq} is a physical constant characteristic for each reaction, so too is $\Delta G'^\circ$ a constant. As we noted in Chapter 6, there is a simple relationship between K'_{eq} and $\Delta G'^\circ$:

$$\Delta G'^\circ = -RT \ln K'_{eq} \qquad (13\text{-}3)$$

The standard free-energy change of a chemical reaction is simply an alternative mathematical way of expressing its equilibrium constant. Table 13-2 shows the relationship between $\Delta G'^\circ$ and K'_{eq}. If the equilibrium constant for a given chemical reaction is 1.0, the standard free-energy change of that reaction is 0.0 (the natural logarithm of 1.0 is zero). If K'_{eq} of a reaction is greater than 1.0, its $\Delta G'^\circ$ is negative. If K'_{eq} is less than 1.0, $\Delta G'^\circ$ is positive. Because the relationship between $\Delta G'^\circ$ and K'_{eq} is exponential, relatively small changes in $\Delta G'^\circ$ correspond to large changes in K'_{eq}.

TABLE 13-2 Relationship between Equilibrium Constants and Standard Free-Energy Changes of Chemical Reactions

K'_{eq}	$\Delta G'^\circ$ (kJ/mol)	$\Delta G'^\circ$ (kcal/mol)[a]
10^3	−17.1	−4.1
10^2	−11.4	−2.7
10^1	−5.7	−1.4
1	0.0	0.0
10^{-1}	5.7	1.4
10^{-2}	11.4	2.7
10^{-3}	17.1	4.1
10^{-4}	22.8	5.5
10^{-5}	28.5	6.8
10^{-6}	34.2	8.2

[a] Although joules and kilojoules are the standard units of energy and are used throughout this text, biochemists and nutritionists sometimes express $\Delta G'^\circ$ values in kilocalories per mole. We have therefore included values in both kilojoules and kilocalories in this table and in Tables 13-4 and 13-6. To convert kilojoules to kilocalories, divide the number of kilojoules by 4.184.

TABLE 13-3 Relationships among K'_{eq}, $\Delta G'^\circ$, and the Direction of Chemical Reactions

When K'_{eq} is . . .	$\Delta G'^\circ$ is . . .	Starting with all components at 1 M, the reaction . . .
>1.0	negative	proceeds forward
1.0	zero	is at equilibrium
<1.0	positive	proceeds in reverse

It may be helpful to think of the standard free-energy change in another way. $\Delta G'^\circ$ is the difference between the free-energy content of the products and the free-energy content of the reactants, under standard conditions. When $\Delta G'^\circ$ is negative, the products contain less free energy than the reactants and the reaction will proceed spontaneously under standard conditions; all chemical reactions tend to go in the direction that results in a decrease in the free energy of the system. A positive value of $\Delta G'^\circ$ means that the products of the reaction contain more free energy than the reactants, and this reaction will tend to go in the reverse direction if we start with 1.0 M concentrations of all components (standard conditions). Table 13-3 summarizes these points.

WORKED EXAMPLE 13-1 Calculation of $\Delta G'^\circ$

Calculate the standard free-energy change of the reaction catalyzed by the enzyme phosphoglucomutase,

$$\text{Glucose 1-phosphate} \rightleftharpoons \text{glucose 6-phosphate}$$

given that, starting with 20 mM glucose 1-phosphate and no glucose 6-phosphate, the final equilibrium mixture at 25 °C and pH 7.0 contains 1.0 mM glucose 1-phosphate and 19 mM glucose 6-phosphate. Does the reaction in the direction of glucose 6-phosphate formation proceed with a loss or a gain of free energy?

Solution: First we calculate the equilibrium constant:

$$K'_{eq} = \frac{[\text{glucose 6-phosphate}]}{[\text{glucose 1-phosphate}]} = \frac{19 \text{ mM}}{1.0 \text{ mM}} = 19$$

We can now calculate the standard free-energy change:

$$\begin{aligned}
\Delta G'^\circ &= -RT \ln K'_{eq} \\
&= -(8.315 \text{ J/mol} \cdot \text{K})(298 \text{ K})(\ln 19) \\
&= -7.3 \text{ kJ/mol}
\end{aligned}$$

Because the standard free-energy change is negative, the conversion of glucose 1-phosphate to glucose 6-phosphate proceeds with a loss (release) of free energy. (For the reverse reaction, $\Delta G'^\circ$ has the same magnitude but the *opposite* sign.)

Table 13-4 gives the standard free-energy changes for some representative chemical reactions. Note that

TABLE 13-4	Standard Free-Energy Changes of Some Chemical Reactions		
		$\Delta G'^\circ$	
Reaction type		(kJ/mol)	(kcal/mol)
Hydrolysis reactions			
Acid anhydrides			
Acetic anhydride + $H_2O \longrightarrow$ 2 acetate		−91.1	−21.8
ATP + $H_2O \longrightarrow$ ADP + P_i		−30.5	−7.3
ATP + $H_2O \longrightarrow$ AMP + PP_i		−45.6	−10.9
PP_i + $H_2O \longrightarrow 2P_i$		−19.2	−4.6
UDP-glucose + $H_2O \longrightarrow$ UMP + glucose 1-phosphate		−43.0	−10.3
Esters			
Ethyl acetate + $H_2O \longrightarrow$ ethanol + acetate		−19.6	−4.7
Glucose 6-phosphate + $H_2O \longrightarrow$ glucose + P_i		−13.8	−3.3
Amides and peptides			
Glutamine + $H_2O \longrightarrow$ glutamate + NH_4^+		−14.2	−3.4
Glycylglycine + $H_2O \longrightarrow$ 2 glycine		−9.2	−2.2
Glycosides			
Maltose + $H_2O \longrightarrow$ 2 glucose		−15.5	−3.7
Lactose + $H_2O \longrightarrow$ glucose + galactose		−15.9	−3.8
Rearrangements			
Glucose 1-phosphate $\longrightarrow$ glucose 6-phosphate		−7.3	−1.7
Fructose 6-phosphate $\longrightarrow$ glucose 6-phosphate		−1.7	−0.4
Elimination of water			
Malate $\longrightarrow$ fumarate + H_2O		3.1	0.8
Oxidations with molecular oxygen			
Glucose + $6O_2 \longrightarrow 6CO_2 + 6H_2O$		−2,840	−686
Palmitate + $23O_2 \longrightarrow 16CO_2 + 16H_2O$		−9,770	−2,338

hydrolysis of simple esters, amides, peptides, and glycosides, as well as rearrangements and eliminations, proceed with relatively small standard free-energy changes, whereas hydrolysis of acid anhydrides is accompanied by relatively large decreases in standard free energy. The complete oxidation of organic compounds such as glucose or palmitate to CO_2 and H_2O, which in cells requires many steps, results in very large decreases in standard free energy. However, standard free-energy changes such as those in Table 13-4 indicate how much free energy is available from a reaction under *standard conditions*. To describe the energy released under the conditions existing in cells, an expression for the *actual free-energy change* is essential.

Actual Free-Energy Changes Depend on Reactant and Product Concentrations

We must be careful to distinguish between two different quantities: the actual free-energy change, ΔG, and the standard free-energy change, $\Delta G'^\circ$. Each chemical reaction has a characteristic standard free-energy change, which may be positive, negative, or zero, depending on the equilibrium constant of the reaction. The standard free-energy change tells us in which direction and how far a given reaction must go to reach equilibrium *when the initial concentration of each component is 1.0 M,* the pH is 7.0, the temperature is 25 °C, and the pressure is 101.3 kPa (1 atm). Thus $\Delta G'^\circ$ is a constant: it has a characteristic, unchanging value for a given reaction. But the *actual* free-energy change, ΔG, is a function of reactant and product concentrations and of the temperature prevailing during the reaction, none of which will necessarily match the standard conditions as defined above. Moreover, the ΔG of any reaction proceeding spontaneously toward its equilibrium is always negative, becomes less negative as the reaction proceeds, and is zero at the point of equilibrium, indicating that no more work can be done by the reaction.

ΔG and $\Delta G'^\circ$ for any reaction $a\text{A} + b\text{B} \rightleftharpoons c\text{C} + d\text{D}$ are related by the equation

$$\Delta G = \Delta G'^\circ + RT \ln \frac{[\text{C}]^c[\text{D}]^d}{[\text{A}]^a[\text{B}]^b} \qquad (13\text{-}4)$$

in which the terms in red are those *actually prevailing* in the system under observation. The concentration terms in this equation express the effects commonly called mass action, and the term $[C]^c[D]^d/[A]^a[B]^b$ is called the **mass-action ratio**, Q. Thus Equation 13-4 can be expressed as $\Delta G = \Delta G'^{\circ} + RT \ln Q$. As an example, let us suppose that the reaction $A + B \rightleftharpoons C + D$ is taking place under the standard conditions of temperature (25 °C) and pressure (101.3 kPa) but that the concentrations of A, B, C, and D are *not* equal and none of the components is present at the standard concentration of 1.0 M. To determine the actual free-energy change, ΔG, under these nonstandard conditions of concentration as the reaction proceeds from left to right, we simply enter the *actual* concentrations of A, B, C, and D in Equation 13-4; the values of R, T, and $\Delta G'^{\circ}$ are the standard values. ΔG is negative and approaches zero as the reaction proceeds, because the actual concentrations of A and B decrease and the concentrations of C and D increase. Notice that when a reaction is at equilibrium—when there is no force driving the reaction in either direction and ΔG is zero—Equation 13-4 reduces to

$$0 = \Delta G = \Delta G'^{\circ} + RT \ln \frac{[C]_{eq}[D]_{eq}}{[A]_{eq}[B]_{eq}}$$

or

$$\Delta G'^{\circ} = -RT \ln K'_{eq}$$

which is the equation relating the standard free-energy change and equilibrium constant (Eqn 13-3).

The criterion for spontaneity of a reaction is the value of ΔG, not $\Delta G'^{\circ}$. A reaction with a positive $\Delta G'^{\circ}$ can go in the forward direction *if ΔG is negative*. This is possible if the term $RT \ln$ ([products]/[reactants]) in Equation 13-4 is negative and has a larger absolute value than $\Delta G'^{\circ}$. For example, the immediate removal of the products of a reaction can keep the ratio [products]/[reactants] well below 1, such that the term $RT \ln$ ([products]/[reactants]) has a large, negative value. $\Delta G'^{\circ}$ and ΔG are expressions of the *maximum* amount of free energy that a given reaction can *theoretically* deliver—an amount of energy that could be realized only if a perfectly efficient device were available to trap or harness it. Given that no such device is possible (some energy is always lost to entropy during any process), the amount of work done by the reaction at constant temperature and pressure is always less than the theoretical amount.

Another important point is that some thermodynamically favorable reactions (that is, reactions for which $\Delta G'^{\circ}$ is large and negative) do not occur at measurable rates. For example, combustion of firewood to CO_2 and H_2O is very favorable thermodynamically, but firewood remains stable for years because the activation energy (see Figs 6-2, 6-3) for the combustion reaction is higher than the energy available at room temperature. If the necessary activation energy is provided (with a lighted match, for example), combustion will begin, converting the wood to the more stable products CO_2 and H_2O and releasing energy as heat and light. The heat released by this exothermic reaction provides the activation energy for combustion of neighboring regions of the firewood; the process is self-perpetuating.

In living cells, reactions that would be extremely slow *if uncatalyzed* are caused to proceed not by supplying additional heat but by lowering the activation energy through use of an enzyme. An enzyme provides an alternative reaction pathway with a lower activation energy than the uncatalyzed reaction, so that at room temperature a large fraction of the substrate molecules have enough thermal energy to overcome the activation barrier, and the reaction rate increases dramatically. *The free-energy change for a reaction is independent of the pathway by which the reaction occurs;* it depends only on the nature and concentration of the initial reactants and the final products. *Enzymes cannot, therefore, change equilibrium constants;* but they can and do increase the *rate* at which a reaction proceeds in the direction dictated by thermodynamics (see Section 6.2).

Standard Free-Energy Changes Are Additive

In the case of two sequential chemical reactions, $A \rightleftharpoons B$ and $B \rightleftharpoons C$, each reaction has its own equilibrium constant and each has its characteristic standard free-energy change, $\Delta G_1'^{\circ}$ and $\Delta G_2'^{\circ}$. As the two reactions are sequential, B cancels out to give the overall reaction $A \rightleftharpoons C$, which has its own equilibrium constant and thus its own standard free-energy change, $\Delta G_{Sum}'^{\circ}$. *The $\Delta G'^{\circ}$ values of sequential chemical reactions are additive.* For the overall reaction $A \rightleftharpoons C$, $\Delta G_{Sum}'^{\circ}$ is the sum of the individual standard free-energy changes, $\Delta G_1'^{\circ}$ and $\Delta G_2'^{\circ}$, of the two reactions: $\Delta G_{Sum}'^{\circ} = \Delta G_1'^{\circ} + \Delta G_2'^{\circ}$.

(1)	$A \longrightarrow \cancel{B}$	$\Delta G_1'^{\circ}$
(2)	$\cancel{B} \longrightarrow C$	$\Delta G_2'^{\circ}$
Sum:	$A \longrightarrow C$	$\Delta G_1'^{\circ} + \Delta G_2'^{\circ}$

This principle of bioenergetics explains how a thermodynamically unfavorable (endergonic) reaction can be driven in the forward direction by coupling it to a highly exergonic reaction through a common intermediate. For example, in many organisms, the synthesis of glucose 6-phosphate is the first step in the utilization of glucose. In principle, the synthesis could be accomplished by this reaction:

Glucose + $P_i \longrightarrow$ glucose 6-phosphate + H_2O
$$\Delta G'^{\circ} = 13.8 \text{ kJ/mol}$$

But the positive value of $\Delta G'^{\circ}$ predicts that under standard conditions the reaction will tend not to proceed spontaneously in the direction written. Another cellular reaction, the hydrolysis of ATP to ADP and P_i, is highly exergonic:

ATP + $H_2O \longrightarrow$ ADP + P_i $\Delta G'^{\circ} = -30.5 \text{ kJ/mol}$

These two reactions share the common intermediates P_i and H_2O and may be expressed as sequential reactions:

(1)   Glucose + ~~P_i~~ $\longrightarrow$ glucose 6-phosphate + ~~H_2O~~

(2)   ATP + ~~H_2O~~ $\longrightarrow$ ADP + ~~P_i~~

Sum: ATP + glucose $\longrightarrow$ ADP + glucose 6-phosphate

The overall standard free-energy change is obtained by adding the $\Delta G'^{\circ}$ values for individual reactions:

$$\Delta G'^{\circ}_{Sum} = 13.8 \text{ kJ/mol} + (-30.5 \text{ kJ/mol}) = -16.7 \text{ kJ/mol}$$

The overall reaction is exergonic. In this case, energy stored in ATP is used to drive the synthesis of glucose 6-phosphate, even though its formation from glucose and inorganic phosphate (P_i) is endergonic. The _pathway_ of glucose 6-phosphate formation from glucose by phosphoryl transfer from ATP is different from reactions (1) and (2) above, but the net result is the same as the sum of the two reactions. In thermodynamic calculations, all that matters is the state of the system at the beginning of the process and its state at the end; the route between the initial and final states is immaterial.

We have said that $\Delta G'^{\circ}$ is a way of expressing the equilibrium constant for a reaction. For reaction (1) above,

$$K'_{eq_1} = \frac{[\text{glucose 6-phosphate}]}{[\text{glucose}][P_i]} = 3.9 \times 10^{-3} \text{ M}^{-1}$$

Notice that H_2O is not included in this expression, as its concentration (55.5 M) is assumed to remain unchanged by the reaction. The equilibrium constant for the hydrolysis of ATP is

$$K'_{eq_2} = \frac{[\text{ADP}][P_i]}{[\text{ATP}]} = 2.0 \times 10^5 \text{ M}$$

The equilibrium constant for the two coupled reactions is

$$K'_{eq_3} = \frac{[\text{glucose 6-phosphate}][\text{ADP}][P_i]}{[\text{glucose}][P_i][\text{ATP}]}$$

$$= (K'_{eq_1})(K'_{eq_2}) = (3.9 \times 10^{-3} \text{ M}^{-1})(2.0 \times 10^5 \text{ M})$$

$$= 7.8 \times 10^2$$

This calculation illustrates an important point about equilibrium constants: although the $\Delta G'^{\circ}$ values for two reactions that sum to a third, overall reaction are _additive_, the K'_{eq} for the overall reaction is the _product_ of the individual K'_{eq} values for the two reactions. Equilibrium constants are _multiplicative_. By coupling ATP hydrolysis to glucose 6-phosphate synthesis, the K'_{eq} for formation of glucose 6-phosphate from glucose has been raised by a factor of about 2×10^5.

This common-intermediate strategy is employed by all living cells in the synthesis of metabolic intermediates and cellular components. Obviously, the strategy works only if compounds such as ATP are continuously available. In the following chapters we consider several of the most important cellular pathways for producing ATP. For more practice in dealing with free-energy changes and

equilibrium constants for coupled reactions, see the worked examples in Chapter 1 (pp. 25–27).

SUMMARY 13.1 Bioenergetics and Thermodynamics

■ Living cells constantly perform work. They require energy for maintaining their highly organized structures, synthesizing cellular components, transporting small molecules and ions across membranes, and generating electric currents.

■ Bioenergetics is the quantitative study of energy relationships and energy conversions in biological systems. Biological energy transformations obey the laws of thermodynamics.

■ All chemical reactions are influenced by two forces: the tendency to achieve the most stable bonding state (for which enthalpy, H, is a useful expression) and the tendency to achieve the highest degree of randomness, expressed as entropy, S. The net driving force in a reaction is ΔG, the free-energy change, which represents the net effect of these two factors: $\Delta G = \Delta H - T\Delta S$.

■ The standard transformed free-energy change, $\Delta G'^{\circ}$, is a physical constant that is characteristic for a given reaction and can be calculated from the equilibrium constant for the reaction: $\Delta G'^{\circ} = -RT \ln K'_{eq}$.

■ The actual free-energy change, ΔG, is a variable that depends on $\Delta G'^{\circ}$ and on the concentrations of reactants and products: $\Delta G = \Delta G'^{\circ} + RT \ln ([\text{products}]/[\text{reactants}])$.

■ When ΔG is large and negative, the reaction tends to go in the forward direction; when ΔG is large and positive, the reaction tends to go in the reverse direction; and when $\Delta G = 0$, the system is at equilibrium.

■ The free-energy change for a reaction is independent of the pathway by which the reaction occurs. Free-energy changes are additive; the net chemical reaction that results from successive reactions sharing a common intermediate has an overall free-energy change that is the sum of the ΔG values for the individual reactions.

13.2 Chemical Logic and Common Biochemical Reactions

The biological energy transductions we are concerned with in this book are chemical reactions. Cellular chemistry does not encompass every kind of reaction learned in a typical organic chemistry course. Which reactions take place in biological systems and which do not is determined by (1) their relevance to that particular metabolic system and (2) their rates. Both considerations play major roles in shaping the metabolic pathways we consider throughout the rest of the book. A relevant reaction is one that makes use of an available substrate and converts it to a useful product. However, even a potentially relevant reaction may not occur. Some chemical transformations are too slow (have activation energies that are too high) to contribute to living systems, even with the aid of powerful enzyme catalysts. The reactions that do occur in cells represent a toolbox that evolution has used to construct metabolic pathways that circumvent the "impossible" reactions. Learning to recognize the plausible reactions can be a great aid in developing a command of biochemistry.

Even so, the number of metabolic transformations taking place in a typical cell can seem overwhelming. Most cells have the capacity to carry out thousands of specific, enzyme-catalyzed reactions: for example, transformation of a simple nutrient such as glucose into amino acids, nucleotides, or lipids; extraction of energy from fuels by oxidation; and polymerization of monomeric subunits into macromolecules.

To study these reactions, some organization is essential. There are patterns within the chemistry of life; you do not need to learn every individual reaction to comprehend the molecular logic of biochemistry. Most of the reactions in living cells fall into one of five general categories: (1) reactions that make or break carbon–carbon bonds; (2) internal rearrangements, isomerizations, and eliminations; (3) free-radical reactions; (4) group transfers; and (5) oxidation-reductions. We discuss each of these in more detail below and refer

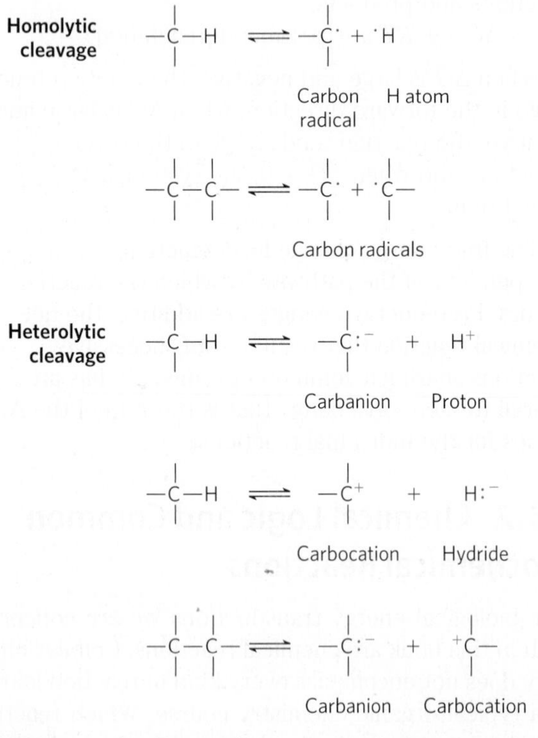

to some examples of each type in later chapters. Note that the five reaction types are not mutually exclusive; for example, an isomerization reaction may involve a free-radical intermediate.

Before proceeding, however, we should review two basic chemical principles. First, a covalent bond consists of a shared pair of electrons, and the bond can be broken in two general ways **(Fig. 13-1)**. In **homolytic cleavage**, each atom leaves the bond as a **radical**, carrying one unpaired electron. In **heterolytic cleavage**, which is more common, one atom retains both bonding electrons. The species most often generated when C—C and C—H bonds are cleaved are illustrated in Figure 13-1. Carbanions, carbocations, and hydride ions are highly unstable; this instability shapes the chemistry of these ions, as we shall see.

The second basic principle is that many biochemical reactions involve interactions between **nucleophiles** (functional groups rich in and capable of donating electrons) and **electrophiles** (electron-deficient functional groups that seek electrons). Nucleophiles combine with and give up electrons to electrophiles. Common biological nucleophiles and electrophiles are shown in **Figure 13-2**.

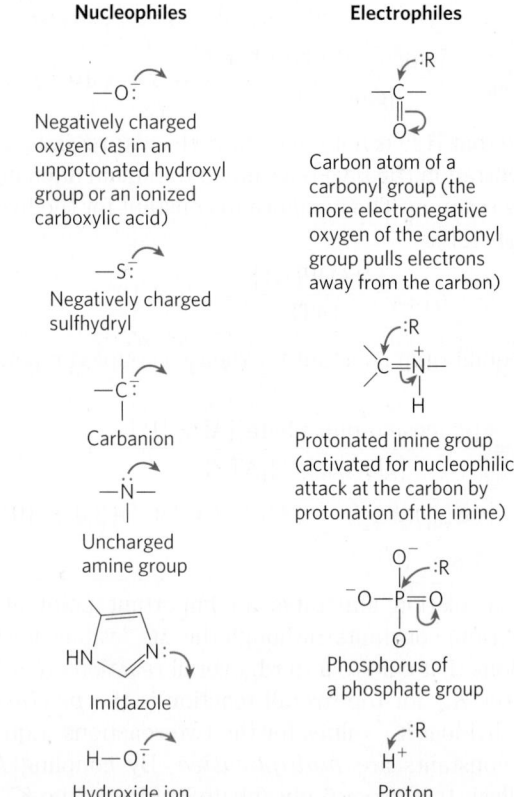

FIGURE 13-2 Common nucleophiles and electrophiles in biochemical reactions. Chemical reaction mechanisms, which trace the formation and breakage of covalent bonds, are communicated with dots and curved arrows, a convention known informally as "electron pushing." A covalent bond consists of a shared pair of electrons. Nonbonded electrons important to the reaction mechanism are designated by dots (:). Curved arrows (⁀) represent the movement of electron pairs. For movement of a single electron (as in a free-radical reaction), a single-headed (fishhook-type) arrow is used (⁀). Most reaction steps involve an unshared electron pair.

FIGURE 13-1 Two mechanisms for cleavage of a C—C or C—H bond. In a homolytic cleavage, each atom keeps one of the bonding electrons, resulting in the formation of carbon radicals (carbons having unpaired electrons) or uncharged hydrogen atoms. In a heterolytic cleavage, one of the atoms retains both bonding electrons. This can result in the formation of carbanions, carbocations, protons, or hydride ions.

Note that a carbon atom can act as either a nucleophile or an electrophile, depending on which bonds and functional groups surround it.

Reactions That Make or Break Carbon–Carbon Bonds

Heterolytic cleavage of a C—C bond yields a **carbanion** and a **carbocation** (Fig. 13-1). Conversely, the formation of a C—C bond involves the combination of a nucleophilic carbanion and an electrophilic carbocation. Carbanions and carbocations are generally so unstable that their formation as reaction intermediates can be energetically unfeasible, even with enzyme catalysts. For the purpose of cellular biochemistry they are impossible reactions—unless chemical assistance is provided in the form of functional groups containing electronegative atoms (O and N) that can alter the electronic structure of adjacent carbon atoms so as to stabilize and facilitate the formation of carbanion and carbocation intermediates.

Carbonyl groups are particularly important in the chemical transformations of metabolic pathways. The carbon of a carbonyl group has a partial positive charge due to the electron-withdrawing property of the carbonyl oxygen, and so is an electrophilic carbon (**Fig. 13-3a**). A carbonyl group can thus facilitate the formation of a carbanion on an adjoining carbon by delocalizing the carbanion's negative charge (Fig. 13-3b). An imine group (see Fig. 1-17) can serve a similar function (Fig. 13-3c). The capacity of carbonyl and imine groups to delocalize electrons can be further enhanced by a general acid catalyst or by a metal ion such as Mg^{2+} (Fig. 13-3d).

The importance of a carbonyl group is evident in three major classes of reactions in which C—C bonds are formed or broken (**Fig. 13-4**): aldol condensations, Claisen ester condensations, and decarboxylations. In each type of reaction, a carbanion intermediate is

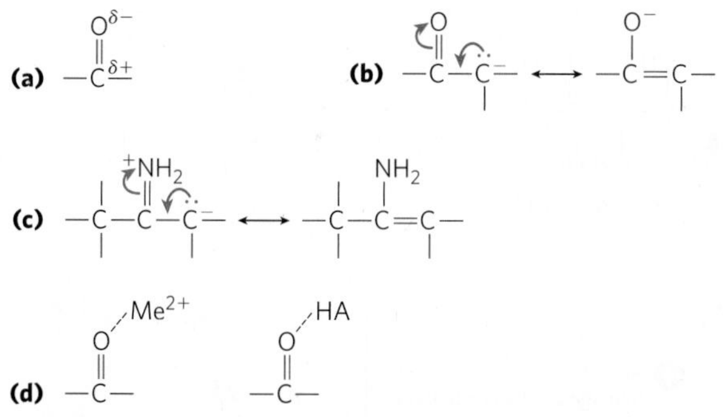

FIGURE 13-3 Chemical properties of carbonyl groups. (a) The carbon atom of a carbonyl group is an electrophile by virtue of the electron-withdrawing capacity of the electronegative oxygen atom, which results in a structure in which the carbon has a partial positive charge. **(b)** Within a molecule, delocalization of electrons into a carbonyl group stabilizes a carbanion on an adjacent carbon, facilitating its formation. **(c)** Imines function much like carbonyl groups in facilitating electron withdrawal. **(d)** Carbonyl groups do not always function alone; their capacity as electron sinks often is augmented by interaction with either a metal ion (Me^{2+}, such as Mg^{2+}) or a general acid (HA).

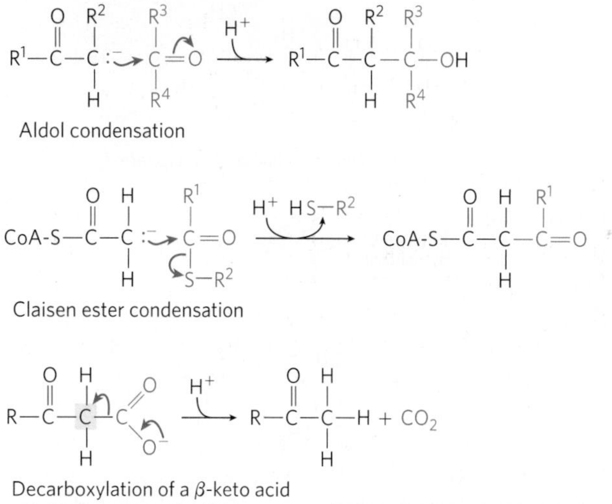

FIGURE 13-4 Some common reactions that form and break C—C bonds in biological systems. For both the aldol condensation and the Claisen condensation, a carbanion serves as nucleophile and the carbon of a carbonyl group serves as electrophile. The carbanion is stabilized in each case by another carbonyl at the adjoining carbon. In the decarboxylation reaction, a carbanion is formed on the carbon shaded blue as the CO_2 leaves. The reaction would not occur at an appreciable rate without the stabilizing effect of the carbonyl adjacent to the carbanion carbon. Wherever a carbanion is shown, a stabilizing resonance with the adjacent carbonyl, as shown in Figure 13-3b, is assumed. An imine (Fig. 13-3c) or other electron-withdrawing group (including certain enzymatic cofactors such as pyridoxal) can replace the carbonyl group in the stabilization of carbanions.

stabilized by a carbonyl group, and in many cases another carbonyl provides the electrophile with which the nucleophilic carbanion reacts.

An **aldol condensation** is a common route to the formation of a C—C bond; the aldolase reaction, which converts a six-carbon compound to two three-carbon compounds in glycolysis, is an aldol condensation in reverse (see Fig. 14-6). In a **Claisen condensation**, the carbanion is stabilized by the carbonyl of an adjacent thioester; an example is the synthesis of citrate in the citric acid cycle (see Fig. 16-9). Decarboxylation also commonly involves the formation of a carbanion stabilized by a carbonyl group; the acetoacetate decarboxylase reaction that occurs in the formation of ketone bodies during fatty acid catabolism provides an example (see Fig. 17-18). Entire metabolic pathways are organized around the introduction of a carbonyl group in a particular location so that a nearby carbon–carbon bond can be formed or cleaved. In some reactions, an imine or a specialized cofactor such as pyridoxal phosphate plays the electron-withdrawing role, instead of a carbonyl group.

The carbocation intermediate occurring in some reactions that form or cleave C—C bonds is generated by the elimination of an excellent leaving group, such as pyrophosphate (see Group Transfer Reactions below). An example is the prenyltransferase reaction (**Fig. 13-5**), an early step in the pathway of cholesterol biosynthesis.

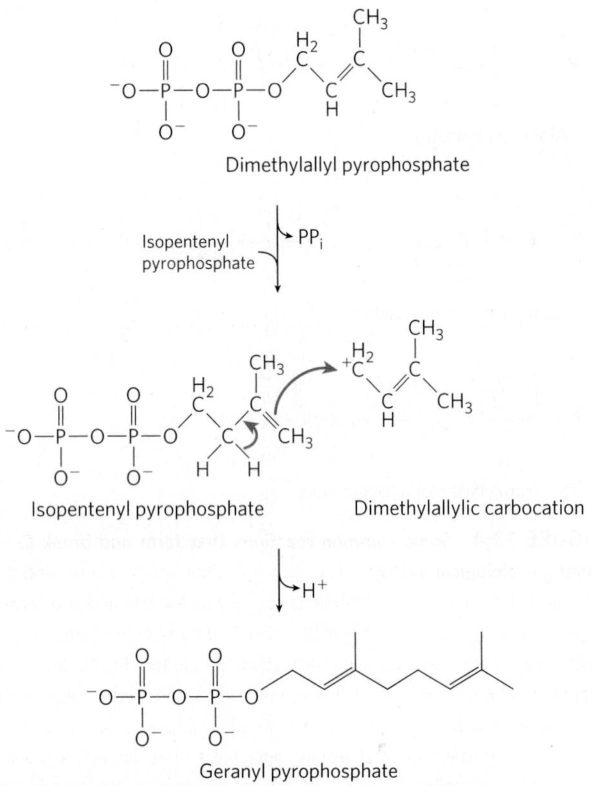

Dimethylallyl pyrophosphate

Isopentenyl pyrophosphate → PP$_i$

Isopentenyl pyrophosphate   Dimethylallylic carbocation

↓ H$^+$

Geranyl pyrophosphate

FIGURE 13-5 Carbocations in carbon–carbon bond formation. In one of the early steps in cholesterol biosynthesis, the enzyme prenyltransferase catalyzes condensation of isopentenyl pyrophosphate and dimethylallyl pyrophosphate to form geranyl pyrophosphate (see Fig. 21-36). The reaction is initiated by elimination of pyrophosphate from the dimethylallyl pyrophosphate to generate a carbocation, stabilized by resonance with the adjacent C=C bond.

Internal Rearrangements, Isomerizations, and Eliminations

Another common type of cellular reaction is an intramolecular rearrangement in which redistribution of electrons results in alterations of many different types without a change in the overall oxidation state of the

molecule. For example, different groups in a molecule may undergo oxidation-reduction, with no net change in oxidation state of the molecule; groups at a double bond may undergo a cis-trans rearrangement; or the positions of double bonds may be transposed. An example of an isomerization entailing oxidation-reduction is the formation of fructose 6-phosphate from glucose 6-phosphate in glycolysis (**Fig. 13-6**; this reaction is discussed in detail in Chapter 14): C-1 is reduced (aldehyde to alcohol) and C-2 is oxidized (alcohol to ketone). Figure 13-6b shows the details of the electron movements in this type of isomerization. A cis-trans rearrangement is illustrated by the prolyl cis-trans isomerase reaction in the folding of certain proteins (see Fig. 4-8). A simple transposition of a C=C bond occurs during metabolism of oleic acid, a common fatty acid (see Fig. 17-10). Some spectacular examples of double-bond repositioning occur in the biosynthesis of cholesterol (see Fig. 21-33).

An example of an elimination reaction that does not affect overall oxidation state is the loss of water from an alcohol, resulting in the introduction of a C=C bond:

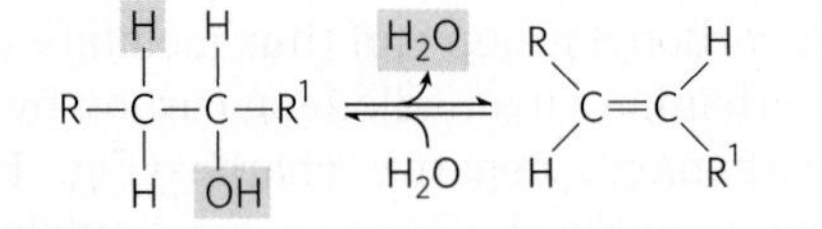

Similar reactions can result from eliminations in amines.

Free-Radical Reactions Once thought to be rare, the homolytic cleavage of covalent bonds to generate free radicals has now been found in a wide range of biochemical processes. These include isomerizations that make use of adenosylcobalamin (vitamin B_{12}) or S-adenosylmethionine, which are initiated with a 5′-deoxyadenosyl radical (see the methylmalonyl-CoA mutase reaction in Box 17-2); certain radical-initiated decarboxylation

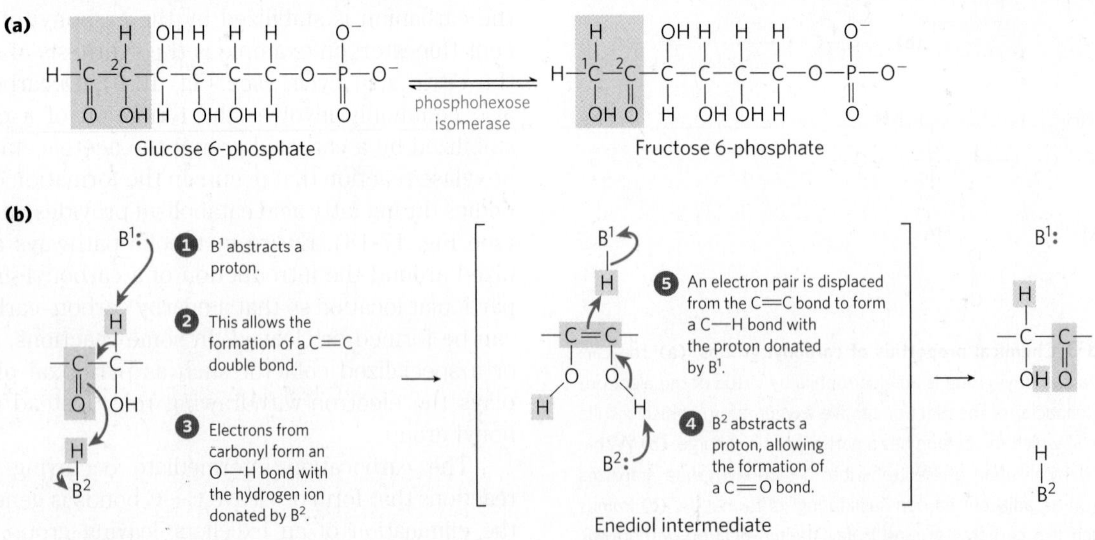

(a)

Glucose 6-phosphate   Fructose 6-phosphate

(b)

① B^1 abstracts a proton.

② This allows the formation of a C=C double bond.

③ Electrons from carbonyl form an O—H bond with the hydrogen ion donated by B^2.

⑤ An electron pair is displaced from the C=C bond to form a C—H bond with the proton donated by B^1.

④ B^2 abstracts a proton, allowing the formation of a C=O bond.

Enediol intermediate

FIGURE 13-6 Isomerization and elimination reactions. (a) The conversion of glucose 6-phosphate to fructose 6-phosphate, a reaction of sugar metabolism catalyzed by phosphohexose isomerase. **(b)** This reaction proceeds through an enediol intermediate. Light red screens follow the

path of oxidation from left to right. B^1 and B^2 are ionizable groups on the enzyme; they are capable of donating and accepting protons (acting as general acids or general bases) as the reaction proceeds.

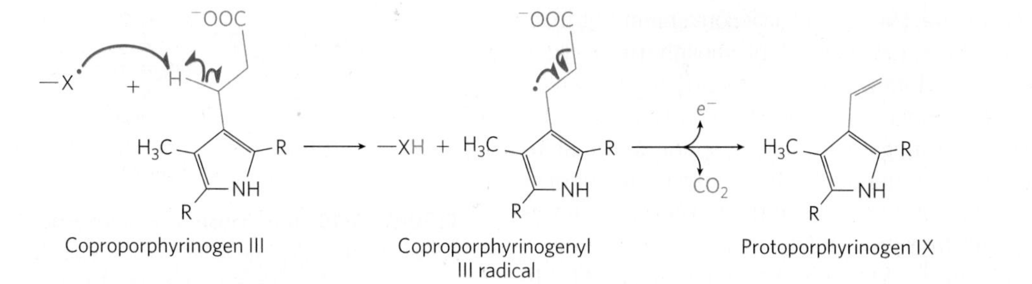

Coproporphyrinogen III   Coproporphyrinogenyl   Protoporphyrinogen IX
                                 III radical

FIGURE 13-7 A free radical–initiated decarboxylation reaction. The biosynthesis of heme (see Fig. 22-26) in *Escherichia coli* includes a decarboxylation step in which propionyl side chains on the coproporphyrinogen III intermediate are converted to the vinyl side chains of protoporphyrinogen IX. When the bacteria are grown anaerobically the enzyme oxygen-independent coproporphyrinogen III oxidase, also called HemN protein, promotes decarboxylation via the free-radical mechanism shown here. The acceptor of the released electron is not known. For simplicity, only the relevant portions of the large coproporphyrinogen III and protoporphyrinogen molecules are shown; the entire structures are given in Figure 22-26. When *E. coli* is grown in the presence of oxygen, this reaction is an oxidative decarboxylation and is catalyzed by a different enzyme. [Source: Information from G. Layer et al., *Curr. Opin. Chem. Biol.* 8:468, 2004, Fig. 4.]

reactions **(Fig. 13-7)**; some reductase reactions, such as that catalyzed by ribonucleotide reductase (see Fig. 22-42); and some rearrangement reactions, such as that catalyzed by DNA photolyase (see Fig. 25-26).

Group Transfer Reactions The transfer of acyl, glycosyl, and phosphoryl groups from one nucleophile to another is common in living cells. Acyl group transfer generally involves the addition of a nucleophile to the carbonyl carbon of an acyl group to form a tetrahedral intermediate:

Tetrahedral intermediate

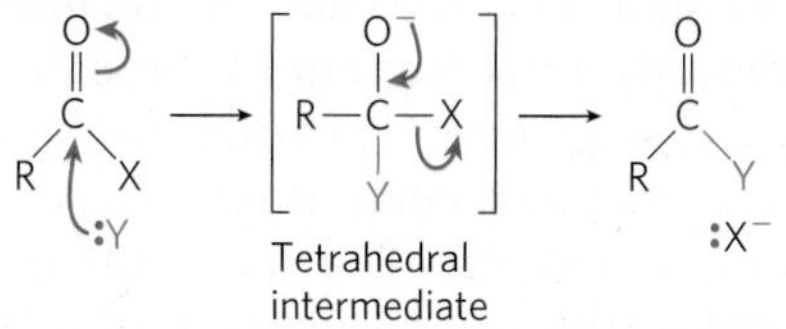

The chymotrypsin reaction is one example of acyl group transfer (see Fig. 6-23). Glycosyl group transfers involve nucleophilic substitution at C-1 of a sugar ring, which is the central atom of an acetal. In principle, the substitution could proceed by an S_N1 or S_N2 pathway, as described in Figure 6-29 for the enzyme lysozyme.

Phosphoryl group transfers play a special role in metabolic pathways, and these transfer reactions are discussed in detail in Section 13.3. A general theme in metabolism is the attachment of a good leaving group to a metabolic intermediate to "activate" the intermediate for subsequent reaction. Among the better leaving groups in nucleophilic substitution reactions are inorganic orthophosphate (the ionized form of H_3PO_4 at neutral pH, a mixture of $H_2PO_4^-$ and HPO_4^{2-}, commonly abbreviated P_i) and inorganic pyrophosphate ($P_2O_7^{4-}$, abbreviated PP_i); esters and anhydrides of phosphoric acid are effectively activated for reaction. Nucleophilic substitution is made more favorable by the attachment of a phosphoryl group to an otherwise poor leaving group such as —OH. Nucleophilic substitutions in which the phosphoryl group ($—PO_3^{2-}$) serves as a leaving group occur in hundreds of metabolic reactions.

Phosphorus can form five covalent bonds. The conventional representation of P_i **(Fig. 13-8a)**, with three

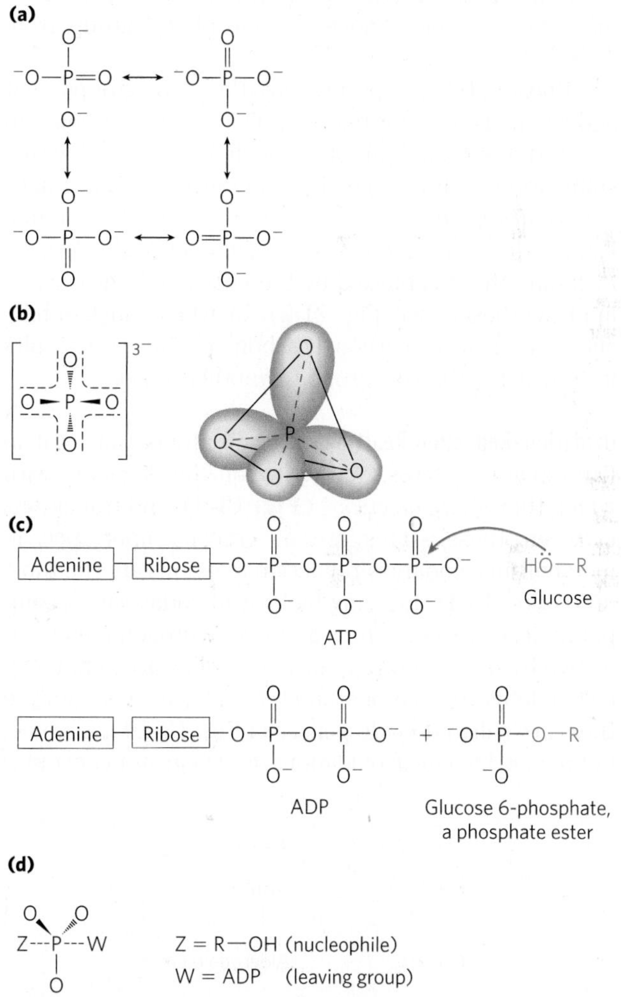

(a)

(b)

(c)

ATP

ADP   Glucose 6-phosphate,
             a phosphate ester

(d)

Z = R—OH (nucleophile)
W = ADP   (leaving group)

FIGURE 13-8 Phosphoryl group transfers: some of the participants. (a) In one (inadequate) representation of P_i, three oxygens are single-bonded to phosphorus, and the fourth is double-bonded, allowing the four different resonance structures shown here. **(b)** The resonance structures of P_i can be represented more accurately by showing all four phosphorus-oxygen bonds with some double-bond character; the hybrid orbitals so represented are arranged in a tetrahedron with P at its center. **(c)** When a nucleophile Z (in this case, the —OH on C-6 of glucose) attacks ATP, it displaces ADP (W). In this S_N2 reaction, a pentacovalent intermediate **(d)** forms transiently.

P—O bonds and one P=O bond, is a convenient but inaccurate picture. In P_i, four equivalent phosphorus–oxygen bonds share some double-bond character, and the anion has a tetrahedral structure (Fig. 13-8b). Because oxygen is more electronegative than phosphorus, the sharing of electrons is unequal: the central phosphorus bears a partial positive charge and can therefore act as an electrophile. In a great many metabolic reactions, a phosphoryl group ($—PO_3^{2-}$) is transferred from ATP to an alcohol, forming a phosphate ester (Fig. 13-8c), or to a carboxylic acid, forming a mixed anhydride. When a nucleophile attacks the electrophilic phosphorus atom in ATP, a relatively stable pentacovalent structure forms as a reaction intermediate (Fig. 13-8d). With departure of the leaving group (ADP), the transfer of a phosphoryl group is complete. The large family of enzymes that catalyze phosphoryl group transfers with ATP as donor are called **kinases** (Greek *kinein*, "to move"). Hexokinase, for example, "moves" a phosphoryl group from ATP to glucose.

Phosphoryl groups are not the only groups that activate molecules for reaction. Thioalcohols (thiols), in which the oxygen atom of an alcohol is replaced with a sulfur atom, are also good leaving groups. Thiols activate carboxylic acids by forming thioesters (thiol esters). In later chapters we discuss several reactions, including those catalyzed by the fatty acyl synthases in lipid synthesis (see Fig. 21-2), in which nucleophilic substitution at the carbonyl carbon of a thioester results in transfer of the acyl group to another moiety.

Oxidation-Reduction Reactions Carbon atoms can exist in five oxidation states, depending on the elements with which they share electrons **(Fig. 13-9)**, and transitions between these states are of crucial importance in metabolism (oxidation-reduction reactions are the topic of Section 13.4). In many biological oxidations, a compound loses two electrons and two hydrogen ions (that is, two hydrogen atoms); these reactions are commonly called dehydrogenations, and the enzymes that catalyze them are called dehydrogenases **(Fig. 13-10)**. In some, but not all, biological oxidations, a carbon atom becomes

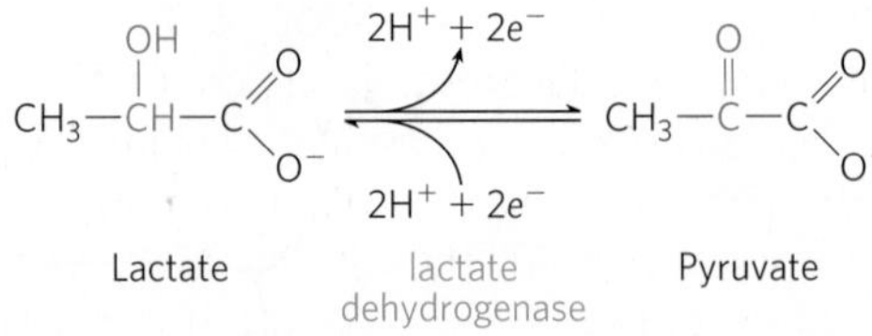

FIGURE 13-10 An oxidation-reduction reaction. Shown here is the oxidation of lactate to pyruvate. In this dehydrogenation, two electrons and two hydrogen ions (the equivalent of two hydrogen atoms) are removed from C-2 of lactate, an alcohol, to form pyruvate, a ketone. In cells the reaction is catalyzed by lactate dehydrogenase and the electrons are transferred to the cofactor nicotinamide adenine dinucleotide (NAD). This reaction is fully reversible; pyruvate can be reduced by electrons transferred from the cofactor.

covalently bonded to an oxygen atom. The enzymes that catalyze these oxidations are generally called oxidases or, if the oxygen atom is derived directly from molecular oxygen (O_2), oxygenases.

Every oxidation must be accompanied by a reduction, in which an electron acceptor acquires the electrons removed by oxidation. Oxidation reactions generally release energy (think of camp fires: the compounds in wood are oxidized by oxygen molecules in the air). Most living cells obtain the energy needed for cellular work by oxidizing metabolic fuels such as carbohydrates or fat (photosynthetic organisms can also trap and use the energy of sunlight). The catabolic (energy-yielding) pathways described in Chapters 14 through 19 are oxidative reaction sequences that result in the transfer of electrons from fuel molecules, through a series of electron carriers, to oxygen. The high affinity of O_2 for electrons makes the overall electron-transfer process highly exergonic, providing the energy that drives ATP synthesis—the central goal of catabolism.

Many of the reactions within these five classes are facilitated by cofactors, in the form of coenzymes and metal ions (vitamin B_{12}, S-adenosylmethionine, folate, nicotinamide, and Fe^{2+} are some examples). Cofactors bind to enzymes—in some cases reversibly, in other cases almost irreversibly—and give them the capacity to promote a particular kind of chemistry (p. 188). Most cofactors participate in a narrow range of closely related reactions. In the following chapters, we introduce and discuss each important cofactor at the point where we first encounter it. The cofactors provide another way to organize the study of biochemical processes, given that the reactions facilitated by a given cofactor generally are mechanistically related.

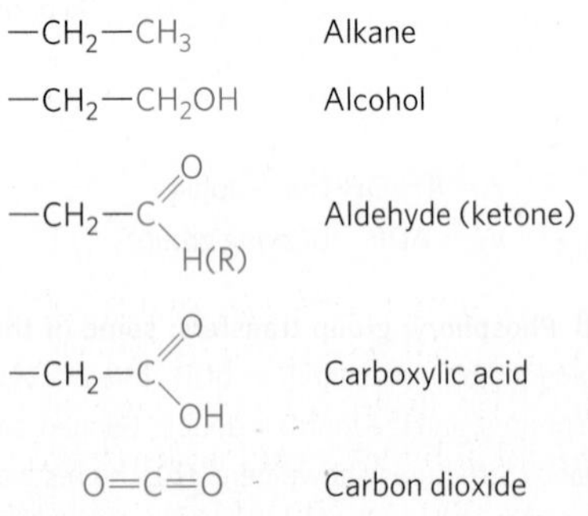

Biochemical and Chemical Equations Are Not Identical

Biochemists write metabolic equations in a simplified way, and this is particularly evident for reactions involving ATP. Phosphorylated compounds can exist in several ionization states and, as we have noted, the different species can bind Mg^{2+}. For example, at pH 7 and 2 mM

FIGURE 13-9 The oxidation levels of carbon in biomolecules. Each compound is formed by oxidation of the carbon shown in red in the compound immediately above. Carbon dioxide is the most highly oxidized form of carbon found in living systems.

Mg^{2+}, ATP exists in the forms ATP^{4-}, $HATP^{3-}$, H_2ATP^{2-}, $MgHATP^-$, and Mg_2ATP. In thinking about the biological role of ATP, however, we are not always interested in all this detail, and so we consider ATP as an entity made up of a sum of species, and we write its hydrolysis as the biochemical equation

$$ATP + H_2O \longrightarrow ADP + P_i$$

where ATP, ADP, and P_i are sums of species. The corresponding standard transformed equilibrium constant, $K'_{eq} = [ADP][P_i]/[ATP]$, depends on the pH and the concentration of free Mg^{2+}. Note that H^+ and Mg^{2+} do not appear in the biochemical equation, because they are held constant. Thus a biochemical equation does not necessarily balance H, Mg, or charge, although it does balance all other elements involved in the reaction (C, N, O, and P in the equation above).

We can write a chemical equation that *does* balance for all elements and for charge. For example, when ATP is hydrolyzed at a pH above 8.5 in the absence of Mg^{2+}, the chemical reaction is represented by

$$ATP^{4-} + H_2O \longrightarrow ADP^{3-} + HPO_4^{2-} + H^+$$

The corresponding equilibrium constant, $K_{eq} = [ADP^{3-}][HPO_4^{2-}][H^+]/[ATP^{4-}]$, depends only on temperature, pressure, and ionic strength.

Both ways of writing a metabolic reaction have value in biochemistry. Chemical equations are needed when we want to account for all atoms and charges in a reaction, as when we are considering the mechanism of a chemical reaction. Biochemical equations are used to determine in which direction a reaction will proceed spontaneously, given a specified pH and $[Mg^{2+}]$, or to calculate the equilibrium constant of such a reaction.

Throughout this book we use biochemical equations, unless the focus is on chemical mechanism, and we use values of $\Delta G'^\circ$ and K'_{eq} as determined at pH 7 and 1 mM Mg^{2+}.

SUMMARY 13.2 Chemical Logic and Common Biochemical Reactions

■ Living systems make use of a large number of chemical reactions that can be classified into five general types.

■ Carbonyl groups play a special role in reactions that form or cleave C—C bonds. Carbanion intermediates are common and are stabilized by adjacent carbonyl groups or, less often, by imines or certain cofactors.

■ A redistribution of electrons can produce internal rearrangements, isomerizations, and eliminations. Such reactions include intramolecular oxidation-reduction, change in cis-trans arrangement at a double bond, and transposition of double bonds.

■ Homolytic cleavage of covalent bonds to generate free radicals occurs in some pathways, such as in

certain isomerization, decarboxylation, reductase, and rearrangement reactions.

■ Phosphoryl transfer reactions are an especially important type of group transfer in cells, required for the activation of molecules for reactions that would otherwise be highly unfavorable.

■ Oxidation-reduction reactions involve the loss or gain of electrons: one reactant gains electrons and is reduced, while the other loses electrons and is oxidized. Oxidation reactions generally release energy and are important in catabolism.

13.3 Phosphoryl Group Transfers and ATP

Having developed some fundamental principles of energy changes in chemical systems and reviewed the common classes of reactions, we can now examine the energy cycle in cells and the special role of ATP as the energy currency that links catabolism and anabolism (see Fig. 1-30). Heterotrophic cells obtain free energy in a chemical form by the catabolism of nutrient molecules, and they use that energy to make ATP from ADP and P_i. ATP then donates some of its chemical energy to endergonic processes such as the synthesis of metabolic intermediates and macromolecules from smaller precursors, the transport of substances across membranes against concentration gradients, and mechanical motion. This donation of energy from ATP generally involves the covalent participation of ATP in the reaction that is to be driven, with the eventual result that ATP is converted to ADP and P_i or, in some reactions, to AMP and 2 P_i. We discuss here the chemical basis for the large free-energy changes that accompany hydrolysis of ATP and other high-energy phosphate compounds, and we show that most cases of energy donation by ATP involve group transfer, not simple hydrolysis of ATP. To illustrate the range of energy transductions in which ATP provides the energy, we consider the synthesis of information-rich macromolecules, the transport of solutes across membranes, and motion produced by muscle contraction.

The Free-Energy Change for ATP Hydrolysis Is Large and Negative

Figure 13-11 summarizes the chemical basis for the relatively large, negative, standard free energy of hydrolysis of ATP. The hydrolytic cleavage of the terminal phosphoric acid anhydride (phosphoanhydride) bond in ATP separates one of the three negatively charged phosphates and thus relieves some of the internal electrostatic repulsion in ATP; the P_i released is stabilized by the formation of several resonance forms not possible in ATP.

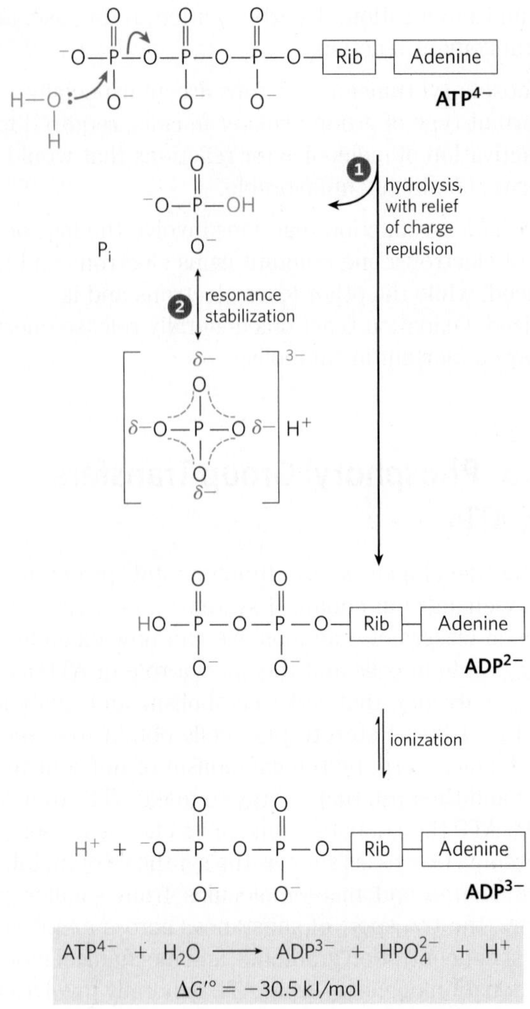

$$ATP^{4-} + H_2O \longrightarrow ADP^{3-} + HPO_4^{2-} + H^+$$
$$\Delta G'^\circ = -30.5 \text{ kJ/mol}$$

FIGURE 13-11 Chemical basis for the large free-energy change associated with ATP hydrolysis. ❶ The charge separation that results from hydrolysis relieves electrostatic repulsion among the four negative charges on ATP. **❷** The product inorganic phosphate (P_i) is stabilized by formation of a resonance hybrid, in which each of the four phosphorus–oxygen bonds has the same degree of double-bond character and the hydrogen ion is not permanently associated with any one of the oxygens. (Some degree of resonance stabilization also occurs in phosphates involved in ester or anhydride linkages, but fewer resonance forms are possible than for P_i.) A third factor (not shown) that favors ATP hydrolysis is the greater degree of solvation (hydration) of the products P_i and ADP relative to ATP, which further stabilizes the products relative to the reactants.

The free-energy change for ATP hydrolysis is –30.5 kJ/mol under standard conditions, but the *actual* free energy of hydrolysis (ΔG) of ATP in living cells is very different: the cellular concentrations of ATP, ADP, and P_i are not identical and are much lower than the 1.0 M of standard conditions (Table 13-5). Furthermore, Mg^{2+} in the cytosol binds to ATP and ADP **(Fig. 13-12)**, and for most enzymatic reactions that involve ATP as phosphoryl group donor, the true substrate is $MgATP^{2-}$. The relevant $\Delta G'^\circ$ is therefore that for $MgATP^{2-}$ hydrolysis. We can calculate ΔG for ATP hydrolysis using data such as those in Table 13-5. The actual free energy of hydrolysis of ATP under intracellular conditions is often called its **phosphorylation potential, ΔG_p.**

Because the concentrations of ATP, ADP, and P_i differ from one cell type to another, ΔG_p for ATP likewise differs among cells. Moreover, in any given cell, ΔG_p can vary from time to time, depending on the metabolic conditions and how they influence the concentrations of ATP, ADP, P_i, and H^+ (pH). We can calculate the actual free-energy change for any given metabolic reaction as it occurs in a cell, providing we know the concentrations of all the reactants and products and other factors (such as pH, temperature, and $[Mg^{2+}]$) that may affect the actual free-energy change.

TABLE 13-5	Total Concentrations of Adenine Nucleotides, Inorganic Phosphate, and Phosphocreatine in Some Cells

	Concentration (mM)[a]				
	ATP	ADP[b]	AMP	P_i	PCr
Rat hepatocyte	3.38	1.32	0.29	4.8	0
Rat myocyte	8.05	0.93	0.04	8.05	28
Rat neuron	2.59	0.73	0.06	2.72	4.7
Human erythrocyte	2.25	0.25	0.02	1.65	0
E. coli cell	7.90	1.04	0.82	7.9	0

[a]For erythrocytes the concentrations are those of the cytosol (human erythrocytes lack a nucleus and mitochondria). In the other types of cells the data are for the entire cell contents, although the cytosol and the mitochondria have very different concentrations of ADP. PCr is phosphocreatine, discussed on p. 516.

[b]This value reflects total concentration; the true value for free ADP may be much lower (p. 509).

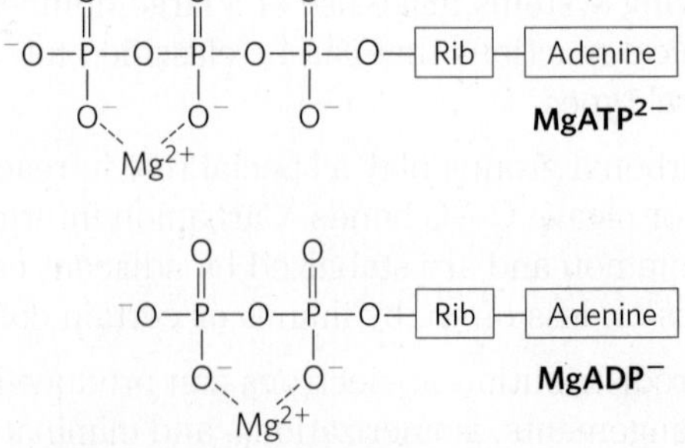

FIGURE 13-12 Mg^{2+} and ATP. Formation of Mg^{2+} complexes partially shields the negative charges and influences the conformation of the phosphate groups in nucleotides such as ATP and ADP.

WORKED EXAMPLE 13-2 Calculation of ΔG_p

Calculate the actual free energy of hydrolysis of ATP, ΔG_p, in human erythrocytes. The standard free energy of hydrolysis of ATP is −30.5 kJ/mol, and the concentrations of ATP, ADP, and P_i in erythrocytes are as shown in Table 13-5. Assume that the pH is 7.0 and the temperature is 37 °C (body temperature). What does this reveal about the amount of energy required to *synthesize* ATP under the same cellular conditions?

Solution: The concentrations of ATP, ADP, and P_i in human erythrocytes are 2.25, 0.25, and 1.65 mM, respectively. The actual free energy of hydrolysis of ATP under these conditions is given by the relationship (see Eqn 13-4)

$$\Delta G_p = \Delta G'^{\circ} + RT \ln \frac{[ADP][P_i]}{[ATP]}$$

Substituting the appropriate values we get

$$\Delta G_p = -30.5 \text{ kJ/mol} + \left[(8.315 \text{ kJ/mol}\cdot\text{K})(310 \text{ K}) \ln \frac{(0.25 \times 10^{-3})(1.65 \times 10^{-3})}{(2.25 \times 10^{-3})} \right]$$

$$= -30.5 \text{ kJ/mol} + (2.58 \text{ kJ/mol}) \ln 1.8 \times 10^{-4}$$

$$= -30.5 \text{ kJ/mol} + (2.58 \text{ kJ/mol})(-8.6)$$

$$= -30.5 \text{ kJ/mol} - 22 \text{ kJ/mol}$$

$$= -52 \text{ kJ/mol}$$

(Note that the final answer has been rounded to the correct number of significant figures (52.5 rounded to 52), following the rule for rounding a number that ends in a 5 to the nearest even number.) Thus ΔG_p, the actual free-energy change for ATP hydrolysis in the intact erythrocyte (−52 kJ/mol), is much larger than the standard free-energy change (−30.5 kJ/mol). By the same token, the free energy required to *synthesize* ATP from ADP and P_i under the conditions prevailing in the erythrocyte would be 52 kJ/mol.

To further complicate the issue, the *total* concentrations of ATP, ADP, and P_i (and H^+) in a cell—such as the values given in Table 13-5—may be substantially higher than the *free* concentrations, which are the thermodynamically relevant values. The difference is due to tight binding of ATP, ADP, and P_i to cellular proteins. For example, the free [ADP] in resting muscle has been variously estimated at between 1 and 37 μM. Using the value 25 μM in Worked Example 13-2, we would get a ΔG_p of −58 kJ/mol. Calculation of the exact value of ΔG_p, however, is perhaps less instructive than the generalization we can make about actual free-energy changes: in vivo, the energy released by ATP hydrolysis is greater than the standard free-energy change, $\Delta G'^{\circ}$.

In the following discussions we use the $\Delta G'^{\circ}$ value for ATP hydrolysis because this allows comparisons, on the same basis, with the energetics of other cellular reactions. Always keep in mind, however, that in living cells ΔG is the relevant quantity—for ATP hydrolysis and all other reactions—and may be quite different from $\Delta G'^{\circ}$.

Here we must make an important point about cellular ATP levels. We have shown (and will discuss further) how the chemical properties of ATP make it a suitable form of energy currency in cells. But it is not merely the molecule's intrinsic chemical properties that

give it this ability to drive metabolic reactions and other energy-requiring processes. Even more important is that, in the course of evolution, there has been a very strong selective pressure for regulatory mechanisms that *hold cellular ATP concentrations far above the equilibrium concentrations* for the hydrolysis reaction. When the ATP level drops, not only does the *amount* of fuel decrease, but the fuel itself *loses its potency:* ΔG for its hydrolysis (that is, its phosphorylation potential, ΔG_p) is diminished. As our discussions of the metabolic pathways that produce and consume ATP will show, living cells have developed elaborate mechanisms—often at what might seem to us the expense of efficiency—to maintain high concentrations of ATP.

Other Phosphorylated Compounds and Thioesters Also Have Large Free Energies of Hydrolysis

Phosphoenolpyruvate (PEP; **Fig. 13-13**) contains a phosphate ester bond that undergoes hydrolysis to yield the enol form of pyruvate, and this direct product can tautomerize to the more stable keto form. Because the reactant (PEP) has only one form (enol) and the product (pyruvate) has two possible forms, the product is stabilized relative to the reactant. This is the

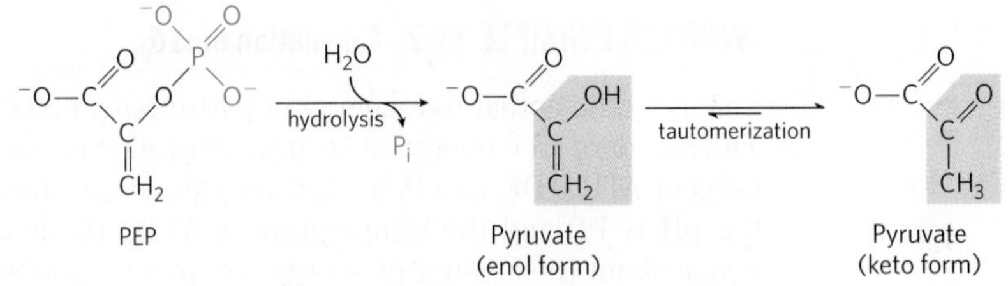

FIGURE 13-13 Hydrolysis of phosphoenolpyruvate (PEP). Catalyzed by pyruvate kinase, this reaction is followed by spontaneous tautomerization of the product, pyruvate. Tautomerization is not possible in PEP, and thus the products of hydrolysis are stabilized relative to the reactants. Resonance stabilization of P_i also occurs, as shown in Figure 13-11.

$$PEP^{3-} + H_2O \longrightarrow \text{pyruvate}^- + HPO_4^{2-}$$
$$\Delta G'^{\circ} = -61.9 \text{ kJ/mol}$$

greatest contributing factor to the high standard free energy of hydrolysis of phosphoenolpyruvate: $\Delta G'^{\circ} = -61.9$ kJ/mol.

Another three-carbon compound, 1,3-bisphosphoglycerate **(Fig. 13-14)**, contains an anhydride bond between the C-1 carboxyl group and phosphoric acid. Hydrolysis of this acyl phosphate is accompanied by a large, negative, standard free-energy change ($\Delta G'^{\circ} = -49.3$ kJ/mol), which can, again, be explained in terms of the structure of reactant and products. When H_2O is added across the anhydride bond of 1,3-bisphosphoglycerate, one of the direct products, 3-phosphoglyceric acid, can lose a proton to give the carboxylate ion, 3-phosphoglycerate, which has two equally probable resonance forms (Fig. 13-14). Removal of the direct product (3-phosphoglyceric acid) and formation of the resonance-stabilized ion favor the forward reaction.

In phosphocreatine **(Fig. 13-15)**, the P—N bond can be hydrolyzed to generate free creatine and P_i. The release of P_i and the resonance stabilization of creatine favor the forward reaction. The standard free-energy change of phosphocreatine hydrolysis is again large, −43.0 kJ/mol.

In all these phosphate-releasing reactions, the several resonance forms available to P_i (Fig. 13-11) stabilize this product relative to the reactant, contributing to an already negative free-energy change. Table 13-6 lists the standard free energies of hydrolysis for some biologically important phosphorylated compounds.

Thioesters, in which a sulfur atom replaces the usual oxygen in the ester bond, also have large, negative, standard free energies of hydrolysis. Acetylcoenzyme A, or acetyl-CoA **(Fig. 13-16)**, is one of many thioesters important in metabolism. The acyl group in these compounds is activated for transacylation, condensation, or oxidation-reduction reactions. Thioesters undergo much less resonance stabilization than do oxygen esters; consequently, the difference in free energy between the reactant and its hydrolysis

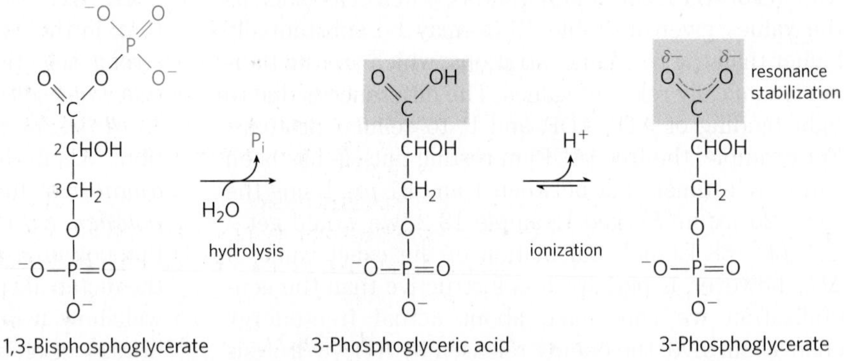

FIGURE 13-14 Hydrolysis of 1,3-bisphosphoglycerate. The direct product of hydrolysis is 3-phosphoglyceric acid, with an undissociated carboxylic acid. Its dissociation allows resonance structures that stabilize the product relative to the reactants. Resonance stabilization of P_i further contributes to the negative free-energy change.

$$1,3\text{-Bisphosphoglycerate}^{4-} + H_2O \longrightarrow 3\text{-phosphoglycerate}^{3-} + HPO_4^{2-} + H^+$$
$$\Delta G'^{\circ} = -49.3 \text{ kJ/mol}$$

FIGURE 13-15 Hydrolysis of phosphocreatine. Breakage of the P—N bond in phosphocreatine produces creatine, which is stabilized by formation of a resonance hybrid. The other product, P_i, is also resonance stabilized.

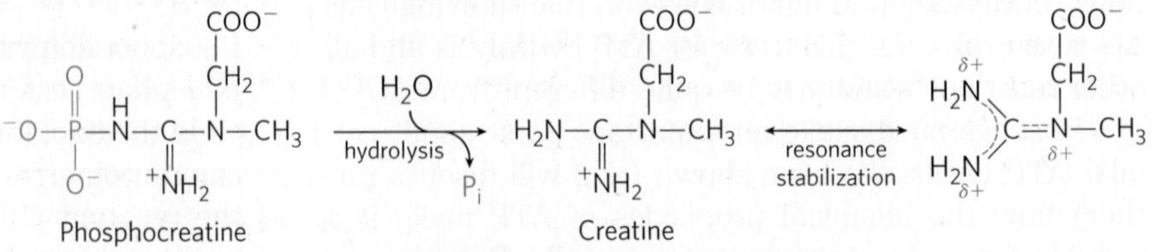

$$\text{Phosphocreatine}^{2-} + H_2O \longrightarrow \text{creatine} + HPO_4^{2-}$$
$$\Delta G'^{\circ} = -43.0 \text{ kJ/mol}$$

TABLE 13-6	Standard Free Energies of Hydrolysis of Some Phosphorylated Compounds and Acetyl-CoA (a Thioester)	
	$\Delta G'^{\circ}$	
	(kJ/mol)	(kcal/mol)
Phosphoenolpyruvate	−61.9	−14.8
1,3-Bisphosphoglycerate ($\to$ 3-phosphoglycerate + P_i)	−49.3	−11.8
Phosphocreatine	−43.0	−10.3
ADP ($\to$ AMP + P_i)	−32.8	−7.8
ATP ($\to$ ADP + P_i)	−30.5	−7.3
ATP ($\to$ AMP + PP_i)	−45.6	−10.9
AMP ($\to$ adenosine + P_i)	−14.2	−3.4
PP_i ($\to 2P_i$)	−19.2	−4.0
Glucose 3-phosphate	−20.9	−5.0
Fructose 6-phosphate	−15.9	−3.8
Glucose 6-phosphate	−13.8	−3.3
Glycerol 3-phosphate	−9.2	−2.2
Acetyl-CoA	−31.4	−7.5

Sources: Data mostly from W. P. Jencks, in *Handbook of Biochemistry and Molecular Biology*, 3rd edn (G. D. Fasman, ed.), *Physical and Chemical Data*, Vol. 1, p. 296, CRC Press, 1976. Value for the free energy of hydrolysis of PP_i from P. A. Frey and A. Arabshahi, *Biochemistry* 34:11,307, 1995.

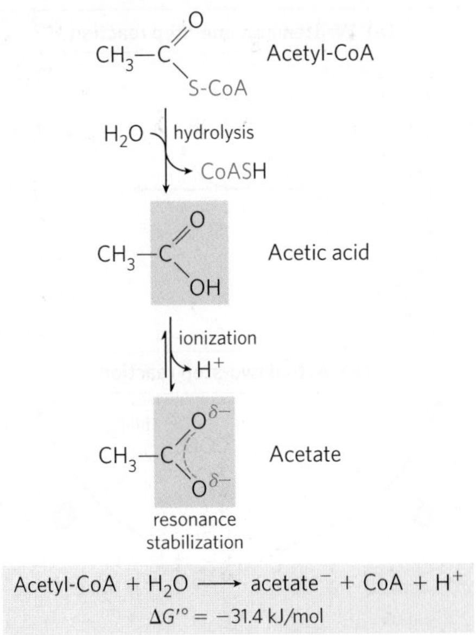

FIGURE 13-16 Hydrolysis of acetyl-coenzyme A. Acetyl-CoA is a thioester with a large, negative, standard free energy of hydrolysis. Thioesters contain a sulfur atom in the position occupied by an oxygen atom in oxygen esters. The complete structure of coenzyme A (CoA, or CoASH) is shown in Figure 8-41.

products, which *are* resonance-stabilized, is greater for thioesters than for comparable oxygen esters **(Fig. 13-17).** In both cases, hydrolysis of the ester generates a carboxylic acid, which can ionize and assume several resonance forms. Together, these factors result in the large, negative $\Delta G'^{\circ}$ (−31.4 kJ/mol) for acetyl-CoA hydrolysis.

To summarize, for hydrolysis reactions with large, negative, standard free-energy changes, the products are more stable than the reactants for one or more of

the following reasons: (1) the bond strain in reactants due to electrostatic repulsion is relieved by charge separation, as for ATP; (2) the products are stabilized by ionization, as for ATP, acyl phosphates, and thioesters; (3) the products are stabilized by isomerization (tautomerization), as for PEP; and/or (4) the products are stabilized by resonance, as for creatine released from phosphocreatine, carboxylate ion released from acyl phosphates and thioesters, and phosphate (P_i) released from anhydride or ester linkages.

ATP Provides Energy by Group Transfers, Not by Simple Hydrolysis

Throughout this book you will encounter reactions or processes for which ATP supplies energy, and the

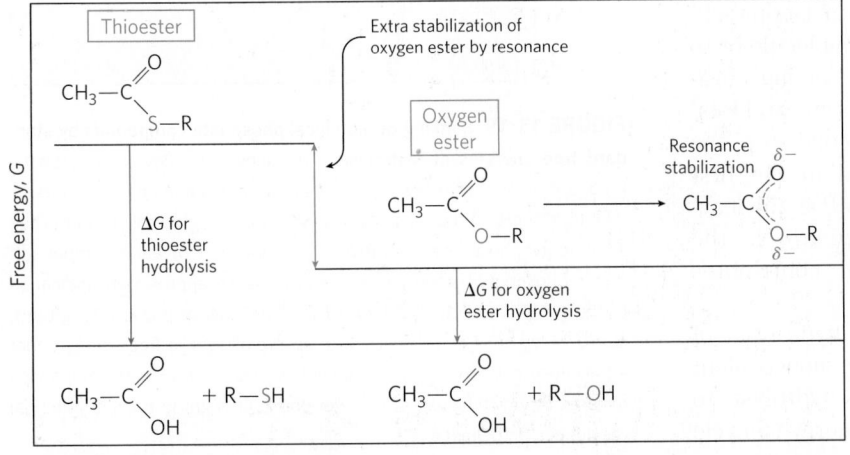

FIGURE 13-17 Free energy of hydrolysis for thioesters and oxygen esters. The *products* of both types of hydrolysis reaction have about the same free-energy content (G), but the thioester has a higher free-energy content than the oxygen ester. Orbital overlap between the O and C atoms allows resonance stabilization in oxygen esters; orbital overlap between S and C atoms is poorer and provides little resonance stabilization.

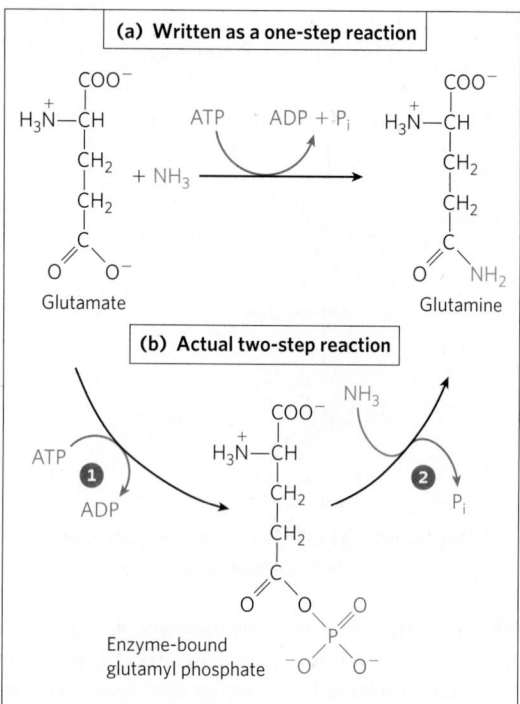

(a) Written as a one-step reaction

Glutamate

Glutamine

(b) Actual two-step reaction

Enzyme-bound
glutamyl phosphate

FIGURE 13-18 ATP hydrolysis in two steps. (a) The contribution of ATP to a reaction is often shown as a single step but is almost always a two-step process. **(b)** Shown here is the reaction catalyzed by ATP-dependent glutamine synthetase. ❶ A phosphoryl group is transferred from ATP to glutamate, then ❷ the phosphoryl group is displaced by NH_3 and released as P_i.

contribution of ATP to these reactions is commonly indicated as in **Figure 13-18a**, with a single arrow showing the conversion of ATP to ADP and P_i (or, in some cases, of ATP to AMP and pyrophosphate, PP_i). When written this way, these reactions of ATP seem to be simple hydrolysis reactions in which water displaces P_i (or PP_i), and one is tempted to say that an ATP-dependent reaction is "driven by the hydrolysis of ATP." This is *not* the case. ATP hydrolysis per se usually accomplishes nothing but the liberation of heat, which cannot drive a chemical process in an isothermal system. A single reaction arrow such as that in Figure 13-18a almost invariably represents a two-step process (Fig. 13-18b) in which part of the ATP molecule, a phosphoryl or pyrophosphoryl group or the adenylate moiety (AMP), is first transferred to a substrate molecule or to an amino acid residue in an enzyme, becoming covalently attached to the substrate or the enzyme and raising its free-energy content. Then, in a second step, the phosphate-containing moiety transferred in the first step is displaced, generating P_i, PP_i, or AMP as the leaving group. Thus ATP participates *covalently* in the enzyme-catalyzed reaction to which it contributes free energy.

Some processes *do* involve direct hydrolysis of ATP (or GTP), however. For example, noncovalent binding of ATP (or GTP), followed by its hydrolysis to ADP (or GDP) and P_i, can provide the energy to cycle

some proteins between two conformations, producing mechanical motion. This occurs in muscle contraction (see Fig. 5-31) and in the movement of enzymes along DNA (see Fig. 25-31) or of ribosomes along messenger RNA (see Fig. 27-31). The energy-dependent reactions catalyzed by helicases, RecA protein, and some topoisomerases (Chapter 25) also involve direct hydrolysis of phosphoanhydride bonds. The AAA+ ATPases involved in DNA replication and other processes described in Chapter 25 use ATP hydrolysis to cycle associated proteins between active and inactive forms. GTP-binding proteins that act in signaling pathways directly hydrolyze GTP to drive conformational changes that terminate signals triggered by hormones or by other extracellular factors (Chapter 12).

The phosphate compounds found in living organisms can be divided, somewhat arbitrarily, into two groups, based on their standard free energies of hydrolysis **(Fig. 13-19)**. "High-energy" compounds have a $\Delta G'^{\circ}$ of hydrolysis more negative than -25 kJ/mol; "low-energy" compounds have a less negative $\Delta G'^{\circ}$. Based on this criterion, ATP, with a $\Delta G'^{\circ}$ of hydrolysis of -30.5 kJ/mol (-7.3 kcal/mol), is a high-energy compound; glucose

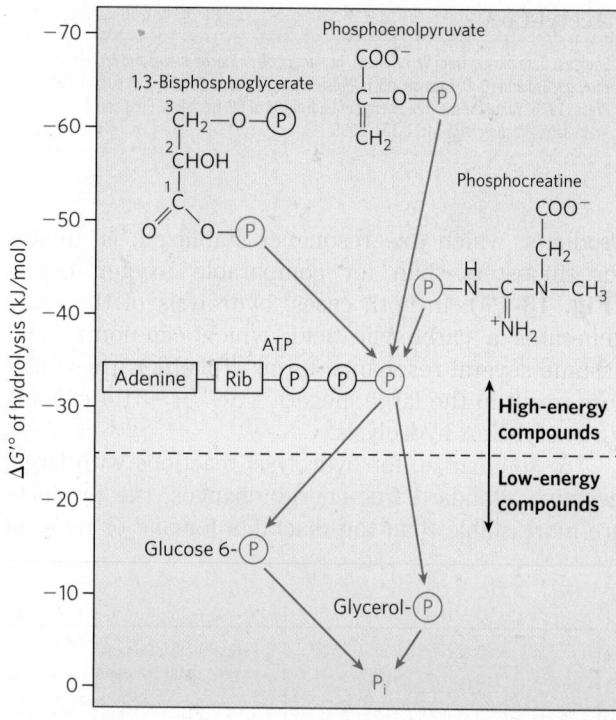

FIGURE 13-19 Ranking of biological phosphate compounds by standard free energies of hydrolysis. This shows the flow of phosphoryl groups, represented by Ⓟ, from high-energy phosphoryl group donors via ATP to acceptor molecules (such as glucose and glycerol) to form their low-energy phosphate derivatives. (The location of each compound's donor phosphoryl group along the scale is an approximate indication of the compound's $\Delta G'^{\circ}$ of hydrolysis.) This flow of phosphoryl groups, catalyzed by kinases, proceeds with an overall loss of free energy under intracellular conditions. Hydrolysis of low-energy phosphate compounds releases P_i, which has an even lower phosphoryl group transfer potential (as defined in the text).

6-phosphate, with a $\Delta G'^\circ$ of hydrolysis of -13.8 kJ/mol (-3.3 kcal/mol), is a low-energy compound.

The term "high-energy phosphate bond," long used by biochemists to describe the P—O bond broken in hydrolysis reactions, is incorrect and misleading, as it wrongly suggests that the bond itself contains the energy. In fact, the breaking of all chemical bonds requires an *input* of energy. The free energy released by hydrolysis of phosphate compounds does not come from the specific bond that is broken; it results from the products of the reaction having a lower free-energy content than the reactants. For simplicity, we sometimes use the term "high-energy phosphate compound" when referring to ATP or other phosphate compounds with a large, negative, standard free energy of hydrolysis.

As is evident from the additivity of free-energy changes of sequential reactions (see Section 13.1), any phosphorylated compound can be synthesized by coupling the synthesis to the breakdown of another phosphorylated compound with a more negative free energy of hydrolysis. For example, because cleavage of P_i from phosphoenolpyruvate releases more energy than is needed to drive the condensation of P_i with ADP, the direct donation of a phosphoryl group from PEP to ADP is thermodynamically feasible:

$$\Delta G'^\circ \text{ (kJ/mol)}$$

(1) $\text{PEP} + H_2O \longrightarrow \text{Pyruvate} + P_i$ -61.9

(2) $\text{ADP} + P_i \longrightarrow \text{ATP} + H_2O$ $+30.5$

Sum: $\text{PEP} + \text{ADP} \longrightarrow \text{Pyruvate} + \text{ATP}$ -31.4

Notice that although the overall reaction is represented as the algebraic sum of the first two reactions, the overall reaction is actually a third, distinct reaction that does not involve P_i; PEP donates a *phosphoryl* group *directly* to ADP. We can describe phosphorylated compounds as having a high or low *phosphoryl group transfer potential*, on the basis of their standard free energies of hydrolysis (as listed in Table 13-6). The phosphoryl group transfer potential of PEP is very high, that of ATP is high, and that of glucose 6-phosphate is low (Fig. 13-19).

Much of catabolism is directed toward the synthesis of high-energy phosphate compounds, but their formation is not an end in itself; they are the means of activating a wide variety of compounds for further chemical transformation. The transfer of a phosphoryl group to a compound effectively puts free energy into that compound, so that it has more free energy to give up during subsequent metabolic transformations. We described above how the synthesis of glucose 6-phosphate is accomplished by phosphoryl group transfer from ATP. In the next chapter we see how this phosphorylation of glucose activates, or "primes," the glucose for catabolic reactions that occur in nearly every living cell. Because of its intermediate position on the scale of group transfer potential, ATP can carry energy from high-energy phosphate compounds produced by catabolism (phosphoenolpyruvate, for

example) to compounds such as glucose, converting them into more reactive species with better leaving groups. ATP thus serves as the universal energy currency in all living cells.

One more chemical feature of ATP is crucial to its role in metabolism: although, in aqueous solution, ATP is thermodynamically unstable and is therefore a good phosphoryl group donor, it is *kinetically* stable. Because of the huge activation energies (200 to 400 kJ/mol) required for uncatalyzed cleavage of its phosphoanhydride bonds, ATP does not spontaneously donate phosphoryl groups to water or to the hundreds of other potential acceptors in the cell. Only when specific enzymes are present to lower the energy of activation does phosphoryl group transfer from ATP proceed. The cell is therefore able to regulate the disposition of the energy carried by ATP by regulating the various enzymes that act on it.

ATP Donates Phosphoryl, Pyrophosphoryl, and Adenylyl Groups

The reactions of ATP are generally S_N2 nucleophilic displacements (see Section 13.2) in which the nucleophile may be, for example, the oxygen of an alcohol or carboxylate, or a nitrogen of creatine or of the side chain of arginine or histidine. Each of the three phosphates of ATP is susceptible to nucleophilic attack **(Fig. 13-20)**, and each position of attack yields a different type of product.

Nucleophilic attack by an alcohol on the γ phosphate (Fig. 13-20a) displaces ADP and produces a new phosphate ester. Studies with ^{18}O-labeled reactants have shown that the bridge oxygen in the new compound is derived from the alcohol, not from ATP; the group transferred from ATP is therefore a phosphoryl ($-PO_3^{2-}$), not a phosphate ($-OPO_3^{2-}$). Phosphoryl group transfer from ATP to glutamate (Fig. 13-18) or to glucose (p. 219) involves attack at the γ position of the ATP molecule.

Attack at the β phosphate of ATP displaces AMP and transfers a pyrophosphoryl (not pyrophosphate) group to the attacking nucleophile (Fig. 13-20b). For example, the formation of 5-phosphoribosyl-1-pyrophosphate (p. 870), a key intermediate in nucleotide synthesis, results from attack of an —OH of the ribose on the β phosphate.

Nucleophilic attack at the α position of ATP displaces PP_i and transfers adenylate (5'-AMP) as an adenylyl group (Fig. 13-20c); the reaction is an **adenylylation** (a-den'-i-li-la'-shun, one of the most ungainly words in the biochemical language). Notice that hydrolysis of the α–β phosphoanhydride bond releases considerably more energy ($\sim$46 kJ/mol) than hydrolysis of the β–γ bond ($\sim$31 kJ/mol) (Table 13-6). Furthermore, the PP_i formed as a byproduct of the adenylylation is hydrolyzed to two P_i by the ubiquitous enzyme **inorganic pyrophosphatase**, releasing 19 kJ/mol and

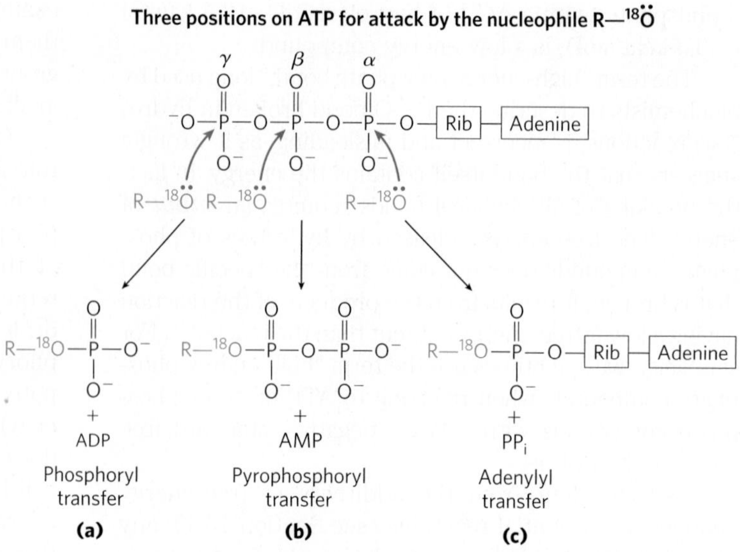

FIGURE 13-20 Nucleophilic displacement reactions of ATP.
Any of the three P atoms (α, β, or γ) may serve as the electrophilic target for nucleophilic attack, in this case by the labeled nucleophile R—^{18}O:. The nucleophile may be an alcohol (ROH), a carboxyl group (RCOO$^-$), or a phosphoanhydride (a nucleoside mono- or diphosphate, for example). **(a)** When the oxygen of the nucleophile attacks the γ position, the bridge oxygen of the product is labeled, indicating that the group transferred from ATP is a phosphoryl (—PO$_3^{2-}$), not a phosphate (—OPO$_3^{2-}$). **(b)** Attack on the β position displaces AMP and leads to the transfer of a pyrophosphoryl (not pyrophosphate) group to the nucleophile. **(c)** Attack on the α position displaces PP$_i$ and transfers the adenylyl group to the nucleophile.

Three positions on ATP for attack by the nucleophile R—^{18}O

ADP — Phosphoryl transfer **(a)**
AMP — Pyrophosphoryl transfer **(b)**
PP$_i$ — Adenylyl transfer **(c)**

thereby providing a further energy "push" for the adenylylation reaction. In effect, both phosphoanhydride bonds of ATP are split in the overall reaction. Adenylylation reactions are therefore thermodynamically very favorable. When the energy of ATP is used to drive a particularly unfavorable metabolic reaction, adenylylation is often the mechanism of energy coupling. Fatty acid activation is a good example of this energy-coupling strategy.

The first step in the activation of a fatty acid—either for energy-yielding oxidation or for use in the synthesis of more complex lipids—is the formation of its thiol ester (see Fig. 17-5). The direct condensation of a fatty acid with coenzyme A is endergonic, but the formation of a fatty acyl–CoA is made exergonic by stepwise removal of *two* phosphoryl groups from ATP. First, adenylate (AMP) is transferred from ATP to the carboxyl group of the fatty acid, forming a mixed anhydride (fatty acyl adenylate) and liberating PP$_i$. The thiol group of coenzyme A then displaces the adenylyl group and forms a thioester with the fatty acid. The sum of these two reactions is energetically equivalent to the exergonic hydrolysis of ATP to AMP and PP$_i$ ($\Delta G'^\circ = -45.6$ kJ/mol) and the endergonic formation of fatty acyl–CoA. The formation of fatty acyl–CoA ($\Delta G'^\circ = -31.4$ kJ/mol) is made energetically favorable by hydrolysis of the PP$_i$ by inorganic pyrophosphatase. Thus, in the activation of a fatty acid, both phosphoanhydride bonds of ATP are broken. The resulting $\Delta G'^\circ$ is the sum of the $\Delta G'^\circ$ values for the breakage of these bonds, or -45.6 kJ/mol + (-19.2) kJ/mol:

$$\text{ATP} + 2\text{H}_2\text{O} \longrightarrow \text{AMP} + 2\text{P}_i \quad \Delta G'^\circ = -64.8 \text{ kJ/mol}$$

The activation of amino acids before their polymerization into proteins (see Fig. 27-19) is accomplished by an analogous set of reactions in which a transfer RNA molecule takes the place of coenzyme A. An interesting use of the cleavage of ATP to AMP and PP$_i$ occurs in the

firefly, which uses ATP as an energy source to produce flashes of light (Box 13-1).

Assembly of Informational Macromolecules Requires Energy

When simple precursors are assembled into high molecular weight polymers with defined sequences (DNA, RNA, proteins), as described in detail in Part III, energy is required both for the condensation of monomeric units and for the creation of *ordered* sequences. The precursors for DNA and RNA synthesis are nucleoside triphosphates, and polymerization is accompanied by cleavage of the phosphoanhydride linkage between the α and β phosphates, with the release of PP$_i$ (Fig. 13-20). The moieties transferred to the growing polymer in these reactions are adenylate (AMP), guanylate (GMP), cytidylate (CMP), or uridylate (UMP) for RNA synthesis, and their deoxy analogs (with TMP in place of UMP) for DNA synthesis. As noted above, the activation of amino acids for protein synthesis involves the donation of adenylyl groups from ATP, and we shall see in Chapter 27 that several steps of protein synthesis on the ribosome are also accompanied by GTP hydrolysis. In all these cases, the exergonic breakdown of a nucleoside triphosphate is coupled to the endergonic process of synthesizing a polymer of a specific sequence.

ATP Energizes Active Transport and Muscle Contraction

ATP can supply the energy for transporting an ion or a molecule across a membrane into another aqueous compartment where its concentration is higher (see Fig. 11-36). Transport processes are major consumers of energy; in human kidney and brain, for example, as much as two-thirds of the energy consumed at rest is used to pump Na$^+$ and K$^+$ across plasma membranes via the Na$^+$K$^+$ ATPase. The transport of Na$^+$ and K$^+$ is

BOX 13-1   Firefly Flashes: Glowing Reports of ATP

Bioluminescence requires considerable amounts of energy. In the firefly, ATP is used in a set of reactions that convert chemical energy into light energy. Males emit a flash of light to attract females, who flash in return to signal their interest. In the 1950s, from many thousands of fireflies collected by children in and around Baltimore, William McElroy and his colleagues at the Johns Hopkins University isolated the principal biochemical components: luciferin, a complex carboxylic acid, and luciferase, an enzyme. The generation of a light flash requires activation of luciferin by an enzymatic reaction involving pyrophosphate cleavage of ATP to form luciferyl adenylate (Fig. 1). In the presence of molecular oxygen and luciferase, the luciferin undergoes a multistep oxidative decarboxylation to oxyluciferin. This process is accompanied by emission of light. The color of the light flash differs from one firefly species to another and seems to be determined by differences in the structure of the luciferase. Luciferin is regenerated from oxyluciferin in a subsequent series of reactions.

In the laboratory, pure firefly luciferin and luciferase are used to measure minute quantities of ATP by the intensity of the light flash produced. As little as a few picomoles (10^{-12} mol) of ATP can be measured in this way. Next-gen pyrosequencing of DNA relies on flashes of light from the luciferin-luciferase reaction to detect the presence of ATP after addition of nucleotides to a growing strand of DNA (see Fig. 8-36).

The firefly, a beetle of the *Lampyridae* family.
[Source: Cathy Keifer/Fotolia.]

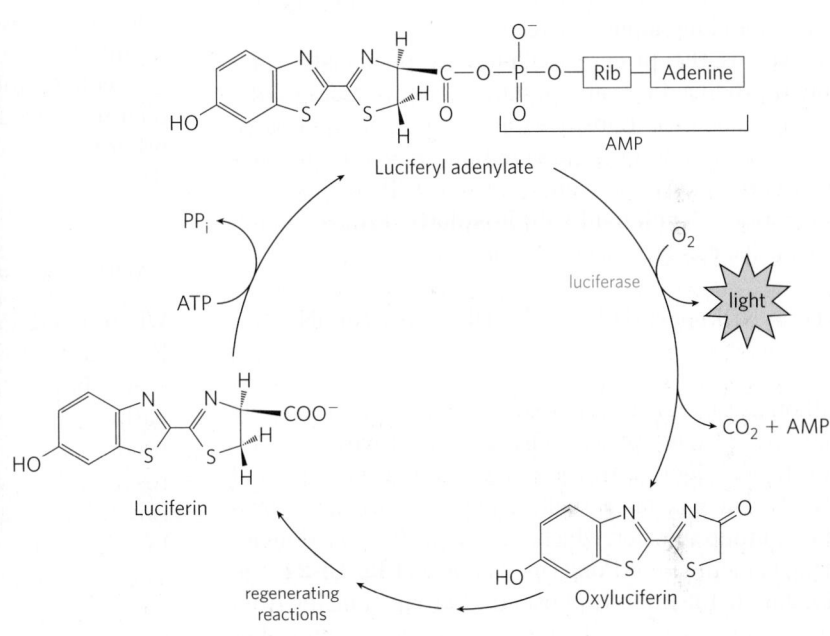

FIGURE 1 Important components in the firefly bioluminescence cycle.

driven by cyclic phosphorylation and dephosphorylation of the transporter protein, with ATP as the phosphoryl group donor. Na^+-dependent phosphorylation of the Na^+K^+ ATPase forces a change in the protein's conformation, and K^+-dependent dephosphorylation favors return to the original conformation. Each cycle in the transport process results in the conversion of ATP to ADP and P_i, and it is the free-energy change of ATP hydrolysis that drives the cyclic changes in protein conformation that result in the electrogenic pumping of Na^+ and K^+. Note that in this case, ATP interacts covalently by phosphoryl group transfer to the *enzyme*, not to the *substrate*.

In the contractile system of skeletal muscle cells, myosin and actin are specialized to transduce the chemical energy of ATP into motion (see Fig. 5-31). ATP binds tightly but noncovalently to one conformation of myosin, holding the protein in that conformation. When myosin catalyzes the hydrolysis of its bound ATP, the ADP and P_i dissociate from the protein, allowing it to relax into a second conformation until another molecule of ATP binds. The binding and subsequent hydrolysis of ATP (by myosin ATPase) provide the energy that forces cyclic changes in the conformation of the myosin head. The change in conformation of many individual myosin molecules results in the sliding of myosin fibrils along actin filaments (see Fig. 5-30), which translates into macroscopic contraction of the muscle fiber. As we noted earlier, this production of mechanical motion at the expense of ATP is one of the few cases in which ATP hydrolysis per se, rather than group transfer from ATP, is the source of the chemical energy in a coupled process.

Transphosphorylations between Nucleotides Occur in All Cell Types

Although we have focused on ATP as the cell's energy currency and donor of phosphoryl groups, all other nucleoside triphosphates (GTP, UTP, and CTP) and all deoxynucleoside triphosphates (dATP, dGTP, dTTP, and dCTP) are energetically equivalent to ATP. The standard free-energy changes associated with hydrolysis of their phosphoanhydride linkages are very nearly identical with those shown in Table 13-6 for ATP. In preparation for their various biological roles, these other nucleotides are generated and maintained as the nucleoside triphosphate (NTP) forms by phosphoryl group transfer to the corresponding nucleoside diphosphates (NDPs) and monophosphates (NMPs).

ATP is the primary high-energy phosphate compound produced by catabolism, in the processes of glycolysis, oxidative phosphorylation, and, in photosynthetic cells, photophosphorylation. Several enzymes then carry phosphoryl groups from ATP to the other nucleotides. **Nucleoside diphosphate kinase**, found in all cells, catalyzes the reaction

$$\text{ATP} + \text{NDP(or dNDP)} \overset{\text{Mg}^{2+}}{\rightleftharpoons} \text{ADP} + \text{NTP(or dNTP)}$$
$$\Delta G'^{\circ} \approx 0$$

Although this reaction is fully reversible, the relatively high [ATP]/[ADP] ratio in cells normally drives the reaction to the right, with the net formation of NTPs and dNTPs. The enzyme actually catalyzes a two-step phosphoryl group transfer, which is a classic case of a double-displacement (Ping-Pong) mechanism (**Fig. 13-21**; see also Fig. 6-13b). First, phosphoryl group transfer from ATP to an active-site His residue produces a phosphoenzyme intermediate; then the phosphoryl group is transferred from the P–His residue to an NDP acceptor. Because the enzyme is nonspecific for the base in the NDP and works equally well on dNDPs and NDPs, it can synthesize all NTPs and dNTPs, given the corresponding NDPs and a supply of ATP.

Phosphoryl group transfers from ATP result in an accumulation of ADP; for example, when muscle is contracting vigorously, ADP accumulates and interferes with ATP-dependent contraction. During periods of intense demand for ATP, the cell lowers the ADP concentration, and at the same time replenishes ATP, by the action of **adenylate kinase**:

$$2\text{ADP} \overset{\text{Mg}^{2+}}{\rightleftharpoons} \text{ATP} + \text{AMP} \qquad \Delta G'^{\circ} \approx 0$$

This reaction is fully reversible, so, after the intense demand for ATP ends, the enzyme can recycle AMP by converting it to ADP, which can then be phosphorylated to ATP in mitochondria. A similar enzyme, guanylate kinase, converts GMP to GDP at the expense of ATP. By pathways such as these, energy conserved in the catabolic production of ATP is used to supply the cell with all required NTPs and dNTPs.

Phosphocreatine (PCr; Fig. 13-15), also called creatine phosphate, serves as a ready source of phosphoryl groups for the quick synthesis of ATP from ADP. The PCr concentration in skeletal muscle is approximately 30 mM, nearly 10 times the concentration of ATP, and in other tissues such as smooth muscle, brain, and kidney, [PCr] is 5 to 10 mM. The enzyme **creatine kinase** catalyzes the reversible reaction

$$\text{ADP} + \text{PCr} \overset{\text{Mg}^{2+}}{\rightleftharpoons} \text{ATP} + \text{Cr} \quad \Delta G'^{\circ} = -12.5 \text{ kJ/mol}$$

When a sudden demand for energy depletes ATP, the PCr reservoir is used to replenish ATP at a rate considerably faster than ATP can be synthesized by catabolic pathways. When the demand for energy slackens, ATP produced by catabolism is used to replenish the PCr reservoir by reversal of the creatine kinase reaction (see Box 23-2). Organisms in the lower phyla employ other PCr-like molecules (collectively called **phosphagens**) as phosphoryl reservoirs.

Inorganic Polyphosphate Is a Potential Phosphoryl Group Donor

Inorganic polyphosphate, denoted by polyP (or (polyP)$_n$, where n is the number of orthophosphate residues), is a linear polymer composed of many tens or hundreds of P$_i$ residues linked through phosphoanhydride bonds.

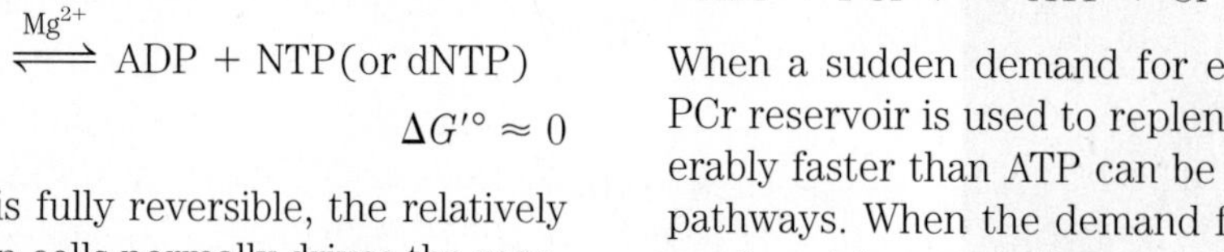

FIGURE 13-21 Ping-Pong mechanism of nucleoside diphosphate kinase. The enzyme binds its first substrate (ATP in our example), and a phosphoryl group is transferred to the side chain of a His residue. ADP departs, another nucleoside (or deoxynucleoside) diphosphate replaces it, and this is converted to the corresponding triphosphate by transfer of the phosphoryl group from the phosphohistidine residue.

This polymer, present in all organisms, may accumulate to high levels in some cells. In yeast, for example, the amount of polyP that accumulates in the vacuoles would represent, if distributed uniformly throughout the cell, a concentration of 200 mM! (Compare this with the concentrations of other phosphoryl group donors listed in Table 13-5.)

$$-O-\overset{\overset{\displaystyle O}{\|}}{\underset{\underset{\displaystyle O^-}{|}}{P}}-O-\overset{\overset{\displaystyle O}{\|}}{\underset{\underset{\displaystyle O^-}{|}}{P}}-O-\overset{\overset{\displaystyle O}{\|}}{\underset{\underset{\displaystyle O^-}{|}}{P}}-O-\overset{\overset{\displaystyle O}{\|}}{\underset{\underset{\displaystyle O^-}{|}}{P}}-O-\overset{\overset{\displaystyle O}{\|}}{\underset{\underset{\displaystyle O^-}{|}}{P}}-O---$$

Inorganic polyphosphate (polyP)

One potential role for polyP is to serve as a phosphagen, a reservoir of phosphoryl groups that can be used to generate ATP, in the same way that creatine phosphate is used in muscle. PolyP has about the same phosphoryl group transfer potential as PP_i. The shortest polyphosphate, PP_i ($n = 2$), can serve as the energy source for active transport of H^+ across the vacuolar membrane in plant cells. For at least one form of the enzyme phosphofructokinase in plants, PP_i is the phosphoryl group donor, a role played by ATP in animals and microbes (p. 540). The finding of high concentrations of polyP in volcanic condensates and steam vents suggests that it could have served as an energy source in prebiotic and early cellular evolution.

In bacteria, the enzyme **polyphosphate kinase-1** (PPK-1) catalyzes the reversible reaction

$$\text{ATP} + (\text{polyP})_n \overset{\text{Mg}^{2+}}{\rightleftharpoons} \text{ADP} + (\text{polyP})_{n+1}$$

$$\Delta G'^\circ = -20 \text{ kJ/mol}$$

by a mechanism involving an enzyme-bound Ⓟ–His intermediate (recall the mechanism of nucleoside diphosphate kinase, described in Fig. 13-21). A second enzyme, **polyphosphate kinase-2** (PPK-2), catalyzes the reversible synthesis of GTP (or ATP) from polyphosphate and GDP (or ADP):

$$\text{GDP(ADP)} + (\text{polyP})_{n+1} \overset{\text{Mg}^{2+}}{\rightleftharpoons} \text{GTP(ATP)} + (\text{polyP})_n$$

PPK-2 is believed to act primarily in the direction of GTP and ATP synthesis, and PPK-1 in the direction of polyphosphate synthesis. PPK-1 and PPK-2 are present in a wide variety of bacteria, including many pathogenic species.

In bacteria, elevated levels of polyP have been shown to promote expression of genes involved in adaptation to conditions of starvation or other threats to survival. In *Escherichia coli*, for example, polyP accumulates when cells are starved for amino acids or P_i, and this accumulation confers a survival advantage. Deletion of the genes for polyphosphate kinases diminishes the ability of certain pathogenic bacteria to invade animal tissues. The enzymes may therefore prove to be suitable targets in the development of new antimicrobial drugs.

No yeast gene encodes a PPK-like protein, but a complex of actin-related proteins in yeast can carry out the synthesis of polyphosphate. The mechanism for polyphosphate synthesis in eukaryotes seems to be different from that in bacteria.

SUMMARY 13.3 Phosphoryl Group Transfers and ATP

■ ATP is the chemical link between catabolism and anabolism. It is the energy currency of the living cell. The exergonic conversion of ATP to ADP and P_i, or to AMP and PP_i, is coupled to many endergonic reactions and processes.

■ Direct hydrolysis of ATP is the source of energy in some processes driven by conformational changes but, in general, it is not ATP hydrolysis but the transfer of a phosphoryl, pyrophosphoryl, or adenylyl group from ATP to a substrate or enzyme that couples the energy of ATP breakdown to endergonic transformations of substrates.

■ Through these group transfer reactions, ATP provides the energy for anabolic reactions, including the synthesis of informational macromolecules, and for the transport of molecules and ions across membranes against concentration gradients and electrical potential gradients.

■ To maintain its high group transfer potential, ATP concentration must be held far above the equilibrium concentration by energy-yielding reactions of catabolism.

■ Cells contain other metabolites with large, negative, free energies of hydrolysis, including phosphoenolpyruvate, 1,3-bisphosphoglycerate, and phosphocreatine. These high-energy compounds, like ATP, have a high phosphoryl group transfer potential. Thioesters also have high free energies of hydrolysis.

■ Inorganic polyphosphate, present in all cells, may serve as a reservoir of phosphoryl groups with high group transfer potential.

13.4 Biological Oxidation-Reduction Reactions

The transfer of phosphoryl groups is a central feature of metabolism. Equally important is another kind of transfer: electron transfer in oxidation-reduction reactions, sometimes referred to as redox reactions. These reactions involve the loss of electrons by one chemical species, which is thereby oxidized, and the gain of electrons by another, which is reduced. The flow of electrons in oxidation-reduction reactions is responsible, directly or indirectly, for all work done by living organisms. In nonphotosynthetic organisms, the sources of electrons are reduced compounds (foods); in photosynthetic organisms, the initial electron donor is a chemical species excited by the absorption of light. The path of electron flow in metabolism is complex. Electrons move from various metabolic intermediates to specialized electron carriers in enzyme-catalyzed reactions. The carriers, in turn, donate electrons to acceptors with higher electron affinities, with the release of energy. Cells possess a variety of molecular energy transducers, which convert the energy of electron flow into useful work.

We begin by discussing how work can be accomplished by an electromotive force (emf), then consider the theoretical and experimental basis for measuring energy changes in oxidation reactions in terms of emf and the relationship between this force, expressed in volts, and the free-energy change, expressed in joules. We also describe the structures and oxidation-reduction chemistry of the most common of the specialized electron carriers, which you will encounter repeatedly in later chapters.

The Flow of Electrons Can Do Biological Work

Every time we use a motor, an electric light or heater, or a spark to ignite gasoline in a car engine, we use the flow of electrons to accomplish work. In the circuit that powers a motor, the source of electrons can be a battery containing two chemical species that differ in affinity for electrons. Electrical wires provide a pathway for electron flow from the chemical species at one pole of the battery, through the motor, to the chemical species at the other pole of the battery. Because the two chemical species differ in their affinity for electrons, electrons flow spontaneously through the circuit, driven by a force proportional to the difference in electron affinity, the **electromotive force, emf**. The emf (typically a few volts) can accomplish work if an appropriate energy transducer—in this case a motor—is placed in the circuit. The motor can be coupled to a variety of mechanical devices to do useful work.

Living cells have an analogous biological "circuit," with a relatively reduced compound such as glucose as the source of electrons. As glucose is enzymatically oxidized, the released electrons flow spontaneously through a series of electron-carrier intermediates to another chemical species, such as O_2. This electron flow is exergonic, because O_2 has a higher affinity for electrons than do the electron-carrier intermediates. The resulting emf provides energy to a variety of molecular energy transducers (enzymes and other proteins) that do biological work. In the mitochondrion, for example, membrane-bound enzymes couple electron flow to the production of a transmembrane pH difference and a transmembrane electrical potential, accomplishing chemiosmotic and electrical work. The proton gradient thus formed has potential energy, sometimes called the proton-motive force by analogy with electromotive force. Another enzyme, ATP synthase in the inner mitochondrial membrane, uses the proton-motive force to do chemical work: synthesis of ATP from ADP and P_i as protons flow spontaneously across the membrane. Similarly, membrane-localized enzymes in *E. coli* convert emf to proton-motive force, which is then used to power

flagellar motion. The principles of electrochemistry that govern energy changes in the macroscopic circuit with a motor and battery apply with equal validity to the molecular processes accompanying electron flow in living cells.

Oxidation-Reductions Can Be Described as Half-Reactions

Although oxidation and reduction must occur together, it is convenient when describing electron transfers to consider the two halves of an oxidation-reduction reaction separately. For example, the oxidation of ferrous ion by cupric ion,

$$Fe^{2+} + Cu^{2+} \rightleftharpoons Fe^{3+} + Cu^+$$

can be described in terms of two half-reactions:

$$(1) \qquad Fe^{2+} \rightleftharpoons Fe^{3+} + e^-$$
$$(2) \qquad Cu^{2+} + e^- \rightleftharpoons Cu^+$$

The electron-donating molecule in an oxidation-reduction reaction is called the reducing agent or reductant; the electron-accepting molecule is the oxidizing agent or oxidant. A given agent, such as an iron cation existing in the ferrous (Fe^{2+}) or ferric (Fe^{3+}) state, functions as a conjugate reductant-oxidant pair (redox pair), just as an acid and corresponding base function as a conjugate acid-base pair. Recall from Chapter 2 that in acid-base reactions we can write a general equation: proton donor $\rightleftharpoons H^+$ + proton acceptor. In redox reactions we can write a similar general equation: electron donor (reductant) $\rightleftharpoons e^-$ + electron acceptor (oxidant). In the reversible half-reaction (1) above, Fe^{2+} is the electron donor and Fe^{3+} is the electron acceptor; together, Fe^{2+} and Fe^{3+} constitute a **conjugate redox pair**. The mnemonic OIL RIG—oxidation *is* *l*osing, *r*eduction *is* *g*aining—may be helpful in remembering what happens to electrons in redox reactions.

The electron transfers in the oxidation-reduction reactions of organic compounds are not fundamentally different from those of inorganic species. Consider the oxidation of a reducing sugar (an aldehyde or ketone) by cupric ion:

$$R-C{\overset{O}{\underset{H}{\big\langle}}} + 4OH^- + 2Cu^{2+} \rightleftharpoons R-C{\overset{O}{\underset{OH}{\big\langle}}} + Cu_2O + 2H_2O$$

This overall reaction can be expressed as two half-reactions:

$$(1) \quad R-C{\overset{O}{\underset{H}{\big\langle}}} + 2OH^- \rightleftharpoons R-C{\overset{O}{\underset{OH}{\big\langle}}} + 2e^- + H_2O$$

$$(2) \quad 2Cu^{2+} + 2e^- + 2OH^- \rightleftharpoons Cu_2O + H_2O$$

Notice that because two electrons are removed from the aldehyde carbon, the second half-reaction (the one-electron reduction of cupric to cuprous ion) must be doubled to balance the overall equation.

Biological Oxidations Often Involve Dehydrogenation

The carbon in living cells exists in a range of oxidation states **(Fig. 13-22)**. When a carbon atom shares an electron pair with another atom (typically H, C, S, N, or O), the sharing is unequal, in favor of the more electronegative atom. The order of increasing electronegativity is H < C < S < N < O. In oversimplified but useful terms, the more electronegative atom "owns" the bonding electrons it shares with another atom. For example, in methane (CH_4), carbon is more electronegative than the four hydrogens bonded to it, and the C atom therefore owns all eight bonding electrons (Fig. 13-22). In ethane, the electrons in the C—C bond are shared equally, so each C atom owns only seven of its eight bonding electrons. In ethanol, C-1 is less electronegative than the oxygen to which it is bonded, and the O atom therefore owns both electrons of the C—O bond, leaving C-1 with only five bonding electrons. With each formal loss of "owned" electrons, the carbon atom has undergone oxidation—even when no oxygen is involved, as in the conversion of an alkane (—CH_2—CH_2—) to an alkene (—CH=CH—). In this case, oxidation (loss of electrons) is coincident with the loss of hydrogen. In biological systems, as we noted earlier in the chapter, oxidation is often synonymous with **dehydrogenation**, and many enzymes that catalyze oxidation reactions are **dehydrogenases**. Notice that the more reduced compounds in Figure 13-22 (top) are richer in hydrogen than in oxygen, whereas the more oxidized compounds (bottom) have more oxygen and less hydrogen.

Not all biological oxidation-reduction reactions involve carbon. For example, in the conversion of molecular nitrogen to ammonia, $6H^+ + 6e^- + N_2 \rightarrow 2NH_3$, the nitrogen atoms are reduced.

Electrons are transferred from one molecule (electron donor) to another (electron acceptor) in one of four ways:

1. Directly as *electrons*. For example, the Fe^{2+}/Fe^{3+} redox pair can transfer an electron to the Cu^+/Cu^{2+} redox pair:

$$Fe^{2+} + Cu^{2+} \rightleftharpoons Fe^{3+} + Cu^+$$

2. As *hydrogen atoms*. Recall that a hydrogen atom consists of a proton (H^+) and a single electron (e^-). In this case we can write the general equation

$$AH_2 \rightleftharpoons A + 2e^- + 2H^+$$

where AH_2 is the hydrogen/electron donor. (Do not mistake the above reaction for an acid dissociation, which involves a proton and no electron.) AH_2 and A together constitute a conjugate redox pair (A/AH_2), which can reduce another compound B (or redox pair, B/BH_2) by transfer of hydrogen atoms:

$$AH_2 + B \rightleftharpoons A + BH_2$$

3. As a *hydride ion* ($:H^-$), which has two electrons. This occurs in the case of NAD-linked dehydrogenases, described below.

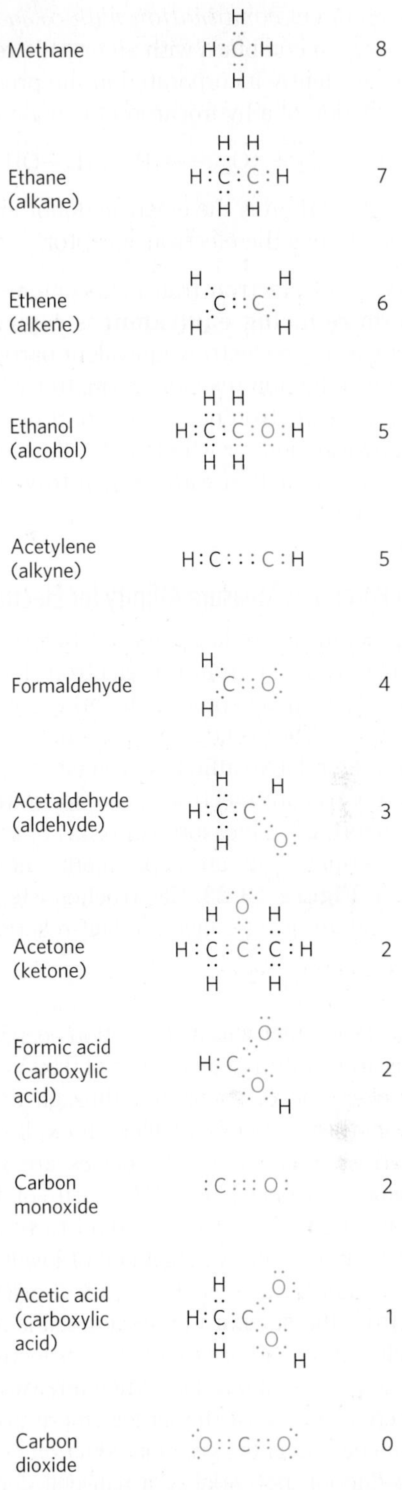

Methane	H:C:H (with H above and below)	8
Ethane (alkane)	H:C:C:H	7
Ethene (alkene)	C::C	6
Ethanol (alcohol)	H:C:C:O:H	5
Acetylene (alkyne)	H:C:::C:H	5
Formaldehyde	C::O	4
Acetaldehyde (aldehyde)	H:C:C	3
Acetone (ketone)	H:C:C:C:H	2
Formic acid (carboxylic acid)	H:C	2
Carbon monoxide	:C:::O:	2
Acetic acid (carboxylic acid)	H:C:C	1
Carbon dioxide	:O::C::O:	0

FIGURE 13-22 Different levels of oxidation of carbon compounds in the biosphere. To approximate the level of oxidation of these compounds, focus on the red carbon atom and its bonding electrons. When this carbon is bonded to the less electronegative H atom, both bonding electrons (red) are assigned to the carbon. When carbon is bonded to another carbon, bonding electrons are shared equally, so one of the two electrons is assigned to the red carbon. When the red carbon is bonded to the more electronegative O atom, the bonding electrons are assigned to the oxygen. The number to the right of each compound is the number of electrons "owned" by the red carbon, a rough expression of the degree of oxidation of that compound. As the red carbon undergoes oxidation (loses electrons), the number gets smaller.

4. Through direct *combination with oxygen*. In this case, oxygen combines with an organic reductant and is covalently incorporated in the product, as in the oxidation of a hydrocarbon to an alcohol:

$$R\text{—}CH_3 + \tfrac{1}{2}O_2 \longrightarrow R\text{—}CH_2\text{—}OH$$

The hydrocarbon is the electron donor and the oxygen atom is the electron acceptor.

All four types of electron transfer occur in cells. The neutral term **reducing equivalent** is commonly used to designate a single electron equivalent participating in an oxidation-reduction reaction, no matter whether this equivalent is an electron per se or is part of a hydrogen atom or a hydride ion, or whether the electron transfer takes place in a reaction with oxygen to yield an oxygenated product.

Reduction Potentials Measure Affinity for Electrons

When two conjugate redox pairs are together in solution, electron transfer from the electron donor of one pair to the electron acceptor of the other may proceed spontaneously. The tendency for such a reaction depends on the relative affinity of the electron acceptor of each redox pair for electrons. The **standard reduction potential, $E°$**, a measure (in volts) of this affinity, can be determined in an experiment such as that described in **Figure 13-23**. Electrochemists have chosen as a standard of reference the half-reaction

$$H^+ + e^- \longrightarrow \tfrac{1}{2}H_2$$

The electrode at which this half-reaction occurs (called a half-cell) is arbitrarily assigned an $E°$ of 0.00 V. When this hydrogen electrode is connected through an external circuit to another half-cell in which an oxidized species and its corresponding reduced species are present at standard concentrations (at 25 °C, each solute at 1 M, each gas at 101.3 kPa), electrons tend to flow through the external circuit from the half-cell of lower $E°$ to the half-cell of higher $E°$. By convention, a half-cell that takes electrons from the standard hydrogen cell is assigned a positive value of $E°$, and one that donates electrons to the hydrogen cell, a negative value. When any two half-cells are connected, that with the larger (more positive) $E°$ will be reduced; it has the greater reduction potential.

The reduction potential of a half-cell depends not only on the chemical species present but also on their activities, approximated by their concentrations. About a century ago, Walther Nernst derived an equation that relates standard reduction potential ($E°$) to the actual reduction potential (E) at any concentration of oxidized and reduced species in a living cell:

$$E = E° + \frac{RT}{nF} \ln \frac{[\text{electron acceptor}]}{[\text{electron donor}]} \quad (13\text{-}5)$$

where R and T have their usual meanings, n is the number of electrons transferred per molecule, and F is the

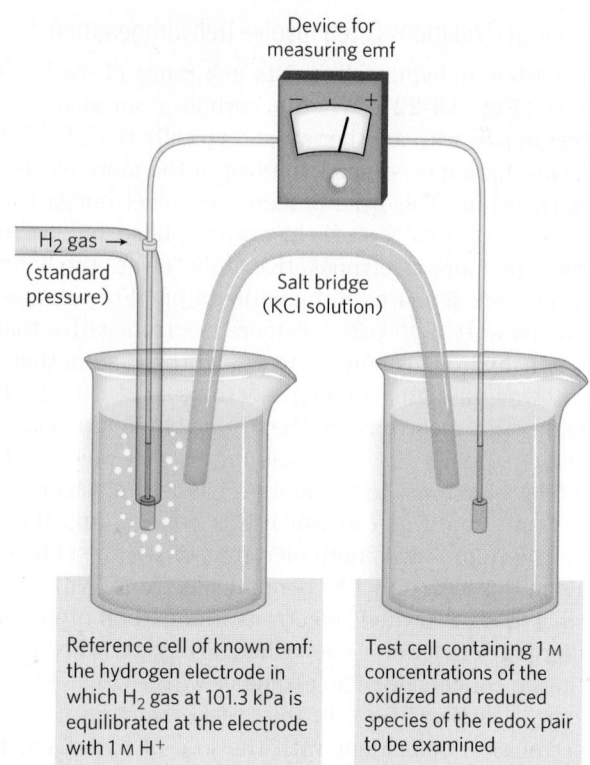

FIGURE 13-23 Measurement of the standard reduction potential ($E'°$) of a redox pair. Electrons flow from the test electrode to the reference electrode, or vice versa. The ultimate reference half-cell is the hydrogen electrode, as shown here, at pH 0. The electromotive force (emf) of this electrode is designated 0.00 V. At pH 7 in the test cell (at 25 °C), $E'°$ for the hydrogen electrode is −0.414 V. The direction of electron flow depends on the relative electron "pressure" or potential of the two cells. A salt bridge containing a saturated KCl solution provides a path for counter-ion movement between the test cell and the reference cell. From the observed emf and the known emf of the reference cell, the experimenter can find the emf of the test cell containing the redox pair. The cell that gains electrons has, by convention, the more positive reduction potential.

Faraday constant, a proportionality constant that converts volts to joules (Table 13-1). At 298 K (25 °C), this expression reduces to

$$E = E° + \frac{0.026 \text{ V}}{n} \ln \frac{[\text{electron acceptor}]}{[\text{electron donor}]} \quad (13\text{-}6)$$

>> **Key Convention:** Many half-reactions of interest to biochemists involve protons. As in the definition of $\Delta G'°$, biochemists define the standard state for oxidation-reduction reactions as pH 7 and express a standard transformed reduction potential, $E'°$, the standard reduction potential at pH 7 and 25 °C. By convention, $\Delta E'°$ for any redox reaction is given as $E'°$ of the electron acceptor minus $E'°$ of the electron donor. <<

The standard reduction potentials given in Table 13-7 and used throughout this book are values for $E'°$ and are

Labels within figure:

Device for measuring emf

H₂ gas → (standard pressure)

Salt bridge (KCl solution)

Reference cell of known emf: the hydrogen electrode in which H₂ gas at 101.3 kPa is equilibrated at the electrode with 1 M H⁺

Test cell containing 1 M concentrations of the oxidized and reduced species of the redox pair to be examined

TABLE 13-7	Standard Reduction Potentials of Some Biologically Important Half-Reactions

Half-reaction	E'° (V)
$\frac{1}{2}O_2 + 2H^+ + 2e^- \longrightarrow H_2O$	0.816
$Fe^{3+} + e^- \longrightarrow Fe^{2+}$	0.771
$NO_3^- + 2H^+ + 2e^- \longrightarrow NO_2^- + H_2O$	0.421
Cytochrome f (Fe^{3+}) + $e^- \longrightarrow$ cytochrome f (Fe^{2+})	0.365
$Fe(CN)_6^{3-}$ (ferricyanide) + $e^- \longrightarrow Fe(CN)_6^{4-}$	0.36
Cytochrome a_3 (Fe^{3+}) + $e^- \longrightarrow$ cytochrome a_3 (Fe^{2+})	0.35
$O_2 + 2H^+ + 2e^- \longrightarrow H_2O_2$	0.295
Cytochrome a (Fe^{3+}) + $e^- \longrightarrow$ cytochrome a (Fe^{2+})	0.29
Cytochrome c (Fe^{3+}) + $e^- \longrightarrow$ cytochrome c (Fe^{2+})	0.254
Cytochrome c_1 (Fe^{3+}) + $e^- \longrightarrow$ cytochrome c_1 (Fe^{2+})	0.22
Cytochrome b (Fe^{3+}) + $e^- \longrightarrow$ cytochrome b (Fe^{2+})	0.077
Ubiquinone + $2H^+ + 2e^- \longrightarrow$ ubiquinol	0.045
Fumarate^{2-} + $2H^+ + 2e^- \longrightarrow$ succinate^{2-}	0.031
$2H^+ + 2e^- \longrightarrow H_2$ (at standard conditions, pH 0)	0.000
Crotonyl-CoA + $2H^+ + 2e^- \longrightarrow$ butyryl-CoA	−0.015
Oxaloacetate^{2-} + $2H^+ + 2e^- \longrightarrow$ malate^{2-}	−0.166
Pyruvate$^-$ + $2H^+ + 2e^- \longrightarrow$ lactate$^-$	−0.185
Acetaldehyde + $2H^+ + 2e^- \longrightarrow$ ethanol	−0.197
FAD + $2H^+ + 2e^- \longrightarrow$ FADH$_2$	−0.219[a]
Glutathione + $2H^+ + 2e^- \longrightarrow$ 2 reduced glutathione	−0.23
S + $2H^+ + 2e^- \longrightarrow H_2S$	−0.243
Lipoic acid + $2H^+ + 2e^- \longrightarrow$ dihydrolipoic acid	−0.29
$NAD^+ + H^+ + 2e^- \longrightarrow$ NADH	−0.320
$NADP^+ + H^+ + 2e^- \longrightarrow$ NADPH	−0.324
Acetoacetate + $2H^+ + 2e^- \longrightarrow$ β-hydroxybutyrate	−0.346
α-Ketoglutarate + $CO_2 + 2H^+ + 2e^- \longrightarrow$ isocitrate	−0.38
$2H^+ + 2e^- \longrightarrow H_2$ (at pH 7)	−0.414
Ferredoxin (Fe^{3+}) + $e^- \longrightarrow$ ferredoxin (Fe^{2+})	−0.432

Source: Data mostly from R. A. Loach, in *Handbook of Biochemistry and Molecular Biology*, 3rd edn (G. D. Fasman, ed.), *Physical and Chemical Data*, Vol. 1, p. 122, CRC Press, 1976.

[a] This is the value for free FAD; FAD bound to a specific flavoprotein (e.g., succinate dehydrogenase) has a different E'° that depends on its protein environment.

therefore valid only for systems at neutral pH. Each value represents the potential difference when the conjugate redox pair, at 1 M concentrations, 25 °C, and pH 7, is connected with the standard (pH 0) hydrogen electrode. Notice in Table 13-7 that when the conjugate pair

$2H^+/H_2$ at pH 7 is connected with the standard hydrogen electrode (pH 0), electrons tend to flow from the pH 7 cell to the standard (pH 0) cell; the measured E'° for the $2H^+/H_2$ pair is −0.414 V.

Standard Reduction Potentials Can Be Used to Calculate Free-Energy Change

Why are reduction potentials so useful to the biochemist? When E values have been determined for any two half-cells, relative to the standard hydrogen electrode, we also know their reduction potentials relative to each other. We can then predict the direction in which electrons will tend to flow when the two half-cells are connected through an external circuit or when components of both half-cells are present in the same solution. Electrons tend to flow to the half-cell with the more positive E, and the strength of that tendency is proportional to ΔE, the difference in reduction potential. The energy made available by this spontaneous electron flow (the free-energy change, ΔG, for the oxidation-reduction reaction) is proportional to ΔE:

$$\Delta G = -nF\Delta E \quad \text{or} \quad \Delta G'^\circ = -nF\Delta E'^\circ \quad (13\text{-}7)$$

where n is the number of electrons transferred in the reaction. With this equation we can calculate the actual free-energy change for any oxidation-reduction reaction from the values of $\Delta E'^\circ$ in a table of reduction potentials (Table 13-7) and the concentrations of reacting species.

WORKED EXAMPLE 13-3   **Calculation of $\Delta G'^\circ$ and ΔG of a Redox Reaction**

Calculate the standard free-energy change, $\Delta G'^\circ$, for the reaction in which acetaldehyde is reduced by the biological electron carrier NADH:

$$\text{Acetaldehyde} + \text{NADH} + H^+ \longrightarrow \text{ethanol} + NAD^+$$

Then calculate the *actual* free-energy change, ΔG, when [acetaldehyde] and [NADH] are 1.00 M, and [ethanol] and [NAD$^+$] are 0.100 M. The relevant half-reactions and their E'° values are:

(1)   Acetaldehyde + $2H^+ + 2e^- \longrightarrow$ ethanol
$$E'^\circ = -0.197 \text{ V}$$

(2) $NAD^+ + 2H^+ + 2e^- \longrightarrow$ NADH + H^+
$$E'^\circ = -0.320 \text{ V}$$

Remember that, by convention, $\Delta E'^\circ$ is E'° of the electron acceptor minus E'° of the electron donor. It represents the difference between the electron affinities of the two half-reactions in the table of reduction potentials (Table 13-7). Note that the more widely separated the two half-reactions in the table, the more energetic the electron-transfer reaction when the two half-reactions occur together. By convention, in tables of reduction potentials, all half-reactions are represented as reductions, but when two half-reactions occur together, one of them must be an oxidation. Although that half-reaction will go in the opposite direction from that shown in Table 13-7, we *do not change the sign* of that half-reaction before calculating $\Delta E'^{\circ}$, because $\Delta E'^{\circ}$ is *defined* as a difference of reduction potentials.

Solution: Because acetaldehyde is accepting electrons ($n = 2$) from NADH, $\Delta E'^{\circ} = -0.197\ V - (-0.320\ V) = 0.123\ V$. Therefore,

$$\Delta G'^{\circ} = -nF\Delta E'^{\circ} = -2\,(96.5\ kJ/V \cdot mol)\,(0.123\ V)$$
$$= -23.7\ kJ/mol$$

This is the free-energy change for the oxidation-reduction reaction at 25 °C and pH 7, when acetaldehyde, ethanol, NAD^+, and NADH are all present at 1.00 M concentrations.

To calculate ΔG when [acetaldehyde] and [NADH] are 1.00 M, and [ethanol] and [NAD^+] are 0.100 M, we can use Equation 13-4 and the standard free-energy change calculated above:

$$\Delta G = \Delta G'^{\circ} + RT \ln \frac{[ethanol][NAD^+]}{[acetaldehyde][NADH]}$$
$$= -23.7\ kJ/mol +$$
$$(8.315\ J/mol \cdot K)(298\ K)\ln\frac{(0.100\ M)(0.100\ M)}{(1.00\ M)(1.00\ M)}$$
$$= -23.7\ kJ/mol + (2.48\ J/mol)\ln 0.01$$
$$= -35.1\ kJ/mol$$

This is the actual free-energy change at the specified concentrations of the redox pairs.

Cellular Oxidation of Glucose to Carbon Dioxide Requires Specialized Electron Carriers

The principles of oxidation-reduction energetics described above apply to the many metabolic reactions that involve electron transfers. For example, in many organisms, the oxidation of glucose supplies energy for the production of ATP. The complete oxidation of glucose:

$$C_6H_{12}O_6 + 6O_2 \longrightarrow 6CO_2 + 6H_2O$$

has a $\Delta G'^{\circ}$ of $-2{,}840\ kJ/mol$. This is a much larger release of free energy than is required for ATP synthesis in cells (50 to 60 kJ/mol; see Worked Example 13-2). Cells convert glucose to CO_2 not in a single, high-energy-releasing reaction but rather in a series of controlled reactions, some of which are oxidations. The free energy released in these oxidation steps is of the same order of magnitude as that required for ATP synthesis from ADP, with some energy to spare. Electrons removed in these oxidation steps are transferred to coenzymes specialized for carrying electrons, such as NAD^+ and FAD (described below).

A Few Types of Coenzymes and Proteins Serve as Universal Electron Carriers

The multitude of enzymes that catalyze cellular oxidations channel electrons from their hundreds of different substrates into just a few types of universal electron carriers. The reduction of these carriers in catabolic processes results in the conservation of free energy released by substrate oxidation. NAD, NADP, FMN, and FAD are water-soluble coenzymes that undergo reversible oxidation and reduction in many of the electron-transfer reactions of metabolism. The nucleotides NAD and NADP move readily from one enzyme to another; the flavin nucleotides FMN and FAD are usually very tightly bound to the enzymes, called flavoproteins, for which they serve as prosthetic groups. Lipid-soluble quinones such as ubiquinone and plastoquinone act as electron carriers and proton donors in the nonaqueous environment of membranes. Iron-sulfur proteins and cytochromes, which have tightly bound prosthetic groups that undergo reversible oxidation and reduction, also serve as electron carriers in many oxidation-reduction reactions. Some of these proteins are water-soluble, but others are peripheral or integral membrane proteins (see Fig. 11-6).

We conclude this chapter by describing some chemical features of nucleotide coenzymes and some of the enzymes (dehydrogenases and flavoproteins) that use them. The oxidation-reduction chemistry of quinones, iron-sulfur proteins, and cytochromes is discussed in Chapters 19 and 20.

NADH and NADPH Act with Dehydrogenases as Soluble Electron Carriers

Nicotinamide adenine dinucleotide (NAD; NAD^+ in its oxidized form) and its close analog nicotinamide adenine dinucleotide phosphate (NADP; $NADP^+$ when oxidized) are composed of two nucleotides joined through their phosphate groups by a phosphoanhydride bond **(Fig. 13-24a)**. Because the nicotinamide ring resembles pyridine, these compounds are sometimes called **pyridine nucleotides**. The vitamin niacin is the source of the nicotinamide moiety in nicotinamide nucleotides.

Both coenzymes undergo reversible reduction of the nicotinamide ring (Fig. 13-24). As a substrate molecule undergoes oxidation (dehydrogenation), giving up two hydrogen atoms, the oxidized form of the nucleotide (NAD^+ or $NADP^+$) accepts a hydride ion (:H^-, the equivalent of a proton and two electrons) and is reduced (to NADH or NADPH). The second proton removed from the substrate is released to the aqueous solvent. The half-reactions for these nucleotide cofactors are

$$NAD^+ + 2e^- + 2H^+ \longrightarrow NADH + H^+$$
$$NADP^+ + 2e^- + 2H^+ \longrightarrow NADPH + H^+$$

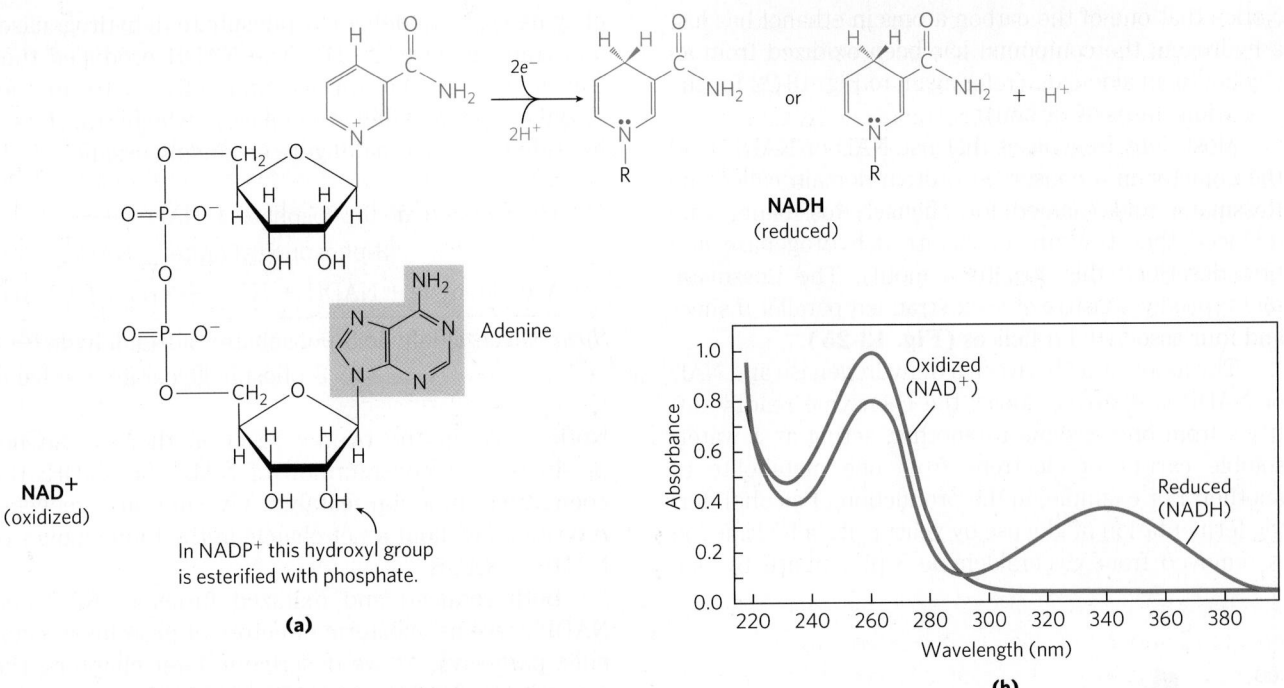

(a)

(b)

FIGURE 13-24 NAD and NADP. (a) Nicotinamide adenine dinucleotide, NAD$^+$, and its phosphorylated analog, NADP$^+$, undergo reduction to NADH and NADPH, accepting a hydride ion (two electrons and one proton) from an oxidizable substrate. The hydride ion is added to either the front or the back of the planar nicotinamide ring. **(b)** The UV absorption spectra of NAD$^+$ and NADH. Reduction of the nicotinamide ring produces a new, broad absorption band with a maximum at 340 nm. The production of NADH during an enzyme-catalyzed reaction can be conveniently followed by observing the appearance of the absorbance at 340 nm (molar extinction coefficient $\varepsilon_{340} = 6{,}200$ M^{-1} cm^{-1}).

Reduction of NAD$^+$ or NADP$^+$ converts the benzenoid ring of the nicotinamide moiety (with a fixed positive charge on the ring nitrogen) to the quinonoid form (with no charge on the nitrogen). The reduced nucleotides absorb light at 340 nm; the oxidized forms do not (Fig. 13-24b). Biochemists use this difference in absorption to assay reactions involving these coenzymes. Note that the plus sign in the abbreviations NAD$^+$ and NADP$^+$ does *not* indicate the net charge on these molecules (in fact, both are negatively charged); rather, it indicates that the nicotinamide ring is in its oxidized form, with a positive charge on the nitrogen atom. In the abbreviations NADH and NADPH, the "H" denotes the added hydride ion. To refer to these nucleotides without specifying their oxidation state, we use NAD and NADP.

The total concentration of NAD$^+$ + NADH in most tissues is about 10^{-5} M; that of NADP$^+$ + NADPH is about 10^{-6} M. In many cells and tissues, the ratio of NAD$^+$ (oxidized) to NADH (reduced) is high, favoring hydride transfer from a substrate *to* NAD$^+$ to form NADH. By contrast, NADPH is generally present at a higher concentration than NADP$^+$, favoring hydride transfer *from* NADPH to a substrate. This reflects the specialized metabolic roles of the two coenzymes: NAD$^+$ generally functions in oxidations—usually as part of a catabolic reaction; NADPH is the usual coenzyme in reductions—nearly always as part of an anabolic

reaction. A few enzymes can use either coenzyme, but most show a strong preference for one over the other. Also, the processes in which these two cofactors function are segregated in eukaryotic cells: for example, oxidations of fuels such as pyruvate, fatty acids, and α-keto acids derived from amino acids occur in the mitochondrial matrix, whereas reductive biosynthetic processes such as fatty acid synthesis take place in the cytosol. This functional and spatial specialization allows a cell to maintain two distinct pools of electron carriers, with two distinct functions.

More than 200 enzymes are known to catalyze reactions in which NAD$^+$ (or NADP$^+$) accepts a hydride ion from a reduced substrate, or NADPH (or NADH) donates a hydride ion to an oxidized substrate. The general reactions are

$$AH_2 + NAD^+ \longrightarrow A + NADH + H^+$$

$$A + NADPH + H^+ \longrightarrow AH_2 + NADP^+$$

where AH$_2$ is the reduced substrate and A is the oxidized substrate. The general name for an enzyme of this type is **oxidoreductase**; they are also commonly called dehydrogenases. For example, alcohol dehydrogenase catalyzes the first step in the catabolism of ethanol, in which ethanol is oxidized to acetaldehyde:

$$\underset{\text{Ethanol}}{CH_3CH_2OH} + NAD^+ \longrightarrow \underset{\text{Acetaldehyde}}{CH_3CHO} + NADH + H^+$$

Notice that one of the carbon atoms in ethanol has lost a hydrogen; the compound has been oxidized from an alcohol to an aldehyde (refer again to Fig. 13-22 for the oxidation states of carbon).

Most dehydrogenases that use NAD or NADP bind the cofactor in a conserved protein domain called the Rossmann fold (named for Michael Rossmann, who deduced the structure of lactate dehydrogenase and first described this structural motif). The Rossmann fold typically consists of a six-stranded parallel β sheet and four associated α helices **(Fig. 13-25)**.

The association between a dehydrogenase and NAD or NADP is relatively loose; the coenzyme readily diffuses from one enzyme to another, acting as a water-soluble carrier of electrons from one metabolite to another. For example, in the production of alcohol during fermentation of glucose by yeast cells, a hydride ion is removed from glyceraldehyde 3-phosphate by one enzyme (glyceraldehyde 3-phosphate dehydrogenase) and transferred to NAD^+. The NADH produced then leaves the enzyme surface and diffuses to another enzyme (alcohol dehydrogenase), which transfers a hydride ion to acetaldehyde, producing ethanol:

(1) Glyceraldehyde 3-phosphate + NAD^+ $\longrightarrow$
$$3\text{-phosphoglycerate} + NADH + H^+$$

(2) Acetaldehyde + NADH + H^+ $\longrightarrow$ ethanol + NAD^+

Sum: Glyceraldehyde 3-phosphate + acetaldehyde $\longrightarrow$
$$3\text{-phosphoglycerate} + \text{ethanol}$$

Notice that in the overall reaction there is no net production or consumption of NAD^+ or NADH; the coenzymes function catalytically and are recycled repeatedly without a net change in the total amount of NAD^+ + NADH.

Both reduced and oxidized forms of NAD and NADP serve as allosteric effectors of proteins in catabolic pathways. As we describe in later chapters, the ratios NAD^+/NADH and $NADP^+$/NADPH serve as sensitive gauges of a cell's fuel supply, allowing rapid, appropriate changes in energy-yielding and energy-dependent metabolism.

NAD Has Important Functions in Addition to Electron Transfer

Some key cellular functions are regulated by enzymes that use NAD^+ not as a redox cofactor but as a substrate in a coupled reaction in which the availability of NAD^+ can be an indicator of the cell's energy status. In DNA replication and repair, the enzyme DNA ligase is adenylylated and then transfers the AMP to a 5′ phosphate in a nicked DNA (see Fig. 25-16); in bacteria, NAD^+ serves as the source of the activating AMP group. A family of proteins called sirtuins regulate the activity of proteins in diverse cellular pathways by deacetylating the ε-amino group of an acetylated Lys residue. The deacetylation is coupled to NAD^+ hydrolysis, yielding O-acetyl-ADP-ribose and nicotinamide. Among the cellular processes regulated by sirtuins are inflammation, apoptosis, aging, and DNA transcription; deacetylation by a sirtuin alters the charge on histones, influencing which genes are expressed (see p. 1149). The availability of NAD^+ for these types of reactions may indicate that the cell is undergoing stress and that pathways designed to respond to stress should be activated.

NAD^+ also plays an important role in cholera infections (see Box 12-1). Cholera toxin has an enzymatic activity that transfers ADP-ribose from NAD^+ to a G protein involved in regulating ion fluxes in the cells lining the gut. This ADP-ribosylation blocks water retention, causing the diarrhea and dehydration characteristic of cholera. ■

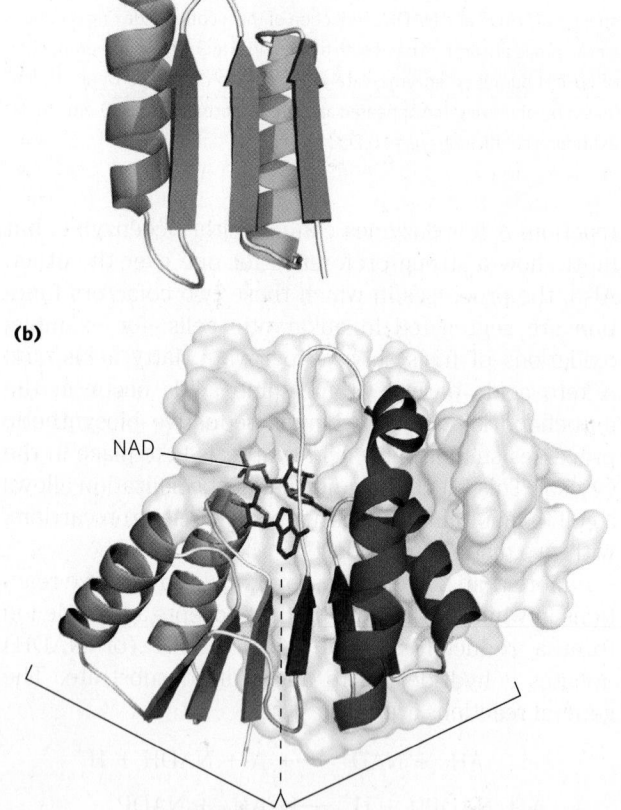

(a)

(b)

NAD

FIGURE 13-25 The Rossmann fold. This structural motif is found in the NAD-binding site of many dehydrogenases. **(a)** It consists of a pair of structurally similar motifs (only one of which is shown here), each having three parallel β sheets and two α helices (β-α-β-α-β). **(b)** The nucleotide-binding domain of the enzyme lactate dehydrogenase with NAD (ball-and-stick structure) bound in an extended conformation through hydrogen bonds and salt bridges to the paired β-α-β-α-β motifs of the Rossmann fold (shades of red and blue). [Source: Derived from PDB ID 3LDH, J. L. White et al., *J. Mol. Biol.* 102:759, 1976.]

Conrad Elvehjem, 1901–1962 [Source: Courtesy of the Department of Biochemistry, University of Wisconsin-Madison.]

FIGURE 13-26 reference structures:

Niacin (nicotinic acid)

Nicotine

Nicotinamide

Tryptophan

FIGURE 13-26 Niacin (nicotinic acid) and its derivative nicotinamide. The biosynthetic precursor of these compounds is tryptophan. In the laboratory, nicotinic acid was first produced by oxidation of the natural product nicotine—thus the name. Both nicotinic acid and nicotinamide cure pellagra, but nicotine (from cigarettes or elsewhere) has no curative activity.

1916. In 1920, Joseph Goldberger showed pellagra to be caused by a dietary insufficiency, and in 1937, Frank Strong, D. Wayne Woolley, and Conrad Elvehjem identified niacin as the curative agent for blacktongue. Supplementation of the human diet with this inexpensive compound has nearly eradicated pellagra in the populations of the developed world, with one significant exception: people who drink excessive amounts of alcohol. In these individuals, intestinal absorption of niacin is much reduced, and caloric needs are often met with distilled spirits that are virtually devoid of vitamins, including niacin. ■

Dietary Deficiency of Niacin, the Vitamin Form of NAD and NADP, Causes Pellagra

Frank Strong, 1908–1993 [Source: Courtesy of the Department of Biochemistry, University of Wisconsin-Madison.]

D. Wayne Woolley, 1914–1966 [Source: Rockefeller Archive Center.]

As we noted in Chapter 6 and will discuss further in later chapters, most coenzymes are derived from the substances we call vitamins. The pyridine-like rings of NAD and NADP are derived from the vitamin **niacin** (nicotinic acid; **Fig. 13-26**), which is synthesized from tryptophan. Humans generally cannot synthesize sufficient quantities of niacin, and this is especially so for individuals with diets low in tryptophan (maize, for example, has a low tryptophan content). Niacin deficiency, which affects all the NAD(P)-dependent dehydrogenases, causes the serious human disease pellagra (Italian for "rough skin") and a related disease in dogs, blacktongue. Pellagra is characterized by the "three Ds": dermatitis, diarrhea, and dementia, followed in many cases by death. A century ago, pellagra was a common human disease; in the southern United States, where maize was a dietary staple, about 100,000 people were afflicted and about 10,000 died as a result of this disease between 1912 and

Flavin Nucleotides Are Tightly Bound in Flavoproteins

Flavoproteins are enzymes that catalyze oxidation-reduction reactions using either flavin mononucleotide (FMN) or flavin adenine dinucleotide (FAD) as coenzyme **(Fig. 13-27)**. These coenzymes, the **flavin nucleotides**, are derived from the vitamin riboflavin. The fused ring structure of flavin nucleotides (the isoalloxazine ring) undergoes reversible reduction, accepting either one or two electrons in the form of one or two hydrogen atoms (each atom an electron plus a proton) from a reduced substrate. The fully reduced forms are abbreviated $FADH_2$ and $FMNH_2$. When a fully oxidized flavin nucleotide accepts only one electron (one hydrogen atom), the semiquinone form of the isoalloxazine ring is produced, abbreviated $FADH^•$ and $FMNH^•$. Because flavin nucleotides have a slightly different chemical specialty from that of the nicotinamide coenzymes—the ability to participate in either one- or two-electron transfers—flavoproteins are involved in a greater diversity of reactions than the NAD(P)-linked dehydrogenases.

Like the nicotinamide coenzymes (Fig. 13-24), the flavin nucleotides undergo a shift in a major absorption band on reduction (again, useful to biochemists who want to monitor reactions involving these coenzymes). Flavoproteins that are fully reduced (two electrons accepted) generally have an absorption maximum near 360 nm. When partially reduced (one electron), they acquire another absorption maximum at about 450 nm; when fully oxidized, the flavin has maxima at 370 and 440 nm.

The flavin nucleotide in most flavoproteins is bound rather tightly to the protein, and in some enzymes, such as succinate dehydrogenase, it is bound covalently. Such tightly bound coenzymes are properly called

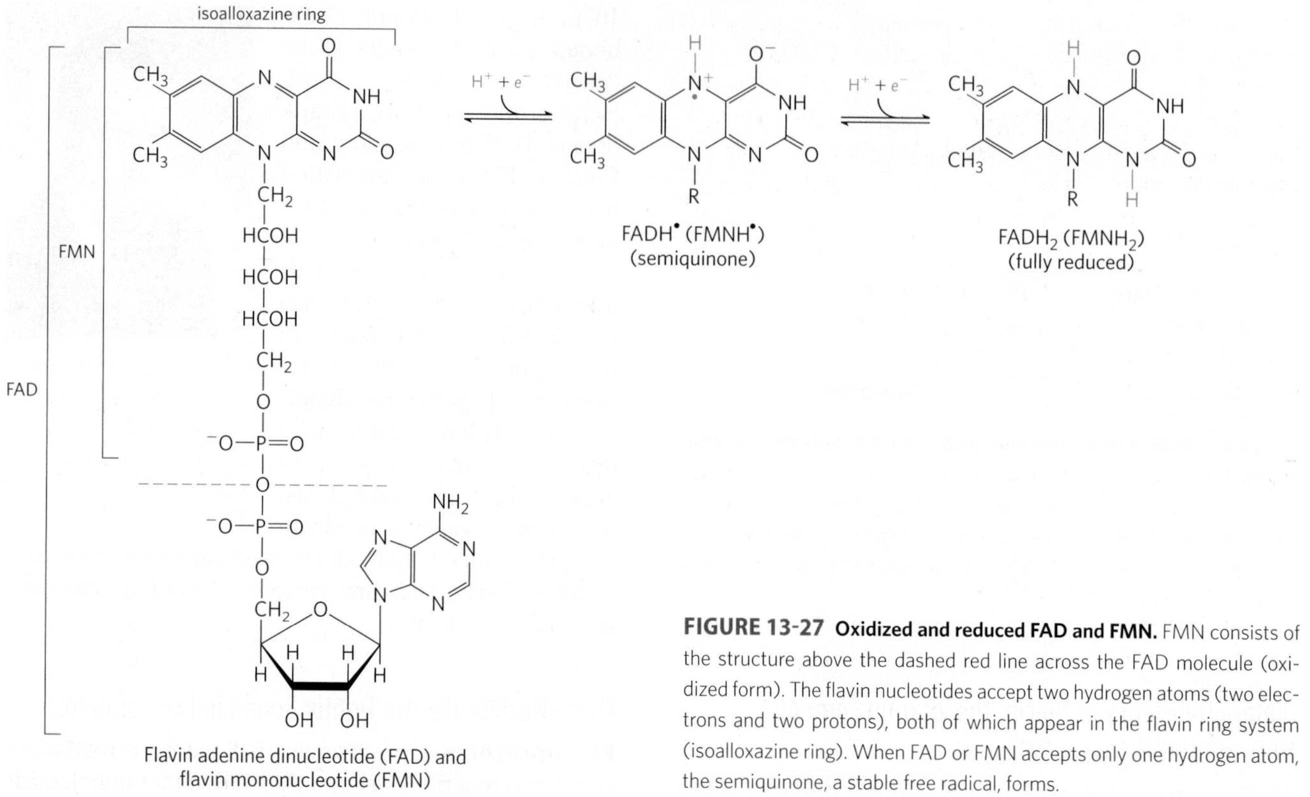

Flavin adenine dinucleotide (FAD) and
flavin mononucleotide (FMN)

FIGURE 13-27 Oxidized and reduced FAD and FMN. FMN consists of the structure above the dashed red line across the FAD molecule (oxidized form). The flavin nucleotides accept two hydrogen atoms (two electrons and two protons), both of which appear in the flavin ring system (isoalloxazine ring). When FAD or FMN accepts only one hydrogen atom, the semiquinone, a stable free radical, forms.

prosthetic groups. They do not transfer electrons by diffusing from one enzyme to another; rather, they provide a means by which the flavoprotein can temporarily hold electrons while it catalyzes electron transfer from a reduced substrate to an electron acceptor. One important feature of the flavoproteins is the variability in the standard reduction potential (E'°) of the bound flavin nucleotide. Tight association between the enzyme and prosthetic group confers on the flavin ring a reduction potential typical of that particular flavoprotein, sometimes quite different from the reduction potential of the free flavin nucleotide. FAD bound to succinate dehydrogenase, for example, has an E'° close to 0.0 V, compared with -0.219 V for free FAD; E'° for other flavoproteins ranges from -0.40 V to $+0.06$ V. Flavoproteins are often very complex; some have, in addition to a flavin nucleotide, tightly bound inorganic ions (iron or molybdenum, for example) capable of participating in electron transfers.

Certain flavoproteins have a distinctly different role, as light receptors. **Cryptochromes** are a family of flavoproteins, widely distributed in the eukaryotic phyla, that mediate the effects of blue light on plant development and the effects of light on mammalian circadian rhythms (oscillations in physiology and biochemistry, with a 24-hour period). The cryptochromes are homologs of another family of flavoproteins, the photolyases. Found in both bacteria and eukaryotes, **photolyases** use the energy of absorbed light to repair chemical defects in DNA.

We examine the function of flavoproteins as electron carriers in Chapters 19 and 20, when we consider their roles in oxidative phosphorylation (in mitochondria) and photophosphorylation (in chloroplasts), and we describe the photolyase reactions in Chapter 25.

SUMMARY 13.4 Biological Oxidation-Reduction Reactions

■ In many organisms, a central energy-conserving process is the stepwise oxidation of glucose to CO_2, in which some of the energy of oxidation is conserved in ATP as electrons are passed to O_2.

■ Biological oxidation-reduction reactions can be described in terms of two half-reactions, each with a characteristic standard reduction potential, E'°.

■ When two electrochemical half-cells, each containing the components of a half-reaction, are connected, electrons tend to flow to the half-cell with the higher reduction potential. The strength of this tendency is proportional to the difference between the two reduction potentials (ΔE) and is a function of the concentrations of oxidized and reduced species.

■ The standard free-energy change for an oxidation-reduction reaction is directly proportional to the difference in standard reduction potentials of the two half-cells: $\Delta G'^\circ = -nF\Delta E'^\circ$.

■ Many biological oxidation reactions are dehydrogenations in which one or two hydrogen atoms ($H^+ + e^-$) are transferred from a substrate to a hydrogen acceptor. Oxidation-reduction reactions in living cells involve specialized electron carriers.

■ NAD and NADP are the freely diffusible coenzymes of many dehydrogenases. Both NAD^+ and $NADP^+$ accept two electrons and one proton. In addition to its role in oxidation-reduction reactions, NAD^+ is the source of AMP in the bacterial DNA ligase reaction and of ADP-ribose in the cholera toxin reaction, and is hydrolyzed in the deacetylation of proteins by some sirtuins.

■ FAD and FMN, the flavin nucleotides, serve as tightly bound prosthetic groups of flavoproteins. They can accept either one or two electrons and one or two protons. Flavoproteins also serve as light receptors in cryptochromes and photolyases.

Key Terms

Terms in bold are defined in the glossary.

autotroph 491	adenylylation 513
heterotroph 491	**inorganic pyrophos-**
metabolism 492	**phatase** 513
metabolic pathways 492	**nucleoside diphosphate**
metabolite 492	**kinase** 516
intermediary	adenylate kinase 516
metabolism 492	creatine kinase 516
catabolism 492	phosphagens 516
anabolism 493	polyphosphate kinase-1,
standard transformed	kinase-2 517
constants 497	electromotive force
homolytic cleavage 502	(emf) 518
radical 502	**conjugate redox pair** 518
heterolytic cleavage 502	dehydrogenation 519
nucleophile 502	**dehydrogenases** 519
electrophile 502	**reducing equivalent** 520
carbanion 503	**standard reduction**
carbocation 503	**potential (E'°)** 520
aldol condensation 503	**pyridine nucleotide** 522
Claisen condensation 503	oxidoreductase 523
kinases 506	**flavoprotein** 525
phosphorylation potential	**flavin nucleotides** 525
(ΔG_p) 508	cryptochrome 526
thioester 510	photolyase 526

Problems

1. Entropy Changes during Egg Development Consider a system consisting of an egg in an incubator. The white and yolk of the egg contain proteins, carbohydrates, and lipids. If fertilized, the egg is transformed from a single cell to a complex organism. Discuss this irreversible process in terms of the entropy changes in the system, surroundings, and universe. Be sure that you first clearly define the system and surroundings.

2. Calculation of $\Delta G'^\circ$ from an Equilibrium Constant Calculate the standard free-energy change for each of the following metabolically important enzyme-catalyzed reactions, using the equilibrium constants given for the reactions at 25 °C and pH 7.0.

(a) Glutamate + oxaloacetate $\underset{\text{aminotransferase}}{\overset{\text{aspartate}}{\rightleftharpoons}}$

$\qquad$ aspartate + α-ketoglutarate $\qquad K'_{eq} = 6.8$

(b) Dihydroxyacetone phosphate $\underset{\text{isomerase}}{\overset{\text{triose phosphate}}{\rightleftharpoons}}$

$\qquad$ glyceraldehyde 3-phosphate $\qquad K'_{eq} = 0.0475$

(c) Fructose 6-phosphate + ATP $\overset{\text{phosphofructokinase}}{\rightleftharpoons}$

$\qquad$ fructose 1,6-bisphosphate + ADP $\qquad K'_{eq} = 254$

3. Calculation of the Equilibrium Constant from $\Delta G'^\circ$ Calculate the equilibrium constant K'_{eq} for each of the following reactions at pH 7.0 and 25 °C, using the $\Delta G'^\circ$ values in Table 13-4.

(a) Glucose 6-phosphate + H_2O $\underset{\text{6-phosphatase}}{\overset{\text{glucose}}{\rightleftharpoons}}$ glucose + P_i

(b) Lactose + H_2O $\overset{\beta\text{-galactosidase}}{\rightleftharpoons}$ glucose + galactose

(c) Malate $\overset{\text{fumarase}}{\rightleftharpoons}$ fumarate + H_2O

4. Experimental Determination of K'_{eq} and $\Delta G'^\circ$ If a 0.1 M solution of glucose 1-phosphate at 25 °C is incubated with a catalytic amount of phosphoglucomutase, the glucose 1-phosphate is transformed to glucose 6-phosphate. At equilibrium, the concentrations of the reaction components are

$$\text{Glucose 1-phosphate} \rightleftharpoons \text{glucose 6-phosphate}$$
$$4.5 \times 10^{-3} \text{ M} \qquad\qquad 9.6 \times 10^{-2} \text{ M}$$

Calculate K'_{eq} and $\Delta G'^\circ$ for this reaction.

5. Experimental Determination of $\Delta G'^\circ$ for ATP Hydrolysis A direct measurement of the standard free-energy change associated with the hydrolysis of ATP is technically demanding because the minute amount of ATP remaining at equilibrium is difficult to measure accurately. The value of $\Delta G'^\circ$ can be calculated indirectly, however, from the equilibrium constants of two other enzymatic reactions having less favorable equilibrium constants:

Glucose 6-phosphate + H_2O $\longrightarrow$ glucose + P_i $\qquad K'_{eq} = 270$
ATP + glucose $\longrightarrow$ ADP + glucose 6-phosphate $\qquad K'_{eq} = 890$

Using this information for equilibrium constants determined at 25 °C, calculate the standard free energy of hydrolysis of ATP.

6. Difference between $\Delta G'^\circ$ and ΔG Consider the following interconversion, which occurs in glycolysis (Chapter 14):

$$\text{Fructose 6-phosphate} \rightleftharpoons \text{glucose 6-phosphate}$$
$$K'_{eq} = 1.97$$

(a) What is $\Delta G'^\circ$ for the reaction (K'_{eq} measured at 25 °C)?

(b) If the concentration of fructose 6-phosphate is adjusted to 1.5 M and that of glucose 6-phosphate is adjusted to 0.50 M, what is ΔG?

(c) Why are $\Delta G'^\circ$ and ΔG different?

7. Free Energy of Hydrolysis of CTP Compare the structure of the nucleoside triphosphate CTP with the structure of ATP.

Cytidine triphosphate (CTP)

Adenosine triphosphate (ATP)

Now predict the K'_{eq} and $\Delta G'^{\circ}$ for the following reaction:

$$\text{ATP} + \text{CDP} \longrightarrow \text{ADP} + \text{CTP}$$

8. Dependence of ΔG on pH The free energy released by the hydrolysis of ATP under standard conditions is -30.5 kJ/mol. If ATP is hydrolyzed under standard conditions except at pH 5.0, is more or less free energy released? Explain.

9. The $\Delta G'^{\circ}$ for Coupled Reactions Glucose 1-phosphate is converted into fructose 6-phosphate in two successive reactions:

Glucose 1-phosphate $\longrightarrow$ glucose 6-phosphate

Glucose 6-phosphate $\longrightarrow$ fructose 6-phosphate

Using the $\Delta G'^{\circ}$ values in Table 13-4, calculate the equilibrium constant, K'_{eq}, for the sum of the two reactions:

Glucose 1-phosphate $\longrightarrow$ fructose 6-phosphate

10. Effect of [ATP]/[ADP] Ratio on Free Energy of Hydrolysis of ATP Using Equation 13-4, plot ΔG against $\ln Q$ (mass-action ratio) at 25 °C for the concentrations of ATP, ADP, and P_i in the table below. $\Delta G'^{\circ}$ for the reaction is -30.5 kJ/mol. Use the resulting plot to explain why metabolism is regulated to keep the ratio [ATP]/[ADP] high.

Concentration (mM)

ATP	5	3	1	0.2	5
ADP	0.2	2.2	4.2	5.0	25
P_i	10	12.1	14.1	14.9	10

11. Strategy for Overcoming an Unfavorable Reaction: ATP-Dependent Chemical Coupling The phosphorylation of glucose to glucose 6-phosphate is the initial step in the catabolism of glucose. The direct phosphorylation of glucose by P_i is described by the equation

Glucose + P_i $\longrightarrow$ glucose 6-phosphate + H_2O

$$\Delta G'^{\circ} = 13.8 \text{ kJ/mol}$$

(a) Calculate the equilibrium constant for the above reaction at 37 °C. In the rat hepatocyte, the physiological concentrations

of glucose and P_i are maintained at approximately 4.8 mM. What is the equilibrium concentration of glucose 6-phosphate obtained by the direct phosphorylation of glucose by P_i? Does this reaction represent a reasonable metabolic step for the catabolism of glucose? Explain.

(b) In principle, at least, one way to increase the concentration of glucose 6-phosphate is to drive the equilibrium reaction to the right by increasing the intracellular concentrations of glucose and P_i. Assuming a fixed concentration of P_i at 4.8 mM, how high would the intracellular concentration of glucose have to be to give an equilibrium concentration of glucose 6-phosphate of 250 μM (the normal physiological concentration)? Would this route be physiologically reasonable, given that the maximum solubility of glucose is less than 1 M?

(c) The phosphorylation of glucose in the cell is coupled to the hydrolysis of ATP; that is, part of the free energy of ATP hydrolysis is used to phosphorylate glucose:

(1)   Glucose + P_i $\longrightarrow$ glucose 6-phosphate + H_2O

$$\Delta G'^{\circ} = 13.8 \text{ kJ/mol}$$

(2)   ATP + H_2O $\longrightarrow$ ADP + P_i $\Delta G'^{\circ} = -30.5$ kJ/mol

Sum: Glucose + ATP $\longrightarrow$ glucose 6-phosphate + ADP

Calculate K'_{eq} at 37 °C for the overall reaction. For the ATP-dependent phosphorylation of glucose, what concentration of glucose is needed to achieve a 250 μM intracellular concentration of glucose 6-phosphate when the concentrations of ATP and ADP are 3.38 mM and 1.32 mM, respectively? Does this coupling process provide a feasible route, at least in principle, for the phosphorylation of glucose in the cell? Explain.

(d) Although coupling ATP hydrolysis to glucose phosphorylation makes thermodynamic sense, we have not yet specified how this coupling is to take place. Given that coupling requires a common intermediate, one conceivable route is to use ATP hydrolysis to raise the intracellular concentration of P_i and thus drive the unfavorable phosphorylation of glucose by P_i. Is this a reasonable route? (Think about the solubility product, K_{sp}, of metabolic intermediates.)

(e) The ATP-coupled phosphorylation of glucose is catalyzed in hepatocytes by the enzyme glucokinase. This enzyme binds ATP and glucose to form a glucose-ATP-enzyme complex, and the phosphoryl group is transferred directly from ATP to glucose. Explain the advantages of this route.

12. Calculations of $\Delta G'^{\circ}$ for ATP-Coupled Reactions From data in Table 13-6, calculate the $\Delta G'^{\circ}$ value for the following reactions:

(a) Phosphocreatine + ADP $\longrightarrow$ creatine + ATP

(b) ATP + fructose $\longrightarrow$ ADP + fructose 6-phosphate

13. Coupling ATP Cleavage to an Unfavorable Reaction To explore the consequences of coupling ATP hydrolysis under physiological conditions to a thermodynamically unfavorable biochemical reaction, consider the hypothetical transformation X $\rightarrow$ Y, for which $\Delta G'^{\circ} = 20.0$ kJ/mol.

(a) What is the ratio [Y]/[X] at equilibrium?

(b) Suppose X and Y participate in a sequence of reactions during which ATP is hydrolyzed to ADP and P_i. The overall reaction is

$$\text{X} + \text{ATP} + H_2O \longrightarrow \text{Y} + \text{ADP} + P_i$$

Calculate [Y]/[X] for this reaction at equilibrium. Assume that the temperature is 25 °C and the equilibrium concentrations of ATP, ADP, and P_i are 1 M.

(c) We know that [ATP], [ADP], and $[P_i]$ are *not* 1 M under physiological conditions. Calculate [Y]/[X] for the ATP-coupled reaction when the values of [ATP], [ADP], and $[P_i]$ are those found in rat myocytes (Table 13-5).

14. Calculations of ΔG at Physiological Concentrations Calculate the actual, physiological ΔG for the reaction

Phosphocreatine + ADP $\longrightarrow$ creatine + ATP

at 37 °C, as it occurs in the cytosol of neurons, with phosphocreatine at 4.7 mM, creatine at 1.0 mM, ADP at 0.73 mM, and ATP at 2.6 mM.

15. Free Energy Required for ATP Synthesis under Physiological Conditions In the cytosol of rat hepatocytes, the temperature is 37 °C and the mass-action ratio, Q, is

$$\frac{[ATP]}{[ADP][P_i]} = 5.33 \times 10^2 \text{ M}^{-1}$$

Calculate the free energy required to synthesize ATP in a rat hepatocyte.

16. Chemical Logic In the glycolytic pathway, a six-carbon sugar (fructose 1,6-bisphosphate) is cleaved to form two three-carbon sugars, which undergo further metabolism (see Fig. 14-6). In this pathway, an isomerization of glucose 6-phosphate to fructose 6-phosphate (shown below) occurs two steps before the cleavage reaction (the intervening step is phosphorylation of fructose 6-phosphate to fructose 1,6-bisphosphate (p. 539)).

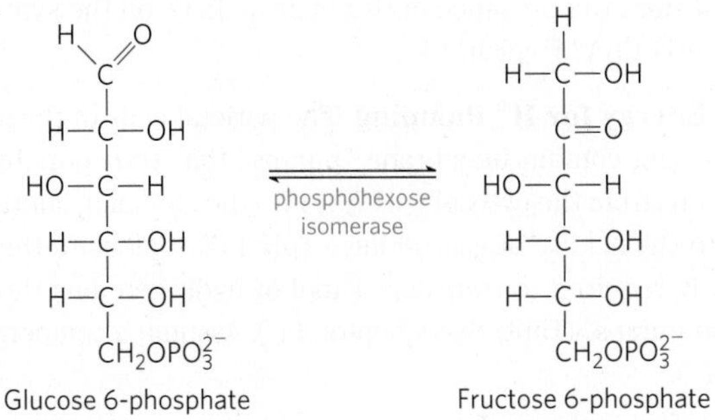

Glucose 6-phosphate   Fructose 6-phosphate

What does the isomerization step accomplish from a chemical perspective? (Hint: Consider what might happen if the C—C bond cleavage were to proceed without the preceding isomerization.)

17. Enzymatic Reaction Mechanisms I Lactate dehydrogenase is one of the many enzymes that require NADH as coenzyme. It catalyzes the conversion of pyruvate to lactate:

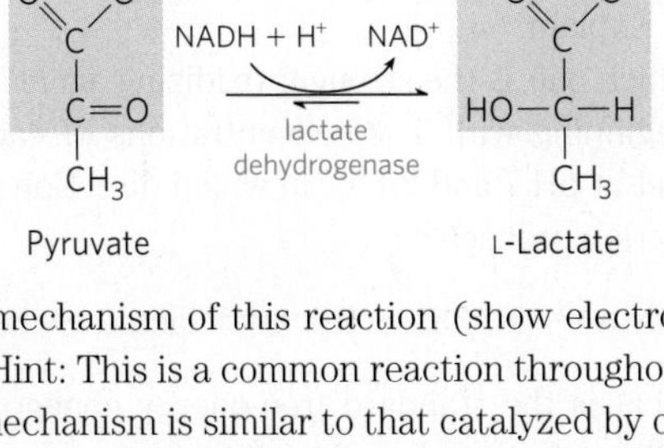

Pyruvate   L-Lactate

Draw the mechanism of this reaction (show electron-pushing arrows). (Hint: This is a common reaction throughout metabolism; the mechanism is similar to that catalyzed by other dehydrogenases that use NADH, such as alcohol dehydrogenase.)

18. Enzymatic Reaction Mechanisms II Biochemical reactions often look more complex than they really are. In the

pentose phosphate pathway (Chapter 14), sedoheptulose 7-phosphate and glyceraldehyde 3-phosphate react to form erythrose 4-phosphate and fructose 6-phosphate in a reaction catalyzed by transaldolase.

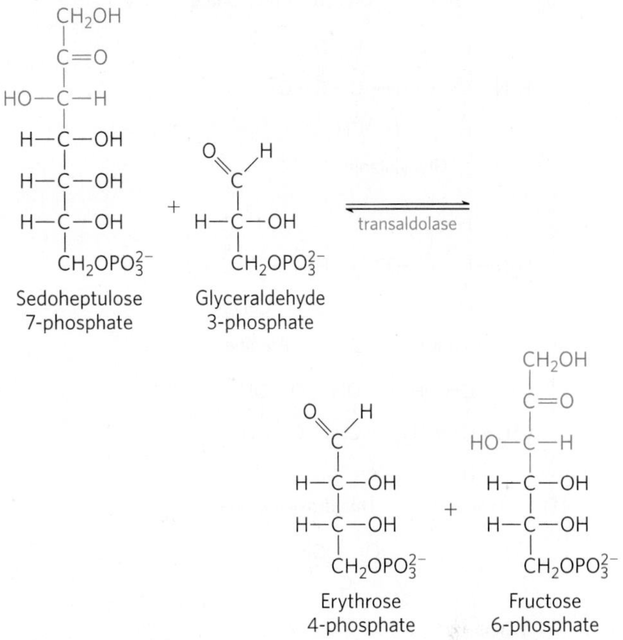

Sedoheptulose 7-phosphate   Glyceraldehyde 3-phosphate

Erythrose 4-phosphate   Fructose 6-phosphate

Draw a mechanism for this reaction (show electron-pushing arrows). (Hint: Take another look at aldol condensations, then consider the name of this enzyme.)

19. Recognizing Reaction Types For the following pairs of biomolecules, identify the type of reaction (oxidation-reduction, hydrolysis, isomerization, group transfer, or internal rearrangement) required to convert the first molecule to the second. In each case, indicate the general type of enzyme and cofactor(s) or reactants that would be required, and any other products that would result.

(a)   *trans*-Δ^2-Enoyl-CoA

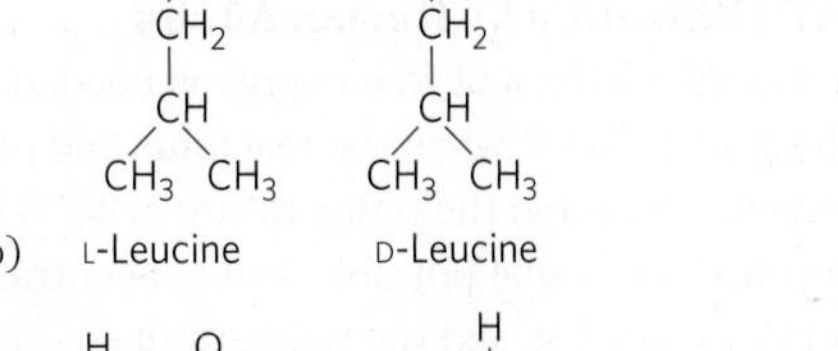

(b)   L-Leucine   D-Leucine

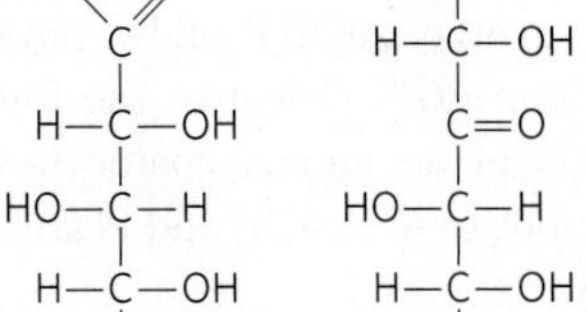

(c)   Glucose   Fructose

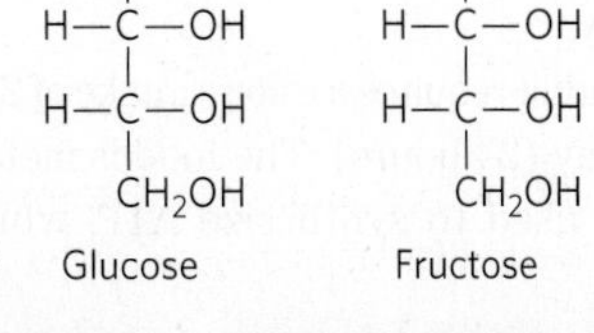

(d) Glycerol   Glycerol 3-phosphate

Glycylalanine

(e) Glycine   Alanine

(f) Glycerol   Dihydroxyacetone

(g) Acetaldehyde   Acetic acid

20. Effect of Structure on Group Transfer Potential
Some invertebrates contain phosphoarginine. Is the standard free energy of hydrolysis of this molecule more similar to that of glucose 6-phosphate or of ATP? Explain your answer.

Phosphoarginine

21. Polyphosphate as a Possible Energy Source
The standard free energy of hydrolysis of inorganic polyphosphate (polyP) is about -20 kJ/mol for each P_i released. We calculated in Worked Example 13-2 that, in a cell, it takes about 50 kJ/mol of energy to synthesize ATP from ADP and P_i. Is it feasible for a cell to use polyphosphate to synthesize ATP from ADP? Explain your answer.

22. Daily ATP Utilization by Human Adults
(a) A total of 30.5 kJ/mol of free energy is needed to synthesize ATP from ADP and P_i when the reactants and products are at 1 M concentrations and the temperature is 25 °C (standard state). Because the actual physiological concentrations of ATP, ADP, and P_i are not 1 M, and the temperature is 37 °C, the free energy required to synthesize ATP under physiological conditions is different from $\Delta G'^{\circ}$. Calculate the free energy required to synthesize ATP in the human hepatocyte when the physiological concentrations of ATP, ADP, and P_i are 3.5, 1.50, and 5.0 mM, respectively.

(b) A 68 kg (150 lb) adult requires a caloric intake of 2,000 kcal (8,360 kJ) of food per day (24 hours). The food is metabolized and the free energy is used to synthesize ATP, which then provides energy for the body's daily chemical and mechanical work. Assuming that the efficiency of converting food energy into ATP is 50%, calculate the weight of ATP used by a human adult in 24 hours. What percentage of the body weight does this represent?

(c) Although adults synthesize large amounts of ATP daily, their body weight, structure, and composition do not change significantly during this period. Explain this apparent contradiction.

23. Rates of Turnover of γ and β Phosphates of ATP
If a small amount of ATP labeled with radioactive phosphorus in the terminal position, $[\gamma\text{-}^{32}\text{P}]\text{ATP}$, is added to a yeast extract, about half of the ^{32}P activity is found in P_i within a few minutes, but the concentration of ATP remains unchanged. Explain. If the same experiment is carried out using ATP labeled with ^{32}P in the central position, $[\beta\text{-}^{32}\text{P}]\text{ATP}$, the ^{32}P does not appear in P_i within such a short time. Why?

24. Cleavage of ATP to AMP and PP_i during Metabolism
Synthesis of the activated form of acetate (acetyl-CoA) is carried out in an ATP-dependent process:

$$\text{Acetate} + \text{CoA} + \text{ATP} \longrightarrow \text{acetyl-CoA} + \text{AMP} + PP_i$$

(a) The $\Delta G'^{\circ}$ for hydrolysis of acetyl-CoA to acetate and CoA is -32.2 kJ/mol and that for hydrolysis of ATP to AMP and PP_i is -30.5 kJ/mol. Calculate $\Delta G'^{\circ}$ for the ATP-dependent synthesis of acetyl-CoA.

(b) Almost all cells contain the enzyme inorganic pyrophosphatase, which catalyzes the hydrolysis of PP_i to P_i. What effect does the presence of this enzyme have on the synthesis of acetyl-CoA? Explain.

25. Energy for H^+ Pumping
The parietal cells of the stomach lining contain membrane "pumps" that transport hydrogen ions from the cytosol (pH 7.0) into the stomach, contributing to the acidity of gastric juice (pH 1.0). Calculate the free energy required to transport 1 mol of hydrogen ions through these pumps. (Hint: See Chapter 11.) Assume a temperature of 37 °C.

26. Standard Reduction Potentials
The standard reduction potential, E'°, of any redox pair is defined for the half-cell reaction:

$$\text{Oxidizing agent} + n \text{ electrons} \longrightarrow \text{reducing agent}$$

The E'° values for the NAD^+/NADH and pyruvate/lactate conjugate redox pairs are -0.32 V and -0.19 V, respectively.

(a) Which redox pair has the greater tendency to lose electrons? Explain.

(b) Which pair is the stronger oxidizing agent? Explain.

(c) Beginning with 1 M concentrations of each reactant and product at pH 7 and 25 °C, in which direction will the following reaction proceed?

$$\text{Pyruvate} + \text{NADH} + H^+ \rightleftharpoons \text{lactate} + NAD^+$$

(d) What is the standard free-energy change ($\Delta G'^{\circ}$) for the conversion of pyruvate to lactate?

(e) What is the equilibrium constant (K'_{eq}) for this reaction?

27. Energy Span of the Respiratory Chain Electron transfer in the mitochondrial respiratory chain may be represented by the net reaction equation

$$\text{NADH} + \text{H}^+ + \tfrac{1}{2}\text{O}_2 \rightleftharpoons \text{H}_2\text{O} + \text{NAD}^+$$

(a) Calculate $\Delta E'^\circ$ for the net reaction of mitochondrial electron transfer. Use E'° values in Table 13-7.

(b) Calculate $\Delta G'^\circ$ for this reaction.

(c) How many ATP molecules can *theoretically* be generated by this reaction if the free energy of ATP synthesis under cellular conditions is 52 kJ/mol?

28. Dependence of Electromotive Force on Concentrations Calculate the electromotive force (in volts) registered by an electrode immersed in a solution containing the following mixtures of NAD^+ and NADH at pH 7.0 and 25 °C, with reference to a half-cell of E'° 0.00 V.

(a) 1.0 mm NAD^+ and 10 mm NADH

(b) 1.0 mm NAD^+ and 1.0 mm NADH

(c) 10 mm NAD^+ and 1.0 mm NADH

29. Electron Affinity of Compounds List the following in order of increasing tendency to accept electrons: (a) α-ketoglutarate + CO_2 (yielding isocitrate); (b) oxaloacetate; (c) O_2; (d) NADP^+.

30. Direction of Oxidation-Reduction Reactions Which of the following reactions would you expect to proceed in the direction shown, under standard conditions, in the presence of the appropriate enzymes?

(a) Malate + $\text{NAD}^+ \longrightarrow$ oxaloacetate + NADH + H^+

(b) Acetoacetate + NADH + $\text{H}^+ \longrightarrow$
β-hydroxybutyrate + NAD^+

(c) Pyruvate + NADH + $\text{H}^+ \longrightarrow$ lactate + NAD^+

(d) Pyruvate + β-hydroxybutyrate $\longrightarrow$
lactate + acetoacetate

(e) Malate + pyruvate $\longrightarrow$ oxaloacetate + lactate

(f) Acetaldehyde + succinate $\longrightarrow$ ethanol + fumarate

Data Analysis Problem

31. Thermodynamics Can Be Tricky Thermodynamics is a challenging area of study and one with many opportunities for confusion. An interesting example is found in an article by Robinson, Hampson, Munro, and Vaney, published in *Science* in 1993. Robinson and colleagues studied the movement of small molecules between neighboring cells of the nervous system through cell-to-cell channels (gap junctions). They found that the dyes Lucifer yellow (a small, negatively charged molecule) and biocytin (a small zwitterionic molecule) moved in only one

direction between two particular types of glia (nonneuronal cells of the nervous system). Dye injected into astrocytes would rapidly pass into adjacent astrocytes, oligodendrocytes, or Müller cells, but dye injected into oligodendrocytes or Müller cells passed slowly if at all into astrocytes. All of these cell types are connected by gap junctions.

Although it was not a central point of their article, the authors presented a molecular model for how this unidirectional transport might occur, as shown in their Figure 3:

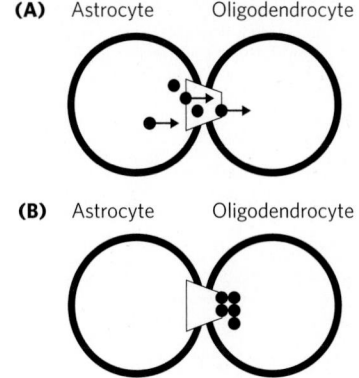

The figure legend reads: "Model of the unidirectional diffusion of dye between coupled oligodendrocytes and astrocytes, based on differences in connection pore diameter. Like a fish in a fish trap, dye molecules (black circles) can pass from an astrocyte to an oligodendrocyte (A) but not back in the other direction (B)."

Although this article clearly passed review at a well-respected journal, several letters to the editor (1994) followed, showing that Robinson and coauthors' model violated the second law of thermodynamics.

(a) Explain how the model violates the second law. Hint: Consider what would happen to the entropy of the system if one started with equal concentrations of dye in the astrocyte and oligodendrocyte connected by the "fish trap" type of gap junctions.

(b) Explain why this model cannot work for small molecules, although it may allow one to catch fish.

(c) Explain why a fish trap *does* work for fish.

(d) Provide two plausible mechanisms for the unidirectional transport of dye molecules between the cells that do not violate the second law of thermodynamics.

References

Letters to the editor. 1994. *Science* 265:1017–1019.

Robinson, S.R., E.C.G.M. Hampson, M.N. Munro, and D.I. Vaney. 1993. Unidirectional coupling of gap junctions between neuroglia. *Science* 262:1072–1074.

CHAPTER 14

Glycolysis, Gluconeogenesis, and the Pentose Phosphate Pathway

Self-study tools that will help you practice what you've learned and reinforce this chapter's concepts are available online.
Go to www.macmillanlearning.com/LehningerBiochemistry7e.

Glucose occupies a central position in the metabolism of plants, animals, and many microorganisms. It is relatively rich in potential energy, and thus a good fuel; the complete oxidation of glucose to carbon dioxide and water proceeds with a standard free-energy change of −2,840 kJ/mol. By storing glucose as a high molecular weight polymer such as starch or glycogen, a cell can stockpile large quantities of hexose units while maintaining a relatively low cytosolic osmolarity. When energy demands increase, glucose can be released from these intracellular storage polymers and used to produce ATP either aerobically or anaerobically.

Glucose is not only an excellent fuel, it is also a remarkably versatile precursor, capable of supplying a huge array of metabolic intermediates for biosynthetic reactions. A bacterium such as *Escherichia coli* can obtain from glucose the carbon skeletons for every amino acid, nucleotide, coenzyme, fatty acid, or other metabolic intermediate it needs for growth. A comprehensive study of the metabolic fates of glucose would encompass hundreds or thousands of transformations. In animals and vascular plants, glucose has four major fates: it may be (1) used in the synthesis of complex polysaccharides destined for the extracellular space; (2) stored in cells (as a polysaccharide or as sucrose); (3) oxidized to a three-carbon compound (pyruvate) via glycolysis to provide ATP and metabolic intermediates; or (4) oxidized via the pentose phosphate (phosphogluconate) pathway to yield ribose 5-phosphate for nucleic acid synthesis and NADPH for reductive biosynthetic processes (**Fig. 14-1**).

Organisms that do not have access to glucose from other sources must make it. Photosynthetic organisms

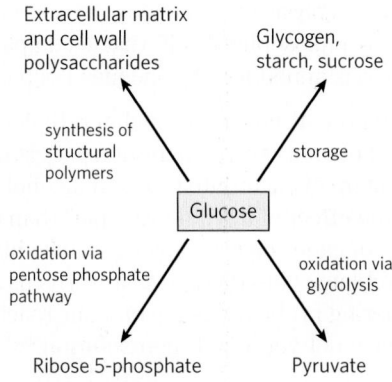

FIGURE 14-1 Major pathways of glucose utilization. Although not the only possible fates for glucose, these four pathways are the most significant in terms of the amount of glucose that flows through them in most cells.

make glucose by first reducing atmospheric CO_2 to trioses, then converting the trioses to glucose. Nonphotosynthetic cells make glucose from simpler three- and four-carbon precursors by the process of gluconeogenesis, effectively reversing glycolysis in a pathway that uses many of the glycolytic enzymes.

In this chapter we describe the individual reactions of glycolysis, gluconeogenesis, and the pentose phosphate pathway and the functional significance of each pathway. We also describe the various metabolic fates of the pyruvate produced by glycolysis. They include the fermentations that are used by many organisms in anaerobic niches to produce ATP and that are exploited industrially as sources of ethanol, lactic acid, and other commercially useful products. And we look at the pathways that feed various sugars from mono-, di-, and polysaccharides into the glycolytic pathway. The discussion of glucose metabolism continues in Chapter 15, where we use the processes of carbohydrate synthesis and degradation to illustrate the many mechanisms by which organisms regulate metabolic pathways. The biosynthetic pathways from glucose to extracellular matrix and cell wall polysaccharides and storage polysaccharides are discussed in Chapter 20.

14.1 Glycolysis

In **glycolysis** (from the Greek *glykys,* "sweet" or "sugar," and *lysis,* "splitting"), a molecule of glucose is degraded in a series of enzyme-catalyzed reactions to yield two molecules of the three-carbon compound pyruvate. During the sequential reactions of glycolysis, some of the free energy released from glucose is conserved in the form of ATP and NADH. Glycolysis was the first metabolic pathway to be elucidated and is probably the best understood. From Eduard Buchner's discovery in 1897 of fermentation in cell-free extracts of yeast until the elucidation of the whole pathway in yeast (by Otto Warburg and Hans von Euler-Chelpin) and in muscle (by Gustav Embden and Otto Meyerhof) in the 1930s, the reactions of glycolysis were a major focus of biochemical research. The philosophical shift that accompanied these discoveries was announced by Jacques Loeb in 1906:

> Through the discovery of Buchner, Biology was relieved of another fragment of mysticism. The splitting up of sugar into CO_2 and alcohol is no more the effect of a "vital principle" than the splitting up of cane sugar by invertase. The history of this problem is instructive, as it warns us against considering problems as beyond our reach because they have not yet found their solution.*

The development of methods of enzyme purification, the discovery and recognition of the importance of coenzymes such as NAD, and the discovery of the

Hans Von Euler-Chelpin, 1873–1964 [Source: Austrian Archives/Corbis.]

Gustav Embden, 1874–1933

Otto Meyerhof, 1884–1951 [Source: Science Source.]

pivotal metabolic role of ATP and other phosphorylated compounds all came out of studies of glycolysis. The glycolytic enzymes of many species have long since been purified and thoroughly studied.

Glycolysis is an almost universal central pathway of glucose catabolism, the pathway with the largest flux of carbon in most cells. The glycolytic breakdown of glucose is the sole source of metabolic energy in some mammalian tissues and cell types (erythrocytes, renal medulla, brain, and sperm, for example). Some plant tissues that are modified to store starch (such as potato tubers) and some aquatic plants (watercress, for example) derive most of their energy from glycolysis; many anaerobic microorganisms are entirely dependent on glycolysis.

Fermentation is a general term for the *anaerobic* degradation of glucose or other organic nutrients to obtain energy, conserved as ATP. Because living organisms first arose in an atmosphere without oxygen, anaerobic breakdown of glucose is probably the most ancient biological mechanism for obtaining energy from organic fuel molecules. And as genome sequencing of a wide variety of organisms has revealed, some archaea and some parasitic microorganisms lack one or more of the enzymes of glycolysis but retain the core of the pathway; they presumably carry out variant forms of glycolysis. In the course of evolution, the chemistry of this reaction sequence has been completely conserved; the glycolytic enzymes of vertebrates are closely similar, in amino acid sequence and three-dimensional structure, to their homologs in yeast and spinach. Glycolysis differs among species only in the details of its regulation and in the subsequent metabolic fate of the pyruvate formed. The thermodynamic principles and the types of regulatory mechanisms that govern glycolysis are common to all pathways of cell metabolism. The glycolytic pathway, of central importance in itself, can also serve as a model for many aspects of the pathways discussed throughout this book.

Before examining each step of the pathway in some detail, we take a look at glycolysis as a whole.

An Overview: Glycolysis Has Two Phases

The breakdown of the six-carbon glucose into two molecules of the three-carbon pyruvate occurs in 10 steps, the first 5 of which constitute the *preparatory phase* (**Fig. 14-2a**). In these reactions, glucose is first

*From J. Loeb, *The Dynamics of Living Matter*, Columbia University Press, New York, 1906.

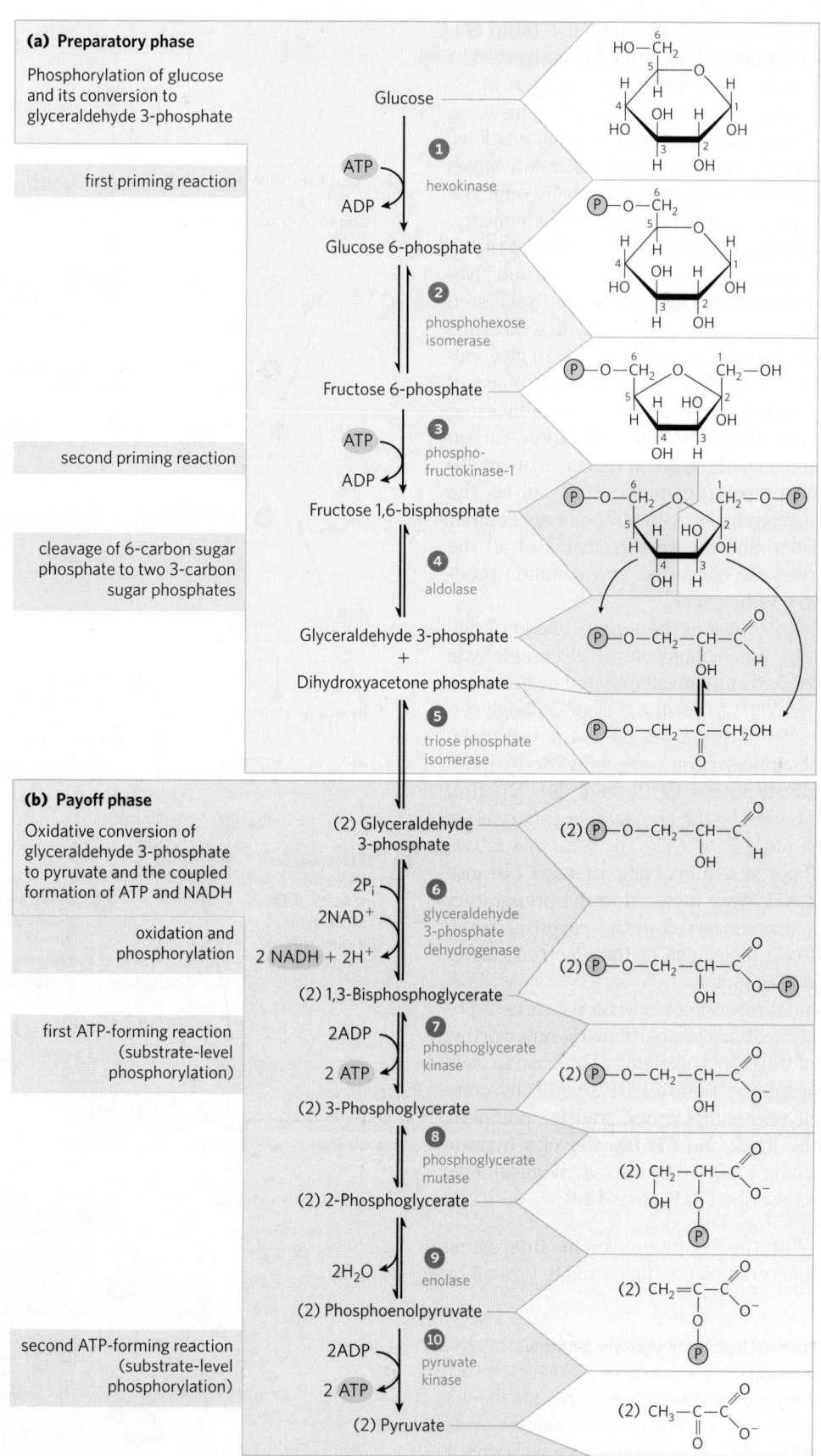

FIGURE 14-2 The two phases of glycolysis. For each molecule of glucose that passes through the preparatory phase **(a)**, two molecules of glyceraldehyde 3-phosphate are formed; both pass through the payoff phase **(b)**. Pyruvate is the end product of the second phase of glycolysis. For each glucose molecule, two ATP are consumed in the preparatory phase and four ATP are produced in the payoff phase, giving a net yield of two ATP per molecule of glucose converted to pyruvate. The numbered reaction steps correspond to the numbered headings in the text discussion. Keep in mind that each phosphoryl group, represented here as ⓟ, has two negative charges ($-PO_3^{2-}$).

phosphorylated at the hydroxyl group on C-6 (step **1**). The D-glucose 6-phosphate thus formed is converted to D-fructose 6-phosphate (step **2**), which is again phosphorylated, this time at C-1, to yield D-fructose 1,6-bisphosphate (step **3**). For both phosphorylations, ATP is the phosphoryl group donor. As all sugar derivatives in glycolysis are the D isomers, we will usually omit the D designation except when emphasizing stereochemistry.

Fructose 1,6-bisphosphate is split to yield two three-carbon molecules, dihydroxyacetone phosphate and glyceraldehyde 3-phosphate (step **4**); this is the "lysis" step that gives the pathway its name. The dihydroxyacetone phosphate is isomerized to a second molecule of glyceraldehyde 3-phosphate (step **5**), ending the first phase of glycolysis. Note that two molecules of ATP are invested before the cleavage of glucose into two three-carbon pieces; there will be a good return on this investment. To summarize: in the preparatory phase of glycolysis the energy of ATP is invested, raising the free-energy content of the intermediates, and the carbon chains of all the metabolized hexoses are converted to a common product, glyceraldehyde 3-phosphate.

The energy gain comes in the *payoff phase* of glycolysis (Fig. 14-2b). Each molecule of glyceraldehyde 3-phosphate is oxidized and phosphorylated by inorganic phosphate (*not* by ATP) to form 1,3-bisphosphoglycerate (step **6**). Energy is then released as the two molecules of 1,3-bisphosphoglycerate are converted to two molecules of pyruvate (steps **7** through **10**). Much of this energy is conserved by the coupled phosphorylation of four molecules of ADP to ATP. The net yield is two molecules of ATP per molecule of glucose used, because two molecules of ATP were invested in the preparatory phase. Energy is also conserved in the payoff phase in the formation of two molecules of the electron carrier NADH per molecule of glucose.

In the sequential reactions of glycolysis, three types of chemical transformations are particularly noteworthy: (1) degradation of the carbon skeleton of glucose to yield pyruvate; (2) phosphorylation of ADP to ATP by compounds with high phosphoryl group transfer potential, formed during glycolysis; and (3) transfer of a hydride ion to NAD^+, forming NADH. The overall chemical logic of the pathway is described in **Figure 14-3**.

Fates of Pyruvate With the exception of some interesting variations in the bacterial realm, the pyruvate formed by

FIGURE 14-3 The chemical logic of the glycolytic pathway. In this simplified version of the pathway, each molecule is shown in a linear form, with carbon and hydrogen atoms not depicted, in order to highlight chemical transformations. Remember that glucose and fructose are present mostly in their cyclized forms in solution, although they are transiently present in linear form at the active sites of some of the enzymes in this pathway.

The preparatory phase, steps **1** to **5**, converts the six-carbon glucose into two three-carbon units, each of them phosphorylated. Oxidation of the three-carbon units is initiated in the payoff phase. To produce pyruvate, the chemical steps must occur in the order shown.

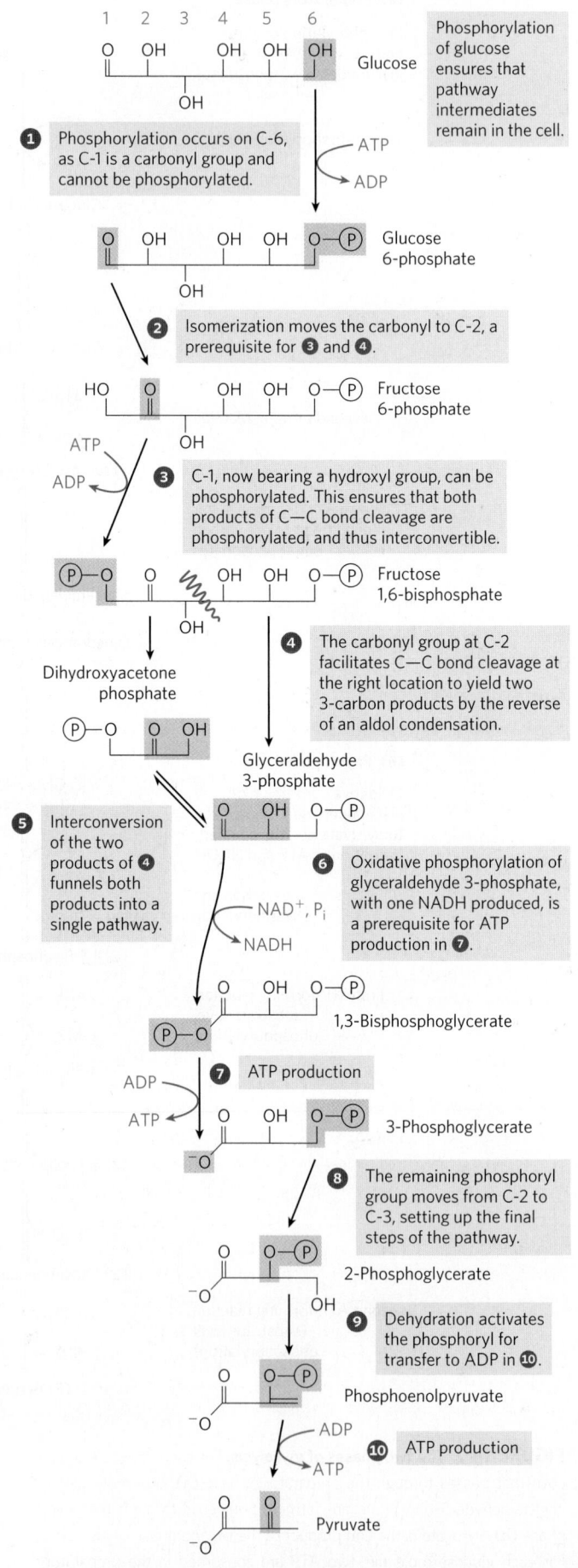

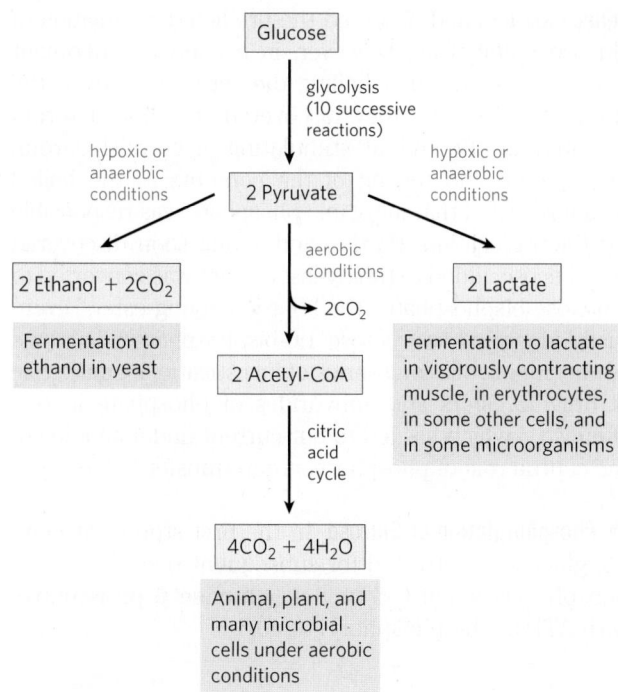

FIGURE 14-4 **Three possible catabolic fates of the pyruvate formed in glycolysis.** Pyruvate also serves as a precursor in many anabolic reactions, not shown here.

glycolysis is further metabolized via one of three catabolic routes. In aerobic organisms or tissues, under aerobic conditions, glycolysis is only the first stage in the complete degradation of glucose **(Fig. 14-4)**. Pyruvate is oxidized, with loss of its carboxyl group as CO_2, to yield the acetyl group of acetyl-coenzyme A; the acetyl group is then oxidized completely to CO_2 by the citric acid cycle (Chapter 16). The electrons from these oxidations are passed to O_2 through a chain of carriers in mitochondria, to form H_2O. The energy from the electron-transfer reactions drives the synthesis of ATP in mitochondria (Chapter 19).

The second route for pyruvate is its reduction to lactate via **lactic acid fermentation**. When vigorously contracting skeletal muscle must function under low-oxygen conditions (**hypoxia**), NADH cannot be reoxidized to NAD^+, but NAD^+ is required as an electron acceptor for the further oxidation of pyruvate. Under these conditions pyruvate is reduced to lactate, accepting electrons from NADH and thereby regenerating the NAD^+ necessary for glycolysis to continue. Certain tissues and cell types (retina and erythrocytes, for example) convert glucose to lactate even under aerobic conditions, and lactate is also the product of glycolysis under anaerobic conditions in some microorganisms (Fig. 14-4).

The third major route of pyruvate catabolism leads to ethanol. In some plant tissues and in certain invertebrates, protists, and microorganisms such as brewer's or baker's yeast, pyruvate is converted under hypoxic or anaerobic conditions to ethanol and CO_2,

a process called **ethanol (alcohol) fermentation** (Fig. 14-4).

The oxidation of pyruvate is an important catabolic process, but pyruvate has anabolic fates as well. It can, for example, provide the carbon skeleton for the synthesis of the amino acid alanine or for the synthesis of fatty acids. We return to these anabolic reactions of pyruvate in later chapters.

ATP and NADH Formation Coupled to Glycolysis During glycolysis some of the energy of the glucose molecule is conserved in ATP, while much remains in the product, pyruvate. The overall equation for glycolysis is

$$\text{Glucose} + 2NAD^+ + 2ADP + 2P_i \longrightarrow$$
$$2 \text{ pyruvate} + 2NADH + 2H^+ + 2ATP + 2H_2O \quad (14\text{-}1)$$

For each molecule of glucose degraded to pyruvate, two molecules of ATP are generated from ADP and P_i, and two molecules of NADH are produced by the reduction of NAD^+. The hydrogen acceptor in this reaction is NAD^+ (see Fig. 13-24), bound to a Rossmann fold as shown in Figure 13-25. The reduction of NAD^+ proceeds by the enzymatic transfer of a hydride ion ($:H^-$) from the aldehyde group of glyceraldehyde 3-phosphate to the nicotinamide ring of NAD^+, yielding the reduced coenzyme NADH. The other hydrogen atom of the substrate molecule is released to the solution as H^+.

We can now resolve the equation of glycolysis into two processes—the conversion of glucose to pyruvate, which is exergonic:

$$\text{Glucose} + 2NAD^+ \longrightarrow 2 \text{ pyruvate} + 2NADH + 2H^+ \quad (14\text{-}2)$$
$$\Delta G_1'^\circ = -146 \text{ kJ/mol}$$

and the formation of ATP from ADP and P_i, which is endergonic:

$$2ADP + 2P_i \longrightarrow 2ATP + 2H_2O \quad (14\text{-}3)$$
$$\Delta G_2'^\circ = 2(30.5 \text{ kJ/mol}) = 61.0 \text{ kJ/mol}$$

The sum of Equations 14-2 and 14-3 gives the overall standard free-energy change of glycolysis, $\Delta G_{Sum}'^\circ$:

$$\Delta G_{Sum}'^\circ = \Delta G_1'^\circ + \Delta G_2'^\circ = -146 \text{ kJ/mol} + 61.0 \text{ kJ/mol}$$
$$= -85 \text{ kJ/mol}$$

Under standard conditions, and under the (nonstandard) conditions that prevail in a cell, glycolysis is an essentially irreversible process, driven to completion by a large net decrease in free energy.

Energy Remaining in Pyruvate Glycolysis releases only a small fraction of the total available energy of the glucose molecule; the two molecules of pyruvate formed by glycolysis still contain most of the chemical potential energy of glucose, energy that can be extracted by oxidative reactions in the citric acid cycle (Chapter 16) and oxidative phosphorylation (Chapter 19).

Importance of Phosphorylated Intermediates Each of the nine glycolytic intermediates between glucose and pyruvate is phosphorylated (Fig. 14-2). The phosphoryl groups seem to have three functions.

1. Because the plasma membrane generally lacks transporters for phosphorylated sugars, the phosphorylated glycolytic intermediates cannot leave the cell. After the initial phosphorylation, no further energy is necessary to retain phosphorylated intermediates in the cell, despite the large difference in their intracellular and extracellular concentrations.

2. Phosphoryl groups are essential components in the enzymatic conservation of metabolic energy. Energy released in the breakage of phosphoanhydride bonds (such as those in ATP) is partially conserved in the formation of phosphate esters such as glucose 6-phosphate. High-energy phosphate compounds formed in glycolysis (1,3-bisphosphoglycerate and phosphoenolpyruvate) donate phosphoryl groups to ADP to form ATP.

3. Binding energy resulting from the binding of phosphate groups to the active sites of enzymes lowers the activation energy and increases the specificity of the enzymatic reactions (Chapter 6). The phosphate groups of ADP, ATP, and the glycolytic intermediates form complexes with Mg^{2+}, and the substrate binding sites of many glycolytic enzymes are specific for these Mg^{2+} complexes. Most glycolytic enzymes require Mg^{2+} for activity.

The Preparatory Phase of Glycolysis Requires ATP

In the preparatory phase of glycolysis, two molecules of ATP are invested and the hexose chain is cleaved into two triose phosphates. The realization that *phosphorylated* hexoses were intermediates in glycolysis came slowly and serendipitously. In 1906, Arthur Harden and William Young tested their hypothesis that inhibitors of proteolytic enzymes would stabilize the glucose-fermenting enzymes in yeast extract. They added blood serum (known to contain inhibitors of proteolytic enzymes) to

Arthur Harden, 1865–1940
[Source: Mary Evans Picture Library/Alamy.]

William Young, 1878–1942
[Source: Courtesy Medical History Museum, The University of Melbourne.]

yeast extracts and observed the predicted stimulation of glucose metabolism. However, in a control experiment intended to show that boiling the serum destroyed the stimulatory activity, they discovered that boiled serum was just as effective at stimulating glycolysis! Careful examination and testing of the contents of the boiled serum revealed that inorganic phosphate was responsible for the stimulation. Harden and Young soon discovered that glucose added to their yeast extract was converted to a hexose bisphosphate (the "Harden-Young ester," eventually identified as fructose 1,6-bisphosphate). This was the beginning of a long series of investigations on the role of organic esters and anhydrides of phosphate in biochemistry, which has led to our current understanding of the central role of phosphoryl group transfer in biology.

❶ Phosphorylation of Glucose In the first step of glycolysis, glucose is activated for subsequent reactions by its phosphorylation at C-6 to yield **glucose 6-phosphate**, with ATP as the phosphoryl donor:

$$\Delta G'^{\circ} = -16.7 \text{ kJ/mol}$$

This reaction, which is irreversible under intracellular conditions, is catalyzed by **hexokinase**. Recall that kinases are enzymes that catalyze the transfer of the terminal phosphoryl group from ATP to an acceptor nucleophile (see Fig. 13-20). Kinases are a subclass of transferases (see Table 6-3). The acceptor in the case of hexokinase is a hexose, normally D-glucose, although hexokinase also catalyzes the phosphorylation of other common hexoses, such as D-fructose and D-mannose, in some tissues.

Hexokinase, like many other kinases, requires Mg^{2+} for its activity, because the true substrate of the enzyme is not ATP^{4-} but the $MgATP^{2-}$ complex (see Fig. 13-12). Mg^{2+} shields the negative charges of the phosphoryl groups in ATP, making the terminal phosphorus atom an easier target for nucleophilic attack by an —OH of glucose. Hexokinase undergoes a profound change in shape, an induced fit, when it binds glucose; two domains of the protein move about 8 Å closer to each other when ATP binds (see Fig. 6-26). This movement brings bound ATP closer to a molecule of glucose also bound to the enzyme and blocks the access of water (from the solvent), which might otherwise enter the active site and attack (hydrolyze) the phosphoanhydride bonds of ATP. Like the other nine enzymes of glycolysis, hexokinase is a soluble, cytosolic protein.

Hexokinase is present in nearly all organisms. The human genome encodes four different hexokinases (I to IV), all of which catalyze the same reaction. Two or more enzymes that catalyze the same reaction but are encoded by different genes are called **isozymes** (see Box 15-2). One of the isozymes present in hepatocytes, hexokinase IV (also called glucokinase), differs from other forms of hexokinase in kinetic and regulatory properties, with important physiological consequences that are described in Section 15.3.

❷ Conversion of Glucose 6-Phosphate to Fructose 6-Phosphate The enzyme **phosphohexose isomerase (phosphoglucose isomerase)** catalyzes the reversible isomerization of glucose 6-phosphate, an aldose, to **fructose 6-phosphate**, a ketose:

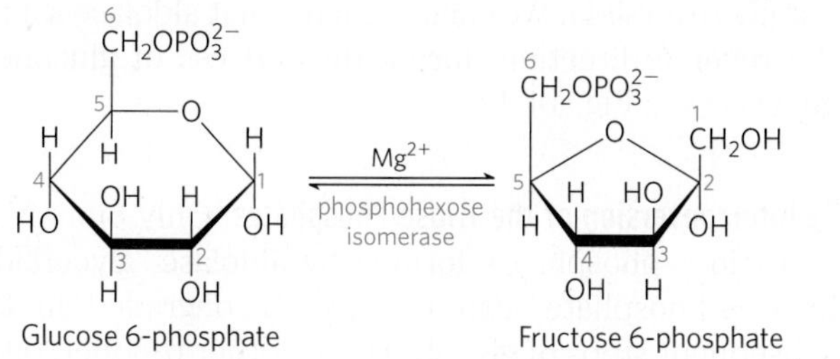

Glucose 6-phosphate   Fructose 6-phosphate

$\Delta G'^{\circ} = 1.7$ kJ/mol

The mechanism for this reaction involves an enediol intermediate (**Fig. 14-5**). The reaction proceeds readily in either direction, as might be expected from the relatively small change in standard free energy.

❸ Phosphorylation of Fructose 6-Phosphate to Fructose 1,6-Bisphosphate In the second of the two priming reactions of glycolysis, **phosphofructokinase-1 (PFK-1)** catalyzes the transfer of a phosphoryl group from ATP to fructose 6-phosphate to yield **fructose 1,6-bisphosphate**:

Fructose 6-phosphate

Fructose 1,6-bisphosphate

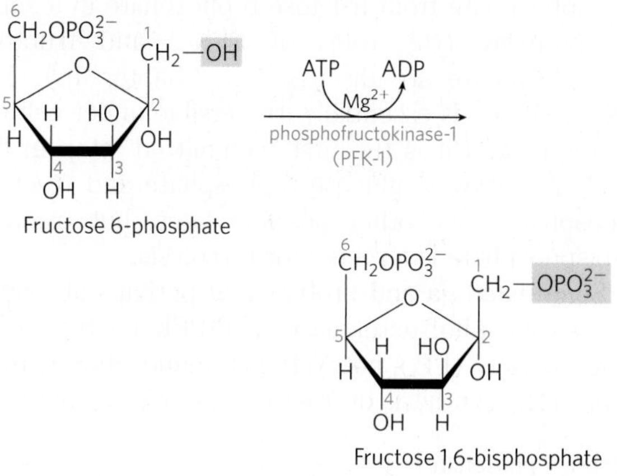

$\Delta G'^{\circ} = -14.2$ kJ/mol

》》 Key Convention: Compounds that contain two phosphate or phosphoryl groups attached at different positions in the molecule are named *bisphosphates* (or *bisphospho* compounds); for example, fructose 1,6-bisphosphate and 1,3-bisphosphoglycerate. Compounds with two phosphates linked together as a pyrophosphoryl group are named *diphosphates;* for example, adenosine diphosphate (ADP). Similar rules apply for the naming of *tris-phosphates* (such as inositol 1,4,5-trisphosphate; see p. 451) and *triphosphates* (such as adenosine triphosphate, ATP). 《

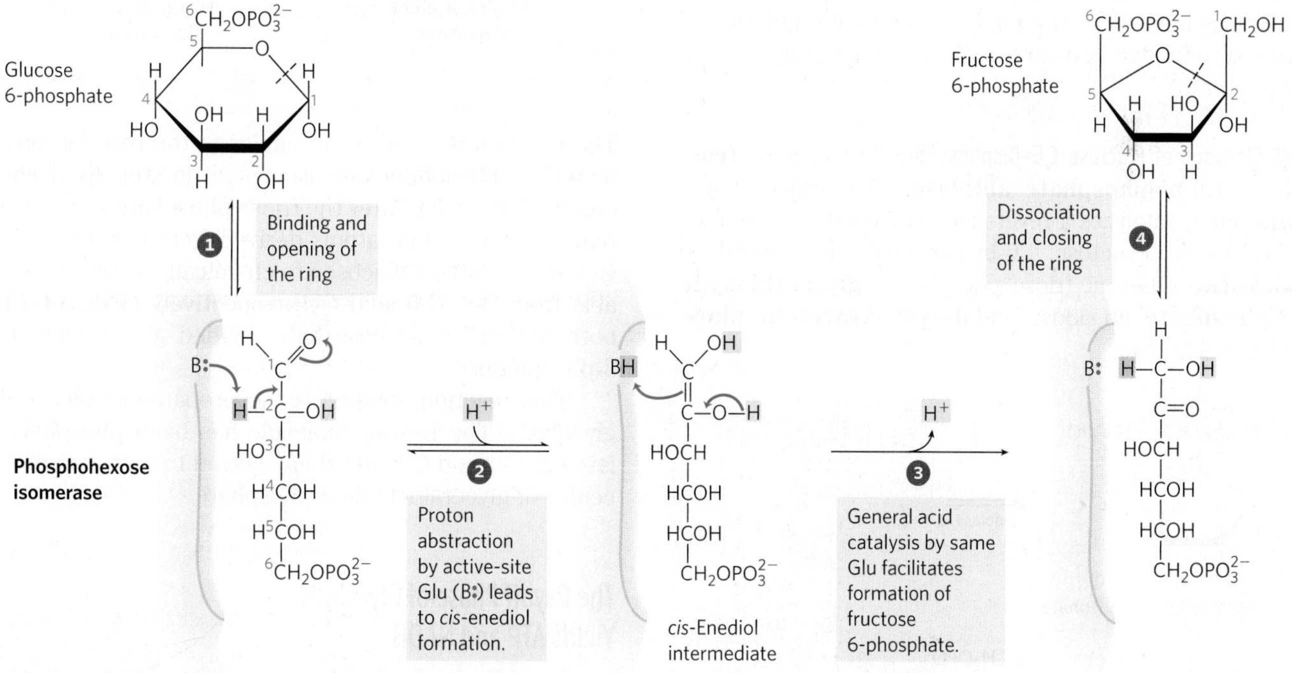

Phosphohexose isomerase

Binding and opening of the ring

Proton abstraction by active-site Glu (B:) leads to *cis*-enediol formation.

cis-Enediol intermediate

General acid catalysis by same Glu facilitates formation of fructose 6-phosphate.

Dissociation and closing of the ring

MECHANISM FIGURE 14-5 The phosphohexose isomerase reaction. The ring opening and closing reactions (steps ❶ and ❹) are catalyzed by an active-site His residue, by mechanisms omitted here for simplicity. The proton (light red) initially at C-2 is made more easily abstractable by electron withdrawal by the adjacent carbonyl and nearby hydroxyl groups.

After its transfer from C-2 to the active-site Glu residue (a weak acid), the proton is freely exchanged with the surrounding solution; that is, the proton abstracted from C-2 in step ❷ is not necessarily the same one that is added to C-1 in step ❸.

The enzyme that forms fructose 1,6-bisphosphate is called PFK-1 to distinguish it from a second enzyme (PFK-2) that catalyzes the formation of fructose 2,6-bisphosphate from fructose 6-phosphate in a separate pathway (the roles of PFK-2 and fructose 2,6-bisphosphate are discussed in Chapter 15). The PFK-1 reaction is essentially irreversible under cellular conditions, and it is the first "committed" step in the glycolytic pathway; glucose 6-phosphate and fructose 6-phosphate have other possible fates, but fructose 1,6-bisphosphate is targeted for glycolysis.

Some bacteria and protists and perhaps all plants have a phosphofructokinase (PP-PFK-1) that uses pyrophosphate (PP_i), not ATP, as the phosphoryl group donor in the synthesis of fructose 1,6-bisphosphate:

$$\text{Fructose 6-phosphate} + PP_i \xrightarrow{Mg^{2+}}$$
$$\text{fructose 1,6-bisphosphate} + P_i$$
$$\Delta G'^\circ = -2.9 \text{ kJ/mol}$$

Phosphofructokinase-1 is subject to complex allosteric regulation; its activity is increased whenever the cell's ATP supply is depleted or when the ATP breakdown products, ADP and AMP (particularly the latter), accumulate. The enzyme is inhibited whenever the cell has ample ATP and is well supplied by other fuels such as fatty acids. In some organisms, fructose 2,6-bisphosphate (not to be confused with the PFK-1 reaction product, fructose 1,6-bisphosphate) is a potent allosteric activator of PFK-1. Ribulose 5-phosphate, an intermediate in the pentose phosphate pathway discussed later in this chapter, also activates phosphofructokinase indirectly. The multiple layers of regulation of this step in glycolysis are discussed in greater detail in Chapter 15.

④ Cleavage of Fructose 1,6-Bisphosphate The enzyme **fructose 1,6-bisphosphate aldolase**, often called simply **aldolase**, catalyzes a reversible aldol condensation (see Fig. 13-4). Fructose 1,6-bisphosphate is cleaved to yield two different triose phosphates, **glyceraldehyde 3-phosphate**, an aldose, and **dihydroxyacetone phosphate**, a ketose:

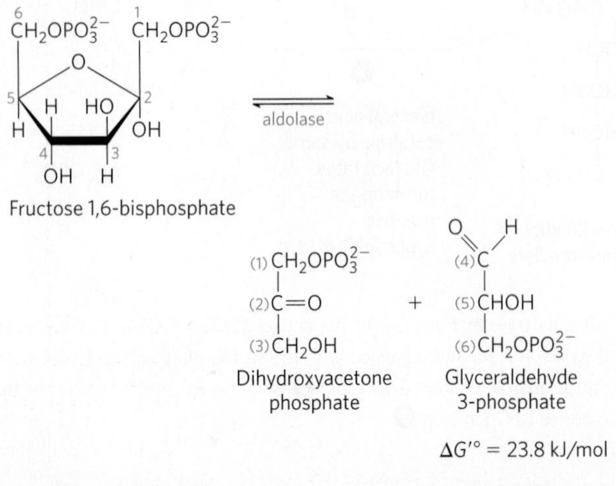

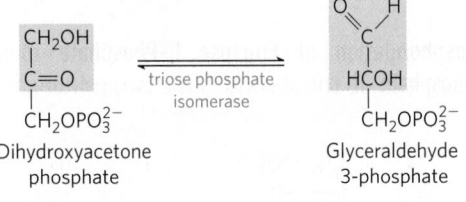

Fructose 1,6-bisphosphate

Dihydroxyacetone phosphate   Glyceraldehyde 3-phosphate

$$\Delta G'^\circ = 23.8 \text{ kJ/mol}$$

There are two classes of aldolases. Class I aldolases, found in animals and plants, use the mechanism shown in **Figure 14-6**. Class II enzymes, in fungi and bacteria, do not form the Schiff base intermediate. Instead, a zinc ion at the active site is coordinated with the carbonyl oxygen at C-2; the Zn^{2+} polarizes the carbonyl group and stabilizes the enolate intermediate created in the C—C bond cleavage step (see Fig. 6-19).

Although the aldolase reaction has a strongly positive standard free-energy change in the direction of fructose 1,6-bisphosphate cleavage, at the lower concentrations of reactants present in cells the actual free-energy change is small and the aldolase reaction is readily reversible. We shall see later that aldolase acts in the reverse direction during the process of gluconeogenesis (see Fig. 14-17).

⑤ Interconversion of the Triose Phosphates Only one of the two triose phosphates formed by aldolase, glyceraldehyde 3-phosphate, can be directly degraded in the subsequent steps of glycolysis. The other product, dihydroxyacetone phosphate, is rapidly and reversibly converted to glyceraldehyde 3-phosphate by the fifth enzyme of the glycolytic sequence, **triose phosphate isomerase**:

$$\Delta G'^\circ = 7.5 \text{ kJ/mol}$$

The reaction mechanism is similar to the reaction promoted by phosphohexose isomerase in step ② of glycolysis (Fig. 14-5). After the triose phosphate isomerase reaction, the carbon atoms derived from C-1, C-2, and C-3 of the starting glucose are chemically indistinguishable from C-6, C-5, and C-4, respectively **(Fig. 14-7)**; both "halves" of glucose have yielded glyceraldehyde 3-phosphate.

This reaction completes the preparatory phase of glycolysis. The hexose molecule has been phosphorylated at C-1 and C-6 and then cleaved to form two molecules of glyceraldehyde 3-phosphate.

The Payoff Phase of Glycolysis Yields ATP and NADH

The payoff phase of glycolysis (Fig. 14-2b) includes the energy-conserving phosphorylation steps in which some of the chemical energy of the glucose molecule is conserved in the form of ATP and NADH. Remember that one molecule of glucose yields two molecules of glyceraldehyde 3-phosphate, and both halves of the

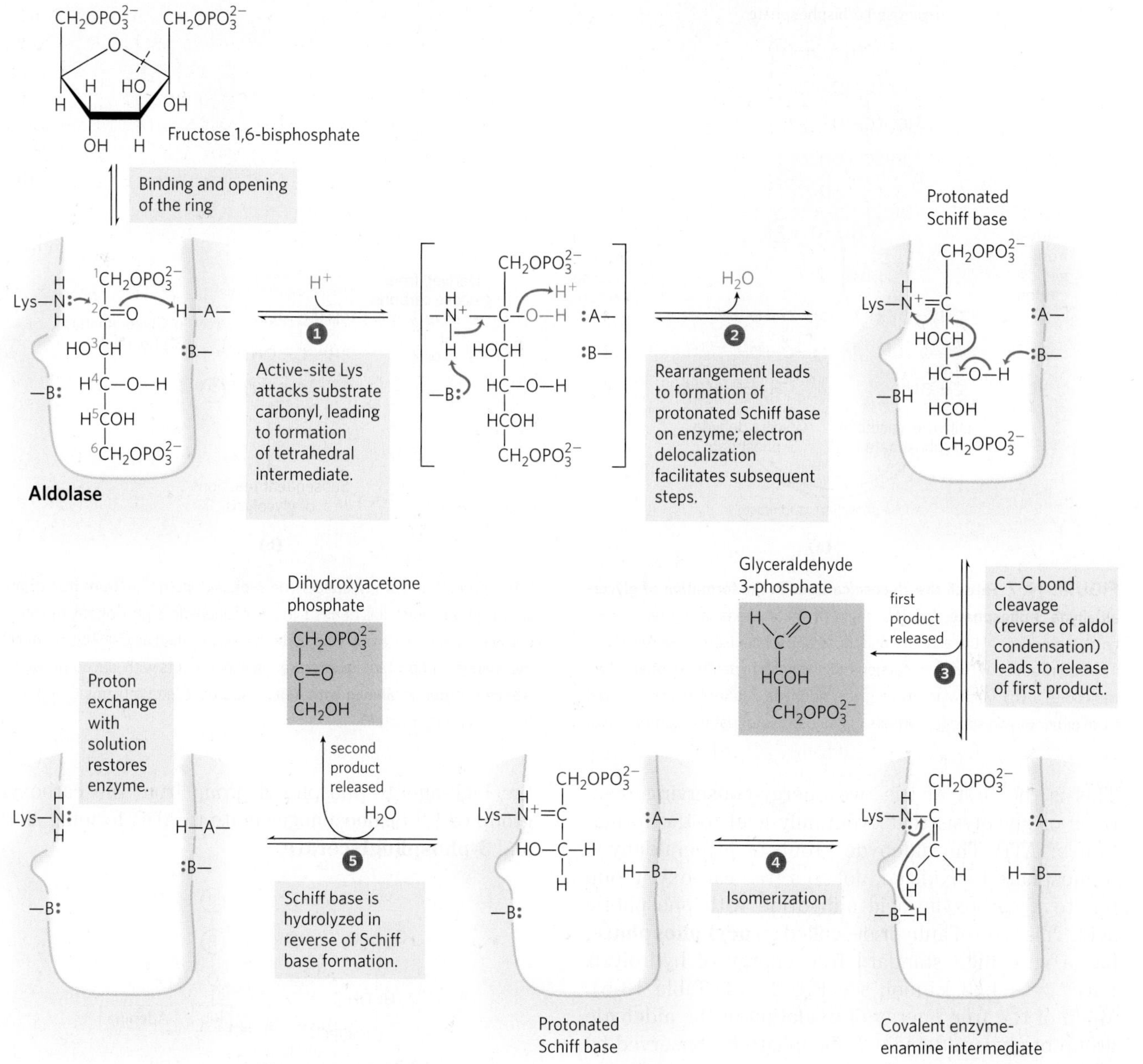

MECHANISM FIGURE 14-6 **The class I aldolase reaction.** The reaction shown here is the reverse of an aldol condensation. Note that cleavage between C-3 and C-4 depends on the presence of the carbonyl group at C-2, which is converted to an imine on the enzyme. A and B represent amino acid residues that serve as general acid (A) or base (B).

glucose molecule follow the same pathway in the second phase of glycolysis. The conversion of two molecules of glyceraldehyde 3-phosphate to two molecules of pyruvate is accompanied by the formation of four molecules of ATP from ADP. However, the net yield of ATP per molecule of glucose degraded is only two, because two ATP were invested in the preparatory phase of glycolysis to phosphorylate the two ends of the hexose molecule.

⑥ Oxidation of Glyceraldehyde 3-Phosphate to 1,3-Bisphosphoglycerate The first step in the payoff phase is the oxidation of glyceraldehyde 3-phosphate to **1,3-bisphosphoglycerate**, catalyzed by **glyceraldehyde 3-phosphate dehydrogenase**:

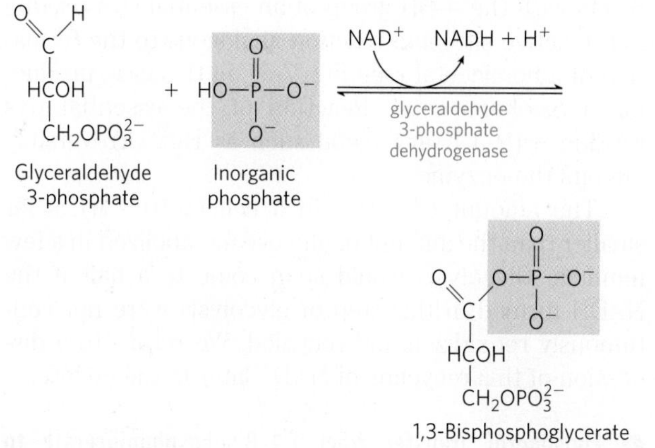

$\Delta G'^\circ = 6.3$ kJ/mol

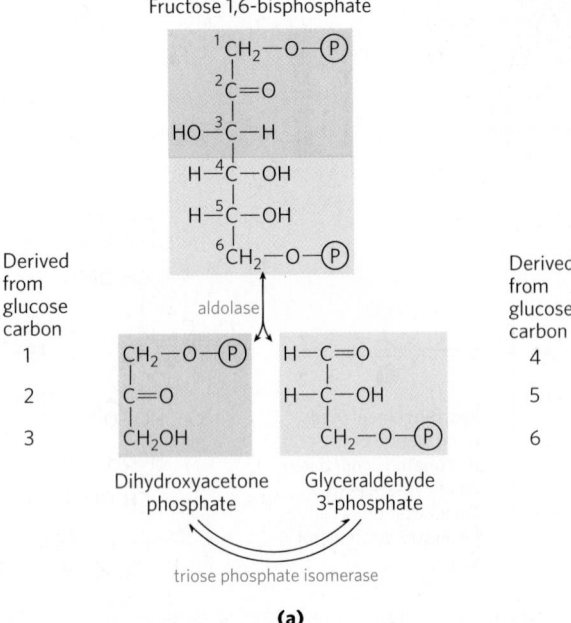

FIGURE 14-7 Fate of the glucose carbons in the formation of glycer-aldehyde 3-phosphate. (a) The origin of the carbons in the two three-carbon products of the aldolase and triose phosphate isomerase reactions. The end product of the two reactions is glyceraldehyde 3-phosphate (two molecules). (b) Each carbon of glyceraldehyde 3-phosphate is derived from either of two specific carbons of glucose. Note that the numbering of the carbon atoms of glyceraldehyde 3-phosphate differs from that of the glucose from which it is derived. In glyceraldehyde 3-phosphate, the most complex functional group (the carbonyl) is specified as C-1. This numbering change is important for interpreting experiments with glucose in which a single carbon is labeled with a radioisotope. (See Problems 6 and 9 at the end of this chapter.)

This is the first of the two energy-conserving reactions of glycolysis that eventually lead to the formation of ATP. The aldehyde group of glyceraldehyde 3-phosphate is oxidized, not to a free carboxyl group but to a carboxylic acid anhydride with phosphoric acid. This type of anhydride, called an **acyl phosphate**, has a very high standard free energy of hydrolysis ($\Delta G'^\circ = -49.3$ kJ/mol; see Fig. 13-14, Table 13-6). Much of the free energy of oxidation of the aldehyde group of glyceraldehyde 3-phosphate is conserved by formation of the acyl phosphate group at C-1 of 1,3-bisphosphoglycerate.

Glyceraldehyde 3-phosphate is covalently bound to the dehydrogenase during the reaction **(Fig. 14-8)**. The aldehyde group of glyceraldehyde 3-phosphate reacts with the —SH group of an essential Cys residue in the active site, in a reaction analogous to the formation of a hemiacetal (see Fig. 7-5), in this case producing a *thio*hemiacetal. Reaction of the essential Cys residue with a heavy metal such as Hg^{2+} irreversibly inhibits the enzyme.

The amount of NAD^+ in a cell ($\leq 10^{-5}$ M) is far smaller than the amount of glucose metabolized in a few minutes. Glycolysis would soon come to a halt if the NADH formed in this step of glycolysis were not continuously reoxidized and recycled. We return to a discussion of this recycling of NAD^+ later in the chapter.

❼ **Phosphoryl Transfer from 1,3-Bisphosphoglycerate to ADP** The enzyme **phosphoglycerate kinase** transfers

the high-energy phosphoryl group from the carboxyl group of 1,3-bisphosphoglycerate to ADP, forming ATP and **3-phosphoglycerate**:

$$\Delta G'^\circ = -18.8 \text{ kJ/mol}$$

Notice that phosphoglycerate kinase is named for the reverse reaction, in which it transfers a phosphoryl

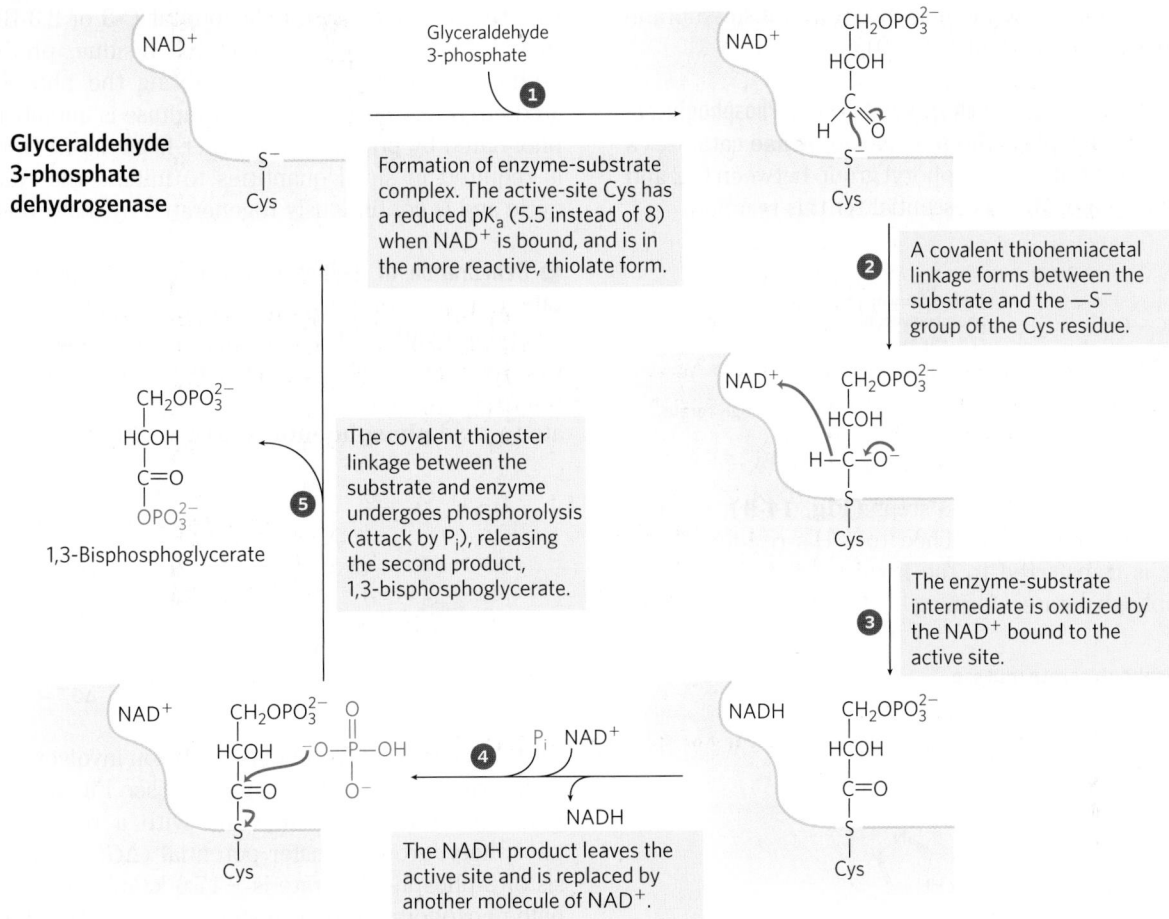

MECHANISM FIGURE 14-8 **The glyceraldehyde 3-phosphate dehydrogenase reaction.**

group from ATP to 3-phosphoglycerate. Like all enzymes, it catalyzes the reaction in both directions. This enzyme acts in the direction suggested by its name during gluconeogenesis (see Fig. 14-17) and during photosynthetic CO_2 assimilation (see Fig. 20-31). In glycolysis, the reaction it catalyzes proceeds as shown above, in the direction of ATP synthesis.

Steps ❻ and ❼ of glycolysis together constitute an energy-coupling process in which 1,3-bisphosphoglycerate is the common intermediate; it is formed in the first reaction (which would be endergonic in isolation), and its acyl phosphate group is transferred to ADP in the second reaction (which is strongly exergonic). The sum of these two reactions is

Glyceraldehyde 3-phosphate + ADP + P_i + NAD^+ $\rightleftharpoons$
$$3\text{-phosphoglycerate} + ATP + NADH + H^+$$
$$\Delta G'^\circ = -12.2 \text{ kJ/mol}$$

Thus the overall reaction is exergonic.

Recall from Chapter 13 that the actual free-energy change, ΔG, is determined by the standard free-energy change, $\Delta G'^\circ$, and the mass-action ratio, Q, which is the ratio [products]/[reactants] (see Eqn 13-4). For step ❻

$$\Delta G = \Delta G'^\circ + RT \ln Q$$

$$= \Delta G'^\circ + RT \ln \frac{[\text{1,3-bisphosphoglycerate}][\text{NADH}]}{[\text{glyceraldehyde 3-phosphate}][P_i][\text{NAD}^+]}$$

Notice that $[H^+]$ is not included in Q. In biochemical calculations, $[H^+]$ is assumed to be a constant (10^{-7} M), and this constant is included in the definition of $\Delta G'^\circ$ (p. 498).

When the mass-action ratio is less than 1.0, its natural logarithm has a negative sign. In the cytosol, where these reactions are taking place, the ratio [NADH]/[NAD$^+$] is a small fraction, contributing to a low Q. Step ❼, by consuming the product of step ❻ (1,3-bisphosphoglycerate), keeps [1,3-bisphosphoglycerate] relatively low in the steady state and thereby keeps Q for the overall energy-coupling process small. When Q is small, the contribution of $\ln Q$ can make ΔG strongly negative. This is simply another way of showing how the two reactions, steps ❻ and ❼, are coupled through a common intermediate.

The outcome of these coupled reactions, both reversible under cellular conditions, is that the energy released on oxidation of an aldehyde to a carboxylate group is conserved by the coupled formation of ATP from ADP and P_i. The formation of ATP by phosphoryl group transfer from a substrate such as 1,3-bisphosphoglycerate is referred to as a **substrate-level phosphorylation**, to distinguish this mechanism from **respiration-linked phosphorylation**. Substrate-level phosphorylations involve soluble enzymes and chemical intermediates (1,3-bisphosphoglycerate in this case). Respiration-linked phosphorylations, on the other hand, involve membrane-bound enzymes and transmembrane gradients of protons (Chapter 19).

❽ Conversion of 3-Phosphoglycerate to 2-Phosphoglycerate The enzyme **phosphoglycerate mutase** catalyzes a reversible shift of the phosphoryl group between C-2 and C-3 of glycerate; Mg^{2+} is essential for this reaction:

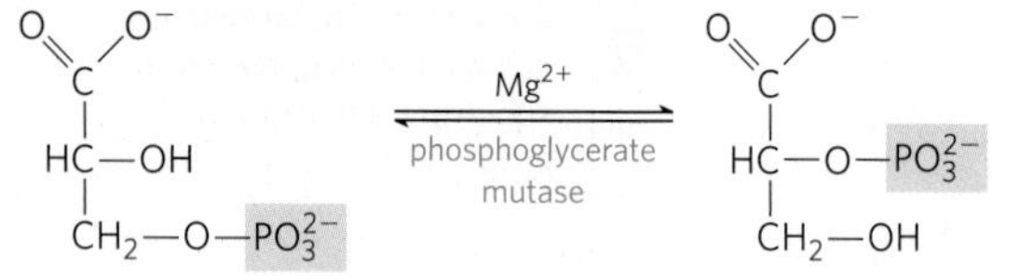

$$\Delta G'^{\circ} = 4.4 \text{ kJ/mol}$$

The reaction occurs in two steps **(Fig. 14-9)**. A phosphoryl group initially attached to a His residue of the mutase is transferred to the hydroxyl group at C-2 of 3-phosphoglycerate, forming 2,3-bisphosphoglycerate (2,3-BPG). The phosphoryl group at C-3 of 2,3-BPG is then transferred to the same His residue, producing 2-phosphoglycerate and regenerating the phosphorylated enzyme. Phosphoglycerate mutase is initially phosphorylated by phosphoryl transfer from 2,3-BPG, which is required in small quantities to initiate the catalytic cycle and is continuously regenerated by that cycle.

❾ Dehydration of 2-Phosphoglycerate to Phosphoenolpyruvate In the second glycolytic reaction that generates a compound with high phosphoryl group transfer potential (the first was step ❻), **enolase** promotes reversible removal of a molecule of water from 2-phosphoglycerate to yield **phosphoenolpyruvate (PEP)**:

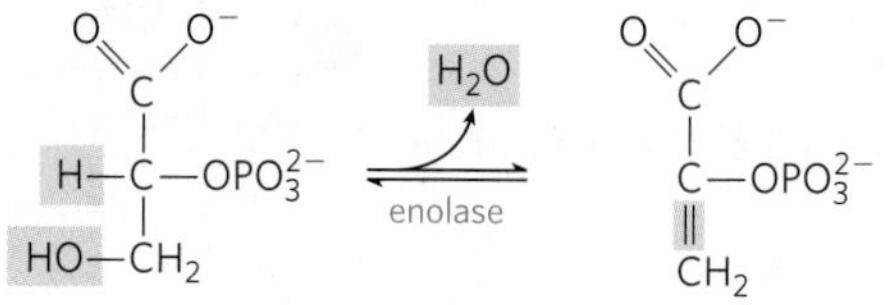

$$\Delta G'^{\circ} = 7.5 \text{ kJ/mol}$$

The mechanism of the enolase reaction involves an enolic intermediate stabilized by Mg^{2+} (see Fig. 6-27). The reaction converts a compound with a relatively low phosphoryl group transfer potential ($\Delta G'^{\circ}$ for hydrolysis of 2-phosphoglycerate is -17.6 kJ/mol) to one with high phosphoryl group transfer potential ($\Delta G'^{\circ}$ for PEP hydrolysis is -61.9 kJ/mol) (see Fig. 13-13, Table 13-6).

❿ Transfer of the Phosphoryl Group from Phosphoenolpyruvate to ADP The last step in glycolysis is the transfer of the phosphoryl group from phosphoenolpyruvate to ADP, catalyzed by **pyruvate kinase**, which requires K^+ and either Mg^{2+} or Mn^{2+}:

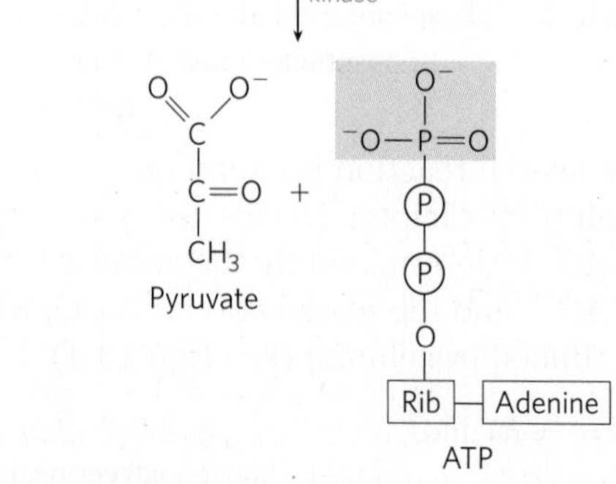

Phosphoglycerate mutase

① Phosphoryl transfer occurs between an active-site His and C-2 (OH) of the substrate. A second active-site His acts as general base catalyst.

② Phosphoryl transfer from C-3 of the substrate to the first active-site His. The second active-site His acts as general acid catalyst.

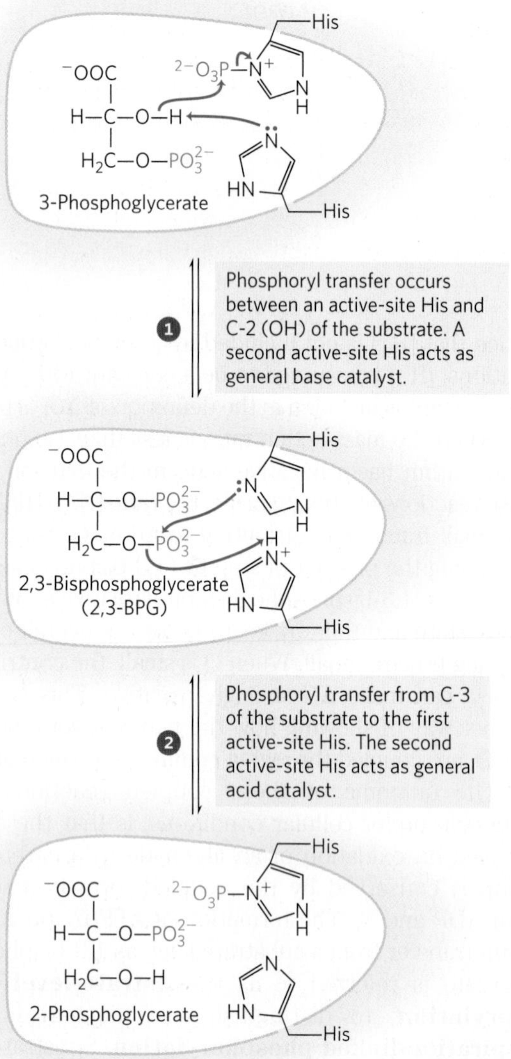

MECHANISM FIGURE 14-9 **The phosphoglycerate mutase reaction.**

$$\Delta G'^{\circ} = -31.4 \text{ kJ/mol}$$

In this substrate-level phosphorylation, the product **pyruvate** first appears in its enol form, then tautomerizes rapidly and nonenzymatically to its keto form, which predominates at pH 7:

[Chemical structures: Pyruvate (enol form) ⇌ Pyruvate (keto form), labeled "tautomerization"]

The overall reaction has a large, negative standard free-energy change, due in large part to the spontaneous conversion of the enol form of pyruvate to the keto form (see Fig. 13-13). About half of the energy released by PEP hydrolysis ($\Delta G'^{\circ} = -61.9$ kJ/mol) is conserved in the formation of the phosphoanhydride bond of ATP ($\Delta G'^{\circ} = -30.5$ kJ/mol), and the rest (-31.4 kJ/mol) constitutes a large driving force pushing the reaction toward ATP synthesis. We discuss the regulation of pyruvate kinase in Chapter 15.

The Overall Balance Sheet Shows a Net Gain of ATP

We can now construct a balance sheet for glycolysis to account for (1) the fate of the carbon skeleton of glucose, (2) the input of P_i and ADP and output of ATP, and (3) the pathway of electrons in the oxidation-reduction reactions. The left side of the following equation shows all the inputs of ATP, NAD^+, ADP, and P_i (consult Fig. 14-2), and the right side shows all the outputs (keep in mind that each molecule of glucose yields two molecules of pyruvate):

$$\text{Glucose} + 2\text{ATP} + 2\text{NAD}^+ + 4\text{ADP} + 2\text{P}_i \longrightarrow$$
$$2 \text{ pyruvate} + 2\text{ADP} + 2\text{NADH} + 2\text{H}^+ + 4\text{ATP} + 2\text{H}_2\text{O}$$

Canceling out common terms on both sides of the equation gives the overall equation for glycolysis under aerobic conditions:

$$\text{Glucose} + 2\text{NAD}^+ + 2\text{ADP} + 2\text{P}_i \longrightarrow$$
$$2 \text{ pyruvate} + 2\text{NADH} + 2\text{H}^+ + 2\text{ATP} + 2\text{H}_2\text{O}$$

The two molecules of NADH formed by glycolysis in the cytosol are, under aerobic conditions, reoxidized to NAD^+ by transfer of their electrons to the electron-transfer chain, which in eukaryotic cells is located in the mitochondria. The electron-transfer chain passes these electrons to their ultimate destination, O_2:

$$2\text{NADH} + 2\text{H}^+ + O_2 \longrightarrow 2\text{NAD}^+ + 2\text{H}_2\text{O}$$

Electron transfer from NADH to O_2 in mitochondria provides the energy for synthesis of ATP by respiration-linked phosphorylation (Chapter 19).

In the overall glycolytic process, one molecule of glucose is converted to two molecules of pyruvate (the pathway of carbon). Two molecules of ADP and two of P_i are converted to two molecules of ATP (the pathway of phosphoryl groups). Four electrons, as two hydride ions, are transferred from two molecules of glyceraldehyde 3-phosphate to two of NAD^+ (the pathway of electrons).

Glycolysis Is under Tight Regulation

During his studies on the fermentation of glucose by yeast, Louis Pasteur discovered that both the rate and the total amount of glucose consumption were many times greater under anaerobic than aerobic conditions. Later studies of muscle showed the same large difference in the rates of anaerobic and aerobic glycolysis. The biochemical basis of this "Pasteur effect" is now clear. The ATP yield from glycolysis under anaerobic conditions (2 ATP per molecule of glucose) is much smaller than that from the complete oxidation of glucose to CO_2 under aerobic conditions (30 or 32 ATP per glucose; see Table 19-5). About 15 times as much glucose must therefore be consumed anaerobically as aerobically to yield the same amount of ATP.

The flux of glucose through the glycolytic pathway is regulated to maintain nearly constant ATP levels (as well as adequate supplies of glycolytic intermediates that serve biosynthetic roles). The required adjustment in the rate of glycolysis is achieved by a complex interplay among ATP consumption, NADH regeneration, and allosteric regulation of several glycolytic enzymes—including hexokinase, PFK-1, and pyruvate kinase—and by second-to-second fluctuations in the concentration of key metabolites that reflect the cellular balance between ATP production and consumption. On a slightly longer time scale, glycolysis is regulated by the hormones glucagon, epinephrine, and insulin, and by changes in the expression of the genes for several glycolytic enzymes. An especially interesting case is the **aerobic glycolysis** in tumors. The German biochemist Otto Warburg first observed in 1928 that tumors of nearly all types carry out glycolysis at a much higher rate than normal tissue, *even when oxygen is available*. This "Warburg effect" is the basis for several methods of detecting and treating cancer (Box 14-1).

Warburg is generally considered the preeminent biochemist of the first half of the twentieth century. He made seminal contributions to many other areas of biochemistry, including respiration, photosynthesis, and the enzymology of intermediary metabolism. Beginning in 1930, Warburg and his associates purified and crystallized seven of the enzymes of glycolysis. They developed an experimental tool that revolutionized biochemical studies of oxidative metabolism: the Warburg manometer, which directly measured the oxygen consumption of tissues by monitoring changes in gas volume, and thus allowed quantitative measurement of any enzyme with oxidase activity.

Otto Warburg, 1883–1970 [Source: Science Photo Library/Science Source.]

BOX 14–1 ✚ MEDICINE   High Rate of Glycolysis in Tumors Suggests Targets for Chemotherapy and Facilitates Diagnosis

In many types of tumors found in humans and other animals, glucose uptake and glycolysis proceed about 10 times faster than in normal, noncancerous tissues. Most tumor cells grow under hypoxic conditions (i.e., with limited oxygen supply) because, at least initially, they lack the capillary network to supply sufficient oxygen. Cancer cells located more than 100 to 200 μm from the nearest capillaries must depend on glycolysis alone (without further oxidation of pyruvate) for much of their ATP production. The energy yield (2 ATP per glucose) is far lower than can be obtained by the complete oxidation of pyruvate to CO_2 in mitochondria (about 30 ATP per glucose; Chapter 19). So, to make the same amount of ATP, tumor cells must take up much more glucose than do normal cells, converting it to pyruvate and then to lactate as they recycle NADH. It is likely that two early steps in the transformation of a normal cell into a tumor cell are (1) the change to dependence on glycolysis for ATP production, and (2) the development of tolerance to a low pH in the extracellular fluid (caused by release of the end product of glycolysis, lactic acid). In general, the more aggressive the tumor, the greater is its rate of glycolysis.

This increase in glycolysis is achieved, at least in part, by increased synthesis of the glycolytic enzymes and of the plasma membrane transporters GLUT1 and GLUT3 (see Table 11-3) that carry glucose into cells. (Recall that GLUT1 and GLUT3 are not dependent on insulin.) The **hypoxia-inducible transcription factor (HIF-1)** is a protein that acts at the level of mRNA synthesis to stimulate the production of at least eight glycolytic enzymes and the glucose transporters when oxygen supply is limited (Fig. 1). With the resulting high rate of glycolysis, the tumor cell can survive anaerobic conditions until the supply of blood vessels has caught up with tumor growth. Another protein induced by HIF-1 is the peptide hormone VEGF (vascular endothelial growth factor), which stimulates the outgrowth of blood vessels (angiogenesis) toward the tumor.

There is also evidence that the tumor suppressor protein p53, which is mutated in most types of cancer (see Section 12.11), controls the synthesis and assembly of mitochondrial proteins essential to the passage of electrons to O_2. Cells with mutant p53 are defective in mitochondrial electron transport and are forced to rely more heavily on glycolysis for ATP production (Fig. 1).

This heavier reliance of tumors than of normal tissue on glycolysis suggests a possibility for anticancer therapy: inhibitors of glycolysis might target and kill tumors by depleting their supply of ATP. Three inhibitors of hexokinase have shown promise as chemotherapeutic agents: 2-deoxyglucose, lonidamine, and 3-bromopyruvate. By preventing the

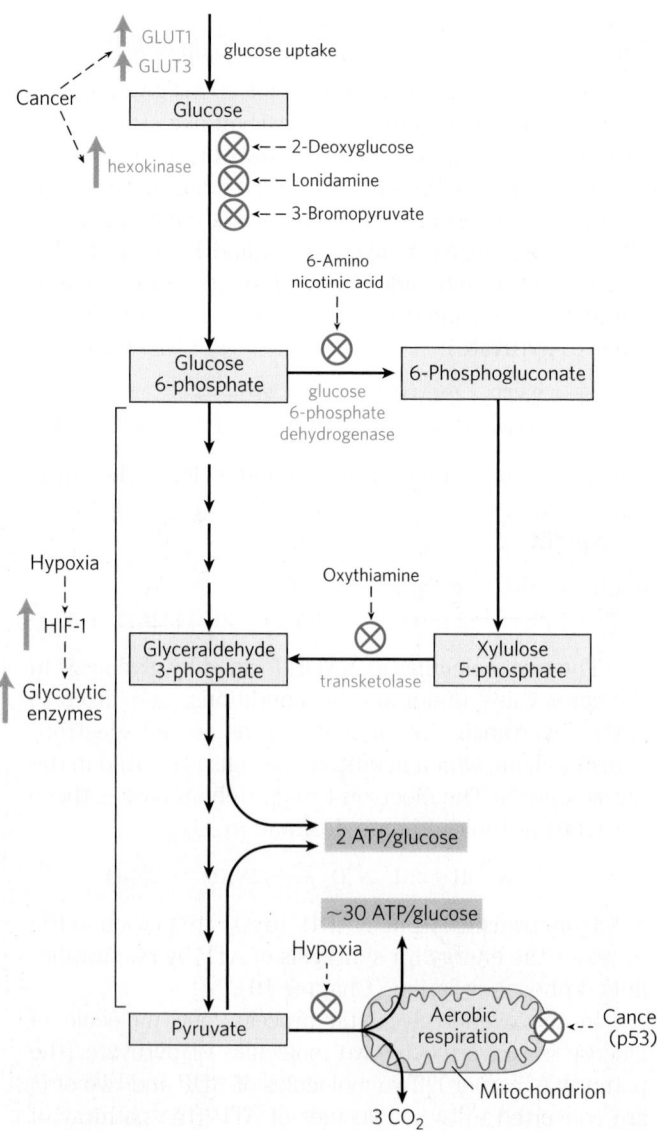

FIGURE 1 The anaerobic metabolism of glucose in tumor cells yields far less ATP (2 per glucose) than the complete oxidation to CO_2 that takes place in healthy cells under aerobic conditions (~30 ATP per glucose), so a tumor cell must consume much more glucose to produce the same amount of ATP. Glucose transporters and most of the glycolytic enzymes are overproduced in tumors. Compounds that inhibit hexokinase, glucose 6-phosphate dehydrogenase, or transketolase block ATP production by glycolysis, thus depriving the cancer cell of energy and killing it.

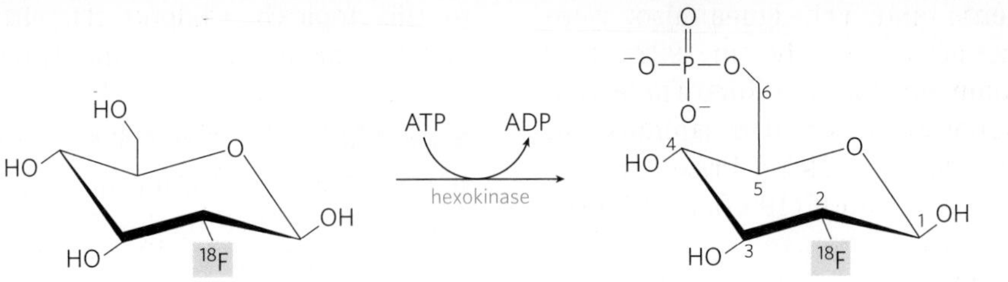

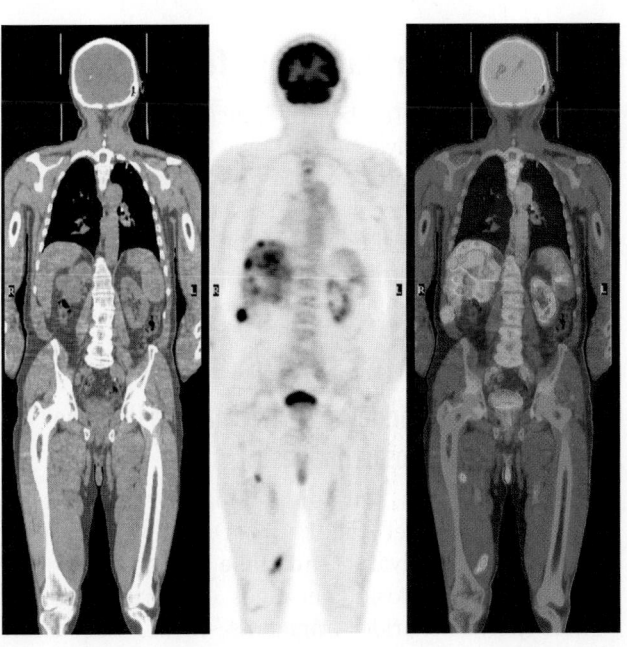

FIGURE 2 Phosphorylation of ^{18}F-labeled 2-fluoro-2-deoxyglucose by hexokinase traps the FdG in cells (as 6-phospho-FdG), where its presence can be detected by positron emission from ^{18}F.

formation of glucose 6-phosphate, these compounds not only deprive tumor cells of glycolytically produced ATP but also prevent the formation of pentose phosphates via the pentose phosphate pathway, which also begins with glucose 6-phosphate. Without pentose phosphates, a cell cannot synthesize the nucleotides essential to DNA and RNA synthesis and thus cannot grow or divide. Another anticancer drug already approved for clinical use is imatinib (Gleevec), described in Box 12-4. It inhibits a specific tyrosine kinase, preventing the increased synthesis of hexokinase normally triggered by that kinase. The thiamine analog oxythiamine, which blocks the action of a transketolase-like enzyme that converts xylulose 5-phosphate to glyceraldehyde 3-phosphate (Fig. 1), is in preclinical trials as an antitumor drug.

The high glycolytic rate in tumor cells also has diagnostic usefulness. The relative rates at which tissues take up glucose can be used in some cases to pinpoint the location of tumors. In positron emission tomography (PET), individuals are injected with a harmless, isotopically labeled glucose analog that is taken up but not metabolized by tissues. The labeled compound is 2-fluoro-2-deoxyglucose (FdG), in which the hydroxyl group at the C-2 of glucose is replaced with ^{18}F (Fig. 2). This compound is taken up via GLUT transporters and is a good substrate for hexokinase, but it cannot be converted to the enediol intermediate in the phosphohexose isomerase reaction (see Fig. 14-5) and therefore accumulates as 6-phospho-FdG. The extent of its accumulation depends on its rate of uptake and phosphorylation, which, as noted above, is typically 10 or more times higher in tumors than in normal tissue. Decay of ^{18}F yields positrons (two per ^{18}F atom) that can be detected by a series of sensitive detectors positioned around the body, which allows accurate localization of accumulated 6-phospho-FdG (Fig. 3).

FIGURE 3 Detection of cancerous tissue by positron emission tomography (PET). The adult male patient had undergone surgical removal of a primary skin cancer (malignant melanoma). The image on the left, obtained by whole-body computed tomography (CT scan), shows the location of the soft tissues and bones. The central panel is a PET scan after the patient had ingested ^{18}F-labeled 2-fluoro-2-deoxyglucose (FdG). Dark spots indicate regions of high glucose utilization. As expected, the brain and bladder are heavily labeled—the brain because it uses most of the glucose consumed in the body, and the bladder because the ^{18}F-labeled 6-phospho-FdG is excreted in the urine. When the intensity of the label in the PET scan is translated into false color (the intensity increases from green to yellow to red) and the image is superimposed on the CT scan, the fused image (right) reveals cancer in the bones of the upper spine, in the liver, and in some regions of muscle, all the result of cancer spreading from the primary malignant melanoma. [Source: ISM/Phototake.]

Trained in carbohydrate chemistry in the laboratory of the great Emil Fischer (who won the Nobel Prize in Chemistry in 1902), Warburg himself won the Nobel Prize in Physiology or Medicine in 1931. Several of Warburg's students and colleagues also were awarded Nobel Prizes: Otto Meyerhof in 1922, Hans Krebs and Fritz Lipmann in 1953, and Hugo Theorell in 1955. Meyerhof's laboratory provided training for Lipmann, and for several other Nobel Prize winners: Severo Ochoa (1959), Andre Lwoff (1965), and George Wald (1967).

Glucose Uptake Is Deficient in Type 1 Diabetes Mellitus

The metabolism of glucose in mammals is limited by the rate of glucose uptake into cells and its phosphorylation by hexokinase. Glucose uptake from the blood is mediated by the GLUT family of glucose transporters (see Table 11-3). The transporters of hepatocytes (GLUT1, GLUT2) and of brain neurons (GLUT3) are always present in plasma membranes. In contrast, the main glucose transporter in the cells of skeletal muscle, cardiac muscle, and adipose tissue (GLUT4) is sequestered in small intracellular vesicles and moves into the plasma membrane only in response to an insulin signal **(Fig. 14-10)**. We discussed this insulin signaling mechanism in Chapter 12 (see Fig. 12-20). Thus in skeletal muscle, heart, and adipose tissue, glucose uptake and metabolism depend on the normal release of insulin by pancreatic β cells in response to elevated blood glucose (see Fig. 23-27).

Individuals with type 1 diabetes mellitus (also called insulin-dependent diabetes) have too few β cells and cannot release sufficient insulin to trigger glucose uptake by the cells of skeletal muscle, heart, or adipose tissue. Thus, after a meal containing carbohydrates, glucose accumulates to abnormally high levels in the blood, a condition known as hyperglycemia. Unable to take up glucose, muscle and fat tissue use the fatty acids of stored triacylglycerols as their principal fuel. In the liver, acetyl-CoA derived from this fatty acid breakdown is converted to "ketone bodies"—acetoacetate and β-hydroxybutyrate—which are exported and carried to other tissues to be used as fuel (Chapter 17). These compounds are especially critical to the brain, which uses ketone bodies as alternative fuel when glucose is unavailable. (Fatty acids cannot pass through the blood-brain barrier and thus are not a fuel for brain neurons.)

In untreated type 1 diabetes, overproduction of acetoacetate and β-hydroxybutyrate leads to their accumulation in the blood, and the consequent lowering of blood pH produces **ketoacidosis**, a life-threatening condition. Insulin injection reverses this sequence of events: GLUT4 moves into the plasma membranes of hepatocytes and adipocytes, glucose is taken up into the cells and phosphorylated, and the blood glucose

level falls, greatly reducing the production of ketone bodies.

Diabetes mellitus has profound effects on the metabolism of both carbohydrates and fats. We return to this topic in Chapter 23, after considering lipid metabolism (Chapters 17 and 21). ∎

SUMMARY 14.1 Glycolysis

∎ Glycolysis is a near-universal pathway by which a glucose molecule is oxidized to two molecules of pyruvate, with energy conserved as ATP and NADH.

∎ All 10 glycolytic enzymes are in the cytosol, and all 10 intermediates are phosphorylated compounds of three or six carbons.

∎ In the preparatory phase of glycolysis, ATP is invested to convert glucose to fructose 1,6-bisphosphate. The bond between C-3 and C-4 is then broken to yield two molecules of triose phosphate.

∎ In the payoff phase, each of the two molecules of glyceraldehyde 3-phosphate derived from glucose undergoes oxidation at C-1; the energy of this oxidation reaction is conserved in the form of one NADH and two ATP per triose phosphate oxidized. The net equation for the overall process is

$$\text{Glucose} + 2\text{NAD}^+ + 2\text{ADP} + 2\text{P}_i \longrightarrow$$
$$2 \text{ pyruvate} + 2\text{NADH} + 2\text{H}^+ + 2\text{ATP} + 2\text{H}_2\text{O}$$

∎ Glycolysis is tightly regulated in coordination with other energy-yielding pathways to ensure a steady supply of ATP.

∎ In type 1 diabetes, defective uptake of glucose by muscle and adipose tissue has profound effects on the metabolism of carbohydrates and fats.

14.2 Feeder Pathways for Glycolysis

Many carbohydrates besides glucose meet their catabolic fate in glycolysis, after being transformed into one of the glycolytic intermediates. The most significant are the storage polysaccharides glycogen and starch, either within cells (endogenous) or obtained in the diet; the disaccharides maltose, lactose, trehalose, and sucrose; and the monosaccharides fructose, mannose, and galactose **(Fig. 14-11)**.

Dietary Polysaccharides and Disaccharides Undergo Hydrolysis to Monosaccharides

For most humans, starch is the major source of carbohydrates in the diet (Fig. 14-11). Digestion begins in the mouth, where salivary **α-amylase** hydrolyzes the internal $(\alpha1\rightarrow4)$ glycosidic linkages of starch, producing short polysaccharide fragments or oligosaccharides. (Note that in this *hydrolysis* reaction, water, not P_i, is

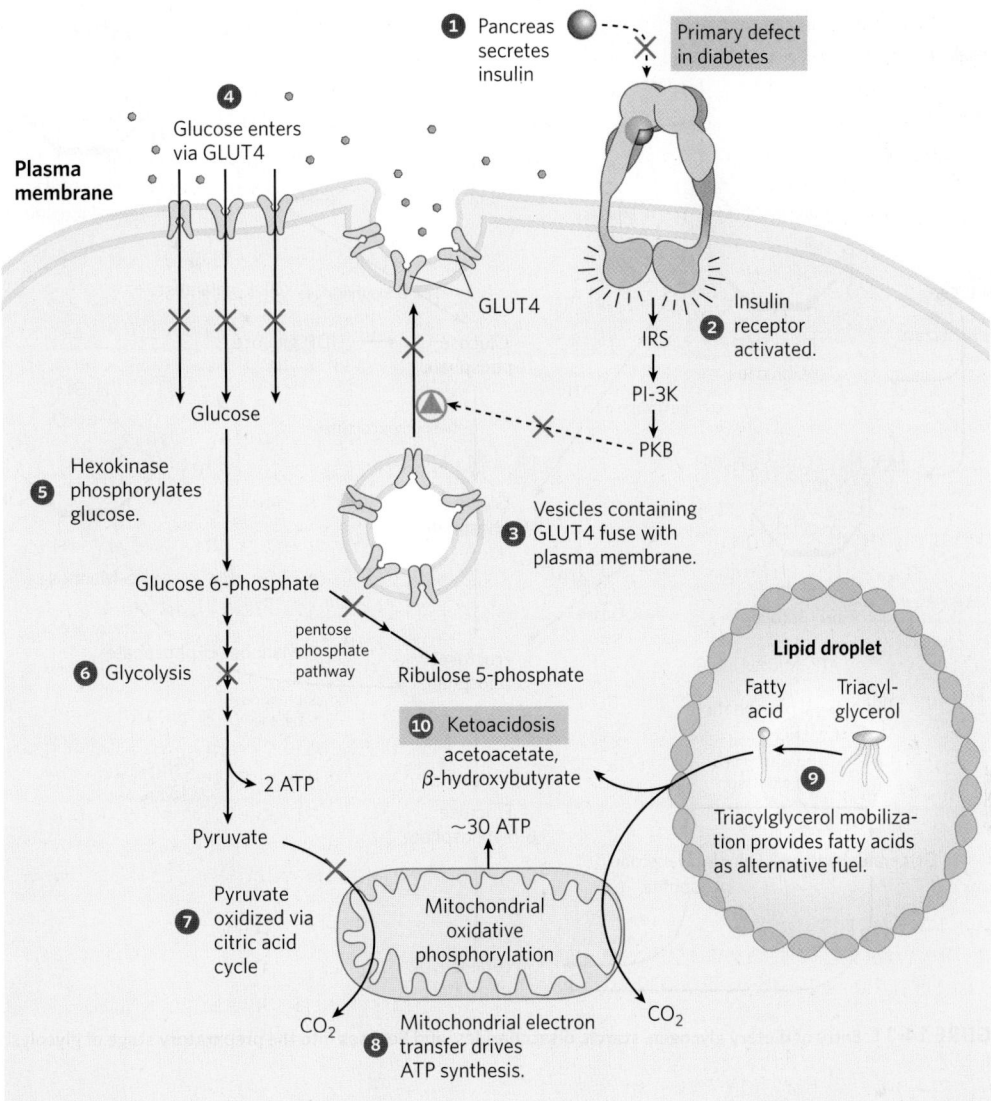

FIGURE 14-10 Effect of type 1 diabetes on carbohydrate and fat metabolism in an adipocyte. Normally, insulin triggers the insertion of GLUT4 transporters into the plasma membrane by the fusion of GLUT4-containing vesicles with the membrane, allowing glucose uptake from the blood. When blood levels of insulin drop, GLUT4 is resequestered in vesicles by endocytosis. In type 1 (insulin-dependent) diabetes mellitus, the insertion of GLUT4 into membranes, as well as other processes normally stimulated by insulin, is inhibited as indicated by **X**. The lack of insulin prevents glucose uptake via GLUT4; as a consequence, cells are deprived of glucose, and blood glucose is elevated.

Lacking glucose for energy supply, adipocytes break down triacylglycerols stored in fat droplets and supply the resulting fatty acids to other tissues for mitochondrial ATP production. Two byproducts of fatty acid oxidation in the liver (acetoacetate and β-hydroxybutyrate, see p. 668) accumulate and are released into the blood, providing fuel for the brain but also decreasing blood pH, causing ketoacidosis. The same sequence of events takes place in muscle, except that myocytes do not store triacylglycerols and instead take up fatty acids that are released into the blood by adipocytes. (The details of insulin signaling are discussed in Section 12.4.)

the attacking species.) In the stomach, salivary α-amylase is inactivated by the low pH, but a second form of α-amylase, secreted by the pancreas into the small intestine, continues the breakdown process. Pancreatic α-amylase yields mainly maltose and maltotriose (the di- and trisaccharides of glucose) and oligosaccharides called limit dextrins, fragments of amylopectin containing ($\alpha1{\rightarrow}6$) branch points. Maltose and dextrins are degraded to glucose by enzymes of the intestinal brush border (the fingerlike microvilli of intestinal

epithelial cells, which greatly increase the area of the intestinal surface). Dietary glycogen has essentially the same structure as starch, and its digestion proceeds by the same pathway.

As we noted in Chapter 7, most animals cannot digest cellulose for lack of the enzyme cellulase, which attacks the ($\beta1{\rightarrow}4$) glycosidic bonds of cellulose. In ruminant animals, the extended stomach includes a chamber in which symbiotic microorganisms that produce cellulase break down cellulose into glucose molecules. These

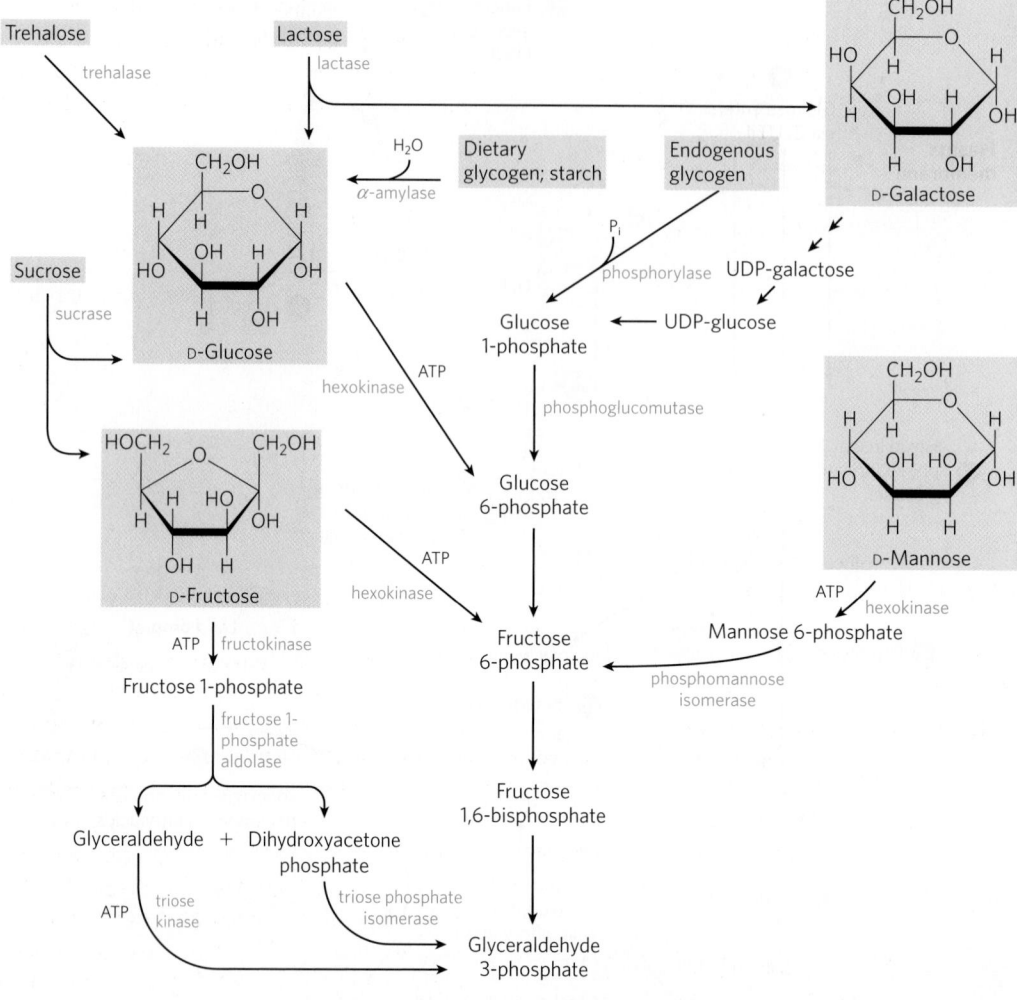

FIGURE 14-11 Entry of dietary glycogen, starch, disaccharides, and hexoses into the preparatory stage of glycolysis.

microorganisms use the resulting glucose in an anaerobic fermentation that produces large quantities of propionate. This propionate serves as the starting material for gluconeogenesis, which produces much of the lactose in milk.

Endogenous Glycogen and Starch Are Degraded by Phosphorolysis

Glycogen stored in animal tissues (primarily liver and skeletal muscle), in microorganisms, or in plant tissues can be mobilized for use within the same cell by a *phosphorolytic* reaction catalyzed by **glycogen phosphorylase** (**starch phosphorylase** in plants) **(Fig. 14-12)**. These enzymes catalyze an attack by P_i on the ($\alpha1{\rightarrow}4$) glycosidic linkage that joins the last two glucose residues at a nonreducing end, generating glucose 1-phosphate and a polymer one glucose unit shorter. *Phosphorolysis* preserves some of the energy of the glycosidic bond in the phosphate ester glucose 1-phosphate. Glycogen phosphorylase (or starch phosphorylase) acts repetitively until it approaches an ($\alpha1{\rightarrow}6$) branch point (see

Fig. 7-13), where its action stops. A **debranching enzyme** removes the branches. The mechanisms and control of glycogen degradation are described in greater detail in Chapter 15.

Glucose 1-phosphate produced by glycogen phosphorylase is converted to glucose 6-phosphate by **phosphoglucomutase**, which catalyzes the reversible reaction

$$\text{Glucose 1-phosphate} \rightleftharpoons \text{glucose 6-phosphate}$$

Phosphoglucomutase employs essentially the same mechanism as phosphoglycerate mutase (Fig. 14-9): both entail a bisphosphate intermediate, and the enzyme is transiently phosphorylated in each catalytic cycle. The general name **mutase** is given to enzymes that catalyze the transfer of a functional group from one position to another in the same molecule. Mutases are a subclass of **isomerases**, enzymes that interconvert stereoisomers or structural or positional isomers (see Table 6-3). The glucose 6-phosphate formed in the phosphoglucomutase reaction can enter glycolysis or another pathway such as the pentose phosphate pathway, described in Section 14.5.

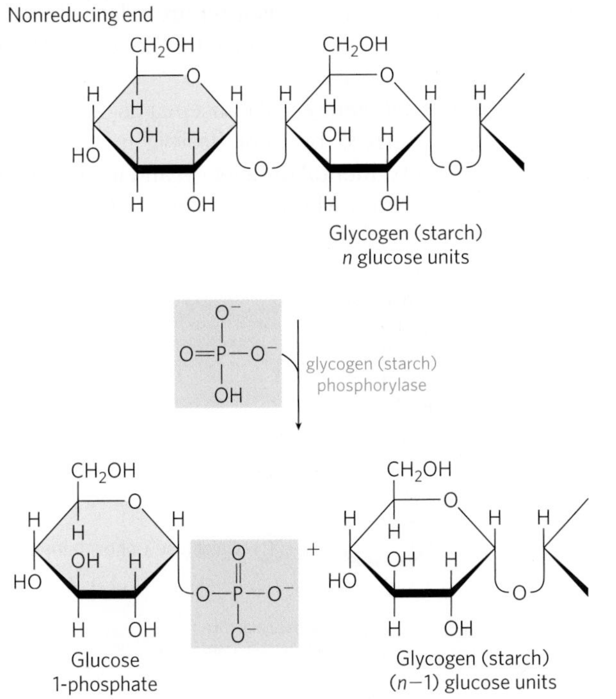

FIGURE 14-12 Breakdown of intracellular glycogen by glycogen phosphorylase. The enzyme catalyzes attack by inorganic phosphate (pink) on the terminal glucosyl residue (blue) at the nonreducing end of a glycogen molecule, releasing glucose 1-phosphate and generating a glycogen molecule shortened by one glucose residue. The reaction is a *phosphorolysis* (not hydrolysis).

WORKED EXAMPLE 14-1 | **Energy Savings for Glycogen Breakdown by Phosphorolysis**

Calculate the energy savings (in ATP molecules per glucose monomer) achieved by breaking down glycogen by *phosphorolysis* rather than *hydrolysis* to begin the process of glycolysis.

Solution: Phosphorolysis produces a phosphorylated glucose (glucose 1-phosphate), which is then converted to glucose 6-phosphate—without expenditure of the cellular energy (1 ATP) needed for formation of glucose 6-phosphate from free glucose. Thus only 1 ATP is consumed per glucose monomer in the preparatory phase, compared with 2 ATP when glycolysis starts with free glucose. The cell therefore gains 3 ATP per glucose monomer (4 ATP produced in the payoff phase minus 1 ATP used in the preparatory phase), rather than 2—a saving of 1 ATP per glucose monomer.

Breakdown of dietary polysaccharides such as glycogen and starch in the gastrointestinal tract by phosphorolysis rather than hydrolysis would produce no energy gain: sugar phosphates are not transported into the cells that line the intestine, but must first be dephosphorylated to the free sugar.

Disaccharides must be hydrolyzed to monosaccharides before entering cells. Intestinal disaccharides and dextrins are hydrolyzed by enzymes attached to the outer surface of the intestinal epithelial cells:

$$\text{Dextrin} + n\text{H}_2\text{O} \xrightarrow{\text{dextrinase}} n \text{ D-glucose}$$

$$\text{Maltose} + \text{H}_2\text{O} \xrightarrow{\text{maltose}} 2 \text{ D-glucose}$$

$$\text{Lactose} + \text{H}_2\text{O} \xrightarrow{\text{lactase}} \text{D-galactose} + \text{D-glucose}$$

$$\text{Sucrose} + \text{H}_2\text{O} \xrightarrow{\text{sucrase}} \text{D-fructose} + \text{D-glucose}$$

$$\text{Trehalose} + \text{H}_2\text{O} \xrightarrow{\text{trehalase}} 2 \text{ D-glucose}$$

The monosaccharides so formed are actively transported into the epithelial cells (see Fig. 11-41), then passed into the blood to be carried to various tissues, where they are phosphorylated and funneled into the glycolytic sequence.

Lactose intolerance, common among adults of most human populations except those originating in Northern Europe and some parts of Africa, is due to the disappearance after childhood of most or all of the lactase activity of the intestinal epithelial cells. Without intestinal lactase, lactose cannot be completely digested and absorbed in the small intestine, and it passes into the large intestine, where bacteria convert it to toxic products that cause abdominal cramps and diarrhea. The problem is further complicated because undigested lactose and its metabolites increase the osmolarity of the intestinal contents, favoring retention of water in the intestine. In most parts of the world where lactose intolerance is prevalent, milk is not used as a food by adults, although milk products predigested with lactase are commercially available in some countries. In certain human disorders, several or all of the intestinal disaccharidases are missing. In these cases, the digestive disturbances triggered by dietary disaccharides can sometimes be minimized by a controlled diet. ■

Other Monosaccharides Enter the Glycolytic Pathway at Several Points

In most organisms, hexoses other than glucose can undergo glycolysis after conversion to a phosphorylated derivative. D-Fructose, present in free form in many fruits and formed by hydrolysis of sucrose in the small intestine of vertebrates, is phosphorylated by hexokinase:

$$\text{Fructose} + \text{ATP} \xrightarrow{\text{Mg}^{2+}} \text{fructose 6-phosphate} + \text{ADP}$$

This is a major pathway of fructose entry into glycolysis in the muscles and kidney. In the liver, fructose enters by a different pathway. The liver enzyme **fructokinase** catalyzes the phosphorylation of fructose at C-1 rather than C-6:

$$\text{Fructose} + \text{ATP} \xrightarrow{\text{Mg}^{2+}} \text{fructose 1-phosphate} + \text{ADP}$$

The fructose 1-phosphate is then cleaved to glyceraldehyde and dihydroxyacetone phosphate by **fructose 1-phosphate aldolase**:

$$CH_2OPO_3^{2-}$$

Fructose 1-phosphate $\rightleftharpoons$ Dihydroxyacetone phosphate + Glyceraldehyde

(via fructose 1-phosphate aldolase)

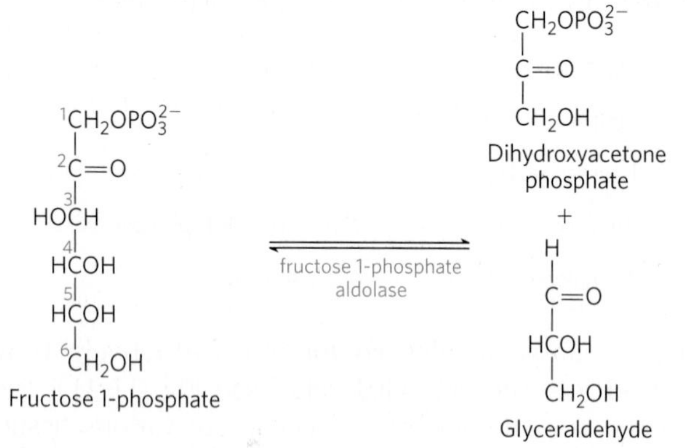

Dihydroxyacetone phosphate is converted to glyceraldehyde 3-phosphate by the glycolytic enzyme triose phosphate isomerase. Glyceraldehyde is phosphorylated by ATP and **triose kinase** to glyceraldehyde 3-phosphate:

$$\text{Glyceradehyde} + \text{ATP} \xrightarrow{Mg^{2+}} \text{glyceraldehyde 3-phosphate} + \text{ADP}$$

Thus both products of fructose 1-phosphate hydrolysis enter the glycolytic pathway as glyceraldehyde 3-phosphate.

D-Galactose, a product of the hydrolysis of lactose (milk sugar), passes in the blood from the intestine to the liver, where it is first phosphorylated at C-1, at the expense of ATP, by the enzyme **galactokinase**:

$$\text{Galactose} + \text{ATP} \xrightarrow{Mg^{2+}} \text{galactose 1-phosphate} + \text{ADP}$$

The galactose 1-phosphate is then converted to its epimer at C-4, glucose 1-phosphate, by a set of reactions in which **uridine diphosphate** (UDP) functions as a coenzyme-like carrier of hexose groups **(Fig. 14-13)**. The epimerization involves first the oxidation of the C-4 —OH group to a ketone, then reduction of the ketone to an —OH, with inversion of the configuration at C-4. NAD is the cofactor for both the oxidation and the reduction.

A defect in any of the three enzymes in this pathway causes **galactosemia** in humans. In galactokinase-deficiency galactosemia, high galactose concentrations are found in blood and urine. Affected individuals develop cataracts in infancy, caused by deposition of the galactose metabolite galactitol in the lens.

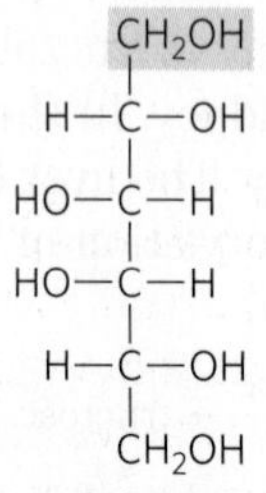

D-Galactitol

The other symptoms in this disorder are relatively mild, and strict limitation of galactose in the diet greatly diminishes their severity.

Transferase-deficiency galactosemia is more serious; it is characterized by poor growth in childhood, speech abnormality, mental deficiency, and liver damage that may be fatal, even when galactose is withheld from

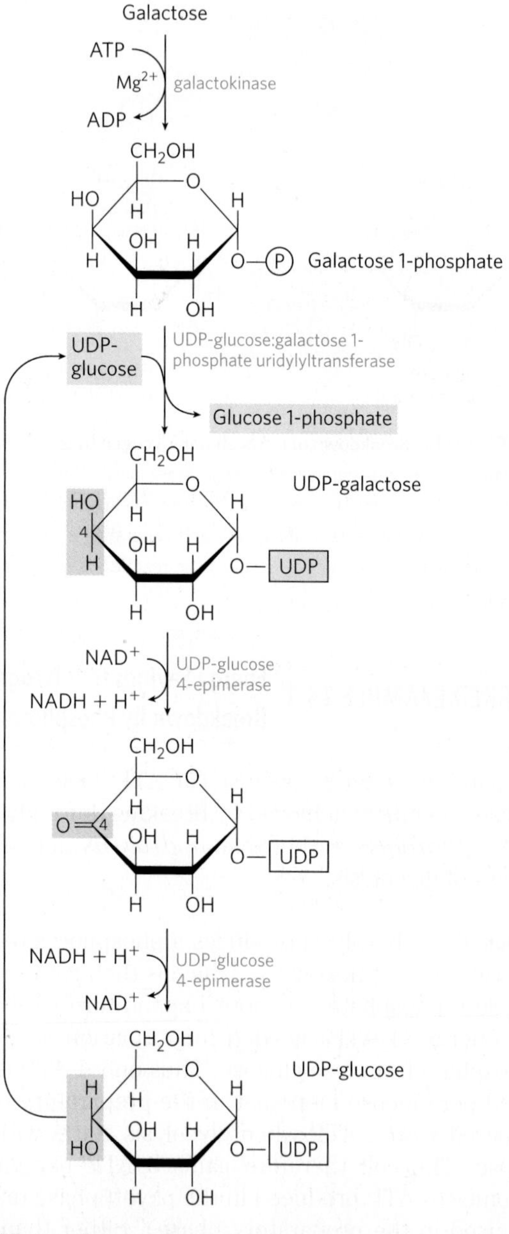

FIGURE 14-13 Conversion of galactose to glucose 1-phosphate. The conversion proceeds through a sugar-nucleotide derivative, UDP-galactose, which is formed when galactose 1-phosphate displaces glucose 1-phosphate from UDP-glucose. UDP-galactose is then converted by UDP-glucose 4-epimerase to UDP-glucose, in a reaction that involves oxidation of C-4 (light red) by NAD^+, then reduction of C-4 by NADH; the result is inversion of the configuration at C-4. The UDP-glucose is recycled through another round of the same reaction. The net effect of this cycle is the conversion of galactose 1-phosphate to glucose 1-phosphate; there is no net production or consumption of UDP-galactose or UDP-glucose.

the diet. Epimerase-deficiency galactosemia leads to similar symptoms, but is less severe when dietary galactose is carefully controlled. ∎

D-Mannose, released in the digestion of various polysaccharides and glycoproteins of foods, can be phosphorylated at C-6 by hexokinase:

$$\text{Mannose} + \text{ATP} \xrightarrow{\text{Mg}^{2+}} \text{mannose 6-phosphate} + \text{ADP}$$

Mannose 6-phosphate is isomerized by **phosphomannose isomerase** to yield fructose 6-phosphate, an intermediate of glycolysis.

SUMMARY 14.2 Feeder Pathways for Glycolysis

■ Endogenous glycogen and starch, storage forms of glucose, enter glycolysis in a two-step process. Phosphorolytic cleavage of a glucose residue from an end of the polymer, forming glucose 1-phosphate, is catalyzed by glycogen phosphorylase or starch phosphorylase. Phosphoglucomutase then converts the glucose 1-phosphate to glucose 6-phosphate, which can enter glycolysis.

■ Ingested polysaccharides and disaccharides are converted to monosaccharides by intestinal hydrolytic enzymes, and the monosaccharides then enter intestinal cells and are transported to the liver or other tissues.

■ A variety of D-hexoses, including fructose, galactose, and mannose, can be funneled into glycolysis. Each is phosphorylated and converted to glucose 6-phosphate, fructose 6-phosphate, or fructose 1-phosphate.

■ Conversion of galactose 1-phosphate to glucose 1-phosphate involves two nucleotide derivatives: UDP-galactose and UDP-glucose. Genetic defects in any of the three enzymes that catalyze conversion of galactose to glucose 1-phosphate result in galactosemias of varying severity.

14.3 Fates of Pyruvate under Anaerobic Conditions: Fermentation

Under aerobic conditions, the pyruvate formed in the final step of glycolysis is oxidized to acetate (acetyl-CoA), which enters the citric acid cycle and is oxidized to CO_2 and H_2O. The NADH formed by dehydrogenation of glyceraldehyde 3-phosphate is ultimately reoxidized to NAD^+ by passage of its electrons to O_2 in mitochondrial respiration. Under hypoxic (low-oxygen) conditions, however—as in very active skeletal muscle, in submerged plant tissues, in solid tumors, or in lactic acid bacteria—NADH generated by glycolysis cannot be reoxidized by O_2. Failure to regenerate NAD^+ would leave the cell with no electron acceptor for the oxidation of glyceraldehyde 3-phosphate, and the energy-yielding reactions of glycolysis would stop. NAD^+ must therefore be regenerated in some other way.

The earliest cells lived in an atmosphere almost devoid of oxygen and had to develop strategies for deriving energy from fuel molecules under anaerobic conditions. Most modern organisms have retained the ability to continually regenerate NAD^+ during anaerobic glycolysis by transferring electrons from NADH to form a reduced end product such as lactate or ethanol.

Pyruvate Is the Terminal Electron Acceptor in Lactic Acid Fermentation

When animal tissues cannot be supplied with sufficient oxygen to support aerobic oxidation of the pyruvate and NADH produced in glycolysis, NAD^+ is regenerated from NADH by the reduction of pyruvate to **lactate**. As mentioned earlier, some tissues and cell types (such as erythrocytes, which have no mitochondria and thus cannot oxidize pyruvate to CO_2) produce lactate from glucose even under aerobic conditions. The reduction of pyruvate in this pathway is catalyzed by **lactate dehydrogenase**, which forms the L isomer of lactate at pH 7:

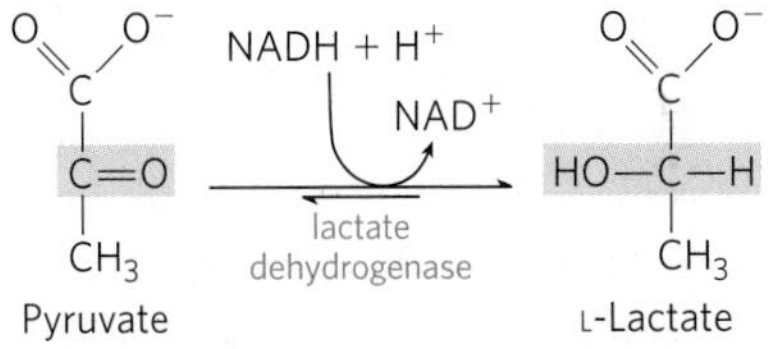

The overall equilibrium of the reaction strongly favors lactate formation, as shown by the large negative standard free-energy change.

In glycolysis, dehydrogenation of the two molecules of glyceraldehyde 3-phosphate derived from each molecule of glucose converts two molecules of NAD^+ to two of NADH. Because the reduction of two molecules of pyruvate to two of lactate regenerates two molecules of NAD^+, there is no net change in NAD^+ or NADH:

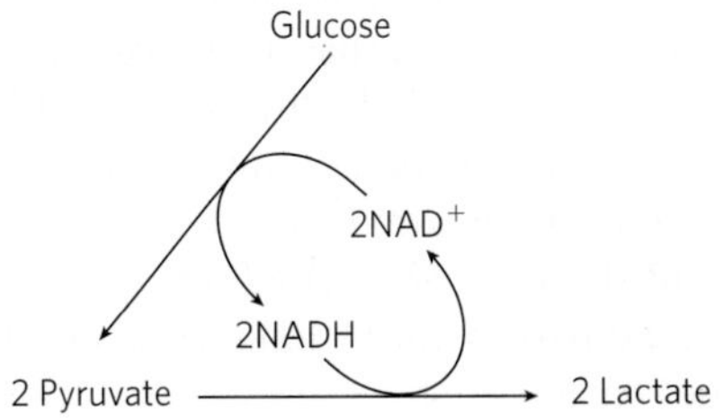

The lactate formed by active skeletal muscles (or by erythrocytes) can be recycled; it is carried in the blood to the liver, where it is converted to glucose during the recovery from strenuous muscular activity. When lactate is produced in large quantities during vigorous muscle contraction (during a sprint, for example), the acidification that results from ionization of lactic acid in muscle and blood limits the period of vigorous activity. The best-conditioned athletes can sprint at top speed for no more than a minute (Box 14-2).

BOX 14-2 Athletes, Alligators, and Coelacanths: Glycolysis at Limiting Concentrations of Oxygen

Most vertebrates are essentially aerobic organisms; they convert glucose to pyruvate by glycolysis, then use molecular oxygen to oxidize the pyruvate completely to CO_2 and H_2O. Anaerobic catabolism of glucose to lactate occurs during short bursts of extreme muscular activity, for example in a 100 m sprint, during which oxygen cannot be carried to the muscles fast enough to oxidize pyruvate. Instead, the muscles use their stored glucose (glycogen) as fuel to generate ATP by fermentation, with lactate as the end product. In a sprint, lactate in the blood builds up to high concentrations. It is slowly converted back to glucose by gluconeogenesis in the liver in the subsequent rest or recovery period, during which oxygen is consumed at a gradually diminishing rate until the breathing rate returns to normal. The excess oxygen consumed in the recovery period represents a repayment of the oxygen debt. This is the amount of oxygen required to supply ATP for gluconeogenesis during recovery respiration, in order to regenerate the glycogen "borrowed" from liver and muscle to carry out intense muscular activity in the sprint. The cycle of reactions that includes glucose conversion to lactate in muscle and lactate conversion to glucose in liver is called the Cori cycle, for Carl and Gerty Cori, whose studies in the 1930s and 1940s clarified the pathway and its role (see Box 15-4).

The circulatory systems of most small vertebrates can carry oxygen to their muscles fast enough to avoid having to use muscle glycogen anaerobically. For example, migrating birds often fly great distances at high speeds without rest and without incurring an oxygen debt. Many running animals of moderate size also maintain an essentially aerobic metabolism in their skeletal muscle. However, the circulatory systems of larger animals, including humans, cannot completely sustain aerobic metabolism in skeletal muscles over long periods of intense muscular activity. These animals generally are slow-moving under normal circumstances and engage in intense muscular activity only in the gravest emergencies, because such bursts of activity require long recovery periods to repay the oxygen debt.

Alligators and crocodiles, for example, are normally sluggish animals. Yet, when provoked, they are capable of lightning-fast charges and dangerous lashings of their powerful tails. Such intense bursts of activity are short and must be followed by long periods of recovery. The fast emergency movements require lactic acid fermentation to generate ATP in skeletal muscles. The stores of muscle glycogen are rapidly expended in intense muscular activity, and

lactate reaches very high concentrations in myocytes and extracellular fluid. Whereas a trained athlete can recover from a 100 m sprint in 30 min or less, an alligator may require many hours of rest and extra oxygen consumption to clear the excess lactate from its blood and regenerate muscle glycogen after a burst of activity.

Other large animals, such as the elephant and rhinoceros, have similar metabolic characteristics, as do diving mammals such as whales and seals. Dinosaurs and other huge, now-extinct animals probably had to depend on lactic acid fermentation to supply energy for muscular activity, followed by very long recovery periods during which they were vulnerable to attack by smaller predators better able to use oxygen and thus better adapted to continuous, sustained muscular activity.

Deep-sea explorations have revealed many species of marine life at great ocean depths, where the oxygen concentration is near zero. For example, the primitive coelacanth, a large fish recovered from depths of 4,000 m or more off the coast of South Africa, has an essentially anaerobic metabolism in virtually all its tissues. It converts carbohydrates to lactate and other products, most of which must be excreted. Some marine vertebrates ferment glucose to ethanol and CO_2 in order to generate ATP.

[Source: John Zocco/Shutterstock.]

Although conversion of glucose to lactate includes two oxidation-reduction steps, there is no net change in the oxidation state of carbon; in glucose ($C_6H_{12}O_6$) and lactic acid ($C_3H_6O_3$), the H:C ratio is the same. Nevertheless, some of the energy of the glucose molecule has been extracted by its conversion to lactate—enough to give a net yield of two molecules of ATP for every glucose molecule consumed. **Fermentation** is the general term for such processes, which extract energy (as ATP) but do not consume oxygen or change the concentrations of NAD^+ or NADH. Fermentations are carried out by a wide range of organisms, many of which occupy anaerobic niches, and they yield a variety of end products, some of which find commercial uses.

Ethanol Is the Reduced Product in Ethanol Fermentation

Yeast and other microorganisms ferment glucose to ethanol and CO_2, rather than to lactate. Glucose is converted to pyruvate by glycolysis, and the pyruvate is converted to ethanol and CO_2 in a two-step process:

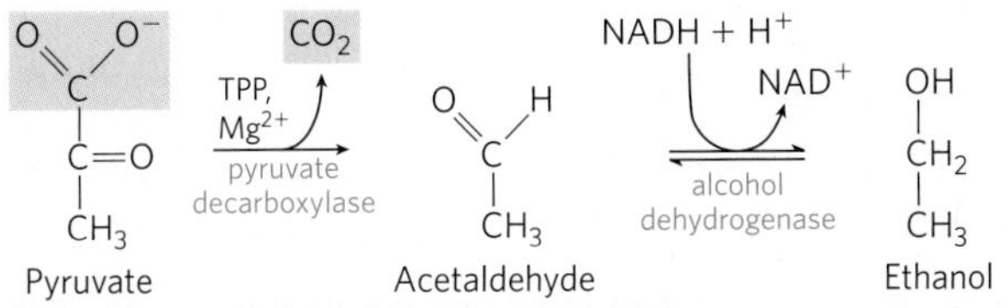

In the first step, pyruvate is decarboxylated in an irreversible reaction catalyzed by **pyruvate decarboxylase**. This reaction is a simple decarboxylation and does not involve the net oxidation of pyruvate. Pyruvate decarboxylase requires Mg^{2+} and has a tightly bound coenzyme, thiamine pyrophosphate, which is discussed below. In the second step, acetaldehyde is reduced to ethanol through the action of **alcohol dehydrogenase**, with the reducing power furnished by NADH derived from the dehydrogenation of glyceraldehyde 3-phosphate. This reaction is a well-studied case of hydride transfer from NADH **(Fig. 14-14)**. Ethanol and CO_2 are thus the end products of ethanol fermentation, and the overall equation is

$$\text{Glucose} + 2\text{ADP} + 2P_i \longrightarrow 2\text{ ethanol} + 2CO_2 + 2\text{ATP} + 2H_2O$$

As in lactic acid fermentation, there is no net change in the ratio of hydrogen to carbon atoms when glucose (H:C ratio = 12/6 = 2) is fermented to two ethanol and two CO_2 (combined H:C ratio = 12/6 = 2). In all fermentations, the H:C ratio of the reactants and products remains the same.

Pyruvate decarboxylase is present in brewer's and baker's yeast (*Saccharomyces cerevisiae*) and in all other organisms that ferment glucose to ethanol, including some plants. The CO_2 produced by pyruvate decarboxylation in brewer's yeast is responsible for the characteristic carbonation of champagne. The ancient art of brewing beer involves several enzymatic processes

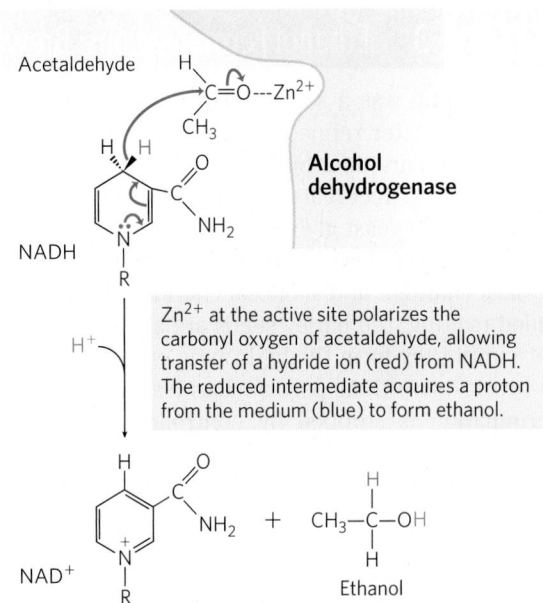

Zn^{2+} at the active site polarizes the carbonyl oxygen of acetaldehyde, allowing transfer of a hydride ion (red) from NADH. The reduced intermediate acquires a proton from the medium (blue) to form ethanol.

MECHANISM FIGURE 14-14 **The alcohol dehydrogenase reaction.**

in addition to the reactions of ethanol fermentation (Box 14-3). In baking, CO_2 released by pyruvate decarboxylase when yeast is mixed with a fermentable sugar causes dough to rise. The enzyme is absent in vertebrate tissues and in other organisms that carry out lactic acid fermentation.

Alcohol dehydrogenase is present in many organisms that metabolize ethanol, including humans. In the liver it catalyzes the oxidation of ethanol, either ingested or produced by intestinal microorganisms, with the concomitant reduction of NAD^+ to NADH. In this case, the reaction proceeds in the direction opposite to that involved in the production of ethanol by fermentation.

Thiamine Pyrophosphate Carries "Active Acetaldehyde" Groups

The pyruvate decarboxylase reaction provides our first encounter with **thiamine pyrophosphate (TPP) (Fig. 14-15)**, a coenzyme derived from vitamin B_1. Lack of vitamin B_1 in the human diet leads to the condition known as beriberi, characterized by an accumulation of body fluids (swelling), pain, paralysis, and ultimately death. ■

Thiamine pyrophosphate plays an important role in the cleavage of bonds adjacent to a carbonyl group, such as the decarboxylation of α-keto acids, and in chemical rearrangements in which an activated acetaldehyde group is transferred from one carbon atom to another (Table 14-1). The functional part of TPP, the thiazolium ring, has a relatively acidic proton at C-2. Loss of this proton produces a carbanion that is the active species in TPP-dependent reactions (Fig. 14-15).

BOX 14-3   Ethanol Fermentations: Brewing Beer and Producing Biofuels

Beer brewing was a science learned early in human history, and later refined for larger-scale production. Brewers prepare beer by ethanol fermentation of the carbohydrates in cereal grains (seeds) such as barley, carried out by yeast glycolytic enzymes. The carbohydrates, largely polysaccharides, must first be degraded to disaccharides and monosaccharides. In a process called malting, the barley seeds are allowed to germinate until they form the hydrolytic enzymes required to break down their polysaccharides, at which point germination is stopped by controlled heating. The product is malt, which contains enzymes that catalyze the hydrolysis of the β linkages of cellulose and other cell wall polysaccharides of the barley husks, and enzymes such as α-amylase and maltase.

The brewer next prepares the wort, the nutrient medium required for fermentation by yeast cells. The malt is mixed with water and then mashed or crushed. This allows the enzymes formed in the malting process to act on the cereal polysaccharides to form maltose, glucose, and other simple sugars, which are soluble in the aqueous medium. The remaining cell matter is then separated, and the liquid wort is boiled with hops to give flavor. The wort is cooled and then aerated.

Now the yeast cells are added. In the aerobic wort the yeast grows and reproduces very rapidly, using energy obtained from available sugars. No ethanol forms during this stage, because the yeast, amply supplied with oxygen, oxidizes the pyruvate formed by glycolysis to CO_2 and H_2O via the citric acid cycle. When all the dissolved oxygen in the vat of wort has been consumed, the yeast cells switch to anaerobic metabolism, and from this point they ferment the sugars into ethanol and CO_2. The fermentation process is controlled in part by the concentration of the ethanol formed, by the pH, and by the amount of remaining sugar. After fermentation has been stopped, the cells are removed and the "raw" beer is ready for final processing.

In the final steps of brewing, the amount of foam (or head) on the beer, which results from dissolved proteins, is adjusted. Normally this is controlled by proteolytic enzymes that arise in the malting process. If these enzymes act on the proteins too long, the beer will have very little head and will be flat; if they do not act long enough, the beer will not be clear when it is cold. Sometimes proteolytic enzymes from other sources are added to control the head.

Much of the technology developed for large-scale production of alcoholic beverages is now finding application to a wholly different problem: the production of ethanol as a renewable fuel. With the continuing depletion of the known stores of fossil fuels and the rising demand for fuel for internal combustion engines, there is increased interest in the use of ethanol as a fuel substitute or extender. The principal advantage of ethanol as a fuel is that it can be produced from relatively *inexpensive* and *renewable* resources rich in sucrose, starch, or cellulose—starch from corn or wheat, sucrose from beets or cane, and cellulose from straw, forest industry waste, or municipal solid waste. Typically, the raw material (feedstock) is first converted chemically to monosaccharides, then fed to a hardy strain of yeast in an industrial-scale fermenter (Fig. 1). The fermentation can yield not only ethanol for fuel but also side products such as proteins that can be used as animal feed.

FIGURE 1 Industrial-scale fermentations to produce biofuel and other products are typically carried out in tanks that hold thousands of liters of medium. [Source: Charles O'Rear/Corbis.]

The carbanion readily adds to carbonyl groups, and the thiazolium ring is thereby positioned to act as an "electron sink" that greatly facilitates reactions such as the decarboxylation catalyzed by pyruvate decarboxylase.

Fermentations Are Used to Produce Some Common Foods and Industrial Chemicals

Our progenitors learned millennia ago to use fermentation in the production and preservation of foods.

Certain microorganisms present in raw food products ferment the carbohydrates and yield metabolic products that give the foods their characteristic forms, textures, and tastes. Yogurt, already known in biblical times, is produced when the bacterium *Lactobacillus bulgaricus* ferments the carbohydrate in milk, producing lactic acid; the resulting drop in pH causes the milk proteins to precipitate, producing the thick texture and sour taste of unsweetened yogurt. Another bacterium, *Propionibacterium freudenreichii*, ferments milk to

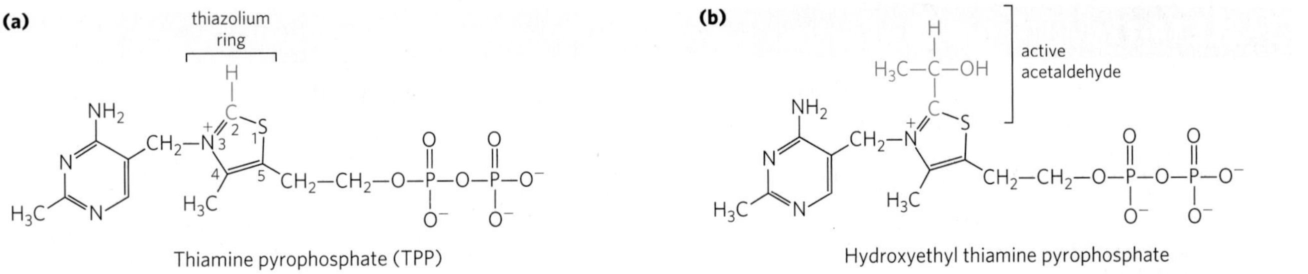

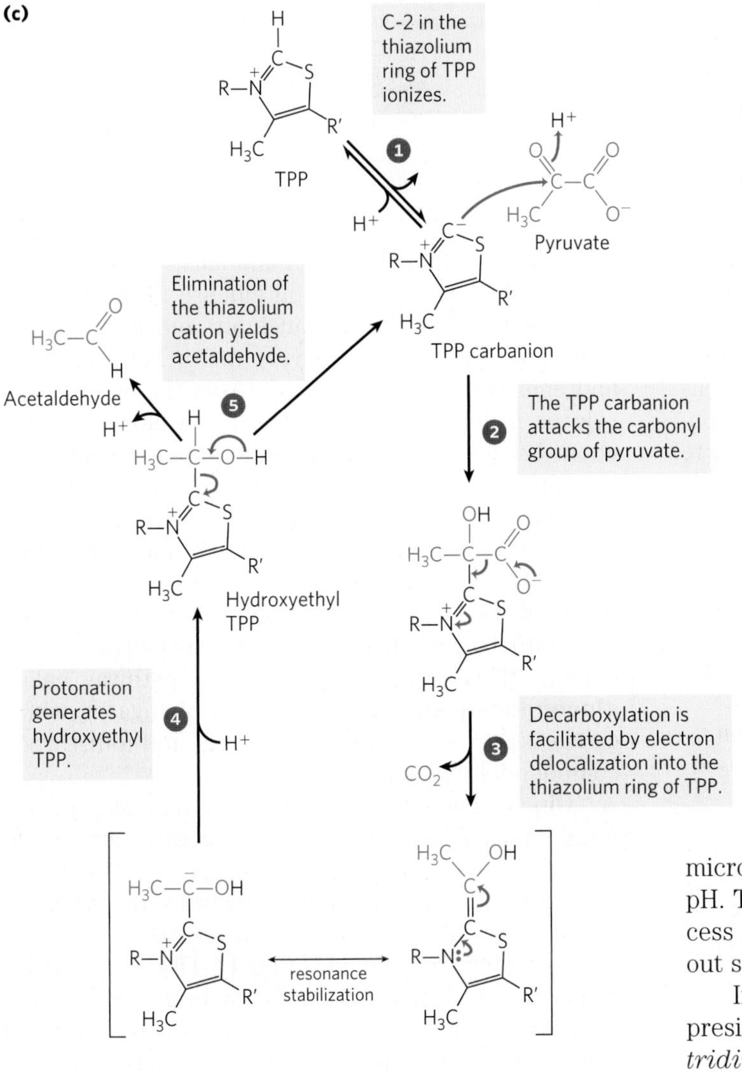

(a)

Thiamine pyrophosphate (TPP)

(b)

Hydroxyethyl thiamine pyrophosphate

(c)

MECHANISM FIGURE 14-15 **Thiamine pyrophosphate (TPP) and its role in pyruvate decarboxylation. (a)** TPP is the coenzyme form of vitamin B_1 (thiamine). The reactive carbon atom in the thiazolium ring of TPP is shown in red. In the reaction catalyzed by pyruvate decarboxylase, two of the three carbons of pyruvate are carried transiently on TPP in the form of a hydroxyethyl, or "active acetaldehyde," group **(b)**, which is subsequently released as acetaldehyde. **(c)** The thiazolium ring of TPP stabilizes carbanion intermediates by providing an electrophilic (electron-deficient) structure into which the carbanion electrons can be delocalized by resonance. Structures with this property, often called "electron sinks," play a role in many biochemical reactions—here, facilitating carbon–carbon bond cleavage.

produce propionic acid and CO_2; the propionic acid precipitates milk proteins, and bubbles of CO_2 cause the holes characteristic of Swiss cheese. Many other food products are the result of fermentations: pickles, sauerkraut, sausage, soy sauce, and a variety of national favorites, such as kimchi (Korea), tempoyak (Indonesia), kefir (Russia), dahi (India), and pozol (Mexico). The drop in pH associated with fermentation also helps to preserve foods, because most of the microorganisms that cause food spoilage cannot grow at low pH. In agriculture, plant byproducts such as corn stalks are preserved for use as animal feed by packing them into a large container (a silo) with limited access to air;

microbial fermentation produces acids that lower the pH. The silage that results from this fermentation process can be kept as animal feed for long periods without spoilage.

In 1910, Chaim Weizmann (later to become the first president of Israel) discovered that the bacterium *Clostridium acetobutyricum* ferments starch to butanol and acetone. This discovery opened the field of industrial fermentations, in which some readily available material rich in carbohydrate (corn starch or molasses, for example) is supplied to a pure culture of a specific microorganism, which ferments it into a product of greater commercial value. The ethanol used to make "gasohol" is produced by microbial fermentation, as are formic, acetic, propionic, butyric, and succinic acids, and glycerol, methanol, isopropanol, butanol, and butanediol. These fermentations are generally carried out in huge closed vats in which temperature and access to air are controlled to favor the multiplication of the desired microorganism and to exclude contaminating organisms. The beauty of industrial fermentations is

TABLE 14-1	Some TPP-Dependent Reactions		
Enzyme	Pathway(s)	Bond cleaved	Bond formed
Pyruvate decarboxylase	Ethanol fermentation	$R^1-\overset{O}{\underset{}{C}}-\overset{O}{\underset{O^-}{C}}$	$R^1-\overset{O}{\underset{H}{C}}$
Pyruvate dehydrogenase α-Ketoglutarate dehydrogenase	Synthesis of acetyl-CoA Citric acid cycle	$R^2-\overset{O}{\underset{}{C}}-\overset{O}{\underset{O^-}{C}}$	$R^2-\overset{O}{\underset{S\text{-}CoA}{C}}$
Transketolase	Carbon-assimilation reactions Pentose phosphate pathway	$R^3-\overset{O}{\underset{}{C}}-\overset{OH}{\underset{H}{C}}-R^4$	$R^3-\overset{O}{\underset{}{C}}-\overset{OH}{\underset{H}{C}}-R^5$

that complicated, multistep chemical transformations are carried out in high yields and with few side products by chemical factories that reproduce themselves—microbial cells. For some industrial fermentations, technology has been developed to immobilize the cells in an inert support, to pass the starting material continuously through the bed of immobilized cells, and to collect the desired product in the effluent—an engineer's dream!

SUMMARY 14.3 Fates of Pyruvate under Anaerobic Conditions: Fermentation

■ The NADH formed in glycolysis must be recycled to regenerate NAD$^+$, which is required as an electron acceptor in the first step of the payoff phase. Under aerobic conditions, electrons pass from NADH to O$_2$ in mitochondrial respiration.

■ Under anaerobic or hypoxic conditions, many organisms regenerate NAD$^+$ by transferring electrons from NADH to pyruvate, forming lactate. Other organisms, such as yeast, regenerate NAD$^+$ by reducing pyruvate to ethanol and CO$_2$. In these anaerobic processes (fermentations), there is no *net* oxidation or reduction of the carbons of glucose.

■ A variety of microorganisms can ferment sugar in fresh foods, resulting in changes in pH, taste, and texture, and preserving food from spoilage. Fermentations are used in industry to produce a wide variety of commercially valuable organic compounds from inexpensive starting materials.

14.4 Gluconeogenesis

The central role of glucose in metabolism arose early in evolution, and this sugar remains the nearly universal fuel and building block in modern organisms, from

microbes to humans. In mammals, some tissues depend almost completely on glucose for their metabolic energy. For the human brain and nervous system, as well as the erythrocytes, testes, renal medulla, and embryonic tissues, glucose from the blood is the sole or major fuel source. The brain alone requires about 120 g of glucose each day—more than half of all the glucose stored as glycogen in muscle and liver. However, the supply of glucose from these stores is not always sufficient; between meals and during longer fasts, or after vigorous exercise, glycogen is depleted. For these times, organisms need a method for synthesizing glucose from noncarbohydrate precursors. This is accomplished by a pathway called **gluconeogenesis** ("new formation of sugar"), which converts pyruvate and related three- and four-carbon compounds to glucose.

Gluconeogenesis occurs in all animals, plants, fungi, and microorganisms. The reactions are essentially the same in all tissues and all species. The important precursors of glucose in animals are three-carbon compounds such as lactate, pyruvate, and glycerol, as well as certain amino acids **(Fig. 14-16)**. In mammals, gluconeogenesis takes place mainly in the liver, and to a lesser extent in renal cortex and in the epithelial cells that line the inside of the small intestine. The glucose produced passes into the blood to supply other tissues. After vigorous exercise, lactate produced by anaerobic glycolysis in skeletal muscle returns to the liver and is converted to glucose, which moves back to muscle and is converted to glycogen—a circuit called the Cori cycle (Box 14-2; see also Fig. 23-21). In plant seedlings, stored fats and proteins are converted, via paths that include gluconeogenesis, to the disaccharide sucrose for transport throughout the developing plant. Glucose and its derivatives are precursors for the synthesis of plant cell walls, nucleotides and coenzymes, and a variety of other essential metabolites. In many microorganisms, gluconeogenesis starts from

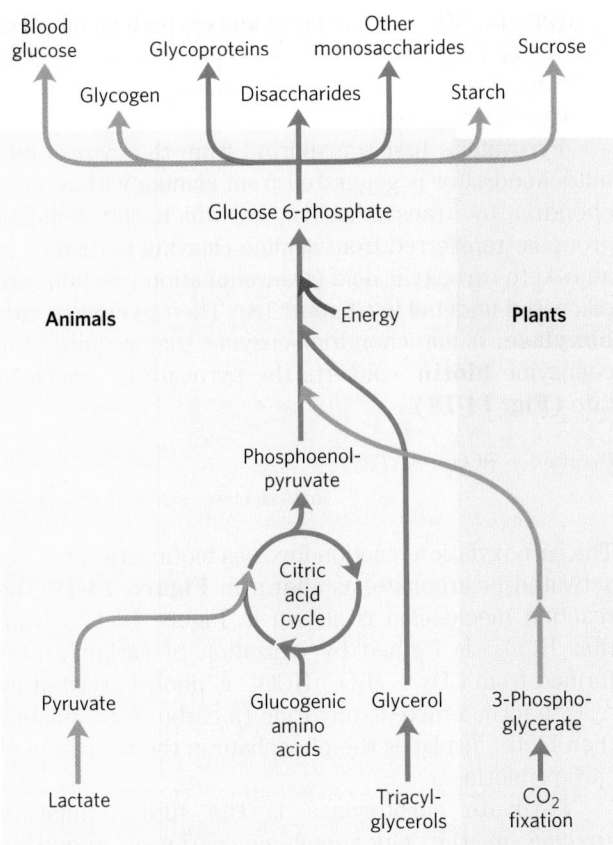

FIGURE 14-16 Carbohydrate synthesis from simple precursors. The pathway from phosphoenolpyruvate to glucose 6-phosphate is common to the biosynthetic conversion of many different precursors of carbohydrates in animals and plants. The path from pyruvate to phosphoenolpyruvate leads through oxaloacetate, an intermediate of the citric acid cycle, which we discuss in Chapter 16. Any compound that can be converted to either pyruvate or oxaloacetate can therefore serve as starting material for gluconeogenesis. This includes alanine and aspartate, which are convertible to pyruvate and oxaloacetate, respectively, and other amino acids that can also yield three- or four-carbon fragments, the so-called glucogenic amino acids (see Table 14-4; see also Fig. 18-15). Plants and photosynthetic bacteria are uniquely able to convert CO_2 to carbohydrates, using the Calvin cycle (see Section 20.5).

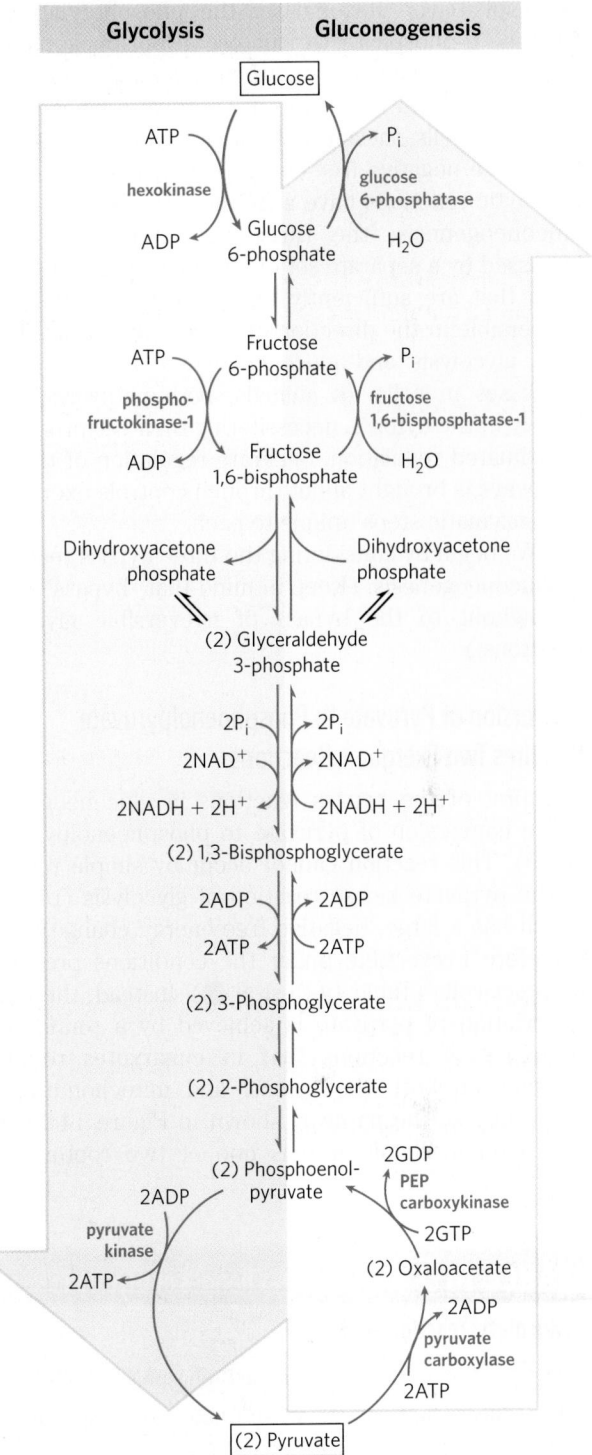

FIGURE 14-17 Opposing pathways of glycolysis and gluconeogenesis in rat liver. The reactions of glycolysis are on the left side, in red; the opposing pathway of gluconeogenesis is on the right, in blue. The major sites of regulation of gluconeogenesis shown here are discussed later in this chapter, and in detail in Chapter 15. Figure 14-20 illustrates an alternative route for oxaloacetate produced in mitochondria.

simple organic compounds of two or three carbons, such as acetate, lactate, and propionate, in their growth medium.

Although the reactions of gluconeogenesis are the same in all organisms, the metabolic context and the regulation of the pathway differ from one species to another and from tissue to tissue. In this section we focus on gluconeogenesis as it occurs in the mammalian liver. In Chapter 20 we show how photosynthetic organisms use this pathway to convert the primary products of photosynthesis into glucose, to be stored as sucrose or starch.

Gluconeogenesis and glycolysis are not identical pathways running in opposite directions, although they do share several steps **(Fig. 14-17)**; 7 of the 10 enzymatic reactions of gluconeogenesis are the reverse of glycolytic reactions. However, three reactions of glycolysis are essentially irreversible in vivo and cannot be used in gluconeogenesis: the conversion of glucose to glucose

6-phosphate by hexokinase, the phosphorylation of fructose 6-phosphate to fructose 1,6-bisphosphate by phosphofructokinase-1, and the conversion of phosphoenolpyruvate to pyruvate by pyruvate kinase (Fig. 14-17). In cells, these three reactions are characterized by a large negative free-energy change, whereas other glycolytic reactions have a ΔG near 0 (Table 14-2). In gluconeogenesis, the three irreversible steps are bypassed by a separate set of enzymes, catalyzing reactions that are sufficiently exergonic to be effectively irreversible in the direction of glucose synthesis. Thus, both glycolysis and gluconeogenesis are irreversible processes in cells. In animals, both pathways occur largely in the cytosol, necessitating their reciprocal and coordinated regulation. Separate regulation of the two pathways is brought about through controls exerted on the enzymatic steps unique to each.

We begin by considering the three bypass reactions of gluconeogenesis. (Keep in mind that "bypass" refers throughout to the bypass of irreversible glycolytic reactions.)

Conversion of Pyruvate to Phosphoenolpyruvate Requires Two Exergonic Reactions

The first of the bypass reactions in gluconeogenesis is the conversion of pyruvate to phosphoenolpyruvate (PEP). This reaction cannot occur by simple reversal of the pyruvate kinase reaction of glycolysis (p. 544), which has a large, negative free-energy change and is therefore irreversible under the conditions prevailing in intact cells (Table 14-2, step ❿). Instead, the phosphorylation of pyruvate is achieved by a roundabout sequence of reactions that in eukaryotes requires enzymes in both the cytosol and mitochondria. As we shall see, the pathway shown in Figure 14-17 and described in detail here is one of two routes from

pyruvate to PEP; it is the predominant path when pyruvate or alanine is the glucogenic precursor. A second pathway, described later, predominates when lactate is the glucogenic precursor.

Pyruvate is first transported from the cytosol into mitochondria or is generated from alanine within mitochondria by transamination, in which the α-amino group is transferred from alanine (leaving pyruvate) to an α-keto carboxylic acid (transamination reactions are discussed in detail in Chapter 18). Then **pyruvate carboxylase**, a mitochondrial enzyme that requires the coenzyme **biotin**, converts the pyruvate to oxaloacetate **(Fig. 14-18)**:

$$\text{Pyruvate} + \text{HCO}_3^- + \text{ATP} \longrightarrow$$
$$\text{oxaloacetate} + \text{ADP} + \text{P}_i \quad (14\text{-}4)$$

The carboxylation reaction involves biotin as a carrier of activated bicarbonate, as shown in **Figure 14-19**; the reaction mechanism is shown in Figure 16-17. (Note that HCO_3^- is formed by ionization of carbonic acid formed from $CO_2 + H_2O$.) HCO_3^- is phosphorylated by ATP to form a mixed anhydride (a carboxyphosphate); then biotin displaces the phosphate in the formation of carboxybiotin.

Pyruvate carboxylase is the first regulatory enzyme in the gluconeogenic pathway, requiring acetyl-CoA as a positive effector. (Acetyl-CoA is produced by fatty acid oxidation (Chapter 17), and its accumulation signals the availability of fatty acids as fuel.) As we shall see in Chapter 16 (see Fig. 16-16), the pyruvate carboxylase reaction can replenish intermediates in another central metabolic pathway, the citric acid cycle.

Because the mitochondrial membrane has no transporter for oxaloacetate, before export to the cytosol the oxaloacetate formed from pyruvate must be reduced to

TABLE 14-2	Free-Energy Changes of Glycolytic Reactions in Erythrocytes		
Glycolytic reaction step		$\Delta G'^\circ$ **(kJ/mol)**	ΔG **(kJ/mol)**
❶ Glucose + ATP $\longrightarrow$ glucose 6-phosphate + ADP		−16.7	−33.4
❷ Glucose 6-phosphate $\rightleftharpoons$ fructose 6-phosphate		1.7	0 to 25
❸ Fructose 6-phosphate + ATP $\longrightarrow$ fructose1,6-bisphosphate + ADP		−14.2	−22.2
❹ Fructose1,6-bisphosphate $\rightleftharpoons$ dihydroxyacetone phosphate + glyceraldehyde 3-phosphate		23.8	−6 to 0
❺ Dihydroxyacetone phosphate $\rightleftharpoons$ glyceraldehyde 3-phosphate		7.5	0 to 4
❻ Glyceraldehyde 3-phosphate + P_i + NAD^+ $\rightleftharpoons$ 1,3-bisphosphoglycerate + NADH + H^+		6.3	−2 to 2
❼ 1,3-Bisphosphoglycerate + ADP $\rightleftharpoons$ 3-phosphoglycerate + ATP		−18.8	0 to 2
❽ 3-Phosphoglycerate $\rightleftharpoons$ 2-phosphoglycerate		4.4	0 to 0.8
❾ 2-Phosphoglycerate $\rightleftharpoons$ phosphoenolpyruvate + H_2O		7.5	0 to 3.3
❿ Phosphoenolpyruvate + ADP $\longrightarrow$ pyruvate + ATP		−31.4	−16.7

Note: $\Delta G'^\circ$ is the standard free-energy change, as defined in Chapter 13 (pp. 497–498). ΔG is the free-energy change calculated from the actual concentrations of glycolytic intermediates present under physiological conditions in erythrocytes, at pH 7. The glycolytic reactions bypassed in gluconeogenesis are shown in red. Biochemical equations are not necessarily balanced for H or charge (p. 506–507).

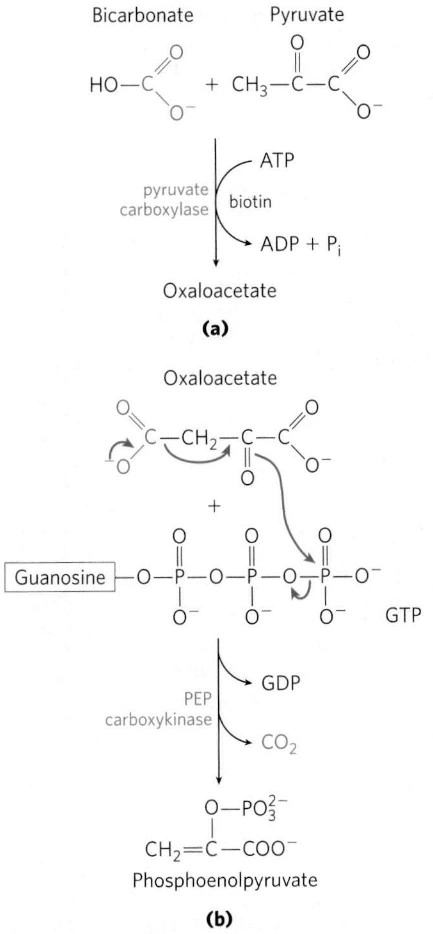

FIGURE 14-18 Synthesis of phosphoenolpyruvate from pyruvate. (a) In mitochondria, pyruvate is converted to oxaloacetate in a biotin-requiring reaction catalyzed by pyruvate carboxylase. **(b)** In the cytosol, oxaloacetate is converted to phosphoenolpyruvate by PEP carboxykinase. The CO_2 incorporated in the pyruvate carboxylase reaction is lost here as CO_2. The decarboxylation leads to a rearrangement of electrons that facilitates attack of the carbonyl oxygen of the pyruvate moiety on the γ phosphate of GTP.

malate by mitochondrial **malate dehydrogenase**, at the expense of NADH:

$$\text{Oxaloacetate} + \text{NADH} + \text{H}^+ \rightleftharpoons \text{L-malate} + \text{NAD}^+ \quad (14\text{-}5)$$

The standard free-energy change for this reaction is quite high, but under physiological conditions (including a very low concentration of oxaloacetate) $\Delta G \approx 0$ and the reaction is readily reversible. Mitochondrial malate dehydrogenase functions in both gluconeogenesis and the citric acid cycle, but the overall flow of metabolites in the two processes is in opposite directions.

Malate leaves the mitochondrion through a specific transporter in the inner mitochondrial membrane (see Fig. 19-31), and in the cytosol it is reoxidized to oxaloacetate, with the production of cytosolic NADH:

$$\text{Malate} + \text{NAD}^+ \longrightarrow \text{oxaloacetate} + \text{NADH} + \text{H}^+ \quad (14\text{-}6)$$

The oxaloacetate is then converted to PEP by **phosphoenolpyruvate carboxykinase** (Fig. 14-18). This Mg^{2+}-dependent reaction requires GTP as the phosphoryl group donor:

$$\text{Oxaloacetate} + \text{GTP} \rightleftharpoons \text{PEP} + \text{CO}_2 + \text{GDP} \quad (14\text{-}7)$$

The reaction is reversible under intracellular conditions; the formation of one high-energy phosphate compound (PEP) is balanced by the hydrolysis of another (GTP). The overall equation for this set of bypass reactions, the sum of Equations 14-4 through 14-7, is

$$\text{Pyruvate} + \text{ATP} + \text{GTP} + \text{HCO}_3^- \longrightarrow$$
$$\text{PEP} + \text{ADP} + \text{GDP} + \text{P}_i + \text{CO}_2 \quad (14\text{-}8)$$
$$\Delta G'^\circ = 0.9 \text{ kJ/mol}$$

Two high-energy phosphate equivalents (one from ATP and one from GTP), each yielding about 50 kJ/mol

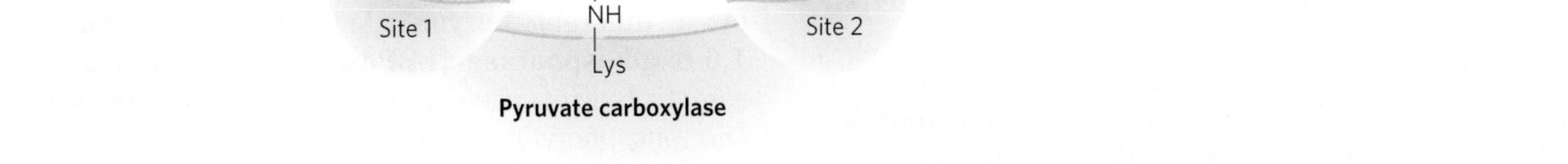

FIGURE 14-19 Role of biotin in the pyruvate carboxylase reaction. The cofactor biotin is covalently attached to the enzyme through an amide linkage to the ε-amino group of a Lys residue, forming a biotinyl-enzyme. The reaction occurs in two phases, which occur at two different sites in the enzyme. At catalytic site 1, bicarbonate ion is converted to CO_2 at the expense of ATP. Then CO_2 reacts with biotin, forming carboxybiotinyl-enzyme. The long arm composed of biotin and the Lys side chain to which it is attached then carry the CO_2 of carboxybiotinyl-enzyme to catalytic site 2 on the enzyme surface, where CO_2 is released and reacts with the pyruvate, forming oxaloacetate and regenerating the biotinyl-enzyme. The general role of flexible arms in carrying reaction intermediates between enzyme active sites is described in Figure 16-18, and the mechanistic details of the pyruvate carboxylase reaction are shown in Figure 16-17. Similar mechanisms occur in other biotin-dependent carboxylation reactions, such as those catalyzed by propionyl-CoA carboxylase (see Fig. 17-12) and acetyl-CoA carboxylase (see Fig. 21-1).

under cellular conditions, must be expended to phosphorylate one molecule of pyruvate to PEP. In contrast, when PEP is converted to pyruvate during glycolysis, only one ATP is generated from ADP. Although the standard free-energy change ($\Delta G'^\circ$) of the two-step path from pyruvate to PEP is 0.9 kJ/mol, the actual free-energy change (ΔG), calculated from measured cellular concentrations of intermediates, is very strongly negative (−25 kJ/mol); this results from the ready consumption of PEP in other reactions such that its concentration remains relatively low. The reaction is thus effectively irreversible in the cell.

Note that the CO_2 added to pyruvate in the pyruvate carboxylase step is the same molecule that is lost in the PEP carboxykinase reaction (Fig. 14-18b). This carboxylation-decarboxylation sequence represents a way of "activating" pyruvate, in that the decarboxylation of oxaloacetate facilitates PEP formation. In Chapter 21 we shall see how a similar carboxylation-decarboxylation sequence is used to activate acetyl-CoA for fatty acid biosynthesis (see Fig. 21-1).

There is a logic to the route of these reactions through the mitochondrion. The [NADH]/[NAD$^+$] ratio in the cytosol is 8×10^{-4}, about 10^5 times lower than in mitochondria. Because cytosolic NADH is consumed in gluconeogenesis (in the conversion of 1,3-bisphosphoglycerate to glyceraldehyde 3-phosphate; Fig. 14-17), glucose biosynthesis cannot proceed unless NADH is available. The transport of malate from the mitochondrion to the cytosol and its reconversion there to oxaloacetate effectively moves reducing equivalents to the cytosol, where they are scarce. This path from pyruvate to PEP therefore provides an important balance between NADH produced and consumed in the cytosol during gluconeogenesis.

A second pyruvate → PEP bypass predominates when lactate is the glucogenic precursor **(Fig. 14-20)**. This pathway makes use of lactate produced by glycolysis in erythrocytes or anaerobic muscle, for example, and it is particularly important in large vertebrates after vigorous exercise (Box 14-2). The conversion of lactate to pyruvate in the cytosol of hepatocytes yields NADH, and the export of reducing equivalents (as malate) from mitochondria is therefore unnecessary. After the pyruvate produced by the lactate dehydrogenase reaction is transported into the mitochondrion, it is converted to oxaloacetate by pyruvate carboxylase, as described above. This oxaloacetate, however, is converted directly to PEP by a mitochondrial isozyme of PEP carboxykinase, and the PEP is transported out of the mitochondrion to continue on the gluconeogenic path. The mitochondrial and cytosolic isozymes of PEP carboxykinase are encoded by separate genes in the nuclear chromosomes, providing another example of two distinct enzymes catalyzing the same reaction but having different cellular locations or metabolic roles (recall the isozymes of hexokinase).

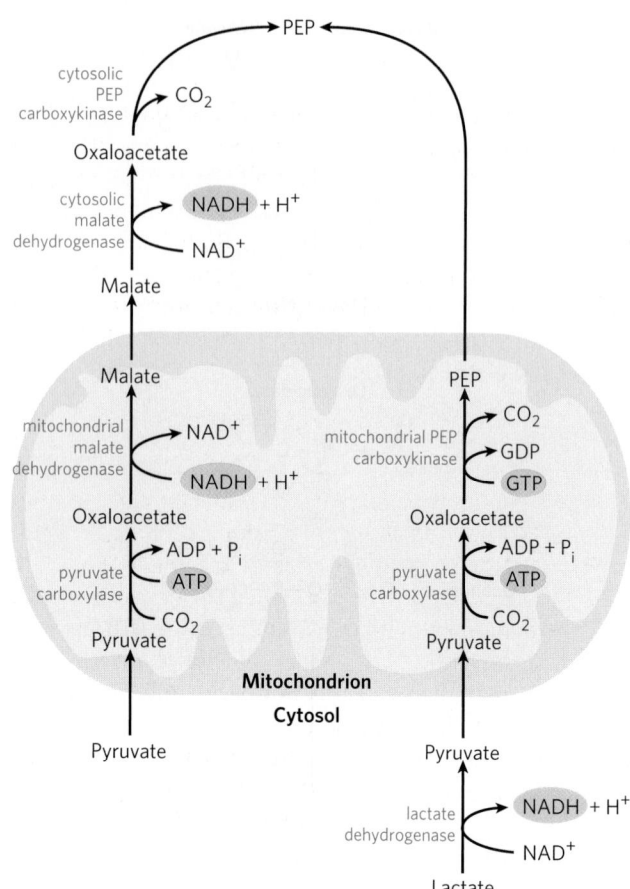

FIGURE 14-20 Alternative paths from pyruvate to phosphoenolpyruvate. The relative importance of the two pathways depends on the availability of lactate or pyruvate and the cytosolic requirements for NADH for gluconeogenesis. The path on the right predominates when lactate is the precursor, because cytosolic NADH is generated in the lactate dehydrogenase reaction and does not have to be shuttled out of the mitochondrion (see text).

Conversion of Fructose 1,6-Bisphosphate to Fructose 6-Phosphate Is the Second Bypass

The second glycolytic reaction that cannot participate in gluconeogenesis is the phosphorylation of fructose 6-phosphate by PFK-1 (Table 14-2, step ③). Because this reaction is highly exergonic and therefore irreversible in intact cells, the generation of fructose 6-phosphate from fructose 1,6-bisphosphate (Fig. 14-17) is catalyzed by a different enzyme, Mg^{2+}-dependent **fructose 1,6-bisphosphatase (FBPase-1)**, which promotes the essentially irreversible *hydrolysis* of the C-1 phosphate (*not* phosphoryl group transfer to ADP):

Fructose 1,6-bisphosphate + H_2O $\longrightarrow$

$$\text{fructose 6-phosphate} + P_i$$
$$\Delta G'^\circ = -16.3\,\text{kJ/mol}$$

FBPase-1 is so named to distinguish it from another, similar enzyme (FBPase-2) with a regulatory role, which we discuss in Chapter 15.

Conversion of Glucose 6-Phosphate to Glucose Is the Third Bypass

The third bypass is the final reaction of gluconeogenesis, the dephosphorylation of glucose 6-phosphate to yield glucose (Fig. 14-17). Reversal of the hexokinase reaction (p. 538) would require phosphoryl group transfer from glucose 6-phosphate to ADP, forming ATP, an energetically unfavorable reaction (Table 14-2, step ❶). The reaction catalyzed by **glucose 6-phosphatase** does not require synthesis of ATP; it is a simple hydrolysis of a phosphate ester:

$$\text{Glucose 6-phosphate} + H_2O \longrightarrow \text{glucose} + P_i$$
$$\Delta G'^{\circ} = -13.8 \text{ kJ/mol}$$

This Mg^{2+}-activated enzyme is found on the lumenal side of the endoplasmic reticulum of hepatocytes, renal cells, and epithelial cells of the small intestine (see Fig. 15-30), but not in other tissues, which are therefore unable to supply glucose to the blood. If other tissues had glucose 6-phosphatase, this enzyme's activity would hydrolyze the glucose 6-phosphate needed within those tissues for glycolysis. Glucose produced by gluconeogenesis in the liver or kidney or ingested in the diet is delivered to these other tissues, including brain and muscle, through the bloodstream.

Gluconeogenesis Is Energetically Expensive, but Essential

The sum of the biosynthetic reactions leading from pyruvate to free blood glucose (Table 14-3) is

$$2 \text{ Pyruvate} + 4ATP + 2GTP + 2NADH + 2H^+ + 4H_2O \longrightarrow$$
$$\text{glucose} + 4ADP + 2GDP + 6P_i + 2NAD^+ \quad (14\text{-}9)$$

For each molecule of glucose formed from pyruvate, six high-energy phosphate groups are required, four from ATP and two from GTP. In addition, two molecules of NADH are required for the reduction of two molecules of 1,3-bisphosphoglycerate. Clearly, Equation 14-9 is not simply the reverse of the equation for conversion of glucose to pyruvate by glycolysis, which would require only two molecules of ATP:

$$\text{Glucose} + 2ADP + 2P_i + NAD^+ \longrightarrow$$
$$2 \text{ pyruvate} + 2ATP + 2NADH + 2H^+ + 2H_2O$$

The synthesis of glucose from pyruvate is a relatively expensive process. Much of this high energy cost is necessary to ensure the irreversibility of gluconeogenesis. Under intracellular conditions, the overall free-energy change of glycolysis is at least -63 kJ/mol. Under the same conditions the overall ΔG of gluconeogenesis is -16 kJ/mol. Thus both glycolysis and gluconeogenesis are essentially irreversible processes in cells. A second advantage to investing energy to convert pyruvate to glucose is that if pyruvate were instead excreted, its considerable potential for ATP production by complete, aerobic oxidation would be lost (more than 10 ATP are produced per pyruvate, as we shall see in Chapter 16).

Citric Acid Cycle Intermediates and Some Amino Acids Are Glucogenic

The biosynthetic pathway to glucose described above allows the net synthesis of glucose not only from pyruvate but also from the four-, five-, and six-carbon intermediates of the citric acid cycle (Chapter 16). Citrate,

TABLE 14-3 Sequential Reactions in Gluconeogenesis Starting from Pyruvate	
Pyruvate + HCO_3^- + ATP $\longrightarrow$ oxaloacetate + ADP + P_i	×2
Oxaloacetate + GTP $\rightleftharpoons$ phosphoenolpyruvate + CO_2 + GDP	×2
Phosphoenolpyruvate + H_2O $\rightleftharpoons$ 2-phosphoglycerate	×2
2-Phosphoglycerate $\rightleftharpoons$ 3-phosphoglycerate	×2
3-Phosphoglycerate + ATP $\rightleftharpoons$ 1,3-bisphosphoglycerate + ADP	×2
1,3-Bisphosphoglycerate + NADH + H^+ $\rightleftharpoons$ glyceraldehyde 3-phosphate + NAD^+ + P_i	×2
Glyceraldehyde 3-phosphate $\rightleftharpoons$ dihydroxyacetone phosphate	
Glyceraldehyde 3-phosphate + dihydroxyacetone phosphate $\rightleftharpoons$ fructose 1,6-bisphosphate	
Fructose 1,6-bisphosphate $\longrightarrow$ fructose 6-phosphate + P_i	
Fructose 6-phosphate $\rightleftharpoons$ glucose 6-phosphate	
Glucose 6-phosphate + H_2O $\longrightarrow$ glucose + P_i	
Sum: 2 Pyruvate + 4ATP + 2GTP + 2NADH + $2H^+$ + $4H_2O$ $\longrightarrow$ glucose + 4ADP + 2GDP + $6P_i$ + $2NAD^+$	

Note: The bypass reactions are in red; all other reactions are reversible steps of glycolysis. The figures at the right indicate that the reaction is to be counted twice, because two three-carbon precursors are required to make a molecule of glucose. The reactions required to replace the cytosolic NADH consumed in the glyceraldehyde 3-phosphate dehydrogenase reaction (the conversion of lactate to pyruvate in the cytosol or the transport of reducing equivalents from mitochondria to the cytosol in the form of malate) are not considered in this summary. Biochemical equations are not necessarily balanced for H and charge (p. 498).

TABLE 14-4	Glucogenic Amino Acids, Grouped by Site of Entry	
Pyruvate		**Succinyl-CoA**
Alanine		Isoleucine[a]
Cysteine		Methionine
Glycine		Threonine
Serine		Valine
Threonine		**Fumarate**
Tryptophan[a]		Phenylalanine[a]
α-Ketoglutarate		Tyrosine[a]
Arginine		**Oxaloacetate**
Glutamate		Asparagine
Glutamine		Aspartate
Histidine		
Proline		

Note: All these amino acids are precursors of blood glucose or liver glycogen, because they can be converted to pyruvate or citric acid cycle intermediates. Of the 20 common amino acids, only leucine and lysine are unable to furnish carbon for net glucose synthesis.

[a]These amino acids are also ketogenic (see Fig. 18-15).

isocitrate, α-ketoglutarate, succinyl-CoA, succinate, fumarate, and malate—all are citric acid cycle intermediates that can undergo oxidation to oxaloacetate (see Fig. 16-7). Some or all of the carbon atoms of most amino acids derived from proteins are ultimately catabolized to pyruvate or to intermediates of the citric acid cycle. Such amino acids can therefore undergo net conversion to glucose and are said to be **glucogenic** (Table 14-4). Alanine and glutamine, the principal molecules that transport amino groups from extrahepatic tissues to the liver (see Fig. 18-9), are particularly important glucogenic amino acids in mammals. After removal of their amino groups in liver mitochondria, the carbon skeletons remaining (pyruvate and α-ketoglutarate, respectively) are readily funneled into gluconeogenesis.

Mammals Cannot Convert Fatty Acids to Glucose

No net conversion of fatty acids to glucose occurs in mammals. As we shall see in Chapter 17, the catabolism of most fatty acids yields only acetyl-CoA. Mammals cannot use acetyl-CoA as a precursor of glucose, because the pyruvate dehydrogenase reaction is irreversible and cells have no other pathway to convert acetyl-CoA to pyruvate. Plants, yeast, and many bacteria do have a pathway (the glyoxylate cycle; see Fig. 20-55) for converting acetyl-CoA to oxaloacetate, so these organisms can use fatty acids as the starting material for gluconeogenesis. This is important during the germination of seedlings, for example;

before leaves develop and photosynthesis can provide energy and carbohydrates, the seedling relies on stored seed oils for energy production and cell wall biosynthesis.

Although mammals cannot convert fatty acids to carbohydrate, they can use the small amount of glycerol produced from the breakdown of fats (triacyl*glycerols*) for gluconeogenesis. Phosphorylation of glycerol by glycerol kinase, followed by oxidation of the central carbon, yields dihydroxyacetone phosphate, an intermediate in gluconeogenesis in liver.

As we shall see in Chapter 21, glycerol phosphate is an essential intermediate in triacylglycerol synthesis in adipocytes, but these cells lack glycerol kinase and so cannot simply phosphorylate glycerol. Instead, adipocytes carry out a truncated version of gluconeogenesis, known as **glyceroneogenesis**: the conversion of pyruvate to dihydroxyacetone phosphate via the early reactions of gluconeogenesis, followed by reduction of the dihydroxyacetone phosphate to glycerol phosphate (see Fig. 21-21).

Glycolysis and Gluconeogenesis Are Reciprocally Regulated

If glycolysis (the conversion of glucose to pyruvate) and gluconeogenesis (the conversion of pyruvate to glucose) were allowed to proceed simultaneously at high rates, the result would be the consumption of ATP and the production of heat. For example, PFK-1 and FBPase-1 catalyze opposing reactions:

$$ATP + \text{fructose 6-phosphate} \xrightarrow{\text{PFK-1}}$$
$$ADP + \text{fructose 1,6-bisphosphate}$$

$$\text{Fructose 1,6-bisphosphate} + H_2O \xrightarrow{\text{FBPase-1}}$$
$$\text{fructose 6-phosphate} + P_i$$

The sum of these two reactions is

$$ATP + H_2O \longrightarrow ADP + P_i + \text{heat}$$

These two enzymatic reactions, and several others in the two pathways, are regulated allosterically and by covalent modification (phosphorylation). In Chapter 15 we take up the mechanisms of this regulation in detail. For now, suffice it to say that the pathways are regulated so that when the flux of glucose through glycolysis goes up, the flux of pyruvate toward glucose goes down, and vice versa.

SUMMARY 14.4 Gluconeogenesis

■ Gluconeogenesis is a ubiquitous multistep process in which glucose is produced from lactate, pyruvate, or oxaloacetate, or any compound (including citric acid cycle intermediates) that can be converted to one of these intermediates. Seven of the steps in gluconeogenesis are catalyzed by the same enzymes used in glycolysis; these are the reversible reactions.

■ Three irreversible steps in glycolysis are bypassed by reactions catalyzed by gluconeogenic enzymes: (1) conversion of pyruvate to PEP via oxaloacetate, catalyzed by pyruvate carboxylase and PEP carboxykinase; (2) dephosphorylation of fructose 1,6-bisphosphate by FBPase-1; and (3) dephosphorylation of glucose 6-phosphate by glucose 6-phosphatase.

■ Formation of one molecule of glucose from pyruvate requires 4 ATP, 2 GTP, and 2 NADH; it is expensive.

■ In mammals, gluconeogenesis in the liver, kidney, and small intestine provides glucose for use by the brain, muscles, and erythrocytes.

■ Pyruvate carboxylase is stimulated by acetyl-CoA, increasing the rate of gluconeogenesis when the cell has adequate supplies of other substrates (fatty acids) for energy production.

■ Animals cannot convert acetyl-CoA derived from fatty acids into glucose; plants and microorganisms can.

■ Glycolysis and gluconeogenesis are reciprocally regulated to prevent wasteful operation of both pathways at the same time.

14.5 Pentose Phosphate Pathway of Glucose Oxidation

In most animal tissues, the major catabolic fate of glucose 6-phosphate is glycolytic breakdown to pyruvate, much of which is then oxidized via the citric acid cycle, ultimately leading to the formation of ATP. Glucose 6-phosphate does have other catabolic fates, however, which lead to specialized products needed by the cell. Of particular importance in some tissues is the oxidation of glucose 6-phosphate to pentose phosphates by the **pentose phosphate pathway** (also called the **phosphogluconate pathway** or the **hexose mono-phosphate pathway; Fig. 14-21**). In this oxidative pathway, $NADP^+$ is the electron acceptor, yielding NADPH. Rapidly dividing cells, such as those of bone marrow, skin, and intestinal mucosa, and those of tumors, use the pentose ribose 5-phosphate to make RNA, DNA, and such coenzymes as ATP, NADH, $FADH_2$, and coenzyme A.

In other tissues, the essential product of the pentose phosphate pathway is not the pentoses but the electron donor NADPH, needed for reductive biosynthesis or to counter the damaging effects of oxygen radicals. Tissues that carry out extensive fatty acid synthesis (liver, adipose, lactating mammary gland) or very active synthesis of cholesterol and steroid

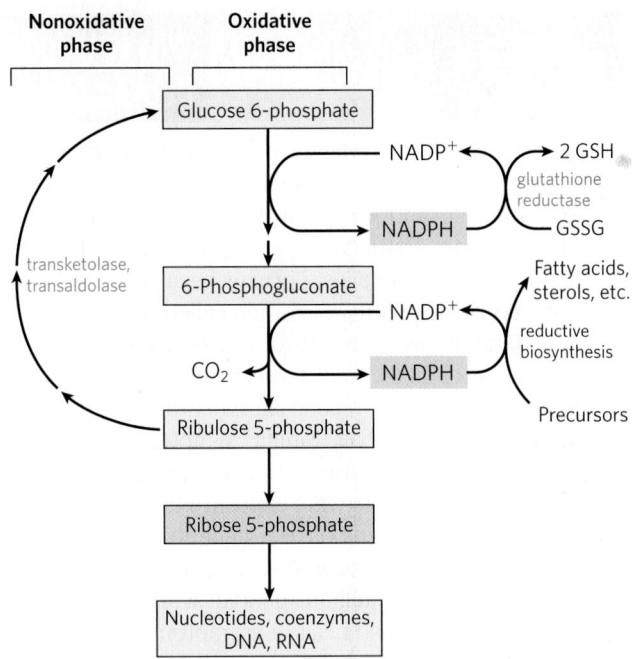

FIGURE 14-21 General scheme of the pentose phosphate pathway. NADPH formed in the oxidative phase is used to reduce glutathione, GSSG (see Box 14-4), and to support reductive biosynthesis. The other product of the oxidative phase is ribose 5-phosphate, which serves as a precursor for nucleotides, coenzymes, and nucleic acids. In cells that are not using ribose 5-phosphate for biosynthesis, the nonoxidative phase recycles six molecules of the pentose into five molecules of the hexose glucose 6-phosphate, allowing continued production of NADPH and converting glucose 6-phosphate (in six cycles) to CO_2.

hormones (liver, adrenal glands, gonads) require the NADPH provided by this pathway. Erythrocytes and the cells of the lens and cornea are directly exposed to oxygen and thus to the damaging free radicals generated by oxygen. By maintaining a reducing atmosphere (a high ratio of NADPH to $NADP^+$ and a high ratio of reduced to oxidized glutathione), such cells can prevent or undo oxidative damage to proteins, lipids, and other sensitive molecules. In erythrocytes, the NADPH produced by the pentose phosphate pathway is so important in preventing oxidative damage that a genetic defect in glucose 6-phosphate dehydrogenase, the first enzyme of the pathway, can have serious medical consequences (Box 14-4). ■

The Oxidative Phase Produces Pentose Phosphates and NADPH

The first reaction of the pentose phosphate pathway **(Fig. 14-22)** is the oxidation of glucose 6-phosphate by **glucose 6-phosphate dehydrogenase (G6PD)** to form 6-phosphoglucono-δ-lactone, an intramolecular ester. $NADP^+$ is the electron acceptor, and the overall equilibrium lies far in the direction of NADPH

BOX 14-4 ⚕ MEDICINE   Why Pythagoras Wouldn't Eat Falafel: Glucose 6-Phosphate Dehydrogenase Deficiency

Fava beans, an ingredient of falafel, have been an important food source in the Mediterranean and Middle East since antiquity. The Greek philosopher and mathematician Pythagoras prohibited his followers from dining on fava beans, perhaps because they make many people sick with a condition called favism, which can be fatal. In favism, erythrocytes begin to lyse 24 to 48 hours after ingestion of the beans, releasing free hemoglobin into the blood. Jaundice and sometimes kidney failure can result. Similar symptoms can occur with ingestion of the antimalarial drug primaquine or of sulfa antibiotics, or following exposure to certain herbicides. These symptoms have a genetic basis: glucose 6-phosphate dehydrogenase (G6PD) deficiency, which affects about 400 million people worldwide. Most G6PD-deficient individuals are asymptomatic; only the combination of G6PD deficiency and certain environmental factors produces the clinical manifestations.

Glucose 6-phosphate dehydrogenase catalyzes the first step in the pentose phosphate pathway (see Fig. 14-22), which produces NADPH. This reductant, essential in many biosynthetic pathways, also protects cells from oxidative damage by hydrogen peroxide (H_2O_2) and superoxide free radicals, highly reactive oxidants generated as metabolic byproducts and through the actions of drugs such as primaquine and natural products such as divicine—the toxic ingredient of fava beans. During normal detoxification, H_2O_2 is converted to H_2O by reduced glutathione and glutathione peroxidase, and the oxidized glutathione is converted back to the reduced form by glutathione reductase and NADPH (Fig. 1). H_2O_2 is also broken down to H_2O and O_2 by catalase, which also requires NADPH. In G6PD-deficient individuals, the NADPH production is diminished and detoxification of H_2O_2 is inhibited. Cellular damage results: lipid peroxidation leading to breakdown of erythrocyte membranes and oxidation of proteins and DNA.

The geographic distribution of G6PD deficiency is instructive. Frequencies as high as 25% occur in tropical Africa, parts of the Middle East, and Southeast Asia, areas where malaria is most prevalent. In addition to such epidemiological observations, in vitro studies show that growth of one malaria parasite, *Plasmodium falciparum*, is inhibited in G6PD-deficient erythrocytes. The parasite is very sensitive to oxidative damage and is killed by a level of oxidative stress that

is tolerable to a G6PD-deficient human host. Because the advantage of resistance to malaria balances the disadvantage of lowered resistance to oxidative damage, natural selection sustains the G6PD-deficient genotype in human populations where malaria is prevalent. Only under overwhelming oxidative stress, caused by drugs, herbicides, or divicine, does G6PD deficiency cause serious medical problems.

An antimalarial drug such as primaquine is believed to act by causing oxidative stress to the parasite. It is ironic that antimalarial drugs can cause human illness through the same biochemical mechanism that provides resistance to malaria. Divicine also acts as an antimalarial drug, and ingestion of fava beans may protect against malaria. By refusing to eat falafel, many Pythagoreans with normal G6PD activity may have unwittingly increased their risk of malaria!

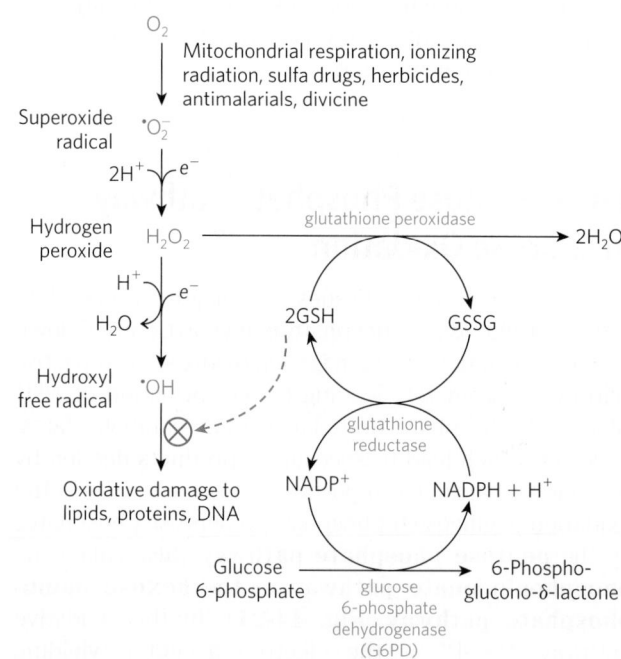

FIGURE 1 Role of NADPH and glutathione in protecting cells against highly reactive oxygen derivatives. Reduced glutathione (GSH) protects the cell by destroying hydrogen peroxide and hydroxyl free radicals. Regeneration of GSH from its oxidized form (GSSG) requires the NADPH produced in the glucose 6-phosphate dehydrogenase reaction.

formation. The lactone is hydrolyzed to the free acid 6-phosphogluconate by a specific **lactonase**, then 6-phosphogluconate undergoes oxidation and decarboxylation by **6-phosphogluconate dehydrogenase**

to form the ketopentose ribulose 5-phosphate; the reaction generates a second molecule of NADPH. (This ribulose 5-phosphate is important in the regulation of glycolysis and gluconeogenesis, as we shall see in

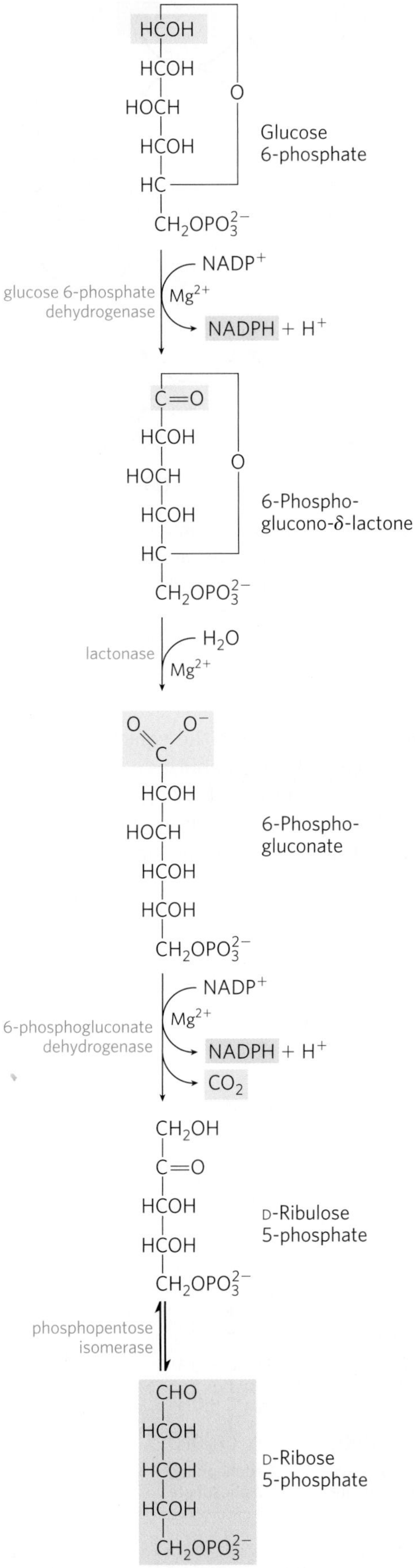

Chapter 15.) **Phosphopentose isomerase** converts ribulose 5-phosphate to its aldose isomer, ribose 5-phosphate. In some tissues, the pentose phosphate pathway ends at this point, and its overall equation is

$$\text{Glucose 6-phosphate} + 2\text{NADP}^+ + \text{H}_2\text{O} \longrightarrow$$
$$\text{ribose 5-phosphate} + \text{CO}_2 + 2\text{NADPH} + 2\text{H}^+$$

The net result is the production of NADPH, a reductant for biosynthetic reactions, and ribose 5-phosphate, a precursor for nucleotide synthesis.

The Nonoxidative Phase Recycles Pentose Phosphates to Glucose 6-Phosphate

In tissues that require primarily NADPH, the pentose phosphates produced in the oxidative phase of the pathway are recycled into glucose 6-phosphate. In this nonoxidative phase, ribulose 5-phosphate is first epimerized to xylulose 5-phosphate:

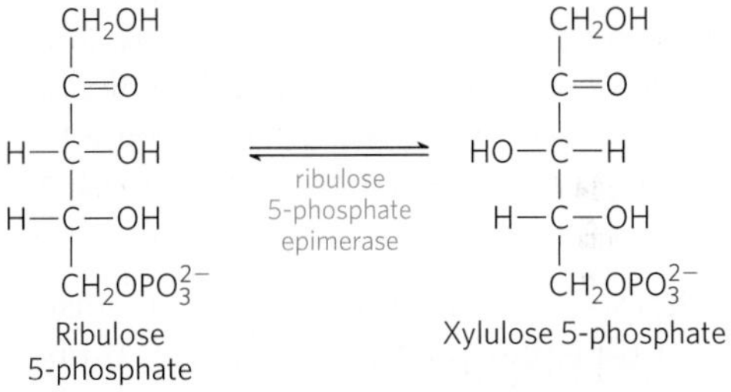

Then, in a series of rearrangements of the carbon skeletons **(Fig. 14-23)**, six five-carbon sugar phosphates are converted to five six-carbon sugar phosphates, completing the cycle and allowing continued oxidation of glucose 6-phosphate with production of NADPH. Continued recycling leads ultimately to the conversion of glucose 6-phosphate to six CO_2. Two enzymes unique to the pentose phosphate pathway act in these interconversions of sugars: transketolase and transaldolase. **Transketolase** catalyzes the transfer of a two carbon fragment from a ketose donor to an aldose acceptor **(Fig. 14-24a)**. In its first appearance in the pentose phosphate pathway, transketolase transfers C-1 and C-2 of xylulose 5-phosphate to ribose 5-phosphate, forming the seven-carbon product sedoheptulose 7-phosphate (Fig. 14-24b). The remaining three-carbon fragment from xylulose is glyceraldehyde 3-phosphate.

Next, **transaldolase** catalyzes a reaction similar to the aldolase reaction of glycolysis: a three-carbon fragment is removed from sedoheptulose 7-phosphate and condensed with glyceraldehyde 3-phosphate, forming fructose 6-phosphate and the tetrose erythrose 4-phosphate **(Fig. 14-25)**. Now transketolase acts again, forming fructose 6-phosphate and glyceraldehyde 3-phosphate from erythrose 4-phosphate and xylulose 5-phosphate **(Fig. 14-26)**. Two molecules of glyceraldehyde 3-phosphate formed by two iterations of these reactions can be converted to a molecule of fructose 1,6-bisphosphate as in gluconeogenesis (Fig. 14-17), and

FIGURE 14-22 Oxidative reactions of the pentose phosphate pathway. The end products are ribose 5-phosphate, CO_2, and NADPH.

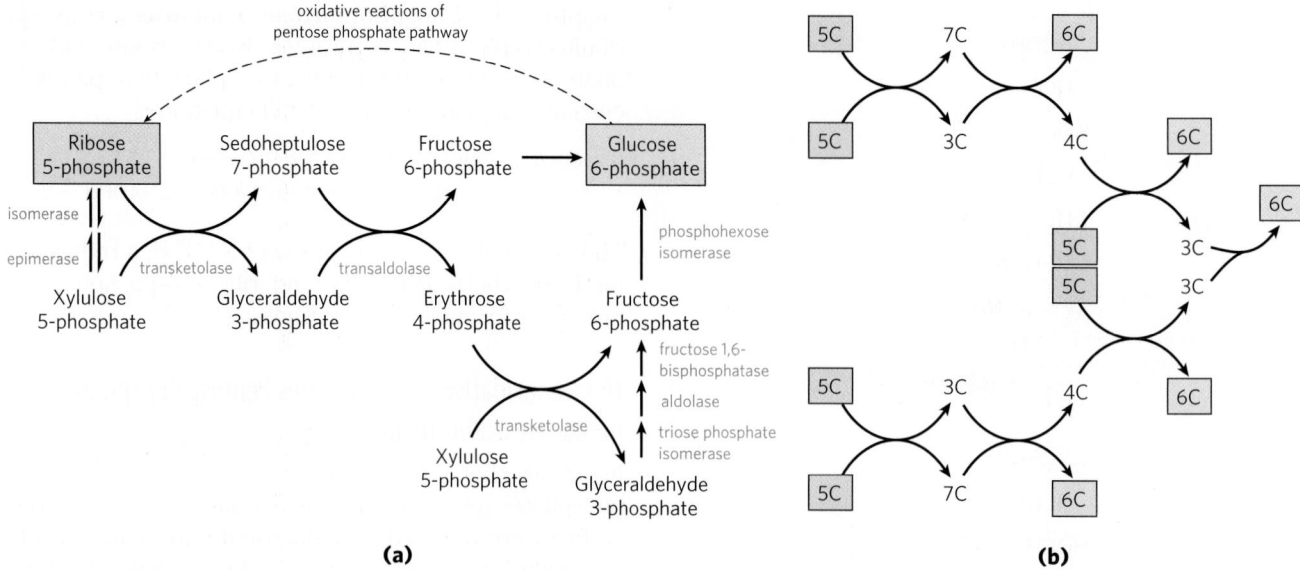

(a)

(b)

FIGURE 14-23 Nonoxidative reactions of the pentose phosphate pathway. (a) These reactions convert pentose phosphates to hexose phosphates, allowing the oxidative reactions (see Fig. 14-22) to continue. Transketolase and transaldolase are specific to this pathway; the other enzymes also serve in the glycolytic or gluconeogenic pathways. **(b)** A schematic diagram showing the pathway from six pentoses (5C) to five hexoses (6C). Note that this involves two sets of the interconversions shown in (a). Every reaction shown here is reversible; unidirectional arrows are used only to make clear the direction of the reactions during continuous oxidation of glucose 6-phosphate. In the light-independent reactions of photosynthesis, the direction of these reactions is reversed (see Fig. 20-37).

finally FBPase-1 and phosphohexose isomerase convert fructose 1,6-bisphosphate to glucose 6-phosphate. Overall, six pentose phosphates have been converted to five hexose phosphates (Fig. 14-23b)—the cycle is now complete.

Transketolase requires the cofactor thiamine pyrophosphate (TPP), which stabilizes a two-carbon carbanion in this reaction **(Fig. 14-27a)**, just as it does in the pyruvate decarboxylase reaction (Fig. 14-15). Transaldolase uses a Lys side chain to form a Schiff base with

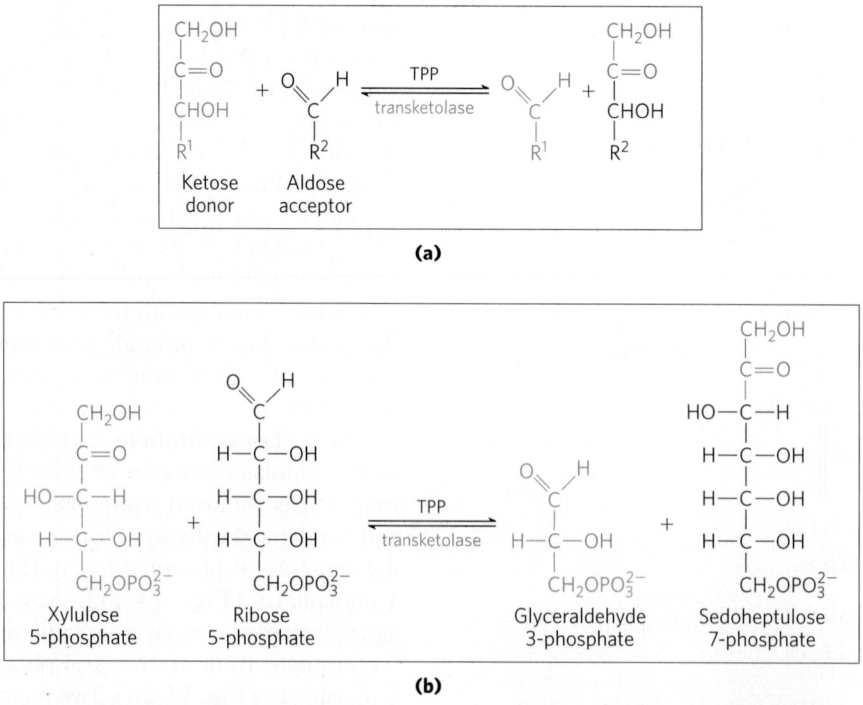

(a)

(b)

FIGURE 14-24 The first reaction catalyzed by transketolase. (a) The general reaction catalyzed by transketolase is the transfer of a two-carbon group, carried temporarily on enzyme-bound TPP, from a ketose donor to an aldose acceptor. **(b)** Conversion of two pentose phosphates to a triose phosphate and a seven-carbon sugar phosphate, sedoheptulose 7-phosphate.

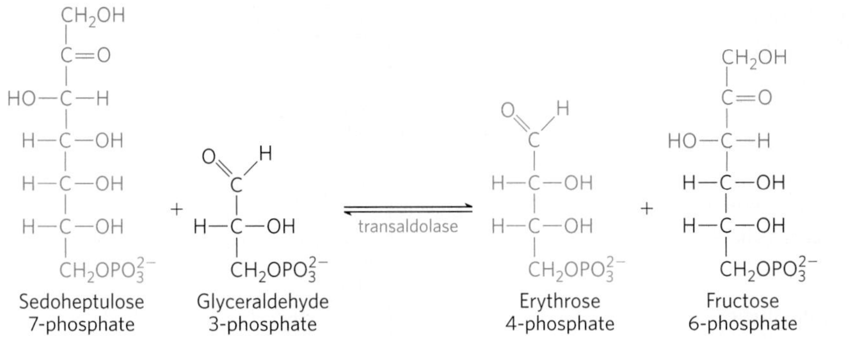

FIGURE 14-25 The reaction catalyzed by transaldolase.

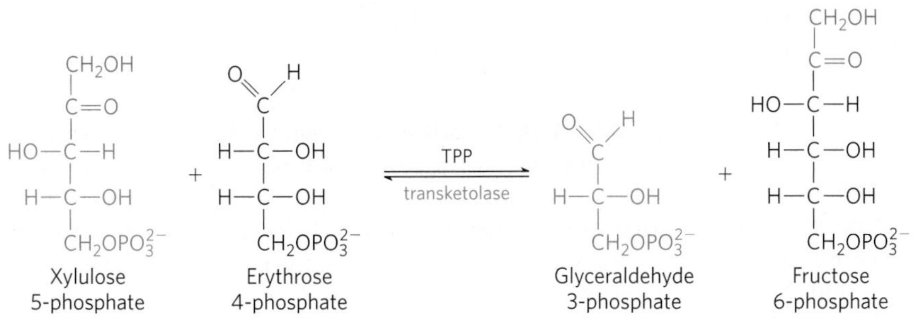

Sedoheptulose 7-phosphate   Glyceraldehyde 3-phosphate   Erythrose 4-phosphate   Fructose 6-phosphate

FIGURE 14-26 The second reaction catalyzed by transketolase.

Xylulose 5-phosphate   Erythrose 4-phosphate   Glyceraldehyde 3-phosphate   Fructose 6-phosphate

the carbonyl group of its substrate, a ketose, thereby stabilizing a carbanion (Fig. 14-27b) that is central to the reaction mechanism.

The process described in Figure 14-22 is known as the **oxidative pentose phosphate pathway**. The first and third steps are oxidations with large, negative standard free-energy changes and are essentially irreversible in the cell. The reactions of the nonoxidative part of the pentose phosphate pathway (Fig. 14-23) are readily reversible and thus also provide a means of converting

hexose phosphates to pentose phosphates. As we shall see in Chapter 20, a process that converts hexose phosphates to pentose phosphates is crucial to the photosynthetic assimilation of CO_2 by plants. That pathway, the **reductive pentose phosphate pathway**, is essentially the reversal of the reactions shown in Figure 14-23 and employs many of the same enzymes.

All the enzymes of the pentose phosphate pathway are located in the cytosol, like those of glycolysis and most of those of gluconeogenesis. In fact, these three pathways are connected through several shared intermediates and enzymes. The glyceraldehyde 3-phosphate formed by the action of transketolase is readily converted to dihydroxyacetone phosphate by the glycolytic enzyme triose phosphate isomerase, and these two trioses can be joined by the aldolase as in gluconeogenesis, forming fructose 1,6-bisphosphate. Alternatively, the triose phosphates can be oxidized to pyruvate by the glycolytic reactions. The fate of the trioses is determined by the cell's relative needs for pentose phosphates, NADPH, and ATP.

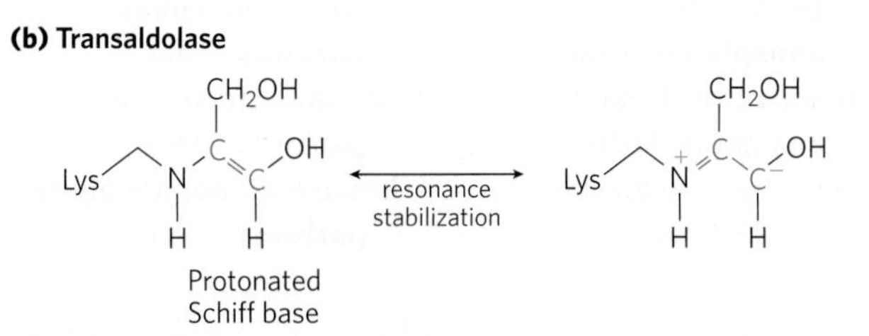

Wernicke-Korsakoff Syndrome Is Exacerbated by a Defect in Transketolase

Wernicke-Korsakoff syndrome is a disorder caused by a severe deficiency of thiamine, a component of TPP. The syndrome is more common among people with alcoholism than in the general population, because chronic, heavy alcohol consumption interferes with the intestinal absorption of thiamine. The syndrome can be exacerbated by a mutation in the gene for transketolase that results in an enzyme with a lowered affinity for TPP—an affinity one-tenth that of the normal enzyme. This defect makes individuals much

(a) Transketolase

TPP

(b) Transaldolase

Protonated Schiff base

FIGURE 14-27 Carbanion intermediates stabilized by covalent interactions with transketolase and transaldolase. (a) The ring of TPP stabilizes the carbanion in the dihydroxyethyl group carried by transketolase; see Figure 14-15 for the chemistry of TPP action. **(b)** In the transaldolase reaction, the protonated Schiff base formed between the ε-amino group of a Lys side chain and the substrate stabilizes the C-3 carbanion formed after aldol cleavage.

more sensitive to a thiamine deficiency: even a moderate thiamine deficiency (tolerable in individuals with an unmutated transketolase) can drop the level of TPP below that needed to saturate the enzyme. The result is a slowing down of the whole pentose phosphate pathway. In people with Wernicke-Korsakoff syndrome this results in a worsening of symptoms, which can include severe memory loss, mental confusion, and partial paralysis. ■

Glucose 6-Phosphate Is Partitioned between Glycolysis and the Pentose Phosphate Pathway

Whether glucose 6-phosphate enters glycolysis or the pentose phosphate pathway depends on the current needs of the cell and on the concentration of $NADP^+$ in the cytosol. Without this electron acceptor, the first reaction of the pentose phosphate pathway (catalyzed by G6PD) cannot proceed. When a cell is rapidly converting NADPH to $NADP^+$ in biosynthetic reductions, the level of $NADP^+$ rises, allosterically stimulating G6PD and thereby increasing the flux of glucose 6-phosphate through the pentose phosphate pathway **(Fig. 14-28)**. When the demand for NADPH slows, the level of $NADP^+$ drops, the pentose phosphate pathway slows, and glucose 6-phosphate is instead used to fuel glycolysis.

SUMMARY 14.5 Pentose Phosphate Pathway of Glucose Oxidation

■ The *oxidative* pentose phosphate pathway (phosphogluconate pathway, or hexose monophosphate pathway) brings about oxidation and decarboxylation at C-1 of glucose 6-phosphate, reducing $NADP^+$ to NADPH and producing pentose phosphates.

■ NADPH provides reducing power for biosynthetic reactions, and ribose 5-phosphate is a precursor for nucleotide and nucleic acid synthesis. Rapidly growing tissues and tissues carrying out active biosynthesis of fatty acids, cholesterol, or steroid hormones send more glucose 6-phosphate through the pentose phosphate pathway than do tissues with less demand for pentose phosphates and reducing power.

■ The first phase of the pentose phosphate pathway consists of two oxidations that convert glucose 6-phosphate to ribulose 5-phosphate and reduce $NADP^+$ to NADPH. The second phase comprises nonoxidative steps that convert pentose phosphates to glucose 6-phosphate, which begins the cycle again.

■ In the second phase, transketolase (with TPP as cofactor) and transaldolase catalyze the interconversion of three-, four-, five-, six-, and seven-carbon sugars, with the reversible conversion of six pentose phosphates to five hexose phosphates. In the carbon-assimilating reactions of photosynthesis, the same enzymes catalyze the reverse process, the *reductive* pentose phosphate pathway: conversion of five hexose phosphates to six pentose phosphates.

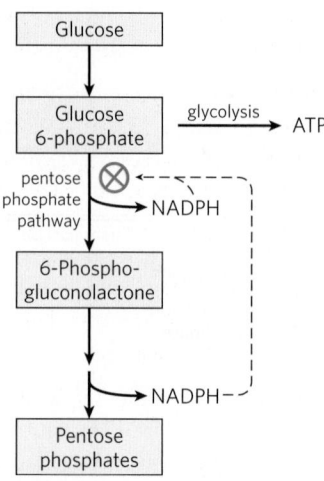

FIGURE 14-28 Role of NADPH in regulating the partitioning of glucose 6-phosphate between glycolysis and the pentose phosphate pathway. When NADPH is forming faster than it is being used for biosynthesis and glutathione reduction (see Fig. 14-21), [NADPH] rises and inhibits the first enzyme in the pentose phosphate pathway. As a result, more glucose 6-phosphate is available for glycolysis.

■ A genetic defect in transketolase that lowers its affinity for TPP exacerbates the Wernicke-Korsakoff syndrome.

■ Entry of glucose 6-phosphate either into glycolysis or into the pentose phosphate pathway is largely determined by the relative concentrations of $NADP^+$ and NADPH.

█ Key Terms

Terms in bold are defined in the glossary.

glycolysis 534	**aerobic glycolysis** 545
fermentation 534	**mutases** 550
lactic acid fermentation 537	**isomerases** 550
hypoxia 537	lactose intolerance 551
ethanol (alcohol) fermentation 537	galactosemia 552
isozymes 539	**thiamine pyrophosphate (TPP)** 555
acyl phosphate 542	**gluconeogenesis** 558
substrate-level phosphorylation 543	**biotin** 560
respiration-linked phosphorylation 543	**pentose phosphate pathway** 565
phosphoenolpyruvate (PEP) 544	**phosphogluconate pathway** 565
	hexose monophosphate pathway 565

█ Problems

1. Equation for the Preparatory Phase of Glycolysis Write balanced biochemical equations for all the reactions in the catabolism of glucose to two molecules of glyceraldehyde 3-phosphate (the preparatory phase of glycolysis), including the standard free-energy change for each reaction. Then write the overall or net equation for the preparatory phase of glycolysis, with the net standard free-energy change.

2. The Payoff Phase of Glycolysis in Skeletal Muscle In working skeletal muscle under anaerobic conditions, glyceraldehyde 3-phosphate is converted to pyruvate (the payoff phase of glycolysis), and the pyruvate is reduced to lactate. Write balanced biochemical equations for all the reactions in this process, with the standard free-energy change for each reaction. Then write the overall or net equation for the payoff phase of glycolysis (with lactate as the end product), including the net standard free-energy change.

3. GLUT Transporters Compare the localization of GLUT4 with that of GLUT2 and GLUT3, and explain why these localizations are important in the response of muscle, adipose tissue, brain, and liver to insulin.

4. Ethanol Production in Yeast When grown anaerobically on glucose, yeast *(S. cerevisiae)* converts pyruvate to acetaldehyde, then reduces acetaldehyde to ethanol using electrons from NADH. Write the equation for the second reaction, and calculate its equilibrium constant at 25 °C, given the standard reduction potentials in Table 13-7.

5. Energetics of the Aldolase Reaction Aldolase catalyzes the glycolytic reaction

Fructose 1,6-bisphosphate $\longrightarrow$
  glyceraldehyde 3-phosphate + dihydroxyacetone phosphate

The standard free-energy change for this reaction in the direction written is +23.8 kJ/mol. The concentrations of the three intermediates in the hepatocyte of a mammal are: fructose 1,6-bisphosphate, 1.4×10^{-5} M; glyceraldehyde 3-phosphate, 3×10^{-6} M; and dihydroxyacetone phosphate, 1.6×10^{-5} M. At body temperature (37 °C), what is the actual free-energy change for the reaction?

6. Pathway of Atoms in Fermentation A "pulse-chase" experiment using ^{14}C-labeled carbon sources is carried out on a yeast extract maintained under strictly anaerobic conditions to produce ethanol. The experiment consists of incubating a small amount of ^{14}C-labeled substrate (the pulse) with the yeast extract just long enough for each intermediate in the fermentation pathway to become labeled. The label is then "chased" through the pathway by the addition of excess unlabeled glucose. The chase effectively prevents any further entry of labeled glucose into the pathway.

(a) If [1-^{14}C]glucose (glucose labeled at C-1 with ^{14}C) is used as a substrate, what is the location of ^{14}C in the product ethanol? Explain.

(b) Where would ^{14}C have to be located in the starting glucose to ensure that all the ^{14}C activity is liberated as ^{14}CO$_2$ during fermentation to ethanol? Explain.

7. Heat from Fermentations Large-scale industrial fermenters generally require constant, vigorous cooling. Why?

8. Fermentation to Produce Soy Sauce Soy sauce is prepared by fermenting a salted mixture of soybeans and wheat with several microorganisms, including yeast, over a period of 8 to 12 months. The resulting sauce (after solids are removed) is rich in lactate and ethanol. How are these two compounds produced? To prevent the soy sauce from having a strong vinegary taste (vinegar is dilute acetic acid), oxygen must be kept out of the fermentation tank. Why?

9. Equivalence of Triose Phosphates ^{14}C-Labeled glyceraldehyde 3-phosphate was added to a yeast extract. After a short time, fructose 1,6-bisphosphate labeled with ^{14}C at C-3 and C-4 was isolated. What was the location of the ^{14}C label in the starting glyceraldehyde 3-phosphate? Where did the second ^{14}C label in fructose 1,6-bisphosphate come from? Explain.

10. Glycolysis Shortcut Suppose you discovered a mutant yeast whose glycolytic pathway was shorter because of the presence of a new enzyme catalyzing the reaction

$$\text{Glyceraldehyde 3-phosphate} + \text{H}_2\text{O} \xrightarrow[\text{NAD}^+ \quad \text{NADH} + \text{H}^+]{} \text{3-phosphoglycerate}$$

Would shortening the glycolytic pathway in this way benefit the cell? Explain.

11. Role of Lactate Dehydrogenase During strenuous activity, the demand for ATP in muscle tissue is vastly increased. In rabbit leg muscle or turkey flight muscle, the ATP is produced almost exclusively by lactic acid fermentation. ATP is formed in the payoff phase of glycolysis by two reactions, promoted by phosphoglycerate kinase and pyruvate kinase. Suppose skeletal muscle were devoid of lactate dehydrogenase. Could it carry out strenuous physical activity; that is, could it generate ATP at a high rate by glycolysis? Explain.

12. Efficiency of ATP Production in Muscle The transformation of glucose to lactate in myocytes releases only about 7% of the free energy released when glucose is completely oxidized to CO$_2$ and H$_2$O. Does this mean that anaerobic glycolysis in muscle is a wasteful use of glucose? Explain.

13. Free-Energy Change for Triose Phosphate Oxidation The oxidation of glyceraldehyde 3-phosphate to 1,3-bisphosphoglycerate, catalyzed by glyceraldehyde 3-phosphate dehydrogenase, proceeds with an unfavorable equilibrium constant ($K'_{eq} = 0.08$; $\Delta G'^{\circ} = 6.3$ kJ/mol), yet the flow through this point in the glycolytic pathway proceeds smoothly. How does the cell overcome the unfavorable equilibrium?

14. Arsenate Poisoning Arsenate is structurally and chemically similar to inorganic phosphate (P$_i$), and many enzymes that require phosphate will also use arsenate. Organic compounds of arsenate are less stable than analogous phosphate compounds, however. For example, acyl *arsenates* decompose rapidly by hydrolysis:

$$\underset{\text{O}^-}{\underset{|}{\text{R}-\overset{\text{O}}{\overset{||}{\text{C}}}-\text{O}-\overset{\text{O}}{\overset{||}{\text{As}}}-\text{O}^-}} + \text{H}_2\text{O} \longrightarrow$$

$$\text{R}-\overset{\text{O}}{\overset{||}{\text{C}}}-\text{O}^- + \underset{\text{O}^-}{\underset{|}{\text{HO}-\overset{\text{O}}{\overset{||}{\text{As}}}-\text{O}^-}} + \text{H}^+$$

On the other hand, acyl *phosphates*, such as 1,3-bisphospho-glycerate, are more stable and undergo further enzyme-catalyzed transformation in cells.

(a) Predict the effect on the net reaction catalyzed by glyceraldehyde 3-phosphate dehydrogenase if phosphate were replaced by arsenate.

(b) What would be the consequence to an organism if arsenate were substituted for phosphate? Arsenate is very toxic to most organisms. Explain why.

15. Requirement for Phosphate in Ethanol Fermentation In 1906 Harden and Young, in a series of classic studies on the fermentation of glucose to ethanol and CO_2 by extracts of brewer's yeast, made the following observations. (1) Inorganic phosphate was essential to fermentation; when the supply of phosphate was exhausted, fermentation ceased before all the glucose was used. (2) During fermentation under these conditions, ethanol, CO_2, and a hexose bisphosphate accumulated. (3) When arsenate was substituted for phosphate, no hexose bisphosphate accumulated, but the fermentation proceeded until all the glucose was converted to ethanol and CO_2.

(a) Why did fermentation cease when the supply of phosphate was exhausted?

(b) Why did ethanol and CO_2 accumulate? Was the conversion of pyruvate to ethanol and CO_2 essential? Why? Identify the hexose bisphosphate that accumulated. Why did it accumulate?

(c) Why did the substitution of arsenate for phosphate prevent the accumulation of the hexose bisphosphate yet allow fermentation to ethanol and CO_2 to go to completion? (See Problem 14.)

16. Role of the Vitamin Niacin Adults engaged in strenuous physical activity require an intake of about 160 g of carbohydrate daily but only about 20 mg of niacin for optimal nutrition. Given the role of niacin in glycolysis, how do you explain the observation?

17. Synthesis of Glycerol Phosphate The glycerol 3-phosphate required for the synthesis of glycerophospholipids can be synthesized from a glycolytic intermediate. Propose a reaction sequence for this conversion.

18. Severity of Clinical Symptoms Due to Enzyme Deficiency The clinical symptoms of two forms of galactosemia—deficiency of galactokinase or of UDP-glucose:galactose 1-phosphate uridylyltransferase—show radically different severity. Although both types produce gastric discomfort after milk ingestion, deficiency of the transferase also leads to liver, kidney, spleen, and brain dysfunction and eventual death. What products accumulate in the blood and tissues with each type of enzyme deficiency? Estimate the relative toxicities of these products from the above information.

19. Muscle Wasting in Starvation One consequence of starvation is a reduction in muscle mass. What happens to the muscle proteins?

20. Pathway of Atoms in Gluconeogenesis A liver extract capable of carrying out all the normal metabolic reactions of the liver is briefly incubated in separate experiments with the following ^{14}C-labeled precursors.

(a) $[^{14}C]$Bicarbonate, $HO-^{14}C\overset{O^-}{\underset{O}{\diagdown}}$

(b) $[1-^{14}C]$Pyruvate, $CH_3-\underset{\underset{O}{\|}}{C}-^{14}COO^-$

Trace the pathway of each precursor through gluconeogenesis. Indicate the location of ^{14}C in all intermediates and in the product, glucose.

21. Energy Cost of a Cycle of Glycolysis and Gluconeogenesis What is the cost (in ATP equivalents) of transforming glucose to pyruvate via glycolysis and back again to glucose via gluconeogenesis?

22. Relationship between Gluconeogenesis and Glycolysis Why is it important that gluconeogenesis is not the exact reversal of glycolysis?

23. Energetics of the Pyruvate Kinase Reaction Explain in bioenergetic terms how the conversion of pyruvate to phosphoenolpyruvate in gluconeogenesis overcomes the large, negative, standard free-energy change of the pyruvate kinase reaction in glycolysis.

24. Glucogenic Substrates A common procedure for determining the effectiveness of compounds as precursors of glucose in mammals is to starve the animal until the liver glycogen stores are depleted and then administer the compound in question. A substrate that leads to a *net* increase in liver glycogen is termed glucogenic, because it must first be converted to glucose 6-phosphate. Show by means of known enzymatic reactions which of the following substances are glucogenic.

(a) Succinate, $^-OOC-CH_2-CH_2-COO^-$

(b) Glycerol, $\underset{\underset{H}{|}}{\overset{OH}{\underset{|}{CH_2}}-\overset{OH}{\underset{|}{C}}-\overset{OH}{\underset{|}{CH_2}}}$

(c) Acetyl-CoA, $CH_3-\underset{\underset{O}{\|}}{C}-S\text{-CoA}$

(d) Pyruvate, $CH_3-\underset{\underset{O}{\|}}{C}-COO^-$

(e) Butyrate, $CH_3-CH_2-CH_2-COO^-$

25. Ethanol Affects Blood Glucose Levels The consumption of alcohol (ethanol), especially after periods of strenuous activity or after not eating for several hours, results in a deficiency of glucose in the blood, a condition known as hypoglycemia. The first step in the metabolism of ethanol by the liver is oxidation to acetaldehyde, catalyzed by liver alcohol dehydrogenase:

$$CH_3CH_2OH + NAD^+ \longrightarrow CH_3CHO + NADH + H^+$$

Explain how this reaction inhibits the transformation of lactate to pyruvate. Why does this lead to hypoglycemia?

26. Blood Lactate Levels during Vigorous Exercise The concentrations of lactate in blood plasma before, during, and after a 400 m sprint are shown in the graph.

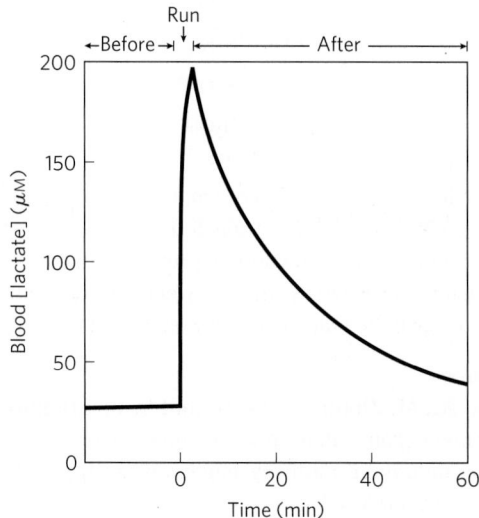

(a) What causes the rapid rise in lactate concentration?

(b) What causes the decline in lactate concentration after completion of the sprint? Why does the decline occur more slowly than the increase?

(c) Why is the concentration of lactate not zero during the resting state?

27. Relationship between Fructose 1,6-Bisphosphatase and Blood Lactate Levels A congenital defect in the liver enzyme fructose 1,6-bisphosphatase results in abnormally high levels of lactate in the blood plasma. Explain.

28. Effect of Phloridzin on Carbohydrate Metabolism Phloridzin, a toxic glycoside from the bark of the pear tree, blocks the normal reabsorption of glucose from the kidney tubule, thus causing blood glucose to be almost completely excreted in the urine. In an experiment, rats fed phloridzin and sodium succinate excreted about 0.5 mol of glucose (made by gluconeogenesis) for every 1 mol of sodium succinate ingested. How is the succinate transformed to glucose? Explain the stoichiometry.

Phloridzin

29. Excess O_2 Uptake during Gluconeogenesis Lactate absorbed by the liver is converted to glucose, with the input of 6 mol of ATP for every mole of glucose produced. The extent of this process in a rat liver preparation can be monitored by administering [^{14}C]lactate and measuring the amount of [^{14}C]

glucose produced. Because the stoichiometry of O_2 consumption and ATP production is known (about 5 ATP per O_2), we can predict the extra O_2 consumption above the normal rate when a given amount of lactate is administered. However, when the extra O_2 used in the synthesis of glucose from lactate is actually measured, it is always higher than predicted by known stoichiometric relationships. Suggest a possible explanation for this observation.

30. Role of the Pentose Phosphate Pathway If the oxidation of glucose 6-phosphate via the pentose phosphate pathway were being used primarily to generate NADPH for biosynthesis, the other product, ribose 5-phosphate, would accumulate. What problems might this cause?

Data Analysis Problem

31. Engineering a Fermentation System Fermentation of plant matter to produce ethanol for fuel is one potential method for reducing the use of fossil fuels and thus the CO_2 emissions that lead to global warming. Many microorganisms can break down cellulose then ferment the glucose to ethanol. However, many potential cellulose sources, including agricultural residues and switchgrass, also contain substantial amounts of arabinose, which is not as easily fermented.

D-Arabinose

Escherichia coli is capable of fermenting arabinose to ethanol, but it is not naturally tolerant of high ethanol levels, thus limiting its utility for commercial ethanol production. Another bacterium, *Zymomonas mobilis,* is naturally tolerant of high levels of ethanol but cannot ferment arabinose. Deanda, Zhang, Eddy, and Picataggio (1996) described their efforts to combine the most useful features of these two organisms by introducing the *E. coli* genes for the arabinose-metabolizing enzymes into *Z. mobilis.*

(a) Why is this a simpler strategy than the reverse: engineering *E. coli* to be more ethanol-tolerant?

Deanda and colleagues inserted five *E. coli* genes into the *Z. mobilis* genome: *araA*, coding for L-arabinose isomerase, which interconverts L-arabinose and L-ribulose; *araB*, L-ribulokinase, which uses ATP to phosphorylate L-ribulose at C-5; *araD*, L-ribulose 5-phosphate epimerase, which interconverts L-ribulose 5-phosphate and L-xylulose 5-phosphate; *talB*, transaldolase; and *tktA*, transketolase.

(b) For each of the three *ara* enzymes, briefly describe the chemical transformation it catalyzes and, where possible, name an enzyme discussed in this chapter that carries out an analogous reaction.

The five *E. coli* genes inserted in *Z. mobilis* allowed the entry of arabinose into the nonoxidative phase of the pentose phosphate pathway (Fig. 14-23), where it was converted to glucose 6-phosphate and fermented to ethanol.

(c) The three *ara* enzymes eventually converted arabinose into which sugar?

(d) The product from part (c) feeds into the pathway shown in Figure 14-23. Combining the five *E. coli* enzymes listed above with the enzymes of this pathway, describe the overall pathway for the fermentation of six molecules of arabinose to ethanol.

(e) What is the stoichiometry of the fermentation of six molecules of arabinose to ethanol and CO_2? How many ATP molecules would you expect this reaction to generate?

(f) *Zymomonas mobilis* uses a slightly different pathway for ethanol fermentation from the one described in this chapter. As a result, the expected ATP yield is only 1 ATP per molecule of arabinose. Although this is less beneficial for the bacterium, it is better for ethanol production. Why?

Another sugar commonly found in plant matter is xylose.

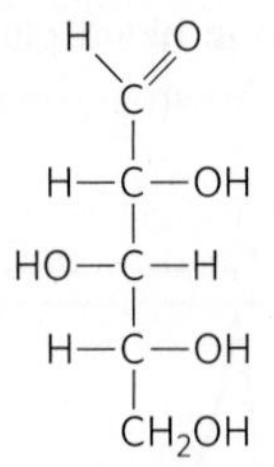

D-Xylose

(g) What additional enzymes would you need to introduce into the modified *Z. mobilis* strain described above to enable it to use xylose as well as arabinose to produce ethanol? You don't need to name the enzymes (they may not even exist in the real world); just give the reactions they would need to catalyze.

Reference

Deanda, K., M. Zhang, C. Eddy, and S. Picataggio. 1996. Development of an arabinose-fermenting *Zymomonas mobilis* strain by metabolic pathway engineering. *Appl. Environ. Microbiol.* 62:4465–4470.

Further Reading is available at www.macmillanlearning.com/LehningerBiochemistry7e.

Principles of Metabolic Regulation

Self-study tools that will help you practice what you've learned and reinforce this chapter's concepts are available online.
Go to www.macmillanlearning.com/LehningerBiochemistry7e.

Metabolic regulation, a central theme in biochemistry, is one of the most remarkable features of living organisms. Of the thousands of enzyme-catalyzed reactions that can take place in a cell, there is probably not one that escapes some form of regulation. This need to regulate every aspect of cellular metabolism becomes clear as one examines the complexity of metabolic reaction sequences. Although it is convenient for the student of biochemistry to divide metabolic processes into "pathways" that play discrete roles in the cell's economy, no such separation exists in the living cell. Rather, every pathway we discuss in this book is inextricably intertwined with all the other cellular pathways in a multidimensional network of reactions **(Fig. 15-1)**. For example, in Chapter 14 we discussed four possible fates for **glucose 6-phosphate** in a hepatocyte: breakdown by glycolysis for the production of ATP, breakdown in the pentose phosphate pathway for the production of NADPH and pentose phosphates, use in the synthesis of complex polysaccharides of the extracellular matrix, or hydrolysis to glucose and phosphate to replenish blood glucose. In fact, glucose 6-phosphate has other possible fates in hepatocytes, too; it may, for example, be used to synthesize other sugars, such as glucosamine, galactose, galactosamine, fucose, and neuraminic acid, for use in protein glycosylation, or it may be partially degraded to provide acetyl-CoA for fatty acid and sterol synthesis. And the bacterium *Escherichia coli* can use glucose to produce the carbon skeleton of *every one* of its several thousand types of molecules. When any cell uses glucose 6-phosphate for one purpose, that "decision" affects all the other pathways for which glucose 6-phosphate is a precursor or intermediate: any change in the allocation of glucose 6-phosphate to one pathway affects, directly or indirectly, the flow of metabolites through all the others.

Such changes in allocation are common in the life of a cell. Louis Pasteur was the first to describe the more than 10-fold increase in glucose consumption by a yeast culture when it was shifted from aerobic to anaerobic conditions. This "Pasteur effect" occurs without a significant change in the concentrations of ATP or most of the hundreds of metabolic intermediates and products derived from glucose. A similar effect occurs in the cells of skeletal muscle when a sprinter leaves the starting blocks. The ability of a cell to carry out all these interconnected metabolic processes simultaneously—obtaining every product in the amount needed and at the right time, in the face of major perturbations from outside, and without generating leftovers—is an *astounding* accomplishment.

In this chapter we use the metabolism of glucose to illustrate some general principles of metabolic regulation. First we look at the general roles of regulation in achieving metabolic homeostasis and introduce metabolic control analysis, a system for analyzing complex

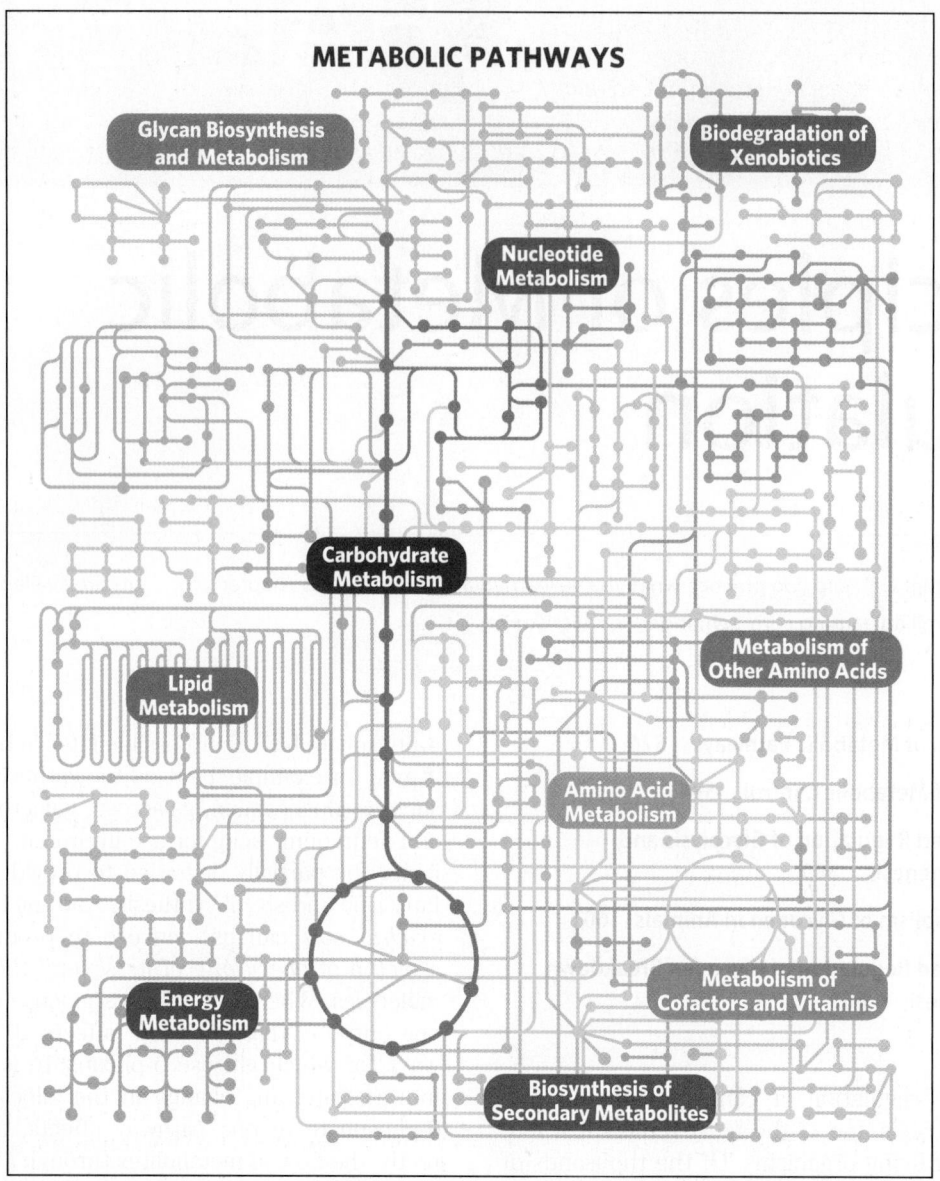

METABOLIC PATHWAYS

Glycan Biosynthesis and Metabolism

Biodegradation of Xenobiotics

Nucleotide Metabolism

Carbohydrate Metabolism

Lipid Metabolism

Metabolism of Other Amino Acids

Amino Acid Metabolism

Energy Metabolism

Metabolism of Cofactors and Vitamins

Biosynthesis of Secondary Metabolites

FIGURE 15-1 Metabolism as a three-dimensional meshwork. A typical eukaryotic cell has the capacity to make about 30,000 different proteins, which catalyze thousands of different reactions involving many hundreds of metabolites, most shared by more than one "pathway." In this much-simplified overview of metabolic pathways, each dot represents an intermediate compound and each connecting line represents an enzymatic reaction. For a more realistic and far more complex diagram of metabolism, see the online KEGG PATHWAY database (www.genome.ad.jp/kegg/pathway/map/map01100.html); in this interactive map, each dot can be clicked to obtain extensive data about the compound and the enzymes for which it is a substrate. [Source: www.genome.ad.jp/kegg/pathway/map/map01100.html.]

metabolic interactions quantitatively. We then describe the specific regulatory properties of the individual enzymes of glucose metabolism; for glycolysis and gluconeogenesis, we described the catalytic activities of the enzymes in Chapter 14. Here we also discuss both the catalytic and regulatory properties of the enzymes of glycogen synthesis and breakdown, one of the best-studied cases of metabolic regulation. Note that in selecting carbohydrate metabolism to illustrate the principles of metabolic regulation, we have artificially separated the metabolism of fats and carbohydrates. In fact, these two activities are very tightly integrated, as we shall see in Chapter 23.

15.1 Regulation of Metabolic Pathways

The pathways of glucose metabolism provide, in the catabolic direction, the energy essential to oppose the forces of entropy and, in the anabolic direction, biosynthetic precursors and a storage form of metabolic energy. These reactions are so important to survival that very complex regulatory mechanisms have evolved to ensure that metabolites move through each pathway in the correct direction and at the correct rate to match exactly the cell's or the organism's changing circumstances. By a variety of mechanisms operating on different time scales, adjustments are made in the rate of metabolite flow through an entire pathway when external circumstances change.

Circumstances do change, sometimes dramatically. For example, the demand for ATP in insect flight muscle increases 100-fold in a few seconds when the insect takes flight. In humans, the availability of oxygen may decrease due to hypoxia (diminished delivery of oxygen to tissues) or ischemia (diminished flow of blood to tissues). Wound healing requires huge amounts of energy and biosynthetic precursors. The relative proportions of carbohydrate, fat, and protein in the diet vary from meal to meal, and the supply of fuels obtained in the diet is intermittent, requiring metabolic adjustments between meals and during periods of starvation.

Cells and Organisms Maintain a Dynamic Steady State

Fuels such as glucose enter a cell, and waste products such as CO_2 leave, but the mass and the gross composition of a typical cell, organ, or adult animal do not change appreciably over time; cells and organisms exist in a dynamic steady state. For each metabolic reaction in a pathway, the substrate is provided by the preceding reaction at the same rate at which it is converted to product. Thus, although the rate (v) of metabolite flow, or **flux**, through this step of the pathway may be high and variable, the concentration of substrate, S, remains constant. So, for the two-step reaction

$$A \xrightarrow{v_1} S \xrightarrow{v_2} P$$

when $v_1 = v_2$, [S] is constant. For example, changes in v_1 for the entry of glucose from various sources into the blood are balanced by changes in v_2 for the uptake of glucose from the blood into various tissues, so the concentration of glucose in the blood ([S]) is held nearly constant at 5 mM. This is **homeostasis** at the molecular level. The failure of homeostatic mechanisms is often at the root of human disease. In diabetes mellitus, for example, the regulation of blood glucose concentration is defective as a result of the lack of or insensitivity to insulin, with profound medical consequences.

When the external perturbation is not merely transient, or when one kind of cell develops into another, the adjustments in cell composition and metabolism can be more dramatic and may require significant and lasting changes in the allocation of energy and synthetic precursors to bring about a new dynamic steady state. Consider, for example, the differentiation of stem cells in the bone marrow into erythrocytes. The precursor cell contains a nucleus, mitochondria, and little or no hemoglobin, whereas the fully differentiated erythrocyte contains prodigious amounts of hemoglobin but has neither nucleus nor mitochondria; the cell's composition has permanently changed in response to external developmental signals, with accompanying changes in metabolism. This **cellular differentiation** requires precise regulation of the levels of cellular proteins.

In the course of evolution, organisms have acquired a remarkable collection of regulatory mechanisms for maintaining homeostasis at the molecular, cellular, and organismal levels, as reflected in the proportion of genes that encode regulatory machinery. In humans, about 2,500 genes (~12% of all genes) encode regulatory proteins, including a variety of receptors, regulators of gene expression, and more than 800 different protein kinases! In many cases, the regulatory mechanisms overlap: one enzyme is subject to regulation by several different mechanisms.

Both the Amount and the Catalytic Activity of an Enzyme Can Be Regulated

The flux through an enzyme-catalyzed reaction can be modulated by changes in the *number* of enzyme molecules or by changes in the *catalytic activity* of each enzyme molecule already present. Such changes occur on time scales from milliseconds to many hours, in response to signals from within or outside the cell. Very rapid allosteric changes in enzyme activity are generally triggered locally, by changes in the local concentration of a small molecule—a substrate of the pathway in which that reaction is a step (say, glucose for glycolysis), a product of the pathway (ATP from glycolysis), or a key metabolite or cofactor (such as NADH) that indicates the cell's metabolic state. Second messengers (such as cyclic AMP and Ca^{2+}) generated intracellularly in response to extracellular signals (hormones, cytokines, and so forth) also mediate allosteric regulation, on a slightly slower time scale set by the rate of the signal-transduction mechanism (see Chapter 12).

Extracellular signals (**Fig. 15-2**, ❶) may be hormonal (insulin or epinephrine, for example) or neuronal (acetylcholine), or may be growth factors or cytokines. The number of molecules of a given enzyme in a cell is a function of the relative rates of synthesis and degradation of that enzyme. The rate of synthesis can be adjusted by the activation (in response to some outside signal) of a transcription factor (Fig. 15-2, ❷; described in more detail in Chapter 28). **Transcription factors** are nuclear proteins that, when activated, bind specific DNA regions (**response elements**) near a gene's promoter (its transcriptional starting point) and activate or repress the transcription of that gene, leading to increased or decreased synthesis of the encoded protein. Activation of a transcription factor is sometimes the result of its binding of a specific ligand and sometimes the result of its phosphorylation or dephosphorylation. Each gene is controlled by one or more response elements that are recognized by specific transcription factors. Genes that have several response elements are therefore controlled by several different transcription factors responding to several different signals. Groups of genes encoding proteins that act together, such as the enzymes of glycolysis or gluconeogenesis, often share common response element sequences, so that a

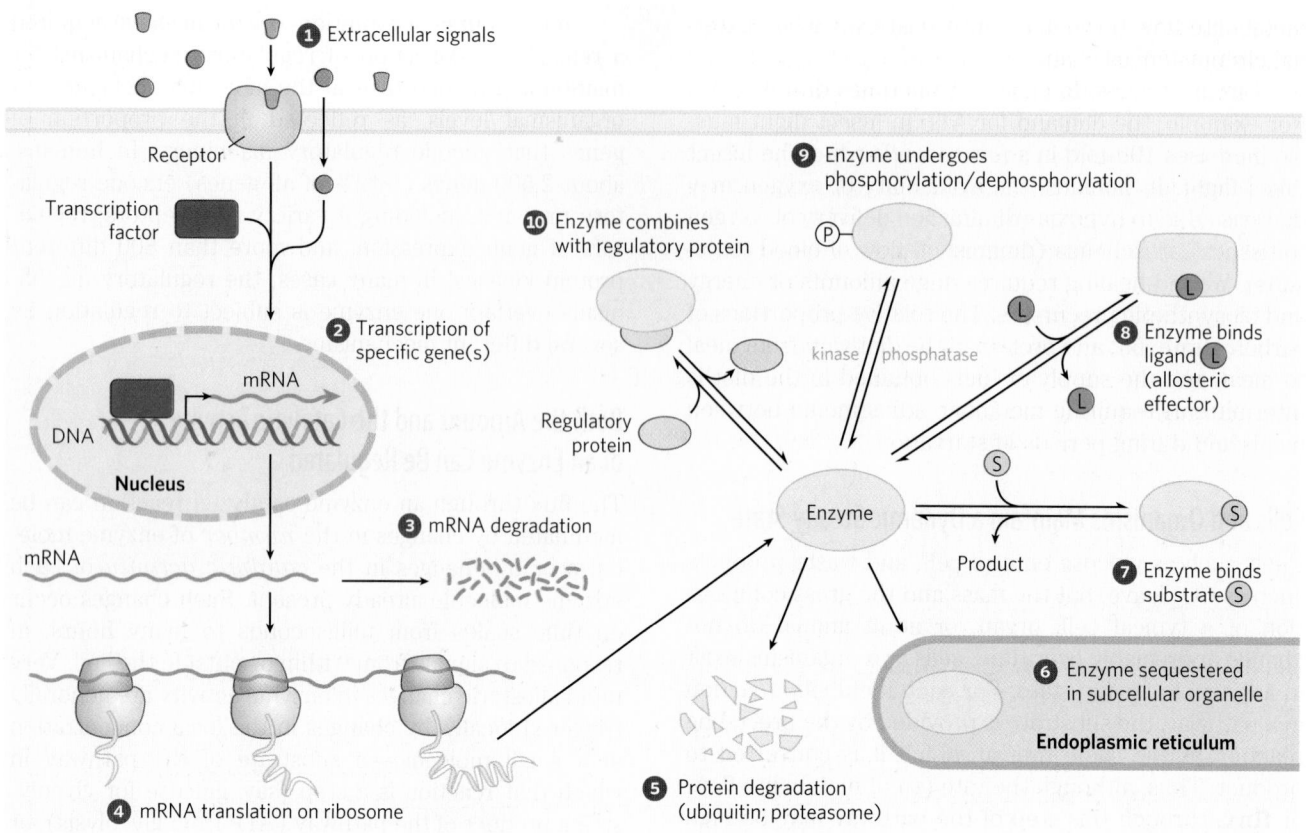

FIGURE 15-2 Factors affecting the activity of enzymes. The total activity of an enzyme can be changed by altering the *number* of its molecules in the cell, or its *effective* activity in a subcellular compartment (**1** through **6**), or by modulating the *activity* of existing molecules (**7** through **10**), as detailed in the text. An enzyme may be influenced by a combination of such factors.

single signal, acting through a particular transcription factor, turns all of these genes on and off together. The regulation of carbohydrate metabolism by specific transcription factors is described in Section 15.3.

The stability of messenger RNAs—their resistance to degradation by cellular ribonucleases (Fig. 15-2, **3**)—varies, and the amount of a given mRNA in the cell is a function of its rates of synthesis and degradation (Chapter 26). The rate at which an mRNA is translated into a protein by ribosomes (Fig. 15-2, **4**) is also regulated, and depends on several factors described in detail in Chapter 27. Note that an n-fold increase in an mRNA does not always mean an n-fold increase in its protein product.

Once synthesized, protein molecules have a finite lifetime, which may range from minutes to many days (Table 15-1). The rate of protein degradation (Fig. 15-2, **5**) differs from one protein to another and depends on the conditions in the cell. Some proteins are tagged by the covalent attachment of ubiquitin for degradation in proteasomes, as discussed in Chapter 27 (see, for example, the case of cyclin, in Fig. 12-35). Rapid **turnover** (synthesis followed by degradation) is energetically expensive, but proteins with a short half-life can reach new steady-state levels much faster than those with a long half-life, and the benefit of this quick responsiveness must balance or outweigh the cost to the cell.

Yet another way to alter the *effective* activity of an enzyme is to sequester the enzyme and its substrate in different compartments (Fig. 15-2, **6**). In muscle, for example, hexokinase cannot act on glucose until the sugar enters the myocyte from the blood, and the rate at which it enters depends on the activity of glucose transporters (see Table 11-3) in the plasma membrane. Within cells, membrane-bounded compartments segregate certain enzymes and enzyme systems, and the transport of substrate across these intracellular membranes may be the limiting factor in enzyme action.

By these several mechanisms for regulating enzyme level, cells can dramatically change their complement of

TABLE 15-1	Average Half-Life of Proteins in Mammalian Tissues
Tissue	**Average half-life (days)**
Liver	0.9
Kidney	1.7
Heart	4.1
Brain	4.6
Muscle	10.7

enzymes in response to changes in metabolic circumstances. In vertebrates, liver is the most adaptable tissue; a change from a high-carbohydrate to a high-lipid diet, for example, affects the transcription of hundreds of genes and thus the levels of hundreds of proteins. These global changes in gene expression can be quantified by the use of DNA microarrays (see Fig. 9-23) that display the entire complement of mRNAs present in a given cell type or organ (the **transcriptome**) or by two-dimensional gel electrophoresis (see Fig. 3-21) that displays the protein complement of a cell type or organ (its **proteome**). Both techniques offer great insights into metabolic regulation. The effect of changes in the proteome is often a change in the total ensemble of low molecular weight metabolites, the **metabolome** (**Fig. 15-3**). The metabolome of *E. coli* growing on glucose is dominated by a few classes of metabolites: glutamate (49%); nucleotides (mainly ribonucleoside triphosphates) (15%); intermediates of glycolysis, the citric acid cycle, and the pentose phosphate pathway (central pathways of carbon metabolism) (15%); and redox cofactors and glutathione (9%).

Once the regulatory mechanisms that involve protein synthesis and degradation have produced a certain number of molecules of each enzyme in a cell, the activity of those enzymes can be further regulated in several other ways: by the concentration of substrate, the presence of allosteric effectors, covalent modifications, or binding of regulatory proteins—all of which can change the activity of an individual enzyme molecule (Fig. 15-2, ❼ to ❿).

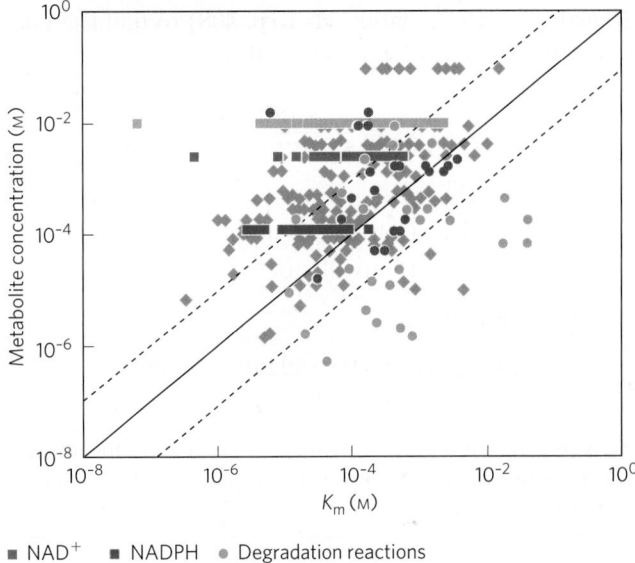

■ NAD$^+$ ■ NADPH   ● Degradation reactions
■ ATP   ◆ Other   ● Central pathways of carbon metabolism

FIGURE 15-4 Comparison of K_m and substrate concentration for some metabolic enzymes. Measured metabolite concentrations for *E. coli* growing on glucose are plotted against the known K_m for enzymes that consume that metabolite. The solid line is the line of unity (where metabolite concentration = K_m), and the dashed lines each denote a tenfold deviation from the line of unity. [Source: Data from B. D. Bennett et al., *Nature Chem. Biol.* 5:593, 2009, Fig. 2.]

All enzymes are sensitive to the concentration of their substrate(s) (Fig. 15-2, ❼). Recall that in the simplest case (an enzyme that follows Michaelis-Menten kinetics), the initial rate of the reaction is half-maximal when the substrate is present at a concentration equal to K_m (that is, when the enzyme is half-saturated with substrate). Activity drops off at lower [S], and when [S] $\ll K_m$, the reaction rate is linearly dependent on [S].

The relationship between [S] and K_m is important because intracellular concentrations of substrate are often in the same range as, or lower than, K_m. The activity of hexokinase, for example, changes with [glucose], and intracellular [glucose] varies with the concentration of glucose in the blood. As we will see, the different forms (isozymes) of hexokinase have different K_m values and are therefore affected differently by changes in intracellular [glucose], in ways that make sense physiologically. For a number of phosphoryl transfers from ATP, and for redox reactions using NADPH or NAD$^+$, the metabolite concentration is well above the K_m (**Fig. 15-4**); these cofactors are not likely to be the limiting factors in such reactions.

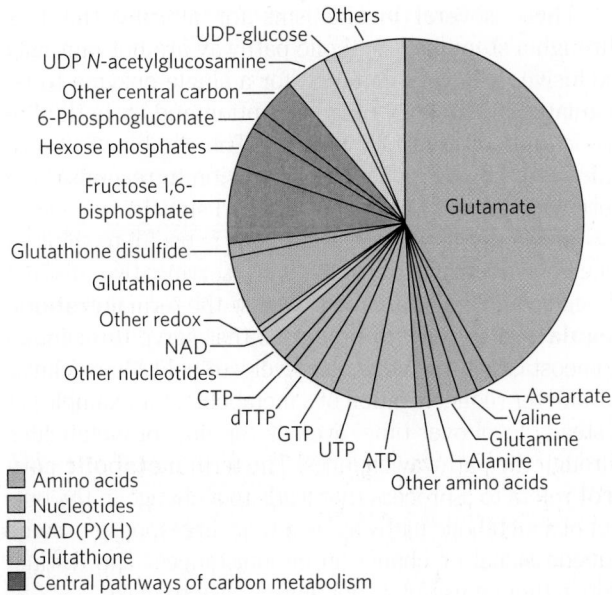

Amino acids
Nucleotides
NAD(P)(H)
Glutathione
Central pathways of carbon metabolism
Other intermediates

FIGURE 15-3 The metabolome of *E. coli* growing on glucose. Summary of the amounts of 103 metabolites as measured by a combination of liquid chromatography and tandem mass spectrometry (LC-MS/MS). [Source: Data from B. D. Bennett et al., *Nature Chem. Biol.* 5:593, 2009, Fig. 1.]

WORKED EXAMPLE 15-1	Activity of a Glucose Transporter

If K_t (the equivalent of K_m) for the glucose transporter in liver (GLUT2) is 40 mM, calculate the effect on the rate of glucose flux into a hepatocyte of increasing the blood glucose concentration from 3 mM to 10 mM.

Solution: We use Equation 11–1 (p. 408) to find the initial velocity (flux) of glucose uptake.

$$V_0 = \frac{V_{max}[S]_{out}}{K_t + [S]_{out}}$$

At 3 mM glucose

$$V_0 = V_{max}(3 \text{ mM})/(40 \text{ mM} + 3 \text{ mM})$$
$$= V_{max}(3 \text{ mM}/43 \text{ mM}) = 0.07 \, V_{max}$$

At 10 mM glucose

$$V_0 = V_{max}(10 \text{ mM})/(40 \text{ mM} + 10 \text{ mM})$$
$$= V_{max}(10 \text{ mM}/50 \text{ mM}) = 0.20 \, V_{max}$$

So a rise in blood glucose from 3 mM to 10 mM increases the rate of glucose influx into a hepatocyte by a factor of 0.20/0.07 ≈ 3.

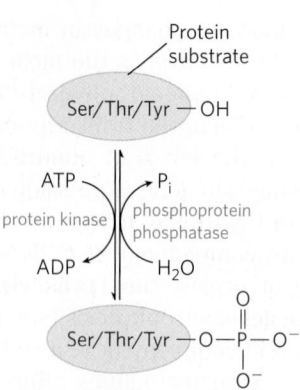

FIGURE 15-5 Protein phosphorylation and dephosphorylation. Protein kinases transfer a phosphoryl group from ATP to a Ser, Thr, or Tyr residue in an enzyme or other protein substrate. Protein phosphatases remove the phosphoryl group as P_i.

Enzyme activity can be either increased or decreased by an allosteric effector (Fig. 15-2, ❽; see Fig. 6-35). Allosteric effectors typically convert hyperbolic kinetics to sigmoid kinetics, or vice versa (see Fig. 15-16b, for example). In the steepest part of the sigmoid curve, a small change in the concentration of substrate, or of allosteric effector, can have a large impact on reaction rate. Recall from Chapter 5 (p. 167) that the cooperativity of an allosteric protein can be expressed as a Hill coefficient, with higher coefficients meaning greater cooperativity. For an allosteric enzyme with a Hill coefficient of 4, activity increases from 10% V_{max} to 90% V_{max} with only a 3-fold increase in [S], compared with the 81-fold rise in [S] needed by an enzyme with no cooperative effects (Hill coefficient of 1; Table 15-2).

Covalent modifications of enzymes or other proteins (Fig. 15-2, ❾) occur within seconds or minutes of a regulatory signal, typically an extracellular signal. By far the most common modifications are phosphorylation and dephosphorylation **(Fig. 15-5)**; up to half the proteins in a eukaryotic cell are phosphorylated under some circumstances. Phosphorylation by a specific protein kinase may alter the electrostatic features of an enzyme's active site, cause movement of an inhibitory region of the enzyme out of the active site, alter the enzyme's interaction with other proteins, or force conformational changes that translate into changes in V_{max} or K_m. For covalent modification to be useful in regulation, the cell must be able to restore the altered enzyme to its original activity state. A family of phosphoprotein phosphatases, at least some of which are themselves under regulation, catalyzes the dephosphorylation of proteins.

Finally, many enzymes are regulated by association with and dissociation from another, regulatory protein (Fig. 15-2, ❿). For example, the cyclic AMP–dependent protein kinase (PKA; see Fig. 12-6) is inactive until cAMP binding separates catalytic from regulatory (inhibitory) subunits of the enzyme.

These several mechanisms for altering the flux through a step in a metabolic pathway are not mutually exclusive. It is very common for a single enzyme to be regulated at the level of transcription and by both allosteric and covalent mechanisms. The combination provides fast, smooth, effective regulation in response to a very wide array of perturbations and signals.

In the discussions that follow, it is useful to think of changes in enzymatic activity as serving two distinct though complementary roles. We use the term **metabolic regulation** to refer to processes that serve to maintain homeostasis at the molecular level—to hold some cellular parameter (concentration of a metabolite, for example) at a steady level over time, even as the flow of metabolites through the pathway changes. The term **metabolic control** refers to a process that leads to a change in the output of a metabolic pathway over time, in response to some outside signal or change in circumstances. The distinction, although useful, is not always easy to make.

Reactions Far from Equilibrium in Cells Are Common Points of Regulation

For some steps in a metabolic pathway the reaction is close to equilibrium, with the cell in its dynamic steady

TABLE 15-2	Relationship between Hill Coefficient and the Effect of Substrate Concentration on Reaction Rate for Allosteric Enzymes
Hill coefficient (n_H)	Required change in [S] to increase V_0 from 10% to 90% V_{max}
0.5	×6,600
1.0	×81
2.0	×9
3.0	×4.3
4.0	×3

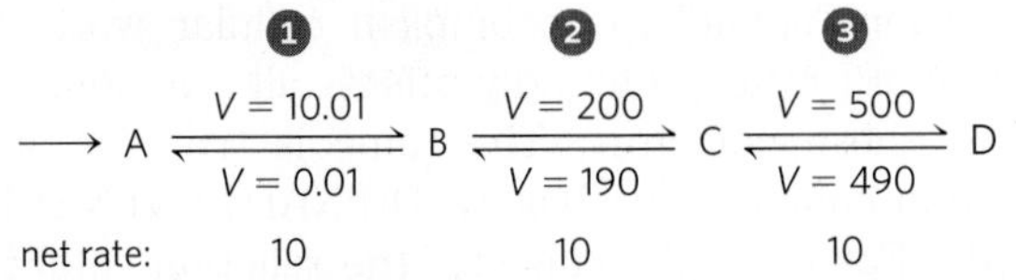

FIGURE 15-6 Near-equilibrium and nonequilibrium steps in a metabolic pathway. Steps ❷ and ❸ of this pathway are near equilibrium in the cell; for each step, the rate (V) of the forward reaction is only slightly greater than the reverse rate, so the net forward rate (10) is relatively low and the free-energy change, ΔG, is close to zero. An increase in [C] or [D] can reverse the direction of these steps. Step ❶ is maintained in the cell far from equilibrium; its forward rate greatly exceeds its reverse rate. The net rate of step ❶ (10) is much larger than the reverse rate (0.01) and is identical to the net rates of steps ❷ and ❸ when the pathway is operating in the steady state. Step ❶ has a large, negative ΔG.

state **(Fig. 15-6)**. The net flow of metabolites through these steps is the small difference between the rates of the forward and reverse reactions, rates that are very similar when a reaction is near equilibrium. Small changes in substrate or product concentration can produce large changes in the net rate, and can even change the direction of the net flow. We can identify these near-equilibrium reactions in a cell by comparing the **mass-action ratio, Q**, with the equilibrium constant for the reaction, K'_{eq}. Recall that for the reaction $A + B \rightarrow C + D$, $Q = [C][D]/[A][B]$. When Q and K'_{eq} are within 1 to 2 orders of magnitude of each other, the reaction is near

equilibrium. This is the case for 6 of the 10 steps in the glycolytic pathway (Table 15-3).

Other reactions are far from equilibrium in the cell. For example, K'_{eq} for the phosphofructokinase-1 (PFK-1) reaction is about 1,000, but Q ([fructose 1,6-bisphosphate][ADP]/[fructose 6-phosphate][ATP]) in a hepatocyte in the steady state is about 0.1 (Table 15-3). It is *because* the reaction is so far from equilibrium that the process is exergonic under cellular conditions and tends to go in the forward direction. The reaction is held far from equilibrium because, under prevailing cellular conditions of substrate, product, and effector concentrations, the rate of conversion of fructose 6-phosphate to fructose 1,6-bisphosphate is limited by the activity of PFK-1, which is itself limited by the number of PFK-1 molecules present and by the actions of allosteric effectors. Thus the net forward rate of the enzyme-catalyzed reaction is equal to the net flow of glycolytic intermediates through other steps in the pathway, and the reverse flow through PFK-1 remains near zero.

The cell *cannot* allow reactions with large equilibrium constants to reach equilibrium. If [fructose 6-phosphate], [ATP], and [ADP] in the cell were held at typical levels (low millimolar concentrations) and the PFK-1 reaction were allowed to reach equilibrium by an increase in [fructose 1,6-bisphosphate], the concentration of fructose 1,6-bisphosphate would rise into the molar range, wreaking osmotic havoc on the cell. Consider another case: if the reaction $ATP \rightarrow ADP + P_i$

TABLE 15-3	Equilibrium Constants, Mass-Action Ratios, and Free-Energy Changes for Enzymes of Carbohydrate Metabolism					
		Mass-action ratio, Q		Reaction near equilibrium in vivo?[a]	$\Delta G'^{\circ}$ (kJ/mol)	ΔG (kJ/mol) in heart
Enzyme	K'_{eq}	Liver	Heart			
Hexokinase	1×10^3	2×10^{-2}	8×10^{-2}	No	-17	-27
PFK-1	1.0×10^3	9×10^{-2}	3×10^{-2}	No	-14	-23
Aldolase	1.0×10^{-4}	1.2×0^{-6}	9×10^{-6}	Yes	$+24$	-6.0
Triose phosphate isomerase	4×10^{-2}	—[b]	2.4×10^{-1}	Yes	$+7.5$	$+3.8$
Glyceraldehyde 3-phosphate dehydrogenase + phosphoglycerate kinase	2×10^3	6×10^2	9.0	Yes	-13	$+3.5$
Phosphoglycerate mutase	1×10^{-1}	1×10^{-1}	1.2×10^{-1}	Yes	$+4.4$	$+0.6$
Enolase	3	2.9	1.4	Yes	-3.2	-0.5
Pyruvate kinase	2×10^4	7×10^{-1}	40	No	-31	-17
Phosphoglucose isomerase	4×10^{-1}	3.1×10^{-1}	2.4×10^{-1}	Yes	$+2.2$	-1.4
Pyruvate carboxylase + PEP carboxykinase	7	1×10^{-3}	—[b]	No	-5.0	-23
Glucose 6-phosphatase	8.5×10^2	1.2×10^2	—[b]	Yes	-17	-5.0

Source: K'_{eq} and Q from E. A. Newsholme and C. Start, *Regulation in Metabolism*, pp. 97, 263, Wiley Press, 1973. ΔG and $\Delta G'^{\circ}$ were calculated from these data.

[a]For simplicity, any reaction for which the absolute value of the calculated ΔG is less than 6 is considered near equilibrium.

[b]Data not available.

were allowed to approach equilibrium in the cell, the actual free-energy change (ΔG) for that reaction (ΔG_p; see Worked Example 13–2, p. 509) would approach zero, and ATP would lose the high phosphoryl group transfer potential that makes it valuable to the cell. It is therefore essential that enzymes catalyzing ATP breakdown and other highly exergonic reactions in a cell be sensitive to regulation, so that when metabolic changes are forced by external circumstances, the flow through these enzymes will be adjusted to ensure that [ATP] remains far above its equilibrium level. When such metabolic changes occur, the activities of enzymes in all interconnected pathways adjust to keep these critical steps away from equilibrium. Thus, not surprisingly, many enzymes (such as PFK-1) that catalyze highly exergonic reactions are subject to a variety of subtle regulatory mechanisms. The multiplicity of these adjustments is so great that we cannot predict by examining the properties of any one enzyme in a pathway whether that enzyme has a strong influence on net flow through the entire pathway. This complex problem can be approached by metabolic control analysis, as described in Section 15.2.

Adenine Nucleotides Play Special Roles in Metabolic Regulation

After the protection of its DNA from damage, perhaps nothing is more important to a cell than maintaining a constant supply and concentration of ATP. Many ATP-using enzymes have K_m values between 0.1 and 1 mM, and the ATP concentration in a typical cell is about 5 to 10 mM (Fig. 15-4). If [ATP] were to drop significantly, these enzymes would be less than fully saturated by their substrate (ATP), and the rates of hundreds of reactions that involve ATP would decrease (**Fig. 15-7**); the cell would probably not survive this *kinetic* effect on so many reactions.

There is also an important *thermodynamic* effect of lowered [ATP]. Because ATP is converted to ADP or

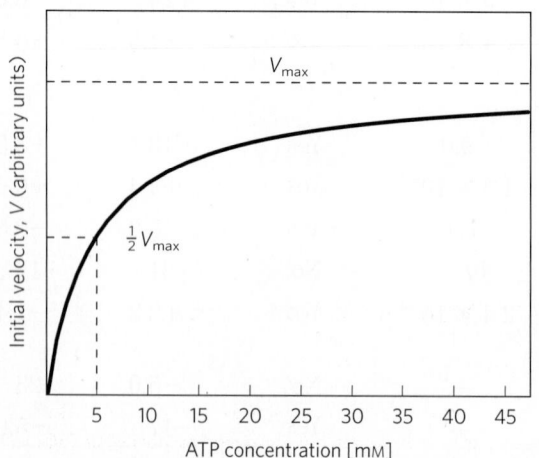

FIGURE 15-7 Effect of ATP concentration on the initial reaction velocity of a typical ATP-dependent enzyme. These experimental data yield a K_m for ATP of 5 mM. The concentration of ATP in animal tissues is ~5 mM.

AMP when "spent" to accomplish cellular work, the [ATP]/[ADP] ratio profoundly affects all reactions that employ these cofactors. The same is true for other important cofactors, such as NADH/NAD$^+$ and NADPH/NADP$^+$. For example, consider the reaction catalyzed by hexokinase:

$$\text{ATP} + \text{glucose} \longrightarrow \text{ADP} + \text{glucose 6-phosphate}$$

$$K'_{eq} = \frac{[\text{ADP}]_{eq}[\text{glucose 6-phosphate}]_{eq}}{[\text{ATP}]_{eq}[\text{glucose}]_{eq}} = 2 \times 10^3$$

Note that this expression holds true *only* when reactants and products are at their *equilibrium* concentrations, where $\Delta G' = 0$. At any other set of concentrations, $\Delta G'$ is not zero. Recall (from Chapter 13) that the ratio of products to substrates (the mass-action ratio, Q) determines the magnitude and sign of $\Delta G'$ and therefore the driving force, $\Delta G'$, of the reaction:

$$\Delta G' = \Delta G'^{\circ} + RT \ln \frac{[\text{ADP}][\text{glucose 6-phosphate}]}{[\text{ATP}][\text{glucose}]}$$

Because an alteration of this driving force profoundly influences every reaction that involves ATP, organisms have evolved under strong pressure to develop regulatory mechanisms responsive to the [ATP]/[ADP] ratio.

AMP concentration is an even more sensitive indicator of a cell's energetic state than is [ATP]. Normally, cells have a far higher concentration of ATP (5 to 10 mM) than of AMP (<0.1 mM). When some process (say, muscle contraction) consumes ATP, AMP is produced in two steps. First, hydrolysis of ATP produces ADP, then the reaction catalyzed by **adenylate kinase** produces AMP:

$$2\text{ADP} \longrightarrow \text{AMP} + \text{ATP}$$

If ATP is consumed such that its concentration drops 10%, the *relative* increase in [AMP] is much greater than that of [ADP] (Table 15-4). It is not surprising, therefore, that many regulatory processes are keyed to changes in [AMP]. Probably the most important mediator of regulation by AMP is **AMP-activated protein kinase (AMPK)**, which responds to an increase in [AMP] by phosphorylating key proteins and thus regulating their activities. (AMPK is not to be confused with the *cyclic* AMP–dependent protein kinase PKA; see Section 15.5.) The rise in [AMP] may be caused by a reduced nutrient supply or by increased exercise. AMP activates AMPK allosterically, causing a conformational change that makes AMPK a good substrate for phosphorylation by another protein kinase, LKB1. Phosphorylation of a specific Thr residue by LKB1 activates AMPK manyfold. LKB1 is itself regulated by many factors, including metabolic stress. AMPK activity increases glucose transport and activates glycolysis and fatty acid oxidation, while suppressing energy-requiring processes such as the synthesis of glycogen, fatty acids, cholesterol, and protein (**Fig. 15-8**). For example, AMPK slows glycogen synthesis by phosphorylating and inhibiting the enzyme glycogen synthase (see below). All of the changes

	TABLE 15-4	Relative Changes in [ATP] and [AMP] When ATP Is Consumed	
Adenine nucleotide	Concentration before ATP depletion (mM)	Concentration after ATP depletion (mM)	Relative change
ATP	5.0	4.5	10%
ADP	1.0	1.0	0
AMP	0.1	0.6	600%

effected by AMPK serve to raise [ATP] and lower [AMP]. In Chapter 23, we discuss the role of AMPK in balancing anabolism and catabolism in the whole organism.

In addition to ATP, hundreds of metabolic intermediates also must be present at appropriate concentrations in the cell. To take just one example: the glycolytic intermediates dihydroxyacetone phosphate and 3-phosphoglycerate are precursors of triacylglycerols and serine, respectively. When these products are needed, the rate of glycolysis must be adjusted to provide them without reducing the glycolytic production of ATP. The same is true for maintaining the levels of other important cofactors, such as NADH and NADPH: changes in their mass-action ratios (that is, in the ratio of reduced to oxidized cofactor) have global effects on metabolism.

Of course, priorities at the *organismal* level have also driven the evolution of regulatory mechanisms. In mammals, the brain has virtually no stored source of energy, depending instead on a constant supply of glucose from the blood. If blood glucose drops from its normal concentration of 4.5 to 5 mM to half that level, mental confusion results, and a fivefold reduction in blood glucose can lead to coma and death. To buffer against changes in blood glucose concentration, release of the hormones insulin and glucagon, elicited by high or low blood glucose, respectively, triggers metabolic changes that tend to return the blood glucose concentration to normal.

Other pressures must also have operated throughout evolution, selecting for regulatory mechanisms that accomplish the following:

1. Maximize the efficiency of fuel utilization by preventing the simultaneous operation of pathways in opposite directions (such as glycolysis and gluconeogenesis).

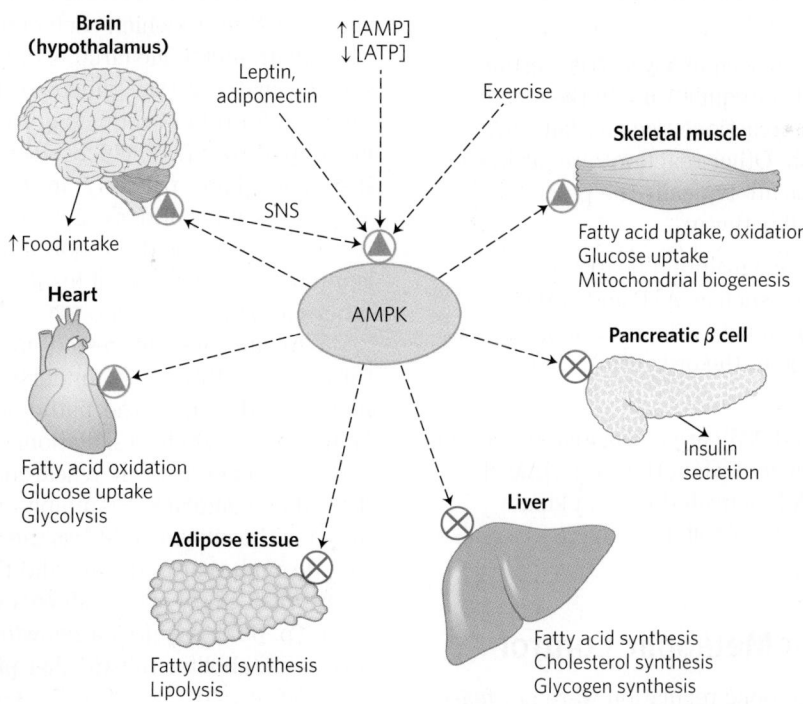

FIGURE 15-8 Role of AMP-activated protein kinase (AMPK) in carbohydrate and fat metabolism. AMPK is activated by elevated [AMP] or decreased [ATP], by exercise, by the sympathetic nervous system (SNS), or by peptide hormones produced in adipose tissue (leptin and adiponectin, described in more detail in Chapter 23). When activated, AMPK phosphorylates target proteins and shifts metabolism in a variety of tissues away from energy-consuming processes such as the synthesis of glycogen, fatty acids, and cholesterol; shifts metabolism in extrahepatic tissues to the use of fatty acids as a fuel; and triggers gluconeogenesis in the liver to provide glucose for the brain. In the hypothalamus, AMPK stimulates feeding behavior to provide more dietary fuel.

2. Partition metabolites appropriately between alternative pathways (such as glycolysis and the pentose phosphate pathway).

3. Draw on the fuel best suited for the immediate needs of the organism (glucose, fatty acids, glycogen, or amino acids).

4. Slow down biosynthetic pathways when their products accumulate.

The remaining chapters of this book present many examples of each kind of regulatory mechanism.

SUMMARY 15.1 Regulation of Metabolic Pathways

■ In a metabolically active cell in a steady state, intermediates are formed and consumed at equal rates. When a transient perturbation alters the rate of formation or consumption of a metabolite, compensating changes in enzyme activities return the system to the steady state.

■ Cells regulate their metabolism by a variety of mechanisms over a time scale ranging from less than a millisecond to days, either by changing the activity of existing enzyme molecules or by changing the number of molecules of a specific enzyme.

■ Various signals activate or inactivate transcription factors, which act in the nucleus to regulate gene expression. Changes in the transcriptome lead to changes in the proteome, and ultimately in the metabolome of a cell or tissue.

■ In multistep processes such as glycolysis, certain reactions are essentially at equilibrium in the steady state; the rates of these reactions rise and fall with substrate concentration. Other reactions are far from equilibrium; these steps are typically the points of regulation of the overall pathway.

■ Regulatory mechanisms maintain nearly constant levels of key metabolites such as ATP and NADH in cells and glucose in the blood, while matching the use or production of glucose to the organism's changing needs.

■ The levels of ATP and AMP are a sensitive reflection of a cell's energy status, and when the [ATP]/[AMP] ratio decreases, the AMP-activated protein kinase (AMPK) triggers a variety of cellular responses to raise [ATP] and lower [AMP].

15.2 Analysis of Metabolic Control

Detailed studies of metabolic regulation were not feasible until the basic chemical steps in a pathway had been clarified and the responsible enzymes characterized. Beginning with Eduard Buchner's discovery (c. 1900) that an extract of broken yeast cells could convert glucose to ethanol and CO_2, a major thrust of biochemical research was to deduce the steps by which this

Eduard Buchner, 1860–1917
[Source: Science Photo Library/Science Source.]

transformation occurred and to purify and characterize the enzymes that catalyzed each step. By the middle of the twentieth century, all 10 enzymes of the glycolytic pathway had been purified and characterized. In the next 50 years much was learned about the regulation of these enzymes by intracellular and extracellular signals, through the kinds of allosteric and covalent mechanisms described in this chapter. The conventional wisdom was that in a linear pathway such as glycolysis, catalysis by one enzyme must be the slowest and must therefore determine the rate of metabolite flow, or flux, through the whole pathway. For glycolysis, PFK-1 was considered the rate-limiting enzyme, because it was known to be closely regulated by fructose 2,6-bisphosphate and other allosteric effectors.

With the advent of genetic engineering technology, it became possible to test this "single rate-determining step" hypothesis by increasing the concentration of the enzyme that catalyzes the "rate-limiting step" in a pathway and determining whether flux through the pathway increases proportionally. Most often it does not; the simple explanation (a single rate-determining step) is wrong. It has now become clear that, in most pathways, the control of flux is distributed among several enzymes, and the extent to which each contributes to the control varies with metabolic circumstances—the supply of the starting material (say, glucose), the supply of oxygen, the need for other products derived from intermediates in the pathway (say, glucose 6-phosphate for the pentose phosphate pathway in cells synthesizing large amounts of nucleotides), the effects of metabolites with regulatory roles, and the hormonal status of the organism (such as the levels of insulin and glucagon), among other factors.

Why are we interested in what limits the flux through a pathway? To understand the action of hormones or drugs, or the pathology that results from a failure of metabolic regulation, we must know where control is exercised. If researchers wish to develop a drug that stimulates or inhibits a pathway, the logical target is the enzyme that has the greatest impact on the flux through that pathway. And the bioengineering of a microorganism to overproduce a product of commercial value (p. 327) requires a knowledge of what limits the flux of metabolites toward that product.

The Contribution of Each Enzyme to Flux through a Pathway Is Experimentally Measurable

There are several ways to determine experimentally how a change in the activity of one enzyme in a pathway affects metabolite flux through that pathway. Consider

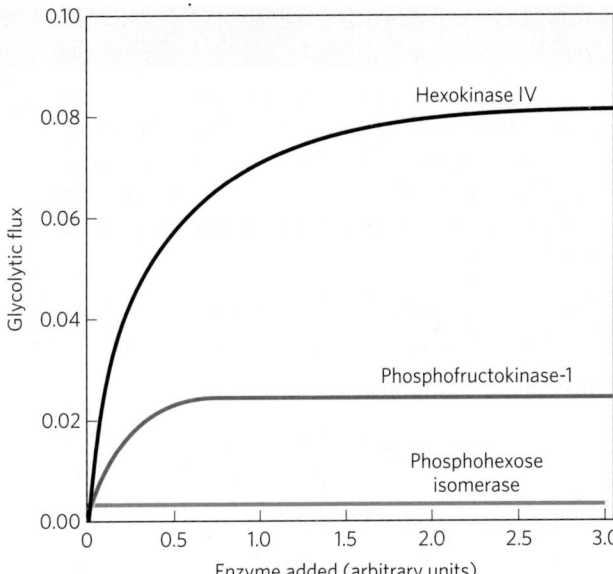

FIGURE 15-9 Dependence of glycolytic flux in a rat liver homogenate on added enzymes. Purified enzymes in the amounts shown on the *x* axis were added to an extract of liver carrying out glycolysis in vitro. The flux through the pathway is shown on the *y* axis. [Source: Data from N. V. Torres et al., *Biochem. J.* 234:169, 1986.]

the experimental results shown in **Figure 15-9**. When a sample of rat liver was homogenized to release all soluble enzymes, the extract carried out the glycolytic conversion of glucose to fructose 1,6-bisphosphate at a measurable rate. (This experiment, for simplicity, focused on just the first part of the glycolytic pathway.) When increasing amounts of purified hexokinase IV (glucokinase) were added to the extract, the rate of glycolysis progressively increased. The addition of purified PFK-1 to the extract also increased the rate of glycolysis, but not as dramatically as did hexokinase. Addition of purified phosphohexose isomerase was without effect. These results suggest that hexokinase and PFK-1 both contribute to setting the flux through the pathway (hexokinase more than PFK-1), and that phosphohexose isomerase does not.

Similar experiments can be done on intact cells or organisms, using specific inhibitors or activators to change the activity of one enzyme while observing the effect on flux through the pathway. The amount of an enzyme can also be altered genetically; bioengineering can produce a cell that makes extra copies of the enzyme under investigation or has a version of the enzyme that is less active than the normal enzyme. Increasing the concentration of an enzyme genetically sometimes has significant effects on flux; sometimes it has no effect.

Three critical parameters, which together describe the responsiveness of a pathway to changes in metabolic circumstances, lie at the center of **metabolic control analysis**. We turn now to a qualitative description of these parameters and their meaning in the context of a living cell. Box 15-1 provides a more rigorous quantitative discussion.

The Flux Control Coefficient Quantifies the Effect of a Change in Enzyme Activity on Metabolite Flux through a Pathway

Quantitative data on metabolic flux, obtained as described in Figure 15-9, can be used to calculate a **flux control coefficient**, C, for each enzyme in a pathway. This coefficient expresses the relative contribution of each enzyme to setting the rate at which metabolites flow through the pathway—that is, the **flux**, J. C can have any value from 0.0 (for an enzyme with no impact on the flux) to 1.0 (for an enzyme that wholly determines the flux). An enzyme can also have a *negative* flux control coefficient. In a branched pathway, an enzyme in one branch, by drawing intermediates away from the other branch, can have a negative impact on the flux through that other branch **(Fig. 15-10)**. C is not a constant, and it is not intrinsic to a single enzyme; it is a function of the whole system of enzymes, and its value depends on the concentrations of substrates and effectors.

When real data from the experiment on glycolysis in a rat liver extract (Fig. 15-9) were subjected to this kind of analysis, investigators found flux control coefficients (for enzymes at the concentrations found in the extract) of 0.79 for hexokinase, 0.21 for PFK-1, and 0.0 for phosphohexose isomerase. It is not just fortuitous that these values add up to 1.0; we can show that for any complete pathway, the sum of the flux control coefficients must equal unity.

The Elasticity Coefficient Is Related to an Enzyme's Responsiveness to Changes in Metabolite or Regulator Concentrations

A second parameter, the **elasticity coefficient**, ε, expresses quantitatively the responsiveness of a single enzyme to changes in the concentration of a metabolite or regulator; it is a function of the enzyme's intrinsic kinetic properties. For example, an enzyme with typical

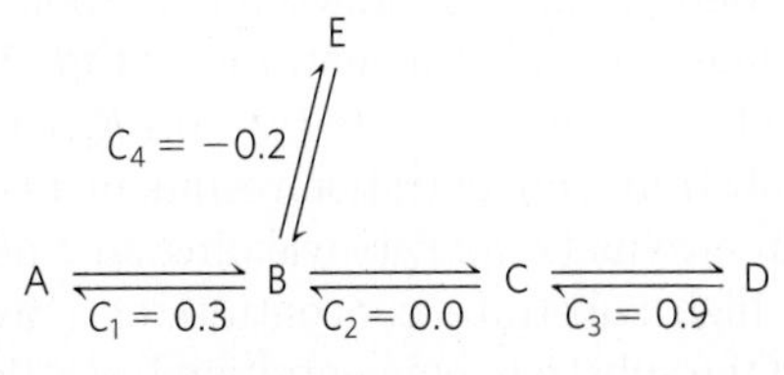

$$C_4 = -0.2 \quad \nearrow E$$

$$A \underset{C_1 = 0.3}{\rightleftharpoons} B \underset{C_2 = 0.0}{\rightleftharpoons} C \underset{C_3 = 0.9}{\rightleftharpoons} D$$

FIGURE 15-10 Flux control coefficient, *C*, in a branched metabolic pathway. In this simple pathway, the intermediate B has two alternative fates. To the extent that reaction B → E draws B away from the pathway A → D, it controls that pathway, which will result in a *negative* flux control coefficient for the enzyme that catalyzes step B → E. Note that the sum of all four coefficients equals 1.0, as it must for any defined system of enzymes.

BOX 15-1 | METHODS | Metabolic Control Analysis: Quantitative Aspects

The factors that influence the flow of intermediates (flux) through a pathway may be determined quantitatively by experiment and expressed in terms useful for predicting the change in flux when a factor involved in the pathway changes. Consider the simple reaction sequence in Figure 1, in which a substrate X (say, glucose) is converted in several steps to a product Z (perhaps pyruvate, formed glycolytically). An enzyme late in the pathway is a dehydrogenase (ydh) that acts on substrate Y. Because the action of a dehydrogenase is easily measured (see Fig. 13-24), we can use the flux (J) through this step (J_{ydh}) to measure the flux through the whole path. We manipulate experimentally the level of an early enzyme in the pathway (xase, which acts on the substrate X) and measure the flux through the path (J_{ydh}) for several levels of the enzyme xase.

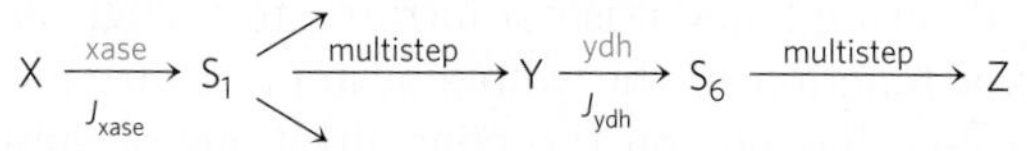

$$X \xrightarrow[J_{xase}]{xase} S_1 \xrightarrow{multistep} Y \xrightarrow[J_{ydh}]{ydh} S_6 \xrightarrow{multistep} Z$$

FIGURE 1 Flux through a hypothetical multienzyme pathway.

The relationship between the flux through the pathway from X to Z in the intact cell and the concentration of each enzyme in the path should be hyperbolic, with virtually no flux at infinitely low enzyme activity and near-maximum flux at very high enzyme activity. In a plot of J_{ydh} against the concentration of xase, E_{xase}, the change in flux with a small change in enzyme level is $\partial J_{ydh}/\partial E_{xase}$, which is simply the slope of the tangent to the curve at any concentration of enzyme, E_{xase}, and which tends toward zero at saturating E_{xase}. At low E_{xase}, the slope is steep; the flux increases with each incremental increase in enzyme activity. At very high E_{xase}, the slope is much smaller; the system is less responsive to added xase because it is already present in excess over the other enzymes in the pathway.

To show quantitatively the dependence of flux through the pathway, ∂J_{ydh}, on ∂E_{xase}, we could use the ratio $\partial J_{ydh}/\partial E_{xase}$. However, its usefulness is limited

because its value depends on the units used to express flux and enzyme activity. By expressing the *fractional* changes in flux and enzyme activity, $\partial J_{ydh}/J_{ydh}$ and $\partial E_{xase}/E_{xase}$, we obtain a unitless expression for the **flux control coefficient**, C, in this case $C_{xase}^{J_{ydh}}$:

$$C_{xase}^{J_{ydh}} \approx \frac{\partial J_{ydh}}{J_{ydh}} \bigg/ \frac{\partial E_{xase}}{E_{xase}} \qquad (1)$$

This can be rearranged to

$$C_{xase}^{J_{ydh}} \approx \frac{\partial J_{ydh}}{\partial E_{xase}} \cdot \frac{E_{xase}}{J_{ydh}}$$

which is mathematically identical to

$$C_{xase}^{J_{ydh}} \approx \frac{\partial \ln J_{ydh}}{\partial \ln E_{xase}}$$

This equation suggests a simple graphical means for determining the flux control coefficient: $C_{xase}^{J_{ydh}}$ is the slope of the tangent to the plot of $\ln J_{ydh}$ versus $\ln E_{xase}$, which can be obtained by replotting the experimental data in Figure 2a to obtain Figure 2b. Notice that $C_{xase}^{J_{ydh}}$ is not a constant; it depends on the starting E_{xase} from which the change in enzyme level takes place. For the cases shown in Figure 2, $C_{xase}^{J_{ydh}}$ is about 1.0 at the lowest E_{xase}, but only about 0.2 at high E_{xase}. A value near 1.0 for $C_{xase}^{J_{ydh}}$ means that the enzyme's concentration wholly determines the flux through the pathway; a value near 0.0 means that the enzyme's concentration does not limit the flux through the path. Unless the flux control coefficient is greater than about 0.5, changes in the activity of the enzyme will not have a strong effect on the flux.

The **elasticity coefficient**, ε, of an enzyme is a measure of how that enzyme's catalytic activity changes when the concentration of a metabolite—substrate, product, or effector—changes. It is obtained from an experimental plot of the rate of the reaction catalyzed by the enzyme versus the concentration of the metabolite, at metabolite concentrations that prevail in the cell. By arguments analogous to those used to

Michaelis-Menten kinetics shows a hyperbolic response to increasing substrate concentration (**Fig. 15-11**). At low concentrations of substrate (say, 0.1 K_m), each increment in substrate concentration results in a comparable increase in enzymatic activity, yielding an ε near 1.0. At relatively high substrate concentrations (say, 10 K_m), increasing the substrate concentration has little effect on

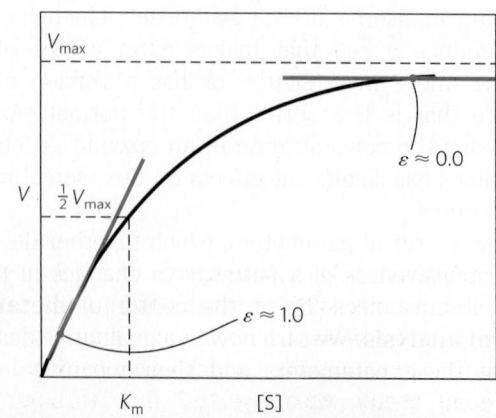

FIGURE 15-11 Elasticity coefficient, ε, of an enzyme with typical Michaelis-Menten kinetics. At substrate concentrations far below the K_m, each increase in [S] produces a correspondingly large increase in the reaction velocity, V. For this region of the curve, the enzyme has an ε of about 1.0. At [S] $\gg K_m$, increasing [S] has little effect on V; ε here is close to 0.0.

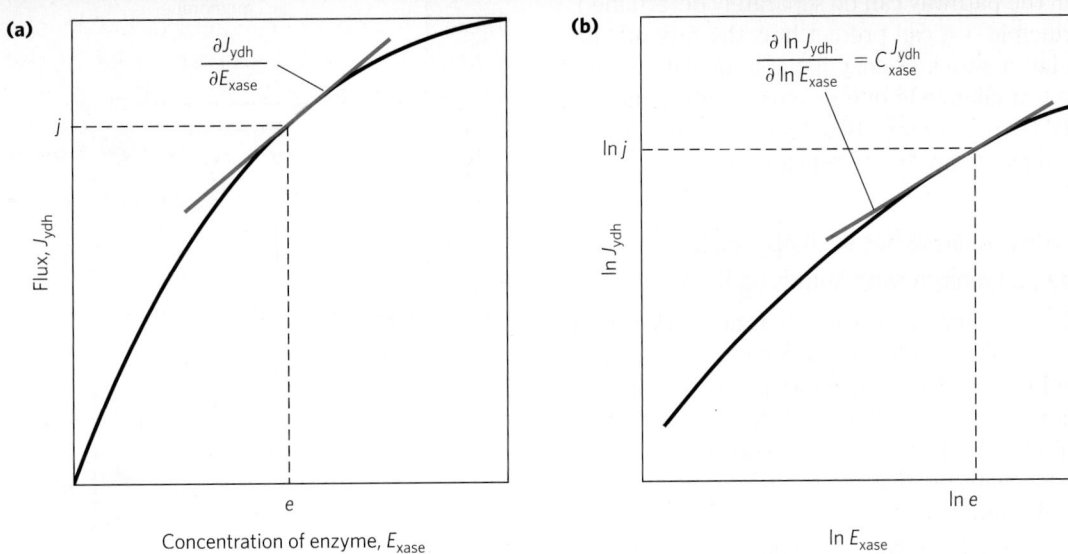

FIGURE 2 The flux control coefficient. **(a)** Typical variation of the pathway flux, J_{ydh}, measured at the step catalyzed by the enzyme ydh, as a function of the amount of the enzyme xase, E_{xase}, which catalyzes an earlier step in the pathway. The flux control coefficient at (e,j) is the product of the slope of the tangent to the curve, $\partial J_{ydh}/\partial E_{xase}$, and the ratio (scaling factor) e/j. **(b)** On a double-logarithmic plot of the same curve, the flux control coefficient is the slope of the tangent to the curve.

derive C, we can show ε to be the slope of the tangent to a plot of $\ln V$ versus $\ln$ [substrate, or product, or effector]:

$$\varepsilon_S^{xase} = \frac{\partial V_{xase}}{\partial S} \cdot \frac{S}{V_{xase}}$$

$$= \frac{\partial \ln |V_{xase}|}{\partial \ln S}$$

For an enzyme with typical Michaelis-Menten kinetics, the value of ε ranges from about 1 at substrate concentrations far below K_m to near 0 as V_{max} is approached. Allosteric enzymes can have elasticities greater than 1.0, but not larger than their Hill coefficient (p. 167).

Finally, the effect of controllers outside the pathway itself (that is, not metabolites) can be measured and expressed as the **response coefficient**, **R**. The change in flux through the pathway is measured for changes in the concentration of the controlling parameter P, and R is defined in a form analogous to that of Equation 1, yielding the expression

$$R_P^{J_{ydh}} = \frac{\partial J_{ydh}}{\partial P} \cdot \frac{P}{J_{ydh}}$$

Using the same logic and graphical methods as described above for determining C, we can obtain R as the slope of the tangent to the plot of $\ln J$ versus $\ln P$.

The three coefficients we have described are related in this simple way:

$$R_P^{J_{ydh}} = C_{xase}^{J_{ydh}} \cdot \varepsilon_P^{xase}$$

Thus the responsiveness of each enzyme in a pathway to a change in an outside controlling factor is a simple function of two things: the control coefficient, a variable that expresses the extent to which that enzyme influences the flux under a given set of conditions, and the elasticity, an intrinsic property of the enzyme that reflects its sensitivity to substrate and effector concentrations.

the reaction rate, because the enzyme is already saturated with substrate. The elasticity in this case approaches zero. For allosteric enzymes that show positive cooperativity, ε may exceed 1.0, but it cannot exceed the Hill coefficient, which is typically between 1.0 and 4.0.

The Response Coefficient Expresses the Effect of an Outside Controller on Flux through a Pathway

We can also derive a quantitative expression for the relative impact of an outside factor (such as a hormone or growth factor), which is neither a metabolite nor an enzyme in the pathway, on the flux through the pathway. The experiment would measure the flux through the pathway (glycolysis, in this case) at various levels of the parameter P (the insulin concentration, for example) to obtain the **response coefficient**, **R**, which expresses the change in pathway flux when P ([insulin]) changes.

The three coefficients C, ε, and R are related in a simple way: the responsiveness (R) of a pathway to an outside factor that affects a certain enzyme is a function of (1) how sensitive the pathway is to changes in the activity of that enzyme (the flux control coefficient, C) and (2) how sensitive that specific enzyme is to changes in the outside controlling factor (the elasticity, ε):

$$R = C \cdot \varepsilon$$

Each enzyme in the pathway can be examined in this way, and the effects of any of several outside factors on flux through the pathway can be separately determined. Thus, in principle, we can predict how the flux of substrate through a series of enzymatic steps will change when there is a change in one or more controlling factors external to the pathway. Box 15-1 shows how these qualitative concepts are treated quantitatively.

Metabolic Control Analysis Has Been Applied to Carbohydrate Metabolism, with Surprising Results

Metabolic control analysis provides a framework within which we can think quantitatively about regulation, interpret the significance of the regulatory properties of each enzyme in a pathway, identify the steps that most affect the flux through the pathway, and distinguish between *regulatory* mechanisms that act to maintain metabolite concentrations and *control* mechanisms that actually alter the flux through the pathway. Analysis of the glycolytic pathway in yeast, for example, has revealed an unexpectedly low flux control coefficient for PFK-1, which, as we have noted, has been viewed as the main point of flux control—the "rate-determining step"—in glycolysis. Experimentally raising the level of PFK-1 fivefold led to a change in flux through glycolysis of less than 10%, suggesting that the real role of PFK-1 regulation is not to control flux through glycolysis but to mediate metabolite homeostasis—to prevent large changes in metabolite concentrations when the flux through glycolysis increases in response to elevated blood glucose or insulin. Recall that the study of glycolysis in a liver extract (Fig. 15-9) also yielded a flux control coefficient that contradicted the conventional wisdom; it showed that hexokinase, not PFK-1, is most influential in setting the flux through glycolysis. We must note here that a liver extract is far from equivalent to a hepatocyte; the ideal way to study flux control is by manipulating one enzyme at a time in the living cell. This is feasible in many cases.

Investigators have used nuclear magnetic resonance (NMR) as a noninvasive means to determine the concentrations of glycogen and metabolites in the five-step pathway from glucose in the blood to glycogen in myocytes **(Fig. 15-12)** of rat and human muscle. They found that the flux control coefficient for glycogen synthase was smaller than that for either the glucose transporter (GLUT4) or hexokinase. (We discuss glycogen synthase and other enzymes of glycogen metabolism in Sections 15.4 and 15.5.) This finding contradicts the conventional wisdom that glycogen synthase is the locus of flux control and suggests that the importance of the phosphorylation/dephosphorylation of glycogen synthase is related instead to the maintenance of metabolite homeostasis—that is, *regulation*, not *control*. Two metabolites in this pathway, glucose and glucose 6-phosphate, are key intermediates in other pathways, including glycolysis, the pentose phosphate pathway, and the synthesis of glucosamine. Metabolic control analysis suggests that when the blood glucose

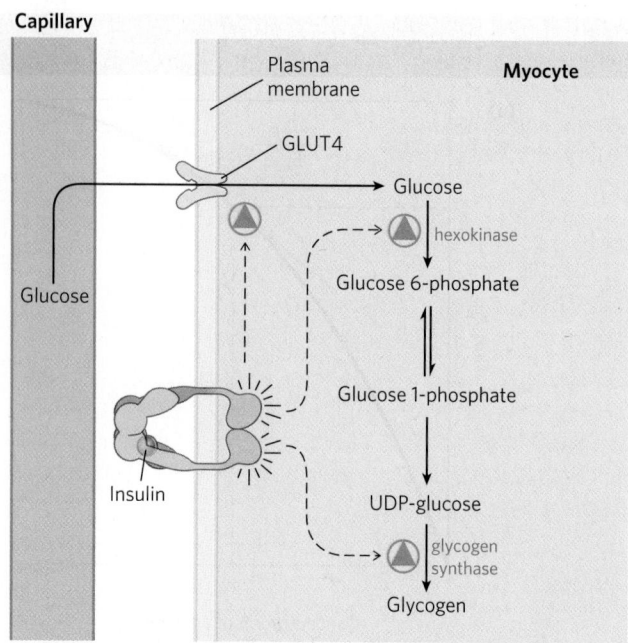

FIGURE 15-12 Control of glycogen synthesis from blood glucose in muscle. Insulin affects three of the five steps in this pathway, but it is the effects on transport and hexokinase activity, not the change in glycogen synthase activity, that increase the flux toward glycogen.

level rises, insulin acts in muscle to (1) increase glucose transport into cells by conveying GLUT4 to the plasma membrane, (2) induce the synthesis of hexokinase, and (3) activate glycogen synthase by covalent alteration (see Fig. 15-41). The first two effects of insulin increase glucose flux through the pathway (control), and the third serves to adapt the activity of glycogen synthase so that metabolite levels (glucose 6-phosphate, for example) will not change dramatically with the increased flux (regulation).

Metabolic Control Analysis Suggests a General Method for Increasing Flux through a Pathway

How could an investigator engineer a cell to increase the flux through one pathway without altering the concentrations of other metabolites or the fluxes through other pathways? More than three decades ago, Henrik Kacser predicted, on the basis of metabolic control analysis, that this could be accomplished by increasing the concentrations of every enzyme in a pathway. The prediction has been confirmed in several experimental tests, and it also fits with the way cells normally control fluxes through a pathway. For example, rats fed a high-protein diet dispose of excess amino groups by converting them to urea in the urea cycle (Chapter 18). After such a dietary shift, the urea output increases fourfold, and the amount of all eight enzymes in the urea cycle increases two- to threefold. Similarly, when increased fatty acid oxidation is triggered by activation of peroxisome proliferator-activated receptor γ (PPARγ, a ligand-activated transcription factor; see Fig. 21-22), synthesis of the *whole set* of fatty acid oxidative enzymes is increased.

With the growing use of DNA microarrays to study the expression of whole sets of genes in response to various perturbations, we should soon learn whether this is the general mechanism by which cells make long-term adjustments in flux through specific pathways.

SUMMARY 15.2 Analysis of Metabolic Control

■ Metabolic control analysis shows that control of the rate of metabolite flux through a pathway is distributed among several of the enzymes in that path.

■ The flux control coefficient, C, is an experimentally determined measure of the effect of an enzyme's concentration on flux through a multienzyme pathway. It is characteristic of the whole system, not intrinsic to the enzyme.

■ The elasticity coefficient, ε, of an enzyme is an experimentally determined measure of its responsiveness to changes in the concentration of a metabolite or regulator molecule.

■ The response coefficient, R, is a measure of the experimentally determined change in flux through a pathway in response to a regulatory hormone or second messenger. It is a function of C and ε: $R = C \cdot \varepsilon$.

■ Some regulated enzymes control the flux through a pathway, while others rebalance the level of metabolites in response to the change in flux. The first activity is *control*; the second, rebalancing activity is *regulation*.

■ Metabolic control analysis predicts, and experiments have confirmed, that flux toward a specific product is most effectively increased by raising the concentration of all enzymes in the pathway.

15.3 Coordinated Regulation of Glycolysis and Gluconeogenesis

In mammals, **gluconeogenesis** occurs primarily in the liver, where its role is to provide glucose for export to other tissues when glycogen stores are exhausted and when no dietary glucose is available. As we discussed in Chapter 14, gluconeogenesis employs several of the enzymes that act in glycolysis, but it is not simply the reversal of glycolysis. Seven of the glycolytic reactions are freely reversible, and the enzymes that catalyze these reactions also function in gluconeogenesis **(Fig. 15-13)**. Three reactions of glycolysis are so exergonic as to be essentially irreversible: those catalyzed by hexokinase, PFK-1, and pyruvate kinase. All three reactions have a large, negative ΔG (Table 15-3 shows the values in heart muscle). Gluconeogenesis uses detours around each of these irreversible steps; for example, the conversion of fructose 1,6-bisphosphate to fructose 6-phosphate is catalyzed by fructose 1,6-bisphosphatase (FBPase-1). Each of these bypass reactions also has a large, negative ΔG.

At each of the three points where glycolytic reactions are bypassed by alternative, gluconeogenic

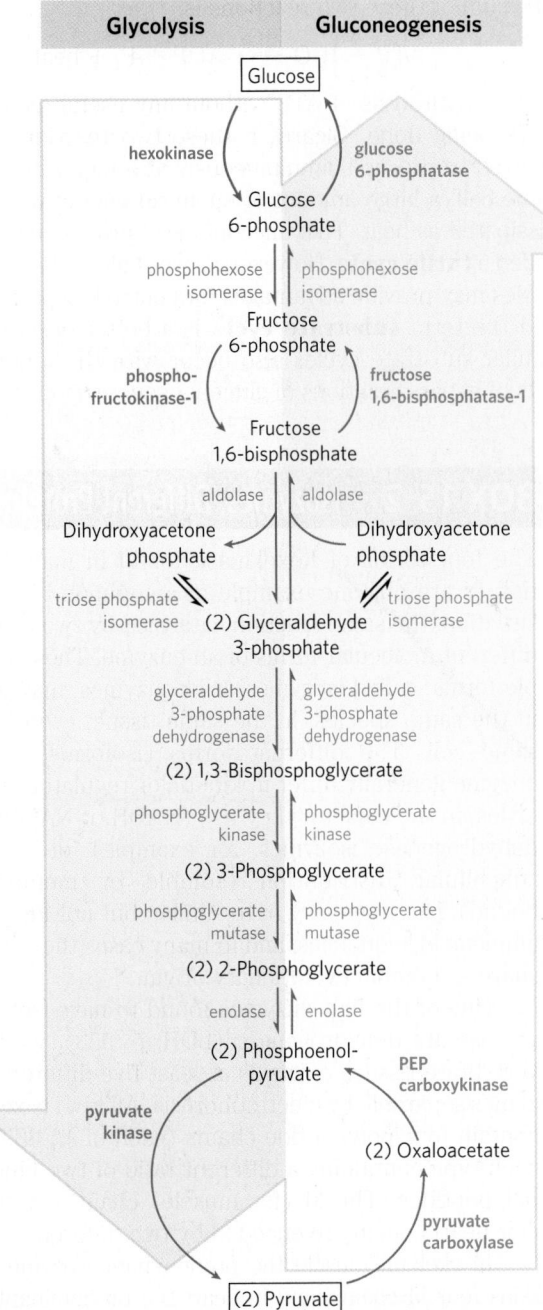

FIGURE 15-13 Glycolysis and gluconeogenesis. Opposing pathways of glycolysis (pink) and gluconeogenesis (blue) in rat liver. Three steps are catalyzed by different enzymes in gluconeogenesis (the "bypass reactions") and glycolysis; seven steps are catalyzed by the same enzymes in the two pathways. Cofactors have been omitted for simplicity.

reactions, simultaneous operation of both pathways would consume ATP without accomplishing any chemical or biological work. For example, PFK-1 and FBPase-1 catalyze opposing reactions:

$$\text{ATP} + \text{fructose 6-phosphate} \xrightarrow{\text{PFK-1}}$$
$$\text{ADP} + \text{fructose 1,6-bisphosphate}$$

$$\text{Fructose 1,6-bisphosphate} + \text{H}_2\text{O} \xrightarrow{\text{FBPase-1}}$$
$$\text{fructose 6-phosphate} + \text{P}_i$$

The sum of these two reactions is:

$$ATP + H_2O \longrightarrow ADP + P_i + heat$$

that is, hydrolysis of ATP without any useful metabolic work being done. Clearly, if these two reactions were allowed to proceed simultaneously at a high rate in the same cell, a large amount of chemical energy would be dissipated as heat. This uneconomical process has been called a **futile cycle**. However, as we shall see later, such cycles may provide advantages for controlling pathways, and the term **substrate cycle** is a better description. Similar substrate cycles also occur with the other two sets of bypass reactions of gluconeogenesis (Fig. 15-13).

We look now in some detail at the mechanisms that regulate glycolysis and gluconeogenesis at the three points where these pathways diverge.

Hexokinase Isozymes of Muscle and Liver Are Affected Differently by Their Product, Glucose 6-Phosphate

Hexokinase, which catalyzes the entry of glucose into the glycolytic pathway, is a regulatory enzyme. Humans have four isozymes (designated I to IV), encoded by four different genes. **Isozymes** are different proteins that catalyze the same reaction (Box 15-2). The predominant hexokinase isozyme of myocytes (**hexokinase II**)

BOX 15-2 Isozymes: Different Proteins That Catalyze the Same Reaction

The four forms of hexokinase found in mammalian tissues are but one example of a common biological situation: the same reaction catalyzed by two or more different molecular forms of an enzyme. These multiple forms, called isozymes or isoenzymes, may occur in the same species, in the same tissue, even in the same cell. The different forms (isoforms) of the enzyme generally differ in kinetic or regulatory properties, in the cofactor they use (NADH or NADPH for dehydrogenase isozymes, for example), or in their subcellular distribution (soluble or membrane-bound). Isozymes may have similar, but not identical, amino acid sequences, and in many cases they clearly share a common evolutionary origin.

One of the first enzymes found to have isozymes was lactate dehydrogenase (LDH; p. 553), which in vertebrate tissues exists as at least five different isozymes separable by electrophoresis. All LDH isozymes contain four polypeptide chains (each of M_r 33,500), each type containing a different ratio of two kinds of polypeptides. The M (for muscle) chain and the H (for heart) chain are encoded by two different genes.

In skeletal muscle the predominant isozyme contains four M chains, and in heart the predominant isozyme contains four H chains. Other tissues have some combination of the five possible types of LDH isozymes:

Type	Composition	Location
LDH$_1$	HHHH	Heart and erythrocyte
LDH$_2$	HHHM	Heart and erythrocyte
LDH$_3$	HHMM	Brain and kidney
LDH$_4$	HMMM	Skeletal muscle and liver
LDH$_5$	MMMM	Skeletal muscle and liver

Differences in the isozyme content of tissues can be used to assess the timing and extent of heart damage due to myocardial infarction (heart attack). Damage to heart tissue results in the release of heart LDH into the blood. Shortly after a heart

attack, the blood level of total LDH increases, and there is more LDH$_2$ than LDH$_1$. After 12 hours the amounts of LDH$_1$ and LDH$_2$ are very similar, and after 24 hours there is more LDH$_1$ than LDH$_2$. This switch in the [LDH$_1$]/[LDH$_2$] ratio, combined with increased concentrations in the blood of another heart enzyme, creatine kinase, is very strong evidence of a recent myocardial infarction. ■

The different LDH isozymes have significantly different values of V_{max} and K_m, particularly for pyruvate. The properties of LDH$_4$ favor rapid reduction of very low concentrations of pyruvate to lactate in skeletal muscle, whereas those of isozyme LDH$_1$ favor rapid oxidation of lactate to pyruvate in the heart.

In general, the distribution of different isozymes of a given enzyme reflects at least four factors:

1. *Different metabolic patterns in different organs.* For glycogen phosphorylase, the isozymes in skeletal muscle and liver have different regulatory properties, reflecting the different roles of glycogen breakdown in these two tissues.

2. *Different locations and metabolic roles for isozymes in the same cell.* The isocitrate dehydrogenase isozymes of the cytosol and the mitochondrion are an example (Chapter 16).

3. *Different stages of development in embryonic or fetal tissues and in adult tissues.* For example, the fetal liver has a characteristic isozyme distribution of LDH, which changes as the organ develops into its adult form. Some enzymes of glucose catabolism in malignant (cancer) cells occur as their fetal, not adult, isozymes.

4. *Different responses of isozymes to allosteric modulators.* This difference is useful in fine-tuning metabolic rates. Hexokinase IV (glucokinase) of liver and the hexokinase isozymes of other tissues differ in their sensitivity to inhibition by glucose 6-phosphate.

has a high affinity for glucose—it is half-saturated at about 0.1 mM. Because glucose entering myocytes from the blood (where the glucose concentration is 4 to 5 mM) produces an intracellular glucose concentration high enough to saturate hexokinase II, the enzyme normally acts at or near its maximal rate. Muscle **hexokinase I** and hexokinase II are allosterically inhibited by their product, glucose 6-phosphate, so whenever the cellular concentration of glucose 6-phosphate rises above its normal level, these isozymes are temporarily and reversibly inhibited, bringing the rate of glucose 6-phosphate formation into balance with the rate of its utilization and reestablishing the steady state.

The different hexokinase isozymes of liver and muscle reflect the different roles of these organs in carbohydrate metabolism: muscle consumes glucose, using it for energy production, whereas liver maintains blood glucose homeostasis by consuming or producing glucose, depending on the prevailing blood glucose concentration. The predominant hexokinase isozyme of liver is **hexokinase IV** (glucokinase), which differs in three important respects from hexokinases I to III of muscle. First, the glucose concentration at which hexokinase IV is half-saturated (about 10 mM) is higher than the usual concentration of glucose in the blood. Because an efficient glucose transporter in hepatocytes **(GLUT2)** rapidly equilibrates the glucose concentrations in cytosol and blood (see Fig. 11-28 for the kinetics of the same transporter, GLUT1, in erythrocytes), the high K_m of hexokinase IV allows its direct regulation by the level of blood glucose **(Fig. 15-14)**. When blood glucose is high, as it is after a meal rich in carbohydrates, excess glucose is transported into hepatocytes, where hexokinase IV converts it to glucose 6-phosphate. Because hexokinase IV is not saturated at 10 mM glucose, its activity continues to increase as the glucose concentration rises to 10 mM or more. Under conditions of low blood glucose, the glucose concentration in a hepatocyte is low relative to the K_m of hexokinase IV, and the glucose generated by gluconeogenesis leaves the cell before being trapped by phosphorylation.

Second, hexokinase IV is not inhibited by glucose 6-phosphate, and it can therefore continue to operate

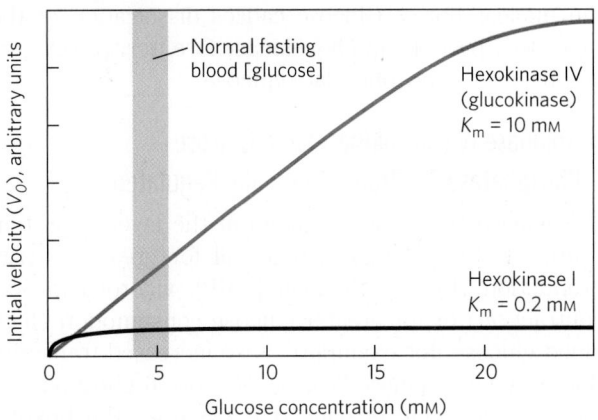

FIGURE 15-14 Comparison of the kinetic properties of hexokinase IV (glucokinase) and hexokinase I. Note the sigmoidicity for hexokinase IV and the much lower K_m for hexokinase I. When blood glucose rises above 5 mM, hexokinase IV activity increases, but hexokinase I is already operating near V_{max} and cannot respond to an increase in glucose concentration. Hexokinases I, II, and III have similar kinetic properties.

when the accumulation of glucose 6-phosphate completely inhibits hexokinases I to III. Finally, hexokinase IV is subject to inhibition by the reversible binding of a regulatory protein specific to liver **(Fig. 15-15)**. The binding is much tighter in the presence of the allosteric effector fructose 6-phosphate. Glucose competes with fructose 6-phosphate for binding and causes dissociation of the regulatory protein from the hexokinase, relieving the inhibition. Immediately after a carbohydrate-rich meal, when blood glucose is high, glucose enters the hepatocyte via GLUT2 and activates hexokinase IV by this mechanism. During a fast, when blood glucose drops below 5 mM, fructose 6-phosphate triggers the inhibition of hexokinase IV by the regulatory protein, so the liver does not compete with other organs for the scarce glucose. The mechanism of inhibition by the regulatory protein is interesting: the protein anchors hexokinase IV inside the nucleus, where it is segregated from the other enzymes of glycolysis in the cytosol (Fig. 15-15). When the glucose concentration in the cytosol rises, it equilibrates with glucose in the nucleus by transport through

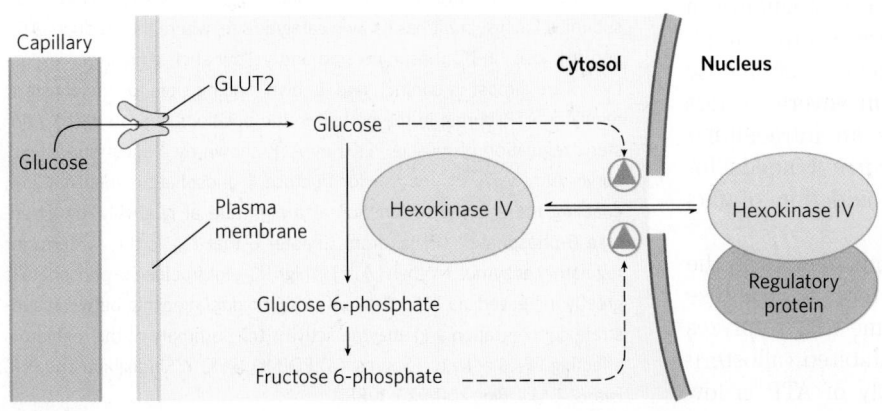

FIGURE 15-15 Regulation of hexokinase IV (glucokinase) by sequestration in the nucleus. The protein inhibitor of hexokinase IV is a nuclear binding protein that draws hexokinase IV into the nucleus when the fructose 6-phosphate concentration in liver is high and releases it to the cytosol when the glucose concentration is high.

the nuclear pores. Glucose causes dissociation of the regulatory protein, and hexokinase IV enters the cytosol and begins to phosphorylate glucose.

Hexokinase IV (Glucokinase) and Glucose 6-Phosphatase Are Transcriptionally Regulated

Hexokinase IV is also regulated at the level of protein synthesis. Circumstances that call for greater energy production (low [ATP], high [AMP], vigorous muscle contraction) or for greater glucose consumption (high blood glucose, for example) cause increased transcription of the hexokinase IV gene. Glucose 6-phosphatase, the gluconeogenic enzyme that bypasses the hexokinase step of glycolysis, is transcriptionally regulated by factors that call for increased production of glucose (low blood glucose, glucagon signaling). The transcriptional regulation of these two enzymes (along with other enzymes of glycolysis and gluconeogenesis) is described below.

Phosphofructokinase-1 and Fructose 1,6-Bisphosphatase Are Reciprocally Regulated

As we have noted, glucose 6-phosphate can flow either into glycolysis or through any of several other pathways, including glycogen synthesis and the pentose phosphate pathway. The metabolically irreversible reaction catalyzed by PFK-1 is the step that commits glucose to glycolysis. In addition to its substrate-binding sites, this complex enzyme has several regulatory sites at which allosteric activators or inhibitors bind.

ATP is not only a substrate for PFK-1 but also an end product of the glycolytic pathway. When high cellular [ATP] signals that ATP is being produced faster than it is being consumed, ATP inhibits PFK-1 by binding to an allosteric site and lowering the affinity of the enzyme for its substrate fructose 6-phosphate **(Fig. 15-16)**. ADP and AMP, which increase in concentration as consumption of ATP outpaces production, act allosterically to relieve this inhibition by ATP. These effects combine to produce higher enzyme activity when ADP or AMP accumulates and lower activity when ATP accumulates.

Citrate (the ionized form of citric acid), a key intermediate in the aerobic oxidation of pyruvate, fatty acids, and amino acids, is also an allosteric regulator of PFK-1; high citrate concentration increases the inhibitory effect of ATP, further reducing the flow of glucose through glycolysis. In this case, as in several others encountered later, citrate serves as an intracellular signal that the cell is meeting its current needs for energy-yielding metabolism by the oxidation of fats and proteins.

The corresponding step in gluconeogenesis is the conversion of fructose 1,6-bisphosphate to fructose 6-phosphate **(Fig. 15-17)**. The enzyme that catalyzes this reaction, FBPase-1, is strongly inhibited (allosterically) by AMP; when the cell's supply of ATP is low

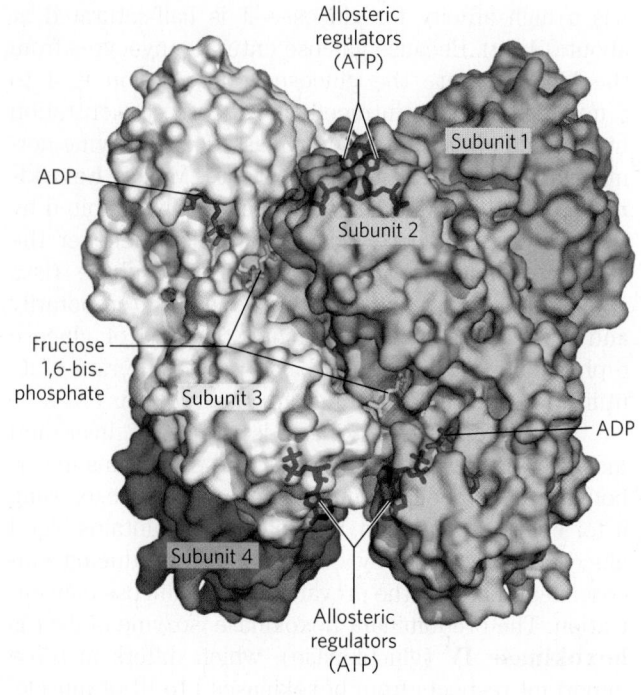

(a)

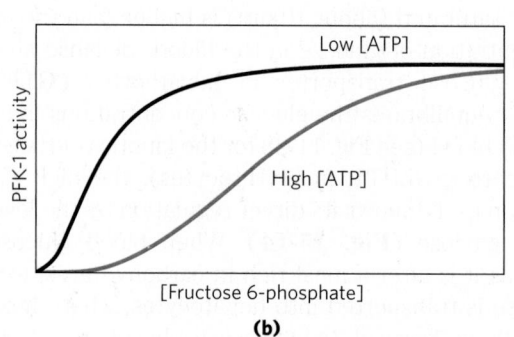

(b)

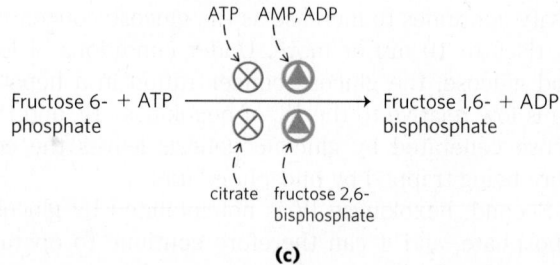

(c)

FIGURE 15-16 Phosphofructokinase-1 (PFK-1) and its regulation. (a) Surface contour image of *E. coli* PFK-1, showing portions of its four identical subunits. Each subunit has its own catalytic site, where the products ADP and fructose 1,6-bisphosphate (red and yellow stick structures, respectively) are almost in contact, and its own binding sites for the allosteric regulator ATP, buried in the protein in the positions indicated. (b) Allosteric regulation of muscle PFK-1 by ATP, shown by a substrate-activity curve. At low [ATP], the $K_{0.5}$ for fructose 6-phosphate is relatively low, enabling the enzyme to function at a high rate at relatively low [fructose 6-phosphate]. (Recall from Chapter 6 that $K_{0.5}$ is the K_m term for regulatory enzymes.) When [ATP] is high, $K_{0.5}$ for fructose 6-phosphate is greatly increased, as indicated by the sigmoid relationship between substrate concentration and enzyme activity. (c) Summary of the regulators affecting PFK-1 activity. [Source: (a) PDB ID 1PFK, Y. Shirakihara and P. R. Evans, *J. Mol. Biol.* 204:973, 1988.]

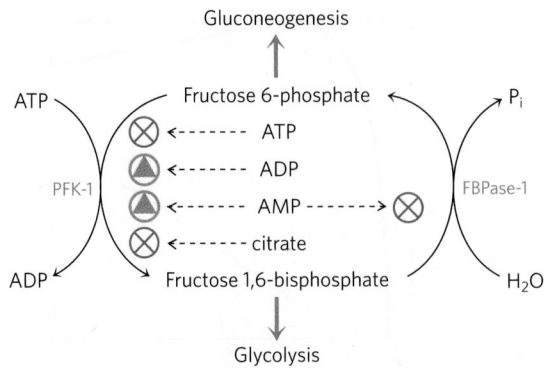

FIGURE 15-17 Regulation of fructose 1,6-bisphosphatase (FBPase-1) and phosphofructokinase-1 (PFK-1). The important role of fructose 2,6-bisphosphate in the regulation of this substrate cycle is detailed in subsequent figures.

(corresponding to high [AMP]), the ATP-requiring synthesis of glucose slows.

Thus these opposing steps in the glycolytic and gluconeogenic pathways—those catalyzed by PFK-1 and FBPase-1—are regulated in a coordinated and reciprocal manner. In general, when sufficient concentrations of acetyl-CoA or citrate (the product of acetyl-CoA condensation with oxaloacetate) are present, or when a high proportion of the cell's adenylate is in the form of ATP, gluconeogenesis is favored. When the level of AMP increases, it promotes glycolysis by stimulating PFK-1 (and, as we shall see in Section 15.5, promotes glycogen degradation by activating glycogen phosphorylase).

Fructose 2,6-Bisphosphate Is a Potent Allosteric Regulator of PFK-1 and FBPase-1

The special role of the liver in maintaining a constant blood glucose level requires additional regulatory mechanisms to coordinate glucose production and consumption. When the blood glucose level decreases, the hormone **glucagon** signals the liver to produce and release more glucose and to stop consuming it for its own needs. One source of glucose is glycogen stored in the liver; another source is gluconeogenesis, using pyruvate, lactate, glycerol, or certain amino acids as starting material. When blood glucose is high, insulin signals the liver to use glucose as a fuel and as a precursor for the synthesis and storage of glycogen and triacylglycerol.

The rapid hormonal regulation of glycolysis and gluconeogenesis is mediated by **fructose 2,6-bisphosphate**, an allosteric effector for the enzymes PFK-1 and FBPase-1:

Fructose 2,6-bisphosphate

When fructose 2,6-bisphosphate binds to its allosteric site on PFK-1, it increases the enzyme's affinity for its substrate fructose 6-phosphate and reduces its affinity for the allosteric inhibitors ATP and citrate **(Fig. 15-18)**. At the physiological concentrations of its substrates, ATP and fructose 6-phosphate, and of its other positive and negative effectors (ATP, AMP, citrate), PFK-1 is virtually inactive in the absence of fructose 2,6-bisphosphate. Fructose 2,6-bisphosphate has the opposite effect on FBPase-1: it reduces its affinity for its substrate (Fig. 15-18b), thereby slowing gluconeogenesis.

The cellular concentration of the allosteric regulator fructose 2,6-bisphosphate is set by the relative rates of its formation and breakdown **(Fig. 15-19a)**. It is formed by phosphorylation of fructose 6-phosphate, catalyzed by **phosphofructokinase-2 (PFK-2)**, and is broken down by **fructose 2,6-bisphosphatase (FBPase-2)**. (Note that these enzymes are distinct from PFK-1 and FBPase-1, which catalyze the formation and breakdown, respectively, of fructose 1,6-bisphosphate.) PFK-2 and FBPase-2 are two separate enzymatic activities of a single, bifunctional protein. The balance of these two activities in the liver, which determines the cellular level of fructose 2,6-bisphosphate, is regulated by glucagon and insulin (Fig. 15-19b).

As we saw in Chapter 12 (p. 449–450), glucagon stimulates the adenylyl cyclase of liver to synthesize 3′,5′-cyclic AMP (cAMP) from ATP. Cyclic AMP then activates cAMP-dependent protein kinase, which transfers a phosphoryl group from ATP to the bifunctional protein PFK-2/FBPase-2. Phosphorylation of this protein enhances its FBPase-2 activity and inhibits its PFK-2 activity. Glucagon thereby lowers the cellular level of fructose 2,6-bisphosphate, inhibiting glycolysis and stimulating gluconeogenesis. The resulting production of more glucose enables the liver to replenish blood glucose in response to glucagon. Insulin has the opposite effect, stimulating the activity of a phosphoprotein phosphatase that catalyzes removal of the phosphoryl group from the bifunctional protein PFK-2/FBPase-2, activating its PFK-2 activity, increasing the level of fructose 2,6-bisphosphate, stimulating glycolysis, and inhibiting gluconeogenesis.

Xylulose 5-Phosphate Is a Key Regulator of Carbohydrate and Fat Metabolism

Another regulatory mechanism also acts by controlling the level of fructose 2,6-bisphosphate. In the mammalian liver, xylulose 5-phosphate (p. 567), a product of the pentose phosphate pathway (hexose monophosphate pathway), mediates the increase in glycolysis that follows ingestion of a high-carbohydrate meal. The xylulose 5-phosphate concentration rises as glucose entering the liver is converted to glucose 6-phosphate and enters both the glycolytic and pentose phosphate pathways. Xylulose 5-phosphate activates phosphoprotein

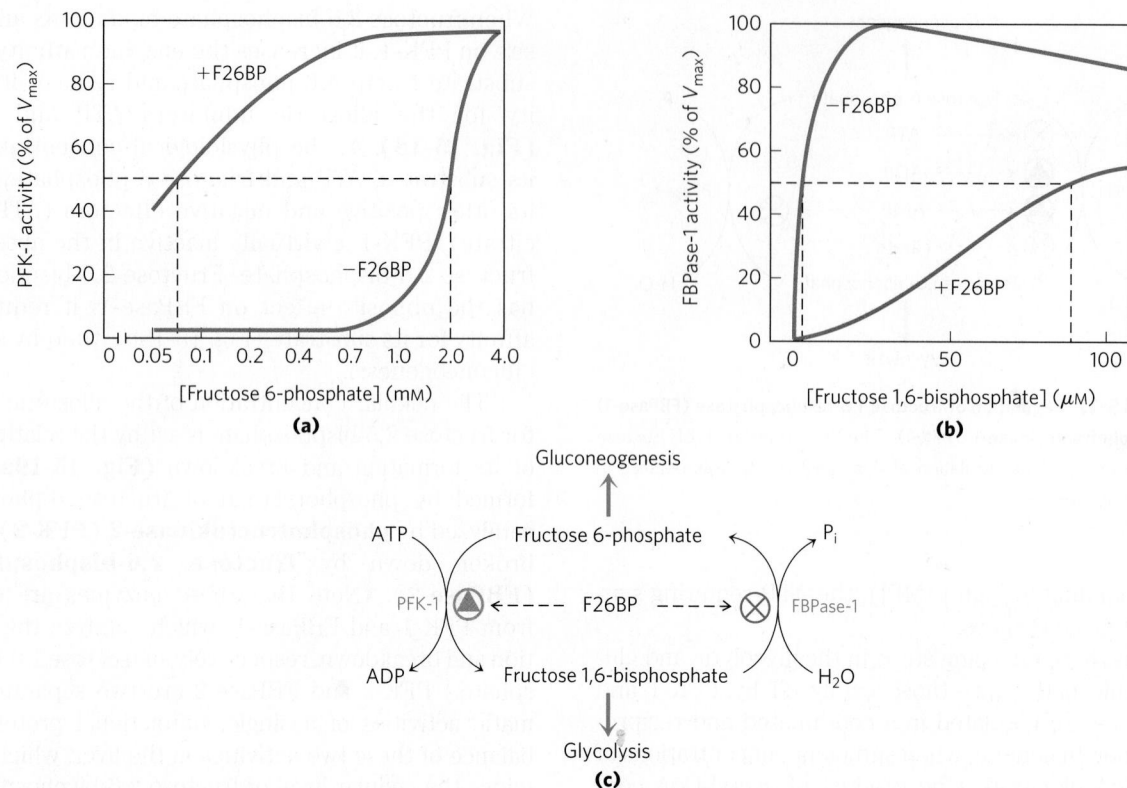

FIGURE 15-18 Role of fructose 2,6-bisphosphate in regulation of glycolysis and gluconeogenesis. Fructose 2,6-bisphosphate (F26BP) has opposite effects on the enzymatic activities of phosphofructokinase-1 (PFK-1, a glycolytic enzyme) and fructose 1,6-bisphosphatase (FBPase-1, a gluconeogenic enzyme). **(a)** PFK-1 activity in the absence of F26BP (blue curve) is half-maximal when the concentration of fructose 6-phosphate is 2 mM (that is, $K_{0.5} = 2$ mM). When 0.13 μM F26BP is present (red curve), the $K_{0.5}$ for fructose 6-phosphate is only 0.08 mM. Thus F26BP activates PFK-1 by increasing its apparent affinity for fructose 6-phosphate (see Fig. 15-16b). **(b)** FBPase-1 activity is inhibited by as little as 1 μM F26BP and is strongly inhibited by 25 μM. In the absence of this inhibitor (blue curve), the $K_{0.5}$ for fructose 1,6-bisphosphate is 5 μM, but in the presence of 25 μM F26BP (red curve), the $K_{0.5}$ is > 70 μM. Fructose 2,6-bisphosphate also makes FBPase-1 more sensitive to inhibition by another allosteric regulator, AMP. **(c)** Summary of regulation by F26BP.

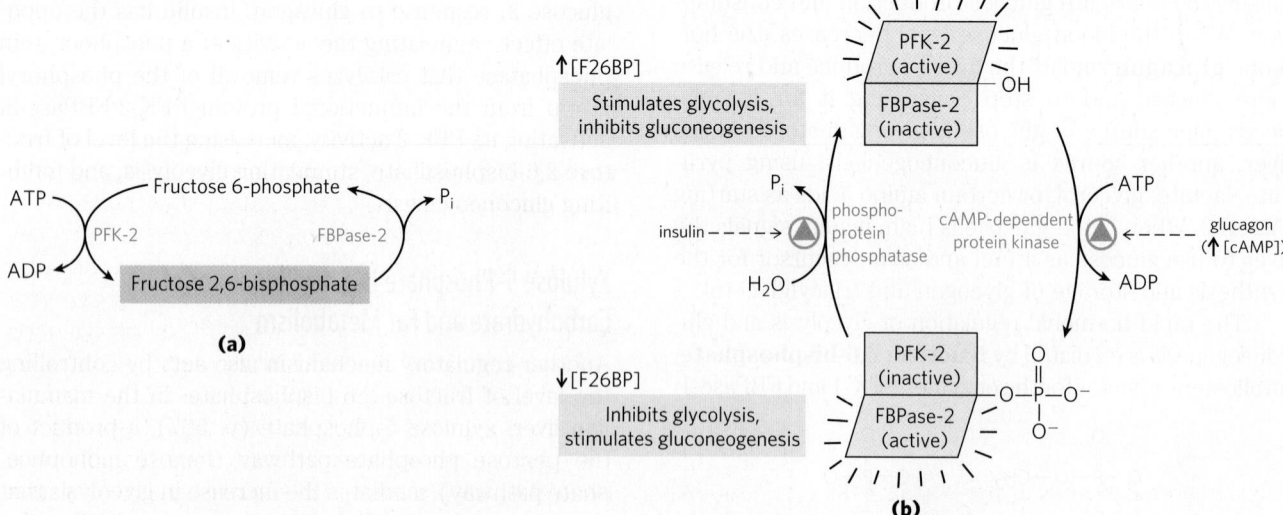

FIGURE 15-19 Regulation of fructose 2,6-bisphosphate level. (a) The cellular concentration of the regulator fructose 2,6-bisphosphate (F26BP) is determined by the rates of its synthesis by phosphofructokinase-2 (PFK-2) and its breakdown by fructose 2,6-bisphosphatase (FBPase-2). **(b)** Both enzyme activities are part of the same polypeptide chain, and they are reciprocally regulated by insulin and glucagon.

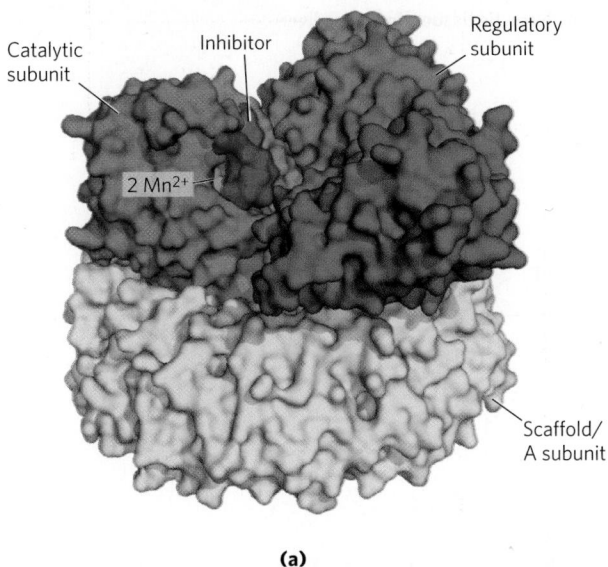

(a)

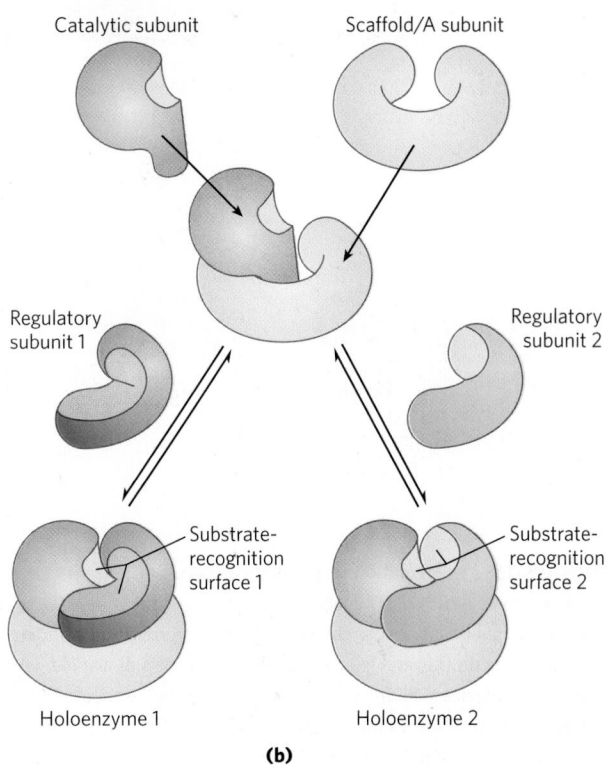

(b)

FIGURE 15-20 Structure and action of phosphoprotein phosphatase 2A (PP2A). (a) The catalytic subunit has two Mn^{2+} ions in its active site, positioned close to the substrate-recognition surface formed by the interface between the catalytic subunit and the regulatory subunit. Microcystin-LR, shown here in red, is a specific inhibitor of PP2A. The catalytic and regulatory subunits rest in a scaffold (the A subunit) that positions them relative to each other and shapes the substrate-recognition site. (b) PP2A recognizes several target proteins, its specificity provided by the regulatory subunit. Each of several regulatory subunits fits the scaffold containing the catalytic subunit, and each regulatory subunit creates its unique substrate-binding site. [Source: (a) PDB ID 2NPP, Y. Xu et al., *Cell* 127:1239, 2006.]

phosphatase 2A (PP2A; **Fig. 15-20**), which dephosphorylates the bifunctional PFK-2/FBPase-2 enzyme (Fig. 15-19). Dephosphorylation activates PFK-2 and inhibits FBPase-2, and the resulting rise in fructose 2,6-bisphosphate concentration stimulates glycolysis and inhibits gluconeogenesis. The increased glycolysis boosts the production of acetyl-CoA, while the increased flow of hexose through the pentose phosphate pathway generates NADPH. Acetyl-CoA and NADPH are the starting materials for fatty acid synthesis, which has long been known to increase dramatically in response to intake of a high-carbohydrate meal. Xylulose 5-phosphate also increases the synthesis of *all* the enzymes required for fatty acid synthesis, meeting the prediction from metabolic control analysis. We return to this effect in our discussion of the integration of carbohydrate and lipid metabolism in Chapter 23.

The Glycolytic Enzyme Pyruvate Kinase Is Allosterically Inhibited by ATP

At least three isozymes of pyruvate kinase are found in vertebrates, differing in their tissue distribution and their response to modulators. High concentrations of ATP, acetyl-CoA, and long-chain fatty acids (signs of abundant energy supply) allosterically inhibit all

isozymes of pyruvate kinase (**Fig. 15-21**). The liver isozyme (L form), but not the muscle isozyme (M form), is subject to further regulation by phosphorylation. When low blood glucose causes glucagon release, cAMP-dependent protein kinase phosphorylates the L isozyme of pyruvate kinase, inactivating it. This slows the use of glucose as a fuel in liver, sparing it for export to the brain and other organs. In muscle, the effect of increased [cAMP] is quite different. In response to epinephrine, cAMP activates glycogen breakdown and glycolysis and provides the fuel needed for the fight-or-flight response.

The Gluconeogenic Conversion of Pyruvate to Phosphoenolpyruvate Is under Multiple Types of Regulation

In the pathway leading from pyruvate to glucose, the first control point determines the fate of pyruvate in the mitochondrion: its conversion either to acetyl-CoA (by the pyruvate dehydrogenase complex) to fuel the citric acid cycle (Chapter 16) or to oxaloacetate (by pyruvate carboxylase) to start the process of gluconeogenesis (**Fig. 15-22**). When fatty acids are readily available as fuels, their breakdown in liver mitochondria yields acetyl-CoA, a signal that further oxidation of glucose for fuel is not necessary. Acetyl-CoA is a positive allosteric modulator of pyruvate carboxylase and a negative modulator of pyruvate dehydrogenase, through stimulation of a protein kinase that inactivates the dehydrogenase. When the cell's energy needs are being met, oxidative phosphorylation slows, [NADH] rises relative to

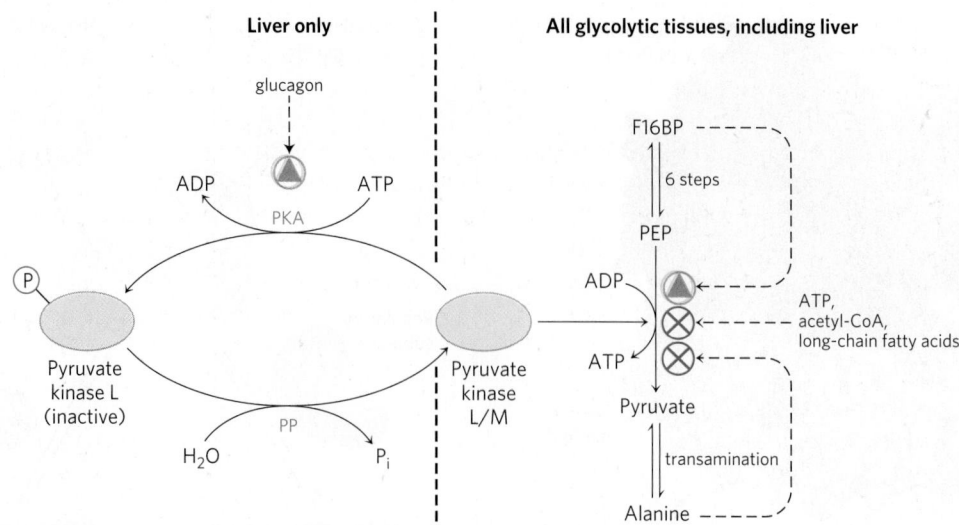

FIGURE 15-21 Regulation of pyruvate kinase. The enzyme is allosterically inhibited by ATP, acetyl-CoA, and long-chain fatty acids (all signs of an abundant energy supply), and the accumulation of fructose 1,6-bisphosphate triggers its activation. Accumulation of alanine, which can be synthesized from pyruvate in one step, allosterically inhibits pyruvate kinase, slowing the production of pyruvate by glycolysis. The liver isozyme (L form) is also regulated hormonally. Glucagon activates cAMP-dependent protein kinase (PKA; see Fig. 15-37), which phosphorylates the pyruvate kinase L isozyme, inactivating it. When the glucagon level drops, a protein phosphatase (PP) dephosphorylates pyruvate kinase, activating it. This mechanism prevents the liver from consuming glucose by glycolysis when blood glucose is low; instead, the liver exports glucose. The muscle isozyme (M form) is not affected by this phosphorylation mechanism.

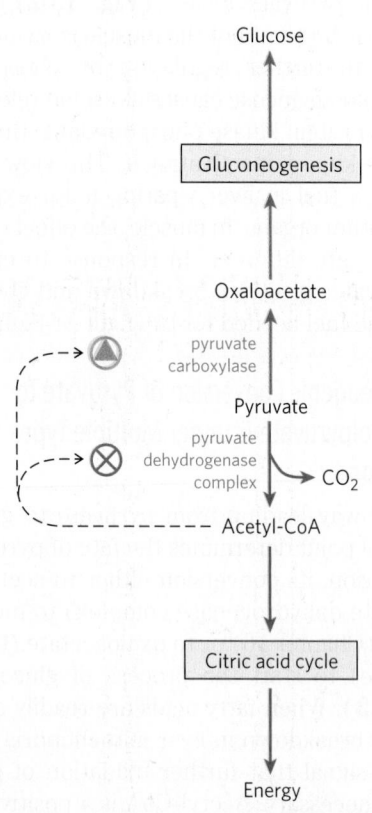

FIGURE 15-22 Two alternative fates for pyruvate. Pyruvate can be converted to glucose and glycogen via gluconeogenesis or oxidized to acetyl-CoA for energy production. The first enzyme in each path is regulated allosterically; acetyl-CoA, produced either by fatty acid oxidation or by the pyruvate dehydrogenase complex, stimulates pyruvate carboxylase and inhibits pyruvate dehydrogenase.

[NAD$^+$] and inhibits the citric acid cycle, and acetyl-CoA accumulates. The increased concentration of acetyl-CoA inhibits the pyruvate dehydrogenase complex, slowing the formation of acetyl-CoA from pyruvate, and stimulates gluconeogenesis by activating pyruvate carboxylase, allowing conversion of excess pyruvate to oxaloacetate (and, eventually, glucose).

Oxaloacetate formed in this way is converted to phosphoenolpyruvate (PEP) in the reaction catalyzed by PEP carboxykinase (Fig. 15-13). In mammals, the regulation of this key enzyme occurs primarily at the level of its synthesis and breakdown, in response to dietary and hormonal signals. Fasting or high glucagon levels act through cAMP to increase the rate of transcription and to stabilize the mRNA. Insulin, or high blood glucose, has the opposite effects. We discuss this transcriptional regulation in more detail below. Generally triggered by a signal from outside the cell (diet, hormones), these changes take place on a time scale of minutes to hours.

Transcriptional Regulation of Glycolysis and Gluconeogenesis Changes the Number of Enzyme Molecules

Most of the regulatory actions discussed thus far are mediated by fast, quickly reversible mechanisms: allosteric effects, covalent alteration (phosphorylation) of the enzyme, or binding of a regulatory protein. Another set of regulatory processes involves changes in the number of molecules of an enzyme in the cell, through changes in the balance of enzyme synthesis and breakdown, and our discussion now turns to regulation of transcription through signal-activated transcription factors.

In Chapter 12 we encountered nuclear receptors and transcription factors in the context of insulin signaling. Insulin acts through its receptor in the plasma membrane to turn on at least two distinct signaling pathways, each involving activation of a protein kinase. The MAP kinase ERK, for example, phosphorylates the transcription factors SRF and Elk1 (see Fig. 12-19), which then stimulate the synthesis of enzymes needed for cell growth and division. Protein kinase B (PKB; also called Akt) phosphorylates another set of transcription factors (PDX1, for example), and these stimulate the synthesis of enzymes that metabolize carbohydrates and the fats formed and stored following excess carbohydrate intake in the diet. In pancreatic β cells, PDX1 also stimulates the synthesis of insulin itself.

More than 150 genes are transcriptionally regulated by insulin; humans have at least seven general types of insulin response elements, each recognized by a subset of transcription factors activated by insulin under various conditions. Insulin stimulates the transcription of the genes that encode hexokinases II and IV, PFK-1, pyruvate kinase, and PFK-2/FBPase-2 (all involved in glycolysis and its regulation); several enzymes of fatty acid synthesis; and glucose 6-phosphate dehydrogenase and 6-phosphogluconate dehydrogenase, enzymes of the pentose phosphate pathway that generate the NADPH required for fatty acid synthesis. Insulin also slows the expression of the genes for two enzymes of gluconeogenesis: PEP carboxykinase and glucose 6-phosphatase (Table 15-5). These changes have the effects of (1) stimulating reactions that consume glucose (synthesis of glycogen, fatty acids, and triacylglycerols) and (2) inhibiting production and release of glucose from the liver into the blood.

One transcription factor important to carbohydrate metabolism is **ChREBP (carbohydrate response element binding protein; Fig. 15-23)**, which is expressed primarily in liver, adipose tissue, and kidney. It coordinates the synthesis of enzymes needed for carbohydrate and fat synthesis. ChREBP in its inactive state is phosphorylated, and is located in the cytosol. When the phosphoprotein phosphatase PP2A (Fig. 15-20) removes a phosphoryl group from ChREBP, the transcription factor can enter the nucleus. Here, nuclear PP2A removes another phosphoryl group, and ChREBP now joins with a partner protein, Mlx, and turns on the synthesis of several enzymes: pyruvate kinase, fatty acid synthase, and acetyl-CoA carboxylase, the first enzyme in the path to fatty acid synthesis.

Controlling the activity of PP2A—and thus, ultimately, the synthesis of this group of metabolic enzymes—is xylulose 5-phosphate, an intermediate of the pentose phosphate pathway (Fig. 14–23). When blood glucose concentration is high, glucose enters the liver and is phosphorylated by hexokinase IV. The glucose 6-phosphate thus formed can enter either the glycolytic pathway or the pentose phosphate pathway. If the latter, two initial oxidations produce xylulose 5-phosphate, which serves as a signal that the glucose-utilizing pathways are well-supplied with substrate. It accomplishes this by allosterically activating PP2A, which then dephosphorylates ChREBP, allowing the transcription factor to turn on the expression of genes for enzymes of glycolysis and fat synthesis (Fig. 15-23).

TABLE 15-5	Some of the Many Genes Regulated by Insulin
Change in gene expression	**Role in glucose metabolism**
Increased expression Hexokinase II Hexokinase IV Phosphofructokinase-1 (PFK-1) PFK-2/FBPase-2 Pyruvate kinase	Essential for glycolysis, which consumes glucose for energy
Glucose 6-phosphate dehydrogenase 6-Phosphogluconate dehydrogenase Malic enzyme	Produce NADPH, which is essential for conversion of glucose to lipids
ATP-citrate lyase Pyruvate dehydrogenase	Produce acetyl-CoA, which is essential for conversion of glucose to lipids
Acetyl-CoA carboxylase Fatty acid synthase complex Stearoyl-CoA dehydrogenase Acyl-CoA–glycerol transferases	Essential for conversion of glucose to lipids
Decreased expression PEP carboxykinase Glucose 6-phosphatase (catalytic subunit)	Essential for glucose production by gluconeogenesis

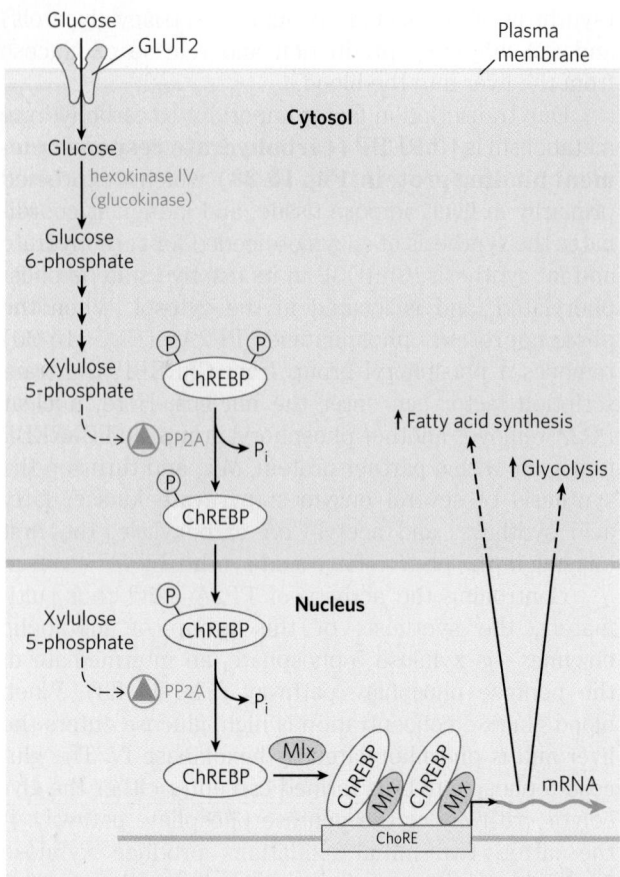

FIGURE 15-23 Mechanism of gene regulation by the transcription factor ChREBP. When ChREBP in the cytosol of a hepatocyte is phosphorylated on a Ser and a Thr residue, it cannot enter the nucleus. Dephosphorylation of P-Ser by protein phosphatase PP2A allows ChREBP to enter the nucleus, where a second dephosphorylation, of P-Thr, activates ChREBP so that it can associate with its partner protein, Mlx. ChREBP-Mlx now binds to the carbohydrate response element (ChoRE) in the promoter and stimulates transcription. PP2A is allosterically activated by xylulose 5-phosphate, an intermediate in the pentose phosphate pathway.

Glycolysis yields pyruvate, and conversion of pyruvate to acetyl-CoA provides the starting material for fatty acid synthesis: acetyl-CoA carboxylase converts acetyl-CoA to malonyl-CoA, the first committed intermediate in the path to fatty acids. The fatty acid synthase complex produces fatty acids for export to adipose tissue and storage as triacylglycerols (Chapter 21). In this way, excess dietary carbohydrate is stored as fat.

Another transcription factor in the liver, **SREBP-1c**, a member of the family of **sterol regulatory element binding proteins** (see Fig. 21-44), turns on the synthesis of pyruvate kinase, hexokinase IV, lipoprotein lipase, acetyl-CoA carboxylase, and the fatty acid synthase complex that will convert acetyl-CoA (produced from pyruvate) into fatty acids for storage in adipocytes. The synthesis of SREBP-1c is stimulated by insulin and depressed by glucagon. SREBP-1c also suppresses the expression of several gluconeogenic enzymes: glucose 6-phosphatase, PEP carboxykinase, and FBPase-1.

The transcription factor **CREB (cyclic AMP response element binding protein)** turns on the synthesis of glucose 6-phosphatase and PEP carboxykinase in response to the increase in [cAMP] triggered by glucagon. In contrast, insulin-stimulated *inactivation* of other transcription factors turns off several gluconeogenic enzymes in the liver: PEP carboxykinase, fructose 1,6-bisphosphatase, the glucose 6-phosphate transporter of the endoplasmic reticulum, and glucose 6-phosphatase. For example, **FOXO1 (forkhead box other)** stimulates the synthesis of gluconeogenic enzymes and suppresses the synthesis of the enzymes of glycolysis, the pentose phosphate pathway, and triacylglycerol synthesis **(Fig. 15-24)**. In its unphosphorylated form, FOXO1 acts as a nuclear transcription factor. In response to insulin, FOXO1 leaves the nucleus and in the cytosol is phosphorylated by PKB, then tagged with ubiquitin and degraded by the proteasome. Glucagon prevents this phosphorylation by PKB, and FOXO1 remains active in the nucleus.

Complicated though the processes outlined above may seem, regulation of the genes encoding enzymes of carbohydrate and fat metabolism is proving far more complex and more subtle than we have shown here. Multiple transcription factors can act on the same gene promoter; multiple protein kinases and phosphatases can activate or inactivate these transcription factors;

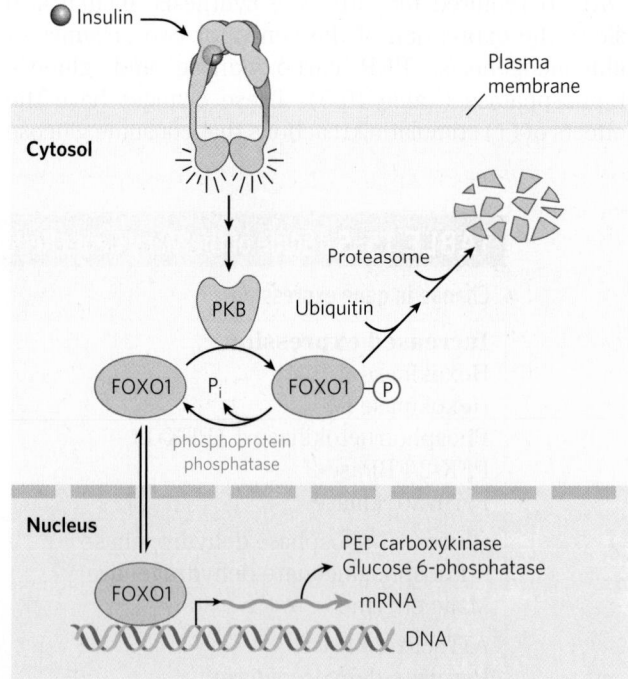

FIGURE 15-24 Mechanism of gene regulation by the transcription factor FOXO1. Insulin activates the signaling cascade shown in Figure 12-20, leading to activation of protein kinase B (PKB). FOXO1 in the cytosol is phosphorylated by PKB, and the phosphorylated transcription factor is tagged by the attachment of ubiquitin for degradation in proteasomes. FOXO1 that remains unphosphorylated or is dephosphorylated can enter the nucleus, bind to a response element, and trigger transcription of the associated genes. Insulin therefore has the effect of turning *off* the expression of these genes, which include PEP carboxykinase and glucose 6-phosphatase.

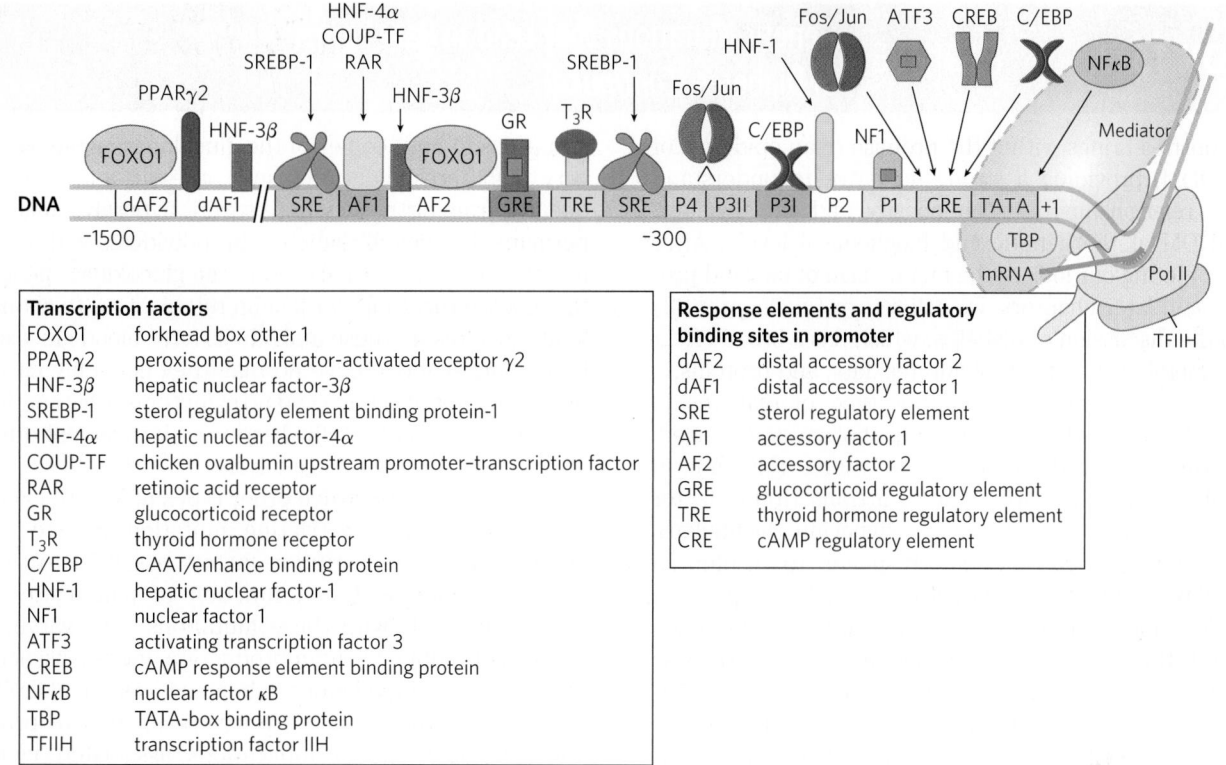

Transcription factors	
FOXO1	forkhead box other 1
PPARγ2	peroxisome proliferator-activated receptor γ2
HNF-3β	hepatic nuclear factor-3β
SREBP-1	sterol regulatory element binding protein-1
HNF-4α	hepatic nuclear factor-4α
COUP-TF	chicken ovalbumin upstream promoter–transcription factor
RAR	retinoic acid receptor
GR	glucocorticoid receptor
T$_3$R	thyroid hormone receptor
C/EBP	CAAT/enhance binding protein
HNF-1	hepatic nuclear factor-1
NF1	nuclear factor 1
ATF3	activating transcription factor 3
CREB	cAMP response element binding protein
NFκB	nuclear factor κB
TBP	TATA-box binding protein
TFIIH	transcription factor IIH

Response elements and regulatory binding sites in promoter	
dAF2	distal accessory factor 2
dAF1	distal accessory factor 1
SRE	sterol regulatory element
AF1	accessory factor 1
AF2	accessory factor 2
GRE	glucocorticoid regulatory element
TRE	thyroid hormone regulatory element
CRE	cAMP regulatory element

FIGURE 15-25 The PEP carboxykinase promoter region, showing the complexity of regulatory input to this gene. This diagram shows the transcription factors (smaller icons, bound to the DNA) known to regulate the transcription of the PEP carboxykinase gene. The extent to which this gene is expressed depends on the combined input affecting all of these factors, which can reflect the availability of nutrients, blood glucose level, and other circumstances that affect the cell's need for this enzyme at any particular time. P1, P2, P3I, P3II, and P4 are protein-binding sites identified by DNase I footprinting (see Box 26-1). The TATA box is the assembly point for the RNA polymerase II (Pol II) transcription complex. [Source: Information from K. Chakravarty, *Crit. Rev. Biochem. Mol. Biol.* 40:129, 2005, Fig. 2.]

and a variety of protein accessory factors modulate the action of the transcription factors. This complexity is apparent, for example, in the gene encoding PEP carboxykinase, a well-studied case of transcriptional control. Its promoter region **(Fig. 15-25)** has 15 or more response elements that are recognized by at least a dozen known transcription factors, with more likely to be discovered. The transcription factors act in combination on this promoter region, and on hundreds of other gene promoters, to fine-tune the levels of hundreds of metabolic enzymes, coordinating their activity in the metabolism of carbohydrates and fats. The critical importance of transcription factors in metabolic regulation is made clear by observing the effects of mutations in their genes. For example, at least five different types of maturity-onset diabetes of the young (MODY) are associated with mutations in specific transcription factors (Box 15-3).

BOX 15-3　☤ MEDICINE　Genetic Mutations That Lead to Rare Forms of Diabetes

The term "diabetes" describes a variety of medical conditions that have in common an excessive production of urine. In Box 11–1 we described diabetes insipidus, in which defective water reabsorption in the kidney results from a mutation in the gene for aquaporin. "Diabetes mellitus" refers specifically to disease in which the ability to metabolize glucose is defective, due either to the failure of the pancreas to produce insulin or to tissue resistance to the actions of insulin.

There are two common types of diabetes mellitus. Type 1, also called insulin-dependent diabetes mellitus (IDDM), is caused by autoimmune attack on the insulin-producing β cells of the pancreas. Individuals with IDDM must take insulin by injection or inhalation to compensate for their missing β cells. IDDM develops in childhood or in the teen years; an older name for the disease is juvenile diabetes. Type 2, also called non-insulin-dependent diabetes mellitus (NIDDM), typically develops in adults over 40 years old. It is far more common than IDDM, and its occurrence in the population is strongly correlated with obesity. The current epidemic of obesity in the more developed

(Continued on next page)

countries brings with it the promise of an epidemic of NIDDM, providing a strong incentive to understand the relationship between obesity and the onset of NIDDM at the genetic and biochemical levels. After completing our look at the metabolism of fats and proteins in later chapters, we will return (in Chapter 23) to the discussion of diabetes, which has a broad effect on metabolism: of carbohydrates, fats, and proteins.

Here we consider another type of diabetes in which carbohydrate and fat metabolism is deranged: maturity-onset diabetes of the young (MODY), in which genetic mutation affects a transcription factor important in carrying the insulin signal into the nucleus, or affects an enzyme that responds to insulin. In MODY2, a mutation in the hexokinase IV (glucokinase) gene affects the liver and pancreas, tissues in which this is the main isoform of hexokinase. The glucokinase of pancreatic β cells functions as a glucose sensor. Normally, when blood glucose rises, so does the glucose level in β cells, and because glucokinase has a relatively high K_m for glucose, its activity increases with rising blood glucose levels. Metabolism of the glucose 6-phosphate formed in this reaction raises the ATP level in β cells, and this triggers insulin release by the mechanism shown in Fig. 23-28. In healthy individuals, blood glucose concentrations of ~5 mM trigger this insulin release. But individuals with inactivating

mutations in both copies of the glucokinase gene have very high thresholds for insulin release, and consequently, from birth, they have severe hyperglycemia—permanent neonatal diabetes. In individuals with one mutated and one normal copy of the glucokinase gene, the glucose threshold for insulin release rises to about 7 mM. As a result, these individuals have blood glucose levels only slightly above normal: they generally have only mild hyperglycemia and no symptoms. This condition (MODY2) is generally discovered by accident during routine blood glucose analysis.

There are at least five other types of MODY, each the result of an inactivating mutation in one or another of the transcription factors essential to the normal development and function of pancreatic β cells. Individuals with these mutations have varying degrees of reduced insulin production and the associated defects in blood glucose homeostasis. In MODY1 and MODY3, the defects are severe enough to produce the long-term complications associated with IDDM and NIDDM—cardiovascular problems, kidney failure, and blindness. MODY4, 5, and 6 are less severe forms of the disease. Altogether, MODY disorders represent a small percentage of NIDDM cases. Also very rare are individuals with mutations in the insulin gene itself; they have defects in insulin signaling of varying severity.

SUMMARY 15.3 Coordinated Regulation of Glycolysis and Gluconeogenesis

■ Gluconeogenesis and glycolysis share seven enzymes, catalyzing the freely reversible reactions of the pathways. For the other three steps, the forward and reverse reactions are catalyzed by different enzymes, and these are the points of regulation of the two pathways.

■ Hexokinase IV (glucokinase) has kinetic properties related to its special role in the liver: releasing glucose to the blood when blood glucose is low, and taking up and metabolizing glucose when blood glucose is high.

■ PFK-1 is allosterically inhibited by ATP and citrate. In most mammalian tissues, including liver, fructose 2,6-bisphosphate is an allosteric activator of this enzyme.

■ Pyruvate kinase is allosterically inhibited by ATP, and the liver isozyme also is inhibited by cAMP-dependent phosphorylation.

■ Gluconeogenesis is regulated at the level of pyruvate carboxylase (which is activated by acetyl-CoA) and FBPase-1 (which is inhibited by fructose 2,6-bisphosphate and AMP).

■ To limit substrate cycling between glycolysis and gluconeogenesis, the two pathways are under reciprocal allosteric control, mainly achieved by the opposing effects of fructose 2,6-bisphosphate on PFK-1 and FBPase-1.

■ Glucagon or epinephrine decreases [fructose 2,6-bisphosphate] by raising [cAMP] and bringing about phosphorylation of the bifunctional enzyme PFK-2/FBPase-2. Insulin increases [fructose 2,6-bisphosphate] by activating a phosphoprotein phosphatase that dephosphorylates and thus activates PFK-2.

■ Xylulose 5-phosphate, an intermediate of the pentose phosphate pathway, activates phosphoprotein phosphatase PP2A, which dephosphorylates several target proteins, including PFK-2/FBPase-2, tilting the balance toward glucose uptake, glycogen synthesis, and lipid synthesis in the liver.

■ Transcription factors including ChREBP, CREB, SREBP, and FOXO1 act in the nucleus to regulate the expression of specific genes coding for enzymes of the glycolytic and gluconeogenic pathways. Insulin and glucagon act antagonistically in activating these transcription factors, thus turning on and off large numbers of genes.

15.4 The Metabolism of Glycogen in Animals

Our discussion of metabolic regulation, using carbohydrate metabolism as the primary example, now turns to the synthesis and breakdown of glycogen. In this section we focus on the metabolic pathways; in Section 15.5 we turn to the regulatory mechanisms.

Excess glucose is converted to polymeric forms for storage—glycogen in vertebrates and many microorganisms, starch in plants. In vertebrates, glycogen may represent up to 10% of the weight of liver and 1% to 2% of the weight of muscle. If this much glucose were dissolved in the cytosol of a hepatocyte, its concentration would be about 0.4 M, enough to dominate the osmotic properties of the cell. When stored as a large polymer (glycogen), however, the same mass of glucose has a concentration of only 0.01 μM. Glycogen is stored in large cytosolic granules. The elementary particle of glycogen, the β-particle, is about 21 nm in diameter and consists of up to 55,000 glucose residues with about 2,000 nonreducing ends. Twenty to 40 of these particles cluster together to form α-rosettes, visible in the electron microscope in tissue samples from well-fed animals **(Fig. 15-26)** but essentially absent after a 24-hour fast.

The glycogen in muscle provides a quick source of energy for either aerobic or anaerobic metabolism. Muscle glycogen can be exhausted in less than an hour during vigorous activity. Liver glycogen serves as a reservoir of glucose for other tissues when dietary glucose is not available (between meals or during a fast); this is especially important for the neurons of the brain, which cannot use fatty acids as fuel. Liver glycogen can be depleted in 12 to 24 hours. In humans, the total amount of energy stored as glycogen is far less than the amount stored as fat (triacylglycerol) (see Table 23-6), but fats cannot be converted to glucose in vertebrates and cannot be catabolized anaerobically.

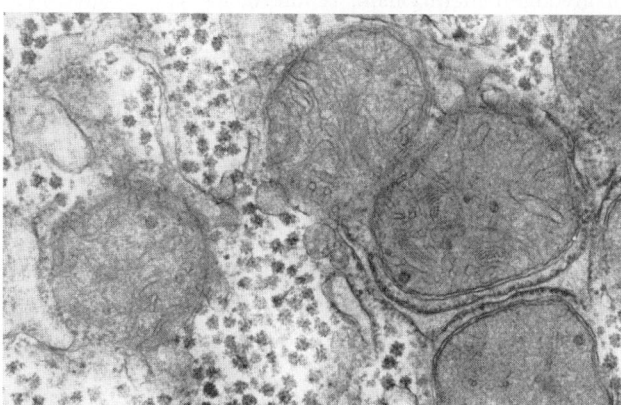

FIGURE 15-26 Glycogen granules in a hepatocyte. Glycogen granules appear as electron-dense particles, often in aggregates or rosettes associated with tubules of the smooth endoplasmic reticulum. Five mitochondria are also evident in this micrograph. [Source: BCC Microimaging. Reproduced with permission.]

Glycogen granules are complex aggregates of glycogen and the enzymes that synthesize it and degrade it, as well as the machinery for regulating these enzymes. The general mechanisms for storing and mobilizing glycogen are the same in muscle and liver, but the enzymes differ in subtle yet important ways that reflect the different roles of glycogen in the two tissues. Glycogen is also obtained in the diet and broken down in the gut, and this involves a separate set of hydrolytic enzymes that convert glycogen to free glucose. (Dietary starch is hydrolyzed in a similar way.) We begin our discussion with the breakdown of glycogen to glucose 1-phosphate **(glycogenolysis)**, then turn to synthesis of glycogen **(glycogenesis)**.

Glycogen Breakdown Is Catalyzed by Glycogen Phosphorylase

In skeletal muscle and liver, the glucose units of the outer branches of glycogen enter the glycolytic pathway through the action of three enzymes: glycogen phosphorylase, glycogen debranching enzyme, and phosphoglucomutase. Glycogen phosphorylase catalyzes the reaction in which an $(\alpha 1 \rightarrow 4)$ glycosidic linkage between two glucose residues at a nonreducing end of glycogen undergoes attack by inorganic phosphate (P_i), removing the terminal glucose residue as α-D-**glucose 1-phosphate (Fig. 15-27)**. This *phosphorolysis* reaction is different from the *hydrolysis* of glycosidic bonds by amylase during intestinal degradation of dietary glycogen and starch. In phosphorolysis, some of the energy of the glycosidic bond is preserved in the formation of the phosphate ester, glucose 1-phosphate (see Section 14.2).

Pyridoxal phosphate is an essential cofactor in the glycogen phosphorylase reaction; its phosphate group acts as a general acid catalyst, promoting attack by P_i on the glycosidic bond. (This is an unusual role for pyridoxal phosphate; its more typical role is as a cofactor in amino acid metabolism; see Fig. 18-6.)

Glycogen phosphorylase acts repetitively on the nonreducing ends of glycogen branches until it reaches a point four glucose residues away from an $(\alpha 1 \rightarrow 6)$ branch point (see Fig. 7-13), where its action stops. Further degradation by glycogen phosphorylase can occur only after the **debranching enzyme**, formally known as **oligo $(\alpha 1 \rightarrow 6)$ to $(\alpha 1 \rightarrow 4)$ glucantransferase**, catalyzes two successive reactions that transfer branches **(Fig. 15-28)**. Once these branches are transferred and the glucosyl residue at C-6 is hydrolyzed, glycogen phosphorylase activity can continue.

Glucose 1-Phosphate Can Enter Glycolysis or, in Liver, Replenish Blood Glucose

Glucose 1-phosphate, the end product of the glycogen phosphorylase reaction, is converted to glucose 6-phosphate by **phosphoglucomutase**, which catalyzes the reversible reaction.

Glucose 1-phosphate $\rightleftharpoons$ glucose 6-phosphate

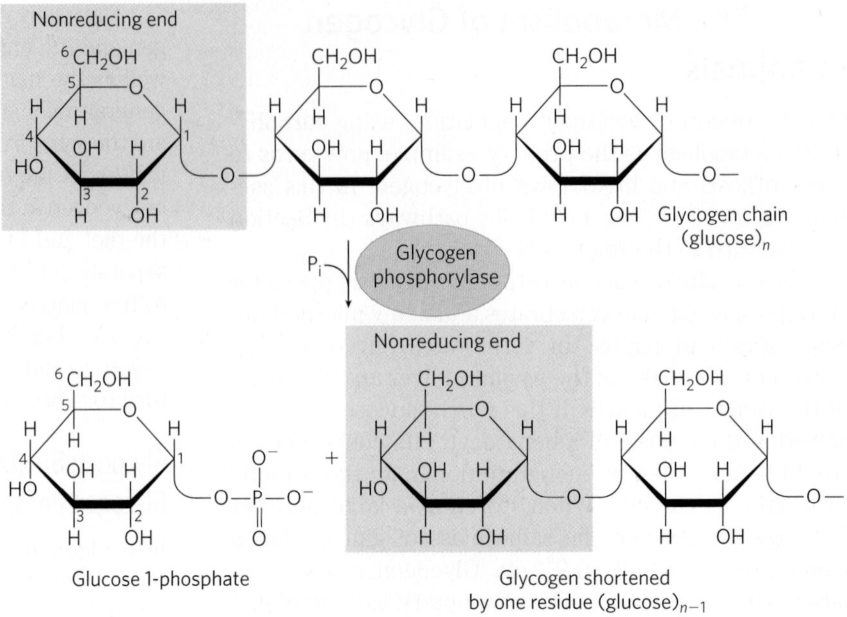

FIGURE 15-27 Removal of a glucose residue from the nonreducing end of a glycogen chain by glycogen phosphorylase. This process is repetitive; the enzyme removes successive glucose residues until it reaches the fourth glucose unit from a branch point (see Fig. 15-28).

Nonreducing end

Glycogen phosphorylase

P_i

Glycogen chain (glucose)$_n$

Glucose 1-phosphate

Nonreducing end

Glycogen shortened by one residue (glucose)$_{n-1}$

Initially phosphorylated at a Ser residue, the enzyme donates a phosphoryl group to C-6 of the substrate, then accepts a phosphoryl group from C-1 **(Fig. 15-29)**.

The glucose 6-phosphate formed from glycogen in skeletal muscle can enter glycolysis and serve as an energy source to support muscle contraction. In liver, glycogen breakdown serves a different purpose: to release glucose into the blood when the blood glucose level drops, as it does between meals. This requires the enzyme glucose 6-phosphatase, present in liver and kidney but not in other tissues. The enzyme is an integral membrane protein of the endoplasmic reticulum, predicted to contain nine transmembrane helices, with its active site on the lumenal side of the ER. Glucose 6-phosphate formed in the cytosol is transported into the ER lumen by a specific transporter (T1) **(Fig. 15-30)** and hydrolyzed at the lumenal surface by glucose 6-phosphatase. The resulting P_i and glucose are thought to be carried back into the cytosol by two different transporters (T2 and T3), and the glucose leaves the hepatocyte via the plasma membrane transporter, GLUT2. Notice that by having the active site of glucose 6-phosphatase inside the ER lumen, the cell separates this reaction from the process of glycolysis, which takes place in the cytosol and would be aborted by the action of glucose 6-phosphatase. Genetic defects in either glucose 6-phosphatase or T1 lead to serious derangement of glycogen metabolism, resulting in type Ia glycogen storage disease (Box 15-4).

Because muscle and adipose tissue lack glucose 6-phosphatase, they cannot convert the glucose 6-phosphate formed by glycogen breakdown to glucose, and these tissues therefore do not contribute glucose to the blood.

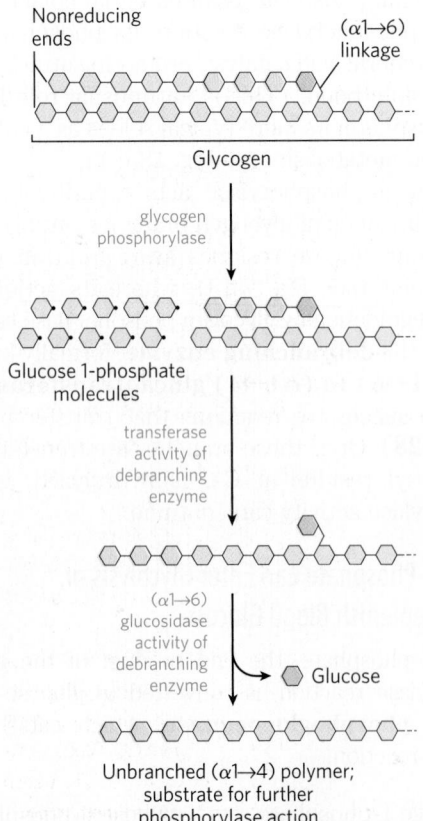

FIGURE 15-28 Glycogen breakdown near an ($\alpha1{\rightarrow}6$) branch point. Following sequential removal of terminal glucose residues by glycogen phosphorylase (see Fig. 15-27), glucose residues near a branch are removed in a two-step process that requires a bifunctional debranching enzyme. First, the transferase activity of the enzyme shifts a block of three glucose residues from the branch to a nearby nonreducing end, to which the segment is reattached in ($\alpha1{\rightarrow}4$) linkage. The single glucose residue remaining at the branch point, in ($\alpha1{\rightarrow}6$) linkage, is then released as free glucose by the ($\alpha1{\rightarrow}6$) glucosidase activity of the debranching enzyme. The glucose residues are shown in shorthand form, which omits the —H, —OH, and —CH$_2$OH groups from the pyranose rings.

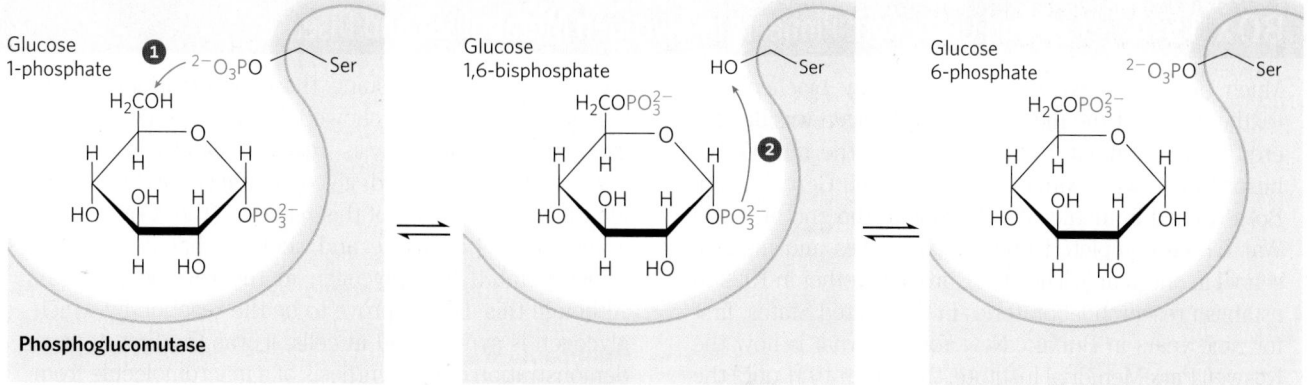

FIGURE 15-29 Reaction catalyzed by phosphoglucomutase. The reaction begins with the enzyme phosphorylated on a Ser residue. In step ❶, the enzyme donates its phosphoryl group (blue) to glucose 1-phosphate,

producing glucose 1,6-bisphosphate. In step ❷, the phosphoryl group at C-1 of glucose 1,6-bisphosphate (red) is transferred back to the enzyme, re-forming the phosphoenzyme and producing glucose 6-phosphate.

The Sugar Nucleotide UDP-Glucose Donates Glucose for Glycogen Synthesis

Luis Leloir, 1906–1987
[Source: AP Photo/John Lindsay.]

Many of the reactions in which hexoses are transformed or polymerized involve **sugar nucleotides**, compounds in which the anomeric carbon of a sugar is activated by attachment to a nucleotide through a phosphate ester linkage. Sugar nucleotides are the substrates for polymerization of monosaccharides into disaccharides, glycogen, starch, cellulose, and more complex extracellular polysaccharides. They are also key intermediates in the production of the aminohexoses and deoxyhexoses found in some of these polysaccharides, and in the synthesis of vitamin C (L-ascorbic acid). The role of sugar nucleotides in the biosynthesis of glycogen and many other carbohydrate derivatives was discovered in 1953 by the Argentine biochemist Luis Leloir.

D-Glucosyl group

UDP-glucose
(a sugar nucleotide)

The suitability of sugar nucleotides for biosynthetic reactions stems from several properties:

1. Their formation is metabolically irreversible, contributing to the irreversibility of the synthetic pathways in which they are intermediates. The condensation of a nucleoside triphosphate with a

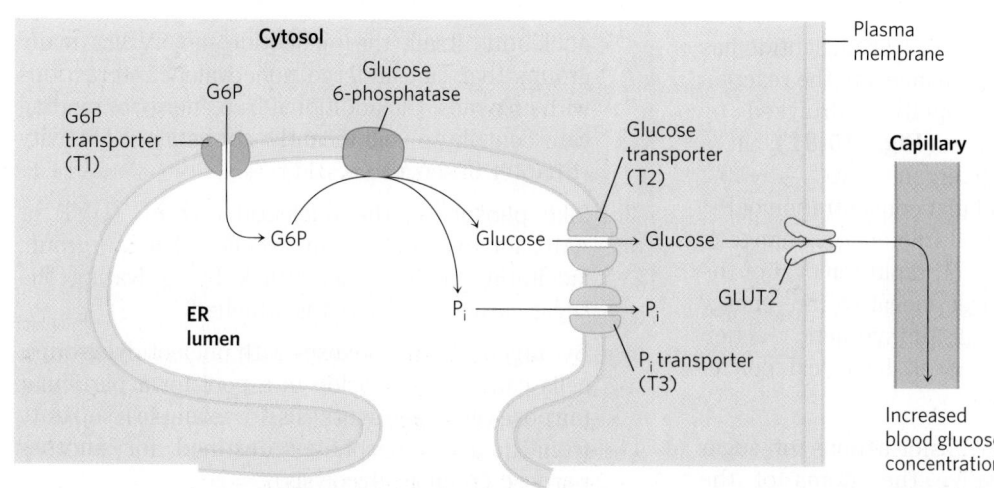

FIGURE 15-30 Hydrolysis of glucose 6-phosphate by glucose 6-phosphatase of the ER. The catalytic site of glucose 6-phosphatase faces the lumen of the ER. A glucose 6-phosphate (G6P) transporter (T1) carries the substrate from the cytosol to the lumen, and the products glucose and P_i pass to the cytosol on specific transporters (T2 and T3). Glucose leaves the cell via the GLUT2 transporter in the plasma membrane.

BOX 15-4 Carl and Gerty Cori: Pioneers in Glycogen Metabolism and Disease

Much of what is written in present-day biochemistry textbooks about the metabolism of glycogen was discovered between about 1925 and 1950 by the remarkable husband and wife team of Carl F. Cori and Gerty T. Cori. Both trained in medicine in Europe at the end of World War I (she completed premedical studies and medical school in one year!). They left Europe together in 1922 to establish research laboratories in the United States, first for nine years in Buffalo, New York, at what is now the Roswell Park Memorial Institute, then from 1931 until the end of their lives at Washington University in St. Louis.

In their early physiological studies of the origin and fate of glycogen in animal muscle, the Coris demonstrated the conversion of glycogen to lactate in tissues, movement of lactate in the blood to the liver, and, in the liver, reconversion of lactate to glycogen—a pathway that came to be known as the Cori cycle

The Coris in Gerty Cori's laboratory, around 1947. [Source: AP Photo.]

(see Fig. 23-21). Pursuing these observations at the biochemical level, they showed that glycogen was mobilized in a phosphorolysis reaction catalyzed by the enzyme they discovered, glycogen phosphorylase. They identified the product of this reaction (the "Cori ester") as glucose 1-phosphate and showed that it could be reincorporated into glycogen in the reverse reaction. Although this did not prove to be the reaction by which glycogen is synthesized in cells, it was the first in vitro demonstration of the synthesis of a macromolecule from simple monomeric subunits, and it inspired others to search for polymerizing enzymes. Arthur Kornberg, discoverer of the first DNA polymerase, said of his experience in the Coris' lab, "Glycogen phosphorylase, not base pairing, was what led me to DNA polymerase."

Gerty Cori became interested in human genetic diseases in which too much glycogen is stored in the liver. She was able to identify the biochemical defect in several of these diseases and to show that the diseases could be diagnosed by assays of the enzymes of glycogen metabolism in small samples of tissue obtained by biopsy. Table 1 summarizes what we now know about 13 genetic diseases of this sort. ■

Carl and Gerty Cori shared the Nobel Prize in Physiology or Medicine in 1947 with Bernardo Houssay of Argentina, who was cited for his studies of hormonal regulation of carbohydrate metabolism. The Cori laboratories in St. Louis became an international center of biochemical research in the 1940s and 1950s, and at least six scientists who trained with the Coris became Nobel laureates: Arthur Kornberg (for DNA synthesis, 1959), Severo Ochoa (for RNA synthesis, 1959), Luis Leloir (for the role of sugar nucleotides in polysaccharide synthesis, 1970), Earl Sutherland (for the discovery of cAMP in the regulation of carbohydrate metabolism, 1971), Christian de Duve (for subcellular fractionation, 1974), and Edwin Krebs (for the discovery of phosphorylase kinase, 1991).

hexose 1-phosphate to form a sugar nucleotide has a small positive free-energy change, but the reaction releases PP_i, which is rapidly hydrolyzed by inorganic pyrophosphatase **(Fig. 15-31)**, in a reaction that is strongly exergonic ($\Delta G'^{\circ} = -19.2$ kJ/mol). This keeps the cellular concentration of PP_i low, ensuring that the actual free-energy change in the cell is favorable. In effect, rapid removal of the product, driven by the large, negative free-energy change of PP_i hydrolysis, pulls the synthetic reaction forward, a common strategy in biological polymerization reactions.

2. Although the chemical transformations of sugar nucleotides do not involve the atoms of the

nucleotide itself, the nucleotide moiety has many groups that can undergo noncovalent interactions with enzymes; the additional free energy of binding can contribute significantly to catalytic activity (Chapter 6; see also p. 311).

3. Like phosphate, the nucleotidyl group (UMP or AMP, for example) is an excellent leaving group, facilitating nucleophilic attack by activating the sugar carbon to which it is attached.

4. By "tagging" some hexoses with nucleotidyl groups, cells can set them aside in a pool for a particular purpose (glycogen synthesis, for example), separate from hexose phosphates destined for another purpose (such as glycolysis).

TABLE 1	Glycogen Storage Diseases of Humans		
Type (name)	**Enzyme affected**	**Primary organ/ cells affected**	**Symptoms**
Type 0	Glycogen synthase	Liver	Low blood glucose, high ketone bodies, early death
Type Ia (von Gierke)	Glucose 6-phosphatase	Liver	Enlarged liver, kidney failure
Type Ib	Microsomal glucose 6-phosphate translocase	Liver	As in type Ia; also high susceptibility to bacterial infections
Type Ic	Microsomal P_i transporter	Liver	As in type Ia
Type II (Pompe)	Lysosomal glucosidase	Skeletal and cardiac muscle	Infantile form: death by age 2; juvenile form: muscle defects (myopathy); adult form: as in muscular dystrophy
Type IIIa (Cori or Forbes)	Debranching enzyme	Liver, skeletal and cardiac muscle	Enlarged liver in infants; myopathy
Type IIIb	Liver debranching enzyme (muscle enzyme normal)	Liver	Enlarged liver in infants
Type IV (Andersen)	Branching enzyme	Liver, skeletal muscle	Enlarged liver and spleen, myoglobin in urine
Type V (McArdle)	Muscle phosphorylase	Skeletal muscle	Exercise-induced cramps and pain; myoglobin in urine
Type VI (Hers)	Liver phosphorylase	Liver	Enlarged liver
Type VII (Tarui)	Muscle PFK-1	Muscle, erythrocytes	As in type V; also hemolytic anemia
Type VIb, VIII, or IX	Phosphorylase kinase	Liver, leukocytes, muscle	Enlarged liver
Type XI (Fanconi-Bickel)	Glucose transporter (GLUT2)	Liver	Failure to thrive, enlarged liver, rickets, kidney dysfunction

Glycogen synthesis takes place in virtually all animal tissues but is especially prominent in the liver and skeletal muscles. The starting point for synthesis of glycogen is glucose 6-phosphate. As we have seen, this can be derived from free glucose in a reaction catalyzed by the isozymes hexokinase I and II in muscle and hexokinase IV (glucokinase) in liver:

$$\text{D-Glucose} + \text{ATP} \longrightarrow \text{D-glucose 6-phosphate} + \text{ADP}$$

However, some ingested glucose takes a more roundabout path to glycogen. It is first taken up by erythrocytes and converted to lactate glycolytically; the lactate is then taken up by the liver and converted to glucose 6-phosphate by gluconeogenesis.

To initiate glycogen synthesis, the glucose 6-phosphate is converted to glucose 1-phosphate in the phosphoglucomutase reaction:

$$\text{Glucose 6-phosphate} \rightleftharpoons \text{glucose 1-phosphate}$$

The product is then converted to UDP-glucose by the action of **UDP-glucose pyrophosphorylase**, in a key step of glycogen biosynthesis:

$$\text{Glucose 1-phosphate} + \text{UTP} \longrightarrow \text{UDP-glucose} + \text{PP}_i$$

Notice that this enzyme is named for the reverse reaction; in the cell, the reaction proceeds in the direction of UDP-glucose formation, because pyrophosphate is rapidly hydrolyzed by inorganic pyrophosphatase (Fig. 15-31).

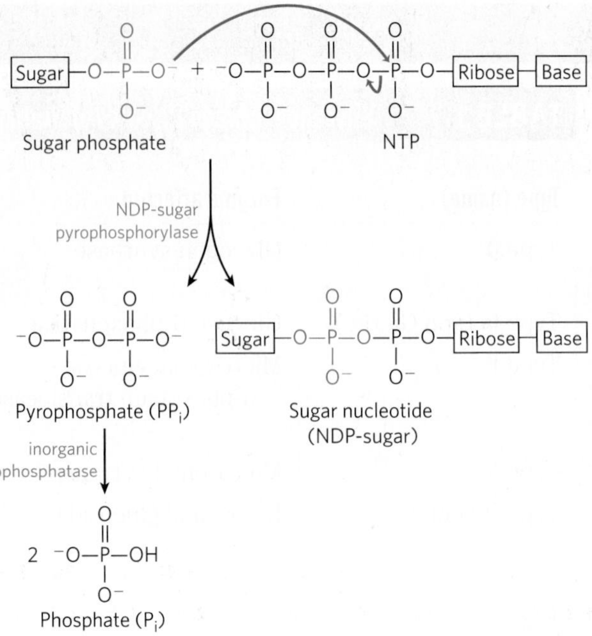

FIGURE 15-31 Formation of a sugar nucleotide. A condensation reaction occurs between a nucleoside triphosphate (NTP) and a sugar phosphate. The negatively charged oxygen on the sugar phosphate serves as a nucleophile, attacking the α phosphate of the nucleoside triphosphate and displacing pyrophosphate. The reaction is pulled in the forward direction by the hydrolysis of PP$_i$ by inorganic pyrophosphatase.

Net reaction: Sugar phosphate + NTP $\longrightarrow$ NDP-sugar + 2P$_i$

UDP-glucose is the immediate donor of glucose residues in the reaction catalyzed by **glycogen synthase**, which promotes the transfer of the glucose residue from UDP-glucose to a nonreducing end of a branched glycogen molecule **(Fig. 15-32)**. The overall equilibrium of the path from glucose 6-phosphate to glycogen lengthened by one glucose unit greatly favors synthesis of glycogen.

Glycogen synthase cannot make the (α1→6) bonds found at the branch points of glycogen; these are formed by the glycogen-branching enzyme, also called **amylo (1→4) to (1→6) transglycosylase**, or glycosyl (4→6) transferase. The glycogen-branching enzyme catalyzes transfer of a terminal fragment of 6 or 7 glucose residues from the nonreducing end of a glycogen branch having at least 11 residues to the C-6 hydroxyl group of a glucose residue at a more interior position of the same or another glycogen chain, thus creating a new branch **(Fig. 15-33)**. Further glucose residues may be added to the new branch by glycogen synthase. The biological effect of branching is to make the glycogen molecule more soluble and to increase the number of nonreducing ends. This increases the number of sites accessible to glycogen phosphorylase and glycogen synthase, both of which act only at nonreducing ends.

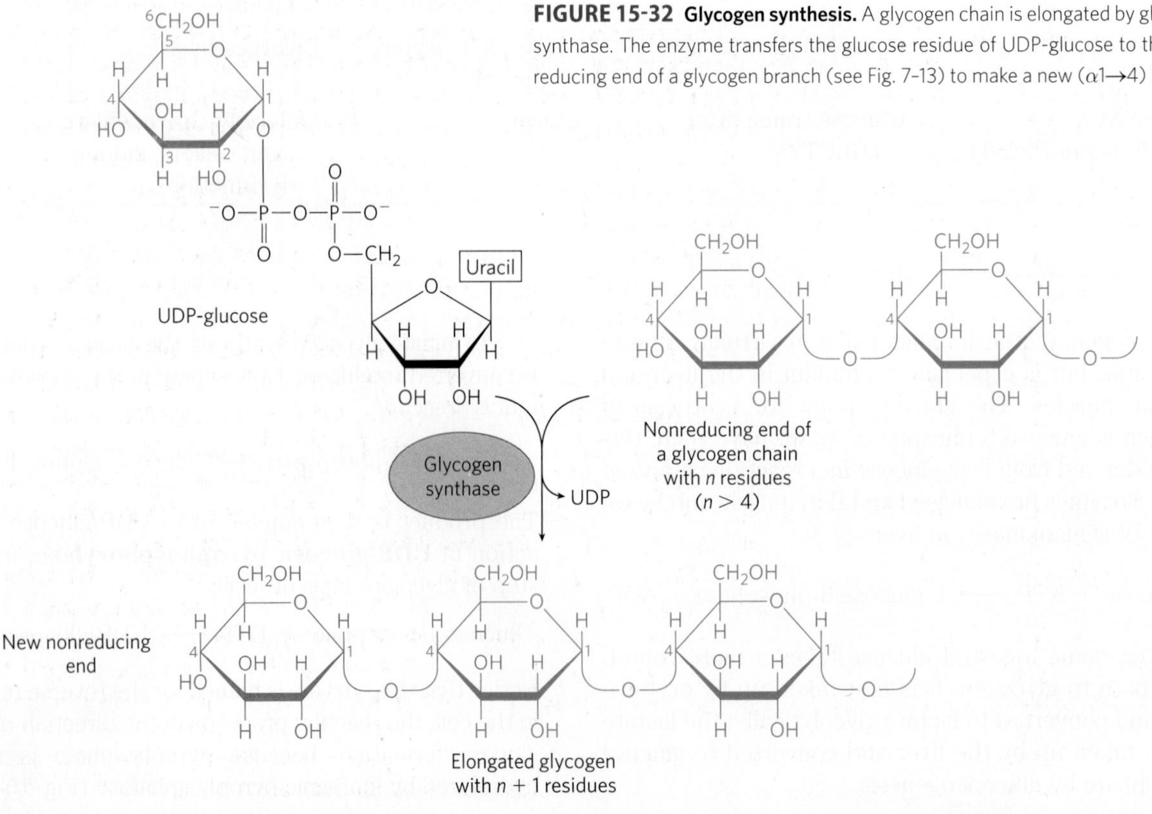

FIGURE 15-32 Glycogen synthesis. A glycogen chain is elongated by glycogen synthase. The enzyme transfers the glucose residue of UDP-glucose to the nonreducing end of a glycogen branch (see Fig. 7-13) to make a new (α1→4) linkage.

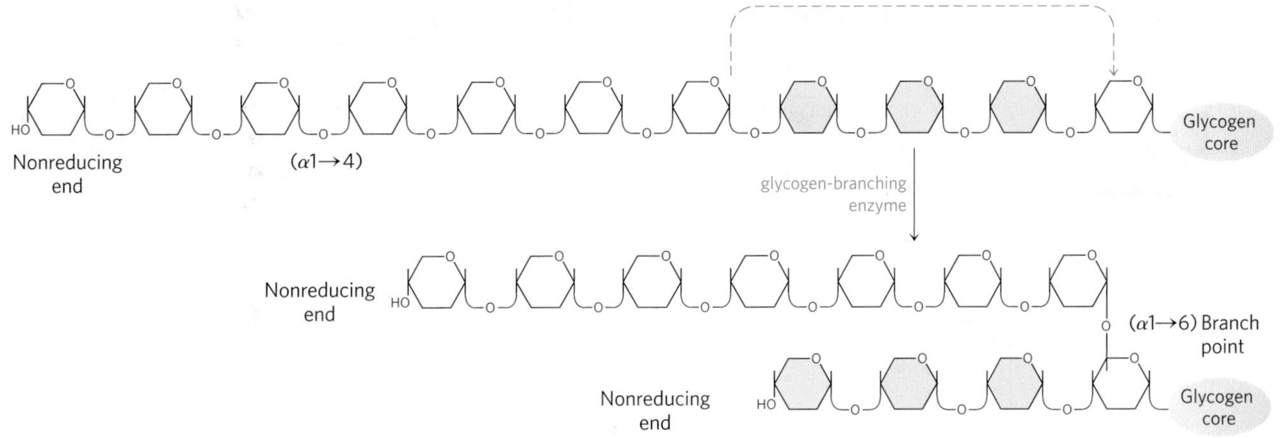

FIGURE 15-33 Branch synthesis in glycogen. The glycogen-branching enzyme (also called amylo (1→4) to (1→6) transglycosylase, or glycosyl-(4→6) transferase) forms a new branch point during glycogen synthesis.

Glycogenin Primes the Initial Sugar Residues in Glycogen

Glycogen synthase cannot initiate a new glycogen chain de novo. It requires a primer, usually a preformed (α1→4) polyglucose chain or branch having at least eight glucose residues. So, how is a *new* glycogen molecule initiated? The intriguing protein **glycogenin (Fig. 15-34)** is both the primer on which new chains are assembled and the enzyme that catalyzes their assembly. The first step in the synthesis of a new glycogen molecule is the transfer of a glucose residue from UDP-glucose to the hydroxyl group of Tyr^{194} of glycogenin, catalyzed by the protein's intrinsic glucosyltransferase activity **(Fig. 15-35)**. The nascent chain is extended by the sequential addition of seven more glucose residues, each derived from UDP-glucose; the reactions are catalyzed by the chain-extending activity of glycogenin. At this point, glycogen synthase takes over, further extending the glycogen chain. Glycogenin remains buried within the β-particle, covalently attached to the single reducing end of the glycogen molecule (Fig. 15-35b). Medical consequences of a mutation in the gene for glycogenin that knocks out that protein's polymerizing activity include muscle weakness and fatigue, depleted glycogen in the liver, and an irregular heartbeat (cardiac arrhythmia).

SUMMARY 15.4 The Metabolism of Glycogen in Animals

■ Glycogen is stored in muscle and liver as large, insoluble particles, which make an insignificant contribution to the osmolarity of the cytosol. Contained within the particles are the enzymes that metabolize glycogen, as well as regulatory enzymes.

■ Glycogen phosphorylase catalyzes phosphorolytic cleavage at the nonreducing ends of glycogen chains, producing glucose 1-phosphate. The debranching enzyme transfers branches onto main chains and releases the residue at the (α1→6) branch as free glucose.

■ Phosphoglucomutase interconverts glucose 1-phosphate and glucose 6-phosphate. Glucose 6-phosphate can enter glycolysis or, in liver, can be converted to free glucose by glucose 6-phosphatase in the endoplasmic reticulum, then released to replenish blood glucose.

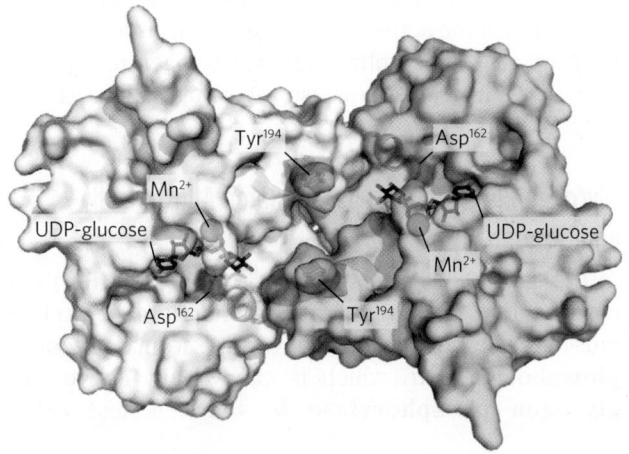

FIGURE 15-34 Glycogenin structure. Muscle glycogenin (M_r 37,000) forms dimers in solution. Humans have a second isoform in liver, glycogenin-2. The substrate, UDP-glucose, is bound to a Rossmann fold near the amino terminus and is some distance from the Tyr^{194} residues—15 Å from the Tyr in the same monomer, 12 Å from the Tyr in the dimeric partner. Each UDP-glucose is bound through its phosphates to a Mn^{2+} ion, which is essential to catalysis. Mn^{2+} is believed to function as an electron-pair acceptor (Lewis acid) to stabilize the leaving group, UDP. The glycosidic bond in the product has the same configuration about the C-1 of glucose as the substrate UDP-glucose, suggesting that the transfer of glucose from UDP to Tyr^{194} occurs in two steps. The first step is probably a nucleophilic attack by Asp^{162}, forming a temporary intermediate with inverted configuration. A second nucleophilic attack by Tyr^{194} then restores the starting configuration. [Source: PDB ID 1LL2, B. J. Gibbons et al., *J. Mol. Biol.* 319:463, 2002.]

(a)

(b)

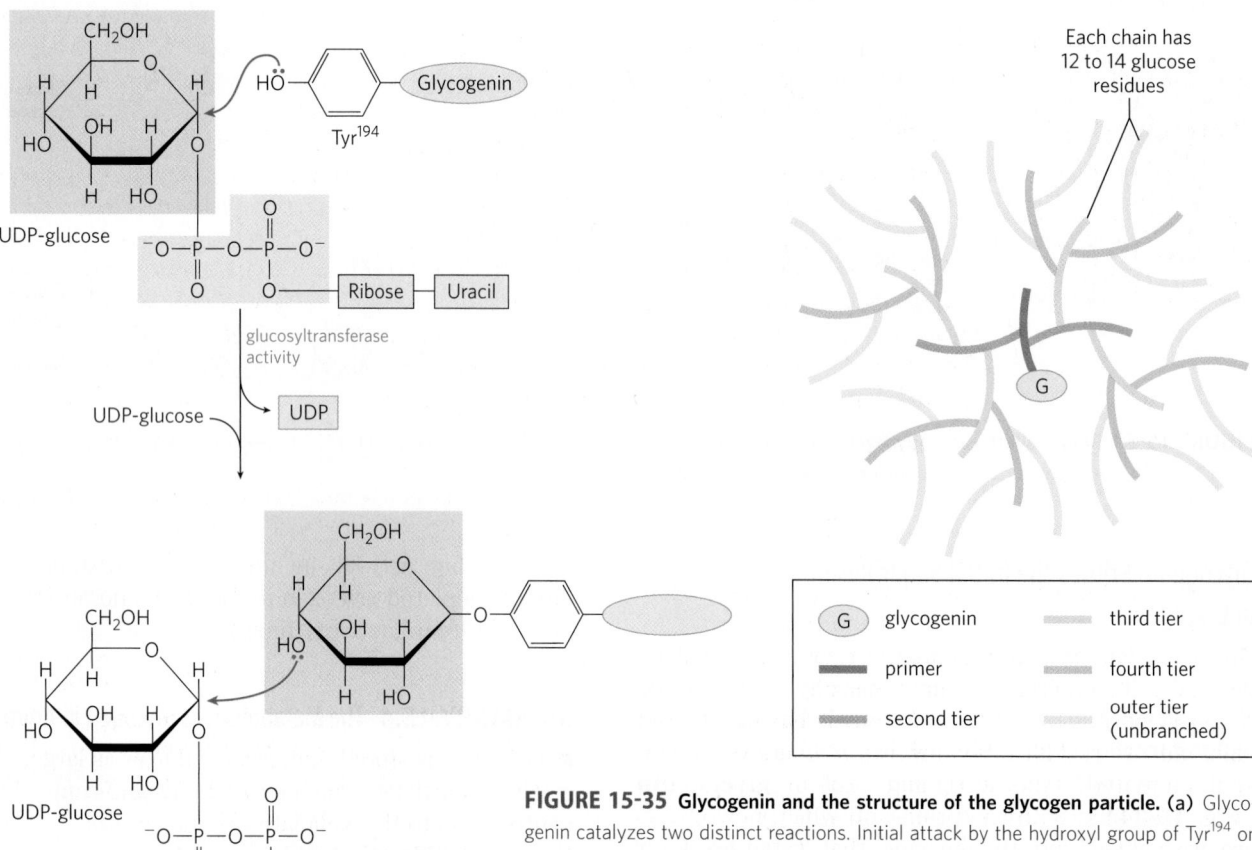

FIGURE 15-35 **Glycogenin and the structure of the glycogen particle. (a)** Glycogenin catalyzes two distinct reactions. Initial attack by the hydroxyl group of Tyr[194] on C-1 of the glucosyl moiety of UDP-glucose results in a glucosylated Tyr residue. The C-1 of another UDP-glucose molecule is now attacked by the C-4 hydroxyl group of the terminal glucose, and this sequence repeats to form a nascent glycogen molecule of eight glucose residues attached by ($\alpha1{\rightarrow}4$) glycosidic linkages. **(b)** Structure of the glycogen particle. Starting at a central glycogenin molecule, glycogen chains (12 to 14 residues) extend in tiers. Inner chains have two ($\alpha1{\rightarrow}6$) branches each. Chains in the outer tier are unbranched. There are 12 tiers in a mature glycogen particle (only 5 are shown here), consisting of about 55,000 glucose residues in a molecule of about 21 nm diameter and $M_r \sim 1 \times 10^7$.

■ The sugar nucleotide UDP-glucose donates glucose residues to the nonreducing end of glycogen in the reaction catalyzed by glycogen synthase. A separate branching enzyme produces the ($\alpha1{\rightarrow}6$) linkages at branch points.

■ New glycogen particles begin with the autocatalytic formation of a glycosidic bond between the glucose of UDP-glucose and a Tyr residue of the protein glycogenin, followed by addition of several glucose residues to form a primer that can be acted on by glycogen synthase.

15.5 Coordinated Regulation of Glycogen Breakdown and Synthesis

As we have seen, the mobilization of stored glycogen is brought about by glycogen phosphorylase, which degrades glycogen to glucose 1-phosphate (Fig. 15-27).

Glycogen phosphorylase provides an especially instructive case of enzyme regulation. It was one of the first known examples of an allosterically regulated enzyme and the first enzyme shown to be controlled by reversible phosphorylation. It was also one of the first allosteric enzymes for which the detailed three-dimensional structures of the active and inactive forms were revealed by x-ray crystallographic studies. Glycogen phosphorylase also illustrates how isozymes play their tissue-specific roles.

Glycogen Phosphorylase Is Regulated Allosterically and Hormonally

In the late 1930s, Carl and Gerty Cori (Box 15-4) discovered that the glycogen phosphorylase of skeletal muscle exists in two interconvertible forms: **glycogen phosphorylase *a***, which is catalytically active, and **glycogen phosphorylase *b***, which is less active

Earl W. Sutherland, Jr.,
1915–1974
[Source: Science Source.]

(**Fig. 15-36**). Subsequent studies by Earl Sutherland showed that phosphorylase *b* predominates in resting muscle, but during vigorous muscular activity, epinephrine triggers phosphorylation of a specific Ser residue in phosphorylase *b*, converting it to its more active form, phosphorylase *a*. (Note that glycogen phosphorylase is often referred to simply as phosphorylase—so honored because it was the first phosphorylase to be discovered; the shortened name has persisted in common usage and in the literature.)

The enzyme (phosphorylase *b* kinase) responsible for activating phosphorylase by transferring a phosphoryl group to its Ser residue is itself activated by epinephrine or glucagon through a series of steps shown in **Figure 15-37**. Sutherland discovered the second messenger cAMP, which increases in concentration in response to stimulation by epinephrine (in muscle) or glucagon (in liver). Elevated [cAMP] initiates an **enzyme cascade**, in which a catalyst activates a catalyst, which activates a catalyst (see Section 12.2). Such cascades allow large amplification of the initial signal (see pink boxes in Fig. 15-37). The rise in [cAMP] activates cAMP-dependent protein kinase, also called protein kinase A (PKA). PKA then phosphorylates and activates **phosphorylase *b* kinase**, which catalyzes the phosphorylation of Ser residues in each of the two identical subunits of glycogen phosphorylase, activating it and thus stimulating glycogen breakdown. In muscle, this provides fuel for glycolysis to sustain muscle contraction for the fight-or-flight response signaled by epinephrine. In liver, glycogen breakdown counters the low blood glucose signaled by glucagon, releasing glucose into the blood. These different roles are reflected in subtle differences in the regulatory mechanisms in muscle and liver. The glycogen phosphorylases of liver and muscle are isozymes, encoded by different genes and differing in their regulatory properties.

In muscle, superimposed on the regulation of phosphorylase by covalent modification are two allosteric control mechanisms (Fig. 15-37). Ca^{2+}, the signal for muscle contraction, binds to and activates phosphorylase *b* kinase, promoting conversion of phosphorylase *b* to the active *a* form. Ca^{2+} binds to phosphorylase *b* kinase through its δ subunit, which is calmodulin (see Fig. 12-12). AMP, which accumulates in vigorously contracting muscle as a result of ATP breakdown, binds to and activates phosphorylase, speeding the release of glucose 1-phosphate from glycogen. When ATP levels are adequate, ATP blocks the allosteric site to which AMP binds, inactivating phosphorylase.

When the muscle returns to rest, a second enzyme, **phosphoprotein phosphatase 1 (PP1)**, removes the phosphoryl groups from phosphorylase *a*, converting it to the less active form, phosphorylase *b*.

Like the enzyme of muscle, the glycogen phosphorylase of liver is regulated hormonally (by phosphorylation/dephosphorylation) and allosterically. The dephosphorylated form is essentially inactive. When the blood glucose level is too low, glucagon (acting through the cascade mechanism shown in Fig. 15-37) activates phosphorylase *b* kinase, which in turn converts phosphorylase *b* to its active *a* form, initiating the release of glucose into the blood. When blood glucose levels return to normal, glucose enters hepatocytes and binds to an inhibitory allosteric site on phosphorylase *a*. This binding also produces a conformational change that exposes the phosphorylated Ser residues to PP1, which catalyzes their dephosphorylation and inactivates the phosphorylase (**Fig. 15-38**). The allosteric site for glucose allows liver glycogen phosphorylase to act as its own glucose sensor and to respond appropriately to changes in blood glucose.

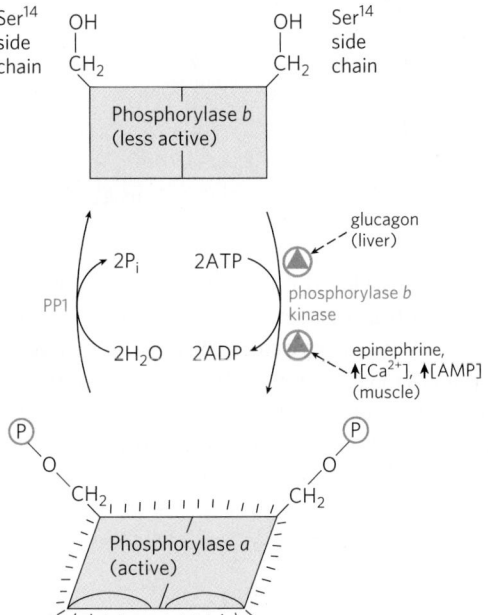

Ser^{14} side chain — OH — CH$_2$

Ser^{14} side chain — OH — CH$_2$

Phosphorylase *b* (less active)

PP1

2P$_i$ ← → 2ATP → glucagon (liver) → phosphorylase *b* kinase

2H$_2$O ← → 2ADP ← epinephrine, ↑[Ca^{2+}], ↑[AMP] (muscle)

℗ — O — CH$_2$ ℗ — O — CH$_2$

Phosphorylase *a* (active)

FIGURE 15-36 Regulation of muscle glycogen phosphorylase by covalent modification. In the more active form of the enzyme, phosphorylase *a*, Ser^{14} residues, one on each subunit, are phosphorylated. Phosphorylase *a* is converted to the less active form, phosphorylase *b*, by enzymatic loss of these phosphoryl groups, catalyzed by phosphoprotein phosphatase 1 (PP1). Phosphorylase *b* can be reconverted (reactivated) to phosphorylase *a* by the action of phosphorylase *b* kinase. (See also Fig. 6-43 on glycogen phosphorylase regulation.)

Glycogen Synthase Is Also Regulated by Phosphorylation and Dephosphorylation

Like glycogen phosphorylase, glycogen synthase can exist in phosphorylated and dephosphorylated forms

FIGURE 15-37 Cascade mechanism of epinephrine and glucagon action. By binding to specific surface receptors, either epinephrine acting on a myocyte (left) or glucagon acting on a hepatocyte (right) activates a GTP-binding protein, $G_{s\alpha}$ (see Fig. 12-7). Active $G_{s\alpha}$ triggers a rise in [cAMP], activating PKA. This sets off a cascade of phosphorylations; PKA activates phosphorylase b kinase, which then activates glycogen phosphorylase. Such cascades effect a large amplification of the initial signal; the figures in pink boxes are probably low estimates of the actual increase in number of molecules at each stage of the cascade. The resulting breakdown of glycogen provides glucose, which in the myocyte can supply ATP (via glycolysis) for muscle contraction and in the hepatocyte is released into the blood to counter the low blood glucose.

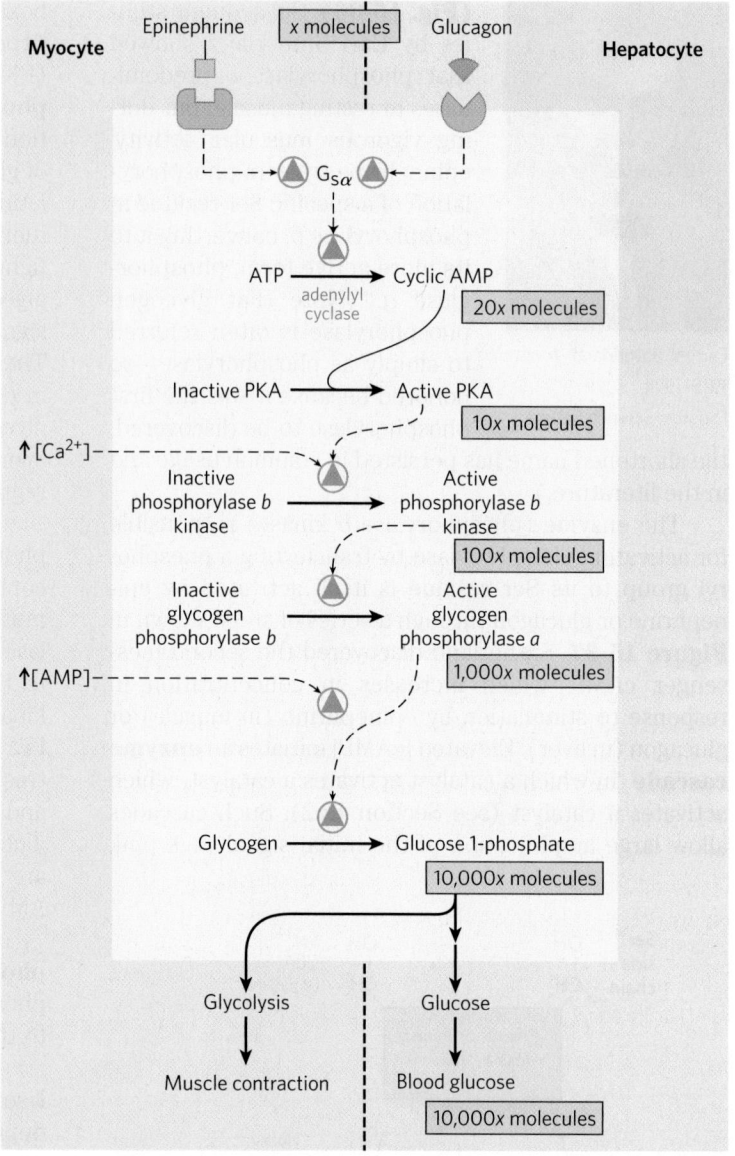

(Fig. 15-39). Its active form, **glycogen synthase a**, is unphosphorylated. Phosphorylation of the hydroxyl side chains of several Ser residues of both subunits converts glycogen synthase a to **glycogen synthase b**, which is inactive unless its allosteric activator, glucose 6-phosphate, is present. Glycogen synthase is remarkable for its ability

to be phosphorylated on various residues by at least 11 different protein kinases. The most important regulatory kinase is **glycogen synthase kinase 3 (GSK3)**, which adds phosphoryl groups to three Ser residues near the carboxyl terminus of glycogen synthase, strongly inactivating it. The action of GSK3 is hierarchical; it cannot

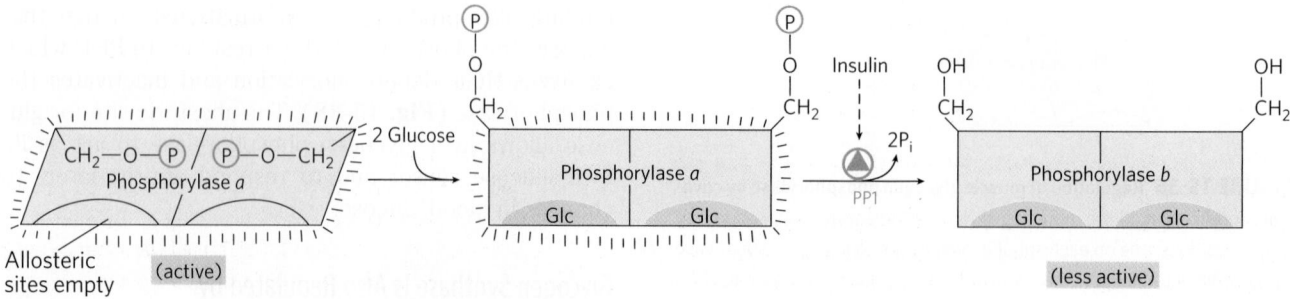

FIGURE 15-38 Glycogen phosphorylase of liver as a glucose sensor. Glucose binding to an allosteric site of the phosphorylase a isozyme of liver induces a conformational change that exposes its phosphorylated Ser residues to the action of phosphoprotein phosphatase 1 (PP1). This phosphatase converts phosphorylase a to phosphorylase b, sharply reducing the activity of phosphorylase and slowing glycogen breakdown in response to high blood glucose. Insulin also acts indirectly to stimulate PP1 and slow glycogen breakdown.

FIGURE 15-39 Effects of GSK3 on glycogen synthase activity. Glycogen synthase *a*, the active form, has three Ser residues near its carboxyl terminus, which are phosphorylated by glycogen synthase kinase 3 (GSK3). This converts glycogen synthase to the inactive (*b*) form. GSK3 action requires prior phosphorylation (priming) by casein kinase (CKII). Insulin triggers activation of glycogen synthase *b* by blocking the activity of GSK3 (see the pathway for this action in Fig. 12-20) and activating phosphoprotein phosphatase 1 (PP1). In muscle, epinephrine activates PKA, which phosphorylates the glycogen-targeting protein G$_M$ (see Fig. 15-42) on a site that causes dissociation of PP1 from glycogen. Glucose 6-phosphate favors dephosphorylation of glycogen synthase by binding to it and promoting a conformation that is a good substrate for PP1. Glucose also promotes dephosphorylation; the binding of glucose to glycogen phosphorylase *a* forces a conformational change that favors dephosphorylation to glycogen phosphorylase *b*, thus allowing the action of PP1 (see Fig. 15-41).

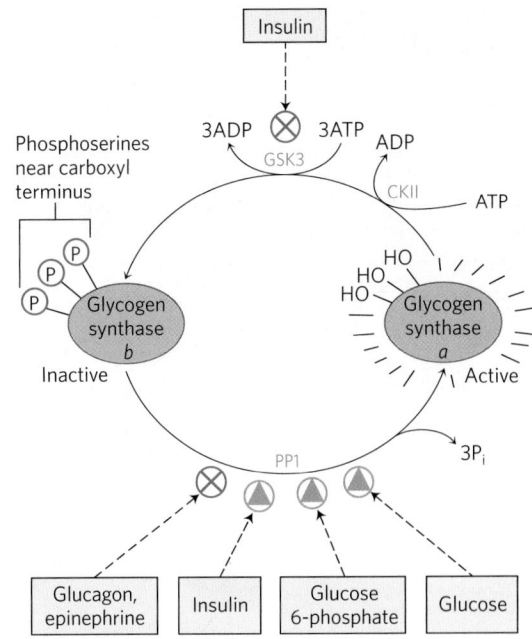

phosphorylate glycogen synthase until another protein kinase, **casein kinase II (CKII)**, has first phosphorylated the glycogen synthase on a nearby residue, an event called **priming (Fig. 15-40a)**. AMP-activated protein kinase (AMPK), which associates with glycogen granules through its carbohydrate-binding domain, also phosphorylates glycogen synthase, inhibiting glycogen synthesis during periods of metabolic stress.

In liver, conversion of glycogen synthase *b* to the active form is promoted by PP1, which is bound to the glycogen particle by G$_L$, a PP1 subunit. PP1 removes the phosphoryl groups from the three Ser residues phosphorylated by GSK3. Glucose 6-phosphate binds to an allosteric site on glycogen synthase *b*, making the enzyme a better substrate for dephosphorylation by PP1 and causing its activation. By analogy with glycogen

phosphorylase, which acts as a glucose sensor, glycogen synthase can be regarded as a glucose 6-phosphate sensor. In muscle, a different phosphatase may have the role played by PP1 in liver, activating glycogen synthase by dephosphorylating it.

Glycogen Synthase Kinase 3 Mediates Some of the Actions of Insulin

As we saw in Chapter 12, one way in which insulin triggers intracellular changes is by activating a protein

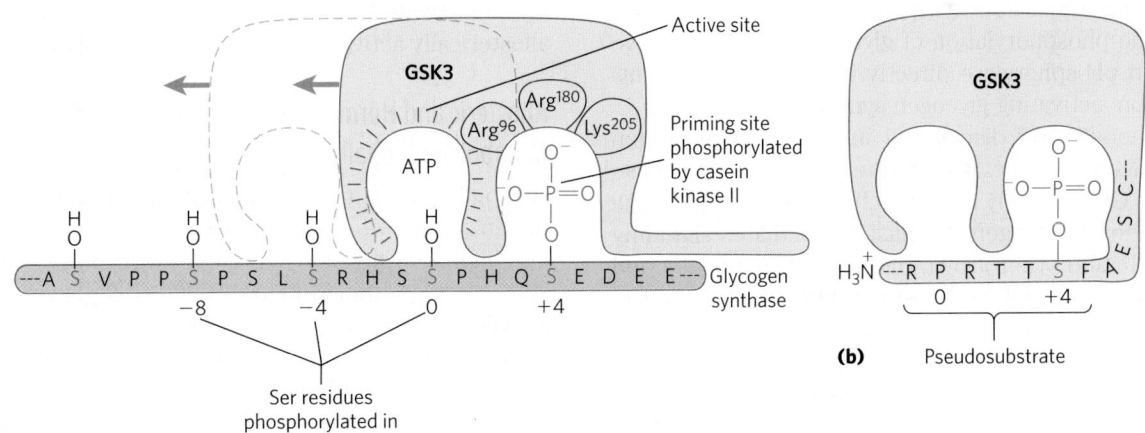

(a)

FIGURE 15-40 Priming of GSK3 phosphorylation of glycogen synthase. (a) Glycogen synthase kinase 3 first associates with its substrate (glycogen synthase) by interaction between three positively charged residues (Arg[96], Arg[180], Lys[205]) and a phosphoserine residue at position +4 in the substrate. (For orientation, the Ser or Thr residue to be phosphorylated in the substrate is assigned the index 0. Residues on the amino-terminal side of this residue are numbered −1, −2, and so forth; residues on the carboxyl-terminal side are numbered +1, +2, and so forth.) This association aligns the active site of the enzyme with a Ser residue at position 0, which it phosphorylates. This creates a new priming site, and the enzyme

moves down the protein to phosphorylate the Ser residue at position −4, and then the Ser at −8. **(b)** GSK3 has a Ser residue near its amino terminus that can be phosphorylated by PKA or PKB (see Fig. 15-41). This produces a "pseudosubstrate" region in GSK3 that folds into the priming site and makes the active site inaccessible to another protein substrate, inhibiting GSK3 until the priming phosphoryl group of its pseudosubstrate region is removed by PP1. Other proteins that are substrates for GSK3 also have a priming site at position +4, which must be phosphorylated by another protein kinase before GSK3 can act on them. (See also Figs 6-38 and 12-25b on glycogen synthase regulation.)

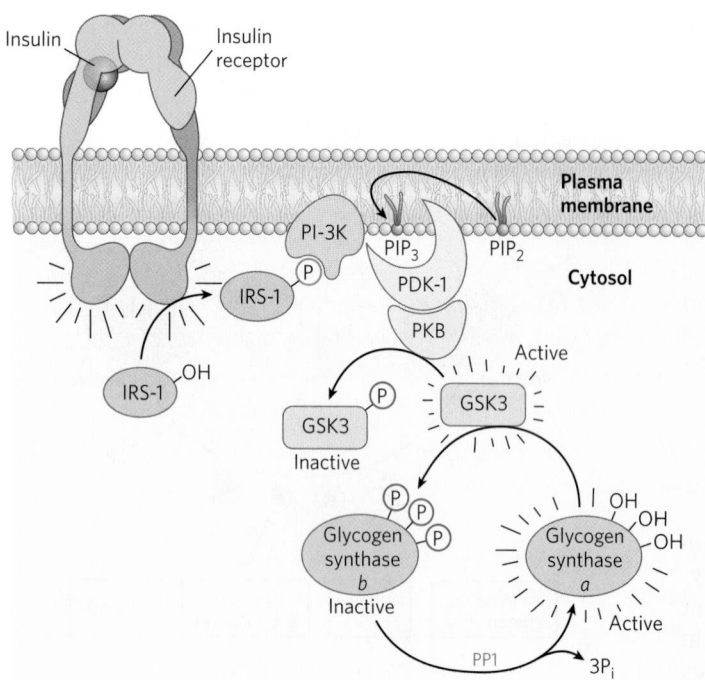

FIGURE 15-41 The path from insulin to GSK3 and glycogen synthase. Insulin binding to its receptor activates a tyrosine protein kinase in the receptor, which phosphorylates insulin receptor substrate-1 (IRS-1). The phosphotyrosine in this protein is then bound by phosphatidylinositol 3-kinase (PI-3K), which converts phosphatidylinositol 4,5-bisphosphate (PIP$_2$) in the membrane to phosphatidylinositol 3,4,5-trisphosphate (PIP$_3$). A protein kinase (PDK-1) that is activated when bound to PIP$_3$ activates a second protein kinase (PKB), which phosphorylates glycogen synthase kinase 3 (GSK3) in its pseudosubstrate region, inactivating it by the mechanism shown in Figure 15-40b. The inactivation of GSK3 allows phosphoprotein phosphatase 1 (PP1) to dephosphorylate and thus activate glycogen synthase. In this way, insulin stimulates glycogen synthesis. (See Fig. 12-20 for more details on insulin action.)

kinase (PKB) that, in turn, phosphorylates and inactivates GSK3 (**Fig. 15-41**; see also Fig. 12-20). Phosphorylation of a Ser residue near the amino terminus of GSK3 converts that region of the protein to a pseudosubstrate, which folds into the site at which the priming phosphorylated Ser residue normally binds (Fig. 15-40b). This prevents GSK3 from binding the priming site of a real substrate, thereby inactivating the enzyme and tipping the balance in favor of dephosphorylation of glycogen synthase by PP1. Glycogen phosphorylase can also affect the phosphorylation of glycogen synthase: active glycogen phosphorylase directly inhibits PP1, preventing it from activating glycogen synthase (Fig. 15-39).

Although first discovered in its role in glycogen metabolism (hence the name glycogen synthase kinase), GSK3 clearly has a much broader role than the regulation of glycogen synthase. It mediates signaling by insulin and other growth factors and nutrients, and it acts in the specification of cell fates during embryonic development. Among its targets are cytoskeletal proteins and proteins essential for mRNA and protein synthesis. These targets, like glycogen synthase, must first undergo a priming phosphorylation by another protein kinase before they can be phosphorylated by GSK3.

Phosphoprotein Phosphatase 1 Is Central to Glycogen Metabolism

A single enzyme, PP1, can remove phosphoryl groups from all three of the enzymes phosphorylated in response to glucagon (liver) and epinephrine (liver and muscle): phosphorylase kinase, glycogen phosphorylase, and glycogen synthase. Insulin stimulates glycogen synthesis by activating PP1 and by inactivating GSK3.

The catalytic subunit of phosphoprotein phosphatase 1 (PP1c) does not exist free in the cytosol, but is tightly bound to its target proteins by one of a family of **glycogen-targeting proteins** that bind glycogen and each of the three enzymes, glycogen phosphorylase, phosphorylase kinase, and glycogen synthase (**Fig. 15-42**). PP1 is itself subject to covalent and allosteric regulation: it is inactivated when phosphorylated by PKA and is allosterically activated by glucose 6-phosphate.

Allosteric and Hormonal Signals Coordinate Carbohydrate Metabolism Globally

Having looked at the mechanisms that regulate individual enzymes, we can now consider the overall shifts in carbohydrate metabolism that occur in the well-fed state, during fasting, and in the fight-or-flight response—signaled by insulin, glucagon, and epinephrine, respectively. We need to contrast two cases in which regulation serves different ends: (1) the role of hepatocytes in supplying glucose to the blood, and (2) the selfish use of carbohydrate fuels by nonhepatic tissues, typified by skeletal muscle (myocytes), to support their own activities.

After ingestion of a carbohydrate-rich meal, the elevation of blood glucose triggers insulin release (**Fig. 15-43**, top). In a hepatocyte, insulin has two immediate effects: it inactivates GSK3, acting through the cascade shown in Figure 15-41, and activates a protein phosphatase, perhaps PP1. These two actions fully activate glycogen synthase. PP1 also inactivates glycogen phosphorylase a and phosphorylase kinase by dephosphorylating both,

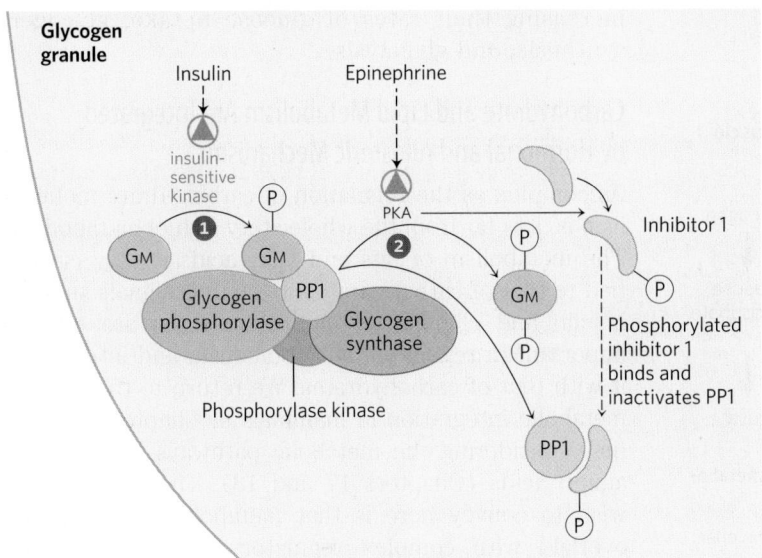

FIGURE 15-42 Glycogen-targeting protein G$_M$. The glycogen-targeting protein G$_M$ is one of a family of proteins that bind other proteins (including PP1) to glycogen particles. G$_M$ can be phosphorylated at two different sites in response to insulin or epinephrine. **1** Insulin-stimulated phosphorylation of G$_M$ site 1 activates PP1, which dephosphorylates phosphorylase kinase, glycogen phosphorylase, and glycogen synthase. **2** Epinephrine-stimulated phosphorylation of G$_M$ site 2 causes dissociation of PP1 from the glycogen particle, preventing its access to glycogen phosphorylase and glycogen synthase. PKA also phosphorylates a protein (inhibitor 1) that, when phosphorylated, inhibits PP1. By these means, insulin inhibits glycogen breakdown and stimulates glycogen synthesis, and epinephrine (or glucagon in the liver) has the opposite effects.

effectively stopping glycogen breakdown. Glucose enters the hepatocyte through the high-capacity transporter GLUT2, always present in the plasma membrane, and the elevated intracellular glucose leads to dissociation of hexokinase IV (glucokinase) from its nuclear regulatory protein (Fig. 15-15). Hexokinase IV enters the cytosol and phosphorylates glucose, stimulating glycolysis and supplying the precursor for glycogen synthesis. Under these conditions, hepatocytes use the excess glucose in the blood to synthesize glycogen, up to the limit of about 10% of the total weight of the liver.

Between meals, or during an extended fast, the drop in blood glucose triggers the release of glucagon, which, acting through the cascade shown in Figure 15-37, activates PKA. PKA mediates all the effects of glucagon (Fig. 15-43, bottom). It phosphorylates phosphorylase kinase, activating it and leading to the activation of glycogen phosphorylase. It phosphorylates glycogen synthase, inactivating it and blocking glycogen synthesis. It phosphorylates PFK-2/FBPase-2, leading to a drop in the concentration of the regulator fructose 2,6-bisphosphate, which has the effect of inactivating the glycolytic enzyme PFK-1 and activating the gluconeogenic enzyme FBPase-1. And it phosphorylates and inactivates the glycolytic enzyme pyruvate kinase. Under these conditions, the liver produces glucose 6-phosphate by glycogen breakdown and by gluconeogenesis, and it stops using glucose to fuel glycolysis or make glycogen, maximizing the amount of glucose it can

FIGURE 15-43 Regulation of carbohydrate metabolism in the liver. Colored arrows indicate causal relationships between the changes they connect. For example, an arrow from ↓A to ↑B means that a decrease in A causes an increase in B. Red arrows connect events that result from high blood glucose; blue arrows connect events that result from low blood glucose.

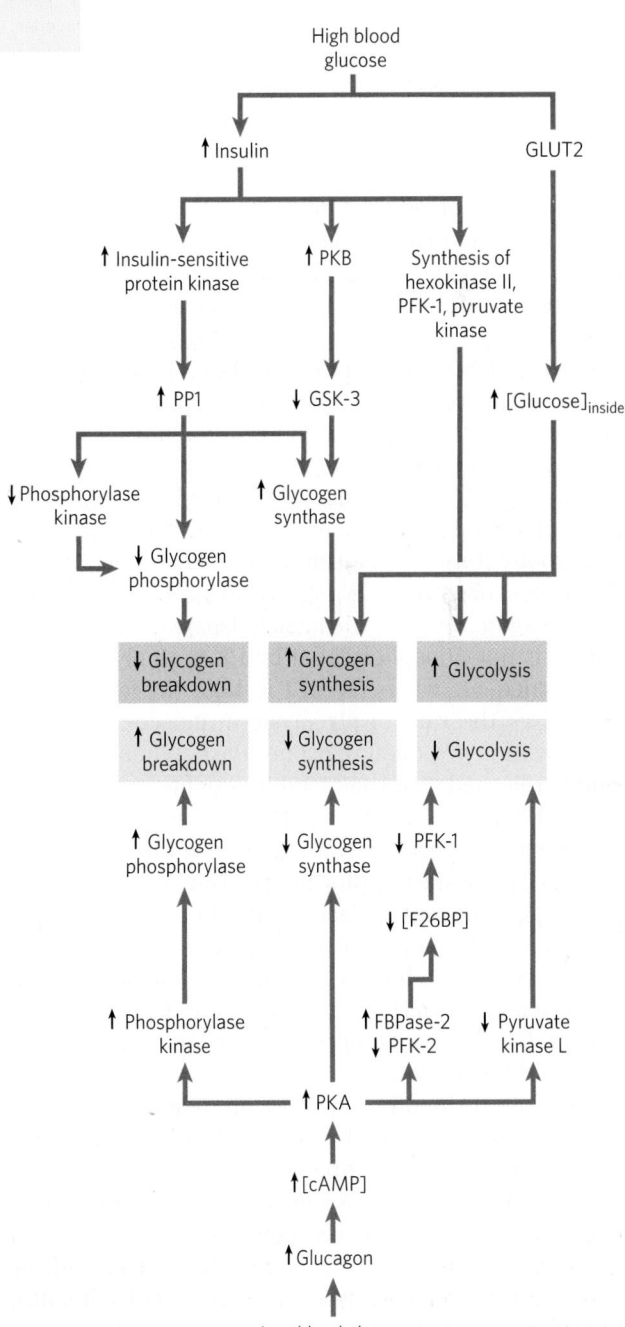

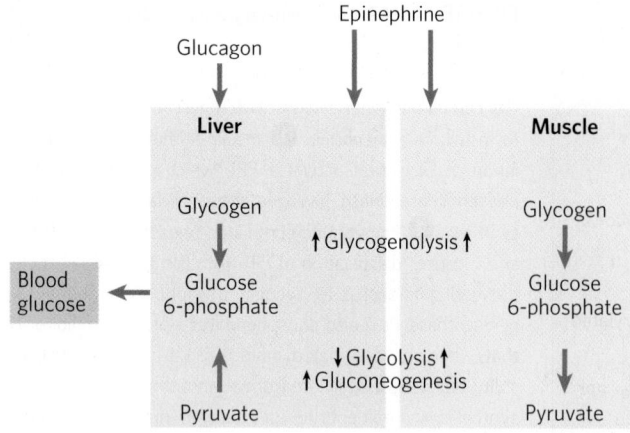

FIGURE 15-44 Difference in the regulation of carbohydrate metabolism in liver and muscle. In liver, either glucagon (indicating low blood glucose) or epinephrine (signaling the need to fight or flee) has the effect of maximizing the output of glucose into the blood. In muscle, epinephrine increases glycogen breakdown and glycolysis, which together provide fuel to produce the ATP needed for muscle contraction.

release to the blood. This release of glucose is possible only in liver and kidney, because other tissues lack glucose 6-phosphatase (Fig. 15-30).

The physiology of skeletal muscle differs from that of liver in three ways important to our discussion of metabolic regulation **(Fig. 15-44)**: (1) muscle uses its stored glycogen only for its own needs; (2) as it goes from rest to vigorous contraction, muscle undergoes very large changes in its demand for ATP, which is supported by glycolysis; (3) muscle lacks the enzymatic machinery for gluconeogenesis. The regulation of carbohydrate metabolism in muscle reflects these differences from liver. First, myocytes lack receptors for glucagon. Second, the muscle isozyme of pyruvate kinase is not phosphorylated by PKA, so glycolysis is not turned off when [cAMP] is high. In fact, cAMP *increases* the rate of glycolysis in muscle, probably by activating glycogen phosphorylase. When epinephrine is released into the blood in a fight-or-flight situation, PKA is activated by the rise in [cAMP] and phosphorylates and activates glycogen phosphorylase kinase. The resulting phosphorylation and activation of glycogen phosphorylase results in faster glycogen breakdown. Epinephrine is not released under low-stress conditions, but with each neuronal stimulation of muscle contraction, cytosolic [Ca^{2+}] rises briefly and activates phosphorylase kinase through its calmodulin subunit.

Elevated insulin triggers increased glycogen synthesis in myocytes by activating PP1 and inactivating GSK3. Unlike hepatocytes, myocytes have a reserve of GLUT4 sequestered in intracellular vesicles. Insulin triggers their movement to the plasma membrane (see Fig. 12-20), where they allow increased glucose uptake. In response to insulin, therefore, myocytes help to lower blood glucose by increasing their rates of glucose uptake, glycogen synthesis, and glycolysis.

Carbohydrate and Lipid Metabolism Are Integrated by Hormonal and Allosteric Mechanisms

As complex as the regulation of carbohydrate metabolism is, it is far from the whole story of fuel metabolism. The metabolism of fats and fatty acids is very closely tied to that of carbohydrates. Hormonal signals such as insulin and changes in diet or exercise are equally important in regulating fat metabolism and integrating it with that of carbohydrates. We return to this overall metabolic integration in mammals in Chapter 23, after first considering the metabolic pathways for fats and amino acids (Chapters 17 and 18). The message we wish to convey here is that metabolic pathways are overlaid with complex regulatory controls that are exquisitely sensitive to changes in metabolic circumstances. These mechanisms act to adjust the flow of metabolites through various metabolic pathways, as needed by the cell and organism, and to do so without causing major changes in the concentrations of intermediates shared with other pathways.

SUMMARY 15.5 Coordinated Regulation of Glycogen Breakdown and Synthesis

■ Glycogen phosphorylase is activated in response to glucagon or epinephrine, which raise [cAMP] and activate PKA. PKA phosphorylates and activates phosphorylase kinase, which converts glycogen phosphorylase *b* to its active *a* form.

■ Phosphoprotein phosphatase 1 (PP1) reverses the phosphorylation of glycogen phosphorylase *a*, inactivating it. Glucose binds to the liver isozyme of glycogen phosphorylase *a*, favoring its dephosphorylation and inactivation.

■ Glycogen synthase *a* is inactivated by phosphorylation catalyzed by GSK3. Insulin blocks GSK3. PP1, which is activated by insulin, reverses the inhibition by dephosphorylating glycogen synthase *b*.

■ Insulin increases glucose uptake into myocytes and adipocytes by triggering movement of the glucose transporter GLUT4 to the plasma membrane.

■ Insulin stimulates the synthesis of hexokinases II and IV, PFK-1, pyruvate kinase, and several enzymes involved in lipid synthesis. Insulin stimulates glycogen synthesis in muscle and liver.

■ In liver, glucagon stimulates glycogen breakdown and gluconeogenesis while blocking glycolysis, thereby sparing glucose for export to the brain and other tissues.

■ In muscle, epinephrine stimulates glycogen breakdown and glycolysis, providing ATP to support contraction.

Key Terms

Terms in bold are defined in the glossary.

glucose 6-phosphate 575
homeostasis 577
cellular differentiation 577
transcription factor 577
response element 577
turnover 578
transcriptome 579
proteome 579
metabolome 579
metabolic regulation 580
metabolic control 580
mass-action ratio, *Q* 581
adenylate kinase 582
AMP-activated protein kinase (AMPK) 582
flux control coefficient, *C* 585
flux, *J* 585
elasticity coefficient, ε 585
response coefficient, *R* 587
gluconeogenesis 589
futile cycle 590
substrate cycle 590
hexokinase II 590
hexokinase I 591
hexokinase IV (glucokinase) 591
GLUT2 591
glucagon 593
fructose 2,6-bisphosphate 593
phosphofructokinase-2 (PFK-2) 593
fructose 2,6-bisphosphatase (FBPase-2) 593
carbohydrate response element binding protein (ChREBP) 597
sterol regulatory element binding protein (SREBP) 598
cyclic AMP response element binding protein (CREB) 598
forkhead box other (FOXO1) 598
glycogenolysis 601
glycogenesis 601
glucose 1-phosphate 601
debranching enzyme 601
oligo (α1→6) to (α1→4) glucantransferase 601
phosphoglucomutase 601
sugar nucleotides 603
UDP-glucose pyrophosphorylase 605
amylo (1→4) to (1→6) transglycosylase 606
glycogenin 607
glycogen phosphorylase *a* 608
glycogen phosphorylase *b* 608
enzyme cascade 609
phosphorylase *b* kinase 609
phosphoprotein phosphatase 1 (PP1) 609
glycogen synthase *a* 610
glycogen synthase *b* 610
glycogen synthase kinase 3 (GSK3) 610
casein kinase II (CKII) 611
priming 611
glycogen-targeting proteins 612

Problems

1. Measurement of Intracellular Metabolite Concentrations Measuring the concentrations of metabolic intermediates in a living cell presents great experimental difficulties—usually a cell must be destroyed before metabolite concentrations can be measured. Yet enzymes catalyze metabolic interconversions very rapidly, so a common problem associated with these types of measurements is that the findings reflect not the physiological concentrations of metabolites but the equilibrium concentrations. A reliable experimental technique requires all enzyme-catalyzed reactions to be instantaneously stopped in the intact tissue so that the metabolic intermediates do not undergo change. This objective is accomplished by rapidly compressing the tissue between large aluminum plates cooled with liquid nitrogen (−190 °C), a process called **freeze-clamping**. After freezing, which stops enzyme action instantly, the tissue is powdered and the enzymes are inactivated by precipitation with perchloric acid. The precipitate is removed by centrifugation, and the clear supernatant extract is analyzed for metabolites. To calculate intracellular concentrations, the intracellular volume is determined from the total water content of the tissue and a measurement of the extracellular volume.

The intracellular concentrations of the substrates and products of the phosphofructokinase-1 reaction in isolated rat heart tissue are given in the table below.

Metabolite	**Concentration (μM)[a]**
Fructose 6-phosphate	87.0
Fructose 1,6-bisphosphate	22.0
ATP	11,400
ADP	1,320

Source: Data from J. R. Williamson, *J. Biol. Chem.* 240:2308, 1965.
[a]Calculated as μmol/mL of intracellular water.

(a) Calculate *Q*, [fructose 1,6-bisphosphate][ADP]/[fructose 6-phosphate][ATP], for the PFK-1 reaction under physiological conditions.

(b) Given a $\Delta G'^{\circ}$ for the PFK-1 reaction of −14.2 kJ/mol, calculate the equilibrium constant for this reaction.

(c) Compare the values of *Q* and K'_{eq}. Is the physiological reaction near or far from equilibrium? Explain. What does this experiment suggest about the role of PFK-1 as a regulatory enzyme?

2. Are All Metabolic Reactions at Equilibrium?

(a) Phosphoenolpyruvate (PEP) is one of the two phosphoryl group donors in the synthesis of ATP during glycolysis. In human erythrocytes, the steady-state concentration of ATP is 2.24 mM, that of ADP is 0.25 mM, and that of pyruvate is 0.051 mM. Calculate the concentration of PEP at 25 °C, assuming that the pyruvate kinase reaction (see Fig. 13-13) is at equilibrium in the cell.

(b) The physiological concentration of PEP in human erythrocytes is 0.023 mM. Compare this with the value obtained in (a). Explain the significance of this difference.

3. Effect of O_2 Supply on Glycolytic Rates The regulated steps of glycolysis in intact cells can be identified by studying the catabolism of glucose in whole tissues or organs. For example, the glucose consumption by heart muscle can be measured by artificially circulating blood through an isolated intact heart and measuring the concentration of glucose before and after the blood passes through the heart. If the circulating blood is deoxygenated, heart muscle consumes glucose at a steady rate. When oxygen is added to the blood, the rate of glucose consumption drops dramatically, then is maintained at the new, lower rate. Explain.

4. Regulation of PFK-1 The effect of ATP on the allosteric enzyme PFK-1 is shown below. For a given concentration of fructose 6-phosphate, the PFK-1 activity increases with increasing concentrations of ATP, but a point is reached beyond which increasing the concentration of ATP inhibits the enzyme.

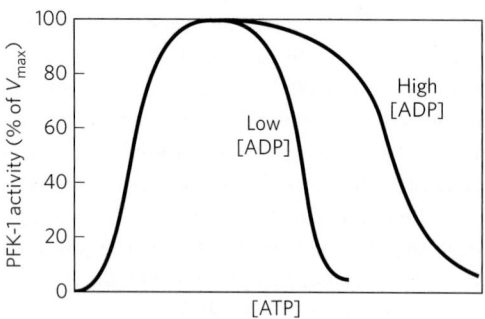

(a) Explain how ATP can be both a substrate and an inhibitor of PFK-1. How is the enzyme regulated by ATP?

(b) In what ways is glycolysis regulated by ATP levels?

(c) The inhibition of PFK-1 by ATP is diminished when the ADP concentration is high, as shown in the graph. How can this observation be explained?

5. Cellular Glucose Concentration The concentration of glucose in human blood plasma is maintained at about 5 mM. The concentration of free glucose inside a myocyte is much lower. Why is the concentration so low in the cell? What happens to glucose after entry into the cell? Glucose is administered intravenously as a food source in certain clinical situations. Given that the transformation of glucose to glucose 6-phosphate consumes ATP, why not administer intravenous glucose 6-phosphate instead?

6. Enzyme Activity and Physiological Function The V_{max} of the glycogen phosphorylase from skeletal muscle is much greater than the V_{max} of the same enzyme from liver tissue.

(a) What is the physiological function of glycogen phosphorylase in skeletal muscle? In liver tissue?

(b) Why does the V_{max} of the muscle enzyme need to be greater than that of the liver enzyme?

7. Glycogen Phosphorylase Equilibrium Glycogen phosphorylase catalyzes the removal of glucose from glycogen. The $\Delta G'^{\circ}$ for this reaction is 3.1 kJ/mol.

(a) Calculate the ratio of $[P_i]$ to [glucose 1-phosphate] when the reaction is at equilibrium. (Hint: The removal of glucose units from glycogen does not change the glycogen concentration.)

(b) The measured ratio $[P_i]$/[glucose 1-phosphate] in myocytes under physiological conditions is more than 100:1. What does this indicate about the direction of metabolite flow through the glycogen phosphorylase reaction in muscle?

(c) Why are the equilibrium and physiological ratios different? What is the possible significance of this difference?

8. Regulation of Glycogen Phosphorylase In muscle tissue, the rate of conversion of glycogen to glucose 6-phosphate is determined by the ratio of phosphorylase a (active) to phosphorylase b (less active). Determine what happens to the rate of glycogen breakdown if a muscle preparation containing glycogen phosphorylase is treated with (a) phosphorylase kinase and ATP; (b) PP1; (c) epinephrine.

9. Glycogen Breakdown in Rabbit Muscle The intracellular use of glucose and glycogen is tightly regulated at four points. To compare the regulation of glycolysis when oxygen is plentiful and when it is depleted, consider the utilization of glucose and glycogen by rabbit leg muscle in two physiological settings: a resting rabbit, with low ATP demands, and a rabbit that sights its mortal enemy, the coyote, and dashes into its burrow. For each setting, determine the relative levels (high, intermediate, or low) of AMP, ATP, citrate, and acetyl-CoA and describe how these levels affect the flow of metabolites through glycolysis by regulating specific enzymes. In periods of stress, rabbit leg muscle produces much of its ATP by anaerobic glycolysis (lactate fermentation) and very little by oxidation of acetyl-CoA derived from fat breakdown.

10. Glycogen Breakdown in Migrating Birds Unlike the rabbit with its short dash, migratory birds require energy for extended periods of time. For example, ducks generally fly several thousand miles during their annual migration. The flight muscles of migratory birds have a high oxidative capacity and obtain the necessary ATP through the oxidation of acetyl-CoA (obtained from fats) via the citric acid cycle. Compare the regulation of muscle glycolysis during short-term intense activity, as in the fleeing rabbit, and during extended activity, as in the migrating duck. Why must the regulation in these two settings be different?

11. Enzyme Defects in Carbohydrate Metabolism Summaries of four clinical case studies follow. For each case determine which enzyme is defective and designate the appropriate treatment, from the lists provided at the end of the problem. Justify your choices. Answer the questions contained in each case study. (You may need to refer to information in Chapter 14.)

Case A The patient develops vomiting and diarrhea shortly after milk ingestion. A lactose tolerance test is administered. (The patient ingests a standard amount of lactose, and the glucose and galactose concentrations in blood plasma are measured at intervals. In individuals with normal carbohydrate metabolism, the levels increase to a maximum in about 1 hour, then decline.) The patient's blood glucose and galactose concentrations do not increase during the test. Why do blood glucose and galactose increase and then decrease during the test in healthy individuals? Why do they fail to rise in the patient?

Case B The patient develops vomiting and diarrhea after ingestion of milk. His blood is found to have a low concentration of glucose but a much higher than normal concentration of reducing sugars. The urine tests positive for galactose. Why is the concentration of reducing sugar in the blood high? Why does galactose appear in the urine?

Case C The patient complains of painful muscle cramps when performing strenuous physical exercise but has no other symptoms. A muscle biopsy indicates a muscle glycogen concentration much higher than normal. Why does glycogen accumulate?

Case D The patient is lethargic, her liver is enlarged, and a biopsy of the liver shows large amounts of excess glycogen. She also has a lower than normal blood glucose level. What is the reason for the low blood glucose in this patient?

Defective Enzyme

(a) Muscle PFK-1

(b) Phosphomannose isomerase

(c) Galactose 1-phosphate uridylyltransferase

(d) Liver glycogen phosphorylase

(e) Triose kinase

(f) Lactase in intestinal mucosa

(g) Maltase in intestinal mucosa

(h) Muscle debranching enzyme

Treatment

1. Jogging 5 km each day

2. Fat-free diet

3. Low-lactose diet

4. Avoiding strenuous exercise

5. Large doses of niacin (the precursor of NAD$^+$)

6. Frequent feedings (smaller portions) of a normal diet

 12. Effects of Insufficient Insulin in a Person with Diabetes A man with insulin-dependent diabetes is brought to the emergency room in a near-comatose state. While vacationing in an isolated place, he lost his insulin medication and has not taken any insulin for two days.

(a) For each tissue listed below, is each pathway faster, slower, or unchanged in this patient, compared with the normal level when he is getting appropriate amounts of insulin?

(b) For each pathway, describe at least one control mechanism responsible for the change you predict.

Tissue and Pathways

1. Adipose: fatty acid synthesis

2. Muscle: glycolysis; fatty acid synthesis; glycogen synthesis

3. Liver: glycolysis; gluconeogenesis; glycogen synthesis; fatty acid synthesis; pentose phosphate pathway

 13. Blood Metabolites in Insulin Insufficiency For the patient described in Problem 12, predict the levels of the following metabolites in his blood *before* treatment in the emergency room, relative to levels maintained during adequate insulin treatment: (a) glucose; (b) ketone bodies; (c) free fatty acids.

14. Metabolic Effects of Mutant Enzymes Predict and explain the effect on glycogen metabolism of each of the following defects caused by mutation: (a) loss of the cAMP-binding site on the regulatory subunit of protein kinase A (PKA); (b) loss of the protein phosphatase inhibitor (inhibitor 1 in Fig. 15-42); (c) overexpression of phosphorylase *b* kinase in liver; (d) defective glucagon receptors in liver.

15. Hormonal Control of Metabolic Fuel Between your evening meal and breakfast, your blood glucose drops and your liver becomes a net producer rather than consumer of glucose. Describe the hormonal basis for this switch, and explain how the hormonal change triggers glucose production by the liver.

16. Altered Metabolism in Genetically Manipulated Mice Researchers can manipulate the genes of a mouse so that a single gene in a single tissue either produces an inactive protein (a "knockout" mouse) or produces a protein that is always (constitutively) active. What effects on metabolism would you predict for mice with the following genetic changes: (a) knockout of glycogen debranching enzyme in the liver;

(b) knockout of hexokinase IV in liver; (c) knockout of FBPase-2 in liver; (d) constitutively active FBPase-2 in liver; (e) constitutively active AMPK in muscle; (f) constitutively active ChREBP in liver?

Data Analysis Problem

17. Optimal Glycogen Structure Muscle cells need rapid access to large amounts of glucose during heavy exercise. This glucose is stored in liver and skeletal muscle in polymeric form as particles of glycogen. The typical glycogen particle contains about 55,000 glucose residues (see Fig. 15-35b). Meléndez-Hevia, Waddell, and Shelton (1993) explored some theoretical aspects of the structure of glycogen, as described in this problem.

(a) The cellular concentration of glycogen in liver is about 0.01 μM. What cellular concentration of free glucose would be required to store an equivalent amount of glucose? Why would this concentration of free glucose present a problem for the cell?

Glucose is released from glycogen by glycogen phosphorylase, an enzyme that can remove glucose molecules, one at a time, from one end of a glycogen chain. Glycogen chains are branched (see Figs 15-28 and 15-35b), and the degree of branching—the number of branches per chain—has a powerful influence on the rate at which glycogen phosphorylase can release glucose.

(b) Why would a degree of branching that was too low (i.e., below an optimum level) reduce the rate of glucose release? (Hint: Consider the extreme case of no branches in a chain of 55,000 glucose residues.)

(c) Why would a degree of branching that was too high also reduce the rate of glucose release? (Hint: Think of the physical constraints.)

Meléndez-Hevia and colleagues did a series of calculations and found that two branches per chain (see Fig. 15-35b) was optimal for the constraints described above. This is what is found in glycogen stored in muscle and liver.

To determine the optimum number of glucose residues per chain, Meléndez-Hevia and coauthors considered two key parameters that define the structure of a glycogen particle: t = the number of tiers of glucose chains in a particle (the molecule in Fig. 15-35b has five tiers); g_c = the number of glucose residues in each chain. They set out to find the values of t and g_c that would maximize three quantities: (1) the amount of glucose stored in the particle (G_T) per unit volume; (2) the number of unbranched glucose chains (C_A) per unit volume (i.e., number of chains in the outermost tier, readily accessible to glycogen phosphorylase); and (3) the amount of glucose available to phosphorylase in these unbranched chains (G_{PT}).

(d) Show that $C_A = 2^{t-1}$. This is the number of chains available to glycogen phosphorylase before the action of the debranching enzyme.

(e) Show that C_T, the total number of chains in the particle, is given by $C_T = 2^t - 1$. Thus $G_T = g_c(C_T) = g_c(2^t - 1)$, the total number of glucose residues in the particle.

(f) Glycogen phosphorylase cannot remove glucose from glycogen chains that are shorter than five glucose residues. Show that $G_{PT} = (g_c - 4)(2^{t-1})$. This is the amount of glucose readily available to glycogen phosphorylase.

(g) Based on the size of a glucose residue and the location of branches, the thickness of one tier of glycogen is $0.12\,g_c$ nm + 0.35 nm. Show that the volume of a particle, V_s, is given by the equation $V_s = {}^{4}\!/_{3}\pi t^3 (0.12g_c + 0.35)^3$ nm^3.

Meléndez-Hevia and coauthors then determined the optimum values of t and g_c—those that gave the maximum value of a quality function, f, that maximizes G_T, C_A, and G_{PT}, while minimizing V_s: $f = \dfrac{G_T C_A G_{PT}}{V_S}$. They found that the optimum value of g_c is independent of t.

(h) Choose a value of t between 5 and 15 and find the optimum value of g_c. How does this compare with the g_c found in liver glycogen (see Fig. 15-35b)? (Hint: You may find it useful to use a spreadsheet program.)

Reference

Meléndez-Hevia, E., T.G. Waddell, and E.D. Shelton. 1993. Optimization of molecular design in the evolution of metabolism: the glycogen molecule. *Biochem. J.* 295:477–483.

Further Reading is available at www.macmillanlearning.com/LehningerBiochemistry7e.

CHAPTER 16

The Citric Acid Cycle

Self-study tools that will help you practice what you've learned and reinforce this chapter's concepts are available online.
Go to www.macmillanlearning.com/LehningerBiochemistry7e.

As we saw in Chapter 14, some cells obtain energy (ATP) by fermentation, breaking down glucose in the absence of oxygen. For most eukaryotic cells and many bacteria, which live under aerobic conditions and oxidize their organic fuels to carbon dioxide and water, glycolysis is but the first stage in the complete oxidation of glucose. Rather than being reduced to lactate, ethanol, or some other fermentation product, the pyruvate produced by glycolysis is further oxidized to H_2O and CO_2. This aerobic phase of catabolism is called **respiration**. In the broader physiological or macroscopic sense, respiration refers to a multicellular organism's uptake of O_2 and release of CO_2. Biochemists and cell biologists, however, use the term in a narrower sense to refer to the molecular processes by which *cells* consume O_2 and produce CO_2—processes more precisely termed **cellular respiration**.

Cellular respiration occurs in three major stages (Fig. 16-1). In the first, organic fuel molecules—glucose, fatty acids, and some amino acids—are oxidized to yield two-carbon fragments in the form of the acetyl group of acetyl-coenzyme A (acetyl-CoA). In the second stage, the acetyl groups are oxidized to CO_2 in the citric acid cycle, and much of the energy of these oxidations is conserved in the reduced electron carriers NADH and $FADH_2$. In the third stage of respiration, these reduced coenzymes are themselves oxidized, giving up protons (H^+) and electrons. The electrons are transferred to O_2 via a series of electron-carrying molecules known as the respiratory chain, resulting in the formation of water (H_2O). In the course of electron transfer, much of the energy available from redox reactions is conserved in the form of ATP, by a process called oxidative phosphorylation (Chapter 19). Respiration is more complex than glycolysis and is believed to have evolved much later, after the rise of oxygen levels in the earth's atmosphere that resulted from the evolution of photosynthesis by cyanobacteria. As discussed in Chapter 1, the increase in oxygen levels was a turning point in evolutionary history.

We consider first the conversion of pyruvate to acetyl groups, then the entry of those groups into the **citric acid cycle**, also called the **tricarboxylic acid (TCA) cycle** or the **Krebs cycle** (after its discoverer, Hans Krebs). We next examine the cycle reactions and the enzymes that catalyze them. Because intermediates of the citric acid cycle are also siphoned off as biosynthetic precursors, we go on to consider some ways in which these intermediates are replenished. The citric acid cycle is a hub in metabolism, with degradative pathways leading in and anabolic pathways leading out, and it is closely regulated in coordination with other pathways.

Hans Krebs, 1900–1981
[Source: Keystone Pictures USA/Alamy.]

16.1 Production of Acetyl-CoA (Activated Acetate)

In aerobic organisms, glucose and other sugars, fatty acids, and most amino acids are ultimately oxidized to CO_2 and H_2O via the citric acid cycle and the respiratory chain. Before entering the citric acid cycle, the carbon skeletons of monosaccharides and fatty acids are degraded

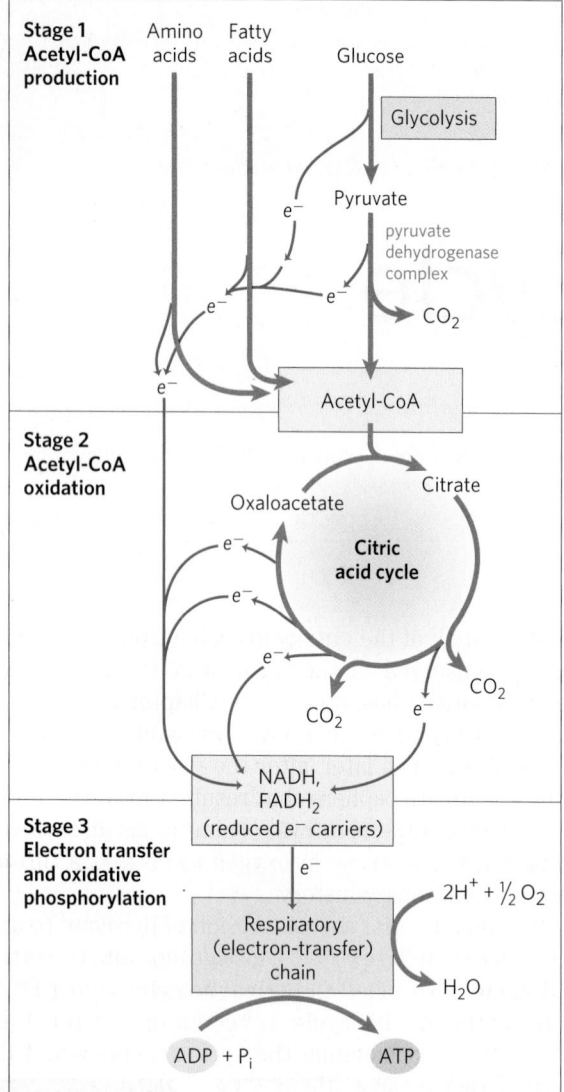

FIGURE 16-1 Catabolism of proteins, fats, and carbohydrates in the three stages of cellular respiration. Stage 1: oxidation of fatty acids, glucose, and some amino acids yields acetyl-CoA. Stage 2: oxidation of acetyl groups in the citric acid cycle includes four steps in which electrons are abstracted. Stage 3: electrons carried by NADH and $FADH_2$ are funneled into a chain of mitochondrial (or, in bacteria, plasma membrane–bound) electron carriers—the respiratory chain—ultimately reducing O_2 to H_2O. This electron flow drives the production of ATP.

to the acetyl group of acetyl-CoA, the form in which the cycle accepts most of its fuel input. Many amino acid carbons also enter the cycle as acetate, although several amino acids are degraded to other cycle intermediates such as succinate and malate, which then enter the cycle.

Pyruvate generated in the cytosol by glycolysis represents a node in the metabolism of carbohydrates, fats, and proteins. Pyruvate that enters mitochondria may be oxidized by the citric acid cycle to generate energy or, after conversion to acetyl-CoA, may be used as the starting material for synthesis of fatty acids and sterols. A third possible fate for pyruvate is as a precursor for the synthesis of amino acids (Chapter 22). In **aerobic glycolysis**,

pyruvate does not undergo further oxidation via the citric acid cycle, even though oxygen is available to support that route; instead, pyruvate is simply reduced to lactate in the cytosol, regenerating NAD^+ for continued ATP production by glycolysis.

We have seen that in many types of cancer, glucose undergoes aerobic glycolysis in the cytosol, with lactate as a side product (the Warburg effect; see Box 14-1). Because the entry of pyruvate into mitochondria (and its subsequent oxidation) slows in tumors, the mechanism by which pyruvate enters mitochondria is of great interest. To enter the matrix, pyruvate first diffuses through large openings in the outer mitochondrial membrane, then is transported across the inner membrane by the **mitochondrial pyruvate carrier (MPC)**, a passive transporter specific for pyruvate. MPC is encoded by two genes, *MPC1* and *MPC2*, which are mutated in a high proportion (80%) of certain cancers, including gliomas (tumors of the glial cells of the brain). These mutations reduce the fraction of cytosolic pyruvate that undergoes oxidation in the citric acid cycle, and this reduction might, in principle, explain the Warburg effect. ∎

In normal cells (not tumors), pyruvate in the mitochondrial matrix is oxidized to acetyl-CoA and CO_2 by the **pyruvate dehydrogenase (PDH) complex**. This highly ordered cluster of enzymes—multiple copies of each of three enzymes—is located in the mitochondria of all eukaryotic cells and in the cytosol of bacteria.

A careful examination of this enzyme complex is rewarding in several respects. The PDH complex is a classic, much-studied example of a multienzyme complex in which a series of chemical intermediates remain bound to the enzyme molecules as a substrate is transformed into the final product. Five cofactors, four derived from vitamins, participate in the reaction mechanism. The regulation of this enzyme complex also illustrates how a combination of covalent modification and allosteric mechanism results in precisely regulated flux through a metabolic step. Finally, the PDH complex is the prototype for two other important enzyme complexes: α-ketoglutarate dehydrogenase, of the citric acid cycle, and the branched-chain α-keto acid dehydrogenase, of the oxidative pathways of several amino acids (see Fig. 18-28). The remarkable similarity in protein structure, cofactor requirements, and reaction mechanisms of these three complexes doubtless reflects a common evolutionary origin; they are paralogs.

Pyruvate Is Oxidized to Acetyl-CoA and CO_2

The overall reaction catalyzed by the pyruvate dehydrogenase complex is an **oxidative decarboxylation**, an irreversible oxidation process in which the carboxyl group is removed from pyruvate as a molecule of CO_2 and the two remaining carbons become the acetyl group of acetyl-CoA **(Fig. 16-2)**. The NADH formed in this reaction gives up a hydride ion (:H^-) to the respiratory

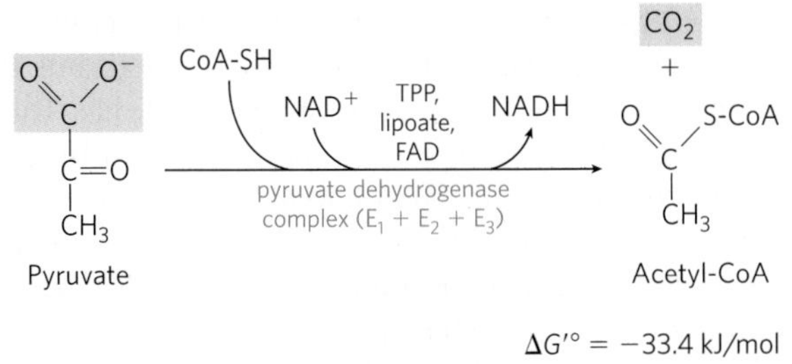

FIGURE 16-2 Overall reaction catalyzed by the pyruvate dehydrogenase complex. The five coenzymes participating in this reaction, and the three enzymes that make up the enzyme complex, are discussed in the text.

chain (Fig. 16-1), which carries the two electrons to oxygen or, in anaerobic microorganisms, to an alternative electron acceptor such as nitrate or sulfate. The transfer of electrons from NADH to oxygen ultimately generates 2.5 molecules of ATP per pair of electrons. The irreversibility of the PDH complex reaction has been demonstrated by isotopic labeling experiments: the complex cannot reattach radioactively labeled CO_2 to acetyl-CoA to yield carboxyl-labeled pyruvate.

The Pyruvate Dehydrogenase Complex Employs Five Coenzymes

The combined dehydrogenation and decarboxylation of pyruvate to the acetyl group of acetyl-CoA (Fig. 16-2) requires the sequential action of three different enzymes and five different coenzymes or prosthetic groups—thiamine pyrophosphate (TPP), flavin adenine dinucleotide (FAD), coenzyme A (CoA, sometimes denoted CoA-SH, to emphasize the role of the —SH group), nicotinamide adenine dinucleotide (NAD), and lipoate. Four different vitamins required in human nutrition are vital components of this system: thiamine (in TPP), riboflavin (in FAD), niacin (in NAD), and pantothenate

(in CoA). We have already described the roles of FAD and NAD as electron carriers (Chapter 13), and we have encountered TPP as the coenzyme of pyruvate decarboxylase (see Fig. 14-15).

Coenzyme A **(Fig. 16-3)** has a reactive thiol (—SH) group that is critical to the role of CoA as an acyl carrier in many metabolic reactions. Acyl groups are covalently linked to the thiol group, forming **thioesters**. Because of their relatively high standard free energies of hydrolysis (see Figs 13-16, 13-17), thioesters have a high acyl group transfer potential—that is, donation of their acyl groups to a variety of acceptor molecules is a favorable reaction. The acyl group attached to coenzyme A may thus be thought of as "activated" for group transfer.

The fifth cofactor of the PDH complex, **lipoate (Fig. 16-4)**, has two thiol groups that can undergo reversible oxidation to a disulfide bond (—S—S—), similar to that between two Cys residues in a protein. Because of its capacity to undergo oxidation-reduction reactions, lipoate can serve both as an electron (hydrogen) carrier and as an acyl carrier, as we shall see.

The Pyruvate Dehydrogenase Complex Consists of Three Distinct Enzymes

The PDH complex contains three enzymes—**pyruvate dehydrogenase** (E_1), **dihydrolipoyl transacetylase** (E_2), and **dihydrolipoyl dehydrogenase** (E_3)—each present in multiple copies. The number of copies of each enzyme and therefore the size of the complex varies among species. The PDH complex isolated from mammals is about 50 nm in diameter—more than five times the size of an entire ribosome and big enough to be visualized with the electron microscope **(Fig. 16-5a)**. In the bovine enzyme, 60 identical copies of E_2 form a pentagonal dodecahedron (the core) with a diameter of about 25 nm (Fig. 16-5b). (The core of the

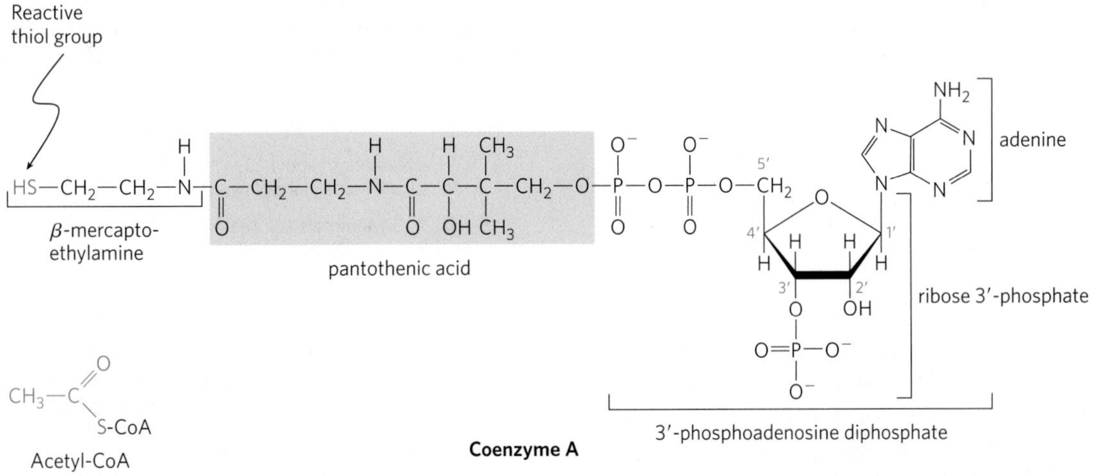

FIGURE 16-3 Coenzyme A (CoA). A hydroxyl group of pantothenic acid is joined to a modified ADP moiety by a phosphate ester bond, and its carboxyl group is attached to β-mercaptoethylamine in amide linkage. The hydroxyl group at the 3' position of the ADP moiety has a phosphoryl group not present in free ADP. The —SH group of the mercaptoethylamine moiety forms a thioester with acetate in acetyl-coenzyme A (acetyl-CoA) (lower left).

Oxidized form Reduced form Acetylated form

Lipoic acid

Lys residue of E$_2$

Polypeptide chain of E$_2$ (dihydrolipoyl transacetylase)

FIGURE 16-4 Lipoic acid (lipoate) in amide linkage with a Lys residue. The lipoyllysyl moiety is the prosthetic group of dihydrolipoyl transacetylase (E$_2$ of the PDH complex). The lipoyl group occurs in oxidized (disulfide) and reduced (dithiol) forms and acts as a carrier of both hydrogen and an acetyl (or other acyl) group.

Escherichia coli enzyme contains 24 copies of E$_2$.) E$_2$ is the point of connection for the prosthetic group lipoate, attached through an amide bond to the ε-amino group of a Lys residue (Fig. 16-4). E$_2$ has three functionally distinct domains (Fig. 16-5c): the amino-terminal *lipoyl domain*, containing the lipoyl-Lys residue(s); the central E$_1$- and E$_3$-*binding domain*; and the inner-core *acyltransferase domain*, which contains the

FIGURE 16-5 The pyruvate dehydrogenase complex. (a) Cryoelectron micrograph of PDH complexes isolated from bovine kidney. In cryoelectron microscopy, biological samples are viewed at extremely low temperatures; this avoids potential artifacts introduced by the usual process of dehydrating, fixing, and staining (see Box 19-1). **(b)** Three-dimensional image of PDH complex, showing the subunit structure: E$_1$, pyruvate dehydrogenase; E$_2$, dihydrolipoyl transacetylase; and E$_3$, dihydrolipoyl dehydrogenase. This image is reconstructed by analysis of a large number of images such as those in (a), combined with crystallographic studies of individual subunits. The core (green) consists of 60 molecules of E$_2$, arranged in 20 trimers to form a pentagonal dodecahedron. The lipoyl domain of E$_2$ (blue) reaches outward to touch the active sites of E$_1$ molecules (yellow) arranged on the E$_2$ core. Several E$_3$ subunits (red) are also bound to the core, where the swinging arm on E$_2$ can reach their active sites. An asterisk marks the site where a lipoyl group is attached to the lipoyl domain of E$_2$. To make the structure clearer, about half of the complex has been cut away from the front. **(c)** E$_2$ consists of three types of domains linked by short polypeptide linkers: a catalytic acyltransferase domain; a binding domain, involved in the binding of E$_2$ to E$_1$ and E$_3$; and one or more (depending on the species) lipoyl domains. [Source: (a) Courtesy of Dr. Richard N. Trelease. (b) Courtesy of Dr. Z. Hong Zhou, Director, Electron Imaging Center for NanoMachines (EICN).]

acyltransferase active site. The yeast PDH complex has a single lipoyl domain with a lipoate attached, but the mammalian complex has two, and *E. coli* has three (Fig. 16-5c). The domains of E$_2$ are separated by linkers, sequences of 20 to 30 amino acid residues, rich in Ala and Pro and interspersed with charged residues; these linkers tend to assume their extended forms, holding the three domains apart.

The active site of E$_1$ has bound TPP, and that of E$_3$ has bound FAD. Two regulatory proteins are also part of the complex: a protein kinase and a phosphoprotein phosphatase, discussed below. This basic E$_1$–E$_2$–E$_3$ structure has been conserved during evolution and used in a number of similar metabolic reactions, including the oxidation of α-ketoglutarate in the citric acid cycle (described below) and the oxidation of α-keto acids derived from the breakdown of the branched-chain amino acids valine, isoleucine, and leucine (see Fig. 18-28). Within a given species, E$_3$ of PDH is identical to E$_3$ of the other two enzyme complexes. The attachment of lipoate to the end of a Lys side chain in E$_2$ produces a long, flexible arm that can move from the active site of E$_1$ to the active sites of E$_2$ and E$_3$, a distance of perhaps 5 nm or more.

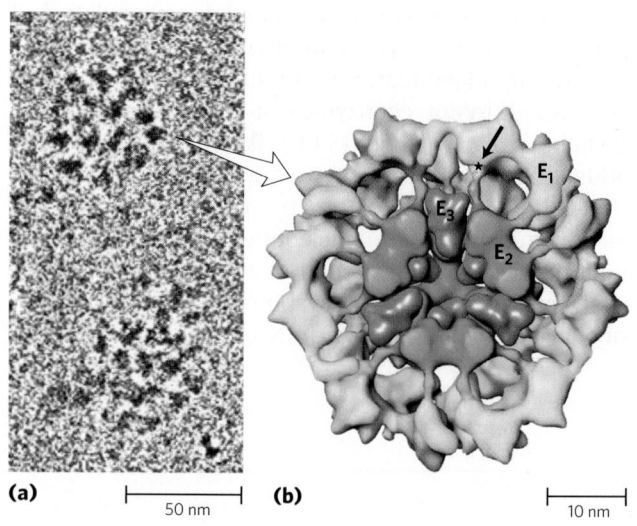

(a) 50 nm **(b)** 10 nm

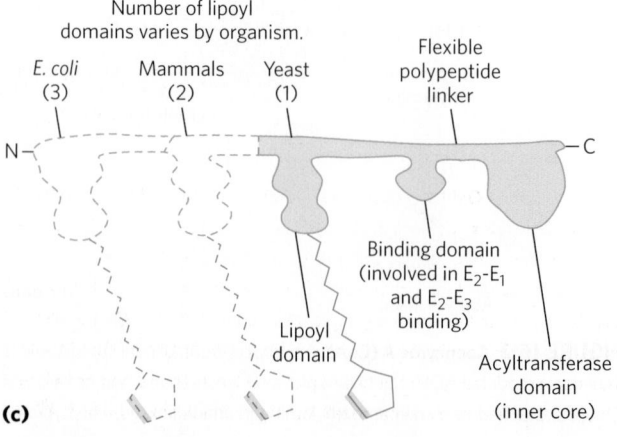

Number of lipoyl domains varies by organism.

E. coli (3) Mammals (2) Yeast (1)

Flexible polypeptide linker

N — C

Lipoyl domain

Binding domain (involved in E$_2$-E$_1$ and E$_2$-E$_3$ binding)

Acyltransferase domain (inner core)

(c)

In Substrate Channeling, Intermediates Never Leave the Enzyme Surface

Figure 16-6 shows schematically how the pyruvate dehydrogenase complex carries out the five consecutive reactions in the decarboxylation and dehydrogenation of pyruvate. Step ❶ is essentially identical to the reaction catalyzed by pyruvate decarboxylase (see Fig. 14-15c); C-1 of pyruvate is released as CO_2, and C-2, which in pyruvate has the oxidation state of an aldehyde, is attached to TPP as a hydroxyethyl group. This first step is the slowest and therefore limits the rate of the overall reaction. It is also the point at which the PDH complex exercises its substrate specificity. In step ❷ the hydroxyethyl group is oxidized to the level of a carboxylic acid (acetate). The two electrons removed in this reaction reduce the —S—S— of a lipoyl group on E_2 to two thiol (—SH) groups. The acetyl moiety produced in this oxidation-reduction reaction is first esterified to one of the lipoyl —SH groups, then transesterified to CoA to form acetyl-CoA (step ❸). Thus the energy of oxidation drives the formation of a high-energy thioester of acetate. The remaining reactions catalyzed by the PDH complex (by E_3, in steps ❹ and ❺) are electron transfers necessary to regenerate the oxidized (disulfide) form of the lipoyl group of E_2, to prepare the enzyme complex for another round of oxidation. The electrons removed from the hydroxyethyl group derived from pyruvate pass through FAD to NAD^+.

Central to the mechanism of the PDH complex are the swinging lipoyllysyl arms of E_2, which accept from E_1 the two electrons and the acetyl group derived from pyruvate, passing them to E_3. All these enzymes and coenzymes are clustered, allowing the intermediates to react quickly without diffusing away from the surface of the enzyme complex. The five-reaction sequence shown in Figure 16-6 is thus an example of **substrate channeling**. The intermediates of the multistep sequence never leave the complex, and the local concentration of the substrate of E_2 is kept very high. Channeling also prevents theft of the activated acetyl group by other enzymes that use this group as substrate. As we shall see, a similar tethering mechanism for the channeling of substrate between active sites is used in some other enzymes, with lipoate, biotin, or a CoA-like moiety serving as cofactors.

As one might predict, mutations in the genes for the subunits of the PDH complex, or a dietary thiamine deficiency, can have severe consequences. Thiamine-deficient animals are unable to oxidize pyruvate normally. This is of particular importance to the brain, which usually obtains all its energy from the aerobic oxidation of glucose in a pathway that necessarily includes the oxidation of pyruvate. Beriberi, a disease that results from thiamine deficiency, is characterized by loss of neural function. This disease occurs primarily in populations that rely on a diet consisting mainly of white (polished) rice, which lacks the hulls in which most of the thiamine of rice is found. People who habitually consume large amounts of alcohol can also develop thiamine deficiency, because much of their dietary intake consists of the vitamin-free "empty calories" of distilled spirits. An elevated level of pyruvate in the blood is often an indicator of defects in pyruvate oxidation due to one of these causes. ■

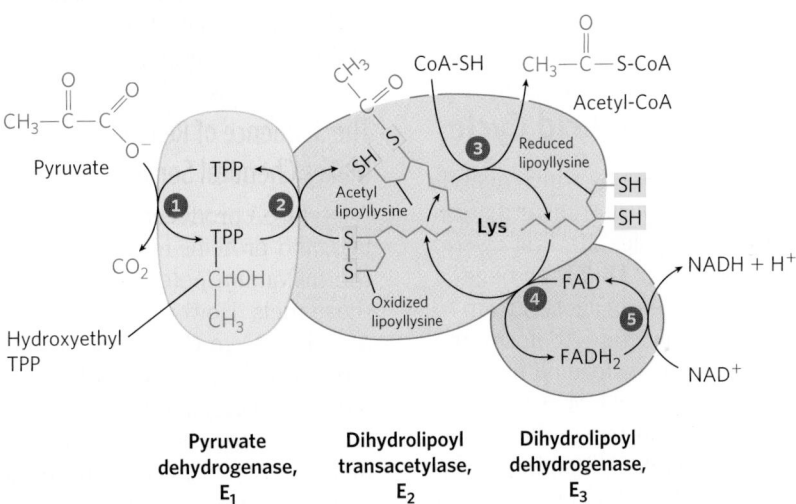

FIGURE 16-6 Oxidative decarboxylation of pyruvate to acetyl-CoA by the PDH complex. The fate of pyruvate is traced in red. In step ❶ pyruvate reacts with the bound thiamine pyrophosphate (TPP) of pyruvate dehydrogenase (E_1) and is decarboxylated to the hydroxyethyl derivative (see Fig. 14-15). Pyruvate dehydrogenase also carries out step ❷, the transfer of two electrons and the acetyl group from TPP to the oxidized form of the lipoyllysyl group of the core enzyme, dihydrolipoyl transacetylase (E_2), to form the acetyl thioester of the reduced lipoyl group. Step ❸ is a transesterification in which the —SH group of CoA replaces the —SH group of E_2 to yield acetyl-CoA and the fully reduced (dithiol) form of the lipoyl group. In step ❹ dihydrolipoyl dehydrogenase (E_3) promotes transfer of two hydrogen atoms from the reduced lipoyl groups of E_2 to the FAD prosthetic group of E_3, restoring the oxidized form of the lipoyllysyl group of E_2. In step ❺ the reduced $FADH_2$ of E_3 transfers a hydride ion to NAD^+, forming NADH. The enzyme complex is now ready for another catalytic cycle. (Subunit colors correspond to those in Fig. 16-5b.)

SUMMARY 16.1 Production of Acetyl-CoA (Activated Acetate)

■ Pyruvate, the product of glycolysis, is transported into the mitochondrial matrix by the mitochondrial pyruvate carrier.

■ Pyruvate is converted to acetyl-CoA, the starting material for the citric acid cycle, by the pyruvate dehydrogenase complex.

■ The PDH complex is composed of multiple copies of three enzymes: pyruvate dehydrogenase, E_1 (with its bound cofactor TPP); dihydrolipoyltransacetylase, E_2 (with its covalently bound lipoyl group); and dihydrolipoyl dehydrogenase, E_3 (with its cofactors FAD and NAD).

■ E_1 catalyzes first the decarboxylation of pyruvate, producing hydroxyethyl-TPP, and then the oxidation of the hydroxyethyl group to an acetyl group. The electrons from this oxidation reduce the disulfide of lipoate bound to E_2, and the acetyl group is transferred into thioester linkage with one —SH group of reduced lipoate.

■ E_2 catalyzes the transfer of the acetyl group to coenzyme A, forming acetyl-CoA.

■ E_3 catalyzes the regeneration of the disulfide (oxidized) form of lipoate; electrons pass first to FAD, then to NAD^+.

■ The long lipoyllysyl arm swings from the active site of E_1 to E_2 to E_3, tethering the intermediates to the enzyme complex to allow substrate channeling.

■ The organization of the PDH complex is very similar to that of the enzyme complexes that catalyze the oxidation of α-ketoglutarate and the branched-chain α-keto acids.

16.2 Reactions of the Citric Acid Cycle

We are now ready to trace the process by which acetyl-CoA undergoes oxidation. This chemical transformation is carried out by the citric acid cycle, the first *cyclic* pathway we have encountered **(Fig. 16-7)**. To begin a turn of the cycle, acetyl-CoA donates its acetyl group to the four-carbon compound oxaloacetate to form the six-carbon citrate. Citrate is then transformed into isocitrate, also a six-carbon molecule, which is dehydrogenated with loss of CO_2 to yield the five-carbon compound α-ketoglutarate (also called oxoglutarate). α-Ketoglutarate undergoes loss of a second molecule of CO_2 and ultimately yields the four-carbon compound succinate. Succinate is enzymatically converted in three steps into the four-carbon oxaloacetate—which is then ready to react with another molecule of acetyl-CoA. In each turn of the cycle, one acetyl group (two carbons) enters as acetyl-CoA and two molecules of CO_2 leave; one molecule of oxaloacetate is used to form citrate and one molecule of oxaloacetate is regenerated. No net removal of oxaloacetate occurs; one

molecule of oxaloacetate can theoretically bring about oxidation of an infinite number of acetyl groups, and, in fact, oxaloacetate is present in cells in very low concentrations. Four of the eight steps in this process are oxidations, in which the energy of oxidation is very efficiently conserved in the form of the reduced coenzymes NADH and $FADH_2$.

As noted earlier, although the citric acid cycle is central to energy-yielding metabolism its role is not limited to energy conservation. Four- and five-carbon intermediates of the cycle serve as precursors for a wide variety of products. To replace intermediates removed for this purpose, cells employ anaplerotic (replenishing) reactions, which are described below.

Eugene Kennedy and Albert Lehninger showed in 1948 that, in eukaryotes, the entire set of reactions of the citric acid cycle takes place in mitochondria. Isolated mitochondria were found to contain not only all the enzymes and coenzymes required for the citric acid cycle, but also all the enzymes and proteins necessary for the last stage of respiration—electron transfer and ATP synthesis by oxidative phosphorylation. As we shall see in later chapters, mitochondria also contain the enzymes for the oxidation of fatty acids and some amino acids to acetyl-CoA, and the oxidative degradation of other amino acids to α-ketoglutarate, succinyl-CoA, or oxaloacetate. Thus, in nonphotosynthetic eukaryotes, the mitochondrion is the site of most energy-yielding oxidative reactions and of the coupled synthesis of ATP. In photosynthetic eukaryotes, mitochondria are the major site of ATP production in the dark, but in daylight, chloroplasts produce most of the organism's ATP. In most bacteria, the enzymes of the citric acid cycle are in the cytosol, and the plasma membrane plays a role analogous to that of the inner mitochondrial membrane in ATP synthesis (Chapter 19).

The Sequence of Reactions in the Citric Acid Cycle Makes Chemical Sense

Acetyl-CoA produced in the breakdown of carbohydrates, fats, and proteins must be completely oxidized to CO_2 if the maximum potential energy is to be extracted from these fuels. However, the direct oxidation of acetate (or acetyl-CoA) to CO_2 is not biochemically feasible. Decarboxylation of this two-carbon acid would yield CO_2 and methane (CH_4). Methane is chemically rather stable, and except for certain methanotrophic bacteria that grow in methane-rich niches, organisms do not have the cofactors and enzymes needed to oxidize methane. Methylene groups (—CH_2—), however, are readily metabolized by enzyme systems present in most organisms. In typical oxidation sequences, two adjacent methylene groups (—CH_2—CH_2—) are involved, at least one of which is adjacent to a carbonyl group. As we noted in Chapter 13 (p. 503), carbonyl groups are particularly important in the chemical transformations of metabolic pathways. The carbon of the carbonyl group has a partial positive charge due to the electron-withdrawing property of the

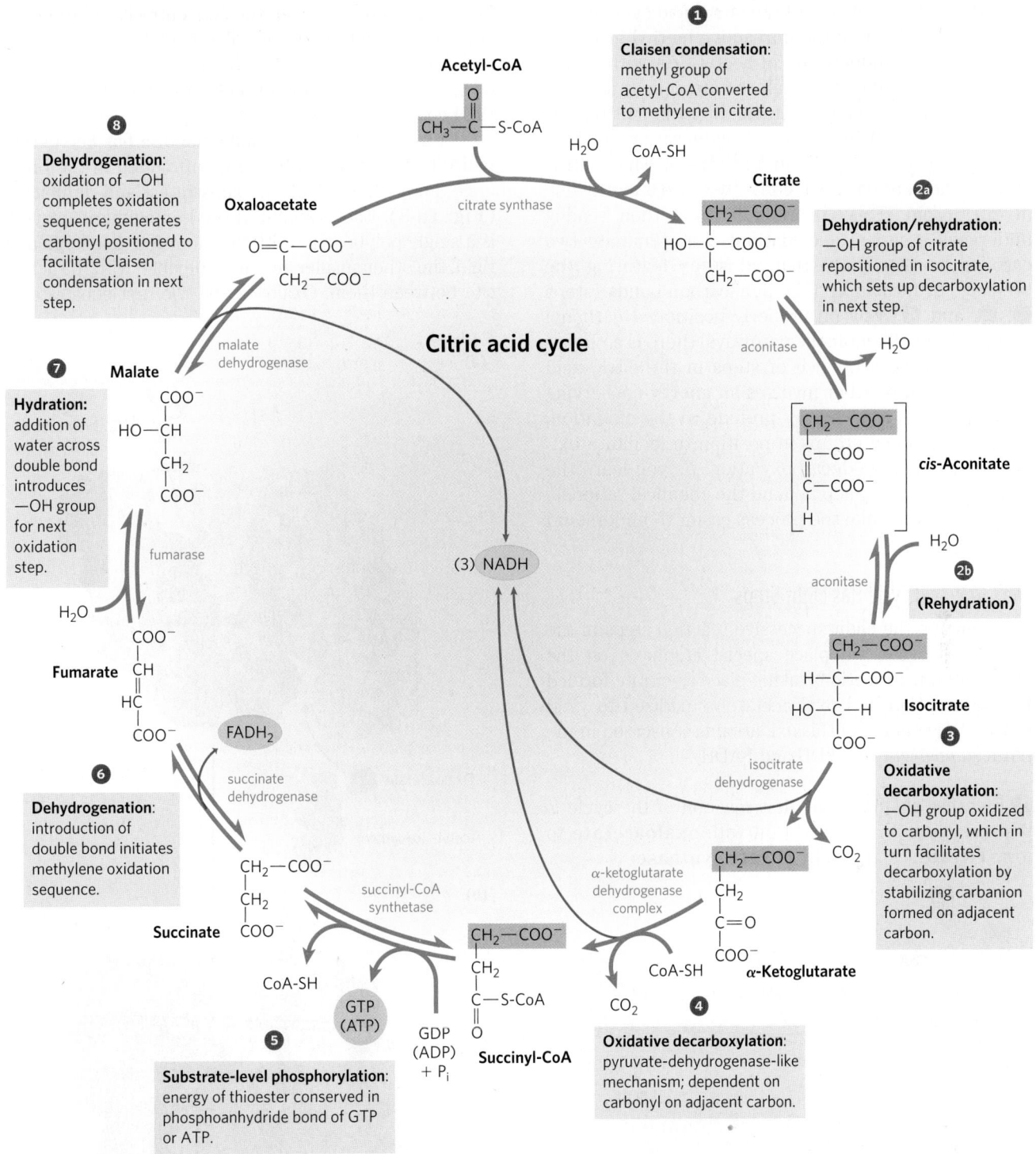

FIGURE 16-7 Reactions of the citric acid cycle. The carbon atoms shaded in pink are those derived from the acetate of acetyl-CoA in the first turn of the cycle; these are *not* the carbons released as CO_2 in the first turn. Note that in succinate and fumarate, the two-carbon group derived from acetate can no longer be specifically denoted; because succinate and fumarate are symmetric molecules, C-1 and C-2 are indistinguishable from C-4 and C-3.

The number beside each reaction step corresponds to a numbered heading on pages 626–633. The red arrows show where energy is conserved by electron transfer to FAD or NAD^+, forming $FADH_2$ or $NADH + H^+$. Steps ❶, ❸, and ❹ are essentially irreversible in the cell; all other steps are reversible. The nucleoside triphosphate product of step ❺ may be either ATP or GTP, depending on which succinyl-CoA synthetase isozyme is the catalyst.

carbonyl oxygen and is therefore an electrophilic center. A carbonyl group can facilitate the formation of a carbanion on an adjoining carbon by delocalizing the carbanion's negative charge. We see an example of the oxidation of a methylene group in the citric acid cycle, as succinate

is oxidized (steps ❻ to ❽ in Fig. 16-7) to form a carbonyl (in oxaloacetate) that is more chemically reactive than either a methylene group or methane.

In short, if acetyl-CoA is to be oxidized efficiently, the methyl group of the acetyl-CoA must be attached to something. The first step of the citric acid cycle neatly solves the problem of the unreactive methyl group by means of the condensation of acetyl-CoA with oxaloacetate. The carbonyl of oxaloacetate acts as an electrophilic center, which is attacked by the methyl carbon of acetyl-CoA in a Claisen condensation (p. 503) to form citrate (step ❶ in Fig. 16-7). The methyl group of acetate has been converted into a methylene in citric acid. This tricarboxylic acid then readily undergoes a series of oxidations that eliminate two carbons as CO_2. Note that all steps featuring the breakage or formation of carbon–carbon bonds (steps ❶, ❸, and ❹) rely on properly positioned carbonyl groups. As in all metabolic pathways, there is a chemical logic to the sequence of steps in the citric acid cycle: each step either involves an energy-conserving oxidation or is a necessary prelude to the oxidation, placing functional groups in position to facilitate oxidation or oxidative decarboxylation. As you learn the steps of the cycle, keep in mind the chemical rationale for each; it will make the process easier to understand and remember.

The Citric Acid Cycle Has Eight Steps

In examining the eight successive reaction steps of the citric acid cycle, we place special emphasis on the chemical transformations taking place as citrate formed from acetyl-CoA and oxaloacetate is oxidized to yield CO_2 and the energy of this oxidation is conserved in the reduced coenzymes NADH and $FADH_2$.

❶ **Formation of Citrate** The first reaction of the cycle is the condensation of acetyl-CoA with **oxaloacetate** to form **citrate**, catalyzed by **citrate synthase**:

$$\Delta G'^{\circ} = -32.2 \text{ kJ/mol}$$

In this reaction, the methyl carbon of the acetyl group is joined to the carbonyl group (C-2) of oxaloacetate. Citroyl-CoA is a transient intermediate formed on the active site of the enzyme (see Fig. 16-9). It rapidly undergoes hydrolysis to free CoA and citrate, which are released from the active site. The hydrolysis of this high-energy thioester intermediate makes the forward reaction highly exergonic. The large, negative standard free-energy change of the forward citrate synthase reaction is essential to the operation of the cycle

because, as noted earlier, the concentration of oxaloacetate is normally very low. The CoA liberated in this reaction is recycled to participate in the oxidative decarboxylation of another molecule of pyruvate by the PDH complex.

Citrate synthase from mitochondria has been crystallized and visualized by x-ray diffraction in the presence and absence of its substrates and inhibitors **(Fig. 16-8)**. Each subunit of the homodimeric enzyme is a single polypeptide with two domains, one large and rigid, the other smaller and more flexible, with the active site between them. Oxaloacetate, the first substrate to

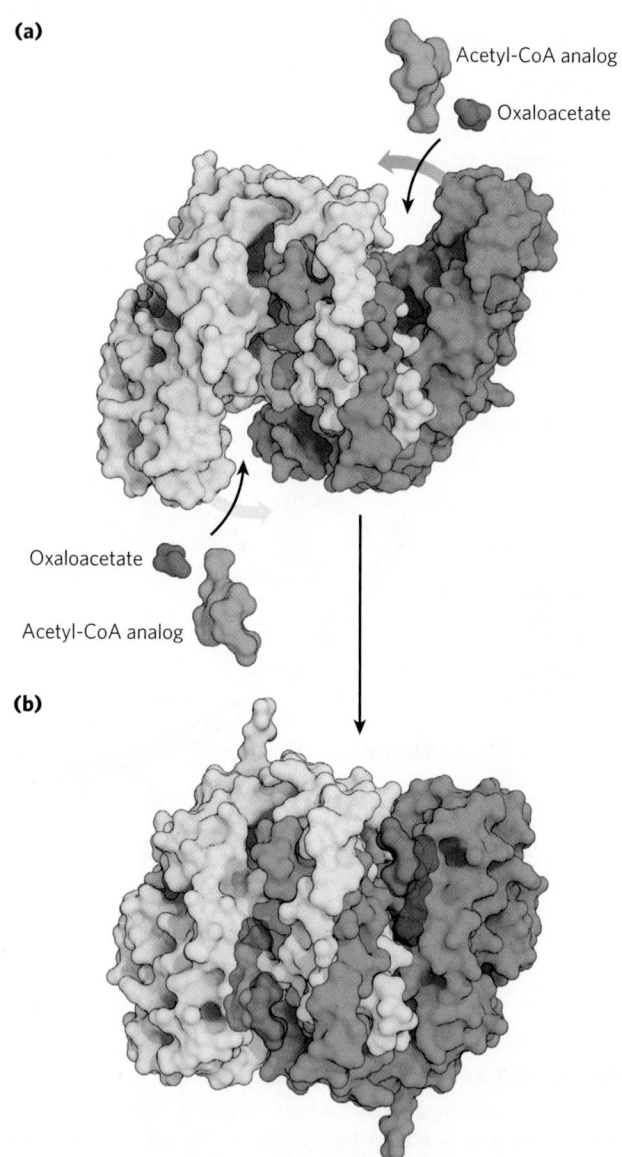

FIGURE 16-8 Structure of citrate synthase. The flexible domain of each subunit undergoes a large conformational change on binding oxaloacetate, creating a binding site for acetyl-CoA. **(a)** Open form of the enzyme alone; **(b)** closed form with bound oxaloacetate and a stable analog of acetyl-CoA (carboxymethyl-CoA). In these representations one subunit is colored tan and one green. [Sources: (a) PDB ID 5CSC, D.-I. Liao et al., *Biochemistry* 30:6031, 1991. (b) Derived from PDB ID 5CTS, M. Karpusas et al., *Biochemistry* 29:2213, 1990.]

bind to the enzyme, induces a large conformational change in the flexible domain, creating a binding site for the second substrate, acetyl-CoA. When citroyl-CoA has formed in the enzyme active site, another conformational change brings about thioester hydrolysis, releasing CoA-SH. This induced fit of the enzyme first to its substrate and then to its reaction intermediate decreases the likelihood of premature and unproductive cleavage of the thioester bond of acetyl-CoA. Kinetic studies of the enzyme are consistent with this ordered bisubstrate mechanism (see Fig. 6-13). The reaction catalyzed by citrate synthase is essentially a Claisen condensation (p. 503), involving a thioester (acetyl-CoA) and a ketone (oxaloacetate) **(Fig. 16-9)**.

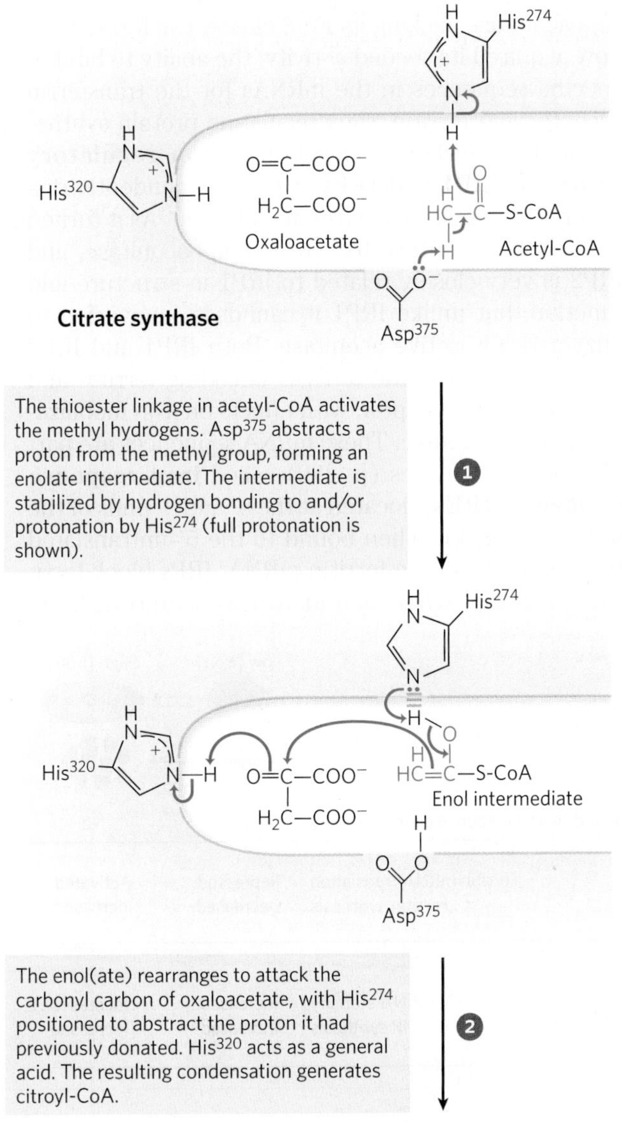
His274
His320
Oxaloacetate
Acetyl-CoA
Citrate synthase
Asp375
The thioester linkage in acetyl-CoA activates the methyl hydrogens. Asp375 abstracts a proton from the methyl group, forming an enolate intermediate. The intermediate is stabilized by hydrogen bonding to and/or protonation by His274 (full protonation is shown).
1
Enol intermediate
The enol(ate) rearranges to attack the carbonyl carbon of oxaloacetate, with His274 positioned to abstract the proton it had previously donated. His320 acts as a general acid. The resulting condensation generates citroyl-CoA.
2

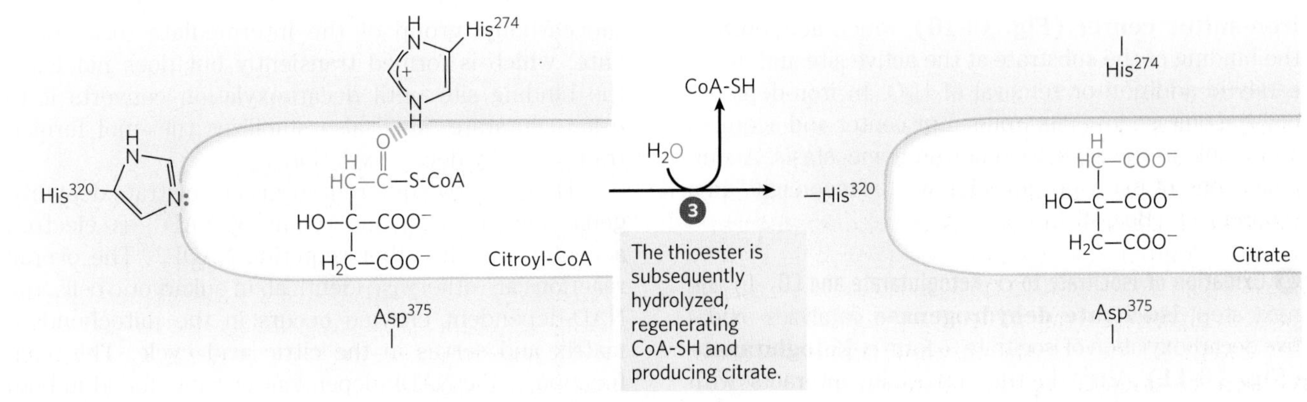
His274
His320
Citroyl-CoA
Asp375
CoA-SH
H2O
3
The thioester is subsequently hydrolyzed, regenerating CoA-SH and producing citrate.
Citrate

MECHANISM FIGURE 16-9 **Citrate synthase.** In the mammalian citrate synthase reaction, oxaloacetate binds first, in a strictly ordered reaction sequence. This binding triggers a conformation change that opens up the binding site for acetyl-CoA. Oxaloacetate is specifically oriented in the active site of citrate synthase by interaction of its two carboxylates with two positively charged Arg residues (not shown here). [Source: Information from S. J. Remington, *Curr. Opin. Struct. Biol.* 2:730, 1992.]

2 **Formation of Isocitrate via *cis*-Aconitate** The enzyme **aconitase** (more formally, **aconitate hydratase**) catalyzes the reversible transformation of citrate to **isocitrate**, through the intermediary formation of the tricarboxylic acid ***cis*-aconitate**, which normally does not dissociate from the active site. Aconitase can promote the reversible addition of H_2O to the double bond of enzyme-bound *cis*-aconitate in two different ways, one leading to citrate and the other to isocitrate:

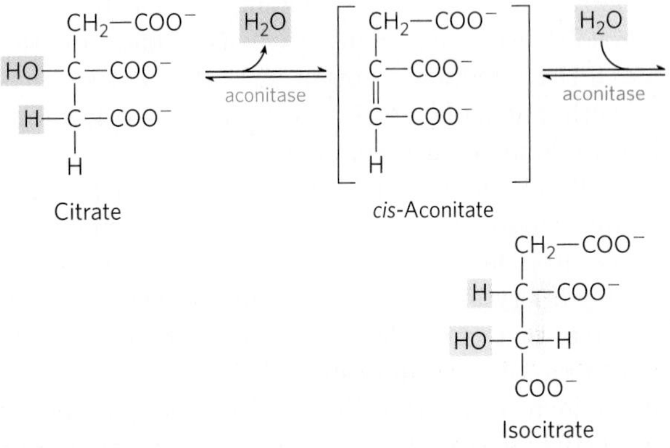
H2O
aconitase
Citrate
cis-Aconitate
Isocitrate

$$\Delta G'^{\circ} = 13.3 \text{ kJ/mol}$$

Although the equilibrium mixture at pH 7.4 and 25 °C contains less than 10% isocitrate, in the cell the reaction is pulled to the right because isocitrate is immediately consumed in the next step of the cycle, lowering its steady-state concentration. Aconitase contains an

BOX 16-1 Moonlighting Enzymes: Proteins with More Than One Job

The "one gene–one enzyme" dictum, put forward by George Beadle and Edward Tatum in 1940 (see Chapter 24), went unchallenged for much of the twentieth century, as did the associated assumption that each protein had only one role. But in recent years, many striking exceptions to this simple formula have been discovered—cases in which a single protein encoded by a single gene clearly is **moonlighting**, doing more than one job in the cell. Aconitase is one such protein: it acts both as an enzyme and as a regulator of protein synthesis.

Eukaryotic cells have two isozymes of aconitase. The mitochondrial isozyme converts citrate to isocitrate in the citric acid cycle. The cytosolic isozyme has two distinct functions. It catalyzes the conversion of citrate to isocitrate, providing the substrate for a cytosolic isocitrate dehydrogenase that generates NADPH as reducing power for fatty acid synthesis and other anabolic processes in the cytosol. But it also has a role in cellular iron homeostasis.

All cells must obtain iron for the activity of the many proteins that require it as a cofactor. In humans, severe iron deficiency results in anemia, an insufficient supply of erythrocytes and a reduced oxygen-carrying capacity that can be life-threatening. Too much iron is also harmful: it accumulates in and damages the liver in hemochromatosis and other diseases. Iron obtained in the diet is carried in the blood by the protein **transferrin** and enters cells via endocytosis mediated by the **transferrin receptor**. Once inside cells, iron is used in the synthesis of hemes, cytochromes, Fe-S proteins, and other Fe-dependent proteins, and excess iron is stored bound to the protein **ferritin**. The levels of transferrin, transferrin receptor, and ferritin are

therefore crucial to cellular iron homeostasis. The synthesis of these three proteins is regulated in response to iron availability—and aconitase, in its moonlighting job, plays a key regulatory role.

Aconitase has an essential Fe-S cluster at its active site (see Fig. 16-10). When a cell is depleted of iron, this Fe-S cluster is disassembled and the enzyme loses its aconitase activity. But the apoenzyme (apoaconitase, lacking its Fe-S cluster) so formed has now acquired its second activity: the ability to bind to specific sequences in the mRNAs for the transferrin receptor and ferritin, thus regulating protein synthesis at the translational level. Two **iron regulatory proteins**, **IRP1** and **IRP2**, were independently discovered as regulators of iron metabolism. As it turned out, IRP1 is identical to cytosolic apoaconitase, and IRP2 is very closely related to IRP1 in structure and function, but unlike IRP1 it cannot be converted to enzymatically active aconitase. Both IRP1 and IRP2 bind to regions in the mRNAs encoding ferritin and the transferrin receptor, with effects on iron mobilization and iron uptake. These mRNA sequences are part of hairpin structures (p. 289) called **iron response elements** (**IREs**), located at the 5′ and 3′ ends of the mRNAs (Fig. 1). When bound to the 5′-untranslated IRE sequence in the ferritin mRNA, IRPs block ferritin synthesis; when bound to the 3′-untranslated

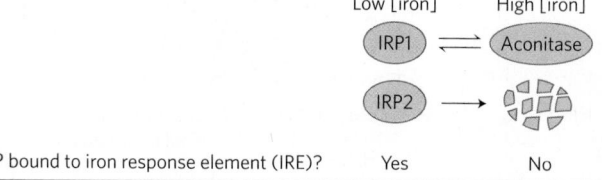

	Low [iron]	High [iron]
IRP bound to iron response element (IRE)?	Yes	No
Ferritin mRNA translation	Repressed	Activated
Ferritin synthesis	Decreased	Increased
TfR mRNA stability	Increased	Decreased
TfR synthesis	Increased	Decreased

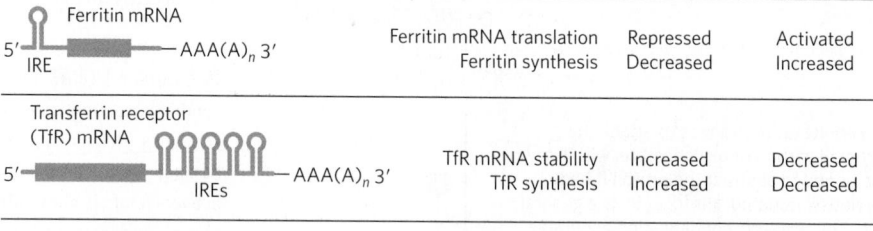

FIGURE 1 Effect of IRP1 and IRP2 on the mRNAs for ferritin and the transferrin receptor. [Source: Information from R. S. Eisenstein, *Annu. Rev. Nutr.* 20:627, 2000, Fig. 1.]

iron-sulfur center (Fig. 16-10), which acts both in the binding of the substrate at the active site and in the catalytic addition or removal of H_2O. In iron-depleted cells, aconitase loses its iron-sulfur center and acquires a new role in the regulation of iron homeostasis. Aconitase is one of many enzymes known to "moonlight" in a second role (Box 16-1).

❸ **Oxidation of Isocitrate to α-Ketoglutarate and CO_2** In the next step, **isocitrate dehydrogenase** catalyzes oxidative decarboxylation of isocitrate to form **α-ketoglutarate** (**Fig. 16-11**). Mn^{2+} in the active site interacts with

the carbonyl group of the intermediate oxalosuccinate, which is formed transiently but does not leave the binding site until decarboxylation converts it to α-ketoglutarate. Mn^{2+} also stabilizes the enol formed transiently by decarboxylation.

There are two different forms of isocitrate dehydrogenase in all cells, one requiring NAD^+ as electron acceptor and the other requiring $NADP^+$. The overall reactions are otherwise identical. In eukaryotic cells, the NAD-dependent enzyme occurs in the mitochondrial matrix and serves in the citric acid cycle. The main function of the NADP-dependent enzyme, found in both

IRE sequence in the transferrin receptor mRNA, they stabilize the mRNA, preventing its degradation and thus allowing the synthesis of more copies of the receptor protein per mRNA molecule. So, in iron-deficient cells, iron uptake becomes more efficient and iron storage (bound to ferritin) is reduced. When cellular iron concentrations return to normal levels, IRP1 is converted to aconitase, and IRP2 undergoes proteolytic degradation, ending the low-iron response.

The enzymatically active aconitase and the moonlighting, regulatory apoaconitase have different structures. As the active aconitase, the protein has two lobes that close around the Fe-S cluster; as IRP1, the two lobes open, exposing the mRNA-binding site (Fig. 2).

Aconitase is just one of a growing list of enzymes known (or believed) to moonlight in a second role. Many of the glycolytic enzymes are included in this group. Pyruvate kinase acts in the nucleus to regulate the transcription of genes that respond to thyroid hormone. Glyceraldehyde 3-phosphate dehydrogenase moonlights both as uracil DNA glycosylase, effecting the repair of damaged DNA, and as a regulator of histone H2B transcription. The crystallins in the lens of the vertebrate eye are several moonlighting glycolytic enzymes, including phosphoglycerate kinase, triose phosphate isomerase, and lactate dehydrogenase. This phenomenon is sometimes called gene sharing: two different proteins, with different functional roles, are encoded by a single gene. Beadle and Tatum's "one gene–one enzyme" dictum, like so many other propositions in science, is only approximately true.

Until recently, the discovery that a protein has more than one function was largely a matter of serendipity: two groups of investigators studying two unrelated questions discovered that "their" proteins had similar properties, compared them carefully, and found them to be identical. With the growth of annotated protein and DNA sequence databases, researchers can now deliberately look for moonlighting proteins by searching the databases for any other protein with the same sequence as the one under study, but with a different function.

This also means that in the databases, a protein annotated as having a given function doesn't necessarily have *only* that function. Protein moonlighting may also explain some puzzling findings: experiments in which a protein with a known function is made inactive by a mutation, and the resulting mutant organisms show a phenotype with no obvious relation to that function.

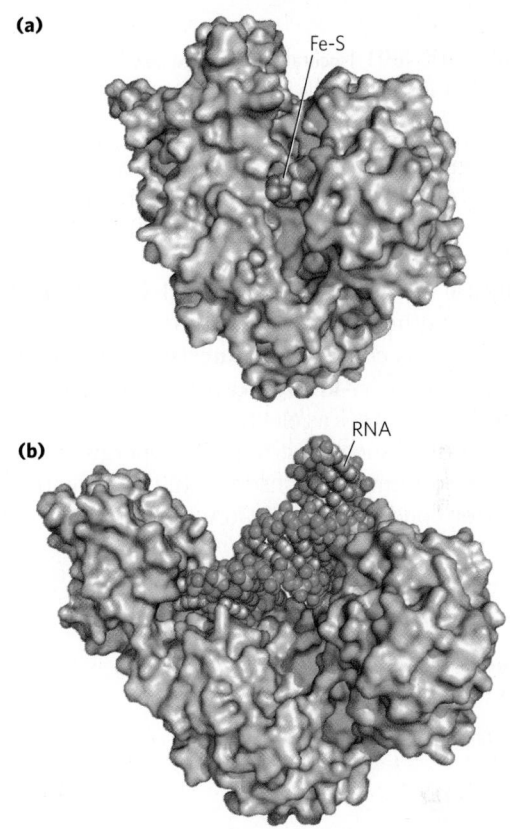

FIGURE 2 Two forms of cytosolic aconitase/IRP1 with two distinct functions. **(a)** In aconitase, the two major lobes are closed and the Fe-S cluster is buried; the protein has been made transparent here to show the Fe-S cluster. **(b)** In IRP1, the lobes open up, exposing a binding site for the mRNA hairpin of the substrate. [Sources: (a) PDB ID 2B3Y, J. Dupuy et al., *Structure* 14:129, 2006. (b) PDB ID 2IPY, W. E. Walden et al., *Science* 314:1903, 2006.]

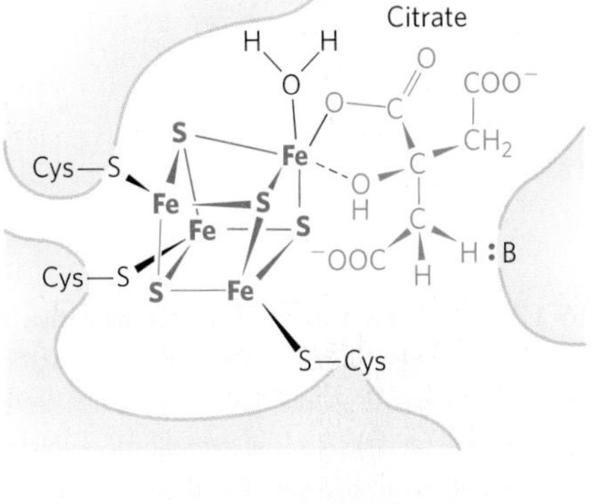

FIGURE 16-10 Iron-sulfur center in aconitase. The iron-sulfur center is in red, the citrate molecule in blue. Three Cys residues of the enzyme bind three iron atoms; the fourth iron is bound to one of the carboxyl groups of citrate and also interacts noncovalently with a hydroxyl group of citrate (dashed bond). A basic residue (:B) in the enzyme helps to position the citrate in the active site. The iron-sulfur center acts in both substrate binding and catalysis. The general properties of iron-sulfur proteins are discussed in Chapter 19 (see Fig. 19-5).

MECHANISM FIGURE 16-11 Isocitrate dehydrogenase. In this reaction, the substrate, isocitrate, loses one carbon by oxidative decarboxylation. See Figure 14-14 for more information on hydride transfer reactions involving NAD^+ and $NADP^+$.

Step ① Isocitrate is oxidized by hydride transfer to NAD^+ or $NADP^+$ (depending on the isocitrate dehydrogenase isozyme).

Step ② Decarboxylation is facilitated by electron withdrawal by the adjacent carbonyl and coordinated Mn^{2+}.

Step ③ Rearrangement of the enol intermediate generates α-ketoglutarate.

(Isocitrate → Oxalosuccinate → α-Ketoglutarate)

the mitochondrial matrix and the cytosol, may be the generation of NADPH, which is essential for reductive anabolic pathways such as fatty acid and sterol synthesis.

④ Oxidation of α-Ketoglutarate to Succinyl-CoA and CO_2 The next step is another oxidative decarboxylation, in which α-ketoglutarate is converted to **succinyl-CoA** and CO_2 by the action of the **α-ketoglutarate dehydrogenase complex**; NAD^+ serves as electron acceptor and CoA as the carrier of the succinyl group. The energy of oxidation of α-ketoglutarate is conserved in the formation of the thioester bond of succinyl-CoA:

(α-Ketoglutarate + CoA-SH + NAD^+ → Succinyl-CoA + CO_2 + NADH, α-ketoglutarate dehydrogenase complex)

$$\Delta G'^\circ = -33.5 \text{ kJ/mol}$$

This reaction is virtually identical to the pyruvate dehydrogenase reaction discussed above and to the reaction sequence responsible for the breakdown of branched-chain amino acids **(Fig. 16-12)**. The α-ketoglutarate dehydrogenase complex closely resembles the PDH complex in both structure and function. It includes three enzymes, homologous to E_1, E_2, and E_3 of the PDH complex, as well as enzyme-bound TPP, bound lipoate, FAD, NAD, and coenzyme A. Both complexes are certainly derived from a common evolutionary ancestor. Although the E_1 components of the two complexes are structurally similar, their amino acid sequences differ and, of course, they have different binding specificities: E_1 of the PDH complex binds pyruvate, and E_1 of the α-ketoglutarate dehydrogenase complex binds α-ketoglutarate. The E_2 components of the two complexes are also very similar, both having covalently bound lipoyl moieties. The subunits of E_3 are identical in the two enzyme complexes. The complex that degrades branched-chain α-keto acids catalyzes the same reaction sequence using the same five cofactors. This is a clear case of gene duplication and subsequent **divergent evolution**, as described in Figure 1-34. For example, if the PDH complex evolved first, duplication of the genes encoding E_1 and E_2 would have provided "extra copies" that could

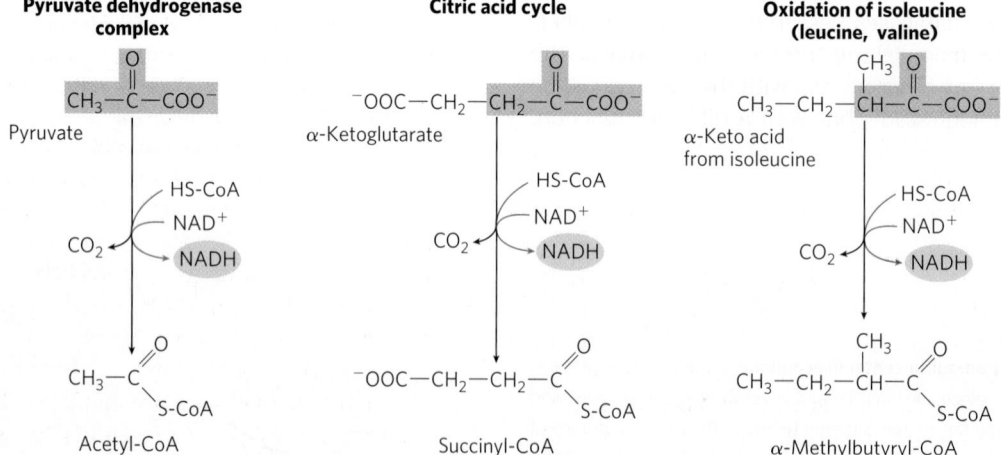

FIGURE 16-12 A conserved mechanism for oxidative decarboxylation. The pathways shown employ the same five cofactors (thiamine pyrophosphate, coenzyme A, lipoate, FAD, and NAD^+), closely similar multienzyme complexes, and the same enzymatic mechanism to carry out oxidative decarboxylations of pyruvate (by the pyruvate dehydrogenase complex), α-ketoglutarate (in the citric acid cycle), and the carbon skeletons of the three branched-chain amino acids, isoleucine (shown here), leucine, and valine. A fourth reaction, catalyzed by glycine decarboxylase, involves a very similar mechanism (see Fig. 20-49).

mutate over time. Mutations eventually led to an enzyme using the same mechanism for oxidative decarboxylation, but with a new substrate specificity (for α-ketoglutarate) that offered a selective advantage. These related enzymes can employ the same E_3 subunit because the substrate for E_3—a reduced lipoate—is the same for both complexes.

⑤ Conversion of Succinyl-CoA to Succinate Succinyl-CoA, like acetyl-CoA, has a thioester bond with a strongly negative standard free energy of hydrolysis ($\Delta G'^\circ \approx$ −36 kJ/mol). In the next step of the citric acid cycle, energy released in the breakage of this bond is used to drive the synthesis of a phosphoanhydride bond in either GTP or ATP, with a net $\Delta G'^\circ$ of only −2.9 kJ/mol. **Succinate** is formed in the process:

$$\Delta G'^\circ = -2.9 \text{ kJ/mol}$$

The enzyme that catalyzes this reversible reaction is called **succinyl-CoA synthetase** or **succinic thiokinase**; both names indicate the participation of a nucleoside triphosphate in the reaction (Box 16-2).

BOX 16-2 Synthases and Synthetases; Ligases and Lyases; Kinases, Phosphatases, and Phosphorylases: Yes, the Names Are Confusing!

Citrate synthase is one of many enzymes that catalyze condensation reactions, yielding a product more chemically complex than its precursors. **Synthases** catalyze condensation reactions in which no nucleoside triphosphate (ATP, GTP, and so forth) is required as an energy source. **Synthetases** catalyze condensations that *do* use ATP or another nucleoside triphosphate as a source of energy for the synthetic reaction. Succinyl-CoA synthetase is such an enzyme. **Ligases** (from the Latin *ligare*, "to tie together") are enzymes that catalyze condensation reactions in which two atoms are joined, using ATP or another energy source. (Thus synthetases are ligases.) DNA ligase, for example, closes breaks in DNA molecules, using energy supplied by either ATP or NAD^+; it is widely used in joining DNA pieces for genetic engineering. Ligases are not to be confused with **lyases**, enzymes that catalyze cleavages (or, in the reverse direction, additions) in which electronic rearrangements occur. The PDH complex, which oxidatively cleaves CO_2 from pyruvate, is a member of the large class of lyases.

The name **kinase** is applied to enzymes that transfer a phosphoryl group from a nucleoside triphosphate such as ATP to an acceptor molecule—a sugar (as in hexokinase and glucokinase), a protein (as in glycogen phosphorylase kinase), another nucleotide (as in nucleoside diphosphate kinase), or a metabolic intermediate such as oxaloacetate (as in PEP carboxykinase). The reaction catalyzed by a kinase is a *phosphorylation*. On the other hand, *phosphorolysis* is a displacement reaction in which phosphate is the attacking species and becomes covalently attached at the point of bond breakage. Such reactions are catalyzed by **phosphorylases**. Glycogen phosphorylase, for example, catalyzes the phosphorolysis of glycogen, producing glucose 1-phosphate. *Dephosphorylation*, the removal of a phosphoryl group from a phosphate ester, is catalyzed by **phosphatases**, with water as the attacking species. Fructose bisphosphatase-1 converts fructose 1,6-bisphosphate to fructose 6-phosphate in gluconeogenesis, and phosphorylase a phosphatase removes phosphoryl groups from phosphoserine in phosphorylated glycogen phosphorylase. Whew!

Unfortunately, these descriptions of enzyme types overlap, and many enzymes are commonly called by two or more names. Succinyl-CoA synthetase, for example, is also called succinate thiokinase; the enzyme is both a synthetase in the citric acid cycle and a kinase when acting in the direction of succinyl-CoA synthesis. This raises another source of confusion in the naming of enzymes. An enzyme may have been discovered by the use of an assay in which, say, A is converted to B. The enzyme is then named for that reaction. Later work may show, however, that in the cell, the enzyme functions primarily in converting B to A. Commonly, the first name continues to be used, although the metabolic role of the enzyme would be better described by naming it for the reverse reaction. The glycolytic enzyme pyruvate kinase illustrates this situation (p. 544). To a beginner in biochemistry, this duplication in nomenclature can be bewildering. International committees have made heroic efforts to systematize the nomenclature of enzymes (see Table 6-3 for a brief summary of the system), but some systematic names have proved too long and cumbersome and are not frequently used in biochemical conversation.

We have tried throughout this book to use the enzyme name most commonly employed by working biochemists and to point out cases in which an enzyme has more than one widely used name. For current information on enzyme nomenclature, refer to the recommendations of the Nomenclature Committee of the International Union of Biochemistry and Molecular Biology (www.chem.qmul.ac.uk/iubmb/nomenclature).

This energy-conserving reaction involves an intermediate step in which the enzyme molecule itself becomes phosphorylated at a His residue in the active site **(Fig. 16-13a)**. This phosphoryl group, which has a high group transfer potential, is transferred to ADP (or GDP) to form ATP (or GTP). Animal cells have two isozymes of succinyl-CoA synthetase, one specific for ADP and the other for GDP. The enzyme has two subunits, α (M_r 32,000), which has the $\textcircled{P}$–His residue (His^{246}) and

the binding site for CoA, and β (M_r 42,000), which confers specificity for either ADP or GDP. The active site is at the interface between subunits. The crystal structure of succinyl-CoA synthetase reveals two "power helices" (one from each subunit), oriented so that their electric dipoles situate partial positive charges close to the negatively charged $\textcircled{P}$–His (Fig. 16-13b), stabilizing the phosphoenzyme intermediate. (Recall the similar role of helix dipoles in stabilizing K^+ ions in the K^+ channel; see Fig. 11-45.)

The formation of ATP (or GTP) at the expense of the energy released by the oxidative decarboxylation of α-ketoglutarate is a substrate-level phosphorylation, like the synthesis of ATP in the glycolytic reactions catalyzed by phosphoglycerate kinase and pyruvate kinase (see Fig. 14-2). The GTP formed by succinyl-CoA synthetase can donate its terminal phosphoryl group to ADP to form ATP, in a reversible reaction catalyzed by **nucleoside diphosphate kinase** (p. 516):

$$GTP + ADP \rightleftharpoons GDP + ATP \qquad \Delta G'^{\circ} = 0 \text{ kJ/mol}$$

Thus the net result of the activity of either isozyme of succinyl-CoA synthetase is the conservation of energy as ATP. There is no change in free energy for the nucleoside diphosphate kinase reaction; ATP and GTP are energetically equivalent.

❻ Oxidation of Succinate to Fumarate The succinate formed from succinyl-CoA is oxidized to **fumarate** by the flavoprotein **succinate dehydrogenase**:

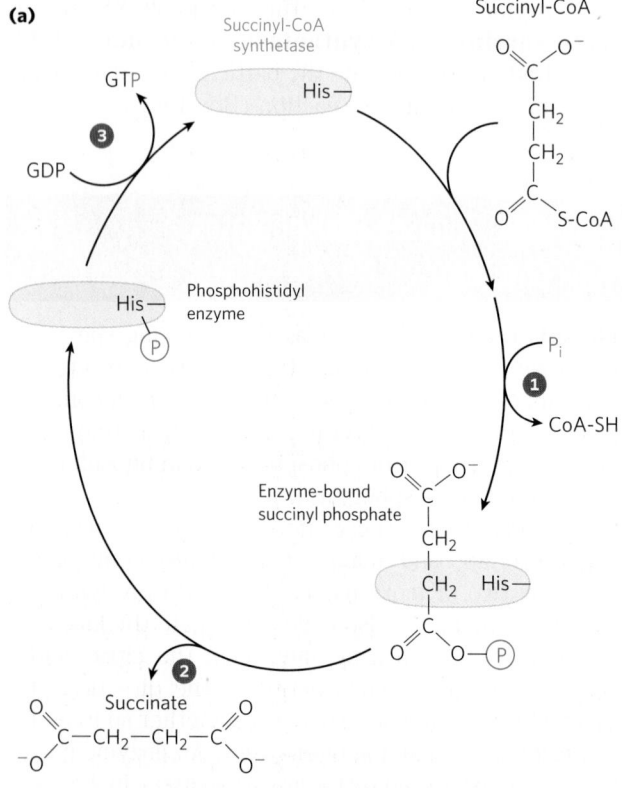

(a)

$$\Delta G'^{\circ} = 0 \text{ kJ/mol}$$

In eukaryotes, succinate dehydrogenase is an integral protein of the mitochondrial inner membrane; in bacteria, of the plasma membrane. The enzyme contains three different iron-sulfur clusters and one molecule of covalently bound FAD (see Fig. 19-9). Electrons pass from succinate through the FAD and iron-sulfur centers before entering the chain of electron carriers in the

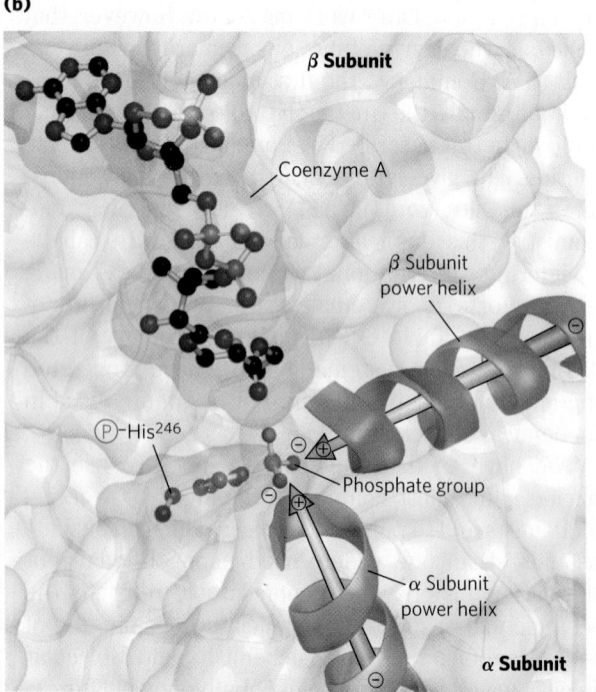

FIGURE 16-13 The succinyl-CoA synthetase reaction. (a) In step ❶ a phosphoryl group replaces the CoA of succinyl-CoA bound to the enzyme, forming a high-energy acyl phosphate. In step ❷ the succinyl phosphate donates its phosphoryl group to a His residue of the enzyme, forming a high-energy phosphohistidyl enzyme. In step ❸ the phosphoryl group is transferred from the His residue to the terminal phosphate of GDP (or ADP), forming GTP (or ATP). **(b)** Active site of succinyl-CoA synthetase of *E. coli*. The active site includes part of both the α (blue) and the β (brown) subunits. The power helices (blue, brown) place the partial positive charges of the helix dipole near the phosphate group of $\textcircled{P}$–His^{246} in the α chain, stabilizing the phosphohistidyl enzyme. The bacterial and mammalian enzymes have similar amino acid sequences and three-dimensional structures. [Source: Derived from PDB ID 1SCU, W. T. Wolodko et al., *J. Biol. Chem.* 269:10,883, 1994.]

mitochondrial inner membrane (the plasma membrane in bacteria). Electron flow from succinate through these carriers to the final electron acceptor, O_2, is coupled to the synthesis of about 1.5 ATP molecules per pair of electrons (respiration-linked phosphorylation). Malonate, an analog of succinate not normally present in cells, is a strong competitive inhibitor of succinate dehydrogenase, and its addition to mitochondria in the laboratory blocks the activity of the citric acid cycle.

Malonate   Succinate

7 Hydration of Fumarate to Malate The reversible hydration of fumarate to **L-malate** is catalyzed by **fumarase** (formally, **fumarate hydratase**). The transition state in this reaction is a carbanion:

Fumarate   Carbanion
                             transition state

L-Malate

$$\Delta G'^\circ = -3.8 \text{ kJ/mol}$$

This enzyme is highly stereospecific; it catalyzes hydration of the trans double bond of fumarate but not the cis double bond of maleate. In the reverse direction (from L-malate to fumarate), fumarase is equally stereospecific: D-malate is not a substrate.

Fumarate   Maleate

L-Malate   D-Malate

8 Oxidation of Malate to Oxaloacetate In the last reaction of the citric acid cycle, **L-malate dehydrogenase** catalyzes the oxidation of L-malate to oxaloacetate, coupled to the reduction of NAD^+ to NADH:

L-Malate   Oxaloacetate

$$\Delta G'^\circ = 29.7 \text{ kJ/mol}$$

The equilibrium of this reaction lies far to the left under standard thermodynamic conditions, but in intact cells oxaloacetate is continually removed by the highly exergonic citrate synthase reaction (step **2** of Fig. 16-7). This keeps the concentration of oxaloacetate in the cell extremely low ($<10^{-6}$ M), pulling the malate dehydrogenase reaction toward the formation of oxaloacetate.

Although the individual reactions of the citric acid cycle were initially worked out in vitro, using minced muscle tissue, the pathway and its regulation have also been studied extensively in vivo. By using precursors isotopically labeled with ^{14}C, researchers have traced the fate of individual carbon atoms through the citric acid cycle. Some of the earliest experiments with ^{14}C produced an unexpected result, however, which aroused considerable controversy about the pathway and mechanism of the citric acid cycle. In fact, these experiments at first seemed to show that citrate was not the first tricarboxylic acid to be formed. Box 16-3 gives some details of this episode in the history of citric acid cycle research. Metabolic flux through the cycle can now be monitored in living tissue by using ^{13}C-labeled precursors and whole-tissue NMR spectroscopy. Because the NMR signal is unique to the compound containing the ^{13}C, biochemists can trace the movement of precursor carbons into each cycle intermediate and into compounds derived from the intermediates. This technique has great promise for studies of regulation of the citric acid cycle and its interconnections with other metabolic pathways such as glycolysis.

The Energy of Oxidations in the Cycle Is Efficiently Conserved

We have now covered one complete turn of the citric acid cycle **(Fig. 16-14)**. A two-carbon acetyl group entered the cycle by combining with oxaloacetate. Two carbon atoms emerged from the cycle as CO_2 from the oxidation of isocitrate and α-ketoglutarate. The energy released by these oxidations was conserved in the reduction of three NAD^+ and one FAD and the production of one ATP or GTP. At the end of the cycle a molecule of oxaloacetate was regenerated. Note that the two carbon atoms appearing as CO_2 are not the same

BOX 16-3 Citrate: A Symmetric Molecule That Reacts Asymmetrically

When compounds enriched in the heavy-carbon isotope ^{13}C and the radioactive carbon isotopes ^{11}C and ^{14}C became available about 60 years ago, they were soon put to use in tracing the pathway of carbon atoms through the citric acid cycle. One such experiment initiated the controversy over the role of citrate. Acetate labeled in the carboxyl group (designated [1-^{14}C] acetate) was incubated aerobically with an animal tissue preparation. Acetate is enzymatically converted to acetyl-CoA in animal tissues, and the pathway of the labeled carboxyl carbon of the acetyl group in the cycle reactions could thus be traced. α-Ketoglutarate was isolated from the tissue after incubation, then degraded by known chemical reactions to establish the position(s) of the isotopic carbon.

Condensation of unlabeled oxaloacetate with carboxyl-labeled acetate would be expected to produce citrate labeled in one of the two primary carboxyl groups. Citrate is a symmetric molecule, its two terminal carboxyl groups being chemically indistinguishable. Therefore, half the labeled citrate molecules were expected to yield α-ketoglutarate labeled in the α-carboxyl group and the other half to yield α-ketoglutarate labeled in the γ-carboxyl group; that is, the α-ketoglutarate isolated was expected to be a mixture of the two types of labeled molecules (Fig. 1,

pathways ❶ and ❷). Contrary to this expectation, the labeled α-ketoglutarate isolated from the tissue suspension contained ^{14}C only in the γ-carboxyl group (Fig. 1, pathway ❶). The investigators concluded that citrate (or any other symmetric molecule) could not be an intermediate in the pathway from acetate to α-ketoglutarate. Rather, an asymmetric tricarboxylic acid, presumably *cis*-aconitate or isocitrate, must be the first product formed from condensation of acetate and oxaloacetate.

In 1948, however, Alexander Ogston pointed out that although citrate has no chiral center (see Fig. 1-21), it has the *potential* to react asymmetrically if an enzyme with which it interacts has an active site that is asymmetric. He suggested that the active site of aconitase may have three points to which the citrate must be bound and that the citrate must undergo a specific three-point attachment to these binding points. As seen in Figure 2, the binding of citrate to three such points could happen in only one way, and this would account for the formation of only one type of labeled α-ketoglutarate. Organic molecules such as citrate that have no chiral center but are potentially capable of reacting asymmetrically with an asymmetric active site are now called **prochiral molecules**.

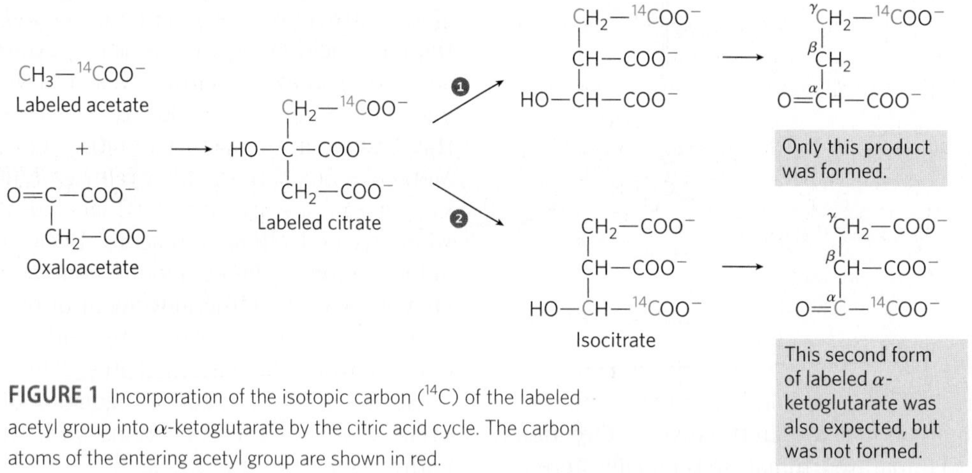

FIGURE 1 Incorporation of the isotopic carbon (^{14}C) of the labeled acetyl group into α-ketoglutarate by the citric acid cycle. The carbon atoms of the entering acetyl group are shown in red.

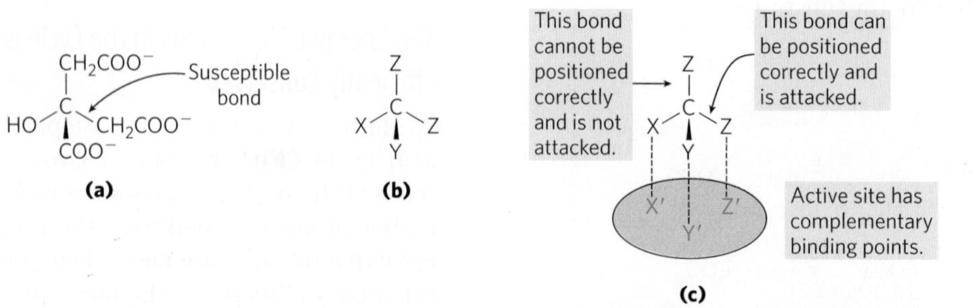

FIGURE 2 The prochiral nature of citrate. (a) Structure of citrate; (b) schematic representation of citrate: X = —OH; Y = —COO$^-$; Z = —CH$_2$COO$^-$. (c) Correct complementary fit of citrate to the binding

site of aconitase. There is only one way in which the three specified groups of citrate can fit on the three points of the binding site. Thus only one of the two —CH$_2$COO$^-$ groups is bound by aconitase.

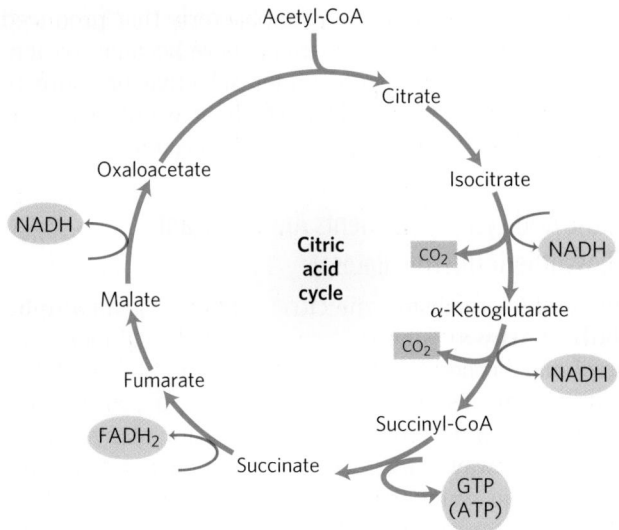

FIGURE 16-14 Products of one turn of the citric acid cycle. At each turn of the cycle, three NADH, one FADH$_2$, one GTP (or ATP), and two CO$_2$ are released in oxidative decarboxylation reactions. Here and in several of the following figures, all cycle reactions are shown as proceeding in one direction only, but keep in mind that most of the reactions are reversible (see Fig. 16-7).

two carbons that entered in the form of the acetyl group; additional turns around the cycle are required to release these carbons as CO$_2$ (Fig. 16-7).

Although the citric acid cycle directly generates only one ATP per turn (in the conversion of succinyl-CoA to succinate), the four oxidation steps in the cycle provide a large flow of electrons into the respiratory chain via NADH and FADH$_2$ and thus lead to formation

of a large number of ATP molecules during oxidative phosphorylation.

We saw in Chapter 14 that the production of two molecules of pyruvate from one molecule of glucose in glycolysis yields 2 ATP and 2 NADH. In oxidative phosphorylation (Chapter 19), passage of two electrons from NADH to O$_2$ drives the formation of about 2.5 ATP, and passage of two electrons from FADH$_2$ to O$_2$ yields about 1.5 ATP. This stoichiometry allows us to calculate the overall yield of ATP from the complete oxidation of glucose. When both pyruvate molecules are oxidized to 6 CO$_2$ via the pyruvate dehydrogenase complex and the citric acid cycle, and the electrons are transferred to O$_2$ via oxidative phosphorylation, as many as 32 ATP are obtained per glucose (Table 16-1). In round numbers, this represents the conservation of 32 × 30.5 kJ/mol = 976 kJ/mol, or 34% of the theoretical maximum of about 2,840 kJ/mol available from the complete oxidation of glucose. These calculations employ the standard free-energy changes; when corrected for the actual free energy required to form ATP within cells (see Worked Example 13-2, p. 509), the calculated efficiency of the process is closer to 65%.

Why Is the Oxidation of Acetate So Complicated?

The eight-step cyclic process for oxidation of simple two-carbon acetyl groups to CO$_2$ may seem unnecessarily cumbersome and not in keeping with the biological principle of maximum economy. The role of the citric acid cycle is not confined to the oxidation of acetate, however. This pathway is the hub of intermediary metabolism. Four- and five-carbon end products of

TABLE 16-1 Stoichiometry of Coenzyme Reduction and ATP Formation in the Aerobic Oxidation of Glucose via Glycolysis, the Pyruvate Dehydrogenase Complex Reaction, the Citric Acid Cycle, and Oxidative Phosphorylation

Reaction	Number of ATP or reduced coenzyme directly formed	Number of ATP ultimately formed[a]
Glucose ⟶ glucose 6-phosphate	−1 ATP	−1
Fructose 6-phosphate ⟶ fructose 1,6-bisphosphate	−1 ATP	−1
2 Glyceraldehyde 3-phosphate ⟶ 2 1,3-bisphosphoglycerate	2 NADH	3 or 5[b]
2 1,3-Bisphosphoglycerate ⟶ 2 3-phosphoglycerate	2 ATP	2
2 Phosphoenolpyruvate ⟶ 2 pyruvate	2 ATP	2
2 Pyruvate ⟶ 2 acetyl-CoA	2 NADH	5
2 Isocitrate ⟶ 2 α-ketoglutarate	2 NADH	5
2 α-Ketoglutarate ⟶ 2 succinyl-CoA	2 NADH	5
2 Succinyl-CoA ⟶ 2 succinate	2 ATP (or 2 GTP)	2
2 Succinate ⟶ 2 fumarate	2 FADH$_2$	3
2 Malate ⟶ 2 oxaloacetate	2 NADH	5
Total		30–32

[a]This is calculated as 2.5 ATP per NADH and 1.5 ATP per FADH$_2$. A negative value indicates consumption.

[b]This number is either 3 or 5, depending on the mechanism used to shuttle NADH equivalents from the cytosol to the mitochondrial matrix; see Figures 19-30 and 19-31.

many catabolic processes feed into the cycle to serve as fuels. Oxaloacetate and α-ketoglutarate, for example, are produced from aspartate and glutamate, respectively, when proteins are degraded. Under some metabolic circumstances, intermediates are drawn out of the cycle to be used as precursors in a variety of biosynthetic pathways.

The citric acid cycle, like all other metabolic pathways, is the product of evolution, and much of this evolution occurred before the advent of aerobic organisms. It does not necessarily represent the *shortest* pathway from acetate to CO_2, but it is the pathway that has, over time, conferred the greatest selective advantage. Early anaerobes most probably used some of the reactions of the citric acid cycle in linear biosynthetic processes. In fact, some modern anaerobic microorganisms use an incomplete citric acid cycle as a source of, not energy, but biosynthetic precursors **(Fig. 16-15)**. These organisms use the first three reactions of the cycle to make α-ketoglutarate, but, lacking α-ketoglutarate dehydrogenase, they cannot carry out the complete set of citric acid cycle reactions. They do have the four enzymes that catalyze the reversible conversion of oxaloacetate to succinyl-CoA and can produce malate, fumarate, succinate, and succinyl-CoA from oxaloacetate in a reversal of the "normal" (oxidative) direction of flow through the cycle. This pathway is a fermentation, with the NADH produced by isocitrate oxidation recycled to NAD^+ by reduction of oxaloacetate to succinate.

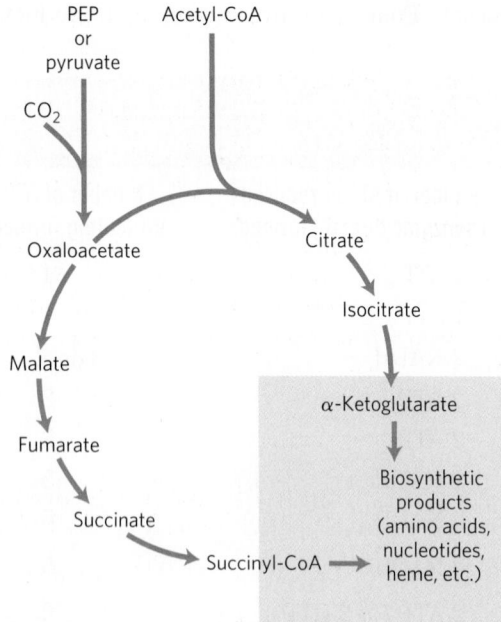

FIGURE 16-15 Biosynthetic precursors produced by an incomplete citric acid cycle in anaerobic bacteria. These anaerobes lack α-ketoglutarate dehydrogenase and therefore cannot carry out the complete citric acid cycle. α-Ketoglutarate and succinyl-CoA serve as precursors in a variety of biosynthetic pathways. (See Fig. 16-14 for the "normal" direction of these reactions in the citric acid cycle.)

With the evolution of cyanobacteria that produced O_2 from water, the earth's atmosphere became oxygenrich and organisms were under selective pressure to develop aerobic metabolism, which, as we have seen, is much more efficient than anaerobic fermentation.

Citric Acid Cycle Components Are Important Biosynthetic Intermediates

In aerobic organisms, the citric acid cycle is an **amphibolic pathway**, one that serves in both catabolic and anabolic processes. Besides its role in the oxidative catabolism of carbohydrates, fatty acids, and amino acids, the cycle provides precursors for many biosynthetic pathways **(Fig. 16-16)**, through reactions that served the same purpose in anaerobic ancestors. α-Ketoglutarate and oxaloacetate can, for example, serve as precursors of the amino acids aspartate and glutamate by simple transamination (Chapter 22). Through aspartate and glutamate, the carbons of oxaloacetate and α-ketoglutarate are then used to build other amino acids, as well as purine and pyrimidine nucleotides. Succinyl-CoA is a central intermediate in the synthesis of the porphyrin ring of heme groups, which serve as oxygen carriers (in hemoglobin and myoglobin) and electron carriers (in cytochromes) (see Fig. 22-25). And oxaloacetate can be converted to glucose via gluconeogenesis (see Fig. 15-13).

Given that the carbon atoms of acetate molecules entering the citric acid cycle appear eight steps later in oxaloacetate, it might seem that the cycle could generate oxaloacetate from acetate, then the oxaloacetate could be used to synthesize glucose. However, the stoichiometry of the citric acid cycle shows that there is no net conversion of acetate to oxaloacetate; for every two carbons that enter the cycle as acetyl-CoA, two leave as CO_2. In many organisms other than vertebrates, and in all vascular plants, another reaction sequence, the **glyoxylate cycle**, serves as a mechanism for converting acetate to carbohydrate. The glyoxylate cycle converts *two* molecules of acetate to one of oxaloacetate in a variant of the citric acid cycle in which the two decarboxylation steps are bypassed (see Fig. 20-55). Thus plants and many simpler organisms can synthesize glucose from fatty acids, but vertebrates cannot.

Anaplerotic Reactions Replenish Citric Acid Cycle Intermediates

As intermediates of the citric acid cycle are removed to serve as biosynthetic precursors, they are replenished by **anaplerotic reactions** (Fig. 16-16; Table 16-2). Under normal circumstances, the reactions by which cycle intermediates are siphoned off into other pathways and those by which they are replenished are in dynamic balance, so that the concentrations of the citric acid cycle intermediates remain almost constant.

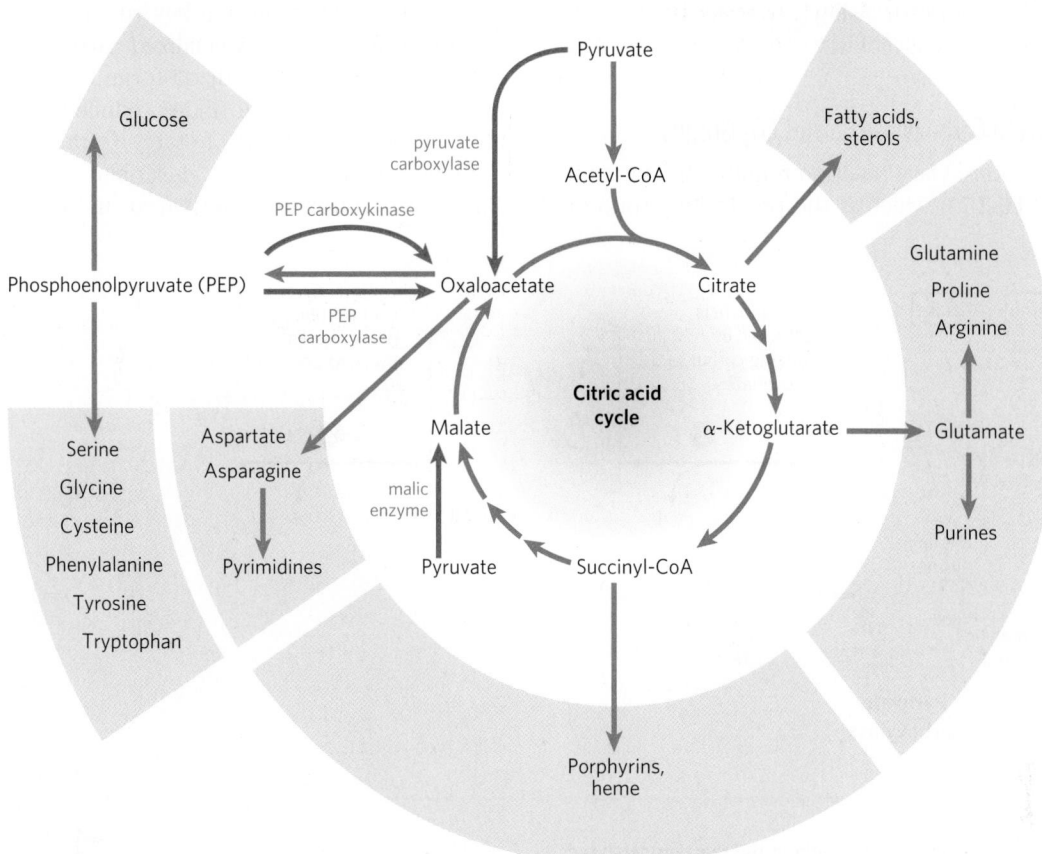

FIGURE 16-16 Role of the citric acid cycle in anabolism. Intermediates of the citric acid cycle are drawn off as precursors in many biosynthetic pathways. Shown in red are four anaplerotic reactions that replenish depleted cycle intermediates (see Table 16-2).

Table 16-2 shows the most common anaplerotic reactions, all of which, in various tissues and organisms, convert either pyruvate or phosphoenolpyruvate to oxaloacetate or malate. The most important anaplerotic reaction in mammalian liver and kidney is the reversible carboxylation of pyruvate by CO_2 to form oxaloacetate, catalyzed by **pyruvate carboxylase**. When the citric acid cycle is deficient in oxaloacetate or any other intermediates, pyruvate is carboxylated to produce more oxaloacetate. The enzymatic addition of a carboxyl group to pyruvate requires energy, which is supplied by ATP—the free energy required to attach a carboxyl group to pyruvate is about equal to the free energy available from ATP.

Pyruvate carboxylase is a regulatory enzyme and is virtually inactive in the absence of acetyl-CoA, its positive allosteric modulator. Whenever acetyl-CoA, the fuel for the citric acid cycle, is present in excess, it stimulates the pyruvate carboxylase reaction to produce more oxaloacetate, enabling the cycle to use more acetyl-CoA in the citrate synthase reaction.

The other anaplerotic reactions shown in Table 16-2 are also regulated to keep the level of intermediates high enough to support the activity of the citric acid cycle. Phosphoenolpyruvate (PEP) carboxylase, for example, is activated by the glycolytic intermediate fructose 1,6-bisphosphate, which accumulates when the

TABLE 16-2	Anaplerotic Reactions	
Reaction		**Tissue(s)/organism(s)**
Pyruvate + HCO_3^- + ATP $\xrightleftharpoons[]{\text{pyruvate carboxylase}}$ oxaloacetate + ADP + P_i		Liver, kidney
Phosphoenolpyruvate + CO_2 + GDP $\xrightleftharpoons[]{\text{PEP carboxykinase}}$ oxaloacetate + GTP		Heart, skeletal muscle
Phosphoenolpyruvate + HCO_3^- $\xrightleftharpoons[]{\text{PEP carboxylase}}$ oxaloacetate + P_i		Higher plants, yeast, bacteria
Pyruvate + HCO_3^- + NAD(P)H $\xrightleftharpoons[]{\text{malic enzyme}}$ malate + NAD(P)$^+$		Widely distributed in eukaryotes and bacteria

citric acid cycle operates too slowly to process all of the pyruvate generated by glycolysis.

Biotin in Pyruvate Carboxylase Carries CO_2 Groups

The pyruvate carboxylase reaction requires the vitamin **biotin (Fig. 16-17)**, which is the prosthetic group of the enzyme. Biotin plays a key role in many carboxylation reactions. It is a specialized carrier of one-carbon groups in their most oxidized form: CO_2. (The transfer of one-carbon groups in more reduced forms is mediated by other cofactors, notably tetrahydrofolate and S-adenosylmethionine, as described in Chapter 18.) Carboxyl groups are activated in a reaction that

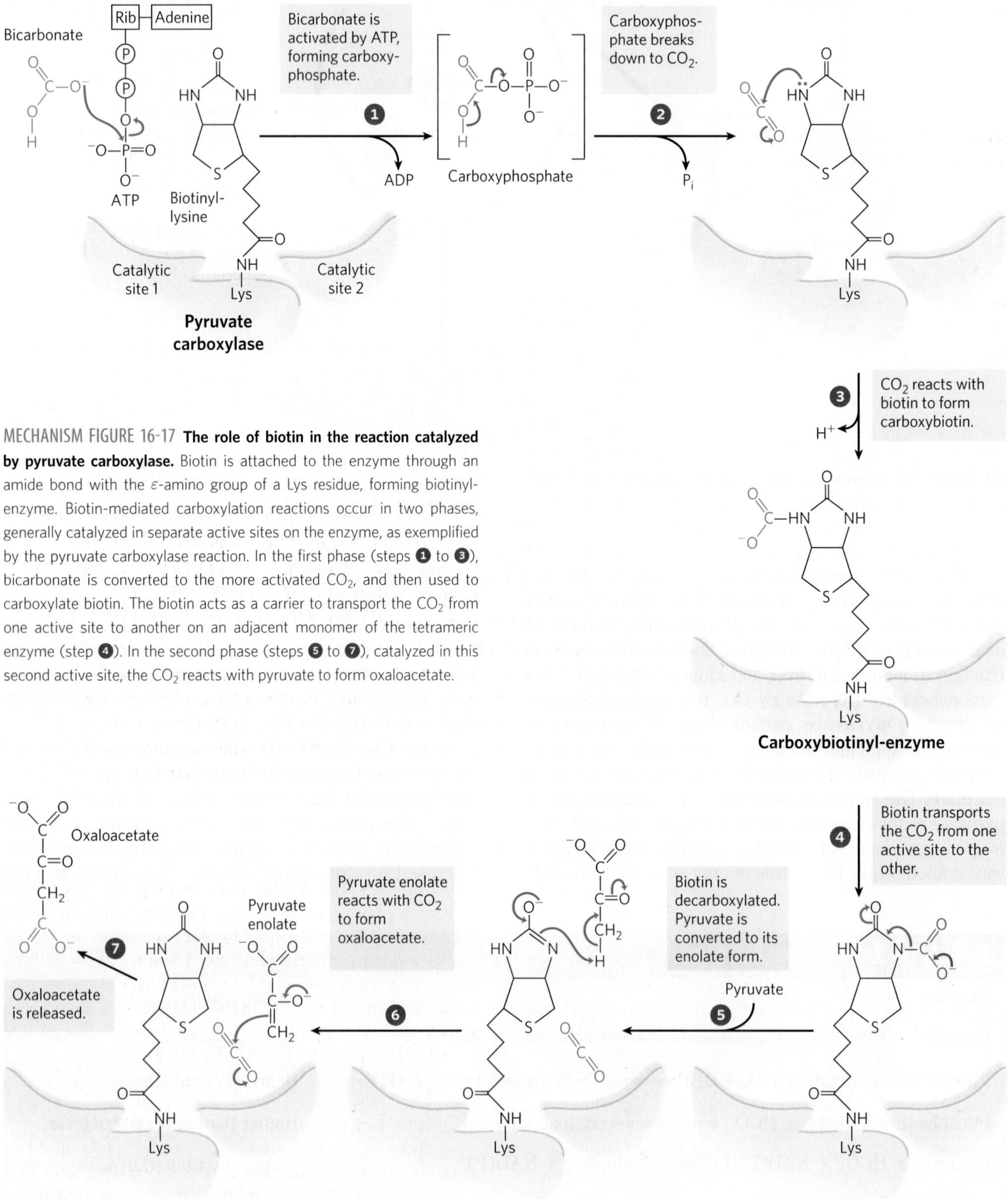

MECHANISM FIGURE 16-17 **The role of biotin in the reaction catalyzed by pyruvate carboxylase.** Biotin is attached to the enzyme through an amide bond with the ε-amino group of a Lys residue, forming biotinyl-enzyme. Biotin-mediated carboxylation reactions occur in two phases, generally catalyzed in separate active sites on the enzyme, as exemplified by the pyruvate carboxylase reaction. In the first phase (steps ❶ to ❸), bicarbonate is converted to the more activated CO_2, and then used to carboxylate biotin. The biotin acts as a carrier to transport the CO_2 from one active site to another on an adjacent monomer of the tetrameric enzyme (step ❹). In the second phase (steps ❺ to ❼), catalyzed in this second active site, the CO_2 reacts with pyruvate to form oxaloacetate.

consumes ATP and joins CO_2 to enzyme-bound biotin. This "activated" CO_2 is then passed to an acceptor (pyruvate in this case) in a carboxylation reaction.

Pyruvate carboxylase has four identical subunits, each containing a molecule of biotin covalently attached through an amide linkage to the ε-amino group of a specific Lys residue in the enzyme active site. Carboxylation of pyruvate proceeds in two steps (Fig. 16-17): first, a carboxyl group derived from HCO_3^- is attached to biotin, then the carboxyl group is transferred to pyruvate to form oxaloacetate. These two steps occur at separate active sites; the long flexible arm of biotin transfers activated carboxyl groups from the first active site (on one monomer of the tetramer) to the second (on the adjacent monomer), functioning much like the long lipoyllysyl arm of E_2 in the PDH complex (Fig. 16-6) and the long arm of the CoA-like moiety in the acyl carrier protein involved in fatty acid synthesis (see Fig. 21-5); these are compared in **Figure 16-18**. Lipoate, biotin, and pantothenate all enter cells on the same transporter; all become covalently attached to proteins by similar reactions; and all provide a flexible tether that allows bound reaction intermediates to move from one active site to another in an enzyme complex, without dissociating from it. That is, all participate in substrate channeling.

Biotin is a vitamin required in the human diet; it is abundant in many foods and is synthesized by intestinal bacteria. Biotin deficiency is rare, but can sometimes be caused by a diet rich in raw eggs. Egg whites contain a large amount of the protein **avidin** (M_r 70,000), which binds very tightly to biotin and prevents its absorption in the intestine. The avidin of egg whites may be a defense mechanism for the potential chick embryo, inhibiting the growth of bacteria. When eggs are cooked, avidin is denatured (and thereby inactivated) along with all other egg white proteins. Purified avidin is a useful reagent in biochemistry and cell biology. A protein that contains covalently bound biotin (derived experimentally or produced in vivo) can be recovered by affinity chromatography (see Fig. 3-17c) based on biotin's strong affinity for avidin. The protein is then eluted from the column with an excess of free biotin. The very high affinity of biotin for avidin is also used in the laboratory in the form of a molecular glue that can hold two structures together (see Fig. 19-27).

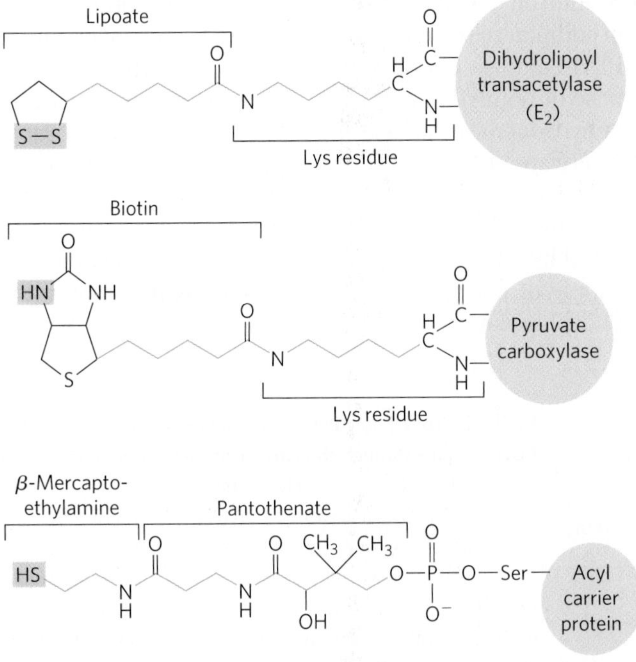

SUMMARY 16.2 Reactions of the Citric Acid Cycle

■ The citric acid cycle (Krebs cycle, TCA cycle) is a nearly universal central catabolic pathway in which compounds derived from the breakdown of carbohydrates, fats, and proteins are oxidized to CO_2, with most of the energy of oxidation temporarily held in the electron carriers $FADH_2$ and NADH. During aerobic metabolism, these electrons are transferred to O_2 and the energy of electron flow is trapped as ATP.

■ Acetyl-CoA enters the citric acid cycle (in the mitochondria of eukaryotes, the cytosol of bacteria) as citrate synthase catalyzes its condensation with oxaloacetate to form citrate.

■ In seven sequential reactions, including two decarboxylations, the citric acid cycle converts citrate to oxaloacetate and releases two CO_2. The pathway is cyclic in that the intermediates of the cycle are not used up; for each oxaloacetate consumed in the path, one is produced.

■ For each acetyl-CoA oxidized by the citric acid cycle, the energy gain consists of three molecules of NADH, one $FADH_2$, and one nucleoside triphosphate (either ATP or GTP).

■ Besides acetyl-CoA, any compound that gives rise to a four- or five-carbon intermediate of the citric acid cycle—for example, the breakdown products of many amino acids—can be oxidized by the cycle.

■ The citric acid cycle is amphibolic, serving in both catabolism and anabolism; cycle intermediates can be drawn off and used as the starting material for a variety of biosynthetic products.

■ Vertebrates cannot synthesize glucose from acetate or from the fatty acids that give rise to acetyl-CoA.

■ When intermediates are shunted from the citric acid cycle to other pathways, they are replenished by several anaplerotic reactions, which produce four-carbon intermediates by carboxylation of three-carbon

FIGURE 16-18 Biological tethers. The cofactors lipoate, biotin, and the combination of β-mercaptoethylamine and pantothenate form long, flexible arms (blue) on the enzymes to which they are covalently bound, acting as tethers that move intermediates from one active site to the next. The group shaded light red is, in each case, the point of attachment of the activated intermediate to the tether.

compounds; these reactions are catalyzed by pyruvate carboxylase, PEP carboxykinase, PEP carboxylase, and malic enzyme.

■ Enzymes that catalyze carboxylations commonly employ biotin to activate CO_2 and to carry it to acceptors such as pyruvate or phosphoenolpyruvate.

16.3 Regulation of the Citric Acid Cycle

As we have seen in Chapter 15, the regulation of key enzymes in metabolic pathways, by allosteric effectors and by covalent modification, ensures the production of intermediates at the rates required to keep the cell in a stable steady state while avoiding wasteful overproduction. The flow of carbon atoms from pyruvate into and through the citric acid cycle is under tight regulation at at least three levels: the transport of pyruvate into mitochondria by the mitochondrial pyruvate carrier (MPC); the conversion of pyruvate to acetyl-CoA, the starting material for the cycle (the pyruvate dehydrogenase complex reaction); and the entry of acetyl-CoA into the cycle (the citrate synthase reaction). Acetyl-CoA is also produced by pathways other than the PDH complex

reaction—most cells produce acetyl-CoA from the oxidation of fatty acids and certain amino acids—and the availability of intermediates from these other pathways is important in the regulation of pyruvate oxidation and of the citric acid cycle. The cycle is also regulated at the isocitrate dehydrogenase and α-ketoglutarate dehydrogenase reactions.

Production of Acetyl-CoA by the Pyruvate Dehydrogenase Complex Is Regulated by Allosteric and Covalent Mechanisms

The PDH complex of mammals is strongly inhibited by ATP and by acetyl-CoA and NADH, the products of the reaction catalyzed by the complex **(Fig. 16-19)**. The allosteric inhibition of pyruvate oxidation is greatly enhanced when long-chain fatty acids are available. AMP, CoA, and NAD^+, all of which accumulate when too little acetate flows into the citric acid cycle, allosterically activate the PDH complex. Thus, this enzyme activity is turned off when ample fuel is available in the form of fatty acids and acetyl-CoA and when the cell's [ATP]/[ADP] and [NADH]/[NAD^+] ratios are high, and it is turned on again when energy demands are high and the cell requires greater flux of acetyl-CoA into the citric acid cycle.

In mammals, these allosteric regulatory mechanisms are complemented by a second level of regulation: covalent protein modification. The PDH complex is inhibited by reversible phosphorylation of a specific Ser residue on one of the two subunits of E_1. As noted earlier, in addition to the enzymes E_1, E_2, and E_3, the mammalian PDH complex contains two proteins with the sole purpose of regulating the activity of the complex. Pyruvate dehydrogenase kinase phosphorylates and thereby inactivates E_1, and a specific phosphoprotein phosphatase removes the phosphoryl group by hydrolysis and thereby activates E_1. The kinase is allosterically activated by ATP: when [ATP] is high (reflecting a sufficient supply of energy), the PDH complex is inactivated by phosphorylation of E_1. When [ATP] declines, kinase

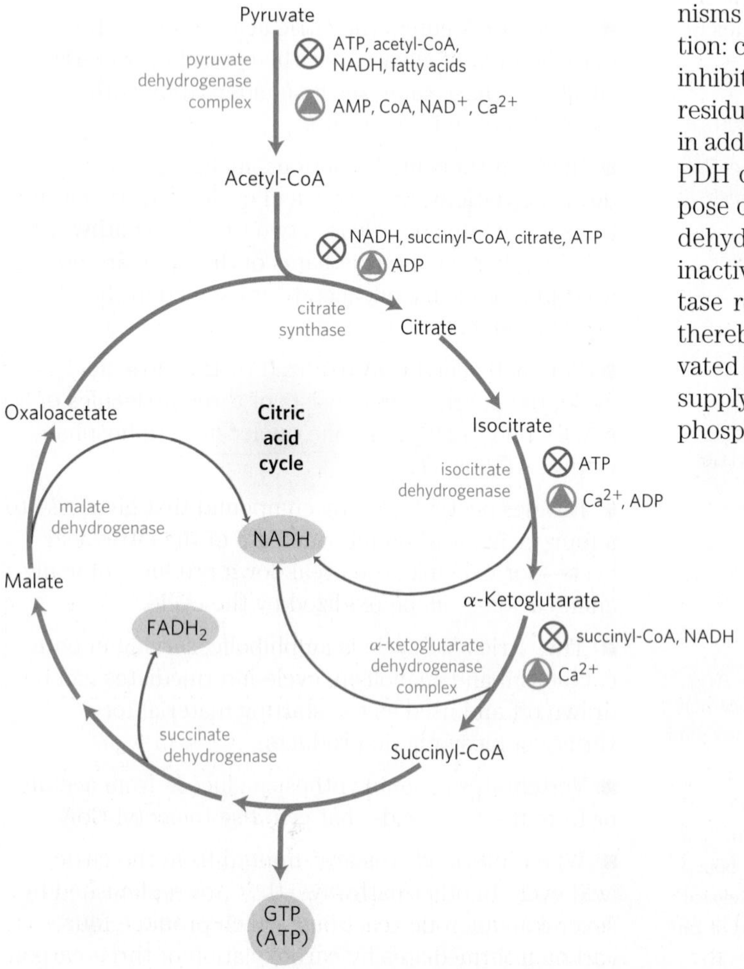

FIGURE 16-19 Regulation of metabolite flow from the PDH complex through the citric acid cycle in mammals. The PDH complex is allosterically inhibited when [ATP]/[ADP], [NADH]/[NAD^+], and [acetyl-CoA]/[CoA] ratios are high, all of which indicate an energy-sufficient metabolic state. When these ratios decrease, allosteric activation of pyruvate oxidation results. The rate of flow through the citric acid cycle can be limited by the availability of the citrate synthase substrates, oxaloacetate and acetyl-CoA, or of NAD^+, which is depleted by its conversion to NADH, slowing the three NAD-dependent oxidation steps. Feedback inhibition by succinyl-CoA, citrate, and ATP also slows the cycle by inhibiting early steps. In muscle tissue, Ca^{2+} stimulates contraction and, as shown here, stimulates energy-yielding metabolism to replace the ATP consumed by contraction.

activity decreases and phosphatase action removes the phosphoryl groups from E_1, activating the complex.

The simple compound dichloroacetate inhibits pyruvate dehydrogenase kinase in the laboratory and so relieves the inhibition of the pyruvate dehydrogenase complex. This stimulates pyruvate oxidation via the citric acid cycle and thus may have a use in directing metabolism in tumor cells away from aerobic glycolysis.

Dichloroacetate

The Citric Acid Cycle Is Regulated at Its Three Exergonic Steps

The flow of metabolites through the citric acid cycle is under stringent regulation. Three factors govern the flux through the cycle: substrate availability, inhibition by accumulating products, and allosteric feedback inhibition of the enzymes that catalyze early steps in the cycle.

Each of the three strongly exergonic steps in the cycle—those catalyzed by citrate synthase, isocitrate dehydrogenase, and α-ketoglutarate dehydrogenase (Fig. 16-19)—can become the rate-limiting step under some circumstances. The availability of the substrates for citrate synthase (acetyl-CoA and oxaloacetate) varies with the metabolic state of the cell and sometimes limits the rate of citrate formation. NADH, a product of isocitrate and α-ketoglutarate oxidation, accumulates under some conditions, and at high [NADH]/[NAD$^+$] both dehydrogenase reactions are severely inhibited by mass action. Similarly, in the cell, the malate dehydrogenase reaction is essentially at equilibrium (that is, it is substrate-limited), and when [NADH]/[NAD$^+$] is high the concentration of oxaloacetate is low, slowing the first step in the cycle. Product accumulation inhibits all three limiting steps of the cycle: succinyl-CoA inhibits α-ketoglutarate dehydrogenase (and also citrate synthase); citrate blocks citrate synthase; and the end product, ATP, inhibits both citrate synthase and isocitrate dehydrogenase. The inhibition of citrate synthase by ATP is relieved by ADP, an allosteric activator of this enzyme. In vertebrate muscle, Ca^{2+}, the signal for contraction and for a concomitant increase in demand for ATP, activates both isocitrate dehydrogenase and α-ketoglutarate dehydrogenase, as well as the PDH complex. In short, the concentrations of substrates and intermediates in the citric acid cycle set the flux through this pathway at a rate that provides optimal concentrations of ATP and NADH.

Under normal conditions, the rates of glycolysis and of the citric acid cycle are integrated so that only as much glucose is metabolized to pyruvate as is needed to supply the citric acid cycle with its fuel, the acetyl groups of acetyl-CoA. Pyruvate, lactate, and acetyl-CoA are normally maintained at steady-state concentrations. The rate of glycolysis is matched to the rate of the citric acid cycle not only through its inhibition by high levels of ATP and NADH, which are common to both the glycolytic and respiratory stages of glucose oxidation, but also by the concentration of citrate. Citrate, the product of the first step of the citric acid cycle, is an important allosteric inhibitor of phosphofructokinase-1 in the glycolytic pathway (see Fig. 15-16).

Substrate Channeling through Multienzyme Complexes May Occur in the Citric Acid Cycle

Although the enzymes of the citric acid cycle are usually described as soluble components of the mitochondrial matrix (except for succinate dehydrogenase, which is membrane-bound), growing evidence suggests that within the mitochondrion these enzymes exist as multienzyme complexes. The classic approach of enzymology—purification of individual proteins from extracts of broken cells—was applied with great success to the citric acid cycle enzymes. However, the first casualty of cell breakage is higher-level organization within the cell—the noncovalent, reversible interaction of one protein with another, or of an enzyme with some structural component such as a membrane, microtubule, or microfilament. When cells are broken open, their contents, including enzymes, are diluted 100- or 1,000-fold **(Fig. 16-20)**.

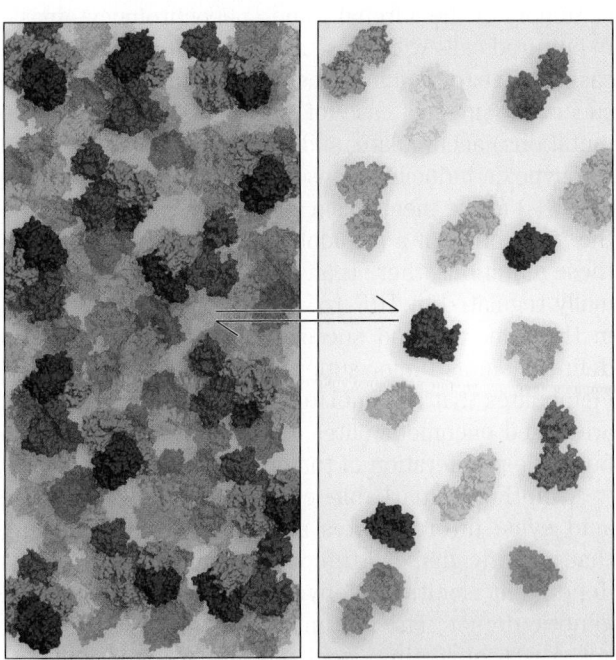

In the cytosol, high concentrations of enzymes 1, 2, and 3 favor their association.

In extract of broken cells, dilution by buffer reduces the concentrations of enzymes 1, 2, and 3, favoring their dissociation.

FIGURE 16-20 Dilution of a solution containing a noncovalently bound protein complex—such as one consisting of three enzymes (illustrated here in red, blue, and green)—favors dissociation of the complex into its constituents.

Several types of evidence suggest that, in cells, multienzyme complexes ensure efficient passage of the product of one enzyme reaction to the next enzyme in the pathway. Such complexes are called **metabolons**. Certain enzymes of the citric acid cycle have been isolated together as supramolecular complexes, or have been found associated with the inner mitochondrial membrane, or have been shown to diffuse in the mitochondrial matrix more slowly than expected for the individual protein in solution. There is strong evidence for substrate channeling through multienzyme complexes in other metabolic pathways, and many enzymes thought of as "soluble" probably function in the cell as highly organized complexes that channel intermediates. We will encounter other examples of channeling when we discuss the biosynthesis of amino acids and nucleotides in Chapter 22.

Some Mutations in Enzymes of the Citric Acid Cycle Lead to Cancer

When the mechanisms for regulating a pathway such as the citric acid cycle are overwhelmed by a major metabolic perturbation, the result can be serious disease. Mutations in citric acid cycle enzymes are very rare in humans and other mammals, but those that do occur are devastating. Genetic defects in the fumarase gene lead to tumors of smooth muscle (leiomas) and kidney; mutations in succinate dehydrogenase lead to tumors of the adrenal gland (pheochromocytomas). In cultured cells with these mutations, fumarate (in the case of fumarase mutations) and, to a lesser extent, succinate (in the case of succinate dehydrogenase mutations) accumulate, and this accumulation induces the hypoxia-inducible transcription factor HIF-1α (see Box 14-1). The mechanism of tumor formation may be the production of a pseudohypoxic state. In cells with these mutations, there is an up-regulation of genes normally regulated by HIF-1α. These effects of mutations in the fumarase and succinate dehydrogenase genes define them as tumor suppressor genes (p. 484). The metabolites that accumulate (fumarate and succinate) are called oncometabolites, because of their ability to favor the proliferation of tumor cells.

Another remarkable connection between citric acid cycle intermediates and cancer is the finding that in many glial cell tumors (gliomas), the NADPH-dependent isocitrate dehydrogenase has an unusual genetic defect. The mutant enzyme loses its normal activity (converting isocitrate to α-ketoglutarate) but *gains* a new activity: it converts α-ketoglutarate to 2-hydroxyglutarate **(Fig. 16-21)**, which accumulates in the tumor cells. α-Ketoglutarate and Fe^{3+} are essential cofactors for a family of histone demethylases that alter gene expression by removing methyl groups from Arg and Lys residues in the histones that organize nuclear DNA. By competing

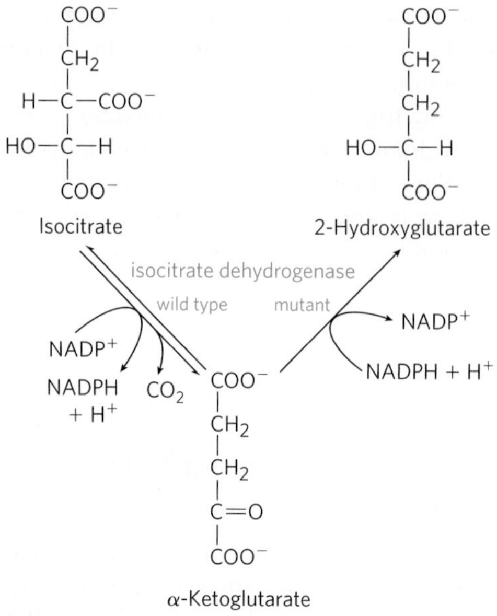

FIGURE 16-21 A mutant isocitrate dehydrogenase acquires a new activity. Wild-type isocitrate dehydrogenase catalyzes the conversion of isocitrate to α-ketoglutarate, but mutations that alter the binding site for isocitrate cause loss of the normal enzymatic activity and gain of a new activity: conversion of α-ketoglutarate to 2-hydroxyglutarate. Accumulation of this product inhibits histone demethylase, altering gene regulation and leading to glial cell tumors in the brain.

with α-ketoglutarate for binding to the histone demethylases, 2-hydroxyglutarate inhibits their activity. Inhibition of the histone demethylases interferes with normal gene regulation, leading to unrestricted glial cell growth. The family of more than 60 dioxygenases that use α-ketoglutarate and Fe^{3+} as cofactors are also competitively inhibited by 2-hydroxyglutarate. Inhibition of one or more of these enzymes could interfere with normal regulation of cell division and thus produce a tumor. ■

SUMMARY 16.3 Regulation of the Citric Acid Cycle

■ The overall rate of the citric acid cycle is controlled by the rate of conversion of pyruvate to acetyl-CoA and by the flux through citrate synthase, isocitrate dehydrogenase, and α-ketoglutarate dehydrogenase. These fluxes are largely determined by the concentrations of substrates and products: the end products ATP and NADH are inhibitory, and the substrates NAD^+ and ADP are stimulatory.

■ The production of acetyl-CoA for the citric acid cycle by the PDH complex is inhibited allosterically by metabolites that signal a sufficiency of metabolic energy (ATP, acetyl-CoA, NADH, and fatty acids) and stimulated by metabolites that indicate a reduced energy supply (AMP, NAD^+, CoA).

■ Complexes of consecutive enzymes in a pathway allow substrate channeling between them.

Key Terms

Terms in bold are defined in the glossary.

respiration 619
cellular respiration 619
citric acid cycle 619
**tricarboxylic acid (TCA)
  cycle** 619
Krebs cycle 619
mitchondrial pyruvate
  carrier (MPC) 620
pyruvate dehydrogenase
  (PDH) complex 620
oxidative
  decarboxylation 620
thioester 621
lipoate 621
**substrate
  channeling** 623
iron-sulfur center 627
**moonlighting
  enzymes** 628

α-ketoglutarate
  dehydrogenase
  complex 630
synthases 631
synthetases 631
ligases 631
lyases 631
kinases 631
phosphorylases 631
phosphatases 631
**nucleoside diphosphate
  kinase** 632
prochiral molecule 634
amphibolic pathway 636
glyoxylate cycle 636
anaplerotic reaction 636
biotin 638
avidin 639
metabolon 642

Problems

1. Balance Sheet for the Citric Acid Cycle The citric acid cycle has eight enzymes: citrate synthase, aconitase, isocitrate dehydrogenase, α-ketoglutarate dehydrogenase, succinyl-CoA synthetase, succinate dehydrogenase, fumarase, and malate dehydrogenase.

(a) Write a balanced equation for the reaction catalyzed by each enzyme.

(b) Name the cofactor(s) required by each enzyme reaction.

(c) For each enzyme determine which of the following describes the type of reaction(s) catalyzed: condensation (carbon–carbon bond formation); dehydration (loss of water); hydration (addition of water); decarboxylation (loss of CO_2); oxidation-reduction; substrate-level phosphorylation; isomerization.

(d) Write a balanced net equation for the catabolism of acetyl-CoA to CO_2.

2. Net Equation for Glycolysis and the Citric Acid Cycle Write the net biochemical equation for the metabolism of a molecule of glucose by glycolysis and the citric acid cycle, including all cofactors.

3. Recognizing Oxidation and Reduction Reactions One biochemical strategy of many living organisms is the stepwise oxidation of organic compounds to CO_2 and H_2O and the conservation of a major part of the energy thus produced in the form of ATP. It is important to be able to recognize oxidation-reduction processes in metabolism. Reduction of an organic molecule results from the hydrogenation of a double bond (Eqn 1, below) or of a single bond with accompanying cleavage (Eqn 2). Conversely, oxidation results from dehydrogenation.

In biochemical redox reactions, the coenzymes NAD and FAD dehydrogenate/hydrogenate organic molecules in the presence of the proper enzymes.

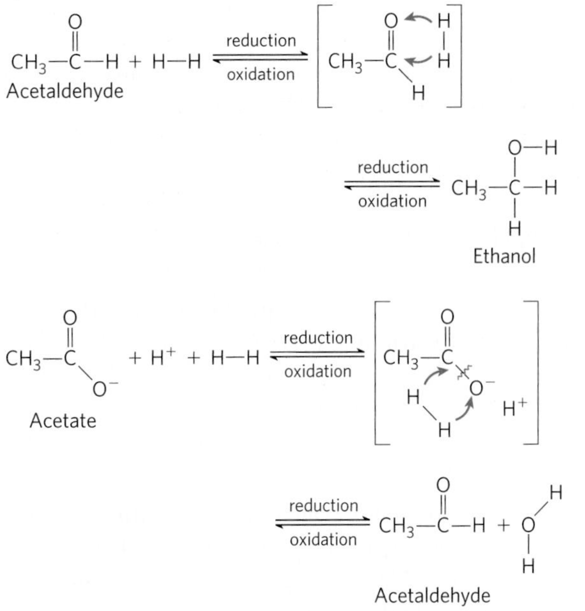

For each of the metabolic transformations in (a) through (h), determine whether oxidation or reduction has occurred. Balance each transformation by inserting H—H and, where necessary, H_2O.

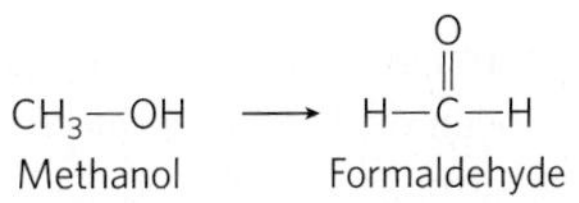

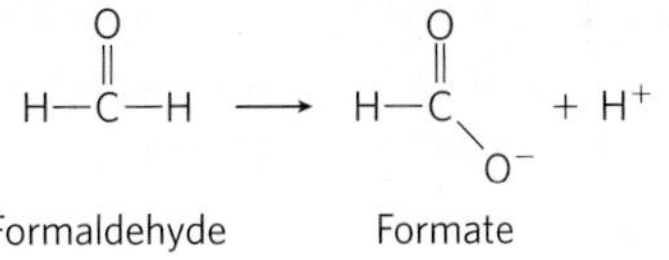

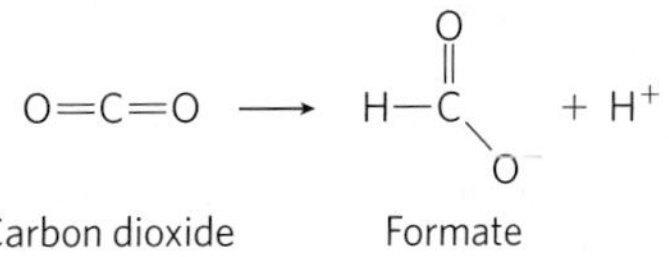

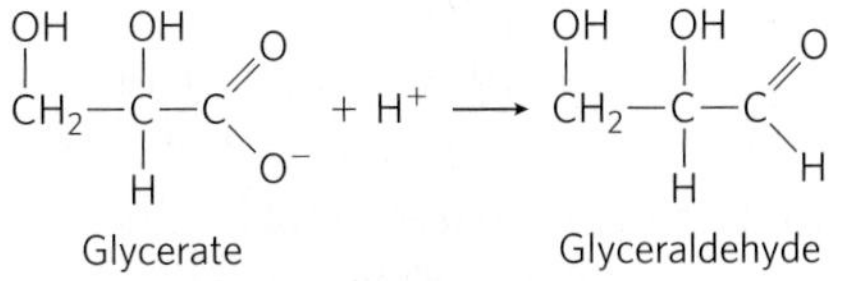

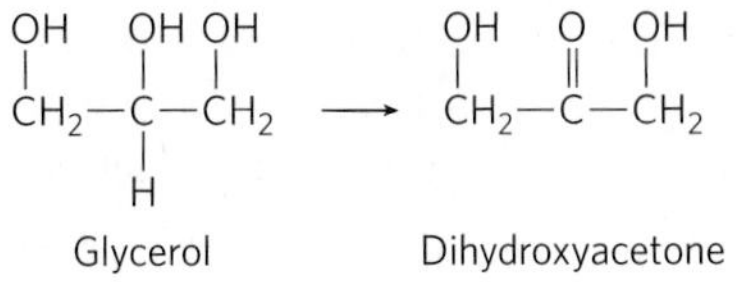

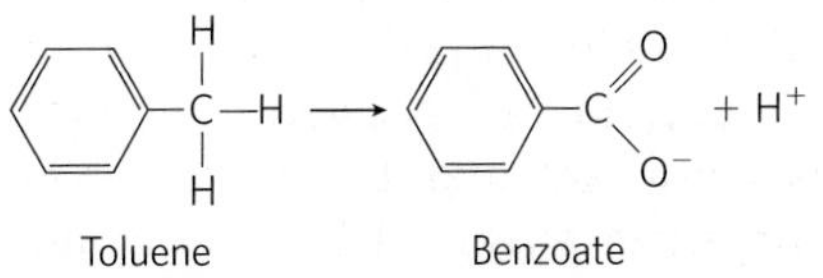

(g)

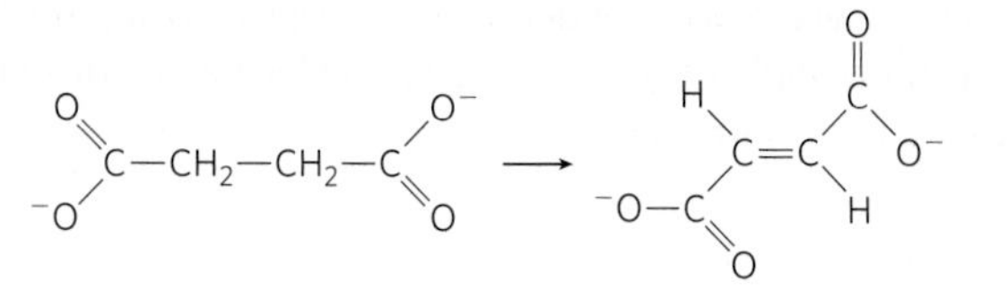

Succinate → Fumarate

(h)

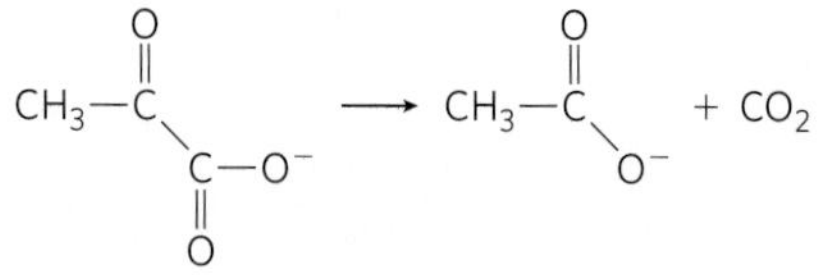

Pyruvate → Acetate + CO_2

4. Relationship between Energy Release and the Oxidation State of Carbon A eukaryotic cell can use glucose ($C_6H_{12}O_6$) and hexanoic acid ($C_6H_{14}O_2$) as fuels for cellular respiration. On the basis of their structural formulas, which substance releases more energy per gram on complete combustion to CO_2 and H_2O?

5. Nicotinamide Coenzymes as Reversible Redox Carriers The nicotinamide coenzymes (see Fig. 13-24) can undergo reversible oxidation-reduction reactions with specific substrates in the presence of the appropriate dehydrogenase. In these reactions, NADH + H^+ serves as the hydrogen source, as described in Problem 3. Whenever the coenzyme is oxidized, a substrate must be simultaneously reduced:

$$\text{Substrate} + \text{NADH} + \text{H}^+ \rightleftharpoons \text{product} + \text{NAD}^+$$
Oxidized   Reduced   Reduced   Oxidized

For each of the reactions in (a) through (f) shown below, determine whether the substrate has been oxidized or reduced or is unchanged in oxidation state (see Problem 3). If a redox change has occurred, balance the reaction with the necessary amount of NAD^+, NADH, H^+, and H_2O. The objective is to recognize when a redox coenzyme is necessary in a metabolic reaction.

(a) 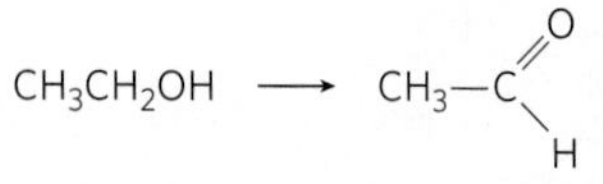 $CH_3CH_2OH \longrightarrow CH_3-C{\overset{O}{\underset{H}{}}}$

Ethanol   Acetaldehyde

(b)

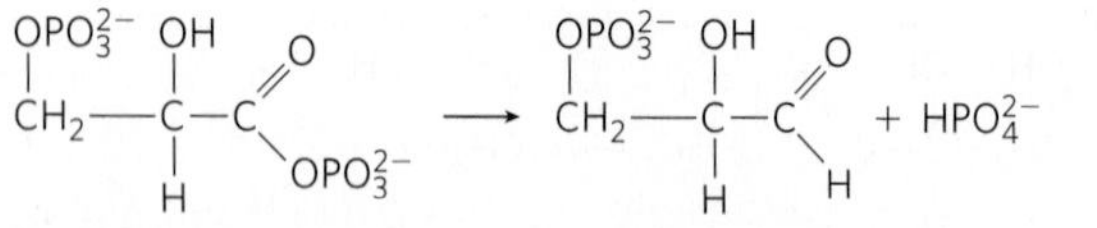

1,3-Bisphosphoglycerate → Glyceraldehyde 3-phosphate + HPO_4^{2-}

(c) 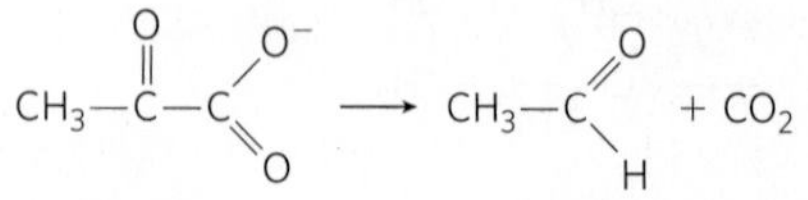 $CH_3-C{\overset{O}{\underset{O^-}{}}} \longrightarrow CH_3-C{\overset{O}{\underset{H}{}}} + CO_2$

Pyruvate   Acetaldehyde

(d) 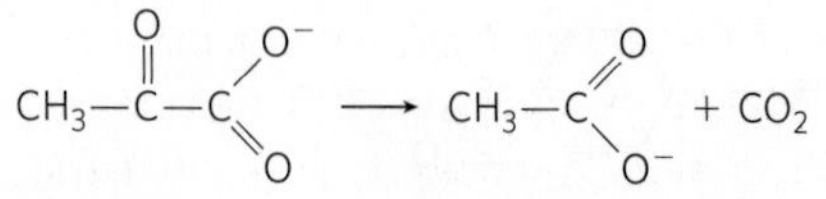 $CH_3-C{\overset{O}{\underset{O^-}{}}} \longrightarrow CH_3-C{\overset{O}{\underset{O^-}{}}} + CO_2$

Pyruvate   Acetate

(e)

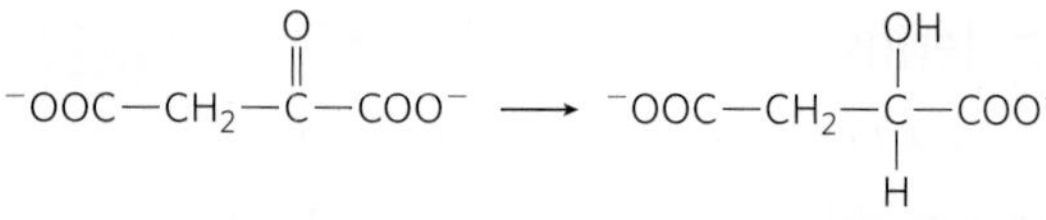

Oxaloacetate → Malate

(f)

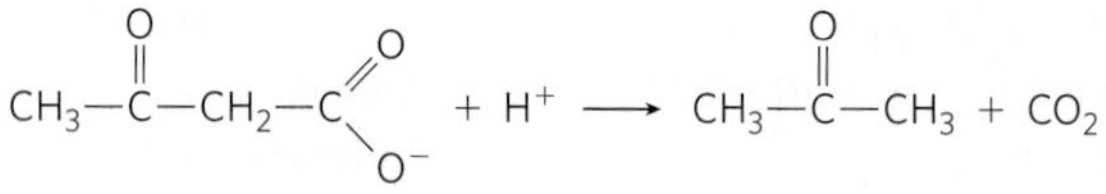

Acetoacetate → Acetone + CO_2

6. Pyruvate Dehydrogenase Cofactors and Mechanism Describe the role of each cofactor involved in the reaction catalyzed by the pyruvate dehydrogenase complex.

7. Thiamine Deficiency Individuals with a thiamine-deficient diet have relatively high levels of pyruvate in their blood. Explain this in biochemical terms.

8. Isocitrate Dehydrogenase Reaction What type of chemical reaction is involved in the conversion of isocitrate to α-ketoglutarate? Name and describe the role of any cofactors. What other reaction(s) of the citric acid cycle are of this same type?

9. Stimulation of Oxygen Consumption by Oxaloacetate and Malate In the early 1930s, Albert Szent-Györgyi reported the interesting observation that the addition of small amounts of oxaloacetate or malate to suspensions of minced pigeon breast muscle stimulated the oxygen consumption of the preparation. Surprisingly, the amount of oxygen consumed was about seven times more than the amount necessary for complete oxidation (to CO_2 and H_2O) of the added oxaloacetate or malate. Why did the addition of oxaloacetate or malate stimulate oxygen consumption? Why was the amount of oxygen consumed so much greater than the amount necessary to completely oxidize the added oxaloacetate or malate?

10. Formation of Oxaloacetate in a Mitochondrion In the last reaction of the citric acid cycle, malate is dehydrogenated to regenerate the oxaloacetate necessary for the entry of acetyl-CoA into the cycle:

L-Malate + $NAD^+ \longrightarrow$ oxaloacetate + NADH + H^+
$$\Delta G'^{\circ} = 30.0 \text{ kJ/mol}$$

(a) Calculate the equilibrium constant for this reaction at 25 °C.

(b) Because $\Delta G'^{\circ}$ assumes a standard pH of 7, the equilibrium constant calculated in (a) corresponds to

$$K'_{eq} = \frac{[\text{oxaloacetate}][\text{NADH}]}{[\text{L-malate}][\text{NAD}^+]}$$

The measured concentration of L-malate in rat liver mitochondria is about 0.20 mM when $[NAD^+]/[NADH]$ is 10. Calculate the concentration of oxaloacetate at pH 7 in these mitochondria.

(c) To appreciate the magnitude of the mitochondrial oxaloacetate concentration, calculate the number of oxaloacetate molecules in a single rat liver mitochondrion. Assume the mitochondrion is a sphere of diameter 2.0 μm.

11. Cofactors for the Citric Acid Cycle Suppose you have prepared a mitochondrial extract that contains all the soluble enzymes of the matrix but has lost (by dialysis) all the low molecular weight cofactors. What must you add to the extract so that the preparation will oxidize acetyl-CoA to CO_2?

12. Riboflavin Deficiency How would a riboflavin deficiency affect the functioning of the citric acid cycle? Explain your answer.

13. Oxaloacetate Pool What factors might decrease the pool of oxaloacetate available for the activity of the citric acid cycle? How can the pool of oxaloacetate be replenished?

14. Energy Yield from the Citric Acid Cycle The reaction catalyzed by succinyl-CoA synthetase produces the high-energy compound GTP. How is the free energy contained in GTP incorporated into the cellular ATP pool?

15. Respiration Studies in Isolated Mitochondria Cellular respiration can be studied in isolated mitochondria by measuring oxygen consumption under different conditions. If 0.01 M sodium malonate is added to actively respiring mitochondria that are using pyruvate as fuel, respiration soon stops and a metabolic intermediate accumulates.

(a) What is the structure of this intermediate?

(b) Explain why it accumulates.

(c) Explain why oxygen consumption stops.

(d) Aside from removal of the malonate, how can this inhibition of respiration be overcome? Explain.

16. Labeling Studies in Isolated Mitochondria The metabolic pathways of organic compounds have often been delineated by using a radioactively labeled substrate and following the fate of the label.

(a) How can you determine whether glucose added to a suspension of isolated mitochondria is metabolized to CO_2 and H_2O?

(b) Suppose you add a brief pulse of $[3\text{-}^{14}C]$pyruvate (labeled in the methyl position) to the mitochondria. After one turn of the citric acid cycle, what is the location of the ^{14}C in the oxaloacetate? Explain by tracing the ^{14}C label through the pathway. How many turns of the cycle are required to release all the $[3\text{-}^{14}C]$pyruvate as CO_2?

17. Pathway of CO_2 in Gluconeogenesis In the first bypass step of gluconeogenesis, the conversion of pyruvate to phosphoenolpyruvate (PEP), pyruvate is carboxylated by pyruvate carboxylase to oxaloacetate, which is subsequently decarboxylated to PEP by PEP carboxykinase (Chapter 14). Because the addition of CO_2 is directly followed by the loss of CO_2, you might expect that in tracer experiments, the ^{14}C of $^{14}CO_2$ would not be incorporated into PEP, glucose, or any intermediates in gluconeogenesis. However, investigators find that when a rat liver preparation synthesizes glucose in the presence of $^{14}CO_2$, ^{14}C slowly appears in PEP and eventually at C-3 and C-4 of glucose. How does the ^{14}C label get into the PEP and glucose? (Hint: During gluconeogenesis in the presence of $^{14}CO_2$, several of the four-carbon citric acid cycle intermediates also become labeled.)

18. $[1\text{-}^{14}C]$Glucose Catabolism An actively respiring bacterial culture is briefly incubated with $[1\text{-}^{14}C]$glucose, and the glycolytic and citric acid cycle intermediates are isolated. Where is the ^{14}C in each of the intermediates listed below? Consider only the initial incorporation of ^{14}C, in the first pass of labeled glucose through the pathways.

(a) Fructose 1,6-bisphosphate

(b) Glyceraldehyde 3-phosphate

(c) Phosphoenolpyruvate

(d) Acetyl-CoA

(e) Citrate

(f) α-Ketoglutarate

(g) Oxaloacetate

19. Role of the Vitamin Thiamine People with beriberi, a disease caused by thiamine deficiency, have elevated levels of blood pyruvate and α-ketoglutarate, especially after consuming a meal rich in glucose. How are these effects related to a deficiency of thiamine?

20. Synthesis of Oxaloacetate by the Citric Acid Cycle Oxaloacetate is formed in the last step of the citric acid cycle by the NAD^+-dependent oxidation of L-malate. Can a net synthesis of oxaloacetate from acetyl-CoA occur using only the enzymes and cofactors of the citric acid cycle, without depleting the intermediates of the cycle? Explain. How is oxaloacetate that is lost from the cycle (to biosynthetic reactions) replenished?

21. Oxaloacetate Depletion Mammalian liver can carry out gluconeogenesis using oxaloacetate as the starting material (Chapter 14). Would the operation of the citric acid cycle be affected by extensive use of oxaloacetate for gluconeogenesis? Explain your answer.

22. Mode of Action of the Rodenticide Fluoroacetate Fluoroacetate, prepared commercially for rodent control, is also produced by a South African plant. After entering a cell, fluoroacetate is converted to fluoroacetyl-CoA in a reaction catalyzed by the enzyme acetate thiokinase:

$$F-CH_2COO^- + CoA\text{-}SH + ATP \longrightarrow$$

$$F-CH_2\underset{\substack{\| \\ O}}{C}-S\text{-}CoA + AMP + PP_i$$

The toxic effect of fluoroacetate was studied in an experiment using intact isolated rat heart. After the heart was perfused with 0.22 mM fluoroacetate, the measured rate of glucose uptake and glycolysis decreased, and glucose 6-phosphate and fructose 6-phosphate accumulated. Examination of the citric acid cycle intermediates revealed that their concentrations were below normal, except for citrate, with a concentration 10 times higher than normal.

(a) Where did the block in the citric acid cycle occur? What caused citrate to accumulate and the other cycle intermediates to be depleted?

(b) Fluoroacetyl-CoA is enzymatically transformed in the citric acid cycle. What is the structure of the end product of fluoroacetate metabolism? Why does it block the citric acid cycle? How might the inhibition be overcome?

(c) In the heart perfusion experiments, why did glucose uptake and glycolysis decrease? Why did hexose monophosphates accumulate?

(d) Why is fluoroacetate poisoning fatal?

23. Synthesis of L-Malate in Wine Making The tartness of some wines is due to high concentrations of L-malate. Write a sequence of reactions showing how yeast cells synthesize L-malate from glucose under anaerobic conditions in the presence of dissolved CO_2 (HCO_3^-). Note that the overall reaction for this fermentation cannot involve the consumption of nicotinamide coenzymes or citric acid cycle intermediates.

24. Net Synthesis of α-Ketoglutarate α-Ketoglutarate plays a central role in the biosynthesis of several amino acids. Write a sequence of enzymatic reactions that could result in the net synthesis of α-ketoglutarate from pyruvate. Your proposed sequence must not involve the net consumption of other citric acid cycle intermediates. Write an equation for the overall reaction and identify the source of each reactant.

25. Amphibolic Pathways Explain, giving examples, what is meant by the statement that the citric acid cycle is amphibolic.

26. Regulation of the Pyruvate Dehydrogenase Complex In animal tissues, the rate of conversion of pyruvate to acetyl-CoA is regulated by the ratio of active, phosphorylated to inactive, unphosphorylated PDH complex. Determine what happens to the rate of this reaction when a preparation of rabbit muscle mitochondria containing the PDH complex is treated with (a) pyruvate dehydrogenase kinase, ATP, and NADH; (b) pyruvate dehydrogenase phosphatase and Ca^{2+}; (c) malonate.

27. Commercial Synthesis of Citric Acid Citric acid is used as a flavoring agent in soft drinks, fruit juices, and many other foods. Worldwide, the market for citric acid is valued at hundreds of millions of dollars per year. Commercial production uses the mold *Aspergillus niger*, which metabolizes sucrose under carefully controlled conditions.

(a) The yield of citric acid is strongly dependent on the concentration of $FeCl_3$ in the culture medium, as indicated in the graph. Why does the yield decrease when the concentration of Fe^{3+} is above or below the optimal value of 0.5 mg/L?

(b) Write the sequence of reactions by which *A. niger* synthesizes citric acid from sucrose. Write an equation for the overall reaction.

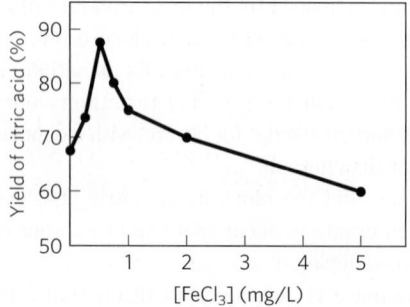

(c) Does the commercial process require the culture medium to be aerated—that is, is this a fermentation or an aerobic process? Explain.

28. Regulation of Citrate Synthase In the presence of saturating amounts of oxaloacetate, the activity of citrate synthase from pig heart tissue shows a sigmoid dependence on the concentration of acetyl-CoA, as shown in the graph below. When succinyl-CoA is added, the curve shifts to the right and the sigmoid dependence is more pronounced.

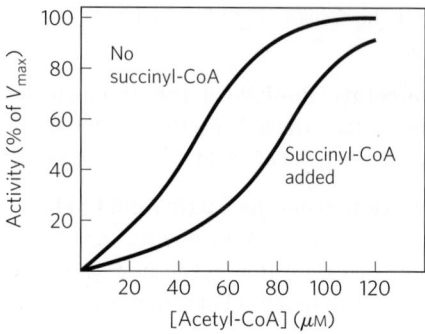

On the basis of these observations, suggest how succinyl-CoA regulates the activity of citrate synthase. (Hint: See Fig. 6-35.) Why is succinyl-CoA an appropriate signal for regulation of the citric acid cycle? How does the regulation of citrate synthase control the rate of cellular respiration in pig heart tissue?

29. Regulation of Pyruvate Carboxylase The carboxylation of pyruvate by pyruvate carboxylase occurs at a very low rate unless acetyl-CoA, a positive allosteric modulator, is present. If you have just eaten a meal rich in fatty acids (triacylglycerols) but low in carbohydrates (glucose), how does this regulatory property shut down the oxidation of glucose to CO_2 and H_2O but increase the oxidation of acetyl-CoA derived from fatty acids?

30. Relationship between Respiration and the Citric Acid Cycle Although oxygen does not participate directly in the citric acid cycle, the cycle operates only when O_2 is present. Why?

31. Effect of [NADH]/[NAD$^+$] on the Citric Acid Cycle How would you expect the operation of the citric acid cycle to respond to a rapid increase in the [NADH]/[NAD$^+$] ratio in the mitochondrial matrix? Why?

32. Thermodynamics of Citrate Synthase Reaction in Cells Citrate is formed by the condensation of acetyl-CoA with oxaloacetate, catalyzed by citrate synthase:

Oxaloacetate + acetyl-CoA + H_2O $\rightleftharpoons$ citrate + CoA + H^+

In rat heart mitochondria at pH 7.0 and 25 °C, the concentrations of reactants and products are: oxaloacetate, 1 μM; acetyl-CoA, 1 μM; citrate, 220 μM; and CoA, 65 μM. The standard free-energy change for the citrate synthase reaction is -32.2 kJ/mol. What is the direction of metabolite flow through the citrate synthase reaction in rat heart cells? Explain.

33. Reactions of the Pyruvate Dehydrogenase Complex Two of the steps in the oxidative decarboxylation of pyruvate (steps ❹ and ❺ in Fig. 16-6) do not involve any of the three carbons of pyruvate yet are essential to the operation of the PDH complex. Explain.

34. Citric Acid Cycle Mutants There are many cases of human disease in which one or another enzyme activity is lacking due to genetic mutation. However, cases in which individuals lack one of the enzymes of the citric acid cycle are extremely rare. Why?

Data Analysis Problem

35. How the Citric Acid Cycle Was Discovered The detailed biochemistry of the citric acid cycle was determined by several researchers over a period of decades. In a 1937 article, Krebs and Johnson summarized their work and the work of others in the first published description of this pathway.

The methods used by these researchers were very different from those of modern biochemistry. Radioactive tracers were not commonly available until the 1940s, so Krebs and other researchers had to use nontracer techniques to work out the pathway. Using freshly prepared samples of pigeon breast muscle, they determined oxygen consumption by suspending minced muscle in buffer in a sealed flask and measuring the volume (in μL) of oxygen consumed under different conditions. They measured levels of substrates (intermediates) by treating samples with acid to remove contaminating proteins, then assaying the quantities of various small organic molecules. The two key observations that led Krebs and colleagues to propose a citric acid *cycle* as opposed to a *linear pathway* (like that of glycolysis) were made in the following experiments.

Experiment I. They incubated 460 mg of minced muscle in 3 mL of buffer at 40 °C for 150 minutes. Addition of *citrate* increased O_2 consumption by 893 μL compared with samples without added citrate. They calculated, based on the O_2 consumed during respiration of other carbon-containing compounds, that the expected O_2 consumption for complete respiration of this quantity of citrate was only 302 μL.

Experiment II. They measured O_2 consumption by 460 mg of minced muscle in 3 mL of buffer when incubated with *citrate* and/or with *1-phosphoglycerol* (glycerol 1-phosphate; this was known to be readily oxidized by cellular respiration) at 40 °C for 140 minutes. The results are shown in the table.

Sample	Substrate(s) added	μL O_2 absorbed
1	No extra	342
2	0.3 mL 0.2 M 1-phosphoglycerol	757
3	0.15 mL 0.02 M citrate	431
4	0.3 mL 0.2 M 1-phosphoglycerol and 0.15 mL 0.02 M citrate	1,385

(a) Why is O_2 consumption a good measure of cellular respiration?

(b) Why does sample 1 (unsupplemented muscle tissue) consume some oxygen?

(c) Based on the results for samples 2 and 3, can you conclude that 1-phosphoglycerol and citrate serve as substrates for cellular respiration in this system? Explain your reasoning.

(d) Krebs and colleagues used the results from these experiments to argue that citrate was "catalytic"—that it helped the muscle tissue samples metabolize 1-phosphoglycerol more completely. How would you use their data to make this argument?

(e) Krebs and colleagues further argued that citrate was not simply consumed by these reactions, but had to be *regenerated*. Therefore, the reactions had to be a *cycle* rather than a linear pathway. How would you make this argument?

Other researchers had found that *arsenate* (AsO_4^{3-}) inhibits α-ketoglutarate dehydrogenase and that *malonate* inhibits succinate dehydrogenase.

(f) Krebs and coworkers found that muscle tissue samples treated with arsenate and citrate would consume citrate only in the presence of oxygen; under these conditions, oxygen was consumed. Based on the pathway in Figure 16-7, what was the citrate converted to in this experiment, and why did the samples consume oxygen?

In their article, Krebs and Johnson further reported the following. (1) In the presence of arsenate, 5.48 mmol of citrate was converted to 5.07 mmol of α-ketoglutarate. (2) In the presence of malonate, citrate was quantitatively converted to large amounts of succinate and small amounts of α-ketoglutarate. (3) Addition of oxaloacetate in the absence of oxygen led to production of a large amount of citrate; the amount was increased if glucose was also added.

Other workers had found the following pathway in similar muscle tissue preparations:

Succinate $\longrightarrow$ fumarate $\longrightarrow$ malate $\longrightarrow$ oxaloacetate $\longrightarrow$ pyruvate

(g) Based only on the data presented in this problem, what is the order of the intermediates in the citric acid cycle? How does this compare with Figure 16-7? Explain your reasoning.

(h) Why was it important to show the *quantitative* conversion of citrate to α-ketoglutarate?

The Krebs and Johnson article also contains other data that filled in most of the missing components of the cycle. The only component left unresolved was the molecule that reacted with oxaloacetate to form citrate.

Reference

Krebs, H.A., and W.A. Johnson. 1937. The role of citric acid in intermediate metabolism in animal tissues. *Enzymologia* 4:148–156. Reprinted in *FEBS Lett.* 117(Suppl.):K2–K10, 1980.

Further Reading is available at www.macmillanlearning.com/LehningerBiochemistry7e.

CHAPTER 17

Fatty Acid Catabolism

> Self-study tools that will help you practice what you've learned and reinforce this chapter's concepts are available online.
> Go to www.macmillanlearning.com/LehningerBiochemistry7e.

The oxidation of long-chain fatty acids to acetyl-CoA is a central energy-yielding pathway in many organisms and tissues. In mammalian heart and liver, for example, it provides as much as 80% of the energetic needs under all physiological circumstances. The electrons removed from fatty acids during oxidation pass through the respiratory chain, driving ATP synthesis; the acetyl-CoA produced from the fatty acids may be completely oxidized to CO_2 in the citric acid cycle, resulting in further energy conservation. In some species and in some tissues, the acetyl-CoA has alternative fates. In liver, acetyl-CoA may be converted to ketone bodies—water-soluble fuels exported to the brain and other tissues when glucose is not available. In vascular plants, acetyl-CoA serves primarily as a biosynthetic precursor, only secondarily as fuel. Although the biological role of fatty acid oxidation differs from organism to organism, the mechanism is essentially the same. The repetitive four-step process by which fatty acids are converted into acetyl-CoA, called **β oxidation**, is the main topic of this chapter.

In Chapter 10 we described the properties of triacylglycerols (also called triglycerides or neutral fats) that make them especially suitable as storage fuels. The long alkyl chains of their constituent fatty acids are essentially hydrocarbons, highly reduced structures with an energy of complete oxidation (~38 kJ/g) more than twice that for the same weight of carbohydrate or protein. This advantage is compounded by the extreme insolubility of lipids in water; cellular triacylglycerols aggregate in lipid droplets, which do not raise the osmolarity of the cytosol, and they are unsolvated. (In storage polysaccharides, by contrast, water of solvation can account for two-thirds of the overall weight of the stored molecules.) And because of their relative chemical inertness, triacylglycerols can be stored in large quantity in cells without the risk of undesired chemical reactions with other cellular constituents.

The properties that make triacylglycerols good storage compounds, however, present problems in their role as fuels. Because they are insoluble in water, ingested triacylglycerols must be emulsified before they can be digested by water-soluble enzymes in the intestine, and triacylglycerols absorbed in the intestine or mobilized from storage tissues must be carried in the blood bound to proteins that counteract their insolubility. Also, to overcome the relative stability of the C—C bonds in a fatty acid, the carboxyl group at C-1 is activated by attachment to coenzyme A, which allows stepwise oxidation of the fatty acyl group at the C-3, or β, position—hence the name β oxidation.

We begin this chapter with a brief discussion of the sources of fatty acids and the routes by which they travel to the site of their oxidation, with special emphasis on the process in vertebrates. We then describe the chemical steps of fatty acid oxidation in mitochondria. The complete oxidation of fatty acids to CO_2 and H_2O takes place in three stages: the oxidation of long-chain fatty acids to two-carbon fragments, in the form of acetyl-CoA (β oxidation); the oxidation of acetyl-CoA to CO_2 in the citric acid cycle (Chapter 16); and the transfer of electrons from reduced electron carriers to the mitochondrial respiratory chain (Chapter 19). In this chapter we focus on the first of these stages. We begin our discussion of β oxidation with the simple case in which a fully saturated fatty acid with an even number of carbon atoms is degraded to acetyl-CoA. We then look briefly at the extra transformations necessary for the degradation of unsaturated fatty acids and fatty acids with an odd number of carbons. Finally, we discuss variations on the β-oxidation theme in specialized organelles—peroxisomes and glyoxysomes—and two less common pathways of fatty acid catabolism, ω and α oxidation. The chapter concludes with a description of an alternative fate for the acetyl-CoA formed by β oxidation in vertebrates: the production of ketone bodies in the liver.

17.1 Digestion, Mobilization, and Transport of Fats

Cells can obtain fatty acid fuels from four sources: fats consumed in the diet, fats stored in cells as lipid droplets, fats synthesized in one organ for export to another, and fats obtained by autophagy (which degrades the cell's own organelles). Some species use all four sources under various circumstances, others use one or two. Vertebrates, for example, obtain fats in the diet, mobilize fats stored in specialized tissue (adipose tissue, consisting of cells called adipocytes), and, in the liver, convert excess dietary carbohydrates to fats for export to other tissues. During starvation, they can recycle lipids by autophagy. On average, 40% or more of the daily energy requirement of humans in highly industrialized countries is supplied by dietary triacylglycerols (although most nutritional guidelines recommend no more than 30% of daily caloric intake from fats). Triacylglycerols provide more than half the energy requirements of some organs, particularly the liver, heart, and resting skeletal muscle. Stored triacylglycerols are virtually the sole source of energy in hibernating animals and migrating birds. Protists obtain fats by consuming organisms lower in the food chain, and some also store fats as cytosolic lipid droplets. Vascular plants mobilize fats stored in seeds during germination, but do not otherwise depend on fats for energy.

Dietary Fats Are Absorbed in the Small Intestine

In vertebrates, before ingested triacylglycerols can be absorbed through the intestinal wall they must be converted from insoluble macroscopic fat particles to finely dispersed microscopic micelles. This solubilization is carried out by bile salts, such as taurocholic acid (p. 374), which are synthesized from cholesterol in the liver, stored in the gallbladder, and released into the small intestine after ingestion of a fatty meal. Bile salts are amphipathic compounds that act as biological detergents, converting dietary fats into mixed micelles of bile salts and triacylglycerols (**Fig. 17-1**, step **❶**).

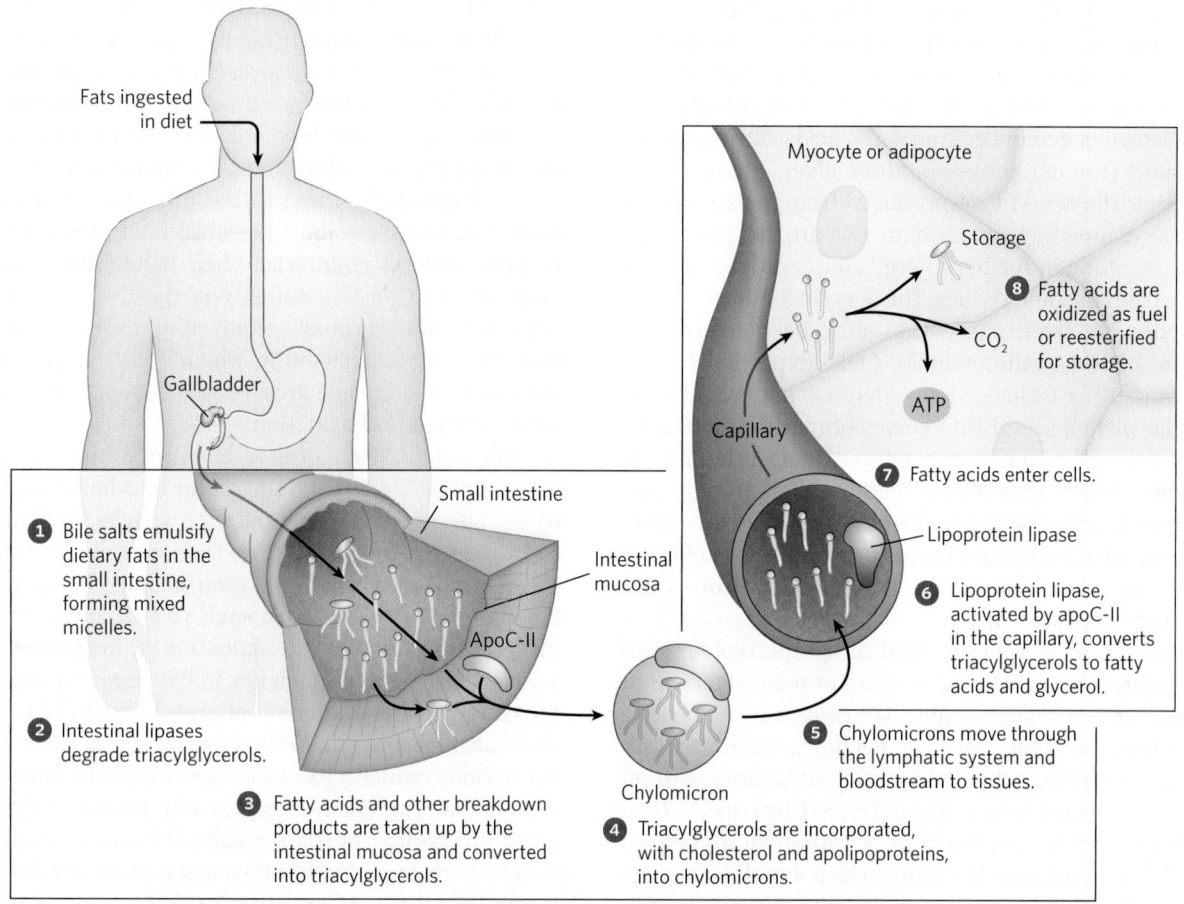

FIGURE 17-1 Processing of dietary lipids in vertebrates. Digestion and absorption of dietary lipids occur in the small intestine, and the fatty acids released from triacylglycerols are packaged and delivered to muscle and adipose tissues. The eight steps are discussed in the text.

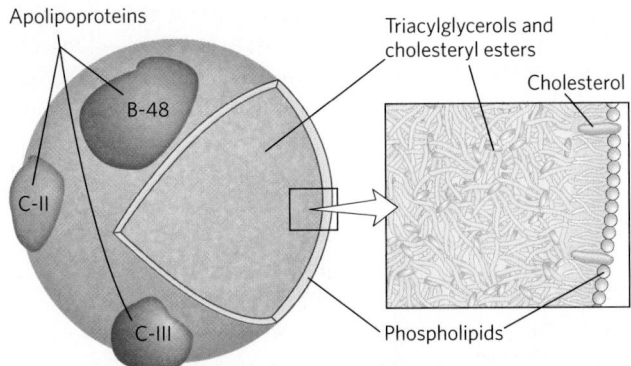

FIGURE 17-2 Molecular structure of a chylomicron. The surface is a layer of phospholipids, with head groups facing the aqueous phase. Triacylglycerols sequestered in the interior (yellow) make up more than 80% of the mass. Several apolipoproteins that protrude from the surface (B-48, C-II, C-III) act as signals in the uptake and metabolism of chylomicron contents. The diameter of chylomicrons ranges from about 100 to 500 nm.

Micelle formation enormously increases the fraction of lipid molecules accessible to the action of water-soluble lipases in the intestine, and lipase action converts triacylglycerols to monoacylglycerols (monoglycerides) and diacylglycerols (diglycerides), free fatty acids, and glycerol (step ❷). These products of lipase action diffuse into the epithelial cells lining the intestinal surface (the intestinal mucosa) (step ❸), where they are reconverted to triacylglycerols and packaged with dietary cholesterol and specific proteins into lipoprotein aggregates called **chylomicrons** (**Fig. 17-2**; see also Fig. 17-1, step ❹).

Apolipoproteins are lipid-binding proteins in the blood that are responsible for the transport of triacylglycerols, phospholipids, cholesterol, and cholesteryl esters between organs. Apolipoproteins ("apo" means "detached" or "separate," designating the protein in its lipid-free form) combine with lipids to form several classes of **lipoprotein** particles, spherical aggregates with hydrophobic lipids at the core and hydrophilic protein side chains and lipid head groups at the surface. Various combinations of lipid and protein produce particles of different densities, ranging from chylomicrons and very-low-density lipoproteins (VLDL) to very-high-density lipoproteins (VHDL), which can be separated by ultracentrifugation. The structures of these lipoprotein particles and their roles in lipid transport are detailed in Chapter 21.

The protein moieties of lipoproteins are recognized by receptors on cell surfaces. In lipid uptake from the intestine, chylomicrons, which contain apolipoprotein C-II (apoC-II), move from the intestinal mucosa into the lymphatic system, and then enter the blood, which carries them to muscle and adipose tissue (Fig. 17-1, step ❺). In the capillaries of these tissues, the extracellular enzyme **lipoprotein lipase**, activated by apoC-II, hydrolyzes triacylglycerols to fatty acids and glycerol

(step ❻), which are taken up by specific transporters in the plasma membranes of cells in the target tissues (step ❼). In muscle, the fatty acids are oxidized for energy; in adipose tissue, they are reesterified for storage as triacylglycerols (step ❽).

The remnants of chylomicrons, depleted of most of their triacylglycerols but still containing cholesterol and apolipoproteins, travel in the blood to the liver, where they are taken up by endocytosis, mediated by receptors for their apolipoproteins. Triacylglycerols that enter the liver by this route may be oxidized to provide energy or to provide precursors for the synthesis of ketone bodies, as described in Section 17.3. When the diet contains more fatty acids than are needed immediately for fuel or as precursors, the liver converts them to triacylglycerols, which are packaged with specific apolipoproteins into VLDLs. These VLDLs are secreted by hepatocytes and transported in the blood to adipose tissue, where the triacylglycerols are removed and stored in lipid droplets within adipocytes.

Hormones Trigger Mobilization of Stored Triacylglycerols

Neutral lipids are stored in adipocytes (and in steroid-synthesizing cells of the adrenal cortex, ovary, and testis) in the form of **lipid droplets**, with a core of triacylglycerols and sterol esters surrounded by a monolayer of phospholipids. The surface of these droplets is coated with **perilipins**, a family of proteins that restrict access to lipid droplets, preventing untimely lipid mobilization. When hormones signal the need for metabolic energy, triacylglycerols stored in adipose tissue are mobilized (brought out of storage) and transported to tissues (skeletal muscle, heart, and renal cortex) in which fatty acids can be oxidized for energy production. The hormones epinephrine and glucagon, secreted in response to low blood glucose levels or a fight-or-flight situation, stimulate the enzyme adenylyl cyclase in the adipocyte plasma membrane (**Fig. 17-3**), which produces the intracellular second messenger cyclic AMP (cAMP; see Fig. 12-4). Cyclic AMP–dependent protein kinase (PKA) triggers changes that open the lipid droplet to the action of three cytosolic lipases, which act on tri-, di-, and monoacylglycerols, releasing fatty acids and glycerol.

The fatty acids thus released (**free fatty acids, FFAs**) pass from the adipocyte into the blood, where they bind to the blood protein **serum albumin** (Fig. 17-3). This protein (M_r 66,000), which makes up about half of the total serum protein, noncovalently binds as many as 10 fatty acids per protein monomer. Bound to this soluble protein, the otherwise insoluble fatty acids are carried to tissues such as skeletal muscle, heart, and renal cortex. In these target tissues, fatty acids dissociate from albumin and are moved by plasma membrane transporters into cells to serve as fuel.

About 95% of the biologically available energy of triacylglycerols resides in their three long-chain fatty

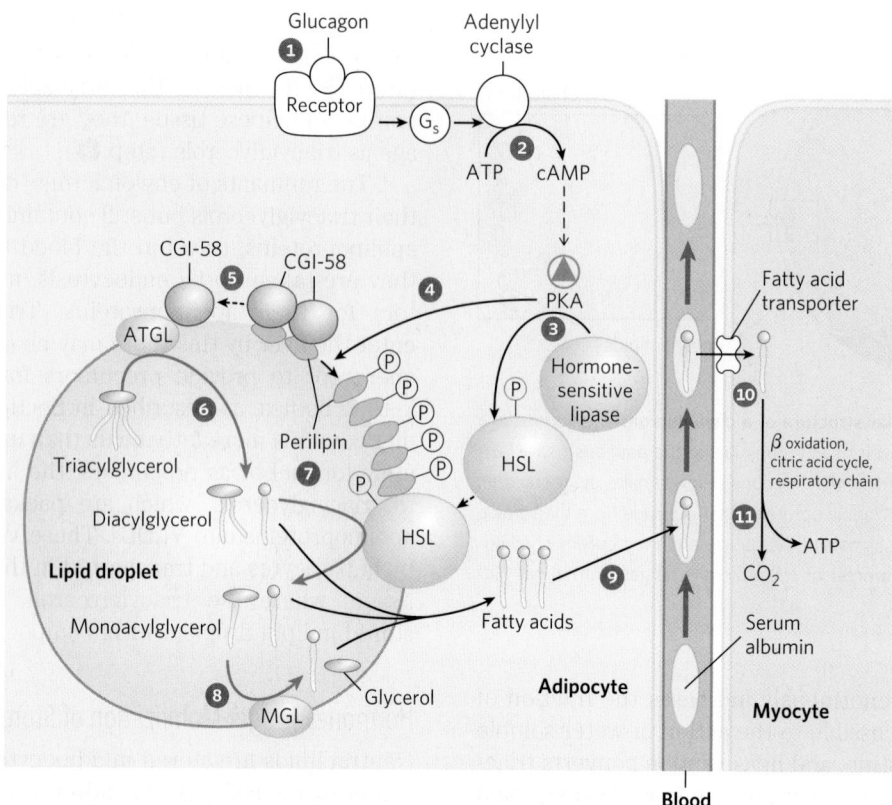

FIGURE 17-3 Mobilization of triacylglycerols stored in adipose tissue. When low levels of glucose in the blood trigger the release of glucagon, ❶ the hormone binds its receptor in the adipocyte membrane and thus ❷ stimulates adenylyl cyclase, via a G protein, to produce cAMP. This activates PKA, which phosphorylates ❸ the hormone-sensitive lipase (HSL) and ❹ perilipin molecules on the surface of the lipid droplet. Phosphorylation of perilipin causes ❺ dissociation of the protein CGI-58 from perilipin. CGI-58 then recruits adipose triacylglycerol lipase (ATGL) to the droplet surface and stimulates its lipase activity. Active ATGL ❻ converts triacylglycerols to diacylglycerols.

The phosphorylated perilipin associates with phosphorylated HSL, allowing it access to the surface of the lipid droplet, where ❼ it converts diacylglycerols to monoacylglycerols. ❽ A third lipase, monoacylglycerol lipase (MGL), hydrolyzes monoacylglycerols. ❾ Fatty acids leave the adipocyte, and are transported in the blood bound to serum albumin. They are released from the albumin and ❿ enter a myocyte via a specific fatty acid transporter. ⓫ In the myocyte, fatty acids are oxidized to CO_2, and the energy of oxidation is conserved in ATP, which fuels muscle contraction and other energy-requiring metabolism in the myocyte.

acids; only 5% is contributed by the glycerol moiety. The glycerol released by lipase action is phosphorylated by **glycerol kinase (Fig. 17-4)**, and the resulting glycerol 3-phosphate is oxidized to dihydroxyacetone phosphate. The glycolytic enzyme triose phosphate isomerase converts this compound to glyceraldehyde 3-phosphate, which is oxidized via glycolysis.

Fatty Acids Are Activated and Transported into Mitochondria

The enzymes of fatty acid oxidation in animal cells are located in the mitochondrial matrix, as demonstrated in 1948 by Eugene P. Kennedy and Albert Lehninger. The fatty acids with chain lengths of 12 or fewer carbons enter mitochondria without the help of membrane transporters. Those with 14 or more carbons, which constitute the majority of the FFAs obtained in the diet or released from adipose tissue, cannot pass directly through the mitochondrial membranes: they must first undergo the three enzymatic reactions of the **carnitine shuttle**. The first

reaction is catalyzed by a family of isozymes of **acyl-CoA synthetase**, each specific for fatty acids having either short, intermediate, or long carbon chains. The isozymes are present in the outer mitochondrial membrane, where they promote the general reaction

$$\text{Fatty acid} + \text{CoA} + \text{ATP} \rightleftharpoons \text{fatty acyl–CoA} + \text{AMP} + \text{PP}_i$$

Thus, acyl–CoA synthetases catalyze the formation of a thioester linkage between the fatty acid carboxyl group and the thiol group of coenzyme A to yield a **fatty acyl–CoA**, coupled to the cleavage of ATP to AMP and PP_i. (Recall the description of this reaction in Chapter 13, to illustrate how the free energy released by cleavage of phosphoanhydride bonds in ATP could be coupled to the formation of a high-energy compound; p. 514.) The reaction occurs in two steps and involves a fatty acyl-adenylate intermediate **(Fig. 17-5)**.

Fatty acyl–CoAs, like acetyl-CoA, are high-energy compounds; their hydrolysis to FFAs and CoA has a large, negative standard free-energy change ($\Delta G'^\circ = -31$ kJ/mol). The formation of a fatty acyl–CoA is made

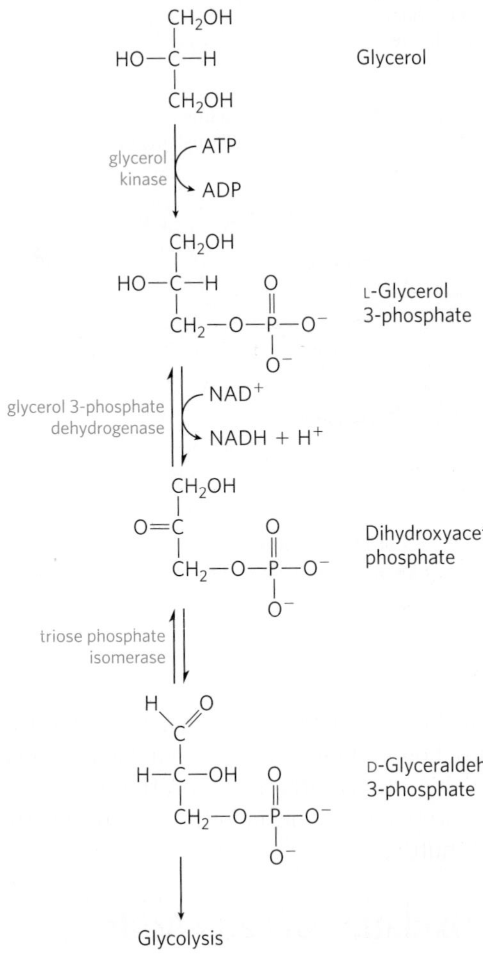

CH$_2$OH
|
HO—C—H   Glycerol
|
CH$_2$OH

glycerol kinase — ATP → ADP

CH$_2$OH
|
HO—C—H   O
| ‖
CH$_2$—O—P—O$^-$ L-Glycerol 3-phosphate
          |
          O$^-$

glycerol 3-phosphate dehydrogenase — NAD$^+$ → NADH + H$^+$

CH$_2$OH
|
O=C   O
| ‖
CH$_2$—O—P—O$^-$ Dihydroxyacetone phosphate
       |
       O$^-$

triose phosphate isomerase

H   O
 \\ //
  C
  |
H—C—OH   O
  | ‖
  CH$_2$—O—P—O$^-$ D-Glyceraldehyde 3-phosphate
          |
          O$^-$

Glycolysis

FIGURE 17-4 Entry of glycerol into the glycolytic pathway.

into the mitochondrion and oxidized to produce ATP, or they can be used in the cytosol to synthesize membrane lipids. Fatty acids destined for mitochondrial oxidation are transiently attached to the hydroxyl group of **carnitine** to form fatty acyl–carnitine—the second reaction of the shuttle.

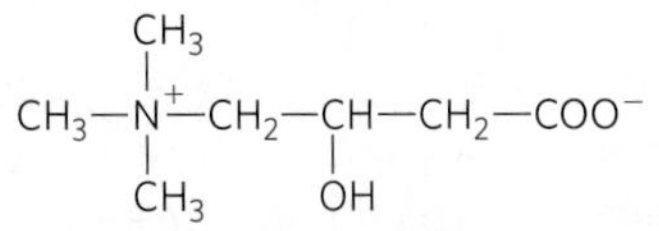

CH$_3$
|
CH$_3$—N$^+$—CH$_2$—CH—CH$_2$—COO$^-$
| |
CH$_3$ OH

Carnitine

This transesterification is catalyzed by **carnitine acyltransferase 1 (also called carnitine palmitoyltransferase 1, CPT1)**, in the outer membrane (**Fig. 17-6**). The acyl-CoA is converted to the carnitine ester as it passes through the outer membrane. The fatty acyl–carnitine ester then diffuses across the intermembrane space and enters the matrix by passive transport through the **acyl-carnitine/carnitine cotransporter** of the inner mitochondrial membrane. This cotransporter moves one molecule of carnitine from the matrix to the intermembrane space as one molecule of fatty acyl–carnitine moves into the matrix.

In the third and final step of the carnitine shuttle, the fatty acyl group is transferred from carnitine to intramitochondrial coenzyme A by **carnitine acyltransferase 2 (also called CPT2)**. This isozyme, located on the inner face of the inner mitochondrial membrane, regenerates fatty acyl–CoA and releases it, along with free carnitine, into the matrix (Fig. 17-6).

This three-step process for transferring fatty acids into the mitochondrion—esterification to CoA, transesterification to carnitine followed by transport, and transesterification back to CoA—links two separate

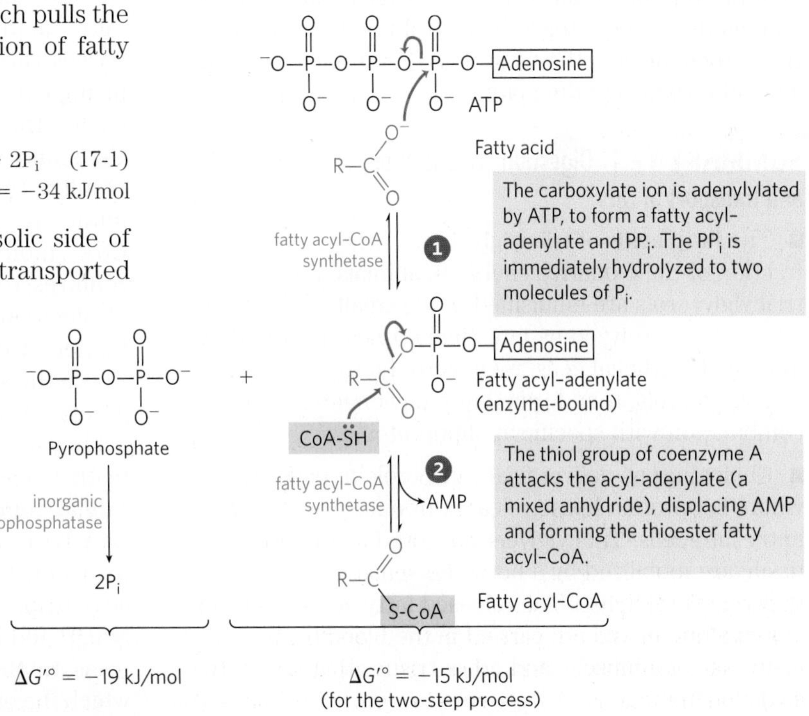

more favorable by the hydrolysis of *two* high-energy bonds in ATP; the pyrophosphate formed in the activation reaction is immediately hydrolyzed by inorganic pyrophosphatase (left side of Fig. 17-5), which pulls the preceding activation reaction in the direction of fatty acyl–CoA formation. The overall reaction is

Fatty acid + CoA + ATP ⇌

fatty acyl–CoA + AMP + 2P$_i$ (17-1)

$$\Delta G'^\circ = -34 \text{ kJ/mol}$$

Fatty acyl–CoA esters formed on the cytosolic side of the outer mitochondrial membrane can be transported

MECHANISM FIGURE 17-5 Activation of a fatty acid by conversion to a fatty acyl-CoA. The conversion is catalyzed by fatty acyl-CoA synthetase and inorganic pyrophosphatase. Formation of the fatty acyl-CoA derivative occurs in two steps. The overall reaction is highly exergonic.

O   O   O
‖   ‖   ‖
$^-$O—P—O—P—O—P—O—[Adenosine]
| | |
O$^-$ O$^-$ O$^-$ ATP

Fatty acid
|
R—C
   \\
    O

fatty acyl-CoA synthetase ①

The carboxylate ion is adenylylated by ATP, to form a fatty acyl-adenylate and PP$_i$. The PP$_i$ is immediately hydrolyzed to two molecules of P$_i$.

O   O
‖   ‖
$^-$O—P—O—P—O$^-$ +
| |
O$^-$ O$^-$

Pyrophosphate

inorganic pyrophosphatase

2P$_i$

$$\Delta G'^\circ = -19 \text{ kJ/mol}$$

O
‖
R—C—O—P—O—[Adenosine]
         |
         O$^-$ Fatty acyl-adenylate (enzyme-bound)

CoA-SH

fatty acyl-CoA synthetase ② → AMP

The thiol group of coenzyme A attacks the acyl-adenylate (a mixed anhydride), displacing AMP and forming the thioester fatty acyl-CoA.

O
‖
R—C—S-CoA   Fatty acyl-CoA

$$\Delta G'^\circ = -15 \text{ kJ/mol}$$
(for the two-step process)

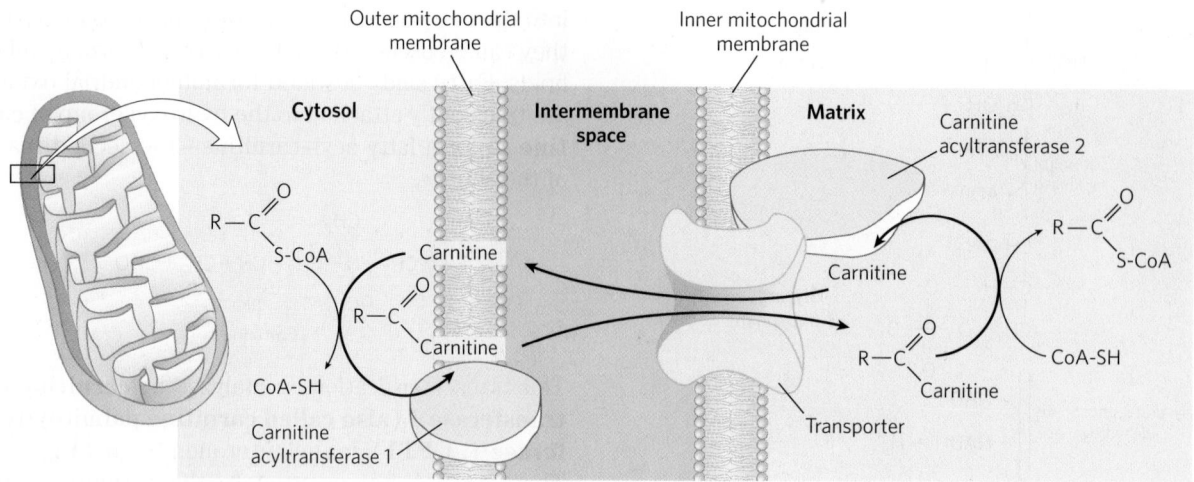

FIGURE 17-6 Fatty acid entry into mitochondria via the acyl-carnitine/carnitine transporter. After fatty acyl–carnitine is formed at the outer membrane or in the intermembrane space, it moves into the matrix by passive transport through the inner membrane. In the matrix, the acyl group is transferred to mitochondrial coenzyme A, freeing carnitine to return to the intermembrane space through the same transporter. Carnitine acyltransferase 1 is inhibited by malonyl-CoA, the first intermediate in fatty acid synthesis (see Fig. 21-2). This inhibition prevents the simultaneous synthesis and degradation of fatty acids.

pools of coenzyme A and of fatty acyl–CoA, one in the cytosol, the other in mitochondria. These pools have different functions. Coenzyme A in the mitochondrial matrix is largely used in oxidative degradation of pyruvate, fatty acids, and some amino acids, whereas cytosolic coenzyme A is used in the biosynthesis of fatty acids (see Fig. 21-10). Fatty acyl–CoA in the cytosolic pool can be used there for membrane lipid synthesis or can be moved into the mitochondrial matrix for oxidation and ATP production. Conversion to the carnitine ester commits the fatty acyl moiety to the oxidative fate.

The carnitine-mediated entry process is the rate-limiting step for oxidation of fatty acids in mitochondria and, as discussed later, is a control point. Once inside the mitochondrion, the fatty acyl–CoA is acted upon by a set of enzymes in the matrix.

SUMMARY 17.1 Digestion, Mobilization, and Transport of Fats

■ The fatty acids of triacylglycerols furnish a large fraction of the oxidative energy in animals. Dietary triacylglycerols are emulsified in the small intestine by bile salts, hydrolyzed by intestinal lipases, absorbed by intestinal epithelial cells, reconverted into triacylglycerols, then formed into chylomicrons by combination with specific apolipoproteins.

■ Chylomicrons deliver triacylglycerols to tissues, where lipoprotein lipase releases free fatty acids for entry into cells. Triacylglycerols stored in adipose tissue are mobilized by a hormone-sensitive triacylglycerol lipase. The released fatty acids bind to serum albumin and are carried in the blood to the heart, skeletal muscle, and other tissues that use fatty acids for fuel.

■ Once inside cells, fatty acids are activated at the outer mitochondrial membrane by conversion to fatty acyl–CoA thioesters. Fatty acyl–CoA that is to be oxidized enters mitochondria in three steps, via the carnitine shuttle.

17.2 Oxidation of Fatty Acids

As noted earlier, mitochondrial oxidation of fatty acids takes place in three stages **(Fig. 17-7)**. In the first stage—β oxidation—fatty acids undergo oxidative removal of successive two-carbon units in the form of acetyl-CoA, starting from the carboxyl end of the fatty acyl chain. For example, the 16-carbon palmitic acid (palmitate at pH 7) undergoes seven passes through the oxidative sequence, in each pass losing two carbons as acetyl-CoA. At the end of seven cycles the last two carbons of palmitate (originally C-15 and C-16) remain as acetyl-CoA. The overall result is the conversion of the 16-carbon chain of palmitate to eight two-carbon acetyl groups of acetyl-CoA molecules. Formation of each acetyl-CoA requires removal of four hydrogen atoms (two pairs of electrons and four H^+) from the fatty acyl moiety by dehydrogenases.

In the second stage of fatty acid oxidation, the acetyl groups of acetyl-CoA are oxidized to CO_2 in the citric acid cycle, which also takes place in the mitochondrial matrix. Acetyl-CoA derived from fatty acids thus enters a final common pathway of oxidation with the acetyl-CoA derived from glucose via glycolysis and pyruvate oxidation (see Fig. 16-1). The first two stages of fatty acid oxidation produce the reduced electron carriers NADH and $FADH_2$, which in the third stage donate electrons to the mitochondrial respiratory chain, through which the electrons pass to oxygen with the concomitant

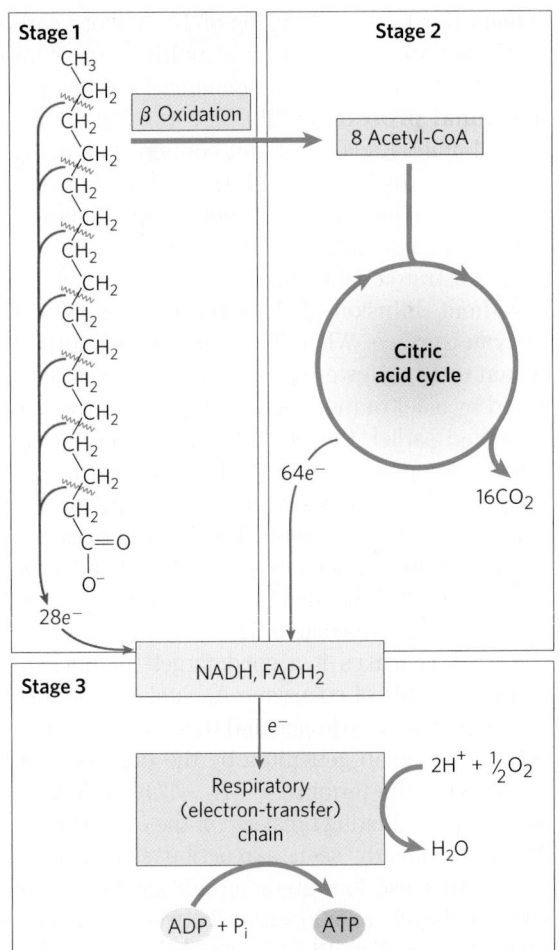

FIGURE 17-7 Stages of fatty acid oxidation. Stage 1: A long-chain fatty acid is oxidized to yield acetyl residues in the form of acetyl-CoA. This process is called β oxidation. Stage 2: The acetyl groups are oxidized to CO_2 via the citric acid cycle. Stage 3: Electrons derived from the oxidations of stages 1 and 2 pass to O_2 via the mitochondrial respiratory chain, providing the energy for ATP synthesis by oxidative phosphorylation.

phosphorylation of ADP to ATP (Fig. 17-7). The energy released by fatty acid oxidation is thus conserved as ATP.

We now take a closer look at the first stage of fatty acid oxidation, beginning with the simple case of a saturated fatty acyl chain with an even number of carbons, then turning to the slightly more complicated cases of unsaturated and odd-number chains. We also consider the regulation of fatty acid oxidation, the β-oxidative

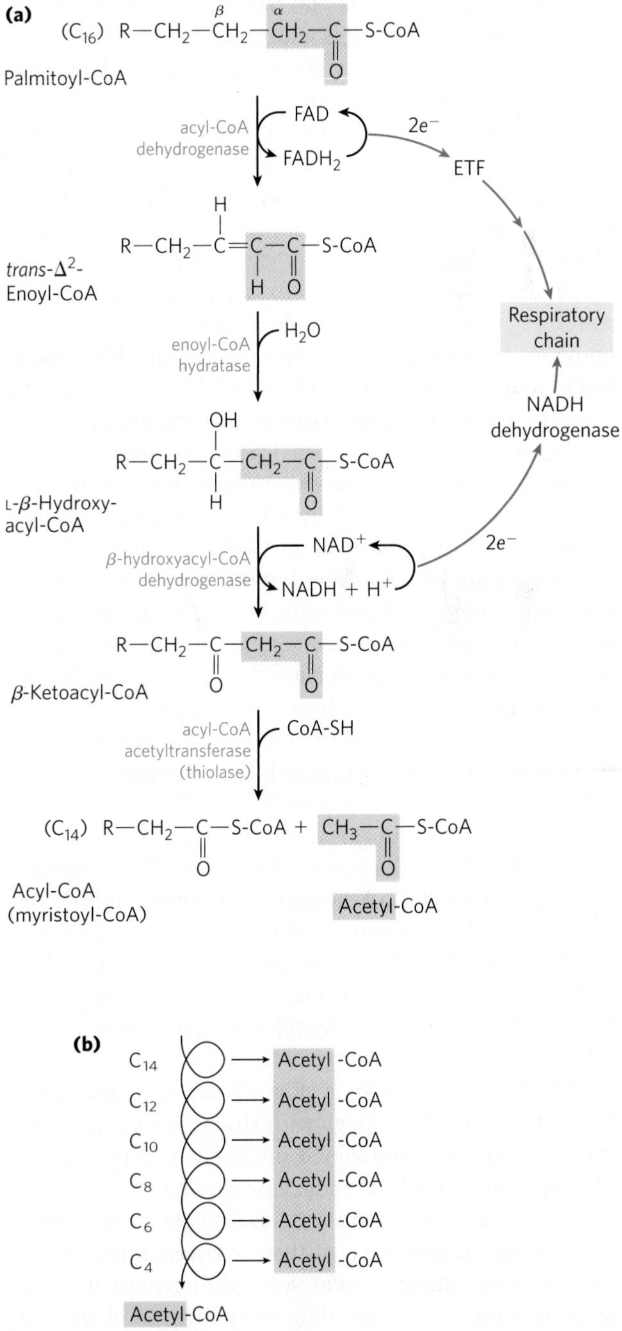

FIGURE 17-8 The β-oxidation pathway. (a) In each pass through this four-step sequence, one acetyl residue (shaded in light red) is removed in the form of acetyl-CoA from the carboxyl end of the fatty acyl chain—in this example palmitate (C_{16}), which enters as palmitoyl-CoA. Electrons from the first oxidation pass through the flavoprotein ETF, and then through a second flavoprotein, into the respiratory chain. Electrons from the second oxidation enter the respiratory chain through NADH dehydrogenase (see Fig. 19-15). **(b)** Six more passes through the β-oxidation pathway yield seven more molecules of acetyl-CoA, the seventh arising from the last two carbon atoms of the 16-carbon chain. Eight molecules of acetyl-CoA are formed in all. The acetyl-CoA may be oxidized in the citric acid cycle, donating more electrons to the respiratory chain.

processes as they occur in organelles other than mitochondria, and, finally, two less-general modes of fatty acid catabolism, α oxidation and ω oxidation.

The β Oxidation of Saturated Fatty Acids Has Four Basic Steps

Four enzyme-catalyzed reactions make up the first stage of fatty acid oxidation **(Fig. 17-8a)**. First, dehydrogenation of fatty acyl–CoA produces a double bond between the α and β carbon atoms (C-2 and C-3), yielding a **trans-Δ^2-enoyl-CoA** (the symbol Δ^2 designates the position of the double bond; you may want to review fatty acid nomenclature, p. 361.) Note that the new double bond has the trans configuration, whereas the double bonds in naturally occurring unsaturated fatty acids are normally in the cis configuration. We consider the significance of this difference later.

This first step is catalyzed by three isozymes of **acyl-CoA dehydrogenase**, each specific for a range of fatty-acyl chain lengths: very-long-chain acyl-CoA dehydrogenase (VLCAD), acting on fatty acids of 12 to 18 carbons; medium-chain (MCAD), acting on fatty acids of 4 to 14 carbons; and short-chain (SCAD), acting on fatty acids of 4 to 8 carbons. VLCAD is in the inner mitochondrial membrane; MCAD and SCAD are in the matrix. All three isozymes are flavoproteins with tightly bound FAD (see Fig. 13-27) as a prosthetic group. The electrons removed from the fatty acyl–CoA are transferred to FAD, and the reduced form of the dehydrogenase immediately donates its electrons to an electron carrier of the mitochondrial respiratory chain, the **electron-transferring flavoprotein (ETF)** (see Fig. 19-15). The oxidation catalyzed by an acyl-CoA dehydrogenase is analogous to succinate dehydrogenation in the citric acid cycle (p. 632); in both reactions the enzyme is bound to the inner membrane, a double bond is introduced into a carboxylic acid between the α and β carbons, FAD is the electron acceptor, and electrons from the reaction ultimately enter the respiratory chain and pass to O_2, with the concomitant synthesis of about 1.5 ATP molecules per electron pair.

In the second step of the β-oxidation cycle (Fig. 17-8a), water is added to the double bond of the $trans$-Δ^2-enoyl-CoA to form the L stereoisomer of β-**hydroxyacyl-CoA (3-hydroxyacyl-CoA)**. This reaction, catalyzed by **enoyl-CoA hydratase**, is formally analogous to the fumarase reaction in the citric acid cycle, in which H_2O adds across an α–β double bond (p. 633).

In the third step, L-β-hydroxyacyl-CoA is dehydrogenated to form β-**ketoacyl-CoA**, by the action of β-**hydroxyacyl-CoA dehydrogenase**; NAD^+ is the electron acceptor. This enzyme is absolutely specific for the L stereoisomer of hydroxyacyl-CoA. The NADH formed in the reaction donates its electrons to **NADH dehydrogenase**, an electron carrier of the respiratory chain, and ATP is formed from ADP as the electrons pass to O_2. The reaction catalyzed by β-hydroxyacyl-CoA dehydrogenase is closely analogous to the malate dehydrogenase reaction of the citric acid cycle (p. 633).

The fourth and last step of the β-oxidation cycle is catalyzed by **acyl-CoA acetyltransferase**, more commonly called **thiolase**, which promotes reaction of β-ketoacyl-CoA with a molecule of free coenzyme A to split off the carboxyl-terminal two-carbon fragment of the original fatty acid as acetyl-CoA. The other product is the coenzyme A thioester of the fatty acid, now shortened by two carbon atoms (Fig. 17-8a). This reaction is called thiolysis, by analogy with the process of hydrolysis, because the β-ketoacyl-CoA is cleaved by reaction with the thiol group of coenzyme A. The thiolase reaction is a reverse Claisen condensation (see Fig. 13-4).

The last three steps of this four-step sequence are catalyzed by either of two sets of enzymes, with the enzymes employed depending on the length of the fatty acyl chain. For fatty acyl chains of 12 or more carbons, the reactions are catalyzed by a multienzyme complex associated with the inner mitochondrial membrane, the **trifunctional protein (TFP)**. TFP is a heterooctamer of $\alpha_4\beta_4$ subunits. Each α subunit contains two activities, the enoyl-CoA hydratase and the β-hydroxyacyl-CoA dehydrogenase; the β subunits contain the thiolase activity. This tight association of three enzymes may allow efficient substrate channeling from one active site to the next, without diffusion of the intermediates away from the enzyme surface. When TFP has shortened the fatty acyl chain to 12 or fewer carbons, further oxidations are catalyzed by a set of four soluble enzymes in the matrix.

As noted earlier, the single bond between methylene ($-CH_2-$) groups in fatty acids is relatively stable. The β-oxidation sequence is an elegant mechanism for destabilizing and breaking these bonds. The first three reactions of β oxidation create a much less stable C—C bond, in which the α carbon (C-2) is bonded to *two* carbonyl carbons (the β-ketoacyl-CoA intermediate). The ketone function on the β carbon (C-3) makes it a good target for nucleophilic attack by the —SH of coenzyme A, catalyzed by thiolase. The acidity of the α hydrogen and the resonance stabilization of the carbanion generated by the departure of this hydrogen make the terminal $-CH_2-CO-S$-CoA a good leaving group, facilitating breakage of the α–β bond.

We have already seen a reaction sequence nearly identical with these four steps of fatty acid oxidation, in the citric acid cycle reaction steps between succinate and oxaloacetate (see Fig. 16-7). A nearly identical reaction sequence occurs again in the pathways by which the branched-chain amino acids (isoleucine, leucine, and valine) are oxidized as fuels (see Fig. 18-28). **Figure 17-9** shows the common features of these three sequences, almost certainly an example of the conservation of a mechanism by gene duplication and evolution of a new specificity in the enzyme products of the duplicated genes.

The Four β-Oxidation Steps Are Repeated to Yield Acetyl-CoA and ATP

In one pass through the β-oxidation sequence, one molecule of acetyl-CoA, two pairs of electrons, and four protons (H^+) are removed from the long-chain fatty acyl–CoA, shortening it by two carbon atoms. The equation for one pass, beginning with the coenzyme A ester of our example, palmitate, is

$$\text{Palmitoyl-CoA} + \text{CoA} + \text{FAD} + NAD^+ + H_2O \longrightarrow$$
$$\text{myristoyl-CoA} + \text{acetyl-CoA} + \text{FADH}_2 + \text{NADH} + H^+$$
$$(17\text{-}2)$$

Following removal of one acetyl-CoA unit from palmitoyl-CoA, the coenzyme A thioester of the shortened fatty acid (now the 14-carbon myristate) remains. The myristoyl-CoA can now go through another set of four β-oxidation reactions, exactly analogous to the first, to yield a second molecule of acetyl-CoA and lauroyl-CoA, the coenzyme A thioester of the 12-carbon laurate. Altogether, seven passes

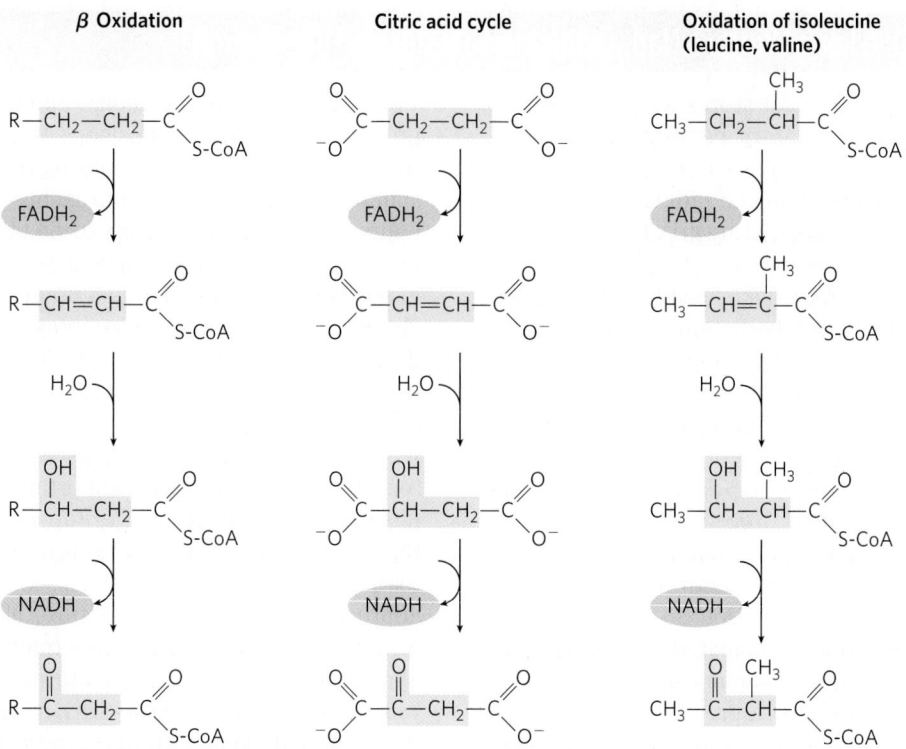

β Oxidation · **Citric acid cycle** · **Oxidation of isoleucine (leucine, valine)**

FIGURE 17-9 A conserved reaction sequence to introduce a carbonyl function on the carbon β to a carboxyl group. The β-oxidation pathway for fatty acyl–CoAs, the pathway from succinate to oxaloacetate in the citric acid cycle, and the pathway by which the deaminated carbon skeletons from isoleucine, leucine, and valine are oxidized as fuels—all use the same reaction sequence.

through the β-oxidation sequence are required to oxidize one molecule of palmitoyl-CoA to eight molecules of acetyl-CoA (Fig. 17-8b). The overall equation is

$$\text{Palmitoyl-CoA} + 7\text{CoA} + 7\text{FAD} + 7\text{NAD}^+ + 7\text{H}_2\text{O} \longrightarrow$$
$$8 \text{ acetyl-CoA} + 7\text{FADH}_2 + 7\text{NADH} + 7\text{H}^+ \quad (17\text{-}3)$$

Each molecule of $FADH_2$ formed during oxidation of the fatty acid donates a pair of electrons to ETF of the respiratory chain, and about 1.5 molecules of ATP are generated during the ensuing transfer of each electron pair to O_2. Similarly, each molecule of NADH formed delivers a pair of electrons to the mitochondrial NADH dehydrogenase, and the subsequent transfer of each pair of electrons to O_2 results in formation of about 2.5 molecules of ATP. Thus four molecules of ATP are formed for each two-carbon unit removed in one pass through the sequence. Note that water is also produced in this process. Each pair of electrons transferred from NADH or $FADH_2$ to O_2 yields one H_2O, referred to as "metabolic water." Reduction of O_2 by NADH also consumes one H^+ per NADH molecule: $\text{NADH} + \text{H}^+ + \frac{1}{2}\text{O}_2 \longrightarrow \text{NAD}^+ + \text{H}_2\text{O}$. In hibernating animals, fatty acid oxidation provides metabolic energy, heat, and water—all essential for survival of an animal that neither eats nor drinks for long periods (Box 17-1). Camels obtain water to supplement the meager supply available in their natural environment by oxidation of fats stored in their hump.

The overall equation for the oxidation of palmitoyl-CoA to eight molecules of acetyl-CoA, including the electron transfers and oxidative phosphorylations, is

$$\text{Palmitoyl-CoA} + 7\text{CoA} + 7\text{O}_2 + 28\text{P}_i + 28\text{ADP} \longrightarrow$$
$$8 \text{ acetyl-CoA} + 28\text{ATP} + 7\text{H}_2\text{O} \quad (17\text{-}4)$$

Acetyl-CoA Can Be Further Oxidized in the Citric Acid Cycle

The acetyl-CoA produced from the oxidation of fatty acids can be oxidized to CO_2 and H_2O by the citric acid cycle. The following equation represents the balance sheet for the second stage in the oxidation of palmitoyl-CoA, together with the coupled phosphorylations of the third stage:

$$8 \text{ Acetyl-CoA} + 16\text{O}_2 + 80\text{P}_i + 80\text{ADP} \longrightarrow$$
$$8\text{CoA} + 80\text{ATP} + 16\text{CO}_2 + 16\text{H}_2\text{O} \quad (17\text{-}5)$$

Combining Equations 17–4 and 17–5, we obtain the overall equation for the complete oxidation of palmitoyl-CoA to carbon dioxide and water:

$$\text{Palmitoyl-CoA} + 23\text{O}_2 + 108\text{P}_i + 108\text{ADP} \longrightarrow$$
$$\text{CoA} + 108\text{ATP} + 16\text{CO}_2 + 23\text{H}_2\text{O} \quad (17\text{-}6)$$

Table 17-1 summarizes the yields of NADH, $FADH_2$, and ATP in the successive steps of palmitoyl-CoA oxidation. Note that because the activation of palmitate to palmitoyl-CoA breaks both phosphoanhydride bonds in ATP (Fig. 17-5), the energetic cost of activating a fatty acid is equivalent to two ATP, and the net gain per molecule of palmitate is 106 ATP. The standard free-energy change for the oxidation of palmitate to CO_2 and H_2O is about 9,800 kJ/mol. Under standard conditions, the energy recovered as the phosphate bond energy of ATP is 106×30.5 kJ/mol = 3,230 kJ/mol, about 33% of the theoretical maximum.

BOX 17-1   A Long Winter's Nap: Oxidizing Fats during Hibernation

Many animals depend on fat stores for energy during hibernation, during migratory periods, and in other situations involving radical metabolic adjustments. One of the most pronounced adjustments of fat metabolism occurs in hibernating grizzly bears. These animals remain in a continuous state of dormancy for as long as seven months. Unlike most hibernating species, the bear maintains a body temperature of about 31 °C, close to the normal (nonhibernating) level (~40 °C). Although expending about 25,000 kJ/day (6,000 kcal/day) while hibernating, the bear does not eat, drink, urinate, or defecate for months at a time. Its heart rate drops from 90 to 8 beats per minute, and its respiration (breathing) drops from 6 to 10 breaths to approximately 1 breath per minute. As we will see in Chapter 19, mitochondrial electron transfer can be uncoupled from ATP production so that all of the energy of fuel oxidation is dissipated as heat, to maintain a body temperature near normal in the face of much lower ambient temperatures.

Experimental studies have shown that hibernating grizzly bears use body fat as their sole fuel. Fat oxidation yields sufficient energy to maintain body temperature, synthesize amino acids and proteins, and carry out other energy-requiring activities, such as membrane transport. Fat oxidation also releases large amounts of water, as described in the text, which replenishes water lost in breathing. The glycerol released by degradation of triacylglycerols is converted into blood glucose by gluconeogenesis. Urea formed during breakdown of amino acids is reabsorbed in the kidneys and recycled, with the amino groups reused to make new amino acids for maintaining body proteins.

Bears store an enormous amount of body fat in preparation for their long sleep. An adult grizzly consumes about 38,000 kJ/day during the late spring and summer, but as winter approaches it feeds for 20 hours a day, consuming up to 84,000 kJ daily. This increase in feeding is a response to a seasonal change in hormone secretion. Large amounts of triacylglycerols are formed from the huge intake of carbohydrates during the fattening-up period. The bear will emerge from hibernation having lost 15% to 40% of its maximum body weight.

The winter sleep of bears is sometimes called torpor, and it differs in important ways from the hibernation behavior of a group of smaller animals that undergo alternating periods of high and low body temperature. In these animals, body temperature approaches ambient temperature, close to 0 °C, for much of the time during hibernation, but rises to almost prehibernation level during brief periods of wakefulness. During these periods, the animals eat, drink, and defecate. In the Arctic ground squirrel (*Urocitellus parryii*), for example, body temperature (37 °C prehibernation) drops to 0 °C during hibernation, and respiration drops to less than 10% of its prehibernation rate.

Studies of hibernation mechanisms may yield insight into several problems in human medicine; for example, slowing the metabolism of organs donated for transplantation might extend the period of their viability. And if humans are to make long trips into space, inducing a torporlike state might relieve the monotony of long missions and conserve on-board resources such as food and oxygen.

A grizzly bear prepares its hibernation nest near the McNeil River in Canada.
[Source: Stouffer Productions/Animals Animals.]

TABLE 17-1 Yield of ATP during Oxidation of One Molecule of Palmitoyl-CoA to CO_2 and H_2O

Enzyme catalyzing the oxidation step	Number of NADH or $FADH_2$ formed	Number of ATP ultimately formed[a]
β Oxidation		
Acyl-CoA dehydrogenase	7 $FADH_2$	10.5
β-Hydroxyacyl-CoA dehydrogenase	7 NADH	17.5
Citric acid cycle		
Isocitrate dehydrogenase	8 NADH	20
α-Ketoglutarate dehydrogenase	8 NADH	20
Succinyl-CoA synthetase		8[b]
Succinate dehydrogenase	8 $FADH_2$	12
Malate dehydrogenase	8 NADH	20
Total		108

[a]These calculations assume that mitochondrial oxidative phosphorylation produces 1.5 ATP per $FADH_2$ oxidized and 2.5 ATP per NADH oxidized.

[b]GTP produced directly in this step yields ATP in the reaction catalyzed by nucleoside diphosphate kinase (p. 516).

However, when the free-energy changes are calculated from actual concentrations of reactants and products under intracellular conditions (see Worked Example 13-2, p. 509), the free-energy recovery is more than 60%; the energy conservation is remarkably efficient.

Oxidation of Unsaturated Fatty Acids Requires Two Additional Reactions

The fatty acid oxidation sequence just described is typical when the incoming fatty acid is saturated (that is, has only single bonds in its carbon chain). However, most of the fatty acids in the triacylglycerols and phospholipids of animals and plants are unsaturated, having one or more double bonds. These bonds are in the cis configuration and cannot be acted upon by enoyl-CoA hydratase, the enzyme catalyzing the addition of H_2O to the trans double bond of the Δ^2-enoyl-CoA generated during β oxidation. Two auxiliary enzymes are needed for β oxidation of the common unsaturated fatty acids: an isomerase and a reductase. We illustrate these auxiliary reactions with two examples.

Oleate is an abundant 18-carbon monounsaturated fatty acid with a cis double bond between C-9 and C-10 (denoted Δ^9). In the first step of oxidation, oleate is converted to oleoyl-CoA and, like the saturated fatty acids, enters the mitochondrial matrix via the carnitine shuttle (Fig. 17-6). Oleoyl-CoA then undergoes three passes through the fatty acid oxidation cycle to yield three molecules of acetyl-CoA and the coenzyme A ester of a Δ^3, 12-carbon unsaturated fatty acid, cis-Δ^3-dodecenoyl-CoA (**Fig. 17-10**). This product cannot serve as a substrate for enoyl-CoA hydratase, which acts only on trans double bonds. The auxiliary enzyme Δ^3,Δ^2-**enoyl-CoA isomerase** isomerizes the cis-Δ^3-enoyl-CoA to the $trans$-Δ^2-enoyl-CoA, which is converted by enoyl-CoA hydratase into

the corresponding L-β-hydroxyacyl-CoA ($trans$-Δ^2-dodecenoyl-CoA). This intermediate is now acted upon by the remaining enzymes of β oxidation to yield acetyl-CoA and the coenzyme A ester of a 10-carbon saturated fatty acid, decanoyl-CoA. The latter undergoes four more passes through the β-oxidation pathway to yield five more molecules of acetyl-CoA. Altogether,

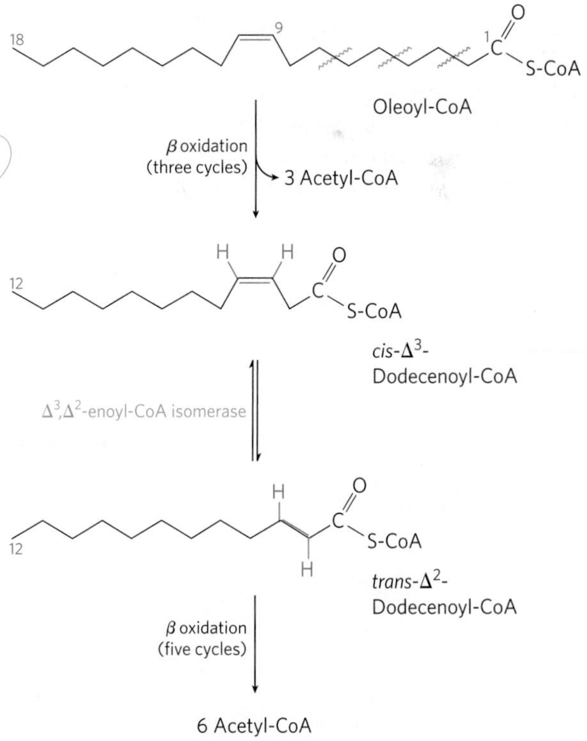

FIGURE 17-10 Oxidation of a monounsaturated fatty acid. Oleic acid, as oleoyl-CoA (Δ^9), is the example used here. Oxidation requires an additional enzyme, enoyl-CoA isomerase, to reposition the double bond, converting the cis isomer to a trans isomer, an intermediate in β oxidation.

nine acetyl-CoAs are produced from one molecule of the 18-carbon oleate.

The other auxiliary enzyme (a reductase) is required for oxidation of polyunsaturated fatty acids—for example, the 18-carbon linoleate, which has a cis-Δ^9,cis-Δ^{12} configuration **(Fig. 17-11)**. Linoleoyl-CoA undergoes three passes through the β-oxidation sequence to yield three molecules of acetyl-CoA and the coenzyme A ester of a 12-carbon unsaturated fatty acid with a cis-Δ^3,cis-Δ^6 configuration. This intermediate cannot be used by the enzymes of the β-oxidation pathway: its double bonds are in the wrong position and have the wrong configuration (cis, not trans). However, the combined action of enoyl-CoA isomerase and **2,4-dienoyl-CoA reductase**, as shown in Figure 17-11, transforms this intermediate into one that can enter the β-oxidation pathway and be degraded to six acetyl-CoAs. The overall result is conversion of linoleate to nine molecules of acetyl-CoA.

Complete Oxidation of Odd-Number Fatty Acids Requires Three Extra Reactions

Although most naturally occurring lipids contain fatty acids with an even number of carbon atoms, fatty acids with an odd number of carbons are common in the lipids of many plants and some marine organisms. Cattle and other ruminant animals form large amounts of the three-carbon **propionate** (CH_3—CH_2—COO^-) during fermentation of carbohydrates in the rumen. The propionate is absorbed into the blood and oxidized by the liver and other tissues.

Long-chain odd-number fatty acids are oxidized in the same pathway as the even-number acids, beginning at the carboxyl end of the chain. However, the substrate for the last pass through the β-oxidation sequence is a fatty acyl–CoA with a five-carbon fatty acid. When this is oxidized and cleaved, the products are acetyl-CoA and **propionyl-CoA**. The acetyl-CoA can be oxidized in the citric acid cycle, of course, but propionyl-CoA enters a different pathway, having three enzymes.

Propionyl-CoA is first carboxylated to form the D stereoisomer of **methylmalonyl-CoA (Fig. 17-12)** by

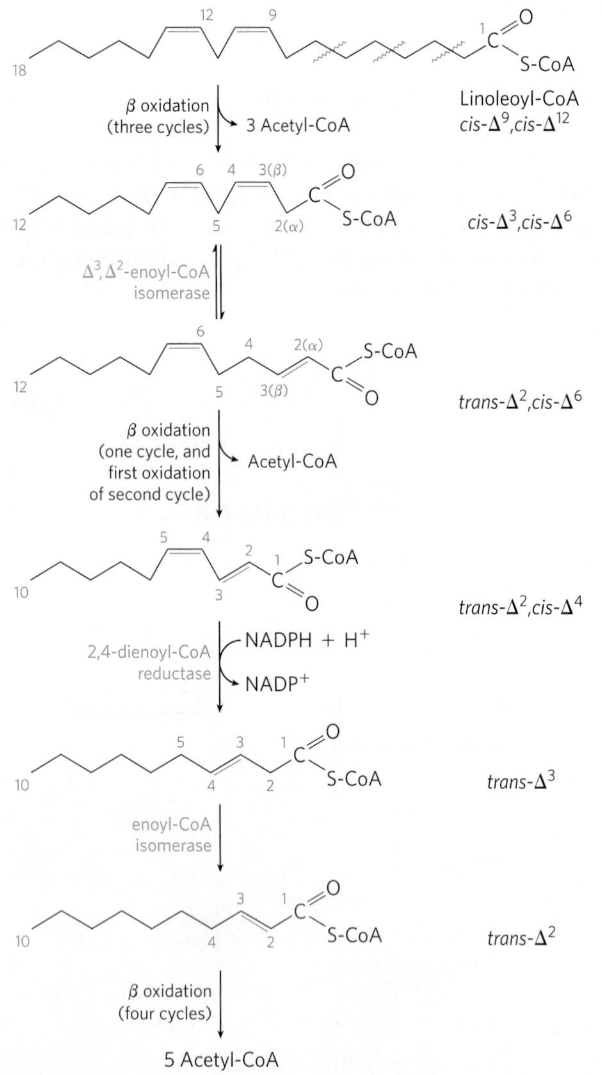

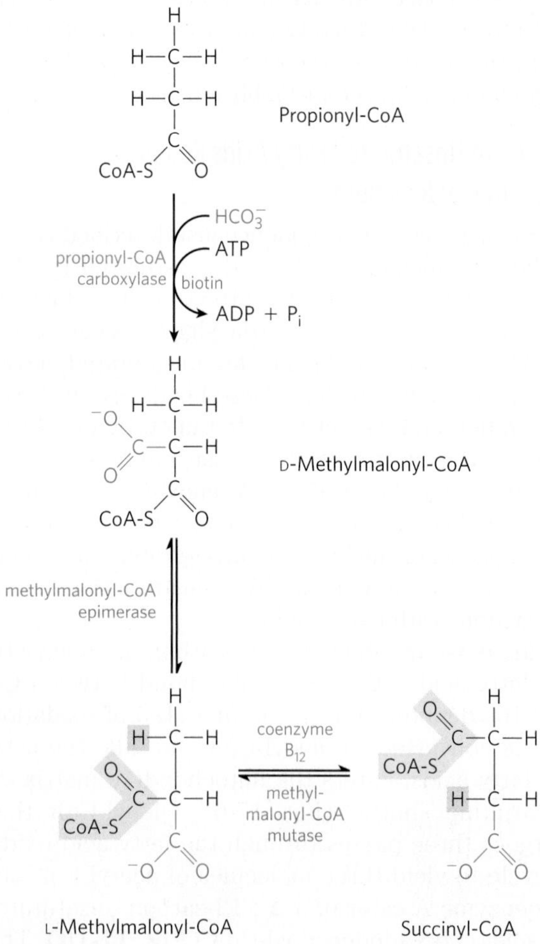

FIGURE 17-11 Oxidation of a polyunsaturated fatty acid. The example here is linoleic acid, as linoleoyl-CoA ($\Delta^{9,12}$). Oxidation requires a second auxiliary enzyme in addition to enoyl-CoA isomerase: NADPH-dependent 2,4-dienoyl-CoA reductase. The combined action of these two enzymes converts a $trans$-Δ^2,cis-Δ^4-dienoyl-CoA intermediate to the $trans$-Δ^2-enoyl-CoA substrate necessary for β oxidation.

FIGURE 17-12 Oxidation of propionyl-CoA produced by β oxidation of odd-number fatty acids. The sequence involves the carboxylation of propionyl-CoA to D-methylmalonyl-CoA and conversion of the latter to succinyl-CoA. This conversion requires epimerization of D- to L-methylmalonyl-CoA, followed by a remarkable reaction in which substituents on adjacent carbon atoms exchange positions (see Box 17-2).

propionyl-CoA carboxylase, which contains the cofactor biotin. In this enzymatic reaction, as in the pyruvate carboxylase reaction (see Fig. 16-17), CO_2 (or its hydrated ion, HCO_3^-) is activated by attachment to biotin before its transfer to the substrate, in this case the propionate moiety. Formation of the carboxybiotin intermediate requires energy, which is provided by ATP. The D-methylmalonyl-CoA thus formed is enzymatically epimerized to its L stereoisomer by **methylmalonyl-CoA epimerase** (Fig. 17-12). The L-methylmalonyl-CoA then undergoes an intramolecular rearrangement to form succinyl-CoA, which can enter the citric acid cycle. This rearrangement is catalyzed by **methylmalonyl-CoA mutase**, which requires as its coenzyme **5′-deoxyadenosylcobalamin**, or **coenzyme B$_{12}$**, which is derived from vitamin B_{12} (cobalamin). Box 17-2 describes the role of coenzyme B_{12} in this remarkable exchange reaction.

Fatty Acid Oxidation Is Tightly Regulated

Oxidation of fatty acids consumes a precious fuel, and it is regulated so as to occur only when the organism's need for energy requires it. In the liver, fatty acyl–CoA formed in the cytosol has two major pathways open to it: (1) β oxidation by enzymes in mitochondria or (2) conversion into triacylglycerols and phospholipids by enzymes in the cytosol. The pathway taken depends on

the rate of transfer of long-chain fatty acyl–CoA into mitochondria. The three-step process (carnitine shuttle) by which fatty acyl groups are carried from cytosolic fatty acyl–CoA into the mitochondrial matrix (Fig. 17-6) is rate-limiting for fatty acid oxidation and is an important point of regulation. Once fatty acyl groups have entered the mitochondrion, they are committed to oxidation to acetyl-CoA.

Malonyl-CoA, the first intermediate in the cytosolic biosynthesis of long-chain fatty acids from acetyl-CoA (see Fig. 21-2), increases in concentration whenever the animal is well supplied with carbohydrate; excess glucose that cannot be oxidized or stored as glycogen is converted in the cytosol into fatty acids for storage as triacylglycerol. The inhibition of carnitine acyltransferase 1 by malonyl-CoA **(Fig. 17-13)** ensures that the oxidation of fatty acids is inhibited whenever the liver is amply supplied with glucose as fuel and is actively making triacylglycerols from excess glucose.

Malonyl-CoA

Two of the enzymes of β oxidation are also regulated by metabolites that signal energy sufficiency. When the $[NADH]/[NAD^+]$ ratio is high, β-hydroxyacyl-CoA

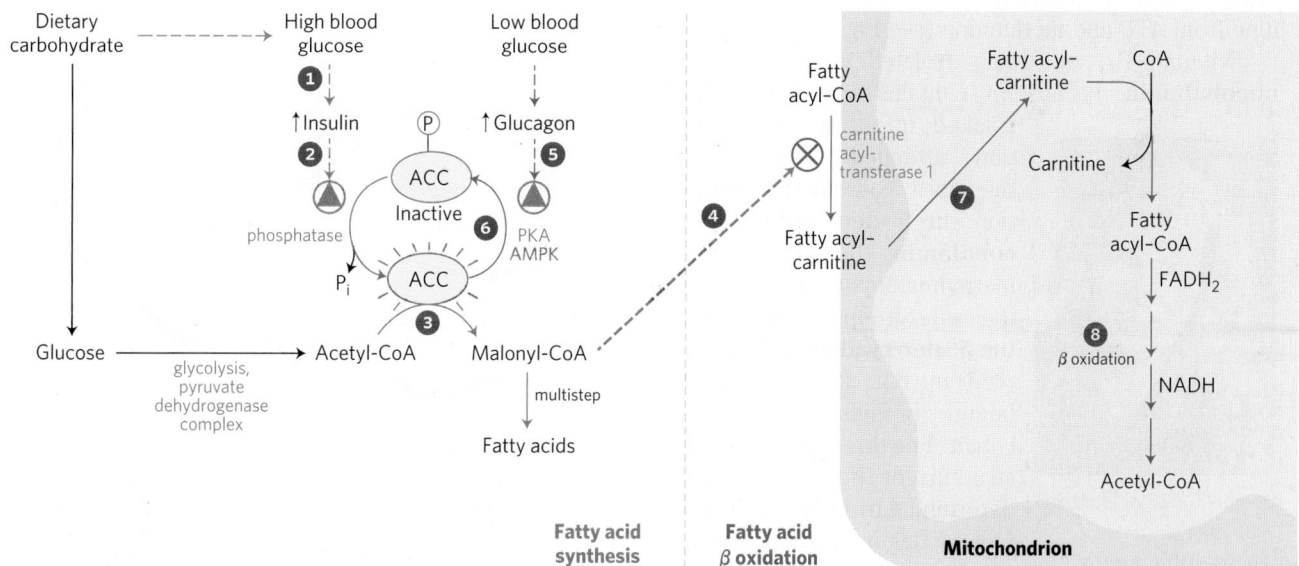

FIGURE 17-13 Coordinated regulation of fatty acid synthesis and breakdown. When the diet provides a ready source of carbohydrate as fuel, β oxidation of fatty acids is unnecessary and is therefore downregulated. Two enzymes are key to the coordination of fatty acid metabolism: acetyl-CoA carboxylase (ACC), the first enzyme in the synthesis of fatty acids (see Fig. 21-1), and carnitine acyltransferase 1, which limits the transport of fatty acids into the mitochondrial matrix for β oxidation (see Fig. 17-6). Ingestion of a high-carbohydrate meal raises the blood glucose level and thus ❶ triggers the release of insulin. ❷ Insulin-dependent protein phosphatase dephosphorylates ACC, activating it. ❸ ACC catalyzes the formation of malonyl-CoA

(the first intermediate of fatty acid synthesis), and ❹ malonyl-CoA inhibits carnitine acyltransferase 1, thereby preventing fatty acid entry into the mitochondrial matrix. When blood glucose levels drop between meals, ❺ glucagon release activates cAMP-dependent protein kinase (PKA), which ❻ phosphorylates and inactivates ACC. When [AMP] rises, AMPK also phosphorylates and inactivates ACC. The concentration of malonyl-CoA falls, the inhibition of fatty acid entry into mitochondria is relieved, and fatty acids ❼ enter the mitochondrial matrix and ❽ become the major fuel. Because glucagon also triggers the mobilization of fatty acids in adipose tissue, a supply of fatty acids begins arriving in the blood.

In the methylmalonyl-CoA mutase reaction (see Fig. 17-12), the group —CO—S-CoA at C-2 of the original propionate exchanges position with a hydrogen atom at C-3 of the original propionate (Fig. 1a). Coenzyme B$_{12}$ is the cofactor for this reaction, as it is for almost all enzymes that catalyze reactions of this general type (Fig. 1b). These coenzyme B$_{12}$–dependent processes are among the very few enzymatic reactions in biology in which there is an exchange of an alkyl or substituted alkyl group (X) with a hydrogen atom on an adjacent carbon, *with no mixing of the transferred hydrogen atom with the hydrogen of the solvent,* H$_2$O. How can the hydrogen atom move between two carbons without mixing with the enormous excess of hydrogen atoms in the solvent?

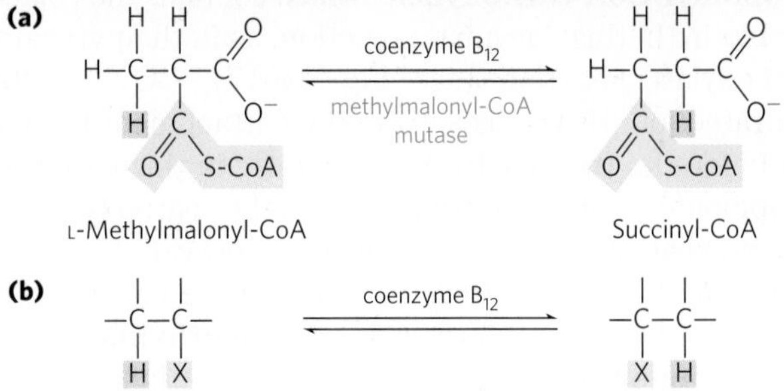

Coenzyme B$_{12}$ is the cofactor form of vitamin B$_{12}$, which is unique among all the vitamins in that it contains not only a complex organic molecule but an essential trace element, cobalt. The complex **corrin ring system** of vitamin B$_{12}$ (colored blue in Fig. 2), to which cobalt (as Co^{3+}) is coordinated, is chemically related to the porphyrin ring system of heme and heme proteins (see Fig. 5-1). A fifth coordination position of cobalt is filled by dimethylbenzimidazole ribonucleotide (shaded yellow), bound covalently by its 3'-phosphate group to a side chain of the corrin ring, through aminoisopropanol. The formation of this complex cofactor occurs in one of only two known reactions in which triphosphate is cleaved from ATP (Fig. 3); the other reaction is the formation of *S*-adenosylmethionine from ATP and methionine (see Fig. 18-18).

Vitamin B$_{12}$ as usually isolated is called **cyanocobalamin**, because it contains a cyano group (picked up during purification) attached to cobalt in the sixth coordination position. In **5'-deoxyadenosylcobalamin**, the cofactor for methylmalonyl-CoA mutase, the cyano group is replaced by the **5'-deoxyadenosyl** group (red in Fig. 2), covalently bound through C-5' to the cobalt. The three-dimensional structure of the cofactor was determined by Dorothy Crowfoot Hodgkin in 1956, using x-ray crystallography.

Dorothy Crowfoot Hodgkin, 1910–1994
[Source: Bettmann/Corbis.]

The key to understanding how coenzyme B$_{12}$ catalyzes hydrogen exchange lies in the properties of the covalent bond between cobalt and C-5' of the deoxyadenosyl group (Fig. 2). This is a relatively weak bond; merely illuminating the compound with visible light is enough to break this Co—C bond. (This extreme photolability probably accounts for the absence of vitamin B$_{12}$ in plants.) Dissociation

(a)

L-Methylmalonyl-CoA ⇌ (coenzyme B$_{12}$, methylmalonyl-CoA mutase) Succinyl-CoA

(b)

⇌ coenzyme B$_{12}$

FIGURE 1

produces a 5'-deoxyadenosyl radical and the Co^{2+} form of the vitamin. The chemical function of 5'-deoxyadenosylcobalamin is to generate free radicals in this way, thus initiating a series of transformations such as that illustrated in Figure 4—a postulated mechanism for the reaction catalyzed by methylmalonyl-CoA mutase and several other coenzyme B$_{12}$–dependent transformations. In this postulated mechanism, the migrating hydrogen atom never exists as a free

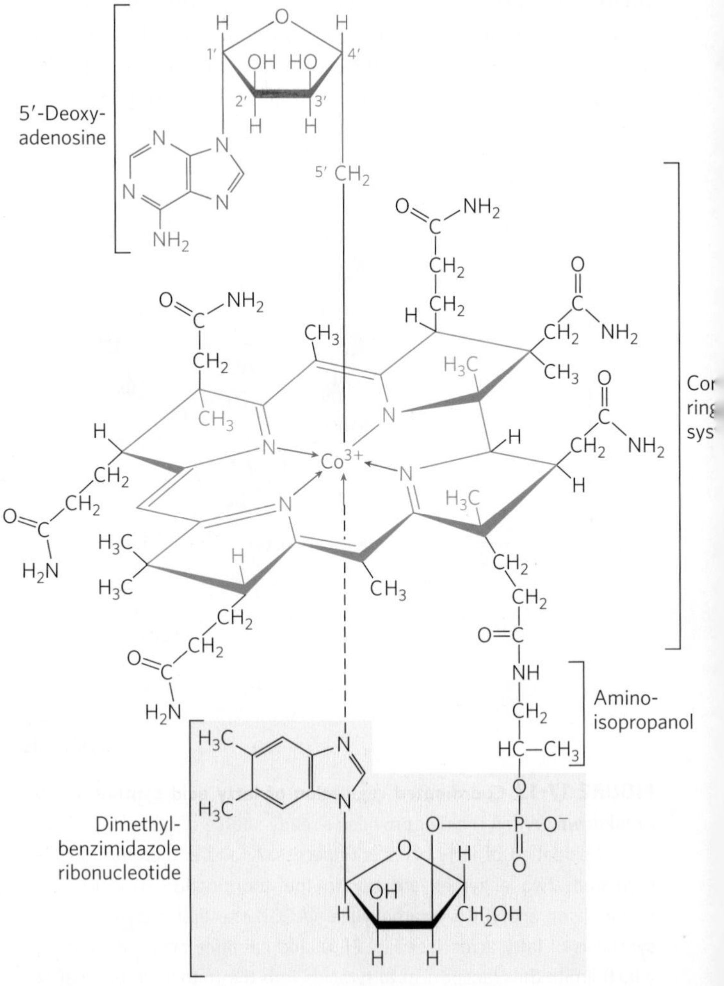

FIGURE 2

species and is thus never free to exchange with the hydrogen of surrounding water molecules.

Vitamin B_{12} deficiency results in serious disease. This vitamin is not made by plants or animals and can be synthesized only by a few species of microorganisms. It is required by healthy people in only minute amounts, about 3 μg/day. The serious disease **pernicious anemia** results from failure to absorb vitamin B_{12} efficiently from the intestine, where it is synthesized by intestinal bacteria or obtained from digestion of meat. Individuals with this disease do not produce sufficient amounts of **intrinsic factor**, a glycoprotein essential to vitamin B_{12} absorption. The pathology in pernicious anemia includes reduced production of erythrocytes, reduced levels of hemoglobin, and severe, progressive impairment of the central nervous system. Administration of large doses of vitamin B_{12} alleviates these symptoms in at least some cases. ■

ATP

Cobalamin

Coenzyme B_{12}

FIGURE 3

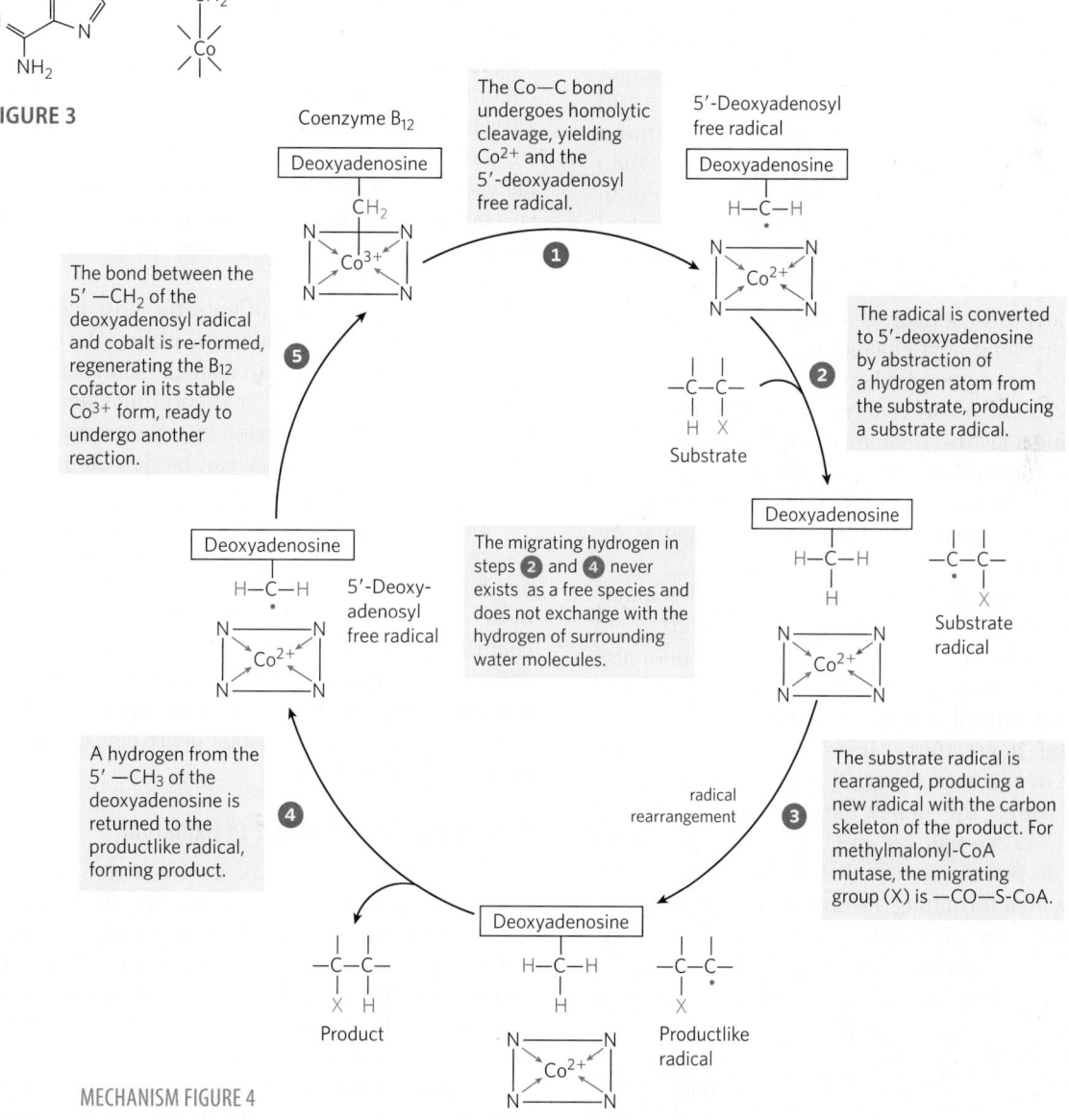

Coenzyme B_{12}

Deoxyadenosine

The Co—C bond undergoes homolytic cleavage, yielding Co^{2+} and the 5'-deoxyadenosyl free radical.

5'-Deoxyadenosyl free radical

Deoxyadenosine

1

The bond between the 5' —CH_2 of the deoxyadenosyl radical and cobalt is re-formed, regenerating the B_{12} cofactor in its stable Co^{3+} form, ready to undergo another reaction.

5

The radical is converted to 5'-deoxyadenosine by abstraction of a hydrogen atom from the substrate, producing a substrate radical.

2

Substrate

Deoxyadenosine

5'-Deoxy-adenosyl free radical

The migrating hydrogen in steps **2** and **4** never exists as a free species and does not exchange with the hydrogen of surrounding water molecules.

Deoxyadenosine

Substrate radical

A hydrogen from the 5' —CH_3 of the deoxyadenosine is returned to the productlike radical, forming product.

4

radical rearrangement

3

The substrate radical is rearranged, producing a new radical with the carbon skeleton of the product. For methylmalonyl-CoA mutase, the migrating group (X) is —CO—S-CoA.

Product

Deoxyadenosine

Productlike radical

MECHANISM FIGURE 4

dehydrogenase is inhibited; in addition, high concentrations of acetyl-CoA inhibit thiolase.

Recall from Chapter 15 that during periods of vigorous muscle contraction or during fasting, the fall in [ATP] and the rise in [AMP] activate AMPK, the AMP-activated protein kinase. AMPK phosphorylates several target enzymes, including acetyl-CoA carboxylase, which catalyzes malonyl-CoA synthesis. This phosphorylation and thus inhibition of acetyl-CoA carboxylase lowers the concentration of malonyl-CoA, relieving the inhibition of fatty acyl–carnitine transport into mitochondria (Fig. 17-13) and allowing β oxidation to replenish the supply of ATP.

Transcription Factors Turn on the Synthesis of Proteins for Lipid Catabolism

In addition to the various short-term regulatory mechanisms that modulate the activity of existing enzymes, transcriptional regulation can change the number of molecules of the enzymes of fatty acid oxidation on a longer time scale—minutes to hours. The **PPAR** family of nuclear receptors are transcription factors that affect many metabolic processes in response to a variety of fatty acid–like ligands. (They were originally recognized as *p*eroxisome *p*roliferator-*a*ctivated *r*eceptors, then were found to function more broadly.) PPARα acts in muscle, adipose tissue, and liver to turn on a set of genes essential for fatty acid oxidation, including the fatty acid transporter, carnitine acyltransferases 1 and 2, the fatty acyl–CoA dehydrogenases for short, medium, long, and very long acyl chains, and related enzymes. This response is triggered when a cell or organism has an increased demand for energy from fat catabolism, such as during a fast between meals or under conditions of longer-term starvation. Glucagon, released in response to low blood glucose, can act through cAMP and the transcription factor CREB to turn on certain genes for lipid catabolism.

Another situation that is accompanied by major changes in the expression of the enzymes of fatty acid oxidation is the transition from fetal to neonatal metabolism in the heart. In the fetus, the principal fuels in heart muscle are glucose and lactate, but in the neonatal heart, fatty acids are the main fuel. At the time of this transition, PPARα is activated and in turn activates the genes essential for fatty acid metabolism. As we will see in Chapter 23, two other transcription factors in the PPAR family also play crucial roles in determining the enzyme complements—and therefore the metabolic activities—of specific tissues at particular times (see Fig. 23-43).

The major sites of fatty acid oxidation, at rest and during exercise, are skeletal and heart muscle. Endurance training increases PPARα expression in muscle, leading to increased levels of fatty acid–oxidizing enzymes and increased oxidative capacity of the muscle.

Genetic Defects in Fatty Acyl–CoA Dehydrogenases Cause Serious Disease

Stored triacylglycerols are typically the chief source of energy for muscle contraction, and an inability to oxidize fatty acids from triacylglycerols has serious consequences for health. The most common genetic defect in fatty acid catabolism in U.S. and northern European populations is due to a mutation in the gene encoding the **medium-chain acyl-CoA dehydrogenase**, mentioned earlier in the chapter. Among northern Europeans, the frequency of carriers (individuals with this recessive mutation on one of the two homologous chromosomes) is about 1 in 40, and about 1 individual in 10,000 has the disease—that is, has two copies of the mutant MCAD allele and is unable to oxidize fatty acids of 6 to 12 carbons. The disease is characterized by recurring episodes of a syndrome that includes fat accumulation in the liver, high blood levels of octanoic acid (8:0), low blood glucose (hypoglycemia), sleepiness, vomiting, and coma. Although individuals may have no symptoms between episodes, the episodes are very serious; mortality due to this disease is 25% to 60% in early childhood. If the genetic defect is detected shortly after birth, the infant can be started on a low-fat, high-carbohydrate diet. With early detection and careful management of the diet—including avoiding long intervals between meals, to prevent the body from turning to its fat reserves for energy—the prognosis for these individuals is good.

To screen newborn infants for possible metabolic defects in fatty acid oxidation, plasma levels of acyl-carnitine are determined in a small blood sample. When fatty acid oxidation fails, acyl-carnitines accumulate in mitochondria and are carried into the cytosol, and then into the blood, where they can be detected by tandem mass spectrometry (p. 101).

More than 20 human genetic defects in fatty acid transport or oxidation have been documented, most much less common than the defect in MCAD. One of the most severe disorders results from loss of the long-chain β-hydroxyacyl-CoA dehydrogenase activity of the trifunctional protein, TFP. Other disorders include defects in the α or β subunits that affect all three activities of TFP and cause serious heart disease and abnormal skeletal muscle. ■

Peroxisomes Also Carry Out β Oxidation

The mitochondrial matrix is the major site of fatty acid oxidation in animal cells, but in certain cells, other compartments also contain enzymes capable of oxidizing fatty acids to acetyl-CoA, by a pathway similar but not identical to that in mitochondria. In plant cells, the major site of β oxidation is not mitochondria but peroxisomes.

In **peroxisomes**, organelles found in both animal and plant cells, the intermediates for β oxidation of fatty acids are coenzyme A derivatives, and the process consists of four steps, as in mitochondrial β oxidation **(Fig. 17-14)**: (1) dehydrogenation, (2) addition of water to the resulting double bond, (3) oxidation of the β-hydroxyacyl-CoA to a ketone, and (4) thiolytic cleavage by coenzyme A. (The identical reactions also occur in glyoxysomes, organelles found only in germinating seeds.)

One difference between the peroxisomal and mitochondrial pathways is in the chemistry of the first step (Fig. 17-4). In peroxisomes, the flavoprotein acyl-CoA oxidase that introduces the double bond passes electrons directly to O_2, producing H_2O_2 (thus the name "peroxisomes"). This strong and potentially damaging oxidant is immediately cleaved to H_2O and O_2 by **catalase**. Recall that in mitochondria, the electrons removed in the first oxidation step pass through the respiratory chain to O_2 to produce H_2O, and this process is accompanied by ATP synthesis. In peroxisomes, the energy released in the first oxidative step of fatty acid breakdown is not conserved as ATP, but is dissipated as heat.

A second important difference between mitochondrial and peroxisomal β oxidation in mammals is in the specificity for fatty acyl-CoAs; the peroxisomal system is much more active on very-long-chain fatty acids such as hexacosanoic acid (26:0) and on branched-chain fatty acids such as phytanic acid and pristanic acid (see Fig. 17-17). These less-common fatty acids are obtained from dietary intake of dairy products, the fat of ruminant animals, meat, and fish. Their catabolism in the peroxisome involves several auxiliary enzymes unique to this organelle. The inability to oxidize these compounds is responsible for several serious human genetic diseases. Individuals with **Zellweger syndrome** are unable to make peroxisomes and therefore lack all the metabolism unique to that organelle. In **X-linked adrenoleukodystrophy (XALD)**, peroxisomes fail to oxidize very-long-chain fatty acids, apparently due to lack of a functional transporter for these fatty acids in the peroxisomal membrane. Both defects lead to accumulation in the blood of very-long-chain fatty acids, especially 26:0. XALD affects young boys before the age of 10 years, causing loss of vision, behavioral disturbances, and death within a few years. ■

In mammals, high concentrations of fats in the diet result in increased synthesis of the enzymes of peroxisomal β oxidation in the liver. Liver peroxisomes do not contain the enzymes of the citric acid cycle and cannot catalyze the oxidation of acetyl-CoA to CO_2. Instead, long-chain or branched fatty acids are catabolized in peroxisomes to shorter-chain products, such as hexanoyl-CoA, which are exported to mitochondria and completely oxidized there. As we will see in Chapter 20, the germinating seeds of plants can synthesize carbohydrates and many other metabolites from acetyl-CoA produced in peroxisomes, using a pathway (the glyoxylate cycle) not present in vertebrates.

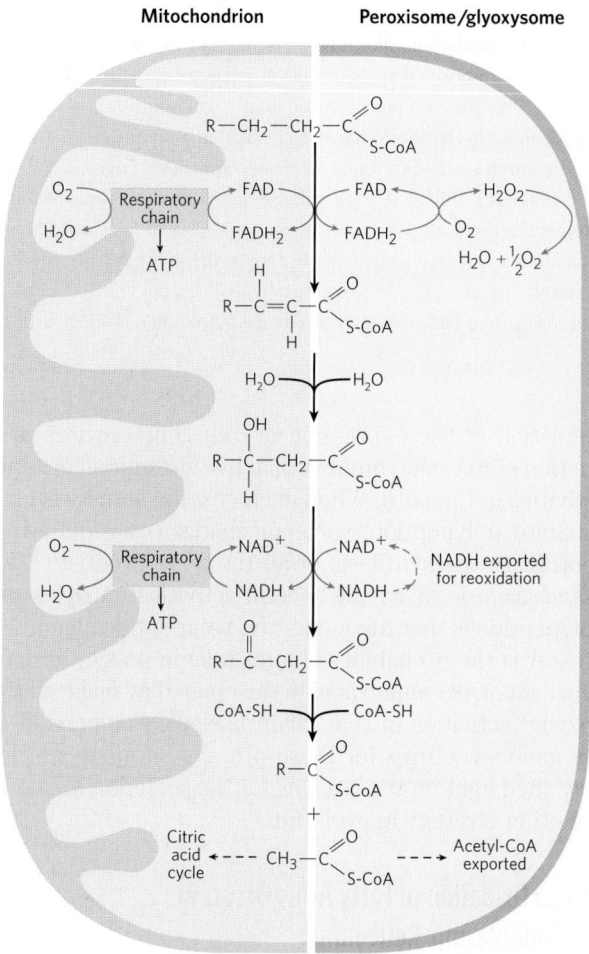

FIGURE 17-14 Comparison of β oxidation in mitochondria and in peroxisomes and glyoxysomes. The peroxisomal/glyoxysomal system differs from the mitochondrial system in three respects: (1) the peroxisomal system prefers very-long-chain fatty acids; (2) in the first oxidative step, electrons pass directly to O_2, generating H_2O_2; and (3) the NADH formed in the second oxidative step cannot be reoxidized in the peroxisome or glyoxysome, so reducing equivalents are exported to the cytosol, eventually entering mitochondria. The acetyl-CoA produced by peroxisomes and glyoxysomes is also exported; the acetate from glyoxysomes (organelles found only in germinating seeds) serves as a biosynthetic precursor (see Fig. 20-55). Acetyl-CoA produced in mitochondria is further oxidized in the citric acid cycle.

The β-Oxidation Enzymes of Different Organelles Have Diverged during Evolution

Although the β-oxidation reactions in mitochondria are essentially the same as those in peroxisomes and glyoxysomes, the enzymes (isozymes) differ significantly between the two types of organelles. The differences apparently reflect an evolutionary divergence that occurred very early, with the separation of gram-positive and gram-negative bacteria (see Fig. 1-7).

In mitochondria, the four β-oxidation enzymes that act on short-chain fatty acyl–CoAs are separate, soluble

(a) Gram-positive bacteria and mitochondrial short-chain-specific system

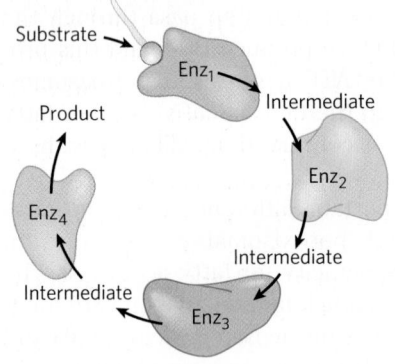

(b) Gram-negative bacteria

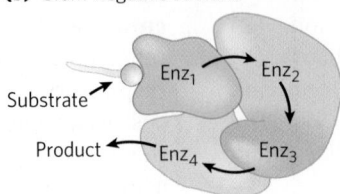

(c) Mitochondrial very-long-chain-specific system

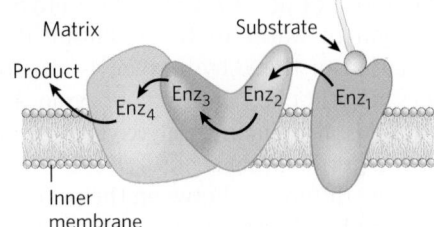

Enz$_1$	Acyl-CoA dehydrogenase
Enz$_2$	Enoyl-CoA hydratase
Enz$_3$	L-β-Hydroxyacyl-CoA dehydrogenase
Enz$_4$	Thiolase
Enz$_5$	D-3-Hydroxyacyl-CoA epimerase
Enz$_6$	Δ^3,Δ^2-Enoyl-CoA isomerase.

(d) Peroxisomal and glyoxysomal systems

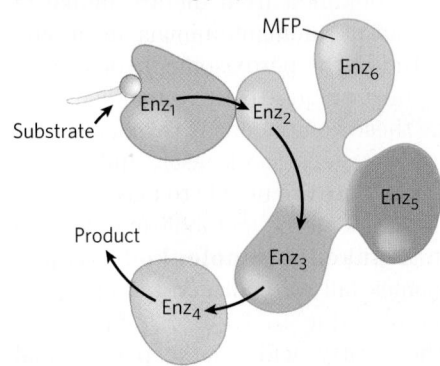

FIGURE 17-15 The enzymes of β oxidation. Shown here are the different subunit structures of the enzymes of β oxidation in gram-positive and gram-negative bacteria, mitochondria, peroxisomes, and glyoxysomes. **(a)** The four enzymes of β oxidation in gram-positive bacteria are separate, soluble entities, as are those of the short-chain-specific system of mitochondria. **(b)** In gram-negative bacteria, the four enzyme activities reside in three polypeptides; Enz$_2$ and Enz$_3$ are parts of a single polypeptide chain. **(c)** The very-long-chain-specific system of mitochondria is also composed of three polypeptides, one of which includes the activities of Enz$_2$ and Enz$_3$; in this case, the system is bound to the inner mitochondrial membrane. **(d)** In the peroxisomal and glyoxysomal β-oxidation systems, Enz$_1$ and Enz$_4$ are separate polypeptides, but Enz$_2$ and Enz$_3$, as well as two auxiliary enzymes (Enz$_5$ and Enz$_6$), are part of a single polypeptide chain: the multifunctional protein, MFP.

proteins (as noted earlier), similar in structure to the analogous enzymes of gram-positive bacteria **(Fig. 17-15a)**. The gram-negative bacteria have four activities in three soluble subunits (Fig. 17-15b), and the eukaryotic enzyme system that acts on long-chain fatty acids—the trifunctional protein, TFP—has three enzyme activities in two subunits that are membrane-associated (Fig. 17-15c). The β-oxidation enzymes of peroxisomes and glyoxysomes, however, form a complex of proteins, one of which contains four enzymatic activities in a single polypeptide chain (Fig. 17-15d). The first enzyme, acyl-CoA oxidase, is a single polypeptide chain; the **multifunctional protein (MFP)** contains the second and third enzyme activities (enoyl-CoA hydratase and hydroxyacyl-CoA dehydrogenase) as well as two auxiliary activities needed for the oxidation of unsaturated fatty acids (D-3-hydroxyacyl-CoA epimerase and Δ^3,Δ^2-enoyl-CoA isomerase); the fourth enzyme, thiolase, is a separate, soluble polypeptide.

The enzymes that catalyze essentially the reversal of β oxidation in the synthesis of fatty acids are also organized differently in bacteria and eukaryotes; in bacteria, the seven enzymes needed for fatty acid synthesis are separate polypeptides, but in mammals, all seven activities are part of a single, huge polypeptide chain (see Fig. 21-3a). One advantage to the cell in having several enzyme activities of the same pathway encoded in a single polypeptide chain is that this solves the problem of regulating the synthesis of enzymes that must interact functionally;

regulation of the expression of *one* gene ensures production of the same number of active sites for all enzyme activities in the path. When each enzyme activity is on a separate polypeptide, some mechanism is required to coordinate the synthesis of all the gene products. The *disadvantage* of having several activities on the same polypeptide is that the longer the polypeptide chain, the greater is the probability of a mistake in its synthesis: a single incorrect amino acid in the chain may make all the enzyme activities in that chain useless. Comparison of the gene structures for these proteins in many species may shed light on the reasons for the selection of one or the other strategy in evolution.

The ω Oxidation of Fatty Acids Occurs in the Endoplasmic Reticulum

Although mitochondrial β oxidation, in which enzymes act at the carboxyl end of a fatty acid, is by far the most important catabolic fate for fatty acids in animal cells, there is another pathway in some species, including vertebrates, that involves oxidation of the ω (omega) carbon—the carbon most distant from the carboxyl group. The enzymes unique to ω **oxidation** are located in the endoplasmic reticulum of liver and kidney, and the preferred substrates are fatty acids of 10 or 12 carbon atoms. In mammals, ω oxidation is normally a minor pathway for fatty acid degradation, but when β oxidation is defective (because of mutation

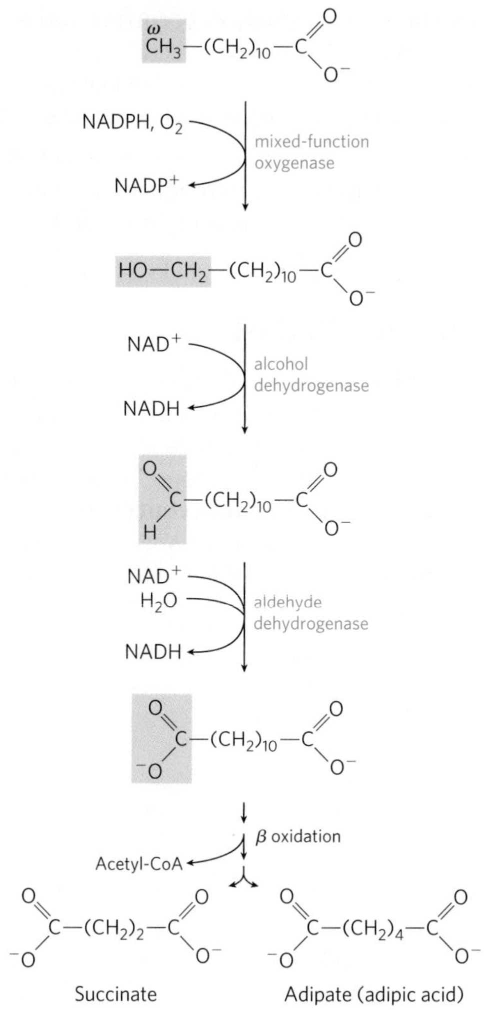

FIGURE 17-16 The ω oxidation of fatty acids in the endoplasmic reticulum. This alternative to β oxidation begins with oxidation of the carbon most distant from the β carbon—the ω (omega) carbon. The substrate is usually a medium-chain fatty acid; shown here is lauric acid (laurate). This pathway is generally not the major route for oxidative catabolism of fatty acids.

or a carnitine deficiency, for example) it becomes more important.

The first step introduces a hydroxyl group onto the ω carbon **(Fig. 17-16)**. The oxygen for this group comes from molecular oxygen (O_2) in a complex reaction that involves cytochrome P450 and the electron donor NADPH. Reactions of this type are catalyzed by **mixed-function oxygenases**, described in Box 21-1. Two more enzymes then act on the ω carbon: **alcohol dehydrogenase** oxidizes the hydroxyl group to an aldehyde, and **aldehyde dehydrogenase** oxidizes the aldehyde group to a carboxylic acid, producing a fatty acid with a carboxyl group at each end. At this point, either end can be attached to coenzyme A, and the molecule can enter the mitochondrion and undergo β oxidation by the usual route, releasing acetyl-CoA and dicarboxylic acids of 12, 10, 8, 6, and 4 carbons. The four-carbon acid (succinic acid) can enter the citric acid cycle and undergo oxidation as a fuel.

Phytanic Acid Undergoes α Oxidation in Peroxisomes

Phytanic acid, a long-chain fatty acid with methyl branches, is derived from the phytol side chain of chlorophyll (see Fig. 20-8). The presence of a methyl group on the β carbon of this fatty acid prevents the formation of a β-keto intermediate, making its β oxidation impossible. Humans obtain phytanic acid in the diet, primarily from dairy products and from the fats of ruminant animals; microorganisms in the rumen of these animals produce phytanic acid as they digest plant chlorophyll. The typical western diet includes 50 to 100 mg of phytanic acid per day.

Phytanic acid is metabolized in peroxisomes by α **oxidation**, in which a single carbon is removed from the carboxyl end of the acid **(Fig. 17-17)**. Phytanoyl-CoA is

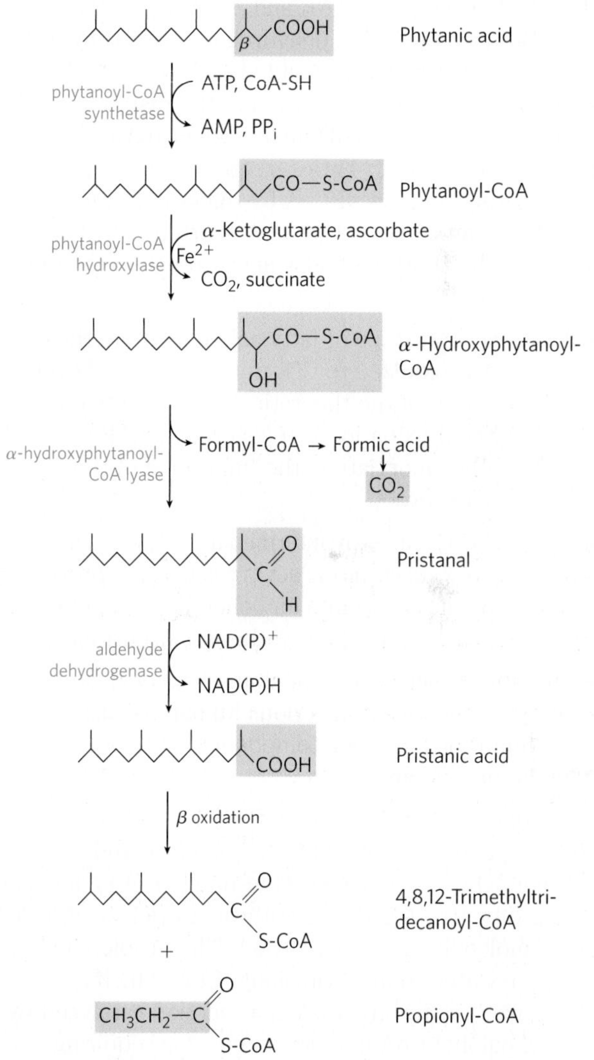

FIGURE 17-17 The α oxidation of a branched-chain fatty acid (phytanic acid) in peroxisomes. Phytanic acid has a methyl-substituted β carbon and therefore cannot undergo β oxidation. The combined action of the enzymes shown here removes the carboxyl carbon of phytanic acid to produce pristanic acid, in which the β carbon is unsubstituted, allowing β oxidation. Notice that β oxidation of pristanic acid releases propionyl-CoA, not acetyl-CoA. This is further catabolized as in Figure 17-12. (The details of the reaction that produces pristanal remain controversial.)

first hydroxylated on its α carbon in a reaction that involves molecular oxygen. The product is decarboxylated to form an aldehyde one carbon shorter, and then oxidized to the corresponding carboxylic acid, which now has no substituent on the β carbon. Further β oxidation produces acetyl-CoA and then propionyl-CoA in successive oxidation cycles. **Refsum disease**, resulting from a genetic defect in phytanoyl-CoA hydroxylase, leads to the accumulation of very high blood levels of phytanic acid, causing (by unknown mechanisms) severe neurological deficits, including blindness and deafness. ■

SUMMARY 17.2 Oxidation of Fatty Acids

■ In the first stage of β oxidation, four reactions remove each acetyl-CoA unit, in turn, from the carboxyl end of a saturated fatty acyl–CoA: (1) dehydrogenation of the α and β carbons (C-2 and C-3) by FAD-linked acyl–CoA dehydrogenases, (2) hydration of the resulting trans-Δ^2 double bond by enoyl-CoA hydratase, (3) dehydrogenation of the resulting L-β-hydroxyacyl-CoA by NAD-linked β-hydroxyacyl-CoA dehydrogenase, and (4) CoA-requiring cleavage of the resulting β-ketoacyl-CoA by thiolase, to form acetyl-CoA and a fatty acyl–CoA shortened by two carbons. The shortened fatty acyl–CoA then reenters the sequence.

■ In the second stage of fatty acid oxidation, the acetyl-CoA is oxidized to CO_2 in the citric acid cycle. A large fraction of the theoretical yield of free energy from fatty acid oxidation is recovered as ATP by oxidative phosphorylation, the final stage of the oxidative pathway.

■ Malonyl-CoA, an early intermediate of fatty acid synthesis, inhibits carnitine acyltransferase 1, preventing the entry of fatty acids into mitochondria. This blocks fatty acid breakdown while synthesis is occurring.

■ Genetic defects in the medium-chain acyl-CoA dehydrogenase result in serious human disease, as do mutations in other components of the β-oxidation system.

■ Oxidation of unsaturated fatty acids requires two additional enzymes: enoyl-CoA isomerase and 2,4-dienoyl-CoA reductase. Odd-number fatty acids are oxidized by the β-oxidation pathway to yield acetyl-CoA and a molecule of propionyl-CoA. The propionyl-CoA is carboxylated to methylmalonyl-CoA, which is isomerized to succinyl-CoA in a reaction catalyzed by methylmalonyl-CoA mutase, an enzyme requiring coenzyme B_{12}.

■ Peroxisomes of plants and animals, and glyoxysomes of plants, carry out β oxidation in four steps similar to those of the mitochondrial pathway. The first oxidation step, however, transfers electrons directly to O_2, generating H_2O_2. Peroxisomes of animal tissues specialize in the oxidation of very-long-chain fatty acids and branched fatty acids.

■ The reactions of ω oxidation, occurring in the endoplasmic reticulum, produce dicarboxylic fatty acyl intermediates, which can undergo β oxidation at either end to yield short dicarboxylic acids such as succinate.

■ The reactions of α oxidation convert branched fatty acids such as phytanic acid to products that can undergo β oxidation, yielding acetyl-CoA and propionyl-CoA.

17.3 Ketone Bodies

In humans and most other mammals, acetyl-CoA formed in the liver during oxidation of fatty acids can either enter the citric acid cycle (stage 2 of Fig. 17-7) or undergo conversion to the "ketone bodies," **acetone**, **acetoacetate**, and D-**β-hydroxybutyrate**, for export to other tissues. (The term "bodies" is an historical artifact; the compounds are soluble in blood and urine, not particulate, and not all of them are ketones.)

Acetone

Acetoacetate

D-β-Hydroxybutyrate

Acetone, produced in smaller quantities than the other ketone bodies, is exhaled. Acetoacetate and D-β-hydroxybutyrate are transported by the blood to tissues other than the liver (extrahepatic tissues), where they are converted to acetyl-CoA and oxidized in the citric acid cycle, providing much of the energy required by tissues such as skeletal and heart muscle and the renal cortex. The brain, which preferentially uses glucose as fuel, can adapt to the use of acetoacetate and D-β-hydroxybutyrate under starvation conditions, when glucose is unavailable. In this situation, the brain cannot use fatty acids as fuel, because they do not cross the blood-brain barrier. The production and export of ketone bodies from the liver to extrahepatic tissues allows continued oxidation of fatty acids in the liver when acetyl-CoA is not being oxidized in the citric acid cycle.

Ketone Bodies, Formed in the Liver, Are Exported to Other Organs as Fuel

The first step in the formation of acetoacetate, occurring in the liver **(Fig. 17-18)**, is the enzymatic condensation of two molecules of acetyl-CoA, catalyzed by thiolase; this is simply the reversal of the last step of β oxidation. The acetoacetyl-CoA then condenses with

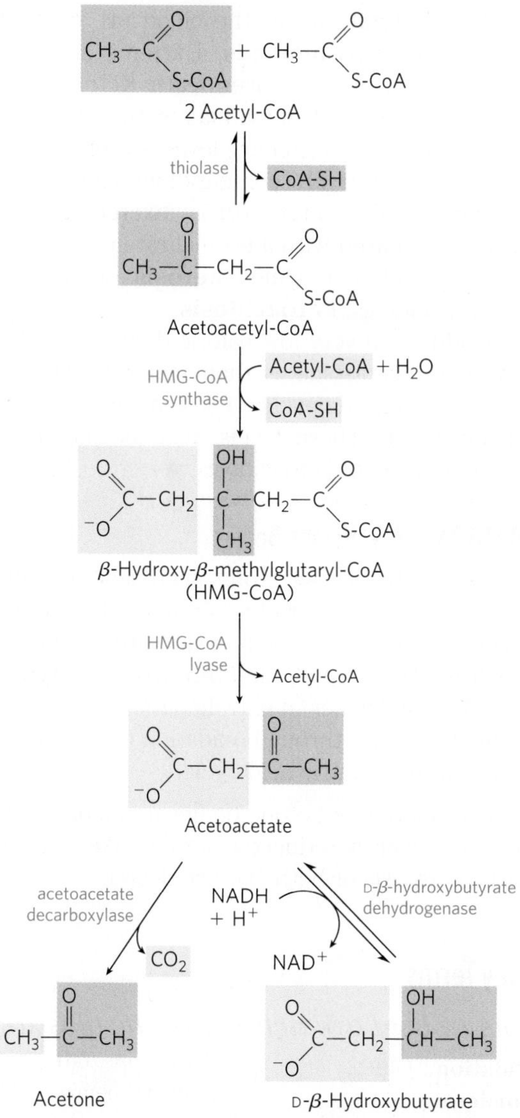

FIGURE 17-18 Formation of ketone bodies from acetyl-CoA. Healthy, well-nourished individuals produce ketone bodies at a relatively low rate. When acetyl-CoA accumulates (as in starvation or untreated diabetes, for example), thiolase catalyzes the condensation of two acetyl-CoA molecules to acetoacetyl-CoA, the parent compound of the three ketone bodies. The reactions of ketone body formation occur in the matrix of liver mitochondria. The six-carbon compound β-hydroxy-β-methylglutaryl-CoA (HMG-CoA) is also an intermediate of sterol biosynthesis, but the enzyme that forms HMG-CoA in that pathway is cytosolic. HMG-CoA lyase is present only in the mitochondrial matrix.

another molecule of acetyl-CoA to form **β-hydroxy-β-methylglutaryl-CoA (HMG-CoA)**, which is cleaved to free acetoacetate and acetyl-CoA. The acetoacetate is reversibly reduced by D-β-hydroxybutyrate dehydrogenase, a mitochondrial enzyme, to D-β-hydroxybutyrate. This enzyme is specific for the D stereoisomer; it does not act on L-β-hydroxyacyl-CoAs and is distinct from L-β-hydroxyacyl-CoA dehydrogenase of the β-oxidation pathway. This difference in stereospecificity of the two enzymes that use β-hydroxyacyl-CoAs as substrates in fatty acid breakdown and fatty acid synthesis means that the cell can maintain

separate pools of β-hydroxyacyl-CoAs, earmarked for either breakdown or synthesis.

In healthy people, acetone is formed in very small amounts from acetoacetate, which is easily decarboxylated, either spontaneously or by the action of **acetoacetate decarboxylase** (Fig. 17-18). Because individuals with untreated diabetes produce large quantities of acetoacetate, their blood contains significant amounts of acetone, which is toxic. Acetone is volatile and imparts a characteristic odor to the breath, which is sometimes useful in diagnosing diabetes. ■

In extrahepatic tissues, D-β-hydroxybutyrate is oxidized to acetoacetate by D-β-hydroxybutyrate dehydrogenase **(Fig. 17-19)**. The acetoacetate is activated to its coenzyme A ester by transfer of CoA from succinyl-CoA, an intermediate of the citric acid cycle (see Fig. 16-7), in a reaction catalyzed by **β-ketoacyl-CoA transferase**, also called thiophorase. The acetoacetyl-CoA is then cleaved by thiolase to yield two molecules of acetyl-CoA, which enter the citric acid cycle. Thus the ketone bodies are used as fuels in all tissues except liver, which lacks β-ketoacyl-CoA transferase. The liver is therefore a producer of ketone bodies for other tissues, but not a consumer.

The production and export of ketone bodies by the liver allows continued oxidation of fatty acids with only minimal oxidation of acetyl-CoA. When intermediates of the citric acid cycle are being siphoned off for glucose

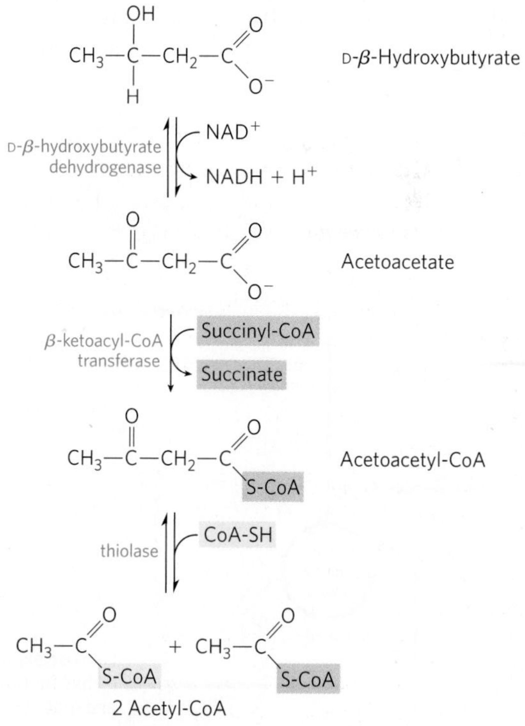

FIGURE 17-19 D-β-Hydroxybutyrate as a fuel. D-β-Hydroxybutyrate, synthesized in the liver, passes into the blood and thus to other tissues, where it is converted in three steps to acetyl-CoA. It is first oxidized to acetoacetate, which is activated with coenzyme A donated from succinyl-CoA, then split by thiolase. The acetyl-CoA thus formed is used for energy production.

synthesis by gluconeogenesis, for example, oxidation of cycle intermediates slows—and so does acetyl-CoA oxidation. Moreover, the liver contains only a limited amount of coenzyme A, and when most of it is tied up in acetyl-CoA, β oxidation slows for want of the free coenzyme. The production and export of ketone bodies frees coenzyme A, allowing continued fatty acid oxidation.

Ketone Bodies Are Overproduced in Diabetes and during Starvation

Starvation and untreated diabetes mellitus lead to overproduction of ketone bodies in the liver, with several adverse effects on health. During starvation, gluconeogenesis depletes citric acid cycle intermediates, diverting acetyl-CoA to ketone body production **(Fig. 17-20)**. In untreated diabetes, when the insulin level is insufficient, extrahepatic tissues cannot take up glucose efficiently from the blood, either for fuel or for conversion to fat. Under these conditions, levels of malonyl-CoA (the starting material for fatty acid synthesis in the liver; see Fig. 21-1) fall, inhibition of carnitine acyltransferase 1 is relieved, and fatty acids enter mitochondria to be degraded to acetyl-CoA—which cannot pass through the citric acid cycle because cycle intermediates have been drawn off for use as substrates in gluconeogenesis. The resulting accumulation of acetyl-CoA accelerates the formation of ketone bodies and their release into the blood beyond the capacity of extrahepatic tissues to oxidize them. The increased blood levels of acetoacetate and

D-β-hydroxybutyrate lower the blood pH, causing the condition known as **acidosis**. Extreme acidosis can lead to coma and in some cases death. Ketone bodies in the blood and urine of individuals with untreated diabetes can reach extraordinary levels—a blood concentration of 90 mg/100 mL (compared with a normal level of <3 mg/100 mL) and urinary excretion of 5,000 mg/24 hr (compared with a normal rate of ≤125 mg/24 hr). This condition is called **ketosis** or, when combined with acidosis, **ketoacidosis**.

Individuals on very low-calorie diets, using the fats stored in adipose tissue as their major energy source, also have increased levels of ketone bodies in their blood and urine. These levels must be monitored to avoid the dangers of ketoacidosis. ■

SUMMARY 17.3 Ketone Bodies

■ The ketone bodies—acetone, acetoacetate, and D-β-hydroxybutyrate—are formed in the liver when fatty acids are the principal fuel supporting whole-body metabolism. Acetoacetate and D-β-hydroxybutyrate serve as fuel molecules in extrahepatic tissues, including the brain, through oxidation to acetyl-CoA and entry into the citric acid cycle.

■ Overproduction of ketone bodies in uncontrolled diabetes or severely reduced calorie intake can lead to acidosis or ketosis or both (ketoacidosis).

Key Terms

Terms in bold are defined in the glossary.

β oxidation 649
chylomicron 651
apolipoprotein 651
lipoprotein 651
perilipin 651
free fatty acids 651
serum albumin 651
carnitine shuttle 652
carnitine
    acyltransferase 1 653
acyl-carnitine/carnitine
    transporter 653
carnitine
    acyltransferase 2 653
trifunctional protein
    (TFP) 656
methylmalonyl-CoA
    mutase 661
coenzyme B$_{12}$ 661

malonyl-CoA 661
pernicious anemia 663
intrinsic factor 663
**PPAR (peroxisome
    proliferator-activated
    receptor)** 664
medium-chain acyl-CoA
    dehydrogenase
    (MCAD) 664
multifunctional protein
    (MFP) 666
ω oxidation 666
**mixed-function
    oxygenases** 667
α oxidation 667
acidosis 670
ketosis 670
ketoacidosis 670

Problems

1. Energy in Triacylglycerols On a per-carbon basis, where does the largest amount of biologically available energy in triacylglycerols reside: in the fatty acid portions or the

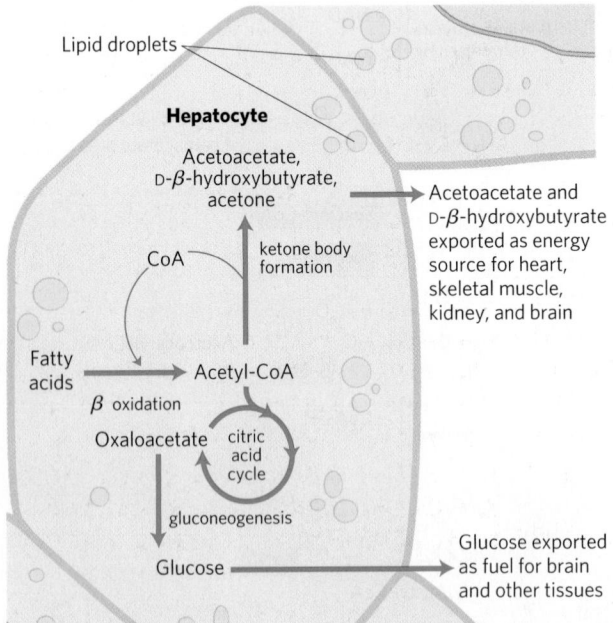

FIGURE 17-20 Ketone body formation and export from the liver. Conditions that promote gluconeogenesis (untreated diabetes, severely reduced food intake) slow the citric acid cycle (by drawing off oxaloacetate) and enhance the conversion of acetyl-CoA to acetoacetate. The released coenzyme A allows continued β oxidation of fatty acids.

glycerol portion? Indicate how knowledge of the chemical structure of triacylglycerols provides the answer.

2. Fuel Reserves in Adipose Tissue Triacylglycerols, with their hydrocarbon-like fatty acids, have the highest energy content of the major nutrients.

(a) If 15% of the body mass of a 70.0 kg adult consists of triacylglycerols, what is the total available fuel reserve, in both kilojoules and kilocalories, in the form of triacylglycerols? Recall that 1.00 kcal = 4.18 kJ.

(b) If the basal energy requirement is approximately 8,400 kJ/day (2,000 kcal/day), how long could this person survive if the oxidation of fatty acids stored as triacylglycerols were the only source of energy?

(c) What would be the weight loss in pounds per day under such starvation conditions (1 lb = 0.454 kg)?

3. Common Reaction Steps in the Fatty Acid Oxidation Cycle and Citric Acid Cycle Cells often use the same enzyme reaction pattern for analogous metabolic conversions. For example, the steps in the oxidation of pyruvate to acetyl-CoA and of α-ketoglutarate to succinyl-CoA, although catalyzed by different enzymes, are very similar. The first stage of fatty acid oxidation follows a reaction sequence closely resembling a sequence in the citric acid cycle. Use equations to show the analogous reaction sequences in the two pathways.

4. β Oxidation: How Many Cycles? How many cycles of β oxidation are required for the complete oxidation of activated oleic acid, $18:1(\Delta^9)$?

5. Chemistry of the Acyl-CoA Synthetase Reaction Fatty acids are converted to their coenzyme A esters in a reversible reaction catalyzed by acyl-CoA synthetase:

$$R-COO^- + ATP + CoA \rightleftharpoons$$
$$\overset{O}{\overset{\|}{R-C-CoA}} + AMP + PP_i$$

(a) The enzyme-bound intermediate in this reaction has been identified as the mixed anhydride of the fatty acid and adenosine monophosphate (AMP), acyl-AMP:

Write two equations corresponding to the two steps of the reaction catalyzed by acyl-CoA synthetase.

(b) The acyl-CoA synthetase reaction is readily reversible, with an equilibrium constant near 1. How can this reaction be made to favor formation of fatty acyl–CoA?

6. Intermediates in Oleic Acid Oxidation What is the structure of the partially oxidized fatty acyl group that is formed when oleic acid, $18:1(\Delta^9)$, has undergone three cycles of β oxidation? What are the next two steps in the continued oxidation of this intermediate?

7. β Oxidation of an Odd-Number Fatty Acid What are the direct products of β oxidation of a fully saturated, straight-chain fatty acid of 11 carbons?

8. Oxidation of Tritiated Palmitate Palmitate uniformly labeled with tritium (^3H) to a specific activity of 2.48×10^8 counts per minute (cpm) per micromole of palmitate is added to a mitochondrial preparation that oxidizes it to acetyl-CoA. The acetyl-CoA is isolated and hydrolyzed to acetate. The specific activity of the isolated acetate is 1.00×10^7 cpm/μmol. Is this result consistent with the β-oxidation pathway? Explain. What is the final fate of the removed tritium?

9. Compartmentation in β Oxidation Free palmitate is activated to its coenzyme A derivative (palmitoyl-CoA) in the cytosol before it can be oxidized in the mitochondrion. If palmitate and $[^{14}C]$coenzyme A are added to a liver homogenate, palmitoyl-CoA isolated from the cytosolic fraction is radioactive, but that isolated from the mitochondrial fraction is not. Explain.

10. Comparative Biochemistry: Energy-Generating Pathways in Birds One indication of the relative importance of various ATP-producing pathways is the V_{max} of certain enzymes of these pathways. The values of V_{max} of several enzymes from the pectoral muscles (chest muscles used for flying) of pigeon and pheasant are listed below.

Enzyme	V_{max} (μmol substrate/min/g tissue)	
	Pigeon	Pheasant
Hexokinase	3.0	2.3
Glycogen phosphorylase	18.0	120.0
Phosphofructokinase-1	24.0	143.0
Citrate synthase	100.0	15.0
Triacylglycerol lipase	0.07	0.01

(a) Discuss the relative importance of glycogen metabolism and fat metabolism in generating ATP in the pectoral muscles of these birds.

(b) Compare oxygen consumption in the two birds.

(c) Judging from the data in the table, which bird is the long-distance flyer? Justify your answer.

(d) Why were these particular enzymes selected for comparison? Would the activities of triose phosphate isomerase and malate dehydrogenase be equally good bases for comparison? Explain.

11. Mutant Carnitine Acyltransferase What changes in metabolic pattern would result from a mutation in the muscle carnitine acyltransferase 1 in which the mutant protein has lost its affinity for malonyl-CoA but not its catalytic activity?

12. Effect of Carnitine Deficiency An individual developed a condition characterized by progressive muscular weakness and aching muscle cramps. The symptoms were aggravated by fasting, exercise, and a high-fat diet. The homogenate of a skeletal muscle specimen from the patient oxidized added oleate more slowly than did control homogenates, consisting of muscle specimens from healthy individuals.

When carnitine was added to the patient's muscle homogenate, the rate of oleate oxidation equaled that in the control homogenates. The patient was diagnosed as having a carnitine deficiency.

(a) Why did added carnitine increase the rate of oleate oxidation in the patient's muscle homogenate?

(b) Why were the patient's symptoms aggravated by fasting, exercise, and a high-fat diet?

(c) Suggest two possible reasons for the deficiency of muscle carnitine in this individual.

13. Fatty Acids as a Source of Water Contrary to legend, camels do not store water in their humps, which actually consist of large fat deposits. How can these fat deposits serve as a source of water? Calculate the amount of water (in liters) that a camel can produce from 1.0 kg of fat. Assume for simplicity that the fat consists entirely of tripalmitoylglycerol.

14. Petroleum as a Microbial Food Source Some microorganisms of the genera *Nocardia* and *Pseudomonas* can grow in an environment where hydrocarbons are the only food source. These bacteria oxidize straight-chain aliphatic hydrocarbons, such as octane, to their corresponding carboxylic acids:

$$CH_3(CH_2)_6CH_3 + NAD^+ + O_2 \rightleftharpoons$$
$$CH_3(CH_2)_6COOH + NADH + H^+$$

How could these bacteria be used to clean up oil spills? What would be some of the limiting factors in the efficiency of this process?

15. Metabolism of a Straight-Chain Phenylated Fatty Acid A crystalline metabolite was isolated from the urine of a rabbit that had been fed a straight-chain fatty acid containing a terminal phenyl group:

A 302 mg sample of the metabolite in aqueous solution was completely neutralized by 22.2 mL of 0.100 M NaOH.

(a) What is the probable molecular weight and structure of the metabolite?

(b) Did the straight-chain fatty acid contain an even or an odd number of methylene ($-CH_2-$) groups (i.e., is n even or odd)? Explain.

16. Fatty Acid Oxidation in Uncontrolled Diabetes When the acetyl-CoA produced during β oxidation in the liver exceeds the capacity of the citric acid cycle, the excess acetyl-CoA forms ketone bodies—acetone, acetoacetate, and D-β-hydroxybutyrate. This occurs in severe, uncontrolled diabetes: because the tissues cannot use glucose, they oxidize large amounts of fatty acids instead. Although acetyl-CoA is not toxic, the mitochondrion must divert the acetyl-CoA to ketone bodies. What problem would arise if acetyl-CoA were not converted to ketone bodies? How does the diversion to ketone bodies solve the problem?

17. Consequences of a High-Fat Diet with No Carbohydrates Suppose you had to subsist on a diet of whale blubber and seal blubber, with little or no carbohydrate.

(a) What would be the effect of carbohydrate deprivation on the utilization of fats for energy?

(b) If your diet were totally devoid of carbohydrate, would it be better to consume odd- or even-number fatty acids? Explain.

18. Even- and Odd-Number Fatty Acids in the Diet In a laboratory experiment, two groups of rats are fed two different fatty acids as their sole source of carbon for a month. The first group gets heptanoic acid (7:0), and the second gets octanoic acid (8:0). After the experiment, a striking difference is seen between the two groups. Those in the first group are healthy and have gained weight, whereas those in the second group are weak and have lost weight as a result of losing muscle mass. What is the biochemical basis for this difference?

19. Metabolic Consequences of Ingesting ω-Fluorooleate The shrub *Dichapetalum toxicarium*, native to Sierra Leone, produces ω-fluorooleate, which is highly toxic to warm-blooded animals.

ω-Fluorooleate

This substance has been used as an arrow poison, and powdered fruit from the plant is sometimes used as a rat poison (hence the plant's common name, ratsbane). Why is this substance so toxic? (Hint: Review Chapter 16, Problem 22.)

20. Mutant Acetyl-CoA Carboxylase What would be the consequences for fat metabolism of a mutation in acetyl-CoA carboxylase that replaced the Ser residue normally phosphorylated by AMPK with an Ala residue? What might happen if the same Ser were replaced by Asp? (Hint: See Fig. 17-13.)

21. Effect of PDE Inhibitor on Adipocytes How would an adipocyte's response to epinephrine be affected by the addition of an inhibitor of cAMP phosphodiesterase (PDE)? (Hint: See Fig. 12-4.)

22. Role of FAD as Electron Acceptor Acyl-CoA dehydrogenase uses enzyme-bound FAD as a prosthetic group to dehydrogenate the α and β carbons of fatty acyl–CoA. What is the advantage of using FAD as an electron acceptor rather than NAD$^+$? Explain in terms of the standard reduction potentials for the Enz-FAD/FADH$_2$ ($E'^\circ = -0.219$ V) and NAD$^+$/NADH ($E'^\circ = -0.320$ V) half-reactions.

23. β Oxidation of Arachidic Acid How many turns of the fatty acid oxidation cycle are required for complete oxidation of arachidic acid (see Table 10-1) to acetyl-CoA?

24. Fate of Labeled Propionate If [3-^{14}C]propionate (^{14}C in the methyl group) is added to a liver homogenate, ^{14}C-labeled oxaloacetate is rapidly produced. Draw a flow chart for the pathway by which propionate is transformed to oxaloacetate, and indicate the location of the ^{14}C in oxaloacetate.

25. Phytanic Acid Metabolism When phytanic acid uniformly labeled with ^{14}C is fed to a mouse, radioactivity can be detected in malate, a citric acid cycle intermediate, within minutes. Draw a metabolic pathway that could account for this. Which of the carbon atoms in malate would contain ^{14}C label?

26. Sources of H₂O Produced in β Oxidation The complete oxidation of palmitoyl-CoA to carbon dioxide and water is represented by the overall equation

$$\text{Palmitoyl-CoA} + 23O_2 + 108P_i + 108ADP \longrightarrow$$
$$\text{CoA} + 16CO_2 + 108ATP + 23H_2O$$

Water is also produced in the reaction

$$\text{ADP} + P_i \longrightarrow \text{ATP} + H_2O$$

but is not included as a product in the overall equation. Why?

27. Biological Importance of Cobalt In cattle, deer, sheep, and other ruminant animals, large amounts of propionate are produced in the rumen through the bacterial fermentation of ingested plant matter. Propionate is the principal source of glucose for these animals, via the route propionate ⟶ oxaloacetate ⟶ glucose. In some areas of the world, notably Australia, ruminant animals sometimes show symptoms of anemia with concomitant loss of appetite and retarded growth, resulting from an inability to transform propionate to oxaloacetate. This condition is due to a cobalt deficiency caused by very low cobalt levels in the soil and thus in plant matter. Explain.

28. Fat Loss during Hibernation Bears expend about 25×10^6 J/day during periods of hibernation, which may last as long as seven months. The energy required to sustain life is obtained from fatty acid oxidation. How much weight loss (in kilograms) has occurred after seven months? How might ketosis be minimized during hibernation? (Assume the oxidation of fat yields 38 kJ/g.)

Data Analysis Problem

29. β Oxidation of Trans Fats Unsaturated fats with trans double bonds are commonly referred to as "trans fats." There has been much discussion about the effects of dietary trans fats on health. In their investigations of the effects of trans fatty acid metabolism on health, Yu and colleagues (2004) showed that a model trans fatty acid was processed differently from its cis isomer. They used three related 18-carbon fatty acids to explore the difference in β oxidation between cis and trans isomers of the same-size fatty acid.

Stearic acid
(octadecenoic acid)

Oleic acid
(cis-Δ⁹-octadecenoic acid)

Elaidic acid
(trans-Δ⁹-octadecenoic acid)

The researchers incubated the coenzyme A derivative of each acid with rat liver mitochondria for 5 minutes, then separated the remaining CoA derivatives in each mixture by HPLC (high-performance liquid chromatography). The results are shown below, with separate panels for the three experiments.

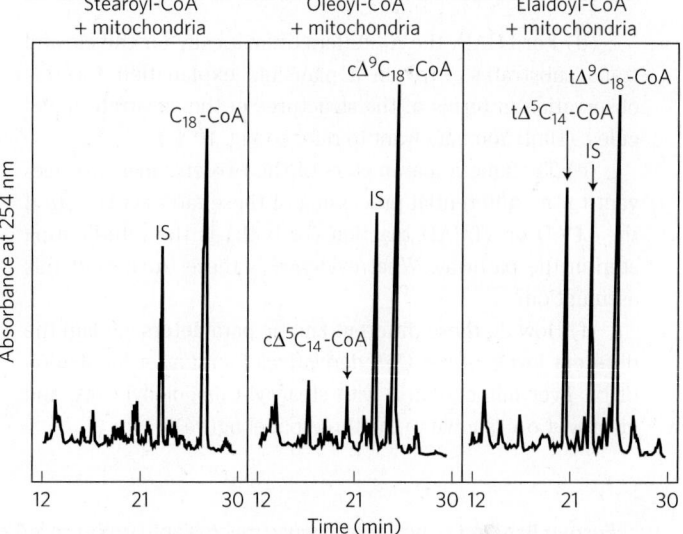

In the figure, IS indicates an internal standard (pentadecanoyl-CoA) added to the mixture, after the reaction, as a molecular marker. The researchers abbreviated the CoA derivatives as follows: stearoyl-CoA, C_{18}-CoA; cis-Δ^5-tetradecenoyl-CoA, $c\Delta^5 C_{14}$-CoA; oleoyl-CoA, $c\Delta^9 C_{18}$-CoA; trans-Δ^5-tetradecenoyl-CoA, $t\Delta^5 C_{14}$-CoA; and elaidoyl-CoA, $t\Delta^9 C_{18}$-CoA.

cis-Δ⁵-Tetradecenoyl-CoA

trans-Δ⁵-Tetradecenoyl-CoA

(a) Why did Yu and colleagues need to use CoA derivatives rather than the free fatty acids in these experiments?

(b) Why were no lower molecular weight CoA derivatives found in the reaction with stearoyl-CoA?

(c) How many rounds of β oxidation would be required to convert the oleoyl-CoA and the elaidoyl-CoA to cis-Δ^5-tetradecenoyl-CoA and trans-Δ^5-tetradecenoyl-CoA, respectively?

Yu and coworkers measured the kinetic parameters of two forms of the enzyme acyl-CoA dehydrogenase: long-chain acyl-CoA dehydrogenase (LCAD) and very-long-chain acyl-CoA dehydrogenase (VLCAD). They used the CoA derivatives of three fatty acids: tetradecanoyl-CoA (C_{14}-CoA), cis-Δ^5-tetradecenoyl-CoA ($c\Delta^5 C_{14}$-CoA), and trans-Δ^5-tetradecenoyl-CoA ($t\Delta^5 C_{14}$-CoA). The results are shown below. (See Chapter 6 for definitions of the kinetic parameters.)

	LCAD			VLCAD		
	C_{14}–CoA	$c\Delta^5 C_{14}$–CoA	$t\Delta^5 C_{14}$–CoA	C_{14}–CoA	$c\Delta^5 C_{14}$–CoA	$t\Delta^5 C_{14}$–CoA
V_{max}	3.3	3.0	2.9	1.4	0.32	0.88
K_m	0.41	0.40	1.6	0.57	0.44	0.97
k_{cat}	9.9	8.9	8.5	2.0	0.42	1.12
k_{cat}/K_m	24	22	5	4	1	1

(d) For LCAD, the K_m differs dramatically for the cis and trans substrates. Provide a plausible explanation for this observation in terms of the structures of the substrate molecules. (Hint: You may want to refer to Fig. 10-1.)

(e) The kinetic parameters of the two enzymes are relevant to the differential processing of these fatty acids *only* if the LCAD or VLCAD reaction (or both) is the rate-limiting step in the pathway. What evidence is there to support this assumption?

(f) How do these different kinetic parameters explain the different levels of the CoA derivatives found after incubation of rat liver mitochondria with stearoyl-CoA, oleoyl-CoA, and elaidoyl-CoA (shown in the three-panel figure)?

Yu and coworkers measured the substrate specificity of rat liver mitochondrial thioesterase, which hydrolyzes acyl-CoA to CoA and free fatty acid. This enzyme was approximately twice as active with C_{14}-CoA thioesters as with C_{18}-CoA thioesters.

(g) Other research has suggested that free fatty acids can pass through membranes. In their experiments, Yu and colleagues found *trans*-Δ^5-tetradecenoic acid outside (i.e., in the medium surrounding) mitochondria that had been incubated with elaidoyl-CoA. Describe the pathway that led to this extramitochondrial *trans*-Δ^5-tetradecenoic acid. Be sure to indicate where in the cell the various transformations take place, as well as the enzymes that catalyze the transformations.

(h) It is often said in the popular press that "trans fats are not broken down by your cells and instead accumulate in your body." In what sense is this statement correct and in what sense is it an oversimplification?

Reference

Yu, W., X. Liang, R. Ensenauer, J. Vockley, L. Sweetman, and H. Schultz. 2004. Leaky β-oxidation of a *trans*-fatty acid. *J. Biol. Chem.* 279:52,160–52,167.

Further Reading is available at www.macmillanlearning.com/LehningerBiochemistry7e.

CHAPTER 18

Amino Acid Oxidation and the Production of Urea

Self-study tools that will help you practice what you've learned and reinforce this chapter's concepts are available online. Go to www.macmillanlearning.com/LehningerBiochemistry7e.

We now turn our attention to the amino acids, the final class of biomolecules that, through their oxidative degradation, make a significant contribution to the generation of metabolic energy. The fraction of metabolic energy obtained from amino acids, whether they are derived from dietary protein or from tissue protein, varies greatly with the type of organism and with metabolic conditions. Carnivores consume primarily protein and thus must obtain most of their energy from amino acids, whereas herbivores may fill only a small fraction of their energy needs by this route. Most microorganisms can scavenge amino acids from their environment and use them as fuel when required by metabolic conditions. Plants, however, rarely if ever oxidize amino acids to provide energy; the carbohydrate produced from CO_2 and H_2O in photosynthesis is generally their sole energy source. Amino acid concentrations in plant tissues are carefully regulated to just meet the requirements for biosynthesis of proteins, nucleic acids, and other molecules needed to support growth. Amino acid catabolism does occur in plants, but its purpose is to produce metabolites for other biosynthetic pathways.

In animals, amino acids undergo oxidative degradation in three different metabolic circumstances:

1. During the normal synthesis and degradation of cellular proteins (protein turnover; Chapter 27),
    some amino acids that are released from protein breakdown and are not needed for new protein synthesis undergo oxidative degradation.

2. When a diet is rich in protein and the ingested amino acids exceed the body's needs for protein synthesis, the surplus is catabolized; amino acids cannot be stored.

3. During starvation or in uncontrolled diabetes mellitus, when carbohydrates are either unavailable or not properly utilized, cellular proteins are used as fuel.

Under all these metabolic conditions, amino acids lose their amino groups to form α-keto acids, the "carbon skeletons" of amino acids. The α-keto acids undergo oxidation to CO_2 and H_2O or, often more importantly, provide three- and four-carbon units that can be converted by gluconeogenesis into glucose, the fuel for brain, skeletal muscle, and other tissues.

The pathways of amino acid catabolism are quite similar in most organisms. The focus of this chapter is on the pathways in vertebrates, because these have received the most research attention. As in carbohydrate and fatty acid catabolism, the processes of amino acid degradation converge on the central catabolic pathways, with the carbon skeletons of most amino acids finding their way to the citric acid cycle. In some cases the reaction pathways of amino acid breakdown closely parallel steps in the catabolism of fatty acids (see Fig. 17-9).

One important feature distinguishes amino acid degradation from other catabolic processes described to

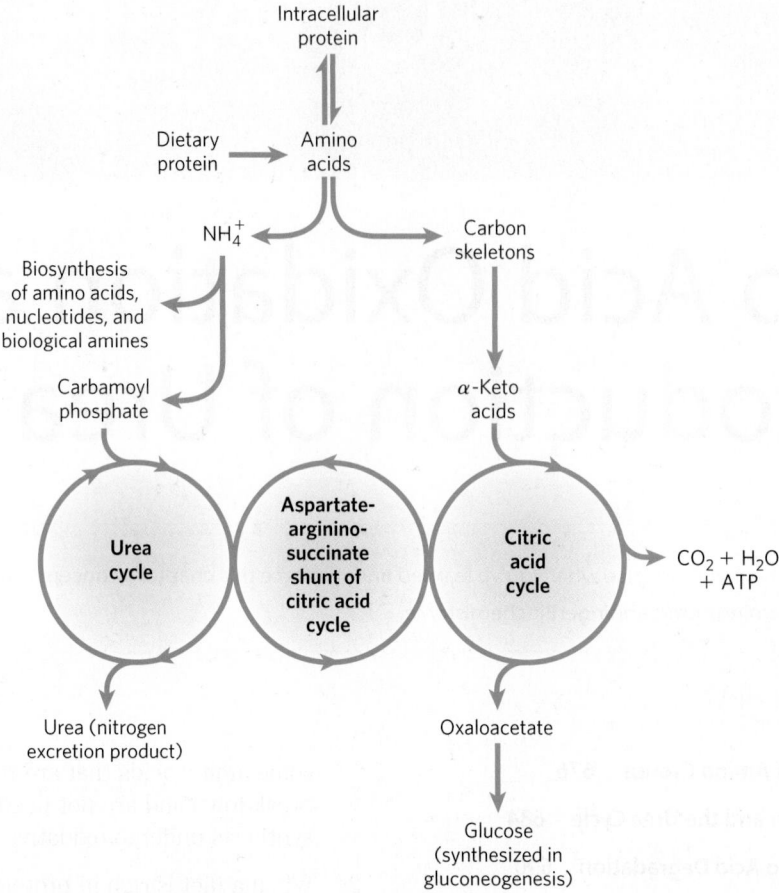

FIGURE 18-1 Overview of amino acid catabolism in mammals. The amino groups and the carbon skeleton take separate but interconnected pathways.

this point: every amino acid contains an amino group, and the pathways for amino acid degradation therefore include a key step in which the α-amino group is separated from the carbon skeleton and shunted into the pathways of amino group metabolism **(Fig. 18-1)**. We deal first with amino group metabolism and nitrogen excretion, then with the fate of the carbon skeletons derived from the amino acids; along the way we see how the pathways are interconnected.

18.1 Metabolic Fates of Amino Groups

Nitrogen, N_2, is abundant in the atmosphere but is too inert for use in most biochemical processes. Because only a few microorganisms can convert N_2 to biologically useful forms such as NH_3 (Chapter 22), amino groups are carefully husbanded in biological systems.

Figure 18-2a provides an overview of the catabolic pathways of ammonia and amino groups in vertebrates. Amino acids derived from dietary protein are the source of most amino groups. Most amino acids are metabolized in the liver. Some of the ammonia generated in this process is recycled and used in a variety of biosynthetic pathways; the excess is either excreted directly or converted to urea or uric acid for excretion, depending on

the organism (Fig. 18-2b). Excess ammonia generated in other (extrahepatic) tissues travels to the liver (in the form of amino groups, as described below) for conversion to the excretory form.

Four amino acids play central roles in nitrogen metabolism: glutamate, glutamine, alanine, and aspartate. The special place of these four amino acids in nitrogen metabolism is not an evolutionary accident. These particular amino acids are the ones most easily converted into citric acid cycle intermediates: glutamate and glutamine to α-ketoglutarate, alanine to pyruvate, and aspartate to oxaloacetate. Glutamate and glutamine are especially important, acting as a kind of general collection point for amino groups. In the cytosol of liver cells (hepatocytes), amino groups from most amino acids are transferred to α-ketoglutarate to form glutamate, which enters mitochondria and gives up its amino group to form NH_4^+. Excess ammonia generated in most other tissues is converted to the amide nitrogen of glutamine, which passes to the liver, then into liver mitochondria. Glutamine or glutamate or both are present in higher concentrations than other amino acids in most tissues.

In skeletal muscle, excess amino groups are generally transferred to pyruvate to form alanine,

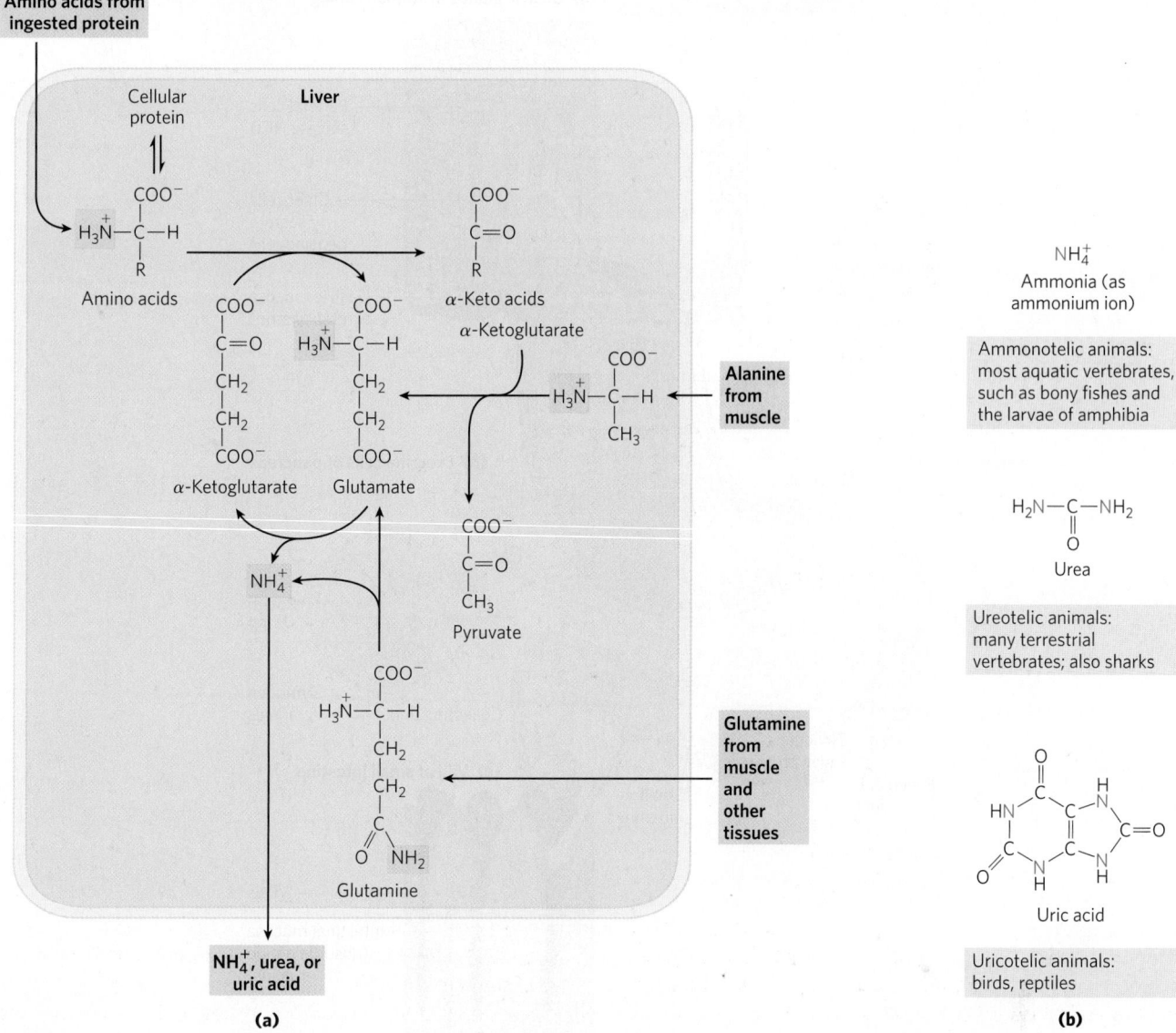

FIGURE 18-2 Amino group catabolism. (a) Overview of catabolism of amino groups (shaded) in vertebrate liver. **(b)** Excretory forms of nitrogen. Excess NH_4^+ is excreted as ammonia (microbes, bony fishes), urea (most terrestrial vertebrates), or uric acid (birds and terrestrial reptiles). Notice that the carbon atoms of urea and uric acid are highly oxidized; the organism discards carbon only after extracting most of its available energy of oxidation.

another important molecule in the transport of amino groups to the liver. We will see in Section 18.2 that aspartate comes into play in the metabolic processes that occur once the amino groups are delivered to the liver.

We begin with a discussion of the breakdown of dietary proteins, then proceed to a general description of the metabolic fates of amino groups.

Dietary Protein Is Enzymatically Degraded to Amino Acids

In humans, the degradation of ingested proteins to their constituent amino acids occurs in the gastrointestinal tract. Entry of dietary protein into the stomach stimulates the gastric mucosa to secrete the hormone **gastrin**, which in turn stimulates the secretion of hydrochloric acid by the parietal cells and pepsinogen by the chief cells of the gastric glands **(Fig. 18-3a)**. The acidic gastric juice (pH 1.0 to 2.5) is both an antiseptic, killing most bacteria and other foreign cells, and a denaturing agent, unfolding globular proteins and rendering their internal peptide bonds more accessible to enzymatic hydrolysis. **Pepsinogen** (M_r 40,554), an inactive precursor, or zymogen (p. 230), is converted to active pepsin (M_r 34,614) by an autocatalytic cleavage (a cleavage mediated by the pepsinogen itself) that occurs only at low pH. In the stomach, pepsin hydrolyzes ingested proteins at peptide bonds on the amino-terminal side of Leu and the aromatic amino acid residues Phe, Trp, and Tyr (see Table 3-6), cleaving long polypeptide chains into a mixture of smaller peptides.

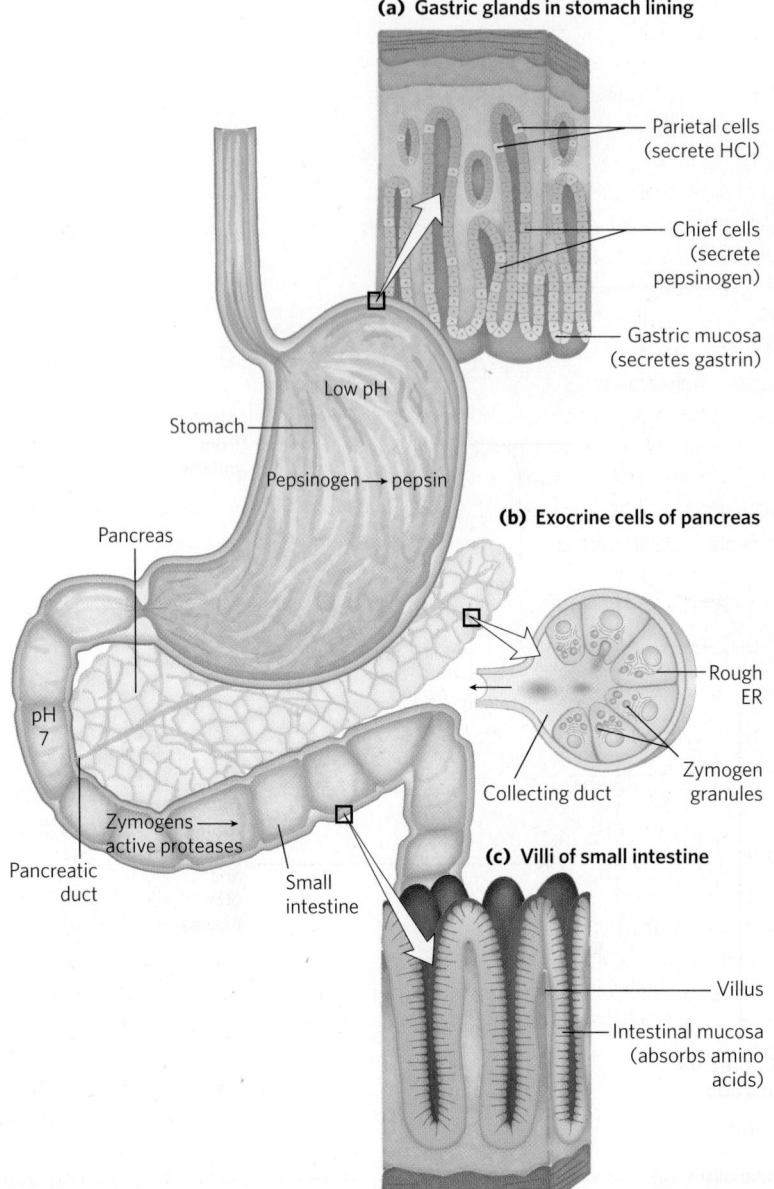

(a) Gastric glands in stomach lining

Parietal cells (secrete HCl)

Chief cells (secrete pepsinogen)

Gastric mucosa (secretes gastrin)

Low pH

Stomach

Pepsinogen → pepsin

Pancreas

pH 7

Zymogens → active proteases

Pancreatic duct

Small intestine

(b) Exocrine cells of pancreas

Rough ER

Collecting duct

Zymogen granules

(c) Villi of small intestine

Villus

Intestinal mucosa (absorbs amino acids)

FIGURE 18-3 Part of the human digestive (gastrointestinal) tract. **(a)** The parietal cells and chief cells of the gastric glands secrete their products in response to the hormone gastrin. Pepsin begins the process of protein degradation in the stomach. **(b)** The cytoplasm of exocrine cells of the pancreas is completely filled with rough endoplasmic reticulum, the site of synthesis of the zymogens of many digestive enzymes. The zymogens are concentrated in membrane-enclosed transport particles called zymogen granules. When an exocrine cell is stimulated, its plasma membrane fuses with the zymogen granule membrane and zymogens are released into the lumen of the collecting duct by exocytosis. The collecting ducts ultimately lead to the pancreatic duct and thence to the small intestine. **(c)** In the small intestine, amino acids are absorbed through the epithelial cell layer (intestinal mucosa) of the villi and enter the capillaries. Recall that the products of lipid hydrolysis in the small intestine enter the lymphatic system after their absorption by the intestinal mucosa (see Fig. 17-1).

As the acidic stomach contents pass into the small intestine, the low pH triggers secretion of the hormone **secretin** into the blood. Secretin stimulates the pancreas to secrete bicarbonate into the small intestine to neutralize the gastric HCl, abruptly increasing the pH to about 7. (All pancreatic secretions pass into the small intestine through the pancreatic duct.) The digestion of proteins now continues in the small intestine. Arrival of amino acids in the upper part of the intestine (duodenum) causes release into the blood of the hormone **cholecystokinin**, which stimulates secretion of several pancreatic enzymes with activity optima at pH 7 to 8. **Trypsinogen**, **chymotrypsinogen**, and **procarboxypeptidases A** and **B**—the zymogens of **trypsin**, **chymotrypsin**, and **carboxypeptidases A** and **B**—are synthesized and secreted by the exocrine cells of the pancreas (Fig. 18-3b). Trypsinogen is converted to its active form, trypsin, by **enteropeptidase**, a proteolytic enzyme secreted by intestinal cells. Free trypsin then catalyzes the conversion of additional trypsinogen to trypsin (see Fig. 6-39). Trypsin also activates chymotrypsinogen, the procarboxypeptidases, and proelastase.

Why this elaborate mechanism for getting active digestive enzymes into the gastrointestinal tract? Synthesis of the enzymes as inactive precursors protects the exocrine cells from destructive proteolytic attack. The pancreas further protects itself against self-digestion by making a specific inhibitor, a protein called **pancreatic trypsin inhibitor** (p. 231). Given the key role of trypsin in proteolytic activation pathways, inhibition of trypsin effectively prevents premature production of active proteolytic enzymes within pancreatic cells.

Trypsin and chymotrypsin further hydrolyze the peptides that were produced by pepsin in the stomach. This stage of protein digestion is accomplished very efficiently, because pepsin, trypsin, and chymotrypsin have different amino acid specificities (see Table 3-6). Degradation of the short peptides in the small intestine is then completed by other intestinal peptidases. These include carboxypeptidases A and B (both of which are zinc-containing enzymes), which remove successive carboxyl-terminal residues from peptides, and an **aminopeptidase** that hydrolyzes successive amino-terminal residues from short peptides. The resulting mixture of free amino acids is transported into the epithelial cells lining the small intestine (Fig. 18-3c), through which the amino acids enter the blood capillaries in the villi and travel to the liver. In humans, most globular proteins from animal sources are almost completely hydrolyzed to amino acids in the gastrointestinal tract, but some fibrous proteins, such as keratin, are only partly digested. In addition, the protein content of some plant foods is protected against breakdown by indigestible cellulose husks.

Acute pancreatitis is a disease caused by obstruction of the normal pathway by which pancreatic secretions enter the intestine. The zymogens of the proteolytic enzymes are converted to their catalytically active forms prematurely, *inside* the pancreatic cells, and attack the pancreatic tissue itself. This causes excruciating pain and damage to the organ that can prove fatal. ∎

Pyridoxal Phosphate Participates in the Transfer of α-Amino Groups to α-Ketoglutarate

The first step in the catabolism of most L-amino acids, once they have reached the liver, is removal of the α-amino groups, promoted by enzymes called **aminotransferases** or **transaminases**. In these **transamination** reactions, the α-amino group is transferred to the α-carbon atom of α-ketoglutarate, leaving behind the corresponding α-keto acid analog of the amino acid **(Fig. 18-4)**. There is no net deamination (loss of amino groups) in these reactions, because the α-ketoglutarate

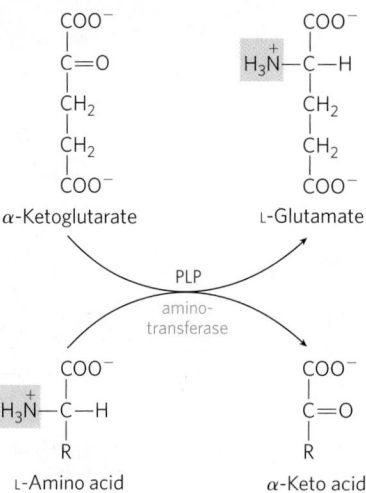

FIGURE 18-4 Enzyme-catalyzed transaminations. In many aminotransferase reactions, α-ketoglutarate is the amino group acceptor. All aminotransferases have pyridoxal phosphate (PLP) as cofactor. Although the reaction is shown here in the direction of transfer of the amino group to α-ketoglutarate, it is readily reversible.

becomes aminated as the α-amino acid is deaminated. The effect of transamination reactions is to collect the amino groups from many different amino acids in the form of L-glutamate. The glutamate then functions as an amino group donor for biosynthetic pathways or for excretion pathways that lead to the elimination of nitrogenous waste products.

Cells contain different types of aminotransferases. Many are specific for α-ketoglutarate as the amino group acceptor but differ in their specificity for the L-amino acid. The enzymes are named for the amino group donor (alanine aminotransferase and aspartate aminotransferase, for example). The reactions catalyzed by aminotransferases are freely reversible, having an equilibrium constant of about 1.0 ($\Delta G' \approx 0$ kJ/mol).

All aminotransferases have the same prosthetic group and the same reaction mechanism. The prosthetic group is **pyridoxal phosphate (PLP)**, the coenzyme form of pyridoxine, or vitamin B_6. We encountered pyridoxal phosphate in Chapter 15, as a coenzyme in the glycogen phosphorylase reaction, but its role in that reaction is not representative of its usual coenzyme function. Its primary role in cells is in the metabolism of molecules with amino groups.

Pyridoxal phosphate functions as an intermediate carrier of amino groups at the active site of aminotransferases. It undergoes reversible transformations between its aldehyde form, pyridoxal phosphate, which can accept an amino group, and its aminated form, pyridoxamine phosphate, which can donate its amino group to an α-keto acid **(Fig. 18-5a)**. Pyridoxal phosphate is generally covalently bound to the enzyme's active site through an aldimine (Schiff base) linkage to the ε-amino group of a Lys residue (Fig. 18-5b, d).

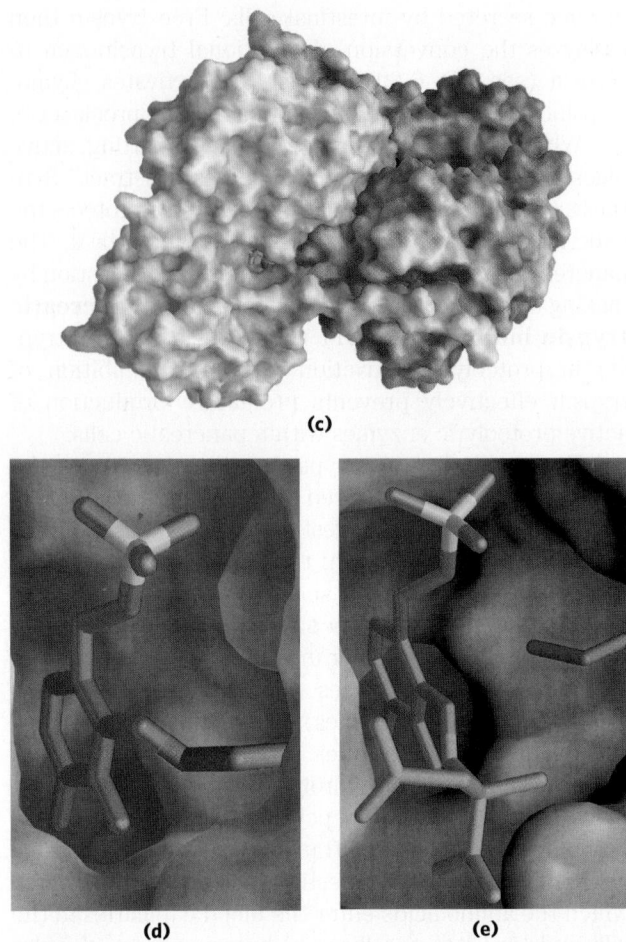

FIGURE 18-5 Pyridoxal phosphate, the prosthetic group of aminotransferases. (a) Pyridoxal phosphate (PLP) and its aminated form, pyridoxamine phosphate, are the tightly bound coenzymes of aminotransferases. The functional groups are shaded. **(b)** Pyridoxal phosphate is bound to the enzyme through noncovalent interactions and a Schiff-base (aldimine) linkage to a Lys residue at the active site. The steps in the formation of a Schiff base from a primary amine and a carbonyl group are detailed in Figure 14-6. **(c)** PLP (red) bound to one of the two active sites of the dimeric enzyme aspartate aminotransferase, a typical aminotransferase; **(d)** close-up view of the active site, with PLP (red, with yellow phosphorus) in aldimine linkage with the side chain of Lys[258] (purple); **(e)** another close-up view of the active site, with PLP linked to the substrate analog 2-methylaspartate (green) via a Schiff base. [Source: (c, d, e) PDB ID 1AJS, S. Rhee et al., *J. Biol. Chem.* 272:17,293, 1997.]

Pyridoxal phosphate participates in a variety of reactions at the α, β, and γ carbons (C-2 to C-4) of amino acids. Reactions at the α carbon **(Fig. 18-6)** include racemizations (interconverting L- and D-amino acids) and decarboxylations, as well as transaminations. Pyridoxal phosphate plays the same chemical role in each of these reactions. A bond to the α carbon of the substrate is broken, removing either a proton or a carboxyl group. The electron pair left behind on the α carbon would form a highly unstable carbanion, but pyridoxal phosphate provides resonance stabilization of this intermediate (Fig. 18-6, inset). The highly conjugated structure of PLP (an electron sink) permits delocalization of the negative charge.

Aminotransferases (Fig. 18-5) are classic examples of enzymes catalyzing bimolecular Ping-Pong reactions (see Fig. 6-13b, d), in which the first substrate reacts and the product must leave the active site before the second substrate can bind. Thus the incoming amino acid binds to the active site, donates its amino group to pyridoxal phosphate, and departs in the form of an α-keto acid. The incoming α-keto acid then binds, accepts the amino group from pyridoxamine phosphate, and departs in the form of an amino acid.

Glutamate Releases Its Amino Group as Ammonia in the Liver

As we have seen, the amino groups from many of the α-amino acids are collected in the liver in the form of the amino group of L-glutamate molecules. These amino groups must next be removed from glutamate to prepare them for excretion. In hepatocytes, glutamate is transported from the cytosol into mitochondria, where it undergoes **oxidative deamination** catalyzed by **L-glutamate dehydrogenase** (M_r 330,000). In mammals, this enzyme is present in the mitochondrial matrix.

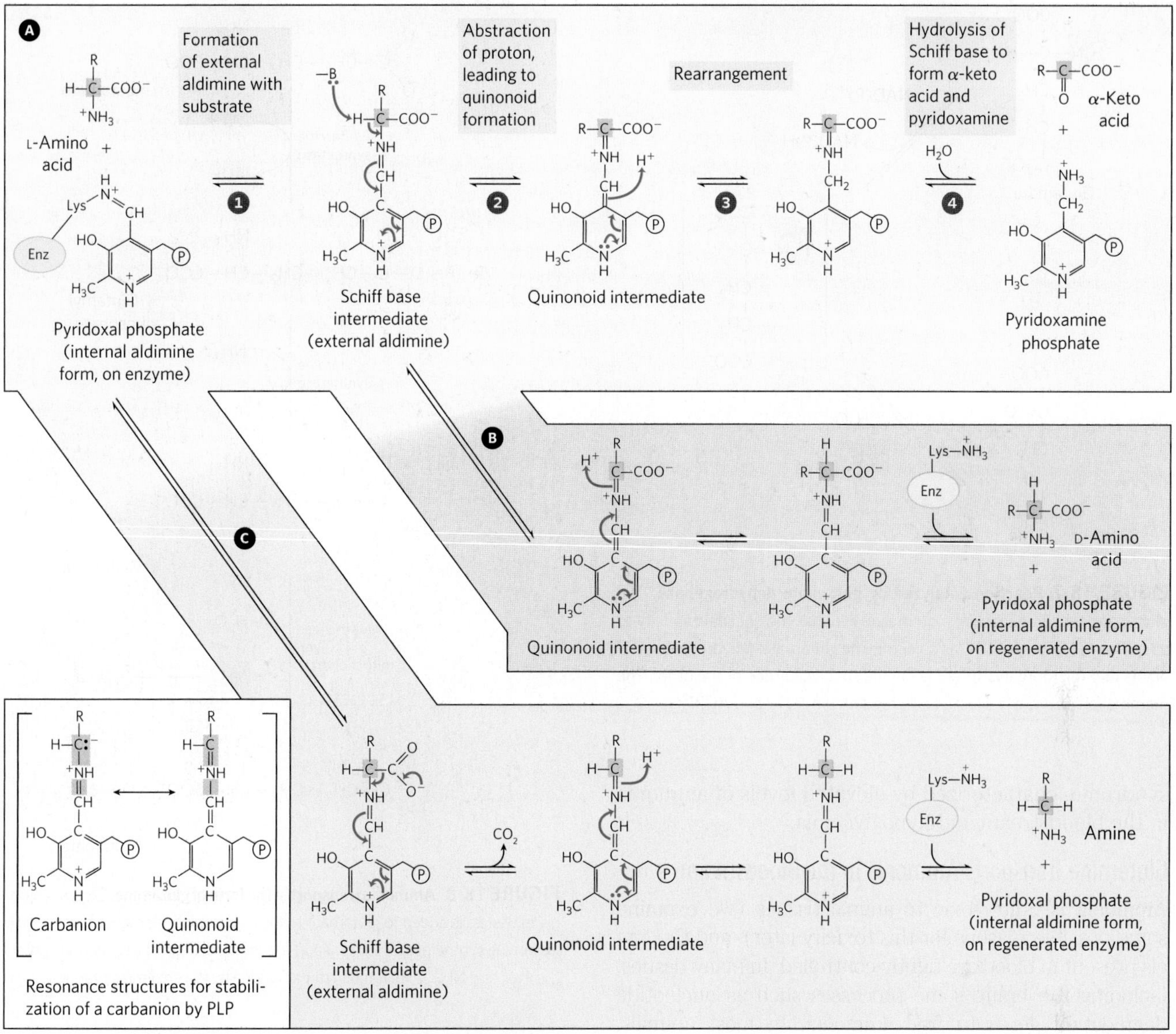

MECHANISM FIGURE 18-6 **Some amino acid transformations at the α carbon that are facilitated by pyridoxal phosphate.** Pyridoxal phosphate is generally bonded to the enzyme through a Schiff base, also called an internal aldimine. This activated form of PLP readily undergoes transimination to form a new Schiff base (external aldimine) with the α-amino group of the substrate amino acid (see Fig. 18-5b, d). Three alternative fates for the external aldimine are shown: **A** transamination, **B** racemization, and **C** decarboxylation. The PLP–amino acid Schiff base is in conjugation with the pyridine ring, an electron sink that permits delocalization of an electron pair to avoid formation of an unstable carbanion on the α carbon (inset). A quinonoid intermediate is involved in all three types of reactions. The transamination route **A** is especially important in the pathways described in this chapter. The pathway highlighted in yellow (shown left to right) represents only part of the overall reaction catalyzed by aminotransferases. To complete the process, a second α-keto acid replaces the one that is released, and this is converted to an amino acid in a reversal of the reaction steps (right to left). Pyridoxal phosphate is also involved in certain reactions at the β and γ carbons of some amino acids (not shown).

It is the only enzyme that can use either NAD^+ or $NADP^+$ as the acceptor of reducing equivalents (**Fig. 18-7**).

The combined action of an aminotransferase and glutamate dehydrogenase is referred to as **transdeamination**. A few amino acids bypass the transdeamination pathway and undergo direct oxidative deamination. The fate of the NH_4^+ produced by any of these deamination processes is discussed in detail in Section 18.2. The α-ketoglutarate formed from glutamate deamination can be used in the citric acid cycle and for glucose synthesis.

Glutamate dehydrogenase operates at an important intersection of carbon and nitrogen metabolism. An allosteric enzyme with six identical subunits, its activity is influenced by a complicated array of allosteric modulators. The best-studied of these are the positive modulator ADP and the negative modulator GTP. The metabolic rationale for this regulatory pattern has not been elucidated in detail. Mutations that alter the allosteric binding site for GTP or otherwise cause permanent activation of glutamate dehydrogenase lead to a human genetic disorder called hyperinsulinism-hyperammonemia

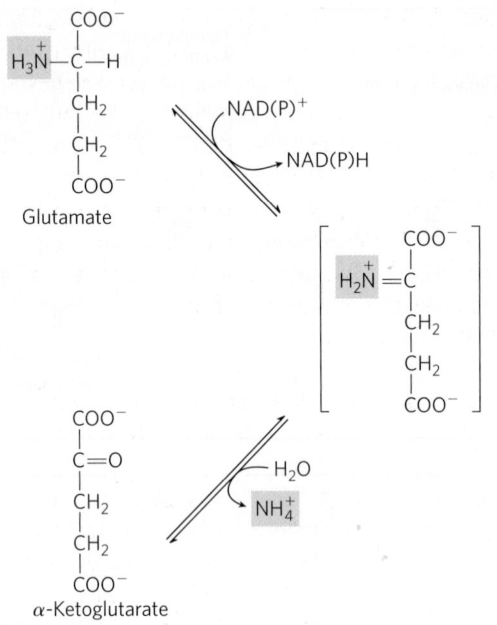

FIGURE 18-7 Reaction catalyzed by glutamate dehydrogenase. The glutamate dehydrogenase of mammalian liver has the unusual capacity to use either NAD$^+$ or NADP$^+$ as cofactor. The glutamate dehydrogenases of plants and microorganisms are generally specific for one or the other. The mammalian enzyme is allosterically regulated by GTP and ADP.

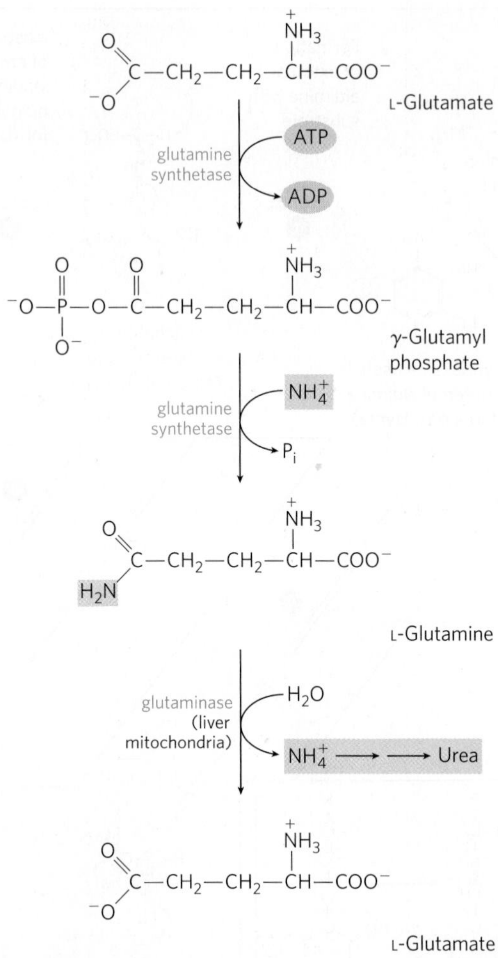

FIGURE 18-8 Ammonia transport in the form of glutamine. Excess ammonia in tissues is added to glutamate to form glutamine, a process catalyzed by glutamine synthetase. After transport in the bloodstream, the glutamine enters the liver, and NH$_4^+$ is liberated in mitochondria by the enzyme glutaminase.

syndrome, characterized by elevated levels of ammonia in the bloodstream and hypoglycemia.

Glutamine Transports Ammonia in the Bloodstream

Ammonia is quite toxic to animal tissues (we examine some possible reasons for this toxicity later), and the levels present in blood are tightly controlled. In many tissues, including the brain, some processes such as nucleotide degradation generate free ammonia. In most animals, much of the free ammonia is converted to a nontoxic compound before export from the extrahepatic tissues into the blood and transport to the liver or kidneys. For this transport function, glutamate, critical to *intracellular* amino group metabolism, is supplanted by L-glutamine. The free ammonia produced in tissues is combined with glutamate to yield glutamine by the action of **glutamine synthetase**. This reaction requires ATP and occurs in two steps **(Fig. 18-8)**. First, glutamate and ATP react to form ADP and a γ-glutamyl phosphate intermediate, which then reacts with ammonia to produce glutamine and inorganic phosphate. Glutamine is a nontoxic transport form of ammonia; it is normally present in blood in much higher concentrations than other amino acids. Glutamine also serves as a source of amino groups in a variety of biosynthetic reactions. Glutamine synthetase is found in all organisms, always playing a central metabolic role. In microorganisms, the enzyme serves as an essential portal for the entry of fixed nitrogen into biological systems. (The roles of glutamine and glutamine synthetase in metabolism are further discussed in Chapter 22.)

In most terrestrial animals, glutamine in excess of that required for biosynthesis is transported in the blood to the intestine, liver, and kidneys for processing. In these tissues, the amide nitrogen is released as ammonium ion in the mitochondria, where the enzyme **glutaminase** converts glutamine to glutamate and NH$_4^+$ (Fig. 18-8). The NH$_4^+$ from intestine and kidney is transported in the blood to the liver. In the liver, the ammonia from all sources is disposed of by urea synthesis. Some of the glutamate produced in the glutaminase reaction may be further processed in the liver by glutamate dehydrogenase, releasing more ammonia and producing carbon skeletons for metabolic fuel. However, most glutamate enters the transamination reactions required for amino acid biosynthesis and other processes (Chapter 22).

In metabolic acidosis (p. 670) there is an increase in glutamine processing by the kidneys. Not all the excess NH$_4^+$ thus produced is released into the bloodstream or converted to urea; some is excreted directly into the urine. In the kidney, the NH$_4^+$ forms salts with metabolic acids, facilitating their removal in the urine. Bicarbonate produced by the decarboxylation of α-ketoglutarate in the citric acid cycle can also serve as a buffer in blood plasma. Taken together, these effects of glutamine metabolism in the kidney tend to counteract acidosis. ∎

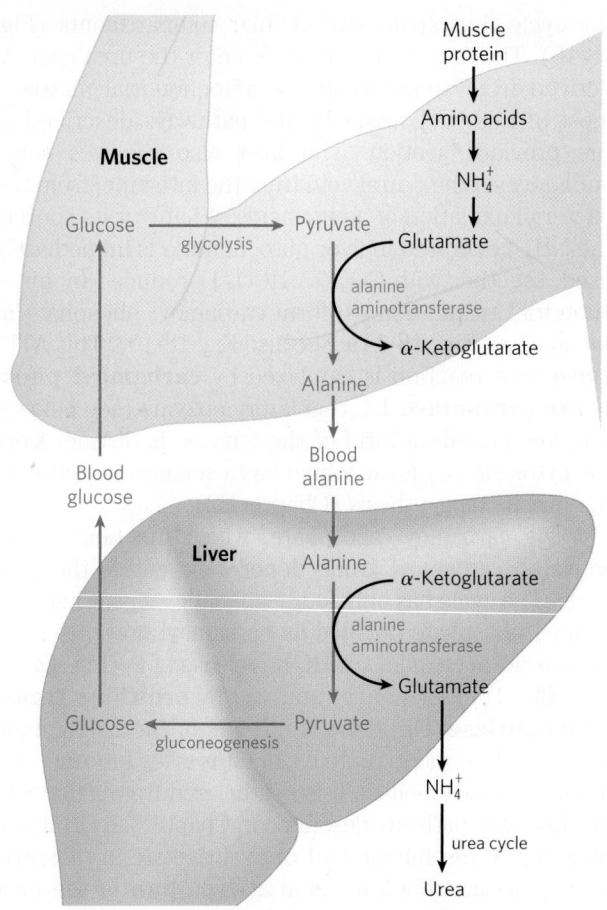

FIGURE 18-9 Glucose-alanine cycle. Alanine serves as a carrier of ammonia and of the carbon skeleton of pyruvate from skeletal muscle to liver. The ammonia is excreted and the pyruvate is used to produce glucose, which is returned to the muscle.

Alanine Transports Ammonia from Skeletal Muscles to the Liver

Alanine also plays a special role in transporting amino groups to the liver in a nontoxic form, via a pathway called the **glucose-alanine cycle (Fig. 18-9)**. In muscle and certain other tissues that degrade amino acids for fuel, amino groups are collected in the form of glutamate by transamination (Fig. 18-2a). Glutamate can be converted to glutamine for transport to the liver, as described above, or it can transfer its α-amino group to pyruvate, a readily available product of muscle glycolysis, by the action of **alanine aminotransferase** (Fig. 18-9). The alanine so formed passes into the blood and travels to the liver. In the cytosol of hepatocytes, alanine aminotransferase transfers the amino group from alanine to α-ketoglutarate, forming pyruvate and glutamate. Glutamate can then enter mitochondria, where the glutamate dehydrogenase reaction releases NH_4^+ (Fig. 18-7), or can undergo transamination with oxaloacetate to form aspartate, another nitrogen donor in urea synthesis, as we shall see.

The use of alanine to transport ammonia from skeletal muscles to the liver is another example of the intrinsic economy of living organisms. Vigorously contracting skeletal muscles operate anaerobically, producing pyruvate and lactate from glycolysis as well as ammonia from protein breakdown. These products must find their way to the liver, where pyruvate and lactate are incorporated into glucose, which is returned to the muscles, and ammonia is converted to urea for excretion. The glucose-alanine cycle, in concert with the Cori cycle (see Box 14-2 and Fig. 23-21), accomplishes this transaction. The energetic burden of gluconeogenesis is thus imposed on the liver rather than the muscle, and all available ATP in muscle is devoted to muscle contraction.

Ammonia Is Toxic to Animals

The catabolic production of ammonia poses a serious biochemical problem, because ammonia is very toxic. The brain is particularly sensitive; damage from ammonia toxicity causes cognitive impairment, ataxia, and epileptic seizures. In extreme cases there is swelling of the brain leading to death. The molecular bases for this toxicity are gradually coming into focus. In the blood, about 98% of ammonia is in the protonated form (NH_4^+), which does not cross the plasma membrane. The small amount of NH_3 present readily crosses all membranes, including the blood-brain barrier, allowing it to enter cells, where much of it becomes protonated and can accumulate inside cells as NH_4^+.

Ridding the cytosol of ammonia requires reductive amination of α-ketoglutarate to glutamate by glutamate dehydrogenase (the reverse of the reaction described earlier; Fig. 18-7) and conversion of glutamate to glutamine by glutamine synthetase. In the brain, only astrocytes—star-shaped cells of the nervous system that provide nutrients, support, and insulation for neurons—express glutamine synthetase. Glutamate and its derivative γ-aminobutyrate (GABA; see Fig. 22-31) are important neurotransmitters; some of the sensitivity of the brain to ammonia may reflect depletion of glutamate in the glutamine synthetase reaction. However, glutamine synthetase activity is insufficient to deal with excess ammonia, or to fully explain its toxicity.

Increased [NH_4^+] also alters the capacity of astrocytes to maintain potassium homeostasis across the membrane. NH_4^+ competes with K^+ for transport into the cell through the Na^+K^+ ATPase, resulting in elevated extracellular [K^+]. The excess extracellular K^+ enters neurons through a symporter, **Na^+-K^+-$2Cl^-$ cotransporter 1 (NKCC1)**, bringing Na^+ and $2Cl^-$ with it. Excess Cl^- in these neurons alters their response when the neurotransmitter GABA interacts with their $GABA_A$ receptors, producing abnormal depolarization and increased neuronal activity that likely account for the neuromuscular incoordination and seizures that often result from ammonia poisoning. If extracellular [NH_4^+] remains elevated, the perturbation of ion and aquaporin channels in astrocytes causes the cells to swell, resulting in fatal brain edema. ■

As we close this discussion of amino group metabolism, note that we have described several processes that deposit excess ammonia in the mitochondria of hepatocytes (Fig. 18-2). We now look at the fate of that ammonia.

SUMMARY 18.1 Metabolic Fates of Amino Groups

■ Humans derive a small fraction of their oxidative energy from the catabolism of amino acids. Amino acids are derived from the normal breakdown (recycling) of cellular proteins, degradation of ingested proteins, and breakdown of body proteins in lieu of other fuel sources during starvation or in uncontrolled diabetes mellitus.

■ Proteases degrade ingested proteins in the stomach and small intestine. Most proteases are initially synthesized as inactive zymogens.

■ An early step in the catabolism of amino acids is the separation of the amino group from the carbon skeleton. In most cases, the amino group is transferred to α-ketoglutarate to form glutamate. This transamination reaction requires the coenzyme pyridoxal phosphate.

■ Glutamate is transported to liver mitochondria, where glutamate dehydrogenase liberates the amino group as ammonium ion (NH_4^+). Ammonia formed in other tissues is transported to the liver as the amide nitrogen of glutamine or, in transport from skeletal muscle, as the amino group of alanine.

■ The pyruvate produced by deamination of alanine in the liver is converted to glucose, which is transported back to muscle as part of the glucose-alanine cycle.

18.2 Nitrogen Excretion and the Urea Cycle

If not reused for the synthesis of new amino acids or other nitrogenous products, amino groups are channeled into a single excretory end product (**Fig. 18-10**). Most aquatic species, such as the bony fishes, are **ammonotelic**, excreting amino nitrogen as ammonia. The toxic ammonia is simply diluted in the surrounding water. Terrestrial animals require pathways for nitrogen excretion that minimize toxicity and water loss. Most terrestrial animals are **ureotelic**, excreting amino nitrogen in the form of urea; birds and reptiles are **uricotelic**, excreting amino nitrogen as uric acid. (The pathway of uric acid synthesis is described in Fig. 22-48.) Plants recycle virtually all amino groups—they excrete nitrogen only under rare circumstances.

In ureotelic organisms, the ammonia deposited in the mitochondria of hepatocytes is converted to urea in the **urea cycle**. This pathway was discovered in 1932 by Hans Krebs (who later also discovered the citric acid cycle) and a medical student associate, Kurt Henseleit. Urea production occurs almost exclusively in the liver and is the fate of most of the ammonia channeled there. The urea passes into the bloodstream and thus to the kidneys and is excreted into the urine. The production of urea now becomes the focus of our discussion.

Urea Is Produced from Ammonia in Five Enzymatic Steps

The urea cycle begins inside liver mitochondria, but three of the subsequent steps take place in the cytosol;

the cycle thus spans two cellular compartments (Fig. 18-10). The first amino group to enter the urea cycle is derived from ammonia in the mitochondrial matrix—most of this NH_4^+ arises by the pathways described in the previous section. The liver also receives some ammonia via the portal vein from the intestine, from the bacterial oxidation of amino acids. Whatever its source, the NH_4^+ generated in liver mitochondria is immediately used, together with CO_2 (as HCO_3^-) produced by mitochondrial respiration, to form carbamoyl phosphate in the matrix (**Fig. 18-11a**; see also Fig. 18-10). This ATP-dependent reaction is catalyzed by **carbamoyl phosphate synthetase I**, a regulatory enzyme (see below). The mitochondrial form of the enzyme is distinct from the cytosolic (II) form, which has a separate function in pyrimidine biosynthesis (Chapter 22).

The carbamoyl phosphate, which functions as an activated carbamoyl group donor, now enters the urea cycle. The cycle has only four enzymatic steps. First, carbamoyl phosphate donates its carbamoyl group to ornithine to form citrulline, with the release of P_i (Fig. 18-10, step ❶). The reaction is catalyzed by **ornithine transcarbamoylase**. Ornithine is not one of the 20 common amino acids found in proteins, but it is a key intermediate in nitrogen metabolism. It is synthesized from glutamate in a five-step pathway described in Chapter 22. Ornithine plays a role resembling that of oxaloacetate in the citric acid cycle, accepting material at each turn of the urea cycle. The citrulline produced in the first step of the urea cycle passes from the mitochondrion to the cytosol.

The next two steps bring in the second amino group. The source is aspartate generated in mitochondria by transamination and transported into the cytosol. A condensation reaction between the amino group of aspartate and the ureido (carbonyl) group of citrulline forms argininosuccinate (step ❷ in Fig. 18-10). This cytosolic reaction, catalyzed by **argininosuccinate synthetase**, requires ATP and proceeds through a citrullyl-AMP intermediate (Fig. 18-11b). The argininosuccinate is then cleaved by **argininosuccinase** (step ❸ in Fig. 18-10) to form free arginine and fumarate, the latter being converted to malate before entering

FIGURE 18-10 The urea cycle and reactions that feed amino groups into the cycle. The enzymes catalyzing these reactions (named in the text) are distributed between the mitochondrial matrix and the cytosol. One amino group enters the urea cycle as carbamoyl phosphate, formed in the matrix; the other enters as aspartate, formed in the matrix by transamination of oxaloacetate and glutamate, catalyzed by aspartate aminotransferase. The urea cycle consists of four steps. ❶ Formation of citrulline from ornithine and carbamoyl phosphate (entry of the first amino group); the citrulline passes into the cytosol. ❷ Formation of argininosuccinate through a citrullyl-AMP intermediate (entry of the second amino group). ❸ Formation of arginine from argininosuccinate; this reaction releases fumarate, which enters the citric acid cycle. ❹ Formation of urea; this reaction also regenerates ornithine. The pathways by which NH_4^+ arrives in the mitochondrial matrix of hepatocytes were discussed in Section 18.1.

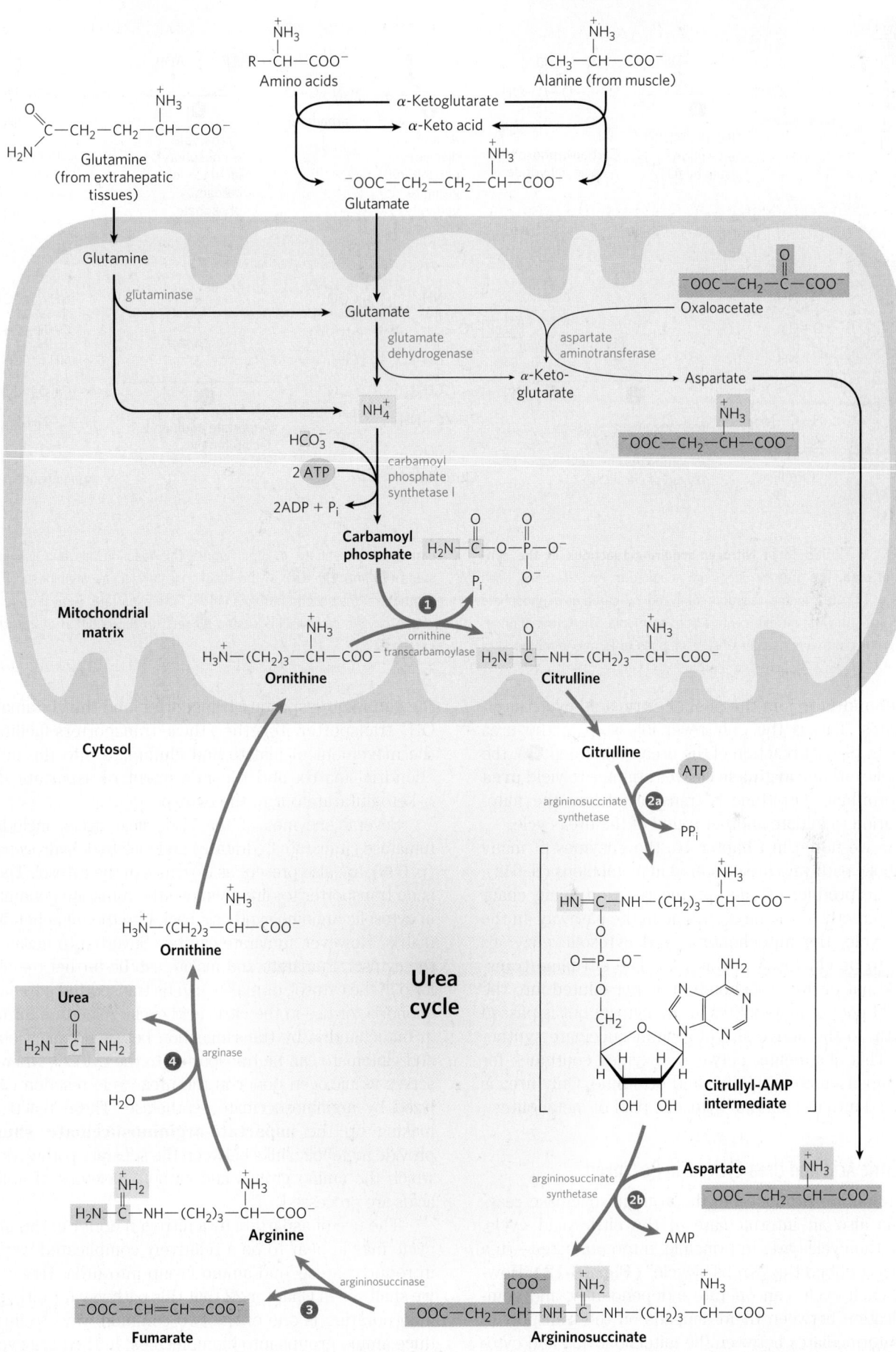

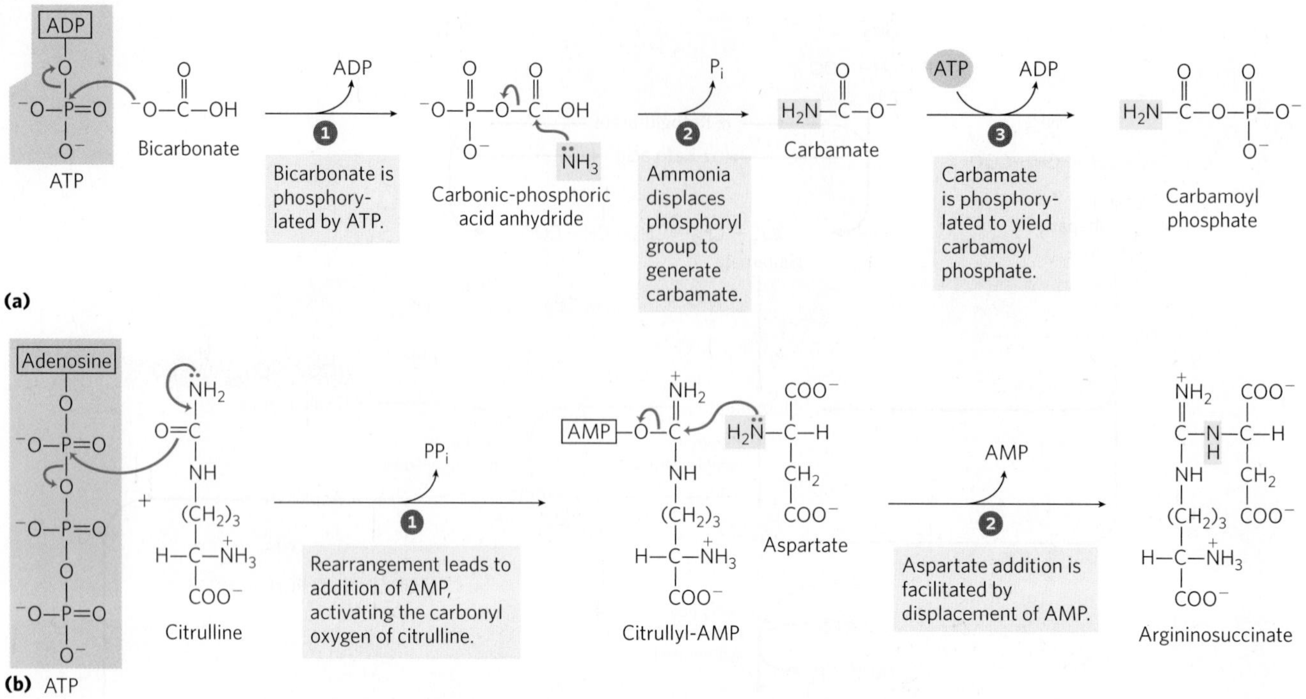

(a)

(b) ATP

MECHANISM FIGURE 18-11 **Nitrogen-acquiring reactions in the synthesis of urea.** The urea nitrogens are acquired in two reactions, each requiring ATP. **(a)** In the reaction catalyzed by carbamoyl phosphate synthetase I, the first nitrogen enters from ammonia. The terminal phosphate groups of two molecules of ATP are used to form one molecule of carbamoyl phosphate. In other words, this reaction has two activation steps (❶ and ❸). **(b)** In the reaction catalyzed by argininosuccinate synthetase, the second nitrogen enters from aspartate. Activation of the ureido oxygen of citrulline in step ❶ sets up the addition of aspartate in step ❷.

mitochondria to join the pool of citric acid cycle intermediates. This is the only reversible step in the urea cycle. In the last reaction of the urea cycle (step ❹), the cytosolic enzyme **arginase** cleaves arginine to yield **urea** and ornithine. Ornithine is transported into the mitochondrion to initiate another round of the urea cycle.

As we noted in Chapter 16, the enzymes of many metabolic pathways are clustered in metabolons (p. 642), with the product of one enzyme reaction being channeled directly to the next enzyme in the pathway. In the urea cycle, the mitochondrial and cytosolic enzymes seem to be clustered in this way. The citrulline transported out of the mitochondrion is not diluted into the general pool of metabolites in the cytosol but is passed directly to the active site of argininosuccinate synthetase. This channeling between enzymes continues for argininosuccinate, arginine, and ornithine. Only urea is released into the general cytosolic pool of metabolites.

The Citric Acid and Urea Cycles Can Be Linked

The fumarate produced in the argininosuccinase reaction is also an intermediate of the citric acid cycle. Thus, the cycles are, in principle, interconnected—in a process dubbed the "Krebs bicycle" **(Fig. 18-12)**. However, each cycle can operate independently, and communication between them depends on the transport of key intermediates between the mitochondrion and cytosol. Major transporters in the inner mitochondrial membrane include the malate-α-ketoglutarate transporter,

the glutamate-aspartate transporter, and the glutamate-OH^- transporter. Together, these transporters facilitate the movement of malate and glutamate into the mitochondrial matrix and the movement of aspartate and α-ketoglutarate out to the cytosol.

Several enzymes of the citric acid cycle, including fumarase (fumarate hydratase) and malate dehydrogenase (p. 633), are also present as isozymes in the cytosol. There is no transporter to directly move the fumarate generated in cytosolic arginine synthesis back into the mitochondrial matrix. However, fumarate can be converted to malate in the cytosol. Fumarate and malate can be further metabolized in the cytosol, or malate can be transported into mitochondria for use in the citric acid cycle. Aspartate formed in mitochondria by transamination between oxaloacetate and glutamate can be transported to the cytosol, where it serves as nitrogen donor in the urea cycle reaction catalyzed by argininosuccinate synthetase. These reactions, making up the **aspartate-argininosuccinate shunt**, provide metabolic links between the separate pathways by which the amino groups and carbon skeletons of amino acids are processed.

The use of aspartate as a nitrogen donor in the urea cycle may appear to be a relatively complicated way to introduce the second amino group into urea. However, we shall see in Chapter 22 that this pathway for nitrogen incorporation is one of the two common ways to introduce amino groups into biomolecules. In the urea cycle, additional pathway interconnections can help explain why aspartate is used as a nitrogen donor. The urea and

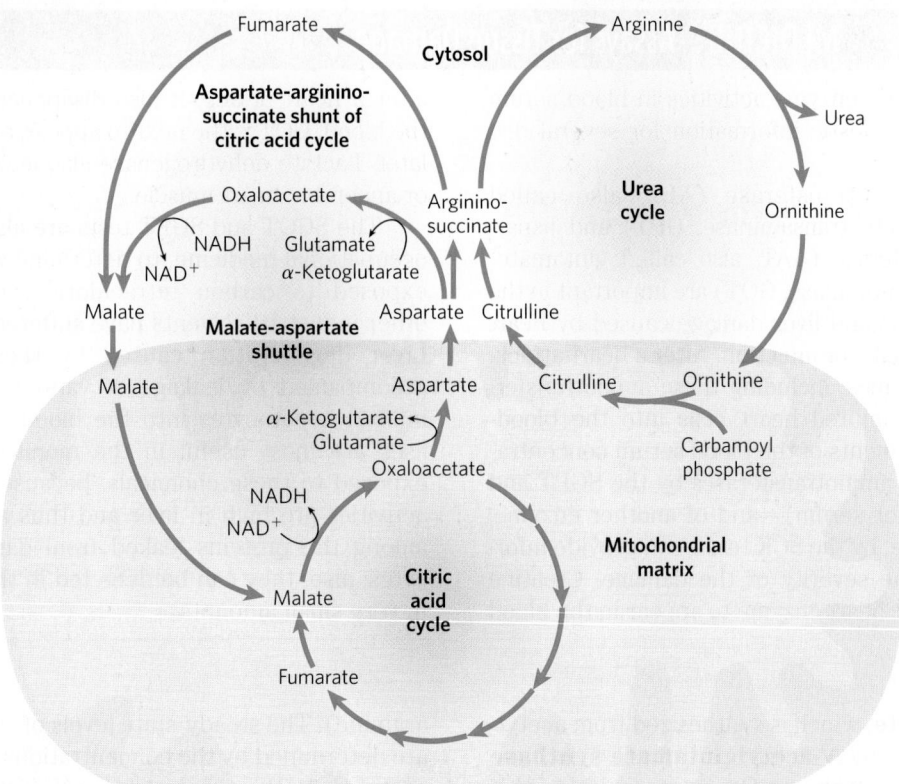

FIGURE 18-12 Links between the urea cycle and citric acid cycle. The interconnected cycles have been called the "Krebs bicycle." The pathways linking the citric acid and urea cycles are known as the aspartate-argininosuccinate shunt; these effectively link the fates of the amino groups and the carbon skeletons of amino acids. The interconnections are quite elaborate. For example, some citric acid cycle enzymes, such as fumarase and malate dehydrogenase, have both cytosolic and mitochondrial isozymes.

Fumarate produced in the cytosol—whether by the urea cycle, purine biosynthesis, or other processes—can be converted to cytosolic malate, which is used in the cytosol or transported into mitochondria to enter the citric acid cycle. These processes are further intertwined with the malate-aspartate shuttle, a set of reactions that brings reducing equivalents into the mitochondrion (see also Fig. 19-31). These different cycles and processes rely on a limited number of transporters in the inner mitochondrial membrane.

citric acid cycles are closely tied to an additional process that brings NADH, in the form of reducing equivalents, into the mitochondrion. As detailed in the next chapter, the NADH produced by glycolysis, fatty acid oxidation, and other processes cannot be transported across the mitochondrial inner membrane. Reducing equivalents are instead brought into the mitochondrion by converting aspartate to oxaloacetate in the cytosol, reducing the oxaloacetate to malate with NADH, and transporting the malate into the mitochondrial matrix via the malate–α-ketoglutarate transporter. Once inside the mitochondrion, the malate can be reconverted to oxaloacetate while generating NADH. The oxaloacetate is converted to aspartate in the matrix and transported out of the mitochondrion by the aspartate-glutamate transporter. This malate-aspartate shuttle completes yet another cycle that functions to keep the mitochondrion supplied with NADH (Fig. 18-12; see also Fig. 19-31).

These processes require that a balance be maintained in the cytosol between the concentrations of glutamate and aspartate. The enzyme that transfers amino groups between these key amino acids is aspartate aminotransferase, AAT (also called glutamate-oxaloacetate transaminase, GOT). This is one of the most active enzymes in hepatocytes and other tissues. When tissue damage occurs, this easily

assayed enzyme and others leak into the blood. Thus, measuring blood levels of liver enzymes is important in diagnosing a variety of medical conditions (Box 18-1).

The Activity of the Urea Cycle Is Regulated at Two Levels

The flux of nitrogen through the urea cycle in an individual animal varies with diet. When the dietary intake is primarily protein, the carbon skeletons of amino acids are used for fuel, producing much urea from the excess amino groups. During prolonged starvation, when breakdown of muscle protein begins to supply much of the organism's metabolic energy, urea production also increases substantially.

These changes in demand for urea cycle activity are met over the long term by regulation of the rates of synthesis of the four urea cycle enzymes and carbamoyl phosphate synthetase I in the liver. All five enzymes are synthesized at higher rates in starving animals and in animals on very-high-protein diets than in well-fed animals eating primarily carbohydrates and fats. Animals on protein-free diets produce lower levels of urea cycle enzymes.

On a shorter time scale, allosteric regulation of at least one key enzyme adjusts the flux through the urea cycle. The first enzyme in the pathway, carbamoyl phosphate synthetase I, is allosterically activated by

BOX 18-1 ⚕ MEDICINE Assays for Tissue Damage

Analyses of certain enzyme activities in blood serum give valuable diagnostic information for several disease conditions.

Alanine aminotransferase (ALT; also called glutamate-pyruvate transaminase, GPT) and aspartate aminotransferase (AAT; also called glutamate-oxaloacetate transaminase, GOT) are important in the diagnosis of heart and liver damage caused by heart attack, drug toxicity, or infection. After a heart attack, a variety of enzymes, including these aminotransferases, leak from injured heart cells into the bloodstream. Measurements of the blood serum concentrations of the two aminotransferases by the SGPT and SGOT tests (S for serum)—and of another enzyme, **creatine kinase**, by the SCK test—can provide information about the severity of the damage. Creatine kinase is the first heart enzyme to appear in the blood

after a heart attack; it also disappears quickly from the blood. GOT is the next to appear, and GPT follows later. Lactate dehydrogenase also leaks from injured or anaerobic heart muscle.

The SGOT and SGPT tests are also important in occupational medicine, to determine whether people exposed to carbon tetrachloride, chloroform, or other industrial solvents have suffered liver damage. Liver degeneration caused by these solvents is accompanied by leakage of various enzymes from injured hepatocytes into the blood. Aminotransferases are most useful in the monitoring of people exposed to these chemicals, because these enzyme activities are high in liver and thus are likely to be among the proteins leaked from damaged hepatocytes; also, they can be detected in the bloodstream in very small amounts.

N-acetylglutamate, which is synthesized from acetyl-CoA and glutamate by *N*-acetylglutamate synthase (Fig. 18-13). In plants and microorganisms, this enzyme catalyzes the first step in the de novo synthesis of arginine from glutamate (see Fig. 22-12), but in mammals, *N*-acetylglutamate synthase activity in the liver has a purely regulatory function (mammals lack the other enzymes needed to convert glutamate to

arginine). The steady-state levels of *N*-acetylglutamate are determined by the concentrations of glutamate and acetyl-CoA (the substrates for *N*-acetylglutamate synthase) and arginine (an activator of *N*-acetylglutamate synthase, and thus an activator of the urea cycle).

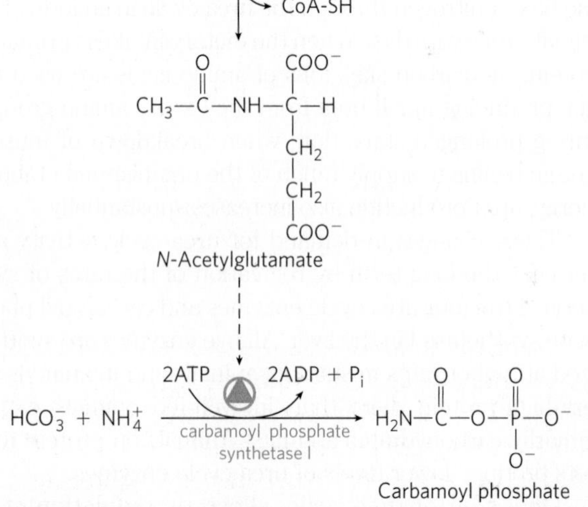

Pathway Interconnections Reduce the Energetic Cost of Urea Synthesis

If we consider the urea cycle in isolation, we see that the synthesis of one molecule of urea requires four high-energy phosphate groups (Fig. 18-10). Two ATP molecules are required to make carbamoyl phosphate, and one ATP to make argininosuccinate—the latter ATP undergoing a pyrophosphate cleavage to AMP and PP$_i$, which is hydrolyzed to two P$_i$. The overall equation of the urea cycle is

$$2NH_4^+ + HCO_3^- + 3ATP^{4-} + H_2O \longrightarrow$$
$$\text{urea} + 2ADP^{3-} + 4P_i^{2-} + AMP^{2-} + 2H^+$$

However, this apparent cost is compensated for by the pathway interconnections detailed above. The fumarate generated by the urea cycle is converted to malate, and the malate is transported into the mitochondrion (Fig. 18-12). Inside the mitochondrial matrix, NADH is generated in the malate dehydrogenase reaction. Each NADH molecule can generate up to 2.5 ATP during mitochondrial respiration (Chapter 19), greatly reducing the overall energetic cost of urea synthesis.

FIGURE 18-13 Synthesis of *N*-acetylglutamate and its activation of carbamoyl phosphate synthetase I.

Genetic Defects in the Urea Cycle Can Be Life-Threatening

⚕ People with genetic defects in any enzyme involved in urea formation cannot tolerate protein-rich diets. Amino acids ingested in excess of the minimum daily requirements for protein synthesis are deaminated in the liver, producing free ammonia that cannot be converted to urea and exported into the bloodstream, and, as we have seen, ammonia is highly toxic. The absence of a urea cycle enzyme can result in hyperammonemia or in the buildup of one or more urea cycle intermediates, depending on the enzyme that is missing. Given that most urea cycle steps are irreversible, the absent enzyme activity can often be identified by determining which cycle intermediate is present in elevated concentration in the blood and/or urine. Although the breakdown of amino acids can have serious health consequences in individuals with urea cycle deficiencies, a protein-free diet is not a treatment option. Humans are incapable of synthesizing half of the 20 common amino acids, and these **essential amino acids** (Table 18-1) must be provided in the diet.

A variety of treatments are available for individuals with urea cycle defects. Careful administration of the aromatic acids benzoate or phenylbutyrate in the diet can help lower the level of ammonia in the blood. Benzoate is converted to benzoyl-CoA, which combines with glycine to form hippurate (**Fig. 18-14**, left). The glycine used up in this reaction must be regenerated, and ammonia is thus taken up in the glycine synthase reaction. Phenylbutyrate is converted to phenylacetate by β oxidation. The phenylacetate is then converted to phenylacetyl-CoA, which combines with glutamine to form phenylacetylglutamine (Fig. 18-14, right). The resulting removal of glutamine triggers its further synthesis by glutamine synthetase (see Eqn 22-1) in a reaction that takes up ammonia. Both hippurate and phenylacetylglutamine are nontoxic compounds that are excreted in the urine. The pathways shown in Figure 18-14 make only minor contributions to normal metabolism, but they become prominent when aromatic acids are ingested.

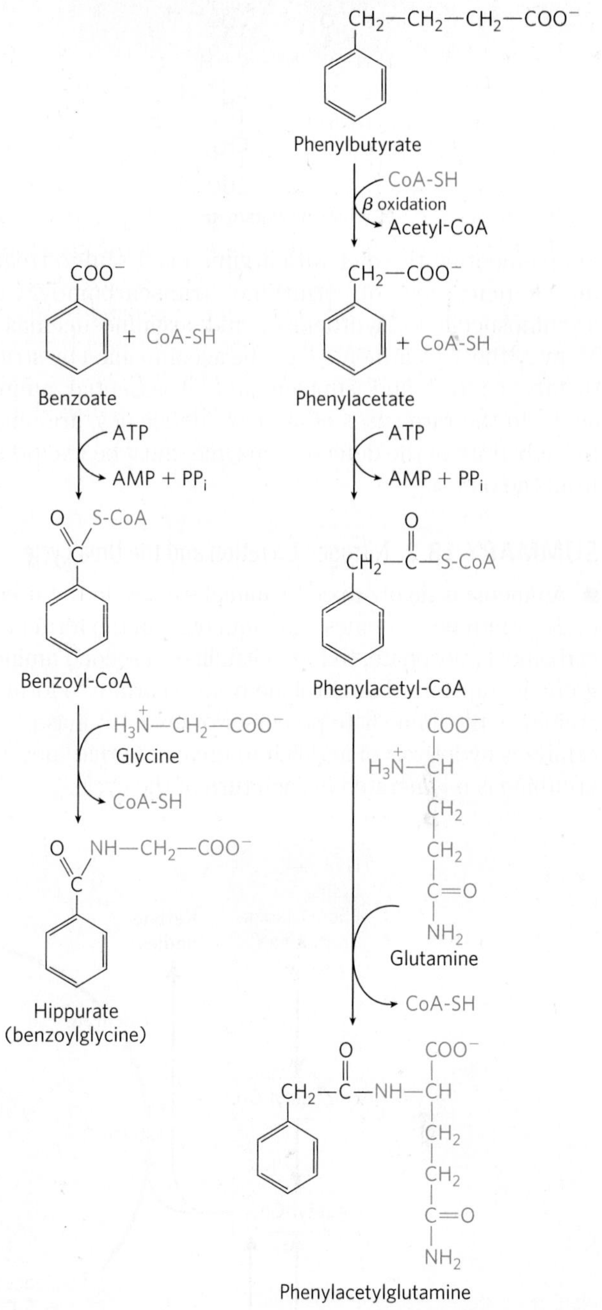

FIGURE 18-14 Treatment for deficiencies in urea cycle enzymes. The aromatic acids benzoate and phenylbutyrate, administered in the diet, are metabolized and combine with glycine and glutamine, respectively. The products are excreted in the urine. Subsequent synthesis of glycine and glutamine to replenish the pool of these intermediates removes ammonia from the bloodstream.

TABLE 18-1	Nonessential and Essential Amino Acids for Humans and the Albino Rat		
Nonessential	**Conditionally essential**[a]	**Essential**	
Alanine	Arginine	Histidine	
Asparagine	Cysteine	Isoleucine	
Aspartate	Glutamine	Leucine	
Glutamate	Glycine	Lysine	
Serine	Proline	Methionine	
	Tyrosine	Phenylalanine	
		Threonine	
		Tryptophan	
		Valine	

[a]Required to some degree in young, growing animals and/or sometimes during illness.

Other therapies are more specific to a particular enzyme deficiency. Deficiency of *N*-acetylglutamate synthase results in the absence of the normal activator of carbamoyl phosphate synthetase I (Fig. 18-13). This condition can be treated by administering carbamoyl glutamate, an analog of *N*-acetylglutamate that is effective in activating carbamoyl phosphate synthetase I.

Carbamoyl glutamate

Supplementing the diet with arginine is useful in treating deficiencies of ornithine transcarbamoylase, argininosuccinate synthetase, and argininosuccinase. Many of these treatments must be accompanied by strict dietary control and supplements of essential amino acids. In the rare cases of arginase deficiency, arginine, the substrate of the defective enzyme, must be excluded from the diet. ■

SUMMARY 18.2 Nitrogen Excretion and the Urea Cycle

■ Ammonia is highly toxic to animal tissues. In the urea cycle, ornithine combines with ammonia, in the form of carbamoyl phosphate, to form citrulline. A second amino group is transferred to citrulline from aspartate to form arginine—the immediate precursor of urea. Arginase catalyzes hydrolysis of arginine to urea and ornithine; ornithine is regenerated in each turn of the cycle.

■ The urea cycle results in a net conversion of oxaloacetate to fumarate, both of which are intermediates in the citric acid cycle. The two cycles are thus interconnected.

■ The activity of the urea cycle is regulated at the level of enzyme synthesis and by allosteric regulation of the enzyme that catalyzes the formation of carbamoyl phosphate.

18.3 Pathways of Amino Acid Degradation

Amino acid catabolism normally accounts for only 10% to 15% of the human body's energy production; these pathways are not nearly as active as glycolysis and fatty acid oxidation. Flux through these catabolic routes also varies greatly, depending on the balance between requirements for biosynthetic processes and the availability of a particular amino acid. The 20 catabolic pathways converge to form only six major products, all of which enter the citric acid cycle **(Fig. 18-15)**. From here the carbon skeletons are diverted to gluconeogenesis or ketogenesis or are completely oxidized to CO_2 and H_2O.

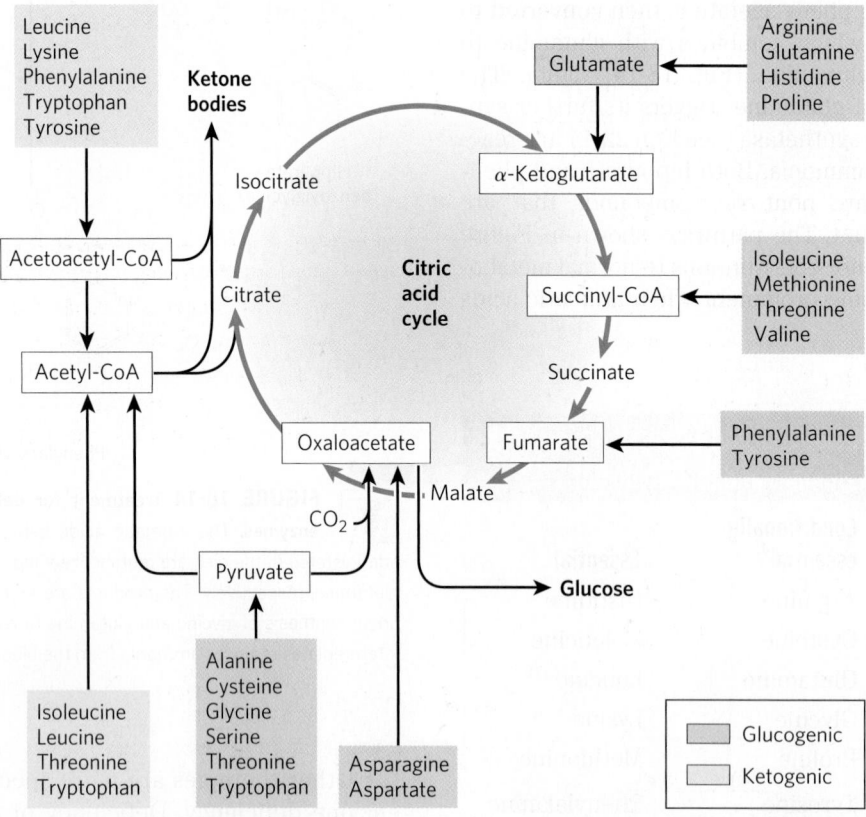

FIGURE 18-15 Summary of amino acid catabolism. Amino acids are grouped according to their major degradative end product. Some amino acids are listed more than once because different parts of their carbon skeletons are degraded to different end products. The figure shows the most important catabolic pathways in vertebrates, but there are minor variations among vertebrate species. Threonine, for instance, is degraded by at least two different pathways (see Figs 18-19, 18-27), and the importance of a given pathway can vary with the organism and its metabolic conditions. The glucogenic and ketogenic amino acids are also delineated in the figure, by color shading. Notice that five of the amino acids are both glucogenic and ketogenic. The amino acids degraded to pyruvate are also potentially ketogenic. Only two amino acids, leucine and lysine, are exclusively ketogenic.

All or part of the carbon skeletons of seven amino acids are ultimately broken down to acetyl-CoA. Five amino acids are converted to α-ketoglutarate, four to succinyl-CoA, two to fumarate, and two to oxaloacetate. Parts or all of six amino acids are converted to pyruvate, which can be converted to either acetyl-CoA or oxaloacetate. We later summarize the individual pathways for the 20 amino acids in flow diagrams, each leading to a specific point of entry into the citric acid cycle. In these diagrams, the carbon atoms that enter the citric acid cycle are shown in color. Note that some amino acids appear more than once, reflecting different fates for different parts of their carbon skeletons. Rather than examining every step of every pathway in amino acid catabolism, we single out for special discussion some enzymatic reactions that are particularly noteworthy for their mechanisms or their medical significance.

Some Amino Acids Are Converted to Glucose, Others to Ketone Bodies

The seven amino acids that are degraded entirely or in part to acetoacetyl-CoA and/or acetyl-CoA—phenylalanine, tyrosine, isoleucine, leucine, tryptophan, threonine, and lysine—can yield ketone bodies in the liver, where acetoacetyl-CoA is converted to acetoacetate and then to acetone and β-hydroxybutyrate (see Fig. 17-18). These are the **ketogenic** amino acids (Fig. 18-15). Their ability to form ketone bodies is particularly evident in uncontrolled diabetes mellitus, in which the liver produces large amounts of ketone bodies from both fatty acids and the ketogenic amino acids.

The amino acids that are degraded to pyruvate, α-ketoglutarate, succinyl-CoA, fumarate, and/or oxaloacetate can be converted to glucose and glycogen by pathways described in Chapters 14 and 15. They are the **glucogenic** amino acids. The division between ketogenic and glucogenic amino acids is not sharp; five amino acids—tryptophan, phenylalanine, tyrosine, threonine, and isoleucine—are both ketogenic and glucogenic. Catabolism of amino acids is particularly critical to the survival of animals with high-protein diets or during starvation. Leucine is an exclusively ketogenic amino acid that is very common in proteins. Its degradation makes a substantial contribution to ketosis under starvation conditions.

Several Enzyme Cofactors Play Important Roles in Amino Acid Catabolism

A variety of interesting chemical rearrangements occur in the catabolic pathways of amino acids. It is useful to begin our study of these pathways by noting the classes of reactions that recur and introducing their enzyme cofactors. We have already considered one important class: transamination reactions requiring pyridoxal phosphate. One-carbon transfers are another common type of reaction in amino acid catabolism. Such transfers usually involve one of three cofactors: biotin, tetrahydrofolate, or S-adenosylmethionine **(Fig. 18-16)**. These cofactors transfer one-carbon groups in different oxidation states: biotin transfers carbon in its most oxidized state, CO_2 (see Fig. 14-19); tetrahydrofolate transfers one-carbon groups in intermediate oxidation states and sometimes as methyl groups; and S-adenosylmethionine transfers methyl groups, the most reduced state of carbon. The latter two cofactors are especially important in amino acid and nucleotide metabolism.

Tetrahydrofolate (H_4 folate), synthesized in bacteria, consists of substituted pterin (6-methylpterin), p-aminobenzoate, and glutamate moieties (Fig. 18-16).

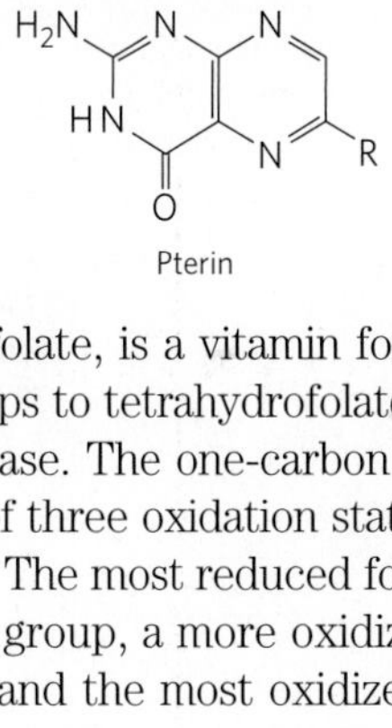

Pterin

The oxidized form, folate, is a vitamin for mammals; it is converted in two steps to tetrahydrofolate by the enzyme dihydrofolate reductase. The one-carbon group undergoing transfer, in any of three oxidation states, is bonded to N-5 or N-10 or both. The most reduced form of the cofactor carries a methyl group, a more oxidized form carries a methylene group, and the most oxidized forms carry a

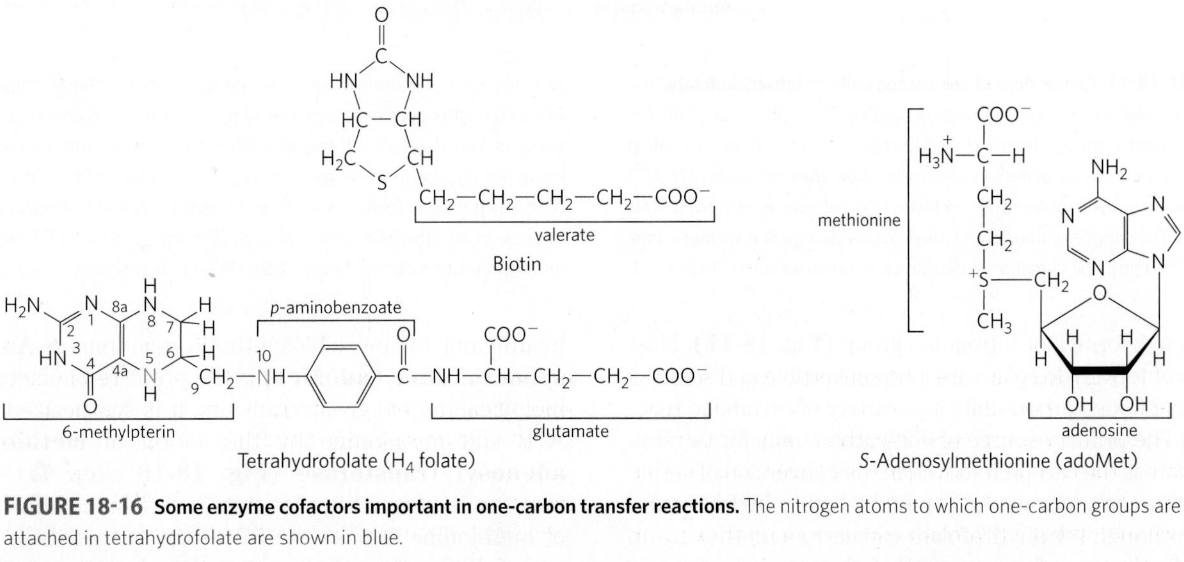

FIGURE 18-16 Some enzyme cofactors important in one-carbon transfer reactions. The nitrogen atoms to which one-carbon groups are attached in tetrahydrofolate are shown in blue.

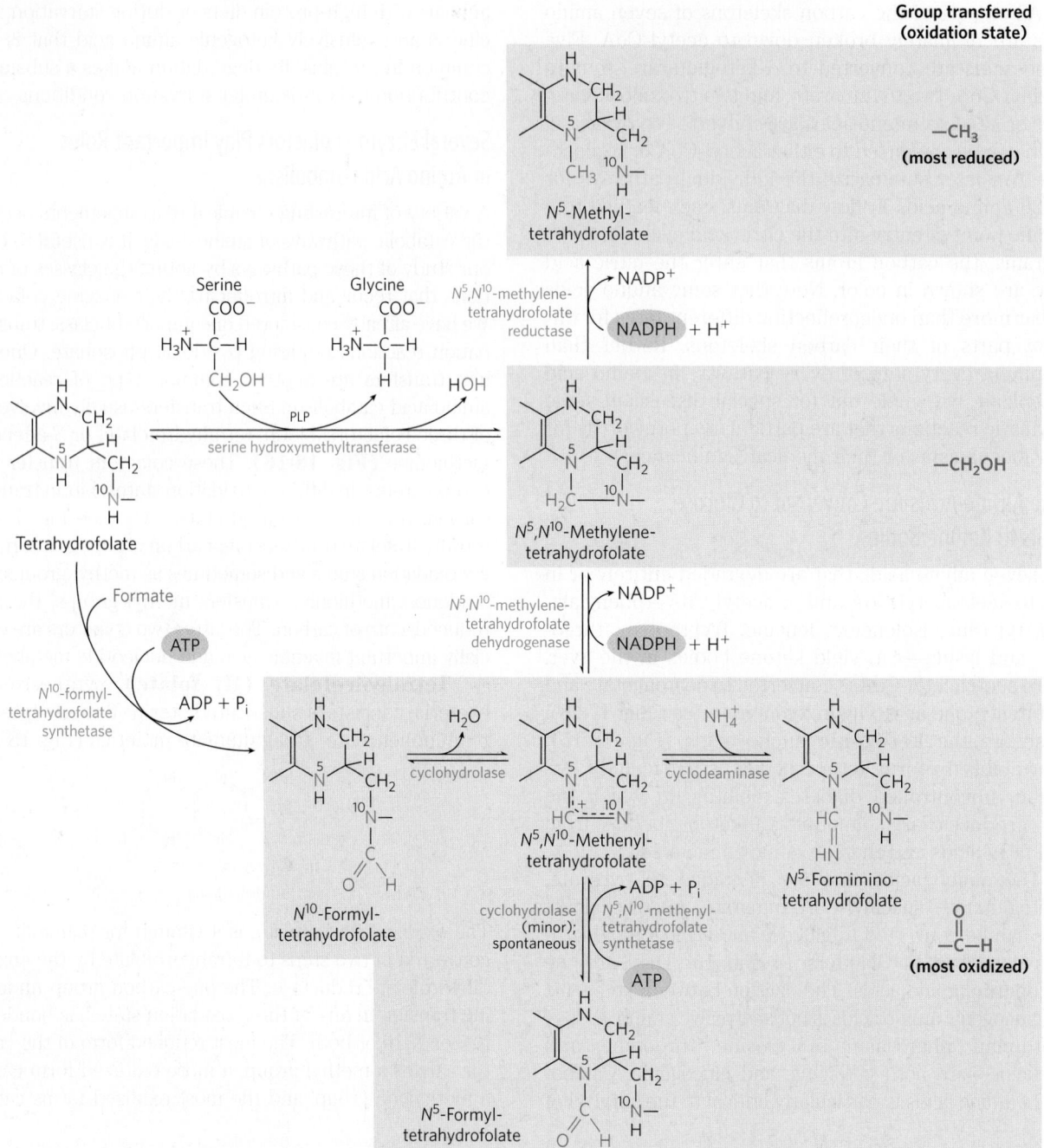

FIGURE 18-17 Conversions of one-carbon units on tetrahydrofolate. The different molecular species are grouped according to oxidation state, with the most reduced at the top and most oxidized at the bottom. All species within a single shaded box are at the same oxidation state. The conversion of N^5,N^{10}-methylenetetrahydrofolate to N^5-methyltetrahydrofolate is effectively irreversible. The enzymatic transfer of formyl groups, as in purine synthesis (see Fig. 22-35) and in the formation of formylmethionine in bacteria (Chapter 27),

generally uses N^{10}-formyltetrahydrofolate rather than N^5-formyltetrahydrofolate. The latter species is significantly more stable and therefore a weaker donor of formyl groups. N^5-Formyltetrahydrofolate is a minor byproduct of the cyclohydrolase reaction, and can also form spontaneously. Conversion of N^5-formyltetrahydrofolate to N^5,N^{10}-methenyltetrahydrofolate requires ATP, because of an otherwise unfavorable equilibrium. Note that N^5-formimino-tetrahydrofolate is derived from histidine in a pathway shown in Figure 18-26.

methenyl, formyl, or formimino group **(Fig. 18-17).** Most forms of tetrahydrofolate are interconvertible and serve as donors of one-carbon units in a variety of metabolic reactions. The primary source of one-carbon units for tetrahydrofolate is the carbon removed in the conversion of serine to glycine, producing N^5,N^{10}-methylenetetrahydrofolate.

Although tetrahydrofolate can carry a methyl group at N-5, the transfer potential of this methyl group is

insufficient for most biosynthetic reactions. **S-Adenosylmethionine (adoMet)** is the preferred cofactor for biological methyl group transfers. It is synthesized from ATP and methionine by the action of **methionine adenosyl transferase (Fig. 18-18,** step **❶**). This reaction is unusual in that the nucleophilic sulfur atom of methionine attacks the 5′ carbon of the ribose moiety of ATP rather than one of the phosphorus atoms.

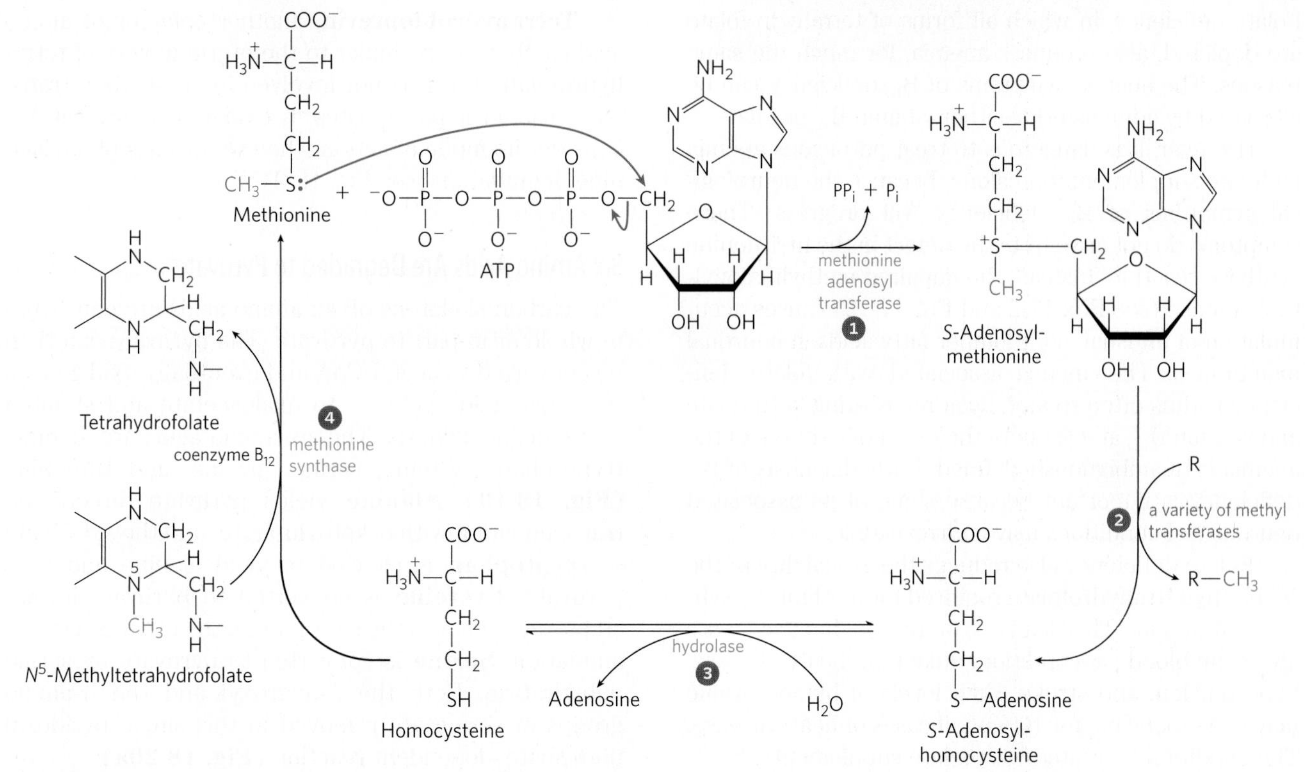

FIGURE 18-18 Synthesis of methionine and *S*-adenosylmethionine in an activated-methyl cycle. The steps are described in the text. In the methionine synthase reaction (step ❹), the methyl group is transferred to cobalamin to form methylcobalamin, which is the methyl donor in the formation of methionine. *S*-Adenosylmethionine, which has a positively charged sulfur (and is thus a sulfonium ion), is a powerful methylating agent in several biosynthetic reactions. The methyl group acceptor (step ❷) is designated R.

Triphosphate is released and is cleaved to P_i and PP_i on the enzyme, and the PP_i is cleaved by inorganic pyrophosphatase; thus three bonds, including two bonds of high-energy phosphate groups, are broken in this reaction. The only other known reaction in which triphosphate is displaced from ATP occurs in the synthesis of coenzyme B_{12} (see Box 17-2, Fig. 3).

S-Adenosylmethionine is a potent alkylating agent by virtue of its destabilizing sulfonium ion. The methyl group is subject to attack by nucleophiles and is about 1,000 times more reactive than the methyl group of N^5-methyltetrahydrofolate.

Transfer of the methyl group from *S*-adenosylmethionine to an acceptor yields **S-adenosylhomocysteine** (Fig. 18-18, step ❷), which is subsequently broken down to homocysteine and adenosine (step ❸). Methionine is regenerated by transfer of a methyl group to homocysteine in a reaction catalyzed by methionine synthase (step ❹), and methionine is reconverted to *S*-adenosylmethionine to complete an activated-methyl cycle.

One form of methionine synthase common in bacteria uses N^5-methyltetrahydrofolate as a methyl donor. Another form of the enzyme present in some bacteria and mammals uses N^5-methyltetrahydrofolate, but the methyl group is first transferred to cobalamin, derived from coenzyme B_{12}, to form methylcobalamin as the methyl donor in methionine formation. This reaction and the rearrangement of L-methylmalonyl-CoA to succinyl-CoA (see Box 17-2, Fig. 1a) are the only known coenzyme B_{12}–dependent reactions in mammals.

The vitamins B_{12} and folate are closely linked in these metabolic pathways. The B_{12} deficiency disease **pernicious anemia** is rare, seen only in individuals who have a defect in the intestinal absorption pathways for this vitamin (see Box 17-2) or in strict vegetarians (B_{12} is not present in plants). The disease progresses slowly, because only small amounts of vitamin B_{12} are required and normal stores of B_{12} in the liver can last three to five years. Symptoms include not only anemia but a variety of neurological disorders.

The anemia can be traced to the methionine synthase reaction. As noted above, the methyl group of methylcobalamin is derived from N^5-methyltetrahydrofolate, and this is the only reaction in mammals that uses N^5-methyltetrahydrofolate. The reaction converting the N^5,N^{10}-methylene form to the N^5-methyl form of tetrahydrofolate is irreversible (Fig. 18-17). Thus, if coenzyme B_{12} is not available for the synthesis of methylcobalamin, metabolic folates become trapped in the N^5-methyl form. The anemia associated with vitamin B_{12} deficiency is called **megaloblastic anemia**. It manifests as a decline in the production of mature erythrocytes (red blood cells) and the appearance in the bone marrow of immature precursor cells, or **megaloblasts**. Erythrocytes are gradually replaced in the blood by smaller numbers of abnormally large erythrocytes called **macrocytes**. The defect in erythrocyte development is a direct consequence of the depletion of the N^5,N^{10}-methylenetetrahydrofolate, which is required for synthesis of the thymidine nucleotides needed for DNA synthesis (see Chapter 22).

Folate deficiency, in which all forms of tetrahydrofolate are depleted, also produces anemia, for much the same reasons. The anemia symptoms of B_{12} deficiency can be alleviated by administering either vitamin B_{12} or folate.

However, it is dangerous to treat pernicious anemia by folate supplementation alone, because the neurological symptoms of B_{12} deficiency will progress. These symptoms do not arise from the defect in the methionine synthase reaction. Instead, the impaired methylmalonyl-CoA mutase (see Box 17-2 and Fig. 17-12) causes accumulation of unusual, odd-number fatty acids in neuronal membranes. The anemia associated with folate deficiency is thus often treated by administering both folate and vitamin B_{12}, at least until the metabolic source of the anemia is unambiguously defined. Early diagnosis of B_{12} deficiency is important because some of its associated neurological conditions may be irreversible.

Folate deficiency also reduces the availability of the N^5-methyltetrahydrofolate required for methionine synthase function. This leads to a rise in homocysteine levels in blood, a condition linked to heart disease, hypertension, and stroke. High levels of homocysteine may be responsible for 10% of all cases of heart disease. The condition is treated with folate supplements. ∎

FIGURE 18-19 Catabolic pathways for alanine, glycine, serine, cysteine, tryptophan, and threonine. The fate of the indole group of tryptophan is shown in Figure 18-21. Details of most of the reactions involving serine and glycine are shown in Figure 18-20. The pathway for threonine degradation shown here accounts for only about a third of threonine catabolism (for the alternative pathway, see Fig. 18-27). Several pathways for cysteine degradation lead to pyruvate. The sulfur of cysteine has several alternative fates, one of which is shown in Figure 22-17. Carbon atoms here and in subsequent figures are color-coded as necessary to trace their fates.

Tetrahydrobiopterin, another cofactor of amino acid catabolism, is similar to the pterin moiety of tetrahydrofolate, but it is not involved in one-carbon transfers; instead it participates in oxidation reactions. We consider its mode of action when we discuss phenylalanine degradation (see Fig. 18-24).

Six Amino Acids Are Degraded to Pyruvate

The carbon skeletons of six amino acids are converted in whole or in part to pyruvate. The pyruvate can then be converted to acetyl-CoA and eventually oxidized via the citric acid cycle, or to oxaloacetate and shunted into gluconeogenesis. The six amino acids are alanine, tryptophan, cysteine, serine, glycine, and threonine **(Fig. 18-19)**. **Alanine** yields pyruvate directly on transamination with α-ketoglutarate, and the side chain of **tryptophan** is cleaved to yield alanine and thus pyruvate. **Cysteine** is converted to pyruvate in two steps; one removes the sulfur atom, the other is a transamination. **Serine** is converted to pyruvate by serine dehydratase. Both the β-hydroxyl and the α-amino groups of serine are removed in this single pyridoxal phosphate–dependent reaction **(Fig. 18-20a)**.

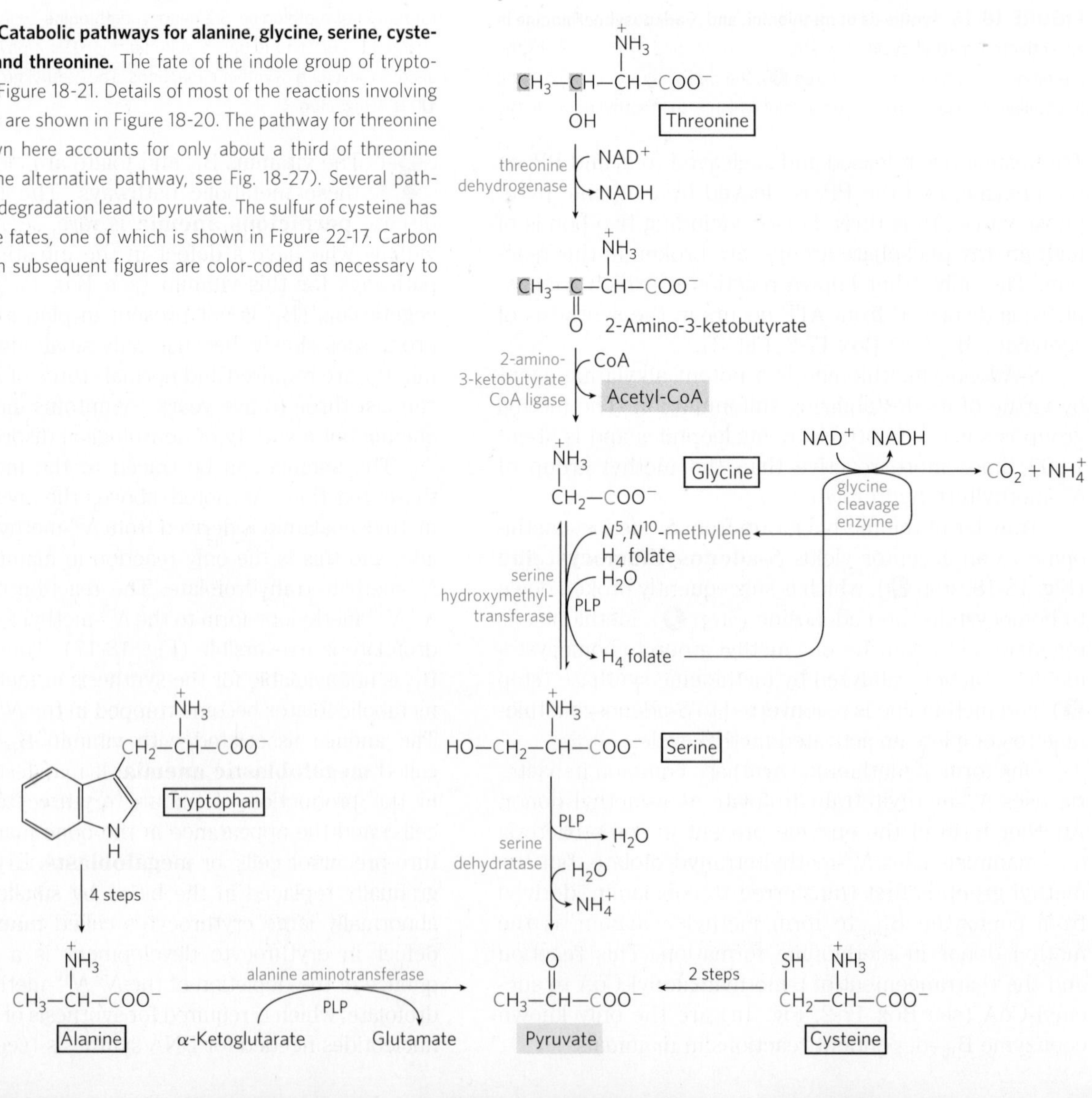

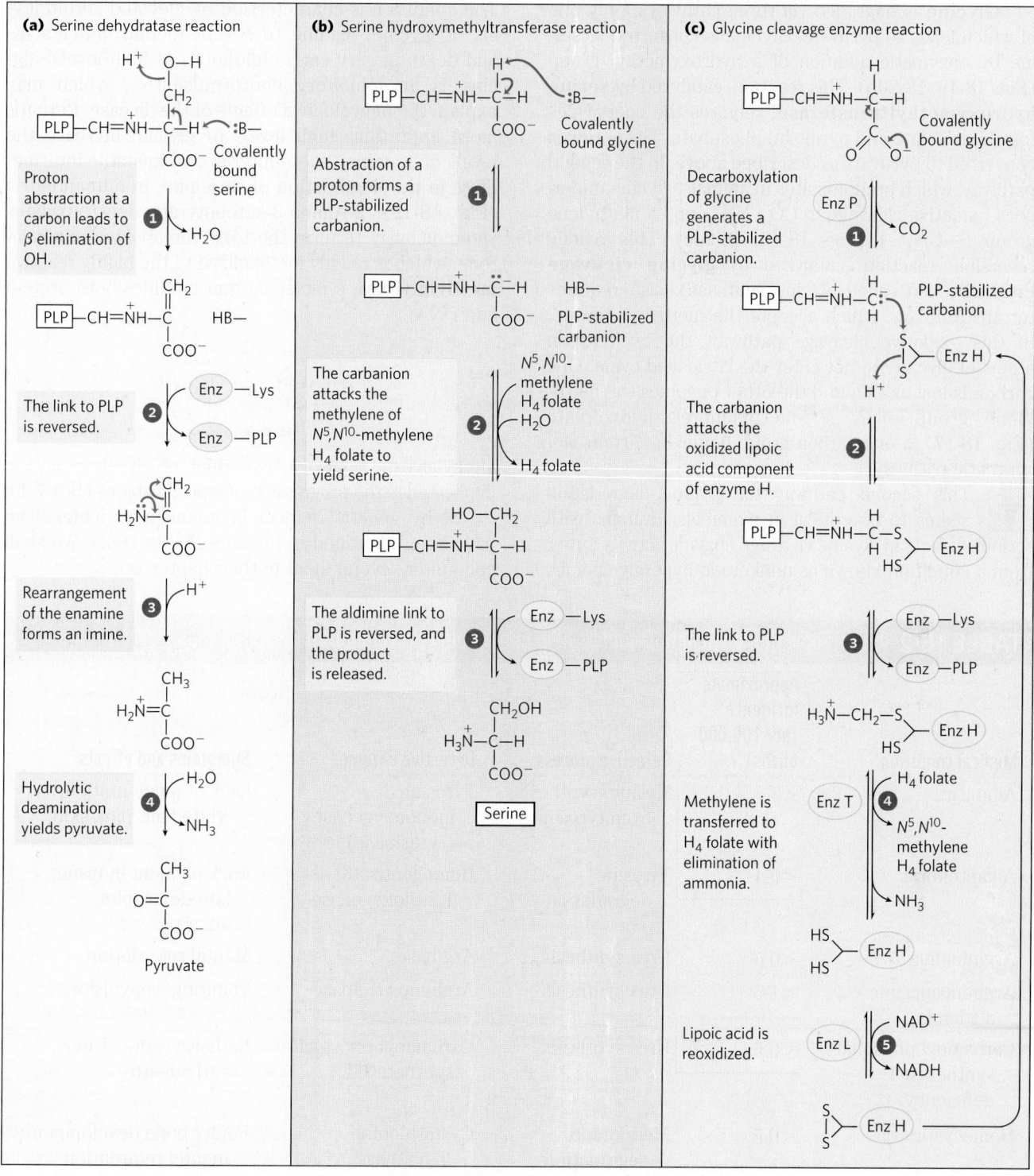

MECHANISM FIGURE 18-20 Interplay of the pyridoxal phosphate and tetrahydrofolate cofactors in serine and glycine metabolism. The first step in each of these reactions (not shown) involves the formation of a covalent imine linkage between enzyme-bound PLP and the substrate amino acid—serine in (a), glycine in (b) and (c). **(a)** A PLP-catalyzed elimination of water in the serine dehydratase reaction (step ❶) begins the pathway to pyruvate. **(b)** In the serine hydroxymethyltransferase reaction, a PLP-stabilized carbanion (product of step ❶) is a key intermediate in the reversible

transfer of the methylene group (as —CH$_2$—OH) from N^5,N^{10}-methylenetetrahydrofolate to form serine. **(c)** The glycine cleavage enzyme is a multienzyme complex, with components P, H, T, and L. The overall reaction, which is reversible, converts glycine to CO$_2$ and NH$_4^+$, with the second glycine carbon taken up by tetrahydrofolate to form N^5,N^{10}-methylenetetrahydrofolate. Pyridoxal phosphate activates the α carbon of amino acids at critical stages in all these reactions, and tetrahydrofolate carries one-carbon units in two of them (see Figs 18-6, 18-17).

Glycine is degraded via three pathways, only one of which leads to pyruvate. Glycine is converted to serine by enzymatic addition of a hydroxymethyl group (Figs 18-19, 18-20b). This reaction, catalyzed by **serine hydroxymethyltransferase**, requires the coenzymes tetrahydrofolate and pyridoxal phosphate. The serine is converted to pyruvate as described above. In the second pathway, which predominates in animals, glycine undergoes oxidative cleavage to CO_2, NH_4^+, and a methylene group ($-CH_2-$) (Figs 18-19, 18-20c). This readily reversible reaction, catalyzed by **glycine cleavage enzyme** (also called glycine synthase), also requires tetrahydrofolate, which accepts the methylene group. In this oxidative cleavage pathway, the two carbon atoms of glycine do not enter the citric acid cycle. One carbon is lost as CO_2 and the other becomes the methylene group of N^5,N^{10}-methylenetetrahydrofolate (Fig. 18-17), a one-carbon group donor in certain biosynthetic pathways.

This second pathway for glycine degradation seems to be critical in mammals. Humans with serious defects in glycine cleavage enzyme activity suffer from a condition known as nonketotic hyperglycinemia.

The condition is characterized by elevated serum levels of glycine, leading to severe mental deficiencies and death in very early childhood. At high levels, glycine is an inhibitory neurotransmitter, which may explain the neurological effects of the disease. Perhaps more important, high levels of glycine increase the levels of 2-amino-3-ketobutyrate, an unstable intermediate in the degradation of threonine in mitochondria (Fig. 18-19). 2-Amino-3-ketobutyrate decarboxylates spontaneously to form the toxic metabolite aminoacetone, which is readily metabolized to the highly reactive **methylglyoxal**, a molecule that modifies both protein and DNA.

Methylglyoxal

Methylglyoxal is also a byproduct of glycolysis and is implicated in the progression of type 2 diabetes (Box 7-1).

Many genetic defects of amino acid metabolism have been identified in humans (Table 18-2). We shall encounter several more in this chapter. ∎

TABLE 18-2 Some Human Genetic Disorders Affecting Amino Acid Catabolism

Medical condition	Approximate incidence (per 100,000 births)	Defective process	Defective enzyme	Symptoms and effects
Albinism	<3	Melanin synthesis from tyrosine	Tyrosine 3-monooxygenase (tyrosinase)	Lack of pigmentation; white hair, pink skin
Alkaptonuria	<0.4	Tyrosine degradation	Homogentisate 1,2-dioxygenase	Dark pigment in urine; late-developing arthritis
Argininemia	<0.5	Urea synthesis	Arginase	Mental retardation
Argininosuccinic acidemia	<1.5	Urea synthesis	Argininosuccinase	Vomiting; convulsions
Carbamoyl phosphate synthetase I deficiency	<0.5	Urea synthesis	Carbamoyl phosphate synthetase I	Lethargy; convulsions; early death
Homocystinuria	<0.5	Methionine degradation	Cystathionine β-synthase	Faulty bone development; mental retardation
Maple syrup urine disease (branched-chain ketoaciduria)	<0.4	Isoleucine, leucine, and valine degradation	Branched-chain α-keto acid dehydrogenase complex	Vomiting; convulsions; mental retardation; early death
Methylmalonic acidemia	<0.5	Conversion of propionyl-CoA to succinyl-CoA	Methylmalonyl-CoA mutase	Vomiting; convulsions; mental retardation; early death
Phenylketonuria	<8	Conversion of phenylalanine to tyrosine	Phenylalanine hydroxylase	Neonatal vomiting; mental retardation

In the third and final pathway of glycine degradation, the achiral glycine molecule is a substrate for the enzyme D-amino acid oxidase. The glycine is converted to glyoxylate, an alternative substrate for lactate dehydrogenase (p. 553). Glyoxylate is oxidized in an NAD^+-dependent reaction to oxalate:

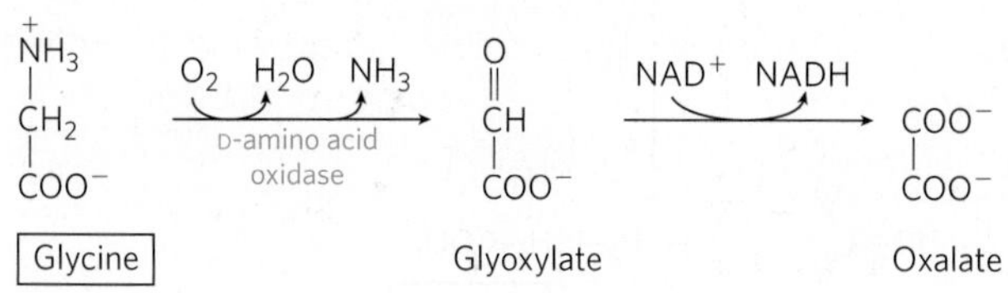

The function of D-amino acid oxidase, present at high levels in the kidney, is thought to be the detoxification of ingested D-amino acids derived from bacterial cell walls and from grilled foodstuffs (high heat causes some spontaneous racemization of the L-amino acids in proteins). Oxalate, whether obtained in foods or produced enzymatically in the kidneys, has medical significance. Crystals of calcium oxalate account for up to 75% of all kidney stones. ∎

There are two significant pathways for **threonine** degradation. One pathway leads to pyruvate via glycine (Fig. 18-19). The conversion to glycine occurs in two steps, with threonine first converted to 2-amino-3-ketobutyrate by the action of threonine dehydrogenase. This pathway is important in a few classes of rapidly dividing human cells, such as embryonic stem cells. The glycine generated by this pathway is broken down primarily by the glycine cleavage enzyme (Fig. 18-19). The N^5,N^{10}-methylenetetrahydrofolate thus generated (Fig. 18-20c) is needed for the synthesis, via pathways described in Chapter 22, of nucleotides used in DNA replication. However, in most human tissues, the degradation of threonine via glycine is a relatively minor pathway, accounting for 10% to 30% of threonine catabolism. It is more important in some other mammals. The major pathway in most human tissues leads to succinyl-CoA, as described later in this chapter.

In the laboratory, serine hydroxymethyltransferase will catalyze the conversion of threonine to glycine and acetaldehyde in one step, but this is not a significant pathway for threonine degradation in mammals.

Seven Amino Acids Are Degraded to Acetyl-CoA

Portions of the carbon skeletons of seven amino acids—**tryptophan**, **lysine**, **phenylalanine**, **tyrosine**, **leucine**, **isoleucine**, and **threonine**—yield acetyl-CoA and/or acetoacetyl-CoA, the latter being converted to acetyl-CoA (**Fig. 18-21**). Some of the final steps in the degradative pathways for leucine, lysine, and tryptophan resemble steps in the oxidation of fatty acids (see Fig. 17-9). Threonine (not shown in Fig. 18-21) yields

some acetyl-CoA via the minor pathway illustrated in Figure 18-19.

The degradative pathways of two of these seven amino acids deserve special mention. Tryptophan breakdown is the most complex of all the pathways of amino acid catabolism in animal tissues; portions of tryptophan (four of its carbons) yield acetyl-CoA via acetoacetyl-CoA. Some of the intermediates in tryptophan catabolism are precursors for the synthesis of other biomolecules (**Fig. 18-22**), including nicotinate, a precursor of NAD and NADP in animals; serotonin, a neurotransmitter in vertebrates; and indoleacetate, a growth factor in plants. Some of these biosynthetic pathways are described in more detail in Chapter 22 (see Figs 22-30, 22-31).

The breakdown of phenylalanine is noteworthy because genetic defects in the enzymes of this pathway lead to several inheritable human diseases (**Fig. 18-23**), as discussed below. Phenylalanine and its oxidation product tyrosine (both with nine carbons) are degraded into two fragments, both of which can enter the citric acid cycle: four of the nine carbon atoms yield free acetoacetate, which is converted to acetoacetyl-CoA and thus acetyl-CoA, and a second four-carbon fragment is recovered as fumarate. Eight of the nine carbons of these two amino acids thus enter the citric acid cycle; the remaining carbon is lost as CO_2. Phenylalanine, after its hydroxylation to tyrosine, is also the precursor of dopamine, a neurotransmitter, and of norepinephrine and epinephrine, hormones secreted by the adrenal medulla (see Fig. 22-31). Melanin, the black pigment of skin and hair, is also derived from tyrosine.

Phenylalanine Catabolism Is Genetically Defective in Some People

Given that many amino acids are either neurotransmitters or precursors or antagonists of neurotransmitters, it is not surprising that genetic defects of amino acid metabolism can cause defective neural development and intellectual deficits. In most such diseases, specific intermediates accumulate. For example, a genetic defect in **phenylalanine hydroxylase**, the first enzyme in the catabolic pathway for phenylalanine (Fig. 18-23), is responsible for the disease **phenylketonuria (PKU)**, the most common cause of elevated levels of phenylalanine in the blood (hyperphenylalaninemia).

Phenylalanine hydroxylase (also called phenylalanine-4-monooxygenase) is one of a general class of enzymes called **mixed-function oxygenases** (see Box 21-1), all of which catalyze simultaneous hydroxylation of a substrate by an oxygen atom of O_2 and reduction of the other oxygen atom to H_2O. Phenylalanine hydroxylase requires the cofactor tetrahydrobiopterin, which carries electrons from NADPH to O_2 and becomes oxidized to dihydrobiopterin in the

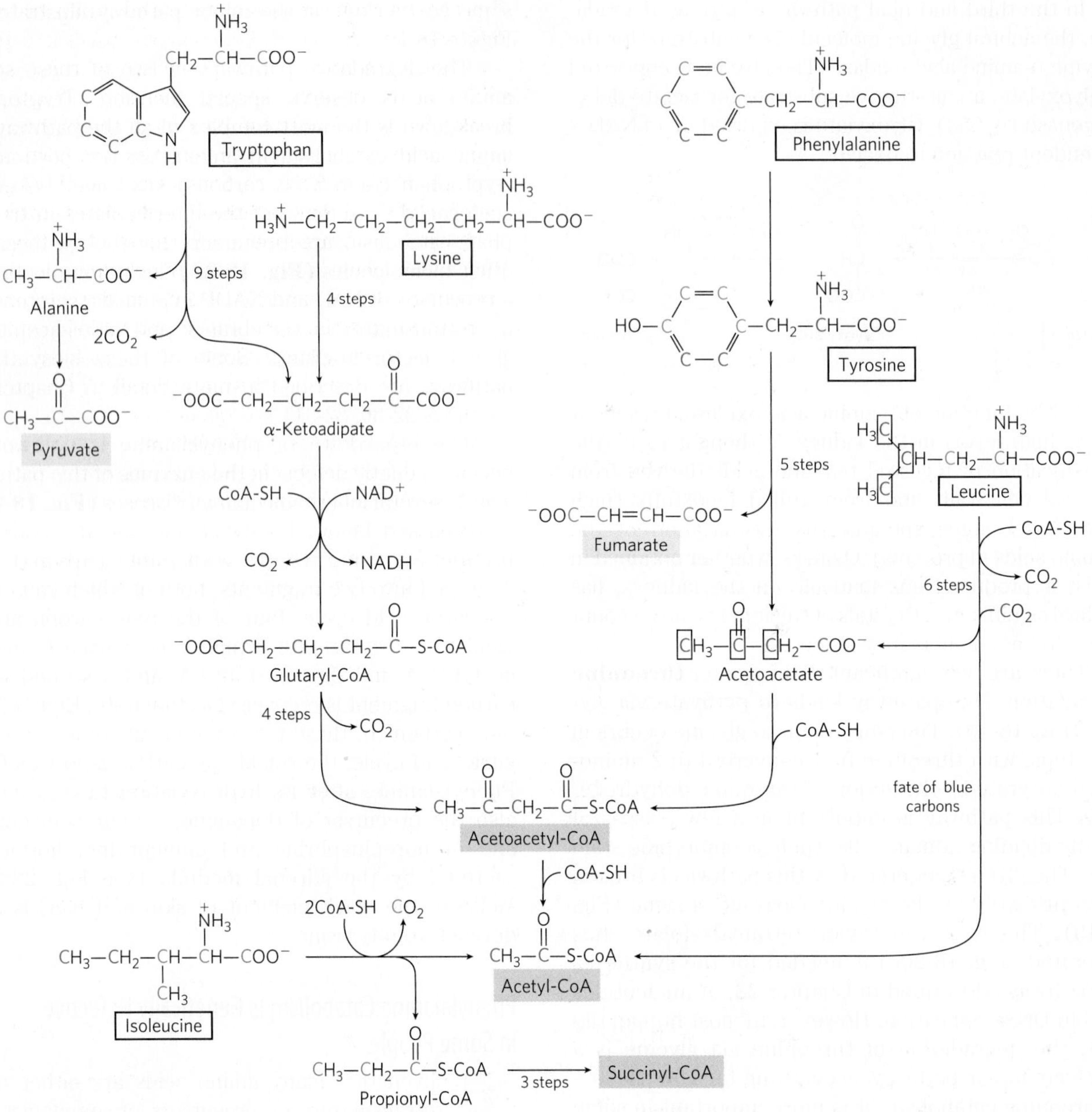

FIGURE 18-21 Catabolic pathways for tryptophan, lysine, phenyl-alanine, tyrosine, leucine, and isoleucine. These amino acids donate some of their carbons (red) to acetyl-CoA. Tryptophan, phenylalanine, tyrosine, and isoleucine also contribute carbons (blue) to pyruvate or citric acid cycle intermediates. The phenylalanine pathway is described in more detail in Figure 18-23. The fate of nitrogen atoms is not traced in this scheme; in most cases they are transferred to α-ketoglutarate to form glutamate.

process **(Fig. 18-24).** It is subsequently reduced by the enzyme **dihydrobiopterin reductase** in a reaction that requires NADPH.

In individuals with PKU, a secondary, normally little-used pathway of phenylalanine metabolism comes into play. In this pathway phenylalanine undergoes transamination with pyruvate to yield **phenylpyruvate (Fig. 18-25).** Phenylalanine and phenylpyruvate accumulate in the blood and tissues and are excreted in the urine—hence the name "phenylketonuria." Much of the phenylpyruvate, rather than being excreted as such, is either decarboxylated to phenylacetate or reduced to phenyllactate. Phenylacetate imparts a characteristic odor to the urine, which nurses have traditionally used to

detect PKU in infants. The accumulation of phenylalanine or its metabolites in early life impairs normal development of the brain, causing severe intellectual deficits. This may be caused by excess phenylalanine competing with other amino acids for transport across the blood-brain barrier, resulting in a deficit of required metabolites.

Phenylketonuria was among the first inheritable metabolic defects discovered in humans. When this condition is recognized early in infancy, mental retardation can be prevented by rigid dietary control. The diet must supply only enough phenylalanine and tyrosine to meet the needs for protein synthesis. Consumption of protein-rich foods must be curtailed. Natural proteins, such as casein of milk, must first be hydrolyzed and much of

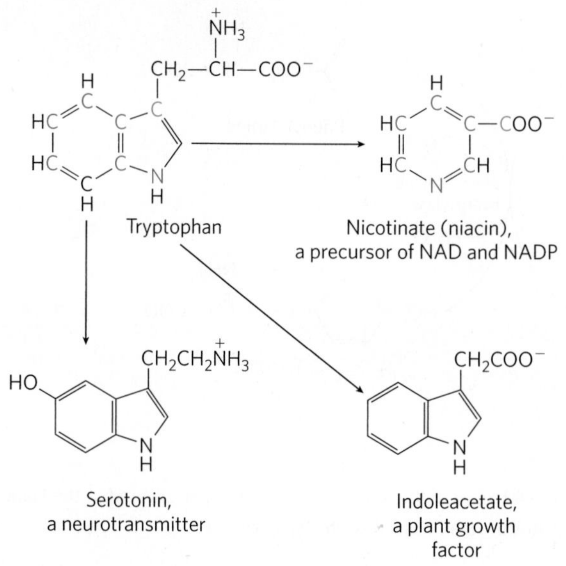

FIGURE 18-22 Tryptophan as precursor. The aromatic rings of tryptophan give rise to nicotinate (niacin), indoleacetate, and serotonin. Colored atoms trace the source of the ring atoms in nicotinate.

the phenylalanine removed to provide an appropriate diet, at least through childhood. Because the artificial sweetener aspartame is a dipeptide of aspartate and the methyl ester of phenylalanine (see Fig. 1-25b), foods sweetened with aspartame bear warnings addressed to individuals on phenylalanine-controlled diets. The dietary restrictions are difficult to follow perfectly for a lifetime, and thus often do not completely eliminate neurological symptoms. A new treatment is being developed in which the enzyme phenylalanine ammonia lyase is taken orally or injected subcutaneously to degrade phenylalanine in proteins ingested as part of a somewhat less restricted diet. Derived from plants, bacteria, and many yeast and fungi, phenylalanine ammonia lyase normally contributes to the biosynthesis of polyphenol compounds such as flavonoids. It degrades phenylalanine to the harmless metabolite *trans*-cinnamic acid and ammonia; the small amounts of ammonia generated are not toxic.

Phenylketonuria can also be caused by a defect in the enzyme that catalyzes the regeneration of tetrahydrobiopterin (Fig. 18-24). The treatment in this case is more

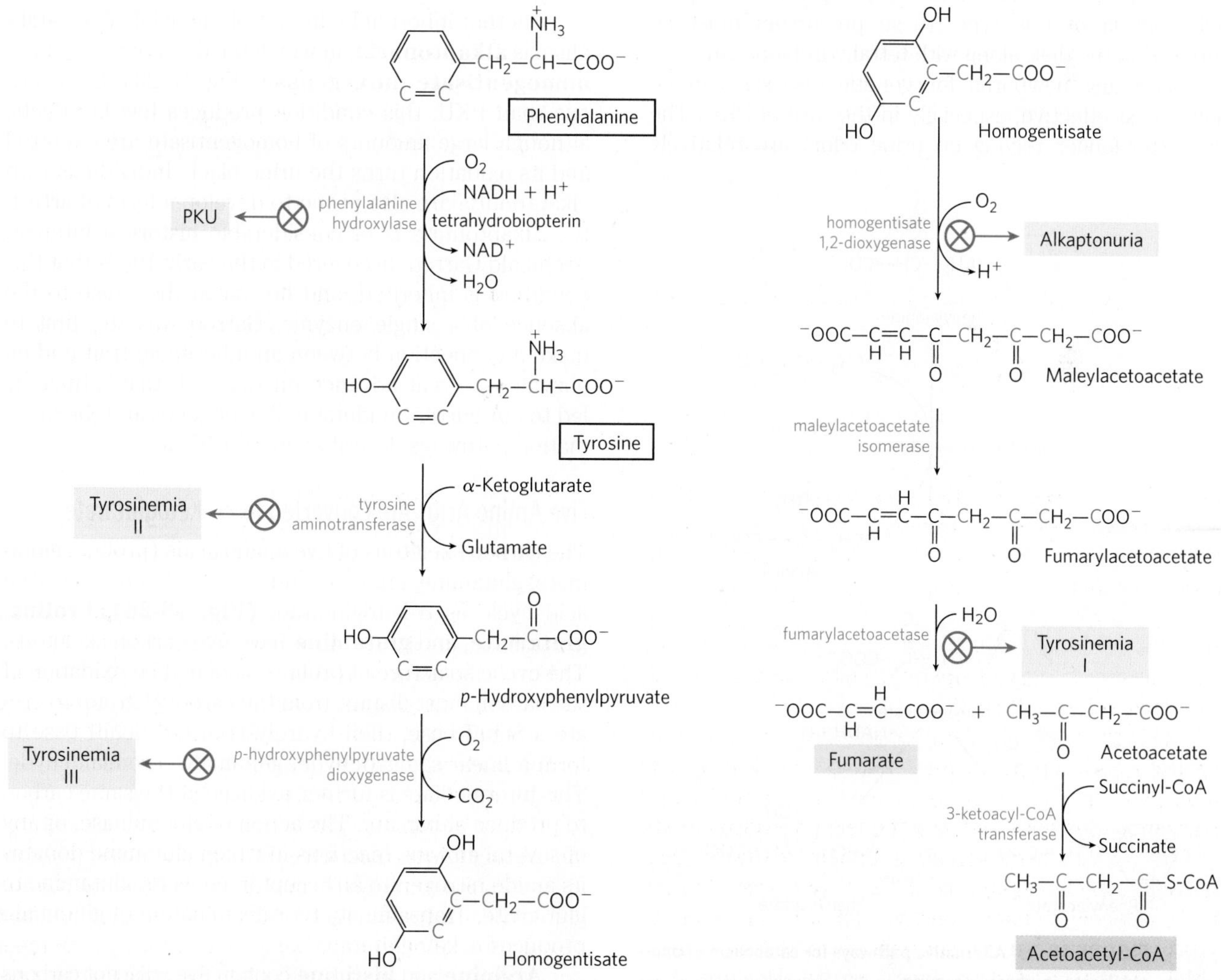

FIGURE 18-23 Catabolic pathways for phenylalanine and tyrosine. In humans these amino acids are normally converted to acetoacetyl-CoA and fumarate. Genetic defects in many of these enzymes cause inheritable human diseases (shaded yellow).

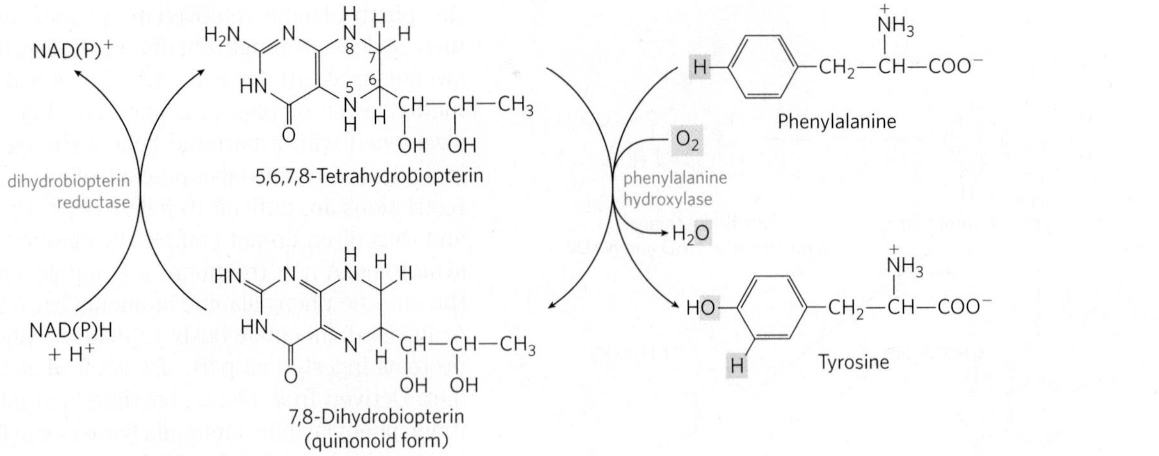

FIGURE 18-24 Role of tetrahydrobiopterin in the phenylalanine hydroxylase reaction. The H atom shaded pink is transferred directly from C-4 to C-3 in the reaction. This feature, discovered at the National Institutes of Health, is called the NIH shift.

complex than restricting the intake of phenylalanine and tyrosine. Tetrahydrobiopterin is also required for the formation of L-3,4-dihydroxyphenylalanine (L-dopa) and 5-hydroxytryptophan—precursors of the neurotransmitters norepinephrine and serotonin, respectively. In phenylketonuria of this type, these precursors must be supplied in the diet, along with tetrahydrobiopterin.

Screening newborns for genetic diseases can be highly cost-effective, especially in the case of PKU. The tests (no longer relying on urine odor) are relatively inexpensive, and the detection and early treatment of PKU in infants (eight to ten cases per 100,000 newborns) saves millions of dollars in later health care costs each year. More importantly, the emotional trauma avoided by early detection with these simple tests is inestimable.

Another inheritable disease of phenylalanine catabolism is **alkaptonuria**, in which the defective enzyme is **homogentisate dioxygenase** (Fig. 18-23). Less serious than PKU, this condition produces few ill effects, although large amounts of homogentisate are excreted and its oxidation turns the urine black. Individuals with alkaptonuria are also prone to develop a form of arthritis. Alkaptonuria is of considerable historical interest. Archibald Garrod discovered in the early 1900s that this condition is inherited, and he traced the cause to the absence of a single enzyme. Garrod was the first to make a connection between an inheritable trait and an enzyme—a great advance on the path that ultimately led to our current understanding of genes and the information pathways described in Part III. ■

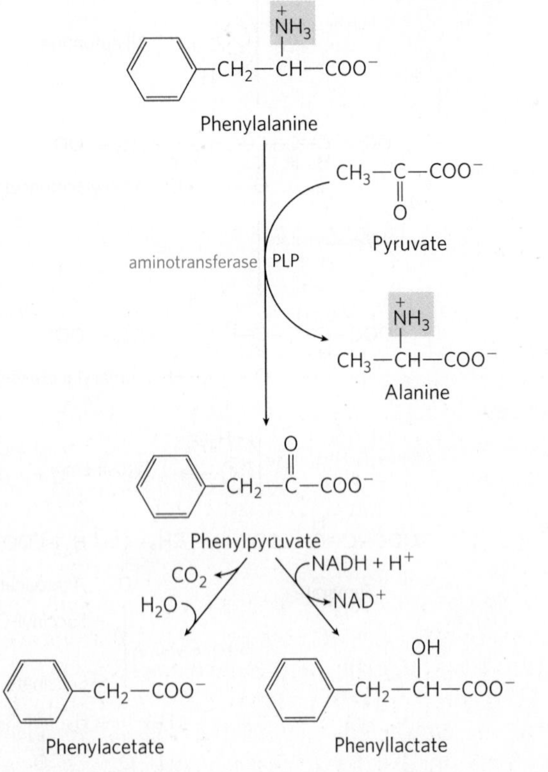

Five Amino Acids Are Converted to α-Ketoglutarate

The carbon skeletons of five amino acids (proline, glutamate, glutamine, arginine, and histidine) enter the citric acid cycle as α-ketoglutarate **(Fig. 18-26)**. **Proline, glutamate,** and **glutamine** have five-carbon skeletons. The cyclic structure of proline is opened by oxidation of the carbon most distant from the carboxyl group to create a Schiff base, then hydrolysis of the Schiff base to form a linear semialdehyde, glutamate γ-semialdehyde. This intermediate is further oxidized at the same carbon to produce glutamate. The action of glutaminase, or any of several enzyme reactions in which glutamine donates its amide nitrogen to an acceptor, converts glutamine to glutamate. Transamination or deamination of glutamate produces α-ketoglutarate.

Arginine and **histidine** contain five adjacent carbons and a sixth carbon attached through a nitrogen atom. The catabolic conversion of these amino acids to glutamate is

FIGURE 18-25 Alternative pathways for catabolism of phenylalanine in phenylketonuria. In PKU, phenylpyruvate accumulates in the tissues, blood, and urine. The urine may also contain phenylacetate and phenyllactate.

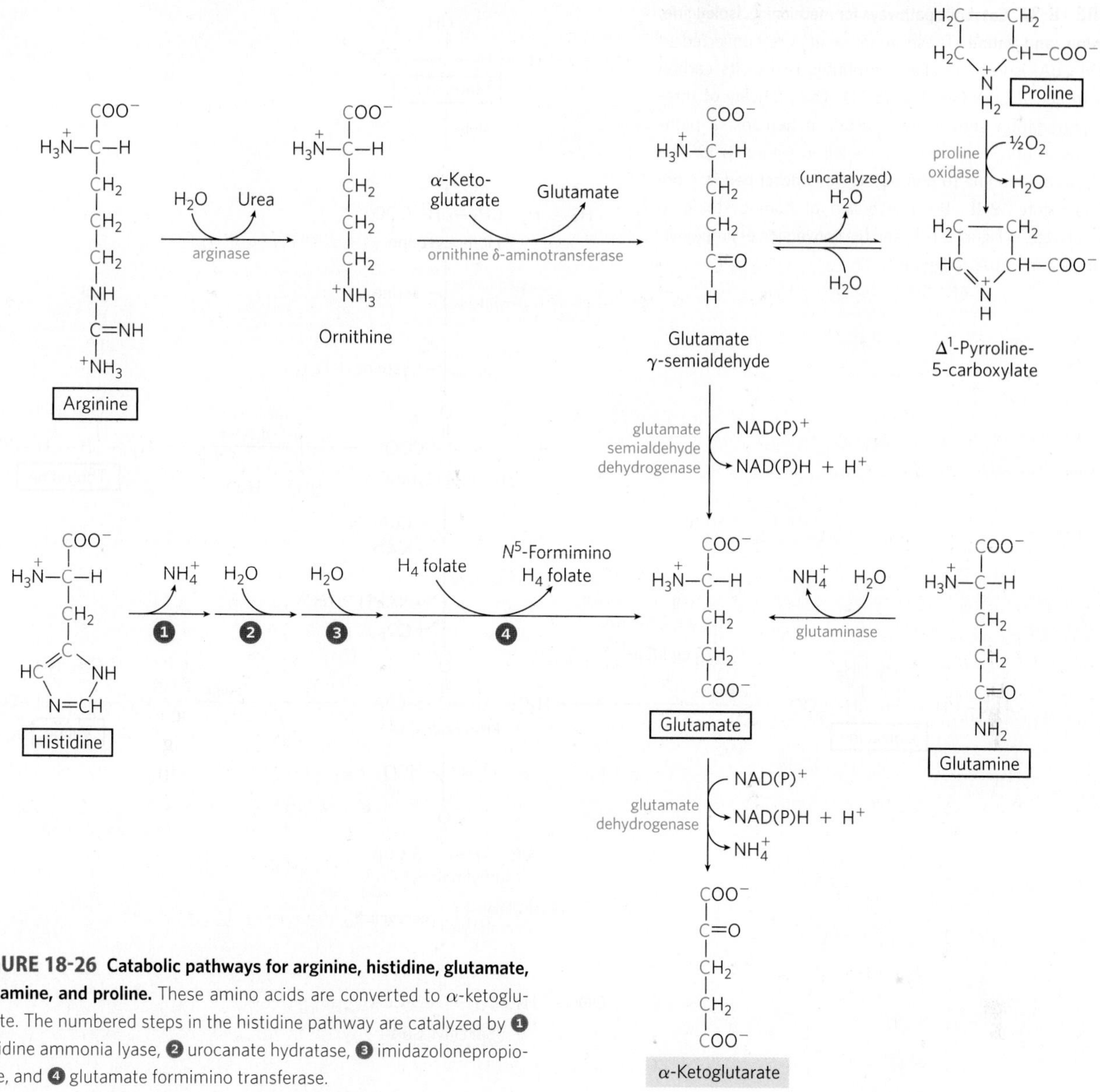

FIGURE 18-26 Catabolic pathways for arginine, histidine, glutamate, glutamine, and proline. These amino acids are converted to α-ketoglutarate. The numbered steps in the histidine pathway are catalyzed by ❶ histidine ammonia lyase, ❷ urocanate hydratase, ❸ imidazolonepropionase, and ❹ glutamate formimino transferase.

therefore slightly more complex than the path from proline or glutamine (Fig. 18-26). Arginine is converted to the five-carbon skeleton of ornithine in the urea cycle (Fig. 18-10), and the ornithine is transaminated to glutamate γ-semialdehyde. Conversion of histidine to the five-carbon glutamate occurs in a multistep pathway; the extra carbon is removed in a step that uses tetrahydrofolate as cofactor.

Four Amino Acids Are Converted to Succinyl-CoA

The carbon skeletons of methionine, isoleucine, threonine, and valine are degraded by pathways that yield succinyl-CoA **(Fig. 18-27)**, an intermediate of the citric acid cycle. **Methionine** donates its methyl group to one of several possible acceptors through *S*-adenosylmethionine, and three of its four remaining carbon atoms are converted to the propionate of propionyl-CoA, a precursor of succinyl-CoA. **Isoleucine** undergoes transamination, followed by oxidative decarboxylation of the resulting α-keto acid. The remaining five-carbon skeleton is further oxidized to

acetyl-CoA and propionyl-CoA. **Valine** undergoes trans-amination and decarboxylation, then a series of oxidation reactions that convert the remaining four carbons to propionyl-CoA. Some parts of the valine and isoleucine degradative pathways closely parallel steps in fatty acid degradation (see Fig. 17-9). In human tissues, **threonine** is also converted in two steps to propionyl-CoA. This is the primary pathway for threonine degradation in humans (see Fig. 18-19 for the alternative pathway). The mechanism of the first step is analogous to that catalyzed by serine dehydratase, and the serine and threonine dehydratases may actually be the same enzyme.

The propionyl-CoA derived from these three amino acids is converted to succinyl-CoA by a pathway described in Chapter 17: carboxylation to methylmalo-nyl-CoA, epimerization of the methylmalonyl-CoA, and conversion to succinyl-CoA by the coenzyme B$_{12}$–dependent methylmalonyl-CoA mutase (see Fig. 17-12). In the rare genetic disease known as methylmalonic

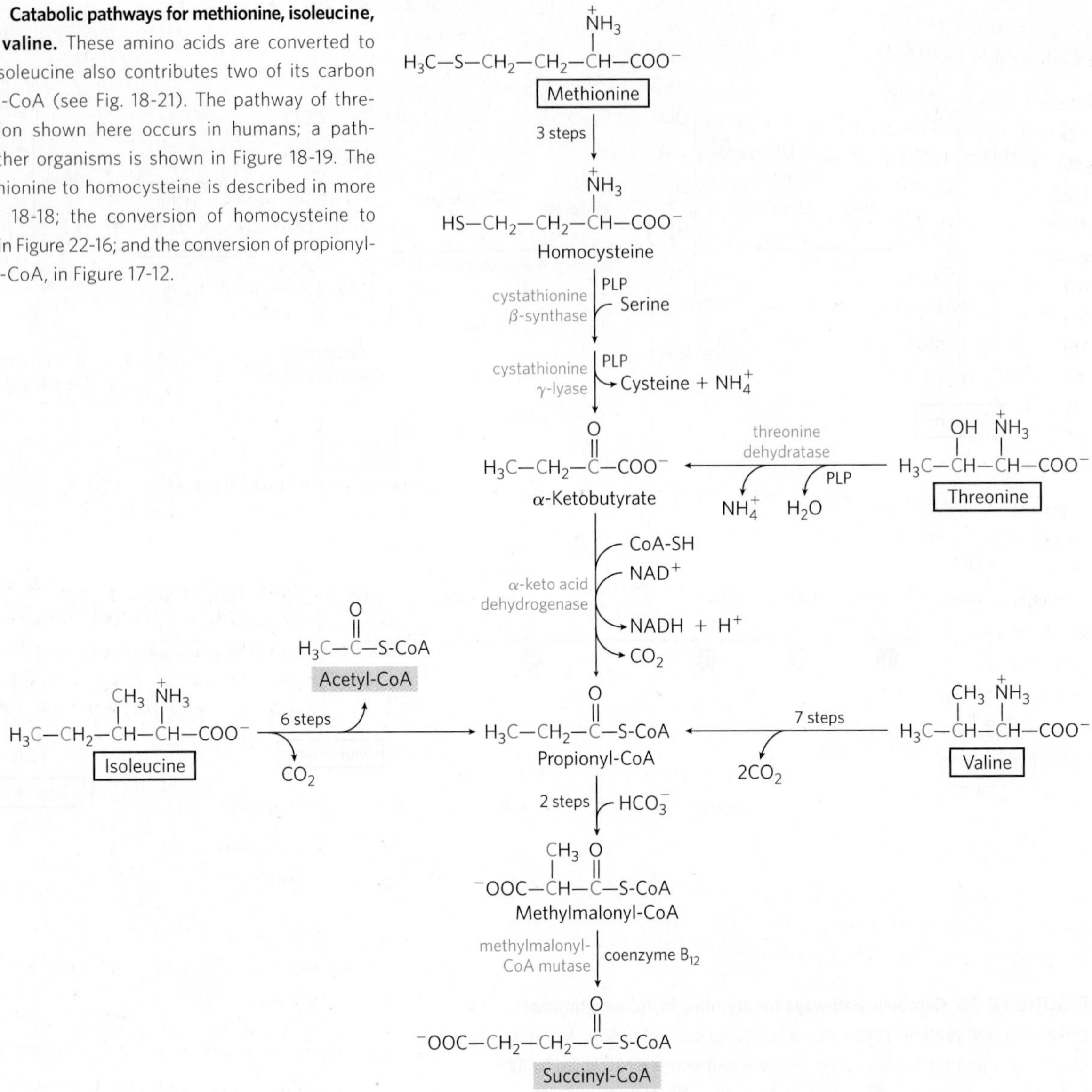

FIGURE 18-27 Catabolic pathways for methionine, isoleucine, threonine, and valine. These amino acids are converted to succinyl-CoA; isoleucine also contributes two of its carbon atoms to acetyl-CoA (see Fig. 18-21). The pathway of threonine degradation shown here occurs in humans; a pathway found in other organisms is shown in Figure 18-19. The route from methionine to homocysteine is described in more detail in Figure 18-18; the conversion of homocysteine to α-ketobutyrate, in Figure 22-16; and the conversion of propionyl-CoA to succinyl-CoA, in Figure 17-12.

acidemia, methylmalonyl-CoA mutase is lacking—with serious metabolic consequences (Table 18-2; Box 18-2).

Branched-Chain Amino Acids Are Not Degraded in the Liver

Although much of the catabolism of amino acids takes place in the liver, the three amino acids with branched side chains (leucine, isoleucine, and valine) are oxidized as fuels primarily in muscle, adipose, kidney, and brain tissue. These extrahepatic tissues contain an aminotransferase, absent in liver, that acts on all three branched-chain amino acids to produce the corresponding α-keto acids **(Fig. 18-28)**. The **branched-chain α-keto acid dehydrogenase complex** then catalyzes oxidative decarboxylation of all three α-keto acids, in each case releasing the carboxyl group as CO_2 and producing the acyl-CoA derivative. This reaction is formally analogous to two other oxidative decarboxylations encountered in Chapter 16: oxidation of pyruvate to

acetyl-CoA by the pyruvate dehydrogenase complex (see Fig. 16-6) and oxidation of α-ketoglutarate to succinyl-CoA by the α-ketoglutarate dehydrogenase complex (p. 630). In fact, all three enzyme complexes are similar in structure and share essentially the same reaction mechanism. Five cofactors (thiamine pyrophosphate, FAD, NAD, lipoate, and coenzyme A) participate, and the three proteins in each complex catalyze homologous reactions. This is clearly a case in which enzymatic machinery that evolved to catalyze one reaction was "borrowed" by gene duplication and further evolved to catalyze similar reactions in other pathways.

Experiments with rats have shown that the branched-chain α-keto acid dehydrogenase complex is regulated by covalent modification in response to the content of branched-chain amino acids in the diet. With little or no excess dietary intake of branched-chain amino acids, the enzyme complex is phosphorylated by a protein kinase and thereby inactivated. Addition of excess branched-chain amino acids to the diet results in

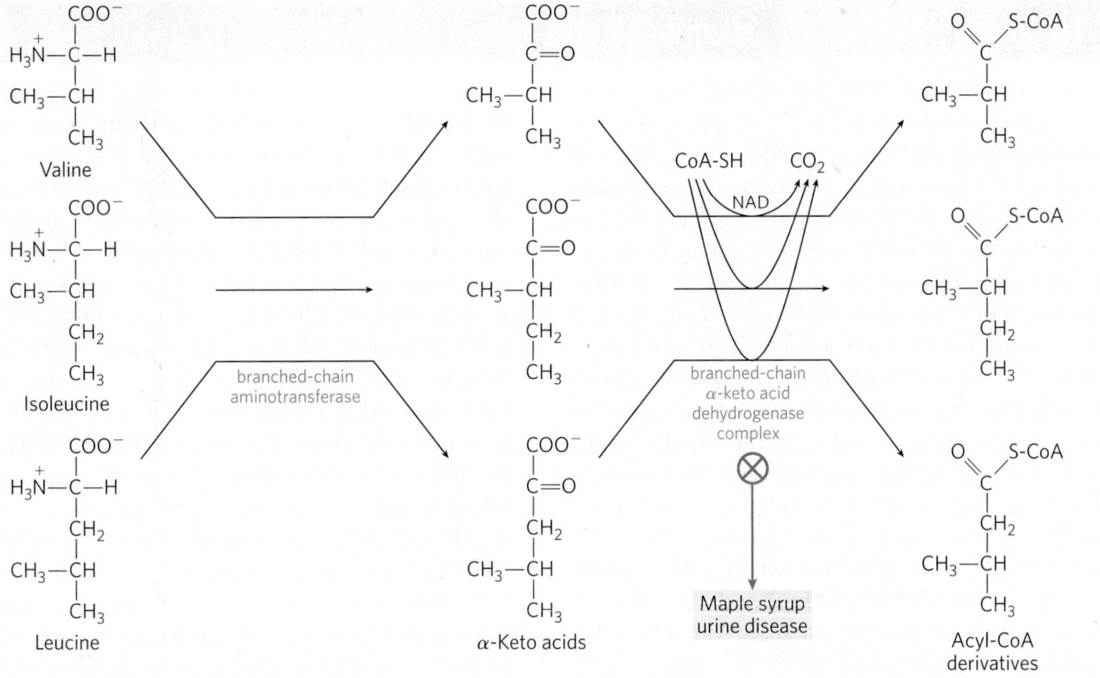

FIGURE 18-28 Catabolic pathways for the three branched-chain amino acids: valine, isoleucine, and leucine. All three pathways occur in extrahepatic tissues and share the first two enzymes, as shown here. The branched-chain α-keto acid dehydrogenase complex is analogous to the pyruvate and α-ketoglutarate dehydrogenase complexes and requires the same five cofactors (some not shown here). This enzyme is defective in people with maple syrup urine disease.

dephosphorylation and consequent activation of the enzyme. Recall that the pyruvate dehydrogenase complex is subject to similar regulation by phosphorylation and dephosphorylation (p. 640).

There is a relatively rare genetic disease in which the three branched-chain α-keto acids (as well as their precursor amino acids, especially leucine) accumulate in the blood and "spill over" into the urine. This condition, called **maple syrup urine disease** because of the characteristic odor imparted to the urine by the α-keto acids, results from a defective branched-chain α-keto acid dehydrogenase complex. Untreated, the disease results in abnormal development of the brain, mental retardation, and death in early infancy. Treatment entails rigid control of the diet, limiting the intake of valine, isoleucine, and leucine to the minimum required to permit normal growth. ■

Asparagine and Aspartate Are Degraded to Oxaloacetate

The carbon skeletons of **asparagine** and **aspartate** ultimately enter the citric acid cycle as malate in mammals or oxaloacetate in bacteria. The enzyme **asparaginase** catalyzes the hydrolysis of asparagine to aspartate, which undergoes transamination with α-ketoglutarate to yield glutamate and oxaloacetate **(Fig. 18-29)**. The oxaloacetate is converted to malate in the cytosol and then transported into the mitochondrial matrix through the malate–α-ketoglutarate transporter. In bacteria, the oxaloacetate produced in the transamination reaction can be used directly in the citric acid cycle.

We have now seen how the 20 common amino acids, after losing their nitrogen atoms, are degraded by dehydrogenation, decarboxylation, and other reactions to yield portions of their carbon backbones in the form of six central metabolites that can enter the citric acid cycle. Those portions degraded to acetyl-CoA are completely oxidized to carbon dioxide and water, with generation of ATP by oxidative phosphorylation.

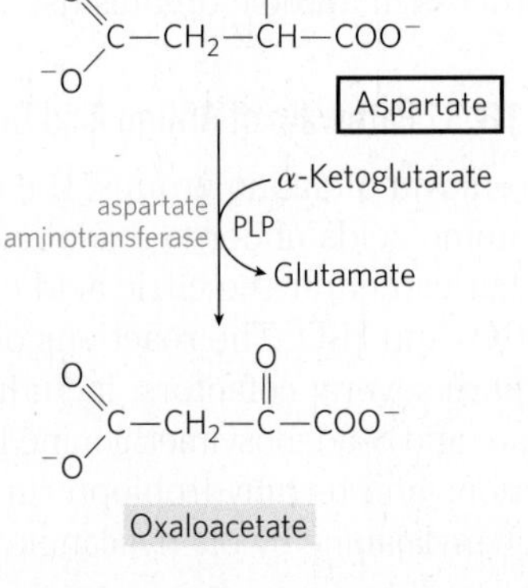

FIGURE 18-29 Catabolic pathway for asparagine and aspartate. Both amino acids are converted to oxaloacetate.

BOX 18-2   ⚕ MEDICINE   Scientific Sleuths Solve a Murder Mystery

Truth can sometimes be stranger than fiction—or at least as strange as a made-for-TV movie. Take, for example, the case of Patricia Stallings. Convicted of the murder of her infant son, she was sentenced to life in prison—but was later found innocent, thanks to the medical sleuthing of three persistent researchers.

The story began in the summer of 1989 when Stallings brought her three-month-old son, Ryan, to the emergency room of Cardinal Glennon Children's Hospital in St. Louis. The child had labored breathing, uncontrollable vomiting, and gastric distress. According to the attending physician, a toxicologist, the child's symptoms indicated that he had been poisoned with ethylene glycol, an ingredient of antifreeze, a conclusion apparently confirmed by analysis at a commercial lab.

After he recovered, the child was placed in a foster home, and Stallings and her husband, David, were allowed to see him in supervised visits. But when the infant became ill, and subsequently died, after a visit in which Stallings had been briefly left alone with him, she was charged with first-degree murder and held without bail. At the time, the evidence seemed compelling, as both the commercial lab and the hospital lab found large amounts of ethylene glycol in the boy's blood and traces of it in a bottle of milk Stallings had fed her son during the visit.

But without knowing it, Stallings had performed a brilliant experiment. While in custody, she learned she was pregnant; she subsequently gave birth to another son, David Stallings Jr., in February 1990. He was placed immediately in a foster home, but within two weeks he started having symptoms similar to

Ryan's. David was eventually diagnosed with a rare metabolic disorder called methylmalonic acidemia (MMA). A recessive genetic disorder of amino acid metabolism, MMA affects about 1 in 48,000 newborns and presents symptoms almost identical with those caused by ethylene glycol poisoning.

Stallings couldn't possibly have poisoned her second son, but the Missouri state prosecutor's office was not impressed by the new developments and pressed forward with her trial anyway. The court wouldn't allow the MMA diagnosis of the second child to be introduced as evidence, and in January 1991 Patricia Stallings was convicted of assault with a deadly weapon and sentenced to life in prison.

Fortunately for Stallings, however, William Sly, chairman of the Department of Biochemistry and Molecular Biology at St. Louis University, and James Shoemaker, head of a metabolic screening lab at the university, got interested in her case when they heard about it from a television broadcast. Shoemaker performed his own analysis of Ryan's blood and didn't detect ethylene glycol. He and Sly then contacted Piero Rinaldo, a metabolic disease expert at Yale University School of Medicine whose lab is equipped to diagnose MMA from blood samples.

When Rinaldo analyzed Ryan's blood serum, he found high concentrations of methylmalonic acid, a breakdown product of the branched-chain amino acids isoleucine and valine, which accumulates in MMA patients because the enzyme that should convert it to the next product in the metabolic pathway is defective (Fig. 1). And particularly telling, he says,

---

As was the case for carbohydrates and lipids, the degradation of amino acids results ultimately in the generation of reducing equivalents (NADH and $FADH_2$) through the action of the citric acid cycle. Our survey of catabolic processes concludes in the next chapter with a discussion of respiration, in which these reducing equivalents fuel the ultimate oxidative and energy-generating process in aerobic organisms.

SUMMARY 18.3 Pathways of Amino Acid Degradation

■ After the removal of amino groups, the carbon skeletons of amino acids undergo oxidation to compounds that can enter the citric acid cycle for oxidation to CO_2 and H_2O. The reactions of these pathways require several cofactors, including tetrahydrofolate and S-adenosylmethionine in one-carbon transfer reactions and tetrahydrobiopterin in the oxidation of phenylalanine by phenylalanine hydroxylase.

■ Depending on their degradative end product, some amino acids can be converted to ketone bodies, some to

glucose, and some to both. Thus amino acid degradation is integrated into intermediary metabolism and can be critical to survival under conditions in which amino acids are a significant source of metabolic energy.

■ The carbon skeletons of amino acids enter the citric acid cycle through five intermediates: acetyl-CoA, α-ketoglutarate, succinyl-CoA, fumarate, and oxaloacetate. Some are also degraded to pyruvate, which can be converted to either acetyl-CoA or oxaloacetate.

■ The amino acids producing pyruvate are alanine, cysteine, glycine, serine, threonine, and tryptophan. Leucine, lysine, phenylalanine, and tryptophan yield acetyl-CoA via acetoacetyl-CoA. Isoleucine, leucine, threonine, and tryptophan also form acetyl-CoA directly.

■ Arginine, glutamate, glutamine, histidine, and proline produce α-ketoglutarate; isoleucine, methionine, threonine, and valine produce succinyl-CoA; four carbon atoms of phenylalanine and tyrosine

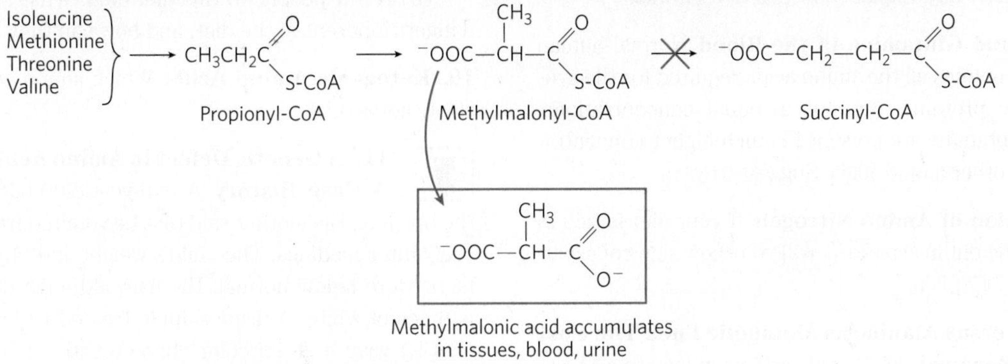

FIGURE 1 Children with a mutation (red X) that inactivates the enzyme methylmalony-CoA mutase cannot degrade isoleucine, methionine, threonine, and valine normally. Instead, a potentially fatal accumulation of methylmalonic acid occurs, with symptoms similar to those of ethylene glycol poisoning.

the child's blood and urine contained massive amounts of ketones, another metabolic consequence of the disease. Like Shoemaker, he did not find any ethylene glycol in a sample of the baby's bodily fluids. The bottle couldn't be tested, since it had mysteriously disappeared. Rinaldo's analyses convinced him that Ryan had died from MMA, but how to account for the results from two labs, indicating that the boy had ethylene glycol in his blood? Could they both be wrong?

When Rinaldo obtained the lab reports, what he saw was, he says, "scary." One lab said that Ryan Stallings' blood contained ethylene glycol, even though the blood sample analysis did not match the lab's own profile for a known sample containing ethylene glycol. "This was not just a matter of questionable interpretation. The quality of their analysis was unacceptable,"

Rinaldo says. And the second laboratory? According to Rinaldo, that lab detected an abnormal component in Ryan's blood and just "assumed it was ethylene glycol." Samples from the bottle had produced nothing unusual, says Rinaldo, yet the lab claimed evidence of ethylene glycol in that, too.

Rinaldo presented his findings to the case's prosecutor, George McElroy, who called a press conference the very next day. "I no longer believe the laboratory data," he told reporters. Having concluded that Ryan Stallings had died of MMA after all, McElroy dismissed all charges against Patricia Stallings on September 20, 1991.

By Michelle Hoffman, Science 253:931, 1991. Copyright 1991 by the American Association for the Advancement of Science.

give rise to fumarate; and asparagine and aspartate produce oxaloacetate.

■ The branched-chain amino acids (isoleucine, leucine, and valine), unlike the other amino acids, are degraded only in extrahepatic tissues.

■ Several serious human diseases can be traced to genetic defects in the enzymes of amino acid catabolism.

Key Terms

Terms in bold are defined in the glossary.

aminotransferases 679
transaminases 679
transamination 679
pyridoxal phosphate (PLP) 679
oxidative deamination 680
L-glutamate dehydrogenase 680

glutamine synthetase 682
glutaminase 682
glucose-alanine cycle 683
ammonotelic 684
ureotelic 684
uricotelic 684
urea cycle 684
urea 686

creatine kinase 688
essential amino acids 689
ketogenic 691
glucogenic 691
tetrahydrofolate 691
S-adenosylmethionine (adoMet) 692

tetrahydrobiopterin 694
phenylketonuria (PKU) 697
mixed-function oxidases 697
alkaptonuria 700
maple syrup urine disease 703

Problems

1. Products of Amino Acid Transamination Name and draw the structure of the α-keto acid resulting when each of the following amino acids undergoes transamination with α-ketoglutarate: (a) aspartate, (b) glutamate, (c) alanine, (d) phenylalanine.

2. Measurement of Alanine Aminotransferase Activity The activity (reaction rate) of alanine aminotransferase is usually measured by including an excess of pure lactate dehydrogenase and NADH in the reaction system. The rate of alanine disappearance is equal to the rate of NADH disappearance measured spectrophotometrically. Explain how this assay works.

3. Alanine and Glutamine in the Blood Normal human blood plasma contains all the amino acids required for the synthesis of body proteins, but not in equal concentrations. Alanine and glutamine are present in much higher concentrations than any other amino acids. Suggest why.

4. Distribution of Amino Nitrogen If your diet is rich in alanine but deficient in aspartate, will you show signs of aspartate deficiency? Explain.

5. Lactate versus Alanine as Metabolic Fuel: The Cost of Nitrogen Removal The three carbons in lactate and alanine have identical oxidation states, and animals can use either carbon source as a metabolic fuel. Compare the net ATP yield (moles of ATP per mole of substrate) for the complete oxidation (to CO_2 and H_2O) of lactate versus alanine when the cost of nitrogen excretion as urea is included.

Lactate   Alanine

6. Ammonia Toxicity Resulting from an Arginine-Deficient Diet In a study conducted some years ago, cats were fasted overnight then given a single meal complete in all amino acids except arginine. Within 2 hours, blood ammonia levels increased from a normal level of 18 μg/L to 140 μg/L, and the cats showed the clinical symptoms of ammonia toxicity. A control group fed a complete amino acid diet or an amino acid diet in which arginine was replaced by ornithine showed no unusual clinical symptoms.

(a) What was the role of fasting in the experiment?

(b) What caused the ammonia levels to rise in the experimental group? Why did the absence of arginine lead to ammonia toxicity? Is arginine an essential amino acid in cats? Why or why not?

(c) Why can ornithine be substituted for arginine?

7. Oxidation of Glutamate Write a series of balanced equations, and an overall equation for the net reaction, describing the oxidation of 2 mol of glutamate to 2 mol of α-ketoglutarate and 1 mol of urea.

8. Transamination and the Urea Cycle Aspartate aminotransferase has the highest activity of all the mammalian liver aminotransferases. Why?

9. The Case against the Liquid Protein Diet A weight-reducing diet heavily promoted some years ago required the daily intake of "liquid protein" (soup of hydrolyzed gelatin), water, and an assortment of vitamins. All other food and drink were to be avoided. People on this diet typically lost 10 to 14 lb in the first week.

(a) Opponents argued that the weight loss was almost entirely due to water loss and would be regained very soon after a normal diet was resumed. What is the biochemical basis for this argument?

(b) A few people on this diet died. What are some of the dangers inherent in the diet, and how can they lead to death?

10. Ketogenic Amino Acids Which amino acids are exclusively ketogenic?

11. A Genetic Defect in Amino Acid Metabolism: A Case History A two-year-old child was taken to the hospital. His mother said that he vomited frequently, especially after feedings. The child's weight and physical development were below normal. His hair, although dark, contained patches of white. A urine sample treated with ferric chloride ($FeCl_3$) gave a green color characteristic of the presence of phenylpyruvate. Quantitative analysis of urine samples gave the results shown in the table.

Substance	Concentration (mM)	
	Patient's urine	Normal urine
Phenylalanine	7.0	0.01
Phenylpyruvate	4.8	0
Phenyllactate	10.3	0

(a) Suggest which enzyme might be deficient in this child. Propose a treatment.

(b) Why does phenylalanine appear in the urine in large amounts?

(c) What is the source of phenylpyruvate and phenyllactate? Why does this pathway (normally not functional) come into play when the concentration of phenylalanine rises?

(d) Why does the boy's hair contain patches of white?

12. Role of Cobalamin in Amino Acid Catabolism Pernicious anemia is caused by impaired absorption of vitamin B_{12}. What is the effect of this impairment on the catabolism of amino acids? Are all amino acids equally affected? (Hint: See Box 17-2.)

13. Vegetarian Diets Vegetarian diets can provide high levels of antioxidants and a lipid profile that can help prevent coronary disease. However, there can be some associated problems. Blood samples were taken from a large group of volunteer subjects who were vegans (strict vegetarians: no animal products), lactovegetarians (vegetarians who eat dairy products), or omnivores (individuals with a varied diet including meat). In each case, the volunteers had followed the diet for several years. The blood levels of both homocysteine and methylmalonate were elevated in the vegan group, somewhat lower in the lactovegetarian group, and much lower in the omnivore group. Explain.

14. Pernicious Anemia Vitamin B_{12} deficiency can arise from a few rare genetic diseases that lead to low B_{12} levels despite a normal diet that includes B_{12}-rich meat and dairy sources. These conditions cannot be treated with dietary B_{12} supplements. Explain.

15. Pyridoxal Phosphate Reaction Mechanisms Threonine can be broken down by the enzyme threonine dehydratase, which catalyzes the conversion of threonine to α-ketobutyrate and ammonia. The enzyme uses PLP as a cofactor. Suggest a mechanism for this reaction, based on the mechanisms in Figure 18-6. Note that this reaction includes an elimination at the β carbon of threonine.

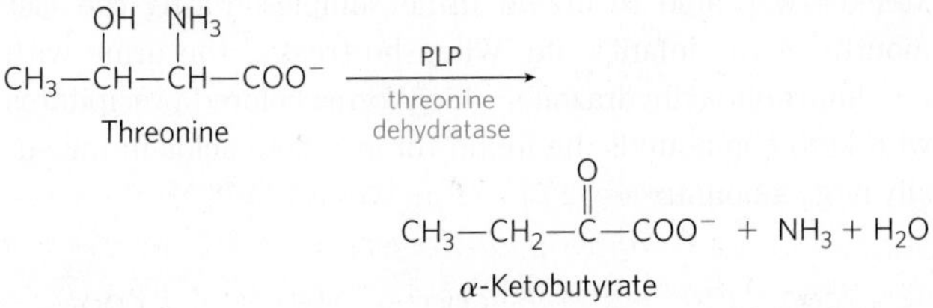

$$\underset{\text{Threonine}}{CH_3-\overset{\overset{OH}{|}}{CH}-\overset{\overset{NH_3}{|}}{CH}-COO^-} \xrightarrow[\substack{\text{threonine}\\\text{dehydratase}}]{PLP} \underset{\alpha\text{-Ketobutyrate}}{CH_3-CH_2-\overset{\overset{O}{\|}}{C}-COO^- + NH_3 + H_2O}$$

16. Pathway of Carbon and Nitrogen in Glutamate Metabolism
When $[2\text{-}^{14}C,^{15}N]$glutamate undergoes oxidative degradation in the liver of a rat, in which atoms of the following metabolites will each isotope be found: (a) urea, (b) succinate, (c) arginine, (d) citrulline, (e) ornithine, (f) aspartate?

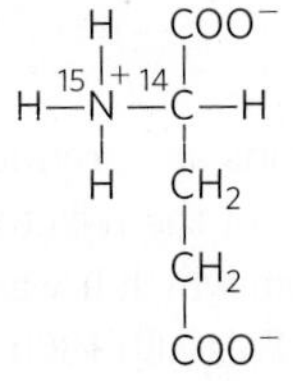

Labeled glutamate

17. Chemical Strategy of Isoleucine Catabolism
Isoleucine is degraded in six steps to propionyl-CoA and acetyl-CoA.

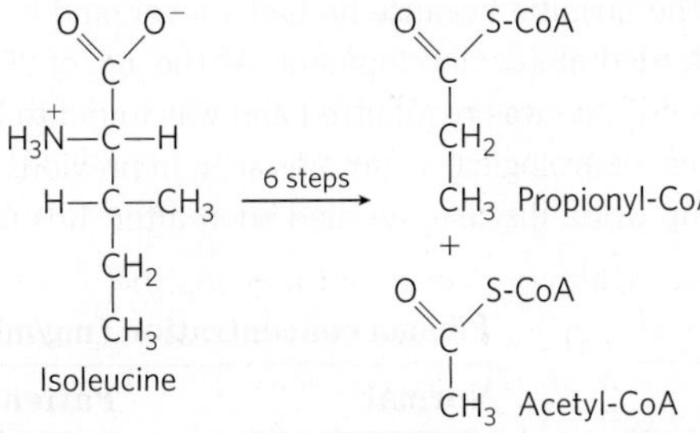

Isoleucine $\xrightarrow{\text{6 steps}}$ Propionyl-CoA + Acetyl-CoA

(a) The chemical process of isoleucine degradation includes strategies analogous to those used in the citric acid cycle and the β oxidation of fatty acids. The intermediates of isoleucine degradation (I to V) shown below are not in the proper order. Use your knowledge and understanding of the citric acid cycle and β-oxidation pathway to arrange the intermediates in the proper metabolic sequence for isoleucine degradation.

I   II   III

IV   V

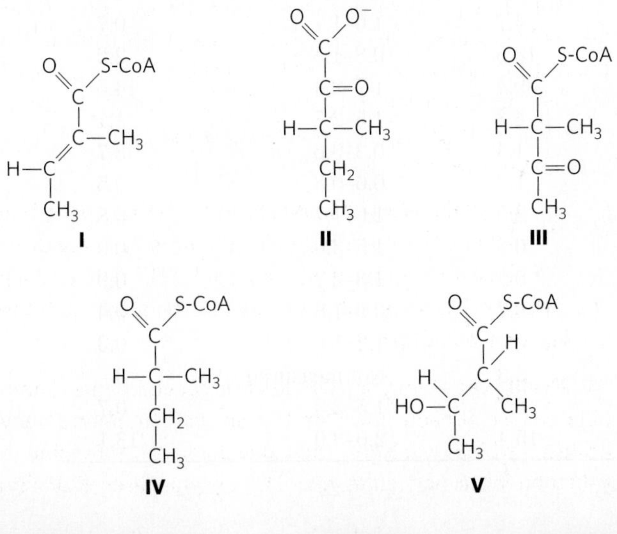

(b) For each step you propose, describe the chemical process, provide an analogous example from the citric acid cycle or β-oxidation pathway (where possible), and indicate any necessary cofactors.

18. Role of Pyridoxal Phosphate in Glycine Metabolism
The enzyme serine hydroxymethyltransferase requires pyridoxal phosphate as cofactor. Propose a mechanism for the reaction catalyzed by this enzyme, in the direction of serine degradation (glycine production). (Hint: See Figs 18-19 and 18-20b.)

19. Parallel Pathways for Amino Acid and Fatty Acid Degradation
The carbon skeleton of leucine is degraded by a series of reactions closely analogous to those of the citric acid cycle and β oxidation. For each reaction, (a) through (f), shown below, indicate its type, provide an analogous example from the citric acid cycle or β-oxidation pathway (where possible), and note any necessary cofactors.

Leucine

(a) $\downarrow$

α-Ketoisocaproate

(b) $\begin{cases} \text{CoA-SH} \\ \text{CO}_2 \end{cases}$

Isovaleryl-CoA

(c) $\downarrow$

β-Methylcrotonyl-CoA

(d) $\left\{ HCO_3^- \right.$

β-Methylglutaconyl-CoA

(e) $\left\{ H_2O \right.$

β-Hydroxy-β-methylglutaryl-CoA

(f) $\downarrow$

$^-OOC-CH_2-\overset{\overset{O}{\|}}{C}-CH_3 + CH_3-\overset{\overset{O}{\|}}{C}-S\text{-}CoA$

Acetoacetate   Acetyl-CoA

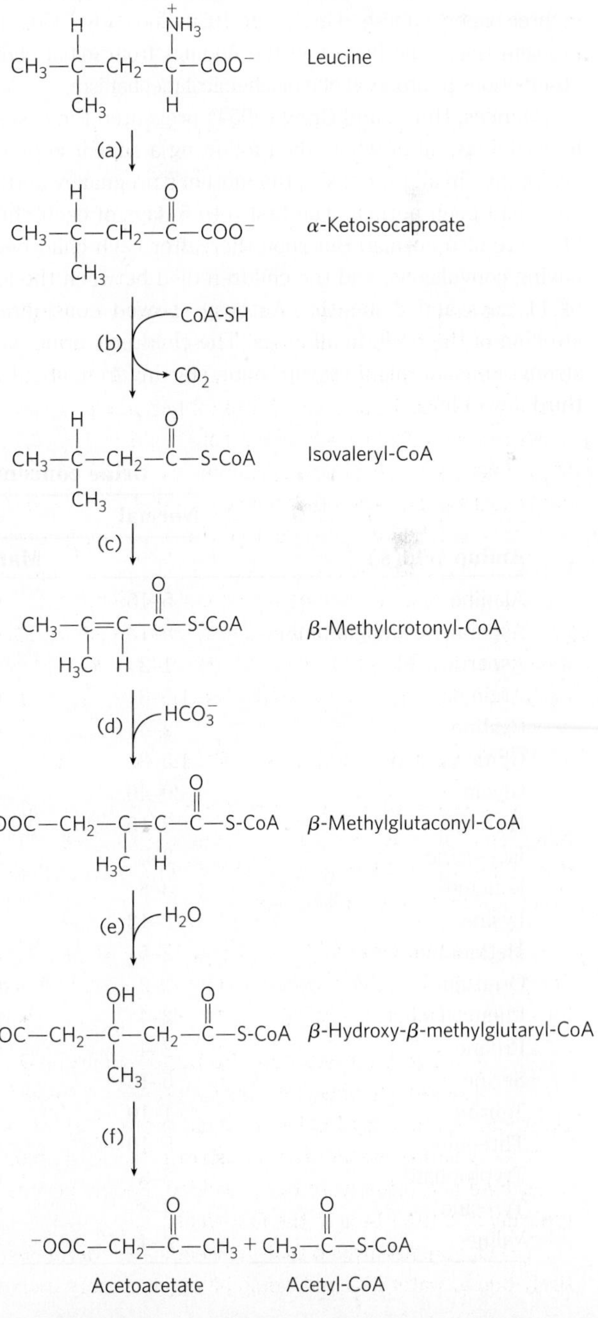

20. Treatments for a Genetic Disease The strict dietary controls required to stem the progress of maple syrup urine disease are difficult to follow for a lifetime, and patients may experience poor metabolic control that leads to neurological symptoms. In these cases, treatment can involve an organ transplant from a suitable donor. Organ transplantation involves considerable risk, but success can greatly alleviate this metabolic disorder and reduce the need for dietary restrictions. Which organ could be transplanted to gain this effect, and why?

Data Analysis Problem

21. Maple Syrup Urine Disease Figure 18-28 shows the pathway for the degradation of branched chain amino acids and the site of the biochemical defect that causes maple syrup urine disease. The initial findings that eventually led to the discovery of the defect in this disease were presented in three papers published in the late 1950s and early 1960s. This problem traces the history of the findings from initial clinical observations to proposal of a biochemical mechanism.

Menkes, Hurst, and Craig (1954) presented the cases of four siblings, all of whom died following a similar course of symptoms. In all four cases, the mother's pregnancy and the birth had been normal. The first 3 to 5 days of each child's life were also normal. But soon thereafter each child began having convulsions, and the children died between the ages of 11 days and 3 months. Autopsy showed considerable swelling of the brain in all cases. The children's urine had a strong, unusual "maple syrup" odor, starting from about the third day of life.

Menkes (1959) reported data collected from six more children. All showed symptoms similar to those described above, and died within 15 days to 20 months of birth. In one case, Menkes was able to obtain urine samples during the last months of the infant's life. When he treated the urine with 2,4-dinitrophenylhydrazone, which forms colored precipitates with keto compounds, he found three α-keto acids in unusually large amounts:

α-Ketoisocaproate α-Ketoisovalerate α-Keto-β-methyl-n-valerate

(a) These α-keto acids are produced by the deamination of amino acids. For each of the α-keto acids above, draw and name the amino acid from which it was derived.

Dancis, Levitz, and Westall (1960) collected further data that led them to propose the biochemical defect shown in Figure 18-28. In one case, they examined a patient whose urine first showed the maple syrup odor when he was 4 months old. At the age of 10 months (March 1956), the child was admitted to the hospital because he had a fever, and he showed grossly retarded motor development. At the age of 20 months (January 1957), he was readmitted and was found to have the degenerative neurological symptoms seen in previous cases of maple syrup urine disease; he died soon after. Results of his

Amino acid(s)	Urine concentration (mg/24 h)			Plasma concentration (mg/mL)	
	Normal	Patient		Normal	Patient
		Mar. 1956	Jan. 1957		Jan. 1957
Alanine	5–15	0.2	0.4	3.0–4.8	0.6
Asparagine and glutamine	5–15	0.4	0	3.0–5.0	2.0
Aspartic acid	1–2	0.2	1.5	0.1–0.2	0.04
Arginine	1.5–3	0.3	0.7	0.8–1.4	0.8
Cystine	2–4	0.5	0.3	1.0–1.5	0
Glutamic acid	1.5–3	0.7	1.6	1.0–1.5	0.9
Glycine	20–40	4.6	20.7	1.0–2.0	1.5
Histidine	8–15	0.3	4.7	1.0–1.7	0.7
Isoleucine	2–5	2.0	13.5	0.8–1.5	2.2
Leucine	3–8	2.7	39.4	1.7–2.4	14.5
Lysine	2–12	1.6	4.3	1.5–2.7	1.1
Methionine	2–5	1.4	1.4	0.3–0.6	2.7
Ornithine	1–2	0	1.3	0.6–0.8	0.5
Phenylalanine	2–4	0.4	2.6	1.0–1.7	0.8
Proline	2–4	0.5	0.3	1.5–3.0	0.9
Serine	5–15	1.2	0	1.3–2.2	0.9
Taurine	1–10	0.2	18.7	0.9–1.8	0.4
Threonine	5–10	0.6	0	1.2–1.6	0.3
Tryptophan	3–8	0.9	2.3	Not measured	0
Tyrosine	4–8	0.3	3.7	1.5–2.3	0.7
Valine	2–4	1.6	15.4	2.0–3.0	13.1

blood and urine analyses are shown in the table on page 708, along with normal values for each component.

(b) The table includes taurine, an amino acid not normally found in proteins. Taurine is often produced as a byproduct of cell damage. Its structure is:

$$\overset{+}{H_3N}-CH_2-CH_2-\overset{\overset{\displaystyle O}{\|}}{\underset{\underset{\displaystyle O}{\|}}{S}}-O^-$$

Based on its structure and the information in this chapter, what is the most likely amino acid precursor of taurine? Explain your reasoning.

(c) Compared with the normal values given in the table, which amino acids showed significantly elevated levels in the patient's blood in January 1957? Which ones in the patient's urine?

Based on their results and their knowledge of the pathway shown in Figure 18-28, Dancis and coauthors concluded that "although it appears most likely to the authors that the primary block is in the metabolic degradative pathway of the branched-chain amino acids, this cannot be considered established beyond question."

(d) How do the data presented here support this conclusion?

(e) Which data presented here do *not* fit this model of maple syrup urine disease? How do you explain these seemingly contradictory data?

(f) What data would you need to collect to be more secure in your conclusion?

References

Dancis, J., M. Levitz, and R. Westall. 1960. Maple syrup urine disease: branched-chain ketoaciduria. *Pediatrics* 25:72–79.

Menkes, J.H. 1959. Maple syrup disease: isolation and identification of organic acids in the urine. *Pediatrics* 23:348–353.

Menkes, J.H., P.L. Hurst, and J.M. Craig. 1954. A new syndrome: progressive familial infantile cerebral dysfunction associated with an unusual urinary substance. *Pediatrics* 14:462–466.

Further Reading is available at www.macmillanlearning.com/LehningerBiochemistry7e.

CHAPTER 19

Oxidative Phosphorylation

Self-study tools that will help you practice what you've learned and reinforce this chapter's concepts are available online.
Go to www.macmillanlearning.com/LehningerBiochemistry7e.

Oxidative phosphorylation is the culmination of energy-yielding metabolism (catabolism) in aerobic organisms. All oxidative steps in the degradation of carbohydrates, fats, and amino acids converge at this final stage of cellular respiration, in which the energy of oxidation drives the synthesis of ATP. Oxidative phosphorylation accounts for most of the ATP synthesized by nonphotosynthetic organisms under most circumstances. In eukaryotes, oxidative phosphorylation occurs in mitochondria and involves huge protein complexes embedded in the mitochondrial membranes. The pathway to ATP synthesis in mitochondria challenged and fascinated biochemists for much of the twentieth century.

Our current understanding of ATP synthesis in mitochondria is based on the theory, introduced by Peter Mitchell in 1961, that transmembrane differences in proton concentration are the reservoir for the energy extracted from biological oxidation reactions. This **chemiosmotic theory** has been accepted as one of the great unifying principles of twentieth-century biology. It provides insight into the processes of oxidative phosphorylation and photophosphorylation in plants, and into such apparently disparate energy transductions as active transport across membranes and the motion of bacterial flagella.

The mechanism of oxidative phosphorylation has three defining components **(Fig. 19-1)**. (1) Electrons flow from electron donors (oxidizable substrates) through a chain of membrane-bound carriers to a final electron acceptor with a large reduction potential (molecular oxygen, O_2).

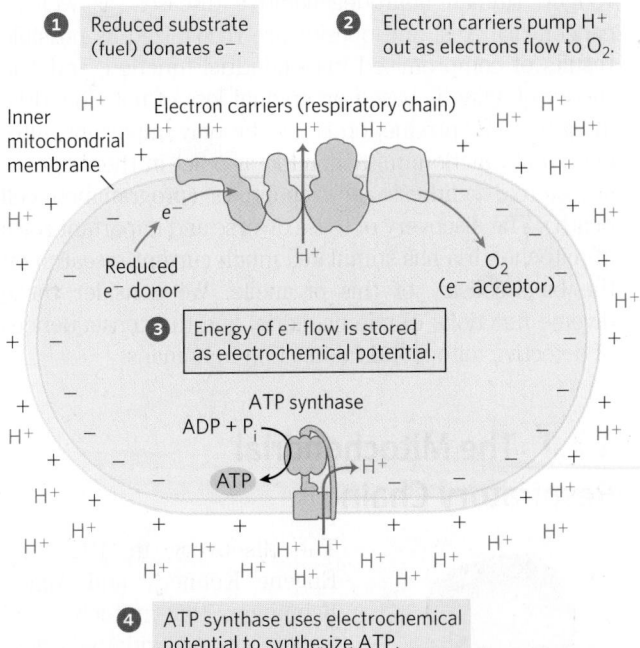

① Reduced substrate (fuel) donates e^-.

② Electron carriers pump H^+ out as electrons flow to O_2.

③ Energy of e^- flow is stored as electrochemical potential.

④ ATP synthase uses electrochemical potential to synthesize ATP.

FIGURE 19-1 The chemiosmotic mechanism for ATP synthesis in mitochondria. Electrons move spontaneously through a chain of membrane-bound carriers, the respiratory chain, driven by the high reduction potential of oxygen and the relatively low reduction potentials of the various reduced substrates (fuels) that undergo oxidation in the mitochondrion. Electron flow creates an electrochemical potential by the transmembrane movement of protons and positive charge. This electrochemical potential drives ATP synthesis by a membrane-bound enzyme, ATP synthase, that is fundamentally similar in mitochondria and chloroplasts, and in bacteria and archaea as well.

(2) The free energy made available by this "downhill" (exergonic) electron flow is coupled to the "uphill" transport of protons across a proton-impermeable membrane, conserving the free energy of fuel oxidation as a transmembrane electrochemical potential (p. 406). (3) The transmembrane flow of protons back down their concentration gradient through specific protein channels provides the free energy for synthesis of ATP, catalyzed by a membrane protein complex (ATP synthase) that couples proton flow to phosphorylation of ADP.

The chapter begins with a description of the components of the mitochondrial electron-transfer chain—the respiratory chain—and their organization into large functional complexes in the inner mitochondrial membrane, the path of electron flow through these complexes, and the proton movements that accompany this flow. We then consider the remarkable enzyme complex that captures, by "rotational catalysis," the energy of proton flow in ATP, and the regulatory mechanisms that coordinate oxidative phosphorylation with the many catabolic pathways by which fuels are oxidized.

The metabolic role of mitochondria is so critical to cellular and organismal function that defects in mitochondrial function have serious medical consequences. Mitochondria are central to neuronal and muscular function, and to the regulation of whole-body energy metabolism and body weight. Human neurodegenerative diseases, as well as cancer, diabetes, and obesity, are recognized as possible results of compromised mitochondrial function, and one theory of aging is based on gradual loss of mitochondrial integrity. ATP production is not the only important mitochondrial function; mitochondria also act in thermogenesis, steroid synthesis, and apoptosis (programmed cell death). The discovery of these diverse and important roles of mitochondria has stimulated much current research on the biochemistry of this organelle. We consider these diverse functions of mitochondria, and the consequences of defective mitochondrial function in humans.

19.1 The Mitochondrial Respiratory Chain

The discovery in 1948, by Eugene Kennedy and Albert Lehninger, that mitochondria are the site of oxidative phosphorylation in eukaryotes marked the beginning of the enzymological studies of biological energy transductions. Mitochondria, like gram-negative bacteria, have two membranes **(Fig. 19-2a)**. The outer mitochondrial membrane is readily permeable to small molecules (M_r <5,000) and ions, which move freely through transmembrane channels formed by a family of integral membrane proteins called porins. The inner membrane is impermeable to most small molecules and ions, including protons (H^+); the only species that cross this membrane do so through specific transporters. The inner membrane bears the components of the respiratory chain and ATP synthase.

The mitochondrial matrix, enclosed by the inner membrane, contains the pyruvate dehydrogenase complex and the enzymes of the citric acid cycle, the fatty acid β-oxidation pathway, and the pathways of amino acid oxidation—all the pathways of fuel oxidation except glycolysis, which takes place in the cytosol. The selectively permeable inner mitochondrial membrane segregates the intermediates and enzymes of cytosolic metabolic pathways from those of metabolic processes occurring in the matrix. However, specific transporters carry pyruvate, fatty acids, and amino acids or their α-keto derivatives into the matrix for access to the machinery of the citric acid cycle. ADP and P_i are specifically transported into the matrix as newly synthesized ATP is transported out. The best current inventory of proteins in mammalian mitochondria lists about 1,100, at least 300 of which have unknown functions.

The bean-shaped representation of a mitochondrion in Figure 19-2a is an oversimplification, derived in part from early studies in which thin sections of cells were observed in the electron microscope. Three-dimensional images obtained either by reconstruction from serial sections or by confocal microscopy reveal great variation in mitochondrial size and shape. In living cells stained with mitochondrion-specific fluorescent dyes, large numbers of variously shaped mitochondria are seen, clustered about the nucleus (Fig. 19-2b).

Tissues with a high demand for aerobic metabolism (brain, skeletal and heart muscle, and eye, for example) contain many hundreds or thousands of mitochondria per cell, and in general, mitochondria of cells with high metabolic activity have more, and more densely packed, cristae (Fig. 19-2c, d). During cell growth and division, mitochondria, like bacteria, divide by fission, and under some circumstances individual mitochondria fuse to form larger, more-extended structures. Stressful conditions, such as the presence of electron-transfer inhibitors or mutations in an electron carrier, trigger mitochondrial fission and sometimes **mitophagy**—breakdown of mitochondria and recycling of the amino acids, nucleotides, and lipids released. As stress is relieved, short, small mitochondria fuse to form long, thin, tubular organelles.

Electrons Are Funneled to Universal Electron Acceptors

Oxidative phosphorylation begins with the entry of electrons into the series of electron carriers called the **respiratory chain**. Most of these electrons arise from the action of dehydrogenases that collect electrons from catabolic pathways and funnel them into universal

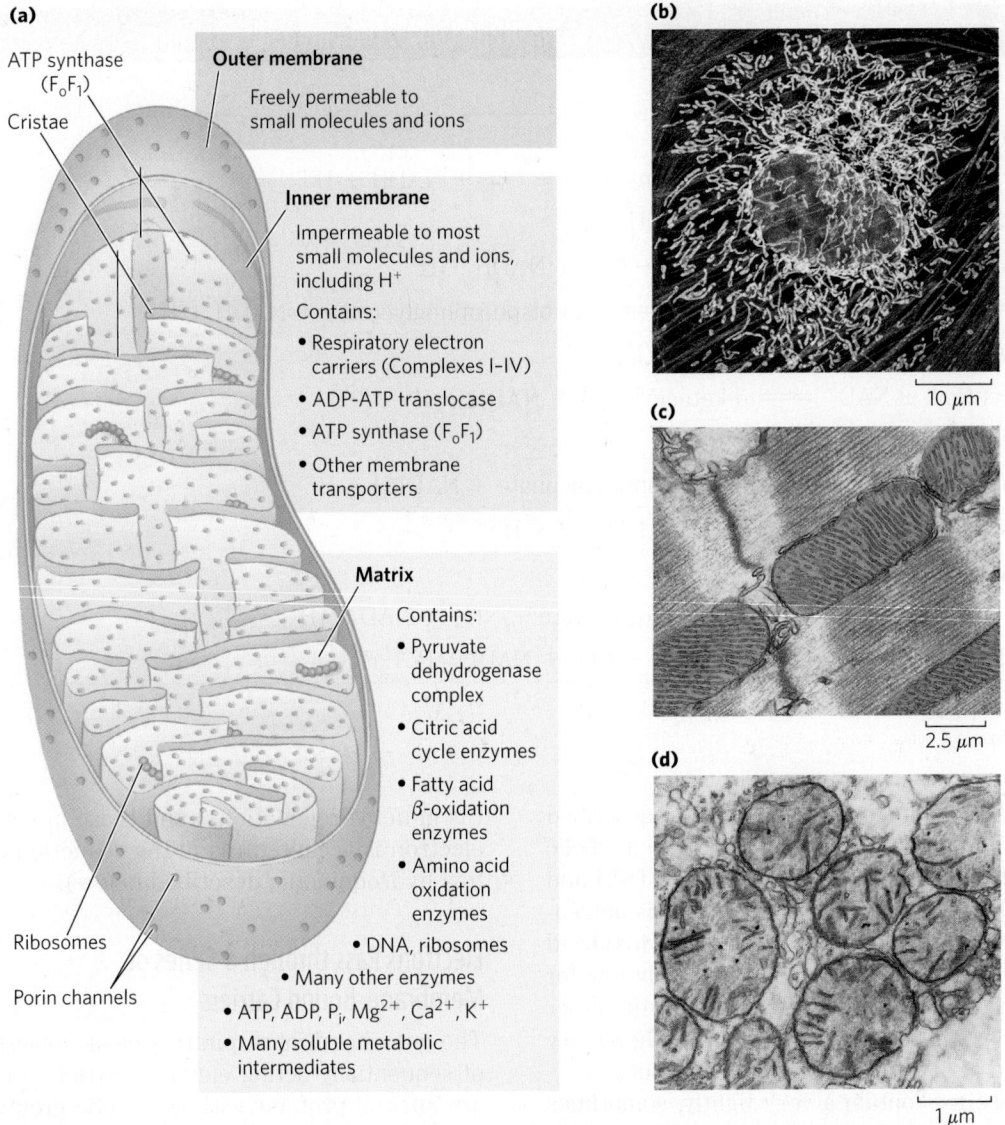

FIGURE 19-2 Biochemical anatomy of a mitochondrion. (a) The outer membrane has pores that make it permeable to small molecules and ions, but not to proteins. The convolutions of the inner membrane, called cristae, provide a very large surface area. The inner membrane of a single liver mitochondrion may have more than 10,000 sets of electron-transfer systems (respiratory chains) and ATP synthase molecules, distributed over the membrane surface. **(b)** A typical animal cell has hundreds or thousands of mitochondria. This endothelial cell from bovine pulmonary artery was stained with fluorescent probes for actin (blue), for DNA (red), and for mitochondria (yellow). Notice the variability in length of the mitochondria.

(c) The mitochondria of heart muscle (blue in this colorized electron micrograph) have more profuse cristae and thus a much larger area of inner membrane, with more than three times as many sets of respiratory chains as **(d)** liver mitochondria. Muscle and liver mitochondria are about the size of a bacterium—1 to 10 μm long. The mitochondria of invertebrates, plants, and microbial eukaryotes are similar to those shown here, but with much variation in size, shape, and degree of convolution of the inner membrane. [Sources: (b) Talley Lambert/Science Source. (c) Thomas Deerinck, NCMIR/Science Source. (d) Biophoto Associates/ Science Source.]

electron acceptors—nicotinamide nucleotides (NAD^+ or $NADP^+$) or flavin nucleotides (FMN or FAD) (see Figs 13-24, 13-27).

Nicotinamide nucleotide–linked dehydrogenases catalyze reversible reactions of the following general types:

$$\text{Reduced substrate} + NAD^+ \rightleftharpoons$$
$$\text{oxidized substrate} + NADH + H^+$$

$$\text{Reduced substrate} + NADP^+ \rightleftharpoons$$
$$\text{oxidized substrate} + NADPH + H^+$$

Most dehydrogenases that act in catabolism are specific for NAD^+ as electron acceptor (Table 19-1). Some are in the cytosol, others are in mitochondria, and still others have mitochondrial and cytosolic isozymes.

NAD-linked dehydrogenases remove two hydrogen atoms from their substrates. One of these is transferred as a hydride ion ($:H^-$) to NAD^+, and the other is released as H^+ in the medium (see Fig. 13-24). NADH and NADPH are water-soluble electron carriers that associate *reversibly* with dehydrogenases. NADH carries electrons from catabolic reactions to their point of entry into the respiratory

TABLE 19-1 Some Important Reactions Catalyzed by NAD(P)H-Linked Dehydrogenases

Reaction[a]	Location[b]
NAD-linked	
α-Ketoglutarate + CoA + NAD$^+$ $\rightleftharpoons$ succinyl-CoA + CO$_2$ + NADH + H$^+$	M
L-Malate + NAD$^+$ $\rightleftharpoons$ oxaloacetate + NADH + H$^+$	M and C
Pyruvate + CoA + NAD$^+$ $\rightleftharpoons$ acetyl-CoA + CO$_2$ + NADH + H$^+$	M
Glyceraldehyde 3-phosphate + P$_i$ + NAD$^+$ $\rightleftharpoons$ 1,3-bisphosphoglycerate + NADH + H$^+$	C
Lactate + NAD$^+$ $\rightleftharpoons$ pyruvate + NADH + H$^+$	C
β-Hydroxyacyl-CoA + NAD$^+$ $\rightleftharpoons$ β-ketoacyl-CoA + NADH + H$^+$	M
NADP-linked	
Glucose 6-phosphate + NADP$^+$ $\rightleftharpoons$ 6-phosphogluconate + NADPH + H$^+$	C
L-Malate + NADP$^+$ $\rightleftharpoons$ pyruvate + CO$_2$ + NADPH + H$^+$	C
NAD- or NADP-linked	
L-Glutamate + H$_2$O + NAD(P)$^+$ $\rightleftharpoons$ α-ketoglutarate + NH$_4^+$ + NAD(P)H	M
Isocitrate + NAD(P)$^+$ $\rightleftharpoons$ α-ketoglutarate + CO$_2$ + NAD(P)H + H$^+$	M and C

[a]These reactions and their enzymes are discussed in Chapters 14 through 18.
[b]M designates mitochondria; C, cytosol.

chain, the NADH dehydrogenase complex described below. NADPH generally supplies electrons to anabolic reactions. Cells maintain separate pools of NADPH and NADH, with different redox potentials. This is accomplished by holding the ratio [reduced form]/[oxidized form] relatively high for NADPH and relatively low for NADH. Neither NADH nor NADPH can cross the inner mitochondrial membrane, but the electrons they carry can be shuttled across indirectly, as we shall see.

Flavoproteins contain a very tightly, sometimes covalently, bound flavin nucleotide, either FMN or FAD (see Fig. 13-27). The oxidized flavin nucleotide can accept either one electron (yielding the semiquinone form) or two (yielding FADH$_2$ or FMNH$_2$). Electron transfer occurs because the flavoprotein has a higher reduction potential than the compound oxidized. Recall that reduction potential is a quantitative measure of the relative tendency of a given chemical species to accept electrons in an oxidation-reduction reaction (p. 520). The standard reduction potential of a flavin nucleotide, unlike that of NAD or NADP, depends on the protein with which it is associated. Local interactions with functional groups in the protein distort the electron orbitals in the flavin ring, changing the relative stabilities of oxidized and reduced forms. The relevant standard reduction potential is therefore that of the particular flavoprotein, not that of isolated FAD or FMN. The flavin nucleotide should be considered part of the flavoprotein's active site rather than a reactant or product in the electron-transfer reaction. Because flavoproteins can participate in either one- or two-electron transfers, they can serve as intermediates between reactions in which two electrons are donated

(as in dehydrogenations) and those in which only one electron is accepted (as in the reduction of a quinone to a hydroquinone, described below).

Electrons Pass through a Series of Membrane-Bound Carriers

The mitochondrial respiratory chain consists of a series of sequentially acting electron carriers, most of which are integral proteins with prosthetic groups capable of accepting and donating either one or two electrons. Three types of electron transfers occur in oxidative phosphorylation: (1) direct transfer of electrons, as in the reduction of Fe^{3+} to Fe^{2+}, (2) transfer as a hydrogen atom (H$^+$ + e^-), and (3) transfer as a hydride ion (:H$^-$), which bears two electrons. The term **reducing equivalent** is used to designate a single electron equivalent transferred in an oxidation-reduction reaction.

In addition to NAD and flavoproteins, three other types of electron-carrying molecules function in the respiratory chain: a hydrophobic quinone called ubiquinone, and two different types of iron-containing proteins, cytochromes and iron-sulfur proteins. **Ubiquinone** (also called **coenzyme Q**, or simply **Q**) is a lipid-soluble benzoquinone with a long isoprenoid side chain **(Fig. 19-3)**. Ubiquinone can accept one electron to become the semiquinone radical ($^\bullet$QH) or two electrons to form ubiquinol (QH$_2$) and, like flavoprotein carriers, it can act at the junction between a two-electron donor and a one-electron acceptor. Because ubiquinone is both small and hydrophobic, it is freely diffusible within the lipid bilayer of the inner mitochondrial membrane and can shuttle reducing equivalents between other, less

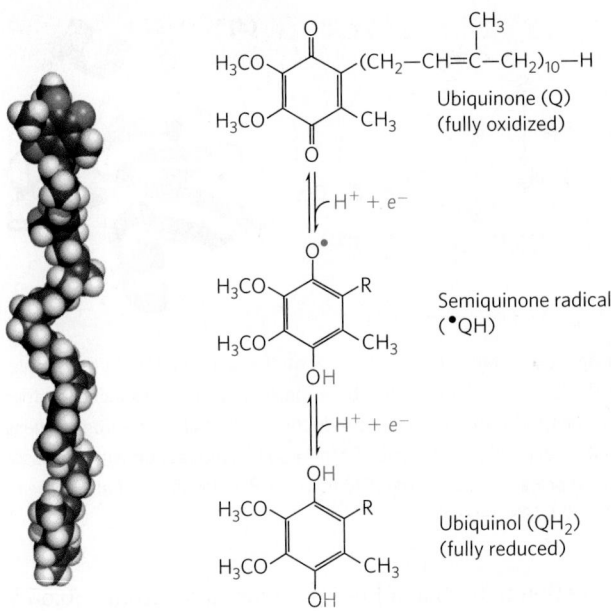

FIGURE 19-3 Ubiquinone (Q, or coenzyme Q). Complete reduction of ubiquinone requires two electrons and two protons, and occurs in two steps through the semiquinone radical intermediate.

mobile electron carriers in the membrane. And because it carries both electrons and protons, it plays a central role in coupling electron flow to proton movement.

The **cytochromes** are proteins with characteristic strong absorption of visible light, due to their iron-containing heme prosthetic groups **(Fig. 19-4a)**. Mitochondria contain three classes of cytochromes, designated a, b, and c, which are distinguished by differences in their light-absorption spectra. Each type of cytochrome in its reduced (Fe^{2+}) state has three absorption bands in the visible range (Fig. 19-4b). The longest-wavelength band is near 600 nm in type a cytochromes, near 560 nm in type b, and near 550 nm in type c. To distinguish among closely related cytochromes of one type, the exact absorption maximum is sometimes used in the names, as in cytochrome b_{562}.

The hemes of a and b cytochromes are tightly, but not covalently, bound to their associated proteins; the hemes of c-type cytochromes are covalently attached through Cys residues (Fig. 19-4). As with the flavoproteins, the standard reduction potential of the heme iron atom of a cytochrome depends on its interaction with

(a)

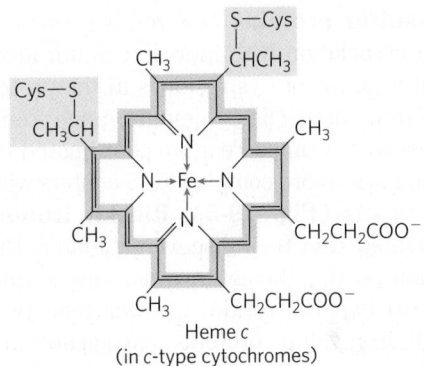

Iron protoporphyrin IX
(in b-type cytochromes)

Heme c
(in c-type cytochromes)

Heme a
(in a-type cytochromes)

(b)

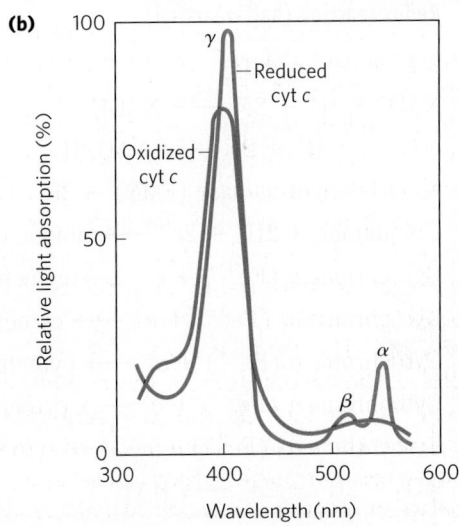

FIGURE 19-4 Prosthetic groups of cytochromes. (a) Each group consists of four five-membered, nitrogen-containing rings in a cyclic structure called a porphyrin. The four nitrogen atoms are coordinated with a central Fe ion, either Fe^{2+} or Fe^{3+}. Iron protoporphyrin IX is found in b-type cytochromes and in hemoglobin and myoglobin (see Fig. 4-17). Heme c is covalently bound to the protein of cytochrome c through thioether bonds to two Cys residues. Heme a, found in a-type cytochromes, has a long isoprenoid tail attached to one of the five-membered rings. The conjugated double-bond system (shaded light red) of the porphyrin ring has delocalized π electrons that are relatively easily excited by photons with the wavelengths of visible light, which accounts for the strong absorption by hemes (and related compounds) in the visible region of the spectrum. **(b)** Absorption spectra of cytochrome c (cyt c) in its oxidized (blue) and reduced (red) forms. The characteristic α, β, and γ bands of the reduced form are labeled.

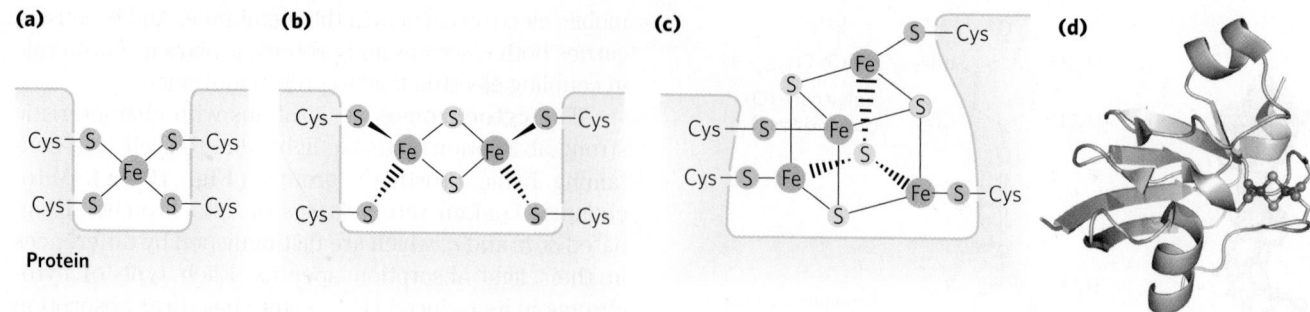

(a) **(b)** **(c)** **(d)**

Protein

FIGURE 19-5 Iron-sulfur centers. The Fe-S centers of iron-sulfur proteins may be as simple as shown in **(a)**, with a single Fe ion surrounded by the S atoms of four Cys residues; Fe is red, inorganic S is yellow, and the S of Cys is orange. Other centers include both inorganic and Cys S atoms, as in **(b)** 2Fe-2S or **(c)** 4Fe-4S centers. **(d)** The ferredoxin of the cyanobacterium *Anabaena* 7120 has one 2Fe-2S center. (Note that in these designations, only the inorganic S atoms are counted. For example, in the 2Fe-2S center (b), each Fe ion is actually surrounded by four S atoms.) The exact standard reduction potential of the iron in these centers depends on the type of center and its interaction with the associated protein. [Source: (d) PDB ID 1FRD, B. L. Jacobson et al., *Biochemistry* 32:6788, 1993.]

protein side chains and is therefore different for each cytochrome. The cytochromes of type *a* and *b* and some of type *c* are integral proteins of the inner mitochondrial membrane. One striking exception is the soluble cytochrome *c* that associates through electrostatic interactions with the outer surface of the inner membrane.

In **iron-sulfur proteins**, the iron is present not in heme but in association with inorganic sulfur atoms or with the sulfur atoms of Cys residues in the protein, or both. These iron-sulfur (Fe-S) centers range from simple structures with a single Fe atom coordinated to four Cys —SH groups to more complex Fe-S centers with two or four Fe atoms **(Fig. 19-5)**. **Rieske iron-sulfur proteins** (named after their discoverer, John S. Rieske) are a variation on this theme, in which one Fe atom is coordinated to two His residues rather than two Cys residues. All iron-sulfur proteins participate in one-electron transfers in which one iron atom of the Fe-S cluster is oxidized or reduced. At least eight Fe-S proteins function in mitochondrial electron transfer. The reduction potential of Fe-S proteins varies from -0.65 V to $+0.45$ V, depending on the microenvironment of the iron within the protein.

In the overall reaction catalyzed by the mitochondrial respiratory chain, electrons move from NADH, succinate, or some other primary electron donor through flavoproteins, ubiquinone, iron-sulfur proteins, and cytochromes, and finally to O_2. A look at the methods used to determine the sequence in which the carriers act is instructive, as the same general approaches have been used to study other electron-transfer chains, such as those of chloroplasts (see Fig. 20-16).

First, the standard reduction potentials of the individual electron carriers have been determined experimentally (Table 19-2). We would expect the carriers to function in order of increasing reduction potential, because electrons tend to flow spontaneously from carriers of lower E'° to carriers of higher E'°. The order of carriers deduced by this method is NADH $\rightarrow$ Q $\rightarrow$ cytochrome *b* $\rightarrow$ cytochrome c_1 $\rightarrow$ cytochrome *c* $\rightarrow$

TABLE 19-2 Standard Reduction Potentials of Respiratory Chain and Related Electron Carriers	
Redox reaction (half-reaction)	**E'° (V)**
$2H^+ + 2e^- \longrightarrow H_2$	-0.414
$NAD^+ + H^+ + 2e^- \longrightarrow NADH$	-0.320
$NADP^+ + H^+ + 2e^- \longrightarrow NADPH$	-0.324
NADH dehydrogenase (FMN) $+ 2H^+ + 2e^- \longrightarrow$ NADH dehydrogenase $(FMNH_2)$	-0.30
Ubiquinone $+ 2H^+ + 2e^- \longrightarrow$ ubiquinol	0.045
Cytochrome *b* $(Fe^{3+}) + e^- \longrightarrow$ cytochrome *b* (Fe^{2+})	0.077
Cytochrome c_1 $(Fe^{3+}) + e^- \longrightarrow$ cytochrome c_1 (Fe^{2+})	0.22
Cytochrome *c* $(Fe^{3+}) + e^- \longrightarrow$ cytochrome *c* (Fe^{2+})	0.254
Cytochrome *a* $(Fe^{3+}) + e^- \longrightarrow$ cytochrome *a* (Fe^{2+})	0.29
Cytochrome a_3 $(Fe^{3+}) + e^- \longrightarrow$ cytochrome a_3 (Fe^{2+})	0.35
$\frac{1}{2}O_2 + 2H^+ + 2e^- \longrightarrow H_2O$	0.817

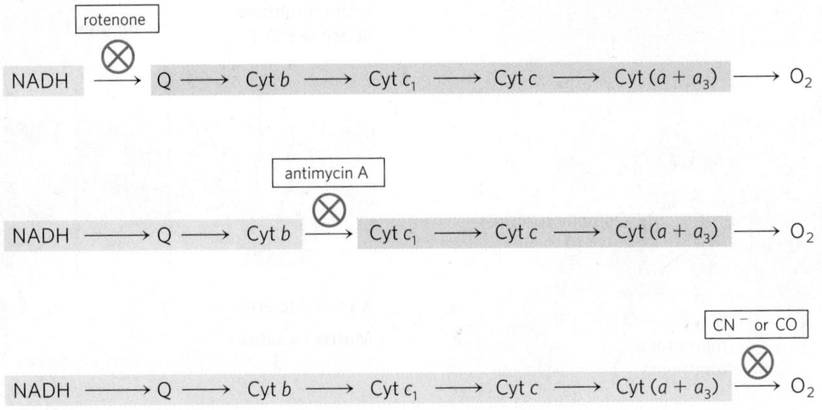

FIGURE 19-6 Method for determining the sequence of electron carriers. This method measures the effects of inhibitors of electron transfer on the oxidation state of each carrier. In the presence of an electron donor and O_2, each inhibitor causes a characteristic pattern of oxidized/reduced carriers: those before the block become reduced (blue), and those after the block become oxidized (light red).

cytochrome $a \rightarrow$ cytochrome $a_3 \rightarrow O_2$. Note, however, that the order of standard reduction potentials is not necessarily the same as the order of *actual* reduction potentials under cellular conditions, which depend on the concentrations of reduced and oxidized forms (see Eqn 13-5, p. 520). A second method for determining the sequence of electron carriers involves reducing the entire chain of carriers experimentally by providing an electron source but no electron acceptor (no O_2). When O_2 is suddenly introduced into the system, the rate at which each electron carrier becomes oxidized, measured spectroscopically, reveals the order in which the carriers function. The carrier nearest O_2 (at the end of the chain) gives up its electrons first, the second carrier from the end is oxidized next, and so on. Such experiments have confirmed the sequence deduced from standard reduction potentials.

In a final confirmation, agents that inhibit the flow of electrons through the chain have been used in combination with measurements of the degree of oxidation of each carrier. In the presence of O_2 and an electron donor, carriers that function before the inhibited step become fully reduced, and those that function after this step are completely oxidized **(Fig. 19-6)**. By using several inhibitors that block different steps in the chain,

investigators have determined the entire sequence; it is the same as deduced in the first two approaches.

Electron Carriers Function in Multienzyme Complexes

The electron carriers of the respiratory chain are organized into membrane-embedded supramolecular complexes that can be physically separated. Gentle treatment of the inner mitochondrial membrane with detergents allows the resolution of four unique electron-carrier complexes, each capable of catalyzing electron transfer through a portion of the chain (Table 19-3; **Fig. 19-7**). Complexes I and II catalyze electron transfer to ubiquinone from two different electron donors: NADH (Complex I) and succinate (Complex II). Complex III carries electrons from reduced ubiquinone to cytochrome c, and Complex IV completes the sequence by transferring electrons from cytochrome c to O_2.

We now look in more detail at the structure and function of each complex of the mitochondrial respiratory chain.

Complex I: NADH to Ubiquinone In mammals, **Complex I**, also called **NADH:ubiquinone oxidoreductase** or **NADH dehydrogenase**, is a large enzyme composed of

TABLE 19-3	The Protein Components of the Mitochondrial Respiratory Chain		
Enzyme complex/protein	Mass (kDa)	Number of subunits[a]	Prosthetic group(s)
I NADH dehydrogenase	850	45 (14)	FMN, Fe-S
II Succinate dehydrogenase	140	4	FAD, Fe-S
III Ubiquinone: cytochrome c oxidoreductase[b]	250	11	Hemes, Fe-S
Cytochrome c[c]	13	1	Heme
IV Cytochrome oxidase[b]	204	13 (3–4)	Hemes; Cu_A, Cu_B

[a]Number of subunits in the bacterial complexes in parentheses.

[b]Mass and subunit data are for the monomeric form.

[c]Cytochrome c is not part of an enzyme complex; it moves between Complexes III and IV as a freely soluble protein.

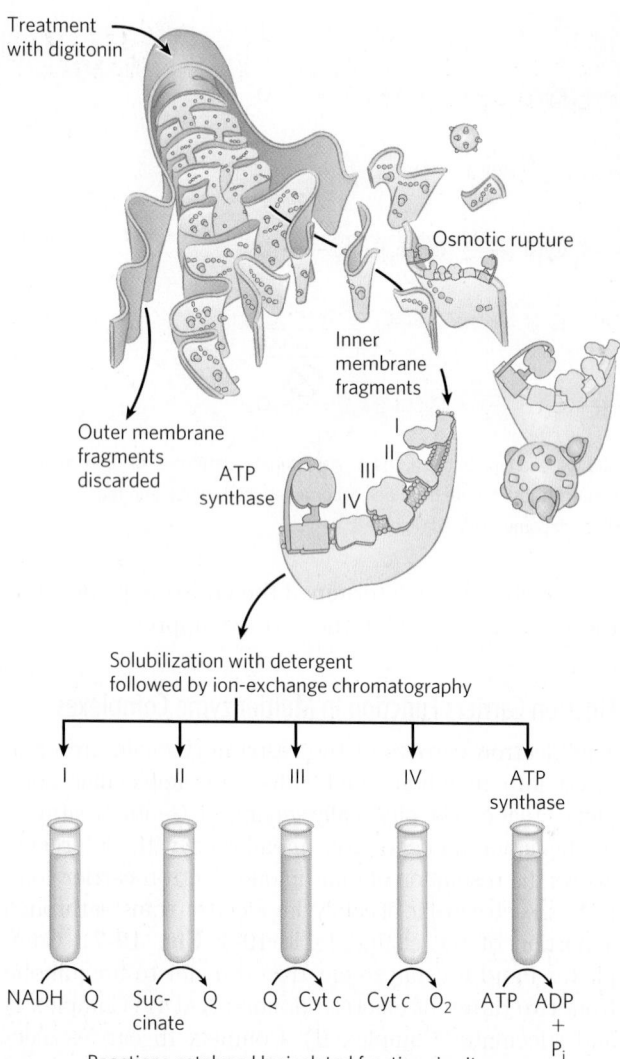

Solubilization with detergent
followed by ion-exchange chromatography

I | II | III | IV | ATP synthase

NADH → Q   Suc- → Q   Q → Cyt c   Cyt c → O₂   ATP → ADP + Pᵢ
cinate

Reactions catalyzed by isolated fractions in vitro

FIGURE 19-7 Separation of functional complexes of the respiratory chain. The outer mitochondrial membrane is first removed by treatment with the detergent digitonin. Fragments of inner membrane are then obtained by osmotic rupture of the membrane, and the fragments are gently dissolved in a second detergent. The resulting mixture of inner membrane proteins is resolved by ion-exchange chromatography into several complexes (I through IV) of the respiratory chain, each with its unique protein composition (see Table 19-3), and the enzyme ATP synthase (sometimes called Complex V). The isolated Complexes I through IV catalyze electron transfers between donors (NADH and succinate), intermediate carriers (Q and cytochrome c), and O₂, as shown. In vitro, isolated ATP synthase has only ATP-hydrolyzing (ATPase), not ATP-synthesizing, activity.

45 different polypeptide chains, including an FMN-containing flavoprotein and at least 8 iron-sulfur centers. Complex I is L-shaped, with one arm embedded in the inner membrane and the other extending into the matrix. Comparative studies of Complex I in bacteria and other organisms show that 7 polypeptides in the membrane arm and 7 in the matrix arm are conserved and essential (**Fig. 19-8**).

Complex I catalyzes two simultaneous and obligately coupled processes: (1) the exergonic transfer to

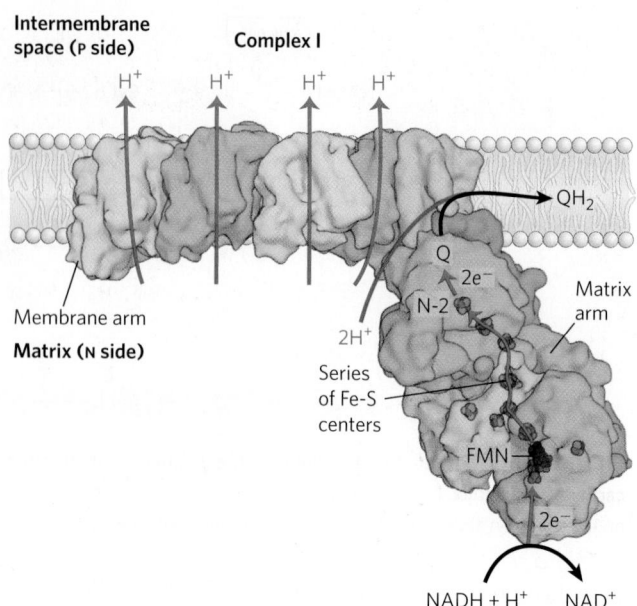

FIGURE 19-8 Structure of Complex I (NADH:ubiquinone oxidoreductase). Complex I (shown here is the crystal structure from the bacterium *Thermus thermophilus*) catalyzes the transfer of a hydride ion from NADH to FMN, and from the FMN, two electrons pass through a series of Fe-S centers to the Fe-S center N-2 in the matrix arm of the complex. Electron transfer from N-2 to ubiquinone on the membrane arm forms QH₂, which diffuses into the lipid bilayer. This electron transfer also drives expulsion from the matrix of four protons per pair of electrons. Proton flux produces an electrochemical potential across the inner mitochondrial membrane (N side negative, P side positive). The detailed mechanism that couples electron and proton transfer in Complex I is not yet known, but three of the membrane subunits are structurally related to a known Na⁺-H⁺ antiporter, and the path of proton movement may be similar in both cases. The fourth putative proton pathway is through an integral subunit closer to the Q-binding site. A long helix (not visible in this view) lying along the surface of the membrane arm may coordinate the action of all four proton pumps when Q is reduced. [Source: PDB ID 4HEA, R. Baradaran et al., *Nature* 494:443, 2013.]

ubiquinone of a hydride ion from NADH and a proton from the matrix, expressed by

$$\text{NADH} + \text{H}^+ + \text{Q} \longrightarrow \text{NAD}^+ + \text{QH}_2 \quad (19\text{-}1)$$

and (2) the endergonic transfer of four protons from the matrix to the intermembrane space. Protons are moved against a transmembrane proton gradient in this process. Complex I is therefore a proton pump driven by the energy of electron transfer, and the reaction it catalyzes is **vectorial**: it moves protons in a specific direction from one location (the matrix, which becomes negatively charged with the departure of protons) to another (the intermembrane space, which becomes positively charged). To emphasize the vectorial nature of the process, the overall reaction is often written with subscripts that indicate the location of the protons: P for the positive side of the inner membrane (the intermembrane space), N for the negative side (the matrix):

$$\text{NADH} + 5\text{H}_N^+ + \text{Q} \longrightarrow \text{NAD}^+ + \text{QH}_2 + 4\text{H}_P^+ \quad (19\text{-}2)$$

TABLE 19-4 Agents That Interfere with Oxidative Phosphorylation

Type of interference	Compound[a]	Target/mode of action
Inhibition of electron transfer	Cyanide Carbon monoxide	Inhibit cytochrome oxidase
	Antimycin A	Blocks electron transfer from cytochrome b to cytochrome c_1
	Myxothiazol Rotenone Amytal Piericidin A	Prevent electron transfer from Fe-S center to ubiquinone
Inhibition of ATP synthase	Aurovertin	Inhibit F_1
	Oligomycin Venturicidin	Inhibit F_o
	DCCD	Blocks proton flow through F_o
Uncoupling of phosphorylation from electron transfer	FCCP DNP	Hydrophobic proton carriers
	Valinomycin	K^+ ionophore
	Uncoupling protein 1	In brown adipose tissue, forms proton-conducting pores in inner mitochondrial membrane
Inhibition of ATP-ADP exchange	Atractyloside	Inhibits adenine nucleotide translocase

[a]DCCD, dicyclohexylcarbodiimide; FCCP, cyanide-p-trifluoromethoxyphenylhydrazone; DNP, 2,4-dinitrophenol.

Amytal (a barbiturate drug), rotenone (a plant product commonly used as an insecticide), and piericidin A (an antibiotic) inhibit electron flow from the Fe-S centers of Complex I to ubiquinone (Table 19-4) and therefore block the overall process of oxidative phosphorylation.

Three of the seven integral protein subunits of the membrane arm are related to a Na^+-H^+ antiporter and are believed to be responsible for pumping three protons; a fourth subunit in the membrane arm, that nearest the Q-binding site, is probably responsible for pumping the fourth proton (Fig. 19-8).

How is the reduction of ubiquinone coupled to proton pumping? Reduction of Q occurs far away from the membrane arm of the protein, where proton pumping occurs, so the coupling is clearly indirect. Presumably, reduction of Q induces a long-range conformational change in Complex I. The high-resolution view of Complex I from crystallographic studies suggests one possible coupling mechanism: a long helix that lies along the membrane arm may transmit allosteric changes throughout that arm. It seems likely that all four protons are pumped simultaneously, so that the energy from a strongly exergonic reaction (Q reduction) is broken into smaller packets, a common strategy employed by living organisms.

Complex II: Succinate to Ubiquinone We encountered **Complex II** in Chapter 16 as **succinate dehydrogenase**, the only membrane-bound enzyme in the citric acid cycle (p. 632). Complex II couples the oxidation of succinate at one site with the reduction of ubiquinone at another site about 40 Å away. Although smaller and simpler than Complex I, Complex II contains five prosthetic groups of two types and four different protein subunits **(Fig. 19-9)**. Subunits C and D are integral membrane proteins, each with three transmembrane helices. They contain a heme group, heme b, and a binding site for Q, the final electron acceptor in the reaction catalyzed by Complex II. Subunits A and B extend into the matrix; they contain three 2Fe-2S centers, bound FAD, and a binding site for the substrate, succinate. Although the overall path of electron transfer is long (from the succinate-binding site to FAD, then through the Fe-S centers to the Q-binding site), none of the individual electron-transfer distances exceeds about 11 Å—a reasonable distance for rapid electron transfer (Fig. 19-9). Electron transfer through Complex II is not accompanied by proton pumping across the inner membrane, although the QH_2 produced by succinate oxidation will be used by Complex III to drive proton transfer. Because Complex II functions in the citric acid cycle, factors that affect its activity (such as the availability of oxidized Q) probably serve to coordinate that cycle with mitochondrial electron transfer.

The heme b of Complex II is apparently not in the direct path of electron transfer; it may serve instead to reduce the frequency with which electrons "leak" out of the system, moving from succinate to molecular oxygen to produce the **reactive oxygen species (ROS)** hydrogen peroxide (H_2O_2) and the **superoxide radical** ($\cdot O_2^-$), as described below. Some individuals with point mutations in Complex II subunits near heme b or the ubiquinone-binding site suffer from hereditary

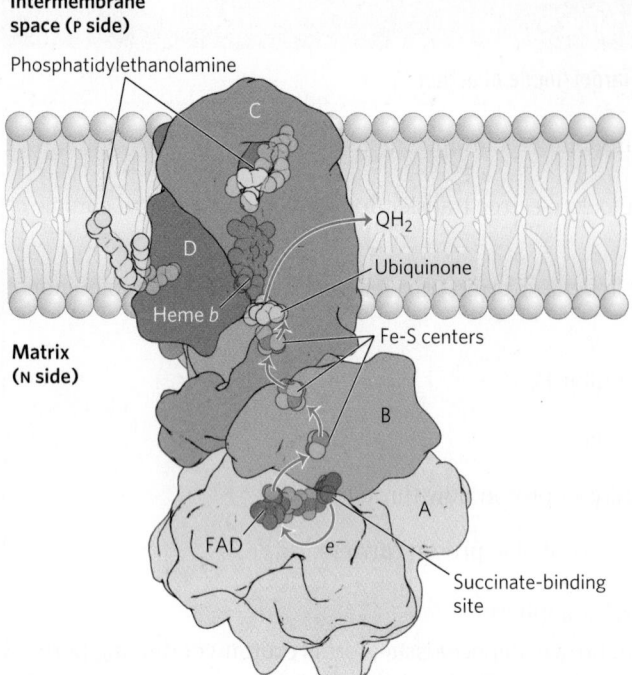

FIGURE 19-9 Structure of Complex II (succinate dehydrogenase). This complex (porcine) has two transmembrane subunits, C and D; subunits A and B extend into the matrix. Just behind the FAD in subunit A is the binding site for succinate. Subunit B has three Fe-S centers, ubiquinone is bound to subunit B, and heme b is sandwiched between subunits C and D. Two phosphatidylethanolamine molecules are so tightly bound to subunit D that they show up in the crystal structure. Electrons move (blue arrows) from succinate to FAD, then through the three Fe-S centers to ubiquinone. The heme b is not on the main path of electron transfer but protects against the formation of reactive oxygen species (ROS) by electrons that go astray. [Source: PDB ID 1ZOY, F. Sun et al., *Cell* 121:1043, 2005.]

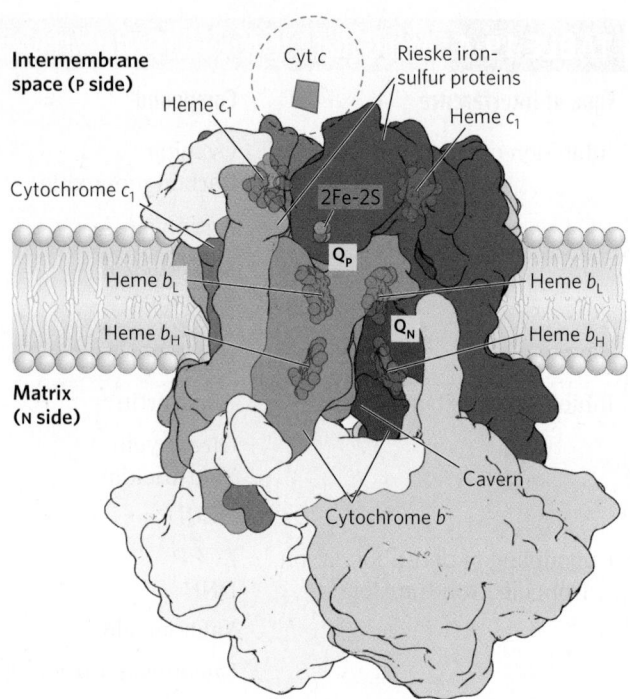

FIGURE 19-10 Structure of Complex III (cytochrome bc_1 complex). The complex (bovine) is a dimer of identical monomers, each with 11 different subunits. The functional core of each monomer consists of three subunits: cytochrome b (green), with its two hemes (b_H and b_L); the Rieske iron-sulfur protein (purple), with its 2Fe-2S centers; and cytochrome c_1 (blue), with its heme. This cartoon illustration shows how cytochrome c_1 and the Rieske iron-sulfur protein project from the P surface and can interact with cytochrome c (not part of the functional complex) in the intermembrane space. The complex has two distinct binding sites for ubiquinone, Q_N and Q_P, which correspond to the sites of inhibition by two drugs that block oxidative phosphorylation. Antimycin A, which blocks electron flow from cytochrome b to cytochrome c_1, specifically from heme b_H to Q, binds at Q_N, close to heme b_H on the N (matrix) side of the membrane. Myxothiazol, which prevents electron flow from QH_2 to the Rieske iron-sulfur protein, binds at Q_P, near the 2Fe-2S center and heme b_L on the P side. The dimeric structure is essential to the function of Complex III. The interface between monomers forms two caverns, each containing a Q_P site from one monomer and a Q_N site from the other. The ubiquinone intermediates move within these sheltered caverns.

Complex III crystallizes in two distinct conformations (not shown). In one, the Rieske Fe-S center is close to its electron acceptor, the heme of cytochrome c_1, but relatively distant from cytochrome b and the QH_2-binding site at which the Rieske Fe-S center receives electrons. In the other, the Fe-S center has moved away from cytochrome c_1 and toward cytochrome b. The Rieske protein is thought to oscillate between these two conformations as it is first reduced, then oxidized. [Source: PDB ID 1BGY, S. Iwata et al., *Science* 281:64, 1998.]

paraganglioma, characterized by benign tumors of the head and neck, commonly in the carotid body, an organ that senses O_2 levels in the blood. These mutations result in greater production of ROS and perhaps greater tissue damage during succinate oxidation. Mutations that affect the succinate-binding region in Complex II may lead to degenerative changes in the central nervous system, and some mutations are associated with tumors of the adrenal medulla. ■

Complex III: Ubiquinone to Cytochrome c Electrons from reduced ubiquinone (ubiquinol, QH_2) pass through two more large protein complexes in the inner mitochondrial membrane before reaching the ultimate electron acceptor, O_2. **Complex III** (also called **cytochrome bc_1 complex** or **ubiquinone:cytochrome c oxidoreductase**) couples the transfer of electrons from ubiquinol to cytochrome c with the vectorial transport of protons from the matrix to the intermembrane space. The functional unit of Complex III **(Fig. 19-10)** is a dimer. Each monomer consists of three proteins central to the action of the complex: cytochrome b, cytochrome c_1, and the Rieske iron-sulfur protein. (Several other

proteins associated with Complex III in vertebrates are not conserved across the phyla and presumably play subsidiary roles.) The two cytochrome b monomers surround a cavern in the middle of the membrane, in which ubiquinone is free to move from the matrix side of the membrane (site Q_N on one monomer) to the intermembrane space (site Q_P on the other monomer) as it shuttles electrons and protons across the inner mitochondrial membrane.

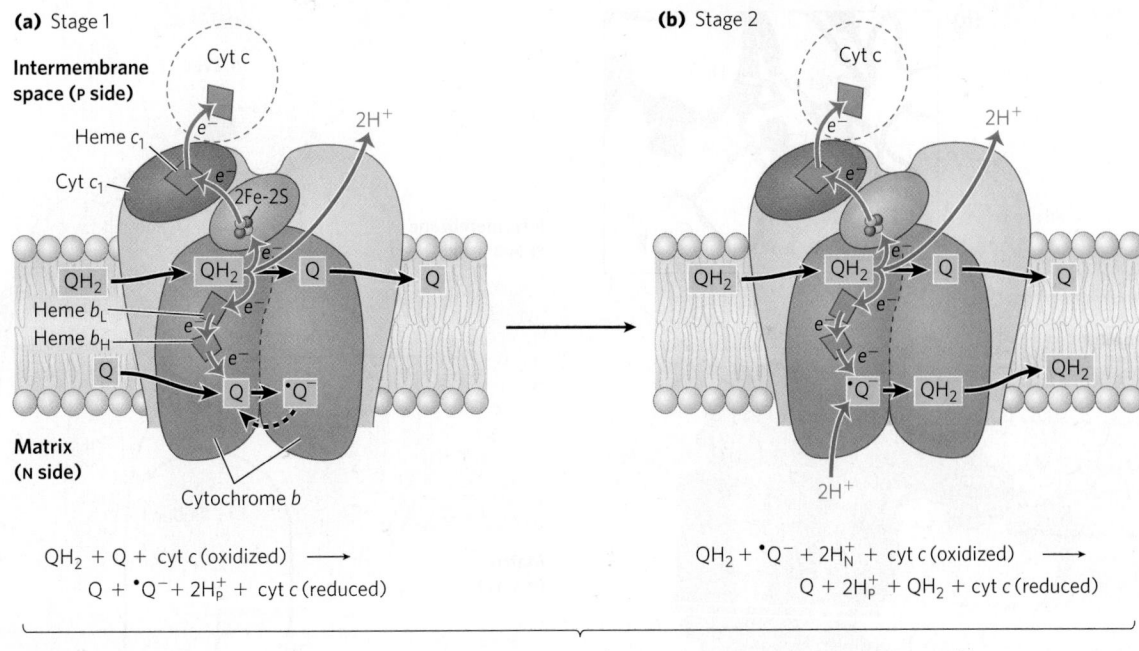

FIGURE 19-11 The Q cycle, shown in two stages. The path of electrons through Complex III is shown with blue arrows; the movement of various forms of ubiquinone, with black arrows. **(a)** In the first stage, Q on the N side is reduced to the semiquinone radical, which moves back into position to accept another electron. **(b)** In the second stage, the semiquinone radical is converted to QH_2. Meanwhile, on the P side of the membrane, two molecules of QH_2 are oxidized to Q, releasing two protons per Q molecule (four protons in all) into the intermembrane space. Each QH_2 donates one electron (via the Rieske Fe-S center) to cytochrome c_1, and one electron (via cytochrome b) to a molecule of Q near the N side, reducing it in two steps to QH_2. This reduction also consumes two protons per Q, which are taken up from the matrix (N side). Reduced cytochrome c_1 passes electrons one at a time to cytochrome c, which dissociates and carries electrons to Complex IV. In each cycle, one reduction of Q at the Q_N site is coupled with two oxidations of QH_2 at the Q_P site by consuming two protons from the matrix and releasing four protons into the intermembrane space.

To account for the role of Q in energy conservation, Mitchell proposed the "Q cycle" **(Fig. 19-11)**. As electrons move from QH_2 through Complex III, QH_2 is oxidized with the release of protons on one side of the membrane (at Q_P), while at the other site (Q_N), Q is reduced and protons are taken up. The product of one catalytic site thus becomes the substrate at the second site, and vice versa. The net equation for the redox reactions of the Q cycle is

$$QH_2 + 2\ \text{cyt}\ c\ (\text{oxidized}) + 2H_N^+ \longrightarrow$$
$$Q + 2\ \text{cyt}\ c\ (\text{reduced}) + 4H_P^+ \quad (19\text{-}3)$$

The Q cycle accommodates the switch between the two-electron carrier ubiquinol (the reduced form of ubiquinone) and the one-electron carriers—hemes b_L and b_H of cytochrome b, and cytochromes c_1 and c—and results in the uptake of two protons on the N side and the release of four protons on the P side, per pair of electrons passing through Complex III to cytochrome c. Two of the protons released on the P side are electrogenic; the other two are electroneutral, balanced by the two charges (electrons) passed to cytochrome c on the P side. Although the path of electrons through this segment of the respiratory chain is complicated, the net effect of the transfer is simple: QH_2 is oxidized to Q, two molecules of cytochrome c are reduced, and two protons are moved from the P side to the N side of the inner mitochondrial membrane.

Cytochrome c is a soluble protein of the intermembrane space, which associates reversibly with the P side of the inner membrane. After its single heme accepts an electron from Complex III, cytochrome c moves in the intermembrane space to Complex IV to donate the electron to a binuclear copper center.

Complex IV: Cytochrome c to O_2 In the final step of the respiratory chain, **Complex IV**, also called **cytochrome oxidase**, carries electrons from cytochrome c to molecular oxygen, reducing it to H_2O. Complex IV is a large, dimeric enzyme of the inner mitochondrial membrane, each monomer having 13 subunits and M_r of 204,000. Bacteria contain a form that is much simpler, with only 3 or 4 subunits per monomer, but still capable of catalyzing both electron transfer and proton pumping. Comparison of the mitochondrial and bacterial complexes suggests that these 3 subunits have been conserved in evolution; in multicellular organisms, the other 10 subunits may contribute to the assembly or stability of Complex IV **(Fig. 19-12)**.

Subunit II of Complex IV contains two Cu ions complexed with the —SH groups of two Cys residues in a binuclear center (Cu_A; Fig. 19-12b) that resembles the 2Fe-2S centers of iron-sulfur proteins. Subunit I contains two heme groups, designated a and a_3, and another copper ion (Cu_B). Heme a_3 and Cu_B form a second binuclear center

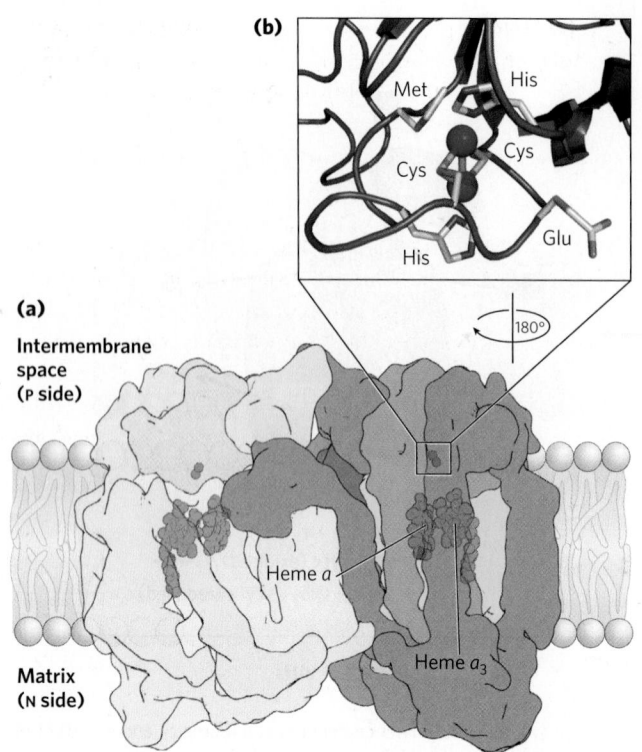

FIGURE 19-12 Structure of Complex IV (cytochrome oxidase). (a) This complex (bovine) has 13 subunits in each identical monomer of its dimeric structure. Subunit I (yellow) has two heme groups, a and a_3, near a single copper ion, Cu_B (not visible here). Heme a_3 and Cu_B form a binuclear Fe-Cu center. Subunit II (purple) contains two Cu ions complexed with the —SH groups of two Cys residues in a binuclear center, Cu_A, that resembles the 2Fe-2S centers of iron-sulfur proteins. This binuclear center and the cytochrome c-binding site are located in a domain of subunit II that protrudes from the P side of the inner membrane (into the intermembrane space). Subunit III (blue) is essential for rapid proton movement through subunit II. The roles of the other 10 subunits in mammalian Complex IV (green) are not fully understood, although some function in assembly or stabilization of the complex. **(b)** The binuclear center of Cu_A. The Cu ions (blue spheres) share electrons equally. When the center is reduced, the ions have the formal charges $Cu^{1+}Cu^{1+}$; when oxidized, $Cu^{1.5+}Cu^{1.5+}$. Six amino acid residues are ligands around the Cu ions: Glu, Met, two His, and two Cys. [Source: PDB ID 1OCC, T. Tsukihara et al., *Science* 272:1136, 1996.]

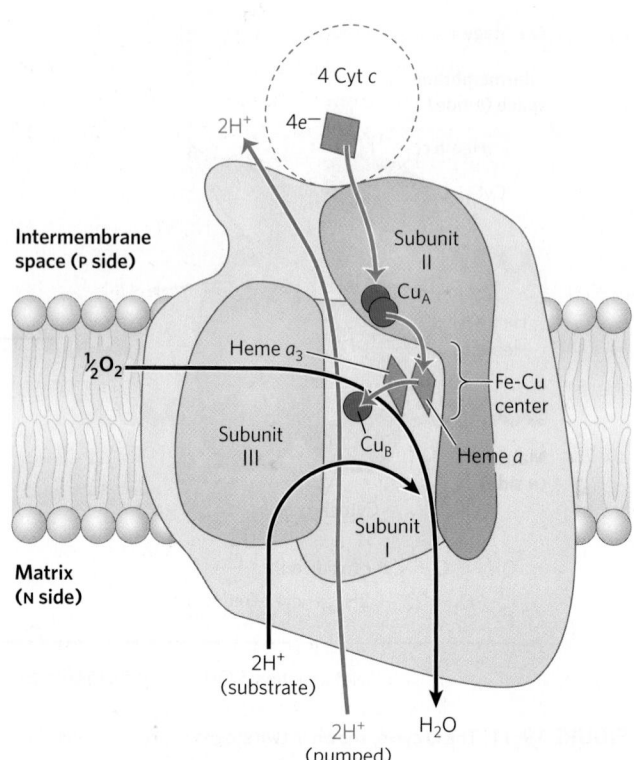

FIGURE 19-13 Path of electrons through Complex IV. For simplicity, only one monomer of the dimeric bovine Complex IV is shown. The three proteins critical to electron flow are subunits I, II, and III. The larger green structure includes the other 10 proteins in each monomer of the dimeric complex. Electron transfer through Complex IV begins with cytochrome c (top). Two molecules of reduced cytochrome c each donate an electron to the binuclear center Cu_A. From here, electrons pass through heme a to the Fe-Cu center (heme a_3 and Cu_B). Oxygen now binds to heme a_3 and is reduced to its peroxy derivative (O_2^{2-}; not shown here) by two electrons from the Fe-Cu center. Delivery of two more electrons from cytochrome c (top), for a total of four electrons, converts the O_2^{2-} to two molecules of water, with consumption of four "substrate" protons from the matrix. At the same time, two protons are pumped from the matrix for each pair of electrons passing through Complex IV, by a mechanism not fully understood. Note that reduction of O_2 to $2H_2O$ requires four electrons, or two pairs of electrons.

that accepts electrons from heme a and transfers them to O_2 bound to heme a_3. The detailed role of subunit III is not clear, but its presence is essential to Complex IV function.

Electron transfer through Complex IV is from cytochrome c to the Cu_A center, to heme a, to the heme a_3–Cu_B center, and finally to O_2 **(Fig. 19-13)**. For every two electrons passing through this complex, the enzyme consumes two "substrate" H^+ from the matrix (N side) in converting $\frac{1}{2}O_2$ to H_2O. It also uses the energy of this redox reaction to pump two protons outward into the intermembrane space (P side) for each pair of electrons that pass through, adding to the electrochemical potential produced by redox-driven proton transport through Complexes I and III. The overall reaction catalyzed by Complex IV is

$$2 \text{ cyt } c(\text{reduced}) + 4H_N^+ + \tfrac{1}{2}O_2 \longrightarrow$$
$$2 \text{ cyt } c(\text{oxidized}) + 2H_P^+ + H_2O \quad (19\text{-}4)$$

This two-electron reduction of $\frac{1}{2}O_2$ requires the oxidation of QH_2, which in turn requires oxidation of NADH or succinate.

At Complex IV, O_2 is reduced at redox centers that carry only one electron at a time. Normally the incompletely reduced oxygen intermediates remain tightly bound to the complex until completely converted to water, but a small fraction of oxygen intermediates escape. These intermediates are reactive oxygen species that can damage cellular components unless eliminated by defense mechanisms described below.

Mitochondrial Complexes Associate in Respirasomes

Although the four electron-transferring complexes can be separated in the laboratory, in the intact mitochondrion, the respiratory complexes tightly associate with each other in the inner membrane to form **respirasomes**, functional combinations of two or more different electron-transferring complexes. For example, when Complex III is gently extracted from mitochondrial membranes, it is found to be associated with Complex I and remains associated during gentle electrophoresis. Purified aggregates of Complexes III and IV can also be isolated, and when viewed using the powerful tool single-particle cryo-electron microscopy (Box 19-1), appear as regular supercomplexes of the right size and contours to accommodate the known crystal structures of both **(Fig. 19-14)**. The kinetics of electron flow through the series of respiratory complexes would be very different in the two extreme cases of tight versus no association between them: (1) if complexes were tightly associated, electron transfers would essentially occur through a solid state, and (2) if the complexes functioned separately, electrons would be carried between them by ubiquinone and cytochrome c. The kinetic evidence supports electron transfer through a solid state, and thus the respirasome model.

Cardiolipin, the lipid that is especially abundant in the inner mitochondrial membrane (see Figs 10-8 and 11-2), may be critical to the integrity of respirasomes; its removal with detergents, or its absence in certain yeast mutants, results in defective mitochondrial electron transfer and a loss of affinity between the

respiratory complexes. Some of the "auxiliary" proteins of the complexes, with no apparent role in electron transfer, may serve to hold respirasomes together.

Other Pathways Donate Electrons to the Respiratory Chain via Ubiquinone

Several other electron-transfer reactions can reduce ubiquinone in the inner mitochondrial membrane **(Fig. 19-15)**. In the first step of the β oxidation of fatty acyl–CoA, catalyzed by the flavoprotein **acyl-CoA dehydrogenase** (see Fig. 17-8), electrons pass from the substrate to the FAD of the dehydrogenase, then to electron-transferring flavoprotein (ETF). ETF passes its electrons to **ETF:ubiquinone oxidoreductase**, which reduces Q in the inner mitochondrial membrane to QH_2. Glycerol 3-phosphate, formed either from glycerol released by triacylglycerol breakdown or by the reduction of dihydroxyacetone phosphate from glycolysis, is oxidized by **glycerol 3-phosphate dehydrogenase** (see Fig. 17-4), a flavoprotein located on the outer face of the inner mitochondrial membrane. The electron acceptor in this reaction is Q; the QH_2 produced enters the pool of QH_2 in the membrane. The important role of glycerol 3-phosphate dehydrogenase in shuttling reducing equivalents from

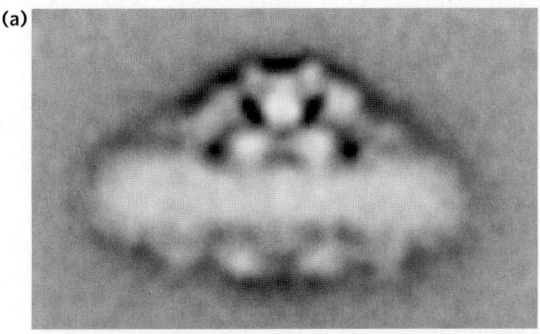

(a)

(b)

FIGURE 19-14 A respirasome composed of Complexes III and IV. **(a)** Purified supercomplexes containing Complexes III and IV (from yeast), visualized by electron microscopy after quick-freezing. The electron densities of hundreds of images were averaged to yield this composite view (see Box 19-1). **(b)** The x-ray–derived structures of one Complex III (red; from yeast) and two Complex IV (green; from bovine heart) could be fitted to the electron-density map to suggest one possible mode of interaction of these complexes in a respirasome. This view is in the plane of the bilayer (yellow). [Source: Courtesy of Egbert Boekema.]

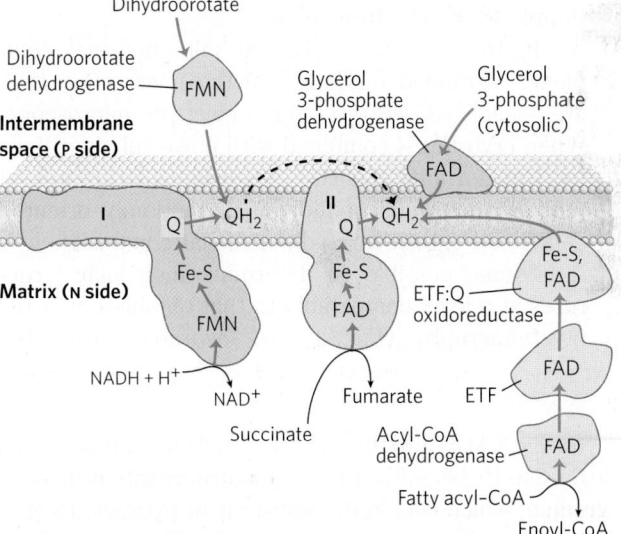

FIGURE 19-15 Paths of electron transfer to ubiquinone in the respiratory chain. Electrons from NADH in the matrix pass through the FMN of a flavoprotein (NADH dehydrogenase) to a series of Fe-S centers (in Complex I) and then to Q. Electrons from succinate oxidation in the citric acid cycle pass through a flavoprotein with several Fe-S centers (Complex II) on the way to Q. Acyl-CoA dehydrogenase, the first enzyme of fatty acid β oxidation, transfers electrons to electron-transferring flavoprotein (ETF), from which they pass to Q via ETF:ubiquinone oxidoreductase. Dihydroorotate, an intermediate in the biosynthetic pathway to pyrimidine nucleotides (see Fig. 22-38), donates two electrons to Q through a flavoprotein (dihydroorotate dehydrogenase). And glycerol 3-phosphate, an intermediate of glycolysis in the cytosol, donates electrons to a flavoprotein (glycerol 3-phosphate dehydrogenase) on the outer face of the inner mitochondrial membrane, from which they pass to Q.

BOX 19-1 METHODS Determining Three-Dimensional Structures of Large Macromolecular Complexes by Single-Particle Cryo-Electron Microscopy

Our understanding of a highly complex process such as mitochondrial oxidative phosphorylation is aided immensely by knowing the detailed molecular structures of the proteins that participate in the process. However, it is often difficult to determine the molecular structure of large, macromolecular complexes such as the respiratory chain complexes or ATP synthase. Integral membrane proteins often resist crystallization once removed from their lipid environment, so their structures cannot be solved by x-ray diffraction. In principle, discrete objects in the 100 to 300 Å diameter range can be visualized by electron microscopy (EM). In practice, the high intensity of the EM beam often damages the specimen before a high-resolution image can be obtained. In **cryo-electron microscopy**, a sample containing many individual copies of the structure of interest is quick-frozen and kept frozen while being observed in two dimensions with the electron microscope, greatly reducing damage of the specimen by the electron beam. Development of the direct electron detector, with greater sensitivity and lower noise, was also a key improvement, allowing shorter exposures of the sample to the electron beam.

Particles such as purified mitochondrial complexes, arranged randomly on the microscope grid, are visualized with the cryo-electron microscope. When cryo-EM is combined with powerful algorithms for transforming the two-dimensional structures of tens of thousands of individual, randomly oriented complexes into a three-dimensional composite, it is sometimes possible to determine molecular structures at a level comparable to that obtained by x-ray crystallography (Fig. 1). In favorable cases, the repetitive aspects can be automated, such as choice of objects to be included in the analysis, imaging of

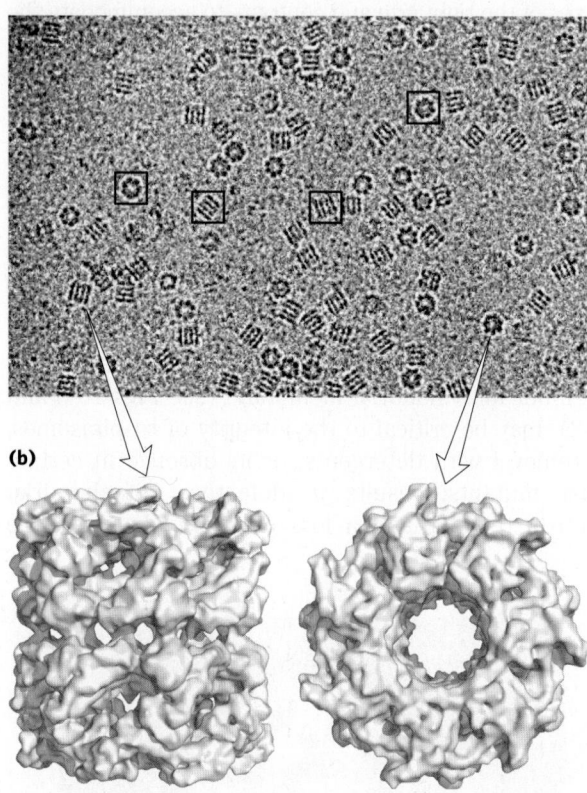

FIGURE 1 Structure of the chaperone protein GroEL as determined by single-particle cryo-EM. **(a)** EM images of many individual GroEL particles. **(b)** Side and top views of the three-dimensional structure derived from analysis of the EM images. [Sources: (a) © Alberto Bartesaghi, PhD. (b) PDB ID 3E76, P. D. Kaiser et al., *Acta Crystallogr.* 65:967, 2009.]

cytosolic NADH into the mitochondrial matrix is described in Section 19.2 (see Fig. 19-32). **Dihydroorotate dehydrogenase**, which acts in the synthesis of pyrimidines (see Fig. 22-38), is also on the outside of the inner mitochondrial membrane and donates electrons to Q in the respiratory chain. The reduced QH_2 passes its electrons through Complex III and ultimately to O_2.

The Energy of Electron Transfer Is Efficiently Conserved in a Proton Gradient

The transfer of two electrons from NADH through the respiratory chain to molecular oxygen can be summarized as

$$NADH + H^+ + \tfrac{1}{2}O_2 \longrightarrow NAD^+ + H_2O \quad (19\text{-}5)$$

This net reaction is highly exergonic. For the redox pair $NAD^+/NADH$, E'° is -0.320 V, and for the pair O_2/H_2O, E'° is 0.816 V. The $\Delta E'^\circ$ for this reaction is therefore 1.14 V, and the standard free-energy change (see Eqn 13-7, p. 521) is

$$
\begin{aligned}
\Delta G'^\circ &= -nF\,\Delta E'^\circ \quad (19\text{-}6)\\
&= -2(96.5 \text{ kJ/V·mol})(1.14 \text{ V})\\
&= -220 \text{ kJ/mol (of NADH)}
\end{aligned}
$$

This *standard* free-energy change is based on the assumption of equal concentrations (1 M) of NADH and NAD^+. In actively respiring mitochondria, the actions of many dehydrogenases keep the actual $[NADH]/[NAD^+]$ ratio well above unity, and the real free-energy change for the reaction shown in Equation 19-5 is therefore substantially greater (more negative) than -220 kJ/mol. A similar calculation for the oxidation of succinate shows that electron transfer from succinate (E'° for fumarate/succinate = 0.031 V) to O_2 has a smaller, but

each object individually, and calculations to produce a three-dimensional structure from the huge number of two-dimensional images. If the structures of individual protein components of a large complex are known from x-ray crystallography, (see Fig. 19-25), atomic models can be fitted into the outlines of the complex to confirm the cryo-EM structure (Fig. 2). The EMDataBank (emdatabank.org) is a unified resource for accessing EM structure maps deposited into data banks and assigned EMD accession codes.

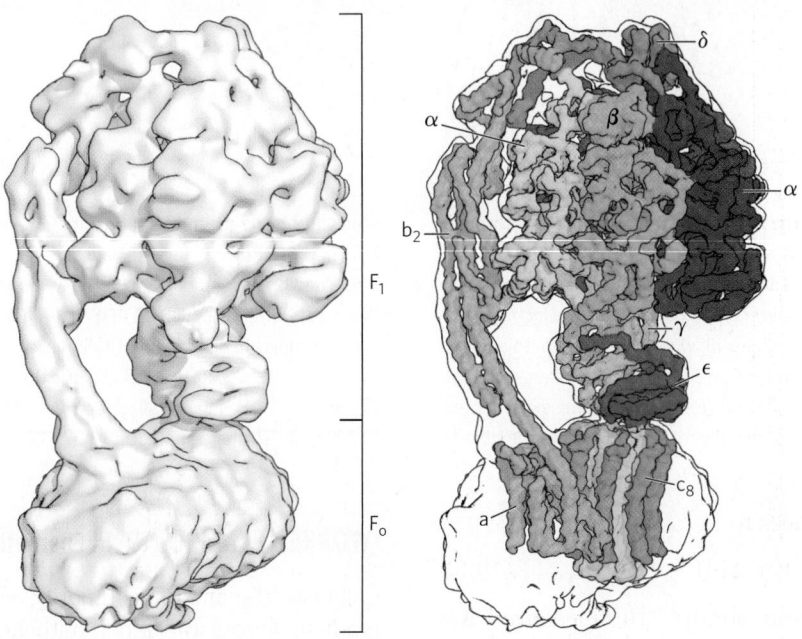

FIGURE 2 (a) Three-dimensional structure of bovine ATP synthase obtained by single-particle cryo-EM. This method revealed more of the structure of the a and b subunits in the membrane-embedded portion of the complex than was available from x-ray crystallography. The structure on the lower left includes subunit a and an extension of subunit b, as well as several other small proteins associated with the F_o structure but not visible in crystallographic structures. **(b)** An atomic model consistent with x-ray crystal structures can be fit to the envelope derived from cryo-EM, yielding a much improved picture of the whole ATP synthase structure. [Sources: (a) EMD-3164 and (b) PDB ID 5ARA, A. Zhou et al., *eLife* 4:e10180, 2015.]

still negative, standard free-energy change of about -150 kJ/mol.

Much of this energy is used to pump protons out of the matrix. For each pair of electrons transferred to O_2, four protons are pumped out by Complex I, four by Complex III, and two by Complex IV **(Fig. 19-16)**. The *vectorial* equation for the process is therefore

$$\text{NADH} + 11\text{H}_\text{N}^+ + \tfrac{1}{2}\text{O}_2 \longrightarrow$$
$$\text{NAD}^+ + 10\text{H}_\text{P}^+ + \text{H}_2\text{O} \quad (19\text{-}7)$$

The electrochemical energy inherent in this difference in proton concentration and separation of charge represents a temporary conservation of much of the energy of electron transfer. The energy stored in such a gradient, termed the **proton-motive force**, has two components: (1) the *chemical potential energy* due to the difference in concentration of a chemical species (H^+) in the two regions separated by the membrane, and (2) the *electrical potential energy* that results from the separation of charge when a proton moves across the membrane without a counterion **(Fig. 19-17)**.

As we showed in Chapter 11, the free-energy change for the creation of an electrochemical gradient by an ion pump is

$$\Delta G = RT \ln(C_2/C_1) + ZF\,\Delta\psi \quad (19\text{-}8)$$

where C_2 and C_1 are the concentrations of an ion in two regions, and $C_2 > C_1$; Z is the absolute value of its electrical charge (1 for a proton); and $\Delta\psi$ is the transmembrane difference in electrical potential, measured in volts.

For protons,

$$\ln(C_2/C_1) = 2.3(\log[\text{H}^+]_\text{P} - \log[\text{H}^+]_\text{N})$$
$$= 2.3(\text{pH}_\text{N} - \text{pH}_\text{P}) = 2.3\,\Delta\text{pH}$$

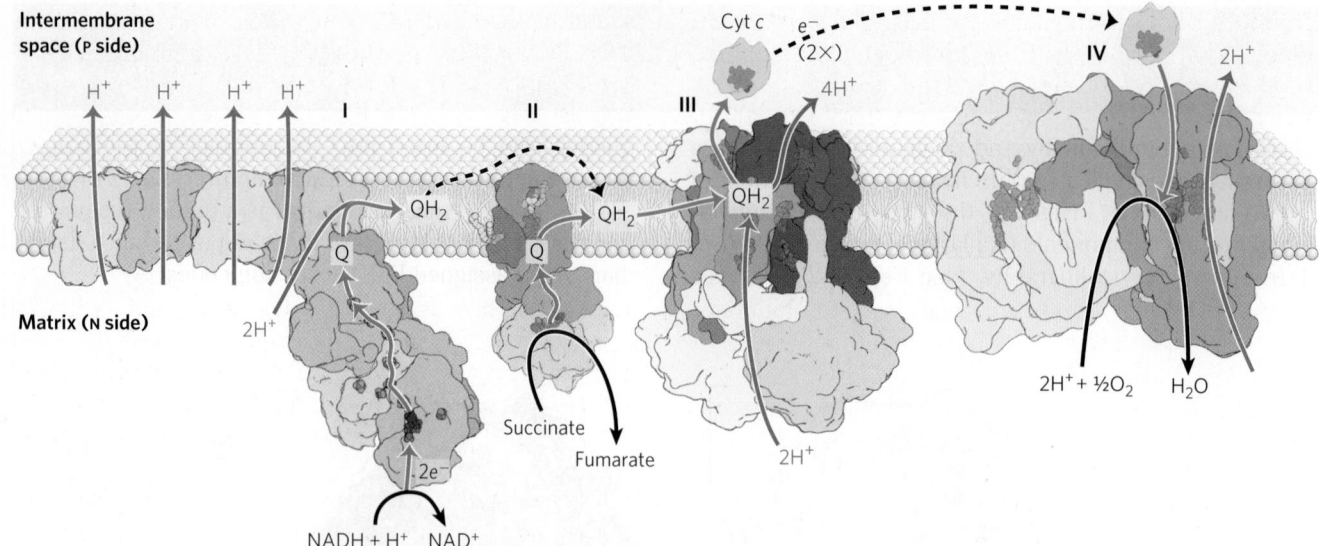

FIGURE 19-16 Summary of the flow of electrons and protons through the four complexes of the respiratory chain. Electrons reach Q through Complexes I and II (as well as through several other paths shown in Fig. 19-15). Reduced Q (QH_2) serves as a mobile carrier of electrons and protons. It passes electrons to Complex III, which passes them to another mobile connecting link, cytochrome c. Complex IV then transfers electrons from reduced cytochrome c to O_2. Electron flow through Complexes I, III, and IV is accompanied by proton efflux from the matrix into the intermembrane space. In bovine heart, the approximate ratios of Complexes I:II:III:IV are 1.1:1.3:3.0:6.7. Broken lines indicate the diffusion of Q in the plane of the inner membrane, and of cytochrome c through the intermembrane space. [Sources: Complex I: PDB ID 4HEA, R. Baradaran et al., *Nature* 494:443, 2013. Complex II: PDB ID 1ZOY, F. Sun et al., *Cell* 121:1043, 2005. Complex III: PDB ID 1BGY, S. Iwata et al., *Science* 281:64, 1998. Cytochrome c: PDB ID 1HRC, G. W. Bushnell et al., *J. Mol. Biol.* 214:585, 1990. Complex IV: PDB ID 1OCC, T. Tsukihara et al., *Science* 272:1136, 1996.]

and Equation 19-8 reduces to

$$\Delta G = 2.3RT\,\Delta pH + F\Delta\psi \qquad (19\text{-}9)$$

In actively respiring mitochondria, the measured $\Delta\psi$ is 0.15 to 0.20 V, and the pH of the matrix is about 0.75 units more alkaline than that of the intermembrane space.

P side
$[H^+]_P = C_2$

Proton pump

N side
$[H^+]_N = C_1$

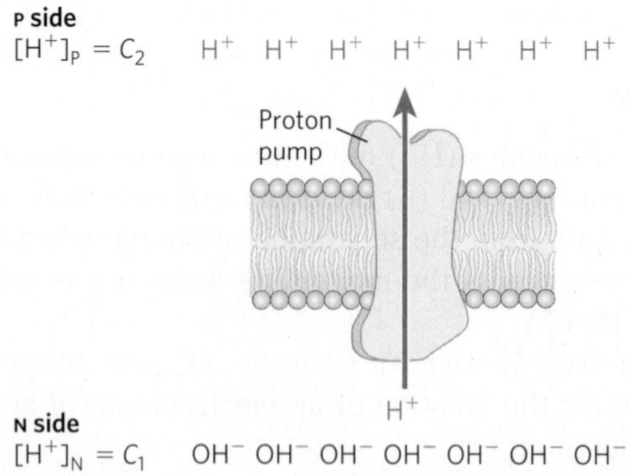

$$\Delta G = RT\ln(C_2/C_1) + ZF\Delta\psi$$
$$= 2.3RT\,\Delta pH + F\Delta\psi$$

FIGURE 19-17 Proton-motive force. The inner mitochondrial membrane separates two compartments of different $[H^+]$, resulting in differences in chemical concentration (ΔpH) and charge distribution ($\Delta\psi$) across the membrane. The net effect is the proton-motive force (ΔG), which can be calculated as shown here. This is explained more fully in the text.

WORKED EXAMPLE 19-1 Energetics of Electron Transfer

Calculate the amount of energy conserved in the proton gradient across the inner mitochondrial membrane per pair of electrons transferred through the respiratory chain from NADH to oxygen. Assume $\Delta\psi$ is 0.15 V and the pH difference is 0.75 unit at body temperature of 37 °C.

Solution: Equation 19-9 gives the free-energy change when *one* mole of protons moves across the inner membrane. Substituting the values of the constants R and F, 310 K for T, and the measured values for ΔpH (0.75 unit) and $\Delta\psi$ (0.15 V) in this equation gives $\Delta G = 19$ kJ/mol (of protons). Because the transfer of two electrons from NADH to O_2 is accompanied by the outward pumping of 10 protons (Eqn 19-7), roughly 190 kJ (of the 220 kJ released by oxidation of 1 mol of NADH) is conserved in the proton gradient.

When protons flow spontaneously *down* their electrochemical gradient, energy is made available to do work. In mitochondria, chloroplasts, and aerobic bacteria, the electrochemical energy in the proton gradient drives the synthesis of ATP from ADP and P_i. We return to the energetics and stoichiometry of ATP synthesis driven by the electrochemical potential of the proton gradient in Section 19.2.

Reactive Oxygen Species Are Generated during Oxidative Phosphorylation

Several steps in the path of oxygen reduction in mitochondria have the potential to produce highly reactive free radicals that can damage cells. The passage of electrons from QH_2 to Complex III and the passage of electrons from Complexes I and II to QH_2 involve the radical $\cdot Q^-$ as an intermediate. The $\cdot Q^-$ radical can, with a low probability, pass an electron to O_2 in the reaction

$$O_2 + e^- \longrightarrow \cdot O_2^-$$

The superoxide free radical thus generated is highly reactive; its formation also leads to production of the even more reactive hydroxyl free radical, $\cdot OH$ **(Fig. 19-18)**.

Reactive oxygen species can wreak havoc, reacting with and damaging enzymes, membrane lipids, and nucleic acids. In actively respiring mitochondria, 0.2% to as much as 2% of the O_2 used in respiration forms $\cdot O_2^-$—more than enough to have lethal effects unless the free radical is quickly disposed of. Factors that slow the flow of electrons through the respiratory chain increase the formation of superoxide, perhaps by prolonging the lifetime of $\cdot O_2^-$ generated in the Q cycle. The formation of ROS is favored when two

conditions are met: (1) mitochondria are not making ATP (for lack of ADP or O_2) and therefore have a large proton-motive force and a high QH_2/Q ratio, and (2) there is a high $NADH/NAD^+$ ratio in the matrix. In these situations, the mitochondrion is under oxidative stress—more electrons are available to enter the respiratory chain than can be immediately passed through to oxygen. When the supply of electron donors (NADH) is matched with that of electron acceptors, there is less oxidative stress, and ROS production is much reduced. Although overproduction of ROS is clearly detrimental, *low* levels of ROS are used by the cell as a signal reflecting the insufficient supply of oxygen (hypoxia), triggering metabolic adjustments (see Fig. 19-34).

To prevent oxidative damage by $\cdot O_2^-$, cells have the enzyme **superoxide dismutase**, which catalyzes the reaction

$$2 \cdot O_2^- + 2H^+ \longrightarrow H_2O_2 + O_2$$

The hydrogen peroxide (H_2O_2) thus generated is rendered harmless by **glutathione peroxidase** (Fig. 19-18). Glutathione reductase recycles the oxidized glutathione to its reduced form, using electrons from the NADPH generated by nicotinamide nucleotide transhydrogenase (in the mitochondrion) or by the pentose phosphate pathway (in the cytosol; see Fig. 14-21). Reduced glutathione also serves to keep protein sulfhydryl groups in their reduced state, preventing some of the deleterious effects of oxidative stress.

Plant Mitochondria Have Alternative Mechanisms for Oxidizing NADH

Plant mitochondria supply the cell with ATP during periods of low illumination or darkness by mechanisms entirely analogous to those used by nonphotosynthetic organisms. In the light, the principal source of mitochondrial NADH is a reaction in which glycine, produced by a process known as photorespiration, is converted to serine (see Fig. 20-48):

2 Glycine + NAD^+ $\longrightarrow$

serine + CO_2 + NH_3 + NADH + H^+

For reasons discussed in Chapter 20, plants must carry out this reaction even when they do not need NADH for ATP production. To regenerate NAD^+ from unneeded NADH, mitochondria of plants (and of some fungi and protists) transfer electrons from NADH directly to ubiquinone and from ubiquinone directly to O_2, bypassing Complexes III and IV and their proton pumps. In this process the energy in NADH is dissipated as heat, which can sometimes be of value to the plant (Box 19-2). *Cyanide-resistant* NADH oxidation is the hallmark of this unique plant electron-transfer pathway; unlike cytochrome oxidase (Complex IV), the alternative QH_2 oxidase is not inhibited by cyanide.

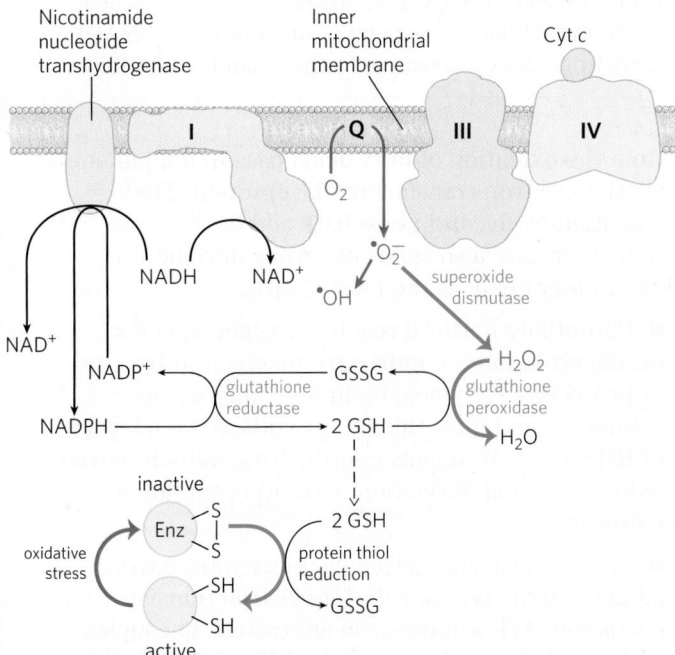

FIGURE 19-18 ROS formation in mitochondria and mitochondrial defenses. When the rate of electron entry into the respiratory chain and the rate of electron transfer through the chain are mismatched, superoxide radical ($\cdot O_2^-$) production increases at Complexes I and III as the partially reduced ubiquinone radical ($\cdot Q^-$) donates an electron to O_2. Superoxide acts on aconitase, a 4Fe-4S protein (not shown), to release Fe^{2+}. In the presence of Fe^{2+}, the Fenton reaction leads to formation of the highly reactive hydroxyl free radical ($\cdot OH$). The reactions shown in blue defend the cell against the damaging effects of superoxide. Reduced glutathione (GSH; see Fig. 22-29) donates electrons for the reduction of H_2O_2 and of the oxidized Cys residues (—S—S—) of enzymes and other proteins; GSH is regenerated from the oxidized form (GSSG) by reduction with NADPH.

BOX 19-2 Hot, Stinking Plants and Alternative Respiratory Pathways

Many flowering plants attract insect pollinators by releasing odorant molecules that mimic an insect's natural food sources or potential egg-laying sites. Plants pollinated by flies or beetles that normally feed on or lay their eggs in dung or carrion sometimes use foul-smelling compounds to attract these insects.

One family of stinking plants is the Araceae, which includes philodendrons, arum lilies, and skunk cabbages. These plants have tiny flowers densely packed on an erect structure, the spadix, surrounded by a modified leaf, the spathe. The spadix releases odors of rotting flesh or dung. Before pollination the spadix also heats up, in some species to as much as 20 to 40 °C above ambient temperatures. Heat production (thermogenesis) helps evaporate odorant molecules for better dispersal, and because rotting flesh and dung are usually warm from the hyperactive metabolism of scavenging microbes, the heat itself might also attract insects. In the case of the eastern skunk cabbage (Fig. 1), which flowers in late winter or early spring when snow still covers the ground, thermogenesis allows the spadix to grow up through the snow.

How does a skunk cabbage heat its spadix? The mitochondria of plants, fungi, and unicellular eukaryotes

FIGURE 1 Eastern skunk cabbage. [Source: Colin Purrington.]

have respiratory chains that are essentially the same as those in animals, but they also have an alternative respiratory pathway. A QH_2 oxidase transfers electrons from the ubiquinone pool directly to oxygen, bypassing the two proton-translocating steps of Complexes III and IV (Fig. 2). Energy that might have been conserved as ATP is instead released as heat. Plant mitochondria also have an alternative NADH dehydrogenase, insensitive to the Complex I inhibitor

SUMMARY 19.1 The Mitochondrial Respiratory Chain

■ Chemiosmotic theory provides the intellectual framework for understanding many biological energy transductions, including oxidative phosphorylation and photophosphorylation. The energy of electron flow is conserved by the concomitant pumping of protons across the membrane, producing an electrochemical gradient, the proton-motive force.

■ In mitochondria, hydride ions removed from substrates (such as α-ketoglutarate and malate) by NAD-linked dehydrogenases donate electrons to the respiratory chain, which transfers the electrons to molecular O_2, reducing it to H_2O.

■ Reducing equivalents from NADH are passed through a series of Fe-S centers to ubiquinone, which transfers the electrons to cytochrome b, the first carrier in Complex III. In this complex, electrons take two separate paths through two b-type cytochromes and cytochrome c_1 to an Fe-S center. The Fe-S center passes electrons, one at a time, through cytochrome c and into Complex IV, cytochrome oxidase. This copper-containing enzyme, which also contains cytochromes a and a_3, accumulates electrons, then passes them to O_2, reducing it to H_2O.

■ Some electrons enter this chain of carriers through alternative paths. Succinate is oxidized by succinate dehydrogenase (Complex II), which contains a flavoprotein that passes electrons through several Fe-S centers to ubiquinone. Electrons derived

from the oxidation of fatty acids pass to ubiquinone via the electron-transferring flavoprotein. The oxidation of glycerol phosphate and of dihydroorotate also sends electrons into the respiratory chain at the level of QH_2.

■ Potentially harmful reactive oxygen species produced in mitochondria are inactivated by a set of protective enzymes, including superoxide dismutase and glutathione peroxidase. Low levels of ROS serve as signals coordinating mitochondrial oxidative phosphorylation with other metabolic pathways.

■ Plants, fungi, and unicellular eukaryotes have, in addition to the typical path for electron transfer coupled to ATP synthesis, an alternative, uncoupled pathway that recycles excess NADH to NAD^+.

19.2 ATP Synthesis

How is a concentration gradient of protons transformed into ATP? We have seen that electron transfer releases, and the proton-motive force conserves, more than enough free energy (about 190 kJ) per "mole" of electron pairs to drive the formation of a mole of ATP, which requires about 50 kJ (p. 509). Mitochondrial oxidative phosphorylation therefore poses no thermodynamic problem. But what is the chemical mechanism that couples proton flux with phosphorylation?

rotenone (see Table 19-4), that transfers electrons from NADH in the matrix directly to ubiquinone, bypassing Complex I and its associated proton pumping. And plant mitochondria have yet another NADH dehydrogenase, on the external face of the inner membrane, that transfers electrons from NADPH or NADH in the intermembrane space to ubiquinone, again bypassing Complex I. Thus when electrons enter the alternative respiratory pathway through the rotenone-insensitive NADH dehydrogenase, the external NADH dehydrogenase, or succinate dehydrogenase (Complex II), and pass to O_2 via the cyanide-resistant alternative oxidase, energy is not conserved as ATP but is released as heat. A skunk cabbage can use the heat to melt snow, produce a foul stench, or attract beetles or flies.

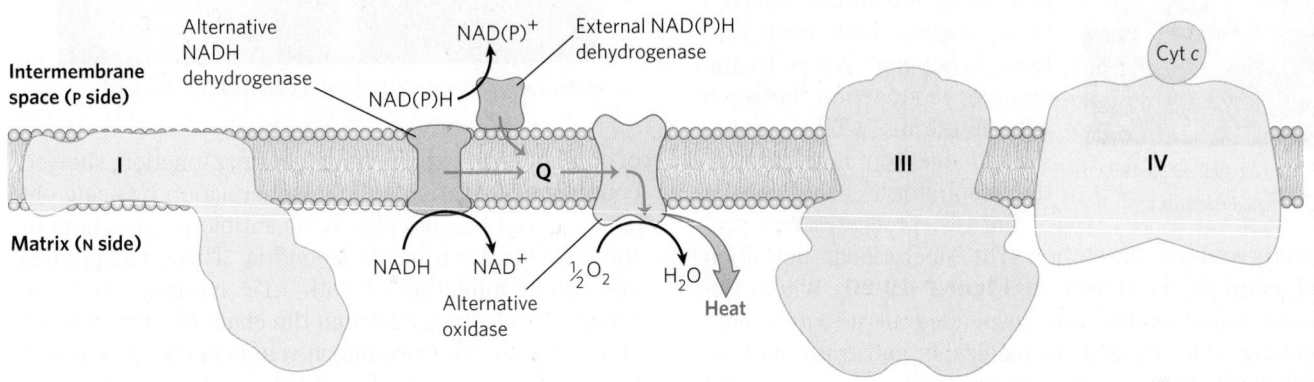

FIGURE 2 Electron carriers of the inner membrane of plant mitochondria. Electrons can flow through Complexes I, III, and IV, as in animal mitochondria, or through plant-specific alternative carriers by the paths shown with blue arrows.

In the Chemiosmotic Model, Oxidation and Phosphorylation Are Obligately Coupled

The **chemiosmotic model**, proposed by Peter Mitchell, is the paradigm for energy coupling. According to the model **(Fig. 19-19)**, the electrochemical energy inherent in the difference in proton concentration and the separation of charge across the inner mitochondrial membrane—the proton-motive force—drives the synthesis of ATP as protons flow passively back

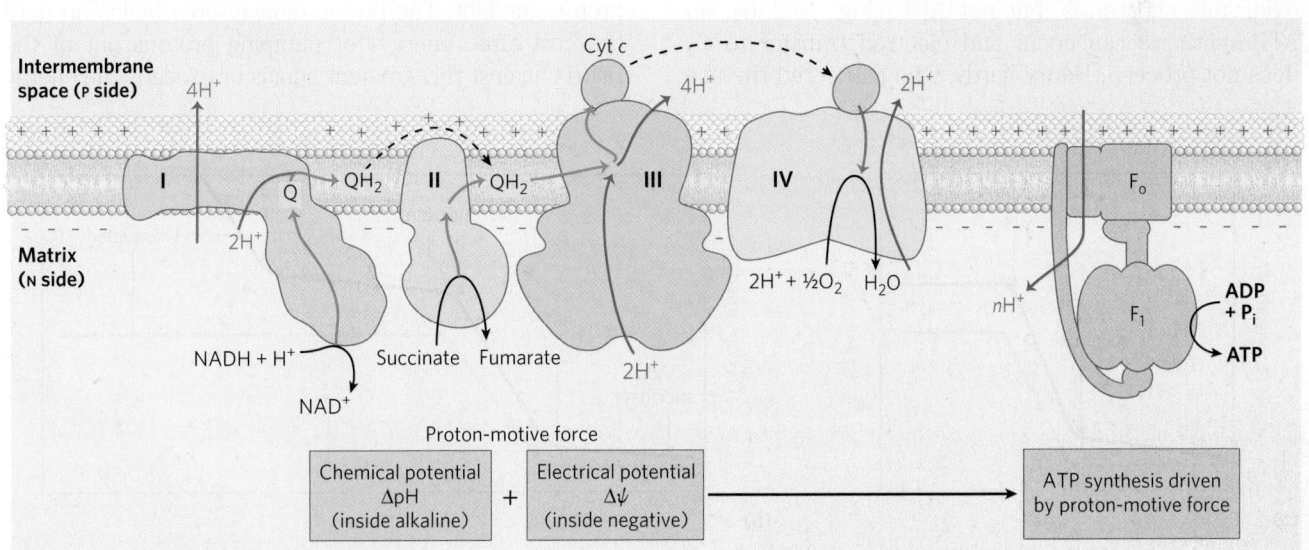

FIGURE 19-19 Chemiosmotic model. In this simple representation of the chemiosmotic theory applied to mitochondria, electrons from NADH and other oxidizable substrates pass through a chain of carriers arranged asymmetrically in the inner membrane. Electron flow is accompanied by proton transfer across the membrane, producing both a chemical gradient (ΔpH) and an electrical gradient ($\Delta\psi$), which, combined, create the proton-motive force. The inner mitochondrial membrane is impermeable to protons; protons can reenter the matrix only through proton-specific channels (F_o). The proton-motive force that drives protons back into the matrix provides the energy for ATP synthesis, catalyzed by the F_1 complex associated with F_o.

into the matrix through a proton pore in **ATP synthase**. To emphasize this crucial role of the proton-motive force, the equation for ATP synthesis is sometimes written

$$\text{ADP} + \text{P}_i + n\text{H}_P^+ \longrightarrow \text{ATP} + \text{H}_2\text{O} + n\text{H}_N^+ \qquad (19\text{-}10)$$

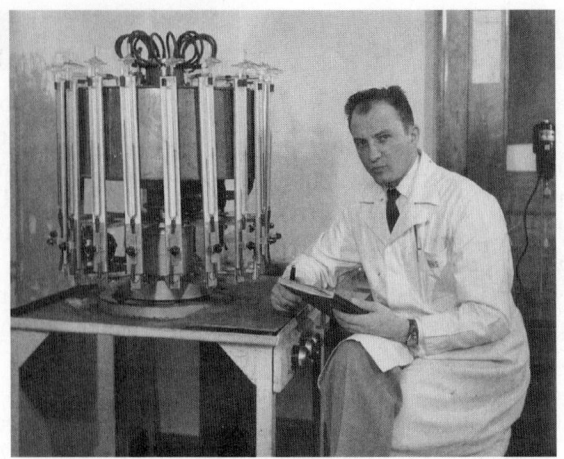

Henry Lardy, 1917–2010 [Source: © Courtesy D. L. Nelson.]

Peter Mitchell, 1920–1992 [Source: AP Photo.]

Mitchell used "chemiosmotic" to describe enzymatic reactions that involve, simultaneously, a chemical reaction and a transport process, and the overall process is sometimes referred to as "chemiosmotic coupling." Here, "coupling" refers to the *obligate* connection between mitochondrial ATP synthesis and electron flow through the respiratory chain; neither of the two processes can proceed without the other. The operational definition of coupling is shown in **Figure 19-20**. When isolated mitochondria are suspended in a buffer containing ADP, P_i, and an oxidizable substrate such as succinate, three easily measured processes occur: (1) the substrate is oxidized (succinate yields fumarate), (2) O_2 is consumed, and (3) ATP is synthesized. Oxygen consumption and ATP synthesis depend on the presence of an oxidizable substrate (succinate in this case) as well as ADP and P_i.

Because substrate oxidation drives ATP synthesis, inhibitors of electron transfer block ATP synthesis (Fig. 19-20a). The converse is also true: inhibition of ATP synthesis blocks electron transfer in intact mitochondria. When isolated mitochondria are given O_2 and oxidizable substrates, but not ADP (Fig. 19-20b), no ATP synthesis can occur and electron transfer to O_2 does not proceed. Henry Lardy, who pioneered the use

of antibiotics to explore mitochondrial function, showed coupling of oxidation and phosphorylation by using oligomycin and venturicidin, toxic antibiotics that bind to the ATP synthase in mitochondria. These compounds are potent inhibitors of both ATP synthesis *and* the transfer of electrons through the chain of carriers to O_2 (Fig. 19-20b). Because oligomycin is known to interact with ATP synthase itself, and not directly with the electron carriers, it follows that electron transfer and ATP synthesis are obligately coupled: neither reaction occurs without the other.

Chemiosmotic theory readily explains the dependence of electron transfer on ATP synthesis in mitochondria. When the flow of protons into the matrix through the proton channel of ATP synthase is blocked (with oligomycin, for example), no path exists for the return of protons to the matrix, and the continued extrusion of protons driven by the activity of the respiratory chain generates a large proton gradient. The proton-motive force builds up until the cost (free energy) of pumping protons out of the matrix against this gradient equals or exceeds the energy

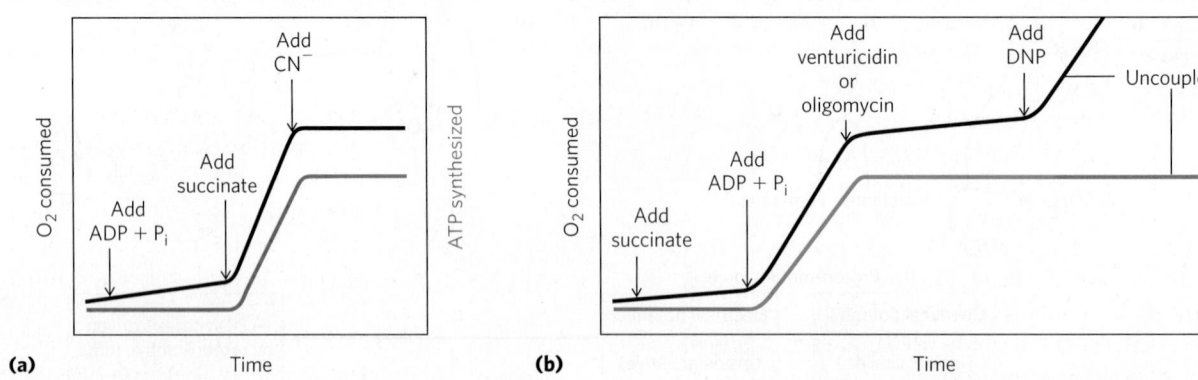

(a) Time **(b)** Time

FIGURE 19-20 Coupling of electron transfer and ATP synthesis in mitochondria. In experiments to demonstrate coupling, mitochondria are suspended in a buffered medium, and an O_2 electrode monitors O_2 consumption. At intervals, samples are removed and assayed for the presence of ATP. **(a)** Addition of ADP and P_i alone results in little or no increase in either respiration (O_2 consumption; black) or ATP synthesis (red). When succinate is added, respiration begins immediately, and ATP

is synthesized. Addition of cyanide (CN^-), which blocks electron transfer between cytochrome oxidase (Complex IV) and O_2, inhibits both respiration and ATP synthesis. **(b)** Mitochondria provided with succinate respire and synthesize ATP only when ADP and P_i are added. Subsequent addition of venturicidin or oligomycin, inhibitors of ATP synthase, blocks both ATP synthesis and respiration. Dinitrophenol (DNP) is an uncoupler, allowing respiration to continue without ATP synthesis.

2,4-Dinitrophenol
(DNP)

Carbonylcyanide-*p*-
trifluoromethoxyphenylhydrazone
(FCCP)

FIGURE 19-21 Two chemical uncouplers of oxidative phosphorylation. Both DNP and FCCP have a dissociable proton (red) and are very hydrophobic. They carry protons across the inner mitochondrial membrane, dissipating the proton gradient. Both also uncouple photophosphorylation.

released by the transfer of electrons from NADH to O_2. At this point electron flow must stop; the free energy for the overall process of electron flow coupled to proton pumping becomes zero, and the system is at equilibrium.

Certain conditions and reagents, however, can uncouple oxidation from phosphorylation. When intact mitochondria are disrupted by treatment with detergent or by physical shear, the resulting membrane fragments can still catalyze electron transfer from succinate or NADH to O_2, but no ATP synthesis is coupled to this respiration. Certain chemical compounds cause uncoupling without physically disrupting mitochondrial structure. Chemical uncouplers include 2,4-dinitrophenol (DNP) and carbonylcyanide-*p*-trifluoromethoxyphenylhydrazone (FCCP) (Table 19-4; **Fig. 19-21**), weak acids with hydrophobic properties that permit them to diffuse readily across mitochondrial membranes. After entering the matrix in the protonated form, they can release a proton, thus dissipating the proton gradient. Resonance stabilization delocalizes the charge on the anionic forms, making them sufficiently hydrophobic to diffuse back across the membrane, where they can pick up a proton and repeat the process. Ionophores such as valinomycin (see Fig. 11-42) allow inorganic ions to pass easily through membranes. Ionophores uncouple electron transfer from oxidative phosphorylation by dissipating the electrical contribution to the electrochemical gradient across the mitochondrial membrane.

A prediction of the chemiosmotic theory is that, because the role of electron transfer in mitochondrial ATP synthesis is simply to pump protons to create the electrochemical potential of the proton-motive force, an artificially created proton gradient should be able to replace electron transfer in driving ATP synthesis. This has been experimentally confirmed **(Fig. 19-22)**. Mitochondria manipulated to impose a difference of proton concentration and a separation of charge across the inner membrane synthesize ATP *in the absence of an oxidizable substrate*; the proton-motive force alone suffices to drive ATP synthesis.

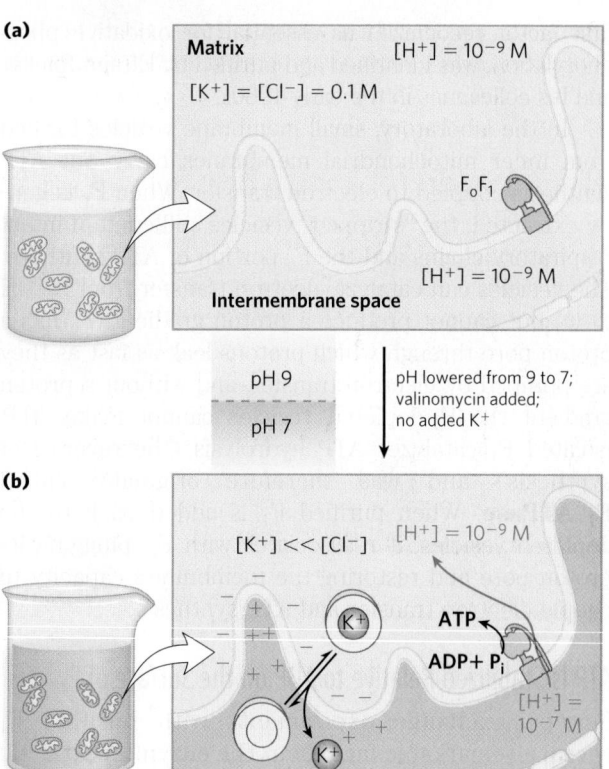

(a)
Matrix $[H^+] = 10^{-9}$ M
$[K^+] = [Cl^-] = 0.1$ M

F_oF_1

$[H^+] = 10^{-9}$ M
Intermembrane space

pH 9
pH 7

pH lowered from 9 to 7;
valinomycin added;
no added K^+

(b)
$[K^+] << [Cl^-]$ $[H^+] = 10^{-9}$ M

ATP
ADP + P_i

$[H^+] = 10^{-7}$ M

FIGURE 19-22 Evidence for the role of a proton gradient in ATP synthesis. An artificially imposed electrochemical gradient can drive ATP synthesis in the absence of an oxidizable substrate as electron donor. In this two-step experiment, **(a)** isolated mitochondria are first incubated in a pH 9 buffer containing 0.1 M KCl. Slow leakage of buffer and KCl into the mitochondria eventually brings the matrix into equilibrium with the surrounding medium. No oxidizable substrates are present. **(b)** Mitochondria are now removed from the pH 9 buffer and resuspended in pH 7 buffer containing valinomycin but no KCl. The change in buffer creates a difference of two pH units across the inner mitochondrial membrane. The outward flow of K^+, carried by valinomycin down the K^+ ion concentration gradient without a counter-ion, creates a charge imbalance across the membrane (matrix negative). The sum of the chemical potential provided by the pH difference and the electrical potential provided by the separation of charges is a proton-motive force large enough to support ATP synthesis in the absence of an oxidizable substrate.

ATP Synthase Has Two Functional Domains, F_0 and F_1

Mitochondrial ATP synthase is an F-type ATPase (see Fig. 11-37b) similar in structure and mechanism to the ATP

Efraim Racker, 1913–1991
[Source: Division of Rare and Manuscript Collections, Cornell University Library.]

synthases of bacteria and (as we will see in Chapter 20) of chloroplasts. This large enzyme complex of the inner mitochondrial membrane catalyzes the formation of ATP from ADP and P_i, driven by the flow of protons from the P to the N side of the membrane (Eqn 19-10). ATP synthase, also called Complex V, has two distinct components: F_1, a peripheral membrane protein, and F_o (*o* denoting oligomycin-sensitive), which is integral to the membrane. F_1, the first factor recognized as essential for oxidative phosphorylation, was identified and purified by Efraim Racker and his colleagues in the early 1960s.

In the laboratory, small membrane vesicles formed from inner mitochondrial membranes carry out ATP synthesis coupled to electron transfer. When F_1 is gently extracted, the "stripped" vesicles still contain intact respiratory chains and the F_o portion of ATP synthase. The vesicles can catalyze electron transfer from NADH to O_2 but cannot produce a proton gradient: F_o has a proton pore through which protons leak as fast as they are pumped by electron transfer, and without a proton gradient the F_1-depleted vesicles cannot make ATP. Isolated F_1 catalyzes ATP hydrolysis (the reversal of synthesis) and was therefore originally called **F_1 ATPase**. When purified F_1 is added back to the depleted vesicles, it reassociates with F_o, plugging its proton pore and restoring the membrane's capacity to couple electron transfer and ATP synthesis.

ATP Is Stabilized Relative to ADP on the Surface of F_1

Isotope exchange experiments with purified F_1 reveal a remarkable fact about the enzyme's catalytic mechanism: on the enzyme surface, the reaction $ADP + P_i \rightleftharpoons ATP + H_2O$ is readily reversible—the free-energy change for ATP synthesis is close to zero. When ATP is hydrolyzed by F_1 in the presence of ^{18}O-labeled water, the P_i released contains an ^{18}O atom. Careful measurement of the ^{18}O content of P_i formed in vitro by F_1-catalyzed hydrolysis of ATP reveals that the P_i has not one but three or four ^{18}O atoms (**Fig. 19-23**). This indicates that the terminal pyrophosphate bond in ATP is cleaved and re-formed repeatedly before P_i leaves the enzyme surface. With P_i free to tumble in its binding site, each hydrolysis inserts ^{18}O randomly at one of the four positions in the molecule. This exchange reaction occurs in unenergized F_oF_1 complexes (with no proton gradient) and with isolated F_1—the exchange does not require the input of energy.

Kinetic studies of the initial rates of ATP synthesis and hydrolysis confirm the conclusion that $\Delta G'^\circ$ for ATP synthesis on the enzyme is near zero. From the measured rates of hydrolysis ($k_1 = 10 \text{ s}^{-1}$) and synthesis ($k_{-1} = 24 \text{ s}^{-1}$), the calculated equilibrium constant for the reaction

$$\text{Enz-ATP} \rightleftharpoons \text{Enz-}(ADP + P_i)$$

is

$$K'_{eq} = \frac{k_{-1}}{k_1} = \frac{24 \text{ s}^{-1}}{10 \text{ s}^{-1}} = 2.4$$

From this K'_{eq}, the calculated apparent $\Delta G'^\circ$ is close to zero. This is much different from the K'_{eq} of about 10^5 ($\Delta G'^\circ = -30.5 \text{ kJ/mol}$) for the hydrolysis of ATP free in solution (i.e., not on the enzyme surface).

What accounts for the huge difference? ATP synthase stabilizes ATP relative to $ADP + P_i$ by binding ATP more tightly, releasing enough energy to counterbalance the cost of making ATP. Careful measurements of the binding constants show that F_oF_1 binds ATP with very high affinity ($K_d \leq 10^{-12} \text{ M}$) and ADP with much lower affinity ($K_d \approx 10^{-5} \text{ M}$). The difference in K_d corresponds to a difference of about 40 kJ/mol in binding energy, and this binding energy drives the equilibrium toward formation of the product ATP.

The Proton Gradient Drives the Release of ATP from the Enzyme Surface

Although ATP synthase equilibrates ATP with $ADP + P_i$, in the absence of a proton gradient the newly synthesized ATP does not leave the surface of the enzyme. It is the proton gradient that causes the enzyme to release the ATP formed on its surface. The reaction coordinate

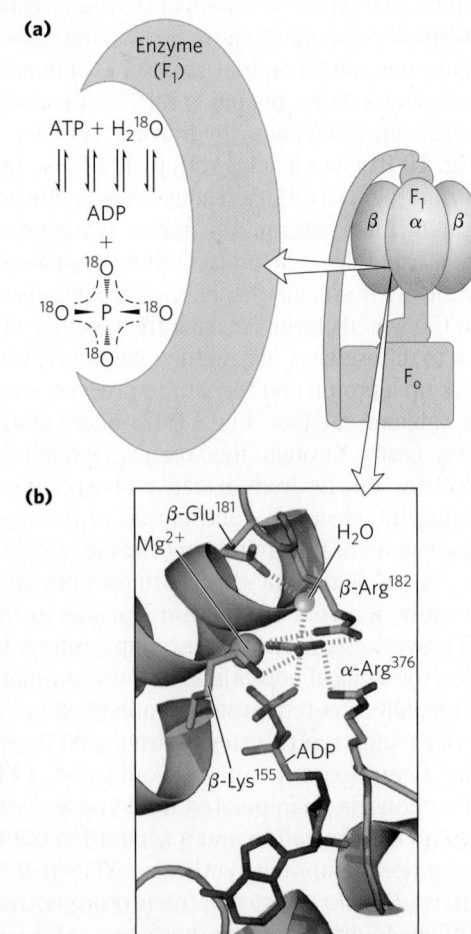

FIGURE 19-23 Catalytic mechanism of F_1. (a) An ^{18}O-exchange experiment. F_1 solubilized from mitochondrial membranes is incubated with ATP in the presence of ^{18}O-labeled water. At intervals, a sample of the solution is withdrawn and analyzed for the incorporation of ^{18}O into the P_i produced from ATP hydrolysis. In minutes, the P_i contains three or four ^{18}O atoms, indicating that both ATP hydrolysis and ATP synthesis have occurred several times during the incubation. **(b)** The likely transition state complex for ATP hydrolysis and synthesis by ATP synthase. The α subunit is shown in gray, β in purple. The positively charged residues β-Arg^{182} and α-Arg^{376} coordinate two oxygens of the pentavalent phosphate intermediate; β-Lys^{155} interacts with a third oxygen, and the Mg^{2+} ion further stabilizes the intermediate. The blue sphere represents the leaving group (H_2O). These interactions result in ready equilibration of ATP and $ADP + P_i$ in the active site. [Source: (b) Derived from PDB ID 1BMF, J. P. Abrahams et al., *Nature* 370:621, 1994.]

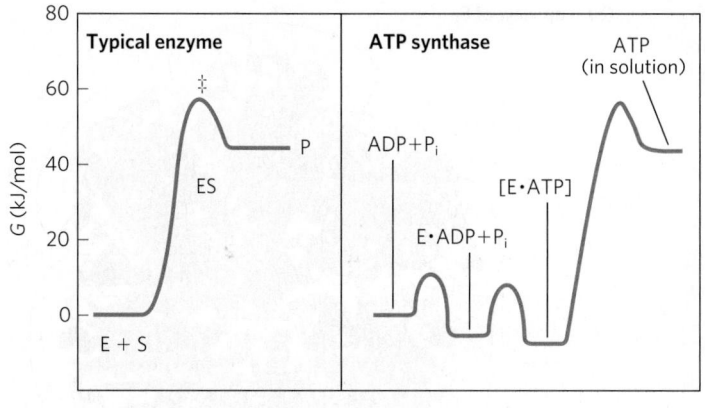

FIGURE 19-24 Reaction coordinate diagrams for ATP synthase and for a more typical enzyme. In a typical enzyme-catalyzed reaction (left), reaching the transition state (‡) between substrate and product is the major energy barrier to overcome. In the reaction catalyzed by ATP synthase (right), release of ATP from the enzyme, not formation of ATP, is the major energy barrier. The free-energy change for the formation of ATP from ADP and P_i in aqueous solution is large and positive, but on the enzyme surface, the very tight binding of ATP provides sufficient binding energy to bring the free energy of the enzyme-bound ATP close to that of ADP + P_i, so the reaction is readily reversible. The equilibrium constant is near 1. The free energy required for the release of ATP is provided by the proton-motive force.

diagram of the process **(Fig. 19-24)** illustrates the difference between the mechanism of ATP synthase and that of many other enzymes that catalyze endergonic reactions.

For the continued synthesis of ATP, the enzyme must cycle between a form that binds ATP very tightly and a form that releases ATP. Chemical and crystallographic studies of the ATP synthase have revealed the structural basis for this alternation in function.

Each β Subunit of ATP Synthase Can Assume Three Different Conformations

John E. Walker [Source: © Findlay Kember/ Associated Press.]

Mitochondrial F_1 has nine subunits of five different types, with the composition $\alpha_3\beta_3\gamma\delta\varepsilon$. Each of the three β subunits has one catalytic site for ATP synthesis. The crystallographic determination of the F_1 structure by John E. Walker and colleagues revealed structural details very helpful in explaining the catalytic mechanism of the enzyme. The knoblike portion of F_1 is a flattened sphere, 8 nm by 10 nm, consisting of alternating α and β subunits arranged like the sections of an orange **(Fig. 19-25a–c)**. The polypeptides that make up the stalk in the F_1 crystal structure are asymmetrically arranged, with one domain of the single γ subunit making up a central shaft that passes through F_1, and another domain of γ associated primarily with one of the three β subunits, designated β-empty (Fig. 19-25b). Although the amino acid sequences of the three β subunits are identical, *their conformations differ*, in part because of the association of the γ subunit with just one of the three.

The conformational differences among β subunits extend to differences in their ATP/ADP-binding sites. When the protein is crystallized in the presence of ADP and App(NH)p, a close structural analog of ATP that cannot be hydrolyzed by the ATPase activity of

F_1, the binding site of one of the three β subunits is filled with App(NH)p, the second is filled with ADP, and the third is empty. The corresponding β subunit conformations are designated β-ATP, β-ADP, and β-empty (Fig. 19-25b). This difference in nucleotide binding among the three subunits is critical to the mechanism of the complex.

App(NH)p (β,γ-imidoadenosine 5'-triphosphate)

The F_o complex, with its proton pore, is composed of three subunits, a, b, and c, in the proportion ab_2c_n, where n ranges from 8 to 15, depending on the species. Subunit c is a small (M_r 8,000), very hydrophobic polypeptide, consisting almost entirely of two transmembrane helices, with a small loop extending from the matrix side of the membrane. The crystal structure of the yeast F_oF_1 shows 10 c subunits, each with two transmembrane helices roughly perpendicular to the plane of the membrane and arranged in two concentric circles. The inner circle is made up of the amino-terminal helices of each c subunit; the outer circle, about 55 Å in diameter, is made up of the carboxyl-terminal helices. *The c subunits in this c ring rotate together as a unit around an axis perpendicular to the membrane.* The ε and γ subunits of F_1 form a leg-and-foot that projects from the bottom (membrane) side of F_1 and stands firmly on the ring of c subunits. The a subunit consists of several hydrophobic helices that span the membrane in close association with one of the c subunits in the c ring. The schematic drawing in Figure 19-25a combines the structural information from studies of bovine F_1, yeast F_oF_1, and the c ring of the strict anaerobic bacterium *Ilyobacter tartaricus*.

FIGURE 19-25 Mitochondrial ATP synthase complex. (a) A cartoon representation of the F_oF_1 complex. **(b)** F_1 viewed from above (that is, from the N side of the membrane), showing the three β (shades of purple) and three α (shades of gray) subunits and the central shaft (γ subunit, green). Each β subunit, near its interface with the neighboring α subunit, has a nucleotide-binding site critical to the catalytic activity. The single γ subunit associates primarily with one of the three $\alpha\beta$ pairs, forcing each of the three β subunits into slightly different conformations, with different nucleotide-binding sites. In the crystalline enzyme, one subunit, β-ADP, has ADP (yellow) in its binding site; the next, β-ATP, has ATP (red); and the third, β-empty, has no bound nucleotide.

(c) The entire enzyme viewed from the side (in the plane of the membrane). The F_1 portion has three α and three β subunits arranged like the segments of an orange around a central shaft, the γ subunit (green). (Two α subunits and one β subunit have been omitted to reveal the γ subunit and the binding sites for ATP and ADP on the β subunits.) The δ subunit confers oligomycin sensitivity on the ATP synthase, and the ε subunit may serve to inhibit the enzyme's ATPase activity under some circumstances. The F_o subunit consists of one a subunit and two b subunits, which anchor the F_oF_1 complex in the membrane and act as a stator, holding the α and β subunits in place. F_o also includes the c ring, made up of a number (8 to 15, depending on the species) of identical c subunits, small, hydrophobic proteins. The c ring and the a subunit interact to provide a transmembrane path for protons. Each of the c subunits in F_o has a critical Asp residue near the middle of the membrane, which undergoes protonation/deprotonation during the catalytic cycle of the ATP synthase. Shown here is the homologous c_{11} ring of the Na^+-ATPase of *Ilyobacter tartaricus*, for which the structure is well established. The Na^+-binding sites, which correspond to the proton-binding sites of the F_oF_1 complex, are shown with their bound Na^+ ions (red spheres). **(d)** A view of F_o perpendicular to the membrane. As in (c), red spheres represent the Na^+- or proton-binding sites in Asp residues. [Sources: F_1: PDB ID 1BMF, J. P. Abrahams et al., *Nature* 370:621, 1994; PDB ID 1JNV, A. C. Hausrath et al., *J. Biol. Chem.* 276:47,227, 2001; PDB ID 2A7U, S. Wilkens et al., *Biochemistry* 44:11,786, 2005; PDB ID 2CLY, V. Kane Dickson et al., *EMBO J.* 25:2911, 2006. F_o: PDB ID 1B9U, O. Dmitriev et al., *J. Biol. Chem.* 274:15,598, 1999. c ring: PDB ID 1YCE, T. Meier et al., *Science* 308:659, 2005.]

(b) Top view of F_1

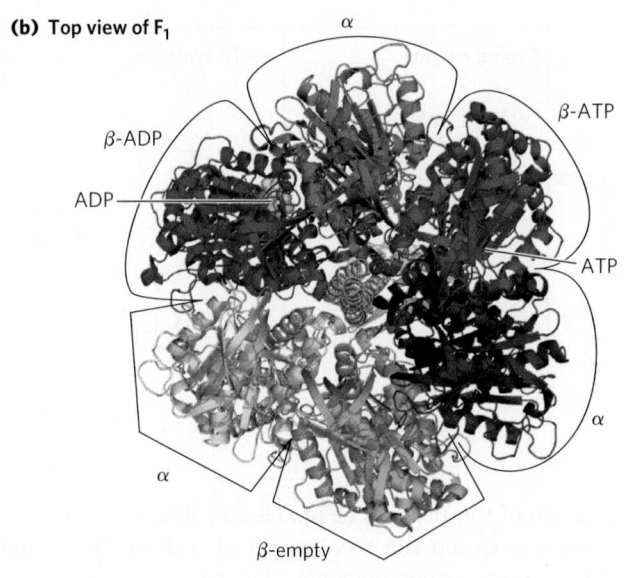

(c) Side view of F_oF_1

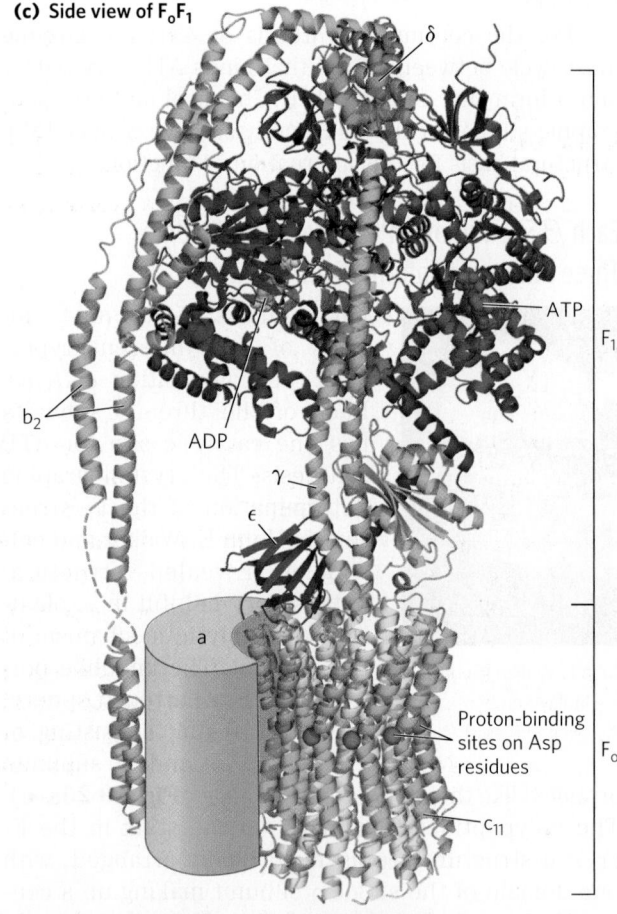

(a)

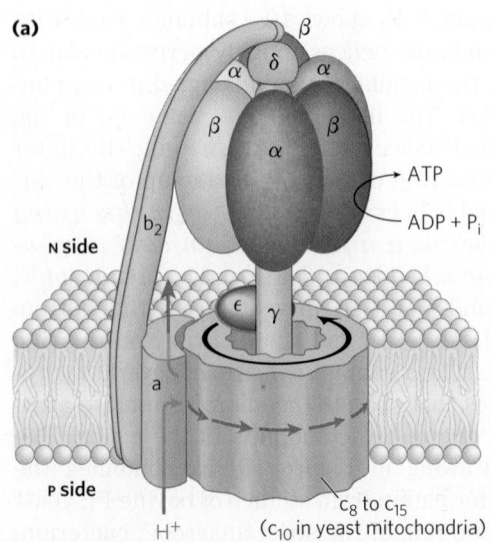

(d) Bottom view of F_o

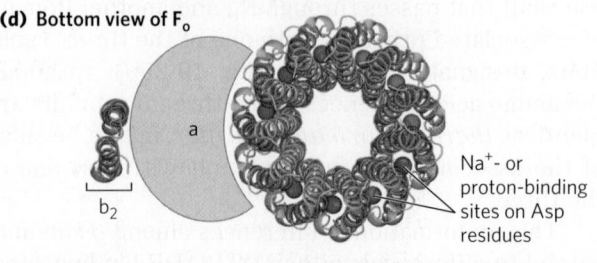

Rotational Catalysis Is Key to the Binding-Change Mechanism for ATP Synthesis

Paul Boyer [Source: Lacy Atkins/Associated Press/ AP Images.]

On the basis of detailed kinetic and binding studies of the reactions catalyzed by F_oF_1, Paul Boyer proposed a **rotational catalysis** mechanism in which the three active sites of F_1 take turns catalyzing ATP synthesis **(Fig. 19-26)**. A given β subunit starts in the β-ADP conformation, which binds ADP and P_i from the surrounding medium. The subunit now changes conformation, assuming the β-ATP form that tightly binds and stabilizes ATP, bringing about the ready equilibration of ADP + P_i with ATP on the enzyme surface. Finally, the subunit changes to the β-empty conformation, which has very low affinity for ATP, and the newly synthesized ATP leaves the enzyme surface. Another round of catalysis begins when this subunit again assumes the β-ADP form and binds ADP and P_i.

The conformational changes central to this mechanism are driven by the passage of protons through the F_o portion of ATP synthase. The streaming of protons through the F_o pore causes the cylinder of c subunits and the attached γ subunit to rotate about the long axis of γ, which is perpendicular to the plane of the membrane. The γ subunit passes through the center of the $\alpha_3\beta_3$ spheroid, which is held stationary relative to the membrane surface by the b_2 and δ subunits (Fig. 19-25a). With each rotation of 120°, γ comes into contact with a different β subunit, and the contact forces that β subunit into the β-empty conformation.

The three β subunits interact in such a way that when one assumes the β-empty conformation, its neighbor to one side *must* assume the β-ADP form, and the other neighbor the β-ATP form. Thus one complete rotation of the γ subunit causes each β subunit to cycle through all three of its possible conformations, and for each rotation, three ATP are synthesized and released from the enzyme surface.

One strong prediction of this **binding-change model** is that the γ subunit should rotate in one direction when F_oF_1 is synthesizing ATP and in the opposite direction when the enzyme is hydrolyzing ATP. This prediction of rotation with ATP hydrolysis was confirmed in elegant experiments in the laboratories of Masasuke Yoshida and Kazuhiko Kinosita, Jr. The rotation of γ in a single F_1 molecule was observed microscopically by attaching a long, thin, fluorescent actin polymer to γ and watching it move relative to $\alpha_3\beta_3$ immobilized on a microscope slide as ATP was hydrolyzed. (The expected reversal of the rotation when ATP is being synthesized could not be tested in this experiment; there is no proton gradient to drive ATP synthesis.) When the entire F_oF_1 complex (not just F_1) was used in a similar experiment, the entire ring of c subunits rotated with γ **(Fig. 19-27)**. The "shaft" rotated in the predicted direction through 360°. The rotation was not smooth, but occurred in three discrete steps of 120°. As calculated from the known rate of ATP hydrolysis by one F_1 molecule and from the frictional drag on the long actin polymer, the efficiency of this mechanism in converting chemical energy into motion is close to 100%. It is, in Boyer's words, "a splendid molecular machine!"

How Does Proton Flow through the F_o Complex Produce Rotary Motion?

One feasible model to explain how proton flow and rotary motion are coupled in the F_o complex is shown in

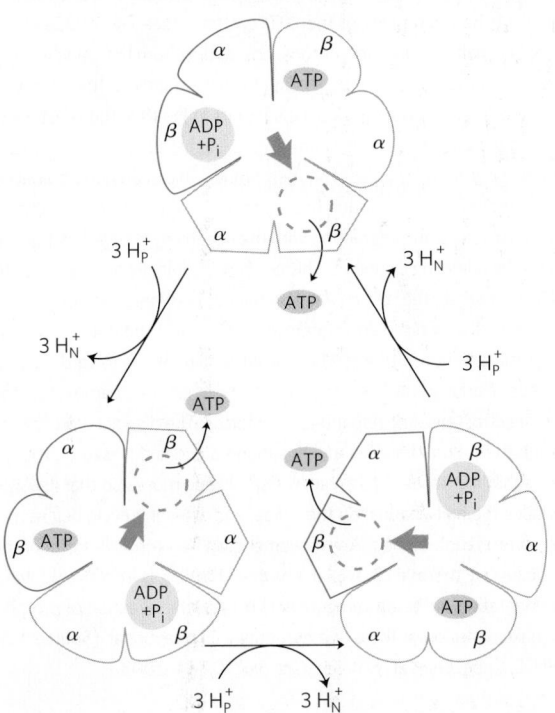

FIGURE 19-26 Binding-change model for ATP synthase. The F_1 complex has three nonequivalent adenine nucleotide–binding sites, one for each pair of α and β subunits. At any given moment, one of these sites is in the β-ATP conformation (which binds ATP tightly), a second is in the β-ADP (loose-binding) conformation, and a third is in the β-empty (very-loose-binding) conformation. In this view from the N side, the proton-motive force causes rotation of the central shaft—the γ subunit, shown as a green arrowhead—which comes into contact with each $\alpha\beta$ subunit pair in succession. This produces a cooperative conformational change in which the β-ATP site is converted to the β-empty conformation, and ATP dissociates; the β-ADP site is converted to the β-ATP conformation, which promotes condensation of bound ADP + P_i to form ATP; and the β-empty site becomes a β-ADP site, which loosely binds ADP + P_i entering from the solvent. This model, based on experimental findings, requires that at least two of the three catalytic sites alternate in activity; ATP cannot be released from one site unless and until ADP and P_i are bound at the other. Note that the direction of rotation reverses when the ATP synthase is acting as an ATPase, as in the experiment depicted in Figure 19-27.

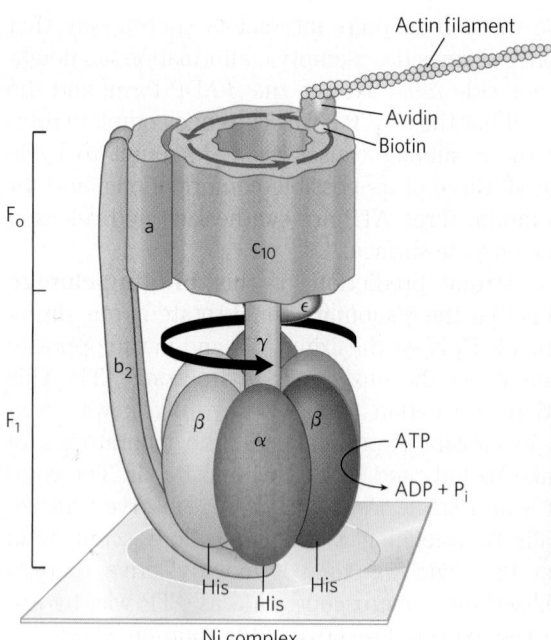

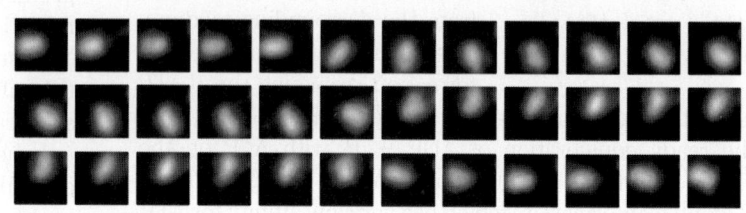

FIGURE 19-27 Experimental demonstration of rotation of F_o and γ. F_1 genetically engineered to contain a run of His residues adheres tightly to a microscope slide coated with a Ni complex; biotin is covalently attached to a c subunit of F_o. The protein avidin, which binds biotin very tightly, is covalently attached to long filaments of actin labeled with a fluorescent probe. Biotin-avidin binding now attaches the actin filaments to the c subunit. When ATP is provided as substrate for the ATPase activity of F_1, the labeled filament is seen to rotate continuously in one direction, proving that the F_o cylinder of c subunits rotates. In another experiment, a fluorescent actin filament was attached directly to the γ subunit. The series of fluorescence micrographs (read left to right) shows the position of the actin filament at intervals of 133 ms. Note that as the filament rotates, it makes a discrete jump about every eleventh frame. Presumably, the cylinder and shaft move as one unit. [Sources: Cartoon: Information from Y. Sambongi et al., *Science* 286:1722, 1999. Micrographs: Courtesy of Ryohei Yasuda and Kazuhiko Kinosita, from Yasuda et al., *Cell* 93:1117, 1998. © Elsevier.]

Figure 19-28. The individual subunits in F_o are arranged in a circle about a central core that is probably filled with membrane lipids. Each c subunit has a critical Asp (or Glu) residue at about the middle of the membrane. Protons cross the membrane through a path made up of both a and c subunits. The a subunit has a proton half-channel leading from the P side (cytosol) to the middle of the membrane, where it ends near the Asp residue of the adjoining c subunit. A proton diffuses from the P side (where the proton concentration is relatively high) down this half-channel and binds to the Asp residue, removing the negative charge on the carboxyl group and thereby displacing a positively charged Arg residue that had been associated with the Asp. This Arg residue swings aside, forming an interaction with the Asp on the adjacent c subunit in the ring and displacing the proton from that Asp; this proton exits through the second half-channel to the N side, where the proton concentration is relatively low, completing the movement of one proton equivalent from outside to inside the matrix. Now another proton has entered the half-channel on the P side, moved to the Asp of the next

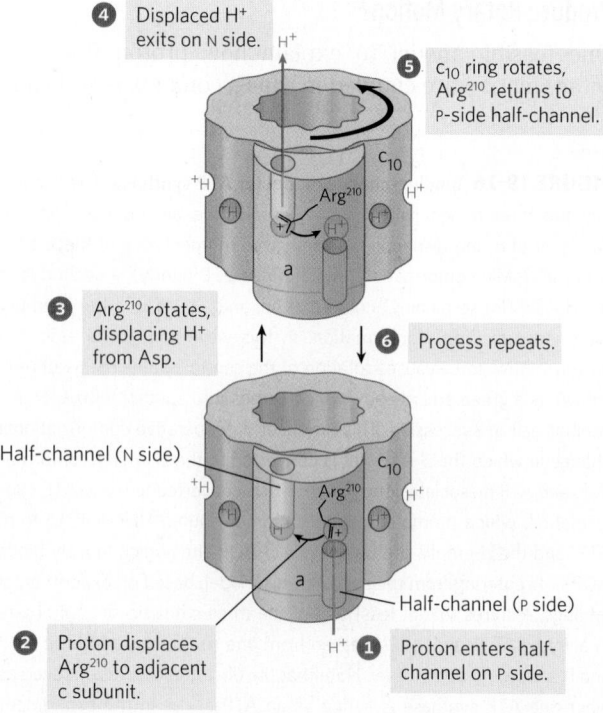

FIGURE 19-28 A model for proton-driven rotation of the c ring. The a subunit of the F_o complex of the ATP synthase (see Fig. 19-25a) has two hydrophilic half-channels for protons, one leading from the P side to the middle of the membrane, the other leading from the middle of the membrane to the N side (matrix). The individual c subunits in F_o (10 in the yeast enzyme) are arranged in a circle about a central core. Each c subunit has a critical Asp residue about midway across the membrane, which can donate or accept a proton (red H^+). The a subunit has a positively charged Arg side chain that forms an electrostatic interaction with the negatively charged carboxylate of Asp on the adjoining c subunit. This c subunit is initially positioned so that a proton that enters the half-channel on the P side (where the proton concentration is relatively high) encounters and protonates the Asp residue, weakening its interaction with Arg. The Arg side chain rotates toward the protonated Asp residue in the next c subunit and displaces its carboxyl proton, as a new electrostatic Arg-Asp interaction forms. The displaced proton moves through the second half-channel in subunit a and is released on the N side. The c subunit with its unprotonated Asp residue moves so that its Arg-Asp pair faces the half-channel on the P side, and a second cycle begins: proton entry, protonation of Asp, Arg movement, and proton exit. Rotation of the ring occurs by thermal (Brownian) motion; the ring is effectively ratcheted. The orientation of the proton gradient dictates the direction of proton flow and makes rotation of the c ring essentially unidirectional. [Source: PDB ID 1OHH, E. Cabezon et al., *Nature Struct. Biol.* 10:744, 2003.]

BOX 19-3 METHODS Atomic Force Microscopy to Visualize Membrane Proteins

In **atomic force microscopy (AFM)**, the sharp tip of a microscopic probe attached to a flexible cantilever is drawn across an uneven surface such as a membrane (Fig. 1). Electrostatic and van der Waals interactions between the tip and the sample produce a force that moves the probe up and down (in the z dimension) as it encounters hills and valleys in the sample. A laser beam reflected from the cantilever detects motions of as little as 1 Å. In one type of atomic force microscope, the force on the probe is held constant (relative to a standard force, on the order of piconewtons) by a feedback circuit that causes the platform holding the sample to rise or fall to keep the force constant. A series of scans in the x and y dimensions (the plane of the membrane) yields a three-dimensional contour map of the surface, with resolution near the atomic scale—0.1 nm in the vertical dimension, 0.5 to 1.0 nm in the lateral dimensions.

In favorable cases, AFM can be used to study single membrane protein molecules, such as the c subunits of the F_o complex (Fig. 19-29c, d).

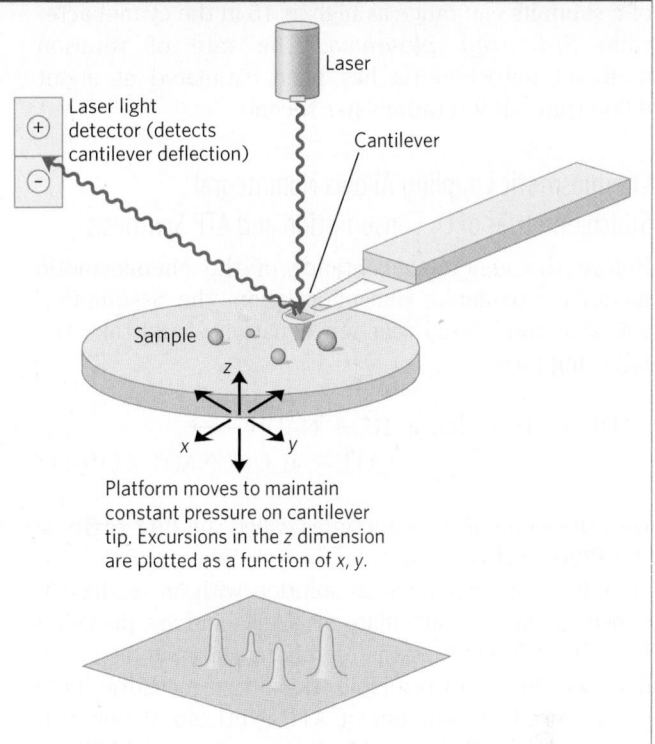

FIGURE 1 The principle of atomic force microscopy.

c subunit, and protonated it, again displacing Arg, which in turn displaces a proton from the next c subunit, and so forth. The rotary movement of the c ring is the result of thermal (Brownian) motion, made unidirectional by the large difference in proton concentration across the membrane. The number of protons that must be transferred to produce one complete rotation of the c ring is equal to the number of c subunits in the ring. Studies of the c ring with atomic force microscopy (Box 19-3) or x-ray diffraction have shown that the number of c subunits differs in different organisms **(Fig. 19-29)**.

FIGURE 19-29 Species differences in number of c subunits in the c ring of the F_o complex. The structures of the c rings from several species have been determined by x-ray crystallography. Each helix in the inner ring is half of a hairpin-shaped c subunit; the outer ring of helices forms the other half of the hairpin structure. The essential Asp residue at position 61 is shown as a red dot. Views of the c ring perpendicular to the membrane show the number of c subunits for **(a)** bovine mitochondria (8 subunits) and **(b)** yeast mitochondria (10). Atomic force microscopy (see Box 19-3) has been used to visualize the c rings of **(c)** a thermophilic bacterium, *Bacillus* species TA2.A1 (13 subunits), and **(d)** spinach (14). According to the model in Figure 19-28, different numbers of c subunits in the c ring should result in different ratios of ATP formed per pair of electrons passing through the respiratory chain (i.e., different P/O ratios). [Sources: (a) PDB ID 1OHH, E. Cabezon et al., *Nature Struct. Biol.* 10:744, 2003. (c) D. Matthies, et al., *J. Mol. Biol.* 388:611, 2009. (d) H. Seelert et al., *Nature* 405:418, 2000. Reprinted by permission from Macmillan Publishers Ltd.]

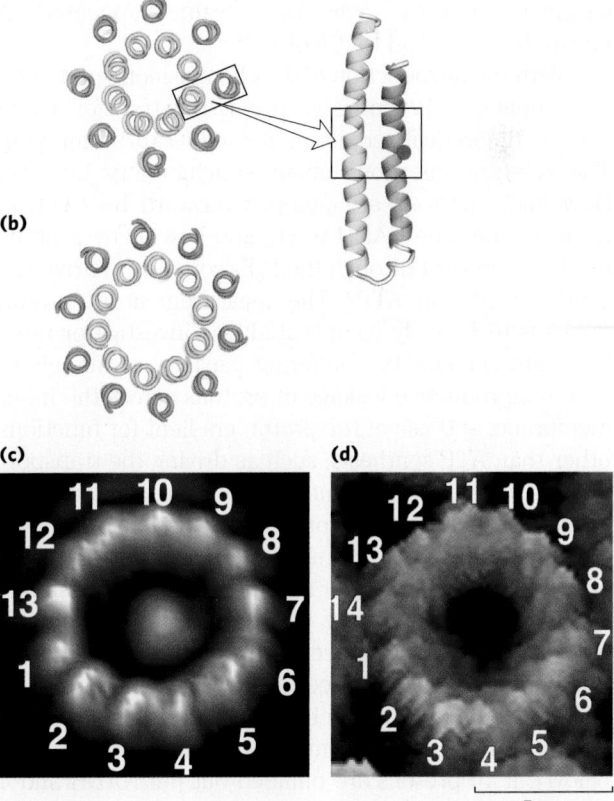

(a) Top view   **Side view**

(b)

(c)   **(d)**

5 nm

In bovine mitochondria the number is 8, in yeast mitochondria and in *Escherichia coli* it is 10, and the number of c subunits can range as high as 15 in the cyanobacterium *Spirulina platensis*. The rate of rotation in intact mitochondria has been estimated at about 6,000 rpm—100 rotations per second.

Chemiosmotic Coupling Allows Nonintegral Stoichiometries of O_2 Consumption and ATP Synthesis

Before the general acceptance of the chemiosmotic model for oxidative phosphorylation, the assumption was that the overall reaction equation would take the following form:

$$x\text{ADP} + x\text{P}_i + \tfrac{1}{2}O_2 + \text{H}^+ + \text{NADH} \longrightarrow$$
$$x\text{ATP} + \text{H}_2\text{O} + \text{NAD}^+ \quad (19\text{-}11)$$

with the value of x—sometimes called the **P/O ratio** or the **P/2e⁻ ratio**—always an integer. When intact mitochondria are suspended in solution with an oxidizable substrate such as succinate or NADH and are provided with O_2, ATP synthesis is readily measurable, as is the decrease in O_2. In principle, these two measurements should yield the number of ATP synthesized per $\tfrac{1}{2}O_2$ consumed, the P/O ratio. Most experiments yielded P/O ($\text{ATP}/\tfrac{1}{2}O_2$) ratios of between 2 and 3 when NADH was the electron donor, and between 1 and 2 when succinate was the donor. Given the assumption that P/O should have an integral value, most experimenters agreed that the P/O ratios must be 3 for NADH and 2 for succinate, and for years those values appeared in research papers and textbooks.

With the introduction of the chemiosmotic paradigm for coupling ATP synthesis to electron transfer, there was no theoretical requirement for P/O to be integral. The relevant questions about stoichiometry became, How many protons are pumped outward by electron transfer from one NADH to O_2, and how many protons must flow inward through the F_oF_1 complex to drive the synthesis of one ATP? The measurement of proton fluxes is technically complicated; the investigator must take into account the buffering capacity of mitochondria, nonproductive leakage of protons across the inner membrane, and use of the proton gradient for functions other than ATP synthesis, such as driving the transport of substrates across the inner mitochondrial membrane (described below). The consensus experimental values for number of protons pumped out per pair of electrons are 10 for NADH and 6 for succinate (which sends electrons into the respiratory chain at the level of ubiquinone). The most widely accepted experimental value for number of protons required to drive the synthesis of an ATP molecule is 4, of which 1 is used in transporting P_i, ATP, and ADP across the mitochondrial membrane (see below). If 10 protons are pumped out per NADH and 4 must flow in to produce 1 ATP, the proton-based P/O ratio is 2.5 for NADH as the electron donor and 1.5 (6/4)

for succinate. However, as we will see in Worked Example 2, the proton stoichiometry of ATP synthesis by ATP synthase depends upon the number of c units in F_o, which ranges from 8 to 15, depending on the species.

The Proton-Motive Force Energizes Active Transport

Although the primary role of the proton gradient in mitochondria is to furnish energy for the synthesis of ATP, the proton-motive force also drives several transport processes essential to oxidative phosphorylation. The inner mitochondrial membrane is generally impermeable to charged species, but two specific systems transport ADP and P_i into the matrix and ATP out to the cytosol **(Fig. 19-30)**.

The **adenine nucleotide translocase**, integral to the inner membrane, binds ADP^{3-} in the intermembrane space and transports it into the matrix in exchange for an ATP^{4-} molecule simultaneously transported outward (see Fig. 13-11 for the ionic forms of ATP and ADP). Because this antiporter moves four negative charges out for every three moved in, its activity is favored by the transmembrane electrochemical gradient, which gives the matrix a net negative charge; the proton-motive

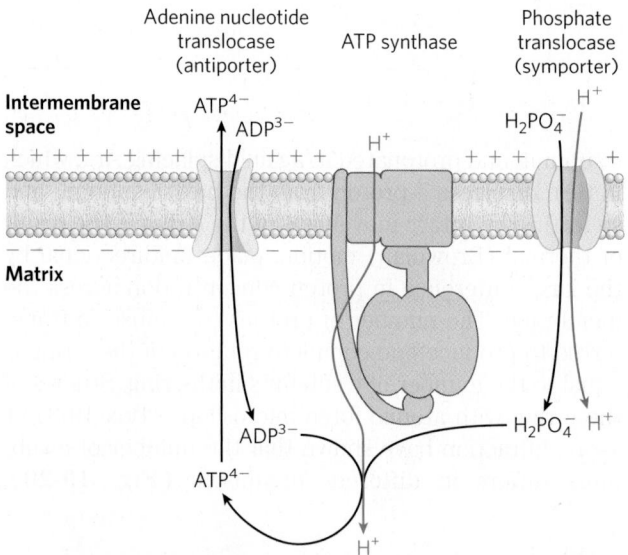

FIGURE 19-30 Adenine nucleotide and phosphate translocases. Transport systems of the inner mitochondrial membrane carry ADP and P_i into the matrix and newly synthesized ATP into the cytosol. The adenine nucleotide translocase is an antiporter; the same protein moves ADP into the matrix and ATP out. The effect of replacing ATP^{4-} with ADP^{3-} in the matrix is the net efflux of one negative charge, which is favored by the charge difference across the inner membrane (outside positive). At pH 7, P_i is present as both HPO_4^{2-} and H_2PO_4^-; the phosphate translocase is specific for H_2PO_4^-. There is no net flow of charge during symport of H_2PO_4^- and H^+, but the relatively low proton concentration in the matrix favors the inward movement of H^+. Thus the proton-motive force is responsible both for providing the energy for ATP synthesis and for transporting substrates (ADP and P_i) into and product (ATP) out of the mitochondrial matrix. All three of these transport systems can be isolated as a single membrane-bound complex (ATP synthasome).

force drives ATP-ADP exchange. Adenine nucleotide translocase is specifically inhibited by atractyloside, a toxic glycoside produced by a species of thistle. If the transport of ADP into and ATP out of mitochondria is inhibited, cytosolic ATP cannot be regenerated from ADP, explaining the toxicity of atractyloside.

A second membrane transport system essential to oxidative phosphorylation is the **phosphate translocase**, which promotes symport of one $H_2PO_4^-$ and one H^+ into the matrix. This transport process, too, is favored by the transmembrane proton gradient (Fig. 19-30). Notice that the process requires movement of one proton from the P to the N side of the inner membrane, consuming some of the energy of electron transfer. A complex of the ATP synthase and both translocases, the **ATP synthasome**, can be isolated from mitochondria by gentle dissection with detergents, suggesting that the functions of these three proteins are very tightly integrated.

ATP and ADP cross the outer mitochondrial membrane via the voltage-dependent anion channel (VDAC), a 19-stranded β barrel (see Fig. 11-11) with an opening about 27 Å wide, connecting the cytosol and the intermembrane space. Each VDAC, when open, can move 10^5 ATP molecules per second. The opening is gated by voltage, as its name indicates, and under some conditions VDAC is closed to ATP.

WORKED EXAMPLE 19-2 | **Stoichiometry of ATP Production: Effect of c Ring Size**

(a) If the ATP synthase of *bovine* mitochondria has 8 c subunits per c ring, what is the predicted ratio of ATP formed per NADH oxidized? (b) What is the predicted value for *yeast* mitochondria, with 10 c subunits per ATP synthase? (c) What are the comparable values for electrons entering the respiratory chain from $FADH_2$?

Solution: (a) The question asks us to determine how many ATP molecules are produced per NADH. This is another way of asking us to calculate the P/O ratio, or x in Equation 19-11. If the c ring has 8 c subunits, then one full rotation will transfer 8 protons to the matrix and produce 3 ATP molecules. But this synthesis also requires the transport of 3 P_i into the matrix, at a cost of 1 proton each, adding 3 more protons to the total number required. This brings the total cost to (11 protons)/(3 ATP) = 3.7 protons/ATP. The generally agreed value for the number of protons pumped out per pair of electrons transferred from NADH is 10 (Eqn 19-7). So, oxidizing 1 NADH produces (10 protons)/(3.7 protons/ATP) = 2.7 ATP.

(b) If the c ring has 10 c subunits, then one full rotation will transfer 10 protons to the matrix and produce 3 ATP molecules. Adding in the 3 protons to transport the 3 P_i into the matrix brings the total cost to (13 protons)/(3 ATP) = 4.3 protons/ATP. Oxidizing 1 NADH produces (10 protons)/(4.3 protons/ATP) = 2.3 ATP.

(c) When electrons enter the respiratory chain from $FADH_2$ (at ubiquinone), only 6 protons are available to drive ATP synthesis. This changes the calculation for bovine mitochondria to (6 protons)/(3.7 protons/ATP) = 1.6 ATP per pair of electrons from $FADH_2$. For yeast mitochondria, the calculation is (6 protons)/(4.3 protons/ATP) = 1.4 ATP per pair of electrons from $FADH_2$.

These calculated values of x, or the P/O ratio, define a range that includes the experimental values of 2.5 ATP/NADH and 1.5 ATP/$FADH_2$, and we therefore use these values throughout this book.

Shuttle Systems Indirectly Convey Cytosolic NADH into Mitochondria for Oxidation

The NADH dehydrogenase of the inner mitochondrial membrane of animal cells can accept electrons only from NADH in the matrix. Given that the inner membrane is not permeable to NADH, how can the NADH generated by glycolysis in the cytosol be reoxidized to NAD^+ by O_2 via the respiratory chain? Special shuttle systems carry reducing equivalents from cytosolic NADH into mitochondria by an indirect route. The most active NADH shuttle, which functions in liver, kidney, and heart mitochondria, is the **malate-aspartate shuttle (Fig. 19-31)**. The reducing equivalents of cytosolic NADH are first transferred to cytosolic oxaloacetate to yield malate, catalyzed by cytosolic malate dehydrogenase. The malate thus formed passes through the inner membrane via the malate–α-ketoglutarate transporter. Within the matrix, the reducing equivalents are passed to NAD^+ by the action of matrix malate dehydrogenase, forming NADH; this NADH can pass electrons directly to the respiratory chain. About 2.5 molecules of ATP are generated as this pair of electrons passes to O_2. Cytosolic oxaloacetate must be regenerated by transamination reactions and the activity of membrane transporters to start another cycle of the shuttle.

Skeletal muscle and brain use a different NADH shuttle, the **glycerol 3-phosphate shuttle (Fig. 19-32)**. It differs from the malate-aspartate shuttle in that it delivers the reducing equivalents from NADH through FAD in glycerol 3-phosphate dehydrogenase to ubiquinone and thus into Complex III, not Complex I (Fig. 19-15), providing enough energy to synthesize only 1.5 ATP molecules per pair of electrons.

The mitochondria of plants have an *externally* oriented NADH dehydrogenase that can transfer electrons directly from cytosolic NADH into the respiratory chain at the level of ubiquinone. Because this pathway bypasses the NADH dehydrogenase of Complex I and the associated proton movement, the yield of ATP from cytosolic NADH is less than that from NADH generated in the matrix (Box 19-2).

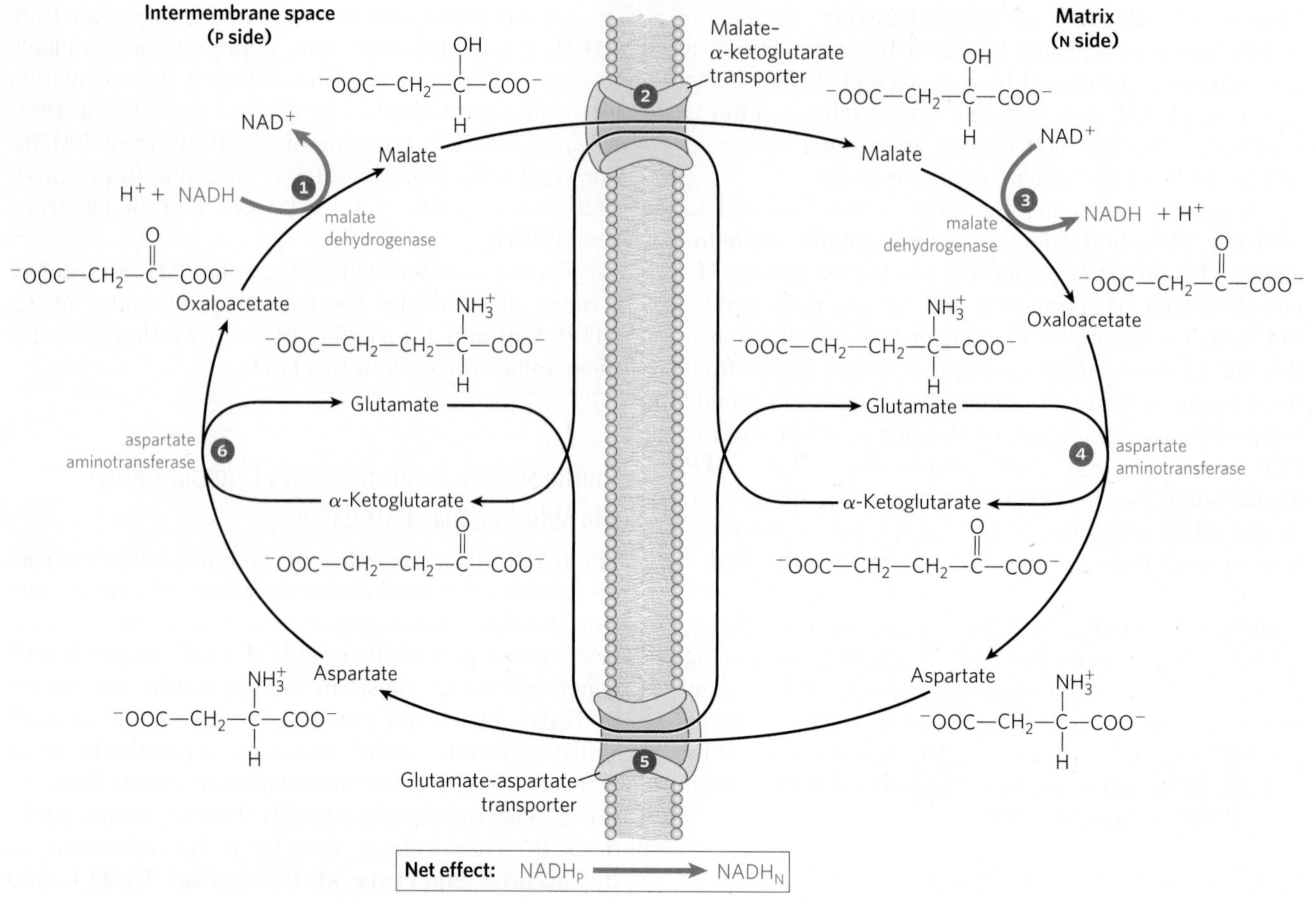

Intermembrane space
(P side)

$^-OOC-CH_2-\overset{\underset{|}{OH}}{\underset{|}{C}}-COO^-$

Malate-
α-ketoglutarate
transporter

Matrix
(N side)

$^-OOC-CH_2-\overset{\underset{|}{OH}}{\underset{|}{C}}-COO^-$

NAD$^+$

❶ Malate

H$^+$ + NADH

malate
dehydrogenase

Malate

NAD$^+$

❸ NADH + H$^+$

malate
dehydrogenase

$^-OOC-CH_2-\overset{\underset{}{O}}{\overset{||}{C}}-COO^-$

Oxaloacetate

$^-OOC-CH_2-CH_2-\overset{\underset{|}{NH_3^+}}{\underset{|}{C}}-COO^-$

Glutamate

$^-OOC-CH_2-CH_2-\overset{\underset{|}{NH_3^+}}{\underset{|}{C}}-COO^-$

Glutamate

$^-OOC-CH_2-\overset{\underset{}{O}}{\overset{||}{C}}-COO^-$

Oxaloacetate

aspartate
aminotransferase ❻

α-Ketoglutarate

$^-OOC-CH_2-CH_2-\overset{\underset{}{O}}{\overset{||}{C}}-COO^-$

α-Ketoglutarate

$^-OOC-CH_2-CH_2-\overset{\underset{}{O}}{\overset{||}{C}}-COO^-$

❹ aspartate
aminotransferase

$^-OOC-CH_2-\overset{\underset{|}{NH_3^+}}{\underset{|}{C}}-COO^-$

Aspartate

Aspartate

$^-OOC-CH_2-\overset{\underset{|}{NH_3^+}}{\underset{|}{C}}-COO^-$

❺

Glutamate-aspartate
transporter

Net effect: NADH$_P$ ⟶ NADH$_N$

FIGURE 19-31 Malate-aspartate shuttle. This shuttle for transporting reducing equivalents from cytosolic NADH into the mitochondrial matrix is used in liver, kidney, and heart. ❶ NADH in the cytosol enters the intermembrane space through openings in the outer membrane (porins), then passes two reducing equivalents to oxaloacetate, producing malate. ❷ Malate crosses the inner membrane via the malate–α-ketoglutarate transporter. ❸ In the matrix, malate passes two reducing

equivalents to NAD$^+$, and the resulting NADH is oxidized by the respiratory chain; the oxaloacetate formed from malate cannot pass directly into the cytosol. ❹ Oxaloacetate is first transaminated to aspartate, and ❺ aspartate can leave via the glutamate-aspartate transporter. ❻ Oxaloacetate is regenerated in the cytosol, completing the cycle, and glutamate produced in the same reaction enters the matrix via the glutamate-aspartate transporter.

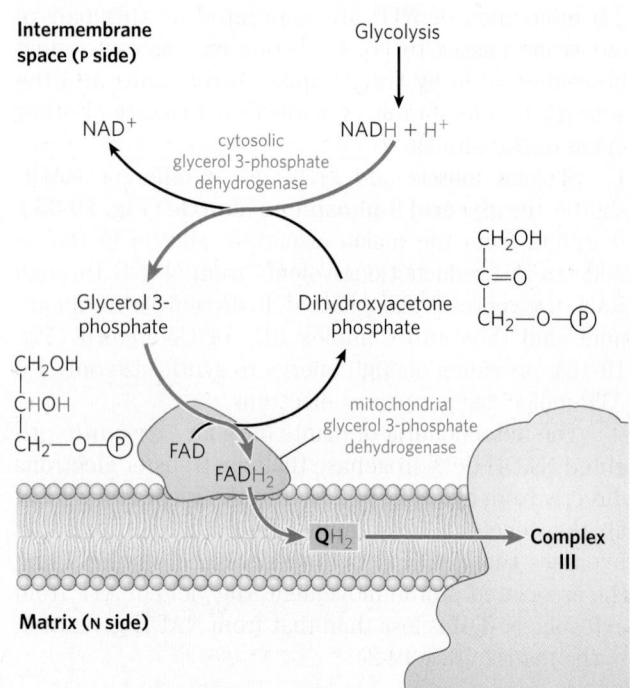

Intermembrane
space (P side)

Glycolysis

NAD$^+$

cytosolic
glycerol 3-phosphate
dehydrogenase

NADH + H$^+$

$\overset{\underset{}{CH_2OH}}{\underset{\underset{}{CH_2-O-Ⓟ}}{\underset{}{C=O}}}$

Glycerol 3-
phosphate

Dihydroxyacetone
phosphate

$\overset{\underset{}{CH_2OH}}{\underset{\underset{}{CH_2-O-Ⓟ}}{\underset{}{CHOH}}}$

FAD

FADH$_2$

mitochondrial
glycerol 3-phosphate
dehydrogenase

QH$_2$ ⟶ **Complex III**

Matrix (N side)

SUMMARY 19.2 ATP Synthesis

■ The flow of electrons through Complexes I, III, and IV results in pumping of protons across the inner mitochondrial membrane, making the matrix alkaline relative to the intermembrane space. This proton gradient provides the energy, in the form of the proton-motive force, for ATP synthesis from ADP and P$_i$ by ATP synthase (F$_o$F$_1$ complex) in the inner membrane.

■ ATP synthase carries out "rotational catalysis," in which the flow of protons through F$_o$ causes each of three nucleotide-binding sites in F$_1$ to cycle from

FIGURE 19-32 Glycerol 3-phosphate shuttle. This alternative means of moving reducing equivalents from the cytosol to the respiratory chain operates in skeletal muscle and the brain. In the cytosol, dihydroxyacetone phosphate accepts two reducing equivalents from NADH in a reaction catalyzed by cytosolic glycerol 3-phosphate dehydrogenase. An isozyme of glycerol 3-phosphate dehydrogenase bound to the outer face of the inner membrane then transfers two reducing equivalents from glycerol 3-phosphate in the intermembrane space to ubiquinone. Note that this shuttle does not involve membrane transport systems.

(ADP + P$_i$)–bound to ATP-bound to empty conformations.

■ ATP formation on the enzyme requires little energy; the role of the proton-motive force is to push ATP from its binding site on the synthase.

■ The ratio of ATP synthesized per ½O$_2$ reduced to H$_2$O (the P/O ratio) is about 2.5 when electrons enter the respiratory chain at Complex I, and 1.5 when electrons enter at ubiquinone. This ratio varies among species, depending on the number of c subunits in the F$_o$ complex.

■ Energy conserved in a proton gradient can drive solute transport uphill across a membrane.

■ The inner mitochondrial membrane is impermeable to NADH and NAD$^+$, but NADH equivalents are moved from the cytosol to the matrix by either of two shuttles. NADH equivalents moved in by the malate-aspartate shuttle enter the respiratory chain at Complex I and yield a P/O ratio of 2.5; those moved in by the glycerol 3-phosphate shuttle enter at ubiquinone and give a P/O ratio of 1.5.

19.3 Regulation of Oxidative Phosphorylation

Oxidative phosphorylation produces most of the ATP made in aerobic cells. Complete oxidation of a molecule of glucose to CO$_2$ yields 30 or 32 ATP (Table 19-5). By comparison, glycolysis under anaerobic conditions (lactate fermentation) yields only 2 ATP per glucose. Clearly, the evolution of oxidative phosphorylation provided a tremendous increase in the energy efficiency of catabolism. Complete oxidation to CO$_2$ of the coenzyme A derivative of palmitate (16:0), which also occurs in the mitochondrial matrix, yields 108 ATP per palmitoyl-CoA (see Table 17-1). A similar calculation can be made for the ATP yield from oxidation of each of the amino acids (Chapter 18). Aerobic oxidative pathways that result in electron transfer to O$_2$ accompanied by oxidative phosphorylation therefore account for the vast majority of the ATP produced in catabolism, so the regulation of ATP production by oxidative phosphorylation to match the cell's fluctuating needs for ATP is absolutely essential.

Oxidative Phosphorylation Is Regulated by Cellular Energy Needs

The rate of respiration (O$_2$ consumption) in mitochondria is tightly regulated; it is generally limited by the availability of ADP as a substrate for phosphorylation. Dependence of the rate of O$_2$ consumption on the availability of the P$_i$ acceptor, ADP (Fig. 19-20b), the **acceptor control** of respiration, can be remarkable. In some animal tissues, the **acceptor control ratio**, the ratio of the maximal rate of ADP-induced O$_2$ consumption to the basal rate in the absence of ADP, is at least 10.

The intracellular concentration of ADP is one measure of the energy status of cells. Another, related measure is the **mass-action ratio** of the ATP-ADP system, [ATP]/([ADP][P$_i$]). Usually this ratio is very high, so the ATP-ADP system is almost fully phosphorylated. When the rate of some energy-requiring process (protein synthesis, for example) increases, the rate of breakdown of ATP to ADP and P$_i$ increases, lowering the mass-action ratio. With more ADP available for oxidative phosphorylation, the rate of respiration increases, causing regeneration of ATP. This continues until the mass-action ratio returns to its normal high level, at which point respiration slows again. The rate of oxidation of cellular fuels is regulated with such sensitivity and precision that the [ATP]/([ADP][P$_i$]) ratio fluctuates only slightly in most tissues, even during extreme variations in energy demand. In short, ATP is formed only as fast as it is used in energy-requiring cellular activities.

An Inhibitory Protein Prevents ATP Hydrolysis during Hypoxia

We have already encountered ATP synthase as an ATP-driven proton pump (see Fig. 11-37), catalyzing the reverse of ATP synthesis. When a cell is hypoxic (deprived of oxygen), as in a heart attack or stroke, electron transfer to oxygen slows, and so does the

TABLE 19-5	ATP Yield from Complete Oxidation of Glucose	
Process	Direct product	Final ATP
Glycolysis	2 NADH (cytosolic)	3 or 5[a]
	2 ATP	2
Pyruvate oxidation (two per glucose)	2 NADH (mitochondrial matrix)	5
Acetyl-CoA oxidation in citric acid cycle (two per glucose)	6 NADH (mitochondrial matrix)	15
	2 FADH$_2$	3
	2 ATP or 2 GTP	2
Total yield per glucose		30 or 32

[a]If the malate/aspartate shuttle is used to transfer reducing equivalents into the mitochondrion, yield is 5 ATP. If the glycerol 3-phosphate shuttle is used, the yield is 3 ATP.

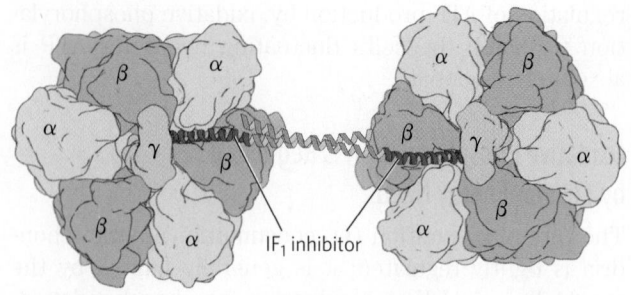

FIGURE 19-33 Structure of bovine F₁-ATPase in a complex with its regulatory protein IF₁. Two F₁ molecules are viewed here from the P side, as in Figure 19-25c. The inhibitor IF₁ (red) binds to the αβ interface of the subunits in the diphosphate (ADP) conformation (α-ADP and β-ADP), freezing the two F₁ complexes and thereby blocking ATP hydrolysis (and synthesis). (Parts of IF₁ that failed to resolve in the crystals of F₁ are shown (light red) as they occur in crystals of isolated IF₁.) This complex is stable only at the low cytosolic pH characteristic of cells that are producing ATP by glycolysis; when aerobic metabolism resumes, the cytosolic pH rises, the inhibitor is destabilized, and ATP synthase becomes active. [Source: Derived from PDB ID 1OHH, E. Cabezon et al., *Nature Struct. Biol.* 10:744, 2003.]

pumping of protons. The proton-motive force soon collapses. Under these conditions, the ATP synthase could operate in reverse, hydrolyzing ATP made by glycolysis to pump protons outward and causing a disastrous drop in ATP levels. This is prevented by a small (84 amino acids) protein inhibitor, IF₁, which simultaneously binds to two ATP synthase molecules, inhibiting their ATPase activity **(Fig. 19-33)**. IF₁ is inhibitory only in its dimeric form, which is favored at pH lower than 6.5. In a cell starved for oxygen, the main source of ATP becomes

glycolysis, and the pyruvic or lactic acid thus formed lowers the pH in the cytosol and the mitochondrial matrix. This favors IF₁ dimerization, leading to inhibition of the ATPase activity of ATP synthase and thereby preventing wasteful hydrolysis of ATP. When aerobic metabolism resumes, production of pyruvic acid slows, the pH of the cytosol rises, the IF₁ dimer is destabilized, and the inhibition of ATP synthase is lifted. IF₁ is an intrinsically disordered protein (p. 138); it acquires a favored conformation only on interaction with ATP synthase.

Hypoxia Leads to ROS Production and Several Adaptive Responses

In hypoxic cells there is an imbalance between the input of electrons from fuel oxidation in the mitochondrial matrix and transfer of electrons to molecular oxygen, leading to increased formation of reactive oxygen species. In addition to the glutathione peroxidase system (Fig. 19-18), cells have two other lines of defense against ROS **(Fig. 19-34)**. One is regulation of pyruvate dehydrogenase (PDH), the enzyme that delivers acetyl-CoA to the citric acid cycle (Chapter 16). Under hypoxic conditions, PDH kinase phosphorylates mitochondrial

FIGURE 19-34 Regulation of gene expression by hypoxia-inducible factor (HIF-1) to reduce ROS formation. Under conditions of low oxygen (hypoxia), HIF-1 is synthesized in greater amounts and acts as a transcription factor, increasing synthesis of the glucose transporter, glycolytic enzymes, pyruvate dehydrogenase kinase (PDH kinase), lactate dehydrogenase, a protease that degrades the cytochrome oxidase subunit COX4-1, and cytochrome oxidase subunit COX4-2. These changes counter the formation of ROS by decreasing the supply of NADH and FADH₂ and making cytochrome oxidase of Complex IV more effective. [Source: Information from D. A. Harris, *Bioenergetics at a Glance*, p. 36, Blackwell Science, 1995.]

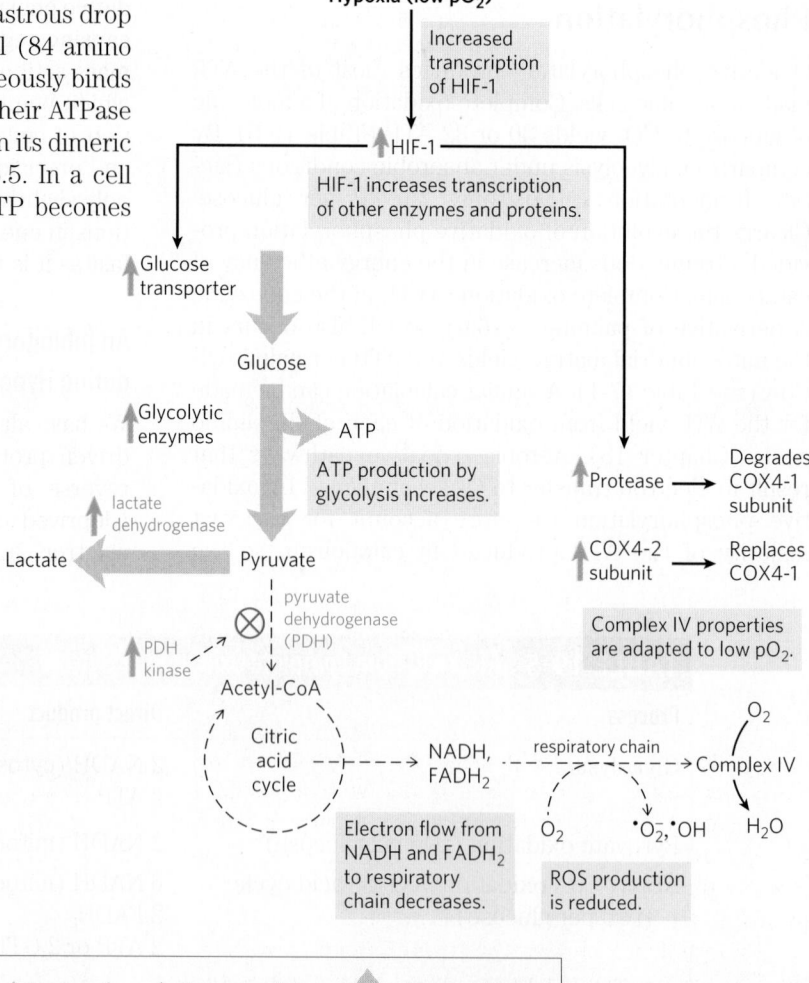

PDH, inactivating it and slowing the delivery of FADH₂ and NADH from the citric acid cycle to the respiratory chain. A second means of preventing ROS formation is the replacement of one subunit of Complex IV, known as COX4-1, with another subunit, COX4-2, that is better suited to hypoxic conditions. With COX4-1, the catalytic properties of Complex IV are optimal for respiration at normal oxygen concentrations; with COX4-2, Complex IV is optimized for operation under hypoxic conditions.

The changes in PDH activity and the COX4-2 content of Complex IV are both mediated by HIF-1, the hypoxia-inducible factor. HIF-1 (another intrinsically disordered protein) accumulates in hypoxic cells and, acting as a transcription factor, triggers increased synthesis of PDH kinase, COX4-2, and a protease that degrades COX4-1. HIF-1 is a master regulator of O₂ homeostasis. Recall that it also mediates the changes in glucose transport and glycolytic enzymes that produce the Warburg effect, the dependence on glycolysis (not mitochondrial respiration) for ATP production, even in the presence of sufficient oxygen (see Box 14-1).

When these mechanisms for dealing with ROS are insufficient, either due to genetic mutation affecting one of the protective proteins or under conditions of very high rates of ROS production, mitochondrial function is compromised. Mitochondrial damage is thought to be involved in aging, heart failure, certain rare cases of diabetes (described below), and several maternally inherited genetic diseases that affect the nervous system. ∎

ATP-Producing Pathways Are Coordinately Regulated

The major catabolic pathways have overlapping and concerted regulatory mechanisms that allow them to function together in an economical and self-regulating manner to produce ATP and biosynthetic precursors. The relative concentrations of ATP and ADP control not only the rates of electron transfer and oxidative phosphorylation but also the rates of the citric acid cycle, pyruvate oxidation, and glycolysis **(Fig. 19-35)**. Whenever ATP consumption increases, the rate of electron transfer and oxidative phosphorylation increases. Simultaneously, the rate of pyruvate oxidation via the citric acid cycle increases, increasing the flow of electrons into the respiratory chain. These events, in turn, can evoke an increased rate of glycolysis, increasing the rate of pyruvate formation. When conversion of ADP to ATP lowers the ADP concentration, acceptor control slows electron transfer and thus oxidative phosphorylation. Glycolysis and the citric acid cycle are also slowed, because ATP is an allosteric inhibitor of the glycolytic enzyme phosphofructokinase-1 (see Fig. 15-16) and of pyruvate dehydrogenase (see Fig. 16-19).

Phosphofructokinase-1 is also inhibited by citrate, the first intermediate of the citric acid cycle. When the cycle is "idling," citrate accumulates within mitochondria, then is transported into the cytosol. When the

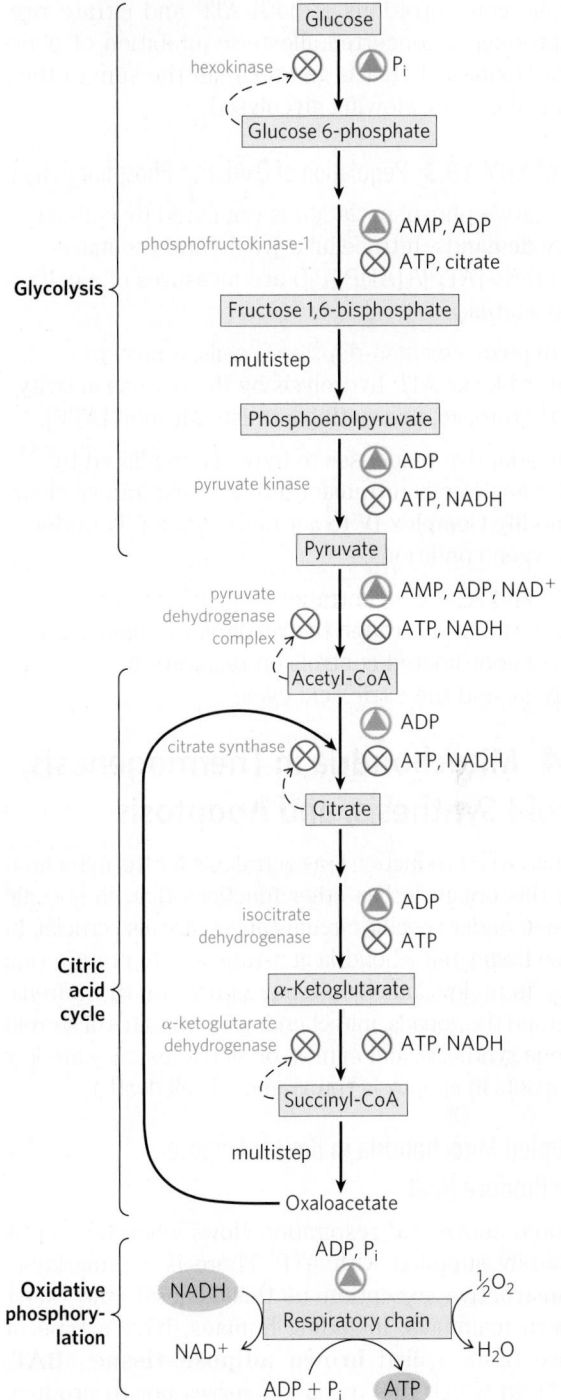

FIGURE 19-35 Regulation of ATP-producing pathways. This diagram shows the coordinated regulation of glycolysis, pyruvate oxidation, the citric acid cycle, and oxidative phosphorylation by the relative concentrations of ATP, ADP, and AMP, and by NADH. High [ATP] (or low [ADP] and [AMP]) produces low rates of glycolysis, pyruvate oxidation, acetate oxidation via the citric acid cycle, and oxidative phosphorylation. All four pathways are accelerated when the use of ATP and the formation of ADP, AMP, and P_i increase. The ability of citrate to inhibit both glycolysis and the citric acid cycle reinforces the action of the adenine nucleotide system. In addition, increased [NADH] and [acetyl-CoA] also inhibit the oxidation of pyruvate to acetyl-CoA, and a high [NADH]/[NAD⁺] ratio inhibits the dehydrogenase reactions of the citric acid cycle (see Fig. 16-19).

cytosolic concentrations of both ATP and citrate rise, they produce a concerted allosteric inhibition of phosphofructokinase-1 that is greater than the sum of their individual effects, slowing glycolysis.

SUMMARY 19.3 Regulation of Oxidative Phosphorylation

■ Oxidative phosphorylation is regulated by cellular energy demands. Intracellular [ADP] and the mass-action ratio $[ATP]/([ADP][P_i])$ are measures of a cell's energy status.

■ In hypoxic (oxygen-deprived) cells, a protein inhibitor blocks ATP hydrolysis by the reverse activity of ATP synthase, preventing a drastic drop in [ATP].

■ The adaptive responses to hypoxia, mediated by HIF-1, slow electron transfer into the respiratory chain and modify Complex IV to act more efficiently under low-oxygen conditions.

■ ATP and ADP concentrations set the rate of electron transfer through the respiratory chain via a series of coordinated controls on respiration, glycolysis, and the citric acid cycle.

19.4 Mitochondria in Thermogenesis, Steroid Synthesis, and Apoptosis

Although ATP production is a central role for the mitochondrion, this organelle has other functions that, in specific tissues or under specific circumstances, are also crucial. In adipose tissue, mitochondria generate heat to protect vital organs from low ambient temperature; in the adrenal glands and the gonads, mitochondria are the sites of steroid hormone synthesis; and in most or all tissues, they are key participants in apoptosis (programmed cell death).

Uncoupled Mitochondria in Brown Adipose Tissue Produce Heat

We noted above that respiration slows when the cell is adequately supplied with ATP. There is a remarkable and instructive exception to this general rule. Most newborn mammals, including humans, have a type of adipose tissue called **brown adipose tissue** (**BAT**; p. 923), in which fuel oxidation serves not to produce ATP but to generate heat to keep the newborn warm. This specialized adipose tissue is brown because of the presence of large numbers of mitochondria and thus high concentrations of cytochromes, with heme groups that are strong absorbers of visible light.

The mitochondria of brown adipocytes are much like those of other mammalian cells, except in having a unique protein in their inner membrane. **Uncoupling protein 1 (UCP1)** provides a path for protons to return to the matrix without passing through the F_oF_1 complex **(Fig. 19-36)**. As a result of this short-circuiting of protons, the energy of oxidation is not conserved by ATP formation but is dissipated as heat, which contributes to maintaining body temperature (see Fig. 23-16).

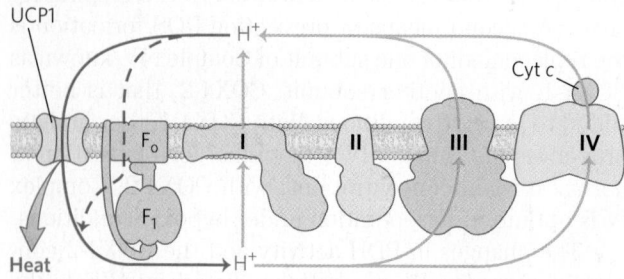

Intermembrane space (P side)

Matrix (N side)

FIGURE 19-36 Heat generation by uncoupled mitochondria. UCP1, the uncoupling protein in the mitochondria of brown adipose tissue, by providing an alternative route for protons to reenter the mitochondrial matrix, causes the energy conserved by proton pumping to be dissipated as heat.

Hibernating animals also depend on the activity of uncoupled BAT mitochondria to generate heat during their long dormancy (see Box 17-1). We will return to the role of UCP1 when we discuss the regulation of body mass in Chapter 23 (pp. 940–941).

Mitochondrial P-450 Monooxygenases Catalyze Steroid Hydroxylations

Mitochondria are the site of biosynthetic reactions that produce steroid hormones, including the sex hormones, glucocorticoids, mineralocorticoids, and vitamin D hormone. These compounds are synthesized from cholesterol or a related sterol in a series of hydroxylations catalyzed by enzymes of the **cytochrome P-450** family (see Box 21-1), all of which have a critical heme group (its absorption at 450 nm gives this family its name). In the hydroxylation reactions, one atom of molecular oxygen is incorporated into the substrate and the second is reduced to H_2O, making cytochrome P-450 enzymes monooxygenases:

$$R-H + O_2 + NADPH + H^+ \longrightarrow R-OH + H_2O + NADP^+$$

In this reaction, two species are oxidized: NADPH and R—H.

There are dozens of P-450 enzymes, all situated in the inner mitochondrial membrane with their catalytic site exposed to the matrix. Steroidogenic cells are packed with mitochondria specialized for steroid synthesis; the mitochondria are generally larger than those in other tissues and have more extensive and highly convoluted inner membranes **(Fig. 19-37)**.

The path of electron flow in the mitochondrial P-450 system is complex, involving a flavoprotein and an iron-sulfur protein that carry electrons from NADPH to the P-450 heme **(Fig. 19-38)**. All P-450 enzymes have a heme that interacts with O_2 and a substrate-binding site that confers specificity.

Another large family of P-450 enzymes is found in the endoplasmic reticulum of hepatocytes. These enzymes catalyze reactions similar to the mitochondrial

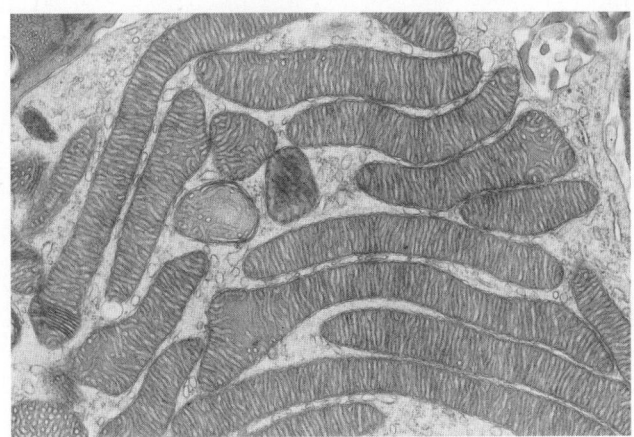

FIGURE 19-37 Mitochondria of adrenal gland, specialized for steroid synthesis. As seen in this electron micrograph of a thin section of adrenal gland, mitochondria are profuse and have extensive cristae, providing a large surface for the P-450 enzymes of the inner membrane. [Source: Don Fawcett/Science Source.]

P-450 reactions, but their substrates include a wide variety of hydrophobic compounds, many of which are **xenobiotics**—compounds not found in nature but synthesized industrially. The P-450 enzymes of the ER have very broad and overlapping substrate specificities. Hydroxylation of the hydrophobic compounds makes them more water-soluble, and they can then be cleared by the kidneys and excreted in urine. Among the substrates for these P-450 oxygenases are many commonly used prescription drugs. Metabolism by P-450 enzymes limits a drug's lifetime in the bloodstream and thus its therapeutic effects. Humans differ in their genetic complement of P-450 enzymes in the ER, as well as in the extent to which certain P-450 enzymes have been induced, such as by a history of ethanol ingestion. In principle, therefore, an individual's genetics and personal history could figure in determinations of therapeutic drug doses. In practice, this precise tailoring of dosage is not yet economically feasible, but it may become so. ■

Mitochondria Are Central to the Initiation of Apoptosis

Apoptosis, also called **programmed cell death**, is a process in which individual cells die for the good of the organism, such as in the course of normal embryonic development, and the organism conserves the cells' molecular components (amino acids, nucleotides, and so forth). Apoptosis may be triggered by an external signal, acting at a plasma membrane receptor, or by internal events such as DNA damage, viral infection, oxidative stress from the accumulation of ROS, or other stress such as a heat shock.

Mitochondria play a critical role in triggering apoptosis. When a stressor gives the signal for cell death, one early consequence is an increase in the permeability of the outer mitochondrial membrane, allowing cytochrome c to escape from the intermembrane space into the cytosol **(Fig. 19-39)**. The increased permeability is due to the opening of the **permeability transition pore complex (PTPC)**, a multisubunit protein in the outer membrane; its opening and closing are affected by several proteins that stimulate or suppress apoptosis. When released into the cytosol, cytochrome c interacts with monomers of the protein **Apaf-1 (apoptosis protease activating factor-1)**, causing the formation of an **apoptosome** composed of seven Apaf-1 and seven cytochrome c molecules. The apoptosome provides the platform on which the proenzyme procaspase-9 is activated to caspase-9, a member of a family of highly specific proteases, called the **caspases**, involved in apoptosis. These proteases share a critical *Cys* residue at their active site, and all cleave proteins only on the carboxyl-terminal side of *Asp* residues, thus the name "caspases." Caspase-9 initiates a cascade of proteolytic activations, with one caspase activating a second, and this in turn activating a third, and so forth (see Fig. 12-40). Note that this role of cytochrome c in apoptosis is a clear case of "moonlighting," in that one protein plays two very different roles in the cell (see Box 16-1).

SUMMARY 19.4 Mitochondria in Thermogenesis, Steroid Synthesis, and Apoptosis

■ In the brown adipose tissue of newborns, electron transfer is uncoupled from ATP synthesis, and the energy of fuel oxidation is dissipated as heat. Hibernating animals use this strategy to avoid freezing.

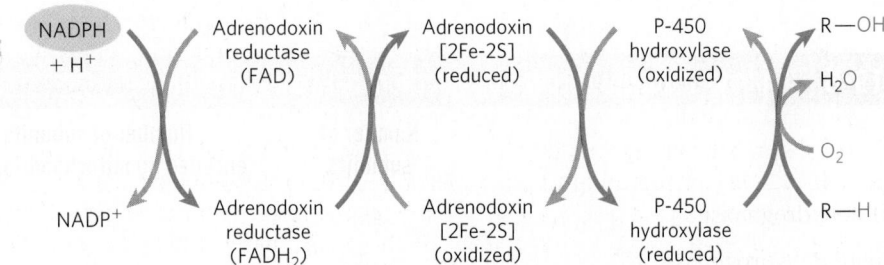

FIGURE 19-38 Path of electron flow in mitochondrial cytochrome P-450 reactions in adrenal gland. Two electrons are transferred from NADPH to the FAD-containing flavoprotein adrenodoxin reductase, which passes the electrons, one at a time, to adrenodoxin, a small, soluble 2Fe-2S protein. Adrenodoxin passes single electrons to the cytochrome P-450 hydroxylase, which interacts directly with O_2 and the substrate (R—H) to form the products, H_2O and R—OH.

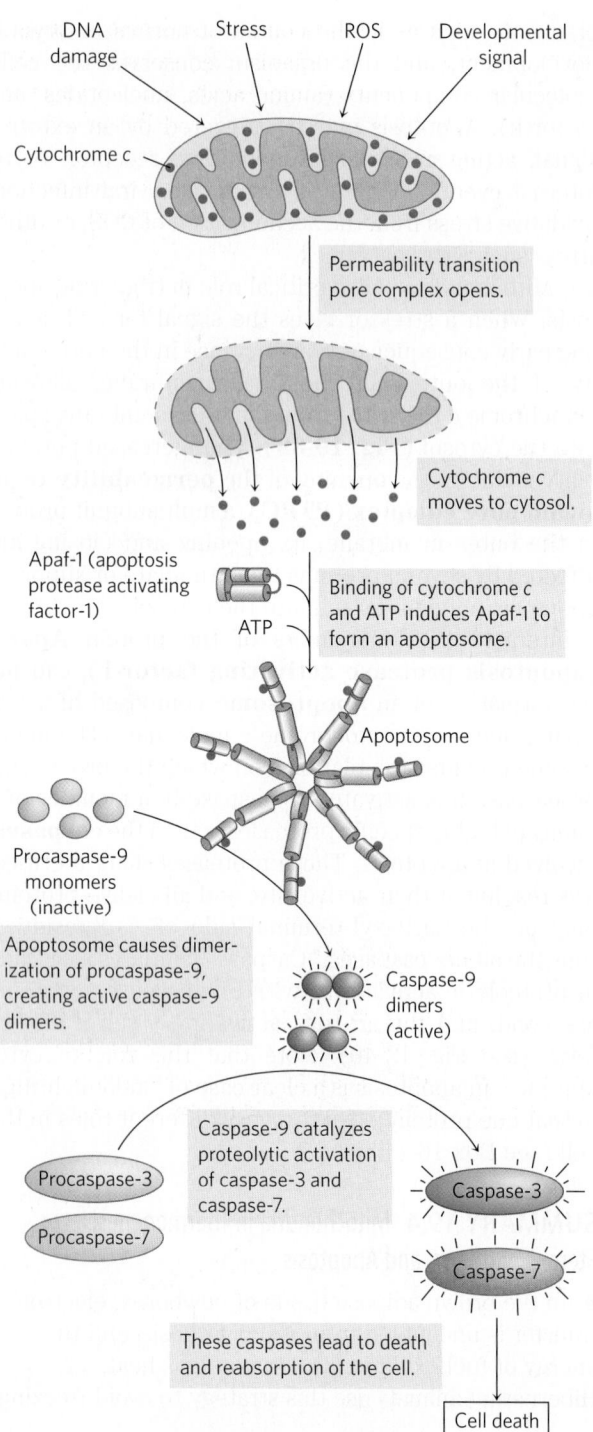

FIGURE 19-39 Role of cytochrome *c* in apoptosis. Cytochrome *c* is a small, soluble, mitochondrial protein, located in the intermembrane space, that carries electrons between Complex III and Complex IV during respiration. In a completely separate role, as outlined here, it acts as a trigger for apoptosis by stimulating the activation of a family of proteases called caspases. [Source: Information from S. J. Riedl and G. S. Salvesen, *Nature Rev. Mol. Cell Biol.* 8:409, 2007, Fig. 3.]

■ Hydroxylation reaction steps in the synthesis of steroid hormones in steroidogenic tissues (adrenal gland, gonads, liver, and kidney) take place in specialized mitochondria.

■ Mitochondrial cytochrome *c*, released into the cytosol, participates in activation of caspase-9, one of the proteases involved in apoptosis.

19.5 Mitochondrial Genes: Their Origin and the Effects of Mutations

Mitochondria contain their own genome, a circular, double-stranded DNA (mtDNA) molecule. Each of the hundreds or thousands of mitochondria in a typical cell has about five copies of this genome. The human mitochondrial chromosome **(Fig. 19-40)** contains 37 genes (16,569 bp), including 13 that encode subunits of proteins of the respiratory chain (Table 19-6); the remaining genes code for rRNA and tRNA molecules essential to the protein-synthesizing machinery of mitochondria. To synthesize these 13 protein subunits, mitochondria have their own ribosomes, distinctly different from those in the cytoplasm. The great majority of mitochondrial proteins—about 1,100 different types—are encoded by nuclear genes, synthesized on cytoplasmic ribosomes, then imported into and assembled in the mitochondria (Chapter 27).

Mitochondria Evolved from Endosymbiotic Bacteria

The existence of mitochondrial DNA, ribosomes, and tRNAs supports the theory of the endosymbiotic origin of mitochondria (see Fig. 1-40), which holds that the first organisms capable of aerobic metabolism, including respiration-linked ATP production, were bacteria. Primitive eukaryotes that lived anaerobically (by fermentation) acquired the ability to carry out oxidative

TABLE 19-6	Respiratory Proteins Encoded by Mitochondrial Genes in Humans		
Complex		Number of subunits	Number of subunits encoded by mitochondrial DNA
I NADH dehydrogenase		45	7
II Succinate dehydrogenase		4	0
III Ubiquinone:cytochrome *c* oxidoreductase		11	1
IV Cytochrome oxidase		13	3
V ATP synthase		8	2

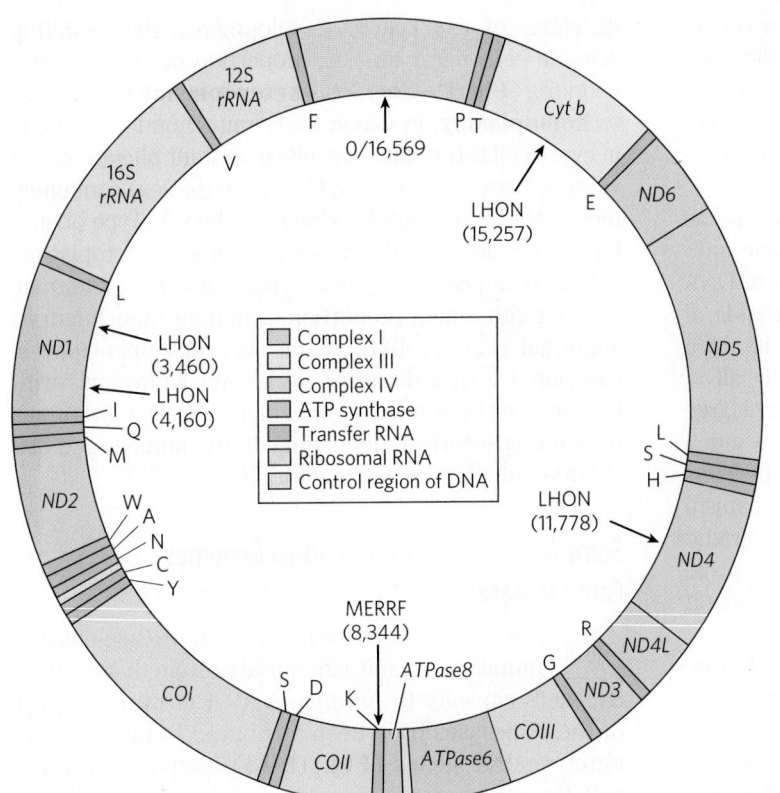

FIGURE 19-40 Mitochondrial genes and mutations. A map of human mitochondrial DNA, showing the genes that encode proteins of Complex I, the NADH dehydrogenase (*ND1* to *ND6*); the cytochrome *b* of Complex III (*Cyt b*); the subunits of cytochrome oxidase, Complex IV (*COI* to *COIII*); and two subunits of ATP synthase (*ATPase6* and *ATPase8*). The colors of the genes correspond to those of the complexes shown in Figure 19-7. Also included here are the genes for ribosomal RNAs (*rRNA*) and for some mitochondrion-specific transfer RNAs; tRNA specificity is indicated by the one-letter codes for amino acids. Arrows indicate the positions of mutations that cause Leber hereditary optic neuropathy (LHON) and myoclonic epilepsy and ragged-red fiber disease (MERRF). Numbers in parentheses indicate the position of the altered nucleotides (nucleotide 1 is at the top of the circle, and numbering proceeds counterclockwise). [Source: Information from M. A. Morris, *J. Clin. Neuroophthalmol.* 10:159, 1990.]

phosphorylation when they established a symbiotic relationship with bacteria living in their cytosol. After a long period of evolution and the movement of many bacterial genes into the nucleus of the "host" eukaryote, the endosymbiotic bacteria eventually became mitochondria.

This hypothesis presumes that early free-living bacteria had the enzymatic machinery for oxidative phosphorylation. And it predicts that their modern bacterial descendants must have respiratory chains closely similar to those of modern eukaryotes. They do. Aerobic bacteria carry out NAD-linked electron transfer from substrates to O_2, coupled to the phosphorylation of cytosolic ADP. The dehydrogenases are located in the bacterial cytosol, and the respiratory chain in the plasma membrane. The electron carriers translocate protons outward across the plasma membrane as electrons are transferred to O_2. Bacteria such as *E. coli* have F_oF_1 complexes in their plasma membranes; the F_1 portion protrudes into the cytosol and catalyzes ATP synthesis from ADP and P_i as protons flow back into the cell through the proton channel of F_o.

The respiration-linked extrusion of protons across the bacterial plasma membrane also provides the driving force for other processes. Certain bacterial transport systems bring about uptake of extracellular nutrients (lactose, for example) against a concentration gradient, in symport with protons (see Fig. 11-39). And the rotary motion of bacterial flagella is provided by "proton turbines," molecular rotary motors driven not by ATP but directly by the transmembrane electrochemical potential generated by respiration-linked proton pumping **(Fig. 19-41)**. It seems likely that the chemiosmotic mechanism evolved early, before the emergence of eukaryotes.

Mutations in Mitochondrial DNA Accumulate throughout the Life of the Organism

The respiratory chain is the major producer of reactive oxygen species in cells, so mitochondrial contents, including the mitochondrial genome, suffer the greatest exposure to, and damage by, ROS. Moreover, the mitochondrial DNA replication system is less effective than the nuclear system at correcting mistakes made during

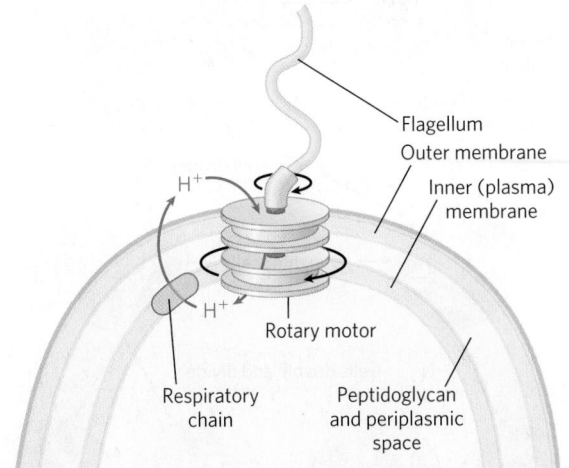

FIGURE 19-41 Rotation of bacterial flagella by proton-motive force. The shaft and rings at the base of the flagellum make up a rotary motor that has been called a "proton turbine." Protons ejected by electron transfer flow back into the cell through the turbine, causing rotation of the shaft of the flagellum. This motion differs fundamentally from the motion of muscle and of eukaryotic flagella and cilia, for which ATP hydrolysis is the energy source.

replication and at repairing DNA damage. As a consequence, defects in mtDNA accumulate over time. One theory of aging is that this gradual accumulation of defects is the primary cause of many of the "symptoms" of aging, which include, for example, progressive weakening of skeletal and heart muscle.

A unique feature of mitochondrial inheritance is the variation among individual cells, and between one individual organism and another, in the effects of a mtDNA mutation. A typical cell has hundreds or thousands of mitochondria, each with multiple copies of its own genome (Fig. 19-2b). Animals inherit essentially all of their mitochondria from the female parent. Eggs are large and contain 10^5 or 10^6 mitochondria; sperm are much smaller and contain far fewer mitochondria—perhaps 100 to 1,000. Furthermore, there is an active mechanism for targeting sperm-derived mitochondria for degradation in the fertilized egg. Just after fertilization, maternal phagosomes migrate to the site of sperm entry, engulf sperm mitochondria, and degrade them.

Suppose that, in a female organism, damage to one mitochondrial genome occurs in a germ cell from which oocytes develop, such that the germ cell contains mainly mitochondria with wild-type genes but one mitochondrion with a mutant gene. During the course of oocyte maturation, as this germ cell and its descendants repeatedly divide, the defective mitochondrion replicates and its progeny, all defective, are randomly distributed to daughter cells. Eventually, the mature egg cells contain different proportions of the defective mitochondria. When an egg cell is fertilized and undergoes the many divisions of embryonic development, the resulting somatic cells differ in their proportion of mutant mitochondria **(Fig. 19-42a)**. This **heteroplasmy** (in contrast to **homoplasmy**, in which every mitochondrial genome in every cell is the same) results in mutant phenotypes of varying degrees of severity. Cells (and tissues) containing mostly wild-type mitochondria have the wild-type phenotype; they are essentially normal. Other heteroplasmic cells have intermediate phenotypes, some almost normal, others (with a high proportion of mutant mitochondria) abnormal (Fig. 19-42b). If the abnormal phenotype is associated with a disease (see below), individuals with the same mtDNA mutation may have disease symptoms of differing severity—depending on the number and distribution of affected mitochondria.

Some Mutations in Mitochondrial Genomes Cause Disease

About 1 in 5,000 people have a disease-causing mutation in a mitochondrial protein that reduces the cell's capacity to produce ATP. A growing number of these diseases have been attributed to mutations in mitochondrial genes **(Fig. 19-43)**. Some tissues and cell types—neurons, myocytes of both skeletal and cardiac muscle, and β cells of the pancreas—are less able than others to tolerate lowered ATP production and are therefore more affected by mutations in mitochondrial proteins.

A group of genetic diseases known as the **mitochondrial encephalomyopathies** affect primarily the

FIGURE 19-42 Heteroplasmy in mitochondrial genomes. (a) When a mature egg cell is fertilized, all of the mitochondria in the resulting diploid cell (zygote) are maternal; none come from the sperm. If some fraction of the maternal mitochondria have a mutant gene, the random distribution of mitochondria during subsequent cell divisions yields some daughter cells with mostly mutant mitochondria, some with mostly wild-type mitochondria, and some in between. Thus daughter cells show a varying degree of heteroplasmy. **(b)** Different degrees of heteroplasmy produce different cellular phenotypes. This section of human muscle tissue is from an individual with defective cytochrome oxidase. The cells were stained so that wild-type cells are blue and cells with mutant cytochrome oxidase are brown. As the micrograph shows, different cells in the same tissue are affected to different degrees by the mitochondrial mutation. [Source: (b) Courtesy of Rob Taylor. Reprinted with permission from R. W. Taylor and D. M. Turnbull, *Nature Rev. Genet.* 6:389, 2005, Fig. 2a.]

Mutant mitochondria
Wild-type mitochondria
Parent cell
Contents double
Cell divides
Daughter cells
Cells double and divide
Largely mutant
Largely wild type
Fully wild type
Heteroplasmy
Homoplasmy

(a)

(b)

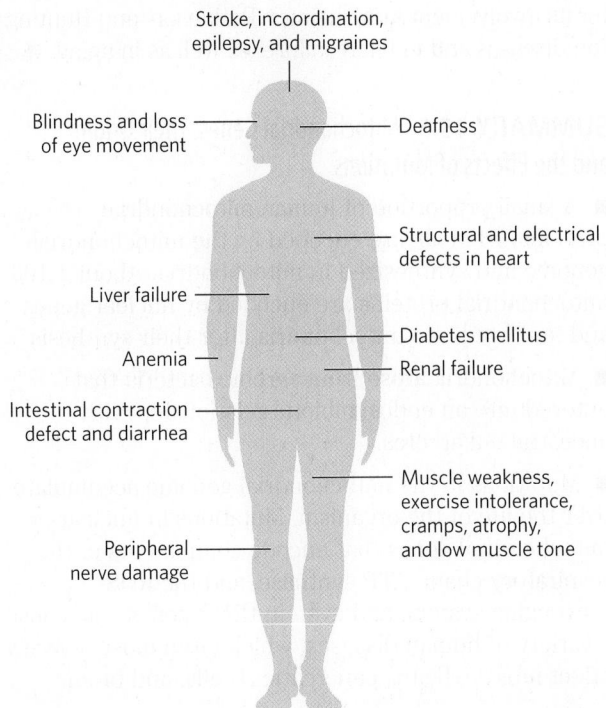

Stroke, incoordination, epilepsy, and migraines

Blindness and loss of eye movement

Deafness

Liver failure

Structural and electrical defects in heart

Anemia

Diabetes mellitus

Renal failure

Intestinal contraction defect and diarrhea

Muscle weakness, exercise intolerance, cramps, atrophy, and low muscle tone

Peripheral nerve damage

FIGURE 19-43 Mutations that cause mitochondrial disorders. Mutations in genes that encode mitochondrial proteins can lead to disorders affecting many organs and organ systems. [Source: Information from S. B Vafai and V. K. Mootha, *Nature* 491:374, 2012.]

brain and skeletal muscle. These diseases are invariably inherited from the mother, because, as noted above, a developing embryo derives all its mitochondria from the egg. The rare disease **Leber hereditary optic neuropathy (LHON)** affects the central nervous system, including the optic nerves, causing bilateral loss of vision in early adulthood. A single base change in the mitochondrial gene *ND4* (Fig. 19-40) changes an Arg residue to a His residue in a polypeptide of Complex I, and the result is mitochondria partially defective in electron transfer from NADH to ubiquinone. Although these mitochondria can produce some ATP by electron transfer from succinate, they apparently cannot supply sufficient ATP to support the very active metabolism of neurons, including the optic nerve. A single base change in the mitochondrial gene for cytochrome *b*, a component of Complex III, also produces LHON, demonstrating that the pathology results from a general reduction of mitochondrial function, not specifically from a defect in electron transfer through Complex I.

A mutation in *ATP6* affects the proton pore in ATP synthase, leading to low rates of ATP synthesis while leaving the respiratory chain intact. Oxidative stress due to the continued supply of electrons from NADH increases the production of ROS, and the damage to mitochondria caused by ROS sets up a vicious cycle. Half of individuals with this mutant gene die within days or months of birth. **Myoclonic epilepsy with ragged-red fiber syndrome (MERRF)** is caused by a mutation in the mitochondrial gene that encodes a tRNA specific for lysine (tRNA^{Lys}). This disease, characterized by uncontrollable muscular jerking, results from defective production of several of the proteins that require mitochondrial tRNAs for their synthesis. Skeletal muscle fibers of individuals with MERRF have abnormally shaped mitochondria that sometimes contain paracrystalline structures **(Fig. 19-44)**. Other mutations in mitochondrial genes are believed to be responsible for the progressive muscular weakness that characterizes mitochondrial myopathy and for enlargement and deterioration of the heart muscle in hypertrophic cardiomyopathy.

If a prospective mother is known to carry a pathogenic mitochondrial gene, the technique of mitochondrial donation can circumvent the passage of that mutant gene to her offspring. A prospective mother's nuclear genes are microscopically transplanted into an enucleated ovum from a donor with healthy mitochondria, then the ovum is fertilized in vitro and the resulting embryo is transplanted into the mother's uterus. This and similar "three-parent baby" procedures, which were approved in the United Kingdom in 2015, raise ethical issues that are being vigorously debated.

Mitochondrial disease can also result from mutations in any of the ~1,100 nuclear genes that encode mitochondrial proteins. For example, a mutation in one of the nuclear-encoded proteins of Complex IV, COX6B1, results in severe defects in brain development and thickened walls of the heart muscle. Other nuclear genes encode proteins essential for the *assembly* of mitochondrial complexes. Mutations in these genes can also lead to serious mitochondrial disease. ∎

A Rare Form of Diabetes Results from Defects in the Mitochondria of Pancreatic β Cells

The mechanism that regulates the release of insulin from pancreatic β cells hinges on the ATP concentration in those cells. When blood glucose is high, β cells take up glucose and oxidize it by glycolysis and

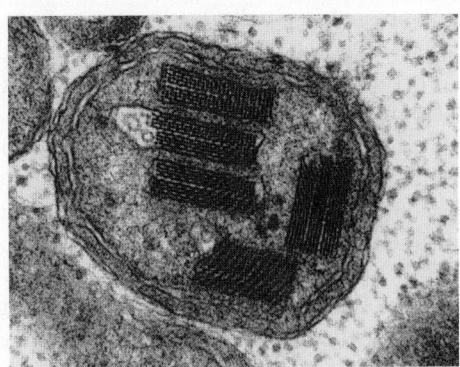

FIGURE 19-44 Paracrystalline inclusions in MERRF mitochondrion. Electron micrograph of an abnormal mitochondrion from the muscle of an individual with MERRF, showing the paracrystalline protein inclusions sometimes present in the mutant mitochondria. [Source: D. Wallace et al., *Cell* 55:601, 1988. © Elsevier.]

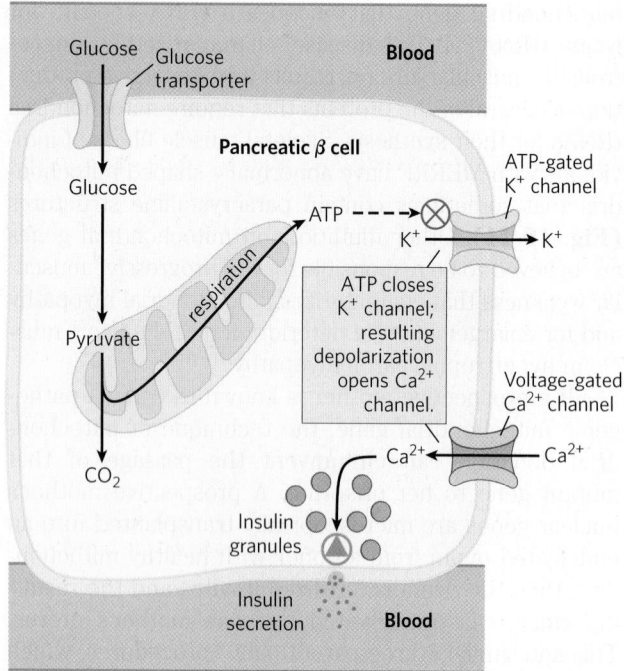

FIGURE 19-45 A mitochondrial defect prevents insulin secretion. In the normal situation, as depicted here, when the blood glucose level rises, production of ATP in β cells increases. ATP, by blocking K⁺ channels, depolarizes the plasma membrane and thus opens voltage-gated Ca²⁺ channels. The resulting influx of Ca²⁺ triggers exocytosis of insulin-containing secretory vesicles, releasing insulin. When oxidative phosphorylation in β cells is defective, [ATP] is never sufficient to trigger this process, and insulin is not released.

the citric acid cycle, raising [ATP] above a threshold level **(Fig. 19-45)**. When [ATP] exceeds this threshold, an ATP-gated K⁺ channel in the plasma membrane closes, depolarizing the membrane and triggering insulin release (see Fig. 23-28). Pancreatic β cells with defects in oxidative phosphorylation cannot increase [ATP] above this threshold, and the resulting failure of insulin release effectively produces diabetes. For example, defects in the gene for glucokinase, the hexokinase IV isozyme present in β cells, lead to a rare form of diabetes called MODY2 (see Box 15-3); low glucokinase activity prevents the generation of above-threshold [ATP], blocking insulin secretion. Mutations in the mitochondrial tRNA^Lys or tRNA^Leu genes also compromise mitochondrial ATP production, and type 2 diabetes mellitus is common among individuals with these defects (although such cases make up a very small fraction of all cases of diabetes).

When nicotinamide nucleotide transhydrogenase, which is part of the mitochondrial defense against ROS (Fig. 19-18), is genetically defective, the accumulation of ROS damages mitochondria, slowing ATP production and blocking insulin release by β cells (Fig. 19-45). Damage caused by ROS, including damage to mtDNA, may also underlie other human diseases; there is some evidence

for its involvement in Alzheimer, Parkinson, and Huntington diseases and in heart failure, as well as in aging. ∎

SUMMARY 19.5 Mitochondrial Genes: Their Origin and the Effects of Mutations

∎ A small proportion of human mitochondrial proteins, 13 in all, are encoded by the mitochondrial genome and synthesized in mitochondria. About 1,100 mitochondrial proteins are encoded by nuclear genes and imported into mitochondria after their synthesis.

∎ Mitochondria arose from aerobic bacteria that entered into an endosymbiotic relationship with ancestral eukaryotes.

∎ Mutations in the mitochondrial genome accumulate over the life of the organism. Mutations in nuclear or mitochondrial genes that encode components of the respiratory chain, ATP synthase, and the ROS-scavenging system, and even in tRNA genes, can cause a variety of human diseases, which often most severely affect muscle, heart, pancreatic β cells, and brain.

∎ It is possible to combine the mitochondria from one woman with the nuclear genes of another to create an ovum free of a mutation that would have led to a mitochondrial disease.

Key Terms

Terms in bold are defined in the glossary.

chemiosmotic theory 711	**proton-motive**
respiratory chain 712	**force** 725
flavoprotein 714	**ATP synthase** 729
reducing equivalent 714	**F₁ ATPase** 732
ubiquinone (coenzyme	rotational catalysis 735
Q, Q) 714	binding-change model 735
cytochromes 715	**P/O ratio** 738
iron-sulfur protein 716	P/2e⁻ ratio 738
Rieske iron-sulfur	malate-aspartate
protein 716	shuttle 739
Complex I 717	glycerol 3-phosphate
NADH dehydrogenase 717	shuttle 739
vectorial 718	**acceptor control** 741
Complex II 719	**mass-action ratio**
succinate dehydrogenase 719	**(Q)** 741
reactive oxygen species	**brown adipose tissue**
(ROS) 719	**(BAT)** 744
superoxide radical	**uncoupling protein 1**
(•O₂⁻) 719	**(UCP1)** 744
Complex III 720	**cytochrome P-450** 744
cytochrome *bc₁* complex 720	xenobiotics 745
Q cycle 721	**apoptosis** 745
Complex IV 721	apoptosome 745
cytochrome oxidase 721	caspase 745
cryo-electron	heteroplasmy 748
microscopy 724	homoplasmy 748

Problems

1. Oxidation-Reduction Reactions Complex I, the NADH dehydrogenase complex of the mitochondrial respiratory chain, promotes the following series of oxidation-reduction reactions, in which Fe^{3+} and Fe^{2+} represent the iron in iron-sulfur centers, Q is ubiquinone, QH_2 is ubiquinol, and E is the enzyme:

(1) $NADH + H^+ + E\text{-}FMN \longrightarrow NAD^+ + E\text{-}FMNH_2$

(2) $E\text{-}FMNH_2 + 2Fe^{3+} \longrightarrow E\text{-}FMN + 2Fe^{2+} + 2H^+$

(3) $2Fe^{2+} + 2H^+ + Q \longrightarrow 2Fe^{3+} + QH_2$

Sum: $NADH + H^+ + Q \longrightarrow NAD^+ + QH_2$

For each of the three reactions catalyzed by Complex I, identify (a) the electron donor, (b) the electron acceptor, (c) the conjugate redox pair, (d) the reducing agent, and (e) the oxidizing agent.

2. All Parts of Ubiquinone Have a Function In electron transfer, only the quinone portion of ubiquinone undergoes oxidation-reduction; the isoprenoid side chain remains unchanged. What is the function of this chain?

3. Use of FAD Rather Than NAD⁺ in Succinate Oxidation All the dehydrogenases of glycolysis and the citric acid cycle use NAD^+ (E'° for $NAD^+/NADH$ is -0.32 V) as electron acceptor except succinate dehydrogenase, which uses covalently bound FAD (E'° for $FAD/FADH_2$ in this enzyme is 0.050 V). Suggest why FAD is a more appropriate electron acceptor than NAD^+ in the dehydrogenation of succinate, based on the E'° values of fumarate/succinate ($E'^{\circ} = 0.031$ V), $NAD^+/NADH$, and the succinate dehydrogenase $FAD/FADH_2$.

4. Degree of Reduction of Electron Carriers in the Respiratory Chain The degree of reduction of each carrier in the respiratory chain is determined by conditions in the mitochondrion. For example, when NADH and O_2 are abundant, the steady-state degree of reduction of the carriers decreases as electrons pass from the substrate to O_2. When electron transfer is blocked, the carriers before the block become more reduced and those beyond the block become more oxidized (see Fig. 19-6). For each of the conditions below, predict the state of oxidation of ubiquinone and cytochromes b, c_1, c, and $a + a_3$.

(a) Abundant NADH and O_2, but cyanide added
(b) Abundant NADH, but O_2 exhausted
(c) Abundant O_2, but NADH exhausted
(d) Abundant NADH and O_2

5. Effect of Rotenone and Antimycin A on Electron Transfer Rotenone, a toxic natural product from plants, strongly inhibits NADH dehydrogenase of insect and fish mitochondria. Antimycin A, a toxic antibiotic, strongly inhibits the oxidation of ubiquinol.

(a) Explain why rotenone ingestion is lethal to some insect and fish species.
(b) Explain why antimycin A is a poison.

(c) Given that rotenone and antimycin A are equally effective in blocking their respective sites in the electron-transfer chain, which would be a more potent poison? Explain.

6. Uncouplers of Oxidative Phosphorylation In normal mitochondria, the rate of electron transfer is tightly coupled to the demand for ATP. When the rate of use of ATP is relatively low, the rate of electron transfer is low; when demand for ATP increases, the electron-transfer rate increases. Under these conditions of tight coupling, the number of ATP molecules produced per atom of oxygen consumed when NADH is the electron donor—the P/O ratio—is about 2.5.

(a) Predict the effect of a relatively low and a relatively high concentration of uncoupling agent on the rate of electron transfer and the P/O ratio.

(b) Ingestion of uncouplers causes profuse sweating and an increase in body temperature. Explain this phenomenon in molecular terms. What happens to the P/O ratio in the presence of uncouplers?

(c) The uncoupler 2,4-dinitrophenol was once prescribed as a weight-reducing drug. How could this agent, in principle, serve as a weight-reducing aid? Uncoupling agents are no longer prescribed, because some deaths occurred following their use. Why might the ingestion of uncouplers cause death?

7. Effects of Valinomycin on Oxidative Phosphorylation When the antibiotic valinomycin (see Fig. 11-42) is added to actively respiring mitochondria, several things happen: the yield of ATP decreases, the rate of O_2 consumption increases, heat is released, and the pH gradient across the inner mitochondrial membrane increases. Does valinomycin act as an uncoupler or as an inhibitor of oxidative phosphorylation? Explain the experimental observations in terms of the antibiotic's ability to transfer K^+ ions across the inner mitochondrial membrane.

8. Cellular ADP Concentration Controls ATP Formation Although both ADP and P_i are required for the synthesis of ATP, the rate of synthesis depends mainly on the concentration of ADP, not P_i. Why?

9. Advantages of Supercomplexes for Electron Transfer There is growing evidence that mitochondrial Complexes I, II, III, and IV are part of a larger supercomplex. What might be the advantage of having all four complexes within a supercomplex?

10. How Many Protons in a Mitochondrion? Electron transfer translocates protons from the mitochondrial matrix to the external medium, establishing a pH gradient across the inner membrane (outside more acidic than inside). The tendency of protons to diffuse back into the matrix is the driving force for ATP synthesis by ATP synthase. During oxidative phosphorylation by a suspension of mitochondria in a medium of pH 7.4, the pH of the matrix has been measured as 7.7.

(a) Calculate [H⁺] in the external medium and in the matrix under these conditions.

(b) What is the outside-to-inside ratio of [H⁺]? Comment on the energy inherent in this concentration difference. (Hint: See Eqn 11-4, p. 413.)

(c) Calculate the number of protons in a respiring liver mitochondrion, assuming its inner matrix compartment is a sphere of diameter 1.5 μm.

(d) From these data, is the pH gradient alone sufficient to generate ATP?

(e) If not, suggest how the necessary energy for synthesis of ATP arises.

11. Rate of ATP Turnover in Rat Heart Muscle Rat heart muscle operating aerobically fills more than 90% of its ATP needs by oxidative phosphorylation. Each gram of tissue consumes O_2 at the rate of 10.0 μmol/min, with glucose as the fuel source.

(a) Calculate the rate at which the heart muscle consumes glucose and produces ATP.

(b) For a steady-state concentration of ATP of 5.0 μmol/g of heart muscle tissue, calculate the time required (in seconds) to completely turn over the cellular pool of ATP. What does this result indicate about the need for tight regulation of ATP production? (Note: Concentrations are expressed as micromoles per gram of muscle tissue because the tissue is mostly water.)

12. Rate of ATP Breakdown in Insect Flight Muscle ATP production in the flight muscle of the fly *Lucilia sericata* results almost exclusively from oxidative phosphorylation. During flight, 187 mL of O_2/h·g of body weight is needed to maintain an ATP concentration of 7.0 μmol/g of flight muscle. Assuming that flight muscle makes up 20% of the weight of the fly, calculate the rate at which the flight-muscle ATP pool turns over. How long would the reservoir of ATP last in the absence of oxidative phosphorylation? Assume that reducing equivalents are transferred by the glycerol 3-phosphate shuttle and that O_2 is at 25°C and 101.3 kPa (1 atm).

13. High Blood Alanine Level Associated with Defects in Oxidative Phosphorylation Most individuals with genetic defects in oxidative phosphorylation are found to have relatively high concentrations of alanine in their blood. Explain this in biochemical terms.

14. Compartmentalization of Citric Acid Cycle Components Isocitrate dehydrogenase is found only in mitochondria, but malate dehydrogenase is found in both the cytosol and mitochondria. What is the role of cytosolic malate dehydrogenase?

15. Transmembrane Movement of Reducing Equivalents Under aerobic conditions, extramitochondrial NADH must be oxidized by the mitochondrial respiratory chain. Consider a preparation of rat hepatocytes containing mitochondria and all the cytosolic enzymes. If [4-³H]NADH is introduced, radioactivity soon appears in the mitochondrial matrix. However, if [7-¹⁴C]NADH is introduced, no radioac-

tivity appears in the matrix. What do these observations reveal about the oxidation of extramitochondrial NADH by the respiratory chain?

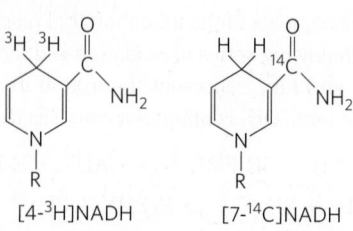

[4-³H]NADH   [7-¹⁴C]NADH

16. NAD Pools and Dehydrogenase Activities Although both pyruvate dehydrogenase and glyceraldehyde 3-phosphate dehydrogenase use NAD^+ as their electron acceptor, the two enzymes do not compete for the same cellular NAD pool. Why?

17. The Malate–α-Ketoglutarate Transport System The transport system that conveys malate and α-ketoglutarate across the inner mitochondrial membrane (see Fig. 19-31) is inhibited by *n*-butylmalonate. Suppose *n*-butylmalonate is added to an aerobic suspension of kidney cells using glucose exclusively as fuel. Predict the effect of this inhibitor on (a) glycolysis, (b) oxygen consumption, (c) lactate formation, and (d) ATP synthesis.

18. Time Scales of Regulatory Events in Mitochondria Compare the likely time scales for the adjustments in respiratory rate caused by (a) increased [ADP] and (b) reduced pO_2. What accounts for the difference?

19. The Pasteur Effect When O_2 is added to an anaerobic suspension of cells consuming glucose at a high rate, the rate of glucose consumption declines greatly as the O_2 is used up, and accumulation of lactate ceases. This effect, first observed by Louis Pasteur in the 1860s, is characteristic of most cells capable of both aerobic and anaerobic glucose catabolism.

(a) Why does the accumulation of lactate cease after O_2 is added?

(b) Why does the presence of O_2 decrease the rate of glucose consumption?

(c) How does the onset of O_2 consumption slow down the rate of glucose consumption? Explain in terms of specific enzymes.

20. Respiration-Deficient Yeast Mutants and Ethanol Production Respiration-deficient yeast mutants (p⁻; "petites") can be produced from wild-type parents by treatment with mutagenic agents. The mutants lack cytochrome oxidase, a deficit that markedly affects their metabolic behavior. One striking effect is that fermentation is not suppressed by O_2—that is, the mutants lack the Pasteur effect (see Problem 19). Some companies are very interested in using these mutants to ferment wood chips to ethanol for energy use. Explain the advantages of using these mutants rather than wild-type yeast for large-scale ethanol production. Why does the absence of cytochrome oxidase eliminate the Pasteur effect?

21. Mitochondrial Disease and Cancer Mutations in the genes that encode certain mitochondrial proteins are associated with a high incidence of some types of cancer. How might defective mitochondria lead to cancer?

22. Variable Severity of a Mitochondrial Disease
Different individuals with a disease caused by the same specific defect in the mitochondrial genome may have symptoms ranging from mild to severe. Explain why.

23. Diabetes as a Consequence of Mitochondrial Defects
Glucokinase is essential in the metabolism of glucose in pancreatic β cells. Humans with two defective copies of the glucokinase gene exhibit a severe, neonatal diabetes, whereas those with only one defective copy of the gene have a much milder form of the disease (mature onset diabetes of the young, MODY2). Explain this difference in terms of the biology of the β cell.

24. Effects of Mutations in Mitochondrial Complex II
Single nucleotide changes in the gene for succinate dehydrogenase (Complex II) are associated with midgut carcinoid tumors. Suggest a mechanism to explain this observation.

Data Analysis Problem

25. Identifying a Protein Central to the Activity of ATP Synthase
Much of our knowledge about the steps in the respiratory chain and the mechanism of ATP synthase came about by dissecting the pathway, using various inhibitors and uncouplers (see Table 19-4) and bacterial mutants. In this problem, we see how Robert Fillingame used dicyclohexylcarbodiimide (DCCD) and *E. coli* mutants resistant to its effects to identify the components that came to be known as the c subunits of the F_o portion of ATP synthase.

DCCD reacts with carboxyl groups in the side chains of Asp and Glu residues. When DCCD is added to a suspension of intact, actively respiring mitochondria, the rate of electron transfer (measured by O_2 consumption) and the rate of ATP production dramatically decrease. If a solution of 2,4-dinitrophenol (DNP) is now added to the preparation, O_2 consumption returns to normal, but ATP production remains inhibited.

(a) Explain the effect of DNP on the inhibited mitochondrial preparation.

(b) Which process is directly affected by DCCD, electron transfer or ATP synthesis?

E. coli carries out oxidative phosphorylation with machinery remarkably similar to that in mammals, and *E. coli* is far more amenable to mutant selection. Addition of DCCD to a culture of wild-type *E. coli* (strain AN180) growing aerobically blocks further growth in a time- and dose-dependent fashion.

Fillingame selected a DCCD-resistant mutant of *E. coli* (RF-7) for which aerobic growth was only slightly diminished in the presence of DCCD. Next, he needed to demonstrate that the DCCD-resistant component in his *E. coli* strains was the ATP synthase. He isolated the membrane fraction from the wild-type and RF-7 strains and assayed them for ATPase activity in the presence and absence of DCCD. He found time- and dose-dependent inhibition of the ATPase activity in the membrane fraction of the wild-type, but not in the RF-7 membrane fraction.

(c) Why did Fillingame assay ATPase activity instead of ATP synthase activity?

(d) Is the DCCD-binding protein missing in the mutant RF-7, or just altered?

Fillingame wanted to know whether the DCCD-sensitive protein was an integral part of the membrane or could be solubilized into the fraction that contained the ATPase activity. He prepared "stripped membrane" and "soluble ATPase" fractions from both wild-type cells and RF-7 mutants, by treating intact membranes with dithiothreitol. He measured the ATPase activity in the native membranes, in the stripped membranes, and in systems reconstituted by mixing the stripped membranes with the soluble fraction from the wild-type or RF-7 mutant strain. The native membranes and reconstituted systems all had ATPase activity; the stripped membrane fractions had very little ATPase activity. Having established that all combinations of reconstituted systems had similar ATPase activity, Fillingame then added DCCD to see which combinations were inhibited.

(e) What results would you expect if the DCCD-binding protein were in the stripped membranes? What would you expect if it were in the soluble fraction?

The results were clear. For the stripped membranes from wild-type cells, the reconstituted ATPase was sensitive to DCCD, regardless of the source of the soluble fraction. For the stripped membranes from mutant cells, the reconstituted ATPase was insensitive to DCCD. So, DCCD sensitivity is due to a protein in the stripped membrane fraction, not to a protein in the fraction solubilized with dithiothreitol.

To identify the DCCD-sensitive protein, Fillingame exposed intact membranes of the wild-type (AN180) and RF-7 *E. coli* to ^{14}C-labeled DCCD, then used SDS-PAGE to separate the proteins. He cut the gel into thin slices from bottom to top and determined the ^{14}C content of each slice, measured as disintegrations per minute (dpm) per 2 mm gel slice, normalized to the amount of protein applied to the gel. The distance migrated is equal to the slice number times 2 mm. The results are plotted below. The arrows denote cytochrome *c*, used as a molecular mass marker; I and II, peaks of interest; and BPB, bromphenol blue, a tracking dye to indicate the front of the sample as it moves through the gel. Many proteins from each sample were labeled with [^{14}C] DCCD (measured in dpm, disintegrations per minute).

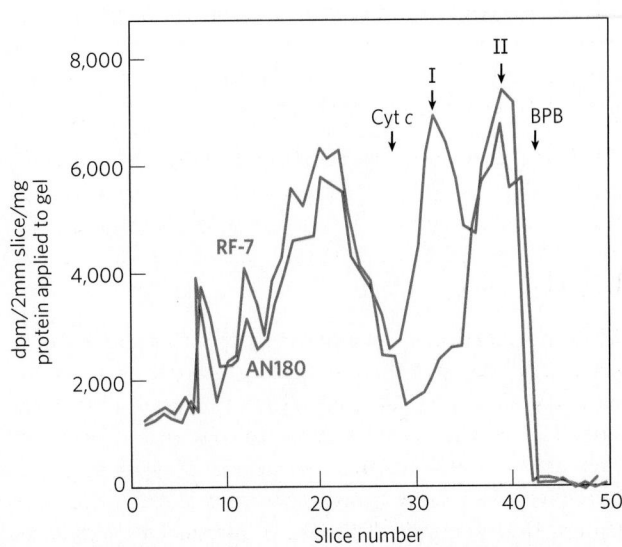

(f) How did Fillingame know which labeled protein(s) was/were of interest?

(g) When he repeated the experiment using "stripped membranes" prepared as before, he found the same protein was specifically labeled in the wild-type fraction. Why was this step necessary?

(h) The protein of interest proved to have an M_r of about 9 kDa, and was readily soluble in a very nonpolar solvent (chloroform/methanol). What could Fillingame deduce about the structure, location, and topology of this protein?

(i) In later work, Fillingame found that the residue that reacts with DCCD in this protein is Asp[61]. An *E. coli* mutant in which this protein has a Ser residue substituted for Ala[21] is much less sensitive to inhibition by DCCD than is the wild type. What explanation can you give for this observation?

(j) Extensive studies of this DCCD-inhibited protein have shown it to be a central part of the F_oF_1 ATP synthase of bacteria, plants, and animals. What is its role in oxidative phosphorylation?

Reference

Fillingame, R.H. 1975. Identification of the dicyclohexylcarbodiimide-reactive protein component of the adenosine 5′-triphosphate energy-transducing system of *Escherichia coli. J. Bacteriol.* 124:870–883.

Further Reading is available at www.macmillanlearning.com/LehningerBiochemistry7e.

CHAPTER 20

Photosynthesis and Carbohydrate Synthesis in Plants

Self-study tools that will help you practice what you've learned and reinforce this chapter's concepts are available online.
Go to www.macmillanlearning.com/LehningerBiochemistry7e.

We have now reached a turning point in our study of cellular metabolism. Thus far in Part II we have described how the major metabolic fuels—carbohydrates, fatty acids, and amino acids—are degraded through converging *catabolic* pathways to enter the citric acid cycle and yield their electrons to the respiratory chain, driving ATP synthesis by *oxidative* phosphorylation. We now turn to ATP synthesis coupled to the *light-driven* flow of electrons to oxygen, and then to the *anabolic* pathways, which use chemical energy in the form of ATP and NADH or NADPH to synthesize cellular components from simple precursor molecules. Anabolic pathways are generally reductive

rather than oxidative. Catabolism and anabolism proceed simultaneously in a dynamic steady state, so the energy-yielding degradation of cellular components is counterbalanced by biosynthetic processes, which create and maintain the intricate orderliness of living cells.

The capture of solar energy by photosynthetic organisms and its conversion to the chemical energy of reduced organic compounds is the ultimate source of nearly all biological energy and organic precursors on Earth. The evolution of oxygenic (oxygen-evolving) photosynthesis about 2.5 billion years ago, and the consequent rise in atmospheric oxygen, shaped the metabolic landscape we have inherited. Photosynthetic and heterotrophic organisms live in a balanced steady state in the biosphere **(Fig. 20-1)**. Photosynthetic organisms trap solar energy and form ATP and NADPH, which they use as energy sources to make carbohydrates and other organic compounds from CO_2 and H_2O; simultaneously, they release O_2 into the atmosphere.

$$CO_2 + H_2O \xrightarrow{\text{light}} (CH_2O) + O_2$$

Aerobic heterotrophs (humans, for example, as well as plants during dark periods) use the O_2 so formed to degrade the energy-rich organic products of photosynthesis to CO_2 and H_2O, generating ATP. The CO_2 returns to the atmosphere, to be used again by photosynthetic organisms. Solar energy thus provides the driving force for the continuous cycling of CO_2 and O_2 through the biosphere and provides the reduced substrates—fuels,

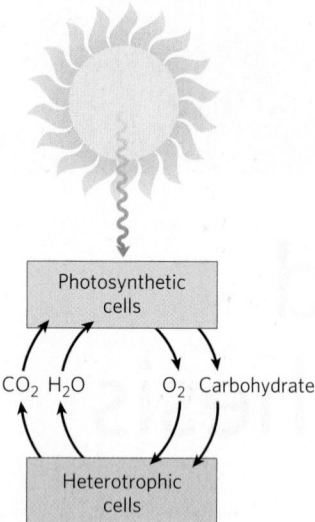

FIGURE 20-1 Solar energy as the ultimate source of all biological energy. Photosynthetic organisms use the energy of sunlight to manufacture glucose and other organic products, which heterotrophic cells use as energy and carbon sources.

such as glucose—on which nonphotosynthetic organisms depend.

In this chapter we begin with **photosynthesis**, which encompasses two processes: the **light-dependent reactions**, in which sunlight provides the energy for the synthesis of ATP and NADPH, and the **carbon-assimilation reactions** (or **carbon-fixation reactions**), in which ATP and NADPH are used to reduce CO_2 to form triose phosphates via a set of reactions known as the Calvin cycle **(Fig. 20-2)**. Photorespiration is an unproductive side reaction during CO_2 fixation, and we look at

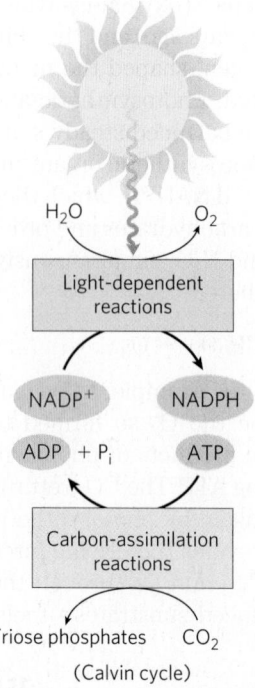

FIGURE 20-2 The light-dependent reactions of photosynthesis generate energy-rich NADPH and ATP at the expense of solar energy. NADPH and ATP are used in the carbon-assimilation reactions, which occur in light or darkness, to reduce CO_2 to form trioses and more complex compounds, such as glucose and sucrose, derived from trioses.

several ways that certain plants can avoid this. We next discuss the conversion of trioses produced in the Calvin cycle to sucrose (for sugar transport) and starch (for energy storage), which is accomplished by mechanisms analogous to those used by animal cells to make glycogen, and describe the synthesis of the cellulose of plant cell walls. Finally, we consider how carbohydrate metabolism is integrated within a plant cell and throughout the plant body.

20.1 Light Absorption

The process of **photophosphorylation** resembles oxidative phosphorylation (Chapter 19) in that electron flow through a series of membrane carriers is coupled to proton pumping, producing the proton motive force that powers ATP formation. In oxidative phosphorylation, the electron donor is NADH and the ultimate electron acceptor is O_2, forming H_2O. In photophosphorylation, electrons flow in the opposite direction: H_2O is the electron *donor* and NADPH is *formed*. How are these endergonic processes possible?

Water is a poor donor of electrons; its standard reduction potential is 0.816 V, compared with -0.320 V for NADH, a good electron donor. Photosynthesis requires the input of energy in the form of light to *create* a good electron donor and a good electron acceptor. Electrons flow from the electron donor through a series of membrane-bound carriers, including cytochromes, quinones, and iron-sulfur proteins, while protons are pumped across a membrane to create an electrochemical potential. Electron transfer and proton pumping are catalyzed by a membrane complex homologous in structure and function to Complex III of mitochondria. The electrochemical potential so produced is the driving force for ATP synthesis from ADP and P_i, catalyzed by a membrane-bound ATP synthase complex closely similar to that of mitochondria and bacteria. The process is summarized in **Figure 20-3**.

Chloroplasts Are the Site of Light-Driven Electron Flow and Photosynthesis in Plants

In photosynthetic eukaryotic cells, both the light-dependent and the carbon-assimilation reactions take place in **chloroplasts (Fig. 20-4)**, organelles that are variable in shape and generally a few micrometers in diameter. Like mitochondria, chloroplasts are surrounded by two membranes: an outer membrane that is permeable to small molecules and ions, and an inner membrane that encloses an internal compartment. This compartment is called the **stroma** in chloroplasts and is analogous to the mitochondrial matrix; it is an aqueous phase containing most of the enzymes required for the carbon-assimilation reactions. Throughout the stroma is a highly convoluted set of internal membranes, analogous to the mitochondrial cristae; these membranes are topologically continuous, forming a single compartment

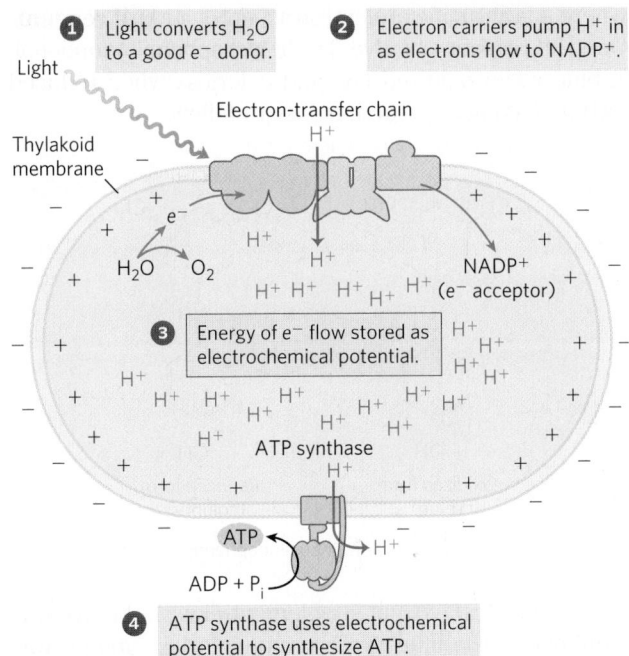

1 Light converts H_2O to a good e^- donor.

2 Electron carriers pump H^+ in as electrons flow to $NADP^+$.

Light

Thylakoid membrane

Electron-transfer chain

H^+

e^-

H_2O O_2

$NADP^+$ (e^- acceptor)

3 Energy of e^- flow stored as electrochemical potential.

ATP synthase

ATP

$ADP + P_i$

4 ATP synthase uses electrochemical potential to synthesize ATP.

FIGURE 20-3 The chemiosmotic mechanism for ATP synthesis in chloroplasts. Movement of electrons through a chain of membrane-bound carriers is driven by the energy of photons absorbed by the green pigment chlorophyll. Electron flow leads to the movement of protons and positive charge across the membrane, creating an electrochemical potential. This electrochemical potential drives ATP synthesis by the membrane-bound ATP synthase, which is fundamentally similar in structure and mechanism to the ATP synthase of mitochondria. Comparison with Figure 19-1 shows the many similarities and significant differences in chemiosmotic synthesis of ATP in chloroplasts and mitochondria.

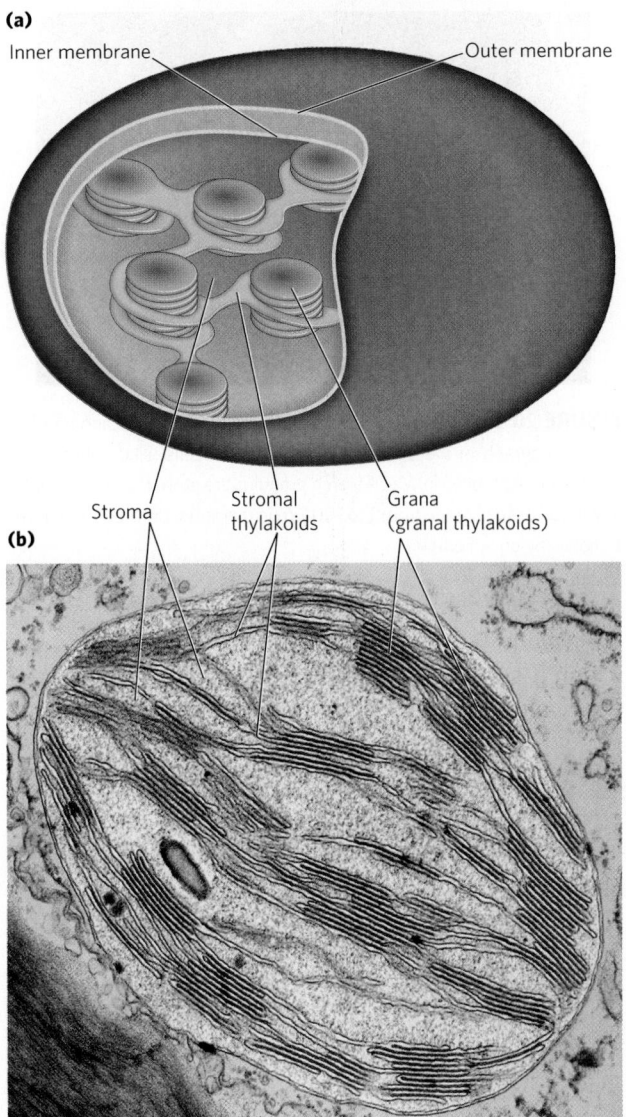

(a)

Inner membrane

Outer membrane

Stroma   Stromal thylakoids   Grana (granal thylakoids)

(b)

FIGURE 20-4 Chloroplast structure. (a) Schematic diagram. **(b)** Electron micrograph at high magnification, showing the highly organized thylakoid membrane structure. [Source: (b) Biophoto Associates/Science Source.]

or lumen. This complex membrane system forms flattened sacks called **thylakoids**. **Granal thylakoids** are disklike pouches arranged in stacks; they are connected by **stromal thylakoids**, which are flatter and spiral around a stack of grana. Embedded in the thylakoid membranes, commonly called **lamellae**, are the photosynthetic pigments and enzyme complexes that carry out the light-dependent reactions and ATP synthesis. The membrane system and its embedded components are constantly being remodeled, adapting to changes in the availability of light and in the plant's demand for energy and reductive power. The inner membranes of chloroplasts, like those of all plastids, are impermeable to polar and charged molecules. Traffic across these membranes is mediated by sets of specific transporters.

Chloroplasts are just one of several types of **plastids,** organelles unique to plants and algae. Plastids are self-reproducing organelles bounded by a double membrane and containing a small genome, the **plastome**, that encodes some of their proteins. Most proteins destined for plastids are encoded in nuclear genes, which are transcribed and translated like other nuclear genes; the proteins are then imported into plastids. Plastids reproduce by binary fission, replicating their

plastome (a single circular DNA molecule) and using their own enzymes and ribosomes to synthesize the proteins encoded by that plastome. **Amyloplasts** are colorless plastids (that is, they lack chlorophyll and other chromophores found in chloroplasts). They have no internal membranes analogous to the thylakoid membranes of chloroplasts, and in plant tissues rich in starch these plastids are packed with starch granules **(Fig. 20-5)**. Chloroplasts can be converted to **proplastids** by the loss of their internal membranes and chlorophyll, and proplastids are interconvertible with amyloplasts **(Fig. 20-6)**. In turn, both amyloplasts and proplastids can develop into chloroplasts. The relative abundance of the plastid types depends on the type of plant tissue and on the intensity of illumination. Cells of green leaves are rich in chloroplasts, whereas amyloplasts dominate in nonphotosynthetic

FIGURE 20-5 Starch-filled amyloplasts. This colorized scanning electron micrograph of a slice of raw potato shows cells filled with amyloplasts, the organelles in which starch granules are stored. Starch granules in various tissues range from 1 to 100 μm in diameter. [Source: Dr. Jeremy Burgess/Science Source.]

tissues that store starch in large quantities, such as potato tubers.

In 1937, Robert Hill found that when leaf extracts containing chloroplasts were illuminated, they (1) evolved O_2 and (2) reduced a nonbiological electron acceptor added to the medium, according to the **Hill reaction**:

$$2H_2O + 2A \xrightarrow{\text{light}} 2AH_2 + O_2$$

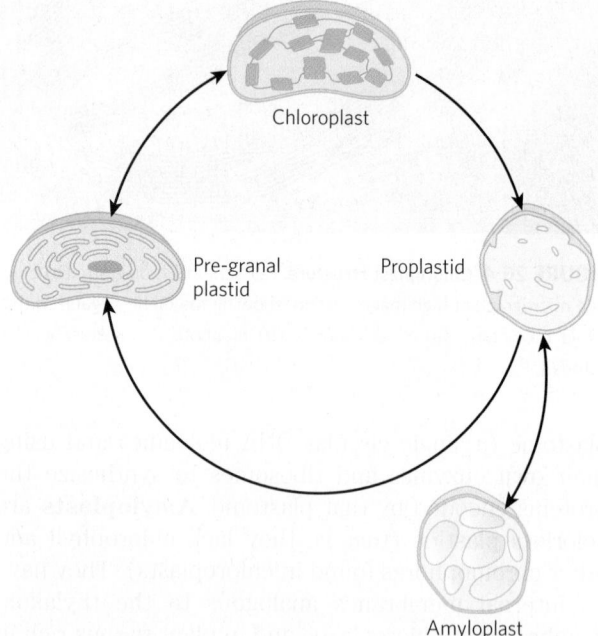

FIGURE 20-6 Plastids: their origins and interconversions. All types of plastids are bounded by a double membrane, and some (notably the mature chloroplast) have extensive internal membranes. Internal membranes can be lost (as a mature chloroplast becomes a proplastid) and resynthesized (as a proplastid gives rise to a pre-granal plastid and then a mature chloroplast). Proplastids in nonphotosynthetic tissues (such as root) give rise to amyloplasts, which contain large quantities of starch. All plant cells have plastids, and these organelles are the sites not only of photosynthesis but of other processes, such as the synthesis of essential amino acids, thiamine, pyridoxal phosphate, flavins, and vitamins A, C, E, and K.

where A is an artificial electron acceptor, or **Hill reagent**. One Hill reagent, the dye 2,6-dichlorophenolindophenol, is blue when oxidized (A) and colorless when reduced (AH_2), making the reaction easy to follow.

Oxidized form (blue)   Reduced form (colorless)

2,6-Dichlorophenolindophenol

When a leaf extract supplemented with the dye was illuminated, the blue dye became colorless and O_2 was evolved. In the dark, no O_2 evolution or dye reduction took place. This was the first evidence that absorbed light energy causes electrons to flow from H_2O to an electron acceptor. Moreover, Hill found that CO_2 was neither required nor reduced to a stable form under these conditions; O_2 evolution could be dissociated from CO_2 reduction. Several years later Severo Ochoa showed that $NADP^+$ is the biological electron acceptor in chloroplasts, according to the equation

$$2H_2O + 2NADP^+ \xrightarrow{\text{light}} 2NADPH + 2H^+ + O_2$$

To understand this photochemical process, we must first consider the more general topic of the effects of light absorption on molecular structure.

Visible light is electromagnetic radiation of wavelengths 400 to 700 nm, a small part of the electromagnetic spectrum **(Fig. 20-7)**, ranging from violet to red. The energy of a single **photon** (a quantum of light) is greater at the violet end of the spectrum than at the red end; shorter wavelength (and higher frequency) corresponds to higher energy. The energy, E, in a single photon of visible light is given by the Planck equation:

$$E = h\nu = hc/\lambda$$

where h is Planck's constant (6.626×10^{-34} J·s), ν is the frequency of the light in cycles/s, c is the speed of light (3.00×10^8 m/s), and λ is the wavelength of the light in meters. The energy of a photon of visible light ranges from 150 kJ/einstein for red light to ~300 kJ/einstein for violet light.

WORKED EXAMPLE 20-1 Energy of a Photon

The light used by vascular plants for photosynthesis has a wavelength of about 700 nm. Calculate the energy in a "mole" of photons (an einstein) of light of this

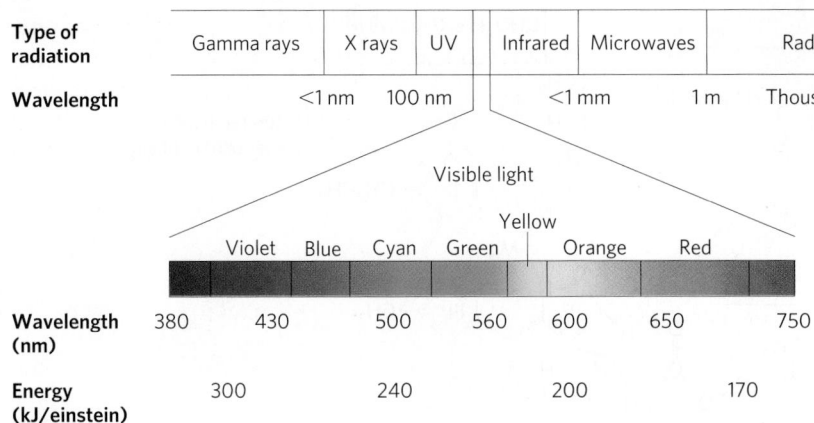

Type of radiation	Gamma rays	X rays	UV		Infrared	Microwaves	Radio waves

FIGURE 20-7 **Electromagnetic radiation.** The spectrum of electromagnetic radiation, and the energy of photons in the visible range. One einstein is 6.022×10^{23} photons.

wavelength, and compare this with the energy needed to synthesize a mole of ATP.

Solution: The energy in a single photon is given by the Planck equation. At a wavelength of 700×10^{-9} m, the energy of a photon is

$$E = hc/\lambda$$
$$= \frac{[(6.626 \times 10^{-34}\,\text{J·s})\,(3.00 \times 10^{8}\,\text{m/s})]}{(7.00 \times 10^{-7}\,\text{m})}$$
$$= 2.84 \times 10^{-19}\,\text{J}$$

An einstein of light is Avogadro's number of photons (6.022×10^{23}); thus the energy of one einstein of photons at 700 nm is given by

$$(2.84 \times 10^{-19}\,\text{J/photon})(6.022 \times 10^{23}\,\text{photons/einstein})$$
$$= 17.1 \times 10^{4}\,\text{J/einstein}$$
$$= 171\,\text{kJ/einstein}$$

So, a "mole" of photons of red light has about five times the energy needed to produce a mole of ATP from ADP and P_i (30.5 kJ/mol).

When a photon is absorbed, an electron in the absorbing molecule (chromophore) is lifted to a higher energy level. This is an all-or-nothing event: to be absorbed, the photon must contain a quantity of energy, called a **quantum**, that exactly matches the energy of the electronic transition. A molecule that has absorbed a photon is in an **excited state**, which is generally unstable. An electron lifted into a higher-energy orbital usually returns rapidly to its lower-energy orbital; that is, the excited molecule decays to the stable **ground state**, giving up the absorbed quantum as light or heat or using it to do chemical work. Light emission accompanying decay of excited molecules, called **fluorescence**, is always at a longer wavelength (lower energy) than that of the absorbed light (see Box 12-2). An alternative mode of decay, central to photosynthesis, involves direct transfer of excitation energy from an excited molecule to a neighboring molecule. Just as the photon is a quantum of light energy, so the **exciton** is a quantum of energy passed from an excited molecule to another molecule in a process called **exciton transfer**.

Chlorophylls Absorb Light Energy for Photosynthesis

The most important light-absorbing pigments in the thylakoid membranes are the **chlorophylls**, green pigments with polycyclic, planar structures resembling the protoporphyrin of hemoglobin (see Fig. 5-1), except that Mg^{2+}, not Fe^{2+}, occupies the central position **(Fig. 20-8a)**. The four inward-oriented nitrogen atoms of chlorophyll are coordinated with the Mg^{2+}. All chlorophylls have a long **phytol** side chain, esterified to a carboxyl-group substituent in ring IV, and chlorophylls also have a fifth five-membered ring not present in heme.

The heterocyclic five-ring system that surrounds the Mg^{2+} has an extended polyene structure, with alternating single and double bonds. Such polyenes characteristically show strong absorption in the visible region of the spectrum **(Fig. 20-9)**; the chlorophylls have unusually high molar extinction coefficients (see Box 3-1) and are therefore particularly well-suited for absorbing visible light during photosynthesis.

Chloroplasts always contain both chlorophyll a and chlorophyll b (Fig. 20-8a). Although both are green, their absorption spectra are sufficiently different (Fig. 20-9) that they complement each other's range of light absorption in the visible region. Most plants contain about twice as much chlorophyll a as chlorophyll b. The pigments in algae and photosynthetic bacteria include chlorophylls that differ only slightly from the plant pigments.

Cyanobacteria and red algae employ **phycobilins** such as phycoerythrobilin and phycocyanobilin (Fig. 20-8b) as their light-harvesting pigments. These open-chain tetrapyrroles have the extended polyene system found in chlorophylls, but not their cyclic structure or central Mg^{2+}. Phycobilins are covalently linked to specific binding proteins, forming **phycobiliproteins**, which associate in highly ordered complexes called phycobilisomes that constitute the primary light-harvesting structures in these microorganisms.

Accessory Pigments Extend the Range of Light Absorption

In addition to chlorophylls, thylakoid membranes contain secondary light-absorbing pigments, or **accessory pigments**, called carotenoids. **Carotenoids** may be yellow,

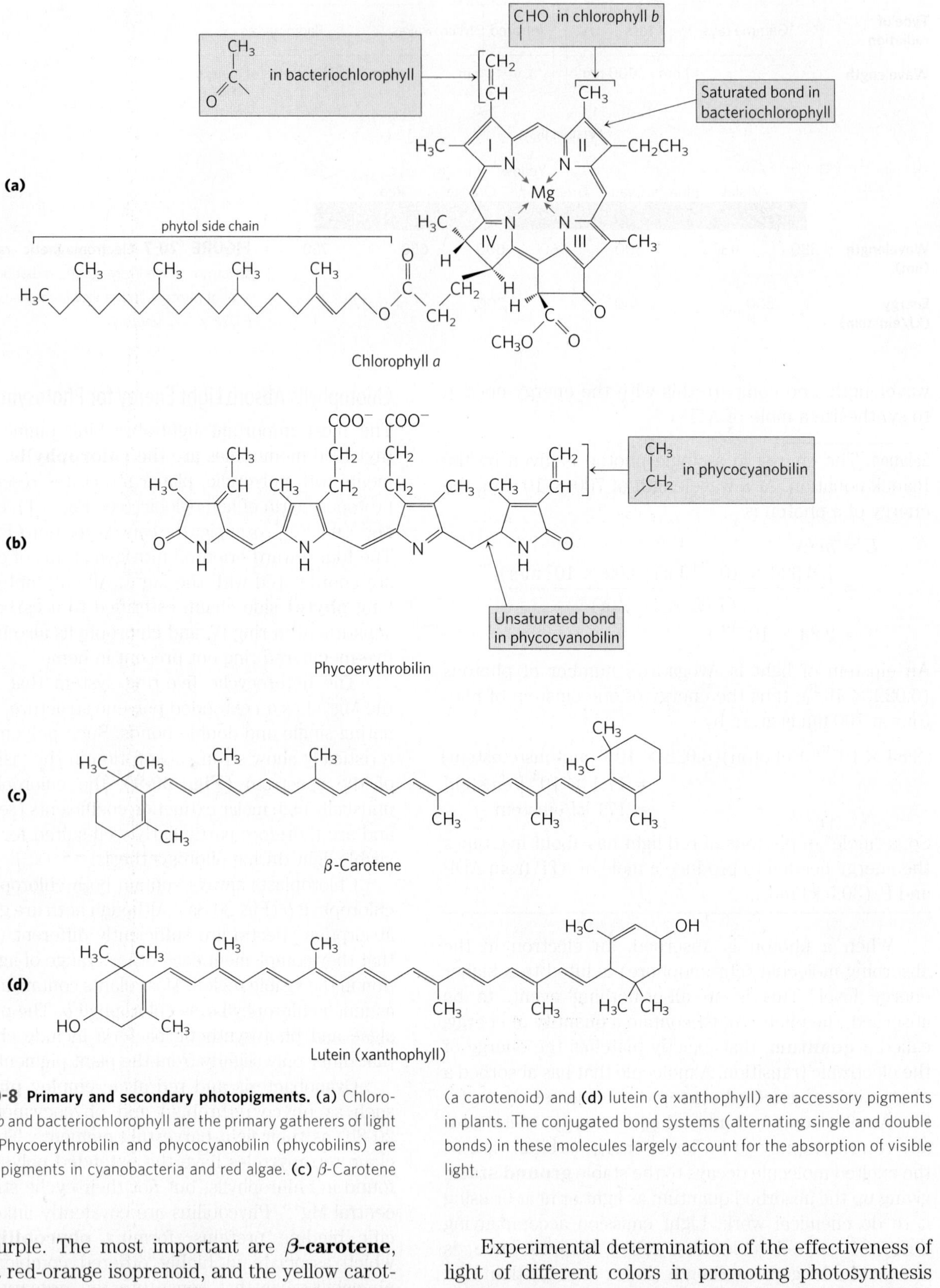

(a)

Chlorophyll *a*

(b)

Phycoerythrobilin

(c)

β-Carotene

(d)

Lutein (xanthophyll)

FIGURE 20-8 Primary and secondary photopigments. (a) Chlorophylls *a* and *b* and bacteriochlorophyll are the primary gatherers of light energy. **(b)** Phycoerythrobilin and phycocyanobilin (phycobilins) are the antenna pigments in cyanobacteria and red algae. **(c)** β-Carotene (a carotenoid) and **(d)** lutein (a xanthophyll) are accessory pigments in plants. The conjugated bond systems (alternating single and double bonds) in these molecules largely account for the absorption of visible light.

red, or purple. The most important are **β-carotene**, which is a red-orange isoprenoid, and the yellow carotenoid **lutein** (Fig. 20-8c, d). The carotenoids absorb light at wavelengths not absorbed by the chlorophylls (Fig. 20-9) and thus are supplementary light receptors. They also protect downstream components from a highly reactive form of oxygen (singlet oxygen) that is formed when intense light exceeds the system's capacity to accept electrons.

Experimental determination of the effectiveness of light of different colors in promoting photosynthesis yields an **action spectrum (Fig. 20-10)**, often useful in identifying the pigment primarily responsible for a biological effect of light. By capturing light in a region of the spectrum not used by other organisms, a photosynthetic organism can claim a unique ecological niche. For example, the phycobilins in red algae and cyanobacteria absorb light in the range 520 to 630 nm (Fig. 20-9),

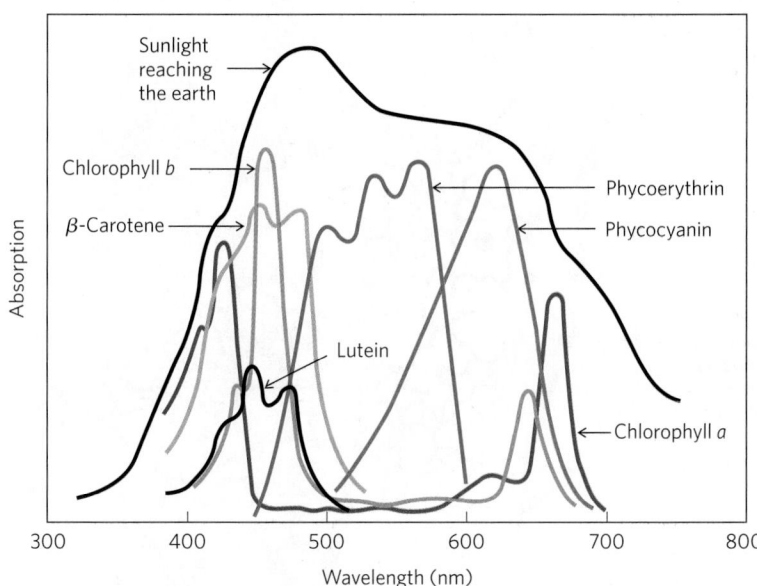

FIGURE 20-9 Absorption of visible light by photopigments. Plants are green because their pigments absorb light from the red and blue regions of the spectrum, leaving primarily green light to be reflected. Compare the absorption spectra of the pigments with the spectrum of sunlight reaching the earth's surface; the combination of chlorophylls (a and b) and accessory pigments enables plants to harvest most of the energy available in sunlight.

The relative amounts of chlorophylls and accessory pigments are characteristic of a particular plant species. Variation in the proportions of these pigments is responsible for the range of colors of photosynthetic organisms, from the deep blue-green of spruce needles, to the greener green of maple leaves, to the red, brown, or purple color of some species of multicellular algae and the leaves of some foliage plants favored by gardeners.

allowing them to occupy niches where light of lower or higher wavelength has been filtered out by the pigments of other organisms living in the water above them, or by the water itself.

Chlorophylls Funnel Absorbed Energy to Reaction Centers by Exciton Transfer

The light-absorbing pigments of thylakoid or bacterial membranes are arranged in functional arrays called

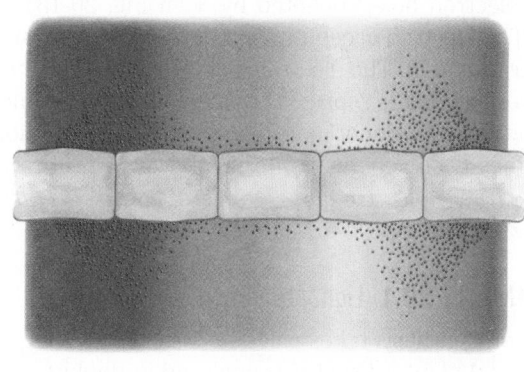

(a)

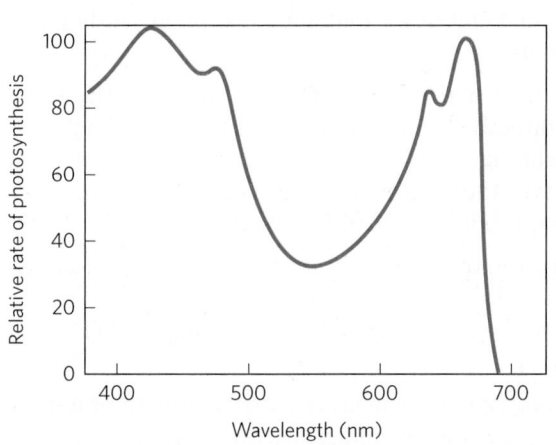

(b)

photosystems. In spinach chloroplasts, for example, each photosystem contains about 200 chlorophyll and 50 carotenoid molecules. All the pigment molecules in a photosystem can absorb photons, but only a few chlorophyll molecules associated with the **photochemical reaction center** are specialized to transduce light into chemical energy. The other pigment molecules in a photosystem serve as **antenna molecules**. They absorb light energy and transmit it rapidly and efficiently to the reaction center (**Fig. 20-11**). Some are part of a core complex around the reaction center. Others form **light-harvesting complexes (LHCs)** around the periphery of the core complex. Chlorophyll and other pigments are always associated with specific binding proteins, which fix the chromophores in relation to each other, to other protein complexes, and to the membrane. For example, each monomer of the trimeric light-harvesting

FIGURE 20-10 Two ways to determine the action spectrum for photosynthesis. (a) Results of a classic experiment performed by T. W. Engelmann in 1882 to determine the wavelength of light that is most effective in supporting photosynthesis. Engelmann placed cells of a filamentous photosynthetic alga on a microscope slide and illuminated them with light from a prism, so that one part of the filament received mainly blue light, another part yellow, another red. To determine which algal cells carried out photosynthesis most actively, Engelmann also placed on the microscope slide bacteria known to migrate toward regions of high O_2 concentration. After a period of illumination, the distribution of bacteria showed highest O_2 levels (produced by photosynthesis) in the regions illuminated with violet and red light.

(b) Results of a similar experiment that used modern techniques (an oxygen electrode) for the measurement of O_2 production. An action spectrum, as shown here, describes the relative rate of photosynthesis for illumination with a constant number of photons of different wavelengths. An action spectrum is useful because, by comparison with absorption spectra (such as those in Fig. 20-9), it suggests which pigments can channel energy into photosynthesis.

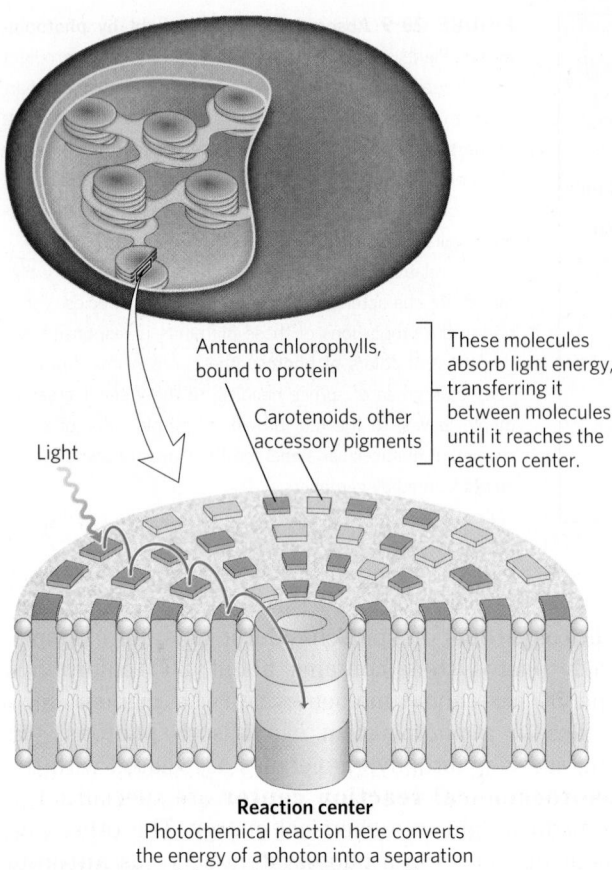

Light

Antenna chlorophylls, bound to protein

Carotenoids, other accessory pigments

These molecules absorb light energy, transferring it between molecules until it reaches the reaction center.

Reaction center
Photochemical reaction here converts the energy of a photon into a separation of charge, initiating electron flow.

FIGURE 20-11 Organization of photosystems in the thylakoid membrane. Photosystems are tightly packed in the thylakoid membrane, with several hundred antenna chlorophylls and accessory pigments surrounding a reaction center. Absorption of a photon by any of the antenna chlorophylls leads to excitation of the reaction center by exciton transfer (red arrow). Also embedded in the thylakoid membrane are the cytochrome b_6f complex and ATP synthase (see Fig. 20-21).

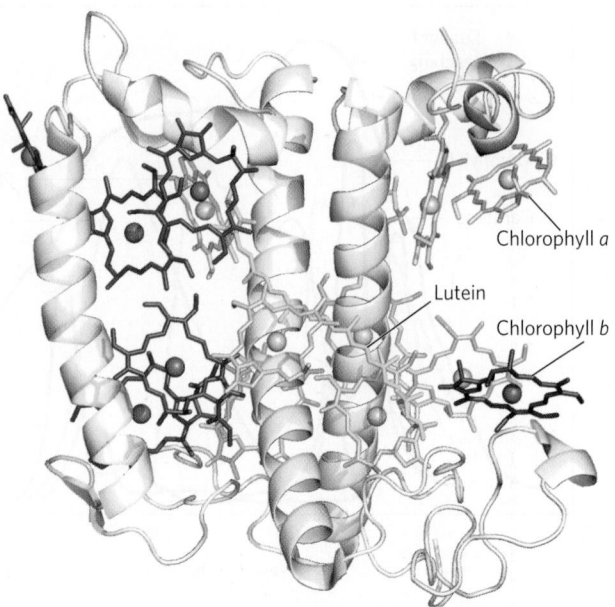

Chlorophyll a

Lutein

Chlorophyll b

FIGURE 20-12 The light-harvesting complex LHCII. The functional unit is a trimer, with 36 chlorophyll and 6 lutein molecules. Shown here is a monomer, viewed in the plane of the membrane, with its three transmembrane α-helical segments, seven chlorophyll a molecules (light green), five chlorophyll b molecules (dark green), and two molecules of lutein (yellow), which form an internal cross-brace. [Source: PDB ID 2BHW, J. Standfuss et al., *EMBO J.* 24:919, 2005.]

complex LHCII **(Fig. 20-12)** contains seven molecules of chlorophyll a, five of chlorophyll b, and two of lutein.

The chlorophyll molecules in light-harvesting complexes and other chlorophyll-binding proteins have light-absorption properties that are subtly different from those of free chlorophyll. When isolated chlorophyll molecules in vitro are excited by light, the absorbed energy is quickly released as fluorescence and heat, but when chlorophyll in intact leaves is excited by visible light **(Fig. 20-13**, step **❶)**, very little fluorescence is observed. Instead, the excited antenna chlorophyll transfers energy directly to a neighboring chlorophyll molecule, which becomes excited as the first molecule returns to its ground state (step **❷**). This transfer of energy, exciton transfer, extends to a third, fourth, or subsequent neighbor, until one of a special pair of chlorophyll a molecules at the photochemical reaction center is excited (step **❸**). In this excited chlorophyll molecule, an electron is promoted to a higher-energy orbital. This electron then passes to a nearby electron acceptor that is part of the electron-transfer chain, leaving the

reaction-center chlorophyll pair with a missing electron (an "electron hole," denoted by $+$ in Fig. 20-13) (step **❹**). The electron acceptor acquires a negative charge in this transaction. The electron lost by the reaction-center chlorophyll pair is replaced by an electron from a neighboring electron-donor molecule (step **❺**), which thereby becomes positively charged. In this way, *excitation by light causes electric charge separation and initiates an oxidation-reduction chain.*

SUMMARY 20.1 Light Absorption

■ Photosynthesis takes place in the chloroplasts of algae and plants, structures enclosed in double membranes and filled with an elaborate system of thylakoid membranes containing the photosynthetic machinery.

■ The light-dependent reactions of photosynthesis are those directly dependent on the absorption of light; the resulting photochemistry takes electrons from H_2O and drives them through a series of membrane-bound carriers, producing NADPH and ATP.

■ A photon of visible light possesses enough energy to bring about photochemical reactions, which in photosynthetic organisms lead eventually to ATP synthesis.

■ Chlorophyll molecules are associated with proteins in light-harvesting complexes arrayed around photochemical reaction centers.

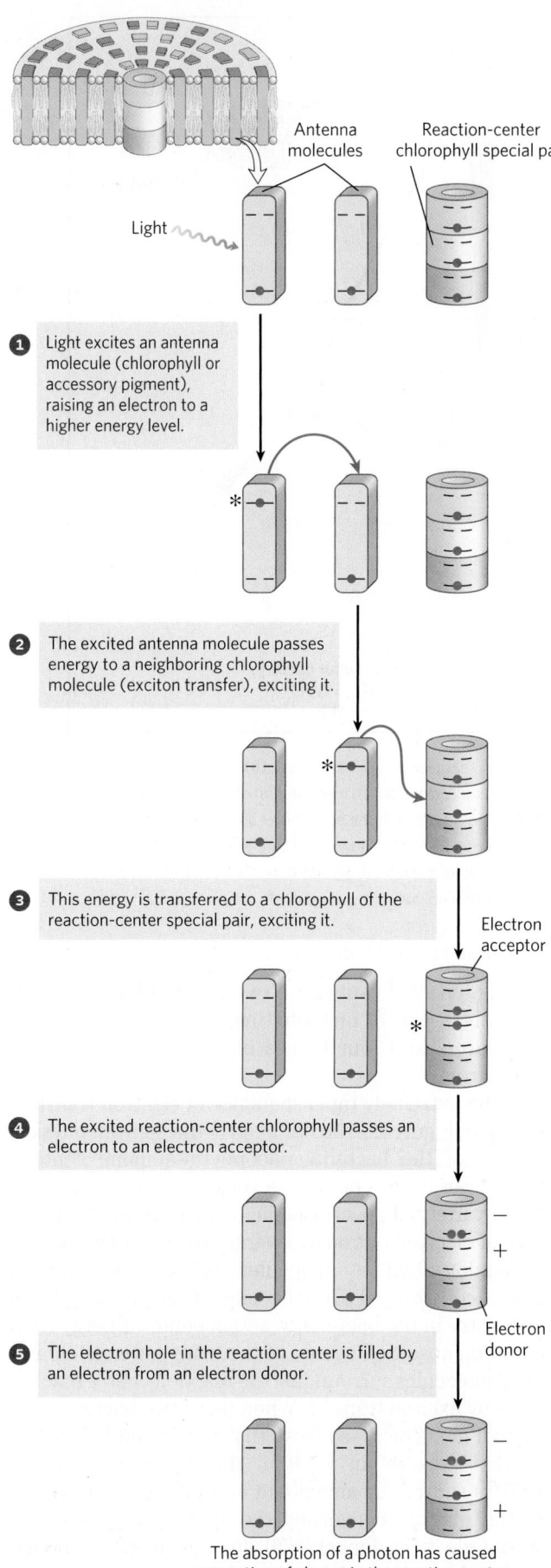

① Light excites an antenna molecule (chlorophyll or accessory pigment), raising an electron to a higher energy level.

② The excited antenna molecule passes energy to a neighboring chlorophyll molecule (exciton transfer), exciting it.

③ This energy is transferred to a chlorophyll of the reaction-center special pair, exciting it.

④ The excited reaction-center chlorophyll passes an electron to an electron acceptor.

⑤ The electron hole in the reaction center is filled by an electron from an electron donor.

The absorption of a photon has caused separation of charge in the reaction center.

FIGURE 20-13 Exciton and electron transfer. This generalized scheme shows conversion of the energy of an absorbed photon into separation of charges at the reaction center. The steps are further described in the text. Note that step **①** may repeat between successive antenna molecules until the exciton reaches the special pair of chlorophylls in the reaction center. An asterisk (*) denotes the excited state of a molecule.

■ In plants, absorption of a photon excites chlorophyll molecules and other (accessory) pigments, which funnel the energy into reaction centers in the thylakoid membranes. In the reaction centers, photoexcitation results in a charge separation that produces a strong electron donor (reducing agent) and a strong electron acceptor.

20.2 Photochemical Reaction Centers

Determining the site and the nature of the light-driven chemistry of photosynthesis proved to be a challenge. One reason for this was the insolubility of the photosynthetic apparatus; the central players are integral membrane proteins, and, like most membrane proteins, they are difficult to solubilize and purify. One major insight came in 1952, when Louis Duysens found that illumination of the photosynthetic membranes of the purple bacterium *Rhodospirillum rubrum* with a pulse of light of a specific wavelength (870 nm) caused a temporary decrease in the absorption of light at that wavelength; a pigment was "bleached" by the 870 nm light. Later studies by Bessel Kok and Horst Witt showed similar bleaching of plant chloroplast pigments by light of 680 and 700 nm. Furthermore, addition of the (nonbiological) electron acceptor $[Fe(CN)_6]^{3-}$ (ferricyanide) caused bleaching at these wavelengths *without prior illumination*. These findings indicated that bleaching of the pigments was due to the loss of an electron from a photochemical reaction center. The pigments were named for the wavelength causing maximum bleaching: P870, P680, and P700. They were later discovered to be a "special pair" of chlorophyll molecules in the reaction center, often designated $(Chl)_2$.

Photosynthetic Bacteria Have Two Types of Reaction Center

Photosynthetic bacteria have relatively simple phototransduction machinery, with one of two general types of reaction center. One type, found in purple bacteria, passes electrons through **pheophytin** (chlorophyll lacking the central Mg^{2+} ion) to a quinone. The other, in green sulfur bacteria, passes electrons through a quinone to an iron-sulfur center. Cyanobacteria and plants have two photosystems, PSI and PSII, one of each type, acting in tandem. Biochemical and biophysical studies have revealed many of the molecular details of reaction centers of bacteria and cyanobacteria, which therefore serve as prototypes for the more complex phototransduction systems of plants.

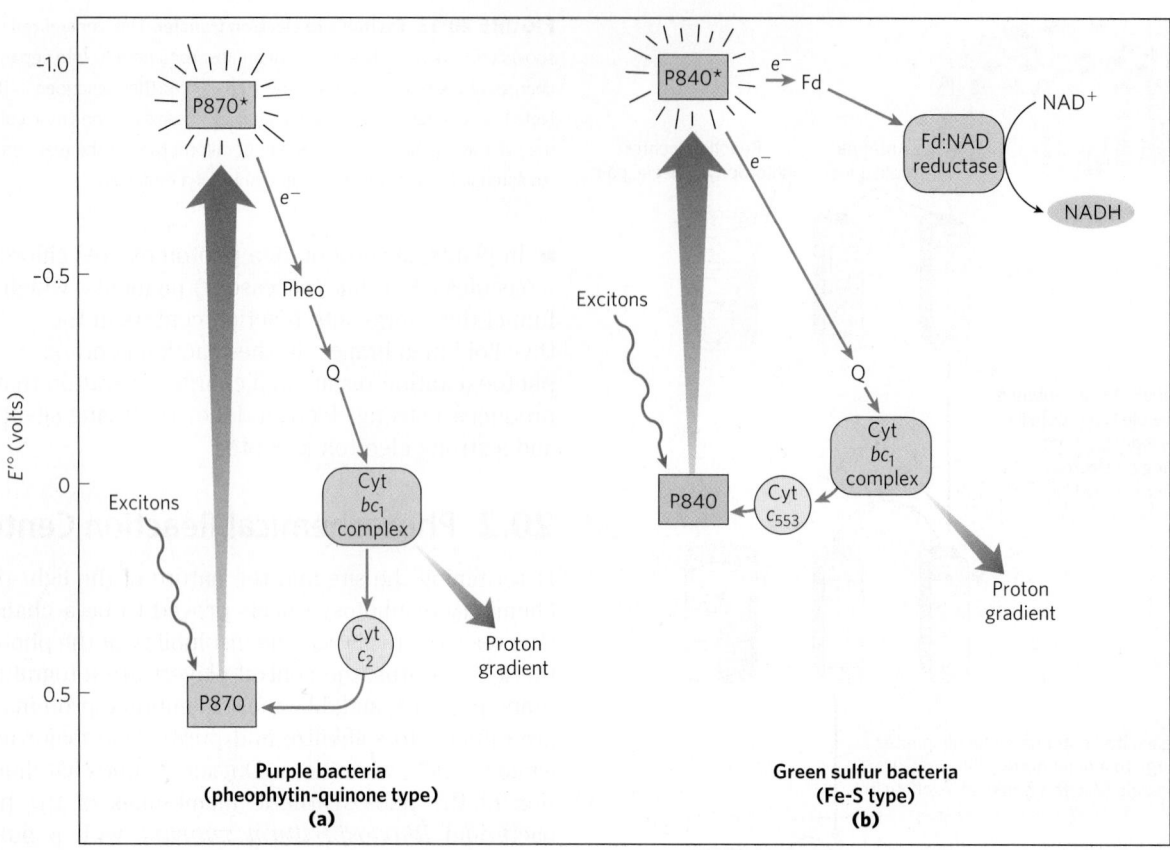

FIGURE 20-14 Functional modules of photosynthetic machinery in purple bacteria and green sulfur bacteria. (a) In purple bacteria, light energy drives electrons from the reaction-center P870 through pheophytin (Pheo), a quinone (Q), and the cytochrome bc_1 complex, then through cytochrome c_2 and thus back to the reaction center. Electron flow through the cytochrome bc_1 complex causes proton pumping, creating an electrochemical potential that powers ATP synthesis. **(b)** Green sulfur bacteria have two routes for electrons driven by excitation of P840: a cyclic route through a quinone to the cytochrome bc_1 complex and back to the reaction center via cytochrome c_{553}, and a noncyclic route from the reaction center through the iron-sulfur protein ferredoxin (Fd), then to NAD^+ in a reaction catalyzed by ferredoxin:NAD reductase.

The Pheophytin-Quinone Reaction Center (Type II Reaction Center) The photosynthetic machinery in purple bacteria consists of three basic modules **(Fig. 20-14)**: a single P870 reaction center; a cytochrome bc_1 electron-transfer complex similar to Complex III of the mitochondrial electron-transfer chain; and an ATP synthase, also similar to that of mitochondria. Illumination drives electrons through pheophytin and a quinone to the cytochrome bc_1 complex; after passing through the complex, electrons flow through cytochrome c_2 back to the reaction center, restoring its preillumination state. This light-driven cyclic flow of electrons provides the energy for proton pumping by the cytochrome bc_1 complex. Powered by the resulting proton gradient, ATP synthase produces ATP, exactly as in mitochondria (see Fig. 19-26).

The three-dimensional structures of the reaction centers of purple bacteria (*Rhodopseudomonas viridis* and *Rhodobacter sphaeroides*), deduced from x-ray crystallography, shed light on how phototransduction takes place in a pheophytin-quinone reaction center. The *R. viridis* reaction center **(Fig. 20-15a)** is a large protein complex containing four polypeptide subunits, two pairs of bacteriochlorophylls, a pair of pheophytins, two quinones, a nonheme iron, and four hemes in the associated c-type cytochrome.

The extremely rapid sequence of electron transfers shown in Figure 20-15b has been deduced from physical studies of the bacterial pheophytin-quinone centers, using brief flashes of light to trigger phototransduction and a variety of spectroscopic techniques to follow the flow of electrons through several carriers. A pair of bacteriochlorophyll *a* molecules $(Chl)_2$—the "special pair"—constitute P870, the site of the initial photochemistry in the bacterial reaction center. Energy from a photon absorbed by one of the many antenna chlorophyll molecules surrounding the reaction center reaches P870 by exciton transfer. When these two bacteriochlorophyll molecules—so close that their bonding orbitals overlap—absorb an exciton, the redox potential of P870 is shifted, by an amount equivalent to the energy of the photon, converting the special pair to a very strong electron donor. P870 donates an electron that passes through a neighboring chlorophyll monomer to pheophytin (Pheo). This produces two radicals, one

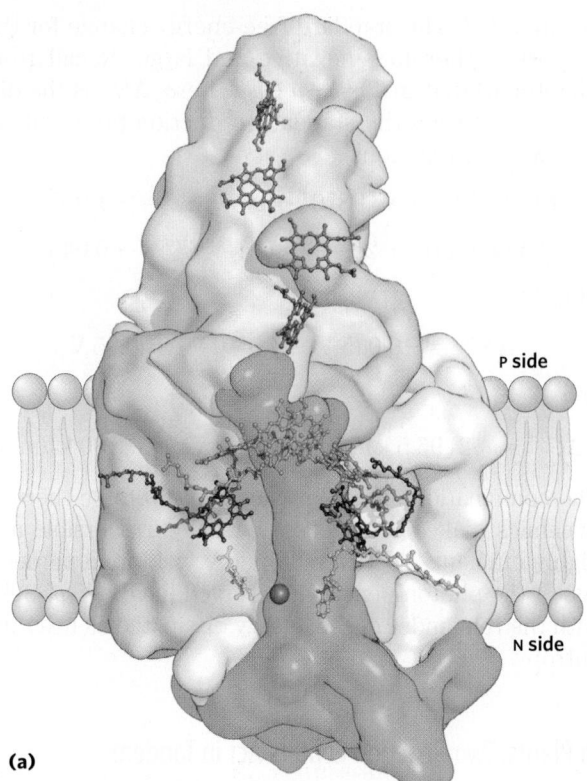

(a)

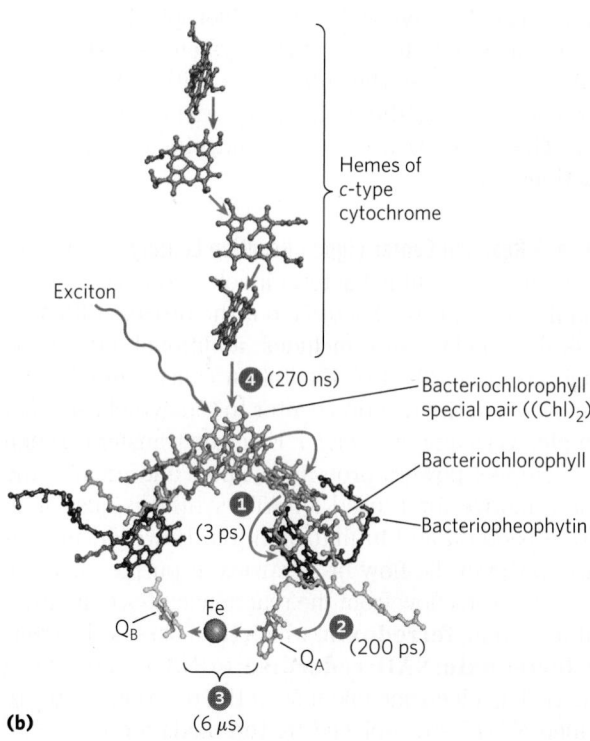

(b)

FIGURE 20-15 Photochemical reaction center of the purple bacterium _Rhodopseudomonas viridis._ (a) The system has four components: three subunits, H, M, and L (brown, lavender, and white, respectively), with a total of 11 transmembrane helical segments, and a fourth protein, cytochrome c (yellow), associated with the complex at the membrane surface. Subunits L and M are paired transmembrane proteins that together form a cylindrical structure with roughly bilateral symmetry about its long axis. Shown on the surface in (a) and separately in (b) are ball-and-stick representations of the embedded prosthetic groups that participate in the photochemical events. Bound to the L and M chains are two pairs of bacteriochlorophyll molecules (green); one of the pairs—the "special pair," $(Chl)_2$—is P870, the site of the first photochemical changes after light absorption. Also incorporated in the system are a pair of pheophytin a (Pheo a) molecules (blue); two quinones, menaquinone (Q_A)

and ubiquinone (Q_B) (orange and yellow), also arranged with bilateral symmetry; and a single nonheme Fe (red) located approximately on the axis of symmetry between the quinones. Shown at the top of the figure are four heme groups (red) associated with the c-type cytochrome of the reaction center.

(b) Sequence of events following excitation of the special pair of bacteriochlorophylls, with the time scale of the electron transfers in parentheses. ❶ The excited special pair passes an electron through a bacteriochlorophyll monomer to pheophytin, ❷ from which the electron moves rapidly to the tightly bound menaquinone, Q_A. ❸ This quinone passes electrons much more slowly to the diffusible ubiquinone, Q_B, through the nonheme Fe. Meanwhile, ❹ the "electron hole" in the special pair is filled by an electron from a heme of cytochrome c. [Source: PDB ID 1PRC, J. Diesenhofer et al., _J. Mol. Biol._ 246:429, 1995.]

positively charged (the special pair, P870) and one negatively charged (the pheophytin):

$$\text{P870} + 1 \text{ exciton} \longrightarrow \text{P870}^* \qquad \text{(excitation)}$$
$$\text{P870}^* + \text{Pheo} \longrightarrow {}^\bullet\text{P870}^+ + {}^\bullet\text{Pheo}^-$$
$$\text{(charge separation)}$$

The pheophytin radical now passes its electron to a tightly bound molecule of quinone (Q_A), converting it to a semiquinone radical, which immediately donates its extra electron to a second, loosely bound quinone (Q_B). Two such electron transfers convert Q_B to its fully reduced form, Q_BH_2, which is free to diffuse in the membrane bilayer, away from the reaction center:

$$2\,{}^\bullet\text{Pheo}^- + 2\text{H}^+ + Q_B \longrightarrow 2\,\text{Pheo} + Q_BH_2$$
$$\text{(quinone reduction)}$$

The hydroquinone (Q_BH_2), carrying in its chemical bonds some of the energy of the photons that originally

excited P870, enters the pool of reduced quinone (QH_2) dissolved in the membrane and moves through the lipid phase of the bilayer to the cytochrome bc_1 complex.

Like the homologous Complex III in mitochondria, the cytochrome bc_1 complex of purple bacteria carries electrons from a quinol donor (QH_2) to an electron acceptor, using the energy of electron transfer to pump protons across the membrane, producing a proton-motive force. The path of electron flow through this complex is believed to be very similar to that through mitochondrial Complex III, involving a Q cycle (see Fig. 19-11) in which protons are consumed on one side of the membrane and released on the other. The ultimate electron acceptor in purple bacteria is the electron-depleted form of P870, denoted ${}^\bullet\text{P870}^+$. Electrons move from the cytochrome bc_1 complex to P870 via a soluble c-type cytochrome, cytochrome c_2 (Fig. 20-14a). The electron-transfer process completes the cycle, returning the reaction center to its unbleached state, ready to absorb another exciton from an antenna chlorophyll.

A remarkable feature of this system is that all the chemistry occurs in the *solid state*, with reacting species held close together in the right orientation for reaction. The result is a very fast and efficient series of reactions.

The Fe-S Reaction Center (Type I Reaction Center) Photosynthesis in green sulfur bacteria involves the same three modules as in purple bacteria, but the process differs in several respects and includes additional enzymatic reactions (Fig. 20-14b). Excitation causes an electron to move from the reaction center to the cytochrome bc_1 complex via a quinone carrier. Electron transfer through this complex powers proton transport and creates the proton-motive force used for ATP synthesis, just as in purple bacteria and in mitochondria. However, in contrast to the cyclic flow of electrons in purple bacteria, some electrons flow from the reaction center to an iron-sulfur protein, **ferredoxin**, which then passes electrons via **ferredoxin:NAD reductase** to NAD^+, producing NADH. The electrons taken from the reaction center to reduce NAD^+ are replaced by the oxidation of H_2S to elemental S, then to SO_4^{2-}, in the reaction that defines the green sulfur bacteria. This oxidation of H_2S by bacteria is chemically analogous to the oxidation of H_2O by oxygenic plants.

Kinetic and Thermodynamic Factors Prevent the Dissipation of Energy by Internal Conversion

The complex construction of reaction centers is the product of evolutionary selection for efficiency in the photosynthetic process. The excited state P870* could, in principle, decay to its ground state by internal conversion, a very rapid process (10 picoseconds; 1 ps = 10^{-12} s) in which the energy of the absorbed photon is converted to heat (molecular motion). Reaction centers are constructed to prevent the inefficiency that would result from internal conversion. The proteins of the reaction center hold the bacteriochlorophylls, bacteriopheophytins, and quinones in a fixed orientation relative to each other. This accounts for the high efficiency and rapidity of the reactions; nothing is left to chance collision or random diffusion. Exciton transfer from an antenna chlorophyll to the special pair of the reaction center is accomplished in less than 100 ps with >90% efficiency. Within 3 ps of the excitation of P870, pheophytin has received an electron and become a negatively charged radical; less than 200 ps later, the electron has reached the quinone Q_B (Fig. 20-15b). The electron-transfer reactions not only are fast but are thermodynamically "downhill"; the excited special pair, P870*, is an excellent electron donor ($E'^\circ = -1$ V), and each successive electron transfer is to an acceptor of substantially less

negative E'°. The standard free-energy change for the process is therefore negative and large. Recall from Chapter 13 that $\Delta G'^\circ = -n\,\Delta E'^\circ$; here, $\Delta E'^\circ$ is the difference between the standard reduction potentials of the two half-reactions

$$(1)\ P870^* \longrightarrow {}^\bullet P870^+ + e^- \quad E'^\circ = -1.0\text{ V}$$

$$(2)\ Q + 2H^+ + 2e^- \longrightarrow QH_2 \quad E'^\circ = -0.045\text{ V}$$

Thus

$$\Delta E'^\circ = -0.045\text{ V} - (-1.0\text{ V}) \approx 0.95\text{ V}$$

and

$$\Delta G'^\circ = -2(96.5\text{ kJ/V}\cdot\text{mol})(0.95\text{ V}) = -180\text{ kJ/mol}$$

The combination of fast kinetics and favorable thermodynamics makes the process virtually irreversible and highly efficient. The overall energy yield (the percentage of the photon's energy conserved in QH_2) is >30%, with the remainder of the energy dissipated as heat and entropy.

In Plants, Two Reaction Centers Act in Tandem

The photosynthetic apparatus of modern cyanobacteria, algae, and vascular plants is more complex than the one-center bacterial systems, and it most likely evolved through the combination of two simpler bacterial photosystems. The thylakoid membranes of chloroplasts have two different kinds of photosystems, each with its own type of photochemical reaction center and set of antenna molecules. The two systems have distinct and complementary functions **(Fig. 20-16)**. **Photosystem II (PSII)** is a pheophytin-quinone type of system (like the single photosystem of purple bacteria) containing roughly equal amounts of chlorophylls a and b. Excitation of the P680 special pair in its reaction center drives electrons through the cytochrome b_6f complex with concomitant movement of protons across the thylakoid membrane. **Photosystem I (PSI)** is structurally and functionally related to the type I reaction center of green sulfur bacteria. It has a P700 reaction center and a high ratio of chlorophyll a to chlorophyll b. The excited P700 passes electrons through a chain of carriers to the Fe-S protein ferredoxin, then to $NADP^+$, producing NADPH. The thylakoid membranes of a single spinach chloroplast have many hundreds of each kind of photosystem.

These two reaction centers in plants act in tandem to catalyze the light-driven movement of electrons from H_2O to $NADP^+$. Electrons are carried between the two photosystems by the soluble protein **plastocyanin**. To replace the electrons that move from PSII through PSI to $NADP^+$, cyanobacteria and plants oxidize H_2O (as green sulfur bacteria oxidize H_2S), producing O_2 (Fig. 20-16, bottom left). This process is called **oxygenic photosynthesis** to distinguish it from the anoxygenic photosynthesis of purple and green sulfur bacteria. All

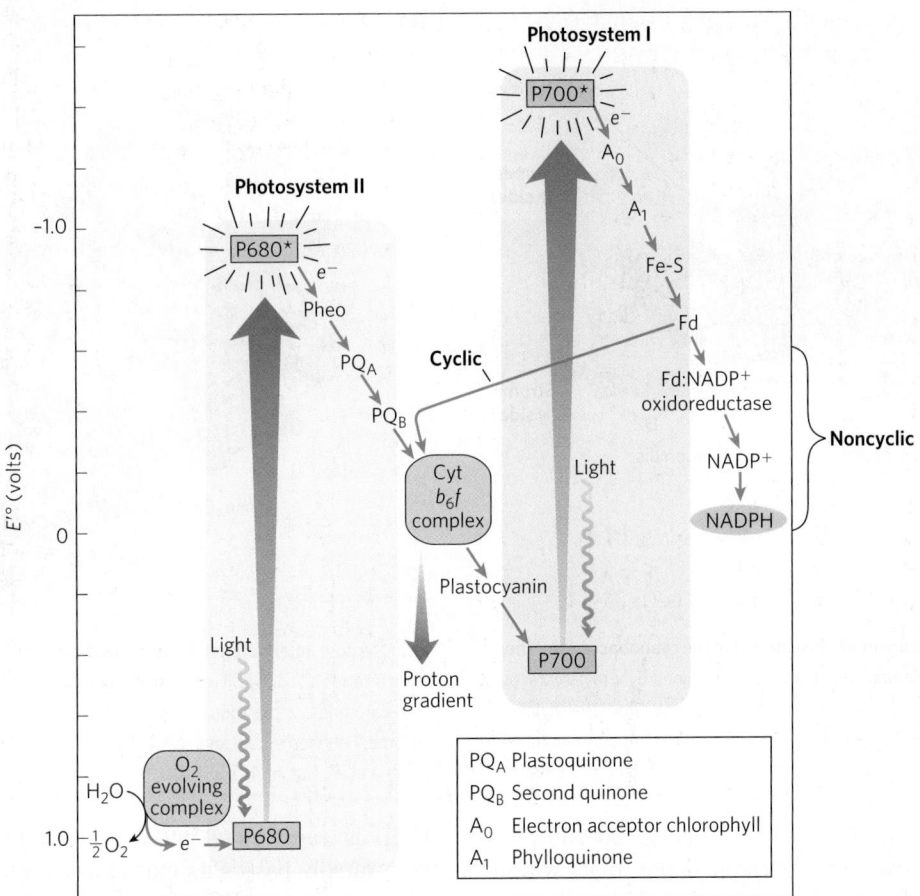

FIGURE 20-16 Integration of photosystems I and II in chloroplasts. This "Z scheme" shows the pathway of electron transfer from H_2O (lower left) to $NADP^+$ (far right) in noncyclic photosynthesis. The position on the vertical scale of each electron carrier reflects its standard reduction potential. To raise the energy of electrons derived from H_2O to the energy level required to reduce $NADP^+$ to NADPH, each electron must be "lifted" twice (heavy arrows) by photons absorbed in PSII and PSI. One photon is required per electron in each photosystem. After excitation, the high-energy electrons flow "downhill" through the carrier chains as shown. Protons move across the thylakoid membrane during the water-splitting reaction and during electron transfer through the cytochrome b_6f complex, producing the proton gradient that is essential to ATP formation. An alternative path of electrons is cyclic electron transfer, in which electrons move from ferredoxin back to the cytochrome b_6f complex, instead of reducing $NADP^+$ to NADPH. The cyclic pathway produces more ATP and less NADPH than the noncyclic.

O_2-evolving photosynthetic cells—those of plants, algae, and cyanobacteria—contain both PSI and PSII; organisms with only one photosystem do not evolve O_2. The diagram in Figure 20-16, often called the **Z scheme** because of its overall form, outlines the pathway of electron flow between the two photosystems and the energy relationships in the light-dependent reactions. The Z scheme thus describes the complete route by which electrons flow from H_2O to $NADP^+$, according to the equation

$$2H_2O + 2NADP^+ + 8 \text{ photons} \longrightarrow$$
$$O_2 + 2NADPH + 2H^+$$

For every two photons absorbed (one by each photosystem), one electron is transferred from H_2O to $NADP^+$. To form one molecule of O_2, which requires transfer of four electrons from two H_2O to two $NADP^+$, a total of eight photons must be absorbed, four by each photosystem.

Photosystem II PSII is dimeric **(Fig. 20-17)**. Each monomer is a huge complex of 19 proteins, including the core complex of P680 reaction-center proteins D1 and D2; two chlorophyll-binding proteins, CP43 and CP47; and associated chromophores, including carotenoids, a non-heme iron, and the inorganic complex Mn_4CaO_5. Sixteen of the proteins in PSII have transmembrane segments, but three are peripheral proteins on the lumenal side that stabilize the Mn_4CaO_5 complex. Surrounding PSII are additional chlorophyll-binding proteins and light-harvesting complexes. When a photon is absorbed by any of these antenna molecules, the resulting exciton moves very rapidly from one to another of the pigments until it reaches the reaction center and excites P680, the special pair of chlorophyll a molecules (Chl $a)_2$, to initiate the photochemistry **(Fig. 20-18)**.

The mechanistic details of the photochemical reactions in PSII are essentially similar to those in the photosystem of purple bacteria, with several important

(a)

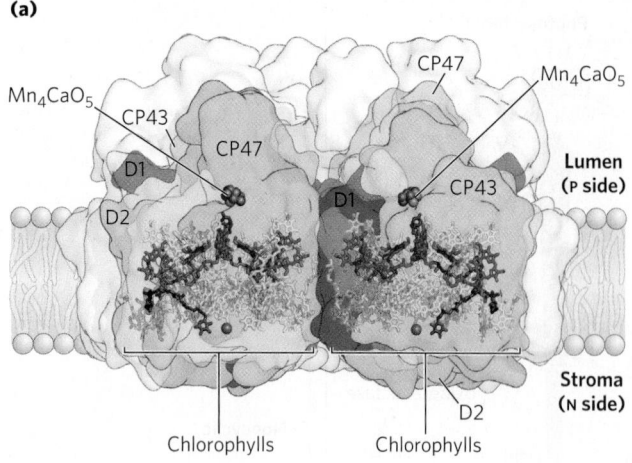

(b)

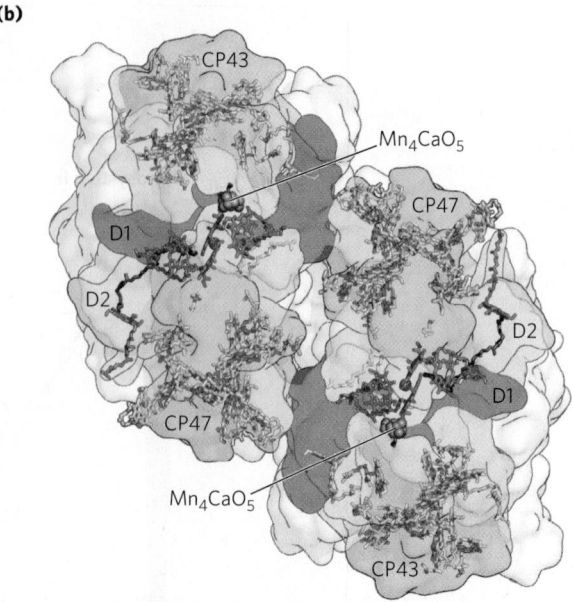

FIGURE 20-17 Structure of photosystem II of the cyanobacterium *Thermosynechococcus vulcanus.* This structure, determined by x-ray crystallography, is viewed **(a)** in the plane of the membrane and **(b)** from the thylakoid lumen (P side). The enormous complex is a dimer; each monomer has its own reaction center. CP43 and CP47 are chlorophyll-binding proteins that form the core antenna, directly associated with the PSII reaction-center proteins D1 and D2. Each PSII monomer contains 35 chlorophylls, 2 pheophytins, 11 β-carotenes, 2 plastoquinones, 1 b-type cytochrome and 1 c-type cytochrome, 1 nonheme iron, and 1 Mn_4CaO_5 complex. [Source: PDB ID 3WU2, Y. Umena et al., *Nature* 473:55, 2011.]

additions. Excitation of P680 in PSII **(Fig. 20-19)** produces P680*, an excellent electron donor that, within picoseconds, transfers an electron to pheophytin, giving it a negative charge (•Pheo⁻). With the loss of its electron,

P680* is transformed into a radical cation, P680⁺. •Pheo⁻ very rapidly passes its extra electron to a protein-bound **plastoquinone**, PQ_A (or Q_A), which in turn passes its electron to another, more loosely bound plastoquinone, PQ_B (or Q_B). When PQ_B has acquired two electrons in two such transfers from PQ_A and two protons from the solvent water, it is in its fully reduced quinol form, PQ_BH_2. The overall reaction initiated by light in PSII is

$$4\ P680 + 4H^+ + 2\ PQ_B + 4\ \text{photons} \longrightarrow$$
$$4\ P680^+ + 2\ PQ_BH_2 \quad (20\text{-}1)$$

Eventually, the electrons in PQ_BH_2 pass through the cytochrome b_6f complex (Fig. 20-16). The electron initially removed from P680 is replaced with an electron obtained from the oxidation of water, as described below. The binding site for plastoquinone is the point of action of many commercial herbicides that kill plants by blocking electron transfer through the cytochrome b_6f complex and preventing photosynthetic ATP production.

Photosystem I PSI and its antenna molecules are part of a supramolecular complex composed of at least 16 proteins, including four chlorophyll-binding proteins arranged around the periphery of the reaction center

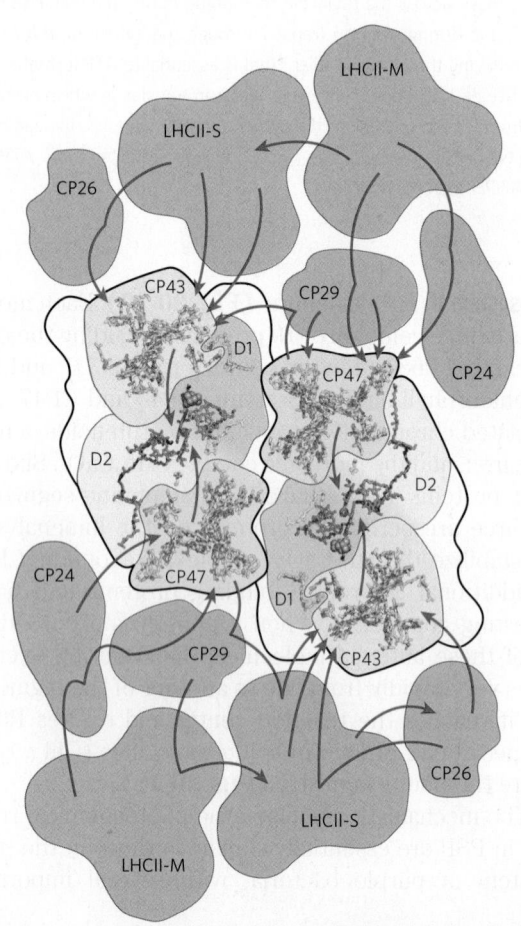

FIGURE 20-18 Path of excitons through photosystem II. The reaction center is surrounded by, and loosely associated with, several antenna complexes, each containing many chlorophyll and other pigment molecules. CP43 and CP47 form the core antenna; CP26, CP29, and CP24 and the trimeric light-harvesting complexes LHCII-M and LHCII-S form the outer antenna. When a photon excites one of these many antenna chlorophylls, the resulting exciton moves (red arrows) from chlorophyll to chlorophyll until it reaches P870, the special pair of chlorophylls of the reaction center in the transmembrane proteins D1 and D2. [Source: PDB ID 3WU2, Y. Umena et al., *Nature* 473:55, 2011.]

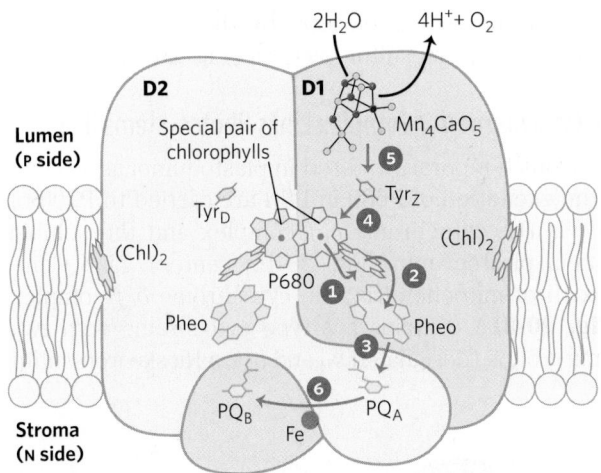

FIGURE 20-19 Electron flow through photosystem II of the cyanobacterium *Synechococcus elongatus*. The monomeric form of the core complex shown here has two major transmembrane proteins, D1 and D2, each with its set of electron carriers. Although the two subunits are nearly symmetric, electron flow occurs through only one of the two branches of electron carriers, that on the right (in D1). The arrows show the path of electron flow from the Mn_4CaO_5 ion cluster of the water-splitting enzyme to plastoquinone PQ_B. The photochemical events occur in the sequence indicated by the step numbers. Notice the close similarity between the positions of electron carriers here and the positions in the bacterial photochemical reaction center shown in Figure 20-15. The role of the Tyr residues and the detailed structure of the Mn_4CaO_5 cluster are discussed below (see Fig. 20-24b).

(Fig. 20-20). The complex also includes 35 carotenoids of several types, three 4Fe-4S clusters, and two phylloquinones. The reaction center's electron carriers are tightly integrated with antenna chlorophylls. As a result, exciton transfer from antenna chlorophylls to the reaction center is rapid and efficient.

The photochemical events that follow the excitation of PSI at the reaction-center P700 (Fig. 20-20c) are formally similar to those occurring in PSII. The excited reaction-center P700* loses an electron to an acceptor, designated A_0 (a chlorophyll *a* molecule, functionally homologous to the pheophytin of PSII), creating A_0^- and P700$^+$. Again, excitation results in charge separation at the

FIGURE 20-20 Structure of photosystem I of a plant and electron flow through the system. Photosystem I is a symmetric trimer. Shown here is one monomer of photosystem I of the garden pea, *Pisum sativum*, viewed from the stromal side in (a) and (b). **(a)** The protein subunits of the reaction center and the light-harvesting complexes. **(b)** The chromophores. The chlorophyll molecules of the core complex are shown in light green, those of LHCI in dark green, and the chlorophylls that span the gap between the light-harvesting complexes and the reaction center in olive green. Lipids are gray and carotenoids orange. **(c)** The path of electrons (blue arrows) through PSI, viewed in the plane of the membrane. When P700, the special pair of chlorophylls, is excited by a photon or exciton, its reduction potential is dramatically reduced, making it a good electron donor. P700 then passes an electron through a nearby chlorophyll (referred to as A_0) to phylloquinone (Q_K). Reduced Q_K is reoxidized as it passes two electrons, one at a time, to an Fe-S center (F_X) near the N side of the membrane. From F_X, electrons move through two more Fe-S centers (F_A and F_B) to ferredoxin in the stroma. Ferredoxin then donates electrons to NADP$^+$ (not shown), reducing it to NADPH, one of the forms in which the energy of photons is trapped in chloroplasts. [Source: (a, b) PDB ID 4RKU, Y. Mazor et al.]

photochemical reaction center. P700$^+$ is a strong oxidizing agent, which quickly acquires an electron from plastocyanin, a soluble Cu-containing electron-transfer protein. A_0^- is an exceptionally strong reducing agent that passes its electron through a chain of carriers that leads to NADP$^+$

(a)

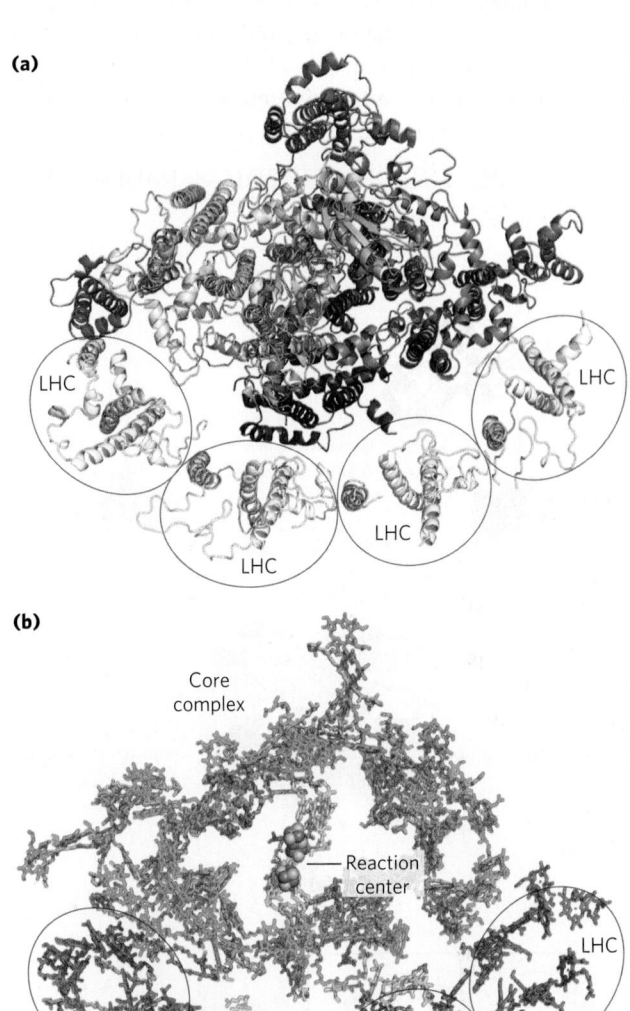

(c)

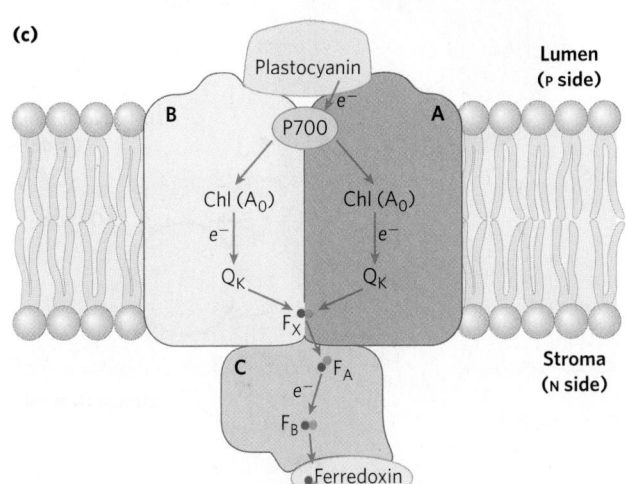

(Fig. 20-16, right side). **Phylloquinone (Q_K)** accepts the electron and passes it to an iron-sulfur protein through three Fe-S centers in PSI. From here, the electron moves to ferredoxin (Fd), another iron-sulfur protein loosely associated with the thylakoid membrane. Recall that ferredoxin contains a 2Fe-2S center (see Fig. 19-5) that undergoes one-electron oxidation and reduction reactions. The fourth electron carrier in the chain is the flavoprotein **ferredoxin:NADP$^+$ oxidoreductase**, which transfers electrons from reduced ferredoxin (Fd$_{red}$) to NADP$^+$:

$$2Fd_{red} + 2H^+ + NADP^+ \longrightarrow 2Fd_{ox} + NADPH + H^+$$

This enzyme is homologous to the ferredoxin:NAD reductase of green sulfur bacteria (Fig. 20-14b).

The Cytochrome b_6f Complex Links Photosystems II and I

Electrons temporarily stored in plastoquinol as a result of the excitation of P680 in PSII are carried to P700 of PSI via the cytochrome b_6f complex and the soluble protein plastocyanin (Fig. 20-16, center). Like Complex III of mitochondria, the cytochrome b_6f complex **(Fig. 20-21)** contains a b-type cytochrome with two heme groups (designated b_H and b_L), a Rieske iron-sulfur

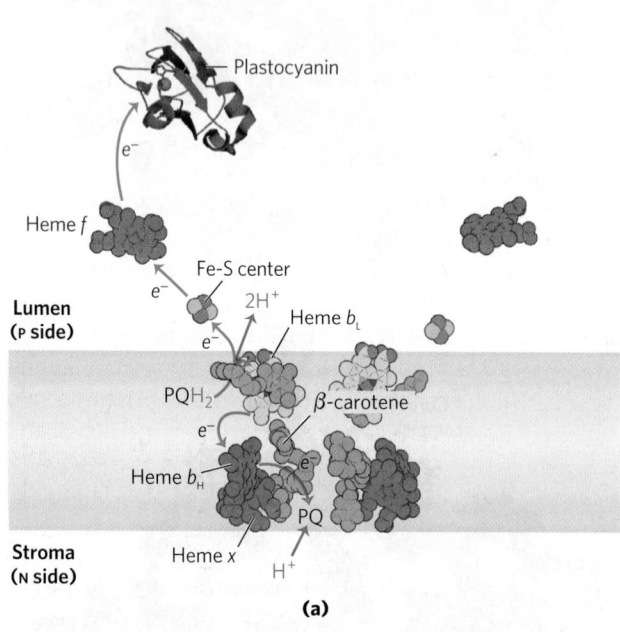

(a)

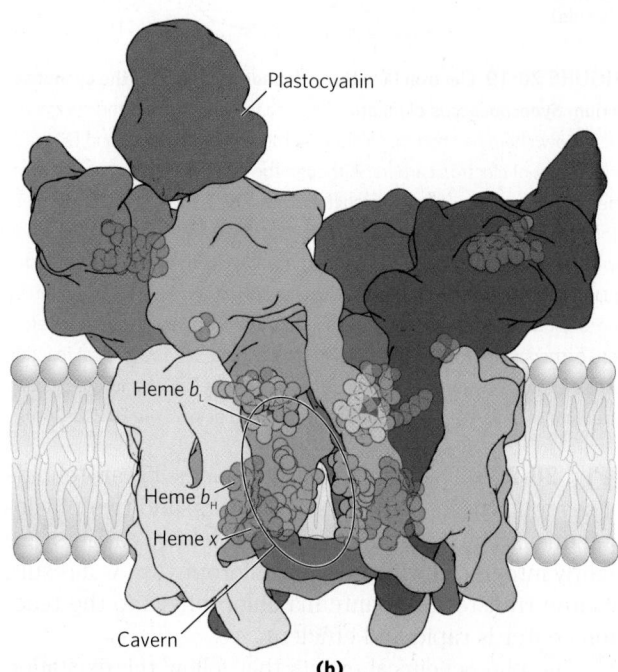

(b)

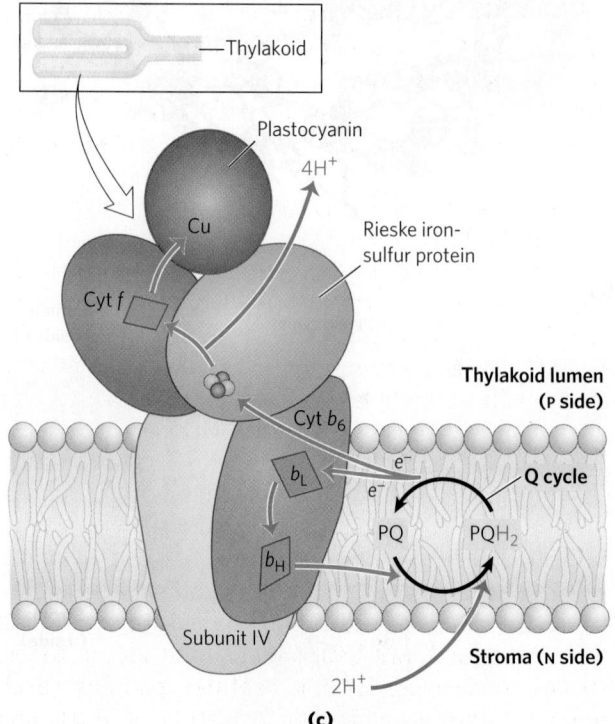

(c)

FIGURE 20-21 Electron and proton flow through the cytochrome b_6f complex. **(a)** The crystal structure of the complex reveals the positions of the electron carriers. In addition to the hemes of cytochrome b (heme b_H and b_L; also called heme b_N and b_P, respectively, because of their proximity to the N and P sides of the bilayer) and cytochrome f (heme f), there is a fourth heme (heme x) near heme b_H; also present is a β-carotene of unknown function. Two sites bind plastoquinone: the PQH$_2$ site near the P side of the bilayer, and the PQ site near the N side. The Fe-S center of the Rieske protein lies just outside the bilayer on the P side, and the heme f site is on a protein domain that extends well into the thylakoid lumen. The electron path is shown for just one of the monomers, but both sets of carriers in the dimer carry electrons to plastocyanin.

(b) The cytochrome b_6f complex is a homodimer arranged to create a cavern connecting the PQH$_2$ and PQ sites. (Compare this with the structure of mitochondrial Complex III, cytochrome bc_1, in Fig. 19-10.) This cavern allows plastoquinone to move between the sites of its oxidation and reduction.

(c) Plastoquinol (PQH$_2$), formed in PSII, is oxidized by the cytochrome b_6f complex in a series of steps like those of the Q cycle in Complex III of mitochondria (see Fig. 19-11). One electron from PQH$_2$ passes to the Fe-S center of the Rieske protein, the other to heme b_L of cytochrome b_6. The net effect is passage of electrons from PQH$_2$ to the soluble protein plastocyanin, which carries them to PSI. [Sources: (a, b) PDB ID 1VF5, G. Kurisu et al., *Science* 302:1009, 2003; PDB ID 2Q5B, Y. S. Bukhman-DeRuyter et al.]

protein (M_r 20,000), and cytochrome f (named for the Latin *frons*, "leaf"). Electrons flow through the cytochrome b_6f complex from PQ_BH_2 to cytochrome f, then to plastocyanin, and finally to $P700^+$, thereby reducing it.

Like Complex III of mitochondria, cytochrome b_6f conveys electrons from a reduced quinone—a mobile, lipid-soluble carrier of two electrons (Q in mitochondria, PQ_B in chloroplasts)—to a water-soluble protein that carries one electron (cytochrome c in mitochondria, plastocyanin in chloroplasts). As in mitochondria, the function of this complex involves a Q cycle (see Fig. 19-11) in which electrons pass, one at a time, from PQ_BH_2 to cytochrome b_6. This cycle results in the pumping of protons across the membrane, from the stromal compartment to the thylakoid lumen. Up to four protons enter the lumen for each pair of electrons that passes through the cytochrome b_6f complex. The result is production of a proton gradient across the thylakoid membrane as electrons pass from PSII to PSI. Because the volume of the flattened thylakoid lumen is small, the influx of a small number of protons has a relatively large effect on lumenal pH. The measured difference in pH between the stroma (pH 8) and the thylakoid lumen (pH 5) represents a 1,000-fold difference in proton concentration—a powerful driving force for ATP synthesis.

Cyclic Electron Flow between PSI and the Cytochrome b_6f Complex Increases the Production of ATP Relative to NADPH

The linear path of electrons from water to NADP produces both a proton gradient, which is used to drive ATP synthesis, and NADPH, which is used in reductive biosynthetic processes. This linear movement of electrons is called **noncyclic electron flow**, to distinguish it from its alternative, **cyclic electron flow**. Cyclic electron flow involves only PSI, not PSII (Fig. 20-16). Electrons passing from P700 to ferredoxin do not continue to $NADP^+$, but move back through the cytochrome b_6f complex to plastocyanin. (This electron path parallels that in green sulfur bacteria, shown in Fig. 20-14b.) Plastocyanin then donates electrons to P700, which transfers them to ferredoxin. In this way, electrons are repeatedly recycled through the cytochrome b_6f complex and the reaction center of PSI, each electron propelled around the cycle by the energy of one photon. Cyclic electron flow is not accompanied by net formation of NADPH or evolution of O_2. However, it *is* accompanied by proton pumping by the cytochrome b_6f complex and by phosphorylation of ADP to ATP, referred to as **cyclic photophosphorylation**. The overall equation for cyclic electron flow and photophosphorylation is simply

$$\text{ADP} + P_i \xrightarrow{\text{light}} \text{ATP} + H_2O$$

By regulating the partitioning of electrons between $NADP^+$ reduction and cyclic photophosphorylation, a plant adjusts the ratio of ATP to NADPH produced in the light-dependent reactions to match its needs for these products in the carbon-assimilation reactions and other biosynthetic processes. As we shall see in Section 20.5, the carbon-assimilation reactions require ATP and NADPH in the ratio 3:2. This regulation of electron-transfer pathways is part of a short-term adaptation to changes in light color (wavelength) and quantity (intensity), as further described below.

State Transitions Change the Distribution of LHCII between the Two Photosystems

Plants are exposed to light of highly variable intensity and wavelength in the course of a day or a season, and, although they can alter their growth patterns somewhat, they cannot uproot themselves and move to optimize their light exposure. Instead, cellular mechanisms have evolved that allow plants to accommodate to changing light conditions. The energy needed to excite PSI (P700) is less (light of longer wavelength, lower energy) than the energy needed to excite PSII (P680). If PSI and PSII were physically contiguous, excitons originating in the antenna system of PSII would migrate to the reaction center of PSI, leaving PSII chronically underexcited and thus interfering with the operation of the two-center system. This imbalance in the supply of excitons is prevented by separating the two photosystems in the thylakoid membrane **(Fig. 20-22)**. PSII is located almost exclusively in the tightly appressed membrane stacks of granal thylakoids; its associated light-harvesting complex (LHCII) mediates the tight association of adjacent membranes in the grana. PSI and the ATP synthase complex are located almost exclusively in the nonappressed membranes of the stromal thylakoids, where they have access to the contents of the stroma, including ADP and $NADP^+$. The cytochrome b_6f complex is present primarily in the granal thylakoids.

The association of LHCII with PSI and PSII depends on light intensity and wavelength, which can change in the short term and lead to **state transitions** in the chloroplast. In state 1, LHCII, PSII, and PSI are poised to maximize the capture of light energy. A critical Thr residue in LHCII is unphosphorylated, and LHCII associates with PSII. Under conditions of intense or blue light, which favor absorption by PSII, that photosystem reduces plastoquinone to plastoquinol (PQH_2) faster than PSI can oxidize it. The resulting accumulation of PQH_2 activates a protein kinase that triggers the transition to state 2 by phosphorylating a Thr residue on LHCII **(Fig. 20-23)**. Phosphorylation weakens the interaction of LHCII with the appressed membrane and with PSII; some LHCII dissociates and moves to the stromal thylakoids. Here it captures photons (excitons) for PSI, speeding the oxidation of PQH_2 and

(a)

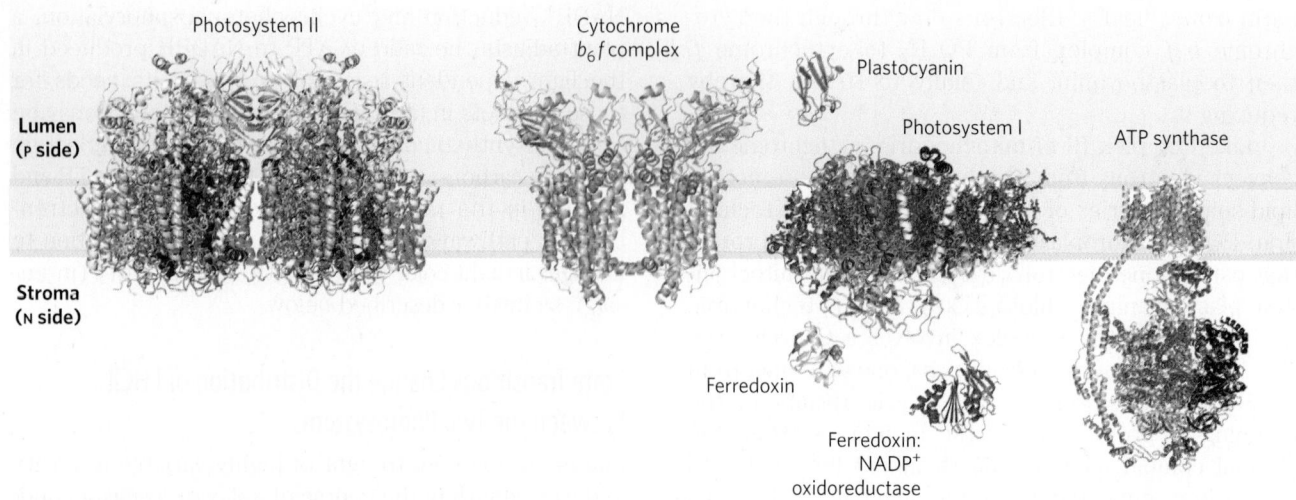

(b)

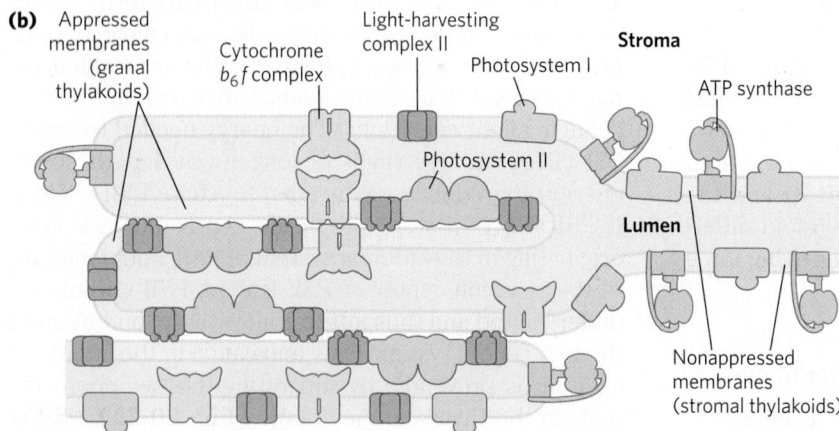

FIGURE 20-22 Localization of PSI and PSII in thylakoid membranes.
(a) Structures of the complexes and soluble proteins of the photosynthetic apparatus of a vascular plant or alga, drawn to the same scale. The bovine ATP synthase is shown. **(b)** Light-harvesting complex LHCII and ATP synthase are located both in appressed regions of the thylakoid membrane (granal thylakoids, in which several membranes are in contact) and in nonappressed regions (stromal thylakoids), and have ready access to ADP and NADP$^+$ in the stroma. PSII is present almost exclusively in the appressed granal regions, and PSI almost exclusively in nonappressed stromal regions. LHCII is the "adhesive" that holds appressed thylakoid membranes together (see Fig. 20-23). [Sources: (a) PSII: PDB ID 3WU2, Y. Umena et al., *Nature* 473:55, 2011; cyt b_6f complex: PDB ID 2E74, E. Yamashita et al., *J. Mol. Biol.* 370:39, 2007; plastocyanin: PDB ID 1AG6, Y. Xue et al., *Protein Sci.* 7:2099, 1998; PSI: PDB ID 4RKU, Y. Mazor et al; ferredoxin: PDB ID 1A70, C. Binda et al., *Acta Crystallogr. D Biol. Crystallogr.* 54:1353, 1998; ferredoxin:NADP reductase: PDB ID 1QG0, Z. Deng et al., *Nature Struct. Biol.* 6:847, 1999; ATP synthase: PDB ID 5ARA, A. Zhou et al., *eLife* 4:e10180, 2015.]

reversing the imbalance between electron flow in PSI and PSII. In less intense light (in the shade, with more red light), PSI oxidizes PQH_2 faster than PSII can make it, and the resulting increase in [PQ] triggers

dephosphorylation of LHCII, reversing the effect of phosphorylation. The state transition in LHCII localization and the transition from cyclic to noncyclic photophosphorylation are coordinately regulated: the path of

FIGURE 20-23 Balancing of electron flow in PSI and PSII by state transition. In granal thylakoids, a hydrophobic domain of LHCII in one membrane inserts into the neighboring membrane and closely appresses the two (state 1). Accumulation of plastoquinol (not shown) stimulates a protein kinase that phosphorylates a Thr residue in the hydrophobic domain of LHCII, which reduces its affinity for the neighboring membrane and converts appressed granal thylakoids to nonappressed stromal thylakoids (state 2). A specific protein phosphatase reverses this regulatory phosphorylation when the [PQ]/[PQH$_2$] ratio increases.

electrons is primarily noncyclic in state 1 and primarily cyclic in state 2.

When light is so intense that the combined activity of PSII and PSI cannot synthesize ATP and NADPH fast enough to keep up with the supply of photons, carotenoids in LHCII absorb the exciton and very rapidly quench the excited chlorophyll before it can create damaging reactive oxygen species (ROS). The trigger for switching from an efficient light-harvesting state to an energy-dissipating state is the lowering of pH in the lumenal space, but the detailed mechanism for this transition is not yet known.

Water Is Split by the Oxygen-Evolving Complex

The ultimate source of the electrons passed to NADPH in plant (oxygenic) photosynthesis is water. Having given up an electron to pheophytin, $P680^+$ (of PSII) must acquire an electron to return to its ground state in preparation for capture of another photon. In principle, the required electron might come from any number of organic or inorganic compounds. Photosynthetic bacteria use a variety of electron donors for this purpose—acetate, succinate, malate, or sulfide—depending on what is available in a particular ecological niche. About 3 billion years ago, evolution of primitive photosynthetic bacteria (progenitors of the modern cyanobacteria) produced a photosystem capable of taking electrons from a donor that is always available: water. Two water molecules are split, yielding four electrons, four protons, and molecular oxygen:

$$2H_2O \longrightarrow 4H^+ + 4e^- + O_2$$

A single photon of visible light does not have enough energy to break the bonds in water; four photons are required in this photolytic cleavage reaction.

The four electrons abstracted from water do not pass directly to $P680^+$, which can accept only one electron at a time. Instead, a remarkable molecular device, the **oxygen-evolving complex** (also called the **water-splitting complex**), passes four electrons *one at a time* to $P680^+$ **(Fig. 20-24a)**. The immediate electron donor to $P680^+$ is a Tyr residue (often designated Z or Tyr_Z) in subunit D1 of the PSII reaction center. The Tyr residue loses both a proton and an electron, generating the electrically neutral Tyr free radical, $^\bullet Tyr$:

$$4 P680^+ + 4 Tyr \longrightarrow 4 P680 + 4\, ^\bullet Tyr \quad (20\text{-}2)$$

The Tyr radical regains its missing electron and proton by oxidizing a cluster of four manganese ions and one calcium ion in the water-splitting complex. With each single-electron transfer, the Mn_4CaO_5 cluster becomes more oxidized; four single-electron transfers, each corresponding to the absorption of one photon, produce a charge of 4+ on the Mn_4CaO_5 cluster (Fig. 20-24):

$$4\, ^\bullet Tyr + [Mn_4CaO_5]^0 \longrightarrow 4 Tyr + [Mn_4CaO_5]^{4+} \quad (20\text{-}3)$$

In this state, the Mn_4CaO_5 cluster can take four electrons from a pair of water molecules, releasing four H^+ and O_2:

$$[Mn_4CaO_5]^{4+} + 2H_2O \longrightarrow$$
$$[Mn_4CaO_5]^0 + 4H^+ + O_2 \quad (20\text{-}4)$$

Because the four protons produced in this reaction are released into the thylakoid lumen, the oxygen-evolving complex acts as a proton pump, driven by electron transfer. The sum of Equations 20-1 (p. 768) through 20-4 is

$$2H_2O + 2PQ_B + 4\ photons \longrightarrow$$
$$O_2 + 2\ PQ_BH_2 \quad (20\text{-}5)$$

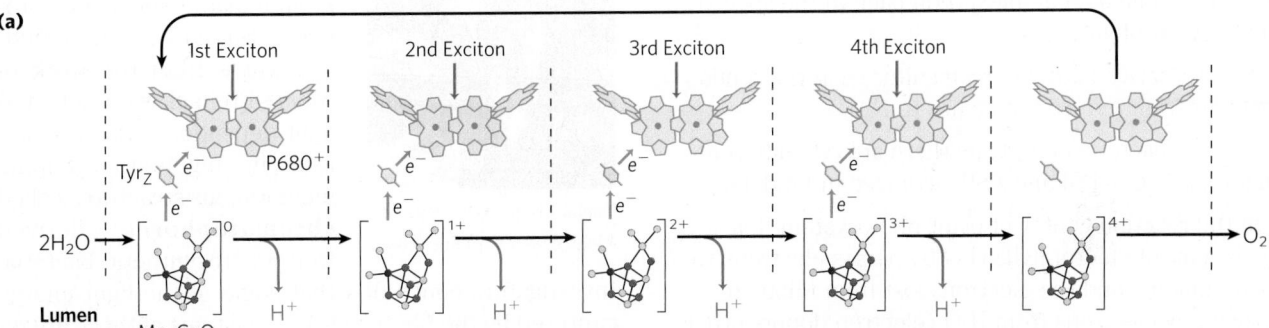

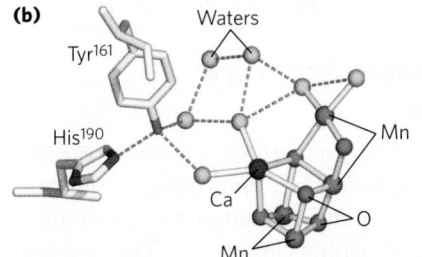

FIGURE 20-24 Water-splitting activity of the oxygen-evolving complex. (a) Shown here is the process that produces a four-electron oxidizing agent—a multinuclear center with four Mn ions, one Ca ion, and five oxygen atoms—in the oxygen-evolving complex of PSII. The sequential absorption of four photons (excitons), each absorption causing the loss of one electron from the Mn_4CaO_5 cluster, produces an oxidizing agent that can remove four electrons from two molecules of water, producing O_2. The electrons lost from the Mn_4CaO_5 cluster pass one at a time to an oxidized Tyr residue in a PSII protein, then to $P680^+$. **(b)** The metallic center of the oxygen-evolving complex, obtained by x-ray crystallography. Tyr^{161}, known to participate in the oxidation of water, is seen hydrogen-bonded to a network of water molecules, including several directly in contact with the Mn_4CaO_5 cluster. This is the site of one of the most important reactions in the biosphere. [Source: (b) PDB ID 3WU2, Y. Umena et al., *Nature* 473:55, 2011.]

The detailed structure of the oxygen-evolving cluster has been obtained by high-resolution x-ray crystallography. The metal cluster takes the shape of a chair (Fig. 20-24b). The seat and legs of the chair are made up of three Mn ions, one Ca ion, and four O atoms; the fourth Mn and another O form the back of the chair. Four water molecules are also seen in the crystal structure, two associated with one of the Mn ions, the other two with the Ca ion. It may be one (or more) of these water molecules that undergoes oxidation to produce O_2. This metal cluster is associated with several peripheral membrane proteins on the lumenal side of the thylakoid membrane that are believed to stabilize the cluster. The Tyr residue designated Z, through which electrons move between water and the PSII reaction center, is connected with a network of hydrogen-bonded water molecules that includes the four associated with the Mn_4CaO_5 cluster. The detailed mechanism of water oxidation by the Mn_4CaO_5 cluster is not known but is under intense investigation. The reaction is central to life on Earth and may involve novel bioinorganic chemistry. Determination of the structure of the polymetallic center has inspired several reasonable and testable hypotheses. Stay tuned.

SUMMARY 20.2 Photochemical Reaction Centers

■ Bacteria have a single photochemical reaction center; in purple bacteria, it is of the pheophytin-quinone type, and in green sulfur bacteria, the Fe-S type.

■ Structural studies of the reaction center of a purple bacterium have provided information about light-driven electron flow from an excited special pair of chlorophyll molecules (P870), through pheophytin, to quinones. Electrons pass from the quinones through the cytochrome bc_1 complex, then back to the special pair of chlorophylls.

■ An alternative path, in green sulfur bacteria, sends electrons from reduced quinones to NAD^+.

■ In cyanobacteria and plants there are two different reaction centers, PSI and PSII, arranged in tandem.

■ In the reaction center of plant photosystem II, a special pair of chlorophylls (P680) passes electrons to plastoquinone, and the electrons lost from P680 are replaced by electrons from H_2O (electron donors other than H_2O are used in other organisms).

■ Plant photosystem I passes electrons from the excited special pair (P700) in its reaction center through a series of carriers to ferredoxin, which then reduces $NADP^+$ to NADPH.

■ Electron flow through the cytochrome b_6f complex drives protons across the plasma membrane, creating a proton-motive force that provides the energy for ATP synthesis by an ATP synthase.

The water-splitting activity also moves protons from the stroma to the thylakoid lumen, contributing to the proton gradient.

■ Linear flow of electrons through the photosystems produces NADPH and ATP. Cyclic electron flow produces only ATP and allows variability in the proportions of NADPH and ATP formed.

■ Localization of PSI and PSII between the granal and stromal thylakoids can change and is indirectly controlled by light intensity, optimizing the distribution of excitons between PSI and PSII for efficient energy capture.

■ The light-driven splitting of H_2O is catalyzed by a Mn_4CaO_5-containing protein complex; O_2 is produced. The reduced plastoquinone carries electrons to the cytochrome b_6f complex; from here they pass to plastocyanin, and then to P700 to replace the electrons lost during its photoexcitation.

20.3 ATP Synthesis by Photophosphorylation

The combined activities of the two plant photosystems move electrons from water to $NADP^+$, conserving some of the energy of absorbed light as NADPH (Fig. 20-16). Simultaneously, protons are pumped across the thylakoid membrane and energy is conserved as an electrochemical potential. We turn now to the process by which this proton gradient drives the synthesis of ATP, the other energy-conserving product of the light-dependent reactions.

Daniel Arnon, 1910–1994
[Source: Reinhard Bachofen.]

In 1954, Daniel Arnon and his colleagues discovered that ATP is generated from ADP and P_i during photosynthetic electron transfer in illuminated spinach chloroplasts. Support for these findings came from the work of Albert Frenkel, who detected light-dependent ATP production in pigment-containing membranous structures called **chromatophores**, derived from photosynthetic bacteria. Investigators concluded that some of the light energy captured by the photosynthetic systems of these organisms is transformed into the phosphate bond energy of ATP in the process of photophosphorylation.

A Proton Gradient Couples Electron Flow and Phosphorylation

Several properties of photosynthetic electron transfer and photophosphorylation in chloroplasts indicate that a proton gradient plays the same role as in mitochondrial oxidative phosphorylation. (1) The reaction centers, electron carriers, and ATP-forming enzymes are located in a proton-impermeable membrane—the thylakoid membrane—which must be intact to support photophosphorylation. (2) Photophosphorylation can be uncoupled from electron flow by reagents that promote the passage of protons through the thylakoid membrane. (3) Photophosphorylation can be blocked by venturicidin and similar agents that inhibit the formation of ATP from ADP and P_i by the mitochondrial ATP synthase. (4) ATP synthesis is catalyzed by CF_oCF_1 complexes, located on the outer surface of the thylakoid membranes, that are very similar in structure and function to the F_oF_1 complexes of mitochondria.

Electron-transferring molecules in the chain of carriers connecting PSII and PSI are oriented asymmetrically in the thylakoid membrane, so photoinduced electron flow results in the net movement of protons across the membrane, from the stromal side to the thylakoid lumen **(Fig. 20-25)**. In 1966, André Jagendorf showed that a pH gradient across the thylakoid membrane (alkaline outside) could furnish the driving force to generate ATP. His early observations provided some of the most important experimental evidence in support of Mitchell's chemiosmotic hypothesis.

André Jagendorf
[Source: Courtesy Professor André Jagendorf.]

Jagendorf incubated chloroplasts, in the dark, in a pH 4 buffer; the buffer slowly penetrated into the lumen of the thylakoids, lowering their internal pH. He added ADP and P_i to the dark suspension of chloroplasts and then suddenly raised the pH of the outer medium to 8, momentarily creating a large pH gradient across the membrane. As protons moved out of the thylakoids into the medium, ATP was generated from ADP and P_i. Because the formation of ATP occurred in the dark (with no input of energy from light), this experiment showed that a pH gradient across the membrane is a high-energy state that, as in mitochondrial oxidative phosphorylation, can mediate the transduction of energy from electron transfer into the chemical energy of ATP.

The Approximate Stoichiometry of Photophosphorylation Has Been Established

As electrons move from water to $NADP^+$ in plant chloroplasts, about 12 protons move from the stroma into the thylakoid lumen per four electrons passed (that is, per O_2 formed). Four of these protons are moved by the oxygen-evolving complex, and up to eight by the cytochrome b_6f complex. The measurable result is a 1,000-fold difference in H^+ concentration across the thylakoid membrane ($\Delta pH = 3$). Recall that the energy stored in a proton gradient (the electrochemical potential) has two components: a proton concentration difference (ΔpH) and an electrical potential ($\Delta \psi$) due to charge separation. In chloroplasts, ΔpH is the dominant component; counterion movement apparently dissipates most of the electrical potential. In illuminated chloroplasts, the energy stored in the proton gradient per mole of protons is

$$\Delta G = 2.3RT\,\Delta pH + ZF\,\Delta \psi = -17\text{ kJ/mol}$$

so the movement of 12 mol of protons across the thylakoid membrane represents conservation of about 200 kJ of energy—enough energy to drive the synthesis of several moles of ATP ($\Delta G'^{\circ} = 30.5$ kJ/mol). Experimental measurements yield values of about 3 ATP per O_2 produced.

At least eight photons must be absorbed to drive four electrons from H_2O to NADPH (one photon per electron at each reaction center). The energy in eight photons of visible light is more than enough for the synthesis of three molecules of ATP.

ATP synthesis is not the only energy-conserving reaction of photosynthesis in plants; the NADPH formed in the final electron transfer is also energetically rich. The overall equation for noncyclic photophosphorylation (a term explained below) is

$$2H_2O + 8 \text{ photons} + 2NADP^+ + \sim 3ADP^+ + \sim 3P_i \longrightarrow$$
$$O_2 + \sim 3ATP + 2NADPH \quad (20\text{-}6)$$

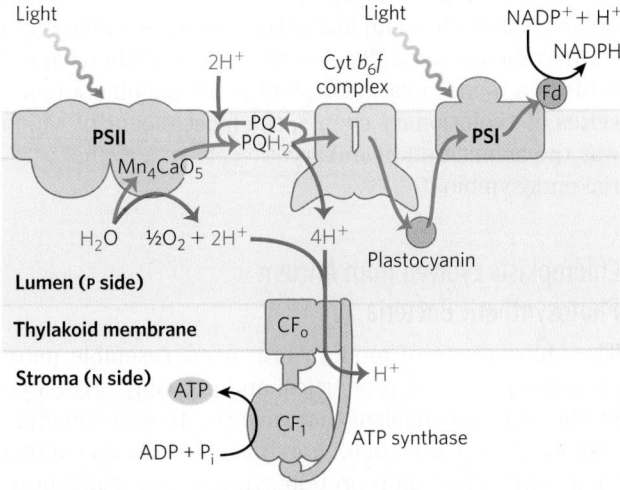

FIGURE 20-25 Proton and electron circuits during photophosphorylation. Electrons (blue arrows) move from H_2O through PSII, through the intermediate chain of carriers, through PSI, and finally to $NADP^+$. Protons (red arrows) are pumped into the thylakoid lumen by the flow of electrons through the carriers linking PSII and PSI, and reenter the stroma through proton channels formed by CF_o of ATP synthase. The CF_1 subunit catalyzes synthesis of ATP.

The ATP Synthase of Chloroplasts Resembles That of Mitochondria

The enzyme responsible for ATP synthesis in chloroplasts is a large complex with two functional components, CF_o and

CF$_1$ (C denoting its location in chloroplasts). CF$_o$ is a transmembrane proton pore composed of several integral membrane proteins and is homologous to mitochondrial F$_o$. CF$_1$ is a peripheral membrane protein complex very similar in subunit composition, structure, and function to mitochondrial F$_1$.

Electron microscopy of sectioned chloroplasts shows ATP synthase complexes as knoblike projections on the *outside* (stromal, or N) surface of thylakoid membranes; these complexes correspond to the ATP synthase complexes that project on the *inside* (matrix, or N) surface of the inner mitochondrial membrane. Thus the relationship between the orientation of the ATP synthase and the direction of proton pumping is the same in chloroplasts and mitochondria. In both cases, the F$_1$ portion of ATP synthase is located on the more alkaline (N) side of the membrane through which protons flow down their concentration gradient; the direction of proton flow relative to F$_1$ is the same in both cases: P to N **(Fig. 20-26)**.

The mechanism of chloroplast ATP synthase is also believed to be essentially identical to that of its mitochondrial analog; ADP and P$_i$ readily condense to form ATP on the enzyme surface, and release of this enzyme-bound ATP requires a proton-motive force. Rotational catalysis sequentially engages each of the three β subunits of the ATP synthase in ATP synthesis, ATP release, and ADP + P$_i$ binding (see Figs 19-26 and 19-27).

The chloroplast ATP synthase of spinach, with 14 c subunits in its CF$_o$ complex, is predicted to have a lower ratio of ATP formed to electrons transferred than do the bovine, yeast, or *E. coli* F$_o$ complexes, with 8, 10, and 10 c subunits, respectively (see Fig. 19-29).

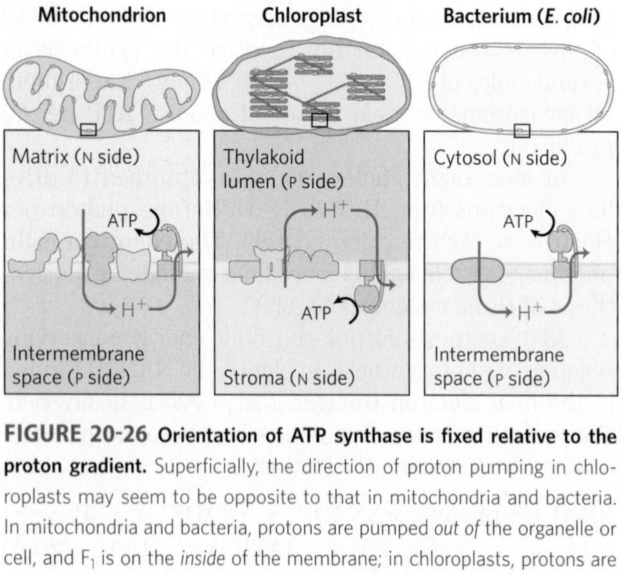

SUMMARY 20.3 ATP Synthesis by Photophosphorylation

■ In plants, both the water-splitting reaction and electron flow through the cytochrome b_6f complex are accompanied by proton pumping across the thylakoid membrane. The proton-motive force thus created drives ATP synthesis by a CF$_o$CF$_1$ complex similar to the mitochondrial F$_o$F$_1$ complex.

■ The catalytic mechanism of CF$_o$CF$_1$ is very similar to that of the ATP synthases of mitochondria and bacteria. Physical rotation driven by the proton gradient is accompanied by ATP synthesis at sites that cycle through three conformations: one with high affinity for ATP, one with high affinity for ADP + P$_i$, and one with low affinity for both nucleotides.

20.4 Evolution of Oxygenic Photosynthesis

The appearance of oxygenic photosynthesis on Earth about 2.5 billion years ago was a crucial event in the evolution of the biosphere. Before that, Earth's atmosphere comprised methane, CO_2, and N_2. The planet was essentially devoid of molecular oxygen and lacked the ozone layer that protects living organisms from solar UV radiation. Oxygenic photosynthesis made available a nearly limitless supply of reducing agent (H_2O) to drive the production of organic compounds by reductive biosynthetic reactions. And mechanisms evolved that allowed organisms to use O_2 as a terminal electron acceptor in highly energetic electron transfers from organic substrates, employing the energy of oxidation to support metabolism. The complex photosynthetic apparatus of a modern vascular plant is the culmination of a series of evolutionary events, the most recent of which was the acquisition by eukaryotic cells of a cyanobacterial endosymbiont.

Chloroplasts Evolved from Ancient Photosynthetic Bacteria

The chloroplasts of modern organisms resemble mitochondria in several properties, and probably originated by the same mechanism that gave rise to mitochondria: endosymbiosis. Like mitochondria, chloroplasts contain their own DNA and protein-synthesizing machinery. Some of the polypeptides of chloroplast proteins are encoded by chloroplast genes and synthesized in the chloroplast; others are encoded by nuclear genes, synthesized outside the chloroplast, and imported (Chapter 27). When plant cells grow and divide, chloroplasts give rise to new chloroplasts by division, during which

FIGURE 20-26 Orientation of ATP synthase is fixed relative to the proton gradient. Superficially, the direction of proton pumping in chloroplasts may seem to be opposite to that in mitochondria and bacteria. In mitochondria and bacteria, protons are pumped *out of* the organelle or cell, and F$_1$ is on the *inside* of the membrane; in chloroplasts, protons are pumped *into* the thylakoid lumen, and CF$_1$ is on the *outside* of the thylakoid membrane. However, exactly the same mechanism of energy conversion (from proton gradient to ATP) occurs in all three cases. ATP is synthesized in the matrix of mitochondria, the stroma of chloroplasts, and the cytosol of bacteria.

their DNA is replicated and divided between daughter chloroplasts. The machinery and mechanisms for light capture, electron flow, and ATP synthesis in modern cyanobacteria are similar in many respects to those in plant chloroplasts. These observations led to the now widely accepted hypothesis that the evolutionary progenitors of modern plant cells were primitive eukaryotes that engulfed photosynthetic cyanobacteria and established stable endosymbiotic relationships with them (see Fig. 1-40).

At least half of the photosynthetic activity on Earth now occurs in microorganisms—algae, other photosynthetic eukaryotes, and photosynthetic bacteria. Cyanobacteria have PSII and PSI in tandem, and the PSII has an associated water-splitting activity resembling that of plants. However, the other groups of photosynthetic bacteria have single reaction centers and do not split H_2O or produce O_2. Many are obligate anaerobes and cannot tolerate O_2; they must use some compound other than H_2O as an electron donor. Some photosynthetic bacteria use inorganic compounds as electron (and hydrogen) donors. For example, green sulfur bacteria use hydrogen sulfide:

$$2H_2S + CO_2 \xrightarrow{\text{light}} (CH_2O) + H_2O + 2S$$

These bacteria, instead of producing molecular O_2, form elemental sulfur as the oxidation product of H_2S. (They further oxidize the S to SO_4^{2-}.) Other photosynthetic bacteria use organic compounds such as lactate as electron donors:

$$2 \text{ Lactate} + CO_2 \xrightarrow{\text{light}} (CH_2O) + H_2O + 2 \text{ pyruvate}$$

The fundamental similarity of photosynthesis in plants and bacteria, despite the differences in the electron donors they employ, becomes more obvious when the equation of photosynthesis is written in the more general form

$$2H_2D + CO_2 \xrightarrow{\text{light}} (CH_2O) + H_2O + 2D$$

in which H_2D is an electron (and hydrogen) donor and D is its oxidized form. H_2D may be water, hydrogen sulfide, lactate, or some other organic compound, depending on the species. Most likely, the bacteria that first developed photosynthetic ability used H_2S as their electron source.

The ancient relatives of modern cyanobacteria probably arose by the combination of genetic material from two types of photosynthetic bacteria, with systems of the type seen in modern purple bacteria (with a PSII-like electron path) and green sulfur bacteria (with a PSI-like path). The bacterium with two independent photosystems may have used one under one set of conditions, the other under different conditions. Over time, a mechanism to connect the two photosystems for simultaneous use evolved, and the PSII-like system acquired the water-splitting capacity found in modern cyanobacteria.

Modern cyanobacteria can synthesize ATP by oxidative phosphorylation or by photophosphorylation, although they have neither mitochondria nor chloroplasts. The enzymatic machinery for both processes is in a highly convoluted plasma membrane **(Fig. 20-27)**. Three protein components function in both processes, giving evidence that the processes have a common evolutionary origin **(Fig. 20-28)**. First, the proton-pumping cytochrome b_6f complex carries electrons from plastoquinone to cytochrome c_6 in photosynthesis, and also carries electrons from ubiquinone to cytochrome c_6 in oxidative phosphorylation—the role played by cytochrome bc_1 in mitochondria. Second, cytochrome c_6, homologous to mitochondrial cytochrome c, carries electrons from Complex III to Complex IV in cyanobacteria; it can also carry electrons from the cytochrome b_6f complex to PSI—a role performed in

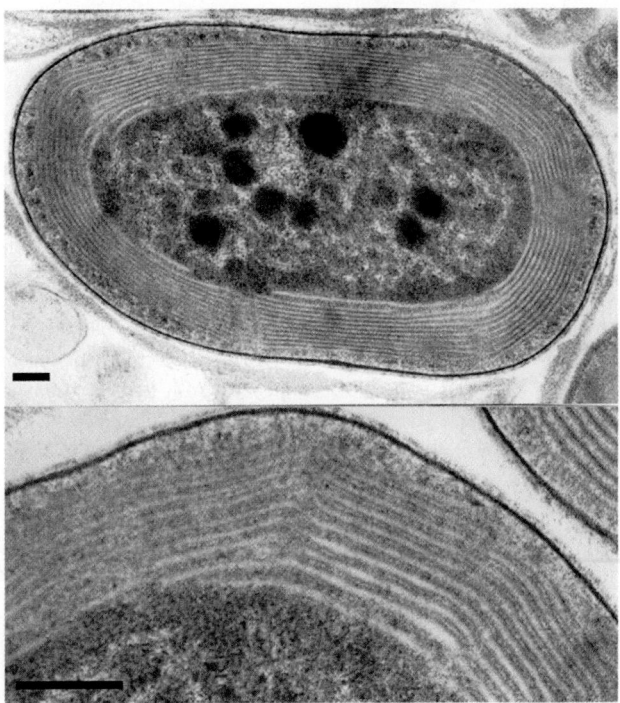

FIGURE 20-27 The photosynthetic membranes of a cyanobacterium. In these thin sections of a cyanobacterium, viewed with a transmission electron microscope, the multiple layers of the internal membranes are seen to fill half the total volume of the cell. The extensive membrane system serves the same role as the thylakoid membranes of vascular plants, providing a large surface area containing all of the photosynthetic machinery. (Bar = 100 nm.) [Source: S. R. Miller et al. Discovery of a free-living chlorophyll d-producing cyanobacterium with a hybrid proteobacterial/cyanobacterial small-subunit rRNA gene. *Proc. Natl. Acad. Sci. USA* 102:850, 2005, Fig. 2. © 2005 National Academy of Sciences.]

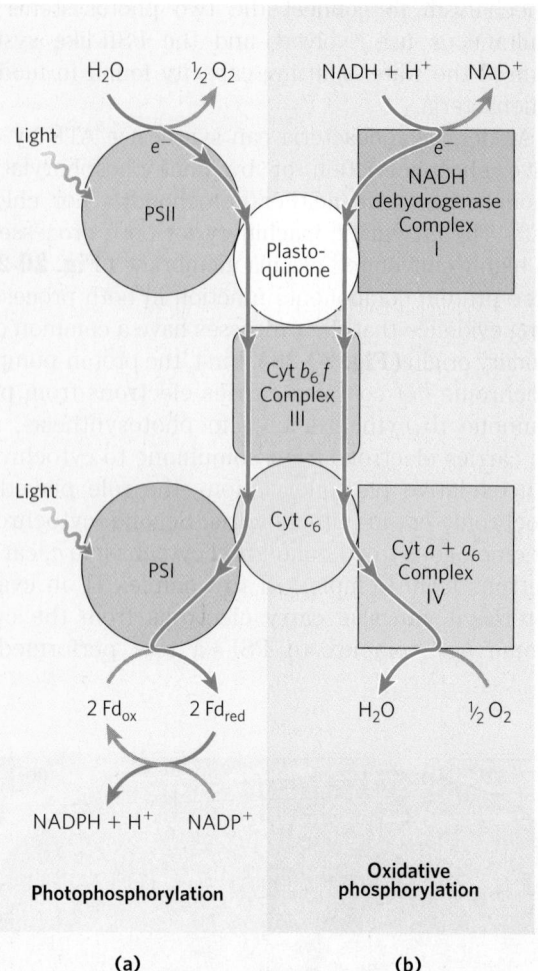

FIGURE 20-28 Dual roles of cytochrome b_6f and cytochrome c_6 in cyanobacteria reflect evolutionary origins. Cyanobacteria use cytochrome b_6f, cytochrome c_6, and plastoquinone for both oxidative phosphorylation and photophosphorylation. **(a)** In photophosphorylation, electrons flow (blue arrows) from water to $NADP^+$. **(b)** In oxidative phosphorylation, electrons flow from NADH to O_2. Both processes are accompanied by proton movement across the membrane, accomplished by a Q cycle.

plants by plastocyanin. We therefore see the functional homology between the cyanobacterial cytochrome b_6f complex and the mitochondrial cytochrome bc_1 complex, and between cyanobacterial cytochrome c_6 and plant plastocyanin. The third conserved component is the ATP synthase, which functions in oxidative phosphorylation and photophosphorylation in cyanobacteria, and in the mitochondria and chloroplasts of photosynthetic eukaryotes. The structure and remarkable mechanism of this enzyme have been strongly conserved throughout evolution.

In *Halobacterium,* a Single Protein Absorbs Light and Pumps Protons to Drive ATP Synthesis

In some modern archaea, a quite different mechanism for converting the energy of light into an electrochemical

gradient has evolved. The halophilic ("salt-loving") archaeon *Halobacterium salinarum* (commonly referred to as a halobacterium) lives only in brine ponds and salt lakes (Great Salt Lake and the Dead Sea, for example), where the high salt concentration—which can exceed 4 M—results from water loss by evaporation. Indeed, halobacteria cannot live in NaCl concentrations lower than 3 M. These organisms are aerobes and normally use O_2 to oxidize organic fuel molecules. However, the solubility of O_2 is so low in brine ponds that sometimes oxidative metabolism must be supplemented by sunlight as an alternative source of energy.

The plasma membrane of *H. salinarum* contains patches of the light-absorbing pigment **bacteriorhodopsin**, which contains retinal (the aldehyde derivative of vitamin A; see Fig. 10-20) as a light-harvesting prosthetic group. When the cells are illuminated, all-*trans*-retinal bound to the bacteriorhodopsin absorbs a photon and undergoes photoisomerization to 13-*cis*-retinal, forcing a conformational change in the protein. The restoration of all-*trans*-retinal is accompanied by outward movement of protons through the plasma membrane. Bacteriorhodopsin, with only 247 amino acid residues, is the simplest light-driven proton pump known. The difference in the three-dimensional structure of bacteriorhodopsin in the dark and after illumination **(Fig. 20-29a)** suggests a pathway by which a concerted series of proton "hops" could effectively move a proton across the membrane. The chromophore retinal is bound through a Schiff-base linkage to the ε-amino group of a Lys residue. In the dark, the nitrogen of this Schiff base is protonated; in the light, photoisomerization of retinal lowers the pK_a of this group and it releases its proton to a nearby Asp residue, triggering a series of proton hops that ultimately result in release of a proton at the outer surface of the membrane (Fig. 20-29b).

The electrochemical potential across the membrane drives protons back into the cell through a membrane ATP synthase complex very similar to that of mitochondria and chloroplasts. Thus, when O_2 is limited, halobacteria can use light to supplement the ATP synthesized by oxidative phosphorylation. Halobacteria do not evolve O_2, nor do they carry out photoreduction of $NADP^+$; their phototransducing machinery is therefore much simpler than that of cyanobacteria or plants. Nevertheless, the proton-pumping mechanism may prove to be prototypical for the many other, more complex, ion pumps.

SUMMARY 20.4 Evolution of Oxygenic Photosynthesis

■ Modern cyanobacteria are derived from an ancient organism that acquired two photosystems, one of the type now found in purple bacteria, the other of the type found in green sulfur bacteria.

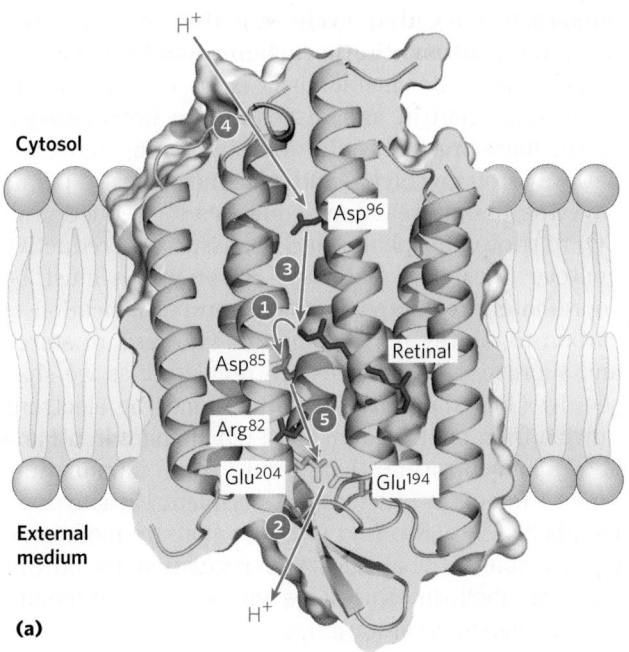

(a)

FIGURE 20-29 A different mechanism for light-driven proton pumping evolved independently in a halophilic archaeon. (a) Bacteriorhodopsin (M_r 26,000) of *Halobacterium halobium* has seven membrane-spanning α helices. The chromophore all-*trans*-retinal (purple) is covalently attached via a Schiff base to the ε-amino group of a Lys residue deep in the membrane interior. Running through the protein are a series of Asp and Glu residues and a series of closely associated water molecules that together provide the transmembrane path for protons (red arrows). Steps ❶ through ❺ indicate proton movements, described below.

(b) In the dark (left panel), the Schiff base is protonated. Illumination (right panel) photoisomerizes the retinal, forcing subtle conformational changes in the protein that alter the distance between the Schiff base and its neighboring amino acid residues. Interaction with these neighbors (Leu[93] and Val[49]) lowers the pK_a of the protonated Schiff base, and the base gives up its proton to a nearby carboxyl group on Asp[85] (step ❶ in (a)). This initiates a series of concerted proton hops between water molecules (see Fig. 2-14) in the interior of the protein, which ends with ❷ release of a proton that was shared by Glu[194] and Glu[204] near the extracellular surface. (Tyr[83] forms a hydrogen bond with Glu[194] that facilitates this proton release.) ❸ The Schiff base reacquires a proton from Asp[96], which ❹ takes up a proton from the cytosol. ❺ Finally, Asp[85] gives up its proton, leading to reprotonation of the Glu[204]-Glu[194] pair. The system is now ready for another round of proton pumping. [Sources: (a) PDB ID 1C8R, H. Luecke et al., *Science* 286:255, 1999. (b) Information from R. B. Gennis and T. G. Ebrey, *Science* 286:252, 1999.]

Dark

Retinal

Thr[89]—OH

Asp[85]

Low pK_a

Protonated Schiff base (high pK_a)

Arg[82]

Proton-release complex (protonated; high pK_a)   Glu[194]   Glu[204]

(b)

Light

Leu[93]   Val[49]

Retinal   Lys[216]

Conformational change lowers pK_a of Schiff base

Asp[85]

Higher pK_a

Proton transfer

Arg[82]

pK_a of proton-release complex lowered

Tyr[83]—OH   Glu[194]   Glu[204]

Proton release H^+

- Many photosynthetic microorganisms obtain electrons for photosynthesis not from water but from donors such as H_2S.

- Cyanobacteria with the tandem photosystems and a water-splitting activity that released oxygen into the atmosphere appeared on Earth about 2.5 billion years ago.

- Chloroplasts, like mitochondria, evolved from bacteria living endosymbiotically in early eukaryotic cells. The ATP synthases of bacteria, cyanobacteria, mitochondria, and chloroplasts share a common evolutionary precursor and a common enzymatic mechanism.

- An entirely different mechanism for converting light energy to a proton gradient has evolved in the modern archaea, in which the light-harvesting pigment is retinal.

20.5 Carbon-Assimilation Reactions

Photosynthetic organisms use the ATP and NADPH produced in the light-dependent reactions of photosynthesis to synthesize all of the thousands of components that make up the organism. This is possible because of certain enzymatic capacities and pathways, not found in chemoheterotrophs (such as humans), that have evolved in photoautotrophs (see Fig. 1-6). Plants (and other autotrophs) can reduce atmospheric CO_2 to trioses, then use the trioses as precursors for the synthesis of cellulose and starch, lipids and proteins, and the many other organic components of plant cells **(Fig. 20-30)**. Lacking these synthetic capacities, humans and other animals are ultimately dependent on photosynthetic organisms to provide the reduced fuels and organic precursors essential to life.

Green plants contain in their chloroplasts the enzymatic machinery that catalyzes the conversion of CO_2 to simple (reduced) organic compounds, a process called **CO_2 assimilation**. This process has also been called **CO_2 fixation** or **carbon fixation**, but we reserve these terms for the specific reaction in which CO_2 is incorporated (fixed) into a three-carbon organic compound, the triose phosphate 3-phosphoglycerate. This simple product of photosynthesis is the precursor of more complex biomolecules, including sugars, polysaccharides, and the metabolites derived from them, all of which are synthesized by metabolic pathways similar to those of animal tissues. Carbon dioxide is assimilated via a cyclic pathway, its key intermediates constantly regenerated. The pathway was elucidated in the early 1950s by Melvin Calvin, Andrew Benson, and James A. Bassham, and is

Melvin Calvin, 1911–1997
[Source: Ted Spiegel/Corbis.]

often called the **Calvin cycle** or, perhaps more descriptively, the **photosynthetic carbon-reduction cycle**.

Carbohydrate metabolism is more complex in plant cells than in animal cells or in nonphotosynthetic microorganisms. In addition to the universal pathways of glycolysis and gluconeogenesis, plants have the unique reaction sequences for reduction of CO_2 to triose phosphates and the associated reductive pentose phosphate pathway—all of which must be coordinately regulated to ensure proper allocation of carbon to energy production and synthesis of starch and sucrose. Key enzymes are regulated, as we shall see, by (1) reduction of disulfide bonds by electrons flowing from photosystem I and (2) changes in pH and Mg^{2+} concentration that result from illumination. When we look at other aspects of plant carbohydrate metabolism, we also find enzymes that are modulated by (3) conventional allosteric regulation by one or more metabolic intermediates and (4) covalent modification (phosphorylation).

Carbon Dioxide Assimilation Occurs in Three Stages

The first stage in the assimilation of CO_2 into biomolecules **(Fig. 20-31)** is the carbon-fixation reaction: condensation of CO_2 with a five-carbon acceptor, **ribulose 1,5-bisphosphate**, to form two molecules of **3-phosphoglycerate**. In the second stage, the 3-phosphoglycerate is reduced to triose phosphates. Overall, three molecules of CO_2 are fixed to three molecules of ribulose 1,5-bisphosphate to form six molecules of glyceraldehyde 3-phosphate (18 carbons) in equilibrium with dihydroxyacetone phosphate. In the third stage, five of the six molecules of triose phosphate (15 carbons) are used to regenerate three molecules of ribulose 1,5-bisphosphate (15 carbons), the starting material. The sixth molecule of triose phosphate, the net product of photosynthesis, can be used to make hexoses for fuel and building materials, sucrose for transport to nonphotosynthetic tissues, or starch for

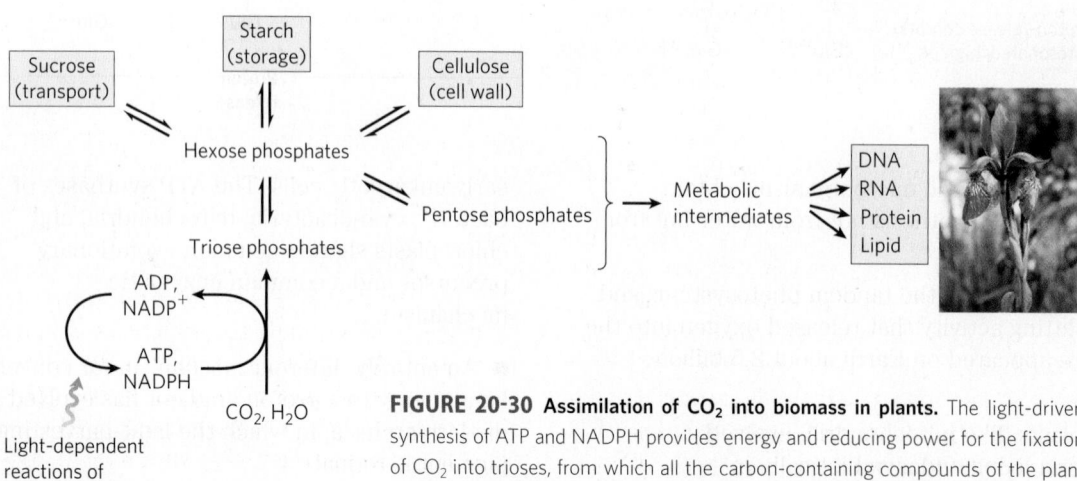

FIGURE 20-30 Assimilation of CO_2 into biomass in plants. The light-driven synthesis of ATP and NADPH provides energy and reducing power for the fixation of CO_2 into trioses, from which all the carbon-containing compounds of the plant cell are synthesized. [Source: Photo from Brzostowska/Shutterstock.]

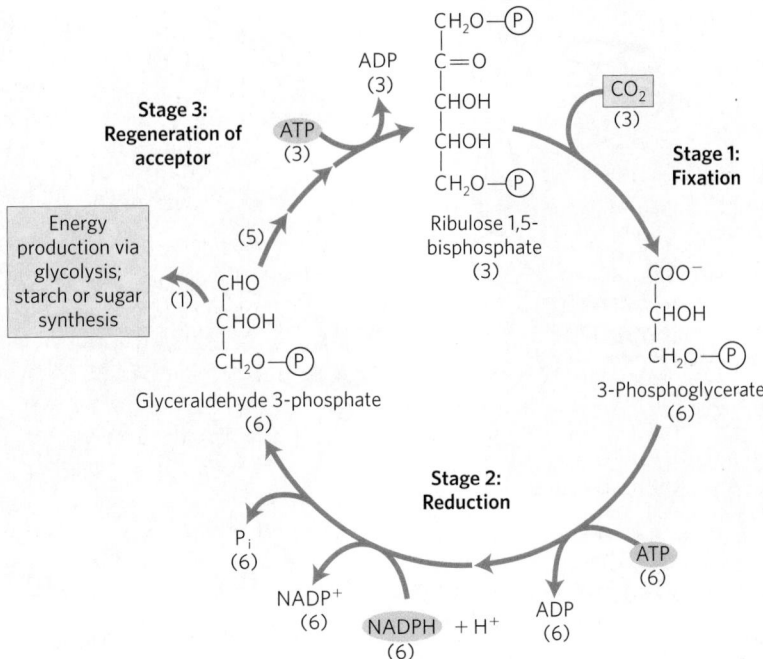

FIGURE 20-31 The three stages of CO_2 assimilation in photosynthetic organisms. Stoichiometries of three key intermediates (numbers in parentheses) reveal the fate of carbon atoms entering and leaving the Calvin cycle, or photosynthetic carbon-reduction cycle. As shown here, three CO_2 are fixed for the net synthesis of one molecule of glyceraldehyde 3-phosphate.

storage. Thus the overall process is cyclical, with the continuous conversion of CO_2 to triose and hexose phosphates. Fructose 6-phosphate is a key intermediate in stage 3 of CO_2 assimilation; it stands at a branch point, leading either to regeneration of ribulose 1,5-bisphosphate or to synthesis of starch. The pathway from hexose phosphate to pentose bisphosphate involves many of the same reactions used in animal cells for the conversion of pentose phosphates to hexose phosphates during the nonoxidative phase of the **pentose phosphate pathway** (see Fig. 14-23). In the photosynthetic assimilation of CO_2, essentially the same set of reactions operates in the other direction, converting hexose phosphates to pentose phosphates. This **reductive pentose phosphate cycle** uses the same enzymes as the oxidative pathway, and several additional enzymes that make the reductive cycle irreversible. All 13 enzymes of the pathway are in the chloroplast stroma.

Stage 1: Fixation of CO_2 into 3-Phosphoglycerate

An important clue to the nature of the CO_2-assimilation mechanisms in photosynthetic organisms came in the late 1940s. Calvin and his associates illuminated a suspension of green algae in the presence of radioactive carbon dioxide ($^{14}CO_2$) for just a few seconds, then quickly killed the cells, extracted their contents, and used chromatographic methods to search for the metabolites in which the labeled carbon first appeared. The first compound that became labeled was 3-phosphoglycerate, with the ^{14}C predominantly located in the carboxyl carbon atom. These experiments strongly suggested that 3-phosphoglycerate is an early intermediate in photosynthesis. The many plants in which this three-carbon compound is the first intermediate are called **C_3 plants**, in contrast to the C_4 plants described below.

The enzyme that catalyzes incorporation of CO_2 into an organic form is **ribulose 1,5-bisphosphate carboxylase/oxygenase**, a name mercifully shortened to **rubisco**. As a carboxylase, rubisco catalyzes the covalent attachment of CO_2 to the five-carbon sugar ribulose 1,5-bisphosphate and cleavage of the unstable six-carbon intermediate to form two molecules of 3-phosphoglycerate, one of which bears the carbon introduced as CO_2 in its carboxyl group (Fig. 20-31). The enzyme's oxygenase activity is discussed in Section 20.6.

There are two distinct forms of rubisco. Form I is found in vascular plants, algae, and cyanobacteria; form II is confined to certain photosynthetic bacteria. Plant rubisco, the crucial enzyme in the production of biomass from CO_2, has a complex form I structure **(Fig. 20-32a)**, with eight identical large subunits (M_r 53,000; encoded in the plastome), each containing a catalytic site, and eight identical small subunits (M_r 14,000; encoded in the nuclear genome) of uncertain function. The form II rubisco of photosynthetic bacteria is simpler in structure, having two subunits that in many respects resemble the large subunits of the plant enzyme (Fig. 20-32b). This similarity is consistent with the endosymbiont hypothesis for the origin of chloroplasts (p. 37). The plant enzyme has an exceptionally low turnover number;

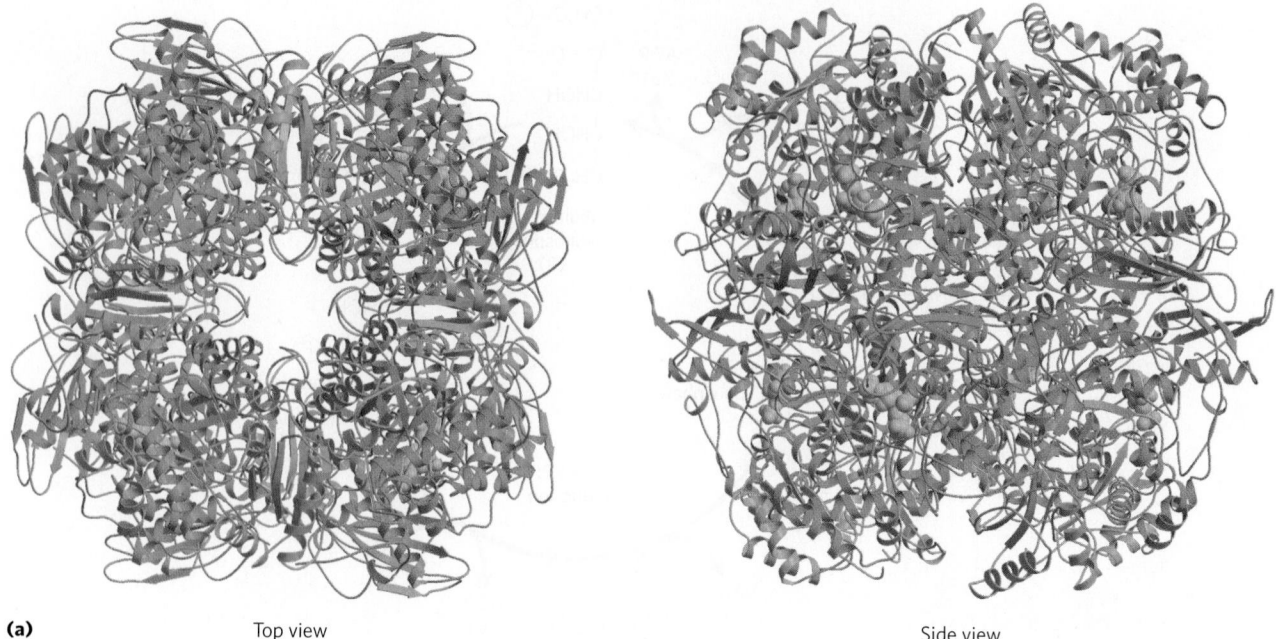

(a) Top view   Side view

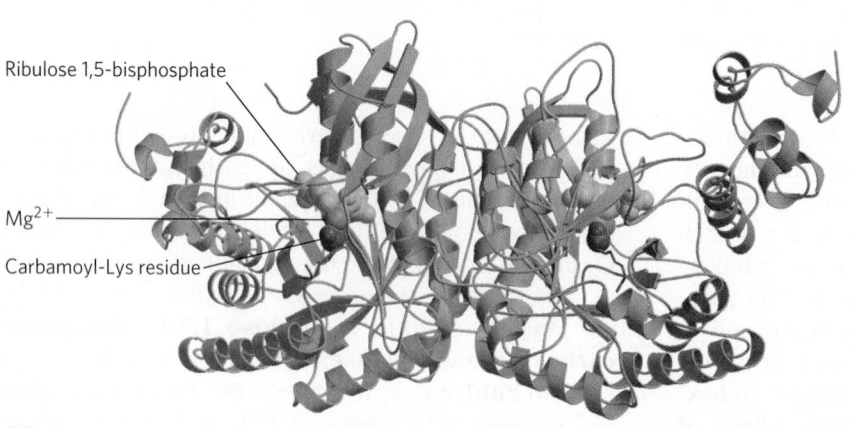

Ribulose 1,5-bisphosphate

Mg^{2+}

Carbamoyl-Lys residue

(b)

FIGURE 20-32 Structure of ribulose 1,5-bisphosphate carboxylase/oxygenase (rubisco). (a) Top and side views of a ribbon model of form I rubisco from spinach. The enzyme has eight large (blue) and eight small (gray) subunits, tightly packed into a structure of M_r >500,000. Rubisco is present at a concentration of about 250 mg/mL in the chloroplast stroma, corresponding to an extraordinarily high concentration of active sites (~4 mM). A transition-state analog, 2-carboxyarabinitol bisphosphate (yellow), is shown bound to each of the eight substrate-binding sites. Mg^{2+} is shown in green. **(b)** Ribbon model of form II rubisco from the bacterium *Rhodospirillum rubrum*. The identical subunits are in gray and blue. [Sources: (a) PDB ID 8RUC, I. Andersson, *J. Mol. Biol.* 259:160, 1996. (b) PDB ID 9RUB, T. Lundqvist and G. Schneider, *J. Biol. Chem.* 266:12,604, 1991.]

only three molecules of CO_2 are fixed per second per molecule of rubisco at 25 °C. To achieve high rates of CO_2 fixation, plants therefore need large amounts of this enzyme. In fact, rubisco makes up almost 50% of soluble protein in chloroplasts and is probably one of the most abundant enzymes in the biosphere.

Central to the proposed mechanism for plant rubisco is a carbamoylated Lys side chain with a bound Mg^{2+} ion. The Mg^{2+} ion brings together and orients the reactants at the active site **(Fig. 20-33)** and polarizes

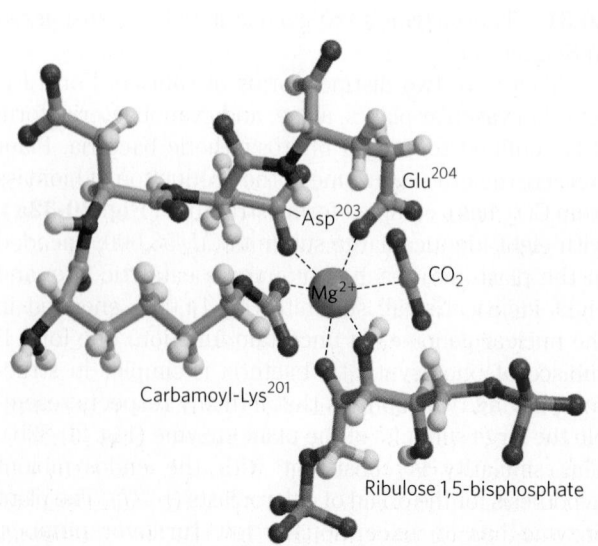

Glu^{204}

Asp^{203}

Mg^{2+}

CO_2

Carbamoyl-Lys^{201}

Ribulose 1,5-bisphosphate

FIGURE 20-33 Central role of Mg^{2+} in the active site of rubisco. Mg^{2+} is coordinated in a roughly octahedral complex with six oxygen atoms: one oxygen in the carbamate on Lys^{201}; two in the carboxyl groups of Glu^{204} and Asp^{203}; two at C-2 and C-3 of the substrate, ribulose 1,5-bisphosphate; and one in the other substrate, CO_2. A water molecule occupies the CO_2-binding site in the crystal structure. In this figure, a CO_2 molecule is modeled in its place. (Residue numbers refer to the spinach enzyme.) [Source: Derived from PDB ID 1RXO, T. C. Taylor and I. Andersson, *J. Mol. Biol.* 265:432, 1997.]

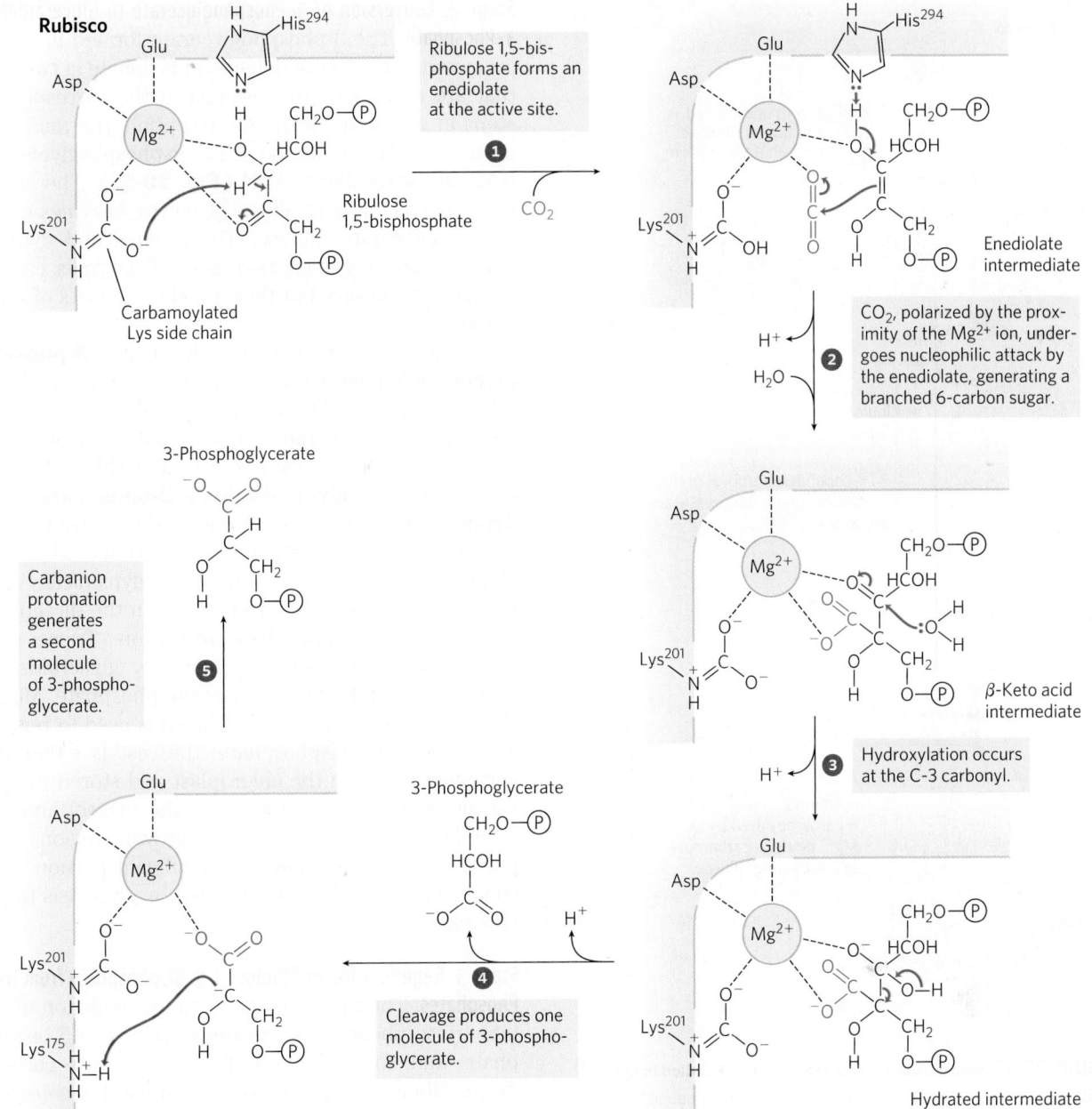

Rubisco

Ribulose 1,5-bis-phosphate forms an enediolate at the active site.

❶

CO_2

Ribulose 1,5-bisphosphate

Carbamoylated Lys side chain

Enediolate intermediate

H^+

H_2O

❷

CO_2, polarized by the proximity of the Mg^{2+} ion, undergoes nucleophilic attack by the enediolate, generating a branched 6-carbon sugar.

3-Phosphoglycerate

Carbanion protonation generates a second molecule of 3-phosphoglycerate.

❺

β-Keto acid intermediate

H^+

❸

Hydroxylation occurs at the C-3 carbonyl.

3-Phosphoglycerate

H^+

❹

Cleavage produces one molecule of 3-phosphoglycerate.

Hydrated intermediate

MECHANISM FIGURE 20-34 First stage of CO_2 assimilation: rubisco's carboxylase activity. The CO_2-fixation reaction is catalyzed by ribulose 1,5-bisphosphate carboxylase/oxygenase. The overall reaction accomplishes the combination of one CO_2 and one ribulose 1,5-bisphosphate to form two molecules of 3-phosphoglycerate, one of which contains the carbon atom from CO_2 (red). Additional proton transfers (not shown), involving Lys^{201}, Lys^{175}, and His^{294}, occur in several of these steps.

the CO_2, opening it to nucleophilic attack by the five-carbon enediolate reaction intermediate formed on the enzyme **(Fig. 20-34)**. The resulting six-carbon intermediate breaks down to yield two molecules of 3-phosphoglycerate.

As the catalyst for the first step of photosynthetic CO_2 assimilation, rubisco is a prime target for regulation. The enzyme is inactive until carbamoylated on the ε-amino group of Lys^{201} **(Fig. 20-35)**. Ribulose 1,5-bisphosphate inhibits carbamoylation by binding tightly to the active site and locking the enzyme in the "closed" conformation, in which Lys^{201} is inaccessible. **Rubisco activase**

overcomes the inhibition by promoting ATP-dependent release of the ribulose 1,5-bisphosphate, exposing the Lys amino group to nonenzymatic carbamoylation by CO_2; this is followed by Mg^{2+} binding, which activates the rubisco. Rubisco activase in some species is activated by light through a redox mechanism similar to that shown in Figure 20-46.

Another regulatory mechanism involves the "nocturnal inhibitor" 2-carboxyarabinitol 1-phosphate, a naturally occurring transition-state analog (p. 210) with a structure similar to that of the β-keto acid intermediate of the rubisco reaction (Fig. 20-34). This compound,

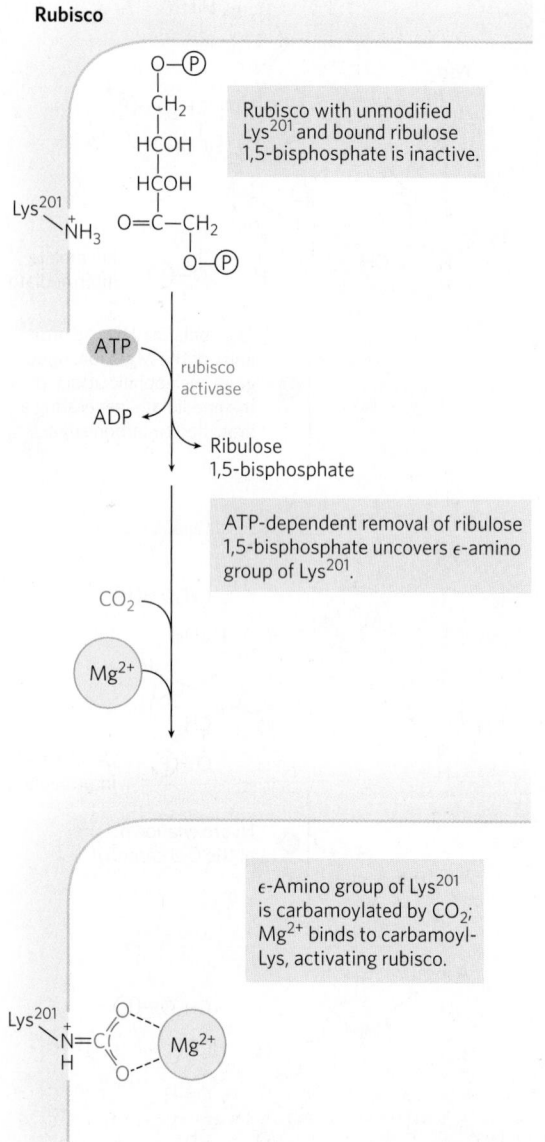

FIGURE 20-35 Role of rubisco activase in carbamoylation of Lys[201] of rubisco. When the substrate ribulose 1,5-bisphosphate is bound to the active site, Lys[201] is not accessible. Rubisco activase couples ATP hydrolysis to expulsion of the bound sugar bisphosphate, exposing Lys[201]; this Lys residue can now be carbamoylated with CO_2 in a reaction that, apparently, is not enzyme-mediated. Mg^{2+} is attracted to and binds to the negatively charged carbamoyl-Lys, and the enzyme is thus activated.

synthesized in the dark in some plants, is a potent inhibitor of carbamoylated rubisco. It is either broken down when light returns or is expelled by rubisco activase, thus activating the rubisco.

2-Carboxyarabinitol 1-phosphate

Stage 2: Conversion of 3-Phosphoglycerate to Glyceraldehyde 3-Phosphate

The 3-phosphoglycerate formed in stage 1 is converted to glyceraldehyde 3-phosphate in two steps that are essentially the reversal of the corresponding steps in glycolysis, with one exception: the nucleotide cofactor for the reduction of 1,3-bisphosphoglycerate is NADPH rather than NADH **(Fig. 20-36)**. The chloroplast stroma contains all the glycolytic enzymes except phosphoglycerate mutase. The stromal and cytosolic enzymes are isozymes; both sets of enzymes catalyze the same reactions, but they are the products of different genes.

In the first step of stage 2, the stromal **3-phosphoglycerate kinase** catalyzes the transfer of a phosphoryl group from ATP to 3-phosphoglycerate, yielding 1,3-bisphosphoglycerate. Next, NADPH donates electrons in a reduction catalyzed by the chloroplast-specific isozyme of **glyceraldehyde 3-phosphate dehydrogenase**, producing glyceraldehyde 3-phosphate and P_i. The high concentrations of NADPH and ATP in the chloroplast stroma allow this thermodynamically unfavorable pair of reactions to proceed in the direction of glyceraldehyde 3-phosphate formation. Triose phosphate isomerase then interconverts glyceraldehyde 3-phosphate and dihydroxyacetone phosphate. Most of the triose phosphate thus produced is used to regenerate ribulose 1,5-bisphosphate; the rest is either converted to starch in the chloroplast and stored for later use or immediately exported to the cytosol and converted to sucrose for transport to growing regions of the plant. In developing leaves, a significant portion of the triose phosphate may be degraded by glycolysis to provide energy.

Stage 3: Regeneration of Ribulose 1,5-Bisphosphate from Triose Phosphates

The first reaction in the assimilation of CO_2 into triose phosphates consumes ribulose 1,5-bisphosphate, and, for continuous flow of CO_2 into carbohydrate, ribulose 1,5-bisphosphate must be constantly regenerated. This is accomplished in a series of reactions **(Fig. 20-37)** that, together with stages 1 and 2, constitute the cyclic pathway shown in Figure 20-31. The product of the first assimilation reaction (3-phosphoglycerate) thus undergoes transformations that regenerate ribulose 1,5-bisphosphate. The intermediates in this pathway include three-, four-, five-, six-, and seven-carbon sugars. In the following discussion, all step numbers refer to Figure 20-37.

Steps ❶ and ❹ are catalyzed by the same enzyme, **aldolase**. It first catalyzes the reversible condensation of glyceraldehyde 3-phosphate with dihydroxyacetone phosphate, yielding fructose 1,6-bisphosphate (step ❶); this is cleaved to fructose 6-phosphate and P_i by fructose 1,6-bisphosphatase (FBPase-1) in step ❷. The reaction is strongly exergonic and essentially irreversible. Step ❸ is catalyzed by **transketolase**, which contains thiamine pyrophosphate (TPP) as its prosthetic group (see Fig. 14-15a) and requires Mg^{2+}. Transketolase catalyzes

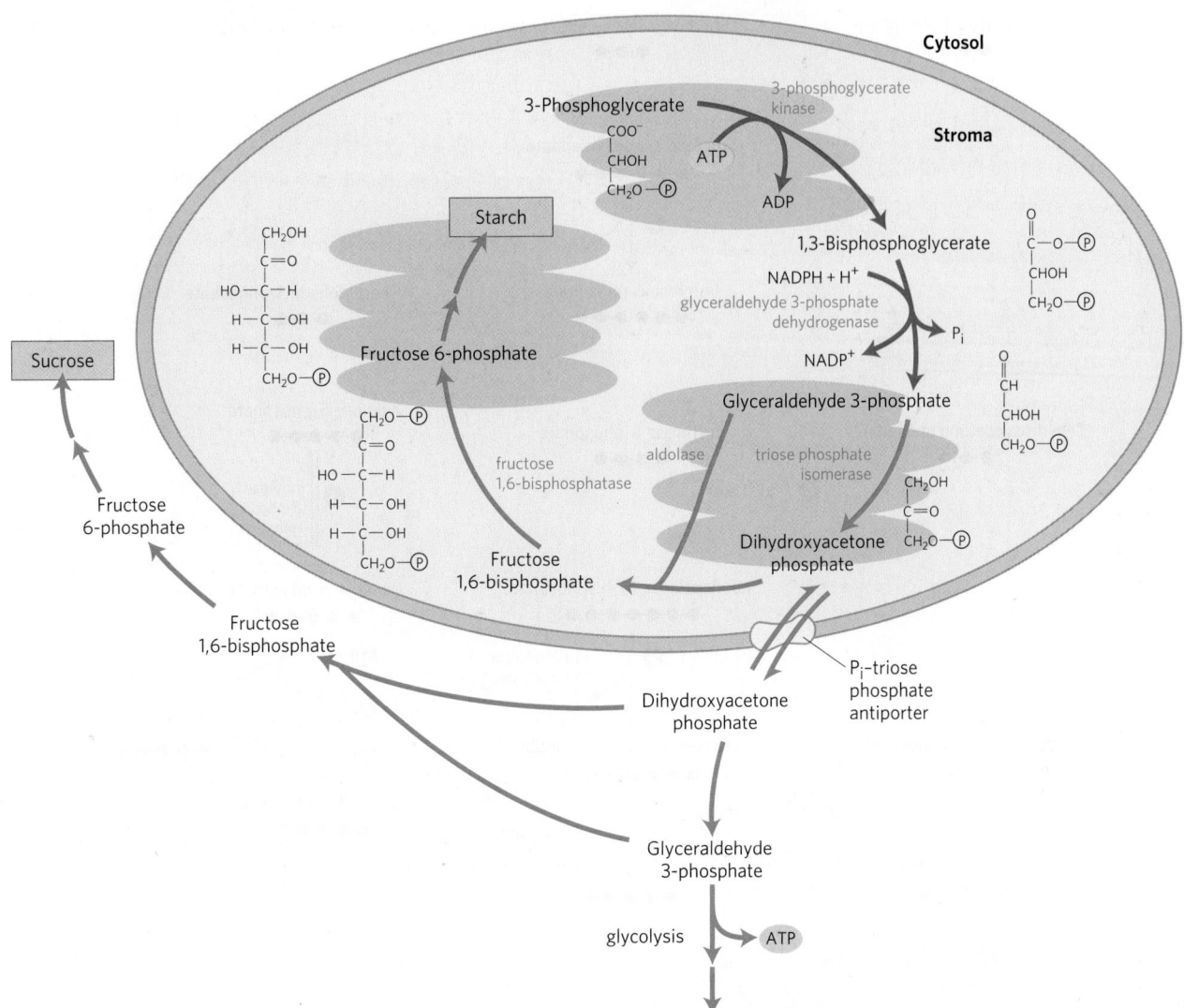

FIGURE 20-36 Second stage of CO_2 assimilation. 3-Phosphoglycerate is converted to glyceraldehyde 3-phosphate (red arrows). The fixed carbon of glyceraldehyde 3-phosphate has several possible fates (blue arrows). Most of the glyceraldehyde 3-phosphate (10 of 12 moles) is recycled to form 6 moles of pentose phosphates, as shown in Figure 20-37. The remaining 2 moles of triose phosphate, the net gain of fixed carbon, is converted to sucrose for transport or is stored in the chloroplast as starch.

In the latter case, glyceraldehyde 3-phosphate condenses with dihydroxyacetone phosphate in the stroma to form fructose 1,6-bisphosphate, a precursor of starch. In other situations, the glyceraldehyde 3-phosphate is converted to dihydroxyacetone phosphate, which leaves the chloroplast via a specific transporter (see Fig. 20-42) and, in the cytosol, can be degraded glycolytically to provide energy or is used to form fructose 6-phosphate and hence sucrose.

the reversible transfer of a two-carbon ketol group (CH_2OH—CO—) from a ketose phosphate donor, fructose 6-phosphate, to an aldose phosphate acceptor, glyceraldehyde 3-phosphate **(Fig. 20-38a, b)**, forming the pentose xylulose 5-phosphate and the tetrose erythrose 4-phosphate. In step ❹, aldolase acts again, combining erythrose 4-phosphate with dihydroxyacetone phosphate to form the seven-carbon **sedoheptulose 1,7-bisphosphate**. An enzyme unique to plastids, sedoheptulose 1,7-bisphosphatase, converts the bisphosphate to sedoheptulose 7-phosphate (step ❺); this is the second irreversible reaction in the pathway. Transketolase now acts again (step ❻), converting sedoheptulose 7-phosphate and glyceraldehyde 3-phosphate to

two pentose phosphates (Fig. 20-38c). **Figure 20-39** shows how a two-carbon fragment is temporarily carried on the transketolase cofactor TPP and is condensed with the three carbons of glyceraldehyde 3-phosphate in step ❻.

The pentose phosphates formed in the transketolase reactions—ribose 5-phosphate and xylulose 5-phosphate—are converted to **ribulose 5-phosphate** (steps ❼ and ❽), which in step ❾, the final reaction of the cycle, is phosphorylated to ribulose 1,5-bisphosphate by ribulose 5-phosphate kinase **(Fig. 20-40)**. This is the third highly exergonic reaction of the pathway, as the phosphate anhydride bond in ATP is swapped for a phosphate ester in ribulose 1,5-bisphosphate.

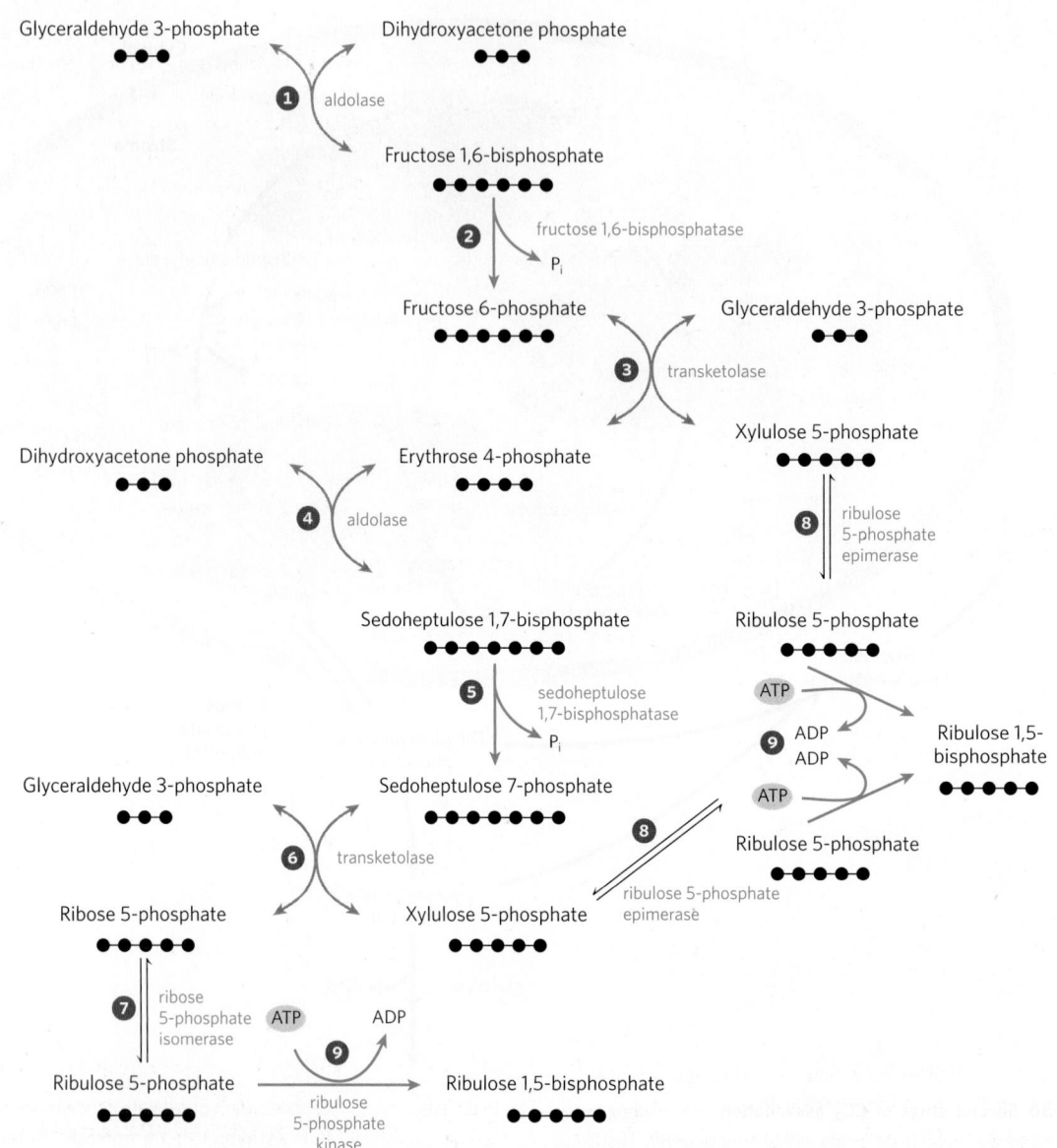

FIGURE 20-37 Third stage of CO₂ assimilation. This schematic diagram shows the interconversions of triose phosphates and pentose phosphates. Black dots represent the number of carbons in each compound. The starting materials are glyceraldehyde 3-phosphate and dihydroxyacetone phosphate. Reactions catalyzed by aldolase (**❶** and **❹**) and transketolase (**❸** and **❻**) produce pentose phosphates that are converted to ribulose 1,5-bisphosphate—ribose 5-phosphate by ribose 5-phosphate isomerase (**❼**) and xylulose 5-phosphate by ribulose 5-phosphate epimerase (**❽**). In step **❾**, ribulose 5-phosphate is phosphorylated, regenerating ribulose 1,5-bisphosphate. The steps with blue arrows are exergonic and make the whole process irreversible: **❷** fructose 1,6-bisphosphatase, **❺** sedoheptulose bisphosphatase, and **❾** ribulose 5-phosphate kinase.

Synthesis of Each Triose Phosphate from CO₂ Requires Six NADPH and Nine ATP

The net result of three turns of the Calvin cycle is the conversion of three molecules of CO_2 and one molecule of phosphate to a molecule of triose phosphate. The stoichiometry of the overall path from CO_2 to triose phosphate, with regeneration of ribulose 1,5-bisphosphate, is shown in **Figure 20-41**. Three molecules of ribulose 1,5-bisphosphate (a total of 15 carbons) condense with three CO_2 (3 carbons) to form six molecules of 3-phosphoglycerate (18 carbons). These six molecules of 3-phosphoglycerate are reduced to six molecules of glyceraldehyde 3-phosphate (which is in equilibrium with dihydroxyacetone phosphate), with the expenditure of six ATP (in the synthesis of 1,3-bisphosphoglycerate) and six NADPH (in the reduction of 1,3-bisphosphoglycerate to glyceraldehyde 3-phosphate). The isozyme of glyceraldehyde 3-phosphate dehydrogenase present in chloroplasts can use NADPH as its electron carrier and normally functions in the direction of 1,3-bisphosphoglycerate reduction. The cytosolic isozyme uses NAD, as does the glycolytic enzyme of animals and other eukaryotes, and in the dark this isozyme acts in glycolysis to oxidize glyceraldehyde 3-phosphate. Both glyceraldehyde

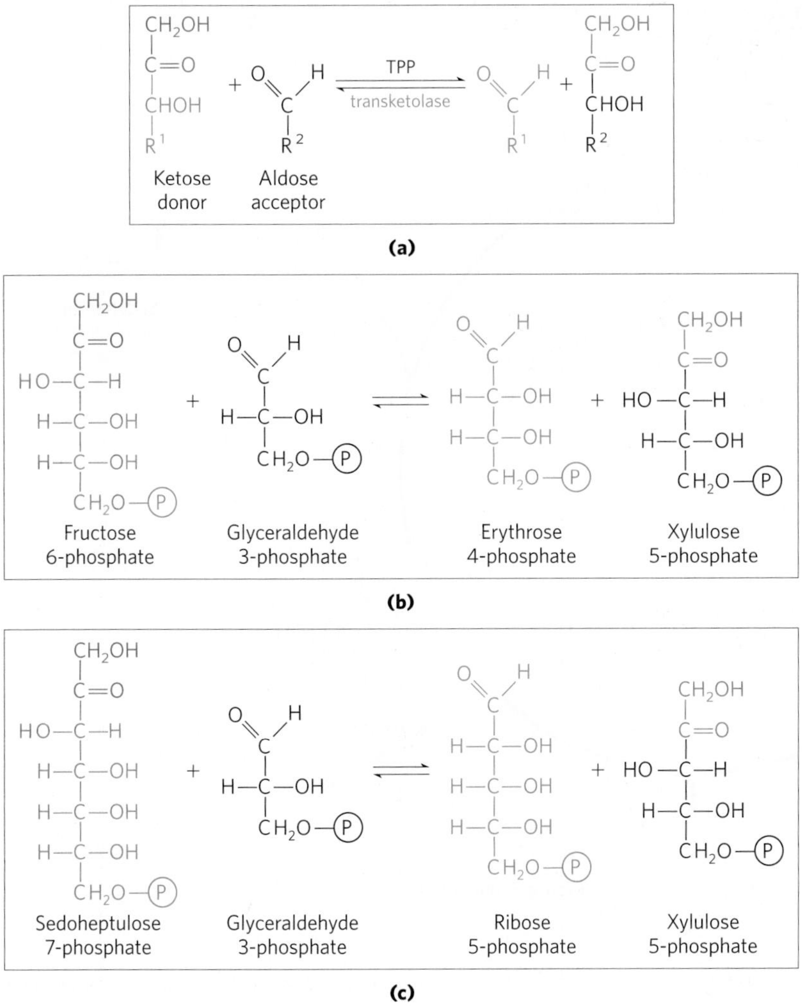

FIGURE 20-38 Transketolase-catalyzed reactions of the Calvin cycle. (a) General reaction catalyzed by transketolase: transfer of a two-carbon group, carried temporarily on enzyme-bound TPP, from a ketose donor to an aldose acceptor. **(b)** Conversion of a hexose and a triose to a four-carbon and a five-carbon sugar (step ❸ in Fig. 20-37). **(c)** Conversion of seven-carbon and three-carbon sugars to two pentoses (step ❻ in Fig. 20-37).

3-phosphate dehydrogenase isozymes, like all enzymes, catalyze the reaction in both directions.

One molecule of glyceraldehyde 3-phosphate is the net product of the carbon-assimilation pathway. The other five triose phosphate molecules (15 carbons) are rearranged in steps ❶ to ❾ of Figure 20-37 to form three molecules of ribulose 1,5-bisphosphate (15 carbons). The last step in this conversion requires one ATP per ribulose 1,5-bisphosphate, or a total of three ATP. Thus, in summary, for every molecule of triose phosphate produced by photosynthetic CO_2 assimilation, six NADPH and nine ATP are required.

NADPH and ATP are produced in the light-dependent reactions of photosynthesis in about the same ratio (2:3) as they are consumed in the Calvin cycle. Nine ATP molecules are converted to ADP and phosphate in the generation of a molecule of triose phosphate; eight of the phosphates are released as P_i and combined with eight ADP to regenerate ATP. The ninth phosphate is incorporated into the triose phosphate itself. To convert the ninth ADP to ATP, a molecule of P_i must be imported from the cytosol, as we shall see.

In the dark, the production of ATP and NADPH by photophosphorylation, and the incorporation of CO_2 into triose phosphate (once referred to as the dark

reactions), cease. The "dark reactions" of photosynthesis were so named to distinguish them from the *primary* light-driven reactions of electron transfer to $NADP^+$ and synthesis of ATP. They do not, in fact, occur at significant rates in the dark and are thus more appropriately called the carbon-assimilation reactions. Later in this section we describe the regulatory mechanisms that turn carbon assimilation on in the light and turn it off in the dark.

The chloroplast stroma contains all the enzymes necessary to convert the triose phosphates produced by CO_2 assimilation (glyceraldehyde 3-phosphate and dihydroxyacetone phosphate) to starch, which is temporarily stored in the chloroplast as insoluble granules. Aldolase condenses the trioses to fructose 1,6-bisphosphate; fructose 1,6-bisphosphatase produces fructose 6-phosphate; phosphohexose isomerase yields glucose 6-phosphate; and phosphoglucomutase produces glucose 1-phosphate, the starting material for starch synthesis (see Section 20.7).

All the reactions of the Calvin cycle except those catalyzed by rubisco, sedoheptulose 1,7-bisphosphatase, and ribulose 5-phosphate kinase also take place in animal tissues. Lacking these three enzymes, animals cannot carry out net conversion of CO_2 to glucose.

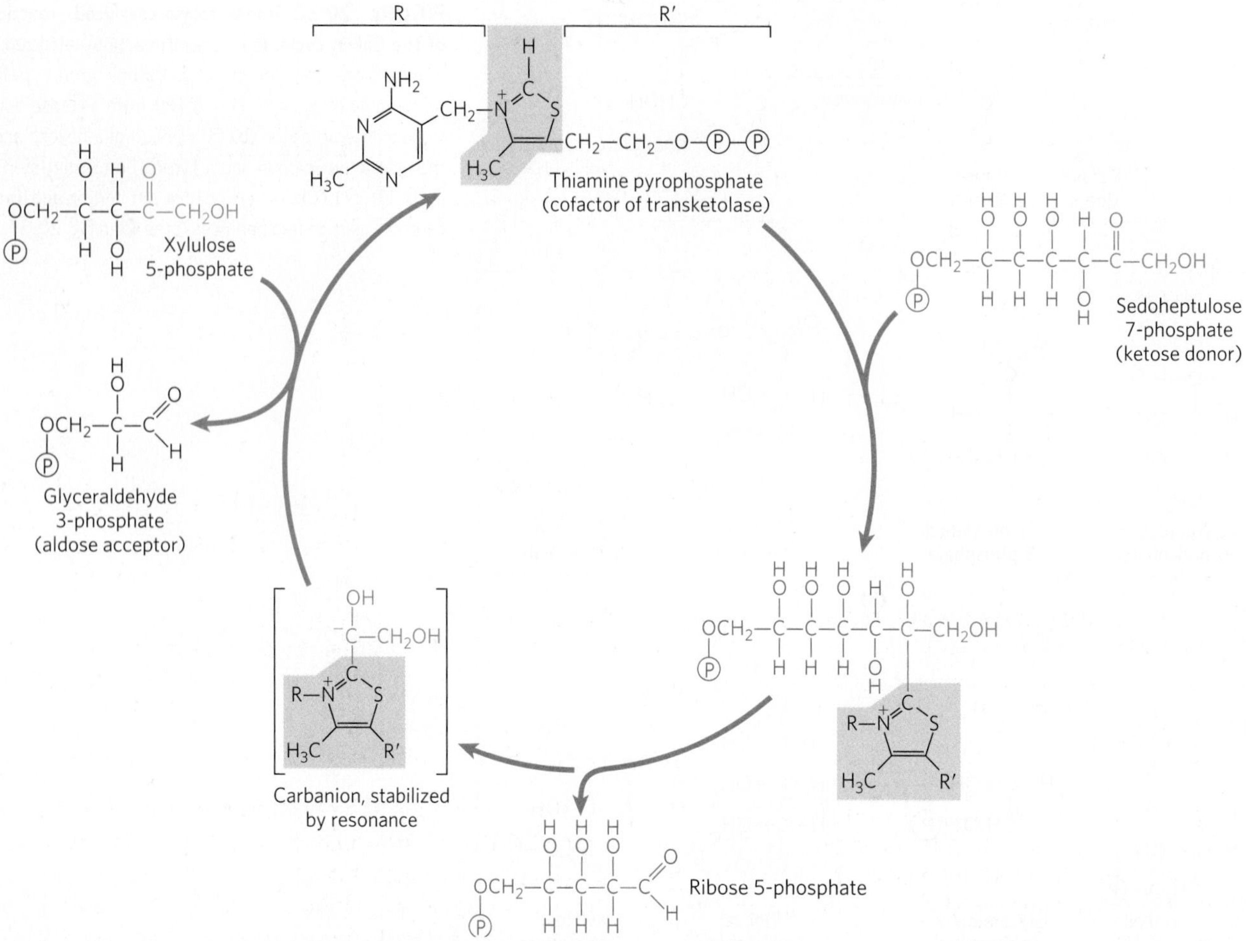

FIGURE 20-39 TPP as a cofactor for transketolase. Transketolase transfers a two-carbon group from sedoheptulose 7-phosphate to glyceraldehyde 3-phosphate, producing two pentose phosphates (step ❻ in Fig. 20-37). Thiamine pyrophosphate serves as a temporary carrier of the two-carbon unit and as an electron sink (see Fig. 14-15) to facilitate the reactions.

A Transport System Exports Triose Phosphates from the Chloroplast and Imports Phosphate

The inner chloroplast membrane is impermeable to most phosphorylated compounds, including fructose 6-phosphate, glucose 6-phosphate, and fructose 1,6-bisphosphate. It does, however, have a specific antiporter that catalyzes the one-for-one exchange of P_i with a triose phosphate, either dihydroxyacetone phosphate or 3-phosphoglycerate (**Fig. 20-42**; see also Fig. 20-36).

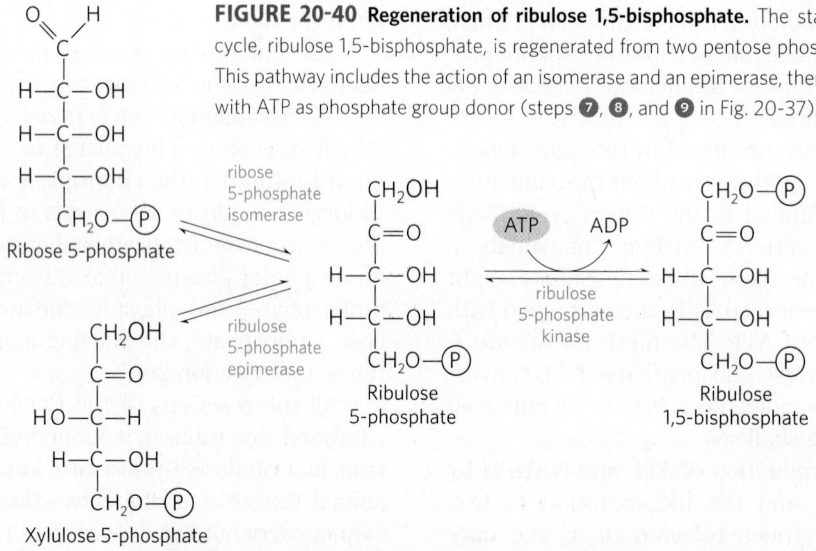

FIGURE 20-40 Regeneration of ribulose 1,5-bisphosphate. The starting material for the Calvin cycle, ribulose 1,5-bisphosphate, is regenerated from two pentose phosphates produced in the cycle. This pathway includes the action of an isomerase and an epimerase, then phosphorylation by a kinase, with ATP as phosphate group donor (steps ❼, ❽, and ❾ in Fig. 20-37).

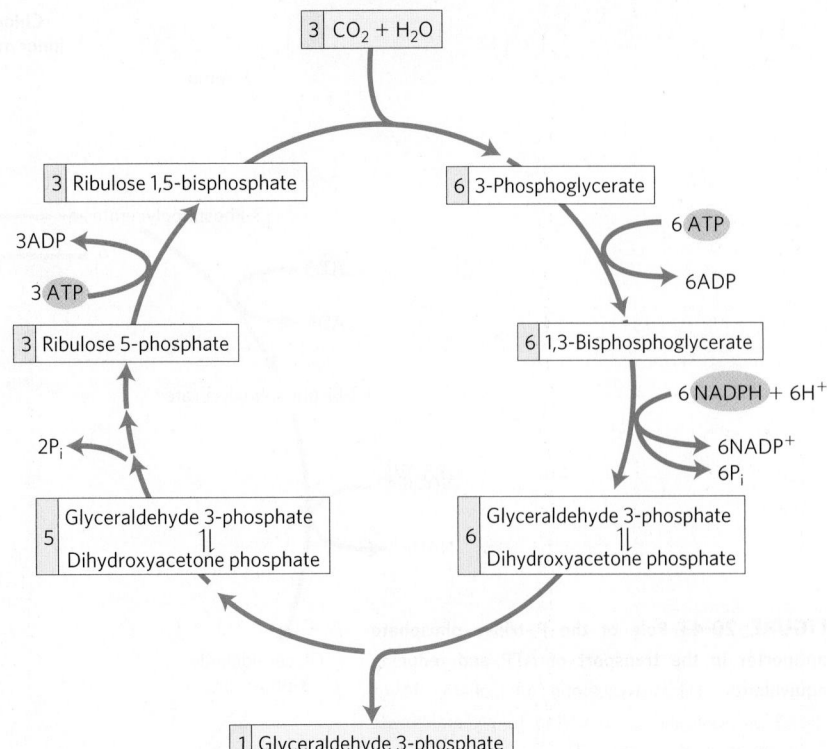

FIGURE 20-41 Stoichiometry of CO_2 assimilation in the Calvin cycle. For every three CO_2 molecules fixed, one molecule of triose phosphate (glyceraldehyde 3-phosphate) is produced and nine ATP and six NADPH are consumed.

This antiporter simultaneously moves P_i into the chloroplast, where it is used in photophosphorylation, and moves triose phosphate into the cytosol, where it can be used to synthesize sucrose, the form in which the fixed carbon is transported to distant plant tissues.

Sucrose synthesis in the cytosol and starch synthesis in the chloroplast are the major pathways by which the excess triose phosphate from photosynthesis is "harvested." Sucrose synthesis (described below) releases four P_i molecules from the four triose phosphates required to make sucrose. For every molecule of triose phosphate

removed from the chloroplast, one P_i is transported into the chloroplast, providing the ninth P_i mentioned above, to be used in regenerating ATP. If this exchange were blocked, triose phosphate synthesis would quickly deplete the available P_i in the chloroplast, slowing ATP synthesis and suppressing assimilation of CO_2 into starch.

The P_i–triose phosphate antiport system serves one additional function. ATP and reducing power are needed in the cytosol for a variety of synthetic and energy-requiring reactions. These requirements are met to an as yet undetermined degree by mitochondria, but a

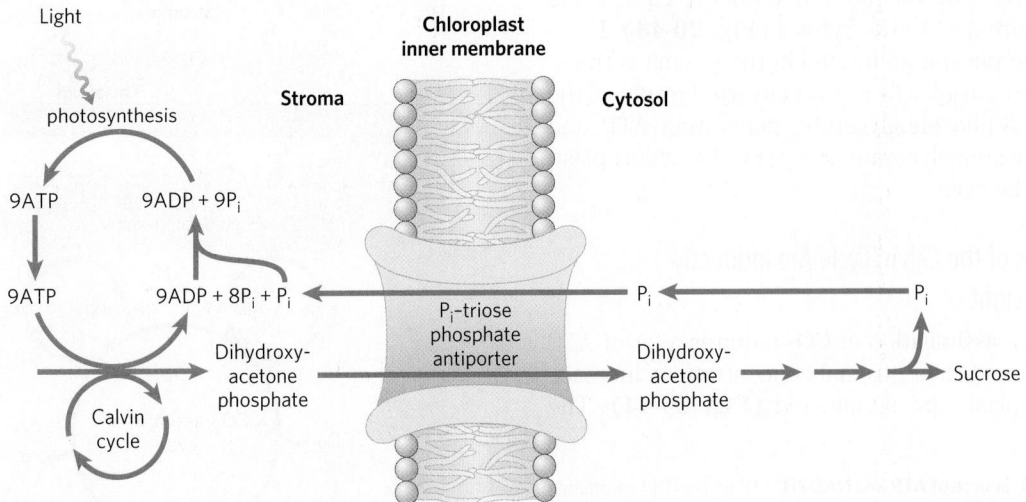

FIGURE 20-42 The P_i–triose phosphate antiport system of the inner chloroplast membrane. This transporter facilitates the exchange of cytosolic P_i for stromal dihydroxyacetone phosphate. The products of photosynthetic carbon assimilation are thus moved into the cytosol, where

they serve as a starting point for sucrose biosynthesis, and P_i required for photophosphorylation is moved into the stroma. This same antiporter can transport 3-phosphoglycerate and acts indirectly in the export of ATP and reducing equivalents (see Fig. 20-43).

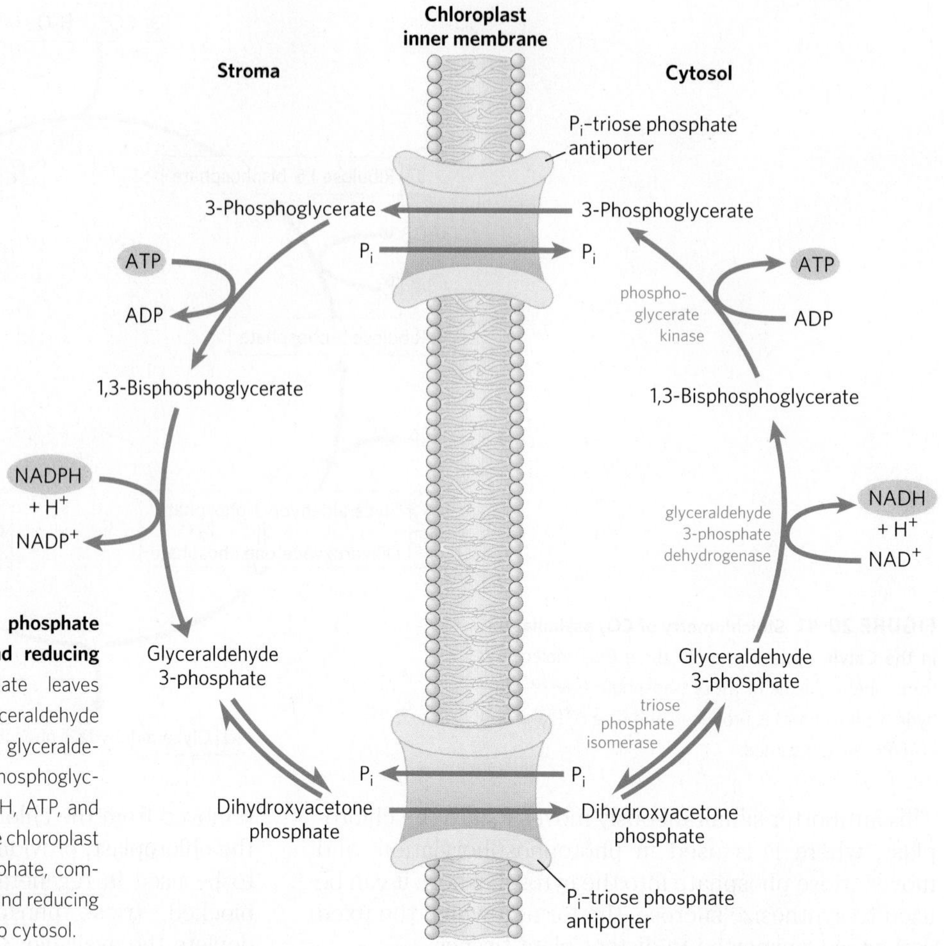

Chloroplast inner membrane

Stroma

Cytosol

P_i–triose phosphate antiporter

3-Phosphoglycerate → 3-Phosphoglycerate

ATP → ADP

P_i → P_i

phospho-glycerate kinase

ATP ← ADP

1,3-Bisphosphoglycerate

NADPH + H⁺ → NADP⁺

1,3-Bisphosphoglycerate

glyceraldehyde 3-phosphate dehydrogenase

NADH + H⁺ ← NAD⁺

Glyceraldehyde 3-phosphate

Glyceraldehyde 3-phosphate

triose phosphate isomerase

P_i ← P_i

Dihydroxyacetone phosphate → Dihydroxyacetone phosphate

P_i–triose phosphate antiporter

FIGURE 20-43 Role of the P_i–triose phosphate antiporter in the transport of ATP and reducing equivalents. Dihydroxyacetone phosphate leaves the chloroplast and is converted to glyceraldehyde 3-phosphate in the cytosol. The cytosolic glyceraldehyde 3-phosphate dehydrogenase and phosphoglycerate kinase reactions then produce NADH, ATP, and 3-phosphoglycerate. The latter reenters the chloroplast and is reduced to dihydroxyacetone phosphate, completing a cycle that effectively moves ATP and reducing equivalents (NAD(P)H) from chloroplast to cytosol.

second potential source of energy is the ATP and NADPH generated in the chloroplast stroma during the light-dependent reactions. However, neither ATP nor NADPH can cross the chloroplast membrane. The P_i–triose phosphate antiport system has the indirect effect of moving ATP equivalents and reducing equivalents from the chloroplast to the cytosol **(Fig. 20-43)**. Dihydroxyacetone phosphate formed in the stroma is transported to the cytosol, where it is converted by glycolytic enzymes to 3-phosphoglycerate, generating ATP and NADH. 3-Phosphoglycerate reenters the chloroplast, completing the cycle.

Four Enzymes of the Calvin Cycle Are Indirectly Activated by Light

The reductive assimilation of CO_2 requires a lot of ATP and NADPH, and their stromal concentrations increase when chloroplasts are illuminated **(Fig. 20-44)**. The

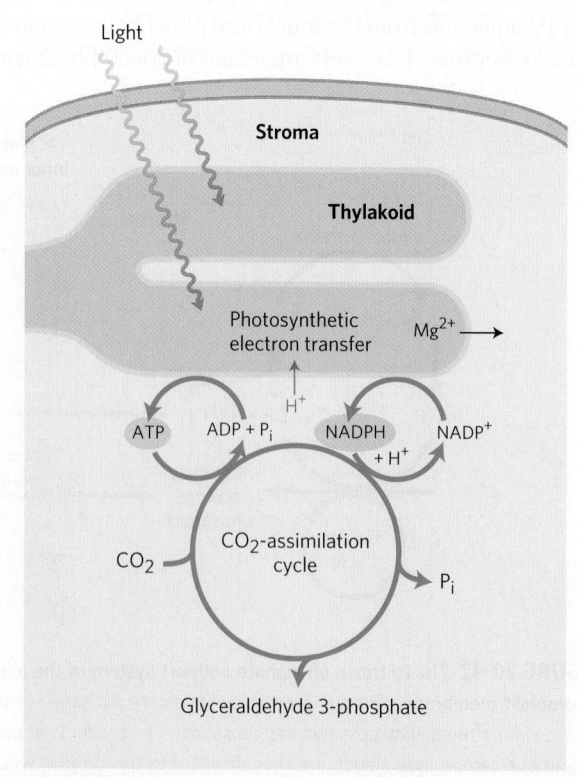

FIGURE 20-44 Source of ATP and NADPH. ATP and NADPH produced by the light-dependent reactions are essential substrates for the reduction of CO_2. The photosynthetic reactions that produce ATP and NADPH are accompanied by movement of protons (red) from the stroma into the thylakoid, creating alkaline conditions in the stroma. Magnesium ions pass from the thylakoid into the stroma, increasing the stromal $[Mg^{2+}]$.

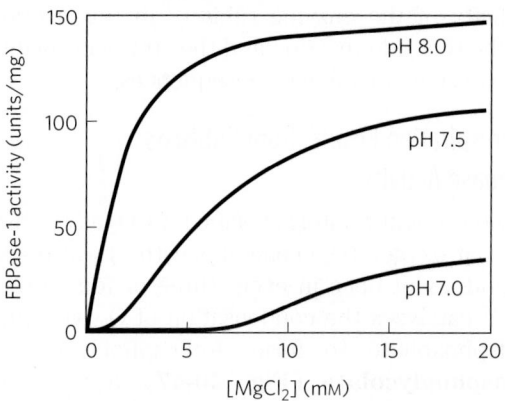

FIGURE 20-45 Activation of chloroplast fructose 1,6-bisphosphatase. Reduced fructose 1,6-bisphosphatase (FBPase-1) is activated by light and by the combination of high pH and high [Mg^{2+}] in the stroma, both of which are results of illumination. [Source: Information from B. Halliwell, *Chloroplast Metabolism: The Structure and Function of Chloroplasts in Green Leaf Cells,* p. 97, Clarendon Press, 1984.]

light-induced transport of protons across the thylakoid membrane also increases the stromal pH from about 7 to about 8, and it is accompanied by a flow of Mg^{2+} from the thylakoid compartment into the stroma, raising the [Mg^{2+}] from 1 to 3 mM to 3 to 6 mM. Several stromal enzymes have evolved to take advantage of these light-induced conditions, which signal the availability of ATP and NADPH: the enzymes are more active in an alkaline environment and at high [Mg^{2+}]. For example, activation of rubisco by formation of the carbamoyllysine is faster at alkaline pH, and high stromal [Mg^{2+}] favors formation of the enzyme's active Mg^{2+} complex. Fructose 1,6-bisphosphatase requires Mg^{2+} and is very dependent on pH **(Fig. 20-45)**; its activity increases more than 100-fold when pH and [Mg^{2+}] rise during chloroplast illumination.

Four Calvin cycle enzymes are subject to a special type of regulation by light. Ribulose 5-phosphate kinase,

fructose 1,6-bisphosphatase, sedoheptulose 1,7-bisphosphatase, and glyceraldehyde 3-phosphate dehydrogenase are activated by light-driven reduction of disulfide bonds between two Cys residues critical to their catalytic activities. When these Cys residues are disulfide-bonded (oxidized), the enzymes are inactive; this is the normal situation in the dark. With illumination, electrons flow from photosystem I to ferredoxin (Fig. 20-16), which passes electrons to a small, soluble, disulfide-containing protein called **thioredoxin (Fig. 20-46)**, in a reaction catalyzed by **ferredoxin-thioredoxin reductase**. Reduced thioredoxin donates electrons for the reduction of the disulfide bonds of the light-activated enzymes, and these reductive cleavage reactions are accompanied by conformational changes that increase enzyme activities. At nightfall, the Cys residues in the four enzymes are reoxidized to their disulfide forms, the enzymes are inactivated, and ATP is not expended in CO_2 assimilation. Instead, starch synthesized and stored during the daytime is degraded to fuel glycolysis at night.

Glucose 6-phosphate dehydrogenase, the first enzyme in the *oxidative* pentose phosphate pathway, is also regulated by this light-driven reduction mechanism, but in the opposite sense. During the day, when photosynthesis produces plenty of NADPH, this enzyme is not needed for NADPH production. Reduction of a critical disulfide bond by electrons from ferredoxin *inactivates* the enzyme.

SUMMARY 20.5 Carbon-Assimilation Reactions

■ Photosynthesis in eukaryotes takes place in chloroplasts. In the CO_2-assimilating reactions (the Calvin cycle), ATP and NADPH are used to reduce CO_2 to triose phosphates. These reactions occur in three stages: the fixation reaction itself, catalyzed by rubisco; reduction of the resulting 3-phosphoglycerate to glyceraldehyde 3-phosphate;

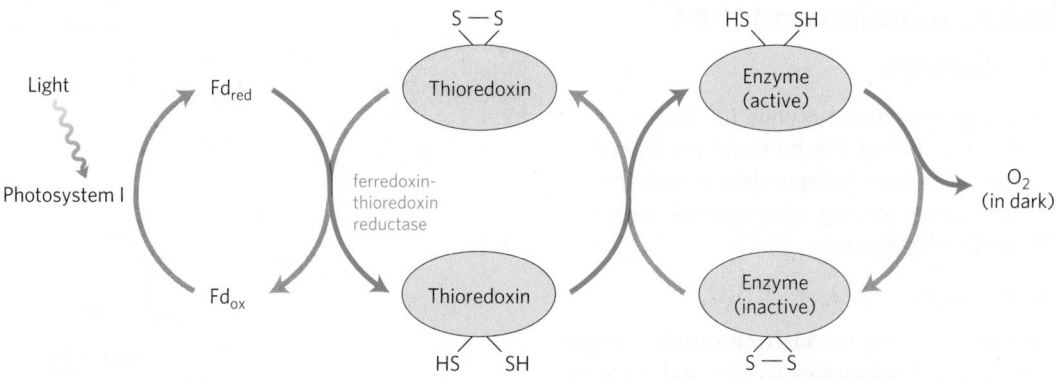

FIGURE 20-46 Light activation of several enzymes of the Calvin cycle. The light activation is mediated by thioredoxin, a small, disulfide-containing protein. In the light, thioredoxin is reduced by electrons moving from photosystem I through ferredoxin (Fd) (blue arrows), then thioredoxin reduces critical disulfide bonds in each of the enzymes sedoheptulose 1,7-bisphosphatase, fructose 1,6-bisphosphatase, ribulose 5-phosphate kinase, and glyceraldehyde 3-phosphate dehydrogenase, activating these enzymes. In the dark, the —SH groups undergo reoxidation to disulfides, inactivating the enzymes.

and regeneration of ribulose 1,5-bisphosphate from triose phosphates.

■ Rubisco condenses CO_2 with ribulose 1,5-bisphosphate, forming an unstable hexose bisphosphate that splits into two molecules of 3-phosphoglycerate. Rubisco is activated by covalent modification (carbamoylation of Lys^{201}) catalyzed by rubisco activase and is inhibited by a natural transition-state analog, the concentration of which rises in the dark and falls during daylight.

■ Stromal isozymes of the glycolytic enzymes catalyze reduction of 3-phosphoglycerate to glyceraldehyde 3-phosphate; reduction of each molecule requires one ATP and one NADPH.

■ Stromal enzymes, including transketolase and aldolase, rearrange the carbon skeletons of triose phosphates to generate intermediates of three, four, five, six, and seven carbons, eventually yielding pentose phosphates. The pentose phosphates are converted to ribulose 5-phosphate, which is phosphorylated to ribulose 1,5-bisphosphate to complete the Calvin cycle.

■ The cost of fixing three CO_2 into one triose phosphate is nine ATP and six NADPH, which are provided by the light-dependent reactions of photosynthesis.

■ An antiporter in the inner chloroplast membrane exchanges P_i in the cytosol for 3-phosphoglycerate or dihydroxyacetone phosphate produced by CO_2 assimilation in the stroma. Oxidation of dihydroxyacetone phosphate in the cytosol generates ATP and NADH, thus moving ATP and reducing equivalents from the chloroplast to the cytosol.

■ Four enzymes of the Calvin cycle are activated indirectly by light and are inactive in the dark, so that hexose synthesis does not compete with glycolysis—which is required to provide energy in the dark.

20.6 Photorespiration and the C_4 and CAM Pathways

As we have seen, photosynthetic cells produce O_2 (by the splitting of H_2O) during the light-driven reactions and use CO_2 during the light-independent processes, so the net gaseous change during photosynthesis is the uptake of CO_2 and release of O_2:

$$CO_2 + H_2O \longrightarrow O_2 + (CH_2O)$$

In the dark, plants also carry out **mitochondrial respiration**, the oxidation of substrates to CO_2 and the conversion of O_2 to H_2O. And there is another process in plants that, like mitochondrial respiration, consumes O_2 and produces CO_2 and, like photosynthesis, is driven by light. This process, **photorespiration**, is a costly side reaction of photosynthesis, a result of the lack of

specificity of the enzyme rubisco. In this section we describe this side reaction and the strategies plants use to minimize its metabolic consequences.

Photorespiration Results from Rubisco's Oxygenase Activity

Rubisco is not absolutely specific for CO_2 as a substrate. Molecular oxygen (O_2) competes with CO_2 at the active site, and about once in every three or four turnovers, rubisco catalyzes the condensation of O_2 with ribulose 1,5-bisphosphate to form 3-phosphoglycerate and **2-phosphoglycolate (Fig. 20-47)**, a metabolically useless product. This is the oxygenase activity referred to in the full name of the enzyme: ribulose 1,5-bisphosphate carboxylase/oxygenase. The reaction with O_2 results in no fixation of carbon and seems to be a net liability to the cell; salvaging the carbons from 2-phosphoglycolate (by the pathway outlined below) consumes significant amounts of cellular energy and releases some previously fixed CO_2.

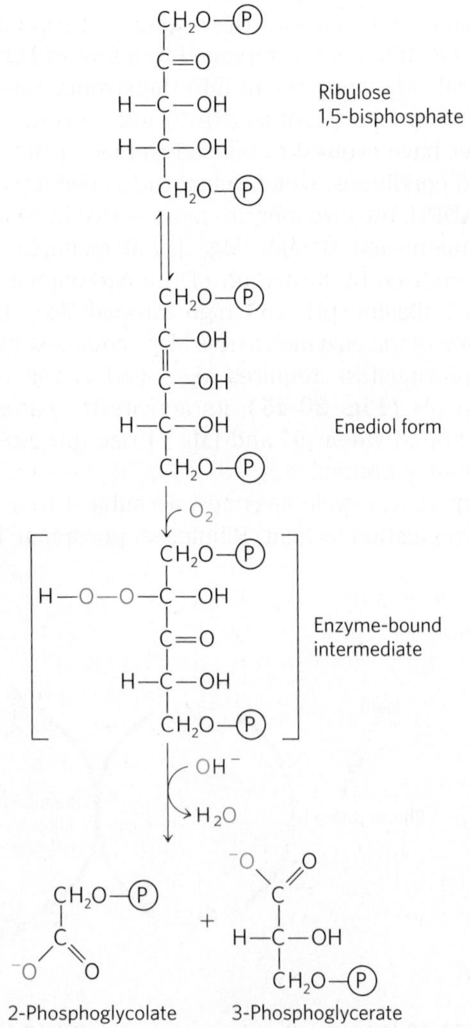

FIGURE 20-47 Oxygenase activity of rubisco. Rubisco can incorporate O_2 rather than CO_2 into ribulose 1,5-bisphosphate. The unstable intermediate thus formed splits into 2-phosphoglycolate (recycled as described in Fig. 20-48) and 3-phosphoglycerate, which can reenter the Calvin cycle.

Given that the reaction with oxygen is deleterious to the organism, why did the evolution of rubisco produce an active site unable to discriminate well between CO_2 and O_2? Perhaps much of this evolution occurred before the time, about 2.5 billion years ago, when production of O_2 by photosynthetic organisms started to raise the oxygen content of the atmosphere. Before that time, there was no selective pressure for rubisco to discriminate between CO_2 and O_2. The K_m for CO_2 is about 9 μM, and that for O_2 is about 350 μM. The modern atmosphere contains about 20% O_2 and only 0.04% CO_2, so an aqueous solution in equilibrium with air at room temperature contains about 250 μM O_2 and 11 μM CO_2—concentrations that allow significant O_2 "fixation" by rubisco and thus a significant waste of energy. The temperature dependence of the solubilities of O_2 and CO_2 is such that at higher temperatures, the ratio of O_2 to CO_2 in solution increases. In addition, the affinity of rubisco for CO_2 decreases with increasing temperature, exacerbating its tendency to catalyze the wasteful oxygenase reaction. And as CO_2 is consumed in the assimilation reactions, the ratio of O_2 to CO_2 in the air spaces of a leaf increases, further favoring the oxygenase reaction.

The Salvage of Phosphoglycolate Is Costly

The **glycolate pathway** converts two molecules of 2-phosphoglycolate to a molecule of serine (three carbons) and a molecule of CO_2 **(Fig. 20-48)**. In the chloroplast, a phosphatase converts 2-phosphoglycolate to glycolate, which is exported to the peroxisome. There, glycolate is oxidized by molecular oxygen, and the resulting aldehyde (glyoxylate) undergoes transamination to glycine. The hydrogen peroxide formed as a side product of glycolate oxidation is rendered harmless by peroxidases in the peroxisome. Glycine passes from the peroxisome to the mitochondrial matrix, where it undergoes oxidative decarboxylation by the glycine decarboxylase complex, an enzyme similar in structure and mechanism to two mitochondrial complexes we have already encountered: the pyruvate dehydrogenase complex and the α-ketoglutarate dehydrogenase complex (Chapter 16). The **glycine decarboxylase complex** oxidizes glycine to CO_2 and NH_3, with the concomitant reduction of NAD^+ to NADH and transfer of the remaining carbon from glycine to the cofactor tetrahydrofolate **(Fig. 20-49)**. The one-carbon unit carried on tetrahydrofolate is then transferred to a second glycine by serine hydroxymethyltransferase, producing serine. The net reaction catalyzed by the glycine decarboxylase complex and serine hydroxymethyltransferase is

$$2 \text{ Glycine} + NAD^+ + H_2O \longrightarrow$$
$$\text{serine} + CO_2 + NH_3 + NADH + H^+$$

The serine is converted to hydroxypyruvate, then to glycerate, and finally to 3-phosphoglycerate, which is used to regenerate ribulose 1,5-bisphosphate, completing the long, expensive cycle (Fig. 20-48).

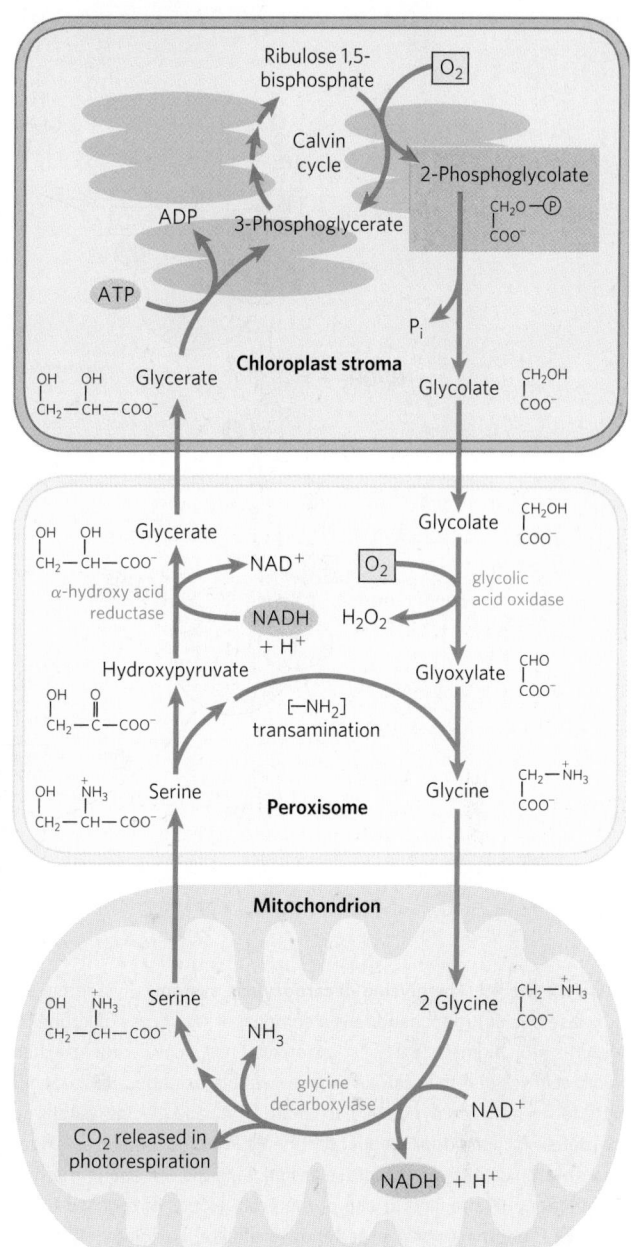

FIGURE 20-48 Glycolate pathway. This pathway, which salvages 2-phosphoglycolate (shaded light red) by converting it to serine and, eventually, 3-phosphoglycerate, involves three cellular compartments. Glycolate formed by dephosphorylation of 2-phosphoglycolate in chloroplasts is oxidized to glyoxylate and transaminated to glycine in peroxisomes. In mitochondria, two glycine molecules condense to form serine and CO_2, released in photorespiration (shaded green). This reaction is catalyzed by glycine decarboxylase, an enzyme present at very high levels in the mitochondria of C$_3$ plants (see text). The serine is converted to hydroxypyruvate and then to glycerate in peroxisomes; glycerate reenters the chloroplasts to be phosphorylated, rejoining the Calvin cycle. Oxygen (shaded blue) is consumed at two steps during photorespiration.

In bright sunlight, the flux through the glycolate salvage pathway can be very high, producing about five times more CO_2 than is typically produced by all the oxidations of the citric acid cycle. To generate this large flux, mitochondria contain prodigious amounts of the glycine

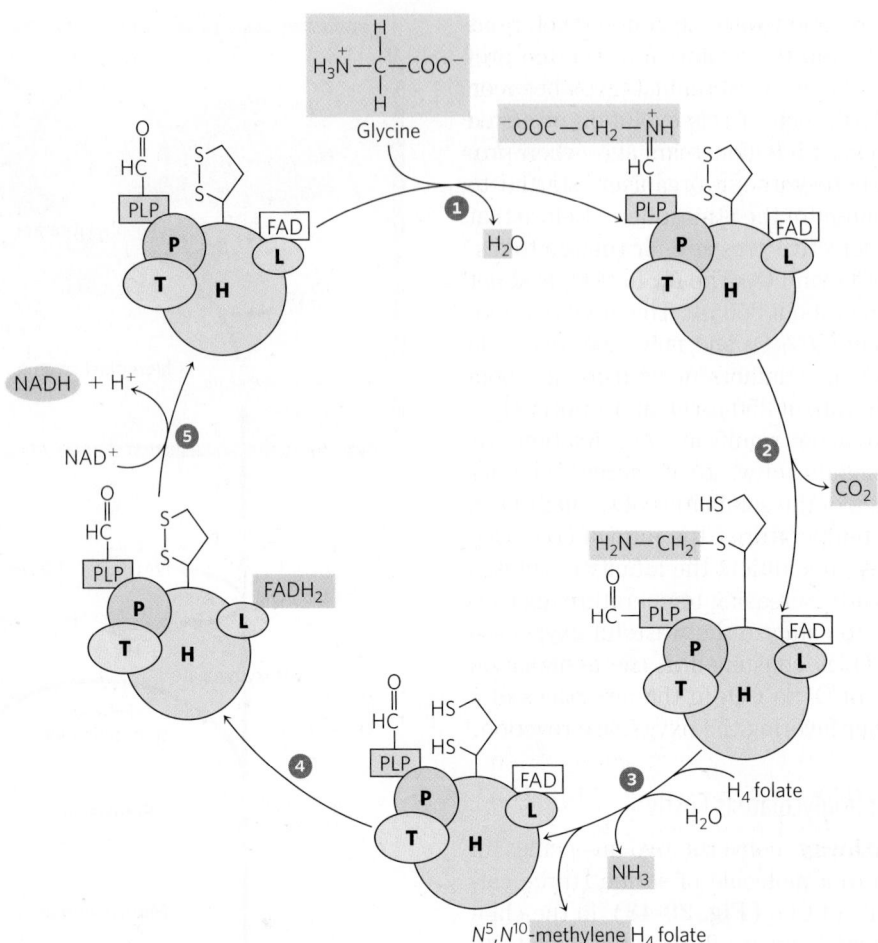

FIGURE 20-49 The glycine decarboxylase system. Glycine decarboxylase in plant mitochondria is a complex of four types of subunits, with the stoichiometry $P_4H_{27}T_9L_2$. Protein H has a covalently attached lipoic acid residue that can undergo reversible oxidation. ❶ A Schiff base forms between pyridoxal phosphate (PLP) and glycine, catalyzed by protein P (named for its bound PLP). ❷ Protein P catalyzes oxidative decarboxylation of glycine, releasing CO_2; the remaining methylamine group is attached to one of the —SH groups of reduced lipoic acid. ❸ Protein T (which uses tetrahydrofolate, H_4 folate, as cofactor) now releases NH_3 from the methylamine moiety and transfers the remaining one-carbon fragment to tetrahydrofolate, producing N^5,N^{10}-methylenetetrahydrofolate. ❹ Protein L oxidizes the two —SH groups of lipoic acid to a disulfide, ❺ passing electrons through FAD to NAD^+ and thus completing the cycle. The N^5,N^{10}-methylenetetrahydrofolate formed in this process is used by serine hydroxymethyltransferase to convert a molecule of glycine to serine, regenerating the tetrahydrofolate that is essential for the reaction catalyzed by protein T. The L subunit of glycine decarboxylase is identical to the dihydrolipoyl dehydrogenase (E_3) of pyruvate dehydrogenase and α-ketoglutarate dehydrogenase (see Fig. 16-6).

decarboxylase complex: the four proteins of the complex make up *half* of all the protein in the mitochondrial matrix in the leaves of pea and spinach plants. In nonphotosynthetic parts of a plant, such as potato tubers, mitochondria have very low concentrations of the glycine decarboxylase complex.

The combined activity of the rubisco oxygenase and the glycolate salvage pathway consumes O_2 and produces CO_2—hence the name photorespiration. This pathway is perhaps better called the **oxidative photosynthetic carbon cycle**, or **C_2 cycle**, names that do not invite comparison with respiration in mitochondria. Unlike mitochondrial respiration, photorespiration does not conserve energy and may actually inhibit net biomass formation as much as 50%. This inefficiency has led to evolutionary adaptations in the carbon-assimilation processes, particularly in plants that have evolved in warm climates. The apparent inefficiency of rubisco, and its effect in limiting biomass production, has inspired efforts to genetically engineer a "better" rubisco, but this goal is not, as yet, within reach (Box 20-1).

In C_4 Plants, CO_2 Fixation and Rubisco Activity Are Spatially Separated

In many plants that grow in the tropics (and in temperate-zone crop plants native to the tropics, such as maize, sugarcane, and sorghum) a mechanism has evolved to circumvent the problem of wasteful photorespiration.

The step in which CO$_2$ is fixed into a three-carbon product, 3-phosphoglycerate, is preceded by several steps, one of which is temporary fixation of CO$_2$ into a four-carbon compound. Plants that use this process are referred to as **C$_4$ plants**, and the assimilation process as **C$_4$ metabolism** or the **C$_4$ pathway**. Plants that use the carbon-assimilation method we have described thus far, in which the *first step* is reaction of CO$_2$ with ribulose 1,5-bisphosphate to form 3-phosphoglycerate, are called C$_3$ plants.

The C$_4$ plants, which typically grow at high light intensity and high temperatures, have several important characteristics: high photosynthetic rates, high growth rates, low photorespiration rates, low rates of water loss, and a specialized leaf structure. Photosynthesis in the leaves of C$_4$ plants involves two cell types: mesophyll and bundle-sheath cells **(Fig. 20-50a)**. There are three variants of C$_4$ metabolism, worked out in the 1960s by Marshall Hatch and Roger Slack (Fig. 20-50b).

In plants of tropical origin, the first intermediate into which ^{14}CO$_2$ is fixed is oxaloacetate, a four-carbon compound. This reaction, which occurs in the cytosol of leaf mesophyll cells, is catalyzed by **phosphoenolpyruvate carboxylase**, for which the substrate is HCO$_3^-$, not CO$_2$. The oxaloacetate thus formed is either reduced to malate at the expense of NADPH (as shown in Fig. 20-50b) or converted to aspartate by transamination:

Oxaloacetate + α-amino acid $\longrightarrow$

L-aspartate + α-keto acid

The malate or aspartate formed in the mesophyll cells then passes into neighboring bundle-sheath cells through plasmodesmata, protein-lined channels that connect two plant cells and provide a path for movement of metabolites and even small proteins between cells. In the bundle-sheath cells, malate is oxidized and decarboxylated to yield pyruvate and CO$_2$ by the action of **malic enzyme**, reducing NADP$^+$. In plants that use aspartate as the CO$_2$ carrier, aspartate arriving in bundle-sheath cells is transaminated to form oxaloacetate and reduced to malate, then the CO$_2$ is released by malic enzyme or PEP carboxykinase. Labeling experiments show that the free CO$_2$ released in the bundle-sheath cells is the same CO$_2$ molecule originally fixed into oxaloacetate in the mesophyll cells. This CO$_2$ is now

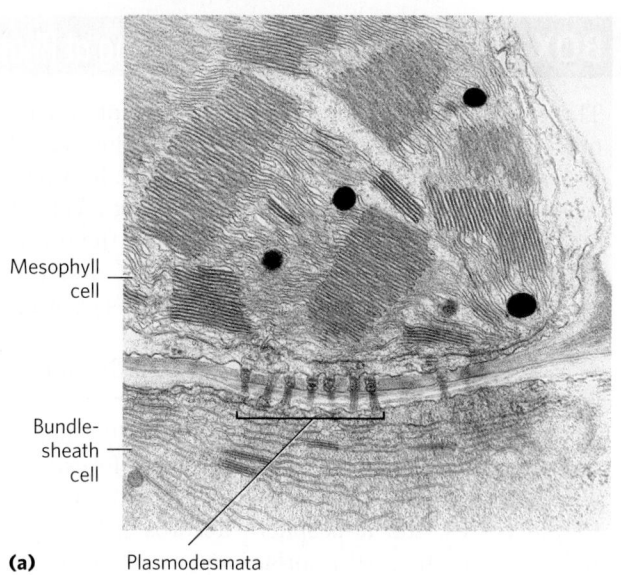

Mesophyll cell

Bundle-sheath cell

(a)   Plasmodesmata

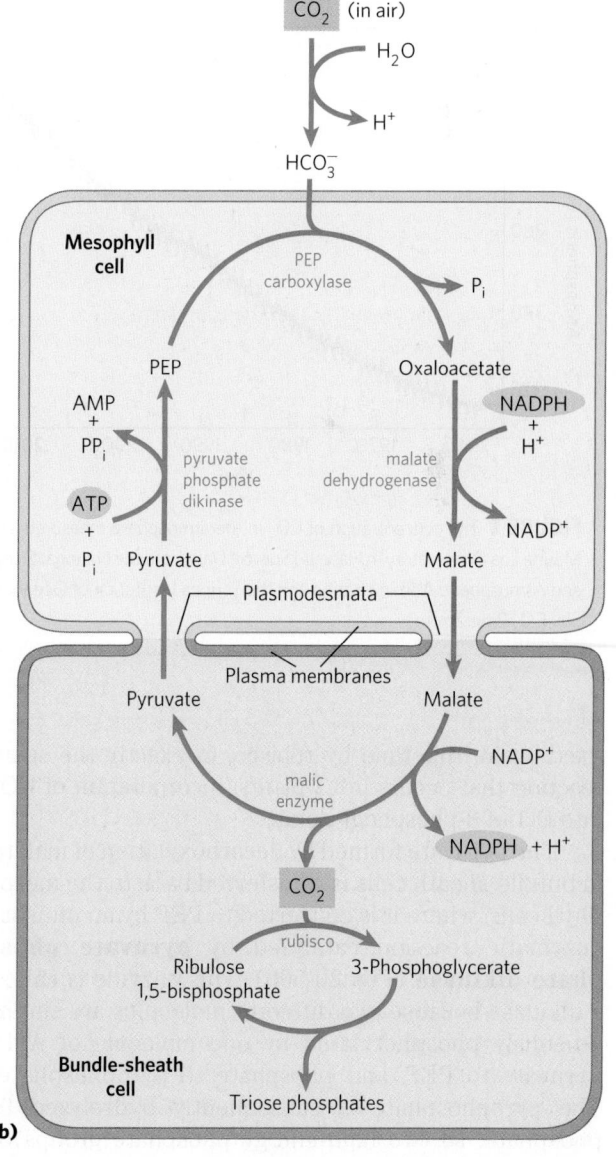

(b)

FIGURE 20-50 Carbon assimilation in C$_4$ plants. The C$_4$ pathway, involving mesophyll cells and bundle-sheath cells, predominates in plants of tropical origin. **(a)** Electron micrograph showing chloroplasts of adjacent mesophyll and bundle-sheath cells. The bundle-sheath cell contains starch granules. Plasmodesmata connecting the two cells are visible. **(b)** The C$_4$ pathway of CO$_2$ assimilation, which occurs through a four-carbon intermediate. [Source: (a) Dr. Ray Evert, University of Wisconsin-Madison, Department of Botany.]

BOX 20-1   Will Genetic Engineering of Photosynthetic Organisms Increase Their Efficiency?

Three pressing world problems have prompted serious attention to the possibility of engineering plants to be more efficient in converting sunlight into biomass: the "greenhouse" effect of increasing levels of atmospheric CO_2 on global climate change, the dwindling supply of oil for generating energy, and the need for more and better food for the world's growing population.

The concentration of CO_2 in the earth's atmosphere has risen steadily over the past 50 years (Fig. 1), a combined effect of the use of fossil fuels for energy and the clearing and burning of tropical forests to allow use of the land for agriculture. As atmospheric CO_2 increases, the atmosphere absorbs more heat radiated from the earth's surface and reradiates more heat toward the surface of the planet (and in all other directions). Retention of heat raises the temperature

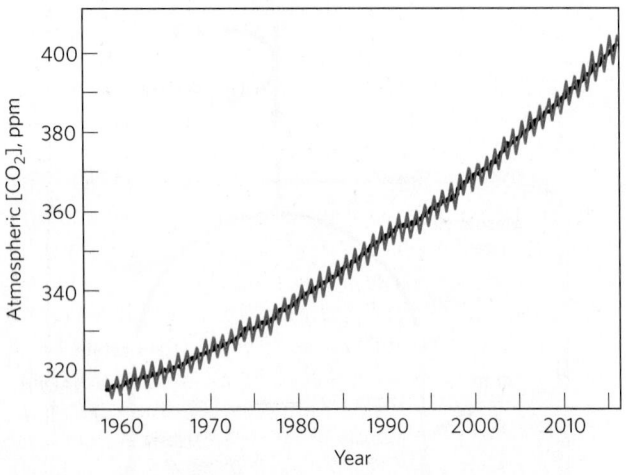

FIGURE 1 The concentration of CO_2 in the atmosphere measured at the Mauna Loa Observatory in Hawaii. [Source: Data from the National Oceanic and Atmospheric Administration and the Scripps Institution of Oceanography CO_2 Program.]

at the surface of the earth; this is the greenhouse effect. One way to limit the increase in atmospheric CO_2 would be to engineer plants or microorganisms with a greater capacity for sequestering CO_2.

The estimated amount of total carbon in all terrestrial systems (atmosphere, soil, biomass) is about 3,200 gigatons (GT), or 3,200 billion metric tons. The atmosphere contains another 760 GT of CO_2.

The flux of carbon through these terrestrial reservoirs (Fig. 2) is largely due to the photosynthetic activities of plants and the degradative activities of microorganisms. Plants fix some 123 GT of carbon annually, then immediately release about half of that to the atmosphere as they respire. Much of the remainder is gradually released to the atmosphere by microbial action on dead plant materials, but biomass is sequestered in woody plants and trees for decades or centuries. Anthropogenic carbon flux, the amount of CO_2 released into the atmosphere by human activities, is 9 GT per year—small compared with total biomass, but enough to tip the balance toward increased CO_2 in the atmosphere. Estimates indicate that the forests of North America sequester 0.7 GT of carbon annually, which represents about a tenth of the annual *global* production of CO_2 from fossil fuels. Clearly, preservation of forests and reforestation are effective ways to limit the flow of CO_2 back into the atmosphere.

A second approach to limiting the increase of atmospheric CO_2, while also addressing the need to replace dwindling fossil fuels, is to use renewable biomass as a source of ethanol to replace fossil fuels in internal combustion engines. This reduces the *unidirectional* movement of carbon from fossil fuels into the atmospheric pool of CO_2, replacing it with the *cyclic* flow of CO_2 from ethanol to CO_2 and back to biomass. When maize, wheat, or switchgrass is fermented to ethanol for fuel, every increase in biomass

fixed again, this time by rubisco, in exactly the same reaction that occurs in C_3 plants: incorporation of CO_2 into C-1 of 3-phosphoglycerate.

The pyruvate formed by decarboxylation of malate in bundle-sheath cells is transferred back to the mesophyll cells, where it is converted to PEP by an unusual enzymatic reaction catalyzed by **pyruvate phosphate dikinase** (Fig. 20-50b). This enzyme is called a dikinase because two different molecules are simultaneously phosphorylated by one molecule of ATP: pyruvate to PEP, and phosphate to pyrophosphate. The pyrophosphate is subsequently hydrolyzed to phosphate, so two high-energy phosphate groups of ATP are used in regenerating PEP. The PEP is now

ready to receive another molecule of CO_2 in the mesophyll cell.

The PEP carboxylase of mesophyll cells has a high affinity for HCO_3^- (which is favored relative to CO_2 in aqueous solution) and can fix CO_2 more efficiently than can rubisco. Unlike rubisco, it does not use O_2 as an alternative substrate, so there is no competition between CO_2 and O_2. The PEP carboxylase reaction, then, serves to fix and concentrate CO_2 in the form of malate. Release of CO_2 from malate in the bundle-sheath cells yields a sufficiently high local concentration of CO_2 for rubisco to function near its maximal rate, and for suppression of the enzyme's oxygenase activity.

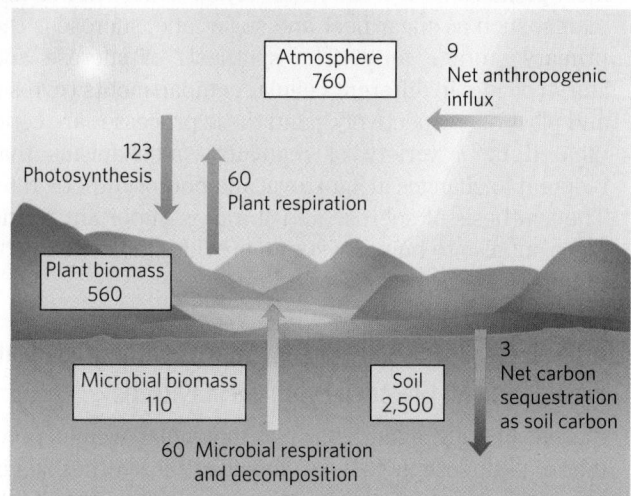

FIGURE 2 The terrestrial carbon cycle. Carbon stocks (boxes) are shown as gigatons (GT), and fluxes (arrows) in GT per year. Animal biomass is negligible here—less than 0.5 GT. [Source: Information from C. Jansson et al., *BioScience* 60:683, 2010, Fig. 1.]

its efficiency in fixing CO_2 and producing biomass. If rubisco could be genetically engineered to turn over faster or to be more selective for CO_2 relative to O_2, would the effect be greater photosynthetic production of biomass and thus greater sequestration of CO_2, greater production of nonfossil fuel, and improved nutrition?

We noted in Chapter 15 that the traditional view of metabolic pathways held that one step in any pathway was the slowest and therefore the limiting factor in material flow through the pathway. However, heroic efforts to engineer cells or organisms to produce more of the "limiting" enzyme in a pathway have often given discouraging results; the organisms often show little or no change in the flux through that pathway. The Calvin cycle is an instructive case in point. Increasing the amount of rubisco in plant cells through genetic engineering has little or no effect on the rate of CO_2 conversion into carbohydrate. Similarly, changes in the levels of enzymes known to be regulated by light and therefore suspected of playing key roles in the regulation of the CO_2-assimilation pathway (fructose 1,6-bisphosphatase, 3-phosphoglycerate kinase, and glyceraldehyde 3-phosphate dehydrogenase) also produce little or no significant improvement in photosynthetic rate. However, changed levels of sedoheptulose 1,7-bisphosphatase, not thought to be a regulatory enzyme, do have a significant impact on photosynthesis. Metabolic control analysis (Section 15.2) suggests that this result is not unexpected; in the living organism, pathways can be limited by more than one step, because every change in one step results in compensating changes in other steps. Careful determination of the flux control coefficient (see Box 15-1) helps to pinpoint which enzymes in a pathway to target for genetic engineering. Clearly, genetic engineers and metabolic control analysts will need to work together on problems like this one.

production brought about by more efficient photosynthesis should result in a corresponding decrease in the use of fossil fuels.

Finally, engineering of food crops to yield more food per acre of land, or per hour of work, could improve human nutrition worldwide.

In principle, these goals might be accomplished by developing a rubisco that didn't also catalyze the wasteful reaction with O_2, or by increasing the turnover number for rubisco, or by increasing the level of rubisco or other enzymes in the pathway for carbon fixation. Rubisco, as we have noted, is an unusually inefficient enzyme, with a turnover number of $3\ s^{-1}$ at 25 °C; most enzymes have turnover numbers orders of magnitude larger. It also catalyzes the wasteful reaction with oxygen, which further reduces

Once CO_2 is fixed into 3-phosphoglycerate in the bundle-sheath cells, the other reactions of the Calvin cycle take place exactly as described earlier. Thus in C₄ plants, mesophyll cells carry out CO_2 assimilation by the C₄ pathway and bundle-sheath cells synthesize starch and sucrose by the C₃ pathway.

Three enzymes of the C₄ pathway are regulated by light, becoming more active in daylight. Malate dehydrogenase is activated by the thioredoxin-dependent reduction mechanism shown in Figure 20-46; PEP carboxylase is activated by phosphorylation of a Ser residue; and pyruvate phosphate dikinase is activated by dephosphorylation. In the latter two cases, the details of how light causes phosphorylation or dephosphorylation are not known.

The pathway of CO_2 assimilation has a greater energy cost in C₄ plants than in C₃ plants. For each molecule of CO_2 assimilated in the C₄ pathway, a molecule of PEP must be regenerated at the expense of two phosphoanhydride bonds in ATP. Thus C₄ plants need five ATP molecules to assimilate one molecule of CO_2, whereas C₃ plants need only three (nine per triose phosphate). As the temperature increases (and the affinity of rubisco for CO_2 decreases, as noted above), a point is reached, at about 28 to 30 °C, at which the gain in efficiency from the elimination of photorespiration more than compensates for this energetic cost. C₄ plants (crabgrass, for example) outgrow most C₃ plants during the summer, as any experienced gardener can attest.

In CAM Plants, CO_2 Capture and Rubisco Action Are Temporally Separated

Succulent plants such as cactus and pineapple, which are native to very hot, very dry environments, have another variation on photosynthetic CO_2 fixation, which reduces loss of water vapor through the pores (stomata) by which CO_2 and O_2 must enter leaf tissue. Instead of separating the initial trapping of CO_2 and its fixation by rubisco across space (as do the C_4 plants), they separate these two events over time. At night, when the air is cooler and moister, the stomata open to allow entry of CO_2, which is then fixed into oxaloacetate by PEP carboxylase. The oxaloacetate is reduced to malate and stored in the vacuoles, to protect cytosolic and plastid enzymes from the low pH produced by malic acid dissociation. During the day the stomata close, preventing the water loss that would result from high daytime temperatures, and the CO_2 trapped overnight in malate is released as CO_2 by the NADP-linked malic enzyme. This CO_2 is now assimilated by the action of rubisco and the Calvin cycle enzymes. Because this method of CO_2 fixation was first discovered in stonecrops, perennial flowering plants of the family Crassulaceae, it is called crassulacean acid metabolism, and the plants are called **CAM plants**.

SUMMARY 20.6 Photorespiration and the C_4 and CAM Pathways

■ When rubisco uses O_2 rather than CO_2 as substrate, the 2-phosphoglycolate so formed is disposed of in an oxygen-dependent pathway. The result is increased consumption of O_2—photorespiration, or, more accurately, the oxidative photosynthetic carbon cycle or C_2 cycle. The 2-phosphoglycolate is converted to glyoxylate, to glycine, and then to serine in a pathway that involves enzymes in the chloroplast stroma, peroxisomes, and mitochondria.

■ In C_4 plants, the carbon-assimilation pathway minimizes photorespiration: CO_2 is first fixed in mesophyll cells into a four-carbon compound, which passes into bundle-sheath cells and releases CO_2 in high concentrations. The released CO_2 is fixed by rubisco, and the remaining reactions of the Calvin cycle occur as in C_3 plants.

■ In CAM plants, CO_2 is fixed into malate in the dark and stored in vacuoles until daylight, when the stomata are closed (minimizing water loss), and the stored malate serves as a source of CO_2 for rubisco.

20.7 Biosynthesis of Starch, Sucrose, and Cellulose

During active photosynthesis in bright light, a plant leaf produces more carbohydrate (as triose phosphates) than it needs for generating energy or synthesizing precursors.

The excess is converted to sucrose and transported to other parts of the plant, to be used as fuel or stored. In most plants, starch is the main storage form, but in a few plants, such as sugar beet and sugarcane, sucrose is the primary storage form. The synthesis of sucrose and starch occurs in different cellular compartments (cytosol and plastids, respectively), and these processes are coordinated by a variety of regulatory mechanisms that respond to changes in light level and photosynthetic rate. The synthesis of sucrose and starch is important to the plant but also to humans: starch provides more than 80% of human dietary calories worldwide.

ADP-Glucose Is the Substrate for Starch Synthesis in Plant Plastids and for Glycogen Synthesis in Bacteria

Starch, like glycogen, is a high molecular weight polymer of D-glucose in ($\alpha 1{\rightarrow}4$) linkage. It is synthesized in chloroplasts for temporary storage as one of the stable end products of photosynthesis, and for long-term storage it is synthesized in amyloplasts of the nonphotosynthetic parts of plants: seeds, roots, and tubers (underground stems).

The mechanism of glucose activation in starch synthesis is similar to that in glycogen synthesis. An activated **nucleotide sugar**, in this case **ADP-glucose**, is formed by condensation of glucose 1-phosphate with ATP in a reaction made essentially irreversible by the presence in plastids of inorganic pyrophosphatase (Fig. 15-31). **Starch synthase** then transfers glucose residues from ADP-glucose to preexisting starch molecules. The monomeric units are almost certainly added to the nonreducing end of the growing polymer, as they are in glycogen synthesis (see Fig. 15-32).

The amylose of starch is unbranched, but amylopectin has numerous ($\alpha 1{\rightarrow}6$)-linked branches (see Fig. 7-13). Chloroplasts contain a branching enzyme, similar to glycogen-branching enzyme (see Fig. 15-33), that introduces the ($\alpha 1{\rightarrow}6$) branches of amylopectin. Taking into account the hydrolysis by inorganic pyrophosphatase of the PP_i produced during ADP-glucose synthesis, the overall reaction for starch formation from glucose 1-phosphate is

$$\text{Starch}_n + \text{glucose 1-phosphate} + \text{ATP} \longrightarrow$$
$$\text{starch}_{n+1} + \text{ADP} + 2P_i$$
$$\Delta G'^\circ = -50 \text{ kJ/mol}$$

Starch synthesis is regulated at the level of ADP-glucose formation, as discussed below.

Many types of bacteria store carbohydrate in the form of glycogen (essentially, highly branched starch), which they synthesize in a reaction analogous to that catalyzed by glycogen synthase in animals. Bacteria, like plant plastids, use ADP-glucose as the activated form of glucose, whereas animal cells use UDP-glucose. Again, the similarity between plastid and bacterial metabolism is consistent with the endosymbiont hypothesis for the origin of organelles (see Fig. 1-40).

UDP-Glucose Is the Substrate for Sucrose Synthesis in the Cytosol of Leaf Cells

Most of the triose phosphate generated by CO_2 fixation in plants is converted to sucrose **(Fig. 20-51)** or starch. In the course of evolution, sucrose may have been selected as the transport form of carbon because of its unusual linkage between the anomeric C-1 of glucose and the anomeric C-2 of fructose. This bond is not hydrolyzed by amylases or other common carbohydrate-cleaving enzymes, and the unavailability of the sucrose molecule's anomeric carbons prevents its reacting nonenzymatically (as does glucose) with amino acids and proteins.

Sucrose is synthesized in the cytosol, beginning with dihydroxyacetone phosphate and glyceraldehyde 3-phosphate exported from the chloroplast. After condensation of two triose phosphates to form fructose 1,6-bisphosphate (catalyzed by aldolase), hydrolysis by fructose 1,6-bisphosphatase yields fructose 6-phosphate. **Sucrose 6-phosphate synthase** then catalyzes the reaction of fructose 6-phosphate with **UDP-glucose** to form **sucrose 6-phosphate** (Fig. 20-51). Finally, **sucrose 6-phosphate phosphatase** removes the phosphate

group, making sucrose available for export to other tissues. The reaction catalyzed by sucrose 6-phosphate synthase is a low-energy process ($\Delta G'^{\circ} = -5.7$ kJ/mol), but the hydrolysis of sucrose 6-phosphate to sucrose is sufficiently exergonic ($\Delta G'^{\circ} = -16.5$ kJ/mol) to make the overall synthesis of sucrose essentially irreversible. Sucrose synthesis is regulated and closely coordinated with starch synthesis, as we shall see.

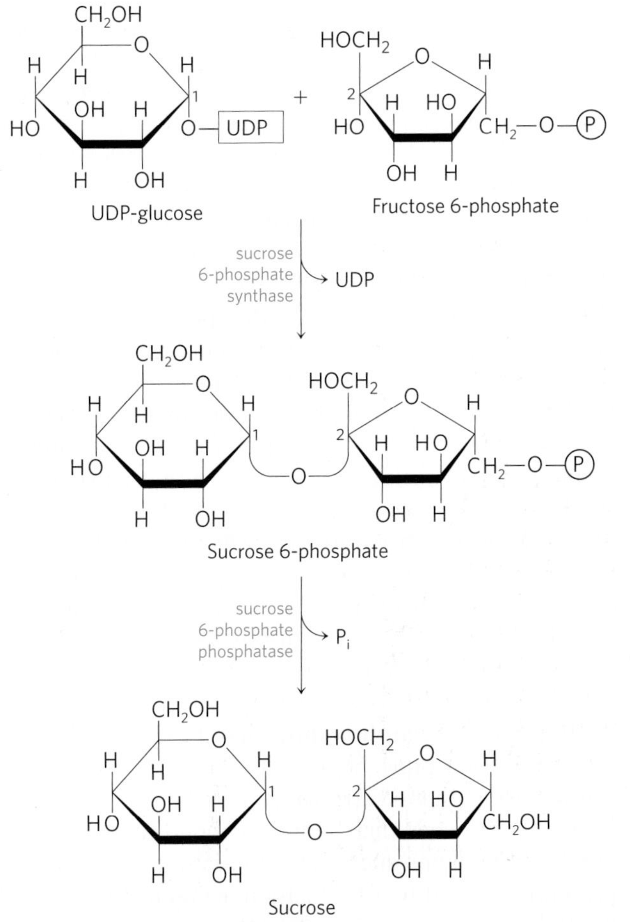

One remarkable difference between the cells of plants and animals is the absence in the plant cell cytosol of the enzyme inorganic pyrophosphatase, which catalyzes the reaction

$$PP_i + H_2O \longrightarrow 2P_i \qquad \Delta G'^{\circ} = -19.2 \text{ kJ/mol}$$

For many biosynthetic reactions that liberate PP_i, pyrophosphatase activity makes the process more favorable energetically, tending to make these reactions irreversible. In plants, this enzyme is present in plastids but absent from the cytosol. As a result, the cytosol of leaf cells contains a substantial concentration of PP_i—enough (~0.3 mm) to make reactions such as that catalyzed by UDP-glucose pyrophosphorylase (see Fig. 15-31) readily reversible. Recall from Chapter 14 (p. 540) that the cytosolic isozyme of phosphofructokinase in plants uses PP_i, not ATP, as the phosphoryl donor.

Conversion of Triose Phosphates to Sucrose and Starch Is Tightly Regulated

Triose phosphates produced by the Calvin cycle in bright sunlight, as we have noted, may be stored temporarily in the chloroplast as starch, or converted to sucrose and exported to nonphotosynthetic parts of the plant, or both. The balance between the two processes is tightly regulated, and both must be coordinated with the rate of carbon fixation. Five-sixths of the triose phosphate formed in the Calvin cycle must be recycled to ribulose 1,5-bisphosphate (Fig. 20-41); if more than one-sixth of the triose phosphate is drawn out of the cycle to make sucrose and starch, the cycle will slow or stop. However, *insufficient* conversion of triose phosphate to starch or sucrose would tie up phosphate, leaving a chloroplast deficient in P_i, which is also essential for operation of the Calvin cycle.

The flow of triose phosphates into sucrose is regulated by the activity of fructose 1,6-bisphosphatase (FBPase-1) and the enzyme that effectively reverses its action, PP_i-dependent phosphofructokinase (PP-PFK-1). These enzymes are therefore critical points for determining the fate of triose phosphates produced by photosynthesis. Both enzymes are regulated by **fructose 2,6-bisphosphate (F26BP)**, which inhibits FBPase-1 and stimulates PP-PFK-1. In vascular plants, the concentration of F26BP varies inversely with the rate of photosynthesis **(Fig. 20-52)**. Phosphofructokinase-2, responsible for F26BP synthesis, is inhibited by dihydroxyacetone phosphate or 3-phosphoglycerate and stimulated by fructose 6-phosphate and P_i. During active photosynthesis, dihydroxyacetone phosphate is produced and P_i is consumed, resulting in inhibition of

FIGURE 20-51 Sucrose synthesis. Sucrose is synthesized from UDP-glucose and fructose 6-phosphate, which are synthesized from triose phosphates in the plant cell cytosol by pathways shown in Figures 15-31 and 20-36. The sucrose 6-phosphate synthase of most plant species is allosterically regulated by glucose 6-phosphate and P_i.

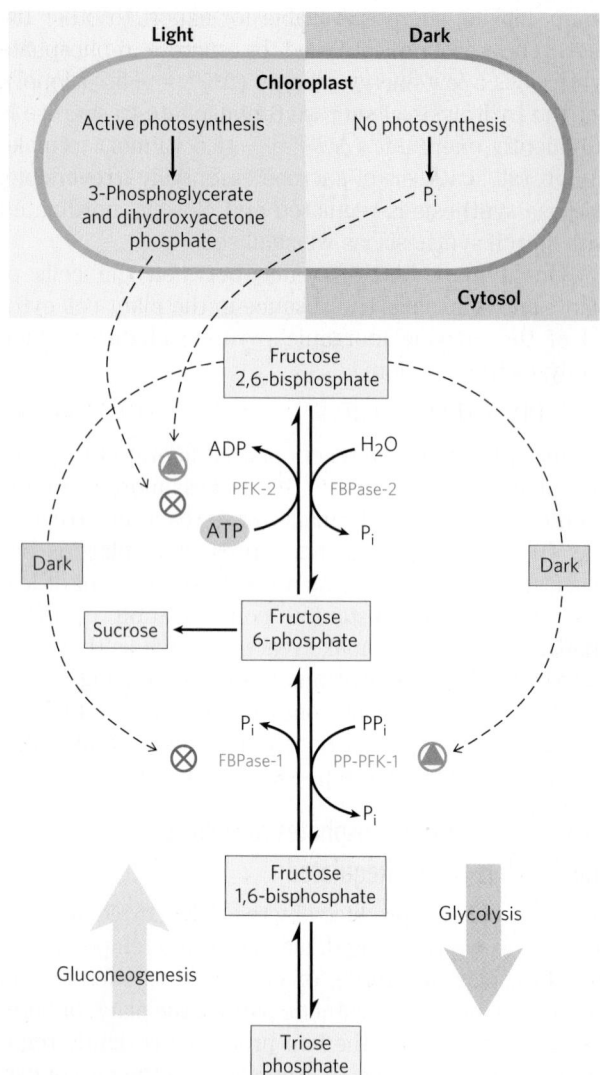

FIGURE 20-52 Fructose 2,6-bisphosphate as regulator of sucrose synthesis. The concentration of the allosteric regulator fructose 2,6-bisphosphate in plant cells is regulated by the products of photosynthetic carbon assimilation and by P_i. Dihydroxyacetone phosphate and 3-phosphoglycerate produced by CO_2 assimilation inhibit phosphofructokinase-2 (PFK-2), the enzyme that synthesizes the regulator; P_i stimulates PFK-2. The concentration of the regulator is therefore inversely proportional to the rate of photosynthesis. In the dark, the concentration of fructose 2,6-bisphosphate increases and stimulates the glycolytic enzyme PP_i-dependent phosphofructokinase-1 (PP-PFK-1), while inhibiting the gluconeogenic enzyme fructose 1,6-bisphosphatase (FBPase-1). When photosynthesis is active (in the light), the concentration of the regulator drops and the synthesis of fructose 6-phosphate and sucrose is favored.

PFK-2 and lowered concentrations of F26BP. This favors greater flux of triose phosphate into fructose 6-phosphate formation and sucrose synthesis. With this regulatory system, sucrose synthesis occurs when the level of triose phosphate produced by the Calvin cycle exceeds that needed to maintain operation of the cycle.

Sucrose synthesis is also regulated at the level of sucrose 6-phosphate synthase, which is allosterically activated by glucose 6-phosphate and inhibited by P_i. This enzyme is further regulated by phosphorylation and dephosphorylation; a protein kinase phosphorylates the

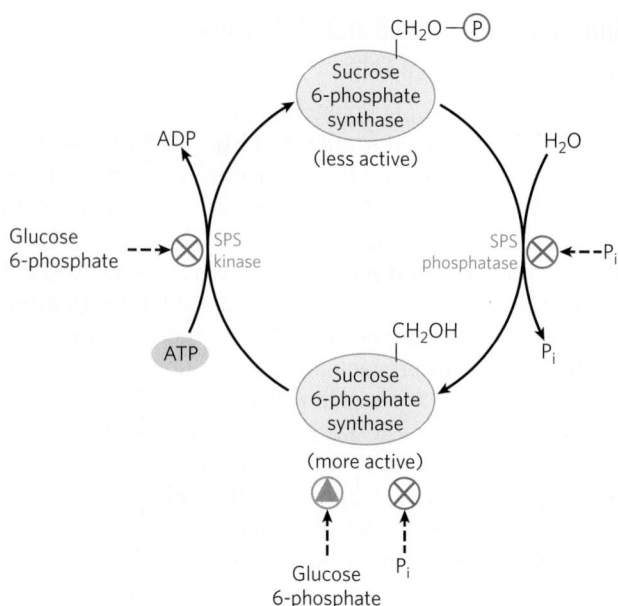

FIGURE 20-53 Regulation of sucrose phosphate synthase by phosphorylation. A protein kinase (SPS kinase) specific for sucrose phosphate synthase (SPS) phosphorylates a Ser residue in SPS, inactivating it; a specific phosphatase (SPS phosphatase) reverses this inhibition. The kinase is inhibited allosterically by glucose 6-phosphate, which also activates SPS allosterically. The phosphatase is inhibited by P_i, which also inhibits SPS directly. Thus, when the concentration of glucose 6-phosphate is high as a result of active photosynthesis, SPS is activated and produces sucrose phosphate. A high P_i concentration, which occurs when photosynthetic conversion of ADP to ATP is slow, inhibits sucrose phosphate synthesis.

enzyme on a specific Ser residue, making it less active, and a phosphatase reverses this inactivation by removing the phosphate **(Fig. 20-53)**. Inhibition of the kinase by glucose 6-phosphate, and of the phosphatase by P_i, enhances the effects of these two compounds on sucrose synthesis. When hexose phosphates are abundant, sucrose 6-phosphate synthase is activated by glucose 6-phosphate; when P_i is elevated (as when photosynthesis is slow), sucrose synthesis is slowed. During active photosynthesis, triose phosphates are converted to fructose 6-phosphate, which is rapidly equilibrated with glucose 6-phosphate by phosphohexose isomerase. Because the equilibrium lies far toward glucose 6-phosphate, as soon as fructose 6-phosphate accumulates, the level of glucose 6-phosphate rises and sucrose synthesis is stimulated.

The key regulatory enzyme in starch synthesis is **ADP-glucose pyrophosphorylase (Fig. 20-54)**; it is activated by 3-phosphoglycerate, which accumulates during active photosynthesis, and inhibited by P_i, which accumulates when light-driven condensation of ADP and P_i slows. When sucrose synthesis slows, 3-phosphoglycerate formed by CO_2 fixation accumulates, activating this enzyme and stimulating the synthesis of starch.

The Glyoxylate Cycle and Gluconeogenesis Produce Glucose in Germinating Seeds

Many plants store lipids (oils) and proteins in their seeds, to be used as sources of energy and as biosynthetic

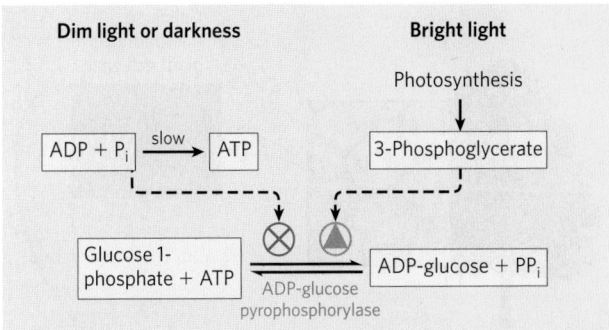

FIGURE 20-54 Regulation of ADP-glucose phosphorylase by 3-phosphoglycerate and P_i. This enzyme, which produces the precursor for starch synthesis, is rate-limiting in starch production. The enzyme is stimulated allosterically by 3-phosphoglycerate (3-PGA) and inhibited by P_i; in effect, the ratio [3-PGA]/[P_i], which rises with increasing rates of photosynthesis, controls starch synthesis at this step.

precursors during germination, before photosynthetic capacity has developed. These stored components are converted to carbohydrates by the combined action of several pathways. Glucogenic amino acids (see Table 14-4) derived from the breakdown of stored seed proteins are transaminated and oxidized to succinyl-CoA, pyruvate, oxaloacetate, fumarate, and α-ketoglutarate (Chapter 18)—all good starting materials for gluconeogenesis. Active gluconeogenesis in germinating seeds provides glucose for the synthesis of sucrose, polysaccharides, and many metabolites derived from hexoses. In plant seedlings, sucrose provides much of the chemical energy needed for initial growth.

Triacylglycerols stored in seeds also provide fuel for the germinating plants. They are hydrolyzed to free fatty acids, which undergo β oxidation to acetyl-CoA in specialized peroxisomes called **glyoxysomes** that develop during seed germination (see Fig. 17-14). The acetyl-CoA formed from seed oils enters the **glyoxylate cycle (Fig. 20-55)**, which brings about the net conversion of acetate to succinate or other four-carbon intermediates of the citric acid cycle:

$$2 \text{ Acetyl-CoA} + \text{NAD}^+ + 2\text{H}_2\text{O} \longrightarrow$$
$$\text{succinate} + 2\text{CoA} + \text{NADH} + \text{H}^+$$

In the glyoxylate cycle, acetyl-CoA condenses with oxaloacetate to form citrate, and citrate is converted to isocitrate, exactly as in the citric acid cycle. The next

FIGURE 20-55 Conversion of stored fatty acids to sucrose in germinating seeds through the glyoxylate cycle. This pathway begins in specialized peroxisomes called glyoxysomes. The citrate synthase, aconitase, and malate dehydrogenase of the glyoxylate cycle are isozymes of the citric acid cycle enzymes; isocitrate lyase and malate synthase are unique to the glyoxylate cycle. Notice that two acetyl groups (shaded light red) enter the cycle and four carbons leave as succinate (shaded blue). Succinate is exported to mitochondria, where it is converted to oxaloacetate by enzymes of the citric acid cycle. Oxaloacetate enters the cytosol and serves as the starting material for gluconeogenesis and for synthesis of sucrose, the transport form of carbon in plants. The glyoxylate cycle was elucidated by Hans Kornberg and Neil Madsen in the laboratory of Hans Krebs.

step, however, is not the breakdown of isocitrate by isocitrate dehydrogenase but the cleavage of isocitrate by **isocitrate lyase**, forming succinate and **glyoxylate**. The glyoxylate then condenses with a second molecule of acetyl-CoA to yield malate, in a reaction catalyzed by **malate synthase**. The malate is subsequently

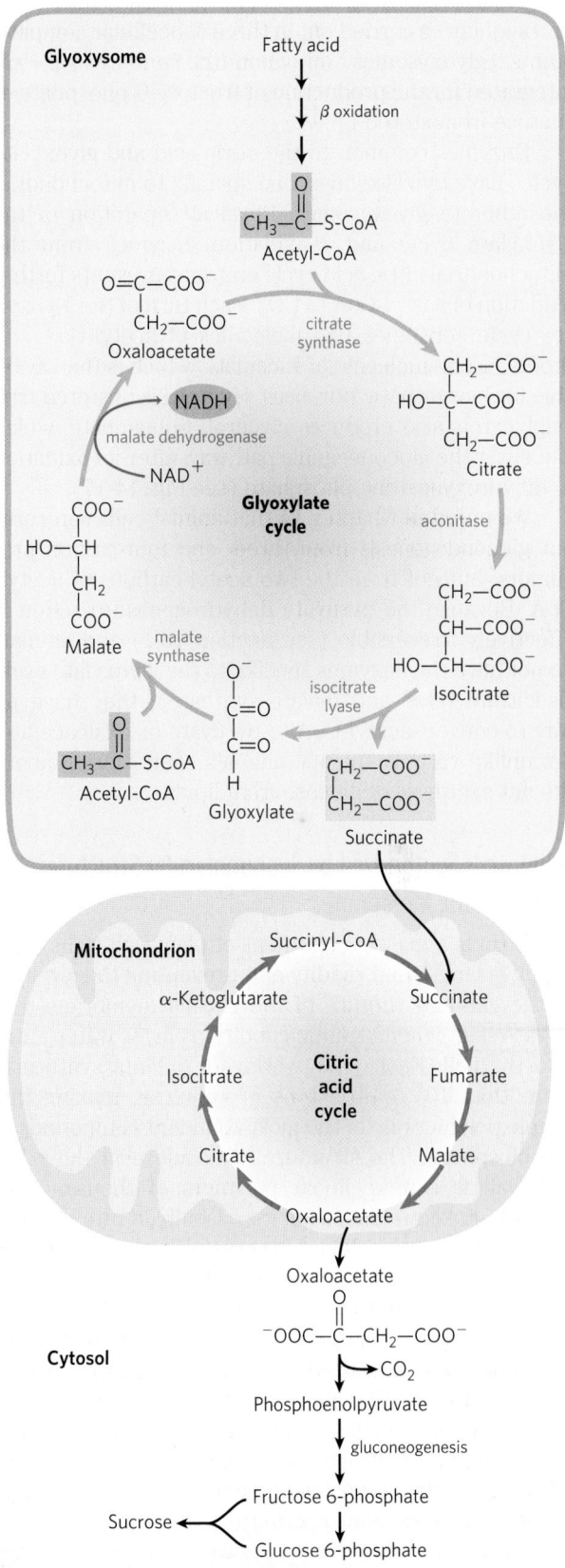

oxidized to oxaloacetate, which can condense with another molecule of acetyl-CoA to start another turn of the cycle. The succinate passes into the mitochondrial matrix, where it is converted by citric acid cycle enzymes to oxaloacetate. The oxaloacetate moves into the cytosol and can be converted to phosphoenolpyruvate by PEP carboxykinase, then to fructose 6-phosphate, the precursor of sucrose, by gluconeogenesis. Thus, reaction sequences carried out in three subcellular compartments (glyoxysomes, mitochondria, and cytosol) are integrated for the production of fructose 6-phosphate or sucrose from stored lipids.

Enzymes common to the citric acid and glyoxylate cycles have two isozymes, one specific to mitochondria, the other to glyoxysomes. Physical separation of the glyoxylate cycle and β-oxidation enzymes from the mitochondrial citric acid cycle enzymes prevents further oxidation of acetyl-CoA to CO_2. Each turn of the glyoxylate cycle consumes two molecules of acetyl-CoA and produces one molecule of succinate, which is then available for biosynthetic purposes. Hydrolysis of stored triacylglycerols also produces glycerol 3-phosphate, which can enter the gluconeogenic pathway, after its oxidation to dihydroxyacetone phosphate (see Fig. 14-17).

We noted in Chapter 14 that animal cells can carry out gluconeogenesis from three- and four-carbon precursors, but not from the two acetyl carbons of acetyl-CoA. Because the pyruvate dehydrogenase reaction is effectively irreversible (see Section 16.1) and animals do not have the enzymes specific to the glyoxylate cycle (isocitrate lyase and malate synthase), they have no way to convert acetyl-CoA to pyruvate or oxaloacetate. So, unlike vascular plants, animals cannot bring about the net synthesis of glucose from lipids.

Cellulose Is Synthesized by Supramolecular Structures in the Plasma Membrane

Cellulose is a major constituent of plant cell walls, providing strength and rigidity and preventing the swelling of the cell and rupture of the plasma membrane that might result when osmotic conditions favor water entry into the cell. Each year, worldwide, plants synthesize more than 10^{11} metric tons of cellulose, making this simple polymer one of the most abundant compounds in the biosphere. The structure of cellulose in the plant cell wall is simple: linear polymers of thousands of ($\beta1{\rightarrow}4$)-linked D-glucose units, assembled into bundles of at least 18 chains, which co-crystallize to form microfibrils, which may in turn be assembled into larger macrofibrils. **(Fig. 20-56)**.

As a major component of the plant cell wall, cellulose must be synthesized from intracellular precursors but deposited and assembled outside the plasma membrane. The enzymatic machinery for initiation, elongation, and export of cellulose chains is therefore more complicated than that used to synthesize starch or glycogen (which are not exported).

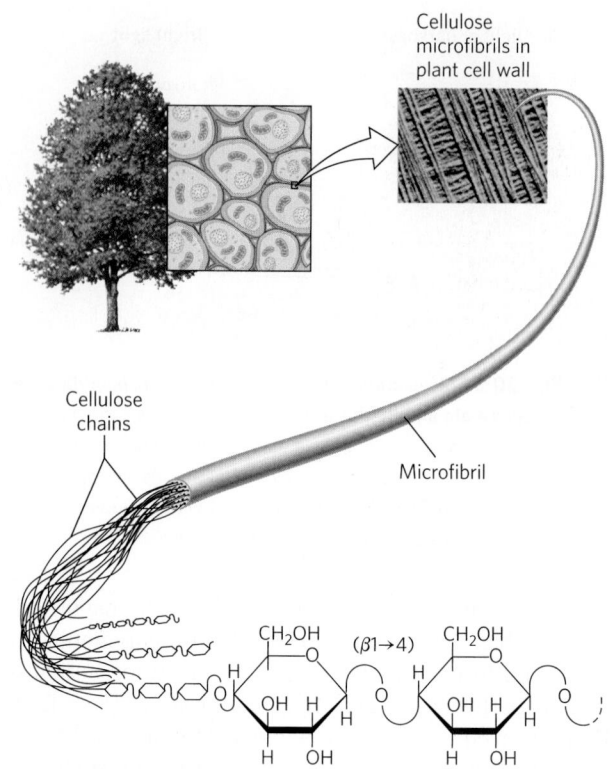

FIGURE 20-56 Cellulose structure. The plant cell wall is made up in part of cellulose molecules arranged side by side to form crystalline arrays—cellulose microfibrils. Several microfibrils may combine to form larger cellulose macrofibrils. The scanning electron microscope shows macrofibrils, 5 to 12 nm in diameter, laid down on the cell surface in several layers distinguishable by the different orientations of the fibrils. [Source: Electron micrograph from Biophoto Associates/Science Source.]

The complex enzymatic machinery that assembles cellulose chains spans the plasma membrane, with one part on the cytoplasmic side positioned to bind the substrate, UDP-glucose, and elongate the chains, and another part extending to the outside, responsible for exporting the cellulose molecules to the extracellular space. Freeze-fracture electron microscopy shows a cellulose synthesis complex, or **rosette**, composed of six large particles arranged in a regular hexagon with a diameter of about 30 nm **(Fig. 20-57a)**. Several proteins, including the catalytic subunit of **cellulose synthase**, make up this structure. Much of the recent progress in understanding cellulose synthesis stems from genetic and molecular genetic studies of *Arabidopsis thaliana*; this small flowering plant is especially amenable to genetic dissection, and its genome has been sequenced. The structure of the plant cellulose synthase is similar to that of the bacterium *Rhodobacter sphaeroides*, which has been determined by x-ray crystallography (Fig. 20-57b).

In one working model of cellulose synthesis, cellulose chains are initiated by the transfer of a glucose residue from UDP-glucose to a "primer" glucose already bound to cellulose synthase on the cytoplasmic side of the plasma membrane, to form a disaccharide. As addition of further glucose residues lengthens the chain, it is

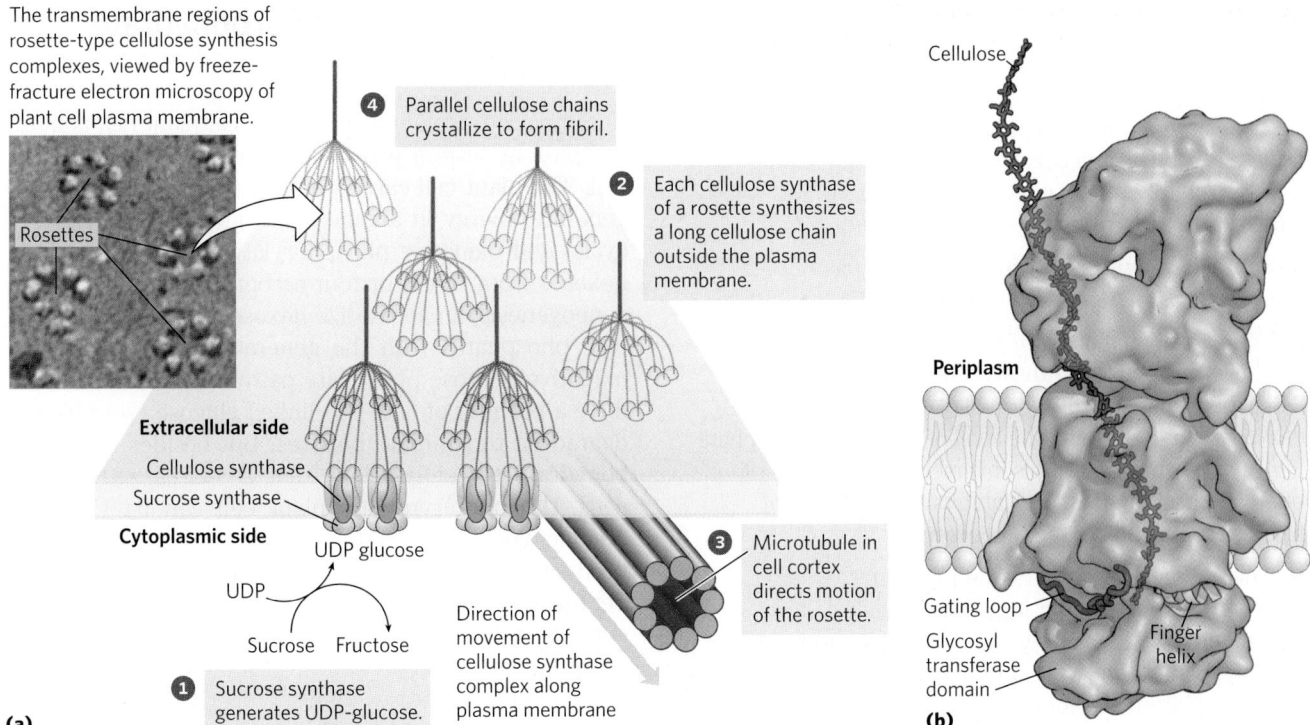

The transmembrane regions of rosette-type cellulose synthesis complexes, viewed by freeze-fracture electron microscopy of plant cell plasma membrane.

Rosettes

4 Parallel cellulose chains crystallize to form fibril.

2 Each cellulose synthase of a rosette synthesizes a long cellulose chain outside the plasma membrane.

Extracellular side

Cellulose synthase

Sucrose synthase

Cytoplasmic side

UDP glucose

UDP

Sucrose   Fructose

1 Sucrose synthase generates UDP-glucose.

Direction of movement of cellulose synthase complex along plasma membrane

3 Microtubule in cell cortex directs motion of the rosette.

(a)

Cellulose

Periplasm

Gating loop

Glycosyl transferase domain

Finger helix

(b)

FIGURE 20-57 A model for the synthesis of cellulose. (a) Schematic derived from a combination of genetic and biochemical studies of *Arabidopsis thaliana* and other vascular plants. (b) The structure of cellulose synthase from the bacterium *Rhodobacter sphaeroides*. The transmembrane part of the protein provides a channel through which the lengthening cellulose polymer (red) is pushed into the periplasm as the chain grows by addition of glucose units on the inside surface of the plasma membrane. Two structures of the enzyme move during the catalytic cycle. The gating loop moves into the substrate-binding site when UDP-glucose binds, then moves out to allow UDP to leave. The finger helix touches the glucose residue at the growing polymer end, then, after a new residue is added, moves so as to touch this new terminal glucose. The glycosyl transferase domain extends into the cytoplasm, where it binds its substrate UDP-glucose. [Sources: (a) Electron micrograph ©courtesy Dr. Candace H. Haigler, North Carolina State University and Dr. Mark Grimson, Texas Tech University. (b) PDB ID 5EJZ, J. L. W. Morgan et al., *Nature* 531:329, 2016. An extension of the cellulose chain was modeled in.]

extruded through a channel formed by the transmembrane helices of cellulose synthase and, on the outer surface of the plasma membrane, joins growing chains from neighboring cellulose synthase molecules to form a cellulose microfibril. Polymers of more than 6 to 8 glucose units are insoluble in water, promoting microfibril crystallization. There is no definite length for a cellulose polymer; synthesis is highly processive, and some polymers are as long as 15,000 glucose units.

The UDP-glucose used for cellulose synthesis (step **1** in Fig. 20-57) is generated from sucrose produced during photosynthesis, in a reaction catalyzed by sucrose synthase (named for the reverse reaction):

$$\text{Sucrose} + \text{UDP} \longrightarrow \text{UDP-glucose} + \text{fructose}$$

A membrane-bound form of sucrose synthase may produce a high local concentration of UDP-glucose for cellulose synthesis.

Each of the six particles of the rosette most likely contains three cellulose synthase molecules, each synthesizing a single cellulose chain (step **2**). The large enzyme complex that catalyzes this process moves along the plasma membrane with directionality often related to the course of microtubules in the cell cortex, the cytoplasmic layer just below the membrane (step **3**). When these microtubules lie perpendicular to the axis of the plant's growth, the cellulose microfibrils are laid down similarly to promote elongation. The motion of the cellulose synthase complexes is believed to be driven by energy released in the polymerization reaction, not by a molecular motor such as kinesin.

The fundamental cellulose microfibril made by one rosette-type cellulose synthesis complex is thought to be composed of 18 chains lying side by side with the same (parallel) orientation of nonreducing and reducing ends. The 18 separate polymers coalesce on the outer surface of the cell and crystallize soon after they are polymerized (step **4**), just prior to integrating into the cell wall.

In UDP-glucose, the glucose is α-linked to the nucleotide, but in cellulose, the glucose residues are ($\beta1\rightarrow4$)-linked, so there is an inversion of configuration at the anomeric carbon (C-1) as the glycosidic bond forms. Glycosyltransferases that invert configuration are generally assumed to use a single-displacement mechanism, with nucleophilic attack by the acceptor species at the anomeric carbon of the donor sugar (in this case, UDP-glucose).

SUMMARY 20.7 Biosynthesis of Starch, Sucrose, and Cellulose

■ Starch synthase in chloroplasts and amyloplasts catalyzes the addition of single glucose residues, donated by ADP-glucose, to the growing polymer chain, probably to the nonreducing end. Branches in amylopectin are introduced by a second enzyme.

■ Sucrose is synthesized in the cytosol from UDP-glucose and fructose 1-phosphate, in two steps.

■ The partitioning of triose phosphates between sucrose synthesis and starch synthesis is regulated by fructose 2,6-bisphosphate (F26BP), an allosteric effector of the enzymes that determine the level of fructose 6-phosphate. F26BP concentration varies inversely with the rate of photosynthesis, and F26BP inhibits the synthesis of fructose 6-phosphate, the precursor of sucrose.

■ The glyoxylate cycle, taking place in the glyoxysomes of germinating seeds of some plants, uses several citric acid cycle enzymes and two additional enzymes: isocitrate lyase and malate synthase. The two decarboxylation steps of the citric acid cycle are bypassed, making possible the *net* formation of succinate, oxaloacetate, and other cycle intermediates from acetyl-CoA. Oxaloacetate thus formed can be used to synthesize glucose (and ultimately sucrose) via gluconeogenesis.

■ In plant cells, cellulose synthesis takes place in rosette-type cellulose synthesis complexes in the plasma membrane, which contain multiple copies of cellulose synthase. This enzyme has a glycosyl transferase activity in its cytoplasmic domain and forms a transmembrane channel through which the growing cellulose chain is extruded. Glucose units are transferred from UDP-glucose to the nonreducing end of the growing chain. Each rosette produces 18 separate chains simultaneously and in parallel. The chains crystallize into microfibrils that integrate into the cell wall. Cortical microtubules orient the developing microfibrils.

20.8 Integration of Carbohydrate Metabolism in Plants

Carbohydrate metabolism in a typical plant cell is more complex in several ways than that in a typical animal cell. The plant cell carries out the same processes that generate energy in animal cells (glycolysis, citric acid cycle, and oxidative phosphorylation); it can generate hexoses from three- or four-carbon compounds by gluconeogenesis; it can oxidize hexose phosphates to pentose phosphates with the generation of NADPH (the oxidative pentose phosphate pathway); and it can produce a polymer of $(\alpha 1{\rightarrow}4)$-linked glucose (starch) and degrade it to generate hexoses. But besides these carbohydrate transformations that it shares with animal cells, the photosynthetic plant cell can fix CO_2 into organic compounds (the rubisco reaction); use the products of fixation to generate trioses, hexoses, and pentoses (the Calvin cycle); and convert acetyl-CoA generated from fatty acid breakdown to four-carbon compounds (the glyoxylate cycle) and the four-carbon compounds to hexoses (gluconeogenesis). These processes, unique to the plant cell, are segregated in several compartments not found in animal cells: the glyoxylate cycle in glyoxysomes, the Calvin cycle in chloroplasts, starch synthesis in amyloplasts, and organic acid storage in vacuoles. The integration of events among these various compartments requires specific transporters in the membranes of each organelle to move products from one organelle to another or into the cytosol.

Pools of Common Intermediates Link Pathways in Different Organelles

Although we have described metabolic transformations in plant cells in terms of individual pathways, these pathways interconnect so completely that we should instead consider pools of metabolic intermediates shared among these pathways and connected by readily reversible reactions **(Fig. 20-58)**. One such

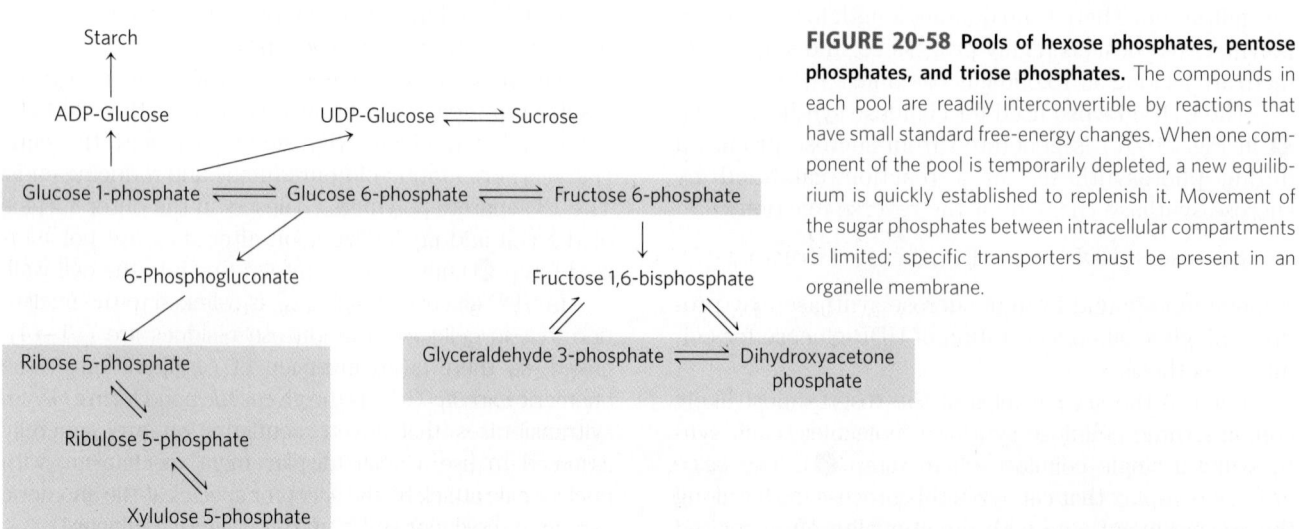

FIGURE 20-58 Pools of hexose phosphates, pentose phosphates, and triose phosphates. The compounds in each pool are readily interconvertible by reactions that have small standard free-energy changes. When one component of the pool is temporarily depleted, a new equilibrium is quickly established to replenish it. Movement of the sugar phosphates between intracellular compartments is limited; specific transporters must be present in an organelle membrane.

metabolite pool includes the hexose phosphates glucose 1-phosphate, glucose 6-phosphate, and fructose 6-phosphate; a second includes the 5-phosphates of the pentoses ribose, ribulose, and xylulose; a third includes the triose phosphates dihydroxyacetone phosphate and glyceraldehyde 3-phosphate. Metabolite fluxes through these pools change in magnitude and direction in response to changes in the circumstances of the plant, and they vary with tissue type. Transporters in the membranes of each organelle move specific compounds in and out, and the regulation of these

transporters presumably influences the degree to which the pools mix.

During daylight hours, triose phosphates produced in photosynthetic leaf tissue ("source" tissues) by the Calvin cycle move out of the chloroplast and into the cytosolic hexose phosphate pool, where they are converted to sucrose for transport via the plant phloem to nonphotosynthetic tissues ("sink" tissues) **(Fig. 20-59)**. In sink tissues such as roots, tubers, and bulbs, sucrose is converted to starch for storage or is used as an energy source via glycolysis.

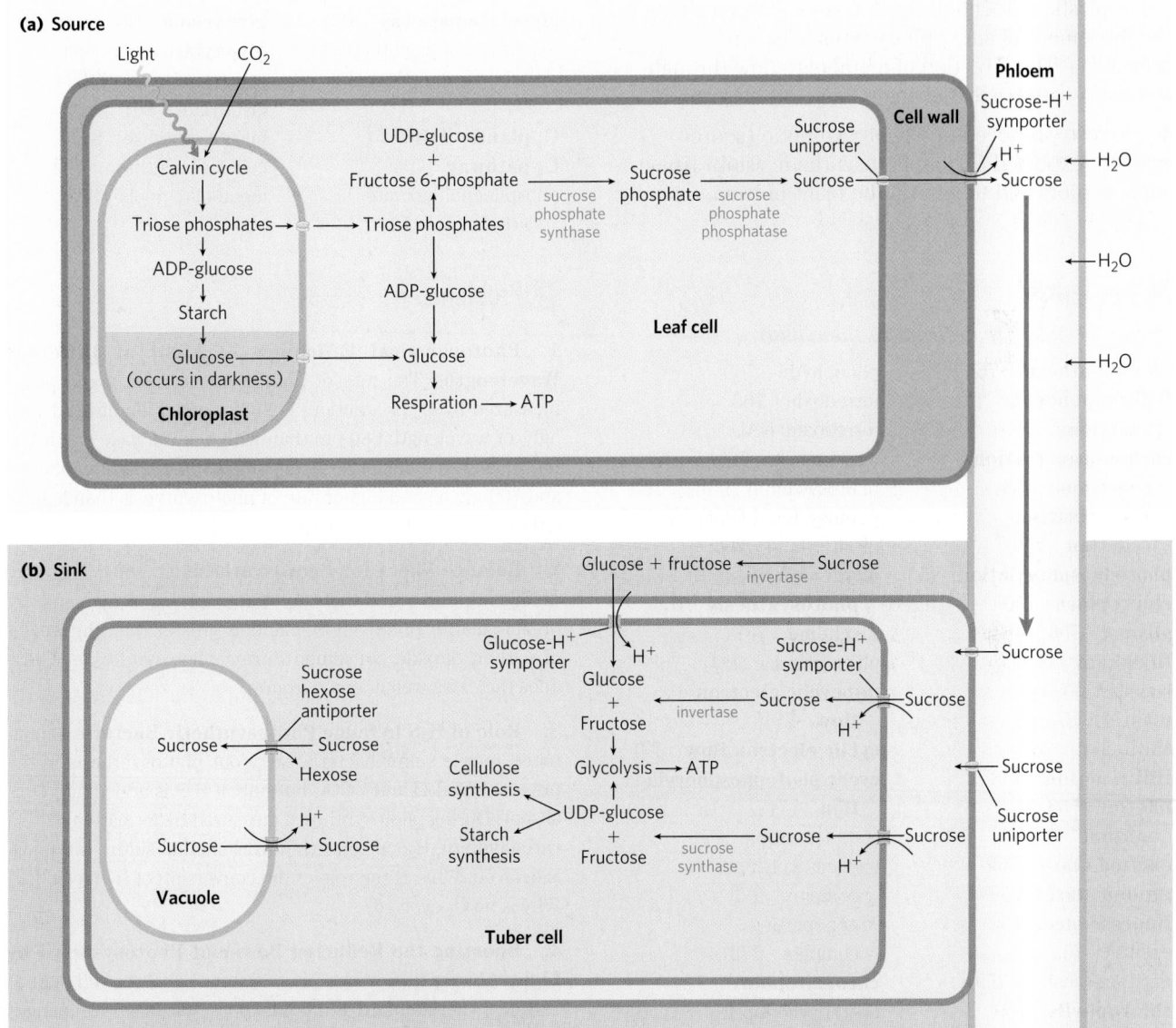

FIGURE 20-59 Movement of sucrose between source and sink tissues. (a) In daylight, photosynthetic leaves (source tissue) fix CO_2 into triose phosphates via the Calvin cycle in chloroplasts. Some of the triose phosphate is used in the chloroplasts to synthesize starch; the rest is exported to the cytosol, where it can be converted via gluconeogenesis to fructose 6-phosphate and glucose 1-phosphate. Sucrose, synthesized from UDP-glucose and fructose, is exported from leaf mesophyll cells to the plant phloem; the resulting high sucrose content draws water into

the phloem by osmosis. The resulting increased turgor pressure (p. 57) pushes the solution in the phloem (sap) toward sink tissues (bulbs, tubers, roots). **(b)** Sucrose moves from the phloem into the sink tissues, where it is converted to starch or cell wall cellulose, or is used as fuel for glycolysis, the citric acid cycle, and oxidative phosphorylation to provide ATP for these nonphotosynthetic tissues. Sugar transport across the plasma membrane and between intracellular compartments is catalyzed by several symporters and antiporters coupled to a proton gradient.

In growing plants, hexose phosphates are also withdrawn from the pool for the synthesis of cell walls. At night, starch is metabolized by glycolysis and oxidative phosphorylation to provide energy for both source and sink tissues.

SUMMARY 20.8 Integration of Carbohydrate Metabolism in Plants

■ The individual pathways of carbohydrate metabolism in plants overlap extensively; they share pools of common intermediates, including hexose phosphates, pentose phosphates, and triose phosphates. Transporters in the membranes of chloroplasts, mitochondria, and amyloplasts mediate the movement of sugar phosphates between organelles. The direction of metabolite flow through the pools within a leaf changes from day to night.

■ Sucrose produced in a photosynthetic (source) tissue is exported to nonphotosynthetic (sink) tissue such as roots and tubers via the plant phloem.

Key Terms

Terms in bold are defined in the glossary.

Problems

1. Photochemical Efficiency of Light at Different Wavelengths The rate of photosynthesis in a green plant, measured by O_2 production, is higher when illuminated with light of wavelength 680 nm than with light of wavelength 700 nm. However, illumination by a combination of light of 680 nm and 700 nm gives a higher rate of photosynthesis than light of either wavelength alone. Explain.

2. Balance Sheet for Photosynthesis In 1804, Theodore de Saussure observed that the total weight of oxygen and dry organic matter produced by plants is greater than the weight of carbon dioxide consumed during photosynthesis. Where does the extra weight come from?

3. Role of H_2S in Some Photosynthetic Bacteria Illuminated purple sulfur bacteria carry out photosynthesis in the presence of H_2O and $^{14}CO_2$, but only if H_2S is added and O_2 is absent. During photosynthesis, measured by formation of $[^{14}C]$ carbohydrate, H_2S is converted to elemental sulfur, but no O_2 is evolved. What is the role of the conversion of H_2S to sulfur? Why is no O_2 evolved?

4. Boosting the Reducing Power of Photosystem I by Light Absorption When photosystem I absorbs red light at 700 nm, the standard reduction potential of P700 changes from 0.40 V to about -1.2 V. What fraction of the absorbed light is trapped in the form of reducing power?

5. Electron Flow through Photosystems I and II Predict how an inhibitor of electron passage through pheophytin would affect electron flow through (a) photosystem II and (b) photosystem I. Explain your reasoning.

6. Limited ATP Synthesis in the Dark In a laboratory experiment, spinach chloroplasts are illuminated in the absence of ADP and P_i, then the light is turned off and ADP and P_i are added. ATP is synthesized for a short time in the dark. Explain this finding.

7. Mode of Action of the Herbicide DCMU When chloroplasts are treated with 3-(3,4-dichlorophenyl)-1,1-dimethylurea (DCMU, or diuron), a potent herbicide, O_2 evolution and photophosphorylation cease. Oxygen evolution, but not photophosphorylation, can be restored by addition of an external electron acceptor, or Hill reagent. How does DCMU act as a weed killer? Suggest a location for the inhibitory action of this herbicide in the scheme shown in Figure 20-16. Explain.

8. Effect of Venturicidin on Oxygen Evolution Venturicidin is a powerful inhibitor of the chloroplast ATP synthase, interacting with CF_o and blocking proton passage through the CF_oCF_1 complex. How would venturicidin affect oxygen evolution in a suspension of well-illuminated chloroplasts? Would your answer change if the experiment were done in the presence of an uncoupling reagent such as 2,4-dinitrophenol (DNP)? Explain.

9. Bioenergetics of Photophosphorylation The steady-state concentrations of ATP, ADP, and P_i in isolated spinach chloroplasts under full illumination at pH 7.0 are 120.0, 6.0, and 700.0 μM, respectively.

(a) What is the free-energy requirement for the synthesis of 1 mol of ATP under these conditions?

(b) The energy for ATP synthesis is furnished by light-induced electron transfer in the chloroplasts. What is the minimum voltage drop necessary (during transfer of a pair of electrons) to synthesize ATP under these conditions? (You may need to refer to Eqn 13-7, p. 521.)

10. Light Energy for a Redox Reaction Suppose you have isolated a new photosynthetic microorganism that oxidizes H_2S and passes the electrons to NAD^+. What wavelength of light would provide enough energy for H_2S to reduce NAD^+ under standard conditions? Assume 100% efficiency in the photochemical event, and use an E'° of -243 mV for H_2S and -320 mV for NAD^+. See Figure 20-7 for the energy equivalents of wavelengths of light.

11. Equilibrium Constant for Water-Splitting Reactions The coenzyme $NADP^+$ is the terminal electron acceptor in chloroplasts, according to the reaction

$$2H_2O + 2NADP^+ \longrightarrow 2NADPH + 2H^+ + O_2$$

Use information in Chapter 19 (Table 19-2) to calculate the equilibrium constant for this reaction at 25 °C. (The relationship between K'_{eq} and $\Delta G'^\circ$ is discussed on p. 498.) How can the chloroplast overcome this unfavorable equilibrium?

12. Energetics of Phototransduction During photosynthesis, eight photons must be absorbed (four by each photosystem) for every O_2 molecule produced:

$$2H_2O + 2NADP^+ + 8\,photons \longrightarrow 2NADPH + 2H^+ + O_2$$

Assuming that these photons have a wavelength of 700 nm (red) and that the light absorption and use of light energy are 100% efficient, calculate the free-energy change for the process.

13. Electron Transfer to a Hill Reagent Isolated spinach chloroplasts evolve O_2 when illuminated in the presence of potassium ferricyanide (a Hill reagent), according to the equation

$$2H_2O + 4Fe^{3+} \longrightarrow O_2 + 4H^+ + 4Fe^{2+}$$

where Fe^{3+} represents ferricyanide and Fe^{2+}, ferrocyanide. Is NADPH produced in this process? Explain.

14. How Often Does a Chlorophyll Molecule Absorb a Photon? The amount of chlorophyll a (M_r 892) in a spinach leaf is about 20 $\mu g/cm^2$ of leaf surface. In noonday sunlight (average energy reaching the leaf is 5.4 $J/cm^2 \cdot min$), the leaf absorbs about 50% of the radiation. How often does a single chlorophyll molecule absorb a photon? Given that the average lifetime of an excited chlorophyll molecule in vivo is 1 ns, what fraction of the chlorophyll molecules are excited at any one time?

15. Effect of Monochromatic Light on Electron Flow The extent to which an electron carrier is oxidized or reduced during photosynthetic electron transfer can sometimes be observed directly with a spectrophotometer. When chloroplasts are illuminated with 700 nm light, cytochrome f, plastocyanin, and plastoquinone are oxidized. When chloroplasts are illuminated with 680 nm light, however, these electron carriers are reduced. Explain.

16. Function of Cyclic Photophosphorylation When the $[NADPH]/[NADP^+]$ ratio in chloroplasts is high, photophosphorylation is predominantly cyclic (see Fig. 20-16). Is O_2 evolved during cyclic photophosphorylation? Is NADPH produced? Explain. What is the main function of cyclic photophosphorylation?

17. Segregation of Metabolism in Organelles What are the advantages to the plant cell of having different organelles to carry out different reaction sequences that share intermediates?

18. Phases of Photosynthesis When a suspension of green algae is illuminated in the absence of CO_2 and then incubated with $^{14}CO_2$ in the dark, $^{14}CO_2$ is converted to $[^{14}C]$glucose for a brief time. What is the significance of this observation with regard to the CO_2-assimilation process, and how is it related to the light-dependent reactions of photosynthesis? Why does the conversion of $^{14}CO_2$ to $[^{14}C]$glucose stop after a brief time?

19. Identification of Key Intermediates in CO_2 Assimilation Calvin and his colleagues used the unicellular green alga *Chlorella* to study the carbon-assimilation reactions of photosynthesis. They incubated $^{14}CO_2$ with illuminated suspensions of algae and followed the time course of appearance of ^{14}C in two compounds, X and Y, under two sets of conditions. Suggest the identities of X and Y, based on your understanding of the Calvin cycle.

(a) Illuminated *Chlorella* were grown with unlabeled CO_2, then the light was turned off and $^{14}CO_2$ was added (vertical dashed line in the graph below). Under these conditions, X was the first compound to become labeled with ^{14}C; Y was unlabeled.

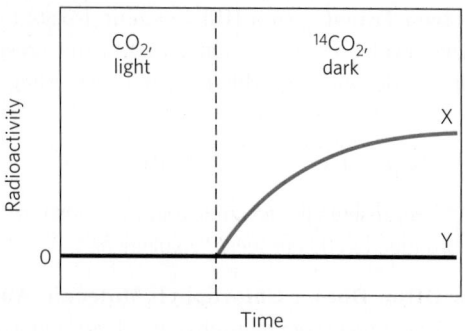

(b) Illuminated *Chlorella* cells were grown with ^{14}CO$_2$. Illumination was continued until all the ^{14}CO$_2$ had disappeared (vertical dashed line in the graph below). Under these conditions, X became labeled quickly but lost its radioactivity with time, whereas Y became more radioactive with time.

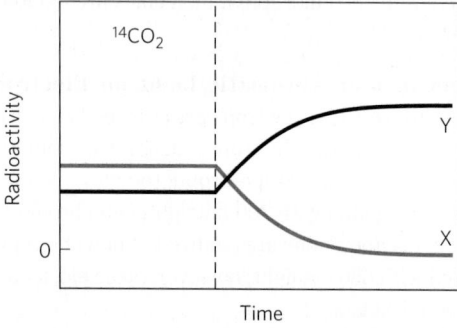

20. Regulation of the Calvin Cycle Iodoacetate reacts irreversibly with the free —SH groups of Cys residues in proteins. Predict which Calvin cycle enzyme(s) would be inhibited by iodoacetate, and explain why.

Iodoacetate   Inactive enzyme

21. Comparison of the Reductive and Oxidative Pentose Phosphate Pathways The *reductive* pentose phosphate pathway generates several intermediates identical to those of the *oxidative* pentose phosphate pathway (Chapter 14). What role does each pathway play in cells where it is active?

22. Photorespiration and Mitochondrial Respiration Compare the oxidative photosynthetic carbon cycle (C$_2$ cycle), also called *photorespiration*, with the *mitochondrial respiration* that drives ATP synthesis. Why are both processes referred to as respiration? Where in the cell do they occur, and under what circumstances? What is the path of electron flow in each?

23. Role of Sedoheptulose 1,7-Bisphosphatase What effect on the cell and the organism might result from a defect in sedoheptulose 1,7-bisphosphatase in (a) a human hepatocyte and (b) the leaf cell of a green plant?

24. Pathway of CO$_2$ Assimilation in Maize If a maize (corn) plant is illuminated in the presence of ^{14}CO$_2$, after about

1 second, more than 90% of all the radioactivity incorporated in the leaves is found at C-4 of malate, aspartate, and oxaloacetate. Only after 60 seconds does ^{14}C appear at C-1 of 3-phosphoglycerate. Explain.

25. Identifying CAM Plants Given some ^{14}CO$_2$ and all the tools typically present in a biochemistry research lab, how would you design a simple experiment to determine whether a plant is a typical C$_4$ plant or a CAM plant?

26. Chemistry of Malic Enzyme: Variation on a Theme Malic enzyme, found in the bundle-sheath cells of C$_4$ plants, carries out a reaction that has a counterpart in the citric acid cycle. What is the analogous reaction? Explain your choice.

27. Differences between C$_3$ and C$_4$ Plants The plant genus *Atriplex* includes some C$_3$ and some C$_4$ species. From the data in the following plots (species 1, black curve; species 2, red curve), identify which is a C$_3$ plant and which is a C$_4$ plant. Justify your answer in molecular terms that account for the data in all three plots.

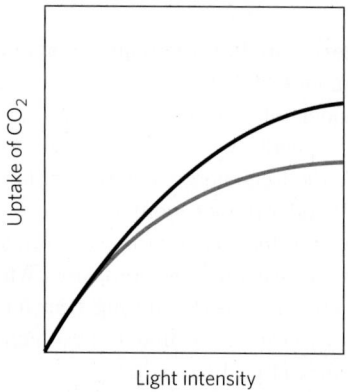

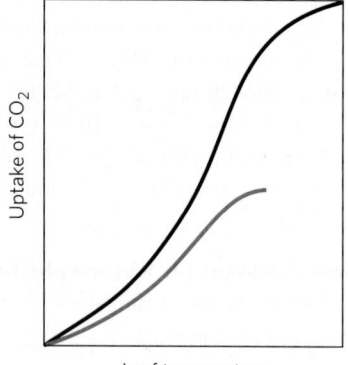

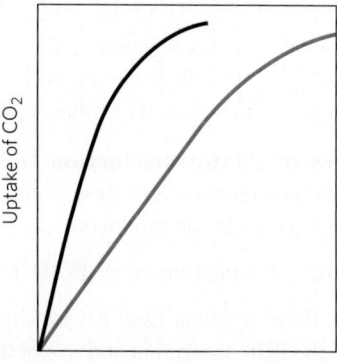

28. C$_4$ Pathway in a Single Cell In typical C$_4$ plants, the initial capture of CO_2 occurs in one cell type and the Calvin cycle reactions occur in another (see Fig. 20-50). Voznesenskaya and colleagues described a plant, *Bienertia cycloptera* (which grows in salty depressions of semidesert in Central Asia), that shows the biochemical properties of a C$_4$ plant, but, unlike typical C$_4$ plants, does not segregate the reactions of CO_2 fixation into two cell types. PEP carboxylase and rubisco are present in the same cell. However, the cells have two types of chloroplasts, which are localized differently. One type, relatively poor in granal thylakoids, is confined to the periphery; the more typical chloroplasts are clustered in the center of the cell, separated from the peripheral chloroplasts by large vacuoles. Thin cytosolic bridges pass through the vacuoles, connecting the peripheral and central cytosol. A micrograph of a *B. cycloptera* leaf cell, with arrows pointing to peripheral chloroplasts, is shown below.

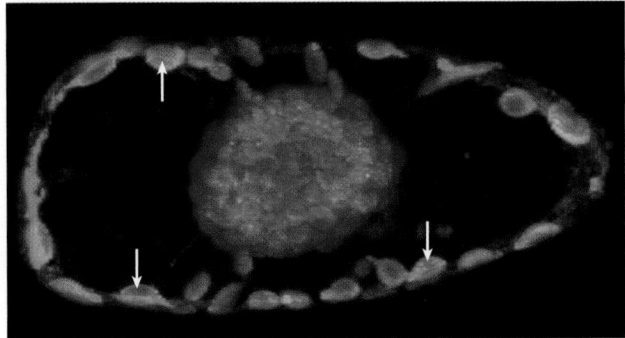

Source: Courtesy of Dr. Gerald Edwards, School of Biological Sciences, Washington State University.

In this plant, where would you expect to find (a) PEP carboxylase, (b) rubisco, and (c) starch granules? Explain your answers with a model for CO_2 fixation in these C$_4$ cells. [Source: Information from E. V. Voznesenskaya et al., *Plant J.* 31:649, 2002.]

29. The Cost of Storing Glucose as Starch Write the sequence of steps and the net reaction required to calculate the cost, in ATP molecules, of converting a molecule of cytosolic glucose 6-phosphate to starch and back to glucose 6-phosphate. What fraction of the maximum number of ATP molecules available from complete catabolism of glucose 6-phosphate to CO_2 and H_2O does this cost represent?

30. Inorganic Pyrophosphatase The enzyme inorganic pyrophosphatase contributes to making many biosynthetic reactions that generate inorganic pyrophosphate essentially irreversible in cells. By keeping the concentration of PP$_i$ very low, the enzyme "pulls" these reactions in the direction of PP$_i$ formation. The synthesis of ADP-glucose in chloroplasts is one such reaction. However, the synthesis of UDP-glucose in the plant cytosol, which also produces PP$_i$, is readily reversible in vivo. How do you reconcile these two facts?

31. Regulation of Starch and Sucrose Synthesis Sucrose synthesis occurs in the cytosol and starch synthesis in the chloroplast stroma, yet the two processes are intricately balanced. What factors shift the reactions in favor of (a) starch synthesis and (b) sucrose synthesis?

32. Regulation of Sucrose Synthesis In the regulation of sucrose synthesis from the triose phosphates produced during photosynthesis, 3-phosphoglycerate and P$_i$ play critical roles (see Fig. 20-52). Explain why the concentrations of these two regulators reflect the rate of photosynthesis.

33. Sucrose and Dental Caries The most prevalent infection in humans worldwide is dental caries, which stems from the colonization and destruction of tooth enamel by a variety of acidifying microorganisms. These organisms synthesize and live within a water-insoluble network of dextrans, called dental plaque, composed of $(\alpha1\rightarrow6)$-linked polymers of glucose with many $(\alpha1\rightarrow3)$ branch points. Polymerization of dextran requires dietary sucrose, and the reaction is catalyzed by a bacterial enzyme, dextran-sucrose glucosyltransferase.

(a) Write the overall reaction for dextran polymerization.

(b) In addition to providing a substrate for the formation of dental plaque, how does dietary sucrose also provide oral bacteria with an abundant source of metabolic energy?

34. Partitioning between the Citric Acid and Glyoxylate Cycles In an organism (such as *E. coli*) that has both the citric acid cycle and the glyoxylate cycle, what determines which of these pathways isocitrate will enter?

Data Analysis Problems

35. Photophosphorylation: Discovery, Rejection, and Rediscovery In the 1930s and 1940s, researchers were beginning to make progress toward understanding the mechanism of photosynthesis. At the time, the role of "energy-rich phosphate bonds" (today, "ATP") in glycolysis and cellular respiration was just becoming known. There were many theories about the mechanism of photosynthesis, especially about the role of light. This problem focuses on what was then called the "primary photochemical process"—that is, on what it is, exactly, that the energy from captured light produces in the photosynthetic cell. Interestingly, one important part of the modern model of photosynthesis was proposed early on, only to be rejected, ignored for several years, then finally revived and accepted.

In 1944, Emerson, Stauffer, and Umbreit proposed that "the function of light energy in photosynthesis is the formation of 'energy-rich' phosphate bonds" (p. 107). In their model (hereafter, the "Emerson model"), the free energy necessary to drive both CO_2 fixation *and* reduction came from these "energy-rich phosphate bonds" (i.e., ATP), produced as a result of light absorption by a chlorophyll-containing protein.

This model was explicitly rejected by Rabinowitch (1945). After summarizing Emerson and coauthors' findings, Rabinowitch stated: "Until more positive evidence is provided, we are inclined to consider as more convincing a general argument against this hypothesis, which can be derived from energy considerations. Photosynthesis is eminently a problem of energy *accumulation*. What good can be served, then, by converting light quanta (even those of red light, which amount to about 43 kcal per Einstein) into 'phosphate quanta' of only 10 kcal per mole? This appears to be a start in the wrong direction—toward *dissipation* rather than toward accumulation of energy" (p. 228). This argument, along with other evidence, led to abandonment of the Emerson model until the 1950s, when it was found to be correct—albeit in a modified form.

For each piece of information from Emerson and coauthors' article presented in (a) through (d), answer the following three questions:

1. How does this information support the Emerson model, in which light energy is used directly by chlorophyll *to make ATP*, and the ATP then provides the energy to drive CO_2 fixation and reduction?

2. How would Rabinowitch explain this information, based on his model (and most other models of the day), in which light energy is used directly by chlorophyll *to make reducing compounds*? Rabinowitch wrote: "Theoretically, there is no reason why *all* electronic energy contained in molecules excited by the absorption of light should not be available for oxidation-reduction" (p. 152). In this model, the reducing compounds are then used to fix and reduce CO_2, and the energy for these reactions comes from the large amounts of free energy released by the reduction reactions.

3. How is this information explained by our modern understanding of photosynthesis?

(a) Chlorophyll contains a Mg^{2+} ion, which is known to be an essential cofactor for many enzymes that catalyze phosphorylation and dephosphorylation reactions.

(b) A crude "chlorophyll protein" isolated from photosynthetic cells showed phosphorylating activity.

(c) The phosphorylating activity of the "chlorophyll protein" was inhibited by light.

(d) The levels of several different phosphorylated compounds in photosynthetic cells changed dramatically in response to light exposure. (Emerson and coworkers were not able to identify the specific compounds involved.)

As it turned out, the Emerson and Rabinowitch models were both partly correct and partly incorrect.

(e) Explain how the two models relate to our current model of photosynthesis.

In his rejection of the Emerson model, Rabinowitch went on to say: "The difficulty of the phosphate storage theory appears most clearly when one considers the fact that, in weak light, eight or ten quanta of light are sufficient to reduce one molecule of carbon dioxide. If each quantum should produce one molecule of high-energy phosphate, the accumulated energy would be only 80–100 kcal per Einstein—while photosynthesis requires *at least* 112 kcal per mole, and probably more, because of losses in irreversible partial reactions" (p. 228).

(f) How does Rabinowitch's value of 8 to 10 photons per molecule of CO_2 reduced compare with the value accepted today?

(g) How would you rebut Rabinowitch's argument, based on our current knowledge about photosynthesis?

References

Emerson, R.L., J.F. Stauffer, and W.W. Umbreit. 1944. Relationships between phosphorylation and photosynthesis in *Chlorella. Am. J. Botany* 31:107–120.

Rabinowitch, E.I. 1945. *Photosynthesis and Related Processes*, Vol. I. Interscience Publishers, New York.

Further Reading is available at www.macmillanlearning.com/LehningerBiochemistry7e.

CHAPTER 21

Lipid Biosynthesis

Self-study tools that will help you practice what you've learned and reinforce this chapter's concepts are available online. Go to www.macmillanlearning.com/LehningerBiochemistry7e.

Lipids play a variety of cellular roles, some only recently recognized. They are the principal form of stored energy in most organisms and major constituents of cellular membranes. Specialized lipids serve as pigments (retinal, carotene), cofactors (vitamin K), detergents (bile salts), transporters (dolichols), hormones (vitamin D derivatives, sex hormones), extracellular and intracellular messengers (eicosanoids, phosphatidylinositol derivatives), and anchors for membrane proteins (covalently attached fatty acids, prenyl groups, phosphatidylinositol). The ability to synthesize a variety of lipids is essential to all organisms. This chapter describes the biosynthetic pathways for some of the most common cellular lipids, illustrating the strategies employed in assembling these water-insoluble products from water-soluble precursors such as acetate. Like other biosynthetic pathways, these reaction sequences are endergonic and reductive. They use ATP as a source of metabolic energy and a reduced electron carrier (usually NADPH) as a reductant.

We first describe the biosynthesis of fatty acids, the primary components of both triacylglycerols and phospholipids, then examine the assembly of fatty acids into triacylglycerols and the simpler membrane phospholipids. Finally, we consider the synthesis of cholesterol, a component of some membranes and the precursor of steroids such as bile acids, sex hormones, and adrenocortical hormones.

21.1 Biosynthesis of Fatty Acids and Eicosanoids

After the discovery that fatty acid oxidation takes place by the oxidative removal of successive two-carbon (acetyl-CoA) units (see Fig. 17-8), biochemists thought the biosynthesis of fatty acids might proceed by a simple reversal of the same enzymatic steps. However, as they were to find out, fatty acid biosynthesis and breakdown occur by different pathways, are catalyzed by different sets of enzymes, and take place in different parts of the cell. Moreover, biosynthesis requires the participation of a three-carbon intermediate, **malonyl-CoA**, that does not appear in the path of fatty acid breakdown.

$$\begin{array}{c} \underset{-O}{\overset{O}{\parallel}}{C} - CH_2 - \underset{S-CoA}{\overset{O}{\parallel}}{C} \\ \text{Malonyl-CoA} \end{array}$$

The pathway of fatty acid synthesis and its regulation now take center stage. We consider the biosynthesis of longer-chain fatty acids, unsaturated fatty acids, and their eicosanoid derivatives at the end of this section.

Malonyl-CoA Is Formed from Acetyl-CoA and Bicarbonate

The formation of malonyl-CoA from acetyl-CoA is an irreversible process, catalyzed by **acetyl-CoA carboxylase**. The bacterial enzyme has three separate polypeptide subunits **(Fig. 21-1)**; in animal cells, all three activities are part of a single multifunctional polypeptide. Plant cells contain both types of acetyl-CoA carboxylase. In all cases, the enzyme contains a biotin prosthetic group covalently bound in amide linkage to the ε-amino group of a Lys residue in one of the three polypeptides or domains of the enzyme molecule. The two-step reaction catalyzed by this enzyme is very similar to other

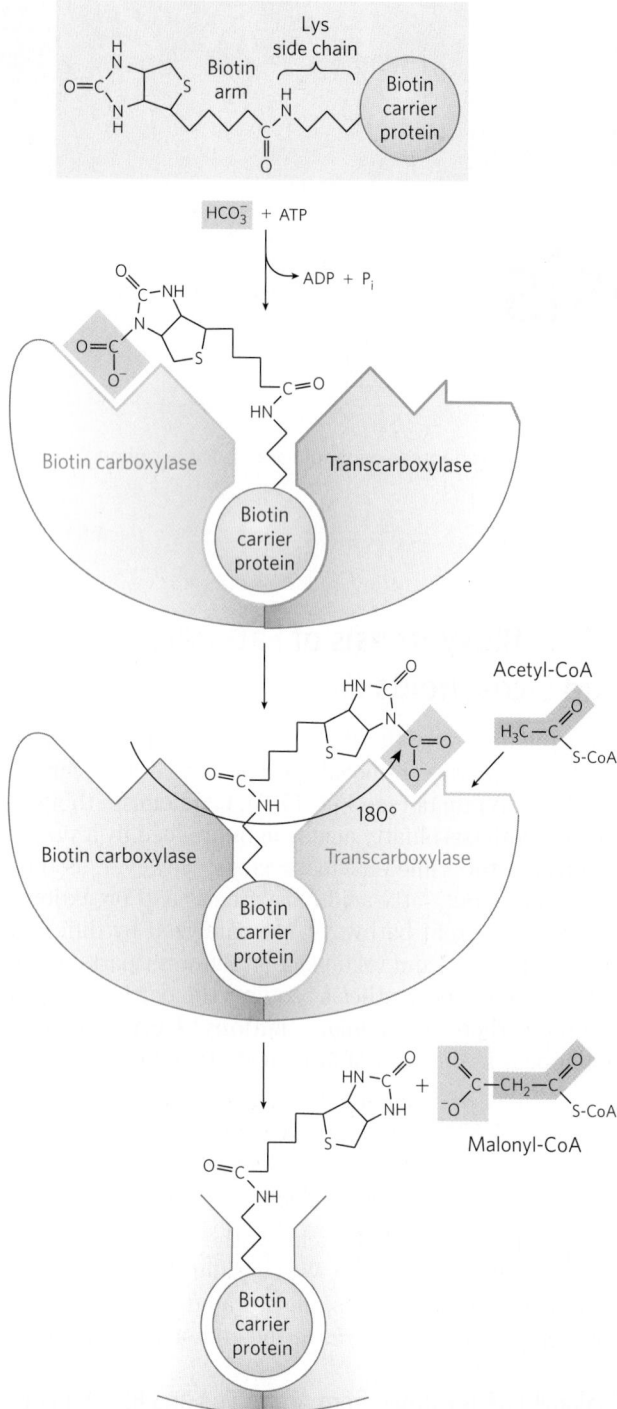

FIGURE 21-1 The acetyl-CoA carboxylase reaction. Acetyl-CoA carboxylase has three functional regions: biotin carrier protein (gray); biotin carboxylase, which activates CO_2 by attaching it to a nitrogen in the biotin ring in an ATP-dependent reaction (see Fig. 16-17); and transcarboxylase, which transfers activated CO_2 (shaded green) from biotin to acetyl-CoA, producing malonyl-CoA. Part of the biotin carrier protein and the long, flexible biotin arm rotate to carry the activated CO_2 from the biotin carboxylase active site to the transcarboxylase active site. The active enzyme in each step is shaded in blue.

to as **fatty acid synthase**. A saturated acyl group produced by each four-step series of reactions becomes the substrate for subsequent condensation with an activated malonyl group. With each passage through the cycle, the fatty acyl chain is extended by two carbons.

Both the electron-carrying cofactor and the activating groups in the reductive anabolic sequence differ from those in the oxidative catabolic process. Recall that in β oxidation, NAD^+ and FAD serve as electron acceptors and the activating group is the thiol (—SH) group of coenzyme A (see Fig. 17-8). By contrast, the reducing agent in the synthetic sequence is NADPH and the activating groups are two different enzyme-bound —SH groups, as described below.

There are two major variants of fatty acid synthase: fatty acid synthase I (FAS I), found in vertebrates and fungi, and fatty acid synthase II (FAS II), found in plants and bacteria. The FAS I found in vertebrates consists of a single multifunctional polypeptide chain (M_r 240,000). The mammalian FAS I is the prototype. Seven active sites for different reactions lie in separate domains **(Fig. 21-3a)**. The mammalian polypeptide functions as a homodimer (M_r 480,000). The subunits seem to function independently. When all the active sites in one subunit are inactivated by mutation, fatty acid synthesis is only moderately reduced. A somewhat different FAS I is found in yeast and other fungi, and is made up of two multifunctional polypeptides that form a complex with an architecture distinct from that of the vertebrate systems (Fig. 21-3b). Three of the seven required active sites are found on the α subunit and four on the β subunit.

With FAS I systems, fatty acid synthesis leads to a single product, and no intermediates are released. When the chain length reaches 16 carbons, that product (palmitate, 16:0; see Table 10-1) leaves the cycle. Carbons C-16 and C-15 of the palmitate are derived from the methyl and carboxyl carbon atoms, respectively, of an acetyl-CoA used directly to prime the system at the outset **(Fig. 21-4)**; the rest of the carbon atoms in the chain are derived from acetyl-CoA via malonyl-CoA.

FAS II, in plants and bacteria, is a dissociated system; each step in the synthesis is catalyzed by a separate and freely diffusible enzyme. Intermediates are also diffusible and may be diverted into other pathways (such as lipoic acid synthesis). Unlike FAS I, FAS II generates a variety of products, including saturated fatty acids of several lengths, as well as unsaturated, branched, and hydroxy fatty acids. An FAS II system is also found in vertebrate mitochondria. The discussion here focuses on the mammalian FAS I.

biotin-dependent carboxylation reactions, such as those catalyzed by pyruvate carboxylase (see Fig. 16-17) and propionyl-CoA carboxylase (see Fig. 17-12). A carboxyl group, derived from bicarbonate (HCO_3^-), is first transferred to biotin in an ATP-dependent reaction. The biotinyl group serves as a temporary carrier of CO_2, transferring it to acetyl-CoA in the second step to yield malonyl-CoA.

Fatty Acid Synthesis Proceeds in a Repeating Reaction Sequence

In all organisms, the long carbon chains of fatty acids are assembled in a repeating four-step sequence **(Fig. 21-2)**, catalyzed by a system collectively referred

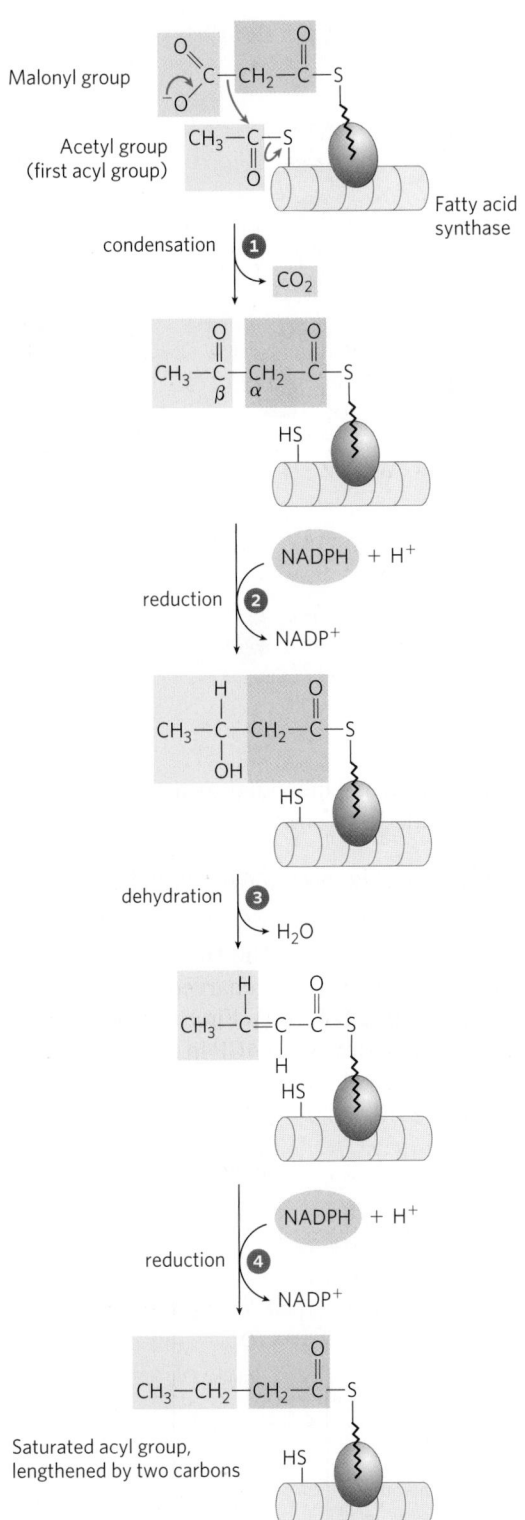

FIGURE 21-2 Addition of two carbons to a growing fatty acyl chain: a four-step sequence. Each malonyl group and acetyl (or longer acyl) group is activated by a thioester that links it to fatty acid synthase, a multienzyme system described below in the text. ❶ Condensation of an activated acyl group (an acetyl group from acetyl-CoA is the first acyl group) and two carbons derived from malonyl-CoA, with elimination of CO_2 from the malonyl group, extends the acyl chain by two carbons. The mechanism of the first step of this reaction is given to illustrate the role of decarboxylation in facilitating condensation. The β-keto product of the condensation is then reduced in three more steps nearly identical to the reactions of β oxidation, but in the reverse sequence: ❷ the β-keto group is reduced to an alcohol, ❸ elimination of H_2O creates a double bond, and ❹ the double bond is reduced to form the corresponding saturated fatty acyl group.

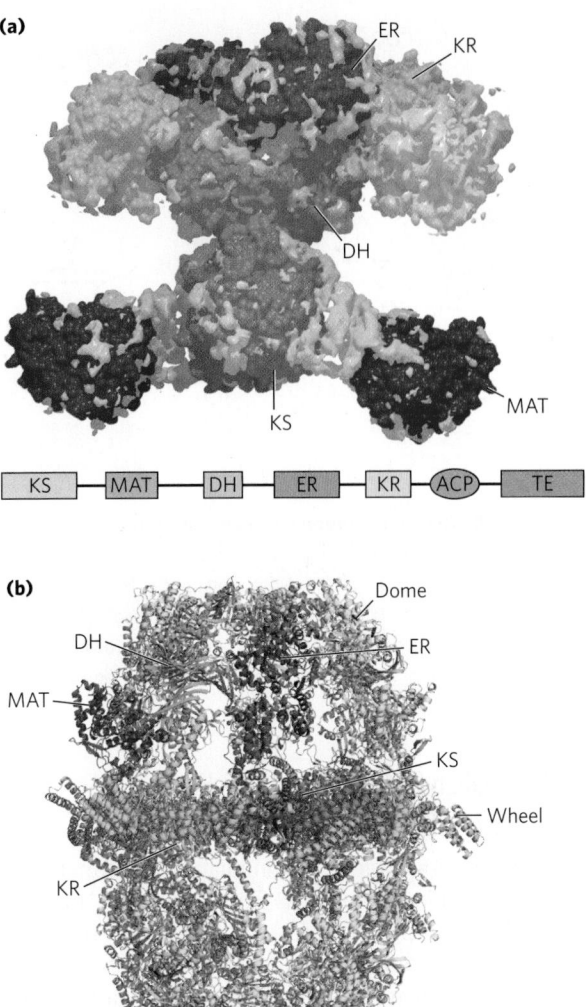

FIGURE 21-3 The structure of fatty acid synthase type I systems. Shown here are low-resolution structures of (a) the mammalian (porcine) and (b) fungal enzyme systems. **(a)** All of the active sites in the mammalian system are located in different domains within a single large polypeptide chain. The different enzymatic activities are: β-ketoacyl-ACP synthase (KS), malonyl/acetyl-CoA–ACP transferase (MAT), β-hydroxyacyl-ACP dehydratase (DH), enoyl-ACP reductase (ER), and β-ketoacyl-ACP reductase (KR). ACP is the acyl carrier protein. The linear arrangement of the domains in the polypeptide is shown below the structure. The seventh domain is a thioesterase (TE) that releases the palmitate product from ACP when synthesis is completed. The ACP and TE domains are disordered in the crystal and are therefore not shown in the structure.

(b) In FAS I from the fungus *Thermomyces lanuginosus*, the same active sites are divided between two multifunctional polypeptide chains that function together. Six copies of each polypeptide are found in the heterododecameric complex. A wheel of six α subunits, which include ACP and the KS and KR active sites, is at the center of the complex. In the wheel, three subunits are located on one face, three on the other. On either side of the wheel are domes formed by trimers of the β subunits, containing the ER and DH active sites, as well as two domains with active sites analogous to MAT in the mammalian enzyme. The domains of one of each type of subunit are colored according to the active site colors of the mammalian enzyme in (a). [Sources: (a) Dimer derived from PDB ID 2CF2, T. Maier et al., *Science* 311:1258, 2006. (b) Derived from PDB IDs 2UV9, 2UVA, 2UVB, and 2UVC, S. Jenni et al., *Science* 316:254, 2007.]

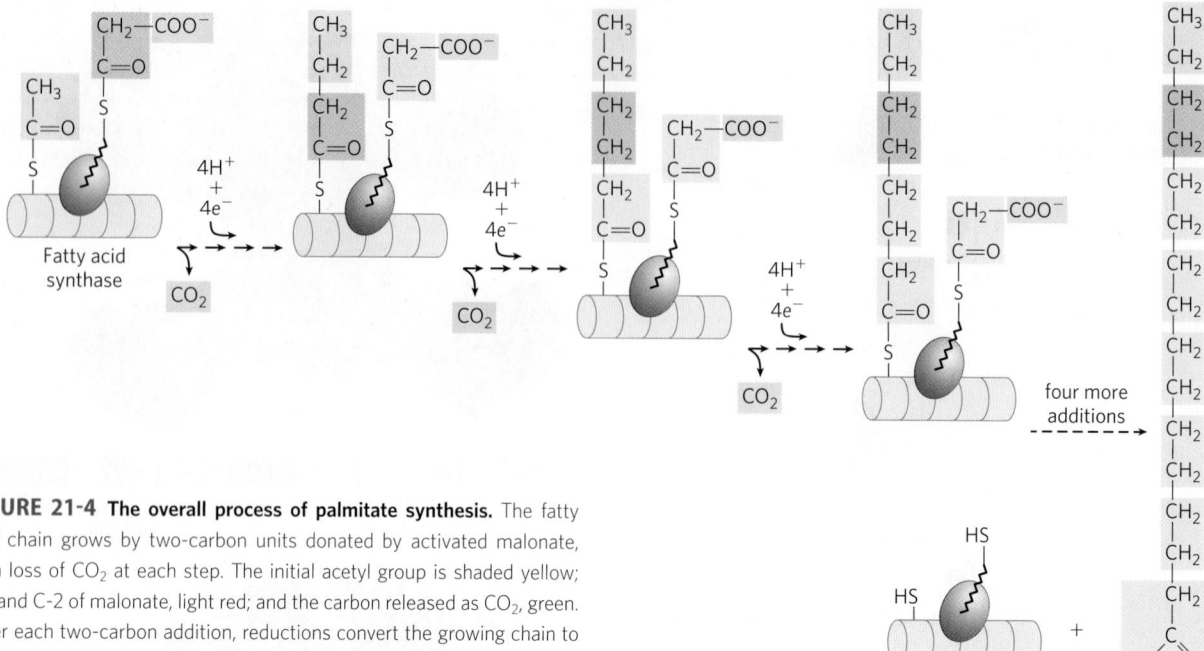

FIGURE 21-4 The overall process of palmitate synthesis. The fatty acyl chain grows by two-carbon units donated by activated malonate, with loss of CO_2 at each step. The initial acetyl group is shaded yellow; C-1 and C-2 of malonate, light red; and the carbon released as CO_2, green. After each two-carbon addition, reductions convert the growing chain to a saturated fatty acid of four, then six, then eight carbons, and so on. The final product is palmitate (16:0).

The Mammalian Fatty Acid Synthase Has Multiple Active Sites

The multiple domains of mammalian FAS I function as distinct but linked enzymes. The active site for each enzyme is found in a separate domain within the larger polypeptide. Throughout the process of fatty acid synthesis, the intermediates remain covalently attached as thioesters to one of two thiol groups. One point of attachment is the —SH group of a Cys residue in one of the synthase domains (β-ketoacyl-ACP synthase; KS); the other is the —SH group of acyl carrier protein, a separate domain of the same polypeptide. Hydrolysis of thioesters is highly exergonic, and the energy released helps to make two steps (❶ and ❺ in Fig. 21-6) in fatty acid synthesis thermodynamically favorable.

Acyl carrier protein (ACP) is the shuttle that holds the system together. The *Escherichia coli* ACP is a small protein (M_r 8,860) containing the prosthetic group **4′-phosphopantetheine** (**Fig. 21-5**; compare this with the panthothenic acid and β-mercaptoethylamine moiety of coenzyme A in Fig. 8-41). The 4′-phosphopantetheine prosthetic group of *E. coli* ACP is believed to serve as a flexible arm, tethering the growing fatty acyl chain to the surface of the fatty acid synthase complex while carrying the reaction intermediates from one enzyme active site to the next. The ACP of mammals has a similar function and the same prosthetic group; as we have

seen, however, it is embedded as a domain in a much larger multifunctional polypeptide.

Fatty Acid Synthase Receives the Acetyl and Malonyl Groups

Before the condensation reactions that build up the fatty acid chain can begin, the two thiol groups on the enzyme complex must be charged with the correct acyl groups (**Fig. 21-6**, top). First, the acetyl group of acetyl-CoA is transferred to ACP in a reaction catalyzed

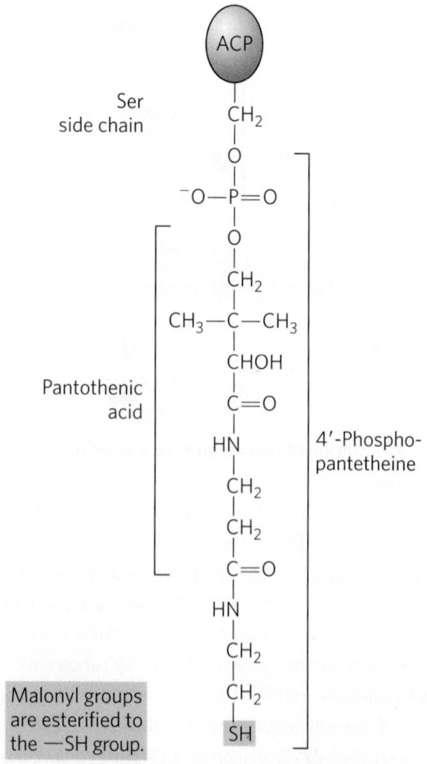

FIGURE 21-5 Acyl carrier protein (ACP). The prosthetic group is 4′-phosphopantetheine, which is covalently attached to the hydroxyl group of a Ser residue in ACP. Phosphopantetheine contains the B vitamin pantothenic acid, also found in the coenzyme A molecule. Its —SH group is the site of entry of malonyl groups during fatty acid synthesis.

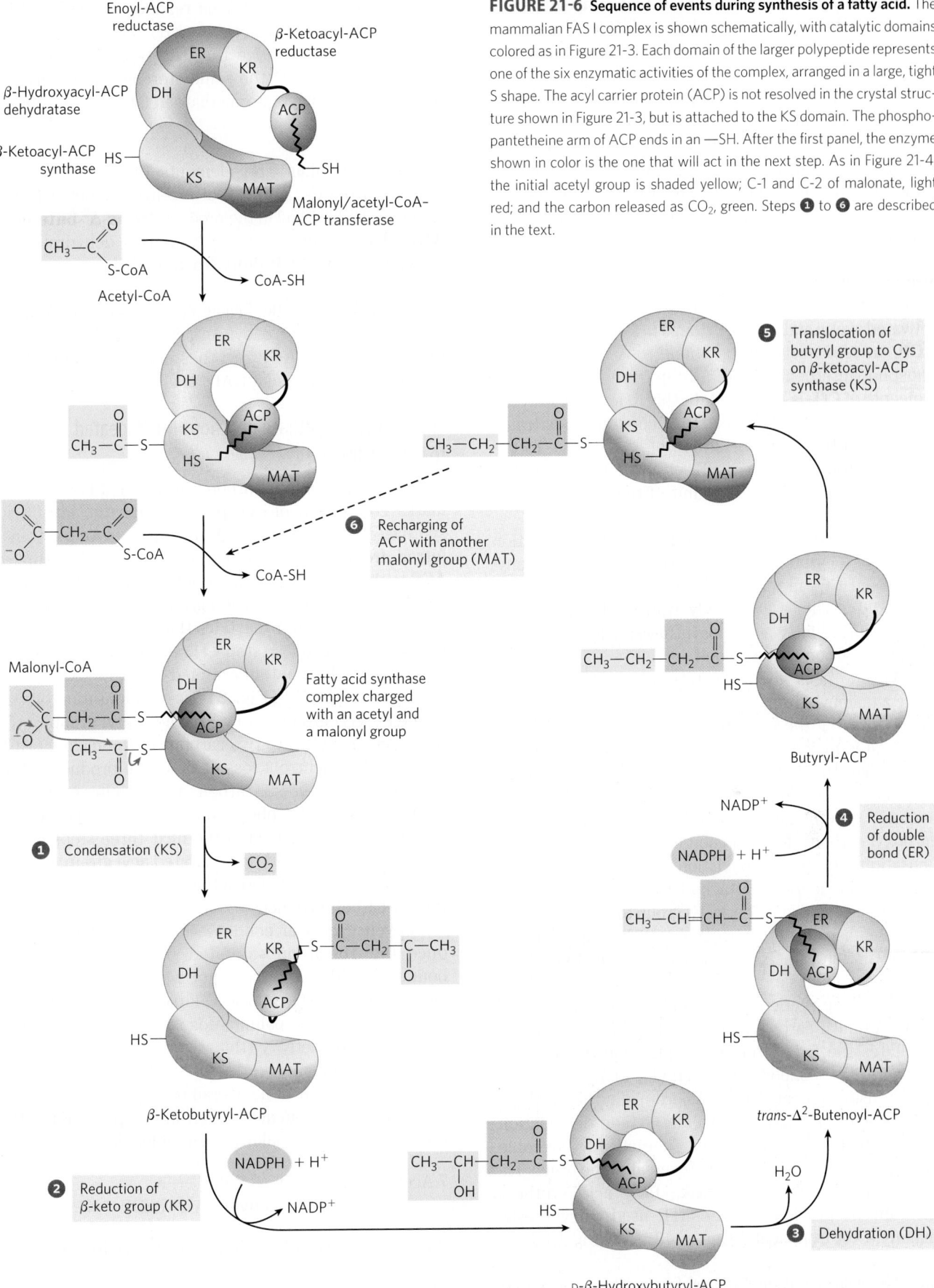

FIGURE 21-6 Sequence of events during synthesis of a fatty acid. The mammalian FAS I complex is shown schematically, with catalytic domains colored as in Figure 21-3. Each domain of the larger polypeptide represents one of the six enzymatic activities of the complex, arranged in a large, tight S shape. The acyl carrier protein (ACP) is not resolved in the crystal structure shown in Figure 21-3, but is attached to the KS domain. The phosphopantetheine arm of ACP ends in an —SH. After the first panel, the enzyme shown in color is the one that will act in the next step. As in Figure 21-4, the initial acetyl group is shaded yellow; C-1 and C-2 of malonate, light red; and the carbon released as CO_2, green. Steps ❶ to ❻ are described in the text.

Enoyl-ACP reductase

β-Ketoacyl-ACP reductase

β-Hydroxyacyl-ACP dehydratase

β-Ketoacyl-ACP synthase

Malonyl/acetyl-CoA–ACP transferase

Acetyl-CoA

CoA-SH

Malonyl-CoA

CoA-SH

Fatty acid synthase complex charged with an acetyl and a malonyl group

❶ Condensation (KS) CO_2

β-Ketobutyryl-ACP

❷ Reduction of β-keto group (KR)   NADPH + H⁺   NADP⁺

D-β-Hydroxybutyryl-ACP

❸ Dehydration (DH) H_2O

trans-Δ²-Butenoyl-ACP

❹ Reduction of double bond (ER)   NADPH + H⁺   NADP⁺

Butyryl-ACP

❺ Translocation of butyryl group to Cys on β-ketoacyl-ACP synthase (KS)

❻ Recharging of ACP with another malonyl group (MAT)

by the **malonyl/acetyl-CoA–ACP transferase** (MAT) domain of the multifunctional polypeptide. The acetyl group is then transferred to the Cys —SH group of the **β-ketoacyl-ACP synthase** (KS). The second reaction, transfer of the malonyl group from malonyl-CoA to the —SH group of ACP, is also catalyzed by malonyl/acetyl-CoA–ACP transferase. In the charged synthase complex, the acetyl and malonyl groups are activated for the chain-lengthening process. We now consider the first four steps of this process in some detail; all step numbers refer to Figure 21-6.

Step ❶ Condensation The first reaction in the formation of a fatty acid chain is a formal Claisen condensation of the activated acetyl and malonyl groups to form **acetoacetyl-ACP**, an acetoacetyl group bound to ACP through the phosphopantetheine —SH group; simultaneously, a molecule of CO_2 is produced. In this reaction, catalyzed by β-ketoacyl-ACP synthase, the acetyl group is transferred from the Cys —SH group of the enzyme to the malonyl group on the —SH of ACP, becoming the methyl-terminal two-carbon unit of the new acetoacetyl group.

The carbon atom of the CO_2 formed in this reaction is the same carbon originally introduced into malonyl CoA from HCO_3^- in the acetyl-CoA carboxylase reaction (Fig. 21-1). Thus CO_2 is only transiently in covalent linkage during fatty acid biosynthesis; it is removed as each two-carbon unit is added.

Why do cells go to the trouble of adding CO_2 to make a malonyl group from an acetyl group, only to lose the CO_2 during the formation of acetoacetate? The use of activated malonyl groups rather than acetyl groups is what makes the condensation reactions thermodynamically favorable. The methylene carbon (C-2) of the malonyl group, sandwiched between carbonyl and carboxyl carbons, forms a good nucleophile. In the condensation step, decarboxylation of the malonyl group facilitates nucleophilic attack of the methylene carbon on the thioester linking the acetyl group to β-ketoacyl-ACP synthase, displacing the enzyme's —SH group. (This is a classic Claisen ester condensation; see Fig. 13-4.) Coupling the condensation to the decarboxylation of the malonyl group renders the overall process highly exergonic. A similar carboxylation-decarboxylation sequence facilitates the formation of phosphoenolpyruvate from pyruvate in gluconeogenesis (see Fig. 14-18).

By using activated malonyl groups in the synthesis of fatty acids and activated acetate in their degradation, the cell makes both processes energetically favorable, although one is effectively the reversal of the other. The extra energy required to make fatty acid synthesis favorable is provided by the ATP used to synthesize malonyl-CoA from acetyl-CoA and HCO_3^- (Fig. 21-1).

Step ❷ Reduction of the Carbonyl Group The acetoacetyl-ACP formed in the condensation step now undergoes reduction of the carbonyl group at C-3 to form D-β-hydroxybutyryl-ACP. This reaction is catalyzed by **β-ketoacyl-ACP reductase** (KR), and the electron donor is NADPH. Notice that the D-β-hydroxybutyryl group does not have the same stereoisomeric form as the L-β-hydroxyacyl intermediate in fatty acid oxidation (see Fig. 17-8).

Step ❸ Dehydration The elements of water are now removed from C-2 and C-3 of D-β-hydroxybutyryl-ACP to yield a double bond in the product, **trans-Δ^2-butenoyl-ACP**. The enzyme that catalyzes this dehydration is **β-hydroxyacyl-ACP dehydratase** (DH).

Step ❹ Reduction of the Double Bond Finally, the double bond of trans-Δ^2-butenoyl-ACP is reduced (saturated) to form **butyryl-ACP** by the action of **enoyl-ACP reductase** (ER); again, NADPH is the electron donor.

The Fatty Acid Synthase Reactions Are Repeated to Form Palmitate

Production of the four-carbon, saturated fatty acyl–ACP marks completion of one pass through the fatty acid synthase complex. In step ❺, the butyryl group is transferred from the phosphopantetheine —SH group of ACP to the Cys —SH group of β-ketoacyl-ACP synthase, which initially bore the acetyl group (Fig. 21-6). To start the next cycle of four reactions that lengthens the chain by two more carbons (step ❻), another malonyl group is linked to the now unoccupied phosphopantetheine —SH group of ACP **(Fig. 21-7)**. Condensation occurs as the butyryl group, acting like the acetyl group in the first cycle, is linked to two carbons of the malonyl-ACP group with concurrent loss of CO_2. The product of this condensation is a six-carbon acyl group, covalently bound to the phosphopantetheine —SH group. Its β-keto group is reduced in the next three steps of the synthase cycle to yield the saturated acyl group, exactly as in the first round of reactions—in this case forming the six-carbon product.

Seven cycles of condensation and reduction produce the 16-carbon saturated palmitoyl group, still bound to ACP. For reasons not well understood, chain elongation by the synthase complex generally stops at this point, and free palmitate is released from the ACP by a hydrolytic activity (thioesterase; TE) in the multifunctional protein.

We can consider the overall reaction for the synthesis of palmitate from acetyl-CoA in two parts. First, the formation of seven malonyl-CoA molecules:

$$7 \text{ Acetyl-CoA} + 7CO_2 + 7ATP \longrightarrow$$
$$7 \text{ malonyl-CoA} + 7ADP + 7P_i \quad (21\text{-}1)$$

then seven cycles of condensation and reduction:

$$\text{Acetyl-CoA} + 7 \text{ malonyl-CoA} + 14NADPH + 14H^+ \longrightarrow$$
$$\text{palmitate} + 7CO_2 + 8CoA + 14NADP^+ + 6H_2O$$
$$(21\text{-}2)$$

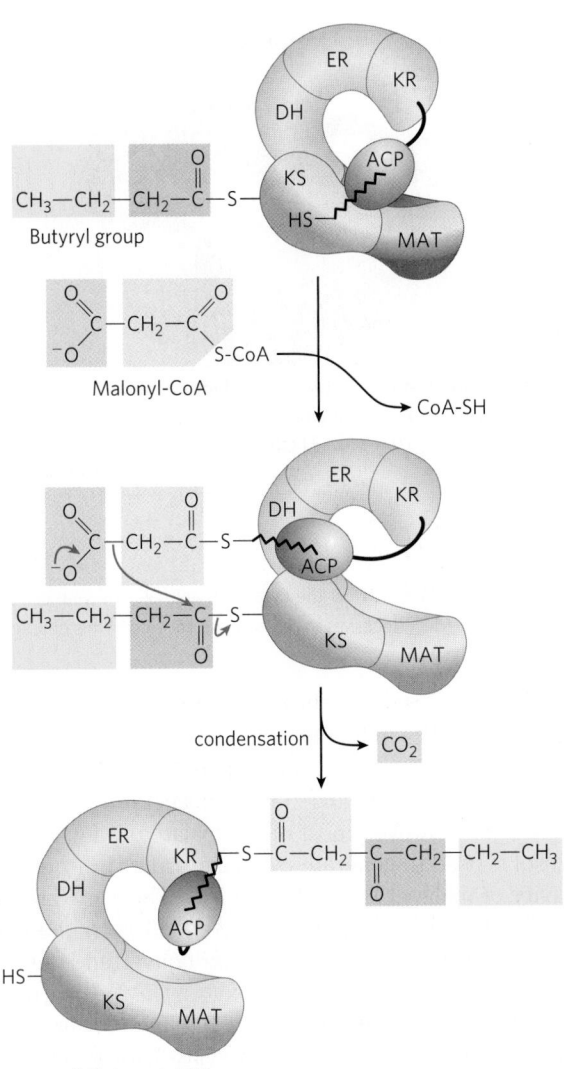

FIGURE 21-7 Beginning of the second round of the fatty acid synthesis cycle. The butyryl group is on the Cys —SH group. The incoming malonyl group is first attached to the phosphopantetheine —SH group. Then, in the condensation step, the entire butyryl group on the Cys —SH is exchanged for the carboxyl group of the malonyl residue, which is lost as CO_2 (green). This step is analogous to step ❶ in Figure 21-6. The product, a six-carbon β-ketoacyl group, now contains four carbons derived from malonyl-CoA and two derived from the acetyl-CoA that started the reaction. The β-ketoacyl group then undergoes steps ❷ through ❹ in Figure 21-6.

Notice that only six net water molecules are produced, because one is used to hydrolyze the thioester linking the palmitate product to the enzyme. The overall process (the sum of Eqns 21-1 and 21-2) is

$$8 \text{ Acetyl-CoA} + 7\text{ATP} + 14\text{NADPH} + 14\text{H}^+ \longrightarrow$$
$$\text{palmitate} + 8\text{CoA} + 7\text{ADP} + 7\text{P}_i + 14\text{NADP}^+ + 6\text{H}_2\text{O}$$
$$(21\text{-}3)$$

The biosynthesis of fatty acids such as palmitate thus requires acetyl-CoA and the input of chemical energy in two forms: the group transfer potential of ATP and the reducing power of NADPH. The ATP is required to attach CO_2 to acetyl-CoA to make malonyl-CoA; the NADPH molecules are required to reduce the β-keto group and the double bond.

In nonphotosynthetic eukaryotes there is an additional cost to fatty acid synthesis, because acetyl-CoA is generated in the mitochondria and must be transported to the cytosol. As we will see, this extra step consumes two ATP per molecule of acetyl-CoA transported, increasing the energetic cost of fatty acid synthesis to three ATP per two-carbon unit.

Fatty Acid Synthesis Is a Cytosolic Process in Many Organisms but Takes Place in the Chloroplasts in Plants

In most higher eukaryotes, the fatty acid synthase complex is found exclusively in the cytosol **(Fig. 21-8)**, as are the biosynthetic enzymes for nucleotides, amino acids, and glucose. This location segregates synthetic processes from degradative reactions, many of which take place in the mitochondrial matrix. There is a corresponding segregation of the electron-carrying cofactors used in anabolism (generally a reductive process) and those used in catabolism (generally oxidative).

Usually, NADPH is the electron carrier for anabolic reactions, and NAD^+ serves in catabolic reactions. In hepatocytes, the $[NADPH]/[NADP^+]$ ratio is very high (~75) in the cytosol, furnishing a strongly reducing environment for the reductive synthesis of fatty acids and other biomolecules. The cytosolic $[NADH]/[NAD^+]$ ratio is much smaller (~8×10^{-4}), so the NAD^+-dependent oxidative catabolism of glucose can take place in the same compartment, and at the same time, as fatty acid synthesis. The $[NADH]/[NAD^+]$ ratio in the mitochondrion is much higher than that in the cytosol, because of the flow of electrons to NAD^+ from the oxidation of fatty acids, amino acids, pyruvate, and acetyl-CoA. This high mitochondrial $[NADH]/[NAD^+]$ ratio favors the reduction of oxygen via the respiratory chain.

In hepatocytes and adipocytes, cytosolic NADPH is largely generated by the pentose phosphate pathway (see Fig. 14-22) and by **malic enzyme (Fig. 21-9a)**. Note that the NADP-linked malic enzyme that operates in the carbon-assimilation pathway of C_4 plants (see Fig. 20-50) is unrelated in function. The pyruvate produced in the reaction shown in Figure 21-9a reenters the mitochondrion. In hepatocytes and in the mammary gland of lactating animals, the NADPH required for fatty acid biosynthesis is supplied primarily by the pentose phosphate pathway (Fig. 21-9b).

In the photosynthetic cells of plants, fatty acid synthesis occurs not in the cytosol but in the chloroplast stroma (Fig. 21-8). This makes sense, given that NADPH is produced in chloroplasts by the light-dependent reactions of photosynthesis:

$$\text{H}_2\text{O} + \text{NADP}^+ \xrightarrow{\text{light}} \tfrac{1}{2}\text{O}_2 + \text{NADPH} + \text{H}^+$$

Acetate Is Shuttled out of Mitochondria as Citrate

In nonphotosynthetic eukaryotes, nearly all the acetyl-CoA used in fatty acid synthesis is formed in mitochondria

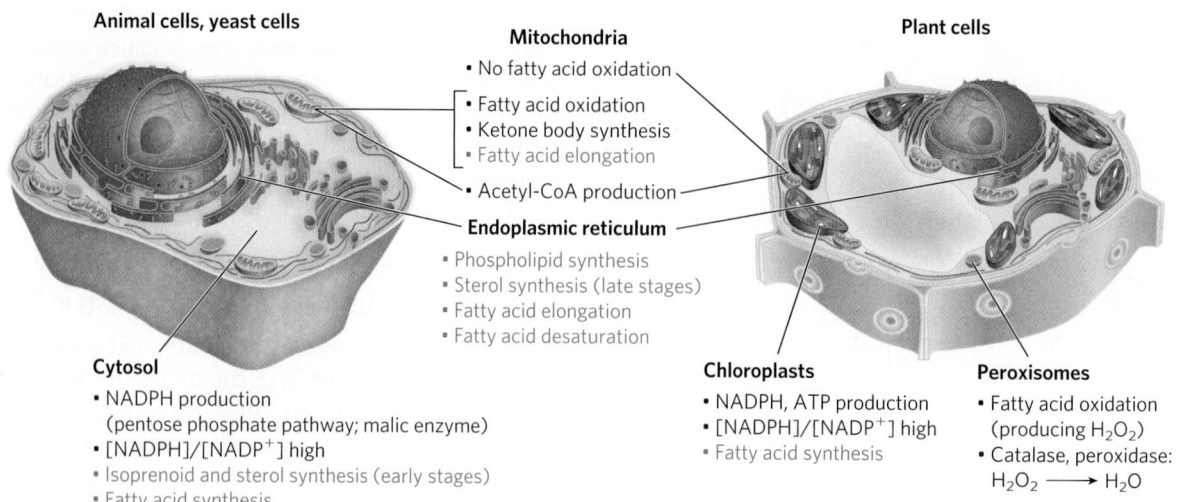

Animal cells, yeast cells

Cytosol
- NADPH production
  (pentose phosphate pathway; malic enzyme)
- [NADPH]/[NADP⁺] high
- Isoprenoid and sterol synthesis (early stages)
- Fatty acid synthesis

Mitochondria
- No fatty acid oxidation
- Fatty acid oxidation
- Ketone body synthesis
- Fatty acid elongation
- Acetyl-CoA production

Endoplasmic reticulum
- Phospholipid synthesis
- Sterol synthesis (late stages)
- Fatty acid elongation
- Fatty acid desaturation

Plant cells

Chloroplasts
- NADPH, ATP production
- [NADPH]/[NADP⁺] high
- Fatty acid synthesis

Peroxisomes
- Fatty acid oxidation
  (producing H_2O_2)
- Catalase, peroxidase:
  $H_2O_2 \longrightarrow H_2O$

FIGURE 21-8 Subcellular localization of lipid metabolism. Yeast and animal cells differ from higher plant cells in the compartmentation of lipid metabolism. Fatty acid synthesis takes place in the compartment in which NADPH is available for reductive synthesis (i.e., where the [NADPH]/[NADP⁺] ratio is high); this is the cytosol in animals and yeast, and the chloroplast in plants. Processes in red type are covered in this chapter.

from pyruvate oxidation and from catabolism of the carbon skeletons of amino acids. Acetyl-CoA arising from the oxidation of fatty acids is not a significant source of acetyl-CoA for fatty acid biosynthesis in animals, because the two pathways are reciprocally regulated, as described below.

The inner mitochondrial membrane is impermeable to acetyl-CoA, so an indirect shuttle transfers acetyl group equivalents across the membrane **(Fig. 21-10)**. Intramitochondrial acetyl-CoA first reacts with oxaloacetate to form citrate, in the citric acid cycle reaction catalyzed by **citrate synthase** (see Fig. 16-9). Citrate then passes through the inner membrane on the **citrate transporter**. In the cytosol, citrate cleavage by **citrate lyase** regenerates acetyl-CoA and oxaloacetate in an ATP-dependent reaction. Oxaloacetate cannot return to the mitochondrial matrix directly, as there is no oxaloacetate transporter. Instead, cytosolic malate

dehydrogenase reduces the oxaloacetate to malate, which can return to the mitochondrial matrix on the malate–α-ketoglutarate transporter, in exchange for citrate. In the matrix, malate is reoxidized to oxaloacetate to complete the shuttle. However, most of the malate produced in the cytosol is used to generate cytosolic NADPH through the activity of malic enzyme (Fig. 21-9a). The pyruvate produced is transported into the mitochondria by the pyruvate transporter (Fig. 21-10), then converted back into oxaloacetate by pyruvate carboxylase in the matrix. In the resulting cycle, two ATP molecules are consumed (by citrate lyase and pyruvate carboxylase) for every molecule of acetyl-CoA delivered to fatty acid synthesis. After citrate cleavage to generate acetyl-CoA, conversion of the four remaining carbons to pyruvate and CO_2 by malic enzyme generates about half the NADPH required for fatty acid synthesis. The pentose phosphate pathway contributes the rest of the needed NADPH.

Fatty Acid Biosynthesis Is Tightly Regulated

When a cell or organism has more than enough metabolic fuel to meet its energy needs, the excess is generally converted to fatty acids and stored as lipids such as triacylglycerols. The reaction catalyzed by acetyl-CoA carboxylase is the rate-limiting step in the biosynthesis of fatty acids, and this enzyme is an important site of regulation. In vertebrates, palmitoyl-CoA, the principal product of fatty acid synthesis, is a feedback inhibitor of the enzyme; citrate is an allosteric activator **(Fig. 21-11a)**, increasing V_{max}. Citrate plays a central role in diverting cellular metabolism from the consumption (oxidation) of metabolic fuel to the storage of fuel as fatty acids. When the concentrations of mitochondrial acetyl-CoA and ATP increase, citrate is transported out of mitochondria; it then becomes both the precursor of cytosolic acetyl-CoA and an allosteric signal for the activation of acetyl-CoA

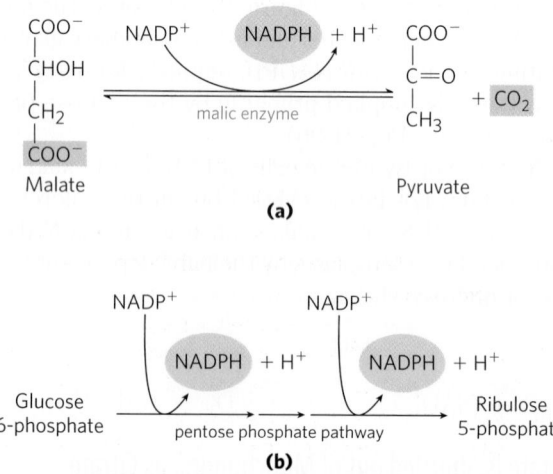

FIGURE 21-9 Production of NADPH. Two routes to NADPH, catalyzed by **(a)** malic enzyme and **(b)** the pentose phosphate pathway.

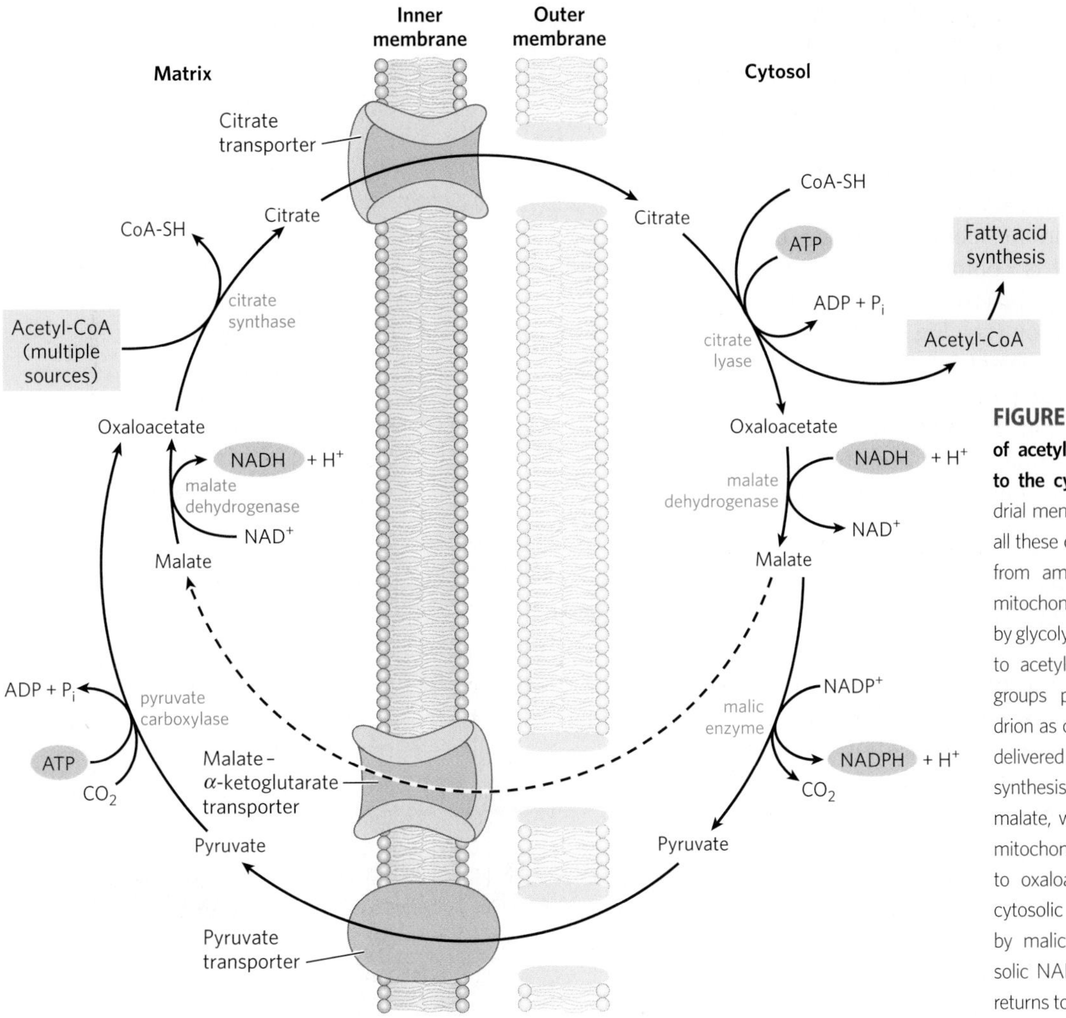

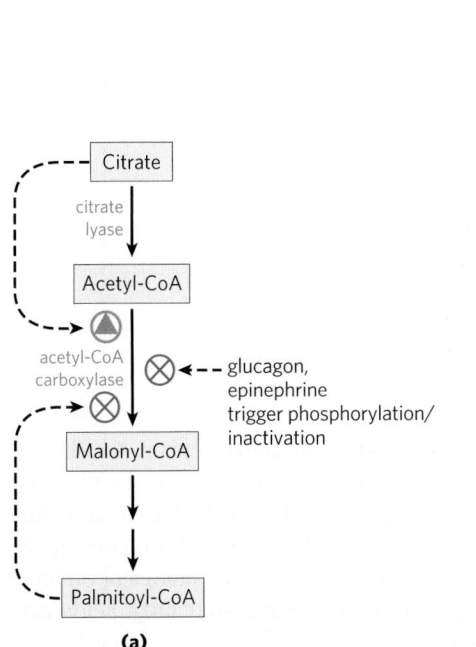

FIGURE 21-10 Shuttle for transfer of acetyl groups from mitochondria to the cytosol. The outer mitochondrial membrane is freely permeable to all these compounds. Pyruvate derived from amino acid catabolism in the mitochondrial matrix, or from glucose by glycolysis in the cytosol, is converted to acetyl-CoA in the matrix. Acetyl groups pass out of the mitochondrion as citrate; in the cytosol they are delivered as acetyl-CoA for fatty acid synthesis. Oxaloacetate is reduced to malate, which can be returned to the mitochondrial matrix and converted to oxaloacetate. The major fate for cytosolic malate, however, is oxidation by malic enzyme to generate cytosolic NADPH; the pyruvate produced returns to the mitochondrial matrix.

carboxylase. At the same time, citrate inhibits the activity of phosphofructokinase-1 (see Fig. 15-16), reducing the flow of carbon through glycolysis.

Acetyl-CoA carboxylase is also regulated by covalent modification. Phosphorylation, triggered by the hormones glucagon and epinephrine, inactivates the enzyme and reduces its sensitivity to activation by citrate, thereby slowing fatty acid synthesis. In its active (dephosphorylated) form, acetyl-CoA carboxylase polymerizes into long filaments (Fig. 21-11b); phosphorylation is accompanied by dissociation into monomeric subunits and loss of activity.

FIGURE 21-11 Regulation of fatty acid synthesis. (a) In the cells of vertebrates, both allosteric regulation and hormone-dependent covalent modification influence the flow of precursors into malonyl-CoA. In plants, acetyl-CoA carboxylase is activated by the changes in [Mg^{2+}] and pH that accompany illumination (not shown here). **(b)** Filaments of acetyl-CoA carboxylase from chicken hepatocytes (the active, dephosphorylated form), as seen with the electron microscope. [Source: (b) Courtesy James M. Ntambi, PhD, Professor of Biochemistry, Steenbock Professor of Nutritional Sciences, University of Wisconsin–Madison.]

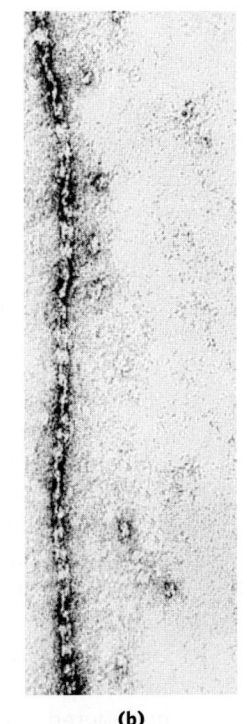

(a)   **(b)**

The acetyl-CoA carboxylase of plants and bacteria is not regulated by citrate or by a phosphorylation-dephosphorylation cycle. Instead, the plant enzyme is activated by an increase in stromal pH and $[Mg^{2+}]$, which occurs on illumination of the plant (see Fig. 20-44). Bacteria do not use triacylglycerols as energy stores. In *E. coli*, the primary role of fatty acid synthesis is to provide precursors for membrane lipids; the regulation of this process is complex, employing guanine nucleotides (such as ppGpp) that coordinate cell growth with membrane formation (see Fig. 8-42).

In addition to the moment-by-moment regulation of enzymatic activity, these pathways are regulated at the level of gene expression. For example, when animals ingest an excess of certain polyunsaturated fatty acids, the expression of genes encoding a wide range of lipogenic enzymes in the liver is suppressed. This gene regulation is mediated by a family of nuclear receptor proteins called PPARs, described in more detail in Chapter 23 (see Fig. 23-43).

If fatty acid synthesis and β oxidation were to proceed simultaneously, the two processes would constitute a futile cycle, wasting energy. We noted earlier (see Fig. 17-13) that β oxidation is blocked by malonyl-CoA, which inhibits carnitine acyltransferase I. Thus, during fatty acid synthesis, production of the first intermediate, malonyl-CoA, shuts down β oxidation at the level of a transport system in the inner mitochondrial membrane. This control mechanism illustrates another advantage of segregating synthetic and degradative pathways in different cellular compartments.

Long-Chain Saturated Fatty Acids Are Synthesized from Palmitate

Palmitate, the principal product of the fatty acid synthase system in animal cells, is the precursor of other long-chain fatty acids **(Fig. 21-12)**. It may be lengthened to form stearate (18:0) or even longer saturated fatty acids by further additions of acetyl groups, through the action of **fatty acid elongation systems** present in the smooth endoplasmic reticulum and in mitochondria. The more active elongation system of the ER extends the 16-carbon chain of palmitoyl-CoA by two carbons, forming stearoyl-CoA. Although different enzyme systems are used, and coenzyme A rather than ACP is the acyl carrier in the reaction, the mechanism of elongation in the ER is otherwise identical to that in palmitate synthesis: donation of two carbons by malonyl-CoA, followed by reduction, dehydration, and reduction to the saturated 18-carbon product, stearoyl-CoA.

Two key products of elongation pathways are linoleate, an omega-6 fatty acid (see Chapter 10 for the alternative nomenclature), and α-linolenate, an omega-3 fatty acid. These are precursors for two extensive families of derivative unsaturated fatty acids, the omega-6 and omega-3 families. Humans cannot synthesize linoleate and α-linolenate and must obtain them in the diet.

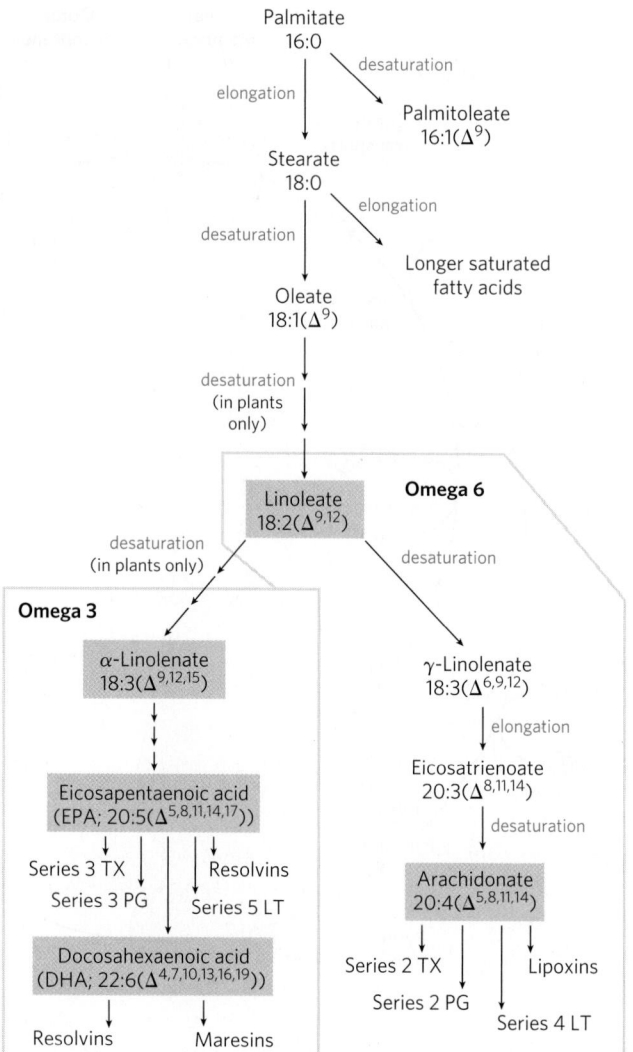

FIGURE 21-12 Routes of synthesis of unsaturated fatty acids and their derivatives. Palmitate is the precursor of stearate and longer-chain saturated fatty acids, as well as the monounsaturated acids palmitoleate and oleate. Mammals cannot convert oleate to linoleate or α-linolenate (shaded light red), which are therefore required in the diet as essential fatty acids. Conversion of linoleate to other polyunsaturated fatty acids and eicosanoids is outlined. Unsaturated fatty acids are symbolized by indicating the number of carbons and the number and position of double bonds, as in Table 10-1. Linoleate and α-linolenate are important omega-6 and omega-3 fatty acids, respectively; they are also precursors for a wide range of unsaturated fatty acids that act as signaling molecules. Two-letter abbreviations specify the eicosanoid prostaglandins (PG), thromboxanes (TX), and leukotrienes (LT). Particular classes of unsaturated fatty acids are further delineated by the number of double bonds, which defines subclasses referred to as series. For example, series 2 TX are thromboxanes with two double bonds in the hydrocarbon chain.

The ratio of omega-6 to omega-3 fatty acids in the diet, if too high, can lead to cardiovascular disease. The importance of this ratio may reflect the multitude of signaling molecules in the omega-6 and omega-3 families (Fig. 21-12), with their equally complex physiological effects. Several of these derivative unsaturated fatty acids are considered below.

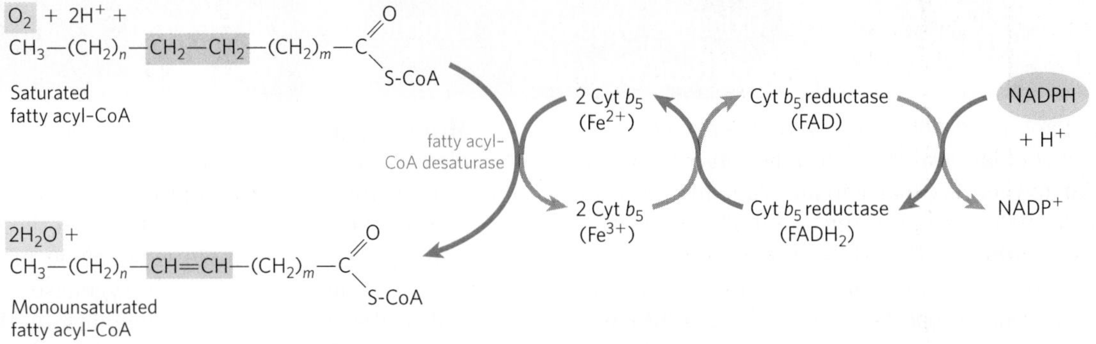

FIGURE 21-13 Electron transfer in the desaturation of fatty acids in vertebrates. Blue arrows show the path of electrons as two substrates—a fatty acyl-CoA and NADPH—undergo oxidation by molecular oxygen. These reactions take place on the lumenal face of the smooth ER. A similar pathway, but with different electron carriers, occurs in plants.

Desaturation of Fatty Acids Requires a Mixed-Function Oxidase

Palmitate and stearate serve as precursors of the two most common monounsaturated fatty acids of animal tissues: palmitoleate, $16:1(\Delta^9)$, and oleate, $18:1(\Delta^9)$; both of these fatty acids have a single cis double bond between C-9 and C-10 (see Table 10-1). The double bond is introduced into the fatty acid chain by an oxidative reaction catalyzed by **fatty acyl–CoA desaturase (Fig. 21-13)**, a **mixed-function oxidase** (Box 21-1). Two different substrates, the fatty acid and NADPH, simultaneously undergo two-electron oxidations. The path of electron flow includes a cytochrome (cytochrome b_5) and a flavoprotein (cytochrome b_5 reductase), both of which, like fatty acyl–CoA desaturase, are in the smooth ER. In plants, oleate is produced by a **stearoyl-ACP desaturase (SCD)** that uses reduced ferredoxin as the electron donor in the chloroplast stroma.

The SCD of animals (as studied in mice) has an important role in the development of obesity and the insulin resistance that often accompanies obesity and precedes development of type 2 diabetes mellitus. Mice have four isozymes, SCD1 through SCD4, of which SCD1 is the best understood. Its synthesis is induced by dietary saturated fatty acids, and also by the action of SREBP and LXR, two protein regulators of lipid metabolism that activate transcription of lipid-synthesizing enzymes (described in Section 21.4). Mice with mutant forms of SCD1 are resistant to diet-induced obesity and do not develop diabetes under conditions that cause both obesity and diabetes in mice with normal SCD1. ■

Mammalian hepatocytes can readily introduce double bonds at the Δ^9 position of fatty acids but cannot introduce additional double bonds between C-10 and the methyl-terminal end. Thus, as noted above, mammals cannot synthesize the omega-6 family precursor linoleate, $18:2(\Delta^{9,12})$, or the omega-3 family precursor α-linolenate, $18:3(\Delta^{9,12,15})$. Plants, however, can synthesize both; the desaturases that introduce double bonds at the Δ^{12} and Δ^{15} positions are located in the ER and in chloroplasts. The ER enzymes act not on free fatty acids but on a phospholipid, phosphatidylcholine, that contains at least one

oleate linked to the glycerol **(Fig. 21-14)**. Both plants and bacteria must synthesize polyunsaturated fatty acids to ensure membrane fluidity at reduced temperatures.

Because they are necessary precursors for the synthesis of other products, linoleate and α-linolenate are **essential fatty acids** for mammals; they must be

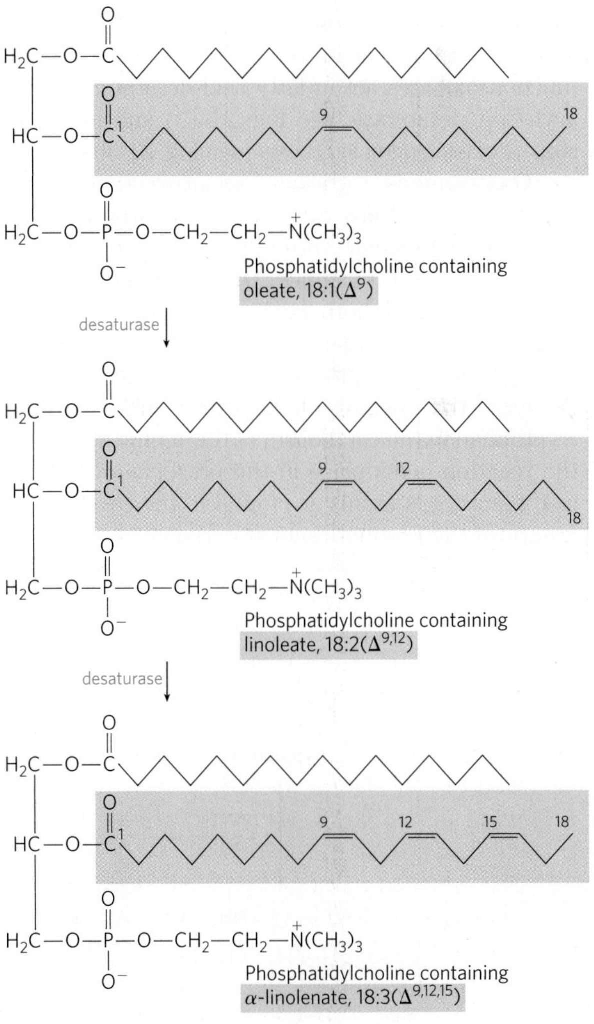

FIGURE 21-14 Action of plant desaturases. Desaturases in plants oxidize phosphatidylcholine-bound oleate to polyunsaturated fatty acids. Some of the products are released from the phosphatidylcholine by hydrolysis.

In this chapter we encounter several enzymes that carry out oxidation-reduction reactions in which molecular oxygen is a participant. The stearoyl-CoA desaturase (SCD) that introduces a double bond into a fatty acyl chain (see Fig. 21-13) is one such enzyme.

The nomenclature for enzymes that catalyze reactions of this general type can be confusing. **Oxidase** is the general name for enzymes that catalyze oxidations in which molecular oxygen is the electron acceptor but oxygen atoms do not appear in the oxidized product. The enzyme that creates a double bond in fatty acyl–CoA during the oxidation of fatty acids in peroxisomes (see Fig. 17-14) is an oxidase of this type; a second example is the cytochrome oxidase of the mitochondrial respiratory chain (see Fig. 19-13). In the first case, the transfer of two electrons to H_2O produces hydrogen peroxide, H_2O_2; in the second, two electrons reduce $\frac{1}{2}O_2$ to H_2O. Many, but not all, oxidases are flavoproteins. **Mixed-function oxidases** oxidize two different substrates simultaneously; again, the molecular oxygen atoms do not appear in the oxidized products. Mixed-function oxidases act in fatty acid desaturation (fatty acyl–CoA desaturase; see Fig. 21-13) and in the last step of plasmalogen synthesis (see Fig. 21-30).

Oxygenases catalyze oxidative reactions in which oxygen atoms *are* directly incorporated into the product molecule, forming a new hydroxyl or carboxyl group, for example. **Dioxygenases** catalyze reactions in which both oxygen atoms of O_2 are incorporated into the organic product. An example of a dioxygenase is tryptophan 2,3-dioxygenase, which catalyzes the opening of the five-membered ring of tryptophan in the catabolism of this amino acid. When the reaction takes place in the presence of $^{18}O_2$, the isotopic oxygen atoms are found in the two carbonyl groups of the product (shown in red):

Tryptophan

N-Formylkynurenine

Monooxygenases, more common and more complex in their action, catalyze reactions in which only one of the two oxygen atoms of O_2 is incorporated into the organic product, the other being reduced to H_2O; an example is squalene monooxygenase (see Fig. 21-37). Monooxygenases require two substrates to serve as reductants of the two oxygen atoms of O_2. The main substrate accepts one of the two oxygen atoms, and a cosubstrate furnishes hydrogen atoms to reduce the other oxygen atom to H_2O. The general reaction equation for monooxygenases is

$$AH + BH_2 + O-O \longrightarrow A-OH + B + H_2O$$

where AH is the main substrate and BH_2 the cosubstrate. Because most monooxygenases catalyze reactions in which the main substrate becomes hydroxylated, they are also called **hydroxylases**. They are also sometimes called **mixed-function oxygenases** to indicate that they oxidize two different substrates simultaneously.

Monooxygenases are divided into several classes, depending on the nature of the cosubstrate. Some use reduced flavin nucleotides ($FMNH_2$ or $FADH_2$), others use NADH or NADPH, and still others use α-ketoglutarate as cosubstrate. The enzyme that hydroxylates the phenyl ring of phenylalanine to form tyrosine is a monooxygenase that uses tetrahydrobiopterin as cosubstrate (see Fig. 18-23). (This is the enzyme that is defective in the human genetic disease phenylketonuria.)

The most numerous and most complex monooxygenation reactions are those employing a type of heme protein called **cytochrome P-450**. Like mitochondrial cytochrome oxidase, enzymes containing a cytochrome P-450 domain can react with O_2 and bind carbon monoxide, but they can be differentiated from cytochrome oxidase because the carbon monoxide complex of their reduced form absorbs light strongly at 450 nm—thus the name P-450.

Cytochrome P-450 enzymes catalyze hydroxylation reactions in which an organic substrate, RH, is hydroxylated to R—OH, incorporating one oxygen atom of O_2; the other oxygen atom is reduced to H_2O by reducing equivalents that are furnished by NADH or NADPH but are usually passed to cytochrome P-450 by an iron-sulfur protein. Figure 1 shows a simplified outline of the action of cytochrome P-450.

One large family of P-450–containing proteins consists of two general types: those highly specific for a single substrate (like typical enzymes) and those with more promiscuous binding sites that accept a

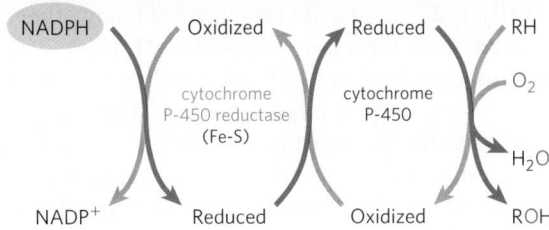

FIGURE 1 Simplified cytochrome P-450 reaction cycle.

variety of substrates, generally similar in being hydrophobic. In the adrenal cortex, for example, a specific cytochrome P-450 participates in the hydroxylation of steroids to yield the adrenocortical hormones (see Fig. 21-49). There are dozens of P-450 enzymes that act on specific substrates in the biosynthetic pathways to steroid hormones and eicosanoids (Fig. 2). Cytochrome P-450 enzymes with broader specificity are important in the hydroxylation of many different drugs, such as barbiturates and other xenobiotics (substances foreign to the organism), particularly if they are hydrophobic and relatively insoluble. The environmental carcinogen benzo[a]pyrene, found in cigarette smoke, undergoes cytochrome P-450–dependent hydroxylation during detoxification. Hydroxylation of xenobiotics, sometimes combined with the attachment of a polar compound such as glucuronic acid to the hydroxyl group, makes them more soluble in water and allows their excretion in urine. Hydroxylation (and glucuronidation) inactivates most drugs, and the rate at which it occurs can determine how long a given dose of a medication remains in the blood at therapeutic levels.

Humans differ in their levels of drug-metabolizing enzymes, both because of their genetics and because past exposure to substrates can induce the synthesis of higher levels of P-450 enzymes. Ethanol and barbiturate drugs share a P-450 enzyme. Long-term heavy drinking induces synthesis of this enzyme. Then, because the barbiturate is inactivated and cleared faster, larger doses are required to get the same therapeutic effect. If an individual takes this larger-than-usual dose of barbiturate and then also drinks alcohol, competition between the alcohol and the barbiturate for the limited amount of enzyme means that both alcohol and barbiturate are cleared more slowly. The resulting high levels of these two central nervous system depressants can be lethal. Similar complications arise when an individual takes two drugs that happen to be inactivated by the same P-450 enzyme; each drug increases the effective dose of the other by slowing its inactivation. It is therefore essential for physicians and pharmacists to know about all of a patient's prescribed and over-the-counter drugs and supplements, as well as a history of heavy drinking, or smoking, or exposure to environmental toxins.

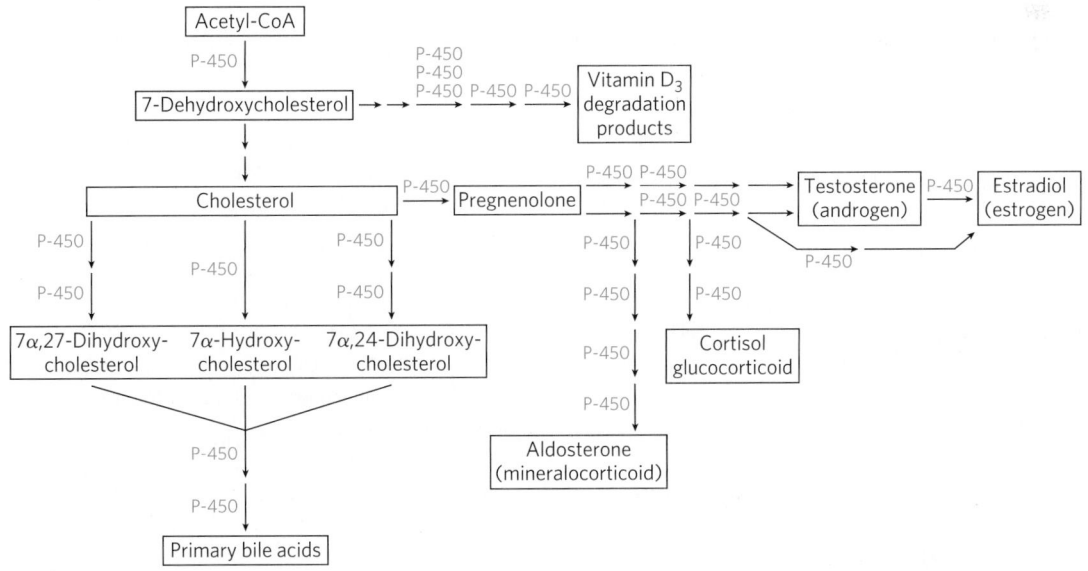

FIGURE 2 Pathways of sterol biosynthesis, showing the steps requiring cytochrome P-450 enzymes.

obtained from dietary plant material. Once ingested, linoleate may be converted to certain other polyunsaturated acids, particularly γ-linolenate, eicosatrienoate, and **arachidonate (eicosatetraenoate)**, all of which can be made only from linoleate (Fig. 21-12). Similarly, α-linolenate is converted to two important derivatives, **eicosapentaenoic acid (EPA)** and **docosahexaenoic acid (DHA)**. Arachidonate, $20:4(\Delta^{5,8,11,14})$, EPA, $20:5(\Delta^{5,8,11,14,17})$, and DHA, $22:6(\Delta^{4,7,10,13,16,19})$ are essential precursors of distinct classes of eicosanoids (Fig. 21-12), lipids with important regulatory functions. The 20- and 22-carbon fatty acids are synthesized from linoleate and α-linolenate by fatty acid elongation reactions analogous to those described on page 820.

Eicosanoids Are Formed from 20- and 22-Carbon Polyunsaturated Fatty Acids

Eicosanoids are a family of very potent biological signaling molecules that act as short-range messengers, affecting tissues near the cells that produce them.

In response to hormonal or other stimuli, phospholipase A_2, present in most types of mammalian cells, attacks membrane phospholipids, releasing arachidonate from the middle carbon of glycerol. Enzymes of the smooth ER then convert arachidonate to **prostaglandins**, beginning with the formation of prostaglandin H_2 (PGH$_2$), the immediate precursor of many other prostaglandins and thromboxanes **(Fig. 21-15a)**. The two reactions that lead to PGH$_2$ are catalyzed by a bifunctional enzyme, **cyclooxygenase (COX)**, also called **prostaglandin H$_2$ synthase**. In the first step, the cyclooxygenase activity introduces molecular oxygen to convert arachidonate to PGG$_2$. The second step, catalyzed by the peroxidase activity of COX, converts PGG$_2$ to PGH$_2$.

>> **Key Convention:** Prostaglandins with different functional groups on the ring are given different letter designations: A, B, C, D, E, F, G, H, and R. The subscript number following the letter, as in PGH$_2$ and PGG$_2$, indicates the number of double bonds. Prostaglandins with two double bonds, all of which are derived from arachidonate, are referred to as series 2 prostaglandins; those with three double bonds, derived from EPA, as series 3 (Fig. 21-12). Similar naming patterns are used for other classes of eicosanoids described below. <<

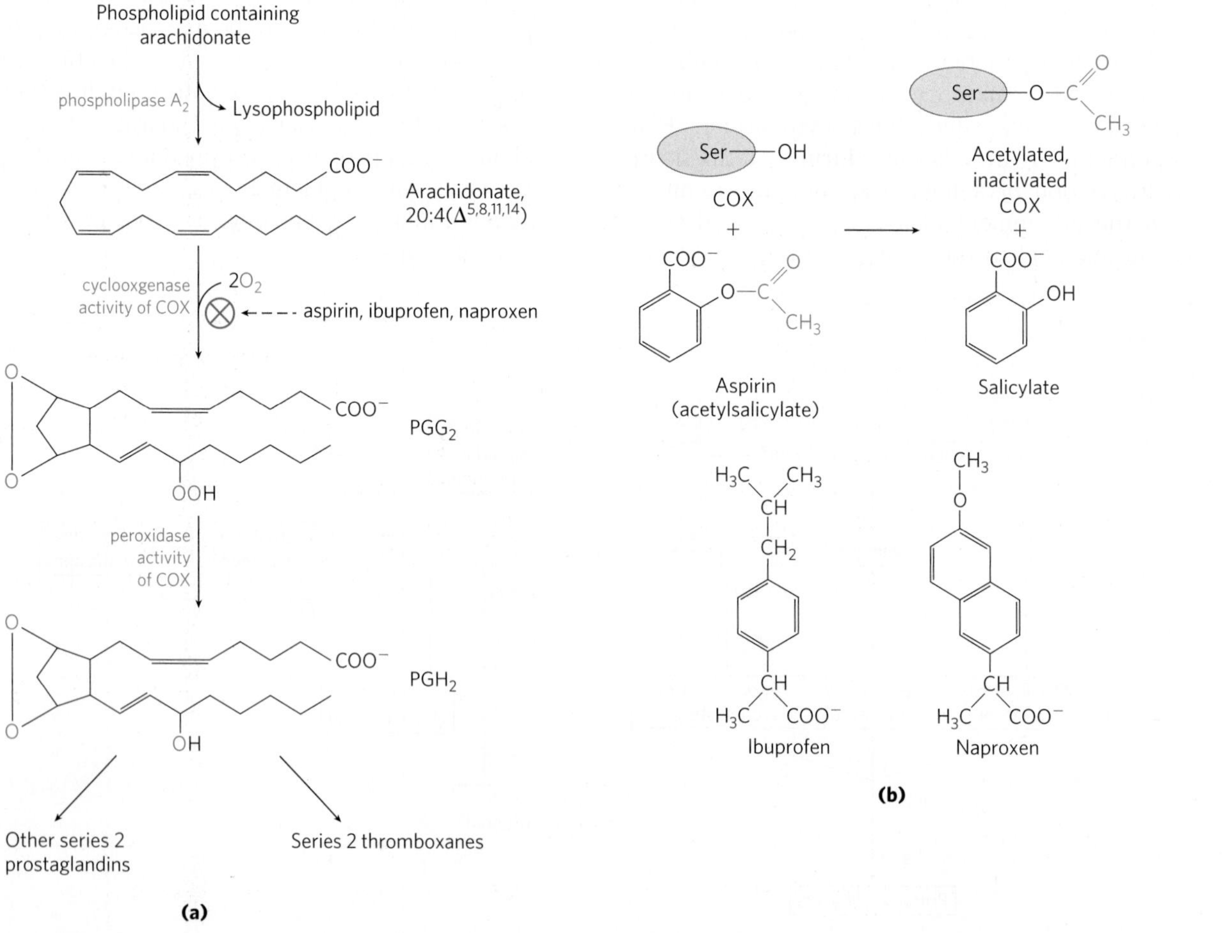

FIGURE 21-15 The "cyclic" pathway from arachidonate to prostaglandins and thromboxanes. (a) After arachidonate is released from phospholipids by the action of phospholipase A_2, the cyclooxygenase and peroxidase activities of COX (also called prostaglandin H$_2$ synthase) catalyze the production of PGH$_2$, the precursor of other prostaglandins and of thromboxanes. **(b)** Aspirin inhibits the first reaction by acetylating an essential Ser residue on the enzyme. Ibuprofen and naproxen inhibit the same step, probably by mimicking the structure of the substrate or an intermediate in the reaction.

Series 2 prostaglandins have important roles in the immediate response to stress or injury, including inflammation, pain, swelling, and dilation of blood vessels. Series 3 prostaglandins, in general, act more slowly and usually moderate the responses associated with series 2 prostaglandins.

Mammals have two isozymes of prostaglandin H_2 synthase, COX-1 and COX-2. These have different functions but closely similar amino acid sequences (60% to 65% sequence identity) and similar reaction mechanisms at both of their catalytic centers. COX-1 is responsible for synthesis of the prostaglandins that regulate the secretion of gastric mucin, and COX-2 for synthesis of the prostaglandins that mediate inflammation, pain, and fever.

Pain can be relieved by inhibiting COX-2. The first drug widely marketed for this purpose was aspirin (acetylsalicylate; Fig. 21-15b). The name "aspirin" (from *a* for acetyl and *spir* for *Spirsaüre*, the German word for the salicylates prepared from the plant *Spiraea ulmaria*) appeared in 1899 when the drug was introduced by the Bayer company. Aspirin irreversibly inactivates the cyclooxygenase activity of both COX isozymes, by acetylating a Ser residue and blocking each enzyme's active site. The synthesis of prostaglandins and thromboxanes is thereby inhibited. Ibuprofen, another widely used *n*onsteroidal *a*nti*i*nflammatory *d*rug (NSAID; Fig. 21-15b), inhibits the same pair of enzymes. However, the inhibition of COX-1 can result in undesired side effects, including stomach irritation and more serious conditions. In the 1990s, NSAID compounds that had a greater specificity for COX-2 were developed as advanced therapies for severe pain. Three of these drugs were approved for use worldwide: rofecoxib (Vioxx), valdecoxib (Bextra), and celecoxib (Celebrex). Though initially considered a success, Vioxx and Bextra were withdrawn as field reports and clinical studies connected the drugs with an increased risk of heart attack and stroke. Celebrex is still on the market but is being used with increased caution. The detailed reasons for the problems with these drugs are still not clear. However, work continues to reveal both new signaling eicosanoids and new modes of action. The problems with the advanced COX-2 inhibitors serve as a cautionary note. We are increasingly aware of the complexity of the web of these signaling interactions, and predicting the consequences of targeting specific components with pharmaceutical agents remains an imperfect process.

Thromboxane synthase, present in blood platelets (thrombocytes), converts PGH_2 to thromboxane A_2, from which other series 2 **thromboxanes** are derived (Fig. 21-15a). The series 2 thromboxanes induce constriction of blood vessels and platelet aggregation, early steps in blood clotting. Low doses of aspirin, taken regularly, reduce the probability of heart attacks and strokes by reducing thromboxane production. ■

Thromboxanes, like prostaglandins, contain a ring of five or six atoms; the pathway from arachidonate to the series 2 prostaglandins and thromboxanes is sometimes called the "cyclic" pathway, to distinguish it from the "linear" pathway that leads from arachidonate to the **leukotrienes**, which are linear compounds **(Fig. 21-16)**. Leukotriene synthesis begins with the action of several lipoxygenases that catalyze the incorporation of molecular oxygen into arachidonate. These enzymes, found in leukocytes and in heart, brain, lung, and spleen, are mixed-function oxidases of the cytochrome P-450 family (see Box 21-1). The various leukotrienes differ in the position of the peroxide group introduced by the lipoxygenases. The linear pathway from arachidonate, unlike the cyclic pathway, is not inhibited by aspirin or other NSAIDs.

Pathogenic organisms, as well as irritants such as air pollution and tobacco smoke, trigger an inflammatory response in the affected tissue, which consists of two phases: initiation and resolution. Eicosanoids of the omega-6 family are critical to initiation— playing key roles in recruiting leukocytes, making blood vessels more permeable, and stimulating chemotaxis and migration of immune system cells. As the source of tissue damage is brought under control, the inflammation must be resolved and the tissue brought back to its normal state. Resolution of inflammation is promoted by several classes of signaling molecules; prominent among these are several leukotrienes and prostaglandins. Many eicosanoids of the omega-3 family (including series 3 prostaglandins and thromboxanes) are

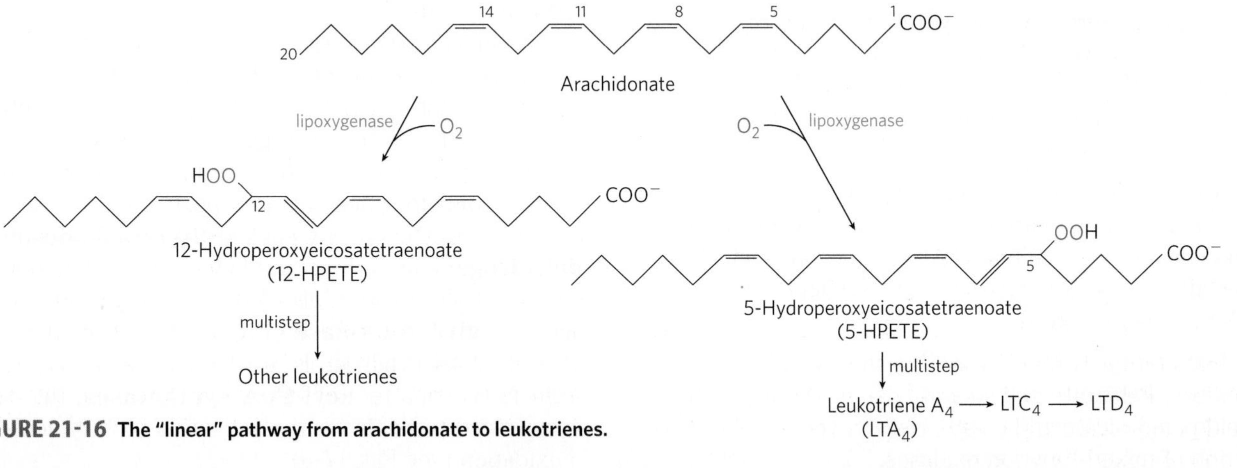

FIGURE 21-16 The "linear" pathway from arachidonate to leukotrienes.

antiinflammatory, although the classification is not absolute; individual eicosanoids can be inflammatory in one tissue and antiinflammatory in another. The resolution phase also uses a set of recently discovered eicosanoids termed **specialized pro-resolving mediators (SPMs)**. The first family of SPMs to be discovered was the lipoxins, followed more recently by resolvins, protectins, and maresins. All SPMs are derived from essential fatty acids (Fig. 21-12). They affect different target cells and tissues in different ways. The sum of their action is to promote removal of debris, microbes, and dead cells, to restore blood vessel integrity, and to regenerate tissue. The SPMs also reduce pain and fever, and are therefore of great interest as pharmaceutical targets. ■

Plants also derive important signaling molecules from fatty acids. As in animals, a key step in the initiation of signaling is activation of a specific phospholipase. In plants, the fatty acid substrate released by phospholipase action is α-linolenate. A lipoxygenase then catalyzes the first step in a pathway that converts α-linolenate to jasmonate, a substance known to have signaling roles in defense against insects, resistance to fungal pathogens, and maturation of pollen. Jasmonate also affects seed germination, root growth, and fruit and seed development.

SUMMARY 21.1 Biosynthesis of Fatty Acids and Eicosanoids

■ Long-chain saturated fatty acids are synthesized from acetyl-CoA by a cytosolic system of six enzymatic activities plus acyl carrier protein (ACP). There are two types of fatty acid synthase. FAS I, found in vertebrates and fungi, consists of multifunctional polypeptides. FAS II is a dissociated system found in bacteria and plants. Both contain two types of —SH groups (one furnished by the phosphopantetheine of ACP, the other by a Cys residue of β-ketoacyl-ACP synthase) that function as carriers of the fatty acyl intermediates.

■ Malonyl-ACP, formed from acetyl-CoA (shuttled out of mitochondria) and CO_2, condenses with an acetyl bound to the Cys —SH to yield acetoacetyl-ACP, with release of CO_2. This is followed by reduction to the D-β-hydroxy derivative, dehydration to the *trans*-Δ^2-unsaturated acyl-ACP, and reduction to butyryl-ACP. NADPH is the electron donor for both reductions. Fatty acid synthesis is regulated at the level of malonyl-CoA formation.

■ Six more molecules of malonyl-ACP react successively at the carboxyl end of the growing fatty acid chain to form palmitoyl-ACP—the end product of the fatty acid synthase reaction. Free palmitate is released by hydrolysis.

■ Palmitate may be elongated to the 18-carbon stearate. Palmitate and stearate can be desaturated to yield palmitoleate and oleate, respectively, by the action of mixed-function oxidases.

■ Mammals cannot make linoleate and must obtain it from plant sources; they convert exogenous linoleate to arachidonate, the parent compound of eicosanoids (prostaglandins, thromboxanes, leukotrienes, and specialized pro-resolving mediators), a family of very potent signaling molecules. The synthesis of prostaglandins and thromboxanes is inhibited by NSAIDs that act on the cyclooxygenase activity of prostaglandin H_2 synthase.

21.2 Biosynthesis of Triacylglycerols

Most of the fatty acids synthesized or ingested by an organism have one of two fates: incorporation into triacylglycerols for the storage of metabolic energy or incorporation into the phospholipid components of membranes. The partitioning between these alternative fates depends on the organism's current needs. During rapid growth, synthesis of new membranes requires the production of membrane phospholipids; when an organism has a plentiful food supply but is not actively growing, it shunts most of its fatty acids into storage fats. Both pathways begin at the same point: the formation of fatty acyl esters of glycerol. In this section we examine the route to triacylglycerols and its regulation, and the production of glycerol 3-phosphate in the process of glyceroneogenesis.

Triacylglycerols and Glycerophospholipids Are Synthesized from the Same Precursors

Animals can synthesize and store large quantities of triacylglycerols, to be used later as fuel (see Box 17-1). Humans can store only a few hundred grams of glycogen in liver and muscle, barely enough to supply the body's energy needs for 12 hours. In contrast, the total amount of stored triacylglycerol in a 70 kg man of average build is about 15 kg, enough to support basal energy needs for as long as 12 weeks (see Table 23-6). Triacylglycerols have the highest energy content of all stored nutrients—more than 38 kJ/g. Whenever carbohydrate is ingested in excess of the organism's capacity to store glycogen, the excess is converted to triacylglycerols and stored in adipose tissue. Plants also manufacture triacylglycerols as an energy-rich fuel, mainly stored in fruits, nuts, and seeds.

In animal tissues, triacylglycerols and glycerophospholipids such as phosphatidylethanolamine share two precursors, fatty acyl–CoA and L-glycerol 3-phosphate, and several biosynthetic steps. The vast majority of the glycerol 3-phosphate is derived from the glycolytic intermediate dihydroxyacetone phosphate (DHAP) by the action of the cytosolic NAD-linked **glycerol 3-phosphate dehydrogenase**; in liver and kidney, a small amount of glycerol 3-phosphate is also formed from glycerol by the action of **glycerol kinase (Fig. 21-17)**. The other precursors of triacylglycerols are fatty acyl–CoAs, formed from fatty acids by **acyl-CoA synthetases**, the same enzymes responsible for the activation of fatty acids for β oxidation (see Fig. 17-5).

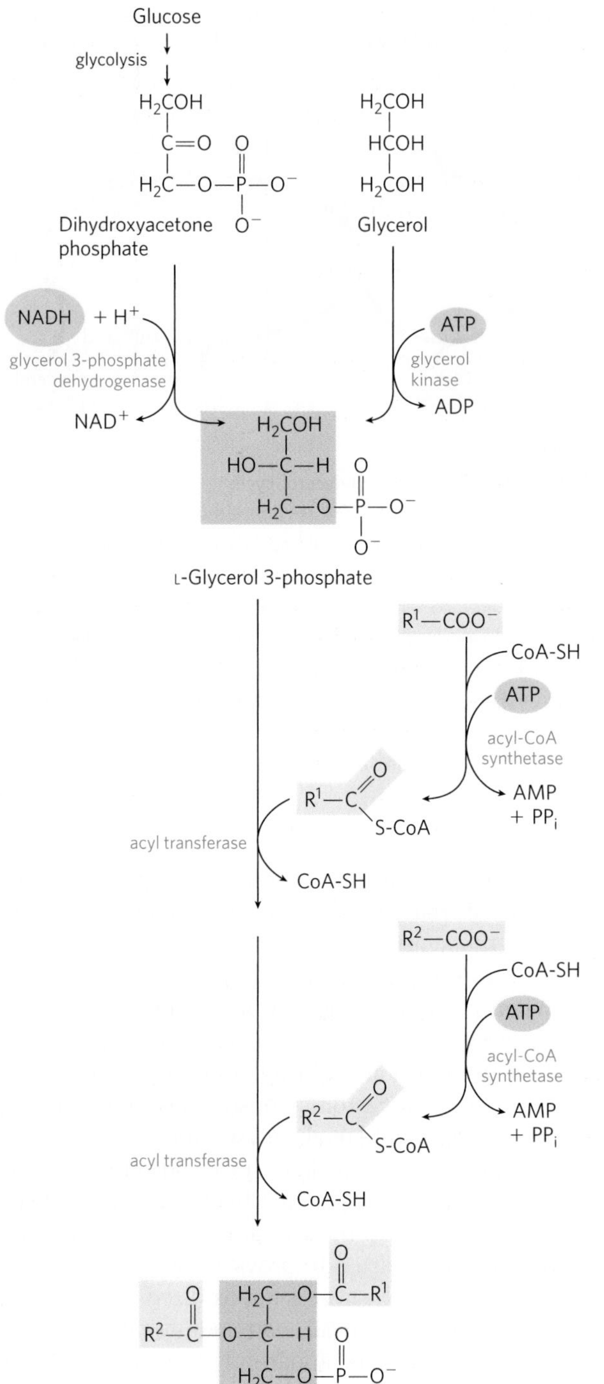

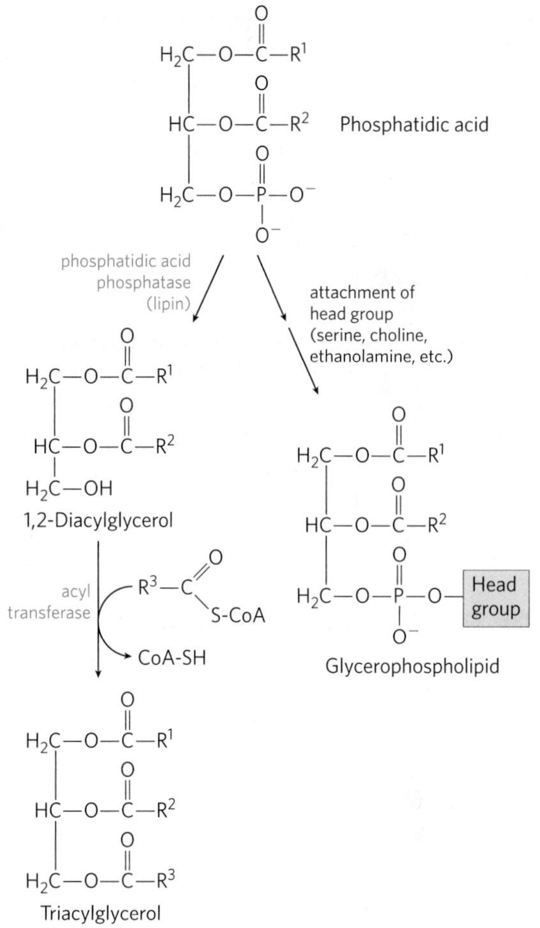

FIGURE 21-18 Phosphatidic acid in lipid biosynthesis. Phosphatidic acid is the precursor of both triacylglycerols and glycerophospholipids. The mechanisms for head-group attachment in phospholipid synthesis are described later in this section.

phosphatidic acid, or phosphatidate (Fig. 21-17). Phosphatidic acid is present in only trace amounts in cells but is a central intermediate in lipid biosynthesis; it can be converted either to a triacylglycerol or to a glycerophospholipid. In the pathway to triacylglycerols, phosphatidic acid is hydrolyzed by **phosphatidic acid phosphatase** (also called lipin) to form a 1,2-diacylglycerol **(Fig. 21-18)**. Diacylglycerols are then converted to triacylglycerols by transesterification with a third fatty acyl–CoA.

Triacylglycerol Biosynthesis in Animals Is Regulated by Hormones

In humans, the amount of body fat stays relatively constant over long periods, although there may be minor short-term changes as caloric intake fluctuates. Carbohydrate, fat, or protein ingested in excess of energy needs is stored in the form of triacylglycerols that can be drawn upon for energy, enabling the body to withstand periods of fasting.

Biosynthesis and degradation of triacylglycerols are regulated such that the favored path depends on the metabolic resources and requirements of the moment. The rate of triacylglycerol biosynthesis is

FIGURE 21-17 Biosynthesis of phosphatidic acid. A fatty acyl group is activated by formation of the fatty acyl–CoA, then transferred to ester linkage with L-glycerol 3-phosphate, formed in either of the two ways shown. Phosphatidic acid is shown here with the correct stereochemistry (L) at C-2 of the glycerol molecule. (The intermediate product with only one esterified fatty acyl group is lysophosphatidic acid.) To conserve space in subsequent figures (and in Fig. 21-14), both fatty acyl groups of glycerophospholipids, and all three acyl groups of triacylglycerols, are shown projecting to the right.

The first stage in the biosynthesis of triacylglycerols is acylation of the two free hydroxyl groups of L-glycerol 3-phosphate by two molecules of fatty acyl–CoA to yield **diacylglycerol 3-phosphate**, more commonly called

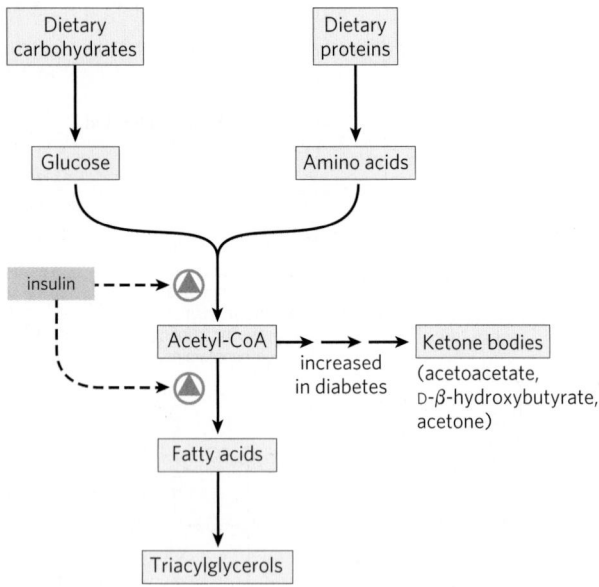

FIGURE 21-19 Regulation of triacylglycerol synthesis by insulin.
Insulin stimulates conversion of dietary carbohydrates and proteins to fat. Individuals with diabetes mellitus either lack insulin or are insensitive to it. This results in diminished fatty acid synthesis, and the acetyl-CoA arising from catabolism of carbohydrates and proteins is shunted instead to ketone body production. People in severe ketosis smell of acetone, so the condition is sometimes mistaken for drunkenness (p. 938).

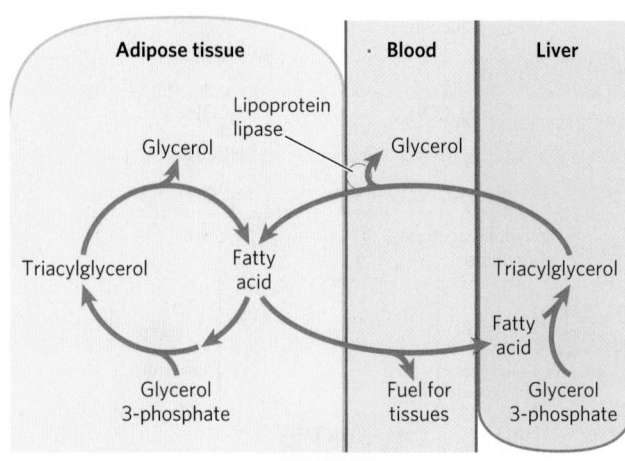

FIGURE 21-20 The triacylglycerol cycle. In mammals, triacylglycerol molecules are broken down and resynthesized in a triacylglycerol cycle during starvation. Some of the fatty acids released by lipolysis of triacylglycerol in adipose tissue pass into the bloodstream, and the remainder are used for resynthesis of triacylglycerol. Some of the fatty acids released into the blood are used for energy (in muscle, for example), and some are taken up by the liver and used in triacylglycerol synthesis. The triacylglycerol formed in the liver is transported in the blood back to adipose tissue, where the fatty acid is released by extracellular lipoprotein lipase, taken up by adipocytes, and reesterified into triacylglycerol.

profoundly altered by the action of several hormones. Insulin, for example, promotes the conversion of carbohydrate to triacylglycerols **(Fig. 21-19)**. People with severe diabetes mellitus, due to failure of insulin secretion or action, not only are unable to use glucose properly but also fail to synthesize fatty acids from carbohydrates or amino acids. If the diabetes is untreated, these individuals have increased rates of fat oxidation and ketone body formation (Chapter 17) and therefore lose weight. ∎

An additional factor in the balance between biosynthesis and degradation of triacylglycerols is that approximately 75% of all fatty acids released by lipolysis are reesterified to form triacylglycerols rather than used for fuel. This ratio persists even under starvation conditions, when energy metabolism is shunted from the use of carbohydrate to the oxidation of fatty acids. Some of this fatty acid recycling takes place in adipose tissue, with the reesterification occurring before release into the bloodstream; some takes place via a systemic cycle in which free fatty acids are transported to the liver, recycled to triacylglycerol, exported again into the blood (transport of lipids in the blood is discussed in Section 21.4), and taken up again by adipose tissue, after release from triacylglycerol by extracellular lipoprotein lipase **(Fig. 21-20**; see also Fig. 17-1). Flux through this **triacylglycerol cycle** between adipose tissue and liver may be low when other fuels are available and the release of fatty acids from adipose tissue is limited, but, as noted above, the proportion of released fatty acids that are reesterified remains roughly

constant at 75% under all metabolic conditions. The level of free fatty acids in the blood thus reflects both the rate of release of fatty acids and the balance between the synthesis and breakdown of triacylglycerols in adipose tissue and liver.

When the mobilization of fatty acids is required to meet energy needs, release from adipose tissue is stimulated by the hormones glucagon and epinephrine (see Figs 17-3, 17-13). Simultaneously, these hormonal signals decrease the rate of glycolysis and increase the rate of gluconeogenesis in the liver (providing glucose for the brain, as further elaborated in Chapter 23). The released fatty acid is taken up by several tissues, including muscle, where it is oxidized to provide energy. Much of the fatty acid taken up by liver is not oxidized but is recycled to triacylglycerol and returned to adipose tissue.

The function of the apparently futile triacylglycerol cycle ("futile" substrate cycles are discussed in Chapter 15) is not well understood, but as we learn more about how the cycle is sustained via metabolism in two separate organs and is coordinately regulated, some possibilities emerge. For example, the excess capacity in the triacylglycerol cycle—the fatty acid that is eventually reconverted to triacylglycerol rather than oxidized as fuel—could represent an energy reserve in the bloodstream during fasting, one that could be more rapidly mobilized in a "fight or flight" emergency than could stored triacylglycerol.

The constant recycling of triacylglycerols in adipose tissue even during starvation raises a second question: what is the source of the glycerol 3-phosphate required for this process? As noted above, glycolysis is suppressed under these conditions by the action of glucagon and epinephrine, so little DHAP is available. And glycerol released during lipolysis cannot be converted directly to glycerol 3-phosphate in adipose tissue, which lacks glycerol kinase (Fig. 21-17). So, how is sufficient glycerol 3-phosphate produced? The answer lies in a pathway discovered more than three decades ago and given little attention until recently, a pathway intimately linked to the triacylglycerol cycle and, in a larger sense, to the balance between fatty acid and carbohydrate metabolism.

Adipose Tissue Generates Glycerol 3-Phosphate by Glyceroneogenesis

Glyceroneogenesis is a shortened version of gluconeogenesis, from pyruvate to DHAP (see Fig. 14-17), followed by conversion of the DHAP to glycerol 3-phosphate by cytosolic NAD-linked glycerol 3-phosphate dehydrogenase **(Fig. 21-21)**. Glycerol 3-phosphate is subsequently used in triacylglycerol synthesis. Glyceroneogenesis was discovered in the 1960s by Lea Reshef, Richard Hanson, and John Ballard, and simultaneously by Eleazar Shafrir and his coworkers, who were intrigued by the presence of two gluconeogenic enzymes, pyruvate carboxylase and phosphoenolpyruvate (PEP) carboxykinase, in adipose tissue, where glucose is not synthesized. After a long period of inattention, interest in this pathway has been renewed by the demonstration of a link between glyceroneogenesis and type 2 diabetes, as we shall see.

Glyceroneogenesis has multiple roles. In adipose tissue, glyceroneogenesis coupled with reesterification of free fatty acids controls the rate of fatty acid release to the blood. In brown adipose tissue, the same pathway

may control the rate at which free fatty acids are delivered to mitochondria for use in thermogenesis. And in fasting humans, glyceroneogenesis in the liver alone supports the synthesis of enough glycerol 3-phosphate to account for up to 65% of fatty acids reesterified to triacylglycerol.

Flux through the triacylglycerol cycle between liver and adipose tissue is controlled to a large degree by the activity of PEP carboxykinase, which limits the rate of both gluconeogenesis and glyceroneogenesis. Glucocorticoid hormones such as cortisol (a biological steroid derived from cholesterol; see Fig. 21-48) and dexamethasone (a synthetic glucocorticoid) regulate the levels of PEP carboxykinase reciprocally in the liver and adipose tissue. Acting through the glucocorticoid receptor, these steroid hormones increase the expression of the gene encoding PEP carboxykinase in the liver, thus increasing gluconeogenesis and glyceroneogenesis **(Fig. 21-22)**.

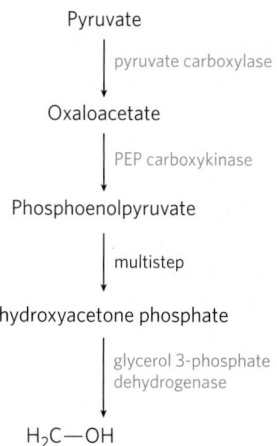

Cortisol

Dexamethasone

Stimulation of glyceroneogenesis leads to an increase in the synthesis of triacylglycerol molecules in the liver and their release into the blood. At the same time, glucocorticoids suppress expression of the gene encoding PEP carboxykinase in adipose tissue. This results in a decrease in glyceroneogenesis in adipose tissue; recycling of fatty acids declines as a result, and more free fatty acids are released into the blood. Thus regulation of glyceroneogenesis in the liver and adipose tissue affects lipid metabolism in opposite ways: a lower rate of glyceroneogenesis in adipose tissue leads to more fatty acid release (rather than recycling), whereas a higher rate in the liver leads to more synthesis and export of triacylglycerols. The net result is an increase in flux through the triacylglycerol cycle. When the glucocorticoids are no longer present, flux through the cycle declines as the expression of PEP carboxykinase increases in adipose tissue and decreases in the liver.

Thiazolidinediones Treat Type 2 Diabetes by Increasing Glyceroneogenesis

The recent attention to glyceroneogenesis, as noted above, has arisen in part from the connection between this pathway and diabetes. High levels of free fatty acids in the blood interfere with glucose utilization in muscle and promote the insulin resistance that leads to type 2 diabetes. A class of drugs called **thiazolidinediones** reduce the levels of fatty acids circulating in

Pyruvate

↓ pyruvate carboxylase

Oxaloacetate

↓ PEP carboxykinase

Phosphoenolpyruvate

↓ multistep

Dihydroxyacetone phosphate

↓ glycerol 3-phosphate dehydrogenase

$$H_2C-OH$$
$$HC-OH \quad O$$
$$H_2C-O-\overset{\displaystyle \|}{\underset{\displaystyle |}{P}}-O^-$$
$$O^-$$

Glycerol 3-phosphate

↓

Triacylglycerol synthesis

FIGURE 21-21 Glyceroneogenesis. The pathway is essentially an abbreviated version of gluconeogenesis, from pyruvate to dihydroxyacetone phosphate (DHAP), followed by conversion of DHAP to glycerol 3-phosphate, which is used for the synthesis of triacylglycerol.

(a)

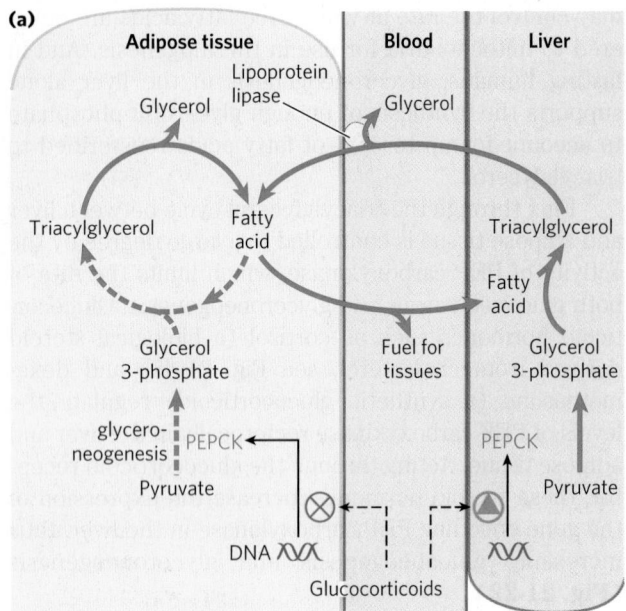

(b)

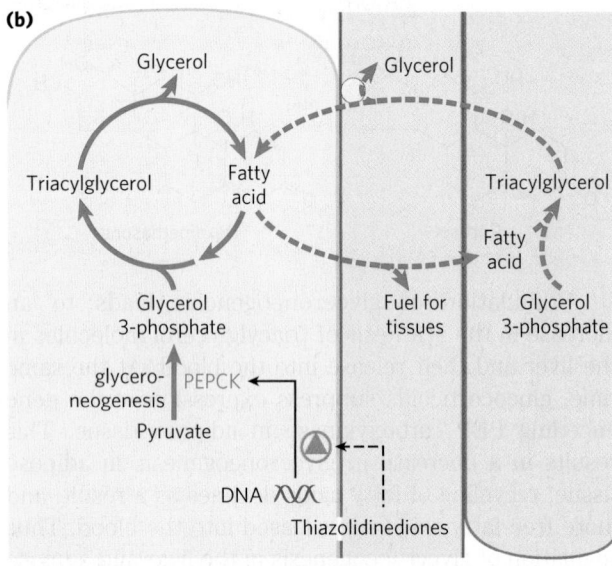

FIGURE 21-22 Regulation of glyceroneogenesis. (a) Glucocorticoid hormones stimulate glyceroneogenesis and gluconeogenesis in the liver, while suppressing glyceroneogenesis in adipose tissue (by reciprocal regulation of the gene expressing PEP carboxykinase (PEPCK) in the two tissues); this increases the flux through the triacylglycerol cycle. The glycerol freed by the breakdown of triacylglycerol in adipose tissue is released to the blood and transported to the liver, where it is primarily converted to glucose, although some is converted to glycerol 3-phosphate by glycerol kinase. **(b)** A class of drugs called thiazolidinediones are now used to treat type 2 diabetes. In this disease, high levels of free fatty acids in the blood interfere with glucose utilization in muscle and promote insulin resistance. Thiazolidinediones activate a nuclear receptor called peroxisome proliferator-activated receptor γ (PPARγ), which induces the activity of PEP carboxykinase. Therapeutically, thiazolidinediones increase the rate of glyceroneogenesis, thus increasing the resynthesis of triacylglycerol in adipose tissue and reducing the amount of free fatty acid in the blood.

the blood and increase sensitivity to insulin. Thiazolidinediones promote the induction of PEP carboxykinase in adipose tissue (Fig. 21-22), leading to increased synthesis of the precursors of glyceroneogenesis. The therapeutic

effect of thiazolidinediones is thus due, at least in part, to the increase in glyceroneogenesis, which in turn increases the resynthesis of triacylglycerol in adipose tissue and reduces the release of free fatty acid from adipose tissue into the blood. ■

SUMMARY 21.2 Biosynthesis of Triacylglycerols

■ Triacylglycerols are formed by reaction of two molecules of fatty acyl–CoA with glycerol 3-phosphate to form phosphatidic acid; this product is dephosphorylated to a diacylglycerol, then acylated by a third molecule of fatty acyl–CoA to yield a triacylglycerol.

■ The synthesis and degradation of triacylglycerols are hormonally regulated.

■ Mobilization and recycling of triacylglycerol molecules result in a triacylglycerol cycle. Triacylglycerols are resynthesized from free fatty acids and glycerol 3-phosphate even during starvation. The dihydroxyacetone phosphate precursor of glycerol 3-phosphate is derived from pyruvate via glyceroneogenesis.

21.3 Biosynthesis of Membrane Phospholipids

In Chapter 10 we introduced two major classes of membrane phospholipids: glycerophospholipids and sphingolipids. Many different phospholipid species can be constructed by combining various fatty acids and polar head groups with the glycerol or sphingosine backbone (see Figs 10-8, 10-12). All the biosynthetic pathways follow a few basic patterns. In general, the assembly of phospholipids from simple precursors requires (1) synthesis of the backbone molecule (glycerol or sphingosine); (2) attachment of fatty acid(s) to the backbone through an ester or amide linkage; (3) addition of a hydrophilic head group to the backbone through a phosphodiester linkage; and, in some cases, (4) alteration or exchange of the head group to yield the final phospholipid product.

In eukaryotic cells, phospholipid synthesis occurs primarily on the surfaces of the smooth ER and the inner mitochondrial membrane. Some newly formed phospholipids remain at the site of synthesis, but most are destined for other cellular locations. The process by which water-insoluble phospholipids move from the site of synthesis to the point of their eventual function is not fully understood, but we discuss some mechanisms that have emerged in recent years.

Cells Have Two Strategies for Attaching Phospholipid Head Groups

The first steps of glycerophospholipid synthesis are shared with the pathway to triacylglycerols (Fig. 21-17): two fatty acyl groups are esterified to C-1 and C-2 of L-glycerol 3-phosphate to form phosphatidic acid.

Commonly, but not invariably, the fatty acid at C-1 is saturated and that at C-2 is unsaturated. A second route to phosphatidic acid is the phosphorylation of a diacylglycerol by a specific kinase.

The polar head group of glycerophospholipids is attached through a phosphodiester bond, in which each of two alcohol hydroxyls (one on the polar head group and one on C-3 of glycerol) forms an ester with phosphoric acid **(Fig. 21-23)**. In the biosynthetic process, one of the hydroxyls is first activated by attachment of a nucleotide, cytidine diphosphate. Cytidine monophosphate is then displaced in a nucleophilic attack by the other hydroxyl **(Fig. 21-24)**. The CDP is attached either to the diacylglycerol, forming the activated phosphatidic acid **CDP-diacylglycerol** (strategy 1), or to the hydroxyl of the head group (strategy 2). Eukaryotic cells employ both strategies, whereas bacteria use only the first. The central importance of cytidine nucleotides in lipid biosynthesis was discovered by Eugene P. Kennedy in the early 1960s, and this pathway is commonly referred to as the Kennedy pathway.

Phospholipid Synthesis in *E. coli* Employs CDP-Diacylglycerol

The first strategy for head-group attachment is illustrated by the synthesis of phosphatidylserine, phosphatidylethanolamine, and phosphatidylglycerol in *E. coli*. The diacylglycerol is activated by condensation of phosphatidic acid with CTP to form CDP-diacylglycerol, with the elimination of pyrophosphate **(Fig. 21-25)**. Displacement of CMP through nucleophilic

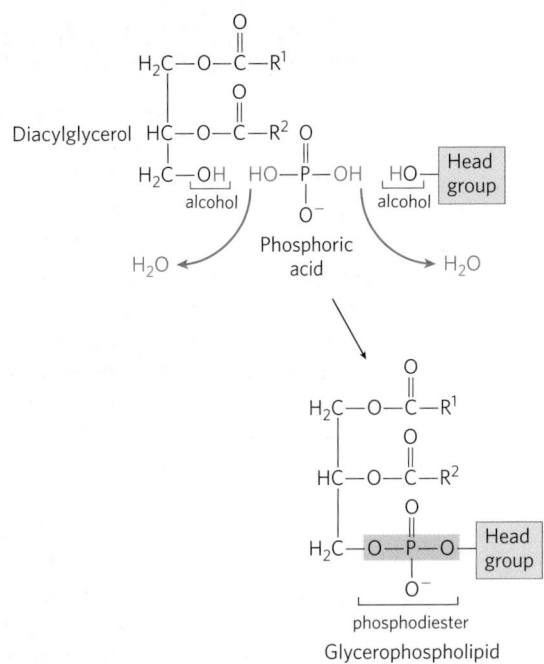

FIGURE 21-23 Head-group attachment. The phospholipid head group is attached to a diacylglycerol by a phosphodiester bond (shaded light red), formed when phosphoric acid condenses with two alcohols, eliminating two molecules of H_2O.

attack by the hydroxyl group of serine or by the C-1 hydroxyl of glycerol 3-phosphate yields **phosphatidylserine** or phosphatidylglycerol 3-phosphate, respectively. The latter is processed further by cleavage of the phosphate monoester (with release of P_i) to yield **phosphatidylglycerol**.

Eugene P. Kennedy, 1919–2011
[Source: Courtesy of EPK Family.]

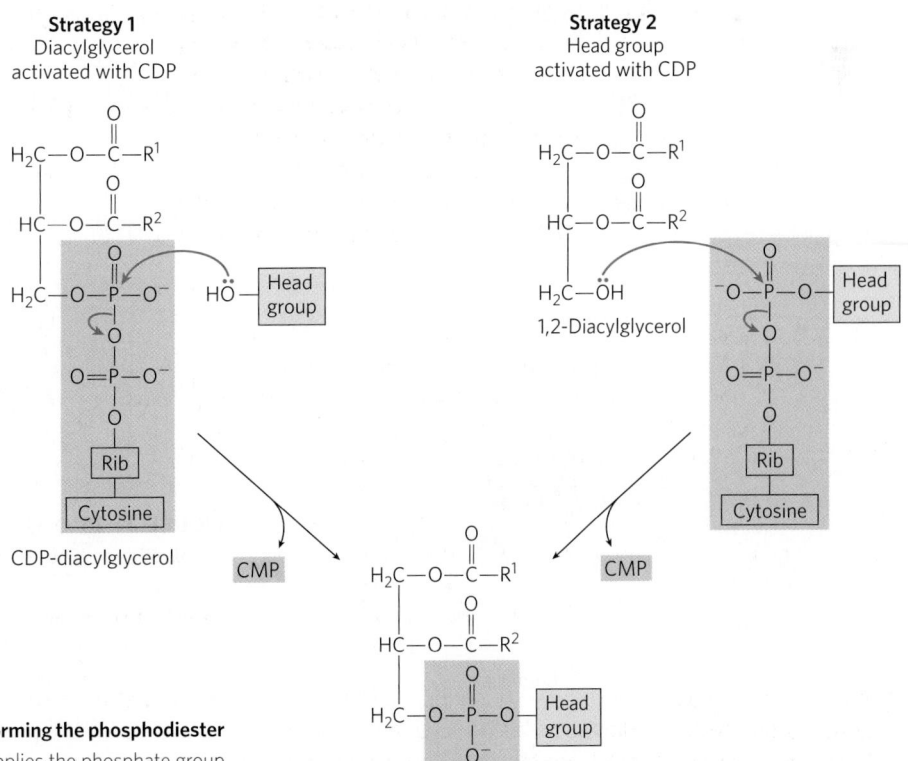

FIGURE 21-24 Two general strategies for forming the phosphodiester bond of phospholipids. In both cases, CDP supplies the phosphate group of the phosphodiester bond.

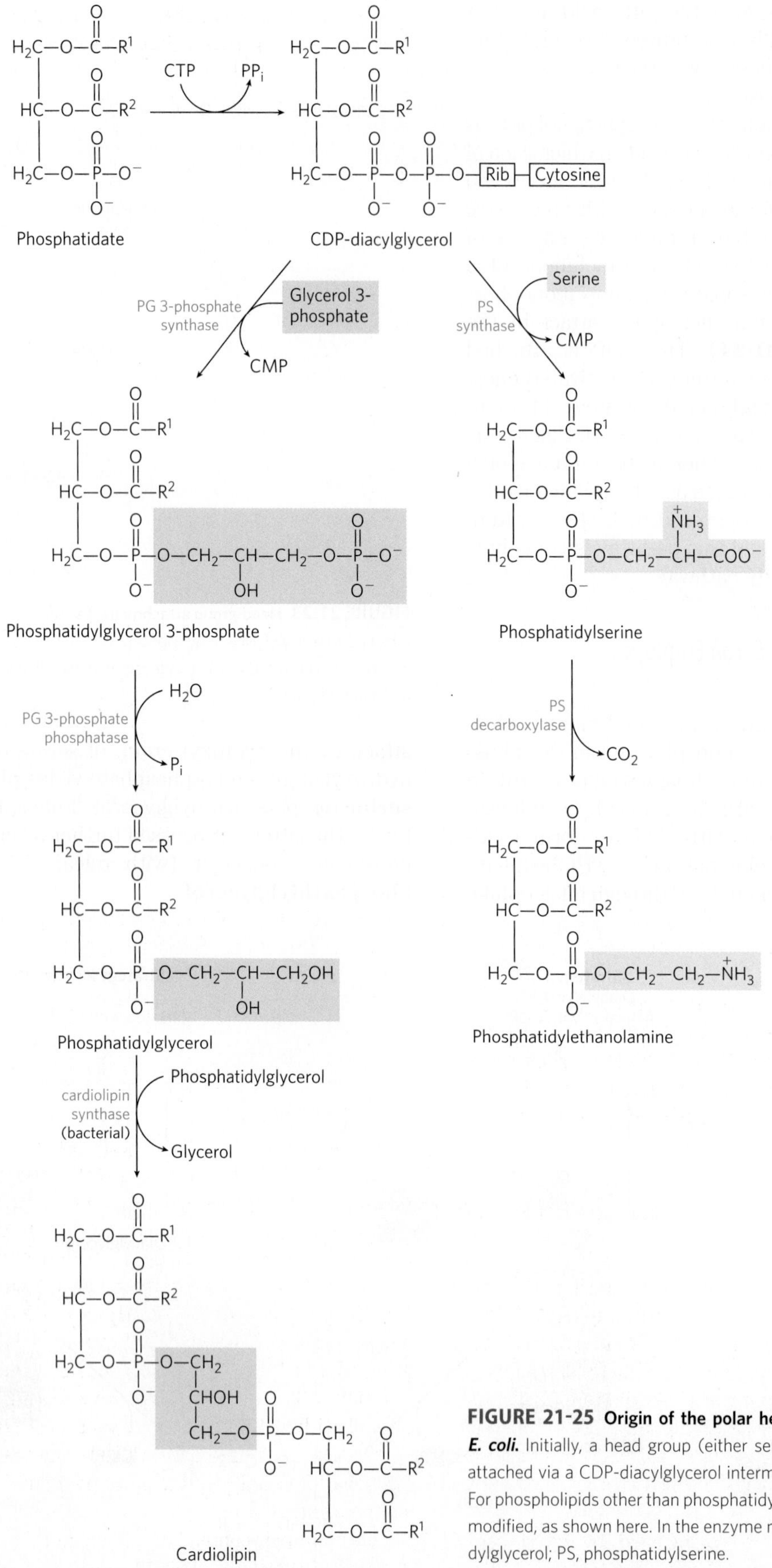

FIGURE 21-25 Origin of the polar head groups of phospholipids in *E. coli.* Initially, a head group (either serine or glycerol 3-phosphate) is attached via a CDP-diacylglycerol intermediate (strategy 1 in Fig. 21-24). For phospholipids other than phosphatidylserine, the head group is further modified, as shown here. In the enzyme names, PG represents phosphatidylglycerol; PS, phosphatidylserine.

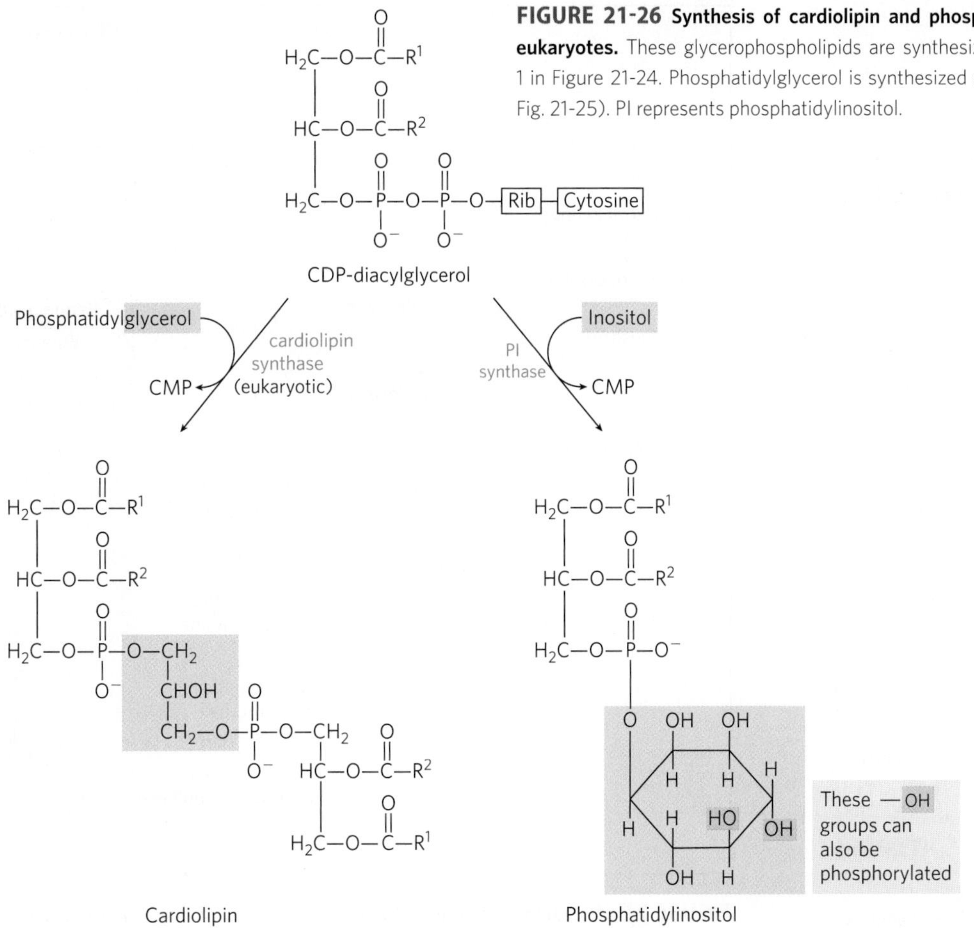

FIGURE 21-26 Synthesis of cardiolipin and phosphatidylinositol in eukaryotes. These glycerophospholipids are synthesized using strategy 1 in Figure 21-24. Phosphatidylglycerol is synthesized as in bacteria (see Fig. 21-25). PI represents phosphatidylinositol.

Phosphatidylserine and phosphatidylglycerol can serve as precursors of other membrane lipids in bacteria (Fig. 21-25). Decarboxylation of the serine moiety in phosphatidylserine, catalyzed by phosphatidylserine decarboxylase, yields **phosphatidylethanolamine**. In *E. coli*, condensation of two molecules of phosphatidylglycerol, with elimination of one glycerol, yields **cardiolipin**, in which two diacylglycerols are joined through a common head group.

Eukaryotes Synthesize Anionic Phospholipids from CDP-Diacylglycerol

In eukaryotes, phosphatidylglycerol, cardiolipin, and phosphatidylinositol (all anionic phospholipids; see Fig. 10-8) are synthesized by the same strategy used for phospholipid synthesis in bacteria. Phosphatidylglycerol is made exactly as in bacteria. Cardiolipin synthesis in eukaryotes differs slightly: phosphatidylglycerol condenses with CDP-diacylglycerol **(Fig. 21-26)**, not with another molecule of phosphatidylglycerol as in *E. coli* (Fig. 21-25).

Phosphatidylinositol is synthesized by condensation of CDP-diacylglycerol with inositol (Fig. 21-26). Specific **phosphatidylinositol kinases** then convert phosphatidylinositol to its phosphorylated derivatives.

Phosphatidylinositol and its phosphorylated products in the plasma membrane play a central role in signal transduction in eukaryotes (see Figs 12-10, 12-11, 12-20).

Eukaryotic Pathways to Phosphatidylserine, Phosphatidylethanolamine, and Phosphatidylcholine Are Interrelated

Yeast, like bacteria, can produce phosphatidylserine by condensation of CDP-diacylglycerol and serine, and can synthesize phosphatidylethanolamine from phosphatidylserine in the reaction catalyzed by phosphatidylserine decarboxylase **(Fig. 21-27)**. Phosphatidylethanolamine may be converted to **phosphatidylcholine** (lecithin) by the addition of three methyl groups to its amino group; *S*-adenosylmethionine is the methyl group donor (see Fig. 18-18) for all three methylation reactions. The pathways to phosphatidylcholine and phosphatidylethanolamine in yeast are summarized in **Figure 21-28**. These pathways are the major sources of phosphatidylethanolamine and phosphatidylcholine in all eukaryotic cells.

In mammals, phosphatidylserine is not synthesized from CDP-diacylglycerol; instead, it is derived from phosphatidylethanolamine or phosphatidylcholine via one of two head-group exchange reactions carried out

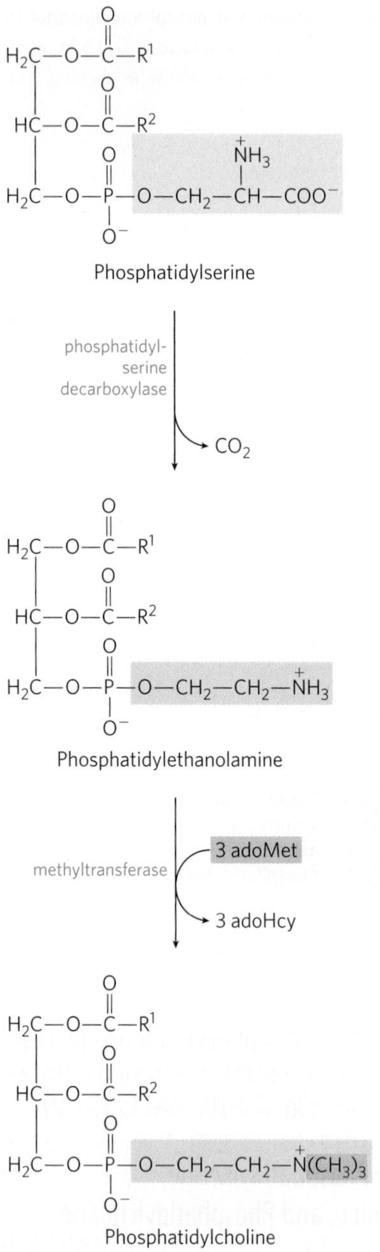

Phosphatidylserine

phosphatidyl-
serine
decarboxylase

CO_2

Phosphatidylethanolamine

methyltransferase

3 adoMet

3 adoHcy

Phosphatidylcholine

FIGURE 21-27 The major path from phosphatidylserine to phosphatidyletha-nolamine and phosphatidylcholine in all eukaryotes. AdoMet is S-adenosylmethio-nine; adoHcy, S-adenosylhomocysteine.

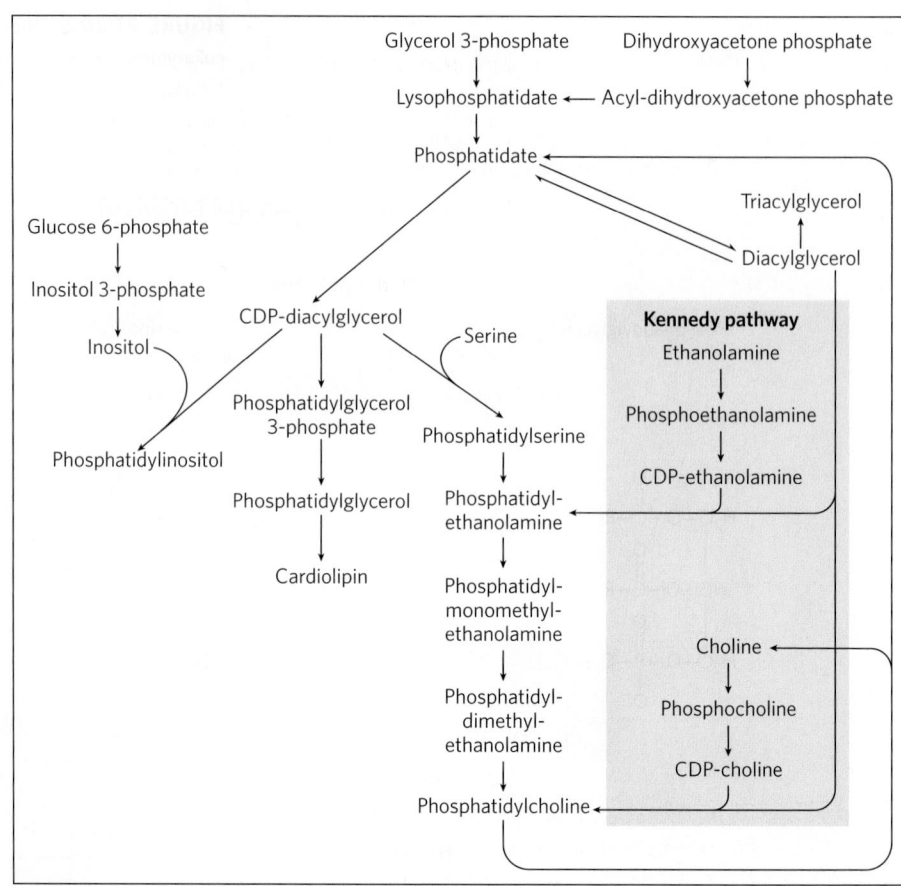

FIGURE 21-28 Summary of the pathways for synthesis of major phospholipids and triacylglycerols in a eukaryote (yeast). Phosphatidic acid is formed by transacylation of L-glycerol 3-phosphate with two fatty acyl groups donated from fatty acyl–CoA. The enzyme phosphatidic acid phosphatase (lipin) converts phosphatidic acid to diacylglycerol, which in the Kennedy pathway condenses with a CDP-activated head group (ethanolamine or choline) to form phosphatidylethanolamine or phosphatidylcholine. Alternatively, phosphatidic acid can be activated with a CDP moiety, which is displaced by condensation with a head group alcohol—inositol, glycerol 3-phosphate, or serine, forming phosphatidylinositol, phosphatidylglycerol, or phosphatidylserine. Decarboxylation of phosphatidylserine yields phosphatidylethanolamine, and meth-ylation of phosphatidylethanolamine produces phosphatidylcholine. Not shown here are the head-group exchange reactions (see Fig. 21-29a) that interconvert phosphatidylethanolamine, phosphatidylserine, and phosphatidylcholine in mammals. Lysophosphatidic acid is phosphatidic acid missing one of the two fatty acyl groups. [Source: Information from G. M. Carman and G.-S. Han, *Annu. Rev. Biochem.* 80:859, 2011, Fig. 2.]

in the endoplasmic reticulum **(Fig. 21-29a)**. Synthesis of phosphatidylethanolamine and phosphatidylcholine in mammals occurs by strategy 2 of Figure 21-24: phos-phorylation and activation of the head group, followed by condensation with diacylglycerol. For example, cho-line is reused ("salvaged") by being phosphorylated then converted to CDP-choline by condensation with CTP. A diacylglycerol displaces CMP from CDP-choline, producing phosphatidylcholine (Fig. 21-29b). An analo-gous salvage pathway converts ethanolamine obtained in the diet to phosphatidylethanolamine. In the liver, phosphatidylcholine is also produced by methylation of

phosphatidylethanolamine (with S-adenosylmethionine, as described above), but in all other tissues phosphati-dylcholine is produced only by condensation of diacyl-glycerol and CDP-choline.

Plasmalogen Synthesis Requires Formation of an Ether-Linked Fatty Alcohol

The biosynthetic pathway to ether lipids, including **plasmalogens** and the **platelet-activating factor** (see Fig. 10-9), requires displacement of an esterified fatty acyl group by a long-chain alcohol to form the ether linkage **(Fig. 21-30)**. Head-group attachment follows, by mechanisms essentially like those used in formation of the common ester-linked phospholipids. Finally, the characteristic double bond of plasmalo-gens (shaded blue in Fig. 21-30) is introduced by the

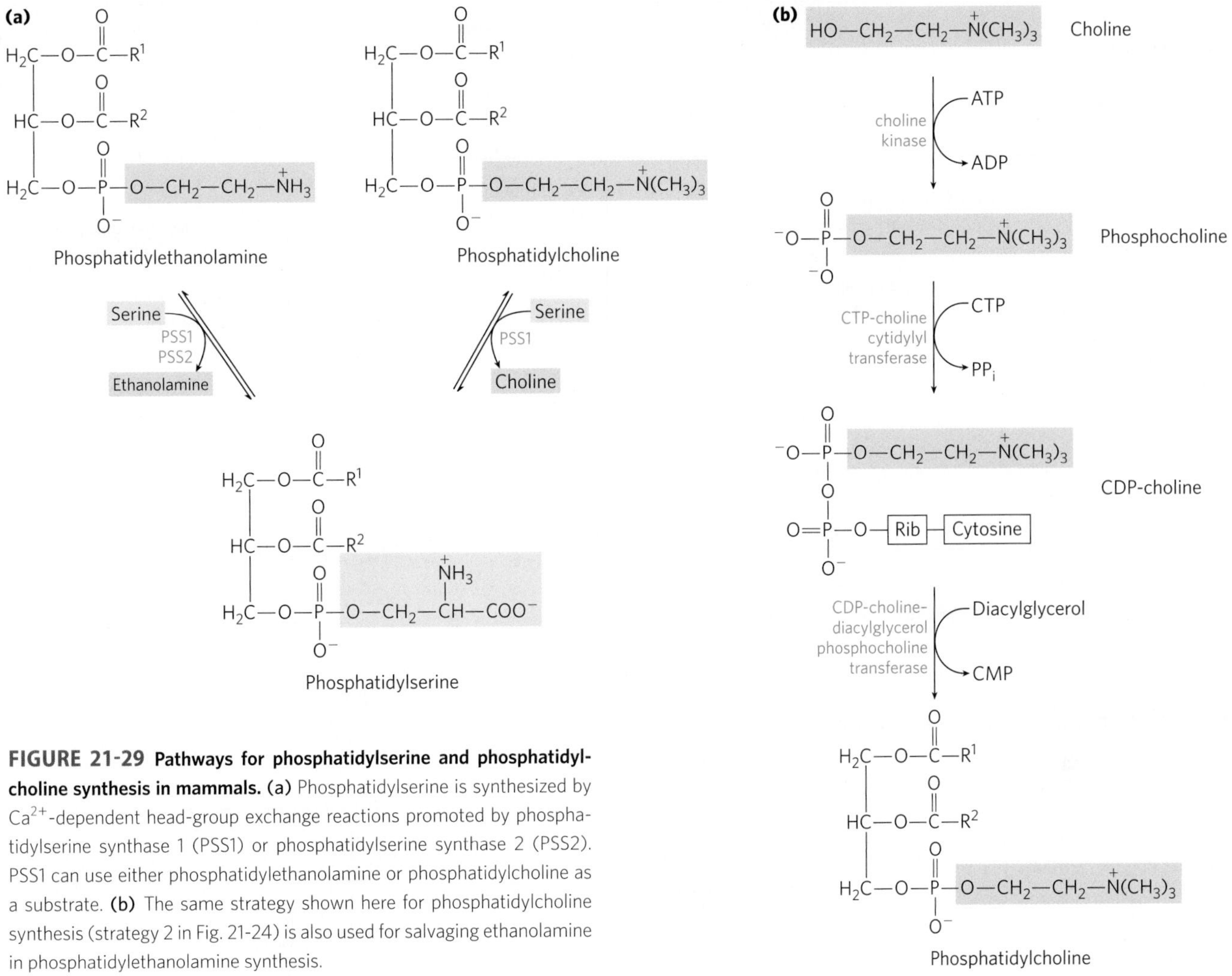

(a)

Phosphatidylethanolamine

Phosphatidylcholine

Phosphatidylserine

(b)

Phosphatidylcholine

FIGURE 21-29 Pathways for phosphatidylserine and phosphatidylcholine synthesis in mammals. (a) Phosphatidylserine is synthesized by Ca^{2+}-dependent head-group exchange reactions promoted by phosphatidylserine synthase 1 (PSS1) or phosphatidylserine synthase 2 (PSS2). PSS1 can use either phosphatidylethanolamine or phosphatidylcholine as a substrate. **(b)** The same strategy shown here for phosphatidylcholine synthesis (strategy 2 in Fig. 21-24) is also used for salvaging ethanolamine in phosphatidylethanolamine synthesis.

action of a mixed-function oxidase similar to that responsible for desaturation of fatty acids (Fig. 21-13). The peroxisome is the primary site of plasmalogen synthesis.

Sphingolipid and Glycerophospholipid Synthesis Share Precursors and Some Mechanisms

The biosynthesis of sphingolipids takes place in four stages: (1) synthesis of the 18-carbon amine **sphinganine** from palmitoyl-CoA and serine; (2) attachment of a fatty acid in amide linkage to yield **N-acylsphinganine**; (3) desaturation of the sphinganine moiety to form **N-acylsphingosine** (ceramide); and (4) attachment of a head group to produce a sphingolipid such as a **cerebroside** or **sphingomyelin (Fig. 21-31)**. The first few steps of this pathway occur in the endoplasmic reticulum; the attachment of head groups in stage 4 occurs in the Golgi complex. The pathway shares several features with the pathways leading to glycerophospholipids: NADPH provides reducing power, and fatty acids enter as their activated CoA derivatives. In cerebroside formation, sugars enter as their activated nucleotide derivatives. Head-group attachment in

sphingolipid synthesis has several novel aspects. Phosphatidylcholine, rather than CDP-choline, serves as the donor of phosphocholine in the synthesis of sphingomyelin.

In glycolipids—the cerebrosides and **gangliosides** (see Fig. 10-12)—the head-group sugar is attached directly to the C-1 hydroxyl of sphingosine in glycosidic linkage rather than through a phosphodiester bond. The sugar donor is a UDP-sugar (UDP-glucose or UDP-galactose).

Polar Lipids Are Targeted to Specific Cellular Membranes

After synthesis on the smooth ER, the polar lipids, including the glycerophospholipids, sphingolipids, and glycolipids, are inserted into specific cellular membranes in specific proportions, by mechanisms not yet understood. Membrane lipids are insoluble in water, so they cannot simply diffuse from their point of synthesis (the ER) to their point of insertion. Instead, they are transported from the ER to the Golgi complex, where additional synthesis can take place. They are then delivered in membrane vesicles that bud from the Golgi complex then move to and fuse with the target membrane.

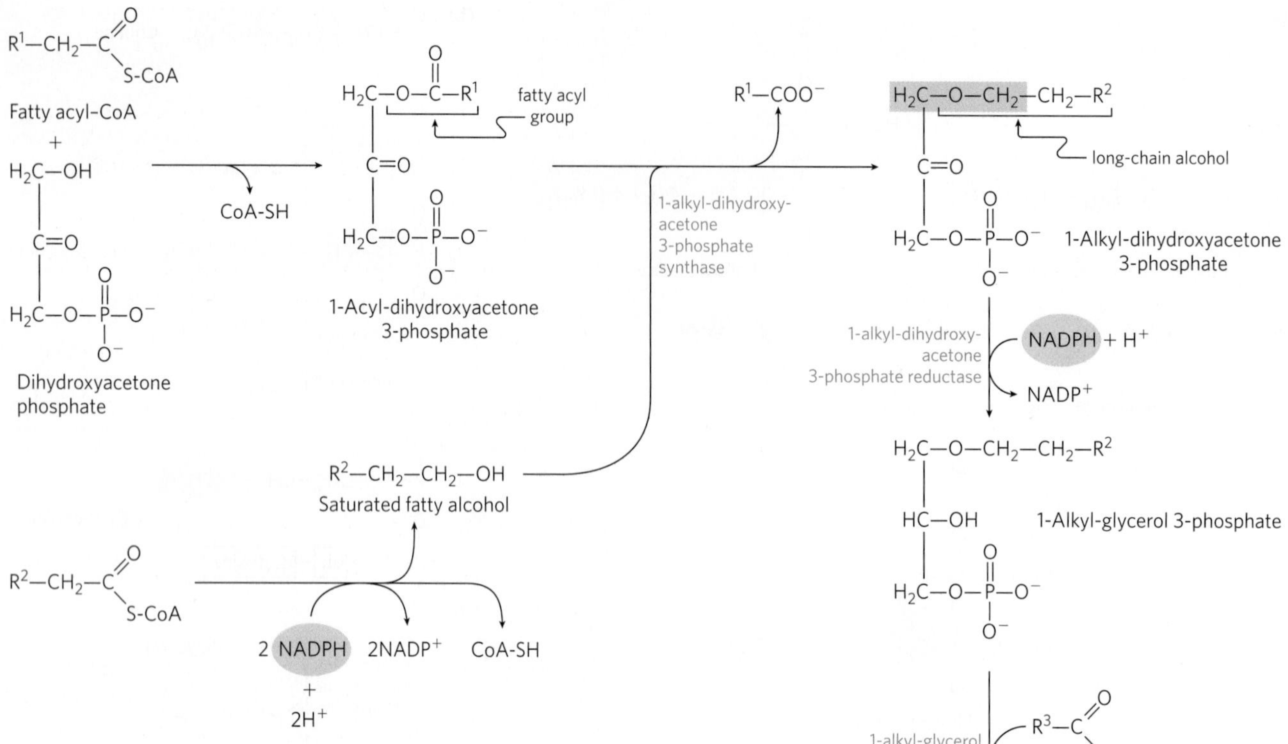

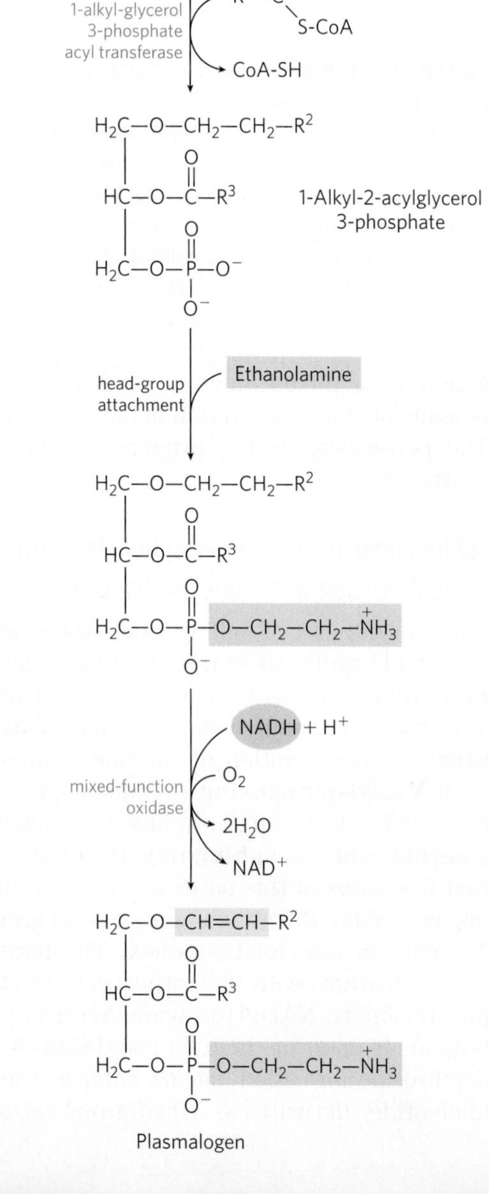

FIGURE 21-30 Synthesis of ether lipids and plasmalogens. The newly formed ether linkage is shaded light red. The intermediate 1-alkyl-2-acyl-glycerol 3-phosphate is the ether analog of phosphatidic acid. Mechanisms for attaching head groups to ether lipids are essentially the same as for their ester-linked analogs. The characteristic double bond of plasmalogens (shaded blue) is introduced in a final step by a mixed-function oxidase system similar to fatty acyl–CoA desaturase, shown in Figure 21-13.

Sphingolipid transfer proteins carry ceramide from the ER to the Golgi complex, where sphingomyelin synthesis occurs. Cytosolic proteins also bind phospholipids and sterols and transport them between cellular membranes. These mechanisms contribute to establishment of the characteristic lipid compositions of organelle membranes (see Fig. 11-2).

SUMMARY 21.3 Biosynthesis of Membrane Phospholipids

- Diacylglycerols are the principal precursors of glycerophospholipids.

- In bacteria, phosphatidylserine is formed by the condensation of serine with CDP-diacylglycerol; decarboxylation of phosphatidylserine produces phosphatidylethanolamine. Phosphatidylglycerol is formed by condensation of CDP-diacylglycerol with glycerol 3-phosphate, followed by removal of the phosphate in monoester linkage.

- Yeast pathways for the synthesis of phosphatidylserine, phosphatidylethanolamine, and phosphatidylglycerol are similar to those in bacteria; phosphatidylcholine is formed by methylation of phosphatidylethanolamine.

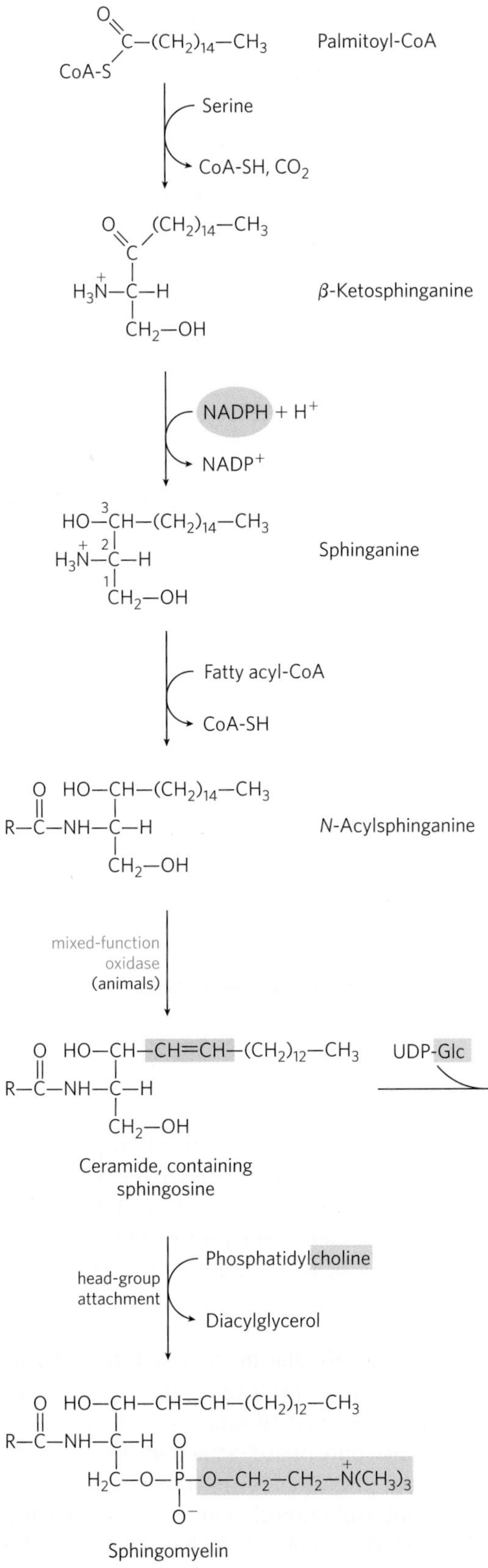

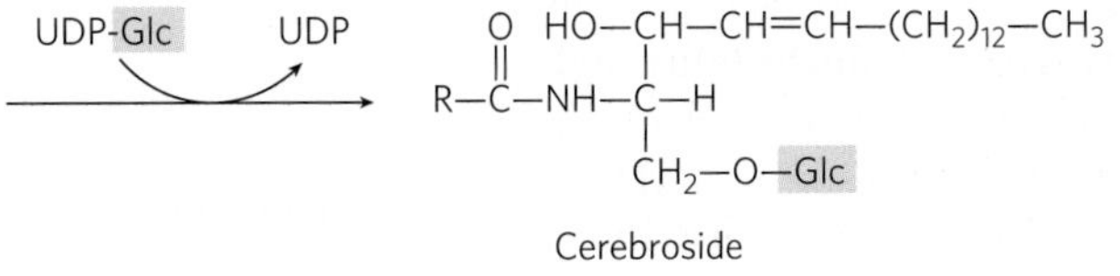

FIGURE 21-31 Biosynthesis of sphingolipids. Condensation of palmitoyl-CoA and serine, forming β-ketosphinganine, followed by reduction with NADPH, yields sphinganine, which is then acylated to *N*-acylsphinganine (a ceramide). In animals, a double bond (shaded light red) is created by a mixed-function oxidase before the final addition of a head group: phosphatidylcholine, to form sphingomyelin, or glucose, to form a cerebroside.

■ Mammalian cells have some pathways similar to those in bacteria, but somewhat different routes for synthesizing phosphatidylcholine and phosphatidylethanolamine. The head-group alcohol (choline or ethanolamine) is activated as the CDP derivative, then condensed with diacylglycerol. Phosphatidylserine is derived only from phosphatidylethanolamine.

■ The characteristic double bond in plasmalogens is introduced by a mixed-function oxidase. The head groups of sphingolipids are attached by unique mechanisms.

■ Phospholipids travel to their intracellular destinations via transport vesicles or specific proteins.

21.4 Cholesterol, Steroids, and Isoprenoids: Biosynthesis, Regulation, and Transport

Cholesterol is doubtless the most publicized lipid, notorious because of the strong correlation between high levels of cholesterol in the blood and the incidence of human cardiovascular diseases. Less well advertised is cholesterol's crucial role as a component of cellular membranes and as a precursor of steroid hormones and bile acids. Cholesterol is an essential molecule in many animals, including humans, but is not required in the mammalian diet—all cells can synthesize it from simple precursors.

The structure of this 27-carbon compound suggests a complex biosynthetic pathway, but all of its carbon atoms are provided by a single precursor—acetate. The **isoprene** units that are the essential intermediates in the pathway from acetate to cholesterol are also precursors to many other natural lipids, and the mechanisms by which isoprene units are polymerized are similar in all these pathways.

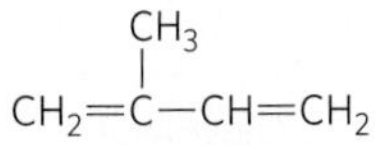

$$CH_2{=}\overset{\displaystyle CH_3}{\underset{\displaystyle}{C}}{-}CH{=}CH_2$$
Isoprene

We begin with an account of the main steps in the biosynthesis of cholesterol from acetate, and then discuss the transport of cholesterol in the blood, its uptake by cells, the normal regulation of cholesterol synthesis, and its regulation in those with defects in cholesterol uptake or transport. We next consider other cellular components derived from cholesterol, such as bile acids and steroid hormones. Finally, an outline of the biosynthetic pathways to some of the many compounds derived from isoprene units, which share early steps with the pathway to cholesterol, illustrates the extraordinary versatility of isoprenoid condensations in biosynthesis.

Cholesterol Is Made from Acetyl-CoA in Four Stages

Cholesterol, like long-chain fatty acids, is made from acetyl-CoA. But the assembly plan of cholesterol is quite different from that of long-chain fatty acids. In early experiments, animals were fed acetate labeled with ^{14}C in either the methyl carbon or the carboxyl carbon. The pattern of labeling in the cholesterol isolated from the two groups of animals **(Fig. 21-32)** provided the blueprint for working out the enzymatic steps in cholesterol biosynthesis.

Synthesis takes place in four stages, as shown in **Figure 21-33**: ❶ condensation of three acetate units to form a six-carbon intermediate, mevalonate; ❷ conversion of mevalonate to activated isoprene units; ❸ polymerization of six 5-carbon isoprene units to form the 30-carbon linear squalene; and ❹ cyclization of squalene to form the four rings of the steroid nucleus, with a further series of changes (oxidations, removal or migration of methyl groups) to produce cholesterol.

Stage ❶ Synthesis of Mevalonate from Acetate The first stage in cholesterol biosynthesis leads to the intermediate **mevalonate (Fig. 21-34)**. Two molecules of acetyl-CoA condense to form acetoacetyl-CoA, which condenses with a third molecule of acetyl-CoA to yield the six-carbon compound β-**hydroxy**-β-**methylglutaryl-CoA (HMG-CoA)**. These first two reactions are catalyzed by **acetyl-CoA acetyl transferase** and **HMG-CoA synthase**, respectively. The cytosolic HMG-CoA synthase in this pathway is distinct from the mitochondrial isozyme that catalyzes HMG-CoA synthesis in ketone body formation (see Fig. 17-18).

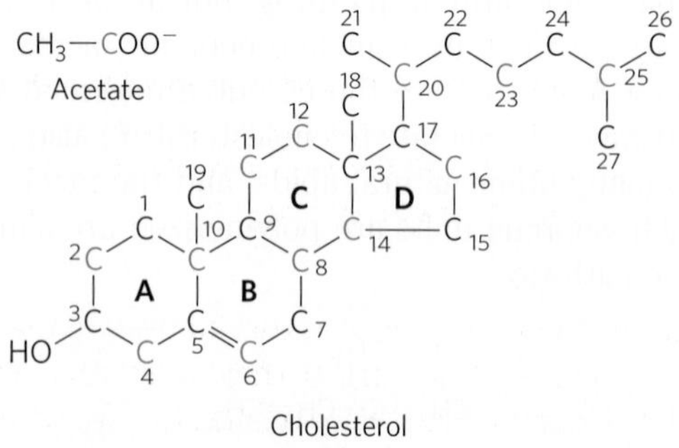

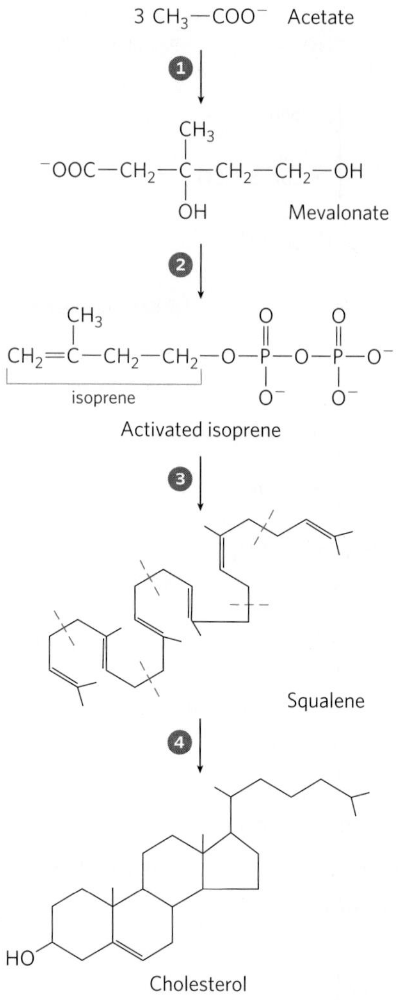

FIGURE 21-33 Summary of cholesterol biosynthesis. The four stages are discussed in the text. Isoprene units in squalene are set off by red dashed lines.

The third reaction is the committed step: reduction of HMG-CoA to mevalonate, for which two molecules of NADPH each donate two electrons. **HMG-CoA reductase**, an integral membrane protein of the smooth ER, is the major point of regulation on the pathway to cholesterol, as we shall see.

Stage ❷ Conversion of Mevalonate to Two Activated Isoprenes In the next stage, three phosphate groups are transferred from three ATP molecules to mevalonate **(Fig. 21-35)**. The phosphate attached to the C-3 hydroxyl group of mevalonate in the intermediate 3-phospho-5-pyrophosphomevalonate is a good leaving group; in the next step, both this phosphate and the nearby carboxyl group leave, producing a double bond in the five-carbon product, Δ^3-**isopentenyl pyrophosphate**. This is the first of the two activated isoprenes central to cholesterol formation. Isomerization of Δ^3-isopentenyl pyrophosphate yields the second activated isoprene, **dimethylallyl pyrophosphate**. Synthesis of isopentenyl pyrophosphate in the cytoplasm of

FIGURE 21-32 Origin of the carbon atoms of cholesterol. This can be deduced from tracer experiments with acetate labeled in the methyl carbon (black) or the carboxyl carbon (red). The individual rings in the fused-ring system are designated A through D.

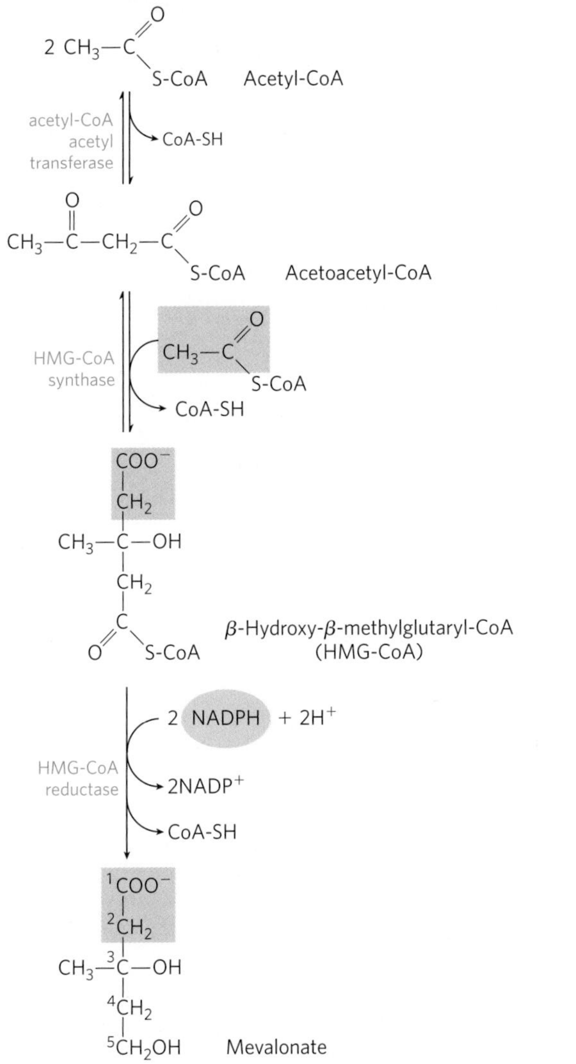

FIGURE 21-34 Formation of mevalonate from acetyl-CoA. The origin of C-1 and C-2 of mevalonate from acetyl-CoA is shaded light red.

plant cells follows the pathway described here. However, plant chloroplasts and many bacteria use a mevalonate-independent pathway. This alternative pathway does not occur in animals, so it is an attractive target for the development of new antibiotics.

Stage ❸ Condensation of Six Activated Isoprene Units to Form Squalene Isopentenyl pyrophosphate and dimethylallyl pyrophosphate now undergo a head-to-tail condensation, in which one pyrophosphate group is displaced and a 10-carbon chain, **geranyl pyrophosphate**, is formed **(Fig. 21-36)**. (The "head" is the end to which pyrophosphate is joined.) Geranyl pyrophosphate undergoes another head-to-tail condensation with isopentenyl pyrophosphate, yielding the 15-carbon intermediate **farnesyl pyrophosphate**. Finally, two molecules of farnesyl pyrophosphate join head to head, with the elimination of both pyrophosphate groups, to form **squalene**. Squalene has 30 carbons, 24 in the main chain and 6 in the form of methyl group branches.

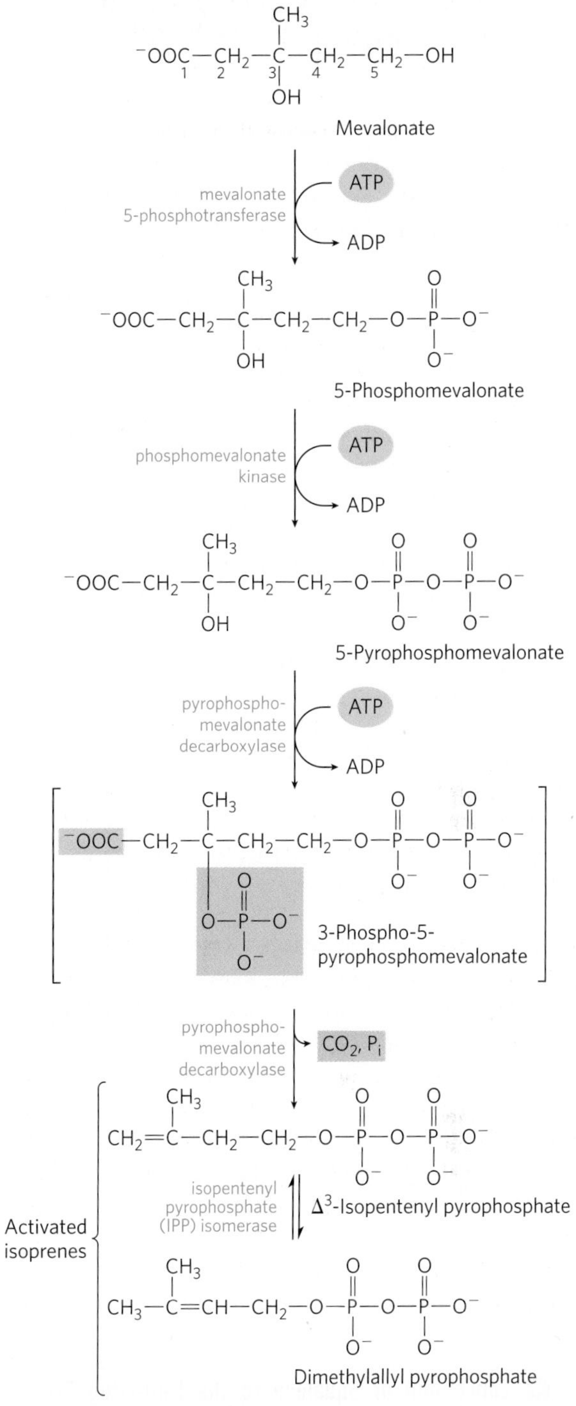

FIGURE 21-35 Conversion of mevalonate to activated isoprene units. Six of these activated units combine to form squalene (see Fig. 21-36). The leaving groups of 3-phospho-5-pyrophosphomevalonate are shaded light red. The bracketed intermediate is hypothetical.

The common names of these intermediates derive from the sources from which they were first isolated. Geraniol, a component of rose oil, has the aroma of geraniums, and farnesol is an aromatic compound found in flowers of the Farnese acacia tree. Many natural scents of plant origin are synthesized from isoprene units. Squalene was first isolated from the liver of sharks (genus *Squalus*).

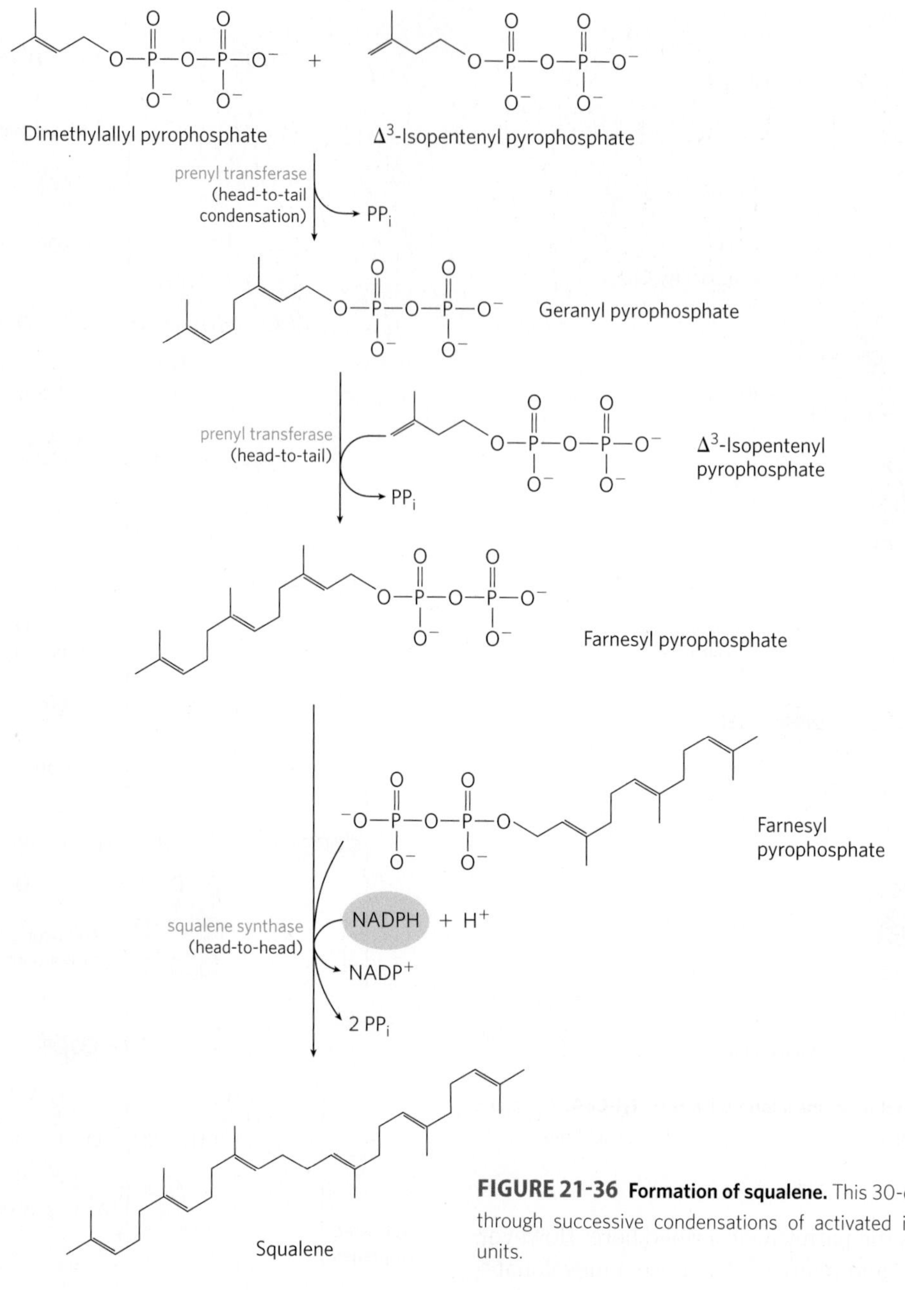

FIGURE 21-36 Formation of squalene. This 30-carbon structure arises through successive condensations of activated isoprene (five-carbon) units.

Stage ④ Conversion of Squalene to the Four-Ring Steroid Nucleus When the squalene molecule is represented as in **Figure 21-37**, the relationship of its linear structure to the cyclic structure of the sterols becomes apparent. All sterols have the four fused rings that form the steroid nucleus, and all are alcohols, with a hydroxyl group at C-3—thus the name "sterol." The action of **squalene monooxygenase** adds one oxygen atom from O_2 to the end of the squalene chain, forming an epoxide. This enzyme is another mixed-function oxidase; NADPH reduces the other oxygen atom of O_2 to H_2O. The double bonds of the product, **squalene 2,3-epoxide**, are positioned so that a remarkable concerted reaction can convert the linear squalene epoxide to a cyclic structure. In animal cells, this cyclization results in the formation of **lanosterol**, which contains the four rings characteristic of the steroid nucleus. Lanosterol is finally converted to cholesterol in a series of about 20 reactions that include the migration of some methyl groups and the removal of others. Elucidation of this extraordinary biosynthetic pathway, one of the most complex known, was accomplished by Konrad Bloch, Feodor Lynen, John Cornforth, and George Popják in the late 1950s.

Cholesterol is the sterol characteristic of animal cells; plants, fungi, and protists make other, closely related sterols instead. They use the same synthetic pathway as far as squalene 2,3-epoxide, at which point the pathways diverge slightly, yielding other sterols, such as stigmasterol in many plants and ergosterol in fungi (Fig. 21-37).

Konrad Bloch, 1912–2000
[Source: AP/Wide World
Photos.]

Feodor Lynen, 1911–1979
[Source: AP/Wide World
Photos.]

John Cornforth, 1917–2013
[Source: Bettmann/Corbis.]

George Popják, 1914–1998
[Source: Professor George
Joseph Popják, MD, DSc, FRS.
Arterioscler. Thromb. Vasc. Biol.
19:830, 1999. ©1999 Wolters
Kluwer Health.]

Squalene

squalene
monooxygenase

NADPH + H$^+$

O$_2$
H$_2$O

NADP$^+$

Squalene 2,3-epoxide

multistep
(plants)

cyclase
(animals)

multistep
(fungi)

Stigmasterol

Ergosterol

cyclase

Lanosterol

multistep

Cholesterol

**FIGURE 21-37 Ring closure converts linear squalene
to the condensed steroid nucleus.** The first step in this
sequence is catalyzed by a mixed-function oxygenase,
for which the cosubstrate is NADPH. The product is an
epoxide, which in the next step is cyclized to the steroid
nucleus. The final product of these reactions in animal
cells is cholesterol; in other organisms, slightly different
sterols are produced, as shown.

WORKED EXAMPLE 21-1 **Energetic Cost of Squalene Synthesis**

What is the energetic cost of the synthesis of squalene from acetyl-CoA, in number of ATPs per molecule of squalene synthesized?

Solution: In the pathway from acetyl-CoA to squalene, ATP is consumed only in the steps that convert mevalonate to the activated isoprene precursors of squalene. Three ATP molecules are used to create each of the six activated isoprenes required to construct squalene, for a total cost of 18 ATP molecules.

Cholesterol Has Several Fates

Most of the cholesterol synthesis in vertebrates takes place in the liver. A small fraction of the cholesterol made there is incorporated into the membranes of hepatocytes, but most of it is exported in one of three forms: as bile acids, biliary cholesterol, or cholesteryl esters **(Fig. 21-38)**. Small quantities of oxysterols such as 25-hydroxycholesterol are formed in the liver and act as regulators of cholesterol synthesis (see below). In other tissues, cholesterol is converted into steroid hormones (in the adrenal cortex and gonads, for example; see Fig. 10-18) or vitamin D hormone (in the liver and kidney; see Fig. 10-19). Such hormones are extremely potent biological signals acting through nuclear receptor proteins.

Bile acids, one of the three forms of cholesterol exported from the liver, are the principal components of bile, a fluid stored in the gallbladder and excreted into the small intestine to aid in the digestion of fat-containing meals. Bile acids and their salts are relatively hydrophilic cholesterol derivatives that serve as emulsifiers in the intestine, converting large particles of fat into tiny micelles and thereby greatly increasing the surface at which digestive lipases can act (see Fig. 17-1). Bile also contains much smaller amounts of cholesterol (biliary cholesterol).

Cholesteryl esters are formed in the liver through the action of **acyl-CoA–cholesterol acyl transferase (ACAT)**. This enzyme catalyzes the transfer of a fatty acid from coenzyme A to the hydroxyl group of cholesterol (Fig. 21-38), converting the cholesterol to a more hydrophobic form and preventing it from entering membranes. Cholesteryl esters are transported in secreted lipoprotein particles to other tissues that use cholesterol, or they are stored in the liver in lipid droplets.

Cholesterol and Other Lipids Are Carried on Plasma Lipoproteins

Cholesterol and cholesteryl esters, like triacylglycerols and phospholipids, are essentially insoluble in water, yet must be moved from the tissue of origin to the tissues in which they will be stored or consumed. They are carried in the blood plasma as **plasma lipoproteins**, macromolecular complexes of specific carrier proteins, called **apolipoproteins**, and various combinations of phospholipids, cholesterol, cholesteryl esters, and triacylglycerols.

Apolipoproteins ("apo" designates the protein in its lipid-free form) combine with lipids to form several classes of lipoprotein particles, spherical complexes with hydrophobic lipids in the core and hydrophilic amino acid side chains at the surface **(Fig. 21-39a)**. Different combinations of lipids and proteins produce particles of different densities, ranging from chylomicrons to high-density lipoproteins. These particles can be separated by ultracentrifugation (Table 21-1) and visualized by electron microscopy (Fig. 21-39b).

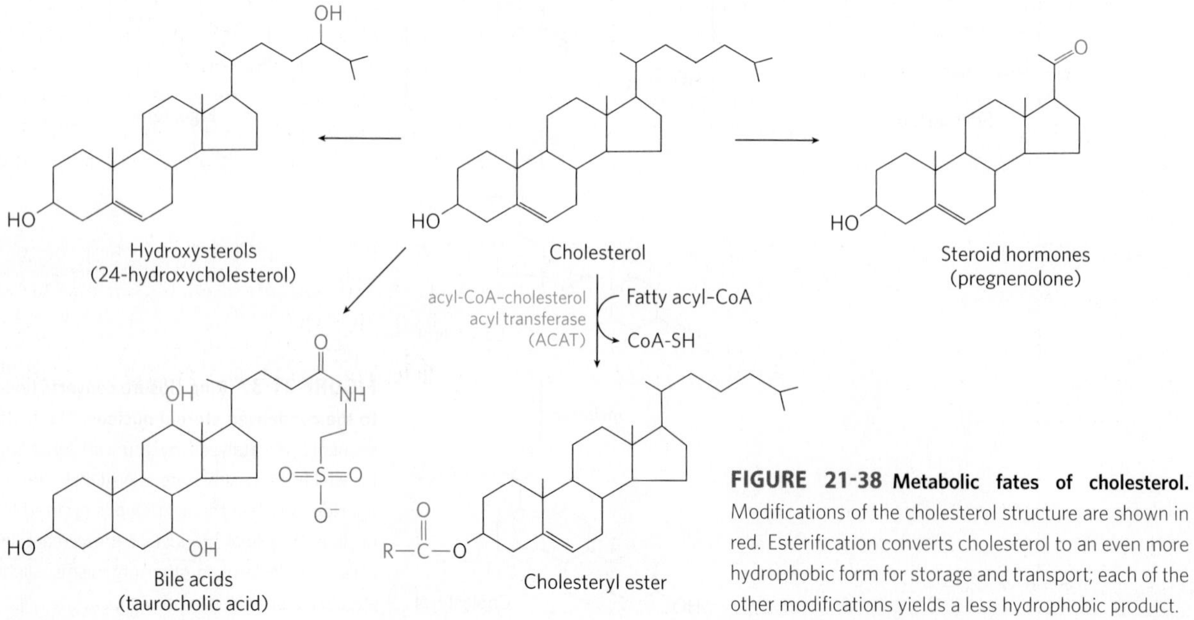

FIGURE 21-38 Metabolic fates of cholesterol. Modifications of the cholesterol structure are shown in red. Esterification converts cholesterol to an even more hydrophobic form for storage and transport; each of the other modifications yields a less hydrophobic product.

Hydroxysterols (24-hydroxycholesterol)

Cholesterol

Steroid hormones (pregnenolone)

acyl-CoA–cholesterol acyl transferase (ACAT)
Fatty acyl-CoA
CoA-SH

Bile acids (taurocholic acid)

Cholesteryl ester

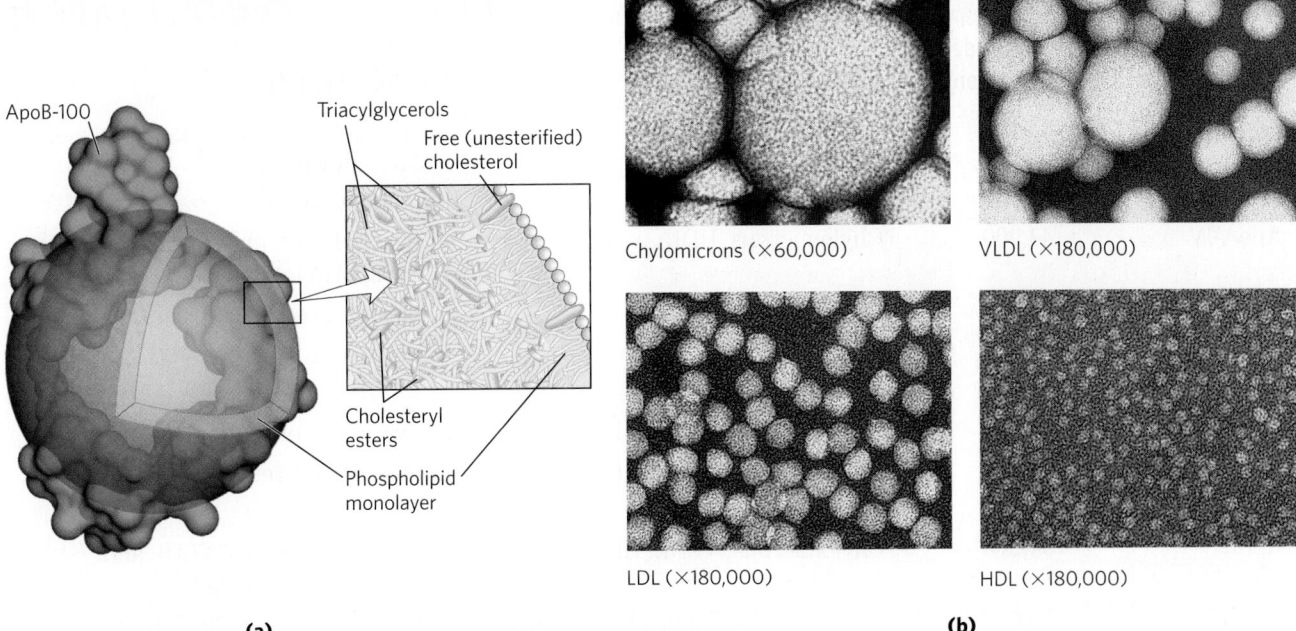

(a)

(b)

FIGURE 21-39 Lipoproteins. (a) Structure of a low-density lipoprotein (LDL). Apolipoprotein B-100 (apoB-100) is one of the largest single polypeptide chains known, with 4,636 amino acid residues (*M*, 512,000). One particle of LDL contains a core with about 1,500 molecules of cholesteryl esters, surrounded by a shell composed of about 500 more molecules of cholesterol, 800 molecules of phospholipids, and one molecule of apoB-100. (b) Four classes of lipoproteins, visualized in the electron microscope after negative staining. Clockwise from top left: chylomicrons, 50 to 200 nm in diameter; VLDL, 28 to 70 nm; HDL, 8 to 11 nm; and LDL, 20 to 25 nm. The particle sizes given are those measured for these samples; particle sizes vary considerably in different preparations. For properties of lipoproteins, see Table 21-1. [Sources: (a) ApoB-100 modeled using data from A. Johs et al., *J. Biol. Chem.* 281:19,732, 2006. (b) Robert Hamilton, Jr., PhD.]

Each class of lipoprotein has a specific function, determined by its point of synthesis, lipid composition, and apolipoprotein content. At least 10 distinct apolipoproteins are found in the lipoproteins of human plasma (Table 21-2), distinguishable by their size, their reactions with specific antibodies, and their characteristic distribution in the lipoprotein classes. These protein components act as signals, targeting lipoproteins to specific tissues or activating enzymes that act on the lipoproteins. They have also been implicated in disease; Box 21-2 describes a link between apoE and Alzheimer disease. **Figure 21-40** provides an overview of the formation and transport of the lipoproteins in mammals. The numbered steps in the following discussion refer to this figure.

Chylomicrons, discussed in Chapter 17 in connection with the movement of dietary triacylglycerols from the intestine to other tissues, are the largest of the lipoproteins and the least dense, containing a high proportion of triacylglycerols (see Fig. 17-2). ❶ Chylomicrons are synthesized from dietary fats in the ER of enterocytes, epithelial cells that line the small intestine. The chylomicrons then move through the lymphatic system and enter the bloodstream via the left subclavian vein. The apolipoproteins of chylomicrons include apoB-48 (unique to this class of lipoproteins), apoE, and apoC-II (Table 21-2). ❷ ApoC-II activates lipoprotein lipase in the capillaries of adipose, heart, skeletal muscle, and lactating mammary tissues, allowing the release of free fatty

TABLE 21-1	Major Classes of Human Plasma Lipoproteins: Some Properties					
		Composition (wt %)				
Lipoprotein	Density (g/mL)	Protein	Phospholipids	Free cholesterol	Cholesteryl esters	Triacylglycerols
Chylomicrons	<1.006	2	9	1	3	85
VLDL	0.95–1.006	10	18	7	12	50
LDL	1.006–1.063	23	20	8	37	10
HDL	1.063–1.210	55	24	2	15	4

Source: Data from D. Kritchevsky, *Nutr. Int.* 2:290, 1986.

TABLE 21-2 Apolipoproteins of the Human Plasma Lipoproteins

Apolipoprotein	Polypeptide molecular weight	Lipoprotein association	Function (if known)
ApoA-I	28,100	HDL	Activates LCAT; interacts with ABC transporter
ApoA-II	17,400	HDL	Inhibits LCAT
ApoA-IV	44,500	Chylomicrons, HDL	Activates LCAT; cholesterol transport/clearance
ApoB-48	242,000	Chylomicrons	Cholesterol transport/clearance
ApoB-100	512,000	VLDL, LDL	Binds to LDL receptor
ApoC-I	7,000	VLDL, HDL	
ApoC-II	9,000	Chylomicrons, VLDL, HDL	Activates lipoprotein lipase
ApoC-III	9,000	Chylomicrons, VLDL, HDL	Inhibits lipoprotein lipase
ApoD	32,500	HDL	
ApoE	34,200	Chylomicrons, VLDL, HDL	Triggers clearance of VLDL and chylomicron remnants
ApoH	50,000	Possibly VLDL, binds phospholipids such as cardiolipin	Roles in coagulation, lipid metabolism, apoptosis, inflammation

Source: Information from D. E. Vance and J. E. Vance (eds), *Biochemistry of Lipids and Membranes*, 5th edn, Elsevier Science Publishing, 2008.

acids (FFA) to these tissues. Chylomicrons thus carry dietary fatty acids to tissues where they will be consumed or stored as fuel. ❸ The remnants of chylomicrons, depleted of most of their triacylglycerols but still containing cholesterol, apoE, and apoB-48, move through the bloodstream to the liver. Receptors in the liver bind to the apoE in the chylomicron remnants and mediate uptake of these remnants by endocytosis. ❹ In the liver, the remnants release their cholesterol and are degraded in lysosomes. This pathway from dietary cholesterol to the liver is the **exogenous pathway** (blue arrows in Fig. 21-40).

When the diet contains more fatty acids and cholesterol than are needed immediately as fuel or precursors to other molecules, ❺ they are converted to triacylglycerols or cholesteryl esters in the liver and packaged with

specific apolipoproteins into **very-low-density lipoprotein (VLDL)**. Excess carbohydrate in the diet can also be converted to triacylglycerols in the liver and exported as VLDL. In addition to triacylglycerols and cholesteryl esters, VLDL contains apoB-100, apoC-I, apoC-II, apoC-III, and apoE (Table 21-2). VLDL is transported in the blood from the liver to muscle and adipose tissue. ❻ In the capillaries of these tissues, apoC-II activates lipoprotein lipase, which catalyzes the release of free fatty acids from triacylglycerols in the VLDL. Adipocytes take up these fatty acids, reconvert them to triacylglycerols, and store the products in intracellular lipid droplets; myocytes, in contrast, primarily oxidize the fatty acids to supply energy. When the insulin level is high (after a meal), VLDL serves primarily to convey

BOX 21-2 ⚕ **MEDICINE** ApoE Alleles Predict Incidence of Alzheimer Disease

There are three common variants, or alleles, of the gene encoding apolipoprotein E in the human population. The most common, accounting for about 78% of human apoE alleles, is *APOE3*; alleles *APOE4* and *APOE2* account for 15% and 7%, respectively. The *APOE4* allele is particularly common in people with Alzheimer disease, and the link is highly predictive. Individuals who inherit *APOE4* have an increased risk of late-onset Alzheimer disease. Those who are homozygous for *APOE4* have a 16-fold increased risk of developing the disease; for those who do, the mean age of onset is just under 70 years. For people who inherit two copies of *APOE3*, by contrast, the mean age of onset of Alzheimer disease exceeds 90 years.

The molecular basis for the association between apoE-4 and Alzheimer disease is not yet known. It is also not clear how apoE-4 might affect growth of the amyloid fibers that appear to be the primary causative agents of the disease (see Fig. 4-32). Speculation has focused on a possible role for apoE in stabilizing the cytoskeletal structure of neurons. The apoE-2 and apoE-3 proteins bind to certain proteins associated with neuronal microtubules, whereas apoE-4 does not. This may accelerate the death of neurons. Whatever the mechanism proves to be, these observations promise to expand our understanding of the biological functions of apolipoproteins.

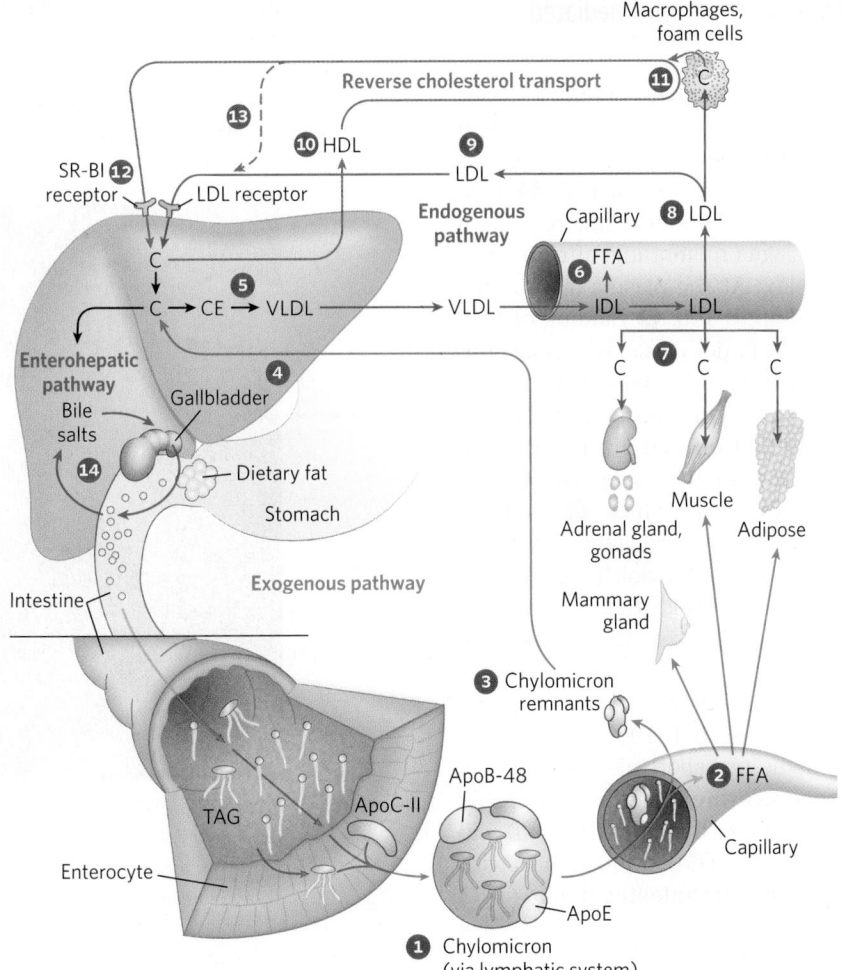

FIGURE 21-40 Lipoproteins and lipid transport. Lipids are transported in the bloodstream as lipoproteins, which exist as several variants that have different functions, different protein and lipid compositions (see Tables 21-1 and 21-2), and thus different densities. Numbered steps are described in the text. In the exogenous pathway (blue arrows), dietary lipids are packaged into chylomicrons; fatty acids from triacylglycerol (TAG) are released by lipoprotein lipase to adipose and muscle tissues, during transport through capillaries. Chylomicron remnants (containing largely protein and cholesterol) are taken up by the liver. Bile salts produced in the liver aid in dispersing dietary fats and are then reabsorbed in the enterohepatic pathway (green arrows). In the endogenous pathway (red arrows), lipids synthesized or packaged in the liver are delivered to peripheral tissues by VLDL. Extraction of lipid from VLDL (along with loss of some apolipoproteins) gradually converts some of it to LDL, which delivers cholesterol to extrahepatic tissues or returns to the liver. Excess cholesterol in extrahepatic tissues is transported back to the liver as HDL in reverse cholesterol transport (purple arrows). C represents cholesterol; CE, cholesteryl ester.

lipids from the diet to adipose tissue for storage. In the fasting state between meals, the fatty acids used to produce VLDL in the liver originate primarily from adipose tissue, and the principal VLDL target is myocytes of the heart and skeletal muscle.

The loss of triacylglycerol converts some VLDL to VLDL remnants, also called intermediate-density lipoprotein (IDL). Further removal of triacylglycerol from IDL (remnants) produces **low-density lipoprotein (LDL)**. Rich in cholesterol and cholesteryl esters, and containing apoB-100 as its major apolipoprotein, ❼ LDL carries cholesterol to extrahepatic tissues such as muscle, adrenal glands, and adipose tissue. These tissues have plasma membrane LDL receptors that recognize apoB-100 and mediate uptake of cholesterol and cholesteryl esters. ❽ LDL also delivers cholesterol to macrophages,

sometimes converting them into foam cells (see Fig. 21-46). ❾ LDL not taken up by peripheral tissues and cells returns to the liver and is taken up via **LDL receptors** in the hepatocyte plasma membrane. Cholesterol that enters hepatocytes by this path may be incorporated into membranes, converted to bile acids, or reesterified by ACAT (Fig. 21-38) for storage within cytosolic lipid droplets. This pathway, from VLDL formation in the liver to LDL return to the liver, is the **endogenous pathway** of cholesterol metabolism and transport (red arrows in Fig. 21-40). Accumulation of excess intracellular cholesterol is prevented by reducing the rate of cholesterol synthesis when sufficient cholesterol is available from LDL in the blood. Regulatory mechanisms to accomplish this are described below. We will return to Figure 21-40 and other pathways of lipoprotein transport after discussing LDL uptake by cells.

Cholesteryl Esters Enter Cells by Receptor-Mediated Endocytosis

Each LDL particle in the bloodstream contains apoB-100, which is recognized by LDL receptors present in the plasma membranes of cells that need to take up cholesterol. **Figure 21-41** shows such a cell. ❶ LDL receptors are synthesized in the endoplasmic reticulum and transported to the plasma membrane, after modification in the Golgi complex. At the plasma membrane, they are available to bind apoB-100. ❷ Binding of LDL to an LDL receptor initiates endocytosis, which ❸ conveys the LDL and its receptor into the cell within an endosome. ❹ The receptor-containing portions of the endosome membrane bud off and are returned to the cell surface, to function again in LDL uptake. ❺ The endosome fuses with a lysosome, which ❻ contains enzymes that hydrolyze the cholesteryl esters, releasing cholesterol and fatty acids into the cytosol. The apoB-100 protein is also degraded to amino acids that are released to the cytosol. ApoB-100 is also present in VLDL, but its receptor-binding domain is not available for binding to the LDL receptor; conversion of VLDL to LDL exposes the receptor-binding domain of apoB-100.

This pathway for the transport of cholesterol in blood and its **receptor-mediated endocytosis** by target tissues was elucidated by Michael Brown and Joseph Goldstein. They discovered that individuals with the genetic disease **familial hypercholesterolemia (FH)** have mutations in the LDL receptor that prevent the normal uptake of LDL by liver and peripheral tissues. The result of defective LDL uptake is very high blood levels of LDL (and of the cholesterol it carries). Individuals with FH have a greatly increased probability of

Michael Brown and Joseph Goldstein
[Source: Mei-Chun Jau. Courtesy of Michael Brown and Joseph Goldstein, University of Texas Southwestern Medical Center.]

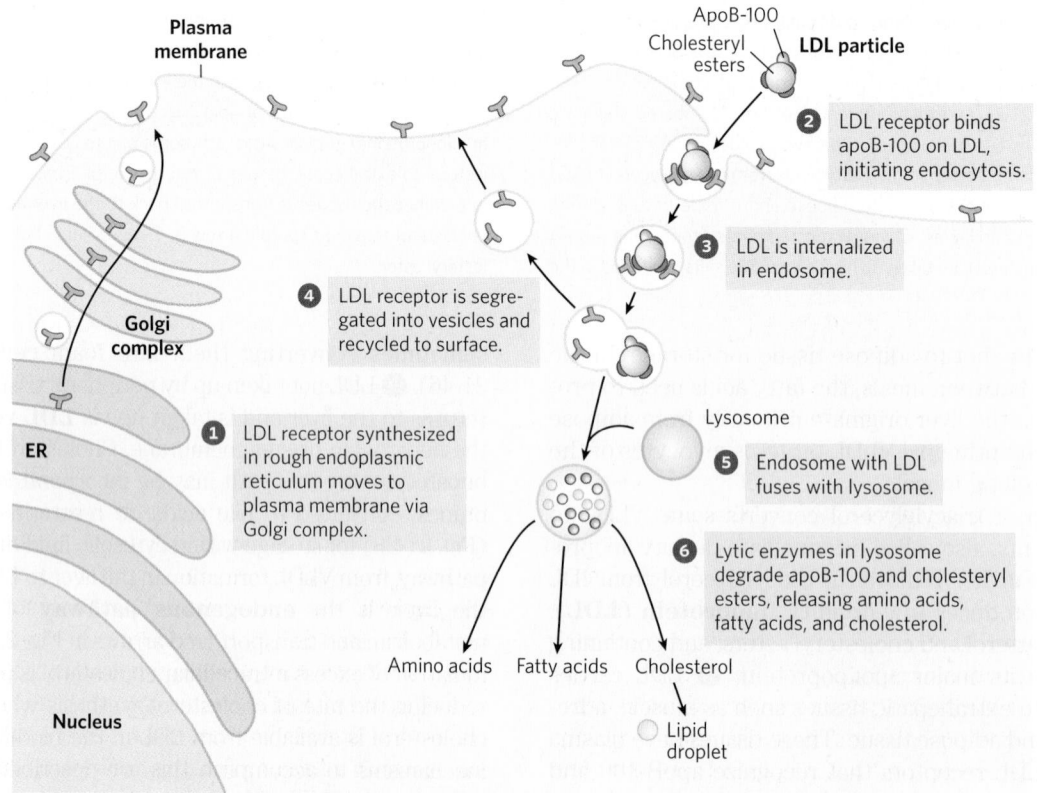

FIGURE 21-41 Uptake of cholesterol by receptor-mediated endocytosis.

developing atherosclerosis, a disease of the cardiovascular system in which blood vessels are occluded by cholesterol-rich plaques (see Fig. 21-46).

Niemann-Pick type-C (NPC) disease is an inherited defect in lipid storage. In this disorder, cholesterol is not transported out of the lysosomes and instead accumulates in lysosomes of liver, brain, and lung, bringing about early death. NPC is the result of a mutation in either of two genes, *NPC1* and *NPC2*, essential to moving cholesterol out of the lysosome and into the cytosol, where it can be further metabolized. *NPC1* encodes a transmembrane lysosomal protein, and *NPC2* encodes a soluble protein. These proteins act in tandem to transfer cholesterol out of the lysosome and into the cytosol for further processing or metabolism. ■

HDL Carries Out Reverse Cholesterol Transport

A fourth major lipoprotein in mammals, **high-density lipoprotein (HDL)**, ❿ originates in the liver and small intestine as small, protein-rich particles that contain relatively little cholesterol and no cholesteryl esters (all step numbers here refer to Fig. 21-40). HDLs contain primarily apoA-I and other apolipoproteins (Table 21-2). They also contain the enzyme **lecithin-cholesterol acyl transferase (LCAT)**, which catalyzes the formation of cholesteryl esters from lecithin (phosphatidylcholine) and cholesterol **(Fig. 21-42)**. LCAT on the surface of nascent (newly forming) HDL particles converts the cholesterol and phosphatidylcholine of chylomicron and VLDL remnants encountered in the bloodstream to cholesteryl esters, which begin to form a core, transforming the disk-shaped nascent HDL to a mature, spherical HDL particle. ⓫ Nascent HDL can also pick up cholesterol from cholesterol-rich extrahepatic cells (including macrophages and foam cells, formed from macrophages; see below). ⓬ Mature HDL then returns to the liver, where the cholesterol is unloaded via the scavenger receptor SR-BI. ⓭ Some of the cholesteryl esters in HDL can also be transferred to LDL by the cholesteryl ester transfer protein. The HDL circuit is **reverse cholesterol transport** (purple arrows in Fig. 21-40). Much of this cholesterol is converted to bile salts by enzymes sequestered in hepatic peroxisomes; the bile salts are stored in the gallbladder and excreted into the intestine when a meal is ingested. ⓮ Bile salts are reabsorbed by the liver and recirculate through the gallbladder in this **enterohepatic circulation** (green arrows in Fig. 21-40).

The unloading of sterols via SR-BI receptors in liver and other tissues does not occur by endocytosis, the mechanism used for LDL uptake. Instead, when HDL binds to SR-BI receptors in the plasma membranes of hepatocytes or steroidogenic tissues such as the adrenal gland, these receptors mediate partial and selective transfer of cholesterol and other lipids in HDL into the cell. Depleted HDL then dissociates to recirculate in the bloodstream and extract more lipids from remnants of

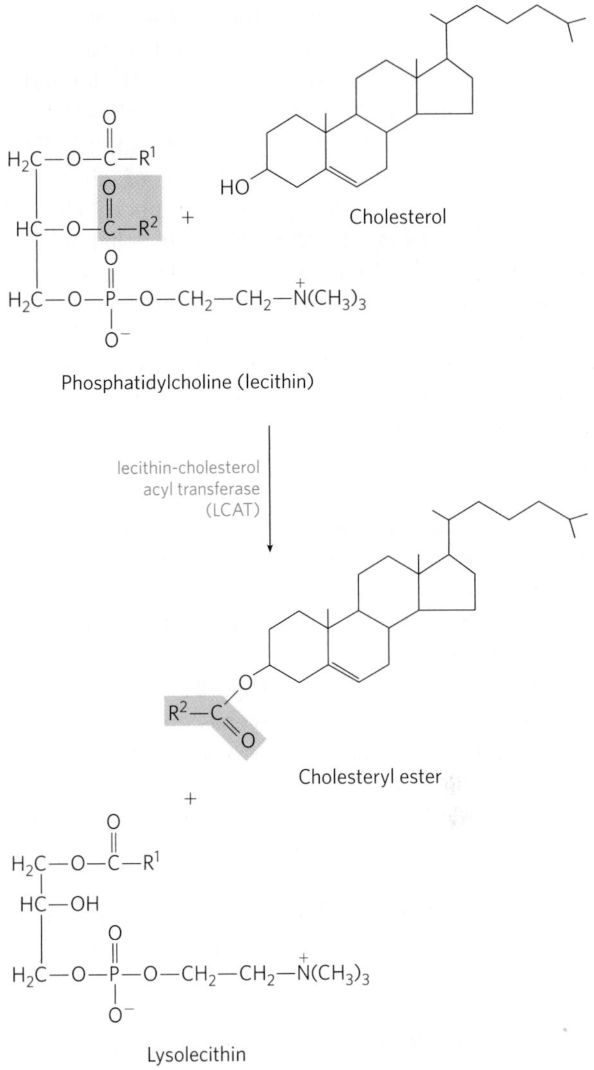

FIGURE 21-42 Reaction catalyzed by lecithin-cholesterol acyl transferase (LCAT). This enzyme is present on the surface of HDL and is stimulated by the HDL component apoA-I. Cholesteryl esters accumulate within nascent HDLs, converting them to mature HDLs.

chylomicrons and VLDL, and from cells overloaded with cholesterol, as described below.

Cholesterol Synthesis and Transport Are Regulated at Several Levels

Cholesterol synthesis is a complex and energy-expensive process. Excess cholesterol cannot be catabolized for use as fuel and must be excreted. Therefore, it is clearly advantageous to an organism to regulate the biosynthesis of cholesterol to complement dietary intake. In mammals, cholesterol production is regulated by intracellular cholesterol concentration, by the supply of ATP, and by the hormones glucagon and insulin. The committed step in the pathway to cholesterol (and a major site of regulation) is the conversion of HMG-CoA to mevalonate (Fig. 21-34), the reaction catalyzed by HMG-CoA reductase.

Short-term regulation of the *activity* of existing HMG-CoA reductase is accomplished by reversible covalent alteration: phosphorylation by the AMP-dependent protein kinase (AMPK), which senses high AMP concentration (indicating low ATP concentration). Thus, when ATP levels drop, the synthesis of cholesterol slows, and catabolic pathways for the generation of ATP are stimulated **(Fig. 21-43)**. Hormones that mediate global regulation of lipid and carbohydrate metabolism also act on HMG-CoA reductase; glucagon stimulates its phosphorylation (inactivation), and insulin promotes dephosphorylation, activating the enzyme and favoring cholesterol synthesis. These covalent regulatory mechanisms are probably not as important, quantitatively, as the mechanisms that affect the synthesis and degradation of the enzyme.

In the longer term, the *number of molecules* of HMG-CoA reductase is increased or decreased in response to cellular concentrations of cholesterol. Regulation of HMG-CoA reductase synthesis by cholesterol is mediated by an elegant system of transcriptional regulation of the HMG-CoA gene **(Fig. 21-44)**. This gene, along with more than 20 other genes encoding enzymes that mediate the uptake and synthesis of cholesterol and unsaturated fatty acids, is controlled by a small family of proteins called **sterol regulatory element-binding proteins (SREBPs)**. When newly synthesized, these proteins are embedded in the ER. Only the soluble regulatory domain fragment of an SREBP functions as a transcription activator, through mechanisms discussed in Chapter 28. When cholesterol and oxysterol levels are high, SREBPs are held in the ER in a complex with another protein called **SREBP cleavage-activating protein (SCAP)**,

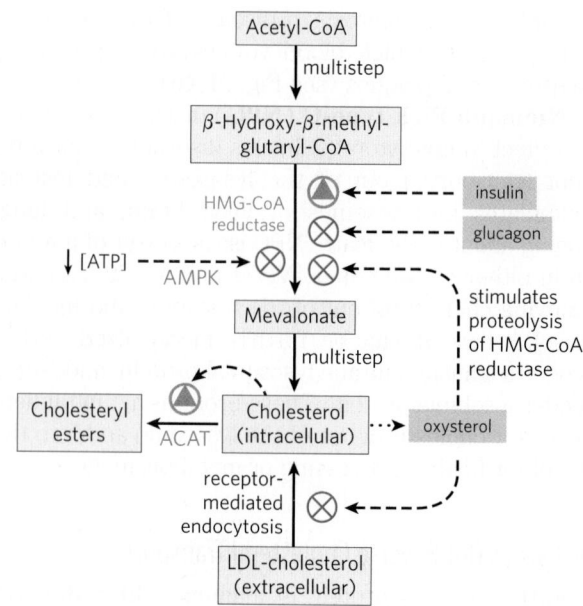

FIGURE 21-43 Regulation of cholesterol formation balances synthesis with dietary uptake and energy state. Insulin promotes dephosphorylation (activation) of HMG-CoA reductase; glucagon promotes its phosphorylation (inactivation); and the AMP-dependent protein kinase AMPK, when activated by low [ATP] relative to [AMP], phosphorylates and inactivates HMG-CoA reductase. Oxysterol metabolites of cholesterol (for example, 24(S)-hydroxycholesterol) stimulate proteolysis of HMG-CoA reductase.

which in turn is anchored in the ER membrane by its interaction with a third membrane protein, **Insig (*insulin-induced gene* protein)** (Fig. 21-44a). SCAP and Insig act as sterol sensors. When sterol levels are high, the

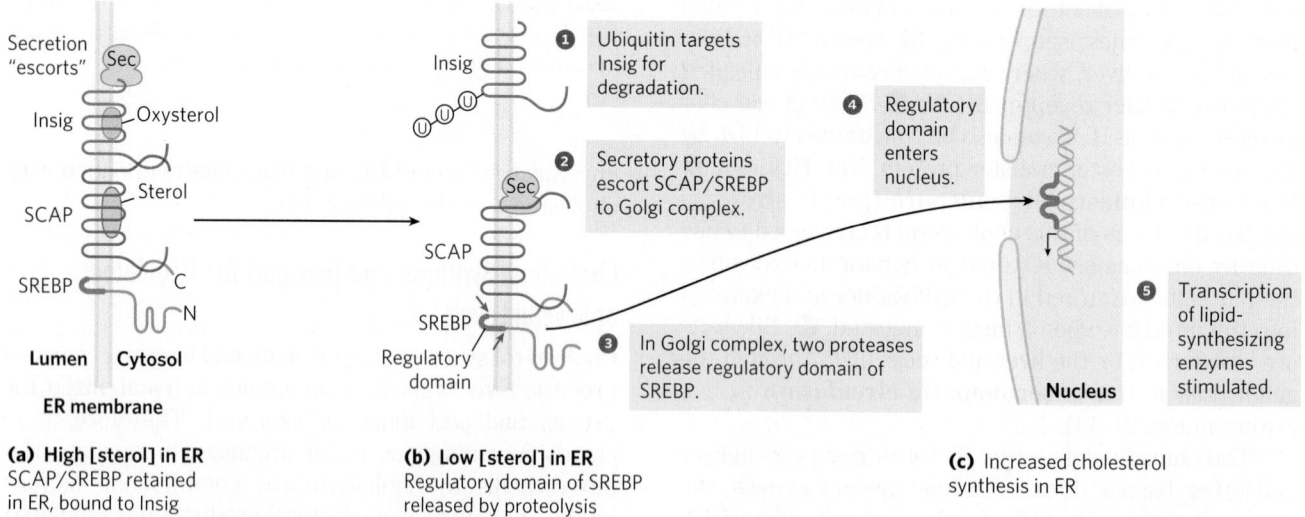

(a) High [sterol] in ER
SCAP/SREBP retained in ER, bound to Insig

(b) Low [sterol] in ER
Regulatory domain of SREBP released by proteolysis

(c) Increased cholesterol synthesis in ER

FIGURE 21-44 Regulation of cholesterol synthesis by SREBP. Sterol regulatory element-binding proteins (SREBPs, shown in green) are embedded in the ER when first synthesized, in a complex with the protein SREBP cleavage-activating protein (SCAP, red), which is in turn bound to Insig (blue). (*N* and *C* represent the amino and carboxyl termini of the proteins.) **(a)** When bound to SCAP and Insig, SREBPs are inactive. **(b)** When sterol levels decline, sterol-binding sites on Insig and SCAP are unoccupied, the complex migrates to the Golgi complex, and SREBP is cleaved (red arrows) to produce a regulatory domain fragment. **(c)** This domain acts in the nucleus to increase the transcription of sterol-regulated genes. Insig is targeted for degradation by the attachment of several ubiquitin molecules. [Source: Information from R. Raghow et al., *Trends Endocrinol. Metab.* 19:65, 2008, Fig. 2.]

Insig-SCAP-SREBP complex is retained in the ER membrane. When the level of sterols in the cell declines (Fig. 21-44b), the SCAP-SREBP complex is escorted by secretory proteins to the Golgi complex. There, two proteolytic cleavages of SREBP release a regulatory fragment, which enters the nucleus and activates transcription of its target genes, including those for HMG-CoA reductase, the LDL receptor protein, and other proteins needed for lipid synthesis. When sterol levels increase sufficiently, the proteolytic release of SREBP amino-terminal domains is again blocked, and proteasome degradation of the existing active domains results in rapid shutdown of the gene targets.

In the long term, the level of HMG-CoA reductase is also regulated by proteolytic degradation of the enzyme itself. High levels of cellular cholesterol are sensed by Insig, which triggers attachment of ubiquitin molecules to HMG-CoA reductase, leading to its degradation by proteasomes (see Fig. 27-50).

Liver X receptor (LXR) is a nuclear transcription factor activated by oxysterol ligands (reflecting high cholesterol levels), which integrates the metabolism of fatty acids, sterols, and glucose. LXRα is expressed primarily in liver, adipose tissue, and macrophages; LXRβ is present in all tissues. When bound to an oxysterol ligand, LXRs form heterodimers with a second type of nuclear receptor, the **retinoid X receptors (RXR)**, and the LXR-RXR dimer activates transcription from a set of genes **(Fig. 21-45)**, including those for acetyl-CoA carboxylase, the first enzyme in fatty acid synthesis; fatty acid synthase; the cytochrome P-450 enzyme CYP7A1, required for sterol conversion to bile acids; apoproteins that participate in cholesterol transport (apoC-I, apoC-II, apoD, and apoE); the ATP-binding cassette (ABC) transporters ABCA1 and ABCG1, required for reverse cholesterol transport (see below); GLUT4, the insulin-stimulated glucose transporter of muscle and adipose tissue; and an SREBP called SREBP1C. The transcriptional regulators LXR and SREBP therefore work together to achieve and maintain cholesterol homeostasis; SREBPs are activated by low levels of cellular cholesterol, and LXRs are activated by high cholesterol levels.

Regulation by LXRs is complemented by the activity of farnesoid X receptor (FXR), which also forms a heterodimer with RXR, with an effect that is often reciprocal to that of LXR-RXR. Although farnesol is a ligand for this receptor, FXR responds primarily to bile acids. High levels of bile acids can be toxic. FXR, expressed mainly in the intestine, liver, kidney, and adrenal glands, provides essential control of bile acid levels by increasing or decreasing expression of multiple genes. For example, LXR activates transcription of the cytochrome P-450 enzyme CYP7A1, which promotes bile acid formation from sterols. FXR represses transcription of CYP7A1 when levels of bile acids are high.

Finally, two other regulatory mechanisms influence cellular cholesterol level: (1) high cellular concentrations of cholesterol activate ACAT, which increases

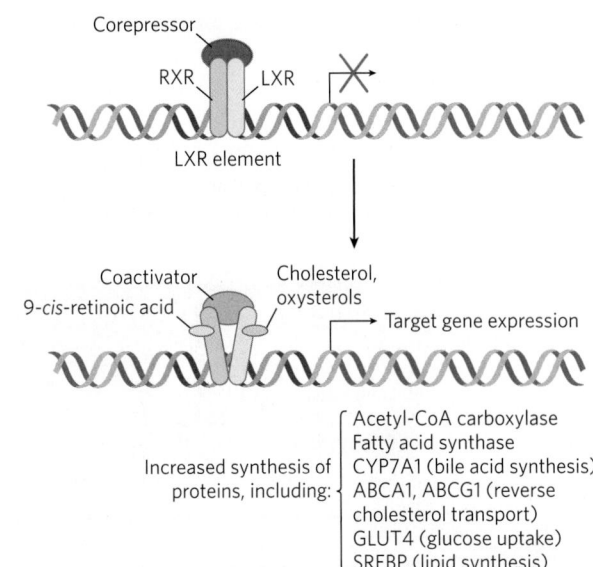

FIGURE 21-45 Action of RXR-LXR dimer on expression of genes for lipid and glucose metabolism. When their ligands are absent, RXR and LXR associate with a corepressor protein, preventing transcription of the genes associated with the LXR element (LXRE). When their respective ligands are present—9-*cis*-retinoic acid for RXR, cholesterol or oxysterols for LXR—the dimer dissociates from the corepressor, then associates with a coactivator protein. This complex binds to the LXR element and turns on expression of the associated genes. Regulation of gene expression is a topic discussed in more detail in Chapter 28. [Source: Information from A. C. Calkin and P. Tontonoz, *Nature Rev. Mol. Cell Biol.* 13:213, 2012, Fig. 1.]

esterification of cholesterol for storage, and (2) high cellular cholesterol levels diminish (via SREBP) transcription of the gene that encodes the LDL receptor, reducing production of the receptor and thus the uptake of cholesterol from the blood.

Dysregulation of Cholesterol Metabolism Can Lead to Cardiovascular Disease

As noted earlier, cholesterol cannot be catabolized by animal cells. Excess cholesterol can only be removed by excretion or by conversion to bile salts. When the sum of cholesterol synthesized and cholesterol obtained in the diet exceeds the amount required for the synthesis of membranes, bile salts, and steroids, pathological accumulations of cholesterol (plaques) can obstruct blood vessels, a condition called **atherosclerosis**. Heart failure due to occluded coronary arteries is a leading cause of death in industrialized societies. Atherosclerosis is linked to high levels of cholesterol in the blood, and particularly to high levels of LDL-cholesterol ("bad cholesterol"); there is a *negative* correlation between HDL ("good cholesterol") levels and arterial disease. Plaque formation in blood vessels is initiated when LDL containing partially oxidized fatty acyl groups adheres to and accumulates in the extracellular matrix of epithelial cells lining

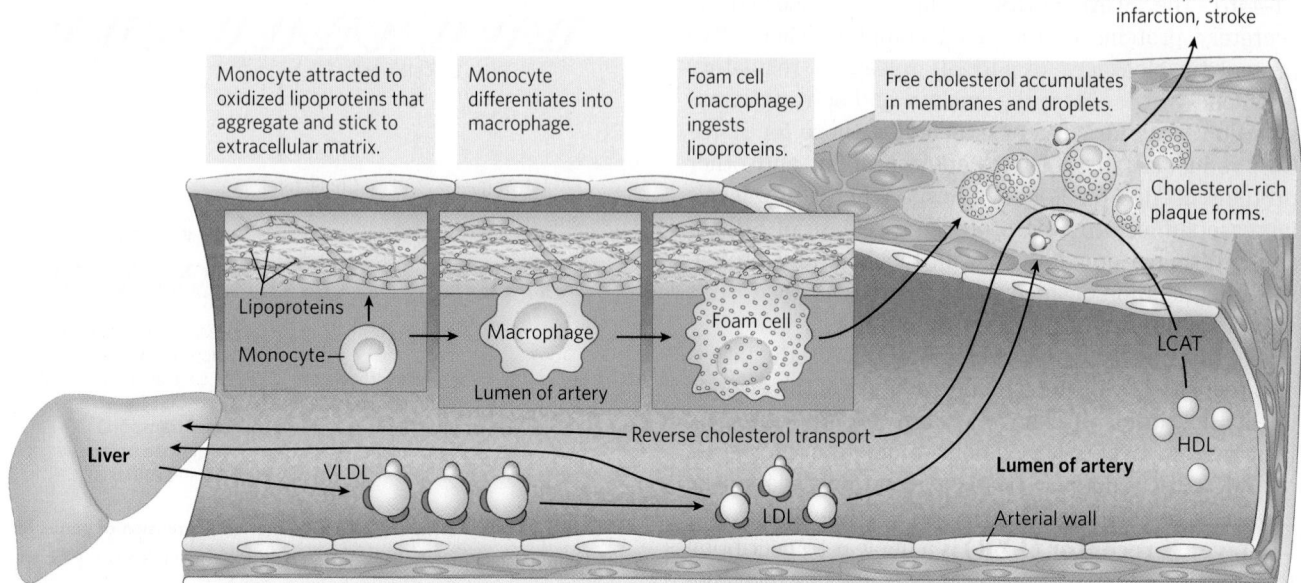

FIGURE 21-46 Formation of atherosclerotic plaques. Excess lipid derived from the diet is deposited on arterial walls, a process facilitated by the conversion of monocytes to foam cells and incorporation of foam cells into growing plaques. Some of this deposition is countered by HDL and reverse cholesterol transport. The LCAT reaction is shown in Figure 21-42.

arteries **(Fig. 21-46)**. Immune cells (monocytes) are attracted to regions with such LDL accumulations, and they differentiate into macrophages, which take up the oxidized LDL and the cholesterol they contain. Macrophages cannot limit their uptake of sterols, and with increasing accumulation of cholesteryl esters and free cholesterol, the macrophages become **foam cells** (they appear foamy when viewed under the microscope). As excess free cholesterol accumulates in foam cells and their membranes, the cells undergo apoptosis. Over long periods of time, arteries become progressively occluded as plaques consisting of extracellular matrix material, scar tissue formed from smooth muscle tissue, and foam cell remnants gradually become larger. Occasionally, a plaque breaks loose from the site of its formation and is carried through the blood to a narrowed region of an artery in the brain or the heart, causing a stroke or a heart attack.

In familial hypercholesterolemia, blood levels of cholesterol are extremely high and severe atherosclerosis develops in childhood. These individuals have a defective LDL receptor and lack receptor-mediated uptake of cholesterol carried by LDL. Consequently, cholesterol is not cleared from the blood; it accumulates in foam cells and contributes to the formation of atherosclerotic plaques. Endogenous cholesterol synthesis continues despite the excessive cholesterol in the blood, because extracellular cholesterol cannot enter cells to regulate intracellular synthesis (Fig. 21-44). A class of drugs called **statins**, some isolated from natural sources and some synthesized industrially, are used to treat patients with elevated serum cholesterol caused

by familial hypercholesterolemia and other conditions. The statins resemble mevalonate (Box 21-3) and are competitive inhibitors of HMG-CoA reductase.

An alternative approach to controlling serum cholesterol levels is to activate LXRs, which has the overall effect of decreasing cholesterol absorption and promoting its excretion. This is the mode of action of a drug called ezetimibe. Because LXR activation also activates SREBP1C, causing the liver to increase its production of fatty acids and triacylglycerols, new classes of drugs that target only intestinal LXRs are being developed. ∎

Reverse Cholesterol Transport by HDL Counters Plaque Formation and Atherosclerosis

HDL plays a critical role in the reverse cholesterol transport pathway **(Fig. 21-47)**, reducing the potential damage from foam cell buildup. Depleted HDL (low in cholesterol) picks up cholesterol stored in extrahepatic tissues, including foam cells at nascent plaques, and carries it to the liver.

Cholesterol movement out of cells requires transporters. The human genome encodes 48 transporters of the ATP-binding cassette (ABC) class, and about half of these promote lipid transport. Two of them transport cholesterol out of cells. In this process, apoA-I interacts with the transporter ABCA1 in a cholesterol-rich cell. ABCA1 transports a load of cholesterol from inside the cell to the outer surface of the plasma membrane, where lipid-free or lipid-poor apoA-I picks it up, then transports it to the liver. Another transporter, ABCG1, interacts with mature HDL, facilitating the movement of cholesterol out

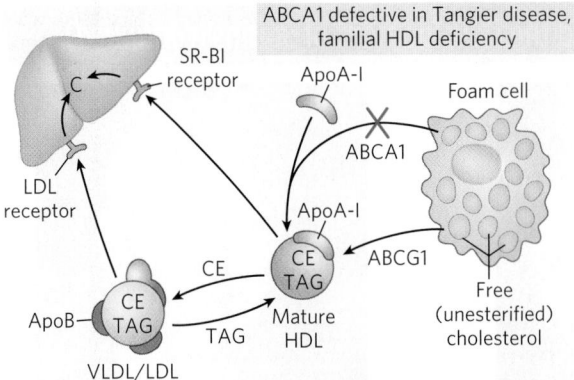

ABCA1 defective in Tangier disease, familial HDL deficiency

FIGURE 21-47 Reverse cholesterol transport. ApoA-I and HDLs pick up excess cholesterol (C) from peripheral cells, with the participation of ABCA1 and ABCG1 transporters, and return it to the liver. In individuals with genetically defective ABCA1, the failure of reverse cholesterol transport leads to severe and early cardiovascular diseases: Tangier disease and familial HDL deficiency disease. CE, cholesteryl esters; TAG, triacylglycerols. [Source: Information from A. R. Tall et al., *Cell Metab.* 7:365, 2008, Fig. 1.]

of the cell and into the HDL. This efflux process is particularly critical to reverse cholesterol transport away from foam cells at the sites of plaques in the blood vessels of individuals with cardiovascular disease.

In **familial HDL deficiency**, HDL levels are very low, and in **Tangier disease** they are almost undetectable (Fig. 21-47). Both genetic disorders are the result of mutations in the ABCA1 protein. ApoA-I in cholesterol-depleted HDL cannot take up cholesterol from cells that lack ABCA1 protein, and apoA-I and cholesterol-poor HDL are rapidly removed from the blood and destroyed. Both familial HDL deficiency and Tangier disease are very rare (worldwide, fewer than 100 families

---

BOX 21-3 ☤ MEDICINE The Lipid Hypothesis and the Development of Statins

Coronary heart disease is the leading cause of death in developed countries. The coronary arteries that bring blood to the heart become narrowed due to the formation of fatty deposits called atherosclerotic plaques, containing cholesterol, fibrous proteins, calcium deposits, blood platelets, and cell debris. Developing the link between artery occlusion (atherosclerosis) and blood cholesterol levels was a project of the twentieth century, triggering a dispute that was resolved only with the development of effective cholesterol-lowering drugs. The Framingham Heart Study, a longitudinal study begun in 1948 and continuing today, was aimed at identifying factors correlated with cardiovascular disease. About 5,000 participants from the city of Framingham, Massachusetts, underwent periodic physical examinations and lifestyle interviews. By 2002, participants of the third generation were included in the study. This monumental study led to the identification of risk factors for cardiovascular disease, including smoking, obesity, physical inactivity, diabetes, high blood pressure, and high blood cholesterol.

In 1913, N. N. Anitschkov, an experimental pathologist in Saint Petersburg, Russia, published a study showing that rabbits fed a diet rich in cholesterol developed lesions very similar to the atherosclerotic plaques seen in aging humans. Anitschkov continued his work over the next few decades, publishing it in prominent western journals. Nevertheless, the work was not accepted as a model for human atherosclerosis, due to a prevailing view that the disease was simply a consequence of aging and could not be prevented. The link between serum cholesterol and atherosclerosis (the lipid hypothesis) was gradually strengthened, however, until researchers in the 1960s openly suggested that therapeutic intervention might be helpful. Controversy persisted until the results of a large study of cholesterol lowering, sponsored by the U.S. National

Institutes of Health, was published in 1984: the Coronary Primary Prevention Trial. This study conclusively showed a statistically significant decrease in heart attacks and strokes as a result of decreasing blood cholesterol level. The study made use of a bile acid–binding resin, cholestyramine, to control cholesterol. The results triggered a search for more effective therapeutic interventions. Some controversy persisted until development of the statins in the late 1980s and 1990s.

Dr. Akira Endo, working at the Sankyo company in Tokyo, discovered the first statin and reported the work in 1976. Endo had been interested in cholesterol metabolism for some time, and speculated in 1971 that the fungi being screened at that time for new antibiotics might also contain an inhibitor of cholesterol synthesis. Over a period of several years, he screened more than 6,000 fungal cultures until a positive result emerged. The compound that resulted was named compactin (Fig. 1). This compound eventually proved

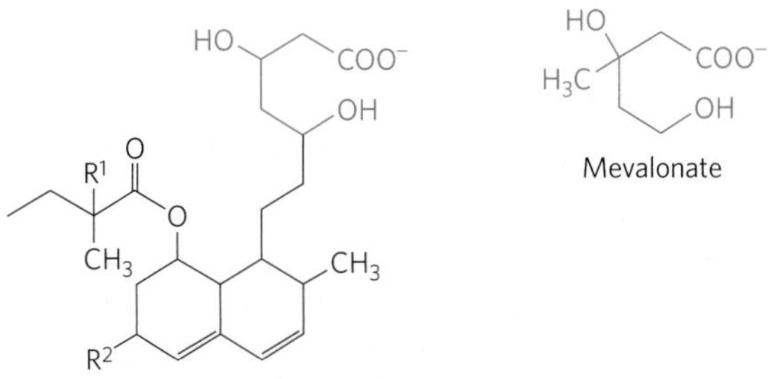

R¹ = H	R² = H	Compactin
R¹ = CH₃	R² = CH₃	Simvastatin (Zocor)
R¹ = H	R² = OH	Pravastatin (Pravachol)
R¹ = H	R² = CH₃	Lovastatin (Mevacor)

FIGURE 1 Statins as inhibitors of HMG-CoA reductase. A comparison of the structures of mevalonate and four pharmaceutical compounds (statins) that inhibit HMG-CoA reductase.

(Continued on next page)

BOX 21-3 ⚕ **MEDICINE** The Lipid Hypothesis and the Development of Statins (*Continued*)

Akira Endo
[Source: Courtesy of Akira Endo, Ph.D.]

Alfred Alberts
[Source: Merck Archives, Merck & Co., Inc. 2016]

P. Roy Vagelos
[Source: Courtesy of P. Roy Vagelos.]

effective in reducing cholesterol levels in dogs and monkeys, and the work came to the attention of Michael Brown and Joseph Goldstein at the University of Texas–Southwestern medical school. Brown and Goldstein began to work with Endo, and confirmed his results. Some dramatic results in the first limited clinical trials convinced several pharmaceutical firms to join the hunt for statins. A team at Merck, led by Alfred Alberts and P. Roy Vagelos, began screening fungal cultures and found a positive result after screening just 18 cultures. The new statin was eventually called lovastatin (Fig. 1). In 1980, a rumor that compactin, at very high doses, was carcinogenic in dogs almost sidelined the race to develop statins, but the benefits to people with familial hypercholesterolemia were already evident. After much consultation with experts around the world and with the U.S. Food and Drug Administration, Merck proceeded carefully to develop lovastatin. Extensive testing over the next two decades revealed no carcinogenic effects from lovastatin, or from the newer generations of statins that have appeared since.

Statins inhibit HMG-CoA reductase, in part, by mimicking the structure of mevalonate (Fig. 1), and thus inhibit cholesterol synthesis. Lovastatin treatment lowers serum cholesterol by as much as 30% in individuals with hypercholesterolemia resulting from one defective copy of the gene for the LDL receptor. When combined with an edible resin that binds bile acids and prevents their reabsorption from the intestine, the statin is even more effective.

Statins are now the most widely used drugs for lowering serum cholesterol levels. Side effects are always a concern with drugs, but in the case of statins, many of the side effects are positive. These drugs can improve blood flow, enhance the stability of atherosclerotic plaques (so they don't rupture and obstruct blood flow), reduce platelet aggregation, and reduce vascular inflammation. In patients taking statins for the first time, some of these effects occur before cholesterol levels drop and may be related to a secondary inhibition of isoprenoid synthesis. Not all effects of statins are positive. A few individuals, usually among those taking statins in combination with other cholesterol-lowering drugs, experience muscle pain or weakness that can become severe and even debilitating. A fairly long list of other side effects has been documented; most are rare. However, for the vast majority of people, the statin-mediated decrease in risks associated with coronary heart disease can be dramatic. As with all medications, statins should be used only in consultation with a physician.

with Tangier disease are known), but the existence of these diseases establishes a role for ABCA1 and ABCG1 proteins in the regulation of plasma HDL levels. ∎

Steroid Hormones Are Formed by Side-Chain Cleavage and Oxidation of Cholesterol

Humans derive all their steroid hormones from cholesterol **(Fig. 21-48)**. Two classes of steroid hormones are synthesized in the cortex of the adrenal gland:

mineralocorticoids, which control the reabsorption of inorganic ions (Na^+, Cl^-, and HCO_3^-) by the kidney, and **glucocorticoids**, which help regulate gluconeogenesis and reduce the inflammatory response. Sex hormones are produced in male and female gonads and the placenta. They include **progesterone**, which regulates the female reproductive cycle, and **androgens** (such as testosterone) and **estrogens** (such as estradiol), which influence the development of secondary sexual characteristics in males and females, respectively. Steroid

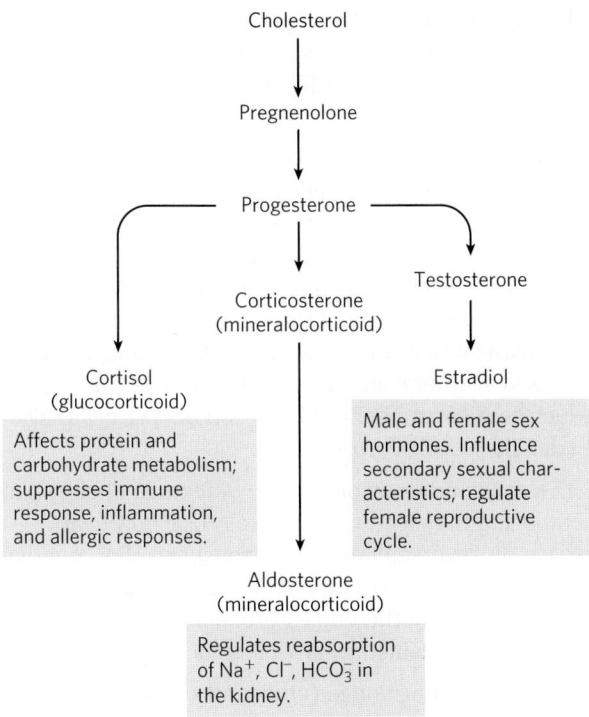

FIGURE 21-48 Some steroid hormones derived from cholesterol. The structures of some of these compounds are shown in Figure 10-18.

hormones are effective at very low concentrations and are therefore synthesized in relatively small quantities. In comparison with the bile salts, their production consumes relatively little cholesterol.

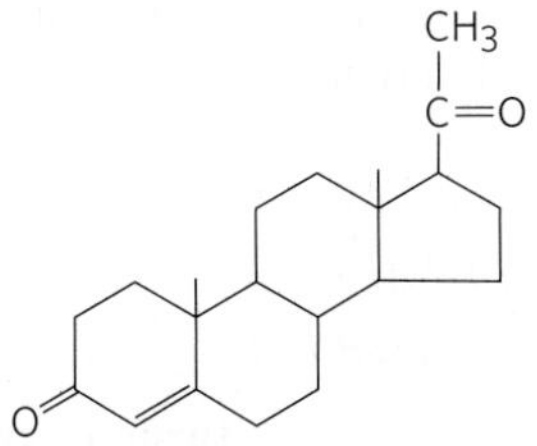

Synthesis of steroid hormones requires removal of some or all of the carbons in the "side chain" on C-17 of the D ring of cholesterol. Side-chain removal takes place in the mitochondria of steroidogenic tissues. Removal requires the hydroxylation of two adjacent carbons in the side chain (C-20 and C-22) followed by cleavage of the bond between them **(Fig. 21-49)**. Formation of the various hormones also requires the introduction of oxygen atoms. All the hydroxylation and oxygenation reactions in steroid biosynthesis are catalyzed by mixed-function oxygenases (see Box 21-1) that use NADPH, O_2, and mitochondrial cytochrome P-450.

Intermediates in Cholesterol Biosynthesis Have Many Alternative Fates

In addition to its role as an intermediate in cholesterol biosynthesis, isopentenyl pyrophosphate is the activated precursor of a huge array of biomolecules with diverse

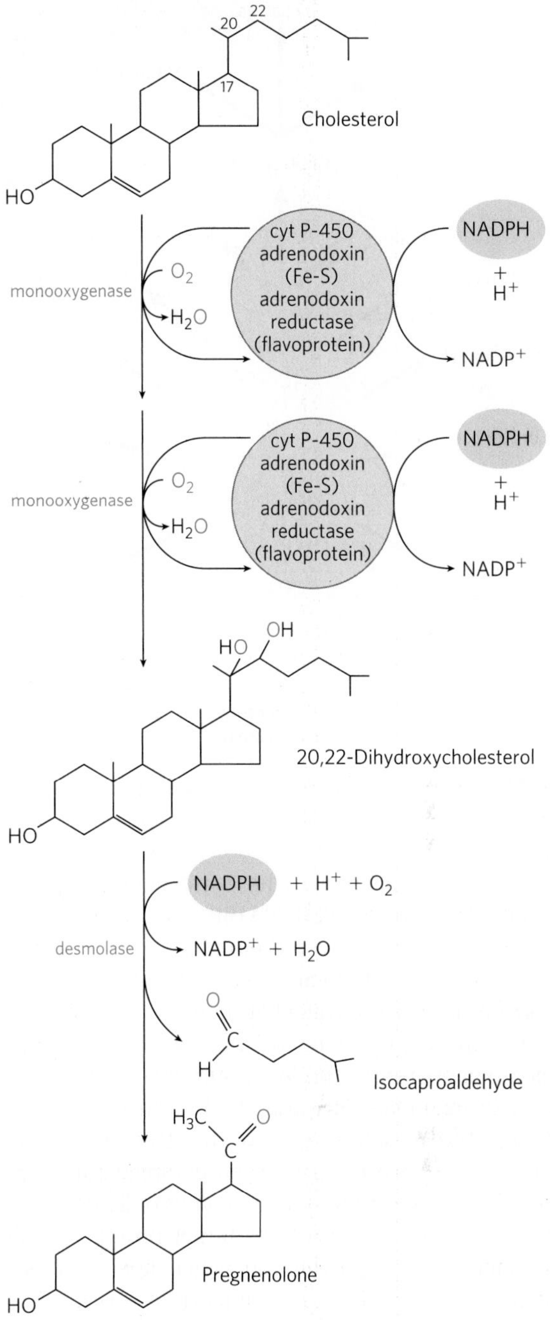

FIGURE 21-49 Side-chain cleavage in the synthesis of steroid hormones. Cytochrome P-450 acts as electron carrier in this monooxygenase system that oxidizes adjacent carbons. The process also requires the electron-transferring proteins adrenodoxin and adrenodoxin reductase. This system for cleaving side chains is found in mitochondria of the adrenal cortex, where active steroid production occurs. Pregnenolone is the precursor of all other steroid hormones (see Fig. 21-48).

biological roles **(Fig. 21-50)**. They include vitamins A, E, and K; plant pigments such as carotene and the phytol chain of chlorophyll; natural rubber; many essential oils, such as the fragrant principles of lemon oil, eucalyptus, and musk; insect juvenile hormone, which controls metamorphosis; dolichols, which serve as lipid-soluble carriers in complex polysaccharide synthesis; and ubiquinone and plastoquinone, electron carriers in mitochondria and

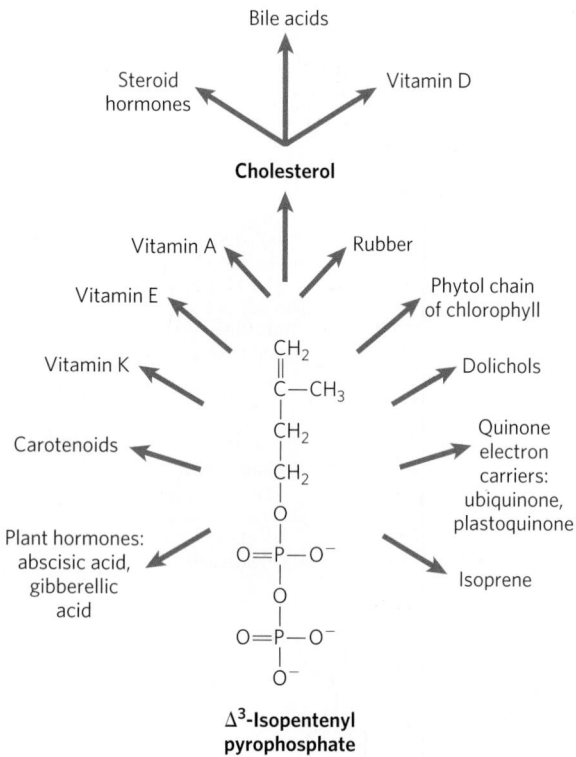

FIGURE 21-50 Overview of isoprenoid biosynthesis. The structures of most of the end products shown here are given in Chapter 10.

chloroplasts. Collectively, these molecules are called isoprenoids. More than 20,000 different isoprenoid molecules have been discovered in nature, and hundreds of new ones are reported each year.

Prenylation, the covalent attachment of an isoprenoid (see Fig. 27-36), is a common mechanism by which proteins are anchored to the inner surface of cellular membranes in mammals (see Fig. 11-13). In some of these proteins, the attached lipid is the 15-carbon farnesyl group; others have the 20-carbon geranylgeranyl group. Different enzymes attach the two types of lipids. It is possible that prenylation reactions target proteins to different membranes, depending on which lipid is attached. Protein prenylation is another important role for the isoprene derivatives of the pathway to cholesterol.

SUMMARY 21.4 Cholesterol, Steroids, and Isoprenoids: Biosynthesis, Regulation and Transport

■ Cholesterol is formed from acetyl-CoA in a complex series of reactions, through the intermediates β-hydroxy-β-methylglutaryl-CoA, mevalonate, and two activated isoprenes, dimethylallyl pyrophosphate and isopentenyl pyrophosphate. Condensation of isoprene units produces the noncyclic squalene, which is cyclized to yield the steroid ring system and side chain.

■ Cholesterol and cholesteryl esters are carried in the blood as plasma lipoproteins. VLDL carries cholesterol, cholesteryl esters, and triacylglycerols from the liver to other tissues, where the triacylglycerols are degraded by lipoprotein lipase, converting VLDL to LDL. The LDL,

rich in cholesterol and its esters, is taken up by receptor-mediated endocytosis, in which the apolipoprotein B-100 of LDL is recognized by receptors in the plasma membrane.

■ Cholesterol synthesis and transport are under complex regulation by hormones, cellular cholesterol content, and energy level (AMP concentration). HMG-CoA reductase is regulated allosterically and by covalent modification. Furthermore, both its synthesis and degradation rates are controlled by a complex of three proteins: Insig, SCAP, and SREBP, which sense cholesterol levels and trigger increased synthesis or degradation of HMG-CoA reductase. The number of LDL receptors per cell is also regulated by cholesterol content.

■ Dietary conditions or genetic defects in cholesterol metabolism may lead to atherosclerosis and heart disease. In reverse cholesterol transport, HDL removes cholesterol from peripheral tissues, carrying it to the liver. By reducing the cholesterol content of foam cells, HDL protects against atherosclerosis.

■ The steroid hormones (glucocorticoids, mineralocorticoids, and sex hormones) are produced from cholesterol by alteration of the side chain and introduction of oxygen atoms into the steroid ring system. In addition to cholesterol, a wide variety of isoprenoid compounds are derived from mevalonate through condensations of isopentenyl pyrophosphate and dimethylallyl pyrophosphate.

■ Prenylation of certain proteins targets them for association with cellular membranes and is essential for their biological activity.

Key Terms

Terms in bold are defined in the glossary.

malonyl-CoA 811	**thromboxane** 825
acetyl-CoA	**leukotriene** 825
carboxylase 811	**specialized pro-resolving**
fatty acid synthase 812	**mediators (SPMs)** 826
acyl carrier protein	**lipoxin** 826
(ACP) 814	glycerol 3-phosphate
fatty acyl–CoA	dehydrogenase 826
desaturase 821	triacylglycerol cycle 828
mixed-function	**glyceroneogenesis** 829
oxidases 821	**thiazolidinediones** 829
stearoyl-ACP desaturase	phosphatidylserine 831
(SCD) 821	phosphatidylglycerol 831
essential fatty acids 821	phosphatidylethanolamine
mixed-function	833
oxygenases 822	**cardiolipin** 833
cytochrome P-450 822	phosphatidylcholine 833
prostaglandin 824	**plasmalogen** 834
cyclooxygenase (COX) 824	platelet-activating
prostaglandin H₂ synthase	factor 834
824	**cerebroside** 835
thromboxane synthase 825	sphingomyelin 835

Problems

1. Pathway of Carbon in Fatty Acid Synthesis Using your knowledge of fatty acid biosynthesis, provide an explanation for the following experimental observations.

(a) Addition of uniformly labeled [^{14}C]acetyl-CoA to a soluble liver fraction yields palmitate uniformly labeled with ^{14}C.

(b) However, addition of a *trace* of uniformly labeled [^{14}C] acetyl-CoA in the presence of an excess of unlabeled malonyl-CoA to a soluble liver fraction yields palmitate labeled with ^{14}C only in C-15 and C-16.

2. Synthesis of Fatty Acids from Glucose After a person has ingested large amounts of sucrose, the glucose and fructose that exceed caloric requirements are transformed to fatty acids for triacylglycerol synthesis. This fatty acid synthesis consumes acetyl-CoA, ATP, and NADPH. How are these substances produced from glucose?

3. Net Equation of Fatty Acid Synthesis Write the net equation for the biosynthesis of palmitate in rat liver, starting from mitochondrial acetyl-CoA and cytosolic NADPH, ATP, and CO_2.

4. Pathway of Hydrogen in Fatty Acid Synthesis Consider a preparation that contains all the enzymes and cofactors necessary for fatty acid biosynthesis from added acetyl-CoA and malonyl-CoA.

(a) If [2-^{2}H]acetyl-CoA (labeled with deuterium, the heavy isotope of hydrogen)

and an excess of unlabeled malonyl-CoA are added as substrates, how many deuterium atoms are incorporated into every molecule of palmitate? What are their locations? Explain.

(b) If unlabeled acetyl-CoA and [2-^{2}H]malonyl-CoA

are added as substrates, how many deuterium atoms are incorporated into every molecule of palmitate? What are their locations? Explain.

5. Energetics of β-Ketoacyl-ACP Synthase In the condensation reaction catalyzed by β-ketoacyl-ACP synthase (see Fig. 21-6), a four-carbon unit is synthesized by the combination of a two-carbon unit and a three-carbon unit, with the release of CO_2. What is the thermodynamic advantage of this process over one that simply combines two two-carbon units?

6. Modulation of Acetyl-CoA Carboxylase Acetyl-CoA carboxylase is the principal regulation point in the biosynthesis of fatty acids. Some of the properties of the enzyme are described below.

(a) Addition of citrate or isocitrate raises the V_{max} of the enzyme as much as 10-fold.

(b) The enzyme exists in two interconvertible forms that differ markedly in their activities:

Protomer (inactive) $\rightleftharpoons$ filamentous polymer (active)

Citrate and isocitrate bind preferentially to the filamentous form, and palmitoyl-CoA binds preferentially to the protomer.

Explain how these properties are consistent with the regulatory role of acetyl-CoA carboxylase in the biosynthesis of fatty acids.

7. Shuttling of Acetyl Groups across the Inner Mitochondrial Membrane The acetyl group of acetyl-CoA, produced by oxidative decarboxylation of pyruvate in the mitochondrion, is transferred to the cytosol by the acetyl group shuttle outlined in Figure 21-10.

(a) Write the overall equation for the transfer of one acetyl group from the mitochondrion to the cytosol.

(b) What is the cost of this process in ATPs per acetyl group?

(c) In Chapter 17 we encountered an acyl group shuttle in the transfer of fatty acyl–CoA from the cytosol to the mitochondrion in preparation for β oxidation (see Fig. 17-6). One result of that shuttle was separation of the mitochondrial and cytosolic pools of CoA. Does the acetyl group shuttle also accomplish this? Explain.

8. Oxygen Requirement for Desaturases The biosynthesis of palmitoleate (see Fig. 21-12), a common unsaturated fatty acid with a cis double bond in the Δ^9 position, uses palmitate as a precursor. Can palmitoleate synthesis be carried out under strictly anaerobic conditions? Explain.

9. Energy Cost of Triacylglycerol Synthesis Use a net equation for the biosynthesis of tripalmitoylglycerol (tripalmitin) from glycerol and palmitate to show how many ATPs are required per molecule of tripalmitin formed.

10. Turnover of Triacylglycerols in Adipose Tissue
When [^{14}C]glucose is added to the balanced diet of adult rats, there is no increase in the total amount of stored triacylglycerols, but the triacylglycerols become labeled with ^{14}C. Explain.

11. Energy Cost of Phosphatidylcholine Synthesis
Write the sequence of steps and the net reaction for the biosynthesis of phosphatidylcholine by the salvage pathway from oleate, palmitate, dihydroxyacetone phosphate, and choline. Starting from these precursors, what is the cost (in number of ATPs) of the synthesis of phosphatidylcholine by the salvage pathway?

12. Salvage Pathway for Synthesis of Phosphatidylcholine A young rat maintained on a diet deficient in methionine fails to thrive unless choline is included in the diet. Explain.

13. Synthesis of Isopentenyl Pyrophosphate If [2-^{14}C] acetyl-CoA is added to a rat liver homogenate that is synthesizing cholesterol, where will the ^{14}C label appear in Δ^3-isopentenyl pyrophosphate, the activated form of an isoprene unit?

14. Activated Donors in Lipid Synthesis In the biosynthesis of complex lipids, components are assembled by transfer of the appropriate group from an activated donor. For example, the activated donor of acetyl groups is acetyl-CoA. For each of the following groups, give the form of the activated donor: (a) phosphate; (b) D-glucosyl; (c) phosphoethanolamine; (d) D-galactosyl; (e) fatty acyl; (f) methyl; (g) the two-carbon group in fatty acid biosynthesis; (h) Δ^3-isopentenyl.

15. Importance of Fats in the Diet When young rats are placed on a completely fat-free diet, they grow poorly, develop a scaly dermatitis, lose hair, and soon die—symptoms that can be prevented if linoleate or plant material is included in the diet. What makes linoleate an essential fatty acid? Why can plant material be substituted?

16. Regulation of Cholesterol Biosynthesis Cholesterol in humans can be obtained from the diet or synthesized de novo. An adult human on a low-cholesterol diet typically synthesizes 600 mg of cholesterol per day in the liver. If the amount of cholesterol in the diet is large, de novo synthesis of cholesterol is drastically reduced. How is this regulation brought about?

17. Lowering Serum Cholesterol Levels with Statins Patients treated with a statin drug generally exhibit a dramatic lowering of serum cholesterol. However, the amount of the enzyme HMG-CoA reductase present in cells can increase substantially. Suggest an explanation for this effect.

18. Roles of Thiol Esters in Cholesterol Biosynthesis Draw a mechanism for each of the three reactions shown in Figure 21-34, detailing the pathway for the synthesis of mevalonate from acetyl-CoA.

19. Potential Side Effects of Treatment with Statins Although clinical trials have not yet been carried out to document benefits or side effects, some physicians

have suggested that patients being treated with statins also take a supplement of coenzyme Q. Suggest a rationale for this recommendation.

Data Analysis Problem

20. Engineering *E. coli* to Produce Large Quantities of an Isoprenoid There is a huge variety of naturally occurring isoprenoids, some of which are medically or commercially important and produced industrially. The production methods include in vitro enzymatic synthesis, which is an expensive and low-yield process. In 1999, Wang, Oh, and Liao reported their experiments to engineer the easily grown bacterium *E. coli* to produce large amounts of astaxanthin, a commercially important isoprenoid.

Astaxanthin is a red-orange carotenoid pigment (an antioxidant) produced by marine algae. Marine animals such as shrimp, lobster, and some fish that feed on the algae get their orange color from the ingested astaxanthin. Astaxanthin is composed of eight isoprene units; its molecular formula is $C_{40}H_{52}O_4$.

Astaxanthin

(a) Circle the eight isoprene units in the astaxanthin molecule. Hint: Use the projecting methyl groups as a guide.

Astaxanthin is synthesized by the pathway shown on the next page, starting with Δ^3-isopentenyl pyrophosphate (IPP). Steps ❶ and ❷ are shown in Figure 21-36, and the reaction catalyzed by IPP isomerase is shown in Figure 21-35.

(b) In step ❹ of the pathway, two molecules of geranylgeranyl pyrophosphate are linked to form phytoene. Is this a head-to-head or a head-to-tail joining? (See Figure 21-36 for details.)

(c) Briefly describe the chemical transformation in step ❺.

(d) The synthesis of cholesterol (Fig. 21-37) includes a cyclization (ring closure) that requires a net oxidation by O_2. Does the cyclization in step ❻ of the astaxanthin synthetic pathway require a net oxidation of the substrate (lycopene)? Explain your reasoning.

E. coli does not make large quantities of many isoprenoids, and does not synthesize astaxanthin. It is known to synthesize small amounts of IPP, DMAPP, geranyl pyrophosphate, farnesyl pyrophosphate, and geranylgeranyl pyrophosphate. Wang and colleagues cloned several of the *E. coli* genes that encode enzymes needed for astaxanthin synthesis, in plasmids that allowed their overexpression. These genes included *idi*, which encodes IPP isomerase, and *ispA*, which encodes a prenyl transferase that catalyzes steps ❶ and ❷.

To engineer an *E. coli* capable of the complete astaxanthin pathway, Wang and colleagues cloned several genes from other bacteria into plasmids that would allow their overexpression in *E. coli*. These genes included *crtE* from *Erwinia uredovora*,

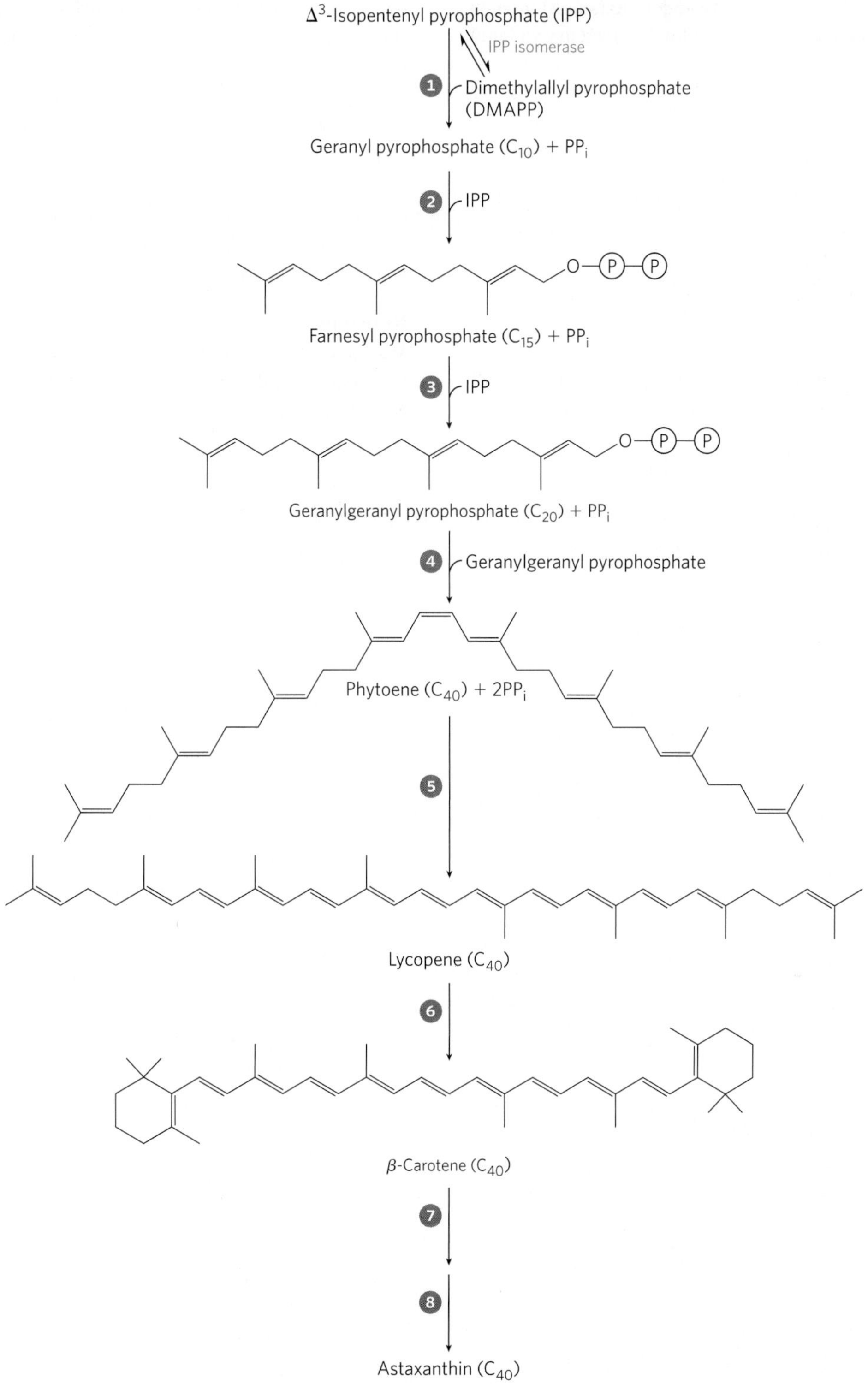

Δ^3-Isopentenyl pyrophosphate (IPP)

IPP isomerase

1 Dimethylallyl pyrophosphate (DMAPP)

Geranyl pyrophosphate (C_{10}) + PP_i

2 IPP

Farnesyl pyrophosphate (C_{15}) + PP_i

3 IPP

Geranylgeranyl pyrophosphate (C_{20}) + PP_i

4 Geranylgeranyl pyrophosphate

Phytoene (C_{40}) + 2PP_i

5

Lycopene (C_{40})

6

β-Carotene (C_{40})

7

8

Astaxanthin (C_{40})

which encodes an enzyme that catalyzes step **3**; and *crtB*, *crtI*, *crtY*, *crtZ*, and *crtW* from *Agrobacterium aurantiacum*, which encode enzymes for steps **4**, **5**, **6**, **7**, and **8**, respectively.

The investigators also cloned the gene *gps* from *Archaeoglobus fulgidus*, overexpressed this gene in *E. coli*, and extracted the gene product. When this extract was reacted with [^{14}C]IPP and DMAPP or geranyl pyrophosphate or farnesyl pyrophosphate, only ^{14}C-labeled geranylgeranyl pyrophosphate was produced in all cases.

(e) Based on these data, which step(s) in the pathway are catalyzed by the enzyme encoded by *gps*? Explain your reasoning.

Wang and coworkers then constructed several *E. coli* strains overexpressing different genes, and measured the orange color of the colonies (wild-type *E. coli* colonies are off-white) and the amount of astaxanthin produced (as measured by its orange color). Their results are shown below (ND indicates not determined).

Strain	Gene(s) overexpressed	Orange color	Astaxanthin yield (μg/g dry weight)
1	crtBIZYW	–	ND
2	crtBIZYW, ispA	–	ND
3	crtBIZYW, idi	–	ND
4	crtBIZYW, idi, ispA	–	ND
5	crtBIZYW, crtE	+	32.8
6	crtBIZYW, crtE, ispA	+	35.3
7	crtBIZYW, crtE, idi	++	234.1
8	crtBIZYW, crtE, idi, ispA	+++	390.3
9	crtBIZYW, gps	+	35.6
10	crtBIZYW, gps, idi	+++	1,418.8

(f) Comparing the results for strains 1 through 4 with those for strains 5 through 8, what can you conclude about the expression level of an enzyme capable of catalyzing step ❸ of the astaxanthin synthetic pathway in wild-type *E. coli*? Explain your reasoning.

(g) Based on the data above, which enzyme is rate-limiting in this pathway, IPP isomerase or the enzyme encoded by *idi*? Explain your reasoning.

(h) Would you expect a strain overexpressing *crtBIZYW*, *gps*, and *crtE* to produce low (+), medium (++), or high (+++) levels of astaxanthin, as measured by its orange color? Explain your reasoning.

Reference

Wang, C.-W., M.-K. Oh, and J.C. Liao. 1999. Engineered isoprenoid pathway enhances astaxanthin production in *Escherichia coli. Biotechnol. Bioeng.* 62:235–241.

Further Reading is available at www.macmillanlearning.com/LehningerBiochemistry7e.

CHAPTER 22

Biosynthesis of Amino Acids, Nucleotides, and Related Molecules

Self-study tools that will help you practice what you've learned and reinforce this chapter's concepts are available online. Go to www.macmillanlearning.com/LehningerBiochemistry7e.

Nitrogen ranks behind only carbon, hydrogen, and oxygen in its contribution to the mass of living systems. Most of this nitrogen is bound up in amino acids and nucleotides. In this chapter we address all aspects of the metabolism of these nitrogen-containing compounds except amino acid catabolism, which is covered in Chapter 18.

Discussing the biosynthetic pathways for amino acids and nucleotides together is a sound approach, not only because both classes of molecules contain nitrogen (which arises from common biological sources) but also because the two sets of pathways are extensively intertwined, with several key intermediates in common. Certain amino acids or parts of amino acids are incorporated into the structure of purines and pyrimidines, and in one case, part of a purine ring is incorporated into an amino acid (histidine). The two sets of pathways also share much common chemistry, in particular a preponderance of reactions involving the transfer of nitrogen or one-carbon groups.

The pathways described here can be intimidating to the beginning biochemistry student. Their complexity arises not so much from the chemistry itself, which in many cases is well understood, but from the sheer number of steps and variety of intermediates. These pathways are best approached by maintaining a focus on metabolic principles we have already discussed, on key intermediates and precursors, and on common classes of reactions. Even a cursory look at the chemistry can be rewarding, for some of the most unusual chemical transformations in biological systems occur in these pathways; for instance, we find prominent examples of the rare biological use of the metals molybdenum, selenium, and vanadium. The effort also offers a practical dividend, especially for students of human or veterinary medicine. Many genetic diseases of humans and animals have been traced to an absence of one or more enzymes of amino acid and nucleotide metabolism, and many pharmaceuticals in common use to combat infectious diseases are inhibitors of enzymes in these pathways—as are a number of the most important agents in cancer chemotherapy.

Regulation is crucial in the biosynthesis of the nitrogen-containing compounds. Because each amino acid and each nucleotide is required in relatively small amounts, the metabolic flow through most of these pathways is not nearly as great as the biosynthetic flow leading to carbohydrate or fat in animal tissues. Because the different amino acids and nucleotides must be made in the correct ratios and at the right time for protein and nucleic acid synthesis, their biosynthetic pathways must be accurately regulated and coordinated with each other. And because amino acids and nucleotides are charged molecules, their levels must be regulated to maintain electrochemical balance in the cell. As discussed in earlier chapters, pathways can be controlled by changes in either the activity or the amounts of specific enzymes. The pathways we encounter in this chapter provide some of the best-understood examples of the regulation of enzyme activity. Control of the *amounts* of different enzymes in a cell (that is, of their synthesis and degradation) is a topic covered in Chapter 28.

22.1 Overview of Nitrogen Metabolism

The biosynthetic pathways leading to amino acids and nucleotides share a requirement for nitrogen. Because soluble, biologically useful nitrogen compounds are generally scarce in natural environments, most organisms maintain strict economy in their use of ammonia, amino acids, and nucleotides. Indeed, as we shall see, free amino acids, purines, and pyrimidines formed during metabolic turnover of proteins and nucleic acids are often salvaged and reused. We first examine the pathways by which nitrogen from the environment is introduced into biological systems.

The Nitrogen Cycle Maintains a Pool of Biologically Available Nitrogen

Although Earth's atmosphere is four-fifths molecular nitrogen (N_2), relatively few species can convert this atmospheric nitrogen into forms useful to living organisms. In the biosphere, the metabolic processes of different species function interdependently to salvage and reuse biologically available nitrogen in a vast **nitrogen cycle (Fig. 22-1)**. The first step in the cycle is **fixation** (reduction) of atmospheric nitrogen by nitrogen-fixing bacteria to yield ammonia (NH_3 or NH_4^+). Although ammonia can be used by most living organisms, soil bacteria that derive their energy by oxidizing ammonia to nitrite (NO_2^-) and ultimately nitrate (NO_3^-) are so abundant and active that nearly all ammonia reaching the soil is oxidized to nitrate. This process is known as **nitrification**. Plants and many bacteria can take up and readily reduce nitrate and nitrite to ammonia through the action of nitrate and nitrite reductases. This ammonia is incorporated into amino acids by plants. Animals then use plants as a source of amino acids to build their proteins. When organisms die, microbial degradation of their proteins returns ammonia to the soil, where nitrifying bacteria again convert it to nitrite and nitrate. A balance is maintained between fixed nitrogen and atmospheric nitrogen by bacteria that reduce nitrate and nitrite to N_2 under anaerobic conditions, a process called **denitrification** (Fig. 22-1). These soil bacteria use NO_3^- or NO_2^- rather than O_2 as the ultimate electron acceptor in a series of reactions that (like oxidative phosphorylation) generates a transmembrane proton gradient, which is used to synthesize ATP.

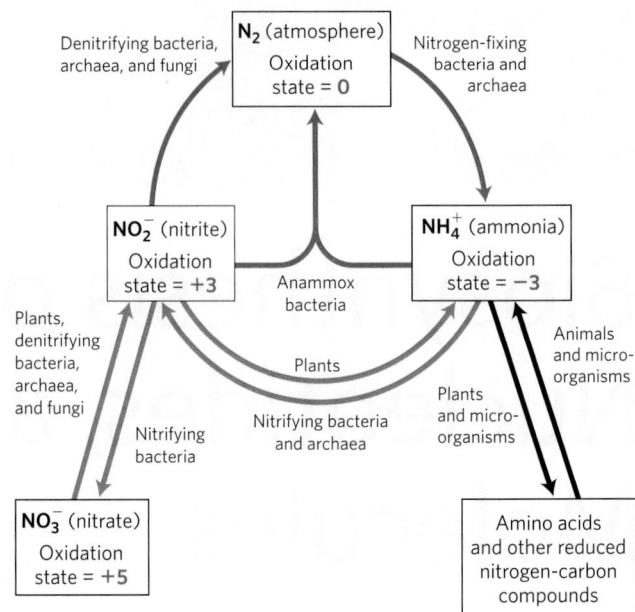

FIGURE 22-1 The nitrogen cycle. The total amount of nitrogen fixed annually in the biosphere exceeds 10^{11} kg. Reactions with red arrows occur largely or entirely in anaerobic environments.

The nitrogen cycle is short-circuited by a group of bacteria that promote anaerobic ammonia oxidation, or **anammox** (Fig. 22-1), a process that converts ammonia and nitrite to N_2. As much as 50% to 70% of the NH_3-to-N_2 conversion in the biosphere may occur through this pathway, which went undetected until the 1980s. The obligate anaerobes that promote anammox are fascinating in their own right and are providing some useful solutions to waste-treatment problems (Box 22-1).

Now let's examine the processes that generate the ammonia that is incorporated into microorganisms, plants, and the animals that eat them.

More than 90% of the NH_4^+ generated by vascular plants, algae, and microorganisms comes from nitrate assimilation, a two-step process. First NO_3^- is reduced to NO_2^- by **nitrate reductase**, then the NO_2^- is reduced to NH_4^+ in a six-electron transfer catalyzed by **nitrite reductase (Fig. 22-2)**. Both reactions involve chains of electron carriers and cofactors we have not yet encountered. Nitrate reductase is a large, soluble protein (M_r 220,000). Within the enzyme, a pair of electrons, donated by NADH, flows through —SH groups of cysteine, FAD, and a cytochrome (cyt b_{557}), then to a novel cofactor containing molybdenum, before reducing the substrate NO_3^- to NO_2^-.

The nitrite reductase of plants is located in the chloroplasts and receives its electrons from ferredoxin (which is reduced in the light-dependent reactions of photosynthesis; see Section 20.2). Six electrons, donated one at a time by ferredoxin, pass through a 4Fe-4S center in the enzyme, then through a novel heme-like molecule (siroheme) before reducing NO_2^- to NH_4^+ (Fig. 22-2). In nonphotosynthetic microbes, NADPH provides the electrons for this reaction.

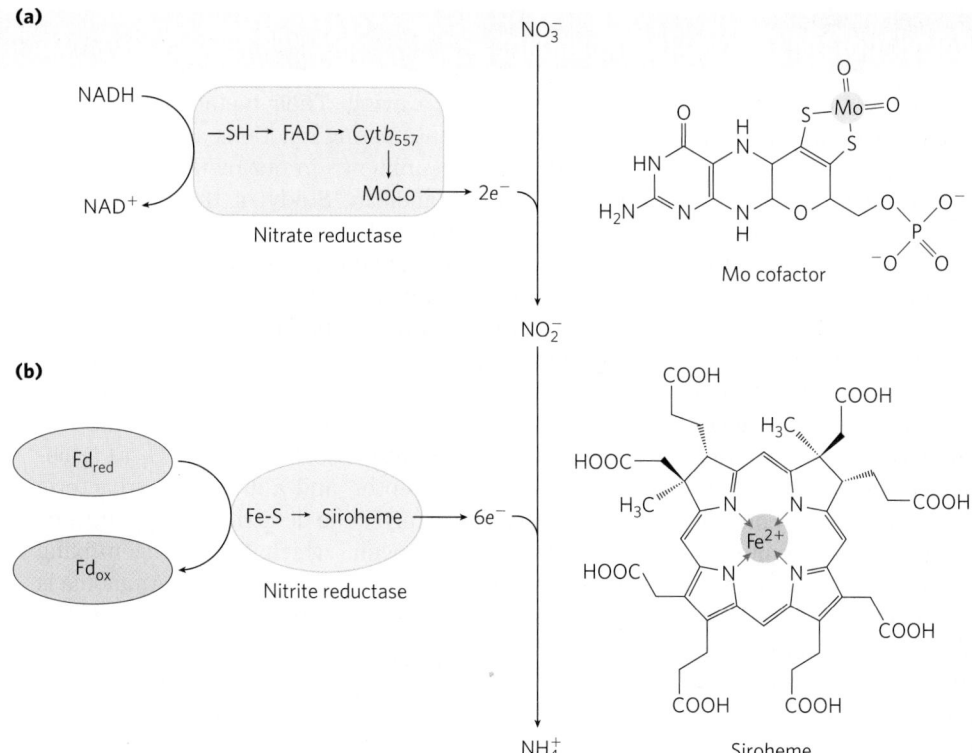

(a)

(b)

FIGURE 22-2 Nitrate assimilation by nitrate reductase and nitrite reductase. (a) Nitrate reductases of plants and bacteria catalyze the two-electron reduction of NO_3^- to NO_2^-, in which a novel Mo-containing cofactor plays a central role. NADH is the electron donor. **(b)** Nitrite reductase converts the product of nitrate reductase into NH_4^+ in a six-electron, eight-proton transfer process in which the metallic center in siroheme carries electrons and the carboxyl groups of siroheme may donate protons. The initial source of electrons is reduced ferredoxin.

Human activity presents an increasing challenge to the global nitrogen balance, and to all life in the biosphere supported by that balance. Fixed nitrogen used in agriculture, primarily in fertilizers, now contributes as much ammonia and other reactive nitrogen species to the biosphere as do natural processes, and industrial activity releases additional reactive nitrogen into the atmosphere. Controlling the damaging effects of agricultural runoff and industrial pollutants will remain an important component of the continuing effort to expand the food supply for a growing human population.

Nitrogen Is Fixed by Enzymes of the Nitrogenase Complex

The availability of fixed nitrogen, an essential nutrient, may have limited the size of the primordial biosphere. As early cells acquired a capacity to fix atmospheric nitrogen, the biosphere expanded. Evidence for biological nitrogen fixation has been found in sedimentary rocks more than 3 billion years old.

In the biosphere of today, only certain bacteria and archaea can fix atmospheric N_2. These organisms, called diazotrophs, include the cyanobacteria of soils and fresh and salt waters, methanogenic archaea (strict anaerobes that obtain energy and carbon by converting H_2 and CO_2 to methane), other kinds of free-living soil bacteria such as *Azotobacter* species, and the nitrogen-fixing bacteria that live as **symbionts** in the root nodules of leguminous plants. The first important product of nitrogen fixation is ammonia, which can be used by all organisms either directly or after its conversion to other soluble compounds such as nitrites, nitrates, or amino acids.

The reduction of nitrogen to ammonia is an exergonic reaction:

$$N_2 + 3H_2 \longrightarrow 2NH_3 \quad \Delta G'^\circ = -33.5 \text{ kJ/mol}$$

The $N\equiv N$ triple bond, however, is very stable, with a bond energy of 930 kJ/mol. Nitrogen fixation therefore has an extremely high activation energy, and atmospheric nitrogen is almost chemically inert under normal conditions. Ammonia is produced industrially by the Haber process (named for its inventor, Fritz Haber), which requires temperatures of 400 to 500 °C and nitrogen and hydrogen at pressures of tens of thousands of kilopascals (several hundred atmospheres) to provide the necessary activation energy.

Biological nitrogen fixation must occur at biological temperatures and at 0.8 atm of nitrogen, and the high activation barrier is overcome by other means. This is accomplished, at least in part, by the binding and hydrolysis of ATP. The overall reaction can be written

$$N_2 + 10H^+ + 8e^- + 16ATP \longrightarrow$$
$$2NH_4^+ + 16ADP + 16P_i + H_2$$

Biological nitrogen fixation is carried out by a highly conserved complex of proteins called the **nitrogenase complex**; its central components are **dinitrogenase reductase** and **dinitrogenase (Fig. 22-3a)**. Dinitrogenase reductase (M_r 60,000) is a dimer of two identical subunits. It contains a single 4Fe-4S redox center (see Fig. 19-5), bound between the subunits, and can be oxidized and reduced by one electron. It also has two binding sites for ATP/ADP (one site on each subunit). Dinitrogenase (M_r 240,000), an $\alpha_2\beta_2$ tetramer, has

BOX 22-1   Unusual Lifestyles of the Obscure but Abundant

Air-breathers that we are, we can easily overlook the bacteria and archaea that thrive in anaerobic environments. Although rarely featured in introductory biochemistry textbooks, these organisms constitute much of the biomass of this planet, and their contributions to the balance of carbon and nitrogen in the biosphere are essential to all forms of life.

As detailed in earlier chapters, the energy used to maintain living systems relies on the generation of proton gradients across membranes. Electrons derived from a reduced substrate are made available to electron carriers in membranes and pass through a series of electron transfers to a final electron acceptor. As a byproduct of this process, protons are released on one side of the membrane, generating the transmembrane proton gradient. The proton gradient is used to synthesize ATP or to drive other energy-requiring processes. For all eukaryotes, the reduced substrate is generally a carbohydrate (glucose or pyruvate) or a fatty acid and the electron acceptor is oxygen.

Many bacteria and archaea are much more versatile. In anaerobic environments such as marine and freshwater sediments, the variety of life strategies is extraordinary. Almost any available redox pair can be an energy source for some specialized organism or group of organisms. For example, a large number of lithotrophic bacteria (a lithotroph is a chemotroph that uses inorganic energy sources; see Fig. 1-6) have a hydrogenase that uses molecular hydrogen to reduce NAD^+:

$$H_2 + NAD^+ \xrightarrow{\text{hydrogenase}} NADH + H^+$$

The NADH is a source of electrons for a variety of membrane-bound electron acceptors, generating the proton gradient needed for ATP synthesis. Other lithotrophs oxidize sulfur compounds (H_2S, elemental sulfur, or thiosulfate) or ferrous iron. A widespread group of archaea called methanogens, all strict anaerobes, extract energy from the reduction of CO_2 to methane. And this is just a small sampling of what anaerobic organisms do for a living. Their metabolic pathways are replete with interesting reactions and highly specialized cofactors unknown in our own world of obligate aerobic metabolism. Study of these organisms can yield practical dividends. It can also provide clues about the origins of life on an early Earth, in an atmosphere that lacked molecular oxygen.

The nitrogen cycle depends on a wide range of specialized bacteria. There are two groups of nitrifying bacteria: those that oxidize ammonia to nitrites and those that oxidize the resulting nitrites to nitrates (see Fig. 22-1). Nitrate is second only to O_2 as a biological electron acceptor, and a great many bacteria and archaea can catalyze the denitrification of nitrates and nitrites to nitrogen, which the nitrogen-fixing bacteria then convert back into ammonia. Ammonia is a major pollutant in sewage and in farm animal waste, and is a byproduct of fertilizer manufacture and oil refining. Waste-treatment plants have generally made use of communities of nitrifying and denitrifying bacteria to convert ammonia waste to atmospheric nitrogen. The process is expensive, requiring inputs of organic carbon and oxygen.

In the 1960s and 1970s, a few articles appeared in the research literature suggesting that ammonia could be oxidized to nitrogen anaerobically, using nitrite as an electron acceptor; this process was called anammox. The reports received little notice until bacteria promoting anammox were discovered in a waste-treatment system in Delft, the Netherlands, in the mid-1980s. A team of Dutch microbiologists led by Gijs Kuenen and Mike Jetten began to study these bacteria, which were soon identified as belonging to an unusual bacterial phylum, Planctomycetes. Some surprises were to follow.

FIGURE 1 The anammox reactions. Ammonia and hydroxylamine are converted to hydrazine and H_2O by hydrazine hydrolase, and the hydrazine is oxidized by hydrazine-oxidizing enzyme, generating N_2 and protons. The protons generate a proton gradient for ATP synthesis. On the anammoxosome exterior, protons are used by the nitrite-reducing enzyme, producing hydroxylamine and completing the cycle. All of the anammox enzymes are embedded in the anammoxosome membrane. [Source: Information from L. A. van Niftrik et al., *FEMS Microbiol. Lett.* 233:10, 2004, Fig. 4.]

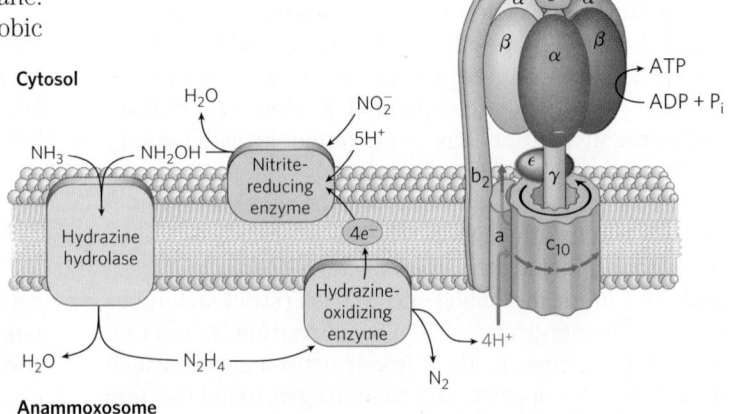

two Fe-containing cofactors that transfer electrons (Fig. 22-3b). One, the **P cluster**, has a pair of 4Fe-4S centers; these share a sulfur atom, making an 8Fe-7S center. The second cofactor in dinitrogenase, the **FeMo** cofactor, is a novel structure composed of 7 Fe atoms, 9 inorganic S atoms, a Cys side chain, and a single carbon atom in the center of the FeS cluster. Also part of the cofactor is a molybdenum atom, with ligands that

(a)

(b)

FIGURE 2 **(a)** Ladderane lipids of the anammoxosome membrane. The mechanism for synthesis of the unstable fused cyclobutane ring structures is unknown. **(b)** Ladderanes can stack to form a very dense, imperme- able, hydrophobic membrane structure, allowing sequestration of the hydrazine produced in the anammox reactions. [Source: Information from L. A. van Niftrik et al., *FEMS Microbiol. Lett.* 233:7, 2004, Fig. 3.]

The biochemistry underlying the anammox process was slowly unraveled (Fig. 1). Hydrazine (N_2H_4), a highly reactive molecule used as a rocket fuel, was an unexpected intermediate. As a small molecule, hydra- zine is both highly toxic and difficult to contain. It read- ily diffuses across typical phospholipid membranes. The anammox bacteria solve this problem by sequestering hydrazine in a specialized organelle, dubbed the **anam- moxosome**. The membrane of this organelle is com- posed of lipids known as **ladderanes** (Fig. 2), never before encountered in biology. The fused cyclobutane rings of ladderanes stack tightly to form a very dense barrier, greatly slowing the release of hydrazine. Cyclobutane rings are strained and difficult to synthe- size; the bacterial mechanisms for synthesizing these lipids are not yet known.

The anammoxosome was a surprising finding. Bacte- rial cells generally do not have compartments, and the lack of a membrane-enclosed nucleus is often cited as the primary distinction between eukaryotes and bacte- ria. One type of organelle in a bacterium was interesting enough, but microbiologists also found that planctomy- cetes have a nucleus: their chromosomal DNA is con- tained within a membrane (Fig. 3). Planctomycetes are an ancient bacterial line with multiple genera, three of which are known to carry out the anammox reactions. Discovery of this subcellular organization has prompted further research to trace the origin of the planctomyce- tes and the evolution of eukaryotic nuclei. Further study of this group may ultimately bring us closer to a key goal of evolutionary biology: a description of the organism affectionately referred to as LUCA—the last universal common ancestor of all life on our planet.

For now, the anammox bacteria offer a major advance in waste treatment, reducing the cost of ammonia removal by as much as 90% (the conven- tional denitrification steps are eliminated completely, and the aeration costs associated with nitrification are lower) and reducing the release of polluting byprod- ucts. Clearly, a greater familiarity with the bacterial underpinnings of the biosphere can pay big dividends as we deal with the environmental challenges of the twenty-first century.

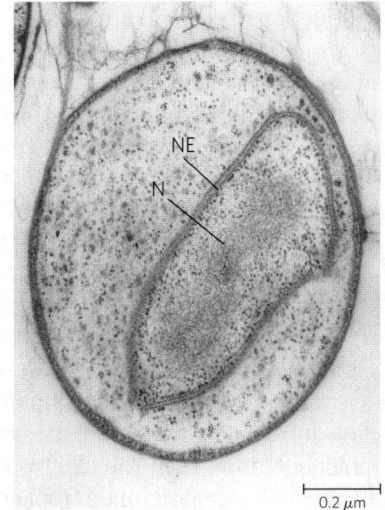

FIGURE 3 Transmission electron micrograph of a cross section through *Gemmata obscuriglobus*, showing the DNA in a nucleus (N) with enclosing nuclear envelope (NE). Bacteria of the *Gemmata* genus (phylum Planctomycetes) do not promote the anammox reactions. [Source: Provided by John Fuerst from R. Lindsay et al., *Arch. Microbiol.* 175:413, 2001, Fig. 6a. © Springer-Verlag, 2001.]

include three inorganic S atoms, a His side chain, and two oxygen atoms from a molecule of homocitrate that is an intrinsic part of the FeMo cofactor. There is also a form of nitrogenase that contains vanadium rather than molybdenum, and some bacterial species can pro- duce both types of enzymes. The vanadium-containing enzyme may be the primary nitrogen-fixing system under some conditions. The vanadium nitrogenase of

(a)

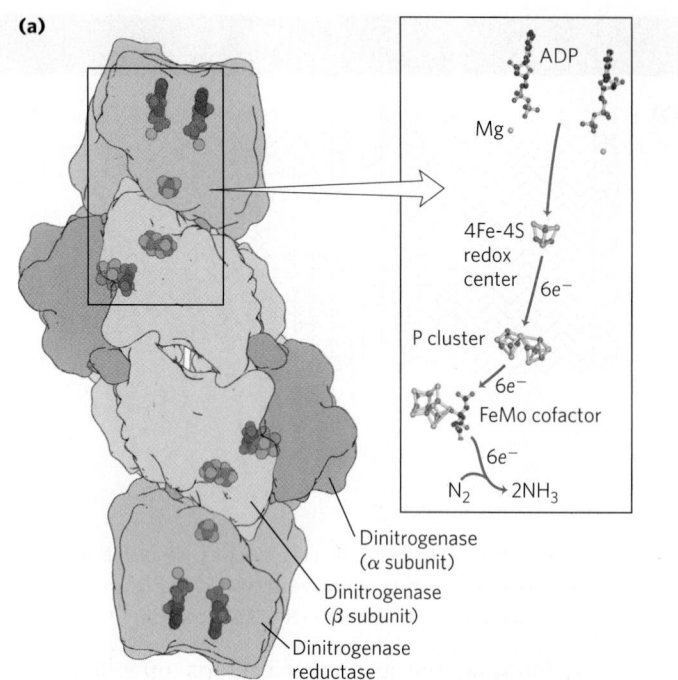

(b)

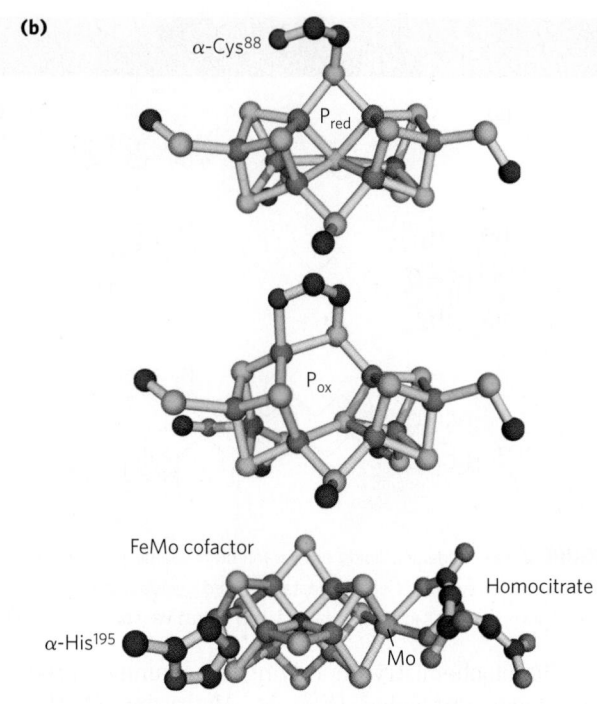

FIGURE 22-3 Enzymes and cofactors of the nitrogenase complex.
(a) The holoenzyme consists of two identical dinitrogenase reductase molecules (green), each with a 4Fe-4S redox center and binding sites for two ATP, and two identical dinitrogenase heterodimers (purple and blue), each with a P cluster (Fe-S center) and an FeMo cofactor. In this structure, ADP is bound in the ATP site, to make the crystal more stable. **(b)** The electron-transfer cofactors. A P cluster is shown here in its reduced (top) and oxidized (middle) forms. The FeMo cofactor (bottom) has a Mo atom with three S ligands, a His ligand, and two oxygen ligands from a molecule of homocitrate. In some organisms, the Mo atom is replaced with a vanadium atom. (Fe is shown in orange, S in yellow.) [Sources: (a) PDB ID 1N2C, H. Schindelin et al., *Nature* 387:370, 1997. (b) P_{red}: PDB ID 3MIN, and P_{ox}: PDB ID 2MIN, J. W. Peters et al., *Biochemistry* 36:1181, 1997; FeMo cofactor: PDB ID 1M1N, O. Einsle et al., *Science* 297:1696, 2002.]

Azotobacter vinelandii has the remarkable capacity to catalyze the reduction of carbon monoxide (CO) to ethylene (C_2H_4), ethane, and propane.

Nitrogen fixation is carried out by a highly reduced form of dinitrogenase and requires eight electrons: six for the reduction of N_2 and two to produce one molecule of H_2. Production of H_2 is an obligate part of the reaction mechanism, but its biological role in the process is not understood.

Dinitrogenase is reduced by the transfer of electrons from dinitrogenase reductase **(Fig. 22-4)**. The dinitrogenase tetramer has two binding sites for the reductase. The required eight electrons are transferred from reductase to dinitrogenase one at a time: a reduced reductase molecule binds to the dinitrogenase and transfers a single electron, then the oxidized reductase dissociates from dinitrogenase, in a repeating cycle. Each turn of the cycle requires the hydrolysis of two ATP molecules by the dimeric reductase. The immediate source of electrons to reduce dinitrogenase reductase varies, with reduced **ferredoxin** (see Section 20.2), reduced flavodoxin, and perhaps other sources playing a role. In at least one species, the ultimate source of electrons to reduce ferredoxin is pyruvate (Fig. 22-4).

The role of ATP in this process is somewhat unusual. Recall that ATP can contribute not only chemical energy, through the hydrolysis of one or more of its phosphoanhydride bonds, but also binding energy

(p. 193), through noncovalent interactions that lower the activation energy. In the reaction carried out by dinitrogenase reductase, both ATP binding and ATP hydrolysis bring about protein conformational changes that help overcome the high activation energy of nitrogen fixation. The binding of two ATP molecules to the reductase shifts the reduction potential (E'°) of this protein from -300 to -420 mV, an enhancement of its reducing power that is required to transfer electrons through dinitrogenase to N_2; the standard reduction potential for the half-reaction $N_2 + 6H^+ + 6e^- \rightarrow 2NH_3$ is -0.34 V. The ATP molecules are then hydrolyzed just before the actual transfer of one electron to dinitrogenase.

ATP binding and hydrolysis change the conformation of nitrogenase reductase in two regions, which are structurally homologous with the switch 1 and switch 2 regions of the GTP-binding proteins involved in biological signaling (see Box 12-1). ATP binding produces a conformational change that brings the 4Fe-4S center of the reductase closer to the P cluster of dinitrogenase (from 18 Å to 14 Å away), which facilitates electron transfer between the reductase and dinitrogenase. The details of electron transfer from the P cluster to the FeMo cofactor, and the means by which eight electrons are accumulated by nitrogenase, are not known. Nor are the intermediates in the reaction known with certainty; two reasonable hypotheses are being tested, both involving the Mo atom as a central player **(Fig. 22-5)**.

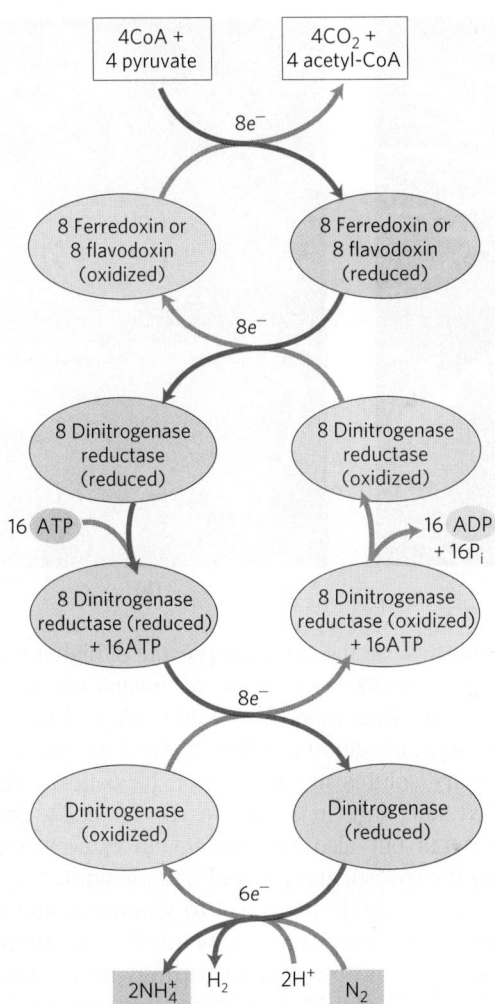

FIGURE 22-4 Electron path in nitrogen fixation by the nitrogenase complex. Electrons are transferred from pyruvate to dinitrogenase via ferredoxin (or flavodoxin) and dinitrogenase reductase. Dinitrogenase reductase reduces dinitrogenase one electron at a time, with at least six electrons required to fix one molecule of N_2. Two additional electrons are used to reduce $2H^+$ to H_2 in a process that obligatorily accompanies nitrogen fixation in anaerobes, making a total of eight electrons required per N_2 molecule. The subunit structures and metal cofactors of the dinitrogenase reductase and dinitrogenase proteins are described in the text and in Figure 22-3.

The nitrogenase complex is remarkably unstable in the presence of oxygen. The reductase is inactivated in air, with a half-life of 30 seconds; dinitrogenase has a half-life of only 10 minutes in air. Free-living bacteria that fix nitrogen cope with this problem in a variety of ways. Some live only anaerobically or repress nitrogenase synthesis when oxygen is present. Some aerobic species, such as *A. vinelandii*, partially uncouple electron transfer from ATP synthesis so that oxygen is burned off as rapidly as it enters the cell (see Box 19-2). When fixing nitrogen, cultures of these bacteria increase in temperature as a result of their efforts to rid themselves of oxygen.

The symbiotic relationship between leguminous plants and the nitrogen-fixing bacteria in their root nodules **(Fig. 22-6)** takes care of both the energy requirements and the oxygen lability of the nitrogenase

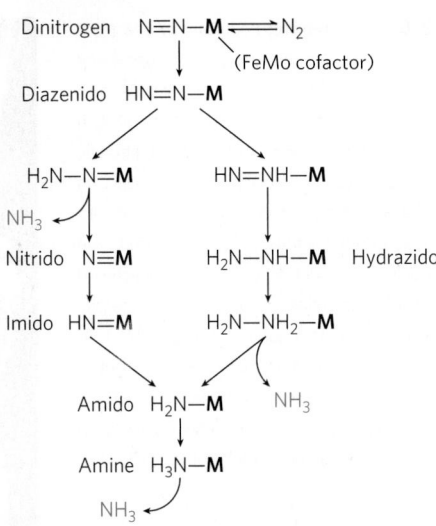

FIGURE 22-5 Two reasonable hypotheses for the intermediates involved in N_2 reduction. In both scenarios, the FeMo cofactor (abbreviated as **M** here) plays a central role, binding directly to one of the nitrogen atoms of N_2 and remaining bound throughout the sequence of reduction steps. [Source: Information from L. C. Seefeldt et al., *Annu. Rev. Biochem.* 78:701, 2009, Fig. 9.]

complex. The energy required for nitrogen fixation was probably the evolutionary driving force for this plant-bacteria association. The bacteria in root nodules have access to a large reservoir of energy in the form of abundant carbohydrate and citric acid cycle intermediates made available by the plant. This may allow the bacteria to fix hundreds of times more nitrogen than do their free-living cousins under conditions generally encountered in soils. To solve the oxygen-toxicity problem, the bacteria in root nodules are bathed in a solution of the oxygen-binding heme protein **leghemoglobin**, produced by the plant (although the heme may be contributed by the bacteria). Leghemoglobin binds all available oxygen so that it cannot interfere with nitrogen fixation, and it efficiently delivers the oxygen to the bacterial electron-transfer system. The benefit to the plant, of course, is a ready supply of reduced nitrogen. In fact, the bacterial symbionts typically produce far more NH_3 than is needed by their symbiotic partner; the excess is released into the soil. The efficiency of the symbiosis between plants and bacteria is evident in the enrichment of soil nitrogen brought about by leguminous plants. This enrichment of NH_3 in the soil is the basis of crop rotation methods, in which plantings of nonleguminous plants (such as maize) that extract fixed nitrogen from the soil are alternated every few years with plantings of legumes such as alfalfa, peas, or clover.

Nitrogen fixation is energetically costly: 16 ATP and 8 electron pairs yield only 2 NH_3. It is therefore not surprising that the process is tightly regulated so that NH_3 is produced only when needed. High [ADP], an indicator of low [ATP], is a strong inhibitor of nitrogenase. NH_4^+ represses the expression of the ~20 nitrogen fixation (*nif*) genes, effectively shutting down the pathway. Covalent alteration of nitrogenase is also used in some diazotrophs to control nitrogen fixation in response to the

FIGURE 22-6 Nitrogen-fixing nodules. (a) Pea plant (*Pisum sativum*) root nodules containing the nitrogen-fixing bacterium *Rhizobium leguminosarum*. The nodules are pink due to the presence of leghemoglobin; this heme protein has a very high binding affinity for oxygen, which strongly inhibits nitrogenase. **(b)** Artificially colorized electron micrograph of a thin section through a pea root nodule. Symbiotic nitrogen-fixing bacteria, or bacteroids (red), live inside the nodule cell, surrounded by the peribacteroid membrane (blue). Bacteroids produce the nitrogenase complex that converts atmospheric nitrogen (N_2) to ammonium (NH_4^+); without the bacteroids, the plant is unable to utilize N_2. (The cell nucleus is shown in yellow/green. Not visible in this micrograph are other organelles of the infected root cell that are normally found in plant cells.) [Source: (a, b) Jeremy Burgess/Science Source.]

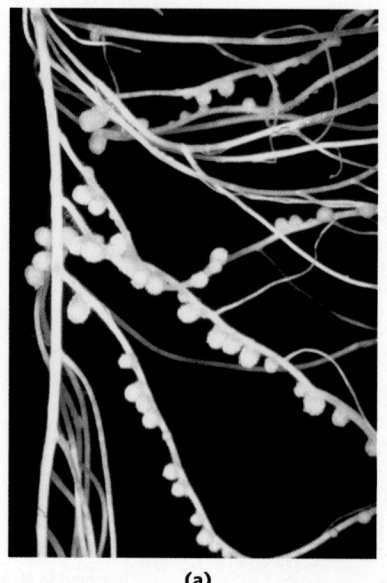
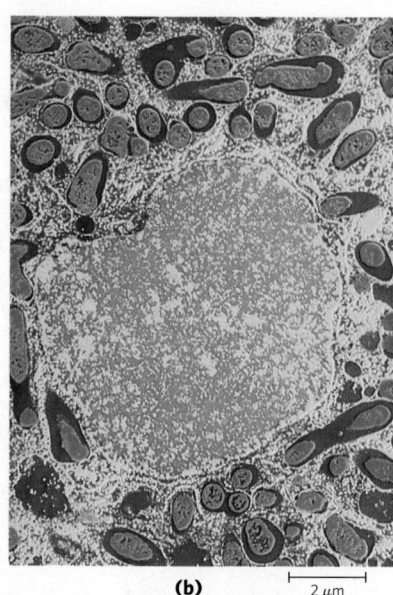

(a) **(b)** 2 μm

availability of NH_4^+ in the surroundings. Transfer of an ADP-ribosyl group from NADH to a specific Arg residue in the nitrogenase reductase shuts down N_2 fixation in *Rhodospirillum*, for example. This is the same covalent modification that we saw in the case of G protein inhibition by the toxins of cholera and pertussis (see Box 12-1).

Nitrogen fixation is the subject of intense study because of its immense practical importance. Industrial production of ammonia for use in fertilizers requires a large and expensive input of energy, and this has spurred a drive to develop recombinant or transgenic organisms that can fix nitrogen. In principle, recombinant DNA techniques (Chapter 9) might be used to transfer the DNA that encodes the enzymes of nitrogen fixation into non-nitrogen-fixing bacteria and plants. However, those genes alone will not suffice. About 20 genes are essential to nitrogenase activity in bacteria, many of them needed for the synthesis, assembly, and insertion of the cofactors. There is also the problem of protecting the enzyme in its new setting from destruction by oxygen. In all, there are formidable challenges in engineering new nitrogen-fixing plants. Success in these efforts will depend on overcoming the problem of oxygen toxicity in any cell that produces nitrogenase.

Ammonia Is Incorporated into Biomolecules through Glutamate and Glutamine

Reduced nitrogen in the form of NH_4^+ is assimilated into amino acids and then into other nitrogen-containing biomolecules. Two amino acids, **glutamate** and **glutamine**, provide the critical entry point. Recall that these same two amino acids play central roles in the catabolism of ammonia and amino groups in amino acid oxidation (Chapter 18). Glutamate is the source of amino groups for most other amino acids, through transamination reactions (the reverse of the reaction shown in Fig. 18-4). The amide nitrogen of glutamine is a source of amino groups in a wide range of biosynthetic processes. In most types of cells, and in extracellular fluids in higher organisms, one

or both of these amino acids are present at higher concentrations—sometimes an order of magnitude or more higher—than other amino acids. An *Escherichia coli* cell requires so much glutamate that this amino acid is one of the primary solutes in the cytosol. Its concentration is regulated not only in response to the cell's nitrogen requirements but also to maintain an osmotic balance between the cytosol and the external medium.

The biosynthetic pathways to glutamate and glutamine are simple, and all or some of the steps occur in most organisms. The most important pathway for the assimilation of NH_4^+ into glutamate requires two reactions. First, **glutamine synthetase** catalyzes the reaction of glutamate and NH_4^+ to yield glutamine. This reaction takes place in two steps, with enzyme-bound γ-glutamyl phosphate as an intermediate (see Fig. 18-8):

(1)   Glutamate + ATP ⟶
$$\text{γ-glutamyl phosphate + ADP}$$

(2)   γ-Glutamyl phosphate + NH_4^+ ⟶
$$\text{glutamine} + P_i + H^+$$

Sum: Glutamate + NH_4^+ + ATP ⟶
$$\text{glutamine + ADP} + P_i + H^+ \quad (22\text{-}1)$$

Glutamine synthetase is found in all organisms. In addition to its importance for NH_4^+ assimilation in bacteria, it has a central role in amino acid metabolism in mammals, converting free NH_4^+, which is toxic, to glutamine for transport in the blood (Chapter 18).

In bacteria and plants, glutamate is produced from glutamine in a reaction catalyzed by **glutamate synthase**. (An alternative name for this enzyme, glutamate:oxoglutarate aminotransferase, yields the acronym GOGAT, by which the enzyme also is known.) α-Ketoglutarate, an intermediate of the citric acid cycle, undergoes reductive amination with glutamine as nitrogen donor:

α-Ketoglutarate + glutamine + NADPH + H^+ ⟶
$$2 \text{ glutamate + NADP}^+ \quad (22\text{-}2)$$

The net reaction of glutamine synthetase and glutamate synthase (Eqns 22-1 and 22-2) is

$$\alpha\text{-Ketoglutarate} + NH_4^+ + NADPH + ATP \longrightarrow$$
$$\text{glutamate} + NADP^+ + ADP + P_i$$

Glutamate synthase is not present in animals, which instead maintain high levels of glutamate by processes such as the transamination of α-ketoglutarate during amino acid catabolism.

Glutamate can also be formed in yet another, albeit minor, pathway: the reaction of α-ketoglutarate and NH_4^+ to form glutamate in one step. This is catalyzed by glutamate dehydrogenase, an enzyme present in all organisms. Reducing power is furnished by NADPH:

$$\alpha\text{-Ketoglutarate} + NH_4^+ + NADPH \longrightarrow$$
$$\text{glutamate} + NADP^+ + H_2O$$

We encountered this reaction in the catabolism of amino acids (see Fig. 18-7). In eukaryotic cells, glutamate dehydrogenase is located in the mitochondrial matrix. The reaction equilibrium favors the reactants, and the K_m for NH_4^+ ($\sim$1 mM) is so high that the reaction probably makes only a modest contribution to NH_4^+ assimilation into amino acids and other metabolites. (Recall that the glutamate dehydrogenase reaction, in reverse (see Fig. 18-10), is one source of NH_4^+ destined for the urea cycle.) Concentrations of NH_4^+ high enough for the glutamate dehydrogenase reaction to make a significant contribution to glutamate levels generally occur only when NH_3 is added to the soil or when organisms are grown in a laboratory in the presence of high NH_3 concentrations. In general, soil bacteria and plants rely on the two-enzyme pathway outlined above (Eqns 22-1, 22-2).

Glutamine Synthetase Is a Primary Regulatory Point in Nitrogen Metabolism

The activity of glutamine synthetase is regulated in virtually all organisms—as expected, given its central metabolic role as an entry point for reduced nitrogen. In enteric bacteria such as *E. coli*, the regulation is unusually complex. Type I enzyme (found in bacteria) has 12 identical subunits of M_r 50,000 **(Fig. 22-7)** and is regulated both allosterically and by covalent modification. (Type II enzyme, in eukaryotes and some bacteria, has 10 identical subunits.) Alanine, glycine, and at least six end products of glutamine metabolism are allosteric inhibitors of the enzyme **(Fig. 22-8)**. Each inhibitor alone produces only partial inhibition, but the effects of multiple inhibitors are more than additive, and all eight together virtually shut down the enzyme. This is an example of cumulative feedback inhibition. This control mechanism provides a constant adjustment of glutamine levels to match immediate metabolic requirements.

Superimposed on the allosteric regulation is inhibition by adenylylation of (addition of AMP to) Tyr^{397}, located near the enzyme's active site **(Fig. 22-9)**. This covalent

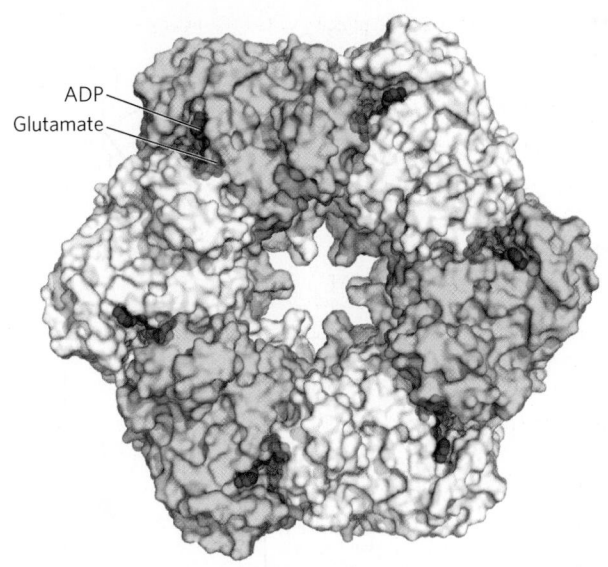

FIGURE 22-7 Subunit structure of bacterial type I glutamine synthetase. This view shows 6 of the 12 identical subunits; a second layer of 6 subunits lies directly beneath those shown. Each of the 12 subunits has an active site, where ATP and glutamate are bound in orientations that favor transfer of a phosphoryl group from ATP to the side-chain carboxyl of glutamate. In this crystal structure, ADP occupies the ATP site. [Source: PDB ID 2GLS, M. M. Yamashita et al., *J. Biol. Chem.* 264:17,681, 1989.]

modification increases sensitivity to the allosteric inhibitors, and activity decreases as more subunits are adenylylated. Both adenylylation and deadenylylation are promoted by **adenylyltransferase** (AT in Fig. 22-9), part of a complex enzymatic cascade that responds to levels of glutamine, α-ketoglutarate, ATP, and P_i. The activity of

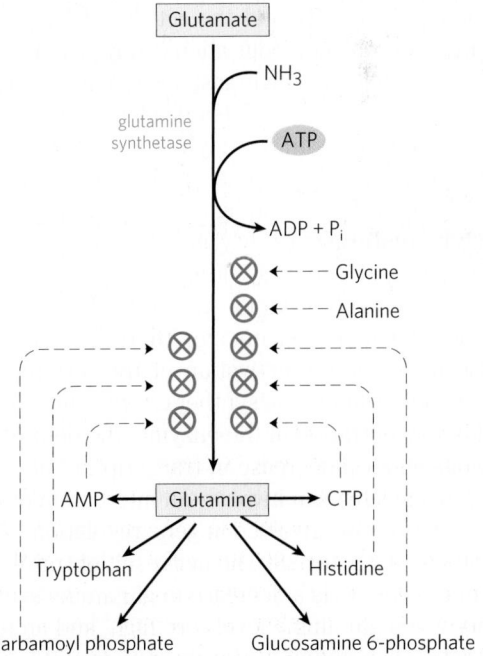

FIGURE 22-8 Allosteric regulation of glutamine synthetase. The enzyme undergoes cumulative regulation by six end products of glutamine metabolism. Alanine and glycine probably serve as indicators of the general status of amino acid metabolism in the cell.

(a)

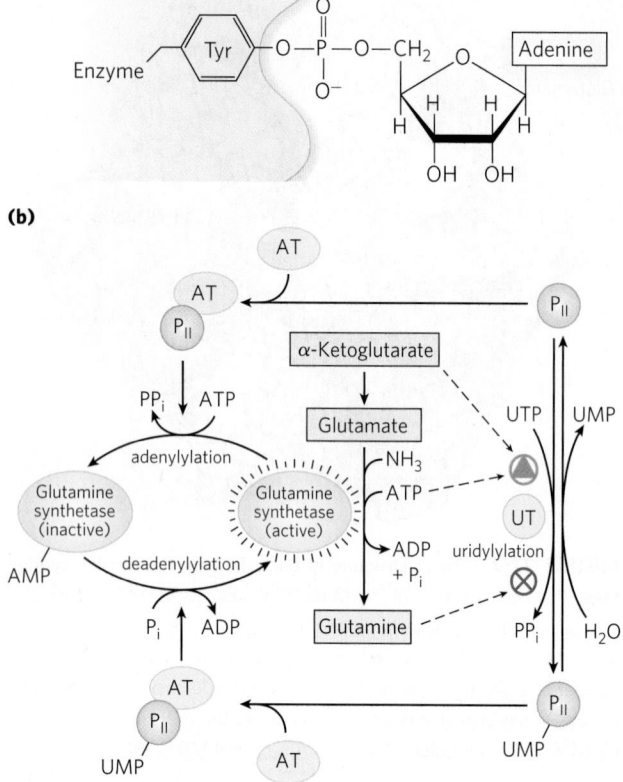

(b)

FIGURE 22–9 Second level of regulation of glutamine synthetase: covalent modifications. (a) An adenylylated Tyr residue. **(b)** Cascade leading to adenylylation (inactivation) of glutamine synthetase. AT represents adenylyltransferase; UT, uridylyltransferase. P_{II} is a regulatory protein, itself regulated by uridylylation. Details of this cascade are discussed in the text.

adenylyltransferase is modulated by binding to a regulatory protein called P_{II}, and the activity of P_{II}, in turn, is regulated by covalent modification (uridylylation), again at a Tyr residue. The adenylyltransferase complex with uridylylated P_{II} (P_{II}-UMP) stimulates deadenylylation, whereas the same complex with deuridylylated P_{II} stimulates adenylylation of glutamine synthetase. Both uridylylation and deuridylylation of P_{II} are brought about by a single enzyme, **uridylyltransferase**. Uridylylation is inhibited by binding of glutamine and P_i to uridylyltransferase and is stimulated by binding of α-ketoglutarate and ATP to P_{II}.

The regulation does not stop there. The uridylylated P_{II} also mediates the activation of transcription of the gene encoding glutamine synthetase, thus increasing the cellular concentration of the enzyme; the deuridylylated P_{II} brings about a decrease in transcription of the same gene. This mechanism involves an interaction of P_{II} with additional proteins involved in gene regulation, of a type described in Chapter 28. The net result of this elaborate system of controls is a decrease in glutamine synthetase activity when glutamine levels are high, and an increase in activity when glutamine levels are low and α-ketoglutarate and ATP (substrates for the synthetase reaction) are available. The multiple layers of regulation permit a sensitive response in which glutamine synthesis is tailored to cellular needs.

Several Classes of Reactions Play Special Roles in the Biosynthesis of Amino Acids and Nucleotides

The pathways described in this chapter include a variety of interesting chemical rearrangements. Several of these recur and deserve special note before we progress to the pathways themselves. These are (1) transamination reactions and other rearrangements promoted by enzymes containing pyridoxal phosphate; (2) transfer of one-carbon groups, with either tetrahydrofolate (usually at the —CHO and —CH$_2$OH oxidation levels) or S-adenosylmethionine (at the —CH$_3$ oxidation level) as cofactor; and (3) transfer of amino groups derived from the amide nitrogen of glutamine. Pyridoxal phosphate (PLP), tetrahydrofolate (H$_4$ folate), and S-adenosylmethionine (adoMet) are described in some detail in Chapter 18 (see Figs 18-6, 18-17, and 18-18). Here we focus on amino group transfer involving the amide nitrogen of glutamine.

More than a dozen known biosynthetic reactions use glutamine as the major physiological source of amino groups, and most of these occur in the pathways outlined in this chapter. As a class, the enzymes catalyzing these reactions are called **glutamine amidotransferases**. All have two structural domains: one binding glutamine, the other binding the second substrate, which serves as amino group acceptor **(Fig. 22-10)**. A conserved Cys residue in the glutamine-binding domain is believed to act as a nucleophile, cleaving the amide bond of glutamine and forming a covalent glutamyl-enzyme intermediate. The NH$_3$ produced in this reaction is not released, but instead is transferred through an "ammonia channel" to a second active site, where it reacts with the second substrate to form the aminated product. The covalent intermediate is hydrolyzed to the free enzyme and glutamate. If the second substrate must be activated, the usual method is the use of ATP to generate an acyl phosphate intermediate (R—OX in Fig. 22-10, with X as a phosphoryl group). The enzyme glutaminase acts in a similar fashion but uses H$_2$O as the second substrate, yielding NH$_4^+$ and glutamate (see Fig. 18-8).

SUMMARY 22.1 Overview of Nitrogen Metabolism

■ The molecular nitrogen that makes up 80% of Earth's atmosphere is unavailable to most living organisms until it is reduced. This fixation of atmospheric N$_2$ takes place in certain free-living bacteria and in symbiotic bacteria in the root nodules of leguminous plants.

■ In soil bacteria and vascular plants, the sequential action of nitrate reductase and nitrite reductase converts NO$_3^-$ to NH$_3$, which can be assimilated into nitrogen-containing compounds.

■ The nitrogen cycle entails formation of ammonia by bacterial fixation of N$_2$, nitrification of ammonia to nitrate by soil organisms, conversion of nitrate to ammonia by higher plants, synthesis of amino acids from ammonia by all organisms, and conversion of

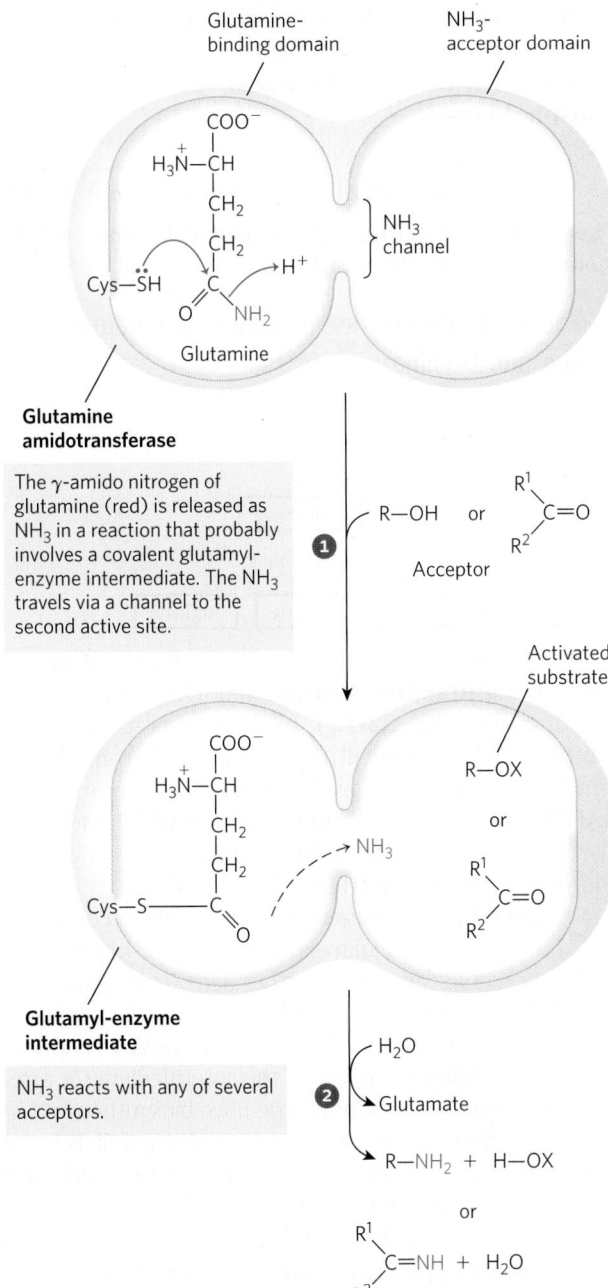

Glutamine amidotransferase

The γ-amido nitrogen of glutamine (red) is released as NH_3 in a reaction that probably involves a covalent glutamyl-enzyme intermediate. The NH_3 travels via a channel to the second active site.

Glutamyl-enzyme intermediate

NH_3 reacts with any of several acceptors.

MECHANISM FIGURE 22-10 Proposed mechanism for glutamine amidotransferases. Each enzyme has two domains. The glutamine-binding domain contains structural elements conserved among many of these enzymes, including a Cys residue required for activity. The NH_3-acceptor (second-substrate) domain varies. Two types of amino acceptors are shown. X represents an activating group, typically a phosphoryl group derived from ATP, that facilitates displacement of a hydroxyl group from R—OH by NH_3.

nitrate and nitrite to N_2 by denitrifying bacteria. The anammox bacteria anaerobically oxidize ammonia to nitrogen, using nitrite as an electron acceptor.

■ Fixation of N_2 as NH_3 is carried out by the nitrogenase complex, in a reaction that requires large investments of ATP and of reducing power. The nitrogenase complex is highly labile in the presence of O_2, and is subject to regulation by the supply of NH_3.

■ In living systems, reduced nitrogen is incorporated first into amino acids and then into a variety of other biomolecules, including nucleotides. The key entry point is the amino acid glutamate. Glutamate and glutamine are the nitrogen donors in a wide range of biosynthetic reactions. Glutamine synthetase, which catalyzes the formation of glutamine from glutamate, is a main regulatory enzyme of nitrogen metabolism.

■ The amino acid and nucleotide biosynthetic pathways make repeated use of the biological cofactors pyridoxal phosphate, tetrahydrofolate, and *S*-adenosylmethionine. Pyridoxal phosphate is required for transamination reactions involving glutamate and for other amino acid transformations. One-carbon transfers require *S*-adenosylmethionine and tetrahydrofolate. Glutamine amidotransferases catalyze reactions that incorporate nitrogen derived from glutamine.

22.2 Biosynthesis of Amino Acids

All amino acids are derived from intermediates in glycolysis, the citric acid cycle, or the pentose phosphate pathway **(Fig. 22-11)**. Nitrogen enters these biosynthetic

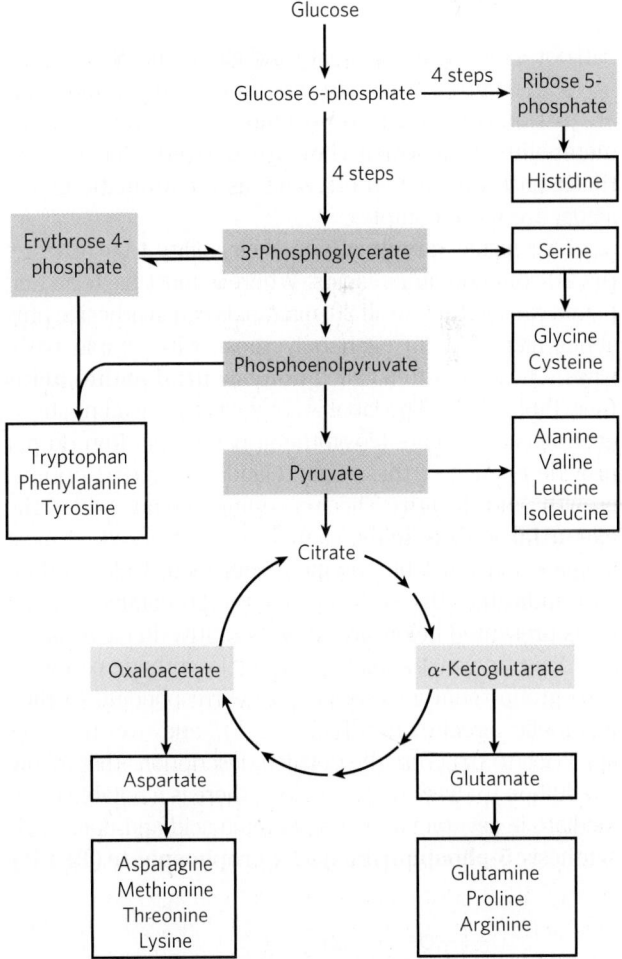

FIGURE 22-11 Overview of amino acid biosynthesis. The carbon skeleton precursors derive from three sources: glycolysis (light red), the citric acid cycle (blue), and the pentose phosphate pathway (purple).

α-Ketoglutarate	**Pyruvate**
Glutamate	Alanine
Glutamine	Valine[a]
Proline	Leucine[a]
Arginine	Isoleucine[a]
3-Phosphoglycerate	**Phosphoenolpyruvate**
Serine	**and erythrose**
Glycine	**4-phosphate**
Cysteine	Tryptophan[a]
Oxaloacetate	Phenylalanine[a]
Aspartate	Tyrosine[b]
Asparagine	**Ribose 5-phosphate**
Methionine[a]	Histidine[a]
Threonine[a]	
Lysine[a]	

TABLE 22-1 Amino Acid Biosynthetic Families, Grouped by Metabolic Precursor

[a]Essential amino acids in mammals.
[b]Derived from phenylalanine in mammals.

pathways by way of glutamate and glutamine. Some pathways are simple, others are not. Ten of the amino acids are just one or several steps removed from the common metabolite from which they are derived. The biosynthetic pathways for others, such as the aromatic amino acids, are more complex.

Organisms vary greatly in their ability to synthesize the 20 common amino acids. Whereas most bacteria and plants can synthesize all 20, mammals can synthesize only about half of them—generally those with simple pathways. These are often called **nonessential amino acids** (see Table 18-1). The label is somewhat misleading, however, because innate biosynthetic pathways often do not provide enough of these amino acids to support optimal growth and health. The remaining amino acids, the **essential amino acids**, cannot be synthesized by most animals and must be obtained from food. Unless otherwise indicated, the pathways for the 20 common amino acids presented below are those operative in bacteria.

A useful way to organize these biosynthetic pathways is to group them into six families corresponding to their metabolic precursors (Table 22-1), and we use this approach to structure the detailed descriptions that follow. In addition to these six precursors, there is a notable intermediate in several pathways of amino acid and nucleotide synthesis: **5-phosphoribosyl-1-pyrophosphate (PRPP)**:

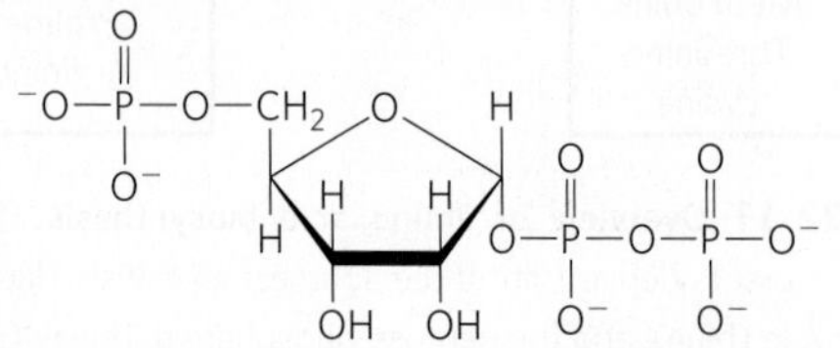

PRPP is synthesized from ribose 5-phosphate derived from the pentose phosphate pathway (see Fig. 14-22), in a reaction catalyzed by **ribose phosphate pyrophosphokinase**:

$$\text{Ribose 5-phosphate} + \text{ATP} \longrightarrow$$
$$\text{5-phosphoribosyl-1-pyrophosphate} + \text{AMP}$$

This enzyme is allosterically regulated by many of the biomolecules for which PRPP is a precursor.

α-Ketoglutarate Gives Rise to Glutamate, Glutamine, Proline, and Arginine

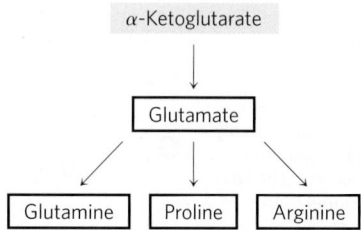

We have already described the biosynthesis of **glutamate** and **glutamine**. **Proline** is a cyclized derivative of glutamate **(Fig. 22-12)**. In the first step of proline synthesis, ATP reacts with the γ-carboxyl group of glutamate to form an acyl phosphate, which is reduced by NADPH or NADH to glutamate γ-semialdehyde. This intermediate undergoes rapid spontaneous cyclization and is then reduced further to yield proline.

Arginine is synthesized from glutamate via ornithine and the urea cycle in animals (Chapter 18). In principle, ornithine could also be synthesized from glutamate γ-semialdehyde by transamination, but the spontaneous cyclization of the semialdehyde in the proline pathway precludes a sufficient supply of this intermediate for ornithine synthesis. Bacteria have a de novo biosynthetic pathway for ornithine (and thus arginine) that parallels some steps of the proline pathway but includes two additional steps that avoid the problem of the spontaneous cyclization of glutamate γ-semialdehyde (Fig. 22-12). In the first step, the α-amino group of glutamate is blocked by an acetylation requiring acetyl-CoA; then, after the transamination step, the acetyl group is removed to yield ornithine.

The pathways to proline and arginine are somewhat different in mammals. Proline can be synthesized by the pathway shown in Figure 22-12, but it is also formed from arginine obtained from dietary or tissue protein. Arginase, a urea cycle enzyme, converts arginine to ornithine and urea (see Figs 18-10, 18-26). The ornithine is converted to glutamate γ-semialdehyde by the enzyme **ornithine δ-aminotransferase (Fig. 22-13)**. The semialdehyde cyclizes to Δ[1]-pyrroline-5-carboxylate, which is then converted to proline (Fig. 22-12). The pathway for arginine synthesis shown in Figure 22-12 is absent in mammals. When arginine from dietary intake or protein turnover is insufficient for protein synthesis, the ornithine δ-aminotransferase reaction operates in the direction of ornithine formation. Ornithine is then converted to citrulline and arginine in the urea cycle.

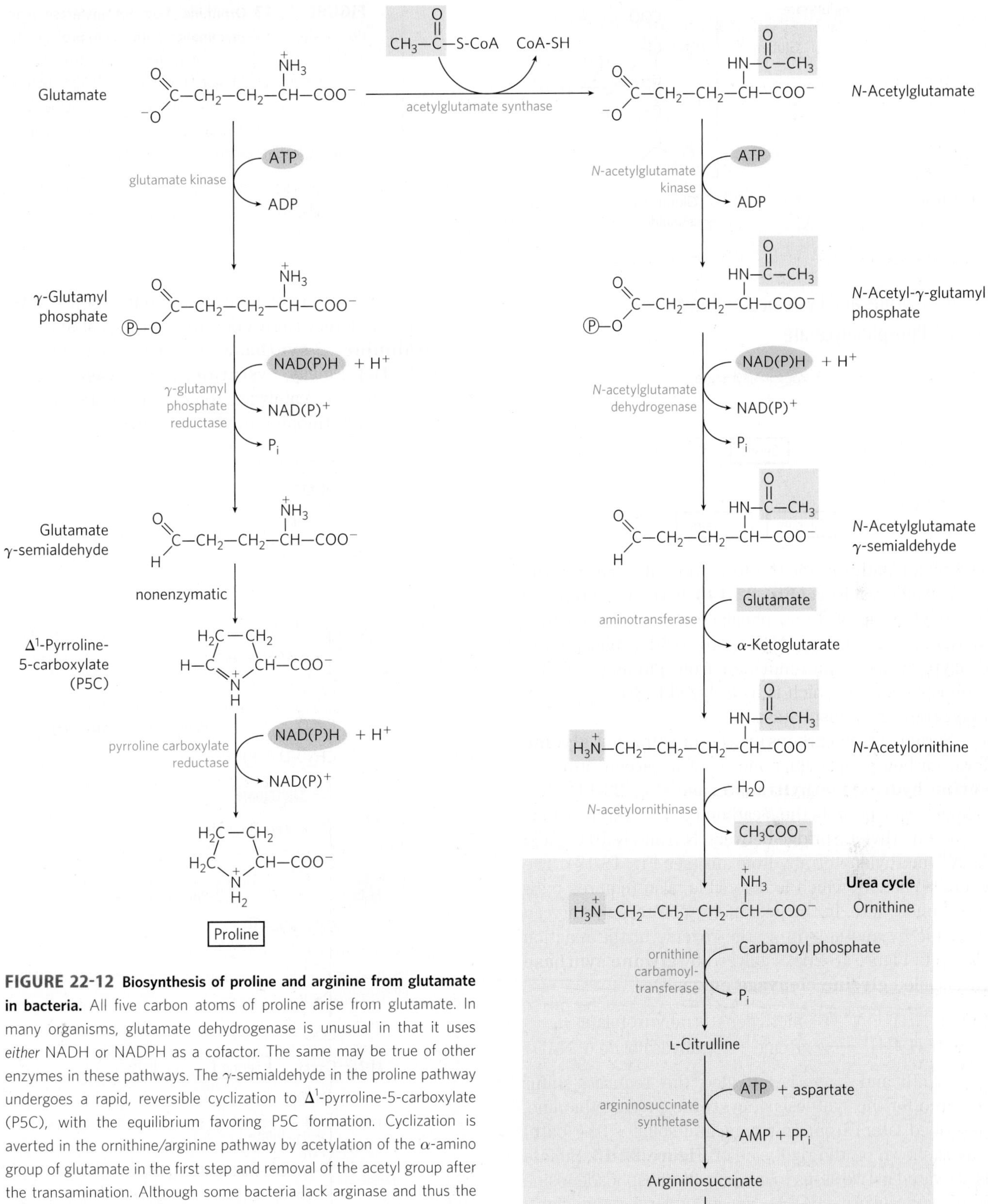

FIGURE 22-12 Biosynthesis of proline and arginine from glutamate in bacteria. All five carbon atoms of proline arise from glutamate. In many organisms, glutamate dehydrogenase is unusual in that it uses *either* NADH or NADPH as a cofactor. The same may be true of other enzymes in these pathways. The γ-semialdehyde in the proline pathway undergoes a rapid, reversible cyclization to Δ^1-pyrroline-5-carboxylate (P5C), with the equilibrium favoring P5C formation. Cyclization is averted in the ornithine/arginine pathway by acetylation of the α-amino group of glutamate in the first step and removal of the acetyl group after the transamination. Although some bacteria lack arginase and thus the complete urea cycle, they can synthesize arginine from ornithine in steps that parallel the mammalian urea cycle, with citrulline and argininosuccinate as intermediates (see Fig. 18-10).

Here, and in subsequent figures in this chapter, the reaction arrows indicate the linear path to the final products, without considering the reversibility of individual steps. For example, the step of the pathway leading to arginine that is catalyzed by *N*-acetylglutamate dehydrogenase is chemically similar to the glyceraldehyde 3-phosphate dehydrogenase reaction in glycolysis (see Fig. 14-8) and is readily reversible.

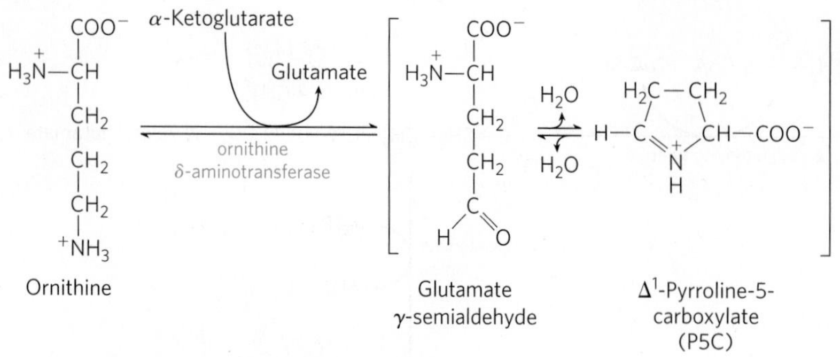

FIGURE 22-13 Ornithine δ-aminotransferase reaction: a step in the mammalian pathway to proline. This enzyme is found in the mitochondrial matrix of most tissues. Although the equilibrium favors P5C formation, the reverse reaction is the only mammalian pathway for synthesis of ornithine (and thus arginine) when arginine levels are insufficient for protein synthesis.

Serine, Glycine, and Cysteine Are Derived from 3-Phosphoglycerate

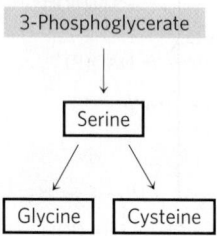

The major pathway for the formation of **serine** is the same in all organisms **(Fig. 22-14)**. In the first step, the hydroxyl group of 3-phosphoglycerate is oxidized by a dehydrogenase (using NAD^+) to yield 3-phosphohydroxypyruvate. Transamination from glutamate yields 3-phosphoserine, which is hydrolyzed to free serine by phosphoserine phosphatase.

Serine (three carbons) is the precursor of **glycine** (two carbons) through removal of a carbon atom by **serine hydroxymethyltransferase** (Fig. 22-14). Tetrahydrofolate accepts the β carbon (C-3) of serine, which forms a methylene bridge between N-5 and N-10 to yield N^5,N^{10}-methylenetetrahydrofolate (see Fig. 18-17). The overall reaction, which is reversible, also requires pyridoxal phosphate. In the liver of vertebrates, glycine can be made by another route: the reverse of the reaction shown in Figure 18-20c, catalyzed by **glycine synthase** (also called **glycine cleavage enzyme**):

$$CO_2 + NH_4^+ + N^5,N^{10}\text{-methylenetetrahydrofolate} +$$
$$NADH + H^+ \longrightarrow \text{glycine} + \text{tetrahydrofolate} + NAD^+$$

Plants and bacteria produce the reduced sulfur required for the synthesis of **cysteine** (and methionine, described later) from environmental sulfates; the pathway is shown on the right side of **Figure 22-15**. Sulfate is activated in two steps to produce 3′-phosphoadenosine 5′-phosphosulfate (PAPS), which undergoes an eight-electron reduction to sulfide. The sulfide is then used in the formation of cysteine from serine in a two-step pathway. Mammals synthesize cysteine from two amino acids: methionine furnishes the sulfur atom, and serine furnishes the carbon skeleton. Methionine is first converted to S-adenosylmethionine (see Fig. 18-18), which can lose its methyl group to any of a number of acceptors to form S-adenosylhomocysteine (adoHcy). This demethylated product is hydrolyzed to free homocysteine, which undergoes a reaction with serine, catalyzed by **cystathionine β-synthase**, to yield cystathionine **(Fig. 22-16)**. Finally, **cystathionine γ-lyase**, a PLP-requiring enzyme, catalyzes removal of ammonia and cleavage of cystathionine to yield free cysteine.

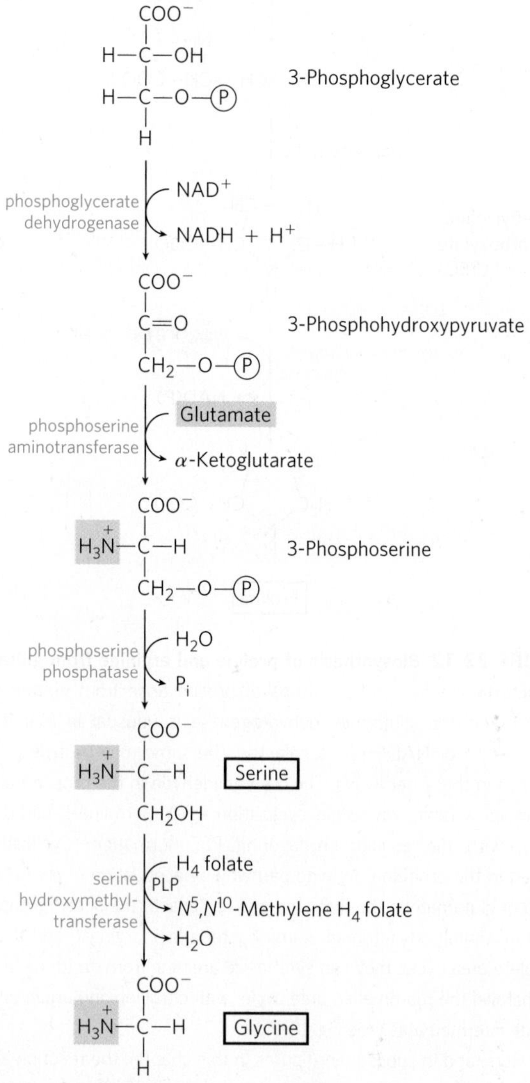

FIGURE 22-14 Biosynthesis of serine from 3-phosphoglycerate and of glycine from serine in all organisms. Glycine is also made from CO_2 and NH_4^+ by the action of glycine synthase, with N^5,N^{10}-methylenetetrahydrofolate as methyl group donor (see text).

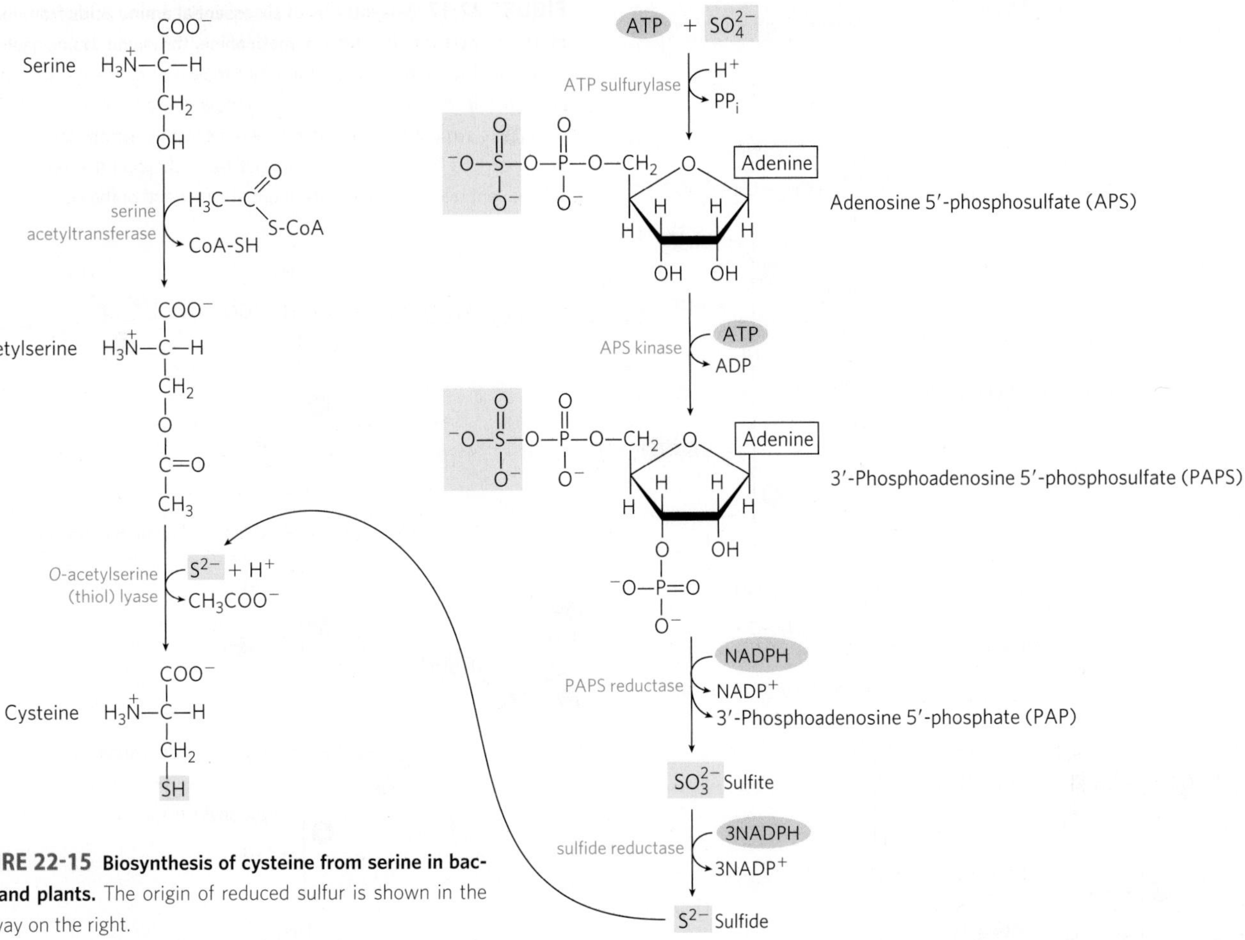

FIGURE 22-15 Biosynthesis of cysteine from serine in bacteria and plants. The origin of reduced sulfur is shown in the pathway on the right.

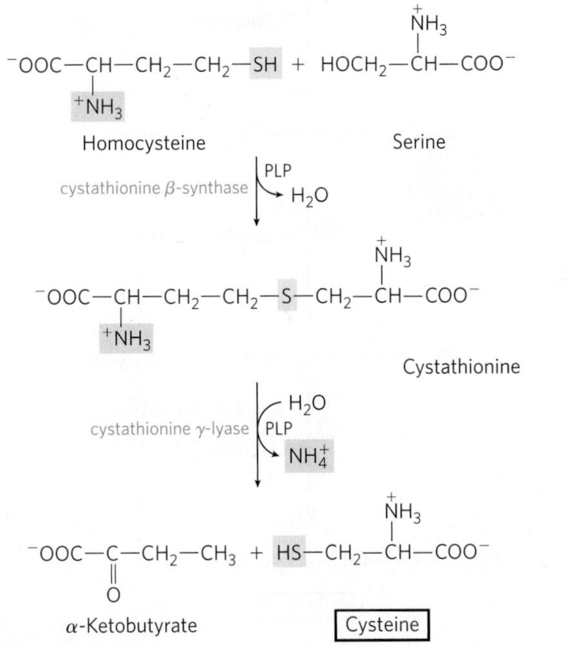

Three Nonessential and Six Essential Amino Acids Are Synthesized from Oxaloacetate and Pyruvate

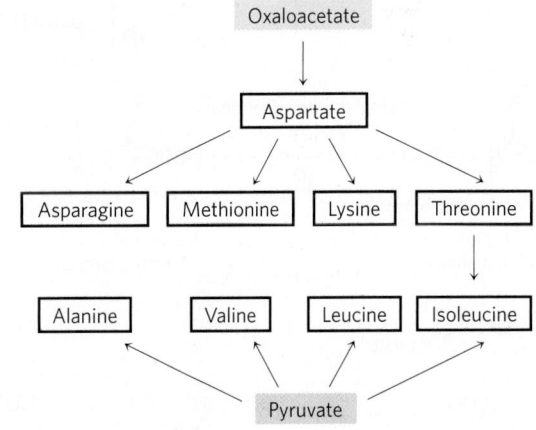

Alanine and **aspartate** are synthesized from pyruvate and oxaloacetate, respectively, by transamination from glutamate. **Asparagine** is synthesized by amidation of aspartate, with glutamine donating the NH_4^+. These are nonessential amino acids, and their simple biosynthetic pathways occur in all organisms.

For reasons incompletely understood, the malignant lymphocytes present in childhood acute lymphoblastic leukemia (ALL) require serum

FIGURE 22-16 Biosynthesis of cysteine from homocysteine and serine in mammals. The homocysteine is formed from methionine, as described in the text.

FIGURE 22-17 Biosynthesis of six essential amino acids from oxaloacetate and pyruvate in bacteria: methionine, threonine, lysine, isoleucine, valine, and leucine. Here, and in other multistep pathways, the enzymes are listed in the key. Note that L,L-α,ε-diaminopimelate, the product of step ⓮, is symmetric. The carbons derived from pyruvate (and the amino group derived from glutamate) are not traced beyond this point, because subsequent reactions may place them at either end of the lysine molecule.

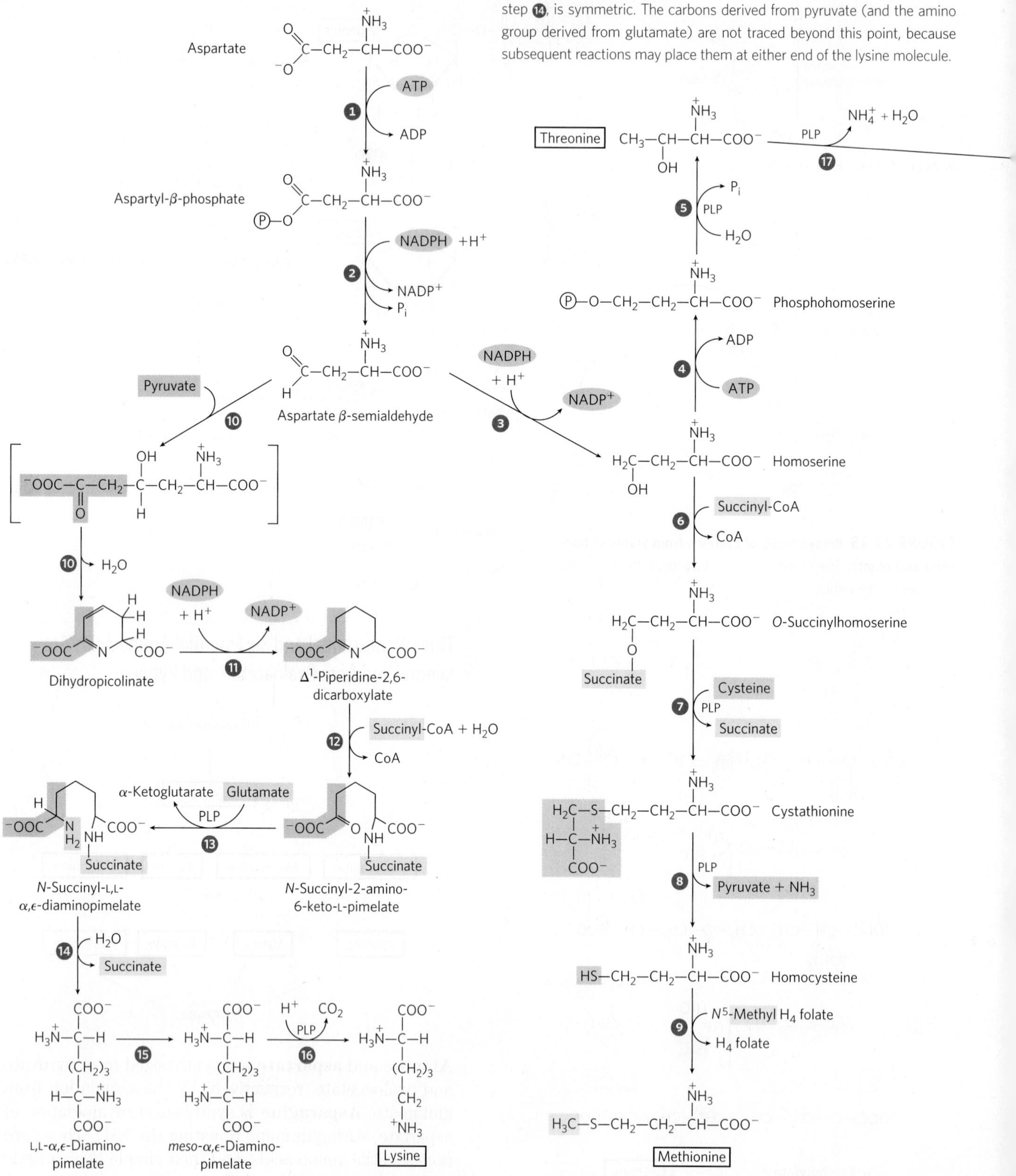

asparagine for growth. The chemotherapy for ALL is administered together with an L-asparaginase derived from bacteria, with the enzyme functioning to reduce serum asparagine. The combined treatment results

in a greater than 95% remission rate in cases of childhood ALL (L-asparaginase treatment alone produces remission in 40% to 60% of cases). However, the asparaginase treatment has some deleterious side

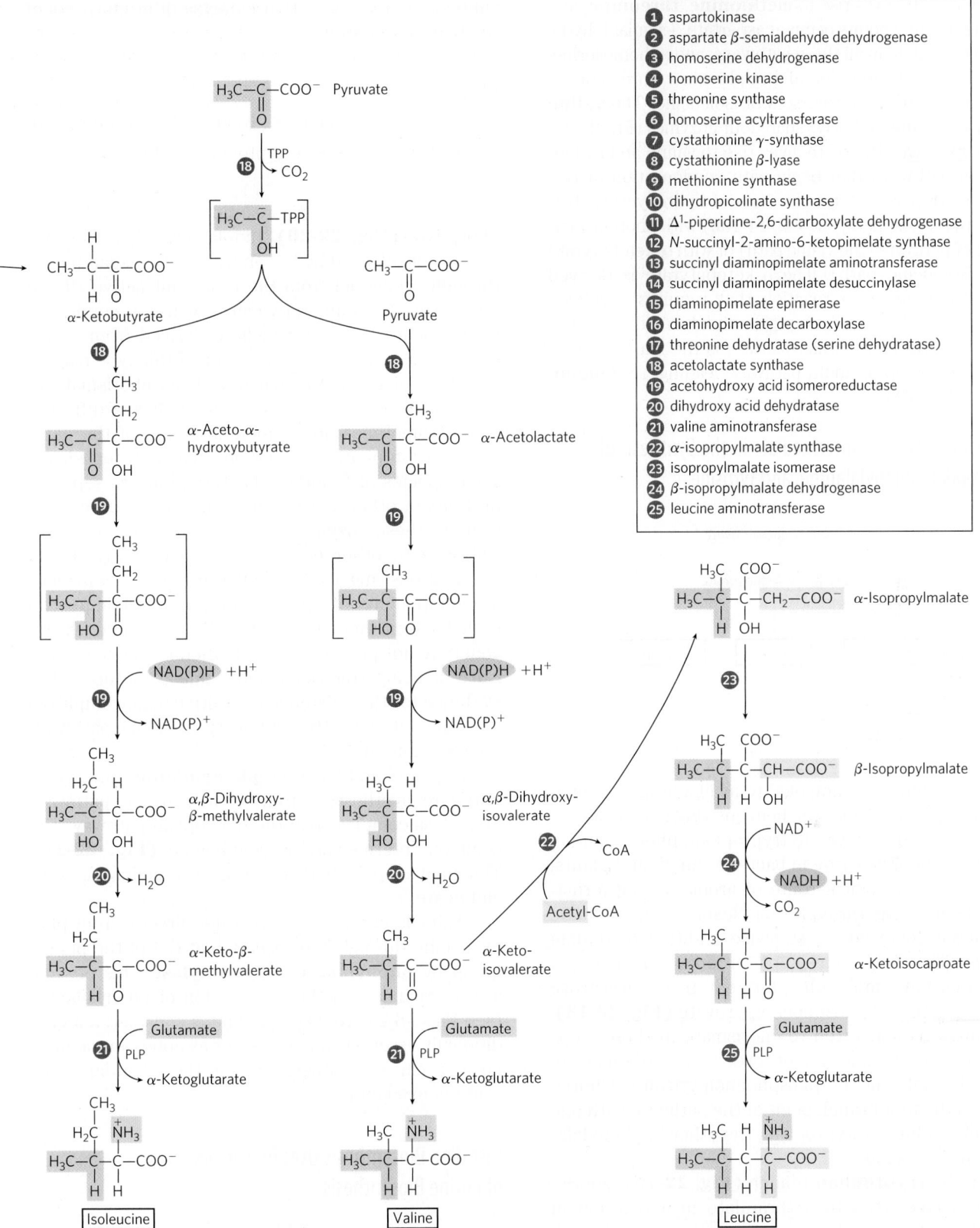

1. aspartokinase
2. aspartate β-semialdehyde dehydrogenase
3. homoserine dehydrogenase
4. homoserine kinase
5. threonine synthase
6. homoserine acyltransferase
7. cystathionine γ-synthase
8. cystathionine β-lyase
9. methionine synthase
10. dihydropicolinate synthase
11. Δ¹-piperidine-2,6-dicarboxylate dehydrogenase
12. N-succinyl-2-amino-6-ketopimelate synthase
13. succinyl diaminopimelate aminotransferase
14. succinyl diaminopimelate desuccinylase
15. diaminopimelate epimerase
16. diaminopimelate decarboxylase
17. threonine dehydratase (serine dehydratase)
18. acetolactate synthase
19. acetohydroxy acid isomeroreductase
20. dihydroxy acid dehydratase
21. valine aminotransferase
22. α-isopropylmalate synthase
23. isopropylmalate isomerase
24. β-isopropylmalate dehydrogenase
25. leucine aminotransferase

effects, and about 10% of patients who achieve remission eventually suffer relapse, with tumors resistant to drug therapy. Researchers are now developing inhibitors of human asparagine synthetase to augment these therapies for childhood ALL. ■

Methionine, threonine, lysine, isoleucine, valine, and leucine are essential amino acids; humans cannot synthesize them. Their biosynthetic pathways in bacteria are complex and interconnected (Fig. 22-17). In some cases, the pathways in bacteria, fungi, and plants differ significantly.

Aspartate gives rise to **methionine**, **threonine**, and **lysine**. Branch points occur at aspartate β-semialdehyde, an intermediate in all three pathways, and at homoserine, a precursor of threonine and methionine. Threonine, in turn, is one of the precursors of isoleucine. The **valine** and **isoleucine** pathways share four enzymes (Fig. 22-17, steps ⑱ to ㉑). Pyruvate gives rise to valine and isoleucine in pathways that begin with condensation of two carbons of pyruvate (in the form of hydroxyethyl thiamine pyrophosphate; see Fig. 14-15b) with another molecule of pyruvate (the valine path) or with α-ketobutyrate (the isoleucine path). The α-ketobutyrate is derived from threonine in a reaction that requires pyridoxal phosphate (Fig. 22-17, step ⑰). An intermediate in the valine pathway, α-ketoisovalerate, is the starting point for a four-step branch pathway leading to **leucine** (steps ㉒ to ㉕).

Chorismate Is a Key Intermediate in the Synthesis of Tryptophan, Phenylalanine, and Tyrosine

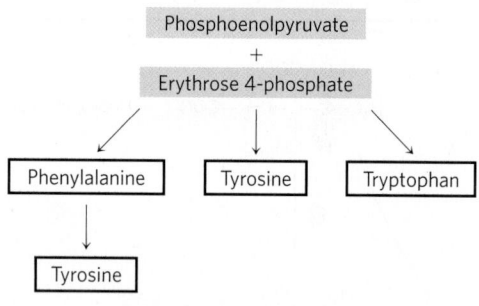

Aromatic rings are not readily available in the environment, even though the benzene ring is very stable. The branched pathway to tryptophan, phenylalanine, and tyrosine, occurring in bacteria, fungi, and plants, is the main biological route of aromatic ring formation. It proceeds through ring closure of an aliphatic precursor followed by stepwise addition of double bonds. The first four steps produce shikimate, a seven-carbon molecule derived from erythrose 4-phosphate and phosphoenolpyruvate **(Fig. 22-18)**. Shikimate is converted to chorismate in three steps that include the addition of three more carbons from another molecule of phosphoenolpyruvate. Chorismate is the first branch point of the pathway, with one branch leading to tryptophan, the other to phenylalanine and tyrosine.

In the **tryptophan** branch **(Fig. 22-19)**, chorismate is converted to anthranilate in a reaction in which glutamine donates the nitrogen that will become part of the indole ring. Anthranilate then condenses with PRPP. The indole ring of tryptophan is derived from the ring carbons and amino group of anthranilate plus two carbons derived from PRPP. The final reaction in the sequence is catalyzed by **tryptophan synthase**. This enzyme has an $\alpha_2\beta_2$ subunit structure and can be dissociated into two α

subunits and a β_2 unit that catalyze different parts of the overall reaction:

$$\text{Indole-3-glycerol phosphate} \xrightarrow{\alpha \text{ subunit}}$$
$$\text{indole + glyceraldehyde 3-phosphate}$$

$$\text{Indole + serine} \xrightarrow{\beta_2 \text{ subunit}} \text{tryptophan + H}_2\text{O}$$

The second part of the reaction requires pyridoxal phosphate **(Fig. 22-20)**. Indole formed in the first part is not released by the enzyme, but instead moves through a channel from the α-subunit active site to one of the β-subunit active sites, where it condenses with a Schiff base intermediate derived from serine and PLP. Intermediate channeling of this type may be a feature of the entire pathway from chorismate to tryptophan. Enzyme active sites catalyzing different steps (sometimes not sequential steps) of the pathway to tryptophan are found on single polypeptides in some species of fungi and bacteria, but are separate proteins in other species. In addition, the activity of some of these enzymes requires a noncovalent association with other enzymes of the pathway. These observations suggest that all the pathway enzymes are components of a large, multienzyme complex in both bacteria and eukaryotes. Such complexes are generally not preserved intact when the enzymes are isolated using traditional biochemical methods, but evidence for the existence of multienzyme complexes is accumulating for this and other metabolic pathways (see Section 16.3).

In plants and bacteria, **phenylalanine** and **tyrosine** are synthesized from chorismate in pathways much less complex than the tryptophan pathway. The common intermediate is prephenate **(Fig. 22-21)**. The final step in both cases is transamination with glutamate.

Animals can produce tyrosine directly from phenylalanine through hydroxylation at C-4 of the phenyl group by **phenylalanine hydroxylase**; this enzyme also participates in the degradation of phenylalanine (see Figs 18-23, 18-24). Tyrosine is considered a conditionally essential amino acid, or as nonessential insofar as it can be synthesized from the essential amino acid phenylalanine.

Histidine Biosynthesis Uses Precursors of Purine Biosynthesis

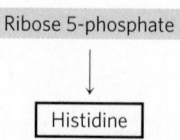

The pathway to **histidine** in all plants and bacteria differs in several respects from other amino acid biosynthetic pathways. Histidine is derived from three precursors

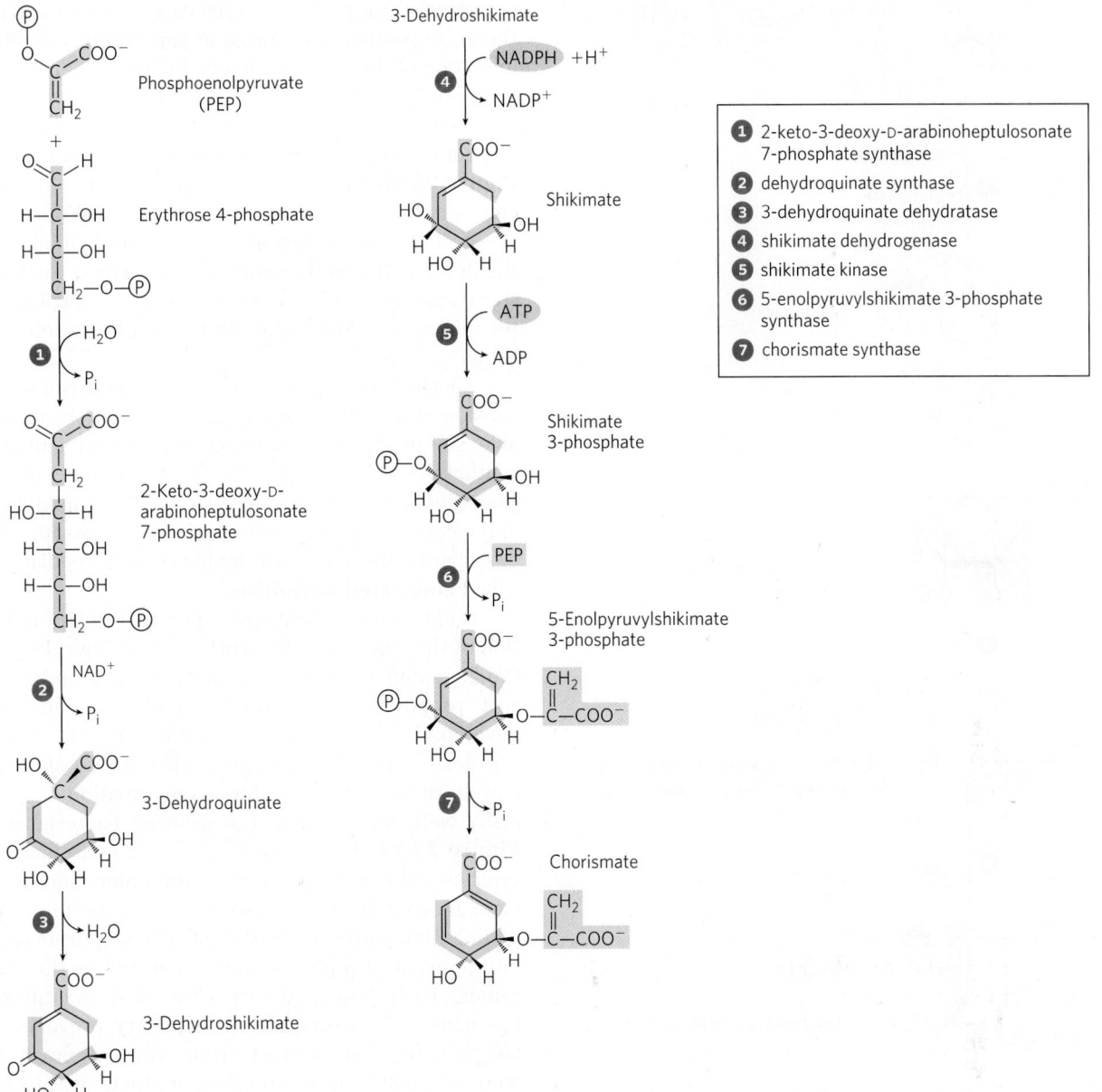

FIGURE 22-18 Biosynthesis of chorismate, an intermediate in the synthesis of aromatic amino acids in bacteria and plants. All carbons are derived from either erythrose 4-phosphate (light purple) or phosphoenolpyruvate (light red). Note that the NAD$^+$ required as a cofactor in step ❷ is released unchanged; it may be transiently reduced to NADH during the reaction, with formation of an oxidized reaction intermediate. Step ❻ is competitively inhibited by glyphosate ($^-$COO—CH$_2$—NH—CH$_2$—PO$_3^{2-}$), the active ingredient in the widely used herbicide Roundup. The herbicide is relatively nontoxic to mammals, which lack this biosynthetic pathway. The intermediates quinate and shikimate are named after the plants in which they have been found to accumulate.

(Fig. 22-22): PRPP contributes five carbons, the purine ring of ATP contributes a nitrogen and a carbon, and glutamine supplies the second ring nitrogen. The key steps are condensation of ATP and PRPP, in which N-1 of the purine ring is linked to the activated C-1 of the ribose of PRPP (step ❶ in Fig. 22-22); purine ring opening that ultimately leaves N-1 and C-2 of adenine linked to the ribose (step ❸); and formation of the imidazole ring, a reaction in which glutamine donates a nitrogen (step ❺). The use of ATP as a metabolite rather than a high-energy cofactor is unusual—but not wasteful, because

it dovetails with the purine biosynthetic pathway. The remnant of ATP that is released after the transfer of N-1 and C-2 is 5-aminoimidazole-4-carboxamide ribonucleotide (AICAR), an intermediate of purine biosynthesis (see Fig. 22-35) that is rapidly recycled to ATP.

Amino Acid Biosynthesis Is under Allosteric Regulation

As detailed in Chapter 15, the control of flux through a metabolic pathway often reflects the activity of multiple

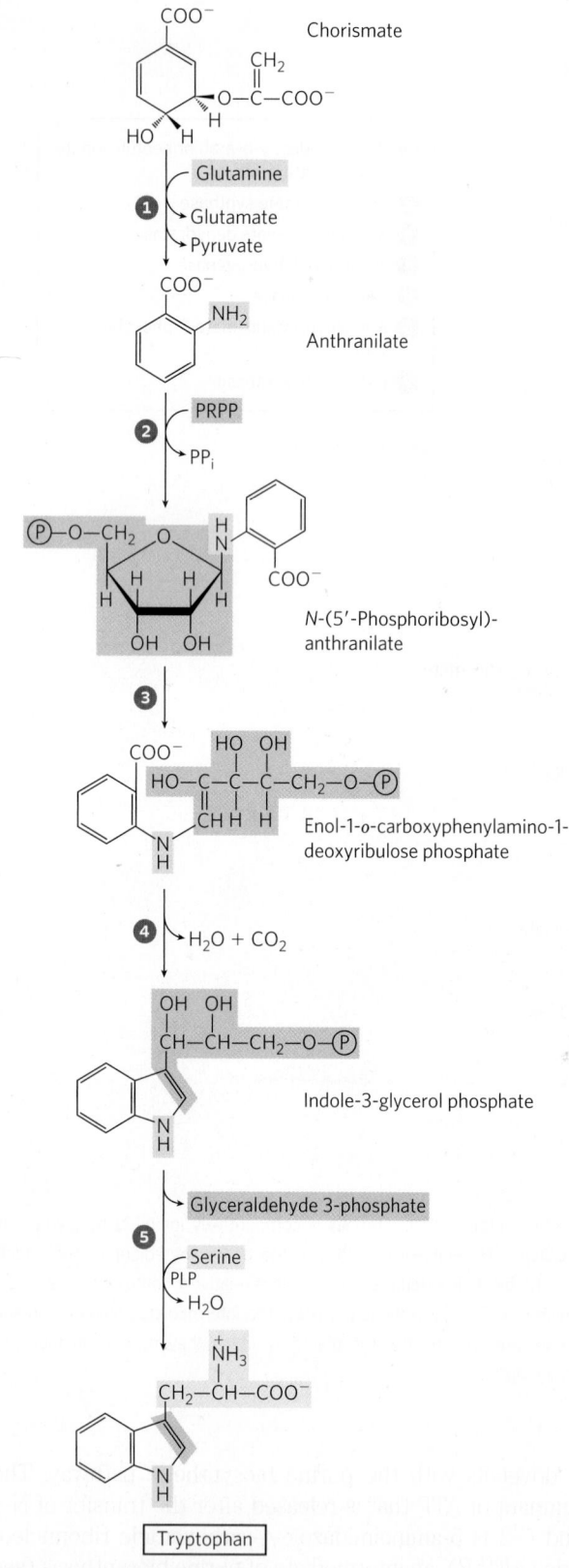

N-(5'-Phosphoribosyl)-anthranilate

Enol-1-o-carboxyphenylamino-1-deoxyribulose phosphate

Indole-3-glycerol phosphate

Tryptophan

1. anthranilate synthase
2. anthranilate phosphoribosyltransferase
3. N-(5'-phosphoribosyl)-anthranilate isomerase
4. indole-3-glycerol phosphate synthase
5. tryptophan synthase

enzymes in that pathway. In the case of amino acid synthesis, regulation takes place in part through feedback inhibition of the first reaction by the end product of the pathway. This first reaction is often catalyzed by an allosteric enzyme that plays an important role in the overall control of flux through that pathway. As an example, **Figure 22-23** shows the allosteric regulation of isoleucine synthesis from threonine (detailed in Fig. 22-17). The end product, isoleucine, is an allosteric inhibitor of the first reaction in the sequence. In bacteria, such allosteric modulation of amino acid synthesis contributes to the minute-to-minute adjustment of pathway activity to cellular needs.

Allosteric regulation of an individual enzyme can be considerably more complex. An example is the remarkable set of allosteric controls exerted on glutamine synthetase of E. coli (Fig. 22-8). Six products derived from glutamine serve as negative feedback modulators of the enzyme, and the overall effects of these and other modulators are more than additive. Such regulation is called **concerted inhibition**.

Additional mechanisms contribute to the regulation of the amino acid biosynthetic pathways. Because the 20 common amino acids must be made in the correct proportions for protein synthesis, cells have developed ways not only of controlling the rate of synthesis of individual amino acids but also of coordinating their formation. Such coordination is especially well developed in fast-growing bacterial cells. **Figure 22-24** shows how E. coli cells coordinate the synthesis of lysine, methionine, threonine, and isoleucine, all made from aspartate. Several important types of inhibition patterns are evident. The step from aspartate to aspartyl-β-phosphate is catalyzed by three isozymes, each independently controlled by different modulators. This **enzyme multiplicity** prevents one biosynthetic end product from shutting down key steps in a pathway when other products of the same pathway are required. The steps from aspartate β-semialdehyde to homoserine and from threonine to α-ketobutyrate (detailed in Fig. 22-17) are also catalyzed by dual, independently controlled isozymes. One isozyme for the conversion of aspartate to aspartyl-β-phosphate is allosterically inhibited by two different modulators, lysine and isoleucine, whose action is more than additive—another example of concerted inhibition. The sequence from aspartate to isoleucine undergoes multiple, overlapping negative feedback inhibitions; for example, isoleucine inhibits the conversion of threonine to α-ketobutyrate (as described above), and threonine inhibits its own formation at three points: from homoserine, from aspartate β-semialdehyde, and

FIGURE 22-19 Biosynthesis of tryptophan from chorismate in bacteria and plants. In E. coli, enzymes catalyzing steps 1 and 2 are subunits of a single complex.

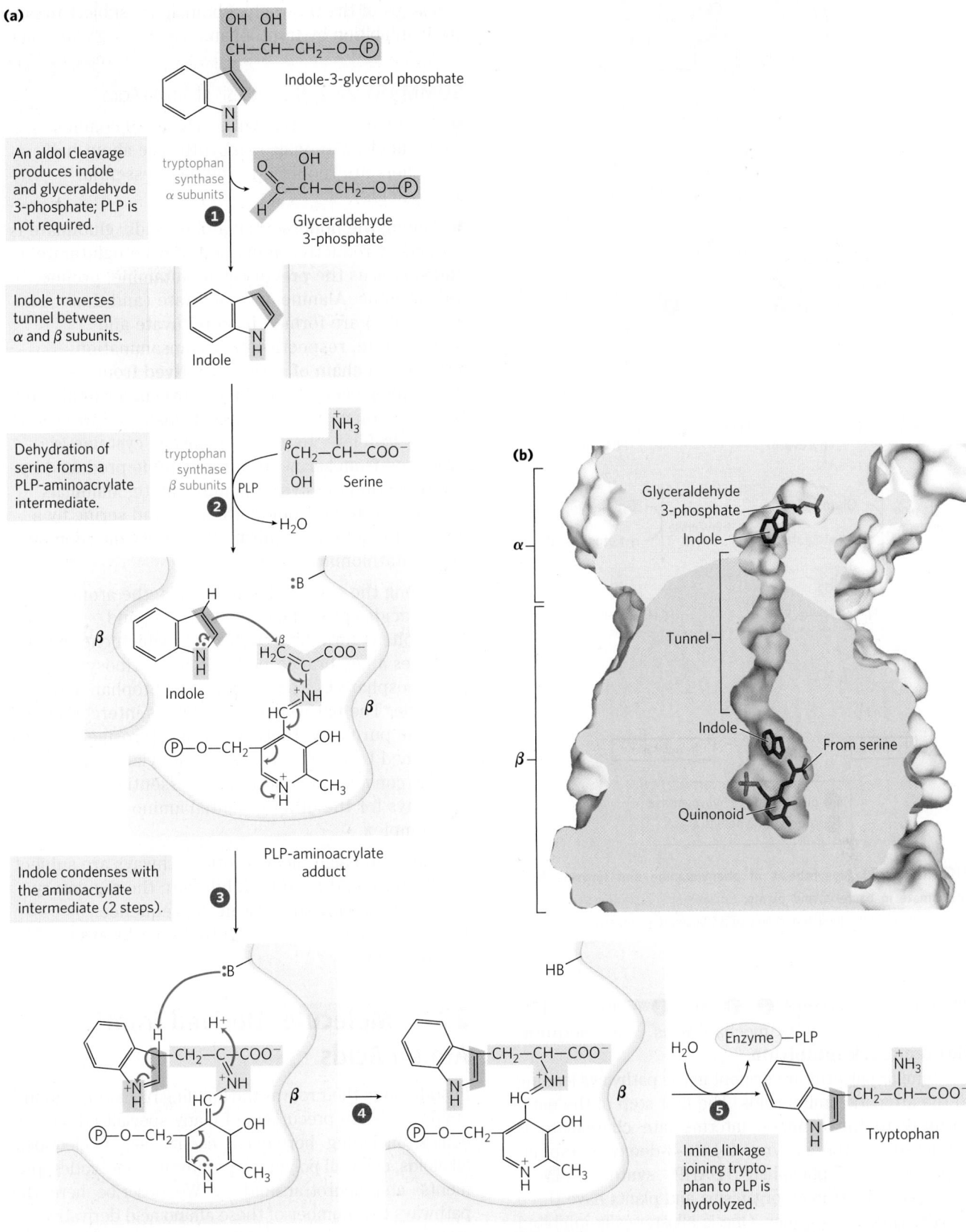

MECHANISM FIGURE 22-20 **Tryptophan synthase reaction. (a)** This enzyme catalyzes a multistep reaction with several types of chemical rearrangements. The PLP-facilitated transformations occur at the β carbon (C-3) of the amino acid, as opposed to the α-carbon reactions described in Figure 18-6. The β carbon of serine is attached to the indole ring system. **(b)** Indole generated on the α subunit (white) moves through a tunnel to the β subunit (blue), where it condenses with the serine moiety. [Source: (b) PDB ID 1KFJ, V. Kulik et al., *J. Mol. Biol.* 324:677, 2002.]

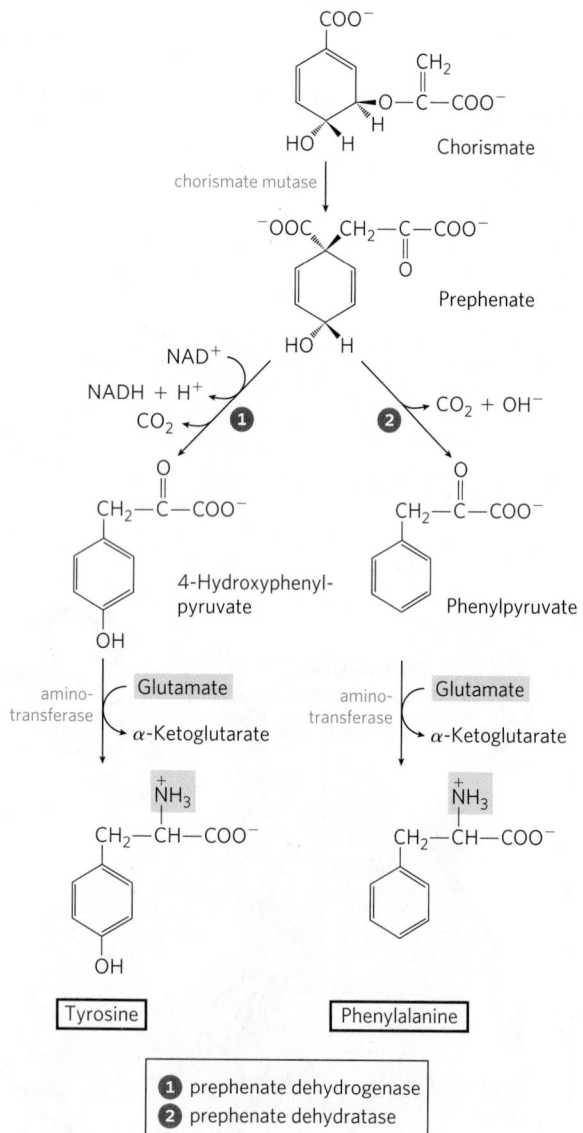

FIGURE 22-21 Biosynthesis of phenylalanine and tyrosine from chorismate in bacteria and plants. Conversion of chorismate to prephenate is a rare biological example of a Claisen rearrangement.

from aspartate (steps ❹, ❸, and ❶ in Fig. 22-17). This overall regulatory mechanism is called **sequential feedback inhibition**.

Similar patterns are evident in the pathways leading to the aromatic amino acids. The first step of the early pathway to the common intermediate chorismate is catalyzed by the enzyme 2-keto-3-deoxy-D-arabino-heptulosonate 7-phosphate (DAHP) synthase (❶ in Fig. 22-18). Most microorganisms and plants have three DAHP synthase isozymes. One is allosterically inhibited (feedback inhibition) by phenylalanine, another by tyrosine, and the third by tryptophan. This scheme helps the overall pathway to respond to cellular requirements for one or more of the aromatic amino acids. Additional regulation takes place after the pathway branches at chorismate. For example, the enzymes catalyzing the first

two steps of the tryptophan branch are subject to allosteric inhibition by tryptophan.

SUMMARY 22.2 Biosynthesis of Amino Acids

■ Plants and bacteria synthesize all 20 common amino acids. Mammals can synthesize about half; the others are required in the diet (essential amino acids).

■ Among the nonessential amino acids, glutamate is formed by reductive amination of α-ketoglutarate and serves as the precursor of glutamine, proline, and arginine. Alanine and aspartate (and thus asparagine) are formed from pyruvate and oxaloacetate, respectively, by transamination. The carbon chain of serine is derived from 3-phosphoglycerate. Serine is a precursor of glycine; the β-carbon atom of serine is transferred to tetrahydrofolate. In microorganisms, cysteine is produced from serine and from sulfide produced by the reduction of environmental sulfate. Mammals produce cysteine from methionine and serine by a series of reactions requiring S-adenosylmethionine and cystathionine.

■ Among the essential amino acids, the aromatic amino acids (phenylalanine, tyrosine, and tryptophan) form by a pathway in which chorismate occupies a key branch point. Phosphoribosyl pyrophosphate is a precursor of tryptophan and histidine. The pathway to histidine is interconnected with the purine synthetic pathway. Tyrosine can also be formed by hydroxylation of phenylalanine (and thus is considered conditionally essential). The pathways for the other essential amino acids are complex.

■ The amino acid biosynthetic pathways are subject to allosteric end-product inhibition; the regulatory enzyme is usually the first in the sequence. Regulation of the various synthetic pathways is coordinated.

22.3 Molecules Derived from Amino Acids

In addition to their role as the building blocks of proteins, amino acids are precursors of many specialized biomolecules, including hormones, coenzymes, nucleotides, alkaloids, cell wall polymers, porphyrins, antibiotics, pigments, and neurotransmitters. We describe here the pathways to a number of these amino acid derivatives.

Glycine Is a Precursor of Porphyrins

The biosynthesis of **porphyrins**, for which glycine is a major precursor, is our first example because of the central importance of the porphyrin nucleus in heme proteins such as hemoglobin and the cytochromes.

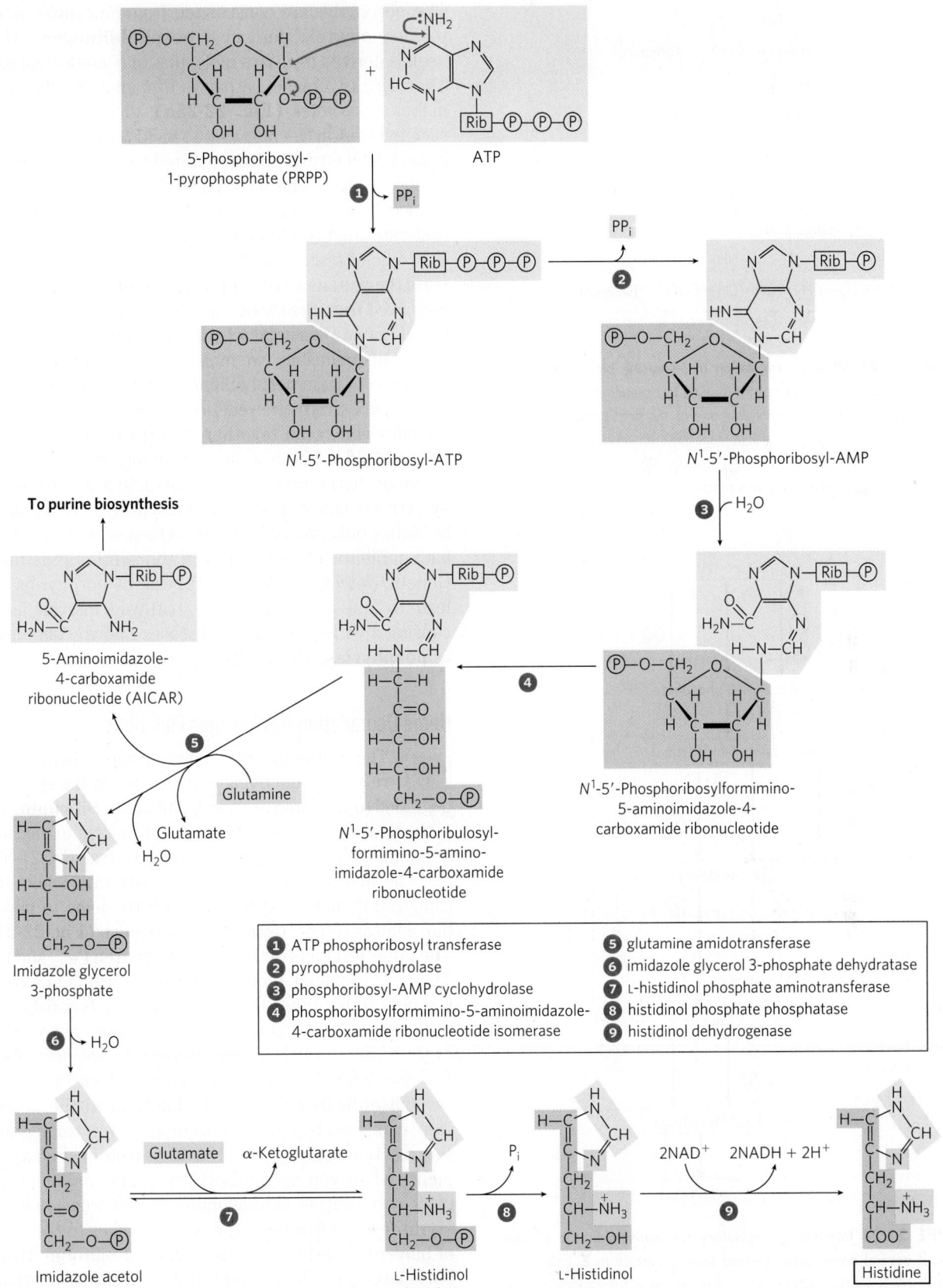

FIGURE 22-22 Biosynthesis of histidine in bacteria and plants. Atoms derived from PRPP and ATP are shaded light red and blue, respectively. Two of the histidine nitrogens are derived from glutamine and glutamate (green). Note that the derivative of ATP remaining after step ❺ (AICAR) is an intermediate in purine biosynthesis (see Fig. 22-35, step ❾), so ATP is rapidly regenerated.

Labels within the figure:

5-Phosphoribosyl-1-pyrophosphate (PRPP)

ATP

N^1-5'-Phosphoribosyl-ATP

N^1-5'-Phosphoribosyl-AMP

To purine biosynthesis

5-Aminoimidazole-4-carboxamide ribonucleotide (AICAR)

N^1-5'-Phosphoribulosyl-formimino-5-amino-imidazole-4-carboxamide ribonucleotide

N^1-5'-Phosphoribosylformimino-5-aminoimidazole-4-carboxamide ribonucleotide

Glutamine

Glutamate

H_2O

Imidazole glycerol 3-phosphate

Imidazole acetol 3-phosphate

Glutamate

α-Ketoglutarate

L-Histidinol phosphate

L-Histidinol

$2NAD^+$ $2NADH + 2H^+$

Histidine

❶ ATP phosphoribosyl transferase
❷ pyrophosphohydrolase
❸ phosphoribosyl-AMP cyclohydrolase
❹ phosphoribosylformimino-5-aminoimidazole-4-carboxamide ribonucleotide isomerase
❺ glutamine amidotransferase
❻ imidazole glycerol 3-phosphate dehydratase
❼ L-histidinol phosphate aminotransferase
❽ histidinol phosphate phosphatase
❾ histidinol dehydrogenase

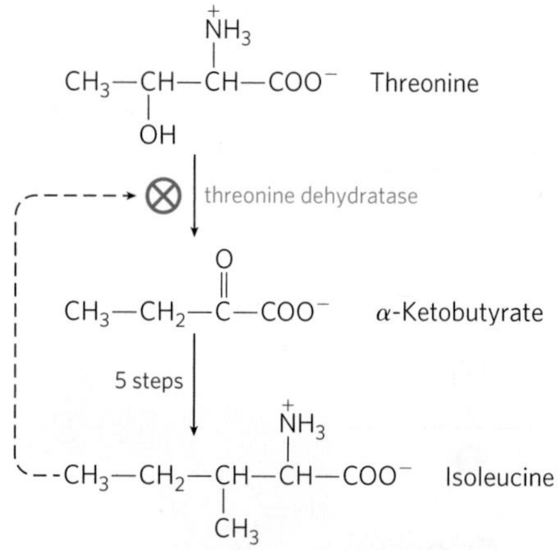

$$CH_3-\overset{\displaystyle OH}{\underset{\displaystyle |}{CH}}-\overset{\displaystyle \overset{+}{N}H_3}{\underset{\displaystyle |}{CH}}-COO^- \quad \text{Threonine}$$

threonine dehydratase

$$CH_3-CH_2-\overset{\displaystyle O}{\overset{\displaystyle \|}{C}}-COO^- \quad \alpha\text{-Ketobutyrate}$$

5 steps

$$CH_3-CH_2-\overset{\displaystyle \overset{+}{N}H_3}{\underset{\displaystyle |}{CH}}-\overset{}{\underset{\displaystyle \underset{\displaystyle CH_3}{|}}{CH}}-COO^- \quad \text{Isoleucine}$$

FIGURE 22-23 Allosteric regulation of isoleucine biosynthesis. The first reaction in the pathway from threonine to isoleucine is inhibited by the end product, isoleucine. This was one of the first examples of allosteric feedback inhibition to be discovered. The steps from α-ketobutyrate to isoleucine correspond to steps **18** through **21** in Figure 22-17 (five steps, because **19** is a two-step reaction).

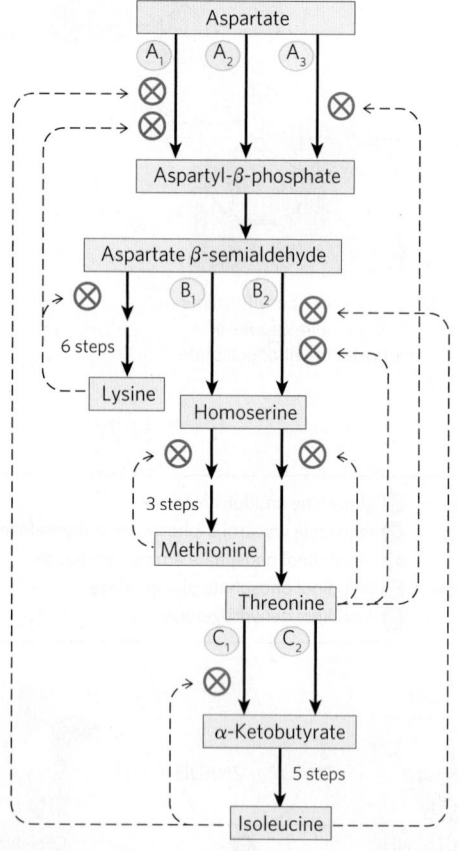

FIGURE 22-24 Interlocking regulatory mechanisms in the biosynthesis of several amino acids derived from aspartate in *E. coli*. Three enzymes (A, B, C) have either two or three isozyme forms, indicated by numerical subscripts. In each case, one isozyme (A_2, B_1, and C_2) has no allosteric regulation; these isozymes are regulated by changes in the amount of enzyme synthesized (Chapter 28). Synthesis of isozymes A_2 and B_1 is repressed when methionine levels are high, and synthesis of isozyme C_2 is repressed when isoleucine levels are high. Enzyme A is aspartokinase; B, homoserine dehydrogenase; C, threonine dehydratase.

The porphyrins are constructed from four molecules of the monopyrrole derivative **porphobilinogen**, which itself is derived from two molecules of δ-aminolevulinate. There are two major pathways to δ-aminolevulinate. In higher eukaryotes **(Fig. 22-25a)**, glycine reacts with succinyl-CoA in the first step to yield α-amino-β-ketoadipate, which is then decarboxylated to δ-aminolevulinate. In plants, algae, and most bacteria, δ-aminolevulinate is formed from glutamate (Fig. 22-25b). The glutamate is first esterified to glutamyl-tRNA^{Glu} (see Chapter 27 on the topic of transfer RNAs); reduction by NADPH converts the glutamate to glutamate 1-semialdehyde, which is cleaved from the tRNA. An aminotransferase converts the glutamate 1-semialdehyde to δ-aminolevulinate.

In all organisms, two molecules of δ-aminolevulinate condense to form porphobilinogen and, through a series of complex enzymatic reactions, four molecules of porphobilinogen come together to form **protoporphyrin (Fig. 22-26)**. The iron atom is incorporated after the protoporphyrin has been assembled, in a step catalyzed by ferrochelatase. Porphyrin biosynthesis is regulated in higher eukaryotes by heme, which serves as a feedback inhibitor of early steps in the synthetic pathway. Genetic defects in the biosynthesis of porphyrins can lead to the accumulation of pathway intermediates, causing a variety of human diseases known collectively as **porphyrias** (Box 22-2).

Heme Degradation Has Multiple Functions

The iron-porphyrin (heme) group of hemoglobin, released from dying erythrocytes in the spleen, is degraded to yield free Fe^{2+} and, ultimately, **bilirubin**. The pathway also contributes the pigment present in mixtures of the bile salts derived from cholesterol (see Fig. 21-38).

The first step in the two-step pathway to bilirubin, catalyzed by heme oxygenase, converts heme to biliverdin, a linear (open) tetrapyrrole derivative **(Fig. 22-27)**. The other products of the reaction are free Fe^{2+} and CO. The Fe^{2+} is quickly bound by ferritin. Carbon monoxide is a poison that binds to hemoglobin (see Box 5-1), and the production of CO by heme oxygenase ensures that, even in the absence of environmental exposure, about 1% of an individual's heme is complexed with CO.

Biliverdin is converted to bilirubin in the second step, catalyzed by biliverdin reductase. You can monitor this reaction colorimetrically in a familiar in situ experiment. When you are bruised, the black and/or purple color results from hemoglobin released from damaged erythrocytes. Over time, the color changes to the green of biliverdin, and then to the yellow of bilirubin. Bilirubin is largely insoluble, and it travels in the bloodstream as a complex with serum albumin. In the liver, bilirubin is transformed to the bile pigment bilirubin diglucuronide. This product is sufficiently water-soluble to be secreted with other components of bile into the small intestine, where microbial enzymes convert it to several products, predominantly urobilinogen. Some urobilinogen

(a)
Succinyl-CoA
Glycine
CoA-SH
δ-aminolevulinate synthase
α-Amino-β-ketoadipate
CO2
δ-aminolevulinate synthase
δ-Aminolevulinate
glutamate 1-semialdehyde aminomutase
(b)
Glutamate
tRNAGlu ATP AMP + PPi
glutamyl-tRNA synthetase
Glutamyl-tRNAGlu
NADPH + H+ NADP+ tRNAGlu
glutamyl-tRNA reductase
Glutamate 1-semialdehyde

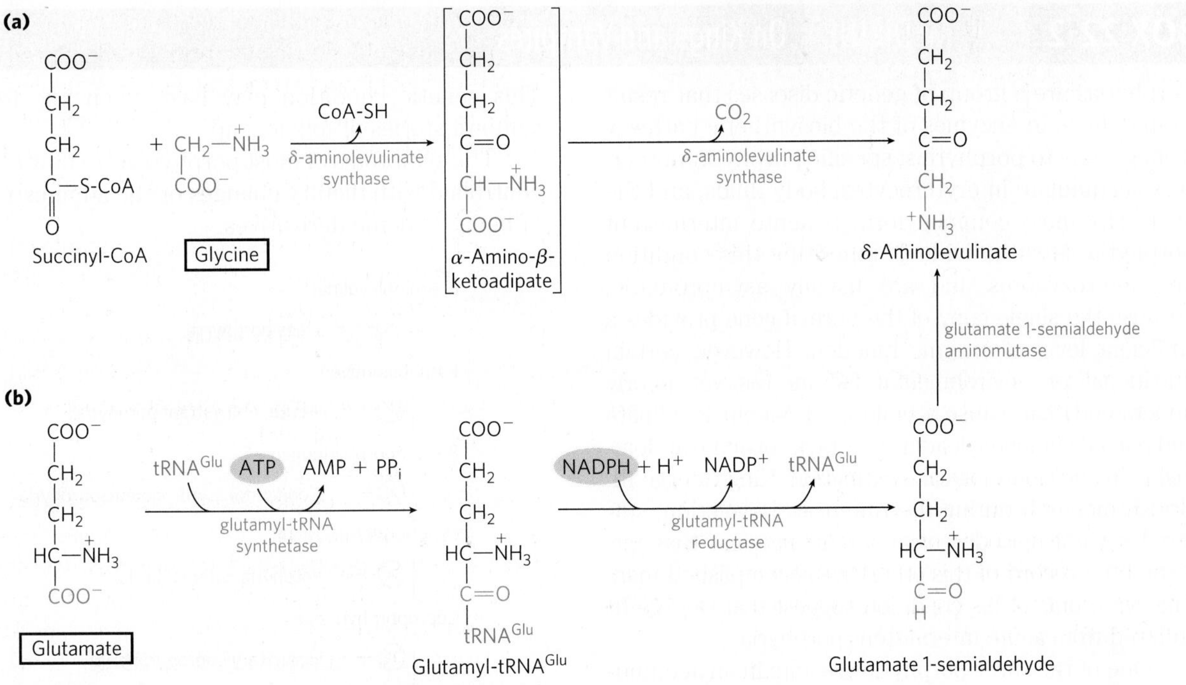

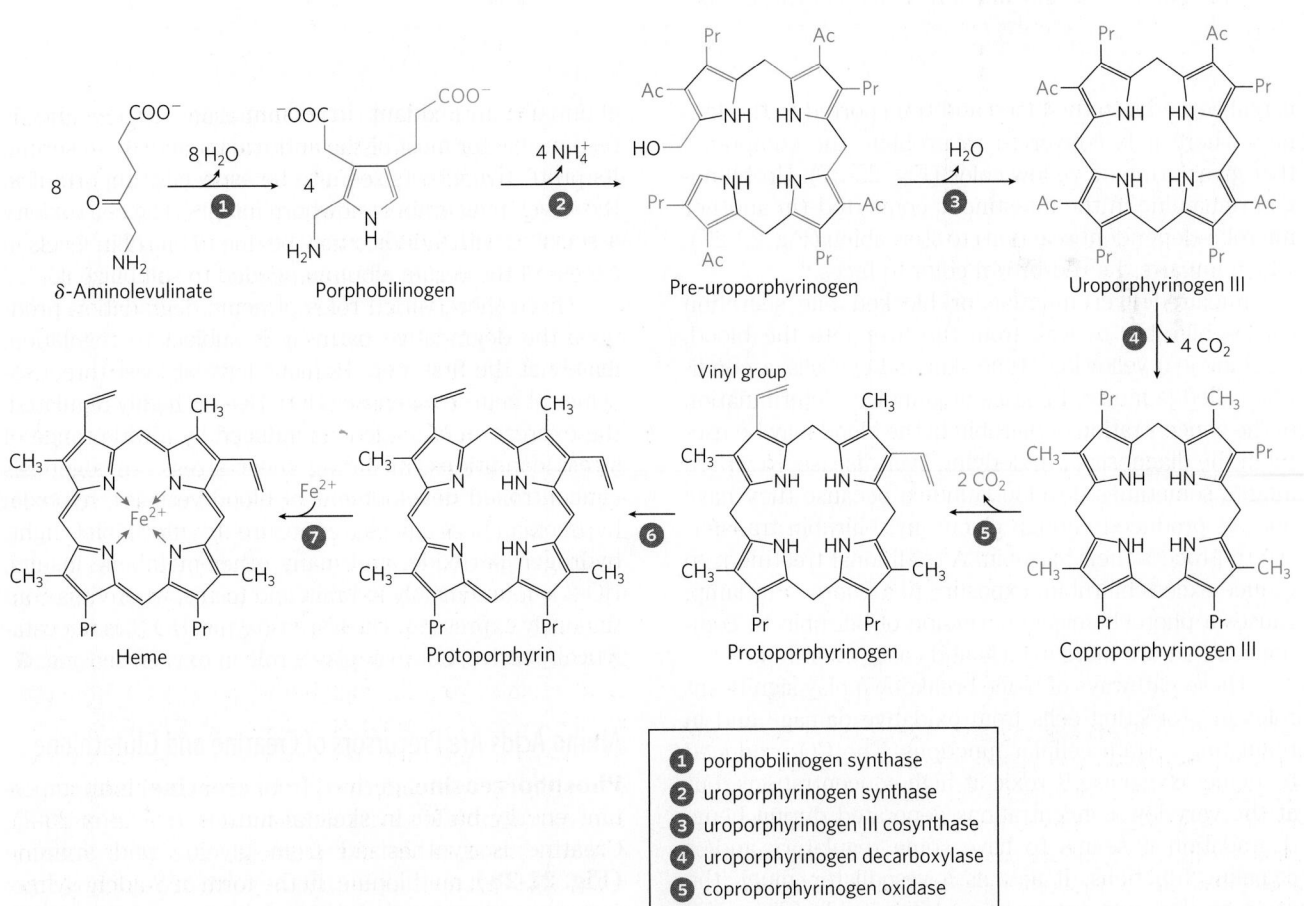

FIGURE 22-25 Biosynthesis of δ-aminolevulinate. (a) In most animals, including mammals, δ-aminolevulinate is synthesized from glycine and succinyl-CoA. The atoms furnished by glycine are shown in red. (b) In bacteria and plants, the precursor of δ-aminolevulinate is glutamate.

δ-Aminolevulinate
8 H2O
Porphobilinogen
4 NH4+
Pre-uroporphyrinogen
H2O
Uroporphyrinogen III
4 CO2
Coproporphyrinogen III
2 CO2
Vinyl group
Protoporphyrinogen
Protoporphyrin
Fe2+
Heme
1 porphobilinogen synthase
2 uroporphyrinogen synthase
3 uroporphyrinogen III cosynthase
4 uroporphyrinogen decarboxylase
5 coproporphyrinogen oxidase
6 protoporphyrinogen oxidase
7 ferrochelatase

FIGURE 22-26 Biosynthesis of heme from δ-aminolevulinate. Ac represents acetyl (—CH_2COO^-); Pr, propionyl (—$CH_2CH_2COO^-$).

BOX 22-2 ⚕ MEDICINE   On Kings and Vampires

Porphyrias are a group of genetic diseases that result from defects in enzymes of the biosynthetic pathway from glycine to porphyrins; specific porphyrin precursors accumulate in erythrocytes, body fluids, and the liver. The most common form is acute intermittent porphyria. Most individuals inheriting this condition are heterozygotes and are usually asymptomatic, because the single copy of the normal gene provides a sufficient level of enzyme function. However, certain nutritional or environmental factors (as yet poorly understood) can cause a buildup of δ-aminolevulinate and porphobilinogen, leading to attacks of acute abdominal pain and neurological dysfunction. King George III, British monarch during the American Revolution, suffered several episodes of apparent madness that tarnished the record of this otherwise accomplished man. The symptoms of his condition suggest that George III suffered from acute intermittent porphyria.

One of the rarer porphyrias results in an accumulation of uroporphyrinogen I, an abnormal isomer of a protoporphyrin precursor. This compound stains the urine red, causes the teeth to fluoresce strongly in ultraviolet light, and makes the skin abnormally sensitive to sunlight. Many individuals with this porphyria are anemic because insufficient heme is synthesized.

This genetic condition may have given rise to the vampire stories of folk legend.

The symptoms of most porphyrias are now readily controlled with dietary changes or the administration of heme or heme derivatives.

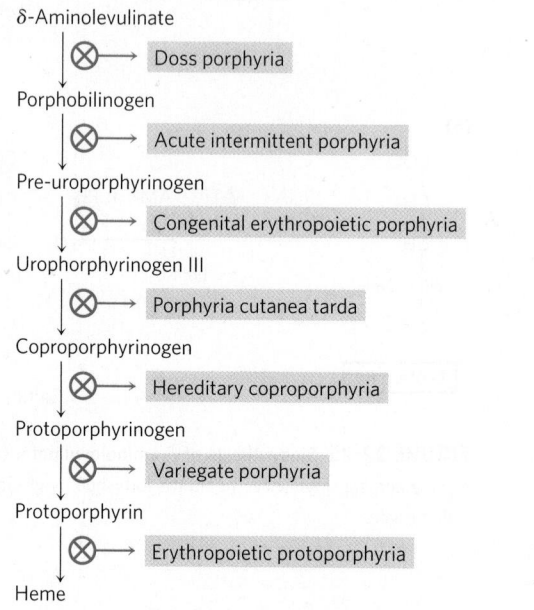

is reabsorbed into the blood and transported to the kidney, where it is converted to urobilin, the compound that gives urine its yellow color (Fig. 22-27). Urobilinogen remaining in the intestine is converted (in another microbe-dependent reaction) to stercobilin (Fig. 22-27), which imparts the red-brown color to feces.

Impaired liver function or blocked bile secretion causes bilirubin to leak from the liver into the blood, resulting in a yellowing of the skin and eyeballs, a condition called jaundice. In cases of jaundice, determination of the concentration of bilirubin in the blood may be useful in the diagnosis of underlying liver disease. Newborn infants sometimes develop jaundice because they have not yet produced enough glucuronyl bilirubin transferase to process their bilirubin. A traditional treatment to reduce excess bilirubin, exposure to a fluorescent lamp, causes a photochemical conversion of bilirubin to compounds that are more soluble and easily excreted.

These pathways of heme breakdown play significant roles in protecting cells from oxidative damage and in regulating certain cellular functions. The CO produced by heme oxygenase is toxic at high concentrations, but at the very low concentrations generated during heme degradation it seems to have some regulatory and/or signaling functions. It acts as a vasodilator, much the same as (but less potent than) nitric oxide (discussed below). Low levels of CO also have some regulatory effects on neurotransmission. Bilirubin is the most

abundant antioxidant in mammalian tissues and is responsible for most of the antioxidant activity in serum. Its protective effects seem to be especially important in the developing brain of newborn infants. The cell toxicity associated with jaundice may be due to bilirubin levels in excess of the serum albumin needed to solubilize it.

Given these varied roles of heme degradation products, the degradative pathway is subject to regulation, mainly at the first step. Humans have at least three isozymes of heme oxygenase (HO). HO-1 is highly regulated; the expression of its gene is induced by a wide range of stress conditions, including shear stress, angiogenesis (uncontrolled development of blood vessels), hypoxia, hyperoxia, heat shock, exposure to ultraviolet light, hydrogen peroxide, and many other metabolic insults. HO-2 is found mainly in brain and testes, where it is continuously expressed. The third isozyme, HO-3, is not catalytically active, but may play a role in oxygen sensing. ∎

Amino Acids Are Precursors of Creatine and Glutathione

Phosphocreatine, derived from **creatine**, is an important energy buffer in skeletal muscle (see Box 23-2). Creatine is synthesized from glycine and arginine **(Fig. 22-28)**; methionine, in the form of *S*-adenosylmethionine, acts as methyl group donor.

Glutathione (GSH), present in plants, animals, and some bacteria, often at high levels, can be thought

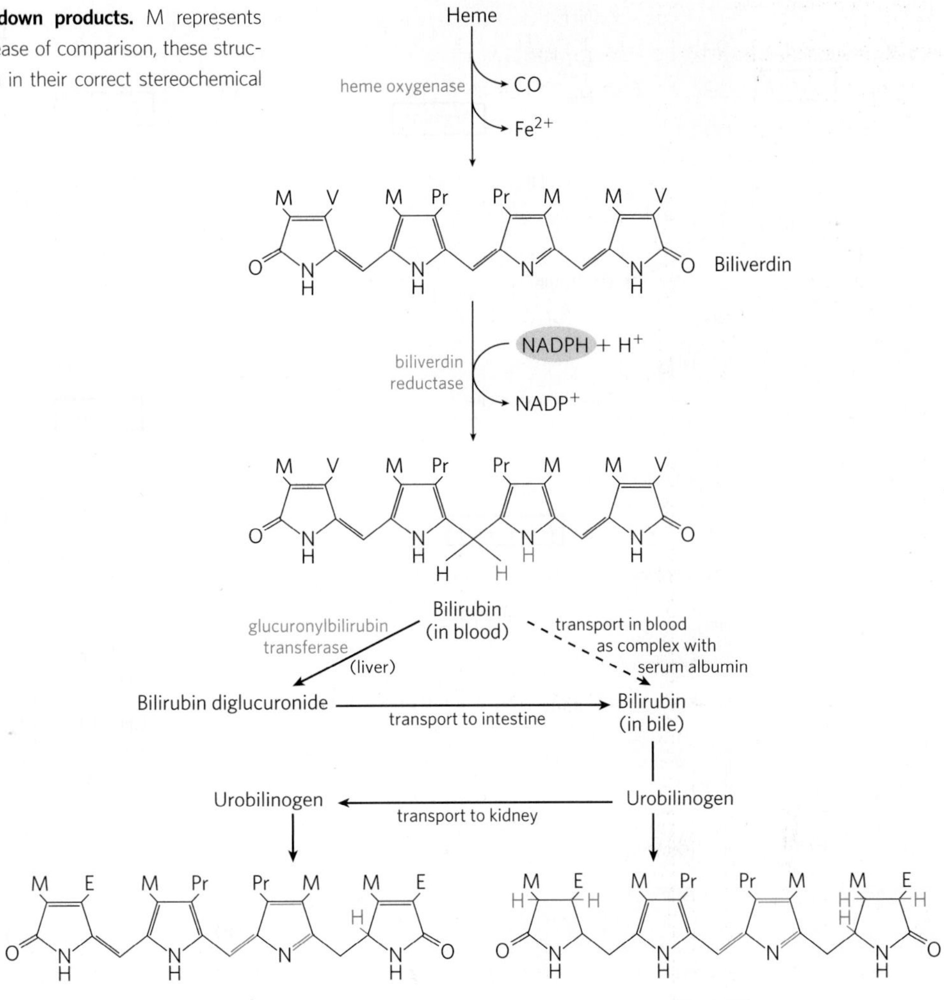

FIGURE 22-27 Bilirubin and its breakdown products. M represents methyl; V, vinyl; Pr, propionyl; E, ethyl. For ease of comparison, these structures are shown in linear form, rather than in their correct stereochemical conformations.

of as a redox buffer. It is derived from glutamate, cysteine, and glycine **(Fig. 22-29)**. The γ-carboxyl group of glutamate is activated by ATP to form an acyl phosphate intermediate, which is then attacked by the α-amino group of cysteine. A second condensation reaction follows, with the α-carboxyl group of cysteine activated to an acyl phosphate to permit reaction with glycine. The oxidized form of glutathione (GSSG), produced in the course of its redox activities, contains two glutathione molecules linked by a disulfide bond.

Glutathione probably helps maintain the sulfhydryl groups of proteins in the reduced state and the iron of heme in the ferrous (Fe^{2+}) state, and it serves as a reducing agent for glutaredoxin in deoxyribonucleotide synthesis (see Fig. 22-41). Its redox function is also used to remove toxic peroxides formed in the normal course of growth and metabolism under aerobic conditions:

$$2\ GSH + R\text{—}O\text{—}O\text{—}H \longrightarrow GSSG + H_2O + R\text{—}OH$$

This reaction is catalyzed by **glutathione peroxidase**, a remarkable enzyme in that it contains a covalently bound selenium (Se) atom in the form of selenocysteine (see Fig. 3-8a), which is essential for its activity.

D-Amino Acids Are Found Primarily in Bacteria

Although D-amino acids do not generally occur in proteins, they do serve some special functions in the structure of bacterial cell walls and peptide antibiotics. Bacterial peptidoglycans (see Fig. 6-30) contain both D-alanine and D-glutamate. D-Amino acids arise directly from the L isomers by the action of amino acid racemases, which have pyridoxal phosphate as cofactor (see Fig. 18-6). Amino acid racemization is uniquely important to bacterial metabolism, and enzymes such as alanine racemase are prime targets for pharmaceutical agents. One such agent, **L-fluoroalanine**, is being tested as an antibacterial drug. Another, **cycloserine**, is used to treat tuberculosis. Because these inhibitors also affect some PLP-requiring human enzymes, however, they have potentially undesirable side effects.

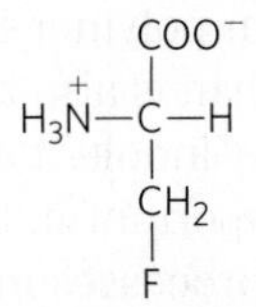

L-Fluoroalanine 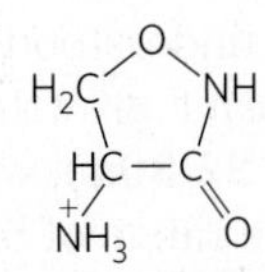   Cycloserine

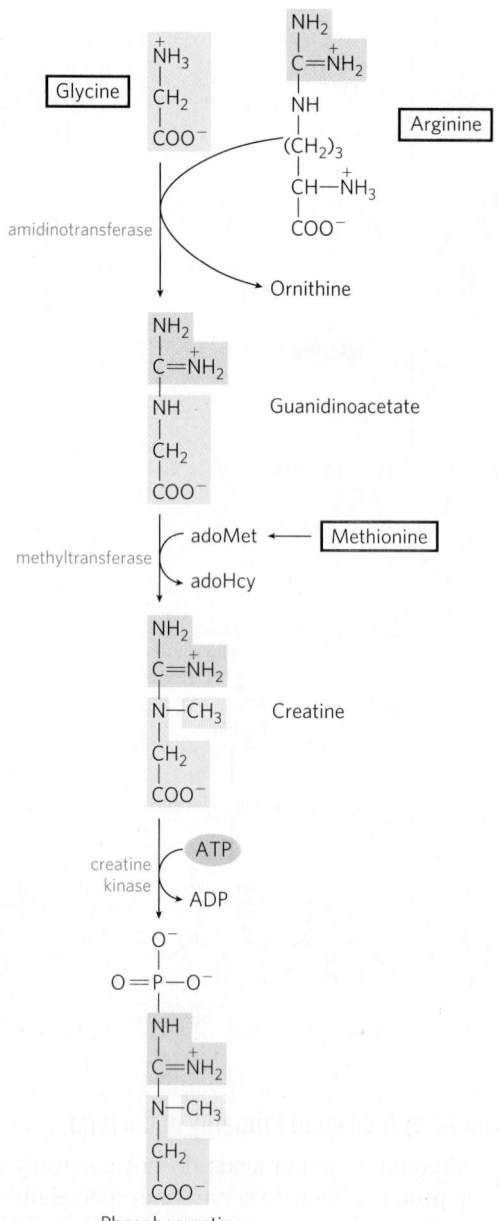

FIGURE 22-28 Biosynthesis of creatine and phosphocreatine. Creatine is made from three amino acids: glycine, arginine, and methionine. This pathway shows the versatility of amino acids as precursors of other nitrogenous biomolecules.

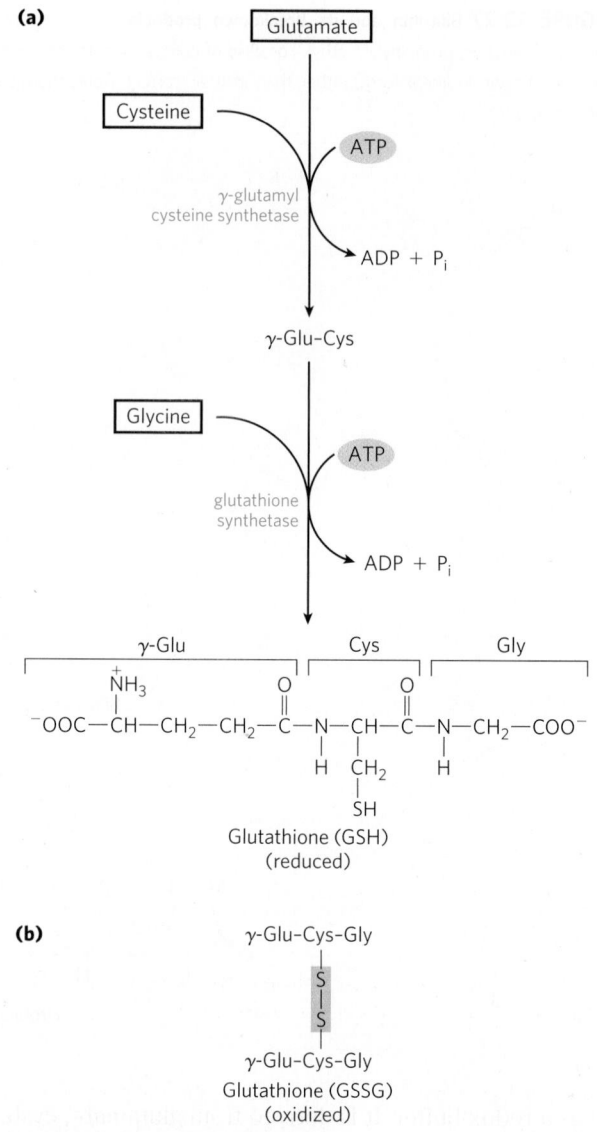

(a)

FIGURE 22-29 Glutathione metabolism. (a) Biosynthesis of glutathione. **(b)** Oxidized form of glutathione.

Aromatic Amino Acids Are Precursors of Many Plant Substances

Phenylalanine, tyrosine, and tryptophan are converted to a variety of important compounds in plants. The rigid polymer **lignin**, derived from phenylalanine and tyrosine, is second only to cellulose in abundance in plant tissues. The structure of the lignin polymer is complex and not well understood. Tryptophan is also the precursor of the plant growth hormone indole-3-acetate, or **auxin (Fig. 22-30a)**, which is important in the regulation of a wide range of biological processes in plants.

Phenylalanine and tyrosine also give rise to many commercially significant natural products, including the tannins that inhibit oxidation in wines; alkaloids such as morphine, which have potent physiological effects; and the flavoring of cinnamon oil (Fig. 22-30b), nutmeg, cloves, vanilla, cayenne pepper, and other products.

Biological Amines Are Products of Amino Acid Decarboxylation

Many important neurotransmitters are primary or secondary amines, derived from amino acids in simple pathways. In addition, some polyamines that form complexes with DNA are derived from the amino acid ornithine, a component of the urea cycle. A common denominator of many of these pathways is amino acid decarboxylation, another PLP-requiring reaction (see Fig. 18-6).

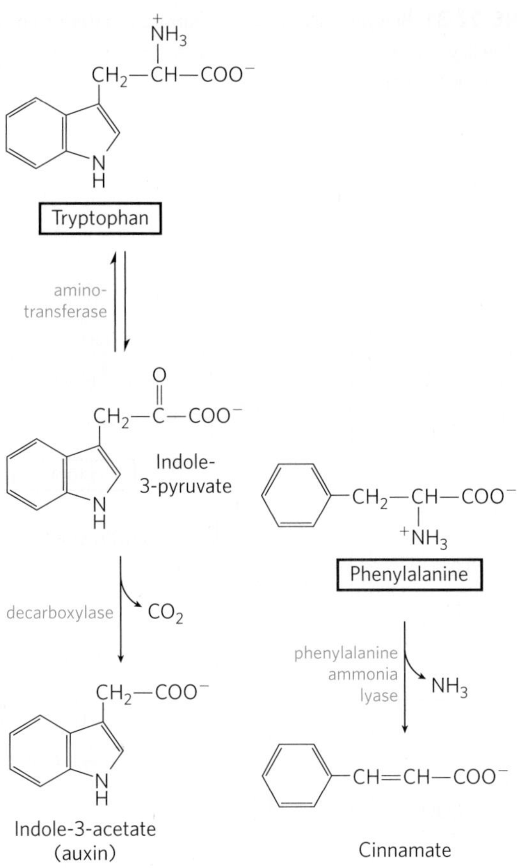

FIGURE 22-30 Biosynthesis of two plant substances from amino acids.
(a) Indole-3-acetate (auxin) and (b) cinnamate (cinnamon flavor).

The synthesis of some neurotransmitters is illustrated in **Figure 22-31**. Tyrosine gives rise to a family of catecholamines that includes **dopamine**, **norepinephrine**, and **epinephrine**. Levels of catecholamines are correlated with, among other things, changes in blood pressure. The neurological disorder Parkinson disease is associated with an underproduction of dopamine, and it has traditionally been treated by administering L-dopa. Overproduction of dopamine in the brain may be linked to psychological disorders such as schizophrenia.

Glutamate decarboxylation gives rise to **γ-aminobutyrate (GABA)**, an inhibitory neurotransmitter. Its underproduction is associated with epileptic seizures. GABA analogs are used in the treatment of epilepsy and hypertension. Levels of GABA can also be increased by administering inhibitors of the GABA-degrading enzyme GABA aminotransferase. Another important neurotransmitter, **serotonin**, is derived from tryptophan in a two-step pathway.

Histidine undergoes decarboxylation to **histamine**, a powerful vasodilator in animal tissues. Histamine is released in large amounts as part of the allergic response, and it also stimulates acid secretion in the stomach. A

growing array of pharmaceutical agents are being designed to interfere with either the synthesis or the action of histamine. A prominent example is the histamine receptor antagonist **cimetidine** (Tagamet), a structural analog of histamine:

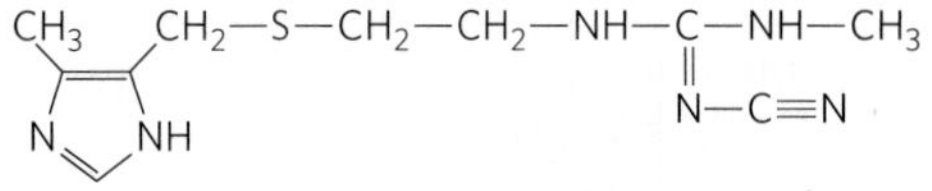

It promotes the healing of duodenal ulcers by inhibiting secretion of gastric acid.

Polyamines such as **spermine** and **spermidine**, involved in DNA packaging, are derived from methionine and ornithine by the pathway shown in **Figure 22-32**. The first step is decarboxylation of ornithine, a precursor of arginine (Fig. 22-12). **Ornithine decarboxylase**, a PLP-requiring enzyme, is the target of several powerful inhibitors used as pharmaceutical agents (see Box 6-3). ■

Arginine Is the Precursor for Biological Synthesis of Nitric Oxide

A surprise finding in the mid-1980s was the role of nitric oxide (NO)—previously known mainly as a component of smog—as an important biological messenger. This simple gaseous substance diffuses readily through membranes, although its high reactivity limits its range of diffusion to about a 1 mm radius from the site of synthesis. In humans NO plays a role in a range of physiological processes, including neurotransmission, blood clotting, and the control of blood pressure. Its mode of action is described in Chapter 12 (see Section 12.5).

Nitric oxide is synthesized from arginine in an NADPH-dependent reaction catalyzed by nitric oxide synthase **(Fig. 22-33)**, a dimeric enzyme structurally related to NADPH cytochrome P-450 reductase (see Box 21-1). The reaction is a five-electron oxidation. Each subunit of the enzyme contains one bound molecule of each of four different cofactors: FMN, FAD, tetrahydrobiopterin, and Fe^{3+} heme. NO is an unstable molecule and cannot be stored. Its synthesis is stimulated by interaction of nitric oxide synthase with Ca^{2+}-calmodulin (see Fig. 12-12).

SUMMARY 22.3 Molecules Derived from Amino Acids

■ Many important biomolecules are derived from amino acids. Glycine is a precursor of porphyrins. Degradation of iron-porphyrin (heme) generates bilirubin, which is converted to bile pigments, with several physiological functions.

■ Glycine and arginine give rise to creatine and phosphocreatine, an energy buffer. Glutathione, formed from three amino acids, is an important cellular reducing agent.

■ Bacteria synthesize D-amino acids from L-amino acids in racemization reactions requiring

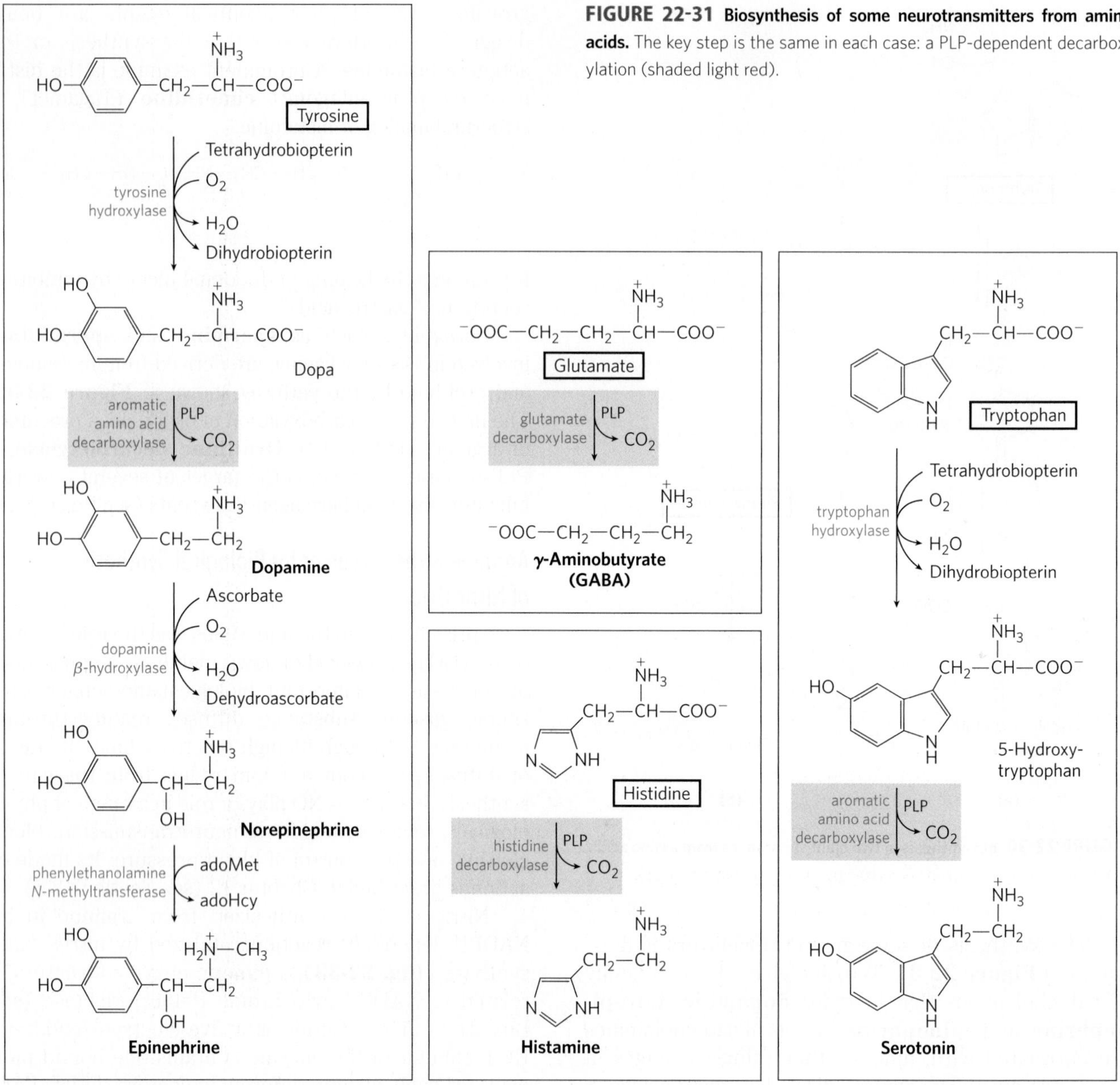

FIGURE 22-31 Biosynthesis of some neurotransmitters from amino acids. The key step is the same in each case: a PLP-dependent decarboxylation (shaded light red).

pyridoxal phosphate. D-Amino acids are commonly found in certain bacterial walls and certain antibiotics.

■ Plants make many substances from aromatic amino acids. The PLP-dependent decarboxylation of some amino acids yields important biological amines, including neurotransmitters.

■ Arginine is the precursor of nitric oxide, a biological messenger.

22.4 Biosynthesis and Degradation of Nucleotides

As discussed in Chapter 8, nucleotides have a variety of important functions in all cells. They are the precursors of DNA and RNA. They are essential carriers of chemical energy—a role primarily of ATP and to some extent GTP. They are components of the cofactors NAD, FAD, S-adenosylmethionine, and coenzyme A, as well as of activated biosynthetic intermediates such as UDP-glucose and CDP-diacylglycerol. Some, such as cAMP and cGMP, are also cellular second messengers.

Two types of pathways lead to nucleotides: the **de novo pathways** and the **salvage pathways**. De novo synthesis of nucleotides begins with their metabolic precursors: amino acids, ribose 5-phosphate, CO_2, and NH_3. Salvage pathways recycle the free bases and nucleosides released from nucleic acid breakdown. Both types of pathways are important in cellular metabolism and both are discussed in this section.

The de novo pathways for purine and pyrimidine biosynthesis seem to be nearly identical in all living organisms. Notably, the free bases guanine, adenine,

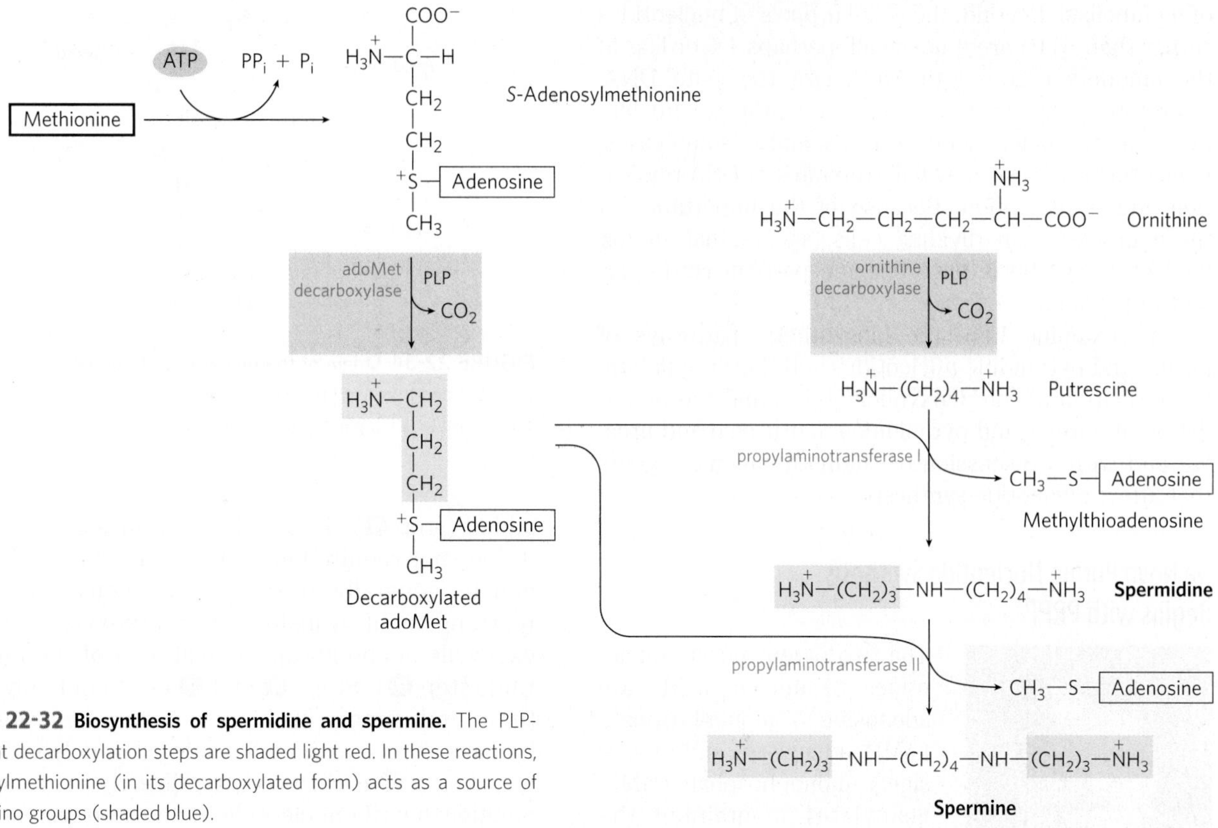

FIGURE 22-32 Biosynthesis of spermidine and spermine. The PLP-dependent decarboxylation steps are shaded light red. In these reactions, S-adenosylmethionine (in its decarboxylated form) acts as a source of propylamino groups (shaded blue).

thymine, cytidine, and uracil are *not* intermediates in these pathways; that is, the bases are not synthesized and then attached to ribose, as might be expected. The purine ring structure is built up one or a few atoms at a time, attached to ribose throughout the process. The pyrimidine ring is synthesized as **orotate**, attached to ribose phosphate, and then converted to the common pyrimidine nucleotides required in nucleic acid synthesis. Although the free bases are not intermediates in the de novo pathways, they are intermediates in some of the salvage pathways.

Several important precursors are shared by the de novo pathways for synthesis of pyrimidines and purines. Phosphoribosyl pyrophosphate (PRPP) is important in

both, and in these pathways the structure of ribose is retained in the product nucleotide, in contrast to its fate in the tryptophan and histidine biosynthetic pathways discussed earlier. An amino acid is an important precursor in each type of pathway: glycine for purines and aspartate for pyrimidines. Glutamine again is the most important source of amino groups—in five different steps in the de novo pathways. Aspartate is also used as the source of an amino group in the purine pathways, in two steps.

Two other features deserve mention. First, there is evidence, especially in the de novo purine pathway, that the enzymes are present as large, multienzyme complexes in the cell, a recurring theme in our discussion

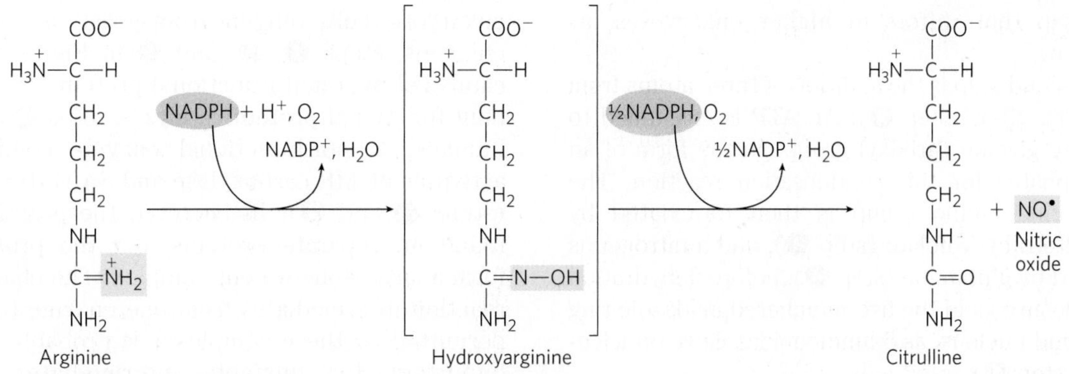

FIGURE 22-33 Biosynthesis of nitric oxide. Both steps are catalyzed by nitric oxide synthase. The nitrogen of the NO is derived from the guanidinium group of arginine.

of metabolism. Second, the cellular pools of nucleotides (other than ATP) are quite small, perhaps 1% or less of the amounts required to synthesize the cell's DNA. Therefore, cells must continue to synthesize nucleotides during nucleic acid synthesis, and in some cases, nucleotide synthesis may limit the rates of DNA replication and transcription. Because of the importance of these processes in dividing cells, agents that inhibit nucleotide synthesis have become particularly important in medicine.

We examine here the biosynthetic pathways of purine and pyrimidine nucleotides and their regulation, the formation of the deoxynucleotides, and the degradation of purines and pyrimidines to uric acid and urea. We end with a discussion of chemotherapeutic agents that affect nucleotide synthesis.

De Novo Purine Nucleotide Synthesis Begins with PRPP

John M. Buchanan, 1917–2007 [Source: Courtesy of MIT Museum.]

The two parent purine nucleotides of nucleic acids are adenosine 5′-monophosphate (AMP; adenylate) and guanosine 5′-monophosphate (GMP; guanylate), containing the purine bases adenine and guanine. **Figure 22-34** shows the origin of the carbon and nitrogen atoms of the purine ring system, as determined by John M. Buchanan using isotopic tracer experiments in birds. The detailed pathway of purine biosynthesis was worked out primarily by Buchanan and G. Robert Greenberg in the 1950s.

In the first committed step of the pathway, an amino group donated by glutamine is attached at C-1 of PRPP **(Fig. 22-35)**. The resulting **5-phosphoribosylamine** is highly unstable, with a half-life of 30 seconds at pH 7.5. The purine ring is subsequently built up on this structure. The pathway described here is identical in all organisms, with the exception of one step that differs in higher eukaryotes, as noted below.

The second step is the addition of three atoms from glycine (Fig. 22-35, step ❷). An ATP is consumed to activate the glycine carboxyl group (in the form of an acyl phosphate) for this condensation reaction. The added glycine amino group is then formylated by N^{10}-formyltetrahydrofolate (step ❸), and a nitrogen is contributed by glutamine (step ❹), before dehydration and ring closure yield the five-membered imidazole ring of the purine nucleus, as 5-aminoimidazole ribonucleotide (AIR; step ❺).

At this point, three of the six atoms needed for the second ring in the purine structure are in place. To complete the process, a carboxyl group is first

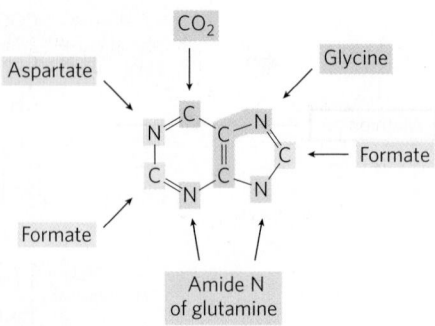

FIGURE 22-34 Origin of the ring atoms of purines. This information was obtained from isotopic experiments with ^{14}C- or ^{15}N-labeled precursors. Formate is supplied in the form of N^{10}-formyltetrahydrofolate.

added (step ❻). This carboxylation is unusual in that it does not require biotin, but instead uses the bicarbonate generally present in aqueous solutions. A rearrangement transfers the carboxylate from the exocyclic amino group to position 4 of the imidazole ring (step ❼). Steps ❻ and ❼ are found only in bacteria and fungi. In higher eukaryotes, including humans, the 5-aminoimidazole ribonucleotide product of step ❺ is carboxylated directly to carboxyaminoimidazole ribonucleotide in one step instead of two (step ❻ₐ). The enzyme catalyzing this reaction is AIR carboxylase.

Aspartate now donates its amino group in two steps (❽ and ❾): formation of an amide bond, followed by elimination of the carbon skeleton of aspartate (as fumarate). (Recall that aspartate plays an analogous role in two steps of the urea cycle; see Fig. 18-10.) The final carbon is contributed by N^{10}-formyltetrahydrofolate (step ❿), and a second ring closure takes place to yield the second fused ring of the purine nucleus (step ⓫). The first intermediate with a complete purine ring is **inosinate (IMP)**.

As in the tryptophan and histidine biosynthetic pathways, the enzymes of IMP synthesis seem to be organized as large, multienzyme complexes in the cell. Once again, evidence comes from the existence of single polypeptides with several functions, some catalyzing nonsequential steps in the pathway. In eukaryotic cells ranging from yeast to fruit flies to chickens, steps ❶, ❸, and ❺ in Figure 22-35 are catalyzed by a multifunctional protein. An additional multifunctional protein catalyzes steps ❿ and ⓫. In humans, a multifunctional enzyme combines the activities of AIR carboxylase and SAICAR synthetase (steps ❻ₐ and ❽). In bacteria, these activities are found on separate proteins, but the proteins may form a large noncovalent complex. The channeling of reaction intermediates from one enzyme to the next permitted by these complexes is probably especially important for unstable intermediates such as 5-phosphoribosylamine.

Conversion of inosinate to adenylate requires the insertion of an amino group derived from aspartate

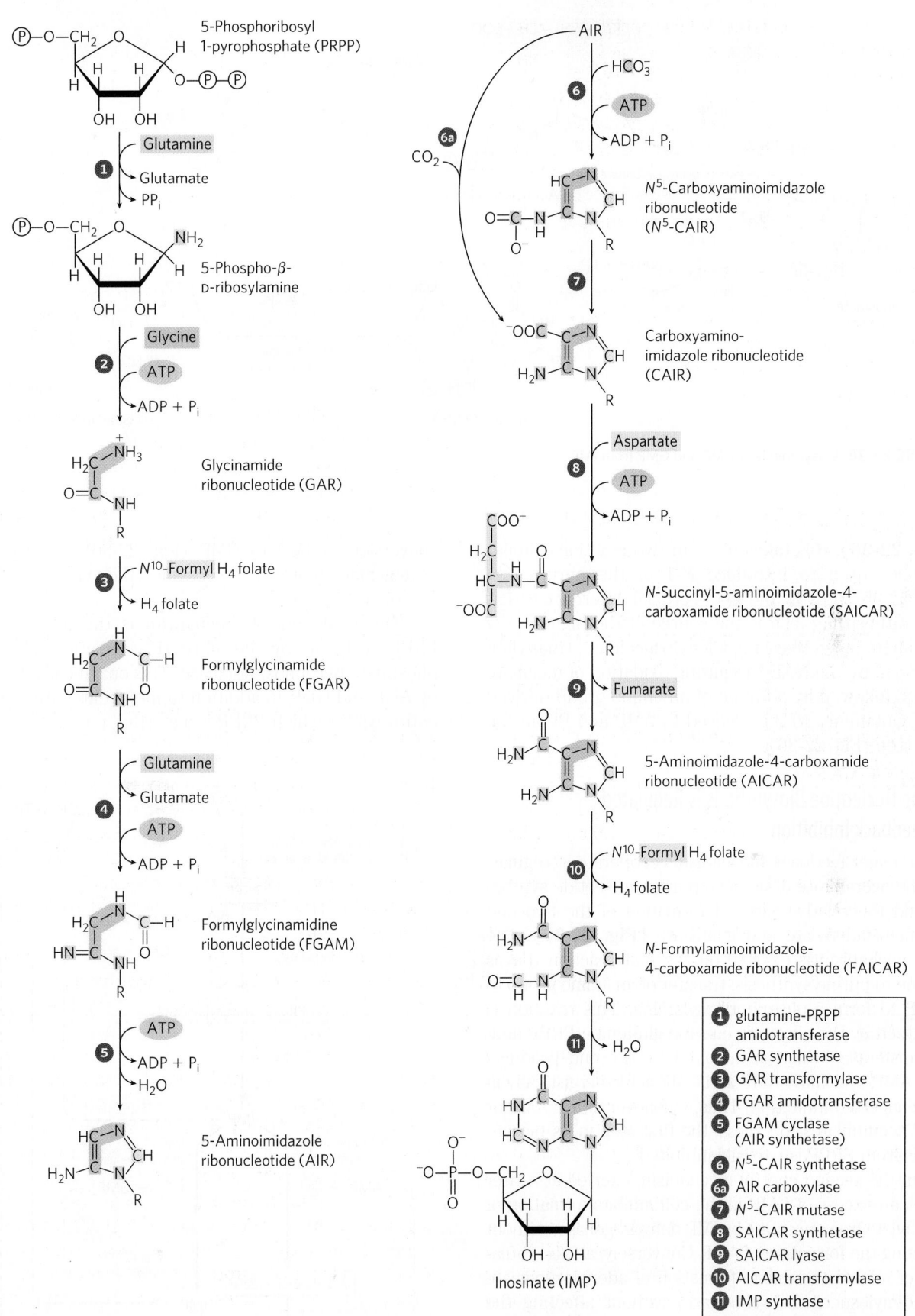

FIGURE 22-35 De novo synthesis of purine nucleotides: construction of the purine ring of inosinate (IMP). Each addition to the purine ring is shaded to match Figure 22-34. After step **2**, R symbolizes the 5-phospho-D-ribosyl group on which the purine ring is built. Formation of 5-phosphoribosylamine (step **1**) is the first committed step in purine synthesis. Note that the product of step **9**, AICAR, is the remnant of ATP released during histidine biosynthesis (see Fig. 22-22, step **5**). Abbreviations are given for most intermediates to simplify the naming of the enzymes. Step **6a** is the alternative path from AIR to CAIR occurring in higher eukaryotes.

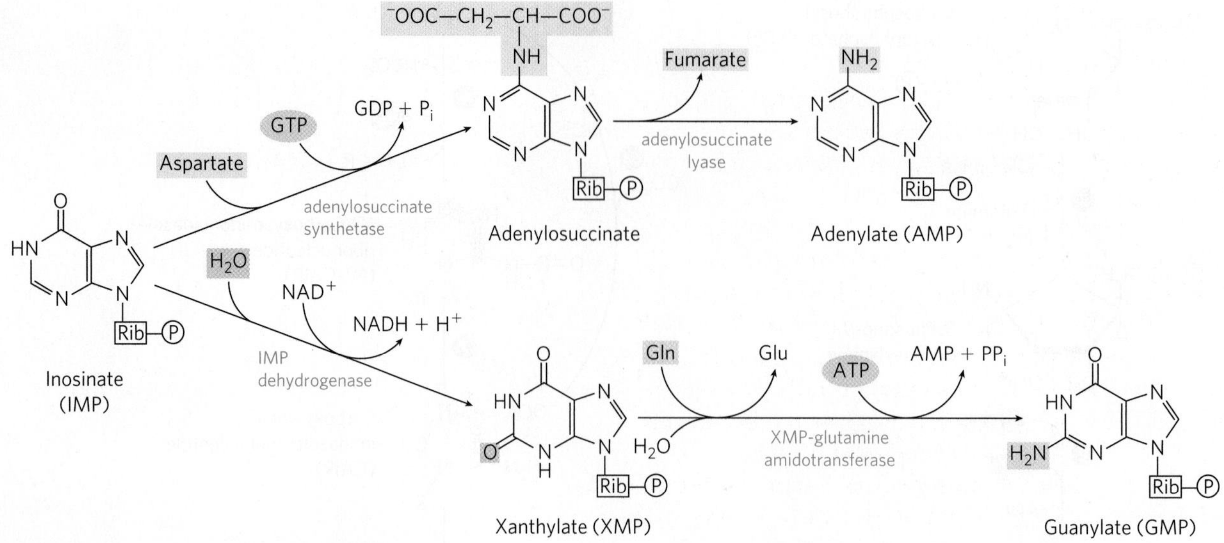

FIGURE 22-36 Biosynthesis of AMP and GMP from IMP.

(Fig. 22-36); this takes place in two reactions similar to those used to introduce N-1 of the purine ring (Fig. 22-35, steps **8** and **9**). A crucial difference is that GTP rather than ATP is the source of the high-energy phosphate in synthesizing adenylosuccinate. Guanylate is formed by the NAD^+-requiring oxidation of inosinate at C-2, followed by addition of an amino group derived from glutamine. ATP is cleaved to AMP and PP_i in the final step (Fig. 22-36).

Purine Nucleotide Biosynthesis Is Regulated by Feedback Inhibition

Three major feedback mechanisms cooperate in regulating the overall rate of de novo purine nucleotide synthesis and the relative rates of formation of the two end products, adenylate and guanylate **(Fig. 22-37)**. The first mechanism is exerted on the first reaction that is unique to purine synthesis: transfer of an amino group to PRPP to form 5-phosphoribosylamine. This reaction is catalyzed by the allosteric enzyme glutamine-PRPP amidotransferase, which is inhibited by the end products IMP, AMP, and GMP. AMP and GMP act synergistically in this concerted inhibition. Thus, whenever either AMP or GMP accumulates to excess, the first step in its biosynthesis from PRPP is partially inhibited.

In the second control mechanism, exerted at a later stage, an excess of GMP in the cell inhibits formation of xanthylate from inosinate by IMP dehydrogenase, without affecting the formation of AMP. Conversely, an accumulation of adenylate inhibits formation of adenylosuccinate by adenylosuccinate synthetase, without affecting the biosynthesis of GMP. When both products are present in sufficient quantities, IMP builds up, and it inhibits an earlier step in the pathway; this is another example of the regulatory strategy called **sequential feedback inhibition**. In the third mechanism, GTP is required in the conversion of IMP to AMP, whereas ATP is required for

conversion of IMP to GMP (Fig. 22-36), a reciprocal arrangement that tends to balance the synthesis of the two ribonucleotides.

The final control mechanism is the inhibition of PRPP synthesis by the allosteric regulation of ribose phosphate pyrophosphokinase. This enzyme is inhibited by ADP and GDP, in addition to metabolites from other pathways for which PRPP is a starting point.

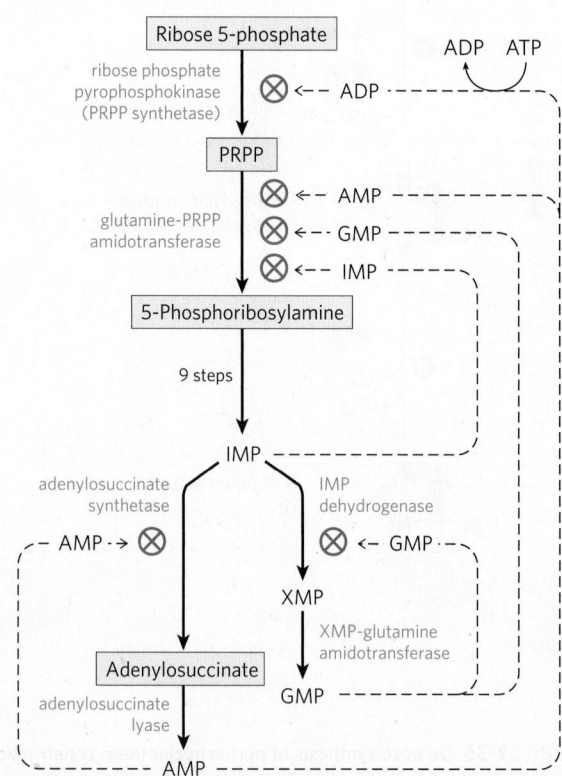

FIGURE 22-37 Regulatory mechanisms in the biosynthesis of adenine and guanine nucleotides in *E. coli.* Regulation of these pathways differs in other organisms.

Pyrimidine Nucleotides Are Made from Aspartate, PRPP, and Carbamoyl Phosphate

The common pyrimidine ribonucleotides are cytidine 5'-monophosphate (CMP; cytidylate) and uridine 5'-monophosphate (UMP; uridylate), which contain the pyrimidines cytosine and uracil. De novo pyrimidine nucleotide biosynthesis **(Fig. 22-38)** proceeds in a somewhat different manner from purine nucleotide synthesis; the six-membered pyrimidine ring is made first and then attached to ribose 5-phosphate. Required in this process is carbamoyl phosphate, also an intermediate in the urea cycle. However, in animals the carbamoyl phosphate required in urea synthesis is made in mitochondria by carbamoyl phosphate synthetase I, whereas the carbamoyl phosphate required in pyrimidine biosynthesis is made in the cytosol by a different form of the enzyme, **carbamoyl phosphate synthetase II**. In bacteria, a single enzyme supplies carbamoyl phosphate for the synthesis of arginine and pyrimidines. The bacterial enzyme has three separate active sites, spaced along a channel nearly 100 Å long **(Fig. 22-39)**. Bacterial carbamoyl phosphate synthetase provides a vivid illustration of the channeling of unstable reaction intermediates between active sites.

Carbamoyl phosphate reacts with aspartate to yield *N*-carbamoylaspartate in the first committed step of pyrimidine biosynthesis (Fig. 22-38). This reaction is catalyzed by **aspartate transcarbamoylase**. In bacteria, this step is highly regulated, and bacterial aspartate transcarbamoylase is one of the most thoroughly studied allosteric enzymes (see below). By removal of water from *N*-carbamoylaspartate, a reaction catalyzed by **dihydroorotase**, the pyrimidine ring is closed to form L-dihydroorotate. This compound is oxidized to the pyrimidine derivative orotate, a reaction in which NAD^+ is the ultimate electron acceptor. In eukaryotes, the first three enzymes in this pathway—carbamoyl phosphate synthetase II, aspartate transcarbamoylase, and dihydroorotase—are part of a single trifunctional protein. The protein, known by the acronym CAD, contains three identical polypeptide chains (each of M_r 230,000), each with active sites for all three reactions. This suggests that large, multienzyme complexes may be the rule in this pathway.

Once orotate is formed, the ribose 5-phosphate side chain, provided once again by PRPP, is attached to yield orotidylate (Fig. 22-38). Orotidylate is then decarboxylated to uridylate, which is phosphorylated to UTP. CTP is formed from UTP by the action of **cytidylate synthetase**, by way of an acyl phosphate intermediate (consuming one ATP). The nitrogen donor is normally glutamine, although the cytidylate synthetases in many species can use NH_4^+ directly.

Pyrimidine Nucleotide Biosynthesis Is Regulated by Feedback Inhibition

Regulation of the rate of pyrimidine nucleotide synthesis in bacteria occurs in large part through aspartate

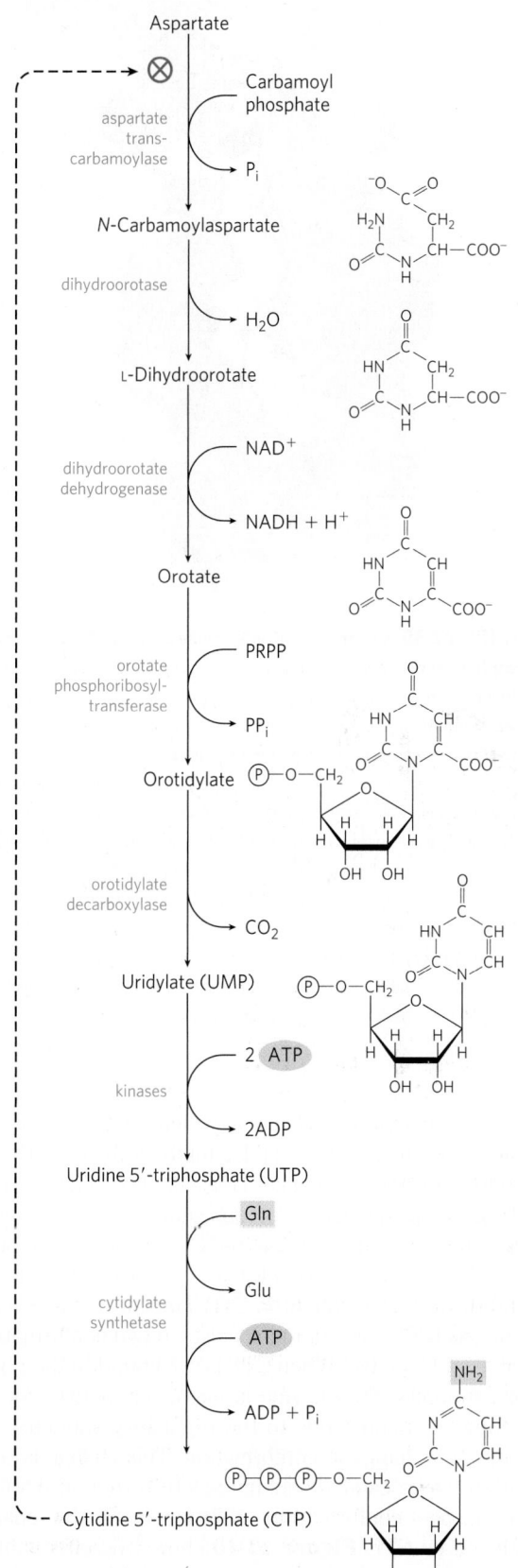

FIGURE 22-38 De novo synthesis of pyrimidine nucleotides: biosynthesis of UTP and CTP via orotidylate. The pyrimidine is constructed from carbamoyl phosphate and aspartate. The ribose 5-phosphate is then added to the completed pyrimidine ring by orotate phosphoribosyltransferase. The first step in this pathway (not shown here; see Fig. 18-11a) is the synthesis of carbamoyl phosphate from CO_2, NH_4^+, and ATP. In eukaryotes, the first step is catalyzed by carbamoyl phosphate synthetase II.

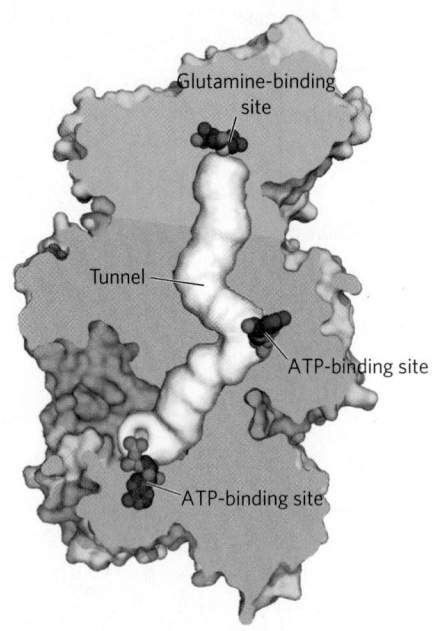

FIGURE 22-39 Channeling of intermediates in bacterial carbamoyl phosphate synthetase. The reaction catalyzed by this enzyme (and its mitochondrial counterpart) is illustrated in Figure 18-11a. The small and large subunits are shown in tan and blue, respectively; the tunnel between active sites (almost 100 Å long) is shown in white. In this reaction, a glutamine molecule binds to the small subunit, donating its amido nitrogen as NH_4^+ in a glutamine amidotransferase–type reaction. The NH_4^+ enters the tunnel, which takes it to a second active site, where it combines with bicarbonate in a reaction requiring ATP. The carbamate then reenters the tunnel to reach the third active site, where it is phosphorylated by ATP to carbamoyl phosphate. To solve this structure, the enzyme was crystallized with ornithine bound to the glutamine-binding site and ADP bound to the ATP-binding sites. [Source: Derived from PDB ID 1M6V, J. B. Thoden et al., *J. Biol. Chem.* 277:39,722, 2002.]

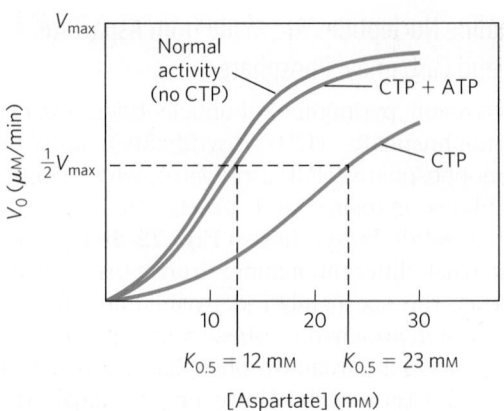

FIGURE 22-40 Allosteric regulation of aspartate transcarbamoylase by CTP and ATP. Addition of 0.8 mM CTP, the allosteric inhibitor of ATCase, increases the $K_{0.5}$ for aspartate (lower curve), thereby reducing the rate of conversion of aspartate to *N*-carbamoylaspartate. ATP at 0.6 mM fully reverses this inhibition by CTP (middle curve).

transcarbamoylase (ATCase), which catalyzes the first reaction in the sequence and is inhibited by CTP, the end product of the sequence (Fig. 22-38). The bacterial ATCase molecule consists of six catalytic subunits and six regulatory subunits (see Fig. 6-34). The catalytic subunits bind the substrate molecules, and the allosteric subunits bind the allosteric inhibitor, CTP. The entire ATCase molecule, as well as its subunits, exists in two conformations, active and inactive. When CTP is not bound to the regulatory subunits, the enzyme is maximally active. As CTP accumulates and binds to the regulatory subunits, they undergo a change in conformation. This change is transmitted to the catalytic subunits, which then also shift to an inactive conformation. ATP prevents the changes induced by CTP. **Figure 22-40** shows the effects of the allosteric regulators on the activity of ATCase.

Nucleoside Monophosphates Are Converted to Nucleoside Triphosphates

Nucleotides to be used in biosynthesis are generally converted to nucleoside triphosphates. The conversion pathways are common to all cells. Phosphorylation of AMP to ADP is promoted by **adenylate kinase**, in the reaction

$$ATP + AMP \rightleftharpoons 2\ ADP$$

The ADP so formed is phosphorylated to ATP by the glycolytic enzymes or through oxidative phosphorylation.

ATP also brings about the formation of other nucleoside diphosphates by the action of a class of enzymes called **nucleoside monophosphate kinases**. These enzymes, which are generally specific for a particular base but nonspecific for the sugar (ribose or deoxyribose), catalyze the reaction

$$ATP + NMP \rightleftharpoons ADP + NDP$$

The efficient cellular systems for rephosphorylating ADP to ATP tend to pull this reaction in the direction of products.

Nucleoside diphosphates are converted to triphosphates by the action of a ubiquitous enzyme, **nucleoside diphosphate kinase**, which catalyzes the reaction

$$NTP_D + NDP_A \rightleftharpoons NDP_D + NTP_A$$

This enzyme is notable in that it is not specific for the base (purines or pyrimidines) or the sugar (ribose or deoxyribose). This nonspecificity applies to both phosphate acceptor (A) and donor (D), although the donor (NTP_D) is almost invariably ATP because it is present in higher concentration than other nucleoside triphosphates under aerobic conditions.

Ribonucleotides Are the Precursors of Deoxyribonucleotides

Deoxyribonucleotides, the building blocks of DNA, are derived from the corresponding ribonucleotides by direct reduction at the 2′-carbon atom of the D-ribose to form the 2′-deoxy derivative. For example, adenosine diphosphate (ADP) is reduced to 2′-deoxyadenosine diphosphate (dADP), and GDP is reduced to dGDP. This reaction is somewhat unusual in that the reduction occurs at a nonactivated carbon; no closely analogous chemical reactions are known. The reaction is catalyzed by **ribonucleotide reductase**, best characterized in *E. coli*, in which its substrates are ribonucleoside diphosphates.

The reduction of the D-ribose portion of a ribonucleoside diphosphate to 2′-deoxy-D-ribose requires a pair of hydrogen atoms, which are ultimately donated by NADPH via an intermediate hydrogen-carrying protein, **thioredoxin**. This ubiquitous protein serves a similar redox function in photosynthesis (see Fig. 20–46) and other processes. Thioredoxin has pairs of —SH groups that carry hydrogen atoms from NADPH to the ribonucleoside diphosphate. Its oxidized (disulfide) form is reduced by NADPH in a reaction catalyzed by **thioredoxin reductase (Fig. 22–41)**, and reduced thioredoxin is then used by ribonucleotide reductase to reduce the nucleoside diphosphates (NDPs) to deoxyribonucleoside diphosphates (dNDPs). A second source of reducing equivalents for ribonucleotide reductase is glutathione (GSH). Glutathione serves as the reductant for a protein closely related to thioredoxin, **glutaredoxin**, which then transfers the reducing power to ribonucleotide reductase (Fig. 22–41).

Ribonucleotide reductase is notable in that its reaction mechanism provides the best-characterized example of the involvement of free radicals in biochemical transformations, once thought to be rare in biological systems.

The enzyme in *E. coli* and most eukaryotes is an $\alpha_2\beta_2$ dimer, with two catalytic subunits, α_2, and two radical-generation subunits, β_2 **(Fig. 22–42)**. Each catalytic subunit contains two kinds of regulatory sites, as described below. The two active sites of the enzyme are formed at the interface between the catalytic (α_2) and radical-generation (β_2) subunits. At each active site, an α subunit contributes two sulfhydryl groups required for activity, and the β_2 subunits contribute a stable tyrosyl radical. The β_2 subunits also have a binuclear iron (Fe^{3+}) cofactor that helps generate and stabilize the Tyr^{122} radical (Fig. 22–42). The tyrosyl radical is too far from the

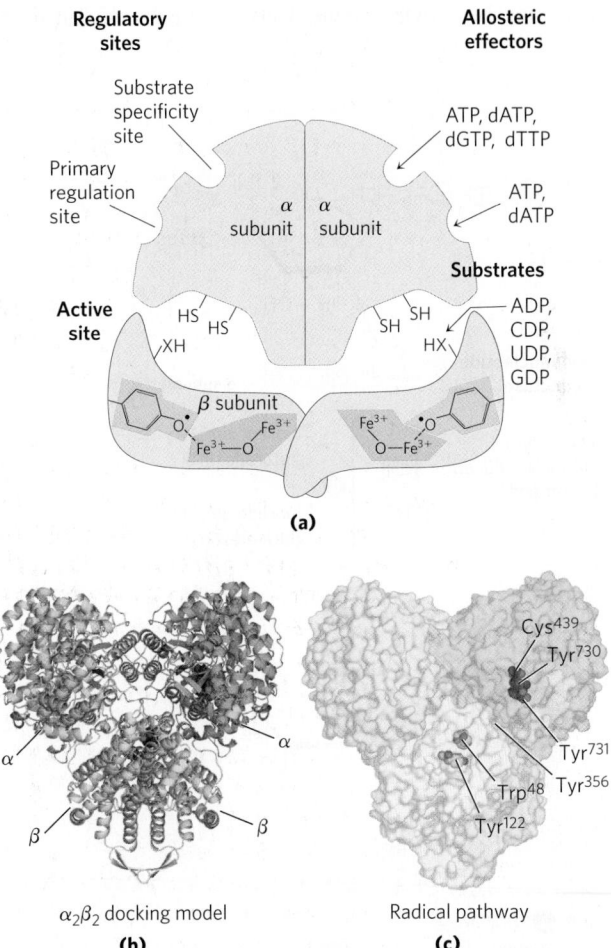

(a)

$\alpha_2\beta_2$ docking model
(b)

Radical pathway
(c)

FIGURE 22-42 Ribonucleotide reductase. (a) A schematic diagram of the subunit structures. Each catalytic subunit (α; also called R1) contains the two regulatory sites described in Figure 22-44 and two Cys residues central to the reaction mechanism. The radical-generation subunits (β; also called R2) each contain a critical Tyr^{122} residue and binuclear iron center. **(b)** The likely structure of $\alpha_2\beta_2$. **(c)** The likely path of radical formation from the initial Tyr^{122} in a β subunit to the active-site Cys^{439}, which is used in the mechanism shown in Figure 22-43. Several aromatic amino acid residues participate in long-range transfer of the radical from the point of its formation at Tyr^{122} to the active site, where the nucleotide substrate is bound. [Sources: (a) Information from L. Thelander and P. Reichard, *Annu. Rev. Biochem.* 48:133, 1979. (b, c) Derived from PDB ID 3UUS, N. Ando et al., *Proc. Natl. Acad. Sci. USA* 108:21,046, 2011.]

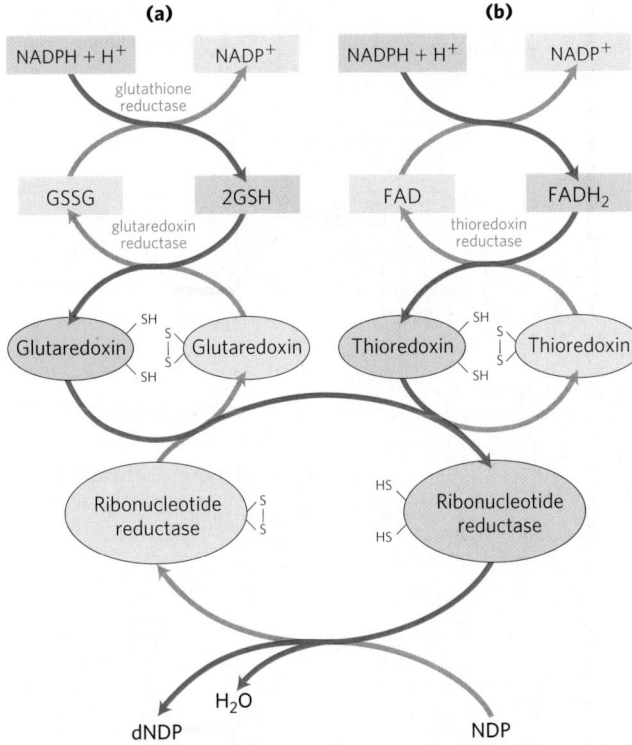

FIGURE 22-41 Reduction of ribonucleotides to deoxyribonucleotides by ribonucleotide reductase. Electrons are transmitted (red arrows) to the enzyme from NADPH via **(a)** glutaredoxin or **(b)** thioredoxin. The sulfide groups in glutaredoxin reductase are contributed by two molecules of bound glutathione (GSH; GSSG indicates oxidized glutathione). Note that thioredoxin reductase is a flavoenzyme, with FAD as prosthetic group.

active site to interact directly with the site, but several aromatic residues form a long-range radical-transfer pathway to the active site (Fig. 22-42c). A likely mechanism for the ribonucleotide reductase reaction is illustrated in **Figure 22-43**. In *E. coli*, the sources of the required reducing equivalents for this reaction are thioredoxin and glutaredoxin, as noted above.

Three classes of ribonucleotide reductase have been reported. Their mechanisms (where known) generally conform to the scheme in Figure 22-43, but they differ in the identity of the group supplying the active-site radical and in the cofactors used to generate it. The *E. coli* enzyme (class I) requires oxygen to regenerate the tyrosyl radical if it is quenched, so this enzyme functions only in an aerobic environment. Class II enzymes, found in

other microorganisms, have 5′-deoxyadenosylcobalamin (see Box 17-2) rather than a binuclear iron center. Class III enzymes have evolved to function in an anaerobic environment. *E. coli* contains a separate class III ribonucleotide reductase when grown anaerobically; this enzyme contains an iron-sulfur cluster (structurally distinct from the binuclear iron center of the class I enzyme) and requires NADPH and *S*-adenosylmethionine for activity. It uses nucleoside triphosphates rather than nucleoside diphosphates as substrates. The evolution of different classes of ribonucleotide reductase for production of DNA precursors in different environments reflects the importance of this reaction in nucleotide metabolism.

Regulation of *E. coli* ribonucleotide reductase is unusual in that not only its *activity* but its *substrate*

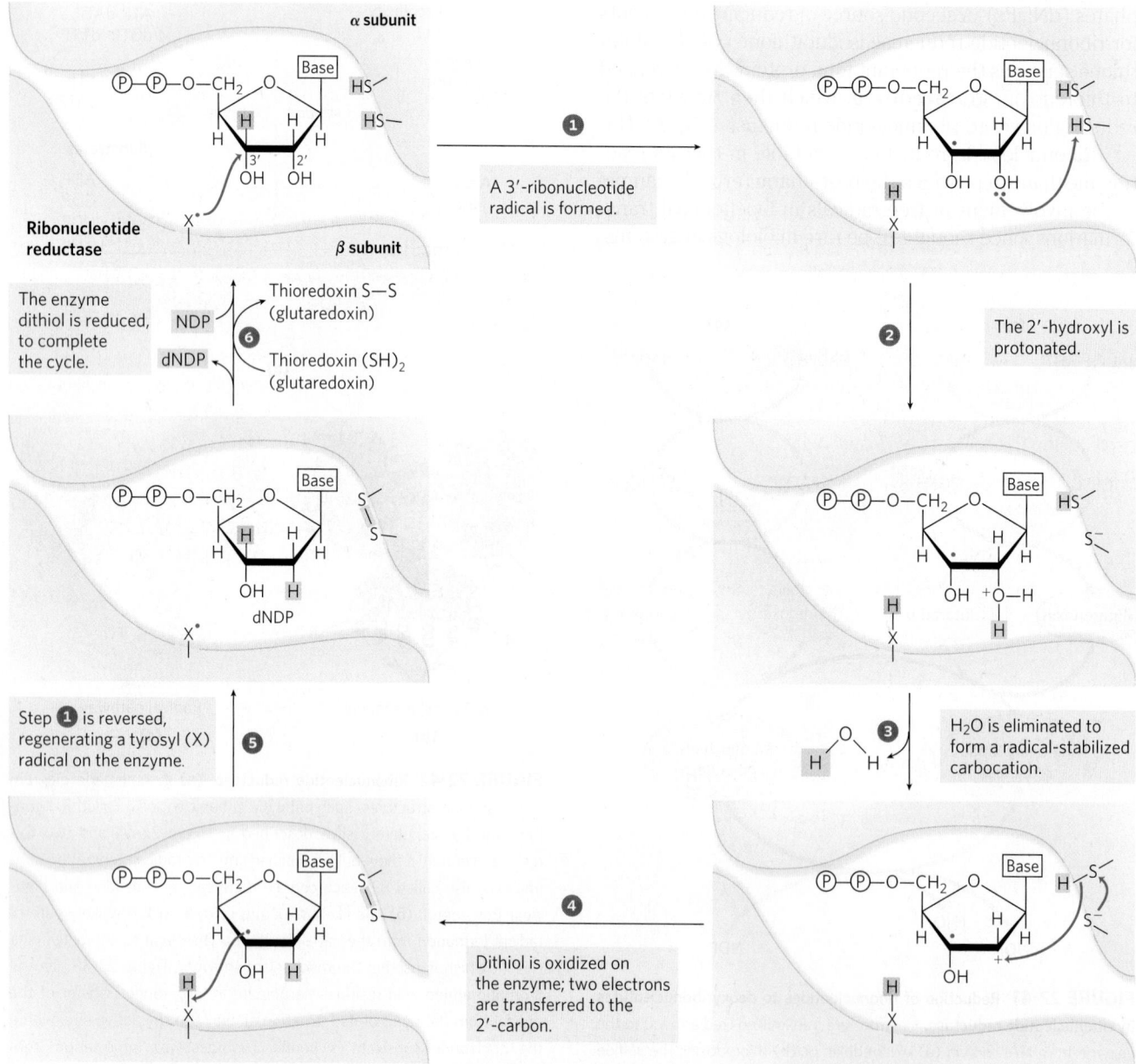

MECHANISM FIGURE 22-43 Proposed mechanism for ribonucleotide reductase. In the enzyme of *E. coli* and most eukaryotes, the active thiol groups are on the α subunit. The active-site radical (—X•) is on the β subunit and in *E. coli* is probably a thiyl radical of Cys[439] (see Fig. 22-42).

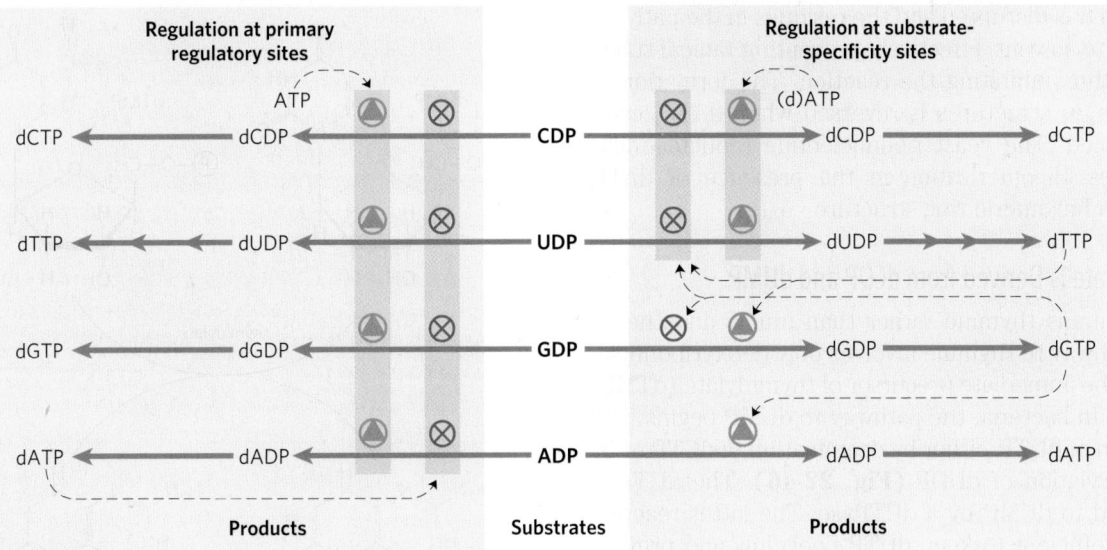

FIGURE 22-44 Regulation of ribonucleotide reductase by deoxynucleoside triphosphates. The overall activity of the enzyme is affected by binding at the primary regulatory site (left). The substrate specificity of the enzyme is affected by the nature of the effector molecule bound at the second type of regulatory site, the substrate-specificity site (right). The diagram indicates inhibition or stimulation of enzyme activity with the four different substrates. The pathway from dUDP to dTTP is described below (see Figs 22-46, 22-47).

specificity is regulated by the binding of effector molecules. Each α subunit has two types of regulatory sites (Fig. 22-42). One type affects overall enzyme activity and binds either ATP, which activates the enzyme, or dATP, which inactivates it. The second type alters substrate specificity in response to the effector molecule—ATP, dATP, dTTP, or dGTP—that is bound there **(Fig. 22-44)**. When ATP or dATP is bound, reduction of UDP and CDP is favored. When dTTP or dGTP is bound, reduction of GDP or ADP, respectively, is stimulated. The scheme is designed to provide a balanced pool of precursors for DNA synthesis. ATP is also a general activator for biosynthesis and ribonucleotide reduction. The presence of dATP in small amounts increases the reduction of pyrimidine nucleotides. An oversupply of the pyrimidine dNTPs is signaled by high levels of dTTP, which shifts the specificity to favor reduction of GDP. High levels of dGTP, in turn, shift the specificity to ADP reduction, and high levels of dATP shut the enzyme down. These effectors are thought to induce several distinct enzyme conformations with altered specificities.

These regulatory effects are accompanied by, and presumably mediated by, large structural rearrangements in the enzyme. When the active form of the *E. coli* enzyme ($\alpha_2\beta_2$) is inhibited by the addition of the allosteric inhibitor dATP, a ringlike $\alpha_4\beta_4$ structure forms, with alternating α_2 and β_2 subunits **(Fig. 22-45)**. In this altered structure, the radical-forming path

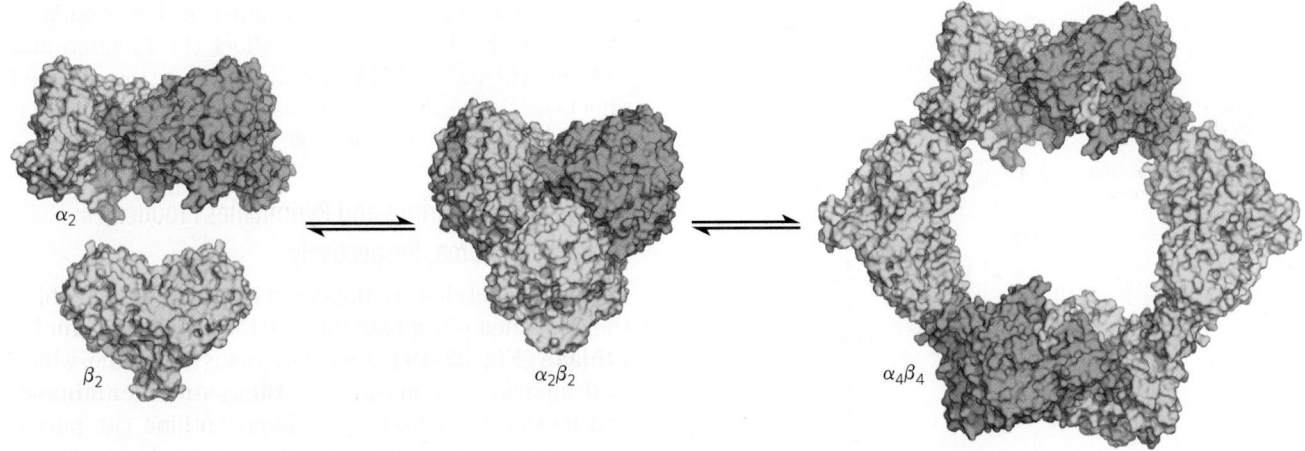

FIGURE 22-45 Oligomerization of ribonucleotide reductase induced by the allosteric inhibitor dATP. At high concentrations of dATP (50 μM), ring-shaped $\alpha_4\beta_4$ structures form. In this conformation, the residues in the radical-forming path are exposed to the solvent, blocking the radical reaction and inhibiting the enzyme. The oligomerization is reversed at lower dATP concentrations. [Source: PDB ID 3UUS, N. Ando et al., *Proc. Natl. Acad. Sci. USA* 108:21,046, 2011.]

from β to α is disrupted and the residues in the path are exposed to solvent, effectively preventing radical transfer and thus inhibiting the reaction. The formation of ringlike $\alpha_4\beta_4$ structures is reversed when dATP levels are reduced. The yeast ribonucleotide reductase also undergoes oligomerization in the presence of dATP, forming a hexameric ring structure, $\alpha_6\beta_6$.

Thymidylate Is Derived from dCDP and dUMP

DNA contains thymine rather than uracil, and the de novo pathway to thymine involves only deoxyribonucleotides. The immediate precursor of thymidylate (dTMP) is dUMP. In bacteria, the pathway to dUMP begins with formation of dUTP, either by deamination of dCTP or by phosphorylation of dUDP **(Fig. 22-46)**. The dUTP is converted to dUMP by a dUTPase. The latter reaction must be efficient to keep dUTP pools low and prevent incorporation of uridylate into DNA.

Conversion of dUMP to dTMP is catalyzed by **thymidylate synthase**. A one-carbon unit at the hydroxymethyl ($-CH_2OH$) oxidation level (see Fig. 18-17) is transferred from N^5,N^{10}-methylenetetrahydrofolate to dUMP, then reduced to a methyl group **(Fig. 22-47)**. The reduction occurs at the expense of oxidation of tetrahydrofolate to dihydrofolate, which is unusual in tetrahydrofolate-requiring reactions. (The mechanism of this reaction is shown in Fig. 22-53.) The dihydrofolate is reduced to tetrahydrofolate by **dihydrofolate reductase**—a regeneration that is essential for the many processes that require tetrahydrofolate. In plants and at least one protist, thymidylate synthase and dihydrofolate reductase reside on a single, bifunctional protein.

About 10% of the human population (and up to 50% of people in impoverished communities) suffers from folic acid deficiency. When the deficiency is severe, the symptoms can include heart disease, cancer, and some types of brain dysfunction. At least some of these symptoms arise from a reduction of thymidylate synthesis, leading to an abnormal incorporation of uracil into DNA. Uracil is recognized by DNA repair pathways (described in Chapter 25) and is cleaved from the DNA.

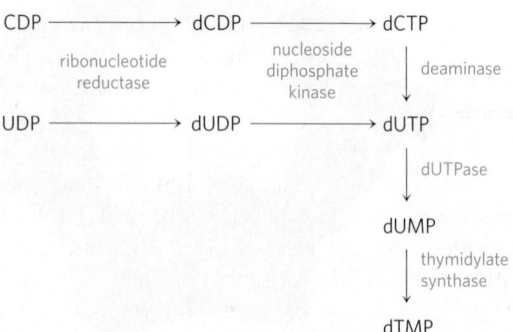

FIGURE 22-46 Biosynthesis of thymidylate (dTMP). The pathways are shown beginning with the reaction catalyzed by ribonucleotide reductase. Figure 22-47 gives details of the thymidylate synthase reaction.

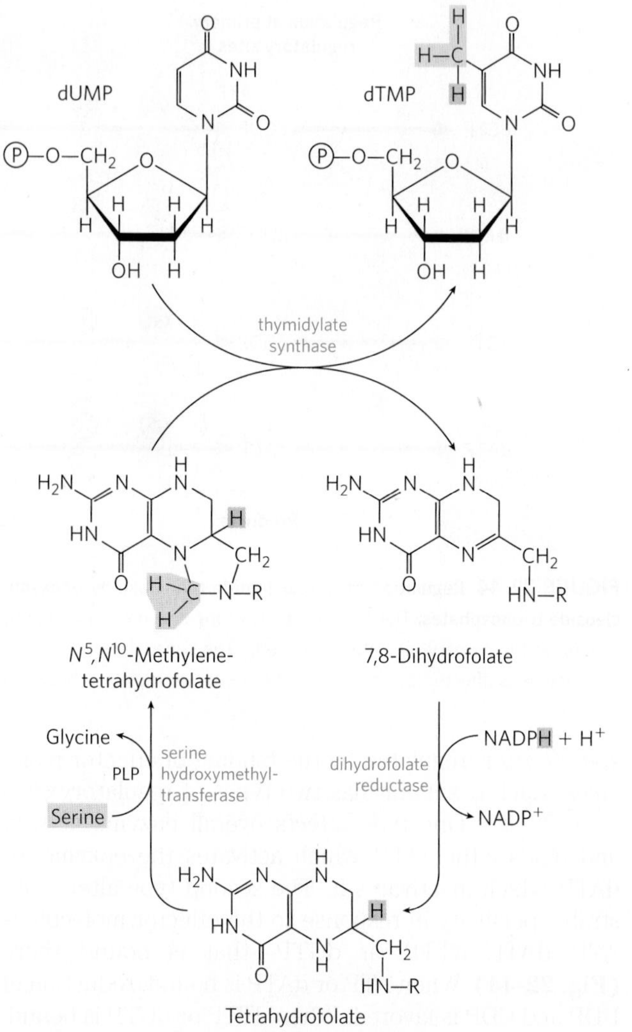

FIGURE 22-47 Conversion of dUMP to dTMP by thymidylate synthase and dihydrofolate reductase. Serine hydroxymethyltransferase is required for regeneration of the N^5,N^{10}-methylene form of tetrahydrofolate. In the synthesis of dTMP, all three hydrogens of the added methyl group are derived from N^5,N^{10}-methylenetetrahydrofolate (light red and gray).

The presence of high levels of uracil in DNA leads to strand breaks that can greatly affect the function and regulation of nuclear DNA, ultimately causing the observed effects on the heart and brain, as well as increased mutagenesis that leads to cancer. ■

Degradation of Purines and Pyrimidines Produces Uric Acid and Urea, Respectively

Purine nucleotides are degraded by a pathway in which they lose their phosphate through the action of **5'-nucleotidase (Fig. 22-48)**. Adenylate yields adenosine, which is deaminated to inosine by **adenosine deaminase**, and inosine is hydrolyzed to hypoxanthine (its purine base) and D-ribose. Hypoxanthine is oxidized successively to xanthine and then uric acid by **xanthine oxidase**, a flavoenzyme with an atom of molybdenum and four iron-sulfur centers in its prosthetic group.

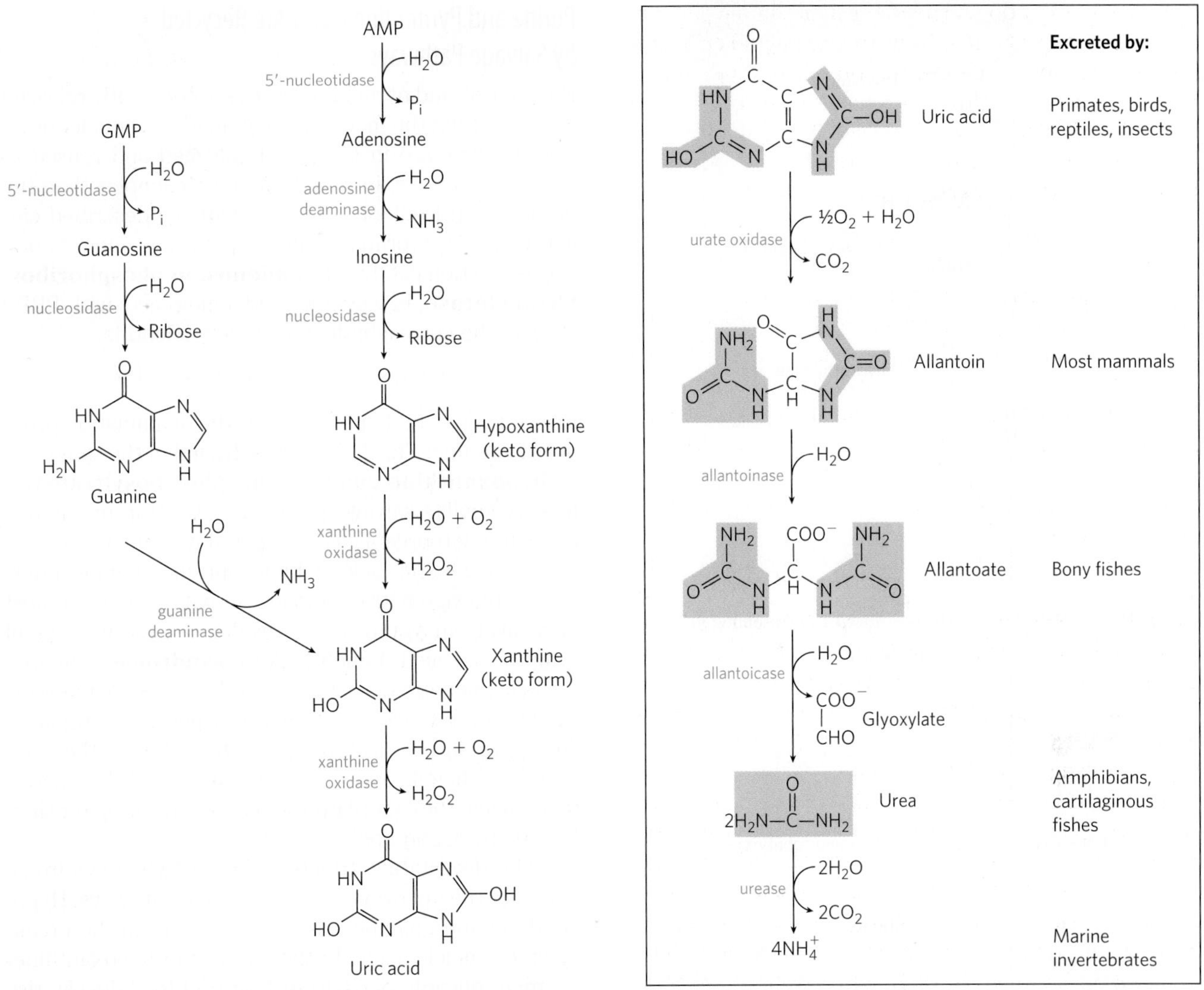

FIGURE 22-48 Catabolism of purine nucleotides. Note that primates excrete much more nitrogen as urea via the urea cycle (Chapter 18) than as uric acid from purine degradation. Similarly, fish excrete much more nitrogen as NH_4^+ than as urea produced by the pathway shown here.

Molecular oxygen is the electron acceptor in this complex reaction.

GMP catabolism also yields uric acid as an end product. GMP is first hydrolyzed to guanosine, which is then cleaved to free guanine. Guanine undergoes hydrolytic removal of its amino group to yield xanthine, which is converted to uric acid by xanthine oxidase (Fig. 22-48).

Uric acid is the excreted end product of purine catabolism in primates, birds, and some other animals. A healthy adult human excretes uric acid at a rate of about 0.6 g/24 h; the excreted product arises in part from ingested purines and in part from turnover of the purine nucleotides of nucleic acids. In most mammals and many other vertebrates, uric acid is degraded to **allantoin** by the action of **urate oxidase**. In other organisms the pathway is further extended, as shown in Figure 22-48.

The pathways for degradation of pyrimidines generally lead to NH_4^+ production and thus to urea synthesis.

Thymine, for example, is degraded to methylmalonylsemialdehyde **(Fig. 22-49)**, an intermediate of valine catabolism. It is further degraded through propionyl-CoA and methylmalonyl-CoA to succinyl-CoA (see Fig. 18-27).

Genetic aberrations in human purine metabolism have been found, some with serious consequences. For example, **adenosine deaminase (ADA) deficiency** leads to severe immunodeficiency disease in which T lymphocytes and B lymphocytes do not develop properly. Lack of ADA leads to a 100-fold increase in the cellular concentration of dATP, a strong inhibitor of ribonucleotide reductase (Fig. 22-44). High levels of dATP produce a general deficiency of other dNTPs in T lymphocytes. The basis for B-lymphocyte toxicity is less clear. Individuals with ADA deficiency lack an effective immune system and do not survive unless treated. Current therapies include bone marrow transplants from a matched donor to replace the hematopoietic

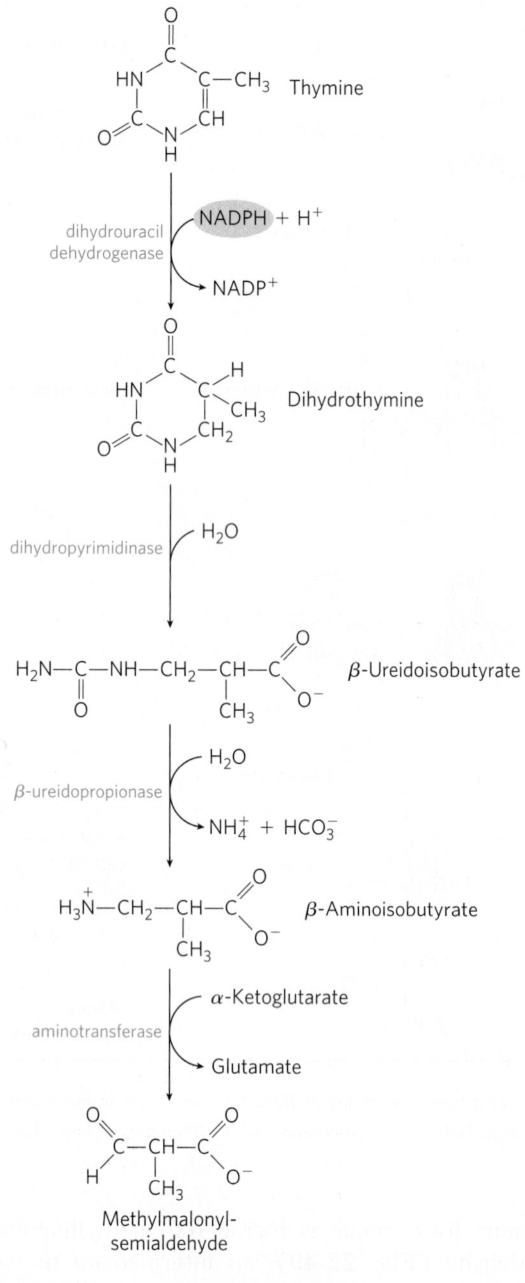

FIGURE 22-49 Catabolism of a pyrimidine. Shown here is the pathway for thymine. The methylmalonylsemialdehyde is further degraded to succinyl-CoA.

stem cells that mature into B and T lymphocytes. However, transplant recipients often suffer a variety of cognitive and physiological problems. Enzyme replacement therapy, requiring once- or twice-weekly intramuscular injection of active ADA, is effective, but the therapeutic benefit often declines after 8 to 10 years and complications arise, including malignancies. For many people, a permanent cure requires replacing the defective gene with a functional one in bone marrow cells. ADA deficiency was one of the first targets of human gene therapy trials (in 1990). Mixed results in early trials have given way to significant successes, and gene therapy is rapidly becoming a viable path for long-term restoration of immune function for these patients. ■

Purine and Pyrimidine Bases Are Recycled by Salvage Pathways

Free purine and pyrimidine bases are constantly released in cells during the metabolic degradation of nucleotides. Free purines are in large part salvaged and reused to make nucleotides, in a pathway much simpler than the de novo synthesis of purine nucleotides described earlier. One of the primary salvage pathways consists of a single reaction catalyzed by **adenosine phosphoribosyltransferase**, in which free adenine reacts with PRPP to yield the corresponding adenine nucleotide:

$$\text{Adenine} + \text{PRPP} \longrightarrow \text{AMP} + \text{PP}_i$$

Free guanine and hypoxanthine (the deamination product of adenine; Fig. 22-48) are salvaged in the same way by **hypoxanthine-guanine phosphoribosyltransferase**. A similar salvage pathway exists for pyrimidine bases in microorganisms, and possibly in mammals.

A genetic lack of hypoxanthine-guanine phosphoribosyltransferase activity, seen almost exclusively in young boys, results in a bizarre set of symptoms called **Lesch-Nyhan syndrome**. Children with this genetic disorder, which becomes manifest by the age of 2 years, are sometimes poorly coordinated and have intellectual deficits. In addition, they are extremely hostile and show compulsive self-destructive tendencies: they mutilate themselves by biting off their fingers, toes, and lips.

The devastating effects of Lesch-Nyhan syndrome illustrate the importance of the salvage pathways. Hypoxanthine and guanine arise constantly from the breakdown of nucleic acids. In the absence of hypoxanthine-guanine phosphoribosyltransferase, PRPP levels rise and purines are overproduced by the de novo pathway, resulting in high levels of uric acid production and gout-like damage to tissue (see below). The brain is especially dependent on the salvage pathways, and this may account for the central nervous system damage in children with Lesch-Nyhan syndrome. This syndrome is another potential target for gene therapy. ■

Excess Uric Acid Causes Gout

Long thought (erroneously) to be due to "high living," gout is a disease of the joints caused by an elevated concentration of uric acid in the blood and tissues. The joints become inflamed, painful, and arthritic, owing to the abnormal deposition of sodium urate crystals. The kidneys are also affected, as excess uric acid is deposited in the kidney tubules. Gout occurs predominantly in males. Its precise cause is not known, but it often involves an underexcretion of urate. A genetic deficiency of one or another enzyme of purine metabolism may also be a factor in some cases.

Gout is effectively treated by a combination of nutritional and drug therapies. Patients exclude foods especially rich in nucleotides and nucleic acids, such as liver or glandular products, from the diet. Major alleviation of

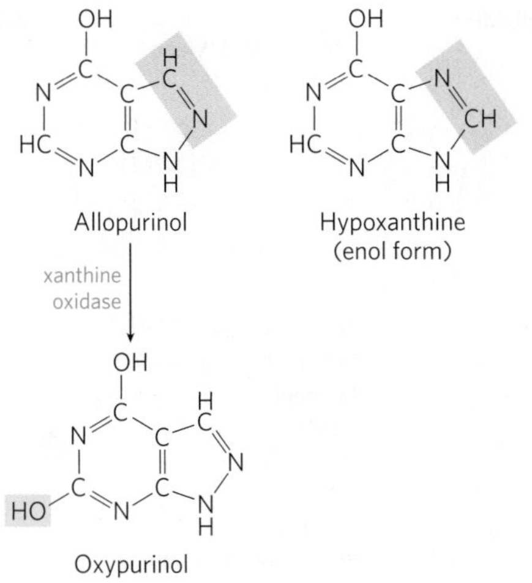

FIGURE 22-50 Allopurinol, an inhibitor of xanthine oxidase. Hypoxanthine is the normal substrate of xanthine oxidase. A slight alteration in the structure of hypoxanthine (shaded light red) yields the medically effective enzyme inhibitor allopurinol. At the active site, allopurinol is converted to oxypurinol, a strong competitive inhibitor that remains tightly bound to the reduced form of the enzyme.

the symptoms is provided by the drug **allopurinol (Fig. 22-50)**, which inhibits xanthine oxidase, the enzyme that catalyzes the conversion of purines to uric acid. Allopurinol is a substrate of xanthine oxidase, which converts allopurinol to oxypurinol (alloxanthine). Oxypurinol inactivates the reduced form of the enzyme by remaining tightly bound in its active site. When xanthine oxidase is inhibited, the excreted products of purine metabolism are xanthine and hypoxanthine, which are more water-soluble than uric acid and less likely to form crystalline deposits. Allopurinol was developed by Gertrude Elion and George Hitchings, who also developed acyclovir, used in treating people with genital and oral herpes infections, and other purine analogs used in cancer chemotherapy. ■

George Hitchings, 1905–1998, and Gertrude Elion, 1918–1999. [Source: Will and Deni Mcintyre/Science Source/Getty Images.]

Many Chemotherapeutic Agents Target Enzymes in Nucleotide Biosynthetic Pathways

The growth of cancer cells is not controlled in the same way as cell growth in most normal tissues. Cancer cells have greater requirements for nucleotides as precursors of DNA and RNA, and consequently are generally more sensitive than normal cells to inhibitors of nucleotide biosynthesis. A growing array of important chemotherapeutic agents—for cancer and other diseases—act by inhibiting one or more enzymes in these pathways. We describe here several well-studied examples that illustrate productive approaches to treatment and help us understand how these enzymes work.

The first set of agents includes compounds that inhibit glutamine amidotransferases. Recall that glutamine is a nitrogen donor in at least half a dozen separate reactions in nucleotide biosynthesis. The binding sites for glutamine and the mechanism by which NH_4^+ is extracted are quite similar in many of these enzymes. Most of the enzymes are strongly inhibited by glutamine analogs such as **azaserine** and **acivicin (Fig. 22-51)**. Azaserine, characterized by John Buchanan in the 1950s, was one of the first examples of a mechanism-based enzyme inactivator (suicide inactivator; p. 210 and Box 6-3). Acivicin shows promise as a cancer chemotherapeutic agent.

Other useful targets for pharmaceutical agents are thymidylate synthase and dihydrofolate reductase, enzymes that provide the only cellular pathway for thymine synthesis **(Fig. 22-52)**. One inhibitor that acts on thymidylate synthase, **fluorouracil**, is an important chemotherapeutic agent. Fluorouracil itself is not the enzyme inhibitor. In the cell, salvage pathways convert it to the deoxynucleoside monophosphate FdUMP, which binds to and inactivates the enzyme. Inhibition by FdUMP **(Fig. 22-53)** is a classic example of mechanism-based enzyme inactivation. Another prominent chemotherapeutic agent, **methotrexate**, is an inhibitor of dihydrofolate reductase. This folate analog acts as a

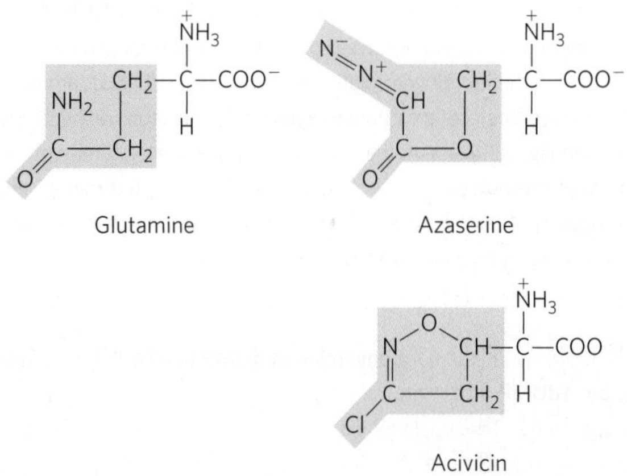

FIGURE 22-51 Azaserine and acivicin, inhibitors of glutamine amidotransferases. These analogs of glutamine interfere in several amino acid and nucleotide biosynthetic pathways.

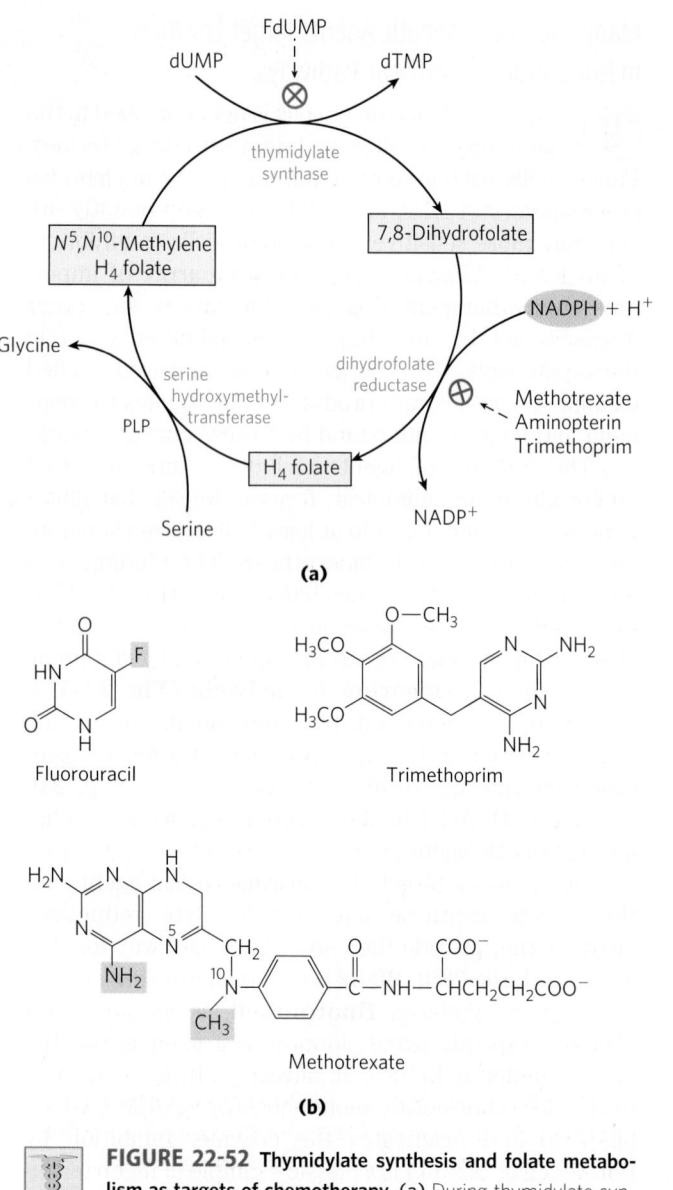

(a)

(b)

FIGURE 22-52 Thymidylate synthesis and folate metabolism as targets of chemotherapy. **(a)** During thymidylate synthesis, N^5,N^{10}-methylenetetrahydrofolate is converted to 7,8-dihydrofolate; the N^5,N^{10}-methylenetetrahydrofolate is regenerated in two steps (see Fig. 22-47). This cycle is a major target of several chemotherapeutic agents. **(b)** Fluorouracil and methotrexate are important chemotherapeutic agents. In cells, fluorouracil is converted to FdUMP, which inhibits thymidylate synthase. Methotrexate, a structural analog of tetrahydrofolate, inhibits dihydrofolate reductase; the shaded amino and methyl groups replace a carbonyl oxygen and a proton, respectively, in folate (see Fig. 22-47). Another important folate analog, aminopterin, is identical to methotrexate except that it lacks the shaded methyl group. Trimethoprim, a tight-binding inhibitor of bacterial dihydrofolate reductase, was developed as an antibiotic.

MECHANISM FIGURE 22-53 Conversion of dUMP to dTMP and its inhibition by FdUMP. The normal reaction mechanism of thymidylate synthase (left). The nucleophilic sulfhydryl group contributed by the enzyme in step ❶ and the ring atoms of dUMP taking part in the reaction are shown in red; :B denotes an amino acid side chain that acts as a base to abstract a proton after step ❸. The hydrogens derived from the methylene group of N^5,N^{10}-methylenetetrahydrofolate are shaded in gray. The 1,3 hydride shift (step ❹), moves a hydride ion (shaded light red) from C-6 of tetrahydrofolate to the methyl group of thymidine,

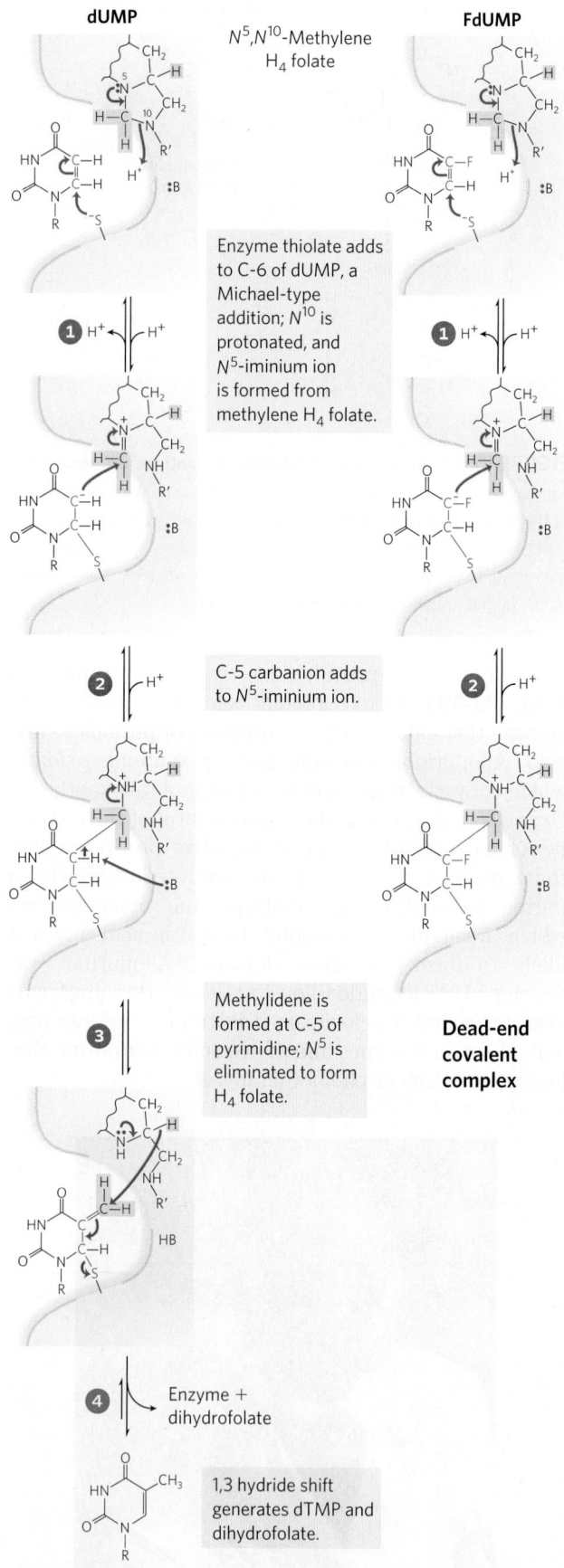

oxidizing tetrahydrofolate to dihydrofolate. This hydride shift is blocked when FdUMP is the substrate (right). Steps ❶ and ❷ proceed normally, but result in a stable complex—consisting of FdUMP linked covalently to the enzyme and to tetrahydrofolate—that inactivates the enzyme.

competitive inhibitor; the enzyme binds methotrexate with about 100 times higher affinity than dihydrofolate. **Aminopterin** is a related compound that acts similarly.

The medical potential of inhibitors of nucleotide biosynthesis is not limited to cancer treatment. All fast-growing cells (including bacteria and protists) are potential targets. **Trimethoprim**, an antibiotic developed by Hitchings and Elion, binds to bacterial dihydrofolate reductase nearly 100,000 times better than to the mammalian enzyme. It is used to treat certain urinary and middle-ear bacterial infections. Parasitic protists, such as the trypanosomes that cause African sleeping sickness (African trypanosomiasis), lack pathways for de novo nucleotide biosynthesis and are particularly sensitive to agents that interfere with their ability to use salvage pathways to scavenge nucleotides from the surrounding environment. Allopurinol (Fig. 22-50) and several similar purine analogs have shown promise for the treatment of African trypanosomiasis and related afflictions. See Box 6-3 for another approach to combating African trypanosomiasis, made possible by advances in our understanding of metabolism and enzyme mechanisms. ■

SUMMARY 22.4 Biosynthesis and Degradation of Nucleotides

■ The purine ring system is built up step by step, beginning with 5-phosphoribosylamine. The amino acids glutamine, glycine, and aspartate furnish all the nitrogen atoms of purines. Two ring-closure steps form the purine nucleus.

■ Pyrimidines are synthesized from carbamoyl phosphate and aspartate, and ribose 5-phosphate is then attached to yield the pyrimidine ribonucleotides.

■ Nucleoside monophosphates are converted to their triphosphates by enzymatic phosphorylation reactions. Ribonucleotides are converted to deoxyribonucleotides by ribonucleotide reductase, an enzyme with novel mechanistic and regulatory characteristics. The thymine nucleotides are derived from dCDP and dUMP.

■ Uric acid and urea are the end products of purine and pyrimidine degradation.

■ Free purines can be salvaged and rebuilt into nucleotides. Genetic deficiencies in certain salvage enzymes cause serious disorders such as Lesch-Nyhan syndrome and ADA deficiency.

■ Accumulation of uric acid crystals in the joints, possibly caused by another genetic deficiency, results in gout.

■ Enzymes of the nucleotide biosynthetic pathways are targets for an array of chemotherapeutic agents used to treat cancer and other diseases.

Key Terms

Terms in bold are defined in the glossary.

Problems

1. ATP Consumption by Root Nodules in Legumes Bacteria residing in the root nodules of the pea plant consume more than 20% of the ATP produced by the plant. Suggest why these bacteria consume so much ATP.

2. Glutamate Dehydrogenase and Protein Synthesis The bacterium *Methylophilus methylotrophus* can synthesize protein from methanol and ammonia. Recombinant DNA techniques have improved the yield of protein by introducing into *M. methylotrophus* the glutamate dehydrogenase gene from *E. coli*. Why does this genetic manipulation increase the protein yield?

3. PLP Reaction Mechanisms Pyridoxal phosphate can help catalyze transformations one or two carbons removed from the α carbon of an amino acid. The enzyme threonine synthase (see Fig. 22-17) promotes the PLP-dependent conversion of phosphohomoserine to threonine. Suggest a mechanism for this reaction.

4. Transformation of Aspartate to Asparagine There are two routes for transforming aspartate to asparagine at the expense of ATP. Many bacteria have an asparagine synthetase that uses ammonium ion as the nitrogen donor. Mammals have an asparagine synthetase that uses glutamine as the nitrogen donor. Given that the latter requires an extra ATP (for the synthesis of glutamine), why do mammals use this route?

5. Equation for the Synthesis of Aspartate from Glucose Write the net equation for the synthesis of aspartate (a nonessential amino acid) from glucose, carbon dioxide, and ammonia.

6. Asparagine Synthetase Inhibitors in Leukemia Therapy Mammalian asparagine synthetase is a glutamine-dependent amidotransferase. Efforts to identify an effective inhibitor of human asparagine synthetase for use in chemotherapy for patients with leukemia have focused not on the amino-terminal glutaminase domain but on the carboxyl-terminal synthetase active site. Explain why the glutaminase domain is not a promising target for a useful drug.

7. Phenylalanine Hydroxylase Deficiency and Diet Tyrosine is normally a nonessential amino acid, but individuals with a genetic defect in phenylalanine hydroxylase require tyrosine in their diet for normal growth. Explain.

8. Cofactors for One-Carbon Transfer Reactions Most one-carbon transfers are promoted by one of three cofactors: biotin, tetrahydrofolate, or S-adenosylmethionine (Chapter 18). S-Adenosylmethionine is generally used as a methyl group donor; the transfer potential of the methyl group in N^5-methyltetrahydrofolate is insufficient for most biosynthetic reactions. However, one example of the use of N^5-methyltetrahydrofolate in methyl group transfer is in methionine formation by the methionine synthase reaction (step ❾ of Fig. 22-17); methionine is the immediate precursor of S-adenosylmethionine (see Fig. 18-18). Explain how the methyl group of S-adenosylmethionine can be derived from N^5-methyltetrahydrofolate, even though the transfer potential of the methyl group in N^5-methyltetrahydrofolate is one-thousandth of that in S-adenosylmethionine.

9. Concerted Regulation in Amino Acid Biosynthesis The glutamine synthetase of *E. coli* is independently modulated by various products of glutamine metabolism (see Fig. 22-8). In this concerted inhibition, the extent of enzyme inhibition is greater than the sum of the separate inhibitions caused by each product. For *E. coli* grown in a medium rich in histidine, what would be the advantage of concerted inhibition?

10. Relationship between Folic Acid Deficiency and Anemia Folic acid deficiency, believed to be the most common vitamin deficiency, causes a type of anemia in which hemoglobin synthesis is impaired and erythrocytes do not mature properly. What is the metabolic relationship between hemoglobin synthesis and folic acid deficiency?

11. Nucleotide Biosynthesis in Amino Acid Auxotrophic Bacteria Wild-type *E. coli* cells can synthesize all 20 common amino acids, but some mutants, called amino acid auxotrophs, are unable to synthesize a specific amino acid and require its addition to the culture medium for optimal growth. Besides their role in protein synthesis, some amino acids are also precursors for other nitrogenous cell products. Consider the three amino acid auxotrophs that are unable to synthesize glycine, glutamine, and aspartate, respectively. For each mutant, what nitrogenous products other than proteins would the cell fail to synthesize?

12. Inhibitors of Nucleotide Biosynthesis Suggest mechanisms for the inhibition of (a) alanine racemase by L-fluoroalanine and (b) glutamine amidotransferases by azaserine.

13. Mode of Action of Sulfa Drugs Some bacteria require p-aminobenzoate in the culture medium for normal growth, and their growth is severely inhibited by the addition of sulfanilamide, one of the earliest sulfa drugs. Moreover, in the presence of this drug, 5-aminoimidazole-4-carboxamide ribonucleotide (AICAR; see Fig. 22-35) accumulates in the culture medium. These effects are reversed by addition of excess p-aminobenzoate.

p-Aminobenzoate   Sulfanilamide

(a) What is the role of p-aminobenzoate in these bacteria? (Hint: See Fig. 18-16.)

(b) Why does AICAR accumulate in the presence of sulfanilamide?

(c) Why are the inhibition and accumulation reversed by addition of excess p-aminobenzoate?

14. Pathway of Carbon in Pyrimidine Biosynthesis Predict the locations of ^{14}C in orotate isolated from cells grown on a small amount of uniformly labeled [^{14}C]succinate. Justify your prediction.

15. Nucleotides as Poor Sources of Energy Under starvation conditions, organisms can use proteins and amino acids as sources of energy. Deamination of amino acids produces carbon skeletons that can enter the glycolytic pathway and the citric acid cycle to produce energy in the form of ATP. Nucleotides are not similarly degraded for use as energy-yielding fuels. What observations about cellular physiology support this statement? What aspect of the structure of nucleotides makes them a relatively poor source of energy?

16. Treatment of Gout Allopurinol (see Fig. 22-50), an inhibitor of xanthine oxidase, is used to treat chronic gout. Explain the biochemical basis for this treatment. Patients treated with allopurinol sometimes develop xanthine stones in the kidneys, although the incidence of kidney damage is much lower than in untreated gout. Explain this observation in the light of the following solubilities in urine: uric acid, 0.15 g/L; xanthine, 0.05 g/L; and hypoxanthine, 1.4 g/L.

17. Inhibition of Nucleotide Synthesis by Azaserine The diazo compound O-(2-diazoacetyl)-L-serine, known also as azaserine (see Fig. 22-51), is a powerful inhibitor of glutamine amidotransferases. If growing cells are treated with azaserine, what intermediates of nucleotide biosynthesis will accumulate? Explain.

Data Analysis Problem

18. Use of Modern Molecular Techniques to Determine the Synthetic Pathway of a Novel Amino Acid Most of the biosynthetic pathways described in this chapter were determined before the development of recombinant DNA technology and genomics, so the techniques were quite different from those that researchers would use today. Here we explore an example of the use of modern molecular techniques to investigate the pathway of synthesis of a novel amino acid, (2S)-4-amino-2-hydroxybutyrate (AHBA). The techniques mentioned here are described in various places in the book; this problem is designed to show how they can be integrated in a comprehensive study.

AHBA is a γ-amino acid that is a component of some aminoglycoside antibiotics, including the antibiotic butirosin. Antibiotics modified by the addition of an AHBA residue are often more resistant to inactivation by bacterial antibiotic-resistance enzymes. As a result, understanding how AHBA is synthesized and added to antibiotics is useful in the design of pharmaceuticals.

In an article published in 2005, Li and coworkers describe how they determined the synthetic pathway of AHBA from glutamate.

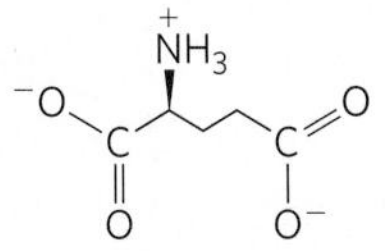

Glutamate 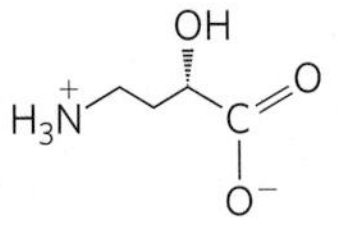   AHBA

(a) Briefly describe the chemical transformations needed to convert glutamate to AHBA. At this point, don't be concerned about the *order* of the reactions.

Li and colleagues began by cloning the butirosin biosynthetic gene cluster from the bacterium *Bacillus circulans*, which makes large quantities of butirosin. They identified five genes that are essential for the pathway: *btrI*, *btrJ*, *btrK*, *btrO*, and *btrV*. They cloned these genes into *E. coli* plasmids that allow overexpression of the genes, producing proteins with "histidine tags" fused to their amino termini to facilitate purification (see p. 332).

The predicted amino acid sequence of the BtrI protein showed strong homology to known acyl carrier proteins (see Fig. 21-5). Using mass spectrometry, Li and colleagues found a molecular mass of 11,812 for the purified BtrI protein (including the His tag). When the purified BtrI was incubated with coenzyme A and an enzyme known to attach CoA to other acyl carrier proteins, the majority molecular species had an M_r of 12,153.

(b) How would you use these data to argue that BtrI can function as an acyl carrier protein with a CoA prosthetic group?

Using standard terminology, Li and coauthors called the form of the protein lacking CoA apo-BtrI and the form with

CoA (linked as in Fig. 21-5) holo-BtrI. When holo-BtrI was incubated with glutamine, ATP, and purified BtrJ protein, the holo-BtrI species of M_r 12,153 was replaced with a species of M_r 12,281, corresponding to the thioester of glutamate and holo-BtrI. Based on these data, the authors proposed the following structure for the M_r 12,281 species, γ-glutamyl-S-BtrI:

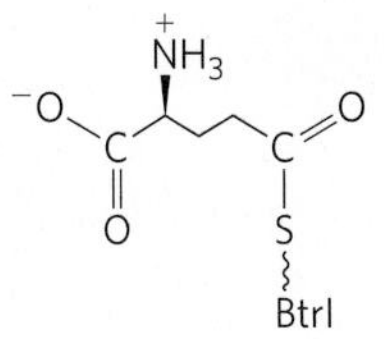

Btrl

(c) What other structure(s) is (are) consistent with the data above?

(d) Li and coauthors argued that the structure shown here (γ-glutamyl-S-BtrI) is likely to be correct because the α-carboxyl group must be removed at some point in the synthetic process. Explain the chemical basis of this argument. (Hint: See Fig. 18-6, reaction C.)

The BtrK protein showed significant homology to PLP-dependent amino acid decarboxylases, and BtrK isolated from *E. coli* was found to contain tightly bound PLP. When γ-glutamyl-S-BtrI was incubated with purified BtrK, a molecular species of M_r 12,240 was produced.

(e) What is the most likely structure of this species?

(f) When the investigators incubated glutamate and ATP with purified BtrI, BtrJ, and BtrK, they found a molecular species of M_r 12,370. What is the most likely structure of this species? Hint: Remember that BtrJ can use ATP to γ-glutamylate nucleophilic groups.

Li and colleagues found that BtrO is homologous to monooxygenase enzymes (see Box 21-1) that hydroxylate alkanes, using FMN as a cofactor, and BtrV is homologous to an NAD(P)H oxidoreductase. Two other genes in the cluster, *btrG* and *btrH*, probably encode enzymes that remove the γ-glutamyl group and attach AHBA to the target antibiotic molecule.

(g) Based on these data, propose a plausible pathway for the synthesis of AHBA and its addition to the target antibiotic. Include the enzymes that catalyze each step and any other substrates or cofactors needed (ATP, NAD, etc.).

Reference

Li, Y., N.M. Llewellyn, R. Giri, F. Huang, and J.B. Spencer. 2005. Biosynthesis of the unique amino acid side chain of butirosin: possible protective-group chemistry in an acyl carrier protein–mediated pathway. *Chem. Biol.* 12:665–675.

Further Reading is available at www.macmillanlearning.com/LehningerBiochemistry7e.

CHAPTER 23

Hormonal Regulation and Integration of Mammalian Metabolism

Self-study tools that will help you practice what you've learned and reinforce this chapter's concepts are available online. Go to www.macmillanlearning.com/LehningerBiochemistry7e.

In Chapters 13 through 22 we have discussed metabolism at the level of the individual cell, emphasizing central pathways common to almost all cells—bacterial, archaeal, and eukaryotic. We have seen how metabolic processes within cells are regulated at the level of individual enzyme reactions by substrate availability, by allosteric mechanisms, and by phosphorylation or other covalent modifications of enzymes.

To fully appreciate the significance of individual metabolic pathways and their regulation, we must view these pathways in the context of the whole organism. An essential characteristic of multicellular organisms is cell differentiation and division of labor. The specialized functions of the tissues and organs of complex organisms such as humans impose characteristic fuel requirements and patterns of metabolism. Hormonal and neuronal signals integrate and coordinate the metabolic activities of different tissues and optimize the allocation of fuels and precursors to each organ.

In this chapter we focus on mammals, looking at the specialized metabolism of several major organs and tissues and the integration of metabolism in the whole organism. We begin by examining the broad range of hormones and hormonal mechanisms, then turn to the tissue-specific functions regulated by these mechanisms. We discuss the distribution of nutrients to various organs, emphasizing the central role of the liver, and the metabolic cooperation among these organs. To illustrate the integrative role of hormones, we describe the interplay of insulin, glucagon, and epinephrine in coordinating fuel metabolism in muscle, liver, and adipose tissue. Other hormones, produced in adipose tissue, muscle, gut, and brain, also play key roles in coordinating metabolism and behavior. We discuss the long-term hormonal regulation of body mass and, finally, the role of obesity in development of metabolic syndrome and type 2 diabetes.

23.1 Hormones: Diverse Structures for Diverse Functions

Hormones are small molecules or proteins that are produced in one tissue, released into the bloodstream, and carried to other tissues, where they act through specific receptors to bring about changes in cellular activities. We also include in this discussion short-lived signals such as

NO, which acts locally, on neighboring cells. Hormones serve to coordinate the metabolic activities of several tissues or organs. Virtually every process in a complex organism is regulated by one or more hormones: maintenance of blood pressure, blood volume, and electrolyte balance; embryogenesis; sexual differentiation, development, and reproduction; hunger, eating behavior, digestion, and fuel allocation—to name but a few. We examine here the methods for detecting and measuring hormones and their interaction with receptors, and we consider a representative selection of hormone types.

The coordination of metabolism in mammals is achieved by the **neuroendocrine system**. Individual cells in one tissue sense a change in the organism's circumstances and respond by secreting a chemical messenger that passes to another cell in the same or different tissue, where the messenger binds to a receptor molecule and triggers a change in this target cell. These chemical messengers may relay information over very short or very long distances. In neuronal signaling **(Fig. 23-1a)**, the chemical messenger is a neurotransmitter (acetylcholine, for example) and may travel only a fraction of a micrometer, across a synaptic cleft to the next neuron in a network. In hormonal signaling, the messengers—hormones—are carried in the bloodstream to neighboring cells or to distant organs and tissues; they may travel a meter or more before encountering their target cell

(Fig. 23-1b). Except for this anatomic difference, these two chemical signaling mechanisms are remarkably similar, and the same molecule can sometimes act as both neurotransmitter and hormone. Epinephrine and norepinephrine, for example, serve as neurotransmitters at certain synapses of the brain and at neuromuscular junctions of smooth muscle and as hormones that regulate fuel metabolism in liver and muscle. The following discussion of cellular signaling emphasizes hormone action, drawing on discussions of fuel metabolism in earlier chapters, but most of the fundamental mechanisms described here also occur in neurotransmitter action.

The Detection and Purification of Hormones Requires a Bioassay

How is a hormone detected and isolated? First, researchers find that a physiological process in one tissue depends on a signal that originates in another tissue. Insulin, for example, was first recognized as a substance that is produced in the pancreas and affects the concentration of glucose in blood and urine (Box 23-1). Once a physiological effect of the putative hormone is discovered, a quantitative bioassay for the hormone can be developed. In the case of insulin, the assay consisted of injecting extracts of pancreas (a crude source of insulin) into experimental animals deficient in insulin, then quantifying the resulting changes in glucose concentration in blood and urine. To isolate a hormone, the biochemist fractionates extracts containing the putative hormone, with the same techniques used to purify other biomolecules (solvent fractionation, chromatography, and electrophoresis), and then assays each fraction for hormone activity. Once the chemical has been purified, its composition and structure can be determined.

This protocol for hormone characterization is deceptively simple. Hormones are extremely potent and are produced in very small amounts. Obtaining sufficient quantities of a hormone to allow its chemical characterization often requires biochemical isolations on a heroic scale. When Andrew Schally and Roger Guillemin independently purified and characterized thyrotropin-releasing hormone (TRH) from the hypothalamus, Schally's group processed about 20 tons of hypothalamus from nearly two million sheep, and Guillemin's group extracted the hypothalamus from about a million pigs. TRH proved to be a simple

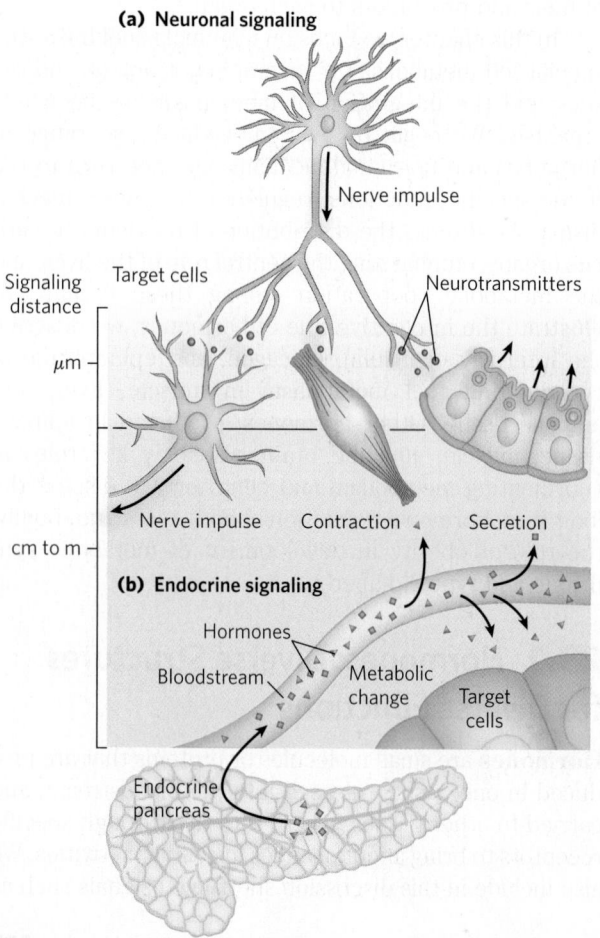

(a) Neuronal signaling

Nerve impulse

Signaling distance

Target cells

Neurotransmitters

μm

Nerve impulse   Contraction   Secretion

cm to m

(b) Endocrine signaling

Hormones

Bloodstream

Metabolic change

Target cells

Endocrine pancreas

FIGURE 23-1 Signaling by the neuroendocrine system. (a) In neuronal signaling, electrical signals (nerve impulses) originate in the cell body of a neuron and travel very rapidly over long distances to the axon tip, where neurotransmitters are released and diffuse to the target cell. The target cell (another neuron, a myocyte, or a secretory cell) is only a fraction of a micrometer or a few micrometers away from the site of neurotransmitter release. **(b)** In endocrine signaling, hormones (such as insulin produced in pancreatic β cells) are secreted into the bloodstream, which carries them throughout the body to target tissues that may be a meter or more away from the secreting cell. Both neurotransmitters and hormones interact with specific receptors on or in their target cells, triggering responses.

BOX 23-1	MEDICINE	How Is a Hormone Discovered? The Arduous Path to Purified Insulin

Millions of people with type 1 diabetes mellitus inject themselves daily with pure insulin to compensate for the lack of production of this critical hormone by their own pancreatic β cells. Insulin injection is not a cure for diabetes, but it allows people who otherwise would have died young to lead long and productive lives. The discovery of insulin, which began with an accidental observation, illustrates the combination of serendipity and careful experimentation that led to the discovery of many hormones.

In 1889, Oskar Minkowski, a young assistant at the Medical College of Strasbourg, and Josef von Mering, at the Hoppe-Seyler Institute in Strasbourg, had a friendly disagreement about whether the pancreas, known to contain lipases, was important in fat digestion in dogs. To resolve the issue, they began an experiment on the digestion of fats. They surgically removed the pancreas from a dog, but before their experiment got any farther, Minkowski noticed that the dog was now producing far more urine than normal (a common symptom of untreated diabetes). Also, the dog's urine had glucose levels far above normal (another symptom of diabetes). These findings suggested that lack of some pancreatic product caused diabetes.

Minkowski tried unsuccessfully to prepare an extract of dog pancreas that would reverse the effect of removing the pancreas—that is, would lower the urinary or blood glucose levels. We now know that insulin is a protein and that the pancreas is very rich in proteases (trypsin and chymotrypsin), normally released directly into the small intestine to aid in digestion. These proteases doubtless degraded the insulin in the pancreatic extracts in Minkowski's experiments.

Despite considerable effort, no significant progress was made in the isolation or characterization of the "antidiabetic factor" until the summer of 1921, when Frederick G. Banting, a young scientist working in the laboratory of J. J. R. MacLeod at the University of Toronto, and a student assistant, Charles Best, took up the problem. By that time, several lines of evidence pointed to a group of specialized cells in the pancreas (the islets of Langerhans; see Fig. 23-27) as the source of the antidiabetic factor, which came to be called insulin (from Latin *insula*, "island").

Taking precautions to prevent proteolysis, Banting and Best (later aided by biochemist J. B. Collip) succeeded in December 1921 in preparing a purified pancreatic extract that cured the symptoms of experimentally induced diabetes in dogs. On January 25, 1922 (just one month later!), their insulin preparation was injected into Leonard Thompson, a 14-year-old boy severely ill with diabetes mellitus. Within days, the levels of ketone bodies and glucose in Thompson's urine dropped dramatically; the extract saved his life and the lives of other seriously ill children who also received these early preparations (Fig. 1). In 1923, Banting and MacLeod won the Nobel Prize for their isolation of insulin. Banting immediately announced that he would share his prize with Best; MacLeod shared his with Collip.

By 1923, pharmaceutical companies were supplying thousands of patients throughout the world with insulin extracted from porcine pancreas. With the development of genetic engineering techniques in the 1980s, it became possible to produce unlimited quantities of human insulin by inserting the cloned human gene for insulin into a microorganism, which was then cultured on an industrial scale. Some people with diabetes are now fitted with implanted insulin pumps, which release adjustable amounts of insulin on demand to meet changing needs at meal times and during exercise. There is a reasonable prospect that, in the future, transplanted pancreatic tissue will provide a source of insulin that responds as well as a normal pancreas, releasing insulin into the bloodstream only when blood glucose rises.

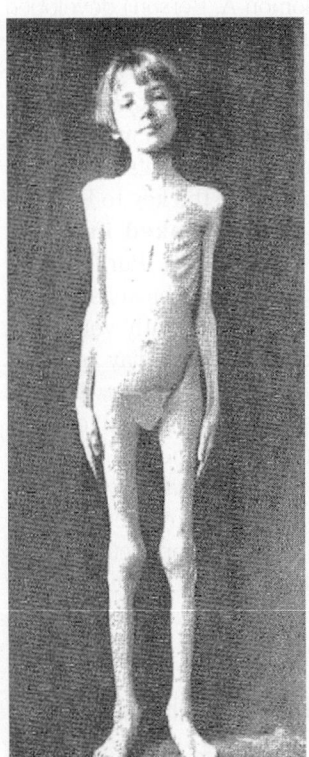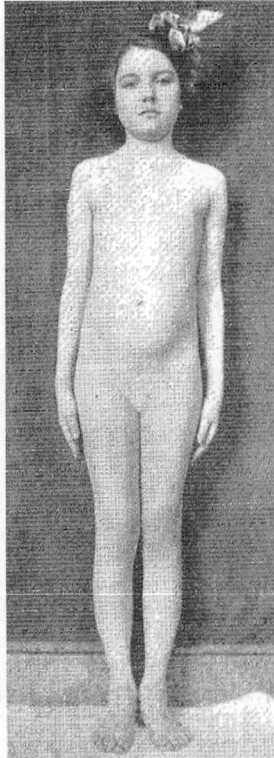

FIGURE 1 A child with type 1 diabetes, before (left) and after (right) three months of treatment with an early preparation of insulin. [Source: H. R. Geyelin et al., *J. Metabol. Res.* 2:767, 1922.]

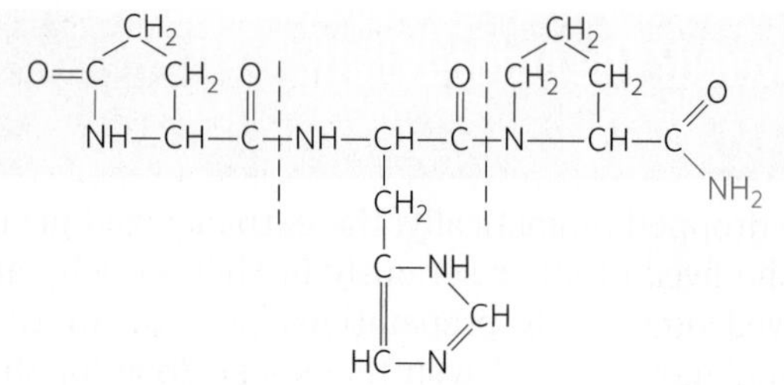

FIGURE 23-2 The structure of thyrotropin-releasing hormone (TRH).
Purified (through heroic efforts) from extracts of hypothalamus, TRH proved to be a derivative of the tripeptide Glu-His-Pro. The side-chain carboxyl group of the amino-terminal Glu forms an amide (red bond) with the residue's α-amino group, creating pyroglutamate, and the carboxyl group of the carboxyl-terminal Pro is converted to an amide (red —NH_2). Such modifications are common among the small peptide hormones. In a typical protein of M_r ~50,000, the charges on the amino- and carboxyl-terminal groups contribute relatively little to the overall charge on the molecule, but in a tripeptide these two charges dominate the properties of the molecule. Formation of the amide derivatives removes these charges.

derivative of the tripeptide Glu–His–Pro **(Fig. 23-2)**. Once the structure of the hormone was known, it could be chemically synthesized in large quantities for use in physiological and biochemical studies. For their work on hypothalamic hormones, Schally and Guillemin shared the Nobel Prize in Physiology or Medicine in 1977, along with Rosalyn Yalow, who (with Solomon A. Berson) developed the extraordinarily sensitive **radioimmunoassay (RIA)** for peptide hormones and used it to study hormone action. This technique revolutionized hormone research by making possible the rapid, quantitative, and specific measurement of hormones in minute amounts.

Hormone-specific antibodies are the key to RIA and its modern equivalent, the **enzyme-linked immuno-sorbent assay** (**ELISA**; see Fig. 5-26b). Purified hormone, injected into rabbits, mice, or chickens, elicits antibodies that bind to the hormone with very high affinity and specificity. These antibodies may be purified and either radioisotopically labeled (for RIA) or conjugated with an enzyme that produces a colored product (for ELISA). The tagged antibodies are then allowed to interact with extracts containing the hormone. The fraction of antibody bound by the hormone in the extract is quantified by radiation detection or photometry. Because of the high affinity of the antibody for the hormone, such assays can be made sensitive to picograms of hormone in a sample.

Hormones Act through Specific High-Affinity Cellular Receptors

As we saw in Chapter 12, all hormones act through highly specific receptors in hormone-sensitive target cells, to which the hormones bind with high affinity. Each cell type has its own combination of hormone receptors, which define the range of its hormone

responsiveness. Moreover, two cell types with the same type of receptor may have different intracellular targets of hormone action and thus may respond differently to the same hormone. The specificity of hormone action results from structural complementarity between the hormone and its receptor; this interaction is extremely selective, so even structurally similar hormones can have different effects if they preferentially bind to different receptors. The high affinity of the interaction allows cells to respond to very low concentrations of hormone. In the design of drugs intended to intervene in hormonal regulation, we need to know the relative specificity and affinity of the drug and the natural hormone.

The intracellular consequences of ligand-receptor interaction are of at least four general types: (1) a second messenger, such as cAMP, cGMP, or inositol trisphosphate, generated inside the cell acts as an allosteric regulator of one or more enzymes; (2) a receptor tyrosine kinase is activated by the extracellular hormone; (3) a change in membrane potential results from the opening or closing of a hormone-gated ion channel; and (4) a steroid or steroidlike molecule causes a change in the level of expression (transcription of DNA into mRNA) of one or more genes, mediated by a nuclear hormone receptor protein (see Fig. 12-2).

Water-soluble peptide and amine hormones, such as insulin and epinephrine, act extracellularly by binding to cell surface receptors that span the plasma membrane **(Fig. 23-3)**. When the hormone binds to its extracellular domain, the receptor undergoes a conformational change analogous to that produced in an allosteric enzyme by binding of an effector molecule. The conformational change triggers the effect of the hormone. With **metabotropic** receptors, the change activates or inhibits an enzyme downstream from the receptor; with **ionotropic** receptors, an ion channel in the plasma membrane opens or closes, resulting in a change in membrane potential (ΔV_m) or in the concentration of an ion such as Ca^{2+}.

A single hormone molecule, in forming a hormone-receptor complex, activates a catalyst that produces many molecules of second messenger, so the receptor serves as both signal transducer and signal amplifier. The signal may be further amplified by a signaling cascade, a series of steps in which a catalyst (such as a protein kinase) activates another catalyst (another protein kinase), resulting in very large amplifications of the original signal. A cascade of this type occurs in the regulation of glycogen synthesis and breakdown by epinephrine (see Fig. 12-7). Epinephrine activates (through its receptor) adenylyl cyclase, which produces many molecules of cAMP for each molecule of receptor-bound hormone. Cyclic AMP in turn activates cAMP-dependent protein kinase (protein kinase A), which activates glycogen phosphorylase *b* kinase, which activates glycogen phosphorylase *b*. The result is signal amplification: one epinephrine molecule causes the production of many thousands or millions of molecules of glucose 1-phosphate from glycogen.

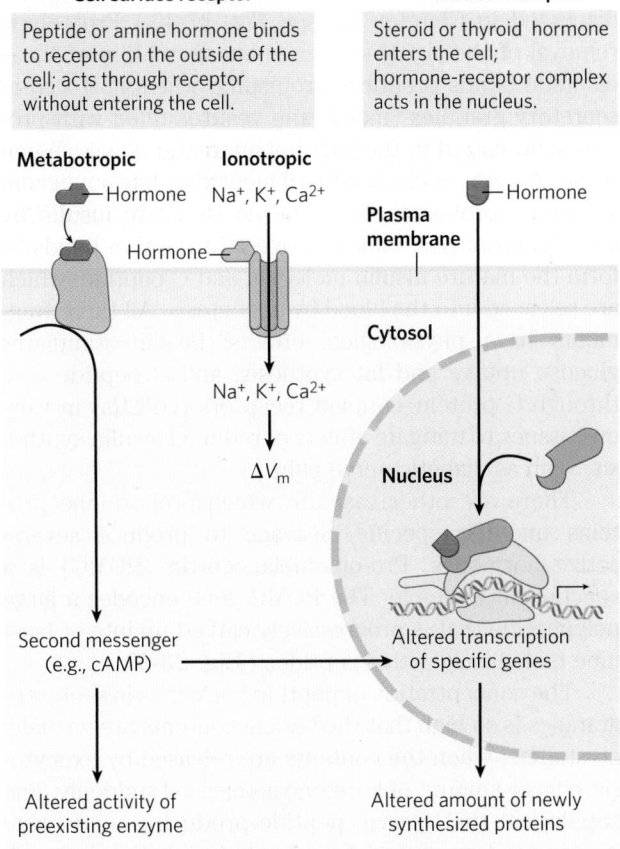

Cell surface receptor

Peptide or amine hormone binds to receptor on the outside of the cell; acts through receptor without entering the cell.

Nuclear receptor

Steroid or thyroid hormone enters the cell; hormone-receptor complex acts in the nucleus.

Metabotropic   Ionotropic

Hormone Na^+, K^+, Ca^{2+}

Hormone

Plasma membrane

Hormone

Cytosol

Na^+, K^+, Ca^{2+}

ΔV_m

Nucleus

Second messenger (e.g., cAMP)

Altered transcription of specific genes

Altered activity of preexisting enzyme

Altered amount of newly synthesized proteins

FIGURE 23-3 Two general mechanisms of hormone action. The peptide and amine hormones are faster acting than steroid and thyroid hormones.

Water-insoluble hormones, including the steroid, retinoid, and thyroid hormones, readily pass through the plasma membrane of their target cells to reach their receptor proteins in the nucleus (Fig. 23-3). The hormone-receptor complex itself carries the message: it interacts with DNA to alter the expression of specific genes, changing the enzyme complement of the cell and thereby changing cellular metabolism (see Fig. 12-30).

Hormones that act through plasma membrane receptors generally trigger very rapid physiological or biochemical responses. Just seconds after the adrenal medulla secretes epinephrine into the bloodstream, skeletal muscle responds by accelerating the breakdown of glycogen. By contrast, the thyroid hormones and the sex (steroid) hormones promote maximal responses in their target tissues only after hours or even days. These differences in response time correspond to different modes of action. In general, the fast-acting hormones lead to a change in the activity of one or more preexisting enzymes in the cell, by allosteric mechanisms or covalent modification. The slower-acting hormones generally alter gene expression, resulting in the synthesis of more (upregulation) or less (downregulation) of the regulated protein(s).

Hormones Are Chemically Diverse

Mammals have several classes of hormones, distinguishable by their chemical structures and their modes of action (Table 23-1). Peptide, catecholamine, and eicosanoid hormones act from outside the target cell via cell surface receptors. Steroid, vitamin D, retinoid, and thyroid hormones enter the cell and act through nuclear receptors. Nitric oxide (a gas) also enters the cell, but activates a cytosolic enzyme, guanylyl cyclase (see Fig. 12-23).

Hormones can also be classified by the way they get from their point of release to their target tissue. **Endocrine** (from the Greek *endon*, "within," and *krinein*, "to release") hormones are released into the blood and carried to target cells throughout the body (insulin and glucagon are examples). **Paracrine** hormones are released into the extracellular space and diffuse to neighboring target cells (the eicosanoid hormones are of this type). **Autocrine** hormones affect the same cell that releases them, binding to receptors on the cell surface.

Mammals are hardly unique in possessing hormonal signaling systems. Insects and nematode worms have highly developed systems for hormonal regulation, with fundamental mechanisms similar to those in mammals.

TABLE 23-1	Classes of Hormones		
Type	**Example**	**Synthetic path**	**Mode of action**
Peptide	Insulin, glucagon	Proteolytic processing of prohormone	Plasma membrane receptors; second messengers
Catecholamine	Epinephrine	From tyrosine	
Eicosanoid	Prostaglandin E_2	From arachidonate (20:4 fatty acid)	
Steroid	Testosterone	From cholesterol	Nuclear receptors; transcriptional regulation
Vitamin D	Calcitriol	From cholesterol	
Retinoid	Retinoic acid	From vitamin A	
Thyroid	Triiodothyronine (T_3)	From Tyr in thyroglobulin	
Nitric oxide	Nitric oxide	From arginine + O_2	Cytosolic receptor (guanylyl cyclase) and second messenger (cGMP)

Plants, too, use hormonal signals to coordinate the activities of their tissues.

To illustrate the structural diversity and range of action of mammalian hormones, we consider representative examples of each major class listed in Table 23-1.

Peptide Hormones The **peptide hormones** vary in size, from 3 to more than 200 amino acid residues. They include the pancreatic hormones insulin, glucagon, and somatostatin; the parathyroid hormone calcitonin; and all the hormones of the hypothalamus and pituitary (described below). These hormones are synthesized on ribosomes in the form of longer, precursor proteins (prohormones), then packaged into secretory vesicles and proteolytically cleaved to form the active peptides. In many peptide hormones the terminal residues are modified, as in TRH (Fig. 23-2).

Insulin is a small protein (M_r 5,800) with two polypeptide chains, A and B, joined by two disulfide bonds. It is synthesized in the pancreas as an inactive single-chain precursor, preproinsulin **(Fig. 23-4)**, with an amino-terminal "signal sequence" that directs its

passage into secretory vesicles. (Signal sequences are discussed in Chapter 27; see Fig. 27-40.) Proteolytic removal of the signal sequence and formation of three disulfide bonds produces proinsulin, which is stored in secretory granules (membrane vesicles filled with protein synthesized in the ER) in pancreatic β cells. When blood glucose is elevated sufficiently to trigger insulin secretion, proinsulin is converted to active insulin by specific proteases, which cleave two peptide bonds to form the mature insulin molecule and C peptide, which are released into the blood by exocytosis. All three fragments have physiological effects: insulin stimulates glucose uptake and fat synthesis, and C peptide acts through G protein–coupled receptors (GPCRs) in various tissues to mitigate effects of reduced insulin synthesis, such as diabetic nerve pain.

There are other cases in which prohormone proteins undergo specific cleavage to produce several active hormones. Pro-opiomelanocortin (POMC) is a spectacular example. The POMC gene encodes a large polypeptide that is progressively carved up into at least nine biologically active peptides **(Fig. 23-5)**.

The concentration of peptide hormones in secretory granules is so high that the vesicle contents are virtually crystalline; when the contents are released by exocytosis, a large amount of hormone is released suddenly. The capillaries that serve peptide-producing endocrine glands are fenestrated (punctuated with tiny holes or "windows"), so the hormone molecules readily enter the bloodstream for transport to target cells elsewhere. As noted earlier, all peptide hormones act by binding to

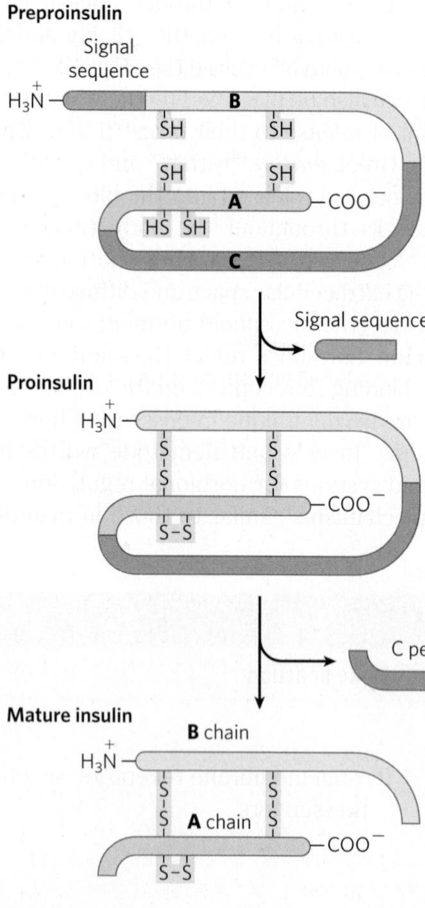

FIGURE 23-4 Insulin. Mature insulin is formed from its larger precursor preproinsulin by proteolytic processing. Removal of a 23 amino acid segment (the signal sequence) at the amino terminus of preproinsulin and formation of three disulfide bonds produce proinsulin. Further proteolytic cuts remove the C peptide from proinsulin to produce mature insulin, composed of A and B chains. The amino acid sequence of bovine insulin is shown in Figure 3-24.

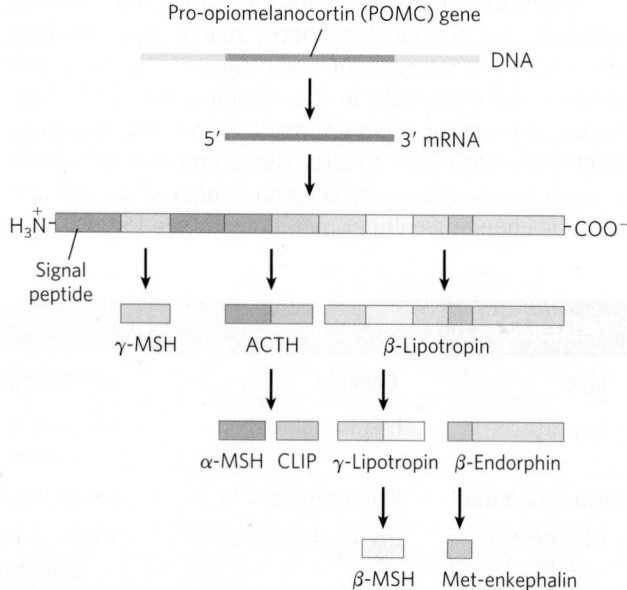

FIGURE 23-5 Proteolytic processing of the pro-opiomelanocortin (POMC) precursor. The initial gene product of the POMC gene is a long polypeptide that undergoes cleavage by a series of specific proteases to produce ACTH (corticotropin), β- and γ-lipotropin, α-, β-, and γ-MSH (melanocyte-stimulating hormone, or melanocortin), CLIP (corticotropin-like intermediary peptide), β-endorphin, and Met-enkephalin. The points of cleavage are pairs of basic residues, Arg-Lys, Lys-Arg, or Lys-Lys.

receptors in the plasma membrane. They cause the generation of a second messenger in the cytosol, which changes the activity of an intracellular enzyme, thereby altering the cell's metabolism.

Catecholamine Hormones The water-soluble compounds **epinephrine (adrenaline)** and **norepinephrine (noradrenaline)** are **catecholamines**, named for the structurally related compound catechol. They are synthesized from tyrosine (see Fig. 22–31).

$$\text{Tyrosine} \longrightarrow \text{L-Dopa} \longrightarrow \text{Dopamine} \longrightarrow$$
$$\text{Norepinephrine} \longrightarrow \text{Epinephrine}$$

Catecholamines produced in the brain and in other neural tissues function as neurotransmitters, but epinephrine and norepinephrine are also hormones, synthesized and secreted by the adrenal glands (adrenals). Like the peptide hormones, catecholamines are highly concentrated in secretory granules and released by exocytosis, and they act through surface receptors to generate intracellular second messengers. They mediate a wide variety of physiological responses to acute stress (see Table 23–7).

Eicosanoid Hormones The **eicosanoid hormones** (prostaglandins, thromboxanes, leukotrienes, and lipoxins) are derived from the 20-carbon polyunsaturated fatty acids arachidonate ($20{:}4(\Delta^{5,8,11,14})$) and eicosapentaenoic acid (EPA; $20{:}5(\Delta^{5,8,11,14,17})$; see Fig. 21–12).

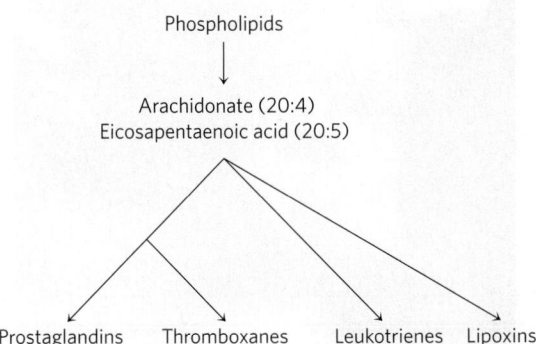

Unlike the hormones described above, they are not synthesized in advance and stored; they are produced when needed. The enzymes of the pathways leading to prostaglandins and thromboxanes are very widely distributed in mammalian tissues; most cells can produce these hormone signals, and cells of many tissues can respond to them through specific plasma membrane receptors. The eicosanoid hormones are paracrine hormones, secreted into the interstitial fluid (not primarily into the blood) and acting on nearby cells.

Some prostaglandins promote the contraction of smooth muscle, including that of the intestine and uterus (and can therefore be used medically to induce labor). They also mediate pain and inflammation in some tissues. Many antiinflammatory drugs act by inhibiting steps in prostaglandin synthetic pathways

(see Fig. 21–15). Thromboxanes regulate platelet function and therefore blood clotting (see Fig. 6–41). Leukotrienes LTC_4 and LTD_4 act through plasma membrane receptors to stimulate contraction of smooth muscle in the intestine, pulmonary airways, and trachea. They are mediators of anaphylaxis, an immune overresponse that can include airway constriction, altered heartbeat, shock, and sometimes death. Lipoxins are short-lived eicosanoid derivatives with potent effects on immune function; they appear in the bloodstream as inflammation ends. ∎

Steroid Hormones The **steroid hormones**—corticosteroid (adrenocortical) hormones and sex hormones—are synthesized from cholesterol in several endocrine tissues.

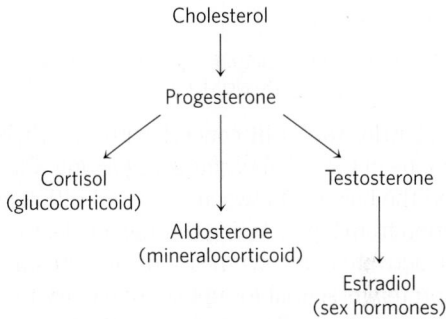

They travel to their target cells through the bloodstream, bound to carrier proteins. More than 50 corticosteroid hormones are produced in the adrenal cortex by reactions that remove the side chain from the D ring of cholesterol and introduce oxygen to form keto and hydroxyl groups. Many of these reactions are catalyzed by cytochrome P-450 enzymes (see Box 21–1). The corticosteroids are of two general types, defined by their actions. Glucocorticoids, such as cortisol, primarily affect the metabolism of carbohydrates; mineralocorticoids, such as aldosterone, regulate the concentrations of electrolytes (K^+, Na^+, Ca^{2+}, Cl^-) in the blood. Two types of sex hormones, androgens (including testosterone) and estrogens (including estradiol; see Fig. 10–18), are synthesized in the testes and ovaries. They affect sexual development, sexual behavior, and a variety of other reproductive and nonreproductive functions. Their synthesis also requires cytochrome P-450 enzymes that cleave the side chain of cholesterol and introduce oxygen atoms.

All steroid hormones act through nuclear receptors to change the level of expression of specific genes (see Fig. 12–30). They can also have more rapid effects, mediated by receptors in the plasma membrane. Humans and other animals are exposed to many exogenous chemicals broadly referred to as "endocrine disruptors," ranging from environmental pollutants such as PCBs (polychlorinated biphenyls), pesticides, and pharmaceuticals to naturally occurring estrogens in plants, such as in soy products. Some endocrine disruptors bind to nuclear steroid receptors and stimulate hormonelike effects; others block the receptors, preventing stimulation by endogenous hormones, or interfere with the normal metabolism of steroid hormones in the liver.

Vitamin D Hormone Calcitriol (1α,25-dihydroxycalcitriol) is produced from vitamin D by enzyme-catalyzed hydroxylation in the liver and kidneys (see Fig. 10-19a). Vitamin D is obtained in the diet or by photolysis of 7-dehydrocholesterol in skin exposed to sunlight.

7-Dehydrocholesterol

↓ UV light

Vitamin D₃
(cholecalciferol)

↓

25-Hydroxycholecalciferol

↓

1α,25-Dihydroxycalcitriol
(calcitriol)

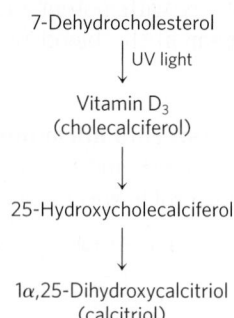 Calcitriol works in concert with parathyroid hormone in Ca^{2+} homeostasis, regulating $[Ca^{2+}]$ in the blood and the balance between Ca^{2+} deposition and Ca^{2+} mobilization from bone. Acting through nuclear receptors, calcitriol activates the synthesis of an intestinal Ca^{2+}-binding protein essential for uptake of dietary Ca^{2+}. Inadequate dietary vitamin D or defects in the biosynthesis of calcitriol result in serious diseases such as rickets, in which bones are weak and malformed (see Fig. 10-19b). ∎

Retinoid Hormones The **retinoid hormones** are potent hormones that regulate the growth, survival, and differentiation of cells via nuclear retinoid receptors. The prohormone retinol is synthesized from β-carotene, primarily in liver (see Fig. 10-20), and many tissues convert retinol to the hormone retinoic acid (RA). RA binds its specific receptor (RAR) in the nucleus, forms a dimer with another nuclear protein, retinoid X receptor (RXR), and alters the rate of expression of genes responsive to RA.

β-Carotene

↓

Vitamin A₁
(retinol)

↓

Retinoic acid

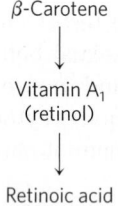 All tissues are retinoid targets, as all cell types have at least one form of nuclear retinoid receptor. In adults, the most significant targets include cornea, skin, epithelia of the lungs and trachea, and the immune system, all of which undergo constant replacement of cells. RA regulates the synthesis of proteins essential for growth or differentiation. Excessive vitamin A (the precursor to retinoid hormones) can cause birth defects, and pregnant women are advised not to use the retinoid creams that have been developed for treatment of severe acne. ∎

Thyroid Hormones The **thyroid hormones** T₄ (thyroxine) and T₃ (triiodothyronine) are synthesized from the precursor protein thyroglobulin (M_r 660,000). Up to 20 Tyr residues in thyroglobulin are enzymatically iodinated in the thyroid gland, then two iodotyrosine residues condense to form the precursor to thyroxine. When needed, thyroxine is released by proteolysis. Condensation of monoiodotyrosine with diiodothyronine produces T₃, which is also an active hormone released by proteolysis.

Thyroglobulin–Tyr

↓

Thyroglobulin–Tyr–I
(iodinated Tyr residues)

↓ proteolysis

Thyroxine (T₄),
triiodothyronine (T₃)

The thyroid hormones act through nuclear receptors to stimulate energy-yielding metabolism, especially in liver and muscle, by increasing the expression of genes encoding key catabolic enzymes. Underproduction of thyroxine slows metabolism and can be the cause of depression. When underproduction is the result of too little iodine in the diet, the thyroid gland enlarges in a futile attempt to produce more thyroxine (**Fig. 23-6**).

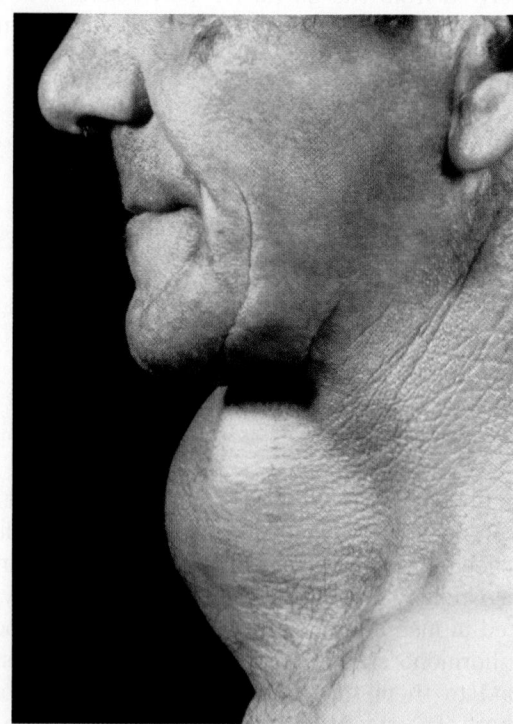

FIGURE 23-6 Goiter. A person with advanced goiter. Goiter is an enlargement of the thyroid gland that can occur in a futile attempt to produce thyroid hormone in the absence of iodine. Iodized salt in the diet prevents most goiter development except in rare cases of genetic defect. Online Mendelian Inheritance in Man (omim.org), an online catalog of human genetic diseases, documents a number of genetic conditions, including mutations in the thyroglobulin gene, that can cause goiter development. [Source: Biophoto Associates/Science Source.]

This condition, called goiter, was once common in regions far from oceans (which provide iodine in the form of fresh seafood) and areas with low-iodine soil (yielding plants with low iodine). Goiter has been almost eliminated in areas where iodine is routinely added to table salt.

Nitric Oxide (NO•) **Nitric oxide** is a relatively stable free radical synthesized from molecular oxygen and the guanidinium nitrogen of arginine (see Fig. 22-33), in a reaction catalyzed by **NO synthase**.

$$\text{Arginine} + 1\tfrac{1}{2}\text{NADPH} + 2\text{O}_2 \longrightarrow$$
$$\text{NO}^{\bullet} + \text{citrulline} + 2\text{H}_2\text{O} + 1\tfrac{1}{2}\text{NADP}^+$$

This enzyme is found in many tissues and cell types: neurons, macrophages, hepatocytes, myocytes of smooth muscle, endothelial cells of the blood vessels, and epithelial cells of the kidney. NO acts near its point of release, entering the target cell and activating the cytosolic enzyme guanylyl cyclase, which catalyzes the formation of the second messenger cGMP (see Fig. 12-23). A cGMP-dependent protein kinase mediates the effects of NO by phosphorylating key proteins and altering their activities. For example, phosphorylation of contractile proteins in the smooth muscle surrounding blood vessels relaxes the muscle, thereby lowering blood pressure.

Hormone Release Is Regulated by a "Top-Down" Hierarchy of Neuronal and Hormonal Signals

The changing levels of specific hormones regulate specific cellular processes, but what regulates the level of each hormone? The brief answer is that the central nervous system receives input from many internal and external sensors—signals about danger, hunger, dietary intake, blood composition and pressure, for example—and orchestrates the production of appropriate hormonal signals by the endocrine tissues. For a more complete answer, we must look at the hormone-producing systems of the human body and some of their functional interrelationships.

Figure 23-7 shows the anatomic locations of the major endocrine glands in humans, and **Figure 23-8** represents the "chain of command" in the hormonal signaling hierarchy. The **hypothalamus**, a small region of the brain **(Fig. 23-9)**, is the coordination center of the endocrine system; it receives and integrates messages from the central nervous system. In response to these messages, the hypothalamus produces regulatory hormones (releasing factors) that pass directly to the nearby pituitary gland through special blood vessels and neurons that connect the two glands (Fig. 23-9b). The pituitary gland has two functionally distinct parts. The **posterior pituitary** contains the axonal endings of many neurons that originate in the hypothalamus. These neurons produce the short peptide hormones oxytocin and vasopressin **(Fig. 23-10)**, which move down the axon to the nerve endings in the pituitary,

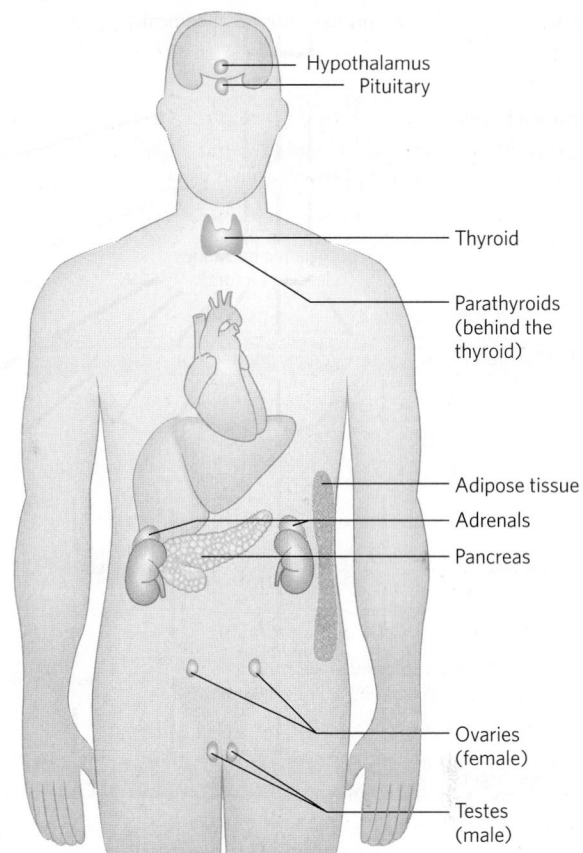

FIGURE 23-7 The major endocrine glands. The glands are shaded pink.

where they are stored in secretory granules to await the signal for their release.

The **anterior pituitary** responds to hypothalamic hormones carried in the blood, producing **tropic hormones**, or **tropins** (from the Greek *tropos*, "turn"). These relatively long polypeptides activate the next rank of endocrine glands (Fig. 23-8), which includes the adrenal cortex, thyroid gland, ovaries, and testes. These glands in turn secrete their specific hormones, which are carried in the bloodstream to target tissues. For example, corticotropin-releasing hormone secreted from the hypothalamus stimulates the anterior pituitary to release corticotropin (ACTH), which travels through the blood to the zona fasciculata of the adrenal cortex and triggers the release of cortisol. Cortisol, the ultimate hormone in this cascade, acts through its receptor in many types of target cells to alter their metabolism. In hepatocytes, one effect of cortisol is to increase the rate of gluconeogenesis.

Hormonal cascades such as those responsible for the release of cortisol and epinephrine result in large amplifications of the initial signal and allow exquisite fine-tuning of the output of the ultimate hormone **(Fig. 23-11)**. At each level in the cascade, a small signal elicits a larger response. For example, the initial electrical signal to the hypothalamus results in the release of a few *nanograms* of corticotropin-releasing hormone, which elicits the release of a few *micrograms* of corticotropin.

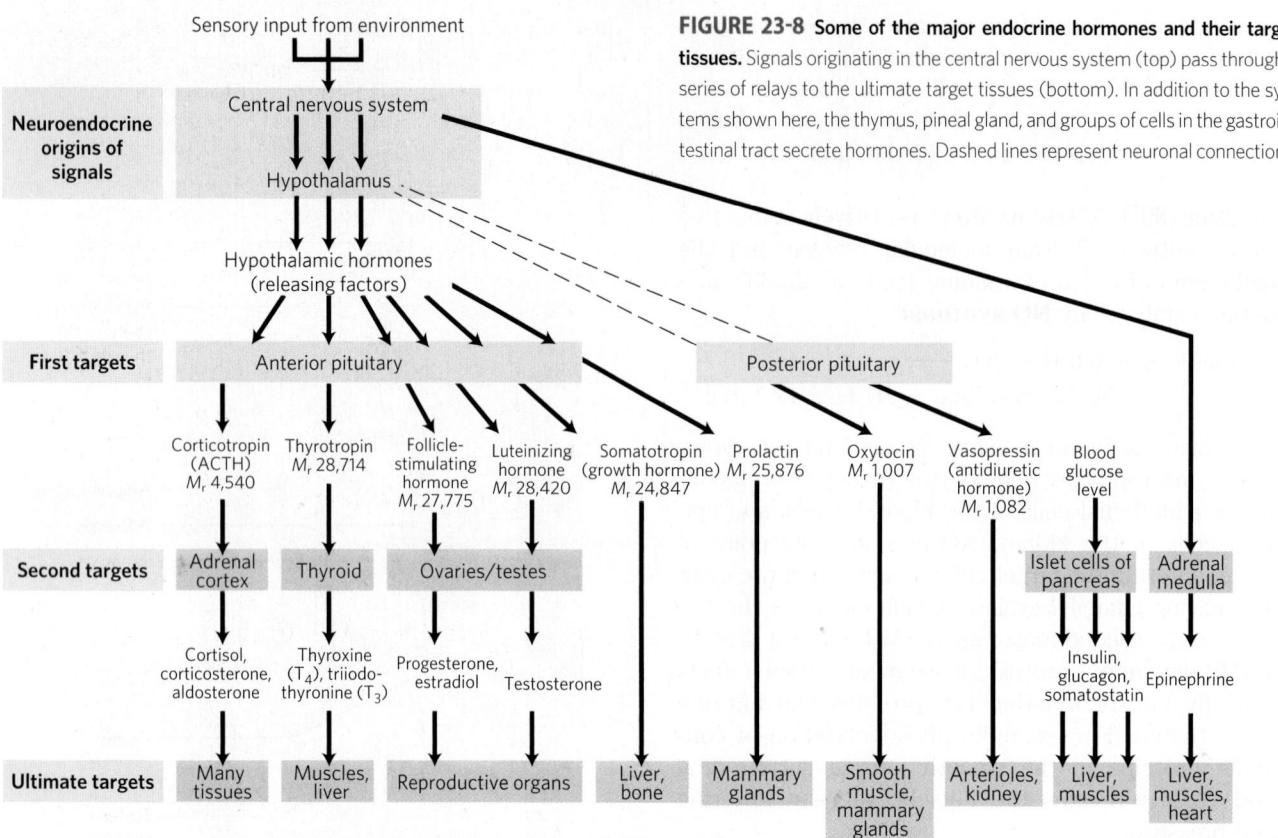

FIGURE 23-8 Some of the major endocrine hormones and their target tissues. Signals originating in the central nervous system (top) pass through a series of relays to the ultimate target tissues (bottom). In addition to the systems shown here, the thymus, pineal gland, and groups of cells in the gastrointestinal tract secrete hormones. Dashed lines represent neuronal connections.

Corticotropin acts on the adrenal cortex to cause the release of *milligrams* of cortisol, for an overall amplification of at least a millionfold.

At each level of a hormonal cascade, feedback inhibition of earlier steps in the cascade is possible; an unnecessarily elevated level of the ultimate hormone or of an intermediate hormone inhibits the release of earlier hormones in the cascade. These feedback mechanisms accomplish the same end as those that limit the output of a biosynthetic pathway (compare Fig. 23-11 with Fig. 22-37): a product is synthesized (or released) only until the necessary concentration is reached.

"Bottom-Up" Hormonal Systems Send Signals Back to the Brain and to Other Tissues

In addition to the top-down hierarchy of hormonal signaling shown in Figure 23-8, some hormones are produced in the digestive tract, muscle, and adipose tissue and communicate the current metabolic state to the

hypothalamus **(Fig. 23-12)**. These signals are integrated in the hypothalamus, and an appropriate neuronal or hormonal response is elicited. The action of the enzyme AMP-activated protein kinase (AMPK) in the hypothalamus is one such integrating mechanism; it

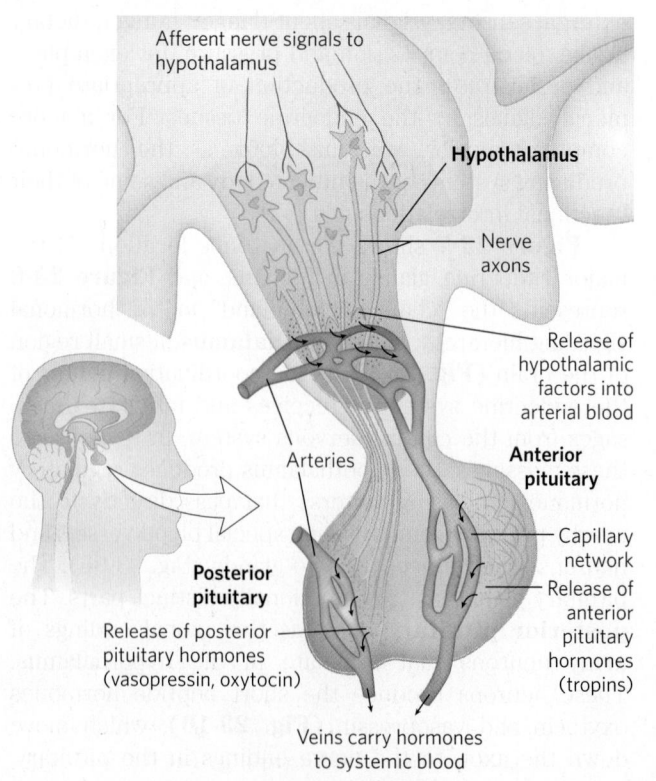

FIGURE 23-9 Neuroendocrine origins of hormone signals. Location of the hypothalamus and pituitary gland and details of the hypothalamus-pituitary system. Signals from connecting neurons stimulate the hypothalamus to secrete releasing factors into a blood vessel that carries the hormones directly to a capillary network in the anterior pituitary. In response to each releasing factor, the anterior pituitary releases the appropriate hormone into the general circulation. Posterior pituitary hormones are synthesized in neurons arising in the hypothalamus; the hormones are transported along axons to nerve endings in the posterior pituitary and stored there until released into the blood in response to a neuronal signal.

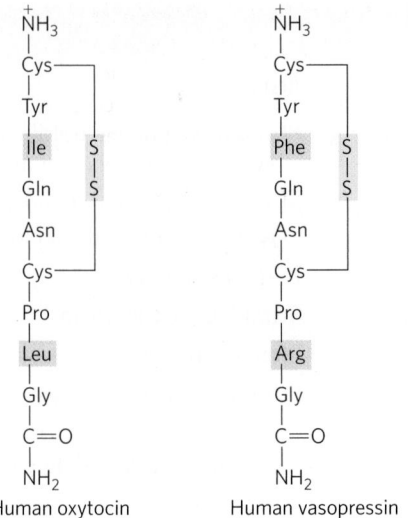

Human oxytocin   Human vasopressin

FIGURE 23-10 Two hormones of the posterior pituitary gland. The carboxyl-terminal residue of both peptides is glycinamide, —NH—CH₂—CONH₂ (as noted in Fig. 23-2, amidation of the carboxyl terminus is common in short peptide hormones). These two hormones, identical in all but two residues (shaded light red), have very different biological effects. Oxytocin acts on the smooth muscle of the uterus and mammary glands, causing uterine contractions during labor and promoting milk release during lactation. Vasopressin (also called antidiuretic hormone) increases water reabsorption in the kidney and promotes the constriction of blood vessels, thereby increasing blood pressure.

sums various inputs and passes on the information by phosphorylating key proteins in the hypothalamus. The known hormonal inputs into this mechanism are numerous, and there are doubtless more to be discovered.

Adipokines, for example, are peptide hormones, produced in adipose tissue, that signal the adequacy of fat reserves. **Leptin,** released when adipose tissue is well-filled with triacylglycerols, acts in the brain to inhibit feeding behavior, whereas **adiponectin** signals depletion of fat reserves and stimulates feeding. **Ghrelin** is produced in the

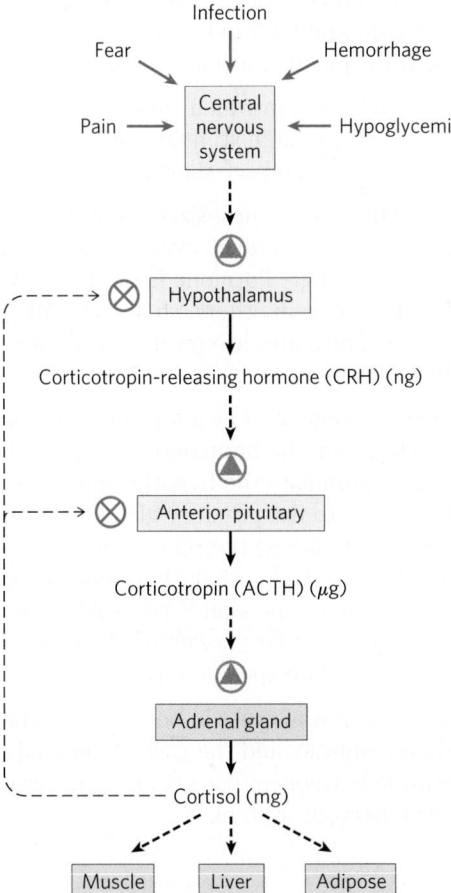

FIGURE 23-11 Cascade of hormone release following central nervous system input to the hypothalamus. Solid black arrows indicate hormone production and release; broken black arrows indicate the action of hormones on target tissues. In each endocrine tissue along the pathway, a stimulus from the level above is received, amplified, and transduced into release of the next hormone in the cascade. The cascade is sensitive to regulation at several levels through feedback inhibition (thin, dashed arrows) by the ultimate hormone (in this case, cortisol). The product therefore regulates its own production, as in feedback inhibition of biosynthetic pathways within a single cell.

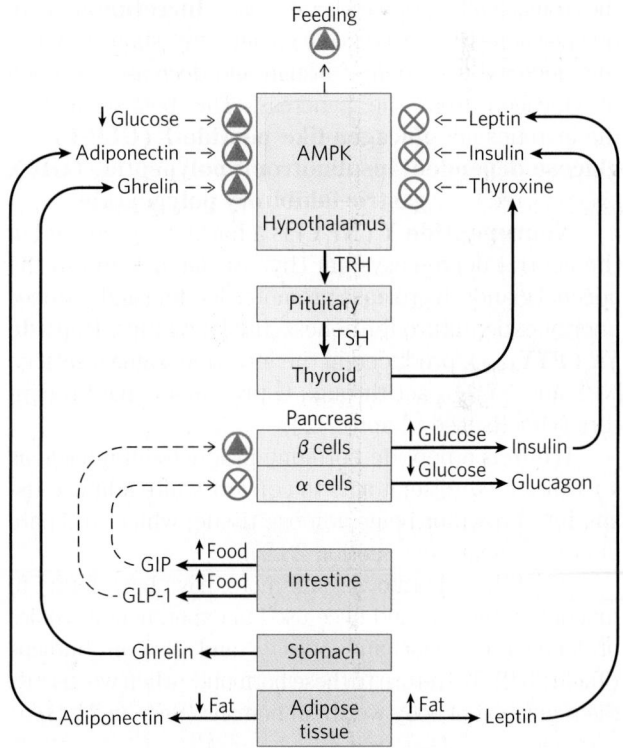

FIGURE 23-12 Regulation of feeding behavior by two-way information flow between tissues and the hypothalamus. When food intake and energy production are adequate, peptide hormones released by the stomach, intestine, and adipose tissue feed back on the hypothalamus to signal satiety and reduce feeding behavior. Other tissue-specific peptide hormones signal inadequate supplies of stored triacylglycerols or low blood glucose levels. All of these signals impinge, directly or indirectly, on AMP-activated protein kinase (AMPK) in the hypothalamus, which integrates these signals and influences feeding behavior and energy-yielding metabolism in the tissues. Nerves carry electrical signals from the brain to the other tissues to complete the information circuit and achieve homeostasis (not shown). TRH, thyrotropin-releasing hormone; GLP-1, glucagon-like peptide-1; GIP, gastric inhibitory polypeptide.

TABLE 23-2	Some Peptide Hormones That Act on Feeding Behavior and Fuel Selection in Mammals		
Hormone	**Production site(s)**	**Target tissue(s)**	**Action(s)**
Insulin	Pancreatic β cells	Muscle, adipose, liver	Stimulates glucose uptake and synthesis of glycogen and fat
Glucagon	Pancreatic α cells	Liver, adipose	Stimulates gluconeogenesis and glucose release to blood
Leptin	Adipose tissue	Hypothalamus	Reduces hunger
Adiponectin	Adipose tissue	Muscle, liver, others	Stimulates catabolism
Ghrelin	Stomach, intestine	Brain	Signals hunger
Incretins: GLP-1, GIP	Intestine	Pancreas	Stimulate insulin release
NPY	Hypothalamus, adrenals	Brain, autonomic nervous system	Stimulates feeding behavior
PYY_{3-36}	Intestine	Brain	Signals satiety
Irisin	Muscle (after exercise)	Adipose	Turns white adipose tissue to beige

gastrointestinal tract when the stomach is empty and acts in the hypothalamus to stimulate feeding behavior; when the stomach fills, ghrelin release ceases. **Incretins** are peptide hormones produced in the gut after ingestion of a meal; they increase secretion of insulin and decrease secretion of glucagon from the pancreas. The best-studied of the incretins are **glucagon-like peptide-1 (GLP-1)** and **glucose-dependent insulinotropic polypeptide (GIP)**, also referred to as **gastric inhibitory polypeptide**.

Neuropeptide Y (NPY) is a hormone produced in the central nervous system (hypothalamus) and in the adrenal glands. Its release promotes feeding and reduces energy expenditure for nonessential activities. **Peptide YY (PYY_{3-36})**, produced in the intestine, signals satiety. NPY and PYY_{3-36} act through G protein–coupled receptors (GPCRs; see Chapter 12).

Irisin is a peptide hormone produced in muscle as a result of exercise; it acts to convert white adipose tissue into brown or beige adipose tissue, which dissipate energy as heat (see Section 23.2).

All of these hormones, and more, have been found to function in the rats and mice used in experimental studies of feeding behavior and obesity, and also in humans (Table 23-2). We return to these hormones when we discuss the regulation of body weight in humans (Section 23.4).

SUMMARY 23.1 Hormones: Diverse Structures for Diverse Functions

■ Hormones are chemical messengers secreted by certain tissues into the blood or interstitial fluid, serving to regulate the activity of other cells or tissues.

■ Radioimmunoassay and ELISA are two highly sensitive techniques for detecting and quantifying hormones.

■ Peptide, catecholamine, and eicosanoid hormones bind to specific receptors in the plasma membrane of target cells, altering the level of an intracellular second messenger, without entering the cell.

■ Steroid, vitamin D, retinoid, and thyroid hormones enter target cells and alter gene expression by interacting with specific nuclear receptors.

■ Hormonal cascades, in which catalysts activate catalysts, amplify an initial stimulus by several orders of magnitude, often in a very short time (seconds).

■ Some hormones are synthesized as prohormones and activated by enzymatic cleavage. In some cases, such as insulin, a single hormone is produced by proteolytic cleavages; in others, such as POMC, several distinct hormones are produced by cleavage of a single prohormone.

■ Hormones are regulated by a top-down hierarchy of interactions between the brain and endocrine glands: nerve impulses stimulate the hypothalamus to send specific hormones to the pituitary gland, thus stimulating (or inhibiting) the release of tropic hormones. The anterior pituitary hormones in turn stimulate other endocrine glands (thyroid, adrenals, pancreas) to secrete their characteristic hormones, which in turn stimulate specific target tissues.

■ Peptide hormones also act in bottom-up signaling: adipose tissue, muscle, and the gastrointestinal tract release peptide hormones that act on other tissues or in the central nervous system.

23.2 Tissue-Specific Metabolism: The Division of Labor

Each tissue of the human body has a specialized function, reflected in its anatomy and metabolic activity (**Fig. 23-13**). Skeletal muscle allows directed motion; adipose tissue stores and distributes energy in the form of fats, which serve as fuel throughout the body and as thermal insulation; in the brain, cells pump ions across

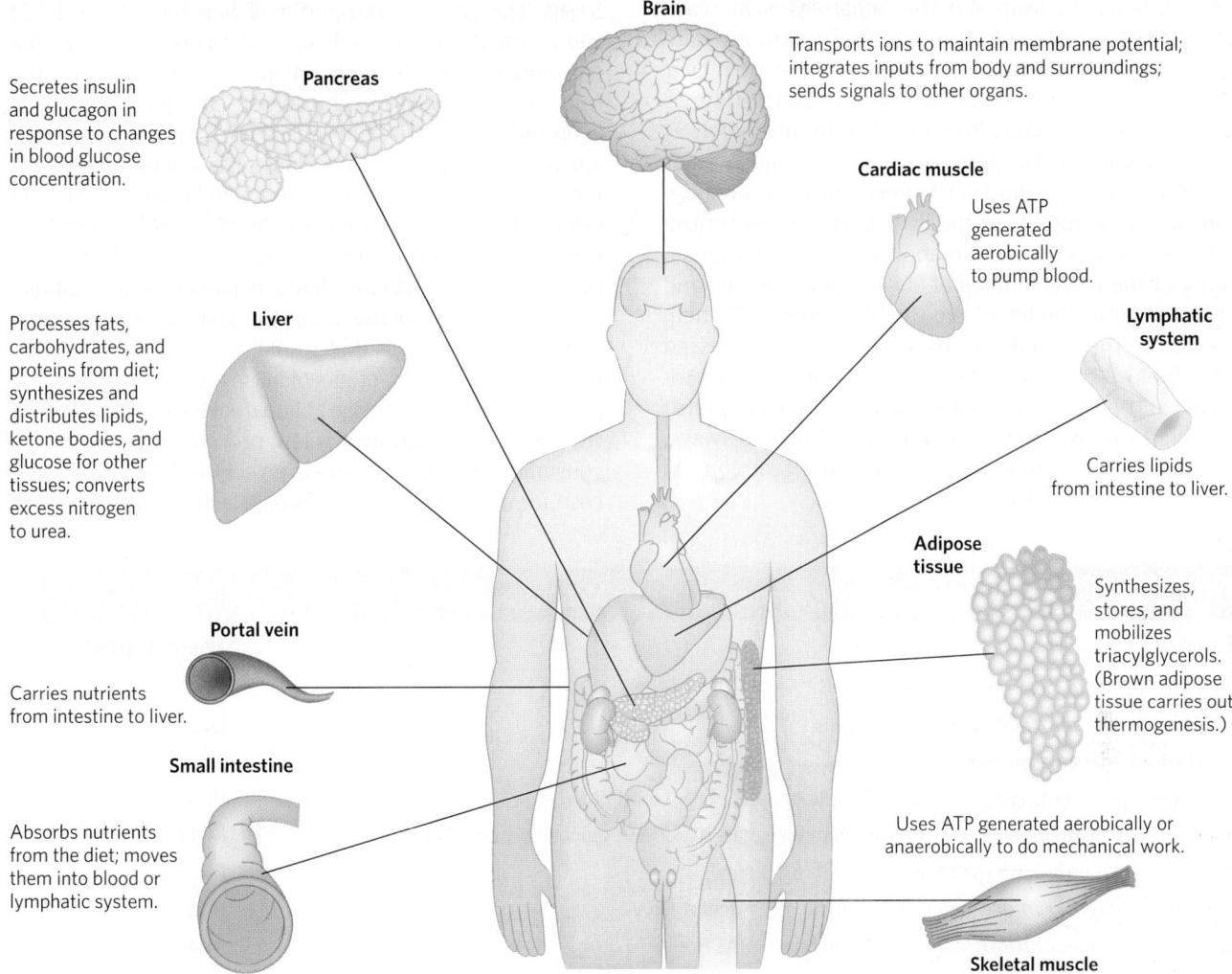

FIGURE 23-13 Specialized metabolic functions of mammalian tissues.

their plasma membranes to produce electrical signals. The liver plays a central processing and distribution role in metabolism and furnishes all other organs and tissues with an appropriate mix of nutrients via the bloodstream. The functional centrality of the liver is indicated by the common reference to all other tissues and organs as "extrahepatic." We therefore begin our discussion of the division of metabolic labor by considering the transformations of carbohydrates, amino acids, and fats in the mammalian liver. This is followed by brief descriptions of the primary metabolic functions of adipose tissue, muscle, brain, and the medium that interconnects all others: the blood.

The Liver Processes and Distributes Nutrients

During digestion in mammals, the three main classes of nutrients (carbohydrates, proteins, and fats) undergo enzymatic hydrolysis into their simple constituents. This breakdown is necessary because the epithelial cells lining the intestinal lumen absorb only relatively small molecules. Many of the fatty acids and monoacylglycerols released by digestion of fats in the intestine are reassembled within these epithelial cells into triacylglycerols (TAGs).

After being absorbed, most sugars and amino acids and some reconstituted TAGs pass from intestinal epithelial cells into blood capillaries and travel in the bloodstream to the liver; the remaining TAGs enter adipose tissue via the lymphatic system. The portal vein (Fig. 23-13) is a direct route from the digestive organs to the liver, and the liver therefore has first access to ingested nutrients. The liver has two main cell types. Kupffer cells are phagocytes, important in immune function. **Hepatocytes**, of primary interest here, transform dietary nutrients into the fuels and precursors required by other tissues and export them via the blood. The kinds and amounts of nutrients supplied to the liver are determined by diet, the time between meals, and several other factors. The demand of extrahepatic tissues for fuels and precursors varies from one organ to another, and with the level of activity and overall nutritional state of the individual.

To meet these changing circumstances, the liver has remarkable metabolic flexibility. For example, when the diet is rich in protein, hepatocytes supply themselves with high levels of enzymes for amino acid catabolism and gluconeogenesis. Within hours after a shift to a high-carbohydrate diet, the levels of these enzymes begin to drop and the hepatocytes increase their synthesis of enzymes essential to carbohydrate metabolism and fat synthesis. Liver enzymes turn over (that is, are synthesized and degraded) at 5 to 10 times the rate of enzyme turnover in other tissues, such as muscle. Extrahepatic tissues also can adjust their metabolism to prevailing conditions, but none of these tissue are as adaptable as the liver, and none so central to the organism's overall metabolism. What follows is a survey of the possible fates of sugars, amino acids, and lipids that enter the liver from the bloodstream. To help you recall the metabolic transformations discussed here, Table 23-3 shows the major pathways and processes, and gives figure numbers (in earlier chapters) for each. Here, we provide summaries of the pathways, referring to the numbered pathways and reactions in Figures 23-14 to 23-16.

Sugars The glucose transporter of hepatocytes (GLUT2) allows rapid, passive diffusion of glucose, so that the concentration of glucose in a hepatocyte is essentially the same as that in the blood. Glucose entering hepatocytes is phosphorylated by glucokinase (hexokinase IV) to yield glucose 6-phosphate. Glucokinase has a much higher K_m for glucose (10 mM) than do the hexokinase isozymes in other cells (pp. 590–591) and, unlike these other isozymes, it is not inhibited by its product, glucose 6-phosphate. The presence of glucokinase allows hepatocytes to continue phosphorylating glucose when the glucose concentration rises well above levels that would overwhelm other hexokinases. The high K_m of glucokinase also ensures that the phosphorylation of glucose in hepatocytes is minimal when the glucose concentration is low, preventing the liver from consuming glucose as fuel via glycolysis. This spares glucose for other tissues. Fructose, galactose, and mannose,

TABLE 23-3 Pathways of Carbohydrate, Amino Acid, and Fat Metabolism Illustrated in Earlier Chapters	
Pathway	**Figure reference(s)**
Citric acid cycle: acetyl-CoA $\longrightarrow$ $2CO_2$	16-7
Oxidative phosphorylation: ATP synthesis	19-19
Carbohydrate catabolism	
Glycogenolysis: glycogen $\longrightarrow$ glucose 1-phosphate $\longrightarrow$ blood glucose	15-27; 15-28
Hexose entry into glycolysis: fructose, mannose, galactose $\longrightarrow$ glucose 6-phosphate	14-11
Glycolysis: glucose $\longrightarrow$ pyruvate	14-2
Pyruvate dehydrogenase reaction: pyruvate $\longrightarrow$ acetyl-CoA	16-2
Lactic acid fermentation: glucose $\longrightarrow$ lactate + 2ATP	14-4
Pentose phosphate pathway: glucose 6-phosphate $\longrightarrow$ pentose phosphates + NADPH	4-22
Carbohydrate anabolism	
Gluconeogenesis: citric acid cycle intermediates $\longrightarrow$ glucose	14-17
Glucose-alanine cycle: glucose $\longrightarrow$ pyruvate $\longrightarrow$ alanine $\longrightarrow$ glucose	18-9
Glycogen synthesis: glucose 6-phosphate $\longrightarrow$ glucose 1-phosphate $\longrightarrow$ glycogen	15-32
Amino acid and nucleotide metabolism	
Amino acid degradation: amino acids $\longrightarrow$ acetyl-CoA, citric acid cycle intermediates	18-15
Amino acid synthesis	22-11
Urea cycle: NH_3 $\longrightarrow$ urea	18-10
Glucose-alanine cycle: alanine $\longrightarrow$ glucose	18-9
Nucleotide synthesis: amino acids $\longrightarrow$ purines, pyrimidines	22-35; 22-38
Hormone and neurotransmitter synthesis	22-31
Fat catabolism	
β Oxidation of fatty acids: fatty acids $\longrightarrow$ acetyl-CoA	17-8
Oxidation of ketone bodies: β-hydroxybutyrate $\longrightarrow$ acetyl-CoA $\longrightarrow$ CO_2 via citric acid cycle	17-19
Fat anabolism	
Fatty acid synthesis: acetyl-CoA $\longrightarrow$ fatty acids	21-6
Triacylglycerol synthesis: acetyl-CoA $\longrightarrow$ fatty acids $\longrightarrow$ triacylglycerol	21-17 to 21-19
Ketone body formation: acetyl-CoA $\longrightarrow$ acetoacetate, β-hydroxybutyrate	17-18
Cholesterol and cholesteryl ester synthesis: acetyl-CoA $\longrightarrow$ cholesterol $\longrightarrow$ cholesteryl esters	21-33 to 21-38
Phospholipid synthesis: fatty acids $\longrightarrow$ phospholipids	21-17; 21-23 to 21-28

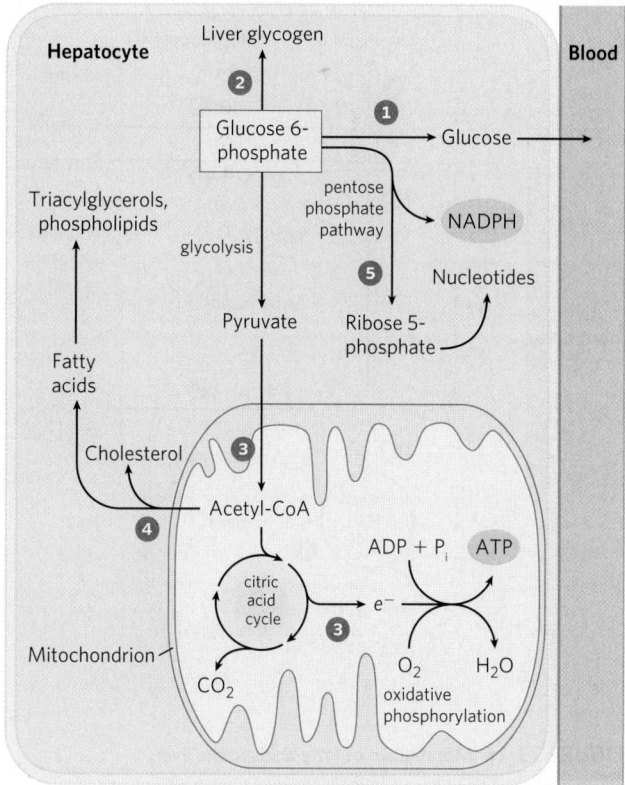

FIGURE 23-14 Metabolic pathways for glucose 6-phosphate in the liver. Here and in Figures 23-15 and 23-16, anabolic pathways are generally shown leading upward, catabolic pathways leading downward, and distribution to other organs horizontally. The numbered processes in each figure are described in the text.

serve as the precursor of fatty acids, which are incorporated into TAGs and phospholipids, and of cholesterol. Much of the lipid synthesized in the liver is transported to other tissues by blood lipoproteins. ❺ Alternatively, glucose 6-phosphate can enter the pentose phosphate pathway, yielding both reducing power (NADPH), needed for the biosynthesis of fatty acids and cholesterol, and D-ribose 5-phosphate, a precursor for nucleotide biosynthesis. NADPH is also an essential cofactor in the detoxification and elimination of many drugs and other xenobiotics metabolized in the liver.

Amino Acids Amino acids that enter the liver follow several important metabolic routes **(Fig. 23-15)**. ❶ They are precursors for protein synthesis, a process discussed in Chapter 27. The liver constantly renews its own proteins, which have a relatively high turnover rate (average half-life of hours to days), and is also the site of biosynthesis of most plasma proteins. ❷ Alternatively, amino acids pass in the bloodstream to other organs to be used in the synthesis

all absorbed from the small intestine, are also converted to glucose 6-phosphate by enzymatic pathways examined in Chapter 14. Glucose 6-phosphate is at the crossroads of carbohydrate metabolism in the liver. It may take any of several major metabolic routes **(Fig. 23-14)**, depending on the current metabolic needs of the organism. By the action of various allosterically regulated enzymes, and through hormonal regulation of enzyme synthesis and activity, the liver directs the flow of glucose into one or more of these pathways.

❶ Glucose 6-phosphate is dephosphorylated by glucose 6-phosphatase to yield free glucose (see Fig. 15-30), which is exported to replenish blood glucose. Export is the predominant pathway when glucose 6-phosphate is in limited supply, because the blood glucose concentration must be kept sufficiently high (4 to 5 mM) to provide adequate energy for the brain and other tissues. ❷ Glucose 6-phosphate not immediately needed to form blood glucose is converted to liver glycogen, or has one of several other fates. Following glycolysis and the pyruvate dehydrogenase reaction, ❸ the acetyl-CoA so formed can be oxidized for ATP production by the citric acid cycle, with ensuing electron transfer and oxidative phosphorylation yielding ATP. (Normally, however, fatty acids are the preferred fuel for ATP production in hepatocytes.) ❹ Acetyl-CoA can also

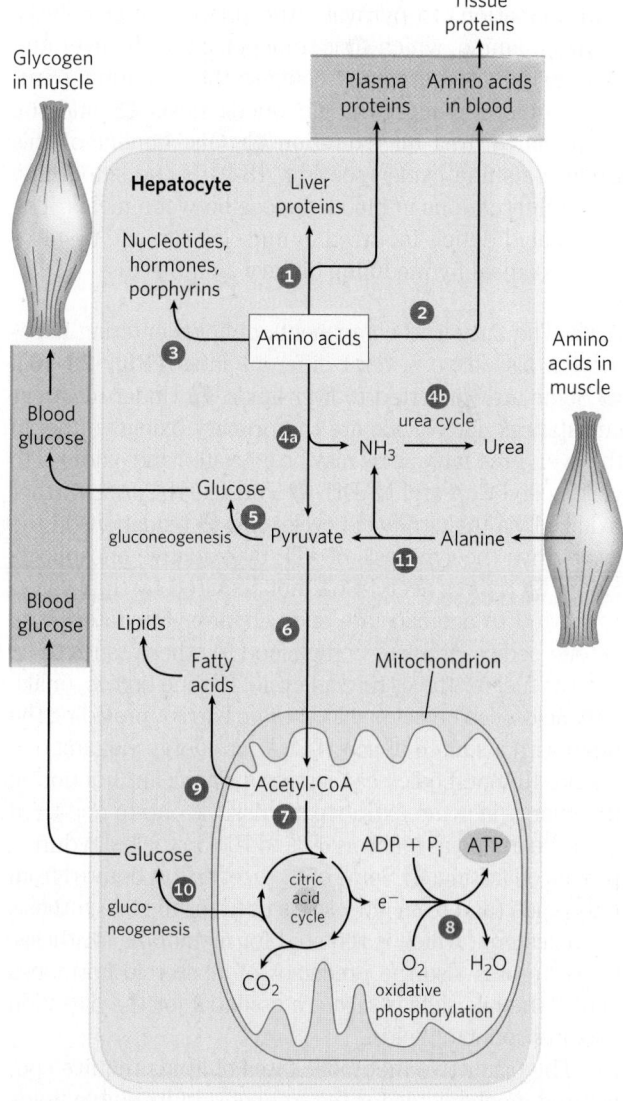

FIGURE 23-15 Metabolism of amino acids in the liver.

of tissue proteins. ❸ Other amino acids are precursors in the biosynthesis of nucleotides, hormones, and other nitrogenous compounds in the liver and other tissues.

❹ₐ Amino acids not needed as biosynthetic precursors are transaminated or deaminated and degraded to yield pyruvate and citric acid cycle intermediates, with various fates; ❹ᵦ the ammonia released is converted to the excretory product urea. ❺ Pyruvate can be converted to glucose and glycogen via gluconeogenesis, or ❻ can be converted to acetyl-CoA, which has several possible fates: ❼ oxidation via the citric acid cycle and ❽ oxidative phosphorylation to produce ATP, or ❾ conversion to lipids for storage. ❿ Citric acid cycle intermediates can be siphoned off into glucose synthesis by gluconeogenesis.

The liver also metabolizes amino acids that arrive intermittently from other tissues. The blood is adequately supplied with glucose just after the digestion and absorption of dietary carbohydrate or, between meals, by the conversion of liver glycogen to blood glucose. During the interval between meals, especially if prolonged, some muscle protein is degraded to amino acids. These amino acids donate their amino groups (by transamination) to pyruvate, the product of glycolysis, to yield alanine, which ⓫ is transported to the liver and deaminated. Hepatocytes convert the resulting pyruvate to blood glucose via gluconeogenesis ❺, and the ammonia to urea for excretion ❹ᵦ. One benefit of this glucose-alanine cycle (see Fig. 18-9) is the smoothing out of fluctuations in blood glucose between meals. The amino acid deficit incurred in muscles is made up after the next meal by incoming dietary amino acids.

Lipids The fatty acid components of lipids entering hepatocytes also have several different fates **(Fig. 23-16)**. ❶ Some are converted to liver lipids. ❷ Under most circumstances, fatty acids are the primary oxidative fuel in the liver. Free fatty acids may be activated and oxidized to yield acetyl-CoA and NADH. ❸ The acetyl-CoA is further oxidized via the citric acid cycle, and ❹ oxidations in the cycle drive the synthesis of ATP by oxidative phosphorylation. ❺ Excess acetyl-CoA, not required by the liver, is converted to acetoacetate and β-hydroxybutyrate; these ketone bodies circulate in the blood to other tissues to be used as fuel for the citric acid cycle. Ketone bodies, unlike fatty acids, can cross the blood-brain barrier, providing the brain with a source of acetyl-CoA for energy-yielding oxidation. Ketone bodies can supply a significant fraction of the energy in some extrahepatic tissues—up to one-third in the heart and as much as 60% to 70% in the brain during prolonged fasting. ❻ Some of the acetyl-CoA derived from fatty acids (and from glucose) is used for the biosynthesis of cholesterol, which is required for membrane synthesis. Cholesterol is also the precursor of all steroid hormones and of the bile salts, which are essential for the digestion and absorption of lipids.

The other two metabolic fates of lipids require specialized mechanisms for the transport of insoluble lipids in blood. ❼ Fatty acids are converted to the phospholipids and TAGs of plasma lipoproteins, which carry

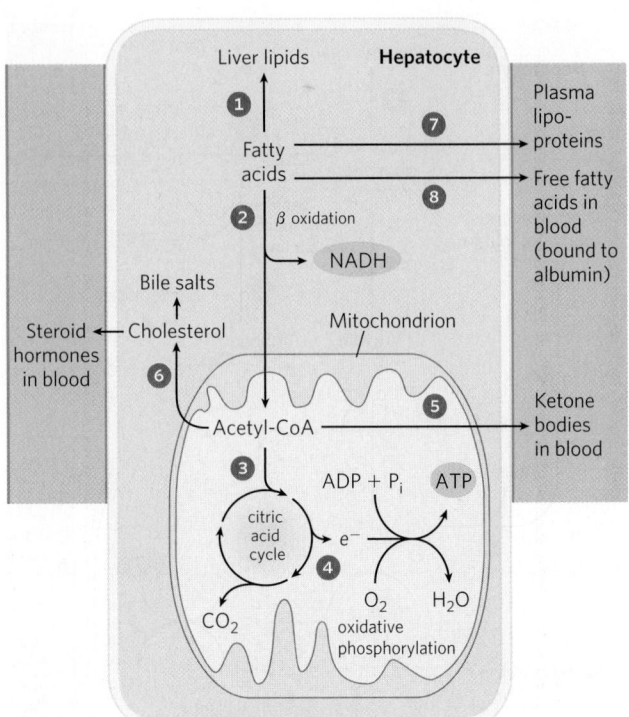

FIGURE 23-16 Metabolism of fatty acids in the liver.

lipids to adipose tissue for storage. ❽ Some free fatty acids are bound to serum albumin and carried to the heart and skeletal muscles, which take up and oxidize free fatty acids as a major fuel. Serum albumin is the most abundant plasma protein; one molecule can carry up to 10 molecules of free fatty acid.

The liver thus serves as the body's distribution center, exporting nutrients in the correct proportions to other organs, smoothing out fluctuations in metabolism caused by intermittent food intake, and processing excess amino groups into urea and other products to be disposed of by the kidneys. Certain nutrients are stored in the liver, including iron ions and vitamin A. The liver also detoxifies foreign organic compounds, such as drugs, food additives, preservatives, and other possibly harmful agents with no food value. Detoxification often includes the cytochrome P-450–dependent hydroxylation of relatively insoluble organic compounds, making them sufficiently soluble for further breakdown and excretion (see Box 21-1).

Adipose Tissues Store and Supply Fatty Acids

There are two types of adipose tissue, white and brown **(Fig. 23-17)**, with different roles, and we focus first on the more abundant of the two. **White adipose tissue (WAT)** is amorphous and widely distributed in the body: under the skin, around deep blood vessels, and in the abdominal cavity. The **adipocytes** of WAT are large (diameter 30 to 70 μm), spherical cells, completely filled with a single large lipid droplet that constitutes about 65% of the cell mass and squeezes the mitochondria and nucleus into a thin layer against the plasma membrane (Fig. 23-17a, c). The lipid droplet contains TAGs and

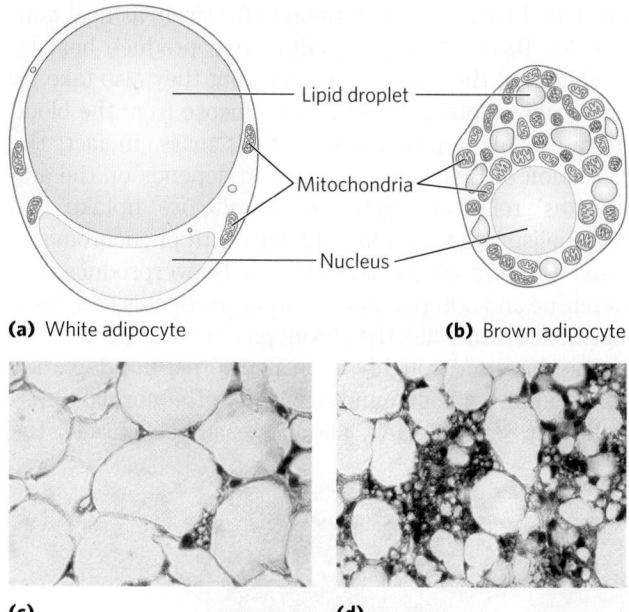

(a) White adipocyte   (b) Brown adipocyte

(c)   (d)

FIGURE 23-17 White and brown adipose tissue. Schematic views of typical mouse adipocytes from **(a)** white adipose tissue (WAT) and **(b)** brown adipose tissue (BAT). White adipocytes are larger and contain a single huge lipid droplet, which squeezes the mitochondria and nucleus against the plasma membrane. In brown adipocytes, mitochondria are much more prominent, the nucleus is near the center of the cell, and multiple small fat droplets are present. Below are light micrographs of **(c)** adipocytes in WAT, stained to show nuclei, and **(d)** a region of mixed white and brown adipocytes, stained with an antibody specific to UCP1. [Source: (c, d) Courtesy of Dr. Patrick Seale.]

sterol esters and is coated with a monolayer of phospholipids, oriented with their head groups facing the cytosol. Specific proteins are associated with the surface of the droplets, including perilipin and the enzymes for synthesis and breakdown of TAGs (see Fig. 17-3). WAT typically makes up about 15% of the mass of a healthy young adult human. Adipocytes are metabolically active, responding quickly to hormonal stimuli in a metabolic interplay with the liver, skeletal muscles, and heart.

Like other cell types, adipocytes have an active glycolytic metabolism, oxidize pyruvate and fatty acids via the citric acid cycle, and carry out oxidative phosphorylation. During periods of high carbohydrate intake, adipose tissue can convert glucose (via pyruvate and acetyl-CoA) to fatty acids, convert the fatty acids to TAGs, and store the TAGs as large lipid droplets—although in humans, much of the fatty acid synthesis occurs in hepatocytes. Adipocytes store TAGs arriving from the liver (carried in the blood as VLDL) and from the intestinal tract (carried in chylomicrons), particularly after meals rich in fat.

When the demand for fuel rises (between meals, for example), lipases in adipocytes hydrolyze stored TAGs to release free fatty acids, which can travel in the bloodstream to skeletal muscle and the heart. The release of fatty acids from adipocytes is greatly accelerated by epinephrine, which stimulates the cAMP-dependent phosphorylation of perilipin and thus gives lipases specific for tri-, di-, and monoacylglycerols access to TAGs in lipid droplets (see Fig. 17-3). Hormone-sensitive lipase is also

stimulated by phosphorylation, but this is not the main cause of increased lipolysis. Insulin counterbalances this effect of epinephrine, decreasing the activity of the lipase.

The breakdown and synthesis of TAGs in adipose tissue constitute a substrate cycle; up to 70% of the fatty acids released by the three lipases are reesterified in adipocytes, re-forming TAGs. Recall from Chapter 15 that such substrate cycles allow fine regulation of the rate and direction of flow of intermediates through a bidirectional pathway. In adipose tissue, glycerol liberated by adipocyte lipases cannot be reused in the synthesis of TAGs, because adipocytes lack glycerol kinase. Instead, the glycerol phosphate required for TAG synthesis is made from pyruvate by glyceroneogenesis, requiring the action of the cytosolic PEP carboxykinase (see Fig. 21-22).

In addition to its central function as a fuel depot, adipose tissue plays an important role as an endocrine organ, producing and releasing hormones that signal the state of energy reserves and coordinate metabolism of fats and carbohydrates throughout the body. We return to this function in Section 23.4 when we discuss the hormonal regulation of body mass.

Brown and Beige Adipose Tissues Are Thermogenic

In small vertebrates and hibernating animals, a significant proportion of the adipose tissue is **brown adipose tissue (BAT)**, distinguished from WAT by its smaller (diameter 20 to 40 μm), differently shaped (polygonal, not round) adipocytes (Fig. 23-17b, d). Like white adipocytes, brown adipocytes store TAGs, but in several smaller lipid droplets per cell rather than as a single central droplet. BAT cells have more mitochondria and a richer supply of capillaries and innervation than WAT cells, and it is the cytochromes of mitochondria and the hemoglobin in capillaries that give BAT its characteristic brown color. A unique feature of brown adipocytes is their production of **uncoupling protein 1 (UCP1)**, also called thermogenin (see Fig. 19-36). This protein is responsible for one of the principal functions of BAT: **thermogenesis**.

In brown adipocytes, fatty acids stored in lipid droplets are released, enter mitochondria, and undergo complete conversion to CO_2 by β oxidation and the citric acid cycle. The reduced $FADH_2$ and NADH so generated pass their electrons through the respiratory chain to molecular oxygen. In WAT, protons pumped out of the mitochondria during electron transfer reenter the matrix through ATP synthase, with the energy of electron transfer conserved in ATP synthesis. In BAT, UCP1 provides an alternative route for the reentry of protons that bypasses ATP synthase. The energy of the proton gradient is thus dissipated as heat, which can maintain the body (especially the nervous system and viscera) at its optimal temperature when the ambient temperature is relatively low.

In the human fetus, differentiation of fibroblast preadipocytes into BAT begins at the twentieth week of gestation, and at the time of birth, BAT represents 1% to 5% of total body mass. The brown fat deposits are located where the heat generated by thermogenesis can ensure that vital tissues—blood vessels to the head, major abdominal blood vessels, and the viscera, including the pancreas, adrenal glands, and kidneys—are not chilled as the newborn enters a world of lower ambient temperature **(Fig. 23-18)**.

At birth, WAT development begins and BAT begins to disappear. Young adult humans have much-diminished deposits of BAT, ranging from 3% of all adipose tissue in males to 7% in females, making up less than 0.1% of body mass. However, adults have, distributed among their WAT cells, significant numbers of adipocytes that can be converted by cold exposure or by β-adrenergic stimulation into cells very similar to brown adipocytes. These **beige adipocytes** have multiple lipid droplets, are richer in mitochondria than white adipocytes, and

produce UCP1, so they function effectively as heat generators. Brown and beige adipocytes produce heat by oxidation of their own fatty acids, but they also take up and oxidize both fatty acids and glucose from the blood at rates out of proportion to their mass. In fact, the detection of BAT by PET scanning depends on the adipocytes' relatively high rate of *glucose* uptake and metabolism (Fig. 23-18b). Humans with pheochromocytoma (tumors of the adrenal gland) overproduce epinephrine and norepinephrine, and one effect is that these hormones stimulate the development of beige adipose tissue, localized roughly as in newborns. In adaptation to warm or cold surroundings, and in the normal differentiation of WAT, BAT, and beige adipose tissue, the

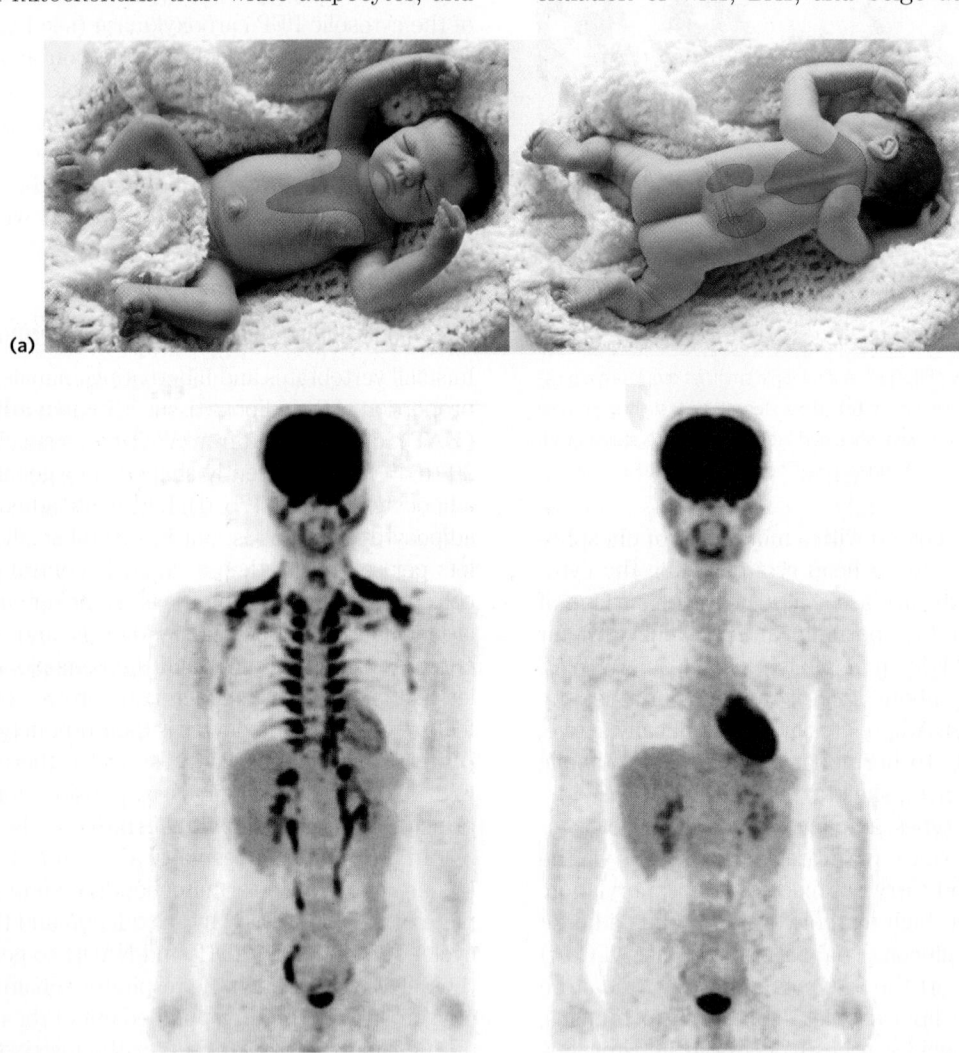

(a)

(b)   After cold exposure   Room-temperature control

FIGURE 23-18 Brown adipose tissue in infants and adults. (a) At birth, human infants have brown fat distributed as shown here, to protect the spine, major blood vessels, and internal organs. **(b)** Positron emission tomography (PET) scanning can show metabolic activity in a living person, in real time. PET scans allow visualization of isotopically labeled glucose in precisely localized regions of the body. A positron-emitting glucose analog, 2-[^{18}F]-fluoro-2-deoxy-D-glucose (FDG), is injected into the bloodstream; a short time later, a PET scan shows how much of the glucose has been taken up by each part of the body—a measure of metabolic activity. On the left is a PET scan of a healthy 25-year-old man who fasted for 12 hours, then stayed for 1 hour in a cold (19 °C) room, with his legs on ice to thoroughly chill him.

At the end of the hour, he was injected with [^{18}F]-FDG, then remained under cold conditions for another hour. Whole-body PET scans were then done at 24 °C. For the control scan, the same man underwent the same PET scan protocol two weeks later, but this time following 2 hours at 27 °C instead of chilling (right). Intense labeling of the brain and heart shows high rates of glucose uptake; labeling of the kidneys and bladder indicates clearance of FDG. In the scan after chilling (left), [^{18}F]-FDG labels BAT in the region above the collarbone and along the vertebrae. [Sources: (a) Adam Steinberg. (b) Reprinted with permission of American Diabetes Association, from M. Saito et al., *Diabetes* 58:1526, 2009, Fig. 1; permission conveyed through Copyright Clearance Center, Inc.]

nuclear transcription factor PPARγ (described later in the chapter) plays a central role. And as we noted above, the peptide hormone irisin, produced in muscle by exercise, triggers the development of beige adipose tissue that continues to burn fuel long after the exercise ends.

Muscles Use ATP for Mechanical Work

Metabolism in skeletal muscle cells—**myocytes**—is specialized to generate ATP as the immediate source of energy for contraction. Moreover, skeletal muscle is adapted to do its mechanical work intermittently, on demand. Sometimes skeletal muscles must work at their maximum capacity for a short time, as in a 100 m sprint; at other times more prolonged work is required, as in running a marathon or in prolonged physical labor.

There are two general classes of muscle tissue, which differ in physiological role and fuel utilization. **Slow-twitch muscle**, also called red muscle, provides relatively low tension but is highly resistant to fatigue. It produces ATP by the relatively slow but steady process of oxidative phosphorylation. Red muscle is very rich in mitochondria and is served by dense networks of blood vessels, which bring the oxygen essential to ATP production. **Fast-twitch muscle**, or white muscle, has fewer mitochondria than red muscle and is less well supplied with blood vessels, but it can develop greater tension and do so faster. White muscle is quicker to fatigue because, when active, it uses ATP faster than it can replace it. There is a genetic component to the proportion of red and white muscle in any individual, but with training, the endurance of fast-twitch muscle can be improved.

Skeletal muscle can use free fatty acids, ketone bodies, or glucose as fuel, depending on the degree of muscular activity **(Fig. 23-19)**. In resting muscle, the primary fuels are free fatty acids from adipose tissue and ketone bodies from the liver. These are oxidized and degraded to yield acetyl-CoA, which enters the citric acid cycle, ultimately yielding the energy for ATP synthesis by oxidative phosphorylation. Moderately active muscle uses blood glucose in addition to fatty acids and ketone bodies. The glucose is phosphorylated, then degraded by glycolysis to pyruvate, which is converted to acetyl-CoA and oxidized via the citric acid cycle and oxidative phosphorylation.

In maximally active fast-twitch muscles, the demand for ATP is so great that the blood flow cannot provide O_2 and fuels fast enough to supply sufficient ATP by aerobic respiration alone. Under these conditions, stored muscle glycogen is broken down to lactate by fermentation (p. 537). Each glucose unit degraded yields three ATP, because phosphorolysis of glycogen produces glucose 6-phosphate (via glucose 1-phosphate), sparing the ATP normally consumed in the hexokinase reaction. Lactic acid fermentation thus responds more quickly than oxidative phosphorylation to an increased need for ATP, supplementing basal ATP production by aerobic oxidation of other fuels via the citric acid cycle and respiratory chain. The use of blood glucose and muscle glycogen as fuels for muscular activity is greatly enhanced by the secretion of epinephrine, which stimulates both the release of glucose from liver glycogen and the breakdown of glycogen in muscle tissue. (Epinephrine mediates the so-called fight-or-flight response, discussed more fully below.)

The relatively small amount of glycogen (about 1% of the total weight of skeletal muscle) limits the glycolytic energy available during all-out exertion. Moreover, the accumulation of lactate and consequent decrease in pH in maximally active muscles reduces their efficiency. Skeletal muscle, however, contains another source of ATP, phosphocreatine (10 to 30 mM), which can rapidly regenerate ATP from ADP by the creatine kinase reaction:

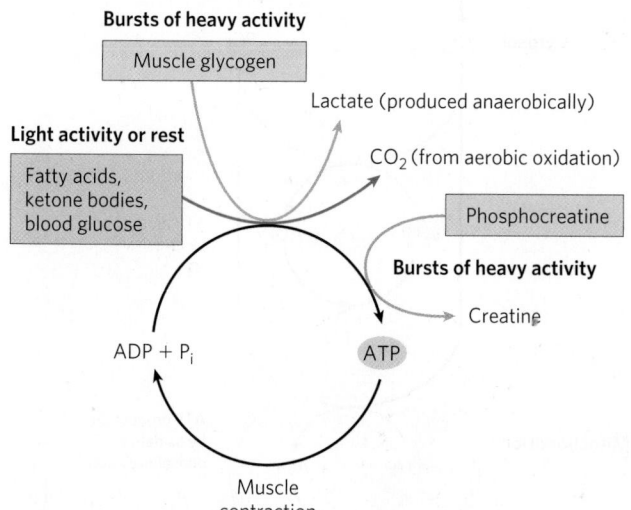

FIGURE 23-19 Energy sources for muscle contraction. Different fuels are used for ATP synthesis during bursts of heavy activity and during light activity or rest. Phosphocreatine can rapidly supply ATP.

During periods of active contraction and glycolysis, this reaction proceeds predominantly in the direction of ATP synthesis; during recovery from exertion, the same enzyme resynthesizes phosphocreatine from creatine and ATP. Because of the relatively high levels of ATP and phosphocreatine in muscle, these compounds can be detected in intact muscle, in real time, by NMR spectroscopy **(Fig. 23-20)**. Creatine serves to shuttle ATP equivalents from the mitochondrion to sites of ATP consumption and can be the limiting factor in the development of new muscle tissue (Box 23-2).

After a period of intense muscular activity, the individual continues breathing heavily for some time, using

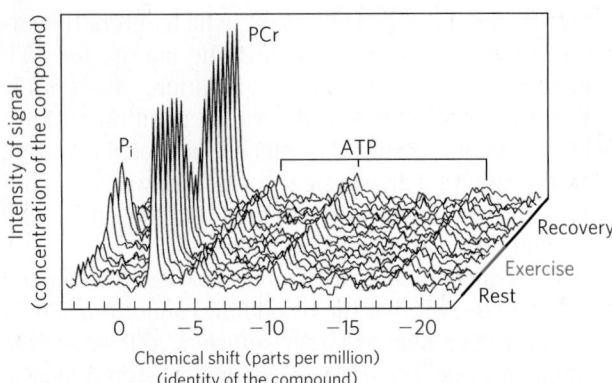

FIGURE 23-20 Phosphocreatine buffers ATP concentration during exercise. A "stack plot" of magnetic resonance spectra (of ^{31}P) shows inorganic phosphate (P$_i$), phosphocreatine (PCr), and ATP (each of its three phosphates giving a signal). The series of plots represents the passage of time, from a period of rest to one of exercise, and then of recovery. Notice that the ATP signal barely changes during exercise, kept high by continued respiration and by the reservoir of phosphocreatine, which diminishes during exercise. During recovery, when ATP production by catabolism is greater than ATP use by the (now resting) muscle, the phosphocreatine reservoir is refilled. [Source: Data from M. L. Blei, K. E. Conley, and M. J. Kushmerick, *J. Physiol.* 465:203, 1993, Fig. 4.]

BOX 23-2   Creatine and Creatine Kinase: Invaluable Diagnostic Aids and the Muscle Builder's Friends

Animal tissues that have a high and fluctuating need for ATP, primarily skeletal muscle, cardiac muscle, and brain, contain several isozymes of creatine kinase. A cytosolic isozyme (cCK) is present in regions of high ATP use (myofibrils and sarcoplasmic reticulum, for example). By converting ADP produced during periods of high ATP use back to ATP, cCK prevents the accumulation of ADP to concentrations that could inhibit ATP-using enzymes by mass action. Another isozyme of creatine kinase is located in regions where the inner and outer membranes of mitochondria come into contact. This mitochondrial isozyme (mCK) probably serves to shuttle ATP equivalents produced in mitochondria to cytosolic sites of ATP use (Fig. 1). The species that diffuses from the mitochondrion to ATP-consuming activities in the cytosol is therefore creatine phosphate, not ATP. The mCK isozyme colocalizes with the adenine nucleotide transporter (in the inner mitochondrial membrane) and porin (in the outer mitochondrial membrane), suggesting that these three components may function together to transport ATP formed in mitochondria into the cytosol.

In knockout mice lacking the mitochondrial isozyme, myocytes compensate by producing more mitochondria, closely associated with myofibrils and sarcoplasmic reticulum, allowing quick diffusion of mitochondrial ATP to the sites of ATP use. Nevertheless, these mice have a reduced capacity for running, indicating a defect in some aspect of energy-supplying metabolism.

FIGURE 1 Mitochondrial creatine kinase (mCK) transfers a phosphoryl group from ATP to creatine (Cr) to form creatine phosphate (PCr), which diffuses to sites of ATP use; at these sites, cytosolic creatine kinase (cCK) passes the phosphoryl group into ATP. Cytosolic CK can also use ATP produced by glycolysis to synthesize PCr. During periods of little ATP demand, the pools of ATP and PCr are replenished in preparation for the next period of intense demand for ATP. In resting muscle, the concentration of PCr is three to five times that of ATP, buffering the cell against rapid depletion of ATP during short bursts of ATP demand. [Source: Information from U. Schlattner et al., *Biochim. Biophys. Acta* 1762:164, 2006, Fig. 1.]

Creatine and phosphocreatine spontaneously break down to form creatinine (Fig. 2). To maintain high creatine levels, these losses have to be compensated for, either by dietary creatine, obtained primarily from meat (muscle) and dairy products, or by de novo synthesis from glycine, arginine, and methionine (see Fig. 22-28), which occurs primarily in liver and kidney. De novo synthesis of creatine is a major consumer of these

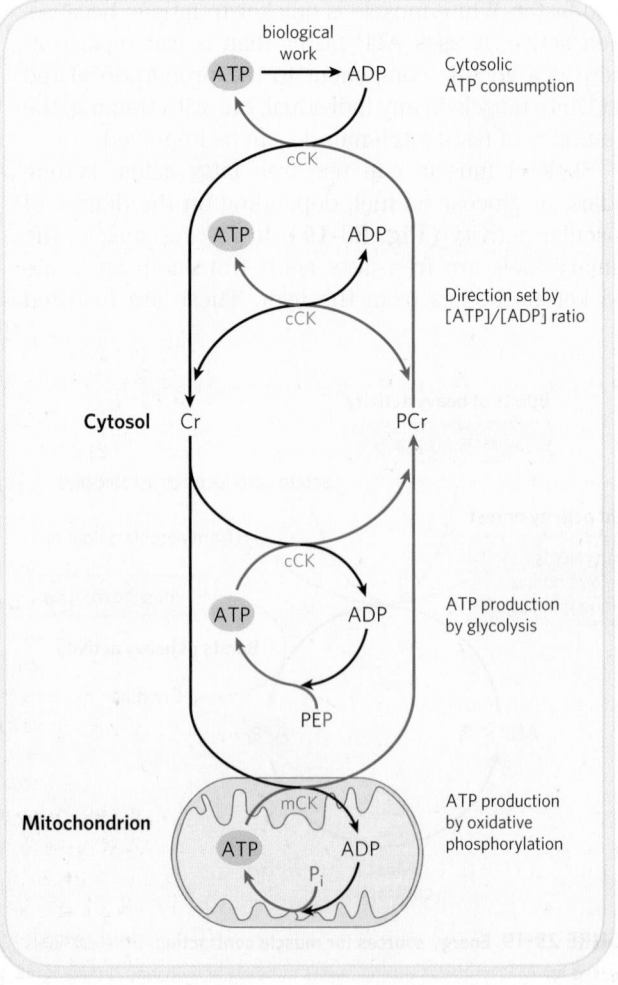

much of the extra O_2 for oxidative phosphorylation in the liver. The ATP produced is used for gluconeogenesis (in the liver) from lactate that has been carried in the blood from the muscles. The glucose thus formed returns to the muscles to replenish their glycogen, completing the Cori cycle (**Fig. 23-21**; see also Box 15-4).

Actively contracting skeletal muscle generates heat as a byproduct of imperfect coupling of the chemical energy of ATP with the mechanical work of contraction. This heat production can be put to good use when ambient temperature is low: skeletal muscle carries out **shivering thermogenesis**, rapidly repeated muscle contraction

that produces heat but little motion, helping to maintain the body at its preferred temperature of 37 °C.

Heart muscle differs from skeletal muscle in that it is continuously active in a regular rhythm of contraction and relaxation, and has a completely aerobic metabolism at all times. Mitochondria are much more abundant in heart muscle than in skeletal muscle, making up almost half the volume of the cells (**Fig. 23-22**). The heart uses mainly free fatty acids as a source of energy, but also some glucose and ketone bodies taken up from the blood; these fuels are oxidized aerobically to generate ATP. Like skeletal muscle, heart muscle does not

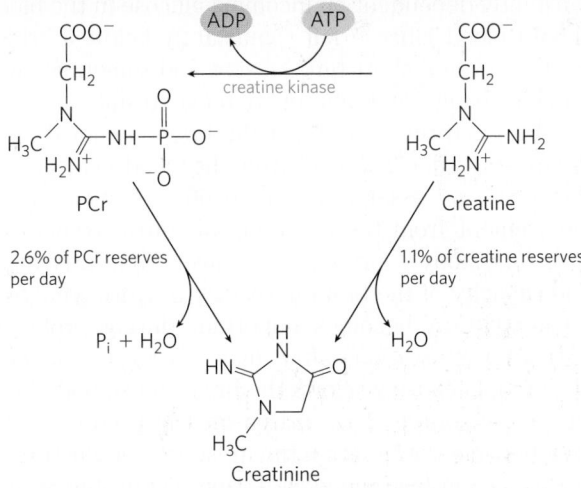

FIGURE 2 Spontaneous (nonenzymatic) formation of creatinine from phosphocreatine or creatine consumes a few percent of the body's total creatine per day, which must be replaced by biosynthesis or from the diet.

FIGURE 3 Many body builders take supplemental creatine to supply creatine phosphate in new muscle tissue. [Source: Steve Williams Photo/Getty Images.]

amino acids, particularly in vegans, for whom this is the only source of creatine; plants do not contain creatine. Muscle tissue has a specific system to take up creatine (exported by liver or kidney) from the blood, against a substantial concentration gradient. Efficient uptake of dietary creatine requires continuous exercise; without exercise, creatine supplementation is of little value.

Heart muscle contains a unique isozyme of creatine kinase (MB), which is not normally found in the blood but appears there when released from heart muscle damaged by a heart attack. The blood level of MB begins to rise within 2 hours of the heart attack, typically peaks 12 to 36 hours after the heart attack, and returns to normal levels in 3 to 5 days. Measurement of the MB isozyme in blood therefore confirms a diagnosis of heart attack and indicates approximately when it occurred.

Children with inborn errors in the enzymes of creatine synthesis or uptake suffer severe intellectual disability and seizures. They have much-reduced levels of

brain creatine as measured by NMR (see Fig. 23-20). Creatine supplementation raises their brain creatine and creatine phosphate concentrations and brings about partial improvement of the symptoms.

In the healthy kidney, creatinine from creatine breakdown is efficiently cleared from the blood into the urine. When renal function is defective, creatinine levels in the blood rise above the normal range of 0.8 to 1.4 mg/dL. Elevated blood creatinine is associated with renal failure in diabetes and other conditions in which renal function is temporarily or permanently compromised. Renal clearance of creatinine varies slightly with age, race, and gender, so correcting the calculation for those factors yields a more sensitive measure of the extent of renal function, the **glomerular filtration rate (GFR)**. ∎

Body builders who are adding muscle mass have a greater need for creatine and commonly take creatine supplements of up to 20 g per day for a few days, followed by lower maintenance doses. The combination of exercise and creatine supplementation increases muscle mass (Fig. 3) and improves performance in high-intensity, short-duration work.

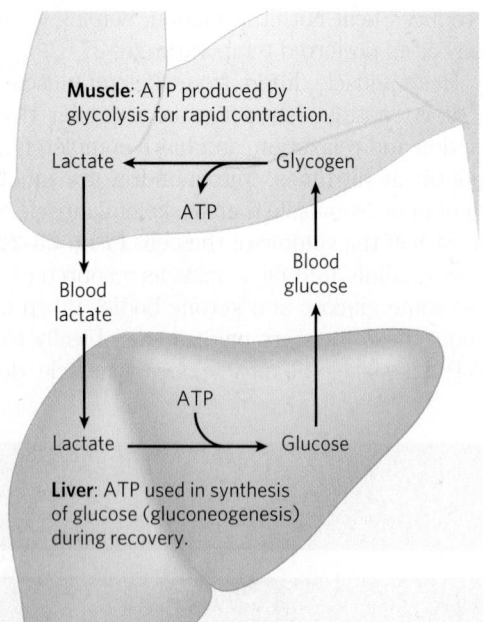

Muscle: ATP produced by glycolysis for rapid contraction.

Lactate ← Glycogen

ATP

Blood lactate

Blood glucose

Lactate → Glucose

ATP

Liver: ATP used in synthesis of glucose (gluconeogenesis) during recovery.

FIGURE 23-21 Metabolic cooperation between skeletal muscle and the liver: the Cori cycle. Extremely active muscles use glycogen as their energy source, generating lactate by glycolysis. During recovery, some of the lactate is transported to the liver and converted to glucose by gluconeogenesis. The glucose is released to the blood and returned to the muscles to replenish their glycogen stores. The overall pathway, glucose → lactate → glucose, constitutes the Cori cycle.

store lipids or glycogen in large amounts. It does have small amounts of reserve energy in the form of phosphocreatine, enough for a few seconds of contraction. Because the heart is normally aerobic and obtains its energy from oxidative phosphorylation, the failure of O_2 to reach part of the heart muscle when the blood vessels are blocked by lipid deposits (atherosclerosis) or blood clots (coronary thrombosis) can cause that region of the heart muscle to die. This is what happens in myocardial infarction, more commonly known as a heart attack. ■

The Brain Uses Energy for Transmission of Electrical Impulses

The metabolism of the brain is remarkable in several respects. The neurons of the adult mammalian brain normally use only glucose as fuel **(Fig. 23-23)**. (Astrocytes, the other major cell type in the brain, can oxidize fatty acids.) The brain, which constitutes about 2% of total body mass, has a very active respiratory metabolism (Fig. 23-18); more than 90% of the ATP produced in the neurons comes from oxidative phosphorylation. The brain uses O_2 at a fairly constant rate, accounting for almost 20% of the total O_2 consumed by the body at rest. Because the brain contains very little glycogen, it is constantly dependent on incoming glucose in the blood. Should blood glucose fall significantly below a critical level for even a short time, severe and sometimes irreversible changes in brain function may result.

Although the neurons of the brain cannot directly use free fatty acids or lipids from the blood as fuels, they can, when necessary, get up to 60% of their energy requirement from the oxidation of β-hydroxybutyrate (a ketone body), formed in the liver from fatty acids. The capacity of the brain to oxidize β-hydroxybutyrate via acetyl-CoA becomes important during prolonged fasting or starvation, after liver glycogen has been depleted, because it allows the brain to use body fat as an energy source. This spares muscle proteins—until they become the brain's ultimate source of glucose, via gluconeogenesis in the liver, during severe starvation.

In neurons, energy is required to create and maintain an electrical potential across the plasma membrane. The

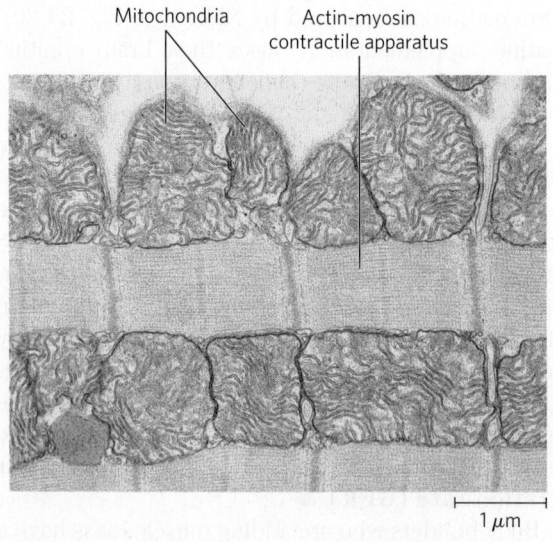

Mitochondria   Actin-myosin contractile apparatus

1 μm

FIGURE 23-22 Electron micrograph of heart muscle. In the profuse mitochondria of heart tissue, pyruvate (from glucose), fatty acids, and ketone bodies are oxidized to drive ATP synthesis. This steady aerobic metabolism allows the human heart to pump blood at a rate of nearly 6 L/min, or about 350 L/h—which amounts to 200×10^6 L of blood over 70 years. [Source: D. W. Fawcett/Science Source.]

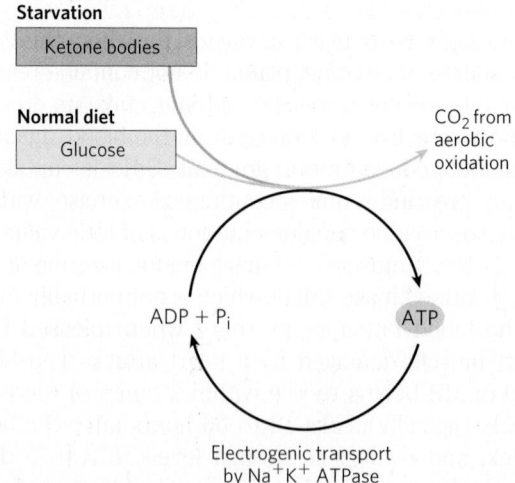

Starvation

Ketone bodies

Normal diet

Glucose

CO_2 from aerobic oxidation

ADP + P_i ATP

Electrogenic transport by Na^+K^+ ATPase

FIGURE 23-23 The fuels that supply ATP in the brain. The energy source used by the brain varies with nutritional state. The ketone body used during starvation is β-hydroxybutyrate. Electrogenic transport by the Na^+K^+ ATPase maintains the transmembrane potential essential to information transfer among neurons.

membrane contains an electrogenic ATP-driven antiporter, the Na^+K^+ ATPase, which simultaneously pumps 2 K^+ ions into and 3 Na^+ ions out of the neuron (see Fig. 11-36). The resulting transmembrane potential changes transiently as an electrical signal, an action potential, sweeps from one end of a neuron to the other (see Fig. 12-29). Action potentials are the chief mechanism of information transfer in the nervous system, so depletion of ATP in neurons would have disastrous effects on all activities coordinated by neuronal signaling.

Blood Carries Oxygen, Metabolites, and Hormones

Blood mediates the metabolic interactions among all tissues. It transports nutrients from the small intestine to the liver and from the liver and adipose tissue to other organs; it also transports waste products from extrahepatic tissues to the liver for processing and to the kidneys for excretion. Oxygen moves in the bloodstream from the lungs to the tissues, and CO_2 generated by tissue respiration returns via the bloodstream to the lungs for exhalation. Blood also carries hormonal signals from one tissue to another. In its role as signal carrier, the circulatory system resembles the nervous system: both regulate and integrate the activities of different organs.

The average adult human has 5 to 6 L of blood. Almost half of this volume is occupied by three types of blood cells **(Fig. 23-24)**: **erythrocytes** (red cells), filled with hemoglobin and specialized for carrying O_2 and CO_2; much smaller numbers of **leukocytes** (white cells) of several types (including **lymphocytes**, also found in lymphatic tissue), which are central to the immune system to defend against infections; and **platelets** (cell fragments), which help to mediate blood clotting. The liquid portion is the **blood plasma**, which is 90% water and 10% solutes. Dissolved or suspended in the plasma are many proteins, lipoproteins, nutrients, metabolites, waste products, inorganic ions, and hormones. More than 70% of the plasma solids are **plasma proteins**, primarily immunoglobulins (circulating antibodies), serum albumin, apolipoproteins (for lipid transport), transferrin (for iron transport), and blood-clotting proteins such as fibrinogen and prothrombin.

The ions and low molecular weight solutes in blood plasma are not fixed components; they are in constant flux between blood and various tissues. Dietary uptake of the inorganic ions that are the predominant electrolytes of blood and cytosol (Na^+, K^+, and Ca^{2+}) is, in general, counterbalanced by their excretion in the urine. For many blood components, something near a dynamic steady state is achieved: the concentration of a component changes little, although a continuous flux occurs between the digestive tract, blood, and urine. The plasma levels of Na^+, K^+, and Ca^{2+} remain close to 140, 5, and 2.5 mM, respectively, with little change in response to dietary intake. Any significant departure from these values can result in serious illness or death. The kidneys play an especially important role in maintaining ion balance by selectively filtering waste products and excess ions out of the blood while preventing the loss of essential nutrients and ions.

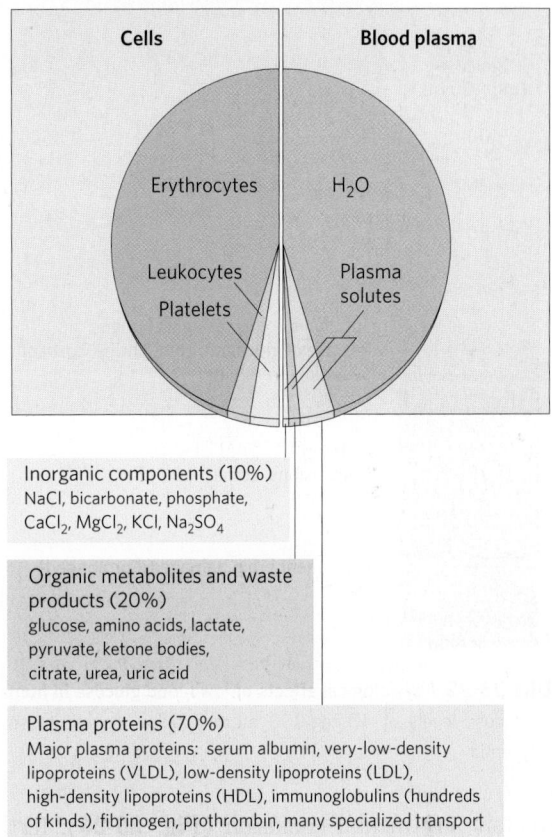

Inorganic components (10%)
NaCl, bicarbonate, phosphate, $CaCl_2$, $MgCl_2$, KCl, Na_2SO_4

Organic metabolites and waste products (20%)
glucose, amino acids, lactate, pyruvate, ketone bodies, citrate, urea, uric acid

Plasma proteins (70%)
Major plasma proteins: serum albumin, very-low-density lipoproteins (VLDL), low-density lipoproteins (LDL), high-density lipoproteins (HDL), immunoglobulins (hundreds of kinds), fibrinogen, prothrombin, many specialized transport proteins such as transferrin

FIGURE 23-24 The composition of blood (by weight). Whole blood can be separated into blood plasma and cells by centrifugation. About 10% of blood plasma is solutes, about 10% of these consisting of inorganic salts, 20% small organic molecules, and 70% plasma proteins. The major dissolved components are listed here. Blood contains many other substances, often in trace amounts. These include other metabolites, enzymes, hormones, vitamins, trace elements, and bile pigments. Measurements of the concentrations of components in blood plasma are important in the diagnosis and treatment of many diseases.

The human erythrocyte loses its nucleus and mitochondria during differentiation. It therefore relies on glycolysis alone for its supply of ATP. The lactate produced by glycolysis returns to the liver, where gluconeogenesis converts it to glucose, to be stored as glycogen or recirculated to peripheral tissues. The erythrocyte has constant access to glucose in the bloodstream.

The concentration of glucose in plasma is subject to tight regulation. We have noted the constant requirement of the brain for glucose and the role of the liver in maintaining blood glucose in the normal range, 60 to 90 mg/100 mL of whole blood (~4.5 mM). (Because erythrocytes make up a significant fraction of blood volume, their removal by centrifugation leaves a supernatant fluid, the plasma, containing the "blood glucose" in a smaller volume. To convert blood glucose to plasma glucose concentration, multiply the blood glucose level by 1.14.) When blood glucose in a human drops to 40 mg/100 mL (the hypoglycemic condition), the person experiences

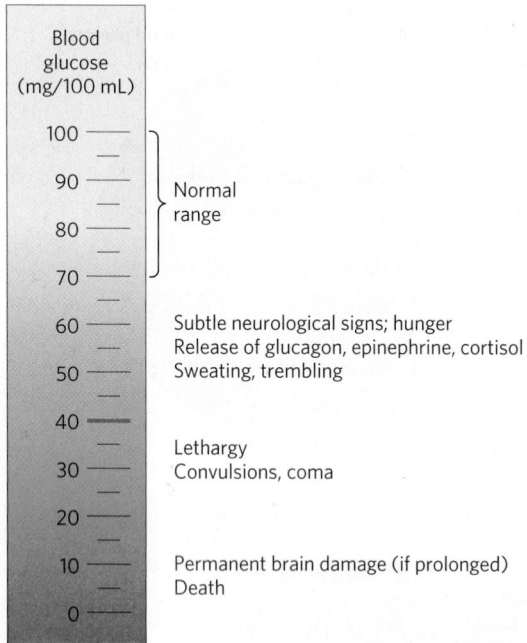

FIGURE 23-25 Physiological effects of low blood glucose in humans. Blood glucose levels of 40 mg/100 mL and below constitute severe hypoglycemia.

discomfort and mental confusion **(Fig. 23-25)**; further reductions lead to coma, convulsions, and, in extreme hypoglycemia, death. Maintaining the normal concentration of glucose in blood is therefore a high priority, and a variety of regulatory mechanisms have evolved to achieve that end. Among the most important regulators of blood glucose are the hormones insulin, glucagon, and epinephrine, as discussed in the next section. ■

SUMMARY 23.2 Tissue-Specific Metabolism: The Division of Labor

■ In mammals there is a division of metabolic labor among specialized tissues and organs. The liver is the central processing and distribution organ for nutrients. Sugars and amino acids produced in digestion cross the intestinal epithelium and enter the blood, which carries them to the liver. Some triacylglycerols derived from ingested lipids also make their way to the liver, where the constituent fatty acids are used in a variety of processes.

■ Glucose 6-phosphate is the key intermediate in carbohydrate metabolism. It may be polymerized into glycogen, dephosphorylated to blood glucose, or converted to fatty acids via acetyl-CoA. It may undergo oxidation by glycolysis, the citric acid cycle, and the respiratory chain to yield ATP, or enter the pentose phosphate pathway to yield pentoses and NADPH.

■ Amino acids are used to synthesize liver and plasma proteins, or their carbon skeletons are converted to glucose and glycogen by gluconeogenesis; the ammonia formed by deamination is converted to urea.

■ The liver converts fatty acids to TAGs, to phospholipids, or to cholesterol and its esters for transport as plasma lipoproteins to adipose tissue for storage. Fatty acids can also be oxidized to yield ATP or to form ketone bodies, which are circulated to other tissues.

■ White adipose tissue stores large reserves of TAGs and releases them into the blood in response to epinephrine or glucagon. Brown adipose tissue is specialized for thermogenesis, the result of fatty acid oxidation in uncoupled mitochondria. Beige adipose tissue expands in response to cold temperature and stimulation by epinephrine. As in brown adipose tissue, the mitochondria of beige adipose tissue have uncoupling protein and are specialized for thermogenesis.

■ Skeletal muscle is specialized to produce and use ATP for mechanical work. During low to moderate muscular activity, oxidation of fatty acids and glucose is the primary source of ATP. During strenuous muscular activity, glycogen is the ultimate fuel, supplying ATP through lactic acid fermentation. During recovery, the lactate is reconverted (through gluconeogenesis) to glycogen and glucose in the liver for use in replenishing muscle glycogen supplies. Phosphocreatine is an immediate source of ATP during active contraction.

■ Heart muscle obtains nearly all its ATP from oxidative phosphorylation, with fatty acids as the primary fuel.

■ The neurons of the brain use only glucose and β-hydroxybutyrate as fuels, the latter being important during fasting or starvation. The brain uses most of its ATP for the active transport of Na^+ and K^+ to maintain the electrical potential across the neuronal membrane.

■ The blood transfers nutrients, waste products, and hormonal signals among tissues and organs. It is made up of cells (erythrocytes, leukocytes, and platelets) and electrolyte-rich water (plasma) containing many proteins.

23.3 Hormonal Regulation of Fuel Metabolism

The minute-by-minute adjustments that keep the blood glucose level near 4.5 mM involve the combined actions of insulin, glucagon, epinephrine, and cortisol on metabolic processes in many body tissues, but especially in liver, muscle, and adipose tissue. Insulin signals these tissues that blood glucose is higher than necessary; as a result, cells take up excess glucose from the blood and convert it to glycogen and triacylglycerols for storage. Glucagon signals that blood glucose is too low, and tissues respond by producing glucose through glycogen breakdown and (in the liver) gluconeogenesis and by oxidizing fats to reduce the need for glucose. Epinephrine is released into the blood to prepare the muscles, lungs, and heart for a burst of activity. Cortisol mediates the body's response to longer-term stresses. We discuss these hormonal regulations in the context of three normal metabolic states—well fed, fasted, and starving—and look at the metabolic consequences of diabetes mellitus, a disorder that results from derangements in the signaling pathways that control glucose metabolism.

TABLE 23-4	Effects of Insulin on Blood Glucose: Uptake of Glucose by Cells and Storage as Triacylglycerols and Glycogen
Metabolic effect	**Target enzyme**
↑ Glucose uptake (muscle, adipose)	↑ Glucose transporter (GLUT4)
↑ Glucose uptake (liver)	↑ Glucokinase (increased expression)
↑ Glycogen synthesis (liver, muscle)	↑ Glycogen synthase
↓ Glycogen breakdown (liver, muscle)	↓ Glycogen phosphorylase
↑ Glycolysis, acetyl-CoA production (liver, muscle)	↑ PFK-1 (by PFK-2) ↑ Pyruvate dehydrogenase complex
↑ Fatty acid synthesis (liver)	↑ Acetyl-CoA carboxylase
↑ Triacylglycerol synthesis (adipose tissue)	↑ Lipoprotein lipase

Insulin Counters High Blood Glucose

Acting through plasma membrane receptors (see Figs 12-19, 12-20), insulin stimulates glucose uptake by muscle and adipose tissue (Table 23-4), where the glucose is converted to glucose 6-phosphate. In the liver,

insulin also activates glycogen synthase and inactivates glycogen phosphorylase, so that much of the glucose 6-phosphate is channeled into glycogen.

Insulin also stimulates the storage of excess fuel as fat in adipose tissue **(Fig. 23-26)**. In the liver, insulin activates both the oxidation of glucose 6-phosphate to

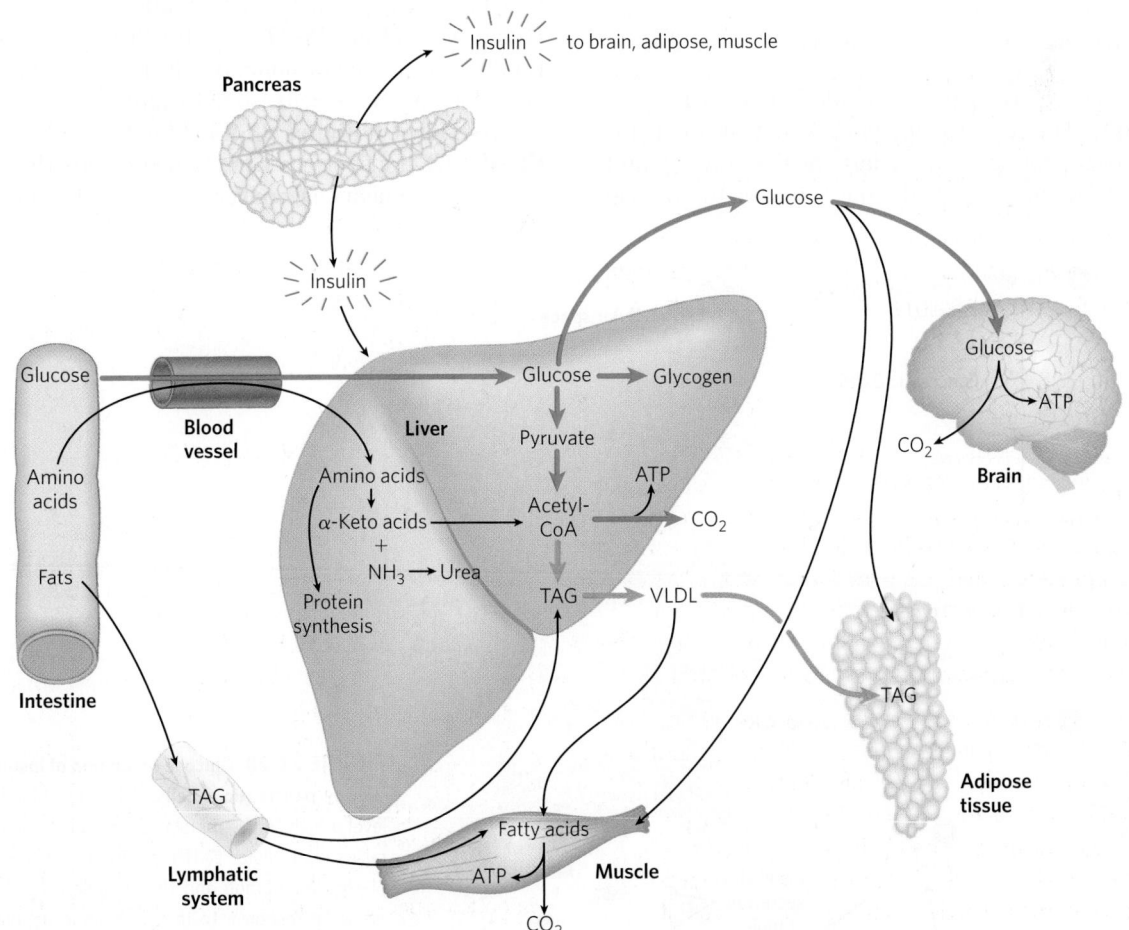

FIGURE 23-26 The well-fed state: the lipogenic liver. Immediately after a calorie-rich meal, glucose, fatty acids, and amino acids enter the liver. Blue arrows follow the path of glucose; orange arrows follow the path of lipids. Insulin released in response to the high blood glucose concentration stimulates glucose uptake by the tissues. Some glucose is exported to the brain for its energy needs, and some goes to adipose and muscle tissue. In the liver, excess glucose is oxidized to acetyl-CoA, which is used to make triacylglycerols for export to adipose and muscle tissue. The NADPH necessary for lipid synthesis is obtained by oxidation of glucose in the pentose phosphate pathway. Excess amino acids are converted to pyruvate and acetyl-CoA, which are also used for lipid synthesis. Dietary fats move from the intestine as chylomicrons, via the lymphatic system, to the liver, muscle, and adipose tissues.

Pancreas

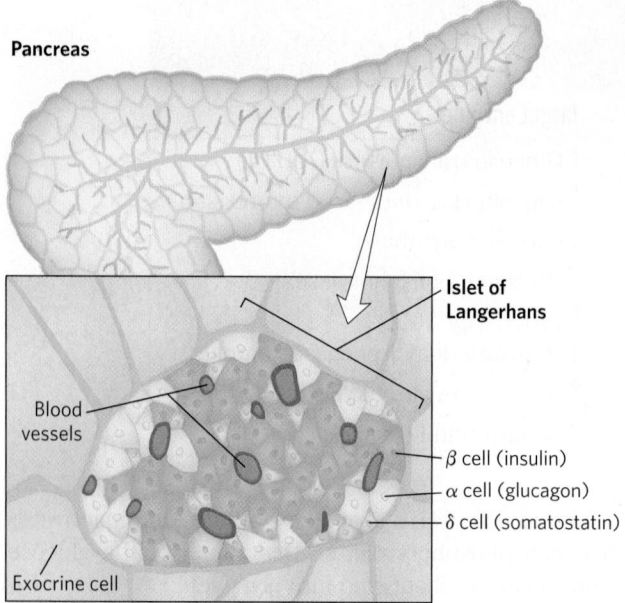

FIGURE 23-27 The endocrine system of the pancreas. The pancreas contains both exocrine cells (see Fig. 18-3b), which secrete digestive enzymes in the form of zymogens, and clusters of endocrine cells, the islets of Langerhans. The islets contain α, β, and δ cells (also known as A, B, and D cells, respectively), each cell type secreting a specific peptide hormone.

pyruvate via glycolysis and the oxidation of pyruvate to acetyl-CoA. The excess acetyl-CoA not needed for energy production is used for fatty acid synthesis, and the fatty acids are exported from the liver to adipose

tissue as the TAGs of plasma lipoproteins (VLDL; see Fig. 21-40). Insulin stimulates the synthesis of TAGs in adipocytes, from fatty acids released from the TAGs of VLDL. These fatty acids are ultimately derived from the excess glucose taken up from blood by the liver. In summary, the effect of insulin is to favor the conversion of excess blood glucose to two storage forms: glycogen (in the liver and muscle) and TAGs (in adipose tissue).

Besides acting directly on muscle and liver to change their metabolism of carbohydrates and fats, insulin can act in the brain to signal these tissues indirectly, as described later.

Pancreatic β Cells Secrete Insulin in Response to Changes in Blood Glucose

When glucose enters the bloodstream from the intestine after a carbohydrate-rich meal, the resulting increase in blood glucose causes increased secretion of insulin (and decreased secretion of glucagon) by the pancreas. Insulin release is largely regulated by the level of glucose in the blood supplying the pancreas. The peptide hormones insulin, glucagon, and somatostatin are produced by clusters of specialized pancreatic cells, the islets of Langerhans **(Fig. 23-27)**. Each cell type of the islets produces a single hormone: α cells produce glucagon; β cells, insulin; and δ cells, somatostatin.

As shown in **Figure 23-28**, when blood glucose rises, ❶ GLUT2 transporters carry glucose into the β cells, where it is immediately converted to glucose 6-phosphate

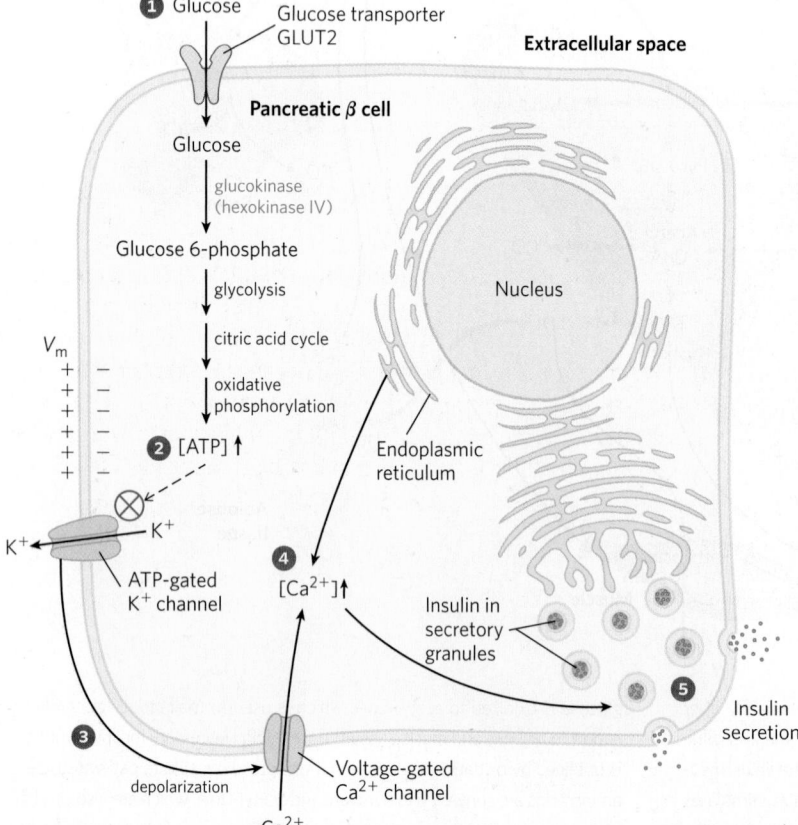

FIGURE 23-28 Glucose regulation of insulin secretion by pancreatic β cells. When the blood glucose level is high, active metabolism of glucose in the β cell raises intracellular [ATP], closing K^+ channels in the plasma membrane and thus depolarizing the membrane. In response to this membrane depolarization, voltage-gated Ca^{2+} channels open, allowing Ca^{2+} to flow into the cell. (Ca^{2+} is also released from the ER, in response to the initial elevation of $[Ca^{2+}]$ in the cytosol.) Cytosolic $[Ca^{2+}]$ is now high enough to trigger insulin release by exocytosis. The numbered processes are discussed in the text.

FIGURE 23-29 ATP-gated K⁺ channels in β cells. (a) The ATP-gated channel, viewed in the plane of the membrane. The channel is formed by four identical Kir6.2 subunits, which are surrounded by four SUR1 (sulfonylurea receptor) subunits. The SUR1 subunits have binding sites for ADP and the drug diazoxide, both of which favor the open channel, and tolbutamide, a sulfonylurea drug that favors the closed channel. The Kir6.2 subunits constitute the channel, and they contain, on the cytosolic side, binding sites for ATP and phosphatidylinositol 4,5-bisphosphate (PIP_2), which favor the closed and the open channel, respectively. **(b)** The structure of the Kir6.2 portion of the channel, viewed in the plane of the membrane. For clarity, only two transmembrane domains and two cytosolic domains are shown. Three K⁺ ions (green) are shown in the region of the selectivity filter. Mutation in certain amino acid residues (shown in red) leads to neonatal diabetes; mutation in others (shown in blue) leads to hyperinsulinism of infancy. This structure was obtained by mapping the known Kir6.2 sequence onto the crystal structures of a bacterial Kir channel (KirBac1.1) and the amino and carboxyl domains of another Kir protein, Kir3.1. Compare this structure with the gated K⁺ channel in Figure 11-46. [Source: (b) KirBac1.1: PDB ID 1P7B, A. Kuo et al., *Science* 300:1922, 2003; Kir3.1: PDB ID 1U4E, S. Pegan et al., *Nature Neurosci.* 8:279, 2005. Coordinates courtesy of Frances M. Ashcroft, Oxford University, used with permission of S. Haider and M. S. P. Sansom to re-create a model published in J. F. Antcliff et al., *EMBO J.* 24:229, 2005.]

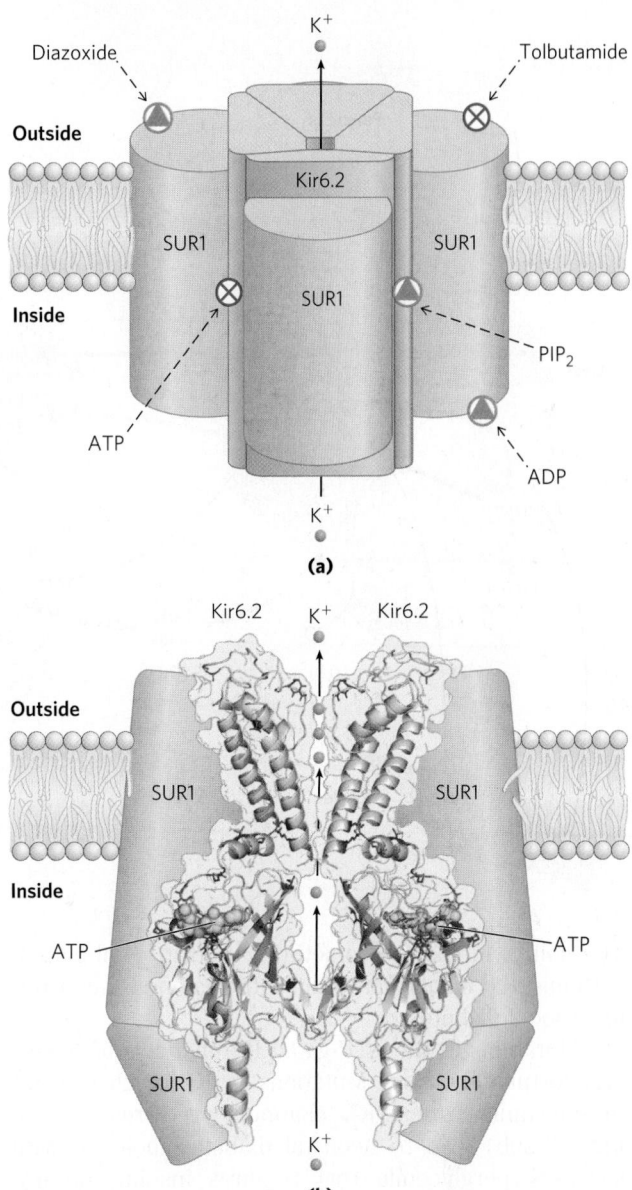

(a)

(b)

by glucokinase and enters glycolysis. With the higher rate of glucose catabolism, ❷ [ATP] increases, causing the closing of **ATP-gated K⁺ channels** in the plasma membrane. ❸ Reduced efflux of K⁺ depolarizes the membrane. (Recall from Section 12.7 that exit of K⁺ through an open K⁺ channel hyperpolarizes the membrane; thus, closing the K⁺ channel effectively depolarizes the membrane.) Membrane depolarization opens voltage-gated Ca^{2+} channels, and ❹ the resulting increase in cytosolic $[Ca^{2+}]$ triggers ❺ the release of insulin by exocytosis. The brain integrates inputs on energy supply and demand, and signals from the parasympathetic and sympathetic nervous systems also affect (stimulate and inhibit, respectively) insulin release. A simple feedback loop limits hormone release: insulin lowers blood glucose by stimulating glucose uptake by the tissues; the reduced blood glucose is detected by the β cell as a diminished flux through the glucokinase reaction; this slows or stops the release of insulin. This feedback regulation holds blood glucose concentration nearly constant despite large fluctuations in dietary intake.

The activity of ATP-gated K⁺ channels is central to the regulation of insulin secretion by β cells. The channels are octamers of four identical Kir6.2 subunits and four identical SUR1 subunits and are constructed along the same lines as the K⁺ channels of bacteria and those of other eukaryotic cells (see Figs 11-45, 11-46). The four Kir6.2 subunits form a cone around the K⁺ channel and function as the selectivity filter and ATP-gating mechanism **(Fig. 23-29)**. When [ATP] rises, indicating increased blood glucose, the K⁺ channels close, thus depolarizing the plasma membrane and triggering insulin release as shown in Figure 23-28.

The **sulfonylurea drugs**, oral medications used in the treatment of type 2 diabetes mellitus, bind to the SUR1 (*sulfonylurea receptor*) subunits of the K⁺ channels, closing the channels and stimulating insulin release. The first generation of these drugs, including tolbutamide, was developed in the 1950s. The second-generation drugs, including glyburide (Micronase), glipizide (Glucotrol), and glimepiride (Amaryl), are more potent and have fewer side effects. The sulfonylurea moiety is screened light red in the following structures.

Glyburide

Glipizide

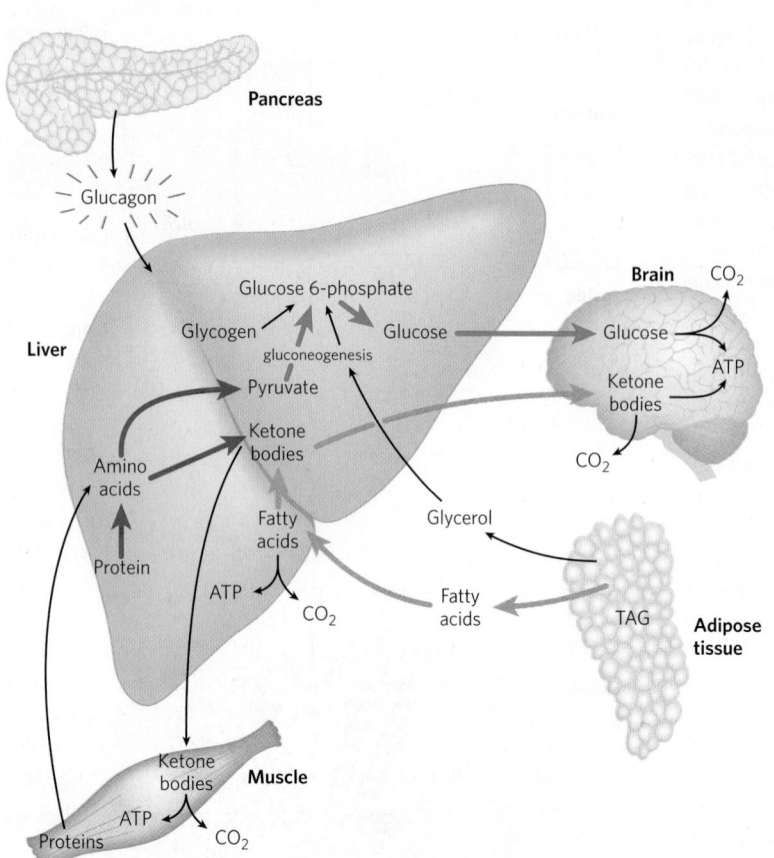

FIGURE 23-30 The fasting state: the glucogenic liver. After some hours without a meal, the liver becomes the principal source of glucose for the brain. Liver glycogen is broken down to glucose 1-phosphate, and this is converted to glucose 6-phosphate, then to free glucose, which is released into the bloodstream. Amino acids from the degradation of proteins in liver and muscle, and glycerol from the breakdown of TAGs in adipose tissue, are used for gluconeogenesis. The liver uses fatty acids as its principal fuel, and excess acetyl-CoA is converted to ketone bodies for export to other tissues; the brain is especially dependent on this fuel when glucose is in short supply (see Fig. 23-23). Blue arrows follow the path of glucose; orange arrows, the path of lipids; and purple arrows, the path of amino acids.

The sulfonylureas are sometimes used in combination with injected insulin but often suffice alone for controlling type 2 diabetes.

Mutations in the ATP-gated K^+ channels of β cells are, fortunately, rare. Mutations in Kir6.2 that result in constantly *open* K^+ channels (red residues in Fig. 23-29b) lead to neonatal diabetes mellitus, with severe hyperglycemia that requires insulin therapy. Other mutations in Kir6.2 or SUR1 (blue residues in Fig. 23-29b) produce permanently *closed* K^+ channels and continuous release of insulin. If untreated, individuals with these mutations develop congenital hyperinsulinemia (hyperinsulinism of infancy); excessive insulin causes severe hypoglycemia (low blood glucose) leading to irreversible brain damage. One effective treatment is surgical removal of part of the pancreas to reduce insulin production. ■

Glucagon Counters Low Blood Glucose

Several hours after the intake of dietary carbohydrate, blood glucose levels fall slightly because of the ongoing oxidation of glucose by the brain and other tissues. Lowered blood glucose triggers secretion of **glucagon** and decreases insulin release **(Fig. 23-30)**.

Glucagon causes an *increase* in blood glucose concentration in several ways (Table 23-5). Like epinephrine,

TABLE 23-5	Effects of Glucagon on Blood Glucose: Production and Release of Glucose by the Liver	
Metabolic effect	**Effect on glucose metabolism**	**Target enzyme**
↑ Glycogen breakdown (liver)	Glycogen ⟶ glucose	↑ Glycogen phosphorylase
↓ Glycogen synthesis (liver)	Less glucose stored as glycogen	↓ Glycogen synthase
↓ Glycolysis (liver)	Less glucose used as fuel in liver	↓ PFK-1
↑ Gluconeogenesis (liver)	Amino acids Glycerol ⟶ glucose Oxaloacetate	↑ FBPase-2 ↓ Pyruvate kinase ↑ PEP carboxykinase
↑ Fatty acid mobilization (adipose tissue)	Less glucose used as fuel by liver, muscle	↑ Hormone-sensitive lipase ↑ PKA (perilipin–ⓅP)
↑ Ketogenesis	Provides alternative to glucose as energy source for brain	↓ Acetyl-CoA carboxylase

it stimulates the net breakdown of liver glycogen by activating glycogen phosphorylase and inactivating glycogen synthase; both effects are the result of phosphorylation of the regulated enzymes, triggered by cAMP. Glucagon inhibits glucose breakdown by glycolysis in the liver and stimulates glucose synthesis by gluconeogenesis. Both effects result from lowering the concentration of fructose 2,6-bisphosphate, an allosteric inhibitor of the gluconeogenic enzyme fructose 1,6-bisphosphatase (FBPase-1) and an activator of the glycolytic enzyme phosphofructokinase-1. Recall that [fructose 2,6-bisphosphate] is ultimately controlled by a cAMP-dependent protein phosphorylation reaction (see Fig. 15-19). Glucagon also inhibits the glycolytic enzyme pyruvate kinase, by promoting its cAMP-dependent phosphorylation, thus blocking the conversion of phosphoenolpyruvate to pyruvate and preventing oxidation of pyruvate via the citric acid cycle. The resulting accumulation of phosphoenolpyruvate favors gluconeogenesis. This effect is augmented by glucagon's stimulation of the synthesis of the gluconeogenic enzyme PEP carboxykinase. By stimulating glycogen breakdown, preventing glycolysis, and promoting gluconeogenesis in hepatocytes, glucagon enables the liver to export glucose, restoring blood glucose to its normal level.

Although its primary target is the liver, glucagon (like epinephrine) also affects adipose tissue, activating TAG breakdown by causing cAMP-dependent phosphorylation of perilipin and hormone-sensitive lipase. The activated lipase liberates free fatty acids, which are exported to the liver and other tissues as fuel, sparing glucose for the brain. The net effect of glucagon is therefore to stimulate glucose synthesis and release by the liver and to mobilize fatty acids from adipose tissue,

to be used instead of glucose by tissues other than the brain. All these effects of glucagon are mediated by cAMP-dependent protein phosphorylation.

During Fasting and Starvation, Metabolism Shifts to Provide Fuel for the Brain

The fuel reserves of a healthy adult human are of three types: glycogen stored in the liver and, in smaller quantities, in muscles; large quantities of TAG in adipose tissues; and tissue proteins, which can be degraded when necessary to provide fuel (Table 23-6).

Two hours after a meal, the blood glucose level is diminished slightly, and tissues receive glucose released from liver glycogen. There is little or no synthesis of TAGs. By four hours after a meal, blood glucose has fallen further, insulin secretion has slowed, and glucagon secretion has increased. These hormonal signals mobilize TAGs from adipose tissue, which now become the primary fuel for muscle and liver. **Figure 23-31** shows the responses to prolonged fasting. ❶ To provide glucose for the brain, the liver degrades certain proteins—those most expendable in an organism not ingesting food. Their nonessential amino acids are transaminated or deaminated (Chapter 18), and ❷ the extra amino groups are converted to urea, which is exported via the bloodstream to the kidneys and excreted in the urine.

Also in the liver, and to some extent in the kidneys, the carbon skeletons of glucogenic amino acids are converted to pyruvate or intermediates of the citric acid cycle. ❸ These intermediates (as well as the glycerol derived from TAGs in adipose tissue) provide the starting materials for gluconeogenesis in the liver,

TABLE 23-6	Available Metabolic Fuels in a Normal-Weight, 70 kg Man and in an Obese, 140 kg Man at the Beginning of a Fast			
Type of fuel	Weight (kg)	Caloric equivalent (thousands of kcal (kJ))		Estimated survival (months)[a]
Normal–weight, 70 kg man				
Triacylglycerols (adipose tissue)	15	140 (590)		
Proteins (mainly muscle)	6	24 (100)		
Glycogen (muscle, liver)	0.23	0.90 (3.8)		
Circulating fuels (glucose, fatty acids, triacylglycerols, etc.)	0.023	0.10 (0.42)		
Total		165 (690)		3
Obese, 140 kg man				
Triacylglycerols (adipose tissue)	80	750 (3,100)		
Proteins (mainly muscle)	8	32 (130)		
Glycogen (muscle, liver)	0.23	0.92 (3.8)		
Circulating fuels	0.025	0.11 (0.46)		
Total		783 (3,200)		14

[a]Survival time is calculated on the assumption of a basal energy expenditure of 1,800 kcal/day.

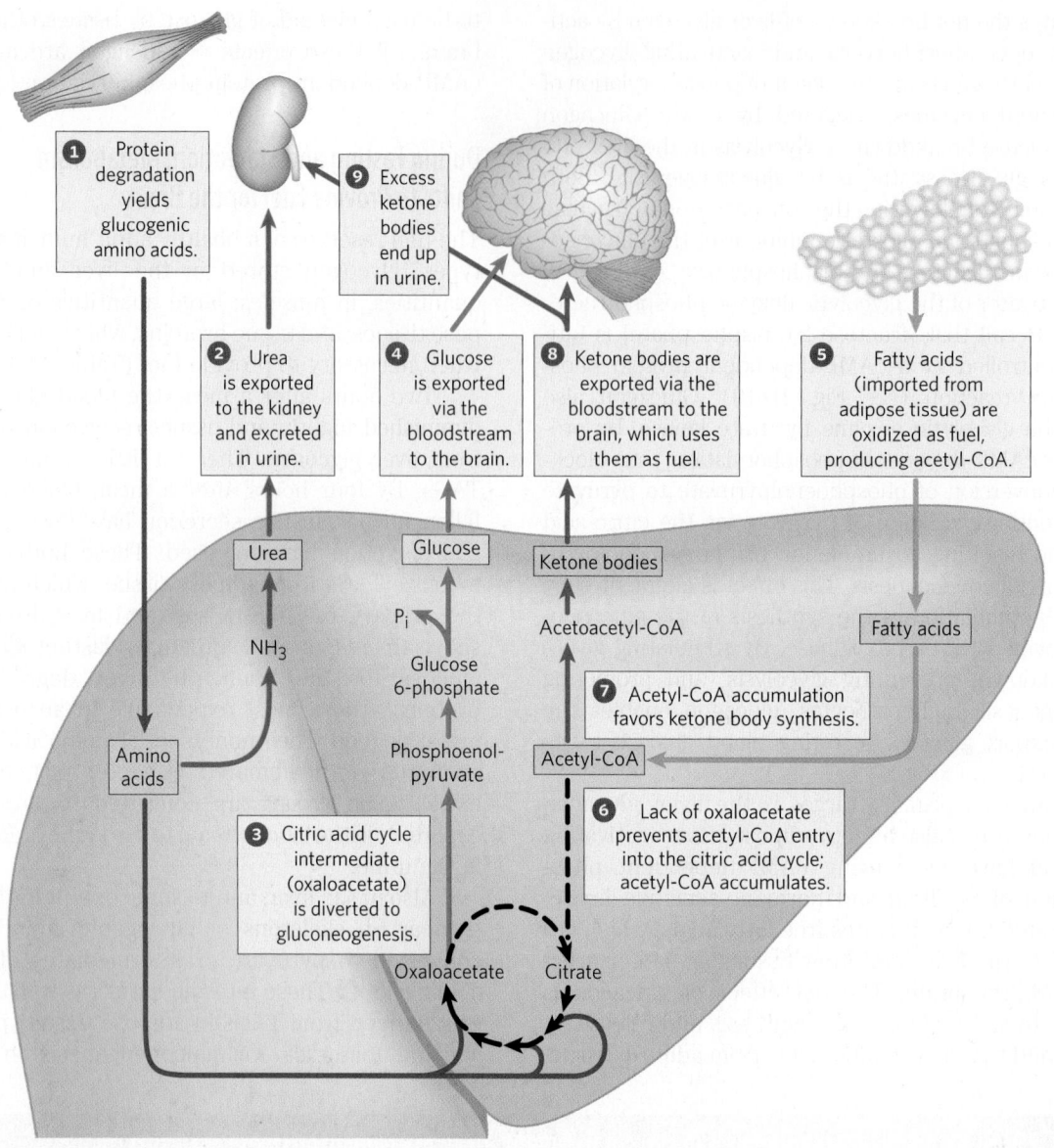

FIGURE 23-31 Fuel metabolism in the liver during prolonged fasting or in uncontrolled diabetes mellitus. The numbered steps are described in the text. After depletion of stored carbohydrates (glycogen), gluconeogenesis in the liver becomes the main source of glucose for the brain (blue arrows). NH_3 from amino acid deamination is converted into urea and excreted (green arrows). Glucogenic amino acids from protein breakdown (purple arrows) provide substrates for gluconeogenesis, and glucose is exported to the brain. Fatty acids from adipose tissue are imported into the liver and oxidized to acetyl-CoA (orange arrows), and acetyl-CoA is the starting material for ketone body formation in the liver and export to brain to serve as energy source (red arrows). When the concentration of ketone bodies in the blood exceeds the ability of the kidneys to reabsorb ketones, these compounds begin to appear in the urine.

❹ yielding glucose for export to the brain. ❺ Fatty acids released from adipose tissue are oxidized to acetyl-CoA in the liver, but as oxaloacetate is depleted by the use of citric acid cycle intermediates for gluconeogenesis, ❻ entry of acetyl-CoA into the cycle is inhibited and acetyl-CoA accumulates. ❼ This favors the formation of acetoacetyl-CoA and ketone bodies. After a few days of fasting, the levels of ketone bodies in the blood rise **(Fig. 23-32)** as they are exported from the liver to the heart, skeletal muscle, and brain, which use these fuels instead of glucose (Fig. 23-31, ❽).

Acetyl-CoA is a critical regulator of the fate of pyruvate: it allosterically inhibits pyruvate dehydrogenase and stimulates pyruvate carboxylase. In these ways, acetyl-CoA prevents its own further production from pyruvate while stimulating the conversion of pyruvate to oxaloacetate, the first step in gluconeogenesis.

Triacylglycerols stored in the adipose tissue of a normal-weight adult could provide enough fuel to maintain a basal rate of metabolism for about three months; a very obese adult has enough stored fuel to endure a fast of more than a year (Table 23-6). When fat reserves are gone, the degradation of *essential* proteins begins; this leads to loss of heart and liver function and, in prolonged starvation, to death. Stored fat can provide adequate energy (calories) during a fast or rigid diet, but

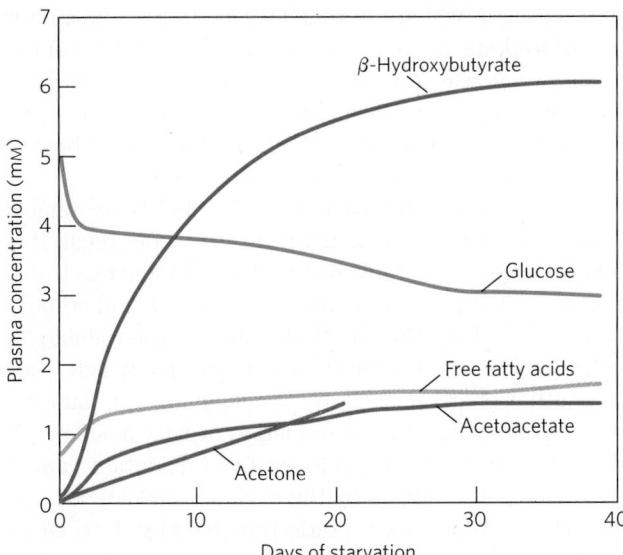

FIGURE 23-32 Plasma concentrations of fatty acids, glucose, and ketone bodies during six weeks of starvation. Despite the hormonal mechanisms for maintaining the glucose level in blood, glucose begins to diminish within 2 days of beginning a fast (blue). The levels of ketone bodies, almost unmeasurable before the fast, rise dramatically after 2 to 4 days of fasting (red), with β-hydroxybutyrate as the major contributor. These water-soluble ketones, acetoacetate and β-hydroxybutyrate, supplement glucose as an energy source for the brain during a long fast. Acetone, a minor ketone body, is not metabolized but is eliminated in the breath. A much smaller rise in blood fatty acids (orange) also occurs, but this does not contribute to energy metabolism in the brain, as fatty acids do not cross the blood-brain barrier. [Data from G. F. Cahill, Jr., *Annu. Rev. Nutr.* 26:1, 2006, Fig. 2.]

vitamins and minerals must be provided, and sufficient dietary glucogenic amino acids are needed to replace those being used for gluconeogenesis. Rations for those on a weight-reduction diet are commonly fortified with vitamins, minerals, and amino acids or proteins.

Epinephrine Signals Impending Activity

When an animal is confronted with a stressful situation that requires increased activity—fighting or fleeing, in the extreme case—neuronal signals from the brain trigger the release of epinephrine and norepinephrine from the adrenal medulla. Both hormones dilate the respiratory passages to facilitate the uptake of O_2, increase the rate and strength of the heartbeat, and raise the blood pressure, thereby promoting the flow of O_2 and fuels to the tissues (Table 23-7). This is the "fight-or-flight" response.

Epinephrine acts primarily on muscle, adipose, and liver tissues. It activates glycogen phosphorylase and inactivates glycogen synthase by cAMP-dependent phosphorylation of the enzymes, thus stimulating the conversion of *liver* glycogen to blood glucose, the fuel for anaerobic muscular work. Epinephrine also promotes the anaerobic breakdown of *muscle* glycogen by lactic acid fermentation, stimulating glycolytic ATP formation. The stimulation of glycolysis is accomplished by raising the concentration of fructose 2,6-bisphosphate, a potent allosteric activator of the key glycolytic enzyme phosphofructokinase-1. Epinephrine also stimulates fat mobilization in adipose tissue, by activating hormone-sensitive lipase and moving aside perilipin (see Fig. 17-3). Finally, epinephrine stimulates glucagon secretion and inhibits insulin secretion, reinforcing its effect of mobilizing fuels and inhibiting fuel storage.

Cortisol Signals Stress, Including Low Blood Glucose

A variety of stressors (anxiety, fear, pain, hemorrhage, infection, low blood glucose, starvation) stimulate release of the glucocorticoid **cortisol** from the adrenal cortex. Cortisol acts on muscle, liver, and adipose tissue to supply the organism with fuel to withstand the stress. Cortisol is a relatively slow-acting hormone that alters metabolism by changing the kinds and amounts of certain enzymes synthesized in its target cells, rather than by regulating the activity of existing enzyme molecules.

TABLE 23-7	Physiological and Metabolic Effects of Epinephrine: Preparation for Action	
Immediate effect		**Overall effect**
Physiological		
↑ Heart rate		
↑ Blood pressure		Increase delivery of O_2 to tissues (muscle)
↑ Dilation of respiratory passages		
Metabolic		
↑ Glycogen breakdown (muscle, liver)		
↓ Glycogen synthesis (muscle, liver)		Increase production of glucose for fuel
↑ Gluconeogenesis (liver)		
↑ Glycolysis (muscle)		Increases ATP production in muscle
↑ Fatty acid mobilization (adipose tissue)		Increases availability of fatty acids as fuel
↑ Glucagon secretion		Reinforce metabolic effects of epinephrine
↓ Insulin secretion		

In adipose tissue, cortisol leads to an increased release of fatty acids from stored TAGs. The exported fatty acids serve as fuel for other tissues, and the glycerol is used for gluconeogenesis in the liver. Cortisol stimulates the breakdown of nonessential muscle proteins and the export of amino acids to the liver, where they serve as precursors for gluconeogenesis. In the liver, cortisol promotes gluconeogenesis by stimulating synthesis of PEP carboxykinase; glucagon has the same effect, whereas insulin has the opposite effect. Glucose produced in this way is stored in the liver as glycogen or exported immediately to tissues that need glucose for fuel. The net effect of these metabolic changes is to restore blood glucose to its normal level and to increase glycogen stores, ready to support the fight-or-flight response commonly associated with stress. The effects of cortisol therefore counterbalance those of insulin. During extended periods of stress, the continued release of cortisol loses its positive adaptive value and begins to cause damage to muscle and bone and to impair endocrine and immune function.

Cushing disease is a medical condition in which cortisol is overproduced by a pituitary tumor. It is treated by surgery to remove the tumor, followed by chemotherapy to kill remaining tumor cells. Addison disease results from underproduction of cortisol, and is treated by administering hydrocortisone (the pharmaceutical name for cortisol). ∎

Diabetes Mellitus Arises from Defects in Insulin Production or Action

Diabetes mellitus is a relatively common disease: about 9% of the U.S. population, and nearly 25% of the U.S. population over the age of 65, show some degree of abnormality in glucose metabolism that is indicative of diabetes or a tendency toward the condition. There are two major clinical classes of diabetes mellitus: **type 1 diabetes**, sometimes referred to as insulin-dependent diabetes mellitus (IDDM), and **type 2 diabetes**, or non-insulin-dependent diabetes mellitus (NIDDM), also called insulin-resistant diabetes.

Type 1 diabetes usually begins early in life, and symptoms quickly become severe. This disease responds to insulin injection, because the metabolic defect stems from an autoimmune destruction of pancreatic β cells and a consequent inability to produce sufficient insulin. Type 1 diabetes requires both insulin therapy and careful, lifelong control of the balance between dietary intake and insulin dose. Characteristic symptoms of type 1 (and type 2) diabetes are excessive thirst and frequent urination (polyuria), leading to the intake of large volumes of water (polydipsia). ("Diabetes mellitus" means "excessive excretion of sweet urine.") These symptoms are due to excretion of large amounts of glucose in the urine, a condition known as **glucosuria**.

Type 2 diabetes is slow to develop (typically in older, obese individuals), and the symptoms are milder and often go unrecognized at first. This is really a group of diseases in which the regulatory activity of insulin is disordered: insulin is produced, but some feature of the insulin-response system is defective. Individuals with this disorder are insulin-resistant. The connection between type 2 diabetes and obesity (discussed below) is an active and promising area of research.

Individuals with either type of diabetes are unable to take up glucose efficiently from the blood; recall that insulin triggers the movement of GLUT4 glucose transporters to the plasma membrane in muscle and adipose tissue (see Fig. 12-20). With glucose unavailable to cells, fatty acids become the principal fuel, which leads to another characteristic metabolic change in diabetes: excessive but incomplete oxidation of fatty acids in the liver. The acetyl-CoA produced by β oxidation cannot be completely oxidized by the citric acid cycle, because the high [NADH]/[NAD$^+$] ratio produced by β oxidation inhibits the cycle (recall that three steps of the cycle convert NAD$^+$ to NADH). Accumulation of acetyl-CoA leads to overproduction of the ketone bodies, acetoacetate and β-hydroxybutyrate, which cannot be used by extrahepatic tissues as fast as they are made in the liver. In addition to β-hydroxybutyrate and acetoacetate, the blood of individuals with diabetes contains small amounts of acetone, which results from the spontaneous decarboxylation of acetoacetate:

$$\underset{\text{Acetoacetate}}{H_3C-\overset{\overset{\displaystyle O}{\|}}{C}-CH_2-COO^-} + H_2O \longrightarrow \underset{\text{Acetone}}{H_3C-\overset{\overset{\displaystyle O}{\|}}{C}-CH_3} + HCO_3^-$$

Acetone is volatile and is exhaled, and in uncontrolled diabetes the breath has a characteristic odor sometimes mistaken for ethanol. An individual with diabetes who is experiencing mental confusion due to high blood glucose is occasionally misdiagnosed as intoxicated, an error that can be fatal. The overproduction of ketone bodies, called **ketosis**, results in greatly increased concentrations of ketone bodies in the blood (ketonemia) and urine (ketonuria).

The ketone bodies are carboxylic acids, which ionize, releasing protons. In uncontrolled diabetes, this acid production can overwhelm the capacity of the blood's bicarbonate buffering system and produce a lowering of blood pH called **acidosis** or, in combination with ketosis, **ketoacidosis**, a potentially life-threatening condition.

Biochemical measurements on blood and urine samples are essential in the diagnosis and treatment of diabetes. A sensitive diagnostic criterion is provided by the **glucose-tolerance test**. The individual fasts overnight, then drinks a test dose of 100 g of glucose dissolved in a glass of water. The blood glucose concentration is measured before the test dose and at 30 min intervals for several hours thereafter. A healthy individual assimilates the glucose readily, the blood glucose rising to no more than about 9 or 10 mM; little or no glucose appears in the urine. In diabetes, individuals assimilate the test dose of glucose poorly; their blood glucose level rises dramatically and returns to the fasting level very slowly. Because the blood glucose levels exceed the kidney threshold (about 10 mM), glucose also appears in the urine. ■

SUMMARY 23.3 Hormonal Regulation of Fuel Metabolism

■ The concentration of glucose in blood is hormonally regulated. Fluctuations in blood glucose (normally 70 to 100 mg/100 mL, or about 4.5 mM) due to dietary intake or vigorous exercise are counterbalanced by a variety of hormonally triggered changes in metabolism in several organs.

■ High blood glucose elicits the release of insulin, which speeds the uptake of glucose by tissues and favors the storage of fuels as glycogen and triacylglycerols while inhibiting fatty acid mobilization in adipose tissue.

■ Low blood glucose triggers release of glucagon, which stimulates glucose release from liver glycogen and shifts fuel metabolism in liver and muscle to fatty acid oxidation, sparing glucose for use by the brain. In prolonged fasting, TAGs become the principal fuel; the liver converts the fatty acids to ketone bodies for export to other tissues, including the brain.

■ Epinephrine prepares the body for increased activity by mobilizing glucose from glycogen and other precursors, releasing the glucose into the blood.

■ Cortisol, released in response to a variety of stressors (including low blood glucose), stimulates gluconeogenesis from amino acids and glycerol in the liver, thus raising blood glucose and counterbalancing the effects of insulin.

■ In diabetes, insulin is either not produced or not recognized by the tissues, and the uptake of blood glucose is compromised. When blood glucose levels are high, glucose is excreted. Tissues then depend on fatty acids for fuel (producing ketone bodies) and degrade cellular proteins to provide glucogenic amino acids for glucose synthesis. Uncontrolled diabetes is characterized by high glucose levels in the blood and urine and the production and excretion of ketone bodies.

23.4 Obesity and the Regulation of Body Mass

In the U.S. population, 35% of adults are obese and another 35% are overweight, as defined in terms of **body mass index (BMI)**, calculated as (weight in kg)/(height in m)^2. A BMI below 25 is considered normal; an individual with a BMI of 25 to 30 is overweight; a BMI greater than 30 indicates **obesity**. Obesity is life-threatening. It significantly increases the likelihood of developing type 2 diabetes, as well as heart attack, stroke, and cancers of the colon, breast, prostate, and endometrium. Consequently, there is great interest in understanding how body mass and the storage of fats in adipose tissue are regulated. ■

To a first approximation, obesity is the result of taking in more calories in the diet than are expended by the body's fuel-consuming activities. The body can deal with an excess of dietary calories in three ways: (1) convert excess fuel to fat and store it in adipose tissue, (2) burn excess fuel by extra exercise, and (3) "waste" fuel by diverting it to heat production (thermogenesis) by uncoupled mitochondria. In mammals, a complex set of hormonal and neuronal signals acts to keep fuel intake and energy expenditure in balance so as to hold the amount of adipose tissue at a suitable level. Dealing effectively with obesity requires understanding how the various checks and balances work under normal conditions and how these homeostatic mechanisms can fail.

Adipose Tissue Has Important Endocrine Functions

One early hypothesis to explain body-mass homeostasis, the "adiposity negative-feedback" model, postulated a mechanism that inhibits eating behavior and increases energy expenditure whenever body weight exceeds a certain value, called the set point; the inhibition is relieved when body weight drops below the set point **(Fig. 23-33)**. This model predicts that a feedback signal originating in adipose tissue influences the brain centers that control eating behavior and metabolic and motor activity. The first such factor to be discovered, in 1994, was leptin, and subsequent research revealed that adipose tissue is an important endocrine organ that produces peptide hormones, known as **adipokines**. Adipokines may act locally (autocrine and paracrine action) or systemically (endocrine action), carrying information about the adequacy of the energy reserves (TAGs) stored in adipose tissue to other tissues and to

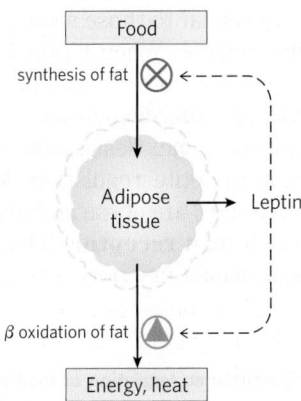

FIGURE 23-33 Set-point model for maintaining constant mass. When the mass of adipose tissue increases (dashed outline), released leptin inhibits feeding and fat synthesis and stimulates oxidation of fatty acids. When the mass of adipose tissue decreases (solid outline), lowered leptin production favors greater food intake and less fatty acid oxidation.

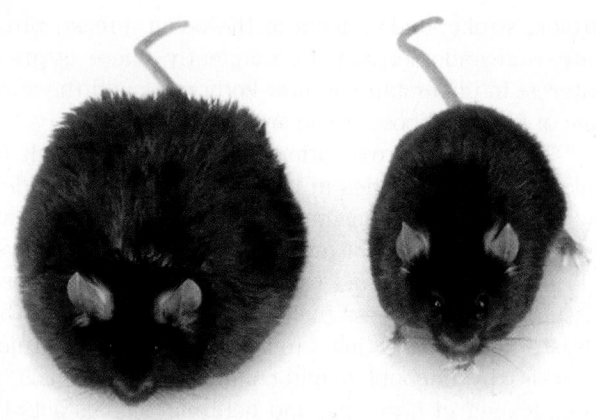

FIGURE 23-34 Obesity caused by defective leptin production. Both of these mice, which are the same age, have defects in the *OB* gene. The mouse on the right was injected daily with purified leptin and weighs 35 g. The mouse on the left got no leptin and consequently ate more food and was less active; it weighs 67 g. [Source: The Rockefeller University/AP Photo.]

the brain. Normally, adipokines produce changes in fuel metabolism and feeding behavior that reestablish adequate fuel reserves and maintain body mass. When adipokines are over- or underproduced, the resulting dysregulation may result in life-threatening disease.

Leptin (Greek *leptos*, "thin") is an adipokine (167 amino acid residues) that, on reaching the brain, acts on receptors in the hypothalamus to curtail appetite. Leptin was first identified as the product of a gene designated *OB* (obese) in laboratory mice. Mice with two defective copies of this gene (*ob/ob* genotype; lowercase letters signify a mutant form of the gene) show the behavior and physiology of animals in a constant state of starvation: their plasma cortisol levels are elevated; they exhibit unrestrained appetite, are unable to stay warm, grow abnormally large, and do not reproduce. As a consequence of unrestrained appetite, they become severely obese, weighing as much as three times more than normal mice **(Fig. 23-34)**. They also have metabolic disturbances similar to those seen in diabetes, and they are insulin-resistant. When leptin is injected into *ob/ob* mice, they eat less, lose weight, and increase their locomotor activity and thermogenesis.

A second mouse gene, designated *DB* (diabetic), also has a role in appetite regulation. Mice with two defective copies (*db/db*) are obese and diabetic. The *DB* gene encodes the **leptin receptor**. When the receptor is defective, the signaling function of leptin is lost.

FIGURE 23-35 Hypothalamic regulation of food intake and energy expenditure. (a) Role of the hypothalamus in its interaction with adipose tissue. The hypothalamus receives input (leptin) from adipose tissue and responds with neuronal signals to adipocytes. **(b)** This signal (norepinephrine) activates protein kinase A, which triggers mobilization of fatty acids from TAG and their uncoupled oxidation in mitochondria, generating heat but not ATP. DAG, diacylglycerol; MAG, monoacylglycerol.

The leptin receptor is expressed primarily in regions of the brain known to regulate feeding behavior—neurons of the **arcuate nucleus** of the hypothalamus **(Fig. 23-35a)**. Leptin carries the message that fat reserves are sufficient, and it promotes reduction of fuel intake and increase in expenditure of energy. Leptin-receptor interaction in the hypothalamus alters the release of neuronal signals to the region of the brain that affects appetite. Leptin also stimulates the sympathetic nervous system, increasing blood pressure, heart rate, and thermogenesis by uncoupling the mitochondria of brown adipocytes (Fig. 23-35b). Recall that the uncoupling protein UCP1 forms a channel in the inner mitochondrial membrane that allows protons to reenter the mitochondrial matrix without passing through the ATP synthase complex. This permits constant oxidation of fuel (fatty acids in a brown or beige adipocyte) without ATP synthesis, dissipating energy as heat and consuming dietary calories or stored fats in potentially very large amounts.

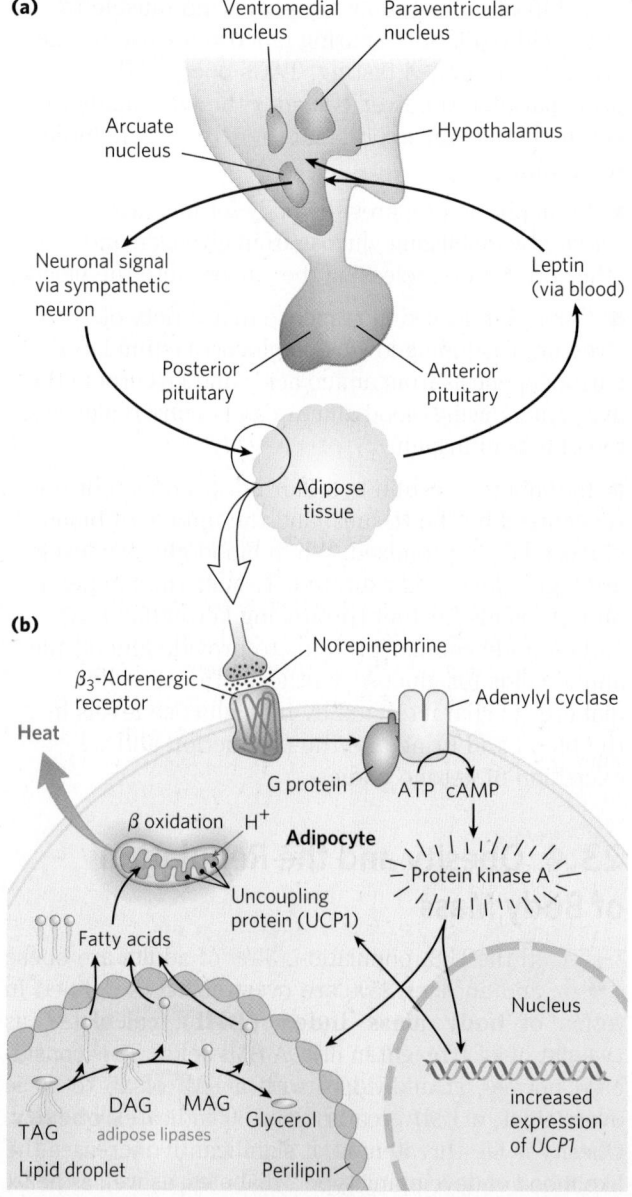

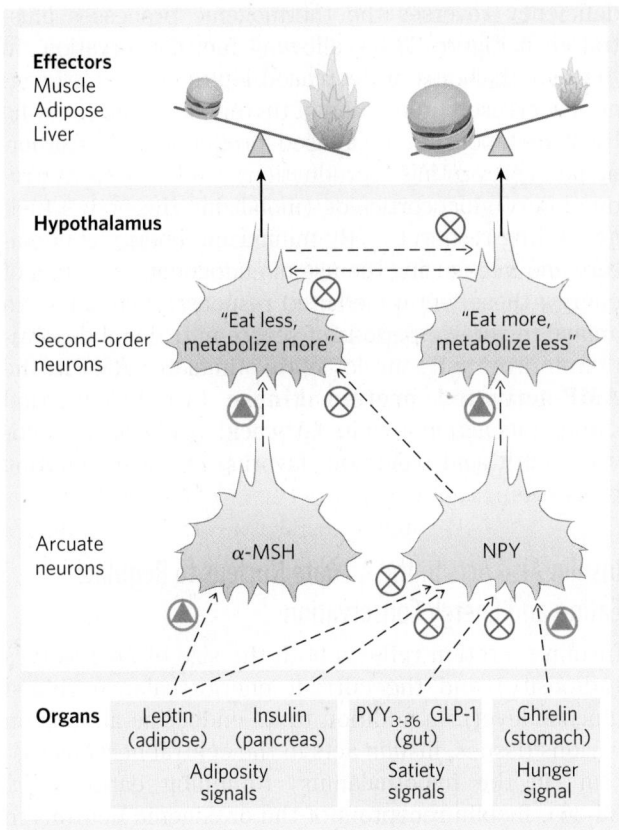

Effectors
Muscle
Adipose
Liver

Hypothalamus

Second-order neurons

"Eat less, metabolize more"

"Eat more, metabolize less"

Arcuate neurons

α-MSH

NPY

Organs

| Leptin (adipose) | Insulin (pancreas) | PYY₃₋₃₆ (gut) | GLP-1 | Ghrelin (stomach) |

Adiposity signals

Satiety signals

Hunger signal

FIGURE 23-36 Hormones that control eating. In the arcuate nucleus of the hypothalamus, two sets of neurosecretory cells receive hormonal input and relay neuronal signals to the cells of muscle, adipose tissue, and liver. Leptin and insulin are released from adipose tissue and the pancreas, respectively, in proportion to the mass of body fat. The two hormones act on anorexigenic neurosecretory cells to trigger release of α-MSH (melanocyte-stimulating hormone); α-MSH carries the signal to second-order neurons in the hypothalamus, which puts out the signals to eat less and metabolize more fuel. Leptin and insulin also act on orexigenic neurosecretory cells to inhibit the release of NPY, reducing the "eat more" signal sent to the tissues. As described later in the text, the gastric hormone ghrelin *stimulates* appetite by activating the NPY-expressing cells; PYY₃₋₃₆, released from the colon, *inhibits* these neurons and decreases appetite. Each of the two types of neurosecretory cells inhibits hormone production by the other, so any stimulus that activates orexigenic cells inactivates anorexigenic cells, and vice versa. This strengthens the effect of stimulatory inputs.

Leptin may also be essential to the normal development of hypothalamic neuronal circuits. In mice, the outgrowth of nerve fibers from the arcuate nucleus during early brain development is slower in the absence of leptin, affecting both the orexigenic and (to a lesser extent) anorexigenic outputs of the hypothalamus. It is possible that the leptin levels *during development* of these circuits determine the details of the hardwiring of this regulatory system.

Leptin Triggers a Signaling Cascade That Regulates Gene Expression

The leptin signal is transduced by a mechanism also used by receptors for interferon and growth factors, the JAK-STAT system **(Fig. 23-37)**. The leptin receptor, which has a single transmembrane segment, dimerizes when leptin binds to the extracellular domains of two monomers. Both monomers are phosphorylated on a Tyr residue of their intracellular domain by a **Janus kinase (JAK)**. The Ⓟ–Tyr residues become docking sites for three proteins that are *s*ignal *t*ransducers and *a*ctivators of *t*ranscription (**STATs** 3, 5, and 6; sometimes called fat-STATS). The docked STATs are then phosphorylated on Tyr residues by the same JAK. After phosphorylation, the STATs dimerize and move to the nucleus, where they bind to specific DNA sequences and stimulate the expression of target genes, including the gene for POMC, from which α-MSH is produced.

The increased catabolism and thermogenesis triggered by leptin are due in part to increased synthesis of the mitochondria in brown and beige adipocytes. Leptin stimulates UCP1 synthesis by altering synaptic transmissions from neurons of the arcuate nucleus to adipose and other tissues via the sympathetic nervous system. The consequent increased release of norepinephrine in these tissues acts through β₃-adrenergic receptors to stimulate transcription of the *UCP1* gene. The resulting uncoupling of electron transfer from oxidative phosphorylation consumes fat and is thermogenic (Fig. 23-35).

Leptin Stimulates Production of Anorexigenic Peptide Hormones

Two types of neurons in the arcuate nucleus control fuel intake and metabolism **(Fig. 23-36)**. The **orexigenic** (appetite-stimulating) neurons stimulate eating by producing and releasing **neuropeptide Y (NPY)**, which causes the next neuron in the circuit to send the signal to the brain: Eat! The blood level of NPY rises during starvation and is elevated in both *ob/ob* and *db/db* mice. The high NPY concentration presumably contributes to the obesity of these mice, who eat voraciously.

The **anorexigenic** (appetite-suppressing) neurons in the arcuate nucleus produce **α-melanocyte–stimulating hormone** (**α-MSH**; also known as melanocortin), formed from its polypeptide precursor pro-opiomelanocortin (POMC; Fig. 23-5). Release of α-MSH causes the next neuron in the circuit to send the signal to the brain: Stop eating!

The amount of leptin released by adipose tissue depends on both the number and the size of adipocytes. When weight loss decreases the mass of lipid tissue, leptin levels in the blood decrease, the production of NPY increases, and the processes in adipose tissue shown in Figure 23-35 are reversed. Uncoupling is diminished, slowing thermogenesis and saving fuel, and fat mobilization slows in response to reduced signaling by cAMP. Consumption of more food, combined with more efficient utilization of fuel, results in replenishment of the fat reserve in adipose tissue, bringing the system back into balance.

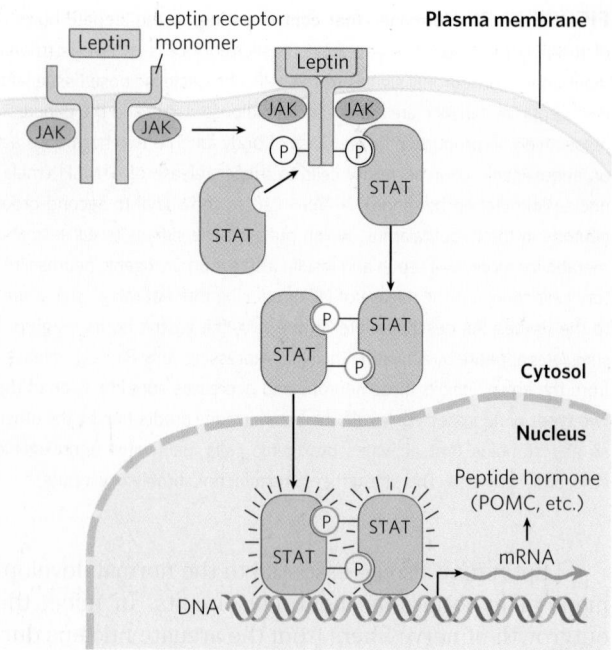

FIGURE 23-37 The JAK-STAT mechanism of leptin signal transduction in the hypothalamus. Leptin binding induces dimerization of the leptin receptor, followed by phosphorylation of specific Tyr residues in the receptor's cytosolic domain, catalyzed by Janus kinase (JAK). STATs bound to the phosphorylated leptin receptor are now phosphorylated on Tyr residues by a separate activity of JAK. The STATs dimerize, binding each other's (P)–Tyr residues, and enter the nucleus. Here, they bind specific regulatory regions in the DNA and alter the expression of certain genes. The products of these genes ultimately influence the organism's feeding behavior and energy expenditure. [Source: Information from J. Auwerx and B. Staels, *Lancet* 351:737, 1998, Fig. 2.]

Might human obesity be the result of insufficient leptin production and therefore be treatable by the injection of leptin? Blood levels of leptin are, in fact, usually much *higher* in obese animals (including humans) than in animals of normal body mass (except, of course, in *ob/ob* mutants, which cannot make leptin). Some downstream element in the leptin response system must be defective in obese individuals, and the elevation in leptin is the result of an (unsuccessful) attempt to overcome the leptin resistance. In those very rare humans with extreme obesity who have a defective leptin gene (*OB*), leptin injection does result in dramatic weight loss. In the vast majority of obese individuals, however, the *OB* gene is intact. In clinical trials, the injection of leptin did not have the weight-reducing effect observed in obese *ob/ob* mice. Clearly, most cases of human obesity involve one or more factors in addition to leptin.

The Leptin System May Have Evolved to Regulate the Starvation Response

The leptin system probably evolved to adjust an animal's activity and metabolism during periods of fasting and starvation, not as a means to restrict weight gain. The *reduction* in leptin level triggered by nutritional

deficiency reverses the thermogenic processes illustrated in Figure 23-35, allowing fuel conservation. In the hypothalamus, a decreased leptin signal also triggers decreased production of thyroid hormone (slowing basal metabolism), decreased production of sex hormones (preventing reproduction), and increased production of glucocorticoids (mobilizing the body's fuel-generating resources). By minimizing energy expenditure and maximizing the use of endogenous reserves of energy, these leptin-mediated responses may allow an animal to survive periods of severe nutritional deprivation. In liver and muscle, leptin stimulates **AMPK**, the **AMP-activated protein kinase** (see below), and through its action inhibits fatty acid synthesis and activates fatty acid oxidation, favoring energy-generating processes.

Insulin Also Acts in the Arcuate Nucleus to Regulate Eating and Energy Conservation

Insulin secretion reflects both the size of fat reserves (adiposity) and the current energy balance (blood glucose level). In addition to its endocrine actions on various tissues, insulin acts in the central nervous system (in the hypothalamus) to inhibit eating (Fig. 23-36). Insulin receptors of the orexigenic neurons in the arcuate nucleus *inhibit* the release of NPY, and insulin receptors of the anorexigenic neurons *stimulate* α-MSH production, thereby decreasing fuel intake and increasing thermogenesis. By mechanisms discussed in Section 23.3, insulin also signals muscle, liver, and adipose tissues to increase the conversion of glucose to acetyl-CoA, providing the starting material for fat synthesis.

Adiponectin Acts through AMPK to Increase Insulin Sensitivity

Adiponectin is a peptide hormone produced almost exclusively in adipose tissue, an adipokine that sensitizes other organs to the effects of insulin. Adiponectin circulates in the blood and powerfully affects the metabolism of fatty acids and carbohydrates in liver and muscle. It increases the uptake of fatty acids from the blood by myocytes and the rate at which fatty acids undergo β oxidation in muscle. It also blocks fatty acid synthesis and gluconeogenesis in hepatocytes, and stimulates glucose uptake and catabolism in muscle and liver.

These effects of adiponectin are indirect and not fully understood, but AMPK clearly mediates many of them. Acting through its GPCR, adiponectin triggers phosphorylation and activation of AMPK. Recall that AMPK is activated by factors that signal the need to shift metabolism toward energy generation and away from energy-requiring biosynthesis (**Fig. 23-38**; see also pp. 582–583). When activated, AMPK profoundly affects the metabolism of individual cells and, through its actions in the brain, the metabolism of the whole animal.

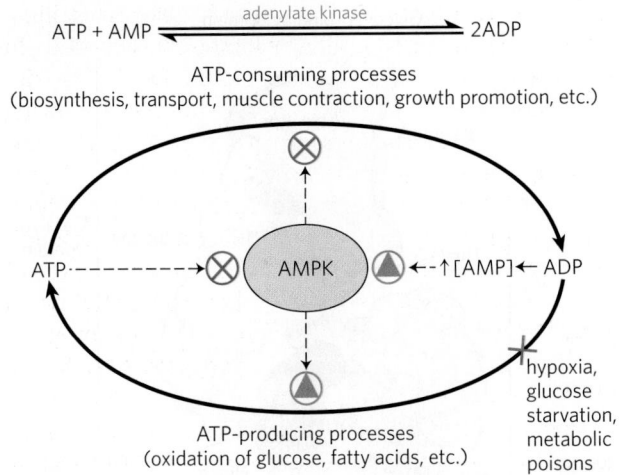

ATP-consuming processes
(biosynthesis, transport, muscle contraction, growth promotion, etc.)

ATP-producing processes
(oxidation of glucose, fatty acids, etc.)

FIGURE 23-38 The role of AMP-activated protein kinase (AMPK) in maintaining energy homeostasis. ADP produced by synthetic reactions is converted to ATP and AMP by adenylate kinase. AMP activates AMPK, which reciprocally regulates ATP-consuming and ATP-generating pathways by phosphorylating key enzymes (see Fig. 23-39). Conditions or agents that inhibit ATP production by catabolic reactions (such as hypoxia, lack of glucose, metabolic poisons) raise [AMP], activate AMPK, and stimulate catabolism. Cellular or organismal activities that consume ATP (muscle contraction, growth) increase [AMP] and stimulate catabolic reactions to replenish ATP. When [ATP] is high, ATP prevents AMP binding to AMPK, thus lowering AMPK activity and slowing catabolism.

AMPK Coordinates Catabolism and Anabolism in Response to Metabolic Stress

AMPK has emerged as a central player in the coordination of metabolic pathways, organism activity, and feeding behavior **(Fig. 23-39)**. This AMP-activated protein kinase monitors the energy and nutrient status *in individual cells* and shifts metabolism toward energy generation when necessary to maintain metabolic homeostasis. Furthermore, by responding to a variety of hormone signals, AMPK in the hypothalamus acts to keep the *whole organism* in energetic balance (Fig. 23-12).

AMPK monitors the energy status of a cell through its response to [AMP]. Many of the energy-consuming

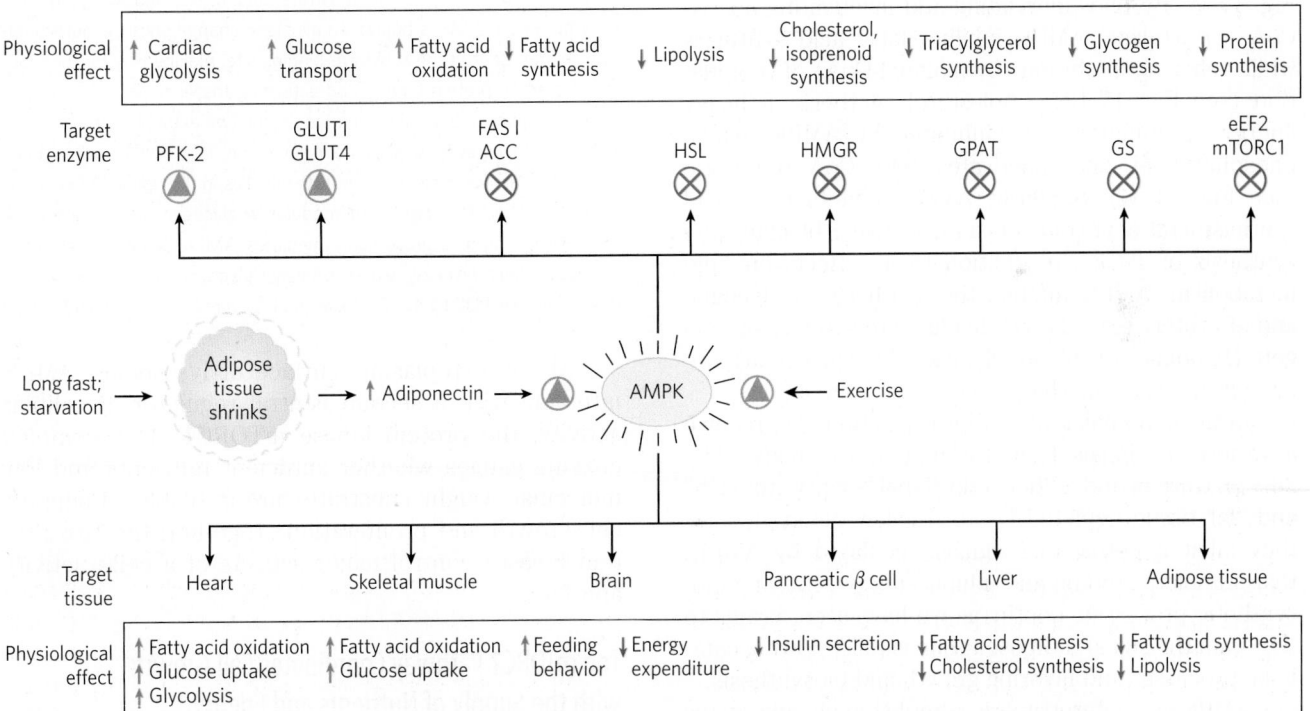

FIGURE 23-39 Formation of adiponectin and its actions through AMPK. Extended fasting or starvation decreases triacylglycerol reserves in adipose tissue, which triggers adiponectin production and release from adipocytes. Adiponectin acts through its plasma membrane receptors in various cell types and organs to inhibit energy-consuming processes and stimulate energy-producing processes. It acts in the brain to stimulate feeding behavior and inhibit energy-consuming physical activity, and in brown fat to inhibit thermogenesis. Adiponectin exerts some of its metabolic effects by activating AMPK, which regulates (by phosphorylation) specific enzymes in key metabolic processes (see Fig. 15-8).

PFK-2, phosphofructokinase-2; GLUT1 and GLUT4, glucose transporters; FAS I, fatty acid synthase I; ACC, acetyl-CoA carboxylase; HSL, hormone-sensitive lipase; HMGR, HMG-CoA reductase; GPAT, an acyl transferase; GS, glycogen synthase; eEF2, eukaryotic elongation factor 2 (required for protein synthesis; see Chapter 27); mTORC1, mammalian target of rapamycin (a protein kinase that regulates protein synthesis on the basis of nutrient availability; see Fig. 23-41). Thiazolidinedione drugs activate the transcription factor PPARγ (see Figs 23-42 and 23-43), which then turns on adiponectin synthesis, indirectly activating AMPK. Exercise, through conversion of ATP to ADP and AMP, also stimulates AMPK.

reactions in cells convert ATP to ADP or AMP. Adenylate kinase catalyzes the reaction 2ADP → AMP + ATP, so [AMP] is a sensitive measure of the cell's energy status (see Table 15-4). AMPK is allosterically activated by AMP binding, and ATP prevents AMP binding, so the enzyme is activated when the cell is energetically depleted (high [AMP]) and inactivated when energy is plentiful (high [ATP], and high [ATP]/[AMP]). AMPK responds to the energetic needs of the whole organism through a second mode of regulation. The enzyme is activated 100-fold by phosphorylation of Thr^{172} by liver kinase B1 (LKB1), which is itself subject to regulation by upstream components, including adiponectin. When activated by phosphorylation and AMP binding, AMPK phosphorylates specific enzymes in metabolic pathways that are crucial to energy homeostasis.

When AMPK senses depletion of ATP in an individual cell, lipid synthesis is inhibited and use of lipid as fuel is stimulated. One enzyme regulated by AMPK in the liver and in white adipose tissue is acetyl-CoA carboxylase, which produces malonyl-CoA, the first intermediate committed to fatty acid synthesis. Malonyl-CoA is a powerful inhibitor of the enzyme carnitine acyltransferase I, which starts the process of β oxidation by transporting fatty acids into the mitochondrion (see Fig. 17-6). By phosphorylating and inactivating acetyl-CoA carboxylase, AMPK inhibits fatty acid synthesis while relieving inhibition (by malonyl-CoA) of β oxidation (see Fig. 17-13). Cholesterol synthesis, a heavy energy consumer, is also inhibited by AMPK, which phosphorylates and inactivates HMG-CoA reductase (see Fig. 21-34). Similarly, AMPK inhibits fatty acid synthase and acyl transferase, effectively blocking the synthesis of TAGs. In addition to its effects on lipid metabolism, AMPK inhibits the synthesis of glycogen and of protein (Fig. 23-39). Inadequate supplies of oxygen (hypoxia) or blood glucose (hypoglycemia) are among the stressors that trigger AMPK activation.

In the hypothalamus, AMPK is positioned to receive a variety of signals from throughout the body (Fig. 23-12). Ghrelin and adiponectin signal "empty stomach" and "fat tissue depleted," and, like low blood glucose, they elicit hypothalamic signals, mediated by AMPK, that stimulate feeding and inhibit energy-requiring biosynthetic processes. Leptin, as we have seen, brings to the brain the signal "adipose tissue is full," slowing catabolic processes and favoring growth and biosynthesis.

AMPK is a heterotrimeric complex in all eukaryotes **(Fig. 23-40)**. In mammals, the several isoforms of each subunit are encoded by different genes. There are at least 12 combinations of the three subunits, some known to be uniquely expressed in one tissue. The catalytic module has the typical protein kinase domain. The regulatory module has four potential AMP-binding sites, at some distance from the catalytic site. A deep substrate-binding cleft includes Thr^{172}, which when phosphorylated, as noted above, activates the enzyme 100-fold.

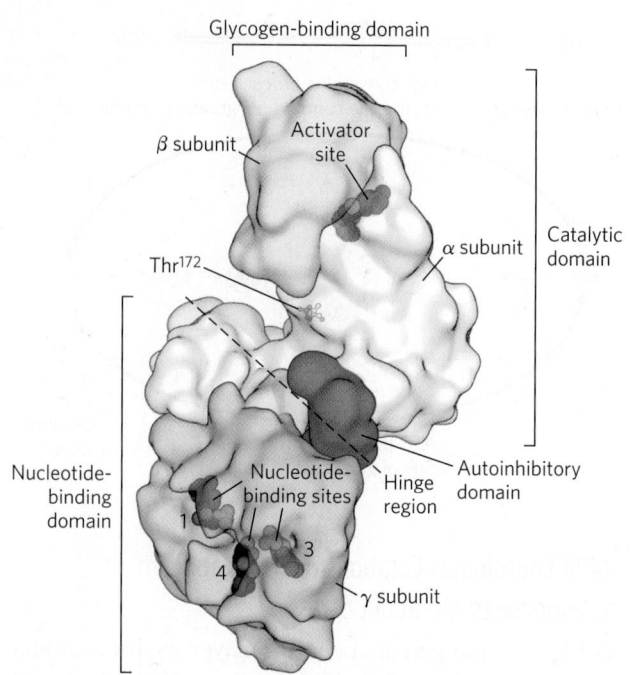

FIGURE 23-40 Structure of AMPK. The enzyme is a heterotrimer with two distinct halves: a catalytic module and a nucleotide-binding module with four nucleotide-binding sites. The hinge region that connects these two modules contains an autoinhibitory domain, which, in the absence of bound AMP, occupies the substrate-binding site and inactivates the enzyme. When AMP is bound, an allosteric change pulls the autoinhibitory segment out of the substrate-binding site, making way for the binding of specific protein targets and activating the enzyme. When ATP is bound to the nucleotide-binding module, the two extra phosphates of the ATP somehow prevent this allosteric activation. Thr^{172}, which, when phosphorylated, activates the enzyme 100-fold, lies in the region between the catalytic and nucleotide-binding modules. A glycogen-binding domain in the catalytic module presumably mediates AMPK binding to glycogen particles, where the enzyme acts to balance glycogen synthesis and breakdown. [Source: PDB ID 4CFE, B. Xiao et al., *Nature Commun.* 4:3017, 2013.]

At the cytoplasmic surface of lysosomes, AMPK interacts with a second central regulator of cellular activity, the protein kinase mTORC1. This complex enzyme gauges whether sufficient nutrients and low molecular weight substrates are available to support cell growth and proliferation. Together, the two protein kinases control major aspects of a cell's activity and fate.

The mTORC1 Pathway Coordinates Cell Growth with the Supply of Nutrients and Energy

The highly conserved, five-subunit Ser/Thr kinase **mTORC1** is activated by intracellular and extracellular signals that indicate abundant energy and nutrient supplies: branched-chain amino acids, insulin, and a high [ATP]/[AMP] ratio. When nutrients are sufficient, one or more of the mTORC1 components hold the entire complex to the outside surface of lysosomes. Once activated, mTORC1 phosphorylates several transcription factors, leading to increased ribosome

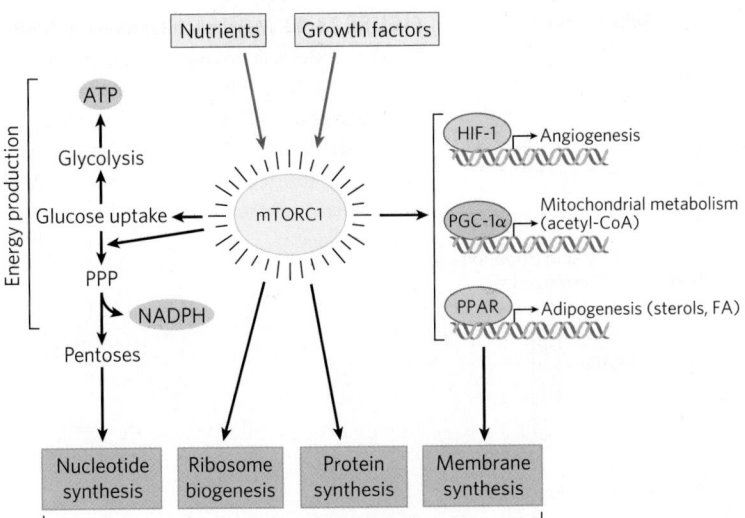

FIGURE 23-41 mTORC1 stimulates cell growth and proliferation when adequate nutrition is available. mTORC1 is a Ser/Thr protein kinase activated by growth factors and metabolites that signal a state of adequate nutrition. By phosphorylating key target proteins, mTORC1 activates energy (ATP and NADPH) production for biosynthesis and stimulates the synthesis of proteins and lipids, allowing cell growth and proliferation. PPP, pentose phosphate pathway; HIF-1, hypoxia-inducible transcription factor; PGC-1α, PPARγ coactivator-1α; PPAR, peroxisome proliferator-activated receptor; FA, fatty acids. [Source: Information from J. L. Yecies and B. D. Manning, *J. Mol. Med.* 89:221, 2011, Fig. 2.]

biogenesis and increased expression of genes encoding the enzymes of lipid synthesis and mitochondrial proliferation **(Fig. 23-41)**. Fasting results in inactivation of mTORC1 by AMPK, leading to increased breakdown of protein and glycogen in the liver and muscle and mobilization of TAGs in adipose tissue. Chronic activation of mTORC1 by overeating results in excess deposition of TAGs in adipose tissue, as well as in liver and muscle, which may contribute to insulin insensitivity and type 2 diabetes (see Section 23.5). Mutations that produce constantly activated mTORC1 are commonly associated with human cancers.

Diet Regulates the Expression of Genes Central to Maintaining Body Mass

Proteins in a family of ligand-activated transcription factors known as **peroxisome proliferator-activated receptors (PPARs)** respond to changes in dietary lipid by altering the expression of genes involved in fat and carbohydrate metabolism. These transcription factors were first recognized for their roles in peroxisome synthesis—thus their name. Their normal ligands are fatty acids or fatty acid derivatives, but they can also bind synthetic agonists and can be activated in the laboratory by genetic manipulation. PPARα, PPARδ, and PPARγ are members of this nuclear receptor superfamily. They act in the nucleus by forming heterodimers with another nuclear receptor, RXR, then binding to regulatory regions of DNA near the genes under their control and changing the rate of transcription of those genes **(Fig. 23-42)**.

PPARγ, expressed in several tissues, including adipose (brown, beige, and white), is required for turning on genes necessary to the differentiation of fibroblasts into adipocytes and genes that encode proteins required for lipid synthesis and storage in adipocytes **(Fig. 23-43)**. PPARγ is activated by the thiazolidinedione drugs that are used to treat type 2 diabetes (discussed below).

PPARα is expressed in liver, kidney, heart, skeletal muscle, and brown adipose tissue. The ligands that activate this transcription factor include eicosanoids, free fatty acids, and the class of drugs called fibrates, such as fenofibrate (TriCor) and ciprofibrate, which are used to treat coronary heart disease by raising HDL and lowering blood TAGs. In hepatocytes, PPARα turns on the genes necessary for the uptake and β oxidation of fatty acids and the formation of ketone bodies during fasting.

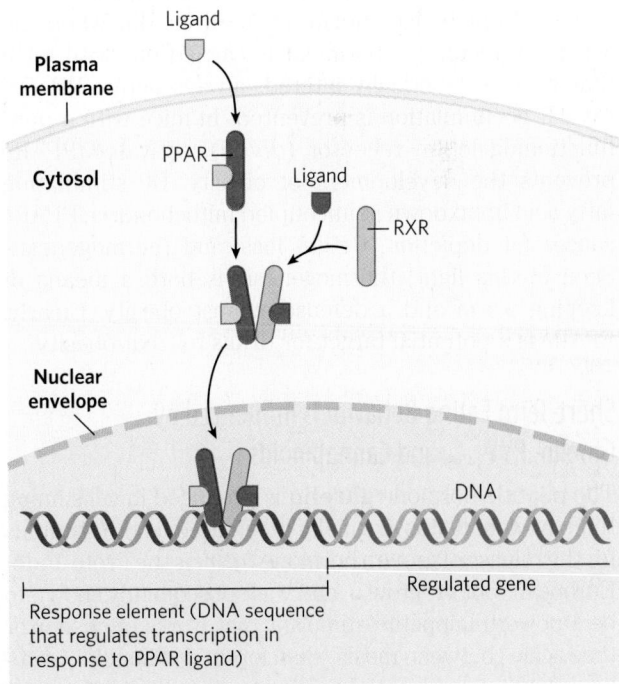

FIGURE 23-42 Mode of action of PPARs. PPARs are transcription factors that, when bound to their cognate ligand, form heterodimers with the nuclear receptor RXR. The dimer binds specific regions of DNA known as response elements, stimulating transcription of genes in those regions. [Source: Information from R. M. Evans et al., *Nature Med.* 10:355, 2004, Fig. 3.]

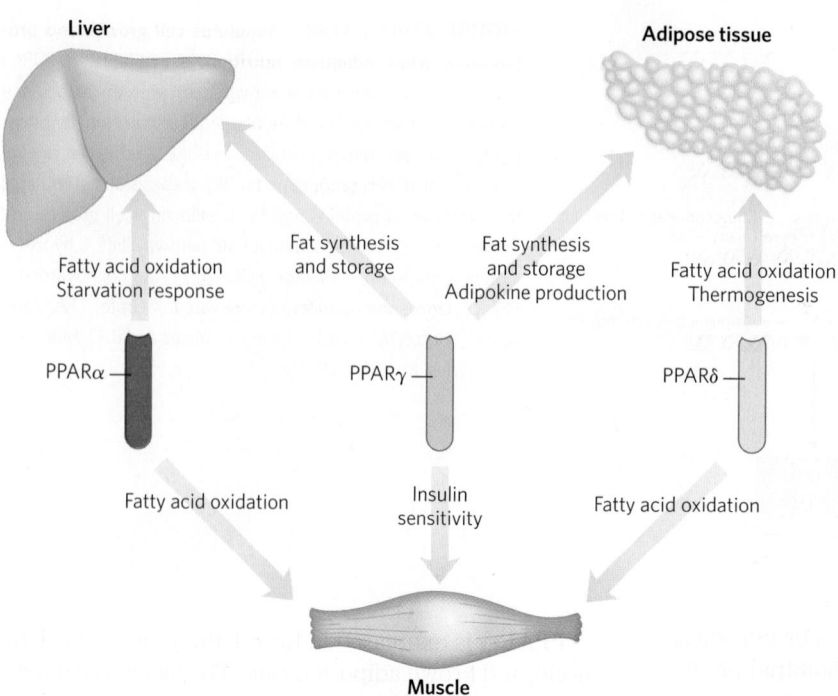

FIGURE 23-43 Metabolic integration by PPARs. The three PPAR isoforms regulate lipid and glucose homeostasis through their coordinated effects on gene expression in liver, muscle, and adipose tissue. PPARα and PPARδ (and its closely related isoform PPARβ) regulate lipid utilization; PPARγ regulates lipid storage and the insulin sensitivity of various tissues.

PPARδ and **PPARβ** are key regulators of fat oxidation, which act by sensing changes in dietary lipid. PPARδ acts in liver and muscle, stimulating the transcription of at least nine genes encoding proteins for β oxidation and for energy dissipation through uncoupling of mitochondria. Normal mice overfed on high-fat diets accumulate massive amounts of white adipose tissue, and fat droplets accumulate in the liver. But when the same overfeeding experiment is carried out with mice that have a genetically altered, always active PPARδ, this fat accumulation is prevented. In mice with a nonfunctioning leptin receptor (*db/db*), activated PPARδ prevents the development of obesity. By stimulating fatty acid breakdown in uncoupled mitochondria, PPARδ causes fat depletion, weight loss, and thermogenesis. Seen in this light, thermogenesis is both a means of keeping warm and a defense against obesity. Clearly, PPARδ is a potential target for drugs to treat obesity.

Short-Term Eating Behavior Is Influenced by Ghrelin, PYY$_{3-36}$, and Cannabinoids

The peptide hormone **ghrelin** is produced in cells lining the stomach. It was originally recognized as the stimulus for the release of growth hormone (*ghre* is the Proto-Indo-European root of "grow"), and was subsequently shown to be a powerful appetite stimulant that works on a shorter time scale (between meals) than leptin and insulin. Ghrelin receptors are located in the pituitary gland, presumably mediating growth hormone release, and in the hypothalamus, affecting appetite, as well as in heart muscle and adipose tissue. Ghrelin acts through a GPCR to generate the second messenger IP$_3$, which mediates the hormone's action. The concentration of ghrelin in the blood fluctuates strikingly throughout the day, peaking just before a meal and dropping sharply just after a meal **(Fig. 23-44)**. Injection of ghrelin into humans produces immediate sensations of intense hunger. Individuals with Prader-Willi syndrome, whose blood levels of ghrelin are exceptionally high, have an uncontrollable appetite, leading to extreme obesity that often results in death before the age of 30.

PYY$_{3-36}$ is a peptide hormone (34 amino acid residues) secreted by endocrine cells in the lining of the small intestine and colon in response to food entering from the stomach. The level of PYY$_{3-36}$ in the blood rises after a meal and remains high for some hours. The hormone is carried in the blood to the arcuate nucleus, where it acts on orexigenic neurons, inhibiting NPY release and reducing hunger (Fig. 23-36). Humans injected with PYY$_{3-36}$ feel little hunger and eat less than normal amounts for about 12 hours.

O-GlcNAc transferase (OGT) is an enzyme that acts to dampen appetite. This enzyme transfers the amino sugar *N*-acetyl glucosamine (p. 247) from the sugar nucleotide UDP-GlcNAc to the hydroxyl group of Ser and Thr side chains. Its targets include synaptic proteins of neurons in the paraventricular nucleus of the hypothalamus. When OGT is inhibited in experimental mice, they eat nearly twice as much as control animals, and triple their adipose mass in two weeks. OGT activity in control mice increases when food is available, leading to a model for satiety in which access to food activates OGT in critical neurons of the paraventricular nucleus. After a lag (the eating period), the posttranslational modification of synaptic proteins by OGT triggers action potentials that carry the message: Stop eating! Without this message, the animal eats twice as much and becomes obese.

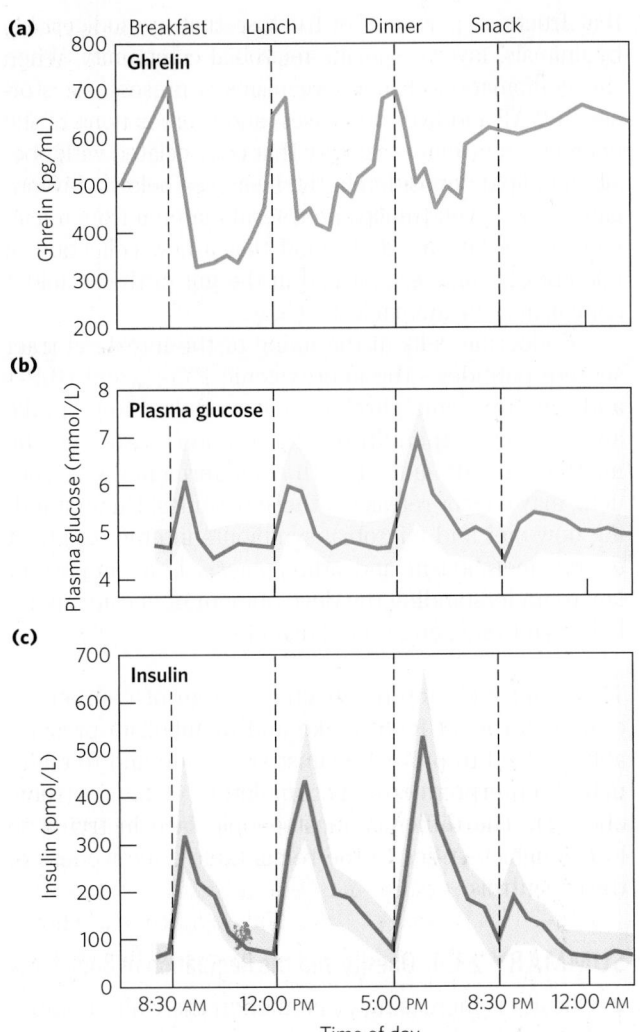

FIGURE 23-44 Variations in blood concentrations of glucose, ghrelin, and insulin relative to meal times. (a) Plasma levels of ghrelin rise sharply just *before* the usual time for meals (7 AM breakfast, 12 noon lunch, 5:30 PM dinner) and drop precipitously just after meals, paralleling subjective feelings of hunger. **(b)** Plasma glucose rises sharply *after* a meal, **(c)** followed immediately by a rise in insulin level in response to the increased blood glucose. [Sources: (a, c) Data from D. E. Cummings et al., *Diabetes* 50:1714, 2001, Fig. 1. (b) Data from M. D. Feher and C. J. Bailey, *Br. J. Diabet. Vasc. Dis.* 4:39, 2004.]

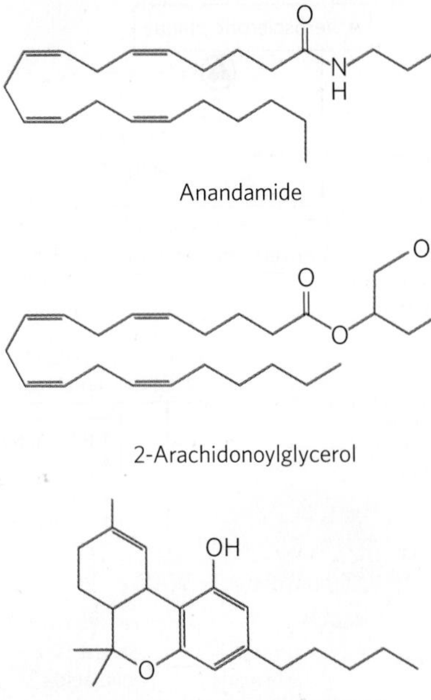

FIGURE 23-45 Cannabinoids. Two endocannabinoids produced by animals, and the psychoactive plant product tetrahydrocannabinol (THC) found in marijuana.

Endocannabinoids (Fig. 23-45) are lipid messengers that act through specific receptors in the brain and throughout the nervous system to increase appetite, heighten the sensory response to food (especially sweet and fatty foods), and elevate mood. When food enters the mouth, neuronal signals travel to the brain, and from there through the vagus nerve to the intestine, which then produces and releases endocannabinoids. The receptors for endocannabinoids are GPCRs that control ion channels of sensory neurons, changing their membrane potentials and sending signals to the brain. Palatable food sensed in this way motivates further consumption of that food. The taste of fats (which are particularly high in caloric content) causes cannabinoid release, which effectively triggers further consumption. Well-conserved across vertebrate species, this system probably evolved to maximize the intake of food and to guard against starvation. In mammals, cannabinoid action stimulates an increase in fat mass and inhibits energy loss by motor activity or thermogenesis. Cannabinoid receptors also mediate the psychoactive effects of tetrahydrocannabinol (Fig. 23-45), one of the active ingredients in marijuana, long known for its stimulating effect on appetite ("the munchies").

Microbial Symbionts in the Gut Influence Energy Metabolism and Adipogenesis

An adult human is host to about 10^{14} microbial cells that inhabit the gut. These microbes function as a major endocrine organ, producing a variety of metabolites with profound effects on host metabolism, feeding behavior, and body mass. Lean and obese individuals harbor different combinations of microbial symbionts in the gut. Investigation of this observation led to the discovery that gut microbes release fermentation products—the short-chain fatty acids acetate, propionate, butyrate, and lactate—that enter the bloodstream and trigger metabolic changes in adipose tissue **(Fig. 23-46)**. Propionate, for example, drives the expansion of WAT by acting on GPCRs (GPR41 and GPR43) in the plasma membranes of several cell types, including adipocytes. These receptors trigger differentiation of precursor cells (preadipocytes) into adipocytes and inhibit lipolysis in existing adipocytes, leading to an increase in WAT mass—that is, obesity. Gut microbes also convert primary bile acids, synthesized in

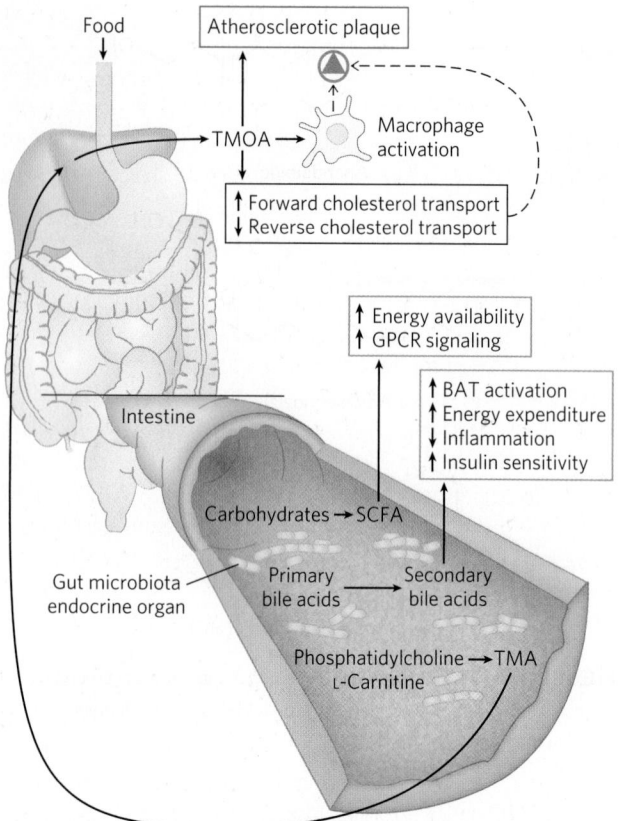

FIGURE 23-46 Effects of gut microbe metabolism on health. The enormous number and diversity of microorganisms (the microbiota) in the colon generate metabolic products that may have significant effects on health, both positive and negative. For example, the metabolism of primary bile acids by microbiota produces secondary products that act through nuclear receptors to stimulate thermogenesis in host brown adipose tissue (BAT) and increase both energy consumption and insulin sensitivity, while reducing inflammation. Metabolism of undigested carbohydrates by microbiota produces short-chain fatty acids (SCFA) that signal expansion of the host's white adipose tissue (WAT), promoting obesity. SCFAs produced by microbiota are also a readily metabolizable energy source for the host. Microbial production of the SCFA propionate prevents lipogenesis in the liver and lowers blood cholesterol, both favorable to health. On the other hand, metabolic conversion of phosphatidylcholine and L-carnitine to trimethylamine (TMA), and its further conversion in liver to trimethylamine-N-oxide (TMAO), results in receptor-mediated changes in cholesterol transport and macrophage activity. The combination of altered sterol transport and increased macrophage activity can lead to formation of atherosclerotic plaque (see Fig. 21-46). Understanding how individuals can achieve and maintain a healthy balance of microbiota is a major challenge for the future.

the liver, into the secondary bile acids deoxycholate and lithocholate, which enter the bloodstream and act through GPCRs and steroid receptors to activate beige adipocytes to produce UCP1 and increase energy expenditure.

These findings raise the possibility of preventing obesity by altering the makeup of the microbial community in the gut. This might be accomplished either by adding, directly to the gut, microbial species (**probiotics**) that disfavor adipogenesis, or by adding to the diet nutrients (**prebiotics**) that favor the dominance of probiotic microbes. For example, experiments in mice have shown

that fructans, polymers of fructose that are indigestible by animals, favor a specific microbial community. When this combination of microorganisms is present, fat storage in WAT and liver decreases, and there is none of the decrease in insulin sensitivity that is associated with obesity and lipid deposition in the liver (see below). Investigators have even transplanted fecal material from a lean mouse to a fat one, and found that a new collection of microbes became established in the gut of the recipient animal, and the animal lost weight.

Endocrine cells in the lining of the intestinal tract secrete peptides—the anorexigenic PYY_{3-36} and GLP-1 and the orexigenic ghrelin—that modulate food intake and energy expenditure. Interaction with specific microbes in the gut, or with their fermentation products, may trigger release of these peptides. Understanding how diet and microbial symbionts interact to affect energy metabolism and adipogenesis is an important key to understanding the development of obesity, metabolic syndrome, and type 2 diabetes.

This exquisitely interconnected system of neuroendocrine controls of food intake and metabolism presumably evolved to protect against starvation and to eliminate counterproductive accumulation of fat (extreme obesity). The difficulty most people face in trying to lose weight testifies to the remarkable effectiveness of these controls.

SUMMARY 23.4 Obesity and the Regulation of Body Mass

■ Obesity is increasingly common in the United States and other developed countries and predisposes people to several chronic, life-threatening conditions, including cardiovascular disease and type 2 diabetes.

■ Adipose tissue produces leptin, a hormone that regulates feeding behavior and energy expenditure so as to maintain adequate reserves of fat. Leptin production and release increase with the number and size of adipocytes.

■ Leptin acts on receptors in the arcuate nucleus of the hypothalamus, causing the release of anorexigenic (appetite-suppressing) peptides, including α-MSH, that act in the brain to inhibit eating. Leptin also stimulates sympathetic nervous system action on adipocytes, leading to uncoupling of mitochondrial oxidative phosphorylation, with consequent thermogenesis.

■ The signal-transduction mechanism for leptin involves phosphorylation of the JAK-STAT system. On phosphorylation by JAK, STATs can bind to regulatory regions in nuclear DNA and alter the expression of genes encoding proteins that set the level of metabolic activity and determine feeding behavior. Insulin acts on receptors in the arcuate nucleus, with results similar to those caused by leptin.

■ The hormone adiponectin stimulates fatty acid uptake and oxidation and inhibits fatty acid synthesis.

It also sensitizes muscle and liver to insulin. The actions of adiponectin are mediated by AMPK, which is also activated by low [AMP] and exercise.

■ Ghrelin, a hormone produced in the stomach, acts on orexigenic (appetite-stimulating) neurons in the arcuate nucleus to produce hunger before a meal. PYY_{3-36}, a peptide hormone of the intestine, acts at the same site to lessen hunger after a meal. Endocannabinoids signal the availability of sweet or fatty food and stimulate its consumption.

■ Microbial symbionts in the gut produce fermentation products and secondary bile acids. They influence release of gut hormones that regulate body mass.

23.5 Obesity, Metabolic Syndrome, and Type 2 Diabetes

In the industrialized world, where the food supply is more than adequate, there is a growing epidemic of obesity and the type 2 diabetes associated with it. As many as 300 million people in the world now have diabetes, and reasonable projections predict a dramatic rise in cases over the next decade, following the world epidemic of obesity. The pathology of diabetes includes cardiovascular disease, renal failure, blindness, neuropathy, and poor healing in the extremities that leads to amputations. In 2014, the global mortality from diabetes was estimated at nearly 4 million, a number that is sure to rise in coming years. Clearly, it is essential to understand type 2 diabetes and its relationship to obesity and to find countermeasures that prevent or reverse the damage done by this disease. ■

In Type 2 Diabetes the Tissues Become Insensitive to Insulin

The hallmark of **type 2 diabetes** is the development of insulin resistance, a state in which more insulin is needed to bring about the biological effects produced by a lower amount of insulin in the normal, healthy state. In the early stages of the disease, pancreatic β cells secrete enough insulin to overcome the lower insulin sensitivity of muscle and liver. But the β cells eventually fail, and the lack of insulin becomes apparent in the body's inability to regulate blood glucose. The intermediate stage, preceding type 2 diabetes mellitus, is sometimes called **metabolic syndrome**, or **syndrome X**. This is typified by obesity, especially in the abdomen; hypertension (high blood pressure); abnormal blood lipids (high TAG and LDL, low HDL); slightly high fasting blood glucose; and a reduced ability to clear glucose in the glucose-tolerance test. Individuals with metabolic syndrome often also show changes in blood proteins, changes that are associated with abnormal clotting (high fibrinogen concentration) or inflammation (high concentrations of the C-reactive peptide, which typically increases with an inflammatory response). About 27% of the adult population in the United States has these symptoms of metabolic syndrome.

What predisposes individuals with metabolic syndrome to develop type 2 diabetes? According to the "lipid toxicity" hypothesis **(Fig. 23-47)**, the action of PPARγ on adipocytes normally keeps the cells ready to synthesize and store triacylglycerols—the adipocytes are insulin-sensitive and produce leptin, which leads to their continued intracellular deposition of TAGs. However, excess caloric intake in obese individuals causes adipocytes to become filled with TAGs, leaving adipose tissue unable to meet any further demand for TAG storage. Lipid-filled adipose tissue releases protein factors that attract macrophages, which infiltrate the tissue and may eventually represent as much as 50% of the adipose tissue by mass. Macrophages trigger the inflammatory response, which impairs TAG deposition in adipocytes and favors release of free fatty acids into the blood. These excess fatty acids enter liver and muscle cells, where they are converted to TAGs that accumulate as lipid droplets. This ectopic (Greek *ektopos*, "out of place") deposition of TAGs leads to insulin insensitivity in liver and muscle, the hallmark of type 2 diabetes.

According to this hypothesis, excess stored fatty acids and TAGs are toxic to liver and muscle. Some individuals are less well equipped genetically to handle this burden of ectopic lipids and are more susceptible to the cellular damage that leads to development of type 2 diabetes. Insulin resistance probably involves impairment of several of the mechanisms by which insulin acts on metabolism, which include changes in protein levels and changes in the activities of signaling enzymes and transcription factors. For example, both adiponectin synthesis in adipocytes and adiponectin levels in the blood decrease with obesity and increase with weight loss.

Several of the drugs that are effective in improving insulin sensitivity in type 2 diabetes are known to act on specific proteins in signaling pathways, and their actions are consistent with the lipotoxicity model. **Thiazolidinediones** (used to treat type 2 diabetes; see Fig. 21-22) bind to PPARγ, turning on a set of adipocyte-specific genes and promoting differentiation of preadipocytes to small adipocytes, thereby increasing the body's capacity for absorbing fatty acids from the diet and storing them as TAGs.

There are clearly genetic factors that predispose toward type 2 diabetes. Although 80% of individuals with type 2 diabetes are obese, most obese individuals do not develop type 2 diabetes. Given the complexity of the regulatory mechanisms we have discussed in this chapter, it is not surprising that the genetics of diabetes is complex, involving interactions between variant genes and environmental factors, including diet and lifestyle. At least 10 genetic loci have been reliably linked to type 2 diabetes; variation in any of these "diabetogenes" alone would cause a relatively small increase in the likelihood of developing type 2 diabetes. For example, people with a variant PPARγ in which an Ala residue replaces Pro at position 12 are at a slightly, though significantly, increased risk of developing type 2 diabetes. ■

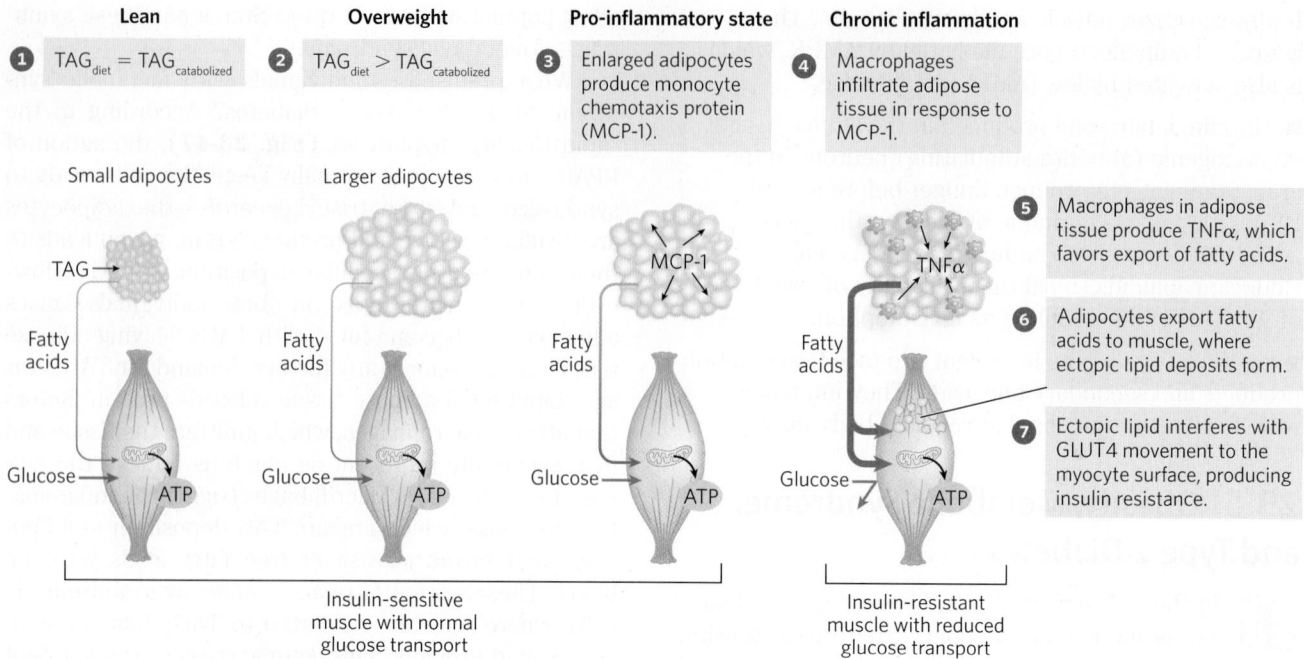

Lean ❶ $TAG_{diet} = TAG_{catabolized}$

Overweight ❷ $TAG_{diet} > TAG_{catabolized}$

Pro-inflammatory state ❸ Enlarged adipocytes produce monocyte chemotaxis protein (MCP-1).

Chronic inflammation ❹ Macrophages infiltrate adipose tissue in response to MCP-1.

❺ Macrophages in adipose tissue produce TNFα, which favors export of fatty acids.

❻ Adipocytes export fatty acids to muscle, where ectopic lipid deposits form.

❼ Ectopic lipid interferes with GLUT4 movement to the myocyte surface, producing insulin resistance.

Insulin-sensitive muscle with normal glucose transport

Insulin-resistant muscle with reduced glucose transport

FIGURE 23-47 Overloading adipocytes with triacylglycerols triggers inflammation in fat tissue and ectopic lipid deposition and insulin resistance in muscle. In an individual of healthy body mass, dietary TAG uptake equals TAG oxidation for energy. In overweight individuals, excess caloric intake results in enlarged adipocytes, engorged with TAG and unable to store more. Enlarged adipocytes secrete MCP-1 (monocyte chemotaxis protein-1), attracting macrophages. Macrophages infiltrate the adipose tissue and produce TNFα (tumor necrosis factor a), which triggers lipid breakdown and release of fatty acids into the blood. The fatty acids enter myocytes, where they accumulate in small lipid droplets. This ectopic lipid storage in muscle somehow causes insulin resistance, perhaps by triggering lipid-activated protein kinases that inactivate some element in the insulin-signaling pathway. GLUT4 glucose transporters leave the myocyte surface, preventing glucose entry into muscle; the myocyte has now become insulin-resistant. It cannot use blood glucose for its fuel, so fatty acids are mobilized from adipose tissue and become the primary fuel. The increased influx of fatty acids into muscle leads to further deposition of ectopic lipids. In some individuals, insulin resistance develops into type 2 diabetes. Other individuals are genetically less susceptible to the deleterious effects of ectopic lipid storage, or are genetically better equipped to manage this storage, and do not develop diabetes. [Source: Information from A. Guilherme et al., *Mol. Cell Biol.* 9:367, 2008, Fig. 1.]

Type 2 Diabetes Is Managed with Diet, Exercise, Medication, and Surgery

Studies show that three factors improve the health of individuals with type 2 diabetes: dietary restriction, regular exercise, and drugs that increase insulin sensitivity or insulin production. Dietary restriction (and accompanying weight loss) reduces the overall burden of handling fatty acids. The lipid composition of the diet influences, through PPARs and other transcription factors, the expression of genes that encode proteins involved in fatty acid oxidation and in burning fat. Exercise activates AMPK, as does adiponectin; AMPK shifts metabolism toward fat oxidation and inhibits fat synthesis.

Exercise consumes calories and in that way directly contributes to weight loss. Exercise also increases release of **irisin** from muscle into the blood. Irisin increases the expression of *UCP1* genes in WAT and also stimulates the development of beige adipocytes, so even after the exercise ends, energy continues to be used in thermogenesis.

Several classes of drugs are used in the management of type 2 diabetes (Table 23-8), some of which we discussed earlier in the chapter. Sulfonylureas act on the ATP-gated K⁺ channels in β cells to stimulate insulin release. Biguanides such as metformin (Glucophage) indirectly activate AMPK. Their direct target is mitochondrial Complex I; its inhibition by metformin reduces ATP production and raises [AMP], which activates AMPK. AMPK triggers the shift to metabolism that conserves energy: increased glucose uptake and oxidation, increased fatty acid mobilization and oxidation, and inhibited fatty acid and sterol synthesis. We have already noted that thiazolidinediones act through PPARγ to increase the concentration of adiponectin in plasma and to stimulate adipocyte differentiation, thereby increasing the capacity for TAG storage.

In cases of extreme obesity and type 2 diabetes, the diabetes can be relieved, or even reversed, by the most effective weight-loss treatment for severe obesity: bariatric surgery, which alters the path of food through the stomach and into the upper region of the small intestine (duodenum) **(Fig. 23-48)**. In Roux-en-Y gastric bypass (RYGBP, named for César Roux, the Swiss surgeon who developed the procedure), the stomach is reduced to a small pouch attached to the esophagus, and the middle region of the small intestine (the jejunum) is attached directly to the pouch. Food bypasses most of the stomach and the duodenum, and goes primarily to the "Roux limb" of the intestine. Stomach acid and digestive

TABLE 23-8 ☤ Treatments for Type 2 Diabetes Mellitus

Intervention/treatment	Direct target	Effect of treatment
Weight loss	Adipose tissue; reduce TAG content	Reduces lipid burden; increases capacity for lipid storage in adipose tissue; restores insulin sensitivity
Exercise	AMPK, activated by increasing [AMP]/[ATP]	Aids weight loss (see Fig. 23-39)
Bariatric surgery	Unknown	Leads to weight loss, better control of blood glucose
Sulfonylureas: glipizide (Glucotrol), glyburide (Micronase), glimepiride (Amaryl)	Pancreatic β cells; K^+ channels blocked	Stimulates insulin secretion by pancreas (see Fig. 23-28)
Biguanides: metformin (Glucophage)	AMPK, activated	Increases glucose uptake by muscle; decreases glucose production in liver
Thiazoladinediones: troglitazone (Rezulin),[a] rosiglitazone (Avandia),[b] pioglitazone (Actos)	PPARγ	Stimulates expression of genes, potentiating the action of insulin in liver, muscle, adipose tissue; increases glucose uptake; decreases glucose synthesis in liver
GLP-1 modulators: exenatide (Byetta), sitagliptin (Januvia)	Glucagon-like peptide-1, dipeptide protease IV	Enhances insulin secretion by pancreas

[a]Voluntarily withdrawn because of side effects.
[b]Prescriptions limited to patients not helped by other treatment, because of possible increased risk of cardiovascular disease.

enzymes travel through bypassed portions of the gut to join the food in the common channel. People who undergo RYGBP surgery not only experience dramatic weight loss but also are less hungry. Remarkably, this surgery also reverses type 2 diabetes in many cases. The explanations for these effects are likely to lie in altered communication among the gut, the brain, and other organs. This may result from changes in the kind and amount of peptide hormones (such as GLP-1 and PYY$_{3-36}$) secreted in the intestine that signal satiety and inhibit feeding behavior. The last word has not been written on this issue. ■

SUMMARY 23.5 Obesity, Metabolic Syndrome, and Type 2 Diabetes

■ Metabolic syndrome, which includes obesity, hypertension, elevated blood lipids, and insulin resistance, is often the prelude to type 2 diabetes.

■ The insulin resistance that characterizes type 2 diabetes may be a consequence of abnormal lipid storage in muscle and liver, in response to a lipid intake that cannot be accommodated by adipose tissue.

■ Expression of the enzymes of lipid synthesis is under tight and complex regulation. PPARs are

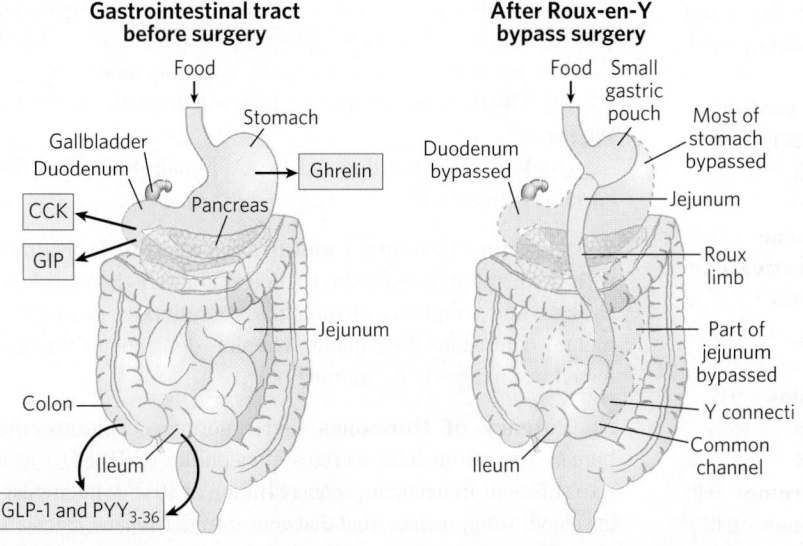

FIGURE 23-48 Bariatric surgery. The digestive system produces peptide hormones—ghrelin, cholecystokinin (CCK), GIP, PYY$_{3-36}$, and GLP-1—that affect energy homeostasis, including hunger and satiety. Shown here are the normal human gastrointestinal tract (top and left) and the GI tract after Roux-en-Y gastric bypass (right). In this surgical procedure, most of the stomach and the upper portion of the small intestine (duodenum) are bypassed. This results in rapid weight loss (as much as 5 lb a week), as well as better blood glucose regulation in individuals with type 2 diabetes. A newer surgical treatment, vertical gastric sleeve surgery (not shown), yields similar results. The biochemical and physiological bases for these effects are not well understood, but presumably include changes in peptide hormone signaling that result from rerouting food through the GI tract.

transcription factors that determine the rate of synthesis of many enzymes involved in lipid metabolism and adipocyte differentiation.

■ Effective treatments for type 2 diabetes include exercise, appropriate diet, and drugs that increase insulin sensitivity or insulin production. Surgical alteration of the digestive tract leads to weight loss and often reverses type 2 diabetes.

Key Terms

Terms in bold are defined in the glossary.

hormone 907	**platelets** 929
neuroendocrine system 908	blood plasma 929
	plasma proteins 929
radioimmunoassay (RIA) 910	ATP-gated K^+ channels 933
enzyme-linked immunosorbent assay (ELISA) 910	**sulfonylurea drugs** 933
	glucagon 934
	cortisol 937
metabotropic 910	**diabetes mellitus** 938
ionotropic 910	type 1 diabetes 938
endocrine 911	type 2 diabetes 938
paracrine 911	glucosuria 938
autocrine 911	**ketosis** 938
peptide hormones 912	**acidosis** 938
insulin 912	**ketoacidosis** 938
epinephrine 913	glucose-tolerance test 938
norepinephrine 913	
catecholamines 913	**body mass index (BMI)** 939
eicosanoid hormones 913	
steroid hormones 913	adipokines 939
vitamin D hormone (calcitriol) 914	leptin 940
	arcuate nucleus 940
retinoid hormones 914	**orexigenic** 941
thyroid hormones 914	neuropeptide Y (NPY) 941
nitric oxide (NO•) 915	**anorexigenic** 941
NO synthase 915	α-melanocyte–stimulating hormone (α-MSH) 941
hypothalamus 915	
posterior pituitary 915	JAK (Janus kinase) 941
anterior pituitary 915	STAT (signal transducer and activator of transcription) 941
tropic hormones (tropins) 915	
hepatocyte 919	**AMP-activated protein kinase (AMPK)** 942
white adipose tissue (WAT) 922	adiponectin 942
adipocyte 922	**mTORC1** 944
brown adipose tissue (BAT) 923	**PPAR (peroxisome proliferator-activated receptor)** 945
uncoupling protein 1 (UCP1) 923	ghrelin 946
thermogenesis 923	PYY$_{3-36}$ 946
beige adipose tissue 924	**endocannabinoids** 947
myocyte 925	**probiotics** 948
erythrocyte 929	**prebiotics** 948
leukocyte 929	**metabolic syndrome** 949
lymphocytes 929	**thiazolidinediones** 949

Problems

1. Peptide Hormone Activity Explain how two peptide hormones as structurally similar as oxytocin and vasopressin can have such different effects (see Fig. 23-10).

2. ATP and Phosphocreatine as Sources of Energy for Muscle During muscle contraction, the concentration of phosphocreatine in skeletal muscle drops, while the concentration of ATP remains fairly constant. However, in a classic experiment, Robert Davies found that if he first treated muscle with 1-fluoro-2,4-dinitrobenzene (FDNB, p. 98), the concentration of ATP declined rapidly while the concentration of phosphocreatine remained unchanged during a series of contractions. Suggest an explanation.

3. Metabolism of Glutamate in the Brain Brain tissue takes up glutamate from the blood, transforms it into glutamine, and releases the glutamine into the blood. What is accomplished by this metabolic conversion? How does it take place? The amount of glutamine produced in the brain can exceed the amount of glutamate entering from the blood. How does this extra glutamine arise? (Hint: You may want to review amino acid catabolism in Chapter 18; recall that NH_4^+ is very toxic to the brain.)

4. Proteins as Fuel during Fasting When muscle proteins are catabolized in skeletal muscle during a fast, what are the fates of the amino acids?

5. Absence of Glycerol Kinase in Adipose Tissue Glycerol 3-phosphate is required for the biosynthesis of triacylglycerols. Adipocytes, specialized for the synthesis and degradation of TAGs, cannot use glycerol directly because they lack glycerol kinase, which catalyzes the reaction

$$\text{Glycerol + ATP} \longrightarrow \text{glycerol 3-phosphate + ADP}$$

How does adipose tissue obtain the glycerol 3-phosphate necessary for TAG synthesis?

6. Oxygen Consumption during Exercise A sedentary adult consumes about 0.05 L of O_2 in 10 seconds. A sprinter, running a 100 m race, consumes about 1 L of O_2 in 10 seconds. After finishing the race, the sprinter continues to breathe at an elevated (but declining) rate for some minutes, consuming an extra 4 L of O_2 above the amount consumed by the sedentary individual.

(a) Why does the need for O_2 increase dramatically during the sprint?

(b) Why does the demand for O_2 remain high after the sprint is completed?

7. Thiamine Deficiency and Brain Function Individuals with thiamine deficiency show some characteristic neurological signs and symptoms, including loss of reflexes, anxiety, and mental confusion. Why might thiamine deficiency be manifested by changes in brain function?

8. Potency of Hormones Under normal conditions, the human adrenal medulla secretes epinephrine ($C_9H_{13}NO_3$) at a rate sufficient to maintain a concentration of 10^{-10} M in circulating blood. To appreciate what that concentration means, calculate the diameter of a round swimming pool, with a water depth of 2.0 m, that would be needed to dissolve 1.0 g (about 1 teaspoon) of epinephrine to a concentration equal to that in blood.

9. Regulation of Hormone Levels in the Blood The half-life of most hormones in the blood is relatively short. For example, when radioactively labeled insulin is injected into an animal, half of the labeled hormone disappears from the blood within 30 min.

(a) What is the importance of the relatively rapid inactivation of circulating hormones?

(b) In view of this rapid inactivation, how is the level of circulating hormone kept constant under normal conditions?

(c) In what ways can the organism make rapid changes in the level of a circulating hormone?

10. Water-Soluble versus Lipid-Soluble Hormones On the basis of their physical properties, hormones fall into one of two categories: those that are very soluble in water but relatively insoluble in lipids (e.g., epinephrine) and those that are relatively insoluble in water but highly soluble in lipids (e.g., steroid hormones). In their role as regulators of cellular activity, most water-soluble hormones do not enter their target cells. The lipid-soluble hormones, by contrast, do enter their target cells and ultimately act in the nucleus. What is the correlation between solubility, the location of receptors, and the mode of action of these two classes of hormones?

11. Metabolic Differences between Muscle and Liver in a "Fight-or-Flight" Situation When an animal confronts a "fight-or-flight" situation, the release of epinephrine promotes glycogen breakdown in the liver, heart, and skeletal muscle. The end product of glycogen breakdown in the liver is glucose; the end product in skeletal muscle is pyruvate.

(a) What is the reason for the different products of glycogen breakdown in the two tissues?

(b) What is the advantage to an animal that must fight or flee of these specific glycogen breakdown routes?

12. Excessive Amounts of Insulin Secretion: Hyperinsulinism Certain malignant tumors of the pancreas cause excessive production of insulin by β cells. Affected individuals exhibit shaking and trembling, weakness and fatigue, sweating, and hunger.

(a) What is the effect of hyperinsulinism on the metabolism of carbohydrates, amino acids, and lipids by the liver?

(b) What are the causes of the observed symptoms? Suggest why this condition, if prolonged, leads to brain damage.

13. Thermogenesis Caused by Thyroid Hormones Thyroid hormones are intimately involved in regulating the basal metabolic rate. Liver tissue of animals given excess thyroxine shows an increased rate of O_2 consumption and increased heat output (thermogenesis), but the ATP concentration in the tissue is normal. Different explanations have been offered for the thermogenic effect of thyroxine. One is that excess thyroxine causes uncoupling of oxidative phosphorylation in mitochondria. How could such an effect account for the observations?

Another explanation suggests that thermogenesis is due to an increased rate of ATP utilization by the thyroxine-stimulated tissue. Is this a reasonable explanation? Why or why not?

14. Function of Prohormones What are the possible advantages of synthesizing hormones as prohormones?

15. Sources of Glucose during Starvation The typical human adult uses about 160 g of glucose per day, 120 g of which is used by the brain. The available reserve of glucose ($\sim$20 g of circulating glucose and $\sim$190 g of glycogen) is adequate for about one day. After the reserve has been depleted during starvation, how would the body obtain more glucose?

16. Parabiotic *ob/ob* Mice By careful surgery, researchers can connect the circulatory systems of two mice so that the same blood circulates through both animals. In these **parabiotic** mice, products released into the blood by one animal reach the other animal via the shared circulation. Both animals are free to eat independently. If a mutant *ob/ob* mouse (both copies of the *OB* gene are defective) and a normal *OB/OB* mouse (two good copies of the *OB* gene) were made parabiotic, what would happen to the weight of each mouse?

17. Calculation of Body Mass Index A portly biochemistry professor weighs 260 lb (118 kg) and is 5 feet 8 inches (173 cm) tall. What is his body mass index? How much weight would he have to lose to bring his body mass index down to 25 (normal)?

18. Insulin Secretion Predict the effects on insulin secretion by pancreatic β cells of exposure to the potassium ionophore valinomycin (see Fig. 11–42). Explain your prediction.

19. Effects of a Deleted Insulin Receptor A strain of mice specifically lacking the insulin receptor of liver is found to have mild fasting hyperglycemia (blood glucose = 132 mg/dL, vs. 101 mg/dL in controls) and a more striking hyperglycemia in the fed state (glucose = 363 mg/dL, vs. 135 mg/dL in controls). The mice have higher than normal levels of glucose 6-phosphatase in the liver and elevated levels of insulin in the blood. Explain these observations.

20. Decisions on Drug Safety The drug Avandia (rosiglitazone) is effective in lowering blood glucose in patients with type 2 diabetes, but a few years after it came into widespread use, it seemed that using the drug came with an increased risk of heart attack. In response, the U.S. Food and Drug Administration severely restricted the conditions under which it could be prescribed. Two years later, after additional studies had been completed, the FDA lifted the restrictions, and today Avandia is available by prescription in the United States, with no special limitations. Many other countries ban it completely. If it were your responsibility to decide whether this drug should remain on the market (labeled with suitable warnings about its side effects) or should be withdrawn, what factors would you weigh in making your decision?

21. Type 2 Diabetes Medication The drugs acarbose (Precose) and miglitol (Glyset), used in the treatment of type 2 diabetes mellitus, inhibit α-glucosidases in the brush border of the small intestine. These enzymes degrade oligosaccharides derived from glycogen or starch to monosaccharides. Suggest a possible mechanism for the salutary effect of these drugs for individuals with diabetes. What side effects, if any, would you expect from these drugs? Why? (Hint: Review lactose intolerance, p. 551.)

Data Analysis Problem

22. Cloning the Sulfonylurea Receptor of the Pancreatic β Cell Glyburide, a member of the sulfonylurea family of drugs shown on p. 933, is used to treat type 2 diabetes. It binds to and closes the ATP-gated K^+ channel shown in Figures 23-28 and 23-29.

(a) Given the mechanism shown in Figure 23-28, would treatment with glyburide result in increased or decreased insulin secretion by pancreatic β cells? Explain your reasoning.

(b) How does treatment with glyburide help reduce the symptoms of type 2 diabetes?

(c) Would you expect glyburide to be useful for treating type 1 diabetes? Explain your answer.

Aguilar-Bryan and coauthors (1995) cloned the gene for the sulfonylurea receptor (SUR) portion of the ATP-gated K^+ channel from hamsters. The research team went to great lengths to ensure that the gene they cloned was, in fact, the SUR-encoding gene. Here we explore how it is possible for researchers to demonstrate that they have cloned the gene of interest rather than another gene.

The first step was to obtain pure SUR protein. As was already known, drugs such as glyburide bind SUR with very high affinity ($K_d < 10$ nM), and SUR has a molecular weight of 140 to 170 kDa. Aguilar-Bryan and coworkers made use of the high-affinity glyburide binding to tag the SUR protein with a radioactive label that would serve as a marker to purify the protein from a cell extract. First, they made a radiolabeled derivative of glyburide, using radioactive iodine (^{125}I):

$[^{125}I]$5-Iodo-2-hydroxyglyburide

(d) In preliminary studies, the ^{125}I-labeled glyburide derivative (hereafter, $[^{125}I]$glyburide) was shown to have the same K_d and binding characteristics as unaltered glyburide. Why was it necessary to demonstrate this? (What alternative possibilities did it rule out?)

Even though $[^{125}I]$glyburide bound to SUR with high affinity, a significant amount of the labeled drug would probably dissociate from the SUR protein during purification. To prevent this, $[^{125}I]$glyburide had to be covalently cross-linked to SUR. There are many methods for covalent cross-linking; Aguilar-Bryan and coworkers used UV light. When aromatic

molecules are exposed to short-wave UV, they enter an excited state and readily form covalent bonds with nearby molecules. By cross-linking the radiolabeled glyburide to the SUR protein, the researchers could simply track the ^{125}I radioactivity to follow SUR through the purification procedure.

The research team treated hamster HIT cells (which express SUR) with $[^{125}I]$glyburide and UV light, purified the ^{125}I-labeled 140 kDa protein, and sequenced its 25 residue amino-terminal segment; they found the sequence PLAFCGTENHSAAYRVDQGVLNNGC. The investigators then generated antibodies that bound to two short peptides in this sequence, one binding to PLAFCGTE and the other to HSAAYRVDQGV, and showed that these antibodies bound the purified ^{125}I-labeled 140 kDa protein.

(e) Why was it necessary to include this antibody-binding step?

Next, the researchers designed PCR primers based on the sequences above, and then cloned a gene from a hamster cDNA library that encoded a protein with these sequences (see Chapter 9 on biotechnology methods). The cloned putative SUR cDNA hybridized to an mRNA of the appropriate length that was present in cells known to contain SUR. The putative SUR cDNA did not hybridize to any mRNA fraction of the mRNAs isolated from hepatocytes, which do not express SUR.

(f) Why was it necessary to include this putative SUR cDNA–mRNA hybridization step?

Finally, the cloned gene was inserted into and expressed in COS cells, which do not normally express the SUR gene. The investigators mixed these cells with $[^{125}I]$glyburide, with or without a large excess of unlabeled glyburide, exposed the cells to UV light, and measured the radioactivity of the 140 kDa protein produced. Their results are shown in the table.

Experiment	Cell type	Added putative SUR cDNA?	Added excess unlabeled glyburide?	^{125}I label in 140 kDa protein
1	HIT	No	No	+++
2	HIT	No	Yes	−
3	COS	No	No	−
4	COS	Yes	No	+++
5	COS	Yes	Yes	−

(g) Why was no ^{125}I-labeled 140 kDa protein found in experiment 2?

(h) How would you use the information in the table to argue that the cDNA encoded SUR?

(i) What other information would you want to collect to be more confident that you had cloned the SUR gene?

Reference

Aguilar-Bryan, L., C.G. Nichols, S.W. Wechsler, J.P. Clement, IV, A.E. Boyd, III, G. González, H. Herrera-Sosa, K. Nguy, J. Bryan, and **D.A. Nelson. 1995.** Cloning of the β cell high-affinity sulfonylurea receptor: a regulator of insulin secretion. *Science* 268:423–426.

Further Reading is available at www.macmillanlearning.com/LehningerBiochemistry7e.

INFORMATION PATHWAYS

The third and final part of this book explores the biochemical mechanisms underlying the apparently contradictory requirements for both genetic continuity and the evolution of living organisms. What is the molecular nature of genetic material? How is genetic information transmitted from one generation to the next with high fidelity? How do the rare changes in genetic material that are the raw material of evolution arise? How is genetic information ultimately expressed in the amino acid sequences of the astonishing variety of protein molecules in a living cell?

Today's understanding of information pathways has arisen from the convergence of genetics, physics, and chemistry in modern biochemistry. This was epitomized by the discovery of the double-helical structure of DNA, postulated by James Watson and Francis Crick in 1953 (see Fig. 8-13). Genetic theory contributed the concept of coding by genes. Physics permitted the determination of molecular structure by x-ray diffraction analysis. Chemistry revealed the composition of DNA. The profound impact of the Watson-Crick hypothesis arose from its ability to account for a wide range of observations derived from studies in these diverse disciplines.

This revolutionized our understanding of the structure of DNA and inevitably stimulated questions about its function. The double-helical structure itself clearly suggested how DNA might be copied so that the information it contains can be transmitted from one generation to the next. Clarification of how the information in DNA is converted into functional proteins came with the discovery of messenger RNA and transfer RNA and with the deciphering of the genetic code.

These and other major advances gave rise to the central dogma of molecular biology, comprising the three major processes in the cellular utilization of genetic information. The first is replication, the copying of parental DNA to form daughter DNA molecules with identical nucleotide sequences. The second is transcription, the process by which parts of the genetic message encoded in DNA are copied precisely into RNA. The third is translation, whereby the genetic message encoded in messenger

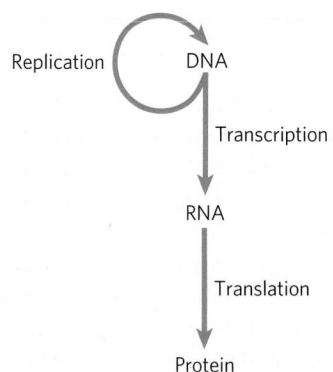

The central dogma of molecular biology, showing the general pathways of information flow via replication, transcription, and translation. The term "dogma" is a misnomer and is retained for historical reasons only. It was introduced by Francis Crick at a time when little evidence supported these ideas, but the "dogma" has become a well-established principle.

RNA is translated on the ribosomes into a polypeptide with a particular sequence of amino acids.

Part III explores these and related processes. In Chapter 24 we examine the structure, topology, and packaging of chromosomes and genes. The processes underlying the central dogma are elaborated in Chapters 25 through 27. Finally, we turn to regulation, examining how the expression of genetic information is controlled (Chapter 28).

A major theme running through these chapters is the added complexity inherent in the biosynthesis of macromolecules that contain information. Assembling nucleic acids and proteins with particular sequences of nucleotides and amino acids represents nothing less than preserving the faithful expression of the template upon which life itself is based. We might expect the formation of phosphodiester bonds in DNA or peptide bonds in proteins to be a trivial feat for cells, given the arsenal of enzymatic and chemical tools described in Part II. However, the framework of patterns and rules established in our examination of metabolic pathways thus far must be enlarged considerably to take into account molecular information. Bonds must be formed between *particular* subunits in informational biopolymers, avoiding either the occurrence or the persistence of sequence errors. This requirement has an enormous impact on the thermodynamics, chemistry, and enzymology of the biosynthetic processes. Formation of a peptide bond requires an energy input of only about 21 kJ per mole of bonds and can be catalyzed by relatively simple enzymes. But to synthesize a bond between two specific amino acids at a particular place in a polypeptide, the cell invests about 125 kJ/mol while making use of more than 200 enzymes, RNA molecules, and specialized proteins. The chemistry involved in peptide bond formation does not change because of this requirement, but additional processes are layered over the basic reaction to ensure that the peptide bond is formed between particular amino acids. Biological information is expensive.

The dynamic interaction between nucleic acids and proteins is another central theme of Part III. Regulatory and catalytic RNA molecules are gradually taking a more prominent place in our understanding of these pathways (discussed in Chapters 26 and 27). However, most of the processes that make up the pathways of cellular information flow are catalyzed and regulated by proteins. An understanding of these enzymes and other proteins can have practical as well as intellectual rewards, because they form the basis of recombinant DNA technology (introduced in Chapter 9).

Evolution again constitutes an overarching theme. Many of the processes outlined in Part III can be traced back billions of years, and a few can be traced to LUCA, the last universal common ancestor. The ribosome, most of the translational apparatus, and some parts of the transcriptional machinery are shared by every living organism on this planet. Genetic information is a kind of molecular clock that can help define ancestral relationships among species. Shared information pathways connect humans to every other species now living on Earth, and to all species that came before. Exploration of these pathways is allowing scientists to slowly open the curtain on the first act—the events that may have heralded the beginning of life on Earth.

CHAPTER 24

Genes and Chromosomes

Self-study tools that will help you practice what you've learned and reinforce this chapter's concepts are available online. Go to www.macmillanlearning.com/LehningerBiochemistry7e.

The size of DNA molecules presents an interesting biological puzzle. Given that these molecules are generally much longer than the cells or viral particles that contain them **(Fig. 24-1)**, how do they fit into their cellular or viral packages? To address this question, we shift our focus from the secondary structure of DNA, considered in Chapter 8, to the extraordinary degree of organization required for the tertiary packaging of DNA into **chromosomes**—the repositories of genetic information. The chapter begins with an examination of the elements that make up viral and cellular chromosomes, and then considers chromosomal size and organization. We then discuss DNA topology, describing the coiling and supercoiling of DNA molecules. Finally, we consider the protein-DNA interactions that organize chromosomes into compact structures.

24.1 Chromosomal Elements

Cellular DNA contains genes and intergenic regions, both of which may serve functions vital to the cell. The more complex genomes, such as those of eukaryotic cells, demand increased levels of chromosomal organization, and this is reflected in the structural features of the chromosomes. We begin by considering the different types of DNA sequences and structural elements within chromosomes.

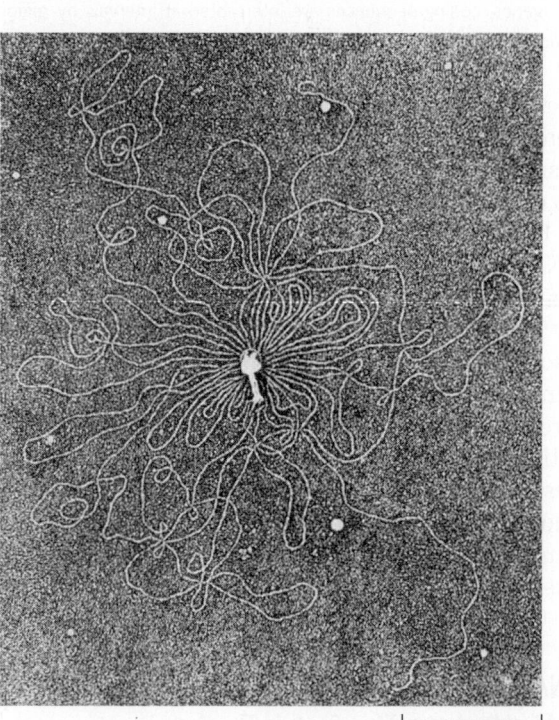

0.5 μm

FIGURE 24-1 Bacteriophage T2 protein coat surrounded by its single, linear molecule of DNA. The DNA was released by lysing the bacteriophage particle in distilled water and allowing the DNA to spread on the water surface. An undamaged T2 bacteriophage particle consists of a head structure that tapers to a tail by which the bacteriophage attaches itself to the outer surface of a bacterial cell. All the DNA shown in this electron micrograph is normally packaged inside the phage head. [Source: Republished with permission of Elsevier, from "Preparation and length measurements of the total deoxyribonucleic acid content of T2 bacteriophages," by A. K. Kleinschmidt, et al. *Biochem. Biophys. Acta* 61, pp. 857–864, 1962; permission conveyed through Copyright Clearance Center, Inc.]

Genes Are Segments of DNA That Code for Polypeptide Chains and RNAs

Our understanding of genes has evolved tremendously over the past century. Classically, a gene was defined as a portion of a chromosome that determines or affects a single character or **phenotype** (visible property), such as eye color. George Beadle and Edward Tatum proposed a molecular definition of a gene in 1940. After exposing spores of the fungus *Neurospora crassa* to x rays and other agents now known to damage DNA and cause alterations in DNA sequence (**mutations**), they detected mutant fungal strains that lacked one or another specific enzyme, sometimes resulting in the failure of an entire metabolic pathway. Beadle and Tatum concluded that a gene is a segment of genetic material that determines, or codes for, one enzyme: the **one gene–one enzyme** hypothesis. Later this concept was broadened to **one gene–one polypeptide**, because many genes code for a protein that is not an enzyme or for one polypeptide of a multisubunit protein.

George W. Beadle, 1903–1989
[Source: National Library of Medicine/Science Photo Library/Science Source.]

Edward L. Tatum, 1909–1975
[Source: Corbis/UPI/ Bettmann/Corbis.]

The modern biochemical definition of a gene is even more precise. A **gene** is all the DNA that encodes the primary sequence of some final gene product, which can be either a polypeptide or an RNA with a structural or catalytic function. DNA also contains other segments or sequences that have a purely regulatory function. **Regulatory sequences** provide signals that may denote the beginning or the end of genes, or influence the transcription of genes, or function as initiation points for replication or recombination (Chapter 28). Some genes can be expressed in different ways to generate multiple gene products from a single segment of DNA; the special transcriptional and translational mechanisms that allow this are described in Chapters 26 through 28.

We can estimate directly the minimum overall size of genes that encode proteins. As described in detail in Chapter 27, each amino acid of a polypeptide chain is coded for by a sequence of three consecutive nucleotides in a single strand of DNA (**Fig. 24-2**), with these "codons" arranged in a sequence that corresponds to

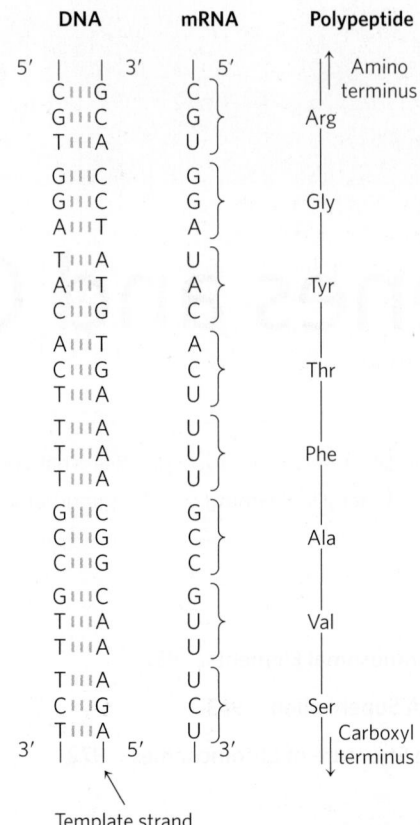

FIGURE 24-2 Colinearity of the coding nucleotide sequences of DNA and mRNA and the amino acid sequence of a polypeptide chain. The triplets of nucleotide units in DNA determine the amino acids in a protein through the intermediary mRNA. One of the DNA strands serves as a template for synthesis of mRNA, which has nucleotide triplets (codons) complementary to those of the DNA. In some bacterial and many eukaryotic genes, coding sequences are interrupted at intervals by regions of noncoding sequences (called introns).

the sequence of amino acids in the polypeptide that the gene encodes. A polypeptide chain of 350 amino acid residues (an average-size chain) corresponds to 1,050 bp of coding DNA. Many genes in eukaryotes and a few in bacteria and archaea are interrupted by noncoding DNA segments and are therefore considerably longer than this simple calculation would suggest.

How many genes are in a single chromosome? The *Escherichia coli* chromosome, one of the bacterial genomes that have been completely sequenced, is a circular DNA molecule (in the sense of an endless loop rather than a perfect circle) with 4,639,675 bp. These base pairs encode about 4,300 genes for proteins and another 157 genes for structural or catalytic RNA molecules. Among eukaryotes, the approximately 3.1 billion base pairs of the human genome include approximately 20,000 genes on the 24 different chromosomes.

DNA Molecules Are Much Longer Than the Cellular or Viral Packages That Contain Them

Chromosomal DNAs are often many orders of magnitude longer than the cells or viruses in which they are

TABLE 24-1	The Sizes of DNA and Viral Particles for Some Bacterial Viruses (Bacteriophages)		
Virus	Size of viral DNA (bp)	Length of viral DNA (nm)	Long dimension of viral particles (nm)
ϕX174	5,386	1,939	25
T7	39,936	14,377	78
λ (lambda)	48,502	17,460	190
T4	168,889	60,800	210

Note: Data on size of DNA are for the replicative form (double-standard). The contour length is calculated assuming that each base pair occupies a length of 3.4 Å (see Fig. 8-13).

located (Fig. 24-1; Table 24-1). This is true of every class of organism or viral parasite.

Viruses Viruses are not free-living organisms; rather, they are infectious parasites that use the resources of a host cell to carry out many of the processes they require to propagate. Many viral particles consist of no more than a genome (usually a single RNA or DNA molecule) surrounded by a protein coat.

Almost all plant viruses and some bacterial and animal viruses have RNA genomes. These genomes tend to be particularly small. For example, the genomes of mammalian retroviruses such as HIV are about 9,000 nucleotides long, and the genome of the bacteriophage Qβ has 4,220 nucleotides. Both types of virus have single-stranded RNA genomes.

The genomes of DNA viruses vary greatly in size (Table 24-1). Many viral DNAs are circular for at least part of their life cycle. During viral replication within a host cell, specific types of viral DNA called **replicative forms** may appear; for example, many linear DNAs become circular and all single-stranded DNAs become double-stranded. A typical medium-size DNA virus is bacteriophage λ (lambda), which infects *E. coli*. In its replicative form inside cells, λ DNA is a circular double helix. This double-stranded DNA contains 48,502 bp and has a contour length of 17.5 μm. Bacteriophage φX174 is a much smaller DNA virus; the DNA in the viral particle is a single-stranded circle, and the double-stranded replicative form contains 5,386 bp. Although viral genomes are small, the contour lengths of their DNAs are typically hundreds of times longer than the long dimensions of the viral particles that contain them (Table 24-1).

Bacteria A single *E. coli* cell contains almost 100 times as much DNA as a bacteriophage λ particle. The chromosome of an *E. coli* cell is a single, double-stranded circular DNA molecule. Its 4,641,652 bp have a contour length of about 1.7 mm, some 850 times the length of the *E. coli* cell **(Fig. 24-3)**. In addition to the very large, circular DNA chromosome in their nucleoid, many bacteria contain one or more small circular DNA molecules that are free in the cytosol. These extrachromosomal elements are called **plasmids** **(Fig. 24-4**; see also p. 324). Most plasmids are only a few thousand base pairs long,

but some contain up to 400,000 bp. They carry genetic information and undergo replication to yield daughter plasmids, which pass into the daughter cells at cell division. Plasmids have been found in yeast and other fungi as well as in bacteria.

In many cases plasmids confer no obvious advantage on their host, and their sole function seems to be self-propagation. However, some plasmids carry genes that are useful to the host bacterium. For example, some plasmid genes make a host bacterium resistant to antibacterial agents. Plasmids carrying the gene for the enzyme β-lactamase confer resistance to β-lactam antibiotics such as penicillin, ampicillin, and amoxicillin (see Fig. 6-32). These and similar plasmids may pass from an antibiotic-resistant cell to an antibiotic-sensitive cell of the same or another bacterial species, making the recipient cell antibiotic-resistant. The extensive use of antibiotics in some human populations has served as a strong selective force, encouraging the spread of antibiotic resistance–coding plasmids (as well as transposable elements, described below, that harbor similar genes) in disease-causing bacteria. Physicians are becoming increasingly reluctant to prescribe antibiotics unless a clear clinical need is confirmed. For similar reasons, the widespread use of antibiotics in animal feeds is being curbed.

Eukaryotes A yeast cell, one of the simplest eukaryotes, has 2.6 times more DNA in its genome than an *E. coli* cell (Table 24-2). Cells of *Drosophila*, the fruit fly used in classical genetic studies, contain more than 35 times as much DNA as *E. coli* cells, and human cells have almost 700 times as much. The cells of many plants and amphibians contain even more. The genetic material of eukaryotic cells is apportioned into chromosomes, the diploid ($2n$) number depending on the species (Table 24-2). A human somatic cell, for example, has 46 chromosomes **(Fig. 24-5)**. Each chromosome of a eukaryotic cell, such as that shown in Figure 24-5a, contains a single, very large, duplex DNA molecule. The DNA molecules in the 24 different types of human chromosomes (22 matching pairs of autosomes plus the X and Y sex chromosomes) vary in length over a 25-fold range. Each type of chromosome in eukaryotes carries a characteristic set of genes.

FIGURE 24-3 A bacterial cell and its DNA. The length of the *E. coli* chromosome (1.7 mm), depicted in linear form, relative to the length of a typical *E. coli* cell (2 μm).

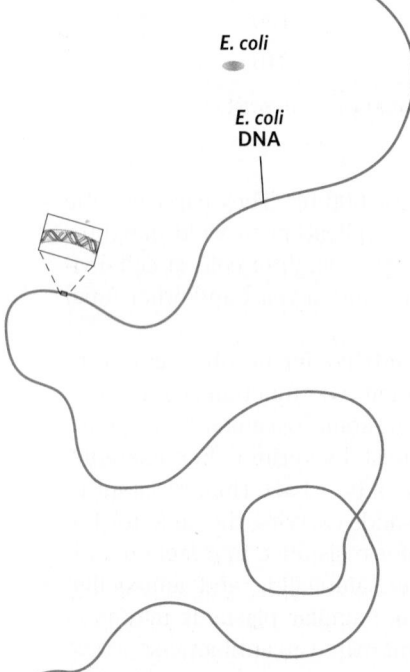

E. coli

E. coli
DNA

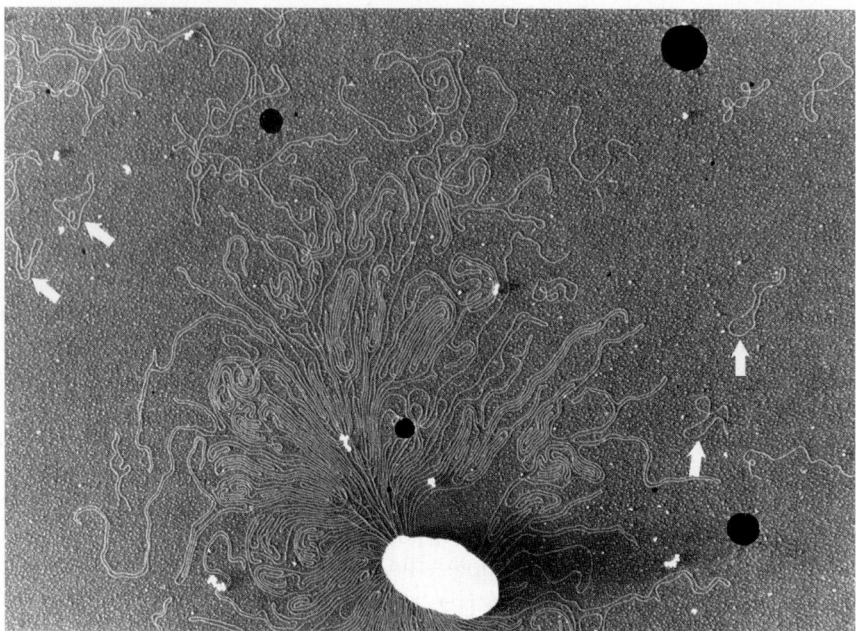

FIGURE 24-4 DNA from a lysed *E. coli* cell. In this electron micrograph, several small, circular plasmid DNAs are indicated by white arrows. The black spots and white specks are artifacts of the preparation. [Source: Huntington Potter, University of Colorado School of Medicine, and David Dressler, Balliol College, Oxford University.]

TABLE 24-2	DNA, Gene, and Chromosome Content in Some Genomes		
	Total DNA (bp)	Number of chromosomes[a]	Approximate number of genes
Escherichia coli K12 (bacterium)	4,641,652	1	4,494[b]
Saccharomyces cerevisiae (yeast)	12,157,105	16[c]	6,340[b]
Caenorhabditis elegans (nematode)	90,269,800	12[d]	23,000
Arabidopsis thaliana (plant)	119,186,200	10	33,000
Drosophila melanogaster (fruit fly)	120,367,260	18	20,000
Oryza sativa (rice)	480,000,000	24	57,000
Mus musculus (mouse)	2,634,266,500	40	27,000
Homo sapiens (human)	3,070,128,600	46	20,000

Note: This information is constantly being refined. For the most current information, consult the websites for the individual genome projects.

[a]The diploid chromosomes number is given for all eukaryotes except yeast.

[b]Includes known RNA-coding genes.

[c]Haploid chromosomes number. Wild yeast strains generally have eight (octoploid) or more sets of these chromosomes.

[d]Number for females, with two X chromosomes. Males have an X but no Y, thus 11 chromosomes in all.

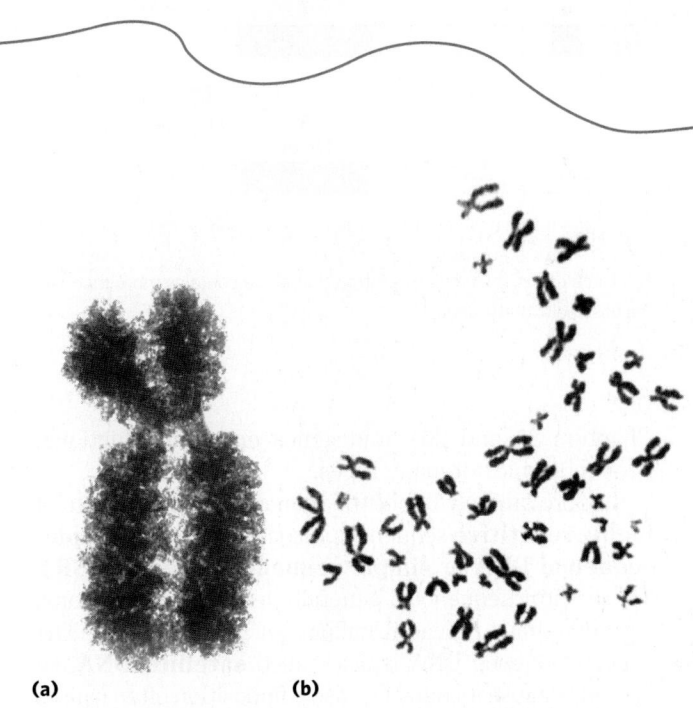

(a)   **(b)**

FIGURE 24-5 **Eukaryotic chromosomes. (a)** A pair of linked and condensed sister chromatids from a Chinese hamster ovary cell. Eukaryotic chromosomes are in this state after replication at metaphase during mitosis. **(b)** A complete set of chromosomes from a leukocyte from one of the authors. There are 46 chromosomes in every normal human somatic cell. [Sources: (a) Don W. Fawcett/Science Source. (b) © Michael M. Cox.]

The DNA molecules of one human genome (22 chromosomes plus X and Y), placed end to end, would extend for about a meter. Most human cells are diploid, so each cell contains a total of 2 m of DNA. An adult human body contains approximately 10^{14} cells and thus a total DNA length of 2×10^{11} km. Compare this with the circumference of the earth (4×10^4 km) or the distance between the earth and the sun (1.5×10^8 km)—a dramatic illustration of the extraordinary degree of DNA compaction in our cells.

Eukaryotic cells also have organelles, mitochondria **(Fig. 24-6)** and chloroplasts, that contain DNA. Mitochondrial DNA (mtDNA) molecules are much smaller than the nuclear chromosomes. In animal cells, mtDNA contains fewer than 20,000 bp (16,569 bp in human mtDNA) and is a circular duplex. Each mitochondrion typically has 2 to 10 copies of this mtDNA molecule, and the number can rise to hundreds in certain cells of an embryo that is undergoing cell differentiation. In a few organisms (trypanosomes, for example), each mitochondrion contains thousands of copies of mtDNA, organized into a complex and interlinked matrix known as a kinetoplast. Plant cell mtDNA ranges in size from 200,000 to 2,500,000 bp. Chloroplast DNA (cpDNA) also exists as circular duplexes and ranges in size from 120,000 to 160,000 bp. Mitochondrial and chloroplast DNAs have an evolutionary origin in the chromosomes of ancient bacteria that gained access to the cytoplasm of host cells

and became the precursors of these organelles (see Fig. 1-40). Mitochondrial DNA codes for the mitochondrial tRNAs and rRNAs and for a few mitochondrial proteins. More than 95% of mitochondrial proteins are encoded by nuclear DNA. Mitochondria and chloroplasts divide when the cell divides. Their DNA is replicated before and during division, and the daughter DNA molecules pass into the daughter organelles.

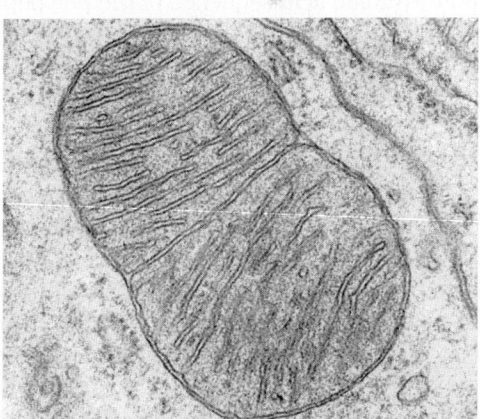

FIGURE 24-6 **A dividing mitochondrion.** Some mitochondrial proteins and RNAs are encoded by one of the copies of the mitochondrial DNA (none of which are visible here). The DNA (mtDNA) is replicated each time the mitochondrion divides, before cell division. [Source: D. W. Fawcett/Science Source.]

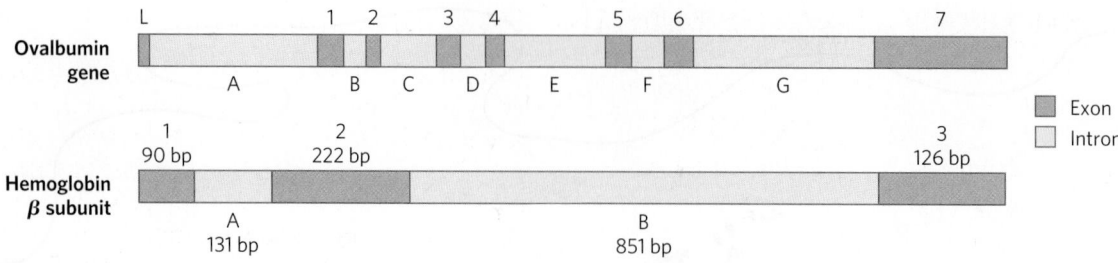

FIGURE 24-7 Introns in two eukaryotic genes. The gene for ovalbumin has seven introns (A to G), splitting the coding sequences into eight exons (L, and 1 to 7). The gene for the β subunit of hemoglobin has two introns and three exons, including one intron that alone contains more than half the base pairs of the gene.

Eukaryotic Genes and Chromosomes Are Very Complex

Many bacterial species have only one chromosome per cell and, in nearly all cases, each chromosome contains only one copy of each gene. A very few genes, such as those for rRNAs, are repeated several times. Genes and regulatory sequences account for almost all the DNA in bacteria. Moreover, almost every gene is precisely colinear with the amino acid sequence (or RNA sequence) it encodes (Fig. 24-2).

The organization of genes in eukaryotic DNA is structurally and functionally much more complex. Studies of eukaryotic chromosome structure and, more recently, the sequencing of entire eukaryotic genomes have yielded many surprises. Many, if not most, eukaryotic genes have a distinctive and puzzling structural feature: their nucleotide sequences contain one or more intervening segments of DNA that do not code for the amino acid sequence of the polypeptide product. These nontranslated inserts interrupt the otherwise colinear relationship between the nucleotide sequence of the gene and the amino acid sequence of the polypeptide it encodes. Such nontranslated DNA segments in genes are called **intervening sequences** or **introns**, and the coding segments are called **exons**. Few bacterial genes contain introns. In higher eukaryotes, the typical gene has much more intron sequence than sequences devoted to exons. For example, in the gene coding for the single polypeptide chain of ovalbumin, an avian egg protein **(Fig. 24-7)**, the introns are much longer than the exons; altogether, the seven introns make up 85% of the gene's DNA. The gene for the muscle protein titin is the intron champion, with 178 introns. Genes for histones seem to have no introns. In most cases the function of introns is not clear. In total, only about 1.5% of human DNA is "coding" or exon DNA, carrying sequence information for protein products. However, when the much larger introns are included in the count, as much as 30% of the human genome consists of protein-coding genes. A great deal of work remains to be done to understand the other genomic sequences. Much of the DNA that is not within genes is made up of repeated sequences of several kinds. These include transposable elements (transposons), molecular parasites that account for nearly half of the DNA in the human genome (see Fig. 9-27 and Chapters 25 and 26), and genes encoding functional RNA molecules of many types.

Approximately 3% of the human genome consists of **highly repetitive** sequences, also referred to as **simple-sequence DNA** or **simple sequence repeats (SSR)**. These short sequences, generally less than 10 bp long, are sometimes repeated millions of times per cell. The simple-sequence DNA is also called **satellite DNA**, so named because its unusual base composition often causes it to migrate as "satellite" bands (separated from the rest of the DNA) when fragmented cellular DNA samples are centrifuged in a cesium chloride density gradient. Studies suggest that simple-sequence DNA does not encode proteins or RNAs. The functional importance of the highly repetitive DNA has been defined in at least some cases. Much of it is associated with two crucial features of eukaryotic chromosomes: centromeres and telomeres.

The **centromere (Fig. 24-8)** is a sequence of DNA that functions during cell division as an attachment point for proteins that link the chromosome to the mitotic spindle. This attachment is essential for the equal and orderly segregation of chromosome sets to daughter cells. The centromeres of *Saccharomyces cerevisiae* have been isolated and studied. The sequences essential to centromere function are about 130 bp long and are very rich in A=T pairs. The centromeric sequences of higher eukaryotes are much longer and, unlike those of yeast, generally consist of thousands of tandem copies of one or several sequences of 5 to 10 bp, in the same orientation.

Telomeres (Greek *telos*, "end") are sequences at the ends of eukaryotic chromosomes that help stabilize the chromosome. Telomeres end with multiple repeated sequences of the form

$$(5')(T_xG_y)_n$$
$$(3')(A_xC_y)_n$$

where x and y are generally between 1 and 4 (Table 24-3). The number of telomere repeats, n, is in the range of

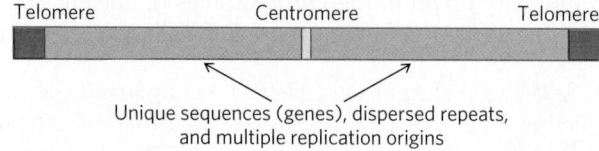

Unique sequences (genes), dispersed repeats, and multiple replication origins

FIGURE 24-8 Important structural elements of a yeast chromosome.

TABLE 24-3	Telomere Sequences
Organism	Telomere repeat sequence
Homo sapiens (human)	$(TTAGGG)_n$
Tetrahymena thermophila (ciliated protozoan)	$(TTGGGG)_n$
Saccharomyces cerevisiae (yeast)	$((TG)_{1-3}(TG)_{2-3})_n$
Arabidopsis thaliana (plant)	$(TTTAGGG)_n$

20 to 100 for most single-celled eukaryotes and is generally more than 1,500 in mammals. The ends of a linear DNA molecule cannot be routinely replicated by the cellular replication machinery (which may be one reason why bacterial DNA molecules are circular). Repeated telomeric sequences are added to eukaryotic chromosome ends primarily by the enzyme telomerase (see Fig. 26-38).

Artificial chromosomes (Chapter 9) have been constructed as a means of better understanding the functional significance of many structural features of eukaryotic chromosomes. A reasonably stable artificial linear chromosome requires only three components: a centromere, a telomere at each end, and sequences that allow the initiation of DNA replication. Yeast artificial chromosomes (YACs; see Fig. 9-6) have been developed as a research tool in biotechnology. Similarly, human artificial chromosomes (HACs) are being developed for the treatment of genetic diseases. These may eventually provide a new path to the intracellular replacement of missing or defective gene products, or somatic gene therapy.

SUMMARY 24.1 Chromosomal Elements

■ Genes are segments of a chromosome that contain the information for a functional polypeptide or RNA molecule. In addition to genes, chromosomes contain a variety of regulatory sequences involved in replication, transcription, and other processes.

■ Genomic DNA and RNA molecules are generally orders of magnitude longer than the viral particles or cells that contain them.

■ Many genes in eukaryotic cells (but few in bacteria and archaea) are interrupted by noncoding sequences, or introns. The coding segments separated by introns are called exons.

■ Only about 1.5% of human genomic DNA encodes proteins. Even when introns are included, less than one-third of human genomic DNA consists of genes. Much of the remainder consists of repeated sequences of various types. Nucleic acid parasites known as transposons account for about half of the human genome.

■ Eukaryotic chromosomes have two important special-function repetitive DNA sequences: centromeres, which are attachment points for the mitotic spindle, and telomeres, located at the ends of chromosomes.

24.2 DNA Supercoiling

Cellular DNA, as we have seen, is extremely compacted, implying a high degree of structural organization. The folding mechanism must not only pack the DNA but also permit access to the information in the DNA. Before considering how this is accomplished in processes such as replication and transcription, we need to examine an important property of DNA structure known as **supercoiling**.

"Supercoiling" means the coiling of a coil. An old-fashioned telephone cord, for example, is typically a coiled wire. The path taken by the wire between the base of the phone and the receiver often includes one or more supercoils **(Fig. 24-9)**. DNA is coiled in the form of a double helix, with both strands of the DNA coiling around an axis. The further coiling of that axis upon itself **(Fig. 24-10)** produces DNA supercoiling. As detailed below, DNA supercoiling is generally a manifestation of structural strain. When there is no net bending of the DNA axis upon itself, the DNA is said to be in a **relaxed** state. Supercoiling affects, and is affected by, replication and transcription, both of which require a separation of DNA strands—a process complicated by the helical interwinding of the strands **(Fig. 24-11)**.

That a DNA molecule would bend on itself and become supercoiled in tightly packaged cellular DNA would seem logical, and perhaps even trivial, were it not for one additional fact: many circular DNA molecules remain highly supercoiled even after they are extracted

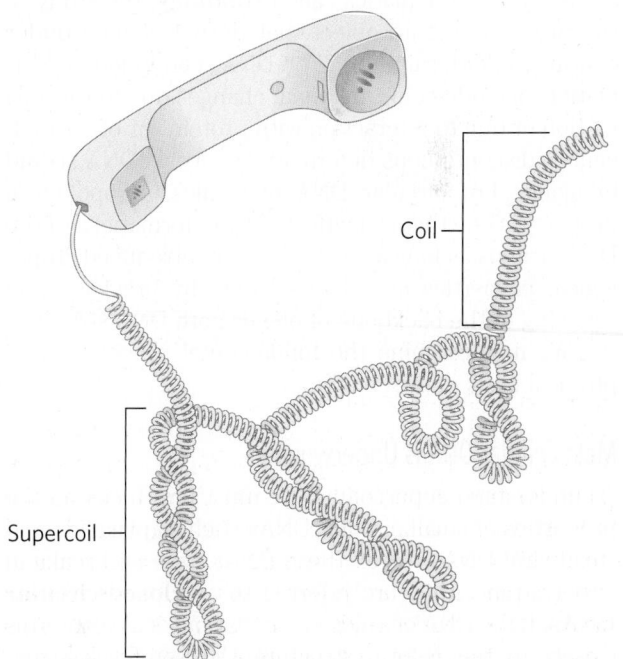

FIGURE 24-9 Supercoils. An old-fashioned phone cord is coiled like a DNA helix, and the coiled cord can itself coil in a supercoil. The illustration is especially appropriate because an examination of phone cords helped lead Jerome Vinograd and his colleagues to the insight that many properties of small circular DNAs can be explained by supercoiling. They first detected DNA supercoiling—in small circular viral DNAs—in 1965.

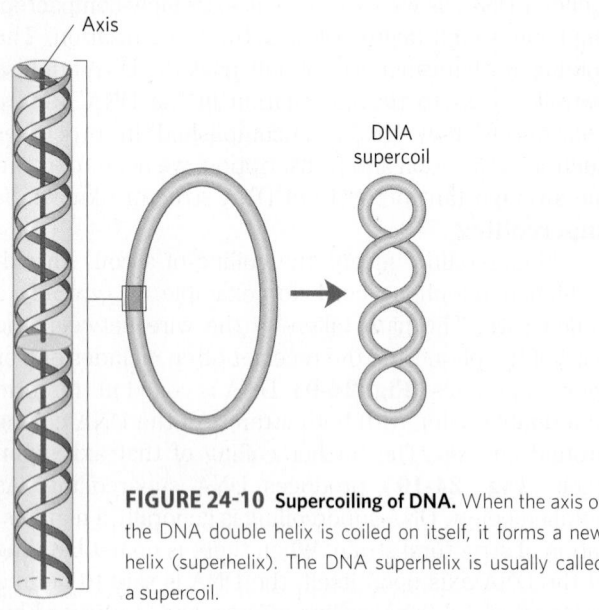

DNA double
helix (coil)

Axis

DNA
supercoil

FIGURE 24-10 Supercoiling of DNA. When the axis of the DNA double helix is coiled on itself, it forms a new helix (superhelix). The DNA superhelix is usually called a supercoil.

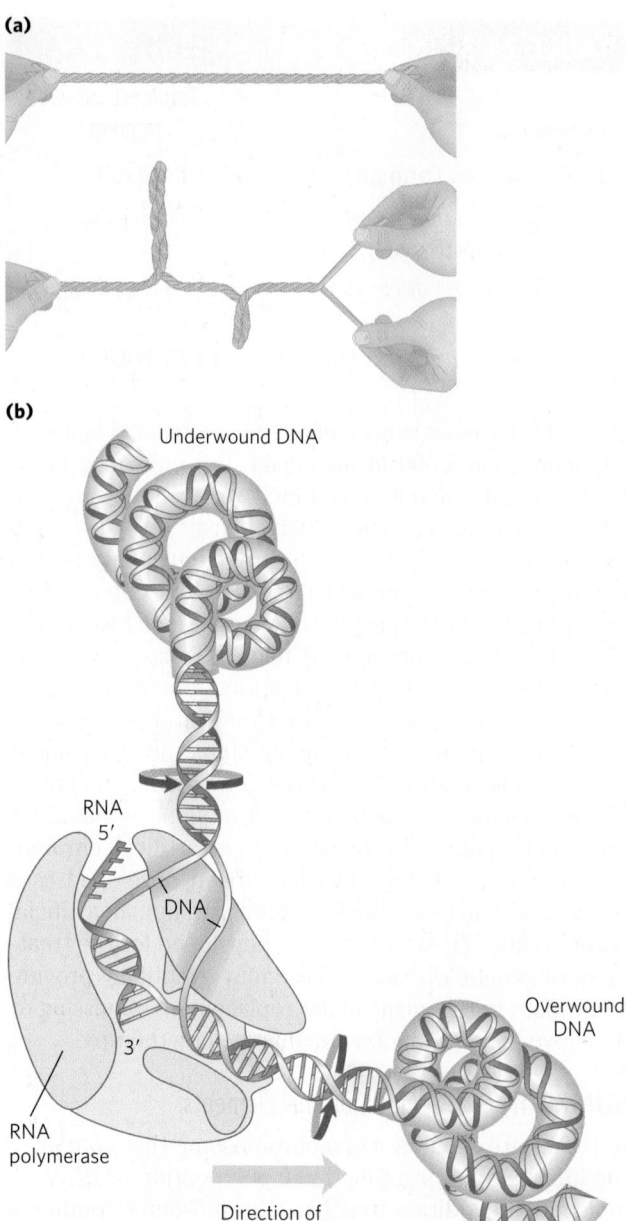

(a)

(b)

Underwound DNA

RNA
5′

DNA

3′

RNA
polymerase

Overwound
DNA

Direction of
transcription

FIGURE 24-11 The effects of replication and transcription on DNA supercoiling. Because DNA is a double-helical structure, strand separation leads to added stress and supercoiling if the DNA is constrained (not free to rotate) ahead of the strand separation. **(a)** The general effect can be illustrated by twisting two strands of a rubber band about each other to form a double helix. If one end is constrained, separating the two strands at the other end will lead to twisting. **(b)** In a DNA molecule, the progress of a DNA polymerase or RNA polymerase (as shown here) along the DNA involves separation of the strands. As a result, the DNA becomes overwound ahead of the enzyme (upstream) and underwound behind it (downstream). Red arrows indicate the direction of winding.

and purified, freed from protein and other cellular components. This indicates that supercoiling is an intrinsic property of DNA tertiary structure. It occurs in all cellular DNAs and is highly regulated by each cell.

Several measurable properties of supercoiling have been established, and the study of supercoiling has provided many insights into DNA structure and function. This work has drawn heavily on concepts derived from a branch of mathematics called **topology**, the study of the properties of an object that do not change under continuous deformations. For DNA, continuous deformations include conformational changes due to thermal motion or due to interaction with proteins or other molecules; discontinuous deformations involve DNA strand breakage. For circular DNA molecules, a topological property is one that is unaffected by deformations of the DNA strands as long as no breaks are introduced. Topological properties are changed only by breakage and rejoining of the backbone of one or both DNA strands.

We now examine the fundamental properties and physical basis of supercoiling.

Most Cellular DNA Is Underwound

To understand supercoiling, we must first focus on the properties of small circular DNAs such as plasmids and small viral DNAs. When these DNAs have no breaks in either strand, they are referred to as **closed-circular DNAs**. If the DNA of a closed-circular molecule conforms closely to the B-form structure (Watson-Crick structure; see Fig. 8-13), with one turn of the double helix per 10.5 bp, the DNA is relaxed rather than supercoiled **(Fig. 24-12)**. Supercoiling results when DNA is subject to some form of structural strain. Purified closed-circular DNA is rarely relaxed, regardless of its biological origin. Furthermore, DNAs derived from a given cellular source have a characteristic degree of supercoiling.

DNA structure is therefore strained in a manner that is regulated by the cell to induce the supercoiling.

In almost every instance, the strain is a result of **underwinding** of the DNA double helix in the closed circle. In other words, the DNA has *fewer* helical turns than would be expected for the B-form structure. The effects of underwinding are summarized in **Figure 24-13**. An 84 bp segment of a circular DNA in the relaxed state

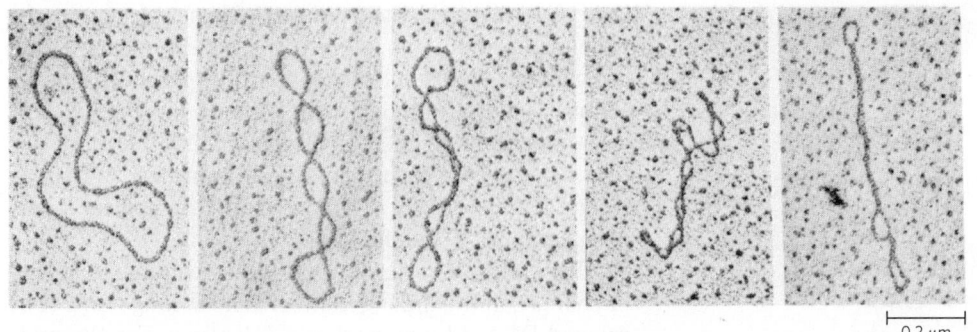

FIGURE 24-12 Relaxed and super-coiled plasmid DNAs. The molecule in the leftmost electron micrograph is relaxed; the degree of supercoiling increases from left to right. [Source: Laurien Polder, from A. Kornberg, *DNA Replication*, p. 29, W. H. Freeman, New York, 1980.]

0.2 μm

would contain eight double-helical turns, one for every 10.5 bp. If one of these turns were removed, there would be (84 bp)/7 = 12.0 bp per turn, rather than the 10.5 found in B-DNA (Fig. 24-13b). This is a deviation from the most stable DNA form, and the molecule is thermodynamically strained as a result. Generally, much of this strain would be accommodated by coiling the axis of the DNA on itself to form a supercoil (Fig. 24-13c; some of the strain in this 84 bp segment would simply become dispersed in the untwisted structure of the larger DNA molecule). In principle, the strain could also be accommodated by separating the two DNA strands over a distance of about 10 bp (Fig. 24-13d). In isolated closed-circular DNA, strain introduced by underwinding is generally accommodated by supercoiling rather than strand separation, because coiling the axis of the DNA usually requires less energy than breaking the hydrogen bonds that stabilize paired bases.

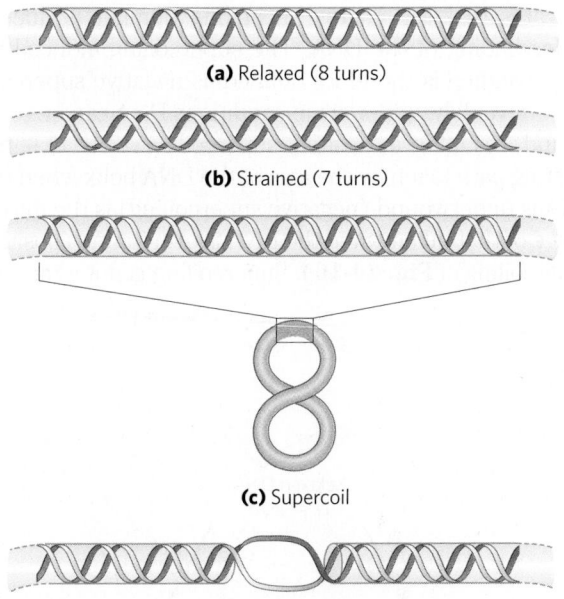

(a) Relaxed (8 turns)

(b) Strained (7 turns)

(c) Supercoil

(d) Strand separation

FIGURE 24-13 Effects of DNA underwinding. (a) A segment of DNA in a closed-circular molecule, 84 bp long, in its relaxed form with eight helical turns. **(b)** Removal of one turn induces structural strain. **(c)** The strain is generally accommodated by formation of a supercoil. **(d)** DNA underwinding also makes the separation of strands somewhat easier. In principle, each turn of underwinding should facilitate strand separation over about 10 bp, as shown here. However, the hydrogen-bonded base pairs would generally preclude strand separation over such a short distance, and the effect becomes important only for longer DNAs and higher levels of DNA underwinding.

Note, however, that the underwinding of DNA in vivo makes separation of the DNA strands easier, facilitating access to the information they contain.

Every cell actively underwinds its DNA with the aid of enzymatic processes (described below), and the resulting strained state represents a form of stored energy. Cells maintain DNA in an underwound state to facilitate its compaction by coiling. The underwinding of DNA is also important to enzymes of DNA metabolism that must bring about strand separation as part of their function.

The underwound state can be maintained only if the DNA is a closed circle or if it is bound and stabilized by proteins so that the strands are not free to rotate about each other. If there is a break in one strand of an isolated, protein-free circular DNA, free rotation at that point will cause the underwound DNA to revert spontaneously to the relaxed state. In a closed-circular DNA molecule, however, the number of helical turns cannot be changed without at least transiently breaking one of the DNA strands. The number of helical turns in a DNA molecule therefore provides a precise description of supercoiling.

DNA Underwinding Is Defined by Topological Linking Number

The field of topology provides some ideas that are useful to the discussion of DNA supercoiling, particularly the concept of **linking number**. Linking number is a topological property of double-stranded DNA, because it does not vary when the DNA is bent or deformed, as long as both DNA strands remain intact. Linking number (Lk) is illustrated in **Figure 24-14**.

Let's begin by visualizing the separation of the two strands of a double-stranded circular DNA. If the two strands are linked as shown in Figure 24-14a, they are effectively joined by what can be described as a topological bond. Even if all hydrogen bonds and base-stacking interactions were abolished such that the strands were not in physical contact, this topological bond would still link the two strands. Visualize one of the circular strands as the boundary of a surface (such as the soap film framed by the loop of a bubble wand before you blow a bubble). The linking number can be defined as the number of times the second strand pierces this surface. For the molecule in Figure 24-14a, $Lk = 1$; for that in Figure 24-14b, $Lk = 6$. The linking number for a closed-circular DNA is always an integer. By convention,

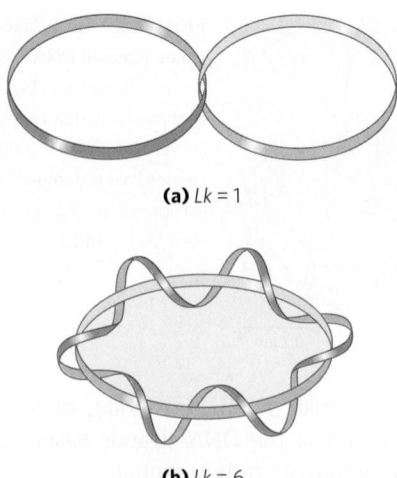

(a) $Lk = 1$

(b) $Lk = 6$

FIGURE 24-14 Linking number, Lk. Here, as usual, each blue ribbon represents one strand of a double-stranded DNA molecule. For the molecule in **(a)**, $Lk = 1$. For the molecule in **(b)**, $Lk = 6$. One of the strands in (b) is kept untwisted for illustrative purposes, to define the border of an imaginary surface (shaded blue). The number of times the twisting strand penetrates this surface provides a rigorous definition of linking number.

if the links between two DNA strands are arranged so that the strands are interwound in a right-handed helix, the linking number is defined as positive $(+)$; for strands interwound in a left-handed helix, the linking number is negative $(-)$. Negative linking numbers are, for all practical purposes, not encountered in DNA.

We can now extend these ideas to a closed-circular DNA with 2,100 bp **(Fig. 24-15a)**. When the molecule is relaxed, the linking number is simply the number of

base pairs divided by the number of base pairs per turn, which is close to 10.5; so in this case, $Lk = 200$. For a circular DNA molecule to have a topological property such as linking number, both strands must be intact, without a break. If there is a break in either strand, the two strands can, in principle, be unraveled and separated completely. In this case, no topological bond exists and Lk is undefined (Fig. 24-15b).

We can now describe DNA underwinding in terms of changes in the linking number. The linking number in relaxed DNA, Lk_0, is used as a reference. For the molecule shown in Figure 24-15a, $Lk_0 = 200$; if two turns are removed from this molecule, $Lk = 198$. The change can be described by the equation

$$\Delta Lk = Lk - Lk_0$$
$$= 198 - 200 = -2 \qquad (24-1)$$

It is often convenient to express the change in linking number in terms of a quantity that is independent of the length of the DNA molecule. This quantity, called the **specific linking difference** or **superhelical density** (σ), is a measure of the number of turns removed relative to the number present in relaxed DNA:

$$\sigma = \frac{\Delta Lk}{Lk_0} \qquad (24-2)$$

In the example in Figure 24-15c, $\sigma = 0.01$, which means that 1% (2 of 200) of the helical turns present in the DNA (in its B form) have been removed. The degree of underwinding in cellular DNAs generally falls in the range of 5% to 7%; that is, $\sigma = -0.05$ to -0.07. The negative sign indicates that the change in linking number is due to underwinding of the DNA. The supercoiling induced by underwinding is therefore defined as negative supercoiling. Conversely, under some conditions DNA can be overwound, resulting in positive supercoiling. Note that the twisting path taken by the axis of the DNA helix when the DNA is underwound (negative supercoiling) is the mirror image of that taken when the DNA is overwound (positive supercoiling) **(Fig. 24-16)**. Supercoiling is not a random

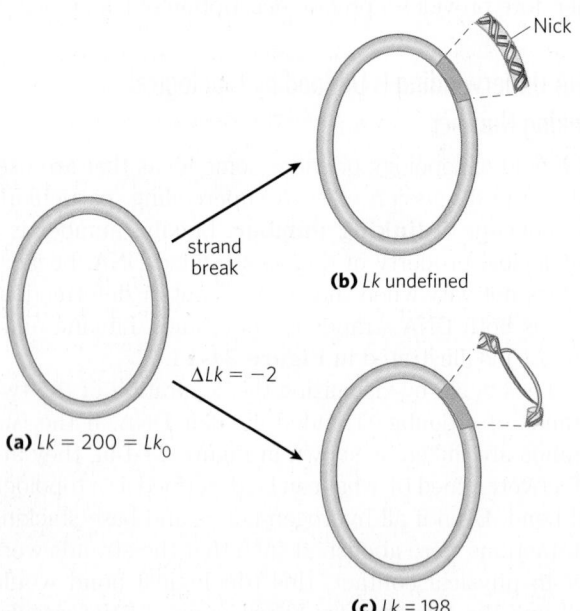

Nick

strand
break

(b) Lk undefined

$\Delta Lk = -2$

(a) $Lk = 200 = Lk_0$

(c) $Lk = 198$

FIGURE 24-15 Linking number applied to closed-circular DNA molecules. A 2,100 bp circular DNA is shown in three forms: **(a)** relaxed, $Lk = 200$; **(b)** relaxed with a nick (break) in one strand, Lk undefined; and **(c)** underwound by two turns, $Lk = 198$. The underwound molecule generally exists as a supercoiled molecule, but underwinding also facilitates the separation of DNA strands.

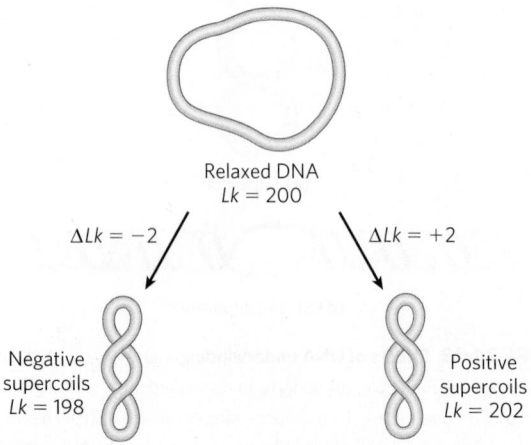

Relaxed DNA
$Lk = 200$

$\Delta Lk = -2$

$\Delta Lk = +2$

Negative
supercoils
$Lk = 198$

Positive
supercoils
$Lk = 202$

FIGURE 24-16 Negative and positive supercoils. For the relaxed DNA molecule of Figure 24-15a, underwinding or overwinding by two helical turns ($Lk = 198$ or 202) will produce negative or positive supercoiling, respectively. Notice that the DNA axis twists in opposite directions in the two cases.

process; the path of the supercoiling is largely prescribed by the torsional strain imparted to the DNA by decreasing or increasing the linking number relative to B-DNA.

Linking number can be changed by ±1 by breaking one DNA strand, rotating one of the ends 360° about the unbroken strand, and rejoining the broken ends. This change has no effect on the number of base pairs or the number of atoms in the circular DNA molecule. Two forms of a circular DNA that differ only in a topological property such as linking number are referred to as **topoisomers**.

WORKED EXAMPLE 24-1   **Calculation of Superhelical Density**

What is the superhelical density (σ) of a closed-circular DNA with a length of 4,200 bp and a linking number (Lk) of 374? What is the superhelical density of the same DNA when $Lk = 412$? Are these molecules negatively or positively supercoiled?

Solution: First, calculate Lk_0 by dividing the length of the closed-circular DNA (in bp) by 10.5 bp/turn: (4,200 bp)/(10.5 bp/turn) = 400. We can now calculate ΔLk from Equation 24-1: $\Delta Lk = Lk - Lk_0 = 374 - 400 = -26$. Substituting the values for ΔLk and Lk_0 into Equation 24-2: $\sigma = \Delta Lk/Lk_0 = -26/400 = -0.065$. Since the superhelical density is negative, this DNA molecule is negatively supercoiled.

When the same DNA molecule has an Lk of 412, $\Delta Lk = 412 - 400 = 12$, and $\sigma = 12/400 = 0.03$. The superhelical density is positive, and the molecule is positively supercoiled.

In addition to causing supercoiling and making strand separation somewhat easier, the underwinding of DNA facilitates structural changes in the molecule. These are of less physiological importance but help illustrate the effects of underwinding. Recall that a cruciform (see Fig. 8-19) generally contains a few unpaired bases; DNA underwinding helps to maintain the required strand separation **(Fig. 24-17)**. Underwinding of a right-handed DNA helix also enables the formation of short stretches of left-handed Z-DNA in regions where the base sequence is consistent with the Z form (see Chapter 8).

Topoisomerases Catalyze Changes in the Linking Number of DNA

DNA supercoiling is a precisely regulated process that influences many aspects of DNA metabolism. Every cell has enzymes with the sole function of underwinding and/or relaxing DNA. The enzymes that increase or decrease the extent of DNA underwinding are **topoisomerases**; the property of DNA that they change is the linking number. These enzymes play an especially important role in processes such as replication and DNA packaging. There

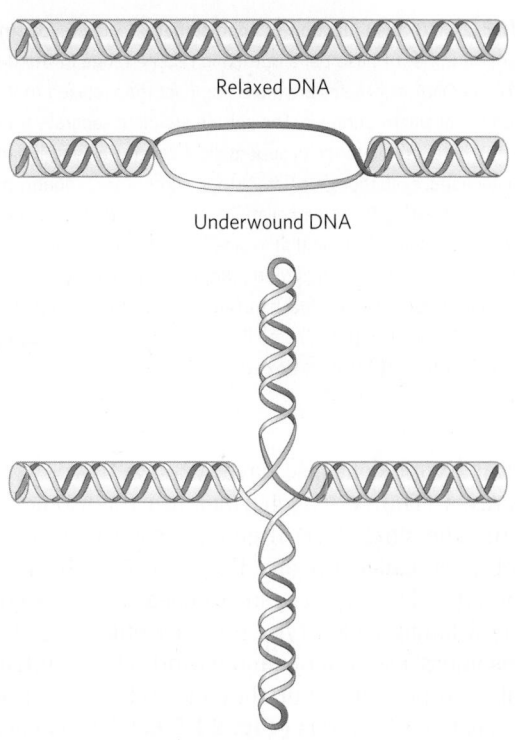

Relaxed DNA

Underwound DNA

Cruciform DNA

FIGURE 24-17 Promotion of cruciform structures by DNA underwinding. In principle, cruciforms can form at palindromic sequences (see Fig. 8-19), but they seldom occur in relaxed DNA because the linear DNA accommodates more paired bases than does the cruciform structure. Underwinding of the DNA facilitates the partial strand separation needed to promote cruciform formation at appropriate sequences.

are two classes of topoisomerases. **Type I topoisomerases** act by transiently breaking one of the two DNA strands, passing the unbroken strand through the break and rejoining the broken ends; they change Lk in increments of 1. **Type II topoisomerases** break both DNA strands and change Lk in increments of 2.

The effects of these enzymes can be demonstrated with agarose gel electrophoresis **(Fig. 24-18)**. A population of identical plasmid DNAs with the same linking number migrates as a discrete band during electrophoresis. Topoisomers with Lk values differing by as little as 1 can be separated by this method, so changes in linking number induced by topoisomerases are readily detected.

E. coli has at least four individual topoisomerases (I through IV). Those of type I (topoisomerases I and III) generally relax DNA by removing negative supercoils (increasing Lk). The way in which bacterial type I topoisomerases change linking number is illustrated in **Figure 24-19**. A bacterial type II enzyme, called either topoisomerase II or DNA gyrase, can introduce negative supercoils (decrease Lk). It uses the energy of ATP to accomplish this. To alter DNA linking number, type II topoisomerases cleave both strands of a DNA molecule and pass another duplex through the break. The degree of supercoiling of bacterial DNA is maintained by regulation of the net activity of topoisomerases I and II.

FIGURE 24-18 Visualization of topoisomers. In this experiment, all DNA molecules have the same number of base pairs but exhibit some range in the degree of supercoiling. Because supercoiled DNA molecules are more compact than relaxed molecules, they migrate more rapidly during gel electrophoresis. The gels shown here separate topoisomers (moving from top to bottom) over a limited range of superhelical density. In lane 1, highly supercoiled DNA migrates in a single band, even though different topoisomers are probably present. Lanes 2 and 3 illustrate the effect of treating the supercoiled DNA with a type I topoisomerase; the DNA in lane 3 was treated for a longer time than that in lane 2. As the superhelical density of the DNA is reduced to the point where it corresponds to the range in which the gel can resolve individual topoisomers, distinct bands appear. Individual bands in the region indicated by the bracket next to lane 3 each contain DNA circles with the same linking number; Lk changes by 1 from one band to the next. [Source: Courtesy of Michael Cox Lab.]

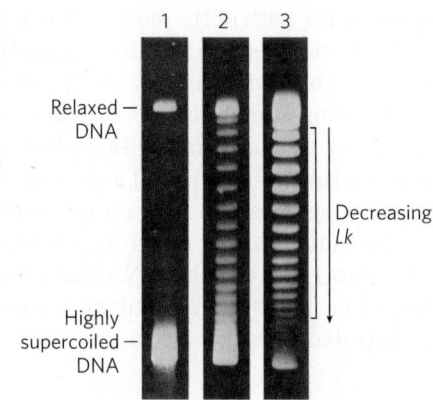

Eukaryotic cells also have type I and type II topoisomerases. The type I enzymes are topoisomerases I and III. The single type II enzyme has two isoforms in vertebrates, called IIα and IIβ. Most type II enzymes, including a DNA gyrase in archaea, are related and define a family called type IIA. The eukaryotic type II topoisomerases cannot underwind DNA (introduce negative supercoils), but they can relax both positive and negative supercoils **(Fig. 24-20a)**. The capacity of type II topoisomerases to pass one duplex DNA segment through a double-strand break in another duplex allows these enzymes to untangle **catenanes**, DNA circles that are topologically linked (Fig. 24-20b). Some topoisomerases are specialized for decatenation functions. For example, bacteria have a separate type II enzyme called topoisomerase IV, which is involved in chromosome untangling during cell division (Chapter 25).

Archaea have an unusual enzyme, topoisomerase VI, which alone defines the type IIB family. The full diversity of DNA topoisomerases is illustrated in Table 24-4.

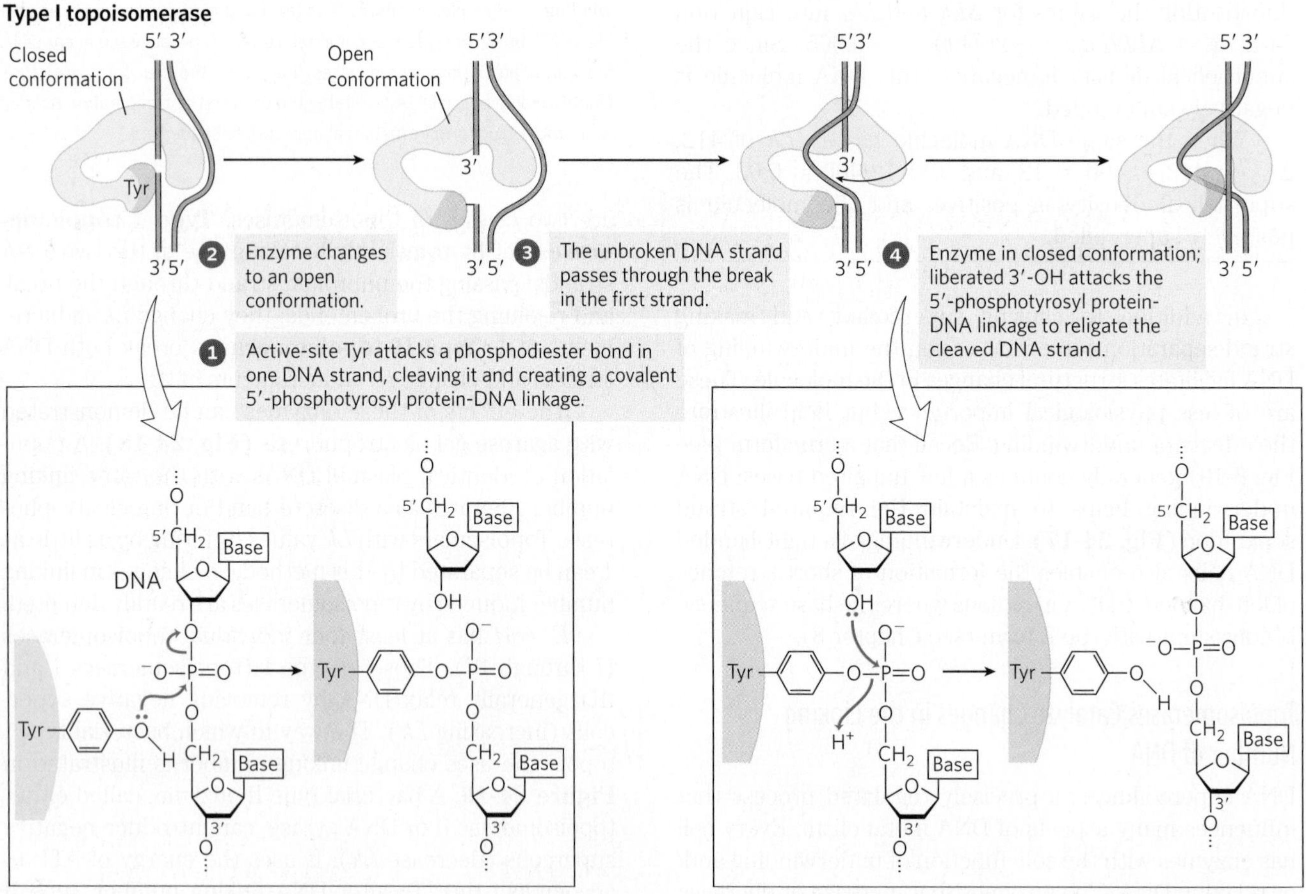

Type I topoisomerase

Closed conformation — Open conformation

2 Enzyme changes to an open conformation.

3 The unbroken DNA strand passes through the break in the first strand.

4 Enzyme in closed conformation; liberated 3'-OH attacks the 5'-phosphotyrosyl protein-DNA linkage to religate the cleaved DNA strand.

1 Active-site Tyr attacks a phosphodiester bond in one DNA strand, cleaving it and creating a covalent 5'-phosphotyrosyl protein-DNA linkage.

MECHANISM FIGURE 24-19 **The type I topoisomerase reaction.** Bacterial topoisomerase I increases Lk by breaking one DNA strand, passing the unbroken strand through the break, then resealing the break. Nucleophilic attack by the active-site Tyr residue breaks one DNA strand. The ends are ligated by a second nucleophilic attack. At each step, one high-energy bond replaces another. [Source: Information from J. J. Champoux, *Annu. Rev. Biochem.* 70:369, 2001, Fig. 3.]

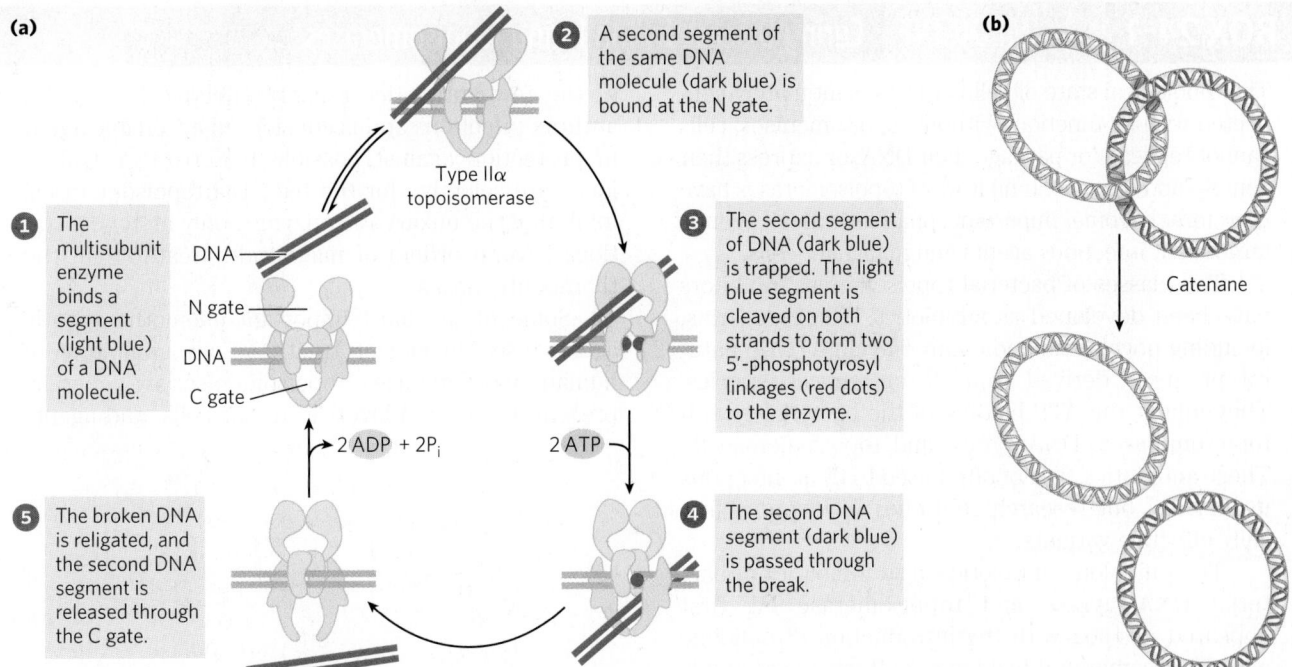

FIGURE 24-20 Alteration of the linking number by a eukaryotic type IIα topoisomerase. (a) The general mechanism features passage of one intact duplex DNA segment through a transient double-strand break in another segment. The DNA segment enters and leaves the topoisomerase through gated cavities, called the N gate and the C gate, above and below the bound DNA. Two ATPs are bound and hydrolyzed during this cycle. The enzyme structure and use of ATP are specific to this reaction. **(b)** When topologically linked as shown, two DNA circles are referred to as a catenane. By cleaving both strands of one circle, and passing a segment of the second circle through the break, a type II topoisomerase can decatenate the circles. [Source: (a) Information from J. J. Champoux, *Annu. Rev. Biochem.* 70:369, 2001, Fig. 11.]

Type	Mechanism	Family (defined by structural class)	Domain(s)	Notes
TABLE 24-4	Diversity in DNA Topoisomerases			
IA	Strand passage[a]	Topoisomerase I	Bacteria, eukaryotes	Relaxes (−)
		Topoisomerase III	Bacteria, eukaryotes	Relaxes (−)
		Reverse gyrase	Archaea, bacteria	Uses ATP to introduce positive supercoils; thermophilic bacteria and archaea only
IB	Swivelase[b]	Topoisomerase IB	Bacteria, eukaryotes	A few bacteria; all eukaryotes
IC	Swivelase	Topoisomerase V	Archaea	*Methanopyrus* only
IIA	Strand passage[c]	Topoisomerase II (DNA gyrase)	Archaea, bacteria	Introduces negative supercoils (ATPase)
		Topoisomerase IIα	Eukaryotes	Relaxes (+ or −)
		Topoisomerase IIβ	Eukaryotes	Relaxes (+ or −)
		Topoisomerase IV	Bacteria	Decatenase[d]
IIB	Strand passage	Topoisomerase VI	Archaea, bacteria, eukaryotes	Among eukaryotes, plants, algae, and protists only

[a]See Figure 24-19.

[b]A nick is made in one strand, and the other strand is allowed to rotate to relieve topological strain.

[c]See Figure 24-20a.

[d]See Figure 24-20b.

BOX 24-1 ⚕ MEDICINE Curing Disease by Inhibiting Topoisomerases

The topological state of cellular DNA is intimately connected with its function. Without topoisomerases, cells cannot replicate or package their DNA, or express their genes—and they die. Inhibitors of topoisomerases have therefore become important pharmaceutical agents, targeted at infectious agents and malignant cells.

Two classes of bacterial topoisomerase inhibitors have been developed as antibiotics. The coumarins, including novobiocin and coumermycin A1, are natural products derived from *Streptomyces* species. They inhibit the ATP binding of the bacterial type II topoisomerases, DNA gyrase and topoisomerase IV. These antibiotics are not often used to treat infections in humans, but research continues to identify clinically effective variants.

The quinolone antibiotics, also inhibitors of bacterial DNA gyrase and topoisomerase IV, first appeared in 1962 with the introduction of nalidixic acid. This compound had limited effectiveness and is no longer used clinically in the United States, but the continued development of this class of drugs eventually led to the introduction of the fluoroquinolones, exemplified by ciprofloxacin (Cipro). The quinolones act by blocking the last step of the topoisomerase reaction, the resealing of the DNA strand breaks. Ciprofloxacin is a broad-spectrum antibiotic. It is one of the few antibiotics reliably effective in treating anthrax infections and is considered a valuable agent in protection against possible bioterrorism. Quinolones are selective for the bacterial topoisomerases, inhibiting the eukaryotic enzymes only at concentrations several orders of magnitude greater than the therapeutic doses.

Some of the most important chemotherapeutic agents used in cancer treatment are inhibitors of human topoisomerases. Topoisomerases are generally present at elevated levels in tumor cells, and agents

Irinotecan

Nalidixic acid

Ciprofloxacin

Topotecan

As we will show in the next few chapters, topoisomerases play a critical role in every aspect of DNA metabolism. As a consequence, they are important drug targets for the treatment of bacterial infections and cancer (Box 24-1).

DNA Compaction Requires a Special Form of Supercoiling

Supercoiled DNA molecules are uniform in several respects. The supercoils are right-handed in a negatively supercoiled DNA molecule (Fig. 24-16), and they tend to be extended and narrow rather than compacted, often with multiple branches **(Fig. 24-21)**. At the superhelical densities normally encountered in cells, the length of the supercoil axis, including branches, is about 40% of the length of the DNA. This type of supercoiling is referred to as **plectonemic** (from the Greek *plektos*, "twisted," and *nema*, "thread"). The term can be applied to any structure with strands intertwined in some simple and regular way, and it is a good description of the general structure of supercoiled DNA in solution.

Plectonemic supercoiling, the form observed in isolated DNAs in the laboratory, does not produce sufficient compaction to package DNA in the cell. A second form of supercoiling, **solenoidal (Fig. 24-22)**, can be adopted by an underwound DNA. Instead of the extended right-handed supercoils characteristic of the plectonemic form, solenoidal supercoiling involves tight left-handed turns, similar to the shape taken up by a garden hose neatly wrapped on a reel. Although their structures are dramatically different, plectonemic and solenoidal supercoiling are two forms of negative supercoiling that can be taken up by the *same* segment of underwound DNA. The two forms are readily interconvertible. Although the plectonemic form is more stable in solution, the solenoidal form can be stabilized by protein binding, as it is in eukaryotic chromosomes; it provides a much greater degree of compaction (Fig. 24-22). Solenoidal supercoiling is the mechanism by which underwinding contributes to DNA compaction in cells.

targeted to these enzymes are much more toxic to the tumors than to most other tissue types. Inhibitors of both type I and type II topoisomerases have been developed as anticancer drugs.

Camptothecin, isolated from a Chinese ornamental tree and first tested clinically in the 1970s, is an inhibitor of eukaryotic type I topoisomerases. Clinical trials indicated limited effectiveness, despite its early promise in preclinical work on mice. However, two effective derivatives, irinotecan (Campto) and topotecan (Hycamtin)—used to treat colorectal cancer and ovarian cancer, respectively—were developed in the 1990s. Additional derivatives are likely to be approved for clinical use in the coming years. All of these drugs act by trapping the topoisomerase-DNA complex in which the DNA is cleaved, inhibiting religation.

The human type II topoisomerases are targeted by a variety of antitumor drugs, including doxorubicin (Adriamycin), etoposide (Etopophos), and ellipticine. Doxorubicin, effective against several kinds of human tumors, is an anthracycline in clinical use. Most of these drugs stabilize the covalent topoisomerase-DNA (cleaved) complex.

All of these anticancer agents generally increase the levels of DNA damage in the targeted, rapidly growing tumor cells. However, noncancerous tissues can also be affected, leading to a more general toxicity and unpleasant side effects that must be managed during therapy. As cancer therapies become more effective and survival statistics for cancer patients improve, the independent appearance of new tumors is becoming a greater problem. In the continuing search for new cancer therapies, topoisomerases are likely to remain prominent targets for research.

Doxorubicin

Etoposide

Ellipticine

SUMMARY 24.2 DNA Supercoiling

■ Most cellular DNAs are supercoiled. Underwinding decreases the total number of helical turns in DNA relative to the relaxed, B form. To maintain an underwound state, DNA must be either a closed circle or bound to protein. Underwinding is quantified by a topological parameter called linking number, Lk.

■ Underwinding is measured in terms of specific linking difference, or superhelical density, σ, which is $(Lk - Lk_0)/Lk_0$. For cellular DNAs, σ is typically

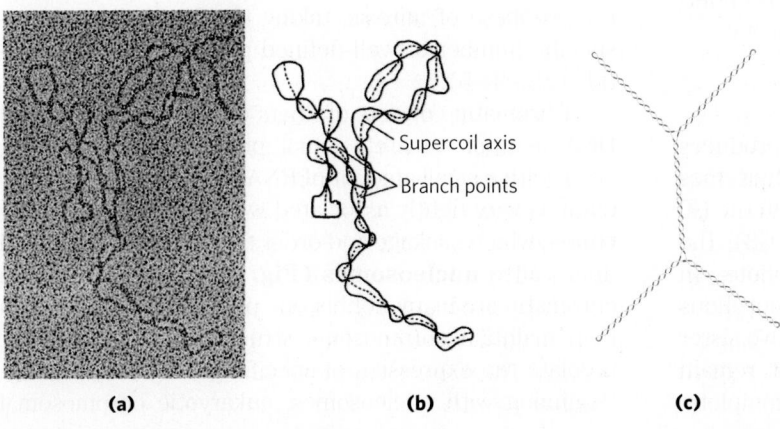

Supercoil axis

Branch points

(a)   **(b)**   **(c)**

FIGURE 24-21 Plectonemic supercoiling. (a) Electron micrograph of plectonemically supercoiled plasmid DNA and **(b)** an interpretation of the observed structure. The dotted lines show the axis of the supercoil; notice the branching of the supercoil. **(c)** An idealized representation of this structure. [Source: Republished with permission of Elsevier, from T. C. Boles et al. (1990) "Structure of plectonemically supercoiled DNA," *J. Mol. Biol.* 213:931–951, Fig. 2; permission conveyed through Copyright Clearance Center, Inc.]

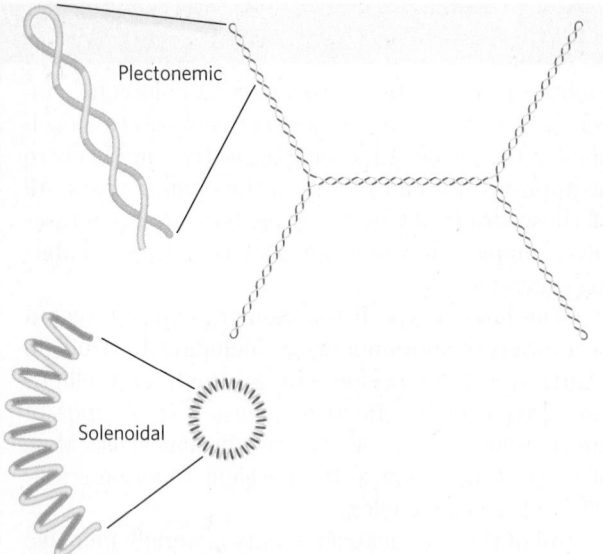

FIGURE 24-22 Plectonemic and solenoidal supercoiling of the same DNA molecule, drawn to scale. Plectonemic supercoiling takes the form of extended right-handed coils. Solenoidal negative supercoiling takes the form of tight left-handed turns about an imaginary tubelike structure. The two forms are readily interconverted, although the solenoidal form is generally not observed unless certain proteins are bound to the DNA. Solenoidal supercoiling provides a much greater degree of compaction.

−0.05 to −0.07, which means that approximately 5% to 7% of the helical turns in the DNA have been removed. DNA underwinding facilitates strand separation by enzymes of DNA metabolism.

■ DNAs that differ only in linking number are called topoisomers. Topoisomerases, enzymes that underwind and/or relax DNA, catalyze changes in linking number. The two classes of topoisomerases, type I and type II, change Lk in increments of 1 or 2, respectively, per catalytic event.

24.3 The Structure of Chromosomes

The term "chromosome" is used to refer to a nucleic acid molecule that is the repository of genetic information in a virus, a bacterium, an archaeon, a eukaryotic cell, or an organelle. It also refers to the densely colored bodies seen in the nuclei of dye-stained eukaryotic cells undergoing mitosis, as visualized using a light microscope.

Chromatin Consists of DNA and Proteins

The eukaryotic cell cycle (see Fig. 12-32) produces remarkable changes in the structure of chromosomes **(Fig. 24-23)**. In nondividing eukaryotic cells (in the G0 phase) and those in interphase (G1, S, and G2), the chromosomal material, **chromatin**, is amorphous. In the S phase of interphase, the DNA in this amorphous state replicates, each chromosome producing two sister chromosomes (called sister chromatids) that remain associated with each other after replication is complete.

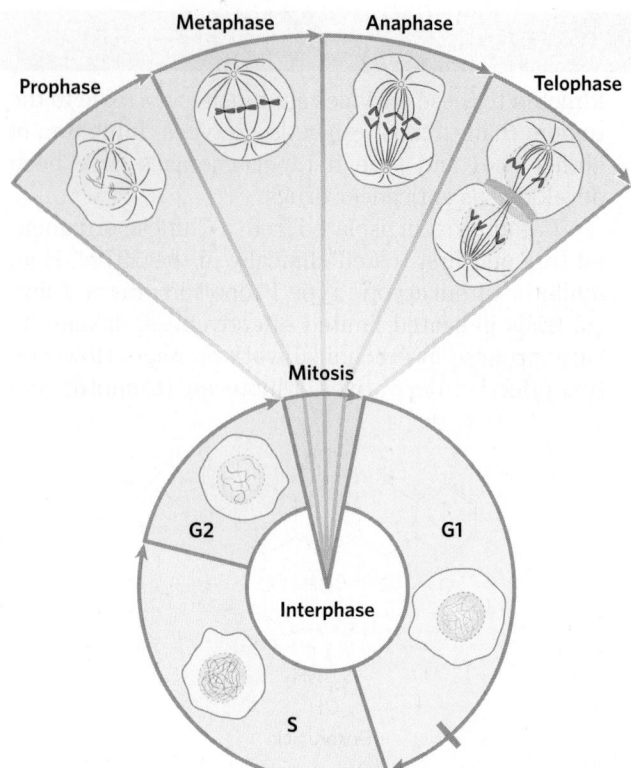

FIGURE 24-23 Changes in chromosome structure during the eukaryotic cell cycle. The relative lengths of the phases shown here are for convenience only. The duration of each phase varies with cell type and with growth conditions (for single-celled organisms) or metabolic state (for multicellular organisms); mitosis is typically the shortest phase. Cellular DNA is uncondensed throughout interphase, as shown in the cartoons of the nucleus. The interphase period can be divided (see Fig. 12-32) into the G1 (gap) phase; the S (synthesis) phase, when the DNA is replicated; and the G2 phase, throughout which the replicated chromosomes (chromatids) cohere to one another. Mitosis can be divided into four stages. The DNA undergoes condensation in prophase. During metaphase, the condensed chromosomes line up in pairs along the plane halfway between the spindle poles. The two chromosomes of each pair are linked to different spindle poles via microtubules that extend between the spindle and the centromere. The sister chromatids separate at anaphase, each drawn toward the spindle pole to which it is connected. The process is completed in telophase. After cell division, the chromosomes decondense and the cycle begins anew.

The chromosomes become much more condensed during prophase of mitosis, taking the form of a species-specific number of well-defined pairs of sister chromatids (Fig. 24-5).

Chromatin consists of fibers containing protein and DNA in approximately equal proportions (by mass), along with a small amount of RNA. The DNA in the chromatin is very tightly associated with proteins called **histones**, which package and order the DNA into structural units called **nucleosomes (Fig. 24-24)**. Also found in chromatin are many nonhistone proteins, some of which help maintain chromosome structure and others that regulate the expression of specific genes (Chapter 28). Beginning with nucleosomes, eukaryotic chromosomal

(a)

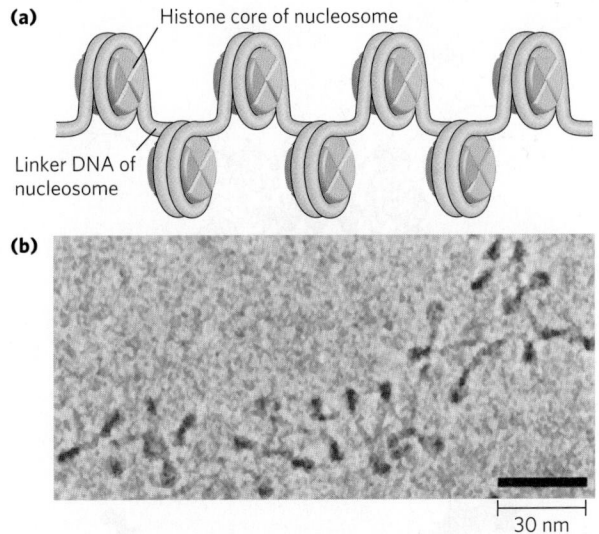

Histone core of nucleosome

Linker DNA of nucleosome

(b)

30 nm

FIGURE 24-24 Nucleosomes. (a) Regularly spaced nucleosomes consist of core histone proteins bound to DNA. **(b)** In this electron micrograph, the DNA-wrapped histone octamer structures are clearly visible. [Source: (b) J. Bednar et al., Nucleosomes, linker DNA, and linker histone form a unique structural motif that directs the higher-order folding and compaction of chromatin, *Proc. Natl. Acad. Sci. USA* vol. 95 no. 24: 14173–14178, November 1998 Cell Biology Fig. 1. © 1998 National Academy of Sciences, U.S.A.]

DNA is packaged into a succession of higher-order structures that ultimately yield the compact chromosome seen with the light microscope. We now turn to a description of this structure in eukaryotes and compare it with the packaging of DNA in bacterial cells.

Histones Are Small, Basic Proteins

Found in the chromatin of all eukaryotic cells, histones have molecular weights between 11,000 and 21,000 and are very rich in the basic amino acids arginine and lysine (together these make up about one-fourth of the amino acid residues). All eukaryotic cells have five major classes of histones, differing in molecular weight and amino acid composition (Table 24-5). The H3 histones

are nearly identical in amino acid sequence in all eukaryotes, as are the H4 histones, suggesting strict conservation of their functions. For example, only 2 of 102 amino acid residues differ between the H4 histone molecules of peas and cows, and only 8 differ between the H4 histones of humans and yeast. Histones H1, H2A, and H2B show less sequence similarity across eukaryotic species.

Each type of histone is subject to enzymatic modification by methylation, acetylation, ADP-ribosylation, phosphorylation, glycosylation, sumoylation, or ubiquitination. Such modifications affect the net electric charge, shape, and other properties of histones, as well as the structural and functional properties of the chromatin, and they play a role in the regulation of transcription.

In addition, eukaryotes generally have several variant forms of certain histones, most notably histones H2A and H3, described in more detail below. The variant forms, along with their modifications, have specialized roles in DNA metabolism.

Nucleosomes Are the Fundamental Organizational Units of Chromatin

The eukaryotic chromosome depicted in Figure 24-5 represents the compaction of a DNA molecule about 10^5 μm long into a cell nucleus that is typically 5 to 10 μm in diameter. This compaction is achieved by means of several levels of highly organized folding. Subjection of chromosomes to treatments that partially unfold them reveals a structure in which the DNA is bound tightly to beads of protein, often regularly spaced. The beads in this "beads-on-a-string" arrangement are complexes of histones and DNA. The bead plus the connecting DNA that leads to the next bead form the nucleosome, the fundamental unit of organization on which the higher-order packing of chromatin is built **(Fig. 24-25)**. The bead of each nucleosome contains eight histone molecules: two copies each of H2A, H2B, H3, and H4. The spacing of the nucleosome beads provides a repeating unit typically of about 200 bp, of which 146 bp are bound tightly around the eight-part histone core and the remainder serve as linker DNA between nucleosome beads. Histone H1 binds to the linker DNA. Brief treatment of chromatin with enzymes that digest DNA causes the linker DNA to degrade preferentially, releasing histone particles containing 146 bp of bound DNA that is protected from digestion. Researchers have crystallized nucleosome cores obtained in this way, and x-ray diffraction analysis reveals a particle made up of the eight histone molecules with the DNA wrapped around the core in the form of a left-handed solenoidal supercoil (Fig. 24-25). Extending out from the nucleosome core are the amino-terminal tails of the histones, which are intrinsically disordered

TABLE 24-5	Types and Properties of the Common Histones			
Histones	Molecular weight	Number of amino acid residues	Content of basic amino acids (% of total)	
			Lys	Arg
H1[a]	21,130	223	29.5	11.3
H2A[a]	13,960	129	10.9	19.3
H2B[a]	13,774	125	16.0	16.4
H3	15,273	135	19.6	13.3
H4	11,236	102	10.8	13.7

[a]The sizes of these histones vary somewhat from species to species. The numbers given here are for bovine histones.

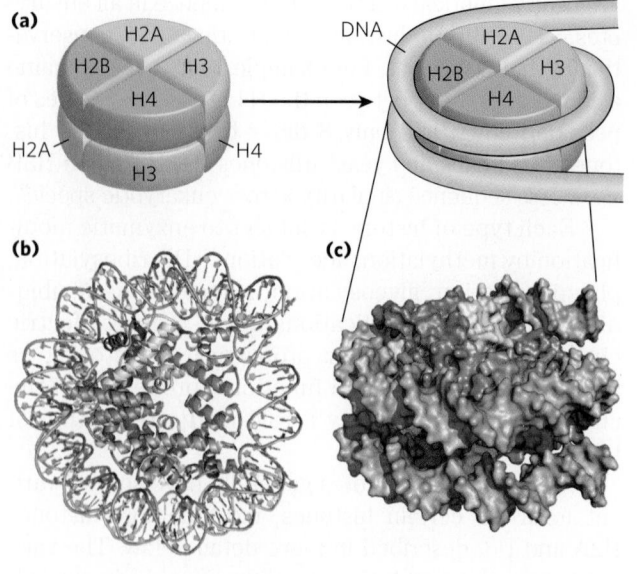

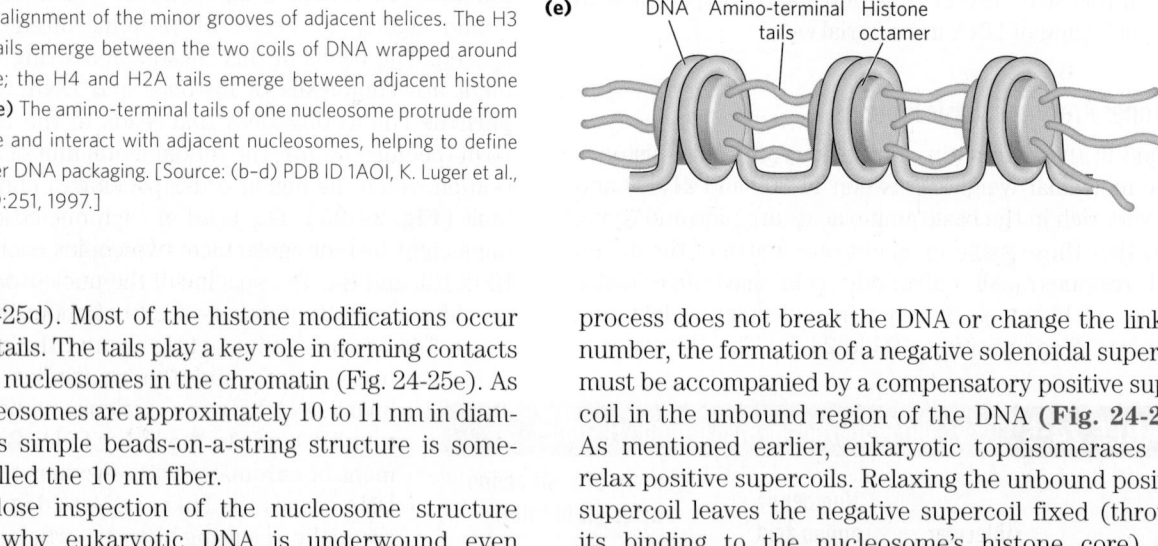

FIGURE 24-25 DNA wrapped around a histone core. (a) The simplified structure of a nucleosome octamer (left), with DNA wrapped around the histone core (right). **(b)** A ribbon representation of the nucleosome from the African frog *Xenopus laevis*. Different colors represent the different histones, matching the colors in (a). **(c)** Surface representation of the nucleosome. The view in (c) is rotated relative to the view in (b) to match the orientation shown in (a). A 146 bp segment of DNA in the form of a left-handed solenoidal supercoil wraps around the histone complex 1.67 times. **(d)** Two views of histone amino-terminal tails protruding from between the two DNA duplexes that supercoil around the nucleosome core. Some tails pass between the supercoils, through holes formed by alignment of the minor grooves of adjacent helices. The H3 and H2B tails emerge between the two coils of DNA wrapped around the histone; the H4 and H2A tails emerge between adjacent histone subunits. **(e)** The amino-terminal tails of one nucleosome protrude from the particle and interact with adjacent nucleosomes, helping to define higher-order DNA packaging. [Source: (b–d) PDB ID 1AOI, K. Luger et al., *Nature* 389:251, 1997.]

(Fig. 24-25d). Most of the histone modifications occur in these tails. The tails play a key role in forming contacts between nucleosomes in the chromatin (Fig. 24-25e). As the nucleosomes are approximately 10 to 11 nm in diameter, this simple beads-on-a-string structure is sometimes called the 10 nm fiber.

A close inspection of the nucleosome structure reveals why eukaryotic DNA is underwound even though eukaryotic cells lack enzymes that underwind DNA. Recall that the solenoidal wrapping of DNA in nucleosomes is but one form of supercoiling that can be taken up by underwound (negatively supercoiled) DNA. The tight wrapping of DNA around the histone core requires the removal of about one helical turn in the DNA. When the protein core of a nucleosome binds in vitro to a relaxed closed-circular DNA, the binding introduces a negative supercoil. Because this binding

process does not break the DNA or change the linking number, the formation of a negative solenoidal supercoil must be accompanied by a compensatory positive supercoil in the unbound region of the DNA **(Fig. 24-26)**. As mentioned earlier, eukaryotic topoisomerases can relax positive supercoils. Relaxing the unbound positive supercoil leaves the negative supercoil fixed (through its binding to the nucleosome's histone core) and results in an overall decrease in linking number. Indeed, topoisomerases have proved necessary for assembling chromatin from purified histones and closed-circular DNA in vitro.

Another factor that affects the binding of DNA to histones in nucleosome cores is the sequence of the bound DNA. Histone cores do not bind at random positions on the DNA; rather, some locations are more likely to be bound than others. This positioning is not fully

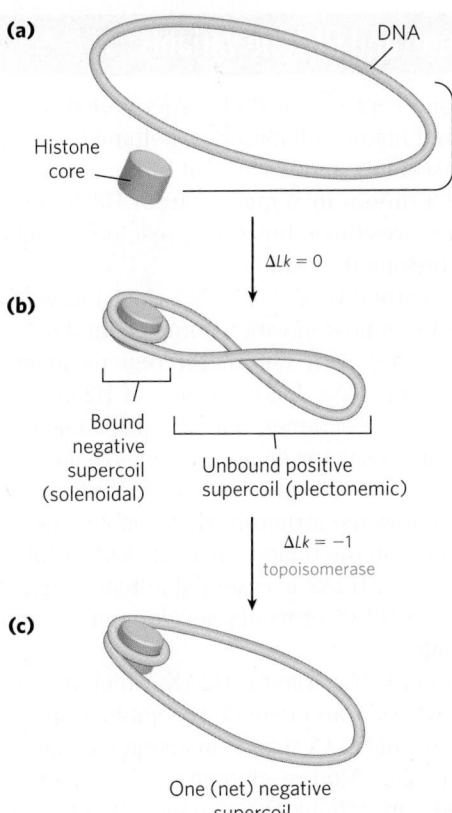

FIGURE 24-26 Chromatin assembly. (a) Relaxed closed-circular DNA. **(b)** Binding of a histone core to form a nucleosome induces one negative supercoil. In the absence of any strand breaks, a positive supercoil must form elsewhere in the DNA ($\Delta Lk = 0$). **(c)** Relaxation of this positive supercoil by a topoisomerase leaves one net negative supercoil ($\Delta Lk = -1$).

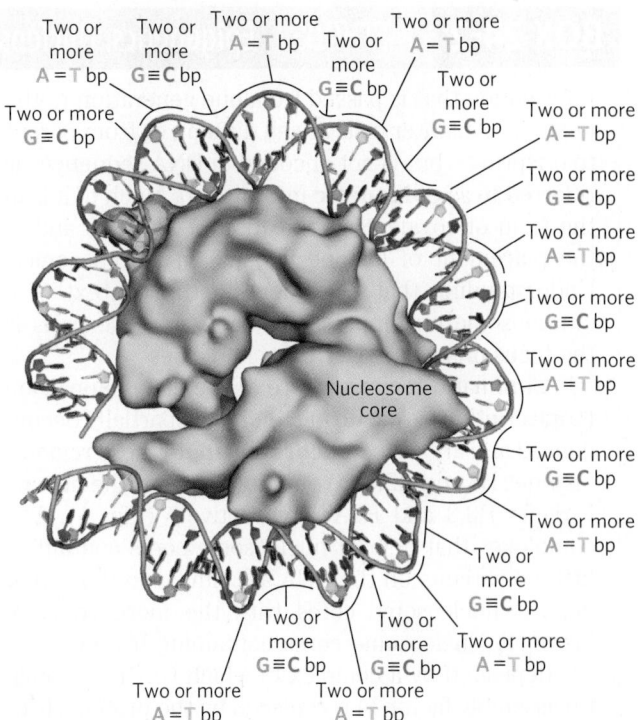

FIGURE 24-27 The effect of DNA sequence on nucleosome binding. Runs of two or more A≡T base pairs facilitate the bending of DNA, while runs of two or more G≡C base pairs have the opposite effect. When spaced at about 10 bp intervals, consecutive A≡T base pairs help bend DNA into a circle. When consecutive G≡C base pairs are spaced 10 bp apart, offset by 5 bp from runs of A≡T base pairs, DNA binding to the nucleosome core is facilitated. [Source: PDB ID 1AOI, K. Luger et al., *Nature* 389:251, 1997.]

understood, but in some cases it seems to depend on a local abundance of A=T base pairs in the DNA helix where it is in contact with the histones **(Fig. 24-27)**. A cluster of two or three A=T base pairs facilitates the compression of the minor groove that is needed for the DNA to wrap tightly around the nucleosome's histone core. Nucleosomes bind particularly well to sequences where AA or AT or TT dinucleotides are staggered at 10 bp intervals, an arrangement that can account for up to 50% of the positions of bound histones in vivo.

Nucleosome cores are deposited on DNA during replication, or following other processes that require a transient displacement of nucleosomes. Other, nonhistone proteins are required for the positioning of some nucleosome cores. In several organisms, certain proteins bind to a specific DNA sequence and facilitate the formation of a nucleosome core nearby. Nucleosome cores seem to be deposited stepwise. A tetramer of two H3 and two H4 histones binds first, followed by H2A-H2B dimers. The incorporation of nucleosomes into chromosomes after chromosomal replication is mediated by a complex of **histone chaperones** that include the proteins CAF1 (chromatin assembly factor 1), RTT106 (regulation of Ty1 transposition), and ASF1

(anti-silencing factor 1). These bind to acetylated variants of histones H3 and H4. The mechanism of deposition is not understood in detail, but parts of the nucleosome complex are known to directly interact with parts of the replication machinery. Some of the same histone chaperones, or different ones, may help assemble nucleosomes after DNA repair, transcription, or other processes. **Histone exchange factors** permit the substitution of histone variants for core histones in some contexts. Proper placement of these variant histones is important. Studies show that mice lacking one of the variant histones die as early embryos (Box 24-2). Precise positioning of nucleosome cores also plays a role in the expression of some eukaryotic genes (Chapter 28).

Nucleosomes Are Packed into Highly Condensed Chromosome Structures

Wrapping of DNA around a nucleosome core compacts the DNA length about sevenfold. The overall compaction in a chromosome, however, is greater than 10,000-fold—ample evidence for even higher orders of structural organization. In vitro, long DNA molecules bound by nucleosomes and histone H1 can fold into a structure

BOX 24-2 METHODS Epigenetics, Nucleosome Structure, and Histone Variants

Information that is passed from one generation to the next—to daughter cells at cell division or from parent to offspring—but is not encoded in DNA sequences is referred to as **epigenetic** information. Much of it is in the form of covalent modification of histones and/or the placement of histone variants in chromosomes. Understanding that placement in the context of a chromosome encompassing millions of base pairs is the focus of some powerful technologies.

Chromatin regions where active gene expression (transcription) is occurring tend to be partially decondensed and are called **euchromatin**. In these regions, histones H3 and H2A are often replaced by the histone variants H3.3 and H2AZ, respectively (Fig. 1). The complexes that deposit nucleosome cores containing histone variants on the DNA are similar to those that deposit nucleosome cores with the more common histones. Nucleosome cores containing histone H3.3 are deposited by a complex in which CAF1 (chromatin assembly factor 1) is replaced by the protein HIRA (a name derived from a class of proteins called HIR,

for *h*istone *r*epressor). Both CAF1 and HIRA can be considered histone chaperones, helping to ensure the proper assembly and placement of nucleosomes. Histone H3.3 differs in sequence from H3 by only four amino acid residues, but these residues all play key roles in histone deposition.

Like histone H3.3, H2AZ is associated with a distinct nucleosome deposition complex, and it is generally associated with chromatin regions involved in active transcription. Incorporation of H2AZ stabilizes the nucleosome octamer, but impedes some cooperative interactions between nucleosomes that are needed to compact the chromosome. This leads to a more open chromosome structure that enables the expression of genes in the region where H2AZ is located. The gene encoding H2AZ is essential in mammals. In fruit flies, loss of H2AZ prevents development beyond the larval stages.

Another H2A variant is H2AX, which is associated with DNA repair and genetic recombination. In mice, the absence of H2AX results in genome instability and male infertility. Modest amounts of H2AX seem to be scattered throughout the genome. When a double-strand break occurs, nearby molecules of H2AX become phosphorylated at Ser[139] in the carboxyl-terminal region. If this phosphorylation is blocked experimentally, formation of the protein complexes necessary for DNA repair is inhibited.

The H3 histone variant known as CENPA is associated with the repeated DNA sequences in centromeres. The chromatin in the centromere region contains the histone chaperones CAF1 and HIRA, and both proteins could be involved in the deposition of nucleosome cores containing CENPA. Elimination of the gene for CENPA is lethal in mice.

The function and positioning of the histone variants can be studied by an application of technologies used in genomics. One useful technology is chromatin immunoprecipitation, or chromatin IP (ChIP). Nucleosomes containing a particular histone variant are precipitated by an antibody that binds specifically to this variant. These nucleosome cores can be studied in isolation from their DNA, but more commonly the DNA associated with them is included in the study to determine where the nucleosome cores of interest bind. The DNA can be labeled and used to probe a microarray (see Fig. 9-23), yielding a map of genomic sequences to which those particular nucleosome cores bind. Because microarrays are often referred to

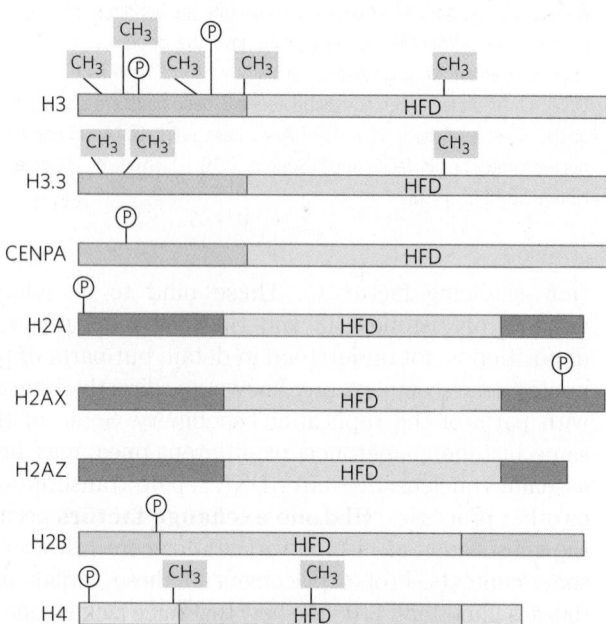

FIGURE 1 Several variants of histones H3, H2A, and H2B are known. Shown here are the standard histones and a few of the known variants. Sites of Lys/Arg residue methylation and Ser phosphorylation are indicated. HFD denotes the histone-fold domain, a structural domain common to all standard histones. Regions denoted in other colors define sequence and structural homologies. [Source: Information from K. Sarma and D. Reinberg, *Nature Rev. Mol. Cell Biol.* 6:139, 2005.]

called the **30 nm fiber (Fig. 24-28a)**. This packing includes one molecule of histone H1 per nucleosome core. A model for the organization of histones and DNA in 30 nm fibers is shown in Figure 24-28b.

The 30 nm fiber—widely studied as a potential second level of chromatin organization—would provide an approximately 100-fold compaction of the DNA. However, close examination of chromatin in vivo has revealed

as chips, this technique is called a ChIP-chip experiment (Fig. 2).

The histone variants, along with the many covalent modifications that histones undergo, help define and isolate the functions of chromatin. They mark the chromatin, facilitating or suppressing specific functions such as chromosome segregation, transcription, and DNA repair. The histone modifications do not disappear at cell division or during meiosis, and thus they become part of the information transmitted from one generation to the next in all eukaryotic organisms.

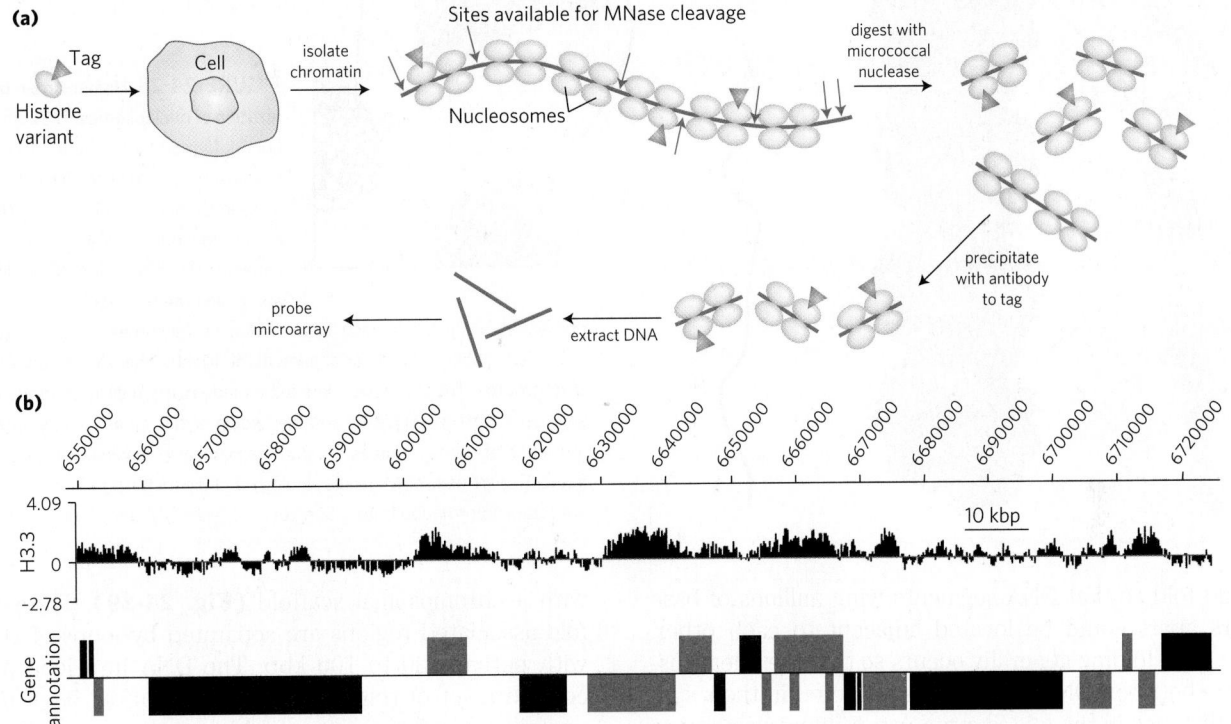

FIGURE 2 A ChIP-chip experiment is designed to reveal the genomic DNA sequences to which a particular histone variant binds. **(a)** A histone variant with an epitope tag (a protein or chemical structure recognized by an antibody; see Chapters 5 and 9) is introduced into a particular cell type, where it is incorporated into nucleosomes. (In some cases, an epitope tag is unnecessary because antibodies are available that bind directly to the histone modification of interest.) Chromatin is isolated from the cells and digested briefly with micrococcal nuclease (MNase). The DNA bound in nucleosomes is protected from digestion, but the linker DNA is cleaved, releasing segments of DNA bound to one or two nucleosomes. An antibody that binds to the epitope tag is added, and the nucleosomes containing the epitope-tagged histone variant are selectively precipitated. The DNA in these nucleosomes is extracted from the precipitate, labeled, and used to probe a microarray representing all or selected parts of the genomic sequences of that particular cell type. **(b)** In this example, the binding of histone H3.3 is characterized in a short segment of chromosome 2L from *Drosophila melanogaster*. Numbers at the top correspond to numbered nucleotide positions in this chromosome arm. Each spot in the microarray represents 100 bp of genomic sequence, so the data here represent more than 1,700 separate spots in the microarray. At each spot, the signal from the labeled DNA that was precipitated with antibody to histone H3.3 is presented as a ratio of that signal relative to the control signal generated when total genomic DNA is isolated without immunoprecipitation, sheared, labeled with a different colored label, and used to probe the same microarray. Signals above the horizontal line indicate genomic positions where histone H3.3 binding is enriched relative to the control. Signals below the line are regions where histone H3.3 is relatively low. Annotated (known) genes in this segment of the genome are shown in the bottom panel of (b), as thickened bars. Bars above the line are genes transcribed 5′ to 3′, left to right, and boxes below the line are transcribed right to left. Red bars are genes where RNA polymerase II is also abundant, indicating active transcription. The histone H3.3 binding is concentrated in and near these genes undergoing active transcription. [Source: Data courtesy of Steve Henikoff. Reprinted with permission from Y. Mito et al., *Nature Genet.* 37:1090, 2005.]

little evidence for 30 nm fibers in cells. Instead, chromosomes seem to condense in a state that could be described as a crumpled globule (Fig. 24-28c). The 10 nm fibers condense by folding onto themselves. The condensation is not as organized as a 30 nm fiber, but is also not random. An average human chromosome of 135 million base pairs is potentially long enough to span a 10 μm nucleus more than 5,000 times. Thus, in principle, chromosomes

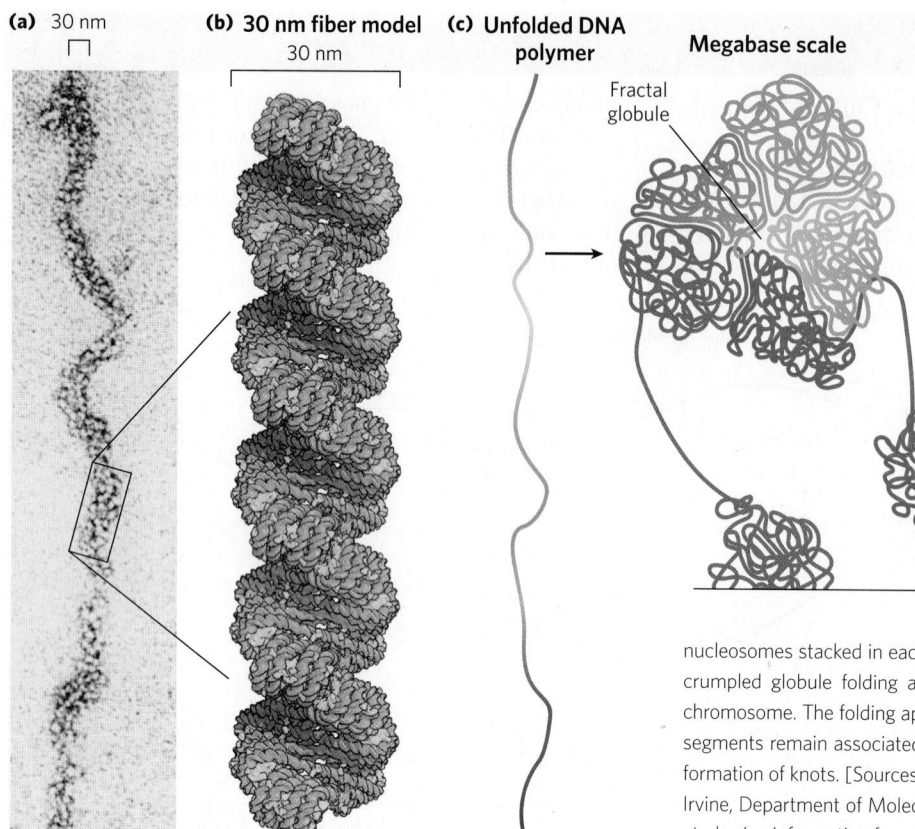

(a) 30 nm

(b) 30 nm fiber model
30 nm

(c) Unfolded DNA polymer

Megabase scale

Fractal globule

FIGURE 24-28 Higher-order organization of nucleosomes. The compact fiber is formed by the tight packing of nucleosomes. **(a)** The 30 nm fiber, as seen by electron microscopy. **(b)** A model for nucleosome organization within a 30 nm fiber. Two 10 nm fibers are coiled around each other, with nucleosomes stacked in each. DNA is blue; nucleosomes are yellow. **(c)** A crumpled globule folding arrangement of 10 nm fibers in a condensed chromosome. The folding appears to be random, but nearby chromosomal segments remain associated and the folding occurs so as to minimize the formation of knots. [Sources: (a) Barbara Hamkalo, University of California, Irvine, Department of Molecular Biology and Biochemistry. (b) Model created using information from F. Song et al., *Science* 344:376, 2014, Fig. 1C.]

could fold so that DNA segments lying millions of base pairs apart could be located adjacent to each other. However, folding generally occurs so that the linear distance between DNA segments is reflected in their spatial distance in the folded structure. Folding also occurs so as to avoid the formation of knots. Transcriptional activity can decrease the level of condensation, producing regions that are relatively open. Transcriptionally inactive regions and regions lacking genes are highly condensed in a form called **heterochromatin**.

The higher levels of folding are not yet fully understood, but certain regions of DNA seem to associate with a chromosomal scaffold **(Fig. 24-29)**. The scaffold-associated regions are separated by loops of DNA with perhaps 20 to 100 kbp. The DNA in a loop may contain a set of related genes. The scaffold itself may contain several proteins, notably topoisomerase II and SMC proteins (described below). The presence of topoisomerase II further emphasizes the relationship between DNA underwinding and chromatin structure. Topoisomerase II is so important to the maintenance of chromatin structure that inhibitors of this enzyme can kill rapidly dividing cells. Several drugs used in cancer chemotherapy are topoisomerase II inhibitors that allow the enzyme

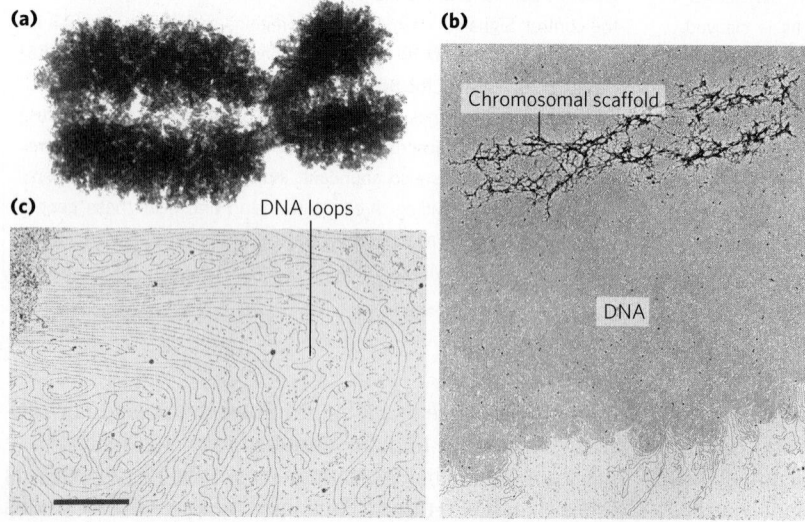

(a)

(b)

(c) DNA loops

Chromosomal scaffold

DNA

FIGURE 24-29 Loops of DNA attached to a chromosomal scaffold. (a) A swollen mitotic chromosome, produced in a buffer of low ionic strength, as seen in the electron microscope. Notice the appearance of chromatin loops at the margins. **(b)** Extraction of the histones leaves a proteinaceous chromosomal scaffold surrounded by naked DNA. **(c)** The DNA appears to be organized in loops attached at their base to the scaffold in the upper left corner; scale bar = 1 μm. The three images are at different magnifications. [Sources: (a, b) Don W. Fawcett/Science Source. (c) U. K. Laemmli et al., "Metaphase chromosome structure: The role of nonhistone proteins," *Cold Spring Harb. Symp. Quant. Biol.* 42:351, 1978. © Cold Spring Harbor Laboratory Press.]

to promote strand breakage but not the resealing of the breaks (see Box 24-1).

There are additional layers of organization in the eukaryotic nucleus. Just before cell division during mitosis, chromosomes can be seen as highly condensed and organized structures **(Figure 24-30a)**. During interphase, chromosomes appear dispersed (Fig. 24-30b, top), but they do not meander randomly in nuclear space (Fig. 24-30b, bottom). Each chromosome is constrained within a subnuclear domain called a **chromosome territory**

(Fig. 24-30c). The exact location of chromosome territories varies from cell to cell in an organism, but some spatial patterns are evident. Some chromosomes have a higher density of genes than others (for example, human chromosomes 1, 16, 17, 19, and 22), and these tend to have territories in the center of the nucleus. Chromosomes with more heterochromatin tend to be located on the nuclear periphery. Spaces between chromosomes are often sites where transcriptional machinery and transcriptionally active genes on adjacent chromosomes are concentrated.

Condensed Chromosome Structures Are Maintained by SMC Proteins

A third major class of chromatin proteins, in addition to the histones and topoisomerases, is the **SMC proteins** (*structural maintenance of chromosomes*). The primary structure of SMC proteins consists of five distinct domains **(Fig. 24-31a)**. The amino- and carboxyl-terminal globular domains, N and C, each of which contains part of an ATP-hydrolytic site, are connected by two regions of α-helical coiled-coil motifs (see Fig. 4-11) joined by a hinge domain. The proteins are generally dimeric, forming a V-shaped complex that is thought to be tied together through the protein's hinge domains (Fig. 24-31b). One N and one C domain come together to form a complete ATP-hydrolytic site at each free end of the V.

Proteins in the SMC family are found in all types of organisms, from bacteria to humans. Eukaryotes have two major types, cohesins and condensins, both of which are bound by regulatory and accessory proteins (Fig. 24-31c). **Cohesins** play a substantial role in linking together sister chromatids immediately after replication and keeping them together as the chromosomes condense to metaphase. This linkage is essential if chromosomes are to segregate properly at cell division. Cohesins, along with a third protein, kleisin, are thought to form a ring around the replicated chromosomes that ties them together until separation is required. The ring may expand and contract in response to ATP hydrolysis. **Condensins** are essential to the condensation of chromosomes as cells enter mitosis. In the laboratory, condensins bind to DNA in a manner that creates positive supercoils; that is, condensin binding causes the DNA to become overwound, in contrast to the underwinding induced by the binding of nucleosomes. A model for the role of condensins in chromatin compaction is presented in **Figure 24-32**. Cohesins and condensins are essential in orchestrating the many changes in chromosome structure during the eukaryotic cell cycle **(Fig. 24-33)**.

Bacterial DNA Is Also Highly Organized

We now turn briefly to the structure of bacterial chromosomes. Bacterial DNA is compacted in a structure called the **nucleoid**, which can occupy a significant

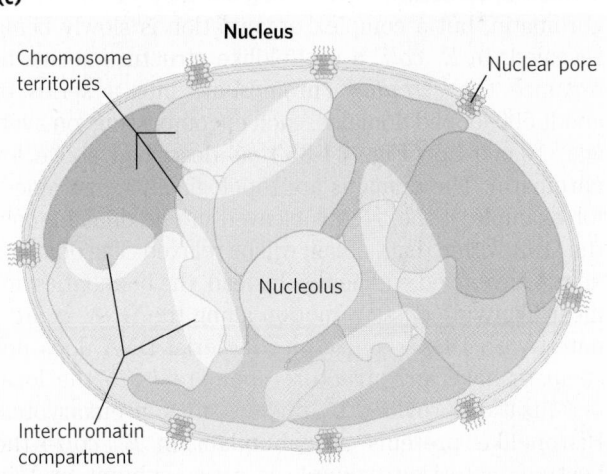

(a)

(b)

(c)

FIGURE 24-30 Chromosomal organization in the eukaryotic nucleus.
(a) Condensed chromosomes at the mitotic anaphase in cells of the bluebell (*Endymion sp.*). **(b)** Interphase nuclei of human breast epithelial cells. The nucleus on the bottom has been treated so that its two copies of chromosome 11 fluoresce green. **(c)** Cartoon showing chromosome territories in a eukaryotic nucleus. The interchromatin compartments are enriched in transcriptional machinery and have abundant actively transcribed genes. The nucleolus is a suborganelle within the nucleus where ribosomes are synthesized and assembled (Chapter 27). [Sources: (a) Pr. G. Giménez-Martín/Science Source. (b) Karen Meaburn and Tom Misteli/National Cancer Institute.]

Nucleus

Chromosome territories

Nuclear pore

Nucleolus

Interchromatin compartment

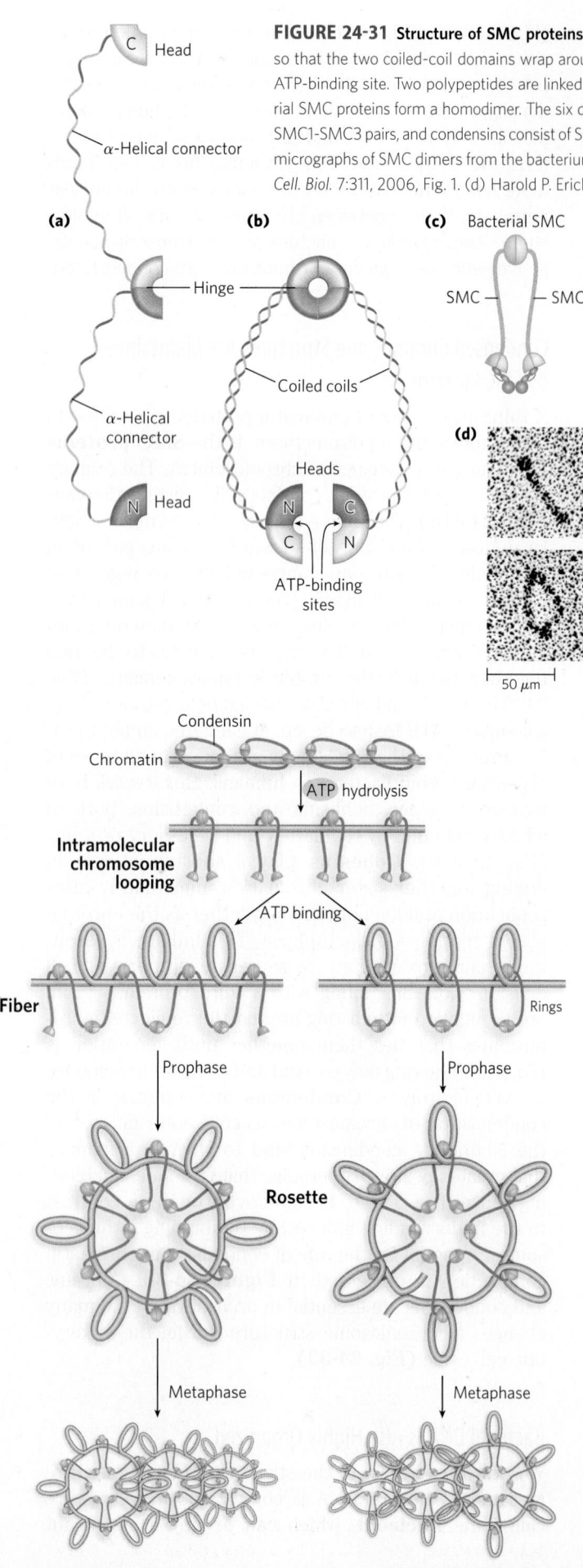

FIGURE 24-31 Structure of SMC proteins. (a) SMC proteins have five domains. **(b)** Each SMC polypeptide is folded so that the two coiled-coil domains wrap around each other and the N and C domains come together to form a complete ATP-binding site. Two polypeptides are linked at the hinge region to form the dimeric V-shaped SMC molecule. **(c)** Bacterial SMC proteins form a homodimer. The six different eukaryotic SMC proteins form heterodimers. Cohesins are made up of SMC1-SMC3 pairs, and condensins consist of SMC2-SMC4 pairs. The SMC5-SMC6 pair is involved in DNA repair. **(d)** Electron micrographs of SMC dimers from the bacterium *Bacillus subtilis*. [Sources: (a–c) Information from T. Hirano, *Nature Rev. Mol. Cell. Biol.* 7:311, 2006, Fig. 1. (d) Harold P. Erickson, Duke University Medical Center, Department of Cell Biology.]

fraction of the cell volume **(Fig. 24-34)**. The DNA seems to be attached at one or more points to the inner surface of the plasma membrane. Much less is known about the structure of the nucleoid than of eukaryotic chromatin, but a complex organization is slowly being revealed. In *E. coli*, a scaffoldlike structure seems to organize the *circular* chromosome into a series of about 500 looped domains, each encompassing, on average, 10,000 bp **(Fig. 24-35)**, as described above for chromatin. The domains are topologically constrained; for example, if the DNA is cleaved in one domain, only the DNA within that domain will be relaxed. The domains do not have fixed end points. Instead, the boundaries are most likely in constant motion along the DNA, coordinated with DNA replication. Bacterial DNA does not seem to have any structure comparable to the local organization provided by nucleosomes in eukaryotes. Histonelike proteins are abundant in *E. coli*—the best-characterized example is a two-subunit protein called HU (M_r 19,000)—but these proteins bind and dissociate within minutes, and no regular, stable

FIGURE 24-32 The possible role of condensins in chromatin condensation. Initially, the DNA is bound at the hinge region of the SMC protein, in the interior of what can become an intramolecular SMC ring. ATP binding leads to head-to-head association, forming supercoiled loops in the bound DNA. Subsequent rearrangement of the head-to-head interactions to form rosettes condenses the DNA. Condensins may organize the looping of the chromosome segments in several ways. Two current models are shown. [Source: Information from T. Hirano, *Nature Rev. Mol. Cell. Biol.* 7:311, 2006, Fig. 6.]

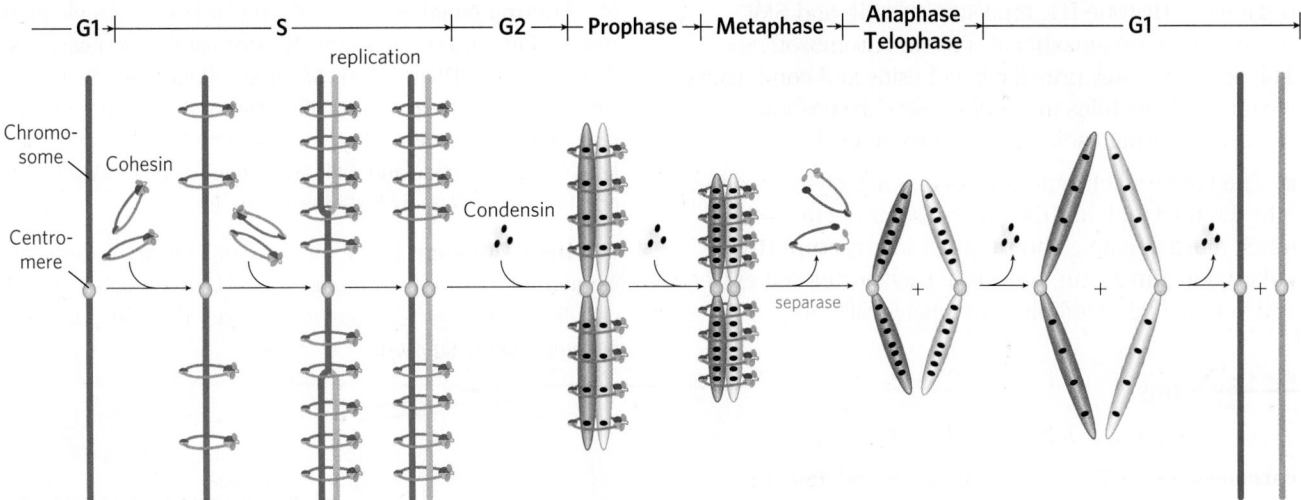

FIGURE 24-33 The roles of cohesins and condensins in the eukaryotic cell cycle. Cohesins are loaded onto the chromosomes during G1 (see Fig. 24-23), tying the sister chromatids together during replication. At the onset of mitosis, condensins bind and maintain the chromatids in a condensed state. During anaphase, the enzyme separase removes the cohesin links. Once the chromatids separate, condensins begin to unload and the daughter chromosomes return to the uncondensed state. [Source: Information from D. P. Bazett-Jones et al., *Mol. Cell* 9:1183, 2002, Fig. 5.]

DNA-histone structure has been found. The dynamic structural changes in the bacterial chromosome may reflect a requirement for more ready access to its genetic information. The bacterial cell division cycle can be as short as 15 min, whereas a typical eukaryotic cell may not divide for hours or even months. In addition, a much greater fraction of bacterial DNA is used to encode RNA and/or protein products. Higher rates of cellular metabolism in bacteria mean that a much higher proportion of the DNA is being transcribed or replicated at a given time than in most eukaryotic cells.

With this overview of the complexity of DNA structure, we are now ready to turn, in the next chapter, to a discussion of DNA metabolism.

SUMMARY 24.3 The Structure of Chromosomes

■ The fundamental unit of organization in the chromatin of eukaryotic cells is the nucleosome, which consists of histones and a 200 bp segment of DNA. A core protein particle containing eight histone molecules (two copies each of histones H2A, H2B, H3, and H4) is encircled by a segment of DNA (about 146 bp) in the form of a left-handed solenoidal supercoil.

■ Nucleosomes can be organized into 30 nm fibers in vitro, but this structure has not been found in cells. Higher-order folding of chromosomes involves attachment to a chromosomal scaffold. Individual chromosomes are constrained within nuclear subdomains called

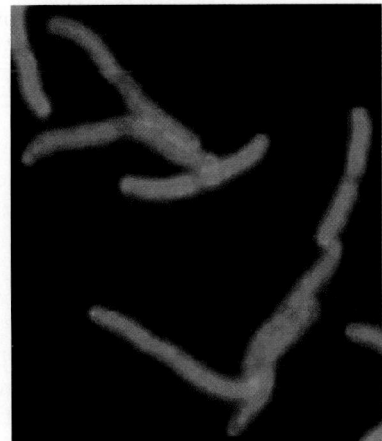

FIGURE 24-34 *E. coli* nucleoids. The DNA of these cells is stained with a dye that fluoresces blue when exposed to UV light. The blue areas define the nucleoids. Notice that some cells have replicated their DNA but have not yet undergone cell division and hence have multiple nucleoids. [Source: Lars Renner.]

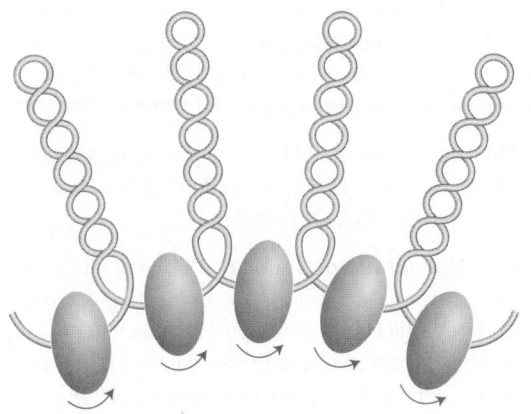

FIGURE 24-35 Looped domains of the *E. coli* chromosome. Each domain is about 10,000 bp long. The domains are not static, but move along the DNA as replication proceeds. Barriers at the boundaries of the domains, of unknown composition, prevent the relaxation of DNA beyond the boundaries of the domain where a strand break occurs. The putative boundary complexes are shown as gray-shaded ovoids. The arrows denote movement of DNA through the boundary complexes.

territories. Histone H1, topoisomerase II, and SMC proteins play organizational roles in chromosomes. The SMC proteins, principally cohesins and condensins, have important roles in keeping the chromosomes organized during each stage of the cell cycle.

■ The bacterial chromosome is extensively compacted into the nucleoid, but the chromosome seems to be much more dynamic and irregular in structure than eukaryotic chromatin, reflecting the shorter cell cycle and very active metabolism of a bacterial cell.

Problems

1. Packaging of DNA in a Virus Bacteriophage T2 has a DNA of molecular weight 120×10^6 contained in a head about 210 nm long. Calculate the length of the DNA (assume the molecular weight of a nucleotide pair is 650) and compare it with the length of the T2 head.

2. The DNA of Phage M13 The base composition of phage M13 DNA is A, 23%; T, 36%; G, 21%; C, 20%. What does this tell you about the DNA of phage M13?

3. The *Mycoplasma* Genome The complete genome of the simplest bacterium known, *Mycoplasma genitalium*, is a circular DNA molecule with 580,070 bp. Calculate the molecular weight and contour length (when relaxed) of this molecule. What is Lk_0 for the *Mycoplasma* chromosome? If $\sigma = -0.06$, what is Lk?

4. Size of Eukaryotic Genes An enzyme isolated from rat liver has 192 amino acid residues and is encoded by a gene with 1,440 bp. Explain the relationship between the number of amino acid residues in the enzyme and the number of nucleotide pairs in its gene.

5. Linking Number A closed-circular DNA molecule in its relaxed form has an Lk of 500. Approximately how many base pairs are in this DNA? How is Lk altered (increases, decreases, doesn't change, becomes undefined) when (a) a protein complex binds to form a nucleosome, (b) one DNA strand is broken, (c) DNA gyrase and ATP are added to the DNA solution, or (d) the double helix is denatured by heat?

6. DNA Topology In the presence of a eukaryotic condensin and a type II topoisomerase, the Lk of a relaxed closed-circular DNA molecule does not change. However, the DNA becomes highly knotted.

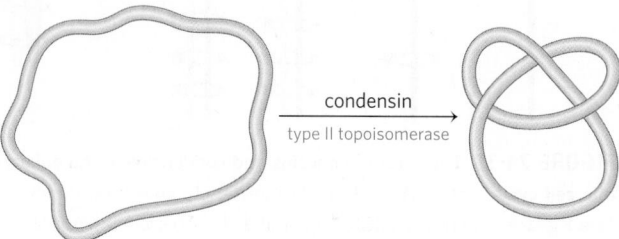

Formation of the knots requires breakage of the DNA, passage of a segment of DNA through the break, and religation by the topoisomerase. Given that every reaction of the topoisomerase would be expected to result in a change in linking number, how can Lk remain the same?

7. Superhelical Density Bacteriophage λ infects *E. coli* by integrating its DNA into the bacterial chromosome. The success of this recombination depends on the topology of the *E. coli* DNA. When the superhelical density (σ) of the *E. coli* DNA is greater than -0.045, the probability of integration is less than 20%; when σ is less than -0.06, the probability is >70%. Plasmid DNA isolated from an *E. coli* culture is found to have a length of 13,800 bp and an Lk of 1,222. Calculate σ for this DNA and predict the likelihood that bacteriophage λ will be able to infect this culture.

8. Altering Linking Number (a) What is the Lk of a 5,000 bp circular duplex DNA molecule with a nick in one strand? (b) What is the Lk of the molecule in (a) when the nick is sealed (relaxed)? (c) How would the Lk of the molecule in (b) be affected by the action of a single molecule of *E. coli* topoisomerase I? (d) What is the Lk of the molecule in (b) after eight enzymatic turnovers by a single molecule of DNA gyrase in the presence of ATP? (e) What is the Lk of the molecule in (d) after four enzymatic turnovers by a single molecule of bacterial type I topoisomerase? (f) What is the Lk of the molecule in (d) after binding of one nucleosome core?

9. Chromatin Early evidence that helped researchers define nucleosome structure is illustrated by the agarose gel below (next page), in which the thick bands represent DNA. This result was generated by briefly treating chromatin with an enzyme that degrades DNA, then removing all protein and subjecting the purified DNA to electrophoresis. Numbers at the side of the gel denote the position to which a linear DNA of the indicated size would migrate. What does this gel tell you about chromatin structure? Why are the DNA bands thick and spread out rather than sharply defined?

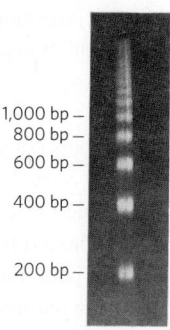

[Source: Courtesy Dr. Roger D. Kornberg, Stanford University School of Medicine.]

10. DNA Structure Explain how the underwinding of a B-DNA helix might facilitate or stabilize the formation of Z-DNA (see Fig. 8-17).

11. Maintaining DNA Structure (a) Describe two structural features required for a DNA molecule to maintain a negatively supercoiled state. (b) List three structural changes that become more favorable when a DNA molecule is negatively supercoiled. (c) What enzyme, with the aid of ATP, can generate negative superhelicity in DNA? (d) Describe the physical mechanism by which this enzyme acts.

12. Yeast Artificial Chromosomes (YACs) YACs are used to clone large pieces of DNA in yeast cells. What three types of DNA sequence are required to ensure proper replication and propagation of a YAC in a yeast cell?

13. Nucleoid Structure in Bacteria In bacteria, the transcription of a subset of genes is affected by DNA topology, with expression increasing or (more often) decreasing when the DNA is relaxed. When a bacterial chromosome is cleaved at a specific site by a restriction enzyme (one that cuts at a long, and thus rare, sequence), only nearby genes (within 10,000 bp) exhibit either an increase or decrease in expression. The transcription of genes elsewhere in the chromosome is unaffected. Explain. (Hint: See Fig. 24-35.)

14. DNA Topology When DNA is subjected to electrophoresis in an agarose gel, shorter molecules migrate faster than longer ones. Closed-circular DNAs of the same size but with different linking numbers also can be separated on an agarose gel: topoisomers that are more supercoiled, and thus more condensed, migrate faster through the gel. In the gel shown below, purified plasmid DNA has migrated from top to bottom. There are two bands, with the faster band much more prominent.

[Source: Courtesy Michael Cox Laboratory.]

(a) What are the DNA species in the two bands? (b) If topoisomerase I were added to a solution of this DNA, what would happen to the upper and lower bands after electrophoresis? (c) If DNA ligase were added to the DNA, would the appearance of the bands change? Explain your answer. (d) If DNA gyrase plus ATP were added to the DNA after the addition of DNA ligase, how would the band pattern change?

15. DNA Topoisomers When DNA is subjected to electrophoresis in an agarose gel, shorter molecules migrate faster than longer ones. Closed-circular DNAs of the same size but different linking number also can be separated on an agarose gel: topoisomers that are more supercoiled, and thus more condensed, migrate faster through the gel—from top to bottom in the two gels shown below. A dye, chloroquine, was added to these gels. Chloroquine intercalates between base pairs and stabilizes a more underwound DNA structure. When the dye binds to a relaxed closed-circular DNA, the DNA is underwound where the dye binds, and unbound regions take on positive supercoils to compensate. In the experiment shown here, topoisomerases were used to make preparations of the same DNA circle with different superhelical densities (σ). Completely relaxed DNA migrated to the position labeled N (for *nicked*), and highly supercoiled DNA (above the limit where individual topoisomers can be distinguished) to the position labeled X.

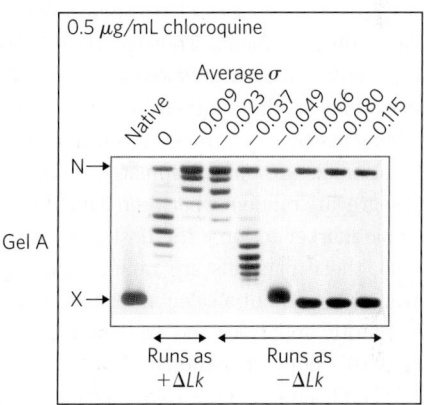

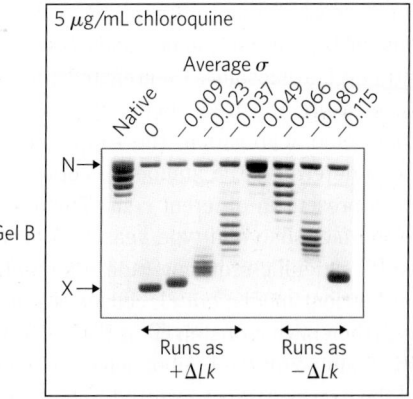

[Source: R. P. Bowater (2005) "Supercoiled DNA: structure," in *Encyclopedia of Life Sciences*, doi:10.1038/npg.els.0006002, John Wiley & Sons, Inc./Wiley InterScience. www.els.net.]

(a) In gel A, why does the $\sigma = 0$ lane (i.e., DNA prepared so that $\sigma = 0$, on average) have multiple bands?

(b) In gel B, is the DNA from the $\sigma = 0$ preparation positively or negatively supercoiled in the presence of the intercalating dye?

(c) In both gels, the $\sigma = -0.115$ lane has two bands, one a highly supercoiled DNA and one relaxed. Propose a reason for the presence of relaxed DNA in these lanes (and others).

(d) The native DNA (leftmost lane in each gel) is the same DNA circle isolated from bacterial cells and untreated. What is the approximate superhelical density of this native DNA?

16. Nucleosomes The human genome comprises just over 3.1 billion base pairs. Assuming it contains nucleosomes that are spaced as described in this chapter, how many molecules of histone H2A are present in one somatic human cell? (Ignore reductions in H2A due to its replacement in some regions by H2A variants.) How would the number change after DNA replication but before cell division?

Data Analysis Problem

17. Defining the Functional Elements of Yeast Chromosomes Figure 24-8 shows the major structural elements of a chromosome of budding yeast (*Saccharomyces cerevisiae*). Heiter, Mann, Snyder, and Davis (1985) determined the properties of some of these elements. They based their study on the finding that in yeast cells, plasmids (which have genes and an origin of replication) act differently from chromosomes (which have these elements plus centromeres and telomeres) during mitosis. The plasmids are not manipulated by the mitotic apparatus and segregate randomly between daughter cells. Without a selectable marker to force the host cells to retain them (see Fig. 9-4), these plasmids are rapidly lost. In contrast, chromosomes, even without a selectable marker, are manipulated by the mitotic apparatus and are lost at a very low frequency (about 10^{-5} per cell division).

Heiter and colleagues set out to determine the important components of yeast chromosomes by constructing plasmids with various parts of chromosomes and observing whether these "synthetic chromosomes" segregated properly during mitosis. To measure the frequencies of different types of failed chromosome segregation, the researchers needed a rapid assay to determine the number of copies of synthetic chromosomes present in different cells. The assay took advantage of the fact that wild-type yeast colonies are white whereas certain adenine-requiring (ade⁻) mutants yield red colonies on nutrient media; *ade2⁻* cells lack functional AIR carboxylase (the enzyme of step ⑥ₐ in Fig. 22-35) and accumulate AIR (5-aminoimidazole ribonucleotide) in their cytoplasm, and the excess AIR is converted to a conspicuous red pigment. The other part of the assay involved the gene *SUP11*, which encodes an ochre suppressor (a type of nonsense suppressor; see Box 27-3) that suppresses the phenotype of some *ade2⁻* mutants.

Heiter and coworkers started with a diploid strain of yeast homozygous for *ade2⁻*; these cells are red. When the mutant cells contain one copy of *SUP11*, the metabolic defect is partly suppressed and the cells are pink. When the cells contain two or more copies of *SUP11*, the defect is completely suppressed and the cells are white.

The researchers inserted one copy of *SUP11* into synthetic chromosomes containing various elements thought to be important in chromosome function, and then observed how well these chromosomes were passed from one generation to the next. These pink cells were plated on nonselective media, and the behavior of the synthetic chromosomes was observed. Heiter and coworkers looked for colonies in which the synthetic chromosomes segregated improperly at the first division after plating, giving rise to a colony that was half one genotype and half the other. Because yeast cells are nonmotile, this would be a sectored colony, with one half one color and the other half another color.

(a) One way for the mitotic process to fail is *nondisjunction*: the chromosome replicates but the sister chromatids fail to separate, so both copies of the chromosome end up in the same daughter cell. Explain how nondisjunction of the synthetic chromosome would give rise to a colony that is half red and half white.

(b) Another way for the mitotic process to fail is *chromosome loss*: the chromosome does not enter the daughter nucleus or is not replicated. Explain how loss of the synthetic chromosome would give rise to a colony that is half red and half pink.

By counting the frequency of the different colony types, Heiter and colleagues could estimate the frequency of these aberrant mitotic events with different types of synthetic chromosome. First, they explored the requirement for centromeric sequences by constructing synthetic chromosomes with DNA fragments of different sizes containing a known centromere. Their results are shown below.

Synthetic chromosome	Size of centromere-containing fragment (kbp)	Chromosome loss (%)	Nondisjunction (%)
1	None	—	>50
2	0.63	1.6	1.1
3	1.6	1.9	0.4
4	3.0	1.7	0.35
5	6.0	1.6	0.35

(c) Based on these data, what can you conclude about the size of the centromere required for normal mitotic segregation? Explain your reasoning.

(d) All the synthetic chromosomes created in these experiments were circular and lacked telomeres. Explain how they could be replicated more-or-less properly.

Heiter and colleagues next constructed a series of linear synthetic chromosomes that included the functional centromeric sequence and telomeres, and measured the total mitotic error frequency (% loss + % nondisjunction) as a function of size.

Synthetic chromosome	Size (kbp)	Total error frequency (%)
6	15	11.0
7	55	1.5
8	95	0.44
9	137	0.14

(e) Based on these data, what can you conclude about the chromosome size required for normal mitotic segregation? Explain your reasoning.

(f) Normal yeast chromosomes are linear, range from 250 kbp to 2,000 kbp in length, and, as noted above, have a mitotic error frequency of about 10^{-5} per cell division. Extrapolating the results from (e), do the centromeric and telomeric sequences used in these experiments explain the mitotic stability of normal yeast chromosomes, or must other elements be involved? Explain your reasoning. (Hint: A plot of log of error frequency vs. length will be helpful.)

Reference

Heiter, P., C. Mann, M. Snyder, and R.W. Davis. 1985. Mitotic stability of yeast chromosomes: a colony color assay that measures nondisjunction and chromosome loss. *Cell* 40:381–392.

Further Reading is available at www.macmillanlearning.com/LehningerBiochemistry7e.

CHAPTER 25

DNA Metabolism

Self-study tools that will help you practice what you've learned and reinforce this chapter's concepts are available online. Go to www.macmillanlearning.com/LehningerBiochemistry7e.

As the repository of genetic information, DNA occupies a unique and central place among biological macromolecules. The nucleotide sequences of DNA encode the primary structures of all cellular RNAs and proteins and, through enzymes, indirectly affect the synthesis of all other cellular constituents. This passage of information from DNA to RNA and protein guides the size, shape, and functioning of every living thing.

DNA is a marvelous device for the stable storage of genetic information. The phrase "stable storage," however, conveys a static and misleading picture. It fails to capture the complexity of processes by which genetic information is preserved in an uncorrupted state and then transmitted from one generation of cells to the next. DNA metabolism comprises both the process that gives rise to faithful copies of DNA molecules (replication) and the processes that affect the inherent structure of the information (repair and recombination). Together, these activities are the focus of this chapter.

The metabolism of DNA is shaped by the requirement for an exquisite degree of accuracy. The chemistry of joining one nucleotide to the next in DNA replication is elegant and simple, almost deceptively so. However, as is the case with all information-containing polymers, forming a covalent link between two monomeric units is just a small part of the biochemical process. As we shall see, complexity arises in the form of enzymatic devices to ensure that the link is formed to the *correct* nucleotide and that genetic information is transmitted intact.

Uncorrected errors that arise during DNA synthesis can have dire consequences, not only because they can permanently affect or eliminate the function of a gene but also because the change may be inheritable.

The enzymes that synthesize DNA may copy DNA molecules that contain millions of bases. They do so with extraordinary fidelity and speed, even though the DNA substrate is highly compacted and bound with other proteins. Formation of phosphodiester bonds to link nucleotides in the backbone of a growing DNA strand is therefore only one part of an elaborate process that requires myriad proteins and enzymes.

Maintaining the integrity of genetic information lies at the heart of DNA repair. As detailed in Chapter 8, DNA is susceptible to many types of damaging reactions. Such reactions are infrequent but significant nevertheless, because of the very low biological tolerance for changes in DNA sequence. DNA is the only macromolecule for which repair systems exist; the number, diversity, and complexity of DNA repair mechanisms reflect the wide range of insults that can harm DNA.

Cells can rearrange their genetic information by processes collectively called recombination—seemingly undermining the principle that the stability and integrity of genetic information are paramount. However, most DNA rearrangements in fact play constructive roles in maintaining genomic integrity, contributing in special ways to DNA replication, DNA repair, and chromosomal segregation.

Special emphasis is given in this chapter to the *enzymes* of DNA metabolism. They merit careful study not only because of their intrinsic biological importance and interest but also for their increasing importance in medicine and for their everyday use as reagents in a wide range of modern biochemical technologies. Many of the seminal discoveries in DNA metabolism have

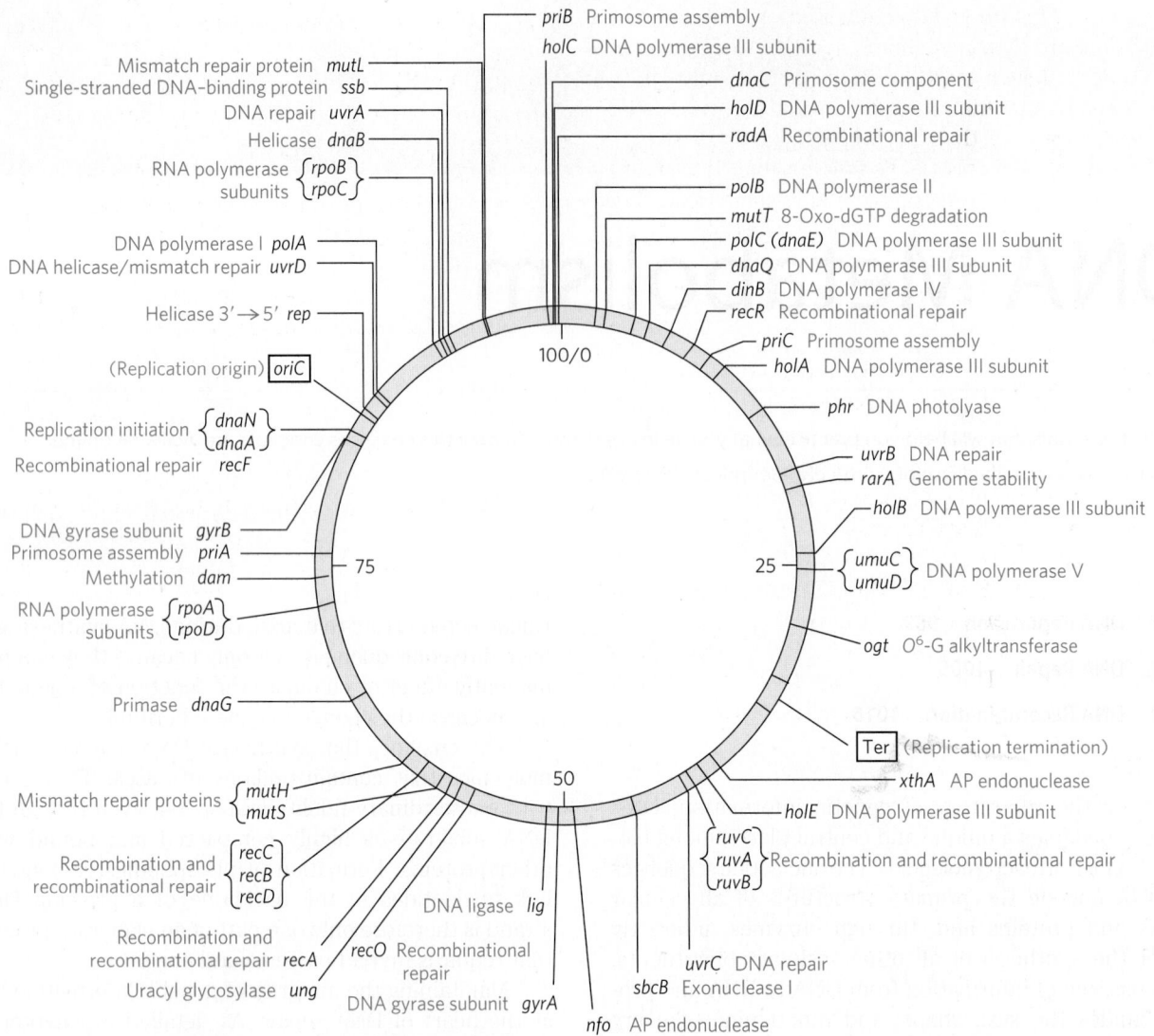

FIGURE 25-1 Map of the *E. coli* chromosome. The map shows the relative positions of genes encoding many of the proteins important in DNA metabolism. The number of genes known to be involved provides a hint of the complexity of these processes. The numbers 0 to 100 inside the circular chromosome denote a genetic measurement called minutes. Each minute corresponds to ~40,000 bp along the DNA molecule. The three-letter names of genes and other elements generally reflect some aspect of their function; for example, *mut*, mutagenesis; *dna*, DNA replication; *pol*, DNA polymerase; *rpo*, RNA polymerase; *uvr*, UV resistance; *rec*, recombination; *dam*, DNA adenine methylation; *lig*, DNA ligase; Ter, termination of replication; and ori, origin of replication (*oriC* in *E. coli*, as shown here).

been made with *Escherichia coli*, so its well-understood enzymes are generally used to illustrate the ground rules. A quick look at some relevant genes on the *E. coli* genetic map **(Fig. 25-1)** provides just a hint of the complexity of the enzymatic systems involved in DNA metabolism.

Before taking a closer look at replication, we must make a short digression into the use of abbreviations in naming genes and proteins—you will encounter many of these in this and later chapters.

▸▸ Key Convention: Bacterial genes generally are named using three italicized, lowercase letters that often reflect a gene's apparent function. For example, the *dna*, *uvr*, and *rec* genes affect *D*NA replication, *r*esistance

to the damaging effects of *UV* radiation, and *rec*ombination, respectively. Where several genes affect the same process, the letters *A*, *B*, *C*, and so forth, are added—as in *dnaA*, *dnaB*, *dnaQ*, for example—usually reflecting their order of discovery rather than the order of their gene products in a reaction sequence. Similar conventions exist for naming eukaryotic genes, although the exact form of the abbreviations may vary with the species, and no single convention applies to all eukaryotic systems. For example, in the budding yeast *Saccharomyces cerevisiae*, gene names are generally three uppercase letters followed by a number, all italicized (for example, the gene *COX1* encodes a subunit of *c*yto-chrome *ox*idase (Chapter 19)). Gene names that pre-date current conventions may differ in format. **◂◂**

The use of abbreviations in naming proteins is less straightforward. During genetic investigations, the protein product of each gene is usually isolated and characterized. Many bacterial genes are identified and named before the roles of their protein products are understood in detail. Sometimes the gene product is found to be a previously isolated protein, and some renaming occurs. Often, however, the product turns out to be an as yet unknown protein, with an activity not easily described by a simple enzyme name.

>> **Key Convention:** Bacterial proteins often retain the name of their genes. When referring to the protein product of an *E. coli* gene, roman type is used and the first letter is capitalized: for example, the *dnaA* and *recA* gene products are the DnaA and RecA proteins, respectively. Conventions for eukaryotic proteins are again complex. For yeast, some proteins have long common names (such as cytochrome oxidase). Others have the same name as the gene, in which case the protein name has one uppercase and two lowercase letters, followed by a number and the letter "p," all in roman type (such as Rad51p). The "p" is to emphasize that this is a protein and to prevent confusion with naming conventions for other organisms. <<

25.1 DNA Replication

Long before the structure of DNA was known, scientists wondered at the ability of organisms to create faithful copies of themselves and, later, at the ability of cells to produce many identical copies of large, complex macromolecules. Speculation about these problems centered around the concept of a **template**, a structure that would allow molecules to be lined up in a specific order and joined to create a macromolecule with a unique sequence and function. The 1940s brought the revelation that DNA was the genetic molecule, but not until James Watson and Francis Crick deduced its structure did the way in which DNA could act as a template for the replication and transmission of genetic information become clear: *one strand is the complement of the other.* The strict base-pairing rules mean that each strand provides the template for a new strand with a predictable and complementary sequence (see Figs 8-14, 8-15).

The fundamental properties of the DNA replication process and the mechanisms used by the enzymes that catalyze it have proved to be essentially identical in all species. This mechanistic unity is a major theme as we proceed from general properties of the replication process, to *E. coli* replication enzymes, and, finally, to replication in eukaryotes.

DNA Replication Follows a Set of Fundamental Rules

Early research on bacterial DNA replication and its enzymes helped to establish several basic properties that have proven applicable to DNA synthesis in every organism.

DNA Replication Is Semiconservative Each DNA strand serves as a template for the synthesis of a new strand, producing two new DNA molecules, each with one new strand and one old strand. This is **semiconservative replication**.

Watson and Crick proposed the hypothesis of semiconservative replication soon after publication of their 1953 paper on the structure of DNA, and the hypothesis was proved by ingeniously designed experiments carried out by Matthew Meselson and Franklin Stahl in 1957. Meselson and Stahl grew *E. coli* cells for many generations in a medium in which the sole nitrogen source (NH_4Cl) contained ^{15}N, the "heavy" isotope of nitrogen, instead of the normal, more abundant "light" isotope, ^{14}N. The DNA isolated from these cells had a density about 1% greater than that of normal [^{14}N]DNA **(Fig. 25-2a)**. Although this is only a small difference, a mixture of heavy [^{15}N]DNA and light [^{14}N]DNA can be separated by centrifugation to equilibrium in a cesium chloride density gradient.

The *E. coli* cells grown in the ^{15}N medium were transferred to a fresh medium containing only the ^{14}N isotope, where they were allowed to grow until the cell population

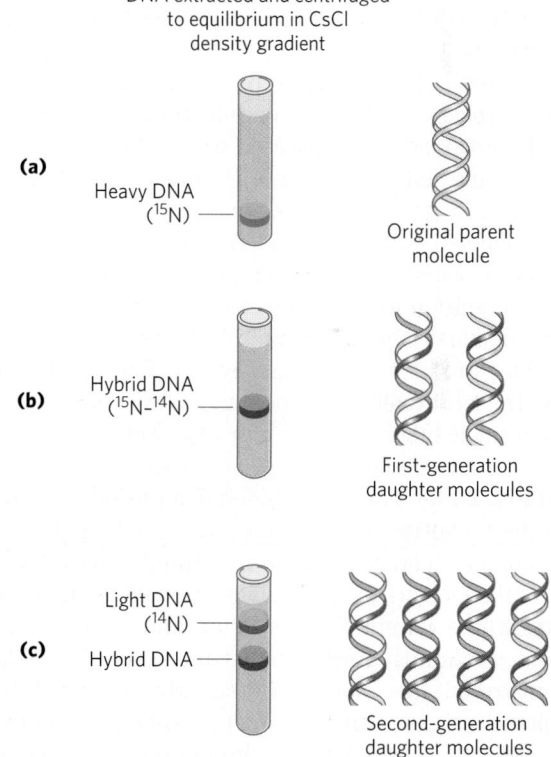

DNA extracted and centrifuged to equilibrium in CsCl density gradient

(a) Heavy DNA (^{15}N) — Original parent molecule

(b) Hybrid DNA (^{15}N–^{14}N) — First-generation daughter molecules

(c) Light DNA (^{14}N) / Hybrid DNA — Second-generation daughter molecules

FIGURE 25-2 The Meselson-Stahl experiment. (a) Cells were grown for many generations in a medium containing only heavy nitrogen, ^{15}N, so that all the nitrogen in their DNA was ^{15}N, as shown by a single band (blue) when centrifuged in a CsCl density gradient. **(b)** Once the cells had been transferred to a medium containing only light nitrogen, ^{14}N, cellular DNA isolated after one generation equilibrated at a higher position in the density gradient (purple band). **(c)** A second cycle of replication yielded a hybrid DNA band (purple) and another band (red), containing only [^{14}N] DNA, confirming semiconservative replication.

had just doubled. The DNA isolated from these first-generation cells formed a *single* band in the CsCl gradient at a position indicating that the double-helical DNA molecules of the daughter cells were hybrids containing one new ^{14}N strand and one parent ^{15}N strand (Fig. 25-2b).

This result argued against conservative replication, an alternative hypothesis, in which one progeny DNA molecule would consist of two newly synthesized DNA strands and the other would contain the two parent strands; this would not yield hybrid DNA molecules in the Meselson-Stahl experiment. The semiconservative replication hypothesis was further supported in the next step of the experiment (Fig. 25-2c). Cells were again allowed to double in number in the ^{14}N medium. The isolated DNA product of this second cycle of replication exhibited *two* bands in the density gradient, one with a density equal to that of light DNA and the other with the density of the hybrid DNA observed after the first cell doubling.

Replication Begins at an Origin and Usually Proceeds Bidirectionally Following the confirmation of a semiconservative mechanism of replication, a host of questions arose. Are the parent DNA strands completely unwound before each is replicated? Does replication begin at random places or at a unique point? After initiation at any point in the DNA, does replication proceed in one direction or both?

An early indication that replication is a highly coordinated process in which the parent strands are simultaneously unwound and replicated was provided by John Cairns, using autoradiography. He made *E. coli* DNA radioactive by growing cells in a medium containing thymidine labeled with tritium (^{3}H). When the DNA was carefully isolated, spread, and overlaid with a photographic emulsion for several weeks, the radioactive thymidine residues generated "tracks" of silver grains in the emulsion, producing an image of the DNA molecule. These tracks revealed that the intact chromosome of *E. coli* is a single huge circle, 1.7 mm long. Radioactive DNA isolated from cells during replication showed an extra loop **(Fig. 25-3)**. Cairns concluded that the loop resulted from the formation of two radioactive daughter strands, each complementary to a parent strand. One or both ends of the loop are dynamic points, termed **replication forks**, where parent DNA is being unwound and the separated strands quickly replicated. Cairns's results demonstrated that both DNA strands are replicated simultaneously, and variations on his experiment showed that replication of bacterial chromosomes is bidirectional: both ends of the loop have active replication forks.

Determination of whether the replication loops originate at a unique point in the DNA required landmarks along the DNA molecule. These were provided by a technique called **denaturation mapping**, developed by Ross Inman and colleagues. Using the 48,502 bp chromosome of bacteriophage λ, Inman showed that DNA could be selectively denatured at sequences unusually rich in A=T base pairs, generating a reproducible

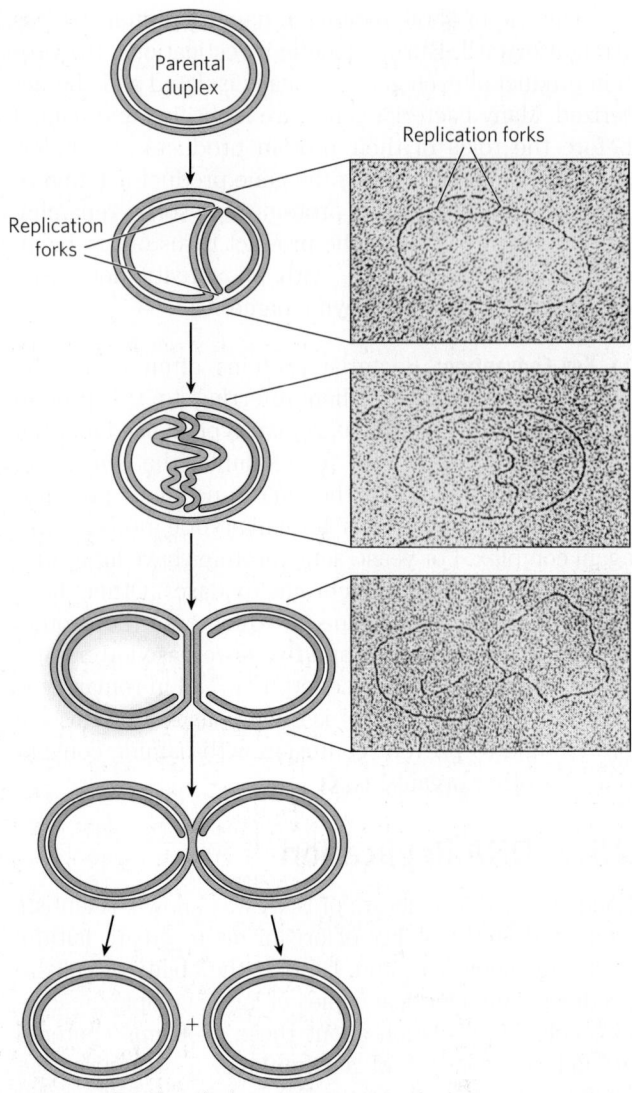

FIGURE 25-3 Visualization of DNA replication. Stages in the replication of circular DNA molecules have been visualized by electron microscopy. Replication of a circular chromosome produces a structure resembling the Greek letter theta, θ, as both strands are replicated simultaneously (new strands shown in light red). The electron micrographs show images of plasmid DNA being replicated from a single replication origin. [Source: Electron micrographs: J. Cairns, *Cold Spring Harb. Symp. Quant. Biol.* 28:44, 1963.]

pattern of single-strand bubbles (see Fig. 8-28). Isolated DNA containing replication loops can be partially denatured in the same way. This allows the position and progress of the replication forks to be measured and mapped, using the denatured regions as points of reference. The technique revealed that in this system, the replication loops always initiate at a unique point, which was termed an **origin**. It also confirmed the earlier observation that replication is usually bidirectional. For circular DNA molecules, the two replication forks meet at a point on the side of the circle opposite to the origin. Specific origins of replication have since been identified and characterized in bacteria and lower eukaryotes.

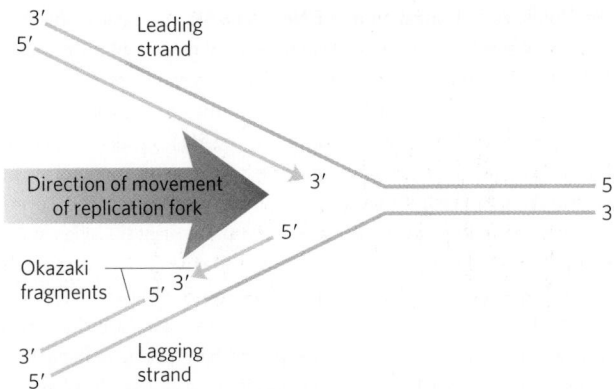

FIGURE 25-4 Defining DNA strands at the replication fork. A new DNA strand (light red) is always synthesized in the 5′→3′ direction. The template is read in the opposite direction, 3′→5′. The leading strand is continuously synthesized in the direction taken by the replication fork. The other strand, the lagging strand, is synthesized discontinuously in short pieces (Okazaki fragments) in a direction opposite to that in which the replication fork moves. The Okazaki fragments are spliced together by DNA ligase. In bacteria, Okazaki fragments are ~1,000 to 2,000 nucleotides long. In eukaryotic cells, they are 150 to 200 nucleotides long.

DNA Synthesis Proceeds in a 5′→3′ Direction and Is Semidiscontinuous A new strand of DNA is always synthesized in the 5′→3′ direction, with the free 3′-OH as the point at which the DNA is elongated. (See p. 283 for definition of the 5′ and 3′ ends of a DNA strand.) Because the two DNA strands are antiparallel, the strand serving as the template is read from its 3′ end toward its 5′ end.

If synthesis always proceeds in the 5′→3′ direction, how can both strands be synthesized simultaneously? If both strands were synthesized *continuously* while the replication fork moved, one strand would have to undergo 3′→5′ synthesis. This problem was resolved by Reiji Okazaki and colleagues in the 1960s. Okazaki found that one of the new DNA strands is synthesized in short pieces, now called **Okazaki fragments**. This work ultimately led to the conclusion that one strand is synthesized continuously and the other discontinuously **(Fig. 25-4)**. The continuous strand, or **leading strand**, is the one for which 5′→3′ synthesis proceeds in the *same* direction that the replication fork moves. The discontinuous strand, or **lagging strand**, is the one in which 5′→3′ synthesis proceeds in the direction *opposite* to the direction of fork movement. Okazaki fragments are typically 100 to 200 nucleotides long in eukaryotes, and 1,000 to 2,000 nucleotides long in bacteria. As we shall see, leading and lagging strand syntheses are tightly coordinated.

DNA Is Degraded by Nucleases

To explain the enzymology of DNA replication, we first introduce the enzymes that degrade DNA rather than synthesize it. These enzymes are known as **nucleases**, or **DNases** if they are specific for DNA rather than RNA. Every cell contains several different nucleases, belonging to two broad classes: exonucleases and

endonucleases. **Exonucleases** degrade nucleic acids from one end of the molecule. Many operate in only the 5′→3′ or the 3′→5′ direction, removing nucleotides only from the 5′ or the 3′ end, respectively, of one strand of a double-stranded nucleic acid or of a single-stranded DNA. **Endonucleases** can begin to degrade at specific internal sites in a nucleic acid strand or molecule, reducing it to smaller and smaller fragments. A few exonucleases and endonucleases degrade only single-stranded DNA. There are a few important classes of endonucleases that cleave only at specific nucleotide sequences (such as the restriction endonucleases that are so important in biotechnology; see Chapter 9, Fig. 9-2). You will encounter many types of nucleases in this and subsequent chapters.

DNA Is Synthesized by DNA Polymerases

Arthur Kornberg, 1918–2007
[Source: World History Archive/Alamy.]

The search for an enzyme that could synthesize DNA began in 1955. Work by Arthur Kornberg and colleagues led to the purification and characterization of a **DNA polymerase** from *E. coli* cells, a single-polypeptide enzyme now called **DNA polymerase I** (M_r 103,000; encoded by the *polA* gene). Much later, investigators found that *E. coli* contains at least four other distinct DNA polymerases, described below.

Detailed studies of DNA polymerase I revealed features of the DNA synthetic process that are now known to be common to all DNA polymerases. The fundamental reaction is a phosphoryl group transfer. The nucleophile is the 3′-hydroxyl group of the nucleotide at the 3′ end of the growing strand. Nucleophilic attack occurs at the α phosphorus of the incoming deoxynucleoside 5′-triphosphate **(Fig. 25-5a)**. Inorganic pyrophosphate is released in the reaction. The general reaction is

$$\underset{\text{DNA}}{(\text{dNMP})_n} + \text{dNTP} \longrightarrow \underset{\substack{\text{Lengthened} \\ \text{DNA}}}{(\text{dNMP})_{n+1}} + \text{PP}_i \quad (25\text{–}1)$$

where dNMP and dNTP are a deoxynucleoside 5′-monophosphate and 5′-triphosphate, respectively. Catalysis by virtually all DNA polymerases prominently involves two Mg^{2+} ions at the active site (Fig. 25-5a). One of these helps to deprotonate the 3′-hydroxyl group, rendering it a more effective nucleophile. The other binds to the incoming dNTP and facilitates departure of the pyrophosphate.

The reaction seems to proceed with only a minimal change in free energy, given that one phosphodiester bond is formed at the expense of a somewhat less stable phosphate anhydride. However, noncovalent base-stacking and base-pairing interactions provide additional stabilization to the lengthened DNA product relative to the free nucleotide. Also, the formation of products is

(a) DNA synthesis

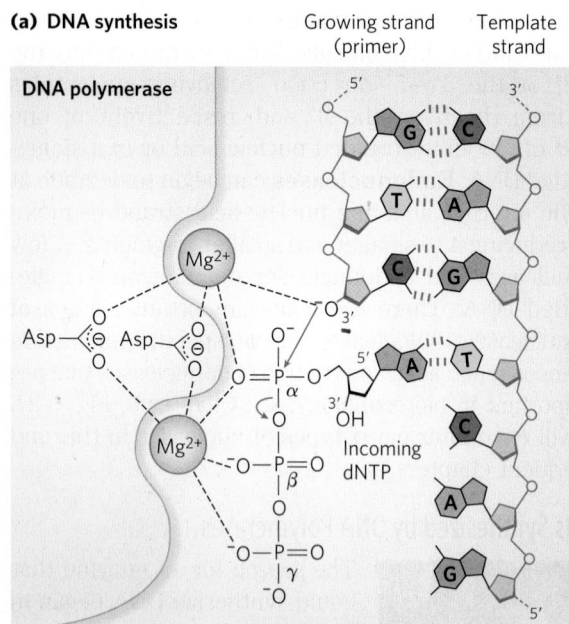

Growing strand (primer) Template strand

DNA polymerase

Incoming dNTP

MECHANISM FIGURE 25-5 Elongation of a DNA chain. (a) The catalytic mechanism for addition of a new nucleotide by DNA polymerase involves two Mg^{2+} ions, coordinated to the phosphate groups of the incoming nucleotide triphosphate, the 3′-hydroxyl group that will act as a nucleophile, and three Asp residues, two of which are highly conserved in all DNA polymerases. The Mg^{2+} ion depicted at the top facilitates attack of the 3′-hydroxyl group of the primer on the α phosphate of the nucleotide triphosphate; the other Mg^{2+} ion facilitates displacement of the pyrophosphate. Both ions stabilize the structure of the pentacovalent transition state. RNA polymerases use a similar mechanism (see Fig. 26-1a). **(b)** DNA polymerase I activity also requires a single unpaired strand to act as template and a primer strand to provide the free hydroxyl group at the 3′ end to which the new nucleotide unit is added. Each incoming nucleotide is selected in part by base-pairing to the appropriate nucleotide in the template strand. The reaction product has a new free 3′ hydroxyl, allowing the addition of another nucleotide. The newly formed base pair translocates to make the active site available to the next pair to be formed. **(c)** The core of most DNA polymerases is shaped like a human hand that wraps around the active site. The structure shown here is DNA polymerase I of *Thermus aquaticus*, bound to DNA. **(d)** A cartoon interpretation of the polymerase structure shows the insertion and postinsertion parts of the active site. The insertion site is where the nucleotide addition occurs, and the postinsertion site is the site to which the newly formed base pair translocates. [Source: (c) PDB ID 4KTQ, Y. Li et al., *EMBO J.* 17:7514, 1998.]

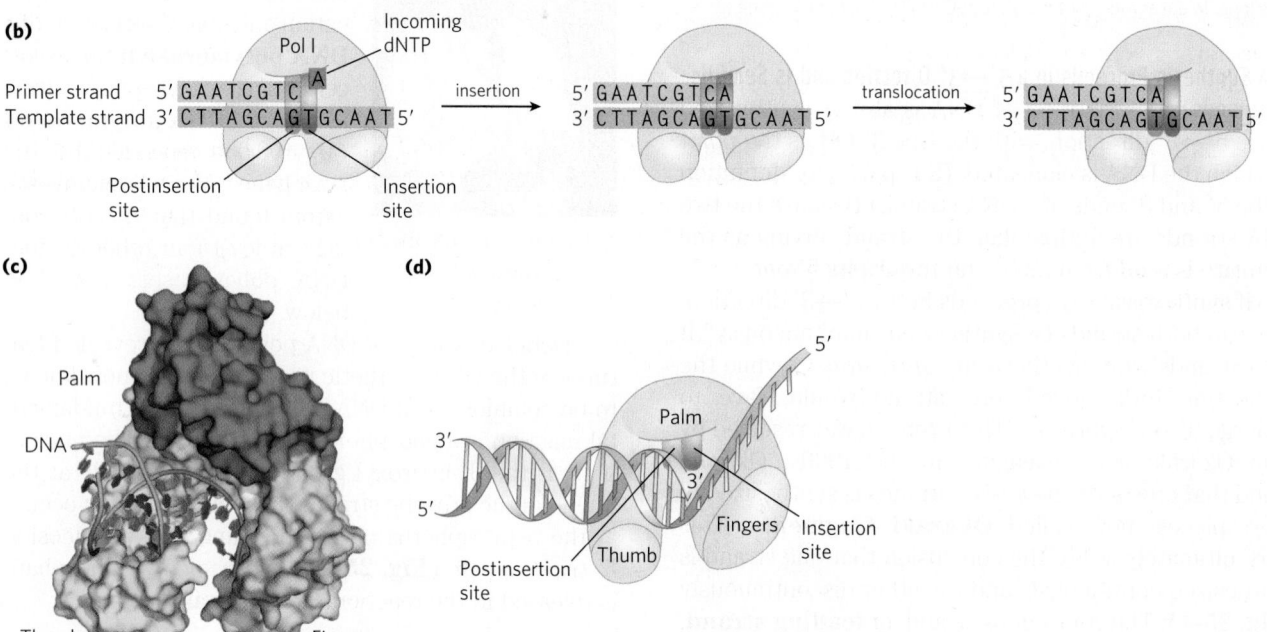

(b)

Pol I   Incoming dNTP

Primer strand 5′ G A A T C G T C
Template strand 3′ C T T A G C A G T G C A A T 5′

Postinsertion site   Insertion site

insertion →

5′ G A A T C G T C A
3′ C T T A G C A G T G C A A T 5′

translocation →

5′ G A A T C G T C A
3′ C T T A G C A G T G C A A T 5′

(c)

Palm
DNA
Thumb   Fingers

(d)

5′
Palm
3′
Fingers   Insertion site
Thumb
Postinsertion site

facilitated in the cell by the 19 kJ/mol generated in the subsequent hydrolysis of the pyrophosphate product by the enzyme pyrophosphatase (p. 513).

Early work on DNA polymerase I led to the definition of two central requirements for DNA polymerization (Fig. 25-5). First, all DNA polymerases require a template. The polymerization reaction is guided by a template DNA strand according to the base-pairing rules predicted by Watson and Crick: where a guanine is present in the template, a cytosine deoxynucleotide is added to the new strand, and so on. This was a particularly important discovery, not only because it provided a chemical basis for accurate semiconservative DNA replication but also because it represented the first example of the use of a template to guide a biosynthetic reaction.

Second, the polymerases require a **primer**. A primer is a strand segment (complementary to the template) with a free 3′-hydroxyl group to which a nucleotide can be added; the free 3′ end of the primer is called the **primer terminus**. In other words, part of the new strand must already be in place: all DNA polymerases can add nucleotides only to a preexisting strand. Many primers are oligonucleotides of RNA rather than DNA, and specialized enzymes synthesize primers when and where they are required.

A DNA polymerase active site has two parts (Fig. 25-5b). The incoming nucleotide is initially positioned in the **insertion site**. Once the phosphodiester bond is formed, the polymerase slides forward on the DNA and the new base pair is positioned in the **postinsertion site**.

These sites are located in a pocket that resembles the palm of a hand (Fig. 25-5c).

After adding a nucleotide to a growing DNA strand, a DNA polymerase either dissociates or moves along the template and adds another nucleotide. Dissociation and reassociation of the polymerase can limit the overall polymerization rate—the process is generally faster when a polymerase adds more nucleotides without dissociating from the template. The average number of nucleotides added before a polymerase dissociates defines its **processivity**. DNA polymerases vary greatly in processivity; some add just a few nucleotides before dissociating, others add many thousands.

Replication Is Very Accurate

Replication proceeds with an extraordinary degree of fidelity. In *E. coli*, a mistake is made only once for every 10^9 to 10^{10} nucleotides added. For the *E. coli* chromosome of $\sim 4.6 \times 10^6$ bp, this means that an error occurs only once per 1,000 to 10,000 replications. During polymerization, discrimination between correct and incorrect nucleotides relies not just on the hydrogen bonds that specify the correct pairing between complementary bases but also on the common geometry of the standard A=T and G≡C base pairs **(Fig. 25-6)**. The active site of DNA polymerase I accommodates only base pairs with this geometry. An incorrect nucleotide may be able to hydrogen-bond with a base in the template, but it generally will not fit into the active site. Incorrect bases can be rejected before the phosphodiester bond is formed.

The accuracy of the polymerization reaction itself, however, is insufficient to account for the high degree of fidelity in replication. Careful measurements in vitro have shown that DNA polymerases insert one incorrect nucleotide for every 10^4 to 10^5 correct ones. These mistakes sometimes occur because a base is briefly in an unusual tautomeric form (see Fig. 8-9), allowing it to hydrogen-bond with an incorrect partner. In vivo, the error rate is reduced by additional enzymatic mechanisms.

One mechanism intrinsic to many DNA polymerases is a separate $3' \rightarrow 5'$ exonuclease activity that double-checks each nucleotide after it is added. This nuclease activity permits the enzyme to remove a newly added nucleotide and is highly specific for mismatched base pairs **(Fig. 25-7)**. If the polymerase has added the wrong nucleotide, translocation of the enzyme to the position where the next nucleotide is to be added is inhibited. This kinetic pause provides the opportunity for a correction. The $3' \rightarrow 5'$ exonuclease activity removes the mispaired nucleotide, and the polymerase begins again. This activity, known as **proofreading**, is not simply the reverse of the polymerization reaction (Eqn 25-1), because pyrophosphate is not involved. The polymerizing

(a) Correct base pairs   **(b) Incorrect base pairs**

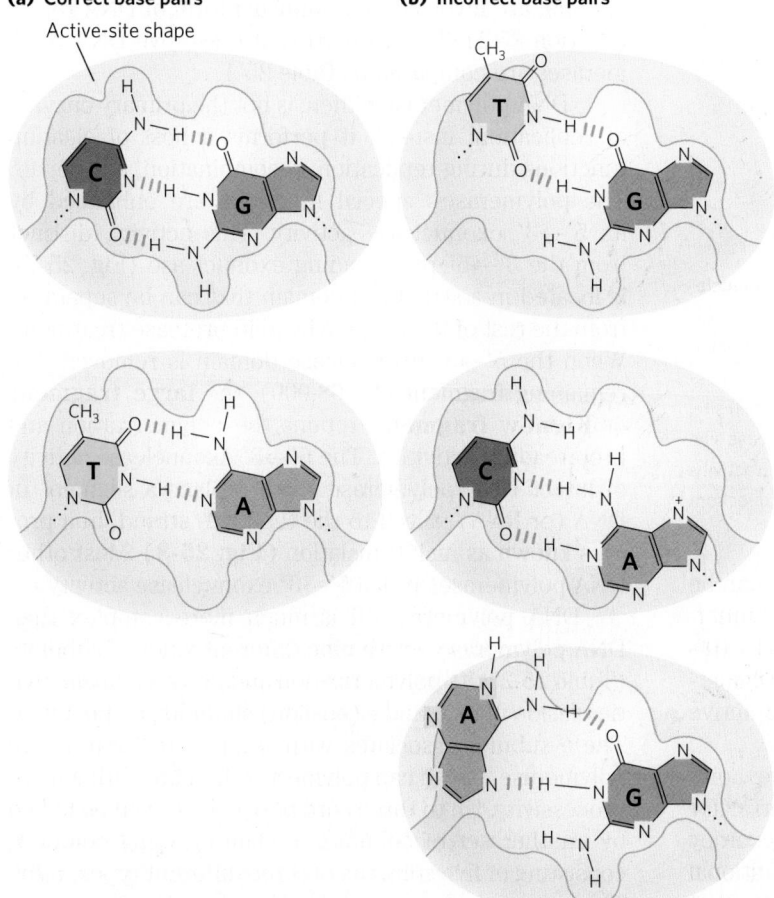

FIGURE 25-6 Contribution of base-pair geometry to the fidelity of DNA replication. (a) The standard A=T and G≡C base pairs have very similar geometries, and an active site sized to fit one will generally accommodate the other. **(b)** The geometry of incorrectly paired bases can exclude them from the active site, as occurs on DNA polymerase.

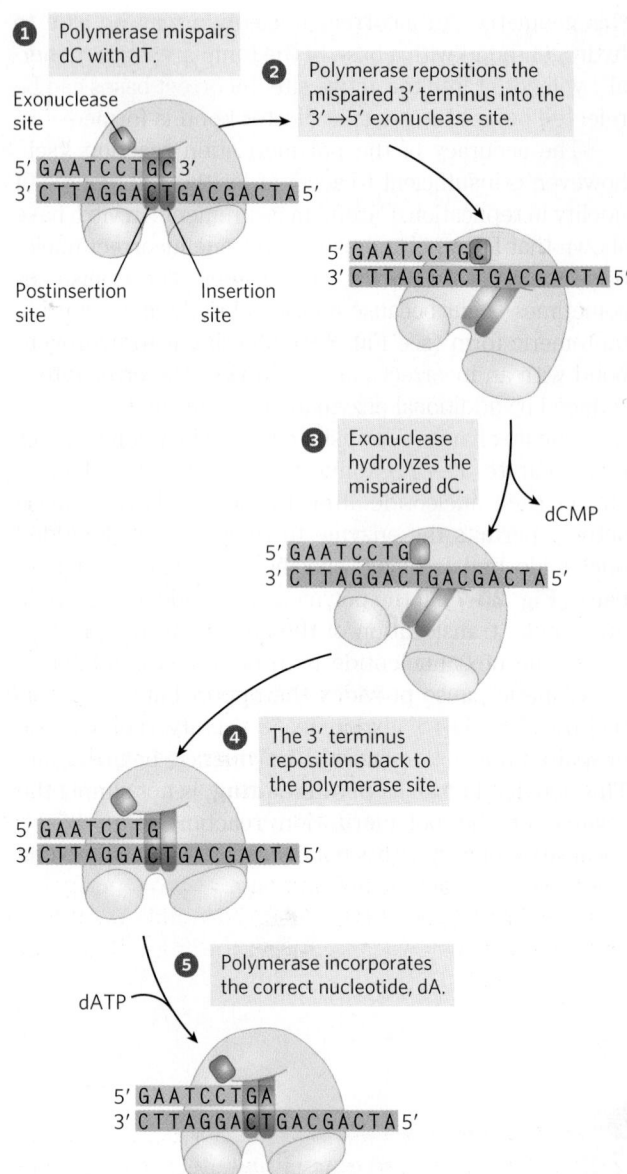

① Polymerase mispairs dC with dT.

Exonuclease site

5′ G A A T C C T G C 3′
3′ C T T A G G A C T G A C G A C T A 5′

② Polymerase repositions the mispaired 3′ terminus into the 3′→5′ exonuclease site.

Postinsertion site   Insertion site

5′ G A A T C C T G C
3′ C T T A G G A C T G A C G A C T A 5′

③ Exonuclease hydrolyzes the mispaired dC.

dCMP

5′ G A A T C C T G
3′ C T T A G G A C T G A C G A C T A 5′

④ The 3′ terminus repositions back to the polymerase site.

5′ G A A T C C T G
3′ C T T A G G A C T G A C G A C T A 5′

⑤ Polymerase incorporates the correct nucleotide, dA.

dATP

5′ G A A T C C T G A
3′ C T T A G G A C T G A C G A C T A 5′

FIGURE 25-7 An example of error correction by the 3′→5′ exonuclease activity of DNA polymerase I. Structural analysis has located the exonuclease activity behind the polymerase activity as the enzyme is oriented in its movement along the DNA. A mismatched base (here, a C–T mismatch) impedes translocation of DNA polymerase I (Pol I) to the next site. The DNA bound to the enzyme slides backward into the exonuclease site, and the enzyme corrects the mistake with its 3′→5′ exonuclease activity. The enzyme then resumes its polymerase activity in the 5′→3′ direction.

and proofreading activities of a DNA polymerase can be measured separately. Proofreading improves the inherent accuracy of the polymerization reaction 10^2- to 10^3-fold. In the monomeric DNA polymerase I, the polymerizing and proofreading activities have separate active sites within the same polypeptide.

When base selection and proofreading are combined, DNA polymerase leaves behind one net error for every 10^6 to 10^8 bases added. Yet the measured accuracy of replication in *E. coli* is higher still. The additional accuracy is provided by a separate enzyme system that

repairs the mismatched base pairs remaining after replication. We describe this mismatch repair, along with other DNA repair processes, in Section 25.2.

E. coli Has at Least Five DNA Polymerases

More than 90% of the DNA polymerase activity observed in *E. coli* extracts can be accounted for by DNA polymerase I. Soon after the isolation of this enzyme in 1955, however, evidence began to accumulate that it is not suited for replication of the large *E. coli* chromosome. First, the rate at which it adds nucleotides (600 nucleotides/min) is too slow (by a factor of 100 or more) to account for the rates at which the replication fork moves in the bacterial cell. Second, DNA polymerase I has a relatively low processivity. Third, genetic studies have demonstrated that many genes, and therefore many proteins, are involved in replication: DNA polymerase I clearly does not act alone. Fourth, and most important, in 1969 John Cairns isolated a bacterial strain with an altered gene for DNA polymerase I that produced an inactive enzyme. Although this strain was abnormally sensitive to agents that damaged DNA, it was nevertheless viable.

A search for other DNA polymerases led to the discovery of *E. coli* **DNA polymerase II** and **DNA polymerase III** in the early 1970s. DNA polymerase II is an enzyme involved in one type of DNA repair (Section 25.3). DNA polymerase III is the principal replication enzyme in *E. coli*. DNA polymerases IV and V, identified in 1999, are involved in an unusual form of DNA repair (Section 25.2). The properties of these five DNA polymerases are compared in Table 25-1.

DNA polymerase I, then, is not the primary enzyme of replication; instead, it performs a host of cleanup functions during replication, recombination, and repair. The polymerase's special functions are enhanced by its 5′→3′ exonuclease activity. This activity, distinct from the 3′→5′ proofreading exonuclease (Fig. 25-7), is located in a structural domain that can be separated from the rest of the enzyme by mild protease treatment. When the 5′→3′ exonuclease domain is removed, the remaining fragment (M_r 68,000), the **large fragment** or **Klenow fragment**, retains the polymerization and proofreading activities. The 5′→3′ exonuclease activity of intact DNA polymerase I can replace a segment of DNA (or RNA) paired to the template strand, in a process known as nick translation (**Fig. 25-8**). Most other DNA polymerases lack a 5′→3′ exonuclease activity.

DNA polymerase III is much more complex than DNA polymerase I, with nine different kinds of subunits (Table 25-2). Its polymerization and proofreading activities reside in its α and ε (epsilon) subunits, respectively. The θ subunit associates with α and ε to form a core polymerase, which can polymerize DNA but with limited processivity. Up to three core polymerases can be linked by another set of subunits, a clamp-loading complex, consisting of five subunits of three different types, $\tau_3\delta\delta'$. The core polymerases are linked through the τ (tau)

TABLE 25-1	Comparison of the Five DNA Polymerases of *E. coli*				
	DNA polymerase				
	I	II[a]	III	IV[a]	V[a]
Structural gene[b]	*polA*	*polB*	*polC (dnaE)*	*dinB*	*umuC*
Subunits (number of different types)	1	7	9	1	3
M_r	103,000	88,000[c]	1,065,400	39,100	110,000
3′→5′ exonuclease (proofreading)	Yes	Yes	Yes	No	No
5′→3′ exonuclease	Yes	No	No	No	No
Polymerization rate (nucleotides/s)	10–20	40	250–1,000	2–3	1
Processivity (nucleotides added before polymerase dissociates)	3–200	1,500	≥500,000	1	6–8

[a]Translesion (mutagenic) DNA polymerases. For DNA polymerase IV, processivity is increased substantially by association with a β clamp. These polymerases are slowed when a DNA lesion is present in the DNA template strand.

[b]For enzymes with more than one subunit, the gene listed here encodes the subunit with polymerization activity. Note that *dnaE* is an earlier designation for the gene now referred to as *polC*.

[c]Polymerization subunit only. DNA polymerase II shares several subunits with DNA polymerase III, including the β, δ, δ′, χ, and ψ subunits (see Table 25-2).

subunits. Two additional subunits, χ (chi) and ψ (psi), are bound to the clamp-loading complex. The entire assembly of 16 protein subunits (eight different types) is called DNA polymerase III* **(Fig. 25-9a)**.

DNA polymerase III* can polymerize DNA, but with a much lower processivity than one would expect for the organized replication of an entire chromosome. The necessary increase in processivity is provided by the addition of the β subunits. The β subunits associate in

pairs to form donut-shaped structures that encircle the DNA and act like clamps (Fig. 25-9b). Each dimer associates with a core subassembly of polymerase III* (one dimeric clamp per active core subassembly) and slides along the DNA as replication proceeds. The β sliding clamp prevents the dissociation of DNA polymerase III from DNA, dramatically increasing processivity—to greater than 500,000 (Table 25-1). The addition of the β subunits converts DNA polymerase III* to DNA polymerase III holoenzyme.

DNA Replication Requires Many Enzymes and Protein Factors

Replication in *E. coli* requires not just a single DNA polymerase but 20 or more different enzymes and proteins, each performing a specific task. The entire complex has been termed the **DNA replicase system** or **replisome**. The enzymatic complexity of replication reflects the constraints imposed by the structure of DNA and by the requirements for accuracy. The main classes of replication enzymes are considered here in terms of the problems they overcome.

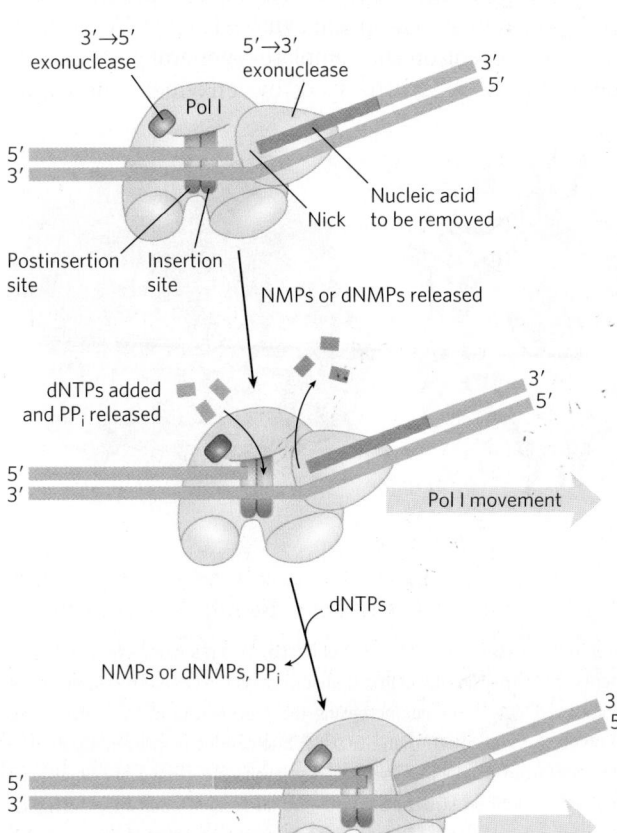

FIGURE 25-8 Nick translation. The bacterial DNA polymerase I has three domains, catalyzing its DNA polymerase, 5′→3′ exonuclease, and 3′→5′ exonuclease activities. The 5′→3′ exonuclease domain is in front of the enzyme as it moves along the DNA and is not shown in Figure 25-5. By degrading the DNA strand ahead of the enzyme and synthesizing a new strand behind, DNA polymerase I can promote a reaction called nick translation, in which a break or nick in the DNA is effectively moved along with the enzyme. This process has a role in DNA repair and in the removal of RNA primers during replication (both described below). The strand of nucleic acid to be removed (either DNA or RNA) is shown in purple, the replacement strand in red. DNA synthesis begins at a nick (a broken phosphodiester bond, leaving a free 3′ hydroxyl and a free 5′ phosphate). A nick remains where DNA polymerase I eventually dissociates, and the nick is later sealed by another enzyme.

TABLE 25-2 Subunits of DNA Polymerase III of *E. coli*

Subunit	Number of subunits per holoenzyme	M_r of subunit	Gene	Function of subunit	
α	3	129,900	*polC* (*dnaE*)	Polymerization activity	Core polymerase
ε	3	27,500	*dnaQ* (*mutD*)	3′→5′ proofreading exonuclease	
θ	3	8,600	*holE*	Stabilization of ε subunit	
τ	3	71,100	*dnaX*	Stable template binding; core enzyme dimerization	Clamp-loading (γ) complex that loads β subunits on lagging strand at each Okazaki fragment[a]
δ	1	38,700	*holA*	Clamp opener	
δ'	1	36,900	*holB*	Clamp loader	
χ	1	16,600	*holC*	Interaction with SSB	
ψ	1	15,200	*holD*	Interaction with τ and χ	
β	6	40,600	*dnaN*	DNA clamp required for optimal processivity	

[a]The clamp-loading complex is also called the γ complex, because of the existence of another version of the complex in which three subunits called γ replace the τ subunits. The γ subunit is encoded by a portion of the gene for the τ subunit (*dnaX*), such that the amino-terminal 66% of the τ subunit has the same amino acid sequence as the γ subunit. The γ subunit is generated by a translational frameshifting mechanism (p. 1085) that leads to premature translational termination. The γ subunit shares the clamp-loading functions of τ but lacks the protein segments that interact with the core polymerase or with DnaB helicase. Clamp-loading complexes incorporating γ subunits may operate independently of the DNA polymerase III holoenzyme, promoting the unloading of β clamps discarded on the lagging strand as the replication fork progresses. They may also promote loading of β clamps for some DNA repair processes that require DNA synthesis away from the replication fork.

Access to the DNA strands that are to act as templates requires separation of the two parent strands. This is generally accomplished by **helicases**, enzymes that move along the DNA and separate the strands, using chemical energy from ATP. Strand separation creates topological stress in the helical DNA structure (see Fig. 24-11), which is relieved by the action of **topoisomerases**. The separated strands are stabilized by **DNA-binding proteins**. As noted earlier, before DNA polymerases can begin synthesizing DNA, primers must be present on the template—generally, short segments of RNA synthesized by enzymes known as

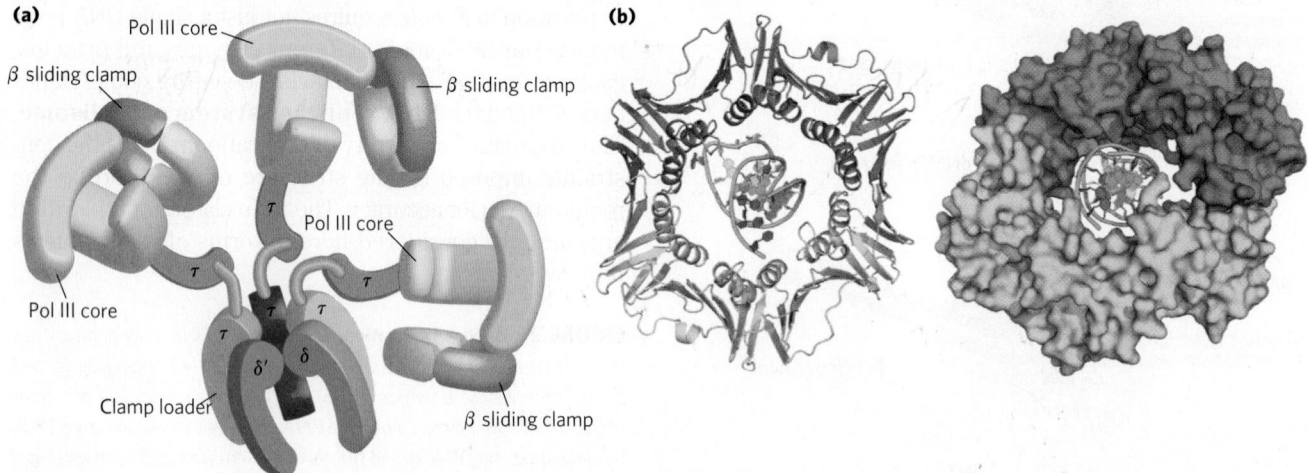

FIGURE 25-9 DNA polymerase III. (a) Architecture of bacterial DNA polymerase III (Pol III). Three core domains, composed of subunits α, ε, and θ, are linked by a five-subunit clamp-loading complex (also known as the γ complex; the name is explained in a footnote to Table 25-2) with the composition $\tau_3\delta\delta'$. The core subunits and clamp-loader complex constitute DNA polymerase III*. The other two subunits of DNA polymerase III*, χ and ψ (not shown), also bind to the clamp-loading complex. Three β clamps interact with the three core subassemblies, each clamp a dimer of the β subunit. The complex interacts with the DnaB helicase (described later in the text) through the τ subunits. **(b)** Two β subunits of *E. coli* polymerase III form a circular clamp that surrounds the DNA. The clamp slides along the DNA molecule, increasing the processivity of the polymerase III holoenzyme to more than 500,000 nucleotides by preventing its dissociation from the DNA. The two β subunits are shown in two shades of purple as ribbon structures (left) and surface contour images (right), surrounding the DNA. [Sources: (a) Information from N. Yao and M. O'Donnell, *Mol. Biosyst.* 4:1075, 2008. (b) PDB ID 2POL, X.-P. Kong et al., *Cell* 69:425, 1992.]

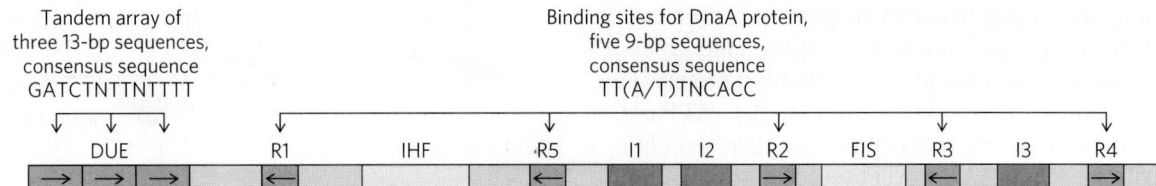

FIGURE 25-10 Arrangement of sequences in the *E. coli* replication origin, *oriC*. Consensus sequences (p. 104) for key repeated elements are shown. N represents any of the four nucleotides. The horizontal arrows indicate the orientations of the nucleotide sequences (left-to-right arrow denotes a sequence in the top strand; right-to-left, in the bottom strand). FIS and IHF are binding sites for proteins described in the text. R sites are bound by DnaA. I sites are additional DnaA-binding sites (with different sequences), bound by DnaA only when the protein is complexed with ATP.

primases. Ultimately, the RNA primers are removed and replaced by DNA; in *E. coli*, this is one of the many functions of DNA polymerase I. A specialized nuclease that degrades RNA in RNA-DNA hybrids, called RNase H1, also removes some RNA primers. After an RNA primer is removed and the gap is filled in with DNA, a nick remains in the DNA backbone in the form of a broken phosphodiester bond. These nicks are sealed by **DNA ligases**. All these processes require coordination and regulation, an interplay best characterized in the *E. coli* system.

Replication of the *E. coli* Chromosome Proceeds in Stages

The synthesis of a DNA molecule can be divided into three stages: initiation, elongation, and termination, distinguished both by the reactions taking place and by the enzymes required. As you will find here and in the next two chapters, synthesis of the major information-containing biological polymers—DNAs, RNAs, and proteins—can be understood in terms of these same three stages, with the stages of each pathway having unique characteristics. The events described below reflect information derived primarily from in vitro experiments using purified *E. coli* proteins, although the principles are highly conserved in all replication systems.

Initiation The *E. coli* replication origin, *oriC*, consists of 245 bp and contains DNA sequence elements that are highly conserved among bacterial replication origins. The general arrangement of the conserved sequences is illustrated in **Figure 25-10**. Two types of sequences are of special interest: five repeats of a 9 bp sequence (R sites) that serve as binding sites for the key initiator protein, DnaA, and a region rich in A=T base pairs called the **DNA unwinding element (DUE)**. There are three additional DnaA-binding sites (I sites), and binding sites for the proteins IHF (integration host factor) and FIS (factor for inversion stimulation). These two proteins were discovered as necessary components of certain recombination reactions described later in this chapter, and their names reflect those roles. Another DNA-binding protein, HU (a histonelike bacterial protein originally dubbed factor U), also participates but does not have a specific binding site.

At least 10 different enzymes or proteins (summarized in Table 25-3) participate in the initiation phase of

TABLE 25-3	Proteins Required to Initiate Replication at the *E. coli* Origin		
Protein	M_r	Number of subunits	Function
DnaA protein	52,000	1	Recognizes *oriC* sequence; opens duplex at specific sites in origin
DnaB protein (helicase)	300,000	6[a]	Unwinds DNA
DnaC protein	174,000	6[a]	Required for DnaB binding at origin
HU	19,000	2	Histonelike protein; DNA-binding protein; stimulates initiation
FIS	22,500	2[a]	DNA-binding protein; stimulates initiation
IHF	22,000	2	DNA-binding protein; stimulates initiation
Primase (DnaG protein)	60,000	1	Synthesizes RNA primers
Single-stranded DNA-binding protein (SSB)	75,600	4[a]	Binds single-stranded DNA
DNA gyrase (DNA topoisomerase II)	400,000	4	Relieves torsional strain generated by DNA unwinding
Dam methylase	32,000	1	Methylates (5′)GATC sequences at *oriC*

[a]Subunits in these cases are identical.

replication. They open the DNA helix at the origin and establish a prepriming complex for subsequent reactions. The crucial component in the initiation process is the DnaA protein, a member of the **AAA+ ATPase** protein family (ATPases *a*ssociated with diverse cellular *a*ctivities). Many AAA+ ATPases, including DnaA, form oligomers and hydrolyze ATP relatively slowly. This ATP hydrolysis acts as a switch that mediates interconversion of the protein between two states. In the case of DnaA, the ATP-bound form is active and the ADP-bound form is inactive.

Eight DnaA protein molecules, all in the ATP-bound state, assemble to form a helical complex encompassing the R and I sites in *oriC* **(Fig. 25-11)**. DnaA has a higher affinity for R sites than I sites, and it binds R sites equally well in its ATP- or ADP-bound form. The I sites, which bind only the ATP-bound DnaA, allow discrimination between the active and inactive forms of DnaA. The tight right-handed wrapping of the DNA around this complex introduces a positive supercoil (see Chapter 24). The associated strain in the nearby DNA, combined with the binding of additional DnaA protein to the DUE region, leads to denaturation in the A=T-rich DUE. The complex formed at the replication origin also includes several DNA-binding proteins—HU, IHF, and FIS—that facilitate DNA bending.

The DnaC protein, another AAA+ ATPase, then loads the DnaB protein onto the separated DNA strands in the denatured region. A hexamer of DnaC, each subunit bound to ATP, forms a tight complex with the ring-shaped, hexameric DnaB helicase. This DnaC-DnaB interaction opens the DnaB ring, the process being aided by a further interaction between DnaB and DnaA. Two of the ring-shaped DnaB hexamers are loaded in the DUE, one onto each DNA strand. The ATP bound to DnaC is hydrolyzed, releasing the DnaC and leaving the DnaB bound to the DNA.

Loading of the DnaB helicase is the key step in replication initiation. As a replicative helicase, DnaB migrates along the single-stranded DNA in the 5'→3' direction, unwinding the DNA as it travels. The DnaB helicases loaded onto the two DNA strands thus travel in opposite directions, creating two potential replication forks. All other proteins at the replication fork are linked directly or indirectly to DnaB. The DNA polymerase III holoenzyme is linked through its τ subunits; additional DnaB interactions are described below. As replication begins and the DNA strands are separated at the fork, many molecules of single-stranded DNA–binding protein (SSB) bind to and stabilize the separated strands, and DNA gyrase (DNA topoisomerase II) relieves the topological stress induced ahead of the fork by the unwinding reaction.

Initiation is the only phase of DNA replication that is known to be regulated, and it is regulated such that replication occurs only once in each cell cycle. The mechanism of regulation is not yet entirely

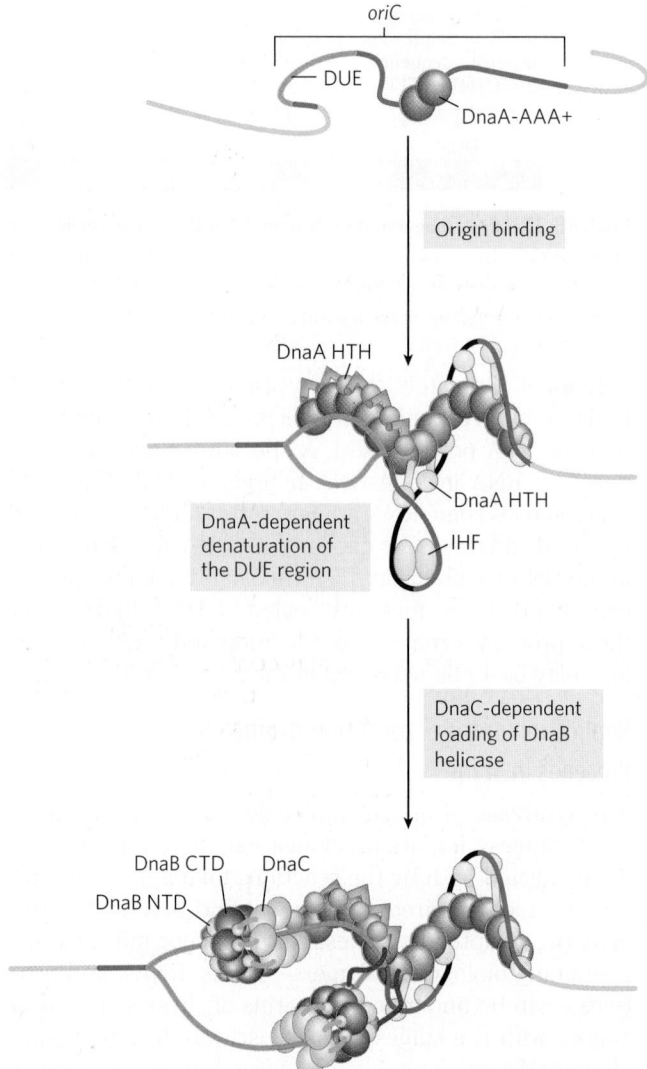

FIGURE 25-11 **Model for initiation of replication at the *E. coli* origin, *oriC*.** Eight DnaA protein molecules, each with a bound ATP, bind at the R and I sites in the origin (see Fig. 25-10). The DNA is wrapped around this complex, which forms a right-handed helical structure that continues into the DUE region. The A=T-rich DUE region is denatured as a result of the strain imparted by the DnaA binding. Formation of the helical DnaA complex is facilitated by the proteins HU, IHF, and FIS. The detailed structural roles of these proteins are not known, but IHF may stabilize a transient DNA loop, as shown here. Hexamers of the DnaB protein bind to each strand, with the aid of DnaC protein. The DnaB helicase activity further unwinds the DNA in preparation for priming and DNA synthesis. [Source: Information from J. P. Erzberger et al., *Nature Struct. Mol. Biol.* 13:676, 2006.]

understood, but genetic and biochemical studies have provided insights into several separate regulatory mechanisms.

Once DNA polymerase III has been loaded onto the DNA, along with the β subunits (signaling completion of the initiation phase), the protein Hda binds to the β subunits and interacts with DnaA to stimulate hydrolysis of its bound ATP. Hda is yet another AAA+ ATPase closely related to DnaA (its name is derived from *homologous to DnaA*). This ATP hydrolysis leads to disassembly of the DnaA complex at the origin. Slow release of ADP by DnaA and rebinding of ATP cycles the protein between its inactive (with bound ADP) and active (with bound ATP) forms on a time scale of 20 to 40 minutes.

The timing of replication initiation is affected by DNA methylation and interactions with the bacterial plasma membrane. The *oriC* DNA is methylated by the Dam methylase (Table 25-3), which methylates the N^6 position of adenine within the palindromic sequence (5′)GATC. (Dam is not a biochemical expletive; it stands for *DNA adenine methylation*.) The *oriC* region of *E. coli* is highly enriched in GATC sequences—it has 11 in its 245 bp; the average frequency of GATC in the *E. coli* chromosome as a whole is 1 in 256 bp.

Immediately after replication, the DNA is hemimethylated: the parent strands have methylated *oriC* sequences but the newly synthesized strands do not. The hemimethylated *oriC* sequences are now sequestered by interaction with the plasma membrane (the mechanism is unknown) and by binding of the protein SeqA. After a time, *oriC* is released from the plasma membrane, SeqA dissociates, and the DNA must be fully methylated by Dam methylase before it can again bind DnaA and initiate a new round of replication.

Elongation The elongation phase of replication includes two distinct but related operations: leading strand synthesis and lagging strand synthesis. Several enzymes at the replication fork are important to the synthesis of both strands. Parent DNA is first unwound by DNA helicases, and the resulting topological stress is relieved by topoisomerases. Each separated strand is then stabilized by SSB. From this point, synthesis of leading and lagging strands is sharply different.

Leading strand synthesis, the more straightforward of the two, begins with the synthesis by primase (DnaG protein) of a short (10 to 60 nucleotide) RNA primer at the replication origin. DnaG interacts with DnaB helicase to carry out this reaction, and the primer is synthesized in the direction opposite to that in which the DnaB helicase is moving. In effect, the DnaB helicase moves along the strand that becomes the lagging strand in DNA synthesis; however, the first primer laid down in the first DnaG-DnaB interaction serves to prime leading strand DNA synthesis in the opposite direction. Deoxyribonucleotides are added to this primer by a DNA polymerase III complex linked to the DnaB helicase tethered to the opposite DNA strand. Leading strand synthesis then proceeds continuously, keeping pace with the unwinding of DNA at the replication fork.

Lagging strand synthesis, as we have noted, is accomplished in short Okazaki fragments **(Fig. 25-12a)**.

First, an RNA primer is synthesized by primase, and, as in leading strand synthesis, DNA polymerase III binds to the RNA primer and adds deoxyribonucleotides (Fig. 25-12b). On this level, the synthesis of each Okazaki fragment seems straightforward, but the reality is quite complex. The complexity lies in the *coordination* of leading and lagging strand synthesis. Both strands are produced by a *single* asymmetric DNA polymerase III dimer; this is accomplished by looping the DNA of the lagging strand as shown in **Figure 25-13**, bringing together the two points of polymerization.

The synthesis of Okazaki fragments on the lagging strand entails some elegant enzymatic choreography. DnaB helicase and DnaG primase constitute a functional unit within the replication complex, the **primosome**. DNA polymerase III uses one set of its core subunits (the core polymerase) to synthesize the leading strand continuously, while the other two sets of core subunits cycle from one Okazaki fragment to the next on the looped lagging strand. In vitro, a DNA polymerase III holoenzyme with only two sets of core subunits can synthesize both leading and lagging strands. However, a third set of core subunits increases the efficiency of lagging strand synthesis as well as the processivity of the overall replisome.

DnaB helicase, bound in front of DNA polymerase III, unwinds the DNA at the replication fork (Fig. 25-13a) as it travels along the lagging strand template in the 5′→3′ direction. DnaG primase occasionally associates with DnaB helicase and synthesizes a short RNA primer (Fig. 25-13b). A new β sliding clamp is then positioned at the primer by the clamp-loading complex of DNA polymerase III (Fig. 25-13c). When synthesis of an Okazaki fragment has been completed, replication halts, and the core subunits of DNA polymerase III dissociate from their β sliding clamp (and from the completed Okazaki fragment) and associate with the new clamp (Fig. 25-13d, e). This initiates synthesis of a new Okazaki fragment. Two sets of core subunits may be engaged in the synthesis of two different Okazaki fragments at the same time. The proteins acting at the replication fork are summarized in Table 25-4.

The clamp-loading complex of DNA polymerase III, consisting of parts of the three τ subunits along with the δ and δ′ subunits, is also an AAA+ ATPase. This complex binds to ATP and to the new β sliding clamp. The binding imparts strain on the dimeric clamp, opening up the ring at one subunit interface **(Fig. 25-14)**. The newly primed lagging strand is slipped into the ring through the resulting break. The clamp loader then hydrolyzes ATP, releasing the β sliding clamp and allowing it to close around the DNA.

The replisome promotes rapid DNA synthesis, adding ~1,000 to 2,000 nucleotides/s to each strand (leading and lagging). Once an Okazaki fragment has been completed, its RNA primer is removed by DNA polymerase I or RNase H1, and replaced with DNA by the

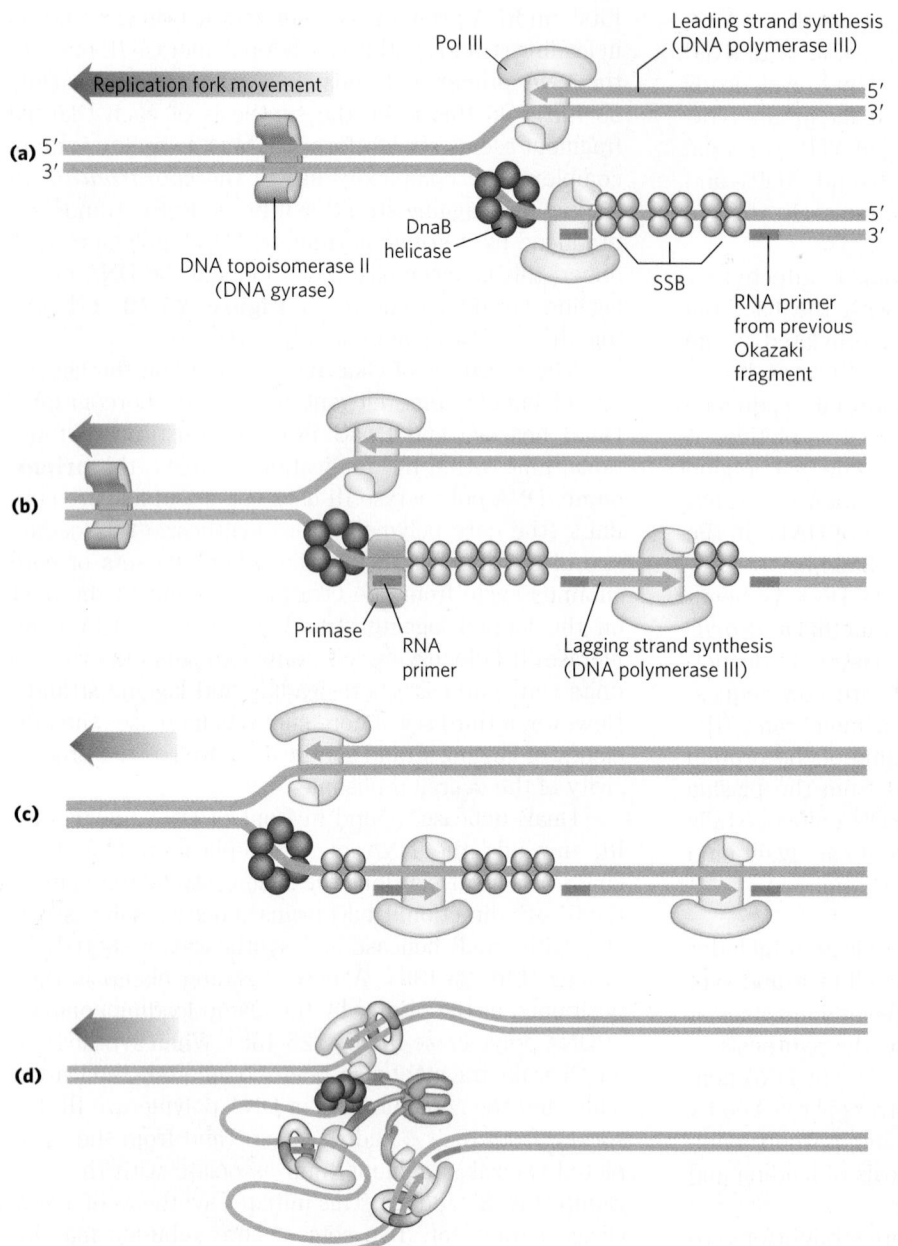

FIGURE 25-12 Synthesis of Okazaki fragments. (a) At intervals, primase synthesizes an RNA primer for a new Okazaki fragment. Notice that if we consider the two template strands as lying side by side, lagging strand synthesis formally proceeds in the opposite direction from fork movement. **(b)** Each primer is extended by DNA polymerase III. **(c)** DNA synthesis continues until the fragment extends as far as the primer of the previously added Okazaki fragment. A new primer is synthesized near the replication fork to begin the process again. **(d)** Each DNA polymerase III holoenzyme has three sets of core subunits, so one or two Okazaki fragments can be synthesized simultaneously, along with the leading strand.

TABLE 25-4	Proteins of the *E. coli* Replisome		
Protein	M_r	Number of subunits	Function
SSB	75,600	4	Binding to single-stranded DNA
DnaB protein (helicase)	300,000	6	DNA unwinding; primosome constituent
Primase (DnaG protein)	60,000	1	RNA primer synthesis; primosome constituent
DNA polymerase III	1,065,400	17	New strand elongation
DNA polymerase I	103,000	1	Filling of gaps; excision of primers
DNA ligase	74,000	1	Ligation
DNA gyrase (DNA topoisomerase II)	400,000	4	Supercoiling

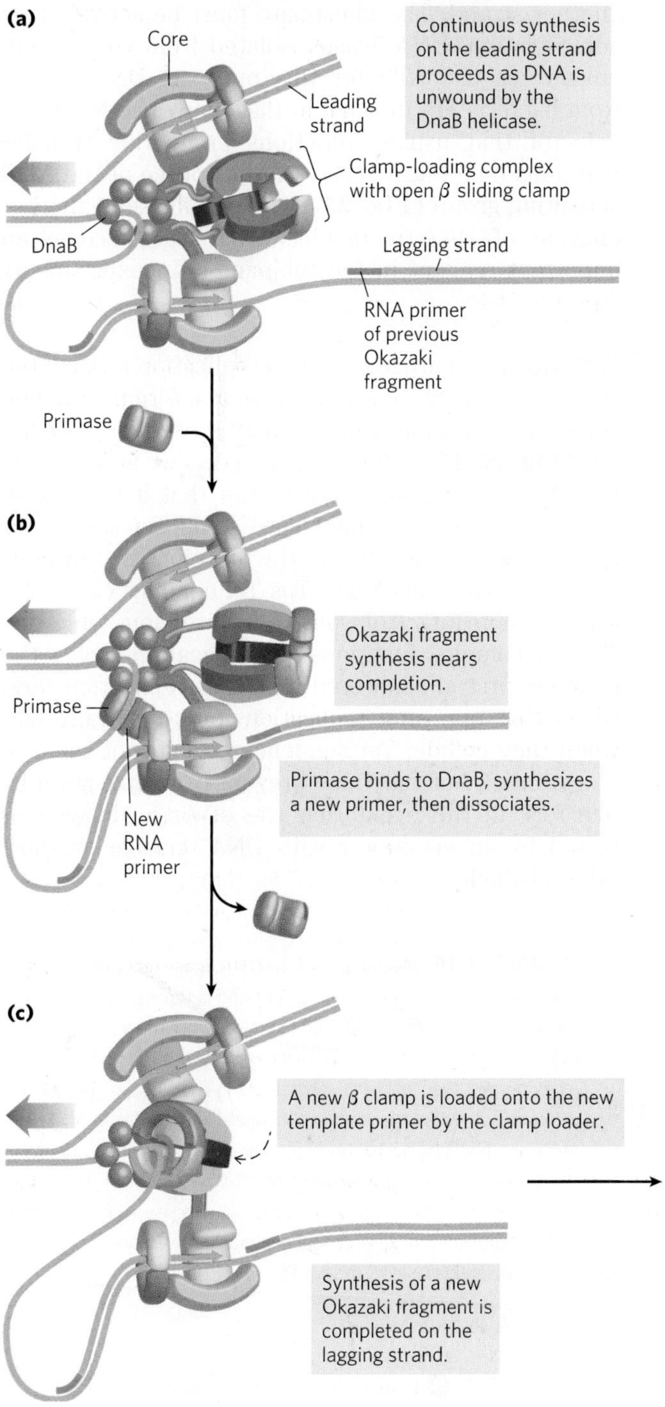

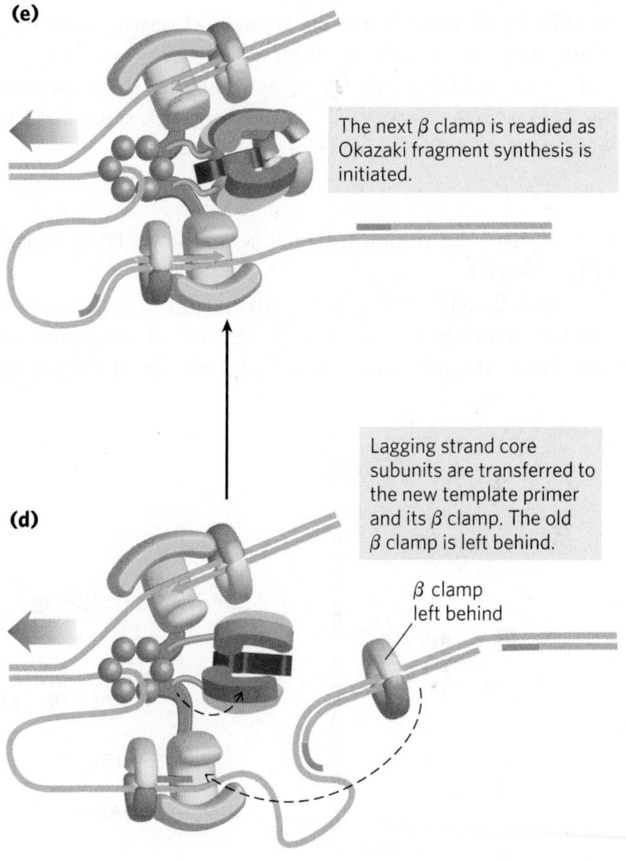

(a)

Core

Leading strand

Clamp-loading complex with open β sliding clamp

Continuous synthesis on the leading strand proceeds as DNA is unwound by the DnaB helicase.

DnaB

Lagging strand

RNA primer of previous Okazaki fragment

Primase

FIGURE 25-13 DNA synthesis on the leading and lagging strands. Events at the replication fork are coordinated by a single DNA polymerase III dimer, in an integrated complex with DnaB helicase. This figure shows the replication process already underway; **(a)** through **(e)** are discussed in the text. Only two sets of polymerase core subunits are shown, to more clearly illustrate the cycling on the lagging strand. The lagging strand is looped so that DNA synthesis proceeds steadily on both the leading and lagging strand templates at the same time. Red arrows indicate the 3′ end of the two new strands and the direction of DNA synthesis. An Okazaki fragment is being synthesized on the lagging strand. The subunit colors and the functions of the clamp-loading complex are explained in Figure 25-14.

(b)

Okazaki fragment synthesis nears completion.

Primase

Primase binds to DnaB, synthesizes a new primer, then dissociates.

New RNA primer

(e)

The next β clamp is readied as Okazaki fragment synthesis is initiated.

Lagging strand core subunits are transferred to the new template primer and its β clamp. The old β clamp is left behind.

β clamp left behind

(c)

A new β clamp is loaded onto the new template primer by the clamp loader.

Synthesis of a new Okazaki fragment is completed on the lagging strand.

(d)

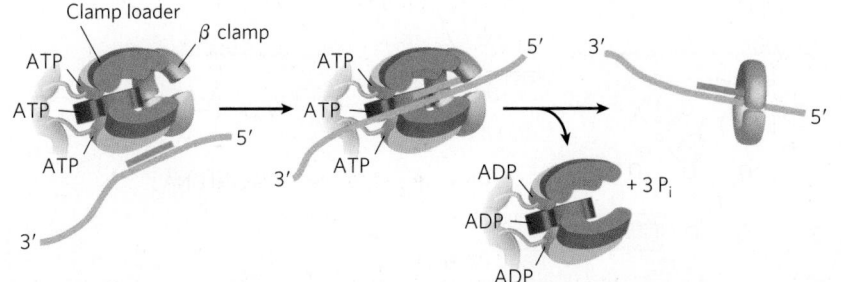

FIGURE 25-14 The DNA polymerase III clamp loader. The five subunits of the clamp-loading (γ) complex are the δ and δ′ subunits and the amino-terminal domain of each of the three τ subunits (see Fig. 25-9). The complex binds to three molecules of ATP and to a dimeric β clamp. This binding forces the β clamp open at one of its two subunit interfaces. Hydrolysis of the bound ATP allows the β clamp to close again around the DNA.

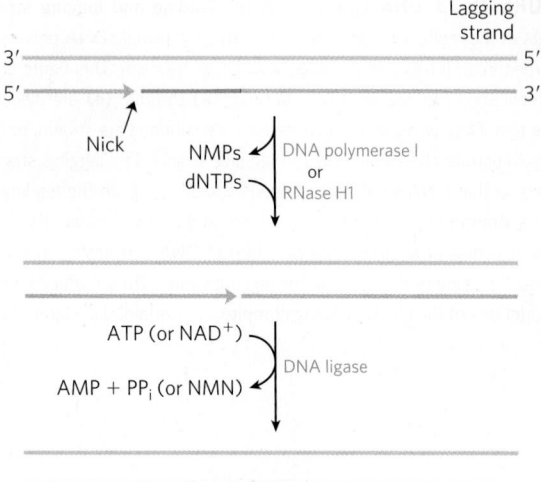

FIGURE 25-15 Final steps in the synthesis of lagging strand segments. RNA primers in the lagging strand are removed by the 5'→3' exonuclease activity of DNA polymerase I or RNase H1, and then replaced with DNA by DNA polymerase I. The remaining nick is sealed by DNA ligase. The role of ATP or NAD⁺ is shown in Figure 25-16.

polymerase; the remaining nick is sealed by DNA ligase **(Fig. 25-15)**.

DNA ligase catalyzes the formation of a phosphodiester bond between a 3' hydroxyl at the end of one DNA strand and a 5' phosphate at the end of another strand. The phosphate must be activated by adenylylation. DNA ligases isolated from viruses and eukaryotes use ATP for this purpose. DNA ligases from bacteria are unusual in that many use NAD⁺—a cofactor that usually functions in hydride transfer reactions (see Fig. 13-24)—as the source of the AMP activating group **(Fig. 25-16)**. DNA ligase is another enzyme of DNA metabolism that has become an important reagent in recombinant DNA experiments (see Fig. 9-1).

Termination Eventually, the two replication forks of the circular *E. coli* chromosome meet at a terminus region containing multiple copies of a 20 bp sequence called Ter **(Fig. 25-17)**. The Ter sequences are arranged on the chromosome to create a trap that a replication fork can enter but cannot leave. The Ter sequences function as binding sites for the protein Tus (terminus utilization substance). The Tus-Ter complex can arrest a replication fork from only one direction. Only one Tus-Ter complex functions per replication cycle—the complex first encountered by either replication fork. Given that opposing replication forks generally halt when they collide, Ter sequences would not seem to be essential, but they may prevent overreplication by one fork in the event that the other is delayed or halted by an encounter with DNA damage or some other obstacle.

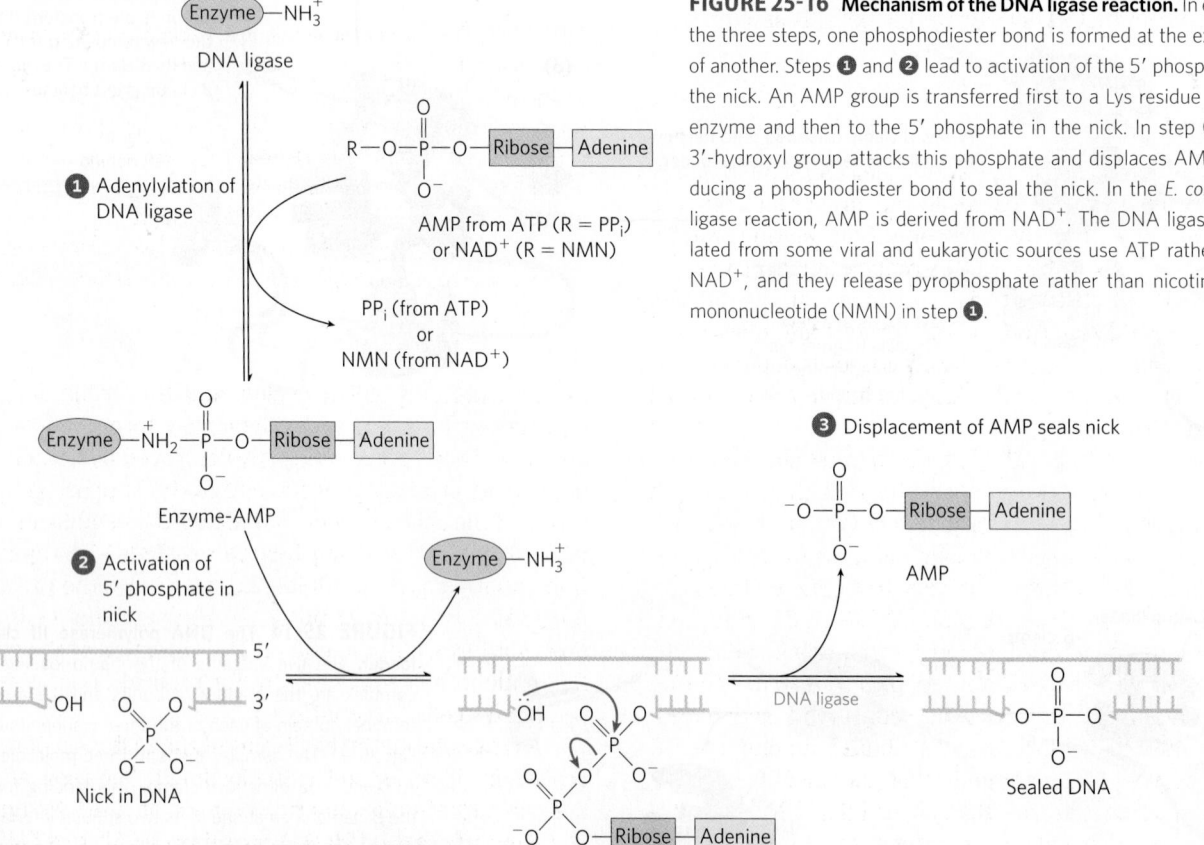

FIGURE 25-16 Mechanism of the DNA ligase reaction. In each of the three steps, one phosphodiester bond is formed at the expense of another. Steps ❶ and ❷ lead to activation of the 5' phosphate in the nick. An AMP group is transferred first to a Lys residue on the enzyme and then to the 5' phosphate in the nick. In step ❸, the 3'-hydroxyl group attacks this phosphate and displaces AMP, producing a phosphodiester bond to seal the nick. In the *E. coli* DNA ligase reaction, AMP is derived from NAD⁺. The DNA ligases isolated from some viral and eukaryotic sources use ATP rather than NAD⁺, and they release pyrophosphate rather than nicotinamide mononucleotide (NMN) in step ❶.

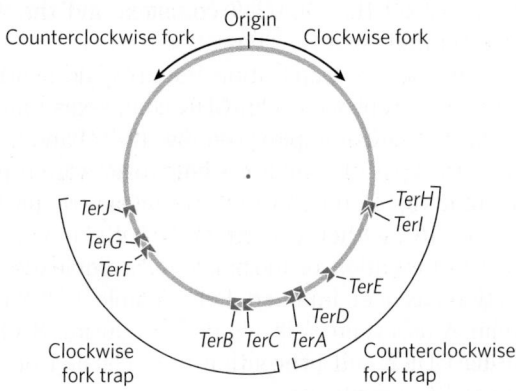

FIGURE 25-17 Termination of chromosome replication in _E. coli._ The Ter sequences (_TerA_ through _TerJ_) are positioned on the chromosome in two clusters with opposite orientations.

So, when either replication fork encounters a functional Tus-Ter complex, it halts; the other fork halts when it meets the first (arrested) fork. The final few hundred base pairs of DNA between these large protein complexes are then replicated (by an as yet unknown mechanism), completing two topologically interlinked (catenated) circular chromosomes **(Fig. 25-18)**. DNA circles linked in this way are known as **catenanes**. Separation of the catenated circles in _E. coli_ requires topoisomerase IV (a type II topoisomerase). The separated chromosomes then segregate into daughter cells at cell division. The terminal phase of replication of other circular chromosomes, including many of the DNA viruses that infect eukaryotic cells, is similar.

Replication in Eukaryotic Cells Is Similar but More Complex

The DNA molecules in eukaryotic cells are considerably larger than those in bacteria and are organized into complex nucleoprotein structures (chromatin; p. 972). The essential features of DNA replication are the same in eukaryotes and bacteria, and many of the protein complexes are functionally and structurally conserved. However, eukaryotic replication is regulated and coordinated with the cell cycle, introducing some additional complexities.

Origins of replication have a well-characterized structure in some lower eukaryotes, but they are much less defined in higher eukaryotes. In vertebrates, a variety of A=T-rich sequences may be used for replication initiation, and the sites may vary from one cell division to the next. Yeast (_S. cerevisiae_) has defined replication origins called autonomously replicating sequences (ARSs), or **replicators**. Yeast replicators span ~150 bp and contain several essential, conserved sequences. About 400 replicators are distributed among the 16 chromosomes of the haploid yeast genome.

Regulation ensures that all cellular DNA is replicated once per cell cycle. Much of this regulation

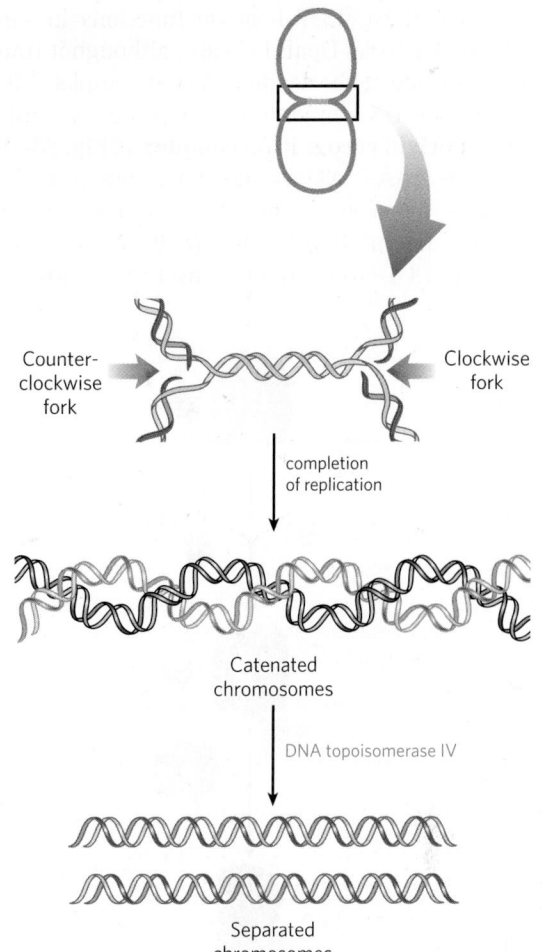

FIGURE 25-18 Role of topoisomerases in replication termination. Replication of the DNA separating opposing replication forks leaves the completed chromosomes joined as catenanes, or topologically interlinked circles. The circles are not covalently linked, but because they are interwound and each is covalently closed, they cannot be separated—except by the action of topoisomerases. In _E. coli_, a type II topoisomerase known as DNA topoisomerase IV plays the primary role in separating catenated chromosomes, transiently breaking both DNA strands of one chromosome and allowing the other chromosome to pass through the break.

involves proteins called cyclins and the cyclin-dependent kinases (CDKs) with which they form complexes (p. 477). The cyclins are rapidly destroyed by ubiquitin-dependent proteolysis at the end of the M phase (mitosis), and the absence of cyclins allows the establishment of **pre-replicative complexes (pre-RCs)** on replication initiation sites. In rapidly growing cells, the pre-RC forms at the end of M phase. In slow-growing cells, it does not form until the end of G1. Formation of the pre-RC renders the cell competent for replication, an event sometimes called **licensing**.

As in bacteria, the key event in the initiation of replication in all eukaryotes is the loading of the replicative helicase, a heterohexameric complex of **minichromosome maintenance (MCM) proteins** (MCM2 to MCM7).

The ring-shaped MCM2–7 helicase functions in some ways like the bacterial DnaB helicase, although it translocates $3' \rightarrow 5'$ along the leading strand template. It is loaded onto the DNA by another six-protein complex, called **ORC (origin recognition complex) (Fig. 25-19)**. ORC has five AAA+ ATPase domains among its subunits and is functionally analogous to the bacterial DnaA. Two other proteins, CDC6 (*cell division cycle*) and CDT1 (*CDC10-dependent transcript 1*), are also

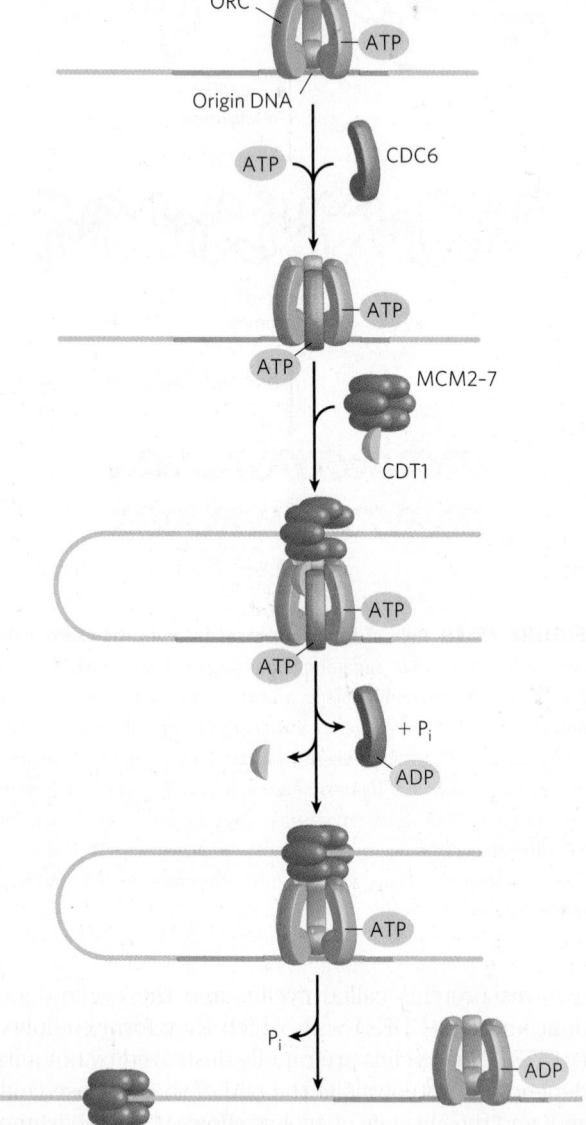

FIGURE 25-19 Assembly of a pre-replicative complex at a eukaryotic replication origin. The initiation site (origin) is bound by ORC, CDC6, and CDT1. These proteins, many of them AAA+ ATPases, promote loading of the replicative helicase, MCM2–7, in a reaction analogous to the loading of the bacterial DnaB helicase by DnaC protein. Loading of the MCM helicase complex onto the DNA forms the pre-replicative complex, or pre-RC, and is the key step in the initiation of replication. [Source: Information from U. Sivaprasad et al., in *DNA Replication and Human Disease* (M. L. DePamphilis, ed.), p. 141, Cold Spring Harbor Laboratory Press, 2006.]

required to load the MCM2–7 complex, and the yeast CDC6 is another AAA+ ATPase.

Commitment to replication requires the synthesis and activity of S-phase cyclin-CDK complexes (such as the cyclin E–CDK2 complex; see Fig. 12-33) and CDC7-DBF4. Both types of complexes help to activate replication by binding to and phosphorylating several proteins in the pre-RC. Other cyclins and CDKs function to inhibit the formation of more pre-RC complexes once replication has been initiated. For example, CDK2 binds to cyclin A as cyclin E levels decline during S phase, inhibiting CDK2 and preventing the licensing of additional pre-RC complexes.

The rate of movement of the replication fork in eukaryotes (~50 nucleotides/s) is only one-twentieth that observed in *E. coli*. At this rate, replication of an average human chromosome proceeding from a single origin would take more than 500 hours. Replication of human chromosomes in fact proceeds bidirectionally from many origins, spaced 30 to 300 kbp apart. Eukaryotic chromosomes are almost always much larger than bacterial chromosomes, so multiple origins are probably a universal feature of eukaryotic cells.

Like bacteria, eukaryotes have several types of DNA polymerases. Some have been linked to particular functions, such as the replication of mitochondrial DNA. The replication of nuclear chromosomes primarily involves three multisubunit DNA polymerases. The highly processive **DNA polymerase ε** synthesizes the leading strand, and **DNA polymerase δ** synthesizes the lagging strand. Both enzymes have $3' \rightarrow 5'$ proofreading exonuclease activities. **DNA polymerase α**, a DNA polymerase/primase, synthesizes RNA primers and also extends them by about 10 nucleotides of DNA. One subunit of DNA polymerase α has a primase activity, and the largest subunit (M_r ~180,000) contains the polymerization activity. However, this polymerase has no proofreading $3' \rightarrow 5'$ exonuclease activity, making it unsuitable for high-fidelity DNA replication.

DNA polymerases ε and δ are associated with and stimulated by proliferating cell nuclear antigen (PCNA; M_r 29,000), a protein found in large amounts in the nuclei of proliferating cells. The three-dimensional structure of PCNA is remarkably similar to that of the β subunit of *E. coli* DNA polymerase III (Fig. 25-9b), although primary sequence homology is not evident. PCNA has a function analogous to that of the β subunit, forming a circular clamp that enhances the processivity of the two polymerases.

Two other protein complexes also function in eukaryotic DNA replication. RPA (replication protein A) is a single-stranded DNA–binding protein, equivalent in function to the *E. coli* SSB protein. RFC (replication factor C) is a clamp loader for PCNA and facilitates the assembly of active replication complexes. The subunits of the RFC complex have significant sequence similarity to the subunits of the bacterial clamp-loading (γ) complex.

Termination of replication on linear eukaryotic chromosomes involves the synthesis of special structures called **telomeres** at the ends of each chromosome, as discussed in the next chapter.

Viral DNA Polymerases Provide Targets for Antiviral Therapy

Many DNA viruses encode their own DNA polymerases, and some of these have become targets for pharmaceuticals. For example, the DNA polymerase of the herpes simplex virus is inhibited by acyclovir, a compound developed by Gertrude Elion and George Hitchings (p. 901). Acyclovir consists of guanine attached to an incomplete ribose ring.

Acyclovir

It is phosphorylated by a virally encoded thymidine kinase; acyclovir binds to this viral enzyme with an affinity 200-fold greater than its binding to the cellular thymidine kinase. This ensures that phosphorylation occurs mainly in virus-infected cells. Cellular kinases convert the resulting acyclo-GMP to acyclo-GTP, which is both an inhibitor and a substrate of DNA polymerases; acyclo-GTP competitively inhibits the herpes DNA polymerase more strongly than cellular DNA polymerases. Because it lacks a 3′ hydroxyl, acyclo-GTP also acts as a chain terminator when incorporated into DNA. Thus viral replication is inhibited at several steps. ■

SUMMARY 25.1 DNA Replication

■ Replication of DNA occurs with very high fidelity and at a designated time in the cell cycle. Replication is semiconservative, each strand acting as template for a new daughter strand. It is carried out in three identifiable phases: initiation, elongation, and termination. The process starts at a single origin in bacteria and usually proceeds bidirectionally.

■ DNA is synthesized in the 5′→3′ direction by DNA polymerases. At the replication fork, the leading strand is synthesized continuously in the same direction as replication fork movement; the lagging strand is synthesized discontinuously as Okazaki fragments, which are subsequently ligated.

■ The fidelity of DNA replication is maintained by (1) base selection by the polymerase, (2) a 3′→5′ proofreading exonuclease activity that is part of most DNA polymerases, and (3) specific repair systems for mismatches left behind after replication.

■ Most cells have several DNA polymerases. In *E. coli*, DNA polymerase III is the primary replication enzyme.

DNA polymerase I is responsible for special functions during replication, recombination, and repair.

■ The separate initiation, elongation, and termination phases of DNA replication involve an array of enzymes and protein factors, many belonging to the AAA+ ATPase family.

■ The major replicative DNA polymerases in eukaryotes are DNA polymerases ε and δ. DNA polymerase α functions to synthesize primers.

25.2 DNA Repair

Most cells have only one or two sets of genomic DNA. Damaged proteins and RNA molecules can be quickly replaced by using information encoded in the DNA, but DNA molecules themselves are irreplaceable. Maintaining the integrity of the information in DNA is a cellular imperative, supported by an elaborate set of DNA repair systems. DNA can become damaged by a variety of processes, some spontaneous, others catalyzed by environmental agents (Chapter 8). Replication itself can very occasionally damage the information content in DNA when errors introduce mismatched base pairs (such as G paired with T).

The chemistry of DNA damage is diverse and complex. The cellular response to this damage includes a wide range of enzymatic systems that catalyze some of the most interesting chemical transformations in DNA metabolism. We first examine the effects of alterations in DNA sequence and then consider specific repair systems.

Mutations Are Linked to Cancer

The best way to illustrate the importance of DNA repair is to consider the effects of *unrepaired* DNA damage (a lesion). The most serious outcome is a change in the base sequence of the DNA, which, if replicated and transmitted to future generations of cells, becomes permanent. A permanent change in the nucleotide sequence of DNA is called a **mutation**. Mutations can involve the replacement of one base pair with another (substitution mutation) or the addition or deletion of one or more base pairs (insertion or deletion mutations). If the mutation affects nonessential DNA or if it has a negligible effect on the function of a gene, it is known as a **silent mutation**. Rarely, a mutation confers some biological advantage. Most nonsilent mutations, however, are neutral or deleterious.

In mammals there is a strong correlation between the accumulation of mutations and cancer. A simple test developed by Bruce Ames measures the potential of a given chemical compound to promote certain easily detected mutations in a specialized bacterial strain (**Fig. 25-20**). Few of the chemicals that we encounter in daily life score as mutagens in this test. However, of the compounds known to be carcinogenic from extensive animal

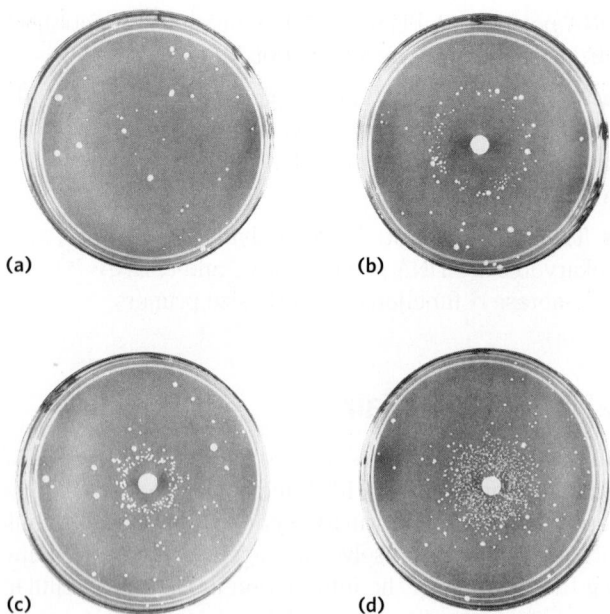

(a) **(b)**

(c) **(d)**

FIGURE 25-20 Ames test for carcinogens, based on their mutagenicity. A strain of *Salmonella typhimurium* having a mutation that inactivates an enzyme of the histidine biosynthetic pathway is plated on a histidine-free medium. Few cells grow. **(a)** The few small colonies of *S. typhimurium* that do grow on a histidine-free medium carry spontaneous back-mutations that permit the histidine biosynthetic pathway to operate. Three identical nutrient plates **(b)**, **(c)**, and **(d)** have been inoculated with an equal number of cells. Each plate then receives a disk of filter paper containing progressively lower concentrations of a mutagen. The mutagen greatly increases the rate of back-mutation and hence the number of colonies. The clear areas around the filter paper indicate where the concentration of mutagen is so high that it is lethal to the cells. As the mutagen diffuses away from the filter paper, it is diluted to sublethal concentrations that promote back-mutation. Mutagens can be compared on the basis of their effect on mutation rate. Because many compounds undergo a variety of chemical transformations after entering cells, compounds are sometimes tested for mutagenicity after first incubating them with a liver extract. Some substances have been found to be mutagenic only after this treatment. [Source: Bruce N. Ames, University of California, Berkeley, Department of Biochemistry and Molecular Biology.]

TABLE 25-5	Types of DNA Repair Systems in *E. coli*
Enzymes/proteins	**Type of damage**
Mismatch repair	
Dam methylase MutH, MutL, MutS proteins DNA helicase II SSB DNA polymerase III Exonuclease I Exonuclease VII RecJ nuclease Exonuclease X DNA ligase	Mismatches
Base-excision repair	
DNA glycosylases AP endonucleases DNA polymerase I DNA ligase	Abnormal bases (uracil, hypoxanthine, xanthine); alkylated bases; in some other organisms, pyrimidine dimers
Nucleotide-excision repair	
ABC excinuclease DNA polymerase I DNA ligase	DNA lesions that cause large structural change (e.g., pyrimidine dimers)
Direct repair	
DNA photolyases	Pyrimidine dimers
O^6-Methylguanine-DNA methyltransferase	O^6-Methylguanine
AlkB protein	1-Methylguanine, 3-methylcytosine

trials, more than 90% are also found to be mutagenic in the Ames test. Because of this strong correlation between mutagenesis and carcinogenesis, the Ames test for bacterial mutagens is widely used as a rapid and inexpensive screen for potential human carcinogens.

The genomic DNA in a typical mammalian cell accumulates many thousands of lesions during a 24-hour period. However, as a result of DNA repair, fewer than 1 in 1,000 become a mutation. DNA is a relatively stable molecule, but in the absence of repair systems, the cumulative effect of many infrequent but damaging reactions would make life impossible. ∎

All Cells Have Multiple DNA Repair Systems

The number and diversity of repair systems reflect both the importance of DNA repair to cell survival and the diverse sources of DNA damage (Table 25-5). Some common types of lesions, such as pyrimidine dimers (see Fig. 8-30), can be repaired by several distinct systems. Many DNA repair processes also seem to be extraordinarily inefficient energetically—an exception to the pattern observed in the vast majority of metabolic pathways, where every ATP is generally accounted for and used optimally. When the integrity of the genetic information is at stake, the amount of chemical energy invested in a repair process seems almost irrelevant.

Accurate DNA repair is possible largely because the DNA molecule consists of two complementary strands. Damaged DNA in one strand can be removed and replaced, without introducing mutations, by using the undamaged complementary strand as a template. We consider here the principal types of repair systems, beginning with those that repair the rare nucleotide mismatches that are left behind by replication.

Mismatch Repair Correction of the rare mismatches left after replication in *E. coli* improves the overall fidelity of replication by an additional factor of 10^2 to 10^3. The mismatches are nearly always corrected to reflect the information in the old (template) strand, which the repair system can distinguish from the newly synthesized strand by the presence of methyl group tags on the template DNA. The mismatch repair system of *E. coli* includes at least 12 protein components (Table 25-5) that function either in strand discrimination or in the repair process itself. The functions of many of these were first worked out by Paul Modrich and colleagues in the 1980s.

The strand discrimination mechanism has not been determined for most bacteria or eukaryotes, but it is well understood for *E. coli* and some closely related bacterial species. In these bacteria, strand discrimination is based on the action of Dam methylase, which, as you will recall, methylates DNA at the N^6 position of all adenines within (5′)GATC sequences. Immediately after passage of the replication fork, there is a short period (a few seconds or minutes) during which the template strand is methylated but the newly synthesized strand is not **(Fig. 25-21)**. The transient unmethylated state of GATC sequences in the newly synthesized strand permits the new strand to be distinguished from the template strand. Replication mismatches in the vicinity of a hemimethylated GATC sequence are then repaired according to the information in the methylated parent (template) strand. If both strands are methylated at a GATC sequence, few mismatches are repaired; if neither strand is methylated, repair occurs but does not favor either strand. The methyl-directed mismatch repair system of *E. coli* efficiently repairs mismatches up to 1,000 bp from a hemimethylated GATC sequence.

How is the mismatch correction process directed by relatively distant GATC sequences? A mechanism is illustrated in **Figure 25-22**. MutL protein forms a complex with MutS protein, and the complex binds to all mismatched base pairs (except C–C). MutH protein binds to MutL and to GATC sequences encountered by the MutL-MutS complex. DNA on both sides of the mismatch is threaded through the MutL-MutS complex, creating a DNA loop; simultaneous movement of both legs of the loop through the complex is equivalent to the complex moving in both directions at once along the DNA. MutH has a site-specific endonuclease activity that is inactive until the complex encounters a

hemimethylated GATC sequence. At this site, MutH catalyzes cleavage of the unmethylated strand on the 5′ side of the G in GATC, which marks the strand for repair. Further steps in the pathway depend on where

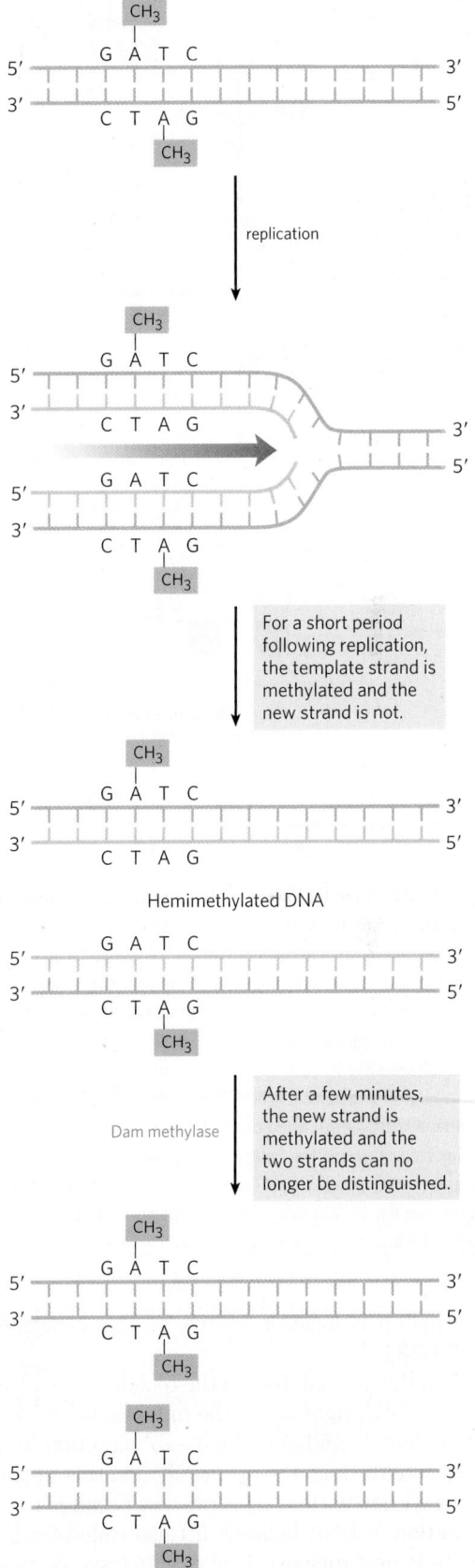

FIGURE 25-21 Methylation and mismatch repair. Methylation of DNA strands can serve to distinguish parent (template) strands from newly synthesized strands in *E. coli* DNA, a function that is critical to mismatch repair (see Fig. 25-22). The methylation occurs at the N^6 of adenines in (5′)GATC sequences. This sequence is a palindrome (see Fig. 8-18), present in opposite orientations on the two strands.

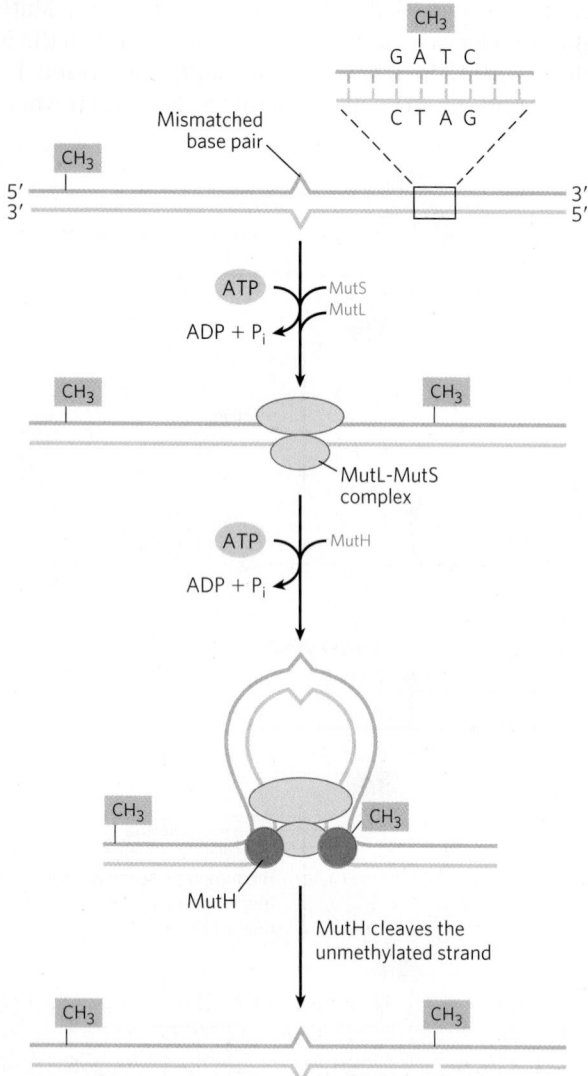

FIGURE 25-22 A model for the early steps of methyl-directed mismatch repair. The proteins involved in this process in *E. coli* have been purified (see Table 25-5). Recognition of the sequence (5′)GATC and of the mismatch are specialized functions of the MutH and MutS proteins, respectively. The MutL protein forms a complex with MutS at the mismatch. DNA is threaded through this complex such that the complex moves simultaneously in both directions along the DNA until it encounters a MutH protein bound at a hemimethylated GATC sequence. MutH cleaves the unmethylated strand on the 5′ side of the G in this sequence. A complex consisting of DNA helicase II and one of several exonucleases then degrades the unmethylated DNA strand from that point toward the mismatch (see Fig. 25-23). [Source: Information from a figure provided by Paul Modrich.]

the mismatch is located relative to this cleavage site **(Fig. 25-23)**.

When the mismatch is on the 5′ side of the cleavage site (Fig. 25-23, right side), the unmethylated strand is unwound and degraded in the 3′→5′ direction from the cleavage site through the mismatch, and this segment is replaced with new DNA. This process requires the combined action of DNA helicase II (also called UvrD helicase), SSB, exonuclease I or exonuclease X (both of

which degrade strands of DNA in the 3′→5′ direction) or exonuclease VII (which degrades single-stranded DNA in either direction), DNA polymerase III, and DNA ligase. The pathway for repair of mismatches on the 3′ side of the cleavage site is similar (Fig. 25-23, left), except that the exonuclease is either exonuclease VII or RecJ nuclease (which degrades single-stranded DNA in the 5′→3′ direction).

Mismatch repair is particularly costly for *E. coli* in terms of energy expended. The mismatch may occur 1,000 or more base pairs from the GATC sequence. The degradation and replacement of a strand segment of this length require an enormous investment in activated deoxynucleotide precursors to repair a *single* mismatched base. This again underscores the importance to the cell of genomic integrity.

Eukaryotic cells also have mismatch repair systems, with several proteins structurally and functionally analogous to the bacterial MutS and MutL (but not MutH) proteins. Alterations in human genes encoding proteins of this type produce some of the most common inherited cancer-susceptibility syndromes (see Box 25-1, p. 1015), further demonstrating the value to the organism of DNA repair systems. The main MutS homologs in most eukaryotes, from yeast to humans, are MSH2 (*MutS h*omolog), MSH3, and MSH6. Heterodimers of MSH2 and MSH6 generally bind to single base-pair mismatches, and bind less well to slightly longer mispaired loops. In many organisms, the longer mismatches (2 to 6 bp) may be bound instead by a heterodimer of MSH2 and MSH3, or are bound by both types of heterodimers in tandem. Homologs of MutL, predominantly a heterodimer of MLH1 (*MutL h*omolog) and PMS1 (*p*ost-*m*eiotic *s*egregation), bind to and stabilize the MSH complexes. Many details of the subsequent events in eukaryotic mismatch repair remain to be worked out. In particular, we do not know how newly synthesized DNA strands are identified, although research reveals that this process does not involve GATC sequences.

Base-Excision Repair Every cell has a class of enzymes called **DNA glycosylases** that recognize particularly common DNA lesions (such as the products of cytosine and adenine deamination; see Fig. 8-29a) and remove the affected base by cleaving the *N*-glycosyl bond. This cleavage creates an apurinic or apyrimidinic site in the DNA, commonly referred to as an **AP site** or **abasic site**. Each DNA glycosylase is generally specific for one type of lesion.

Uracil DNA glycosylases, for example, found in most cells, specifically remove from DNA the uracil that results from spontaneous deamination of cytosine. Mutant cells that lack this enzyme have a high rate of G≡C to A=T mutations. This glycosylase does not remove uracil residues from RNA or thymine residues from DNA. The capacity to distinguish thymine from uracil, the product of cytosine deamination—necessary for the selective repair of the latter—may be one reason why DNA evolved to contain thymine instead of uracil (p. 297).

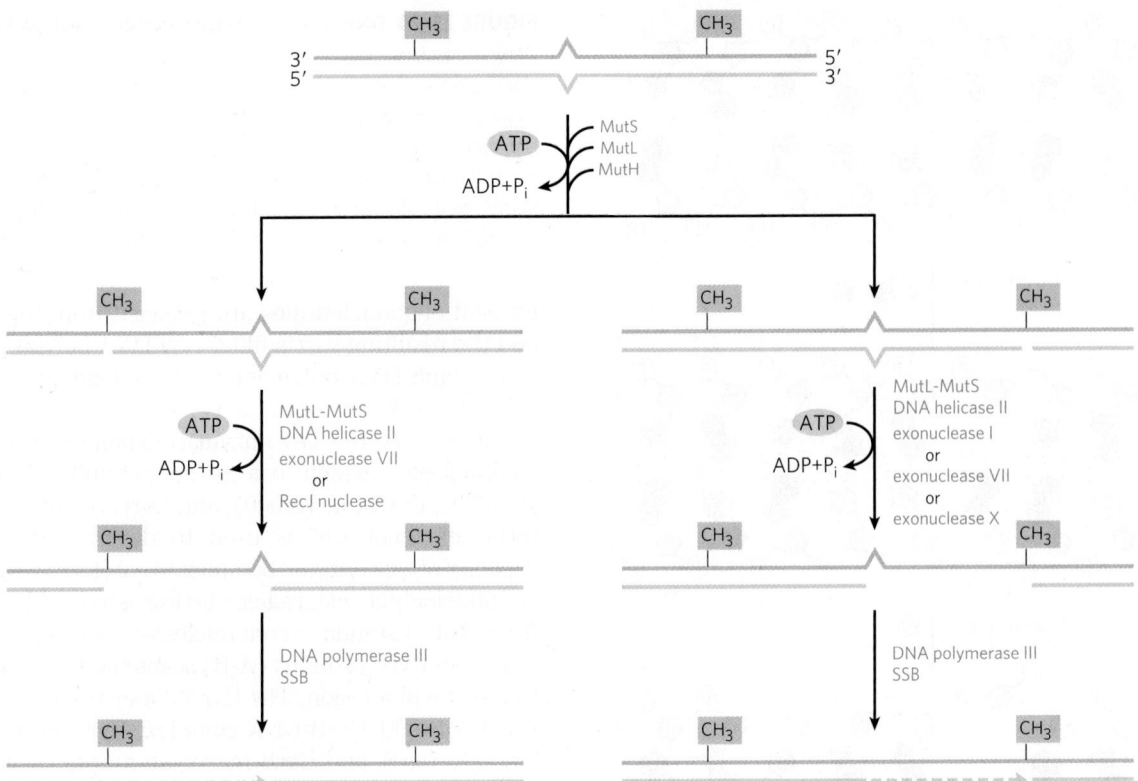

FIGURE 25-23 Completion of methyl-directed mismatch repair. The combined action of DNA helicase II, SSB, and one of four different exonucleases removes a segment of the new strand between the MutH cleavage site and a point just beyond the mismatch. The particular exonuclease depends on the location of the cleavage site relative to the mismatch, as shown by the alternative pathways here. The resulting gap is filled in (dashed line) by DNA polymerase III, and the nick is sealed by DNA ligase (not shown).

Most bacteria have just one type of uracil DNA glycosylase, whereas humans have at least four types, with different specificities—an indicator of the importance of removing uracil from DNA. The most abundant human uracil glycosylase, UNG, is associated with the replisome, where it eliminates the occasional U residue inserted in place of a T during replication. The deamination of C residues is 100-fold faster in single-stranded DNA than in double-stranded DNA, and humans have an enzyme, hSMUG1, that removes any U residues occurring in single-stranded DNA during replication or transcription. Two other human DNA glycosylases, TDG and MBD4, remove either U or T residues paired with G, which are generated by deamination of cytosine or 5-methylcytosine, respectively.

Other DNA glycosylases recognize and remove a variety of damaged bases, including formamidopyrimidine and 8-hydroxyguanine (both arising from purine oxidation), hypoxanthine (from adenine deamination), and alkylated bases such as 3-methyladenine and 7-methylguanine. Glycosylases that recognize other lesions, including pyrimidine dimers, have also been identified in some classes of organisms. Remember that AP sites also arise from slow, spontaneous hydrolysis of the N-glycosyl bonds in DNA (see Fig. 8-29b).

Once an AP site has been formed by a DNA glycosylase, another type of enzyme must repair it. The repair is *not* made by simply inserting a new base and re-forming the N-glycosyl bond. Instead, the deoxyribose 5′-phosphate left behind is removed and replaced with a new nucleotide. This process begins with one of the **AP endonucleases**, enzymes that cut the DNA strand containing the AP site. The position of the incision relative to the AP site (5′ or 3′ to the site) depends on the type of AP endonuclease. A segment of DNA including the AP site is then removed, DNA polymerase I replaces the DNA, and DNA ligase seals the remaining nick **(Fig. 25-24)**. In eukaryotes, nucleotide replacement is carried out by specialized polymerases, as described below.

Nucleotide-Excision Repair DNA lesions that cause large distortions in the helical structure of DNA generally are repaired by the nucleotide-excision system, a repair pathway critical to the survival of all free-living organisms. In nucleotide-excision repair **(Fig. 25-25)**, a multisubunit enzyme (excinuclease) hydrolyzes two phosphodiester bonds, one on either side of the distortion caused by the lesion. In *E. coli* and other bacteria, the enzyme system hydrolyzes the fifth phosphodiester bond on the 3′ side and the eighth phosphodiester bond on the 5′ side to generate a fragment of 12 to 13 nucleotides (depending on whether the lesion involves one or two bases). In humans and other eukaryotes,

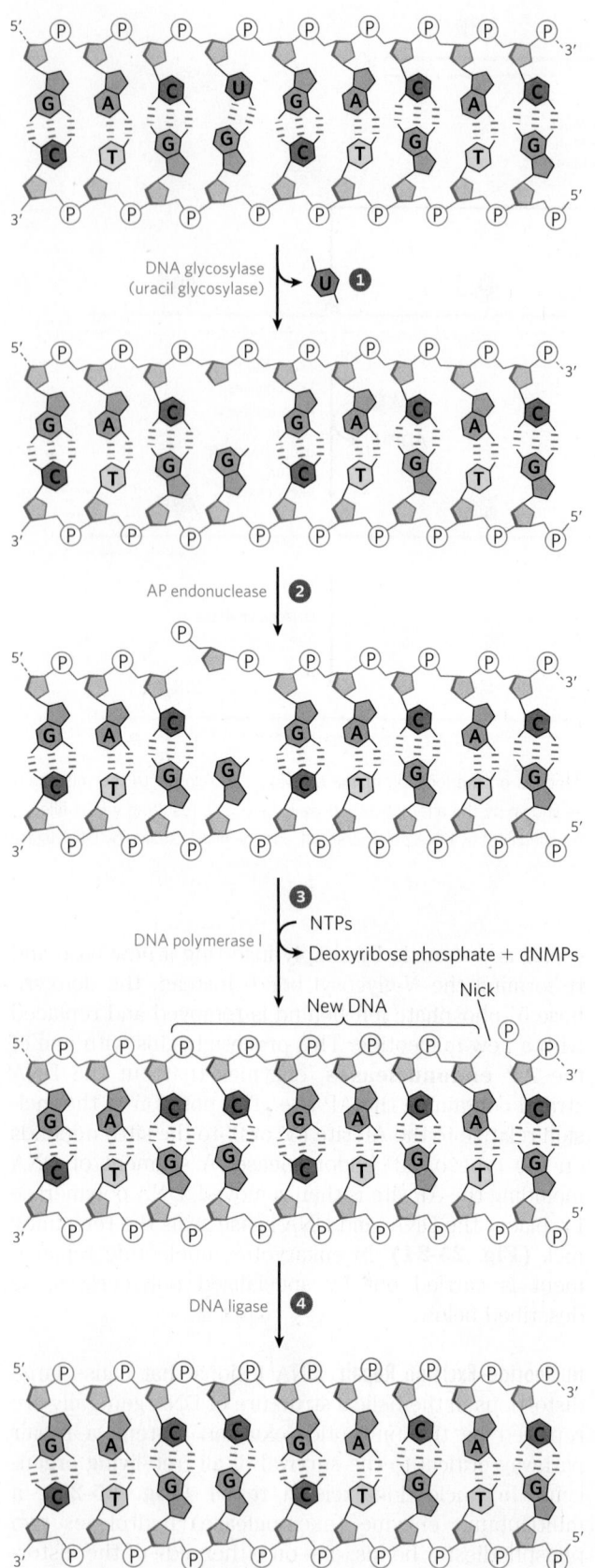

FIGURE 25-24 DNA repair by the base-excision repair pathway. ❶ A DNA glycosylase recognizes a damaged base (in this case, a uracil) and cleaves between the base and deoxyribose in the backbone. ❷ An AP endonuclease cleaves the phosphodiester backbone near the AP site. ❸ DNA polymerase I initiates repair synthesis from the free 3′ hydroxyl at the nick, removing (with its 5′→3′ exonuclease activity) and replacing a portion of the damaged strand. ❹ The nick remaining after DNA polymerase I has dissociated is sealed by DNA ligase.

excised oligonucleotides are released from the duplex and the resulting gap is filled—by DNA polymerase I in *E. coli* and DNA polymerase ε in humans. DNA ligase seals the nick.

In *E. coli*, the key enzymatic complex is the ABC excinuclease, which has three subunits, UvrA (M_r 104,000), UvrB (M_r 78,000), and UvrC (M_r 68,000). The term "excinuclease" is used to describe the unique capacity of this enzyme complex to catalyze two specific endonucleolytic cleavages, distinguishing this activity from that of standard endonucleases. A complex of the UvrA and UvrB proteins (A_2B) scans the DNA and binds to the site of a lesion. The UvrA dimer then dissociates, leaving a tight UvrB-DNA complex. UvrC protein then binds to UvrB, and UvrB makes an incision at the fifth phosphodiester bond on the 3′ side of the lesion. This is followed by a UvrC-mediated incision at the eighth phosphodiester bond on the 5′ side. The resulting 12 to 13 nucleotide fragment is removed by UvrD helicase. The short gap thus created is filled in by DNA polymerase I and DNA ligase. This pathway (Fig. 25-25, left) is a primary repair route for many types of lesions, including cyclobutane pyrimidine dimers, 6-4 photoproducts (see Fig. 8-30), and several other types of base adducts, including benzo[a]pyrene-guanine, which is formed in DNA by exposure to cigarette smoke. The nucleolytic activity of the ABC excinuclease is novel in the sense that two cuts are made in the DNA.

The mechanism of eukaryotic excinucleases is quite similar to that of the bacterial enzyme, although at least 16 polypeptides with no similarity to the *E. coli* excinuclease subunits are required for the dual incision. Some of the nucleotide-excision repair and base-excision repair in eukaryotes is closely tied to transcription (see Chapter 26). Genetic deficiencies in nucleotide-excision repair in humans give rise to a variety of serious diseases (see Box 25-1).

Direct Repair Several types of damage are repaired without removing a base or nucleotide. The best-characterized example is direct photoreactivation of cyclobutane pyrimidine dimers, a reaction promoted by **DNA photolyases**. Pyrimidine dimers result from a UV-induced reaction. Through a mechanism worked out by Aziz Sancar and colleagues, photolyases use energy derived from absorbed light to reverse the damage **(Fig. 25-26)**. Photolyases generally contain two cofactors that serve as light-absorbing agents,

the enzyme system hydrolyzes the sixth phosphodiester bond on the 3′ side and the twenty-second phosphodiester bond on the 5′ side, producing a fragment of 27 to 29 nucleotides. Following the dual incision, the

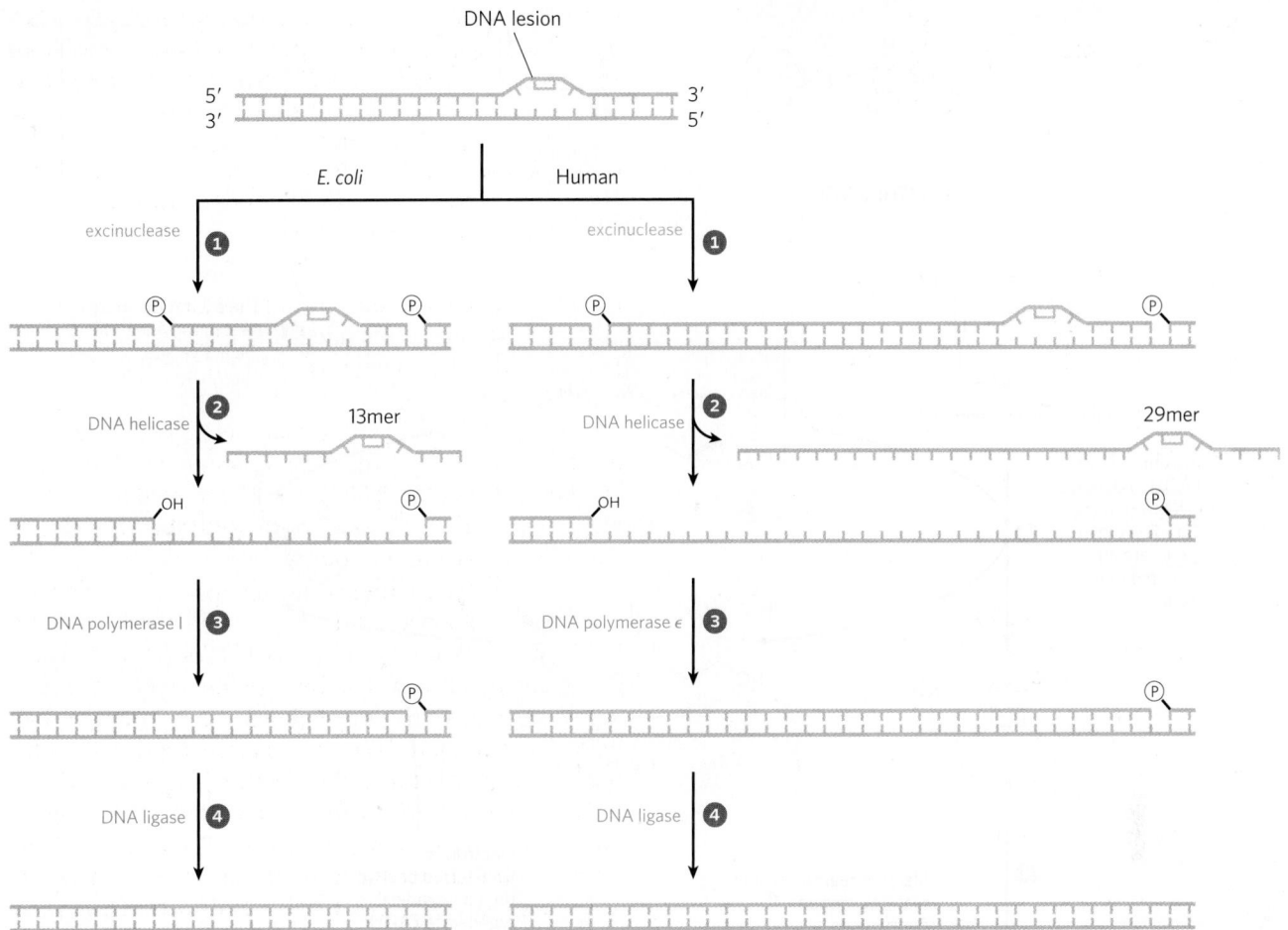

FIGURE 25-25 Nucleotide-excision repair in *E. coli* and humans. The general pathway of nucleotide-excision repair is similar in all organisms. ❶ An excinuclease binds to DNA at the site of a bulky lesion and cleaves the damaged DNA strand on either side of the lesion. ❷ The DNA segment—of 13 nucleotides (13mer) or 29 nucleotides (29mer)—is removed with the aid of a helicase. ❸ The gap is filled in by DNA polymerase, and ❹ the remaining nick is sealed with DNA ligase. [Source: Information from a figure provided by Aziz Sancar.]

or chromophores: in all organisms, one is FADH$_2$; in *E. coli* and yeast, the other is a folate. The reaction mechanism entails the generation of free radicals. DNA photolyases are not found in humans and other placental mammals.

Additional examples are seen in the repair of nucleotides with alkylation damage. The modified nucleotide O^6-methylguanine forms in the presence of alkylating agents and is a common and highly mutagenic lesion (p. 299). It tends to pair with thymine rather than cytosine during replication, and therefore causes G≡C to A=T mutations **(Fig. 25-27)**. Direct repair of O^6-methylguanine is carried out by O^6-methylguanine-DNA methyltransferase, a protein that catalyzes transfer of the methyl group of O^6-methylguanine to one of its own Cys residues. This methyltransferase is not strictly an enzyme, because a single methyl transfer event permanently methylates the protein, inactivating it in this pathway. The consumption of an entire protein molecule to correct a single damaged base is another vivid

illustration of the priority given to maintaining the integrity of cellular DNA.

O^6-Methylguanine nucleotide   Guanine nucleotide

A very different but equally direct mechanism is used to repair 1-methyladenine and 3-methylcytosine. The amino groups of A and C residues are sometimes methylated when the DNA is single-stranded, and the

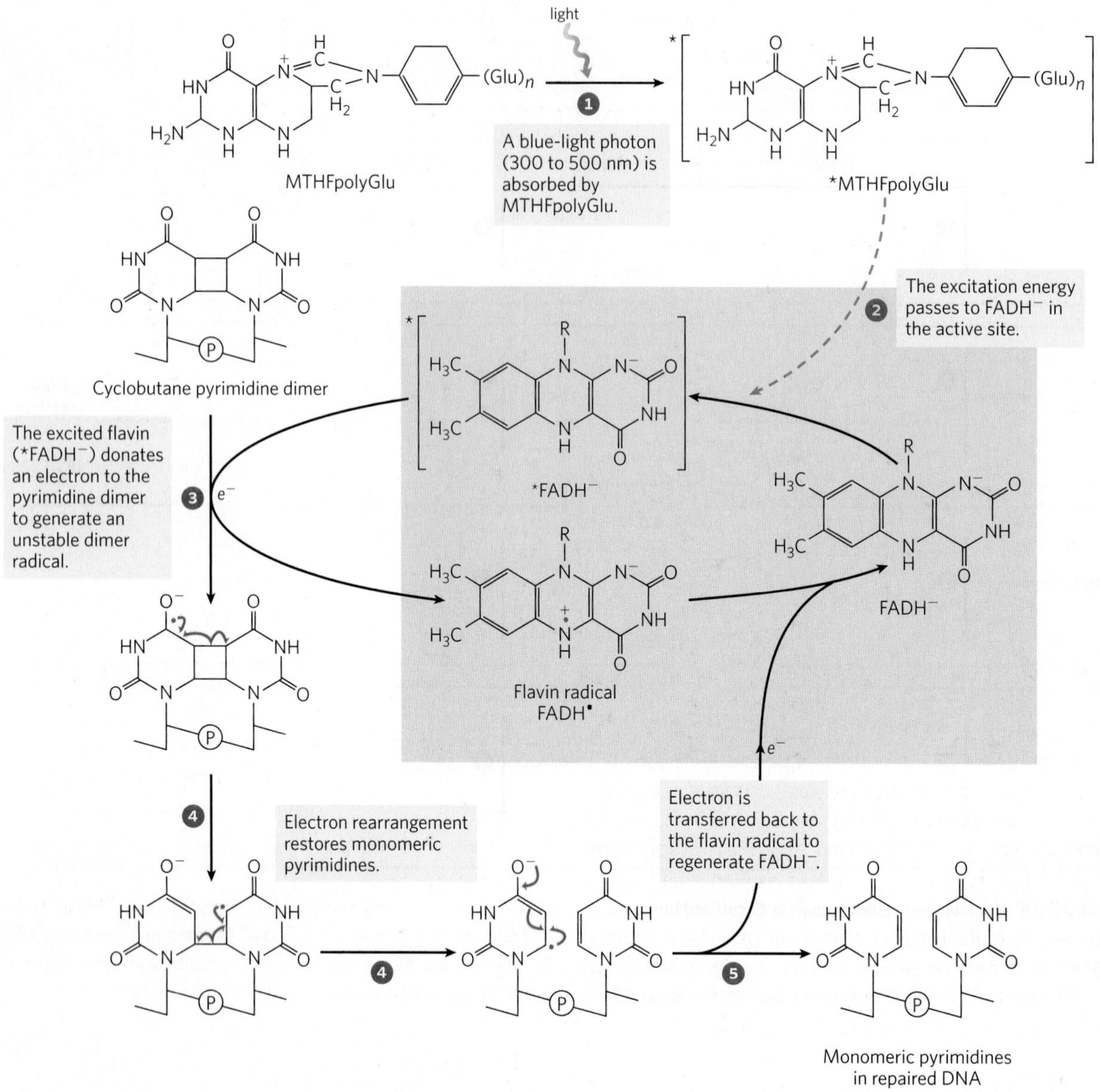

MECHANISM FIGURE 25-26 Repair of pyrimidine dimers with photolyase. Energy derived from absorbed light is used to reverse the photoreaction that caused the lesion. The two chromophores in *E. coli* photolyase (M_r 54,000), N^5,N^{10}-methenyltetrahydrofolylpolyglutamate (MTHFpolyGlu) and FADH⁻, perform complementary functions. MTHFpolyGlu functions as a photoantenna to absorb blue-light photons. The excitation energy passes to FADH⁻, and the excited flavin (*FADH⁻) donates an electron to the pyrimidine dimer, leading to the rearrangement as shown.

methylation directly affects proper base pairing. In *E. coli*, oxidative demethylation of these alkylated nucleotides is mediated by the AlkB protein, a member of the α-ketoglutarate-Fe^{2+}–dependent dioxygenase superfamily **(Fig. 25-28)**. (See Box 4-3 for a description of another member of this enzyme family.)

The Interaction of Replication Forks with DNA Damage Can Lead to Error-Prone Translesion DNA Synthesis

The repair pathways considered to this point generally work only for lesions in double-stranded DNA, the undamaged strand providing the correct genetic information to restore the damaged strand to its original state. However, in certain types of lesions, such as double-strand breaks, double-strand cross-links, or lesions in a single-stranded DNA, the complementary strand is itself damaged or is absent. Double-strand breaks and lesions in single-stranded DNA most often arise when a replication fork encounters an unrepaired DNA lesion **(Fig. 25-29)**. Such lesions and DNA cross-links can also result from ionizing radiation and oxidative reactions.

At a stalled bacterial replication fork, there are two avenues for repair. In the absence of a second strand, the information required for accurate repair must come from a separate, homologous chromosome. The repair system thus involves homologous genetic recombination. This

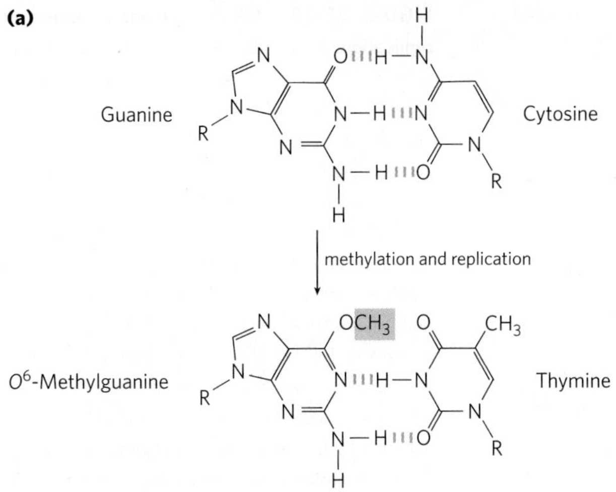

FIGURE 25-27 Example of how DNA damage results in mutations.
(a) The methylation product O^6-methylguanine pairs with thymine rather than cytosine residues. (b) If not repaired, this leads to a G≡C to A=T mutation after replication.

recombinational DNA repair is considered in detail in Section 25.3. Under some conditions, a second repair pathway, **error-prone translesion DNA synthesis** (often abbreviated TLS), becomes available. When this pathway is active, DNA repair is significantly less accurate and a high mutation rate can result. In bacteria, error-prone translesion DNA synthesis is part of a cellular stress response to extensive DNA damage known, appropriately enough, as the **SOS response**. Some SOS proteins, such as the UvrA and UvrB proteins already described, are normally present in the cell but are induced to higher levels as part of the SOS response (Table 25-6). Additional SOS proteins participate in the pathway for error-prone repair; these include the UmuC

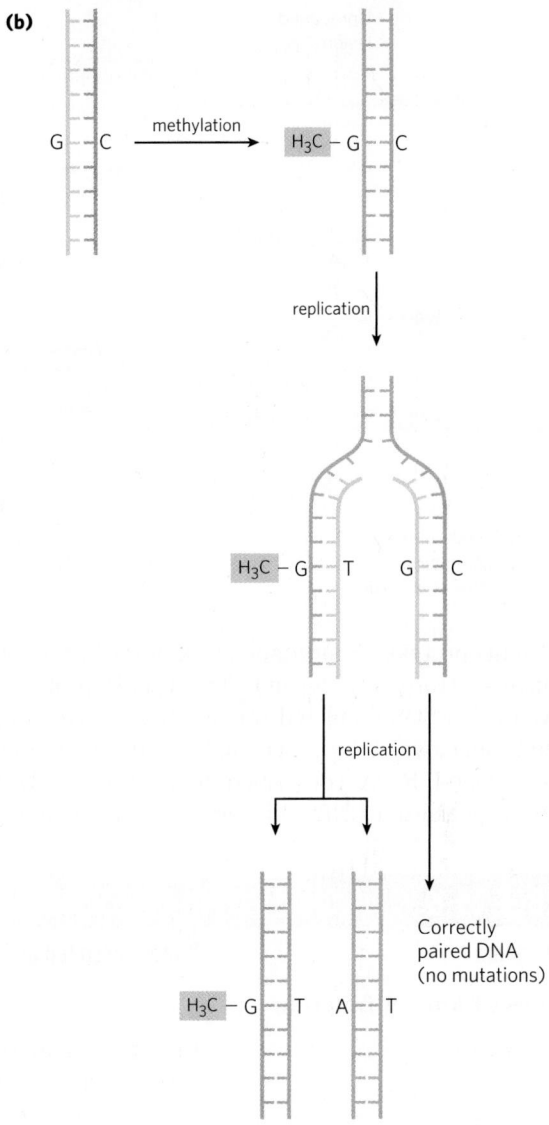

FIGURE 25-28 Direct repair of alkylated bases by AlkB. The AlkB protein is an α-ketoglutarate-Fe^{2+}–dependent hydroxylase (see Box 4-3). It catalyzes the oxidative demethylation of 1-methyladenine and 3-methylcytosine residues.

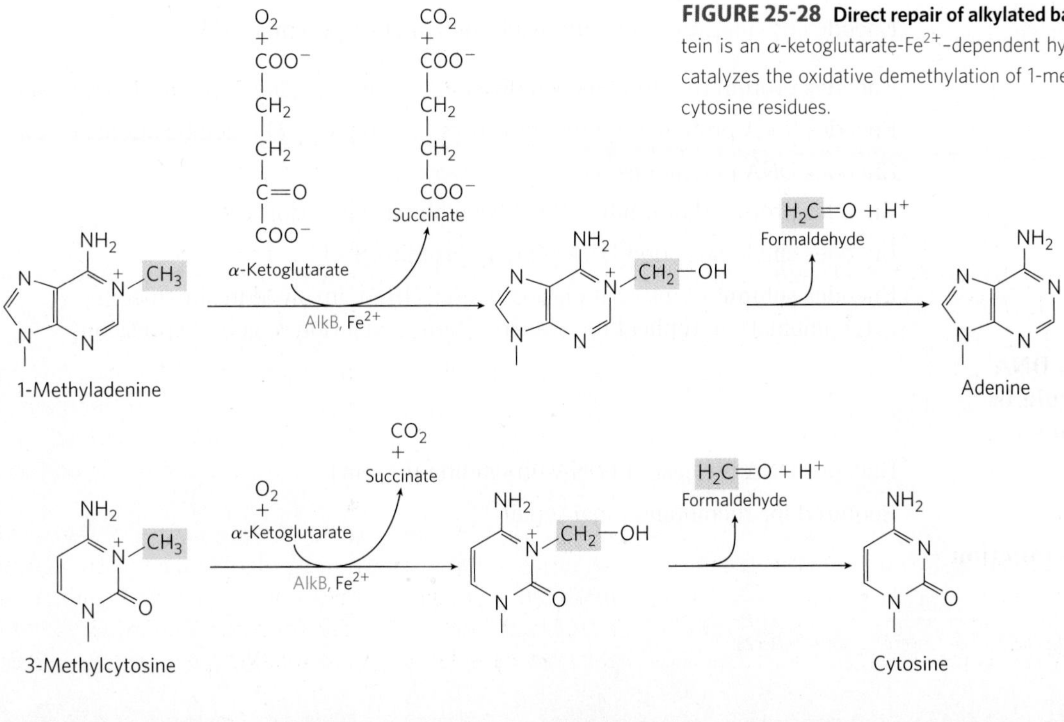

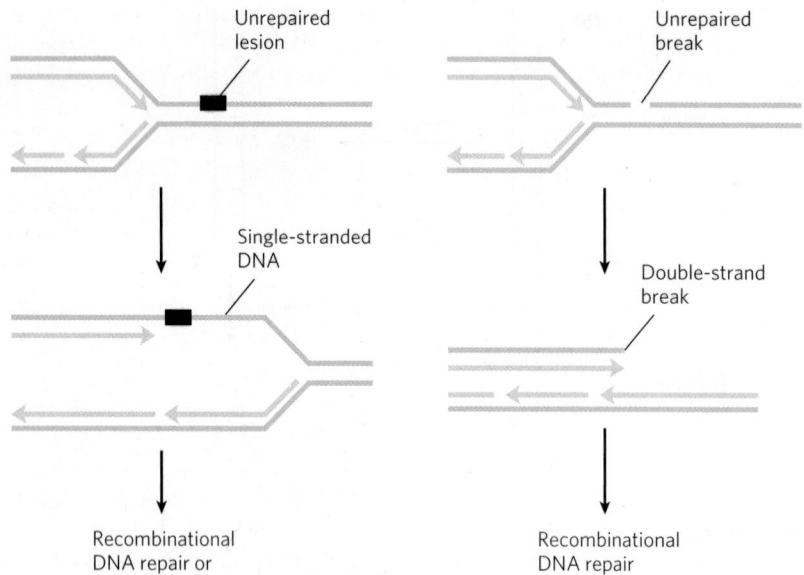

Single-stranded DNA

Unrepaired lesion

Recombinational DNA repair or error-prone repair

Unrepaired break

Double-strand break

Recombinational DNA repair

FIGURE 25-29 DNA damage and its effect on DNA replication. If the replication fork encounters an unrepaired lesion or strand break, replication generally halts and the fork may collapse. A lesion is left behind in an unreplicated, single-stranded segment of the DNA (left); a strand break becomes a double-strand break (right). In each case, the damage to one strand cannot be repaired by mechanisms described earlier in this chapter, because the complementary strand required to direct accurate repair is damaged or absent. There are two possible avenues for repair: recombinational DNA repair (one pathway is described in Fig. 25-30) or, when lesions are unusually numerous, error-prone repair. The latter mechanism involves a novel DNA polymerase (DNA polymerase V, encoded by the *umuC* and *umuD* genes and activated by the RecA protein) that can replicate, albeit inaccurately, over many types of lesions. The repair mechanism is "error-prone" because mutations often result.

and UmuD proteins (*unmutable*; lack of the *umu* gene eliminates error-prone repair). The UmuD protein is cleaved in an SOS-regulated process to a shorter form called UmuD′, which forms a complex with UmuC and a protein called RecA (described in Section 25.3) to create a specialized DNA polymerase, DNA polymerase V (UmuD′$_2$UmuCRecA), which can replicate past many of the DNA lesions that would normally block replication. Proper base pairing is often impossible at the site of such a lesion, so this translesion replication is error-prone.

Given the emphasis on the importance of genomic integrity throughout this chapter, the existence of a

TABLE 25-6 Genes Induced as Part of the SOS Response in *E. coli*

Gene name	Protein encoded and/or role in DNA repair
Genes of known function	
polB (dinA)	Encodes polymerization subunit of DNA polymerase II, required for replication restart in recombinational DNA repair
uvrA *uvrB*	Encode ABC excinuclease subunits UvrA and UvrB
umuC *umuD*	Encode core and polymerization subunits of DNA polymerase V
sulA	Encodes protein that inhibits cell division, possibly to allow time for DNA repair
recA	Encodes RecA protein, required for error-prone repair and recombinational repair
dinB	Encodes DNA polymerase IV
dinD	Encodes protein that inhibits RecA recombinase function
ssb	Encodes single-stranded DNA-binding protein (SSB)
himA	Encodes subunit of integration host factor (IHF), involved in site-specific recombination, replication, transposition, regulation of gene expression
Genes involved in DNA metabolism, but role in DNA repair unknown	
uvrD	Encodes DNA helicase II (DNA-unwinding protein)
recN	Required for recombinational repair
Gene of unknown function	
dinF	

Note: Some of these genes and their functions are further discussed in Chapter 28.

system that increases the rate of mutation may seem incongruous. However, we can think of this system as a desperation strategy. The *umuC* and *umuD* genes are fully induced only late in the SOS response, and they are not activated for translesion synthesis initiated by UmuD cleavage unless the levels of DNA damage are particularly high and all replication forks are blocked. The mutations resulting from DNA polymerase V–mediated replication kill some cells and create deleterious mutations in others, but this is the biological price a species pays to overcome an otherwise insurmountable barrier to replication, as it permits at least a few mutant daughter cells to survive.

Yet another DNA polymerase, DNA polymerase IV, is also induced during the SOS response. Replication by DNA polymerase IV, a product of the *dinB* gene, is also highly error-prone. The bacterial DNA polymerases IV and V (Table 25-1) are part of a family of TLS polymerases found in all organisms. These enzymes lack a proofreading exonuclease and have a more open active site than other DNA polymerases, one that accommodates damaged template nucleotides. With these enzymes, the fidelity of base selection during replication may be reduced by a factor of 10^2, lowering overall replication fidelity to one error in ~1,000 nucleotides.

Mammals have many low-fidelity DNA polymerases of the TLS polymerase family. However, the presence of these enzymes does not necessarily translate into an unacceptable mutational burden, because most of the enzymes also have specialized functions in DNA repair. DNA polymerase η (eta), for example, found in all eukaryotes, promotes translesion synthesis primarily across cyclobutane T–T dimers. Few mutations result, because the enzyme preferentially inserts two A residues across from the linked T residues. Several other low-fidelity polymerases, including DNA polymerases β, ι (iota), and λ, have specialized roles in eukaryotic base-excision repair. Each of these enzymes has a 5′-deoxyribose phosphate lyase activity in addition to its polymerase activity. After base removal by a glycosylase and backbone cleavage by an AP endonuclease, these polymerases remove the abasic site (a 5′-deoxyribose phosphate) and fill in the very short gap. The frequency of mutation due to DNA polymerase η activity is minimized by the short lengths (often one nucleotide) of DNA synthesized.

What emerges from research into cellular DNA repair systems is a picture of a DNA metabolism that maintains genomic integrity with multiple and often redundant systems. In the human genome, more than 130 genes encode proteins dedicated to the repair of DNA. In many cases, the loss of function of one of these proteins results in genomic instability and an increased occurrence of oncogenesis (Box 25-1). These repair systems are often integrated with the DNA replication systems and are complemented by recombination systems, which we turn to next.

BOX 25-1   ✚ MEDICINE   DNA Repair and Cancer

Human cancers develop when genes that regulate normal cell division (oncogenes and tumor suppressor genes; see Chapter 12) fail to function, are activated at the wrong time, or are altered. As a consequence, cells may grow out of control and form a tumor. The genes controlling cell division can be damaged by spontaneous mutation or overridden by the invasion of a tumor virus (Chapter 26). Not surprisingly, alterations in DNA repair genes that result in a higher rate of mutation can greatly increase an individual's susceptibility to cancer. Defects in the genes encoding the proteins involved in nucleotide-excision repair, mismatch repair, recombinational repair, and error-prone translesion DNA synthesis have all been linked to human cancers. Clearly, DNA repair can be a matter of life and death.

Nucleotide-excision repair requires a larger number of proteins in humans than in bacteria, although the overall pathways are very similar. Genetic defects that inactivate nucleotide-excision repair have been associated with several genetic diseases, the best-studied of which is xeroderma pigmentosum (XP). Because nucleotide-excision repair is the sole repair pathway for pyrimidine dimers in humans, people with XP are extremely sensitive to light and readily develop sunlight-induced skin cancers. Most people with XP also have neurological abnormalities, presumably because of their inability to repair certain lesions caused by the high rate of oxidative metabolism in neurons. Defects in the genes encoding any of at least seven different protein components of the nucleotide-excision repair system can result in XP, giving rise to seven different genetic groups, denoted XPA to XPG. Several of these proteins (notably XPB, XPD, and XPG) also play roles in transcription-coupled base-excision repair of oxidative lesions, described in Chapter 26. Note that XPC and XPE are parts of complexes that recognize damaged DNA, whereas XPA, XPB, XPD, XPF, and XPG are all components of a much larger multisubunit complex that represents the human excinuclease depicted in Figure 25-25. These proteins are involved in making the DNA incisions and removing the 29mer segment of DNA.

Most microorganisms have redundant pathways for the repair of cyclobutane pyrimidine dimers—making use of DNA photolyases and sometimes base-excision repair as alternatives to nucleotide-excision repair—but humans and other placental mammals do not. This lack of a backup for nucleotide-excision repair for removing pyrimidine dimers has led to speculation that

(Continued on next page)

BOX 25-1 ⚕ MEDICINE DNA Repair and Cancer (*Continued*)

early mammalian evolution involved small, furry, nocturnal animals with little need to repair UV damage. However, mammals do have a pathway for the translesion bypass of cyclobutane pyrimidine dimers, which involves DNA polymerase η. This enzyme preferentially inserts two A residues opposite a T–T pyrimidine dimer, minimizing mutations. People with a genetic condition in which DNA polymerase η function is missing exhibit an XP-like illness known as XP-variant, or XP-V. Clinical manifestations of XP-V are similar to those of the classic XP diseases, although mutation levels are higher in XP-V when cells are exposed to UV light. Apparently, the nucleotide-excision repair system works in concert with DNA polymerase η in normal human cells, repairing and/or bypassing pyrimidine dimers as needed to keep cell growth and DNA replication going. Exposure to UV light introduces a heavy load of pyrimidine dimers, and some must be bypassed by translesion synthesis to keep replication on track. When one system is missing, it is partly compensated for by the other. A loss of DNA polymerase η activity leads to stalled replication forks and bypass of UV lesions by different, more mutagenic, translesion synthesis (TLS) polymerases. As when other DNA repair systems are absent, the resulting increase in mutations often leads to cancer.

One of the most common inherited cancer-susceptibility syndromes is hereditary nonpolyposis colon cancer (HNPCC). This syndrome has been traced to defects in mismatch repair. Human and other eukaryotic cells have several proteins analogous to the bacterial MutL and MutS proteins (see Fig. 25-22). Defects in at least five different mismatch repair genes can give rise to HNPCC. The most prevalent are defects in the *hMLH1* (*h*uman *MutL homolog 1*) and *hMSH2* (*h*uman *MutS homolog 2*) genes. In individuals with HNPCC, cancer generally develops at an early age, with colon cancers being most common.

Most human breast cancer occurs in women with no known predisposition. However, about 10% of cases are associated with inherited defects in two genes, *BRCA1* and *BRCA2*. Human BRCA1 and BRCA2 are large proteins (1,834 and 3,418 amino acid residues, respectively) that interact with a wide range of other proteins involved in transcription, chromosome maintenance, DNA repair, and control of the cell cycle. BRCA2 has been implicated in the recombinational DNA repair of double-strand breaks. One of the key roles of BRCA2 is to load the human RecA homolog, called Rad51, onto DNA at the sites of double-strand breaks. BRCA1 has as yet imperfectly defined roles in the repair of double-strand breaks, transcription, and some other processes of DNA metabolism. Women with defects in either the *BRCA1* or *BRCA2* gene have a greater than 80% chance of developing breast cancer.

SUMMARY 25.2 DNA Repair

■ Cells have many systems for DNA repair. Mismatch repair in *E. coli* is directed by transient nonmethylation of (5′)GATC sequences on the newly synthesized strand.

■ Base-excision repair systems recognize and repair damage caused by environmental agents (such as radiation and alkylating agents) and spontaneous reactions of nucleotides. Some repair systems recognize and excise only damaged or incorrect bases, leaving an AP (abasic) site in the DNA. This is repaired by excision and replacement of the DNA segment containing the AP site.

■ Nucleotide-excision repair systems recognize and remove a variety of bulky lesions and pyrimidine dimers. They excise a segment of the DNA strand including the lesion, leaving a gap that is filled by DNA polymerase and ligase activities.

■ Some DNA damage is repaired by direct reversal of the reaction causing the damage: pyrimidine dimers are directly converted to monomeric pyrimidines by a photolyase, and the methyl group of O^6-methylguanine is removed by a methyltransferase.

■ In bacteria, error-prone translesion DNA synthesis, involving TLS DNA polymerases, occurs in response to

very heavy DNA damage. In eukaryotes, similar polymerases have specialized roles in DNA repair that minimize the introduction of mutations.

25.3 DNA Recombination

The rearrangement of genetic information within and among DNA molecules encompasses a variety of processes, collectively placed under the heading of genetic recombination. The practical applications of DNA rearrangements in altering the genomes of increasing numbers of organisms are now being explored (Chapter 9).

Genetic recombination events fall into at least three general classes. **Homologous genetic recombination** (also called general recombination) involves genetic exchanges between any two DNA molecules (or segments of the same molecule) that share an extended region of nearly identical sequence. The actual sequence of bases is irrelevant, as long as it is similar in the two DNAs. In **site-specific recombination**, the exchanges occur only at a *particular* DNA sequence. **DNA transposition** is distinct from both other classes in that it usually involves a short segment of DNA with the remarkable capacity to move from one location in a chromosome to another. These "jumping genes" were first observed in maize in the 1940s by Barbara

Barbara McClintock,
1902-1992
[Source: AP Photo.]

McClintock. There is, in addition, a wide range of unusual genetic rearrangements for which no mechanism or purpose has yet been proposed. Here we focus on the three general classes.

Homologous genetic recombination is largely a pathway to repair double-strand breaks in DNA. An alternative process for double-strand break repair, called **nonhomologous end joining (NHEJ)**, is also described here. The functions of genetic recombination systems are as varied as their mechanisms. They include roles in specialized DNA repair systems, specialized activities in DNA replication, regulation of expression of certain genes, facilitation of proper chromosome segregation during eukaryotic cell division, maintenance of genetic diversity, and implementation of programmed genetic rearrangements during embryonic development. In most cases, genetic recombination is closely integrated with other processes in DNA metabolism, and this becomes a theme of our discussion.

Bacterial Homologous Recombination Is a DNA Repair Function

In bacteria, homologous genetic recombination is primarily a DNA repair process and in this context (as noted in Section 25.2) is referred to as **recombinational DNA repair**. It is usually directed at the reconstruction of replication forks that have stalled or collapsed at the site of DNA damage. Homologous genetic recombination can also occur during conjugation (mating), when chromosomal DNA is transferred from one bacterial cell (donor) to another (recipient). Recombination during conjugation, although rare in wild bacterial populations, contributes to genetic diversity.

An example of what happens when a replication fork encounters DNA damage is shown in **Figure 25-30**.

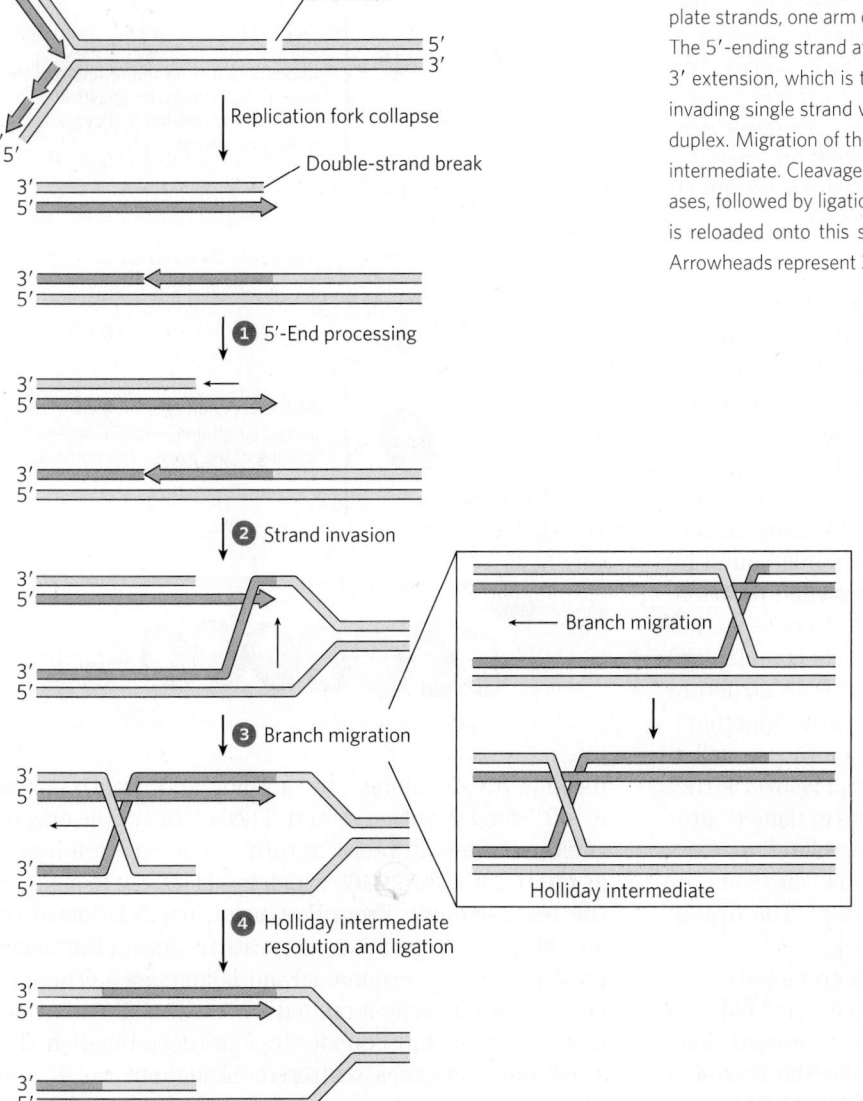

FIGURE 25-30 Recombinational DNA repair at a collapsed replication fork. When a replication fork encounters a break in one of the template strands, one arm of the fork is lost and the replication fork collapses. The 5'-ending strand at the break is degraded to create a single-stranded 3' extension, which is then used in a strand invasion process, pairing the invading single strand with its complementary strand within the adjacent duplex. Migration of the branch (shown in the box) can create a Holliday intermediate. Cleavage of the Holliday intermediate by specialized nucleases, followed by ligation, restores a viable replication fork. The replisome is reloaded onto this structure (not shown), and replication continues. Arrowheads represent 3' ends.

(a)

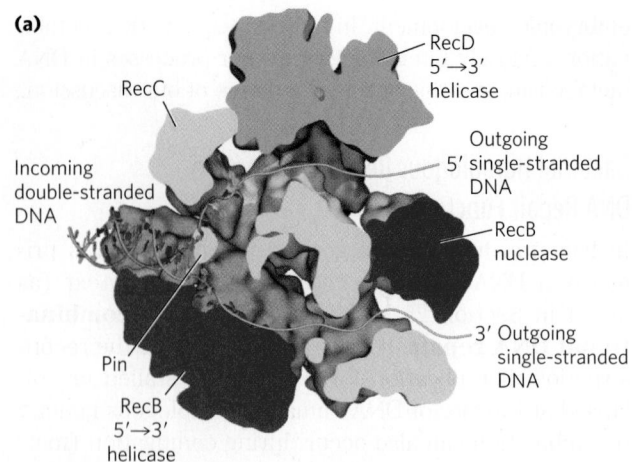

(b)

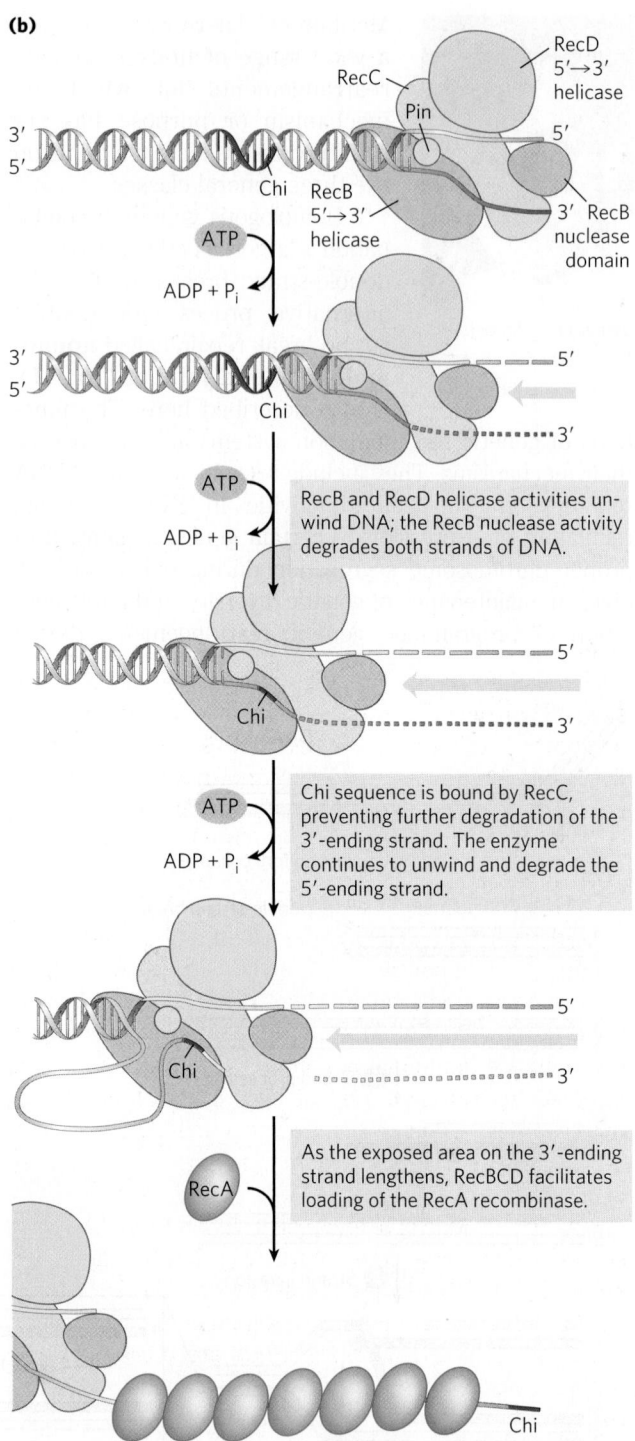

FIGURE 25-31 The RecBCD helicase/nuclease. (a) A cutaway view of the RecBCD enzyme structure as it is bound to DNA. The subunits are shown in different colors; the DNA is entering from the left, and the unwound DNA strands (not part of the solved structure) are shown exiting to the right. A bulbous protein structure called a pin, part of the RecC subunit, facilitates the separation of strands. **(b)** Activities of the RecBCD enzyme at a DNA end. The RecB and RecD subunits are helicases, molecular motors that propel the complex along the DNA, a process that requires ATP. The single nuclease domain of RecB degrades both strands as the complex travels, cleaving the 3′-ending strand more often than the 5′-ending strand. When a chi site is encountered on the 3′-ending strand, the RecC subunit binds to it, halting the advance of this strand through the complex. The 5′-ending strand continues to be degraded as the 3′-ending strand is looped out, eventually creating a 3′-single-stranded extension. As the 3′-single-stranded extension lengthens, RecA protein is finally loaded onto the processed DNA by the RecBCD enzyme. [Source: (a) PDB ID 1W36, M. R. Singleton et al., *Nature* 432:187, 2004.]

A common feature of the DNA repair pathways illustrated in Figures 25-22 to 25-25 is that they introduce a transient break into one of the DNA strands. If a replication fork encounters a damaged site under repair near a break in one of the template strands, one arm of the replication fork becomes disconnected by a double-strand break and the fork collapses. The end of that break is processed by degrading the 5′-ending strand. The resulting 3′ single-stranded extension is bound by a recombinase that uses it to promote strand invasion: the 3′ end invades the intact duplex DNA connected to the other arm of the fork and pairs with its complementary sequence. This creates a branched DNA structure (a point where three DNA segments come together). The DNA branch can be moved in a process called **branch migration** to create an X-like crossover structure known as a **Holliday intermediate**, named after researcher Robin Holliday, who first postulated its existence. The Holliday intermediate is cleaved, or "resolved," by a special class of nucleases. The overall process reconstructs the replication fork.

In *E. coli*, the DNA end-processing is promoted by the RecBCD nuclease/helicase. The RecBCD enzyme binds to linear DNA at a free (broken) end and moves inward along the double helix, unwinding and degrading the DNA in a reaction coupled to ATP hydrolysis **(Fig. 25-31)**. The RecB and RecD subunits are helicase motors, with RecB

moving 3′→5′ along one strand, and RecD moving 5′→3′ along the other strand. The activity of the enzyme is altered when it interacts with a sequence referred to as **chi**, (5′)GCTGGTGG, which binds tightly to a site on the RecC subunit. From that point, degradation of the strand with a 3′ terminus is greatly reduced, but degradation of the 5′-terminal strand is increased. This process creates a single-stranded DNA with a 3′ end, which is used during subsequent steps in recombination. The 1,009 chi sequences scattered throughout the *E. coli* genome enhance the frequency of recombination about 5- to 10-fold within 1,000 bp of the chi site. The

FIGURE 25-32 RecA protein filaments. RecA and other recombinases in this class function as filaments of nucleoprotein. **(a)** Filament formation proceeds in discrete nucleation and extension steps. Nucleation is the addition of the first few RecA subunits. Extension occurs by adding RecA subunits so that the filament grows in the 5'→3' direction. When disassembly occurs, subunits are subtracted from the trailing end. **(b)** Colorized electron micrograph of a RecA filament bound to DNA. **(c)** Segment of a RecA filament with four helical turns (24 RecA subunits). Notice the bound double-stranded DNA in the center. The core domain of RecA is structurally related to the motor domains of helicases. [Sources: (b) By permission of the Estate of Ross Inman. Special thanks to Kim Voss. (c) Modified from PDB ID 3CMX, Z. Chen et al., *Nature* 453:489, 2008.]

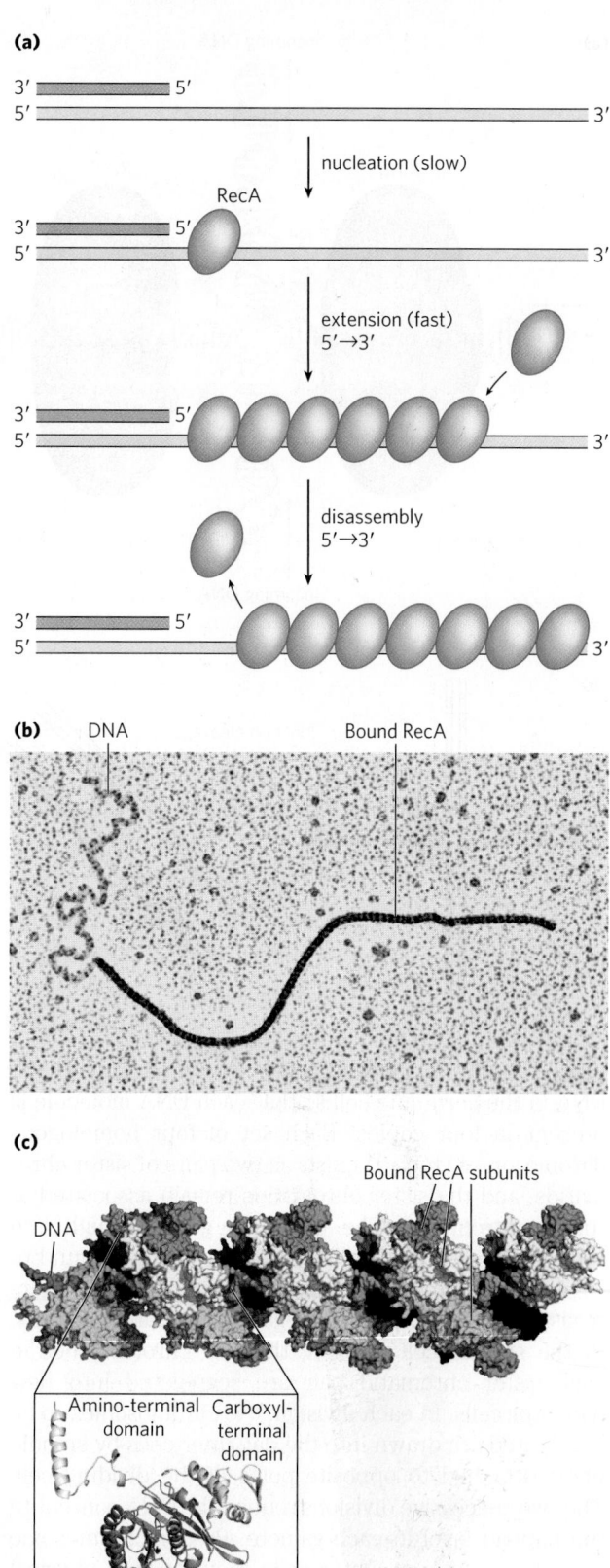

enhancement declines as the distance from chi increases. Sequences that enhance recombination frequency have also been identified in several other organisms.

The bacterial recombinase is the RecA protein. RecA is unusual among the proteins of DNA metabolism in that its active form is an ordered, helical filament of up to several thousand subunits that assemble cooperatively on DNA **(Fig. 25-32)**. This filament usually forms on single-stranded DNA, such as that produced by the RecBCD enzyme. Its formation is not as straightforward as shown in Figure 25-32, because the single-stranded DNA–binding protein (SSB) is normally present and specifically impedes the binding of the first few subunits to DNA (filament nucleation). The RecBCD enzyme acts directly as a RecA loader, facilitating the nucleation of a RecA filament on single-stranded DNA that is coated with SSB. The filaments assemble and disassemble predominantly in a 5'→3' direction. Many other bacterial proteins regulate the formation and disassembly of RecA filaments. RecA protein promotes the central steps of homologous recombination, including the DNA strand invasion step of Figure 25-30, as well as other strand exchange reactions occurring in vitro.

After strand invasion, branch migration is promoted by a complex called RuvAB **(Fig. 25-33a)**. Once a Holliday intermediate has been created, it can be cleaved by a specialized nuclease called RuvC (Fig. 25-33b), and nicks are sealed by DNA ligase. A viable replication fork structure is thus reconstructed, as outlined in Figure 25-30.

After the recombination steps are completed, the replication fork reassembles in a process called **origin-independent restart of replication.** Four proteins (PriA, PriB, PriC, and DnaT) act with DnaC to load the DnaB helicase onto the reconstructed replication fork. The DnaG primase then synthesizes an RNA primer, and DNA polymerase III reassembles on DnaB to restart DNA synthesis. A complex of PriA, PriB, PriC, and DnaT, along with DnaB, DnaC, and DnaG, is called the **replication restart primosome**. In this way, the process of recombination is tightly intertwined with replication. One process of DNA metabolism supports the other.

Eukaryotic Homologous Recombination Is Required for Proper Chromosome Segregation during Meiosis

In eukaryotes, homologous genetic recombination can have several roles in replication and cell division, including the repair of stalled replication forks. Recombination occurs with the highest frequency during **meiosis**, the process by which diploid germ-line cells with two sets of chromosomes divide to produce haploid gametes (sperm cells or ova) in animals (haploid spores in plants)—each gamete having only one member of each chromosome

(a)

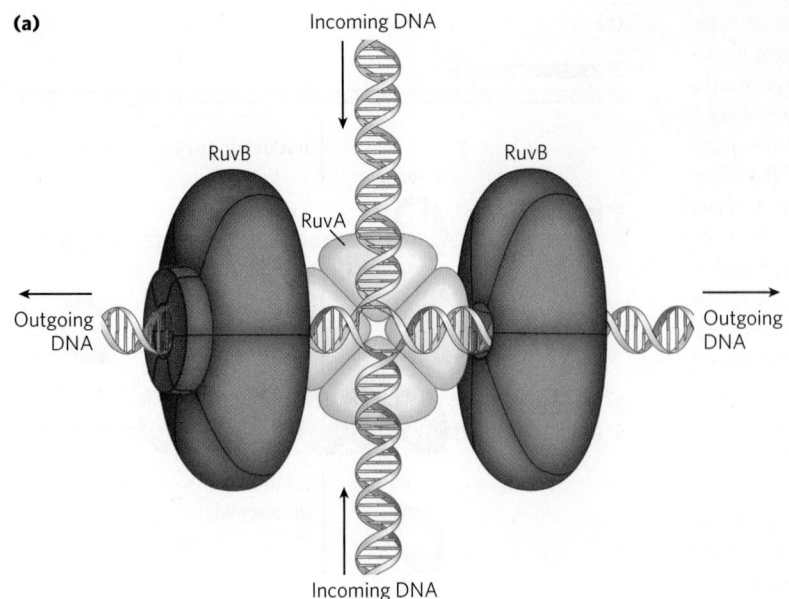

Incoming DNA

RuvB

RuvA

RuvB

Outgoing DNA

Outgoing DNA

Incoming DNA

FIGURE 25-33 Catalysis of DNA branch migration and resolution of a Holliday intermediate by the RuvA, RuvB, and RuvC proteins. (a) The RuvA protein binds directly to a Holliday intermediate where the four DNA arms come together. The hexameric RuvB protein is a DNA translocase. Two hexamers bind to opposing arms of the Holliday intermediate and propel the DNA outward in a reaction coupled to ATP hydrolysis. The branch thus moves. **(b)** RuvC is a specialized nuclease that binds to the RuvAB complex and cleaves the Holliday intermediate on opposing sides of the junction (red arrows), so that two contiguous DNA arms remain in each product.

(b)

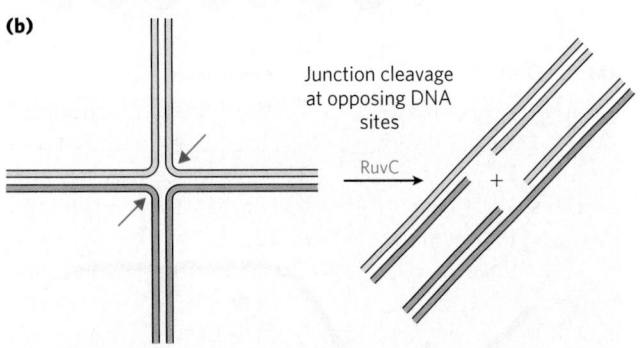

Junction cleavage at opposing DNA sites

RuvC

+

pair **(Fig. 25-34)**. Meiosis begins with replication of the DNA in the germ-line cell so that each DNA molecule is present in four copies. Each set of four homologous chromosomes (tetrad) exists as two pairs of sister chromatids, and the sister chromatids remain associated at their centromeres. The cell then goes through two rounds of cell division without an intervening round of DNA replication. In the first cell division, the two pairs of sister chromatids are segregated into daughter cells. In the second cell division, the two chromosomes in each sister chromatid pair are segregated into new daughter cells. In each division, the chromosomes to be segregated are drawn into the daughter cells by spindle fibers attached to opposite poles of the dividing cell. The two successive divisions reduce the DNA content to the haploid level in each gamete. Proper chromosome segregation into daughter cells requires that physical links exist between the homologous chromosomes to be segregated. As the spindle fibers attach to the centromeres of chromosomes and start to pull, the links between homologous chromosomes create tension. This tension, sensed by cellular mechanisms not yet understood, signals that this pair of chromosomes or sister chromatids is properly aligned for segregation. Once the tension is sensed, the links are gradually dissolved and

segregation proceeds. If improper spindle fiber attachment occurs (e.g., if the centromeres of a chromosome pair are attached to the same cellular pole), a cellular kinase senses the lack of tension and activates a system that removes the spindle attachments, allowing the cell to try again.

During the second meiotic division, the centromeric attachments between sister chromatids, augmented by cohesins deposited during replication (see Fig. 24-33), provide the needed physical links to guide segregation. However, during the first meiotic cell division, the two pairs of sister chromatids to be segregated are not related by a recent replication event and are not linked by cohesins or any other physical association. Instead, the homologous pairs of sister chromatids are aligned and new links are created by recombination, a process involving the breakage and rejoining of DNA **(Fig. 25-35)**. This exchange, also referred to as crossing over, can be observed with the light microscope. Crossing over links the two pairs of sister chromatids together at points called chiasmata (singular, chiasma). Also during crossing over, genetic material is exchanged between the pairs of sister chromatids. These exchanges increase genetic diversity in the resulting gametes. The importance of meiotic recombination to

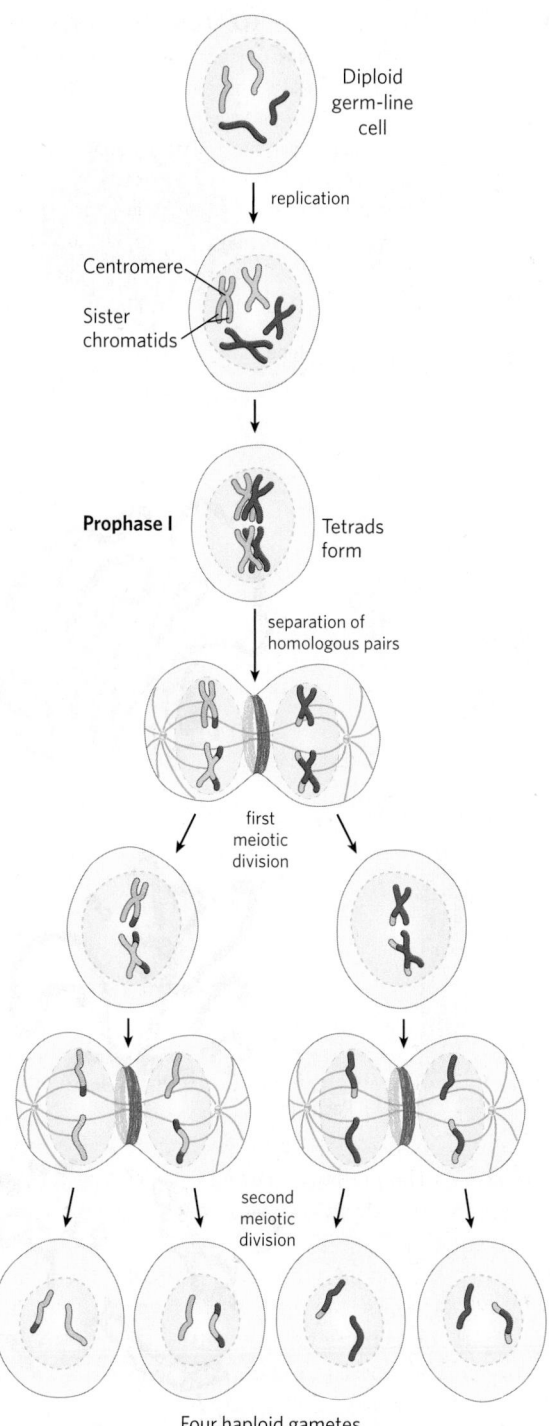

Diploid germ-line cell

replication

Centromere

Sister chromatids

Prophase I

Tetrads form

separation of homologous pairs

first meiotic division

second meiotic division

Four haploid gametes

FIGURE 25-34 Meiosis in animal germ-line cells. The chromosomes of a hypothetical diploid germ-line cell (four chromosomes; two homologous pairs) replicate and are held together at their centromeres. Each replicated double-stranded DNA molecule is called a chromatid (sister chromatid). In prophase I, just before the first meiotic division, the two homologous sets of chromatids align to form tetrads, held together by covalent links at homologous junctions (chiasmata). Crossovers occur within the chiasmata (see Fig. 25-35). These transient associations between homologs ensure that the two tethered chromosomes segregate properly in the next step, when attached spindle fibers pull them toward opposite poles of the dividing cell in the first meiotic division. The products of this division are two daughter cells each with two pairs of different sister chromatids. The pairs now line up across the equator of the cell in preparation for separation of the chromatids (now called chromosomes). The second meiotic division produces four haploid daughter cells that can serve as gametes. Each has two chromosomes, half the number of the diploid germ-line cell. The chromosomes have re-sorted and recombined.

two points on a chromosome is roughly proportional to the distance between the points, and this allows determination of the relative positions of and distances between different genes. The independent assortment of unlinked genes on different chromosomes **(Fig. 25-36)** makes another major contribution to the genetic diversity of gametes. These genetic realities guide many of the modern applications of genomics, such as defining haplotypes (see Fig. 9-28) or searching for disease genes in the human genome (see Fig. 9-32).

Homologous recombination thus serves at least three identifiable functions in eukaryotes: (1) it contributes to the repair of several types of DNA damage; (2) it provides, in eukaryotic cells, a transient physical link between chromatids that promotes the orderly segregation of chromosomes at the first meiotic cell division; and (3) it enhances genetic diversity in a population.

Recombination during Meiosis Is Initiated with Double-Strand Breaks

A likely pathway for homologous recombination during meiosis is outlined in Figure 25-35a. The model has four key features. First, homologous chromosomes align. Second, a double-strand break in a DNA molecule is created, and the exposed ends are processed by an exonuclease, leaving a single-stranded extension with a free 3'-hydroxyl group at the broken end (step ❶). Third, the exposed 3' ends invade the intact duplex DNA of the homolog, and this is followed by branch migration and/or replication to create a pair of Holliday intermediates (steps ❷ to ❹). Fourth, cleavage of the two crossovers creates either of two pairs of complete recombinant products (step ❺). Notice the similarity of these steps to the bacterial recombinational repair processes outlined in Figure 25-30.

In this **double-strand break repair model** for recombination, the 3' ends are used to initiate the genetic exchange. Once paired with the complementary strand on the intact homolog, a region of hybrid DNA is created

proper chromosome segregation is well illustrated by the physiological and societal consequences of their failure (Box 25-2).

Crossing over is not an entirely random process, and "hot spots" have been identified on many eukaryotic chromosomes. However, the assumption that crossing over can occur with equal probability at almost any point along the length of two homologous chromosomes remains a reasonable approximation in many cases, and it is this assumption that permits the genetic mapping of genes on a particular chromosome. The frequency of homologous recombination in any region separating

(a)

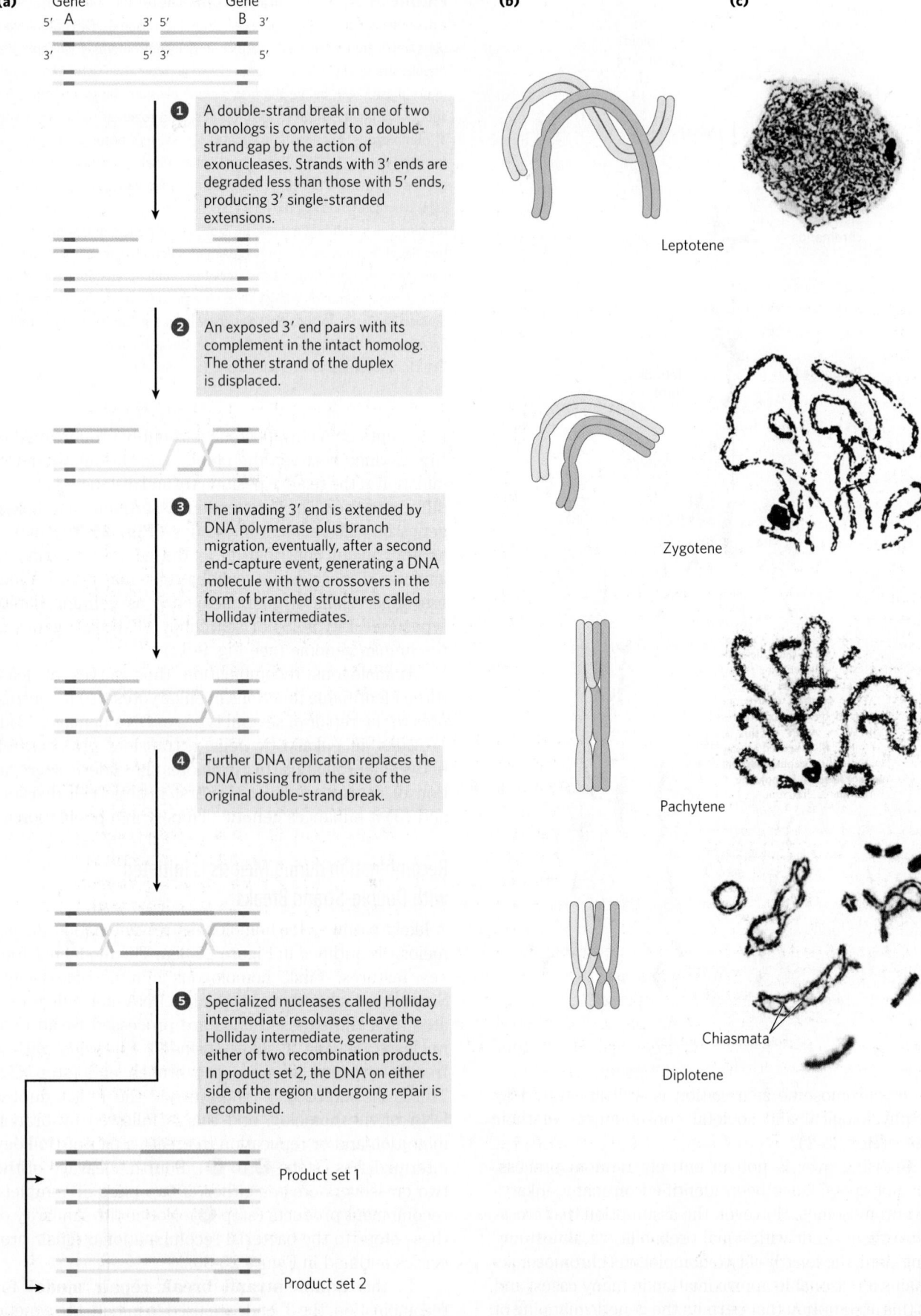

Gene A   Gene B

1 A double-strand break in one of two homologs is converted to a double-strand gap by the action of exonucleases. Strands with 3′ ends are degraded less than those with 5′ ends, producing 3′ single-stranded extensions.

2 An exposed 3′ end pairs with its complement in the intact homolog. The other strand of the duplex is displaced.

3 The invading 3′ end is extended by DNA polymerase plus branch migration, eventually, after a second end-capture event, generating a DNA molecule with two crossovers in the form of branched structures called Holliday intermediates.

4 Further DNA replication replaces the DNA missing from the site of the original double-strand break.

5 Specialized nucleases called Holliday intermediate resolvases cleave the Holliday intermediate, generating either of two recombination products. In product set 2, the DNA on either side of the region undergoing repair is recombined.

Product set 1

Product set 2

(b)

(c)

Leptotene

Zygotene

Pachytene

Chiasmata

Diplotene

FIGURE 25-35 Recombination during prophase I in meiosis. (a) A model of double-strand break repair for homologous genetic recombination. The two homologous chromosomes (one shown in red, the other blue) involved in this recombination event have identical or very nearly identical sequences. Each of the two genes shown has different alleles on the two chromosomes. The steps are described in the text. **(b)** Crossing over occurs during prophase of meiosis I. The several stages of prophase I are aligned with the recombination processes shown in (a). Double-strand breaks are introduced and processed in the leptotene stage. The strand invasion and completion of crossover occur later. As homologous sequences in the two pairs of sister chromatids are aligned in the zygotene stage, synaptonemal complexes form and strand invasion occurs. The homologous chromosomes are tightly aligned by the pachytene stage. **(c)** Homologous chromosomes of a grasshopper, viewed at successive stages of meiotic prophase I. The chiasmata become visible in the diplotene stage. [Source: (c) B. John, *Meiosis*, Figs 2.1a, 2.2a, 2.2b, 2.3a, Cambridge University Press, 1990. Reprinted with the permission of Cambridge University Press.]

BOX 25-2 🜔 MEDICINE Why Proper Segregation of Chromosomes Matters

When chromosomal alignment and recombination are not correct and complete in meiosis I, segregation of chromosomes can go awry. One result may be aneuploidy, a condition in which a cell has the wrong number of chromosomes. The haploid products of meiosis (gametes or spores) may have no copies or two copies of a chromosome. When a gamete having two copies of a chromosome joins with a gamete having one copy of a chromosome during fertilization, cells in the resulting embryo have three copies of that chromosome (they are trisomic).

In *S. cerevisiae*, aneuploidy resulting from errors in meiosis occurs at a rate of about 1 in 10,000 meiotic events. In fruit flies, the rate is about 1 in a few thousand. Rates of aneuploidy in mammals are considerably higher. In mice, the rate is 1 in 100, and it is even higher in other mammals. The rate of aneuploidy in fertilized human eggs has been estimated as 10% to 30%; most of these aneuploid cells are monosomies (they have a single copy of a chromosome) or trisomies. This is almost certainly an underestimate. Most trisomies are lethal, and many result in miscarriage long before the pregnancy is detected. Almost all monosomies are fatal in the early stages of fetal development. Aneuploidy is the leading cause of pregnancy loss. The few trisomic fetuses that survive to birth generally have three copies of chromosome 13, 18, or 21 (trisomy 21 is Down syndrome). Abnormal complements of the sex chromosomes are also found in the human population. The societal consequences of aneuploidy in humans are considerable. Aneuploidy is the leading genetic cause of developmental and mental disabilities. At the heart of these high rates is a feature of meiosis in female mammals that has special significance for the human species.

In a human male, germ-line cells begin to undergo meiosis at puberty, and each meiotic event requires a relatively short time. In contrast, meiosis in the germ-line cells of human females is a highly protracted process. The production of an egg begins before a female is born, with the onset of meiosis in the fetus, at 12 to 13 weeks of gestation. Meiosis is initiated in all the developing fetal germ-line cells over a period of a few weeks. The cells proceed through much of meiosis I. Chromosomes line up and generate crossovers, continuing just beyond the pachytene phase (see Fig. 25-35)—and then the process stops. The chromosomes enter an arrested phase called the dictyate stage, with the crossovers in place, a kind of suspended animation where they remain as the female matures—so, typically remaining in this phase for anywhere from about 13 to 50 years. At sexual maturity, individual germ-line cells continue through the two meiotic cell divisions to produce egg cells.

Between the onset of the dictyate stage and the final completion of meiosis, something may happen that disrupts or damages the crossovers linking homologous chromosomes in the germ-line cells. As a woman ages, the rate of trisomy in the egg cells she produces increases, dramatically so as she approaches menopause (Fig. 1). There are many hypotheses on why this occurs, and several different factors may play a role. However, most of the hypotheses are centered on recombination crossovers in meiosis I and their stability over the protracted dictyate stage.

It is not yet clear what medical steps could be taken to reduce the incidence of aneuploidy in women of child-bearing age. What is revealed is the inherent importance of recombination and generation of crossovers in human meiosis.

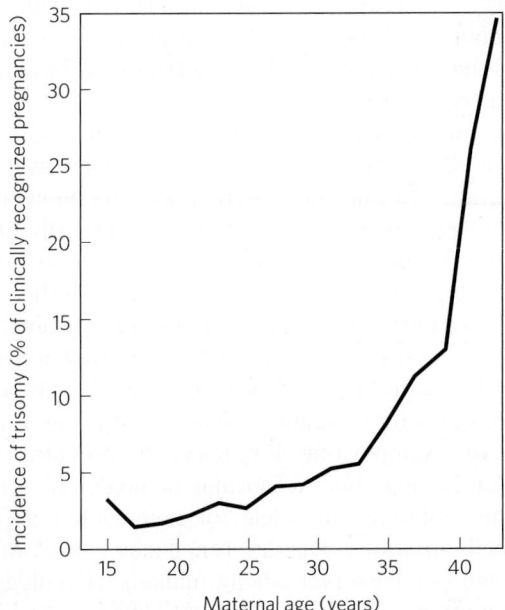

FIGURE 1 The increasing incidence of human trisomy with increasing age of the mother. [Source: Data from T. Hassold and P. Hunt, *Nature Rev. Genet.* 2:280, 2001, Fig. 6.]

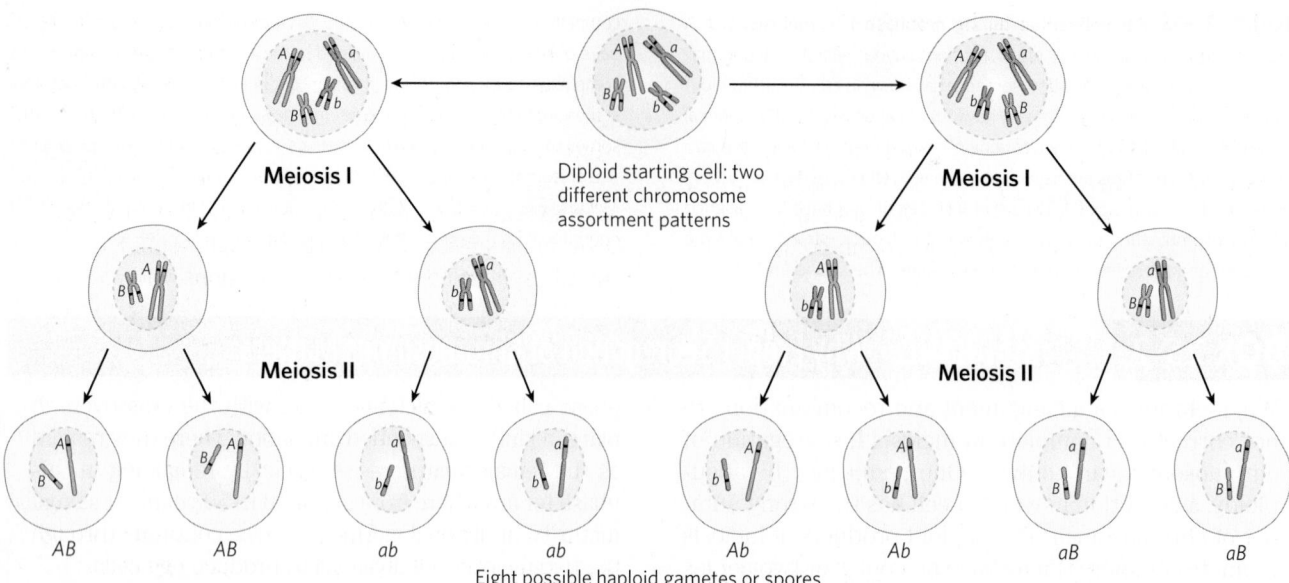

AB   AB   ab   ab   Ab   Ab   aB   aB

Eight possible haploid gametes or spores

FIGURE 25-36 The contribution of independent assortment to genetic diversity. In this example, the two chromosomes have already been replicated to create two pairs of sister chromatids. Blue and red distinguish the sister chromatids of each pair. One gene on each chromosome is highlighted, with different alleles (*A* or *a*, *B* or *b*) in the homologs. Independent assortment can lead to gametes with any combination of the alleles present on the two different chromosomes. Crossing over (not shown here; see Fig. 25-34) would also contribute to genetic diversity in a typical meiotic sequence.

that contains complementary strands from two different parent DNAs (the product of step ❷ in Fig. 25-35a). Each of the 3′ ends can then act as a primer for DNA replication. Meiotic homologous recombination can vary in many details from one species to another, but most of the steps outlined above are generally present in some form. There are two ways to resolve the Holliday intermediate with a RuvC-like nuclease so that the two products carry genes in the same linear order as in the substrates—the original, un-recombined chromosomes (step ❺). If cleaved one way, the DNA flanking the region containing the hybrid DNA is not recombined; if cleaved the other way, the flanking DNA is recombined. Both outcomes are observed in vivo.

The homologous recombination illustrated in Figure 25-35 is an elaborate process that is essential to accurate chromosome segregation. Its molecular consequences for the generation of genetic diversity are subtle. To understand how this process contributes to diversity, we should keep in mind that the two homologous chromosomes that undergo recombination are not necessarily *identical*. The linear array of genes may be the same, but the base sequences in some of the genes may differ slightly (in different alleles). In a human, for example, one chromosome may contain the allele for hemoglobin A (normal hemoglobin) while the other contains the allele for hemoglobin S (the sickle cell mutation). The difference may consist of no more than one base pair among millions. Homologous recombination does not change the linear array of genes, but it can determine which alleles become linked on a single chromosome and are passed to the next generation together. The independent assortment of different chromosomes (Fig. 25-36) determines which gene alleles from different chromosomes are inherited together.

Some Double-Strand Breaks Are Repaired by Nonhomologous End Joining

Double-strand breaks sometimes occur when recombinational DNA repair is not feasible, such as during phases of the cell cycle when no replication is occurring and no sister chromatids are present. At these times, another path is needed to avoid the cell death that would result from a broken chromosome. That alternative is provided by **nonhomologous end joining (NHEJ)**. The broken chromosome ends are simply processed and ligated back together.

Nonhomologous end joining is an important pathway for double-strand break repair in all eukaryotes and has also been detected in some bacteria. The importance of NHEJ increases with genomic complexity, and the process accounts for most double-strand break repair outside meiosis in mammals. In yeast, most double-strand breaks are repaired by recombination, and only a few by NHEJ. NHEJ is a mutagenic process, and a smaller genome, such as that of yeast, has relatively little tolerance for the loss of information. The small genomic alterations may be tolerable in mammalian somatic cells, because they are balanced by the undamaged information on the homolog in each diploid cell, and in these non-germ-line cells, the mutations are not inherited. In vertebrates, a loss of the genes encoding NHEJ function can produce a predisposition to cancer.

Unlike homologous recombinational repair, NHEJ does not conserve the original DNA sequence. The pathway in eukaryotes is illustrated in **Figure 25-37**. The reaction is initiated at the broken ends of a double-strand break by the binding of a heterodimer consisting of the proteins Ku70 and Ku80 ("KU" being the initials of the individual with scleroderma whose serum autoantibodies were used to identify this protein complex; the numbers refer to the approximate molecular weights of the subunits). The Ku proteins are conserved in almost all eukaryotes and act as a kind of molecular scaffold to assemble the other protein components. Ku70-Ku80 interacts with another protein complex containing a protein kinase called DNA-PKcs and a nuclease known as Artemis. Once the complex is assembled, the two broken DNA ends are synapsed (held together). DNA-PKcs autophosphorylates in several locations and also phosphorylates Artemis. Artemis, when phosphorylated, acquires an endonuclease function that can remove 5′ or 3′ single-stranded extensions or hairpins that might be present at the ends. The DNA ends are then separated with the aid of a helicase, and strands from the two different ends are annealed at locations where short regions of complementarity are encountered. Artemis cleaves any unpaired DNA segments that are created. Small DNA gaps are filled by a DNA polymerase, Pol μ or Pol λ. Finally, the nicks are sealed by a protein complex consisting of XRCC4 (*x-ray cross complementation group*), XLF (*XRCC4-like factor*), and DNA ligase IV.

DNA ends are not joined randomly by NHEJ. Instead, when a double-strand break occurs, the ends are generally constrained by the structure of chromatin and thus remain close together. Very rare events linking end sequences that are normally far apart in the chromosome, or are on different chromosomes, may be responsible for occasional dramatic and usually deleterious genomic rearrangements.

Site-Specific Recombination Results in Precise DNA Rearrangements

Homologous genetic recombination can involve any two homologous sequences. The second general type of recombination, site-specific recombination, is a very different type of process: recombination is limited to specific sequences. Recombination reactions of this type occur in virtually every cell, filling specialized roles that vary greatly from one species to another. Examples include regulation of the expression of certain genes and promotion of programmed DNA rearrangements in embryonic development or in the replication cycles of some viral and plasmid DNAs. Each site-specific recombination system consists of an enzyme called a recombinase and a short (20 to 200 bp), unique DNA sequence where the recombinase acts (the recombination site). One or more auxiliary proteins may regulate the timing or outcome of the reaction.

There are two general classes of site-specific recombination systems, which rely on either Tyr or Ser residues in the active site. In vitro studies of many site-specific recombination systems in the tyrosine class have elucidated some general principles, including the fundamental reaction pathway **(Fig. 25-38a)**. Several of these enzymes have been crystallized, revealing structural details of the reaction. A separate recombinase recognizes and binds to each of two recombination sites on two different DNA molecules or within the same DNA.

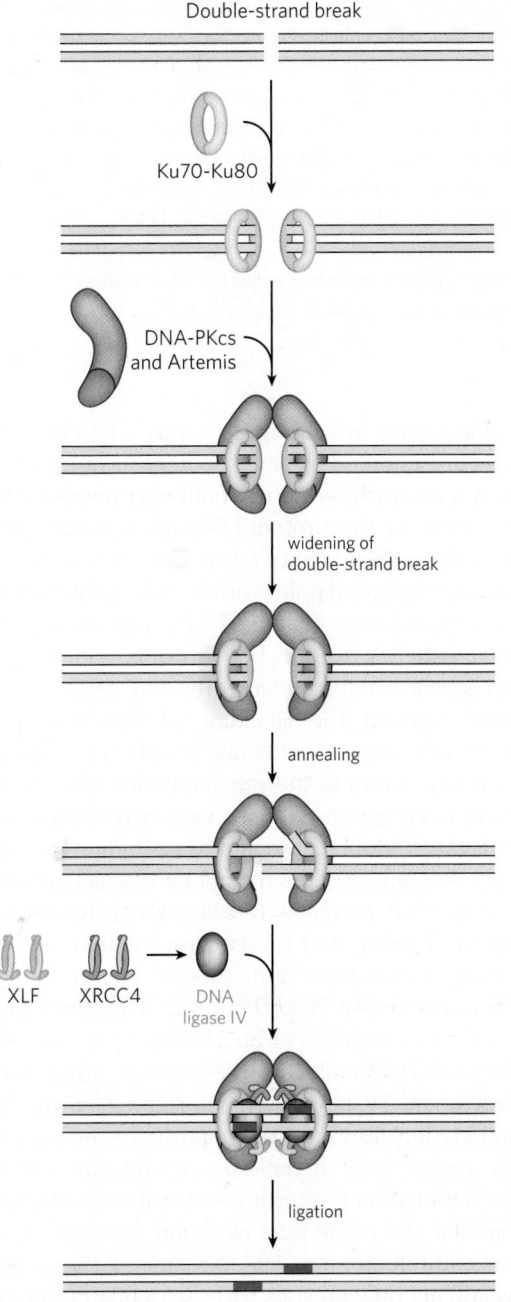

FIGURE 25-37 Nonhomologous end joining. The Ku70-Ku80 complex is the first to bind the DNA ends, followed by a complex including DNA-PKcs and the nuclease Artemis. These proteins then recruit a complex consisting of XRCC4, XLF, and DNA ligase IV. Either of two DNA polymerases, Pol μ or Pol λ (not shown), subsequently extends the annealed DNA strands, as needed, before ligation. [Source: Information from J. M. Sekiguchi and D. O. Ferguson, *Cell* 124:260, 2006, Fig. 1.]

Figure labels:
- Double-strand break
- Ku70-Ku80
- DNA-PKcs and Artemis
- widening of double-strand break
- annealing
- XLF   XRCC4   DNA ligase IV
- ligation

(a)

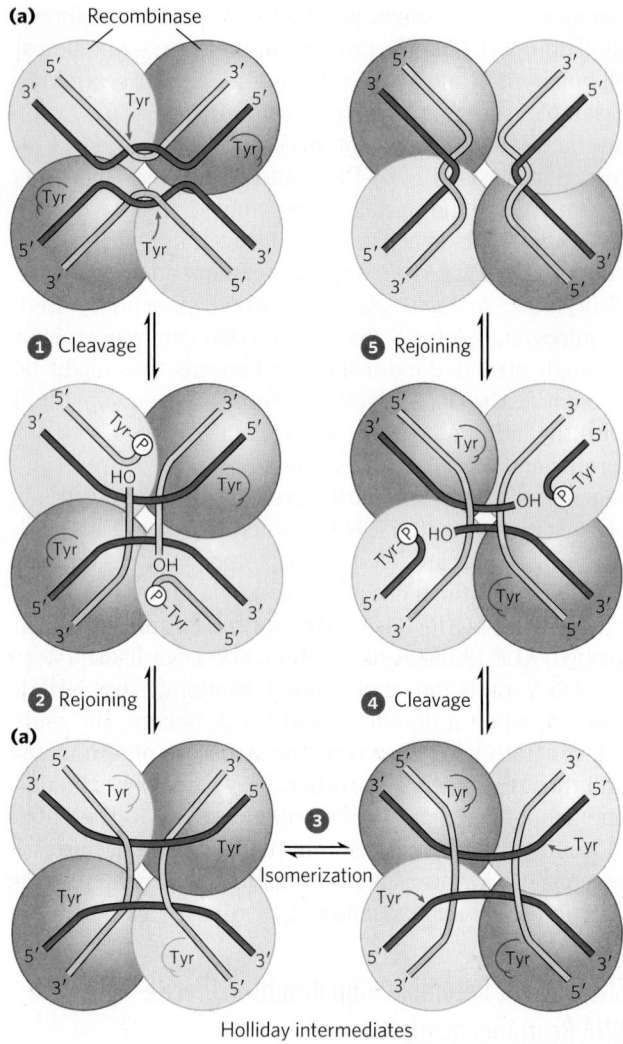

① Cleavage

⑤ Rejoining

② Rejoining

④ Cleavage

(a)

③ Isomerization

Holliday intermediates

(b)

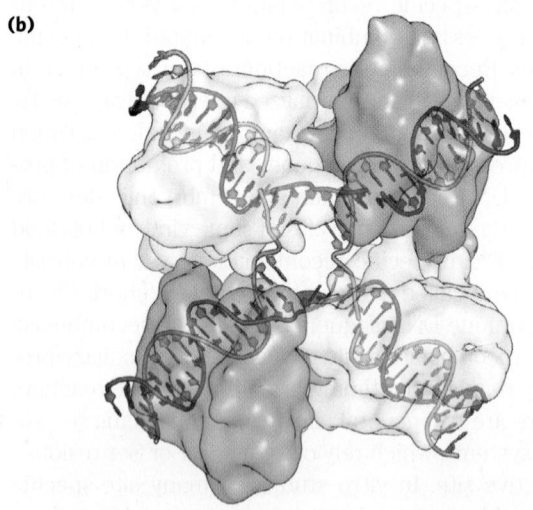

One DNA strand in each site is cleaved at a specific point within the site, and the recombinase becomes covalently linked to the DNA at the cleavage site through a phosphotyrosine bond (step **①**). The transient protein-DNA linkage preserves the phosphodiester bond that is lost in cleaving the DNA, so high-energy cofactors such as ATP

FIGURE 25-38 A site-specific recombination reaction. (a) The reaction shown here is for a common class of site-specific recombinases called integrase-class recombinases (named after bacteriophage λ integrase, the first recombinase characterized). These enzymes use Tyr residues as nucleophiles at the active site. The reaction is carried out within a tetramer of identical subunits. Recombinase subunits bind to a specific sequence, the recombination site. **①** One strand in each DNA is cleaved at particular points in the sequence. The nucleophile is the —OH group of an active-site Tyr residue, and the product of rejoining **②** is a covalent phosphotyrosine link between protein and DNA. After isomerization **③**, the cleaved strands join to new partners, producing a Holliday intermediate. Steps **④** and **⑤** complete the reaction by a process similar to the first two steps. The original sequence of the recombination site is regenerated after recombining the DNA flanking the site. These steps occur within a complex of multiple recombinase subunits that sometimes includes other proteins not shown here. (b) Surface contour model of a four-subunit integrase-class recombinase called the FLP recombinase, bound to a Holliday intermediate (shown with light blue and dark blue helix strands). The protein has been rendered transparent so that the bound DNA is visible. Another group of recombinases, called the resolvase/invertase family, use a Ser residue as nucleophile at the active site. [Source: (b) PDB ID 1P4E, P. A. Rice and Y. Chen, *J. Biol. Chem.* 278:24,800, 2003.]

are unnecessary in subsequent steps. The cleaved DNA strands are rejoined to new partners to form a Holliday intermediate, with new phosphodiester bonds created at the expense of the protein-DNA linkage (step **②**). An isomerization then occurs (step **③**), and the process is repeated at a second point within each of the two recombination sites (steps **④** and **⑤**). In systems that employ an active-site Ser residue, both strands of each recombination site are cut concurrently and rejoined to new partners without the Holliday intermediate. In both types of systems, the exchange is always reciprocal and precise, regenerating the recombination sites when the reaction is complete. We can view a recombinase as a site-specific endonuclease and ligase in one package.

The sequences of the recombination sites recognized by site-specific recombinases are partially asymmetric (nonpalindromic), and the two recombining sites align in the same orientation during the recombinase reaction. The outcome depends on the location and orientation of the recombination sites **(Fig. 25-39)**. If the two sites are on the same DNA molecule, the reaction either inverts or deletes the intervening DNA, determined by whether the recombination sites have the opposite or the same orientation, respectively. If the sites are on different DNAs, the recombination is intermolecular; if one or both DNAs are circular, the result is an insertion. Some recombinase systems are highly specific for one of these reaction types and act only on sites with particular orientations.

Complete chromosomal replication can require site-specific recombination. Recombinational DNA repair of a circular bacterial chromosome, while essential, sometimes generates deleterious byproducts. The resolution of a Holliday intermediate at a replication fork by a nuclease such as RuvC, followed by completion of replication, can give rise to one of two products: the usual

Inversion

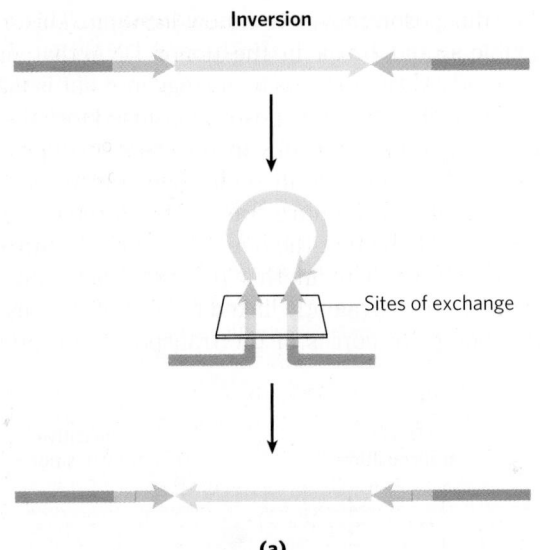

Sites of exchange

(a)

Deletion and insertion

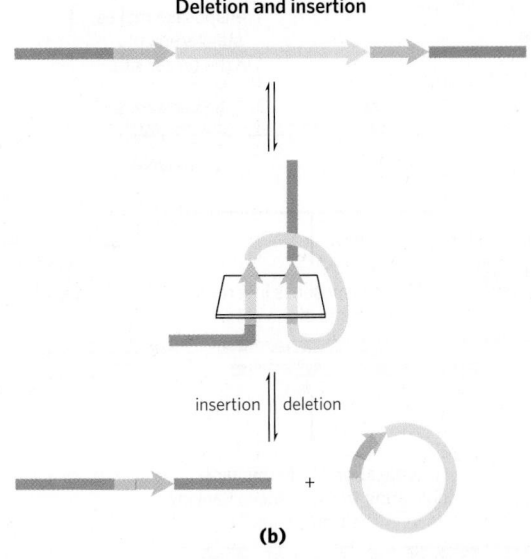

insertion ‖ deletion

+

(b)

FIGURE 25-39 Effects of site-specific recombination. The outcome of site-specific recombination depends on the location and orientation of the recombination sites (red and green) in a double-stranded DNA molecule. Orientation here (shown by arrowheads) refers to the order of nucleotides in the recombination site, *not* the 5'→3' direction. **(a)** Recombination sites with opposite orientation in the same DNA molecule. The result is an inversion. **(b)** Recombination sites with the same orientation, either on one DNA molecule, producing a deletion, or on two DNA molecules, producing an insertion.

two monomeric chromosomes or a contiguous dimeric chromosome **(Fig. 25-40)**. In the latter case, the covalently linked chromosomes cannot be segregated to daughter cells at cell division, and the dividing cells become "stuck." A specialized site-specific recombination system in *E. coli*, the XerCD system, converts the dimeric chromosomes to monomeric chromosomes so that cell division can proceed. The reaction is a site-specific deletion (Fig. 25-39b). This is another example of the close coordination between DNA recombination processes and other aspects of DNA metabolism.

Transposable Genetic Elements Move from One Location to Another

We now consider the third general type of recombination system: recombination that allows the movement of transposable elements, or **transposons**. These segments of DNA, found in virtually all cells, move, or "jump," from one place on a chromosome (the donor site) to another on the same or a different chromosome (the target site). DNA sequence homology is not usually required for this movement, called **transposition**; the new location is determined more or less randomly. Insertion of a transposon in an essential gene could kill the cell, so transposition is tightly regulated and usually very infrequent. Transposons are perhaps the simplest of molecular parasites, adapted to replicate passively within the chromosomes of host cells. In some cases

FIGURE 25-40 DNA deletion to undo a deleterious effect of recombinational DNA repair. The resolution of a Holliday intermediate during recombinational DNA repair (if cut at the points indicated by the red arrows) can generate a contiguous dimeric chromosome. A specialized site-specific recombinase in *E. coli*, XerCD, converts the dimer to monomers, allowing chromosome segregation and cell division to proceed.

they carry genes that are useful to the host cell, and thus exist in a kind of symbiosis with the host.

Bacteria have two classes of transposons. **Insertion sequences** (simple transposons) contain only the sequences required for transposition and the genes for the proteins (transposases) that promote the

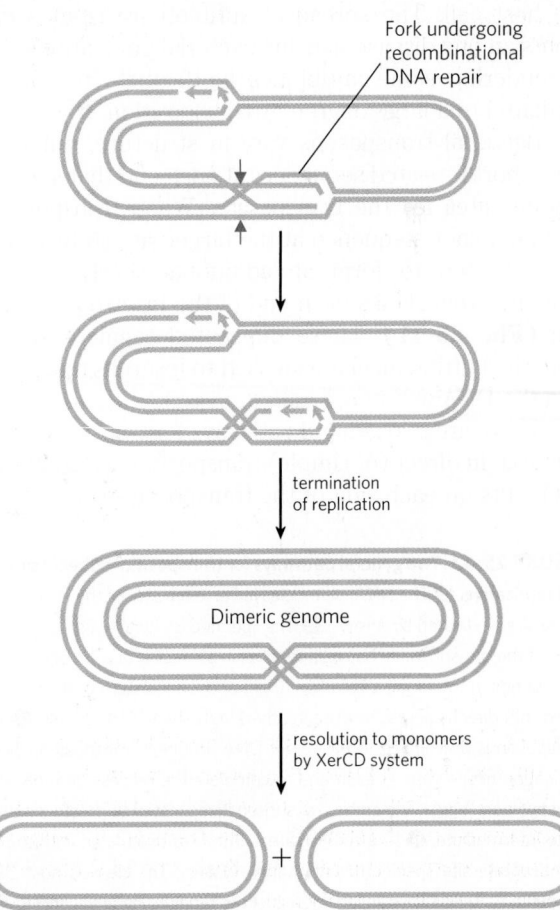

Fork undergoing recombinational DNA repair

termination of replication

Dimeric genome

resolution to monomers by XerCD system

+

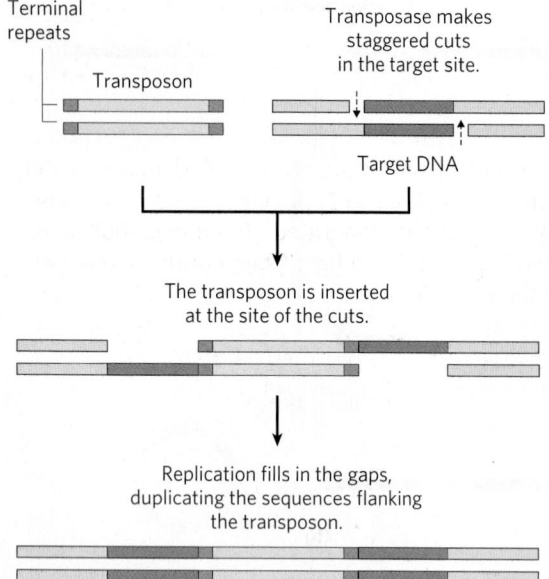

FIGURE 25-41 Duplication of the DNA sequence at a target site when a transposon is inserted. The sequences duplicated following transposon insertion are shown in red. These sequences are generally only a few base pairs long, so their size relative to that of a typical transposon is greatly exaggerated in this drawing.

process. **Complex transposons** contain one or more genes in addition to those needed for transposition. These extra genes might, for example, confer resistance to antibiotics and thus enhance the survival chances of the host cell. The spread of antibiotic-resistance elements among disease-causing bacterial populations that is rendering some antibiotics ineffectual (p. 959) is mediated to a large degree by transposition. ∎

Bacterial transposons vary in structure, but most have short repeated sequences at each end that serve as binding sites for the transposase. When transposition occurs, a short sequence at the target site (5 to 10 bp) is duplicated to form an additional short repeated sequence that flanks each end of the inserted transposon **(Fig. 25-41)**. These duplicated segments result from the cutting mechanism used to insert a transposon into the DNA at a new location.

There are two general pathways for transposition in bacteria. In direct (or simple) transposition **(Fig. 25-42, left)**, cuts on each side of the transposon excise it, and

FIGURE 25-42 Two general pathways for transposition: direct (simple) and replicative. ❶ The DNA is first cleaved on each side of the transposon, at the sites indicated by arrows. ❷ The liberated 3′-hydroxyl groups at the ends of the transposon act as nucleophiles in a direct attack on phosphodiester bonds in the target DNA. The target phosphodiester bonds are staggered (not directly across from each other) in the two DNA strands. ❸ The transposon is now linked to the target DNA. In direct transposition (left), replication fills in gaps at each end to complete the process. In replicative transposition (right), the entire transposon is replicated to create a cointegrate intermediate. ❹ The cointegrate is often resolved later, with the aid of a separate site-specific recombination system. The cleaved host DNA left behind after direct transposition is either repaired by DNA end-joining or degraded (not shown); the latter outcome can be lethal to the organism.

the transposon moves to a new location. This leaves a double-strand break in the donor DNA that must be repaired. At the target site, a staggered cut is made (as in Fig. 25-41), the transposon is inserted into the break, and DNA replication fills in the gaps to duplicate the target-site sequence. In replicative transposition (Fig. 25-42, right), the entire transposon is replicated, leaving a copy behind at the donor location. A **cointegrate** is an intermediate in this process, consisting of the donor region covalently linked to DNA at the target site. Two complete copies of the transposon are present in

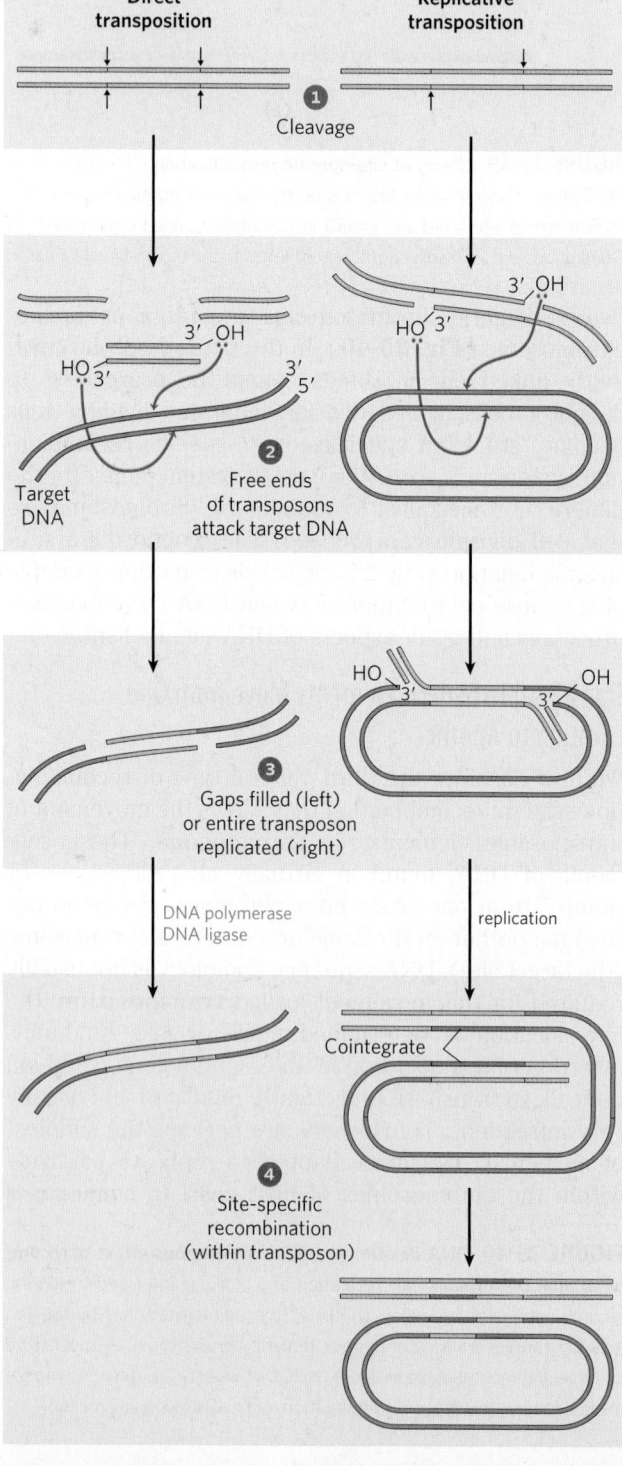

the cointegrate, both having the same relative orientation in the DNA. In some well-characterized transposons, the cointegrate intermediate is converted to products by site-specific recombination, in which specialized recombinases promote the required deletion reaction.

Eukaryotes also have transposons, structurally similar to bacterial transposons, and some use similar transposition mechanisms. In other cases, however, the mechanism of transposition seems to involve an RNA intermediate. Evolution of these transposons is intertwined with the evolution of certain classes of RNA viruses. Both are described in the next chapter. As illustrated in Figure 9-27, nearly half of the human genome is made up of various types of transposon elements.

Immunoglobulin Genes Assemble by Recombination

Some DNA rearrangements are a programmed part of development in eukaryotic organisms. An important example is the generation of complete immunoglobulin genes from separate gene segments in vertebrate genomes. A human (like other mammals) is capable of producing *millions* of different immunoglobulins (antibodies) with distinct binding specificities, even though the human genome contains only ~20,000 genes. Recombination allows an organism to produce an extraordinary

diversity of antibodies from a limited DNA-coding capacity. Studies of the recombination mechanism reveal a close relationship to DNA transposition and suggest that this system for generating antibody diversity may have evolved from an ancient cellular invasion by transposons.

We can use the human genes that encode proteins of the immunoglobulin G (IgG) class to illustrate how antibody diversity is generated. Immunoglobulins consist of two heavy and two light polypeptide chains (see Fig. 5-21). Each chain has two regions, a variable region, with a sequence that differs greatly from one immunoglobulin to another, and a region that is virtually constant within a class of immunoglobulins. There are also two distinct families of light chains, kappa and lambda, which differ somewhat in the sequences of their constant regions. For all three types of polypeptide chains (heavy chain, and kappa and lambda light chains), diversity in the variable regions is generated by a similar mechanism. The genes for these polypeptides are divided into segments, and the genome contains clusters with multiple versions of each segment. The joining of one version of each gene segment creates a complete gene.

Figure 25-43 depicts the organization of the DNA encoding the kappa light chains of human IgG and shows how a mature kappa light chain is generated. In

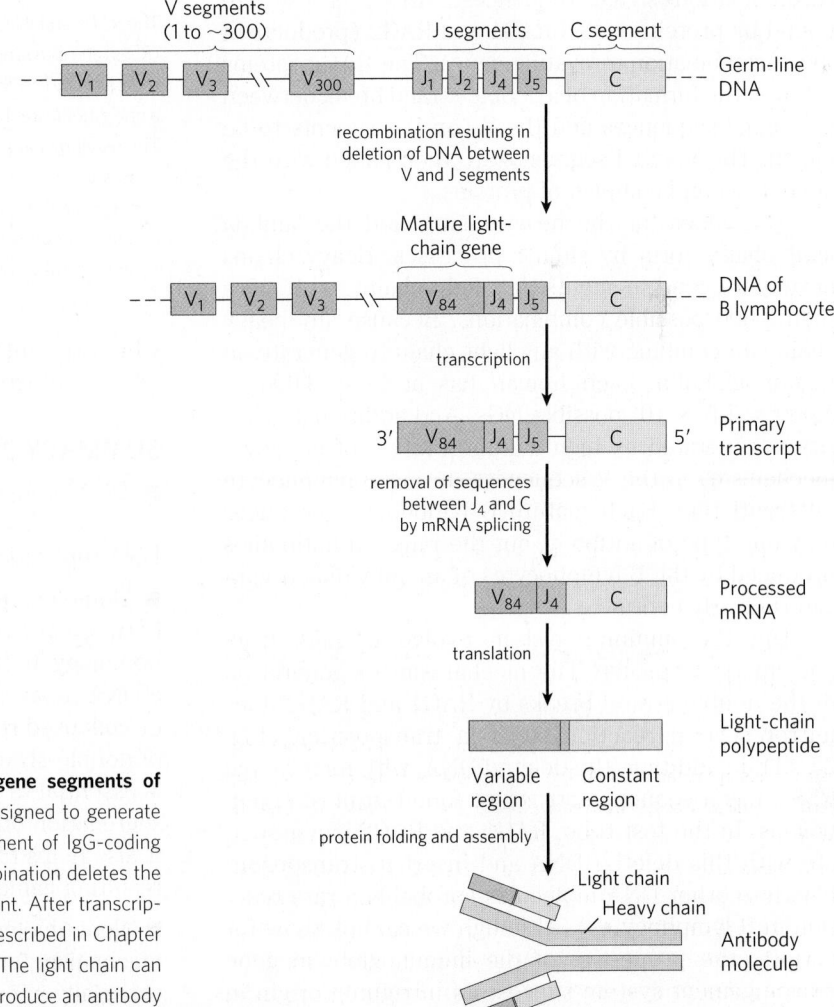

FIGURE 25-43 Recombination of the V and J gene segments of the human IgG kappa light chain. This process is designed to generate antibody diversity. At the top is shown the arrangement of IgG-coding sequences in a stem cell of the bone marrow. Recombination deletes the DNA between a particular V segment and a J segment. After transcription, the transcript is processed by RNA splicing, as described in Chapter 26; translation produces the light-chain polypeptide. The light chain can combine with any of 5,000 possible heavy chains to produce an antibody molecule.

undifferentiated cells, the coding information for this polypeptide chain is separated into three segments. The V (variable) segment encodes the first 95 amino acid residues of the variable region, the J (joining) segment encodes the remaining 12 residues of the variable region, and the C segment encodes the constant region. The genome contains ~300 different V segments, 4 different J segments, and 1 C segment.

As a stem cell in the bone marrow differentiates to form a mature B lymphocyte, one V segment and one J segment are brought together by a specialized recombination system (Fig. 25-43). During this programmed DNA deletion, the intervening DNA is discarded. There are about $300 \times 4 = 1{,}200$ possible V–J combinations. The recombination process is not as precise as the site-specific recombination described earlier, so additional variation occurs in the sequence at the V–J junction. This increases the overall variation by a factor of at least 2.5, so the cells can generate about $2.5 \times 1{,}200 = 3{,}000$ different V–J combinations. The final joining of the V–J combination to the C region is accomplished by an RNA-splicing reaction after transcription, a process described in Chapter 26.

The recombination mechanism for joining the V and J segments is illustrated in **Figure 25-44**. Just beyond each V segment and just before each J segment lie recombination signal sequences (RSSs). These are bound by proteins called RAG1 and RAG2 (products of the *r*ecombination *a*ctivating *g*ene). The RAG proteins catalyze the formation of a double-strand break between the signal sequences and the V (or J) segments to be joined. The V and J segments are then joined with the aid of a second complex of proteins.

The genes for the heavy chains and the lambda light chains form by similar processes. Heavy chains have more gene segments than light chains, with more than 5,000 possible combinations. Because any heavy chain can combine with any light chain to generate an immunoglobulin, each human has at least $3{,}000 \times 5{,}000 = 1.5 \times 10^7$ possible IgGs. And additional diversity is generated by high mutation rates (of unknown mechanism) in the V sequences during B-lymphocyte differentiation. Each mature B lymphocyte produces only one type of antibody, but the range of antibodies produced by the B lymphocytes of an individual organism is clearly enormous.

Did the immune system evolve in part from ancient transposons? The mechanism for generation of the double-strand breaks by RAG1 and RAG2 does mirror several reaction steps in transposition (Fig. 25-44). In addition, the deleted DNA, with its terminal RSSs, has a sequence structure found in most transposons. In the test tube, RAG1 and RAG2 can associate with this deleted DNA and insert it, transposon-like, into other DNA molecules (probably a rare reaction in B lymphocytes). Although we cannot know for certain, the properties of the immunoglobulin gene rearrangement system suggest an intriguing origin in

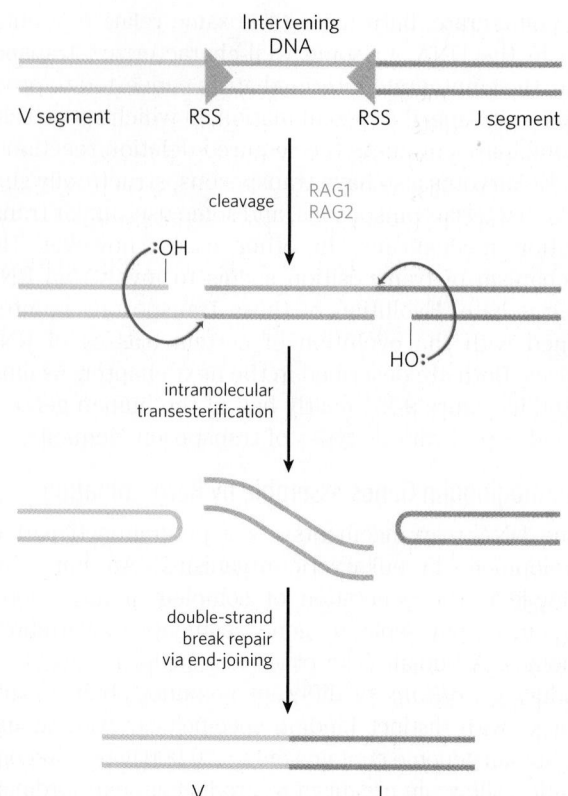

FIGURE 25-44 Mechanism of immunoglobulin gene rearrangement. The RAG1 and RAG2 proteins bind to the recombination signal sequences (RSSs) and cleave one DNA strand between the RSS and the V (or J) segments to be joined. The liberated 3′ hydroxyl then acts as a nucleophile, attacking a phosphodiester bond in the other strand to create a double-strand break. The resulting hairpin bends on the V and J segments are cleaved, and the ends are covalently linked by a complex of proteins specialized for end-joining repair of double-strand breaks. The steps in the generation of the double-strand break catalyzed by RAG1 and RAG2 are chemically related to steps in transposition reactions.

which the distinction between host and parasite has become blurred by evolution.

SUMMARY 25.3 DNA Recombination

■ DNA sequences are rearranged in recombination reactions, usually in processes tightly coordinated with DNA replication or repair.

■ Homologous genetic recombination can take place between any two DNA molecules that share sequence homology. In bacteria, recombination serves mainly as a DNA repair process, focused on reactivating stalled or collapsed replication forks or on the general repair of double-strand breaks. In eukaryotes, recombination is essential to ensure accurate chromosome segregation during the first meiotic cell division. It also helps to create genetic diversity in the resulting gametes.

■ Nonhomologous end joining provides an alternative mechanism for the repair of double-strand breaks, especially in eukaryotic cells.

■ Site-specific recombination occurs only at specific target sequences, and this process can also involve a Holliday intermediate. Recombinases cleave the DNA at specific points and ligate the strands to new partners. This type of recombination is found in virtually all cells, and its many functions include DNA integration and regulation of gene expression.

■ In almost all cells, transposons use recombination to move within or between chromosomes. In vertebrates, a programmed recombination reaction related to transposition joins immunoglobulin gene segments to form immunoglobulin genes during B-lymphocyte differentiation.

Key Terms

Terms in bold are defined in the glossary.

template 989	ORC (origin recognition
semiconservative	complex) 1004
replication 989	DNA polymerase ε 1004
replication fork 990	DNA polymerase δ 1004
origin 990	DNA polymerase α 1004
Okazaki fragment 991	base-excision repair 1008
leading strand 991	DNA glycosylases 1008
lagging strand 991	AP site 1008
nucleases 991	AP endonucleases 1009
exonucleases 991	DNA photolyases 1010
endonucleases 991	error-prone translesion DNA
DNA polymerases 991	synthesis 1013
DNA polymerase I 991	**SOS response** 1013
primer 992	**homologous genetic**
primer terminus 992	**recombination** 1016
processivity 993	**site-specific**
proofreading 993	**recombination** 1016
DNA polymerase III 994	DNA transposition 1016
replisome 995	**recombinational DNA**
helicases 996	**repair** 1017
topoisomerases 996	**branch migration** 1018
primases 997	**Holliday**
DNA ligases 997	**intermediate** 1018
DNA unwinding element	**meiosis** 1019
(DUE) 997	double-strand break repair
AAA+ ATPases 998	model 1021
primosome 999	**nonhomologous end**
catenane 1003	**joining (NHEJ)** 1024
pre-replicative complex	**transposon** 1027
(pre-RC) 1003	**transposition** 1027
licensing 1003	**insertion sequence** 1027
minichromosome maintenance	**cointegrate** 1028
(MCM) protein 1003	

Problems

1. Conclusions from the Meselson-Stahl Experiment The Meselson-Stahl experiment (see Fig. 25-2) proved that DNA undergoes semiconservative replication in *E. coli*. In the "dispersive" model of DNA replication, the parent DNA strands are cleaved into pieces of random size, then joined with pieces of newly replicated DNA to yield daughter duplexes. Explain how the results of Meselson and Stahl's experiment ruled out such a model.

2. Heavy Isotope Analysis of DNA Replication A culture of *E. coli* growing in a medium containing $^{15}NH_4Cl$ is switched to a medium containing $^{14}NH_4Cl$ for three generations (an eightfold increase in population). What is the molar ratio of hybrid DNA ($^{15}N–^{14}N$) to light DNA ($^{14}N–^{14}N$) at this point?

3. Replication of the *E. coli* Chromosome The *E. coli* chromosome contains 4,641,652 bp.

(a) How many turns of the double helix must be unwound during replication of the *E. coli* chromosome?

(b) From the data in this chapter, how long would it take to replicate the *E. coli* chromosome at 37 °C if two replication forks proceeded from the origin? Assume replication occurs at a rate of 1,000 bp/s. Under some conditions *E. coli* cells can divide every 20 min. How might this be possible?

(c) In the replication of the *E. coli* chromosome, about how many Okazaki fragments would be formed? What factors guarantee that the numerous Okazaki fragments are assembled in the correct order in the new DNA?

4. Base Composition of DNAs Made from Single-Stranded Templates Predict the base composition of the total DNA synthesized by DNA polymerase on templates provided by an equimolar mixture of the two complementary strands of bacteriophage $\phi X174$ DNA (a circular DNA molecule). The base composition of one strand is A, 24.7%; G, 24.1%; C, 18.5%; and T, 32.7%. What assumption is necessary to answer this problem?

5. DNA Replication Kornberg and his colleagues incubated soluble extracts of *E. coli* with a mixture of dATP, dTTP, dGTP, and dCTP, all labeled with ^{32}P in the α-phosphate group. After a time, the incubation mixture was treated with trichloroacetic acid, which precipitates the DNA but not the nucleotide precursors. The precipitate was collected, and the extent of precursor incorporation into DNA was determined from the amount of radioactivity present in the precipitate.

(a) If any one of the four nucleotide precursors were omitted from the incubation mixture, would radioactivity be found in the precipitate? Explain.

(b) Would ^{32}P be incorporated into the DNA if only dTTP were labeled? Explain.

(c) Would radioactivity be found in the precipitate if ^{32}P labeled the β or γ phosphate rather than the α phosphate of the deoxyribonucleotides? Explain.

6. The Chemistry of DNA Replication All DNA polymerases synthesize new DNA strands in the $5' \rightarrow 3'$ direction. In some respects, replication of the antiparallel strands of duplex DNA would be simpler if there were also a second type of polymerase, one that synthesized DNA in the $3' \rightarrow 5'$ direction. The two types of polymerase could, in principle, coordinate DNA synthesis without the complicated mechanics required for lagging strand replication. However, no such $3' \rightarrow 5'$-synthesizing enzyme has been found. Suggest two possible mechanisms for $3' \rightarrow 5'$ DNA synthesis. Pyrophosphate should be one product of both proposed reactions. Could one or both mechanisms be supported in a cell? Why or why not? (Hint: You may suggest the use of DNA precursors not actually present in extant cells.)

7. Activities of DNA Polymerases You are characterizing a new DNA polymerase. When the enzyme is incubated with ^{32}P-labeled DNA and no dNTPs, you observe the release of [^{32}P]dNMPs. This release is prevented by adding unlabeled dNTPs. Explain the reactions that most likely underlie these observations. What would you expect to observe if you added pyrophosphate instead of dNTPs?

8. Leading and Lagging Strands Prepare a table that lists the names and compares the functions of the precursors, enzymes, and other proteins needed to make the leading strand versus the lagging strand during DNA replication in *E. coli*.

9. Function of DNA Ligase Some *E. coli* mutants contain defective DNA ligase. When these mutants are exposed to ^{3}H-labeled thymine and the DNA produced is sedimented on an alkaline sucrose density gradient, two radioactive bands appear. One corresponds to a high molecular weight fraction, the other to a low molecular weight fraction. Explain.

10. Fidelity of Replication of DNA What factors promote the fidelity of replication during synthesis of the leading strand of DNA? Would you expect the lagging strand to be made with the same fidelity? Give reasons for your answers.

11. Importance of DNA Topoisomerases in DNA Replication DNA unwinding, such as that occurring in replication, affects the superhelical density of DNA. In the absence of topoisomerases, the DNA would become overwound ahead of a replication fork as the DNA is unwound behind it. A bacterial replication fork will stall when the superhelical density (σ) of the DNA ahead of the fork reaches $+0.14$ (see Chapter 24).

Bidirectional replication is initiated at the origin of a 6,000 bp plasmid in vitro, in the absence of topoisomerases. The plasmid initially has a σ of -0.06. How many base pairs will be unwound and replicated by each replication fork before the forks stall? Assume that both forks travel at the same rate and that each includes all components necessary for elongation except topoisomerase.

12. The Ames Test In a nutrient medium that lacks histidine, a thin layer of agar containing $\sim10^9$ *Salmonella typhimurium* histidine auxotrophs (mutant cells that require histidine to survive) produces ~13 colonies over a two-day incubation period at $37\ °C$ (see Fig. 25-20). How do these colonies arise in the absence of histidine? The experiment is repeated in the presence of 0.4 μg of 2-aminoanthracene. The number of colonies produced over two days exceeds 10,000.

What does this indicate about 2-aminoanthracene? What can you surmise about its carcinogenicity?

13. DNA Repair Mechanisms Vertebrate and plant cells often methylate cytosine in DNA to form 5-methylcytosine (see Fig. 8-5a). In these same cells, a specialized repair system recognizes G–T mismatches and repairs them to G≡C base pairs. How might this repair system be advantageous to the cell? (Explain in terms of the presence of 5-methylcytosine in the DNA.)

14. The Energetic Cost of Mismatch Repair In an *E. coli* cell, DNA polymerase III makes a rare error and inserts a G opposite an A residue at a position 650 bp away from the nearest GATC sequence. The mismatch is accurately repaired by the mismatch repair system. How many phosphodiester bonds derived from deoxynucleotides (dNTPs) are expended during this repair? ATP molecules are also used in this process. Which enzyme(s) consume the ATP?

15. DNA Repair in People with Xeroderma Pigmentosum The condition known as xeroderma pigmentosum (XP) arises from mutations in at least seven different human genes (see Box 25-1). The deficiencies are generally in genes encoding enzymes involved in some part of the pathway for human nucleotide-excision repair. The various types of XP are denoted A through G (XPA, XPB, etc.), with a few additional variants lumped under the label XP-V.

Cultures of fibroblasts from healthy individuals and from patients with XPG are irradiated with ultraviolet light. The DNA is isolated and denatured, and the resulting single-stranded DNA is characterized by analytical ultracentrifugation.

(a) Samples from the normal fibroblasts show a significant reduction in the average molecular weight of the single-stranded DNA after irradiation, but samples from the XPG fibroblasts show no such reduction. Why might this be?

(b) If you assume that a nucleotide-excision repair system is operative in fibroblasts, which step might be defective in the cells from the patients with XPG? Explain.

16. Direct Repair The lesion O^6-meG is normally repaired by direct transfer of the methyl group to the protein O^6-methylguanine-DNA methyltransferase. For the nucleotide sequence AAC(O^6-meG)TGCAC, with a damaged (methylated) G residue, what would be the sequence of both strands of each double-stranded DNA resulting from replication in the following situations?

(a) Replication occurs before repair.
(b) Replication occurs after repair.
(c) Two rounds of replication occur, followed by repair.

17. Holliday Intermediates How does the formation of Holliday intermediates in homologous genetic recombination differ from their formation in site-specific recombination?

18. Cleavage of Holliday Intermediates A Holliday intermediate is formed between two homologous chromosomes, at a point between genes A and B, as shown below. The chromosomes have different alleles of the two genes (A and a, B and b). Where would the Holliday intermediate have to be cleaved (points X and/or Y) to generate a chromosome that would carry (a) an *Ab* genotype or (b) an *ab* genotype?

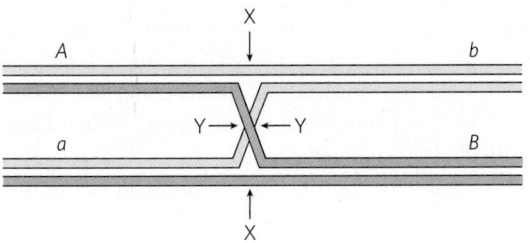

19. A Connection between Replication and Site-Specific Recombination Most wild strains of *S. cerevisiae* have multiple copies of the circular DNA plasmid 2μ (named for its contour length of about 2 μm), which has ~6,300 bp. For its replication, the plasmid uses the host replication system, under the same strict control as the host cell chromosomes, replicating only once per cell cycle. Replication of the plasmid is bidirectional, with both replication forks initiating at a single, well-defined origin. However, one replication cycle of a 2μ plasmid can result in more than two copies of the plasmid, allowing amplification of the plasmid copy number (number of plasmid copies per cell) whenever plasmid segregation at cell division leaves one daughter cell with fewer than the normal complement of plasmid copies. Amplification requires a site-specific recombination system encoded by the plasmid, which serves to invert one part of the plasmid relative to the other. Explain how a site-specific inversion event could result in amplification of the plasmid copy number. (Hint: Consider the situation when replication forks have duplicated one recombination site but not the other.)

Data Analysis Problem

20. Mutagenesis in *Escherichia coli* Many mutagenic compounds act by alkylating the bases in DNA. The alkylating agent R7000 (7-methoxy-2-nitronaphtho[2,1-*b*]furan) is an extremely potent mutagen.

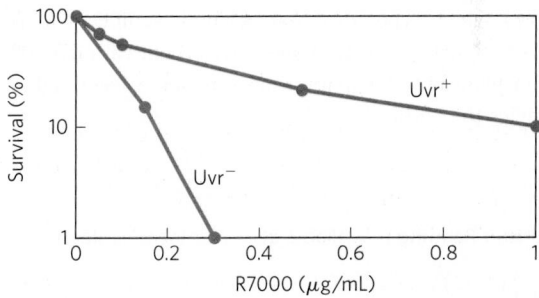

R7000

In vivo, R7000 is activated by the enzyme nitroreductase, and this more reactive form covalently attaches to DNA—primarily, but not exclusively, to G≡C base pairs.

In a 1996 study, Quillardet, Touati, and Hofnung explored the mechanisms by which R7000 causes mutations in *E. coli*. They compared the genotoxic activity of R7000 in two strains of *E. coli*: the wild-type (Uvr⁺) and mutants lacking *uvrA* activity (Uvr⁻; see Table 25-6). They first measured rates of mutagenesis. Rifampicin is an inhibitor of RNA polymerase (see Chapter 26). In its presence, cells will not grow unless certain mutations occur in the gene encoding RNA polymerase; the appearance of rifampicin-resistant colonies thus provides a useful measure of mutagenesis rates.

The effects of different concentrations of R7000 were determined, with the results shown in the graph below.

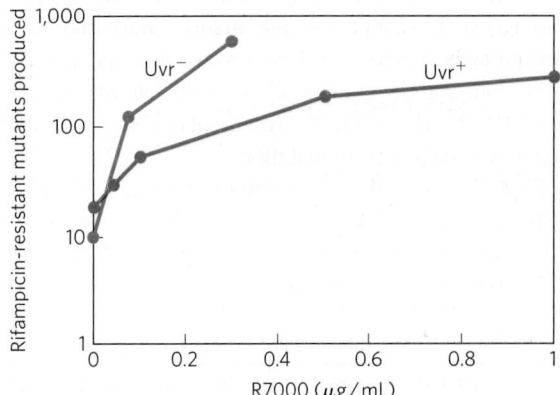

(a) Why are some mutants produced even when no R7000 is present?

Quillardet and colleagues also measured the survival rate of bacteria treated with different concentrations of R7000, with the following results.

(b) Explain how treatment with R7000 is lethal to cells.

(c) Explain the differences in the mutagenesis curves and in the survival curves for the two types of bacteria, Uvr⁺ and Uvr⁻, as shown in the graphs.

The researchers went on to measure the amount of R7000 covalently attached to the DNA in Uvr⁺ and Uvr⁻ *E. coli*. They incubated bacteria with [³H]R7000 for 10 or 70 minutes, extracted the DNA, and measured its ³H content in counts per minute (cpm) per microgram of DNA.

	³H in DNA (cpm/μg)	
Time (min)	**Uvr⁺**	**Uvr⁻**
10	76	159
70	69	228

(d) Explain why the amount of ³H drops over time in the Uvr⁺ strain and rises over time in the Uvr⁻ strain.

Quillardet and colleagues then examined the particular DNA sequence changes caused by R7000 in the Uvr⁺ and Uvr⁻ bacteria. For this, they used six different strains of *E. coli*, each with a different point mutation in the *lacZ* gene, which encodes β-galactosidase (this enzyme catalyzes the same reaction as lactase; see Fig. 14-11). Cells with any of these mutations have a nonfunctional β-galactosidase and are unable to metabolize lactose (i.e., a Lac$^-$ phenotype). Each type of point mutation required a specific reverse mutation to restore *lacZ* gene function and Lac$^+$ phenotype. By plating cells on a medium containing lactose as the sole carbon source, it was possible to select for these reverse-mutated, Lac$^+$ cells. And by counting the number of Lac$^+$ cells following mutagenesis of a particular strain, the researchers could measure the frequency of each type of mutation.

First, they looked at the mutation spectrum in Uvr$^-$ cells. The following table shows the results for the six strains, CC101 through CC106 (with the point mutation required to produce Lac$^+$ cells indicated in parentheses).

Number of Lac$^+$ cells (average ± SD)

R7000 (μg/mL)	CC101 (A=T to C≡G)	CC102 (G≡C to A=T)	CC103 (G≡C to C≡G)	CC104 (G≡C to T=A)	CC105 (A=T to T=A)	CC106 (A=T to G≡C)
0	6 ± 3	11 ± 9	2 ± 1	5 ± 3	2 ± 1	1 ± 1
0.075	24 ± 19	34 ± 3	8 ± 4	82 ± 23	40 ± 14	4 ± 2
0.15	24 ± 4	26 ± 2	9 ± 5	180 ± 71	130 ± 50	3 ± 2

(e) Which types of mutation show significant increases above the background rate due to treatment with R7000? Provide a plausible explanation for why some have higher frequencies than others.

(f) Can all of the mutations you listed in (e) be explained as resulting from covalent attachment of R7000 to a G≡C base pair? Explain your reasoning.

(g) Figure 25-27b shows how methylation of guanine residues can lead to a G≡C to A=T mutation. Using a similar pathway, show how an R7000–G adduct could lead to the G≡C to A=T or T=A mutations shown above. Which base pairs with the R7000–G adduct?

The results for the Uvr$^+$ bacteria are shown in the table below.

Number of Lac$^+$ cells (average ± SD)

R7000 (μg/mL)	CC101 (A=T to C≡G)	CC102 (G≡C to A=T)	CC103 (G≡C to C≡G)	CC104 (G≡C to T=A)	CC105 (A=T to T=A)	CC106 (A=T to G≡C)
0	2 ± 2	10 ± 9	3 ± 3	4 ± 2	6 ± 1	0.5 ± 1
1	7 ± 6	21 ± 9	8 ± 3	23 ± 15	13 ± 1	1 ± 1
5	4 ± 3	15 ± 7	22 ± 2	68 ± 25	67 ± 14	1 ± 1

(h) Do these results show that all mutation types are repaired with equal fidelity? Provide a plausible explanation for your answer.

Reference

Quillardet, P., E. Touati, and M. Hofnung. 1996. Influence of the *uvr*-dependent nucleotide excision repair on DNA adducts formation and mutagenic spectrum of a potent genotoxic agent: 7-methoxy-2-nitronaphtho[2,1-*b*]furan (R7000). *Mutat. Res.* 358:113–122.

Further Reading is available at www.macmillanlearning.com/LehningerBiochemistry7e.

CHAPTER 26

RNA Metabolism

Self-study tools that will help you practice what you've learned and reinforce this chapter's concepts are available online.
Go to www.macmillanlearning.com/LehningerBiochemistry7e.

Expression of the information in a gene generally involves production of an RNA molecule transcribed from a DNA template. Strands of RNA and DNA may seem similar at first glance, differing only in that RNA has a hydroxyl group at the 2′ position of the aldopentose, and uracil usually replaces thymine. However, unlike DNA, most RNAs carry out their functions as single strands, strands that fold back on themselves and have the potential for much greater structural diversity than DNA (Chapter 8). RNA is thus suited to a variety of cellular functions.

RNA is the only macromolecule known to have a role both in the storage and transmission of information and in catalysis, which has led to much speculation about its possible role as an essential chemical intermediate in the development of life on this planet. The discovery of catalytic RNAs, or ribozymes, has changed the very definition of an enzyme, extending it beyond the domain of proteins. Proteins nevertheless remain essential to RNA and its functions. In the biosphere of today, all nucleic acids, including RNAs, are complexed with proteins. Some of these complexes are quite elaborate, and RNA can assume both structural and catalytic roles within complicated biochemical machines.

All RNA molecules except the RNA genomes of certain viruses are derived from information permanently stored in DNA. During **transcription**, an enzyme system converts the genetic information in a segment of double-stranded DNA into an RNA strand with a base sequence complementary to one of the DNA strands. Three major kinds of RNA are produced. **Messenger RNAs (mRNAs)** encode the amino acid sequence of one or more polypeptides specified by a gene or set of genes. **Transfer RNAs (tRNAs)** read the information encoded in the mRNA and transfer the appropriate amino acid to a growing polypeptide chain during protein synthesis. **Ribosomal RNAs (rRNAs)** are constituents of ribosomes, the intricate cellular machines that synthesize proteins. Many additional, specialized RNAs have regulatory or catalytic functions or are precursors to the three main classes of RNA. These special-function RNAs are no longer thought of as minor species in the catalog of cellular RNAs. In vertebrates, RNAs that do not fit into one of the classical categories (mRNA, tRNA, rRNA) seem to vastly outnumber those that do.

During replication the entire chromosome is usually copied, but transcription is more selective. Only particular genes or groups of genes are transcribed at any one time, and some portions of the DNA genome are never transcribed. The cell restricts the expression of genetic information to the formation of gene products needed at any particular moment. Specific regulatory sequences mark the beginning and end of the DNA segments to be transcribed and designate which strand in duplex DNA is to be used as the template. The transcript itself may interact with other RNA molecules as part of the overall regulatory program. The regulation of transcription is described in detail in Chapter 28.

The sum of all the RNA molecules produced in a cell under a given set of conditions is called the cellular **transcriptome**. Given the relatively small fraction of the human genome devoted to protein-coding genes, we might expect that only a small part of the human genome is transcribed. This is not the case. Modern analyses of transcription patterns have revealed that much of the genome of humans and other mammals is transcribed into RNA. The products are predominantly not mRNAs, tRNAs, or rRNAs, but rather special-function RNAs, a host of which are being discovered. Many of these seem to be involved in regulation of gene expression; however, the rapid pace of discovery has forced us to realize that we do not yet know what many of these RNAs do.

In this chapter we examine the synthesis of RNA on a DNA template and the postsynthetic processing and turnover of RNA molecules. In doing so, we encounter many of the specialized functions of RNA, including catalytic functions. Interestingly, the substrates for RNA enzymes are often other RNA molecules. We also describe systems in which RNA is the template and DNA the product, rather than vice versa. The information pathways thus come full circle and reveal that template-dependent nucleic acid synthesis has standard rules, regardless of the nature of template or product (RNA or DNA). This examination of the biological interconversion of DNA and RNA as information carriers leads inevitably to a discussion of the evolutionary origin of biological information.

26.1 DNA-Dependent Synthesis of RNA

Our discussion of RNA synthesis begins with a comparison between transcription and DNA replication (Chapter 25). Transcription resembles replication in its fundamental chemical mechanism, its polarity (direction of synthesis), and its use of a template. And like replication, transcription has initiation, elongation, and termination phases. In the literature on transcription, initiation is further divided into discrete phases of DNA binding and initiation of RNA synthesis. Transcription differs from replication in that it does not require a primer and, generally, involves only limited segments of a DNA molecule. Additionally, within transcribed segments, only one DNA strand serves as a template for a particular RNA molecule.

RNA Is Synthesized by RNA Polymerases

The discovery of DNA polymerase and its dependence on a DNA template spurred a search for an enzyme that synthesizes RNA complementary to a DNA strand. By 1960, four research groups had independently detected an enzyme in cellular extracts that could form an RNA polymer from ribonucleoside 5'-triphosphates. Subsequent work on the purified *Escherichia coli* RNA polymerase helped to define the fundamental properties of transcription (**Fig. 26-1**). **DNA-dependent RNA polymerase** requires, in addition to a DNA template, all four ribonucleoside 5'-triphosphates (ATP, GTP, UTP, and CTP) as precursors of the nucleotide

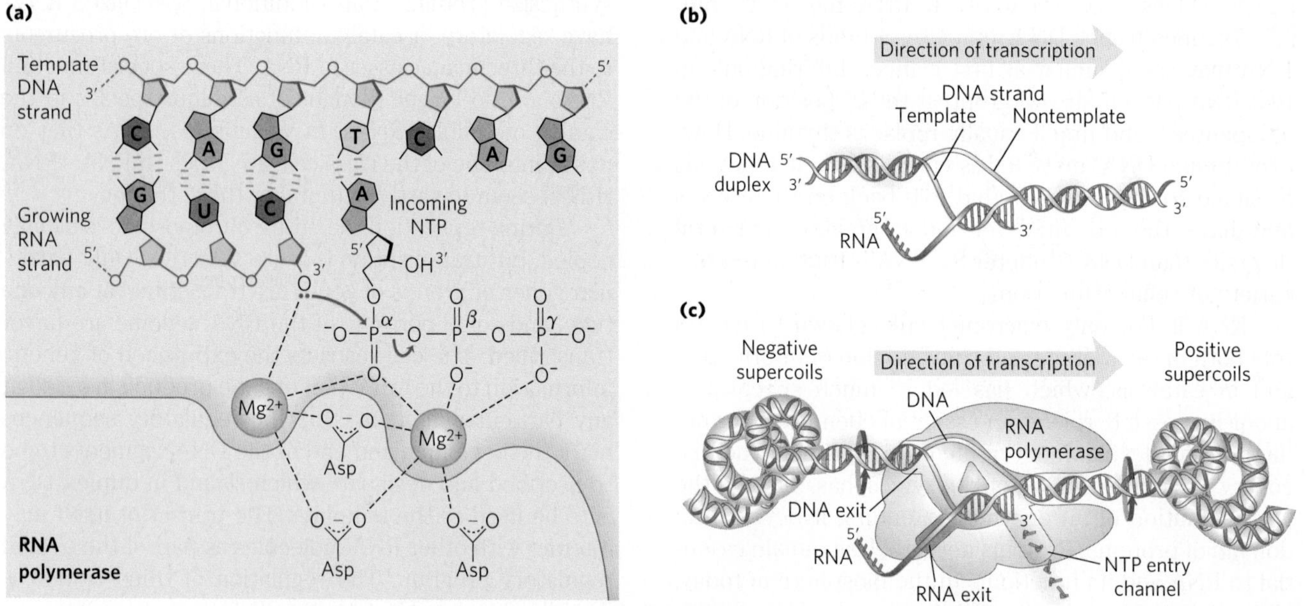

(a)

(b)

(c)

MECHANISM FIGURE 26-1 **Transcription by RNA polymerase in *E. coli*.** For synthesis of an RNA strand complementary to one of two DNA strands in a double helix, the DNA is transiently unwound. **(a)** Catalytic mechanism of RNA synthesis by RNA polymerase. Notice that this is essentially the same mechanism used by DNA polymerases (see Fig. 25-5a). The reaction involves two Mg^{2+} ions, coordinated to the phosphate groups of the incoming nucleoside triphosphates (NTPs) and to three Asp residues, which are highly conserved in the RNA polymerases of all species. One Mg^{2+} ion facilitates attack by the 3'-hydroxyl group on the α phosphate of the NTP; the other Mg^{2+} ion facilitates displacement of the pyrophosphate. Both metal ions stabilize the pentacovalent transition state.

(b) About 17 bp of DNA are unwound at any given time. RNA polymerase and the transcription bubble move from left to right along the DNA as shown, facilitating RNA synthesis. The DNA is unwound ahead and rewound behind as RNA is transcribed. As the DNA is rewound, the RNA-DNA hybrid is displaced and the RNA strand is extruded.

(c) Movement of an RNA polymerase along DNA tends to create positive supercoils (overwound DNA) ahead of the transcription bubble and negative supercoils (underwound DNA) behind it. The RNA polymerase is in close contact with the DNA ahead of the transcription bubble as well as with the separated DNA strands and the RNA within and immediately behind the bubble. A channel in the protein funnels new NTPs to the polymerase active site. The polymerase footprint encompasses about 35 bp of DNA during elongation.

(5') C G C T A T A G C G T T T (3')   DNA nontemplate (coding) strand

(3') G C G A T A T C G C A A A (5')   DNA template strand

(5') C G C U A U A G C G U U U (3')   RNA transcript

FIGURE 26-2 Template and nontemplate (coding) DNA strands. The two complementary strands of DNA are defined by their function in transcription. The RNA transcript is synthesized on the template strand and is identical in sequence (with U in place of T) to the nontemplate strand, or coding strand.

units of RNA, as well as Mg^{2+}. The protein also binds one Zn^{2+}. The chemistry and mechanism of RNA synthesis closely resemble those used by DNA polymerases (see Fig. 25-5a). RNA polymerase elongates an RNA strand by adding ribonucleotide units to the 3′-hydroxyl end, building RNA in the 5′→3′ direction. The 3′-hydroxyl group acts as a nucleophile, attacking the α phosphate of the incoming ribonucleoside triphosphate (Fig. 26-1a) and releasing pyrophosphate. The overall reaction is

$$\underset{\text{RNA}}{(NMP)_n} + NTP \longrightarrow \underset{\text{Lengthened RNA}}{(NMP)_{n+1}} + PP_i$$

RNA polymerase requires DNA for activity and is most active when bound to a double-stranded DNA. As noted above, only one of the two DNA strands serves as a template. The template DNA strand is copied in the 3′→5′ direction (antiparallel to the new RNA strand), just as in DNA replication. Each nucleotide in the newly formed RNA is selected by Watson-Crick base-pairing interactions: U residues are inserted in the RNA to pair with A residues in the DNA template, G residues are inserted to pair with C residues, and so on. Base-pair geometry (see Fig. 25-6) may also play a role in base selection.

Unlike DNA polymerase, RNA polymerase does not require a primer to initiate synthesis. Initiation occurs when RNA polymerase binds at specific DNA sequences called promoters (described below). The 5′-triphosphate group of the first residue in a nascent (newly formed) RNA molecule is not cleaved to release PP_i, but instead remains intact throughout the transcription process. During the elongation phase of transcription, the growing end of the new RNA strand base-pairs temporarily with the DNA template to form a short hybrid RNA-DNA double helix, about 8 bp long (Fig. 26-1b). The

RNA in this hybrid duplex "peels off" shortly after its formation, and the DNA duplex re-forms.

To enable RNA polymerase to synthesize an RNA strand complementary to one of the DNA strands, the DNA duplex must unwind over a short distance, forming a transcription "bubble." During transcription, the *E. coli* RNA polymerase generally keeps about 17 bp unwound. The 8 bp RNA-DNA hybrid occurs in this unwound region. Elongation of a transcript by *E. coli* RNA polymerase proceeds at a rate of 50 to 90 nucleotides/s. Because DNA is a helix, movement of a transcription bubble requires considerable strand rotation of the nucleic acid molecules. DNA strand rotation is restricted in most DNAs by DNA-binding proteins and other structural barriers. As a result, a moving RNA polymerase generates waves of positive supercoils ahead of the transcription bubble and negative supercoils behind (Fig. 26-1c). This has been observed both in vitro and in vivo (in bacteria). In the cell, the topological problems caused by transcription are relieved through the action of topoisomerases (Chapter 24).

>> Key Convention: The two complementary DNA strands have different roles in transcription. The strand that serves as template for RNA synthesis is called the **template strand**. The DNA strand complementary to the template, the **nontemplate strand**, or **coding strand**, is identical in base sequence to the RNA transcribed from the gene, with U in the RNA in place of T in the DNA **(Fig. 26-2)**. The coding strand for a particular gene may be located in either strand of a given chromosome (as shown in **Fig. 26-3** for a virus). By convention, the regulatory sequences that control transcription (described later in this chapter) are designated by the sequences in the coding strand. **<<**

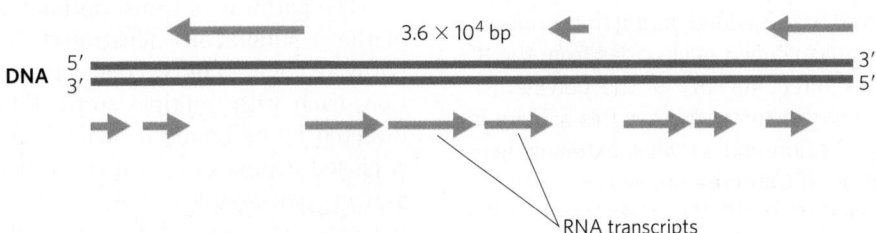

FIGURE 26-3 Organization of coding information in the adenovirus genome. The genetic information of the adenovirus genome (a conveniently simple example) is encoded by a double-stranded DNA molecule of 36,000 bp, both strands of which encode proteins. The information for most proteins is encoded by (that is, identical to) the top strand—by convention, the strand oriented 5′ to 3′ from left to right. The bottom strand acts as template for these transcripts. However, a few proteins are encoded by the bottom strand, which is transcribed in the opposite direction (and uses the top strand as template). Synthesis of mRNAs in adenovirus is much more complex than shown here. Many of the mRNAs derived using the upper strand as template are initially synthesized as a single, long transcript (25,000 nucleotides), which is then extensively processed to produce the separate mRNAs. Adenovirus causes upper respiratory tract infections in some vertebrates.

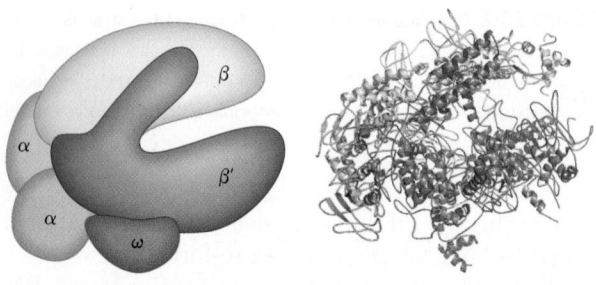

FIGURE 26-4 Structure of the RNA polymerase holoenzyme of the bacterium _Thermus aquaticus._ The overall structure of this enzyme is very similar to that of the _E. coli_ RNA polymerase; no DNA or RNA is shown here. The several subunits of the bacterial RNA polymerase give the enzyme the shape of a crab claw. The pincers are formed by the large β and β' subunits. The subunits are shown in the same colors in the schematic and the ribbon structure. [Source: Ribbon structure: Information from G. Zhang et al., _Cell_ 98:811, 1999, based on PDB ID 1HQM, L. Minakhin et al., _Proc. Natl. Acad. Sci. USA_ 98:892, 2001.]

The DNA-dependent RNA polymerase of _E. coli_ is a large, complex enzyme with five core subunits ($\alpha_2\beta\beta'\omega$; M_r 390,000) and a sixth subunit, one of a group designated σ, with variants designated by size (molecular weight). The σ subunit binds transiently to the core and directs the enzyme to specific binding sites on the DNA (described below). These six subunits constitute the RNA polymerase holoenzyme **(Fig. 26-4)**. The RNA polymerase holoenzyme of _E. coli_ thus exists in several forms, depending on the type of σ subunit. The most common subunit is σ^{70} (M_r 70,000), and the upcoming discussion focuses on the corresponding RNA polymerase holoenzyme.

RNA polymerases lack a separate proofreading $3'\rightarrow5'$ exonuclease active site (such as that of many DNA polymerases), and the error rate for transcription is higher than that for chromosomal DNA replication—approximately one error for every 10^4 to 10^5 ribonucleotides incorporated into RNA. Because many copies of an RNA are generally produced from a single gene, and all RNAs are eventually degraded and replaced, a mistake in an RNA molecule is of less consequence to the cell than a mistake in the permanent information stored in DNA. Many RNA polymerases, including bacterial RNA polymerase and the eukaryotic RNA polymerase II (discussed below), do pause when a mispaired base is added during transcription, and they can remove mismatched nucleotides from the 3' end of a transcript by direct reversal of the polymerase reaction. But we do not yet know whether this activity is a true proofreading function and to what extent it may contribute to the fidelity of transcription.

RNA Synthesis Begins at Promoters

Initiation of RNA synthesis at random points in a DNA molecule would be an extraordinarily wasteful process. Instead, an RNA polymerase binds to specific sequences in the DNA called **promoters**, which direct the transcription of adjacent segments of DNA (genes). The sequences where RNA polymerases bind are variable,

and much research has focused on identifying the particular sequences that are critical to promoter function.

In _E. coli_, RNA polymerase binding occurs within a region stretching from about 70 bp before the transcription start site to about 30 bp beyond it. By convention, the DNA base pairs that correspond to the beginning of an RNA molecule are given positive numbers, and those preceding the RNA start site are given negative numbers. The promoter region thus extends between positions -70 and $+30$. Analyses and comparisons of the most common class of bacterial promoters (those recognized by an RNA polymerase holoenzyme containing σ^{70}) have revealed similarities in two short sequences centered about positions -10 and -35 **(Fig. 26-5)**. These sequences are important interaction sites for the σ^{70} subunit. Although the sequences are not identical for all bacterial promoters in this class, certain nucleotides that are particularly common at each position form a **consensus sequence** (recall the _E. coli oriC_ consensus sequence; see Fig. 25-10). The consensus sequence at the -10 region is (5')TATAAT(3'); at the -35 region it is (5')TTGACA(3'). A third AT-rich recognition element, called the UP (upstream promoter) element, occurs between positions -40 and -60 in the promoters of certain highly expressed genes. The UP element is bound by the α subunit of RNA polymerase. The efficiency with which an RNA polymerase containing σ^{70} binds to a promoter and initiates transcription is determined in large measure by these sequences, the spacing between them, and their distance from the transcription start site.

Many independent lines of evidence attest to the functional importance of the sequences in the -35 and -10 regions. Mutations that affect the function of a given promoter often occur in a base pair in these regions. Variations in the consensus sequence also affect the efficiency of RNA polymerase binding and transcription initiation. A change in only one base pair can decrease the rate of binding by several orders of magnitude. The promoter sequence thus establishes a basal level of expression that can vary greatly from one _E. coli_ gene to the next. A method that provides information about the interaction between RNA polymerase and promoters is illustrated in Box 26-1.

The pathway of transcription initiation and the fate of the σ subunit are illustrated in **Figure 26-6**. The pathway consists of two major parts, binding and initiation, each with multiple steps. First, the polymerase, directed by its bound σ factor, binds to the promoter. A closed complex (in which the bound DNA is intact) and an open complex (in which the bound DNA is intact but partially unwound near the -10 sequence) form in succession. Second, transcription is initiated within the complex, leading to a conformational change that converts the complex to the elongation form, followed by movement of the transcription complex away from the promoter (promoter clearance). Any of these steps can be affected by the specific makeup of the promoter sequences. The σ subunit dissociates stochastically

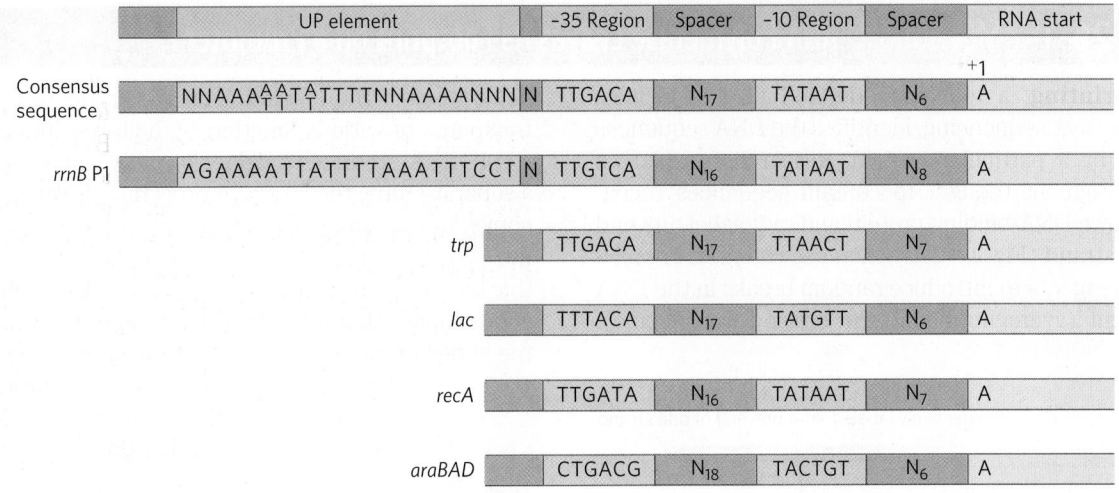

FIGURE 26-5 Typical E. coli promoters recognized by an RNA polymerase holoenzyme containing σ^{70}. Sequences of the nontemplate strand are shown, read in the 5'→3' direction, as is the convention for representations of this kind. The sequences differ from one promoter to the next, but comparisons of many promoters reveal similarities, particularly in the −10 and −35 regions. The sequence element UP, not present in all E. coli promoters, is shown in the P1 promoter for the highly expressed rRNA gene rrnB. UP elements, generally occurring in the region between −40 and −60, strongly stimulate transcription at the promoters that contain them. The UP element in the rrnB P1 promoter encompasses the region between −38 and −59. The consensus sequence for E. coli promoters recognized by σ^{70} is shown second from the top. Spacer regions contain slightly variable numbers of nucleotides (N). Only the first nucleotide coding the RNA transcript (at position +1) is shown.

(at random) as the polymerase enters the elongation phase of transcription. The protein NusA (M_r 54,430) binds to the elongating RNA polymerase, competitively with the σ subunit. Once transcription is complete, NusA dissociates from the enzyme, the RNA polymerase dissociates from the DNA, and a σ factor (σ^{70} or another) can again bind to the enzyme to initiate transcription.

As described further in Chapter 27, transcription of mRNAs and their translation are tightly coupled in bacteria. As a protein-coding gene is being transcribed, ribosomes rapidly bind to and begin to translate the mRNA before its synthesis is complete. Another protein, NusG, binds directly to both the ribosome and RNA polymerase, linking the two complexes. The rate of translation directly affects the rate of transcription.

E. coli has other classes of promoters, bound by RNA polymerase holoenzymes with different σ subunits (Table 26-1). An example is the promoters of the heat shock genes. The products of this set of genes are made at higher levels when the cell has received an insult, such as a sudden increase in temperature. RNA polymerase binds to the promoters of these genes only when σ^{70} is replaced with the σ^{32} (M_r 32,000) subunit, which is specific for the heat shock promoters (see Fig. 28-3). By using different σ subunits, the cell can coordinate the expression of sets of genes, permitting major changes in cell physiology. Which sets of genes are expressed is determined by the availability of the various σ subunits, which is determined by several factors: regulated rates of synthesis and degradation, postsynthetic modifications that switch individual σ subunits between active and inactive forms, and a specialized class of anti-σ proteins, each type binding to and sequestering a particular σ subunit (rendering it unavailable for transcription initiation).

Transcription Is Regulated at Several Levels

Requirements for any gene product vary with cellular conditions or developmental stage, and transcription of each gene is carefully regulated to form gene products only in the proportions needed. Regulation can occur at any step in transcription, including elongation and termination. However, much of the regulation is directed at the polymerase binding and transcription initiation steps outlined in Figure 26-6. Differences in promoter sequences are just one of several levels of control.

The binding of proteins to sequences both near to and distant from the promoter can also affect levels of gene expression. Protein binding can *activate* transcription by facilitating either RNA polymerase binding or steps farther along in the initiation process, or it can *repress* transcription by blocking the activity of the polymerase. In *E. coli*, one protein that activates transcription is the **cAMP receptor protein (CRP)**, which increases the transcription of genes coding for enzymes that metabolize sugars other than glucose when cells are grown in the absence of glucose. **Repressors** are proteins that block the synthesis of RNA at specific genes. In the case of the Lac repressor (Chapter 28), transcription of the genes for the enzymes of lactose metabolism is blocked when lactose is unavailable.

Transcription is the first step in the complicated and energy-intensive pathway of protein synthesis, so much of the regulation of protein levels in both bacterial and eukaryotic cells is directed at transcription, particularly its early stages. In Chapter 28 we describe many mechanisms by which this regulation is accomplished.

BOX 26-1 METHODS RNA Polymerase Leaves Its Footprint on a Promoter

Footprinting, a technique derived from principles used in DNA sequencing, identifies the DNA sequences bound by a particular protein. Researchers isolate a DNA fragment thought to contain sequences recognized by a DNA-binding protein and radiolabel one end of one strand (Fig. 1). They then use chemical or enzymatic reagents to introduce random breaks in the DNA fragment (averaging about one break per molecule).

Separation of the labeled cleavage products (broken fragments of various lengths) by high-resolution electrophoresis produces a ladder of radioactive bands. In a separate tube, the cleavage procedure is repeated on copies of the same DNA fragment in the presence of the DNA-binding protein. The researchers then subject the two sets of cleavage products to electrophoresis and compare them side by side. A gap ("footprint") in the series of radioactive bands derived from the DNA-protein sample, attributable to protection of the DNA by the bound protein, identifies the sequences that the protein binds.

The precise location of the protein-binding site can be determined by directly sequencing (see Fig. 8-35) copies of the same DNA fragment and including the sequencing lanes (not shown here) on the same gel with the footprint. Figure 2 shows footprinting results for the binding of RNA polymerase to a DNA fragment containing a promoter. The polymerase covers 60 to 80 bp; protection by the bound enzyme includes the -10 and -35 regions.

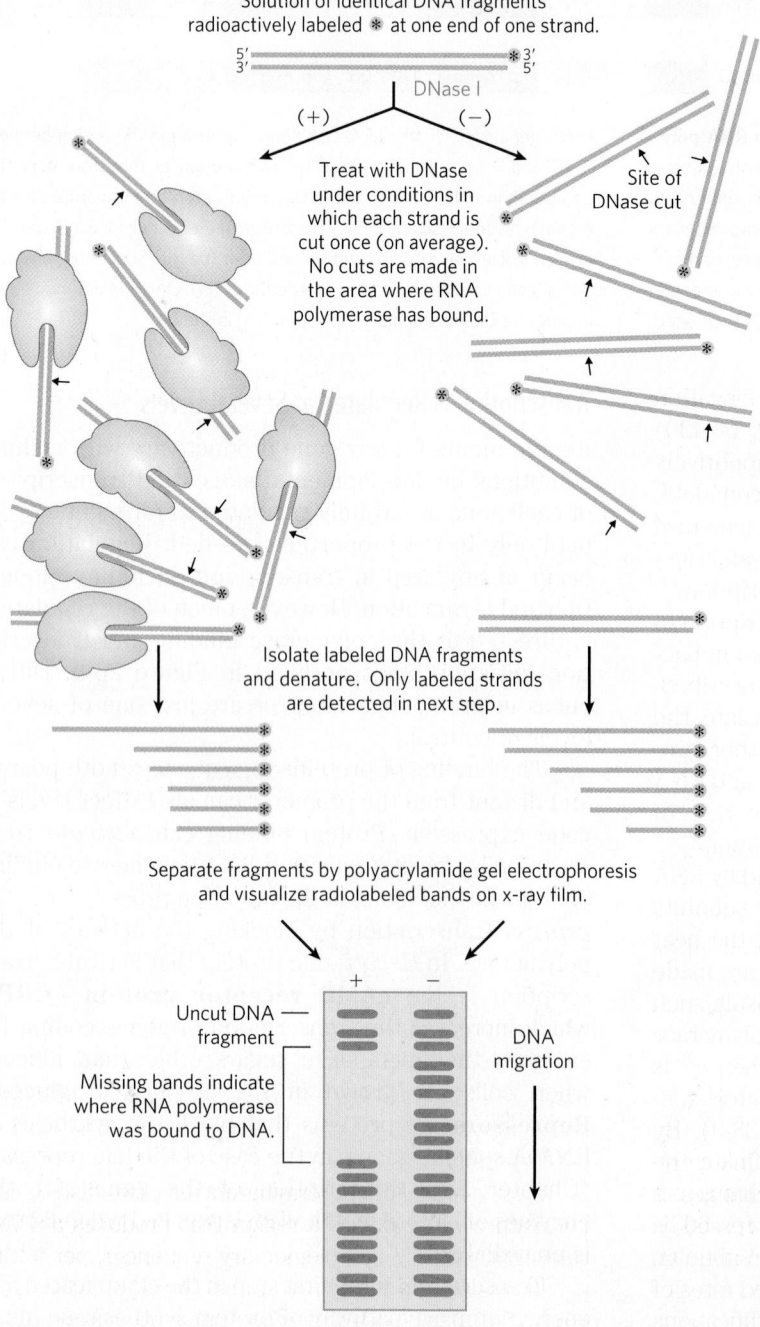

FIGURE 1 Footprint analysis of the RNA polymerase–binding site on a DNA fragment. Separate experiments are carried out in the presence (+) and absence (−) of the polymerase.

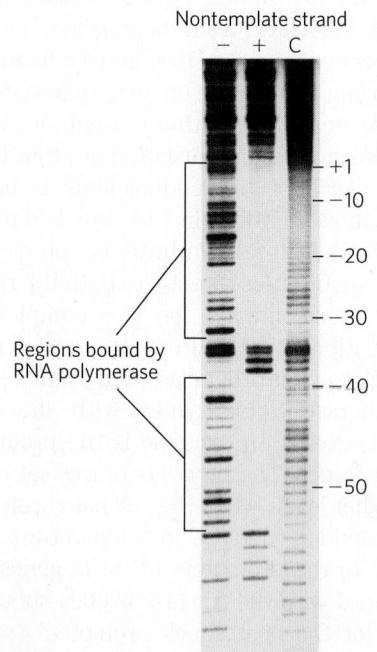

FIGURE 2 Footprinting results of RNA polymerase binding to the *lac* promoter (see Fig. 26-5). In this experiment, the 5′ end of the nontemplate strand was radioactively labeled. Lane C is a control in which the labeled DNA fragments were cleaved with a chemical reagent that produces a more uniform banding pattern. [Source: © Courtesy of Carol Gross Laboratory.]

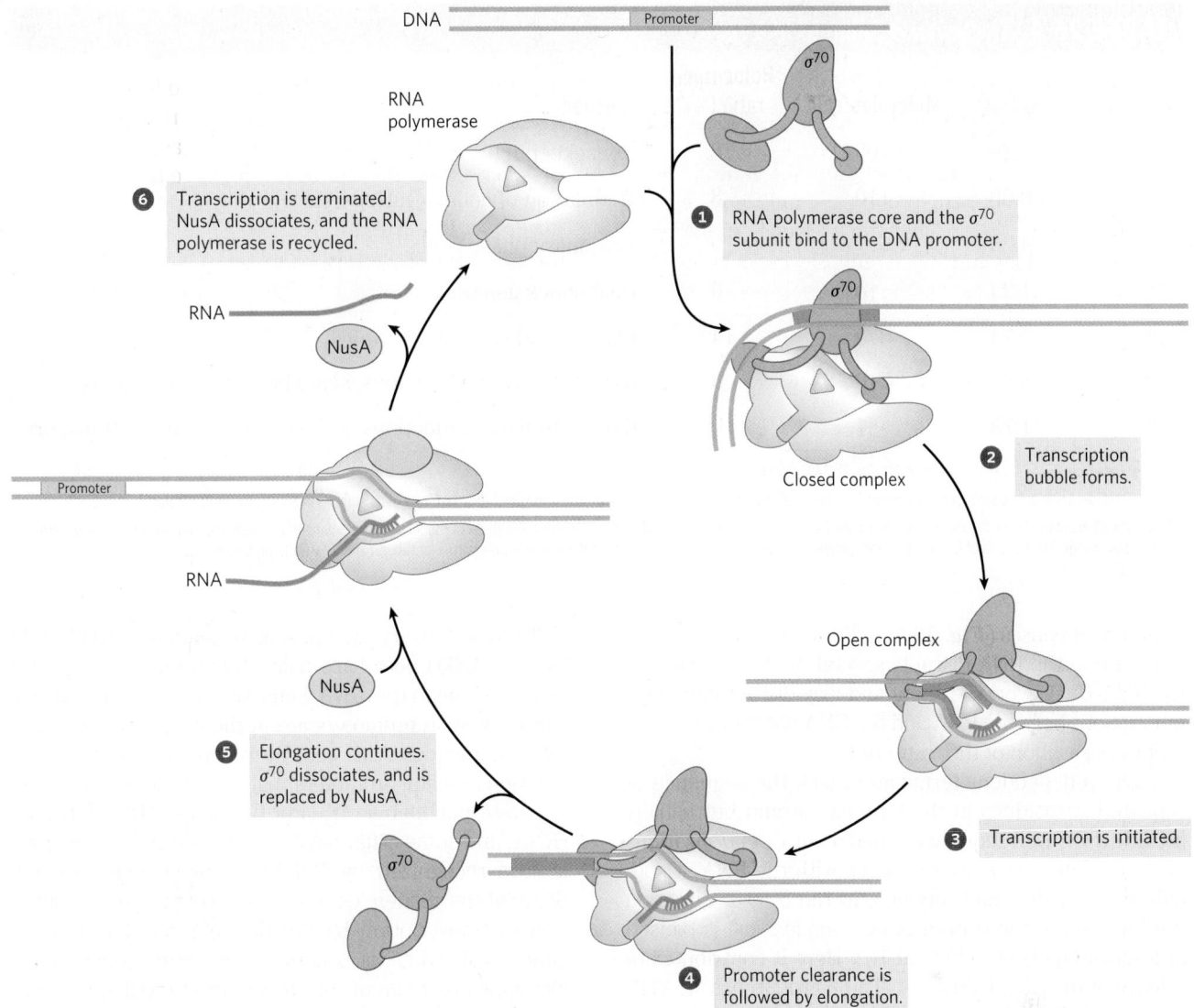

FIGURE 26-6 Transcription initiation and elongation by *E. coli* RNA polymerase. Initiation of transcription requires several steps generally divided into two phases, binding and initiation. In the binding phase, the initial interaction of the RNA polymerase with the promoter leads to formation of a closed complex, in which the promoter DNA is stably bound but not unwound. A 12 to 15 bp region of DNA—from within the −10 region to position +2 or +3—is then unwound to form an open complex. Additional intermediates (not shown) have been detected in the pathways leading to the closed and open complexes, along with several changes in protein conformation. The initiation phase encompasses transcription initiation and promoter clearance (steps ❶ through ❹ here). Once elongation commences, the σ subunit is released and is replaced by the protein NusA. The polymerase leaves the promoter and becomes committed to elongation of the RNA (step ❺). When transcription is complete, the RNA is released, the NusA protein dissociates, and the RNA polymerase dissociates from the DNA (step ❻). Another σ subunit binds to the RNA polymerase and the process begins again.

Specific Sequences Signal Termination of RNA Synthesis

RNA synthesis is processive; that is, the RNA polymerase introduces a large number of nucleotides into a growing RNA molecule before dissociating (p. 993). This is necessary because, if the polymerase released an RNA transcript prematurely, it could not resume synthesis of the same RNA and would have to start again. However, an encounter with certain DNA sequences results in a pause in RNA synthesis, and at some of these sequences transcription is terminated. Our focus here is again on the well-studied systems in bacteria. *E. coli* has at least two classes of termination signals: one class relies on a protein factor called ρ (rho), and the other is ρ-independent.

Most ρ-independent terminators have two distinguishing features. The first is a region that produces an RNA transcript with self-complementary sequences, permitting the formation of a hairpin structure (see Fig. 8-19a) centered 15 to 20 nucleotides before the projected end of the RNA strand. The second feature is a highly conserved string of three A residues in the template strand that are transcribed into U residues near the 3′ end of the hairpin. When a polymerase arrives at a termination site with this

TABLE 26-1 The Seven σ Subunits of *E. coli*

σ subunit	K_d (nM)	Molecules/cell[a]	Holoenzyme ratio (%)[a]	Function
σ^{70}	0.26	700	78	Housekeeping
σ^{54}	0.30	110	8	Modulation of cellular nitrogen levels
σ^{38}	4.26	<1	0	Stationary phase genes
σ^{32}	1.24	<10	0	Heat shock genes
σ^{28}	0.74	370	14	Flagella and chemotaxis genes
σ^{24}	2.43	<10	0	Extracytoplasmic functions; some heat shock functions
σ^{18}	1.73	<1	0	Extracytoplasmic functions, including ferric citrate transport

Source: Data from H. Maeda et al., *Nucleic Acids Res.* 28:3497, 2000.

Note: σ factors are widely distributed in bacteria; the number varies from a single σ factor in *Mycoplasma genitalium* to 63 distinct σ factors in *Streptomyces coelicolor*.

[a]Approximate number of each σ subunit per cell and the fraction of RNA polymerase holoenzyme complexed with each σ subunit during exponential growth. The numbers change as growth conditions change. The fraction of RNA polymerase complexed with each σ subunit reflects both the amount of the particular subunit and its affinity for the enzyme.

structure, it pauses **(Fig. 26-7a)**. Formation of the hairpin structure in the RNA disrupts several A=U base pairs in the RNA-DNA hybrid segment and may disrupt important interactions between RNA and the RNA polymerase, facilitating dissociation of the transcript.

The ρ-dependent terminators lack the sequence of repeated A residues in the template strand but usually include a CA-rich sequence called a *rut* (*r*ho *ut*ilization) element. The ρ protein associates with the RNA at specific binding sites and migrates in the 5'→3' direction until it reaches the transcription complex that is paused at a termination site (Fig. 26-7b). Here it contributes to release of the RNA transcript. The ρ protein has an ATP-dependent RNA-DNA helicase activity that promotes translocation of the protein along the RNA, and ATP is hydrolyzed by the ρ protein during the termination process. The detailed mechanism by which the protein promotes the release of the RNA transcript is not known.

Eukaryotic Cells Have Three Kinds of Nuclear RNA Polymerases

The transcriptional machinery in the nucleus of a eukaryotic cell is much more complex than that in bacteria. Eukaryotes have three RNA polymerases, designated I, II, and III, which are distinct complexes but have certain subunits in common. Each polymerase has a specific function and is recruited to a specific promoter sequence.

RNA polymerase I (Pol I) is responsible for the synthesis of only one type of RNA, a transcript called preribosomal RNA (or pre-rRNA), which contains the precursor for the 18S, 5.8S, and 28S rRNAs (see Fig. 26-24). Pol I promoters differ greatly in sequence from one species to another. The principal function of RNA polymerase II (Pol II) is synthesis of mRNAs and some specialized RNAs. This enzyme can recognize thousands of promoters that vary greatly in sequence. Some Pol II promoters have a few sequence features in common, including a

TATA box (eukaryotic consensus sequence TATA(A/T)A(A/T)(A/G)) near base pair −30 and an Inr sequence (initiator) near the RNA start site at +1 **(Fig. 26-8)**. However, such promoters are in the minority, and elaborate interactions with regulatory proteins guide Pol II function at many promoters that lack these features.

RNA polymerase III (Pol III) makes tRNAs, the 5S rRNA, and some other small specialized RNAs. The promoters recognized by Pol III are well characterized. Some of the sequences required for the regulated initiation of transcription by Pol III are located within the gene itself, whereas others are in more conventional locations upstream of the RNA start site (Chapter 28).

RNA Polymerase II Requires Many Other Protein Factors for Its Activity

RNA polymerase II is central to eukaryotic gene expression and has been studied extensively. Although this polymerase is strikingly more complex than its bacterial counterpart, the complexity masks a remarkable conservation of structure, function, and mechanism. Pol II isolated from either yeast or human cells is a 12 subunit enzyme with an aggregate molecular weight of more than 510,000. The largest subunit (RBP1) exhibits a high degree of homology to the β' subunit of bacterial RNA polymerase. Another subunit (RBP2) is structurally similar to the bacterial β subunit, and two others (RBP3 and RBP11) show some structural homology to the two bacterial α subunits. Pol II must function with genomes that are more complex and with DNA molecules more elaborately packaged than in bacteria. The need for protein-protein contacts with the numerous other protein factors required to navigate this labyrinth accounts in large measure for the added complexity of the eukaryotic polymerase.

The largest subunit of Pol II (RBP1) also has an unusual feature, a long carboxyl-terminal tail consisting of many repeats of a consensus heptad amino acid

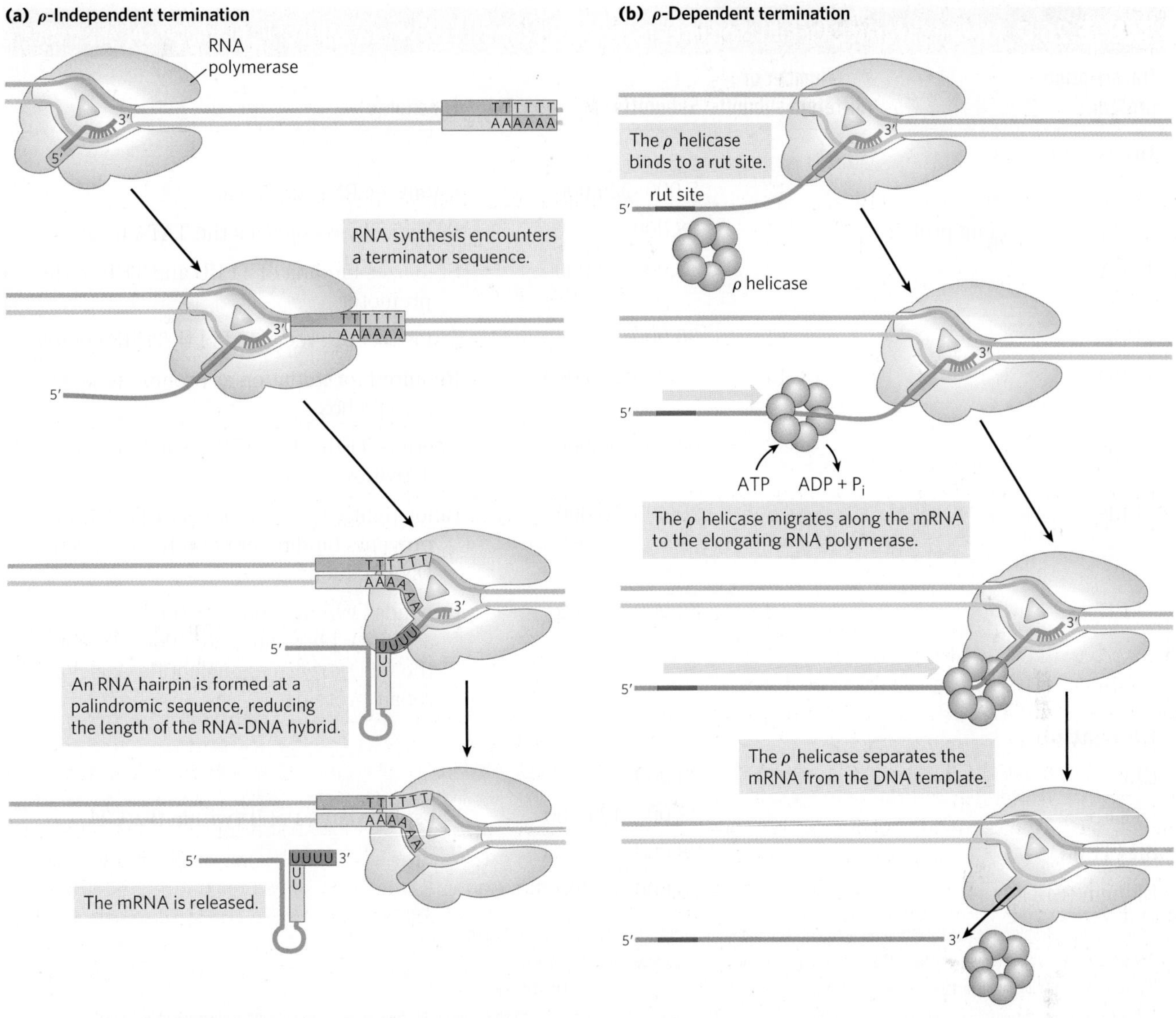

(a) ρ-Independent termination

RNA polymerase

RNA synthesis encounters a terminator sequence.

An RNA hairpin is formed at a palindromic sequence, reducing the length of the RNA-DNA hybrid.

The mRNA is released.

(b) ρ-Dependent termination

The ρ helicase binds to a rut site.

rut site

ρ helicase

ATP   ADP + P_i

The ρ helicase migrates along the mRNA to the elongating RNA polymerase.

The ρ helicase separates the mRNA from the DNA template.

FIGURE 26-7 Termination of transcription in _E. coli_. (a) ρ-Independent termination. RNA polymerase pauses at a variety of DNA sequences, some of which are terminators. One of two outcomes is then possible: either the polymerase bypasses the site and continues on its way, or the complex undergoes a conformational change (isomerization). In the latter case, intramolecular pairing of complementary sequences in the newly formed RNA transcript may form a hairpin that disrupts the RNA-DNA hybrid or the interactions between RNA and polymerase, or both,

resulting in isomerization. An A=U hybrid region at the 3′ end of the new transcript is relatively unstable, and the RNA dissociates from the complex completely, leading to termination. This is the usual outcome at terminators. At other pause sites, the complex may escape after the isomerization step to continue RNA synthesis. (b) ρ-Dependent termination. RNAs that include a rut site (purple) recruit the ρ helicase. The ρ helicase migrates along the mRNA in the 5′→3′ direction and separates it from the polymerase.

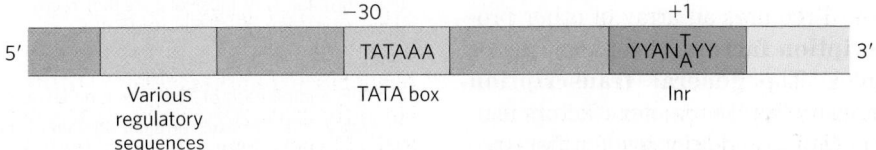

FIGURE 26-8 Some common sequences in promoters recognized by eukaryotic RNA polymerase II. The TATA box is the major assembly point for the proteins of the preinitiation complexes of Pol II. The DNA is unwound at the initiator sequence (Inr), and the transcription start site is usually within or very near this sequence. In the Inr consensus sequence shown here, N represents any nucleotide; Y, a pyrimidine nucleotide. Many additional sequences serve as binding sites for a wide variety of proteins that affect the activity of Pol II. These sequences are important in regulating Pol II promoters and differ greatly in type and number; in general,

the eukaryotic promoter is much more complex than suggested here (see Fig. 15-25). Many of the sequences are located within a few hundred base pairs of the TATA box on the 5′ side; others may be thousands of base pairs away. The sequence elements summarized here are more variable among the Pol II promoters of eukaryotes than among the _E. coli_ promoters (see Fig. 26-5). The majority of Pol II promoters lack a TATA box or a consensus Inr element or both. Additional sequences around the TATA box and downstream (to the right as shown here) of Inr may be recognized by one or more transcription factors.

TABLE 26-2 Proteins Required for Initiation of Transcription at the RNA Polymerase II (Pol II) Promoters of Eukaryotes

Transcription protein	Number of different subunits	Subunit(s) M_r[a]	Function(s)
Initiation			
Pol II	12	7,000–220,000	Catalyzes RNA synthesis
TBP (TATA-binding protein)	1	38,000	Specifically recognizes the TATA box
TFIIA	2	13,000, 42,000	Stabilizes binding of TFIIB and TBP to the promoter
TFIIB	1	35,000	Binds to TBP; recruits Pol II–TFIIF complex
TFIID[b]	13–14	14,000–213,000	Required for initiation at promoters lacking a TATA box
TFIIE	2	33,000, 50,000	Recruits TFIIH; has ATPase and helicase activities
TFIIF	2–3	29,000–58,000	Binds tightly to Pol II; binds to TFIIB and prevents binding of Pol II to nonspecific DNA sequences
TFIIH	10	35,000–89,000	Unwinds DNA at promoter (helicase activity); phosphorylates Pol II (within the CTD); recruits nucleotide-excision repair proteins
Elongation[c]			
ELL[d]	1	80,000	
pTEFb	2	43,000, 124,000	Phosphorylates Pol II (within the CTD)
SII (TFIIS)	1	38,000	
Elongin (SIII)	3	15,000, 18,000, 110,000	

[a] M_r reflects the subunits present in the complexes of human cells. Some components differ somewhat in size in yeast.

[b] The presence of multiple copies of some TFIID subunits brings the total subunit composition of the complex to 21–22.

[c] The function of all elongation factors is to suppress the pausing or arrest of transcription by the Pol II–TFIIF complex.

[d] Name derived from *e*leven-nineteen *l*ysine-rich *l*eukemia. The gene for ELL is the site of chromosomal recombination events frequently associated with acute myeloid leukemia.

sequence, –YSPTSPS–. There are 27 repeats in the yeast enzyme (18 exactly matching the consensus) and 52 (21 exact) in the mouse and human enzymes. This carboxyl-terminal domain (CTD) is separated from the main body of the enzyme by an inherently unstructured linker sequence. The CTD has many important roles in Pol II function, as outlined below.

RNA polymerase II requires an array of other proteins, called **transcription factors**, to form the active transcription complex. The **general transcription factors** required at every Pol II promoter (factors usually designated TFII with an additional identifier) are highly conserved in all eukaryotes (Table 26-2). The process of transcription by Pol II can be described in terms of several phases—assembly, initiation, elongation, termination—each associated with characteristic proteins **(Fig. 26-9)**. The step-by-step pathway described below leads to active transcription in vitro. In the cell, many of the proteins may be present in larger, preassembled complexes, simplifying the pathways for assembly on promoters. As you read about this process, consult Figure 26-9 and Table 26-2 to help keep track of the many participants.

FIGURE 26-9 Transcription at RNA polymerase II promoters. (a) TBP (often with TFIIA and sometimes with TFIID) and TFIIB bind sequentially to a promoter. TFIIF plus Pol II are then recruited to that complex. The further addition of TFIIE and TFIIH results in a closed complex. Within the complex, the DNA is unwound at the Inr region by the helicase activity of TFIIH and perhaps of TFIIE, creating an open complex that completes assembly. The carboxyl-terminal domain of the largest Pol II subunit is phosphorylated by TFIIH, and the polymerase then escapes the promoter and initiates transcription. Elongation is accompanied by the release of many transcription factors and is also enhanced by elongation factors (see Table 26-2). After termination, Pol II is released, dephosphorylated, and recycled. **(b)** Human TBP bound to DNA. The DNA is bent in this complex, opening the minor groove to allow specific hydrogen-bonding between protein and DNA. **(c)** A cutaway view of transcription elongation promoted by the Pol II core enzyme. [Source: (b) PDB ID 1TGH, Z. S. Juo et al., *J. Mol. Biol.* 261:239, 1996.]

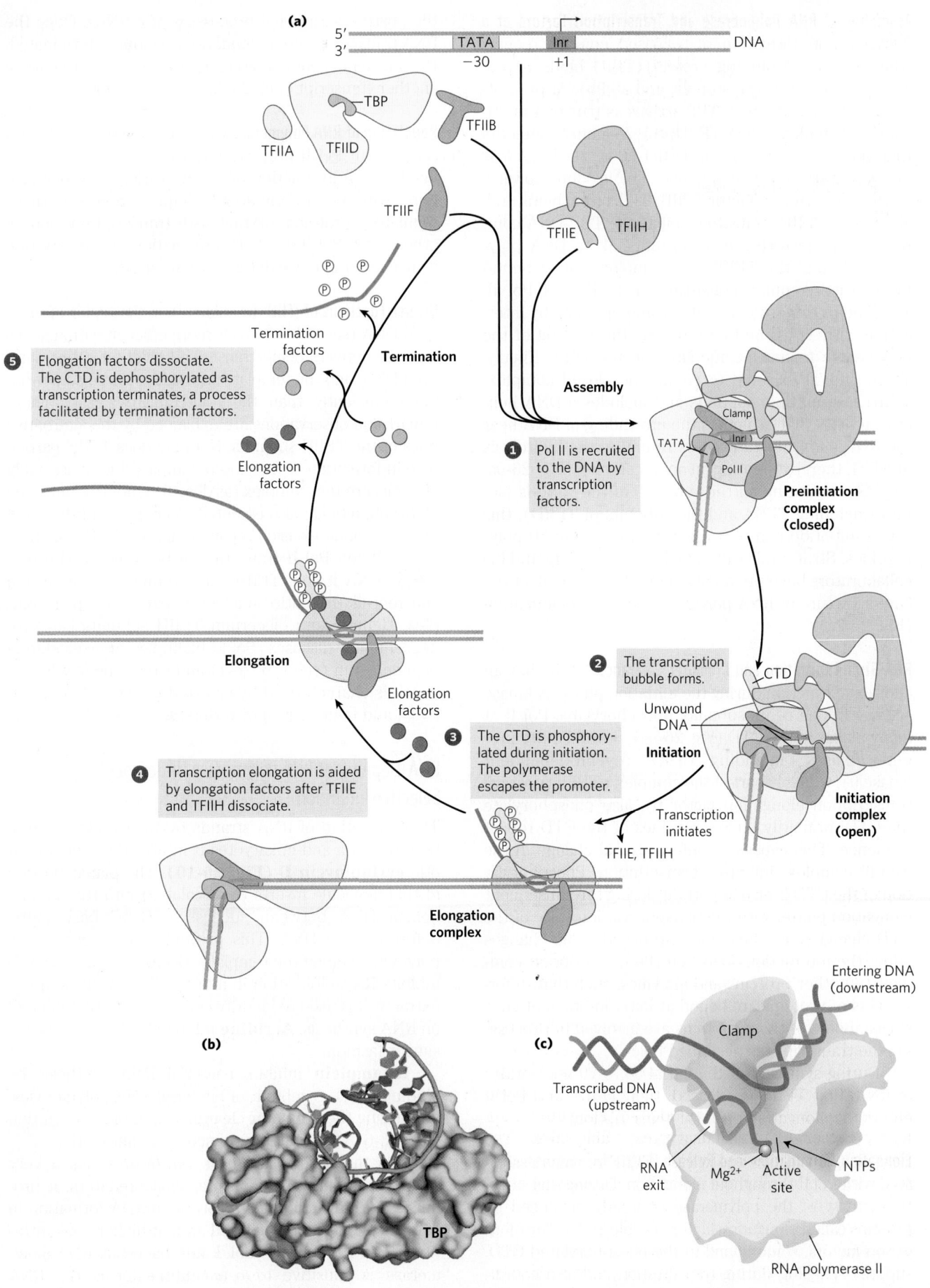

(a)

5′ — TATA — Inr — DNA
3′ —
−30   +1

TBP
TFIIA   TFIID   TFIIB
TFIIF
TFIIE   TFIIH

Assembly

❶ Pol II is recruited to the DNA by transcription factors.

Clamp
TATA   Inr
Pol II

Preinitiation complex (closed)

❺ Elongation factors dissociate. The CTD is dephosphorylated as transcription terminates, a process facilitated by termination factors.

Termination factors

Termination

Elongation factors

CTD
Unwound DNA

Initiation

❷ The transcription bubble forms.

Initiation complex (open)

Transcription initiates

TFIIE, TFIIH

Elongation

Elongation factors

❹ Transcription elongation is aided by elongation factors after TFIIE and TFIIH dissociate.

Elongation factors

❸ The CTD is phosphorylated during initiation. The polymerase escapes the promoter.

Elongation complex

(b)

TBP

(c)

Entering DNA (downstream)

Clamp

Transcribed DNA (upstream)

RNA exit   Mg²⁺   Active site   NTPs

RNA polymerase II

Assembly of RNA Polymerase and Transcription Factors at a Promoter The formation of a closed complex begins when the TATA-binding protein (TBP) binds to the TATA box (Figs 26-9a, step ❶, and 26-9b). At promoters lacking a TATA box, TBP arrives as part of a multisubunit complex called TFIID; the sequence elements that direct the binding of TFIID at these TATA-less promoters are poorly understood. TBP is bound, in turn, by the transcription factor TFIIB. TFIIA then binds and, along with TFIIB, helps to stabilize the TBP-DNA complex. TFIIB provides an important link to DNA polymerase II, and the TFIIB-TBP complex is next bound by another complex consisting of TFIIF and Pol II. TFIIF helps target Pol II to its promoters, both by interacting with TFIIB and by reducing the binding of the polymerase to nonspecific sites on the DNA. Finally, TFIIE and TFIIH bind to create the closed complex. TFIIH has multiple subunits and includes a DNA helicase activity that promotes the unwinding of DNA near the RNA start site (a process requiring the hydrolysis of ATP), thereby creating an open complex (Fig. 26-9a, step ❷). Counting all the subunits of the various factors (including TFIIA and the subunits of TFIID), this active initiation complex can have more than 50 polypeptides. Structural studies by Roger Kornberg and his collaborators have provided a more detailed look at the core structure of RNA polymerase II during elongation (Fig. 26-9c).

RNA Strand Initiation and Promoter Clearance TFIIH has an additional function during the initiation phase. A kinase activity in one of its subunits phosphorylates Pol II at many places in the CTD (Fig. 26-9a). Several other protein kinases, including CDK9 (cyclin-dependent kinase 9), which is part of the complex pTEFb (positive transcription elongation factor b), also phosphorylate the CTD, primarily on Ser residues of the CTD repeat sequence. This causes a conformational change in the overall complex, initiating transcription. Phosphorylation of the CTD is also important during the subsequent elongation phase, with the phosphorylation state of the CTD changing as transcription proceeds. The changes affect the interactions between the transcription complex and other proteins and enzymes, such that different sets of proteins are bound at initiation than at later stages. Some of these proteins are involved in processing the transcript (as described below).

During synthesis of the initial 60 to 70 nucleotides of RNA, first TFIIE is released, then TFIIH, and Pol II enters the elongation phase of transcription.

Elongation, Termination, and Release TFIIF remains associated with Pol II throughout elongation. During this stage, the activity of the polymerase is greatly enhanced by proteins called elongation factors (Table 26-2). The elongation factors, some bound to the phosphorylated CTD, suppress pausing during transcription and also coordinate interactions between protein complexes involved in

the posttranscriptional processing of mRNAs. Once the RNA transcript is completed, transcription is terminated. Pol II is dephosphorylated and recycled, ready to initiate another transcript (Fig. 26-9a, steps ❸ to ❺).

Regulation of RNA Polymerase II Activity Regulation of transcription at Pol II promoters is an elaborate process. It involves the interaction of a wide variety of other proteins with the preinitiation complex. Some of these regulatory proteins interact with transcription factors, others with Pol II itself. The regulation of transcription is described in more detail in Chapter 28.

Diverse Functions of TFIIH In eukaryotes, the repair of damaged DNA (see Table 25-5) is more efficient within genes that are actively being transcribed than for other damaged DNA, and the template strand is repaired somewhat more efficiently than the nontemplate strand. These remarkable observations are explained by the alternative roles of the TFIIH subunits. Not only does TFIIH participate in formation of the closed complex during assembly of a transcription complex (as described above), but some of its subunits are also essential components of the separate nucleotide-excision repair complex (see Fig. 25-25).

When Pol II transcription halts at the site of a DNA lesion, TFIIH can interact with the lesion and recruit the entire nucleotide-excision repair complex. Genetic loss of certain TFIIH subunits can produce human diseases. Some examples are xeroderma pigmentosum (see Box 25-1) and Cockayne syndrome, which is characterized by arrested growth, photosensitivity, and neurological disorders. ∎

DNA-Dependent RNA Polymerase Undergoes Selective Inhibition

The elongation of RNA strands by RNA polymerase in both bacteria and eukaryotes is inhibited by the antibiotic **actinomycin D** (**Fig. 26-10**). The planar portion of this molecule inserts (intercalates) into the double-helical DNA between successive G≡C base pairs, deforming the DNA. This prevents movement of the polymerase along the template. Because actinomycin D inhibits RNA elongation in intact cells as well as in cell extracts, it is used to identify cell processes that depend on RNA synthesis. **Acridine** inhibits RNA synthesis in a similar fashion.

Rifampicin inhibits bacterial RNA synthesis by binding to the β subunit of bacterial RNA polymerases, preventing the promoter clearance step of transcription (Fig. 26-6). It is sometimes used as an antibiotic.

The mushroom *Amanita phalloides* has a very effective defense mechanism against predators. It produces **α-amanitin**, which disrupts mRNA formation in animal cells by blocking Pol II and, at higher concentrations, Pol III. Neither Pol I nor bacterial RNA polymerase is sensitive to α-amanitin—nor is the RNA polymerase II of *A. phalloides* itself!

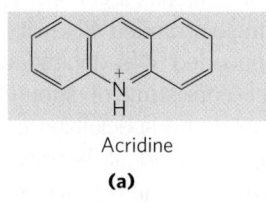

Actinomycin D

Acridine

(a)

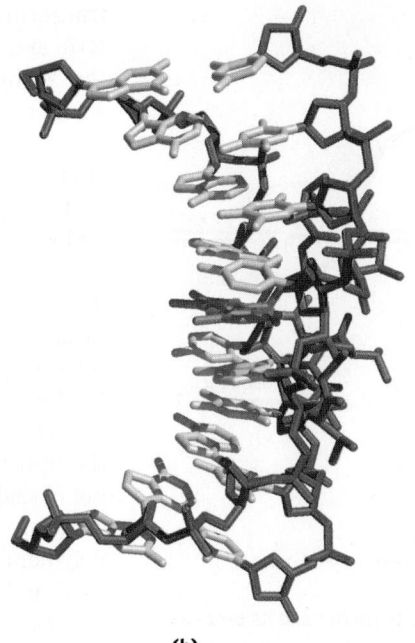

(b)

FIGURE 26-10 Actinomycin D and acridine, inhibitors of DNA transcription. (a) The shaded portion of actinomycin D is planar and intercalates between two successive G≡C base pairs in duplex DNA. The two cyclic peptide moieties of actinomycin D bind to the minor groove of the double helix. Sarcosine (Sar) is *N*-methylglycine; meVal, methylvaline. Acridine also acts by intercalation in DNA. **(b)** A complex of actinomycin D with DNA. The DNA backbone is shown in blue, the bases are white, the intercalated part of actinomycin (shaded in (a)) is orange, and the remainder of the actinomycin is red. The DNA is bent as a result of the actinomycin binding. [Source: (b) PDB ID 1DSC, C. Lian et al., *J. Am. Chem. Soc.* 118:8791, 1996.]

SUMMARY 26.1 DNA-Dependent Synthesis of RNA

■ Transcription is catalyzed by DNA-dependent RNA polymerases, which use ribonucleoside 5'-triphosphates to synthesize RNA complementary to the template strand of duplex DNA. Transcription occurs in several phases: binding of RNA polymerase to a DNA site called a promoter, initiation of transcript synthesis, elongation, and termination.

■ Bacterial RNA polymerase requires a special subunit to recognize the promoter. As the first committed step in transcription, binding of RNA polymerase to the promoter and initiation of transcription are closely regulated. Transcription stops at sequences called terminators.

■ Eukaryotic cells have three types of RNA polymerases. Binding of RNA polymerase II to its promoters requires an array of proteins called transcription factors. Elongation factors participate in the elongation phase of transcription. The largest subunit of Pol II has a long carboxyl-terminal domain that is phosphorylated during the initiation and elongation phases.

26.2 RNA Processing

Many of the RNA molecules in bacteria and virtually all RNA molecules in eukaryotes are processed to some degree after synthesis. Some of the most interesting molecular events in RNA metabolism occur during this postsynthetic processing. Intriguingly, several of the enzymes that catalyze these reactions consist of RNA rather than protein. The discovery of these catalytic RNAs, or **ribozymes**, has brought a revolution in thinking about RNA function and about the origin of life.

A newly synthesized RNA molecule is called a **primary transcript**. Perhaps the most extensive processing of primary transcripts occurs in eukaryotic mRNAs and in the tRNAs of both bacteria and eukaryotes. Special-function RNAs are also processed.

The primary transcript for a eukaryotic mRNA typically contains sequences encompassing one gene, although the sequences encoding the polypeptide may not be contiguous. Noncoding tracts that break up the coding region of the transcript are called introns, and the coding segments are called exons (see the discussion of introns and exons in DNA in Chapter 24). In a process called **RNA splicing**, the introns are removed from the primary transcript, and the exons are joined to form a continuous sequence that specifies a functional polypeptide. Eukaryotic mRNAs are also modified at each end. A modified residue called a 5' cap is added at the 5' end. The 3' end is cleaved, and 80 to 250 A residues are added to create a poly(A) "tail." The sometimes elaborate protein complexes that carry out each of these three mRNA-processing reactions do not operate independently. They seem to be organized in association with each other and with the phosphorylated CTD of Pol II; each complex affects the function of the others. Proteins involved in mRNA transport to the cytoplasm are also associated with the mRNA in the nucleus, and the processing of the transcript is coupled to its transport. In effect, a eukaryotic mRNA, as it is synthesized, is ensconced in an elaborate complex comprising dozens of proteins. The composition of the complex changes as the primary transcript is processed, transported to the cytoplasm, and delivered to the ribosome for translation. The associated proteins modulate all aspects of the function and fate of the mRNA. These processes

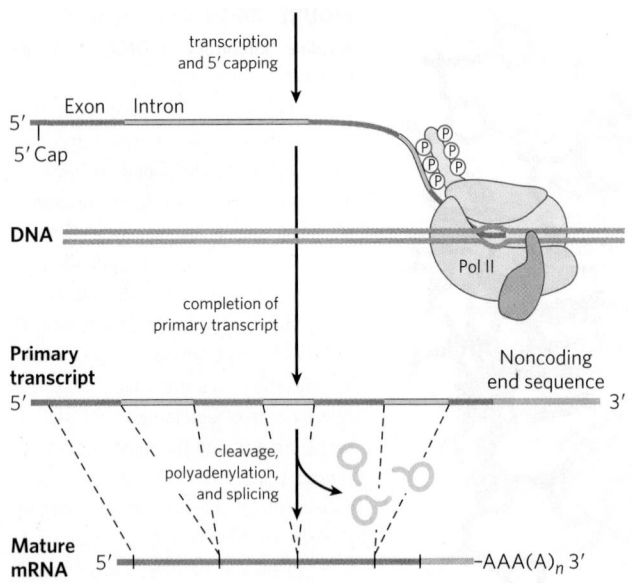

FIGURE 26-11 Formation of the primary transcript and its processing during maturation of mRNA in a eukaryotic cell. The 5′ cap (red) is added before synthesis of the primary transcript is complete. A noncoding end sequence (intron) following the last exon is shown in orange. Splicing can occur either before or after the cleavage and polyadenylation steps. All the processes shown here take place in the nucleus.

are outlined in **Figure 26-11** and described in more detail below.

The primary transcripts of bacterial and eukaryotic tRNAs are processed by the removal of sequences from each end (cleavage) and, in a few cases, by the removal of introns (splicing). Many bases and sugars in tRNAs are also modified; mature tRNAs are replete with unusual bases not found in other nucleic acids (see Fig. 26-22). Many of the special-function RNAs also undergo elaborate processing, often involving the removal of segments from one or both ends.

The ultimate fate of any RNA is its complete and regulated degradation. The rate of turnover of RNAs plays a critical role in determining their steady-state levels and the rate at which cells can shut down expression of a gene when its product is no longer needed. During the development of multicellular organisms, for example, certain proteins must be expressed at one stage only, and the mRNA encoding such a protein must be made and destroyed at the appropriate times.

Eukaryotic mRNAs Are Capped at the 5′ End

Most eukaryotic mRNAs have a **5′ cap**, a residue of 7-methylguanosine linked to the 5′-terminal residue of the mRNA through an unusual 5′,5′-triphosphate linkage **(Fig. 26-12)**. The 5′ cap helps protect mRNA from ribonucleases. It also binds to a specific cap-binding complex of proteins and participates in binding of the mRNA to the ribosome to initiate translation (Chapter 27).

The 5′ cap is formed by condensation of a molecule of GTP with the triphosphate at the 5′ end of the transcript. The guanine is subsequently methylated at N-7, and additional methyl groups are often added at the 2′ hydroxyls of the first and second nucleotides adjacent to the cap (Fig. 26-12a). The methyl groups are derived from S-adenosylmethionine. All these reactions (Fig. 26-12b) occur very early in transcription, after the first 20 to 30 nucleotides of the transcript have been added. All four of the enzymes in the cap-synthesizing complex, and through them the 5′ end of the transcript itself, are associated with the RNA polymerase II CTD until the cap is synthesized. The capped 5′ end is then released from the cap-synthesizing complex and bound by the cap-binding complex (Fig. 26-12c).

The 5′ cap does not provide complete protection of the transcript. The influenza virus has a genome consisting of eight segments of single-stranded RNA. Its genes are transcribed by a virally encoded RNA-dependent RNA polymerase, a heterotrimer consisting of subunits PA, PB1, and PB2. The virus needs no specialized enzymes for the synthesis of 5′ caps; instead, it borrows these structures from host-cell transcripts in a process termed "cap-snatching." A capped host transcript is bound by the viral PB2 polymerase subunit and cleaved by an endonuclease in the PA subunit. The PB1 subunit uses the resulting capped oligonucleotide to prime viral RNA synthesis.

Both Introns and Exons Are Transcribed from DNA into RNA

In bacteria, a polypeptide chain is generally encoded by a DNA sequence that is colinear with the amino acid sequence, continuing along the DNA template without interruption until the information needed to specify the polypeptide is complete. However, the notion that *all* genes are continuous was disproved in 1977 when Phillip Sharp and Richard Roberts independently discovered that many genes for polypeptides in eukaryotes are interrupted by noncoding sequences (introns).

The vast majority of genes in vertebrates contain introns; among the few exceptions are those that encode histones. The occurrence of introns in other eukaryotes varies. Many genes of the yeast *Saccharomyces cerevisiae* lack introns, but introns are more common in some other yeast species. Introns are also found in a few bacterial and archaeal genes. Introns in DNA are transcribed along with the rest of the gene by RNA polymerases. The introns in the primary RNA transcript are then spliced, and the exons are joined to form a mature, functional RNA. In eukaryotic mRNAs, most exons are less than 1,000 nucleotides long, with many in the 100 to 200 nucleotide size range, encoding stretches of 30 to 60 amino acids within a longer polypeptide. Introns vary in size from 50 to more than 700,000 nucleotides, with a median length of about 1,800. Genes of higher eukaryotes, including humans, typically have much more DNA devoted to introns than to exons. The ~20,000 genes of the human genome include more than 200,000 introns.

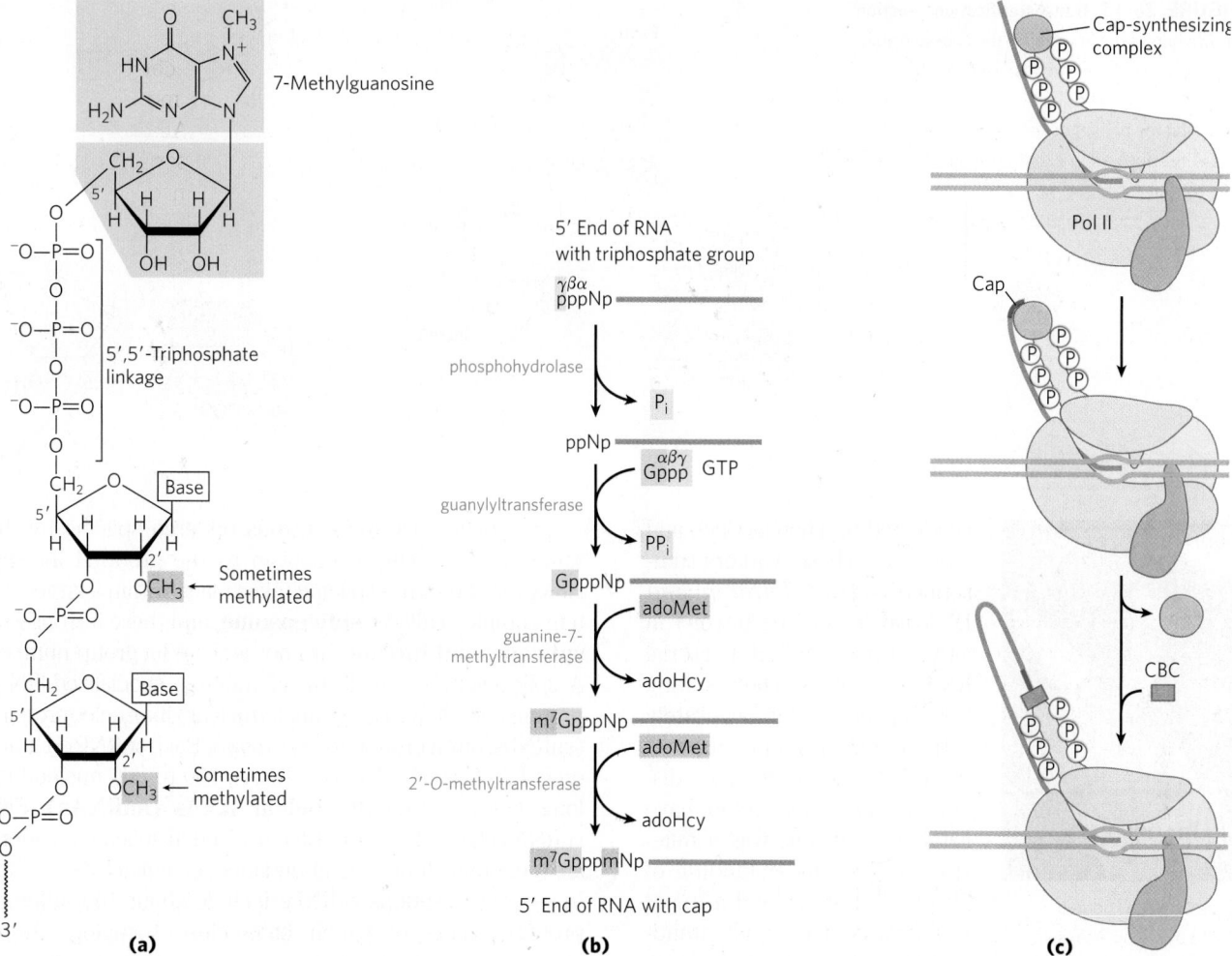

FIGURE 26-12 The 5′ cap of mRNA. (a) 7-Methylguanosine (m⁷G) is joined to the 5′ end of almost all eukaryotic mRNAs in an unusual 5′,5′-triphosphate linkage. Methyl groups (light red) are often found at the 2′ position of the first and second nucleotides. RNAs in yeast cells lack the 2′-methyl groups. The 2′-methyl group on the second nucleotide is generally found only in RNAs from vertebrate cells. **(b)** Generation of the 5′ cap requires four separate steps (adoHcy is *S*-adenosylhomocysteine). **(c)** Synthesis of the cap is carried out by enzymes tethered to the CTD of Pol II. The cap remains tethered to the CTD through an association with the cap-binding complex (CBC).

RNA Catalyzes the Splicing of Introns

There are four classes of introns. The first two, the group I and group II introns, differ in the details of their splicing mechanisms but share one surprising characteristic: they are *self-splicing*—no protein enzymes are involved. Group I introns are found in some nuclear, mitochondrial, and chloroplast genes that code for rRNAs, mRNAs, and tRNAs. Group II introns are generally found in the primary transcripts of mitochondrial or chloroplast mRNAs in fungi, algae, and plants. Group I and group II introns are also found among the rare examples of introns in bacteria. Neither class requires a high-energy cofactor (such as ATP) for splicing. The splicing mechanisms in both groups involve two transesterification reaction steps **(Fig. 26-13)**, in which a ribose 2′- or 3′-hydroxyl group makes a nucleophilic attack on a phosphorus, and a new phosphodiester bond is formed at the expense of the old, maintaining the balance of energy. These reactions are very similar to the DNA breaking and rejoining reactions promoted by topoisomerases (see Fig. 24-19) and site-specific recombinases (see Fig. 25-38).

The group I splicing reaction requires a guanine nucleoside or nucleotide cofactor, but the cofactor is not used as a source of energy; instead, the 3′-hydroxyl group of guanosine is used as a nucleophile in the first step of the splicing pathway. The guanosine 3′-hydroxyl group forms a normal 3′,5′-phosphodiester bond with the 5′ end of the intron **(Fig. 26-14)**. The 3′ hydroxyl of the exon that is displaced in this step then acts as a nucleophile in a similar reaction at the 3′ end of the intron. The result is precise excision of the intron and ligation of the exons.

In group II introns the reaction pattern is similar, except for the nucleophile in the first step, which in this case is the 2′-hydroxyl group of an A residue *within* the intron **(Fig. 26-15)**. A branched lariat structure is formed as an intermediate.

Self-splicing of introns was first revealed in 1982 in studies of the splicing mechanism of the group I rRNA intron from the ciliated protozoan *Tetrahymena thermophila*,

FIGURE 26-13 Transesterification reaction.
Shown here is the first step in the two-step splicing of group I introns. In this example, the 3′ OH
of a guanosine molecule acts as nucleophile,
attacking the phosphodiester linkage between U
and A residues at an exon-intron junction of an
mRNA molecule (see Fig. 26-14).

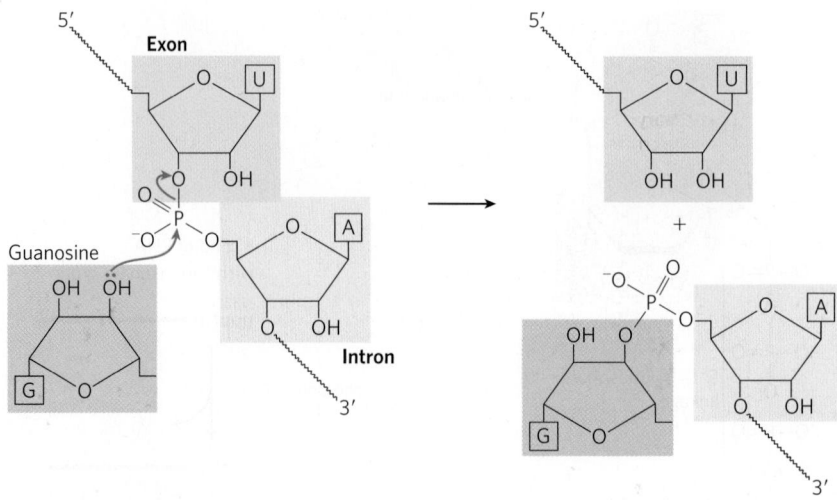

conducted by Thomas Cech and colleagues. These workers transcribed isolated *Tetrahymena* DNA (including the intron) in vitro, using purified bacterial RNA polymerase. The resulting RNA spliced itself accurately without any protein enzymes from *Tetrahymena*. The discovery that RNAs could have catalytic functions was a milestone in our understanding of biological systems and a major step forward in the understanding of how life probably evolved.

In eukaryotes, most introns undergo splicing by the same lariat-forming mechanism as the group II introns. However, the intron splicing takes place within a large protein complex called a **spliceosome**, and these introns, the **spliceosomal introns**, are not assigned a group number. A spliceosome is made up of multiple specialized RNA-protein complexes called small *nuclear ribonucleoproteins* (snRNPs, often pronounced *snurps*). Each snRNP contains one of a class of eukaryotic RNAs, 100 to 200 nucleotides long, known as **small nuclear RNAs (snRNAs)**. Five snRNAs (U1, U2, U4, U5, U6) involved in splicing reactions are generally found in abundance in eukaryotic nuclei. In yeast, the various snRNPs include about 100 different proteins, most of which have close homologs in all other eukaryotes. In humans, these conserved protein

Thomas Cech [Source: Photo by Glenn Asakawa/ University of Colorado.]

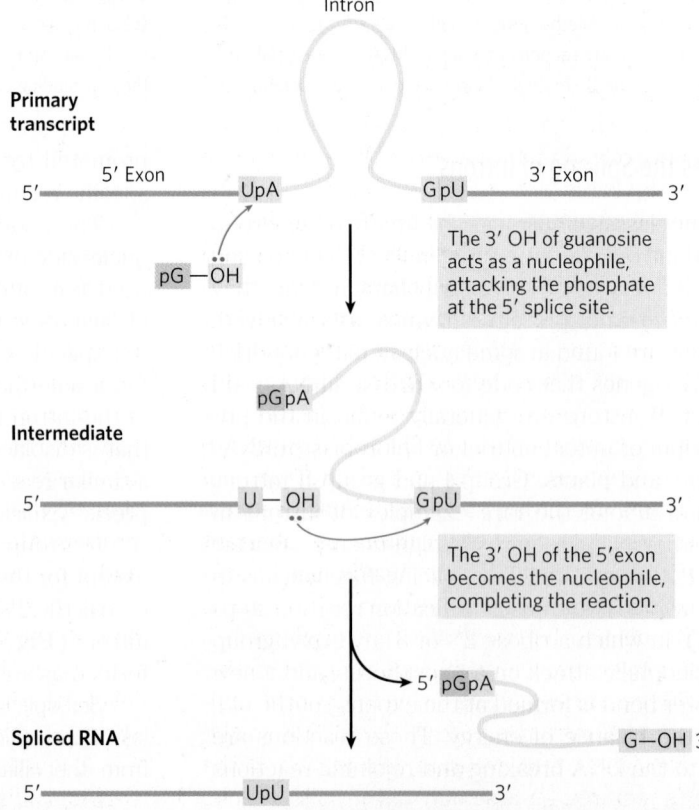

FIGURE 26-14 Splicing mechanism of group I introns. The nucleophile in the first step may be guanosine, GMP, GDP, or GTP. The spliced intron is eventually degraded.

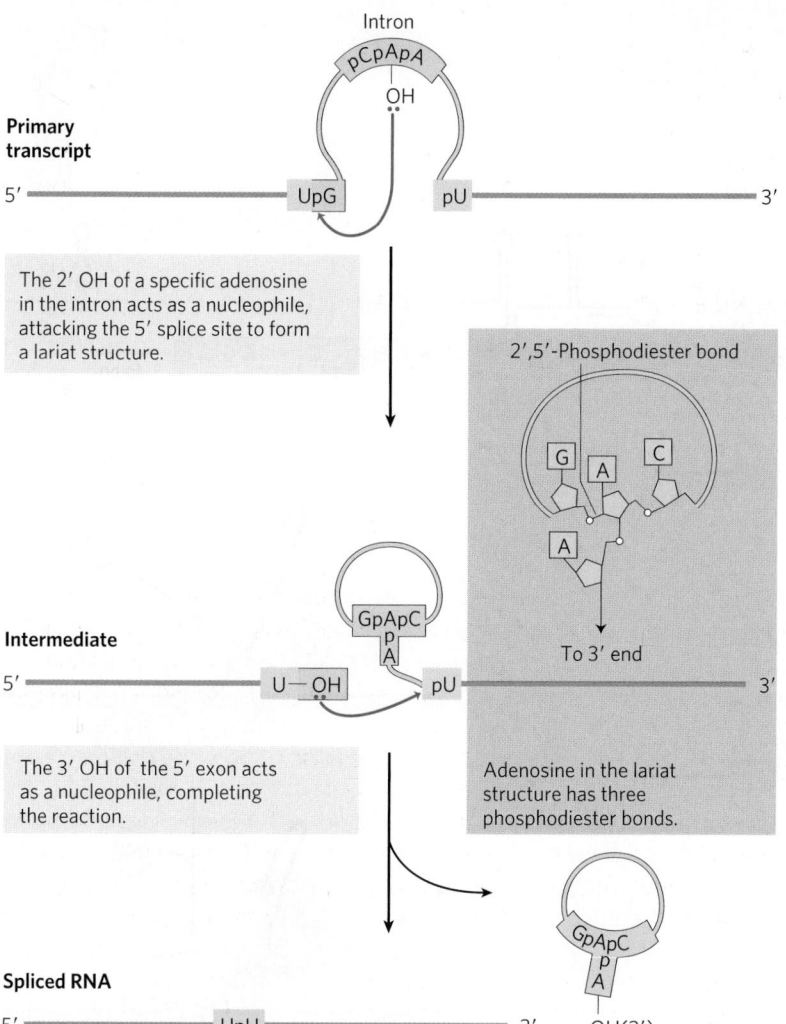

components are augmented by more than 200 additional proteins. Spliceosomes are thus among the most complex macromolecular machines in any eukaryotic cell. The RNA components of a spliceosome are the catalysts of the various splicing steps. The overall complex can be considered a highly flexible nucleoprotein chaperone that can adapt to the great diversity in size and sequence of nuclear mRNAs.

Spliceosomal introns generally have the dinucleotide sequence GU at the 5' end and AG at the 3' end, and these sequences mark the sites where splicing occurs. The U1 snRNA contains a sequence complementary to sequences near the 5' splice site of nuclear mRNA introns **(Fig. 26-16a)**, and the U1 snRNP binds to this region in the primary transcript. A U2 snRNP binds to the 3' end. Addition of the U4, U5, and U6 snRNPs leads to formation of the spliceosome (Fig. 26-16b). Key parts of the splicing active site found in U6 are initially sequestered by base pairing to parts of U4 to prevent aberrant cleavage of nontarget phosphodiester bonds. The U6 and U4 snRNAs must be unwound and separated to expose the active site needed for the first step in splicing. All steps in the process are reversible. Individual proteins associated with the various snRNPs sometimes have multiple functions: splicing, transport of the mRNA to the cytoplasm, translation, and eventual degradation of the mRNA. ATP is required for assembly of the spliceosome,

but the RNA cleavage-ligation reactions do not require ATP. Some mRNA introns are spliced by a less common type of spliceosome, in which the U1 and U2 snRNPs are replaced by the U11 and U12 snRNPs. Whereas U1- and U2-containing spliceosomes remove introns with (5')GU and AG(3') terminal sequences, as shown in Figure 26-16, the U11- and U12-containing spliceosomes remove a rare class of introns that have (5')AU and AC(3') terminal sequences to mark the splice sites.

Some components of the splicing apparatus are tethered to the CTD of RNA polymerase II, indicating that splicing, like other RNA processing reactions, is tightly coordinated with transcription (Fig. 26-16c). As the first splice junction is synthesized, it is bound by a tethered spliceosome. The second splice junction is then captured by this complex as it passes, facilitating juxtaposition of the intron ends and the subsequent splicing process. After splicing, the intron remains in the nucleus and is eventually degraded.

The spliceosomes used in nuclear RNA splicing almost certainly evolved from more ancient group II introns, with the snRNPs contributing much greater levels of catalytic flexibility and regulation relative to their self-splicing ancestors.

A fourth and final class of introns, found in certain tRNAs, is distinguished from the group I and II introns

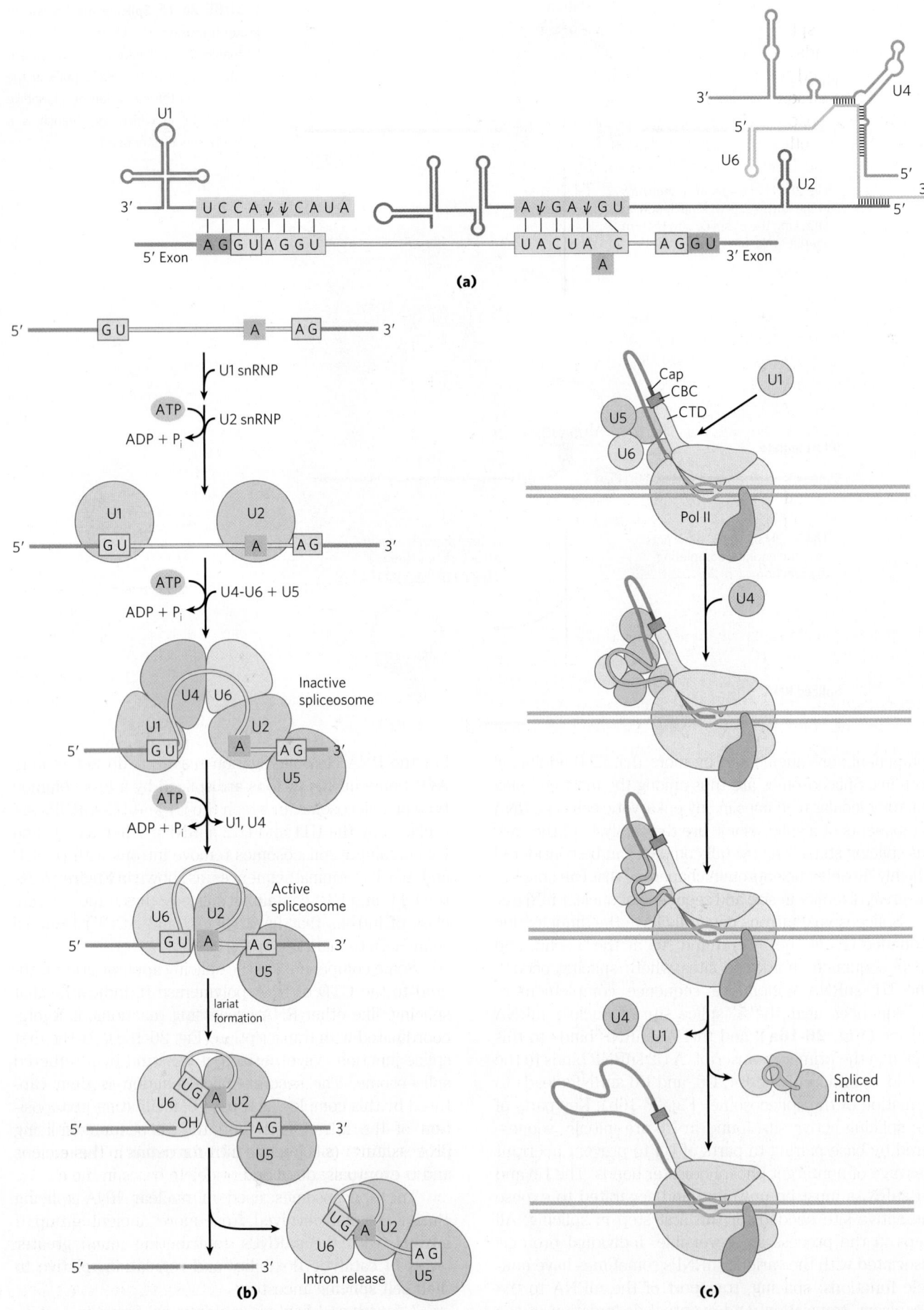

(a)

(b)

(c)

in that the splicing reaction requires ATP and an endo-nuclease. The splicing endonuclease cleaves the phosphodiester bonds at both ends of the intron, and the two exons are joined by a mechanism similar to the DNA ligase reaction (see Fig. 25-16).

Although spliceosomal introns seem to be limited to eukaryotes, the other three intron classes are not. Genes with group I and II introns have now been found in both bacteria and bacterial viruses. Bacteriophage T4, for example, has several protein-coding genes with group I introns. Introns may be more common in archaea than in bacteria.

Eukaryotic mRNAs Have a Distinctive 3′ End Structure

At their 3′ end, most eukaryotic mRNAs undergoing translation in the cell cytoplasm have a string of A residues, about 30 residues in yeast and 50 to 100 in animals, called the **poly(A) tail**. This tail serves as a binding site for one or more specific proteins. The poly(A) tail and its associated proteins have a variety of roles in coordinating transcription and translation, and may help protect mRNA from enzymatic destruction. Many bacterial mRNAs also acquire poly(A) tails, but these tails stimulate decay of mRNA rather than protecting it from degradation.

The poly(A) tail is added in a multistep process. The transcript is extended beyond the site where the poly(A) tail is to be added, then is cleaved at the poly(A) addition site by an endonuclease component of a large enzyme complex, again associated with the CTD of RNA polymerase II **(Fig. 26-17)**. The mRNA site where cleavage occurs is marked by two sequence elements: the highly conserved sequence (5′)AAUAAA(3′), 10 to 30 nucleotides on the 5′ side (upstream) of the cleavage site, and a less well-defined sequence rich in G and U residues, 20 to 40 nucleotides downstream of the cleavage site. Cleavage generates the free 3′-hydroxyl group that defines the end of the mRNA, to which A

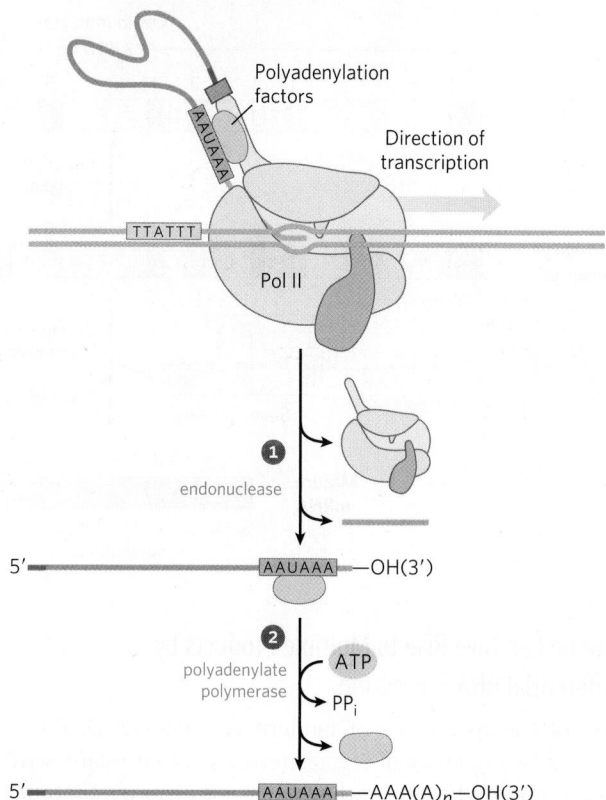

FIGURE 26-17 Addition of the poly(A) tail to the primary RNA transcript of eukaryotes. Pol II synthesizes RNA beyond the segment of the transcript containing the cleavage signal sequences, including the highly conserved upstream sequence (5′)AAUAAA. This cleavage signal sequence is bound by an enzyme complex that includes an endonuclease, a polyadenylate polymerase, and several other multisubunit proteins involved in sequence recognition, stimulation of cleavage, and regulation of the length of the poly(A) tail; all of these proteins are tethered to the CTD. ❶ The RNA is cleaved by the endonuclease at a point 10 to 30 nucleotides 3′ to (downstream of) the sequence AAUAAA. ❷ The polyadenylate polymerase synthesizes a poly(A) tail 80 to 250 nucleotides long, beginning at the cleavage site.

FIGURE 26-16 Splicing mechanism in mRNA primary transcripts. **(a)** RNA pairing interactions in the formation of spliceosome complexes. The U1 snRNA has a sequence near its 5′ end that is complementary to the splice site at the 5′ end of the intron. Base pairing of U1 to this region of the primary transcript helps define the 5′ splice site during spliceosome assembly (ψ is pseudouridine; see Fig. 26-22). U2 is paired to the intron at a position encompassing the A residue (shaded light red) that becomes the nucleophile during the splicing reaction. Base pairing of U2 snRNA causes a bulge that displaces and helps to activate the adenylate, the 2′ OH of which will form the lariat structure through a 2′,5′-phosphodiester bond. **(b)** Assembly of spliceosomes. All steps are reversible, but are shown proceeding in the forward direction for simplicity. The U1 and U2 snRNPs bind, then the remaining snRNPs (the U4-U6 complex and U5) bind to form an inactive spliceosome. Internal rearrangements convert this species to an active spliceosome in which U1 and U4 have been expelled and U6 is paired with both the 5′ splice site and U2. This is followed by the catalytic steps, which parallel those of the splicing of group II introns (see Fig. 26-15). The active spliceosome complex is illustrated on the cover of this book. **(c)** Coordination of splicing and transcription brings the two splice sites together. See the text for details. All steps are reversible. The spliceosome is more than twice the size of RNA polymerase II.

residues are immediately added by **polyadenylate polymerase**, which catalyzes the reaction

$$\text{RNA} + n\text{ATP} \longrightarrow \text{RNA} - (\text{AMP})_n + n\text{PP}_i$$

where $n = 80$ to 250. This enzyme does not require a template but does require the cleaved mRNA as a primer. These longer poly(A) tails are added in the nucleus, and then shortened significantly after the mRNA is transported to the cytoplasm.

The overall processing of a typical eukaryotic mRNA is summarized in **Figure 26-18**. In some cases the polypeptide-coding region of the mRNA is also modified by RNA "editing" (see Section 27.1 for details). This editing includes processes that add or delete bases in the coding regions of primary transcripts or that change the sequence (such as by enzymatic deamination of a C residue to create a U residue). A particularly dramatic example occurs in trypanosomes, which are parasitic protists: large regions of an mRNA are synthesized without any uridylate, and the U residues are inserted later by RNA editing.

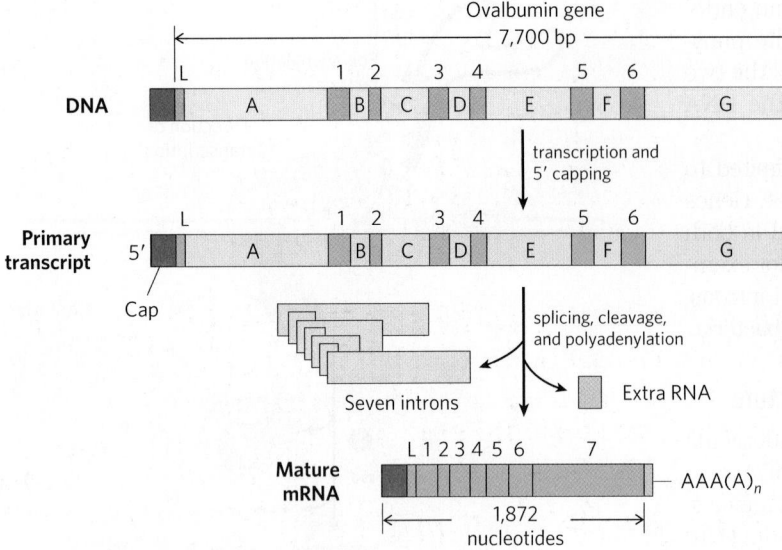

FIGURE 26-18 Overview of the processing of a eukaryotic mRNA. The ovalbumin gene, shown here, has introns A to G and exons 1 to 7 and L (L encodes a signal peptide sequence that targets the protein for export from the cell; see Fig. 27-40). About three-quarters of the RNA is removed during processing. Pol II extends the primary transcript well beyond the cleavage and polyadenylation site ("extra RNA") before terminating transcription. Termination signals for Pol II have not yet been defined.

A Gene Can Give Rise to Multiple Products by Differential RNA Processing

One of the paradoxes of modern genomics is that the apparent complexity of organisms does not correlate with the number of protein-coding genes, or even the amount of genomic DNA. Some eukaryotic mRNA transcripts can be processed in more than one way to produce *different* mRNAs and thus different polypeptides. Much of the variability in processing is the result of **alternative splicing**, in which a particular exon may or may not be incorporated into the mature mRNA transcript. Alternative splicing occurs in a relatively small number of genes in yeast, but in more than 95% of human genes.

Figure 26-19a illustrates how alternative splicing patterns can produce more than one protein from a

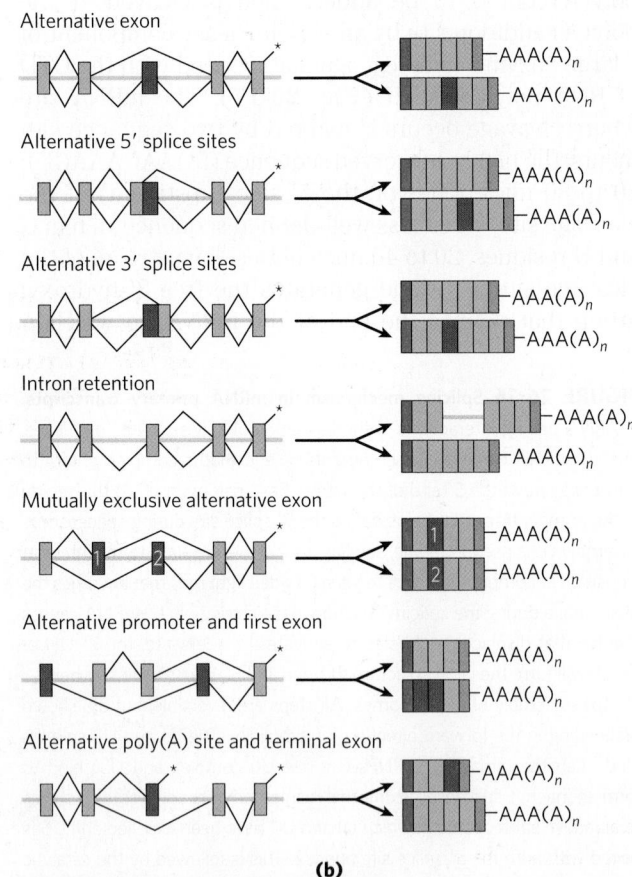

FIGURE 26-19 Alternative splicing in eukaryotes. (a) Alternative splicing patterns. Two different 3′ splice sites are shown. Different mature mRNAs are produced from the same primary transcript. **(b)** Summary of splicing patterns. Exons are shown in shades of green, and introns/untranslated regions as yellow lines. Positions where polyadenosine is to be added are marked with asterisks.

Exons joined in a particular splicing scheme are linked with black lines. For each transcript, the alternative linkage patterns shown above and below the transcript produce the top and bottom spliced mRNA, respectively. In the products, 5′ caps are represented by red boxes, 3′ untranslated regions by orange boxes. [Source: (b) Information from B. J. Blencowe, *Cell* 126:37, 2006, Fig. 2.]

common primary transcript. The primary transcript contains molecular signals for all the alternative processing pathways, and the pathway favored in a given cell or metabolic situation is determined by processing factors, RNA-binding proteins that promote one particular path. For example, the protein composition of the snRNPs involved in splicing may vary somewhat in the spliceosomes that participate in the processing of different genes, and may change further so that the processing of a particular gene is altered at different stages of animal development. As one example, such alternative processing produces three different forms of the myosin heavy chain at different stages of fruit fly development. There are many additional patterns of alternative splicing (Fig. 26-19b).

Complex transcripts can also have more than one site where poly(A) tails can form. If there are two or more sites for cleavage and polyadenylation, use of the one closest to the 5′ end will remove more of the primary transcript sequence **(Fig. 26-20)**. This mechanism, called **poly(A) site choice**, generates diversity in the variable domains of immunoglobulin heavy chains (see Fig. 25-43).

Both alternative splicing and poly(A) site choice come into play in the processing of many genes. For example, a single RNA transcript is processed using both mechanisms to produce two different hormones: the calcium-regulating hormone calcitonin in rat thyroid and calcitonin-gene-related peptide (CGRP) in rat brain **(Fig. 26-21)**. Together, alternative splicing and poly(A) site choice greatly increase the number of different proteins generated from the genomes of higher eukaryotes.

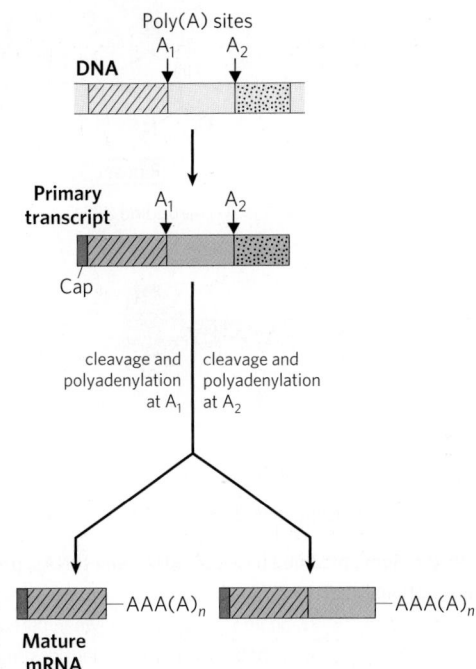

FIGURE 26-20 Poly(A) site choice. Two alternative cleavage and polyadenylation sites, A₁ and A₂, are shown.

Ribosomal RNAs and tRNAs Also Undergo Processing

Posttranscriptional processing is not limited to mRNA. Ribosomal RNAs of bacterial, archaeal, and eukaryotic cells are made from longer precursors called **preribosomal RNAs**, or pre-rRNAs. Transfer RNAs are similarly derived from longer precursors. These RNAs may also

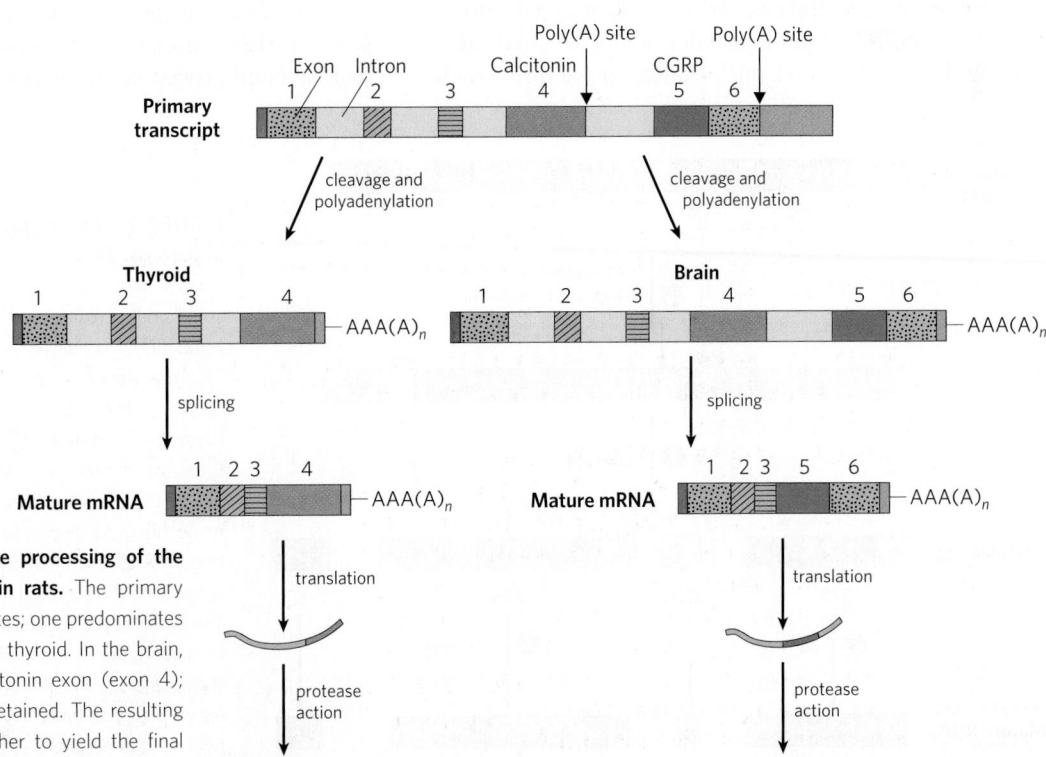

FIGURE 26-21 Alternative processing of the calcitonin gene transcript in rats. The primary transcript has two poly(A) sites; one predominates in the brain, the other in the thyroid. In the brain, splicing eliminates the calcitonin exon (exon 4); in the thyroid, this exon is retained. The resulting peptides are processed further to yield the final hormone products: calcitonin-gene-related peptide (CGRP) in the brain and calcitonin in the thyroid.

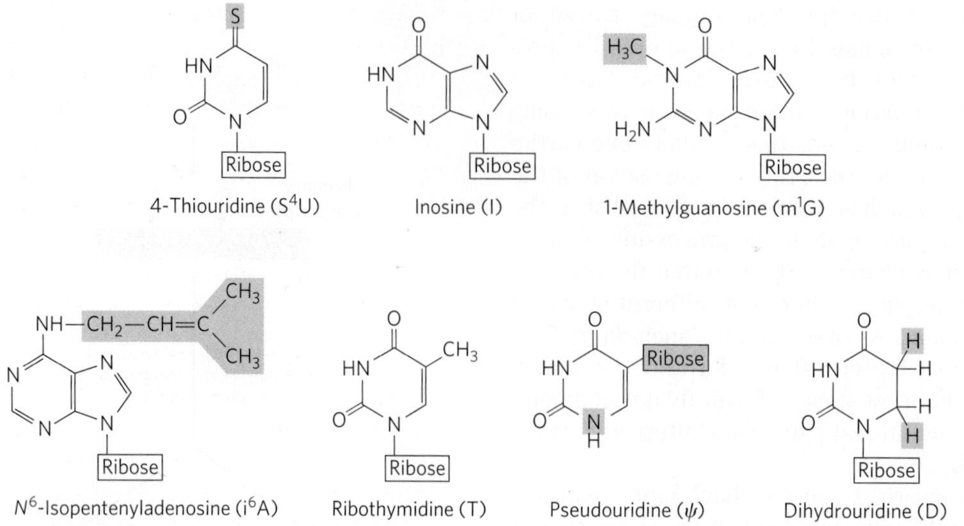

4-Thiouridine (S⁴U)   Inosine (I)   1-Methylguanosine (m¹G)

N⁶-Isopentenyladenosine (i⁶A)   Ribothymidine (T)   Pseudouridine (ψ)   Dihydrouridine (D)

FIGURE 26-22 Some modified bases of rRNAs and tRNAs, produced in posttranscriptional reactions. Notice the unusual ribose attachment point in pseudouridine. The standard symbols are shown in parentheses. This is just a small sampling of the 96 modified nucleosides known to occur in different RNA species, with 81 different types known in tRNAs and 30 observed to date in rRNAs. A complete listing of these modified bases can be found in the Modomics database of RNA modification pathways (http://modomics.genesilico.pl).

contain a variety of modified nucleosides; some examples are shown in **Figure 26-22**.

Ribosomal RNAs In bacteria, 16S, 23S, and 5S rRNAs (and some tRNAs, although most tRNAs are encoded elsewhere) arise from a single 30S RNA precursor of about 6,500 nucleotides. RNA at both ends of the 30S precursor and segments between the rRNAs are removed during processing **(Fig. 26-23)**. The 16S and 23S rRNAs contain modified nucleosides. In *E. coli*, the 11 modifications in the 16S rRNA include a pseudouridine and 10 nucleosides methylated on the base or the 2′-hydroxyl group or both. The 23S rRNA has 10 pseudouridines, 1 dihydrouridine, and 12 methylated nucleosides. In bacteria, each modification is generally catalyzed by a distinct enzyme. Methylation reactions use *S*-adenosylmethionine as cofactor. No cofactor is required for pseudouridine formation.

The genome of *E. coli* encodes seven pre-rRNA molecules. All of these genes have essentially identical rRNA-coding regions, but they differ in the segments between these regions. The segment between the 16S and 23S rRNA genes generally encodes one or two tRNAs, with different tRNAs produced from different pre-rRNA transcripts. Coding sequences for tRNAs are also found on the 3′ side of the 5S rRNA in some precursor transcripts.

The situation in eukaryotes is more complicated. A 45S pre-rRNA transcript is synthesized by RNA polymerase I and processed in the nucleolus to form the 18S,

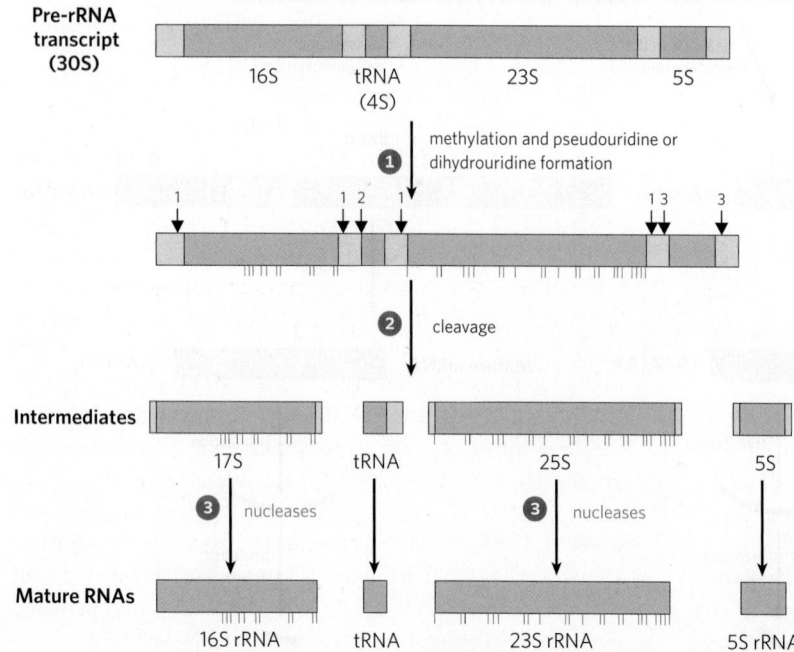

FIGURE 26-23 Processing of pre-rRNA transcripts in bacteria. ❶ Before cleavage, the 30S RNA precursor is methylated at specific bases (red tick marks), and some uridine residues are converted to pseudouridine (blue ticks) or dihydrouridine (black tick) residues. The methylation reactions are of multiple types, some occurring on bases and some on 2′-hydroxyl groups. ❷ Cleavage liberates precursors of rRNAs and tRNA(s). Cleavage at the points labeled 1, 2, and 3 is carried out by the enzymes RNase III, RNase P, and RNase E, respectively. As discussed later in the text, RNase P is a ribozyme. ❸ The final 16S, 23S, and 5S rRNA products result from the action of a variety of specific nucleases. The seven copies of pre-rRNA gene in the *E. coli* chromosome differ in the number, location, and identity of tRNAs included in the primary transcript. Some copies of the gene have additional tRNA gene segments between the 16S and 23S rRNA segments and at the far 3′ end of the primary transcript.

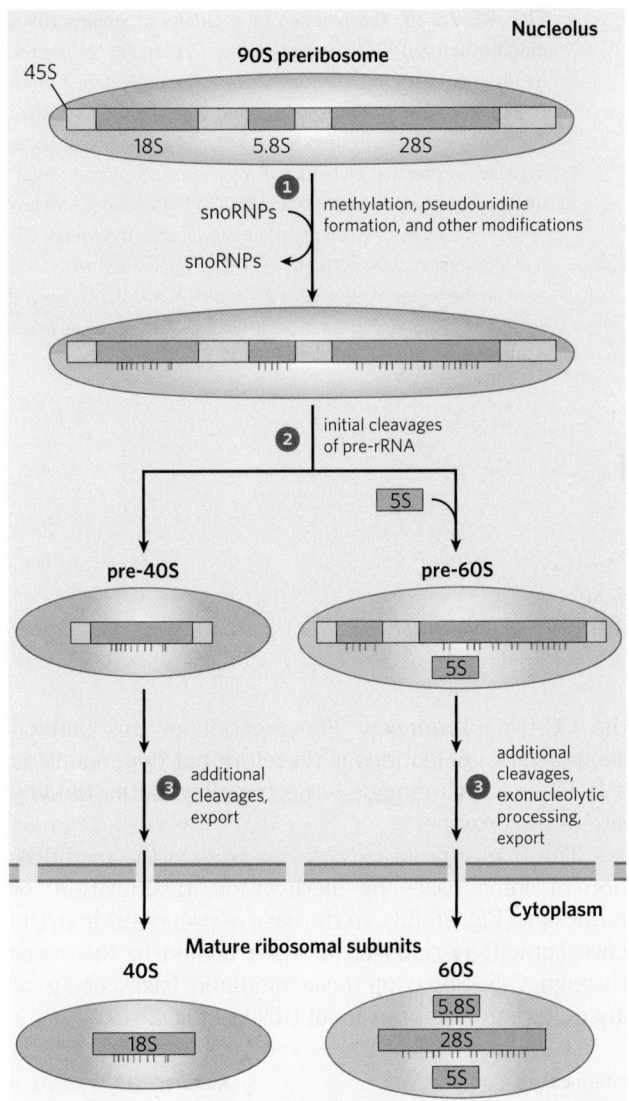

FIGURE 26-24 Processing of pre-rRNA transcripts in vertebrates. During transcription, the 45S primary transcript is incorporated into a nucleolar 90S preribosomal complex, in which rRNA processing and ribosome assembly are tightly coupled. ❶ The 45S precursor is methylated at more than 100 of its 14,000 nucleotides, either on the bases or on the 2'-OH groups (red tick marks); some uridines are converted to pseudouridine (blue ticks), and a few other modifications occur (green ticks are dihydrouridine). ❷, ❸ A series of enzymatic cleavages of the 45S precursor produces the 18S, 5.8S, and 28S rRNAs, and the ribosomal subunits gradually take shape with the assembling ribosomal proteins. The cleavage reactions and all of the modifications require small nucleolar RNAs (snoRNAs), found in protein complexes (snoRNPs) in the nucleolus that are reminiscent of spliceosomes. The 5S rRNA is produced separately.

28S, and 5.8S rRNAs characteristic of eukaryotic ribosomes **(Fig. 26-24).** As in bacteria, the processing includes cleavage reactions mediated by endo- or exoribonucleases and nucleoside modification reactions. Some pre-rRNAs also include introns that must be spliced. The entire process is initiated in the nucleolus, in large complexes that assemble on the rRNA precursor as it is synthesized by Pol I. There is a tight coupling between rRNA

transcription, rRNA maturation, and ribosome assembly in the nucleolus. Each complex includes the ribonucleases that cleave the rRNA precursor, the enzymes that modify particular bases, large numbers of **small nucleolar RNAs**, or **snoRNAs**, that guide nucleoside modification and some cleavage reactions, and ribosomal proteins. In yeast, the entire process involves the pre-rRNA, more than 170 nonribosomal proteins, snoRNAs for each nucleoside modification (about 70 in all, since some snoRNAs guide two types of modification), and the 78 ribosomal proteins. Humans have an even greater number of modified nucleosides, about 200, and a greater number of associated snoRNAs. The composition of the complexes may change as the ribosomes are assembled, and many of the intermediate complexes may rival the ribosome itself, and the snRNPs, in complexity. The 5S rRNA of most eukaryotes is made as a completely separate transcript by a different polymerase (Pol III).

The most common nucleoside modifications in eukaryotic rRNAs are, again, conversion of uridine to pseudouridine and adoMet-dependent nucleoside methylation (often at 2'-hydroxyl groups). These reactions rely on snoRNA-protein complexes, or **snoRNPs**, each consisting of a snoRNA and four or five proteins, which include the enzyme that carries out the modification. There are two classes of snoRNPs, both defined by key conserved sequence elements referred to as lettered boxes. The box H/ACA snoRNPs function in pseudouridylylation, and box C/D snoRNPs in 2'-O-methylations. Unlike the situation in bacteria, the same enzyme may participate in modifications at many sites, guided by the snoRNAs.

The snoRNAs are 60 to 300 nucleotides long. Many are encoded within the introns of other genes and cotranscribed with those genes. Each snoRNA includes a 10 to 21 nucleotide sequence that is perfectly complementary to some site on an rRNA. The conserved sequence elements in the remainder of the snoRNA fold into structures that are bound by the snoRNP proteins **(Fig. 26-25).**

Transfer RNAs Most cells have 40 to 50 distinct tRNAs, and eukaryotic cells have multiple copies of many of the tRNA genes. Transfer RNAs are derived from longer RNA precursors by enzymatic removal of nucleotides from the 5' and 3' ends **(Fig. 26-26).** In eukaryotes, introns are present in a few tRNA transcripts and must be excised. Where two or more different tRNAs are contained in a single primary transcript, they are separated by enzymatic cleavage. The endonuclease RNase P, found in all organisms, removes RNA at the 5' end of tRNAs. This enzyme contains both protein and RNA. The RNA component is essential for activity, and in bacterial cells it can carry out its processing function with precision even without the protein component. RNase P is another example of a catalytic RNA, as described in more detail below. The 3' end of tRNAs is processed by one or more nucleases, including the exonuclease RNase D.

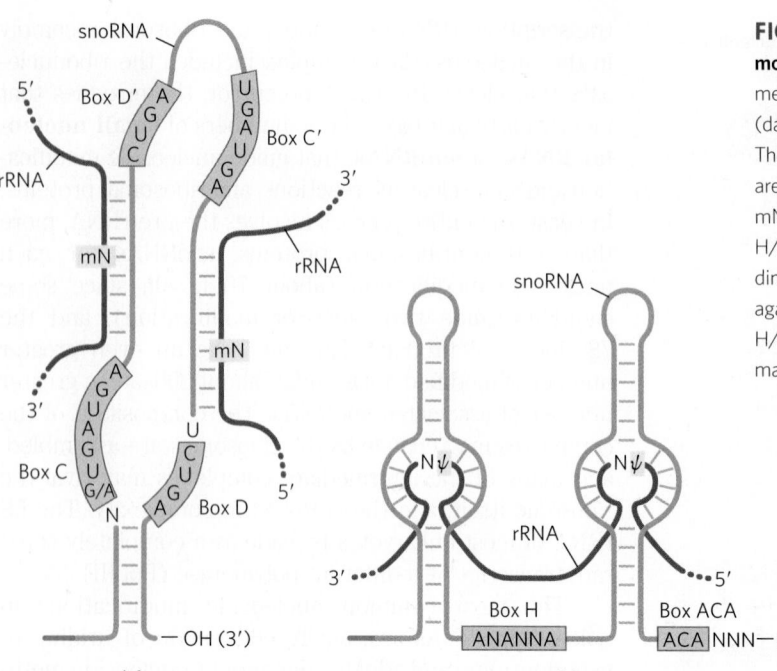

FIGURE 26-25 The function of snoRNAs in guiding rRNA modification. (a) RNA pairing with box C/D snoRNAs to guide methylation reactions. The methylation sites in the target rRNA (dark green) are in the regions paired with the C/D snoRNA. The highly conserved C and D (and C' and D') box sequences are binding sites for proteins that make up the larger snoRNP. mN denotes a methylated nucleotide. (b) RNA pairing with box H/ACA snoRNAs to guide pseudouridylylations. The pseudouridine conversion sites in the target rRNA (green segments) are again in the regions paired with the snoRNA, and the conserved H/ACA box sequences are protein-binding sites. [Source: Information from T. Kiss, *Cell* 109:145, 2002.]

Transfer RNA precursors may undergo further posttranscriptional processing. The 3'-terminal trinucleotide CCA(3') to which an amino acid is attached during protein synthesis (Chapter 27) is absent from some bacterial and all eukaryotic tRNA precursors and is added during processing (Fig. 26-26). This addition is carried out by tRNA nucleotidyltransferase, an unusual enzyme that binds the three ribonucleoside triphosphate precursors in separate active sites and catalyzes formation of the phosphodiester bonds to produce the CCA(3') sequence. The creation of this defined sequence of nucleotides is therefore not dependent on a DNA or RNA template—the template is the binding site of the enzyme.

The final type of tRNA processing is the modification of some bases by methylation, deamination, or reduction (Fig. 26-22). In the case of pseudouridine, the base (uracil) is removed and reattached to the sugar through C-5. Some of these modified bases occur at characteristic positions in all tRNAs (Fig. 26-26).

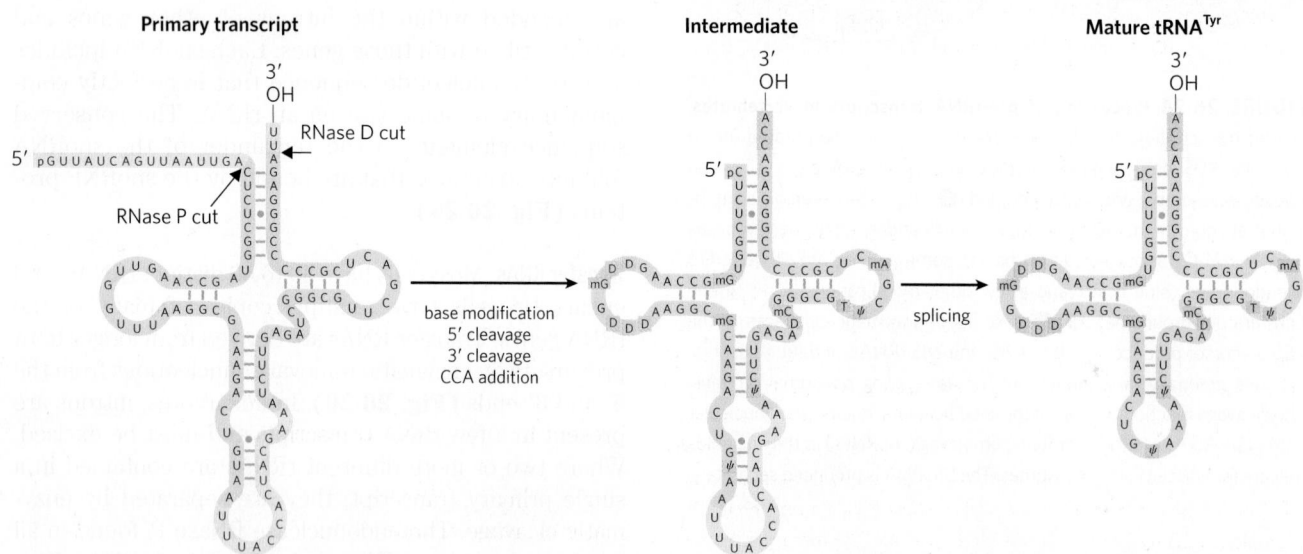

FIGURE 26-26 Processing of tRNAs in bacteria and eukaryotes. The yeast tRNA^Tyr (the tRNA specific for tyrosine binding; see Chapter 27) is used to illustrate the important steps. Short blue lines represent normal base pairing; blue dots indicate G–U base pairs. The nucleotide sequences shown in yellow are removed from the primary transcript. The ends are processed first, the 5' end before the 3' end. CCA is then added to the 3' end, a necessary step in processing all eukaryotic tRNAs and for those bacterial tRNAs that lack this sequence in the primary transcript. While the ends are being processed, specific bases in the rest of the transcript are modified (see Fig. 26-22). For the eukaryotic tRNA shown here, the final step is splicing of the 14 nucleotide intron. Introns are found in some eukaryotic tRNAs but not in bacterial tRNAs.

Special-Function RNAs Undergo Several Types of Processing

The number of known classes of special-function RNAs is expanding rapidly, as is the variety of functions known to be associated with them. Many of these RNAs undergo processing.

The snRNAs and snoRNAs not only facilitate RNA processing reactions but are themselves synthesized as larger precursors and then processed. Many snoRNAs are encoded within the introns of other genes. As the introns are spliced from the pre-mRNA, the snoRNP proteins bind to the snoRNA sequences, and ribonucleases remove the extra RNA at the 5' and 3' ends. The snRNAs destined for spliceosomes are synthesized as pre-snRNAs by RNA polymerase II, and ribonucleases remove the extra RNA at each end. Particular nucleosides in snRNAs are also subject to 11 types of modification, with 2'-O-methylation and conversion of uridine to pseudouridine predominating.

MicroRNAs (miRNAs) are a special class of RNAs involved in gene regulation. They are noncoding RNAs, about 22 nucleotides long, complementary in sequence to particular regions of mRNAs. Found in plants and in animals, from worms to mammals, they promote mRNA degradation and suppress translation. About 1,500 human genes encode miRNAs, and one or more of these miRNAs affect the expression of most protein-coding genes.

The miRNAs are synthesized from much larger precursors, in several steps **(Fig. 26-27)**. The primary transcripts for miRNAs (pri-miRNAs) vary greatly in size; some are encoded in the introns of other genes and are coexpressed with these host genes. Their roles in gene regulation are detailed in Chapter 28.

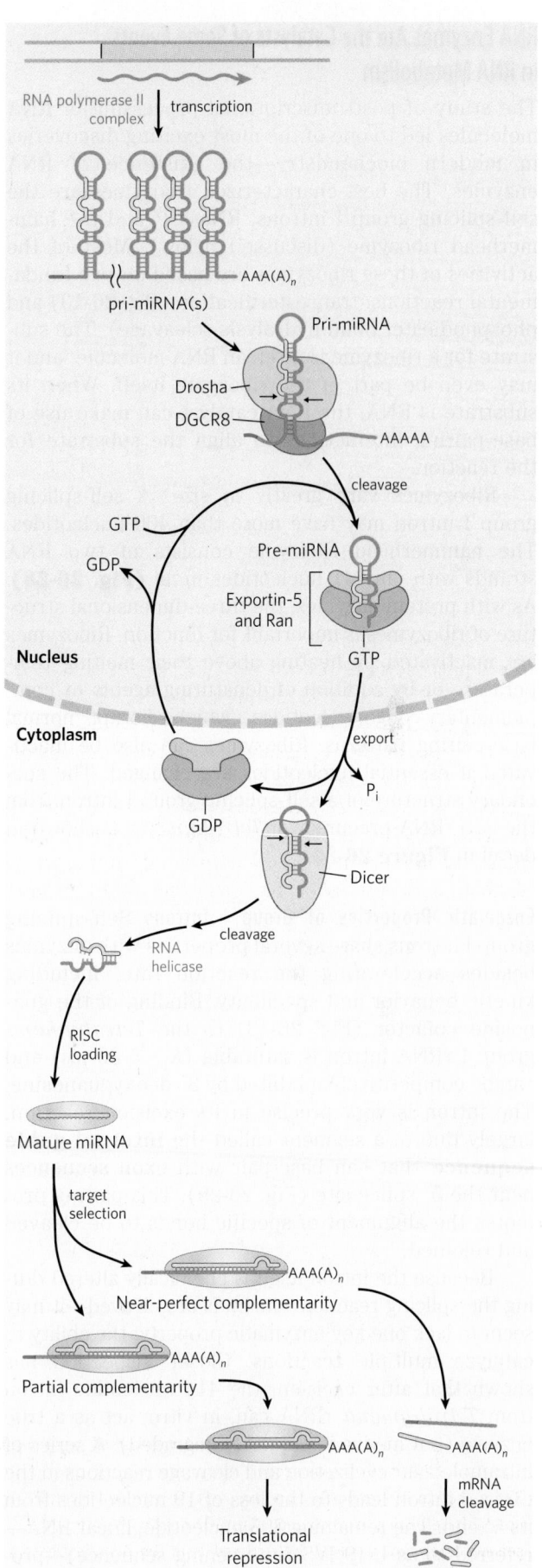

FIGURE 26-27 Synthesis and processing of miRNAs. The primary transcript of miRNAs is a larger RNA of variable length, called pri-miRNA. Much of its processing is mediated by two endoribonucleases in the RNase III family, Drosha and Dicer. First, in the nucleus, the pri-miRNA is reduced to a 70 to 80 nucleotide precursor miRNA (pre-miRNA) by a protein complex including Drosha and another protein, DGCR8. The pre-miRNA is then exported to the cytoplasm in a complex with two proteins, exportin-5 and the Ran GTPase (see Fig. 27-44). In the cytoplasm, Ran hydrolyzes the GTP, then exportin-5 and the pre-miRNA are released. The Ran-GDP and exportin-5 proteins are transported back into the nucleus. The pre-miRNA is acted on by Dicer to produce the nearly mature miRNA paired with a short RNA complement. The complement is removed by an RNA helicase, and the mature miRNA is incorporated into protein complexes, such as the RNA-induced silencing complex (RISC), which then bind a target mRNA. If the complementarity between miRNA and its target is nearly perfect, the target mRNA is cleaved. If the complementarity is only partial, the complex blocks translation of the target mRNA. [Source: Information from E. Wienholds and R. H. A. Plasterk, *FEBS Lett.* 579:5911, 2005; V. N. Kim et al., *Nature Rev. Mol. Cell Biol.* 10:126, 2009, Figs 2–4.]

RNA Enzymes Are the Catalysts of Some Events in RNA Metabolism

The study of posttranscriptional processing of RNA molecules led to one of the most exciting discoveries in modern biochemistry—the existence of RNA enzymes. The best-characterized ribozymes are the self-splicing group I introns, RNase P, and the hammerhead ribozyme (discussed below). Most of the activities of these ribozymes are based on two fundamental reactions: transesterification (Fig. 26-13) and phosphodiester bond hydrolysis (cleavage). The substrate for a ribozyme is often an RNA molecule, and it may even be part of the ribozyme itself. When its substrate is RNA, the RNA catalyst can make use of base-pairing interactions to align the substrate for the reaction.

Ribozymes vary greatly in size. A self-splicing group I intron may have more than 400 nucleotides. The hammerhead ribozyme consists of two RNA strands with only 41 nucleotides in all **(Fig. 26-28)**. As with protein enzymes, the three-dimensional structure of ribozymes is important for function. Ribozymes are inactivated by heating above their melting temperature or by addition of denaturing agents or complementary oligonucleotides, which disrupt normal base-pairing patterns. Ribozymes can also be inactivated if essential nucleotides are changed. The secondary structure of a self-splicing group I intron from the 26S rRNA precursor of *Tetrahymena* is shown in detail in **Figure 26-29**.

Enzymatic Properties of Group I Introns

Self-splicing group I introns share several properties with enzymes besides accelerating the reaction rate, including kinetic behavior and specificity. Binding of the guanosine cofactor (Fig. 26-13) to the *Tetrahymena* group I rRNA intron is saturable ($K_m < 30 \ \mu M$) and can be competitively inhibited by 3′-deoxyguanosine. The intron is very precise in its excision reaction, largely due to a segment called the **internal guide sequence** that can base-pair with exon sequences near the 5′ splice site (Fig. 26-29). This pairing promotes the alignment of specific bonds to be cleaved and rejoined.

Because the intron itself is chemically altered during the splicing reaction—its ends are cleaved—it may seem to lack one key enzymatic property: the ability to catalyze multiple reactions. Closer inspection has shown that after excision, the 414 nucleotide intron from *Tetrahymena* rRNA can, in vitro, act as a true enzyme (but in vivo it is quickly degraded). A series of intramolecular cyclization and cleavage reactions in the excised intron leads to the loss of 19 nucleotides from its 5′ end. The remaining 395 nucleotide, linear RNA—referred to as L-19 IVS (intervening sequence)—promotes nucleotidyl transfer reactions in which some oligonucleotides are lengthened at the expense of others

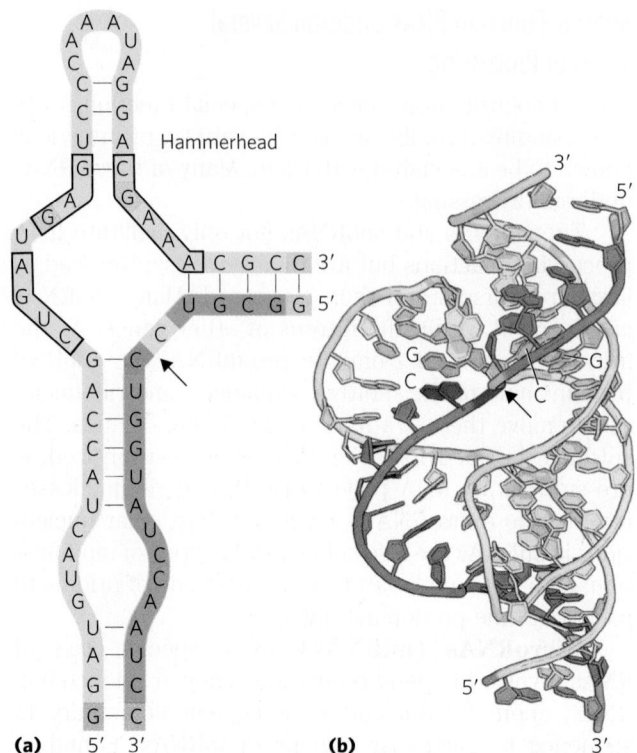

FIGURE 26-28 Hammerhead ribozyme. Certain viruslike elements, or virusoids, have small RNA genomes and usually require another virus to assist in their replication or packaging or both. Some virusoid RNAs include small segments that promote site-specific RNA cleavage reactions associated with replication. These segments are called hammerhead ribozymes, because their secondary structures are shaped like the head of a hammer. Hammerhead ribozymes have been defined and studied separately from the much larger viral RNAs. **(a)** The minimal sequences required for catalysis by the ribozyme. The boxed nucleotides are highly conserved and are required for catalytic function. Guanine nucleotides shaded pink form part of the active site. The arrow indicates the site of self-cleavage. **(b)** Three-dimensional structure (see Fig. 8-25b for a space-filling view of another hammerhead ribozyme). The strands are colored as in (a). The hammerhead ribozyme is a metalloenzyme; Mg^{2+} ions are required for activity in vivo. The phosphodiester bond at the site of self-cleavage is indicated by an arrow. [Source: (b) PDB ID 3ZD5, M. Martick and W. G. Scott, *Cell* 126:309, 2006.]

(Fig. 26-30). The best substrates are oligonucleotides, such as a synthetic $(C)_5$ oligomer, that can base-pair with the same guanylate-rich internal guide sequence that held the 5′ exon in place for self-splicing.

The enzymatic activity of the L-19 IVS ribozyme results from a cycle of transesterification reactions mechanistically similar to self-splicing. Each ribozyme molecule can process about 100 substrate molecules per hour and is not altered in the reaction—that is, the intron acts as a catalyst. It follows Michaelis-Menten kinetics, is specific for RNA oligonucleotide substrates, and can be competitively inhibited. The k_{cat}/K_m (specificity constant) is $10^3 \ M^{-1}s^{-1}$, lower than that of many enzymes, but the ribozyme accelerates hydrolysis by a factor of 10^{10} relative to the uncatalyzed reaction. It makes use of

(a)

(b)

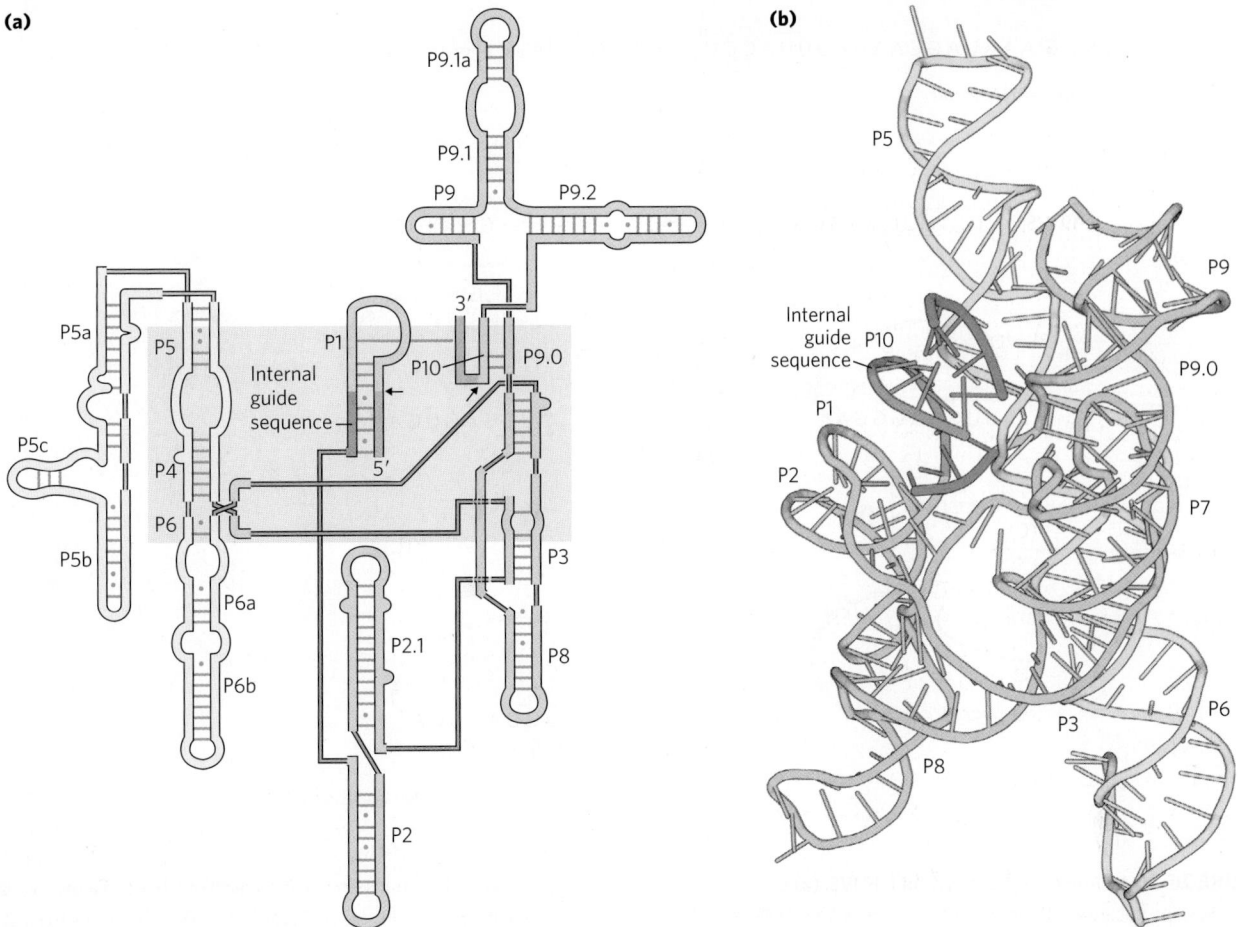

FIGURE 26-29 Secondary structure of the self-splicing rRNA intron of _Tetrahymena_. (a) A two-dimensional representation of secondary structure immediately prior to the initiation of the reaction. Intron sequences are shaded yellow and light red; flanking exon sequences are green; the internal guide sequences that help to align reacting segments at the active site are purple. Each thin, light-red line represents a bond between adjacent nucleotides in a continuous sequence (a device necessitated by showing this complex molecule in two dimensions). Short blue lines represent normal base pairing; blue dots indicate G–U base pairs. All nucleotides are shown. The catalytic core of the self-splicing activity is shaded in gray. Some base-paired regions are labeled (P1, P3, P2.1, P5a, and so forth) according to an established convention for this RNA molecule. The P1 region, which contains the internal guide sequence (purple), is the location of the 5′ splice site (black arrow). Part of the internal guide sequence pairs with the end of the 3′ exon, bringing the 5′ and 3′ splice sites (black arrows) into close proximity. (b) Three-dimensional structure of a reaction intermediate of the same intron, after guanosine-mediated cleavage (Fig. 26-14) and prior to exon ligation. Segments are colored as in (a). [Sources: (a) PDB ID 1GID, J. H. Cate et al., _Science_ 273:1678, 1996. (b) PDB ID 1U6B, P. L. Adams et al., _Nature_ 430:45, 2004.]

substrate orientation, covalent catalysis, and metal-ion catalysis—strategies used by protein enzymes.

Characteristics of Other Ribozymes _E. coli_ RNase P has both an RNA component (the M1 RNA, with 377 nucleotides) and a protein component (M_r 17,500). In 1983, Sidney Altman and Norman Pace and their coworkers discovered that under some conditions, the M1 RNA alone is capable of catalysis, cleaving tRNA precursors at the correct position. The protein component apparently serves to stabilize the RNA or facilitate its function in vivo. The RNase P ribozyme recognizes the three-dimensional shape of its pre-tRNA substrate, along with the CCA sequence, and thus can cleave the 5′ leaders from diverse tRNAs (Fig. 26-26).

The known catalytic repertoire of ribozymes continues to expand. Some virusoids, small RNAs associated with plant RNA viruses, include a structure that promotes a self-cleavage reaction; the hammerhead ribozyme illustrated in Figure 26-28 is in this class, catalyzing the hydrolysis of an internal phosphodiester bond. There are at least nine structural classes of ribozymes that engage in self-cleavage; all use general acid and base catalysis (Fig. 6-8) to promote the attack of a 2′-hydroxyl group on an adjacent phosphodiester bond. The splicing reaction that occurs in a spliceosome relies on a catalytic center formed by the U2, U5, and U6 snRNAs (Fig. 26-16). And, as we shall see in Chapter 27, an RNA component of ribosomes catalyzes the synthesis of proteins.

Exploring catalytic RNAs has provided new insights into catalytic function in general and has important implications for our understanding of the origin and evolution of life on this planet, a topic discussed in Section 26.3.

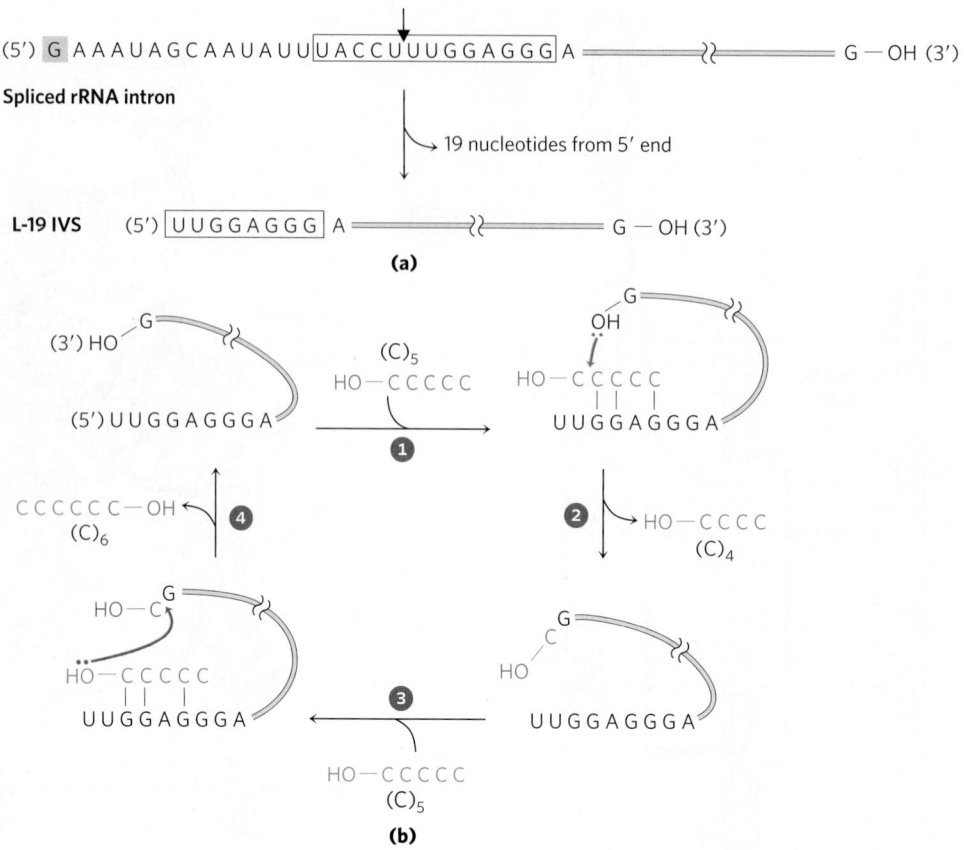

FIGURE 26-30 In vitro catalytic activity of L-19 IVS. (a) L-19 IVS is generated by the autocatalytic removal of 19 nucleotides from the 5′ end of the spliced *Tetrahymena* intron. The cleavage site is indicated by the arrow in the internal guide sequence (boxed). The G residue (shaded light red) added in the first step of the splicing reaction (see Fig. 26-14) is part of the removed sequence. A portion of the internal guide sequence remains at the 5′ end of L-19 IVS. **(b)** L-19 IVS lengthens some RNA oligonucleotides at the expense of others in a cycle of transesterification reactions (steps **1** through **4**). The 3′ OH of the G residue at the 3′ end of L-19 IVS plays a key role in this cycle (note that this is *not* the G residue added in the splicing reaction). (C)₅ is one of the ribozyme's better substrates because it can base-pair with the guide sequence remaining in the intron. Although this catalytic activity is probably irrelevant to the cell, it has important implications for current hypotheses on evolution, discussed at the end of this chapter.

Cellular mRNAs Are Degraded at Different Rates

The expression of genes is regulated at many levels. A crucial factor governing a gene's expression is the cellular concentration of its associated mRNA. The concentration of any molecule depends on two factors: its rate of synthesis and its rate of degradation. When synthesis and degradation of an mRNA are balanced, the concentration of the mRNA remains in a steady state. A change in either rate will lead to net accumulation or depletion of the mRNA. Degradative pathways ensure that mRNAs do not build up in the cell and direct the synthesis of unnecessary proteins.

The rates of degradation vary greatly for mRNAs from different eukaryotic genes. For a gene product that is needed only briefly, the half-life of its mRNA may be only minutes or even seconds. Gene products needed constantly by the cell may have mRNAs that are stable over many cell generations. The average half-life of the mRNAs of a vertebrate cell is about 3 hours, with the pool of each type of mRNA turning over about 10 times per cell generation. The half-life of bacterial mRNAs is much shorter—only about 1.5 min—perhaps because of regulatory requirements.

Messenger RNA is degraded by ribonucleases present in all cells. In *E. coli*, the process begins with one or several cuts by an endoribonuclease, followed by 3′→5′ degradation by exoribonucleases. In lower eukaryotes, the major pathway involves first shortening the poly(A) tail, then decapping the 5′ end and degrading the mRNA in the 5′→3′ direction. A 3′→5′ degradative pathway also exists and may be the major path in higher eukaryotes. All eukaryotes have a complex of up to 10 conserved 3′→5′ exoribonucleases, called the **exosome**, which takes part in the processing of the 3′ end of rRNAs, tRNAs, and some special-function RNAs (including snRNAs and snoRNAs), as well as the degradation of mRNAs.

A hairpin structure in bacterial mRNAs with a ρ-independent terminator (Fig. 26-7) confers stability against degradation. Similar hairpin structures can make some parts of a primary transcript more stable, leading to nonuniform degradation of transcripts. In eukaryotic cells, both the 3′ poly(A) tail and the 5′ cap are important to the stability of many mRNAs.

Polynucleotide Phosphorylase Makes Random RNA-like Polymers

In 1955, Marianne Grunberg-Manago and Severo Ochoa discovered the bacterial enzyme **polynucleotide phosphorylase**, which in vitro catalyzes the reaction

$$(NMP)_n + NDP \rightleftharpoons \underset{\substack{\text{Lengthened}\\\text{polynucleotide}}}{(NMP)_{n+1}} + P_i$$

Polynucleotide phosphorylase was the first nucleic acid–synthesizing enzyme discovered (Arthur Kornberg's discovery of DNA polymerase followed soon thereafter). The reaction catalyzed by polynucleotide phosphorylase differs fundamentally from the polymerase activities discussed so far in that it is not template-dependent. The enzyme uses the 5′-diphosphates of ribonucleosides as substrates and cannot act on the homologous 5′-triphosphates or on deoxyribonucleoside 5′-diphosphates. The RNA polymer formed by polynucleotide phosphorylase contains the usual 3′,5′-phosphodiester linkages, which can be hydrolyzed by ribonuclease. The reaction is readily reversible and can be pushed in the direction of breakdown of the polyribonucleotide by increasing the phosphate concentration. The probable function of this enzyme in the cell is the degradation of mRNAs to nucleoside diphosphates.

 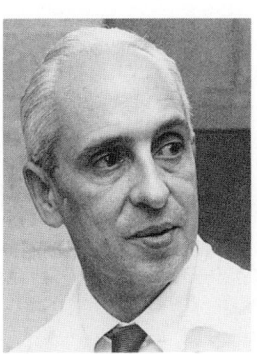

Marianne Grunberg-Manago, 1921-2013 [Source: Philippe Eranian/Sygma/Corbis.]

Severo Ochoa, 1905-1993 [Source: AP Images.]

Because the polynucleotide phosphorylase reaction does not use a template, the polymer it forms does not have a specific base sequence. The reaction proceeds equally well with any or all of the four nucleoside diphosphates, and the base composition of the resulting polymer reflects nothing more than the relative concentrations of the 5′-diphosphate substrates in the medium.

Polynucleotide phosphorylase can be used in the laboratory to prepare RNA polymers with many different base sequences and frequencies. Use of synthetic RNA polymers of this sort was critical in deducing the genetic code for the amino acids (Chapter 27).

SUMMARY 26.2 RNA Processing

■ Eukaryotic mRNAs are modified by addition of a 7-methylguanosine residue at the 5′ end and by cleavage and polyadenylation at the 3′ end to form a long poly(A) tail.

■ Many primary mRNA transcripts contain introns (noncoding regions), which are removed by splicing. Excision of the group I introns found in some rRNAs requires a guanosine cofactor. Some group I and group II introns are capable of self-splicing; no protein enzymes are required. Nuclear mRNA precursors have a third (the largest) class of introns, which are spliced with the aid of RNA-protein complexes called snRNPs that are assembled in spliceosomes. A fourth class of introns, found in some tRNAs, consists of the only introns known to be spliced by protein enzymes.

■ The function of many eukaryotic mRNAs is regulated by complementary microRNAs (miRNAs). The miRNAs are themselves derived from larger precursors through a series of processing reactions.

■ Ribosomal RNAs and transfer RNAs are derived from longer precursor RNAs that are trimmed by nucleases. Some bases are modified enzymatically during the maturation process. Some nucleoside modifications are guided by snoRNAs, within protein complexes called snoRNPs.

■ The self-splicing introns and the RNA component of RNase P (which cleaves the 5′ end of tRNA precursors) are two examples of ribozymes. These biological catalysts have the properties of true enzymes. They generally promote hydrolytic cleavage and transesterification, using RNA as substrate. Combinations of these reactions can be promoted by the excised group I intron of *Tetrahymena* rRNA, resulting in a type of RNA polymerization reaction.

■ Polynucleotide phosphorylase reversibly forms RNA-like polymers from ribonucleoside 5′-diphosphates, adding or removing ribonucleotides at the 3′-hydroxyl end of the polymer. In vivo, the enzyme degrades RNA.

26.3 RNA-Dependent Synthesis of RNA and DNA

In our discussion of DNA and RNA synthesis up to this point, the role of the template strand has been reserved for DNA. However, some enzymes use an RNA template for nucleic acid synthesis. With the important exception of viruses with an RNA genome, these enzymes play only a modest role in information pathways. RNA viruses are the source of most RNA-dependent polymerases characterized so far.

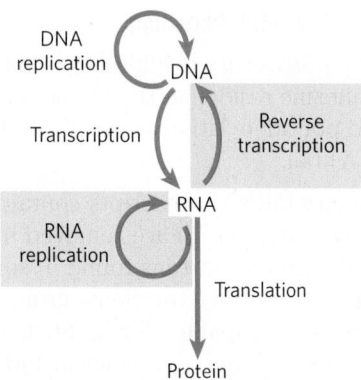

FIGURE 26-31 Extension of the central dogma to include RNA-dependent synthesis of RNA and DNA.

The existence of RNA replication requires an elaboration of the central dogma (**Fig. 26-31**; contrast this with the diagram on p. 955). The enzymes of the RNA replication process have profound implications for investigations into the nature of self-replicating molecules that may have existed in prebiotic times.

Reverse Transcriptase Produces DNA from Viral RNA

Certain RNA viruses that infect animal cells carry within the viral particle an RNA-dependent DNA polymerase called **reverse transcriptase**. On infection, the single-stranded RNA viral genome (~10,000 nucleotides) and the enzyme enter the host cell. The reverse transcriptase first catalyzes the synthesis of a DNA strand complementary to the viral RNA (**Fig. 26-32**), then degrades the RNA strand of the viral RNA-DNA hybrid and replaces it with DNA. The resulting duplex DNA often becomes incorporated into the genome of the eukaryotic host cell. These integrated

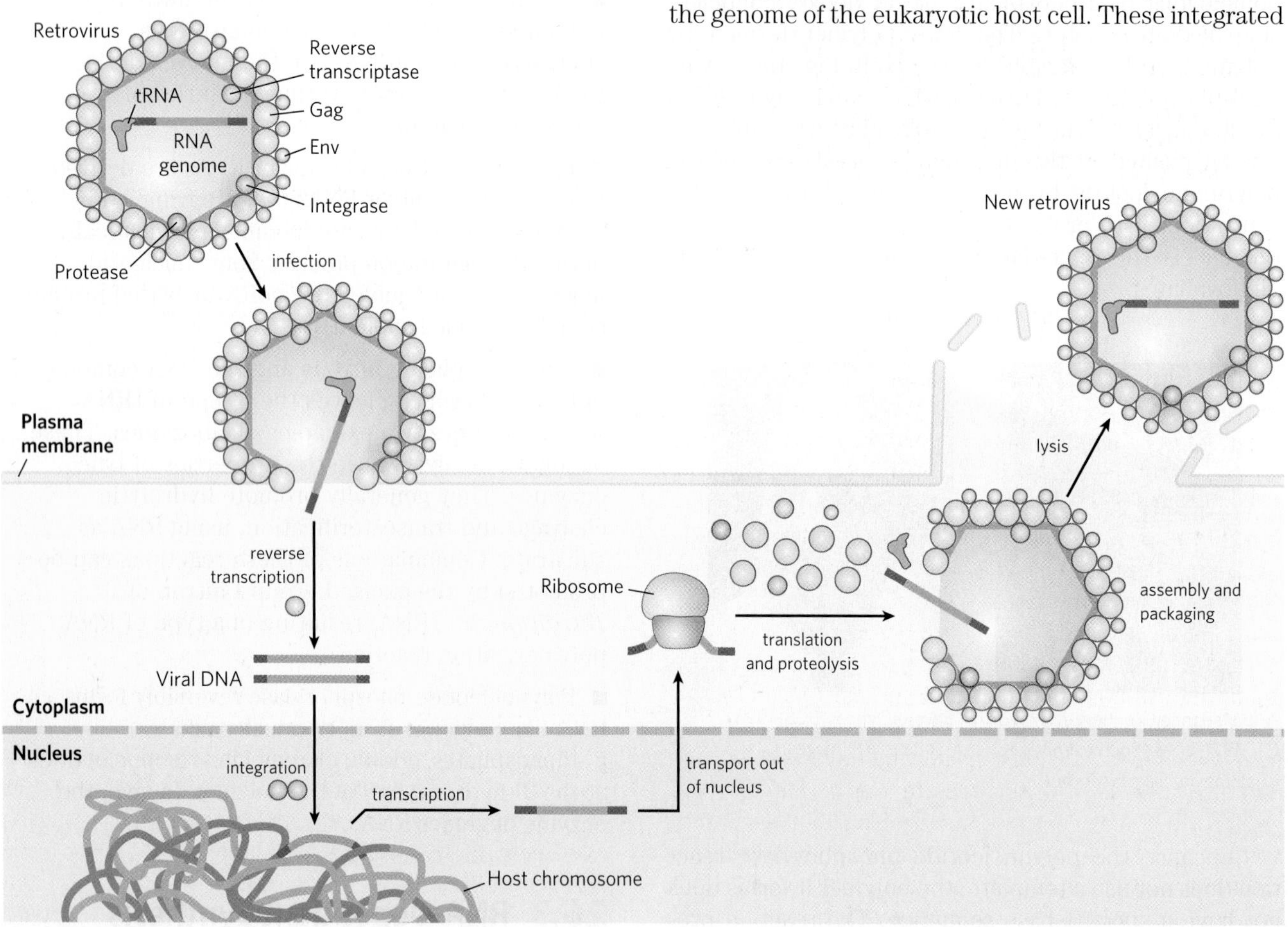

FIGURE 26-32 Retroviral infection of a mammalian cell and integration of the retrovirus into the host chromosome. Viral particles entering the host cell carry viral reverse transcriptase and a cellular tRNA (picked up from a former host cell) already base-paired to the viral RNA. The purple segments represent the long terminal repeats on the viral RNA. The tRNA facilitates immediate conversion of viral RNA to double-stranded DNA by the action of reverse transcriptase, as described in the text. The double-stranded DNA enters the nucleus and is integrated into the host genome. The integration is catalyzed by a virally encoded integrase. Integration of viral DNA into host DNA is mechanistically similar to the

insertion of transposons in bacterial chromosomes (see Fig. 25-42). For example, a few base pairs of host DNA become duplicated at the site of integration, forming short repeats of 4 to 6 bp at each end of the inserted retroviral DNA (not shown). On transcription and translation of the integrated viral DNA, new viruses are formed and released by cell lysis (right). In the viruses, the viral RNA is enclosed by capsid proteins called Gag and outer envelope proteins called Env. Additional viral proteins (reverse transcriptase, integrase, and a viral protease needed for posttranslational processing of viral proteins) are packaged within the virus particle with the RNA.

(and dormant) viral genes can be activated and transcribed, and the gene products—viral proteins and the viral RNA genome itself—are packaged as new viruses. The RNA viruses that contain reverse transcriptases are known as **retroviruses** (*retro* is the Latin prefix for "backward").

The existence of reverse transcriptases in RNA viruses was predicted by Howard Temin in 1962, and the enzymes were ultimately detected by Temin and, independently, by David Baltimore in 1970. Their discovery aroused much attention as dogma-shaking proof that genetic information can flow "backward" from RNA to DNA.

Howard Temin, 1934–1994
[Source: Corbis/UPI/Bettmann.]

David Baltimore
[Source: AP Photo.]

Retroviruses typically have three genes: *gag* (a name derived from the historical designation group associated antigen), *pol*, and *env* (**Fig. 26-33**). The transcript that contains *gag* and *pol* is translated into a long "polyprotein," a single large polypeptide that is cleaved into six proteins with distinct functions. The proteins derived from the *gag* gene make up the interior core of the viral particle. The *pol* gene encodes the protease that cleaves the long polypeptide, an integrase that inserts the viral DNA into the host chromosomes, and reverse transcriptase. Many reverse transcriptases have two subunits, α and β. The *pol* gene specifies the β subunit (M_r 90,000), and the α subunit (M_r 65,000) is simply a proteolytic fragment of the β subunit. The *env* gene encodes the proteins of the viral envelope. At each end of the linear RNA genome are long terminal repeat (LTR) sequences of a few hundred nucleotides. Transcribed into the duplex DNA, these sequences facilitate integration of the viral chromosome into the host DNA and contain promoters for viral gene expression.

Reverse transcriptases catalyze three different reactions: (1) RNA-dependent DNA synthesis, (2) RNA degradation, and (3) DNA-dependent DNA synthesis. Like many DNA and RNA polymerases, reverse transcriptases contain Zn^{2+}. Each transcriptase is most active with the RNA of its own virus, but each can be used experimentally to make DNA complementary to a variety of RNAs. The DNA and RNA synthesis and RNA degradation activities use separate active sites on the protein. For DNA synthesis to begin, the reverse transcriptase

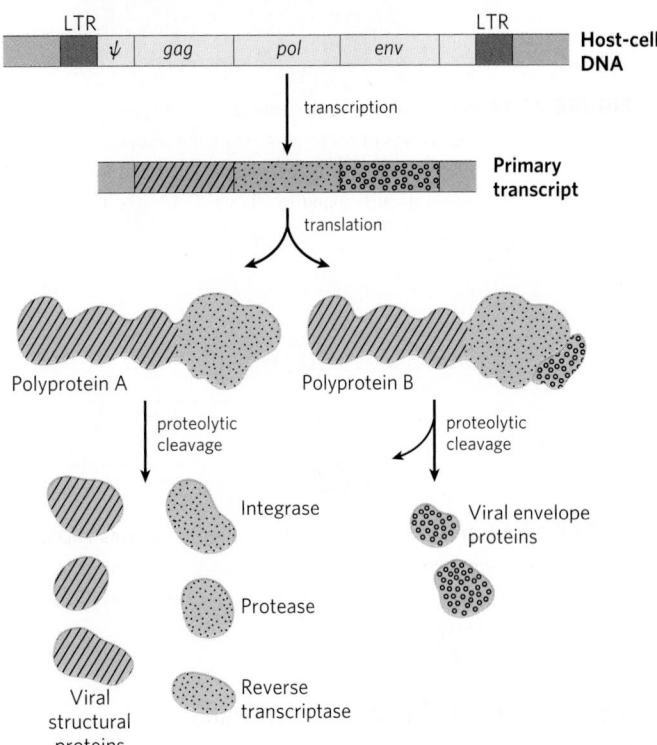

FIGURE 26-33 Structure and gene products of an integrated retroviral genome. The long terminal repeats (LTRs) have sequences needed for the regulation and initiation of transcription. The sequence denoted ψ is required for packaging of retroviral RNAs into mature viral particles. Transcription of the retroviral DNA produces a primary transcript encompassing the *gag*, *pol*, and *env* genes. Translation (Chapter 27) produces a polyprotein, a single long polypeptide derived from the *gag* and *pol* genes, which is cleaved into six distinct proteins. Splicing of the primary transcript yields an mRNA derived largely from the *env* gene, which is also translated into a polyprotein, then cleaved to generate viral envelope proteins.

requires a primer, a cellular tRNA obtained during an earlier infection and carried in the viral particle. This tRNA is base-paired at its 3' end with a complementary sequence in the viral RNA. The new DNA strand is synthesized in the 5'→3' direction, as in all RNA and DNA polymerase reactions. Reverse transcriptases, like RNA polymerases, do not have 3'→5' proofreading exonucleases. They generally have error rates of about 1 per 20,000 nucleotides added. An error rate this high is extremely unusual in DNA replication and seems to be a characteristic of most enzymes that replicate the genomes of RNA viruses. A consequence is a higher mutation rate and faster rate of viral evolution, which is a factor in the frequent appearance of new strains of disease-causing retroviruses.

Reverse transcriptases have become important reagents in the study of DNA-RNA relationships and in DNA cloning techniques. They make possible the synthesis of DNA complementary to an mRNA template, and synthetic DNA prepared in this manner, called **complementary DNA (cDNA)**, can be used to clone cellular genes (see Fig. 9-14).

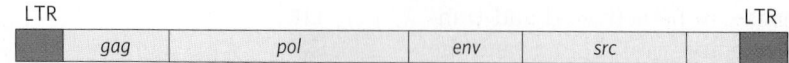

FIGURE 26-34 Rous sarcoma virus genome. The *src* gene encodes a tyrosine kinase, one of a class of enzymes that function in systems affecting cell division, cell-cell interactions, and intercellular communication (Chapter 12). The same gene is found in chicken DNA (the usual host for this virus) and in the genomes of many other eukaryotes, including humans. When associated with the Rous sarcoma virus, this oncogene is often expressed at abnormally high levels, contributing to unregulated cell division and cancer.

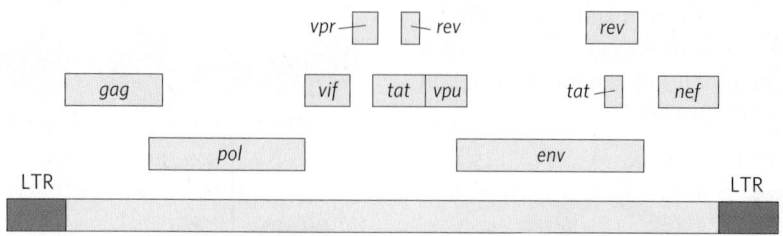

FIGURE 26-35 The genome of HIV, the virus that causes AIDS. In addition to the typical retroviral genes, HIV contains several small genes with a variety of functions (not identified here and not all known). Some of these genes overlap. Alternative splicing mechanisms produce many different proteins from this small (9.7×10^3 nucleotides) genome.

Some Retroviruses Cause Cancer and AIDS

Retroviruses have featured prominently in recent advances in the molecular understanding of cancer. Most retroviruses do not kill their host cells but remain integrated in the cellular DNA, replicating when the cell divides. Some retroviruses, classified as RNA tumor viruses, contain an oncogene that can cause the cell to grow abnormally. The first retrovirus of this type to be studied was the Rous sarcoma virus (also called avian sarcoma virus; **Fig. 26-34**), named for F. Peyton Rous, who studied chicken tumors now known to be caused by this virus. Since the initial discovery of oncogenes by Harold Varmus and Michael Bishop, many dozens of such genes have been found in retroviruses.

The human immunodeficiency virus (HIV), which causes acquired immune deficiency syndrome (AIDS), is a retrovirus. Identified in 1983, HIV has an RNA genome with standard retroviral genes along with several other unusual genes **(Fig. 26-35)**. Unlike many other retroviruses, HIV kills many of the cells it infects (principally T lymphocytes) rather than causing tumor formation. This gradually leads to suppression of the immune system in the host organism. The reverse transcriptase of HIV is even more error-prone than other known reverse transcriptases—10 times more so—resulting in high mutation rates in this virus. One or more errors are generally made every time the viral genome is replicated, so any two viral RNA molecules are likely to differ.

Many modern vaccines for viral infections consist of one or more coat proteins of the virus, produced by methods described in Chapter 9. These proteins are not infectious on their own but stimulate the immune system to recognize and resist subsequent viral invasions (Chapter 5). Because of the high error rate of the HIV reverse transcriptase, the *env* gene in this virus (along with the rest of the genome) undergoes very rapid mutation, complicating the development of an effective vaccine. However, repeated cycles of cell invasion and replication are needed to propagate an HIV infection, so inhibition of viral enzymes offers the most effective therapy currently available. The HIV protease is targeted by a class of drugs called protease inhibitors (see Fig. 6-25). Reverse transcriptase is the target of some additional drugs widely used to treat HIV-infected individuals (Box 26-2). ∎

Many Transposons, Retroviruses, and Introns May Have a Common Evolutionary Origin

Some well-characterized eukaryotic DNA transposons from sources as diverse as yeast and fruit flies have a structure very similar to that of retroviruses; these are sometimes called retrotransposons **(Fig. 26-36)**. Retrotransposons encode an enzyme homologous to the retroviral reverse transcriptase, and their coding regions are flanked by LTR sequences. They transpose from one position to another in the cellular genome by means of an RNA intermediate, using reverse transcriptase to

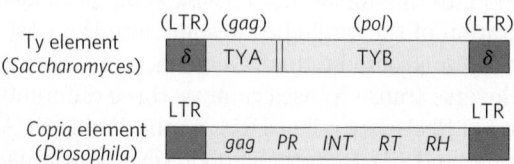

FIGURE 26-36 Eukaryotic transposons. The Ty element of the yeast *Saccharomyces* and the *copia* element of the fruit fly *Drosophila* are examples of eukaryotic retrotransposons, which often have a structure similar to retroviruses but lack the *env* gene. The δ sequences of the Ty element are functionally equivalent to retroviral LTRs. In the *copia* element, *INT* and *RT* are homologous to the integrase and reverse transcriptase segments, respectively, of the *pol* gene.

BOX 26-2 ⚕ MEDICINE Fighting AIDS with Inhibitors of HIV Reverse Transcriptase

Research into the chemistry of template-dependent nucleic acid biosynthesis, combined with modern techniques of molecular biology, has elucidated the life cycle and structure of the human immunodeficiency virus, the retrovirus that causes AIDS. A few years after the isolation of HIV, this research resulted in the development of drugs capable of prolonging the lives of people infected by HIV.

The first drug to be approved for clinical use was AZT, a structural analog of deoxythymidine. AZT was first synthesized in 1964 by Jerome P. Horwitz. It failed as an anticancer drug (the purpose for which it was made), but in 1985 it was found to be a useful treatment for AIDS. AZT is taken up by T lymphocytes, immune system cells that are particularly vulnerable to HIV infection, and converted to AZT triphosphate. (AZT triphosphate taken directly

3'-Azido-2',3'-dideoxy-
thymidine (AZT)

2',3'-Dideoxyinosine (DDI)

would be ineffective because it cannot cross the plasma membrane.) HIV's reverse transcriptase has a higher affinity for AZT triphosphate than for dTTP, and binding of AZT triphosphate to this enzyme competitively inhibits dTTP binding. When AZT is added to the 3' end of the growing DNA strand, lack of a 3' hydroxyl means that the DNA strand is terminated prematurely and viral DNA synthesis grinds to a halt.

AZT triphosphate is not as toxic to the T lymphocytes themselves because *cellular* DNA polymerases have a lower affinity for this compound than for dTTP. At concentrations of 1 to 5 μM, AZT affects HIV reverse transcription but not most cellular DNA replication. Unfortunately, AZT seems to be toxic to the bone marrow cells that are the progenitors of erythrocytes, and many individuals taking AZT develop anemia. AZT can increase the survival time of people with advanced AIDS by about a year, and it delays the onset of AIDS in those who are still in the early stages of HIV infection. Some other AIDS drugs, such as dideoxyinosine (DDI), have a similar mechanism of action. Newer drugs target and inactivate the HIV protease. Because of the high error rate of HIV reverse transcriptase and the resulting rapid evolution of HIV, the most effective treatments of HIV infection use a combination of drugs directed at both the protease and the reverse transcriptase.

---

make a DNA copy of the RNA, followed by integration of the DNA at a new site. Most transposons in eukaryotes use this mechanism for transposition, distinguishing them from bacterial transposons, which move as DNA directly from one chromosomal location to another (see Fig. 25–42).

Retrotransposons lack an *env* gene and so cannot form viral particles. They can be thought of as defective viruses, trapped in cells. Comparisons between retroviruses and eukaryotic transposons suggest that reverse transcriptase is an ancient enzyme that predates the evolution of multicellular organisms.

Many group I and group II introns are also mobile genetic elements. In addition to their self-splicing activities, they encode DNA endonucleases that promote their movement. During genetic exchanges between cells of the same species, or when DNA is introduced into a cell by parasites or by other means, these endonucleases promote insertion of the intron into an identical site in another DNA copy of a homologous gene that does not contain the intron, in a process termed **homing (Fig. 26-37)**. Whereas group I intron homing is DNA-based, group II intron homing

occurs through an RNA intermediate. The endonucleases of the group II introns have associated reverse transcriptase activity. The proteins can form complexes with the intron RNAs themselves, after the introns are spliced from the primary transcripts. Because the homing process involves insertion of the RNA intron into DNA and reverse transcription of the intron, the movement of these introns has been called retrohoming. Over time, every copy of a particular gene in a population may acquire the intron. Much more rarely, the intron may insert itself into a new location in an unrelated gene. If this event does not kill the host cell, it can lead to the evolution and distribution of an intron in a new location. The structures and mechanisms used by mobile introns support the idea that at least some introns originated as molecular parasites whose evolutionary past can be traced to retroviruses and transposons.

Telomerase Is a Specialized Reverse Transcriptase

Telomeres, the structures at the ends of linear eukaryotic chromosomes (see Fig. 24–8), generally consist of many tandem copies of a short oligonucleotide sequence.

FIGURE 26-37 Introns that move: homing and retrohoming. Certain introns include a gene (shown in red) for enzymes that promote homing (some group I introns) or retrohoming (some group II introns). **(a)** The gene in the spliced intron is bound by a ribosome and translated. Group I homing introns specify a site-specific endonuclease, called a homing endonuclease. Group II retrohoming introns specify a protein with both endonuclease and reverse transcriptase activities (not shown here).

(b) Homing. Allele *a* of a gene *X* containing a group I homing intron is present in a cell containing allele *b* of the same gene, which lacks the intron. The homing endonuclease produced by *a* cleaves *b* at the position corresponding to the intron in *a*, and double-strand break repair (recombination with allele *a*; see Fig. 25-35) then creates a new copy of the intron in *b*. **(c)** Retrohoming. Allele *a* of gene *Y* contains a retrohoming group II intron; allele *b* lacks the intron. The spliced intron inserts itself into the coding strand of *b* in a reaction that is the reverse of the splicing that excised the intron from the primary transcript (see Fig. 26-15), except that here the insertion is into DNA rather than RNA. The noncoding DNA strand of *b* is then cleaved by the intron-encoded endonuclease/reverse transcriptase. This same enzyme uses the inserted RNA as a template to synthesize a complementary DNA strand. The RNA is then degraded by cellular ribonucleases and replaced with DNA.

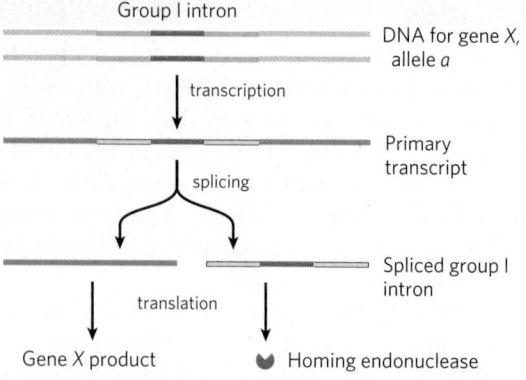

(a) Production of homing endonuclease

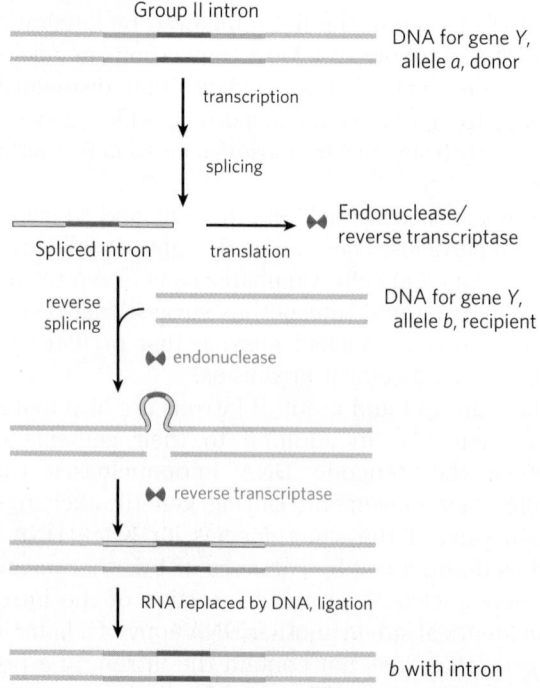

(b) Homing

(c) Retrohoming

This sequence usually has the form T_xG_y in one strand and C_yA_x in the complementary strand, where x and y are typically in the range of 1 to 4 (p. 962). Telomeres vary in length from a few dozen base pairs in some ciliated protozoans to tens of thousands of base pairs in mammals. The TG strand is longer than its complement, leaving a region of single-stranded DNA of up to a few hundred nucleotides at the 3′ end.

The ends of a linear chromosome are not readily replicated by cellular DNA polymerases. DNA replication requires a template and primer, and beyond the end of a linear DNA molecule no template is available for the pairing of an RNA primer. Without a special mechanism for replicating the ends, chromosomes would be shortened somewhat in each cell generation. The enzyme **telomerase**, discovered by Carol Greider and Elizabeth Blackburn, solves this problem by adding telomeres to chromosome ends.

Carol Greider [Source: Courtesy of Carol Greider.]

Elizabeth Blackburn [Source: Elisabeth Fall/Fallfoto.com.]

The discovery and purification of this enzyme provided insight into a reaction mechanism that is remarkable and unprecedented. Telomerase, like some other enzymes described in this chapter, contains both RNA and protein components. The RNA component is about 150 nucleotides long and contains about 1.5 copies of the appropriate C_yA_x telomere repeat. This region of the RNA acts as a template for synthesis of the T_xG_y strand of the telomere. Telomerase thereby acts as a cellular reverse transcriptase that provides the active site for RNA-dependent DNA synthesis. Unlike retroviral reverse transcriptases, telomerase copies only a small segment of RNA that it carries within itself. Telomere synthesis requires the 3′ end of a chromosome as primer and proceeds in the usual 5′→3′ direction. Having synthesized one copy of the repeat, the enzyme repositions to resume extension of the telomere **(Fig. 26-38a)**.

After extension of the T_xG_y strand by telomerase, the complementary C_yA_x strand is synthesized by cellular DNA polymerases, starting with an RNA primer (see Fig. 25-12). The single-stranded region is protected by specific binding proteins in many lower eukaryotes, especially those species with telomeres of less than a few hundred base pairs. In higher eukaryotes (including mammals) with telomeres many thousands of base pairs long, the single-stranded end is sequestered in a specialized structure called a **T loop** (Fig. 26-38b). The single-stranded end is folded back and paired with its complement in the double-stranded portion of the telomere. The formation of a T loop involves invasion of the 3′ end of the telomere's single

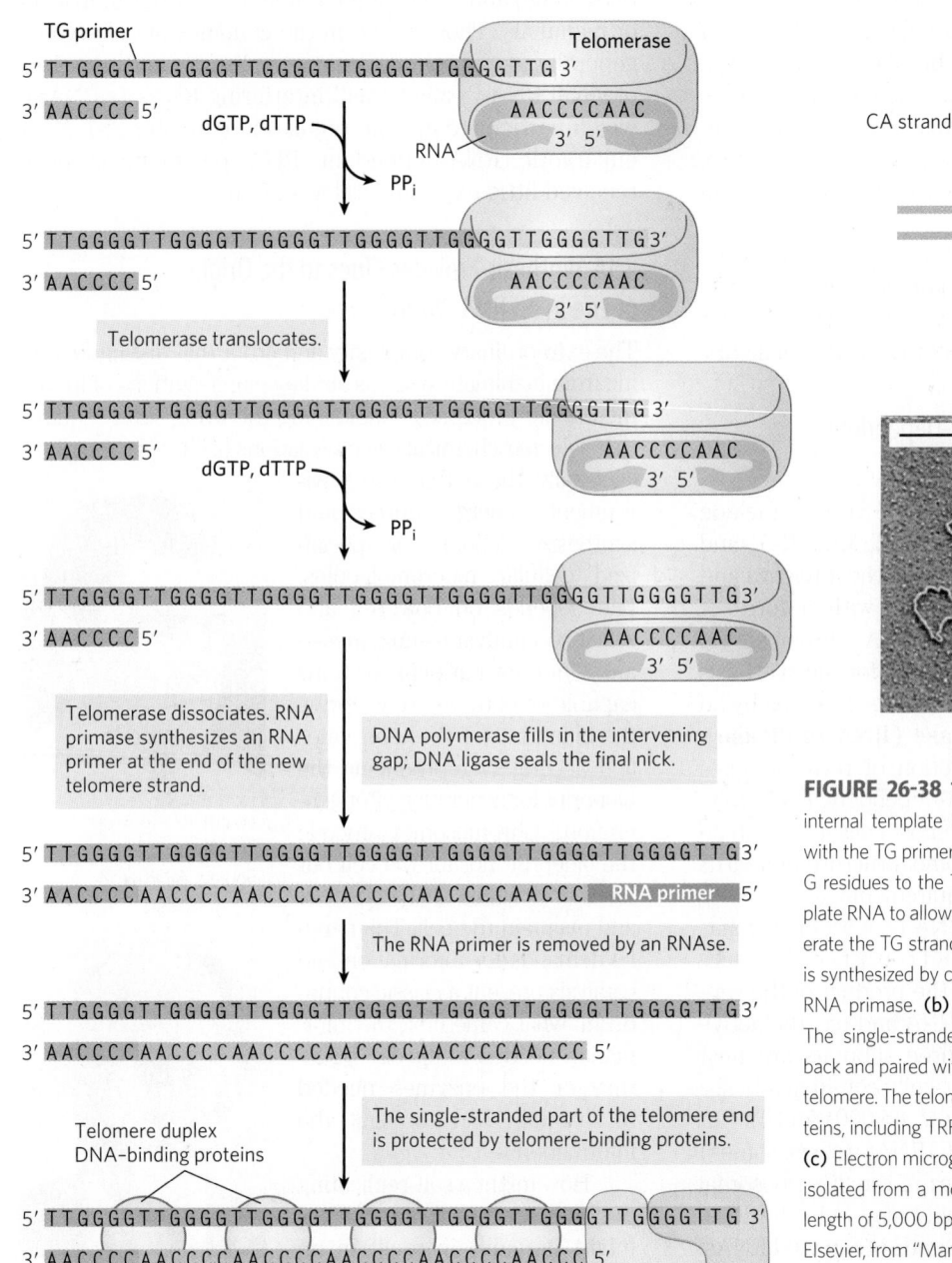

FIGURE 26-38 Telomere synthesis and structure. (a) The internal template RNA of telomerase binds to and base-pairs with the TG primer (T_xG_y) of DNA. Telomerase adds more T and G residues to the TG primer, then repositions the internal template RNA to allow the addition of more T and G residues to generate the TG strand of the telomere. The complementary strand is synthesized by cellular DNA polymerases, after priming by an RNA primase. **(b)** Proposed structure of T loops in telomeres. The single-stranded tail synthesized by telomerase is folded back and paired with its complement in the duplex portion of the telomere. The telomere is bound by several telomere-binding proteins, including TRF1 and TRF2 (*telomere repeat binding factors*). **(c)** Electron micrograph of a T loop at the end of a chromosome isolated from a mouse hepatocyte. The scale bar represents a length of 5,000 bp. [Source: (c) Republished with permission of Elsevier, from "Mammalian telomeres end in a large duplex loop" by J. D. Griffith, et al., *Cell*, 97(4): 503–514, May 1999; permission conveyed through Copyright Clearance Center, Inc.]

strand into the duplex DNA, perhaps by a mechanism similar to the initiation of homologous genetic recombination (see Fig. 25-35). In mammals, the looped DNA is bound by two proteins, TRF1 and TRF2, with the latter protein involved in formation of the T loop. T loops protect the 3' ends of chromosomes, making them inaccessible to nucleases and the enzymes that repair double-strand breaks.

In protozoans (such as *Tetrahymena*), loss of telomerase activity results in a gradual shortening of telomeres with each cell division, ultimately leading to the death of the cell line. A similar link between telomere length and cell senescence (cessation of cell division) has been observed in humans. In germ-line cells, which contain telomerase activity, telomere lengths are maintained; in somatic cells, which lack telomerase, they are not. There is a linear, inverse relationship between the length of telomeres in cultured fibroblasts and the age of the individual from whom the fibroblasts were taken: telomeres in human somatic cells gradually shorten as an individual ages. If the telomerase reverse transcriptase is introduced into human somatic cells in vitro, telomerase activity is restored and the cellular life span increases markedly.

Is the gradual shortening of telomeres a key to the aging process? Is our natural life span determined by the length of the telomeres we are born with? Further research in this area should yield some fascinating insights.

Some RNAs Are Replicated by RNA-Dependent RNA Polymerase

Apart from the retroviruses, the RNA viruses include some *E. coli* bacteriophages, such as f2, MS2, R17, and Qβ, as well as eukaryotic viruses such as the influenza and Sindbis viruses (the latter associated with a form of encephalitis). The single-stranded RNA chromosomes of these viruses also function as mRNAs for the synthesis of viral proteins. They are replicated in the host cell by an **RNA-dependent RNA polymerase (RNA replicase)**. All RNA viruses—with the exception of retroviruses—must encode a protein with RNA-dependent RNA polymerase activity, either because the host cells lack such an enzyme or because the RNA genome structure of a virus imposes specialized enzymatic requirements.

The RNA replicase of most RNA bacteriophages has a molecular weight of ~210,000 and consists of four subunits. One subunit (M_r 65,000) is the product of the replicase gene encoded by the viral RNA and has the active site for replication. The other three subunits are host proteins normally involved in host-cell protein synthesis: the *E. coli* elongation factors Tu (M_r 45,000) and Ts (M_r 34,000) (which ferry amino acyl–tRNAs to ribosomes) and the protein S1 (an integral part of the 30S ribosomal subunit). These three host proteins may help the RNA replicase locate and bind to the 3' ends of the viral RNAs.

The RNA replicase isolated from Qβ-infected *E. coli* cells catalyzes the formation of an RNA complementary to the viral RNA, in a reaction equivalent to that catalyzed by DNA-dependent RNA polymerases. New RNA strand synthesis proceeds in the 5'→3' direction by a chemical mechanism identical to that used in all other nucleic acid synthetic reactions that require a template. RNA replicase requires RNA as its template and will not function with DNA. It lacks a separate proofreading endonuclease activity and has an error rate similar to that of RNA polymerase. Unlike the DNA and RNA polymerases, RNA replicases are specific for the RNA of their own virus; the RNAs of the host cell are generally not replicated. This explains how RNA viruses are preferentially replicated in the host cell, which contains many other types of RNA.

RNA-dependent RNA polymerases are not limited to viruses. Enzymes of this type are found in plants, protists, fungi, and some simpler animals, but not in insects or mammals. Those found in the genomes of eukaryotes generally play a role in the metabolism of another class of small RNAs, called small interfering RNAs (siRNAs), which participate in gene regulation (Chapter 28). Most eukaryotic RNA-dependent RNA polymerases have received little experimental attention.

RNA Synthesis Provides Clues to the Origin of Life in an RNA World

The extraordinary complexity and order that distinguish living from inanimate systems are key manifestations of fundamental life processes. Maintaining the living state requires that *selected* chemical transformations occur very rapidly— especially those that use environmental energy sources and synthesize elaborate or specialized cellular macromolecules. Life depends on powerful and selective catalysts—enzymes— and on informational systems capable of both securely storing the blueprint for these enzymes and accurately reproducing the blueprint for generation after generation. Chromosomes encode the blueprint not for the cell but for the enzymes that construct and maintain the cell. The parallel demands for information and catalysis present a classic conundrum: what came first, the information needed to specify structure or the enzymes needed to maintain and transmit the information?

Carl Woese
[Source: AP Photo.]

How might a self-replicating polymer come to be? How might it maintain itself in an environment where the precursors for polymer synthesis

Francis Crick, 1916–2004
[Source: AP Images.]

Leslie Orgel, 1927–2007
[Source: © Courtesy of The Salk Institute for Biological Studies.]

are scarce? How could evolution progress from such a polymer to the modern DNA-protein world? These difficult questions can be addressed by careful experimentation, providing clues about how life on Earth began and evolved.

The unveiling of the structural and functional complexity of RNA led Carl Woese, Francis Crick, and Leslie Orgel to propose in the 1960s that this macromolecule might serve as both information carrier and catalyst. Since that time, at least six lines of evidence have given increasing substance to their **RNA world hypothesis**.

1. Prebiotic Chemistry Experiments The probable origin of purine and pyrimidine bases is suggested by experiments designed to test hypotheses about prebiotic chemistry (pp. 33–34). Beginning with simple molecules thought to be present in the early atmosphere (CH_4, NH_3, H_2O, H_2), electrical discharges mimicking lightning generate, first, more reactive molecules such as HCN and aldehydes, then an array of amino acids and organic acids (see Fig. 1-35). When molecules such as HCN become abundant, purine and pyrimidine bases are synthesized in detectable amounts. Remarkably, a concentrated solution of ammonium cyanide, refluxed for a few days, generates adenine in yields of up to 0.5% **(Fig. 26-39)**. Adenine may well have been the first and most abundant nucleotide constituent to appear on Earth. Intriguingly, most enzyme cofactors contain adenosine as part of their structure, although it plays no direct role in the cofactor function (see Fig. 8-41). This may suggest an evolutionary relationship. Based on the simple synthesis of adenine from cyanide, adenine may simply have been abundant and available.

2. The Existence of Catalytic RNAs In an "RNA world," RNAs, not proteins, act as catalysts. Perhaps more than anything else, the discovery of catalytic RNAs (ribozymes) in the early 1980s gave life to the RNA world hypothesis and led to widespread speculation that an RNA world might have been important in the transition from prebiotic chemistry to life (see Fig. 1-37). The parent of all

life on this planet, in the sense that it could reproduce itself across the generations from the origin of life to the present, might have been a self-replicating RNA, or a polymer with equivalent chemical characteristics.

3. The Expanding Catalytic Repertoire of Ribozymes A self-replicating polymer would quickly use up available supplies of precursors provided by the relatively slow processes of prebiotic chemistry. Thus, from an early stage in evolution, metabolic pathways would be required to generate precursors efficiently, with the synthesis of precursors presumably catalyzed by ribozymes. The extant ribozymes found in nature have a limited repertoire of catalytic functions, and of the ribozymes that may once have existed, no trace is left. To explore the RNA world hypothesis more deeply, we need to know whether RNA has the potential to catalyze the many different reactions needed in a primitive system of metabolic pathways.

The search for RNAs with new catalytic functions has been aided by the development of a method that rapidly searches pools of random polymers of RNA and extracts those with particular activities; known as **SELEX,** this is nothing less than accelerated evolution in a test tube (Box 26-3). It has been used to generate RNA molecules that bind to amino acids, organic dyes, nucleotides, cyanocobalamin, and other molecules. Researchers have isolated ribozymes that catalyze ester and amide bond formation, S_N2 reactions, metallation of (addition of metal ions to) porphyrins, and carbon–carbon bond formation. The evolution of enzymatic cofactors with nucleotide "handles" that facilitate their binding to ribozymes might have further expanded the repertoire of chemical processes available to primitive metabolic systems.

4. The Structure of the Ribosome As we shall see in Chapter 27, some natural RNA molecules, components of ribosomes, catalyze the formation of peptide bonds, offering a glimpse of how the RNA world might have been transformed by the greater catalytic potential of proteins. The evolution of a capacity to synthesize proteins would have been a major event in the RNA world, allowing the generation of polymers that could greatly stabilize complex RNA structures. However, the onset of peptide synthesis would also have hastened the demise of the RNA world. Proteins are simply better catalysts. The information-carrying role of RNA may have passed to DNA because DNA is chemically more stable. RNA replicase and reverse transcriptase may be modern versions of enzymes that once played important roles in making the transition to the modern DNA-based system.

5. Extant Vestiges of an RNA World The known functions of RNA continue to multiply with each decade. Retroviruses, other RNA viruses, and retrotransposons inhabit a semi-independent universe, maintaining a parasitic existence within the biosphere. For evolutionary biologists, these almost-living entities provide a window on

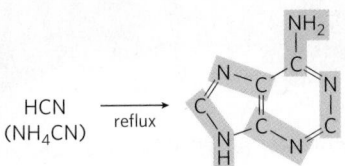

HCN
(NH_4CN) →(reflux)

FIGURE 26-39 Possible prebiotic synthesis of adenine from ammonium cyanide. Adenine is derived from five molecules of cyanide, denoted by shading.

BOX 26-3 METHODS The SELEX Method for Generating RNA Polymers with New Functions

SELEX (*systematic evolution of ligands by exponential enrichment*) is used to generate **aptamers**, oligonucleotides selected to tightly bind a specific molecular target. The process is generally automated to allow rapid identification of one or more aptamers with the desired binding specificity.

Figure 1 illustrates how SELEX is used to select an RNA species that binds tightly to ATP. In step ❶, a random mixture of RNA polymers is subjected to "unnatural selection" by passing it through a resin to which ATP is attached. The practical limit for the complexity of an RNA mixture is about 10^{15} different sequences, which allows complete randomization of 25 nucleotides ($4^{25} = 10^{15}$). For longer RNAs, the RNA pool used to initiate the search does not include all possible sequences. RNA polymers that pass through the column are discarded (step ❷); those that bind to ATP are washed from the column with salt solution and collected (step ❸). In step ❹, the collected RNA polymers are amplified by reverse transcriptase to make many DNA complements to the selected RNAs, then an RNA polymerase makes many RNA complements of the resulting DNA molecules. Finally, in step ❺, this new pool of RNA is subjected to the same selection procedure, and the cycle is repeated a dozen or more times. At the end, only a few aptamers—in this case, RNA sequences with considerable affinity for ATP—remain.

Critical sequence features of an RNA aptamer that binds ATP are shown in Figure 2; molecules with this general structure bind ATP (and other adenosine nucleotides) with $K_d < 50\ \mu M$. Figure 3 presents the three-dimensional structure of a 36 nucleotide RNA aptamer (shown as a complex with AMP) generated by SELEX. This RNA has the backbone structure shown in Figure 2.

In addition to its use in exploring the potential functionality of RNA, SELEX has an important practical side in identifying short RNAs with pharmaceutical uses. Finding an aptamer that binds specifically to every potential therapeutic target may be impossible, but the capacity of SELEX to rapidly select and amplify a specific oligonucleotide sequence from a highly complex pool of sequences makes this a promising approach for generating new therapies. For example, one could select an RNA that binds tightly to a receptor protein prominent in the plasma membrane of cells in a particular cancerous tumor. Blocking the activity of the receptor, or targeting a toxin to the tumor cells by attaching it to the aptamer, would kill the cells. SELEX also has been used to select DNA aptamers that detect anthrax spores. Many other promising applications are under development. ■

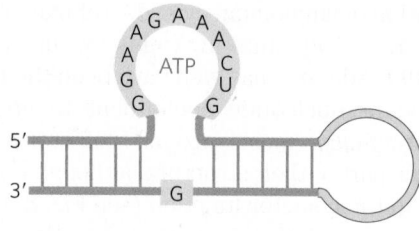

FIGURE 2 RNA aptamer that binds ATP. The shaded nucleotides are those required for the binding activity.

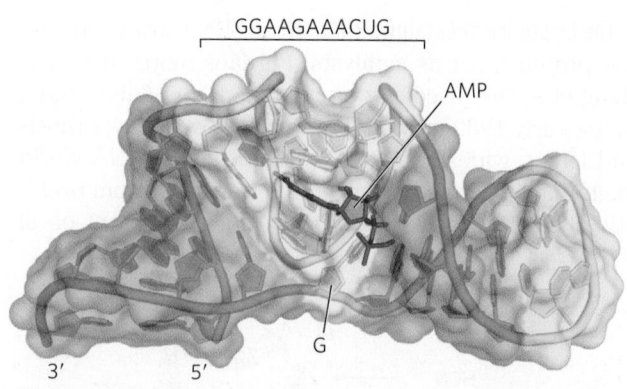

FIGURE 3 RNA aptamer bound to AMP. The bases of the conserved nucleotides (forming the binding pocket) are pale green; the bound AMP is red. [Source: Information from PDB ID 1RAW, T. Dieckmann et al., *RNA* 2:628, 1996.]

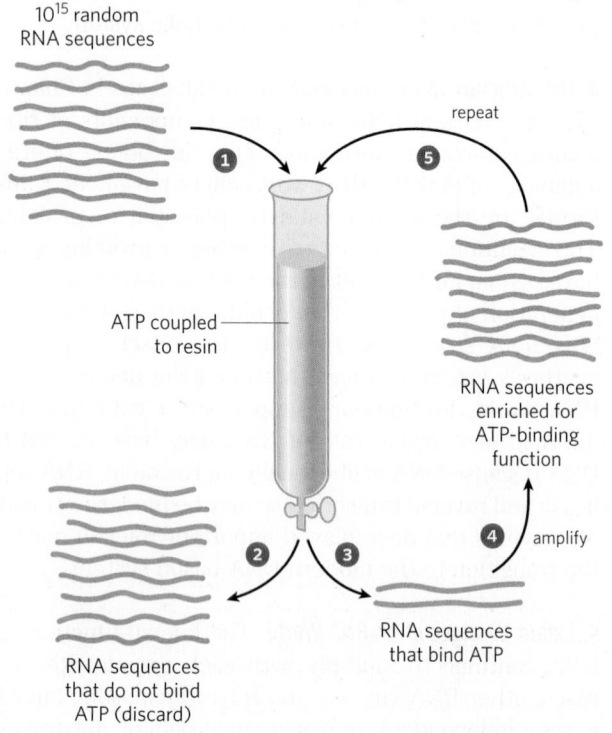

FIGURE 1 The SELEX procedure.

key steps in the evolution of life. Transposons may have been an early innovation in an RNA world. With the appearance of the first, inefficient self-replicators, transposition could have been an important alternative to replication as a strategy for successful reproduction and survival. Early parasitic RNAs would simply hop into a self-replicating molecule via catalyzed transesterification, then passively undergo replication. Natural selection would have driven transposition to become site-specific, targeting sequences that did not interfere with the catalytic activities of the host RNA. Replicators and RNA transposons could have existed in a primitive symbiotic relationship, each contributing to the evolution of the other. Modern introns, retroviruses, and transposons may all be vestiges of a "piggyback" strategy pursued by early parasitic RNAs. These elements continue to make major contributions to the evolution of their hosts.

6. Progress in the Search for an RNA Replicator The RNA world hypothesis requires a nucleotide polymer to reproduce itself. Can a ribozyme bring about its own synthesis in a template-directed manner? Researchers are getting closer to finding such a ribozyme or ribozyme system. For example, Gerald Joyce and colleagues, in 2009, reported on the first set of two ribozymes that could cross-catalyze each other's formation **(Fig. 26-40)**. One ribozyme, E, catalyzes the joining of two oligonucleotides (A′ and B′) to create a second, complementary ribozyme called E′. E′ could then catalyze the joining of two other oligonucleotides (A and B) to form another molecule of E. In this system, the formation of E and E′ was templated, and the amounts grew exponentially as long as substrates were available and proteins were absent. The system evolved so that more-efficient enzymes appeared in the population. A more general

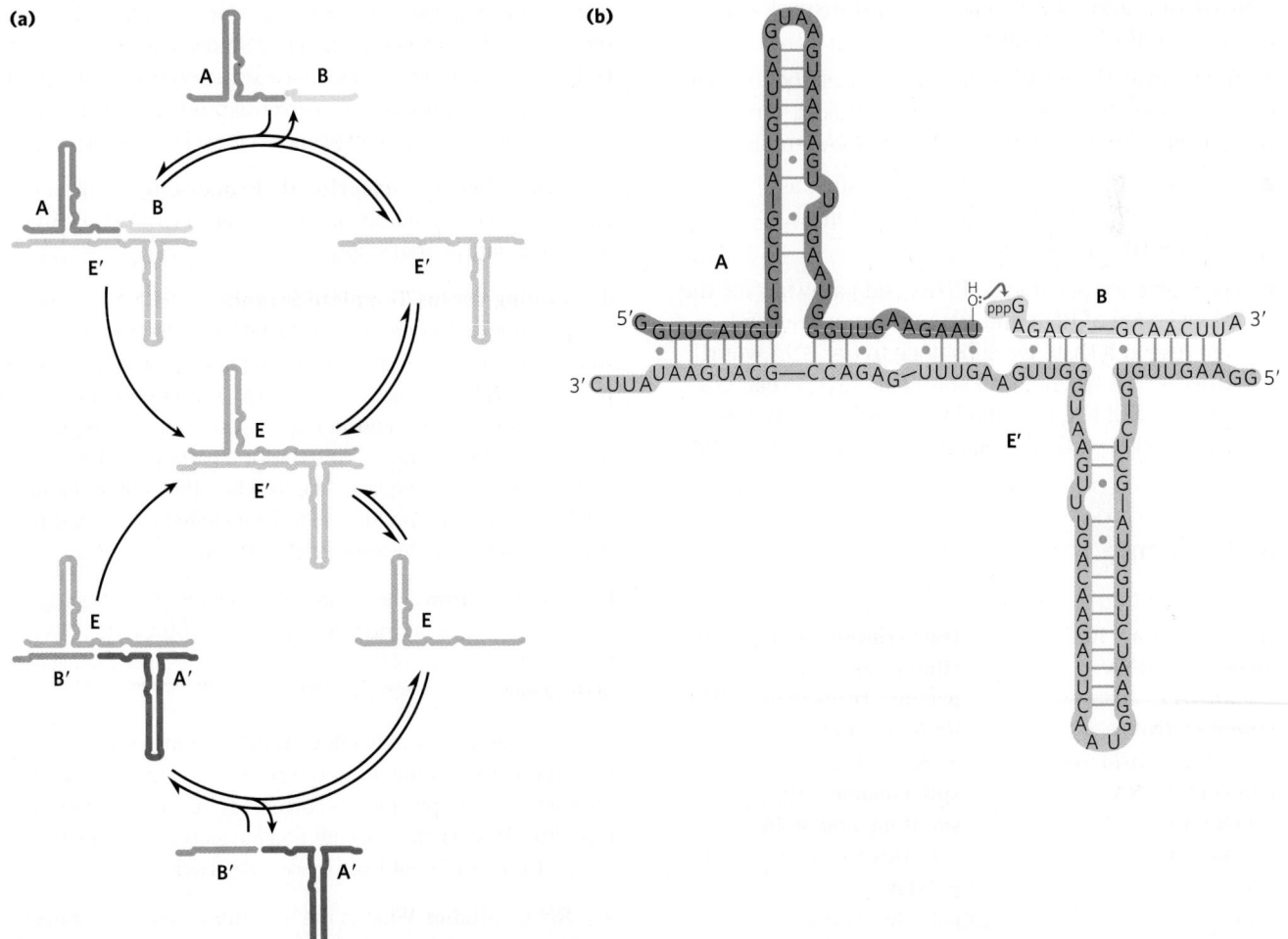

FIGURE 26-40 Self-sustained replication of an RNA enzyme. This system has many of the properties of a living system. The RNA molecules incorporate information and catalytic function, and the reactions produce an exponential increase in the product RNAs. When variants of the RNA substrates are introduced, the system undergoes natural selection such that the best replicators eventually dominate the population. **(a)** A possible reaction scheme. Oligoribonucleotides A and B anneal to ribozyme E′ and are ligated catalytically to form ribozyme E. The joining of oligoribonucleotides A′ and

B′ is similarly catalyzed by ribozyme E. The levels of E and E′ grow exponentially, with a doubling time of about one hour at 42°C, as long as there is a supply of the precursors A, B, A′, and B′. **(b)** The ligation reaction involves attack of the 3′ OH of one oligoribonucleotide on the α phosphate of the 5′-triphosphate of the other oligoribonucleotide. Pyrophosphate is released. Base pairing of the substrates with the ribozyme plays a key role in aligning the substrates for the reaction. [Source: Information from T. A. Lincoln and G. F. Joyce, *Science* 323:1229, 2009.]

RNA-polymerase-like ribozyme was described in 2011 by Philipp Holliger and colleagues.

Although the RNA world remains a hypothesis, with many gaps yet to be explained, experimental evidence supports a growing list of its key elements. Further experimentation should increase our understanding. Important clues to the puzzle will be found in the workings of fundamental chemistry, in living cells, and perhaps on other planets.

SUMMARY 26.3 RNA-Dependent Synthesis of RNA and DNA

■ RNA-dependent DNA polymerases, also called reverse transcriptases, were first discovered in retroviruses, which must convert their RNA genomes into double-stranded DNA as part of their life cycle. These enzymes transcribe the viral RNA into DNA, a process that can be used experimentally to form complementary DNA.

■ Many eukaryotic transposons are related to retroviruses, and their mechanism of transposition includes an RNA intermediate.

■ Telomerase, the enzyme that synthesizes the telomere ends of linear chromosomes, is a specialized reverse transcriptase that contains an internal RNA template.

■ RNA-dependent RNA polymerases, such as the replicases of RNA bacteriophages, are template-specific for the viral RNA.

■ The existence of catalytic RNAs and pathways for the interconversion of RNA and DNA are among several lines of research that give substance to the RNA world hypothesis. The biochemical potential of RNAs can be explored by SELEX, a method for rapidly selecting RNA sequences with particular binding or catalytic properties.

Key Terms

Terms in bold are defined in the glossary.

transcription 1035	**transcription factors** 1044
messenger RNA (mRNA) 1035	**ribozymes** 1047
transfer RNA (tRNA) 1035	**primary transcript** 1047
ribosomal RNA (rRNA) 1035	**RNA splicing** 1047
transcriptome 1035	**5′ cap** 1048
DNA-dependent RNA polymerase 1036	**spliceosome** 1050
template strand 1037	**small nuclear RNA (snRNA)** 1050
nontemplate strand 1037	**poly(A) tail** 1053
coding strand 1037	polyadenylate polymerase 1053
promoter 1038	**alternative splicing** 1054
consensus sequence 1038	**poly(A) site choice** 1055
cAMP receptor protein (CRP) 1039	**small nucleolar RNA (snoRNA)** 1057
repressor 1039	snoRNP 1057
footprinting 1040	**microRNA (miRNA)** 1059
	internal guide sequence 1060

exosome 1062	homing 1067
polynucleotide phosphorylase 1063	telomerase 1068
reverse transcriptase 1064	T loop 1069
retrovirus 1065	RNA-dependent RNA polymerase (RNA replicase) 1070
complementary DNA (cDNA) 1065	**SELEX** 1071
	aptamer 1072

Problems

1. RNA Polymerase (a) How long would it take for the *E. coli* RNA polymerase to synthesize the primary transcript for the *E. coli* genes encoding the enzymes for lactose metabolism, the 5,300 bp *lac* operon (considered in Chapter 28)? (b) How far along the DNA would the transcription "bubble" formed by RNA polymerase move in 10 seconds?

2. Error Correction by RNA Polymerases DNA polymerases are capable of editing and error correction, whereas the capacity for error correction in RNA polymerases seems to be limited. Given that a single base error in either replication or transcription can lead to an error in protein synthesis, suggest a possible biological explanation for this difference.

3. RNA Posttranscriptional Processing Predict the likely effects of a mutation in the sequence (5′)AAUAAA in a eukaryotic mRNA transcript.

4. Coding versus Template Strands The RNA genome of phage Qβ is the nontemplate strand, or coding strand, and when introduced into the cell, it functions as an mRNA. Suppose the RNA replicase of phage Qβ synthesized primarily template-strand RNA and uniquely incorporated this, rather than nontemplate strands, into the viral particles. What would be the fate of the template strands when they entered a new cell? What enzyme would have to be included in the viral particles for successful invasion of a host cell?

5. Transcription The gene encoding the *E. coli* enzyme β-galactosidase begins with the sequence ATGACCATGATTACG. What is the sequence of the RNA transcript specified by this part of the gene?

6. The Chemistry of Nucleic Acid Biosynthesis Describe three properties common to the reactions catalyzed by DNA polymerase, RNA polymerase, reverse transcriptase, and RNA replicase. How is the enzyme polynucleotide phosphorylase similar to and different from these four enzymes?

7. RNA Splicing What is the minimum number of transesterification reactions needed to splice an intron from an mRNA transcript? Explain.

8. RNA Processing If the splicing of mRNA in a vertebrate cell is blocked, the rRNA modification reactions are also blocked. Suggest a reason for this.

9. RNA Genomes The RNA viruses have relatively small genomes. For example, the single-stranded RNAs of retroviruses have about 10,000 nucleotides, and the Qβ RNA is only

4,220 nucleotides long. Given the properties of reverse transcriptase and RNA replicase described in this chapter, can you suggest a reason for the small size of these viral genomes?

10. Screening of RNAs by SELEX The practical limit for the number of different RNA sequences that can be screened in a SELEX experiment is 10^{15}. (a) Suppose you are working with oligonucleotides 32 nucleotides long. How many sequences exist in a randomized pool containing every sequence possible? (b) What percentage of these can be screened in a SELEX experiment? (c) Suppose you wish to select an RNA molecule that catalyzes the hydrolysis of a particular ester. From what you know about catalysis, propose a SELEX strategy that might allow you to select the appropriate catalyst.

11. Slow Death The death cap mushroom, *Amanita phalloides*, contains several dangerous substances, including the lethal α-amanitin. This toxin blocks RNA elongation in consumers of the mushroom by binding to eukaryotic RNA polymerase II with very high affinity; it is deadly in concentrations as low as 10^{-8} M. The initial reaction to ingestion of the mushroom is gastrointestinal distress (caused by some of the other toxins). These symptoms disappear, but about 48 hours later, the mushroom-eater dies, usually from liver dysfunction. Speculate on why it takes this long for α-amanitin to kill.

12. Detection of Rifampicin-Resistant Strains of Tuberculosis Rifampicin is an important antibiotic used to treat tuberculosis and other mycobacterial diseases. Some strains of *Mycobacterium tuberculosis*, the causative agent of tuberculosis, are resistant to rifampicin. These strains become resistant through mutations that alter the *rpoB* gene, which encodes the β subunit of the RNA polymerase. Rifampicin cannot bind to the mutant RNA polymerase and so is unable to block the initiation of transcription. DNA sequences from a large number of rifampicin-resistant *M. tuberculosis* strains have been found to have mutations in a specific 69 bp region of *rpoB*. One well-characterized rifampicin-resistant strain has a single base pair alteration in *rpoB* that results in a His residue being replaced by an Asp residue in the β subunit.

(a) Based on your knowledge of protein chemistry, suggest a technique that would allow detection of the rifampicin-resistant strain containing this particular mutant protein.

(b) Based on your knowledge of nucleic acid chemistry, suggest a technique to identify the mutant form of *rpoB*.

Biochemistry Online

13. The Ribonuclease Gene Human pancreatic ribonuclease has 128 amino acid residues.

(a) What is the minimum number of nucleotide pairs required to code for this protein?

(b) The mRNA expressed in human pancreatic cells was copied with reverse transcriptase to create a "library" of human DNA. The sequence of the mRNA coding for human pancreatic

ribonuclease was determined by sequencing the complementary DNA (cDNA) from this library that included an open reading frame for the protein. Use the nucleotide database at NCBI (www.ncbi.nlm.nih.gov/nucleotide) to find the published sequence of this mRNA. (Search for accession number D26129.) What is the length of this mRNA?

(c) How can you account for the discrepancy between the size you calculated in (a) and the actual length of the mRNA?

Data Analysis Problem

14. A Case of RNA Editing The AMPA (α-amino-3-hydroxy-5-methyl-4-isoxazolepropionic acid) receptor is an important component of the human nervous system. It is present in several forms, in different neurons, and some of this variety results from posttranscriptional modification. This problem explores research on the mechanism of this RNA editing.

An initial report by Sommer and coauthors (1991) looked at the sequence encoding a key Arg residue in the AMPA receptor. The sequence of the cDNA (see Fig. 9-14) for the AMPA receptor showed a CGG (Arg; see Fig. 27-7) codon for this amino acid. Surprisingly, the genomic DNA showed a CAG (Gln) codon at this position.

(a) Explain how this result is consistent with posttranscriptional modification of the AMPA receptor mRNA.

Rueter and colleagues (1995) explored this mechanism in detail. They first developed an assay to differentiate between edited and unedited transcripts, based on the Sanger method of DNA sequencing (see Fig. 8-34). They modified the technique to determine whether the base in question was an A (as in CAG) or not. They designed two DNA primers based on the genomic DNA sequence of this region of the AMPA gene. These primers, and the genomic DNA sequence of the nontemplate strand for the relevant region of the AMPA receptor gene, are shown at the bottom of the page; the A residue that is edited is in red.

To detect whether this A was present or had been edited to another base, Rueter and coworkers used the following procedure:

1. Prepared cDNA complementary to the mRNA, using primer 1, reverse transcriptase, dATP, dGTP, dCTP, and dTTP.
2. Removed the mRNA.
3. Annealed ^{32}P-labeled primer 2 to the cDNA and reacted this with DNA polymerase, dGTP, dCTP, dTTP, and ddATP (dideoxy ATP; see Fig. 8-34).
4. Denatured the resulting duplexes and separated them with polyacrylamide gel electrophoresis (see Fig. 3-18).
5. Detected the ^{32}P-labeled DNA species with autoradiography.

They found that edited mRNA produced a 22 nucleotide [^{32}P] DNA, whereas unedited mRNA produced a 19 nucleotide [^{32}P] DNA.

(b) Using the sequences below, explain how the edited and unedited mRNAs resulted in these different products.

(5′) ...GTCTCTGGTTTTCCTTGGGTGCCTTTATGCAGCAAGGATGCGATATTTCGCCAAG...
Primer 1:   CGTTCCTACGCTATAAAGCGGTTC (5′)
Primer 2:   (5′)CCTTGGGTGCCTTTA

Using the same procedure, this time to measure the fraction of transcripts edited under different conditions, the researchers found that extracts of cultured epithelial cells (a common cell line called HeLa) could edit the mRNA at a high level. To determine the nature of the editing machinery, they pretreated an active HeLa cell extract as described in the following table and measured its ability to edit AMPA mRNA. Proteinase K degrades only proteins; micrococcal nuclease, only DNA.

Sample	Pretreatment	% mRNA edited
1	None	18
2	Proteinase K	5
3	Heat to 65 °C	3
4	Heat to 85 °C	3
5	Micrococcal nuclease	17

(c) Use these data to argue that the editing machinery consists of protein. What is a key weakness in this argument?

To determine the exact nature of the edited base, Rueter and colleagues used the following procedure:

1. Produced mRNA, using [α-^{32}P]ATP in the reaction mixture.
2. Edited the labeled mRNA by incubating with HeLa extract.
3. Hydrolyzed the edited mRNA with nuclease P1 to produce single nucleotide monophosphates.
4. Separated the nucleotide monophosphates with thin-layer chromatography (TLC; see Fig. 10-25b).
5. Identified the resulting ^{32}P-labeled nucleotide monophosphates with autoradiography.

In unedited mRNA, they found only [^{32}P]AMP; in edited mRNA, they found mostly [^{32}P]AMP with some [^{32}P]IMP (inosine monophosphate; see Fig. 22-36).

(d) Why was it necessary to use [α-^{32}P]ATP rather than [β-^{32}P]ATP or [γ-^{32}P]ATP in this experiment?

(e) Why was it necessary to use [α-^{32}P]ATP rather than [α-^{32}P]GTP, [α-^{32}P]CTP, or [α-^{32}P]UTP?

(f) How does the result exclude the possibility that the entire A nucleotide (sugar, base, and phosphate) was removed and replaced by an I nucleotide during the editing process?

The researchers next edited mRNA that was labeled with [2,8-^{3}H]ATP and repeated the above procedure. The only ^{3}H-labeled mononucleotides produced were AMP and IMP.

(g) How does this result exclude removal of the A base (leaving the sugar-phosphate backbone intact) followed by replacement with an I base as a mechanism of editing? What, then, is the most likely mechanism of editing in this case?

(h) How does changing an A to an I residue in the mRNA explain the Gln to Arg change in protein sequence in the two forms of AMPA receptor protein? (Hint: See Fig. 27-8.)

References

Rueter, S.M., C.M. Burns, S.A. Coode, P. Mookherjee, and R.B. Emeson. 1995. Glutamate receptor RNA editing in vitro by enzymatic conversion of adenosine to inosine. *Science* 267:1491–1494.

Sommer, B., M. Köhler, R. Sprengel, and P.H. Seeburg. 1991. RNA editing in brain controls a determinant of ion flow in glutamate-gated channels. *Cell* 67:11–19.

Further Reading is available at www.macmillanlearning.com/LehningerBiochemistry7e.

CHAPTER 27

Protein Metabolism

Self-study tools that will help you practice what you've learned and reinforce this chapter's concepts are available online.
Go to www.macmillanlearning.com/LehningerBiochemistry7e.

Proteins are the end products of most information pathways. A typical cell requires thousands of different proteins at any given moment. These must be synthesized in response to the cell's current needs, transported (targeted) to their appropriate cellular locations, and degraded when no longer needed. Many of the fundamental components and mechanisms used by the protein biosynthetic machinery are remarkably well conserved in all life-forms from bacteria to higher eukaryotes, indicating that they were present in the last universal common ancestor (LUCA) of all extant organisms.

An understanding of protein synthesis, the most complex biosynthetic process, has been one of the greatest challenges in biochemistry. Eukaryotic protein synthesis requires more than 70 different ribosomal proteins; 20 or more enzymes to activate the amino acid precursors; a dozen or more auxiliary enzymes and other protein factors for the initiation, elongation, and termination of polypeptides; perhaps 100 additional enzymes for the final processing of different proteins; and 40 or more kinds of transfer and ribosomal RNAs. Overall, almost 300 different macromolecules cooperate to synthesize polypeptides. Many of these macromolecules are among the most abundant to be found in any cell. Some are organized into the complex three-dimensional structure of the ribosome.

To appreciate the central importance of protein synthesis, consider the cellular resources devoted to this process. Protein synthesis can account for up to 90% of the chemical energy used by a cell for all biosynthetic reactions. Every bacterial, archaeal, and eukaryotic cell contains from several to thousands of copies of many different proteins and RNAs. The 15,000 ribosomes, 100,000 molecules of protein synthesis–related protein factors and enzymes, and 200,000 tRNA molecules in a typical bacterial cell can account for more than 35% of the cell's dry weight.

Despite the great complexity of protein synthesis, proteins are made at exceedingly high rates. A polypeptide of 100 residues is synthesized in an *Escherichia coli* cell (at 37 °C) in about 5 seconds. Synthesis of the thousands of different proteins in a cell is tightly regulated, so that just enough copies are made to match the current metabolic circumstances. To maintain the appropriate mix and concentration of proteins, the targeting and degradative processes must keep pace with synthesis. Research is gradually uncovering the finely coordinated cellular choreography that guides each protein to its proper cellular location and selectively degrades it when it is no longer required.

The study of protein synthesis offers another important reward: a look at a world of RNA catalysts that may have existed before the dawn of life "as we know it." Elucidation of the three-dimensional structures of ribosomes, beginning in 2000, has given us an increasingly detailed look at the mechanics of protein synthesis. It has also confirmed a hypothesis first put forward by Harry Noller two decades earlier: proteins are synthesized by a gigantic RNA enzyme.

Harry Noller
[Source: Courtesy of Harry Noller.]

27.1 The Genetic Code

Paul Zamecnik, 1912-2009
[Source: Archives and Special Collections, Massachusetts General Hospital.]

Three major advances set the stage for our present knowledge of protein biosynthesis. First, in the early 1950s, Paul Zamecnik and his colleagues designed a set of experiments to investigate where in the cell proteins are synthesized. They injected radioactive amino acids into rats and, at different time intervals after the injection, removed the liver, homogenized it, fractionated the homogenate by centrifugation, and examined the subcellular fractions for the presence of radioactive protein. When hours or days were allowed to elapse after injection of the labeled amino acids, *all* the subcellular fractions contained labeled proteins. However, when only minutes had elapsed, labeled protein appeared only in a fraction containing small ribonucleoprotein particles. These particles, visible in animal tissues by electron microscopy, were therefore identified as the site of protein synthesis from amino acids, and later were named ribosomes **(Fig. 27-1)**.

The second key advance was made by Mahlon Hoagland and Zamecnik when they found that amino acids were "activated" for protein synthesis when incubated with ATP and the cytosolic fraction of liver cells. The amino acids became attached to a heat-stable soluble RNA of the type that had been discovered and characterized by Robert Holley, and later called transfer RNA (tRNA), to form **aminoacyl-tRNAs**. The enzymes that catalyze this process are the **aminoacyl-tRNA synthetases**.

The third advance resulted from Francis Crick's reasoning on how the genetic information encoded in the 4-letter language of nucleic acids could be translated into the 20-letter language of proteins. A small nucleic acid (perhaps an RNA) could serve the role of an adaptor, with one part of the adaptor molecule binding a specific amino acid and another part recognizing the nucleotide sequence encoding that amino acid in an mRNA **(Fig. 27-2)**. This idea was soon verified. The tRNA adaptor, the same molecule that activates the amino acid for peptide bond formation, also "translates" the nucleotide sequence of an mRNA into the amino acid sequence of a polypeptide. The overall process of mRNA-guided protein synthesis is often referred to simply as **translation**.

These three developments soon led to recognition of the major stages of protein synthesis and ultimately to elucidation of the genetic code that specifies each amino acid.

The Genetic Code Was Cracked Using Artificial mRNA Templates

By the 1960s, it was apparent that at least three nucleotide residues of DNA are necessary to encode each amino acid. The four code letters of DNA (A, T, G, and C) in groups of two can yield only $4^2 = 16$ different combinations, insufficient to encode 20 amino acids. Groups of three, however, yield $4^3 = 64$ different combinations.

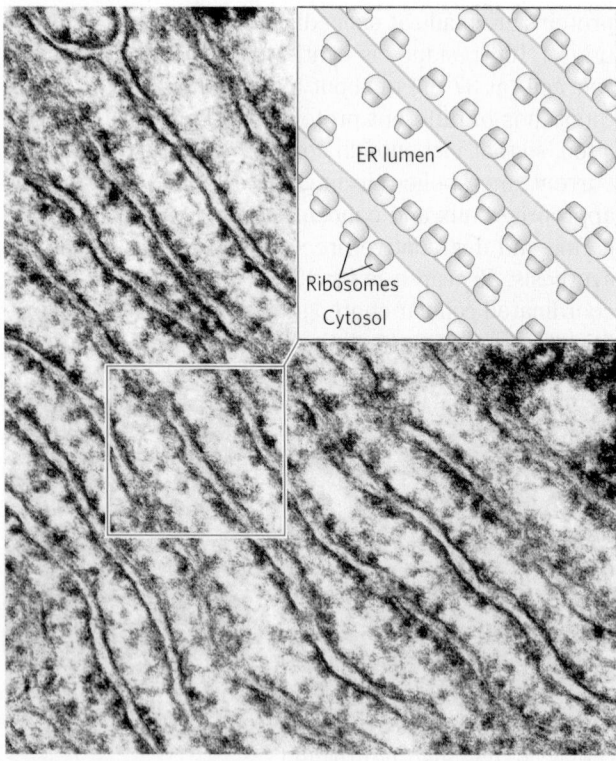

FIGURE 27-1 Ribosomes and endoplasmic reticulum. Electron micrograph and schematic drawing of a portion of a pancreatic cell, showing ribosomes attached to the outer (cytosolic) face of the endoplasmic reticulum (ER). The ribosomes are the numerous small dots bordering the parallel layers of membranes. [Source: Joseph F. Gennaro Jr./Science Source.]

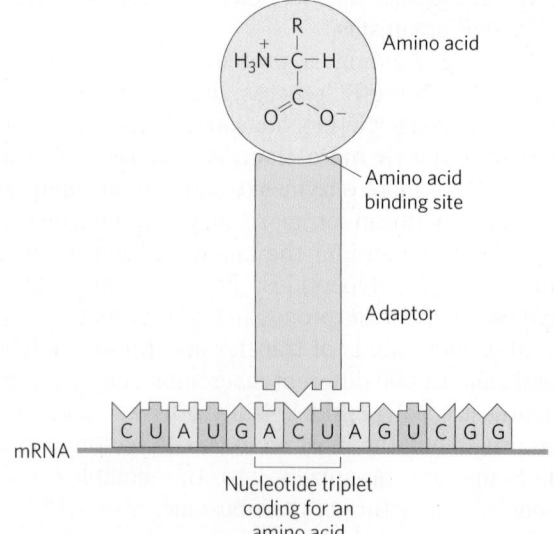

FIGURE 27-2 Crick's adaptor hypothesis. Today we know that the amino acid is covalently bound at the 3' end of a tRNA molecule and that a specific nucleotide triplet elsewhere in the tRNA interacts with a particular triplet codon in mRNA through hydrogen bonding of complementary bases.

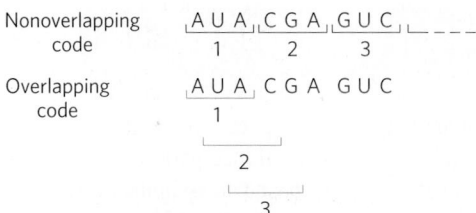

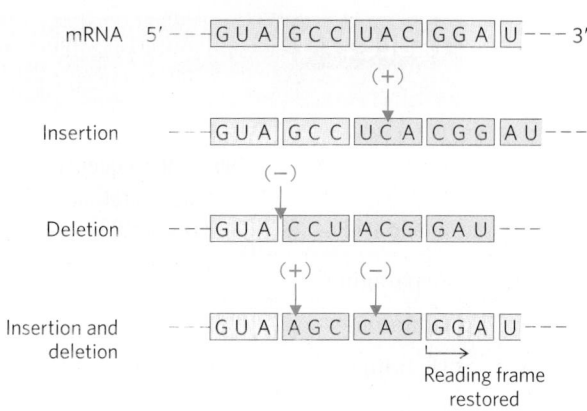

FIGURE 27-3 Overlapping versus nonoverlapping genetic codes. In a nonoverlapping code, codons (numbered consecutively) do not share nucleotides. In an overlapping code, some nucleotides in the mRNA are shared by different codons. In a triplet code with maximum overlap, many nucleotides, such as the third nucleotide from the left (A), are shared by three codons. Note that in an overlapping code, the triplet sequence of the first codon limits the possible sequences for the second codon. A nonoverlapping code provides much more flexibility in the triplet sequence of neighboring codons and therefore in the possible amino acid sequences designated by the code. The genetic code used in all living systems is now known to be nonoverlapping.

FIGURE 27-4 The triplet, nonoverlapping code. Evidence for the general nature of the genetic code came from many types of experiments, including genetic experiments on the effects of deletion and insertion mutations. Inserting or deleting one base pair (shown here in the mRNA transcript) alters the sequence of triplets in a nonoverlapping code; all amino acids coded by the mRNA following the change are affected. Combining insertion and deletion mutations affects some amino acids but can eventually restore the correct amino acid sequence. Adding or subtracting three nucleotides (not shown) leaves the remaining triplets intact, providing evidence that a codon has three, rather than four or five, nucleotides. The triplet codons shaded in gray are those transcribed from the original gene; codons shaded in blue are new codons resulting from the insertion or deletion mutations.

Several key properties of the genetic code were established in early genetic studies **(Figs 27-3, 27-4)**. A **codon** is a triplet of nucleotides that codes for a specific amino acid. Translation occurs in such a way that these nucleotide triplets are read in a successive, nonoverlapping fashion. A specific first codon in the sequence establishes the **reading frame**, in which a new codon begins every three nucleotide residues. There is no punctuation between codons for successive amino acid residues. The amino acid sequence of a protein is defined by a linear sequence of contiguous triplets. In principle, any given single-stranded DNA or mRNA sequence has three possible reading frames. Each reading frame gives a different sequence of codons **(Fig. 27-5)**, but only one is likely to encode a given protein. A key question remained: what were the three-letter code words for each amino acid?

Marshall Nirenberg, 1927–2010
[Source: AP Photo.]

In 1961, Marshall Nirenberg and Heinrich Matthaei reported the first breakthrough. They incubated synthetic polyuridylate, poly(U), with an *E. coli* extract, GTP, ATP, and a mixture of the 20 amino acids in 20 different tubes, each tube containing a different radioactively labeled amino acid. Because poly(U) mRNA is made up of many successive UUU triplets, it should promote the synthesis of a polypeptide containing only the amino acid encoded by UUU. A radioactive polypeptide was indeed formed in only one of the 20 tubes, the one containing radioactive phenylalanine. Nirenberg and Matthaei therefore concluded that the triplet codon UUU encodes phenylalanine. The same approach soon revealed that polycytidylate, poly(C), encodes a polypeptide containing only proline (polyproline), and polyadenylate, poly(A), encodes polylysine. Polyguanylate did not generate any polypeptide in this experiment because it spontaneously forms tetraplexes (see Fig. 8-20d) that cannot be bound by ribosomes.

The synthetic polynucleotides used in such experiments were prepared by using polynucleotide phosphorylase (p. 1063), which catalyzes the formation of RNA polymers starting from ADP, UDP, CDP, and GDP. This enzyme, discovered by Severo Ochoa, requires no template and makes polymers with a base composition that directly reflects the relative concentrations of the nucleoside 5′-diphosphate precursors in the medium. If polynucleotide phosphorylase is presented with UDP only, it makes only poly(U). If it is presented with a mixture of five parts ADP and one part CDP, it makes a

FIGURE 27-5 Reading frames in the genetic code. In a triplet, nonoverlapping code, all mRNAs have three potential reading frames, shaded here in different colors. The triplets, and hence the amino acids specified, are different in each reading frame.

TABLE 27-1 Incorporation of Amino Acids into Polypeptides in Response to Random Polymers of RNA

Amino acid	Observed frequency of incorporation (Lys = 100)	Tentative assignment for nucleotide composition of corresponding codon[a]	Expected frequency of incorporation based on assignment (Lys = 100)
Asparagine	24	A_2C	20
Glutamine	24	A_2C	20
Histidine	6	AC_2	4
Lysine	100	AAA	100
Proline	7	AC_2, CCC	4.8
Threonine	26	A_2C, AC_2	24

Note: Presented here is a summary of data from one of the early experiments designed to elucidate the genetic code. A synthetic RNA containing only A and C residues in 5:1 ratio directed polypeptide synthesis, and both the identity and the quantity of incorporated amino acids were determined. Based on the relative abundance of A and C residues in the synthetic RNA, and assigning the codon AAA (the most likely codon) a frequency of 100, there should be three different codons of composition A_2C, each at a relative frequency of 20; three of composition AC_2, each at a relative frequency of 4.0; and CCC at a relative frequency of 0.8. The CCC assignment was based on information derived from prior studies with poly (C). Where two tentative codon assignments are made, both are proposed to code for the same amino acid.

[a]These designations of nucleotide composition contain no information on nucleotide sequence (except, of course, AAA and CCC).

polymer in which about five-sixths of the residues are adenylate and one-sixth are cytidylate. This random polymer is likely to have many triplets of the sequence AAA, smaller numbers of AAC, ACA, and CAA triplets, relatively few ACC, CCA, and CAC triplets, and very few CCC triplets (Table 27-1). Using a variety of artificial mRNAs made by polynucleotide phosphorylase from different starting mixtures of ADP, GDP, UDP, and CDP, the Nirenberg and Ochoa groups soon identified the base compositions of the triplets coding for almost all the amino acids. Although these experiments revealed the base composition of the coding triplets, they usually could not reveal the sequence of the bases.

>> Key Convention: Much of the following discussion deals with tRNAs. The amino acid specified by a tRNA is indicated by a superscript, such as tRNA^{Ala}, and the aminoacylated tRNA by a hyphenated name: alanyl-tRNA^{Ala} or Ala-tRNA^{Ala}. **<<**

In 1964, Nirenberg and Philip Leder achieved another experimental breakthrough. Isolated *E. coli* ribosomes would bind a specific aminoacyl-tRNA in the presence of the corresponding synthetic polynucleotide messenger. For example, ribosomes incubated with poly(U) and phenylalanyl-tRNA^{Phe} (Phe-tRNA^{Phe}) bind both RNAs, but if the ribosomes are incubated with poly(U) and some other aminoacyl-tRNA, the aminoacyl-tRNA is not bound, because it does not recognize the UUU triplets in poly(U) (Table 27-2). Even trinucleotides could promote specific binding of appropriate tRNAs, so these experiments could be carried out with chemically synthesized small oligonucleotides. With this technique, researchers determined which aminoacyl-tRNA bound to 54 of the 64 possible triplet codons. For some codons,

either no aminoacyl-tRNA or more than one would bind. Another method was needed to complete and confirm the entire genetic code.

At about this time, a complementary approach was provided by H. Gobind Khorana, who developed chemical methods to synthesize polyribonucleotides with defined, repeating sequences of two to four bases. The polypeptides produced by these mRNAs had one or a few amino acids in repeating patterns. These patterns, when combined with information from the random polymers used by Nirenberg and colleagues, permitted

H. Gobind Khorana, 1922–2011 [Source: Courtesy of Archives, University of Wisconsin–Madison.]

TABLE 27-2 Trinucleotides That Induce Specific Binding of Aminoacyl-tRNAs to Ribosomes

Trinucleotide	Relative increase in ^{14}C-labeled aminoacyl-tRNA bound to ribosome[a]		
	Phe-tRNA^{Phe}	Lys-tRNA^{Lys}	Pro-tRNA^{Pro}
UUU	4.6	0	0
AAA	0	7.7	0
CCC	0	0	3.1

Source: Information from M. Nirenberg and P. Leder, *Science* 145:1399, 1964.

[a]Each number represents the factor by which the amount of bound ^{14}C increased when the indicated trinucleotide was present, relative to a control with no trinucleotide.

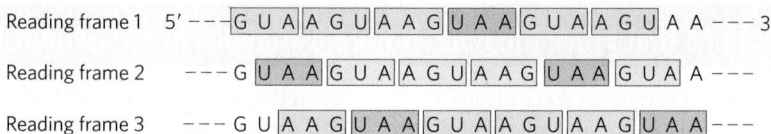

FIGURE 27-6 Effect of a termination codon in a repeating tetranucleotide. Termination codons (light red) are encountered every fourth codon in three different reading frames (shown in different colors). Dipeptides or tripeptides are synthesized, depending on where the ribosome initially binds.

unambiguous codon assignments. The copolymer $(AC)_n$, for example, has alternating ACA and CAC codons: ACACACACACACACA. The polypeptide synthesized on this messenger contained equal amounts of threonine and histidine. Given that a histidine codon has one A and two Cs (Table 27-1), CAC must code for histidine and ACA for threonine.

Consolidation of the results from many experiments permitted assignment of 61 of the 64 possible codons. The other three were identified as termination codons, in part because they disrupted amino acid coding patterns when they occurred in a synthetic RNA polymer **(Fig. 27-6)**. Meanings for all the triplet codons (tabulated in **Fig. 27-7**) were established by 1966 and have been verified in many different ways.

The cracking of the genetic code is regarded as one of the most important scientific discoveries of the twentieth century.

Codons are the key to the translation of genetic information, directing the synthesis of specific proteins. The reading frame is set when translation of an mRNA molecule begins, and it is maintained as the synthetic machinery reads sequentially from one triplet to the next. If the initial reading frame is off by one or two bases, or if translation somehow skips a nucleotide in the mRNA, all the subsequent codons will be out of register; the result is usually a "missense" protein with a garbled amino acid sequence.

Several codons serve special functions (Fig. 27-7). The **initiation codon** AUG is the most common signal for the beginning of a polypeptide in all cells, in addition to coding for Met residues in internal positions of polypeptides. The **termination codons** (UAA, UAG, and UGA), also called stop codons or nonsense codons, normally signal the end of polypeptide synthesis and do not code for any known amino acids. Some deviations from these rules are discussed in Box 27-1.

As described in Section 27.2, initiation of protein synthesis in the cell is an elaborate process that relies on initiation codons and other signals in the mRNA. In retrospect, the experiments of Nirenberg, Khorana, and others to identify codon function should not have worked in the absence of initiation codons. Serendipitously, experimental conditions caused the normal initiation requirements for protein synthesis to be relaxed. Diligence combined with chance to produce a breakthrough—a common occurrence in the history of biochemistry.

In a random sequence of nucleotides, 1 in every 20 codons in each reading frame is, on average, a termination codon. In general, a reading frame without a termination codon among 50 or more consecutive codons is referred to as an **open reading frame (ORF)**. Long open reading frames usually correspond to genes that encode proteins. In the analysis of sequence databases, sophisticated programs are used to search for open reading frames in order to find genes among the often huge background of nongenic DNA. An uninterrupted gene coding for a typical protein with a molecular weight of 60,000 would require an open reading frame with 500 or more codons.

A striking feature of the genetic code is that an amino acid may be specified by more than one codon, so the code is described as **degenerate**. This does *not* suggest that the code is flawed: although an amino acid

First letter of codon (5′ end)

Second letter of codon

	U	C	A	G
U	UUU Phe	UCU Ser	UAU Tyr	UGU Cys
	UUC Phe	UCC Ser	UAC Tyr	UGC Cys
	UUA Leu	UCA Ser	UAA Stop	UGA Stop
	UUG Leu	UCG Ser	UAG Stop	UGG Trp
C	CUU Leu	CCU Pro	CAU His	CGU Arg
	CUC Leu	CCC Pro	CAC His	CGC Arg
	CUA Leu	CCA Pro	CAA Gln	CGA Arg
	CUG Leu	CCG Pro	CAG Gln	CGG Arg
A	AUU Ile	ACU Thr	AAU Asn	AGU Ser
	AUC Ile	ACC Thr	AAC Asn	AGC Ser
	AUA Ile	ACA Thr	AAA Lys	AGA Arg
	AUG Met	ACG Thr	AAG Lys	AGG Arg
G	GUU Val	GCU Ala	GAU Asp	GGU Gly
	GUC Val	GCC Ala	GAC Asp	GGC Gly
	GUA Val	GCA Ala	GAA Glu	GGA Gly
	GUG Val	GCG Ala	GAG Glu	GGG Gly

FIGURE 27-7 "Dictionary" of amino acid code words in mRNAs. The codons are written in the 5′→3′ direction. The third base of each codon (in bold type) plays a lesser role in specifying an amino acid than the first two. The three termination codons are shaded in light red, the initiation codon AUG in green. All the amino acids except methionine and tryptophan have more than one codon. In most cases, codons that specify the same amino acid differ only at the third base.

BOX 27-1   Exceptions That Prove the Rule: Natural Variations in the Genetic Code

In biochemistry, as in other disciplines, exceptions to general rules can be problematic for instructors and frustrating for students. At the same time, though, they teach us that life is complex and inspire us to search for more surprises. Understanding the exceptions can even reinforce the original rule in unpredictable ways.

One would expect little room for variation in the genetic code. Even a single amino acid substitution can have profoundly deleterious effects on the structure of a protein. Nevertheless, variations in the code do occur in some organisms, and they are both interesting and instructive. The types of variation and their rarity provide powerful evidence for a common evolutionary origin of all living things.

To alter the code, changes must occur in the gene(s) encoding one or more tRNAs, with the obvious target for alteration being the anticodon. Such a change would lead to the systematic insertion of an amino acid at a codon that, according to the standard code (see Fig. 27-7), does not specify that amino acid. The genetic code, in effect, is defined by two elements: (1) the anticodons on tRNAs, which determine where an amino acid is placed in a growing polypeptide, and (2) the specificity of the enzymes—the aminoacyl-tRNA synthetases—that charge the tRNAs, which determines the identity of the amino acid attached to a given tRNA.

Most sudden changes in the code would have catastrophic effects on cellular proteins, so code alterations are more likely to persist where relatively few proteins would be affected—such as in small genomes encoding only a few proteins. The biological consequences of a code change could also be limited by restricting changes to the three termination codons, which do not generally occur *within* genes (see Box 27-3 for exceptions to *this* rule). This pattern is, in fact, observed.

Of the very few variations in the genetic code that we know of, most occur in mitochondrial DNA (mtDNA), which encodes only 10 to 20 proteins. Mitochondria have their own tRNAs, so their code variations do not affect the much larger cellular genome. The most common changes in mitochondria involve termination codons. These changes affect termination in the products of only a subset of genes, and sometimes the effects are minor, because the genes have multiple (redundant) termination codons.

Vertebrate mtDNAs have genes that encode 13 proteins, 2 rRNAs, and 22 tRNAs (see Fig. 19-40).

Given the small number of codon reassignments, along with an unusual set of wobble rules (p. 1084), the 22 tRNAs are sufficient to decode the protein-coding genes, as opposed to the 32 tRNAs required for the standard code. In mitochondria, these changes can be viewed as a kind of genomic streamlining, as a smaller genome confers a replication advantage on the organelle. Four codon families (in which the amino acid is determined entirely by the first two nucleotides) are decoded by a single tRNA with a U residue in the first (or wobble) position in the anticodon. Either the U pairs somehow with any of the four possible bases in the third position of the codon or a "two out of three" mechanism is used—that is, no base pairing is needed at the third position. Other tRNAs recognize codons with either A or G in the third position, and yet others recognize U or C, so that virtually all the tRNAs recognize either two or four codons.

In the standard code, only two amino acids are specified by single codons: methionine and tryptophan (see Table 27-3). If all mitochondrial tRNAs recognize two codons, we would expect additional Met and Trp codons in mitochondria. And we find that the single most common code variation is UGA, usually a termination codon, specifying tryptophan. The tRNA^{Trp} recognizes and inserts a Trp residue at either UGA or the usual Trp codon, UGG. The second most common variation is conversion of AUA from an Ile codon to a Met codon; the usual Met codon is AUG, and a single tRNA recognizes both codons. The known coding variations in mitochondria are summarized in Table 1.

Turning to the much rarer changes in the codes for cellular (as distinct from mitochondrial) genomes, we find that the only known variation in a bacterium is again the use of UGA to encode Trp residues, occurring in the simplest free-living cell, *Mycoplasma capricolum*. Among eukaryotes, rare extramitochondrial coding changes occur in a few species of ciliated protists, in which both termination codons UAA and UAG can specify glutamine. There are also rare but interesting cases in which stop codons have been adapted to encode amino acids that are not among the standard 20, as detailed in Box 27-2.

Changes in the code need not be absolute; a codon might not always encode the same amino acid. For example, in many bacteria—including *E. coli*—GUG (Val) is sometimes used as an initiation codon that specifies Met. This occurs only for those genes in which the GUG is properly located relative to

may have two or more codons, each codon specifies only one amino acid. The degeneracy of the code is not uniform. Whereas methionine and tryptophan have single codons, for example, three amino acids (Arg, Leu, Ser)

have six codons, five amino acids have four, isoleucine has three, and nine amino acids have two (Table 27-3).

The genetic code is nearly universal. With the intriguing exception of a few minor variations in

TABLE 1	Known Variant Codon Assignments in Mitochondria				
	Codons[a]				
	UGA	AUA	AGA AGG	CUN	CGG
Normal (cellular) code assignment	Stop	Ile	Arg	Leu	Arg
Animals					
Vertebrates	Trp	Met	Stop	+	+
Drosophila	Trp	Met	Ser	+	+
Yeasts					
Saccharomyces cerevisiae	Trp	Met	+	Thr	+
Torulopsis glabrata	Trp	Met	+	Thr	?
Schizosaccharomyces pombe	Trp	+	+	+	+
Filamentous fungi	Trp	+	+	+	+
Trypanosomes	Trp	+	+	+	+
Higher plants	+	+	+	+	Trp
Chlamydomonas reinhardtii	?	+	+	+	?

[a]N indicates any nucleotide; +, codon has the same meaning as in the cellular code; ?, codon not observed in this mitochondrial genome.

particular mRNA sequences that affect the initiation of translation (as discussed in Section 27.2).

The most surprising alteration in the genetic code occurs in some fungal species of the genus *Candida*, as originally discovered for *Candida albicans*. *C. albicans* is an organism of high genomic complexity, yet its genetic code has undergone a dramatic change: the CUG codon, which usually encodes Leu residues, encodes Ser instead. The natural selection pressure for this change is completely unknown. Furthermore, Ser and Leu are quite different in chemical structure. However, even this change can be understood based on the properties of a universal code. When several codons encode the same amino acid and use multiple tRNAs, not all of the codons are used with equal frequency. In a phenomenon called **codon bias**, some codons for a particular amino acid are used more frequently (sometimes much more frequently) than others. The tRNAs for the frequently used codons are often present at much higher concentrations than the tRNAs required for the rarely used codons. Code degeneracy leads to the presence of six codons for Leu. In bacteria, CUG often encodes Leu. However, in fungi of genera that are very closely related to *Candida* but do not have the coding change, CUG only

rarely encodes Leu and is often entirely absent in highly expressed proteins. A change in the coding sense of CUG would thus have a much smaller effect on fungal cell metabolism than might be expected if all codons were used equally. The coding change may have occurred by a gradual loss of CUG codons in genes and of the tRNA that recognizes CUG as a Leu codon, followed by a capture event—a mutation in the anticodon of a tRNA^Ser that allowed it to recognize CUG. Alternatively, there may have been an intermediate stage in which CUG was recognized as encoding both Leu and Ser, perhaps with contextual signals in the mRNAs that helped one tRNA or another recognize specific CUG codons (see Box 27-2). Phylogenetic analysis indicates that the reassignment of CUG as a Ser codon occurred in *Candida* ancestors about 150 to 170 million years ago.

These variations tell us that the code is not quite as universal as once believed, but that its flexibility is severely constrained. The variations are obviously derivatives of the cellular code, and no example of a completely different code has been found. The limited scope of code variants strengthens the principle that all life on this planet evolved on the basis of a single (slightly flexible) genetic code.

mitochondria, some bacteria, and some single-celled eukaryotes (Box 27-1), amino acid codons are identical in all species examined so far. Human beings, *E. coli*, tobacco plants, amphibians, and viruses share the same

genetic code. This suggests that all life-forms have a common evolutionary ancestor, whose genetic code has been preserved throughout biological evolution. Even the variations reinforce this theme.

TABLE 27-3	Degeneracy of the Genetic Code		
Amino acid	Number of codons	Amino acid	Number of codons
Met	1	Tyr	2
Trp	1	Ile	3
Asn	2	Ala	4
Asp	2	Gly	4
Cys	2	Pro	4
Gln	2	Thr	4
Glu	2	Val	4
His	2	Arg	6
Lys	2	Leu	6
Phe	2	Ser	6

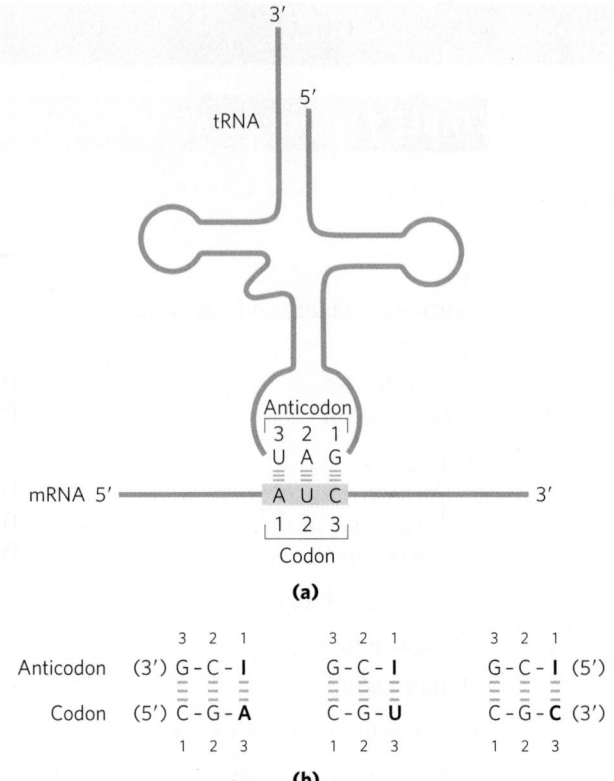

FIGURE 27-8 Pairing relationship of codon and anticodon. (a) Alignment of the two RNAs is antiparallel. The tRNA is shown in the traditional cloverleaf configuration. (b) Three different codon pairing relationships are possible when the tRNA anticodon contains inosinate.

Wobble Allows Some tRNAs to Recognize More than One Codon

When several different codons specify one amino acid, the difference between them usually lies at the third base position (at the 3′ end). For example, alanine is encoded by the triplets GCU, GCC, GCA, and GCG. The codons for most amino acids can be symbolized by XY^A_G or XY^U_C. The first two letters of each codon are the primary determinants of specificity, a feature that has some interesting consequences.

Transfer RNAs base-pair with mRNA codons at a three-base sequence on the tRNA called the **anticodon**. The first base of the codon in mRNA (read in the 5′→3′ direction) pairs with the third base of the anticodon **(Fig. 27-8a)**. If the anticodon triplet of a tRNA recognized only one codon triplet through Watson-Crick base pairing at all three positions, cells would have a different tRNA for each amino acid codon. This is not the case, however, because the anticodons in some tRNAs include the nucleotide inosinate (designated I), which contains the uncommon base hypoxanthine (see Fig. 8-5b). Inosinate can form hydrogen bonds with three different nucleotides (U, C, and A; Fig. 27-8b), although these pairings are much weaker than the hydrogen bonds of Watson-Crick base pairs G≡C and A=U. In yeast, one tRNA^{Arg} has the anticodon (5′)ICG, which recognizes three arginine codons: (5′)CGA, (5′)CGU, and (5′)CGC. The first two bases are identical (CG) and form strong Watson-Crick base pairs with the corresponding bases of the anticodon, but the third base (A, U, or C) forms rather weak hydrogen bonds with the I residue at the first position of the anticodon.

Examination of these and other codon-anticodon pairings led Crick to conclude that the third base of most codons pairs rather loosely with the corresponding base of its anticodon; to use his picturesque word, the third base of such codons (and the first base of their corresponding anticodons) "wobbles." Crick proposed a set of four relationships called the **wobble hypothesis**:

1. The first two bases of an mRNA codon always form strong Watson-Crick base pairs with the corresponding bases of the tRNA anticodon and confer most of the coding specificity.

2. The first base of the anticodon (reading in the 5′→3′ direction; this pairs with the third base of the codon) determines the number of codons recognized by the tRNA. When the first base of the anticodon is C or A, base pairing is specific and only one codon is recognized by that tRNA. When the first base is U or G, binding is less specific and two different codons may be read. When inosine (I) is the first (wobble) nucleotide of an anticodon, three different codons can be recognized—the maximum number for any tRNA. These relationships are summarized in Table 27-4.

3. When an amino acid is specified by several different codons, the codons that differ in either of the first two bases require different tRNAs.

4. A minimum of 32 tRNAs are required to translate all 61 codons (31 to encode the amino acids, 1 for initiation).

TABLE 27-4 How the Wobble Base of the Anticodon Determines the Number of Codons a tRNA Can Recognize

1. One codon recognized:

Anticodon	$(3')$ X–Y–**C** $(5')$	$(3')$ X–Y–**A** $(5')$
Codon	$(5')$ X'–Y'–**G** $(3')$	$(5')$ X'–Y'–**U** $(3')$

2. Two codons recognized:

Anticodon	$(3')$ X–Y–**U** $(5')$	$(3')$ X–Y–**G** $(5')$
Codon	$(5')$ X'–Y'–$^{\text{A}}_{\text{G}}$ $(3')$	$(5')$ X'–Y'–$^{\text{C}}_{\text{U}}$ $(3')$

3. Three codons recognized:

Anticodon	$(3')$ X–Y–**I** $(5')$
Codon	$(5')$ X'–Y'–$^{\text{A}}_{\text{U}}_{\text{C}}$ $(3')$

Note: X and Y denote bases complementary to and capable of strong Watson-Crick base pairing with X' and Y', respectively. Wobble bases—in the 3' position of codons and 5' position of anticodons—are shaded in white.

The wobble (or third) base of the codon contributes to specificity, but because it pairs only loosely with its corresponding base in the anticodon, it permits rapid dissociation of the tRNA from its codon during protein synthesis. If all three bases of a codon engaged in strong Watson-Crick pairing with the three bases of the anticodon, tRNAs would dissociate too slowly and this would limit the rate of protein synthesis. Codon-anticodon interactions balance the requirements for accuracy and speed.

The Genetic Code Is Mutation-Resistant

The genetic code plays an interesting role in safeguarding the genomic integrity of every living organism. Evolution did not produce a code in which codon assignments appeared at random. Instead, the code is strikingly resistant to the deleterious effects of the most common kinds of mutations—**missense mutations**, in which a single new base pair replaces another. In the third, or wobble, position of the codon, single base substitutions produce a change in the encoded amino acid only about 25% of the time. Most such changes are thus **silent mutations**, in which the nucleotide is different but the encoded amino acid remains the same.

Due to the types of spontaneous DNA damage that affect genomes (see Chapter 8), the most frequent missense mutation is a **transition mutation**, in which a purine is replaced by a purine, or a pyrimidine by a pyrimidine (for example, G≡C changed to A=T). All three codon positions have evolved so that there is some resistance to transition mutations. A mutation in the first position of the codon will usually produce an amino acid coding change, but the change often results in an amino acid with similar chemical properties. This is especially true for the hydrophobic amino acids that dominate the first column of the code shown in Figure 27-7. Consider the Val codon GUU. A change to AUU would substitute Ile for Val. A change to CUU would replace Val with Leu. The resulting changes in the structure and/or function of the protein encoded by that gene would often (but not always) be small.

Computational studies have shown that alternative genetic codes, delineated at random, are almost always less resistant to mutation than the existing code. The results indicate that the code underwent considerable streamlining before the appearance of LUCA, the ancestral cell.

The genetic code tells us how protein sequence information is stored in nucleic acids and provides some clues about how that information is translated into protein.

Translational Frameshifting and RNA Editing Affect How the Code Is Read

Once the reading frame has been set during protein synthesis, codons are translated without overlap or punctuation until the ribosomal complex encounters a termination codon. The other two possible reading frames usually contain no useful genetic information, but a few genes are structured so that ribosomes "hiccup" at a certain point in the translation of their mRNAs, changing the reading frame from that point on. This seems to be a mechanism either to allow two or more related but distinct proteins to be produced from a single transcript or to regulate the synthesis of a protein.

One of the best-documented examples of **translational frameshifting** occurs during translation of the mRNA for the overlapping *gag* and *pol* genes of the Rous sarcoma virus (see Fig. 26-34). The reading frame for *pol* is offset to the left by one base pair (–1 reading frame) relative to the reading frame for *gag* **(Fig. 27-9)**.

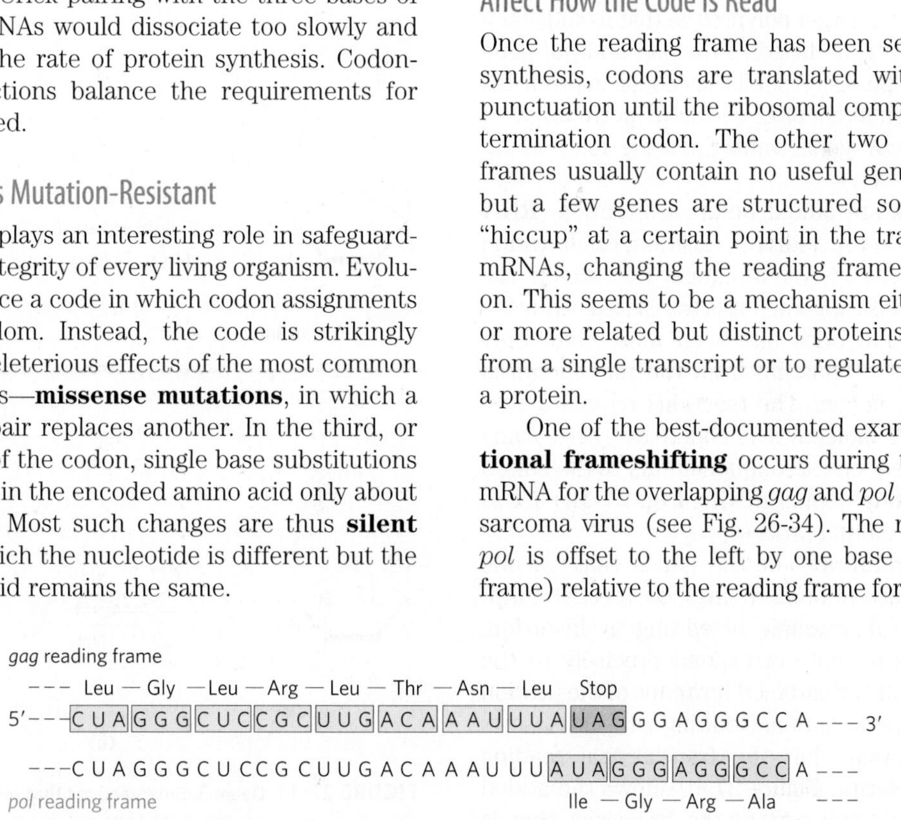

FIGURE 27-9 Translational frameshifting in a retroviral transcript. The *gag-pol* overlap region in Rous sarcoma virus RNA is shown.

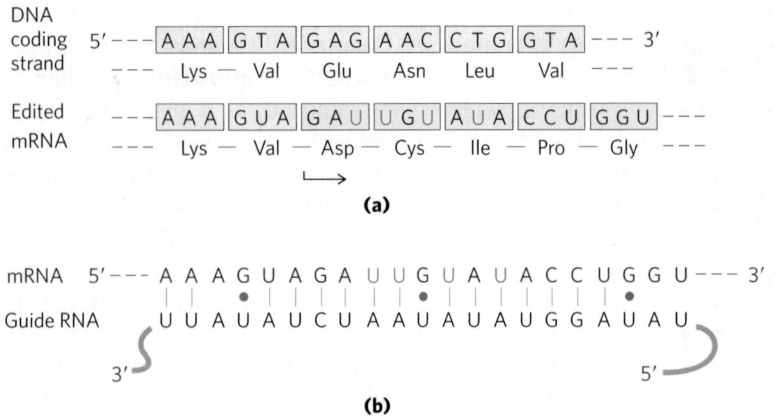

DNA
coding   5′ --- AAA|GTA|GAG|AAC|CTG|GTA --- 3′
strand   ---   Lys — Val — Glu — Asn — Leu — Val ---

Edited   --- AAA|GUA|GAU|UGU|AUA|CCU|GGU ---
mRNA   ---   Lys — Val — Asp — Cys — Ile — Pro — Gly ---

                          ⟶

(a)

mRNA   5′ --- A A A G U A G A U U G U A U A C C U G G U --- 3′
                | | • | | | | | • | | | | | | •
Guide RNA   U U A U A U C U A A U A U A U G G A U A U

           3′   5′

(b)

The product of the *pol* gene (reverse transcriptase) is translated as a larger polyprotein, on the same mRNA that is used for the Gag protein alone (see Fig. 26-33). The polyprotein, or Gag-Pol protein, is then trimmed to the mature reverse transcriptase by proteolytic digestion. Production of the polyprotein requires a translational frameshift in the overlap region to allow the ribosome to bypass the UAG termination codon at the end of the *gag* gene (shaded light red in Fig. 27-9).

Frameshifts occur during about 5% of translations of this mRNA, and the Gag-Pol polyprotein (and ultimately reverse transcriptase) is synthesized at about one-twentieth the frequency of the Gag protein, a level that suffices for efficient reproduction of the virus. In some retroviruses, another translational frameshift allows translation of an even larger polyprotein that includes the product of the *env* gene fused to the *gag* and *pol* gene products (see Fig. 26-33). A similar mechanism produces both the τ and γ subunits of *E. coli* DNA polymerase III from a single *dnaX* gene transcript (see footnote to Table 25-2).

Some mRNAs are edited before translation. **RNA editing** can involve the addition, deletion, or alteration of nucleotides in the RNA in a manner that affects the meaning of the transcript when it is translated. Addition or deletion of nucleotides has been most commonly observed in RNAs originating from the mitochondrial and chloroplast genomes. The reactions require a special class of RNA molecules encoded by these same organelles, with sequences complementary to the edited mRNAs. These guide RNAs (gRNAs; **Fig. 27-10**) act as templates for the editing process.

The initial transcripts of the genes that encode cytochrome oxidase subunit II in some protist mitochondria provide an example of editing by insertion. These transcripts do not correspond precisely to the sequence needed at the carboxyl terminus of the protein product. A posttranscriptional editing process inserts four U residues that shift the translational reading frame of the transcript. Figure 27-10 shows the added U residues in the small part of the transcript that is affected by editing. Note that the base pairing between the initial transcript and the guide RNA includes

several G═U base pairs (blue dots), which are common in RNA molecules.

RNA editing by alteration of nucleotides most commonly involves the enzymatic deamination of adenosine or cytidine residues, forming inosine or uridine, respectively **(Fig. 27-11)**, although other base changes have been described. Inosine is interpreted as a G residue during translation. The adenosine deamination reactions are carried out by *a*denosine *d*eaminases that *a*ct on *R*NA **(ADARs)**. The cytidine deaminations are carried out by the *apoB* mRNA *e*diting *c*atalytic peptide **(APOBEC)** family of enzymes, which includes the *a*ctivation-*i*nduced *d*eaminase **(AID)** enzymes. Both the ADAR and APOBEC groups of deaminase enzymes have a homologous zinc-coordinating catalytic domain.

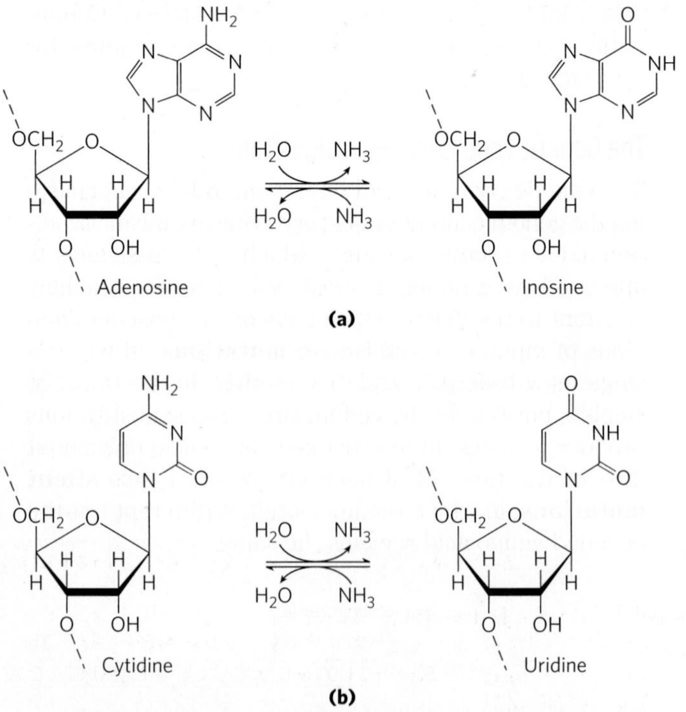

(a)

(b)

FIGURE 27-11 Deamination reactions that result in RNA editing. (a) Conversion of adenosine nucleotides to inosine nucleotides is catalyzed by ADAR enzymes. (b) Cytidine-to-uridine conversions are catalyzed by the APOBEC family of enzymes.

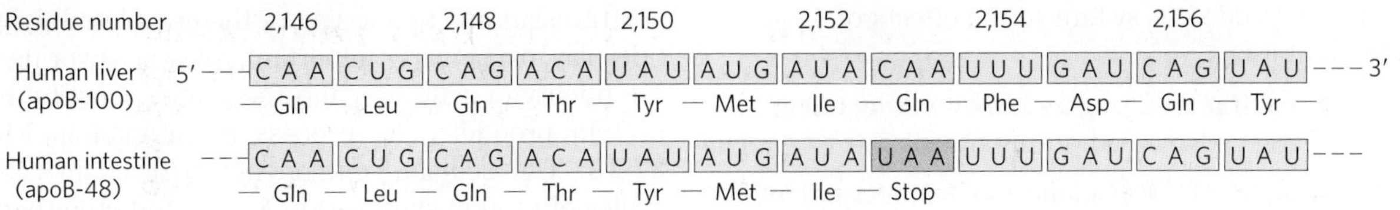

| Residue number | 2,146 | 2,148 | 2,150 | 2,152 | 2,154 | 2,156 |

Human liver
(apoB-100)

5′ - - - |CAA|CUG|CAG|ACA|UAU|AUG|AUA|CAA|UUU|GAU|CAG|UAU| - - - 3′

— Gln — Leu — Gln — Thr — Tyr — Met — Ile — Gln — Phe — Asp — Gln — Tyr —

Human intestine
(apoB-48)

- - - |CAA|CUG|CAG|ACA|UAU|AUG|AUA|UAA|UUU|GAU|CAG|UAU| - - -

— Gln — Leu — Gln — Thr — Tyr — Met — Ile — Stop

FIGURE 27-12 RNA editing of the transcript of the gene for the apoB-100 component of LDL. Deamination, which occurs only in the intestine, converts a specific cytidine to uridine, changing a Gln codon to a stop codon and producing a truncated protein.

The ADAR-promoted A-to-I editing of RNA transcripts is particularly common in primates. Most of the editing occurs in Alu elements, a subset of short interspersed elements (SINEs), eukaryotic transposons that are particularly common in mammalian genomes. Human DNA contains more than a million 300 bp Alu elements, making up about 10% of the genome. These elements are concentrated near protein-coding genes, often in introns and untranslated regions at the 3′ and 5′ ends of transcripts. When it is first synthesized (before processing), the *average* human mRNA includes 10 to 20 Alu elements. Certain microRNAs are also targeted by ADARs. The miRNA alterations generally reduce expression and/or function.

The ADAR enzymes bind to and promote A-to-I editing only in duplex regions of RNA. The abundant Alu elements offer many opportunities for intramolecular base pairing within the transcripts, providing the duplex targets required by ADARs. Some of the editing affects the coding sequences of genes. Defects in ADAR function have been associated with a variety of human neurological conditions, including amyotrophic lateral sclerosis (ALS), epilepsy, and major depression.

The genomes of all vertebrates are replete with SINEs, and many different types of SINES are present in most of these organisms. The Alu elements predominate only in primates. Careful screening of genes and transcripts indicates that A-to-I editing is 30 to 40 times more prevalent in humans than in mice, largely due to the abundance of Alu elements. Large-scale A-to-I editing and an increased level of alternative splicing (see Fig. 26-19b) are two features that set primate genomes apart from those of other mammals. It is not yet clear whether these reactions are incidental or have played key roles in the evolution of primates and, ultimately, humans.

There are six general classes of APOBEC cytidine deaminases: APOBEC1–5 and AID. Most of these enzymes act on DNA. The AID proteins function in increasing antibody diversity during immunoglobulin gene maturation (see Figs 25-43 and 25-44). The APOBEC proteins have a range of functions, including suppression of pathogenic retroviral and retrotransposon gene expression, and roles in innate immunity. APOBEC1 and some of the APOBEC3 proteins (seven APOBEC paralogs are encoded by the human genome) edit mRNAs. A well-studied example of RNA editing by APOBEC1-mediated deamination occurs in the gene for the apolipoprotein B component of low-density lipoprotein in vertebrates.

One form of apolipoprotein B, apoB-100 (M_r 513,000), is synthesized in the liver; a second form, apoB-48 (M_r 250,000), is synthesized in the intestine. Both are encoded by an mRNA produced from the gene for apoB-100. An APOBEC cytidine deaminase found only in the intestine binds to the mRNA at the codon for amino acid residue 2,153 (CAA = Gln) and converts the C to a U to create the termination codon UAA. The apoB-48 produced in the intestine from this modified mRNA is simply an abbreviated form (corresponding to the amino-terminal half) of apoB-100 **(Fig. 27-12)**. This reaction permits tissue-specific synthesis of two different proteins from one gene.

The functions of APOBEC2, 4, and 5 are poorly understood. However, their ability to cause genomic mutations can make them a liability to the cell. One or more APOBEC enzymes are often overexpressed in tumor cells, and their mutagenic ability can contribute to the formation of tumors. They also provide a mechanism for introducing multiple mutations into a targeted segment of a chromosome, leading to selective and more rapid evolution of that DNA region.

SUMMARY 27.1 The Genetic Code

■ The particular amino acid sequence of a protein is constructed through the translation of information encoded in mRNA. This process is carried out by ribosomes.

■ Amino acids are specified by mRNA codons consisting of nucleotide triplets. Translation requires adaptor molecules, the tRNAs, that recognize codons and insert amino acids into their appropriate sequential positions in the polypeptide.

■ The base sequences of the codons were deduced from experiments using synthetic mRNAs of known composition and sequence.

■ The codon AUG signals initiation of translation. The triplets UAA, UAG, and UGA are signals for termination.

■ The genetic code is degenerate: it has multiple codons for almost every amino acid.

■ The standard genetic code is universal in all species, with some minor deviations in mitochondria and a few single-celled organisms. The deviations occur in a pattern that reinforces the concept of a universal code.

■ The third position in each codon is much less specific than the first and second and is said to wobble.

■ The genetic code is resistant to the effects of missense mutations.

■ Translational frameshifting and RNA editing affect how the genetic code is read during translation.

■ RNA editing by ADARs (adenosine deaminases) and APOBECs (cytidine deaminases) also alters the coding sequence of some mRNAs. Many APOBEC enzymes target DNA, where they function in facilitating antibody diversity and suppression of retroviruses and retrotransposons.

27.2 Protein Synthesis

As we have seen for DNA and RNA (Chapters 25 and 26), the synthesis of polymeric biomolecules can be considered in terms of initiation, elongation, and termination stages. These fundamental processes are typically bracketed by two additional stages: activation of precursors before synthesis and postsynthetic processing of the completed polymer. Protein synthesis follows the same pattern. The activation of amino acids before their incorporation into polypeptides and the

posttranslational processing of the completed polypeptide play particularly important roles in ensuring both the fidelity of synthesis and the proper function of the protein product. The process is outlined in **Figure 27-13**. The cellular components involved in the five stages of protein synthesis in *E. coli* and other bacteria are listed in Table 27-5; the requirements in eukaryotic cells are similar, although the components are usually more numerous. An initial overview of the stages of protein synthesis provides a useful outline for the discussion that follows.

Protein Biosynthesis Takes Place in Five Stages

Stage 1: Activation of Amino Acids For the synthesis of a polypeptide with a defined sequence, two fundamental chemical requirements must be met: (1) the carboxyl group of each amino acid must be activated to facilitate formation of a peptide bond, and (2) a link must be established between each new amino acid and the information in the mRNA that encodes it. Both these requirements are met by attaching the amino acid to a tRNA in the

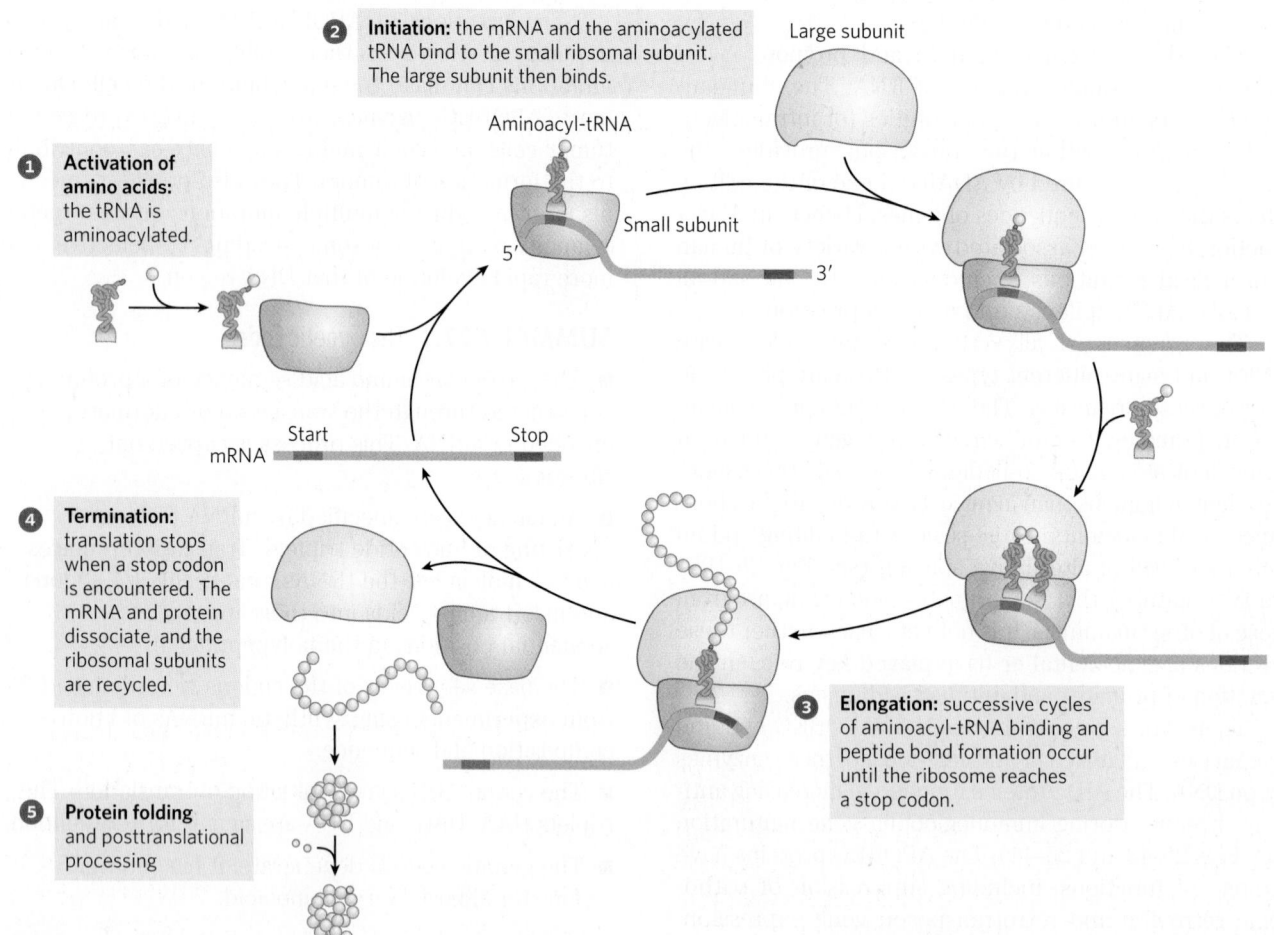

FIGURE 27-13 An overview of the five stages of protein synthesis. ❶ The tRNAs are aminoacylated. ❷ Translation is initiated when an mRNA and an aminoacylated tRNA are bound to the ribosome. ❸ In elongation, the ribosome moves along the mRNA, matching tRNAs to each codon and catalyzing peptide bond formation. ❹ Translation is terminated at a stop

codon, and the ribosomal subunits are released and recycled for another round of protein synthesis. ❺ Following synthesis, the protein must fold into its active conformation and ribosome components are recycled. Most proteins are processed after synthesis. Some amino acids may be removed; others can undergo any of hundreds of known chemical modifications.

TABLE 27-5 Components Required for the Five Major Stages of Protein Synthesis in *E. Coli*

Stage	Essential components
1. Activation of amino acids	20 amino acids 20 aminoacyl-tRNA syntheses 32 or more tRNAs ATP Mg^{2+}
2. Initiation	mRNA *N*-Formylmethionyl-tRNA^{fMet} Initiation codon in mRNA (AUG) 30S ribosomal subunit 50S ribosomal subunit Initiation factors (IF1, IF2, IF3) GTP Mg^{2+}
3. Elongation	Functional 70S ribosomes (initiation complex) Aminoacyl-tRNAs specified by codons Elongation factors (EF-Tu, EF-Ts, EF-G) GTP Mg^{2+}
4. Termination and ribosome recycling	Termination codon in mRNA Release factors (RF1, RF2, RF3, RRF) EF-G IF3
5. Folding and posttranslational processing	Chaperones and folding enzymes (PPI, PDI); specific enzymes, cofactors, and other components for removal of initiating residues and signal sequences, additional proteolytic processing, modification of terminal residues, and attachment of acetyl, phosporyl, methyl, carboxyl, carbohydrate, or prosthetic groups

first stage of protein synthesis. Attaching the right amino acid to the right tRNA is critical. This reaction takes place in the cytosol, not on the ribosome. Each of the 20 amino acids is covalently attached to a specific tRNA at the expense of ATP energy, through the action of Mg^{2+}-dependent activating enzymes known as aminoacyl-tRNA synthetases. When attached to their amino acid (aminoacylated), the tRNAs are said to be "charged."

Stage 2: Initiation The mRNA bearing the code for the polypeptide to be synthesized binds to the smaller of two ribosomal subunits and to the initiating aminoacyl-tRNA. The large ribosomal subunit then binds to form an initiation complex. The initiating aminoacyl-tRNA base-pairs with the mRNA codon AUG that signals the beginning of the polypeptide. This process, which requires GTP, is promoted by cytosolic proteins called initiation factors.

Stage 3: Elongation The nascent polypeptide is lengthened by covalent attachment of successive amino acid units, each carried to the ribosome and correctly positioned by its tRNA, which base-pairs to its corresponding codon in the mRNA. Elongation requires cytosolic proteins known

as elongation factors. The binding of each incoming aminoacyl-tRNA and the movement of the ribosome along the mRNA are facilitated by the hydrolysis of GTP as each residue is added to the growing polypeptide.

Stage 4: Termination and Ribosome Recycling Completion of the polypeptide chain is signaled by a termination codon in the mRNA. The new polypeptide is released from the ribosome, aided by proteins called release factors, and the ribosome is recycled for another round of synthesis.

Stage 5: Folding and Posttranslational Processing To achieve its biologically active form, the new polypeptide must fold into its proper three-dimensional conformation. Before or after folding, the new polypeptide may undergo enzymatic processing, including removal of one or more amino acids (usually from the amino terminus); addition of acetyl, phosphoryl, methyl, carboxyl, or other groups to certain amino acid residues; proteolytic cleavage; and/or attachment of oligosaccharides or prosthetic groups.

Before looking at these five stages in detail, we must examine two key components of protein biosynthesis: the ribosome and tRNAs.

TABLE 27-6	RNA and Protein Components of the *E. Coli* Ribosome			
Subunit	Number of different proteins	Total number of proteins	Protein designations	Number and type of rRNAs
30S	21	21	S1–S21	1 (16S rRNA)
50S	33	36	L1–L36[a]	2 (5S and 23S rRNAs)

[a]The L1 to L36 protein designations do not correspond to 36 different proteins. The protein originally designated L7 is a modified form of L12, and L8 is a complex of three other proteins. Also, L26 proved to be the same protein as S20 (and not part of the 50S subunit). This gives 33 different proteins in the large subunit. There are four copies of the L7/L12 protein, with the three extra copies bringing the total protein count to 36.

The Ribosome Is a Complex Supramolecular Machine

Each *E. coli* cell contains 15,000 or more ribosomes, which comprise nearly a quarter of the dry weight of the cell. Bacterial ribosomes contain about 65% rRNA and 35% protein; they have a diameter of about 18 nm and are composed of two unequal subunits with sedimentation coefficients of 30S and 50S and a combined sedimentation coefficient of 70S. Both subunits contain dozens of ribosomal proteins (r-proteins) and at least one large rRNA (Table 27-6).

Following Zamecnik's discovery that ribosomes are the complexes responsible for protein synthesis, and following elucidation of the genetic code, the study of ribosomes accelerated. In the late 1960s Masayasu Nomura and colleagues demonstrated that both ribosomal subunits can be broken down into their RNA and protein components, then reconstituted in vitro. Under appropriate experimental conditions, the RNA and protein spontaneously reassemble to form 30S or 50S subunits nearly identical in structure and activity to native subunits. This breakthrough fueled decades of research into the function and structure of ribosomal RNAs and proteins. At the same time, increasingly sophisticated structural methods revealed more and more details about ribosome structure.

The dawn of a new millennium illuminated the first high-resolution structures of bacterial ribosomal subunits

Masayasu Nomura, 1927–2011 [Source: Courtesy of Archives, University of Wisconsin–Madison.]

by Venki Ramakrishnan, Thomas Steitz, Ada Yonath, Harry Noller, and others. This work yielded a wealth of surprises **(Fig. 27-14a)**. First, a traditional focus on the protein components of ribosomes was shifted. The ribosomal subunits are huge RNA molecules. In the 50S subunit, the 5S and 23S rRNAs form the structural core. The proteins are secondary elements in the complex, decorating the surface. Second, and most important, there is no protein within 18 Å of the active site for peptide bond formation. The high-resolution structure thus confirms what Noller had predicted much earlier: the ribosome is a ribozyme. In addition to the insight that the detailed structures of the ribosome and its subunits provide into the mechanism of protein synthesis (as elaborated below), these findings have stimulated a new look at the evolution of life (Section 26.3). The ribosomes of eukaryotic cells have also yielded to structural analysis (Fig. 27-14b).

The bacterial ribosome is complex, with a combined molecular weight of ~2.7 million. The two irregularly shaped ribosomal subunits fit together to form a cleft through which the mRNA passes as the ribosome moves along it during translation (Fig. 27-14a). The 57 proteins in bacterial ribosomes vary enormously in size and structure. Molecular weights range from about 6,000 to 75,000. Most of the proteins have globular domains arranged on the ribosome surface. Some also have snakelike extensions that protrude into the rRNA core of the ribosome, stabilizing its structure. The functions of some of these proteins have not yet been elucidated in detail, although a structural role seems evident for many of them.

The sequences of the rRNAs of many organisms are now known. Each of the three single-stranded rRNAs of *E. coli* has a specific three-dimensional conformation with extensive intrachain base pairing. The folding patterns of the rRNAs are highly conserved in all organisms, particularly the regions implicated in key functions **(Fig. 27-15)**. The predicted secondary structure of the rRNAs has largely been confirmed by structural analysis, but fails to convey the extensive network of tertiary interactions apparent in the complete structure.

The ribosomes of eukaryotic cells (other than mitochondrial and chloroplast ribosomes) are larger and more complex than bacterial ribosomes **(Fig. 27-16;** compare Fig. 27-14b**)**, with a diameter of

Venkatraman Ramakrishnan [Source: © Alastair Grant/AP Photo.]

Thomas A. Steitz [Source: © Lucas Jackson/ Reuters/Corbis.]

Ada E. Yonath [Source: © Yin Bogu/Xinhua Press/Corbis.]

(a)

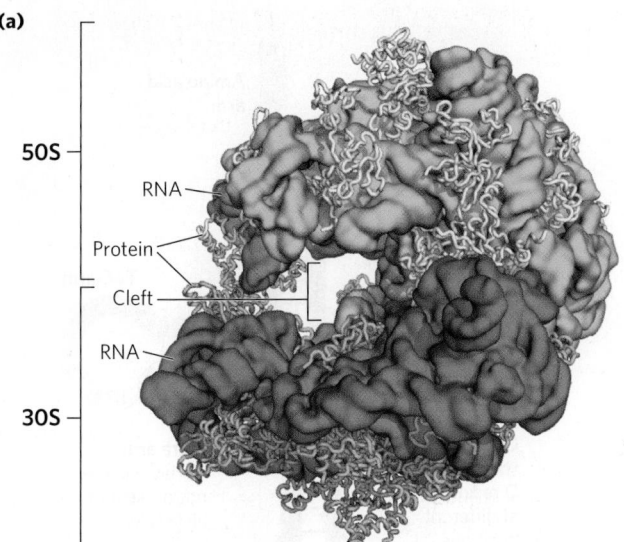

(b)

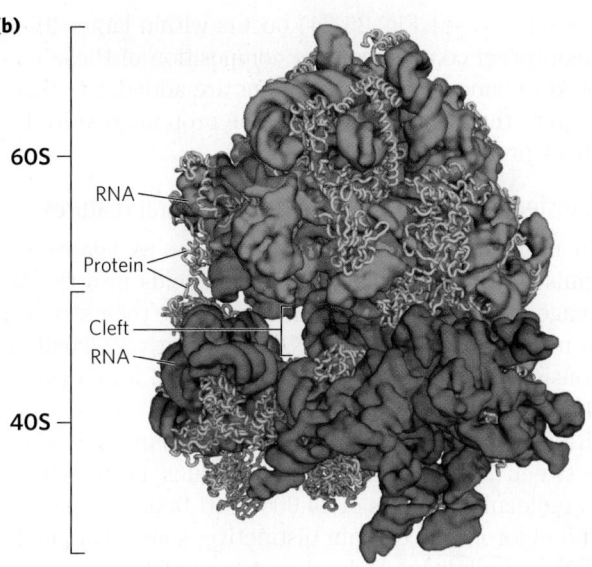

FIGURE 27-14 The structure of ribosomes. Our understanding of ribosome structure has been greatly enhanced by multiple high-resolution images of the ribosomes from bacteria and yeast. **(a)** The bacterial ribosome. The 50S and 30S subunits come together to form the 70S ribosome. The cleft between them is where protein synthesis occurs. **(b)** The yeast ribosome has a similar structure with somewhat increased complexity. [Sources: (a) Derived from PDB ID 4V4I, A. Korostelev et al., *Cell* 126:1065, 2006. (b) Derived from PDB ID 4V7R, A. Ben-Shem et al., *Science* 330:1203, 2010.]

about 23 nm and a sedimentation coefficient of about 80S. They also have two subunits, which vary in size among species but on average are 60S and 40S. Altogether, eukaryotic ribosomes contain more than 80 different proteins. The ribosomes of mitochondria and chloroplasts are somewhat smaller and simpler than bacterial ribosomes. Nevertheless, ribosomal structure and function are strikingly similar in all organisms and organelles.

In both bacteria and eukaryotes, ribosomes are assembled through a hierarchical incorporation of r-proteins as the rRNAs are synthesized. Much of the processing of

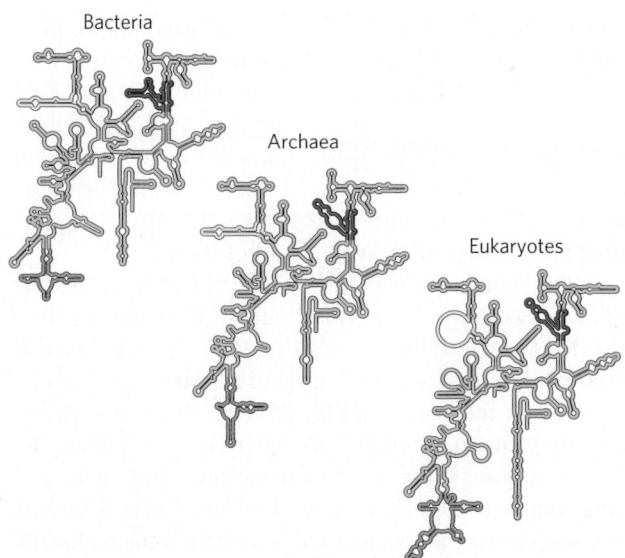

FIGURE 27-15 Conservation of secondary structure in the small subunit rRNAs from the three domains of life. The red, yellow, and purple indicate areas where the structures of the rRNAs from bacteria, archaea, and eukaryotes have diverged. Conserved regions are shown in green. [Source: Information from www.rna.icmb.utexas.edu.]

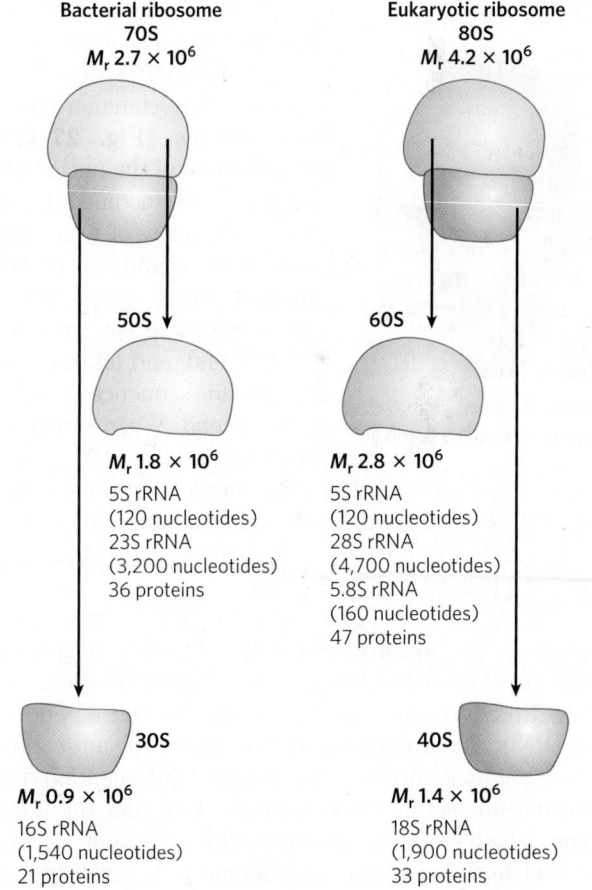

FIGURE 27-16 Summary of the composition and mass of ribosomes in bacteria and eukaryotes. Ribosomal subunits are identified by their S (Svedberg unit) values, sedimentation coefficients that refer to their rate of sedimentation in a centrifuge. The S values are not additive when subunits are combined, because S values are approximately proportional to the two-thirds power of molecular weight and are also slightly affected by shape.

pre-rRNAs (see Fig. 26-24) occurs within large ribonucleoprotein complexes. The composition of these complexes changes as new r-proteins are added, the rRNAs acquire their final form, and some proteins required for rRNA processing dissociate.

Transfer RNAs Have Characteristic Structural Features

To understand how tRNAs can serve as adaptors in translating the language of nucleic acids into the language of proteins, we must first examine their structure in more detail. Transfer RNAs are relatively small and consist of a single strand of RNA folded into a precise three-dimensional structure (see Fig. 8-25a). The tRNAs of bacteria and in the cytosol of eukaryotes have between 73 and 93 nucleotide residues, corresponding to molecular weights of 24,000 to 31,000. Mitochondria and chloroplasts contain distinctive, somewhat smaller tRNAs. Cells have at least one kind of tRNA for each amino acid; at least 32 tRNAs are required to recognize all the amino acid codons (some recognize more than one codon), but some cells use more than 32.

Yeast alanine tRNA (tRNA^{Ala}) was the first nucleic acid to be completely sequenced, by Robert Holley in 1965. It contains 76 nucleotide residues, 10 of which have modified bases. Comparisons of tRNAs from various species have revealed many common structural features **(Fig. 27-17)**. Eight or more of the nucleotide residues have modified bases and sugars, many of which are methylated derivatives of the principal bases. Most tRNAs have a guanylate (pG) residue at the 5' end, and all have the trinucleotide sequence CCA(3') at the 3' end. When drawn in two dimensions, all tRNAs have a hydrogen-bonding pattern that forms a cloverleaf structure with four arms; the longer tRNAs have a short fifth arm, or extra arm. In three dimensions, a tRNA has the form of a twisted L **(Fig. 27-18)**.

Robert W. Holley, 1922–1993
[Source: Bettmann/Corbis.]

Two of the arms of a tRNA are critical for its adaptor function. The **amino acid arm** can carry a specific amino acid esterified by its carboxyl group to the 2'- or 3'-hydroxyl group of the A residue at the 3' end of the tRNA. The **anticodon arm** contains the anticodon. The other major arms are the **D arm**, which contains the unusual nucleotide dihydrouridine (D), and the **TψC arm**, which contains ribothymidine (T), not usually present in RNAs, and pseudouridine (ψ), which has an unusual carbon–carbon bond between the base and ribose (see Fig. 26-22). The D and TψC arms contribute important interactions for the overall folding of tRNA molecules, and the TψC arm interacts with the large-subunit rRNA.

Having looked at the structures of ribosomes and tRNAs, we now consider in detail the five stages of protein synthesis.

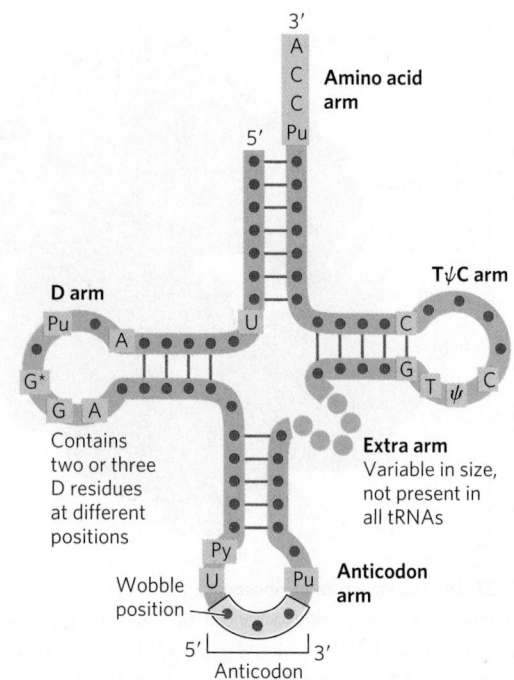

FIGURE 27-17 General cloverleaf secondary structure of tRNAs. The large dots on the backbone represent nucleotide residues; the blue lines represent base pairs. Characteristic and/or invariant residues common to all tRNAs are shaded in light red. Transfer RNAs vary in length from 73 to 93 nucleotides. Extra nucleotides occur in the extra arm or in the D arm. At the end of the anticodon arm is the anticodon loop, which always contains seven unpaired nucleotides. The D arm contains two or three D (5,6-dihydrouridine) residues, depending on the tRNA. In some tRNAs, the D arm has only three hydrogen-bonded base pairs. Pu represents purine nucleotide; Py, pyrimidine nucleotide; ψ, pseudouridylate; G*, guanylate or 2'-O-methylguanylate.

Stage 1: Aminoacyl-tRNA Synthetases Attach the Correct Amino Acids to Their tRNAs

During the first stage of protein synthesis, taking place in the cytosol, aminoacyl-tRNA synthetases esterify the 20 amino acids to their corresponding tRNAs. Each enzyme is specific for one amino acid and one or more corresponding tRNAs. Most organisms have one aminoacyl-tRNA synthetase for each amino acid. For amino acids with two or more corresponding tRNAs, the same enzyme usually aminoacylates all of them.

The structures of all the aminoacyl-tRNA synthetases of *E. coli* have been determined. Researchers have divided them into two classes (Table 27-7), based on substantial differences in primary and tertiary structure and in reaction mechanism **(Fig. 27-19)**; these two classes are the same in all organisms. There is no evidence that the two classes share a common ancestor, and the biological, chemical, or evolutionary reasons for two enzyme classes for essentially identical processes remain obscure.

The reaction catalyzed by an aminoacyl-tRNA synthetase is

$$\text{Amino acid} + \text{tRNA} + \text{ATP} \xrightarrow{\text{Mg}^{2+}}$$
$$\text{aminoacyl-tRNA} + \text{AMP} + \text{PP}_i$$

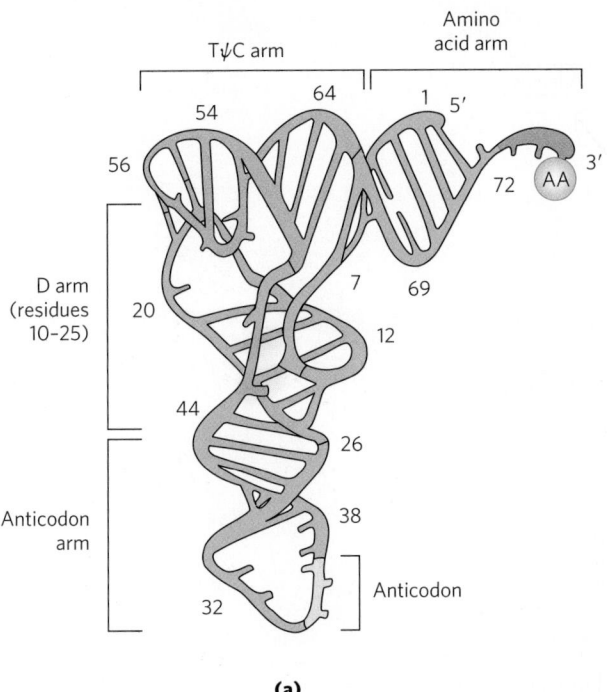

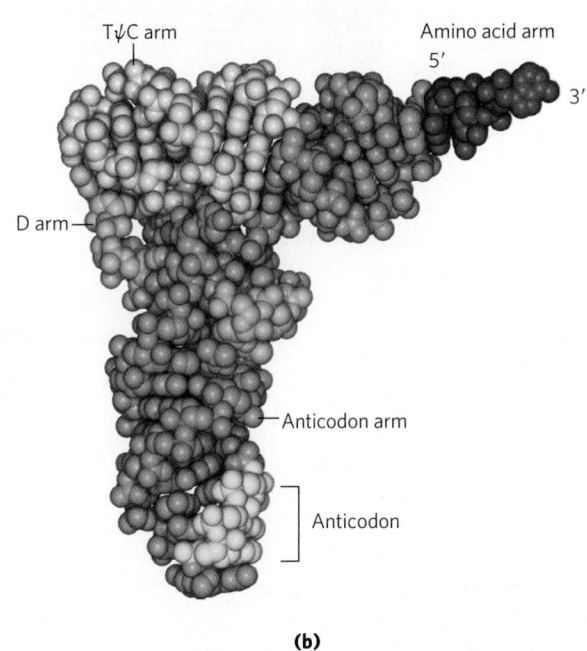

FIGURE 27-18 Three-dimensional structure of yeast tRNA^{Phe} deduced from x-ray diffraction analysis. The shape resembles a twisted L. **(a)** Schematic diagram with the various arms identified in Figure 27-17 shaded in different colors. **(b)** A space-filling model, with the same color coding. The CCA sequence at the 3′ end (purple) is the attachment point for the amino acid. [Source: (b) PDB ID 4TRA, E. Westhof et al., *Acta Crystallogr. A* 44:112, 1988.]

This reaction occurs in two steps in the enzyme's active site. In step ❶ (Fig. 27-19), an enzyme-bound intermediate, aminoacyl adenylate (aminoacyl-AMP), is formed. In the second step, the aminoacyl group is transferred from enzyme-bound aminoacyl-AMP to its corresponding specific tRNA. The course of this second step depends on the class to which the enzyme belongs, as shown by pathways ❷ⓐ and ❷ⓑ in Figure 27-19. The resulting ester linkage between the amino acid and the tRNA **(Fig. 27-20)** has a highly negative standard free energy of hydrolysis ($\Delta G'^{\circ} = -29$ kJ/mol). The pyrophosphate formed in the activation reaction undergoes hydrolysis to phosphate by inorganic pyrophosphatase. Thus *two* high-energy phosphate bonds are ultimately expended for each amino acid molecule activated, rendering the overall reaction for amino acid activation essentially irreversible:

$$\text{Amino acid} + \text{tRNA} + \text{ATP} \xrightarrow{\text{Mg}^{2+}}$$
$$\text{aminoacyl-tRNA} + \text{AMP} + 2\text{P}_i$$
$$\Delta G'^{\circ} \approx -29 \text{ kJ/mol}$$

Proofreading by Aminoacyl-tRNA Synthetases The aminoacylation of tRNA accomplishes two ends: (1) it activates an amino acid for peptide bond formation and (2) it ensures appropriate placement of the amino acid in a growing polypeptide. The identity of the amino acid attached to a tRNA is not checked on the ribosome, so attachment of the correct amino acid to the tRNA is essential to the fidelity of protein synthesis.

As you will recall from Chapter 6, enzyme specificity is limited by the binding energy available from enzyme-substrate interactions. Discrimination between two similar amino acid substrates has been studied in detail in the case of Ile-tRNA synthetase, which distinguishes between valine and isoleucine, amino acids that differ by only a single methylene group (—CH_2—):

TABLE 27-7	The Two Classes of Aminoacyl-tRNA Synthetases		
Class I		**Class II**	
Arg	Leu	Ala	Lys
Cys	Met	Asn	Phe
Gln	Trp	Asp	Pro
Glu	Tyr	Gly	Ser
Ile	Val	His	Thr

Note: Here, Arg represents arginyl-tRNA synthetase, and so forth. The classification applies to all organisms for which tRNA synthetases have been analyzed and is based on protein structural distinctions and on the mechanistic distinction outlined in Figure 27-19.

Valine   Isoleucine

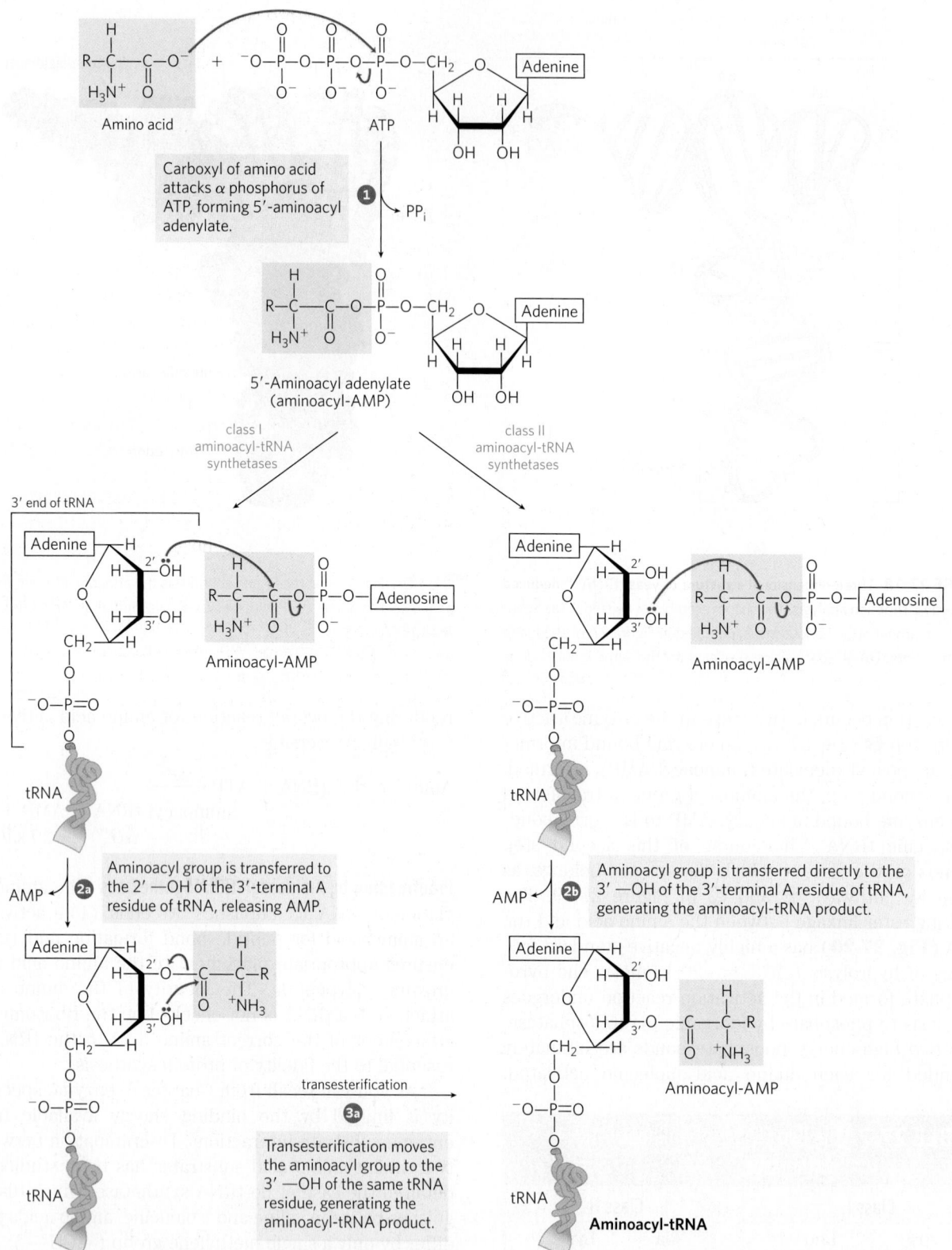

MECHANISM FIGURE 27-19 Aminoacylation of tRNA by aminoacyl-tRNA synthetases. Step **❶** is formation of an aminoacyl adenylate, which remains bound to the active site. In the second step, the aminoacyl group is transferred to the tRNA. The mechanism of this step is somewhat different for the two classes of aminoacyl-tRNA synthetases (see Table 27-7). For class I enzymes, **❷ₐ** the aminoacyl group is transferred first to the 2'-hydroxyl group of the 3'-terminal A residue, then **❸ₐ** to the 3'-hydroxyl group by a transesterification reaction. For class II enzymes, **❷ᵦ** the aminoacyl group is transferred directly to the 3'-hydroxyl group of the terminal adenylate.

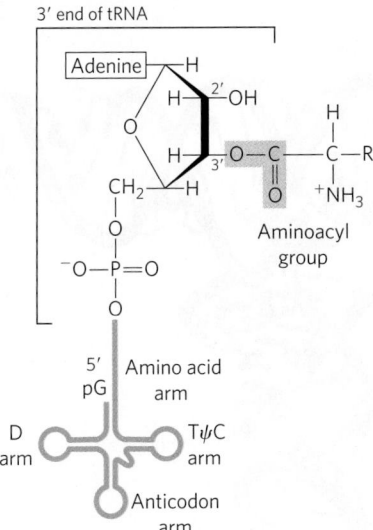

3′ end of tRNA

Adenine

Aminoacyl group

5′ pG   Amino acid arm

D arm   TψC arm

Anticodon arm

FIGURE 27-20 General structure of aminoacyl-tRNAs. The aminoacyl group is esterified to the 3′ position of the terminal A residue. The ester linkage that both activates the amino acid and joins it to the tRNA is shaded light red.

Ile-tRNA synthetase favors activation of isoleucine (to form Ile-AMP) over valine by a factor of 200—as we would expect, given the amount by which a methylene group (in Ile) could enhance substrate binding. Yet valine is erroneously incorporated into proteins in positions normally occupied by an Ile residue at a frequency of only about 1 in 3,000. How is this greater than 10-fold increase in accuracy brought about? Ile-tRNA synthetase, like some other aminoacyl-tRNA synthetases, has a proofreading function.

Recall a general principle from the discussion of proofreading by DNA polymerases (see Fig. 25-7): if available binding interactions do not provide sufficient discrimination between two substrates, the necessary specificity can be achieved by substrate-specific binding in *two successive* steps. The effect of forcing the system through two successive filters is multiplicative. In the case of Ile-tRNA synthetase, the first filter is the initial binding of the amino acid to the enzyme and its activation to aminoacyl-AMP. The second is the binding of any *incorrect* aminoacyl-AMP products to a separate active site on the enzyme; a substrate that binds in this second active site is hydrolyzed. The R group of valine is slightly smaller than that of isoleucine, so Val-AMP fits the hydrolytic (proofreading) site of the Ile-tRNA synthetase, but Ile-AMP does not. Thus Val-AMP is hydrolyzed to valine and AMP in the proofreading active site, and tRNA bound to the synthetase does not become aminoacylated to the wrong amino acid.

In addition to proofreading after formation of the aminoacyl-AMP intermediate, most aminoacyl-tRNA synthetases can hydrolyze the ester linkage between amino acids and tRNAs in the aminoacyl-tRNAs. This hydrolysis is greatly accelerated for incorrectly charged tRNAs, providing yet a third filter to enhance the fidelity of the overall process. The few aminoacyl-tRNA synthetases that activate amino acids with no close structural relatives (Cys-tRNA synthetase, for example) demonstrate little or no proofreading activity; in these cases, the active site for aminoacylation can sufficiently discriminate between the proper substrate and any incorrect amino acid.

The overall error rate of protein synthesis (~1 mistake per 10^4 amino acids incorporated) is not nearly as low as that of DNA replication. Because flaws in a protein are eliminated when the protein is degraded and are not passed on to future generations, they have less biological significance. The degree of fidelity in protein synthesis is sufficient to ensure that most proteins contain no mistakes and that the large amount of energy required to synthesize a protein is rarely wasted. One defective protein molecule is usually unimportant when many correct copies of the same protein are present.

Interaction between an Aminoacyl-tRNA Synthetase and a tRNA: A "Second Genetic Code" An individual aminoacyl-tRNA synthetase must be specific not only for a single amino acid but for certain tRNAs as well. Discriminating among dozens of tRNAs is just as important for the overall fidelity of protein biosynthesis as is distinguishing among amino acids. The interaction between aminoacyl-tRNA synthetases and tRNAs has been referred to as the "second genetic code," reflecting its critical role in maintaining the accuracy of protein synthesis. The "coding" rules appear to be more complex than those in the "first" code.

Figure 27-21 summarizes what we know about the nucleotides involved in recognition by some aminoacyl-tRNA synthetases. Some nucleotides are conserved in all tRNAs and therefore cannot be used for discrimination. By observing changes in nucleotides that alter substrate specificity, researchers have identified nucleotide positions necessary for discrimination by the aminoacyl-tRNA synthetases. These nucleotide positions seem to be concentrated in the amino acid arm and the anticodon arm, including the nucleotides of the anticodon itself, but are also located in other parts of the tRNA molecule. Determination of the crystal structures of aminoacyl-tRNA synthetases complexed with their cognate tRNAs and ATP has added a great deal to our understanding of these interactions **(Fig. 27-22).**

Ten or more specific nucleotides may be involved in recognition of a tRNA by its specific aminoacyl-tRNA synthetase. But in a few cases the recognition mechanism is quite simple. Across a range of organisms from bacteria to humans, the primary determinant of tRNA recognition by the Ala-tRNA synthetases is a single G≡U base pair in the amino acid arm of tRNA^{Ala} **(Fig. 27-23a).** A short synthetic RNA with as few as 7 bp arranged in a simple hairpin minihelix is efficiently aminoacylated by the Ala-tRNA synthetase, as long as the RNA contains the critical G≡U (Fig. 27-23b). This relatively simple alanine system

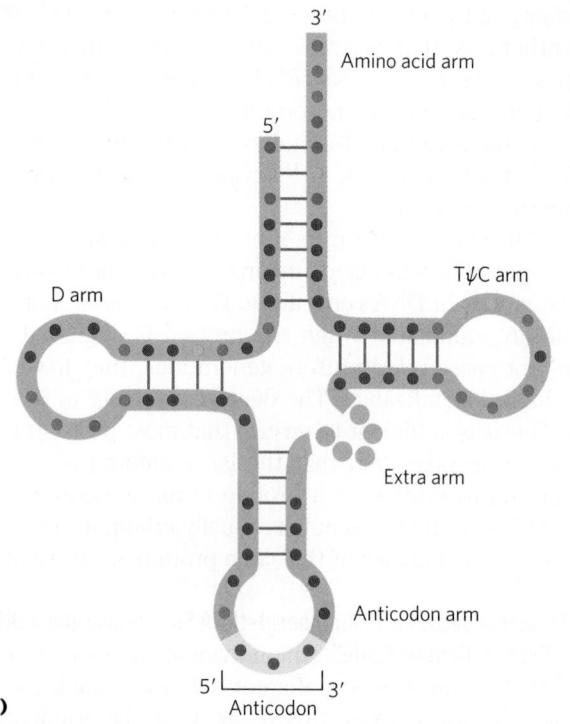

(a)

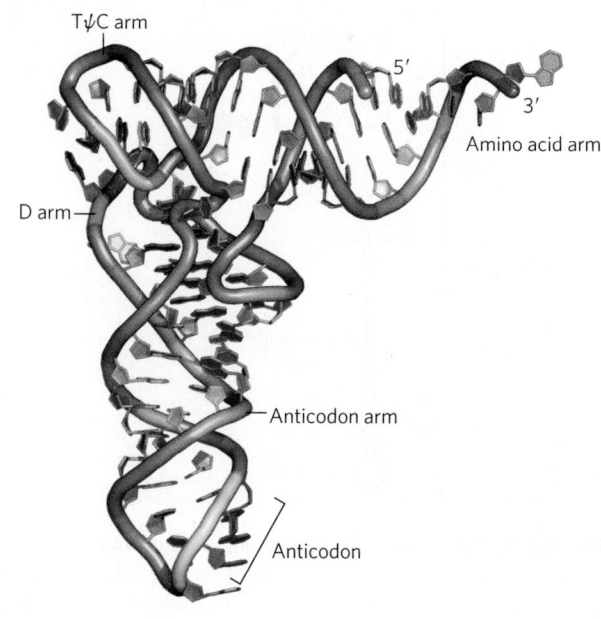

(b)

FIGURE 27-21 Nucleotide positions in a tRNA that are recognized by aminoacyl-tRNA synthetases. (a) Some positions (purple dots) are the same in all tRNAs and therefore cannot be used to discriminate one from another. Other positions are known recognition points for one (orange) or more (blue) aminoacyl-tRNA synthetases. Structural features other than sequence are important for recognition by some of the synthetases. **(b)** The same structural features are shown in three dimensions, with the orange and blue residues again representing positions recognized by one or more aminoacyl-tRNA synthetases, respectively. [Source: PDB ID 1EHZ, H. Shi and P. B. Moore, *RNA* 6:1091, 2000.]

may be an evolutionary relic of a period when RNA oligonucleotides, ancestors to tRNA, were aminoacylated in a primitive system for protein synthesis.

The interaction of aminoacyl-tRNA synthetases and their cognate tRNAs is critical to accurate reading of the genetic code. Any expansion of the code to include new amino acids would necessarily require a new aminoacyl-tRNA synthetase–tRNA pair. A limited expansion of the genetic code has been observed in nature; a more extensive expansion has been accomplished in the laboratory (Box 27-2).

Stage 2: A Specific Amino Acid Initiates Protein Synthesis

Protein synthesis begins at the amino-terminal end and proceeds by the stepwise addition of amino acids to the carboxyl-terminal end of the growing polypeptide, as determined by Howard Dintzis in 1961 **(Fig. 27-24)**. The AUG initiation codon thus specifies an *amino-terminal* methionine residue. Although methionine has only one codon, (5′)AUG, all organisms have two tRNAs for methionine. One is used exclusively when (5′)AUG is the initiation codon for protein synthesis. The other is

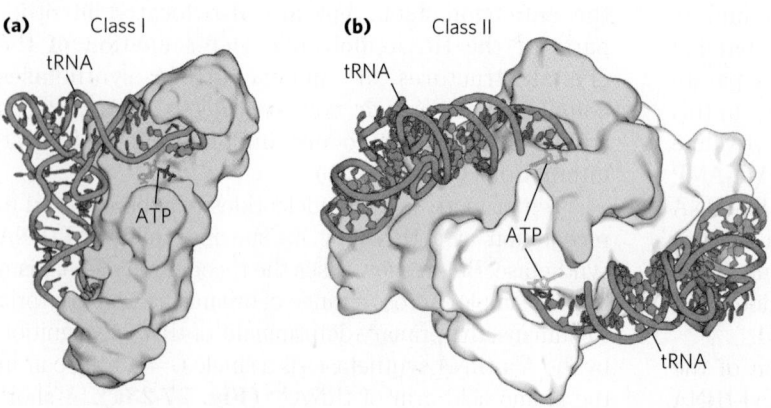

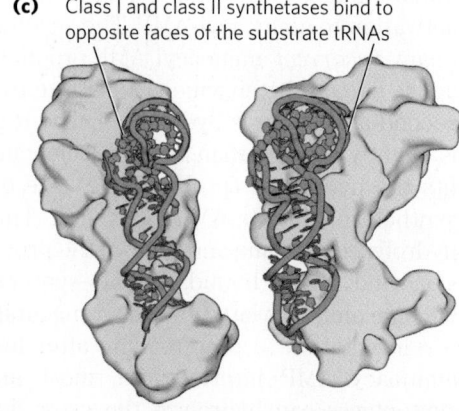

FIGURE 27-22 Aminoacyl-tRNA synthetases. The synthetases are complexed with their cognate tRNAs (green). Bound ATP (red) pinpoints the active site near the end of the aminoacyl arm. **(a)** Gln-tRNA synthetase of *E. coli*, a typical monomeric class I synthetase. **(b)** Asp-tRNA synthetase of yeast, a typical dimeric class II synthetase. **(c)** The two classes of aminoacyl-tRNA synthetases recognize different faces of their tRNA substrates. [Sources: (a, c (left)) PDB ID 1QRT, J. G. Arnez and T. A. Steitz, *Biochemistry* 35:14,725, 1996. (b, c (right)) PDB ID 1ASZ, J. Cavarelli et al., *EMBO J.* 13:327, 1994.]

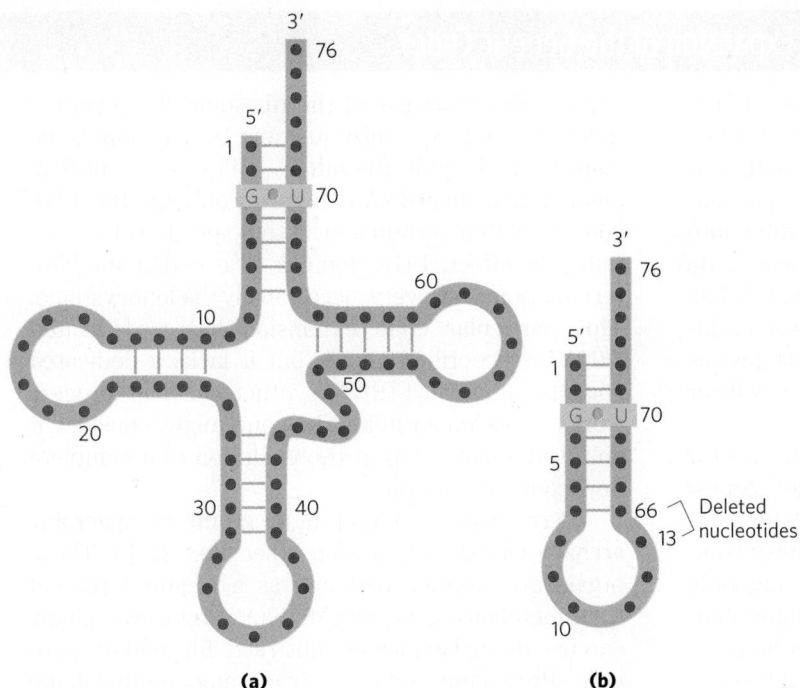

FIGURE 27-23 Structural elements of tRNA^{Ala} that are required for recognition by Ala-tRNA synthetase. **(a)** The tRNA^{Ala} structural elements recognized by the Ala-tRNA synthetase are unusually simple. A single G≡U base pair (light red) is the only element needed for specific binding and aminoacylation. **(b)** A short synthetic RNA minihelix, with the critical G≡U base pair but lacking most of the remaining tRNA structure. This is aminoacylated specifically with alanine almost as efficiently as the complete tRNA^{Ala}.

used to code for a Met residue in an internal position in a polypeptide.

The distinction between an initiating (5′)AUG and an internal one is straightforward. In bacteria, the two types of tRNA specific for methionine are designated tRNA^{Met} and tRNA^{fMet}. The amino acid incorporated in response

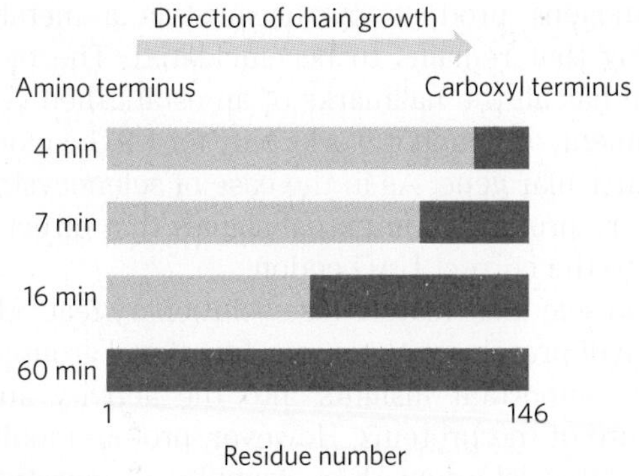

to the (5′)AUG initiation codon is *N*-formylmethionine (fMet). It arrives at the ribosome as *N*-formylmethionyl-tRNA^{fMet} (fMet-tRNA^{fMet}), which is formed in two successive reactions. First, methionine is attached to tRNA^{fMet} by the Met-tRNA synthetase (which in *E. coli* aminoacylates both tRNA^{fMet} and tRNA^{Met}):

$$\text{Methionine} + \text{tRNA}^{\text{fMet}} + \text{ATP} \longrightarrow$$
$$\text{Met-tRNA}^{\text{fMet}} + \text{AMP} + \text{PP}_i$$

Next, a transformylase transfers a formyl group from N^{10}-formyltetrahydrofolate to the amino group of the Met residue:

$$N^{10}\text{-Formyltetrahydrofolate} + \text{Met-tRNA}^{\text{fMet}} \longrightarrow$$
$$\text{tetrahydrofolate} + \text{fMet-tRNA}^{\text{fMet}}$$

The transformylase is more selective than the Met-tRNA synthetase; it is specific for Met residues attached to tRNA^{fMet}, presumably recognizing some unique structural feature of that tRNA. By contrast, Met-tRNA^{Met} inserts methionine in interior positions in polypeptides.

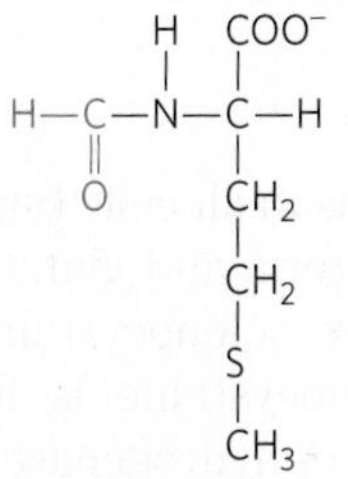

$$
\begin{array}{ccc}
 & \text{H} & \text{COO}^- \\
 & | & | \\
\text{H—C—N—C—H} \\
\| & & | \\
\text{O} & & \text{CH}_2 \\
 & & | \\
 & & \text{CH}_2 \\
 & & | \\
 & & \text{S} \\
 & & | \\
 & & \text{CH}_3
\end{array}
$$

N-Formylmethionine

Addition of the *N*-formyl group to the amino group of methionine by the transformylase prevents fMet from entering interior positions in a polypeptide while also allowing fMet-tRNA^{fMet} to be bound at a specific ribosomal initiation site that accepts neither Met-tRNA^{Met} nor any other aminoacyl-tRNA.

Direction of chain growth →

Amino terminus   Carboxyl terminus

4 min

7 min

16 min

60 min

1   146

Residue number

FIGURE 27-24 Proof that polypeptides grow by addition of amino acid residues to the carboxyl end: the Dintzis experiment. Reticulocytes (immature erythrocytes) actively synthesizing hemoglobin were incubated with radioactive leucine (selected because it occurs frequently in both the α- and β-globin chains). Samples of completed α chains were isolated from the incubating reticulocytes at various times, and the distribution of radioactivity determined. The dark red zones show the portions of completed α-globin chains containing radioactive Leu residues. At 4 min, only a few residues at the carboxyl end of α-globin were labeled, because the only *complete* globin chains with incorporated label after 4 min were those that had nearly completed synthesis at the time the label was added. With longer incubation times, successively longer segments of the polypeptide contained labeled residues, always in a block at the carboxyl end of the chain. The unlabeled end of the polypeptide (the amino terminus) was thus defined as the initiating end, which means that polypeptides grow by successive addition of amino acids to the carboxyl end.

BOX 27-2   Natural and Unnatural Expansion of the Genetic Code

As we have seen, the 20 standard amino acids found in proteins offer only limited chemical functionality. Living systems generally overcome these limitations by using enzymatic cofactors or by modifying particular amino acids after they have been incorporated into proteins. In principle, expansion of the genetic code to introduce new amino acids into proteins offers another route to new functionality, but it is a very difficult route to exploit. Such a change might just as easily result in inactivation of thousands of cellular proteins.

Expanding the genetic code to include a new amino acid requires several cellular changes. A new aminoacyl-tRNA synthetase must generally be present, along with a cognate tRNA. Both of these components must be highly specific, interacting only with each other and the new amino acid. Significant concentrations of the new amino acid must be present in the cell, which may entail the evolution of new metabolic pathways. As outlined in Box 27-1, the anticodon on the tRNA would most likely pair with a codon that usually specifies termination. Making all of this work in a living cell seems unlikely, but it has happened both in nature and in the laboratory.

There are actually 22 rather than 20 amino acids specified by the known genetic code. The two extra ones are selenocysteine and pyrrolysine, each found in only very few proteins but both offering a glimpse into the intricacies of code evolution.

Selenocysteine

Pyrrolysine

A few proteins in all cells (such as formate dehydrogenase in bacteria and glutathione peroxidase in mammals) require selenocysteine for their activity. In *E. coli*, selenocysteine is introduced into the enzyme formate dehydrogenase during translation, in response to an in-frame UGA codon. A special type of Ser-tRNA, present at lower levels than other Ser-tRNAs, recognizes UGA and no other codons. This tRNA is charged with serine by the normal serine aminoacyl-tRNA synthetase, and the serine is enzymatically converted to selenocysteine by a separate

enzyme before its use at the ribosome. The charged tRNA does not recognize just any UGA codon; some contextual signal in the mRNA, still to be identified, ensures that this tRNA recognizes only the few UGA codons, within certain genes, that specify selenocysteine. In effect, UGA doubles as a codon for both termination and (very occasionally) selenocysteine. This particular code expansion has a dedicated tRNA, as described above, but it lacks a dedicated cognate aminoacyl-tRNA synthetase. The process works for selenocysteine, but one might consider it an intermediate step in the evolution of a complete new codon definition.

Pyrrolysine is found in a group of anaerobic archaea called methanogens (see Box 22-1). These organisms produce methane as a required part of their metabolism, and the Methanosarcinaceae family can use methylamines as substrates for methanogenesis. Producing methane from monomethylamine requires the enzyme monomethylamine methyltransferase. The gene encoding this enzyme has an in-frame UAG termination codon. The structure of the methyltransferase was elucidated in 2002, revealing the presence of the novel amino acid pyrrolysine at the position specified by the UAG codon. Subsequent experiments demonstrated that—unlike selenocysteine—pyrrolysine is attached directly to a dedicated tRNA by a cognate pyrrolysyl-tRNA synthetase. These methanogens produce pyrrolysine via a metabolic pathway that remains to be elucidated. The overall system has all the hallmarks of an established codon assignment, although it works only for UAG codons in this particular gene. As in the case of selenocysteine, there are probably contextual signals that direct this tRNA to the correct UAG codon.

Can scientists match this evolutionary feat? Modification of proteins with various functional groups can provide important insights into the activity and/or structure of the proteins. However, protein modification is often laborious. For example, an investigator who wishes to attach a new group to a particular Cys residue will have to somehow block other Cys residues that may be present on the same protein. If one could instead adapt the genetic code to enable a cell to insert a modified amino acid at a particular location in a protein, the process could be rendered much more convenient. Peter Schultz and coworkers have done just that.

To develop a new codon assignment, one again needs a new aminoacyl-tRNA synthetase and a novel cognate tRNA, both adapted to work only with a particular new amino acid. Efforts to create such an "unnatural" code expansion initially focused on *E. coli*. The codon UAG was chosen as the best target for encoding a new amino acid. UAG is the least used of

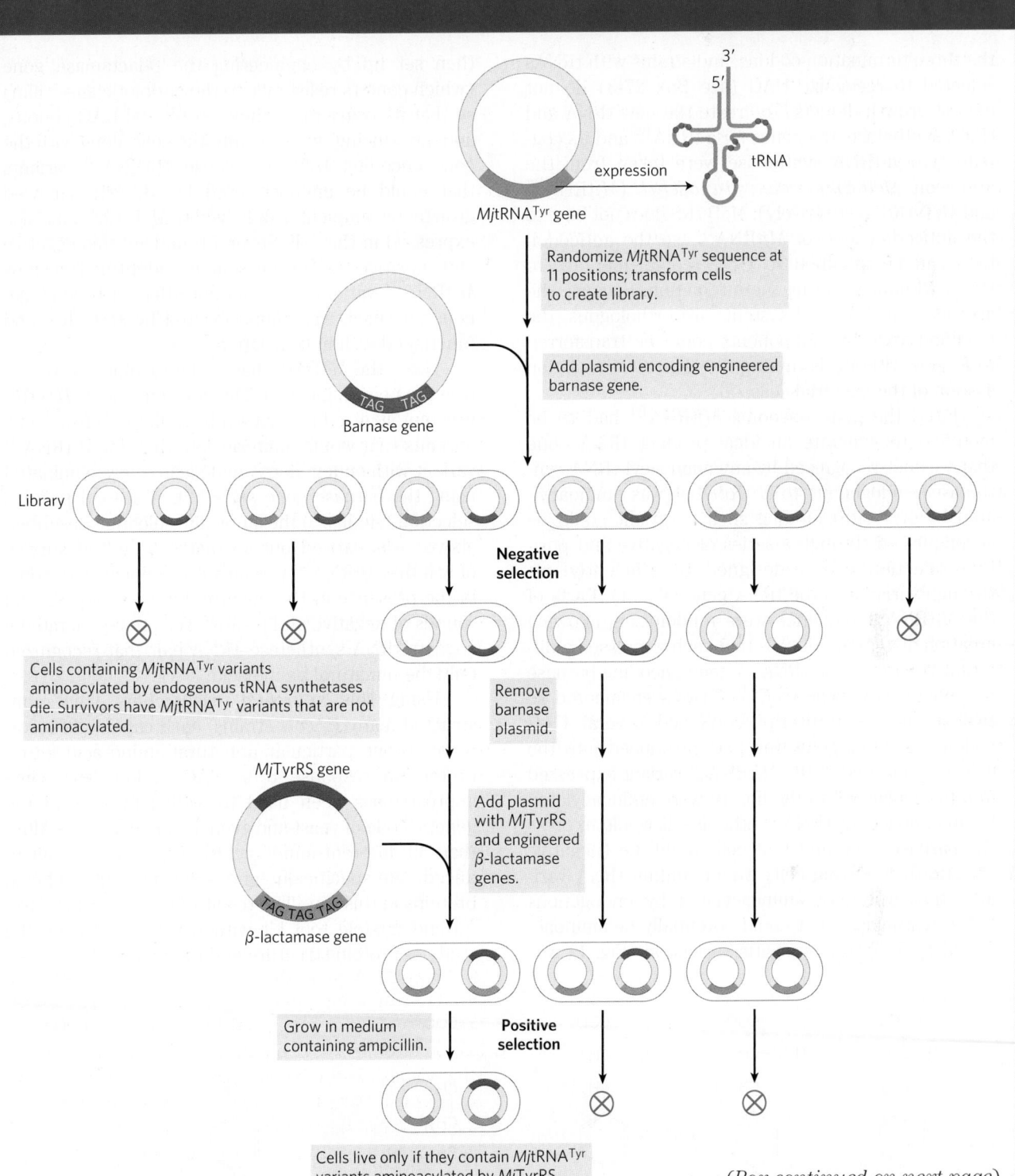

FIGURE 1 Selecting *Mjt*RNA^Tyr variants that function only with the tyrosyl-tRNA synthetase *Mj*TyrRS. The sequence of the gene encoding *Mjt*RNA^Tyr, on a plasmid, is randomized at 11 positions that do not interact with *Mj*TyrRS (red dots). The mutagenized plasmids are introduced into *E. coli* cells to create a library of millions of *Mjt*RNA^Tyr variants, represented by the six cells shown here. The toxic barnase gene, engineered to include the sequence TAG so that its transcript includes UAG codons, is present on a separate plasmid, providing a negative selection. If this gene is expressed, the cells die. It can be successfully expressed only if the *Mjt*RNA^Tyr variant expressed by that particular cell is aminoacylated by endogenous (*E. coli*) aminoacyl-tRNA synthetases, inserting an amino acid instead of stopping translation. Another gene, encoding β-lactamase, and also engineered with TAG sequences to produce UAG stop codons, is provided on yet another plasmid that also expresses the gene encoding *Mj*TyrRS. This serves as a means of positive selection for the remaining *Mjt*RNA^Tyr variants. Those variants that are aminoacylated by *Mj*TyrRS allow expression of the β-lactamase gene, so these cells can grow on ampicillin. Multiple rounds of negative and positive selection yield the best *Mjt*RNA^Tyr variants that are aminoacylated uniquely by *Mj*TyrRS and used efficiently in translation.

(Box continued on next page)

BOX 27-2   Natural and Unnatural Expansion of the Genetic Code (*continued*)

the three termination codons, and strains with tRNAs selected to recognize UAG (see Box 27-3) do not exhibit growth defects. To create the new tRNA and tRNA synthetase, the genes for a tRNA^{Tyr} and its cognate tyrosyl-tRNA synthetase were taken from the archaeon *Methanococcus jannaschii* (*Mj*tRNA^{Tyr} and *Mj*TyrRS, respectively). *Mj*TyrRS does not bind to the anticodon loop of *Mj*tRNA^{Tyr}, so the anticodon loop can be modified to CUA (complementary to UAG) without affecting the interaction. Because the archaeal and bacterial systems are orthologous, the modified archaeal components could be transferred to *E. coli* without disrupting the intrinsic translation system of the bacterial cells.

First, the gene encoding *Mj*tRNA^{Tyr} had to be modified to generate an ideal product tRNA—one that was not recognized by any aminoacyl-tRNA synthetases endogenous to *E. coli* but was aminoacylated by *Mj*TyrRS. Finding such a variant could be accomplished through a series of negative and positive selection cycles designed to efficiently sift through variants of the tRNA gene (Fig. 1). Parts of the *Mj*tRNA^{Tyr} sequence were randomized, allowing creation of a library of cells that each expressed a different version of the tRNA. A gene encoding barnase (a ribonuclease toxic to *E. coli*) was engineered so that its mRNA transcript contained several UAG codons, and this gene was also introduced into the cells on a plasmid. If the *Mj*tRNA^{Tyr} variant expressed in a particular cell in the library were aminoacylated by an endogenous tRNA synthetase, it would express the barnase gene and that cell would die (negative selection). Surviving cells would contain tRNA variants that were not aminoacylated by endogenous tRNA synthetases, but could potentially be aminoacylated by *Mj*TyrRS. A positive selection (Fig. 1) was

then set up by engineering the β-lactamase gene (which confers resistance to the antibiotic ampicillin) so that its transcript contained several UAG codons, and introducing this gene into the cells along with the gene encoding *Mj*TyrRS. Those *Mj*tRNA^{Tyr} variants that could be aminoacylated by *Mj*TyrRS allowed growth on ampicillin only when *Mj*TyrRS was also expressed in the cell. Several rounds of this negative and positive selection scheme identified a new *Mj*tRNA^{Tyr} variant that was not affected by endogenous enzymes, was aminoacylated by *Mj*TyrRS, and functioned well in translation.

Next, the *Mj*TyrRS had to be modified to recognize the new amino acid. The gene encoding *Mj*TyrRS was mutagenized to create a large library of variants. Variants that would aminoacylate the new *Mj*tRNA^{Tyr} variant with endogenous amino acids were eliminated using the barnase gene selection. A second positive selection (similar to the ampicillin selection described above) was carried out so that cells would survive only if the *Mj*tRNA^{Tyr} variant were aminoacylated only in the presence of the unnatural amino acid. Several rounds of negative and positive selection generated a cognate tRNA synthetase–tRNA pair that recognized only the unnatural amino acid.

Using this approach, researchers have constructed many *E. coli* strains, each capable of incorporating one particular unnatural amino acid into a protein in response to a UAG codon. The same approach has been used to artificially expand the genetic code of yeast and even mammalian cells. More than 30 different amino acids (Fig. 2) can be introduced site-specifically and efficiently into cloned proteins in this way. The result is an increasingly useful and flexible tool kit with which to advance the study of protein structure and function.

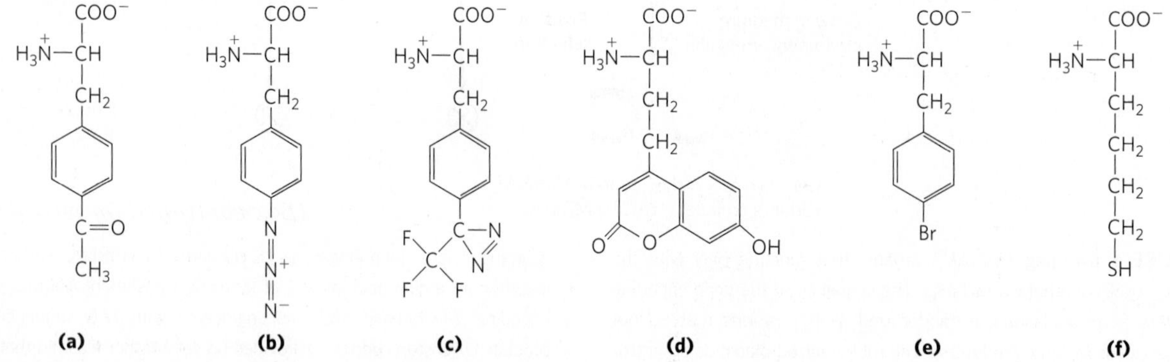

FIGURE 2 A sampling of unnatural amino acids that have been added to the genetic code. These unnatural amino acids add uniquely reactive chemical groups such as **(a)** a ketone; **(b)** an azide; **(c)** a photocrosslinker, a functional group designed to form a covalent bond with a nearby group when activated by light; **(d)** a highly fluorescent amino acid; **(e)** an amino acid with a heavy atom (Br) for use in crystallography; and **(f)** a long-chain cysteine analog that can form extended disulfide bonds. [Source: Information from J. Xie and P. G. Schultz, *Nature Rev. Mol. Cell Biol.* 7:775, 2006.]

In eukaryotic cells, all polypeptides synthesized by cytosolic ribosomes begin with a Met residue (rather than fMet), but, again, the cell uses a specialized initiating tRNA that is distinct from the tRNA^{Met} used at (5′) AUG codons at interior positions in the mRNA. Polypeptides synthesized by mitochondrial and chloroplast ribosomes, however, begin with *N*-formylmethionine. This strongly supports the view that mitochondria and chloroplasts originated from bacterial ancestors that were symbiotically incorporated into precursor eukaryotic cells at an early stage of evolution (see Fig. 1-40).

How can the single (5′)AUG codon determine whether a starting *N*-formylmethionine (or methionine, in eukaryotes) or an interior Met residue is ultimately inserted? The details of the initiation process provide the answer.

The Three Steps of Initiation The **initiation** of polypeptide synthesis in bacteria requires (1) the 30S ribosomal subunit, (2) the mRNA coding for the polypeptide to be made, (3) the initiating fMet-tRNA^{fMet}, (4) a set of three proteins called initiation factors (IF1, IF2, and IF3), (5) GTP, (6) the 50S ribosomal subunit, and (7) Mg^{2+}. Formation of the initiation complex takes place in three steps **(Fig. 27-25)**.

In step ❶, the 30S ribosomal subunit binds two initiation factors, IF1 and IF3. Factor IF3 prevents the 30S and 50S subunits from combining prematurely. The mRNA then binds to the 30S subunit. The initiating (5′) AUG is guided to its correct position by the **Shine-Dalgarno sequence** (named for Australian researchers John Shine and Lynn Dalgarno, who identified it) in the mRNA. This consensus sequence is an initiation signal of four to nine purine residues, 8 to 13 bp to the 5′ side of the initiation codon **(Fig. 27-26a)**. The sequence base-pairs with a complementary pyrimidine-rich sequence near the 3′ end of the 16S rRNA of the 30S ribosomal subunit (Fig. 27-26b). This mRNA-rRNA interaction positions the initiating (5′)AUG sequence of the mRNA in the precise position on the 30S subunit where it is required for initiation of translation. The particular (5′) AUG where fMet-tRNA^{fMet} is to be bound is distinguished from other methionine codons by its proximity to the Shine-Dalgarno sequence in the mRNA.

Bacterial ribosomes have three sites that bind tRNAs, the **aminoacyl (A) site**, the **peptidyl (P) site**, and the **exit (E) site**. The A and P sites bind aminoacyl-tRNAs, whereas the E site binds only uncharged tRNAs that have completed their task on the ribosome. Both the 30S and the 50S subunits contribute to the characteristics of the A and P sites, whereas the E site is largely confined to the 50S subunit. The initiating (5′) AUG is positioned at the P site, the only site to which fMet-tRNA^{fMet} can bind (Fig. 27-25). The fMet-tRNA^{fMet} is the only aminoacyl-tRNA that binds first to the P site; during the subsequent elongation stage, all other incoming aminoacyl-tRNAs (including the Met-tRNA^{Met} that binds to interior AUG codons) bind first to the A site and only subsequently to the P and E sites. The E site is

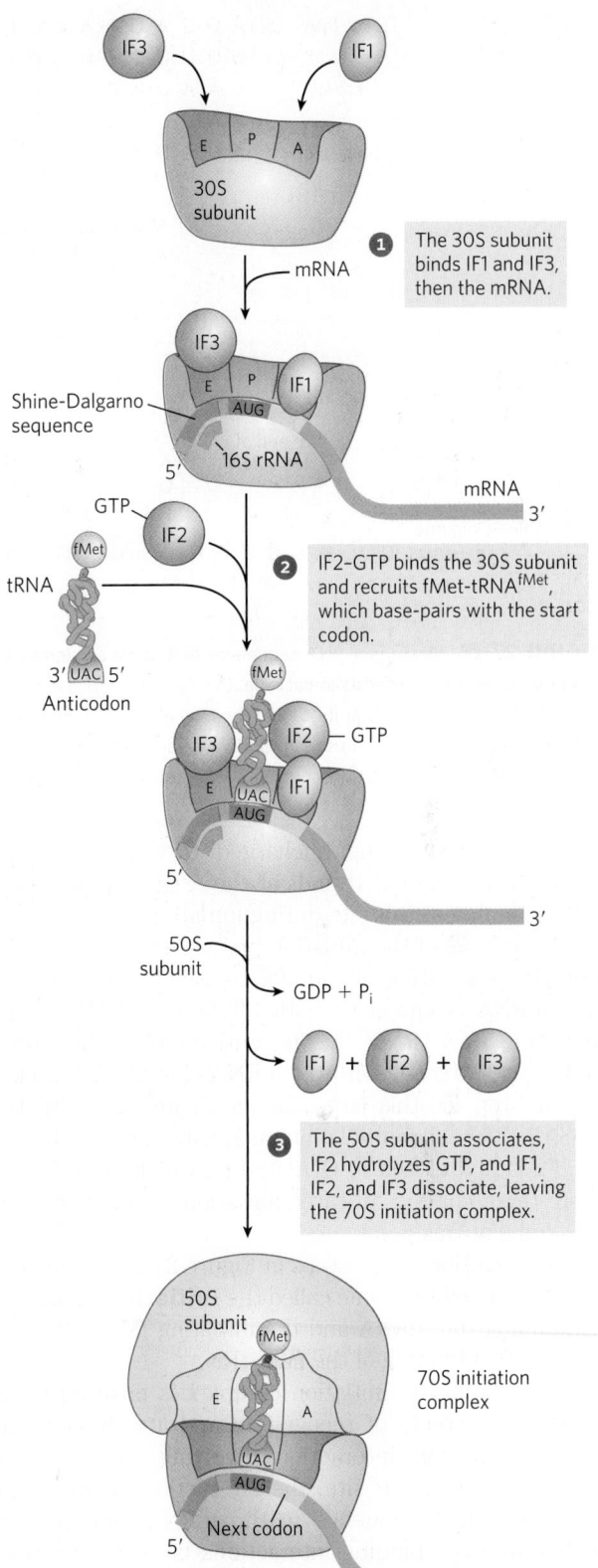

FIGURE 27-25 Formation of the initiation complex in bacteria. The complex forms in three steps (described in the text) at the expense of the hydrolysis of GTP to GDP and P$_i$. IF1, IF2, and IF3 are initiation factors. E designates the exit site; P, the peptidyl site; and A, the aminoacyl site. Here the anticodon of the tRNA is oriented 3′ to 5′, left to right, as in Figure 27-8 but opposite to the orientation in Figures 27-21 and 27-23.

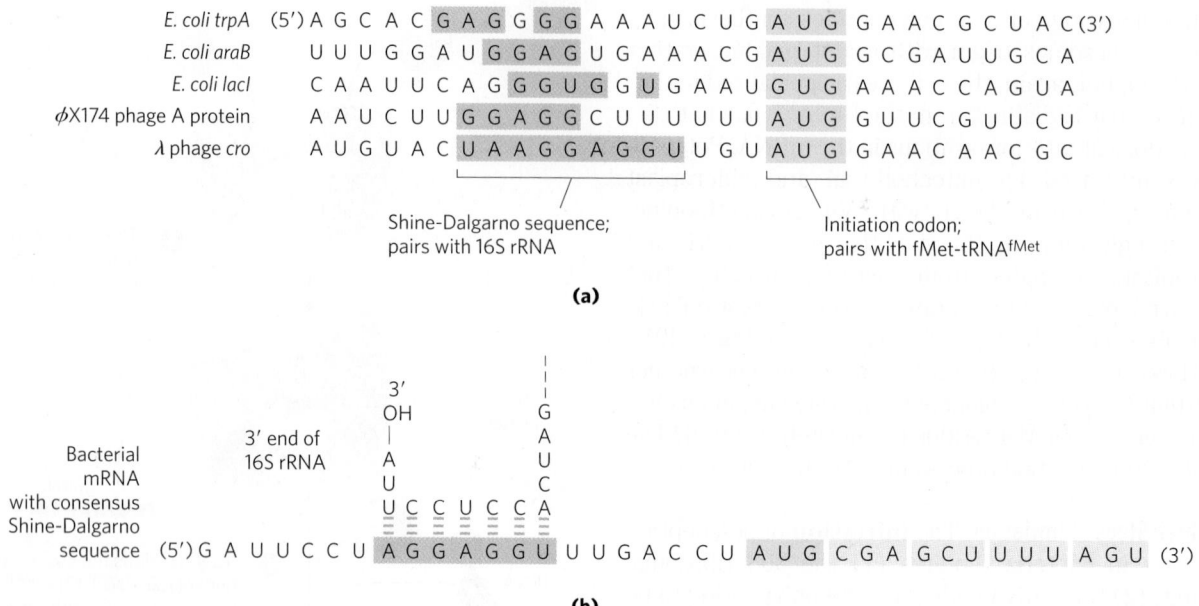

FIGURE 27-26 Messenger RNA sequences that serve as signals for initiation of protein synthesis in bacteria. (a) Alignment of the initiating AUG (shaded in green) at its correct location on the 30S ribosomal subunit depends in part on upstream Shine-Dalgarno sequences (light red). Portions of the mRNA transcripts of five bacterial genes are shown.

Note the unusual example of the *E. coli* LacI protein, which initiates with a GUG (Val) codon (see Box 27-1). In *E. coli*, AUG is the start codon in approximately 91% of genes, with GUG (7%) and UUG (2%) assuming this role more rarely. **(b)** The Shine-Dalgarno sequence of the mRNA pairs with a sequence near the 3′ end of the 16S rRNA.

the site from which the "uncharged" tRNAs leave during elongation. Factor IF1 binds at the A site and prevents tRNA binding at this site during initiation.

In step ❷ of the initiation process (Fig. 27-25), the complex consisting of the 30S ribosomal subunit, IF3, and mRNA is joined by both GTP-bound IF2 and the initiating fMet-tRNA^{fMet}. The anticodon of this tRNA now pairs correctly with the mRNA's initiation codon.

In step ❸, this large complex combines with the 50S ribosomal subunit; simultaneously, the GTP bound to IF2 is hydrolyzed to GDP and P_i, which are released from the complex. All three initiation factors leave the ribosome at this point.

Completion of the steps in Figure 27-25 produces a functional 70S ribosome called the **initiation complex**, containing the mRNA and the initiating fMet-tRNA^{fMet}. The correct binding of the fMet-tRNA^{fMet} to the P site in the complete 70S initiation complex is ensured by at least three points of recognition and attachment: the codon-anticodon interaction involving the initiation AUG fixed in the P site, the interaction between the Shine-Dalgarno sequence in the mRNA and the 16S rRNA, and the binding interactions between the ribosomal P site and the fMet-tRNA^{fMet}. The initiation complex is now ready for elongation.

Initiation in Eukaryotic Cells

Translation is generally similar in eukaryotic and bacterial cells; most of the significant differences are in the number of components and the mechanistic details. The initiation process in eukaryotes is outlined in **Figure 27-27**. Eukaryotic mRNAs are bound to the ribosome as a complex with a number of

specific binding proteins. Eukaryotic cells have at least 12 initiation factors. Initiation factors eIF1A and eIF3 are the functional homologs of the bacterial IF1 and IF3, binding to the 40S subunit in step ❶, blocking tRNA binding to the A site and premature joining of the large and small ribosomal subunits, respectively. The factor eIF1 binds to the E site. The charged initiator tRNA is bound by the initiation factor eIF2, which also has bound GTP. In step ❷, this ternary complex binds to the 40S ribosomal subunit, along with two other proteins involved in later steps, eIF5 (not shown in Fig. 27-27) and eIF5B. This creates a 43S preinitiation complex. The mRNA binds to the eIF4F complex, which, in step ❸, mediates its association with the 43S preinitiation complex. The eIF4F complex is made up of eIF4E (binding to the 5′ cap), eIF4A (an ATPase and RNA helicase), and eIF4G (a linker protein). The eIF4G protein binds to eIF3 and eIF4E to provide the first link between the 43S preinitiation complex and the mRNA. The eIF4G also binds to the poly(A) binding protein (PABP) at the 3′ end of the mRNA, circularizing the mRNA **(Fig. 27-28)** and facilitating the translational regulation of gene expression, as described in Chapter 28.

Addition of the mRNA and its associated factors creates a 48S complex. This complex scans the bound mRNA, starting at the 5′ cap, until an AUG codon is encountered. The scanning process (step ❹ in Fig. 27-27) may be facilitated by the RNA helicase of eIF4A and by another bound factor, eIF4B (not shown in Fig. 27-27), whose precise molecular activity is not understood.

Once the initiating AUG site is encountered, the 60S ribosomal subunit associates with the complex in step ❺,

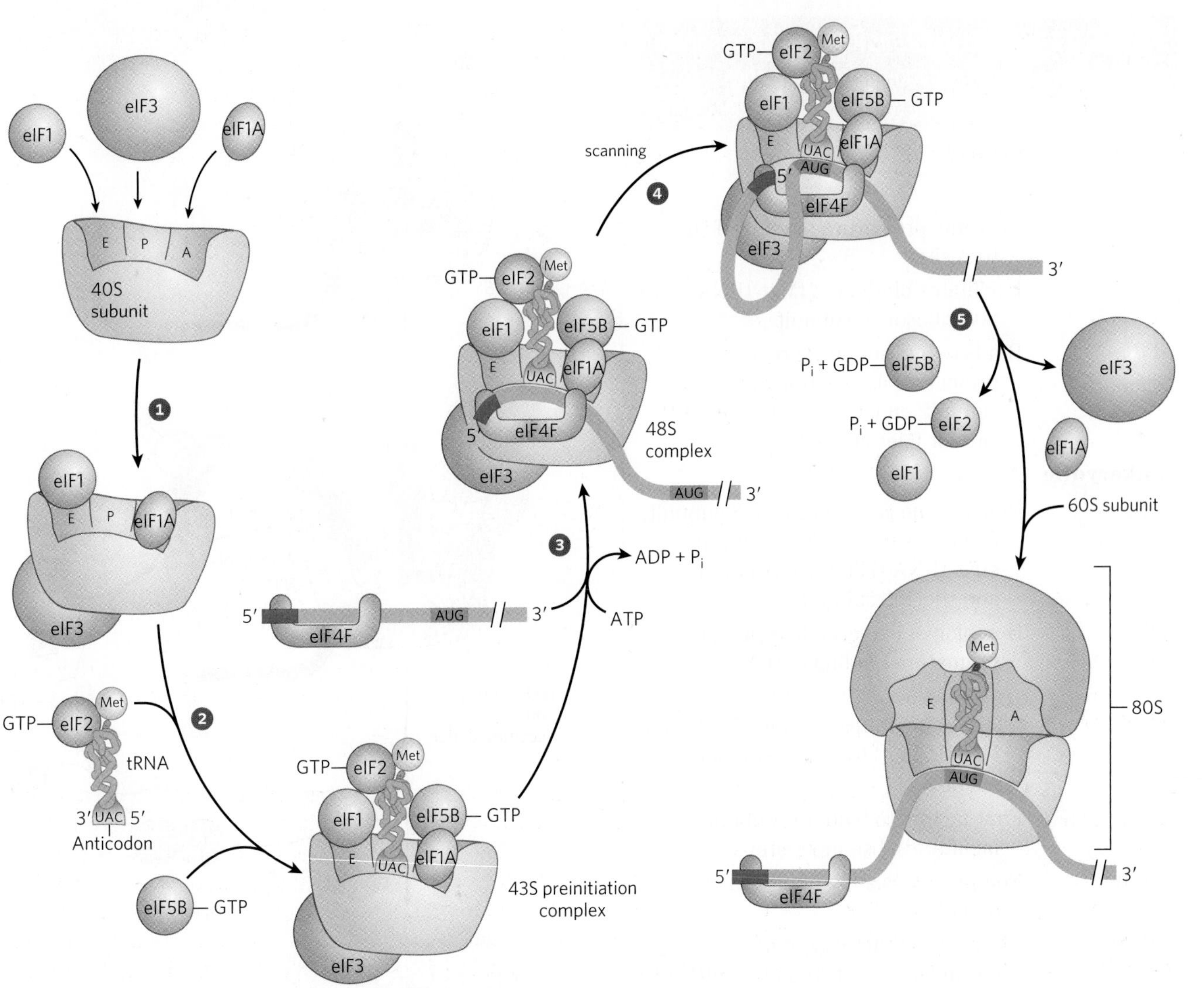

FIGURE 27-27 panel

FIGURE 27-27 Initiation of protein synthesis in eukaryotes. The five steps are described in the text. Eukaryotic initiation factors mediate the association of, first, the charged initiator tRNA to form a 43S preinitiation complex, and then the mRNA (with the 5′ cap shown in red) to form a 48S complex. The final 80S initiation complex is formed as the 60S subunit associates, coupled with release of most of the initiation factors.

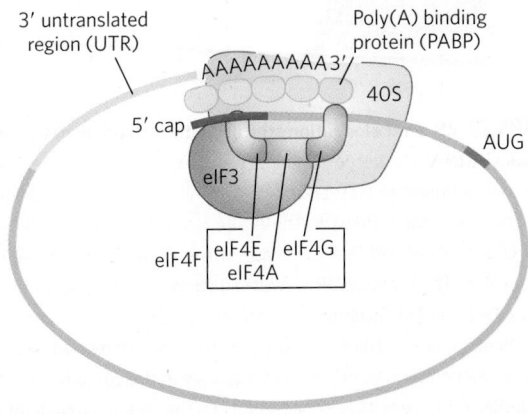

FIGURE 27-28 Circularization of mRNA in the eukaryotic initiation complex. The 3′ and 5′ ends of eukaryotic mRNAs are linked by the eIF4F complex of proteins. The eIF4E subunit binds to the 5′ cap, and the eIF4G protein binds to the poly(A) binding protein (PABP) at the 3′ end of the mRNA. The eIF4G protein also binds to eIF3, linking the circularized mRNA to the 40S subunit of the ribosome.

which is accompanied by release of many of the initiation factors. This requires the activity of eIF5 and eIF5B. The eIF5 protein promotes the GTPase activity of eIF2, producing an eIF2-GDP complex with reduced affinity for the initiator tRNA. The eIF5B protein is homologous to the bacterial IF2. It hydrolyzes its bound GTP and triggers dissociation of eIF2-GDP and other initiation factors, followed closely by association of the 60S subunit. This completes formation of the initiation complex.

The roles of the various bacterial and eukaryotic initiation factors in the overall process are summarized in Table 27-8. The mechanism by which these proteins act is an important area of investigation.

Stage 3: Peptide Bonds Are Formed in the Elongation Stage

The third stage of protein synthesis is **elongation**. Again, we begin with bacterial cells. Elongation requires (1) the

TABLE 27-8	Protein Factors Required for Initiation of Translation in Bacterial and Eukaryotic Cells
Factor	**Function**
Bacterial	
IF1	Prevents premature binding of tRNAs to A site
IF2	Facilitates binding of fMet-tRNA^{fMet} to 30S ribosomal subunit
IF3	Binds to 30S subunit; prevents premature association of 50S subunit; enhances specificity of P site for fMet-rRNA^{fMet}
Eukaryotic	
eIF1	Binds to the E site of the 40S subunit; facilitates interaction between eIF2-tRNA-GTP ternary complex and the 40S subunit
eIF1A	Homolog of bacterial IF1; prevents premature binding of tRNAs to A site
eIF2	GTPase; facilitates binding of initiating Met-tRNA^{Met} to 40S ribosomal subunit
eIF2B[a], eIF3	First factors to bind 40S subunit; facilitate subsequent steps
eIF4F	Complex consisting of eIF4E, eIF4A, and eIF4G
eIF4A	RNA helicase activity; removes secondary structure in the mRNA to permit binding to 40S subunit; part of the eIF4F complex
eIF4B	Binds to mRNA; facilitates scanning of mRNA to locate the first AUG
eIF4E	Binds to the 5' cap of mRNA; part of the eIF4F complex
eIF4G	Binds to eIF4E and to poly (A) binding protein (PABP); part of the eIF4F complex
eIF5[a]	Promotes dissociation of several other initiation factors from 40S subunit as a prelude to association of 60S subunit to form 80S initiation complex
eIF5b	GTPase homologous to bacterial IF2; promotes dissociation of initiation factors before final ribosome assembly

[a]Not shown in Figure 27-27.

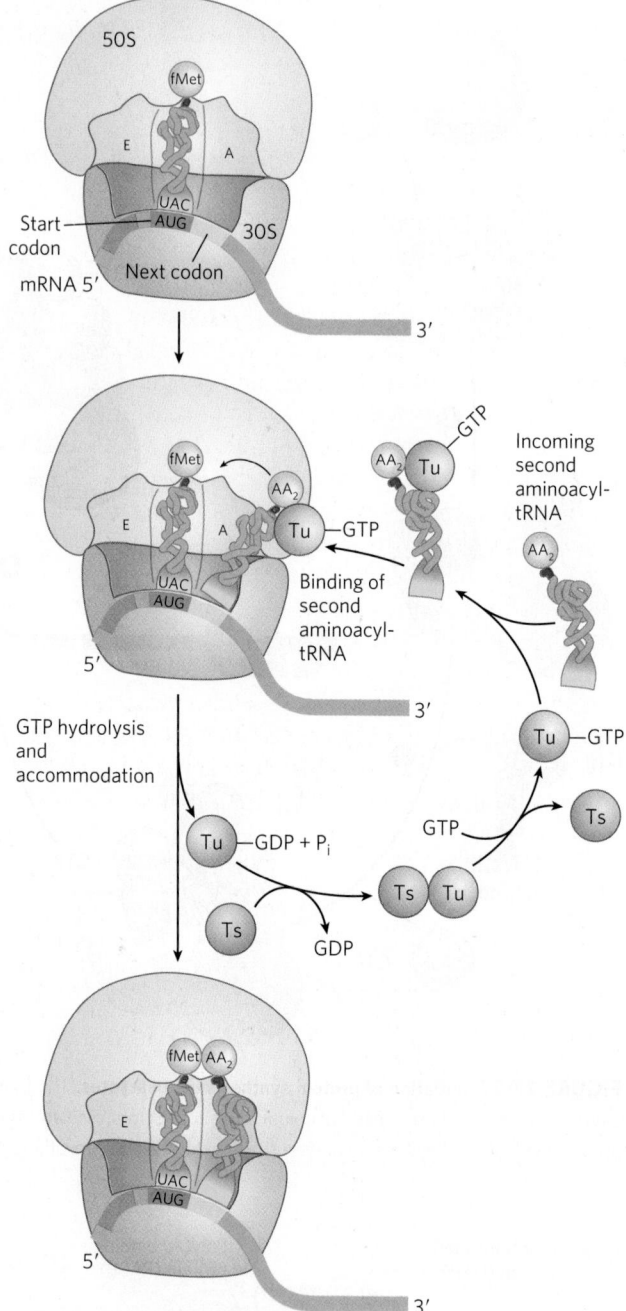

FIGURE 27-29 First elongation step in bacteria: binding of the second aminoacyl-tRNA. The second aminoacyl-tRNA (AA$_2$) enters the A site of the ribosome bound to GTP-bound EF-Tu (shown here as Tu). Binding of the second aminoacyl-tRNA to the A site is accompanied by hydrolysis of the GTP to GDP and P$_i$ and release of the EF-Tu–GDP complex from the ribosome. The bound GDP is released when the EF-Tu–GDP complex binds to EF-Ts, and EF-Ts is subsequently released when another molecule of GTP binds to EF-Tu. This recycles EF-Tu and makes it available to repeat the cycle. "Accommodation" involves a change in the conformation of the second tRNA that pulls its aminoacyl end into the peptidyl transferase site.

initiation complex described above, (2) aminoacyl-tRNAs, (3) a set of three soluble cytosolic proteins called **elongation factors** (EF-Tu, EF-Ts, and EF-G in bacteria), and (4) GTP. Cells use three steps to add each amino

acid residue, and the steps are repeated as many times as there are residues to be added.

Elongation Step 1: Binding of an Incoming Aminoacyl-tRNA In the first step of the elongation cycle **(Fig. 27-29),**

the appropriate incoming aminoacyl-tRNA binds to a complex of GTP-bound EF-Tu. The resulting aminoacyl-tRNA–EF-Tu–GTP complex binds to the A site of the 70S initiation complex. The GTP is hydrolyzed and an EF-Tu–GDP complex is released from the 70S ribosome. The EF-Tu–GTP complex is regenerated in a process requiring EF-Ts and GTP.

Elongation Step 2: Peptide Bond Formation A peptide bond is now formed between the two amino acids bound by their tRNAs to the A and P sites on the ribosome. This occurs by transfer of the initiating *N*-formylmethionyl group from its tRNA to the amino group of the second amino acid, now in the A site **(Fig. 27-30)**. The α-amino group of the amino acid in the A site acts as a nucleophile, displacing the tRNA in the P site to form a peptide bond. This reaction produces a dipeptidyl-tRNA in the A site, and the now "uncharged" (deacylated) tRNA^{fMet} remains bound to the P site. The tRNAs then shift to a hybrid binding state, with elements of each spanning two different sites on the ribosome, as shown in Figure 27-30.

The enzymatic activity that catalyzes peptide bond formation has historically been referred to as **peptidyl transferase** and was widely assumed to be intrinsic to one or more of the proteins in the large ribosomal subunit.

We now know that this reaction is catalyzed by the 23S rRNA, adding to the known catalytic repertoire of ribozymes. This discovery has interesting implications for the evolution of life (Chapter 26).

Elongation Step 3: Translocation In the final step of the elongation cycle, **translocation**, the ribosome moves one codon toward the 3′ end of the mRNA **(Fig. 27-31a)**. This movement shifts the anticodon of the dipeptidyl-tRNA, which is still attached to the second codon of the mRNA, from the A site to the P site, and shifts the deacylated tRNA from the P site to the E site, from where the tRNA is released into the cytosol. The third codon of the mRNA now lies in the A site and the second codon in the P site. Movement of the ribosome along the mRNA requires EF-G (also known as translocase) and the energy provided by hydrolysis of another molecule of GTP. A change in the three-dimensional conformation of the entire ribosome results in its movement along the mRNA. Because the structure of EF-G mimics the structure of the EF-Tu–tRNA complex (Fig. 27-31b), EF-G can bind the A site and, presumably, displace the peptidyl-tRNA.

After translocation, the ribosome, with its attached dipeptidyl-tRNA and mRNA, is ready for the next elongation cycle and attachment of a third amino

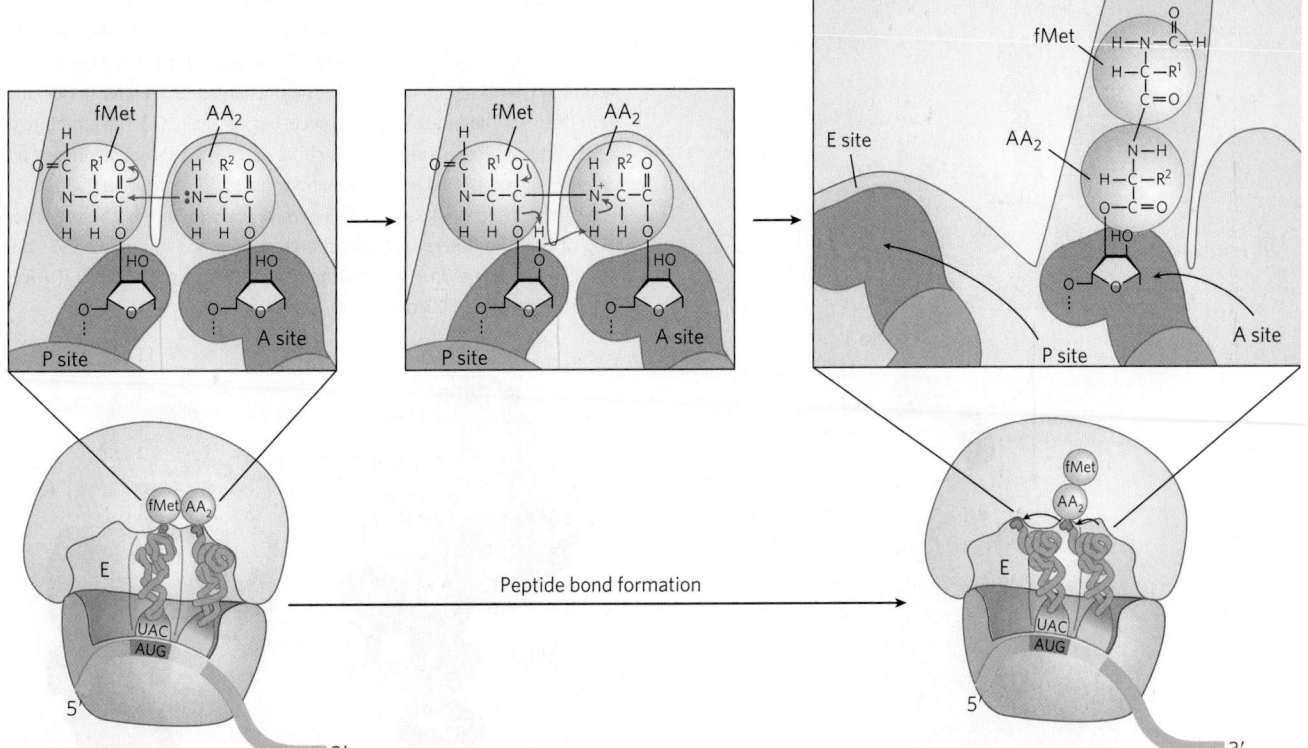

FIGURE 27-30 Second elongation step in bacteria: formation of the first peptide bond. The peptidyl transferase catalyzing this reaction is the 23S rRNA ribozyme. The *N*-formylmethionyl group is transferred to the amino group of the second aminoacyl-tRNA in the A site, forming a dipeptidyl-tRNA. At this stage, both tRNAs bound to the ribosome shift position in the 50S subunit to take up a hybrid binding state. The uncharged tRNA shifts so that its 3′ and 5′ ends are in the E site. Similarly, the 3′ and 5′ ends of the peptidyl-tRNA shift to the P site. The anticodons remain in the P and A sites. Note the involvement of the 2′-hydroxyl group of the 3′-terminal adenosine as a general acid-base catalyst in this reaction.

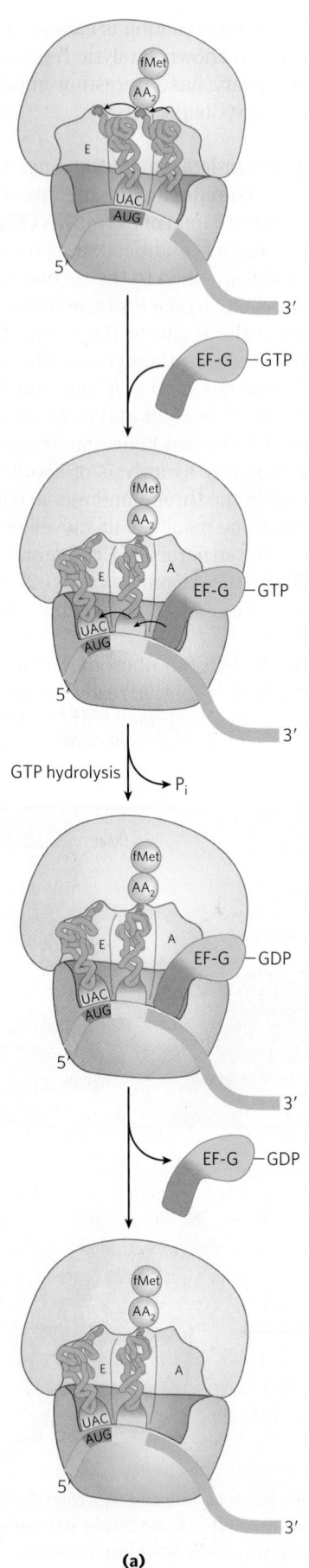

(a)

acid residue. This process occurs in the same way as addition of the second residue (as shown in Figs 27-29, 27-30, and 27-31). For each amino acid residue correctly added to the growing polypeptide, two GTPs are hydrolyzed to GDP and P$_i$ as the ribosome moves from codon to codon along the mRNA toward the 3′ end.

The polypeptide remains attached to the tRNA of the most recent amino acid to be inserted. This association maintains the functional connection between the information in the mRNA and its decoded polypeptide output. At the same time, the ester linkage between this tRNA and the carboxyl terminus of the growing polypeptide activates the terminal carboxyl group for nucleophilic attack by the incoming amino acid to form a new peptide bond (Fig. 27-30). As the existing ester linkage between the polypeptide and tRNA is broken during peptide bond formation, the linkage between the polypeptide and the information in the mRNA persists, because each newly added amino acid is still attached to its tRNA.

The elongation cycle in eukaryotes is similar to that in bacteria. Three eukaryotic elongation factors (eEF1α, eEF1$\beta\gamma$, and eEF2) have functions analogous to those of the bacterial elongation factors (EF-Tu, EF-Ts, and EF-G, respectively). When a new aminoacyl-tRNA binds to the A site, an allosteric interaction leads to ejection of the uncharged tRNA from the E site.

FIGURE 27-31 Third elongation step in bacteria: translocation. (a) The ribosome moves one codon toward the 3′ end of the mRNA, using energy provided by hydrolysis of GTP bound to EF-G (translocase). The dipeptidyl-tRNA is now entirely in the P site, leaving the A site open for an incoming (third) aminoacyl-tRNA. The uncharged tRNA later dissociates from the E site, and the elongation cycle begins again. **(b)** The structure of EF-G mimics the structure of EF-Tu complexed with tRNA. Shown here are (left) EF-Tu complexed with tRNA and (right) EF-G complexed with GDP. The carboxyl-terminal part of EF-G mimics the structure of the anticodon loop of tRNA in both shape and charge distribution. [Sources: (b) (left) PDB ID 1B23, P. Nissen et al., *Structure* 7:143, 1999; (right) PDB ID 1DAR, S. al-Karadaghi et al., *Structure* 4:555, 1996.]

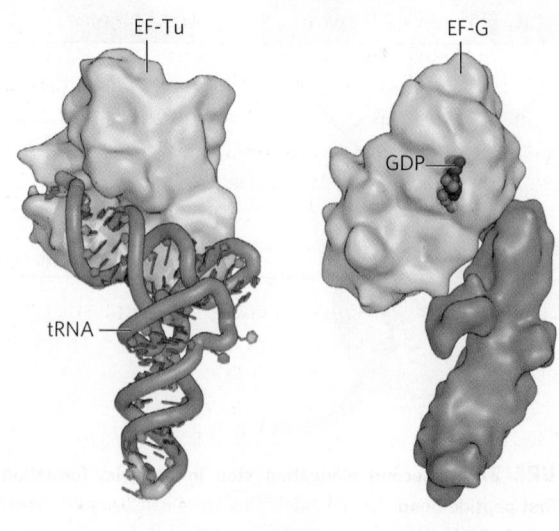

(b)

Proofreading on the Ribosome The GTPase activity of EF-Tu during the first step of elongation in bacterial cells (Fig. 27-29) makes an important contribution to the rate and fidelity of the overall biosynthetic process. Both the EF-Tu–GTP and EF-Tu–GDP complexes exist for a few milliseconds before they dissociate. These two intervals provide opportunities for the codon-anticodon interactions to be proofread. Incorrect aminoacyl-tRNAs normally dissociate from the A site during one of these periods. If the GTP analog guanosine 5′-O-(3-thiotriphosphate) (GTPγS) is used in place of GTP, hydrolysis is slowed, improving the fidelity (by increasing the proofreading intervals) but reducing the rate of protein synthesis.

Guanosine 5′-O-(3-thiotriphosphate)
(GTPγS)

The process of protein synthesis (including the characteristics of codon-anticodon pairing already described) has clearly been optimized through evolution to balance the requirements for speed and fidelity. Improved fidelity might diminish speed, whereas increases in speed would probably compromise fidelity. And, recall that the proofreading mechanism on the ribosome establishes only that the proper codon-anticodon pairing has taken place, not that the correct amino acid is attached to the tRNA. If a tRNA is successfully aminoacylated with the wrong amino acid (as can be done experimentally), this incorrect amino acid is efficiently incorporated into a protein in response to whatever codon is normally recognized by the tRNA.

Stage 4: Termination of Polypeptide Synthesis Requires a Special Signal

Elongation continues until the ribosome adds the last amino acid coded by the mRNA. **Termination**, the fourth stage of polypeptide synthesis, is signaled by the presence of one of three termination codons in the mRNA (UAA, UAG, UGA), immediately following the final coded amino acid. Mutations in a tRNA anticodon that allow an amino acid to be inserted at a termination codon are generally deleterious to the cell (Box 27-3). In bacteria, once a termination codon occupies the ribosomal A site, three **termination factors**, or **release factors**—the proteins RF1, RF2, and RF3—contribute to (1) hydrolysis of the terminal peptidyl-tRNA bond;

BOX 27-3   Induced Variation in the Genetic Code: Nonsense Suppression

When a mutation produces a termination codon in the interior of a gene, translation is prematurely halted and the incomplete polypeptide is usually inactive. These are called nonsense mutations. The gene can be restored to normal function if a second mutation either (1) converts the misplaced termination codon to a codon specifying an amino acid or (2) suppresses the effects of the termination codon. Such restorative mutations are called **nonsense suppressors**; they generally involve mutations in tRNA genes to produce altered (suppressor) tRNAs that can recognize the termination codon and insert an amino acid at that position. Most known suppressor tRNAs have single base substitutions in their anticodons.

Suppressor tRNAs constitute an experimentally induced variation in the genetic code to allow the reading of what are usually termination codons, much like the naturally occurring code variations described in Box 27-1. Nonsense suppression does not completely disrupt normal information transfer in a cell, because the cell usually has several copies of each tRNA gene; some of these duplicate genes are weakly expressed and account for only a minor part of the cellular pool of a particular tRNA. Suppressor mutations usually affect a "minor" tRNA, leaving the major tRNA to read its codon normally.

For example, E. coli has three identical genes for tRNA^{Tyr}, each producing a tRNA with the anticodon (5′)GUA. One of these genes is expressed at relatively high levels, and thus its product represents the major tRNA^{Tyr} species; the other two genes are transcribed in only small amounts. A change in the anticodon of the tRNA product of one of these duplicate tRNA^{Tyr} genes, from (5′)GUA to (5′)CUA, produces a minor tRNA^{Tyr} species that will insert tyrosine at UAG stop codons. This insertion of tyrosine at UAG is carried out inefficiently, but it can produce enough full-length protein from a gene with a nonsense mutation to allow the cell to survive. The major tRNA^{Tyr} continues to translate the genetic code normally for the majority of proteins.

The mutation that leads to creation of a suppressor tRNA does not always occur in the anticodon. The suppression of UGA nonsense codons generally involves the tRNA^{Trp} that normally recognizes UGG. The alteration that allows it to read UGA (and insert Trp residues at these positions) is a G-to-A change at position 24 (in an arm of the tRNA somewhat removed from the anticodon); this tRNA can now recognize both UGG and UGA. A similar change is found in tRNAs affected by the most common naturally occurring variation in the genetic code (UGA = Trp; see Box 27-1).

(2) release of the free polypeptide and the last tRNA, now uncharged, from the P site; and (3) dissociation of the 70S ribosome into its 30S and 50S subunits, ready to start a new cycle of polypeptide synthesis **(Fig. 27-32)**. RF1 recognizes the termination codons UAG and UAA, and RF2 recognizes UGA and UAA. Either RF1 or RF2 (depending on which codon is present) binds at a termination codon and induces peptidyl transferase to transfer the growing polypeptide to a water molecule rather than to another amino acid. The release factors have domains thought to mimic the structure of tRNA, as shown for the elongation factor EF-G in Figure 27-31b. The specific function of RF3 has not been firmly established, although it is thought to release the ribosomal subunit. In eukaryotes, a single release factor, eRF, recognizes all three termination codons.

Ribosome recycling leads to dissociation of the translation components. The release factors dissociate from the posttermination complex (with an uncharged tRNA in the P site) and are replaced by EF-G and a protein called ribosome recycling factor (RRF; M_r 20,300). Hydrolysis of GTP by EF-G leads to dissociation of the 50S subunit from the 30S–tRNA–mRNA complex. EF-G and RRF are replaced by IF3, which promotes dissociation of the tRNA. The mRNA is then released. The complex of IF3 and the 30S subunit is then ready to initiate another round of protein synthesis (Fig. 27-25).

Ribosome Rescue Ribosomes may stall during protein biosynthesis, especially while translating an mRNA that is damaged or incomplete. When the ribosome encounters the end of an mRNA before encountering a stop codon, the translocation step leads to formation of a stable "non-stop complex," in which the A site has no mRNA that can interact with a new charged tRNA. The non-stop complex cannot be recycled by the normal termination factors. Instead, the ribosome is rescued by a process called trans-translation **(Fig. 27-33)**. In virtually all bacteria, the rescue system consists of an RNA called transfer-messenger RNA (tmRNA) and a very small protein, small protein B (SmpB). These bind to the stalled complex in such a way that the tmRNA is positioned in the empty A site so that the ribosome can continue translation until it encounters a stop codon embedded in the tmRNA. The ribosome is then recycled, and both the defective mRNA and the polypeptide translated from it are degraded. Similar systems exist in eukaryotes.

FIGURE 27-32 Termination of protein synthesis in bacteria. Synthesis is terminated in response to a termination codon in the A site. First, a release factor, RF (RF1 or RF2, depending on which termination codon is present), binds to the A site. This leads to hydrolysis of the ester linkage between the nascent polypeptide and the tRNA in the P site and release of the completed polypeptide. Finally, the mRNA, deacylated tRNA, and release factor leave the ribosome, which dissociates into its 30S and 50S subunits, aided by ribosome recycling factor (RRF), IF3, and energy provided by EF-G–mediated GTP hydrolysis. The 30S subunit complex with IF3 is ready to begin another cycle of translation (see Fig. 27-25).

Energy Cost of Fidelity in Protein Synthesis Synthesis of a protein true to the information specified in its mRNA requires energy. Formation of each aminoacyl-tRNA uses two high-energy phosphate groups. An additional ATP is consumed each time an incorrectly activated amino acid is hydrolyzed by the deacylation activity of an aminoacyl-tRNA synthetase as part of its proofreading activity. A GTP is cleaved to GDP and P_i during the first elongation step, and another during the translocation step. Thus, on average, the energy derived from the hydrolysis of more than four NTPs to NDPs is required for the formation of each peptide bond of a polypeptide.

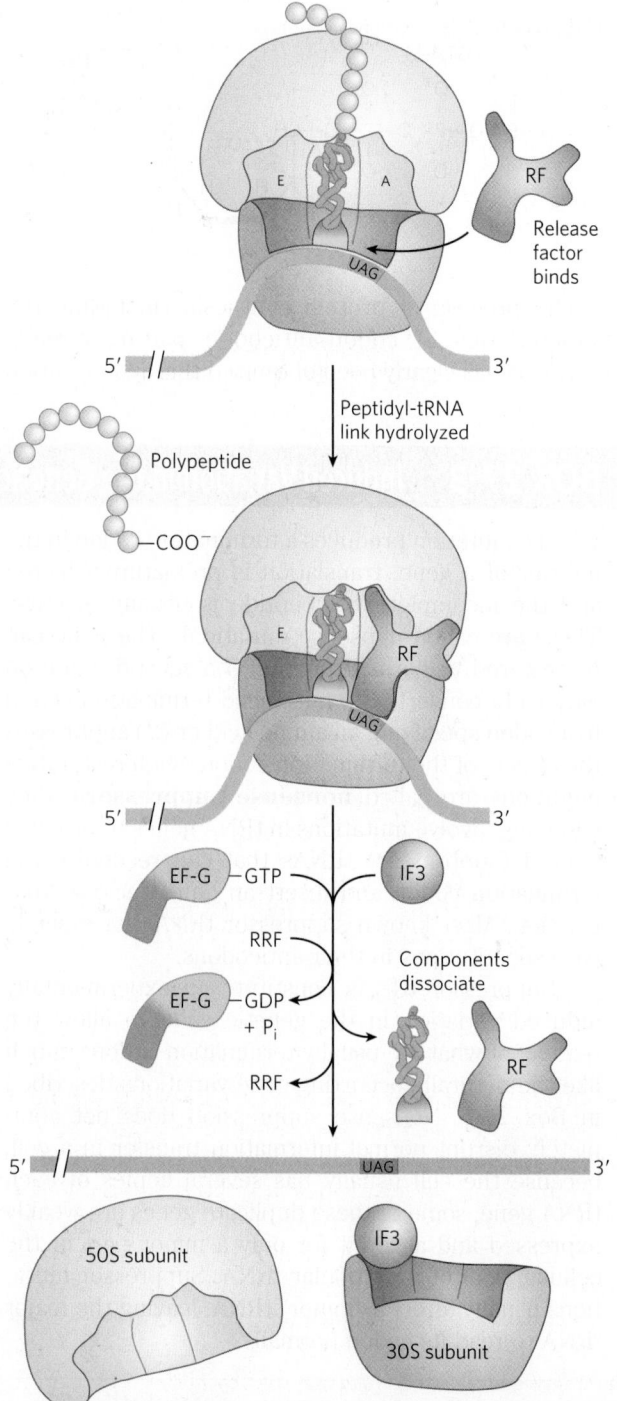

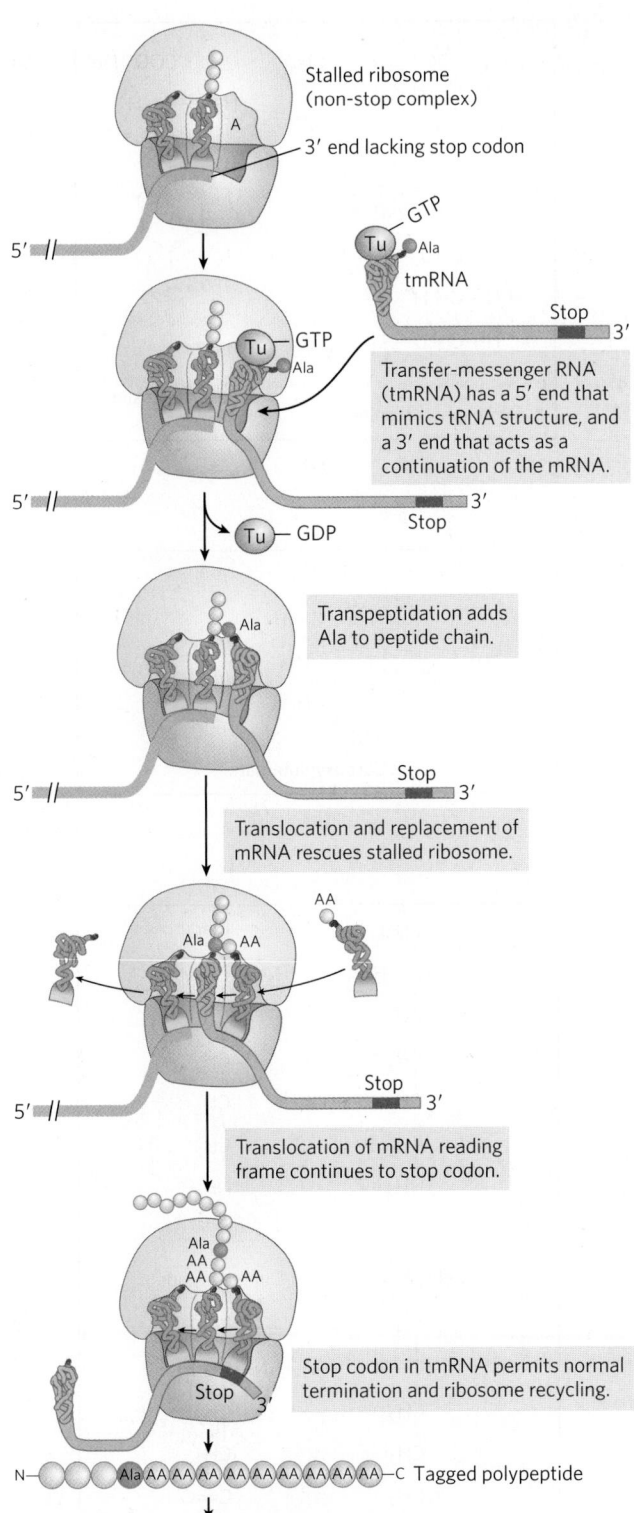

FIGURE 27-33 Rescue of stalled bacterial ribosomes by tmRNA. In bacteria, tmRNA rescues stalled ribosomes by mimicking both tRNA and mRNA. The multistep pathway permits release and degradation of the damaged mRNA and marks the truncated polypeptide for degradation.

This represents an exceedingly large thermodynamic "push" in the direction of synthesis: at least 4×30.5 kJ/mol $= 122$ kJ/mol of phosphodiester bond energy to generate a peptide bond, which has a standard free energy of hydrolysis of only about -21 kJ/mol. The net free-energy change during peptide bond synthesis

is thus -101 kJ/mol. Proteins are information-containing polymers. The biochemical goal is not simply the formation of a peptide bond but the formation of a peptide bond between two *specified* amino acids. Each of the high-energy phosphate compounds expended in this process plays a critical role in maintaining proper alignment between each new codon in the mRNA and its associated amino acid at the growing end of the polypeptide. This energy permits very high fidelity in the biological translation of the genetic message of mRNA into the amino acid sequence of proteins.

Rapid Translation of a Single Message by Polysomes Large clusters of 10 to 100 ribosomes that are very active in protein synthesis can be isolated from both eukaryotic and bacterial cells. Electron micrographs show a fiber between adjacent ribosomes in the cluster, which is called a **polysome (Fig. 27-34a)**. The connecting strand

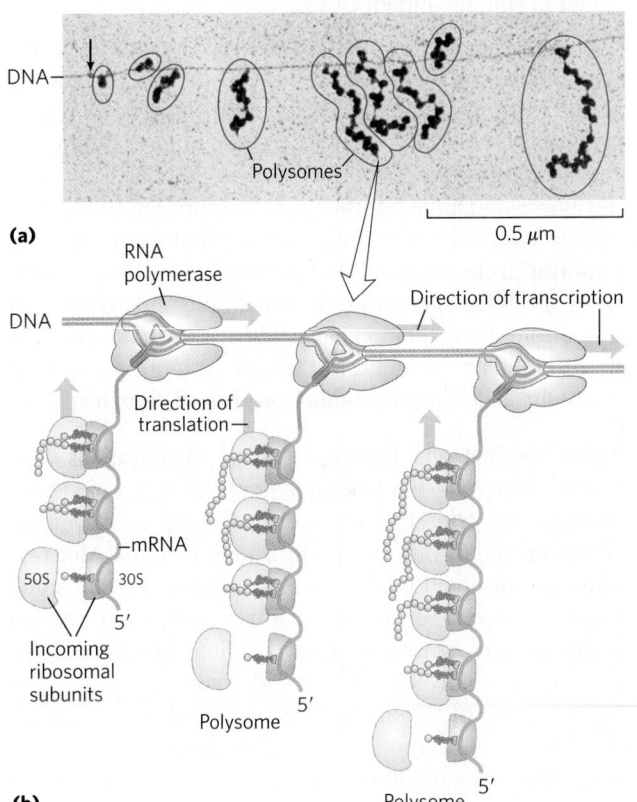

FIGURE 27-34 Coupling of transcription and translation in bacteria.
(a) Electron micrograph of polysomes forming during transcription of a segment of DNA from *E. coli*. Each mRNA is being translated by many ribosomes simultaneously. The nascent polypeptide chains are difficult to see under the conditions used to prepare the sample shown here. The arrow marks the approximate beginning of the gene that is being transcribed. **(b)** Each mRNA is translated by ribosomes while it is still being transcribed from DNA by RNA polymerase. This is possible because the mRNA in bacteria does not have to be transported from a nucleus to the cytoplasm before encountering ribosomes. In this schematic diagram the ribosomes are depicted as smaller than the RNA polymerase. In reality, the ribosomes (M_r 2.7×10^6) are an order of magnitude larger than the RNA polymerase (M_r 3.9×10^5). [Source: (a) O. L. Miller, Jr., et al. *Science* 169:392, 1970, Fig. 3. © 1970 American Association for the Advancement of Science.]

is a single molecule of mRNA that is being translated simultaneously by many closely spaced ribosomes, allowing the highly efficient use of the mRNA.

In bacteria, transcription and translation are tightly coupled. Messenger RNAs are synthesized and translated in the same $5'\rightarrow3'$ direction. Ribosomes begin translating the 5′ end of the mRNA before transcription is complete (Fig. 27-34b). The situation is quite different in eukaryotic cells, where newly transcribed mRNAs must leave the nucleus before they can be translated.

Bacterial mRNAs generally exist for just a few minutes (p. 1062) before they are degraded by nucleases. To maintain high rates of protein synthesis, the mRNA for a given protein or set of proteins must be made continuously and translated with maximum efficiency. The short lifetime of mRNAs in bacteria allows a rapid cessation of synthesis when the protein is no longer needed.

Stage 5: Newly Synthesized Polypeptide Chains Undergo Folding and Processing

In the final stage of protein synthesis, the nascent polypeptide chain is folded and processed into its biologically active form. During or after its synthesis, the polypeptide progressively assumes its native conformation, with the formation of appropriate hydrogen bonds and van der Waals and ionic interactions, and through the hydrophobic effect. Protein chaperones (Chapter 4) play an important role in correct folding in all cells. Some newly made proteins, bacterial, archaeal, and eukaryotic, do not attain their final biologically active conformation until they have been altered by one or more processing reactions called **posttranslational modifications**.

Amino-Terminal and Carboxyl-Terminal Modifications The first residue inserted in all polypeptides is *N*-formylmethionine (in bacteria) or methionine (in eukaryotes). However, the formyl group, the amino-terminal Met residue, and often additional amino-terminal (and, in some cases, carboxyl-terminal) residues may be removed enzymatically in formation of the final functional protein. In as many as 50% of eukaryotic proteins, the amino group of the amino-terminal residue is *N*-acetylated after translation. Carboxyl-terminal residues are also sometimes modified.

Loss of Signal Sequences As we shall see in Section 27.3, the 15 to 30 residues at the amino-terminal end of some proteins play a role in directing the protein to its ultimate destination in the cell. Such **signal sequences** are eventually removed by specific peptidases.

Modification of Individual Amino Acid Residues The hydroxyl groups of certain Ser, Thr, and Tyr residues of some proteins are enzymatically phosphorylated by ATP **(Fig. 27-35a)**; the phosphate groups add negative charges to these polypeptides. The functional significance of this modification varies from one protein to the next. For example, the milk protein casein has

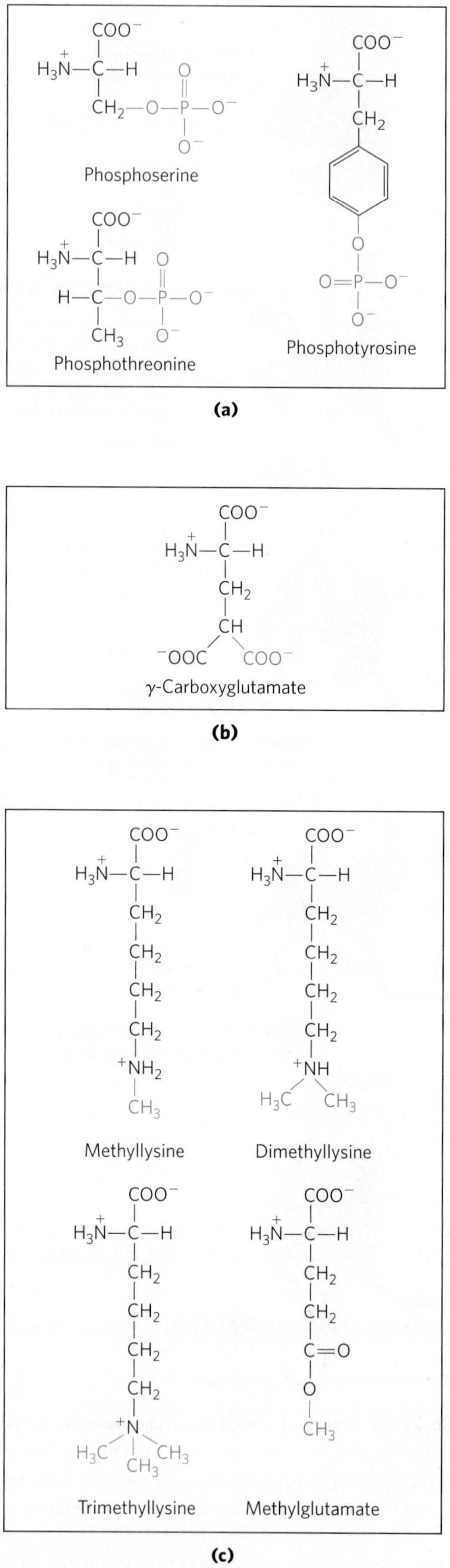

FIGURE 27-35 Some modified amino acid residues. (a) Phosphorylated amino acids. **(b)** A carboxylated amino acid. **(c)** Some methylated amino acids.

many phosphoserine groups that bind Ca^{2+}. Calcium, phosphate, and amino acids are all valuable to suckling young, so casein efficiently provides three essential nutrients. And as we have seen in numerous instances, phosphorylation-dephosphorylation cycles regulate the activity of many enzymes and regulatory proteins.

Extra carboxyl groups may be added to Glu residues of some proteins. For example, the blood-clotting protein prothrombin contains γ-carboxyglutamate residues (Fig. 27-35b) in its amino-terminal region; the γ-carboxyl groups are introduced by an enzyme that requires vitamin K. These carboxyl groups bind Ca^{2+}, which is required to initiate the clotting mechanism.

Monomethyl- and dimethyllysine residues (Fig. 27-35c) occur in some muscle proteins and in cytochrome c. The calmodulin of most species contains one trimethyllysine residue at a specific position. In other proteins, the carboxyl groups of some Glu residues undergo methylation, removing their negative charge.

Attachment of Carbohydrate Side Chains The carbohydrate side chains of glycoproteins are attached covalently during or after synthesis of the polypeptide. In some glycoproteins, the carbohydrate side chain is attached enzymatically to Asn residues (N-linked oligosaccharides), in others to Ser or Thr residues (O-linked oligosaccharides) (see Fig. 7-30). Many proteins that function extracellularly, as well as the lubricating proteoglycans that coat mucous membranes, contain oligosaccharide side chains (see Fig. 7-28).

Addition of Isoprenyl Groups Some eukaryotic proteins are modified by the addition of groups derived from isoprene (isoprenyl groups). A thioether bond is formed between the isoprenyl group and a Cys residue of the protein (see Fig. 11-13). The isoprenyl groups are derived from pyrophosphorylated intermediates of the cholesterol biosynthetic pathway (see Fig. 21-35), such as farnesyl pyrophosphate **(Fig. 27-36)**. Proteins modified in this way include the Ras proteins (small G proteins), which are products of the *ras* oncogenes and proto-oncogenes, and the trimeric G proteins (both discussed in Chapter 12), as well as lamins, proteins found in the nuclear matrix. The isoprenyl group helps to anchor the protein in a membrane. The transforming (carcinogenic) activity of the *ras* oncogene is lost when isoprenylation of the Ras protein is blocked, a finding that has stimulated interest in identifying inhibitors of this posttranslational modification pathway for use in cancer chemotherapy.

Addition of Prosthetic Groups Many proteins require for their activity covalently bound prosthetic groups. Two examples are the biotin molecule of acetyl-CoA carboxylase and the heme group of hemoglobin or cytochrome c.

Proteolytic Processing Many proteins are initially synthesized as large, inactive precursor polypeptides that are proteolytically trimmed to form their smaller, active

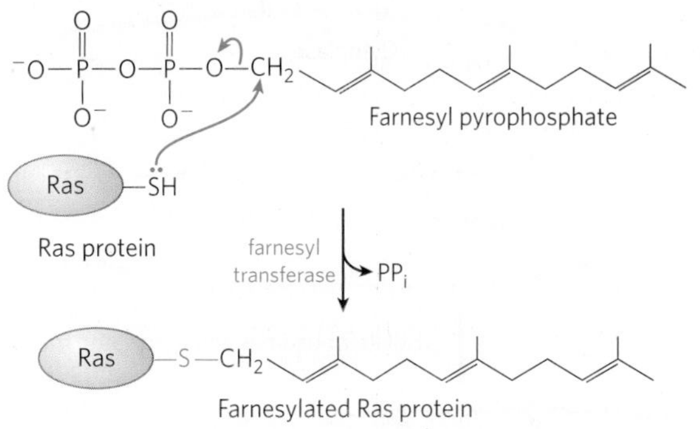

FIGURE 27-36 Farnesylation of a Cys residue. The thioether linkage is shown in red. The Ras protein is the product of the *ras* oncogene.

forms. Examples include proinsulin, some viral proteins, and proteases such as chymotrypsinogen and trypsinogen (see Fig. 6-39).

Formation of Disulfide Cross-Links After folding into their native conformations, some proteins form intrachain or interchain disulfide bridges between Cys residues. In eukaryotes, disulfide bonds are common in proteins to be exported from cells. The cross-links formed in this way help to protect the native conformation of the protein molecule from denaturation in the extracellular environment, which can differ greatly from intracellular conditions and is generally oxidizing.

Ribosome Profiling Provides a Snapshot of Cellular Translation

Modern DNA sequencing methods can be applied in a variety of creative ways to allow researchers to study information pathways. One application, called **ribosome profiling**, identifies the mRNA sequences that are being translated at a particular moment in a cell **(Fig. 27-37)**. Researchers harvest the cells or tissue and rapidly isolate the ribosomes, with their bound mRNAs. Ribonucleases are used to remove any RNA that is not bound by the ribosomes, and the RNA bound and protected by the ribosomes is then isolated. This RNA is converted into DNA by reverse transcriptase (see Chapter 26), then subjected to deep sequencing (p. 310). Not only do the resulting sequences reveal the parts of genes that are being translated at a particular moment, but the relative number of reads for each segment indicates the relative proportions in which the segments are being translated. As an example, all of the eight different subunits of the bacterial F_oF_1-ATPase (see Fig. 19-25) are encoded in a single operon and translated from a single, polycistronic mRNA. We might assume, then, that all the subunit proteins, translated from this mRNA, would be synthesized at similar levels. But the subunits are not present in equal numbers in the final complex; the bacterial F_oF_1-ATPase has 10 c subunits, 3 each of the α and β subunits, and 1 or 2 of the other subunits. Ribosome profiling shows that the eight genes within the mRNA are translated in the ratio 1:1:1:1:2:3:3:10. Translation of

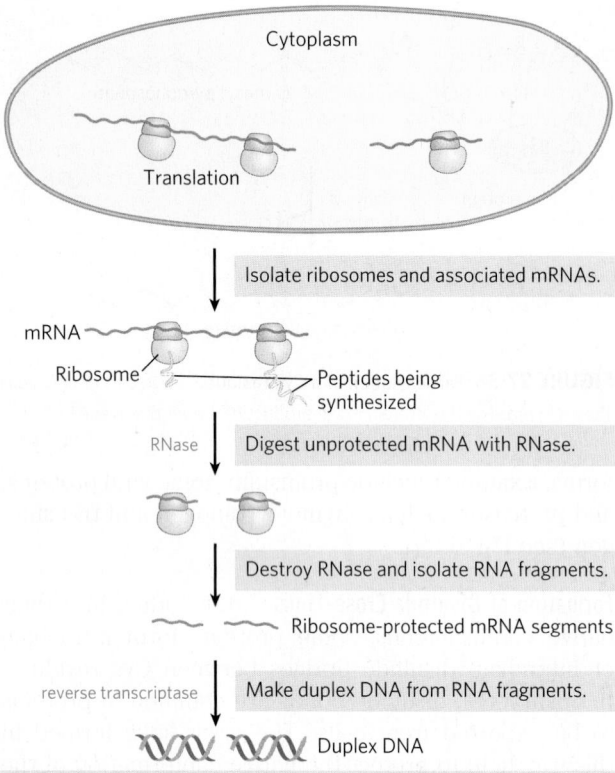

Cytoplasm

Translation

Isolate ribosomes and associated mRNAs.

mRNA

Ribosome

Peptides being synthesized

RNase → Digest unprotected mRNA with RNase.

Destroy RNase and isolate RNA fragments.

Ribosome-protected mRNA segments

reverse transcriptase → Make duplex DNA from RNA fragments.

Duplex DNA

Deep DNA sequencing reveals mRNA segments undergoing translation.

FIGURE 27-37 Ribosome profiling. This technique makes use of modern DNA sequencing methods to determine the mRNAs in a cell that are being translated at a given time. After ribosomes and associated RNA are isolated, ribonucleases are used to remove all RNA that is not bound to and thus protected by the ribosomes. The protected RNA segments are then separated from the ribosomes and converted to DNA by reverse transcriptase, and the DNA is subjected to deep sequencing (see Chapter 8).

the individual genes is thus adjusted to produce the amounts of each subunit that correspond precisely to the subunit ratios in the final complex. Similar results have been obtained for many different macromolecular protein complexes in a variety of organisms.

Protein Synthesis Is Inhibited by Many Antibiotics and Toxins

Protein synthesis is a central function in cellular physiology and is the primary target of many naturally occurring antibiotics and toxins. Except as noted otherwise, these antibiotics inhibit protein synthesis in bacteria. The differences between bacterial and eukaryotic protein synthesis, though in some cases subtle, are such that most of the compounds discussed below are relatively harmless to eukaryotic cells. Natural selection has favored the evolution of compounds that exploit minor differences in order to affect bacterial systems selectively, so that these biochemical weapons are synthesized by some microorganisms and are extremely toxic to others. Because nearly every step in protein synthesis can be specifically inhibited by one antibiotic or another, antibiotics have become valuable tools in the study of protein biosynthesis.

Puromycin, made by the mold *Streptomyces alboniger*, is one of the best-understood inhibitory antibiotics. Its structure is very similar to the 3′ end of an aminoacyl-tRNA, enabling it to bind to the ribosomal A site and participate in peptide bond formation, producing peptidylpuromycin **(Fig. 27-38)**. However, because puromycin resembles only the 3′ end of the tRNA, it does not engage in translocation and dissociates from the ribosome shortly after it is linked to the carboxyl terminus of the peptide. This prematurely terminates polypeptide synthesis.

Tetracyclines inhibit protein synthesis in bacteria by blocking the A site on the ribosome, preventing the binding of aminoacyl-tRNAs. **Chloramphenicol** inhibits protein synthesis by bacterial (and mitochondrial and chloroplast) ribosomes by blocking peptidyl transfer; it does not affect cytosolic protein synthesis in eukaryotes. Conversely, **cycloheximide** blocks the peptidyl transferase of 80S eukaryotic ribosomes but not that of 70S bacterial (and mitochondrial and chloroplast) ribosomes. **Streptomycin**, a basic trisaccharide, causes misreading of the genetic code (in bacteria) at relatively low concentrations and inhibits initiation at higher concentrations.

Tetracycline

Chloramphenicol

Cycloheximide

Streptomycin

Several other inhibitors of protein synthesis are notable because of their toxicity to humans and other mammals. **Diphtheria toxin** (M_r 58,330) catalyzes the ADP-ribosylation of a diphthamide (a modified histidine) residue of eukaryotic elongation factor eEF2, thereby inactivating it. **Ricin** (M_r 29,895), an extremely toxic protein of the castor bean, inactivates the 60S subunit of eukaryotic ribosomes by depurinating a specific adenosine residue in 28S rRNA. Ricin was used in the infamous 1978 murder of BBC journalist and Bulgarian dissident Georgi Markov, presumably by the Bulgarian secret police. Using a syringe hidden at the end of an umbrella, a member of the secret police injected Markov in the leg with a ricin-infused pellet. He died four days later.

SUMMARY 27.2 Protein Synthesis

■ Protein synthesis occurs on the ribosomes, which consist of protein and rRNA. Bacteria have 70S ribosomes, with a large (50S) and a small (30S) subunit. Eukaryotic ribosomes are significantly larger (80S) and contain more proteins.

■ Transfer RNAs have 73 to 93 nucleotide residues, some of which have modified bases. Each tRNA has an amino acid arm with the terminal sequence CCA(3′) to which an amino acid is esterified, an anticodon arm, a TψC arm, and a D arm; some tRNAs have a fifth arm. The anticodon is responsible for the specificity of interaction between the aminoacyl-tRNA and the complementary mRNA codon.

■ The growth of polypeptides on ribosomes begins with the amino-terminal amino acid and proceeds by successive additions of new residues to the carboxyl-terminal end.

■ Protein synthesis occurs in five stages.

1. Amino acids are activated by specific aminoacyl-tRNA synthetases in the cytosol. These enzymes catalyze the formation of aminoacyl-tRNAs, with simultaneous cleavage of ATP to AMP and PP$_i$. The fidelity of protein synthesis depends on the accuracy of this reaction, and some of these enzymes carry out proofreading steps at separate active sites.

2. In bacteria, the initiating aminoacyl-tRNA in all proteins is N-formylmethionyl-tRNA^{fMet}. Initiation of protein synthesis involves formation of a complex between the 30S ribosomal subunit, mRNA, GTP, fMet-tRNA^{fMet}, three initiation factors, and the 50S subunit; GTP is hydrolyzed to GDP and P$_i$.

3. In the elongation steps, GTP and elongation factors are required for binding the incoming aminoacyl-tRNA to the A site on the ribosome. In the first peptidyl transfer reaction, the fMet residue is transferred to the amino group of the incoming aminoacyl-tRNA. Movement of the ribosome along

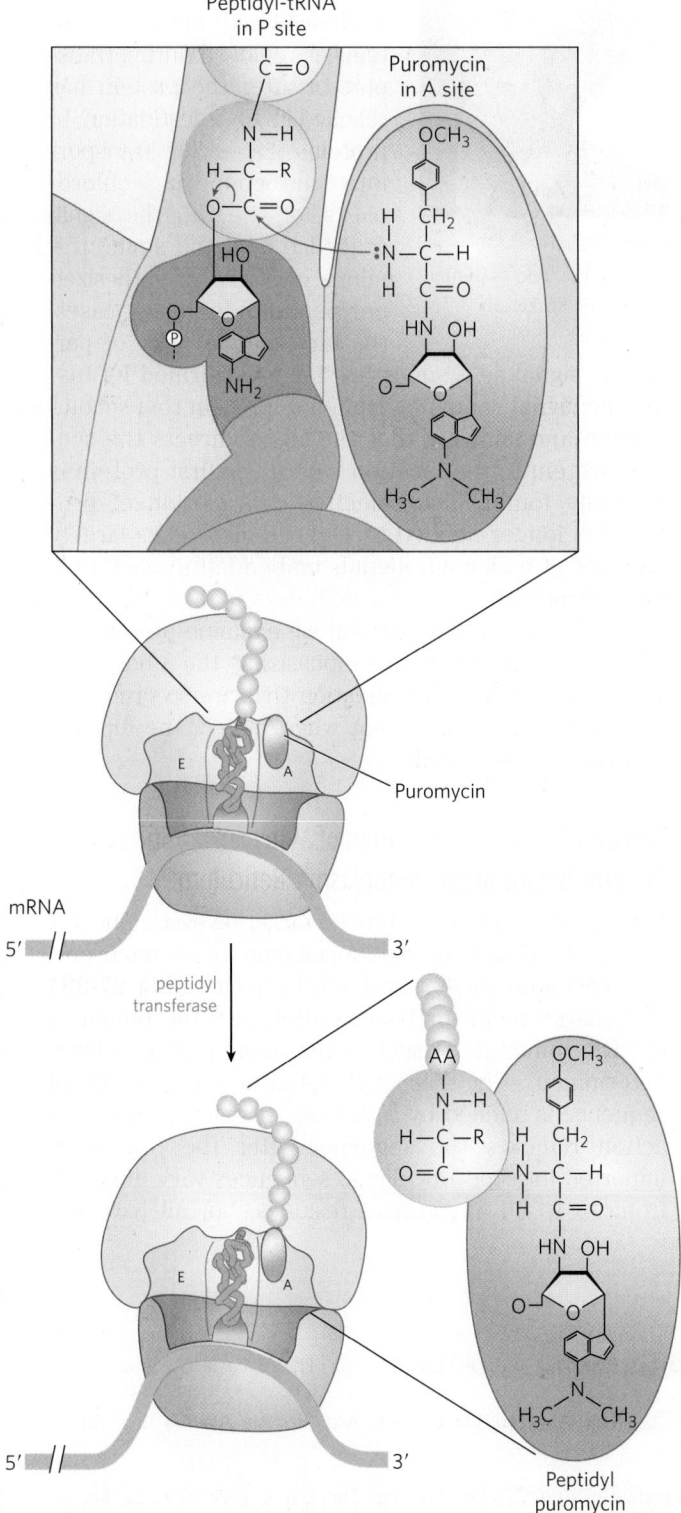

FIGURE 27-38 Disruption of peptide bond formation by puromycin. The antibiotic puromycin resembles the aminoacyl end of a charged tRNA, and it can bind to the ribosomal A site and participate in peptide bond formation. The product of this reaction, peptidyl puromycin, is not translocated to the P site. Instead, it dissociates from the ribosome, causing premature chain termination.

the mRNA then translocates the dipeptidyl-tRNA from the A site to the P site, a process requiring hydrolysis of GTP. Deacylated tRNAs dissociate from the ribosomal E site.

4. After many such elongation cycles, synthesis of the polypeptide is terminated with the aid of release factors. At least four high-energy phosphate equivalents (from ATP and GTP) are required to generate each peptide bond, an energy investment required to guarantee fidelity of translation.

5. Polypeptides fold into their active, three-dimensional forms. Many proteins are further processed by posttranslational modification reactions.

■ Ribosome profiling allows investigators to determine which gene sequences are being translated at any particular moment.

■ Many well-studied antibiotics and toxins inhibit some aspect of protein synthesis.

27.3 Protein Targeting and Degradation

The eukaryotic cell is made up of many structures, compartments, and organelles, each with specific functions that require distinct sets of proteins and enzymes. These proteins (with the exception of those produced in mitochondria and plastids) are synthesized on ribosomes in the cytosol, so how are they directed to their final cellular destinations?

We are now beginning to understand this complex and fascinating process. Proteins destined for secretion, integration in the plasma membrane, or inclusion in lysosomes generally share the first few steps of a pathway that begins in the endoplasmic reticulum. Proteins destined for mitochondria, chloroplasts, or the nucleus use three separate mechanisms. And proteins destined for the cytosol simply remain where they are synthesized.

The most important element in many of these targeting pathways is a short sequence of amino acids called a **signal sequence**, whose function was first

Günter Blobel
[Source: Courtesy of Günter Blobel, The Rockefeller University.]

postulated by Günter Blobel and colleagues in 1970. The signal sequence directs a protein to its appropriate location in the cell and, for many proteins, is removed during transport or after the protein has reached its final destination. In proteins slated for transport into mitochondria, chloroplasts, or the ER, the signal sequence is at the amino terminus of a newly synthesized polypeptide. In many cases, the targeting capacity of particular signal sequences has been confirmed by fusing the signal sequence from one protein to a second protein and showing that the signal directs the second protein to the location where the first protein is normally found. The selective degradation of proteins no longer needed by the cell also relies largely on a set of molecular signals embedded in each protein's structure.

In this concluding section we examine protein targeting and degradation, emphasizing the underlying signals and molecular regulation that are so crucial to cellular metabolism. Except where noted, the focus is now on eukaryotic cells.

Posttranslational Modification of Many Eukaryotic Proteins Begins in the Endoplasmic Reticulum

Perhaps the best-characterized targeting system begins in the ER. Most lysosomal, membrane, or secreted proteins have an amino-terminal signal sequence **(Fig. 27-39)** that marks them for translocation into the lumen of the ER; hundreds of such signal sequences have been determined. The carboxyl terminus of the signal sequence is defined by a cleavage site, where protease action removes the sequence after the protein is imported into the ER. Signal sequences vary in length from 13 to 36 amino acid residues, but all have the

Human influenza
virus A
Met Lys Ala Lys Leu Leu Val Leu Leu Tyr Ala Phe Val Ala Gly Asp Gln --

Human
preproinsulin
Met Ala Leu Trp Met Arg Leu Leu Pro Leu Leu Ala Leu Leu Ala Leu Trp Gly Pro Asp Pro Ala Ala Ala Phe Val --

Bovine growth hormone
Met Met Ala Ala Gly Pro Arg Thr Ser Leu Leu Leu Ala Phe Ala Leu Leu Cys Leu Pro Trp Thr Gln Val Val Gly Ala Phe --

Bee promellitin
Met Lys Phe Leu Val Asn Val Ala Leu Val Phe Met Val Val Tyr Ile Ser Tyr Ile Tyr Ala Ala Pro --

Drosophila glue protein
Met Lys Leu Leu Val Val Ala Val Ile Ala Cys Met Leu Ile Gly Phe Ala Asp Pro Ala Ser Gly Cys Lys --

cleavage site

FIGURE 27-39 Amino-terminal signal sequences of some eukaryotic proteins that direct their translocation into the ER. The hydrophobic core (yellow) is preceded by one or more basic residues (blue). Polar and short-side-chain residues immediately precede (to the left, as shown here) the cleavage sites (indicated by red arrows).

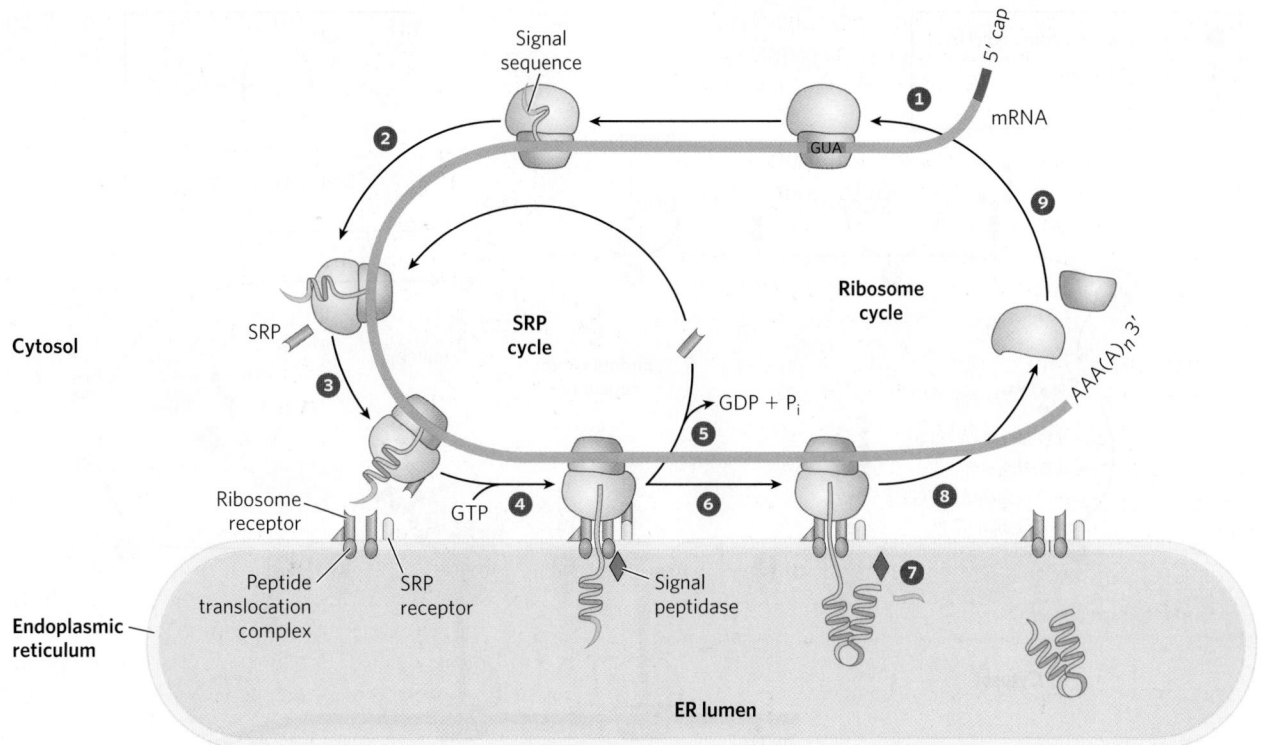

FIGURE 27-40 Directing eukaryotic proteins with the appropriate signals to the endoplasmic reticulum. This process involves the SRP cycle and the translocation and cleavage of the nascent polypeptide. The steps are described in the text. SRP is a rod-shaped complex containing a 300 nucleotide RNA (7SL-RNA) and six different proteins (combined M_r 325,000). One protein subunit of SRP binds directly to the signal sequence, obstructing elongation by sterically blocking the entry of aminoacyl-tRNAs and inhibiting peptidyl transferase. Another protein subunit binds and hydrolyzes GTP. The SRP receptor is a heterodimer of α (M_r 69,000) and β (M_r 30,000) subunits, both of which bind and hydrolyze multiple GTP molecules during this process.

following features: (1) about 10 to 15 hydrophobic amino acid residues; (2) one or more positively charged residues, usually near the amino terminus, preceding the hydrophobic sequence; and (3) a short sequence at the carboxyl terminus (near the cleavage site) that is relatively polar, typically having amino acid residues with short side chains (especially Ala) at the positions closest to the cleavage site.

As originally demonstrated by George Palade, proteins with these signal sequences are synthesized on ribosomes attached to the ER. The signal sequence itself helps to direct the ribosome to the ER, as illustrated in **Figure 27-40**. The targeting pathway begins in step **1**, with initiation of protein synthesis on free ribosomes. The signal sequence appears early in the synthetic process (step **2**), because it is at the amino terminus, which, as we have seen, is synthesized first. As it emerges from the ribosome (step **3**), the signal sequence—and the ribosome itself—is bound by the large **signal recognition particle (SRP)**; SRP then binds GTP

and halts elongation of the polypeptide when it is about 70 amino acids long and the signal sequence has completely emerged from the ribosome. In step **4**, the GTP-bound SRP directs the ribosome (still bound to the mRNA) and the incomplete polypeptide to GTP-bound SRP receptors in the cytosolic face of the ER; the nascent polypeptide is delivered to a **peptide translocation complex** in the ER, which interacts directly with the ribosome. In step **5**, SRP dissociates from the ribosome, accompanied by hydrolysis of GTP in both SRP and the SRP receptor. Elongation of the polypeptide now resumes (step **6**), with the ATP-driven translocation complex feeding the growing polypeptide into the ER lumen until the complete protein has been synthesized. In step **7**, the signal sequence is removed by a signal peptidase within the ER lumen. The ribosome dissociates (step **8**) and is recycled (step **9**).

Glycosylation Plays a Key Role in Protein Targeting

In the ER lumen, newly synthesized proteins are further modified in several ways. Following the removal of signal sequences, polypeptides are folded, disulfide bonds formed, and many proteins glycosylated to form glycoproteins. In many glycoproteins, the linkage to their

George Palade, 1912–2008
[Source: AP Photo.]

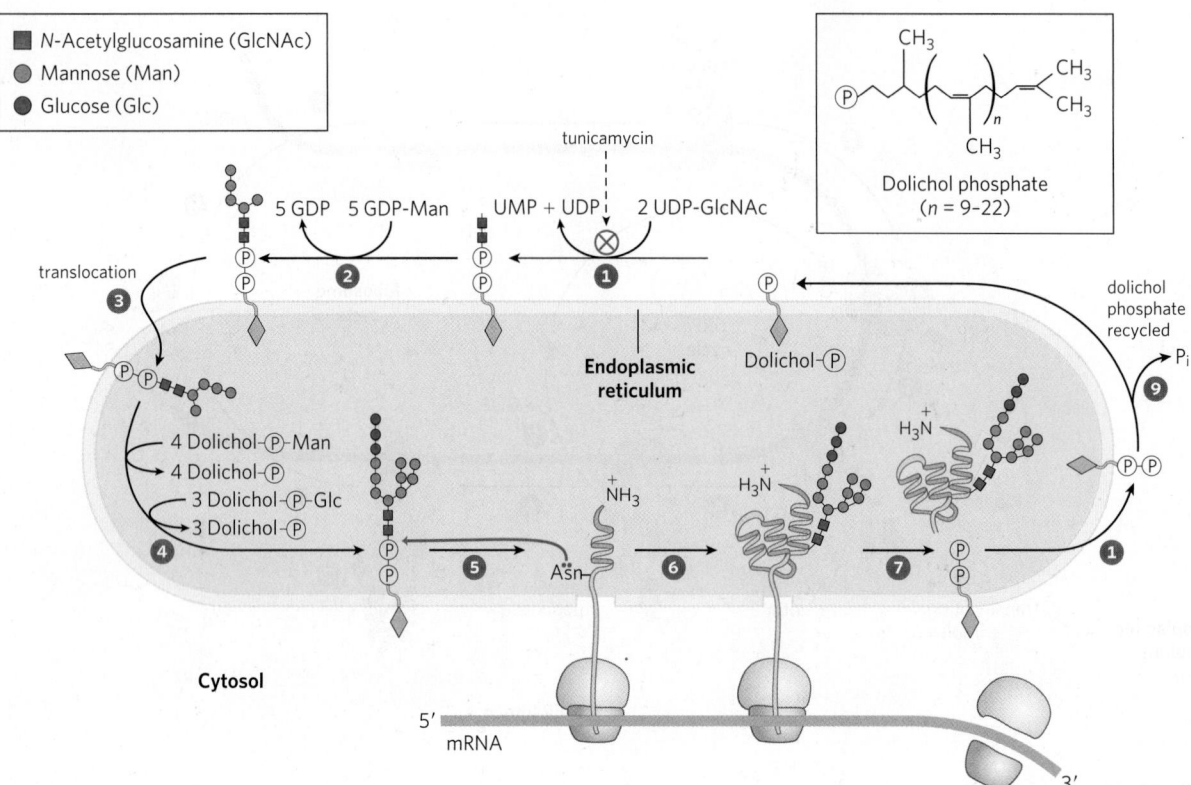

FIGURE 27-41 Synthesis of the core oligosaccharide of glycoproteins. The core oligosaccharide is built up by the successive addition of monosaccharide units. ❶, ❷ The first steps occur on the cytosolic face of the ER. ❸ Translocation moves the incomplete oligosaccharide across the membrane (mechanism not shown), and ❹ completion of the core oligosaccharide occurs within the lumen of the ER. The precursors that contribute additional mannose and glucose residues to the growing oligosaccharide in the lumen are dolichol phosphate derivatives. In the first step in construction of the N-linked oligosaccharide moiety of a glycoprotein, ❺, ❻ the core oligosaccharide is transferred from dolichol phosphate to an Asn residue of the protein within the ER lumen. The core oligosaccharide is then further modified in the ER and the Golgi complex in pathways that differ for different proteins. The five sugar residues shown surrounded by a beige screen, after step ❼, are retained in the final structure of all N-linked oligosaccharides. ❽ The released dolichol pyrophosphate is again translocated so that the pyrophosphate is on the cytosolic face of the ER, then ❾ a phosphate is hydrolytically removed to regenerate dolichol phosphate.

oligosaccharides is through Asn residues. These N-linked oligosaccharides are diverse (Chapter 7), but the pathways by which they form have a common first step. A 14 residue core oligosaccharide is built up stepwise, then transferred from a dolichol phosphate donor molecule to certain Asn residues in the protein **(Fig. 27-41)**. The transferase is on the lumenal face of the ER and thus cannot catalyze glycosylation of cytosolic proteins. After transfer, the core oligosaccharide is trimmed and elaborated in different ways on different proteins, but all N-linked oligosaccharides retain a pentasaccharide core derived from the original 14 residue oligosaccharide. Several antibiotics act by interfering with one or more steps in this process and have aided in elucidating the steps of protein glycosylation. The best characterized is **tunicamycin**, which mimics the structure of UDP-N-acetylglucosamine and blocks the first step of the process (Fig. 27-41, step ❶). A few proteins are O-glycosylated in the ER, but most O-glycosylation occurs in the Golgi complex or in the cytosol (for proteins that do not enter the ER).

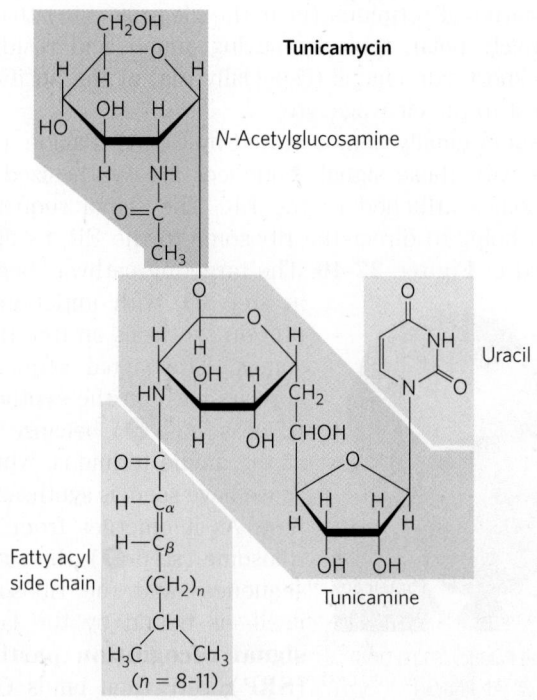

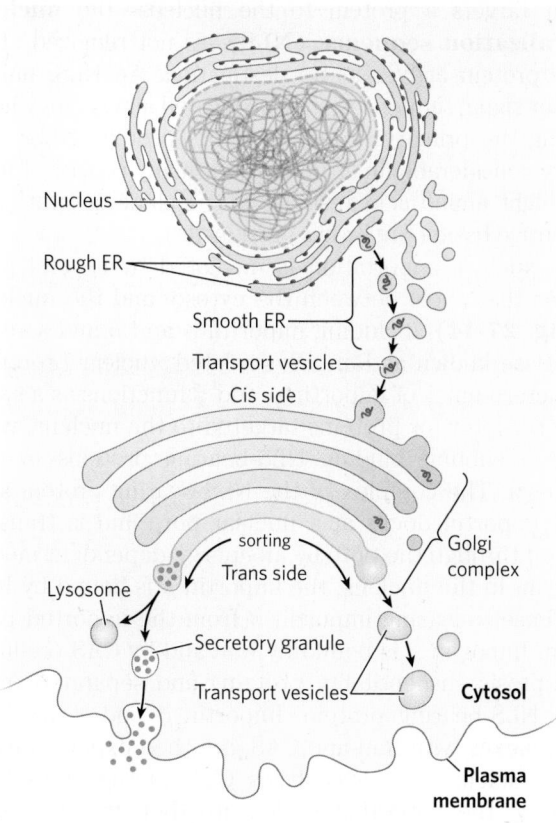

Suitably modified proteins can now be moved to a variety of intracellular destinations. Proteins travel from the ER to the Golgi complex in transport vesicles **(Fig. 27-42)**. In the Golgi complex, oligosaccharides are *O*-linked to some proteins, and *N*-linked oligosaccharides are further modified. By mechanisms not yet fully understood, the Golgi complex also sorts proteins and sends them to their final destinations. The processes that segregate proteins targeted for secretion from those targeted for the plasma membrane or lysosomes must distinguish among these proteins on the basis of structural features other than signal sequences, which were removed in the ER lumen.

This sorting process is best understood in the case of hydrolases destined for transport to lysosomes. On arrival of a hydrolase (a glycoprotein) in the Golgi complex, an as yet undetermined feature (sometimes called a signal patch) of the three-dimensional structure of the hydrolase is recognized by a phosphotransferase, which phosphorylates terminal mannose residues in the oligosaccharides **(Fig. 27-43)**. The presence of one or

FIGURE 27-43 Phosphorylation of mannose residues on lysosome-targeted enzymes. *N*-Acetylglucosamine phosphotransferase recognizes some as yet unidentified structural feature of hydrolases destined for lysosomes.

more mannose 6-phosphate residues in its *N*-linked oligosaccharide is the structural signal that targets a protein to lysosomes. A receptor protein in the membrane of the Golgi complex recognizes the mannose 6-phosphate signal and binds the hydrolase so marked. Vesicles containing these receptor-hydrolase complexes bud from the trans side of the Golgi complex and make their way to sorting vesicles. Here, the receptor-hydrolase complex dissociates in a process facilitated by the lower pH in the vesicle and by phosphatase-catalyzed removal of phosphate groups from the mannose 6-phosphate residues. The receptor is then recycled to the Golgi complex, and vesicles containing the hydrolases bud from the sorting vesicles and move to the lysosomes. In cells treated with tunicamycin (Fig. 27-41, step **①**), hydrolases that should be targeted to lysosomes are instead secreted, confirming that the *N*-linked oligosaccharide plays a key role in targeting these enzymes to lysosomes.

The pathways that target proteins to mitochondria and chloroplasts also rely on amino-terminal signal sequences. Although mitochondria and chloroplasts contain DNA, most of their proteins are encoded by nuclear DNA and must be targeted to the appropriate organelle. Unlike other targeting pathways, however, the mitochondrial and chloroplast pathways begin only *after* a precursor protein has been completely synthesized and released from the ribosome. Precursor proteins destined for mitochondria or chloroplasts are bound by cytosolic chaperone proteins and delivered to receptors on the exterior surface of the target organelle. Specialized translocation mechanisms then transport the protein to its final destination in the organelle, after which the signal sequence is removed.

Signal Sequences for Nuclear Transport Are Not Cleaved

Molecular communication between the nucleus and the cytosol requires the movement of macromolecules through nuclear pores. RNA molecules synthesized in the nucleus are exported to the cytosol. Ribosomal proteins synthesized on cytosolic ribosomes are imported into the nucleus and assembled into 60S and 40S ribosomal subunits in the nucleolus; completed subunits are then exported back to the cytosol. A variety of nuclear proteins (RNA and DNA polymerases, histones, topoisomerases, proteins that regulate gene expression, and so forth) are synthesized in the cytosol and imported into the nucleus. This traffic is modulated by a complex system of molecular signals and transport proteins that is gradually being elucidated.

In most multicellular eukaryotes, the nuclear envelope breaks down at each cell division, and once division is completed and the nuclear envelope reestablished, the dispersed nuclear proteins must be reimported. To allow this repeated nuclear importation, the signal sequence

that targets a protein to the nucleus—the **nuclear localization sequence (NLS)**—is not removed after the protein arrives at its destination. An NLS, unlike other signal sequences, may be located almost anywhere along the primary sequence of the protein. NLSs can vary considerably in structure, but many consist of four to eight amino acid residues and include several consecutive basic (Arg or Lys) residues.

Nuclear importation is mediated by several proteins that cycle between the cytosol and the nucleus **(Fig. 27-44)**, including importin α and β and a small GTPase known as Ran (*R*as-related *n*uclear protein). A heterodimer of importin α and β functions as a soluble receptor for proteins targeted to the nucleus, with the α subunit binding NLS-bearing proteins in the cytosol. The complex of the NLS-bearing protein and the importin docks at a nuclear pore and is translocated through the pore by an energy-dependent mechanism. In the nucleus, the importin β is bound by Ran GTPase, releasing importin β from the imported protein. Importin β is bound by Ran and by CAS (*c*ellular *a*poptosis *s*usceptibility protein) and separated from the NLS-bearing protein. Importin α and β, in their complexes with Ran and CAS, are then exported from the nucleus. Ran hydrolyzes GTP in the cytosol to release the importins, which are then free to begin another importation cycle. Ran itself is also cycled back into the nucleus by the binding of Ran-GDP to nuclear transport factor 2 (NTF2). Inside the nucleus, the GDP bound to Ran is replaced with GTP through the action of Ran guanosine nucleotide–exchange factor (RanGEF; see Box 12-1).

During mitosis, when the nuclear envelope transiently breaks down, the Ran GTPase and the importins play additional roles. The Ran GTPase–importin β complex helps to position the spindle microtubules on the cell perimeter to facilitate chromosome segregation as the cell divides, and this complex also regulates microtubule interaction with other cellular structures.

Bacteria Also Use Signal Sequences for Protein Targeting

Bacteria can target proteins to their inner or outer membranes, to the periplasmic space between these membranes, or to the extracellular medium. They use signal sequences at the amino terminus of the proteins **(Fig. 27-45)**, much like those on eukaryotic proteins targeted to the ER, mitochondria, and chloroplasts.

Most proteins exported from *E. coli* make use of the pathway shown in **Figure 27-46**. Following translation, a protein to be exported may fold only slowly, the amino-terminal signal sequence impeding the folding. The soluble chaperone protein SecB binds to the protein's signal sequence or other

(a)

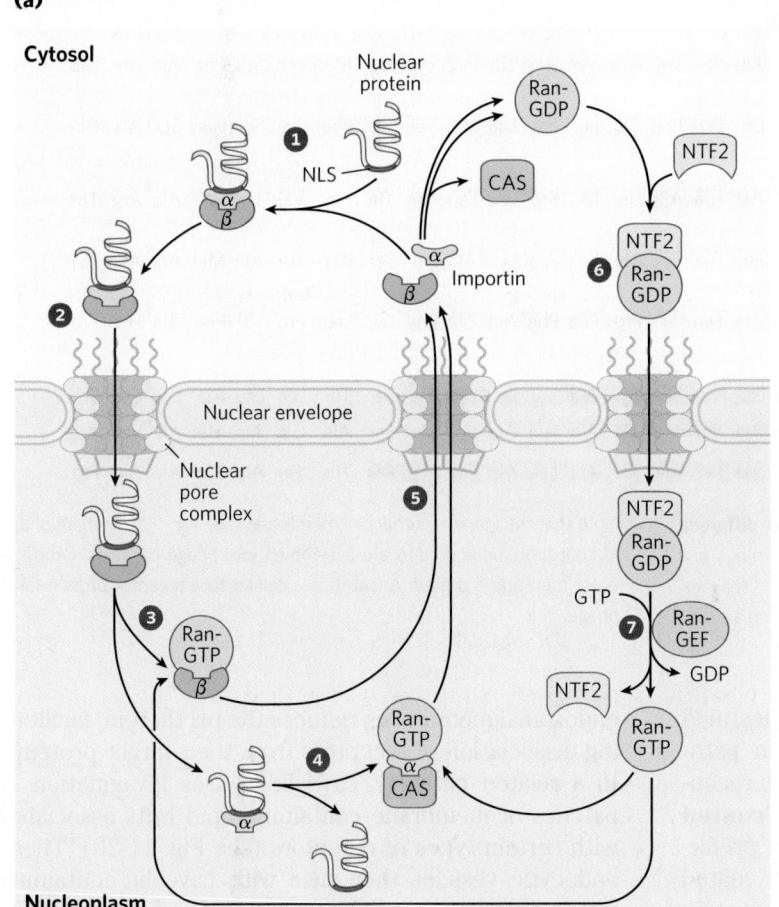

(b)

0.2 μm

FIGURE 27-44 Targeting of nuclear proteins. (a) ❶ A protein with an appropriate nuclear localization signal (NLS) is bound by a complex of importins α and β. ❷ The resulting complex binds to a nuclear pore and translocates. ❸ Inside the nucleus, dissociation of importin β is promoted by the binding of Ran-GTP. ❹ Importin α binds to Ran-GTP and CAS (cellular apoptosis susceptibility protein), releasing the nuclear protein. ❺ Importins α and β and CAS are transported out of the nucleus and recycled. They are released in the cytosol when Ran hydrolyzes its bound GTP. ❻ Ran-GDP is bound by NTF$_2$, and transported back into the nucleus. ❼ RanGEF promotes the exchange of GDP for GTP in the nucleus, and Ran-GTP is ready to process another NLS-bearing protein-importin complex. **(b)** Scanning electron micrograph of the surface of the nuclear envelope, showing numerous nuclear pores. The nuclear pore complex is one of the largest molecular aggregates in the cell (M_r ~5 × 10^7). It is made up of multiple copies of more than 30 different proteins. [Sources: (a) Information from C. Strambio-De-Castillia et al., *Nature Rev. Mol. Cell Biol.* 11:490, 2010, Fig. 1. (b) Don W. Fawcett/Science Source.]

features of its incompletely folded structure. The bound protein is then delivered to SecA, a protein associated with the inner surface of the plasma membrane. SecA acts as both a receptor and a translocating ATPase. Released from SecB and bound to SecA, the protein is delivered to a translocation complex in the membrane, made up of SecY, E, and G, and is translocated stepwise through the membrane at the SecYEG complex in lengths of about 20 amino acid residues. Each step requires the hydrolysis of ATP, catalyzed by SecA.

An exported protein is thus pushed through the membrane by a SecA protein located on the cytoplasmic surface, not pulled through the membrane by a protein on the periplasmic surface. This difference may simply reflect the need for the translocating ATPase to be where the ATP is. The transmembrane electrochemical potential can also provide energy for translocation of the protein, by an as yet unknown mechanism.

Although most exported bacterial proteins use this pathway, some follow an alternative pathway that uses signal recognition and receptor proteins homologous to components of the eukaryotic SRP and SRP receptor (Fig. 27-40).

Cells Import Proteins by Receptor-Mediated Endocytosis

Some proteins are imported into eukaryotic cells from the surrounding medium; examples include low-density lipoprotein (LDL), the iron-carrying protein transferrin,

Inner membrane proteins

cleavage site

Phage fd, major coat protein: Met Lys Lys Ser Leu Val Leu Lys Ala Ser Val Ala Val Ala Thr Leu Val Pro Met Leu Ser Phe Ala↓Ala Glu --

Phage fd, minor coat protein: Met Lys Lys Leu Leu Phe Ala Ile Pro Leu Val Val Pro Phe Tyr Ser His Ser↓Ala Glu --

Periplasmic proteins

Alkaline phosphatase: Met Lys Gln Ser Thr Ile Ala Leu Ala Leu Leu Pro Leu Leu Phe Thr Pro Val Thr Lys Ala↓Arg Thr --

Leucine-specific binding protein: Met Lys Ala Asn Ala Lys Thr Ile Ile Ala Gly Met Ile Ala Leu Ala Ile Ser His Thr Ala Met Ala↓Asp Asp --

β-Lactamase of pBR322: Met Ser Ile Gln His Phe Arg Val Ala Leu Ile Pro Phe Phe Ala Ala Phe Cys Leu Pro Val Phe Ala↓His Pro --

Outer membrane proteins

Lipoprotein: Met Lys Ala Thr Lys Leu Val Leu Gly Ala Val Ile Leu Gly Ser Thr Leu Leu Ala Gly↓Cys Ser --

LamB: Leu Arg Lys Leu Pro Leu Ala Val Ala Val Ala Ala Gly Val Met Ser Ala Gln Ala Met Ala↓Val Asp --

OmpA: Met Met Ile Thr Met Lys Lys Thr Ala Ile Ala Ile Ala Val Ala Leu Ala Gly Phe Ala Thr Val Ala Gln Ala↓Ala Pro --

FIGURE 27-45 Signal sequences that target proteins to different locations in bacteria. Basic amino acids near the amino terminus are highlighted in blue, hydrophobic core amino acids in yellow. Cleavage sites marking the ends of the signal sequences are indicated by red arrows. Note that the inner bacterial cell membrane (see Fig. 1-7) is where phage fd coat proteins and DNA are assembled into phage particles. OmpA is outer membrane protein A; LamB is a cell surface receptor protein for λ phage.

peptide hormones, and circulating proteins destined for degradation. There are several importation pathways **(Fig. 27-47)**. In one path, proteins bind to receptors in invaginations of the membrane called **coated pits**, which concentrate endocytic receptors in preference to other cell-surface proteins. The pits are coated on their cytosolic side with a lattice of the protein **clathrin**, which forms closed polyhedral structures **(Fig. 27-48)**. The clathrin lattice grows as more receptors are occupied by target proteins. Eventually, a complete membrane-bounded endocytic vesicle is pinched off the plasma membrane with the aid of the large GTPase **dynamin**, and enters the cytoplasm. The clathrin is quickly removed by uncoating enzymes, and the vesicle fuses with an endosome. ATPase activity in the endosomal membranes reduces the pH therein, facilitating dissociation of receptors from their target proteins. In a related pathway, caveolin causes invagination of patches of membrane containing lipid rafts associated with certain types of receptors (see Fig. 11-20). These endocytic vesicles then fuse with caveolin-containing internal structures called caveosomes, where the internalized molecules are sorted and redirected to other parts of the cell and the caveolins are prepared for recycling to the membrane surface. There are also clathrin- and caveolin-independent pathways; some make use of dynamin and others do not.

The imported proteins and receptors then go their separate ways, their fates varying with the cell and protein type. Transferrin and its receptor are eventually

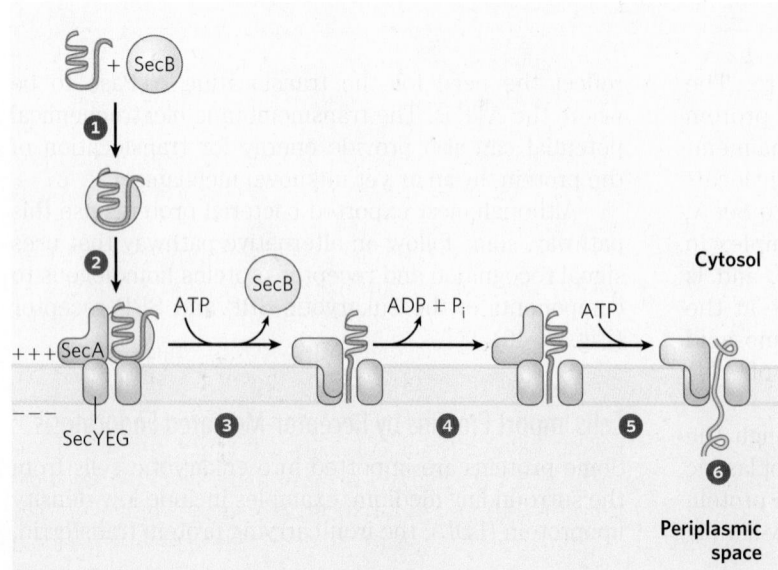

FIGURE 27-46 Model for protein export in bacteria. ❶ A newly translated polypeptide binds to the cytosolic chaperone protein SecB, which **❷** delivers it to SecA, a protein associated with the translocation complex (SecYEG) in the bacterial cell membrane. **❸** SecB is released, and SecA inserts itself into the membrane, forcing about 20 amino acid residues of the protein to be exported through the translocation complex. **❹** Hydrolysis of an ATP by SecA provides the energy for a conformational change that causes SecA to withdraw from the membrane, releasing the polypeptide. **❺** SecA binds another ATP, and the next stretch of 20 amino acid residues is pushed across the membrane through the translocation complex. Steps **❹** and **❺** are repeated until **❻** the entire protein has passed through and is released to the periplasm. The electrochemical potential across the membrane (denoted by + and −) also provides some of the driving force required for protein translocation.

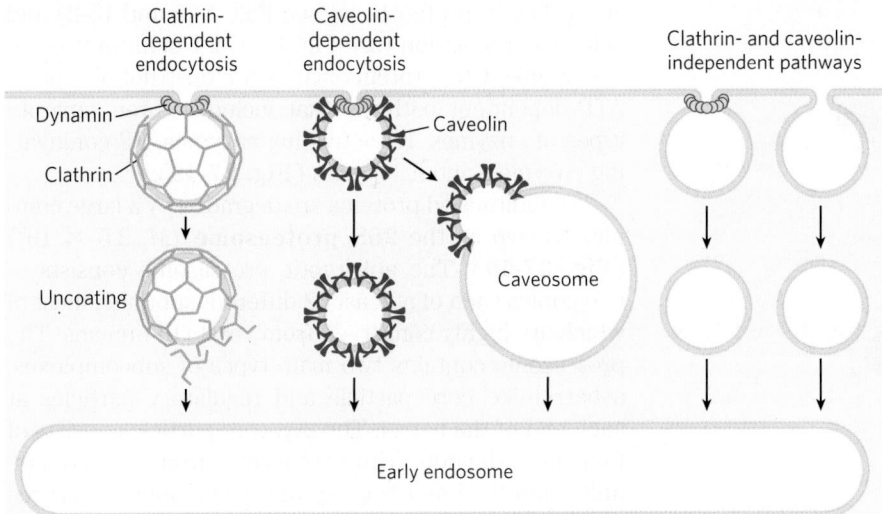

FIGURE 27-47 Summary of endocytosis pathways in eukaryotic cells. Pathways dependent on clathrin or caveolin make use of the GTPase dynamin to pinch vesicles from the plasma membrane. Some pathways do not use clathrin or caveolin; some of these make use of dynamin and some do not. [Source: Information from S. Mayor and R. E. Pagano, *Nature Rev. Mol. Cell Biol.* 8:603, 2007.]

recycled. Some hormones, growth factors, and immune complexes, after eliciting the appropriate cellular response, are degraded along with their receptors. LDL is degraded after the associated cholesterol has been delivered to its destination, but the LDL receptor is recycled (see Fig. 21-41).

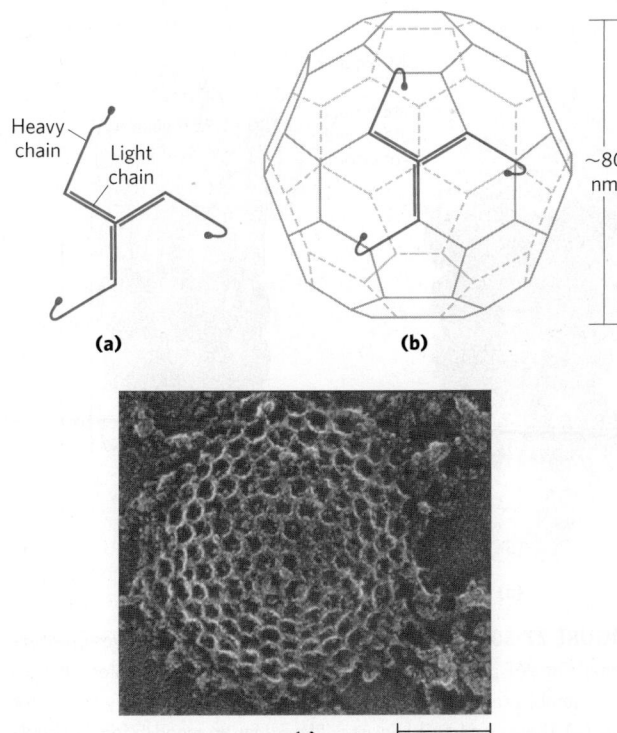

FIGURE 27-48 Clathrin. (a) Three light (L) chains (M_r 35,000) and three heavy (H) chains (M_r 180,000) of the $(HL)_3$ clathrin unit, organized as a three-legged structure called a triskelion. **(b)** Triskelions tend to assemble into polyhedral lattices. **(c)** Electron micrograph of a coated pit on the cytosolic face of the plasma membrane of a fibroblast. [Source: (c) ©1980 Heuser. The Rockefeller University Press. J. Heuser, *J. Cell Biol.* 84:560, 1980. doi:10.1083/jcb.84.3.560.]

Receptor-mediated endocytosis is exploited by some toxins and viruses to gain entry to cells. Influenza virus, diphtheria toxin, and cholera toxin all enter cells in this way.

Protein Degradation Is Mediated by Specialized Systems in All Cells

Protein degradation is critical to overall cellular proteostasis, preventing the buildup of abnormal or unwanted proteins and permitting the recycling of amino acids. The half-lives of eukaryotic proteins vary from 30 seconds to many days. Most proteins turn over rapidly relative to the lifetime of a cell, although a few (such as hemoglobin) can last for the life of the cell (about 110 days for an erythrocyte). Rapidly degraded proteins include those that are defective because of incorrectly inserted amino acids or because of damage accumulated during normal functioning. And enzymes that act at key regulatory points in metabolic pathways often turn over rapidly.

Defective proteins and those with characteristically short half-lives are generally degraded in both bacterial and eukaryotic cells by selective ATP-dependent cytosolic systems. A second system in vertebrates, operating in lysosomes, recycles the amino acids of membrane proteins, extracellular proteins, and proteins with characteristically long half-lives.

In *E. coli*, many proteins are degraded by one of several proteolytic systems that contain AAA+ ATPases (see Chapter 25), including Lon (the name refers to the "long form" of proteins, observed only when this protease is absent), ClpXP, ClpAP, ClpCP, ClpYQ, and FtsH. Each system targets particular proteins distinguished by their structure or subcellular location or both. Typically, ATP hydrolysis is used to maneuver a target protein through a pore into a proteolytic chamber, unfolding the protein in the process. Proteins are cleaved within the chamber. Once a protein has been reduced to

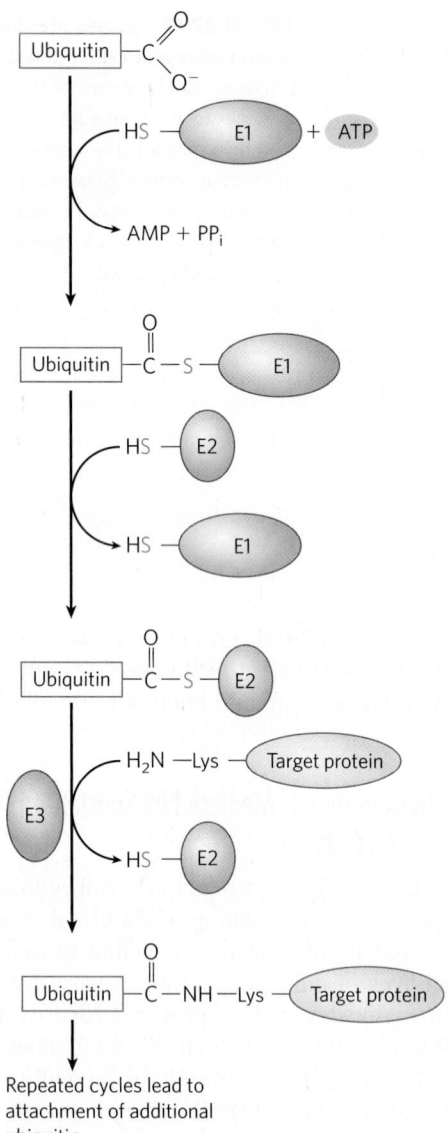

Repeated cycles lead to attachment of additional ubiquitin.

FIGURE 27-49 Three-step pathway by which ubiquitin is attached to a protein. The pathway includes two different enzyme-ubiquitin intermediates. First, the free carboxyl group of ubiquitin's carboxyl-terminal Gly residue becomes linked to an E1-class activating enzyme through a thioester. The ubiquitin is then transferred to an E2 conjugating enzyme. An E3 ligase ultimately catalyzes transfer of the ubiquitin from E2 to the target, linking ubiquitin through an amide (isopeptide) bond to the ε-amino group of a Lys residue in the target protein. Additional cycles produce polyubiquitin, a covalent polymer of ubiquitin subunits that targets the attached protein for destruction in eukaryotes. Multiple pathways of this sort, with different protein targets, are present in most eukaryotic cells.

small, inactive peptides, other ATP-independent proteases complete their degradation.

The ATP-dependent pathway in eukaryotic cells is quite different, involving the protein **ubiquitin**, which, as its name suggests, occurs throughout the eukaryotic kingdoms. One of the most highly conserved proteins known, ubiquitin (76 amino acid residues) is essentially identical in organisms as different as yeasts and humans

and is key to proteostasis (see Figs 4-25 and 15-2) and cell cycle regulation (see Fig. 12-35). Ubiquitin is covalently linked to proteins slated for destruction via an ATP-dependent pathway that includes three separate types of enzymes: E1 activating enzymes, E2 conjugating enzymes, and E3 ligases **(Fig. 27-49)**.

Ubiquitinated proteins are degraded by a large complex known as the **26S proteasome** (M_r 2.5×10^6) **(Fig. 27-50)**. The eukaryotic proteasome consists of two copies each of at least 32 different subunits, most of which are highly conserved from yeasts to humans. The proteasome contains two main types of subcomplexes, a barrel-like core particle and regulatory particles at each end of the barrel. The 20S core particle consists of four rings; the outer rings are formed from seven α subunits, and the inner rings from seven β subunits. Three of the seven subunits in each β ring have protease activities, each with different substrate specificity. The stacked rings of the core particle form the barrel-like structure within which target proteins are degraded. The 19S regulatory particle on each end of the core particle contains approximately 18 subunits, including some that recognize and bind to ubiquitinated proteins. Six of the subunits are AAA+ ATPases that probably function in unfolding the ubiquitinated proteins and translocating the unfolded polypeptide into the core

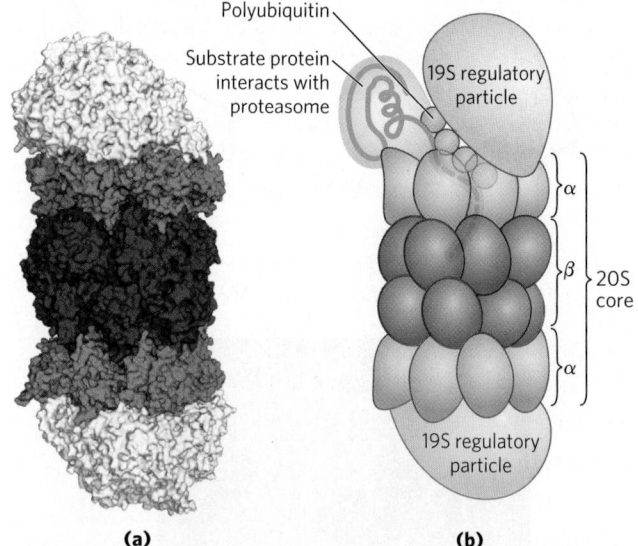

(a)   (b)

FIGURE 27-50 Three-dimensional structure of the eukaryotic proteasome. The 26S proteasome is highly conserved in all eukaryotes. The two subassemblies are the 20S core particle and the 19S regulatory particle, or cap. **(a)** The core particle consists of four rings arranged to form a barrel-like structure. Each of the inner rings has seven different β subunits (dark brown), three of which have protease activities. The outer rings each have seven different α subunits (light brown). A regulatory particle forms a cap on each end of the core particle (gray). **(b)** The regulatory particle binds ubiquitinated proteins, unfolds them, and translocates them into the core particle, where they are degraded to peptides of 3 to 25 amino acid residues. [Source: (a) PDB ID 3L5Q, K. Sadre-Bazzaz et al., *Mol. Cell* 37:728, 2010.]

particle for degradation. The 19S particle also deubiquitinates the proteins as they are degraded in the proteasome. Most cells have additional regulatory complexes that can replace the 19S particle. These alternative regulators do not hydrolyze ATP and do not bind to ubiquitin, but they are important for the degradation of particular cellular proteins. The 26S proteasome can be effectively "accessorized," with regulatory complexes changing with changing cellular conditions.

Although we do not yet understand all the signals that trigger ubiquitination, one simple signal has been found. For many proteins, the identity of the first residue that remains after removal of the amino-terminal Met residue, and any other posttranslational proteolytic processing of the amino-terminal end, has a profound influence on half-life (Table 27-9). These amino-terminal signals have been conserved over billions of years of evolution and are the same in bacterial protein degradation systems and in the human ubiquitination pathway. More complex signals, such as the destruction box discussed in Chapter 12 (see Fig. 12-35), are also being identified.

Ubiquitin-dependent proteolysis is as important for the regulation of cellular processes as for the elimination of defective proteins. Many proteins required at only one stage of the eukaryotic cell cycle are rapidly degraded by the ubiquitin-dependent pathway after fulfilling their function. Ubiquitin-dependent destruction of cyclin is critical to cell-cycle regulation. The E1, E2, and E3 components of the ubiquitination pathway (Fig. 27-49) are large families of proteins. Different E1, E2, and E3 enzymes exhibit different specificities for target proteins and thus regulate different cellular processes. Some of these enzymes are highly localized in certain cellular compartments, reflecting a specialized function.

Not surprisingly, defects in the ubiquitination pathway have been implicated in a wide range of disease states. An inability to degrade certain proteins that activate cell division (the products of oncogenes) can lead to tumor formation, and a too-rapid degradation of proteins that act as tumor suppressors can have the same effect. The ineffective or overly rapid degradation of cellular proteins also seems to play a role in a range of other conditions: renal diseases; asthma; neurodegenerative disorders such as Alzheimer and Parkinson diseases, associated with the formation of characteristic proteinaceous structures in neurons; cystic fibrosis, caused in some cases by a too-rapid degradation of a chloride ion channel, with resultant loss of function (see Box 11-2); Liddle syndrome, in which a sodium channel in the kidney is not degraded, leading to excessive Na^+ absorption and early-onset hypertension—and many other disorders. Drugs designed to inhibit proteasome function are being developed as potential treatments for some of these conditions. In a changing metabolic environment, protein degradation is as important to a cell's survival as is protein synthesis, and much remains to be learned about these interesting pathways. ■

SUMMARY 27.3 Protein Targeting and Degradation

■ After synthesis, many proteins are directed to particular locations in the cell. One targeting mechanism involves a peptide signal sequence, generally found at the amino terminus of a newly synthesized protein.

■ In eukaryotic cells, one class of signal sequences is recognized by the signal recognition particle (SRP), which binds the signal sequence as soon as it appears on the ribosome and transfers the entire ribosome and incomplete polypeptide to the ER. Polypeptides with these signal sequences are moved into the ER lumen as they are synthesized; once in the lumen, they may be modified and moved to the Golgi complex, then sorted and sent to lysosomes, the plasma membrane, or transport vesicles.

■ Proteins targeted to mitochondria and chloroplasts in eukaryotic cells, and those destined for export in bacteria, also make use of an amino-terminal signal sequence.

■ Proteins targeted to the nucleus have an internal signal sequence that, unlike other signal sequences, is not cleaved once the protein is successfully targeted.

■ Some eukaryotic cells import proteins by receptor-mediated endocytosis.

■ All cells eventually degrade proteins, using specialized proteolytic systems. Defective proteins and those slated for rapid turnover are generally degraded by an ATP-dependent system. In eukaryotic cells, the proteins are first tagged by linkage to ubiquitin, a highly conserved protein. Ubiquitin-dependent proteolysis is carried out by proteasomes, also highly conserved, and is critical to the regulation of many cellular processes.

TABLE 27-9	Relationship between Protein Half-Life and Amino-Terminal Amino Acid Residue	
Amino-terminal residue	**Half-life[a]**
Stabilizing	
Ala, Gly, Met, Ser, Thr, Val	>20 h
Destabilizing	
Gln, Ile	~30 min
Glu, Tyr	~10 min
Pro	~7 min
Asp, Leu, Lys, Phe	~3 min
Arg	~2 min

Source: Information from A. Bachmair et al., *Science* 234:179, 1986.

[a]Half-lives were measured in yeast for the β-galactosidase protein modified so that in each experiment it had a different amino-terminal residue. Half-lives may vary for different proteins and in different organisms, but this general pattern appears to hold for all organisms.

Key Terms

Terms in bold are defined in the glossary.

aminoacyl-tRNA 1078	termination 1107
aminoacyl-tRNA	**release factors** 1107
synthetases 1078	**nonsense**
translation 1078	**suppressor** 1107
codon 1079	**polysome** 1109
reading frame 1079	**posttranslational**
initiation codon 1081	**modification** 1110
termination	**ribosome profiling** 1111
codons 1081	**puromycin** 1112
open reading frame	tetracycline 1112
(ORF) 1081	chloramphenicol 1112
anticodon 1084	cycloheximide 1112
wobble 1084	streptomycin 1112
translational	diphtheria toxin 1113
frameshifting 1085	ricin 1113
RNA editing 1086	**signal sequence** 1114
initiation 1101	signal recognition particle
Shine-Dalgarno	**(SRP)** 1115
sequence 1101	peptide translocation
aminoacyl (A) site 1101	complex 1115
peptidyl (P) site 1101	tunicamycin 1116
exit (E) site 1101	nuclear localization sequence
initiation complex 1102	**(NLS)** 1118
elongation 1103	coated pits 1120
elongation factors 1104	clathrin 1120
peptidyl	dynamin 1120
transferase 1105	**ubiquitin** 1122
translocation 1105	**proteasome** 1122

Problems

1. Messenger RNA Translation Predict the amino acid sequences of peptides formed by ribosomes in response to the following mRNA sequences, assuming that the reading frame begins with the first three bases in each sequence.

(a) GGUCAGUCGCUCCUGAUU

(b) UUGGAUGCGCCAUAAUUUGCU

(c) CAUGAUGCCUGUUGCUAC

(d) AUGGACGAA

2. How Many Different mRNA Sequences Can Specify One Amino Acid Sequence? Write all the possible mRNA sequences that can code for the simple tripeptide segment Leu–Met–Tyr. Your answer will give you some idea of the number of possible mRNAs that can code for one polypeptide.

3. Can the Base Sequence of an mRNA Be Predicted from the Amino Acid Sequence of Its Polypeptide Product? A given sequence of bases in an mRNA will code for one and only one sequence of amino acids in a polypeptide, if the reading frame is specified. From a given sequence of amino acid residues in a protein such as cytochrome *c*, can we predict the base sequence of the unique mRNA that encoded it? Give reasons for your answer.

4. Coding of a Polypeptide by Duplex DNA The template strand of a segment of double-helical DNA contains the sequence

(5′)CTTAACACCCCTGACTTCGCGCCGTCG(3′)

(a) What is the base sequence of the mRNA that can be transcribed from this strand?

(b) What amino acid sequence could be coded by the mRNA in (a), starting from the 5′ end?

(c) If the complementary (nontemplate) strand of this DNA were transcribed and translated, would the resulting amino acid sequence be the same as in (b)? Explain the biological significance of your answer.

5. Methionine Has Only One Codon Methionine is one of two amino acids with only one codon. How does the single codon for methionine specify both the initiating residue and interior Met residues of polypeptides synthesized by *E. coli*?

6. The Genetic Code in Action Translate the following mRNA, starting at the first 5′ nucleotide, assuming that translation occurs in an *E. coli* cell. If all tRNAs make maximum use of wobble rules but do not contain inosine, how many distinct tRNAs are required to translate this RNA?

(5′)AUGGGUCGUGAGUCAUCGUUAAUUG
UAGCUGGAGGGGAGGAAUGA(3′)

7. Synthetic mRNAs The genetic code was elucidated through the use of polyribonucleotides synthesized either enzymatically or chemically in the laboratory. Given what we now know about the genetic code, how would you make a polyribonucleotide that could serve as an mRNA coding predominantly for many Phe residues and a small number of Leu and Ser residues? What other amino acid(s) would be encoded by this polyribonucleotide, but in smaller amounts?

8. Energy Cost of Protein Biosynthesis Determine the minimum energy cost, in terms of ATP equivalents expended, for the biosynthesis of the β-globin chain of hemoglobin (146 residues), starting from a pool including all necessary amino acids, ATP, and GTP. Compare your answer with the direct energy cost of the biosynthesis of a linear glycogen chain of 146 glucose residues in ($\alpha 1 \rightarrow 4$) linkage, starting from a pool including glucose, UTP, and ATP (Chapter 15). From your data, what is the *extra* energy cost of making a protein, in which all the residues are ordered in a specific sequence, compared with the cost of making a polysaccharide containing the same number of residues but lacking the informational content of the protein?

In addition to the direct energy cost for the synthesis of a protein, there are indirect energy costs—those required for the cell to make the necessary enzymes for protein synthesis. Compare the magnitude of the indirect costs to a eukaryotic cell of the biosynthesis of linear ($\alpha 1 \rightarrow 4$) glycogen chains and the biosynthesis of polypeptides, in terms of the enzymatic machinery involved.

9. Predicting Anticodons from Codons Most amino acids have more than one codon and attach to more than one tRNA, each with a different anticodon. Write all possible anticodons for the four codons of glycine: (5′)GGU, GGC, GGA, and GGG.

(a) From your answer, which of the positions in the anticodons are primary determinants of their codon specificity in the case of glycine?

(b) Which of these anticodon-codon pairings has/have a wobbly base pair?

(c) In which of the anticodon-codon pairings do all three positions exhibit strong Watson-Crick hydrogen bonding?

10. Effect of Single-Base Changes on Amino Acid Sequence Much important confirmatory evidence on the genetic code has come from assessing changes in the amino acid sequence of mutant proteins after a single base has been changed in the gene that encodes the protein. Which of the following amino acid replacements would be consistent with the genetic code if the replacements were caused by a single base change? Which cannot be the result of a single-base mutation? Why?

    (a) Phe → Leu   (e) Ile → Leu

    (b) Lys → Ala   (f) His → Glu

    (c) Ala → Thr   (g) Pro → Ser

    (d) Phe → Lys

11. Resistance of the Genetic Code to Mutation The following RNA sequence represents the beginning of an open reading frame. What changes (if any) can occur at each position without generating a change in the encoded amino acid residue?

        (5′)AUGAUAUUGCUAUCUUGGACU

12. Basis of the Sickle Cell Mutation Sickle cell hemoglobin has a Val residue at position 6 of the β-globin chain instead of the Glu residue found in normal hemoglobin A. Can you predict what change took place in the DNA codon for glutamate to account for replacement of the Glu residue by Val?

13. Proofreading by Aminoacyl-tRNA Synthetases The isoleucyl-tRNA synthetase has a proofreading function that ensures the fidelity of the aminoacylation reaction, but the histidyl-tRNA synthetase lacks such a proofreading function. Explain.

14. Importance of the "Second Genetic Code" Some aminoacyl-tRNA synthetases do not recognize and bind the anticodon of their cognate tRNAs but instead use other structural features of the tRNAs to impart binding specificity. The tRNAs for alanine apparently fall into this category.

    (a) What features of tRNA^{Ala} are recognized by Ala-tRNA synthetase?

    (b) Describe the consequences of a C→G mutation in the third position of the anticodon of tRNA^{Ala}.

    (c) What other kinds of mutations might have similar effects?

    (d) Mutations of these types are never found in natural populations of organisms. Why? (Hint: Consider what might happen both to individual proteins and to the organism as a whole.)

15. Rate of Protein Synthesis A bacterial ribosome can synthesize about 20 peptide bonds per minute. If the average bacterial protein is approximately 260 amino acid residues long, how many proteins can the ribosomes in an *E. coli* cell synthesize in 20 minutes if all ribosomes are functioning at maximum rates?

16. The Role of Translation Factors A researcher isolates mutant variants of the bacterial translation factors IF2, EF-Tu, and EF-G. In each case, the mutation allows proper folding of the protein and the binding of GTP but does not allow GTP hydrolysis. At what stage would translation be blocked by each mutant protein?

17. Maintaining the Fidelity of Protein Synthesis The chemical mechanisms used to avoid errors in protein synthesis are different from those used during DNA replication. DNA polymerases use a 3′→5′ exonuclease proofreading activity to remove mispaired nucleotides incorrectly inserted into a growing DNA strand. There is no analogous proofreading function on ribosomes, and, in fact, the identity of an amino acid attached to an incoming tRNA and added to the growing polypeptide is never checked. A proofreading step that hydrolyzed the previously formed peptide bond after insertion of an incorrect amino acid into a growing polypeptide (analogous to the proofreading step of DNA polymerases) would be impractical. Why? (Hint: Consider how the link between the growing polypeptide and the mRNA is maintained during elongation; see Figs 27-29 and 27-30.)

18. Ribosome Profiling In the ribosome profiling method (Fig. 27-37), segments of mRNA that are bound to and protected by ribosomes are isolated and converted into DNA for sequencing. However, the mRNA segments are usually only part of the overall RNA protected by ribosomes and isolated after ribonuclease treatment. What other RNA species would be isolated by this protocol?

19. Predicting the Cellular Location of a Protein The gene for a eukaryotic polypeptide 300 amino acid residues long is altered so that a signal sequence recognized by SRP occurs at the polypeptide's amino terminus and a nuclear localization signal (NLS) occurs internally, beginning at residue 150. Where is the protein likely to be found in the cell?

20. Requirements for Protein Translocation across a Membrane The secreted bacterial protein OmpA has a precursor, ProOmpA, which has the amino-terminal signal sequence required for secretion. If purified ProOmpA is denatured with 8 M urea and the urea is then removed (such as by running the protein solution rapidly through a gel filtration column), the protein can be translocated across isolated bacterial inner membranes in vitro. However, translocation becomes impossible if ProOmpA is first allowed to incubate for a few hours in the absence of urea. Furthermore, the capacity for translocation is maintained for an extended period if ProOmpA is first incubated in the presence of another bacterial protein called trigger factor. Describe the probable function of this factor.

21. Protein-Coding Capacity of a Viral DNA The 5,386 bp genome of bacteriophage ϕX174 includes genes for 10 proteins, designated A to K (omitting "I"), with sizes given in the table below. How much DNA would be required to encode these 10 proteins? How can you reconcile the size of the ϕX174 genome with its protein-coding capacity?

Protein	Number of amino acid residues	Protein	Number of amino acid residues
A	455	F	427
B	120	G	175
C	86	H	328
D	152	J	38
E	91	K	56

Data Analysis Problem

22. Designing Proteins by Using Randomly Generated Genes Studies of the amino acid sequence and corresponding three-dimensional structure of wild-type or mutant proteins have led to significant insights into the principles that govern protein folding. An important test of this understanding would be to *design* a protein based on these principles and see whether it folds as expected.

Kamtekar and colleagues (1993) used aspects of the genetic code to generate random protein sequences with defined patterns of hydrophilic and hydrophobic residues. Their clever approach combined knowledge about protein structure, amino acid properties, and the genetic code to explore the factors that influence protein structure.

The researchers set out to generate a set of proteins with the simple four-helix bundle structure shown below, with α helices (shown as cylinders) connected by segments of random coil (light red).

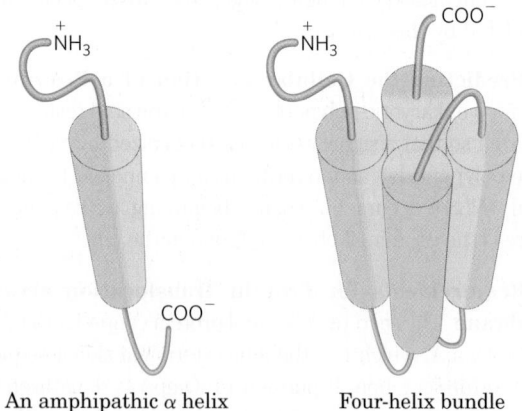

An amphipathic α helix   Four-helix bundle

Each α helix is amphipathic—the R groups on one side of the helix are exclusively hydrophobic (yellow), and those on the other side are exclusively hydrophilic (blue). A protein consisting of four of these helices separated by short segments of random coil would be expected to fold so that the hydrophilic sides of the helices face the solvent.

(a) What forces or interactions hold the four α helices together in this bundled structure?

Figure 4-4a shows a segment of α helix consisting of 10 amino acid residues. With the gray central rod as a divider, four of the R groups (purple spheres) extend from the left side of the helix, and six extend from the right.

(b) Number the R groups in Figure 4-4a, from top (amino terminus; 1) to bottom (carboxyl terminus; 10). Which R groups extend from the left side and which from the right?

(c) Suppose you wanted to design this 10 amino acid segment to be an amphipathic helix, with the left side hydrophilic and the right side hydrophobic. Give a sequence of 10 amino

acids that could potentially fold into such a structure. There are many possible correct answers.

(d) Give one possible double-stranded DNA sequence that could encode the amino acid sequence you chose for (c). (It is an internal portion of a protein, so you do not need to include start or stop codons.)

Rather than designing proteins with specific sequences, Kamtekar and colleagues designed proteins with partially random sequences, with hydrophilic and hydrophobic amino acid residues placed in a controlled pattern. They did this by taking advantage of some interesting features of the genetic code to construct a library of synthetic DNA molecules with partially random sequences arranged in a particular pattern.

To design a DNA sequence that would encode random hydrophobic amino acid sequences, the researchers began with the degenerate codon NTN, where N can be A, G, C, or T. They filled each N position by including an equimolar mixture of A, G, C, and T in the DNA synthesis reaction to generate a mixture of DNA molecules with different nucleotides at that position (see Fig. 8-32). Similarly, to encode random polar amino acid sequences, they began with the degenerate codon NAN and used an equimolar mixture of A, G, and C (but in this case, no T) to fill the N positions.

(e) Which amino acids can be encoded by the NTN triplet? Are all amino acids in this set hydrophobic? Does the set include *all* the hydrophobic amino acids?

(f) Which amino acids can be encoded by the NAN triplet? Are all of these polar? Does the set include *all* the polar amino acids?

(g) In creating the NAN codons, why was it necessary to leave T out of the reaction mixture?

Kamtekar and coworkers cloned this library of random DNA sequences into plasmids, selected 48 that produced the correct patterning of hydrophilic and hydrophobic amino acids, and expressed these in *E. coli.* The next challenge was to determine whether the proteins folded as expected. It would be very time-consuming to express each protein, crystallize it, and determine its complete three-dimensional structure. Instead, the investigators used the *E. coli* protein-processing machinery to screen out sequences that led to highly defective proteins. In this initial screening, they kept only those clones that resulted in a band of protein with the expected molecular weight on SDS polyacrylamide gel electrophoresis (see Fig. 3-18).

(h) Why would a grossly misfolded protein fail to produce a band of the expected molecular weight on electrophoresis?

Several proteins passed this initial test, and further exploration showed that they had the expected four-helix structure.

(i) Why didn't all of the random-sequence proteins that passed the initial screening test produce four-helix structures?

Reference

Kamtekar, S., J.M. Schiffer, H. Xiong, J.M. Babik, and M.H. Hecht. 1993. Protein design by binary patterning of polar and nonpolar amino acids. *Science* 262:1680–1685.

CHAPTER 28

Regulation of Gene Expression

Self-study tools that will help you practice what you've learned and reinforce this chapter's concepts are available online.
Go to www.macmillanlearning.com/LehningerBiochemistry7e.

f the 4,000 or so genes in the typical bacterial genome, or the 20,000 genes in the human genome, only a fraction are expressed in a cell at any given time. Some gene products are present in very large amounts: the elongation factors required for protein synthesis, for example, are among the most abundant proteins in bacteria, and ribulose 1,5-bisphosphate carboxylase/oxygenase (rubisco) of plants and photosynthetic bacteria is one of the most abundant enzymes in the biosphere. Other gene products occur in much smaller amounts; for instance, a cell may contain only a few molecules of the enzymes that repair rare DNA lesions. Requirements for some gene products change over time. The need for enzymes in certain metabolic pathways may wax and wane as food sources change or are depleted. During development of a multicellular organism, some proteins that influence cellular differentiation are present for just a brief time in only a few cells. Specialization of cellular function can greatly affect the need for various gene products; an example is the uniquely high concentration of a single protein—hemoglobin—in erythrocytes. Given the high cost of protein synthesis, regulation of gene expression is essential to making optimal use of available energy.

The cellular concentration of a protein is determined by a delicate balance of at least seven processes, each having several potential points of regulation:

1. Synthesis of the primary RNA transcript (transcription)

2. Posttranscriptional modification of mRNA

3. Degradation of mRNA

4. Protein synthesis (translation)

5. Posttranslational modification of proteins

6. Protein targeting and transport

7. Degradation of protein

These processes are summarized in **Figure 28-1**. We have examined several of these mechanisms in previous chapters. Posttranscriptional modification of mRNA, by processes such as alternative splicing patterns (see Fig. 26-19b) or RNA editing (see Figs 27-10 and 27-12), can affect which proteins are produced from an mRNA transcript and in what amounts. A variety of nucleotide sequences in an mRNA can affect the rate of its degradation (p. 1014). Many factors affect the rate at which an mRNA is translated into a protein, as well as the posttranslational modification, targeting, and eventual degradation of that protein (Chapter 27).

Of the regulatory processes illustrated in Figure 28-1, those operating at the level of transcription initiation are particularly well-documented. These processes are a major focus of this chapter, although we also consider other mechanisms. As noted in earlier chapters, the complexity of an organism is not reflected in the number of its protein-coding genes. Instead, as complexity increases from bacteria to mammals, mechanisms of gene regulation become more elaborate, and posttranscriptional and translational regulation play greater roles. For many genes, the regulatory processes can require a considerable investment of chemical energy.

Control of transcription initiation permits the synchronized regulation of multiple genes encoding products with interdependent activities. For example, when their DNA is heavily damaged, bacterial cells require a coordinated increase in the levels of the many DNA repair enzymes. And perhaps the most sophisticated

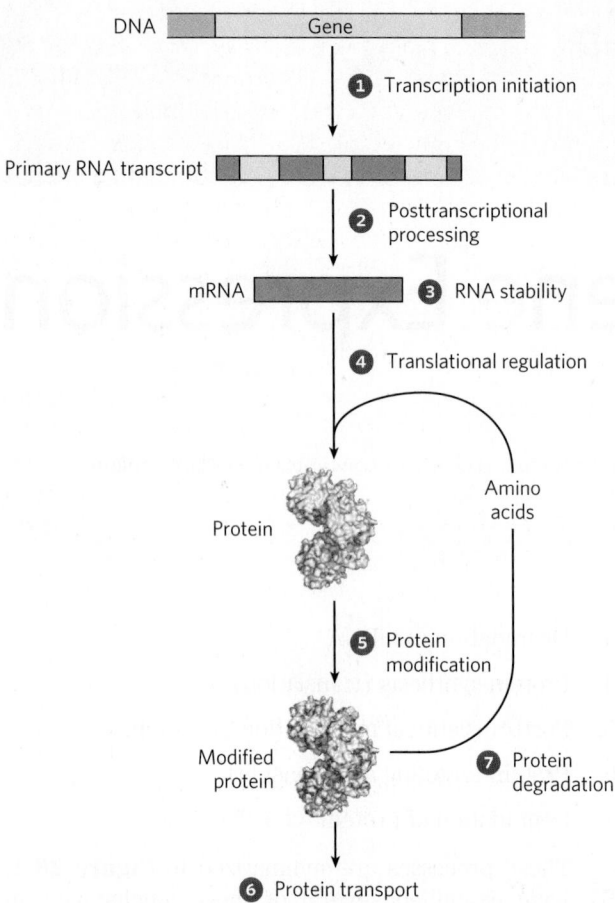

DNA [____Gene____]

❶ Transcription initiation

Primary RNA transcript

❷ Posttranscriptional processing

mRNA   ❸ RNA stability

❹ Translational regulation

Protein

Amino acids

❺ Protein modification

Modified protein   ❼ Protein degradation

❻ Protein transport

FIGURE 28-1 Seven processes that affect the steady-state concentration of a protein. Each process has several potential points of regulation.

form of coordination occurs in the complex regulatory circuits that guide the development of multicellular eukaryotes, which can include many types of regulatory mechanisms.

We begin by examining the interactions between proteins and DNA that are the key to transcriptional regulation. We next discuss the specific proteins that influence the expression of specific genes, first in bacterial and then in eukaryotic cells. Information about posttranscriptional and translational regulation is included in the discussion, where relevant, to provide a more complete overview of the rich complexity of regulatory mechanisms.

28.1 Principles of Gene Regulation

Genes for products that are required at all times, such as those for the enzymes of central metabolic pathways, are expressed at a more or less constant level in

virtually every cell of a species or organism. Such genes are often referred to as **housekeeping genes**. Unvarying expression of a gene is called **constitutive gene expression**.

For other gene products, cellular levels rise and fall in response to molecular signals; this is **regulated gene expression**. Gene products that increase in concentration under particular molecular circumstances are referred to as **inducible**; the process of increasing their expression is **induction**. The expression of many of the genes encoding DNA repair enzymes, for example, is induced by a system of regulatory proteins that responds to high levels of DNA damage. Conversely, gene products that decrease in concentration in response to a molecular signal are referred to as **repressible**, and the process is called **repression**. For example, in bacteria, ample supplies of tryptophan lead to repression of the genes for the enzymes that catalyze tryptophan biosynthesis.

Transcription is mediated and regulated by protein-DNA interactions, especially those involving the protein components of RNA polymerase (Chapter 26). We first consider how the activity of RNA polymerase is regulated, and proceed to a general description of the proteins participating in this regulation. We then examine the molecular basis for the recognition of specific DNA sequences by DNA-binding proteins.

RNA Polymerase Binds to DNA at Promoters

RNA polymerases bind to DNA and initiate transcription at promoters (see Fig. 26-5), sites generally found near points at which RNA synthesis begins on the DNA template. The regulation of transcription initiation often entails changes in how RNA polymerase interacts with a promoter.

The nucleotide sequences of promoters vary considerably, affecting the binding affinity of RNA polymerases and thus the frequency of transcription initiation. Some *Escherichia coli* genes are transcribed once per second, others less than once per cell generation. Much of this variation is due to differences in promoter sequence. In the absence of regulatory proteins, differences in promoter sequence may affect the frequency of transcription initiation by a factor of 1,000 or more. Most *E. coli* promoters have a sequence close to a consensus **(Fig. 28-2)**. Mutations that result in a shift away from the consensus sequence usually decrease the function of bacterial promoters; conversely, mutations toward consensus usually enhance promoter function.

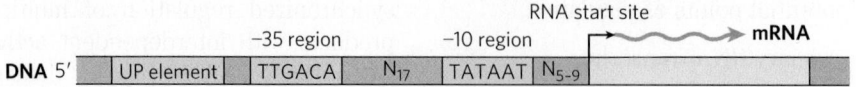

RNA start site

DNA 5′ [UP element | TTGACA | N_{17} | TATAAT | N_{5-9}] → mRNA

−35 region   −10 region

FIGURE 28-2 Consensus sequence for many *E. coli* promoters. Most base substitutions in the −10 and −35 regions have a negative effect on promoter function. Some promoters also include the UP (upstream promoter) element (see Fig. 26-5).

>> **Key Convention:** By convention, DNA sequences are shown as they exist in the nontemplate strand, with the 5′ terminus on the left. Nucleotides are numbered from the transcription start site, with positive numbers to the right (in the direction of transcription) and negative numbers to the left. N indicates any nucleotide. <<

Although housekeeping genes are expressed constitutively, the cellular concentrations of the proteins they encode vary widely. For these genes, the RNA polymerase–promoter interaction strongly influences the rate of transcription initiation; differences in promoter sequence allow the cell to synthesize the appropriate level of each housekeeping gene product.

The basal rate of transcription initiation at the promoters of nonhousekeeping genes is also determined by the promoter sequence, but expression of these genes is further modulated by regulatory proteins. Many of these proteins work by enhancing or interfering with the interaction between RNA polymerase and the promoter.

The sequences of eukaryotic promoters are much more variable than their bacterial counterparts. The three eukaryotic RNA polymerases usually require an array of general transcription factors in order to bind to a promoter, and these can heavily influence basal transcription rates. Yet, as with bacterial gene expression, the basal level of transcription is determined in part by the effect of promoter sequences on the function of RNA polymerase and its associated transcription factors.

Transcription Initiation Is Regulated by Proteins and RNAs

At least three types of proteins regulate transcription initiation by RNA polymerase: **specificity factors** alter the specificity of RNA polymerase for a given promoter or set of promoters, **repressors** impede access of RNA polymerase to the promoter, and **activators** enhance the RNA polymerase–promoter interaction.

As our understanding of the roles of protein regulators slowly matures, many new roles for gene regulation by RNAs are also beginning to emerge. Among these are the **long, noncoding RNAs (lncRNAs)**, generally defined as RNAs more than 200 nucleotides long that lack an open reading frame that encodes a protein—thus distinguishing them from the small functional RNAs (miRNA, snoRNA, snRNA, etc.) described in Chapter 26. The lncRNAs are found in all types of organisms, with tens of thousands expressed in mammalian cells. Known functions of lncRNAs include regulation of nucleosome

positioning and chromatin structure, control of DNA methylation and posttranscriptional histone modifications, transcriptional gene silencing, multiple roles in transcriptional activation and repression, and much more.

We introduced bacterial protein specificity factors in Chapter 26 (see Fig. 26-5), although we did not refer to them by that name. The σ subunit of the *E. coli* RNA polymerase holoenzyme is a specificity factor that mediates promoter recognition and binding. Most *E. coli* promoters are recognized by a single σ subunit (M_r 70,000), σ^{70}. Under some conditions, some of the σ^{70} subunits are replaced by one of six other specificity factors. One notable case arises when bacteria are subjected to heat stress, leading to the replacement of σ^{70} by σ^{32} (M_r 32,000). When bound to σ^{32}, RNA polymerase is directed to a specialized set of promoters with a different consensus sequence **(Fig. 28-3)**. These promoters control the expression of a set of genes that encode proteins, including some protein chaperones (p. 146), that are part of a stress-induced system called the heat shock response. Thus, through changes in the binding affinity of the polymerase that direct the enzyme to different promoters, a set of genes involved in related processes is coordinately regulated. In eukaryotic cells, some of the general transcription factors, in particular the TATA-binding protein (TBP; see Fig. 26-9), may be considered specificity factors.

Repressors bind to specific sites on the DNA. In bacterial cells, such binding sites, called **operators**, are generally near a promoter. RNA polymerase binding, or its movement along the DNA after binding, is blocked when the repressor is present. Regulation by means of a repressor protein that blocks transcription is referred to as **negative regulation**. Repressor binding to DNA is regulated by a molecular signal, or **effector**, usually a small molecule or a protein that binds to the repressor and causes a conformational change. The interaction between repressor and signal molecule either increases or decreases transcription. In some cases, the conformational change results in dissociation of a DNA-bound repressor from the operator **(Fig. 28-4a)**. Transcription initiation can then proceed unhindered. In other cases, interaction between an inactive repressor and the signal molecule causes the repressor to bind to the operator (Fig. 28-4b). In eukaryotic cells, gene regulation by a repressor is less common. Where it does occur (more often in lower eukaryotes such as yeast), the binding site for a repressor may be some distance from the promoter.

FIGURE 28-3 Consensus sequence for promoters that regulate expression of the *E. coli* heat shock genes. This system responds to temperature increases as well as some other environmental stresses, resulting in the induction of a set of proteins. Binding of RNA polymerase to heat shock promoters is mediated by a specialized σ subunit of the polymerase, σ^{32}, which replaces σ^{70} in the RNA polymerase initiation complex.

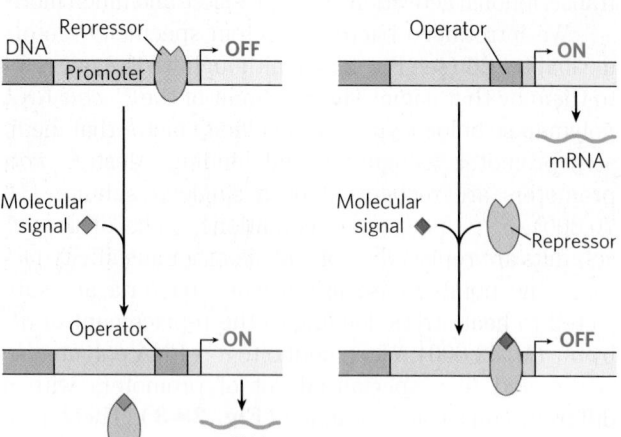

(a) Negative regulation
Molecular signal causes dissociation of repressor from DNA, inducing transcription.

(b) Negative regulation
Molecular signal causes binding of repressor to DNA, inhibiting transcription.

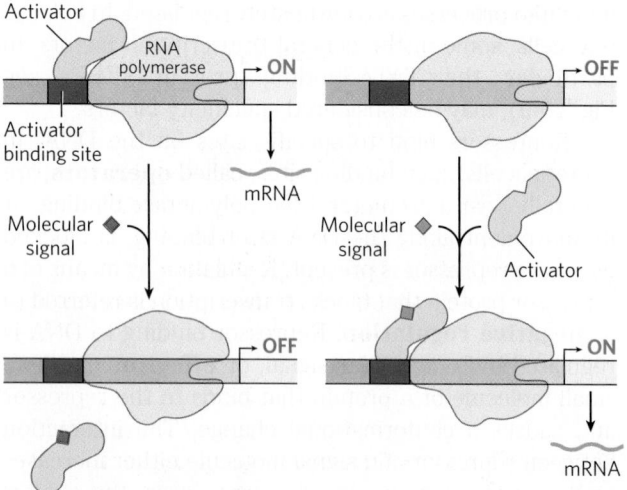

(c) Positive regulation
Molecular signal causes dissociation of activator from DNA, inhibiting transcription.

(d) Positive regulation
Molecular signal causes binding of activator to DNA, inducing transcription.

FIGURE 28-4 Common patterns of regulation of transcription initiation. Two types of negative regulation are illustrated. **(a)** Repressor binds to the operator in the absence of the molecular signal; the external signal causes dissociation of the repressor to permit transcription. **(b)** Repressor binds in the presence of the signal; the repressor dissociates, and transcription ensues when the signal is removed. Positive regulation is mediated by gene activators. Again, two types are shown. **(c)** Activator binds in the absence of the molecular signal and transcription proceeds; when the signal is added, the activator dissociates and transcription is inhibited. **(d)** Activator binds in the presence of the signal; it dissociates only when the signal is removed. Note that "positive" and "negative" regulation refer to the type of regulatory protein involved: the bound protein either facilitates or inhibits transcription. In either case, addition of the molecular signal may increase or decrease transcription, depending on its effect on the regulatory protein.

Binding of these repressors to their binding sites has the same effect as in bacterial cells: inhibiting the assembly or activity of a transcription complex at the promoter.

Activators provide a molecular counterpoint to repressors; they bind to DNA and *enhance* the activity

of RNA polymerase at a promoter; this is **positive regulation**. In bacteria, activator-binding sites are often adjacent to promoters that are bound weakly or not at all by RNA polymerase alone, such that little transcription occurs in the absence of the activator. Some activators are usually bound to DNA, enhancing transcription until dissociation of the activator is triggered by the binding of a signal molecule (Fig. 28-4c). In other cases the activator binds to DNA only after interaction with a signal molecule (Fig. 28-4d). Signal molecules can therefore increase or decrease transcription, depending on how they affect the activator.

Positive regulation by activators is particularly common in eukaryotes. Many eukaryotic activators bind to DNA sites, called enhancers, that are distant from the promoter, affecting the rate of transcription at a promoter that may be located thousands of base pairs away.

The distance between a promoter and the binding site of an activator or repressor is bridged by looping out of the DNA between the two sites **(Fig. 28-5)**.

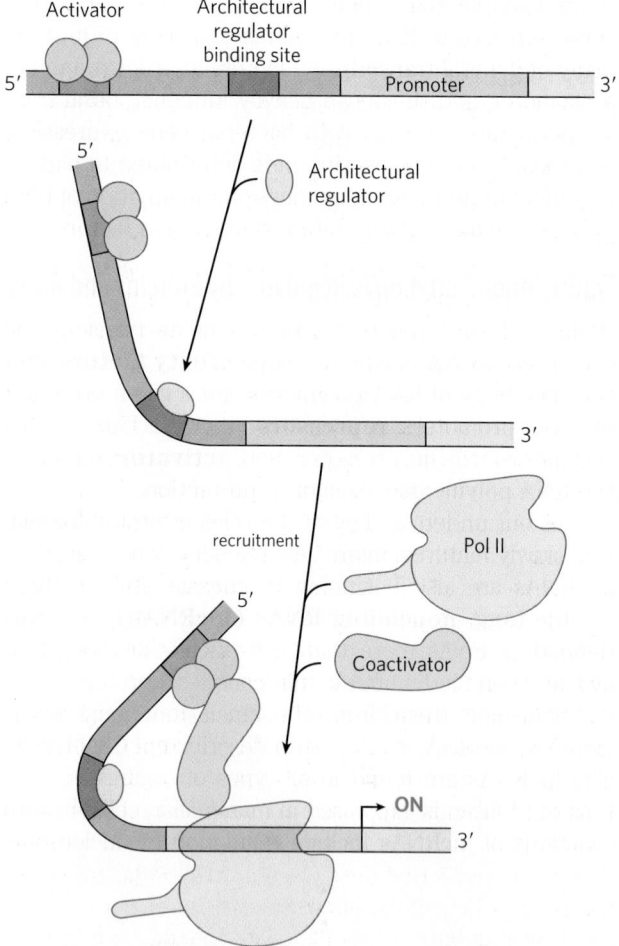

FIGURE 28-5 Interaction between activators/repressors and RNA polymerase in eukaryotes. Eukaryotic activators and repressors frequently bind sites thousands of base pairs distant from the promoters they regulate. DNA looping, often facilitated by architectural regulators, brings the sites together. The interaction between activators and RNA polymerase may be mediated by coactivators, as shown. Repression is sometimes mediated by repressors (described later) that bind to activators, thereby preventing the activating interaction with RNA polymerase.

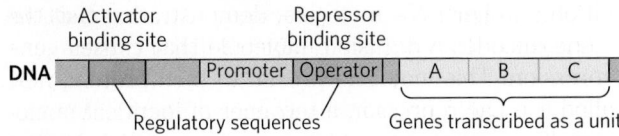

FIGURE 28-6 Representative bacterial operon. Genes A, B, and C are transcribed on one polycistronic mRNA. Typical regulatory sequences include binding sites for proteins that either activate or repress transcription from the promoter.

The looping is facilitated in some cases by proteins called **architectural regulators** that bind to intervening sites. Interaction between activators and the RNA polymerase at the promoter is often mediated by intermediary proteins called coactivators. In some instances, protein repressors may take the place of coactivators, binding to the activators and preventing the activating interaction.

Many Bacterial Genes Are Clustered and Regulated in Operons

Bacteria have a simple general mechanism for coordinating the regulation of multiple genes: these genes are clustered on the chromosome and are transcribed together. Many bacterial mRNAs are polycistronic—multiple genes on a single transcript—and the single promoter that initiates transcription of the cluster is the site of regulation for expression of all the genes in the cluster. The gene cluster and promoter, plus additional sequences that function together in regulation, are called an **operon (Fig. 28-6)**. Operons that include two to six genes transcribed as a unit are common; some operons contain 20 or more genes. The identity and order of the genes in an operon are not random. In many cases, genes in the same operon encode subunits of a larger protein complex, and cotranslation directly enables assembly of the complex. Some operons organize genes involved in related processes that require coordinated regulation. In other cases, the genes may seem to be unrelated, but they encode products required by the cell under similar conditions.

Many of the principles of bacterial gene expression were first defined by studies of lactose metabolism in *E. coli*, which can use lactose as its sole carbon source. In 1960, François Jacob and Jacques Monod published a short paper in the *Proceedings of the French Academy of Sciences* that described how two adjacent genes involved in lactose metabolism were coordinately regulated by a genetic element located at one end of the gene cluster. The genes were those for β-galactosidase, which cleaves lactose to galactose and glucose, and for galactoside permease (lactose permease, p. 418), which transports lactose into the cell **(Fig. 28-7)**. The terms "operon" and "operator" were first introduced in this paper. With the operon model, gene regulation could, for the first time, be considered in molecular terms.

François Jacob, 1920–2013
[Source: Corbis/Bettmann.]

Jacques Monod, 1910–1976
[Source: Corbis/Bettmann.]

The *lac* Operon Is Subject to Negative Regulation

The lactose (*lac*) operon **(Fig. 28-8a)** includes the genes for β-galactosidase (*Z*), galactoside permease (*Y*), and thiogalactoside transacetylase (*A*). The last of these enzymes seems to modify toxic galactosides to

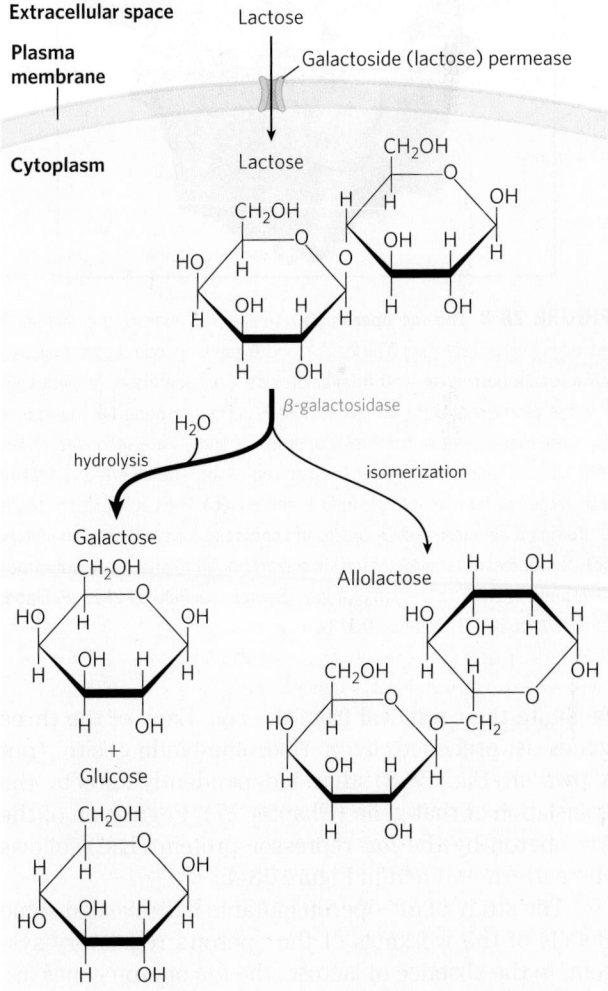

FIGURE 28-7 Lactose metabolism in *E. coli*. Uptake and metabolism of lactose require the activities of galactoside (lactose) permease and β-galactosidase. Conversion of lactose to allolactose by transglycosylation is a minor reaction also catalyzed by β-galactosidase.

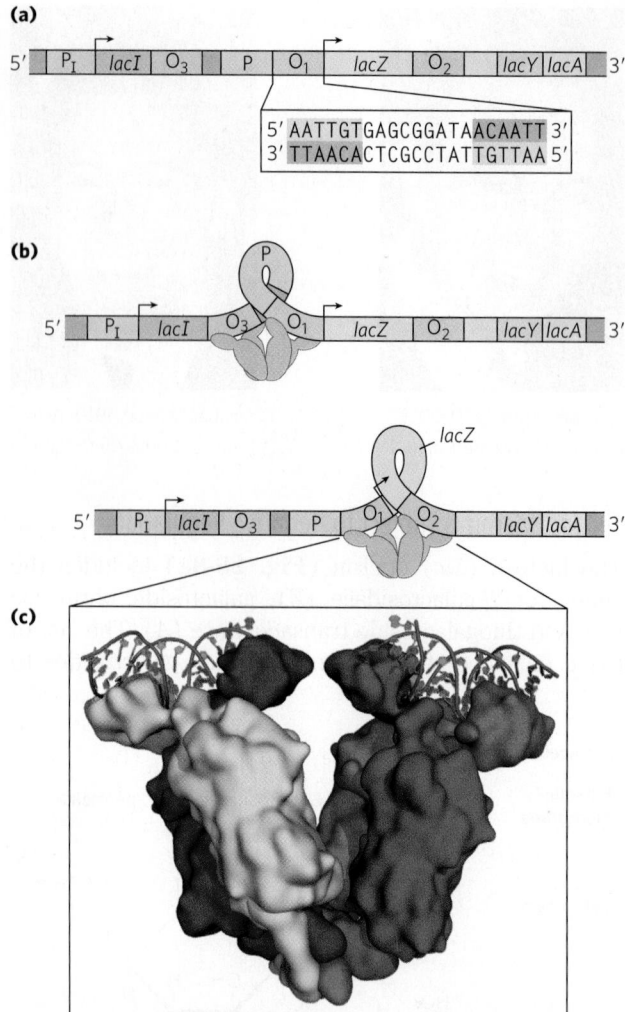

(a)

5′ | P_I | lacI | O_3 | | P | O_1 | | lacZ | O_2 | | lacY | lacA | 3′

5′ AATTGTGAGCGGATAACAATT 3′
3′ TTAACACTCGCCTATTGTTAA 5′

(b)

5′ | P_I | lacI | O_3 | | P | | lacZ | O_2 | | lacY | lacA | 3′

P

(c)

lacZ

5′ | P_I | lacI | O_3 | | P | O_1 | O_2 | | lacY | lacA | 3′

FIGURE 28-8 The lac operon. (a) In the *lac* operon, the *lacI* gene encodes the Lac repressor. The *lac Z, Y,* and *A* genes encode β-galactosidase, galactoside permease, and thiogalactoside transacetylase, respectively. P is the promoter for the *lac* genes, and P_I is the promoter for the *I* gene. O_1 is the main operator for the *lac* operon; O_2 and O_3 are secondary operator sites of lesser affinity for the Lac repressor. The inverted repeat to which the Lac repressor binds in O_1 is shown. **(b)** The Lac repressor binds to the main operator and O_2 or O_3, and seems to form a loop in the DNA. **(c)** Lac repressor (shades of red) is shown bound to short, discontinuous segments of DNA (blue and orange). [Source: (c) PDB ID 2PE5, R. Daber et al., *J. Mol. Biol.* 370:609, 2007.]

facilitate their removal from the cell. Each of the three genes is preceded by a ribosome-binding site (not shown in Fig. 28-8) that independently directs the translation of that gene (Chapter 27). Regulation of the *lac* operon by the *lac* repressor protein (Lac) follows the pattern outlined in Figure 28-4a.

The study of *lac* operon mutants has revealed some details of the workings of the operon's regulatory system. In the absence of lactose, the *lac* operon genes are repressed. Mutations in the operator or in another gene, the *I* gene, result in constitutive synthesis of the gene products. When the *I* gene is defective, repression can be restored by introducing a functional *I* gene into the

cell on another DNA molecule, demonstrating that the *I* gene encodes a diffusible molecule that causes gene repression. This molecule proved to be a protein, now called the Lac repressor, a tetramer of identical monomers. The operator to which it binds most tightly (O_1) abuts the transcription start site (Fig. 28-8a). The *I* gene is transcribed from its own promoter (P_I) independent of the *lac* operon genes. The *lac* operon has two secondary binding sites for the Lac repressor, O_2 and O_3. O_2 is centered near position +410, within the gene encoding β-galactosidase (Z); O_3 is near position −90, within the *I* gene. To repress the operon, the Lac repressor seems to bind to both the main operator and one of the two secondary sites, with the intervening DNA looped out (Fig. 28-8b, c). Either binding arrangement blocks transcription initiation.

Despite this elaborate binding complex, repression is not absolute. Binding of the Lac repressor reduces the rate of transcription initiation by a factor of 10^3. If the O_2 and O_3 sites are eliminated by deletion or mutation, the binding of repressor to O_1 alone reduces transcription by a factor of about 10^2. Even in the repressed state, each cell has a few molecules of β-galactosidase and galactoside permease, presumably synthesized on the rare occasions when the repressor transiently dissociates from the operators. This basal level of transcription is essential to operon regulation.

When cells are provided with lactose, the *lac* operon is induced. An inducer (signal) molecule binds to a specific site on the Lac repressor, causing a conformational change that results in dissociation of the repressor from the operator. The inducer in the *lac* operon system is not lactose itself but allolactose, an isomer of lactose (Fig. 28-7). After entry into the *E. coli* cell (via the few existing molecules of lactose permease), lactose is converted to allolactose by one of the few existing β-galactosidase molecules. Release of the operator by Lac repressor, triggered as the repressor binds to allolactose, allows expression of the *lac* operon genes and leads to a 10^3-fold increase in the concentration of β-galactosidase.

Several β-galactosides structurally related to allolactose are inducers of the *lac* operon but are not substrates for β-galactosidase; others are substrates but not inducers. One particularly effective and nonmetabolizable inducer of the *lac* operon that is often used experimentally is isopropylthiogalactoside (IPTG).

Isopropyl-β-D-thiogalactoside
(IPTG)

An inducer that cannot be metabolized allows researchers to explore the physiological function of lactose as a carbon source for growth, separate from its function in the regulation of gene expression.

In addition to the multitude of operons now known in bacteria, a few polycistronic operons have been found in the cells of lower eukaryotes. In the cells of higher eukaryotes, however, almost all protein-coding genes are transcribed separately.

The mechanisms by which operons are regulated can vary significantly from the simple model presented in Figure 28-8. Even the *lac* operon is more complex than indicated here, with an activator also contributing to the overall scheme, as we shall see in Section 28.2. Before any further discussion of the layers of regulation of gene expression, however, we examine the critical molecular interactions between DNA-binding proteins (such as repressors and activators) and the DNA sequences to which they bind.

Regulatory Proteins Have Discrete DNA-Binding Domains

Regulatory proteins generally bind to specific DNA sequences. Their affinity for these target sequences is roughly 10^4 to 10^6 times higher than their affinity for any other DNA sequence. Most regulatory proteins have discrete DNA-binding domains containing substructures that interact closely and specifically with the DNA. These binding domains usually include one or

more of a relatively small group of recognizable and characteristic structural motifs.

To bind specifically to DNA sequences, regulatory proteins must recognize surface features on the DNA. Most of the chemical groups that differ among the four bases and thus permit discrimination between base pairs are hydrogen-bond donor and acceptor groups exposed in the major groove of DNA **(Fig. 28-9)**, and most of the protein-DNA contacts that impart specificity are hydrogen bonds. A notable exception is the nonpolar surface near C-5 of pyrimidines, where thymine is readily distinguished from cytosine by its protruding methyl group. Protein-DNA contacts are also possible in the minor groove of the DNA, but the hydrogen-bonding patterns there generally do not allow ready discrimination between base pairs.

Within regulatory proteins, the amino acid side chains most often hydrogen-bonding to bases in the DNA are those of Asn, Gln, Glu, Lys, and Arg residues. Is there a simple recognition code in which a particular amino acid always pairs with a particular base? The two hydrogen bonds that can form between Gln or Asn and the N^6 and N-7 positions of adenine cannot form with any other base. And an Arg residue can form two hydrogen bonds with N-7 and O^6 of guanine **(Fig. 28-10)**. Examination of the structure of many DNA-binding proteins, however, has shown that a protein can recognize each base pair in more than one way, leading to the

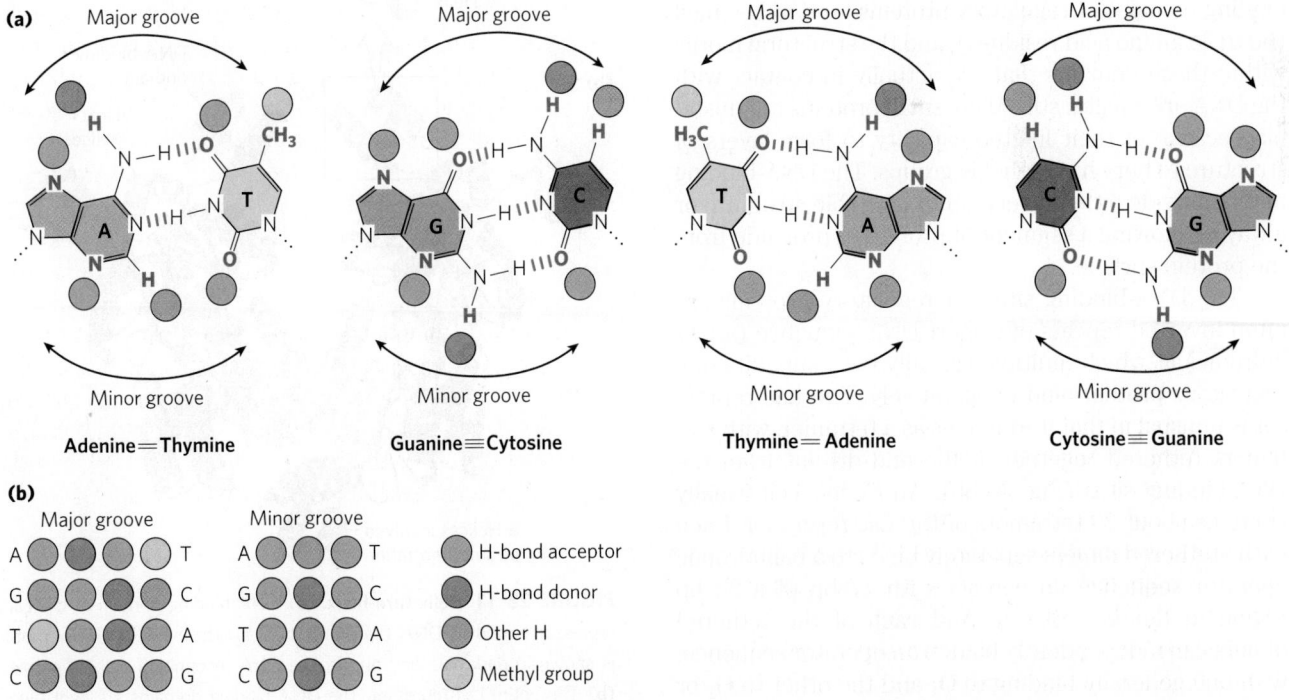

FIGURE 28-9 Groups in DNA available for protein binding. (a) Shown here are functional groups on all four base pairs that are displayed in the major and minor grooves of DNA (see Fig. 8-13). Hydrogen-bond acceptor and donor atoms are marked by blue and red disks, respectively. Other hydrogen atoms are marked with purple disks, and methyl groups with yellow disks.

(b) Recognition patterns for each base pair (from left to right). The much greater variation in the patterns for the major groove gives rise to a much greater discriminatory power in the major groove relative to the minor groove. [Source: Information from J. L. Huret, *Atlas Genet. Cytogenet. Oncol. Haematol.*, 2006, http://atlasgeneticsoncology.org/Educ/DNAEngID30001ES.html.]

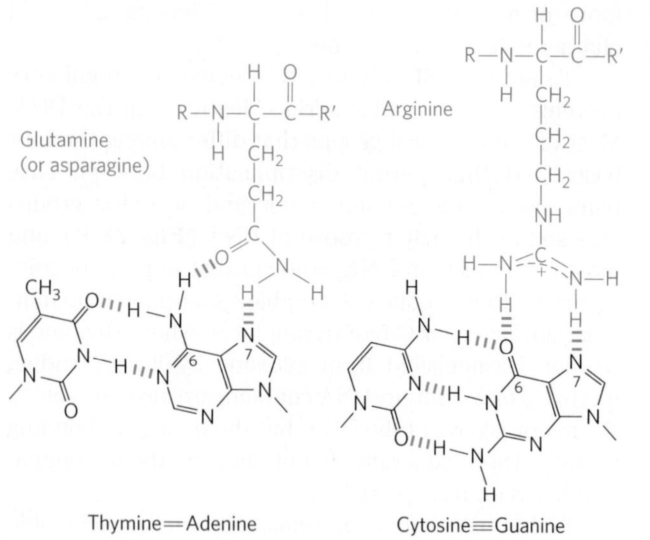

Glutamine
(or asparagine)

Arginine

Thymine═Adenine

Cytosine≡Guanine

FIGURE 28-10 Specific amino acid residue–base pair interactions. The two examples shown have been observed in DNA-protein binding.

conclusion that there is no simple amino acid–base code. For some proteins, the Gln-adenine interaction can specify A═T base pairs, but in others a van der Waals pocket for the methyl group of thymine can recognize A═T base pairs. Researchers cannot yet examine the structure of a DNA-binding protein and infer the DNA sequence to which it binds.

To interact with bases in the major groove of DNA, a protein requires a relatively small substructure that can stably protrude from the protein surface. The DNA-binding domains of regulatory proteins tend to be small (60 to 90 amino acid residues), and the structural motifs within these domains that are actually in contact with the DNA are smaller still. Many small proteins are unstable because of their limited capacity to form layers of structure to bury hydrophobic groups. The DNA-binding motifs provide either a very compact stable structure or a way of allowing a segment of protein to protrude from the protein surface.

The DNA-binding sites for regulatory proteins are often inverted repeats of a short DNA sequence (a palindrome) at which multiple (usually two) subunits of a regulatory protein bind cooperatively. The Lac repressor is unusual in that it functions as a tetramer, with two dimers tethered together at the end distant from the DNA-binding sites (Fig. 28-8b). An *E. coli* cell usually contains about 20 tetramers of the Lac repressor. Each of the tethered dimers separately binds to a palindromic operator sequence, in contact with 17 bp of a 22 bp region in the *lac* operon. And each of the tethered dimers can independently bind to an operator sequence, with one generally binding to O_1 and the other to O_2 or O_3 (as in Fig. 28-8b). The symmetry of the O_1 operator sequence corresponds to the twofold axis of symmetry of two paired Lac repressor subunits. The tetrameric Lac repressor binds to its operator sequences in vivo with an estimated dissociation constant of 10^{-10} M.

The repressor discriminates between the operators and other sequences by a factor of about 10^6, so binding to these few base pairs among the 4.6 million or so of the *E. coli* chromosome is highly specific.

Several DNA-binding motifs have been described, but here we focus on two that play prominent roles in the binding of DNA by regulatory proteins from all domains of life: the helix-turn-helix and the zinc finger. We also consider two other types of such motifs: the homeodomain and the RNA recognition motif, which, as its name implies, also binds RNA; both motifs play prominent roles in some eukaryotic regulatory proteins.

Helix-Turn-Helix The **helix-turn-helix** motif is crucial to the interaction of many regulatory proteins with DNA in bacteria, and similar motifs occur in some eukaryotic regulatory proteins. The helix-turn-helix comprises about 20 amino acid residues in two short α-helical segments, each 7 to 9 residues long, separated by a β turn **(Fig. 28-11)**. This structure generally is not stable by itself; it is simply the reactive portion of a somewhat larger DNA-binding domain. One of the two α-helical segments is called the recognition helix, because it usually contains

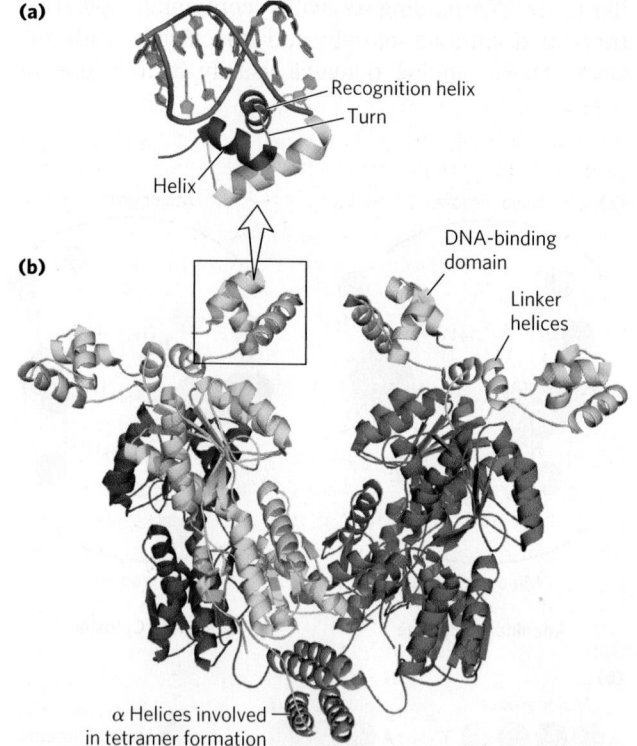

(a)

Recognition helix
Turn
Helix

DNA-binding domain

Linker helices

(b)

α Helices involved
in tetramer formation

FIGURE 28-11 Helix-turn-helix. (a) DNA-binding domain of the Lac repressor bound to DNA (blue and orange). The helix-turn-helix motif is shown in dark blue and purple; the DNA recognition helix is purple. **(b)** The entire Lac repressor. The DNA-binding domains are light blue; the α helices involved in tetramer formation are green. The remainder of the protein (shades of red) has the binding sites for allolactose. The allolactose-binding domains are linked to the DNA-binding domains through linker helices (yellow). [Source: PDB ID 2PE5, R. Daber et al., *J. Mol. Biol.* 370:609, 2007.]

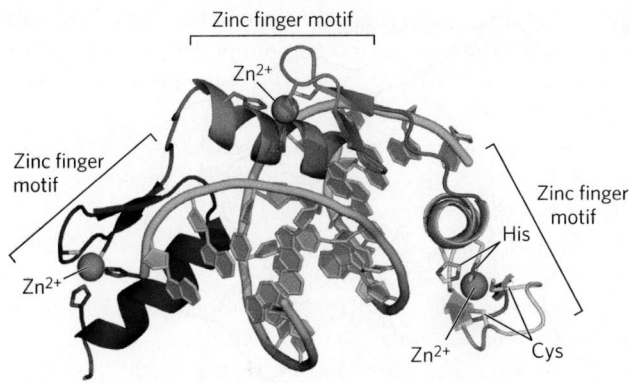

FIGURE 28-12 Zinc fingers. Three zinc fingers (shades of red) of the regulatory protein Zif268, complexed with DNA (blue). Each Zn^{2+} coordinates with two His and two Cys residues. [Source: PDB ID 1ZAA, N. P. Pavletich and C. O. Pabo, *Science* 252:809, 1991.]

many of the amino acids that interact with DNA in a sequence-specific way. This α helix is stacked on other segments of the protein structure so that it protrudes from the protein surface. When bound to DNA, the recognition helix is positioned in or nearly in the major groove. The Lac repressor has this DNA-binding motif (Fig. 28-11).

Zinc Finger In a **zinc finger**, about 30 amino acid residues form an elongated loop held together at the base by a single Zn^{2+} ion, which is coordinated to four of the residues (four Cys, or two Cys and two His). The zinc does not itself interact with DNA; rather, the coordination of zinc with the amino acid residues stabilizes this small structural motif. Several hydrophobic side chains in the core of the structure also lend stability. **Figure 28-12** shows the interaction between DNA and three zinc fingers of a single polypeptide from the mouse regulatory protein Zif268.

Many eukaryotic DNA-binding proteins contain zinc fingers. The interaction of a single zinc finger with DNA is typically weak, and many DNA-binding proteins, like Zif268, have multiple zinc fingers that substantially enhance binding by interacting simultaneously with the DNA. One DNA-binding protein of the frog *Xenopus* has 37 zinc fingers. There are few known examples of the zinc finger motif in bacterial proteins.

The precise manner in which proteins with zinc fingers bind to DNA differs from one protein to the next. Some zinc fingers contain the amino acid residues that are important in sequence discrimination, whereas others seem to bind DNA nonspecifically (the amino acids required for specificity are located elsewhere in the protein). Zinc fingers can also function as RNA-binding motifs, such as in certain proteins that bind eukaryotic mRNAs and act as translational repressors. We discuss this role later (Section 28.3).

Homeodomain Another type of DNA-binding domain has been identified in some proteins that function as transcriptional regulators, especially during eukaryotic

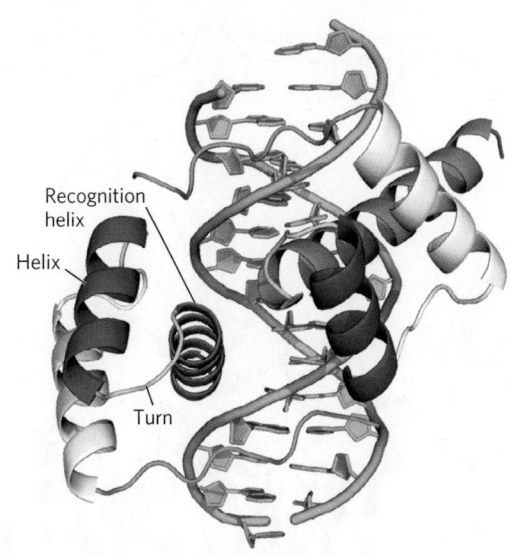

FIGURE 28-13 Homeodomains. Shown here are two homeodomains bound to DNA. In each homeodomain, one of the α helices (purple), layered on two others (dark blue and gray), can be seen protruding into the major groove. This is only a small part of a larger regulatory protein from a class called Pax, active in the regulation of development in fruit flies (see Section 28.3). [Source: PDB ID 1FJL, D. S. Wilson et al., *Cell* 82:709, 1995.]

development. This domain of 60 amino acid residues—called the **homeodomain**, because it was discovered in homeotic genes (genes that regulate the development of body patterns)—is highly conserved and has now been identified in proteins from a wide variety of organisms, including humans **(Fig. 28-13)**. The DNA-binding segment of the domain is related to the helix-turn-helix motif. The DNA sequence that encodes this domain is known as the **homeobox**.

RNA Recognition Motif An RNA-binding domain is not out of place in this discussion. **RNA recognition motifs (RRMs)** are found in some eukaryotic gene activators, where they may do double duty in binding DNA and RNA. When bound to specific binding sites in DNA, these activators induce transcription. The same activators are sometimes regulated in part by specific lncRNAs that compete with DNA binding and decrease gene transcription. Other proteins with RRM motifs bind to mRNA, rRNA, or any of a range of other smaller, noncoding RNAs. The RRM consists of 90 to 100 amino acid residues, arranged in a four-strand antiparallel β-sheet sandwiched against two α helices, with a β_1-α_1-β_2-β_3-α_2-β_4 topology **(Fig. 28-14)**. This motif may be present as part of DNA-binding regulatory proteins that also have other DNA-binding motifs, or may occur in proteins that bind uniquely to RNA.

Regulatory Proteins Also Have Protein-Protein Interaction Domains

Regulatory proteins contain domains not only for DNA binding but also for protein-protein interactions—with

(a)

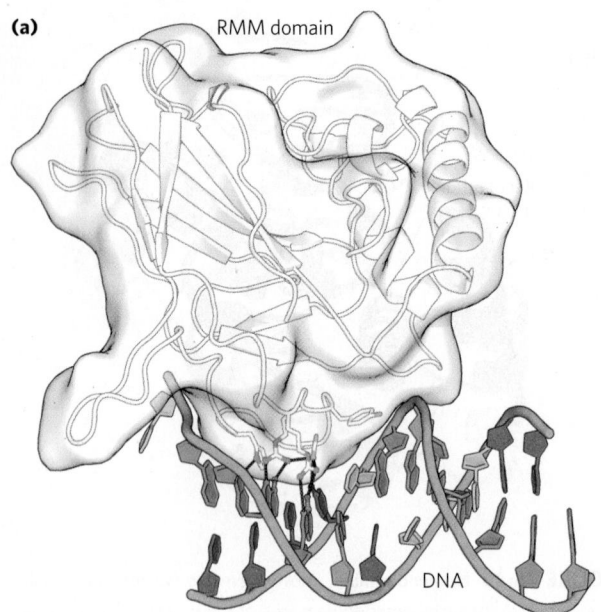

(b)

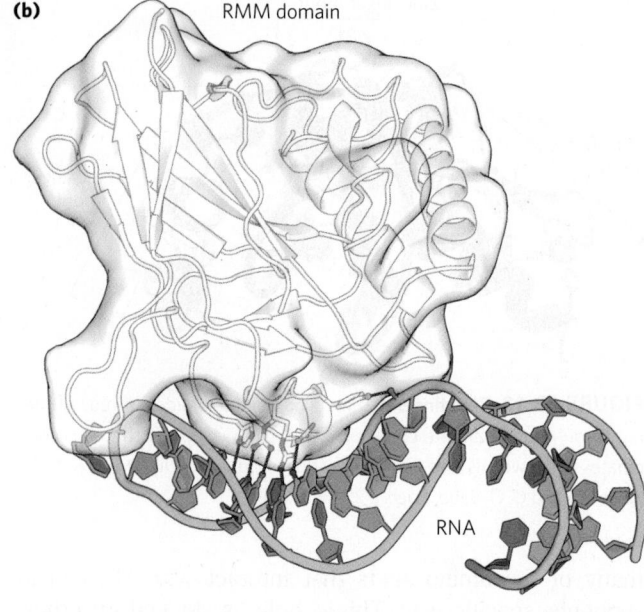

FIGURE 28-14 RNA recognition motifs (RRMs). An RRM from the p50 subunit of regulatory protein NF-κB is shown, bound to **(a)** DNA and **(b)** RNA. Hydrogen-bonding interactions between particular amino acid residues and bases in the DNA or RNA are shown with black lines. NF-κB is the name of a family of structurally related eukaryotic transcription factors that regulate processes ranging from immune and inflammatory responses to cell growth and apoptosis. [Sources: (a) PDB ID 1OOA, D. B. Huang et al., *Proc. Natl. Acad. Sci. USA* 100:9268, 2003. (b) PDB ID 1VKX, F. E. Chen et al., *Nature* 391:410, 1998.]

RNA polymerase, other regulatory proteins, or other subunits of the same regulatory protein. Examples include many eukaryotic transcription factors that function as gene activators, which often bind as dimers to the DNA through DNA-binding domains that contain zinc fingers. Some structural domains are devoted to the interactions required for dimer formation, which is generally a prerequisite for DNA binding. Like DNA-binding motifs, the structural motifs that mediate protein-protein interactions tend to fall within one of a few common categories. Two important examples are the leucine zipper and the basic helix-loop-helix. Structural motifs such as these are the basis for classifying some regulatory proteins into structural families.

Leucine Zipper The **leucine zipper** is an amphipathic α helix with a series of hydrophobic amino acid residues concentrated on one side **(Fig. 28-15)**, with the hydrophobic surface forming the area of contact between the two polypeptides of a dimer. A striking feature of these α helices is the occurrence of Leu residues at every seventh position, forming a straight line along the hydrophobic surface. Although researchers initially thought the Leu residues interdigitated (hence the name "zipper"), we now know that they line up side by side as the interacting α helices coil around each other (forming a coiled coil; Fig. 28-15b). Regulatory proteins with leucine zippers often have a separate DNA-binding domain with a high concentration of basic (Lys or Arg) residues that can interact with the negatively charged phosphates of the DNA backbone. Leucine zippers

have been found in many eukaryotic and a few bacterial proteins.

Basic Helix-Loop-Helix Another common structural motif, the **basic helix-loop-helix**, occurs in some eukaryotic regulatory proteins implicated in the control of gene expression during development of multicellular organisms. These proteins share a conserved region of about 50 amino acid residues important in both DNA binding and protein dimerization. This region can form two short amphipathic α helices linked by a loop of variable length, the helix-loop-helix (distinct from the helix-turn-helix motif associated with DNA binding). The helix-loop-helix motifs of two polypeptides interact to form dimers **(Fig. 28-16)**. In these proteins, DNA binding is mediated by an adjacent short amino acid sequence rich in basic residues, similar to the separate DNA-binding region in proteins containing leucine zippers.

Protein-Protein Interactions in Eukaryotic Regulatory Proteins In eukaryotes, most genes are regulated by activators, and most genes are monocistronic. If a different activator were required for each gene, the number of activators (and genes encoding them) would need to be equivalent to the number of regulated genes. However, in yeast, about 300 transcription factors (many of them activators) are responsible for the regulation of many thousands of genes. Many of the transcription factors regulate the induction of multiple genes, but most genes are subject to regulation by multiple transcription factors

(a)

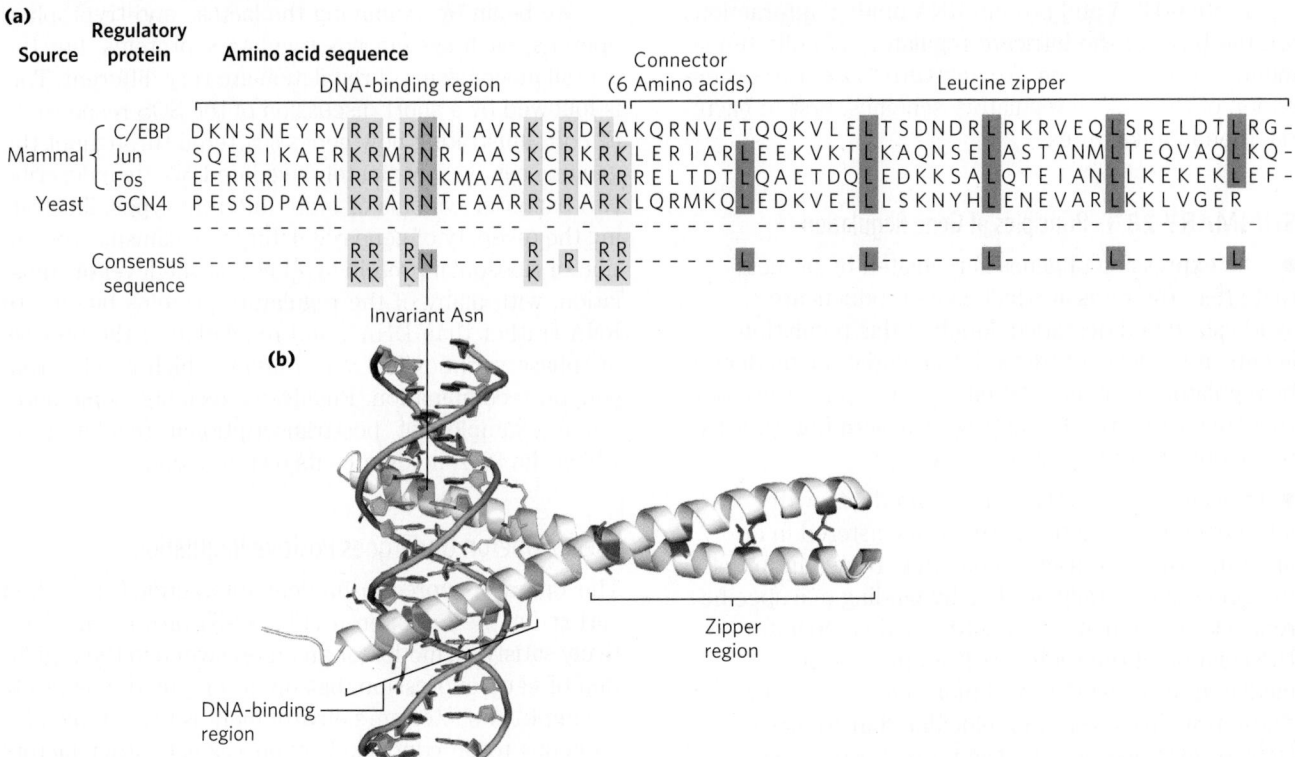

FIGURE 28-15 Leucine zippers. (a) Comparison of amino acid sequences of several leucine zipper proteins. Notice the Leu (L) residues (red) at every seventh position in the zipper region, and the number of Lys (K) and Arg (R) residues in the DNA-binding region (yellow). (b) Leucine zipper from the yeast activator protein GCN4. Only the "zippered" α helices (gray), derived from different subunits of the dimeric protein, are shown. The two helices wrap around each other in a gently coiled coil. The interacting Leu side chains and the conserved residues in the DNA-binding region are colored to correspond to the sequence in (a). [Sources: (a) Information from S. L. McKnight, *Sci. Am.* 264 (April):54, 1991. (b) PDB ID 1YSA, T. E. Ellenberger et al., *Cell* 71:1223, 1992.]

(for example, see Fig. 15-25). Appropriate regulation of different genes is accomplished by different combinations of a limited repertoire of transcription factors at each gene, a mechanism referred to as **combinatorial control**.

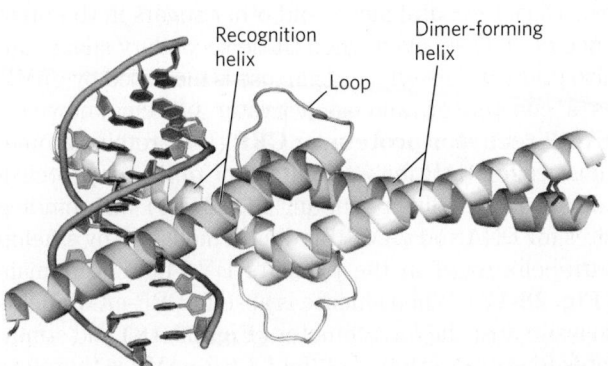

FIGURE 28-16 Helix-loop-helix. The human transcription factor Max, bound to its DNA target site. The protein is dimeric; one subunit is colored. The recognition helix (pink) is linked via the loop to the dimer-forming helix (lavender), which merges with the carboxyl-terminal end of the subunit. Interaction of the carboxyl-terminal helices of the two subunits describes a coiled coil very similar to that of a leucine zipper (see Fig. 28-15b), but with only one pair of interacting Leu residues (red side chains at the right) in this example. The overall structure is sometimes called a helix-loop-helix/leucine zipper motif. [Source: PDB ID 1HLO, P. Brownlie et al., *Structure* 5:509, 1997.]

Combinatorial control is accomplished in part by mixing and matching the variants within a regulatory protein family to form a series of different active protein dimers. Several families of eukaryotic transcription factors have been defined on the basis of close structural similarities. Within each family, dimers can sometimes form between two identical proteins (a homodimer) or between two different members of the family (a heterodimer). A hypothetical family of four different leucine-zipper proteins could thus form up to 10 different dimeric species. In many cases, the different combinations have distinct regulatory and functional properties and regulate different genes. As we shall see, multiple regulatory proteins of this kind function in the regulation of most eukaryotic genes, further contributing to combinatorial control.

In addition to having structural domains devoted to DNA binding and protein dimerization, directing a particular protein dimer to a particular gene, many regulatory proteins have domains that interact with RNA polymerase, with regulatory RNAs, with unrelated regulatory proteins, or with some combination of the three. At least three types of additional domains for protein-protein interaction have been characterized (primarily in eukaryotes): glutamine-rich, proline-rich, and acidic domains, the names reflecting the amino acid residues that are especially abundant.

Protein-DNA and protein-RNA binding interactions are the basis of the intricate regulatory circuits fundamental to gene function. We now turn to a closer examination of these gene regulatory schemes, first in bacteria, then in eukaryotes.

SUMMARY 28.1 Principles of Gene Regulation

■ The expression of genes is regulated by processes that affect the rates at which gene products are synthesized and degraded. Much of this regulation occurs at the level of transcription initiation, mediated by regulatory proteins that either repress transcription (negative regulation) or activate transcription (positive regulation) at specific promoters.

■ In bacteria, genes that encode products with interdependent functions are often clustered in an operon, a single transcriptional unit. Transcription of the genes is generally blocked by binding of a specific repressor protein at a DNA site called an operator. Dissociation of the repressor from the operator is mediated by a specific small molecule, an inducer. These principles were first elucidated in studies of the lactose (*lac*) operon. The Lac repressor dissociates from the *lac* operator when the repressor binds to its inducer, allolactose.

■ Regulatory proteins are DNA-binding proteins that recognize specific DNA sequences; most have distinct DNA-binding domains. Within these domains, common structural motifs that bind DNA (and/or RNA) are the helix-turn-helix, zinc finger, homeodomain, and RNA recognition motif.

■ Regulatory proteins also contain domains for protein-protein interactions, including the leucine zipper and helix-loop-helix, which are involved in dimerization, and other motifs required for activation of transcription. Mixing and matching of protein family variants in dimeric transcription factors provides for more efficient and responsive regulation through combinatorial control.

28.2 Regulation of Gene Expression in Bacteria

As in many other areas of biochemical investigation, the study of the regulation of gene expression advanced earlier and faster in bacteria than in other experimental organisms. The examples of bacterial gene regulation presented here are chosen from among scores of well-studied systems, partly for their historical significance, but primarily because they provide a good overview of the range of regulatory mechanisms in bacteria. Many of the principles of bacterial gene regulation are also relevant to understanding gene expression in eukaryotic cells.

We begin by examining the lactose and tryptophan operons; each system has regulatory proteins, but the overall mechanisms of regulation are very different. This is followed by a short discussion of the SOS response in *E. coli*, illustrating how genes scattered throughout the genome can be coordinately regulated. We then describe two bacterial systems of quite different types, illustrating the diversity of gene regulatory mechanisms: regulation of ribosomal protein synthesis at the level of translation, with many of the regulatory proteins binding to RNA (rather than DNA), and regulation of the process of "phase variation" in *Salmonella*, which results from genetic recombination. Finally, we examine some additional examples of posttranscriptional regulation in which the RNA modulates its own function.

The *lac* Operon Undergoes Positive Regulation

The operator-repressor-inducer interactions described earlier for the *lac* operon (Fig. 28-8) provide an intuitively satisfying model for an on/off switch in the regulation of gene expression, but operon regulation is rarely so simple. A bacterium's environment is too complex for its genes to be controlled by one signal. Other factors besides lactose affect the expression of the *lac* genes, such as the availability of glucose. Glucose, metabolized directly by glycolysis, is the preferred energy source in *E. coli*. Other sugars can serve as the main or sole nutrient, but extra enzymatic steps are required to prepare them for entry into glycolysis, necessitating the synthesis of additional enzymes. Clearly, expressing the genes for proteins that metabolize sugars such as lactose or arabinose is wasteful when glucose is abundant.

What happens to the expression of the *lac* operon when both glucose and lactose are present? A regulatory mechanism known as **catabolite repression** restricts expression of the genes required for catabolism of lactose, arabinose, and other sugars in the presence of glucose, even when these secondary sugars are also present. The effect of glucose is mediated by cAMP, as a coactivator, and an activator protein known as **cAMP receptor protein**, or **CRP** (the protein is sometimes called CAP, for *catabolite gene activator protein*). CRP is a homodimer (subunit M_r 22,000) with binding sites for DNA and cAMP. Binding is mediated by a helix-turn-helix motif in the protein's DNA-binding domain (**Fig. 28-17**). When glucose is absent, CRP-cAMP binds to a site near the *lac* promoter (**Fig. 28-18**) and stimulates RNA transcription 50-fold. CRP-cAMP is therefore a positive regulatory element responsive to glucose levels, whereas the Lac repressor is a negative regulatory element responsive to lactose. The two act in concert. CRP-cAMP has little effect on the *lac* operon when the Lac repressor is blocking transcription, and dissociation of the repressor from the *lac* operator has little effect on transcription of the *lac* operon unless CRP-cAMP is present to facilitate transcription; when CRP is not bound, the wild-type *lac* promoter is a relatively weak

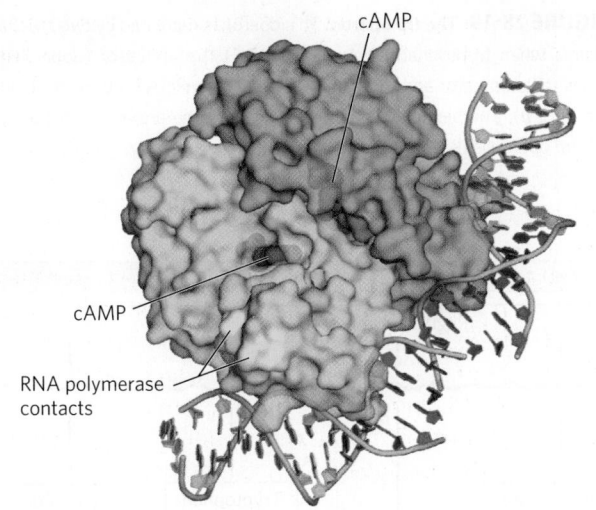

FIGURE 28-17 CRP homodimer with bound cAMP. Note the bending of the DNA around the protein. The region that interacts with RNA polymerase is indicated (yellow). [Source: PDB ID 1RUN, G. Parkinson et al., *Nature Struct. Biol.* 3:837, 1996.]

(a) Glucose high, cAMP low, lactose absent

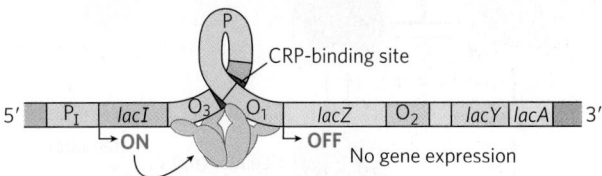

(b) Glucose low, cAMP high, lactose absent

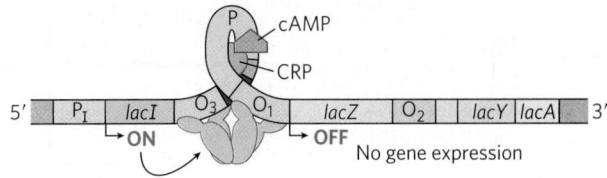

(c) Glucose high, cAMP low, lactose present

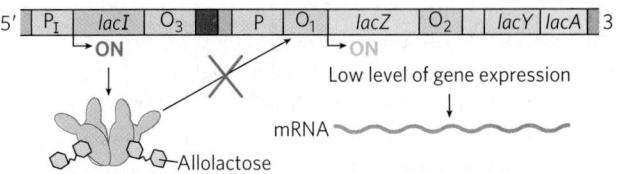

(d) Glucose low, cAMP high, lactose present

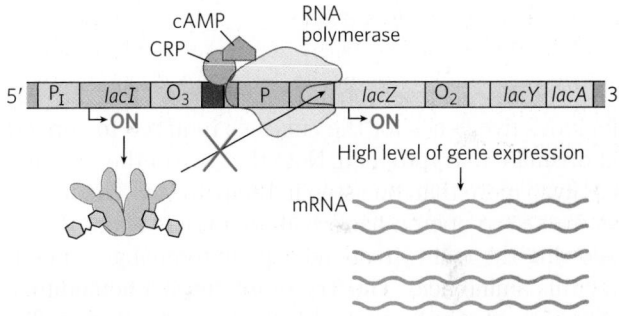

FIGURE 28-18 Positive regulation of the *lac* operon by CRP. The binding site for CRP-cAMP is near the promoter. The combined effects of glucose and lactose availability on *lac* operon expression are shown. When lactose is absent, the repressor binds to the operator and prevents transcription of the *lac* genes. It does not matter whether glucose is **(a)** present or **(b)** absent. **(c)** If lactose is present, the repressor dissociates from the operator. However, if glucose is also available, low cAMP levels prevent CRP-cAMP formation and DNA binding. RNA polymerase may occasionally bind and initiate transcription, resulting in a very low level of *lac* genes transcription. **(d)** When lactose is present and glucose levels are low, cAMP levels rise. The CRP-cAMP complex forms and facilitates robust binding of RNA polymerase to the *lac* promoter and high levels of transcription.

promoter (Fig. 28-18a, c). The open complex of RNA polymerase and the promoter (see Fig. 26-6) does not form readily unless CRP-cAMP is present. CRP interacts directly with RNA polymerase (at the region shown in Fig. 28-17) through the polymerase's α subunit.

The effect of glucose on CRP is mediated by the cAMP interaction (Fig. 28-18). CRP binds to DNA most avidly when cAMP concentrations are high. In the presence of glucose, the synthesis of cAMP is inhibited and efflux of cAMP from the cell is stimulated. As [cAMP] declines, CRP binding to DNA declines, thereby decreasing the expression of the *lac* operon. Strong induction of the *lac* operon therefore requires both lactose (to inactivate the *lac* repressor) and a lowered concentration of glucose (to trigger an increase in [cAMP] and increased binding of cAMP to CRP).

CRP and cAMP participate in the coordinated regulation of many operons, primarily those that encode enzymes for the metabolism of secondary sugars such as lactose and arabinose. A network of operons with a common regulator is called a **regulon**. This arrangement, which allows coordinated shifts in cellular functions that can require the action of hundreds of genes, is a major theme in the regulated expression of dispersed networks of genes in eukaryotes. Other bacterial regulons include the heat shock gene system that responds to changes in temperature (p. 1039) and the genes induced in *E. coli* as part of the SOS response to DNA damage, described later.

Many Genes for Amino Acid Biosynthetic Enzymes Are Regulated by Transcription Attenuation

The 20 common amino acids are required in large amounts for protein synthesis, and *E. coli* can synthesize

all of them. The genes for the enzymes needed to synthesize a given amino acid are generally clustered in an operon and are expressed whenever existing supplies of that amino acid are inadequate for cellular requirements. When the amino acid is abundant, the biosynthetic enzymes are not needed and the operon is repressed.

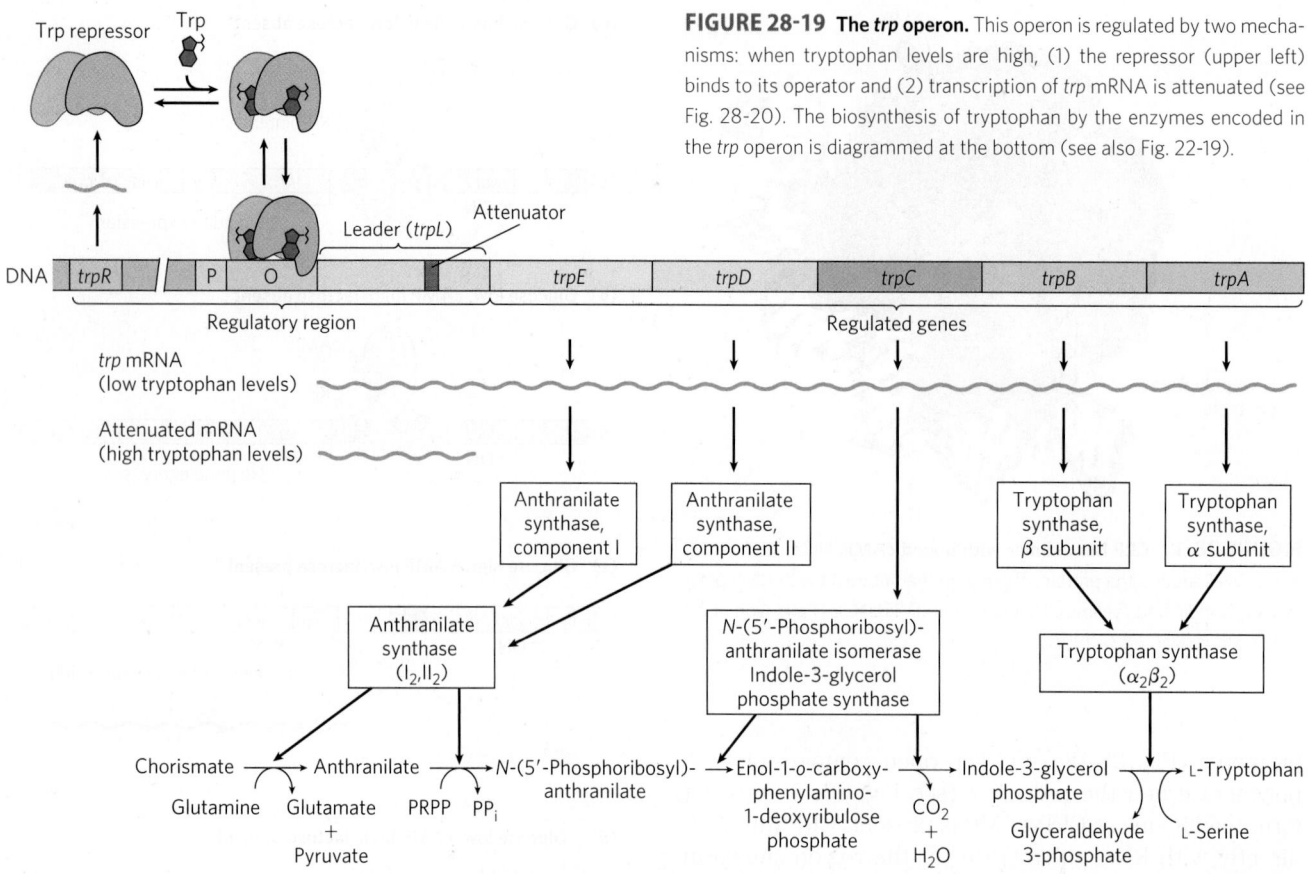

FIGURE 28-19 The *trp* operon. This operon is regulated by two mechanisms: when tryptophan levels are high, (1) the repressor (upper left) binds to its operator and (2) transcription of *trp* mRNA is attenuated (see Fig. 28-20). The biosynthesis of tryptophan by the enzymes encoded in the *trp* operon is diagrammed at the bottom (see also Fig. 22-19).

The *E. coli* tryptophan (*trp*) operon **(Fig. 28-19)** includes five genes for the enzymes required to convert chorismate to tryptophan. Note that two of the enzymes catalyze more than one step in the pathway. The mRNA from the *trp* operon has a half-life of only about 3 min, allowing the cell to respond rapidly to changing needs for this amino acid. The Trp repressor is a homodimer. When tryptophan is abundant, it binds to the Trp repressor, causing a conformational change that permits the repressor to bind to the *trp* operator and inhibit expression of the *trp* operon. The *trp* operator site overlaps the promoter, so binding of the repressor blocks binding of RNA polymerase.

Once again, this simple on/off circuit mediated by a repressor is not the entire regulatory story. Different cellular concentrations of tryptophan can vary the rate of synthesis of the biosynthetic enzymes over a 700-fold range. Once repression is lifted and transcription begins, the rate of transcription is fine-tuned to cellular tryptophan requirements by a second regulatory process, called **transcription attenuation**, in which transcription is initiated normally but is abruptly halted *before* the operon genes are transcribed. The frequency with which transcription is attenuated is regulated by the availability of tryptophan and relies on the very close coupling of transcription and translation in bacteria.

The *trp* operon attenuation mechanism uses signals encoded in four sequences within a 162 nucleotide **leader** region at the 5′ end of the mRNA, preceding the initiation codon of the first gene **(Fig. 28-20a)**. The leader contains a region known as the **attenuator**, made up of sequences 3 and 4. These sequences base-pair to form a G≡C-rich stem-and-loop structure closely followed by a series of U residues. The attenuator structure acts as a transcription terminator (Fig. 28-20b). Sequence 2 is an alternative complement for sequence 3 (Fig. 28-20c). If sequences 2 and 3 base-pair, the attenuator structure cannot form and transcription continues into the *trp* biosynthetic genes; the loop formed by the pairing of sequences 2 and 3 does not obstruct transcription.

Regulatory sequence 1 is crucial for a tryptophan-sensitive mechanism that determines whether sequence 3 pairs with sequence 2 (allowing transcription to continue) or with sequence 4 (attenuating transcription). Formation of the attenuator stem-and-loop structure depends on events that occur during *translation* of regulatory sequence 1, which encodes a leader peptide (so called because it is encoded by the leader region of the mRNA) of 14 amino acids, two of which are Trp residues. The leader peptide has no other known cellular function; its synthesis is simply an operon regulatory device. This peptide is translated immediately after it is transcribed, by a ribosome that follows closely behind RNA polymerase as transcription proceeds.

When tryptophan concentrations are high, concentrations of charged tryptophan tRNA (Trp-tRNA^{Trp})

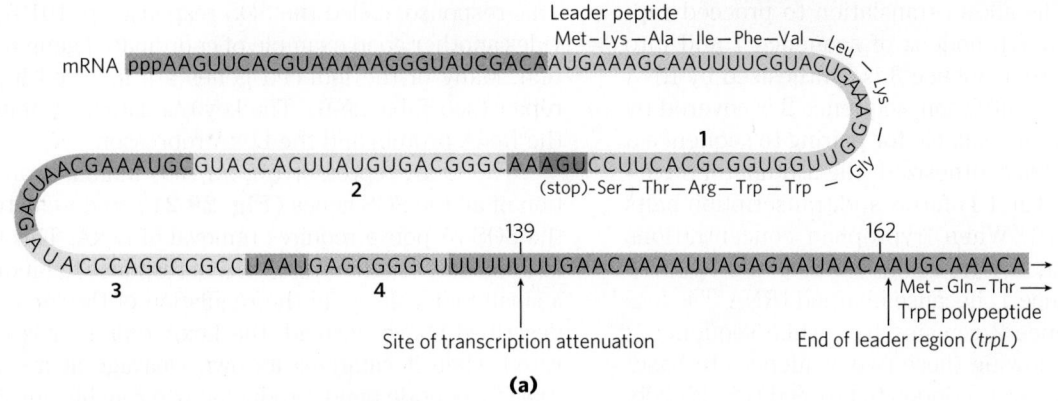

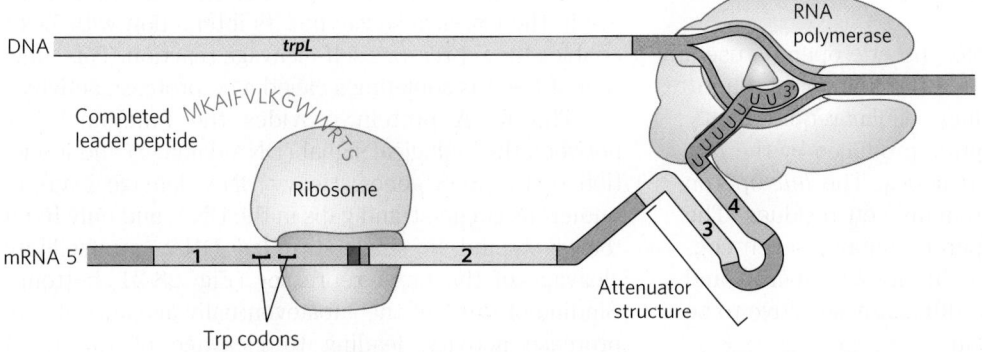

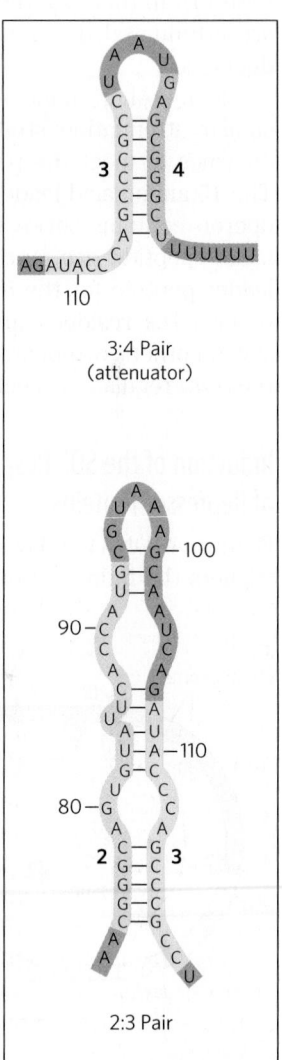

When tryptophan levels are high, the ribosome quickly translates sequence 1 (open reading frame encoding leader peptide) and blocks sequence 2 before sequence 3 is transcribed. Continued transcription leads to attenuation at the terminator-like attenuator structure formed by sequences 3 and 4.

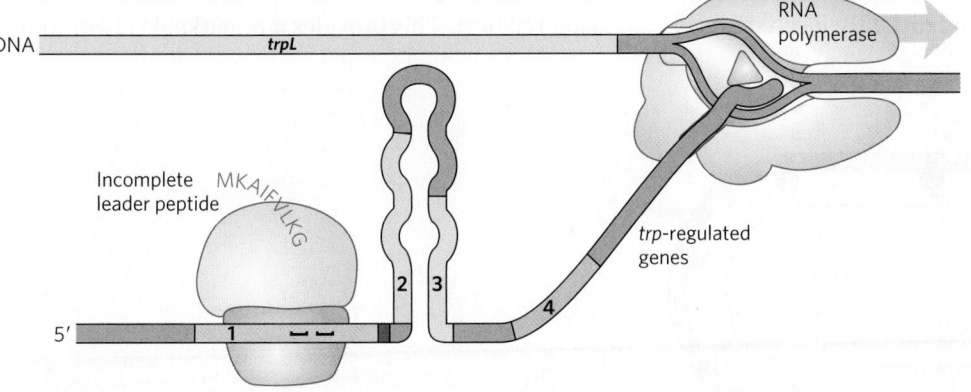

When tryptophan levels are low, the ribosome pauses at the Trp codons in sequence 1. Formation of the paired structure between sequences 2 and 3 prevents attenuation, because sequence 3 is no longer available to form the attenuator structure with sequence 4. The 2:3 structure, unlike the 3:4 attenuator, does not prevent transcription.

(b)

(c)

FIGURE 28-20 Transcriptional attenuation in the *trp* operon. Transcription is initiated at the beginning of the 162 nucleotide mRNA leader encoded by a DNA region called *trpL* (see Fig. 28-19). A regulatory mechanism determines whether transcription is attenuated at the end of the leader or continues into the structural genes. **(a)** The *trp* mRNA leader (*trpL*). The attenuation mechanism in the *trp* operon involves sequences 1 to 4 (highlighted). **(b)** Sequence 1 encodes a small peptide, the leader peptide, containing two Trp residues (W); it is translated immediately after transcription begins. Sequences 2 and 3 are complementary, as are sequences 3 and 4. The attenuator structure forms by the pairing of sequences 3 and 4 (top). Its structure and function are similar to those of a transcription terminator (see Fig. 26-7a). Pairing of sequences 2 and 3 (bottom) prevents the attenuator structure from forming. Note that the leader peptide has no other cellular function. Translation of its open reading frame has a purely regulatory role that determines which complementary sequences (2 and 3, or 3 and 4) are paired. **(c)** Base-pairing schemes for the complementary regions of the *trp* mRNA leader.

are also high. This allows translation to proceed rapidly past the two Trp codons of sequence 1 and into sequence 2, before sequence 3 is synthesized by RNA polymerase. In this situation, sequence 2 is covered by the ribosome and unavailable for pairing to sequence 3 when sequence 3 is synthesized; the attenuator structure (sequences 3 and 4) forms and transcription halts (Fig. 28-20b, top). When tryptophan concentrations are low, however, the ribosome stalls at the two Trp codons in sequence 1, because charged tRNA^{Trp} is less available. Sequence 2 remains free while sequence 3 is synthesized, allowing these two sequences to base-pair and permitting transcription to proceed (Fig. 28-20b, bottom). In this way, the proportion of transcripts that are attenuated declines as tryptophan concentration declines.

Many other amino acid biosynthetic operons use a similar attenuation strategy to fine-tune biosynthetic enzymes to meet the prevailing cellular requirements. The 15 amino acid leader peptide produced by the *phe* operon contains seven Phe residues. The *leu* operon leader peptide has four contiguous Leu residues. The leader peptide for the *his* operon contains seven contiguous His residues. In fact, in the *his* operon and several others, attenuation is sufficiently sensitive to be the *only* regulatory mechanism.

Induction of the SOS Response Requires Destruction of Repressor Proteins

Extensive DNA damage in the bacterial chromosome triggers the induction of many distantly located genes.

This response, called the SOS response (p. 1013), provides another good example of coordinated gene regulation. Many of the induced genes are involved in DNA repair (see Table 25-6). The key regulatory proteins are the RecA protein and the LexA repressor.

The LexA repressor (M_r 22,700) inhibits transcription of all the SOS genes **(Fig. 28-21)**, and induction of the SOS response requires removal of LexA. This is not a simple dissociation from DNA in response to binding of a small molecule, as in the regulation of the *lac* operon described above. Instead, the LexA repressor is inactivated when it catalyzes its own cleavage at a specific Ala–Gly peptide bond, producing two roughly equal protein fragments. At physiological pH, this autocleavage reaction requires the RecA protein. RecA is not a protease in the classical sense, but its interaction with LexA enables the repressor's self-cleavage reaction. This function of RecA is sometimes called a co-protease activity.

The RecA protein provides the functional link between the biological signal (DNA damage) and induction of the SOS genes. Heavy DNA damage leads to numerous single-strand gaps in the DNA, and only RecA that is bound to single-stranded DNA can facilitate cleavage of the LexA repressor (Fig. 28-21, bottom). Binding of RecA at the gaps eventually activates its co-protease activity, leading to cleavage of the LexA repressor and SOS induction.

During induction of the SOS response in a severely damaged cell, RecA also cleaves and thus inactivates the repressors that otherwise allow propagation of certain viruses in a dormant lysogenic state within the bacterial host. This provides a remarkable illustration of

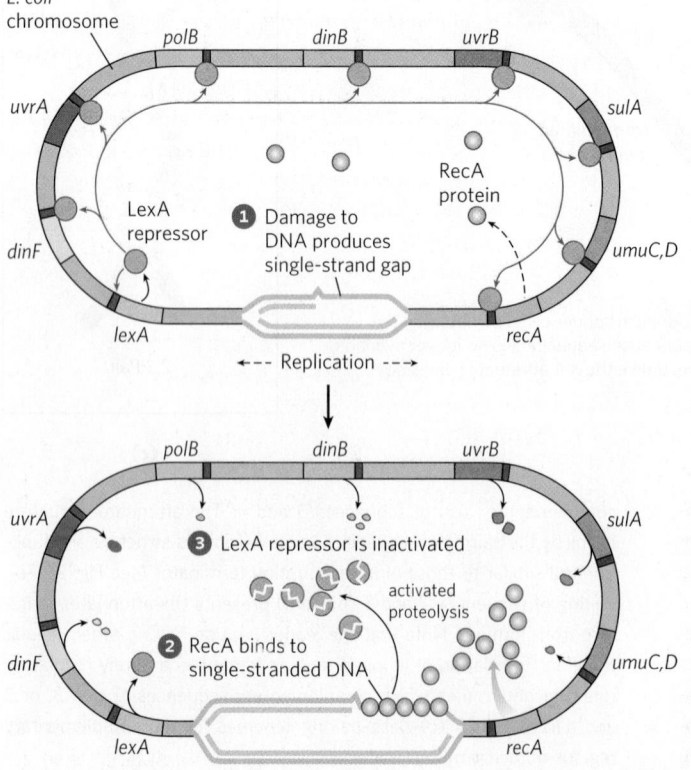

FIGURE 28-21 SOS response in *E. coli*. See Table 25-6 for the functions of many of these proteins. The LexA protein is the repressor in this system, which has an operator site (red) near each gene. Because the *recA* gene is not entirely repressed by the LexA repressor, the normal cell contains about 1,000 RecA monomers. ❶ When DNA is extensively damaged (such as by UV light), DNA replication is halted and the number of single-strand gaps in the DNA increases. ❷ RecA protein binds to this damaged, single-stranded DNA, activating the protein's coprotease activity. ❸ While bound to DNA, the RecA protein facilitates cleavage and inactivation of the LexA repressor. When the repressor is inactivated, the SOS genes, including *recA*, are induced; RecA levels increase 50- to 100-fold.

evolutionary adaptation. These repressors, like LexA, undergo self-cleavage at a specific Ala–Gly peptide bond, so induction of the SOS response permits replication of the virus and lysis of the cell, releasing new viral particles. Thus the bacteriophage can make a hasty exit from a compromised bacterial host cell.

Synthesis of Ribosomal Proteins Is Coordinated with rRNA Synthesis

In bacteria, an increased cellular demand for protein synthesis is met by increasing the number of ribosomes rather than altering the activity of individual ribosomes. In general, the number of ribosomes increases as the cellular growth rate increases. At high growth rates, ribosomes make up approximately 45% of the cell's dry weight. The proportion of cellular resources devoted to making ribosomes is so large, and the function of ribosomes so important, that cells must coordinate the synthesis of the ribosomal components: the ribosomal proteins (r-proteins) and RNAs (rRNAs). This regulation is distinct from the mechanisms described so far: it occurs largely at the level of *translation*.

The 52 genes that encode the r-proteins are distributed across at least 20 operons, each with 1 to 11 genes. Some of these operons also contain the genes for the subunits of DNA primase, RNA polymerase, and protein synthesis elongation factors—revealing the close coupling of replication, transcription, and protein synthesis during bacterial cell growth.

The r-protein operons are regulated primarily through a translational feedback mechanism. One r-protein encoded by each operon also functions as a **translational repressor**, which binds to the mRNA transcribed from that operon and blocks translation of all the genes the messenger encodes **(Fig. 28-22)**. In general, the r-protein that plays the role of repressor also binds directly to an rRNA. Each translational repressor r-protein binds with higher affinity to the appropriate rRNA than to its mRNA, so the mRNA is bound and translation repressed only when the level of the r-protein exceeds that of the rRNA. This ensures that translation of the mRNAs encoding r-proteins is repressed only when synthesis of these r-proteins exceeds that needed to make functional ribosomes. In this way, the rate of r-protein synthesis is kept in balance with rRNA availability.

The mRNA-binding site for the translational repressor is near the translational start site of one of the genes in the operon, usually the first gene (Fig. 28-22). In other operons this would affect only that one gene, because in bacterial polycistronic mRNAs, most genes have independent translation signals. In the r-protein operons, however, the translation of one gene depends on the translation of all the others. The mechanism of this translational coupling is not yet understood in detail. However, in some cases, the translation of multiple genes seems to be blocked by folding of the mRNA into an elaborate three-dimensional structure that is stabilized

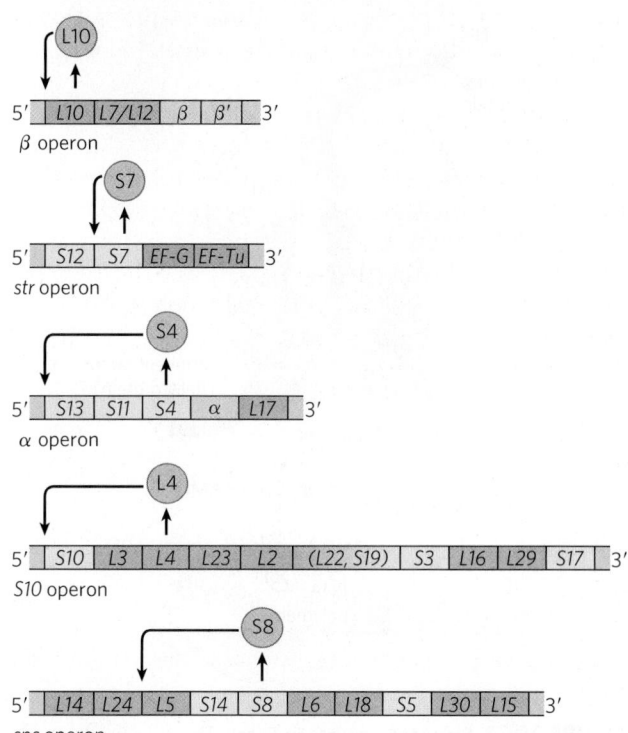

FIGURE 28-22 Translational feedback in some ribosomal protein operons. The r-proteins that act as translational repressors are shaded light red. Each translational repressor blocks the translation of all genes in that operon by binding to the indicated site on the mRNA. Genes that encode subunits of RNA polymerase are shown in purple; genes that encode elongation factors are blue. The r-proteins of the large (50S) ribosomal subunit are designated L1 to L34; those of the small (30S) subunit, S1 to S21.

both by internal base pairing and by binding of the translational repressor protein. When the translational repressor is absent, ribosome binding and translation of one or more of the genes disrupts the folded structure of the mRNA and allows all the genes to be translated.

Because the synthesis of r-proteins is coordinated with the availability of rRNA, the regulation of ribosome production reflects the regulation of rRNA synthesis. In *E. coli*, rRNA synthesis from the seven rRNA operons responds to cellular growth rate and to changes in the availability of crucial nutrients, particularly amino acids. The regulation coordinated with amino acid concentrations is known as the **stringent response (Fig. 28-23)**. When amino acid concentrations are low, rRNA synthesis is halted. Amino acid starvation leads to the binding of uncharged tRNAs to the ribosomal A site; this triggers a sequence of events that begins with the binding of an enzyme called **stringent factor** (RelA protein) to the ribosome. When bound to the ribosome, stringent factor catalyzes formation of the unusual nucleotide guanosine tetraphosphate (ppGpp); it adds pyrophosphate to the 3′ position of GTP, in the reaction

$$GTP + ATP \rightarrow pppGpp + AMP$$

then a phosphohydrolase cleaves off one phosphate to convert some pppGpp to ppGpp. The abrupt rise in

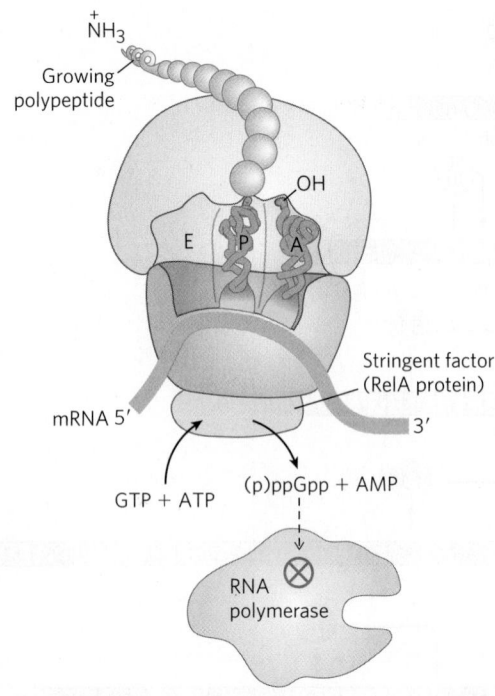

FIGURE 28-23 Stringent response in *E. coli*. This response to amino acid starvation is triggered by binding of an uncharged tRNA in the ribosomal A site. A protein called stringent factor binds to the ribosome and catalyzes the synthesis of pppGpp, which is converted by a phosphohydrolase to ppGpp. The signal ppGpp reduces transcription of some genes and increases that of others, in part by binding to the β subunit of RNA polymerase and altering the enzyme's promoter specificity. Synthesis of rRNA is reduced when ppGpp levels increase.

pppGpp and ppGpp levels in response to amino acid starvation results in a great reduction in rRNA synthesis, mediated at least in part by the binding of ppGpp to RNA polymerase.

The nucleotides pppGpp and ppGpp, along with cAMP, belong to a class of modified nucleotides that act as cellular second messengers. In *E. coli*, these two nucleotides serve as starvation signals; they cause large changes in cellular metabolism by increasing or decreasing the transcription of hundreds of genes. In eukaryotic cells, similar nucleotide second messengers also have multiple regulatory functions. The coordination of cellular metabolism with cell growth is highly complex, and further regulatory mechanisms undoubtedly remain to be discovered.

The Function of Some mRNAs Is Regulated by Small RNAs in Cis or in Trans

As described throughout this chapter, proteins play an important and well-documented role in regulating gene expression. But RNA also has a crucial role—one that is becoming increasingly recognized as more examples of regulatory RNAs are discovered. Once an mRNA is synthesized, its functions can be controlled by RNA-binding proteins, as seen for the r-protein operons just

described, or by an RNA. A separate RNA molecule may bind to the mRNA "in trans" and affect its activity. Alternatively, a portion of the mRNA itself may regulate its own function. When part of a molecule affects the function of another part of the same molecule, it is said to act "in cis."

A well-characterized example of RNA regulation in trans is regulation of the mRNA of the gene *rpoS* (*RNA polymerase sigma factor*), which encodes σ^{38}, one of the seven *E. coli* sigma factors (see Table 26-1). The cell uses this specificity factor in certain stress situations, such as when it enters the stationary phase (a state of no growth, necessitated by lack of nutrients) and σ^{38} is needed to transcribe large numbers of stress response genes. The σ^{38} mRNA is present at low levels under most conditions but is not translated, because a large hairpin structure upstream of the coding region inhibits ribosome binding **(Fig. 28-24)**. Under certain stress

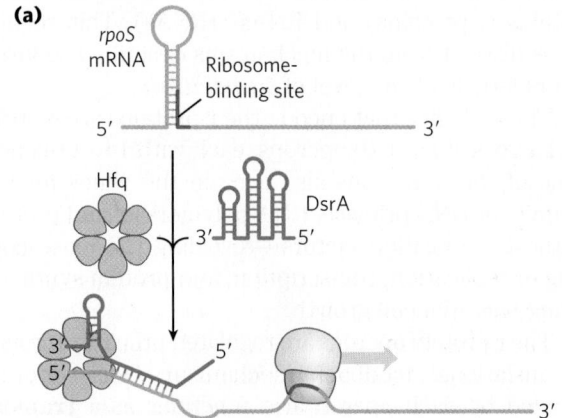

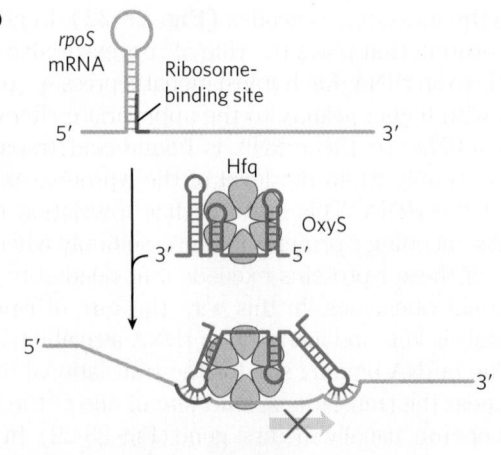

FIGURE 28-24 Regulation of bacterial mRNA function in trans by sRNAs. Several sRNAs (small RNAs)—DsrA, RprA, and OxyS—participate in regulation of the *rpoS* gene. All require the protein Hfq, an RNA chaperone that facilitates RNA-RNA pairing. Hfq has a toroid structure, with a pore in the center. **(a)** DsrA promotes translation by pairing with one strand of a stem-loop structure that otherwise blocks the ribosome-binding site. RprA (not shown) acts in a similar way. **(b)** OxyS blocks translation by pairing with the ribosome-binding site. [Source: Information from M. Szymański and J. Barciszewski, *Genome Biol.* 3:reviews0005.1, 2002.]

conditions, one or both of two small special-function RNAs, DsrA (*downstream region A*) and RprA (*rpoS regulator RNA A*), are induced. Both can pair with one strand of the hairpin in the σ^S mRNA, disrupting the hairpin and thus allowing translation of *rpoS*. Another small RNA, OxyS (*oxidative stress gene S*), is induced under conditions of oxidative stress and inhibits the translation of *rpoS*, probably by pairing with and blocking the ribosome-binding site on the mRNA. OxyS is expressed as part of a system that responds to a different type of stress (oxidative damage) than does the *rpoS* RNA, and its task is to prevent expression of unneeded repair pathways. DsrA, RprA, and OxyS are all relatively small bacterial RNA molecules (less than 300 nucleotides), designated sRNAs (*s* for small; there are, of course, other "small" RNAs with other designations in eukaryotes). All sRNAs require for their function a protein called Hfq, an RNA chaperone that facilitates RNA-RNA pairing. The known bacterial genes regulated in this way are few in number, just a few dozen in a typical bacterial species. However, these examples provide good model systems for understanding patterns present in the more complex and numerous examples of RNA-mediated regulation in eukaryotes.

Regulation in cis involves a class of RNA structures known as **riboswitches**. As described in Box 26-3, aptamers are RNA molecules, generated in vitro, that are capable of specific binding to a particular ligand. As one might expect, such ligand-binding RNA domains are also present in nature—in riboswitches—in a significant number of bacterial mRNAs (and even in some eukaryotic mRNAs). These natural aptamers are structured domains found in untranslated regions at the 5′ ends of certain bacterial mRNAs. Some riboswitches also regulate the transcription of certain noncoding RNAs. Binding of an mRNA's riboswitch to its appropriate ligand results in a conformational change in the mRNA, and transcription is inhibited by stabilization of a premature transcription termination structure, or translation is inhibited (in cis) by occlusion of the ribosome-binding site **(Fig. 28-25)**. In most cases, the riboswitch acts in a kind of feedback loop. Most genes regulated in this way are involved in the synthesis or transport of the ligand that is bound by the riboswitch; thus, when the ligand is present in high concentrations, the riboswitch inhibits expression of the genes needed to replenish this ligand.

Each riboswitch binds only one ligand. Distinct riboswitches have been detected that respond to more than a dozen different ligands, including thiamine pyrophosphate (TPP, vitamin B_1), cobalamin (vitamin B_{12}), flavin mononucleotide, lysine, *S*-adenosylmethionine (adoMet), purines, *N*-acetylglucosamine 6-phosphate, glycine, and some metal cations such as Mn^{2+}. It is likely that many more remain to be discovered. The riboswitch that responds to TPP seems to be the most widespread; it is found in many bacteria, fungi, and some plants. The bacterial TPP riboswitch inhibits translation

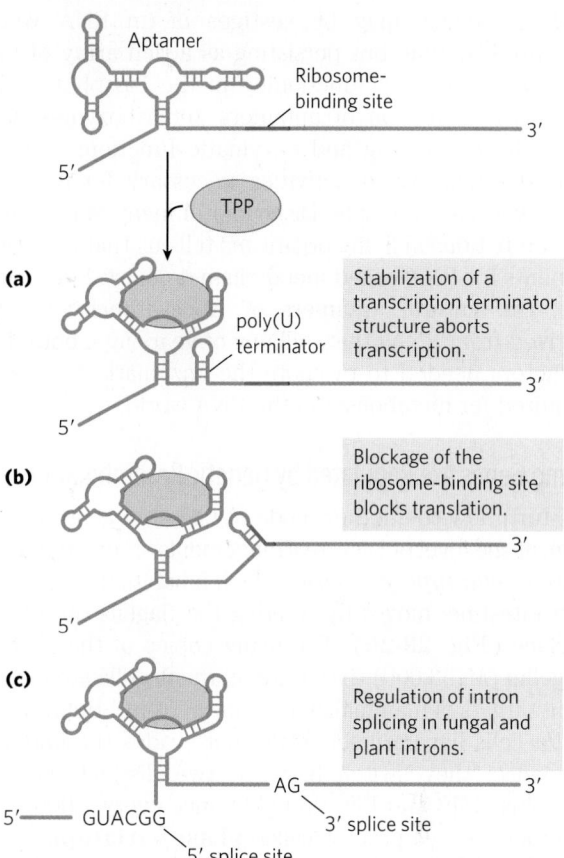

FIGURE 28-25 Regulation of bacterial mRNA function in cis by riboswitches. The known modes of action are illustrated by several different riboswitches, based on a widespread natural aptamer that binds thiamine pyrophosphate. TPP binding to the aptamer leads to a conformational change that produces the varied results illustrated in **(a)**, **(b)**, and **(c)** in several different systems in which the aptamer is utilized. [Source: Information from W. C. Winkler and R. R. Breaker, *Annu. Rev. Microbiol.* 59:487, 2005.]

in some species and induces premature transcription termination in others (Fig. 28-25). The eukaryotic TPP riboswitch is found in the introns of certain genes and modulates the alternative splicing of those genes (see Fig. 26-19b). It is not yet clear how common riboswitches are. However, estimates suggest that more than 4% of the genes of *Bacillus subtilis* are regulated by riboswitches.

As riboswitches become better understood, researchers are finding medical applications. For example, most of the riboswitches described to date, including the one that responds to adoMet, have been found only in bacteria. A drug that bound to and activated the adoMet riboswitch would shut down the genes encoding the enzymes that synthesize and transport adoMet, effectively starving the bacterial cells of this essential cofactor. Drugs of this type are being sought for use as a new class of antibiotics. ■

The pace of discovery of functional RNAs shows no signs of abating and continues to enrich the hypothesis that RNA played a special role in the evolution of life (Chapter 26). The sRNAs and riboswitches, like ribozymes and ribosomes, may be vestiges of an RNA world obscured by time but persisting as a rich array of biological devices still functioning in the biosphere. The laboratory selection of aptamers and ribozymes with novel ligand-binding and enzymatic functions tells us that the RNA-based activities necessary for a viable RNA world are possible. Discovery of many of the same RNA functions in living organisms tells us that key components for RNA-based metabolism do exist. For example, the natural aptamers of riboswitches may be derived from RNAs that, billions of years ago, bound to cofactors needed to promote the enzymatic processes required for metabolism in the RNA world.

Some Genes Are Regulated by Genetic Recombination

We turn now to another mode of bacterial gene regulation, at the level of DNA rearrangement—recombination. *Salmonella typhimurium*, which inhabits the mammalian intestine, moves by rotating the flagella on its cell surface **(Fig. 28-26)**. The many copies of the protein flagellin (M_r 53,000) that make up the flagella are prominent targets of mammalian immune systems. But *Salmonella* cells have a mechanism that evades the immune response: they switch between two distinct flagellin proteins (FljB and FliC) roughly once every 1,000 generations, using a process called **phase variation**.

The switch is accomplished by periodic inversion of a segment of DNA containing the promoter for a flagellin gene. The inversion is a site-specific recombination reaction (see Fig. 25-38) mediated by the Hin recombinase at specific 14 bp sequences (*hix* sequences) at each end of the DNA segment. When the DNA segment is in one orientation, the gene for FljB flagellin and the gene encoding

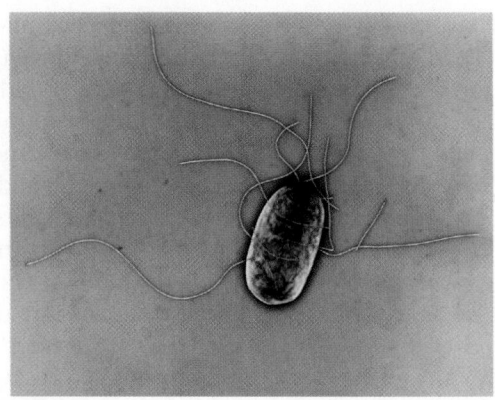

FIGURE 28-26 *Salmonella typhimurium.* The appendages emanating from the top are flagella. [Source: Eye of Science/Science Source.]

a repressor, FljA, are expressed **(Fig. 28-27a)**; the repressor shuts down expression of the gene for FliC flagellin. When the DNA segment is inverted (Fig. 28-27b), the *fljA* and *fljB* genes are no longer transcribed, and the *fliC* gene is induced as the repressor becomes depleted. The Hin recombinase, encoded by the *hin* gene in the DNA segment that undergoes inversion, is expressed when the DNA segment is in either orientation, so the cell can always switch from one state to the other.

This type of regulatory mechanism has the advantage of being absolute: gene expression is impossible when the gene is physically separated from its promoter (note the position of the *fljB* promoter in Fig. 28-27b). An absolute on/off switch may be important in this system (even though it affects only one of the two flagellin genes) because a flagellum with just one copy of the wrong flagellin might be vulnerable to host antibodies

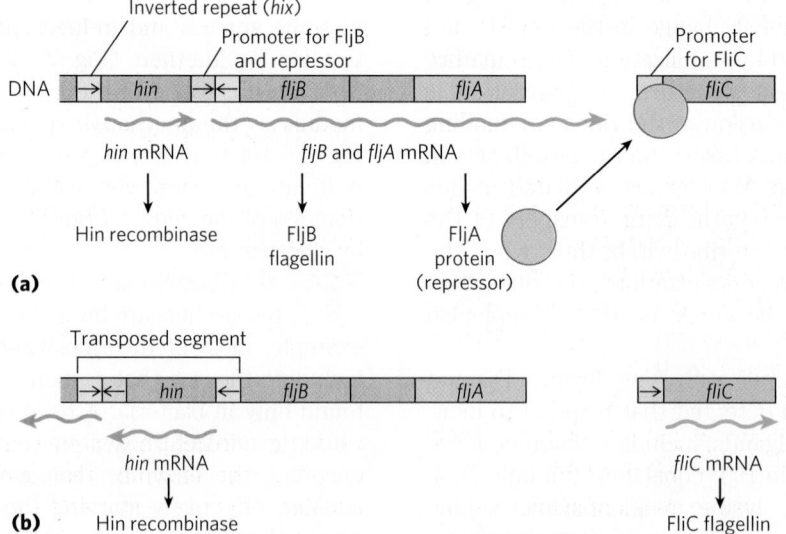

FIGURE 28-27 Regulation of flagellin genes in *Salmonella*: phase variation. The products of genes *fliC* and *fljB* are different flagellins. The *hin* gene encodes the recombinase that catalyzes inversion of the DNA segment containing the *fljB* promoter and the *hin* gene. The recombination sites (inverted repeats) are called *hix* (yellow). **(a)** In one orientation, *fljB* is expressed along with a repressor protein (product of the *fljA* gene) that represses transcription of the *fliC* gene. **(b)** In the opposite orientation, only the *fliC* gene is expressed; the *fljA* and *fljB* genes cannot be transcribed. The interconversion between these two states, known as phase variation, also requires two other nonspecific DNA-binding proteins (not shown), HU and FIS.

TABLE 28-1	Examples of Gene Regulation by Recombination		
System	Recombinase/ recombination site	Type of recombination	Function
Phase variation (*Salmonella*)	Hin/*hix*	Site-specific	Alternative expression of two flagellin genes allows evasion of host immune response.
Host range (bacteriophage μ)	Gin/*gix*	Site-specific	Alternative expression of two sets of tail fiber genes affects host range.
Mating-type switch (yeast)	HO endonuclease, RAD52 protein, other proteins/*MAT*	Nonreciprocal gene conversion[a]	Alternative expression of two mating types of yeast, a and α, creates cells of different mating types that can mate and undergo meiosis.
Antigenic variation (trypanosomes)[b]	Varies	Nonreciprocal gene conversion[a]	Successive expression of different genes encoding the variable surface glycoproteins (VSGs) allows evasion of host immune response.

[a]In nonreciprocal gene conversion (a class of recombination events not discussed in Chapter 25), genetic information is moved from one part of the genome (where it is silent) to another (where it is expressed). The reaction is similar to replicative transposition (see Fig. 25-42).

[b]Trypanosomes cause African sleeping sickness and other diseases (see Box 6-3). The outer surface of a trypanosome is made up of multiple copies of a single VSG, the major surface antigen. A cell can change surface antigens to more than 100 different forms, precluding an effective defense by the host immune system.

against that protein. The *Salmonella* system is by no means unique. Similar regulatory systems occur in some other bacteria and in some bacteriophages, and recombination systems with similar functions have been found in eukaryotes (Table 28-1). Gene regulation by DNA rearrangements that move genes and/or promoters is particularly common in pathogens that benefit by changing their host range or by changing their surface proteins, thereby staying ahead of host immune systems.

SUMMARY 28.2 Regulation of Gene Expression in Bacteria

■ In addition to repression by the Lac repressor, the *E. coli lac* operon undergoes positive regulation by the cAMP receptor protein (CRP). When [glucose] is low, [cAMP] is high and CRP-cAMP binds to a specific site on the DNA, stimulating transcription of the *lac* operon and production of lactose-metabolizing enzymes. The presence of glucose depresses [cAMP], decreasing expression of *lac* and other genes involved in metabolism of secondary sugars. A group of coordinately regulated operons is referred to as a regulon.

■ Operons that produce the enzymes of amino acid synthesis have a regulatory circuit called attenuation, which uses a transcription termination site, called the attenuator, in the mRNA. Formation of the attenuator is modulated by a mechanism that couples transcription and translation while responding to small changes in amino acid concentration.

■ In the SOS system, multiple unlinked genes repressed by a single repressor are induced simultaneously when DNA damage triggers RecA protein–facilitated autocatalytic proteolysis of the repressor.

■ In the synthesis of ribosomal proteins, one protein in each r-protein operon acts as a translational repressor. The mRNA is bound by the repressor, and translation is blocked only when the r-protein is present in excess of available rRNA.

■ Posttranscriptional regulation of some mRNAs is mediated by sRNAs that act in trans or by riboswitches, part of the mRNA structure itself, that act in cis.

■ Some genes are regulated by genetic recombination processes that move promoters relative to the genes being regulated. Regulation can also take place at the level of translation.

28.3 Regulation of Gene Expression in Eukaryotes

Initiation of transcription is a crucial regulation point for gene expression in all organisms. Although eukaryotes and bacteria use some of the same regulatory mechanisms, the regulation of transcription in the two systems is fundamentally different.

We can define a transcriptional ground state as the inherent activity of promoters and transcriptional machinery in vivo in the absence of regulatory sequences. In bacteria, RNA polymerase generally has access to every promoter and can bind and initiate transcription at some level of efficiency in the absence of activators or repressors. In eukaryotes, however, strong promoters are generally inactive in vivo in the absence of regulatory proteins. This fundamental difference gives rise to at least five important features that distinguish the regulation of gene expression at eukaryotic promoters from that observed in bacteria.

First, access to eukaryotic promoters is restricted by the structure of chromatin, and activation of transcription is associated with many changes in chromatin structure in the transcribed region. Second, although eukaryotic cells have both positive and negative regulatory mechanisms, positive mechanisms are more prominent. Almost every eukaryotic gene requires activation to be transcribed. Third, regulatory mechanisms involving lncRNAs are more common in eukaryotic transcriptional regulation. Fourth, eukaryotic cells have larger, more complex multimeric regulatory proteins than do bacteria. Finally, transcription in the eukaryotic nucleus is separated from translation in the cytoplasm in both space and time.

The complexity of regulatory circuits in eukaryotic cells is extraordinary, as is evident from the following discussion. The section ends with an illustrated description of one of the most elaborate circuits: the regulatory cascade that controls development in fruit flies.

Transcriptionally Active Chromatin Is Structurally Distinct from Inactive Chromatin

The effects of chromosome structure on gene regulation in eukaryotes have no clear parallel in bacteria. In the eukaryotic cell cycle, interphase chromosomes appear, at first viewing, to be dispersed and amorphous (see Fig. 24-23). Nevertheless, several forms of chromatin can be found along these chromosomes. About 10% of the chromatin in a typical eukaryotic cell is in a more condensed form than the rest of the chromatin. This form, **heterochromatin**, is transcriptionally inactive. Heterochromatin is generally associated with particular chromosome structures—the centromeres, for example. The remaining, less condensed chromatin is called **euchromatin**.

Transcription of a eukaryotic gene is strongly repressed when its DNA is condensed within heterochromatin. Some, but not all, of the euchromatin is transcriptionally active. Transcriptionally active chromosomal regions are distinguished from heterochromatin in at least three ways: the positioning of nucleosomes, the presence of histone variants, and the covalent modification of nucleosomes. These transcription-associated structural changes in chromatin are collectively called **chromatin remodeling**. The remodeling employs a set of enzymes that promote these changes (Table 28-2).

Four known families of chromatin remodeling complexes, distinguished by their structural features, act directly to alter nucleosome composition in transcribed regions. They may unwrap, translocate, remove, or exchange nucleosomes on the DNA, hydrolyzing ATP in the process (Table 28-2; see the table footnote for an explanation of the abbreviated names of enzyme complexes described here). In some cases, the enzymes catalyze the exchange of pairs of histones within nucleosomes to alter nucleosome composition. The multitude of different complexes are specialized to function at particular genes or chromosomal regions. There are two related complexes in the **SWI/SNF** family in all eukaryotic cells, both of which remodel chromatin so that nucleosomes become more irregularly spaced. They also stimulate the binding of transcription factors. Each complex includes a component called a bromodomain near the carboxyl terminus of the active ATPase subunit, which interacts with acetylated histone tails. The two distinct complexes generally function at different sets of genes. Most of the **ISWI** family complexes optimize nucleosome spacing to allow chromatin assembly and transcriptional silencing. There are generally 9 or 10 different **CHD** family complexes in eukaryotic cells, separated into three subfamilies. The different family members have specialized roles, either ejecting nucleosomes to activate transcription or assembling chromatin to repress transcription. The **INO80** family complexes have a variety of roles in remodeling chromatin for transcriptional activation and DNA repair. One family member, SWR1, promotes subunit exchange in nucleosomes to introduce histone variants such as H2AZ (see Box 24-2), found in transcriptionally active regions. The action of these complexes is incompletely understood, but we know they are essential for transcriptional activation.

The covalent modification of histones is altered dramatically within transcriptionally active chromatin. The core histones of nucleosome particles (H2A, H2B, H3, H4; see Fig. 24-25) are modified by methylation of Lys or Arg residues, phosphorylation of Ser or Thr residues, acetylation (see below), ubiquitination (see Fig. 27-49), or sumoylation. Each of the core histones has two distinct structural domains. A central domain is involved in histone-histone interaction and the wrapping of DNA around the nucleosome. A lysine-rich amino-terminal domain is generally positioned near the exterior of the assembled nucleosome particle; the covalent modifications occur at specific residues concentrated in this amino-terminal domain. The patterns of modification have led some researchers to propose the existence of a histone code, in which modification patterns are recognized by enzymes that alter the structure of chromatin. Indeed, some of the modifications are essential for interactions with proteins that play key roles in transcription.

The acetylation and methylation of histones figure prominently in the processes that activate chromatin for transcription. During transcription, histone H3 is methylated (by specific histone methylases) at Lys^4 in nucleosomes near the 5′ end of the coding region and at Lys^{36} within the coding region. These methylations enable the binding of **histone acetyltransferases (HATs)**, enzymes that acetylate particular Lys residues.

TABLE 28-2 Some Enzyme Complexes That Catalyze Chromatin Structural Changes Associated with Transcription

Enzyme complex[a]	Oligomeric structure (number of polypeptides)	Source	Activities
Histone movement, replacement, or editing, requiring ATP			
SWI/SNF family	8–17 $M_r > 10^6$	Eukaryotes	Nucleosome remodeling; transcriptional activation
ISWI family	2–4	Eukaryotes	Nucleosome remodeling; transcriptional repression; transcriptional activation in some cases
CHD family	1–10	Eukaryotes	Nucleosome remodeling; nucleosome ejection for transcriptional activation; some have repressive roles
INO80 family	>10	Eukaryotes	Nucleosome remodeling and transcriptional activation; family member SWR1 engages in replacement of H2A-H2B with H2AZ-H2B
Histone modification			
GCN5-ADA2-ADA3	3	Yeast	GCN5 has type A HAT activity
SAGA/PCAF	>20	Eukaryotes	Includes GCN5-ADA2-ADA3; acetylates residues in H3, H2B, H2AZ
NuA4	≥12	Eukaryotes	EsaI component has HAT activity; acetylates H4, H2A, and H2AZ
Histone chaperones not requiring ATP			
HIRA	1	Eukaryotes	Deposition of H3.3 during transcription

[a]The abbreviations for eukaryotic genes and proteins are often more confusing or obscure than those used for bacteria. SWI (*switching*) was discovered as a protein required for expression of certain genes involved in mating-type switching in yeast, and SNF (sucrose *non*fermenting) as a factor for expression of the yeast gene for sucrase. Subsequent studies revealed multiple SWI and SNF proteins that act in a complex. The SWI/SNF complex has a role in expression of a wide range of genes and has been found in many eukaryotes, including humans. ISWI is *i*mitation *SWI*. CHD is *c*hromodomain, *h*elicase, *D*NA binding; INO80, *ino*sitol-requiring *80*; and SWR1, *SWi2/Snf2-r*elated ATPase 1. The complex of GCN5 (*g*eneral *c*ontrol *n*onderepressible) and ADA (*a*lteration/*d*eficiency *a*ctivation) proteins was discovered during investigation of the regulation of nitrogen metabolism genes in yeast. These proteins can be part of the larger SAGA (*S*PF, *A*DA2,3, *G*CN5, *a*cetyltransferase) complex in yeasts. The equivalent of SAGA in humans is PCAF (*p*300/*C*BP-*a*ssociated *f*actor). NuA4 is *nu*cleosome *a*cetyltransferase of H*4*; ESA1, essential *S*AS2-related acetyltransferase. HIRA is *hi*stone *r*egulator *A*.

Cytosolic (type B) HATs acetylate newly synthesized histones before the histones are imported into the nucleus. The subsequent assembly of the histones into chromatin after replication is facilitated by histone chaperones: CAF1 for H3 and H4 (see Box 24-2), and NAP1 for H2A and H2B.

Where chromatin is being activated for transcription, the nucleosomal histones are further acetylated by nuclear (type A) HATs. The acetylation of multiple Lys residues in the amino-terminal domains of histones H3 and H4 can reduce the affinity of the entire nucleosome for DNA. Acetylation of particular Lys residues is critical for the interaction of nucleosomes with other proteins. When transcription of a gene is no longer required, the extent of acetylation of nucleosomes in that vicinity is reduced by the activity of **histone deacetylases (HDACs)**, as part of a general gene-silencing process that restores the chromatin to a transcriptionally inactive state. The deacetylases include SIRT1, SIRT2, SIRT6, and SIRT7, NAD$^+$-dependent enzymes in the sirtuin family (SIRT1-7 in mammals). These deacetylate specific Lys residues in histones and other, cytoplasmic targets. In addition to the removal of certain acetyl groups, new covalent modification of histones marks chromatin as transcriptionally inactive. For example, Lys^9 of histone H3 is often methylated in heterochromatin.

The net effect of chromatin remodeling in the context of transcription is to make a segment of the chromosome more accessible and to "label" (chemically modify) it so as to facilitate the binding and activity of transcription factors that regulate expression of the gene or genes in that region.

Most Eukaryotic Promoters Are Positively Regulated

As already noted, eukaryotic RNA polymerases have little or no intrinsic affinity for their promoters; initiation of transcription is almost always dependent on the action of multiple activator proteins. One important reason for the apparent predominance of positive regulation seems obvious: the storage of DNA within chromatin effectively renders most promoters inaccessible,

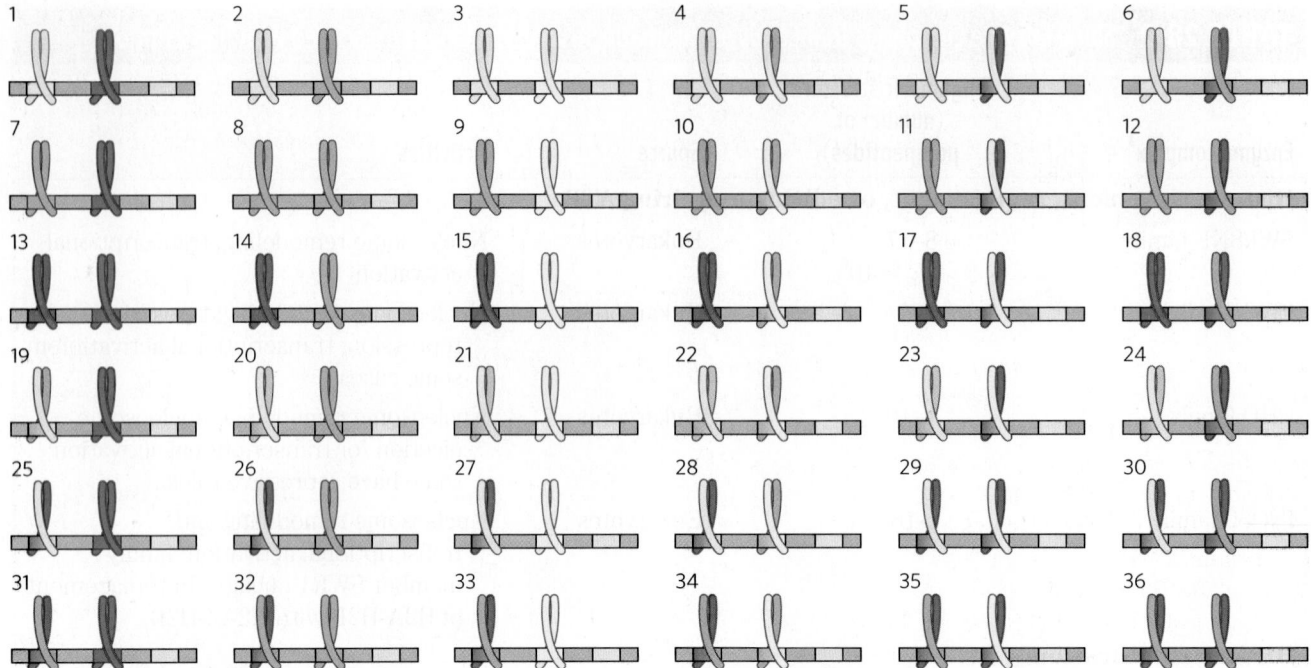

FIGURE 28-28 The advantages of combinatorial control. Combinatorial control allows specific regulation of many genes using a limited repertoire of regulatory proteins. Consider the possibilities inherent in regulation by two different families of leucine zipper proteins (red and green). If each regulatory gene family had three members (as shown here, in dark, medium, and light shades, each binding to a different DNA sequence) that could freely form either homo- or heterodimers, there would be six possible dimeric species in each family and each dimer would recognize a different bipartite regulatory DNA sequence. If a gene had a regulatory site for each protein family, 36 different regulatory combinations would be possible, using just the six proteins from these two families. With six or more sites used in the regulation of a typical eukaryotic gene, the number of possible variants is much greater than this example suggests.

so genes are silent in the absence of other regulation. The structure of chromatin affects access to some promoters more than others, but repressors that bind to DNA so as to preclude access of RNA polymerase (negative regulation) would often be simply redundant. Other factors must be at play in the use of positive regulation, and speculation generally centers around two: the large size of eukaryotic genomes and the greater efficiency of positive regulation.

First, nonspecific DNA binding of regulatory proteins becomes a more important problem in the much larger genomes of higher eukaryotes. And the chance that a single specific binding sequence will occur randomly at an inappropriate site also increases with genome size. Combinatorial control thus becomes important in a large genome **(Fig. 28-28)**. Specificity for transcriptional activation can be improved if each of several positive-regulatory proteins must bind specific DNA sequences to activate a gene. The average number of regulatory sites for a gene in a multicellular organism is six, and genes that are regulated by a dozen such sites are common. The requirement for binding of several positive-regulatory proteins to specific DNA sequences vastly reduces the probability of the random occurrence of a functional juxtaposition of all the necessary binding sites. In addition, the number of regulatory proteins that must be encoded by a genome to regulate all of its genes can be reduced (Fig. 28-28). Thus, a new regulator is not

needed for every gene, although regulation is complex enough in higher eukaryotes that regulatory proteins may represent 5% to 10% of all protein-coding genes.

In principle, a similar combinatorial strategy could be used by multiple negative-regulatory elements, but this brings us to the second reason for the use of positive regulation: it is simply more efficient. If the ~20,000 genes in the human genome were negatively regulated, each cell would have to synthesize, at all times, all of the different repressors in concentrations sufficient to permit specific binding to each "unwanted" gene. In positive regulation, most of the genes are usually inactive (that is, RNA polymerases do not bind to the promoters) and the cell synthesizes only the activator proteins needed to promote transcription of the subset of genes required in the cell at that time.

These arguments notwithstanding, there are examples of negative regulation in eukaryotes, from yeasts to humans, as we shall see. Some of that negative regulation involves lncRNAs, which are more economical to synthesize than repressor proteins.

DNA-Binding Activators and Coactivators Facilitate Assembly of the Basal Transcription Factors

To continue our exploration of the regulation of gene expression in eukaryotes, we return to the interactions between promoters and RNA polymerase II (Pol II), the enzyme responsible for the synthesis of eukaryotic mRNAs. Although many (but not all) Pol II promoters include the TATA box and Inr (initiator) sequences, with their standard spacing (see Fig. 26-8), they vary greatly in both the number and the location of additional sequences required for the regulation of transcription.

The additional regulatory sequences, generally bound by transcription activators, are usually called **enhancers** in higher eukaryotes and **upstream activator sequences (UASs)** in yeast. A typical enhancer may be found hundreds or even thousands of base pairs upstream from the transcription start site, or may even be downstream, within the gene itself. When bound by the appropriate regulatory proteins, an enhancer increases transcription at nearby promoters regardless of its orientation in the DNA. The UASs of yeast function in a similar way, although generally they must be positioned upstream and within a few hundred base pairs of the transcription start site.

Successful binding of the active Pol II holoenzyme at one of its promoters usually requires the combined action of proteins of five types: (1) **transcription activators**, which bind to enhancers or UASs and facilitate transcription; (2) **architectural regulators**, which facilitate DNA looping; (3) **chromatin modification and remodeling proteins**, described above; (4) **coactivators**; and (5) **basal transcription factors**, also called general transcription factors (see Fig. 26-9, Table 26-2), required at most Pol II promoters **(Fig. 28-29)**. The coactivators are required for essential communication between activators and the complex composed of Pol II and the

basal transcription factors. Coactivators also play a direct role in assembly of the preinitiation complex (PIC). Furthermore, a variety of repressor proteins can interfere with communication between Pol II and the activators, resulting in repression of transcription

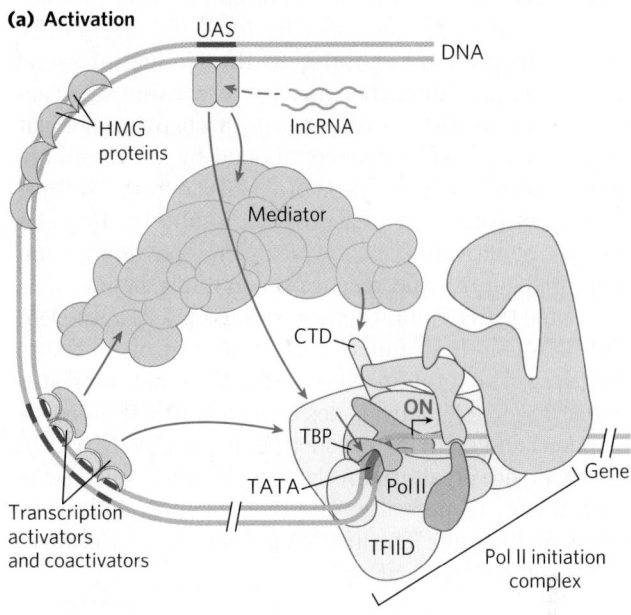

(a) Activation

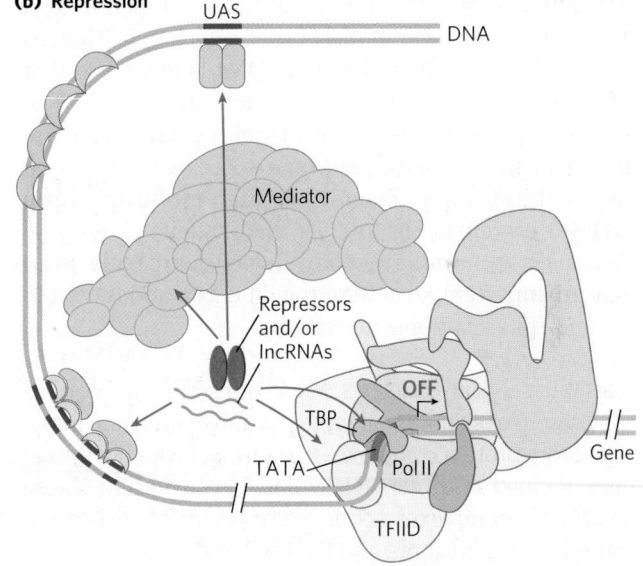

(b) Repression

FIGURE 28-29 Eukaryotic promoters and regulatory proteins. RNA polymerase II and its associated basal (general) transcription factors form a preinitiation complex at the TATA box and Inr site of the cognate promoters, a process facilitated by transcription activators, acting through coactivators (Mediator, TFIID, or both). **(a)** A composite promoter with typical sequence elements and protein complexes found in both yeast and higher eukaryotes. The carboxyl-terminal domain (CTD) of Pol II (see Fig. 26-9) is an important point of interaction with Mediator and other protein complexes. Histone modification enzymes (not shown) catalyze methylation and acetylation; remodeling enzymes alter the content and placement of nucleosomes. The transcription activators have distinct DNA-binding domains and activation domains. In some cases, their function is affected by interaction with lncRNAs. Green arrows indicate common modes of interaction often required for the activation of transcription, as discussed in the text. The HMG proteins are a common type of architectural regulator (see Fig. 28-5), allowing the looping of the DNA required to bring together system components bound at distant binding sites. **(b)** Eukaryotic transcriptional repressors function through a range of mechanisms. Some bind directly to DNA, displacing a protein complex required for activation (not shown); many others interact with various parts of the transcription or activation protein complexes to prevent activation. Possible points of interaction are indicated with red arrows. **(c)** The structure of an HMG protein complex with DNA shows how HMG proteins facilitate DNA looping. The binding is relatively nonspecific, although DNA sequence preferences have been identified for many HMG proteins. Shown here is the HMG domain of the protein HMG-D of *Drosophila*, bound to DNA. [Source: (c) PDB ID 1QRV, F. V. Murphy IV et al., *EMBO J.* 18:6610, 1999.]

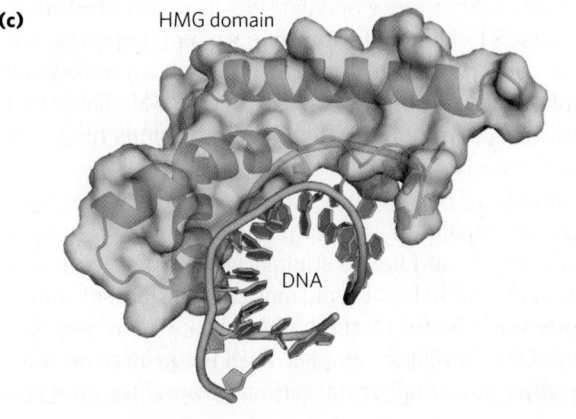

(c)

(Fig. 28-29b). Here we focus on the protein complexes shown in Figure 28-29 and how they interact to activate transcription.

Transcription Activators The requirements for activators vary greatly from one promoter to another. A few are known to activate transcription at hundreds of promoters, whereas others are specific for a few promoters. Many activators are sensitive to the binding of signal molecules, providing the capacity to activate or deactivate transcription in response to a changing cellular environment. Some enhancers bound by activators are quite distant from the promoter's TATA box. Multiple enhancers (often six or more) are bound by a similar number of activators for a typical gene, providing combinatorial control and response to multiple signals.

Some transcription activators can bind to both DNA and RNA, and their function is affected by one or more lncRNAs. The protein NF-κB, for example, activates transcription of many genes involved in the immune response and cytokine production. It can bind to a DNA enhancer site or, alternatively, to an lncRNA called *lethe* (Fig. 28-14; named after the river of forgetfulness in Greek mythology). The lncRNA reduces transcription of genes controlled by NF-κB.

Architectural Regulators How do activators function at a distance? The answer in most cases seems to be that, as indicated earlier, the intervening DNA is looped so that the various protein complexes can interact directly. The looping is promoted by architectural regulators that are abundant in chromatin and bind to DNA with limited specificity. Most prominently, the **high mobility group (HMG)** proteins (Fig. 28-29c; "high mobility" refers to their electrophoretic mobility in polyacrylamide gels) play an important structural role in chromatin remodeling and transcriptional activation.

Coactivator Protein Complexes Most transcription requires the presence of additional protein complexes. Some major regulatory protein complexes that interact with Pol II have been defined both genetically and biochemically. These coactivator complexes act as intermediaries between the transcription activators and the Pol II complex.

A major eukaryotic coactivator, a complex consisting of 20 to 30 or more polypeptides is called **Mediator** (Fig. 28-29). Many of the 20 core polypeptides are highly conserved from fungi to humans. An additional complex of four subunits can interact with Mediator and inhibit transcription initiation. Mediator binds tightly to the carboxyl-terminal domain (CTD) of the largest subunit of Pol II. The Mediator complex is required for both basal and regulated transcription at many promoters used by Pol II, and it also stimulates phosphorylation of the CTD by TFIIH (a basal transcription factor). Transcription activators interact with one or more components of the Mediator complex, with the precise interaction sites differing from one activator to another.

Coactivator complexes function at or near the promoter's TATA box.

Additional coactivators, functioning with one or a few genes, have also been described. Some of these operate in conjunction with Mediator, and some may act in systems that do not employ Mediator.

TATA-Binding Protein and Basal Transcription Factors The first component to bind in the assembly of a **preinitiation complex (PIC)** at the TATA box of a typical Pol II promoter is the **TATA-binding protein (TBP)**. At promoters lacking a TATA box, TBP is usually delivered as part of a larger complex (13 to 14 subunits) called TFIID. The complete complex also includes the basal transcription factors TFIIB, TFIIE, TFIIF, TFIIH; Pol II; and perhaps TFIIA. This minimal PIC, however, is often insufficient for initiation of transcription and generally does not form at all if the promoter is obscured within chromatin. Positive regulation, leading to transcription, is imposed by the activators and coactivators. Mediator interacts directly with TFIIH and TFIIE, allowing their recruitment to the PIC.

Choreography of Transcriptional Activation We can now begin to piece together the sequence of transcriptional activation events at a typical Pol II promoter **(Fig. 28-30)**. The exact order of binding of some components may vary, but the model in Figure 28-30 illustrates the principles of activation as well as one common path. Many transcription activators have significant affinity for their binding sites even when the sites are within condensed chromatin. The binding of activators is often the event that triggers subsequent activation of the promoter. Binding of one activator may enable the binding of others, gradually displacing some nucleosomes.

Crucial remodeling of the chromatin then takes place in stages, facilitated by interactions between activators and HATs or enzyme complexes such as SWI/SNF, or both. In this way, a bound activator can draw in other components necessary for further chromatin remodeling to permit transcription of specific genes. The bound activators interact with the large Mediator complex. Mediator, in turn, provides an assembly surface for the binding of, first, TBP (or TFIID), then TFIIB, and then other components of the PIC, including Pol II. Mediator stabilizes the binding of Pol II and its associated transcription factors and greatly facilitates formation of the PIC. Complexity in these regulatory circuits is the rule rather than the exception, with multiple DNA-bound activators promoting transcription.

The script can change from one promoter to another. For example, many promoters have a different set of recognition sequences and may not have a TATA box, and in multicellular eukaryotes the subunit composition of factors such as TFIID can vary from one tissue to another. However, most promoters seem to require a precisely ordered assembly of components to initiate transcription. The assembly process is not always fast.

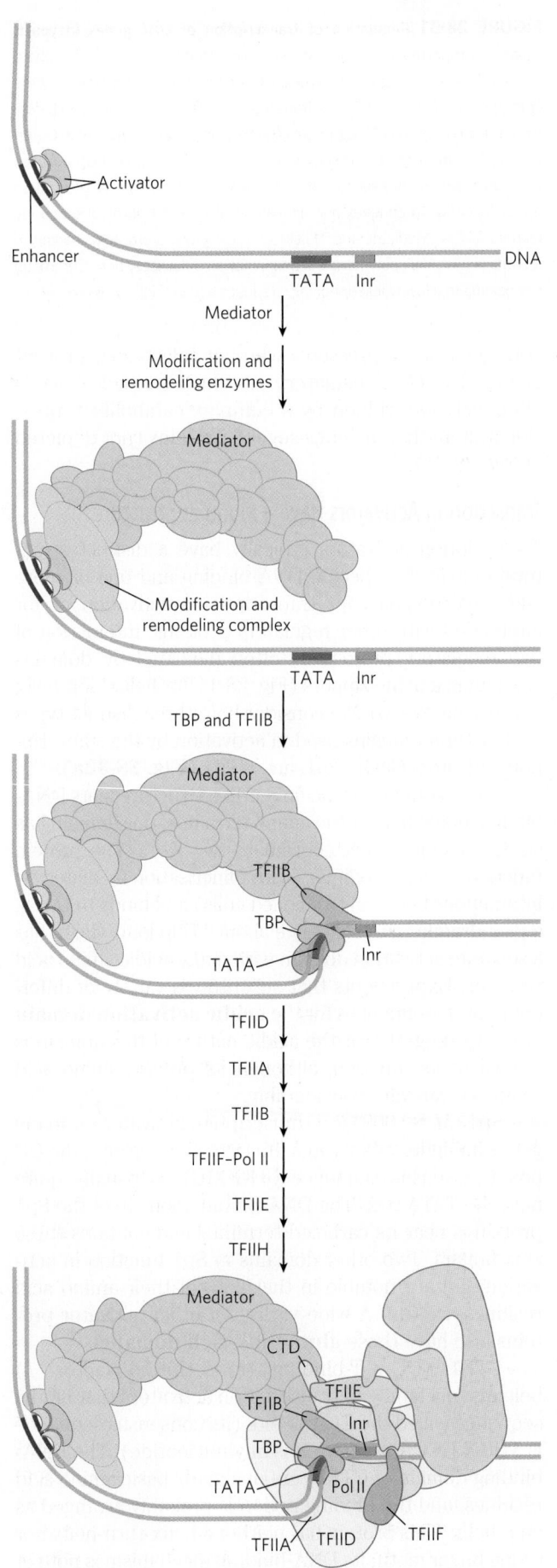

FIGURE 28-30 The components of transcriptional activation. Activators bind the DNA first. The activators recruit the histone modification/ nucleosome remodeling complexes and a coactivator such as Mediator. Mediator facilitates the binding of TBP (or TFIID) and TFIIB, and the other basal transcription factors and Pol II then bind. Phosphorylation of the carboxyl-terminal domain (CTD) of Pol II leads to transcription initiation (not shown). [Source: Information from J. A. D'Alessio et al., *Mol. Cell* 36:924, 2009.]

For some genes it may take minutes; for certain genes of higher eukaryotes, the process can take days.

Reversible Transcriptional Activation Although rarer, some eukaryotic regulatory proteins that bind to Pol II promoters or that interact with transcription activators can act as repressors, inhibiting the formation of active PICs (Fig. 28-30). Some activators can adopt different conformations, enabling them to serve as transcription activators or as repressors. For example, some steroid hormone receptors (described later) function in the nucleus as transcription activators, stimulating the transcription of certain genes when a particular steroid hormone signal is present. When the hormone is absent, the receptor proteins revert to a repressor conformation, *preventing* the formation of PICs. In some cases, this repression involves interaction with histone deacetylases and other proteins that help restore the surrounding chromatin to its transcriptionally inactive state. Mediator, when it includes the inhibitory subunits, may also block transcription initiation. This may be a regulatory mechanism to ensure ordered assembly of the PIC, by delaying transcriptional activation until all required factors are present, or it may be a mechanism that helps deactivate promoters when transcription is no longer required.

The Genes of Galactose Metabolism in Yeast Are Subject to Both Positive and Negative Regulation

Some of the general principles described above can be illustrated by one well-studied eukaryotic regulatory circuit **(Fig. 28-31)**. The enzymes required for the importation and metabolism of galactose in yeast are encoded by genes scattered over several chromosomes (Table 28-3). Each of the *GAL* genes is transcribed separately, and yeast cells have no operons like those in bacteria. However, all the *GAL* genes have similar promoters and are regulated coordinately by a common set of proteins. The promoters for the *GAL* genes consist of the TATA box and Inr sequences, as well as an upstream activator sequence (UAS$_G$) recognized by the transcription activator Gal4 protein (Gal4p). Regulation of gene expression by galactose entails an interplay between Gal4p and two other proteins, Gal80p and Gal3p (Fig. 28-31). Gal80p forms a complex with Gal4p, preventing Gal4p from functioning as an activator of the *GAL* promoters. When galactose is present, it binds Gal3p, which then interacts with the Gal80p-Gal4p complex and allows Gal4p to function as an activator at the *GAL* promoters. As the various

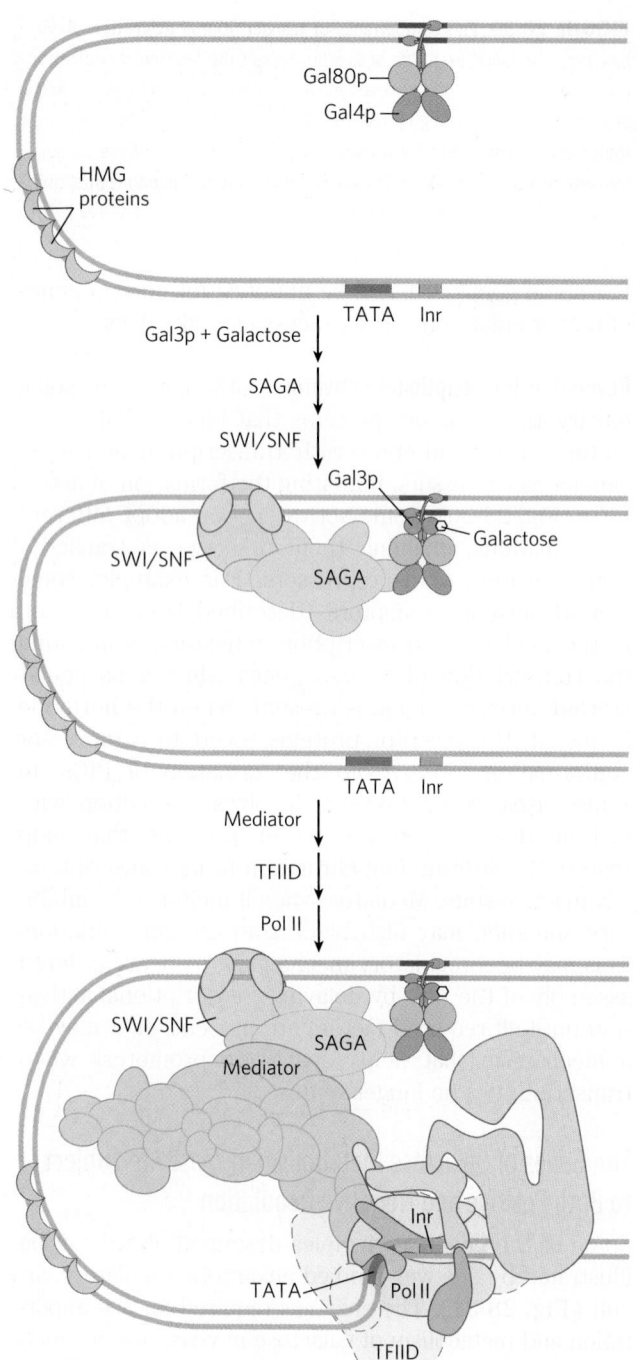

FIGURE 28-31 Regulation of transcription of *GAL* genes in yeast.
Galactose imported into the yeast cell is converted to glucose 6-phosphate by a pathway involving five enzymes, whose genes are scattered over three chromosomes (see Table 28-3). Transcription of these genes is regulated by the combined actions of the proteins Gal4p, Gal80p, and Gal3p, with Gal4p playing the central role of transcription activator. The Gal4p-Gal80p complex is inactive. Binding of galactose to Gal3p leads to interaction of Gal3p with the Gal80p-Gal4p complex and activates Gal4p. The Gal4p subsequently recruits SAGA, Mediator, and TFIID to the galactose promoters, leading to recruitment of RNA polymerase II and initiation of transcription. Chromatin remodeling to allow transcription also requires a SWI/SNF complex.

GAL genes are repressed—whether galactose is present or not. The *GAL* regulatory system described above is effectively overridden by a complex catabolite repression system that includes several proteins (not depicted in Fig. 28-31).

Transcription Activators Have a Modular Structure

Transcription activators typically have a distinct structural domain for specific DNA binding and one or more additional domains for transcriptional activation or for interaction with other regulatory proteins. Interaction of two regulatory proteins is often mediated by domains containing leucine zippers (Fig. 28-15) or helix-loop-helix motifs (Fig. 28-16). We consider here three distinct types of structural domains used in activation by the transcription activators Gal4p, Sp1, and CTF1 **(Fig. 28-32a)**.

Gal4p contains a zinc finger–like structure in its DNA-binding domain, near the amino terminus; this domain has six Cys residues that coordinate two Zn^{2+}. The protein functions as a homodimer (with dimerization mediated by interactions between two coiled coils) and binds to UAS_G, a palindromic DNA sequence about 17 bp long. Gal4p has a separate activation domain with many acidic amino acid residues. Experiments that substitute a variety of different peptide sequences for the **acidic activation domain** of Gal4p suggest that the acidic nature of this domain is critical to its function, although its precise amino acid sequence can vary considerably.

Sp1 (M_r 80,000) is a transcription activator for many genes in higher eukaryotes. Its DNA-binding site, the GC box (consensus sequence GGGCGG), is usually quite near the TATA box. The DNA-binding domain of the Sp1 protein is near its carboxyl terminus and contains three zinc fingers. Two other domains in Sp1 function in activation and are notable in that 25% of their amino acid residues are Gln. A wide variety of other activator proteins also have these **glutamine-rich domains**.

CTF1 (*CCAAT*-binding *t*ranscription *f*actor 1) belongs to a family of transcription activators that bind a sequence called the CCAAT site (its consensus sequence is $TGGN_6GCCAA$, where N is any nucleotide). The DNA-binding domain of CTF1 contains many basic amino acid residues, and the binding region is probably arranged as an α helix. This protein has neither a helix-turn-helix nor a zinc finger motif; its DNA-binding mechanism is not yet clear. CTF1 has a **proline-rich activation domain**,

galactose genes are induced and their products build up, Gal3p may be replaced with Gal1p (a galactokinase needed for galactose metabolism that also acts as a regulator) for sustained activation of the regulatory circuit.

Other protein complexes also have a role in activating transcription of the *GAL* genes. These include the SAGA complex for histone acetylation and chromatin remodeling, the SWI/SNF complex for chromatin remodeling, and Mediator. The Gal4 protein is responsible for recruitment of these additional factors needed for transcriptional activation. SAGA may be the first and primary recruitment target for Gal4p.

Glucose is the preferred carbon source for yeast, as it is for bacteria. When glucose is present, most of the

TABLE 28-3	Genes of Galactose Metabolism in Yeast					
				Relative protein expression in different carbon sources		
Gene	Protein function	Chromosomal location	Protein size (number of residues)	Glucose	Glycerol	Galactose
Regulated genes						
GAL1	Galactokinase	II	528	−	−	+++
GAL2	Galactose permease	XII	574	−	−	+++
PGM2	Phosphoglucomutase	XIII	569	+	+	++
GAL7	Galactose 1-phosphate uridylyltransferase	II	365	−	−	+++
GAL10	UDP-glucose 4-epimerase	II	699	−	−	+++
MEL1	α-Galactosidase	II	453	−	+	++
Regulatory genes						
GAL3	Inducer	IV	520	−	+	++
GAL4	Transcriptional activator	XVI	881	+/−	+	+
GAL80	Transcriptional inhibitor	XIII	435	+	+	++

Source: Information from R. Reece and A. Platt, *Bioessays* 19:1001, 1997.

with Pro accounting for more than 20% of the amino acid residues.

The discrete activation and DNA-binding domains of regulatory proteins often act completely independently, as has been demonstrated in "domain-swapping" experiments. Genetic engineering techniques (Chapter 9) can join the proline-rich activation domain of CTF1 to the DNA-binding domain of Sp1 to create a protein that, like intact Sp1, binds to GC boxes on the DNA and activates transcription at a nearby promoter (as in Fig. 28-32b). The DNA-binding domain of Gal4p has similarly been replaced experimentally with the DNA-binding domain of the *E. coli* LexA repressor (of the SOS response; Fig. 28-21). This chimeric protein neither binds at UAS$_G$ nor activates the yeast *GAL* genes (as would intact Gal4p) unless the UAS$_G$ sequence in the DNA is replaced by the LexA recognition site.

Eukaryotic Gene Expression Can Be Regulated by Intercellular and Intracellular Signals

The effects of steroid hormones (and of thyroid and retinoid hormones, which have a similar mode of action)

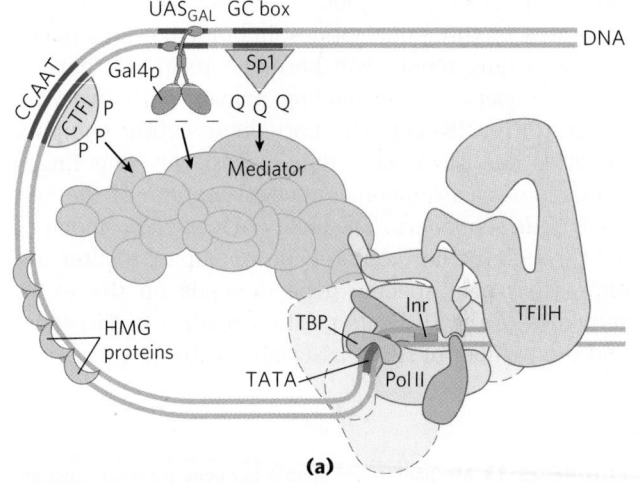

(a)

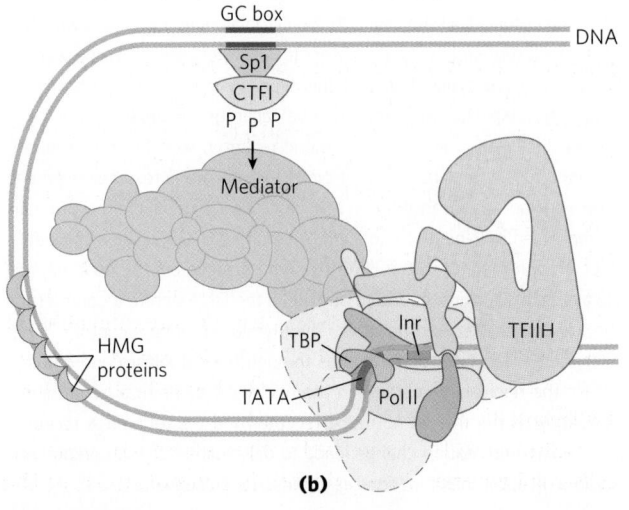

(b)

FIGURE 28-32 Transcription activators. (a) Typical activators such as CTF1, Gal4p, and Sp1 have a DNA-binding domain and an activation domain. The nature of the activation domain is indicated by symbols: − − −, acidic; Q Q Q, glutamine-rich; P P P, proline-rich. These proteins generally activate transcription by interacting with coactivator complexes such as Mediator. Note that the binding sites illustrated here are not generally found together near a single gene. **(b)** A chimeric protein containing the DNA-binding domain of Sp1 and the activation domain of CTF1 activates transcription if a GC box is present.

provide additional well-studied examples of the modulation of eukaryotic regulatory proteins by direct interaction with molecular signals (see Fig. 12-30). Unlike other types of hormones, steroid hormones do not have to bind to plasma membrane receptors. Instead, they can interact with intracellular receptors that are transcription activators. Steroid hormones too hydrophobic to dissolve readily in the blood (estrogen, progesterone, and cortisol, for example) travel on specific carrier proteins from their point of release to their target tissues. In the target tissue, the hormone passes through the plasma membrane by simple diffusion. Once inside the cell, the hormone interacts with one of two types of steroid-binding nuclear receptor **(Fig. 28-33)**. In both cases, the hormone-receptor complex acts by binding to highly specific DNA sequences called **hormone response elements (HREs)**, thereby altering gene expression. Acting at these sites, the receptors act as transcription activators, recruiting coactivators and Pol II (plus its associated transcription factors) to trigger transcription of the gene.

The DNA sequences (HREs) to which hormone-receptor complexes bind are similar in length and arrangement for the various steroid hormones, but differ in sequence. Each receptor has a consensus HRE sequence (Table 28-4) to which the hormone-receptor complex binds well, with each consensus consisting of two six-nucleotide sequences, either contiguous or separated by three nucleotides, in tandem or in a palindromic arrangement. The hormone receptors have a highly conserved DNA-binding domain with two zinc fingers **(Fig. 28-34)**. The hormone-receptor complex binds to the DNA as a dimer, with the zinc finger domains of each monomer recognizing one of the six-nucleotide sequences. The ability of a given hormone to act through the hormone-receptor complex to alter the expression of a specific gene depends on the exact sequence of the HRE, its position relative to the gene, and the number of HREs associated with the gene.

The ligand-binding region of the receptor protein—always at the carboxyl terminus—is specific to the particular receptor. For example, in the ligand-binding region, the glucocorticoid receptor is only 30% similar to the estrogen receptor and 17% similar to the thyroid

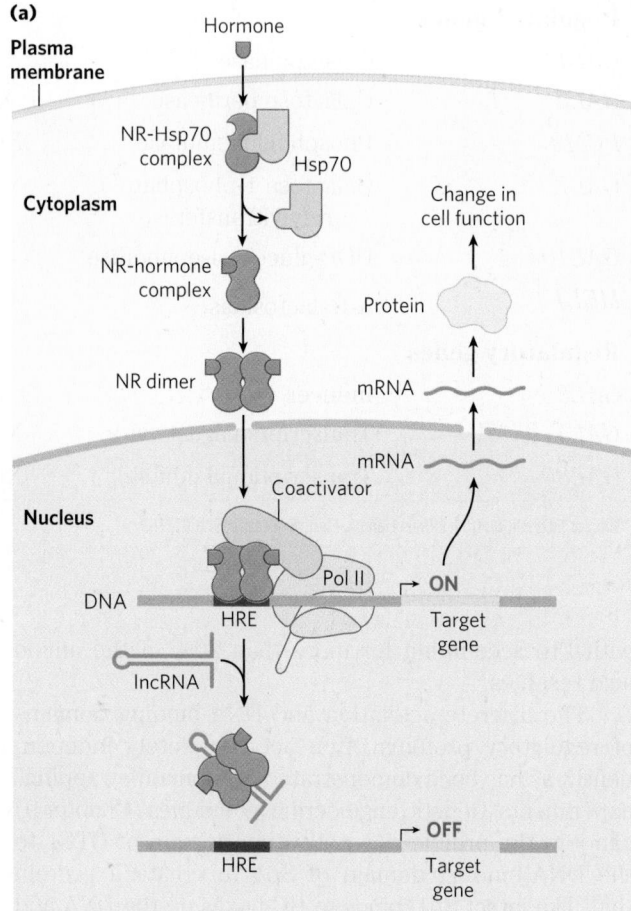

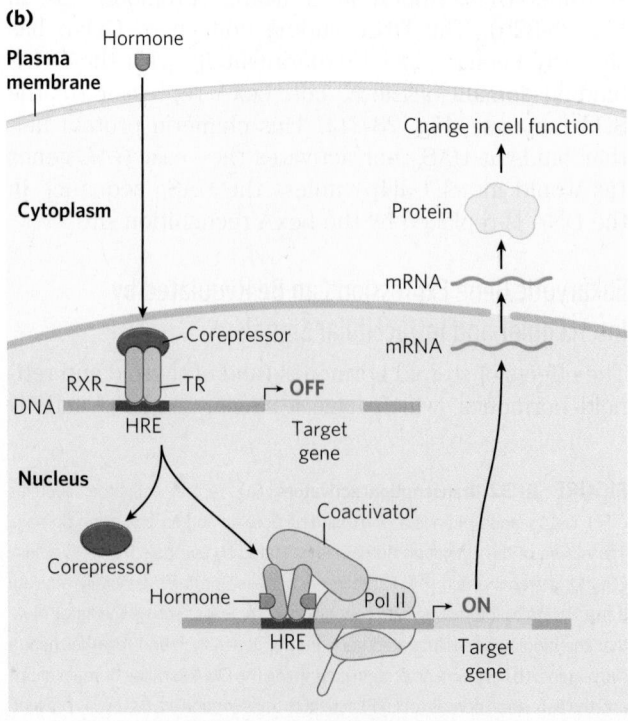

FIGURE 28-33 Mechanisms of steroid hormone receptor function.
There are two types of steroid-binding nuclear receptors. **(a)** Monomeric type I receptors (NR) are found in the cytoplasm, in a complex with the heat shock protein Hsp70. Receptors for estrogen, progesterone, androgens, and glucocorticoids are of this type. When the steroid hormone binds, the Hsp70 dissociates and the receptor dimerizes, exposing a nuclear localization signal. The dimeric receptor, with hormone bound, migrates to the nucleus, where it binds to a hormone response element (HRE) and acts as a transcription activator. The activity of the receptor can be repressed by binding to an lncRNA (such as GAS5), which competes directly with binding to the HRE. **(b)** Type II receptors, by contrast, are always in the nucleus, bound to an HRE in the DNA and to a corepressor that renders the receptor inactive. The thyroid hormone receptor (TR) is of this type. The hormone migrates through the cytoplasm and diffuses across the nuclear membrane. In the nucleus it binds to a heterodimer consisting of the thyroid hormone receptor and the retinoid X receptor (RXR). A conformation change leads to dissociation of the corepressor, and the receptor then functions as a transcription activator.

TABLE 28-4	Hormone Response Elements (HREs) Bound by Steroid-Type Hormone Receptors
Receptors	**HRE consensus sequence bound**[a]
Androgen	GG(A/T)ACAN₂TGTTCT
Glucocorticoid	GGTACAN₃TGTTCT
Retinoic acid (some)	AGGTCAN₅AGGTCA
Vitamin D	AGGTCAN₃AGGTCA
Thyroid hormone	AGGTCAN₃AGGTCA
RX[b]	AGGTCANAGGTCANAG GTCANAGGTCA

[a]N represents any nucleotide.
[b]Forms a dimer with the retinoic acid receptor or vitamin D receptor.

hormone receptor. The size of the ligand-binding region varies dramatically; in the vitamin D receptor it has only 25 amino acid residues, whereas in the mineralocorticoid receptor it has 603 residues. Mutations that change one amino acid residue in these regions can result in loss of responsiveness to a specific hormone. Some humans unable to respond to cortisol, testosterone, vitamin D, or thyroxine have mutations of this type.

The lncRNAs introduce another dimension to regulation by hormone receptors. An lncRNA called GAS5 (*g*rowth *a*rrest *s*pecific *5*) inhibits transcriptional activation by the glucocorticoid receptor by directly competing with DNA for receptor binding. GAS5 also inhibits activity of the closely related androgen, progesterone, and mineralocorticoid receptors. In addition, GAS5

interacts with and sequesters an miRNA called miR-21, which interacts with and inhibits the activity of some regulatory proteins that act as tumor suppressors. Expression of GAS5 is suppressed in a wide range of tumors, resulting in increased expression of steroid hormones, higher levels of active miR-21, and faster tumor growth. Low GAS5 levels thus correlate with worsened outcomes for cancer patients, making this lncRNA a subject of intense ongoing investigation.

Some hormone receptors, including the human progesterone receptor, activate transcription with the aid of a different lncRNA of ~700 nucleotides that acts as a coactivator—**steroid receptor RNA activator (SRA)**. SRA is part of a ribonucleoprotein complex, but it is the RNA component that is required for transcription coactivation. The detailed set of interactions between SRA and other components of the regulatory systems for these genes remains to be worked out.

Regulation Can Result from Phosphorylation of Nuclear Transcription Factors

We noted in Chapter 12 that the effects of insulin on gene expression are mediated by a series of steps leading ultimately to the activation of a protein kinase in the nucleus that phosphorylates specific DNA-binding proteins, thereby altering their ability to act as transcription factors (see Fig. 12-19). This general mechanism mediates the effects of many nonsteroid hormones. For example, the β-adrenergic pathway that leads to elevated levels of cytosolic cAMP, which acts as a second messenger in both eukaryotes and bacteria (Fig. 28-18), also affects the transcription of a set of genes, each of which is located near a specific DNA sequence called a **cAMP response element (CRE)**. The catalytic subunit of protein kinase A, released when cAMP levels rise (see Fig. 12-6), enters the nucleus and phosphorylates a nuclear protein, the **CRE-binding protein (CREB)**. When phosphorylated, CREB binds to CREs near certain genes and acts as a transcription factor, turning on expression of these genes.

Many Eukaryotic mRNAs Are Subject to Translational Repression

Regulation at the level of translation assumes a much more prominent role in eukaryotes than in bacteria and is observed in a range of cellular situations. In contrast to the tight coupling of transcription and translation in bacteria, the transcripts generated in a eukaryotic nucleus must be processed and transported to the cytoplasm before translation. This can impose a significant delay on the appearance of a protein. When a rapid increase in protein production is needed, a translationally repressed mRNA already in the cytoplasm can be activated for translation without delay. Translational regulation may play an especially important role in regulating certain very long eukaryotic genes (a few are measured in the millions of base pairs), for which transcription and mRNA processing can require many

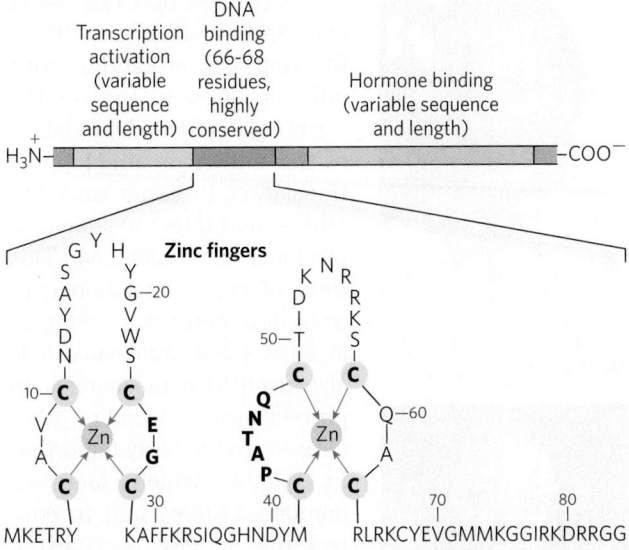

FIGURE 28-34 Typical steroid hormone receptors. These receptor proteins have a binding site for the hormone, a DNA-binding domain, and a region that activates transcription of the regulated gene. The highly conserved DNA-binding domain has two zinc fingers. The sequence shown here is that for the estrogen receptor, but the residues in bold type are common to all steroid hormone receptors.

hours. Some genes are regulated at both the transcriptional and translational stages, with the latter playing a role in the fine-tuning of cellular protein levels. In some non-nucleated cells, such as reticulocytes (immature erythrocytes), transcriptional control is entirely unavailable and translational control of stored mRNAs becomes essential. As described below, translational controls can also have spatial significance during development, when the regulated translation of prepositioned mRNAs creates a local gradient of the protein product.

Eukaryotes have at least four main mechanisms of translational regulation.

1. Translation initiation factors are subject to phosphorylation by protein kinases. The phosphorylated forms are often less active and cause a general depression of translation in the cell.

2. Some proteins bind directly to mRNA and act as translational repressors, many of them binding at specific sites in the 3′ untranslated region (3′UTR). So positioned, these proteins interact with other translation initiation factors bound to the mRNA, or with the 40S ribosomal subunit, to prevent translation initiation **(Fig. 28-35)**.

3. Binding proteins, present in eukaryotes from yeast to mammals, disrupt the interaction between eIF4E and eIF4G (see Fig. 27-28). The mammalian versions are known as 4E-BPs (eIF4E binding proteins). When cell growth is slow, these proteins limit translation by binding to the site on eIF4E that normally interacts with eIF4G. When cell growth resumes or increases in response to growth factors or other stimuli, the binding proteins are inactivated by protein kinase–dependent phosphorylation.

4. RNA-mediated regulation of gene expression often occurs at the level of translational repression.

The variety of translational regulation mechanisms provides flexibility, allowing focused repression of a few mRNAs or global regulation of all cellular translation.

Translational regulation has been particularly well studied in reticulocytes. One such mechanism in these cells involves eIF2, the initiation factor that binds to the initiator tRNA and conveys it to the ribosome; when Met-tRNA has bound to the P site, the factor eIF2B binds to eIF2, recycling it with the aid of GTP binding and hydrolysis. The maturation of reticulocytes includes destruction of the cell nucleus, leaving behind a plasma membrane packed with hemoglobin. Messenger RNAs deposited in the cytoplasm before the loss of the nucleus allow for the replacement of hemoglobin. When reticulocytes become deficient in iron or heme, the translation of globin mRNAs is repressed. A protein kinase called **HCR (hemin-controlled repressor)** is then activated, catalyzing the phosphorylation of eIF2. When phosphorylated, eIF2 forms a stable complex with eIF2B that sequesters the eIF2, making it unavailable for participation in translation. In this way, the reticulocyte coordinates the synthesis of globin with the availability of heme.

Many additional examples of translational regulation have been found in studies of the development of multicellular organisms, as discussed in more detail below.

Posttranscriptional Gene Silencing Is Mediated by RNA Interference

In higher eukaryotes, including nematodes, fruit flies, plants, and mammals, microRNAs (miRNAs) mediate the silencing of many genes. In a phenomenon first described and explained by Craig Mello and Andrew Fire, the RNAs function by interacting with mRNAs, often in the 3′UTR, resulting in either degradation of the mRNA or inhibition of translation. In either case, the mRNA, and thus the gene that produces it, is silenced. This form of gene regulation controls developmental timing in at least some organisms. It is also used as a mechanism to protect against invading RNA viruses (particularly important in plants, which lack an immune system) and to control the activity of transposons. In addition, small RNA molecules may play a critical (as yet undefined) role in the formation of heterochromatin.

Many miRNAs are present only transiently during

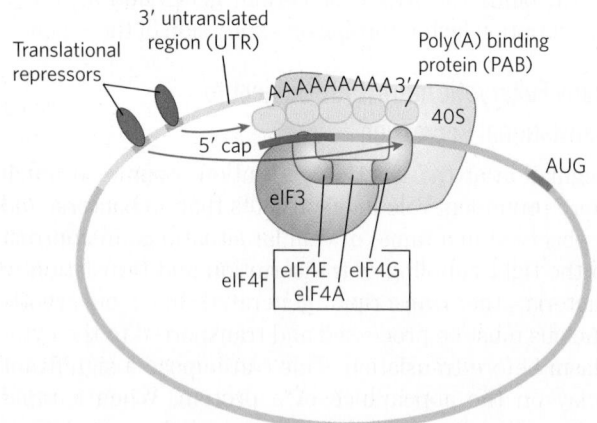

FIGURE 28-35 Translational regulation of eukaryotic mRNA. One of the most important mechanisms for translational regulation in eukaryotes is the binding of translational repressors (RNA-binding proteins) to specific sites in the 3′ untranslated region (3′UTR) of the mRNA. These proteins interact with eukaryotic initiation factors or with the ribosome to prevent or slow translation.

Craig Mello
[Source: Courtesy of Craig Mello.]

Andrew Fire
[Source: Linda A. Cicero/ Stanford News Service.]

development, and these are sometimes referred to as **small temporal RNAs (stRNAs)**. Thousands of different miRNAs have been identified in higher eukaryotes, and they may affect the regulation of a third of mammalian genes. They are transcribed as precursor RNAs ~70 nucleotides long, with internally complementary sequences that form hairpinlike structures. Details of the pathway for processing of miRNAs were described in Chapter 26 (see Fig. 26-27). The precursors are cleaved by endonucleases such as Drosha and Dicer to form short duplexes of 20 to 25 nucleotides. One strand of the processed miRNA is transferred to the target mRNA (or to a viral or transposon RNA), leading to inhibition of translation or degradation of the mRNA **(Fig. 28-36)**. Some miRNAs bind to and affect a single mRNA and thus affect expression of only one gene. Others interact with multiple mRNAs and form the mechanistic core of regulons that coordinate the expression of multiple genes.

This gene regulation mechanism has an interesting and very useful practical side. If an investigator introduces into an organism a duplex RNA molecule corresponding in sequence to virtually any mRNA, Dicer cleaves the duplex into short segments, called **small interfering RNAs (siRNAs)**. These bind to the mRNA and silence it (Fig. 28-36b). The process is known as **RNA interference (RNAi)**. In plants, almost any gene

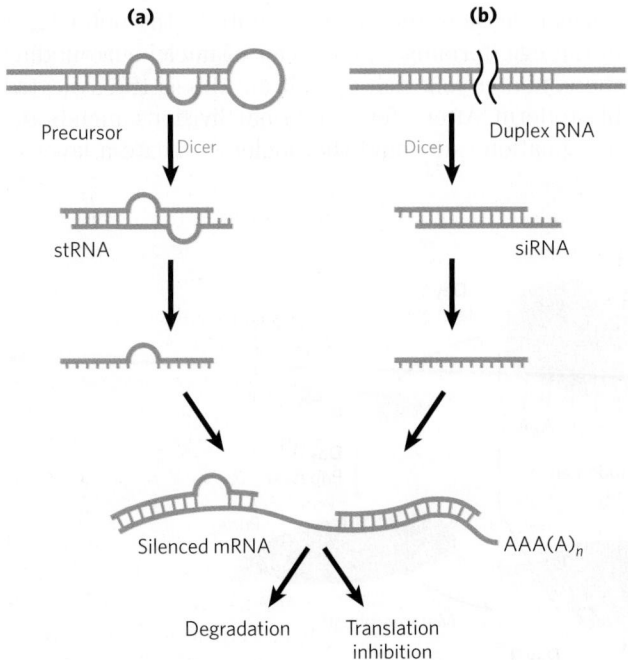

FIGURE 28-36 Gene silencing by RNA interference. (a) Small temporal RNAs (stRNAs, a class of miRNAs) are generated by Dicer-mediated cleavage of longer precursors that fold to create duplex regions. The stRNAs then bind to mRNAs, leading to degradation of mRNA or inhibition of translation. **(b)** Double-stranded RNAs can be constructed and introduced into a cell. Dicer processes the duplex RNAs into small interfering RNAs (siRNAs), which interact with the target mRNA. Again, either the mRNA is degraded or translation is inhibited.

can be effectively shut down in this way. Nematodes can readily ingest entire functional RNAs, and simply introducing the duplex RNA into the worm's diet produces very effective suppression of the target gene. The technique is an important tool in the ongoing efforts to study gene function, because it can disrupt gene function without creating a mutant organism. The procedure can be applied to humans as well. Laboratory-produced siRNAs have been used to block HIV and poliovirus infections in cultured human cells for a week or so at a time. The wider application of RNAi-based pharmaceuticals was initially stymied by the difficulty inherent in delivering RNAi molecules to their required target, given the many nucleases that degrade RNA in human tissues. With recent advances in delivery methods, there are now more than a dozen RNAi pharmaceuticals in advanced clinical trials to treat a range of conditions, from familial amyloidotic polyneuropathy to viral infections and cancer.

RNA-Mediated Regulation of Gene Expression Takes Many Forms in Eukaryotes

The special-function RNAs in eukaryotes include miRNAs, described above; snRNAs, involved in RNA splicing (see Fig. 26-16); snoRNAs, involved in rRNA modification (see Fig. 26-25), and lncRNAs, already encountered in this chapter. These RNAs, like all RNAs (regardless of their length) that do not encode proteins, including rRNAs and tRNAs, come under the general designation of **ncRNAs** (noncoding RNAs). Mammalian genomes seem to encode more ncRNAs than coding mRNAs. Not surprisingly, additional functional classes of ncRNAs are still being discovered. Here we describe a few more examples of ncRNAs that participate in gene regulation, which are designated lncRNAs when their length exceeds 200 nucleotides.

Heat shock factor 1 (HSF1) is an activator protein that, in nonstressed cells, exists as a monomer bound by the chaperone Hsp90. Under stress conditions, HSF1 is released from Hsp90 and trimerizes. The HSF1 trimer binds to DNA and activates transcription of genes encoding products required to deal with the stress. An lncRNA called HSR1 (heat shock RNA 1; ~600 nucleotides) stimulates HSF1 trimerization and DNA binding. HSR1 does not act alone; it functions in a complex with the translation elongation factor eEF1A.

Additional RNAs affect transcription in a variety of ways. A 331 nucleotide lncRNA called 7SK, abundant in mammals, binds to the Pol II transcription elongation factor pTEFb (see Table 26-2) and represses transcript elongation. The ncRNA B2 (~178 nucleotides) binds directly to Pol II during heat shock and represses transcription. The B2-bound Pol II assembles into stable PICs, but transcription is blocked. The B2 RNA thus halts the transcription of many genes during heat shock; the mechanism that allows HSF1-responsive genes to be expressed in the presence of B2 remains to be worked out.

The recognized roles of ncRNAs in gene expression and in many other cellular processes are rapidly expanding. At the same time, the study of the biochemistry of gene regulation is becoming much less protein-centric.

Development Is Controlled by Cascades of Regulatory Proteins

For sheer complexity and intricacy of coordination, the patterns of gene regulation that bring about development of a zygote into a multicellular animal or plant have no peer. Development requires transitions in morphology and protein composition that depend on tightly coordinated changes in expression of the genome. More genes are expressed during early development than in any other part of the life cycle. For example, in the sea urchin, an oocyte has about 18,500 *different* mRNAs, compared with about 6,000 different mRNAs in the cells of a typical differentiated tissue. The mRNAs in the oocyte give rise to a cascade of events that regulate the expression of many genes across both space and time.

Several organisms have emerged as important model systems for the study of development, because they are easy to maintain in a laboratory and have relatively short generation times. These include nematodes, fruit flies, zebra fish, mice, and the plant *Arabidopsis*. This discussion focuses on the development of fruit flies. Our understanding of the molecular events during development of *Drosophila*

melanogaster is particularly well advanced and can be used to illustrate patterns and principles of general significance.

The life cycle of the fruit fly includes complete metamorphosis during its progression from an embryo to an adult **(Fig. 28-37)**. Among the most important characteristics of the embryo are its **polarity** (the anterior and posterior parts of the animal are readily distinguished, as are its dorsal and ventral surfaces) and its **metamerism** (the embryo body is made up of serially repeating segments, each with characteristic features). During development, these segments become organized into a head, thorax, and abdomen. Each segment of the adult thorax has a different set of appendages. Development of this complex pattern is under genetic control, and a variety of pattern-regulating genes have been discovered that greatly affect the organization of the body.

The *Drosophila* egg, along with 15 nurse cells, is surrounded by a layer of follicle cells **(Fig. 28-38)**. As the egg cell forms (before fertilization), mRNAs and proteins originating in the nurse and follicle cells are deposited in the egg cell, where some play a critical role in development. Once a fertilized egg is laid, its nucleus divides and the nuclear descendants continue to divide in synchrony every 6 to 10 min. Plasma membranes are not formed around the nuclei, which are distributed within the egg cytoplasm, forming a syncytium. Between the eighth and eleventh rounds of nuclear division, the nuclei migrate to the outer layer of the egg, forming a monolayer of nuclei surrounding the common yolk-rich cytoplasm; this is the syncytial blastoderm. After a few additional divisions, membrane invaginations surround the nuclei to create a layer of

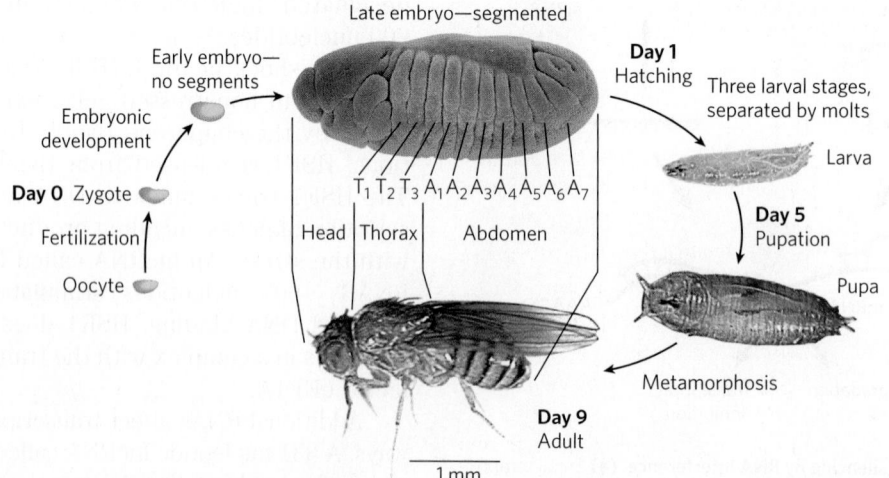

FIGURE 28-37 Life cycle of the fruit fly *Drosophila melanogaster*. *Drosophila* undergoes a complete metamorphosis, which means that the adult insect is radically different in form from its immature stages, a transformation that requires extensive alterations during development. By the late embryonic stage, segments have formed, each containing specialized structures from which the various appendages and other features of the adult fly will develop. [Sources: Later embryo: Courtesy of F. R. Turner, Department of Biology, University of Indiana, Bloomington. Other photos: Courtesy of Prof. Dr. Christian Klambt, Westfälische Wilhelms-Universität Münster, Institut für Neuro- und Verhaltensbiologie.]

FIGURE 28-38 Early development in *Drosophila*. During development of the egg, maternal mRNAs (including the *bicoid* and *nanos* gene transcripts, discussed in the text) and proteins are deposited in the developing oocyte (unfertilized egg cell) by nurse cells and follicle cells. After fertilization, the two nuclei of the egg divide in synchrony within the common cytoplasm (syncytium), then migrate to the periphery. Membrane invaginations surround the nuclei to create a monolayer of cells at the periphery; this is the cellular blastoderm stage. During the early nuclear divisions, several nuclei at the far posterior become pole cells, which later become the germline cells.

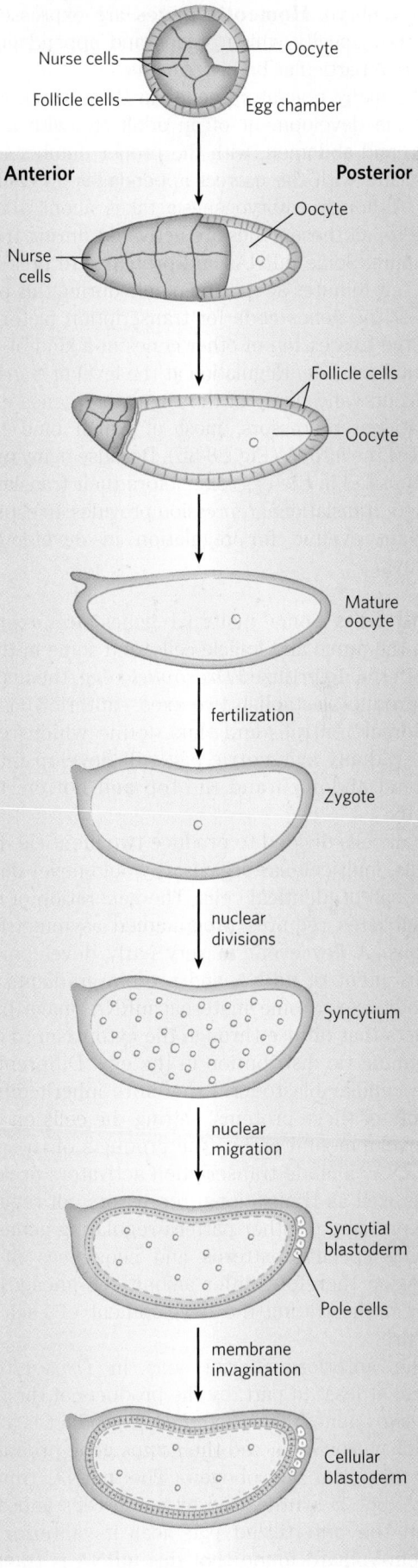

cells that form the cellular blastoderm. At this stage, the mitotic cycles in the various cells lose their synchrony. The developmental fate of the cells is determined by the mRNAs and proteins originally deposited in the egg by the nurse and follicle cells.

Proteins that, through changes in local concentration or activity, cause the surrounding tissue to take up a particular shape or structure are sometimes referred to as **morphogens**; they are the products of pattern-regulating genes. As defined by Christiane Nüsslein-Volhard, Edward B. Lewis, and Eric F. Wieschaus, three major classes of pattern-regulating genes—maternal, segmentation, and homeotic genes—function in successive stages of development to specify the basic features of the *Drosophila* embryo body. **Maternal genes** are expressed in the unfertilized egg, and the resulting **maternal mRNAs** remain dormant until fertilization. These provide most of the proteins needed in very early development, until the cellular blastoderm is formed. Some of the proteins encoded by maternal mRNAs direct the spatial organization of the developing embryo at early stages, establishing its polarity. **Segmentation genes**, transcribed after fertilization, direct the formation of the proper number of body segments. At least three subclasses of segmentation genes act at successive stages: **gap genes** divide the developing embryo into several broad regions; **pair-rule genes**, together with **segment polarity genes**, define 14 stripes that become the 14 segments of a

Christiane Nüsslein-Volhard
[Source: L'Oréal/Micheline Pelletier.]

Edward B. Lewis, 1918–2004
[Source: CalTech Archives, California Institute of Technology.]

Eric F. Wieschaus
[Source: Courtesy of Eric F. Wieschaus.]

normal embryo. **Homeotic genes** are expressed still later; they specify which organs and appendages will develop in particular body segments.

The many regulatory genes in these three classes direct the development of an adult fly, with a head, thorax, and abdomen, with the proper number of segments, and with the correct appendages on each segment. Although embryogenesis takes about a day to complete, all these genes are activated during the first four hours. Some mRNAs and proteins are present for only a few minutes at specific points during this period. Some of the genes code for transcription factors that affect the expression of other genes in a kind of developmental cascade. Regulation at the level of translation also occurs, and many of the regulatory genes encode translational repressors, most of which bind to the 3′UTR of the mRNA (Fig. 28–35). Because many mRNAs are deposited in the egg long before their translation is required, translational repression provides an especially important avenue for regulation in developmental pathways.

Maternal Genes Some maternal genes are expressed within the nurse and follicle cells, and some in the egg itself. In the unfertilized *Drosophila* egg, the maternal gene products establish two axes—anterior-posterior and dorsal-ventral—and thus define which regions of the radially symmetric egg will develop into the head and abdomen and the top and bottom of the adult fly.

If all cells divided to produce two identical daughter cells, multicellular organisms would never be more than a ball of identical cells. The generation of different cell fates requires programmed asymmetric cell divisions. A key event in very early development is establishment of mRNA and protein gradients along the body axes. Some maternal mRNAs have protein products that diffuse through the cytoplasm to create an asymmetric distribution in the egg. Different cells in the cellular blastoderm therefore inherit different amounts of these proteins, setting the cells on different developmental paths. The products of the maternal mRNAs include transcription activators or repressors as well as translational repressors, all regulating the expression of other pattern-regulating genes. The resulting specific patterns and sequences of gene expression therefore differ among cell lineages, ultimately orchestrating the development of each adult structure.

The anterior-posterior axis in *Drosophila* is defined, at least in part, by the products of the *bicoid* and *nanos* genes. The *bicoid* gene product is a major anterior morphogen, and the *nanos* gene product is a major posterior morphogen. The mRNA from the *bicoid* gene is synthesized by nurse cells and deposited in the unfertilized egg near its anterior pole. Nüsslein-Volhard found that this mRNA is translated soon after fertilization, and the Bicoid protein diffuses

through the cell to create, by the seventh nuclear division, a concentration gradient radiating out from the anterior pole (**Fig. 28–39a**). The Bicoid protein is a transcription factor that activates the expression of several segmentation genes; the protein contains a homeodomain (p. 1135). Bicoid is also a translational repressor that inactivates certain mRNAs. The amounts of Bicoid protein in various parts of the embryo affect the subsequent expression of other genes in a threshold-dependent manner. Genes are transcriptionally activated or translationally repressed only where the Bicoid protein concentration exceeds the threshold. Changes in the shape of the Bicoid concentration gradient have dramatic effects on the body pattern. Lack of Bicoid protein results in development of an embryo with two abdomens but neither head nor thorax (Fig. 28–39b); however, embryos without Bicoid will develop normally if an adequate amount of *bicoid* mRNA is injected into the egg at the appropriate end. The *nanos* gene has an analogous role, but its mRNA is deposited at the posterior end of the egg, and the anterior-posterior protein gradient peaks at the posterior pole. The Nanos protein is a translational repressor.

A broader look at the effects of maternal genes reveals the outline of a developmental circuit. In addition to the *bicoid* and *nanos* mRNAs, which are deposited in the egg asymmetrically, several other maternal mRNAs are deposited uniformly throughout the egg cytoplasm. Three of these mRNAs encode the Pumilio, Hunchback, and Caudal proteins, all affected by *nanos* and *bicoid* (**Fig. 28–40**). Caudal and Pumilio are involved in development of the posterior end of the fly. Caudal is a transcription activator with a homeodomain; Pumilio is a translational repressor. Hunchback protein plays an important role in development of the anterior end and is also a transcriptional regulator of a variety of genes, in some cases a positive regulator, in other cases negative. Bicoid suppresses translation of *caudal* at the anterior end and also acts as a transcription activator of *hunchback* in the cellular blastoderm. Because *hunchback* is expressed both in maternal mRNAs and in genes in the developing egg, it is considered both a maternal and a segmentation gene. The result of the activities of Bicoid is an increased concentration of Hunchback at the anterior end of the egg. The Nanos and Pumilio proteins act as translational repressors of *hunchback*, suppressing synthesis of its protein near the posterior end of the egg. Pumilio does not function in the absence of the Nanos protein, and the gradient of Nanos expression confines the activity of both proteins to the posterior region. Translational repression of the *hunchback* gene leads to degradation of *hunchback* mRNA near the posterior end. However, lack of Bicoid in the posterior leads to expression of *caudal*. In this way, the Hunchback and Caudal proteins become asymmetrically distributed in the egg.

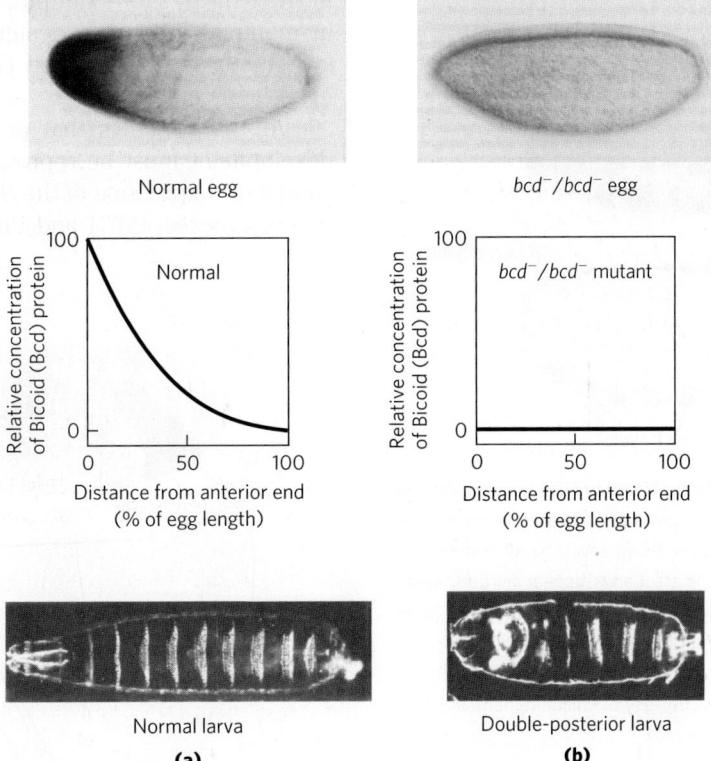

FIGURE 28-39 Distribution of a maternal gene product in a _Drosophila_ egg. (a) Micrograph of an immunologically stained egg (top), showing distribution of the _bicoid_ (_bcd_) gene product. The graph shows stain intensity along the length of the egg. This distribution is essential for normal development of the anterior structures in the larva (bottom). **(b)** If the _bcd_ gene is not expressed by the mother (_bcd⁻/bcd⁻_ mutant) and thus no _bicoid_ mRNA is deposited in the egg, the resulting larva has two posteriors (and soon dies). [Source: Republished with permission of Elsevier, from "The bicoid protein determines position in the _Drosophila_ embryo in a concentration-dependent manner" by Wolfgang Driever and Christiane Nüsslein-Volhard, _Cell_ 54:83–93, July 1, 1988; permission conveyed through Copyright Clearance Center, Inc.]

Segmentation Genes Gap genes, pair-rule genes, and segment polarity genes, three subclasses of segmentation genes in _Drosophila_, are activated at successive stages of embryonic development. Expression of the gap genes is generally regulated by the products of one or more maternal genes. At least some gap genes encode transcription factors that affect the expression of other segmentation or (later) homeotic genes.

One well-characterized segmentation gene is _fushi tarazu_ (_ftz_), of the pair-rule subclass. When _ftz_ is deleted, the embryo develops 7 segments instead of the normal 14, each segment twice the normal width. The Fushi-tarazu protein (Ftz) is a transcription activator with a homeodomain. The mRNAs and proteins

FIGURE 28-40 Regulatory circuits of the anterior-posterior axis in a _Drosophila_ egg. The _bicoid_ and _nanos_ mRNAs are localized near the anterior and posterior poles, respectively. The _caudal, hunchback,_ and _pumilio_ mRNAs are distributed evenly throughout the egg cytoplasm. The gradients of Bicoid (Bcd) and Nanos proteins affect expression of the _caudal_ and _hunchback_ mRNAs as shown, leading to accumulation of Hunchback protein in the anterior and Caudal protein in the posterior of the egg. Because Pumilio protein requires Nanos protein for its activity as a translational repressor of _hunchback,_ it functions only at the posterior end.

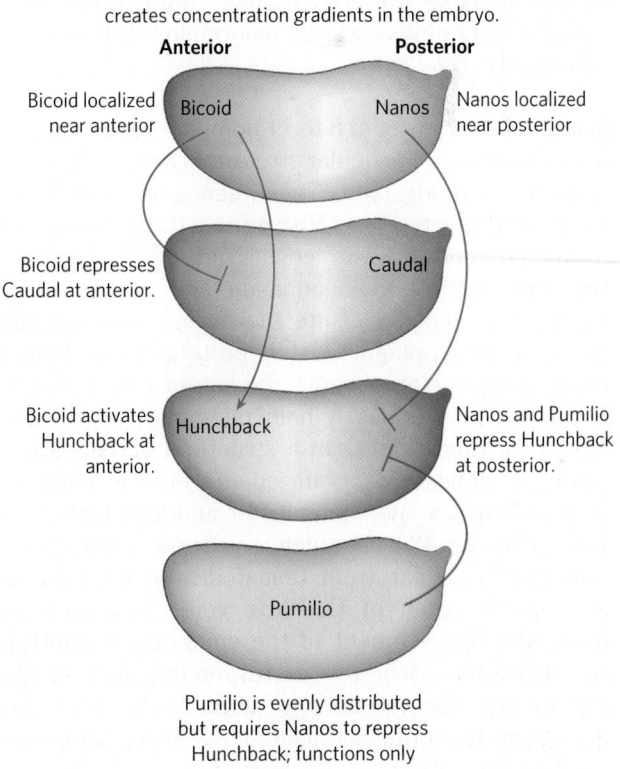

Translation of mRNA and diffusion of products creates concentration gradients in the embryo.

Anterior   **Posterior**

Bicoid localized near anterior — Bicoid   Nanos — Nanos localized near posterior

Bicoid represses Caudal at anterior.   Caudal

Bicoid activates Hunchback at anterior.   Hunchback   Nanos and Pumilio repress Hunchback at posterior.

Pumilio

Pumilio is evenly distributed but requires Nanos to repress Hunchback; functions only at posterior.

(a) Side view

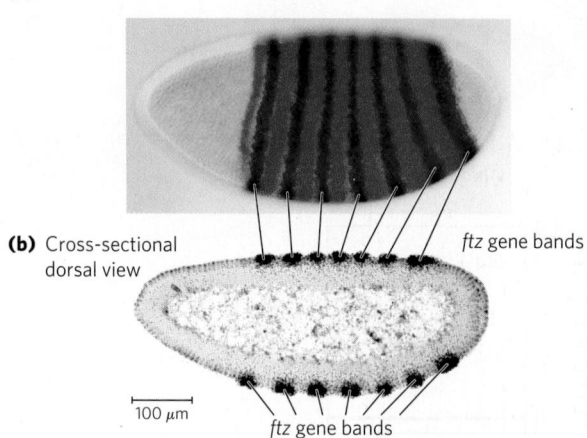

(b) Cross-sectional dorsal view

ftz gene bands

100 μm

ftz gene bands

FIGURE 28-41 Distribution of the *fushi tarazu* (*ftz*) gene product in early *Drosophila* embryos. **(a)** Following a gene-specific staining procedure, the gene product can be detected in seven bands around the circumference of the embryo. **(b)** These bands appear as dark spots (generated by a radioactive label) in a cross-sectional autoradiograph and demarcate the anterior margins of the segments that will appear in the late embryo. [Sources: (a) Courtesy of Stephen J. Small, Department of Biology, New York University. (b) Courtesy of Phillip Ingham, Imperial College London.]

derived from the normal *ftz* gene accumulate in a striking pattern of seven stripes that encircle the posterior two-thirds of the embryo **(Fig. 28-41)**. The stripes demarcate the positions of segments that develop later; these segments are eliminated if *ftz* function is lost. The Ftz protein and a few similar regulatory proteins directly or indirectly regulate the expression of vast numbers of genes in the continuing developmental cascade.

Homeotic Genes A set of 8 to 11 homeotic genes directs the formation of particular structures at specific locations in the body plan. These genes are now more commonly referred to as **Hox genes**, the term derived from "homeobox," the conserved gene sequence that encodes the homeodomain and is present in all of these genes. Despite the name, these are not the only development-related proteins to include a homeodomain (for example, the *bicoid* gene product, described above, has a homeodomain), and "Hox" is more a functional than a structural classification. The *Hox* genes are organized in genomic clusters. *Drosophila* has one such cluster and mammals have four **(Fig. 28-42)**. The genes in these clusters are remarkably similar from nematodes to humans. In *Drosophila*, each of the *Hox* genes is expressed in a particular segment of the embryo and controls the development of the corresponding part of the mature fly. The terminology used to describe *Hox* genes can be confusing. They have historical names

in the fruit fly (for example, *ultrabithorax*), whereas in mammals they are designated by two competing systems based on lettered (A, B, C, D) or numbered (1, 2, 3, 4) clusters.

All of the genes that promote different stages of development must be repressed after their function is complete. Expression of the *Hox* genes is suppressed by two complexes, PRC1 and PRC2 (polycomb repressive

(a)

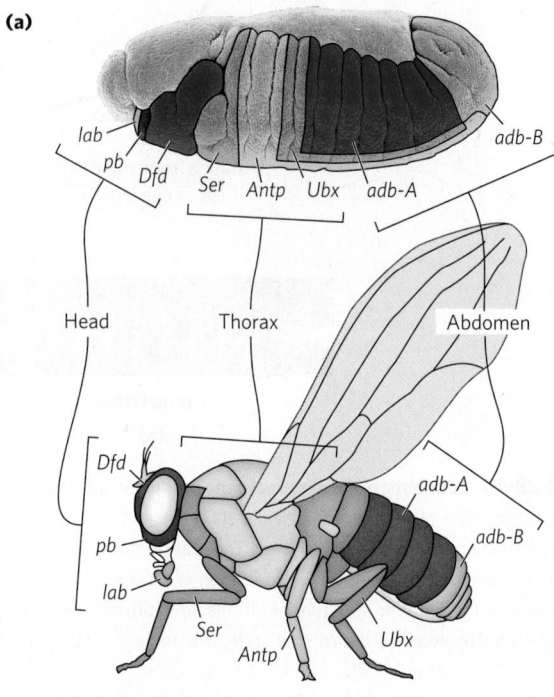

(b)

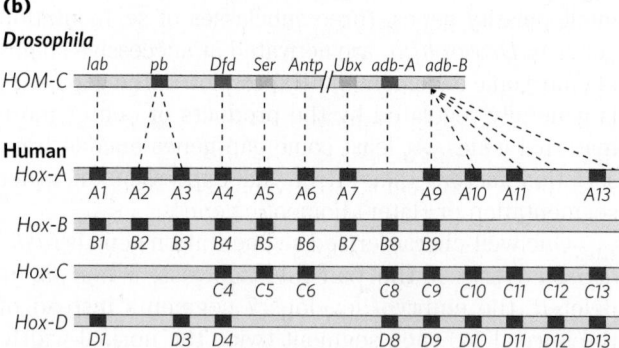

FIGURE 28-42 The *Hox* gene clusters and their effects on development. **(a)** Each *Hox* gene in the fruit fly is responsible for the development of structures in a defined part of the body and is expressed in defined regions of the embryo, as shown here with color coding. **(b)** *Drosophila* has one *Hox* gene cluster; the human genome has four. Many of these genes are highly conserved in multicellular animals. Evolutionary relationships, as indicated by sequence alignments, between genes in the fruit fly *Hox* gene cluster and those in the mammalian *Hox* gene clusters are shown by dashed lines. Similar relationships among the four sets of mammalian *Hox* genes are indicated by vertical alignment. [Source: (a) F. R. Turner, University of Indiana, Department of Biology.]

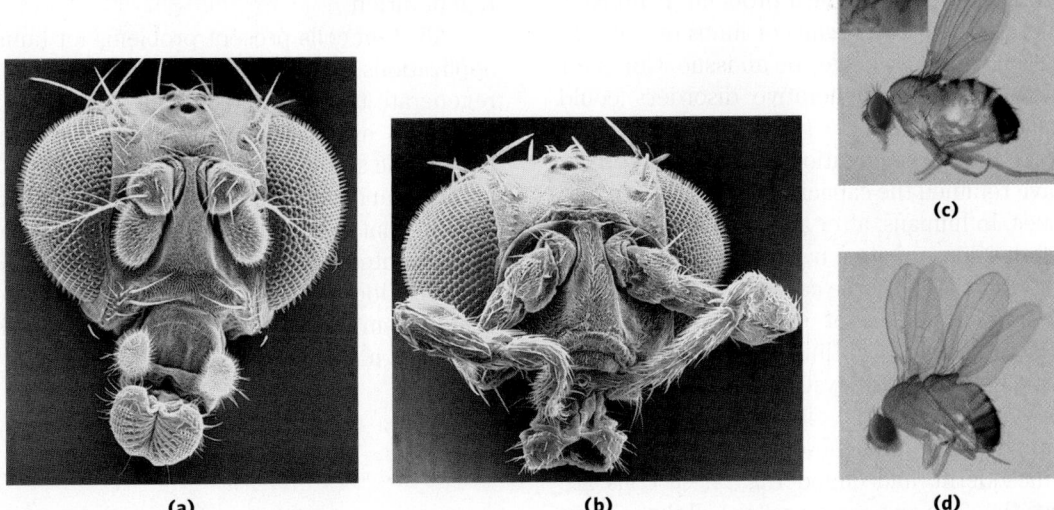

FIGURE 28-43 Effects of mutations in *Hox* genes in *Drosophila*.
(a) Normal head. **(b)** Homeotic mutant (*antennapedia*) in which antennae are replaced by legs. **(c)** Normal body structure. **(d)** Homeotic mutant (*bithorax*) in which a segment has developed incorrectly to
produce an extra set of wings. [Sources: (a, b) F. R. Turner, University of
Indiana, Department of Biology. (c, d) Courtesy of Nipam Patel, Patel Lab,
Berkeley, California.]

complexes). PRC1 ubiquitinates histone H2A; PRC2
methylates Lys^{27} in histone H3. These modifications
lead to structural changes in chromatin that lower transcription levels. Certain lncRNAs play a role in targeting PRC1 and PRC2 to chromosomal regions where
gene silencing is required. Defects in PRC1 or PRC2
can lead to uncontrolled cell growth, and the function
of one or both of these complexes is often perturbed in
tumor cells.

Loss of *Hox* genes in fruit flies, by mutation
or deletion, causes the development of a normal
appendage or body structure at an inappropriate
body position. An important example is the *ultrabithorax (ubx)* gene. When Ubx function is lost, the
first abdominal segment develops incorrectly, having
the structure of the third thoracic segment. Other
known homeotic mutations cause the formation of
an extra set of wings, or two legs at the position in
the head where the antennae are normally found
(Fig. 28-43). The *Hox* genes often span long regions
of DNA. The *ubx* gene, for example, is 77,000 bp long.
More than 73,000 bp of this gene are in introns, one
of which is more than 50,000 bp long. Transcription
of the *ubx* gene takes nearly an hour. The delay this
imposes on *ubx* gene expression is believed to be a
timing mechanism involved in the temporal regulation of subsequent steps in development. Many *Hox*
genes are further regulated by miRNAs encoded by
intergenic regions of the *Hox* gene clusters. All of
the *Hox* gene products are transcription factors that
regulate the expression of an array of downstream
genes.

Many of the principles of development outlined
above apply to other eukaryotes, from nematodes to
humans. Some of the regulatory proteins are conserved.
For example, the products of the homeobox-containing
genes *HOXA7* in mouse and *antennapedia* in fruit fly
differ in only one amino acid residue. Of course,
although the molecular regulatory mechanisms may
be similar, many of the ultimate developmental events
are not conserved (humans do not have wings or
antennae). The different outcomes are brought about
by differences in the downstream target genes controlled by the *Hox* genes. The discovery of structural
determinants with identifiable molecular functions is
the first step in understanding the molecular events
underlying development. As more genes and their
protein products are discovered, the biochemical side
of this vast puzzle will be elucidated in increasingly
rich detail.

Stem Cells Have Developmental Potential That Can Be Controlled

If we can understand development, and the mechanisms of gene regulation behind it, we can control it. An
adult human has many different types of tissues. Many
of the cells are terminally differentiated and no longer
divide. If an organ malfunctions due to disease, or a limb
is lost in an accident, the tissues are not readily
replaced. Most cells, because of the regulatory processes in place, or even because of the loss of some or
all of the genomic DNA, are not easily reprogrammed.
Medical science has made organ transplants possible,

but organ donors are a limited resource and organ rejection remains a major medical problem. If humans could regenerate their own organs or limbs or nervous tissue, rejection would no longer be an issue. Cures for kidney failure or neurodegenerative disorders could become reality.

The key to tissue regeneration lies in **stem cells**—cells that have retained the capacity to differentiate into various tissues. In humans, after an egg is fertilized, the first few cell divisions create a ball of **totipotent** cells, called the morula, that have the capacity to differentiate individually into any tissue or even into a complete organism **(Fig. 28-44)**. Continued cell division produces a hollow ball, the blastocyst. The outer cells of the blastocyst eventually form the placenta. The inner layers form the germ layers of the developing fetus—the ectoderm, mesoderm, and endoderm. These cells are **pluripotent**: they can give rise to cells of all three germ layers and can differentiate into many types of tissues. However, they cannot differentiate into a complete organism. Some of these cells are **unipotent**: they can develop into only one type of cell and/or tissue. It is the pluripotent cells of the blastocyst, the **embryonic stem cells**, that are currently used in embryonic stem cell research.

Stem cells have two functions: to replenish themselves and, at the same time, provide cells that can differentiate. These tasks are accomplished in multiple ways **(Fig. 28-45a)**. All or parts of the stem cell population can, in principle, be involved in replenishment, differentiation, or both.

Other types of stem cells can potentially be used for medical benefit. In the adult organism, **adult stem cells**, as products of additional differentiation, have a more limited potential for further development than do embryonic stem cells. For example, the hematopoietic stem cells of bone marrow can give rise to many types of blood cells and also to cells with the capacity to regenerate bone. They are referred to as **multipotent**. However, these cells cannot differentiate into a liver or kidney or neuron. Adult stem cells are often said to have a **niche**, a microenvironment that promotes stem cell maintenance while allowing differentiation of some daughter cells as replacements for cells in the tissue they serve (Fig. 28-45b). Hematopoietic stem cells in the bone marrow occupy a niche in which signaling from neighboring cells and other cues maintain the stem cell lineage. At the same time, some daughter cells differentiate to provide needed blood cells. Understanding the niche in which stem cells operate, and the signals the niche provides, is essential in

efforts to harness the potential of stem cells for tissue regeneration.

All stem cells present problems for human medical applications. Adult stem cells have a limited capacity to regenerate tissues, are generally present in small numbers, and are hard to isolate from an adult human. Embryonic stem cells have much greater differentiation potential and can be cultured to generate large numbers of cells, but their use is accompanied by ethical concerns related to the necessary destruction of human embryos. Identifying a source of plentiful and medically useful stem cells that does not raise such concerns remains a major goal of medical research.

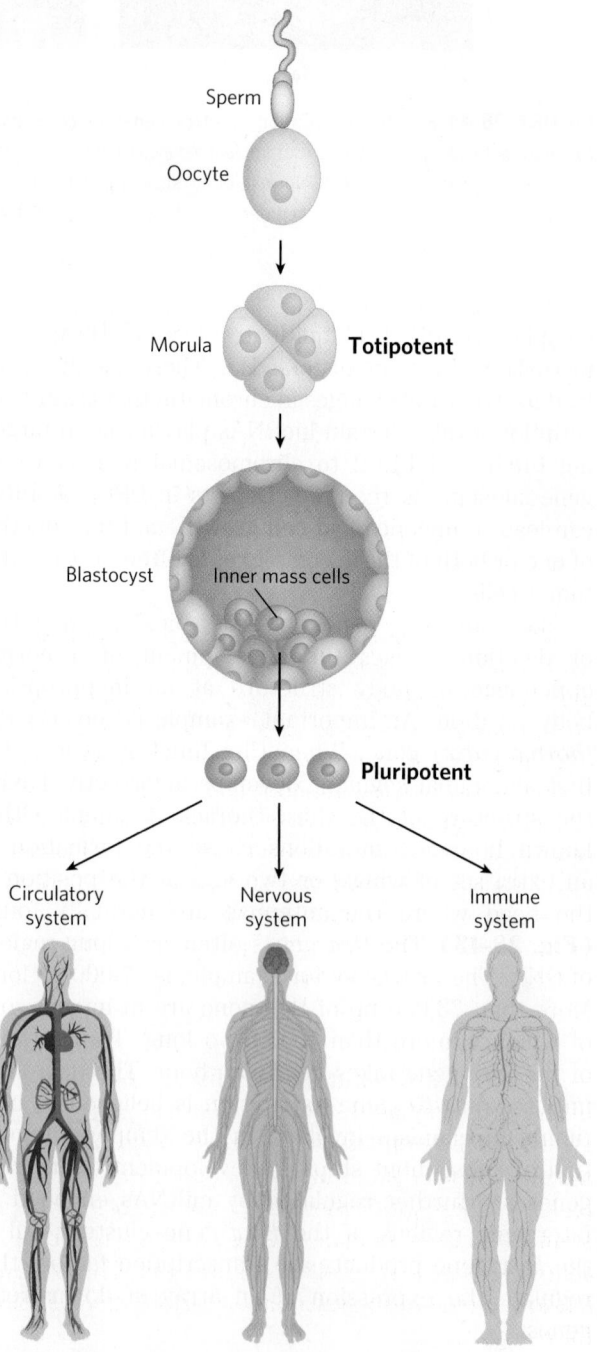

FIGURE 28-44 Totipotent and pluripotent stem cells. Cells at the morula stage are totipotent and have the capacity to differentiate into a complete organism. The source of pluripotent embryonic stem cells is the inner mass cells of the blastocyst. Pluripotent cells give rise to many tissue types but cannot form complete organisms.

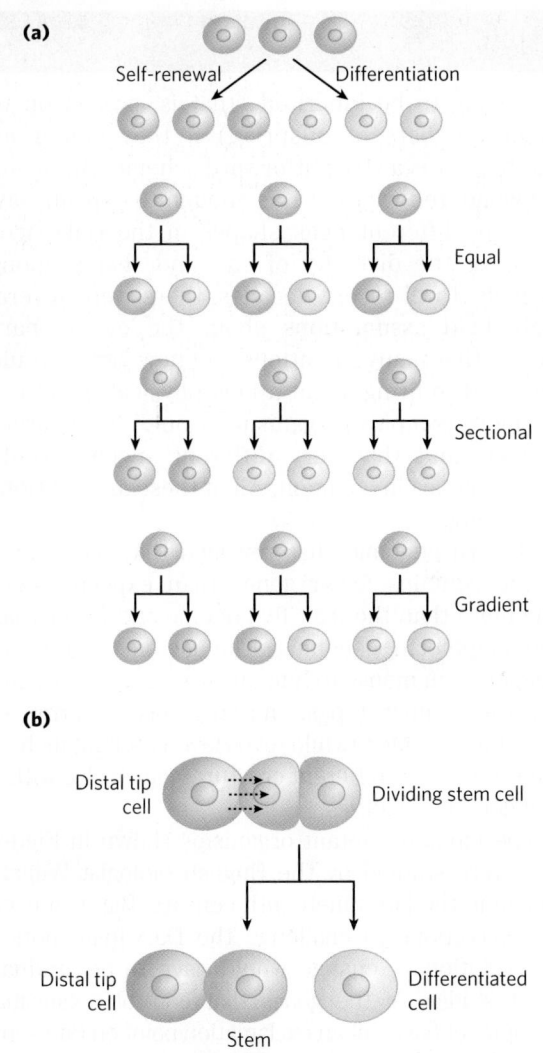

James Thomson
[Source: Courtesy of James Thomson.]

FIGURE 28-45 Stem cell proliferation versus differentiation and development. Stem cells must strike a balance between self-renewal and differentiation. **(a)** Some possible cell division patterns that allow the replenishment of stem cells and production of some differentiated cells. Each cell may produce one stem cell and one differentiated cell, or two differentiated cells, or two stem cells in defined parts of the tissue or culture. Or a gradient of growth conditions can be established, with cell fates differing from one end of the gradient to the other. **(b)** Establishing a developmental niche through stem cell contact with a cell or group of cells. Molecular signals provided by the niche cells (in this case, in plants, a distal tip cell) help orient the mitotic spindle for stem cell division and ensure that one daughter cell retains stem cell properties.

Our ability to culture stem cells (i.e., maintain them in an undifferentiated state), and to manipulate them to grow and differentiate into particular tissues, is very much a function of our understanding of developmental biology. The identification and culturing of pluripotent stem cells from human blastocysts was reported by James Thomson and colleagues in 1998. This advance led to the long-term availability of established cell lines for research.

Thus far, mouse and human embryonic stem cells have been used for most research. Although both types of stem cells are pluripotent, they require very different culture conditions, optimized to allow cell division indefinitely without differentiation. Mouse embryonic stem cells are grown on a layer of gelatin and require the presence of leukemia inhibitory factor (LIF). Human embryonic stem cells are grown on a feeder layer of mouse embryonic fibroblasts and require basic fibroblast growth factor (bFGF, or FGF2). The use of a feeder cell layer implies that the mouse cells are providing a diffusible product or some surface signal, not yet known, that is needed by human stem cells to either promote cell division or prevent differentiation.

A significant advance, reported in 2007, centers on success in reversing differentiation. In effect, skin cells—first from mice, then from humans—have been reprogrammed to take on the characteristics of pluripotent stem cells. The reprogramming involves manipulations to get the cells to express at least four transcription factors, Oct4, Sox2, Nanog, and Lin28, all of which are known to help maintain the stem cell–like state. Gradual improvements in this technology may make the harvesting of embryonic stem cells unnecessary and provide a source of stem cells that is genetically matched to a prospective patient.

Our discussion of developmental regulation and stem cells brings us full circle, back to a biochemical beginning. Evolution appropriately provides the first and last words of this book. If evolution is to generate the kind of changes in an organism that would render it a different species, it is the developmental program that must be affected. Developmental and evolutionary processes are closely allied, each informing the other (Box 28-1). The continuing study of biochemistry has everything to do with enriching the future of humanity and understanding our origins.

SUMMARY 28.3 Regulation of Gene Expression in Eukaryotes

■ In eukaryotes, positive regulation is more common than negative regulation, and transcription is accompanied by large changes in chromatin structure.

■ Promoters for Pol II typically have a TATA box and Inr sequence, as well as multiple binding sites for transcription activators. The latter sites, sometimes located hundreds or thousands of base pairs away from the TATA box, are called upstream activator sequences in yeast and enhancers in higher eukaryotes.

BOX 28-1   Of Fins, Wings, Beaks, and Things

South America has several species of seed-eating finches, commonly called grassquits. About 3 million years ago, a small group of grassquits, of a single species, took flight from the continent's Pacific coast. Perhaps driven by a storm, they lost sight of land and traveled nearly 1,000 km. Small birds such as these might easily have perished on such a journey, but the smallest of chances brought this group to a newly formed volcanic island in an archipelago later to be known as the Galápagos. It was a virgin landscape with untapped plant and insect food sources, and the newly arrived finches survived. Over the years, new islands formed and were colonized by new plants and insects—and by the finches. The birds exploited the new resources on the islands, and groups of birds gradually specialized and diverged into new species. By the time Charles Darwin stepped onto the islands in 1835, many different finch species were to be found on the various islands of the archipelago, feeding on seeds, fruits, insects, pollen, or even blood.

The diversity of living creatures was a source of wonder for humans long before scientists sought to understand its origins. The extraordinary insight handed down to us by Darwin, inspired in part by his encounter with the Galápagos finches, provided a broad explanation for the existence of organisms with a vast array of appearances and characteristics. It also gave rise to many questions about the mechanisms underlying evolution. Answers to those questions have started to appear, first through the study of genomes and nucleic acid metabolism in the last half of the twentieth century, and more recently through an emerging field nicknamed evo-devo—a blend of evolutionary and developmental biology.

In its modern synthesis, the theory of evolution has two main elements: mutations in a population generate genetic diversity; natural selection then acts on this diversity to favor individuals with more useful genomic tools and to disfavor others. Mutations occur at significant rates in every individual's genome, in every cell (see Section 8.3). Advantageous mutations in single-celled organisms or in the germ line of multicellular organisms can be inherited, and they are more likely to be inherited (that is, passed on to greater numbers of offspring) if they confer an advantage. It is a straightforward scheme. But many have wondered whether it is enough to explain, say, the many different beak shapes in the Galápagos finches or the diversity of size and shape among mammals. Until recent decades, there were several widely held assumptions about the evolutionary process: that many mutations and new genes would be needed to bring about a new physical structure, that more-complex organisms would have larger genomes, and that very different species would have few genes in common. All of these assumptions were wrong.

Modern genomics has revealed that the human genome contains fewer genes than expected—not many more than the fruit fly genome and fewer than some amphibian genomes. The genomes of every mammal, from mouse to human, are surprisingly similar in the number, types, and chromosomal arrangement of genes. Meanwhile, evo-devo is telling us how complex and very different creatures can evolve within these genomic realities.

The kinds of mutant organisms shown in Figure 28-43 were studied by the English biologist William Bateson in the late nineteenth century. Bateson used his observations to challenge the Darwinian notion that evolutionary change would have to be gradual. Recent studies of the genes that control organismal development have put an exclamation point on Bateson's ideas. Subtle changes in regulatory patterns during development, reflecting just one or a few mutations, can result in startling physical changes and fuel surprisingly rapid evolution.

The Galápagos finches provide a wonderful example of the link between evolution and development. There are at least 14 (some specialists list 15) species of Galápagos finches, distinguished in large measure by their beak structure. The ground finches, for example, have broad, heavy beaks adapted to crushing large, hard seeds. The cactus finches have longer, slender beaks ideal for probing cactus fruits and flowers (Fig. 1). Clifford Tabin and colleagues

---

■ Large complexes of proteins are generally required to regulate transcriptional activity. The effects of transcription activators on Pol II are mediated by coactivator protein complexes such as Mediator. The modular structures of the activators have distinct activation and DNA-binding domains. Other protein complexes, including histone acetyltransferases and ATP-dependent complexes such as SWI/SNF and ISWI, reversibly remodel and modify chromatin structure.

■ Hormones affect the regulation of gene expression in one of two ways. Steroid hormones interact directly with intracellular receptors that are DNA-binding regulatory proteins; binding of the hormone has either positive or negative effects on the transcription of targeted genes. Nonsteroid hormones bind to cell surface receptors, triggering a signaling pathway that can lead to phosphorylation of a regulatory protein, affecting its activity.

carefully surveyed a set of genes expressed during avian craniofacial development. They identified a single gene, *Bmp4*, whose expression level correlated with formation of the more robust beaks of the ground finches. More-robust beaks were also formed in chicken embryos when high levels of Bmp4 were artificially expressed in the appropriate tissues, confirming the importance of *Bmp4*. In a similar study, the formation of long, slender beaks was linked to the expression of calmodulin (see Fig. 12-12) in particular tissues at appropriate developmental stages. Thus, major changes in the shape and function of the beak can be brought about by subtle changes in the expression of just two genes involved in developmental regulation. Very few mutations are required, and the needed mutations affect regulation. New genes are *not* required.

The system of regulatory genes that guides development is remarkably conserved among all vertebrates. Elevated expression of *Bmp4* in the right tissue at the right time leads to more-robust jaw parts in zebrafish. The same gene plays a key role in tooth development in mammals. The development of eyes is triggered by the expression of a single gene, *Pax6*, in fruit flies and in mammals. The mouse *Pax6* gene will trigger the development of fruit fly eyes in the fruit fly, and the fruit fly *Pax6* gene will trigger the development of mouse eyes in the mouse. In each organism, these genes are part of the much larger regulatory cascade that ultimately creates the correct structures in the correct locations in each organism. The cascade is ancient; for example, the *Hox* genes (described in the text) have been part of the developmental program of multicellular eukaryotes for more than 500 million years. Subtle changes in the cascade can have large effects on development, and thus on the ultimate appearance, of the organism. These same subtle changes can fuel remarkably rapid evolution. For example, the 400 to 500 described species of cichlids (spiny-finned fish) in Lake Malawi and Lake Victoria on the African continent are all derived from one or a few populations that colonized each lake in the past 100,000 to 200,000 years. The Galápagos finches simply followed a path of evolution and change that living creatures have been traveling for billions of years.

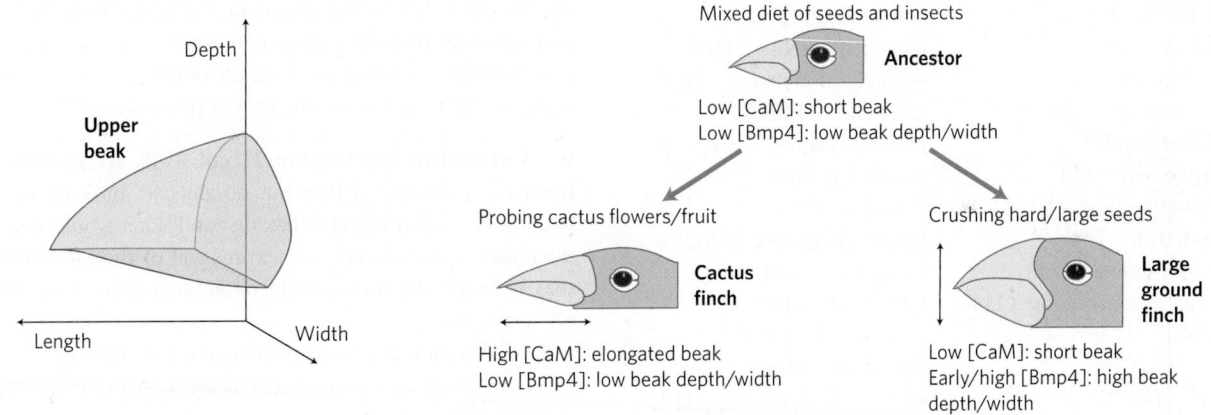

FIGURE 1 Evolution of new beak structures to exploit new food sources. In the Galápagos finches, the different beak structures of the cactus finch and the large ground finch, which feed on different, specialized food sources, were produced to a large extent by a few mutations that altered the timing and level of expression of just two genes: those encoding calmodulin (CaM) and Bmp4. [Source: Information from A. Abzhanov et al., *Nature* 442:563, 2006, Fig. 4.]

■ Regulation mediated by ncRNAs plays an important role in eukaryotic gene expression, with the range of known mechanisms expanding to include interactions with proteins, mRNA, and other ncRNAs.

■ Development of a multicellular organism presents the most complex regulatory challenge. The fate of cells in the early embryo is determined by establishment of anterior-posterior and dorsal-ventral gradients of proteins that act as transcription activators or translational repressors, regulating the genes required for development of structures appropriate to a particular part of the organism. Sets of regulatory genes operate in temporal and spatial succession, transforming given areas of an egg cell into predictable structures in the adult organism.

■ The differentiation of stem cells into functional tissues can be controlled by extracellular signals and conditions.

Key Terms

Terms in bold are defined in the glossary.

Problems

1. Effect of mRNA and Protein Stability on Regulation
E. coli cells are growing in a medium with glucose as the sole carbon source. Tryptophan is suddenly added. The cells continue to grow, and divide every 30 min. Describe (qualitatively) how the amount of tryptophan synthase activity in the cells changes with time under the following conditions:

(a) The *trp* mRNA is stable (degraded slowly over many hours).

(b) The *trp* mRNA is degraded rapidly, but tryptophan synthase is stable.

(c) The *trp* mRNA and tryptophan synthase are both degraded rapidly.

2. The Lactose Operon A researcher engineers a *lac* operon on a plasmid but inactivates all parts of the *lac* operator (*lacO*) and the *lac* promoter, replacing them with the binding site for the LexA repressor (which acts in the SOS response) and a promoter regulated by LexA. The plasmid is introduced into *E. coli* cells that have a *lac* operon with an inactive *lacZ* gene. Under what conditions will these transformed cells produce β-galactosidase?

3. Negative Regulation Describe the probable effects on gene expression in the *lac* operon of a mutation in (a) the *lac* operator that deletes most of O_1, (b) the *lacI* gene that inactivates the repressor, and (c) the promoter that alters the region around position −10.

4. Specific DNA Binding by Regulatory Proteins A typical bacterial repressor protein discriminates between its specific DNA-binding site (operator) and nonspecific DNA by a factor of 10^4 to 10^6. About 10 molecules of repressor per cell are sufficient to ensure a high level of repression. Assume that a very similar repressor existed in a human cell, with a similar specificity for its binding site. How many copies of the repressor would be required to elicit a level of repression similar to that in the bacterial cell? (Hint: The *E. coli* genome contains about 4.6 million bp; the human haploid genome has about 3.2 billion bp.)

5. Repressor Concentration in *E. coli* The dissociation constant for a particular repressor-operator complex is very low, about 10^{-13} M. An *E. coli* cell (volume 2×10^{-12} mL) contains 10 copies of the repressor. Calculate the cellular concentration of the repressor protein. How does this value compare with the dissociation constant of the repressor-operator complex? What is the significance of this answer?

6. Catabolite Repression *E. coli* cells are growing in a medium containing lactose but no glucose. Indicate whether each of the following changes or conditions would increase, decrease, or not change the expression of the *lac* operon. It may be helpful to draw a model depicting what is happening in each situation.

(a) Addition of a high concentration of glucose

(b) A mutation that prevents dissociation of the Lac repressor from the operator

(c) A mutation that completely inactivates β-galactosidase

(d) A mutation that completely inactivates galactoside permease

(e) A mutation that prevents binding of CRP to its binding site near the *lac* promoter

7. Transcription Attenuation How would transcription of the *E. coli trp* operon be affected by the following manipulations of the leader region of the *trp* mRNA?

(a) Increasing the distance (number of bases) between the leader peptide gene and sequence 2

(b) Increasing the distance between sequences 2 and 3

(c) Removing sequence 4

(d) Changing the two Trp codons in the leader peptide gene to His codons

(e) Eliminating the ribosome-binding site for the gene that encodes the leader peptide

(f) Changing several nucleotides in sequence 3 so that it can base-pair with sequence 4 but not with sequence 2

8. Repressors and Repression How would the SOS response in *E. coli* be affected by a mutation in the *lexA* gene that prevented autocatalytic cleavage of the LexA protein?

9. Regulation by Recombination In the phase variation system of *Salmonella*, what would happen to the cell if the Hin recombinase became more active and promoted recombination (DNA inversion) several times in each cell generation?

10. Initiation of Transcription in Eukaryotes A new RNA polymerase activity is discovered in crude extracts of cells derived from an exotic fungus. The RNA polymerase initiates transcription only from a single, highly specialized promoter. As the polymerase is purified its activity declines, and the purified enzyme is completely inactive unless crude extract is added to the reaction mixture. Suggest an explanation for these observations.

11. Functional Domains in Regulatory Proteins A biochemist replaces the DNA-binding domain of the yeast Gal4 protein with the DNA-binding domain from the Lac repressor and finds that the engineered protein no longer regulates transcription of the *GAL* genes in yeast. Draw a diagram of the different functional domains you would expect to find in the Gal4 protein and in the engineered protein. Why does the engineered protein no longer regulate transcription of the *GAL* genes? What might be done to the DNA-binding site recognized by this chimeric protein to make it functional in activating transcription of *GAL* genes?

12. Nucleosome Modification during Transcriptional Activation To prepare genomic regions for transcription, certain histones in the resident nucleosomes are acetylated and methylated at specific locations. Once transcription is no longer needed, these modifications need to be reversed. In mammals, the methylation of Arg residues in histones is reversed by peptidylarginine deiminases (PADIs). The reaction promoted by these enzymes does not yield unmethylated arginine. Instead, it produces citrulline residues in the histone. What is the other product of the reaction? Suggest a mechanism for this reaction.

13. Gene Repression in Eukaryotes Explain why repression of a eukaryotic gene by an RNA might be more efficient than repression by a protein repressor.

14. Inheritance Mechanisms in Development A *Drosophila* egg that is *bcd⁻/bcd⁻* may develop normally, but the adult fruit fly will not be able to produce viable offspring. Explain.

Data Analysis Problem

15. Engineering a Genetic Toggle Switch in *Escherichia coli* Gene regulation is often described as an "on or off" phenomenon: a gene is either fully expressed or not expressed at all. In fact, repression and activation of a gene involve ligand-binding reactions, so genes can show intermediate levels of

expression when intermediate levels of regulatory molecules are present. For example, for the *E. coli lac* operon, consider the binding equilibrium of the Lac repressor, operator DNA, and inducer (see Fig. 28-8). Although this is a complex, cooperative process, it can be approximately modeled by the following reaction (R is repressor; IPTG is the inducer isopropyl-β-D-thiogalactoside):

$$R + IPTG \xrightleftharpoons{K_d = 10^{-4} M} R \cdot IPTG$$

Free repressor, R, binds to the operator and prevents transcription of the *lac* operon; the R · IPTG complex does not bind to the operator and thus transcription of the *lac* operon can proceed.

(a) Using Equation 5-8, we can calculate the relative expression level of the proteins of the *lac* operon as a function of [IPTG]. Use this calculation to determine over what range of [IPTG] the expression level would vary from 10% to 90%.

(b) Describe qualitatively the level of *lac* operon proteins present in an *E. coli* cell before, during, and after induction with IPTG. You need not give the amounts at exact times—just indicate the general trends.

Gardner, Cantor, and Collins (2000) set out to make a "genetic toggle switch"—a gene-regulatory system with two key characteristics, A and B, of a light switch. (A) *It has only two states:* it is either fully on or fully off; it is not a dimmer switch. In biochemical terms, the target gene or gene system (operon) is either fully expressed or not expressed at all; it cannot be expressed at an intermediate level. (B) *Both states are stable:* although you must use a finger to flip the light switch from one state to the other, once you have flipped it and removed your finger, the switch stays in that state. In biochemical terms, exposure to an inducer or some other signal changes the expression state of the gene or operon, and it remains in that state once the signal is removed.

(c) Explain how the *lac* operon lacks both characteristics A and B.

To make their "toggle switch," Gardner and coworkers constructed a plasmid from the following components:

ori   An origin of replication

*amp*ᴿ   A gene conferring resistance to the antibiotic ampicillin

OP*lac*   The operator-promoter region of the *E. coli lac* operon

OP*λ*   The operator-promoter region of λ phage

lacI   The gene encoding the *lac* repressor protein, LacI. In the absence of IPTG, this protein strongly represses OP*lac*; in the presence of IPTG, it allows full expression from OP*lac*.

*rep*ᵗˢ   The gene encoding a temperature-sensitive mutant λ repressor protein, rept*ˢ*. At 37 °C this protein strongly represses OP*λ*; at 42 °C it allows full expression from OP*λ*.

GFP   The gene for green fluorescent protein (GFP), a highly fluorescent reporter protein (see Fig. 9-16)

T   Transcription terminator

The investigators arranged these components, as shown in the following figure, so that the two promoters were reciprocally repressed: OP*lac* controlled expression of *rep*^ts, and OPλ controlled expression of *lacI*. The state of this system was reported by the expression level of *GFP*, which was also under the control of OP*lac*.

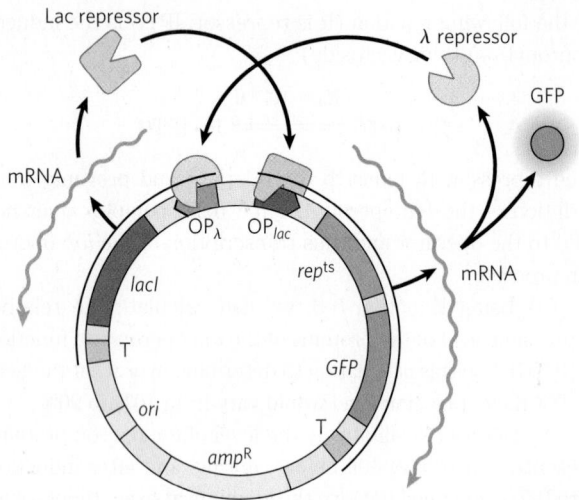

(d) The constructed system has two states: GFP-on (high level of expression) and GFP-off (low level of expression). For each state, describe which proteins are present and which promoters are being expressed.

(e) Treatment with IPTG would be expected to toggle the system from one state to the other. From which state to which? Explain your reasoning.

(f) Treatment with heat (42 °C) would be expected to toggle the system from one state to the other. From which state to which? Explain your reasoning.

(g) Why would this plasmid be expected to have characteristics A and B as described above?

To confirm that their construct did indeed exhibit these characteristics, Gardner and colleagues first showed that, once switched, the GFP expression level (high or low) was stable for long periods of time (characteristic B). Next, they measured the GFP level at different concentrations of the inducer IPTG, with the following results.

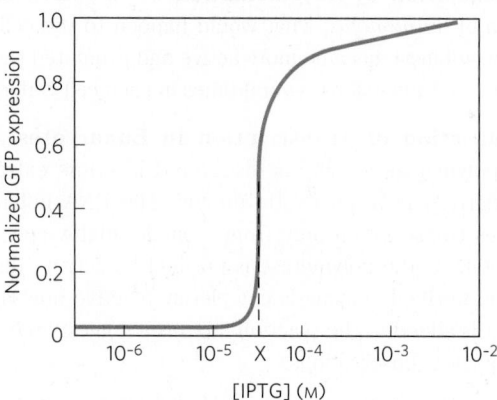

They noticed that the average GFP expression level was intermediate at concentration X of IPTG. However, when they measured the GFP expression level *in individual cells* at [IPTG] = X, they found either a high level or a low level of GFP—no cells showed an intermediate level.

(h) Explain how this finding demonstrates that the system has characteristic A. What is happening to cause the bimodal distribution of expression levels at [IPTG] = X?

Reference

Gardner, T.S., C.R. Cantor, and J.J. Collins. 2000. Construction of a genetic toggle switch in *Escherichia coli*. *Nature* 403:339–342.

Further Reading is available at www.macmillanlearning.com/LehningerBiochemistry7e.

Abbreviated Solutions to Problems

Fuller solutions to all chapter problems are published in the *Absolute Ultimate Guide to Lehninger Principles of Biochemistry*. For all numerical problems, answers are expressed with the correct number of significant figures.

Chapter 1

1. (a) Diameter of magnified cell = 500 mm **(b)** 2.7×10^{12} actin molecules **(c)** 36,000 mitochondria **(d)** 3.9×10^{10} glucose molecules **(e)** 50 glucose molecules per hexokinase molecule

2. (a) 1×10^{-12} g = 1 pg **(b)** 10% **(c)** 5%

3. (a) 1.6 mm; 800 times longer than the cell; DNA must be tightly coiled. **(b)** 4,000 proteins

4. (a) Metabolic rate is limited by diffusion, which is limited by surface area. **(b)** 12 μm^{-1} for the bacterium; 0.04 μm^{-1} for the amoeba; surface-to-volume ratio 300 times higher in the bacterium.

5. 2×10^{6} s (about 23 days)

6. The vitamin molecules from the two sources are identical; the body cannot distinguish the source; only associated impurities might vary with the source.

7. (a)

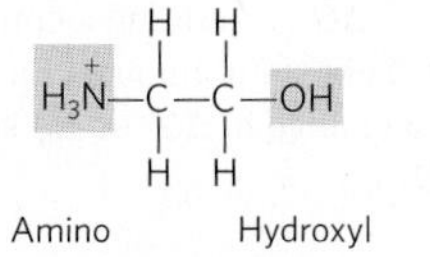

Amino   Hydroxyl

(b)

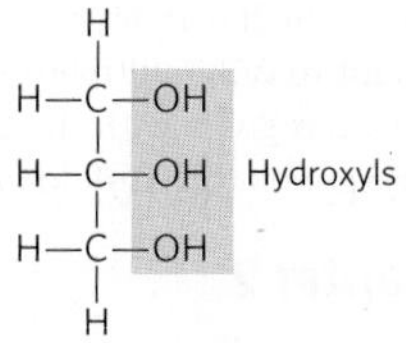

Hydroxyls

(c)

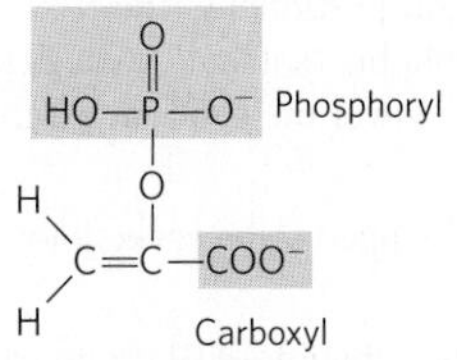

Phosphoryl

Carboxyl

(d)

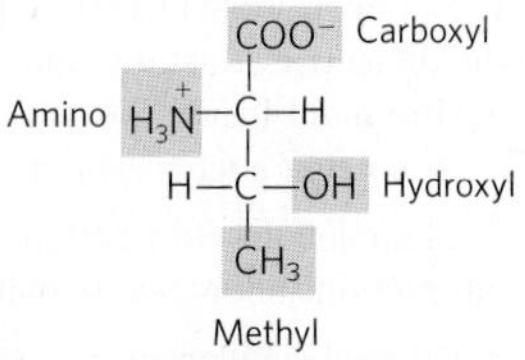

Carboxyl

Amino

Hydroxyl

Methyl

(e)

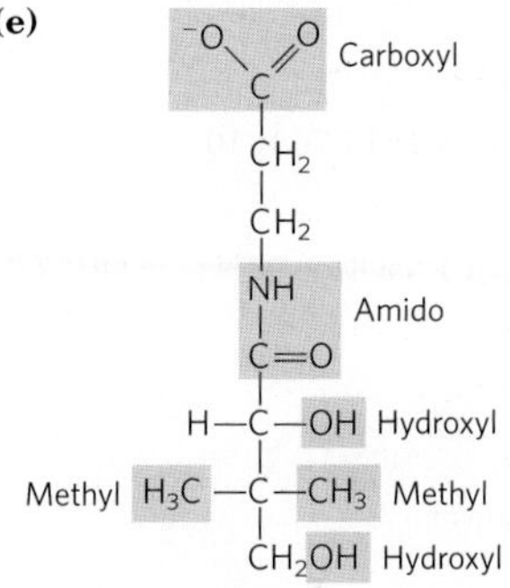

Carboxyl

Amido

Hydroxyl

Methyl   Methyl

Hydroxyl

(f)

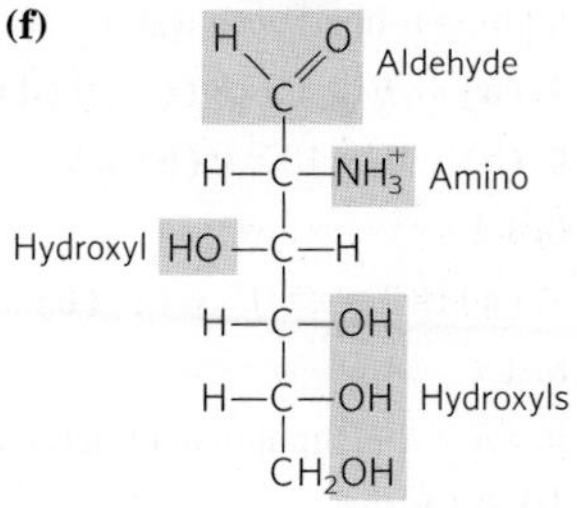

Aldehyde

Amino

Hydroxyl

Hydroxyls

8.

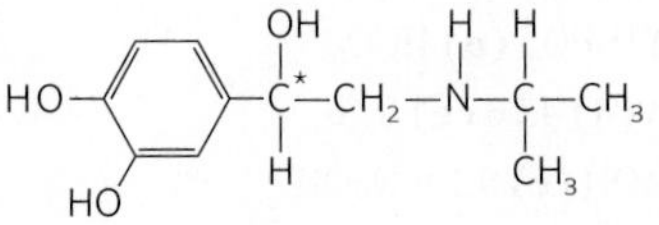

The two enantiomers have different interactions with a chiral biological "receptor" (a protein).

9. (a) Only the amino acids have amino groups; separation could be based on the charge or binding affinity of these groups. Fatty acids are less soluble in water than amino acids, and the two types of molecules also differ in size and shape—either of these property differences could

be the basis for separation. **(b)** Glucose is a smaller molecule than a nucleotide; separation could be based on size. The nitrogenous base and/or the phosphate group also endow nucleotides with characteristics (solubility, charge) that could be used for separation from glucose.

10. It is improbable that silicon could serve as the central organizing element for life, especially in an O_2-containing atmosphere such as that of Earth. Long chains of silicon atoms are not readily synthesized; the polymeric macromolecules necessary for more complex functions would not readily form. Oxygen disrupts bonds between silicon atoms, and silicon–oxygen bonds are extremely stable and difficult to break, preventing the breaking and making of bonds that is essential to life processes.

11. Only one enantiomer of the drug was physiologically active. Dexedrine consisted of the single enantiomer; Benzedrine consisted of a racemic mixture.

12. (a) 3 Phosphoric acid groups; α-D-ribose; guanine **(b)** Choline; phosphoric acid; glycerol; oleic acid; palmitic acid **(c)** Tyrosine; 2 glycines; phenylalanine; methionine

13. (a) CH_2O; $C_3H_6O_3$

(b)

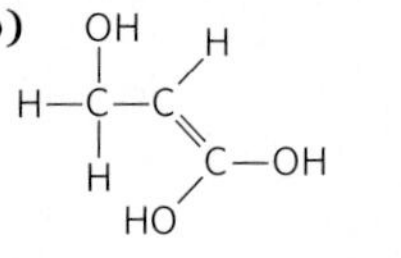

1 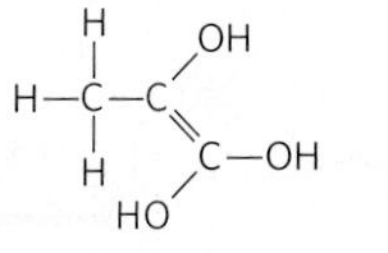   2 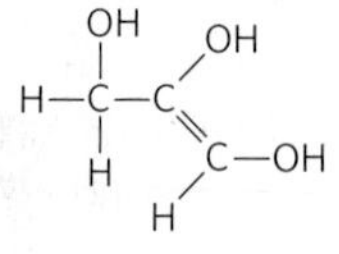   3

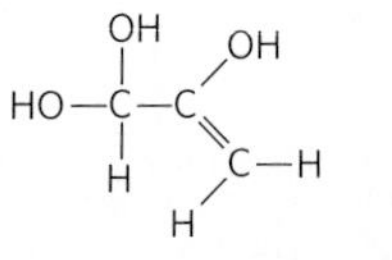

4 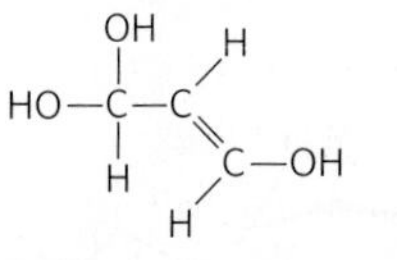   5 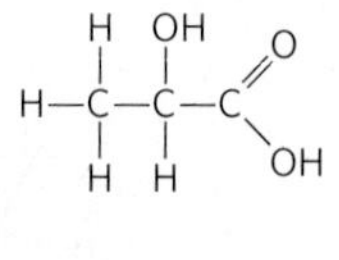   6

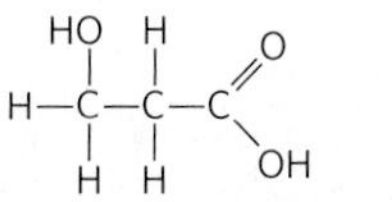

7 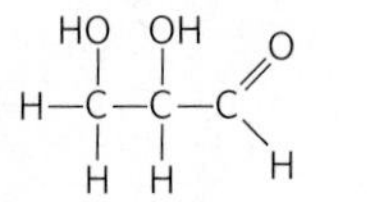   8 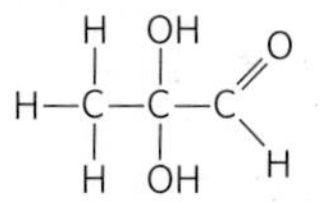   9

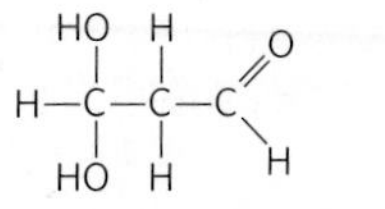

10 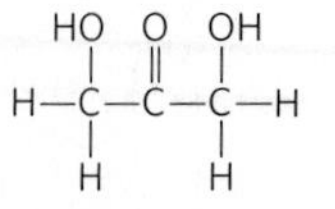   11 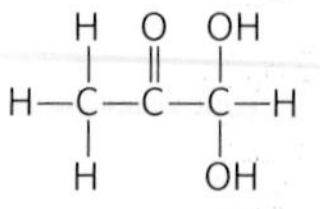   12

(c) X contains a chiral center; eliminates all but **6** and **8**. **(d)** X contains an acidic functional group; eliminates **8**; structure **6** is consistent with all data. **(e)** Structure **6**; we cannot distinguish between the two possible enantiomers.

14. The compound shown is (R)-propranolol; the carbon bearing the hydroxyl group is the chiral carbon. (S)-Propranolol has the structure:

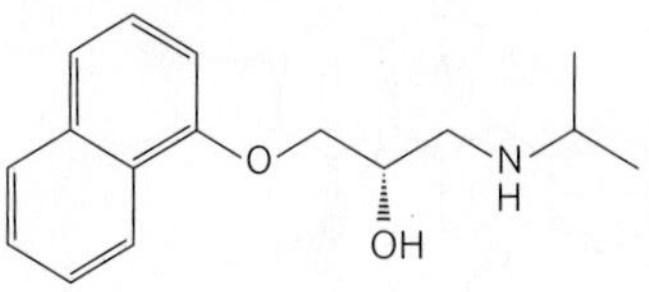

15. The compound shown is (*S,S*)-methylphenidate. (*R,R*)-methylphenidate has the structure:

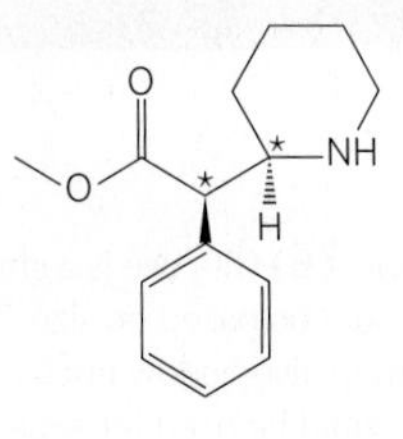

The chiral carbons are indicated with asterisks.

16. (a) A more negative $\Delta G°$ corresponds to a larger K_{eq} for the binding reaction, so the equilibrium is shifted more toward products and tighter binding—and thus greater sweetness and higher MRS. **(b)** Animal-based sweetness assays are time-consuming. A computer program to predict sweetness, eve n if not always completely accurate, would allow chemists to design effective sweeteners much faster. Candidate molecules could then be tested in the conventional assay. **(c)** The range 0.25 to 0.4 nm corresponds to about 1.5 to 2.5 single-bond lengths. The figure below can be used to construct an approximate ruler; any atoms in the light red rectangle are between 0.25 and 0.4 nm from the origin of the ruler.

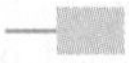

There are many possible AH-B groups in the molecules; a few are shown here.

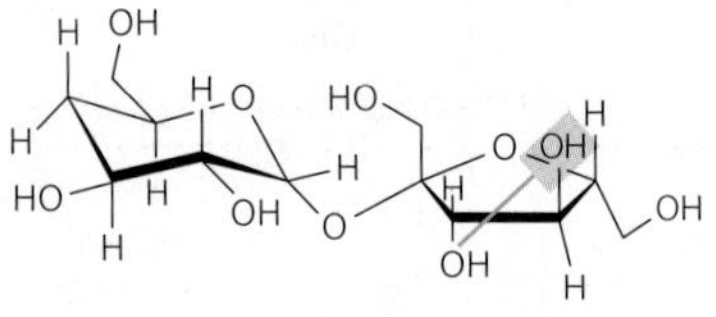

Deoxysucrose

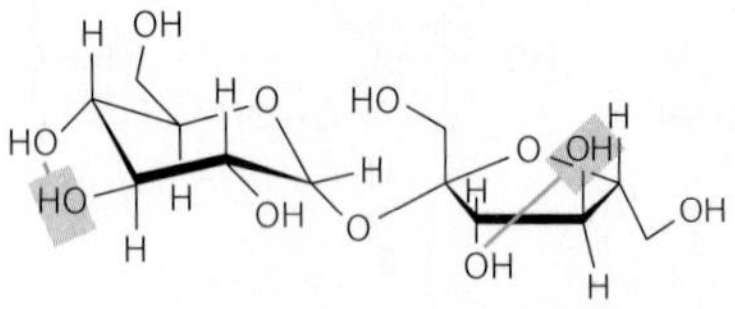

Sucrose

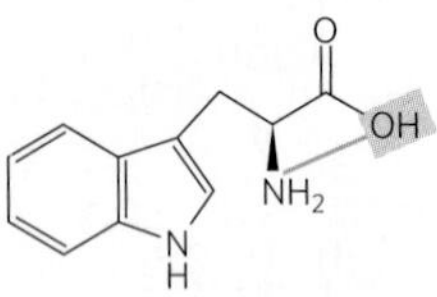

D-Tryptophan

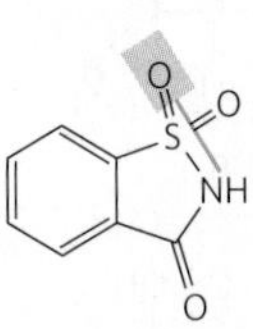

Saccharin

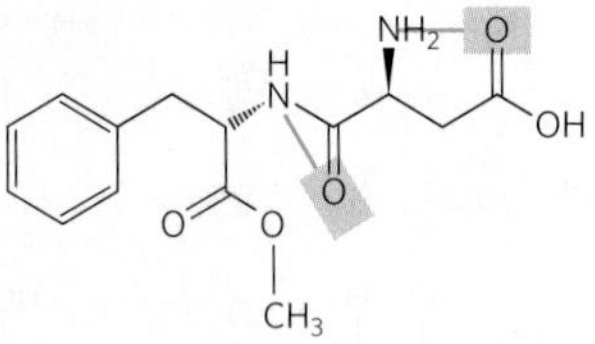

Aspartame

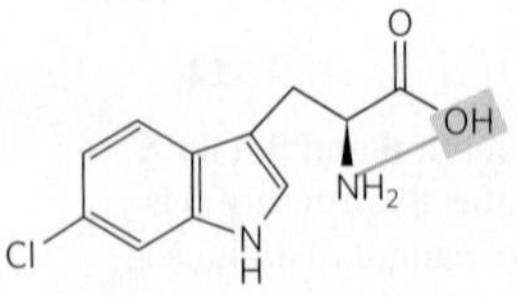

6-Chloro-D-tryptophan

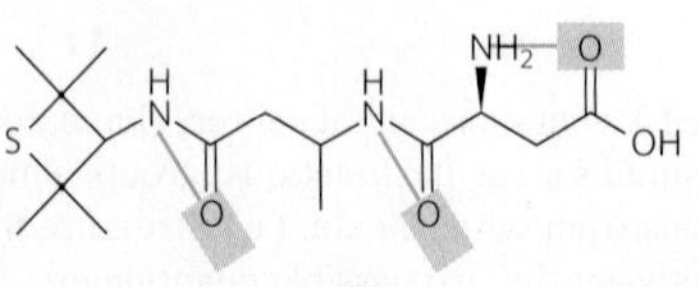

Alitame

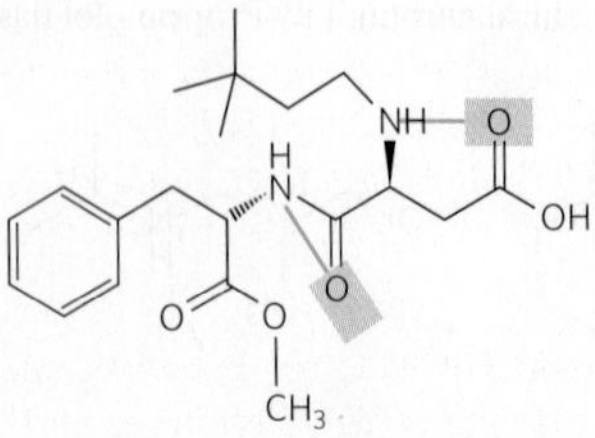

Neotame

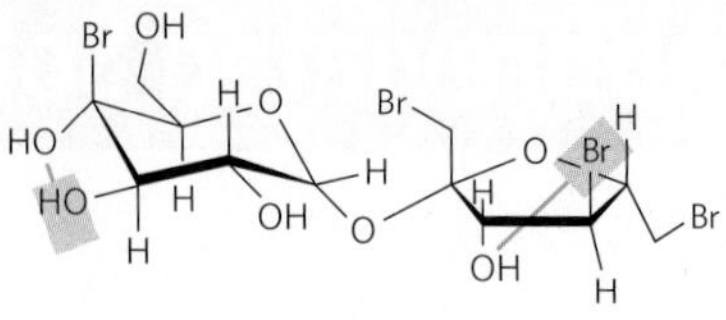

Tetrabromosucrose

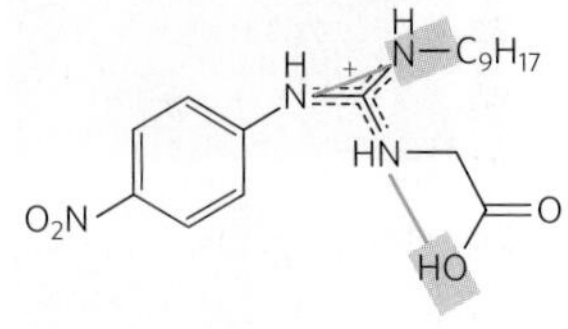

Sucronic acid

(d) First, each molecule has multiple AH-B groups, so it is difficult to know which is the important one. Second, because the AH-B motif is very simple, many nonsweet molecules will have this group. **(e)** Sucrose and deoxysucrose. Deoxysucrose lacks one of the AH-B groups present in sucrose and has a slightly lower MRS than sucrose— as is expected if the AH-B groups are important for sweetness. **(f)** There are many such examples; here are a few: (1) D-Tryptophan and 6-chloro-D-tryptophan have the same AH-B group but very different MRS values. (2) Aspartame and neotame have the same AH-B groups but very different MRS values. (3) Neotame has two AH-B groups and alitame has three, yet neotame is more than five times sweeter than alitame. (4) Bromine is less electronegative than oxygen and thus is expected to weaken an AH-B group, yet tetrabromosucrose is much sweeter than sucrose. **(g)** Given enough "tweaking" of parameters, any model can be made to fit a defined dataset. Because the objective was to create a model to predict $\Delta G°$ for molecules not tested in vivo, the researchers needed to show that the model worked well for molecules it had not been trained on. The degree of inaccuracy with the test set could give researchers an idea of how the model would behave for novel molecules. **(h)** MRS is related to K_{eq}, which is related exponentially to $\Delta G°$, so adding a constant amount to $\Delta G°$ multiplies the MRS by a constant amount. Based on the values given with the structures, a change in $\Delta G°$ of 1.3 kcal/mol corresponds to a 10-fold change in MRS.

Chapter 2

1. Stronger; ionic attractive force is proportional to the *inverse* of the dielectric constant, and a hydrophobic "solvent" such as the environment inside the protein has a lower dielectric constant than a polar solvent such as water.

2. Biomolecular interactions generally need to be reversible; weak interactions allow reversibility.

3. Ethanol is polar; ethane is not. The ethanol —OH group can hydrogen-bond with water.

4. (a) 4.76 **(b)** 9.19 **(c)** 4.0 **(d)** 4.82

5. (a) 1.51×10^{-4} M **(b)** 3.02×10^{-7} M **(c)** 7.76×10^{-12} M

6. 1.1

7. (a) $HCl \rightleftharpoons H^+ + Cl^-$ **(b)** 3.3 **(c)** $NaOH \rightleftharpoons Na^+ + OH^-$ **(d)** 9.8

8. 1.1

9. 1.7×10^{-9} mol of acetylcholine

10. 0.1 M HCl

11. (a) Greater **(b)** Higher **(c)** Lower

12. 3.3 mL

13. (a) $RCOO^-$ **(b)** RNH_2 **(c)** $H_2PO_4^-$ **(d)** HCO_3^-

14. (a) 5.06 **(b)** 4.28 **(c)** 5.46 **(d)** 4.76 **(e)** 3.76

15. (a) 0.1 M HCl **(b)** 0.1 M NaOH **(c)** 0.1 M NaOH

16. (d) Bicarbonate, a weak base, titrates —OH to —O⁻, making the compound more polar and more water-soluble.

17. Stomach; the neutral form of aspirin present at the lower pH is less polar and passes through the membrane more easily.

18. 9

19. 7.4

20. (a) pH 8.6 to 10.6 **(b)** 4/5 **(c)** 10 mL **(d)** $pH = pK_a - 2$

21. 8.9

22. 2.4

23. 6.9

24. 1.4

25. $NaH_2PO_4 \cdot H_2O$, 5.8 g/L; Na_2HPO_4, 8.2 g/L

26. $[A^-]/[HA] = 0.10$

27. Mix 150 mL of 0.10 M sodium acetate and 850 mL of 0.10 M acetic acid.

28. Acetic acid; its pK_a is closest to the desired pH.

29. (a) 4.6 **(b)** 0.1 pH unit **(c)** 4 pH units

30. 4.3

31. 0.13 M acetate and 0.07 M acetic acid

32. 1.7

33. 7

34. (a)

Fully protonated   Fully deprotonated

(b) Fully protonated **(c)** Zwitterion **(d)** Zwitterion **(e)** Fully deprotonated

35. (a) Blood pH is controlled by the carbon dioxide–bicarbonate buffer system, $CO_2 + H_2O \rightleftharpoons H^+ + HCO_3^-$. During *hypoventilation*, $[CO_2]$ increases in the lungs and arterial blood, driving the equilibrium to the right, raising $[H^+]$ and lowering pH. **(b)** During *hyperventilation*, $[CO_2]$ decreases in the lungs and arterial blood, reducing $[H^+]$ and increasing pH above the normal 7.4 value. **(c)** Lactate is a moderately strong acid, completely dissociating under physiological conditions and thus lowering the pH of blood and muscle tissue. Hyperventilation removes H^+, raising the pH of blood and tissues in anticipation of the acid buildup.

36. 7.4

37. Dissolving more CO_2 in the blood increases $[H^+]$ in blood and extracellular fluids, lowering pH: $CO_2(d) + H_2O \rightleftharpoons H_2CO_3 \rightleftharpoons H^+ + HCO_3^-$

38. (a) Use the substance in its surfactant form to emulsify the spilled oil, collect the emulsified oil, then switch to the nonsurfactant form. The oil and water will separate and the oil can be collected for further use. **(b)** The equilibrium lies strongly to the right. The stronger acid (lower pK_a), H_2CO_3, donates a proton to the conjugate base of the weaker acid (higher pK_a), amidine. **(c)** The strength of a surfactant depends on the hydrophilicity of its head groups: the more hydrophilic, the more powerful the surfactant. The amidinium form of s-surf is much more hydrophilic than the amidine form, so it is a more powerful surfactant. **(d)** *Point A:* amidinium; the CO_2 has had plenty of time to react with the amidine to produce the amidinium form. *Point B:* amidine; Ar has removed CO_2 from the solution, leaving the amidine form. **(e)** The conductivity rises as uncharged amidine reacts with CO_2 to produce the charged amidinium form. **(f)** The conductivity falls as Ar removes CO_2, shifting the equilibrium to the uncharged amidine form. **(g)** Treat s-surf with CO_2 to produce the surfactant amidinium form and use this to emulsify the spill. Treat the emulsion with Ar to remove the CO_2 and produce the nonsurfactant amidine form. The oil will separate from the water and can be recovered.

Chapter 3

1. L; determine the absolute configuration at the α carbon and compare it with D- and L-glyceraldehyde.

2. (a) I **(b)** II **(c)** IV **(d)** II **(e)** IV **(f)** II and IV **(g)** III **(h)** III **(i)** V **(j)** III **(k)** V **(l)** II **(m)** III **(n)** V **(o)** I, III, and V

3. (a) pI > pK_a of the α-carboxyl group and pI < pK_a of the α-amino group, so both groups are charged (ionized). **(b)** 1 in 2.19×10^7.

The pI of alanine is 6.01. From Table 3-1 and the Henderson-Hasselbalch equation, 1/4,680 carboxyl groups and 1/4,680 amino groups are uncharged. The fraction of alanine molecules with both groups uncharged is the product of these fractions.

4. (a), (b), (c)

pH	Structure	Net charge	Migrates toward
1	1	+2	Cathode
4	2	+1	Cathode
8	3	0	Does not migrate
12	4	−1	Anode

5. (a) Asp **(b)** Met **(c)** Glu **(d)** Gly **(e)** Ser

6. (a) 2 **(b)** 4 **(c)**

7. (a) Structure at pH 7:

$pK_2 = 8.03$ $pK_1 = 3.39$

(b) Electrostatic interaction between the carboxylate anion and the protonated amino group of the alanine zwitterion favorably affects ionization of the carboxyl group. This favorable electrostatic interaction decreases as the length of the poly(Ala) increases, resulting in an increase in pK_1. **(c)** Ionization of the protonated amino group destroys the favorable electrostatic interaction noted in (b). With increasing distance between the charged groups, removal of the proton from the amino group in poly(Ala) becomes easier and thus pK_2 is lower. The intramolecular effects of the amide (peptide bond) linkages keep pK_a values lower than they would be for an alkyl-substituted amine.

8. 75,000

9. (a) 32,000. The elements of water are lost when a peptide bond forms, so the molecular weight of a Trp residue is not the same as the molecular weight of free tryptophan. **(b)** 2

10. The protein has four subunits, with molecular masses of 160, 90, 90, and 60 kDa. The two 90 kDa subunits (possibly identical) are linked by one or more disulfide bonds.

11. (a) at pH 3, +2; at pH 8, 0; at pH 11, −1 **(b)** pI = 7.8

12. pI ≈ 1; carboxylate groups; Asp and Glu

13. Lys, His, Arg; negatively charged phosphate groups in DNA interact with positively charged side groups in histones.

14. (a) (Glu)$_{20}$ **(b)** (Lys–Ala)$_3$ **(c)** (Asn–Ser–His)$_5$ **(d)** (Asn–Ser–His)$_5$

15. (a) Specific activity after step 1 is 200 units/mg; step 2, 600 units/mg; step 3, 250 units/mg; step 4, 4,000 units/mg; step 5, 15,000 units/mg; step 6, 15,000 units/mg. **(b)** Step 4 **(c)** Step 3 **(d)** Yes. Specific activity did not increase in step 6; SDS polyacrylamide gel electrophoresis.

16. (a) [NaCl] = 0.5 mM **(b)** [NaCl] = 0.05 mM.

17. C elutes first, B second, A last.

18. Tyr–Gly–Gly–Phe–Leu

19.

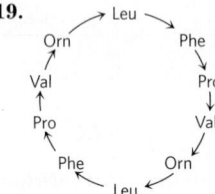

The arrows correspond to the orientation of the peptide bonds, —CO → NH—.

20. 88%, 97%. The percentage (x) of correct amino acid residues released in cycle n is x_n/x. All residues released in the first cycle are correct, even though the efficiency of cleavage is not perfect.

21. (a) Y (1), F (7), and R (9) **(b)** Positions 4 and 9; K (Lys) is more common at 4, R (Arg) is invariant at 9. **(c)** Positions 5 and 10; E (Glu) is more common at both positions. **(d)** Position 2; S (Ser)

22. (a) Peptide 2 **(b)** Peptide 1 **(c)** Peptide 2 **(d)** Peptide 3

23. (a) Any linear polypeptide chain has only two kinds of free amino groups: a single α-amino group at the amino terminus, and an ε-amino group on each Lys residue. These amino groups react with FDNB to form a DNP–amino acid derivative. Insulin gave two different α-amino-DNP derivatives, suggesting that it has two amino termini and thus two polypeptide chains—one with an amino-terminal Gly and the other with an amino-terminal Phe. Because the DNP-lysine product is ε-DNP-lysine, the Lys is not at an amino terminus. **(b)** Yes. The A chain has amino-terminal Gly; the B chain has amino-terminal Phe; and (nonterminal) residue 29 in the B chain is Lys. **(c)** Phe–Val–Asp–Glu–. Peptide B1 shows that the amino-terminal residue is Phe. Peptide B2 also includes Val, but since no DNP-Val is formed, Val is not at the amino terminus; it must be on the carboxyl side of Phe. Thus the sequence of B2 is DNP-Phe–Val. Similarly, the sequence of B3 must be DNP-Phe–Val–Asp, and the sequence of the A chain must begin Phe–Val–Asp–Glu–. **(d)** No. The known amino-terminal sequence of the A chain is Phe–Val–Asn–Gln–. The Asn and Gln appear in Sanger's analysis as Asp and Glu because the vigorous hydrolysis in step 7 hydrolyzed the amide bonds in Asn and Gln (as well as the peptide bonds), forming Asp and Glu. Sanger et al. could not distinguish Asp from Asn or Glu from Gln at this stage in their analysis. **(e)** The sequence exactly matches that in Fig. 3-24. Each peptide in the table gives specific information about which Asx residues are Asn or Asp and which Glx residues are Glu or Gln.

Ac1: residues 20–21. This is the only Cys–Asx sequence in the A chain; there is ~1 amido group in this peptide, so it must be Cys–Asn:

N–Gly–Ile–Val–Glx–Glx–Cys–Cys–Ala–Ser–Val–
      1   5   10
Cys–Ser–Leu–Tyr–Glx–Leu–Glx–Asx–Tyr–Cys–**Asn**–C
           15   20

Ap15: residues 14–15–16. This is the only Tyr–Glx–Leu in the A chain; there is ~1 amido group, so the peptide must be Tyr–Gln–Leu:

N–Gly–Ile–Val–Glx–Glx–Cys–Cys–Ala–Ser–Val–
      1   5   10
Cys–Ser–Leu–Tyr–**Gln**–Leu–Glx–Asx–Tyr–Cys–Asn–C
           15   20

Ap14: residues 14–15–16–17. It has ~1 amido group, and we already know that residue 15 is Gln, so residue 17 must be Glu:

N–Gly–Ile–Val–Glx–Glx–Cys–Cys–Ala–Ser–Val–
      1   5   10
Cys–Ser–Leu–Tyr–Gln–Leu–**Glu**–Asx–Tyr–Cys–Asn–C
           15   20

Ap3: residues 18–19–20–21. It has ~2 amido groups, and we know that residue 21 is Asn, so residue 18 must be Asn:

N–Gly–Ile–Val–Glx–Glx–Cys–Cys–Ala–Ser–Val–
      1   5   10
Cys–Ser–Leu–Tyr–Gln–Leu–Glu–**Asn**–Tyr–Cys–Asn–C
           15   20

Ap1: residues 17–18–19–20–21, which is consistent with residues 18 and 21 being Asn.

Ap5pa1: residues 1–2–3–4. It has ~0 amido group, so residue 4 must be Glu:

N–Gly–Ile–Val–**Glu**–Glx–Cys–Cys–Ala–Ser–Val–
      1   5   10
Cys–Ser–Leu–Tyr–Gln–Leu–Glu–Asn–Tyr–Cys–Asn–C
           15   20

Ap5: residues 1 through 13. It has ~1 amido group, and we know that residue 4 is Glu, so residue 5 must be Gln:

N–Gly–Ile–Val–Glu–**Gln**–Cys–Cys–Ala–Ser–Val–
      1   5   10
Cys–Ser–Leu–Tyr–Gln–Leu–Glu–Asn–Tyr–Cys–Asn–C
           15   20

Chapter 4

1. (a) Shorter bonds have a higher bond order (are multiple rather than single) and are stronger. The peptide C—N bond is stronger than a single bond and is midway between a single and a double bond in character. **(b)** Rotation about the peptide bond is difficult at physiological temperatures because of its partial double-bond character.

2. (a) The principal structural units in the wool fiber polypeptide (α-keratin) are successive turns of the α helix, at 5.4 Å intervals; coiled coils produce the 5.2 Å spacing. Steaming and stretching the fiber yields an extended polypeptide chain with the β conformation, with a distance between adjacent R groups of about 7.0 Å. As the polypeptide reassumes an α-helical structure, the fiber shortens. **(b)** Wool shrinks in the presence of moist heat, as polypeptide chains are converted from an extended β conformation to the native α-helix conformation. The structure of silk—β sheets, with their small, closely packed amino acid side chains—is more stable than that of wool.

3. ~42 peptide bonds per second

4. At pH > 6, the carboxyl groups of poly(Glu) are deprotonated; repulsion among negatively charged carboxylate groups leads to unfolding. Similarly, at pH 7, the amino groups of poly(Lys) are protonated; repulsion among these positively charged groups also leads to unfolding.

5. (a) Disulfide bonds are covalent bonds, which are much stronger than the noncovalent interactions that stabilize most proteins. They cross-link protein chains, increasing their stiffness, mechanical strength, and hardness. **(b)** Cystine residues (disulfide bonds) prevent the complete unfolding of the protein.

6. φ = (f) and ψ = (e)

7. (a) Bends are most likely at residues 7 and 19; Pro residues in the cis configuration accommodate turns well. **(b)** The Cys residues at positions 13 and 24 can form disulfide bonds. **(c)** External surface: polar and charged residues (Asp, Gln, Lys); interior: nonpolar and aliphatic residues (Ala, Ile); Thr, though polar, has a hydropathy index near zero and thus can be found either on the external surface or in the interior of the protein.

8. 30 amino acid residues; 0.87

9. Myoglobin is all three. The folded structure, the "globin fold," is a motif found in all globins. The polypeptide folds into a single domain, which for this protein represents the entire three-dimensional structure.

10. Protein (a), a β barrel, is described by Ramachandran plot (c), which shows most of the allowable conformations in the upper left quadrant where the bond angles characteristic of the β conformation are concentrated. Protein (b), a series of α helices, is described by plot (d), where most of the allowable conformations are in the lower left quadrant.

11. The bacterial enzyme is a collagenase; it destroys the connective tissue barrier of the host, allowing the bacterium to invade the tissues. Bacteria do not contain collagen.

12. (a) The number of moles of DNP-valine formed per mole of protein equals the number of amino termini and thus the number of polypeptide chains. **(b)** 4 **(c)** Different chains would probably run as discrete bands on an SDS polyacrylamide gel.

13. Peptide (a); it has more amino acid residues that favor an α-helical structure (see Table 4-1).

14. (a) Aromatic residues seem to play an important role in stabilizing amyloid fibrils. Thus, molecules with aromatic substituents may inhibit amyloid formation by interfering with the stacking or association of the aromatic side chains. **(b)** Amyloid is formed in the pancreas in association with type 2 diabetes, and is formed in the brain in Alzheimer disease. Although the amyloid fibrils in the two diseases involve different proteins, the fundamental structure of the amyloid is similar and is similarly stabilized in both, so they are potential targets for similar drugs designed to disrupt this structure.

15. (a) NFκB transcription factor, also called RelA transforming factor **(b)** No. You will obtain similar results, but with additional related proteins listed. **(c)** The protein has two subunits. There are multiple variants of the subunits, with the best characterized being 50, 52, or 65 kDa. These pair with each other to form a variety of homodimers and heterodimers. The structures of a number of different variants can be found in the PDB. **(d)** The NFκB transcription factor is a dimeric protein that binds specific DNA sequences, enhancing transcription of nearby genes. One such gene is the immunoglobulin κ (kappa) light chain, from which the transcription factor gets its name.

16. (a) Aba is a suitable replacement because Aba and Cys have side chains of approximately the same size and are similarly hydrophobic. However, Aba cannot form disulfide bonds, so it will not be a suitable replacement if these are required. **(b)** There are many important differences between the synthesized protein and HIV protease produced by a human cell, any of which could result in an inactive synthetic enzyme. (1) Although Aba and Cys have a similar size and hydrophobicity, Aba may not be similar enough for the protein to fold properly. (2) HIV protease may require disulfide bonds for proper functioning. (3) Many proteins synthesized by ribosomes fold while being produced; the protein in this study folded only after the chain was complete. (4) Proteins synthesized by ribosomes may interact with the ribosomes as they fold; this is not possible for the protein in the study. (5) Cytosol is a more complex solution than the buffer used in the study; some proteins may require specific, unknown proteins for proper folding. (6) Proteins synthesized in cells often require chaperones for proper folding; these are not present in the study buffer. (7) In cells, HIV protease is synthesized as part of a larger chain that is then proteolytically processed; the protein in the study was synthesized as a single molecule. **(c)** Because the enzyme *is* functional when Aba is substituted for Cys, disulfide bonds do not play an important role in the structure of HIV protease. **(d)** *Model 1:* It would fold like the L-protease. *Argument for:* The covalent structure is the same (except for chirality), so it should fold like the L-protease. *Argument against:* Chirality is not a trivial detail; three-dimensional shape is a key feature of biological molecules. The synthetic enzyme will not fold like the L-protease. *Model 2:* It would fold to the mirror image of the L-protease. *For:* Because the individual components are mirror images of those in the biological protein, it will fold in the mirror-image shape. *Against:* The interactions involved in protein folding are very complex, so the synthetic protein will most likely fold in another form. *Model 3:* It would fold to something else. *For:* The interactions involved in protein folding are very complex, so the synthetic protein will most likely fold in another form. *Against:* Because the individual components are mirror images of those in the biological protein, it will fold in the mirror-image shape. **(e)** *Model 1.* The enzyme is active, but with the enantiomeric form of the biological substrate, and it is inhibited by the enantiomeric form of the biological inhibitor. This is consistent with the D-protease being the mirror image of the L-protease. **(f)** Evans blue is achiral; it binds to both forms of the enzyme. **(g)** No. Because proteases contain only L-amino acids and recognize only L-peptides, chymotrypsin would not digest the D-protease. **(h)** Not necessarily. Depending on the individual enzyme, any of the problems listed in (b) could result in an inactive enzyme.

Chapter 5

1. Protein B has a higher affinity for ligand X; it will be half-saturated at a much lower concentration of X than will protein A. Protein A has $K_a = 10^6 \text{ M}^{-1}$; protein B has $K_a = 10^9 \text{ M}^{-1}$.

2. (a), (b), (c) $n_H < 1.0$. Apparent negative cooperativity in ligand binding can be caused by the presence of two or more types of ligand-binding sites with different affinities for the ligand on the same or different proteins in the same solution. Apparent negative cooperativity is also commonly observed in heterogeneous protein preparations. There are few well-documented cases of true negative cooperativity.

3. (a) Decreases **(b)** Increases **(c)** Decreases **(d)** Increases

4. $k_d = 8.9 \times 10^{-5} \text{ s}^{-1}$.

5. (a) 0.13 pM **(b)** 0.6 pM **(c)** 7.6 μM

6. (a) 0.5 nM (shortcut: the K_d is equivalent to the ligand concentration where $Y = 0.5$). **(b)** Protein 2 has the highest affinity, as it has the lowest K_d.

7. The cooperative behavior of hemoglobin arises from subunit interactions.

8. (a) The observation that HbA (maternal) is about 60% saturated when the pO$_2$ is 4 kPa, whereas HbF (fetal) is more than 90% saturated under the same physiological conditions, indicates that HbF has a higher O$_2$ affinity than HbA. **(b)** The higher O$_2$ affinity of HbF ensures that oxygen will flow from maternal blood to fetal blood in the placenta. Fetal blood approaches full saturation where the O$_2$ affinity of HbA is low. **(c)** The observation that the O$_2$-saturation curve of HbA undergoes a larger shift on BPG binding than that of HbF suggests that HbA binds BPG more tightly than does HbF. Differential binding of BPG to the two hemoglobins may determine the difference in their O$_2$ affinities.

9. (a) Hb Memphis **(b)** HbS, Hb Milwaukee, Hb Providence, possibly Hb Cowtown **(c)** Hb Providence

10. More tightly. An inability to form tetramers would limit the cooperativity of these variants, and the binding curve would become more hyperbolic. Also, the BPG-binding site would be disrupted. Oxygen binding would probably be tighter, because the default state in the absence of bound BPG is the tight-binding R state.

11. (a) 1×10^{-8} M **(b)** 5×10^{-8} M **(c)** 8×10^{-8} M
(d) 2×10^{-7} M. Note that a rearrangement of Eqn 5-8 gives $[L] = YK_d/(1 - Y)$.

12. The epitope is likely to be a structure that is buried when G-actin polymerizes to F-actin.

13. Many pathogens, including HIV, have mechanisms for repeatedly altering the surface proteins to which immune system components initially bind. Thus the host organism regularly faces new antigens and requires time to mount an immune response to each one. As the immune system responds to one variant, new variants are created.

14. Binding of ATP to myosin triggers dissociation of myosin from the actin thin filament. In the absence of ATP, actin and myosin bind tightly to each other.

15.

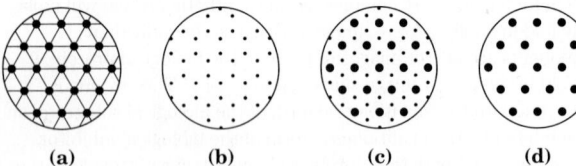

(a)　　　**(b)**　　　**(c)**　　　**(d)**

16. (a) Chain L is the light chain and chain H is the heavy chain of the Fab fragment. Chain Y is lysozyme. **(b)** β structures are predominant in the variable and constant regions of the fragment. **(c)** Fab heavy-chain fragment: 218 amino acid residues; light-chain fragment: 214; lysozyme: 129. Less than 15% of the lysozyme molecule is in contact with the Fab fragment. **(d)** Residues that seem to be in contact with lysozyme include, in the H chain: Gly^{31}, Tyr^{32}, Arg^{99}, Asp^{100}, and Tyr^{101}; in the L chain: Tyr^{32}, Tyr^{49}, Tyr^{50}, and Trp^{92}. In lysozyme, residues Asn^{19}, Gly^{22}, Tyr^{23}, Ser^{24}, Lys^{116}, Gly^{117}, Thr^{118}, Asp^{119}, Gln^{121}, and Arg^{125} seem to be situated at the antigen-antibody interface. Not all these residues are adjacent in the primary structure. Folding of the polypeptide into higher levels of structure brings nonconsecutive residues together to form the antigen-binding site.

17. (a) Two **(b)** No time at all. Suitable antibodies are almost always present before any challenge from the virus. **(c)** $>10^8$. **(d)** Many possible answers

18. (a)

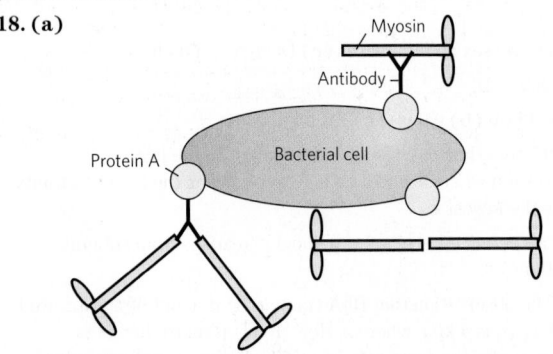

The drawing is not to scale; any given cell would have many more myosin molecules on its surface. **(b)** ATP is needed to provide the chemical energy to drive the motion (see Chapter 13). **(c)** An antibody that bound to the myosin tail, the actin-binding site, would block actin binding and prevent movement. An antibody that bound to actin would also prevent actin-myosin interaction and thus movement. **(d)** There are two possible explanations: (1) Trypsin cleaves only at Lys and Arg residues (see Table 3-6), so would not cleave at many sites in the protein. (2) Not all Arg or Lys residues are equally accessible to trypsin; the most-exposed sites would be cleaved first. **(e)** The S1 model. The hinge model predicts that bead-antibody-HMM complexes (with the hinge) would move, but bead-antibody-SHMM complexes (no hinge) would not. The S1 model predicts that because both complexes include S1, both would move. The finding that the beads move with SHMM (no hinge) is consistent only with the S1 model. **(f)** With fewer myosin molecules bound, the

beads could temporarily fall off the actin as a myosin let go of it. The beads would then move more slowly, as time is required for a second myosin to bind. At higher myosin density, as one myosin lets go, another quickly binds, leading to faster motion. **(g)** Above a certain density, what limits the rate of movement is the intrinsic speed with which myosin molecules move the beads. The myosin molecules are moving at a maximum rate, and adding more will not increase speed. **(h)** Because the force is produced in the S1 head, damaging the S1 head would probably inactivate the resulting molecule, and SHMM would be incapable of producing movement. **(i)** The S1 head must be held together by noncovalent interactions that are strong enough to retain the active shape of the molecule.

Chapter 6

1. The activity of the enzyme that converts sugar to starch is destroyed by heat denaturation.

2. 2.4×10^{-6} M

3. 9.5×10^8 years

4. The enzyme-substrate complex is more stable than the enzyme alone.

5. (a) 190 Å **(b)** Three-dimensional folding of the enzyme brings the amino acid residues into proximity.

6. The reaction rate can be measured by following the decrease in absorption by NADH (at 340 nm) as the reaction proceeds. Determine the K_m value; using substrate concentrations well above the K_m, measure initial rate (rate of NADH disappearance with time, measured spectrophotometrically) at several known enzyme concentrations, and plot initial rate versus concentration of enzyme. The plot should be linear, with a slope that provides a measure of LDH concentration.

7. (b), (e), (g)

8. (a) 1.7×10^{-3} M **(b)** 0.33, 0.67, 0.91 **(c)** The upper curve corresponds to enzyme B ($[X] > K_m$ for this enzyme); the lower curve, enzyme A.

9. (a) 0.2 μM s^{-1} **(b)** 0.6 μM s^{-1} **(c)** 0.9 μM s^{-1}

10. (a) 2,000 s^{-1} **(b)** Measured $V_{max} = 1$ μM s^{-1}; $K_m = 2$ μM

11. (a) 400 s^{-1} **(b)** 10 μM **(c)** $\alpha = 2$, $\alpha' = 3$ **(d)** Mixed inhibitor

12. (a) 24 nM **(b)** 4 μM (V_0 is exactly half V_{max}, so $[A] = K_m$) **(c)** 40 μM (V_0 is exactly half V_{max}, so $[A] = 10$ times K_m in the presence of inhibitor) **(d)** No. $k_{cat}/K_m = (0.33$ s$^{-1})/(4 \times 10^{-6}$ M$) = 8.25 \times 10^4$ M^{-1} s^{-1}, well below the diffusion-controlled limit.

13. $V_{max} \approx 140$ μM/min; $K_m \approx 1 \times 10^{-5}$ M

14. (a) $V_{max} = 51.5$ mM/min; $K_m = 0.59$ mM **(b)** Competitive inhibition

15. $V_{max} = 0.50$ μmol/min; $K_m = 2.2$ mM

16. Curve A

17. 2.0×10^7 min^{-1}

18. The basic assumptions of the Michaelis-Menten equation still hold. The reaction is at steady state, and the rate is determined by $V_0 = k_2[ES]$. The equations needed to solve for $[ES]$ are

$$[E_t] = [E] + [ES] + [EI] \quad \text{and} \quad [EI] = \frac{[E][I]}{K_I}$$

$[E]$ can be obtained by rearranging Eqn 6-19. The rest follows the pattern of the Michaelis-Menten equation derivation in the text.

19. 29,000. The calculation assumes that there is only one essential Cys residue per enzyme molecule.

20. Activity of the prostate enzyme equals total phosphatase activity in a blood sample minus phosphatase activity in the presence of enough tartrate to completely inhibit the prostate enzyme.

21. The inhibition is mixed. Because K_m seems not to change appreciably, this could be the special case of mixed inhibition called noncompetitive.

22. The [S] at which $V_0 = V_{max}/2\alpha'$ is obtained when all terms except V_{max} on the right side of Eqn 6-30—that is, $[S]/(\alpha K_m + \alpha'[S])$—equal $\frac{1}{2}\alpha'$. Begin with $[S]/(\alpha K_m + \alpha'[S]) = \frac{1}{2}\alpha'$ and solve for [S].

23. The optimum activity occurs when Glu[35] is protonated and Asp[52] is unprotonated.

24. (a) In the wild-type enzyme, the substrate is held in place by a hydrogen bond and an ion-dipole interaction between the charged side chain of Arg[109] and the polar carbonyl of pyruvate. During catalysis, the charged Arg[109] side chain also stabilizes the polarized carbonyl transition state. In the mutant, the binding is reduced to just a hydrogen bond, substrate binding is weaker, and ionic stabilization of the transition state is lost, reducing catalytic activity. **(b)** Because Lys and Arg are roughly the same size and have a similar positive charge, they probably have very similar properties. Furthermore, because pyruvate binds to Arg[171] by (presumably) an ionic interaction, an Arg-to-Lys mutation would be expected to have little effect on substrate binding. **(c)** The "forked" arrangement aligns two positively charged groups of Arg residues with the negatively charged oxygens of pyruvate and facilitates two combined hydrogen-bond and ion-dipole interactions. When Lys is present, only one such combined hydrogen-bond and ion-dipole interaction is possible, thus reducing the strength of the interaction. The positioning of the substrate is less precise. **(d)** Ile[250] interacts with the ring of NADH through the hydrophobic effect. This type of interaction is not possible with the hydrophilic side chain of Gln. **(e)** The structure is shown below. **(f)** The mutant enzyme rejects pyruvate because pyruvate's hydrophobic methyl group will not interact with the highly hydrophilic guanidinium group of Arg[102]. The mutant binds oxaloacetate because of the strong ionic interaction between the Arg[102] side chain and the carboxyl of oxaloacetate. **(g)** The protein must be flexible enough to accommodate the added bulk of the side chain and the larger substrate.

Chapter 7

1. With reduction of the carbonyl oxygen to a hydroxyl group, the chemistry at C-1 and C-3 is the same; the glycerol molecule is not chiral.

2. Epimers differ by the configuration about only *one* carbon. **(a)** D-altrose (C-2), D-glucose (C-3), D-gulose (C-4) **(b)** D-idose (C-2), D-galactose (C-3), D-allose (C-4) **(c)** D-arabinose (C-2), D-xylose (C-3)

3. Osazone formation destroys the configuration around C-2 of aldoses, so aldoses differing only at the C-2 configuration give the same derivative, with the same melting point.

4. To convert α-D-glucose to β-D-glucose, the bond between C-1 and the hydroxyl on C-5 (as in Fig. 7-6); to convert D-glucose to D-mannose, either the —H or —OH bond on C-2. Conversion between chair conformations does not require bond breakage; this is the critical distinction between configuration and conformation.

5. No. Glucose and galactose differ at C-4.

6. (a) Both are polymers of D-glucose, but they differ in the glycosidic linkage: $(\beta1\rightarrow4)$ for cellulose, $(\alpha1\rightarrow4)$ for glycogen. **(b)** Both are hexoses, but glucose is an aldohexose, fructose a ketohexose. **(c)** Both are disaccharides, but maltose has two $(\alpha1\rightarrow4)$-linked D-glucose units, and sucrose has $(\alpha1\leftrightarrow2\beta)$-linked D-glucose and D-fructose.

7.

8. A hemiacetal is formed when an aldose or ketose condenses with an alcohol; a glycoside is formed when a hemiacetal condenses with an alcohol (see Fig. 7-5).

9. Fructose cyclizes to either the pyranose or the furanose structure. Increasing the temperature shifts the equilibrium in the direction of the furanose, the less sweet form.

10. The rate of mutarotation is sufficiently high that, as the enzyme consumes β-D-glucose, more α-D-glucose is converted to the β form and, eventually, all the glucose is oxidized. Glucose oxidase is specific for glucose and does not detect other reducing sugars (such as galactose) that react with Fehling's reagent.

11. Boiling a solution of sucrose in water hydrolyzes some of the sucrose to invert sugar. Hydrolysis is accelerated and occurs at lower temperatures with the addition of a small amount of acid (lemon juice or cream of tartar, for example).

12. Prepare a slurry of sucrose and water for the core; add a small amount of sucrase (invertase); immediately coat with chocolate.

13. Sucrose has no free anomeric carbon to undergo mutarotation.

14.

Yes; yes

15. N-Acetyl-β-D-glucosamine is a reducing sugar; its C-1 can be oxidized (pp. 249–250). D-Gluconate is not a reducing sugar; its C-1 is already at the oxidation state of a carboxylic acid. GlcN($\alpha1\leftrightarrow1\alpha$) Glc is not a reducing sugar; the anomeric carbons of both monosaccharides are involved in the glycosidic bond.

16. Humans lack cellulase in the gut and cannot break down cellulose.

17. Native cellulose consists of glucose units linked by $(\beta1\rightarrow4)$ glycosidic bonds, which force the polymer chain into an extended conformation. Parallel series of these extended chains form intermolecular hydrogen bonds, aggregating into long, tough, insoluble fibers. Glycogen consists of glucose units linked by $(\alpha1\rightarrow4)$ glycosidic bonds, which cause bends in the chain and prevent formation of long fibers. In addition, glycogen is highly branched and, because many of its hydroxyl groups are exposed to water, is highly hydrated and disperses in water.

Cellulose is a structural material in plants, consistent with its side-by-side aggregation into insoluble fibers. Glycogen is a storage fuel in animals. Highly hydrated glycogen granules with their many nonreducing ends can be rapidly hydrolyzed by glycogen phosphorylase to release glucose 1-phosphate.

18. Cellulose is several times longer; it assumes an extended conformation, whereas amylose has a helical structure.

19. 6,000 residues/s

20. 11 s

21. The ball-and-stick model of the disaccharide in Fig. 7-18b shows no steric interactions, but a space-filling model, showing atoms with their correct relative sizes, would show several strong steric hindrances in the $-170°$, $-170°$ conformer that are not present in the $30°$, $-40°$ conformer.

22. The negative charges on chondroitin sulfate repel each other and force the molecule into an extended conformation. The polar molecule attracts many water molecules, increasing the molecular volume. In the dehydrated solid, each negative charge is counterbalanced by a positive ion, and the molecule condenses.

23. Positively charged amino acid residues would bind the highly negatively charged groups on heparin. In fact, Lys residues of antithrombin III interact with heparin.

24. 8 possible sequences, 144 possible linkages, and 64 stereochemical possibilities, for a total of 73,728 permutations!

25.

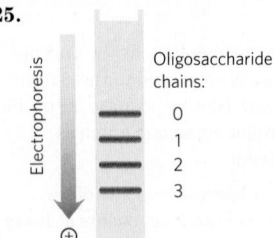

26. Oligosaccharides; their subunits can be combined in more ways than the amino acid subunits of oligopeptides. Each hydroxyl group can participate in glycosidic bonds, and the configuration of each glycosidic bond can be either $α$ or $β$. The polymer can be linear or branched.

27. **(a)** Branch-point residues yield 2,3-di-O-methylglucose; unbranched residues yield 2,3,6-tri-O-methylglucose. **(b)** 3.75%

28. Chains of (1→6)-linked D-glucose residues with occasional (1→3)-linked branches, with one branch every 20 or so residues.

29. **(a)** The tests involve trying to dissolve only part of the sample in a variety of solvents, then analyzing both dissolved and undissolved materials to see whether their compositions differ. **(b)** For a pure substance, all molecules are the same and any dissolved fraction will have the same composition as any undissolved fraction. An impure substance is a mixture of more than one compound. When treated with a particular solvent, more of one component may dissolve, leaving more of the other component(s) behind. As a result, the dissolved and undissolved fractions will have different compositions. **(c)** A quantitative assay allows researchers to be sure that none of the activity has been lost through degradation. When determining the structure of a molecule, it is important that the sample under analysis consist only of intact (undegraded) molecules. If the sample is contaminated with degraded material, this will give confusing and perhaps uninterpretable structural results. A qualitative assay would detect the presence of activity, even if it had become significantly degraded. **(d)** Results 1 and 2. Result 1 is consistent with the known structure, because type B antigen has three molecules of galactose; types A and O each have only two.

Result 2 is also consistent, because type A has two amino sugars (N-acetylgalactosamine and N-acetylglucosamine); types B and O have only one (N-acetylglucosamine). Result 3 is *not* consistent with the known structure: for type A, the glucosamine:galactosamine ratio is 1:1; for type B, it is 1:0. **(e)** The samples were probably impure and/or partly degraded. The first two results were correct possibly because the method was only roughly quantitative and thus not as sensitive to inaccuracies in measurement. The third result is more quantitative and thus more likely to differ from predicted values because of impure or degraded samples. **(f)** An exoglycosidase. If it were an endoglycosidase, one of the products of its action on O antigen would include galactose, N-acetylglucosamine, or N-acetylgalactosamine, and at least one of those sugars would be able to inhibit the degradation. Given that the enzyme is not inhibited by any of these sugars, it must be an exoglycosidase, removing only the terminal sugar from the chain. The terminal sugar of O antigen is fucose, so fucose is the only sugar that could inhibit the degradation of O antigen. **(g)** The exoglycosidase removes N-acetylgalactosamine from A antigen and galactose from B antigen. Because fucose is not a product of either reaction, it will not prevent removal of these sugars, and the resulting substances will no longer be active as A or B antigen. However, the products should be active as O antigen, because degradation stops at fucose. **(h)** All the results are consistent with Fig. 10-14. (1) D-Fucose and L-galactose, which would protect against degradation, are not present in any of the antigens. (2) The terminal sugar of A antigen is N-acetylgalactosamine, and this sugar alone protects this antigen from degradation. (3) The terminal sugar of B antigen is galactose, which is the only sugar capable of protecting this antigen.

Chapter 8

1. N-3 and N-7

2. (5')GCGCAATATTTTGAGAAATATTGCGC(3'); it contains a palindrome. The individual strands can form hairpin structures; the two strands can form a cruciform.

3. $9.4 × 10^{-4}$ g

4. **(a)** 40° **(b)** 0°

5. The RNA helix is in the A conformation; the DNA helix is generally in the B conformation.

6. In eukaryotic DNA, about 5% of C residues are methylated. 5-Methylcytosine can spontaneously deaminate to form thymine; the resulting G–T pair is one of the most common mismatches in eukaryotic cells.

7. Higher

8. Without the base, the ribose ring can be opened to generate the noncyclic aldehyde form. This, and the loss of base-stacking interactions, could contribute significant flexibility to the DNA backbone.

9. CGCGCGTGCGCGCGCG

10. Base stacking in nucleic acids tends to reduce the absorption of UV light. Denaturation involves loss of base stacking, and UV absorption increases.

11. 0.35 mg/mL

12.

Deoxyribose   Guanine   Phosphate

Solubilities: phosphate > deoxyribose > guanine. The highly polar phosphate groups and sugar moieties are on the outside of the double helix, exposed to water; the hydrophobic bases are in the interior of the helix.

13. Primer 1: CCTCGAGTCAATCGATGCTG

Primer 2: CGCGCACATCAGACGAACCA

Recall that all DNA sequences are written in the 5′→3′ direction, left to right; that DNA polymerase synthesizes DNA in the 5′→3′ direction; that the two strands of a DNA molecule are antiparallel; and that both PCR primers must target the end sequences so that their 3′ ends are oriented toward the segment to be amplified.

14. The primers can be used to probe libraries containing long genomic clones to identify contig ends that lie close to each other. If the contigs flanking the gap are close enough, the primers can be used in PCR to directly amplify the intervening DNA separating the contigs, which can then be cloned and sequenced.

15. The 3′-H would prevent addition of any subsequent nucleotides, so the sequence for each cluster would end after the first nucleotide addition.

16. If dCTP is omitted, when the first G residue is encountered in the template, ddCTP will be added, and polymerization will halt. Only one band will be seen in the sequencing gel.

17.

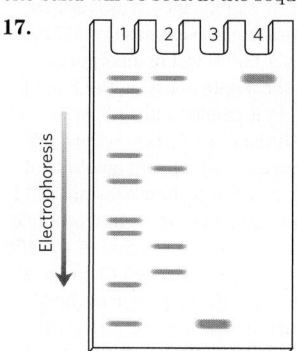

18.

(5′)P—GCGCCAUUGC(3′)—OH
(5′)P—GCGCCAUUG(3′)—OH
(5′)P—GCGCCAUU(3′)—OH
(5′)P—GCGCCAU(3′)—OH
(5′)P—GCGCCA(3′)—OH
(5′)P—GCGCC(3′)—OH
(5′)P—GCGC(3′)—OH
(5′)P—GCG(3′)—OH
(5′)P—GC(3′)—OH

and the nucleoside 5′-phosphates

19. (a) Water is a participant in most biological reactions, including those that cause mutations. The low water content in endospores reduces the activity of mutation-causing enzymes and slows the rate of nonenzymatic depurination reactions, which are hydrolysis reactions. **(b)** UV light induces formation of cyclobutane pyrimidine dimers. Because *B. subtilis* is a soil organism, spores can be lofted to the top of the soil or into the air, where they may be subject to prolonged UV exposure.

20. DMT is a blocking group that prevents reaction of the incoming base with itself.

21. (a) Right-handed. The base at one 5′ end is adenine; at the other 5′ end, cytosine. **(b)** Left-handed **(c)** If you cannot see the structures in stereo, see additional tips in the *Absolute Ultimate Guide for Lehninger Principles of Biochemistry* or use a search engine to find tips online.

22. (a) It would not be easy! The data for different samples from the same organism show significant variation, and the recovery is never 100%. The numbers for C and T show much more consistency than those for A and G, so for C and T it is much easier to make the case that samples from the same organism have the same composition. But even with the less consistent values for A and G, (1) the range of values for different tissues does overlap substantially; (2) the difference between different preparations of the same tissue is about the same as the difference between samples from different tissues; and (3) in samples for which recovery is high, the numbers are more consistent.

(b) This technique would not be sensitive enough to detect a difference between normal and cancerous cells. Cancer is caused by mutations, but these changes in DNA—a few base pairs out of several billion—would be too small to detect with these techniques. **(c)** The ratios of A:G and T:C vary widely among different species. For example, in the bacterium *Serratia marcescens*, both ratios are 0.4, meaning that the DNA contains mostly G and C. In *Haemophilus influenzae*, by contrast, the ratios are 1.74 and 1.54, meaning that the DNA is mostly A and T. **(d)** Conclusion 4 has three requirements. (1) A = T: The table shows an A:T ratio very close to 1 in all cases. Certainly, the variation in this ratio is substantially less than the variation in the A:G and T:C ratios. (2) G = C: Again, the G:C ratio is very close to 1, and the other ratios vary widely. (3) A + G = T + C: This is the purine:pyrimidine ratio, which also is very close to 1. **(e)** The different "core" fractions represent different regions of the wheat germ DNA. If the DNA were a monotonous repeating sequence, the base composition of all regions would be the same. Because different core regions have different sequences, the DNA sequence must be more complex.

Chapter 9

1.

(a) (5′) – – – G(3′)   and   (5′)AATTC – – – (3′)
(3′) – – – CTTAA(5′)   (3′)   G – – – (5′)

(b) (5′) – – – GAATT(3′)   and   (5′)AATTC – – – (3′)
(3′) – – – CTTAA(5′)   (3′)TTAAG – – – (5′)

(c) (5′) – – – GAATTAATTC – – – (3′)
(3′) – – – CTTAATTAAG – – – (5′)

(d) (5′) – – – G(3′)   and   (5′)C – – – (3′)
(3′) – – – C(5′)   (3′)G – – – (5′)

(e) (5′) – – – GAATTC – – – (3′)
(3′) – – – CTTAAG – – – (5′)

(f) (5′) – – – CAG(3′)   and   (5′)CTG – – – (3′)
(3′) – – – GTC(5′)   (3′)GAC – – – (5′)

(g) (5′) – – – CAGAATTC – – – (3′)
(3′) – – – GTCTTAAG – – – (5′)

(h) Method 1: Cut the DNA with EcoRI as in (a), then treat the DNA as in (b) or (d), and then ligate a synthetic DNA fragment with the BamHI recognition sequence between the two resulting blunt ends. Method 2 (more efficient): Synthesize a DNA fragment with the structure

(5′)AATTGGATCC(3′)
(3′)CCTAGGTTAA(5′)

This would ligate efficiently to the sticky ends generated by EcoRI cleavage, would introduce a BamHI site, but would not regenerate the EcoRI site. **(i)** The four fragments (with N = any nucleotide), in order of discussion in the problem, are

(5′)AATTCNNNNCTGCA(3′)
(3′)GNNNNG(5′)

(5′)AATTCNNNNGTGCA(3′)
(3′)GNNNNC(5′)

(5′)AATTGNNNNCTGCA(3′)
(3′)CNNNNG(5′)

(5′)AATTGNNNNGTGCA(3′)
(3′)CNNNNC(5′)

2. Yeast artificial chromosomes (YACs) are not stable in a cell unless they have two telomere-containing ends and a large DNA segment cloned into the chromosome. YACs less than 10,000 bp long are soon lost during continued mitosis and cell division.

3. (a) Plasmids in which the original pBR322 was regenerated without insertion of a foreign DNA fragment; these would retain resistance to ampicillin. Also, two or more molecules of pBR322 might be ligated together with or without insertion of foreign DNA. **(b)** The clones in lanes 1 and 2 each have one DNA fragment inserted in different orientations. The clone in lane 3 has two DNA fragments, ligated such that the EcoRI proximal ends are joined.

4. (5′)GAAAGTCCGCGTTATAGGCATG(3′)
(3′)ACGTCTTTCAGGCGCAATATCCGTACTTAA(5′)

5. Your test would require DNA primers, a heat-stable DNA polymerase, deoxynucleoside triphosphates, and a PCR machine (thermal cycler). The primers would be designed to amplify a DNA segment encompassing the CAG repeat. The DNA strand shown is the coding strand, oriented 5′→3′, left to right. The primer targeted to DNA to the left of the repeat would be identical to any 25-nucleotide sequence shown in the region to the left of the CAG repeat. The primer on the right side must be complementary and antiparallel to a 25-nucleotide sequence to the right of the CAG repeat. Using the primers, DNA including the CAG repeat would be amplified by PCR, and its size would be determined by comparison with size markers after electrophoresis. The length of the DNA would reflect the length of the CAG repeat, providing a simple test for the disease.

6. Design PCR primers that are complementary to the DNA in the deleted segment but would direct DNA synthesis away from each other. No PCR product is generated unless the ends of the deleted segment are joined to create a circle.

7. The plant expressing firefly luciferase must take up luciferin, the substrate of luciferase, before it can "glow" (albeit weakly). The plant expressing green fluorescent protein glows without requiring any other compound.

8.

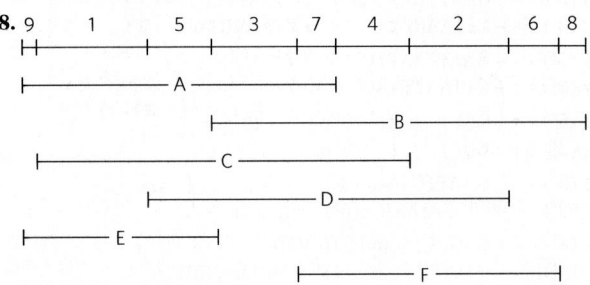

9. The production of labeled antibodies is difficult and expensive, and the labeling of every antibody to every protein target would be impractical. By labeling one antibody preparation for binding to all antibodies of a particular class, the same labeled antibody preparation can be used in many different immunofluorescence experiments.

10. Express the protein in yeast strain 1 as a fusion protein with one of the domains of Gal4p—say, the DNA-binding domain. Using yeast strain 2, make a library in which essentially every protein of the fungus is expressed as a fusion protein with the interaction domain of Gal4p. Mate strain 1 with the strain 2 library, and look for colonies that are colored due to expression of the reporter gene. These colonies will generally arise from mated cells containing a fusion protein that interacts with your target protein.

11. Cover spot 4, add solution containing activated T, irradiate, wash.

1. A–T 2. G–T 3. A–T 4. G–C

Cover spots 2 and 4, add solution containing activated G, irradiate, wash.

1. A–T–G 2. G–T 3. A–T–G 4. G–C

Cover spot 3, add solution containing activated C, irradiate, wash.

1. A–T–G–C 2. G–T–C 3. A–T–G 4. G–C–C

Cover spots 1, 3, and 4, add solution containing activated C, irradiate, wash.

1. A–T–G–C 2. G–T–C–C 3. A–T–G 4. G–C–C

Cover spots 1 and 2, add solution containing activated G, irradiate, wash.

1. A–T–G–C 2. G–T–C–C 3. A–T–G–G 4. G–C–C–G

12. ATSAAG**W**DEWEGGK**V**LIHL**DG**KLQNRGALLELDIGAV

13. The pattern of haplotypes in the Aleut and Eskimo populations suggests that their ancestors' migration into the American Arctic regions was separate from the migrations that eventually populated the rest of North and South America.

14. Interbreeding between the Denisovans and *Homo sapiens* must have occurred in Asia, some time in the many millennia during which humans migrated from Africa to Asia and then to Australia and Melanesia.

15. The same disease condition can be caused by defects in two or more genes that are on different chromosomes.

16. (a) DNA solutions are highly viscous because the very long molecules are tangled in solution. Shorter molecules tend to tangle less and form a less viscous solution, so decreased viscosity corresponds to shortening of the polymers—as caused by nuclease activity. **(b)** An endonuclease. An exonuclease removes single nucleotides from the 5′ or 3′ end and would produce TCA-soluble ^{32}P-labeled nucleotides. An endonuclease cuts DNA into oligonucleotide fragments and produces little or no TCA-soluble ^{32}P-labeled material. **(c)** The 5′ end. If the phosphate were left on the 3′ end, the kinase would incorporate significant ^{32}P as it added phosphate to the 5′ end; treatment with the phosphatase would have no effect on this. In this case, samples A and B would incorporate significant amounts of ^{32}P. When the phosphate is left on the 5′ end, the kinase does not incorporate any ^{32}P: it cannot add a phosphate if one is already present. Treatment with the phosphatase removes 5′ phosphate, and the kinase then incorporates significant amounts of ^{32}P. Sample A will have little or no ^{32}P, and B will show substantial ^{32}P incorporation—as was observed. **(d)** Random breaks would produce a distribution of fragments of random size. The production of specific fragments indicates that the enzyme is site-specific. **(e)** Cleavage at the site of recognition. This produces a specific sequence at the 5′ end of the fragments. If cleavage occurred near but not within the recognition site, the sequence at the 5′ end of the fragments would be random. **(f)** The results are consistent with two recognition sequences, as shown below, cleaved where shown by the arrows:

$$\downarrow$$
(5′) – – – GTT AAC – – – (3′)
(3′) – – – CAA TTG – – – (5′)
$$\uparrow$$

which gives the (5′)pApApC and (3′)TpTp fragments; and

$$\downarrow$$
(5′) – – – GTC GAC – – – (3′)
(3′) – – – CAG CTG – – – (5′)
$$\uparrow$$

which gives the (5′)pGpApC and (3′)CpTp fragments.

Chapter 10

1. The term "lipid" does not specify a particular chemical structure. Compounds are categorized as lipids based on their greater solubility in organic solvents than in water.

2.

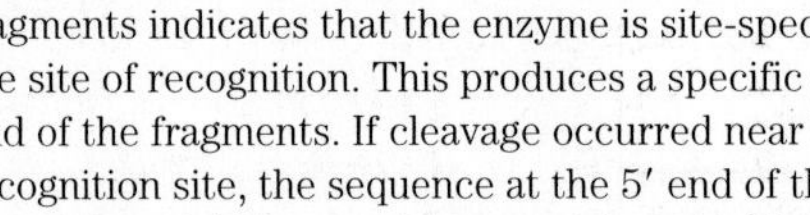

3. (a) The number of cis double bonds. Each cis double bond causes a bend in the hydrocarbon chain, lowering the melting temperature. **(b)** Six different triacylglycerols can be constructed, in order of increasing melting points:

OOO < OOP = OPO < PPO = POP < PPP

where O = oleic and P = palmitic acid. The greater the content of saturated fatty acid, the higher is the melting point. **(c)** Branched-chain fatty acids increase the fluidity of membranes because they decrease the extent of membrane lipid packing.

4. It reduces double bonds, which increases the melting point of lipids containing the fatty acids.

5. Long, saturated acyl chains, nearly solid at air temperatures, form a hydrophobic layer in which a polar compound such as H_2O cannot dissolve or diffuse.

6. Spearmint is (R)-carvone; caraway is (S)-carvone.

7.

COOH
|
H — C — $\overset{+}{N}H_3$
|
CH₃

(R)-2-Aminopropanoic acid

COOH
|
$H_3\overset{+}{N}$ — C — H
|
CH₃

(S)-2-Aminopropanoic acid

OH
|
H_3C — C — COOH
|
H

(R)-2-Hydroxypropanoic acid

COOH
|
H_3C — C — OH
|
H

(S)-2-Hydroxypropanoic acid

8. *Hydrophobic units:* (a) 2 fatty acids; (b), (c), and (d) 1 fatty acid and the hydrocarbon chain of sphingosine; (e) the steroid nucleus and acyl side chain. *Hydrophilic units:* (a) phosphoethanolamine; (b) phosphocholine; (c) D-galactose; (d) several sugar molecules; (e) alcohol group (—OH).

9. It could only be a sphingolipid (sphingomyelin).

10.

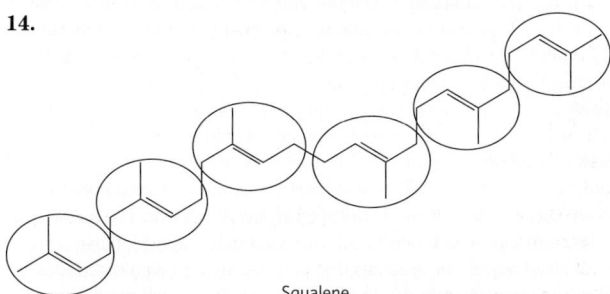

Phosphatidylserine

11. The part of the membrane lipid that determines blood type is the oligosaccharide in the head group of the membrane sphingolipids (see Fig. 10-14). This same oligosaccharide is attached to certain membrane glycoproteins, which also serve as points of recognition by the antibodies that distinguish blood groups.

12. (a) The free —OH group on C-2 and the phosphocholine head group on C-3 are hydrophilic; the fatty acid on C-1 of lysolecithin is hydrophobic. **(b)** Certain steroids such as prednisone inhibit the action of phospholipase A_2, inhibiting the release of arachidonic acid from C-2. Arachidonic acid is converted to a variety of eicosanoids, some of which cause inflammation and pain. **(c)** Phospholipase A_2 releases arachidonic acid, a precursor of other eicosanoids with vital protective functions in the body; it also breaks down dietary glycerophospholipids.

13. Diacylglycerol is hydrophobic and remains in the membrane. Inositol 1,4,5-trisphosphate is highly polar, very soluble in water, and more readily diffusible in the cytosol. Both are second messengers.

14.

Squalene

15. (a) Glycerol and the sodium salts of palmitic and stearic acids **(b)** D-Glycerol 3-phosphocholine and the sodium salts of palmitic and oleic acids

16. Solubility in water: monoacylglycerol > diacylglycerol > triacylglycerol

17. First eluted to last eluted: cholesteryl palmitate and triacylglycerol; cholesterol and *n*-tetradecanol; phosphatidylcholine

and phosphatidylethanolamine; sphingomyelin; phosphatidylserine and palmitate

18. (a) Subject acid hydrolysates of each compound to chromatography (GC or silica gel TLC) and compare the result with known standards. *Sphingomyelin hydrolysate:* sphingosine, fatty acids, phosphocholine, choline, and phosphate; *cerebroside hydrolysate:* sphingosine, fatty acids, sugars, but no phosphate. **(b)** Strong alkaline hydrolysis of sphingomyelin yields sphingosine; phosphatidylcholine yields glycerol. Detect hydrolysate components on thin-layer chromatograms by comparing with standards or by their differential reaction with FDNB (only sphingosine reacts to form a colored product). Treatment with phospholipase A_1 or A_2 releases free fatty acids from phosphatidylcholine, but not from sphingomyelin.

19. Phosphatidylethanolamine and phosphatidylserine

20. (a) GM1 and globoside. Both glucose and galactose are hexoses, so "hexose" in the molar ratio refers to glucose + galactose. The ratios for the four gangliosides are: GM1, 1:3:1:1; GM2, 1:2:1:1; GM3, 1:2:0:1; globoside, 1:3:1:0. **(b)** Yes. The ratio matches GM2, the ganglioside expected to build up in Tay-Sachs disease (see Box 10-1, Fig. 1). **(c)** This analysis is similar to that used by Sanger to determine the amino acid sequence of insulin. The analysis of each fragment reveals only its *composition*, not its *sequence*, but because each fragment is formed by sequential removal of one sugar, we can draw conclusions about sequence. The structure of the normal asialoganglioside is ceramide–glucose–galactose–galactosamine–galactose, consistent with Box 10-1 (excluding Neu5Ac, removed before hydrolysis). **(d)** The Tay-Sachs asialoganglioside is ceramide–glucose–galactose–galactosamine, consistent with Box 10-1. **(e)** The structure of the normal asialoganglioside, GM1, is: *ceramide–glucose* [2 —OH involved in glycosidic links; 1 —OH involved in ring structure; 3 —OH (2, 3, 6) free for methylation]–*galactose* [2 —OH in links; 1 —OH in ring; 3 —OH (2, 4, 6) free for methylation]–*galactosamine* [2 —OH in links; 1 —OH in ring; 1 —NH₂ instead of —OH; 2 —OH (4, 6) free for methylation]–*galactose* [1 —OH in link; 1 —OH in ring; 4 —OH (2, 3, 4, 6) free for methylation]. **(f)** Two key pieces of information are missing: What are the linkages between the sugars? Where is Neu5Ac attached?

Chapter 11

1. The area per molecule would be calculated from the known amount (number of molecules) of lipid used and the area occupied by a monolayer when it begins to resist compression (when the required force increases dramatically, as shown in the plot of force vs. area).

2. The data for dog erythrocytes support a bilayer of lipid: a single cell, with surface area 98 μm^2, has a lipid monolayer area of 200 μm^2. In the case of sheep and human erythrocytes, the data suggest a monolayer, not a bilayer. In fact, significant experimental errors occurred in these early experiments; recent, more accurate measurements support a bilayer in all cases.

3. 63

4. (a) Lipids that form bilayers are amphipathic molecules: they contain a hydrophilic and a hydrophobic region. To minimize the hydrophobic area exposed to the water surface, these lipids form two-dimensional sheets, with the hydrophilic regions exposed to water and the hydrophobic regions buried in the interior of the sheet. Furthermore, to avoid exposing the hydrophobic edges of the sheet to water, lipid bilayers close on themselves. **(b)** These sheets form the closed membrane surfaces that envelop cells and compartments within cells (organelles).

5. 2 nm. Two palmitates placed end to end span about 4 nm, approximately the thickness of a typical bilayer.

6. Salt extraction indicates a peripheral location, and inaccessibility to protease in intact cells indicates an internal location. X seems to be a peripheral protein on the cytosolic face of the membrane.

7. Construct a hydropathy plot; hydrophobic regions of 20 or more residues suggest transmembrane segments. Determine whether the protein in intact erythrocytes reacts with a membrane-impermeant reagent specific for primary amines; if it does, the transporter's amino terminus is on the outside of the cell.

8. ~1%; estimated by calculating the surface area of the cell and of 10,000 transporter molecules (using the dimensions of hemoglobin (5.5 nm diameter, p. 163) as a model globular protein).

9. ~22. To estimate the fraction of membrane surface covered by phospholipids, you would need to know (or estimate) the average cross-sectional area of a phospholipid molecule in a bilayer (e.g., from an experiment such as that diagrammed in Problem 1 in this chapter) and the average cross-sectional area of a 50 kDa protein.

10. Rate of diffusion would decrease. Movement of individual lipids in bilayers occurs much faster at 37 °C, when the lipids are in the "fluid" phase, than at 10 °C, when they are in the "solid" phase. This effect is more pronounced than the usual decrease in Brownian motion with decreased temperature.

11. Interactions among membrane lipids are due to the hydrophobic effect, noncovalent and reversible, allowing membranes to spontaneously reseal.

12. The temperature of body tissues at the extremities is lower than that of tissues closer to the center of the body. If lipid is to remain fluid at this lower temperature, it must contain a higher proportion of unsaturated fatty acids; unsaturated fatty acids lower the melting point of lipid mixtures.

13. The energetic cost of moving the highly polar, sometimes charged, head group through the hydrophobic interior of the bilayer is prohibitive.

14. At pH 7, tryptophan bears a positive and a negative charge, but indole is uncharged. The movement of the less polar indole through the hydrophobic core of the bilayer is energetically more favorable.

15.

The amino acids with the greatest hydropathy index (V, L, F, and C) are clustered on one side of the helix. This amphipathic helix is likely to dip into the lipid bilayer along its hydrophobic surface while exposing its other surface to the aqueous phase. Alternatively, a group of helices may cluster with their polar surfaces in contact with one another and their hydrophobic surfaces facing the lipid bilayer.

16. 35 kJ/mol, neglecting the effects of transmembrane electrical potential; 0.60 mol

17. 13 kJ/mol

18. Most of the O_2 consumed by a tissue is for oxidative phosphorylation, the source of most of the ATP. Therefore, about two-thirds of the ATP synthesized by the kidney is used for pumping K^+ and Na^+.

19. No. The symporter may carry more than one equivalent of Na^+ for each mole of glucose transported.

20. Treat a suspension of cells with unlabeled NEM in the presence of excess lactose, remove the lactose, then add radiolabeled NEM. Use SDS-PAGE to determine the M_r of the radiolabeled band (the transporter).

21. The leucine transporter is specific for the L isomer, but the binding site can accommodate either L-leucine or L-valine. Reduction of V_{max} in the absence of Na^+ indicates that leucine (or valine) is transported by symport with Na^+.

22. V_{max} is reduced; K_t is unaffected.

23. 3×10^{-2} s

24. (a) Glycophorin A: 1 transmembrane segment; myoglobin: no segments long enough to cross a membrane (not a membrane protein); aquaporin: 6 transmembrane segments (may be a membrane channel or receptor protein) **(b)** The 15-residue window provides a better signal-to-noise ratio. **(c)** A narrower window reduces the impact of "edge effects" when a transmembrane sequence occurs near either end of the protein.

25. (a) The rise per residue for an α helix (Chapter 4) is about 1.5 Å = 0.15 nm. To span a 4 nm bilayer, an α helix must contain about 27 residues; thus for seven spans, about 190 residues are required. A protein of M_r 64,000 has about 580 residues. **(b)** A hydropathy plot is used to locate transmembrane regions. **(c)** Because about half of this portion of the receptor consists of charged residues, it probably represents an intracellular loop that connects two adjacent membrane-spanning regions of the protein. **(d)** Because this helix is composed mostly of hydrophobic residues, this portion of the receptor is probably one of the membrane-spanning regions of the protein.

26. (a) *Model A:* supported. The two dark lines are either the protein layers or the phospholipid heads, and the clear space is either the bilayer or the hydrophobic core, respectively. *Model B:* not supported. This model requires a more-or-less uniformly stained band surrounding the cell. *Model C:* supported, with one reservation. The two dark lines are the phospholipid heads; the clear zone is the tails. This assumes that the membrane proteins are not visible, because they do not stain with osmium or do not happen to be in the sections viewed. **(b)** *Model A:* supported. A "naked" bilayer (4.5 nm) + two layers of protein (2 nm) sums to 6.5 nm, which is within the observed range of thickness. *Model B:* neither. This model makes no predictions about membrane thickness. *Model C:* unclear. The result is hard to reconcile with this model, which predicts a membrane as thick as, or slightly thicker than (due to the projecting ends of embedded proteins), a "naked" bilayer. The model is supported only if the smallest values for membrane thickness are correct or if a substantial amount of protein projects from the bilayer. **(c)** *Model A:* unclear. The result is hard to reconcile with this model. If the proteins are bound to the membrane by ionic interactions, the model predicts that the proteins contain a high proportion of charged amino acids, in contrast to what was observed. Also, because the protein layer must be very thin (see (b)), there would not be much room for a hydrophobic protein core, so hydrophobic residues would be exposed to the solvent. *Model B:* supported. The proteins have a mixture of hydrophobic residues (interacting with lipids) and charged residues (interacting with water). *Model C:* supported. The proteins have a mixture of hydrophobic residues (anchoring in the membrane) and charged residues (interacting with water). **(d)** *Model A:* unclear. The result is hard to reconcile with this model, which predicts a ratio of exactly 2.0; this would be hard to achieve under physiologically relevant pressures. *Model B:* neither. This model makes no predictions about amount of lipid in the membrane. *Model C:* supported. Some membrane surface area is taken up with proteins, so the ratio would be less than 2.0, as was observed under more physiologically relevant conditions. **(e)** *Model A:* unclear. The model predicts proteins in extended conformations rather than globular, so is supported only if one assumes that proteins layered on the surfaces include helical segments. *Model B:* supported. The model predicts mostly globular proteins (containing some helical segments). *Model C:* supported. The model predicts mostly globular proteins. **(f)** *Model A:* unclear. The phosphorylamine head groups are protected by the protein layer, but only if the proteins completely cover the surface will the phospholipids be completely protected from phospholipase. *Model B:* supported. Most head groups are accessible to phospholipase. *Model C:* supported. All head groups are accessible to phospholipase. **(g)** *Model A:* not supported. Proteins are entirely accessible to trypsin digestion, and virtually all will undergo multiple cleavage, with no protected hydrophobic segments. *Model B:* not supported. Virtually all proteins are in the bilayer and inaccessible to trypsin. *Model C:* supported. Segments of protein that penetrate or span the bilayer are protected from trypsin; those exposed at the surfaces will be cleaved. The trypsin-resistant portions have a high proportion of hydrophobic residues.

Chapter 12

1. X is cAMP; its production is stimulated by epinephrine. **(a)** Centrifugation sediments adenylyl cyclase (which catalyzes cAMP formation) in the particulate fraction. **(b)** Added cAMP stimulates glycogen phosphorylase. **(c)** cAMP is heat-stable; it can be prepared by treating ATP with barium hydroxide.

2. Unlike cAMP, dibutyryl cAMP passes readily through the plasma membrane.

3. (a) It increases [cAMP]. **(b)** cAMP regulates Na^+ permeability. **(c)** Replace lost body fluids and electrolytes.

4. (a) The mutation makes R unable to bind and inhibit C, so C is constantly active. **(b)** The mutation prevents cAMP binding to R, leaving C inhibited by bound R.

5. Albuterol raises [cAMP], leading to relaxation and dilation of the bronchi and bronchioles. Because β-adrenergic receptors control many other processes, this drug would have undesirable side effects. To minimize these effects, find an agonist specific for the subtype of β-adrenergic receptors found in bronchial smooth muscle.

6. Hormone degradation; hydrolysis of GTP bound to a G protein; degradation, metabolism, or sequestration of second messenger; receptor desensitization; removal of receptor from the cell surface

7. Fuse CFP to β-arrestin and YFP to the cytoplasmic domain of the β-adrenergic receptor, or vice versa. In either case, illuminate at 433 nm and observe fluorescence at both 476 and 527 nm. If the interaction occurs, emitted light intensity will decrease at 476 nm and increase at 527 nm on addition of epinephrine to cells expressing the fusion proteins. If the interaction does not occur, the wavelength of emitted light will remain at 476 nm. Some reasons why this might fail: The fusion proteins (1) are inactive or otherwise unable to interact, (2) are not translocated to their normal subcellular location, or (3) are not stable to proteolytic breakdown.

8. Vasopressin acts by elevating cytosolic $[Ca^{2+}]$ to 10^{-6} M, activating protein kinase C. EGTA injection blocks vasopressin action but should not affect the response to glucagon, which uses cAMP, *not* Ca^{2+}, as second messenger.

9. Amplification results as one molecule of a catalyst activates many molecules of another catalyst, in an amplification cascade involving, in order, insulin receptor, IRS-1, Raf, MEK, ERK; ERK activates a transcription factor, which stimulates mRNA production.

10. A mutation in *ras* that inactivated the Ras GTPase activity would create a protein that, once activated by the binding of GTP, would continue to give, through Raf, the insulin-response signal.

11. *Shared properties of Ras and G_s:* Both bind either GDP or GTP; both are activated by GTP; both, when active, activate a downstream enzyme; both have intrinsic GTPase activity that shuts them off after a short period of activation. *Differences:* Ras is a small, monomeric protein; G_s is heterotrimeric. *Functional difference between G_s and G_i:* G_s activates adenylyl cyclase; G_i inhibits it.

12. *Kinase (factor in parentheses):* PKA (cAMP); PKG (cGMP); PKC (Ca^{2+}, DAG); Ca^{2+}/CaM kinase (Ca^{2+}, CaM); cyclin-dependent kinase (cyclin); protein Tyr kinase (ligand for the receptor, such as insulin); MAPK (Raf); Raf (Ras); glycogen phosphorylase kinase (PKA).

13. G_s remains in its activated form when the nonhydrolyzable analog is bound. The analog therefore prolongs the effect of epinephrine on the injected cell.

14. (a) Use the α-bungarotoxin–bound beads for affinity purification (see Fig. 3-17c) of AChR. Extract proteins from the electric organs and pass the mixture through the chromatography column; the AChR binds selectively to the beads. Elute the AChR with a solute that

weakens its interaction with α-bungarotoxin. **(b)** Use binding of [^{125}I] α-bungarotoxin as a *quantitative assay* for AChR during purification by various techniques. At each step, assay AChR by measuring [^{125}I] α-bungarotoxin binding to the proteins in the sample. Optimize purification for the highest specific activity of AChR (counts/min of bound [^{125}I]α-bungarotoxin per mg of protein) in the final material.

15. Hyperpolarization results in the closing of voltage-dependent Ca^{2+} channels in the presynaptic region of the rod cell. The resulting decrease in $[Ca^{2+}]_{in}$ diminishes release of an inhibitory neurotransmitter that suppresses activity in the next neuron of the visual circuit. When this inhibition is removed in response to a light stimulus, the circuit becomes active and visual centers in the brain are excited.

16. Individuals with Oguchi disease might have a defect in rhodopsin kinase or in arrestin.

17. Rod cells would no longer show any change in membrane potential in response to light. This experiment has been done. Illumination did activate PDE, but the enzyme could not significantly reduce the 8-Br-cGMP level, which remained well above that needed to keep the gated ion channels open. Thus, light had no impact on membrane potential.

18. (a) On exposure to heat, TRPV1 channels open, causing an influx of Na^+ and Ca^{2+} into the sensory neuron. This depolarizes the neuron, triggering an action potential. When the action potential reaches the axon terminus, neurotransmitter is released, signaling the nervous system that heat has been sensed. **(b)** Capsaicin mimics the effects of heat by opening TRPV1 at low temperature, leading to the false sensation of heat. The extremely low EC_{50} indicates that even very small amounts of capsaicin will have dramatic sensory effects. **(c)** At low levels, menthol should open the TRPM8 channel, leading to a sensation of cool; at high levels, both TRPM8 and TRPV3 open, leading to a mixed sensation of cool and heat, such as you may have experienced with very strong peppermints.

19. (a) These mutations might lead to permanent activation of the PGE_2 receptor, leading to unregulated cell division and tumor formation. **(b)** The viral gene might encode a constitutively active form of the receptor, causing a constant signal for cell division and thus tumor formation. **(c)** E1A protein might bind to pRb and prevent E2F from binding, so E2F is constantly active and cells divide uncontrollably. **(d)** Lung cells do not normally respond to PGE_2 because they do not express the PGE_2 receptor; mutations resulting in a constitutively active PGE_2 receptor would not affect lung cells.

20. A normal tumor suppressor gene encodes a protein that restrains cell division. A mutant form of the protein fails to suppress cell division, but if either of the two alleles in an individual encodes normal protein, normal function will continue. A normal oncogene encodes a regulator protein that triggers cell division, but only when an appropriate signal (growth factor) is present. The mutant version of the oncogene product constantly sends the signal to divide, whether or not growth factors are present.

21. In a child who develops multiple tumors in both eyes, every retinal cell had a defective copy of the *Rb* gene at birth. Early in the child's life, several cells independently underwent a second mutation that damaged the one good *Rb* allele, producing a tumor. A child who develops a single tumor had, at birth, two good copies of the *Rb* gene in every cell; mutation in both *Rb* alleles in one cell (extremely rare) caused the single tumor.

22. Two cells expressing the same surface receptor may have different complements of target proteins for protein phosphorylation.

23. (a) The cell-based model, which predicts different receptors present on different cells **(b)** This experiment addresses the issue of the independence of different taste sensations. Even though the receptors for sweet and/or umami are missing, the animals' other taste sensations are normal; thus, pleasant and unpleasant taste sensations are independent. **(c)** Yes. Loss of either T1R1 or T1R3 subunits abolishes umami taste sensation. **(d)** Both models. With either model, removing one receptor would abolish that taste sensation. **(e)** Yes. Loss of either the T1R2 or T1R3 subunits almost completely abolishes the sweet taste sensation; complete elimination of sweet taste requires deletion of both subunits. **(f)** At very high sucrose concentrations,

T1R2 and, to a lesser extent, T1R3 receptors, as homodimers, can detect sweet taste. **(g)** The results are consistent with either model of taste encoding, but do strengthen the researchers' conclusions. Ligand binding can be completely separated from taste sensation. If the ligand for the receptor in "sweet-tasting cells" binds a molecule, mice prefer that molecule as a sweet compound.

Chapter 13

1. Consider the developing chick as the system; the nutrients, egg shell, and outside world are the surroundings. Transformation of the single cell into a chick drastically reduces the entropy of the system. Initially, the parts of the egg outside the embryo (the surroundings) contain complex fuel molecules (a low-entropy condition). During incubation, some of these complex molecules are converted to large numbers of CO_2 and H_2O molecules (high entropy). This increase in the entropy of the surroundings is larger than the decrease in entropy of the chick (the system).

2. (a) -4.8 kJ/mol **(b)** 7.56 kJ/mol **(c)** -13.7 kJ/mol

3. (a) 262 **(b)** 608 **(c)** 0.30

4. $K'_{eq} = 21$; $\Delta G'^{\circ} = -7.6$ kJ/mol

5. -31 kJ/mol

6. (a) -1.68 kJ/mol **(b)** -4.4 kJ/mol **(c)** At a given temperature, the value of $\Delta G'^{\circ}$ for any reaction is fixed and is defined for standard conditions (here, both fructose 6-phosphate and glucose 6-phosphate at 1 M). In contrast, ΔG is a variable that can be calculated for any set of reactant and product concentrations.

7. $K'_{eq} \approx 1$; $\Delta G'^{\circ} \approx 0$

8. Less. The overall equation for ATP hydrolysis can be approximated as

$$ATP^{4-} + H_2O \rightarrow ADP^{3-} + HPO_4^{2-} + H^+$$

(This is only an approximation because the ionized species shown here are the major, but not the only, forms present.) Under standard conditions ([ATP] = [ADP] = [P$_i$] = 1 M), the concentration of water is 55 M and does not change during the reaction. Because H^+ ions are produced in the reaction, at a higher [H^+] (pH 5.0) the equilibrium would be shifted to the left and less free energy would be released.

9. 10

10.

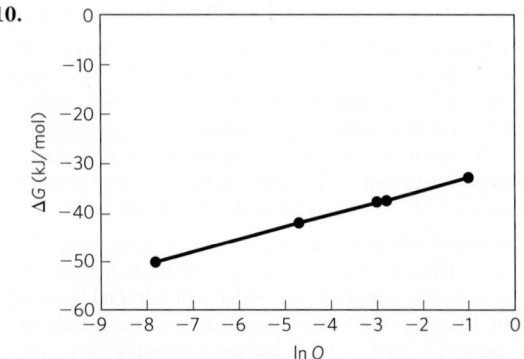

ΔG for ATP hydrolysis is lower when [ATP]/[ADP] is low ($\ll 1$) than when [ATP]/[ADP] is high. The energy available to the cell from a given [ATP] is lower when the [ATP]/[ADP] ratio falls and greater when it rises.

11. (a) 3.85×10^{-3} M^{-1}; [glucose 6-phosphate] $= 8.9 \times 10^{-8}$ M. No. The cellular [glucose 6-phosphate] is much greater than this, favoring the reverse reaction. **(b)** 14 M. No. The maximum solubility of glucose is less than 1 M. **(c)** 837 ($\Delta G'^{\circ} = -16.7$ kJ/mol); [glucose] $= 1.2 \times 10^{-7}$ M. Yes. This reaction path can occur with a concentration of glucose that is readily soluble and does not produce a large osmotic force. **(d)** No. This would require such high [P$_i$] that the phosphate salts of divalent cations would precipitate. **(e)** By directly transferring the phosphoryl group from ATP to glucose, the phosphoryl group transfer potential ("tendency" or "pressure") of ATP is utilized without generating high concentrations of intermediates. The essential part of this transfer is, of course, the enzymatic catalysis.

12. (a) -12.5 kJ/mol **(b)** -14.6 kJ/mol

13. (a) 3×10^{-4} **(b)** 68.7 **(c)** 7.4×10^4

14. -13 kJ/mol

15. 46.7 kJ/mol

16. Isomerization moves the carbonyl group from C-1 to C-2, setting up a carbon–carbon bond cleavage between C-3 and C-4. Without isomerization, bond cleavage would occur between C-2 and C-3, generating one two-carbon and one four-carbon compound.

17. The mechanism is the same as that of the alcohol dehydrogenase reaction (see Fig. 14-14).

18. The first step is the reverse of an aldol condensation (see the aldolase mechanism, Fig. 14-6); the second step is an aldol condensation (see Fig. 13-4).

19. (a) Oxidation-reduction, dehydrogenase with NAD cofactor; NADH + H$^+$ also produced **(b)** Isomerization, isomerase **(c)** Internal rearrangement, isomerase **(d)** Phosphoryl group transfer, kinase and ATP; ADP produced **(e)** Hydrolysis, protease or peptidase and H_2O **(f)** Oxidation-reduction, dehydrogenase with NAD cofactor; NADH + H$^+$ also produced **(g)** Oxidation-reduction, dehydrogenase with NAD cofactor and H_2O; NADH + H$^+$ also produced

20. ATP; the products of phosphoarginine hydrolysis are stabilized by resonance forms not available in the intact molecule.

21. Yes. If [ADP] and [polyphosphate] are kept high, and [ATP] is kept low, the actual free-energy change would be negative.

22. (a) 46 kJ/mol **(b)** 46 kg; 68% **(c)** ATP is synthesized as it is needed, then broken down to ADP and P$_i$; its concentration is maintained in a steady state.

23. The ATP system is in a dynamic steady state; [ATP] remains constant because the rate of ATP consumption equals its rate of synthesis. ATP consumption involves release of the terminal (γ) phosphoryl group; synthesis of ATP from ADP involves replacement of this group. Hence the terminal phosphoryl undergoes rapid turnover. In contrast, the central (β) phosphoryl undergoes only relatively slow turnover.

24. (a) 1.7 kJ/mol **(b)** Inorganic pyrophosphatase catalyzes the hydrolysis of pyrophosphate and drives the net reaction toward the synthesis of acetyl-CoA.

25. 36 kJ/mol

26. (a) NAD$^+$/NADH **(b)** Pyruvate/lactate **(c)** Lactate formation **(d)** -26.1 kJ/mol **(e)** 3.63×10^4

27. (a) 1.14 V **(b)** -220 kJ/mol **(c)** ~ 4

28. (a) -0.35 V **(b)** -0.320 V **(c)** -0.29 V

29. In order of increasing tendency: (a), (d), (b), (c)

30. (c) and (d)

31. (a) The lowest-energy, highest-entropy state occurs when the dye concentration is the same in both cells. If a "fish trap" gap junction allowed unidirectional transport, more of the dye would end up in the oligodendrocyte and less in the astrocyte. This would be a higher-energy, lower-entropy state than the starting state, violating the second law of thermodynamics. The model proposed by Robinson et al. requires an impossible spontaneous decrease in entropy. In terms of energy, the model entails a spontaneous change from a lower-energy to a higher-energy state without an energy input—again, thermodynamically impossible. **(b)** Molecules, unlike fish, do not exhibit *directed behavior*; they move randomly by Brownian motion. Diffusion results in *net* movement of molecules from a region of higher concentration to a region of lower concentration simply because it is more likely that a molecule on the high-concentration side will enter the connecting channel. Look at this as a pathway with a rate-limiting step: the narrow end of the channel. The narrower end limits the rate at which molecules pass through because random motion of the molecules is less likely to move them through the smaller cross section. The wide end of the channel does *not* act like a funnel for molecules, although it may for fish, because molecules are not "crowded" by the sides of the narrowing funnel as fish would be. The narrow end limits the rate of movement equally in both directions. When the concentrations on both sides are equal, the rates of movement in both directions are equal and there will be no change in concentration. **(c)** Fish exhibit *nonrandom behavior*, adjusting their actions in response to the environment. Fish that enter the large opening of the channel tend to move forward because fish have behavior that tends to make them prefer forward movement, and they experience "crowding" as they move through the narrowing channel. It is easy for fish to enter the large opening, but they don't move out of the trap as readily because they are less likely to enter the small opening. **(d)** There are many possible explanations, some of which were proposed by the letter-writers who criticized the article. Here are two: (1) *The dye could bind to a molecule in the oligodendrocyte*. Binding effectively removes the dye from the bulk solvent, so it doesn't "count" as a solute for thermodynamic considerations yet remains visible in the fluorescence microscope. (2) *The dye could be sequestered in a subcellular organelle of the oligodendrocyte*, either actively pumped in at the expense of ATP or drawn in by its attraction to other molecules in that organelle.

Chapter 14

1. Net equation: Glucose + 2ATP → 2 glyceraldehyde 3-phosphate + 2ADP + 2H$^+$; $\Delta G'^\circ$ = 2.1 kJ/mol

2. Net equation: 2 Glyceraldehyde 3-phosphate + 4ADP + 2P$_i$ → 2 lactate + 2NAD$^+$; $\Delta G'^\circ$ = −114 kJ/mol

3. GLUT2 (and GLUT1) is found in liver and is always present in the plasma membrane of hepatocytes. GLUT3 is always present in the plasma membrane of certain brain cells. GLUT4 is normally sequestered in vesicles in cells of muscle and adipose tissue, and enters the plasma membrane only in response to insulin. Thus, liver and brain can take up glucose from blood regardless of insulin level, but muscle and adipose tissue take up glucose only when insulin levels are elevated in response to high blood glucose.

4. CH$_3$CHO + NADH + H$^+$ ⇌ CH$_3$CH$_2$OH + NAD$^+$

$$K'_{eq} = 1.45 \times 10^4$$

5. −8.6 kJ/mol

6. (a) ^{14}CH$_3$CH$_2$OH **(b)** [3-^{14}C]glucose or [4-^{14}C]glucose

7. Fermentation releases energy, some conserved in the form of ATP but much of it dissipated as heat. Unless the fermenter contents are cooled, the temperature would become high enough to kill the microorganisms.

8. Soybeans and wheat contain starch, a polymer of glucose. The microorganisms break down starch to glucose, glucose to pyruvate via glycolysis, and—because the process is carried out in the absence of O$_2$ (i.e., it is a fermentation)—pyruvate to lactic acid and ethanol. If O$_2$ were present, pyruvate would be oxidized to acetyl-CoA, then to CO$_2$ and H$_2$O. Some of the acetyl-CoA, however, would also be hydrolyzed to acetic acid (vinegar) in the presence of oxygen.

9. C-1. This experiment demonstrates the reversibility of the aldolase reaction. The C-1 of glyceraldehyde 3-phosphate is equivalent to C-4 of fructose 1,6-bisphosphate (see Fig. 14-7). The starting glyceraldehyde 3-phosphate must have been labeled at C-1. The C-3 of dihydroxyacetone phosphate becomes labeled through the triose phosphate isomerase reaction, thus giving rise to fructose 1,6-bisphosphate labeled at C-3.

10. No. There would be no anaerobic production of ATP; aerobic ATP production would be diminished only slightly.

11. No. Lactate dehydrogenase is required to recycle NAD$^+$ from the NADH formed during the oxidation of glyceraldehyde 3-phosphate.

12. The transformation of glucose to lactate occurs when myocytes are low in oxygen, and it provides a means of generating ATP under O$_2$-deficient conditions. Because lactate can be oxidized to pyruvate, glucose is not wasted; pyruvate is oxidized by aerobic reactions when O$_2$ becomes plentiful. This metabolic flexibility gives the organism a greater capacity to adapt to its environment.

13. The cell rapidly removes the 1,3-bisphosphoglycerate in a favorable subsequent step, catalyzed by phosphoglycerate kinase.

14. (a) 3-Phosphoglycerate is the product. **(b)** In the presence of arsenate there is no net ATP synthesis under anaerobic conditions.

15. (a) Ethanol fermentation requires 2 mol of P$_i$ per mole of glucose. **(b)** Ethanol is the reduced product formed during reoxidation of NADH to NAD$^+$, and CO$_2$ is the byproduct of the conversion of pyruvate to ethanol. Yes. Pyruvate must be converted to ethanol, to produce a continuous supply of NAD$^+$ for the oxidation of glyceraldehyde 3-phosphate. Fructose 1,6-bisphosphate accumulates; it is formed as an intermediate in glycolysis. **(c)** Arsenate replaces P$_i$ in the glyceraldehyde 3-phosphate dehydrogenase reaction to yield an acyl arsenate, which spontaneously hydrolyzes. This prevents formation of ATP, but 3-phosphoglycerate continues through the pathway.

16. Dietary niacin is used to synthesize NAD$^+$. Oxidations carried out by NAD$^+$ are part of cyclic processes, with NAD$^+$ as electron carrier (reducing agent); one molecule of NAD$^+$ can oxidize many thousands of molecules of glucose, and thus the dietary requirement for the precursor vitamin (niacin) is relatively small.

17. Dihydroxyacetone phosphate + NADH + H$^+$ → glycerol 3-phosphate + NAD$^+$ (catalyzed by a dehydrogenase)

18. *Galactokinase deficiency:* galactose (less toxic); *UDP-glucose:galactose 1-phosphate uridylyltransferase deficiency:* galactose 1-phosphate (more toxic)

19. The proteins are degraded to amino acids and used for gluconeogenesis.

20. (a) In the pyruvate carboxylase reaction, ^{14}CO$_2$ is added to pyruvate, but PEP carboxykinase removes the *same* CO$_2$ in the next step. Thus, ^{14}C is not (initially) incorporated into glucose.

(b)

21. 4 ATP equivalents per glucose molecule

22. Gluconeogenesis would be highly endergonic, and it would be impossible to separately regulate gluconeogenesis and glycolysis.

23. The cell "spends" 1 ATP and 1 GTP in converting pyruvate to PEP.

24. (a), (b), (d) are glucogenic; (c) (e) are not.

25. Consumption of alcohol forces competition for NAD^+ between ethanol metabolism and gluconeogenesis. The problem is compounded by strenuous exercise and lack of food, because at these times the level of blood glucose is already low.

26. (a) The rapid increase in glycolysis; the rise in pyruvate and NADH results in a rise in lactate. **(b)** Lactate is transformed to glucose via pyruvate; this is a slower process, because formation of pyruvate is limited by NAD^+ availability, the lactate dehydrogenase equilibrium is in favor of lactate, and conversion of pyruvate to glucose is energy-requiring. **(c)** The equilibrium for the lactate dehydrogenase reaction is in favor of lactate formation.

27. Lactate is transformed to glucose in the liver by gluconeogenesis (see Figs 14-16, 14-17). A defect in FBPase-1 would prevent entry of lactate into the gluconeogenic pathway in hepatocytes, causing lactate to accumulate in the blood.

28. Succinate is transformed to oxaloacetate, which passes into the cytosol and is converted to PEP by PEP carboxykinase. Two moles of PEP are then required
to produce a mole of glucose by the route outlined in Fig. 14-17.

29. If the catabolic and anabolic pathways of glucose metabolism are operating simultaneously, unproductive cycling of ADP and ATP occurs, with extra O_2 consumption.

30. At the very least, accumulation of ribose 5-phosphate would tend to force this reaction in the reverse direction by mass action (see Eqn 13-4). It might also affect other metabolic reactions that involve ribose 5-phosphate as a substrate or product—such as the pathways of nucleotide synthesis.

31. (a) Ethanol tolerance is likely to involve many more genes, and thus the engineering would be a much more involved project.
(b) L-Arabinose isomerase (the *araA* enzyme) converts an aldose to a ketose by moving the carbonyl of a nonphosphorylated sugar from C-1 to C-2. No analogous enzyme is discussed in this chapter; all the enzymes described here act on phosphorylated sugars. An enzyme that carries out a similar transformation with phosphorylated sugars is phosphohexose isomerase. L-Ribulokinase (*araB*) phosphorylates a sugar at C-5 by transferring the γ phosphate from ATP. Many such reactions are described in this chapter, including the hexokinase reaction. L-Ribulose 5-phosphate epimerase (*araD*) switches the —H and —OH groups on a chiral carbon of a sugar. No analogous reaction is described in the chapter, but it is described in Chapter 20 (see Fig. 20-40). **(c)** The three *ara* enzymes would convert arabinose to xylulose 5-phosphate by the following pathway: Arabinose $\xrightarrow{\text{L-arabinose isomerase}}$ L-ribulose $\xrightarrow{\text{L-ribulokinase}}$ L-ribulose 5-phosphate $\xrightarrow{\text{epimerase}}$ xylulose 5-phosphate. **(d)** The arabinose is converted to xylulose 5-phosphate as in (c), which enters the pathway in Fig. 14-23; the glucose 6-phosphate product is then fermented to ethanol and CO_2. **(e)** 6 molecules of arabinose + 6 molecules of ATP are converted to 6 molecules of xylulose 5-phosphate, which feed into the pathway in Fig. 14-23 to yield 5 molecules of glucose 6-phosphate, each of which is fermented to yield 3 ATP (they enter as glucose 6-phosphate, not glucose)—15 ATP in all. Overall, you would expect a yield of 15 ATP − 6 ATP = 9 ATP from the 6 arabinose molecules. The other products are 10 molecules of ethanol and 10 molecules of CO_2. **(f)** Given the lower ATP yield, for an amount of growth (i.e., of available ATP) equivalent to growth without the added genes, the engineered *Z. mobilis* must ferment more arabinose, and thus it produces more ethanol. **(g)** One way to allow the use of xylose would be to add the genes for two enzymes: an analog of the *araD* enzyme that converts xylose to ribose by switching the —H and —OH on C-3, and an analog of the *araB* enzyme that phosphorylates ribose at C-5. The resulting ribose 5-phosphate would feed into the existing pathway.

Chapter 15

1. (a) 0.0293 **(b)** 308 **(c)** No. Q is much lower than K'_{eq}, indicating that the PFK-1 reaction is far from equilibrium in cells; this reaction is slower than the subsequent reactions in glycolysis. Flux through the glycolytic pathway is largely determined by the activity of PFK-1.

2. (a) 1.4×10^{-9} M **(b)** The physiological concentration (0.023 mM) is 16,000 times the equilibrium concentration; this reaction does not reach equilibrium in the cell. Many reactions in the cell are not at equilibrium.

3. In the absence of O_2, the ATP needs are met by anaerobic glucose metabolism (fermentation to lactate). Because aerobic oxidation of glucose produces far more ATP than does fermentation, less glucose is needed to produce the same amount of ATP.

4. (a) There are two binding sites for ATP: a catalytic site and a regulatory site. Binding of ATP to a regulatory site inhibits PFK-1, by reducing V_{max} or increasing K_m for ATP at the catalytic site.
(b) Glycolytic flux is reduced when ATP is plentiful. **(c)** The graph indicates that increased [ADP] suppresses the inhibition by ATP. Because the adenine nucleotide pool is fairly constant, consumption of ATP leads to an increase in [ADP]. The data show that the activity of PFK-1 may be regulated by the [ATP]/[ADP] ratio.

5. The phosphate group of glucose 6-phosphate is completely ionized at pH 7, giving the molecule an overall negative charge. Because membranes are generally impermeable to electrically charged molecules, glucose 6-phosphate cannot pass from the bloodstream into cells and hence cannot enter the glycolytic pathway and generate ATP. (This is why glucose, once phosphorylated, cannot escape from the cell.)

6. (a) *In muscle:* Glycogen breakdown supplies energy (ATP) via glycolysis. Glycogen phosphorylase catalyzes the conversion of stored glycogen to glucose 1-phosphate, which is converted to glucose 6-phosphate, an intermediate in glycolysis. During strenuous activity, skeletal muscle requires large quantities of glucose 6-phosphate. *In the liver:* Glycogen breakdown maintains a steady level of blood glucose between meals (glucose 6-phosphate is converted to free glucose). **(b)** In actively working muscle, ATP flux requirements are very high and glucose 1-phosphate must be produced rapidly, requiring a high V_{max}.

7. (a) 3.3/1 **(b), (c)** The value of this ratio in the cell (>100:1) indicates that [glucose 1-phosphate] is far below the equilibrium value. The rate at which glucose 1-phosphate is removed (through entry into glycolysis) is greater than its rate of production (by the glycogen phosphorylase reaction), so metabolite flow is from glycogen to glucose 1-phosphate. The glycogen phosphorylase reaction is probably the regulatory step in glycogen breakdown.

8. (a) Increases **(b)** Decreases **(c)** Increases

9. *Resting:* [ATP] high; [AMP] low; [acetyl-CoA] and [citrate] intermediate. *Running:* [ATP] intermediate; [AMP] high; [acetyl-CoA] and [citrate] low. Glucose flux through glycolysis increases during the anaerobic sprint because (1) the ATP inhibition of glycogen phosphorylase and PFK-1 is partially relieved, (2) AMP stimulates both enzymes, and (3) lower citrate and acetyl-CoA levels relieve their inhibitory effects on PFK-1 and pyruvate kinase, respectively.

10. The migrating bird relies on the highly efficient aerobic oxidation of fats, rather than the anaerobic metabolism of glucose used by a sprinting rabbit. The bird reserves its muscle glycogen for short bursts of energy during emergencies.

11. *Case A:* (f), (3); *Case B:* (c), (3); *Case C:* (h), (4); *Case D:* (d), (6)

12. (a) (1) Adipose: fatty acid synthesis slower. (2) Muscle: glycolysis, fatty acid synthesis, and glycogen synthesis slower. (3) Liver: glycolysis faster; gluconeogenesis, glycogen synthesis, and fatty acid synthesis slower; pentose phosphate pathway unchanged.
(b) (1) Adipose and (3) liver: fatty acid synthesis slower because lack of insulin results in inactive acetyl-CoA carboxylase, the first enzyme of fatty acid synthesis. Glycogen synthesis inhibited by cAMP-dependent phosphorylation (thus activation) of glycogen

synthase. (2) Muscle: glycolysis slower because GLUT4 is inactive, so glucose uptake is inhibited. (3) Liver: glycolysis slower because the bifunctional PFK-2/FBPase-2 is converted to the form with active FBPase-2, decreasing [fructose 2,6-bisphosphate], which allosterically stimulates phosphofructokinase and inhibits FBPase-1; this also accounts for the stimulation of gluconeogenesis.

13. (a) Elevated **(b)** Elevated **(c)** Elevated

14. (a) PKA cannot be activated in response to glucagon or epinephrine, and glycogen phosphorylase is not activated. **(b)** PP1 remains active, allowing it to dephosphorylate glycogen synthase (activating it) and glycogen phosphorylase (inhibiting it). **(c)** Phosphorylase remains phosphorylated (active), increasing the breakdown of glycogen. **(d)** Gluconeogenesis cannot be stimulated when blood glucose is low, leading to dangerously low blood glucose during periods of fasting.

15. The drop in blood glucose triggers release of glucagon by the pancreas. In the liver, glucagon activates glycogen phosphorylase by stimulating its cAMP-dependent phosphorylation and stimulates gluconeogenesis by lowering [fructose 2,6-bisphosphate], thus stimulating FBPase-1.

16. (a) Reduced capacity to mobilize glycogen; lowered blood glucose between meals **(b)** Reduced capacity to lower blood glucose after a carbohydrate meal; elevated blood glucose **(c)** Reduced fructose 2,6-bisphosphate (F26BP) in liver, stimulating glycolysis and inhibiting gluconeogenesis **(d)** Reduced F26BP, stimulating gluconeogenesis and inhibiting glycolysis **(e)** Increased uptake of fatty acids and glucose; increased oxidation of both **(f)** Increased conversion of pyruvate to acetyl-CoA; increased fatty acid synthesis

17. (a) Given that each particle contains about 55,000 glucose residues, the equivalent free glucose concentration would be $55,000 \times 0.01 \ \mu M = 550 \ mM$, or 0.55 M. This would present a serious osmotic challenge for the cell! (Body fluids have a substantially lower osmolarity.) **(b)** The lower the number of branches, the lower the number of free ends available for glycogen phosphorylase activity, and the slower the rate of glucose release. With no branches, there would be just one site for phosphorylase to act. **(c)** The outer tier of the particle would be too crowded with glucose residues for the enzyme to gain access to cleave bonds and release glucose. **(d)** The number of chains doubles in each succeeding tier: tier 1 has one chain (2^0), tier 2 has two (2^1), tier 3 has four (2^2), and so on. Thus, for t tiers, the number of chains in the outermost tier, C_A, is 2^{t-1}. **(e)** The total number of chains is $2^0 + 2^1 + 2^2 + \ldots 2^{t-1} = 2^t - 1$. Each chain contains g_c glucose molecules, so the total number of glucose molecules, G_T, is $g_c(2^t - 1)$. **(f)** Glycogen phosphorylase can release all but four of the glucose residues in a chain of length g_c. Therefore, from each chain in the outer tier it can release $(g_c - 4)$ glucose molecules. Given that there are 2^{t-1} chains in the outer tier, the number of glucose molecules the enzyme can release, G_{PT}, is $(g_c - 4)(2^{t-1})$. **(g)** The volume of a sphere is $\frac{4}{3}\pi r^3$. In this case, r is the thickness of one tier times the number of tiers, or $(0.12g_c + 0.35)t$ nm. Thus $V_s = \frac{4}{3}\pi t^3(0.12g_c + 0.35)^3$ nm^3. **(h)** You can show algebraically that the value of g_c that maximizes f is independent of t. Choosing $t = 7$:

g_c	C_A	G_T	G_{PT}	V_s	f
5	64	635	64	1,232	2,111
6	64	762	128	1,760	3,547
7	64	889	192	2,421	4,512
8	64	1,016	256	3,230	5,154
9	64	1,143	320	4,201	5,572
10	64	1,270	384	5,350	5,834
11	64	1,397	448	6,692	5,986
12	64	1,524	512	8,240	6,060
13	64	1,651	576	10,011	6,079
14	64	1,778	640	12,019	6,059
15	64	1,905	704	14,279	6,011
16	64	2,032	768	16,806	5,943

The optimum value of g_c (i.e., at maximum f) is 13. In nature, g_c varies from 12 to 14, which corresponds to f values very close to the optimum. If you choose another value for t, the numbers will differ but the optimal g_c will still be 13.

Chapter 16

1. (a)
❶ *Citrate synthase:*
Acetyl-CoA + oxaloacetate + H_2O → citrate + CoA
❷ *Aconitase:*
Citrate → isocitrate
❸ *Isocitrate dehydrogenase:*
Isocitrate + NAD$^+$ → α-ketoglutarate + CO_2 + NADH
❹ *α-Ketoglutarate dehydrogenase:*
α-Ketoglutarate + NAD$^+$ + CoA → succinyl-CoA + CO_2 + NADH
❺ *Succinyl-CoA synthetase:*
Succinyl-CoA + P$_i$ + GDP → succinate + CoA + GTP
❻ *Succinate dehydrogenase:*
Succinate + FAD → fumarate + FADH$_2$
❼ *Fumarase:*
Fumarate + H_2O → malate
❽ *Malate dehydrogenase:*
Malate + NAD$^+$ → oxaloacetate + NADH + H$^+$

(b), (c) ❶ CoA, condensation; ❷ none, isomerization; ❸ NAD$^+$, oxidative decarboxylation; ❹ NAD$^+$, CoA, and thiamine pyrophosphate, oxidative decarboxylation; ❺ CoA, substrate-level phosphorylation; ❻ FAD, oxidation; ❼ none, hydration; ❽ NAD$^+$, oxidation

(d) Acetyl-CoA + 3NAD$^+$ + FAD + GDP + P$_i$ + 2H_2O →
$\qquad$ 2CO_2 + CoA + 3NADH + FADH$_2$ + GTP + 2H$^+$

2. Glucose + 4ADP + 4P$_i$ + 10NAD$^+$ + 2FAD →
$\qquad$ 4ATP + 10NADH + 2FADH$_2$ + 6CO_2

3. (a) Oxidation; methanol → formaldehyde + [H—H]
(b) Oxidation; formaldehyde → formate + [H—H]
(c) Reduction; CO_2 + [H—H] → formate + H$^+$
(d) Reduction; glycerate + H$^+$ + [H—H] → glyceraldehyde + H_2O
(e) Oxidation; glycerol → dihydroxyacetone + [H—H]
(f) Oxidation; 2H_2O + toluene → benzoate + H$^+$ + 3[H—H]
(g) Oxidation; succinate → fumarate + [H—H]
(h) Oxidation; pyruvate + H_2O → acetate + CO_2 + [H—H]

4. From the structural formulas, we see that the carbon-bound H/C ratio of hexanoic acid (11/6) is higher than that of glucose (7/6). Hexanoic acid is more reduced and yields more energy on complete combustion to CO_2 and H_2O.

5. (a) Oxidized; ethanol + NAD$^+$ → acetaldehyde + NADH + H$^+$
(b) Reduced; 1,3-bisphosphoglycerate + NADH + H$^+$ →
$\qquad$ glyceraldehyde 3-phosphate + NAD$^+$ + HPO$_4^{2-}$
(c) Unchanged; pyruvate + H$^+$ → acetaldehyde + CO_2
(d) Oxidized; pyruvate + NAD$^+$ → acetate + CO_2 + NADH + H$^+$
(e) Reduced; oxaloacetate + NADH + H$^+$ → malate + NAD$^+$
(f) Unchanged; acetoacetate + H$^+$ → acetone + CO_2

6. *TPP:* thiazolium ring adds to α carbon of pyruvate, then stabilizes the resulting carbanion by acting as an electron sink. *Lipoic acid:* oxidizes pyruvate to level of acetate (acetyl-CoA), and activates acetate as a thioester. *CoA-SH:* activates acetate as thioester. *FAD:* oxidizes lipoic acid. *NAD$^+$:* oxidizes FAD.

7. Lack of TPP, caused by thiamine deficiency, inhibits pyruvate dehydrogenase; pyruvate accumulates.

8. Oxidative decarboxylation; NAD$^+$ or NADP$^+$; α-ketoglutarate dehydrogenase reaction

9. Oxygen consumption is a measure of the activity of the first two stages of cellular respiration: glycolysis and the citric acid cycle. The addition of oxaloacetate or malate stimulates the citric acid cycle and thus stimulates respiration. The added oxaloacetate or malate serves a catalytic role: it is regenerated in the latter part of the citric acid cycle.

10. (a) 5.6×10^{-6} **(b)** 1.1×10^{-8} M **(c)** 28 molecules

11. ADP (or GDP), P_i, CoA-SH, TPP, NAD^+; *not* lipoic acid, which is covalently attached to the isolated enzymes that use it

12. The flavin nucleotides, FMN and FAD, would not be synthesized. Because FAD is required in the citric acid cycle, flavin deficiency would strongly inhibit the cycle.

13. Oxaloacetate might be withdrawn for aspartate synthesis or for gluconeogenesis. Oxaloacetate is replenished by the anaplerotic reactions catalyzed by PEP carboxykinase, PEP carboxylase, malic enzyme, or pyruvate carboxylase (see Fig. 16-16).

14. The terminal phosphoryl group of GTP can be transferred to ADP in a reaction catalyzed by nucleoside diphosphate kinase, with an equilibrium constant of 1.0: $GTP + ADP \rightarrow GDP + ATP$.

15. (a) $^-OOC-CH_2-CH_2-COO^-$ (succinate) **(b)** Malonate is a competitive inhibitor of succinate dehydrogenase. **(c)** A block in the citric acid cycle stops NADH formation, which stops electron transfer, which stops respiration. **(d)** A large excess of succinate (substrate) overcomes the competitive inhibition.

16. (a) Add uniformly labeled [^{14}C]glucose and check for the release of $^{14}CO_2$. **(b)** Equally distributed in C-2 and C-3 of oxaloacetate; an infinite number of turns

17. Oxaloacetate equilibrates with succinate, in which C-1 and C-4 are equivalent. Oxaloacetate derived from succinate is labeled at C-1 and C-4, and the PEP derived from it is labeled at C-1, which gives rise to C-3 and C-4 of glucose.

18. (a) C-1 **(b)** C-3 **(c)** C-3 **(d)** C-2 (methyl group) **(e)** C-4 **(f)** C-4 **(g)** Equally distributed in C-2 and C-3

19. Thiamine is required for the synthesis of TPP, a prosthetic group in the pyruvate dehydrogenase and α-ketoglutarate dehydrogenase complexes. A thiamine deficiency reduces the activity of these enzyme complexes and causes the observed accumulation of precursors.

20. No. For every two carbons that enter as acetate, two leave the cycle as CO_2; thus there is no net synthesis of oxaloacetate. Net synthesis of oxaloacetate occurs by the carboxylation of pyruvate, an anaplerotic reaction.

21. Yes. The citric acid cycle would be inhibited. Oxaloacetate is present at relatively low concentrations in mitochondria, and removing it for gluconeogenesis would tend to shift the equilibrium for the citrate synthase reaction toward oxaloacetate.

22. (a) Inhibition of aconitase **(b)** Fluorocitrate; competes with citrate; by a large excess of citrate **(c)** Citrate and fluorocitrate are inhibitors of PFK-1. **(d)** All catabolic processes necessary for ATP production are shut down.

23. *Glycolysis:*
Glucose + $2P_i$ + 2ADP + $2NAD^+$ →
$$2 \text{ pyruvate} + 2ATP + 2NADH + 2H^+ + 2H_2O$$

Pyruvate carboxylase reaction:
2 Pyruvate + $2CO_2$ + 2ATP + $2H_2O$ →
$$2 \text{ oxaloacetate} + 2ADP + 2P_i + 4H^+$$

Malate dehydrogenase reaction:
2 Oxaloacetate + 2NADH + $2H^+$ → 2 L-malate + $2NAD^+$

This sequence recycles nicotinamide coenzymes under anaerobic conditions. The overall reaction is: glucose + $2CO_2$ → 2 L-malate + $4H^+$. Four H^+ are produced per glucose, increasing the acidity and thus the tartness of the wine.

24. Pyruvate + ATP + CO_2 + H_2O → oxaloacetate + ADP + P_i + H^+
Pyruvate + CoA + NAD^+ → acetyl-CoA + CO_2 + NADH + H^+
Oxaloacetate + acetyl-CoA → citrate + CoA
Citrate → isocitrate
Isocitrate + NAD^+ → α-ketoglutarate + CO_2 + NADH + H^+
Net reaction: 2 Pyruvate + ATP + $2NAD^+$ + H_2O →
$$\alpha\text{-ketoglutarate} + CO_2 + ADP + P_i + 2NADH + 3H^+$$

25. The cycle participates in catabolic and anabolic processes. For example, it generates ATP by substrate oxidation, but also provides precursors for amino acid synthesis (see Fig. 16-16).

26. (a) Decreases **(b)** Increases **(c)** Decreases

27. (a) Citrate is produced through the action of citrate synthase on oxaloacetate and acetyl-CoA. Citrate synthase can be used for net synthesis of citrate when (1) there is a continuous influx of new oxaloacetate and acetyl-CoA and (2) isocitrate synthesis is restricted, as in a medium low in Fe^{3+}. Aconitase requires Fe^{3+}, so an Fe^{3+}-restricted medium restricts the synthesis of aconitase.

(b) Sucrose + H_2O → glucose + fructose
Glucose + $2P_i$ + 2ADP + $2NAD^+$ →
$$2 \text{ pyruvate} + 2ATP + 2NADH + 2H^+ + 2H_2O$$
Fructose + $2P_i$ + 2ADP + $2NAD^+$ →
$$2 \text{ pyruvate} + 2ATP + 2NADH + 2H^+ + 2H_2O$$
2 Pyruvate + $2NAD^+$ + 2CoA →
$$2 \text{ acetyl-CoA} + 2NADH + 2H^+ + 2CO_2$$
2 Pyruvate + $2CO_2$ + 2ATP + $2H_2O$ →
$$2 \text{ oxaloacetate} + 2ADP + 2P_i + 4H^+$$
2 Acetyl-CoA + 2 oxaloacetate + $2H_2O$ → 2 citrate + 2CoA

The overall reaction is

Sucrose + H_2O + $2P_i$ + 2ADP + $6NAD^+$ →
$$2 \text{ citrate} + 2ATP + 6NADH + 10H^+$$

(c) The overall reaction consumes NAD^+. Because the cellular pool of this oxidized coenzyme is limited, it must be regenerated from NADH by the electron-transfer chain, with consumption of O_2. Consequently, the overall conversion of sucrose to citric acid is an aerobic process and requires molecular oxygen.

28. Succinyl-CoA is an intermediate of the citric acid cycle; its accumulation signals reduced flux through the cycle, calling for reduced entry of acetyl-CoA into the cycle. Citrate synthase, by regulating the primary oxidative pathway of the cell, regulates the supply of NADH and thus the flow of electrons from NADH to O_2.

29. Fatty acid catabolism increases [acetyl-CoA], which stimulates pyruvate carboxylase. The resulting increase in [oxaloacetate] stimulates acetyl-CoA consumption by the citric acid cycle, and [citrate] rises, inhibiting glycolysis at the level of PFK-1. In addition, increased [acetyl-CoA] inhibits the pyruvate dehydrogenase complex, slowing the utilization of pyruvate from glycolysis.

30. Oxygen is needed to recycle NAD^+ from the NADH produced by the oxidative reactions of the citric acid cycle. Reoxidation of NADH occurs during mitochondrial oxidative phosphorylation.

31. Increased [NADH]/[NAD^+] inhibits the citric acid cycle by mass action at the three NAD^+-reducing steps; high [NADH] shifts equilibrium toward NAD^+.

32. Toward citrate; ΔG for the citrate synthase reaction under these conditions is about -8 kJ/mol.

33. Steps ❹ and ❺ are essential in the reoxidation of the enzyme's reduced lipoamide cofactor.

34. The citric acid cycle is so central to metabolism that a serious defect in any cycle enzyme would probably be lethal to the embryo.

35. (a) The only reaction in muscle tissue that consumes significant amounts of oxygen is cellular respiration, so O_2 consumption is a good proxy for respiration. **(b)** Freshly prepared muscle tissue contains some residual glucose; O_2 consumption is due to oxidation of this glucose. **(c)** Yes. Because the amount of O_2 consumed increased when citrate or 1-phosphoglycerol was added, both can serve as substrate for cellular respiration in this system. **(d)** *Experiment I:* Citrate is causing much more O_2 consumption than would be expected from its complete oxidation. Each molecule of citrate seems to be acting as though it were more than one molecule. The only possible explanation is that each molecule of citrate functions more than once in the reaction—which is how a catalyst operates.

Experiment II: The key is to calculate the excess O_2 consumed by each sample compared with the control (sample 1).

Sample	Substrate(s) added	μL O_2 absorbed	Excess μL O_2 consumed
1	No extra	342	0
2	0.3 mL 0.2 M 1-phosphoglycerol	757	415
3	0.15 mL 0.02 M citrate	431	89
4	0.3 mL 0.2 M 1-phosphoglycerol + 0.15 mL 0.02 M citrate	1,385	1,043

If both citrate and 1-phosphoglycerol were simply substrates for the reaction, you would expect the excess O_2 consumption by sample 4 to be the sum of the individual excess consumptions by samples 2 and 3 (415 μL + 89 μL = 504 μL). However, the excess consumption when both substrates are present is roughly twice this amount (1,043 μL). Thus citrate increases the ability of the tissue to metabolize 1-phosphoglycerol. This behavior is typical of a catalyst. Both experiments (I and II) are required to make this case convincing. Based on experiment I only, citrate is somehow accelerating the reaction, but it is not clear whether it acts by helping substrate metabolism or by some other mechanism. Based on experiment II only, it is not clear which molecule is the catalyst, citrate or 1-phosphoglycerol. Together, the experiments show that citrate is acting as a "catalyst" for the oxidation of 1-phosphoglycerol. **(e)** Given that the pathway can consume citrate (see sample 3), if citrate is to act as a catalyst it must be regenerated. If the set of reactions first consumes then regenerates citrate, it must be a circular rather than a linear pathway. **(f)** When the pathway is blocked at α-ketoglutarate dehydrogenase, citrate is converted to α-ketoglutarate but the pathway goes no further. Oxygen is consumed by reoxidation of the NADH produced by isocitrate dehydrogenase.

(g)

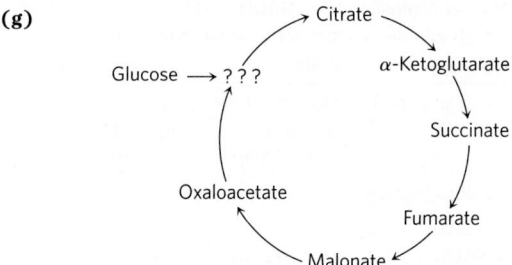

This differs from Fig. 16-7 in that it does not include *cis*-aconitate and isocitrate (between citrate and α-ketoglutarate), or succinyl-CoA, or acetyl-CoA. **(h)** Establishing a quantitative conversion was essential to rule out a branched or other, more complex pathway.

Chapter 17

1. The fatty acid portion; the carbons in fatty acids are more reduced than those in glycerol.

2. (a) 4.0×10^5 kJ (9.6×10^4 kcal) **(b)** 48 days **(c)** 0.48 lb/day

3. The first step in fatty acid oxidation is analogous to the conversion of succinate to fumarate; the second step, to the conversion of fumarate to malate; the third step, to the conversion of malate to oxaloacetate.

4. 8 cycles; the last releases 2 acetyl-CoA.

5. (a) R—COO$^-$ + ATP $\rightarrow$ acyl-AMP + PP$_i$
Acyl-AMP + CoA $\rightarrow$ acyl-CoA + AMP

(b) Irreversible hydrolysis of PP$_i$ to 2P$_i$ by cellular inorganic pyrophosphatase

6. *cis*-Δ^3-Dodecanoyl-CoA; it is converted to *cis*-Δ^2-dodecanoyl-CoA, then β-hydroxydodecanoyl-CoA.

7. 4 acetyl-CoA and 1 propionyl-CoA

8. Yes. Some of the tritium is removed from palmitate during the dehydrogenation reactions of β oxidation. The removed tritium appears as tritiated water.

9. Fatty acyl groups condensed with CoA in the cytosol are first transferred to carnitine, releasing CoA, then transported into the mitochondrion, where they are again condensed with CoA. The cytosolic and mitochondrial pools of CoA are thus kept separate, and no radioactive CoA from the cytosolic pool enters the mitochondrion.

10. (a) In the pigeon, β oxidation predominates; in the pheasant, anaerobic glycolysis of glycogen predominates. **(b)** Pigeon muscle would consume more O_2. **(c)** Fat contains more energy per gram than glycogen does. In addition, the anaerobic breakdown of glycogen is limited by the tissue's tolerance to lactate buildup. Thus the pigeon, using the oxidative catabolism of fats, is the long-distance flyer. **(d)** These enzymes are the regulatory enzymes of their respective pathways and thus limit ATP production rates.

11. Malonyl-CoA would no longer inhibit fatty acid entry into the mitochondrion and β oxidation, so there might be a futile cycle of simultaneous fatty acid synthesis in the cytosol and fatty acid breakdown in mitochondria.

12. (a) The carnitine-mediated entry of fatty acids into mitochondria is the rate-limiting step in fatty acid oxidation. Carnitine deficiency slows fatty acid oxidation; added carnitine increases the rate. **(b)** All increase the metabolic need for fatty acid oxidation. **(c)** Carnitine deficiency might result from a deficiency of lysine, its precursor, or from a defect in one of the enzymes in the biosynthesis of carnitine.

13. Oxidation of fats releases metabolic water; 1.4 L of water per kg of tripalmitoylglycerol (ignores the small contribution of glycerol to the mass).

14. The bacteria can be used to completely oxidize hydrocarbons to CO_2 and H_2O. However, contact between hydrocarbons and bacterial enzymes may be difficult to achieve. Bacterial nutrients such as nitrogen and phosphorus may be limiting, which would inhibit growth.

15. (a) M_r 136; phenylacetic acid **(b)** Even; removal of two carbons at a time from odd-number chains would leave phenylpropionate.

16. Because the mitochondrial pool of CoA is small, CoA must be recycled from acetyl-CoA via the formation of ketone bodies. This allows the operation of the β-oxidation pathway, necessary for energy production.

17. (a) Glucose yields pyruvate via glycolysis, and pyruvate is the main source of oxaloacetate. Without glucose in the diet, [oxaloacetate] drops and the citric acid cycle slows. **(b)** Odd-numbered; propionate conversion to succinyl-CoA provides intermediates for the citric acid cycle and four-carbon precursors for gluconeogenesis.

18. For the odd-number heptanoic acid, β oxidation produces propionyl-CoA, which can be converted in several steps to oxaloacetate, a starting material for gluconeogenesis. The even-number fatty acid cannot support gluconeogenesis, because it is entirely oxidized to acetyl-CoA.

19. β Oxidation of ω-fluorooleate forms fluoroacetyl-CoA, which enters the citric acid cycle and produces fluorocitrate, a powerful inhibitor of aconitase. Inhibition of aconitase shuts down the citric acid cycle. Without reducing equivalents from the citric acid cycle, oxidative phosphorylation (ATP synthesis) is fatally slowed.

20. Ser to Ala: blocks β oxidation in mitochondria. Ser to Asp: blocks fatty acid synthesis, stimulates β oxidation.

21. Response to glucagon or epinephrine would be prolonged, giving a greater mobilization of fatty acids in adipocytes.

22. Enz-FAD, having a more positive standard reduction potential, is a better electron acceptor than NAD$^+$, and the reaction is driven in the direction of fatty acyl–CoA oxidation. This more favorable equilibrium is obtained at the cost of 1 ATP; only 1.5 ATP are produced per FADH$_2$ oxidized in the respiratory chain (vs. 2.5 per NADH).

23. 9 turns; arachidic acid, a 20-carbon saturated fatty acid, yields 10 molecules of acetyl-CoA, the last two formed in the ninth turn.

24. See Fig. 17-12. [3-^{14}C]Succinyl-CoA is formed, which gives rise to oxaloacetate labeled at C-2 and C-3.

25. Phytanic acid → pristanic acid → propionyl-CoA → → → succinyl-CoA → succinate → fumarate → malate. All malate carbons would be labeled, but C-1 and C-4 would have only half as much label as C-2 and C-3.

26. ATP hydrolysis in the energy-requiring reactions of a cell takes up water in the reaction ATP + H_2O → ADP + P_i; thus, in the steady state, there is no *net* production of H_2O.

27. Methylmalonyl-CoA mutase requires the cobalt-containing cofactor formed from vitamin B_{12}.

28. Mass lost per day is about 0.66 kg, or about 140 kg in seven months. Ketosis could be avoided by degradation of nonessential body proteins to supply amino acid skeletons for gluconeogenesis.

29. (a) Fatty acids are converted to their CoA derivatives by enzymes in the cytoplasm; the acyl-CoAs are then imported into mitochondria for oxidation. Given that the researchers were using isolated mitochondria, they had to use CoA derivatives. **(b)** Stearoyl-CoA was rapidly converted to 9 acetyl-CoA by the β-oxidation pathway. All intermediates reacted rapidly, and none were detectable at significant levels. **(c)** Two rounds. Each round removes two carbon atoms, thus two rounds convert an 18-carbon to a 14-carbon fatty acid and 2 acetyl-CoA. **(d)** The K_m is higher for the trans isomer than for the cis, so a higher concentration of trans isomer is required for the same rate of breakdown. Roughly speaking, the trans isomer binds less well than the cis, probably because differences in shape, even though not at the target site for the enzyme, affect substrate binding to the enzyme. **(e)** The substrate for LCAD/VLCAD builds up differently, depending on the particular substrate; this is expected for the rate-limiting step in a pathway. **(f)** The kinetic parameters show that the trans isomer is a poorer substrate than the cis for LCAD, but there is little difference for VLCAD. Because it is a poorer substrate, the trans isomer accumulates to higher levels than the cis. **(g)** One possible pathway is shown below (indicating "inside" and "outside" mitochondria).

Elaidoyl-CoA $\xrightarrow{\text{carnitine acyltransferase 1}}$ elaidoyl-carnitine $\xrightarrow{\text{transport}}$
(outside)   (outside)

elaidoyl-carnitine $\xrightarrow{\text{carnitine acyltransferase 2}}$ elaidoyl-CoA $\xrightarrow{\text{2 rounds of }\beta\text{ oxidation}}$
(inside)   (inside)

5-*trans*-tetradecenoyl-CoA $\xrightarrow{\text{thioesterase}}$ 5-*trans*-tetradecanoic acid $\xrightarrow{\text{diffusion}}$
(inside)   (inside)

5-*trans*-tetradecanoic acid
(outside)

(h) It is correct insofar as trans fats are broken down less efficiently than cis fats, and thus trans fats may "leak" out of mitochondria. It is incorrect to say that trans fats are not broken down by cells; they are broken down, but at a slower rate than cis fats.

Chapter 18

1.

(a) $^-OOC-CH_2-\overset{\overset{\displaystyle O}{\|}}{C}-COO^-$ Oxaloacetate

(b) $^-OOC-CH_2-CH_2-\overset{\overset{\displaystyle O}{\|}}{C}-COO^-$ α-Ketoglutarate

(c) $CH_3-\overset{\overset{\displaystyle O}{\|}}{C}-COO^-$ Pyruvate

(d) (phenyl)$-CH_2-\overset{\overset{\displaystyle O}{\|}}{C}-COO^-$ Phenylpyruvate

2. This is a coupled-reaction assay. The product of the slow transamination (pyruvate) is rapidly consumed in the subsequent "indicator reaction" catalyzed by lactate dehydrogenase, which consumes NADH. Thus the rate of disappearance of NADH is a measure of the rate of the aminotransferase reaction. The indicator reaction is monitored by observing the decrease in absorption of NADH at 340 nm with a spectrophotometer.

3. Alanine and glutamine play special roles in the transport of amino groups from muscle and from other nonhepatic tissues, respectively, to the liver.

4. No. The nitrogen in alanine can be transferred to oxaloacetate via transamination, to form aspartate.

5. 15 mol of ATP per mole of lactate; 13 mol of ATP per mole of alanine, when nitrogen removal is included

6. (a) Fasting resulted in low blood glucose; subsequent administration of the experimental diet led to rapid catabolism of glucogenic amino acids. **(b)** Oxidative deamination caused the rise in NH_3 levels; the absence of arginine (an intermediate in the urea cycle) prevented conversion of NH_3 to urea; arginine is not synthesized in sufficient quantities in the cat to meet the needs imposed by the stress of the experiment. This suggests that arginine is an essential amino acid in the cat's diet. **(c)** Ornithine is converted to arginine by the urea cycle.

7. H_2O + glutamate + NAD^+ →
$$\alpha\text{-Ketoglutarate} + NH_4^+ + NADH + H^+$$
$NH_4^+ + 2ATP + H_2O + CO_2 →$
$$\text{carbamoyl phosphate} + 2ADP + P_i + 3H^+$$
Carbamoyl phosphate + ornithine → citrulline + P_i + H^+
Citrulline + aspartate + ATP →
$$\text{argininosuccinate} + AMP + PP_i + H^+$$
Argininosuccinate → arginine + fumarate
Fumarate + H_2O → malate
Malate + NAD^+ → oxaloacetate + NADH + H^+
Oxaloacetate + glutamate → aspartate + α-ketoglutarate
Arginine + H_2O → urea + ornithine

---

2 Glutamate + CO_2 + $4H_2O$ + $2NAD^+$ + 3ATP →
$$2 \alpha\text{-ketoglutarate} + 2NADH + 7H^+ + \text{urea} +$$
$$2ADP + AMP + PP_i + 2P_i \quad (1)$$

Additional reactions that need to be considered:

AMP + ATP → 2ADP   (2)
$O_2 + 8H^+ + 2NADH + 6ADP + 6P_i → 2NAD^+ + 6ATP + 8H_2O$ (3)
$H_2O + PP_i → 2P_i + H^+$ (4)

Summing equations (1) through (4) gives

2 Glutamate + CO_2 + O_2 + 2ADP + $2P_i$ →
$$2 \alpha\text{-ketoglutarate} + \text{urea} + 3H_2O + 2ATP$$

8. The second amino group introduced into urea is transferred from aspartate, which is generated during the transamination of glutamate to oxaloacetate, a reaction catalyzed by aspartate aminotransferase. Approximately one-half of all the amino groups excreted as urea must pass through the aspartate aminotransferase reaction, making this the most highly active aminotransferase.

9. (a) A person on a diet consisting only of protein must use amino acids as the principal source of metabolic fuel. Because the catabolism of amino acids requires the removal of nitrogen as urea, the process consumes abnormally large quantities of water to dilute and excrete the urea in the urine. Furthermore, electrolytes in the "liquid protein" must be diluted with water and excreted. If the daily water loss through the kidney is not balanced by a sufficient water intake, a net loss of body water results. **(b)** When considering the nutritional benefits of protein, one must keep in mind the total amount of amino acids needed for protein synthesis and the distribution of amino acids in the dietary protein. Gelatin contains a nutritionally unbalanced distribution of amino acids. As large amounts of gelatin are ingested and the excess amino acids are catabolized, the capacity of the urea cycle may be exceeded, leading to ammonia toxicity. This is further complicated by the dehydration that may result from excretion of large quantities of urea. A combination of these two factors could produce coma and death.

10. Lysine and leucine

11. (a) Phenylalanine hydroxylase; a low-phenylalanine diet **(b)** The normal route of phenylalanine metabolism via hydroxylation to tyrosine is blocked, and phenylalanine accumulates. **(c)** Phenylalanine is transformed to phenylpyruvate by transamination, and then to phenyllactate by reduction. The transamination reaction has an equilibrium constant of 1.0, and phenylpyruvate is formed in significant amounts when phenylalanine accumulates. **(d)** Because of the deficiency in production of tyrosine, which is a precursor of melanin, the pigment normally present in hair

12. Not all amino acids are affected. Catabolism of the carbon skeletons of valine, methionine, and isoleucine is impaired because a functional methylmalonyl-CoA mutase (a coenzyme B_{12} enzyme) is absent. The physiological effects of loss of this enzyme are described in Table 18-2 and Box 18-2.

13. The vegan diet lacks vitamin B_{12}, leading to the increase in homocysteine and methylmalonate (reflecting the deficiencies in methionine synthase and methylmalonic acid mutase, respectively) in individuals on the diet for several years. Dairy products provide some vitamin B_{12} in the lactovegetarian diet.

14. The genetic forms of pernicious anemia generally arise as a result of defects in the pathway that mediates absorption of dietary vitamin B_{12} (see Box 17-2). Because dietary supplements are not absorbed in the intestine, these conditions are treated by injecting supplementary B_{12} directly into the bloodstream.

15. The mechanism is identical to that for serine dehydratase (see Fig. 18-20a), except that the extra methyl group of threonine is retained, yielding α-ketobutyrate instead of pyruvate.

16. (a) $^{15}NH_2-CO-^{15}NH_2$

(b) $^-OO^{14}C-CH_2-CH_2-^{14}COO^-$

(c) $R-NH-\underset{\|}{\overset{^{15}NH}{C}}-^{15}NH_2$

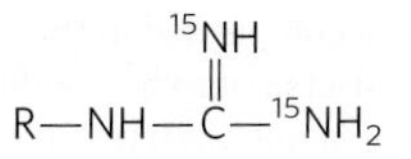

(d) $R-NH-\underset{\|}{\overset{O}{C}}-^{15}NH_2$

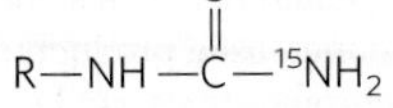

(e) No label

(f) $^-OO^{14}C-\underset{H}{\overset{^{15}NH_2}{C}}-CH_2-^{14}COO^-$

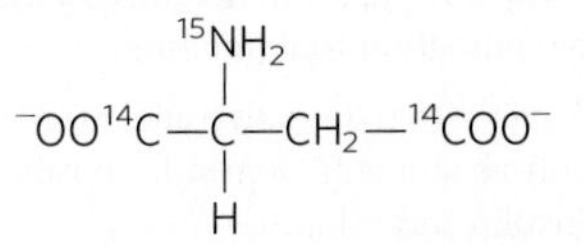

17. (a) Isoleucine $\overset{❶}{\longrightarrow}$ II $\overset{❷}{\longrightarrow}$ IV $\overset{❸}{\longrightarrow}$ I $\overset{❹}{\longrightarrow}$ V $\overset{❺}{\longrightarrow}$ III $\overset{❻}{\longrightarrow}$ acetyl-CoA + propionyl-CoA **(b)** Step ❶ transamination, no analogous reaction, PLP; ❷ oxidative decarboxylation, analogous to the pyruvate dehydrogenase reaction, NAD⁺, TPP, lipoate, FAD; ❸ oxidation, analogous to the succinate dehydrogenase reaction, FAD; ❹ hydration, analogous to the fumarase reaction, no cofactor; ❺ oxidation, analogous to the malate dehydrogenase reaction, NAD⁺; ❻ thiolysis (reverse aldol condensation), analogous to the thiolase reaction, CoA

18. A likely mechanism is:

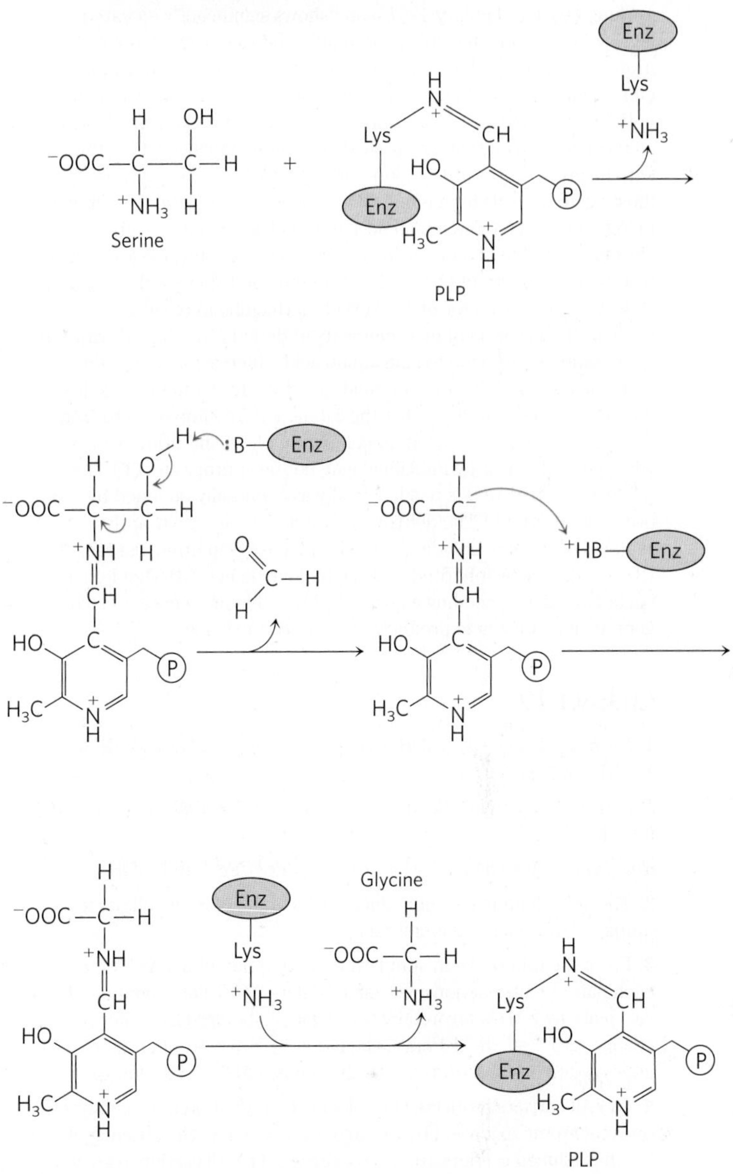

The formaldehyde (HCHO) produced in the second step reacts rapidly with tetrahydrofolate at the enzyme active site to produce N^5,N^{10}-methylenetetrahydrofolate (see Fig. 18-17).

19. (a) Transamination; no analogies; PLP **(b)** Oxidative decarboxylation; analogous to oxidative decarboxylation of pyruvate to acetyl-CoA prior to entry into the citric acid cycle, and of α-ketoglutarate to succinyl-CoA in the citric acid cycle; NAD⁺, FAD, lipoate, and TPP **(c)** Dehydrogenation (oxidation); analogous to dehydrogenation of succinate to fumarate in the citric acid cycle, and of fatty acyl–CoA to enoyl-CoA in β oxidation; FAD **(d)** Carboxylation; no analogies in citric acid cycle or β oxidation; ATP and biotin **(e)** Hydration; analogous to hydration of fumarate to malate in the citric acid cycle, and of enoyl-CoA to 3-hydroxyacyl-CoA in β oxidation; no cofactors **(f)** Reverse aldol reaction; analogous to reverse of citrate synthase reaction in the citric acid cycle; no cofactors

20. Most amino acid catabolism occurs in the liver, including the key steps that are blocked in maple syrup urine disease. A liver transplant from a suitable donor with a normally functioning branched-chain α-keto acid dehydrogenase complex could alleviate disease symptoms.

21. (a) Leucine; valine; isoleucine. **(b)** Cysteine (derived from cystine). If cysteine were decarboxylated as shown in Fig. 18-6, it would yield $H_3N—CH_2—CH_2—SH$, which could be oxidized to taurine. **(c)** The January 1957 blood shows significantly elevated levels of isoleucine, leucine, methionine, and valine; the January 1957 urine, significantly elevated isoleucine, leucine, taurine, and valine. **(d)** All patients had high levels of isoleucine, leucine, and valine in both blood and urine, suggesting a defect in the breakdown of these amino acids. Given that the urine also contained high levels of the keto forms of these three amino acids, the block in the pathway must occur after deamination but before dehydrogenation (as shown in Fig. 18-28). **(e)** The model does not explain the high levels of methionine in blood and taurine in urine. The high taurine levels may be due to the death of brain cells during the end stage of the disease. However, the reason for high levels of methionine in blood are unclear; the pathway of methionine degradation is not linked with the degradation of branched-chain amino acids. Increased methionine could be a secondary effect of buildup of the other amino acids. It is important to keep in mind that the January 1957 samples were from an individual who was dying, so comparing blood and urine results with those of a healthy individual may not be appropriate. **(f)** The following information is needed (and was eventually obtained by other workers): (1) The dehydrogenase activity is significantly reduced or missing in individuals with maple syrup urine disease. (2) The disease is inherited as a single-gene defect. (3) The defect occurs in a gene encoding all or part of the dehydrogenase. (4) The genetic defect leads to production of inactive enzyme.

Chapter 19

1. *Reaction 1:* (a), (d) NADH; (b), (e) E-FMN; (c) NAD^+/NADH and E-FMN/$FMNH_2$

Reaction 2: (a), (d) E-$FMNH_2$; (b), (e) Fe^{3+}; (c) E-FMN/ $FMNH_2$ and Fe^{3+}/Fe^{2+}

Reaction 3: (a), (d) Fe^{2+}; (b), (e) Q; (c) Fe^{3+}/Fe^{2+} and Q/QH_2

2. The side chain makes ubiquinone soluble in lipids and allows it to diffuse in the semifluid membrane.

3. From the difference in standard reduction potential ($\Delta E'^\circ$) for each pair of half-reactions, we can calculate $\Delta G'^\circ$. The oxidation of succinate by FAD is favored by the negative free-energy change ($\Delta G'^\circ = -3.7$ kJ/mol). Oxidation by NAD^+ would require a large, positive, standard free-energy change ($\Delta G'^\circ = 68$ kJ/mol).

4. (a) All carriers reduced; CN^- blocks the reduction of O_2 catalyzed by cytochrome oxidase. **(b)** All carriers reduced; in the absence of O_2, the reduced carriers are not reoxidized. **(c)** All carriers oxidized **(d)** Early carriers more reduced; later carriers more oxidized

5. (a) Inhibition of NADH dehydrogenase by rotenone decreases the rate of electron flow through the respiratory chain, which in turn decreases the rate of ATP production. If this reduced rate is unable to meet the organism's ATP requirements, the organism dies. **(b)** Antimycin A strongly inhibits the oxidation of Q in the respiratory chain, reducing the rate of electron transfer and leading to the consequences described in (a). **(c)** Because antimycin A blocks *all* electron flow to oxygen, it is a more potent poison than rotenone, which blocks electron flow from NADH but not from $FADH_2$.

6. (a) The rate of electron transfer necessary to meet the ATP demand increases, and thus the P/O ratio decreases. **(b)** High concentrations of uncoupler produce P/O ratios near zero. The P/O ratio decreases, and more fuel must be oxidized to generate the same amount of ATP. The extra heat released by this oxidation raises the body temperature. **(c)** Increased activity of the respiratory chain in the presence of an uncoupler requires the degradation of additional fuel. By oxidizing more fuel (including fat reserves) to produce the same amount of ATP, the body loses weight. When the P/O ratio approaches zero, the lack of ATP results in death.

7. Valinomycin acts as an uncoupler. It combines with K^+ to form a complex that passes through the inner mitochondrial membrane,

dissipating the membrane potential. ATP synthesis decreases, which causes the rate of electron transfer to increase. This results in an increase in the H^+ gradient, O_2 consumption, and amount of heat released.

8. The steady-state concentration of P_i in the cell is much higher than that of ADP. The P_i released by ATP hydrolysis changes total [P_i] very little.

9. More-efficient electron transfer between complexes

10. (a) External medium: 4.0×10^{-8} M; matrix: 2.0×10^{-8} M **(b)** [H^+] gradient contributes 1.7 kJ/mol toward ATP synthesis. **(c)** 21 **(d)** No **(e)** From the overall transmembrane potential

11. (a) 0.91 μmol/s·g **(b)** 5.5 s; to provide a constant level of ATP, regulation of ATP production must be tight and rapid.

12. 53 μmol/s·g. With a steady state [ATP] of 7.0 μmol/g, this is equivalent to 10 turnovers of the ATP pool per second; the reservoir would last about 0.13 s.

13. The citric acid cycle is stalled for lack of an acceptor of electrons from NADH. Pyruvate produced by glycolysis cannot enter the cycle as acetyl-CoA; accumulated pyruvate is transaminated to alanine and exported to the liver.

14. Cytosolic malate dehydrogenase plays a key role in the transport of reducing equivalents across the inner mitochondrial membrane via the malate-aspartate shuttle.

15. The inner mitochondrial membrane is impermeable to NADH, but the reducing equivalents of NADH are transferred (shuttled) through the membrane indirectly: they are transferred to oxaloacetate in the cytosol, the resulting malate is transported into the matrix, and mitochondrial NAD^+ is reduced to NADH.

16. Pyruvate dehydrogenase is located in mitochondria; glyceraldehyde 3-phosphate dehydrogenase, in the cytosol. The NAD pools are separated by the inner mitochondrial membrane.

17. (a) Glycolysis becomes anaerobic. **(b)** Oxygen consumption ceases. **(c)** Lactate formation increases. **(d)** ATP synthesis decreases to 2 ATP/glucose.

18. The response to (a), increased [ADP], is faster because the response to (b), reduced pO_2, requires protein synthesis.

19. (a) NADH is reoxidized via electron transfer instead of lactic acid fermentation. **(b)** Oxidative phosphorylation is more efficient. **(c)** The high mass-action ratio of the ATP system inhibits phosphofructokinase-1.

20. Fermentation to ethanol could be accomplished in the presence of O_2, which is an advantage because strict anaerobic conditions are difficult to maintain. The Pasteur effect is not observed, because the citric acid cycle and respiratory chain are inactive.

21. Reactive oxygen species react with macromolecules, including DNA. If a mitochondrial defect leads to increased production of ROS, proto-oncogenes in the nuclear chromosomes can be damaged, producing oncogenes and leading to unregulated cell division and cancer (see Section 12.11).

22. Different extents of heteroplasmy for the defective gene produce different degrees of defective mitochondrial function.

23. Complete lack of glucokinase (two defective alleles) makes it impossible to carry out glycolysis at a sufficient rate to raise [ATP] to the threshold required for insulin secretion.

24. Defects in Complex II result in increased production of ROS, damage to DNA, and mutations that lead to unregulated cell division (cancer; see Section 12.11). It is not clear why the cancer tends to occur in the midgut.

25. (a) DNP is an uncoupler; it shuttles protons across membranes, preventing development of a large proton gradient, which would otherwise prevent further electron flow. O_2 can still be reduced. **(b)** ATP synthase, not the respiratory chain, is affected, because electron transfer occurs in the presence of the uncoupler (DNP) and the inhibitor (DCCD). **(c)** In broken membrane fragments there is no proton gradient to drive ATP synthesis; the enzyme acts in the reverse direction, splitting ATP and pumping protons. **(d)** Just altered. If the DCCD-sensitive protein were altogether missing, we would expect the mutant to grow poorly, if at all, under aerobic conditions and to have little or no ATPase activity. **(e)** Look for DCCD-insensitive ATPase activity in hybrid reconstituted systems; whichever component was supplied by the RF-7 mutant is the source of the protein. (Conversely, look for DCCD-sensitive ATPase activity in hybrid reconstituted systems; whichever component was supplied by the wild-type is the source of the protein.) **(f)** He looked for fractions (gel slices) that were labeled (reacted with DCCD) in the wild-type samples but not in the mutant. **(g)** Getting the same result when using "stripped membranes" allowed him to confirm that the DCCD-sensitive protein was not in the soluble fraction with the ATPase activity. It was tightly associated with the membrane, confirming the results of the hybrid reconstitution experiments. **(h)** A 9 kDa protein would be expected to have about 80 residues. Its solubility in chloroform/methanol indicates a very high content of hydrophobic residues, suggesting it is mostly embedded in the membrane. About 20 residues are required for one transmembrane α helix, so an 80-residue protein would probably have fewer than four transmembrane helices. **(i)** With 40 residues separating Ala^{21} and Asp^{61}, these two residues could lie close to each other on separate transmembrane helices. The substitution of a polar Ser side chain at residue 21 could interfere with the attack of DCCD on Asp^{61}. **(j)** The DCCD-sensitive protein was the c subunit that makes up the rotary motor in the F_o complex.

Chapter 20

1. For the maximum photosynthetic rate, PSI (which absorbs light of 700 nm) and PSII (which absorbs light of 680 nm) must be operating simultaneously.

2. From water consumed in the overall reaction

3. H_2S is the hydrogen donor in photosynthesis. No O_2 is evolved, because the single photosystem lacks the water-splitting complex.

4. 0.44

5. (a) Stops **(b)** Slows; some electron flow continues by the cyclic pathway.

6. During illumination, a proton gradient is established. When ADP and P_i are added, ATP synthesis is driven by the gradient, which becomes exhausted in the absence of light.

7. DCMU blocks electron transfer between PSII and the first site of ATP production.

8. Venturicidin blocks proton movement through the CF_oCF_1 complex; electron flow (O_2 evolution) continues only until the free-energy cost of pumping protons against the rising proton gradient equals the free energy available in a photon. DNP, by dissipating the proton gradient, restores electron flow and O_2 evolution.

9. (a) 56 kJ/mol **(b)** 0.29 V

10. From the difference in reduction potentials, you can calculate $\Delta G'^\circ = 15$ kJ/mol for the redox reaction. Fig. 20-7 shows that the energy of photons in any region of the visible spectrum is more than sufficient to drive this endergonic reaction.

11. 1.35×10^{-77}; the reaction is highly unfavorable! In chloroplasts, the input of light energy overcomes this barrier.

12. -920 kJ/mol

13. No. The electrons from H_2O flow to the artificial electron acceptor Fe^{3+}, not to $NADP^+$.

14. About once every 0.1 s; 1 in 10^8 excited

15. Light of 700 nm excites PSI but not PSII; electrons flow from P700 to $NADP^+$, but no electrons flow from P680 to replace them. When light of 680 nm excites PSII, electrons tend to flow to PSI, but the electron carriers between the two photosystems quickly become completely reduced.

16. No. The excited electron from P700 returns to refill the electron "hole" created by illumination. PSII is not needed to supply electrons, and no O_2 is evolved from H_2O. No NADPH is formed, because the excited electron returns to P700.

17. In organelles, concentrations of specific enzymes and metabolites are elevated; separate pools of cofactors and intermediates are maintained; regulatory mechanisms affect only one set of enzymes and pools.

18. ATP and NADPH are generated in the light and are essential for CO_2 fixation; conversion stops as the supply of ATP and NADPH becomes exhausted. Some enzymes are switched off in the dark.

19. X is 3-phosphoglycerate; Y is ribulose 1,5-bisphosphate.

20. Ribulose 5-phosphate kinase, fructose 1,6-bisphosphatase, sedoheptulose 1,7-bisphosphatase, and glyceraldehyde 3-phosphate dehydrogenase; all are activated by reduction of a critical disulfide bond to a pair of sulfhydryls; iodoacetate reacts irreversibly with free sulfhydryls.

21. The reductive pentose phosphate pathway regenerates ribulose 1,5-bisphosphate from triose phosphates produced during photosynthesis. The oxidative pentose phosphate pathway provides NADPH for reductive biosynthesis and pentose phosphates for nucleotide synthesis.

22. Both types of "respiration" occur in plants, consume O_2, and produce CO_2. (Mitochondrial respiration also occurs in animals.) Mitochondrial respiration takes places continuously, though primarily at night or on cloudy days; electrons derived from various fuels are passed through a chain of carriers in the inner mitochondrial membrane to O_2. Photorespiration takes place in chloroplasts, peroxisomes, and mitochondria, during the daytime, when photosynthetic carbon fixation is occurring. Electron flow in photorespiration is shown in Fig. 20-48; that for mitochondrial respiration, in Fig. 19-19.

23. (a) Without production of NADPH by the pentose phosphate pathway, cells would be unable to synthesize lipids and other reduced products. **(b)** Without generation of ribulose 1,5-bisphosphate, the Calvin cycle would effectively be blocked.

24. In maize, CO_2 is fixed by the C_4 (Hatch-Slack) pathway, in which PEP is carboxylated rapidly to oxaloacetate (some of which undergoes transamination to aspartate) and reduced to malate. Only after subsequent decarboxylation does the CO_2 enter the Calvin cycle.

25. Measure the rate of fixation of $^{14}CO_2$ in the light (daytime) and the dark. Greater fixation in the dark identifies the CAM plant. You could also determine the titratable acidity; acids stored in the vacuole during the night can be quantified in this way.

26. Isocitrate dehydrogenase reaction

27. Species 1, C_4; species 2, C_3

28. (a) Peripheral chloroplasts **(b), (c)** Central region (red)

29. Storage consumes 1 mol of ATP per mole of glucose 6-phosphate, or 3.3% of the total ATP available from glucose 6-phosphate metabolism (i.e., the efficiency of storage is 96.7%).

30. $[PP_i]$ is high in the cytosol because the cytosol lacks inorganic pyrophosphatase.

31. (a) Low $[P_i]$ in the cytosol and high [triose phosphate] in the chloroplast **(b)** High [triose phosphate] in the cytosol

32. 3-Phosphoglycerate is the primary product of photosynthesis; $[P_i]$ rises when light-driven synthesis of ATP from ADP and P_i slows.

33. (a) Sucrose + (glucose)$_n$ → (glucose)$_{n+1}$ + fructose **(b)** Fructose generated in the synthesis of dextran is readily imported and metabolized by the bacteria.

34. The first enzyme in each path is under reciprocal allosteric regulation. Inhibition of one path shunts isocitrate into the other path.

35. (a) (1) The presence of Mg^{2+} supports the hypothesis that chlorophyll is directly involved in catalysis of the phosphorylation reaction, ADP + P_i → ATP. (2) Many enzymes (or other proteins)

that contain Mg^{2+} are not phosphorylating enzymes, so the presence of Mg^{2+} in chlorophyll does not prove its role in phosphorylation reactions. (3) The presence of Mg^{2+} is essential to chlorophyll's photochemical properties: light absorption and electron transfer. **(b)** (1) Enzymes catalyze reversible reactions, so an isolated enzyme that can, under certain lab conditions, catalyze removal of a phosphoryl group could probably, under different conditions (such as in cells), catalyze addition of a phosphoryl group. So, chlorophyll could be involved in the phosphorylation of ADP. (2) There are two possible explanations: the chlorophyll protein is a phosphatase only and does not catalyze ADP phosphorylation under cellular conditions, or the crude preparation contains a contaminating phosphatase activity that is unconnected to the photosynthetic reactions. (3) The preparation was probably contaminated with a nonphotosynthetic phosphatase activity. **(c)** (1) This light inhibition would be expected if the chlorophyll protein catalyzed the reaction ADP + P_i + light $\rightarrow$ ATP. Without light, the reverse reaction, a dephosphorylation, would be favored. In the presence of light, energy is provided and the equilibrium would shift to the right, reducing the phosphatase activity. (2) This inhibition must be an artifact of the isolation or assay methods. (3) The crude preparation methods in use at the time were unlikely to preserve intact chloroplast membranes, so the inhibition must be an artifact. **(d)** In the presence of light, (1) ATP is synthesized and other phosphorylated intermediates are consumed; (2) glucose is produced and is metabolized by cellular respiration to produce ATP, with changes in the levels of phosphorylated intermediates; (3) ATP is produced and other phosphorylated intermediates are consumed. **(e)** Light energy is used to produce ATP (as in the Emerson model) *and* is used to produce reducing power (as in the Rabinowitch model). **(f)** The approximate stoichiometry for photophosphorylation is that 8 photons yield 2 NADPH and about 3 ATP. Two NADPH and 3 ATP are required to reduce 1 CO_2. Thus, at a minimum, 8 photons are required per CO_2 molecule reduced, in good agreement with Rabinowitch's value. **(g)** Because the energy of light is used to produce *both* ATP and NADPH, each photon absorbed contributes more than just 1 ATP for photosynthesis. The process of energy extraction from light is more efficient than Rabinowitch supposed, and plenty of energy is available for this process—even with red light.

Chapter 21

1. (a) The 16 carbons of palmitate are derived from 8 acetyl groups of 8 acetyl-CoA molecules. The ^{14}C-labeled acetyl-CoA gives rise to malonyl-CoA labeled at C-1 and C-2. **(b)** The metabolic pool of malonyl-CoA, the source of all palmitate carbons except C-16 and C-15, does not become labeled with small amounts of ^{14}C-labeled acetyl-CoA. Hence, only [15,16-^{14}C] palmitate is formed.

2. Both glucose and fructose are degraded to pyruvate in glycolysis. Pyruvate is converted to acetyl-CoA by the pyruvate dehydrogenase complex. Some of this acetyl-CoA enters the citric acid cycle, which produces reducing equivalents, NADH and NADPH. Mitochondrial electron transfer to O_2 yields ATP.

3. 8 Acetyl-CoA + 15ATP + 14NADPH + 9H_2O $\rightarrow$
$$\text{palmitate} + 8\text{CoA} + 15\text{ADP} + 15P_i + 14\text{NADP}^+ + 2H^+$$

4. (a) 3 deuteriums per palmitate; all located on C-16; all other two-carbon units are derived from unlabeled malonyl-CoA. **(b)** 7 deuteriums per palmitate; located on all *even*-numbered carbons except C-16

5. By using the three-carbon unit malonyl-CoA, the activated form of acetyl-CoA (recall that malonyl-CoA synthesis requires ATP), metabolism is driven in the direction of fatty acid synthesis by the exergonic release of CO_2.

6. The rate-limiting step in fatty acid synthesis is carboxylation of acetyl-CoA, catalyzed by acetyl-CoA carboxylase. High [citrate] and [isocitrate] indicate that conditions are favorable for fatty acid synthesis: an active citric acid cycle is providing a plentiful supply of ATP, reduced pyridine nucleotides, and acetyl-CoA. Citrate stimulates (increases the V_{max} of) acetyl-CoA carboxylase (a). Because citrate binds more tightly to the filamentous (active) form of the enzyme,

high [citrate] drives the protomer $\rightleftharpoons$ filament equilibrium in the direction of the active form (b). In contrast, palmitoyl-CoA (the end product of fatty acid synthesis) drives the equilibrium in the direction of the inactive (protomer) form. Hence, when the end product of fatty acid synthesis accumulates, the biosynthetic path slows.

7. (a) Acetyl-CoA$_{(mit)}$ + ATP + CoA$_{(cyt)}$ $\rightarrow$ acetyl-CoA$_{(cyt)}$ + ADP + P_i + CoA$_{(mit)}$ **(b)** 1 ATP per acetyl group **(c)** Yes

8. The double bond in palmitoleate is introduced by an oxidation catalyzed by fatty acyl–CoA desaturase, a mixed-function oxidase that requires O_2 as a cosubstrate.

9. 3 Palmitate + glycerol + 7ATP + 4H_2O $\rightarrow$
$$\text{tripalmitin} + 7\text{ADP} + 7P_i + 7H^+$$

10. In adult rats, stored triacylglycerols are maintained at a steady-state level through a balance of rates of degradation and biosynthesis. Hence, triacylglycerols of adipose (fat) tissue are constantly turned over, which explains the incorporation of ^{14}C label from dietary glucose.

11. Net reaction:

Dihydroxyacetone phosphate + NADH + palmitate + oleate + 3ATP + CTP + choline + 4H_2O $\rightarrow$
$$\text{phosphatidylcholine} + \text{NAD}^+ + 2\text{AMP} + \text{ADP} + H^+ + \text{CMP} + 5P_i$$

7ATP per molecule of phosphatidylcholine

12. Methionine deficiency reduces the level of adoMet, which is required for de novo synthesis of phosphatidylcholine. The salvage pathway does not employ adoMet, but uses available choline. Thus phosphatidylcholine can be synthesized even when the diet is deficient in methionine, as long as choline is available.

13. ^{14}C label appears in three places in the activated isoprene:

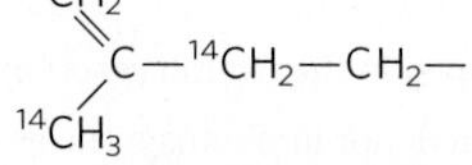

14. (a) ATP **(b)** UDP-glucose **(c)** CDP-ethanolamine **(d)** UDP-galactose **(e)** Fatty acyl–CoA **(f)** S-Adenosylmethionine **(g)** Malonyl-CoA **(h)** Δ^3-Isopentenyl pyrophosphate

15. Linoleate is required in the synthesis of prostaglandins. Animals cannot transform oleate to linoleate, so linoleate is an essential fatty acid. Plants can convert oleate to linoleate, and they provide animals with the required linoleate (see Fig. 21-12).

16. The rate-determining step in the biosynthesis of cholesterol is the synthesis of mevalonate, catalyzed by HMG-CoA reductase. This enzyme is allosterically regulated by mevalonate and cholesterol derivatives. High intracellular [cholesterol] also reduces transcription of the gene encoding HMG-CoA reductase.

17. When cholesterol levels decline because of treatment with a statin, cells attempt to compensate by increasing expression of the gene encoding HMG-CoA reductase; however, statins are good competitive inhibitors of HMG-CoA reductase activity and reduce overall production of cholesterol.

18. Note: In the absence of detailed knowledge of the literature on this enzyme, students might propose several plausible alternatives. *Thiolase reaction:* Begins with nucleophilic attack of an active-site Cys residue on the first acetyl-CoA substrate, displacing —S-CoA and forming a covalent thioester link between Cys and the acetyl group. A base on the enzyme then extracts a proton from the methyl group of the second acetyl-CoA, leaving a carbanion that attacks the carbonyl carbon of the thioester formed in the first step. The sulfhydryl of the Cys residue is displaced, creating the product acetoacetyl-CoA. *HMG-CoA synthase reaction:* Begins in the same way, with a covalent thioester link formed between the enzyme's Cys residue and the acetyl group of acetyl-CoA, with displacement of —S-CoA. The —S-CoA dissociates as CoA-SH, and acetoacetyl-CoA binds to the enzyme. A proton is abstracted from the methyl group of the enzyme-linked acetyl, forming a carbanion that attacks the ketone carbonyl of the acetoacetyl-CoA substrate. The carbonyl is converted to a hydroxyl ion in this reaction, which is protonated to create —OH. The thioester link with the enzyme is then cleaved hydrolytically to generate the HMG-CoA product. *HMG-CoA reductase reaction:* Two successive hydride ions derived from NADPH first displace the —S-CoA, and then reduce the aldehyde to a hydroxyl group.

19. Statins inhibit HMG-CoA reductase, an enzyme in the pathway to the synthesis of activated isoprenes, which are precursors of cholesterol and a wide range of isoprenoids, including coenzyme Q (ubiquinone). Hence, statins might reduce the levels of coenzyme Q available for mitochondrial respiration. Ubiquinone is obtained in the diet as well as by direct biosynthesis, but it is not yet clear how much is required and how well dietary sources can substitute for reduced synthesis. Reductions in the levels of particular isoprenoids may account for some side effects of statins.

20. (a)

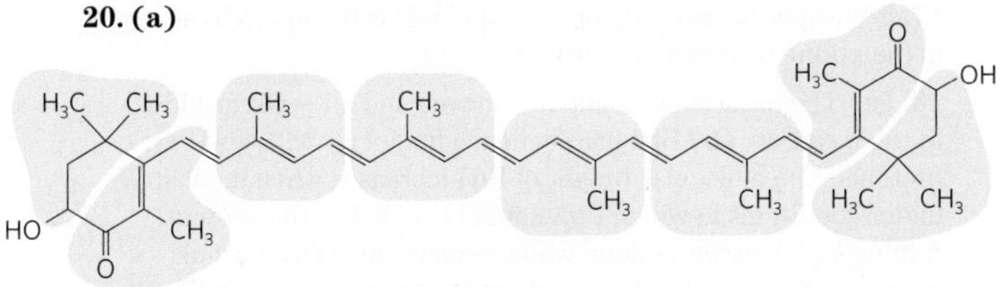

Astaxanthin

(b) Head-to-head. There are two ways to look at this. First, the "tail" of geranylgeranyl pyrophosphate has a branched dimethyl structure, as do both ends of phytoene. Second, no free —OH is formed by the release of PPᵢ, indicating that the two —O—Ⓟ—Ⓟ "heads" are linked to form phytoene. **(c)** Four rounds of dehydrogenation convert four single bonds to double bonds. **(d)** No. A count of single and double bonds in the reaction below shows that one double bond is replaced by two single bonds—so, there is no net oxidation or reduction:

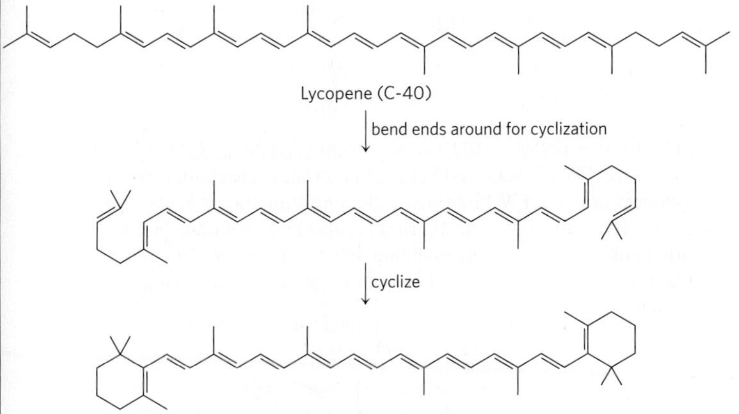

Lycopene (C-40)

↓ bend ends around for cyclization

↓ cyclize

β-Carotene (C-40)

(e) Steps ❶ through ❸. The enzyme can convert IPP and DMAP to geranylgeranyl pyrophosphate, but catalyzes no further reactions in the pathway, as confirmed by results with the other substrates. **(f)** Strains 1 through 4 lack *crtE* and have much lower astaxanthin production than strains 5 through 8, all of which overexpress *crtE*. Thus, overexpression of *crtE* leads to a substantial increase in astaxanthin production. Wild-type *E. coli* has some step ❸ activity, but this conversion of farnesyl pyrophosphate to geranylgeranyl pyrophosphate is strongly rate-limiting. **(g)** IPP isomerase. Comparing strains 5 and 6 shows that adding *ispA*, which catalyzes steps ❶ and ❷, has little effect on astaxanthin production, so these steps are not rate-limiting. However, comparing strains 5 and 7 shows that adding *idi* substantially increases astaxanthin production, so IPP isomerase must be the rate-limiting step when *crtE* is overexpressed. **(h)** A low (+) level, comparable to that of strains 5, 6, and 9. Without overexpression of *idi*, production of astaxanthin is limited by low IPP isomerase activity and the resulting limited supply of IPP.

Chapter 22

1. In their symbiotic relationship with the plant, bacteria supply ammonium ion by reducing atmospheric nitrogen, a process that requires large quantities of ATP.

2. Transfer of nitrogen from NH_3 to carbon skeletons can be catalyzed by (1) glutamine synthetase and (2) glutamate dehydrogenase. The latter enzyme produces glutamate, the amino group donor in all transamination reactions, necessary to the formation of amino acids for protein synthesis.

3. A link is formed between enzyme-bound PLP and the phosphohomoserine substrate, with rearrangement to generate the ketimine at the α carbon of the substrate. This activates the β carbon for proton abstraction, leading to displacement of the phosphate and formation of a double bond between the β and γ carbons. A rearrangement (beginning with proton abstraction at the pyridoxal carbon adjacent to the substrate amino nitrogen) moves the α–β double bond and converts the ketimine to the aldimine form of PLP. Attack of water at the β carbon is then facilitated by the linked pyridoxal, followed by hydrolysis of the imine link between PLP and the product, to generate threonine.

4. In the mammalian route, toxic ammonium ions are transformed to glutamine, reducing toxic effects on the brain.

5. Glucose + $2CO_2$ + $2NH_3 \rightarrow 2$ aspartate + $2H^+$ + $2H_2O$

6. The amino-terminal glutaminase domain is similar in *all* glutamine amidotransferases. A drug that targeted this active site would probably inhibit many enzymes and produce many more side effects than a more specific inhibitor that targets the unique carboxyl-terminal synthetase active site.

7. If phenylalanine hydroxylase is defective, the biosynthetic route to tyrosine is blocked and tyrosine must be obtained from the diet.

8. In adoMet synthesis, triphosphate is released from ATP. Hydrolysis of the triphosphate renders the reaction thermodynamically more favorable.

9. If the inhibition of glutamine synthase were not concerted, saturating concentrations of histidine would shut down the enzyme and cut off production of glutamine, which the bacterium needs to synthesize other products.

10. Folic acid is a precursor of tetrahydrofolate (see Fig. 18-16), required in the biosynthesis of glycine (see Fig. 22-14), a precursor of porphyrins. A folic acid deficiency therefore impairs hemoglobin synthesis.

11. *Glycine auxotrophs:* adenine and guanine; *glutamine auxotrophs:* adenine, guanine, and cytosine; *aspartate auxotrophs:* adenine, guanine, cytosine, and uridine

12. (a) See Fig. 18-6, step ❷, for the reaction mechanism of amino acid racemization. The F atom of fluoroalanine is an excellent leaving group. Fluoroalanine causes irreversible (covalent) inhibition of alanine racemase. One plausible mechanism (where Nuc denotes any nucleophilic amino acid side chain in the enzyme active site) is:

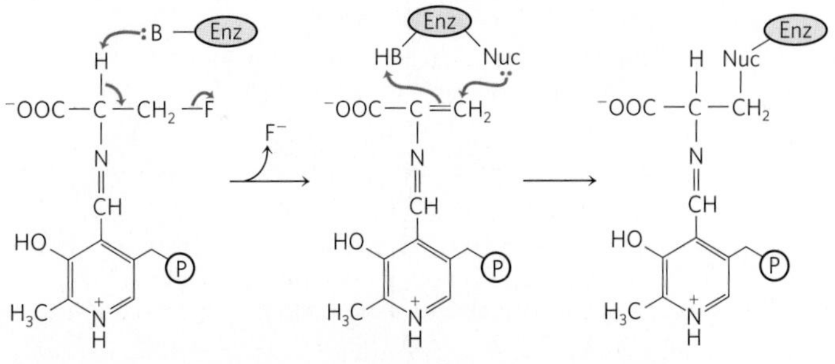

(b) Azaserine (see Fig. 22-51) is an analog of glutamine. The diazoacetyl group is highly reactive and forms covalent bonds with nucleophiles at the active site of a glutamine amidotransferase.

13. (a) As shown in Fig. 18-16, *p*-aminobenzoate is a component of N^5,N^{10}-methylenetetrahydrofolate, the cofactor involved in the transfer of one-carbon units. **(b)** In the presence of sulfanilamide, a structural analog of *p*-aminobenzoate, bacteria are unable to synthesize tetrahydrofolate, a cofactor necessary for converting AICAR to FAICAR; thus AICAR accumulates. **(c)** The competitive inhibition by sulfanilamide of the enzyme involved in tetrahydrofolate biosynthesis is overcome by the addition of excess substrate (p-aminobenzoate).

14. The ^{14}C-labeled orotate arises from the following pathway (the first three steps are part of the citric acid cycle):

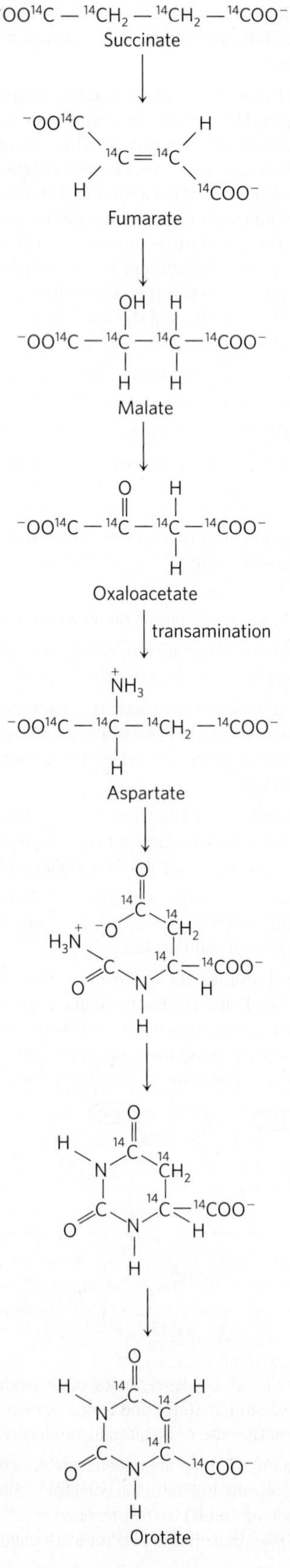

$$^-OO^{14}C - {}^{14}CH_2 - {}^{14}CH_2 - {}^{14}COO^-$$
Succinate

Fumarate

Malate

Oxaloacetate

transamination

Aspartate

Orotate

15. Organisms do not store nucleotides to be used as fuel, and they do not completely degrade them, but rather hydrolyze them to release the bases, which can be recovered in salvage pathways. The low C:N ratio of nucleotides makes them poor sources of energy.

16. Treatment with allopurinol has two consequences. (1) It inhibits conversion of hypoxanthine to uric acid, causing accumulation of hypoxanthine, which is more soluble and more readily excreted; this alleviates the clinical problems associated with AMP degradation. (2) It inhibits conversion of guanine to uric acid, causing accumulation of xanthine, which is less soluble than uric acid; this is the source of xanthine stones. Because the amount of GMP degradation is low relative to AMP degradation, the kidney damage caused by xanthine stones is less than the damage caused by untreated gout.

17. 5-Phosphoribosyl-1-pyrophosphate; this is the first NH_3 acceptor in the purine biosynthetic pathway.

18. (a) The α-carboxyl group is removed and an —OH is added to the γ carbon. **(b)** BtrI has sequence homology with acyl carrier proteins. The molecular weight of BtrI increases when incubated under conditions in which CoA could be added to the protein. Adding CoA to a Ser residue would replace an —OH, formula weight (FW) 17, with a 4′-phosphopantetheine group (see Fig. 21-5; formula $C_{11}H_{21}N_2O_7PS$), FW 356. Thus, $11,182 - 17 + 356 = 12,151$, which is very close to the observed M_r of 12,153. **(c)** The thioester could form with the α-carboxyl group. **(d)** In the most common reaction for removing the α-carboxyl group of an amino acid (see Fig. 18-6, reaction **C**), the carboxyl group must be free. Furthermore, it is difficult to imagine a decarboxylation reaction starting with a carboxyl group in its thioester form. **(e)** $12,240 - 12,281 = 41$, close to the M_r of CO_2 (44). Given that BtrK is probably a decarboxylase, the most likely structure is the decarboxylated form:

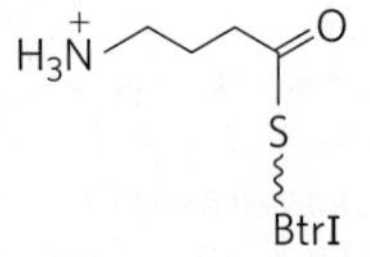

BtrI

(f) $12,370 - 12,240 = 130$. Glutamic acid ($C_5H_9NO_4$; M_r 147), minus the —OH (FW 17) removed in the glutamylation reaction, leaves a glutamyl group of FW 130; thus, γ-glutamylating the molecule shown above would add 130 to its M_r. BtrJ is capable of γ-glutamylating other substrates, so it may γ-glutamylate this molecule. The most likely site for this is the free amino group, giving the following structure:

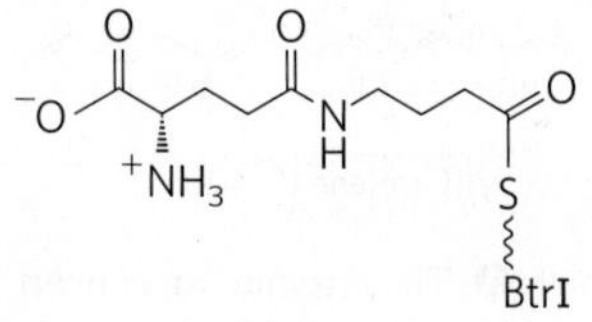

BtrI

(g)

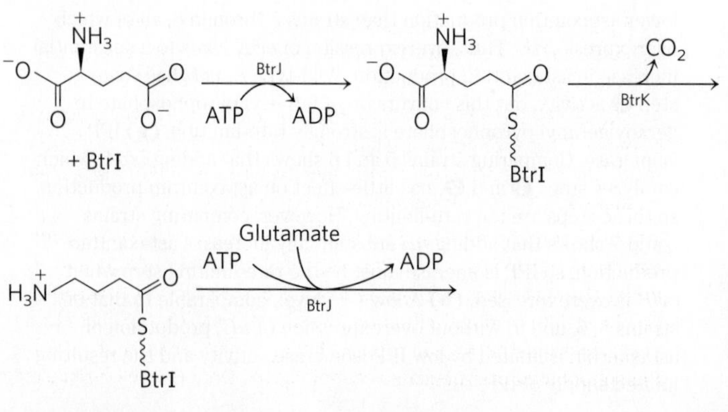

O_2 H_2O

BtrO

$FMNH_2$ FMN

BtrV

NAD^+ NADH

$^+NH_3$ BtrI

Antibiotic

BtrG + BtrH

Glutamate + BtrI

OH

$^+NH_3$ BtrI

OH

$H_3\overset{+}{N}$ Antibiotic

Chapter 23

1. They are recognized by two different receptors, typically found in different cell types, and are coupled to different downstream effects.

2. Steady-state levels of ATP are maintained by phosphoryl group transfer to ADP from phosphocreatine. 1-Fluoro-2,4-dinitrobenzene inhibits creatine kinase.

3. Ammonia is highly toxic to nervous tissue, especially the brain. In healthy individuals, excess NH_3 is removed by transformation of glutamate to glutamine, which travels to the liver and is subsequently transformed to urea. The additional glutamine arises from conversion of glucose to α-ketoglutarate, transamination of α-ketoglutarate to glutamate, and conversion of glutamate to glutamine.

4. Glucogenic amino acids are used to make glucose for the brain; others are deaminated then oxidized in mitochondria via the citric acid cycle.

5. From glucose, by the following route: Glucose $\rightarrow$ dihydroxyacetone phosphate (in glycolysis); dihydroxyacetone phosphate + NADH + H^+ $\rightarrow$ glycerol 3-phosphate + NAD^+ (glycerol 3-phosphate dehydrogenase reaction)

6. **(a)** Increased muscular activity increases the demand for ATP, which is met by increased O_2 consumption. **(b)** After the sprint, lactate produced by anaerobic glycolysis is converted to glucose and glycogen, which requires ATP and therefore O_2.

7. Glucose is the primary fuel for the brain. TPP-dependent oxidative decarboxylation of pyruvate to acetyl-CoA is essential to complete glucose metabolism.

8. 190 m

9. **(a)** Inactivation provides a rapid means to change the concentration of active hormone. **(b)** Insulin level is maintained by equal rates of synthesis and degradation. **(c)** Changes in rate of release from storage, rate of transport, and rate of conversion from prohormone to active hormone

10. Water-soluble hormones bind to receptors on the outer surface of the cell, triggering formation of a second messenger (e.g., cAMP) inside the cell. Lipid-soluble hormones can pass through the plasma membrane to act on target molecules or receptors directly.

11. **(a)** Heart and skeletal muscle lack glucose 6-phosphatase. Any glucose 6-phosphate produced enters the glycolytic pathway and, under O_2-deficient conditions, is converted to lactate via pyruvate. **(b)** In a "fight-or-flight" situation, the concentration of glycolytic precursors must be high in preparation for muscular activity. Phosphorylated intermediates cannot escape from the cell, because

the membrane is not permeable to charged species, and glucose 6-phosphate is not exported on the glucose transporter. The liver, by contrast, must release the glucose necessary to maintain blood glucose level; glucose is formed from glucose 6-phosphate and enters the bloodstream.

12. **(a)** Excessive uptake and use of blood glucose by the liver, leading to hypoglycemia; shutdown of amino acid and fatty acid catabolism **(b)** Little circulating fuel is available for ATP requirements. Brain damage results because glucose is the main source of fuel for the brain.

13. Thyroxine acts as an uncoupler of oxidative phosphorylation. Uncouplers lower the P/O ratio, and the tissue must increase respiration to meet normal ATP demands. Thermogenesis could also be due to the increased rate of ATP use by the thyroid-stimulated tissue, as increased ATP demands are met by increased oxidative phosphorylation and thus increased respiration.

14. Because prohormones are inactive, they can be stored in quantity in secretory granules. Rapid activation is achieved by enzymatic cleavage in response to an appropriate signal.

15. In animals, glucose can be synthesized from many precursors (see Fig. 14-16). In humans, the principal precursors are glycerol from TAGs and glucogenic amino acids from protein.

16. The *ob/ob* mouse, which is initially obese, will lose weight. The *OB/OB* mouse will retain its normal body weight.

17. BMI = 39.3. For BMI of 25, weight must be 75 kg; he must lose 43 kg, or 95 lb.

18. Reduced insulin secretion. Valinomycin has the same effect as opening the K^+ channel, allowing K^+ exit and consequent hyperpolarization.

19. The liver does not receive the insulin message and therefore continues to have high levels of glucose 6-phosphatase and gluconeogenesis, increasing blood glucose both during a fast and after a glucose-containing meal. The elevated blood glucose triggers insulin release from pancreatic β cells, hence the high level of insulin in the blood.

20. Some things to consider: What do the data show about the frequency of heart attack attributable to the drug among people taking the drug? How does this frequency compare with the data on individuals spared the long-term consequences of type 2 diabetes? Are other, equally effective treatment options with fewer adverse effects available?

21. Without intestinal glucosidase activity, absorption of glucose from dietary glycogen and starch is reduced, blunting the usual rise in blood glucose after a meal. The undigested oligosaccharides are fermented by intestinal bacteria, and the gases released cause intestinal discomfort.

22. **(a)** Increased; closing the ATP-gated K^+ channel would depolarize the membrane, increasing insulin release. **(b)** Type 2 diabetes results from decreased sensitivity to insulin, not a deficit of insulin production; increasing circulating insulin levels will reduce the symptoms associated with this disease. **(c)** Individuals with type 1 diabetes have deficient pancreatic β cells, so glyburide will have no beneficial effect. **(d)** Iodine, like chlorine (the atom it replaces in the labeled glyburide), is a halogen, but it is a larger atom and has slightly different chemical properties. The iodinated glyburide might not bind to SUR. If it bound to another molecule instead, the experiment would result in cloning of the gene for this other, incorrect protein. **(e)** Although a protein has been "purified," the "purified" preparation might be a mixture of several proteins that co-purify under those experimental conditions. In this case, the amino acid sequence could be that of a protein that co-purifies with SUR. Using antibody binding to show that the peptide sequences are present in SUR excludes this possibility. **(f)** Although the cloned gene does encode the 25 amino acid sequence found in SUR, it could be a gene that, coincidentally, encodes the same sequence in another protein. In this case, this other gene would most likely be expressed in different cells than the *SUR* gene. The mRNA hybridization results are consistent with the putative *SUR* cDNA actually encoding SUR. **(g)** The excess unlabeled glyburide competes with labeled glyburide for the binding site on SUR. As a result, there is significantly less binding of labeled glyburide, so little or no radioactivity is detected in the 140 kDa protein. **(h)** In the absence of excess unlabeled glyburide, labeled 140 kDa protein is found only in the presence of the putative *SUR* cDNA. Excess unlabeled glyburide competes with the labeled glyburide, and no ^{125}I-labeled 140 kDa protein is detected. This shows that the cDNA produces a glyburide-binding protein of the same molecular weight as SUR—strong evidence that the cloned gene encodes the SUR protein. **(i)** Several additional steps are possible, such as: (1) Express the putative *SUR* cDNA in CHO (Chinese hamster ovary) cells and show that the transformed cells have ATP-gated K^+ channel activity. (2) Show that HIT cells with mutations in the putative *SUR* gene lack ATP-gated K^+ channel activity. (3) Show that humans or experimental animals with mutations in the putative *SUR* gene are unable to secrete insulin.

Chapter 24

1. 6.1×10^4 nm; 290 times longer than the T2 phage head

2. The number of A residues does not equal the number of T residues, nor does the number of G equal the number of C, so the DNA is not a base-paired double helix; the M13 DNA is single-stranded.

3. $M_r = 3.8 \times 10^8$; length $= 200$ μm; $Lk_0 = 55,200$; $Lk = 51,900$

4. The exons contain 3 bp/amino acid $\times$ 192 amino acids = 576 bp. The remaining 864 bp are in introns, possibly in a leader or signal sequence, and/or in other noncoding DNA.

5. 5,000 bp. **(a)** Doesn't change; Lk cannot change without breaking and re-forming the covalent backbone of the DNA. **(b)** Becomes undefined; a circular DNA with a break in one strand has, by definition, no Lk. **(c)** Decreases; in the presence of ATP, gyrase underwinds DNA. **(d)** Doesn't change; this assumes that neither of the DNA strands is broken in the heating process.

6. For Lk to remain unchanged, the topoisomerase must introduce the same number of positive and negative supercoils.

7. $\sigma = -0.067$; >70% probability

8. (a) Undefined; the strands of a nicked DNA could be separated and thus have no Lk. **(b)** 476 **(c)** The DNA is already relaxed, so the topoisomerase does not cause a net change; $Lk = 476$. **(d)** 460; gyrase plus ATP reduces the Lk in increments of 2. **(e)** 464; eukaryotic type I topoisomerases increase the Lk of underwound or negatively supercoiled DNA in increments of 1. **(f)** 460; nucleosome binding does not break any DNA strands and thus cannot change Lk.

9. A fundamental structural unit in chromatin repeats about every 200 bp; the DNA is accessible to the nuclease only at 200 bp intervals. The brief treatment was insufficient to cleave the DNA at every accessible point, so a ladder of DNA bands is created in which the DNA fragments are multiples of 200 bp. The thickness of the DNA bands suggests that the distance between cleavage sites varies somewhat. For instance, not all the fragments in the lowest band are exactly 200 bp long.

10. A right-handed helix has a positive Lk; a left-handed helix (such as Z-DNA) has a negative Lk. Decreasing the Lk of a closed circular B-DNA by underwinding facilitates formation of regions of Z-DNA within certain sequences. (See Chapter 8, p. 289, for a description of sequences that permit the formation of Z-DNA.)

11. (a) Both strands must be covalently closed, and the molecule must be either circular or constrained at both ends. **(b)** Formation of cruciforms, left-handed Z-DNA, plectonemic or solenoidal supercoils, and unwinding of the DNA are favored. **(c)** *E. coli* DNA topoisomerase II or DNA gyrase **(d)** It binds the DNA at a point where it crosses on itself, cleaves both strands of one of the crossing segments, passes the other segment through the break, then reseals the break. The result is a change in Lk of -2.

12. Centromere, telomeres, and an autonomous replicating sequence or replication origin

13. The bacterial nucleoid is organized into domains approximately 10,000 bp long. Cleavage by a restriction enzyme relaxes the DNA within a domain, but not outside the domain. Any gene in the cleaved domain for which expression is affected by DNA topology will be affected by the cleavage; genes outside the domain will not.

14. (a) The lower, faster-migrating band is negatively supercoiled plasmid DNA. The upper band is nicked, relaxed DNA. **(b)** DNA topoisomerase I would relax the supercoiled DNA. The lower band would disappear, and all of the DNA would converge on the upper band. **(c)** DNA ligase would produce little change in the pattern. Some minor additional bands might appear near the upper band, due to the trapping of topoisomers not quite perfectly relaxed by the ligation reaction. **(d)** The upper band would disappear, and all of the DNA would be in the lower band. The supercoiled DNA in the lower band might become even more supercoiled and migrate somewhat faster.

15. (a) When DNA ends are sealed to create a relaxed, closed circle, some DNA species are completely relaxed but others are trapped in slightly underwound or overwound states. This gives rise to a distribution of topoisomers centered on the most relaxed species. **(b)** Positively supercoiled **(c)** The DNA that is relaxed despite the addition of dye is DNA with one or both strands broken. DNA isolation procedures inevitably introduce small numbers of strand breaks in some of the closed-circular molecules. **(d)** -0.05. This is determined by simply comparing native DNA with samples of known σ. In both gels, the native DNA migrates most closely with the sample of $\sigma = -0.049$.

16. 62 million (the genome refers to the haploid genetic content of the cell; the cell is actually diploid, so the number of nucleosomes is doubled). The number is obtained by dividing 3.1 billion bp by 200 bp/nucleosome (giving 15.5 million nucleosomes), multiplying by 2 copies of H2A per nucleosome, and again multiplying by 2 to account for the diploid state of the cell. The 62 million would double upon replication.

17. (a) In nondisjunction, one daughter cell and all of its descendants get two copies of the synthetic chromosome and are white; the other daughter cell and all of its descendants get no copies of the synthetic chromosome and are red. This gives rise to a half-white, half-red colony. **(b)** In chromosome loss, one daughter cell and all of its descendants get one copy of the synthetic chromosome and are pink; the other daughter cell and all of its descendants get no copies of the synthetic chromosome and are red. This gives rise to a half-pink, half-red colony. **(c)** The minimum functional centromere must be smaller than 0.63 kbp, because all fragments of this size or larger confer relative mitotic stability. **(d)** Telomeres are required to fully replicate only linear DNA; a circular molecule can replicate without them. **(e)** The larger the chromosome, the more faithfully it is segregated. The data show neither a minimum size below which the synthetic chromosome is completely unstable nor a maximum size above which stability no longer changes.

(f)

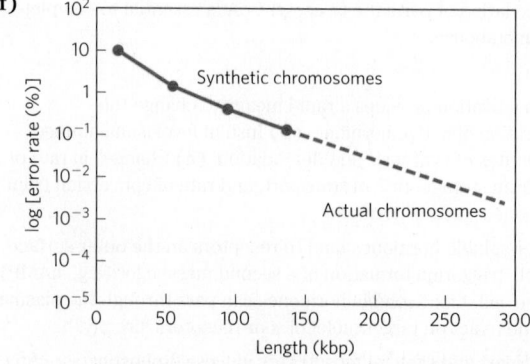

As shown in the graph, even if the synthetic chromosomes were as long as the normal yeast chromosomes, they would not be as stable. This suggests that other, as yet undiscovered, elements are required for stability.

Chapter 25

1. In random, dispersive replication, in the second generation, all the DNAs would have the same density and would appear as a single band, not the two bands observed in the Meselson-Stahl experiment.

2. In this extension of the Meselson-Stahl experiment, after three generations the molar ratio of ^{15}N–^{14}N DNA to ^{14}N–^{14}N DNA is 2/6 = 0.33.

3. (a) 4.42×10^5 turns **(b)** 40 min. In cells dividing every 20 min, a replicative cycle is initiated every 20 min, each cycle beginning before the prior one is complete. **(c)** 2,000 to 5,000 Okazaki fragments. The fragments are 1,000 to 2,000 nucleotides long and are firmly bound to the template strand by base pairing. Each fragment is quickly joined to the lagging strand, thus preserving the correct order of the fragments.

4. A, 28.7%; G, 21.3%; C, 21.3%; T, 28.7%. The DNA strand made from the template strand: A, 32.7%; G, 18.5%; C, 24.1%; T, 24.7%; the DNA strand made from the complementary template strand: A, 24.7%; G, 24.1%; C, 18.5%; T, 32.7%. This assumes that the two template strands are replicated completely.

5. (a) No. Incorporation of ^{32}P into DNA results from the synthesis of new DNA, which requires the presence of *all four* nucleotide precursors. **(b)** Yes. Although all four nucleotide precursors must be present for DNA synthesis, only one of them has to be radioactive for radioactivity to appear in the new DNA. **(c)** No. Radioactivity is incorporated only if the ^{32}P label is in the α phosphate; DNA polymerase cleaves off pyrophosphate—that is, the β- and γ-phosphate groups.

6. *Mechanism 1:* 3′-OH group of an incoming dNTP attacks the α phosphate of the triphosphate at the 5′ end of the growing DNA strand, displacing pyrophosphate. This mechanism uses normal dNTPs, and the growing end of the DNA always has a triphosphate on the 5′ end.

Mechanism 2: This uses a new type of precursor, nucleotide 3′-triphosphates. The growing end of the DNA strand has a 5′-OH group, which attacks the α phosphate of an incoming deoxynucleoside 3′-triphosphate, displacing pyrophosphate. Note that this mechanism would require the evolution of new metabolic pathways to supply the needed deoxynucleoside 3′-triphosphates.

7. The DNA polymerase contains a 3′→5′ exonuclease activity that degrades DNA to produce [^{32}P]dNMPs. The activity is not a 5′→3′ exonuclease, because the addition of unlabeled dNTPs inhibits the production of [^{32}P]dNMPs (polymerization activity would suppress a proofreading exonuclease but not an exonuclease operating downstream of the polymerase). Addition of pyrophosphate would generate [^{32}P]dNTPs through reversal of the polymerase reaction.

8. *Leading strand:* Precursors: dATP, dGTP, dCTP, dTTP (also needs a template DNA strand and DNA primer); enzymes and other proteins: DNA gyrase, helicase, single-stranded DNA–binding protein, DNA polymerase III, topoisomerases, and pyrophosphatase. *Lagging strand:* Precursors: ATP, GTP, CTP, UTP, dATP, dGTP, dCTP, dTTP (also needs an RNA primer); enzymes and other proteins: DNA gyrase, helicase, single-stranded DNA–binding protein, primase, DNA polymerase III, DNA polymerase I, DNA ligase, topoisomerases, and pyrophosphatase. NAD$^+$ is also required as a cofactor for DNA ligase.

9. Mutants with defective DNA ligase produce a DNA duplex in which one of the strands remains in pieces (as Okazaki fragments). When this duplex is denatured, sedimentation results in one fraction containing the intact single strand (the high molecular weight band) and one fraction containing the unspliced fragments (the low molecular weight band).

10. Watson-Crick base pairing between template and leading strand; proofreading and removal of wrongly inserted nucleotides by the 3′-exonuclease activity of DNA polymerase III. Yes—perhaps. Because the factors ensuring fidelity of replication are operative in both the leading and the lagging strands, the lagging strand would probably be made with the same fidelity. However, the greater number of distinct chemical operations involved in making the lagging strand might provide a greater opportunity for errors to arise.

11. ~1,200 bp (600 in each direction)

12. A small fraction (13 of 10^9 cells) of the histidine-requiring mutants spontaneously undergo back-mutation and regain their capacity to synthesize histidine. 2-Aminoanthracene increases the rate of back-mutations about 1,800-fold and is therefore mutagenic. Since most carcinogens are mutagenic, 2-aminoanthracene is probably carcinogenic.

13. Spontaneous deamination of 5-methylcytosine (see Fig. 8-29a) produces thymine, and thus a G–T mismatched pair. These are among the most common mismatches in the DNA of eukaryotes. The specialized repair system restores the G≡C pair.

14. ~1,950 (650 in the DNA degraded between the mismatch and GATC, plus 650 in DNA synthesis to fill the resulting gap, plus 650 in degradation of the pyrophosphate products to inorganic phosphate). ATP is hydrolyzed by the MutL-MutS complex and by the UvrD helicase.

15. (a) UV irradiation produces pyrimidine dimers; in normal fibroblasts these are excised by cleavage of the damaged strand by a special excinuclease. Thus the denatured single-stranded DNA contains the many fragments created by the cleavage, and the average molecular weight is lowered. These fragments of single-stranded DNA are absent from the XPG samples, as indicated by the unchanged average molecular weight. **(b)** The absence of fragments in the single-stranded DNA from the XPG cells after irradiation suggests the special excinuclease is defective or missing.

16. Using G* to represent O^6-meG:

(a) (5′)AACG*TGCAC
       TTG T ACGTG

   (5′)AACGTGCAC
       TTGCACGTG

(b) (5′)AACGTGCAC
       TTGCACGTG

   (5′)AACGTGCAC
       TTGCACGTG

(c) (5′)AACG*TGCAC
       TTG T ACGTG

   (5′)AACATGCAC
       TTGTACGTG

2× (5′)AACGTGCAC
       TTGCACGTG

17. During homologous genetic recombination, a Holliday intermediate may be formed almost anywhere within the two paired, homologous chromosomes; the branch point of the intermediate can move extensively by branch migration. In site-specific recombination, the Holliday intermediate is formed between two specific sites, and branch migration is generally restricted by heterologous sequences on either side of the recombination sites.

18. (a) Points Y **(b)** Points X

19. Once replication has proceeded from the origin to a point where one recombination site has been replicated but the other has not, site-specific recombination not only inverts the DNA between the recombination sites but also changes the direction of one replication fork relative to the other. The forks will chase each other around the DNA circle, generating many tandem copies of the plasmid. The multimeric circle can be resolved to monomers by additional site-specific recombination events.

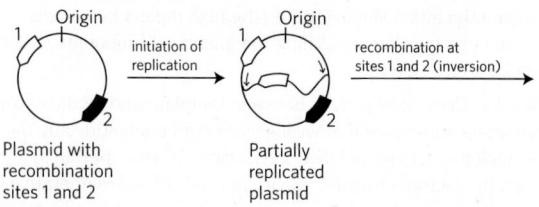

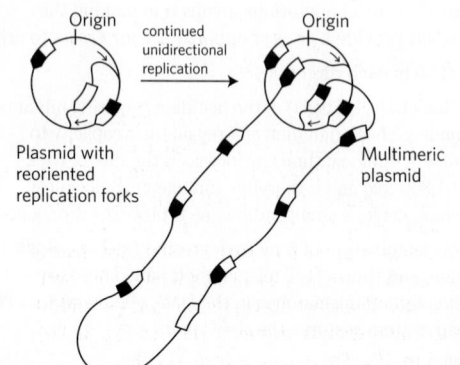

20. (a) Even in the absence of an added mutagen, background mutations occur due to radiation, cellular chemical reactions, and so forth. **(b)** If the DNA is sufficiently damaged, a substantial fraction of gene products are nonfunctional and the cell is nonviable. **(c)** Cells with reduced DNA repair capability are more sensitive to mutagens. Because they less readily repair lesions caused by R7000, Uvr⁻ bacteria have an increased mutation rate and increased chance of lethal effects. **(d)** In the Uvr⁺ strain, the excision-repair system removes DNA bases with attached [³H]R7000, decreasing the amount of ³H in these cells over time. In the Uvr⁻ strain, the DNA is not repaired and the ³H level increases as [³H]R7000 continues to react with the DNA. **(e)** All mutations listed in the table except A≡T to G≡C show significant increases over background. Each type of mutation results from a different type of interaction between R7000 and DNA. Because different types of interactions are not equally likely (due to differences in reactivity, steric constraints, etc.), the resulting mutations occur with different frequencies. **(f)** No. Only those that start with a G≡C base pair are explained by this model. Thus A≡T to C≡G and A≡T to T≡A must be due to R7000 attaching to an A or a T. **(g)** R7000–G pairs with A. First, R7000 adds to G≡C to give R7000–G≡C. (Compare this with what happens with the CH₃-G in Fig. 25-27b.) If this is not repaired, one strand is replicated as R7000-G≡A, which is repaired to T≡A. The other strand is wild-type. If the replication produces R7000-G≡T, a similar pathway leads to an A≡T base pair. **(h)** No. Compare data in the two tables, and keep in mind that different mutations occur at different frequencies.

A≡T to C≡G: moderate in both strains; but better repair in Uvr⁺

G≡C to A≡T: moderate in both; no real difference

G≡C to C≡G: higher in Uvr⁺; certainly less repair!

G≡C to T≡A: high in both; no real difference

A≡T to T≡A: high in both; no real difference

A≡T to G≡C: low in both; no real difference

Certain adducts may be more readily recognized by the repair apparatus than others, and these are repaired more rapidly and result in fewer mutations.

Chapter 26

1. (a) 60 to 100 s **(b)** 500 to 900 nucleotides

2. A single base error in DNA replication, if not corrected, would cause one of the two daughter cells, and all its progeny, to have a mutated chromosome. A single base error in RNA transcription would not affect the chromosome; it would lead to formation of some defective copies of one protein, but because mRNAs turn over rapidly, most copies of the protein would not be defective. The progeny of this cell would be normal.

3. Normal posttranscriptional processing at the 3′ end (cleavage and polyadenylation) would be inhibited or blocked.

4. Because the template-strand RNA does not encode the enzymes needed to initiate viral infection, it would probably be inert or simply degraded by cellular ribonucleases. Replication of the template-strand RNA and propagation of the virus could occur only if intact RNA replicase (RNA-dependent RNA polymerase) were introduced into the cell along with the template strand.

5. AUGACCAUGAUUACG

6. (1) Use of a template strand of nucleic acid; (2) synthesis in the 5′→3′ direction; (3) use of nucleoside triphosphate substrates, with formation of a phosphodiester bond and displacement of PP$_i$. Polynucleotide phosphorylase forms phosphodiester bonds but differs in all other listed properties.

7. Generally two: one to cleave the phosphodiester bond at one intron-exon junction, the other to link the resulting free exon end to the exon at the other end of the intron. If the nucleophile in the first step were water, this step would be a hydrolysis, and only one transesterification step would be required to complete the splicing process.

8. Many snRNAs, required for rRNA modification reactions, are encoded in introns. If splicing does not occur, snoRNAs are not produced.

9. These enzymes lack a $3'{\rightarrow}5'$ proofreading exonuclease and have a high error rate; the likelihood of a replication error that would inactivate the virus is much lower in a small genome than in a large one.

10. (a) $4^{32} = 1.8 \times 10^{19}$ **(b)** 0.006% **(c)** For the "unnatural selection" step, use a chromatographic resin with a bound molecule that is a transition-state analog of the ester hydrolysis reaction.

11. Though RNA synthesis is quickly halted by α-amanitin toxin, it takes several days for the critical mRNAs and proteins in the liver to degrade, causing liver dysfunction and death.

12. (a) After lysis of the cells and partial purification of the contents, the protein extract could be subjected to isoelectric focusing. The β subunit could be detected by an antibody-based assay. The difference in amino acid residues between the normal β subunit and the mutated form (i.e., the different charges on the amino acids) would alter the electrophoretic mobility of the mutant protein in an isoelectric focusing gel, relative to that of the protein from a nonresistant strain. **(b)** Direct DNA sequencing (by the Sanger method)

13. (a) 384 **(b)** 1,620 nucleotide pairs **(c)** Most of the nucleotides are untranslated regions at the $3'$ and $5'$ ends of the mRNA. Also, most mRNAs code for a signal sequence (Chapter 27) in their protein products, which is eventually removed to produce the mature, functional protein.

14. (a) cDNA is produced by reverse transcription of mRNA; thus, the mRNA sequence is probably CGG. Because the genomic DNA transcribed to make the mRNA has the sequence CAG, the primary transcript most likely also has CAG, which is posttranscriptionally modified to CGG. **(b)** The unedited mRNA sequence is the same as that of the DNA (except for U replacing T). Unedited mRNA has the following sequence (* indicates site of editing):

<pre>
                                                    *
(5')---GUCUCUGGUUUUCCUUGGGUGCCUUUAUGCAGCAAGGAUGCGAUAUUUCGCCAAG---(3')
</pre>

In step 1, primer 1 anneals as shown:

<pre>
                                                    *
(5')---GUCUCUGGUUUUCCUUGGGUGCCUUUAUGCAGCAAGGAUGCGAUAUUUCGCCAAG---(3')
                                  ||||||||||||||||||||||||||
              Primer 1: (3')CGTTCCTACGCTATAAAGCGGTTC(5')
</pre>

cDNA (underlined) is synthesized from right to left:

<pre>
                                                    *
(5')---GUCUCUGGUUUUCCUUGGGUGCCUUUAUGCAGCAAGGAUGCGAUAUUUCGCCAAG---(3')
       |||||||||||||||||||||||||||||||||||||||||||||||||
(3')---CAGAGACCAAAAGGAACCCACGGAAATACGTCGTTCCTACGCTATAAAGCGGTTC(5')
</pre>

Then step 2 yields just the cDNA:

<pre>
                                         *
(3')---CAGAGACCAAAAGGAACCCACGGAAATACGTCGTTCCTACGCTATAAAGCGGTTC(5')
</pre>

In step 3, primer 2 anneals to the cDNA:

<pre>
       Primer 2:   (5')CCTTGGGTGCCTTTA(3')
                       |||||||||||||||
(3')---CAGAGACCAAAAGGAACCCACGGAAATACGTCGTTCCTACGCTATAAAGCGGTTC(5')
                                         *
</pre>

DNA polymerase adds nucleotides to the $3'$ end of this primer. Moving from left to right, it inserts T, G, C, and A. However, because the A from ddATP lacks the $3'$-OH group needed to attach the next nucleotide, the chain is not elongated past this point. This A is shown in *italic*; the new DNA is underlined:

<pre>
       Primer 2:   (5')CCTTGGGTGCCTTTA<u>TGC<i>A</i></u>
                       ||||||||||||||||||||
(3')---CAGAGACCAAAAGGAACCCACGGAAATACGTCGTTCCTACGCTATAAAGCGGTTC(5')
                                         *
</pre>

This yields a 19 nucleotide fragment for the unedited transcript. In the edited transcript, the *A is changed to G; in the cDNA this corresponds to C. At the start of step 3:

<pre>
       Primer 2:   (5')CCTTGGGTGCCTTTA(3')
                       |||||||||||||||
(3')---CAGAGACCAAAAGGAACCCACGGAAATACGCCGTTCCTACGCTATAAAGCGGTTC(5')
                                         *
</pre>

In this case, DNA polymerase can elongate past the edited base and will stop at the next T in the cDNA. The dideoxy A is *italic*; the new DNA is underlined:

<pre>
       Primer 2:   (5')CCTTGGGTGCCTTTA<u>TGCGGC<i>A</i></u>
                       ||||||||||||||||||||||||||
(3')---CAGAGACCAAAAGGAACCCACGGAAATACGCCGTTCCTACGCTATAAAGCGGTTC(5')
</pre>

This gives the 22 nucleotide product. **(c)** Treatments (proteases, heat) known to disrupt protein function inhibit the editing activity, whereas treatments (nuclease) that do not affect proteins have little or no effect on editing. A key weakness of this argument is that the protein-disrupting treatments do not completely abolish editing. There could be some background editing or degradation of the mRNA even without the enzyme, or some of the enzyme might survive the treatments. **(d)** Only the α phosphate of NTPs is incorporated into polynucleotides. If the researchers had used the other types of [^{32}P] NTPs, none of the products would have been labeled. **(e)** Because only an A is being edited, only the fate of any A in the sequence is of interest. **(f)** Given that only ATP was labeled, if the entire nucleotide were removed, all radioactivity would have been removed from the mRNA, so only unmodified [^{32}P]AMP would be present on the chromatography plate. **(g)** If the base were removed and replaced, one would expect to see only [^{3}H]AMP. The presence of [^{3}H]IMP indicates that the A to I change occurs without removal of H at positions 2 and 8. The most likely mechanism is chemical modification of A to I by hydrolytic deamination (see Fig. 22-36). **(h)** CAG is changed to CIG. This codon is read as CGG.

Chapter 27

1. (a) Gly–Gln–Ser–Leu–Leu–Ile **(b)** Leu–Asp–Ala–Pro **(c)** His–Asp–Ala–Cys–Cys–Tyr **(d)** Met–Asp–Glu in eukaryotes; fMet–Asp–Glu in bacteria

2. UUAAUGUAU, UUGAUGUAU, CUUAUGUAU, CUCAUGUAU, CUAAUGUAU, CUGAUGUAU, UUAAUGUAC, UUGAUGUAC, CUUAUGUAC, CUCAUGUAC, CUAAUGUAC, CUGAUGUAC

3. No. Because nearly all the amino acids have more than one codon (e.g., Leu has six), any given polypeptide can be encoded by several different base sequences. However, some amino acids are encoded by only one codon, and those with multiple codons often share the same nucleotide at two of the three positions, so *certain parts* of the mRNA sequence encoding a protein of known amino acid sequence can be predicted with high certainty.

4. (a) (5′)CGACGGCGCGAAGUCAGGGGUGUUAAG(3′) **(b)** Arg–Arg–Arg–Glu–Val–Arg–Gly–Val–Lys **(c)** No. The complementary antiparallel strands in double-helical DNA do not have the same base sequence in the 5′→3′ direction. RNA is transcribed from only one specific strand of duplex DNA. The RNA polymerase must therefore recognize and bind to the correct strand.

5. There are two tRNAs for methionine: tRNA^{fMet}, which is the initiating tRNA, and tRNA^{Met}, which can insert a Met residue in interior positions in a polypeptide. Only fMet-tRNA^{fMet} is recognized by the initiation factor IF2 and is aligned with the initiating AUG positioned at the ribosomal P site in the initiation complex. AUG codons in the interior of the mRNA can bind and incorporate only Met-tRNA^{Met}.

6. (5′)AUG-GGU-CGU-GAG-UCA-UCG-UUA-AUU-GUA-GCU-GGA-GGG-GAG-GAA-UGA(3′) is translated as Met–Gly–Arg–Glu–Ser–Ser–Leu–Ile–Val–Ala–Gly–Gly–Glu–Glu–Trp. The peptide is 15 amino acids long, instead of 14. Ten tRNAs are needed, one for each type of amino acid.

7. Allow polynucleotide phosphorylase to act on a mixture of UDP and CDP in which UDP has, say, five times the concentration of CDP. The result would be a synthetic RNA polymer with many UUU triplets (coding for Phe), a smaller number of UUC (Phe), UCU (Ser), and CUU (Leu), a much smaller number of UCC (Ser), CUC (Leu), and CCU (Pro), and an even smaller number of CCC (Pro).

8. A minimum of 583 ATP equivalents (based on 4 per amino acid residue added, except that there are only 145 translocation steps). Correction of each error requires 2 ATP equivalents. For glycogen synthesis, 292 ATP equivalents are required. The extra energy cost for β-globin synthesis reflects the cost of the information content of the protein. At least 20 activating enzymes, 70 ribosomal proteins, 4 rRNAs, 32 or more tRNAs, an mRNA, and 10 or more auxiliary enzymes must be made by the eukaryotic cell in order to synthesize a protein from amino acids. The synthesis of an (α1→4) chain of glycogen from glucose requires only 4 or 5 enzymes (Chapter 15).

9.

Glycine codons	Anticodons
(5′)GGU	(5′)ACC, GCC, ICC
(5′)GGC	(5′)GCC, ICC
(5′)GGA	(5′)UCC, ICC
(5′)GGG	(5′)CCC, UCC

(a) The 3′ and middle position **(b)** Pairings with anticodons (5′) GCC, ICC, and UCC **(c)** Pairings with anticodons (5′)ACC and CCC

10. (a), (c), (e), and (g) only; (b), (d), and (f) cannot be the result of single-base mutations, because (b) and (f) would require substitutions of two bases, and (d) would require substitutions of all three bases.

11. (5′)AUGAUAUUGCUAUCUUGGACU

Changes:

CC	AU	U	C	C
U		C	A	A
		G	G	G

14 of 63 possible one-base changes would result in no coding change.

12. The two DNA codons for Glu are GAA and GAG, and the four DNA codons for Val are GTT, GTC, GTA, and GTG. A single base change in GAA to form GTA or in GAG to form GTG could account for the Glu → Val replacement in sickle-cell hemoglobin. Much less likely are two-base changes, from GAA to GTG, GTT, or GTC; and from GAG to GTA, GTT, or GTC.

13. Isoleucine is similar in structure to several other amino acids, particularly valine. Distinguishing between valine and isoleucine in the aminoacylation process requires a second filter—a proofreading function. Histidine has a structure unlike that of any other amino acid, providing opportunities for binding specificity adequate to ensure accurate aminoacylation of the cognate tRNA.

14. (a) The Ala-tRNA synthetase recognizes the G^3–U^{70} base pair in the amino acid arm of tRNA^{Ala}. **(b)** The mutant tRNA^{Ala} would insert Ala residues at codons encoding Pro. **(c)** A mutation that might have similar effects is an alteration in tRNA^{Pro} that allowed it to be recognized and aminoacylated by Ala-tRNA synthetase. **(d)** Most of the proteins in the cell would be inactivated, so these would be lethal mutations and hence never observed. This represents a powerful selective pressure for maintaining the genetic code.

15. The 15,000 ribosomes in an *E. coli* cell can synthesize more than 23,000 proteins in 20 minutes.

16. *IF2:* The 70S ribosome would form, but initiation factors would not be released and elongation could not start. *EF-Tu:* The second aminoacyl-tRNA would bind to the ribosomal A site, but no peptide bond would form. *EF-G:* The first peptide bond would form, but the ribosome would not move along the mRNA to vacate the A site for binding of a new EF-Tu-tRNA.

17. The amino acid most recently added to a growing polypeptide chain is the only one covalently attached to a tRNA and thus is the only link between the polypeptide and the mRNA encoding it. A proofreading activity would sever this link, halting synthesis of the polypeptide and releasing it from the mRNA.

18. Ribosomal RNA (rRNA) is part of the structure of ribosomes and will be a major component of the RNA isolated in the early steps of ribosome profiling. Some protocols include steps to selectively remove, or reduce the amount of, the rRNA segments in the sample before continuing to the steps needed to convert the RNA into DNA and then sequence it.

19. The protein would be directed into the ER, and from there the targeting would depend on additional signals. SRP binds the amino-terminal signal early in protein synthesis and directs the nascent polypeptide and ribosome to receptors in the ER. Because the protein is translocated into the lumen of the ER as it is synthesized, the NLS is never accessible to the proteins involved in nuclear targeting.

20. Trigger factor is a molecular chaperone that stabilizes an unfolded and translocation-competent conformation of ProOmpA.

21. DNA with a minimum of 5,784 bp; some of the coding sequences must be nested or overlapping.

22. (a) The helices associate through the hydrophobic effect and van der Waals interactions. **(b)** R groups 3, 6, 7, and 10 extend to the left; 1, 2, 4, 5, 8, and 9 extend to the right. **(c)** One possible sequence is

$$1 \quad 2 \quad 3 \quad 4 \quad 5 \quad 6 \quad 7 \quad 8 \quad 9 \quad 10$$

N–Phe–Ile–Glu–Val–Met–Asn–Ser–Ala–Phe–Gln–*C*

(d) One possible DNA sequence for the amino acid sequence in (c) is

Nontemplate strand

$(5')$TTTATTGAAGTAATGAATAGTGCATTCCAG$(3')$
|||||||||||||||||||||||||||||||
$(3')$AAATAACTTCATTACTTATCACGTAAGGTC$(5')$

Template strand

(e) Phe, Leu, Ile, Met, and Val. All are hydrophobic, but the set does not include *all* the hydrophobic amino acids; Trp, Pro, and Ala are missing. **(f)** Tyr, His, Gln, Asn, Lys, Asp, and Glu. All of these are hydrophilic, although Tyr is less hydrophilic than the others. The set does not include *all* the hydrophilic amino acids; Ser, Thr, and Arg are missing. **(g)** Omitting T from the mixture excludes codons starting or ending with T—thus excluding Tyr, which is not very hydrophilic, and, more importantly, excluding the two possible stop codons (TAA and TAG). No other amino acids in the NAN set are excluded by omitting T. **(h)** Misfolded proteins are often degraded in the cell. Therefore, if a synthetic gene has produced a protein that forms a band on the SDS gel, it is likely that this protein is folded properly. **(i)** Protein folding depends on more than the hydrophobic effect and van der Waals interactions. There are many reasons why a synthesized random-sequence protein might not fold into the four-helix structure. For example, hydrogen bonds between hydrophilic side chains could disrupt the structure. Also, not all sequences have an equal propensity to form an α helix.

Chapter 28

1. (a) Tryptophan synthase levels remain high despite the presence of tryptophan. **(b)** Levels again remain high. **(c)** Levels rapidly decrease, preventing wasteful synthesis of tryptophan.

2. The *E. coli* cells will produce β-galactosidase when they are subjected to high levels of a DNA-damaging agent such as UV light. Under such conditions, RecA binds to single-stranded chromosomal DNA and facilitates autocatalytic cleavage of the LexA repressor, releasing LexA from its binding site and allowing transcription of downstream genes.

3. (a) Constitutive, low-level expression of the operon; most mutations in the operator would make the repressor less likely to bind. **(b)** Either constitutive expression, as in (a), or constant repression, if the mutation destroyed the capability to bind to lactose and related compounds, thus destroying the response to inducers. **(c)** Either increased or decreased expression of the operon (under conditions in which it is induced), depending on whether the mutation made the promoter more or less similar, respectively, to the consensus *E. coli* promoter.

4. 7,000 copies

5. 8×10^{-9} M, about 10^5 times greater than the dissociation constant. With 10 copies of active repressor in the cell, the operator site is always bound by the repressor molecule.

6. (a) through **(e)**. Each condition decreases expression of *lac* operon genes.

7. (a) Less attenuation. The ribosome completing the translation of sequence 1 would no longer overlap and block sequence 2; sequence 2 would always be available to pair with sequence 3, preventing formation of the attenuator structure. **(b)** More attenuation. Sequence 2 would pair less efficiently with sequence 3; the attenuator structure would be formed more often, even when sequence 2 was not blocked by a ribosome. **(c)** No attenuation. The only regulation would be that afforded by the Trp repressor. **(d)** Attenuation loses its sensitivity to Trp tRNA. It might become sensitive to His tRNA. **(e)** Attenuation would rarely, if ever, occur. Sequences 2 and 3 always block formation of the attenuator. **(f)** Constant attenuation. Attenuator always forms, regardless of the availability of tryptophan.

8. Induction of the SOS response could not occur, making the cells more sensitive to high levels of DNA damage.

9. Each *Salmonella* cell would have flagella made up of both types of flagellar protein, and the cell would be vulnerable to antibodies generated in response to either protein.

10. A dissociable factor necessary for activity (e.g., a specificity factor similar to the σ subunit of the *E. coli* enzyme) may have been lost during purification of the polymerase.

11.

Gal4 protein

Gal4p DNA-binding domain	Gal4p activator domain

Engineered protein

Lac repressor DNA-binding domain	Gal4p activator domain

The engineered protein cannot bind to the Gal4p-binding site in the *GAL* gene (UAS$_G$) because it lacks the Gal4p DNA-binding domain. Modify the Gal4p-binding site in the DNA to give it the nucleotide sequence to which the Lac repressor normally binds (using methods described in Chapter 9).

12. Methylamine. The reaction proceeds with attack of water on the guanidinium carbon of the modified arginine.

13. Synthesis of a protein first requires the synthesis of an mRNA long enough to encode the protein and to include any necessary regulatory sequences, with one ribonucleoside triphosphate used up for every nucleotide residue included in the mRNA. Then, the mRNA must be translated—one of the most energy-intensive processes in the cell (Chapter 27). To maintain repression, the repressor protein would need to be synthesized repeatedly. With the use of RNA as repressor, the RNA can be shorter than a protein repressor–encoding mRNA, and no translation step is required.

14. The *bcd* mRNA needed for development is contributed to the egg by the mother. The fertilized egg develops normally even if its genotype is *bcd⁻/bcd⁻*, as long as the mother has one normal *bcd* gene and the *bcd⁻* allele is recessive. However, the adult *bcd⁻/bcd⁻* female will be sterile because she has no normal *bcd* mRNA to contribute to her eggs.

15. (a) For 10% expression (90% repression), 10% of the repressor has bound inducer and 90% is free and available to bind the operator. The calculation uses Eqn 5-8 (p. 161), with $Y = 0.1$ and $K_d = 10^{-4}$ M:

$$Y = \frac{[\text{IPTG}]}{[\text{IPTG}] + K_d} = \frac{[\text{IPTG}]}{[\text{IPTG}] + 10^{-4}\,\text{M}}$$

$$0.1 = \frac{[\text{IPTG}]}{[\text{IPTG}] + 10^{-4}\,\text{M}}$$

so

$$0.9[\text{IPTG}] = 10^{-5} \quad \text{or} \quad [\text{IPTG}] = 1.1 \times 10^{-5}\,\text{M}$$

For 90% expression, 90% of the repressor has bound inducer, so $Y = 0.9$. Entering the values for Y and K_d in Eqn 5-8 gives $[\text{IPTG}] = 9 \times 10^{-4}$ M. Thus, gene expression varies 10-fold over a roughly 10-fold [IPTG] range. **(b)** You would expect the protein levels to be low before induction, rise during induction, and then decay as synthesis stops and the proteins are degraded. **(c)** As shown in (a), the *lac* operon has more levels of expression than just on or off; thus it does not have characteristic A. As shown in (b), expression of the *lac* operon subsides once the inducer is removed; thus it lacks characteristic B. **(d)** *GFP-on:* rep^{ts} (designating the protein product of the *rep^{ts}* gene) and GFP are expressed at high levels; rep^{ts} represses OP$_\lambda$, so no LacI is produced. *GFP-off:* LacI is expressed at a high level; LacI represses OP_{lac}, so rep^{ts} and GFP are not produced. **(e)** IPTG treatment switches the system from GFP-off to GFP-on. IPTG has an effect only when LacI is present, so affects only the GFP-off state. Adding IPTG relieves the repression of OP_{lac}, allowing high-level expression of rep^{ts}, which turns off expression of LacI, and high-level expression of GFP. **(f)** Heat treatment switches the system from GFP-on to GFP-off. Heat has an effect only when rep^{ts} is present, so affects only the GFP-on state. Heat inactivates rep^{ts} and relieves the repression of OP_{λ}, allowing high-level expression of LacI. LacI then acts at OP_{lac} to repress synthesis of rep^{ts} and GFP. **(g)** *Characteristic A:* The system is not stable in the intermediate state. At some point, one repressor will act more strongly than the other due to chance fluctuations in expression; this shuts off expression of the other repressor and locks the system in one state. *Characteristic B:* Once one repressor is expressed, it prevents the synthesis of the other; thus the system remains in one state even after the switching stimulus has been removed. **(h)** At no time does any cell express an intermediate level of GFP—this is a confirmation of characteristic A. At the intermediate concentration (X) of inducer, some cells have switched to GFP-on while others have not yet made the switch and remain in the GFP-off state; none are in between. The bimodal distribution of expression levels at [IPTG] = X is caused by the mixed population of GFP-on and GFP-off cells.

Glossary

a

ABC transporters: Plasma membrane proteins with sequences that make up *ATP-binding cassettes*; serve to transport a large variety of substrates, including inorganic ions, lipids, and nonpolar drugs, out of the cell, using ATP as the energy source.

absolute configuration: The configuration of four different substituent groups around an asymmetric carbon atom, in relation to D- and L-glyceraldehyde.

absorption: Transport of the products of digestion from the intestinal tract into the blood.

acceptor control: Regulation of the rate of respiration by the availability of ADP as phosphate group acceptor.

accessory pigments: Visible light–absorbing pigments (carotenoids, xanthophylls, and phycobilins) in plants and photosynthetic bacteria that complement chlorophylls in trapping energy from sunlight.

acid dissociation constant: The dissociation constant (K_a) of an acid, describing its dissociation into its conjugate base and a proton.

acidosis: A metabolic condition in which the capacity of the body to buffer H^+ is diminished; usually accompanied by decreased blood pH.

actin: A protein that makes up the thin filaments of muscle; also an important component of the cytoskeleton of many eukaryotic cells.

action spectrum: A plot of the efficiency of light at promoting a light-dependent process such as photosynthesis as a function of wavelength.

activation energy ($\Delta G^\ddagger$): The amount of energy (in joules) required to convert all the molecules in 1 mol of a reacting substance from the ground state to the transition state.

activator: (1) A DNA-binding protein that positively regulates the expression of one or more genes; that is, transcription rates increase when an activator is bound to the DNA. (2) A positive modulator of an allosteric enzyme.

active site: The region of an enzyme surface that binds the substrate molecule and catalytically transforms it; also known as the catalytic site.

active transporter: Membrane protein that moves a solute across a membrane against an electrochemical gradient in an energy-requiring reaction.

activity: The true thermodynamic activity or potential of a substance, as distinct from its molar concentration.

acyl phosphate: Any molecule with the general chemical form $R-\overset{\displaystyle O}{\underset{\displaystyle \|}{C}}-O-OPO_3^{2-}$.

adaptor proteins: Signaling proteins, generally lacking enzymatic activities, that have binding sites for two or more cellular components and serve to bring those components together.

adenosine 3′,5′-cyclic monophosphate: *See* cyclic AMP.

S-adenosylmethionine (adoMet): An enzymatic cofactor involved in methyl group transfers.

adipocyte: An animal cell specialized for the storage of fats (triacylglycerols).

adipose tissue: Connective tissue specialized for the storage of large amounts of triacylglycerols. *See also* beige adipose tissue; brown adipose tissue; white adipose tissue.

ADP (adenosine diphosphate): A ribonucleoside 5′-diphosphate serving as phosphate group acceptor in the cell energy cycle.

aerobe: An organism that lives in air and uses oxygen as the terminal electron acceptor in respiration.

aerobic: Requiring or occurring in the presence of oxygen.

aerobic glycolysis: Cellular energy generation by glycolysis alone (without subsequent oxidation of pyruvate) even though oxygen is available. *See* glycolysis.

agonist: A compound, typically a hormone or neurotransmitter, that elicits a physiological response when it binds to its specific receptor.

alcohol fermentation: *See* ethanol fermentation.

aldose: A simple sugar in which the carbonyl carbon atom is an aldehyde; that is, the carbonyl carbon is at one end of the carbon chain.

alkalosis: A metabolic condition in which the capacity of the body to buffer OH^- is diminished; usually accompanied by an increase in blood pH.

allosteric enzyme: A regulatory enzyme with catalytic activity modulated by the noncovalent binding of a specific metabolite at a site other than the active site.

allosteric protein: A protein (generally with multiple subunits) with multiple ligand-binding sites, such that ligand binding at one site affects ligand binding at another.

allosteric site: The specific site on the surface of an allosteric protein molecule to which the modulator or effector molecule binds.

α helix: A helical conformation of a polypeptide chain, usually right-handed, with maximal intrachain hydrogen bonding; one of the most common secondary structures in proteins.

α oxidation: An alternative path for the oxidation of β-methyl fatty acids in peroxisomes.

alternative splicing: A process in which nonconsecutive exons are selectively spliced (linked) in alternative ways to generate different mature mRNAs.

Ames test: A simple bacterial test for carcinogenicity, based on the assumption that carcinogens are mutagens.

amino acid activation: ATP-dependent enzymatic esterification of the carboxyl group of an amino acid to the 3′-hydroxyl group of its corresponding tRNA.

amino acids: α-Amino–substituted carboxylic acids, the building blocks of proteins.

aminoacyl-tRNA: An aminoacyl ester of a tRNA.

aminoacyl-tRNA synthetases: Enzymes that catalyze synthesis of an aminoacyl-tRNA at the expense of ATP energy.

amino-terminal residue: The only amino acid residue in a polypeptide chain that has a free α-amino group; defines the amino terminus of the polypeptide.

aminotransferases: Enzymes that catalyze the transfer of amino groups from α-amino to α-keto acids; also called transaminases.

ammonotelic: Excreting excess nitrogen in the form of ammonia.

AMP-activated protein kinase (AMPK): A protein kinase activated by 5′-adenosine monophosphate (AMP). AMPK action generally shifts metabolism away from biosynthesis toward energy production.

amphibolic pathway: A metabolic pathway used in both catabolism and anabolism.

amphipathic: Containing both polar and nonpolar domains.

amphitropic proteins: Proteins that associate reversibly with the membrane and thus can be found in the cytosol, in the membrane, or in both places.

ampholyte: A substance that can act as either a base or an acid.

amphoteric: Capable of donating and accepting protons, thus able to serve as an acid or a base.

AMPK: *See* AMP-activated protein kinase.

amyloidoses: A variety of progressive diseases characterized by abnormal deposits of misfolded proteins in one or more organs or tissues.

anabolism: The phase of intermediary metabolism concerned with the energy-requiring biosynthesis of cell components from smaller precursors.

anaerobe: An organism that lives without oxygen. Obligate anaerobes die when exposed to oxygen.

anaerobic: Occurring in the absence of air or oxygen.

analyte: A molecule to be analyzed by mass spectrometry.

anammox: Anaerobic oxidation of ammonia to N_2, using nitrite as electron acceptor; carried out by specialized chemolithotrophic bacteria.

anaplerotic reaction: An enzyme-catalyzed reaction that can replenish the supply of intermediates in the citric acid cycle.

angstrom (Å): A unit of length (10^{-8} cm) used to indicate molecular dimensions.

anhydride: The product of the condensation of two carboxyl or phosphate groups in which the elements of water are eliminated to form a compound with the general structure R—X—O—X—R, where X is either carbon or phosphorus.

anion-exchange resin: A polymeric resin with fixed cationic groups, used in the chromatographic separation of anions.

anomeric carbon: The carbon atom in a sugar at the new chiral center formed when a sugar cyclizes to form a hemiacetal. This is the carbonyl carbon of aldehydes and ketones.

anomers: Two stereoisomers of a given sugar that differ only in the configuration about the carbonyl (anomeric) carbon atom.

anorexigenic: Tending to decrease appetite and food consumption. *Compare* orexigenic.

antagonist: A compound that interferes with the physiological action of another substance (the agonist), usually at a hormone or neurotransmitter receptor.

antibiotic: One of many different organic compounds that are formed and secreted by various species of microorganisms and plants, are toxic to other species, and presumably have a defensive function.

antibody: A defense protein synthesized by the immune system of vertebrates. *See also* immunoglobulin.

anticodon: A specific sequence of three nucleotides in a tRNA, complementary to a codon for an amino acid in an mRNA.

antigen: A molecule capable of eliciting the synthesis of a specific antibody in vertebrates.

antiparallel: Describes two linear polymers that are opposite in polarity or orientation.

antiport: Cotransport of two solutes across a membrane in opposite directions.

apoenzyme: The protein portion of an enzyme, exclusive of any organic or inorganic cofactors or prosthetic groups that might be required for catalytic activity.

apolipoprotein: The protein component of a lipoprotein.

apoprotein: The protein portion of a protein, exclusive of any organic or inorganic cofactors or prosthetic groups that might be required for activity.

apoptosis: (app′-a-toe′-sis) Programmed cell death in which a cell brings about its own death and lysis, in response to a signal from outside or programmed in its genes, by systematically degrading its own macromolecules.

aptamer: An oligonucleotide that binds specifically to one molecular target, usually selected by an iterative cycle of affinity-based enrichment (SELEX).

aquaporins (AQPs): A family of integral membrane proteins that mediate the flow of water across membranes.

archaea: Members of Archaea, one of the three domains of living organisms; include many species that thrive in extreme environments of high ionic strength, high temperature, or low pH.

arcuate nucleus: A group of neurons in the hypothalamus that function in regulation of hunger and feeding behavior.

asymmetric carbon atom: A carbon atom that is covalently bonded to four different groups and thus may exist in two different tetrahedral configurations.

ATP (adenosine triphosphate): A ribonucleoside 5′-triphosphate functioning as a phosphate group donor in the cellular energy cycle; carries chemical energy between metabolic pathways by serving as a shared intermediate coupling endergonic and exergonic reactions.

ATPase: An enzyme that hydrolyzes ATP to yield ADP and phosphate, usually coupled to a process requiring energy.

ATP synthase: An enzyme complex that forms ATP from ADP and phosphate during oxidative phosphorylation in the inner mitochondrial membrane or the bacterial plasma membrane, and during photophosphorylation in chloroplasts.

attenuator: An RNA sequence involved in regulating the expression of certain genes; functions as a transcription terminator.

autophagy: Catabolic lysosomal degradation of cellular proteins and other components.

autophosphorylation: Strictly, the phosphorylation of an amino acid residue in a protein that is catalyzed by the same protein molecule; often extended to include phosphorylation of one subunit of a homodimer by the other subunit.

autotroph: An organism that can synthesize its own complex molecules from very simple carbon and nitrogen sources, such as carbon dioxide and ammonia.

auxin: A plant growth hormone.

auxotrophic mutant (auxotroph): A mutant organism defective in the synthesis of a particular biomolecule, which must therefore be supplied for the organism's growth.

Avogadro's number (N): The number of molecules in a gram molecular weight (a mole) of any compound (6.02×10^{23}).

b

bacteria: Members of Bacteria, one of the three domains of living organisms; they have a plasma membrane but no internal organelles or nucleus.

bacteriophage: A virus capable of replicating in a bacterial cell; also called a phage.

baculovirus: Any of a group of double-stranded DNA viruses that infect invertebrates, particularly insects; widely used for protein expression in biotechnology.

basal metabolic rate: An animal's rate of oxygen consumption when at complete rest, long after a meal.

base pair: Two nucleotides in nucleic acid chains that are paired by hydrogen bonding of their bases; for example, A with T or U, and G with C.

BAT: *See* brown adipose tissue.

B cell: *See* B lymphocyte.

beige adipose tissue: Thermogenic adipose tissue activated by cooling of the individual; expresses the uncoupling protein UCP1 (thermogenin) at a high level. *Compare* brown adipose tissue; white adipose tissue.

β conformation: An extended, zigzag arrangement of a polypeptide chain; a common secondary structure in proteins.

β oxidation: Oxidative degradation of fatty acids into acetyl-CoA by successive oxidations at the β-carbon atom; as distinct from ω oxidation.

β turn: A type of protein secondary structure consisting of four amino acid residues arranged in a tight turn so that the polypeptide turns back on itself.

bilayer: A double layer of oriented amphipathic lipid molecules, forming the basic structure of biological membranes. The hydrocarbon tails face inward to form a continuous nonpolar phase.

bile acids: Polar derivatives of cholesterol, secreted by the liver into the intestine, that serve to emulsify dietary fats, facilitating lipase action.

bile salts: Amphipathic steroid derivatives with detergent properties, participating in digestion and absorption of lipids.

binding energy: The energy derived from noncovalent interactions between enzyme and substrate or receptor and ligand.

binding site: The crevice or pocket on a protein in which a ligand binds.

bioassay: A method for measuring the amount of a biologically active substance (such as a hormone) in a sample by quantifying the biological response to aliquots of that sample.

bioinformatics: The computerized analysis of biological data, using methods derived from statistics, linguistics, mathematics, chemistry, biochemistry, and physics. The data are often nucleic acid or protein sequences or structural data, but can also include other experimental data, patient statistics, and materials in the scientific literature. Bioinformatics research focuses on methods for data storage, retrieval, and analysis.

biosphere: All the living matter on or in the earth, the seas, and the atmosphere.

biotin: A vitamin; an enzymatic cofactor in carboxylation reactions.

B lymphocyte (B cell): One of a class of blood cells (lymphocytes), responsible for the production of circulating antibodies.

body mass index (BMI): A measure of obesity, calculated as weight in kilograms divided by $(\text{height in meters})^2$. A BMI of more than 27.5 is defined as overweight; more than 30, as obese.

bond energy: The energy required to break a bond.

branch migration: Movement of the branch point in a branched DNA formed from two DNA molecules with identical sequences. *See also* Holliday intermediate.

brown adipose tissue (BAT): Thermogenic adipose tissue rich in mitochondria that contain the uncoupling protein UCP1 (thermogenin), which uncouples electron transfer through the respiratory chain from ATP synthesis. *Compare* beige adipose tissue; white adipose tissue.

buffer: A system capable of resisting changes in pH, consisting of a conjugate acid-base pair in which the ratio of proton acceptor to proton donor is near unity.

C

calorie: The amount of heat required to raise the temperature of 1.0 g of water from 14.5 to 15.5 °C. One calorie (cal) equals 4.18 joules (J).

Calvin cycle: The cyclic pathway in plants that fixes carbon dioxide and produces triose phosphates.

CAM plants: Succulent plants of hot, dry climates, in which CO_2 is fixed into oxaloacetate in the dark, then fixed by rubisco in the light when stomata close to exclude O_2.

cAMP: *See* cyclic AMP.

cAMP receptor protein (CRP): In bacteria, a specific regulatory protein that controls initiation of transcription of the genes that produce the enzymes required to use some other nutrient when glucose is lacking; also called catabolite gene activator protein (CAP).

CAP: *See* cAMP receptor protein.

capsid: The protein coat of a virion, or virus particle.

carbanion: A negatively charged carbon atom.

carbocation: A positively charged carbon atom; also called a carbonium ion.

carbohydrate: A polyhydroxy aldehyde or ketone, or substance that yields such a compound on hydrolysis. Many carbohydrates have the empirical formula $(CH_2O)_n$; some also contain nitrogen, phosphorus, or sulfur.

carbon-assimilation reactions: A reaction sequence in which atmospheric CO_2 is converted into organic compounds.

carbon-fixation reactions: The reactions, catalyzed by rubisco during photosynthesis or by other carboxylases, in which atmospheric CO_2 is initially incorporated (fixed) into an organic compound.

carbonium ion: *See* carbocation.

carboxyl-terminal residue: The only amino acid residue in a polypeptide chain that has a free α-carboxyl group; defines the carboxyl terminus of the polypeptide.

cardiolipin: A membrane phospholipid in which two phosphatidic acid moieties share a single glycerol head group.

carnitine shuttle: A mechanism for moving fatty acids from the cytosol to the mitochondrial matrix as fatty esters of carnitine.

carotenoids: Lipid-soluble photosynthetic pigments made up of isoprene units.

cascade: *See* enzyme cascade; regulatory cascade.

catabolism: The phase of intermediary metabolism concerned with the energy-yielding degradation of nutrient molecules.

catabolite gene activator protein (CAP): *See* cAMP receptor protein.

catalic site: *See* active site.

catecholamines: Hormones, such as epinephrine, that are amino derivatives of catechol.

catenane: Two or more circular polymeric molecules interlinked by one or more noncovalent topological links, resembling the links of a chain.

cation-exchange resin: An insoluble polymer with fixed negative charges, used in the chromatographic separation of cationic substances.

CD spectroscopy: *See* circular dichroism spectroscopy.

cDNA: *See* complementary DNA.

cDNA library: A collection of cloned DNA fragments derived entirely from the complement of mRNA being expressed in a particular organism or cell type under a defined set of conditions.

cellular differentiation: The process in which a precursor cell becomes specialized to carry out a particular function, by acquiring a new complement of proteins and RNA.

central dogma: The organizing principle of molecular biology: genetic information flows from DNA to RNA to protein.

centromere: A specialized site in a chromosome, serving as the attachment point for the mitotic or meiotic spindle.

cerebroside: A sphingolipid containing one sugar residue as a head group.

channeling: The direct transfer of a reaction product (common intermediate) from the active site of an enzyme to the active site of the enzyme catalyzing the next step in a pathway.

chaperone: Any of several classes of proteins or protein complexes that catalyze the accurate folding of proteins in all cells.

chaperonin: One of two major classes of chaperones in virtually all organisms; a complex of proteins that functions in protein folding: GroES/GroEL in bacteria; Hsp60 in eukaryotes.

chemiosmotic coupling: Coupling of ATP synthesis to electron transfer by a transmembrane difference in charge and pH.

chemiosmotic theory: The theory that energy derived from electron transfer reactions is temporarily stored as a transmembrane difference in charge and pH, which subsequently drives formation of ATP in oxidative phosphorylation and photophosphorylation.

chemotaxis: A cell's sensing of and movement toward or away from a specific chemical agent.

chemotroph: An organism that obtains energy by metabolizing organic compounds derived from other organisms.

chiral center: An atom with substituents arranged so that the molecule is not superposable on its mirror image.

chiral compound: A compound that contains an asymmetric center (chiral atom or chiral center) and thus can occur in two nonsuperposable mirror-image forms (enantiomers).

chlorophylls: A family of green pigments that function as receptors of light energy in photosynthesis; magnesium-porphyrin complexes.

chloroplast: A chlorophyll-containing photosynthetic organelle in some eukaryotic cells.

chondroitin sulfate: One of a family of sulfated glycosaminoglycans, a major component of the extracellular matrix.

chromatin: A filamentous complex of DNA, histones, and other proteins, constituting the eukaryotic chromosome.

chromatography: A process in which complex mixtures of molecules are separated by many repeated partitionings between a flowing (mobile) phase and a stationary phase.

chromatophore: A compound or moiety (natural or synthetic) that absorbs visible or ultraviolet light.

chromosome: A single large DNA molecule and its associated proteins, containing many genes; stores and transmits genetic information.

chromosome territory: A region of the nucleus preferentially occupied by a particular chromosome.

chylomicron: A plasma lipoprotein consisting of a large droplet of triacylglycerols stabilized by a coat of protein and phospholipid; carries lipids from the intestine to tissues.

circular dichroism (CD) spectroscopy: A method used to characterize the degree of folding in a protein, based on differences in the absorption of right-handed versus left-handed circularly polarized light.

cis and trans isomers: *See* geometric isomers.

cistron: A unit of DNA or RNA corresponding to one gene.

citric acid cycle: A cyclic pathway for the oxidation of acetyl residues to carbon dioxide, in which formation of citrate is the first step; also known as the Krebs cycle or tricarboxylic acid cycle.

Cleland nomenclature: A shorthand notation developed by W. W. Cleland for describing the progress of enzymatic reactions with multiple substrates and products.

clones: The descendants of a single cell.

cloning: The production of large numbers of identical DNA molecules, cells, or organisms from a single, ancestral DNA molecule, cell, or organism.

closed system: A system that exchanges neither matter nor energy with the surroundings. *See also* system.

cobalamin: *See* coenzyme B$_{12}$.

coding strand: In DNA transcription, the DNA strand identical in base sequence to the RNA transcribed from it, with U in the RNA in place of T in the DNA; as distinct from the template strand. Also called the nontemplate strand.

codon: A sequence of three adjacent nucleotides in a nucleic acid that codes for a specific amino acid.

coenzyme: An organic cofactor required for the action of certain enzymes; often has a vitamin component.

coenzyme A: A pantothenic acid–containing coenzyme that serves as an acyl group carrier in certain enzymatic reactions.

coenzyme B$_{12}$: An enzymatic cofactor derived from the vitamin cobalamin, involved in certain types of carbon skeletal rearrangements.

cofactor: An inorganic ion or a coenzyme required for enzyme activity.

cognate: Describes two biomolecules that normally interact; for example, an enzyme and its usual substrate, or a receptor and its usual ligand.

cohesive ends: *See* sticky ends.

cointegrate: An intermediate in the migration of certain DNA transposons in which the donor DNA and target DNA are covalently attached.

colligative properties: The properties of a solution that depend on the number of solute particles per unit volume; for example, freezing-point depression.

combinatorial control: The use of combinations of a limited repertoire of regulatory proteins to provide gene-specific regulation of many individual genes.

competitive inhibition: A type of enzyme inhibition reversed by increasing the substrate concentration; a competitive inhibitor generally competes with the normal substrate or ligand for a protein's binding site.

complementary: Having a molecular surface with chemical groups arranged to interact specifically with chemical groups on another molecule.

complementary DNA (cDNA): A DNA complementary to a specific mRNA, used in DNA cloning; usually made through the use of reverse transcriptase.

condensation: A reaction type in which two compounds are joined with the elimination of water.

configuration: The spatial arrangement of an organic molecule conferred by the presence of (1) double bonds, about which there is no freedom of rotation, or (2) chiral centers, around which substituent groups are arranged in a specific sequence. Configurational isomers cannot be interconverted without breaking one or more covalent bonds.

conformation: A spatial arrangement of substituent groups that are free to assume different positions in space, without breaking any bonds, because of the freedom of bond rotation.

conjugate acid-base pair: A proton donor and its corresponding deprotonated species; for example, acetic acid (donor) and acetate (acceptor).

conjugated protein: A protein containing one or more prosthetic groups.

conjugate redox pair: An electron donor and its corresponding electron acceptor; for example, Cu$^+$ (donor) and Cu^{2+} (acceptor), or NADH (donor) and NAD$^+$ (acceptor).

consensus sequence: A DNA or amino acid sequence consisting of the residues that most commonly occur at each position in a set of similar sequences.

conservative substitution: Replacement of an amino acid residue in a polypeptide by another residue with similar properties; for example, substitution of Glu by Asp.

constitutive enzymes: Enzymes required at all times by a cell and present at some constant level; for example, many enzymes of the central metabolic pathways. Sometimes called housekeeping enzymes.

contig: A series of overlapping clones or a continuous sequence defining an uninterrupted section of a chromosome.

contour length: The length of a nucleic acid molecule as measured along its helical axis.

cooperativity: The characteristic of an enzyme or other protein in which binding of the first molecule of a ligand changes the affinity for the second molecule. In positive cooperativity, the affinity for the second ligand molecule increases; in negative cooperativity, it decreases.

cotransport: The simultaneous transport, by a single transporter, of two solutes across a membrane. *See also* antiport; symport.

coupled reactions: Two chemical reactions that have a common intermediate and thus a means of energy transfer from one to the other.

covalent bond: A chemical bond that involves sharing of electron pairs.

C$_4$ pathway: The metabolic pathway in which CO$_2$ is first added to phosphoenolpyruvate by the enzyme PEP carboxylase to produce the four-carbon compound within mesophyll cells that is later transported to the bundle-sheath cells, where the CO$_2$ is released for use in the Calvin cycle.

C$_4$ plants: Plants (generally tropical) in which CO$_2$ is first fixed into a four-carbon compound (oxaloacetate or malate) before entering the Calvin cycle via the rubisco reaction.

CRISPR/Cas: Bacterial systems that evolved to provide a defense against bacteriophage infection. CRISPR stands for clustered, *regularly interspaced short palindromic repeats*. Cas stands for *CRISPR-associated*. Engineered CRISPR/Cas systems are used for efficient, targeted genome editing in a wide range of organisms.

cristae: Infoldings of the inner mitochondrial membrane.

CRP: *See* cAMP receptor protein.

cruciform: A secondary structure in double-stranded RNA or DNA in which the double helix is denatured at palindromic repeat sequences in each strand, and each separated strand is paired internally to form opposing hairpin structures. *See also* hairpin.

cryo-electron microscopy (cryo-EM): A technique for determining the structure of proteins or protein complexes; individual molecules are spread on a grid in random orientations, quick-frozen to preserve the integrity of the sample, and visualized by EM. Images of individual molecules are computationally rotated to optimize their coincidence, yielding three-dimensional structures.

cyclic AMP (cAMP; adenosine 3′,5′-cyclic monophosphate): A second messenger; its formation in a cell by adenylyl cyclase is stimulated by certain hormones or other molecular signals.

cyclic electron flow: In chloroplasts, the light-induced flow of electrons originating from and returning to photosystem I.

cyclic photophosphorylation: ATP synthesis driven by cyclic electron flow through photosystem I.

cyclin: One of a family of proteins that activate cyclin-dependent protein kinases and thereby regulate the cell cycle.

cytochrome P-450: A family of heme-containing enzymes, with a characteristic absorption band at 450 nm, that participate in biological hydroxylations with O_2.

cytochromes: Heme proteins serving as electron carriers in respiration, photosynthesis, and other oxidation-reduction reactions.

cytokine: One of a family of small secreted proteins (such as interleukins and interferons) that activate cell division or differentiation by binding to plasma membrane receptors in target cells.

cytokinesis: The final separation of daughter cells following mitosis.

cytoplasm: The portion of a cell's contents outside the nucleus but within the plasma membrane; includes organelles such as mitochondria.

cytoskeleton: The filamentous network that provides structure and organization to the cytoplasm; includes actin filaments, microtubules, and intermediate filaments.

cytosol: The continuous aqueous phase of the cytoplasm, with its dissolved solutes; excludes the organelles such as mitochondria.

d

dalton: Unit of atomic or molecular weight; 1 dalton (Da) is the weight of a hydrogen atom (1.66×10^{-24} g).

dark reactions: *See* carbon-assimilation reactions.

deamination: The enzymatic removal of amino groups from biomolecules such as amino acids or nucleotides.

degenerate code: A code in which a single element in one language is specified by more than one element in a second language.

dehydrogenases: Enzymes that catalyze the removal of pairs of hydrogen atoms from substrates.

deletion mutation: A mutation resulting from deletion of one or more nucleotides from a gene or chromosome.

ΔG: *See* free-energy change.

ΔG‡: *See* activation energy.

ΔG′°: *See* standard free-energy change.

denaturation: Partial or complete unfolding of the specific native conformation of a polypeptide chain, protein, or nucleic acid such that the function of the molecule is lost.

denatured protein: A protein that has lost enough of its native conformation by exposure to a destabilizing agent such as heat or detergent that its function is lost.

de novo pathway: A pathway for the synthesis of a biomolecule, such as a nucleotide, from simple precursors; as distinct from a salvage pathway.

deoxyribonucleic acid: *See* DNA.

deoxyribonucleotides: Nucleotides containing 2-deoxy-D-ribose as the pentose component.

desaturases: Enzymes that catalyze the introduction of double bonds into the hydrocarbon portion of fatty acids.

desensitization: A universal process by which sensory mechanisms cease to respond after prolonged exposure to the specific stimulus they detect.

desolvation: In aqueous solution, the release of bound water surrounding a solute.

diabetes mellitus: A group of metabolic diseases with symptoms that result from a deficiency in insulin production or utilization; characterized by an inability to transport glucose from the blood into cells at normal glucose concentrations.

dialysis: Removal of small molecules from a solution of a macromolecule by their diffusion through a semipermeable membrane into a suitably buffered solution.

dideoxy sequencing: *See* Sanger sequencing.

differential centrifugation: Separation of cell organelles or other particles of different size by their different rates of sedimentation in a centrifugal field.

differentiation: Specialization of cell structure and function during growth and development.

diffusion: Net movement of molecules in the direction of lower concentration.

digestion: Enzymatic hydrolysis of major nutrients in the gastrointestinal system to yield their simpler components.

diploid: Having two sets of genetic information; describes a cell with two chromosomes of each type. *Compare* haploid.

disaccharide: A carbohydrate consisting of two covalently joined monosaccharide units.

dissociation constant (K_d): An equilibrium constant for the dissociation of a complex of two or more biomolecules into its components; for example, dissociation of a substrate from an enzyme.

disulfide bond: A covalent bond resulting from the oxidative linkage of two Cys residues, from the same or different polypeptide chains, forming a cystine residue.

DNA (deoxyribonucleic acid): A polynucleotide with a specific sequence of deoxyribonucleotide units covalently joined through 3′,5′-phosphodiester bonds; serves as the carrier of genetic information.

DNA chimera: DNA containing genetic information derived from two different species.

DNA chip: An informal term for a DNA microarray, referring to the small size of typical microarrays.

DNA cloning: *See* cloning.

DNA library: A collection of cloned DNA fragments.

DNA ligases: Enzymes that create a phosphodiester bond between the 3′ end of one DNA segment and the 5′ end of another.

DNA looping: The interaction of proteins bound at distant sites on a DNA molecule so that the intervening DNA forms a loop.

DNA microarray: A collection of DNA sequences immobilized on a solid surface, with individual sequences laid out in patterned arrays that can be probed by hybridization.

DNA polymerases: Enzymes that catalyze template-dependent synthesis of DNA from its deoxyribonucleoside 5′-triphosphate precursors. The bacterium *Escherichia coli* has five DNA polymerases, numbered I through V; eukaryotes have a larger number.

DNA sequencing technologies: A rapidly growing set of modern automated technologies that allow rapid, inexpensive sequencing of DNA, ranging from parts of single genes to entire genomes. *See* ion semiconductor sequencing; pyrosequencing; reversible terminator sequencing; Sanger sequencing; single-molecule real-time (SMRT) sequencing.

DNA supercoiling: The coiling of DNA upon itself, generally as a result of bending, underwinding, or overwinding of the DNA helix.

DNA transposition: *See* transposition.

domain: A distinct structural unit of a polypeptide; domains may have separate functions and may fold as independent, compact units.

double helix: The natural coiled conformation of two complementary, antiparallel DNA chains.

double-reciprocal plot: A plot of $1/V_0$ versus $1/[S]$, which allows a more accurate determination of V_{max} and K_m than a plot of V_0 versus [S]; also called the Lineweaver-Burk plot.

e

$E'°$: *See* standard reduction potential.

ECM: *See* extracellular matrix.

electrochemical gradient: The resultant of the gradients of concentration and of electric charge of an ion across a membrane; the driving force for oxidative phosphorylation and photophosphorylation.

electrochemical potential: The energy required to maintain a separation of charge and of concentration across a membrane.

electrogenic: Contributing to an electrical potential across a membrane.

electron acceptor: A substance that receives electrons in an oxidation-reduction reaction.

electron carrier: A protein, such as a flavoprotein or a cytochrome, that can reversibly gain and lose electrons; functions in the transfer of electrons from organic nutrients to oxygen or some other terminal acceptor.

electron donor: A substance that donates electrons in an oxidation-reduction reaction.

electron transfer: Movement of electrons from electron donor to electron acceptor; especially, from substrates to oxygen via the carriers of the respiratory (electron-transfer) chain.

electrophile: An electron-deficient group with a strong tendency to accept electrons from an electron-rich group (nucleophile).

electrophoresis: Movement of charged solutes in response to an electrical field; often used to separate mixtures of ions, proteins, or nucleic acids.

ELISA: *See* enzyme-linked immunosorbent assay.

elongation factors: (1) Proteins that function in the elongation phase of eukaryotic transcription. (2) Specific proteins required in the elongation of polypeptide chains by ribosomes.

eluate: The effluent from a chromatographic column.

enantiomers: Stereoisomers that are nonsuperposable mirror images of each other.

endergonic reaction: A chemical reaction that consumes energy (i.e., for which ΔG is positive).

endocannabinoid: An endogenous substance capable of binding to and functionally activating cannabinoid receptors.

endocrine: Pertaining to cellular secretions that enter the bloodstream and have their effects on distant tissues.

endocytosis: The uptake of extracellular material by its inclusion in a vesicle (endosome) formed by invagination of the plasma membrane.

endonucleases: Enzymes that hydrolyze the interior phosphodiester bonds of a nucleic acid—that is, act at bonds other than the terminal bonds.

endoplasmic reticulum: An extensive system of double membranes in the cytoplasm of eukaryotic cells; it encloses secretory channels and is often studded with ribosomes (rough endoplasmic reticulum).

endothermic reaction: A chemical reaction that takes up heat (i.e., for which ΔH is positive).

end-product inhibition: *See* feedback inhibition.

enhancers: DNA sequences that facilitate the expression of a given gene; may be located a few hundred, or even thousand, base pairs away from the gene.

enthalpy (*H***):** The heat content of a system.

enthalpy change (Δ*H*): For a reaction, approximately equal to the difference between the energy used to break bonds and the energy gained by formation of new bonds.

entropy (*S***):** The extent of randomness or disorder in a system.

enzyme: A biomolecule, either protein or RNA, that catalyzes a specific chemical reaction. It does not affect the equilibrium of the catalyzed reaction; it enhances the rate of the reaction by providing a reaction path with a lower activation energy.

enzyme cascade: A series of reactions, often involved in regulatory events, in which one enzyme activates another (often by phosphorylation), which activates a third, and so on. The effect of a catalyst activating a catalyst is a large amplification of the signal that initiated the cascade. *See also* regulatory cascade.

enzyme-linked immunosorbent assay (ELISA): A sensitive immunoassay that uses an enzyme linked to an antibody or antigen to detect a specific protein.

epigenetic: Describes any inherited characteristic of a living organism that is acquired by means that do not involve the nucleotide sequence of the parental chromosomes; for example, covalent modifications of histones.

epimerases: Enzymes that catalyze the reversible interconversion of two epimers.

epimers: Two stereoisomers differing in configuration at one asymmetric center in a compound having two or more asymmetric centers.

epithelial cell: Any cell that forms part of the outer covering of an organism or organ.

epitope: An antigenic determinant; the particular chemical group or groups in a macromolecule (antigen) to which a given antibody binds.

epitope tag: A protein sequence or domain bound by a well-characterized antibody.

equilibrium: The state of a system in which no further net change is occurring; the free energy is at a minimum.

equilibrium constant (K_{eq}): A constant, characteristic for each chemical reaction, that relates the specific concentrations of all reactants and products at equilibrium at a given temperature and pressure.

erythrocyte: A cell containing large amounts of hemoglobin and specialized for oxygen transport; a red blood cell.

essential amino acids: Amino acids that cannot be synthesized by humans and must be obtained from the diet.

essential fatty acids: The group of polyunsaturated fatty acids produced by plants, but not by humans; required in the human diet.

ethanol fermentation: The anaerobic conversion of glucose to ethanol via glycolysis; also called alcohol fermentation. *See also* fermentation.

euchromatin: The regions of interphase chromosomes that are more open (less condensed), where genes are being actively expressed. *Compare* heterochromatin.

eukaryotes: Members of Eukarya, one of the three domains of living organisms; unicellular or multicellular organisms with cells having a membrane-bounded nucleus, multiple chromosomes, and internal organelles.

excited state: An energy-rich state of an atom or molecule, produced by absorption of light energy; as distinct from ground state.

exergonic reaction: A chemical reaction that proceeds with the release of free energy (i.e., for which ΔG is negative).

exocytosis: The fusion of an intracellular vesicle with the plasma membrane, releasing the vesicle contents to the extracellular space.

exon: The segment of a eukaryotic gene that encodes a portion of the final product of the gene; a segment of RNA that remains after posttranscriptional processing and is transcribed into a protein or incorporated into the structure of an RNA. *See also* intron.

exonucleases: Enzymes that hydrolyze only those phosphodiester bonds that are in the terminal positions of a nucleic acid.

exothermic reaction: A chemical reaction that releases heat (i.e., for which ΔH is negative).

expression vector: A vector incorporating sequences that allow the transcription and translation of a cloned gene. *See* vector.

extracellular matrix (ECM): An interwoven combination of glycosaminoglycans, proteoglycans, and proteins, just outside the plasma membrane, that provides cell anchorage, positional recognition, and traction during cell migration.

extrahepatic: Describes all tissues outside the liver; implies the centrality of the liver in metabolism.

f

FAD (flavin adenine dinucleotide): The coenzyme of some oxidation-reduction enzymes; contains riboflavin.

F_1 ATPase: The multiprotein subunit of ATP synthase that has the ATP-synthesizing catalytic sites. It interacts with the F_o subunit of ATP synthase, coupling proton movement to ATP synthesis.

fatty acid: A long-chain aliphatic carboxylic acid in natural fats and oils; also a component of membrane phospholipids and glycolipids.

feedback inhibition: Inhibition of an allosteric enzyme at the beginning of a metabolic sequence by the end product of the sequence; also known as end-product inhibition.

fermentation: Energy-yielding anaerobic breakdown of a nutrient molecule, such as glucose, without net oxidation; yields lactate, ethanol, or some other simple product.

fibrin: A protein factor that forms the cross-linked fibers in blood clots.

fibrinogen: The inactive precursor protein of fibrin.

fibroblast: A cell of the connective tissue that secretes connective tissue proteins such as collagen.

fibrous proteins: Insoluble proteins that serve a protective or structural role; contain polypeptide chains that generally share a common secondary structure.

first law of thermodynamics: The law stating that, in all processes, the total energy of the universe remains constant.

Fischer projection formulas: A method for representing molecules that shows the configuration of groups around chiral centers; also known as projection formulas.

5′ end: The end of a nucleic acid that lacks a nucleotide bound at the 5′ position of the terminal residue.

flagellum: A cell appendage used in propulsion. Bacterial flagella have a much simpler structure than eukaryotic flagella, which are similar to cilia.

flavin-linked dehydrogenases: Dehydrogenases requiring one of the riboflavin coenzymes, FMN or FAD.

flavin nucleotides: Nucleotide coenzymes (FMN and FAD) containing riboflavin.

flavoprotein: An enzyme containing a flavin nucleotide as a tightly bound prosthetic group.

flippases: Membrane proteins in the ABC transporter family that catalyze movement of phospholipids from the extracellular leaflet (monolayer) to the cytosolic leaflet of a membrane bilayer.

floppases: Membrane proteins in the ABC transporter family that catalyze movement of phospholipids from the cytosolic leaflet (monolayer) to the extracellular leaflet of a membrane bilayer.

fluid mosaic model: A model describing biological membranes as a fluid lipid bilayer with embedded proteins; the bilayer exhibits both structural and functional asymmetry.

fluorescence: Emission of light by excited molecules as they revert to the ground state.

fluorescence recovery after photobleaching: *See* FRAP.

fluorescence resonance energy transfer: *See* FRET.

FMN (flavin mononucleotide): Riboflavin phosphate, a coenzyme of certain oxidation-reduction enzymes.

fold: *See* motif.

footprinting: A technique for identifying the nucleic acid sequence bound by a DNA- or RNA-binding protein.

fraction: A portion of a biological sample that has been subjected to a procedure designed to separate macromolecules based on a property such as solubility, net charge, molecular weight, or function.

fractionation: The process of separating the proteins or other components of a complex molecular mixture into fractions based on differences in properties such as solubility, net charge, molecular weight, or function.

frame shift: A mutation caused by insertion or deletion of one or more paired nucleotides, changing the reading frame of codons during protein synthesis; the polypeptide product has a garbled amino acid sequence beginning at the mutated codon.

FRAP (fluorescence recovery after photobleaching): A technique used to quantify the diffusion of membrane components (lipids or proteins) in the plane of the bilayer.

free energy (G): The component of the total energy of a system that can do work at constant temperature and pressure.

free energy of activation ($\Delta G^{\ddagger}$): *See* activation energy.

free-energy change (ΔG): The amount of free energy released (negative ΔG) or absorbed (positive ΔG) in a reaction at constant temperature and pressure.

free radical: *See* radical.

FRET (fluorescence resonance energy transfer): A technique for estimating the distance between two proteins or two domains of a protein by measuring the nonradiative transfer of energy between reporter chromophores when one protein or domain is excited and the fluorescence emitted from the other is quantified.

functional group: The specific atom or group of atoms that confers a particular chemical property on a biomolecule.

furanose: A simple sugar containing the five-membered furan ring.

fusion protein: (1) One of a family of proteins that facilitate membrane fusion. (2) The protein product of a gene created by the fusion of two distinct genes or portions of genes.

futile cycle: *See* substrate cycle.

g

G_i: *See* inhibitory G protein.

G_s: *See* stimulatory G protein.

gametes: Reproductive cells with a haploid gene content; sperm or egg cells.

ganglioside: A sphingolipid containing a complex oligosaccharide as a head group; especially common in nervous tissue.

GAPs: *See* GTPase activator proteins.

GEFs: *See* guanosine nucleotide–exchange factors.

gel filtration: *See* size-exclusion chromatography.

gene: A chromosomal segment that codes for a single functional polypeptide chain or RNA molecule.

gene expression: Transcription, and, in the case of proteins, translation, to yield the product of a gene; a gene is expressed when its biological product is present and active.

gene fusion: The enzymatic attachment of one gene, or part of a gene, to another.

general acid-base catalysis: Catalysis involving proton transfer(s) to or from a molecule other than water.

genetic code: The set of triplet code words in DNA (or mRNA) coding for the amino acids of proteins.

genetic engineering: Any process by which genetic material, particularly DNA, is altered by a molecular biologist.

genetic map: A diagram showing the relative sequence and position of specific genes along a chromosome.

genome: All the genetic information encoded in a cell or virus.

genome annotation: The process of assigning actual or likely functions to genes discovered during genomic DNA sequencing projects.

genomic library: A DNA library containing DNA segments that represent all (or most) of the sequences in an organism's genome.

genomics: A science devoted broadly to the understanding of cellular and organism genomes.

genotype: The genetic constitution of an organism, as distinct from its physical characteristics, or phenotype.

geometric isomers: Isomers related by rotation about a double bond; also called cis and trans isomers.

germ-line cell: A type of animal cell that is formed early in embryogenesis and may multiply by mitosis or produce, by meiosis, cells that develop into gametes (egg or sperm cells).

GFP: *See* green fluorescent protein.

globular proteins: Soluble proteins with a globular (somewhat rounded) shape.

glucogenic: Capable of being converted into glucose or glycogen by the process of gluconeogenesis.

gluconeogenesis: The biosynthesis of a carbohydrate from simpler, noncarbohydrate precursors such as oxaloacetate or pyruvate.

GLUT: Designation for a family of membrane proteins that transport glucose.

glycan: A polymer of monosaccharide units joined by glycosidic bonds; polysaccharide.

glyceroneogenesis: The synthesis in adipocytes of glycerol 3-phosphate from pyruvate for use in triacylglycerol synthesis.

glycerophospholipid: An amphipathic lipid with a glycerol backbone; fatty acids are ester-linked to C-1 and C-2 of the glycerol, and a polar alcohol is attached through a phosphodiester linkage to C-3.

glycoconjugate: A compound containing a carbohydrate component bound covalently to a protein or lipid, forming a glycoprotein or glycolipid.

glycogenesis: The process of converting glucose to glycogen.

glycogenin: The protein that both primes the synthesis of new glycogen chains and catalyzes the polymerization of the first few sugar residues of each chain before glycogen synthase continues the extension.

glycogenolysis: The enzymatic breakdown of stored (not dietary) glycogen.

glycolate pathway: The metabolic pathway in photosynthetic organisms that converts glycolate produced during photorespiration into 3-phosphoglycerate.

glycolipid: A lipid containing a carbohydrate group.

glycolysis: The catabolic pathway by which a molecule of glucose is broken down into two molecules of pyruvate. *Compare* aerobic glycolysis.

glycome: The full complement of carbohydrates and carbohydrate-containing molecules of a cell or tissue under a particular set of conditions.

glycomics: The systematic characterization of the glycome.

glycoprotein: A protein containing a carbohydrate group.

glycosaminoglycan: A heteropolysaccharide of two alternating units: one is either *N*-acetylglucosamine or *N*-acetylgalactosamine; the other is a uronic acid (usually glucuronic acid). Formerly called a mucopolysaccharide.

glycosidic bonds: *See O*-glyosidic bonds.

glycosphingolipid: An amphipathic lipid with a sphingosine backbone to which are attached a long-chain fatty acid and a polar alcohol.

glyoxylate cycle: A variant of the citric acid cycle, for the net conversion of acetate into succinate and, eventually, new carbohydrate; present in bacteria and some plant cells.

glyoxysome: A specialized peroxisome containing the enzymes of the glyoxylate cycle; found in cells of germinating seeds.

glypican: A heparan sulfate proteoglycan attached to a membrane through a glycosyl phosphatidylinositol (GPI) anchor.

Golgi complex: A complex membranous organelle of eukaryotic cells; functions in the posttranslational modification of proteins and their secretion from the cell or incorporation into the plasma membrane or organellar membranes.

GPCRs: *See* G protein–coupled receptors.

GPI-anchored protein: A protein held to the outer monolayer of the plasma membrane by its covalent attachment through a short oligosaccharide chain to a phosphatidylinositol molecule in the membrane.

G protein–coupled receptor kinases (GRKs): A family of protein kinases that phosphorylate Ser and Thr residues near the carboxyl terminus of G protein–coupled receptors, initiating their internalization.

G protein–coupled receptors (GPCRs): A large family of membrane receptor proteins with seven transmembrane helical segments, often associating with G proteins to transduce an extracellular signal into a change in cellular metabolism.

G proteins: A large family of GTP-binding proteins that act in intracellular signaling pathways and in membrane trafficking. Active when GTP is bound, they self-inactivate by converting GTP to GDP. Also called guanosine nucleotide–binding proteins.

gram molecular weight: For a compound, the weight in grams that is numerically equal to its molecular weight; the weight of one mole.

grana: Stacks of thylakoids, flattened membranous sacs or disks, in chloroplasts.

green fluorescent protein (GFP): A small protein from a marine organism that produces bright fluorescence in the green region of the visible spectrum. Fusion proteins with GFP are commonly used to determine the subcellular location of the fused protein by fluorescence microscopy.

GRKs: *See* G protein–coupled receptor kinases.

gRNA: *See* guide RNA.

ground state: The normal, stable form of an atom or molecule, as distinct from the excited state.

group transfer potential: A measure of the ability of a compound to donate an activated group (such as a phosphate or acyl group); generally expressed as the standard free energy of hydrolysis.

growth factors: Proteins or other molecules that act from outside a cell to stimulate cell growth and division.

GTPase activator proteins (GAPs): Regulatory proteins that bind activated G proteins and stimulate their intrinsic GTPase activity, speeding their self-inactivation.

guanosine nucleotide–binding proteins: *See* G proteins.

guanosine nucleotide–exchange factors (GEFs): Regulatory proteins that bind to and activate G proteins by stimulating the exchange of bound GDP for GTP.

guide RNA (gRNA): An RNA found in CRISPR systems that has sequences complementary to those in a target DNA such as a phage DNA.

h

hairpin: Secondary structure in single-stranded RNA or DNA, in which complementary parts of a palindromic repeat fold back and pair to form an antiparallel duplex helix closed at one end.

half-life: The time required for the disappearance or decay of one-half of a given component in a system.

haploid: Having a single set of genetic information; describes a cell with one chromosome of each type. *Compare* diploid.

haplotype: A combination of alleles of different genes located sufficiently close together on a chromosome that they tend to be inherited together.

hapten: A small molecule that, when linked to a larger molecule, elicits an immune response.

Haworth perspective formulas: A method for representing cyclic chemical structures so as to define the configuration of each substituent group; commonly used for representing sugars.

helicases: Enzymes that catalyze separation of strands in a DNA molecule before replication.

heme: The iron-porphyrin prosthetic group of heme proteins.

heme protein: A protein containing a heme as prosthetic group.

hemoglobin: A heme protein in erythrocytes; functions in oxygen transport.

Henderson-Hasselbalch equation: An equation relating pH, pK_a, and ratio of the concentrations of proton-acceptor (A^-) and proton-donor (HA) species in a solution:

$$pH = pK_a + \log \frac{[A^-]}{[HA]}.$$

heparan sulfate: A sulfated polymer of alternating *N*-acetylglucosamine and a uronic acid, either glucuronic or iduronic acid; typically found in the extracellular matrix.

hepatocyte: The major cell type of liver tissue.

heterochromatin: The regions of chromosomes that are condensed, in which gene expression is generally suppressed. *Compare* euchromatin.

heteropolysaccharide: A polysaccharide containing more than one type of sugar.

heterotroph: An organism that requires complex nutrient molecules, such as glucose, as a source of energy and carbon.

heterotropic: Describes an allosteric modulator that is distinct from the normal ligand.

heterotropic enzyme: An allosteric enzyme requiring a modulator other than its substrate.

hexose: A simple sugar with a backbone containing six carbon atoms.

hexose monophosphate pathway: *See* pentose phosphate pathway.

high-performance liquid chromatography (HPLC): Chromatographic procedure, often conducted at relatively high pressures using automated equipment, which permits refined and highly reproducible profiles.

Hill coefficient: A measure of cooperative interaction between protein subunits.

Hill reaction: The evolution of oxygen and photoreduction of an artificial electron acceptor by a chloroplast preparation in the absence of carbon dioxide.

histones: The family of basic proteins that associate tightly with DNA in the chromosomes of all eukaryotic cells.

Holliday intermediate: An intermediate in genetic recombination in which two double-stranded DNA molecules are joined by a reciprocal crossover involving one strand of each molecule.

holoenzyme: A catalytically active enzyme, including all necessary subunits, prosthetic groups, and cofactors.

homeobox: A conserved DNA sequence of 180 bp that encodes a protein domain found in many proteins with a regulatory role in development.

homeodomain: The protein domain encoded by the homeobox; a regulatory unit that determines the segmentation of a body plan.

homeostasis: The maintenance of a dynamic steady state by regulatory mechanisms that compensate for changes in external circumstances.

homeotic genes: Genes that regulate development of the pattern of segments in the *Drosophila* body plan; similar genes are found in most vertebrates.

homologs: Genes or proteins that possess a clear sequence and functional relationship to each other.

homologous genetic recombination: Recombination between two DNA molecules of similar sequence, taking place in all cells; occurs during meiosis and mitosis in eukaryotes.

homologous proteins: Proteins having similar sequences and functions in different species; for example, the hemoglobins.

homopolysaccharide: A polysaccharide made up of one type of monosaccharide unit.

homotropic: Describes an allosteric modulator that is identical to the normal ligand.

homotropic enzyme: An allosteric enzyme that uses its substrate as a modulator.

hormone: A chemical substance, synthesized in small amounts by an endocrine tissue, that is carried in the blood to another tissue, or diffuses to a nearby cell, where it acts as a messenger to regulate the function of the target tissue or organ.

hormone receptor: A protein in, or on the surface of, target cells that binds a specific hormone and initiates the cellular response.

hormone response element (HRE): A short (12 to 20 bp) DNA sequence that binds receptors for steroid, retinoid, thyroid, and vitamin D hormones, altering expression of the contiguous genes. Each hormone has a consensus sequence preferred by the cognate receptor.

housekeeping genes: Genes that encode products (such as the enzymes of the central energy-yielding pathways) needed by cells at all times; also called constitutive genes because they are expressed under all conditions.

HPLC: *See* high-performance liquid chromatography.

HRE: *See* hormone response element.

hyaluronan: A high molecular weight, acidic polysaccharide typically composed of the alternating disaccharide GlcUA(β1→3) GlcNAc; major component of the extracellular matrix, forming larger complexes (proteoglycans) with proteins and other acidic polysaccharides. Also called hyaluronic acid.

hydrogen bond: A weak electrostatic attraction between one electronegative atom (such as oxygen or nitrogen) and a hydrogen atom covalently linked to a second electronegative atom.

hydrolases: Enzymes (e.g., proteases, lipases, phosphatases, nucleases) that catalyze hydrolysis reactions.

hydrolysis: Cleavage of a bond, such as an anhydride or peptide bond, by addition of the elements of water, yielding two or more products.

hydronium ion: The hydrated hydrogen ion (H_3O^+).

hydropathy index: A scale that expresses the relative hydrophobic and hydrophilic tendencies of a chemical group.

hydrophilic: Polar or charged; describes molecules or groups that associate with (dissolve easily in) water.

hydrophobic: Nonpolar; describes molecules or groups that are insoluble in water.

hydrophobic effect: The aggregation of nonpolar molecules in aqueous solution, excluding water molecules; caused largely by an entropic effect related to the hydrogen-bonding structure of the surrounding water.

hyperchromic effect: The large increase in light absorption at 260 nm as a double-helical DNA unwinds (melts).

hypoxia: The metabolic condition in which the supply of oxygen is severely limited.

i

immune response: The capacity of a vertebrate to generate antibodies to an antigen, a macromolecule foreign to the organism.

immunoblotting: A technique using antibodies to detect the presence of a protein in a biological sample after the proteins have been separated by gel electrophoresis, transferred to a membrane, and immobilized; also called Western blotting.

immunoglobulin: An antibody protein generated against, and capable of binding specifically to, an antigen.

induced fit: A change in the conformation of an enzyme in response to substrate binding that renders the enzyme catalytically active; also denotes changes in the conformation of any macromolecule in response to ligand binding, such that the binding site better conforms to the shape of the ligand.

inducer: A signal molecule that, when bound to a regulatory protein, increases the expression of a given gene.

induction: An increase in the expression of a gene in response to a change in the activity of a regulatory protein.

informational macromolecules: Biomolecules containing information in the form of specific sequences of different monomers; for example, many proteins, lipids, polysaccharides, and nucleic acids.

inhibitory G protein (G$_i$): A trimeric GTP-binding protein that, when activated by an associated plasma membrane receptor, inhibits a neighboring membrane enzyme such as adenylyl cyclase. *Compare* stimulatory G protein (G$_s$).

initiation codon: AUG (sometimes GUG or, more rarely, UUG in bacteria and archaea); codes for the first amino acid in a polypeptide sequence: *N*-formylmethionine in bacteria; methionine in archaea and eukaryotes.

initiation complex: A complex of a ribosome with an mRNA and the initiating Met-tRNA^{Met} or fMet-tRNA^{fMet}, ready for the elongation steps.

inorganic pyrophosphatase: An enzyme that hydrolyzes a molecule of inorganic pyrophosphate to yield two molecules of (ortho) phosphate; also known as pyrophosphatase.

insertion mutation: A mutation caused by insertion of one or more extra bases, or a mutagen, between successive bases in DNA.

insertion sequence: Specific base sequences at either end of a transposable segment of DNA.

in situ: "In position"; that is, in its natural position or location.

integral proteins: Proteins firmly bound to a membrane by interactions resulting from the hydrophobic effect; as distinct from peripheral proteins.

integrin: One of a large family of heterodimeric transmembrane proteins that mediate adhesion of cells to other cells or to the extracellular matrix.

intercalation: Insertion between stacked aromatic or planar rings; for example, insertion of a planar molecule between two successive bases in a nucleic acid.

intermediary metabolism: In cells, the enzyme-catalyzed reactions that extract chemical energy from nutrient molecules and use it to synthesize and assemble cell components.

intrinsically disordered proteins: Proteins, or segments of proteins, that lack a definable three-dimensional structure in solution.

intron: A sequence of nucleotides in a gene that is transcribed but excised before the gene is translated; also called intervening sequence. *See also* exon.

in vitro: "In glass"; that is, in the test tube.

in vivo: "In life"; that is, in the living cell or organism.

ion channels: Integral proteins that provide for the regulated transport of a specific ion, or ions, across a membrane.

ion-exchange chromatography: A process for separating complex mixtures of ionic compounds by many repeated partitionings between a flowing (mobile) phase and a stationary phase consisting of a polymeric resin that contains fixed charged groups.

ionizing radiation: A type of radiation, such as x rays, that causes loss of electrons from some organic molecules, thus making them more reactive.

ionophore: A compound that binds one or more metal ions and is capable of diffusing across a membrane, carrying the bound ion.

ionotropic: Describes a membrane receptor that acts as a ligand-gated ion channel. *Compare* metabotropic.

ion product of water (K$_w$): The product of the concentrations of H$^+$ and OH$^-$ in pure water; $K_w = [H^+][OH^-] = 1 \times 10^{-14}$ at 25 °C.

ion semiconductor sequencing: DNA sequencing technology in which nucleotide additions are detected by measuring electrons released.

iron-sulfur protein: One of a large family of electron-transfer proteins in which the electron carrier is one or more iron ions associated with two or more sulfur atoms of Cys residues or of inorganic sulfide.

isoelectric focusing: An electrophoretic method for separating macromolecules on the basis of isoelectric pH.

isoelectric pH (isoelectric point, pI): The pH at which a solute has no net electric charge and thus does not move in an electric field.

isoenzymes: *See* isozymes.

isomerases: Enzymes that catalyze the transformation of compounds into their positional isomers.

isomers: Any two molecules with the same molecular formula but a different arrangement of molecular groups.

isoprene: The hydrocarbon 2-methyl-1,3-butadiene, a recurring structural unit of terpenoids.

isoprenoid: Any of a large number of natural products synthesized by enzymatic polymerization of two or more isoprene units; also called terpenoid.

isozymes: Multiple forms of an enzyme that catalyze the same reaction but differ in amino acid sequence, substrate affinity, V_{max}, and/or regulatory properties; also called isoenzymes.

k

K$_a$: *See* acid dissociation constant.

K$_d$: *See* dissociation constant.

K$_{eq}$: *See* equilibrium constant.

K$_m$: *See* Michaelis constant.

K$_t$ (K$_{transport}$): A kinetic parameter for a membrane transporter, analogous to the Michaelis constant, K_m, for an enzymatic reaction. The rate of substrate uptake is half-maximal when the substrate concentration equals K_t.

K$_w$: *See* ion product of water.

ketoacidosis: A pathological condition sometimes experienced by people with untreated diabetes in which the ketone bodies acetoacetate and D-β-hydroxybutyrate reach extraordinary levels in tissues, urine, and blood (ketosis), which lowers the blood pH (acidosis).

ketogenic: Yielding acetyl-CoA, a precursor for ketone body formation, as a breakdown product.

ketone bodies: Acetoacetate, D-β-hydroxybutyrate, and acetone; water-soluble fuels normally exported by the liver but overproduced during fasting or in untreated diabetes mellitus.

ketose: A simple monosaccharide in which the carbonyl group is a ketone.

ketosis: A condition in which the concentration of ketone bodies in the blood, tissues, and urine is abnormally high.

kinases: Enzymes that catalyze phosphorylation of certain molecules by ATP.

kinetics: The study of reaction rates.

Krebs cycle: *See* citric acid cycle.

l

lagging strand: The DNA strand that, during replication, must be synthesized in the direction opposite to that in which the replication fork moves.

law of mass action: The law stating that the rate of any given chemical reaction is proportional to the product of the activities (or concentrations) of the reactants.

leader: A short sequence near the amino terminus of a protein or the 5′ end of an RNA that has a specialized targeting or regulatory function.

leading strand: The DNA strand that, during replication, is synthesized in the same direction in which the replication fork moves.

leaky mutant: A mutant gene that gives rise to a product with a detectable level of biological activity.

leaving group: The departing or displaced molecular group in a unimolecular elimination or bimolecular substitution reaction.

lectin: A protein that binds a carbohydrate, commonly an oligosaccharide, with very high affinity and specificity, mediating cell-cell interactions.

lethal mutation: A mutation that inactivates a biological function essential to the life of the cell or organism.

leucine zipper: A protein structural motif involved in protein-protein interactions in many eukaryotic regulatory proteins; consists of two interacting α helices in which Leu residues at every seventh position are a prominent feature of the interacting surfaces.

leukocyte: A white blood cell; involved in the immune response in mammals.

leukotriene: Any of a class of noncyclic eicosanoid signaling lipids with three conjugated double bonds; they mediate inflammatory responses, including smooth muscle activity.

ligand: A small molecule that binds specifically to a larger one; for example, a hormone is the ligand for its specific protein receptor.

ligases: Enzymes that catalyze condensation reactions in which two atoms are joined, using the energy of ATP or another energy-rich compound.

light-dependent reactions: The reactions of photosynthesis that require light and cannot occur in the dark; also known as light reactions.

Lineweaver-Burk equation: An algebraic transform of the Michaelis-Menten equation, allowing determination of V_{max} and K_m by extrapolation of [S] to infinity:
$$\frac{1}{V_0} = \frac{K_m}{V_{max}[S]} + \frac{1}{V_{max}}.$$

linking number: The number of times one closed circular DNA strand is wound about another; the number of topological links holding the circles together.

lipases: Enzymes that catalyze the hydrolysis of triacylglycerols.

lipid: A small water-insoluble biomolecule generally containing fatty acids, sterols, or isoprenoid compounds.

lipidome: The full complement of lipid-containing molecules in a cell, organ, or tissue under a particular set of conditions.

lipidomics: The systematic characterization of the lipidome.

lipoate (lipoic acid): A vitamin for some microorganisms; an intermediate carrier of hydrogen atoms and acyl groups in α-keto acid dehydrogenases.

lipoprotein: A lipid-protein aggregate that carries water-insoluble lipids in the blood. The protein component alone is an apolipoprotein.

liposome: A small, spherical vesicle composed of a phospholipid bilayer, forming spontaneously when phospholipids are suspended in an aqueous buffer.

lipoxin: One of a class of hydroxylated linear derivatives of arachidonate that act as potent antiinflammatory agents.

long noncoding RNA (lncRNA): An RNA generally more than 200 nucleotides long that lacks an open reading frame encoding a protein; lncRNAs have numerous and varied functions in gene regulation and structural maintenance of chromatin.

lyases: Enzymes that catalyze removal of a group from a molecule to form a double bond, or addition of a group to a double bond.

lymphocytes: A subclass of leukocytes involved in the immune response. *See also* B lymphocytes; T lymphocytes.

lysis: Destruction of a plasma membrane or (in bacteria) cell wall, releasing the cellular contents and killing the cell.

lysosome: An organelle of eukaryotic cells; contains many hydrolytic enzymes and serves as a degrading and recycling center for unneeded cellular components.

m

macromolecule: A molecule having a molecular weight in the range of a few thousand to many millions.

mass-action ratio (Q): For the reaction $aA + bB \rightleftharpoons cC + dD$, the ratio $[C]^c[D]^d/[A]^a[B]^b$.

matrix: The space enclosed by the inner membrane of the mitochondrion.

mechanistic target of rapamycin complex 1: *See* mTORC1.

meiosis: A type of cell division in which diploid cells give rise to haploid cells destined to become gametes or spores.

membrane potential (V_m): The difference in electrical potential across a biological membrane, commonly measured by insertion of a microelectrode; typical values range from -25 mV (by convention, the negative sign indicates that the inside is negative relative to the outside) to more than -100 mV across some plant vacuolar membranes.

membrane transport: Movement of a polar solute across a membrane via a specific membrane protein (a transporter).

messenger RNA (mRNA): A class of RNA molecules, each of which is complementary to one strand of DNA; carries the genetic message from the chromosome to the ribosomes.

metabolic control: The mechanisms by which flux through a metabolic pathway is changed to reflect a cell's altered circumstances.

metabolic regulation: The mechanisms by which a cell resists changes in the concentrations of individual metabolites that would otherwise occur when metabolic control mechanisms alter flux through a pathway.

metabolic syndrome: A combination of medical conditions that together predispose to cardiovascular disease and type 2 diabetes; includes high blood pressure, high concentrations of LDL and triacylglycerol in the blood, slightly elevated fasting blood glucose concentration, and obesity.

metabolism: The entire set of enzyme-catalyzed transformations of organic molecules in living cells; the sum of anabolism and catabolism.

metabolite: A chemical intermediate in the enzyme-catalyzed reactions of metabolism.

metabolome: The complete set of small-molecule metabolites (metabolic intermediates, signals, secondary metabolites)

present in a given cell or tissue under specific conditions.

metabolomics: The systematic characterization of the metabolome of a cell or tissue.

metabolon: A supramolecular assembly of sequential metabolic enzymes.

metabotropic: Describes a membrane receptor that acts through a second messenger. *Compare* ionotropic.

metalloprotein: A protein with a metal ion as its prosthetic group.

metamerism: Division of the body into segments, as in insects, for example.

micelle: An aggregate of amphipathic molecules in water, with the nonpolar portions in the interior and the polar portions at the exterior surface, exposed to water.

Michaelis constant (K_m): The substrate concentration at which an enzyme-catalyzed reaction proceeds at one-half its maximum velocity.

Michaelis-Menten equation: The equation describing the hyperbolic dependence of the initial reaction velocity, V_0, on substrate concentration, [S], in many enzyme-catalyzed reactions: $V_0 = \dfrac{V_{max}[S]}{K_m + [S]}$.

Michaelis-Menten kinetics: A kinetic pattern in which the initial rate of an enzyme-catalyzed reaction exhibits a hyperbolic dependence on substrate concentration.

microRNA (miRNA): A class of small RNA molecules (20 to 25 nucleotides after processing is complete) involved in gene silencing by inhibiting translation and/or promoting degradation of particular mRNAs.

microsomes: Membranous vesicles formed by fragmentation of the endoplasmic reticulum of eukaryotic cells; recovered by differential centrifugation.

miRNA: *See* microRNA.

mismatch: A base pair in a nucleic acid that cannot form a normal Watson-Crick pair.

mismatch repair: An enzymatic system for repairing base mismatches in DNA.

mitochondrion: An organelle of eukaryotic cells; contains the enzyme systems required for the citric acid cycle, fatty acid oxidation, respiratory electron transfer, and oxidative phosphorylation.

mitosis: In eukaryotic cells, the multistep process that results in chromosome replication and cell division.

mixed-function oxidases: Enzymes that catalyze reactions in which two different substrates are oxidized by molecular oxygen simultaneously, but the oxygen atoms do not appear in the oxidized product.

mixed-function oxygenases: Enzymes (e.g., a monooxygenase) that catalyze reactions in which two reductants—one generally

NADPH, the other the substrate—are oxidized by molecular oxygen, with one oxygen atom incorporated into the product and the other reduced to H_2O; often use cytochrome P-450 to carry electrons from NADPH to O_2. Oxygenases, unlike oxidases, promote reactions in which at least one oxygen atom is incorporated into the final product.

mixed inhibition: The reversible inhibition pattern resulting when an inhibitor molecule can bind to either the free enzyme or the enzyme-substrate complex (not necessarily with the same affinity).

modulator: A metabolite that, when bound to the allosteric site of an enzyme, alters its kinetic characteristics.

molar solution: One mole of solute dissolved in water to give a total volume of 1,000 mL.

mole: One gram molecular weight of a compound. *See also* Avogadro's number.

monocistronic mRNA: An mRNA that can be translated into only one protein.

monoclonal antibodies: Antibodies produced by a cloned hybridoma cell, which are therefore identical and directed against the same epitope of the antigen. (Hybridoma cells are stable, antibody-producing cell lines that grow well in tissue culture; created by fusing an antibody-producing B lymphocyte with a myeloma cell.)

monosaccharide: A carbohydrate consisting of a single sugar unit.

moonlighting enzymes: Enzymes that play two distinct roles, at least one of which is catalytic; the other may be catalytic, regulatory, or structural.

motif: Any distinct folding pattern for elements of secondary structure, observed in one or more proteins. A motif can be simple or complex, and can represent all or just a small part of a polypeptide chain. Also called a fold or supersecondary structure.

mRNA: *See* messenger RNA.

mTORC1 (mechanistic target of rapamycin complex 1): A multiprotein complex of mTOR (mechanistic target of rapamycin) and several regulatory subunits, which together act as a Ser/Thr protein kinase; stimulated by nutrients and energy-sufficient conditions, it triggers cell growth and proliferation.

mucopolysaccharide: *See* glycosaminoglycan.

multidrug transporters: Plasma membrane transporters in the ABC transporter family that expel several commonly used antitumor drugs, thereby interfering with antitumor therapy.

multienzyme system: A group of related enzymes participating in a given metabolic pathway.

mutarotation: The change in specific rotation of a pyranose or furanose sugar or glycoside that accompanies equilibration of its α- and β-anomeric forms.

mutases: Enzymes that catalyze the transposition of functional groups.

mutation: An inheritable change in the nucleotide sequence of a chromosome.

myocyte: A muscle cell.

myofibril: A unit of thick and thin filaments of muscle fibers.

myosin: A contractile protein; the major component of the thick filaments of muscle and other actin-myosin systems.

n

NAD, NADP (nicotinamide adenine dinucleotide, nicotinamide adenine dinucleotide phosphate): Nicotinamide-containing coenzymes that function as carriers of hydrogen atoms and electrons in some oxidation-reduction reactions.

Na^+K^+ ATPase: The electrogenic ATP-driven active transporter in the plasma membrane of most animal cells that pumps three Na^+ outward for every two K^+ moved inward.

native conformation: The biologically active conformation of a macromolecule.

ncRNA (noncoding RNA): Any RNA that does not encode instructions for a protein product.

negative cooperativity: A property of some multisubunit enzymes or proteins in which binding of a ligand or substrate to one subunit impairs binding to another subunit.

negative feedback: Regulation of a biochemical pathway in which a reaction product inhibits an earlier step in the pathway.

neuron: A cell of nervous tissue specialized for transmission of a nerve impulse.

neurotransmitter: A low molecular weight compound (usually containing nitrogen) secreted from the axon terminal of a neuron and bound by a specific receptor on the next neuron or on a myocyte; transmits a nerve impulse.

NHEJ: *See* nonhomologous end joining.

nitrogenase complex: A system of enzymes capable of reducing atmospheric nitrogen to ammonia in the presence of ATP.

nitrogen cycle: The cycling of various forms of biologically available nitrogen through the plant, animal, and microbial worlds, and through the atmosphere and geosphere.

nitrogen fixation: Conversion of atmospheric nitrogen (N_2) into a reduced, biologically available form by nitrogen-fixing organisms.

NMR: *See* nuclear magnetic resonance spectroscopy.

noncoding RNA: *See* ncRNA.

noncyclic electron flow: The light-induced flow of electrons from water to $NADP^+$ in oxygen-evolving photosynthesis; involves photosystems I and II.

nonessential amino acids: Amino acids that can be made by humans from simpler precursors and are thus not required in the diet.

nonheme iron proteins: Proteins, usually acting in oxidation-reduction reactions, that contain iron but no porphyrin groups.

nonhomologous end joining (NHEJ): A process in which a double-strand break in a chromosome is repaired by end-processing and ligation, often creating mutations at the ligation site.

nonpolar: Hydrophobic; describes molecules or groups that are poorly soluble in water.

nonsense codon: A codon that does not specify an amino acid, but signals termination of a polypeptide chain.

nonsense mutation: A mutation that results in premature termination of a polypeptide chain.

nonsense suppressor: A mutation, usually in the gene for a tRNA, that causes an amino acid to be inserted into a polypeptide in response to a termination codon.

nontemplate strand: *See* coding strand.

nuclear magnetic resonance (NMR) spectroscopy: A technique that utilizes certain quantum mechanical properties of atomic nuclei to study the structure and dynamics of the molecules of which the nuclei are a part.

nucleases: Enzymes that hydrolyze the internucleotide (phosphodiester) linkages of nucleic acids.

nucleic acids: Biologically occurring polynucleotides in which the nucleotide residues are linked in a specific sequence by phosphodiester bonds; DNA and RNA.

nucleoid: In bacteria, the nuclear zone that contains the chromosome but has no surrounding membrane.

nucleolus: In eukaryotic cells, a densely staining structure in the nucleus; the site of rRNA synthesis and ribosome formation.

nucleophile: An electron-rich group with a strong tendency to donate electrons to an electron-deficient nucleus (electrophile); the entering reactant in a bimolecular substitution reaction.

nucleoplasm: The portion of a eukaryotic cell's contents that is enclosed by the nuclear membrane.

nucleoside: A compound consisting of a purine or pyrimidine base covalently linked to a pentose.

nucleoside diphosphate kinase: An enzyme that catalyzes transfer of the terminal phosphate of a nucleoside 5'-triphosphate to a nucleoside 5'-diphosphate.

nucleoside diphosphate sugar: A coenzymelike carrier of a sugar molecule, functioning in the enzymatic synthesis of polysaccharides and sugar derivatives.

nucleoside monophosphate kinase: An enzyme that catalyzes transfer of the terminal phosphate of ATP to a nucleoside 5'-monophosphate.

nucleosome: In eukaryotes, a structural unit for packaging chromatin; consists of a DNA strand wound around a histone core.

nucleotide: A nucleoside phosphorylated at one of its pentose hydroxyl groups.

nucleus: In eukaryotes, an organelle that contains chromosomes.

O

***O*-glycosidic bonds:** Bonds between a sugar and another molecule (typically an alcohol, purine, pyrimidine, or sugar) through an intervening oxygen.

oligomer: A short polymer, usually of amino acids, sugars, or nucleotides; the definition of "short" is somewhat arbitrary, but usually fewer than 50 subunits.

oligomeric protein: A multisubunit protein having two or more polypeptide chains.

oligonucleotide: A short polymer of nucleotides (usually fewer than 50).

oligopeptide: A short polymer of amino acids joined by peptide bonds.

oligosaccharide: Several monosaccharide groups joined by glycosidic bonds.

ω oxidation: An alternative mode of fatty acid oxidation in which the initial oxidation is at the carbon most distant from the carboxyl carbon; as distinct from β oxidation.

oncogene: A cancer-causing gene; any of several mutant genes that cause cells to exhibit rapid, uncontrolled proliferation. *See also* proto-oncogene.

open reading frame (ORF): A group of contiguous, nonoverlapping nucleotide codons in a DNA or RNA molecule that does not include a termination codon.

open system: A system that exchanges matter and energy with its surroundings. *See also* system.

operator: A region of DNA that interacts with a repressor protein to control the expression of a gene or group of genes.

operon: A unit of genetic expression consisting of one or more related genes and the operator and promoter sequences that regulate their transcription.

opsin: The protein portion of the visual pigment, which becomes rhodopsin with addition of the chromophore retinal.

optical activity: The capacity of a substance to rotate the plane of plane-polarized light.

optimum pH: The characteristic pH at which an enzyme has maximal catalytic activity.

orexigenic: Tending to increase appetite and food consumption. *Compare* anorexigenic.

ORF: *See* open reading frame.

organelles: Membrane-bounded structures found in eukaryotic cells; contain enzymes and other components required for specialized cell functions.

origin: The nucleotide sequence or site in DNA at which replication is initiated.

orthologs: Genes or proteins from different species that possess a clear sequence and functional relationship to each other.

osmosis: Bulk flow of water through a semipermeable membrane into another aqueous compartment containing solute at a higher concentration.

osmotic pressure: Pressure generated by the osmotic flow of water through a semipermeable membrane into an aqueous compartment containing solute at a higher concentration.

oxidases: Enzymes that catalyze oxidation reactions in which molecular oxygen serves as the electron acceptor, but neither of the oxygen atoms is incorporated into the product. *Compare* oxygenases.

oxidation: The loss of electrons from a compound.

oxidation-reduction reaction: A reaction in which electrons are transferred from a donor to an acceptor molecule; also called a redox reaction.

oxidative phosphorylation: The enzymatic phosphorylation of ADP to ATP coupled to electron transfer from a substrate to molecular oxygen.

oxidizing agent (oxidant): The acceptor of electrons in an oxidation-reduction reaction.

oxygenases: Enzymes that catalyze reactions in which oxygen atoms are directly incorporated into the product, forming a hydroxyl or carboxyl group. In a monooxygenase reaction, only one oxygen atom is incorporated; the other is reduced to H_2O. In a dioxygenase reaction, both oxygens are incorporated into the product. *Compare* oxidases.

oxygenic photosynthesis: Light-driven ATP and NADPH synthesis in organisms that use water as the electron source, producing O_2.

P

palindrome: A segment of duplex DNA in which the base sequences of the two strands exhibit twofold rotational symmetry about an axis.

paradigm: In biochemistry, an experimental model or example.

paralogs: Genes or proteins present in the same species that possess a clear sequence and functional relationship to each other.

partition coefficient: A constant that expresses the ratio in which a given solute will be partitioned or distributed between two given immiscible liquids at equilibrium.

passive transporter: A membrane protein that increases the rate of movement of a solute across the membrane along its electrochemical gradient without the input of energy.

pathogenic: Disease-causing.

PCR: *See* polymerase chain reaction.

PDB (Protein Data Bank): An international database (www.pdb.org) that archives data describing the three-dimensional structure of nearly all macromolecules for which structures have been published.

pentose: A simple sugar with a backbone containing five carbon atoms.

pentose phosphate pathway: A pathway, present in most organisms, that interconverts hexoses and pentoses and is a source of reducing equivalents (NADPH) and pentoses for biosynthetic processes; it begins with glucose 6-phosphate and includes 6-phosphogluconate as an intermediate. Also called the phosphogluconate pathway and hexose monophosphate pathway.

peptidases: Enzymes that hydrolyze peptide bonds.

peptide: Two or more amino acids covalently joined by peptide bonds.

peptide bond: A substituted amide linkage between the α-amino group of one amino acid and the α-carboxyl group of another, with elimination of the elements of water.

peptidoglycan: A major component of bacterial cell walls; generally consists of parallel heteropolysaccharides cross-linked by short peptides.

peptidyl transferase: The enzyme activity that synthesizes the peptide bonds of proteins; a ribozyme, part of the rRNA of the large ribosomal subunit.

peripheral proteins: Proteins loosely or reversibly bound to a membrane by hydrogen bonds or electrostatic forces; generally water-soluble once released from the membrane. *Compare* integral proteins.

permeases: *See* transporters.

peroxisome: An organelle of eukaryotic cells; contains peroxide-forming and peroxide-destroying enzymes.

peroxisome proliferator-activated receptor: *See* PPAR.

pH: The negative logarithm of the hydrogen ion concentration of an aqueous solution.

phage: *See* bacteriophage.

phenotype: The observable characteristics of an organism.

phosphatases: Enzymes that cleave phosphate esters by hydrolysis, the addition of the elements of water.

phosphodiester linkage: A chemical grouping that contains two alcohols esterified to one molecule of phosphoric acid, which thus serves as a bridge between them.

phosphogluconate pathway: *See* pentose phosphate pathway.

phospholipid: A lipid containing one or more phosphate groups.

phosphoprotein phosphatases: *See* protein phosphatases.

phosphorolysis: Cleavage of a compound with phosphate as the attacking group; analogous to hydrolysis.

phosphorylases: Enzymes that catalyze phosphorolysis.

phosphorylation: Formation of a phosphate derivative of a biomolecule, usually by enzymatic transfer of a phosphoryl group from ATP.

phosphorylation potential (ΔG_p): The actual free-energy change of ATP hydrolysis under the nonstandard conditions prevailing in a cell.

photochemical reaction center: The part of a photosynthetic complex where the energy of an absorbed photon causes charge separation, initiating electron transfer.

photon: The ultimate unit (a quantum) of light energy.

photophosphorylation: The enzymatic formation of ATP from ADP coupled to the light-dependent transfer of electrons in photosynthetic cells.

photoreduction: The light-induced reduction of an electron acceptor in photosynthetic cells.

photorespiration: Oxygen consumption occurring in illuminated temperate-zone plants that is largely due to oxidation of phosphoglycolate.

photosynthesis: The use of light energy to produce carbohydrates from carbon dioxide and a reducing agent such as water. *Compare* oxygenic photosynthesis.

photosynthetic phosphorylation: *See* photophosphorylation.

photosystem: In photosynthetic cells, a functional set of light-absorbing pigments and its reaction center, where the energy of an absorbed photon is transduced into a separation of electric charges.

phototroph: An organism that can use the energy of light to synthesize its own fuels from simple molecules such as carbon dioxide, oxygen, and water; as distinct from a chemotroph.

pI: *See* isoelectric pH.

pK_a: The negative logarithm of an equilibrium constant.

plasmalogen: A phospholipid with an alkenyl ether substituent on C-1 of glycerol.

plasma membrane: The exterior membrane enclosing the cytoplasm of a cell.

plasma proteins: The proteins present in blood plasma.

plasmid: An extrachromosomal, independently replicating, small circular DNA molecule; commonly employed in genetic engineering.

plastid: In plants, a self-replicating organelle; may differentiate into a chloroplast or amyloplast.

platelets: Small, enucleated cells that initiate blood clotting; they arise from bone marrow cells called megakaryocytes. Also known as thrombocytes.

pleated sheet: The side-by-side, hydrogen-bonded arrangement of polypeptide chains in the extended β conformation.

plectonemic: Describes a structure in a molecular polymer that has a net twisting of strands about each other in a simple and regular way.

PLP: *See* pyridoxal phosphate.

polar: Hydrophilic, or "water-loving"; describes molecules or groups that are soluble in water.

polarity: (1) In chemistry, the nonuniform distribution of electrons in a molecule; polar molecules are usually soluble in water. (2) In molecular biology, the distinction between the 5′ and 3′ ends of nucleic acids.

poly(A) site choice: The strand cleavage and addition of a 3′-poly(A) tract at alternative locations within an mRNA transcript to generate different mature mRNAs.

poly(A) tail: A length of adenosine residues added to the 3′ end of many mRNAs in eukaryotes (and sometimes in bacteria).

polycistronic mRNA: A contiguous mRNA with more than two genes that can be translated into proteins.

polyclonal antibodies: A heterogeneous pool of antibodies produced in an animal by different B lymphocytes in response to an antigen. Different antibodies in the pool recognize different parts of the antigen.

polylinker: A short, often synthetic, fragment of DNA containing recognition sequences for several restriction endonucleases.

polymerase chain reaction (PCR): A repetitive laboratory procedure that results in geometric amplification of a specific DNA sequence.

polymorphic: Describes a protein for which amino acid sequence variants are present in a population of organisms, but the variations do not destroy the protein's function.

polynucleotide: A covalently linked sequence of nucleotides in which the 3′ hydroxyl of the pentose of one nucleotide residue is joined by a phosphodiester bond to the 5′ hydroxyl of the pentose of the next residue.

polypeptide: A long chain of amino acids linked by peptide bonds; the molecular weight is generally less than 10,000.

polyribosome: *See* polysome.

polysaccharide: A linear or branched polymer of monosaccharide units linked by glycosidic bonds.

polysome: A complex of an mRNA molecule and two or more ribosomes; also called a polyribosome.

polyunsaturated fatty acid (PUFA): A fatty acid with more than one double bond, generally nonconjugated.

P/O ratio: The number of moles of ATP formed in oxidative phosphorylation per $\frac{1}{2}O_2$ reduced (thus, per pair of electrons passed to O_2). Experimental values used in this text are 2.5 for passage of electrons from NADH to O_2, and 1.5 for passage of electrons from FADH to O_2.

porphyria: An inherited disease resulting from the lack of one or more enzymes required to synthesize porphyrins.

porphyrin: A complex nitrogenous compound containing four substituted pyrroles covalently joined in a ring; often complexed with a central metal atom.

positive cooperativity: A property of some multisubunit enzymes or proteins in which binding of a ligand or substrate to one subunit facilitates binding to another subunit.

positive-inside rule: The general observation that most plasma membrane proteins are oriented so that most of their positively charged residues (Lys and Arg) are on the cytosolic face.

posttranscriptional processing: The enzymatic processing of the primary RNA transcript to produce functional RNAs, including mRNAs, tRNAs, rRNAs, and many other classes of RNAs.

posttranslational modification: The enzymatic processing of a polypeptide chain after translation from its mRNA.

PPAR (peroxisome proliferator-activated receptor): A family of nuclear transcription factors, activated by lipidic ligands, that alter the expression of specific genes, including those encoding enzymes of lipid synthesis and breakdown.

prebiotic: A selectively fermented, nondigestible food ingredient that supports the growth of health-promoting bacteria.

pre–steady state: In an enzyme-catalyzed reaction, the period preceding establishment of the steady state, often encompassing just the first enzymatic turnover.

primary structure: A description of the covalent backbone of a polymer (macromolecule), including the sequence of monomeric subunits and any interchain and intrachain covalent bonds.

primary transcript: The immediate RNA product of transcription before any posttranscriptional processing reactions.

primases: Enzymes that catalyze formation of RNA oligonucleotides used as primers by DNA polymerases.

primer: A short oligomer (e.g., of sugars or nucleotides) to which an enzyme adds additional monomeric subunits.

primer terminus: The end of a primer to which monomeric subunits are added.

priming: (1) In protein phosphorylation, the phosphorylation of an amino acid residue that becomes the binding site and point of reference for phosphorylation of other residues in the same protein. (2) In DNA replication, the synthesis of a short oligonucleotide to which DNA polymerases can add additional nucleotides.

primosome: An enzyme complex that synthesizes the primers required for lagging strand DNA synthesis.

probiotic: A live microorganism that, administered in adequate amounts, confers a health benefit on the host.

processivity: For any enzyme that catalyzes the synthesis of a biological polymer, the property of adding multiple subunits to the polymer without dissociating from the substrate.

prochiral molecule: A symmetric molecule that can react asymmetrically with an enzyme having an asymmetric active site, generating a chiral product.

projection formulas: *See* Fischer projection formulas.

prokaryote: A term used historically to refer to any species in the domains Bacteria and Archaea. The differences between bacteria (formerly "eubacteria") and archaea are sufficiently great that this term is of marginal usefulness. A tendency to use "prokaryote" when referring only to bacteria is common and misleading; "prokaryote" also implies an ancestral relationship to eukaryotes, which is incorrect. We do not use "prokaryote" and "prokaryotic" in this text.

promoter: A DNA sequence at which RNA polymerase may bind, leading to initiation of transcription.

proofreading: The correction of errors in the synthesis of an information-containing biopolymer by removing incorrect monomeric subunits after they have been covalently added to the growing polymer.

prostaglandin: One of a class of polyunsaturated, cyclic eicosanoid lipids that act as paracrine hormones.

prosthetic group: A metal ion or organic compound (other than an amino acid) covalently bound to a protein and essential to its activity.

proteases: Enzymes that catalyze the hydrolytic cleavage of peptide bonds in proteins.

proteasome: A supramolecular assembly of enzyme complexes that function in the degradation of damaged or unneeded cellular proteins.

protein: A macromolecule composed of one or more polypeptide chains, each with a characteristic sequence of amino acids linked by peptide bonds.

Protein Data Bank: *See* PDB.

protein kinases: Enzymes that transfer the terminal phosphoryl group of ATP or another nucleoside triphosphate to a Ser, Thr, Tyr, Asp, or His side chain in a target protein, thereby regulating the activity or other properties of that protein.

protein phosphatases: Enzymes that hydrolyze a phosphate ester or anhydride bond on a protein, releasing inorganic phosphate, P_i. Also called phosphoprotein phosphatases.

protein targeting: The process by which newly synthesized proteins are sorted and transported to their proper locations in the cell.

proteoglycan: A hybrid macromolecule consisting of a heteropolysaccharide joined to a polypeptide; the polysaccharide is the major component.

proteome: The full complement of proteins expressed in a given cell, or the complete complement of proteins that can be expressed by a given genome.

proteomics: Broadly, the study of the protein complement of a cell or organism.

proteostasis: The maintenance of a cellular steady-state collection of proteins that are required for cell functions under a given set of conditions.

protomer: A general term describing any repeated unit of one or more stably associated protein subunits in a larger protein structure. In a protomer with multiple subunits, the subunits may be identical or different.

proton acceptor: An anionic compound capable of accepting a proton from a proton donor; that is, a base.

proton donor: The donor of a proton in an acid-base reaction; that is, an acid.

proton-motive force: The electrochemical potential inherent in a transmembrane gradient of H^+ concentration; used in oxidative phosphorylation and photophosphorylation to drive ATP synthesis.

proto-oncogene: A cellular gene, usually encoding a regulatory protein, that can be converted into an oncogene by mutation.

PUFA: *See* polyunsaturated fatty acid.

purine: A nitrogenous heterocyclic base that is a component of nucleotides and nucleic acids; contains fused pyrimidine and imidazole rings.

puromycin: An antibiotic that inhibits polypeptide synthesis by incorporating into a growing polypeptide chain, causing its premature termination.

pyranose: A simple sugar containing the six-membered pyran ring.

pyridine nucleotide: A nucleotide coenzyme containing the pyridine derivative nicotinamide; NAD or NADP.

pyridoxal phosphate (PLP): A coenzyme containing the vitamin pyridoxine (vitamin B_6); functions in amino group transfer reactions.

pyrimidine: A nitrogenous heterocyclic base that is a component of nucleotides and nucleic acids.

pyrimidine dimer: A covalently joined dimer of two adjacent pyrimidine residues in DNA, induced by absorption of UV light; most commonly derived from two adjacent thymines (a thymine dimer).

pyrophosphatase: *See* inorganic pyrophosphatase.

pyrosequencing: A DNA sequencing technology in which each nucleotide addition by DNA polymerase triggers a series of reactions ending in a luciferase-generated flash of light.

q

Q: *See* mass-action ratio.

quantitative PCR (qPCR): A PCR procedure that allows determination of how much of an amplified template was in the original sample; also called real-time PCR.

quantum: The ultimate unit of energy.

quaternary structure: The three-dimensional structure of a multisubunit protein, particularly the manner in which the subunits fit together.

r

racemic mixture (racemate): An equimolar mixture of the D and L stereoisomers of an optically active compound.

radical: An atom or group of atoms possessing an unpaired electron; also called a free radical.

radioactive isotope: An isotopic form of an element with an unstable nucleus that stabilizes itself by emitting ionizing radiation.

radioimmunoassay (RIA): A sensitive, quantitative method for detecting trace amounts of a biomolecule, based on its capacity to displace a radioactive form of the molecule from combination with its specific antibody.

Ras superfamily of G proteins: Small (M_r ~20,000), monomeric guanosine nucleotide–binding proteins that regulate signaling and membrane trafficking pathways; inactive with GDP bound, activated by displacement of the GDP by GTP, then inactivated by their intrinsic GTPase. Also called small G proteins.

rate constant: The proportionality constant that relates the velocity of a chemical reaction to the concentration(s) of the reactant(s).

rate-limiting step: (1) Generally, the step in an enzymatic reaction with the greatest activation energy or with the transition state of highest free energy. (2) The slowest step in a metabolic pathway.

reaction intermediate: Any chemical species in a reaction pathway that has a finite chemical lifetime.

reactive oxygen species (ROS): Highly reactive products of the partial reduction of O_2, including hydrogen peroxide (H_2O_2), superoxide ($^{\bullet}O^{2-}$), and hydroxyl free radical ($^{\bullet}OH$); minor byproducts of oxidative phosphorylation.

reading frame: A contiguous, nonoverlapping set of three-nucleotide codons in DNA or RNA.

receptor Tyr kinase (RTK): A large family of plasma membrane proteins with a ligand-binding site on the extracellular domain, a single transmembrane helix, and a cytoplasmic domain with protein Tyr kinase activity controlled by the extracellular ligand.

recombinant DNA: DNA formed by the joining of genes into new combinations.

recombination: Any enzymatic process by which the linear arrangement of a nucleic acid sequence in a chromosome is altered by cleavage and rejoining.

recombinational DNA repair: Recombinational processes directed at the repair of DNA strand breaks or cross-links, especially at inactivated replication forks.

redox pair: An electron donor and its corresponding oxidized form; for example, NADH and NAD^+.

redox reaction: *See* oxidation-reduction reaction.

reducing agent (reductant): The electron donor in an oxidation-reduction reaction.

reducing end: The end of a polysaccharide that has a terminal sugar with a free anomeric carbon; the terminal residue can act as a reducing sugar.

reducing equivalent: A general term for an electron or an electron equivalent in the form of a hydrogen atom or a hydride ion.

reducing sugar: A sugar in which the carbonyl (anomeric) carbon is not involved in a glycosidic bond and can therefore undergo oxidation.

reduction: The gain of electrons by a compound or ion.

regulator of G protein signaling (RGS): A protein structural domain that stimulates the GTPase activity of heterotrimeric G proteins.

regulatory cascade: A multistep regulatory pathway in which a signal leads to activation of a series of proteins in succession, with each protein in the succession catalytically activating the next, such that the original signal is amplified exponentially. *See also* enzyme cascade.

regulatory enzyme: An enzyme with a regulatory function, through its capacity to undergo a change in catalytic activity by allosteric mechanisms or by covalent modification.

regulatory gene: A gene that gives rise to a product involved in regulation of the expression of another gene; for example, a gene encoding a repressor protein.

regulatory sequence: A DNA sequence involved in regulating the expression of a gene; for example, a promoter or operator.

regulon: A group of genes or operons that are coordinately regulated, even though some, or all, may be spatially distant in the chromosome or genome.

relaxed DNA: Any DNA that exists in its most stable, unstrained structure, typically the B form under most cellular conditions.

release factors: Protein factors in the cytosol required for the release of a completed polypeptide chain from a ribosome; also known as termination factors.

renaturation: The refolding of an unfolded (denatured) globular protein so as to restore its native structure and function.

replication: Synthesis of daughter nucleic acid molecules identical to the parental nucleic acid.

replication fork: The Y-shaped structure generally found at the point where DNA is being synthesized.

replicative form: Any of the full-length structural forms of a viral chromosome that serve as distinct replication intermediates.

replisome: The multiprotein complex that promotes DNA synthesis at the replication fork.

repressible enzyme: In bacteria, an enzyme for which synthesis is inhibited when its reaction product is readily available to the cell.

repression: A decrease in the expression of a gene in response to a change in the activity of a regulatory protein.

repressor: The protein that binds to the regulatory sequence or operator for a gene, blocking its transcription.

residue: A single unit in a polymer; for example, an amino acid in a polypeptide chain. The term reflects the fact that sugars, nucleotides, and amino acids lose a few atoms (generally the elements of water) when incorporated in their respective polymers.

respiration: Any metabolic process that leads to the uptake of oxygen and release of CO_2.

respiration-linked phosphorylation: ATP formation from ADP and P_i, driven by electron flow through a series of membrane-bound carriers, with a proton gradient as the direct source of energy driving rotational catalysis by ATP synthase.

respiratory chain: The electron-transfer chain; a sequence of electron-carrying proteins that transfers electrons from substrates to molecular oxygen in aerobic cells.

response element: A region of DNA, near (upstream from) a gene, bound by specific proteins that influence the rate of transcription of the gene.

restriction endonucleases: Site-specific endodeoxyribonucleases that cleave both strands of DNA at points in or near the specific site recognized by the enzyme; important tools in genetic engineering.

restriction fragment: A segment of double-stranded DNA produced by the action of a restriction endonuclease on a larger DNA.

retinal: A 20-carbon isoprene aldehyde derived from carotene, which serves as the light-sensitive component of the visual pigment rhodopsin. Illumination converts 11-*cis*-retinal to all-*trans*-retinal.

retrovirus: An RNA virus containing a reverse transcriptase.

reverse transcriptase: An RNA-directed DNA polymerase of retroviruses; capable of making DNA complementary to an RNA.

reversible inhibition: Inhibition by a molecule that binds reversibly to the enzyme, such that the enzyme activity returns when the inhibitor is no longer present.

reversible terminator sequencing: A DNA sequencing technology in which nucleotide additions are detected and scored by the color of fluorescence displayed when a labeled nucleotide with a removable sequence terminator is added.

R group: (1) Formally, an abbreviation denoting any alkyl group. (2) Occasionally, used in a more general sense to denote virtually any organic substituent (e.g., the R groups of amino acids).

RGS: *See* regulator of G protein signaling.

rhodopsin: The visual pigment, composed of the protein opsin and the chromophore retinal.

RIA: *See* radioimmunoassay.

ribonuclease: A nuclease that catalyzes the hydrolysis of certain internucleotide linkages of RNA.

ribonucleic acid: *See* RNA.

ribonucleotide: A nucleotide containing D-ribose as its pentose component.

ribosomal RNA (rRNA): A class of RNA molecules serving as components of ribosomes.

ribosome: A supramolecular complex of rRNAs and proteins, approximately 18 to 22 nm in diameter; the site of protein synthesis.

ribosome profiling: A technique employing next-generation DNA sequencing of cDNA fragments derived from RNA bound to cellular ribosomes, to determine what mRNAs are being translated at a given moment.

riboswitch: A structured segment of an mRNA that binds to a specific ligand and affects translation or processing of the mRNA.

ribozymes: Ribonucleic acid molecules with catalytic activities; RNA enzymes.

ribulose 1,5-bisphosphate carboxylase/oxygenase (rubisco): The enzyme that fixes CO_2 into organic form (3-phosphoglycerate) in organisms (plants and some microorganisms) capable of CO_2 fixation.

Rieske iron-sulfur protein: A type of iron-sulfur protein in which two of the ligands to the central iron ion are His side chains; act in many electron-transfer sequences, including oxidative phosphorylation and photophosphorylation.

RNA (ribonucleic acid): A polyribonucleotide of a specific sequence linked by successive 3′,5′-phosphodiester bonds.

RNA editing: Posttranscriptional modification of an mRNA that alters the meaning of one or more codons during translation.

RNA polymerase: An enzyme that catalyzes formation of RNA from ribonucleoside 5′-triphosphates, using a strand of DNA or RNA as a template.

RNA recognition motif (RRM): A DNA- or RNA-binding motif consisting of a four-strand antiparallel β sheet with two α helices on one face.

RNA splicing: Removal of introns and joining of exons in a primary transcript.

ROS: *See* reactive oxygen species.

rRNA: *See* ribosomal RNA.

RTK: *See* receptor Tyr kinase.

rubisco: *See* ribulose 1,5-bisphosphate carboxylase/oxygenase.

S

salvage pathway: A pathway for synthesis of a biomolecule, such as a nucleotide, from intermediates in the degradative pathway for the biomolecule; a recycling pathway, as distinct from a de novo pathway.

Sanger sequencing: A DNA sequencing method based on the use of dideoxynucleoside triphosphates, developed by Frederick Sanger; also called dideoxy sequencing.

sarcomere: A functional and structural unit of the muscle contractile system.

satellite DNA: Highly repeated, nontranslated segments of DNA in eukaryotic chromosomes; most often associated with the centromeric region. Its function is unknown. Also called simple-sequence DNA.

saturated fatty acid: A fatty acid containing a fully saturated alkyl chain.

scaffold proteins: Noncatalytic proteins that nucleate formation of multienzyme complexes by providing two or more specific binding sites for those proteins.

scramblases: Membrane proteins that catalyze movement of phospholipids across the membrane bilayer, leading to uniform distribution of a lipid between the two membrane leaflets (monolayers).

secondary metabolism: Pathways that lead to specialized products not found in every living cell.

secondary structure: The local spatial arrangement of the main-chain atoms in a segment of a polymer (polypeptide or polynucleotide) chain.

second law of thermodynamics: The law stating that, in any chemical or physical process, the entropy of the universe tends to increase.

second messenger: An effector molecule synthesized in a cell in response to an external signal (first messenger) such as a hormone.

sedimentation coefficient: A physical constant specifying the rate of sedimentation of a particle in a centrifugal field under specified conditions.

selectins: A large family of membrane proteins that bind oligosaccharides on other cells tightly and specifically and carry signals across the plasma membrane.

SELEX: A method for rapid experimental identification of nucleic acid sequences (usually RNA) that have particular catalytic or ligand-binding properties.

sequence polymorphisms: Any alterations in genomic sequence (base-pair changes, insertions, deletions, rearrangements) that help distinguish subsets of individuals in a population or distinguish one species from another.

sequencing depth: The number of times, on average, that a given genomic nucleotide is included in sequenced DNA segments.

serine proteases: One of four major classes of proteases, having a reaction mechanism in which an active-site Ser residue acts as a covalent catalyst.

sgRNA: *See* single guide RNA.

Shine-Dalgarno sequence: A sequence in an mRNA that is required for binding bacterial ribosomes.

short tandem repeat (STR): A short (typically 3 to 6 bp) DNA sequence, repeated many times in tandem at a particular location in a chromosome.

SH2 domain: A protein domain that binds tightly to a phosphotyrosine residue in certain proteins, such as the receptor Tyr kinases, initiating formation of a multiprotein complex that acts in a signaling pathway.

shuttle vector: A recombinant DNA vector that can be replicated in two or more different host species. *See also* vector.

sickle cell anemia: A human disease characterized by defective hemoglobin molecules in individuals homozygous for a mutant allele coding for the β chain of hemoglobin.

σ: (1) A subunit of the bacterial RNA polymerase that confers specificity for certain promoters; usually designated by a superscript indicating its size (e.g., σ^{70} has a molecular weight of 70,000). (2) *See* superhelical density.

signal sequence: An amino acid sequence, often at the amino terminus, that signals the cellular fate or destination of a newly synthesized protein.

signal transduction: The process by which an extracellular signal (chemical, mechanical, or electrical) is amplified and converted to a cellular response.

silent mutation: A gene mutation that causes no detectable change in the biological characteristics of the gene product.

simple diffusion: Movement of solute molecules across a membrane to a region of lower concentration, unassisted by a protein transporter.

simple protein: A protein yielding only amino acids on hydrolysis.

simple-sequence DNA: *See* satellite DNA.

single guide RNA (sgRNA): A combination of gRNA and tracrRNA that allows one RNA to both activate Cas nucleases (particularly Cas9) and target the system to a particular DNA sequence. Parts of the RNA can be engineered to target the system to any desired DNA sequence.

single-molecule real-time (SMRT) sequencing: A DNA sequencing technology in which nucleotide additions are detected as flashes of fluorescent colored light, with sensitivity enhanced so that long DNA molecules (up to 15,000 nucleotides) can be sequenced and the results displayed in real time.

single nucleotide polymorphism (SNP): A genomic base-pair change that helps distinguish one species from another or one subset of individuals in a population.

site-directed mutagenesis: A set of methods used to create specific alterations in the sequence of a gene.

site-specific recombination: A type of genetic recombination that occurs only at specific sequences.

size-exclusion chromatography: A procedure for separation of molecules by size, based on the capacity of porous polymers to exclude solutes above a certain size; also called gel filtration.

small G proteins: *See* Ras superfamily of G proteins.

small nuclear RNA (snRNA):   A class of short RNAs, typically 100 to 200 nucleotides long, present in the nucleus; involved in the splicing of eukaryotic mRNAs.

small nucleolar RNA (snoRNA):   A class of short RNAs, generally 60 to 300 nucleotides long, that guide modification of rRNAs in the nucleolus.

SMRT sequencing: *See* single-molecule real-time sequencing.

SNP: *See* single nucleotide polymorphism.

somatic cells: All body cells except the germ-line cells.

SOS response: In bacteria, a coordinated induction of a variety of genes in response to high levels of DNA damage.

Southern blot: A DNA hybridization procedure in which one or more specific DNA fragments are detected in a larger population by hybridization to a complementary, labeled nucleic acid probe.

specialized pro-resolving mediator (SPM):   One of several eicosanoids, derived from essential fatty acids, that promote the resolution phase of the inflammatory response.

specific acid-base catalysis: Acid or base catalysis involving the constituents of water (hydroxide or hydronium ions).

specific activity: The number of micromoles (μmol) of a substrate transformed by an enzyme preparation per minute per milligram of protein at 25 °C; a measure of enzyme purity.

specificity: The ability of an enzyme or receptor to discriminate among competing substrates or ligands.

specific rotation: The rotation, in degrees, of the plane of plane-polarized light (D-line of sodium) by an optically active compound at 25 °C, with a specified concentration and light path.

sphingolipid: An amphipathic lipid with a sphingosine backbone to which are attached a long-chain fatty acid and a polar alcohol.

spliceosome: A complex of RNAs and proteins involved in the splicing of mRNAs in eukaryotic cells.

splicing: *See* RNA splicing.

SPM: *See* specialized pro-resolving mediator.

standard free-energy change ($\Delta G°$):   The free-energy change for a reaction occurring under a set of standard conditions: temperature, 298 K; pressure, 1 atm (101.3 kPa); and all solutes at 1 M concentration. $\Delta G'°$ denotes the standard free-energy change at pH 7.0 in 55.5 M water.

standard reduction potential ($E'°$):   The electromotive force exhibited at an electrode by 1 M concentrations of a reducing agent and its oxidized form at 25 °C and pH 7.0; a measure of the relative tendency of the reducing agent to lose electrons.

statin: Any of a class of drugs used to reduce blood cholesterol in humans; acts by inhibiting the enzyme HMG-CoA reductase, an early step in sterol synthesis.

steady state: A nonequilibrium state of a system through which matter is flowing and in which all components remain at a constant concentration.

stem cells: The common, self-regenerating cells in bone marrow that give rise to differentiated blood cells such as erythrocytes and lymphocytes.

stereoisomers: Compounds that have the same composition and the same order of atomic connections but different molecular arrangements.

sterol: A lipid containing the steroid nucleus.

sticky ends: Two DNA ends in the same DNA molecule, or in different molecules, with short overhanging single-stranded segments that are complementary to one another, facilitating ligation of the ends; also known as cohesive ends.

stimulatory G protein (G_s):   A trimeric regulatory GTP-binding protein that, when activated by an associated plasma membrane receptor, stimulates a neighboring membrane enzyme such as adenylyl cyclase; its effects oppose those of G_i.

stop codons: *See* termination codons.

STR: *See* short tandem repeat.

stroma: The space and aqueous solution enclosed within the inner membrane of a chloroplast, not including the contents of the thylakoid membranes.

structural gene: A gene coding for a protein or RNA molecule; as distinct from a regulatory gene.

substitution mutation: A mutation caused by replacement of one base by another.

substrate: The specific compound acted upon by an enzyme.

substrate channeling: Movement of the chemical intermediates in a series of enzyme-catalyzed reactions from the active site of one enzyme to that of the next enzyme in the pathway, without leaving the surface of the protein complex that includes both enzymes.

substrate cycle: A cycle of enzyme-catalyzed reactions that results in release of thermal energy by the hydrolysis of ATP; sometimes referred to as a futile cycle.

substrate-level phosphorylation: Phosphorylation of ADP or some other nucleoside 5'-diphosphate coupled to dehydrogenation of an organic substrate; independent of the respiratory chain.

suicide inactivator: A relatively inert molecule that is transformed by an enzyme, at its active site, into a reactive substance that irreversibly inactivates the enzyme.

sulfonylurea drugs: A group of oral medications used in the treatment of type 2 diabetes; act by closing K^+ channels in pancreatic β cells, stimulating insulin secretion.

supercoil: The twisting of a helical (coiled) molecule on itself; a coiled coil.

supercoiled DNA: DNA that twists upon itself because it is under- or overwound (and thereby strained) relative to B-form DNA.

superhelical density (σ):   In a helical molecule such as DNA, the number of supercoils (superhelical turns) relative to the number of coils (turns) in the relaxed molecule.

supersecondary structure: *See* motif.

suppressor mutation: A mutation, at a site different from that of a primary mutation, that totally or partially restores a function lost by the primary mutation.

Svedberg (S):   A unit of measure of the rate at which a particle sediments in a centrifugal field.

symbionts: Two or more organisms that are mutually interdependent and usually living in physical association.

symport: Cotransport of solutes across a membrane in the same direction.

syndecan: A heparan sulfate proteoglycan with a single transmembrane domain and an extracellular domain bearing three to five chains of heparan sulfate and, in some cases, chondroitin sulfate.

synteny: Conserved gene order along the chromosomes of different species.

synthases: Enzymes that catalyze condensation reactions that do not require nucleoside triphosphate as an energy source.

synthetases: Enzymes that catalyze condensation reactions using ATP or another nucleoside triphosphate as an energy source.

system: An isolated collection of matter; all other matter in the universe apart from the system is called the surroundings.

systems biology: The study of complex biochemical systems, integrating information from genomics, proteomics, and metabolomics.

t

tag: An extra segment of protein that is fused to a protein of interest by modification of its gene, usually for purposes of purification.

T cell: *See* T lymphocyte.

telomere: A specialized nucleic acid structure at the ends of linear eukaryotic chromosomes.

template: A macromolecular mold or pattern for the synthesis of an informational macromolecule.

template strand: A strand of nucleic acid used by a polymerase as a template to synthesize a complementary strand, as distinct from the coding strand.

terminal transferase: An enzyme that catalyzes addition of nucleotide residues of a single kind to the 3′ end of DNA chains.

termination codons: UAA, UAG, and UGA; in protein synthesis, these codons signal termination of a polypeptide chain. Also known as stop codons.

termination factors: *See* release factors.

termination sequence: A DNA sequence, at the end of a transcriptional unit, that signals the end of transcription.

tertiary structure: The three-dimensional conformation of a polymer in its native, folded state.

tetrahydrobiopterin: The reduced coenzyme form of biopterin.

tetrahydrofolate: The reduced, active coenzyme form of the vitamin folate.

thermogenesis: The biological generation of heat by muscle activity (shivering), uncoupled oxidative phosphorylation, or the operation of substrate (futile) cycles.

thermogenin: *See* uncoupling protein 1.

thiamine pyrophosphate (TPP): The active coenzyme form of vitamin B_1; involved in aldehyde transfer reactions.

thiazolidinediones: A class of medications used in the treatment of type 2 diabetes; act to reduce circulating fatty acids and increase sensitivity to insulin. Also known as glitazones.

thioester: An ester of a carboxylic acid with a thiol or mercaptan.

3′ end: The end of a nucleic acid that lacks a nucleotide bound at the 3′ position of the terminal residue.

thrombocytes: *See* platelets.

thromboxane: Any of a class of eicosanoid lipids with a six-membered ether-containing ring; involved in platelet aggregation during blood clotting.

thylakoid: A closed, continuous system of flattened disks, formed by the pigment-bearing internal membranes of chloroplasts.

thymine dimer: *See* pyrimidine dimer.

tissue culture: A method by which cells derived from multicellular organisms are grown in liquid media.

titration curve: A plot of pH versus the equivalents of base added during titration of an acid.

T lymphocyte (T cell): One of a class of blood cells (lymphocytes) of thymic origin, involved in cell-mediated immune reactions.

tocopherol: Any of several forms of vitamin E.

topoisomerases: Enzymes that introduce positive or negative supercoils in closed, circular duplex DNA.

topoisomers: Different forms of a covalently closed, circular DNA molecule that differ only in linking number.

topology: The study of the properties of an object that do not change under continuous deformations such as twisting or bending.

topology diagram: A structural representation in which the connections between elements of secondary structure are depicted in two dimensions.

TPP: *See* thiamine pyrophosphate.

trace element: A chemical element required by an organism in only trace amounts.

tracrRNA: *See* trans-activating CRISPR RNA.

trans-activating CRISPR RNA (tracrRNA): A bacterially encoded RNA required for the activation and function of the relatively simple CRISPR/Cas system in the human pathogen *Streptococcus pyogenes*.

transaminases: *See* aminotransferases.

transamination: Enzymatic transfer of an amino group from an α-amino acid to an α-keto acid.

transcription: The enzymatic process whereby the genetic information contained in one strand of DNA is used to specify a complementary sequence of bases in an mRNA.

transcriptional control: Regulation of the synthesis of a protein by regulation of the formation of its mRNA.

transcription factor: In eukaryotes, a protein that affects the regulation and transcription initiation of a gene by binding to a regulatory sequence near or within the gene and interacting with RNA polymerase and/or other transcription factors.

transcriptome: The entire complement of RNA transcripts present in a given cell or tissue under specific conditions.

transduction: (1) Generally, the conversion of energy or information from one form to another. (2) The transfer of genetic information from one cell to another by means of a viral vector.

transfer RNA (tRNA): A class of RNA molecules (M_r 25,000 to 30,000), each of which combines covalently with a specific amino acid as the first step in protein synthesis.

transformation: Introduction of an exogenous DNA into a cell, causing the cell to acquire a new phenotype.

transgenic: Describes an organism that has genes from another organism incorporated in its genome as a result of recombinant DNA procedures.

transition state: An activated form of a molecule in which the molecule has undergone a partial chemical reaction; the highest point on the reaction coordinate.

transition-state analog: A stable molecule that resembles the transition state of a particular reaction and therefore binds the enzyme that catalyzes the reaction more tightly than does the substrate in the enzyme-substrate complex.

translation: The process in which the genetic information in an mRNA molecule specifies the sequence of amino acids during protein synthesis.

translational control: Regulation of the synthesis of a protein by regulation of the rate of its translation on the ribosome.

translational frameshifting: A programmed change in the reading frame during translation of an mRNA on a ribosome, occurring by any of several mechanisms.

translational repressor: A repressor that binds to an mRNA, blocking translation.

translocase: An enzyme that causes movement, such as movement of a ribosome along an mRNA.

transpiration: Passage of water from the roots of a plant to the atmosphere via the vascular system and the stomata of leaves.

transporters: Proteins that span a membrane and transport specific nutrients, metabolites, ions, or proteins across the membrane; sometimes called permeases.

transposition: Movement of a gene or set of genes from one site in the genome to another.

transposon (transposable element): A segment of DNA that can move from one position in the genome to another.

triacylglycerol: An ester of glycerol with three molecules of fatty acid; also called a triglyceride or neutral fat.

tricarboxylic acid (TCA) cycle: *See* citric acid cycle.

trimeri G proteins: Members of the G protein family with three subunits; function in a variety of signaling pathways. They are inactive with GDP bound, activated by associated receptors as the GDP is displaced by GTP, then inactivated by their intrinsic GTPase activity.

triose: A simple sugar with a backbone containing three carbon atoms.

tRNA: *See* transfer RNA.

tropic hormones (tropins): Peptide hormones that stimulate a specific target gland to secrete its hormone; for example, thyrotropin produced by the pituitary stimulates secretion of thyroxine by the thyroid.

t-SNAREs: Protein receptors in a targeted membrane (typically the plasma membrane) that bind to v-SNAREs in the membrane of a secretory vesicle, mediating fusion of the vesicle and target membranes.

tumor suppressor gene: One of a class of genes that encode proteins that normally regulate the cell cycle by suppressing cell division. Mutation of one copy of the gene is usually without effect, but when both copies are defective, the cell continues dividing without limitation, producing a tumor.

turnover number: The number of times an enzyme molecule transforms a substrate molecule per unit time, under conditions giving maximal activity at saturating substrate concentrations.

two-component signaling systems: Signal-transducing systems of bacteria and plants; composed of a receptor His kinase that phosphorylates an internal His residue when occupied by its ligand, then catalyzes phosphoryl transfer to a response regulator, which then alters the output of the signaling system.

type 2 diabetes mellitus: A metabolic disorder characterized by insulin resistance and poorly regulated blood glucose level; also known as adult-onset diabetes or noninsulin-dependent diabetes (NIDD).

U

ubiquitin: A small, highly conserved eukaryotic protein that targets an intracellular protein for degradation by proteasomes. Several ubiquitin molecules are covalently attached in tandem to a Lys residue of the target protein by a ubiquitinating enzyme.

ultraviolet (UV) radiation: Electromagnetic radiation in the region of 200 to 400 nm.

uncompetitive inhibition: The reversible inhibition pattern resulting when an inhibitor molecule can bind to the enzyme-substrate complex but not to the free enzyme.

uncoupling protein 1 (UCP1): A protein of the inner mitochondrial membrane in brown and beige adipose tissue that allows transmembrane movement of protons, short-circuiting the normal use of protons to drive ATP synthesis and dissipating the energy of substrate oxidation as heat. Also called thermogenin.

uniport: A transport system that carries only one solute, as distinct from cotransport.

unsaturated fatty acid: A fatty acid containing one or more double bonds.

urea cycle: A cyclic metabolic pathway in vertebrate liver that synthesizes urea from amino groups and carbon dioxide.

ureotelic: Excreting excess nitrogen in the form of urea.

uricotelic: Excreting excess nitrogen in the form of urate (uric acid).

V

V_m: *See* membrane potential.

V_{max}: The maximum velocity of an enzymatic reaction when the binding site is saturated with substrate.

van der Waals interaction: Weak intermolecular forces between molecules as a result of each inducing polarization in the other.

vector: A DNA molecule known to replicate autonomously in a host cell, to which a segment of DNA may be spliced to allow its replication in a cell; for example, a plasmid or an artificial chromosome.

vectorial: Describes an enzymatic reaction or transport process in which the protein has a specific orientation in a biological membrane such that the substrate is moved from one side of the membrane to the other as it is converted into product.

vectorial metabolism: Metabolic transformations in which the location (not the chemical composition) of a substrate changes relative to the plasma membrane or organellar membrane; for example, the action of transporters and the proton pumps of oxidative phosphorylation and photophosphorylation.

vesicle: A small, spherical, membrane-bounded particle with an internal aqueous compartment that contains components such as hormones or neurotransmitters to be moved within or out of a cell.

viral vector: A viral DNA altered so that it can act as a vector for recombinant DNA.

virion: A virus particle.

virus: A self-replicating, infectious, nucleic acid–protein complex that requires an intact host cell for its replication; its genome is DNA or RNA.

vitamin: An organic substance required in small quantities in the diet of some species;

generally functions as a component of a coenzyme.

v-SNAREs: Protein receptors in the membrane of a secretory vesicle that bind to t-SNAREs in a targeted membrane (typically the plasma membrane) and mediate fusion of the vesicle and target membranes.

W

Western blotting: *See* immunoblotting.

white adipose tissue (WAT): Nonthermogenic adipose tissue rich in triacylglycerols, stored and mobilized in response to hormonal signals. Transfer of electrons in the respiratory chain of WAT mitochondria is tightly coupled to ATP synthesis. *Compare* beige adipose tissue; brown adipose tissue.

wild type: The normal (unmutated) genotype or phenotype.

wobble: The relatively loose base pairing between the base at the 3′ end of a codon and the complementary base at the 5′ end of the anticodon.

X

x-ray crystallography: The analysis of x-ray diffraction patterns of a crystalline compound, used to determine the molecule's three-dimensional structure.

Z

zinc finger: A specialized protein motif of some DNA-binding proteins, involved in DNA recognition; characterized by a single atom of zinc coordinated to four Cys residues or to two His and two Cys residues.

Z scheme: In oxygenic photosynthesis, the path of electrons from water through photosystem II and the cytochrome b_6f complex to photosystem I and finally to NADPH. When the sequence of electron carriers is plotted against their reduction potentials, the path of electrons looks like a sideways Z.

zwitterion: A dipolar ion with spatially separated positive and negative charges.

zymogen: An inactive precursor of an enzyme; for example, pepsinogen, the precursor of pepsin.

Index

Key: b = boxed material; f = figures; s = structural formulas; t = tables; **boldface** = boldfaced terms

A

A. *See* adenine
A. *See* absorbance
A bands, **180**, 181, 181f
A kinase anchoring proteins (AKAPs), **450**, 451f
A site. *See* aminoacyl site
AAA+ ATPase, **998**, 999
AAT. *See* aspartate aminotransferase
abasic site, 297, 297f, **1008**
ABC excinuclease, 1010
ABC transporters. *See* ATP-binding cassette transporters
Abl gene, 482b
absolute configuration, **78**
of amino acids, 78, 78f
absorbance (A), **80b**
absorption, of dietary fats, 650–651, 650f, 651f
ACAT. *See* acyl-CoA–cholesterol acyl transferase
ACC. *See* acetyl-CoA carboxylase
acceptor control, **741**
acceptor control ratio, **741**
accessory pigments, **759**
acetic acid–acetate buffer system, 64, 64f
acetoacetate, **668**, 668s
in diabetes mellitus, 938
formation and use of, 668–670, 669f, 670f
acetoacetate decarboxylase, **669**, 669f
acetoacetyl-ACP, 815f, **816**
acetone, **668**, 668s
in diabetes mellitus, 938
formation and use of, 668–670, 669f, 670f
acetylation, 228, 229f
of histones, 1148–1149, 1149t
acetylcholine, 428s
in neuronal signaling, 472–473, 472f
acetylcholine receptors, 427–428, 473
membrane patches of, 400
acetyl-CoA. *See* acetyl-coenzyme A
acetyl-CoA acetyl transferase, **838**, 839f
acetyl-CoA carboxylase (ACC), **811**
malonyl-CoA formation by, 811–812, 812f
regulation of, 818–820, 819f
in regulation of fatty acid oxidation, 661, 661f, 664
acetyl-coenzyme A (acetyl-CoA), 15f, 15s
amino acid degradation to, 690f, 697, 698f, 699f
in cholesterol synthesis, 838–842, 838f, 839f, 840f, 841f
fatty acid oxidation to. *See* fatty acid oxidation
in fatty acid synthesis, 814, 815f, 816–817, 817f
shuttling of, 817–818, 819f
free-energy of hydrolysis of, 510–511, 511f, 511t
in glyoxylate cycle, 801–802, 801f
ketone bodies from, 668–670, 669f, 670f
in liver, 921, 921f, 922, 922f
malonyl-CoA formation from, 811–812, 812f
oxidation of. *See* citric acid cycle
pyruvate kinase regulation by, 595, 596f
pyruvate oxidation to, 619
allosteric and covalent regulation of, 640–641, 640f
oxidative decarboxylation reaction, 620–621, 621f
PDH complex coenzymes, 621, 621f, 622f
PDH complex enzymes, 621–622, 622f
PDH complex substrate channeling, 623, 623f
N-acetylglucosamine, in chitin, 255–256, 256f
N-acetylglutamate, **688**
carbamoyl phosphate synthetase I regulation by, 688, 688f
N-acetylglutamate synthase, **688**, 688f
N-acetylmuramic acid, 258, 258f

N-acetylneuraminic acid (Neu5Ac), 247, 247f, 268
in gangliosides, 370
acid dissociation constants (K_a), 61, 61f, **62**
acid-base catalysis
chymotrypsin, 215, 215f
enzymatic, 197, 197f
acidic activation domain, **1154**, 1155f
acidic R groups, 79f, 81
acidosis, 61, **670**
in diabetes mellitus, 67–68, 67f, 68b, 670, 938
effects of, 68b
ketone bodies causing, 670, 938
metabolic, 682
acids
amino acids as, 81–85, 83f, 84f, 85f
as buffers, 64–67, 64f, 65f, 66f
ionization of
acid dissociation constants and, 61–62, 61f
equilibrium constants and, 59–60
pH scale and, 60–61, 60f, 60t
pure water, 58–59, 59f
titration curves and, 62–63, 62f, 63f
acivicin, **901**, 901f
AcMNPV. *See Autographa californica* multicapsid nucleopolyhedrovirus
aconitase, **627**
in citric acid cycle, 625f, 627–628, 628b–629b, 629f
cis-aconitate, **627**
in citric acid cycle, 625f, 627–628, 628b–629b, 629f
aconitate hydratase, **627**
in citric acid cycle, 625f, 627–628, 628b–629b, 629f
ACP. *See* acyl carrier protein
acquired immune deficiency syndrome (AIDS)
protease inhibitors for, 215, 218, 218f, 219f
retrovirus as cause of, 1066, 1066f
reverse transcriptase inhibitors for, 1067b
T_H cells in, 174
acridine, **1046**, 1047f
ACTH. *See* adrenocorticotropic hormone
actin, 7, 9f, **179**, 180f
myosin thick filament interactions with, 182–183, 182f, 183f
α-actinin, **181**
actinomycin D, **1046**, 1047f
action potentials, voltage-gated ion channels producing, 472–473, 472f
action spectrum, **760**, 761f
activation energy ($\Delta G^\ddagger$), **28**, **191**
of ATP hydrolysis, 513
in enzymatic reactions, 191–192, 191f
binding energy and, 195, 195f
rate constant relationship to, 192
of transmembrane passage, 406–407, 407f
activation-induced deaminases (AIDs), **1086**, 1087
activators, **1129**, 1130, 1130f
DNA-binding, in transcription factor assembly, 1150–1153, 1151f, 1153f
modular structure of, 1154–1155, 1155f
active site, **158**, **190**
enzyme catalysis at, 190, 190f
of ribosomes, 1090, 1091f
active transport, **406**, 406f
ATP energy for, 514–516
against concentration or electrochemical gradient, 412–413, 412f
ion gradients providing energy for, 418–422, 418t, 420f, 421f, 422f
proton-motive force driving, 738–739, 738f
active transporters, **406**
ABC transporters, 417–418, 417f, 418t, 419b–420b
P-type ATPases, 413–416, 414f, 415f
V-type and F-type ATPases, 416, 416f
activity, **95**

of enzymes, 95–96, 96f
flux effects of, 585, 585f, 586b–587b
regulation of, 577–580, 578f, 578t, 579f, 580f, 580t
Actos. *See* pioglitazone
acute intermittent porphyria, 884b
acute lymphoblastic leukemia (ALL), 873–875
acute myeloid leukemia, 482b
acute pancreatitis, **679**
acyclovir, 1005, 1005s
acyl carrier protein (ACP), **814**, 814f
acetyl and malonyl group transfer to, 814, 815f, 816
acyl groups
in lipid bilayer, ordered states of, 397–398, 398f, 398t
transfer of, 505–506
acyl phosphate, **542**
acylation, in chymotrypsin mechanism, 213–215, 214f, 215f, 216f–217f
acyl-carnitine/carnitine cotransporter, **653**, 654f
acyl-CoA acetyltransferase, **656**
in fatty acid oxidation, 655f, 656
in ketone body formation, 668–669, 669f
acyl-CoA dehydrogenase, 655f, **656**, **723**
electron transfer by, 723, 723f
genetic defects in, 664
acyl-CoA synthetases, **652**
in triacylglycerol and glycerophospholipid synthesis, 826, 827f
acyl-CoA–cholesterol acyl transferase (ACAT), **842**, 842f
N-acylsphinganine, **835**, 837f
N-acylsphingosine, **835**, 837f
ADA. *See* adenosine deaminase
adaptor proteins, **450**
multivalent, 467–470, 468f, 469f, 470f
ADARs. *See* adenosine deaminases that act on RNA
Addison disease, 938
adenine (A), **280**, 280s, 280t
base pairing of, 284–285, 285f
biosynthesis of, 890–892, 890f, 891f, 892f
deamination of, 297, 297f
degradation of, 898–901, 899f, 900f, 901f
methylation of, 299
origin of, 1071, 1071f
adenine nucleotide translocase, **738**, 738f, 739
adenosine, 281s
in enzyme cofactors, 311, 312f
adenosine 3′,5′-cyclic monophosphate (cyclic AMP, cAMP), **312**
in epinephrine and glucagon action, 609, 610f
FRET studies of, 452b–453b
as regulatory molecule, 312, 313f
as second messenger
for β-adrenergic receptors, 441–447, 441f, 442f, 443f, 444b–446b, 447f, 447t
Ca^{2+} crosstalk with, 455–456
other regulatory molecules using, 449–451, 450t, 451f, 452b–453b
removal of, 442f, 448
in triacylglycerol mobilization, 651, 652f
adenosine deaminase (ADA), **898**, 899f
adenosine deaminase (ADA) deficiency, **899**, 900
adenosine deaminases that act on RNA (ADARs), **1086**, 1086f, 1087
adenosine diphosphate (ADP)
ATP synthesis from, 516
in coordinated regulation of cellular respiration pathways, 743–744, 743f
in glycolytic pathway, 536f, 542–545
in metabolic regulation, 582–584, 582f, 583f, 583t
oxidative phosphorylation regulation by, 741, 743–744, 743f
PFK-1 and FBPase-1 regulation by, 592–593, 592f, 593f
as signaling molecule, 312–313

Abbreviations for Amino Acids

A	Ala	Alanine	N	Asn	Asparagine
B	Asx	Asparagine or aspartate	P	Pro	Proline
			Q	Gln	Glutamine
C	Cys	Cysteine	R	Arg	Arginine
D	Asp	Aspartate	S	Ser	Serine
E	Glu	Glutamate	T	Thr	Threonine
F	Phe	Phenylalanine	V	Val	Valine
G	Gly	Glycine	W	Trp	Tryptophan
H	His	Histidine	X	—	Unknown or nonstandard amino acid
I	Ile	Isoleucine			
K	Lys	Lysine	Y	Tyr	Tyrosine
L	Leu	Leucine	Z	Glx	Glutamine or glutamate
M	Met	Methionine			

Asx and Glx are used in describing the results of amino acid analytical procedures in which Asp and Glu are not readily distinguished from their amide counterparts, Asn and Gln.

The Standard Genetic Code

UUU	Phe	UCU	Ser	UAU	Tyr	UGU	Cys
UUC	Phe	UCC	Ser	UAC	Tyr	UGC	Cys
UUA	Leu	UCA	Ser	UAA	Stop	UGA	Stop
UUG	Leu	UCG	Ser	UAG	Stop	UGG	Trp
CUU	Leu	CCU	Pro	CAU	His	CGU	Arg
CUC	Leu	CCC	Pro	CAC	His	CGC	Arg
CUA	Leu	CCA	Pro	CAA	Gln	CGA	Arg
CUG	Leu	CCG	Pro	CAG	Gln	CGG	Arg
AUU	Ile	ACU	Thr	AAU	Asn	AGU	Ser
AUC	Ile	ACC	Thr	AAC	Asn	AGC	Ser
AUA	Ile	ACA	Thr	AAA	Lys	AGA	Arg
AUG	Met*	ACG	Thr	AAG	Lys	AGG	Arg
GUU	Val	GCU	Ala	GAU	Asp	GGU	Gly
GUC	Val	GCC	Ala	GAC	Asp	GGC	Gly
GUA	Val	GCA	Ala	GAA	Glu	GGA	Gly
GUG	Val	GCG	Ala	GAG	Glu	GGG	Gly

*AUG also serves as the initiation codon in protein synthesis.

1 H 1.008																	2 He 4.003
3 Li 6.94	4 Be 9.01											5 B 10.81	6 C 12.011	7 N 14.01	8 O 16.00	9 F 19.00	10 Ne 20.18
11 Na 22.99	12 Mg 24.31											13 Al 26.98	14 Si 28.09	15 P 30.97	16 S 32.06	17 Cl 35.45	18 Ar 39.95
19 K 39.10	20 Ca 40.08	21 Sc 44.96	22 Ti 47.90	23 V 50.94	24 Cr 52.00	25 Mn 54.94	26 Fe 55.85	27 Co 58.93	28 Ni 58.71	29 Cu 63.55	30 Zn 65.37	31 Ga 69.72	32 Ge 72.59	33 As 74.92	34 Se 78.96	35 Br 79.90	36 Kr 83.30
37 Rb 85.47	38 Sr 87.62	39 Y 88.91	40 Zr 91.22	41 Nb 92.91	42 Mo 95.94	43 Te 98.91	44 Ru 101.07	45 Rh 102.91	46 Pd 106.4	47 Ag 107.87	48 Cd 112.40	49 In 114.82	50 Sn 118.69	51 Sb 121.75	52 Te 126.70	53 I 126.90	54 Xe 131.30
55 Cs 132.91	56 Ba 137.34	57–70 *	71 Lu 174.97	72 Hf 178.49	73 Ta 180.95	74 W 183.85	75 Re 186.2	76 Os 190.2	77 Ir 192.2	78 Pt 195.09	79 Au 196.97	80 Hg 200.59	81 Tl 204.37	82 Pb 207.19	83 Bi 208.98	84 Po (209)	85 At (210)	86 Rn (222)
87 Fr (223)	88 Ra 226.03	89–102 **	103 Lr 262.11	104 Rf 261.11	105 Db 262.11	106 Sg 263.12	107 Bh 264.12	108 Hs 265.13	109 Mt 268	110 Ds 281	111 Rg 281	112 Cn 285	113 Nh 286	114 Fl 289	115 Mc 289	116 Lv 293	117 Ts 293	118 Og 294

*Lanthanides

57 La 138.91	58 Ce 140.12	59 Pr 140.91	60 Nd 144.24	61 Pm 144.91	62 Sm 150.36	63 Eu 151.96	64 Gd 157.25	65 Tb 158.93	66 Dy 162.50	67 Ho 164.93	68 Er 167.26	69 Tm 168.93	70 Yb 173.04

**Actinides

89 Ac 227.03	90 Th 232.04	91 Pa 231.04	92 U 238.03	93 Np 237.05	94 Pu 244.06	95 Am 243.06	96 Cm 247.07	97 Bk 247.07	98 Cf 251.08	99 Es 252.08	100 Fm 257.10	101 Md 258.10	102 No 259.10

National and international bioinformatics resources that provide access to databases, tools, and literature critical to practitioners of biochemistry

National Center for Biotechnology Information (NCBI)	www.ncbi.nlm.nih.gov
UniProt	www.uniprot.org
ExPASy Bioinformatics Resource Portal	www.expasy.org
GenomeNet	www.genome.jp

Some useful structure databases

Protein Data Bank (PDB)	www.pdb.org
EMDataBank Unified Resource for 3DEM	www.emdatabank.org
National Center for Biomedical Glycomics	http://glycomics.ccrc.uga.edu
LIPIDMAPS Lipidomics Gateway	www.lipidmaps.org
Nucleic Acid Database (NDB)	http://ndbserver.rutgers.edu
Modomics database of RNA modification pathways	http://modomics.genesilico.pl

Some other resources and tools mentioned in this text

Structural Classification of Proteins database (SCOP2)	http://scop2.mrc-lmb.cam.ac.uk
PROSITE Sequence logo	http://prosite.expasy.org/sequence_logo.html
ProtScale hydrophobicity and other profiles of amino acids	http://web.expasy.org/protscale
Predictor of Natural Disordered Regions (PONDR)	www.pondr.com
Enzyme nomenclature	www.chem.qmul.ac.uk/iubmb/enzyme
Ensembl genome databases	www.ensembl.org
PANTHER (Protein ANalysis THrough Evolutionary Relationships) Classification System	www.pantherdb.org
Basic Local Alignment Search Tool (BLAST)	https://blast.ncbi.nlm.nih.gov/Blast.cgi
Kyoto Encyclopedia of Genes and Genomes (KEGG)	www.genome.jp/kegg
KEGG pathway maps	www.genome.ad.jp/kegg/pathway/map/map01100.html
Biochemical nomenclature	www.chem.qmul.ac.uk/iubmb/nomenclature
Online Mendelian Inheritance in Man	www.omim.org